QUANTUM MECHANICS
WITH APPLICATIONS TO NANOTECHNOLOGY AND INFORMATION SCIENCE

QUANTUM MECHANICS
WITH APPLICATIONS TO NANOTECHNOLOGY AND INFORMATION SCIENCE

Yehuda B. Band
Department of Chemistry, Department of Electro-Optics and Department of Physics, and Ilse Katz Institute for Nanoscale Science and Technology
Ben-Gurion University
Beer-Sheva, Israel

and

Yshai Avishai
Department of Physics, and Ilse Katz Institute for Nanoscale Science and Technology
Ben-Gurion University
Beer-Sheva, Israel

AMSTERDAM • WALTHAM • HEIDELBERG • LONDON
NEW YORK • OXFORD • PARIS • SAN DIEGO
SAN FRANCISCO • SINGAPORE • SYDNEY • TOKYO
Academic Press is an imprint of Elsevier

Academic Press is an imprint of Elsevier
The Boulevard, Langford Lane, Kidlington, Oxford OX5 1GB, UK
Radarweg 29, PO Box 211, 1000 AE Amsterdam, The Netherlands

First edition 2013

Copyright © 2013 Elsevier Ltd. All rights reserved.

No part of this publication may be reproduced, stored in a retrieval system or transmitted in any form or by any means electronic, mechanical, photocopying, recording or otherwise without the prior written permission of the publisher.

Permissions may be sought directly from Elsevier's Science & Technology Rights Department in Oxford, UK: phone (+44) (0) 1865 843830; fax (+44) (0) 1865 853333; email: permissions@elsevier.com. Alternatively you can submit your request online by visiting the Elsevier web site at *http://elsevier.com/locate/permissions*, and selecting *Obtaining permission to use Elsevier material*.

Notice
No responsibility is assumed by the publisher for any injury and/or damage to persons or property as a matter of products liability, negligence or otherwise, or from any use or operation of any methods, products, instructions or ideas contained in the material herein. Because of rapid advances in the medical sciences, in particular, independent verification of diagnoses and drug dosages should be made.

British Library Cataloguing-in-Publication Data
A catalogue record for this book is available from the British Library.

Library of Congress Cataloging-in-Publication Data
A catalog record for this book is available from the Library of Congress.

ISBN: 978-0-444-53786-7

For information on all Academic Press publications
visit our website at *www.store.elsevier.com*

Printed and bound in UK
14 13 12 11 10 9 8 7 6 5 4 3 2 1

Working together to grow
libraries in developing countries

www.elsevier.com | www.bookaid.org | www.sabre.org

ELSEVIER | BOOK AID International | Sabre Foundation

CONTENTS

Preface xvii

CHAPTER 1 INTRODUCTION TO QUANTUM MECHANICS 1

- **1.1** What is Quantum Mechanics? 2
  - 1.1.1 A Brief Early History of Quantum Mechanics 2
  - 1.1.2 Energy Quantization 2
  - 1.1.3 Waves, Light, and Blackbody Radiation 6
  - 1.1.4 Wave–Particle Duality 8
  - 1.1.5 Angular Momentum Quantization 12
  - 1.1.6 Tunneling 13
  - 1.1.7 Photoelectric Effect 15
- **1.2** Nanotechnology and Information Technology 17
  - 1.2.1 STM and AFM Microscopies 18
  - 1.2.2 Molecular Electronics 18
  - 1.2.3 Quantum Dots, Wires and Wells, and Nanotubes 19
  - 1.2.4 Bio-Nanotechnology 21
  - 1.2.5 Information Technology 21
- **1.3** A First Taste of Quantum Mechanics 21
  - 1.3.1 Quantum States and Probability Distributions 22
  - 1.3.2 Observable Operators 24
  - 1.3.3 Quantum Entanglement 26
  - 1.3.4 The Postulates of Quantum Mechanics 28
  - 1.3.5 Time-Dependent and -Independent Schrödinger Equations 29
  - 1.3.6 Momentum, Energy, and Angular Momentum 32
  - 1.3.7 Dirac Delta Functions 36
  - 1.3.8 Position and Momentum States, $|\mathbf{x}\rangle$ and $|\mathbf{p}\rangle$ 38
  - 1.3.9 Ehrenfest's Theorem 39
  - 1.3.10 One-Dimensional Wave Equations 41
  - 1.3.11 Particle-in-a-Box and Piecewise-Constant Potentials 42
  - 1.3.12 The Delta Function Potential 52
  - 1.3.13 Wave Packets 53
  - 1.3.14 The Linear Potential and Quantum Tunneling 54
  - 1.3.15 The Harmonic Oscillator 55

CHAPTER 2 THE FORMALISM OF QUANTUM MECHANICS 61

- **2.1** Hilbert Space and Dirac Notation   61
  - 2.1.1 Position and Momentum Representations   63
  - 2.1.2 Basis-State Expansions   64
- **2.2** Hermitian and Anti-Hermitian Operators   66
  - 2.2.1 Compatible Operators and Degeneracy   66
- **2.3** The Uncertainty Principle   67
- **2.4** The Measurement Problem   69
- **2.5** Mixed States: Density Matrix Formulation   70
  - 2.5.1 Many-Particle Systems: Correlation Functions   75
  - 2.5.2 Purity and von Neumann Entropy   77
  - 2.5.3 Distance Between States   77
  - 2.5.4 The Measurement Problem Revisited   78
- **2.6** The Wigner Representation   79
- **2.7** Schrödinger and Heisenberg Representations   85
  - 2.7.1 Interaction Representation   87
  - 2.7.2 Harmonic Oscillator Raising–Lowering Operators   88
  - 2.7.3 Coherent States and Squeezed States   92
- **2.8** The Correspondence Principle and the Classical Limit   97
- **2.9** Symmetry and Conservation Laws in Quantum Mechanics   98
  - 2.9.1 Exchange Symmetry   98
  - 2.9.2 Inversion Symmetry   100
  - 2.9.3 Time-Reversal Symmetry   100
  - 2.9.4 Additional Generators of Galilean Transformations   102

CHAPTER 3 ANGULAR MOMENTUM AND SPHERICAL SYMMETRY 105

- **3.1** Angular Momentum in Quantum Mechanics   105
  - 3.1.1 Angular Momentum Raising and Lowering Operators   107
  - 3.1.2 Electron Spin: $j = 1/2$ 110
  - 3.1.3 Angular Momentum in Spherical Coordinates   111
  - 3.1.4 Spherical Harmonics   112
- **3.2** Spherically Symmetric Systems   116
  - 3.2.1 Angular Momentum Decomposition of Plane Waves   119
  - 3.2.2 Spherical Quantum Dot   120
  - 3.2.3 The 3D Harmonic Oscillator   121
  - 3.2.4 The Morse Oscillator   122
  - 3.2.5 van der Waals and Lennard-Jones Potentials   123
  - 3.2.6 The Hydrogen Atom   124
- **3.3** Rotations and Angular Momentum   132
  - 3.3.1 Euler angles α, β, γ and the Rotation Matrix   133
  - 3.3.2 Rotation and D Functions   134
  - 3.3.3 Rigid-Rotor Eigenfunctions   137
- **3.4** Addition (Coupling) of Angular Momenta   139
  - 3.4.1 Clebsch–Gordan Coefficients and $3j$ Symbols   140
  - 3.4.2 Clebsch–Gordan Series   143

- **3.5** Tensor Operators 144
  - 3.5.1 Irreducible Representations of the Density Matrix 146
  - 3.5.2 Vector Fields 148
  - 3.5.3 Spinor Fields 149
  - 3.5.4 Multipole Expansions 150
- **3.6** Symmetry Considerations 151
  - 3.6.1 Selection Rules 152
  - 3.6.2 Inversion Symmetry 152
  - 3.6.3 Time-Reversal Symmetry 153
  - 3.6.4 Wigner–Eckart Theorem 153
  - 3.6.5 $6j$ and Higher Coefficients 156

CHAPTER 4 SPIN 159

- **4.1** Spin Angular Momentum 159
- **4.2** Spinors 160
  - 4.2.1 Pauli Matrices 161
  - 4.2.2 Rotation of Spinors 164
  - 4.2.3 Spin-Orbitals 165
- **4.3** Electron in a Magnetic Field 166
  - 4.3.1 Charged Particle in a Magnetic Field: Orbital Effects 169
- **4.4** Time-Reversal Properties of Spinors 172
- **4.5** Spin–Orbit Interaction in Atoms 175
- **4.6** Hyperfine Interaction 178
  - 4.6.1 Electric Quadrupole Hyperfine Interaction 181
  - 4.6.2 Zeeman Splitting of Hyperfine States 182
- **4.7** Spin-Dipolar Interactions 183
- **4.8** Introduction to Magnetic Resonance 185
  - 4.8.1 The Rotating-Wave Approximation 187
  - 4.8.2 Spin Relaxation and the Bloch Equation 188
  - 4.8.3 Nuclear Spin Hamiltonian 189
  - 4.8.4 Chemical Shifts 189
  - 4.8.5 Fourier Transform NMR 190

CHAPTER 5 QUANTUM INFORMATION 193

- **5.1** Classical Computation and Classical Information 194
  - 5.1.1 Information and Entropy 194
  - 5.1.2 Shannon Entropy 196
  - 5.1.3 Data Compression 198
  - 5.1.4 Classical Computers and Gates 200
  - 5.1.5 Classical Cryptography 202
  - 5.1.6 Computational Complexity 204
- **5.2** Quantum Information 205
  - 5.2.1 Qubits 205
  - 5.2.2 Quantum Entanglement and Bell States 207
  - 5.2.3 Quantum Gates 213

		5.2.4	No-Cloning Theorem 218

- 5.2.4 No-Cloning Theorem 218
- 5.2.5 Dense Coding 219
- 5.2.6 Data Compression of Quantum Information 220
- 5.2.7 Quantum Teleportation 221
- 5.2.8 Quantum Cryptography 222
- 5.2.9 Quantum Circuits 223
- 5.2.10 Quantum Computing Despite Measurement 224

5.3 Quantum Computing Algorithms 224
- 5.3.1 Deutsch and Deutsch–Jozsa Algorithms 225
- 5.3.2 The Grover Search Algorithm 228
- 5.3.3 Quantum Fourier Transform 232
- 5.3.4 Shor Factorization Algorithm 234
- 5.3.5 Quantum Simulation 239

5.4 Decoherence 240
5.5 Quantum Error Correction 240
5.6 Experimental Implementations 243
- 5.6.1 Ion Traps 243
- 5.6.2 Neutral Atoms in Optical Lattices 245
- 5.6.3 Cavity Based Quantum Computing 246
- 5.6.4 Nuclear Magnetic Resonance Systems 247
- 5.6.5 All-Optical Quantum Computers 248
- 5.6.6 Solid-State Qubits 248

5.7 The EPR Paradox 250
5.8 Bell's Inequalities 251
- 5.8.1 Bell's Inequalities and the EPR Paradox 252
- 5.8.2 Bell's Analysis using Hidden Variables 253
- 5.8.3 General Aspects of Bell's Inequalities 256

CHAPTER 6 QUANTUM DYNAMICS AND CORRELATIONS 259

6.1 Two-Level Systems 259
- 6.1.1 Two-Level Dynamics (Spin Dynamics) 261
- 6.1.2 The Bloch Sphere Picture 262
- 6.1.3 Coupling to a Bath: Decoherence 266
- 6.1.4 Periodically Driven Two-Level System 268
- 6.1.5 Atoms in an Electromagnetic Field: Dispersion and Absorption 273
- 6.1.6 Doppler Cooling of Atoms 275
- 6.1.7 Optical Trapping of Atoms 278
- 6.1.8 Two or More Correlated "Spins" 279
- 6.1.9 The N-Two-Level System Bloch Sphere 282
- 6.1.10 Ramsey Fringe Spectroscopy 285

6.2 Three-Level Systems 287
- 6.2.1 Two or More Three-Level Correlated Systems 288

6.3 Classification of Correlation and Entanglement 290
- 6.3.1 Entanglement Witness Operators 292

6.4 Three-Level System Dynamics 293
6.5 Continuous-Variable Systems 295

Contents

6.6 Wave Packet Dynamics 297
6.7 Time-Dependent Hamiltonians 299
6.8 Quantum Optimal Control Theory 300

CHAPTER 7 APPROXIMATION METHODS 303

7.1 Basis-State Expansions 303
    7.1.1 Time-Dependent Basis Set Expansions 303
7.2 Semiclassical Approximations 304
    7.2.1 The WKB Approximation 304
    7.2.2 Semiclassical Treatment of Dynamics 309
    7.2.3 Semiclassical Hamilton–Jacobi Expansion 309
7.3 Perturbation Theory 309
    7.3.1 Nondegenerate Perturbation Theory 310
    7.3.2 Degenerate Perturbation Theory 313
    7.3.3 Time-Dependent Perturbation Theory 315
7.4 Dynamics in an Electromagnetic Field 321
    7.4.1 Spontaneous and Stimulated Emission of Radiation 323
    7.4.2 Electric Dipole and Multipole Radiation 324
    7.4.3 Thomson, Rayleigh, Raman, and Brillouin Transitions 326
    7.4.4 Decay Width 329
    7.4.5 Doppler Shift 330
7.5 Exponential and Nonexponential Decay 331
7.6 The Variational Method 331
7.7 The Sudden Approximation 333
7.8 The Adiabatic Approximation 335
    7.8.1 Chirped Pulse Adiabatic Passage 336
    7.8.2 Stimulated Raman Adiabatic Passage 338
    7.8.3 The Landau–Zener Problem 339
    7.8.4 Generalized Displacements and Forces 343
    7.8.5 Berry Phase 344
7.9 Linear Response Theory 349
    7.9.1 Susceptibilities 350
    7.9.2 Kubo Formulas 352
    7.9.3 Onsager Reciprocal Relations 360
    7.9.4 Fluctuation–Dissipation Theorem 362

CHAPTER 8 IDENTICAL PARTICLES 367

8.1 Permutation Symmetry 367
    8.1.1 The Symmetric Group S_N 369
    8.1.2 Young Tableaux 370
8.2 Exchange Symmetry 371
    8.2.1 Symmetrization Postulate 372
8.3 Permanents and Slater Determinants 374
8.4 Simple Two- and Three-Electron States 375
8.5 Exchange Symmetry for Two Two-Level Systems 377
8.6 Many-Particle Exchange Symmetry 378

CHAPTER 9 **ELECTRONIC PROPERTIES OF SOLIDS**   381

- **9.1** The Free Electron Gas   381
  - 9.1.1 Density of States in 2D and 1D Systems   386
  - 9.1.2 Fermi–Dirac Distribution   388
- **9.2** Elementary Theories of Conductivity   391
  - 9.2.1 Drude Theory of Conductivity   392
  - 9.2.2 Thermal Conductivity of Metals   397
  - 9.2.3 Sommerfeld Theory of Transport in Metals   399
- **9.3** Crystal Structure   400
  - 9.3.1 Bravais Lattices and Crystal Systems   401
  - 9.3.2 The Reciprocal Lattice   405
  - 9.3.3 Quasicrystals   412
- **9.4** Electrons in a Periodic Potential   414
  - 9.4.1 From Atomic Orbits to Band Structure   414
  - 9.4.2 Band Structure and Electron Transport   415
  - 9.4.3 Periodic Potential and Band Formation   416
  - 9.4.4 Bloch Wave Functions and Energy Bands   418
  - 9.4.5 Schrödinger Equation in Reciprocal Lattice Space   422
  - 9.4.6 Sinusoidal Potential: Mathieu Functions   423
  - 9.4.7 Tight-Binding Model   427
  - 9.4.8 Wannier Functions   429
  - 9.4.9 Electric Field Effects   430
- **9.5** Magnetic Field Effects   434
  - 9.5.1 Electron in a Magnetic Field   435
  - 9.5.2 Aharonov–Bohm Effect   436
  - 9.5.3 The Hall Effect and Magnetoresistance   443
  - 9.5.4 Landau Quantization   445
  - 9.5.5 2D Electron Gas in a Perpendicular Magnetic Field   448
  - 9.5.6 Electron Subject to Periodic Potential and Magnetic Field   454
  - 9.5.7 de Haas–van Alphen and Shubnikov–de Haas Effects   459
  - 9.5.8 The Quantum Hall Effect   462
  - 9.5.9 Paramagnetism and Diamagnetism   468
  - 9.5.10 Magnetic Order   475
- **9.6** Semiconductors   481
  - 9.6.1 Semiconductor Band Structure   482
  - 9.6.2 Charge Carrier Density   484
  - 9.6.3 Extrinsic Semiconductors   486
  - 9.6.4 Inhomogeneous Semiconductors: p-n Junctions   490
  - 9.6.5 Excitons   500
  - 9.6.6 Spin–Orbit Coupling in Solids   504
  - 9.6.7 $\mathbf{k} \cdot \mathbf{p}$ Perturbation Theory   506
  - 9.6.8 Spin Hall Effect   510
  - 9.6.9 Photon Induced Processes in Semiconductors   512
- **9.7** Spintronics   515
  - 9.7.1 Tools for Manipulating Spins   516
  - 9.7.2 Tunneling Magnetoresistance   520
  - 9.7.3 Spintronic Devices   523

Contents

- **9.8** Low-Energy Excitations 526
  - 9.8.1 Phonons 527
  - 9.8.2 Plasmons 531
  - 9.8.3 Magnons 535
  - 9.8.4 Polarons 538
  - 9.8.5 Polaritons 541
- **9.9** Insulators 541
  - 9.9.1 Defining Insulators 542
  - 9.9.2 Classification of Insulators 543

CHAPTER 10 ELECTRONIC STRUCTURE OF MULTIELECTRON SYSTEMS 545

- **10.1** The Multielectron System Hamiltonian 545
- **10.2** Slater and Gaussian Type Atomic Orbitals 546
- **10.3** Term Symbols for Atoms 547
- **10.4** Two-Electron Systems 547
  - 10.4.1 The Helium Atom 548
  - 10.4.2 The Hartree Method: Helium 548
- **10.5** Hartree Approximation for Multielectron Systems 551
- **10.6** The Hartree–Fock Method 552
  - 10.6.1 Hartree–Fock for Helium 557
- **10.7** Koopmans' Theorem 558
- **10.8** Atomic Radii 560
- **10.9** Multielectron Fine Structure: Hund's Rules 560
- **10.10** Electronic Structure of Molecules 562
  - 10.10.1 H_2^+: Molecular Orbitals 563
  - 10.10.2 The Hydrogen Molecule 565
  - 10.10.3 The Hückel Approximation 569
- **10.11** Hartree–Fock for Metals 572
- **10.12** Electron Correlation 574
  - 10.12.1 Configuration Interaction 574
  - 10.12.2 Moller–Plesset Many-Body Perturbation Theory 576
  - 10.12.3 Coupled Cluster Method 577

CHAPTER 11 MOLECULES 579

- **11.1** Molecular Symmetries 579
  - 11.1.1 Molecular Orbitals and Group Theory 580
  - 11.1.2 Character Tables and Mulliken Symbols 581
- **11.2** Diatomic Electronic States 582
  - 11.2.1 Interatomic Potentials at Large Internuclear Distances 585
  - 11.2.2 Hund's Coupling 586
  - 11.2.3 Hyperfine Interactions in Diatomic Molecules 589
- **11.3** The Born-Oppenheimer Approximation 589
  - 11.3.1 Potential Energy Crossings and Pseudocrossings 590
  - 11.3.2 Born–Oppenheimer Nuclear Derivative Coupling 591
  - 11.3.3 The Hellman–Feynman Theorem 592
- **11.4** Rotational and Vibrational Structure 593

11.5 Vibrational Modes and Symmetry 595
11.6 Selection Rules for Optical Transitions 596
    11.6.1 Selection Rules for Diatomic Molecules 597
11.7 The Franck–Condon Principle 600
    11.7.1 Bound-Free Matrix Elements 602

CHAPTER 12 SCATTERING THEORY 605

12.1 Classical Scattering Theory 605
12.2 Quantum Scattering 609
    12.2.1 Time-Dependent and Stationary Approaches 609
    12.2.2 Preparation of the Initial State 611
    12.2.3 Time-Dependent Formulation 612
12.3 Stationary Scattering Theory 616
    12.3.1 Cross-Sections 617
    12.3.2 Two-Body Collisions 618
    12.3.3 From Wave Functions to Cross-Sections 622
    12.3.4 Green's Function 623
12.4 Aspects of Formal Scattering Theory 627
    12.4.1 The Transition Operator (The T Matrix) 628
    12.4.2 The S Matrix and Möller Operators 633
12.5 Central Potentials 635
    12.5.1 Central Potentials and Spin 635
    12.5.2 Axial Symmetry 635
    12.5.3 Partial Wave Analysis 636
    12.5.4 Phase Shift Analysis 647
    12.5.5 Scattering from a Coulomb Potential 659
    12.5.6 Scattering of Two Identical Particles 664
12.6 Resonance Scattering 666
    12.6.1 Influence of Bound States 666
    12.6.2 Resonance Cross-Sections 668
    12.6.3 Feshbach Resonance 669
    12.6.4 Fano Resonance 677
12.7 Approximation Methods 682
    12.7.1 Born Approximation 683
    12.7.2 WKB Approximation 684
    12.7.3 The Variational Principle 688
    12.7.4 Eikonal Approximation 690
12.8 Particles with Internal Degrees of Freedom 693
    12.8.1 Spin 693
    12.8.2 Composite Particles 693
    12.8.3 Channels 693
    12.8.4 Asymptotic States and Cross-Sections 695
    12.8.5 Möller Operators 696
    12.8.6 The Multichannel S Matrix 699
    12.8.7 Scattering from Two Potentials 703
    12.8.8 Scattering of Particles with Spin 704

Contents

    12.8.9   Inelastic Scattering and Scattering Reactions   715
    12.8.10   Scattering from a Collection of Identical Particles   722
12.9   Scattering in Low-Dimensional Systems   724
    12.9.1   Scattering in Two Dimensions   724
    12.9.2   1D Scattering: S Matrix   731
    12.9.3   1D Scattering: Anderson Localization   737
    12.9.4   Scattering in Quasi-One-Dimensional Systems   741

CHAPTER 13 LOW-DIMENSIONAL QUANTUM SYSTEMS   749

13.1   Mesoscopic Systems   750
    13.1.1   Low-Dimensional Nanostructures   752
    13.1.2   Quantum Wells   752
    13.1.3   Quantum Wires   753
    13.1.4   Quantum Dots   754
    13.1.5   Heterojunctions and Superlattices   754
    13.1.6   Quantum Point Contacts   755
13.2   The Landauer Conductance Formula   755
    13.2.1   Aharonov–Bohm Interferometer   757
    13.2.2   Multiport Landauer Formula   760
    13.2.3   Conductance of Quantum Point Contacts   762
13.3   Properties of Quantum Dots   764
    13.3.1   Equilibrium Properties of Quantum Dots   765
    13.3.2   Charge Transport through Quantum Dots   770
13.4   Disorder in Mesoscopic Systems   773
    13.4.1   Disorder in Quantum Dots   773
    13.4.2   Disordered Systems and Random Matrices   774
    13.4.3   Application of Random Matrix Theory to Disordered Mesoscopic Systems   777
13.5   Kondo Effect in Quantum Dots   783
    13.5.1   The Dot Kondo Hamiltonian   784
    13.5.2   The Dot Kondo Temperature   785
    13.5.3   Abrikosov–Suhl Resonance and Kondo Conductance   785
13.6   Graphene   787
    13.6.1   Carbon Nanotubes   788
    13.6.2   Lattice Structure and Dirac Cones   790
    13.6.3   Dirac Equation and its Relevance to Graphene   792
    13.6.4   Tight-Binding Model for Graphene   797
    13.6.5   Continuum Theory   800
    13.6.6   Landau Levels in Graphene   804
    13.6.7   Potential Scattering in Graphene   806
13.7   Inventory of Recently Discovered Low-Dimensional Phenomena   810
    13.7.1   Persistent Currents   811
    13.7.2   Weak Localization   811
    13.7.3   Shot Noise   813
    13.7.4   Strongly Correlated Low-Dimensional Systems   815
    13.7.5   Wigner Crystals   815
    13.7.6   The Fractional Quantum Hall Effect   816

- 13.7.7 High-Temperature Superconductivity and 2D Magnetism 819
- 13.7.8 1D Correlated Electron Systems 820
- 13.7.9 1D Spin Chains 821
- 13.7.10 Quantum Spin Hall Effect 822

CHAPTER 14 MANY-BODY THEORY 825

14.1 Second Quantization 825
- 14.1.1 Interacting Identical Particles: A Generic Many-Body Problem 826
- 14.1.2 A Basis for Many-Body Wave Functions 827
- 14.1.3 Mapping onto Fock Space 830
- 14.1.4 Creation and Annihilation Operators 832
- 14.1.5 The Hamiltonian in Fock Space 837
- 14.1.6 Field Operators 843
- 14.1.7 Electromagnetic Field Quantization: Photons 850
- 14.1.8 Quantization of Lattice Vibrations: Phonons 853
- 14.1.9 Systems with Two Kinds of Particles 856

14.2 Statistical Mechanics in Second Quantization 857

14.3 The Electron Gas 860
- 14.3.1 Electrons in the Jellium Model 861

14.4 Mean-Field Theory 866
- 14.4.1 Mean-Field Equations in Second Quantization 867

CHAPTER 15 DENSITY FUNCTIONAL THEORY 871

15.1 The Hohenberg–Kohn Theorems 872

15.2 The Thomas–Fermi Approximation 875

15.3 The Kohn–Sham Equations 878
- 15.3.1 Local Density Approximation 881

15.4 Spin DFT and Magnetic Systems 882
- 15.4.1 Spin DFT Local Density Approximation 883

15.5 The Gap Problem in DFT 884

15.6 Time-Dependent DFT 885
- 15.6.1 The Runge–Gross Theorem 886
- 15.6.2 Time-Dependent Kohn–Sham Equations 888
- 15.6.3 Adiabatic LDA 888

15.7 DFT Computer Packages 888

APPENDIX A LINEAR ALGEBRA 891

A.1 Vector Spaces 891
- A.1.1 Dirac Notation 892
- A.1.2 Inner Product Spaces 893

A.2 Operators and Matrices 898
- A.2.1 Outer Product 899
- A.2.2 Determinants and Permanents 900
- A.2.3 Normal, Unitary, Hermitian-Conjugate and Hermitian Operators 901
- A.2.4 Trace and Projection 902
- A.2.5 Antilinear and Antiunitary Operators 903

Contents

APPENDIX B SOME ORDINARY DIFFERENTIAL EQUATIONS   905

APPENDIX C VECTOR ANALYSIS   913
- **C.1** Scalar and Vector Products   913
- **C.2** Differential Operators   914
- **C.3** Divergence and Stokes Theorems   915
- **C.4** Curvilinear Coordinates   916

APPENDIX D FOURIER ANALYSIS   921
- **D.1** Fourier Series   922
  - D.1.1 Fourier Series of Functions of a Discrete Variable   924
- **D.2** Fourier Integrals   925
- **D.3** Fourier Series and Integrals in Three-Space Dimensions   927
  - D.3.1 3D Fourier Integrals   927
- **D.4** Fourier Integrals of Time-Dependent Functions   928
- **D.5** Convolution   928
- **D.6** Fourier Expansion of Operators   929
- **D.7** Fourier Transforms   929
- **D.8** FT for Solving Differential and Integral Equations   931

APPENDIX E SYMMETRY AND GROUP THEORY   933
- **E.1** Group Theory Axioms   933
- **E.2** Group Multiplication Tables   933
- **E.3** Examples of Groups   934
  - E.3.1 Point Groups   935
  - E.3.2 Space Groups   936
  - E.3.3 Continuous Groups   936
- **E.4** Some Properties of Groups   936
- **E.5** Group Representations   937
  - E.5.1 Irreducible Representations   938
  - E.5.2 Group Orthogonality Theorem   939
  - E.5.3 Characters and Character Tables   939
  - E.5.4 Constructing Irreducible Representations   942

Bibliography   943

Index   953

PREFACE

Quantum mechanics transcends and supplants classical mechanics at the atomic and subatomic levels. It provides the underlying framework for many subfields of physics, chemistry, and the engineering sciences. It is the only framework for understanding the structure of materials, from the semiconductors in our computers to the metals in our automobiles. It is also the support structure for much of nanotechnology and the promising paradigm of quantum information theory. Moreover, it is the foundation of condensed matter physics, atomic physics, molecular physics, quantum chemistry, materials design, elementary particle, and nuclear physics.

The purpose of this book is to present the fundamentals of quantum theory within a modern perspective, with emphasis on applications to nanoscience and nanotechnology, and information science and information technology. As the frontiers of science have advanced, the sort of curriculum adequate for students in the science and engineering 20 years ago is no longer satisfactory today. Hence, the emphasis is on new topics that are not included in previous books on quantum mechanics [1–11]. These topics include quantum information theory, decoherence and dissipation, quantum measurement theory, disordered systems, and nanotechnology, including spintronics, and reduced dimensional systems such as quantum dots, wires and wells.

The intended readers of this book comprise scientists and engineers, including undergraduate and graduate students in physics, chemistry, materials science, electrical engineering, computer and information science, and nanotechnology. This book can serve as a textbook for a number of courses, including a one-semester undergraduate course in quantum mechanics, a two-semester quantum mechanics undergraduate course, a graduate level quantum mechanics course, an engineering quantum mechanics course, a quantum information and quantum computing course, a nanotechnology course, and a quantum chemistry course. Table 1 specifies the appropriate chapters for each of these courses.

A web page for this book, `https://sites.google.com/site/thequantumbook/`, contains links to interesting web sites related to the subject matter, a link to a list of errors that were found, color versions of the figures of this book, and a means for reporting errors that you find. Please use the e-mail addresses on the web page of the book to contact us with any comments and suggestions regarding this book.

Although we assume the reader is familiar with material ordinarily presented in first-year physics and first-year calculus courses, as well as in linear algebra, Appendices that review the requisite mathematical background material are provided, and a review of classical mechanics is presented in Chapter 16, see `https://sites.google.com/site/thequantumbook/QM_Classical.pdf`.

The layout of this book is as follows. Chapter 1 contains an introduction to quantum mechanics. It explains what quantum mechanics is and why it is essential to properly describe matter and radiation. It includes a brief introductory of nanotechnology and information science and then provides a first taste of quantum mechanics. [Readers who are not well versed in classical mechanics, may need to read Chp_16_Classical_Mechanics.pdf, which is linked to the web page of the book. It presents classical mechanics, including the Lagrangian and Hamiltonian formulation of mechanics; it provides a contrast with quantum mechanics, yet introduces some concepts that are carried over to quantum mechanics.] Chapter 2 presents the formalism of quantum mechanics, including the mathematical notation required for Hilbert Spaces, Dirac notation, and the various representations of quantum mechanics. Chapter 3 presents angular momentum and spherical symmetry, and Chapter 4 covers spin angular momentum, fine structure, hyperfine structure, and magnetic resonance. Chapter 5 considers quantum information, after briefly introducing some concepts from classical information theory. This chapter also introduces the Einstein–Podolsky–Rosen paradox and the Bell's inequalities. The quantum dynamics and quantum correlations of two-level systems (including spin systems), three-level systems, and multi-level systems, as well as wave packet dynamics, and quantum optical control theory are the topics discussed in Chapter 6. The approximation methods described in Chapter 7 include basis-state expansions, semiclassical approximations time-independent and time-dependent perturbation theory, variational methods, sudden and adiabatic approximations, the Berry phase concept, and linear response theory. Chapter 8 on identical particles discusses exchange symmetry of bosons and fermions. Chapter 9 presents the electronic properties of solids, starting from the treatment of the free electron gas and electrons in a periodic potential and then describes metals, semiconductors, and insulators. In Chapter 10, mean-field theories to treat multi-electron systems such as atoms, molecules, and also condensed phase systems are introduced, including Hartree–Fock and configuration interaction, which goes beyond mean field. Some topics for describing molecules are introduced in Chapter 11, including point groups, the Born–Oppenheimer approximation, and the Franck–Condon principle. Scattering theory is presented in Chapter 12, including scattering in one dimension and two dimensions, and scattering in disordered systems. Chapter 13 introduces the quantum mechanics needed to treat low-dimensional systems such as quantum dots, wires, and wells, and other low-dimensional systems. Many-body theory is the topic of Chapter 14, including the basic formulation of second quantization, its application to statistical mechanics, and mean-field theory methods. Finally, density functional theory, the most widely used method for calculating ground-state electronic structure, is the topic of Chapter 15. Appendices on linear algebra and Dirac notation for vectors in Hilbert space (Chapter A), some simple ordinary differential equations required in our treatment of quantum mechanics (B), vector analysis (C), Fourier analysis (D), and group theory (E) are presented at the end of the book.

Because of space limitations, a number of chapters that were originally planned to be part of the book will not appear in the printed version but will be linked to the web page of the book (shortly after printing). Chp_16_Classical_Mechanics.pdf deals with classical mechanics (see above), Chapter_17_Decoherence_Dissipation.pdf considers decoherence and dissipation phenomena and covers the spin-boson model, the Caldeira–Leggett model, master equations, and more generally, the field of *open system dynamics*. Chp_18_Many_Body_Th_Applications.pdf presents field theory methods to treat Landau–Fermi liquid theory, superconductivity, Bose–Einstein condensation, superfluidity, the Hubard model, and the Kondo effect. Finally, Chp_19_Insulators.pdf gives additional detail regarding insulating materials. A list of errors and typos entitled QM_errors_typos.pdf will also be linked to the book web page.

ACKNOWLEDGMENTS

We are deeply grateful to the following people for comments and suggestions regarding the material in this book: Roi Baer, S. I. Ben-Avraham, Dmitry Budker, Doron Cohen, Claude Cohen-Tannoudji, Jean N. Fuchs, Konstantin Kikoin, Leeor Kronik, Pier Mello, William P. Reinhardt, Michael Revzen, Moshe Schechter, Piotr Szańkowski, Richard Tasgal, and Marek Trippenbach. They and the students in our classes (too numerous to mention) have helped make this a better book.

We are grateful to our families for their patience and understanding over the many years of writing. We dedicate this book to Eliezer, Sarah, Renee, Alisa, David, Miriam, and Sharon and also to Carmella, Tirza, Tomer, and Tuval.

Table 1 Possible courses based on *Quantum Mechanics with Applications to Nanotechnology and Quantum Information Science* and recommended book chapters.

	Undergrad Quantum Mechanics 1	Undergrad Quantum Mechanics 2	Quantum Information and Quantum Computing	Nano-technology	Engineering Quantum Mechanics	Quantum Chemistry (2 Semesters)	Condensed Matter (Solid State)	Graduate Quantum Mechanics
1 Intro. to Quantum Mechanics	x		x		x	x		
2 Formalism of Quantum Mechanics	x		x		x	x		
3 Angular Momentum and Spherical Symmetry	x		Sec. 4.1		x	x		
4 Spin	x		x		x	x		
5 Quantum Information			x					x
6 Quantum Dynamics and Correlations		Optional	x					x
7 Approximation Methods		x			x	x		
8 Identical Particles		x		Optional	x	x		
9 Electronic Properties of Solids		Optional		x	x		x	
10 Electronic Structure of Multi-Electron Systems		x		x	x	x	x	

(*Continued*)

	Undergrad Quantum Mechanics 1	Undergrad Quantum Mechanics 2	Quantum Information and Quantum Computing	Nano-technology	Engineering Quantum Mechanics	Quantum Chemistry (2 Semesters)	Condensed Matter (Solid State)	Graduate Quantum Mechanics
11 Molecules		x				x		
12 Scattering Theory				Optional				x
13 Low-Dimensional Systems			Optional	x			x	Optional
14 Many-Body Theory				Optional		Optional	x	x
15 Density Functional Theory				x		Optional	Optional	Optional
Mathematical Appendices	x					x		
16 Review of Classical Mechanics	x					x		
17 Decoherence and Dissipation			x				x	x
18 Field Theory Applications				Optional		Optional	x	x
19 Insulators				Optional			x	x

Introduction to Quantum Mechanics 1

As thou knowest not what is the way of the spirit, ..., even so thou knowest not the works of God who maketh all.
Ecclesiastes 11: 5.

Quantum mechanics determines the properties of physical systems such as atoms, molecules, condensed phase materials, light, etc. It developed out of the failure of classical mechanics, i.e., the failure of Newton's laws and classical electromagnetism, to properly describe such systems. The failure of classical mechanics is particularly acute for systems on the nanoscale, hence the critical need for quantum mechanics in nanotechnology. But even for such macroscopic objects as metal or semiconductor materials, classical mechanics fails in describing their electronic and physical properties because the properties of the electrons in these systems is not properly accounted for.

In this chapter, we shall take the first steps in our study of quantum mechanics. We will treat a number of physical phenomena that cannot be described classically. These phenomena clearly show that a theory other than classical mechanics is necessary for atomic and subatomic phenomena. We shall consider Energy Quantization, Blackbody Radiation, Wave–Particle Duality, Angular Momentum Quantization, Quantum Mechanical Tunneling, and Quantum Entanglement in Sec. 1.1. Then, in Sec. 1.2, we present a brief overview of nanotechnology and information science, and detail why quantum mechanics is so vital for these fields. Today we are able to manipulate matter, atom by atom, and sometimes even electron by electron. But this ability is rather recent. Although it was dreamed of as early as the late 1950s,[1] it is only in the last several decades that this dream has become a reality. Nanoscience and nanotechnology are the science and technology (and perhaps the art) of manipulating materials on an atomic and molecular scale. The hope is that nanoscience and nanotechnology will evolve to the point that we will be able to build submicroscopic size devices, and completely control the structure of matter with molecular precision, so as to build complex microscopic objects. Information technology is also entering a regime where quantum mechanics plays a role. By information technology we mean technology for managing and processing information. As computer memory and processor devices get smaller, quantum mechanics begins to play a role in their behavior. Moreover, serious consideration is being given to new types of information technology devices based upon quantum bits (quantum two-level systems) rather than normal bits (classical devices that can be in either of two states typically called "0" and "1"). Such devices are inherently quantum mechanical in their behavior. Although we are slowly improving our ability to manipulate matter at the atomic level, there is a lot of room for improvement. The better we understand quantum mechanics, the better will be our ability to advance nanoscience and nanotechnology. Section 1.3 introduces some of the most basic concepts of quantum mechanics, such as the superposition principle of quantum states, operators that act on quantum states, the nature of measurement in quantum mechanics, the concept of an entangled quantum state, and propagation of quantum states in time. Then we develop the solution to a few simple one-dimensional quantum problems, including a particle in a box, reflection and transmission, barrier penetration and 1D quantum tunneling, 1D bound states, resonance states, and the quantum harmonic oscillator.

The Appendices are meant to help bring readers up to the knowledge level in mathematics required for understanding quantum mechanics: appendices on linear algebra and Dirac notation for vectors in Hilbert space, some simple ordinary differential equations, vector analysis, Fourier analysis and group theory are provided. If you find yourself having

---

[1] See, e.g., Nobel prize winner Richard Feynman's 1959 lecture "There's Plenty of Room at the Bottom" [17], in which he said, "The principles of physics, as far as I can see, do not speak against the possibility of maneuvering things atom by atom. It is not an attempt to violate any laws; it is something, in principle, that can be done; but in practice, it has not been done because we are too big."

trouble with the mathematics used in the ensuing chapters, you should refer to the appendices, and to the references provided therein. Specifically, Appendix A on linear algebra and Dirac notation contains material that is directly relevant and intimately connected with the formulation of quantum mechanics and should be studied *before* beginning Sec. 1.3 which presents some of the main concepts of quantum mechanics and Chapter 2 which presents the formalism of quantum mechanics. Readers without any background in probability theory should consult a source containing at least the rudiments of probability theory [12–16] *before* beginning Sec. 1.3.

Let us begin.

1.1 WHAT IS QUANTUM MECHANICS?

Classical mechanics is an excellent approximation to describe phenomena involving systems with large masses and systems that are not confined to very small volumes (e.g., a rock thrown in the earth's gravitational field, a system of planets orbiting around a sun, a spinning top, or a heavy charged ion in an electrical potential). However, it fails totally at the atomic level. Quantum mechanics is the only theory that properly describes atomic and subatomic phenomena; it and only it explains why an atom or molecule, or even a solid body, can exist, and it allows us to determine the properties of such systems. Quantum mechanics allows us to predict and understand the structure of atoms and molecules, atomic-level structure of bulk crystals and interfaces, equations of state, phase diagrams of materials and the nature of phase transitions, melting points, elastic moduli, defect formation energies, tensile and shear strengths of materials, fracture energies, phonon spectra (i.e., the vibrational frequencies of condensed phase materials), specific heats of materials, thermal expansion coefficients, thermal conductivities, electrical conductivities and conductances, magnetic properties, surface energies, diffusion and reaction energetics, etc.

At around the turn of the twentieth century, it became clear that the laws of classical physics were incapable of describing atoms and molecules. Moreover, classical laws could not properly treat light fields emanating from the sun or from a red-hot piece of metal. The laws of quantum mechanics were put on firm footing in the late 1920s after a quarter of a century of great turmoil in which an *ad hoc* set of hypotheses were added to classical mechanics in an attempt to patch it up so it can describe systems that are inherently quantum in nature. We shall review the nature of the crisis that developed in science at around the turn of the twentieth century in some detail to better understand the need for quantum mechanics, i.e., the need to replace classical mechanics.

1.1.1 A BRIEF EARLY HISTORY OF QUANTUM MECHANICS

The history of the early discoveries that led to the development of quantum mechanics, and some of its early successes, is summarized in Table 1.1. We shall discuss these discoveries in the beginning sections of this chapter, and throughout this book. A rapid growth in the number of discoveries of quantum phenomena began in the mid 1930s, and continues to this day.

1.1.2 ENERGY QUANTIZATION

In classical mechanics, a mechanical system can be in a state of every possible energy, with the proviso that the energy is bounded from below by the minimum of the potential. Not so in quantum mechanics; only specific bound state energies exist. Let us take the hydrogen atom as an example. The spectrum of the light emitted by an excited hydrogen atom is shown in Fig. 1.1.[2] As we shall see shortly, light can be described as being made up of particles called photons, and light of frequency ν is made up of photons with energy $E = h\nu$, where h is a dimensional constant called the *Planck constant*

---

[2] The spectrum in the figure is plotted in terms of wavelength λ rather than frequency. The wavelength and frequency of a photon are related by the relation $\nu = c/\lambda$, where c is the speed of light in vacuum, $c = 2.99792458 \times 10^8$ m/s.

1.1 What is Quantum Mechanics?

Table 1.1 *The early history of quantum mechanics.*

Year	Discoverer	Discovery
1888	Heinrich Rudolf Hertz	Observation of the photoelectric effect
1896	Henrik A. Lorentz Pieter Zeeman	Explanation of normal Zeeman effect (splitting of spectral lines in a magnetic field)
1896	Antoine-Henri Becquerel	Discovery of penetrating radiation (radioactivity, α rays)
1897	Joseph John Thomson	Discovery of electrons
1900	Max Planck	Blackbody radiation law
1905	Albert Einstein	Explanation of the photoelectric effect
1905	Albert Einstein	Explanation of Brownian motion
1910	Max von Laue, William L. Bragg William H. Bragg	Diffraction of x-rays from crystals; x-ray spectrometer developed
1913	Niels H. D. Bohr	Bohr semiclassical theory of the quantization of energy levels
1914	James Franck Gustav Hertz	Franck–Hertz experiment showed quantized atomic energy levels
1922	Otto Stern Walther Gerlach	Demonstration of atomic magnetic moments that give rise to magnetic phenomena
1923	Arthur H. Compton	Compton Effect (scattering of photons by free electrons)
1924	Louis-Victor de Broglie	Wave–particle duality
1924	Wolfgang Pauli	Postulated the existence of spin angular momentum
1925	Samuel Goudsmit George Uhlenbeck	Postulated that electrons have spin angular momentum
1925	W. Heisenberg	Developed the matrix form of quantum mechanics
1926	Erwin Schrödinger	Developed the wave equation for matter, i.e., the Schrödinger equation
1925	Wolfgang Pauli	Pauli exclusion principle for electrons
1926	Max Born	Statistical interpretation of wave mechanics
1926	Llewellyn H. Thomas	Thomas precession factor in the spin–orbit Hamiltonian
1926	Enrico Fermi	Fermi statistics and Fermi distribution of fermions

(Continued)

Table 1.1 (*Continued*)

Year	Discoverer	Discovery
1927	Clinton Davisson, Lester Germer	Electron matter-waves diffracting off crystals, confirming existence of de Broglie waves
1927	W. Heisenberg	Formulation of the uncertainty principle
1927	Paul A. M. Dirac	Quantum theory of radiation
1927	Robert Mulliken, Friedrich Hund	Development of molecular orbital theory
1928	Paul A. M. Dirac	Relativistic theory of the electron – Dirac equation
1928	George Gamow	Decay of nuclei via quantum tunneling of α particles proposed
1928	Chandrasekhara Venkata Raman	Raman scattering, i.e., inelastic light scattering observed in liquids
1929	W. Heisenberg, Wolfgang Pauli	General quantum theory of fields
1930	Douglas Hartree, Vladimir Fock	The Hartree–Fock method (mean-field approximation to many-body quantum mechanics)
1931	Maria Göppert-Mayer	Calculation of two-photon absorption cross section
1932	John von Neumann	Quantum theory put into operator form
1932	James Chadwick	Discovery of the neutron
1934	Enrico Fermi	Theory of weak interactions (beta decay)
1935	Albert Einstein, Boris Podolsky, Nathan Rosen	Formulation of the EPR paradox

with units J s in SI (International System of Units), $h = 6.62606878 \times 10^{-34}$ J s. The energy of a photon emitted in the decay of a hydrogenic state of energy E_i to a state of lower energy E_f is equal to the energy difference $E_i - E_f$,

$$h\nu = E_i - E_f. \tag{1.1}$$

Figure 1.1 shows a *discrete* spectrum, i.e., it is composed of well-defined frequencies. Hence, energies of the hydrogen atom are discrete. This discrete nature of the energies of an electron around a proton is not understandable from a classical mechanics perspective, wherein states of the hydrogen atom should be able to take on all possible energy values. The "quantization" of the observed energies as determined from the emission spectrum just doesn't make sense from a classical mechanics point of view. This situation of discrete energies exists not only for hydrogen atoms, but for all atoms and molecules, and in fact for all bound states of quantum systems. This said, we further note that a continuum of energies is possible for unbound states (in the case of the hydrogen atom, these correspond to states where the electron is not bound to the proton – they are scattering states with positive energy, as opposed to the bound states that have negative energy relative to a proton and an electron at rest and infinitely separated in distance.

1.1 What is Quantum Mechanics?

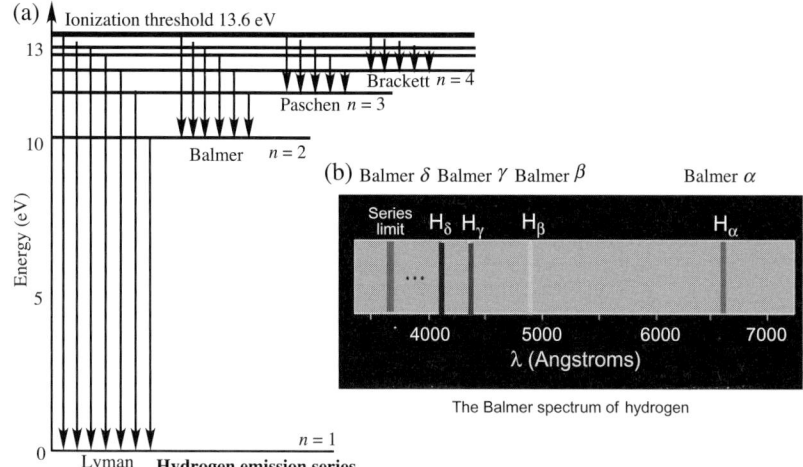

FIG 1.1 The emission spectrum of hydrogen. (a) Lyman, Balmer, Paschen, and Brackett transitions in hydrogen. (b) Two views of the Balmer spectrum in hydrogen versus wavelength. The $n' \to n$ transitions, where n, n' are principal quantum numbers, are called "Balmer" when $n = 2$. Balmer lines with $n' = 3$ are called α, β for $n' = 4$, γ for $n' = 5$, etc. The H in H$_\alpha$, H$_\beta$, etc., stands for hydrogen. (For the color version of this figure, please refer the color plate section at the end of book.)

For bound states, the potential energy of the electron is negative and larger in magnitude than the kinetic energy of the electron. In general, a quantum system has both a discrete set of bound states and a continuous set of unbound states.

As an aside, we note that sometimes the frequency of a photon is given as the angular frequency ω in units of radians per second, which is related to the frequency ν by the relation, $\omega = 2\pi\nu$, i.e., ω [rad/s] $= 2\pi \times \nu$ [cycles/s]. The conservation of energy condition (1.1) for photoemission, $h\nu = E_i - E_f$, can therefore be written as $\hbar\omega = E_i - E_f$, where $\hbar$ (pronounced "h-bar") is Planck's constant divided by 2π, $\hbar \equiv \frac{h}{2\pi} = 1.0545716 \times 10^{-34}$ J s.

The energies of the bound states of a hydrogen atom are given (to an excellent approximation) by the formula $E_n = -\alpha^2 (m_e c^2) \frac{1}{2n^2}$, where the dimensionless constant $\alpha = 1/137.03599976 = 7.297352533 \times 10^{-3}$ is called the *fine structure constant*, m_e is the mass of the electron, and n is an integer, $n = 1, 2, \ldots$, called the *principal quantum number*. The fine structure constant is given in terms of the electron charge $(-e)$, $\hbar$, and the speed of light, c, by $\alpha \equiv \frac{e^2}{4\pi\hbar c}$ in SI units ($\alpha \equiv \frac{e^2}{\hbar c}$ in Gaussian units – see Sec. 3.2.6 for a full discussion of atomic units). It is a small number, since the strength of the electromagnetic interaction is small. The product of α^2 and the rest mass energy $m_e c^2$ (recall the famous Einstein formula $E = mc^2$ for the rest mass energy of a particle) sets the scale of the hydrogen atom energies. The lowest energy of a hydrogen atom is obtained with $n = 1$, $E_1 = -\alpha^2 m_e c^2/2$, and bound states exist for every integer value of n (we shall consider the hydrogen atom in detail in Sec. 3.2.6 – here, simply note that bound states of a hydrogen atom exist only at very special values of energy).

The quantized nature of atomic states was a complete puzzle at the turn of the twentieth century. After Rutherford proposed a model of the atom wherein electrons orbit an atomic nucleus like planets round the Sun in 1911, he assigned his graduate student Neils Bohr the task of explaining the empirical spectral behavior being studied by others with his nuclear model. Bohr combined Einstein's idea of photons that were used to explain the photoelectric effect (1905) (see Sec. 1.1.7) and Balmer's empirical formula for the spectra of atoms (1885) to produce a revolutionary quantum theory of atomic energy levels. Bohr's theory (1913) began with two assumptions: (1) There exist stationary orbits for electrons orbiting the nucleus and the electrons in these orbits do not radiate energy. Electrons do not spiral into the nucleus (i.e., do not lose energy E via photoemission, as would be predicted by the Larmor formula, $dE/dt = -(2e^2/3c^3)|d\nu/dt|^2$, which says that the energy loss rate, dE/dt, is proportional to the square of the acceleration) because they have quantized angular momentum. (2) Electrons can gain (lose) energy upon absorption (emission) of a photon, thereby going to another orbit with higher (lower) energy. This energy change is quantized according to Planck's relationship $h\nu = E_i - E_f$.

In 1914, James Franck and Gustav Hertz performed an experiment that conclusively demonstrated the existence of quantized excited states in mercury atoms, thereby helping to confirm the Bohr quantum theory developed a year earlier. Electrons were accelerated by a voltage toward a positively charged grid in a glass tube filled with mercury vapor. Behind the grid was a collection plate held at a small negative voltage with respect to the grid. When the accelerating voltage provided enough energy to the free electrons, inelastic collisions of these electrons with an atom in the vapor could force it into an excited state, with a concomitant energy loss of the free electron equal to the excitation energy of the atom. A series of dips in the measured current at constant volt increments (of 4.9 volts) showed that a specific amount of energy (4.9 eV) was being lost by the electrons and imparted to the atoms. Franck and Hertz won the Nobel Prize in 1925 for proving that energies of atomic states are quantized.

1.1.3 WAVES, LIGHT, AND BLACKBODY RADIATION

Isaac Newton thought light consisted of particles. These particles could bounce back upon reflection from a mirror or a pool of water. But it became clear from the work of Christian Huygens (the Huygens principle – 1670), Leonhard Euler (wave theory used to predict construction of achromatic lenses), Thomas Young (principle of interference[3] – 1801), Augustin Jean Fresnel (partial refraction and reflection from interface – 1801), and Josef Fraunhofer (diffraction[4] gratings – 1801) among many others, that light behaves as a wave and shows *interference* and *diffraction* phenomena. Optics is integrated into electromagnetic theory, which is a wave theory. The wave equation in vacuum for electromagnetic fields, i.e., electric fields $\mathbf{E}(\mathbf{r},t)$ and magnetic fields $\mathbf{H}(\mathbf{r},t)$, is given by

$$\left(\nabla^2 - \frac{1}{c^2}\frac{\partial^2}{\partial t^2}\right)\mathbf{E}(\mathbf{r},t) = 0, \tag{1.2}$$

with an identical equation for the magnetic field $\mathbf{H}(\mathbf{r},t)$. (Readers not comfortable with the Laplacian operator, ∇^2, or differential operators in general, please see Appendix C). These wave equations describe all propagation phenomena for light in vacuum [18].

Solutions to (1.2) can be formed from plane waves,

$$\mathbf{E}(\mathbf{r},t) = \mathbf{E}_{\mathbf{k},\omega} e^{i(\mathbf{k}\cdot\mathbf{r}-\omega t)}, \tag{1.3}$$

which are solutions to (1.2) as long as $k^2 = \omega^2/c^2$, for any vector amplitude $\mathbf{E}_{\mathbf{k},\omega}$, as can be easily verified by substituting (1.3) into Eq. (1.2). Any superposition (i.e., linear combination) of these solutions is also a solution (just as a superposition of water waves in a lake that originate from two people throwing a stone into the lake co-exist, and propagate through one another), since the wave equation (1.2) is a linear equation. The waves in (1.3) are called plane waves, because their wave fronts (the surface of points in physical space having the same phase) are planes perpendicular to the vector $\mathbf{k}$. The vector $\mathbf{k}$ is called the *wave vector* and the relation $k^2 = \omega^2/c^2$, which is required for (1.3) to be a solution to (1.2), is called the dispersion relation; it relates the photon momentum $\hbar\mathbf{k}$ to the photon energy $\hbar\omega$ (as discussed below).

Blackbody radiation, the electromagnetic radiation of a body that absorbs all radiation that impinges upon it (and therefore looks black at very low temperatures) could not be explained by electromagnetic theory at the turn of the twentieth century. When matter is in thermal equilibrium with the electromagnetic radiation surrounding it, the radiation emitted by the body is completely determined in terms of the temperature of the body. Such matter is called a blackbody, and therefore the radiation is called blackbody radiation. In order to explain the spectrum of blackbody radiation, Max Planck suggested the hypothesis of the quantization of energy (1900): for an electromagnetic wave of angular frequency ω, the energy of the radiation was taken to be proportional to $\hbar\omega$. Einstein generalized this hypothesis to obtain a particle picture of electromagnetic radiation (1905): light consists of a beam of photons, each possessing an energy $\hbar\omega$, and the energy density is given by the product of the density of photons of angular frequency ω times $\hbar\omega$. Einstein then showed how the introduction of photons made it possible to understand the unexplained characteristics of the photoelectric effect (see below). In 1923, Arthur Holly Compton showed that photons actually exist by discovering the Compton effect

---

[3] *Interference* is the addition of two or more waves that results in a new wave pattern. The new wave, which also satisfies the wave equation, is the superposition (see Appendix A.1.1) of the component waves. The superposition can be constructive, meaning the amplitude of the superposition is larger than that of its individual components, or destructive, in which case their is cancellation of the component waves.
[4] *Diffraction* is the change in the direction and intensity of a wave after passing an obstacle or an aperture.

1.1 What is Quantum Mechanics?

(the scattering of photons by free electrons). The scattering behavior showed that a photon had an energy $E = \hbar\omega$ and a momentum $\mathbf{p} = (\hbar\omega/c)\mathbf{u}$, where $\mathbf{u}$ is a unit vector in the direction of the propagation of the photon (for comparison with matter-waves, see the discussion of de Broglie waves in the next section). The *dispersion relation* (i.e., the relation between energy and momentum) for photons can therefore be written as linear relation,

$$E = pc. \tag{1.4}$$

The blackbody radiation law was developed by Max Planck in 1900. Let $u_\omega(T)d\omega$ be the mean energy per unit volume in the frequency range between ω and $\omega + d\omega$. Planck's blackbody radiation law can be written as follows:

$$u_\omega(T) = \frac{\hbar}{\pi^2 c^3} \left(\frac{k_B T}{\hbar}\right)^3 \frac{x^3}{e^x - 1}. \tag{1.5}$$

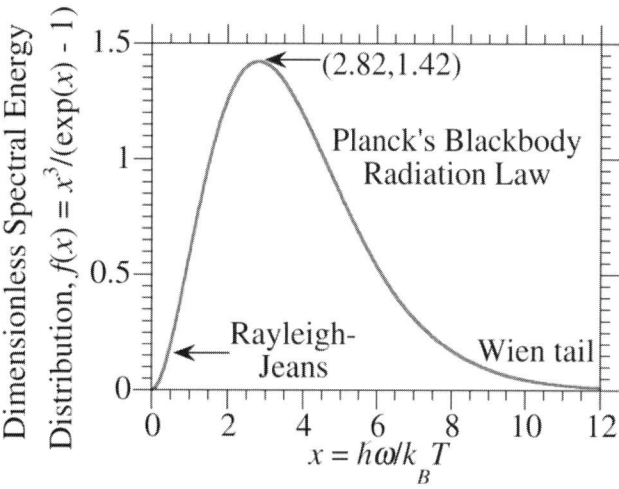

FIG 1.2 Planck's blackbody radiation law, $f(x) = x^3/(e^x - 1)$, where the dimensionless variable $x \equiv \hbar\omega/k_B T$.

Here x is a dimensionless variable, $x \equiv \hbar\omega/k_B T$, and k_B is the Boltzmann constant, $k_B = 1.3806503 \times 10^{-23}$ J K^{-1}. The blackbody radiation law states that the energy density per unit frequency is a universal function of one dimensionless parameter x, $f(x) = x^3/(e^x - 1)$. The factor $(e^x - 1)^{-1}$ appearing on the RHS of Eq. (1.5) arises due to the Bose–Einstein distribution for integer-spin particles (in this case, photons). A plot of the function $f(x) = x^3/(e^x - 1)$ versus x that appears on the RHS of Eq. (1.5) is shown in Fig. 1.2. The low frequency ($x \ll 1$) dependence of the Planck blackbody energy density goes as ω^2, as is clear from expanding the exponent in the denominator of Eq. (1.5), whereas the high frequency dependence is exponential, $u_\omega(T) \sim \exp(-x)$ for $x \gg 1$, as you will show in Problem 1.1. The high frequency behavior (the exponential tail) is called the *Wien tail*, and was understood in terms of statistical mechanics (the Boltzmann distribution). The low frequency limit of the blackbody spectrum, $u_\omega(T) \sim x^2$ for $x \ll 1$, is known as the *Rayleigh–Jeans limit*. The connection between the high and low frequency limits was not understood until the ideas of Planck and Einstein were introduced at the turn of the twentieth century. The frequency at which $u_\omega(T)$ is a maximum can be easily determined by setting the derivative $du_\omega(T)/d\omega$ equal to zero, and is numerically determined to be $\omega_{\max} \approx 2.82\, k_B T/\hbar$. This result is sometimes called Wien's formula or the Wien displacement law. You will derive the Planck blackbody radiation law, Eq. (1.5), in Chapter 9, where you will multiply the following three factors to obtain $u_\omega(T)$: the energy of a photon of angular frequency ω, $\hbar\omega$, the thermal occupation probability given by the Bose–Einstein distribution, $(e^x - 1)^{-1}$, and the density of photon states per unit energy per unit volume, which is proportional to ω^2.

Problem 1.1

Expand Eq. (1.5) for $x \ll 1$ and $x \gg 1$ to derive the Rayleigh–Jeans limit and the Wien tail.

The total energy density in all frequencies, $u(T)$, is given by integrating Eq. (1.5) over frequency (i.e., over $xk_B T/\hbar$). The integral can be evaluated analytically and one finds, $u(T) = 4\sigma T^4/c$, where the constant σ is given by $\sigma = \frac{\pi^2 k_B^4}{60\hbar^3 c^2} = 5.6697 \times 10^{-5}$ erg s^{-1} cm^{-2} K^{-4}, and is called the *Stefan–Boltzmann constant*. The T^4 temperature dependence of the total energy emitted and the numerical value agree with experimental observations. The energy per unit time per unit area, i.e., the intensity, emitted by the blackbody is given by $I(T) = cu(T)/4 = \sigma T^4$. Note that Planck's constant is an integral part of the discussion of blackbody radiation. Even the Stefan–Boltzmann constant involves $\hbar$, hence, arguably, involves quantum mechanics.

1.1.4 WAVE–PARTICLE DUALITY

In the previous section we discussed the fact that light can have both particle and wave properties. Figure 1.3(a) shows a schematic representation of an experiment to look for the interference in the intensity of the light on a screen, a distance d behind an opaque wall with two narrow slits cut into it is obtained when a monochromatic plane wave light field with angular frequency ω impinges upon the opaque wall. Such an experiment was first carried out with light by Thomas Young in 1801. It seemed amazing at the start of the quantum era that mono-energetic particles show the same type of interference pattern, but now it is well known that matter-waves (i.e., particles with mass, which also behave as waves) can experience interference and diffraction just like light. We shall introduce these two concepts in this section. In what follows, we consider only a scalar field (electric field of light is a vector field), as is the case for (spinless) matter-waves.

Waves in 3D emanating from a point have intensities that fall off as the inverse of the distance squared from the source, $1/r^2$ (since the surface area of a sphere is $4\pi r^2$ and the integrated intensity is constant on the surface of a sphere, no matter its size). In the Young double-slit experiment, each point along each slit serves as a source for light. The fields from the two slits are to be added together coherently, and then the resulting field is squared to obtain the intensity. The resulting intensity pattern has interference fringe patterns as is shown in Fig. 1.3(c).

Let us first consider the field emanating from two holes located at positions $(0, 0, \pm a/2)$ before considering the Young double-slit experiment in which light emanates from all the points along the slits $\{(0, y, \pm a/2)\}$. The intensity at a point $(d, 0, z)$ on the screen as a function of wall-to-screen distance d, the distance between the holes a, and the z coordinate, $I(d, z)$, is given by:

$$I(d, z) = C \left| \frac{e^{i\mathbf{k}_1 \cdot \mathbf{r}_1}}{r_1} + \frac{e^{i\mathbf{k}_2 \cdot \mathbf{r}_2}}{r_2} \right|^2. \quad (1.6)$$

Here C is a constant; since we calculate only the relative intensity, we do not need to determine C. The length of the vectors $\mathbf{k}_1$ and $\mathbf{k}_2$ is $k = \omega/c$ (recall that the wavelength–wavenumber relation is $\lambda = 2\pi/k$), and the wave vector $\mathbf{k}_i$ points from the ith hole to the point $(d, 0, z)$ on the screen, as shown in Fig. 1.3(a) with y taken to be zero because now we are considering the case of two holes located at $(0, 0, \pm a/2)$. The distances r_1 and r_2 are the distances from the holes to the point $(d, 0, z)$ on the screen, i.e., $r_1 = \sqrt{d^2 + (z - a/2)^2}$, $r_2 = \sqrt{d^2 + (z + a/2)^2}$. In order to evaluate the right hand side (RHS) of Eq. (1.6), the arguments of the exponentials must be determined, and they are simply $\mathbf{k}_1 \cdot \mathbf{r}_1 = kr_1$ and $\mathbf{k}_2 \cdot \mathbf{r}_2 = kr_2$. The intensity falls off with d as $1/d^2$ at large d. Figure 1.3(b) shows the interference pattern obtained in the calculated relative intensity at the screen as a function of z for two values of the distance between the slits, $a = 100$ and 533 µm, when $d = 1000$ µm and $kd = 100$. For $a = 100$, only a few interference fringes are seen, but for $a = 533$, a dense pattern is obtained. The intensity fall off with increasing z as can be seen clearly for $|z| > a$. If we neglect the difference between r_1 and r_2 in the denominators on the RHS of Eq. (1.6), and also in the exponentials, the intensity can be written in the asymptotic form $I(d, z) = \frac{2C}{d^2 + z^2} [1 + \cos((\mathbf{k}_2 - \mathbf{k}_1) \cdot \mathbf{r})]$. Expanding the argument of the cosine for large d, i.e., $d \gg a, z$, the relative intensity can be written as $I(d, z) = \frac{2C}{d^2} [1 + \cos(ak \sin \theta)]$, where $\sin \theta \approx z/d$.

Problem 1.2

Consider two fields emanating from points $i = 1, 2$ in Fig. 1.3(a) with $y = 0$, having field amplitudes, $E_i(\mathbf{r}, t) = E_0 \frac{e^{i\mathbf{k}_i \cdot \mathbf{r}_i}}{r_i} e^{-i\omega t}$, where the wave vectors $\mathbf{k}_i$ are in the direction from the points to $(d, 0, z)$ on the screen.

(a) Show that $I(d, z) = \frac{2C}{d^2} [1 + \cos(ak \sin \theta)]$ results from Eq. (1.6) when the screen is far away from the opaque wall, $d \gg a, z$, where $\sin \theta \approx z/d$.
(b) From the form of the intensity in (a), find the angles θ for which the intensity vanish.

1.1 What is Quantum Mechanics?

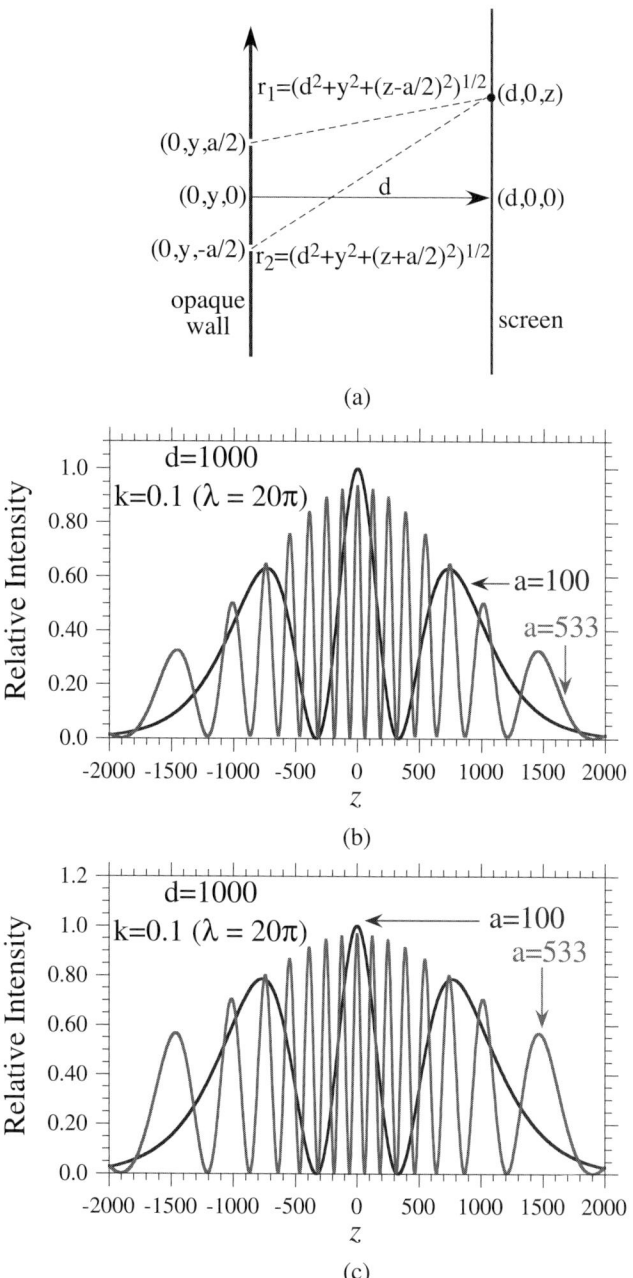

FIG 1.3 Interference pattern in a Young double-slit experiment. (a) Schematic of the geometry of the double-slit setup. (b) Relative intensity at the screen from two holes in the opaque wall situated at $(0, 0, \pm a/2)$ versus z for two values of a. (c) Relative intensity at the screen from two slits in the opaque wall versus z for two values of a.

Let us now return to the Young's double-slit case. We can think of the slits as a continuous set of holes at $(0, y, \pm a/2)$. The distances r_1 and r_2 are given by $r_1 = \sqrt{d^2 + y^2 + (z - a/2)^2}$, $r_2 = \sqrt{d^2 + y^2 + (z + a/2)^2}$, and the electric field arising at point $(d, 0, z)$ is given by the integral over y of the electric fields from each of the "holes" running along the slits. Hence, the intensity is given by

$$I(d, z) = C \left| \int_{-\infty}^{\infty} dy \left(\frac{e^{ik_1 r_1}}{r_1} + \frac{e^{ik_2 r_2}}{r_2} \right) \right|^2 \tag{1.7}$$

$$= C\pi^2 \left| H_0^{(1)}\left(k\sqrt{d^2 + (z - \frac{a}{2})^2}\right) + H_0^{(1)}\left(k\sqrt{d^2 + (z + \frac{a}{2})^2}\right) \right|^2,$$

where $H_0^{(1)}$ is the first Hankel function of order zero (discussed in Appendix B). Figure 1.3(c) shows the intensity pattern at the screen as a function of z for two values of the distance between the slits, $a = 100$ and 533. The intensity fall off with increasing z is now somewhat slower than in Fig. 1.3(b). Asymptotically, $H_n^{(1)}(\zeta) \xrightarrow[\zeta \to \infty]{} \sqrt{\frac{2}{\pi z}} e^{i(\zeta - n\pi/2 - \pi/4)}$, and therefore the asymptotic intensity pattern for large d is given by $I(z, d) = \frac{4C}{\pi k d} [1 + \cos(ak \sin \theta)]$, where $\sin \theta \approx z/d$. The $1/d$ intensity dependence (rather than $1/d^2$ as above) arises because the circumference of a circle of radius d is $2\pi d$. Otherwise, the Young double-slit case is not all that different from the double hole case discussed above.

Another wave property of light is seen in the diffraction pattern of the light intensity from a single finite-width slit (i.e., cover one of the slits in the Young double-slit experiment so the light can go through only one of the slits and look at the intensity on a screen sufficiently far behind the opaque wall containing the single-slit). The intensity on the screen a distance d behind the wall versus position x takes the form

$$I(x, d) = I_0 \frac{kL_x}{4\pi d} \left(\frac{\sin(kL_x x/2d)}{(kL_x x/2d)^2} \right)^2, \tag{1.8}$$

where again $k = \omega/c$ is the wave vector of the light, d is the distance from the opaque wall to the screen, and L_x is the width of the slit. The diffraction pattern from a rectangle of dimensions L_x and L_y cut into the opaque wall is given by the expression

$$I(x, y, d) = I_0 \frac{k^2 L_x L_y}{16\pi^2 d^2} \left(\frac{\sin(kL_x x/2d)}{(kL_x x/2d)} \right)^2 \left(\frac{\sin^2(kL_y y/2d)}{(kL_y y/2d)^2} \right)^2.$$

The intensity pattern from a square aperture, $L_x = L_y$, is shown in Fig. 1.4(a). One can also consider a hole of circular aperture cut into the wall; Figure 1.4(b) shows the interference pattern from a circular aperture of radius a (i.e., a round pinhole),

$$I(r, d) = I_0 \frac{a^2 k^2}{16\pi^2 d^2} \left(\frac{2J_1(kar/d)}{(kar/d)} \right)^2,$$

where $J_1(\cdot)$ is the Bessel function of the first kind of order unity. Again, a distinctive diffraction pattern is obtained. We conclude from the interference and diffraction patterns in Figs 1.3 and 1.4 and the discussion of the previous section that light propagation phenomena can be fully understood only by considering *both* the wave and the particle aspects of light. This seems to be a paradox, but the paradox is resolved in terms of a fully quantum theory of light.

Today we know that beams of mono-energetic particles having mass (e.g., electrons, atoms, C_{60} molecules, etc.) also show interference and diffraction phenomena, just like light. In 1924, Louis de Broglie, as part of his Ph.D. thesis, reconsidered the theory associated with Compton scattering experiments and hypothesized that matter (specifically electrons, but quite generally any matter), could exhibit wave properties. At that time, this "matter-wave" hypothesis sent a shock wave through the scientific community. de Broglie showed that mono-energetic particles having a momentum $\mathbf{p} = m\mathbf{v}$

1.1 What is Quantum Mechanics?

propagating in the direction $\hat{\mathbf{u}}$ should behave as a wave with wavelength λ specified by the relation

$$\boxed{\mathbf{p} = \frac{2\pi\hbar}{\lambda}\hat{\mathbf{u}},} \qquad (1.9)$$

where λ is the wavelength of the matter-wave. This hypothesis justified Bohr's assumption, made in 1913, that electrons maintained stable orbitals at special designated radii and did not spiral into the nucleus because they had quantized angular momentum (see next section). In 1927, Davisson and Germer designed an experiment to measure the energies of electrons scattered from a metal surface. Electrons were accelerated by a voltage drop and allowed to strike the surface of nickel metal. The electron beam scattered off the metal according to the Bragg's law of scattering (see paragraph below) that was already known for scattering of photons off crystals. This confirmed de Broglie's matter wave hypothesis. In 1929, de Broglie was awarded the Nobel Prize in Physics for his discovery of the wave nature of electrons. So, we conclude that both particles and light show "wave–particle duality".

Before concluding this section, we mention one more important wave phenomenon, this time in the context of wave scattering of light off a periodic potential. The structure of crystals is studied by x-ray diffraction, a technique first developed by Max von Laue, William L. Bragg, and (his father) William H. Bragg, who developed the first x-ray spectrometer around 1910. Figure 1.5(a) shows a schematic diagram of an x-ray spectrometer for investigating crystal structure. The condition for Bragg scattering is that the path length difference between waves that scattered

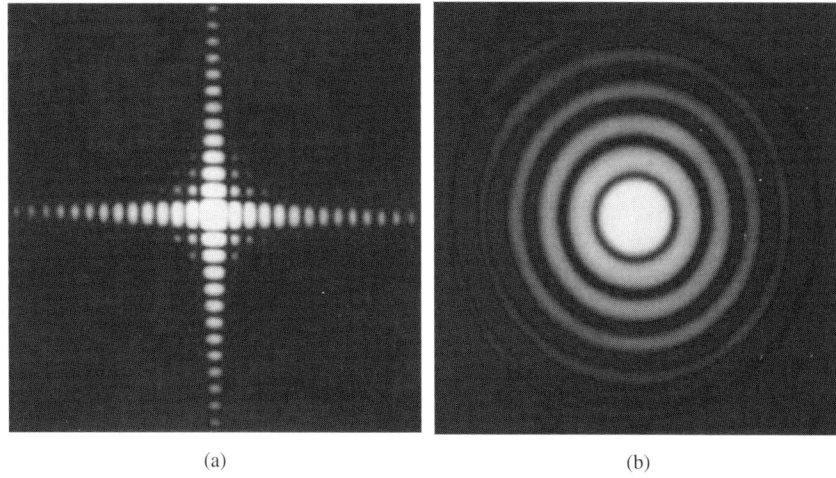

FIG 1.4 (a) Diffraction pattern from a square aperture in an opaque wall. (b) Diffraction pattern for a circular aperture. The intensity patterns are measured on a screen sufficiently far behind the wall.

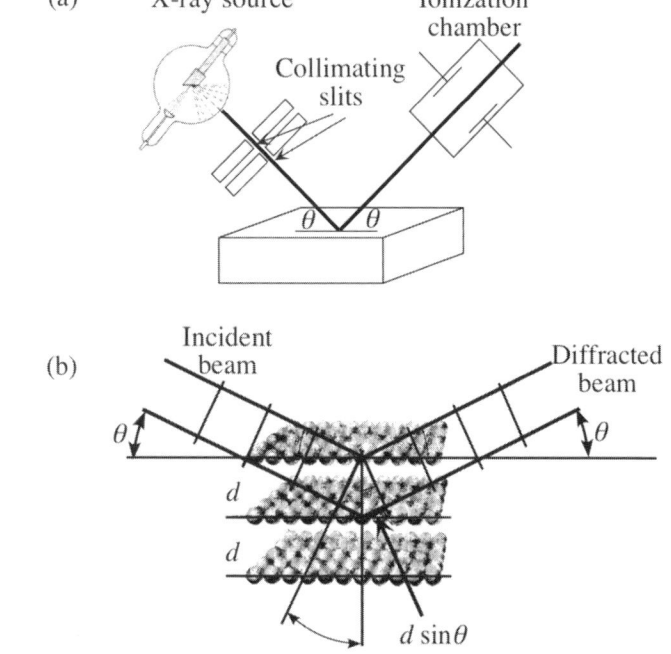

FIG 1.5 (a) Diagram of an x-ray spectrometer for investigating crystal structure. (b) Spectral reflection by Bragg scattering. (Adapted from Pauling, *General Chemistry* [19].)

off two different atomic planes of atoms separated by a distance d in the crystal, $2d \sin \theta$ [see Fig. 1.5(b)] equal an integral multiple of the x-ray wavelength λ,

$$2d \sin \theta = n\lambda, \tag{1.10}$$

or $\mathbf{k} \cdot \mathbf{d} = n\pi$, where $\mathbf{k} = \mathbf{p}/\hbar = (2\pi/\lambda)\,\hat{\mathbf{u}}$. Then constructive interference of these waves result. The reason x-rays are used is that the wavelength λ of the light should be comparable to d so that condition (1.10) can be satisfied for small integers n. This type of scattering off periodic structures is called *Bragg scattering*. It is the basis for much of our understanding of crystal structure in solid-state physics. Not only x-rays Bragg scatter off crystals, but any wave with a wavelength comparable to the crystal period as long as the wave interacts with the crystal, e.g., high energy electron beams (with de Broglie wavelengths on the order of d), as we have seen in the previous paragraph.

1.1.5 ANGULAR MOMENTUM QUANTIZATION

Otto Stern and Walther Gerlach carried out an experiment in 1922 that showed that spin angular momentum is quantized. This experiment is now known as the Stern–Gerlach experiment, and is depicted schematically in Fig. 1.6. They passed silver atoms,[5] through an inhomogeneous magnetic field, and the resulting force deflected the atoms into two opposite directions. We now understand the fact that the deflection pattern had two oppositely displaced components (and no component which had zero displacement), as indicating a half-integer angular momentum (in units of $\hbar$, see below) of the silver atoms. The deflection depends on the projection of the magnetic moment (which is proportional to the angular momentum) of the atom on the magnetic field axis, and only two projections are possible for spin 1/2 (projection 1/2 and $-1/2$).

A particle that possesses a nonvanishing angular momentum, also has a nonvanishing magnetic moment, $\boldsymbol{\mu}$, and the magnetic energy U_{mag} of such a particle in a magnetic field $\mathbf{H}$ is given by[6]

$$U_{\text{mag}} = -\boldsymbol{\mu} \cdot \mathbf{H}. \tag{1.11}$$

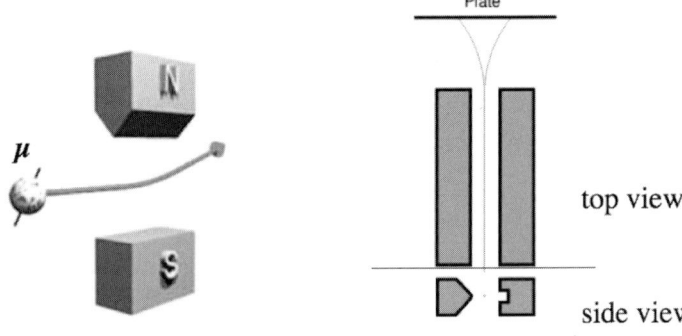

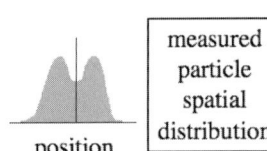

FIG 1.6 The Stern–Gerlach experiment. A beam of particles with magnetic moment $\boldsymbol{\mu}$ passes through an inhomogeneous magnetic field. A force on the particles results. Particles in different spin states experience different forces. For spin 1/2 particles, a bimodal distribution of particle deflections is observed.

---

[5] The ground state electronic configuration of silver atoms is, $\ldots 5s\ ^2S_{1/2}$, meaning that a silver atom has zero orbital angular momentum but a total angular momentum of 1/2.
[6] Sometimes, Eqs (1.11) and (1.12) are written in terms of the magnetic induction field $\mathbf{B}$. In Gaussian units, the $\mathbf{B}$ and $\mathbf{H}$ fields are equal in vacuum (see Table 3.2), but in SI units, they have different units and magnitudes since $\mathbf{B} = \mu_0 \mathbf{H}$, where $\mu_0 = 4\pi \times 10^{-7}$ N A^{-2} is the permeability of free space.

1.1 What is Quantum Mechanics?

The force on the particle is simply given by the gradient of this potential energy by

$$\mathbf{F}_{\text{mag}} = -\nabla U_{\text{mag}} = \nabla(\boldsymbol{\mu} \cdot \mathbf{H}). \qquad (1.12)$$

The magnetic moment of a particle is known to be proportional to its angular momentum $\mathbf{J}$,

$$\boldsymbol{\mu} \propto \mathbf{J}. \qquad (1.13)$$

Equation (1.12) states that when the magnetic field $\mathbf{H}(\mathbf{r})$ is inhomogeneous, the particle experiences a force. For example, consider a magnetic field with a z component that depends upon z, $(\mathbf{H}(\mathbf{r}))_z = H_z(z)$. The magnetic force in the z direction is given by $\mathbf{F}_{\text{mag},z} = \mu_z \frac{\partial H_z(z)}{\partial z}$. In classical mechanics, all possible angles θ between the vectors $\boldsymbol{\mu}$ and $\mathbf{H}(\mathbf{r})$ are possible, and the force depends on the angle, $\mathbf{F}_{\text{mag},z} = (\mu \cos\theta) \frac{\partial H_z(z)}{\partial z}$, so why should only two displacements corresponding to two central values of the force be observed in a Stern–Gerlach experiment? According to classical mechanics, the atoms should be deflected in a manner that depends on the angle between $\boldsymbol{\mu}$ and $\mathbf{H}(\mathbf{r})$, rather than having the bimodal deflection actually observed. The answer to this question came only with the development of quantum mechanics. In quantum mechanics, the projection of the angular momentum, and hence of the magnetic moment, on any axis (in this case, on the magnetic field direction) is quantized. Therefore it cannot take on any arbitrary value; in the case of a spin 1/2 particle, only two values of the projection are possible, $\hbar/2$ and $-\hbar/2$. Moreover, the length of the angular momentum vector, $|\mathbf{J}|$, is found to be given by $\sqrt{3/4}\,\hbar$ for a spin 1/2 particle. This is all very strange and difficult to interpret for the classically trained scientist! It will become clear, even simple, once we learn the quantum theory of angular momentum.

The angular momentum of silver atoms arises from the angular momentum (actually, spin) of the electrons comprising the atom, but this connection of the angular momentum of the atoms with the spin of the electrons contained in the atoms was not made until after 1925 when Samuel A. Goudsmit and George E. Uhlenbeck, under the guidance of their supervisor Paul Ehrenfest at the University of Leiden, proposed that the electron has its own intrinsic spin angular momentum $\mathbf{S}$ and intrinsic magnetic moment $\boldsymbol{\mu}$. Some additional history of spin is discussed in the beginning of Chapter 4. A classical analogy to the spin of an electron orbiting a nucleus in an atom is the rotation of the earth as it orbits around the sun. The spin of the electron is like the spin of the earth around its axis (which takes 24 hours to complete a turn). In quantum mechanics, apart from the spatial degrees of freedom of elementary particles, an inner degree of freedom called "spin" exists. We say that a particle has spin s if it can have $2s+1$ projections of its spin on an external axis. For example, "spin 1/2" can be described by a two-dimensional vector where the two components are for spin-up and spin-down; i.e., the spin states $|\uparrow\rangle$ and $|\downarrow\rangle$ can be represented by two-component vectors

$$|\uparrow\rangle \equiv \begin{pmatrix} 1 \\ 0 \end{pmatrix}, \quad |\downarrow\rangle \equiv \begin{pmatrix} 0 \\ 1 \end{pmatrix}. \qquad (1.14)$$

Similarly, "spin 1" states can be represented by vectors of dimension 3, i.e., the projection of the angular momentum on an external axis (say, the z-axis) can take the values $+1$, 0, and -1 (in units of $\hbar$), and these three states can be represented by three-dimensional vectors, as will be discussed in Chapter 3.

Photons have an internal angular momentum that is associated with their polarization state. We know that light propagating along a given direction, say along the z-axis, can be linearly polarized along a given axis perpendicular to the z-axis. The polarized light along this axis can be decomposed into right and left circular polarizations. A right circularly polarized photon has angular momentum given by $1\hbar$ along its direction of propagation, whereas the angular momentum of a left circularly polarized photon is $-1\hbar$. So, not only matter-waves have a spin angular momentum, but light waves also do, and this angular momentum is quantized in units of $\hbar$.

1.1.6 TUNNELING

In 1928, George Gamow proposed that some unstable nuclei decay via quantum mechanical tunneling of alpha particles, i.e., $^{4}_{2}\text{He}$ particles,[7] out of the nucleus. The alpha particles penetrate through the nuclear Coulomb potential barrier

---

[7] In the notation used here, the superscript, 4, is the number of nucleons and the subscript, 2, is the number of protons in the nucleus.

(resulting from the combination of the attractive nuclear forces and the repulsive Coulomb force — the remaining nucleus and the alpha particle are both positively charged, hence, a repulsive Coulomb potential exists between them, see Fig. 1.7), and manage to leave the nucleus even though their energy is not sufficient to classically surmount the barrier by a process called *nuclear fission*. An analogy is a ball with an insufficient initial velocity (hence, kinetic energy) to roll over a mound, yet having a finite probability to make it over the mound. An example of alpha decay is the process:

$$^{238}_{92}\text{U} \rightarrow {}^{234}_{90}\text{Th} + {}^{4}_{2}\text{He}. \quad (1.15)$$

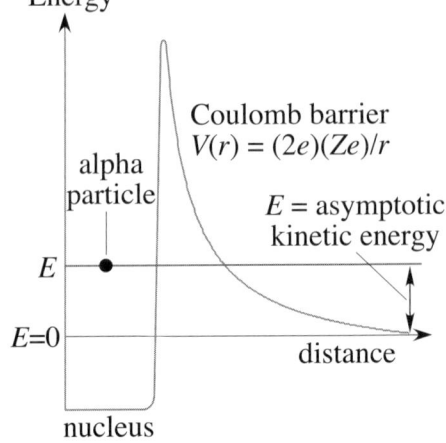

FIG 1.7 Alpha particle decay of a nucleus. An alpha particle, at the energy indicated, can tunnel out of the nucleus and penetrate through the repulsive Coulomb potential. The asymptotic kinetic energy E of the alpha particle is indicated.

Figure 1.7 schematically represents the alpha decay of a nucleus in terms of a potential between the alpha particle and the remaining nucleus, which includes the short-range attractive potential between the alpha particle and the remaining nucleus, due to the strong attractive nuclear force between nucleons in the nucleus, and the long-range repulsive Coulomb potential between the alpha particle and the remaining nucleus. The alpha particle tunnels out of the nucleus through the repulsive Coulomb potential.

The phenomenon of quantum tunneling is used extensively in nanotechnology. Here, we briefly mention only two applications: field-effect transistors and scanning tunneling microscopy. At this point, these are applications that are difficult to describe, since we have not yet developed the background knowledge required; we nevertheless do so, simply to underscore the application to which quantum tunneling is put in modern-day instruments.

Field-effect transistors are solid state devices made out of semiconductors; they were first envisioned by William Shockley in 1945 and first developed based on Shockley's original field-effect theories by Bell Labs scientist John Atalla in 1960. Electrons emitted from the cathode pass through a potential barrier created and controlled by a variable electric field. The electric field controls the shape and height of the tunneling barrier, and therefore the current flowing in the transistor. Most field-effect transistors in use today are metal-oxide-semiconductor field-effect transistors (MOSFETs). Figure 1.8 shows a schematic diagram of a MOSFET. They have four terminals, source, drain, gate, and body. Commonly, the source terminal is connected to the body terminal. The voltage applied between the gate and source terminals modulates the current between the source and drain terminals. If no positive voltage is applied between gate and source the MOSFET is nonconducting. A positive voltage applied to the gate sets up an electric field between it and the rest of the transistor. The positive gate voltage pushes away the holes (effectively positively charged particles) inside the p-type substrate and attracts the moveable electrons in the n-type regions under the source and drain electrodes. This produces a layer just under the gate insulator, in the p-doped Si region, through which electrons can propagate from source to drain. Increasing the positive gate voltage pushes the holes further away

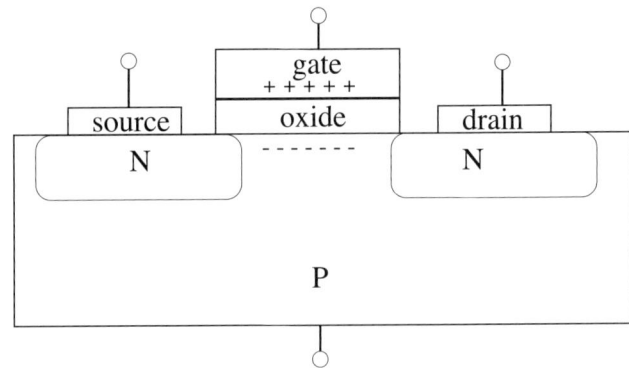

FIG 1.8 Schematic illustration of a metal-oxide-semiconductor field-effect transistor (MOSFET). See text for explanation.

1.1 What is Quantum Mechanics?

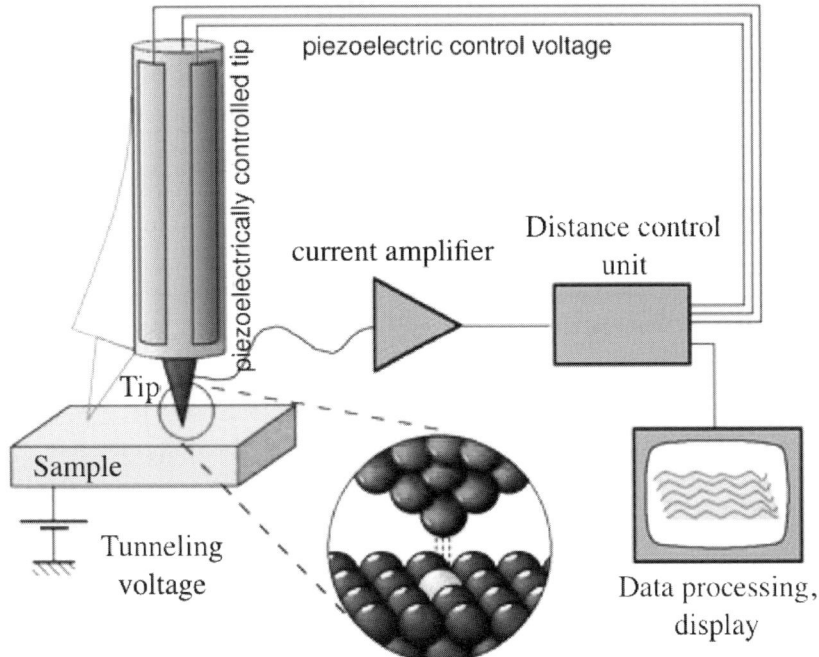

FIG 1.9 Schematic representation of a scanning tunneling microscope (STM), an instrument for imaging surfaces at atomic resolution. Reproduced from http://en.wikipedia.org/wiki/Scanning_tunneling_microscope

and enlarges the thickness of the channel. Based on quantum tunneling arguments (that will be elucidated in Sec. 1.3.11), the current flowing between source and drain is expected to depend exponentially in the gate voltage.

Scanning tunneling microscopy (STM), invented in 1981 by Gerd Binnig and Heinrich Rohrer (Nobel Prize in Physics, 1986), is a commonly used technique for viewing surface structure of conducting materials on a nanoscale. It relies on quantum tunneling of electrons from the atomically sharp voltage-biased microscope tip to the sample (or vice versa). The current between tip and surface is controlled by means of the voltage difference applied between the tip and surface. Without introducing a potential difference, there is a potential barrier for electrons to go between tip and surface (or vice versa) and the current is exponentially dependent on the distance between probe and the surface. Controlling the potential changes the potential barrier, hence the extent of the tunneling and thus the current magnitude. A 3D map of the surface of a conducting material can be built up from STM measurements. Figure 1.9 shows a schematic diagram of an STM apparatus. Insulators cannot be scanned by STM since electrons have no available energy states to tunnel into or out of in the completely filled bands in insulators, but atomic force microscopes (see Sec. 1.2.1) can be used to look at the surfaces of insulators.

1.1.7 PHOTOELECTRIC EFFECT

In 1888, Heinrich Rudolf Hertz carried out an experiment that kicked off the quantum revolution. He illuminated a surface of a metal with narrow bandwidth (i.e., single frequency) ultraviolet light and the radiation was absorbed in the metal. When the frequency of the radiation was above a given threshold frequency, v_0 (specific to the metal being used), a current of electrons was produced. He measured the current as a function of the frequency of radiation, intensity of radiation, and potential at which the surface of the metal was held relative to ground. Figure 1.10(a) shows a schematic of the experiment and illustrates photons in a ultra-violet (uv) light beam impinging on the metal and liberating electrons that leave the metal: (1) without a potential applied to the metal, (2) with a potential drop between metal and electron detector, and (3) with a potential barrier applied. Fig. 1.10(b) shows the energy levels of electrons in the metal, and the external potential applied across the system. Hertz found that no electrons were emitted for radiation with a frequency below that of the threshold v_0, independent of the intensity of the radiation. This result was not understood, and could not be understood based on the physics known at the time. In 1905, Einstein proposed his explanation of Hertz's experiment; he was awarded the Nobel prize in physics (1921) for this work. Einstein explained that the light photons in the beam have a characteristic energy hv given by their frequency v, where h is Planck's constant. If the photon energy hv is larger than the work function $W = hv_0$, defined as the difference of the potential energy outside the metal and the energy, E_F, of the highest state populated by electrons in the metal [see Fig. 1.10(b)], there will be electrons that are ejected from the material. If v is too low (below the frequency v_0), there are no electrons that are able to escape the surface of the metal. Increasing the light beam intensity does not change the energy of the constituent photons (although the number of photons will increase in proportion to the light intensity), and thus the *energy* of the emitted electrons does not depend on the intensity of the incoming light. All the energy of a photon must be absorbed upon its absorption, and this energy is used to liberate one electron from the metal, if its energy is large enough; otherwise the electron cannot get out of the metal. If the photon energy is larger than the work function of the metal, the liberated photoelectron will have a maximum final kinetic

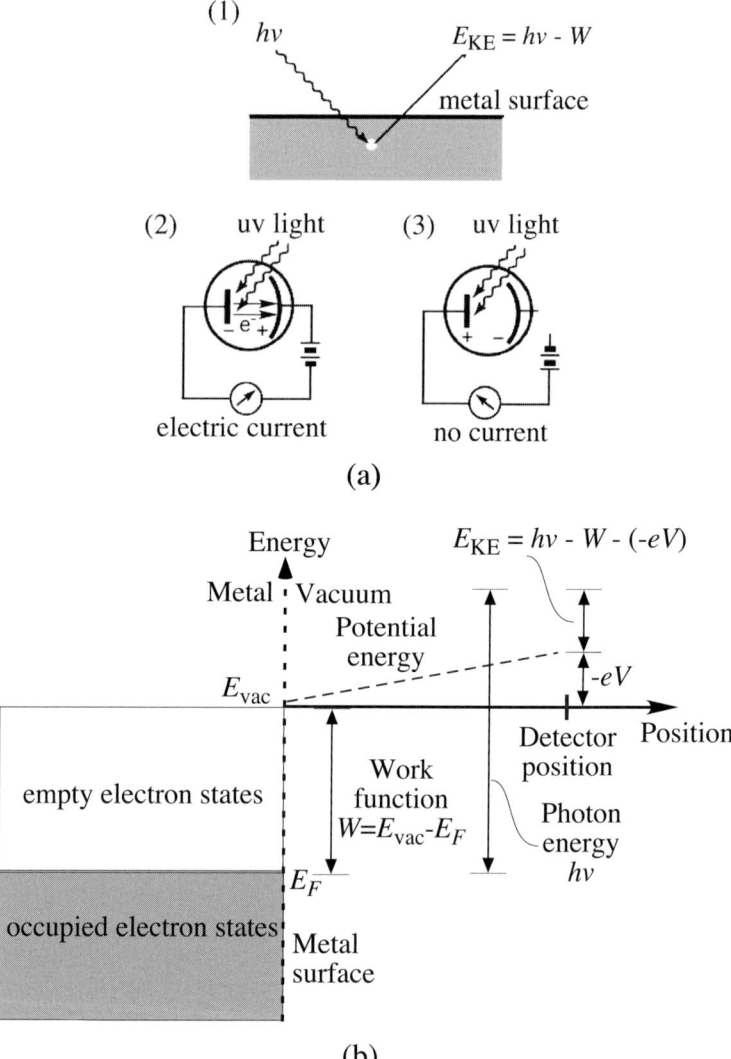

FIG 1.10 Schematic representation of a photoelectric effect experiment. (a-1): Metal surface subjected to a photon beam with photons of frequency v. (a-2): Experimental setup with a potential drop between metal and electron detector. (a-3): Experimental setup with a potential barrier between metal and electron detector. (b) Metal–vacuum interface showing occupied electron states in the metal, work function W, photon energy hv, and maximum electron kinetic energy leaving the metal, $E_{KE} = hv - W - (-eV)$.

energy as a free particle given by $E_{KE} = hv - W$. However, if an external potential V is applied to the surface of the metal relative to the anode that the free electrons strike, the *maximum* electron kinetic energy will equal $E_{KE} = hv - W - (-eV)$. Figure 1.11 shows the linear relation between the maximum kinetic energy of the electrons (measured in units of GHz, the energy in Joules is given by $E\,[J] = h\,[J\,s] \times E\,[GHz] \times 10^9$), versus photon frequency. Note that in the units chosen, the slope of the line in the figure is unity. The electron current versus frequency, the maximum electron kinetic energy versus photon intensity, and the electron current versus photon intensity are shown in Figs 1.11(a–d).

1.2 Nanotechnology and Information Technology

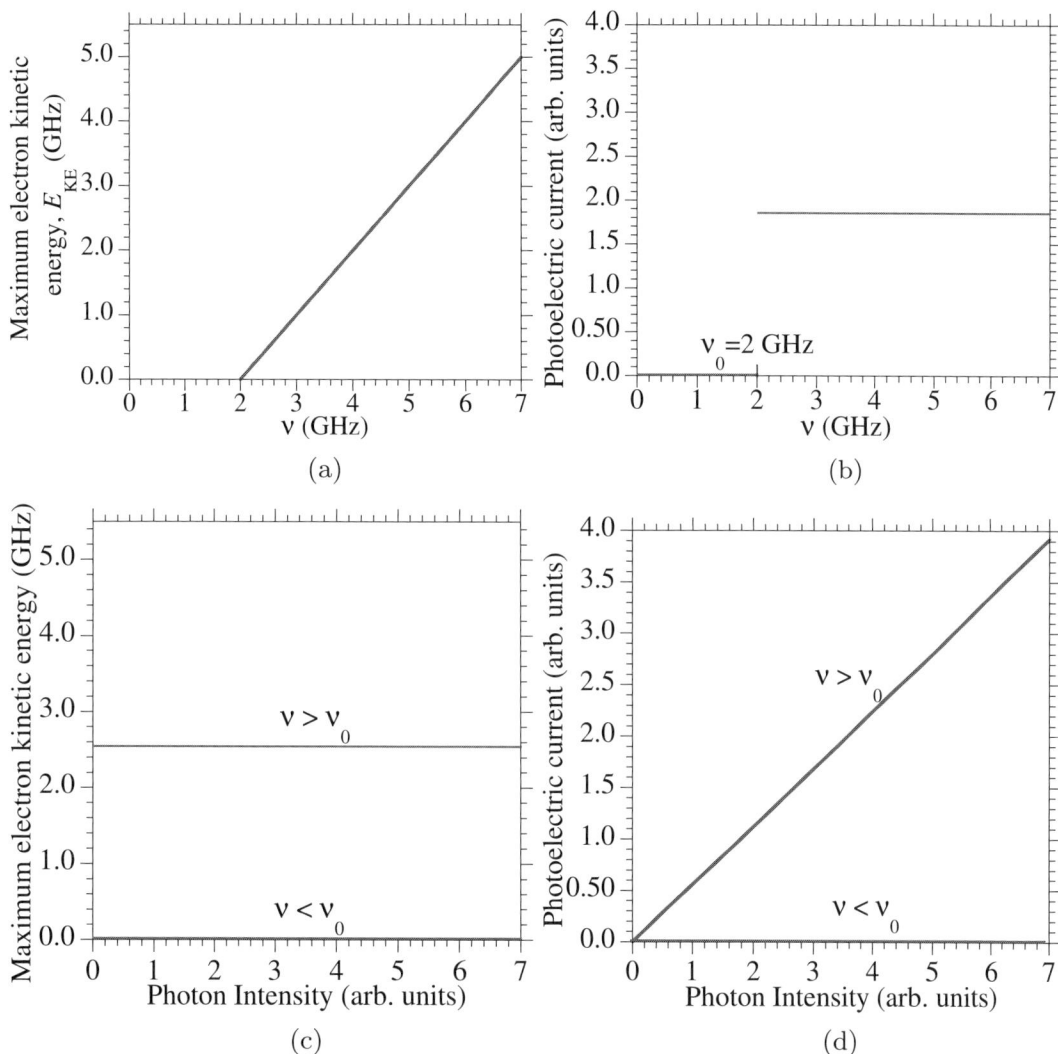

FIG 1.11 Photoelectric characteristics. (a) E_{KE} versus ν. (b) Electron photo-current versus ν. (c) E_{KE} versus intensity I. (d) Photo-current versus I.

Hopefully, the discussion in the last few sections has given you a hint of some of the strange and wonderful phenomena that occur in the quantum world. In the next section we discuss how quantum phenomena come into play in nanotechnology and information technology.

1.2 NANOTECHNOLOGY AND INFORMATION TECHNOLOGY

Nanoscience and nanotechnology can be defined as the science and technology of manipulating materials on an atomic scale. Nanotechnology is used to develop molecular devices such as molecular gears and other molecular machines, nanometer-scale electronic systems, nanocomputers, quantum computers based on qubits (quantum bits), microscopic size atomic clocks, and protein-based molecular devices. For the latter, it might be possible to exploit a host of examples of natural self-replicating machines (e.g., bacteria, viruses) to make such devices efficiently and cheaply.

Nanotechnology makes use of existing micromanipulation techniques, such as the scanning tunneling microscopy (STM) and atomic force microscopy (AFM), deep ultraviolet lithography, electron beam lithography, focused ion beam machining, nanoimprint lithography, atomic layer deposition, molecular vapor deposition, molecular beam epitaxy, and molecular self-assembly techniques such as those employing diblock copolymers.[8] Although some of these techniques were developed before the onset of the nano-era, they are all an integral part of the field. They will undoubtedly evolve and improve, even as new techniques are developed. In recent years, the quest for new devices capable of manipulating electron spins via magnetic fields, in a similar manner to electron charge controlled by electric fields, are beginning to be developed; this new area is called *spintronics*. Conventional electronic devices rely on the transport of electrical charge carriers – electrons or holes – in semiconductors (e.g., silicon). Exploiting the spin of the electron rather than (or in addition to) its charge is expected to lead to a remarkable new generation of spintronic devices that will be smaller, more versatile, and more robust than those currently used in silicon chips and circuit elements. Moreover, an electron spin is a natural candidate for a qubit, and hence, spintronics might be the ultimate scientific basis for quantum information devices. The main idea is that information can be stored in electron spins in a particular spin orientation (up or down), and can then be carried along a wire, and eventually read at a terminal. Spin orientation of conduction electrons survives for a relatively long time (nanoseconds, compared to femtoseconds for electron momentum decay). This makes spintronic devices particularly attractive for memory storage and magnetic sensors applications, and, potentially for quantum computing and quantum information, where an electron spin can serve as a qubit.

In Chapter 5 we take up the subject of information science and technology, first in a classical setting, and then in the quantum information context. In Chapter 9 we introduce some of the important solid-state and condensed matter concepts that are so important to nanotechnolgy, and in Chapter 13 we shall study Quantum Dots, Quantum Wires, Quantum Wells, and Nanotubes. This section is meant only to briefly introduce these topics.

1.2.1 STM AND AFM MICROSCOPIES

Scanning tunneling microscopy (STM) and atomic force microscopy (AFM) are two of the important techniques used to characterize surfaces of materials on a nanometer scale. We have already briefly discussed STM at the end of Sec. 1.1.6 (see Fig. 1.9 for a schematic of an STM setup) and we shall have more to say about it later in this chapter in Sec. 1.3.14. For nonconducting materials STM cannot be used, since electrons cannot flow in insulators, as already discussed, but AFM can. Gerd Binnig, Calvin Quate, and Christoph Gerber developed AFM in 1986. AFM measures the forces that act between the tip and a surface, and uses this information to produce atomic-scale images of the surface. Mechanical contact forces, long-range van der Waals forces, capillary forces, electrostatic forces such as the image-dipole force between a charge and a surface, magnetic forces, Casimir forces (due to vacuum fluctuations – see Chapter 14), can be measured by AFM, and in general, more than one of these forces contributes to the total force measured. An atomic force microscope consists of a microscopic scale cantilever with a sharp tip to scan a surface, as shown schematically in Fig. 1.12. The cantilever is typically silicon or silicon nitride and the tip is nanometer size. When the tip is brought near a sample surface, forces between the tip and the sample deflect the cantilever in proportion to the force exerted on it. The deflection can be measured using reflection of a laser beam from the top of the cantilever into a photodiode array, or by piezoelectric probes. A feedback mechanism is employed to adjust the tip to sample distance to maintain a constant force between tip and sample. AFMs can be operated in a number of modes, including static and oscillating cantilever modes. In the latter, the oscillation amplitude, phase and resonance frequency are modified by tip–sample interaction. The changes in oscillation with respect to the external reference oscillation provide information about the sample.

1.2.2 MOLECULAR ELECTRONICS

Nanoelectronics and molecular electronics are two rapidly developing research areas in nanotechnology. Electronic conduction in *mesoscopic systems* ("meso" comes from the Greek and means middle or intermediate, and refers here to systems that are larger than atoms or molecules, yet smaller than micron scale systems) is now understood in terms of a

---

[8] Block copolymers are molecules composed of long sequences (blocks) of monomer units, covalently bound together. A diblock copolymer is composed of two unlike units, say A and B, bound together in the form ABABAB...

1.2 Nanotechnology and Information Technology

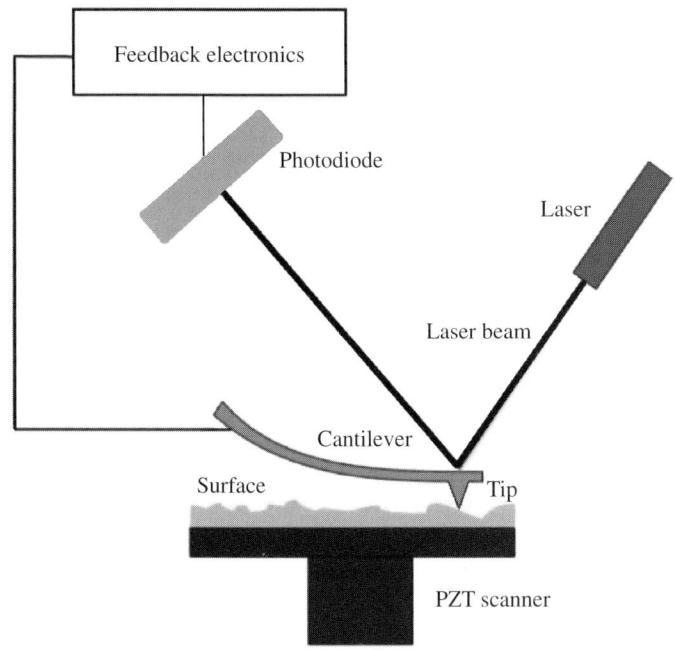

FIG 1.12 Schematic illustration of an atomic force microscope. Reproduced from http://en.wikipedia.org/wiki/Atomic_force_microscope

general quantum scattering theory approach. The conductance (and hence the resistance) of mesoscopic systems can be calculated in terms of what are called *Landauer formulas*, named after Rolf Landauer who pioneered the quantum scattering approach applied to electronic conductance. Quantized conductance of electronic waveguide like mesoscopic systems (i.e., "point contacts") is now well understood. The Coulomb blockade transport regime wherein mesoscopic structures are charged with a finite number of electrons and their capacitance determines transport of further electrons through the structures is also well understood. When impurities or dislocations or a random potential are present, localization of the electron wave functions occurs if the degree of randomness of the impurities or defects is sufficiently large. Dephasing by coupling to the environment is also an important mechanism that affects electron transport. Our knowledge of the interaction of atoms and molecules with surfaces, and of the principles of operation of scanning tunneling microscopy and atomic force microscopy are crucial to further progress in nano-electronics.

A basic building block of all electronics today is the transistor, a semiconductor device that uses a small amount of voltage or electrical current to control a larger voltage or current. For example, a transistor can be used to set the voltage on a wire to be either high or low, representing a binary 1 or 0, respectively. Transistors can have very fast response (as fast as 10^{-11} s, or 100 GHz), and are used in a very wide variety of applications: amplification, switching, signal modulation, etc. Transistors are used for both digital and analog electronic circuits. Transistors govern the operation of computers, TVs, cellular phones, and practically all other modern electronics. Two important types of transistors used today are *field-effect transistors* (FETs), sometimes called unipolar transistors, in which an input voltage controls a output voltage or current, and *bipolar junction transistors* (BJT), in which an input current controls an output current.

Molecular electronics involves the use of molecular building blocks for the fabrication of electronic components for passive elements such as wires and for active elements such as transistors. The hope is that molecular electronics will change the size scale of today's electronic devices, which are on the order of micrometers, to that of tens of nanometers. At such small scales, electronic transport should be described quantum mechanically, but the fact that the molecular wires and transistors interact with the "outside world" makes this an *open quantum system* problem, which is also a nonequilibrium problem since there is a potential across the molecular device and current flows through it. Such problems are hard to treat, unless significant approximations are introduced, but methods are beginning to be developed for such problems (e.g., time-dependent density functional theory).

1.2.3 QUANTUM DOTS, WIRES AND WELLS, AND NANOTUBES

Most quantum mechanical-based devices require the confinement of electrons such that they are prevented from moving in specified directions. If the number of these forbidden (or blocked) directions is three, two, or one, the system is said to be zero-, one-, or two-dimensional, respectively. Such devices are called *low-dimensional systems*. Here we briefly describe the most common devices realizing this important physics.

A *quantum dot* is a mesoscopic device that confines electrons in a small volume in all three dimensions. It can be attached to metallic electrodes and serve as an element in an electrical circuit. When an external potential is applied across the dot, it serves as a gate controlling the number of electrons in the dot and the electron passage through the dot. The device can be so effective that electrons can be added one by one, and for that reason, a quantum dot is sometimes referred to as *single electron transistor*. Geometrically, a quantum dot is a *zero-dimensional system*. A quantum wire confines the electrons in two directions; the electrons are free to propagate in the third direction, whose size is very much bigger than the confinement size in the two confined directions. A nanotube is an example of a quantum wire. The physics of interacting electrons in 1D is rather rich and exposes some spectacular phenomena referred to as *Luttinger liquid* behavior. A quantum well effectively confines electrons to a plane; the confinement direction is so small that only one mode in this direction can be populated. The study of electron properties in 2D systems (quantum wells), also expose some spectacular phenomena, such as the *quantum Hall effect* which will be discussed in Sec. 9.5.8.

Figure 1.13 shows 0D, 1D, 2D, and 3D structures made of carbon atoms. The 0D case shows the carbon fullerene molecule C_{60} named Buckminsterfullerene, "buckyball" for short, which is a naturally occurring quantum dot (see also Fig. 13.31). It was named after Richard Buckminster Fuller, who developed the geodesic dome (a spherical shell structure based on a network of great circles, i.e., geodesics). Its physical appearance is like that of a soccer ball, except its size is smaller by a factor of about 10^7. This molecule was discovered by Robert Curl, Harold Kroto, and Richard Smalley (Nobel Prize laureates in chemistry, 1996); it is a prototypical quantum dot.

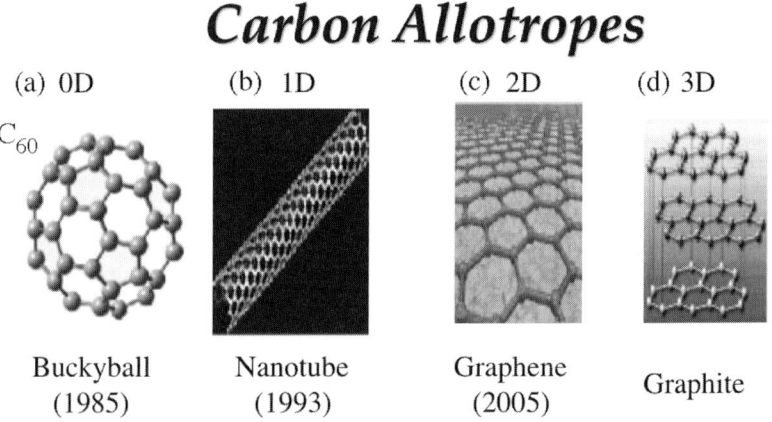

FIG 1.13 Schematic illustration of 0D, 1D, 2D, and 3D quantum structures made of carbon atoms.

Another example of quantum dot systems are semiconductors grown using controlled solution precipitation methods. For example, CdSe quantum dots of sizes in the range of 4–5 nm are relatively easy to grow. Upon illuminating such particles with ultraviolet light, the particles fluoresce with wavelengths that depend sensitively on the size of the quantum dot.

Quantum wires and nanotubes [see Fig. 1.13(b)] can be conducting and can then be used as a conveyor of electrons. It is possible to make quantum wires out of carbon nanotubes, but inorganic nanotubes can also be fabricated and even DNA nanotubes have been produced. The radius of the wire determines the degree of confinement; the smaller the radius, the more important the quantum effects in the wire. Nanotubes that are 10,000 times thinner than a human hair have been made. Nanotubes is one of the hot topics in nanotechnology. An important spin-off of the fullerene research that led to the discovery of the C_{60} buckyball molecule are nanotubes based on carbon or other elements, e.g., WS_2 and MoS_2. These systems consist of graphitic layers wrapped into cylinders. Figure 1.13(b) shows a single-walled carbon nanotube, which serves as a quantum wire and restricts motion of conduction electrons to be in the direction of the wire. They are only few nanometers in diameter, but can be up to a millimeter long. Their properties (e.g., specific heat, heat transport, thermal conductivity), synthesis, and characterization have been extensively studied.

Quantum confinement in quantum wells begins when the quantum well thickness becomes comparable to the de Broglie wavelength of the carriers. Quantum wells can be formed using semiconductor structures that can be grown by molecular beam epitaxy or chemical vapor deposition with monolayer thickness control. Moreover, the 2D sheet of carbon known as graphene [see Fig. 1.13(c)] is another example of a quantum system confined to 2D.

1.2.4 BIO-NANOTECHNOLOGY

Bio-nanotechnology is an interdisciplinary field that deals with biological and medical applications of biomaterials, biosensors, drugs and drug delivery systems. It also involves understanding the structure and function of biological devices at the nanoscale, from the level of single molecules up to complex molecular machines. The latter important subtopic in bio-nanotechnology involves molecular motors and intelligent molecular machines that can be applied to bio-sensing for health care applications. If we learn to integrate electrical and optical biomolecules to produce active devices, networks and bio-sensors, and develop the skills to produce DNA-based nanostructures and machines, the promise of designer-devices tailored to specific tasks will be fulfilled, and the potential of bio-nanotechnology will come to fruition. As the scale of such devices get smaller, classical descriptions fail and quantum mechanics is required. Although much of bio-nanotechnology is adequately described by classical mechanics, many exceptions exist, e.g., the photosynthesis process involves the absorption of a photon and the transfer of energy to an electron, a process that requires a quantum description. Another possible example involves the folding of proteins which requires that the protein move from conformational local minima of the multidimensional protein potential to one or a few lowest local minima: this process might involve quantum tunneling over barriers on the potential, and this would occur in a dissipative and decohering environment.

1.2.5 INFORMATION TECHNOLOGY

Information science and information technology today encompass many aspects of computing, communications, data storage and data security technologies. These involve networking, computer design, database and software design, cryptography, hardware devices, etc.

There are many devices that are used today to create, transmit, transform, and interact with information in electronic form. Understanding how such devices work requires knowledge of physics, including quantum physics. Moreover, quite a few of these devices operate close to fundamental physical limits, including the quantum limit.

Information is physical; it is stored in bits (or qubits) that are physical, and is transported and communicated in ways that directly involve physics. Even the measure of information (the determination of the amount of information contained in a particular message or file) involves physical concepts. Bounds on information storage and information transfer are determined by physical arguments. In short, information is part of physics (as much so as force or energy).

Until recently, only classical protocols for information storage, transfer, retrieval and processing were studied. But slowly, the study of quantum information began in the middle of the twentieth century, and has blossomed into an important branch of atomic, molecular, and optical physics. Quantum information was suddenly popularized by the discovery of a quantum mechanical algorithm for factoring numbers by Peter Shor in 1997. If quantum computers capable of running the Shor algorithm were available, the encryption systems (that are based on the *RSA algorithm* for factoring large numbers) that we use to secure information in banking systems, and on the web, would be rendered useless. But, what quantum mechanics takes away, it gives back; protocols that ensure information security by virtue of quantum mechanical laws have been developed. Therefore, quantum information has become a very hot topic. We take up this subject in Chapter 5.

1.3 A FIRST TASTE OF QUANTUM MECHANICS

We shall now begin to describe some of the most basic features of quantum mechanics so as to learn the language of quantum theory. In so doing, we highlight the probabilistic nature of quantum mechanics, describe the superposition principle of quantum states, discuss the operators that act on quantum states, detail the nature of the operators that can be measured, introduce the concept of an entangled quantum state, introduce the postulates of quantum mechanics, and describe how quantum states develop in time (i.e., propagate). We then run through some of the simplest quantum paradigms: one-dimensional quantum systems, a particle in a box and a particle in a piecewise-constant potential, quantum tunneling, and the quantum harmonic oscillator. These topics are treated here, in this introductory chapter, so the reader can become acquainted with the basic quantum mechanical concepts <u>before</u> starting to deal with some of the more formal aspects of the theory, such as the uncertainty principle, the correspondence principle and mixed (impure) quantum states. Appendix A, which introduces the mathematics of vector spaces, the notion of inner product and additional linear algebra concepts, should be studied in conjunction with Secs. 1.3.1–1.3.6. Please turn now to Appendix A for a quick review of linear algebra and Dirac notation. Then, continue reading this chapter. Furthermore, after you finish studying Chapter 2, you may want to come back and re-read this first-taste material; it will then seem really easy!

Quantum mechanics is the best verified scientific theory we ever had; experimental measurements have verified the theoretical predictions of quantum mechanics to unbelievable accuracy. Nevertheless, there are problems, like its incompatibility with general relativity. And there are conceptual issues regarding quantum measurement theory and decoherence phenomena. These problems should be kept in mind. The reader should also keep in mind that scientific theories cannot be proven right; they can only be proven wrong (if their predictions do not square with measurements).

1.3.1 QUANTUM STATES AND PROBABILITY DISTRIBUTIONS

States in quantum mechanics are denoted by vectors in a vector space (for a discussion of vector spaces, see Appendix A). In what is now known as *Dirac notation* for vectors in a vector space, a state vector is denoted by the symbol $|\ \rangle$, and the state vector of a given state ψ is denoted by the symbol $|\psi\rangle$. In quantum mechanics, a specific preparation of a state of a system does not explicitly determine the outcome of a subsequent measurement of the system, but rather only the probabilities of the various possible outcomes. That is, the preparation determines the probability distributions for all possible measurements that can be performed following the preparation.

For example, after preparing a spin 1/2 particle (e.g., an electron) in a given spin state, we can measure the spin projection along a certain axis, say, the z-axis, using a Stern–Gerlach apparatus. If the particle is in spin state $|\uparrow\rangle$, and we perform a measurement of its spin projection along the z-axis, we will get $+1/2$ (in units of $\hbar$) for the spin projection. If the particle is in spin state $|\downarrow\rangle$, and we perform the same measurement, we will get $-1/2$ (in units of $\hbar$). However, as we shall see shortly, if we perform a measurement on an arbitrary state of the electron and we get $+1/2$, this does not mean that the state before the measurement was $|\uparrow\rangle$. It could have been the superposition state, e.g., $|\psi\rangle = \alpha|\uparrow\rangle + \beta|\downarrow\rangle$ with nonvanishing α and β, and $|\alpha|^2 + |\beta|^2 = 1$. If this were indeed the state of the particle, and if we have many identical copies of the particle in this state, and we perform a measurement on each one of them, we would measure $+1/2$ with probability $|\alpha|^2$, and $-1/2$ with probability $|\beta|^2 = 1 - |\alpha|^2$. In any case, if we measure $+1/2$, the state immediately after the measurement is $|\uparrow\rangle$ (this is one of the postulates of quantum mechanics, as explained in Sec. 1.3.4). Below we shall represent the states $|\uparrow\rangle$ and $|\downarrow\rangle$ as the two-dimensional vectors $\binom{1}{0}$ and $\binom{0}{1}$ respectively. The superposition state $|\psi\rangle = \alpha|\uparrow\rangle + \beta|\downarrow\rangle$ will then be represented as $\alpha\binom{1}{0} + \beta\binom{0}{1} = \binom{\alpha}{\beta}$.

An *observable* is a property of the system that can be measured, hence every measurement of the system specifies a given observable. The position, momentum, momentum squared, and energy of the system, are all examples of observables that can be measured, as is the projection of the spin of a particle on any axis. Observables are represented in quantum mechanics by operators that operate on state vectors, as we shall see in what follows. The quantum state of the system specifies the probability distributions obtained in all such measurements.

In a particular coordinate system with coordinates q, the state $|\psi\rangle$ specifies a function $\psi(q)$ of the coordinates, $\langle q|\psi\rangle \equiv \psi(q)$ [q could be three-dimensional for a system consisting of a single particle in three dimensions, $q \equiv \mathbf{r} = (x, y, z)$, or could in general be n-dimensional, e.g., $n = 3N$ for N particles in three dimensions]. The quantity $\psi(q)$ is called the *wave function* of the state ψ in coordinate (configuration) space. It is the projection of the state $|\psi\rangle$ onto $|q\rangle$ (which is the state with the particle located at position q). The quantity $\langle q|\psi\rangle$ is a complex number and can be viewed as a complex valued function of the coordinates q.[9] Moreover, the quantities $\langle q|\psi\rangle$ and $\langle p|\psi\rangle$ can be interpreted as the inner product of the state vector $|\psi\rangle$ with state vectors $|q\rangle$ and $|p\rangle$, respectively, as explained in detail in Sec. 1.3.8 entitled "Position and Momentum States, $|\mathbf{x}\rangle$ and $|\mathbf{p}\rangle$". When the wave function is properly normalized (see next paragraph), the square of the modulus of this function determines the *probability distribution* of the state as a function of the coordinates, i.e., the quantity $|\psi(q)|^2\, dq$ is the probability that a measurement will find the system in the

---

[9] As explained below, $\langle q|\psi\rangle$ is the wave function in position space, and the wave function in momentum space, $\langle p|\psi\rangle \equiv \psi(p)$, can also be defined, and is related to the wave function in coordinate space through a Fourier transform relation.

1.3 A First Taste of Quantum Mechanics

element dq of configuration space about the point q, $P(q)\,dq = |\psi(q)|^2\,dq$. The probability of the possible outcomes of any measurements can be calculated using the wave function $\psi(q)$, as explained below. If a particle has a spin, the full wave function describing the state of the particle must also have a spin part; often it is possible to write the full wave function as a product of a spatial wave function and a spin wave function (this will be detailed in Sec. 4.2).

In order to insure a probability interpretation of $|\psi|^2$, we can normalize the wave function ψ to ensure that it has unit length. Upon doing so, the wave function at position q is given by $\Psi(q) = \frac{\psi(q)}{\sqrt{\int |\psi(q)|^2\,dq}}$, and therefore $\int |\Psi(q)|^2\,dq = 1$, as you can easily verify. This is what we mean by the wave function being normalized. The square of the length of the state vector $|\psi\rangle$ is $\langle \psi | \psi \rangle \equiv \int \psi^*(q)\psi(q)\,dq$, since $\langle \psi | \psi \rangle = \int dq\, \langle \psi | q \rangle \langle q | \psi \rangle$ because $\int dq\, |q\rangle\langle q| = 1$ (this is a statement of the completeness of the states $|q\rangle$). The square of the length of $|\Psi\rangle$, $\langle \Psi | \Psi \rangle = \frac{\langle \psi | \psi \rangle}{\langle \psi | \psi \rangle}$, is unity (hence the length $\|\Psi\| \equiv \sqrt{\langle \Psi | \Psi \rangle} = 1$).

Two state vectors $|\psi_1\rangle$ and $|\psi_2\rangle$ (or two wave functions ψ_1 and ψ_2) are said to be *orthogonal* if $\langle \psi_1 | \psi_2 \rangle \equiv \int \psi_1^*(q)\psi_2(q)\,dq = 0$. For the discussion of the mathematics required to calculate the length of a state vector, and the orthogonality properties of state vectors, see Appendix A.

The quantum mechanical state of a system, and therefore its wave function, can vary with time, i.e., the state $|\psi\rangle$ depends on time, $|\psi(t)\rangle$, and the wave function is a function of time, $\psi(q,t)$. If the wave function is known at some initial instant, then it is in principle determined at every succeeding instant of time by the time-dependent Schrödinger equation [as we shall see in Secs. 1.3.5 and 1.3.4, $i\hbar \frac{\partial}{\partial t}\psi(t) = H\psi(t)$, where H is the *Hamiltonian operator*, or in Dirac notation for state vectors, $i\hbar \frac{\partial}{\partial t}|\psi(t)\rangle = \hat{H}|\psi(t)\rangle$].

The *superposition principle* is valid in any linear mathematical construct, such as linear vector spaces and linear differential equations. In optics, the superposition principle states that the sum of two optical waves is also an optical wave. Interference phenomena in optics result due to this superposition. In quantum mechanics, it is the wave functions, or the state vectors, that obey the superposition principle; arbitrary linear combinations of state vectors or wave functions can be added to one another. Such linear combinations occur when the state of the system simultaneously "possesses" two or more values for an observable quantity (e.g., a state having more than one value of the energy).

As an example, let us consider a hydrogen atom that was originally in its ground state, $|\psi_1\rangle$ (i.e., its wave function is the ground state wave function $\psi_1(q)$), and in a given experiment, the atom suffers a collision so its state can change to the excited state $|\psi_2\rangle$; let us assume that only these two states are populated in the experiment. When one repeats this same experiment many times, one finds that there is a probability P_1 for finding the hydrogen atom in the ground state, and a probability $P_2 = 1 - P_1$ for finding it in a given excited state $|\psi_2\rangle$. The probabilities, of course, depend on the details of the type of collision, but the sum of the probabilities must equal unity, $P_1 + P_2 = 1$. In quantum mechanics, one writes the state of the hydrogen atom after the collision as $|\psi\rangle = c_1|\psi_1\rangle + c_2|\psi_2\rangle$, where $|c_1|^2 = P_1$ and $|c_2|^2 = P_2$, and the wave function of the hydrogen atom after the collision is written as $\psi = c_1\psi_1 + c_2\psi_2$. We say that the state of the hydrogen atom after the collision, $|\psi\rangle$, is in a superposition of states $|\psi_1\rangle$ and $|\psi_2\rangle$ with *probability amplitudes* c_1 and c_2, respectively. Alternatively, we say that the wave function of the hydrogen atom is in a superposition of wave functions ψ_1 and ψ_2 with probability amplitudes c_1 and c_2, respectively. P_1 and P_2 specify the amplitudes c_1 and c_2 only up to a phase. In quantum mechanics the amplitudes $c_i = |c_i|\,e^{i\theta_i}$ can be complex numbers; their phases are not specified by their magnitudes. Hence, c_1 and c_2 completely specify the probabilities P_1 and P_2, but not vice versa. The possibility of combining quantum states in coherent superpositions that are qualitatively different from their individual components introduces a nonclassical nature to the concept of a state of a system.

As another example, consider the spin $1/2$ particle case discussed at the beginning of this section, wherein the particle can be in spin state $|\uparrow\rangle$ or in state $|\downarrow\rangle$ [another notation that is often used for the spin-up and spin-down states is $|+\rangle$ and $|-\rangle$, respectively (moreover, several additional notations are also commonly used – see Sec. 4.2)]. Experimentally, it is possible to put the spin $1/2$ particle into a *superposition state* $|\psi\rangle$ given by

$$|\psi\rangle = \alpha|\uparrow\rangle + \beta|\downarrow\rangle \quad [\text{or } |\psi\rangle = \alpha|+\rangle + \beta|-\rangle],$$

$$\text{or } |\psi\rangle = \alpha \begin{pmatrix} 1 \\ 0 \end{pmatrix} + \beta \begin{pmatrix} 0 \\ 1 \end{pmatrix} = \begin{pmatrix} \alpha \\ \beta \end{pmatrix}, \tag{1.16}$$

with amplitudes α and β such that $|\alpha|^2 + |\beta|^2 = 1$. This state $|\psi\rangle$ does not have a well-specified projection of angular momentum along the z-axis. If a measurement is done to determine the projection of the angular momentum, the probability of finding the particle in spin-up, $P_{+1/2}$, is $|\alpha|^2$ and the probability $P_{-1/2}$ is $|\beta|^2$. That is, if measurements are performed on identically prepared spin 1/2 particles, with all of them put into the state $|\psi\rangle$, the probability of measuring spin-up will be $P_{+1/2} = |\alpha|^2$, and spin-down will be $P_{-1/2} = |\beta|^2$. In future chapters we shall see that any two-level system can be mapped onto the spin 1/2 particle case, and vice versa, so this example is really no different than that in the previous paragraph.

In classical optics, *coherence* refers to the condition of phase stability that is necessary for interference to be observable. In quantum theory, the concept of coherence is related to the amplitudes c_i of the basis functions ϕ_i used in describing the wave function of the system in a coherent superposition,

$$\psi = \sum_i c_i \phi_i, \quad \left(\text{or } |\psi\rangle = \sum_i c_i |\phi_i\rangle \right), \tag{1.17}$$

and the necessity that they be well determined and stable. In other words, the amplitudes c_i of the basis states $|\phi_i\rangle$ in the expansion $|\psi\rangle = \sum_i c_i |\phi_i\rangle$ should be completely determined (the mathematical definition of basis states and completeness are discussed in Appendix A). [It is possible to put a system into an incoherent state, i.e., into what is known as a mixed state, wherein the phases of the amplitudes are not well determined. We shall learn more about mixed states in Sec. 2.5, and about decoherence processes that can cause incoherent states to develop in Chapter 17, linked to the book web page.]

One can ask whether it is possible to put a macroscopic system (or if you like, a macroscopic object) into a superposition state. In Schrödinger's famous *cat paradox*, the question asked is whether one can put a cat into a superposition state $|\psi\rangle_{\text{cat}} = \frac{1}{\sqrt{2}}(|\psi_{\text{alive}}\rangle + |\psi_{\text{dead}}\rangle)$. In this state, the cat is neither alive nor dead, but is in a linear combination of the two (until we perform a measurement and collapse the cat's wave function, see Sec. 1.3.4; if we find the cat to be dead, we would be responsible for its death because we carried out the measurement). It turns out to be exceedingly difficult to put macroscopic objects into such a superposition state (often called a *Schrödinger cat state* because of decoherence, as will be discussed in Sec. 5.4 and Chapter 17.

1.3.2 OBSERVABLE OPERATORS

A physical variable that can, in principle, be measured is called an *observable*. In quantum mechanics, observables are represented as operators that are applied to state vectors (alternatively, to wave functions). Upon applying an operator $\hat{\mathcal{O}}$ to a state $|\psi\rangle$ one obtains another state $|\phi\rangle$: $|\phi\rangle = \hat{\mathcal{O}}|\psi\rangle$. Equivalently, upon applying an operator $\mathcal{O}$ to a wave function ψ one obtains another wave function ϕ: $\phi = \mathcal{O}\psi$ (see Sec. A.2 in Appendix A). The mean value (often called the average or *expectation value*) of an operator $\mathcal{O}$ in a given state $|\psi\rangle$ (wave function ψ) is given by

$$\langle \mathcal{O} \rangle_\psi \equiv \langle \psi | \hat{\mathcal{O}} | \psi \rangle = \int dq\, \psi^*(q) \mathcal{O} \psi(q). \tag{1.18}$$

The expectation value $\langle \psi | \hat{\mathcal{O}} | \psi \rangle$ is in general a complex number, but if the operator is an observable [i.e., if the operator is *Hermitian* (sometimes called *self-adjoint*)], it must be real (measurable quantities must be real). When the system is in state $|\psi\rangle$, the average over many measurements of a physically observable quantity corresponding to the operator $\mathcal{O}$ should be compared with the theoretical value given by the mean value of $\mathcal{O}$ within the state vector $|\psi\rangle$, $\langle \psi | \hat{\mathcal{O}} | \psi \rangle$. We shall sometimes write $\langle \psi | \mathcal{O} | \psi \rangle$ in order to simplify notation, i.e., we shall often leave off the hat in writing operators when it is clear that we are talking about operators. We also sometimes blur the distinction between an operator that acts on $|\psi\rangle$ in the vector (Hilbert) space and an operator acting on the wave function $\psi(q)$ in coordinate (configuration) space.

An operator is said to be linear if it has the properties $\mathcal{O}(\psi_1 + \psi_2) = \mathcal{O}\psi_1 + \mathcal{O}\psi_2$ and $\mathcal{O}(c\psi) = c\mathcal{O}\psi$, where ψ_1, ψ_2, and ψ are arbitrary wave functions and c is an arbitrary (real or complex) number. In Dirac notation, $\hat{\mathcal{O}}(|\psi_1\rangle + |\psi_2\rangle) = \hat{\mathcal{O}}|\psi_1\rangle + \hat{\mathcal{O}}|\psi_2\rangle$ and $\hat{\mathcal{O}}(c|\psi\rangle) = c\hat{\mathcal{O}}|\psi\rangle$. If we expand the wave function ψ in a set of basis states as in Eq. (1.17), then $\mathcal{O}\psi = \sum_i c_i \mathcal{O}\phi_i$ (in Dirac notation, $\hat{\mathcal{O}}|\psi\rangle = \sum_i c_i \hat{\mathcal{O}}|\phi_i\rangle$). The mean value (i.e., the expectation value) of $\mathcal{O}$ can

1.3 A First Taste of Quantum Mechanics

then be easily expanded using the basis set expansion as follows:

$$\langle\psi|\hat{\mathcal{O}}|\psi\rangle = \sum_{i,j} c_i^* c_j \int dq\, \phi_i^*(q)\mathcal{O}\phi_j(q). \tag{1.19}$$

For any operator $\mathcal{O}$ there are special functions, let us denote them as ϕ_j, such that when the operator acts on any one of these functions, it simply returns the same function times a (in general, complex) number o_j:

$$\mathcal{O}\phi_j = o_j\phi_j, \quad (\text{or } \hat{\mathcal{O}}|\phi_j\rangle = o_j|\phi_j\rangle). \tag{1.20}$$

These special functions are called *eigenfunctions* (eigen comes from the German word meaning self), and the numbers o_j are called *eigenvalues*. The eigenfunctions ϕ_j of the operator $\mathcal{O}$ are the solutions of the equation (1.20) where the eigenvalues o_j are the values of the constant for which this equation has solutions. For example, the eigenvalues and eigenvectors of the z projection of the spin operator, S_z, for a spin 1/2 particle are $o_1 = \hbar/2$ and $o_2 = -\hbar/2$, and $|\phi_1\rangle = |\uparrow\rangle$ and $|\phi_2\rangle = |\downarrow\rangle$, respectively. We have already seen that the states $|\uparrow\rangle$ and $|\downarrow\rangle$ are represented by $\binom{1}{0}$ and $\binom{0}{1}$. We shall see in future chapters that the operator S_z is represented by the matrix $(\hbar/2)\begin{pmatrix}1 & 0\\0 & -1\end{pmatrix}$, whose eigenvalues and eigenvectors are as stated in the previous two sentences.

The eigenvalues of an operator corresponding to a real physical quantity, i.e., of an observable operator, and the mean value of such an operator in every state $|\psi\rangle$, $\langle\psi|\hat{\mathcal{O}}|\psi\rangle$, must be real. This imposes a restriction on the corresponding observable operators. Operators that are Hermitian (self-adjoint) [see Appendix A, Eq. (A.40)] must have real eigenvalues, and it is a necessary condition that observable operators be Hermitian. In order to define a Hermitian operator, we must first define the Hermitian conjugate of an operator. The *Hermitian conjugate* of an operator $\mathcal{O}$, $\mathcal{O}^\dagger$, is defined as the operator that satisfies the following equation for any vectors $|\psi_1\rangle$ and $|\psi_2\rangle$:

$$\langle\psi_2|\hat{\mathcal{O}}^\dagger|\psi_1\rangle = \langle\psi_1|\hat{\mathcal{O}}|\psi_2\rangle^*. \tag{1.21}$$

In wave function notation, (1.21) is $\int dq\, \psi_2^*(q)\mathcal{O}^\dagger \psi_1(q) = \left[\int dq\, \psi_1^*(q)\mathcal{O}\psi_2(q)\right]^*$. An operator $\mathcal{H}$ is *Hermitian* if it is equal to its Hermitian conjugate:

$$\mathcal{H} = \mathcal{H}^\dagger. \tag{1.22}$$

The eigenvalues $\{h_i\}$ of a Hermitian operator $\mathcal{H}$ are real, and their eigenvectors are orthogonal. That is,

$$\mathcal{H}\phi_i = h_i\phi_i, \quad (\text{or } \hat{\mathcal{H}}|\phi_i\rangle = h_i|\phi_i\rangle), \tag{1.23}$$

with $h_i^* = h_i$ (real eigenvalues) and

$$\langle\phi_i|\phi_j\rangle \equiv \int \phi_i^*(q)\phi_j(q)\,dq = 0 \quad \text{if } i \neq j. \tag{1.24}$$

These properties are proved in Sec. A.2 of Appendix A [see proof near Eq. (A.41)].

We shall see that operators in quantum mechanics can be represented by matrices. The Hermitian conjugate of a matrix O, $O^\dagger$, is defined such that

$$O^\dagger \equiv \left(O^t\right)^*, \tag{1.25}$$

where O^t indicates transpose of the matrix O, i.e., in terms of *matrix elements*,

$$\left(O^\dagger\right)_{i,j} = O_{j,i}^*. \tag{1.26}$$

A matrix H is Hermitian if

$$H^\dagger = H, \quad \text{i.e.,} \quad \left(H^\dagger\right)_{i,j} \equiv H_{j,i}^* = H_{i,j}. \tag{1.27}$$

Not only are the eigenvalues $\{h_i\}$ of a Hermitian matrix H real, but its eigenvectors are orthogonal. That is,

$$\sum_l H_{k,l} \phi_l^{(i)} = h_i \phi_k^{(i)}, \tag{1.28}$$

where h_i is real, $\phi_k^{(i)}$ is the kth component of the ith eigenvector (and corresponds to the state vector $|\phi_i\rangle$ in Hilbert space), and $\sum_l \phi_l^{(i)*} \phi_l^{(j)} = \delta_{i,j}$ (in Dirac notation, $\langle \phi_i | \phi_j \rangle = \delta_{i,j}$). Here the eigenvectors have been normalized so that they are not only orthogonal, but they are orthonormal.

Operators representing real physical quantities (observables) must be Hermitian. The definition of an observable operator is typically broadened in quantum mechanics to be any self-adjoint (Hermitian) operator on the quantum vector space of states, without reference to whether it could, in practice, be measured.

The expectation value of a Hermitian operator $\mathcal{O}$ in state ψ [see (1.19)] is particularly easy to calculate when ψ is expanded using the eigenstates of $\mathcal{O}$:

$$\langle \psi | \hat{\mathcal{O}} | \psi \rangle = \sum_{i,j} c_i^* c_j \int dq\, \phi_i^*(q) \mathcal{O} \phi_j(q) = \sum_{i,j} c_i^* c_j o_j \int dq\, \phi_i^*(q) \phi_j(q) = \sum_i |c_i|^2 o_i. \tag{1.29}$$

Here we have used the orthonormality property $\int dq\, \phi_i^*(q) \phi_j(q) = \delta_{i,j}$. For example, for the spin 1/2 particle example in the superposition state (1.16), $|\psi\rangle = \alpha|\uparrow\rangle + \beta|\downarrow\rangle$, the mean value of the z projection of the spin operator, S_z, is given by $\langle \psi | S_z | \psi \rangle = (\alpha^* \langle \uparrow | + \beta^* \langle \downarrow |) S_z (\alpha | \uparrow \rangle + \beta | \downarrow \rangle)$, which can be written as

$$\langle \psi | S_z | \psi \rangle = \frac{\hbar}{2} \begin{pmatrix} \alpha^* & \beta^* \end{pmatrix} \begin{pmatrix} 1 & 0 \\ 0 & -1 \end{pmatrix} \begin{pmatrix} \alpha \\ \beta \end{pmatrix} = \frac{\hbar}{2} \left[|\alpha|^2 - |\beta|^2 \right]. \tag{1.30}$$

Since the states $|\uparrow\rangle$ and $|\downarrow\rangle$ are eigenstates of S_z with eigenvalues $o_1 = (\hbar/2)$, $o_2 = (-\hbar/2)$, and $\langle \phi_i | \phi_j \rangle = \delta_{i,j}$, the inner product in (1.30) is particularly simple to evaluate; this will be further detailed in Sec. 4.2.

1.3.3 QUANTUM ENTANGLEMENT

Entangled states of a system consisting of two or more particles are purely quantum in nature; no classical analog exists. These specially correlated states have features that can be very disorienting if considered within the laws of classical mechanics. We shall first explain what is meant by an entangled state, and then discuss the disorienting features of these states.

An entangled multiparticle state is one that cannot be written as a product of single particle states. This feature of quantum mechanics was originally called "Verschränkung" (loosely translated as interconnection) by Schrödinger, and it underlies some important relations between subsystems of a compound quantum system. Let us give two examples of entangled states for two-particle systems.

The first example is the *singlet spin-state* for two spin 1/2 particles:

$$|\text{singlet}\rangle = \frac{1}{\sqrt{2}} (|\uparrow\rangle|\downarrow\rangle - |\downarrow\rangle|\uparrow\rangle). \tag{1.31}$$

Here, the left state refers to particle 1 and the right one to particle 2, i.e., $|\text{singlet}\rangle = \frac{1}{\sqrt{2}} (|\uparrow\rangle_1 |\downarrow\rangle_2 - |\downarrow\rangle_1 |\uparrow\rangle_2)$. Clearly, this state cannot be written as a product of single particle spin states in the form $|\chi_\alpha\rangle|\chi_\beta\rangle$; it is a superposition of $|\chi_\alpha\rangle|\chi_\beta\rangle$ and $|\chi_\beta\rangle|\chi_\alpha\rangle$ with a well-defined phase relation between them. The state of the first particle is entangled with that of the second particle, and vice versa. In the case of (1.31), if the first particle is spin-up the second is spin down, whereas if the first particle is spin-down, the second is spin-up. More generally, states of the form $(c_1|\uparrow\rangle|\downarrow\rangle + c_2|\downarrow\rangle|\uparrow\rangle)$ with nonvanishing coefficients c_1 and c_2 are entangled (typically, we consider normalized states so $|c_1|^2 + |c_2|^2 = 1$). Clearly the *triplet spin-state*,

$$|\text{triplet}\rangle = \frac{1}{\sqrt{2}} (|\uparrow\rangle|\downarrow\rangle + |\downarrow\rangle|\uparrow\rangle), \tag{1.32}$$

1.3 A First Taste of Quantum Mechanics

is also entangled.[10] If a measurement is performed on either the singlet or triplet, even after the spins are well-separated in physical space, and particle 1 is determined experimentally to be in, say, state $|\uparrow\rangle$, particle 2 is in state $|\downarrow\rangle$ with unit probability. Similarly, if 1 is in state $|\downarrow\rangle$, 2 is in state $|\uparrow\rangle$ with unit probability. Additional examples of entangled spin states are $\frac{1}{\sqrt{2}}(|\uparrow\rangle|\uparrow\rangle + |\downarrow\rangle|\downarrow\rangle)$, and, more generally, the states $(d_1|\uparrow\rangle|\uparrow\rangle + d_2|\downarrow\rangle|\downarrow\rangle)$.

It is important to stress the nonclassical nature of these states. Let us consider a particle of zero spin that is at rest and disintegrates into two particles, which, by conservation of linear momentum go in opposite directions, each with spin 1/2 so that the state of the two spin 1/2 particles is the spin singlet state. After the disintegration, we measure the spins of the particles with two Stern–Gerlach apparatuses located on opposite sides of the initial particle at rest. Many identical systems are prepared and the same measurement is performed on all of the identical systems. The probability of measuring spin-up for particle 1 is 50%, but having measured spin-up, the probability of measuring spin-down for particle 2 is 100%. Moreover, if instead, we were to measure the spin of particle 2, the probability of measuring spin-up would be 50%, but having measured spin-up for particle 2, the probability of measuring spin-down for particle 1 is 100%. This type of behavior is not possible in a classical world. We shall discuss such nonclassical behavior further in Sec. 5.7 in connection with an experiment suggested by Einstein, Podolsky, and Rosen; here we simply set the stage for the argument that will be discussed there. The famous paradox involving entangled states, developed originally by Einstein, Podolsky, and Rosen (EPR) in 1935 [20], criticized aspects of quantum mechanics, and highlighted the disorienting features of entangled states. EPR tried to answer the question: "Can the quantum-mechanical description of 'reality' be considered complete?," where a "complete theory" is one that has the property that "every element of physical reality must have a counterpart in the physical theory." Reality (or realism) means that all measurement outcomes depend on preexisting properties of objects that are independent of the measurement. The EPR argument about the incompleteness of quantum mechanics was not, however, universally accepted. For example, Bohr viewed the EPR argument as a demonstration of the inapplicability of classical descriptions to quantum phenomena. Today, most scientists accept Bohr's view, but Einstein was never convinced. In 1964, Bell investigated the EPR conclusion – that the quantum description of physical reality is not complete – by using it as a working hypothesis and quantified the EPR idea of a deterministic world. In a deterministic world, (1) measurement results are determined by properties the particles carry prior to, and independent of, the measurement (this is what is called "realism"), and (2) results obtained at one location are independent of any actions performed at space-like separation (this is called "locality" – the prohibition of any influences between events in space-like separated regions). Recent experiments show that quantum mechanics does properly predict the results of experiments that violate EPR's criteria of reality and locality. We shall take up the EPR paradox and its generalization, developed by John S. Bell, in Secs. 5.7–5.8. For the moment, we simply observe that the nature of entanglement is one of the most nonclassical aspects of quantum mechanics.

Upon "throwing away" (i.e., determining not to measure) particle 2 of the singlet state, and probing (i.e., performing a measurement on) the state of particle 1, having discarded particle 2, we find that it is not in a *pure state*, given by a wave function, but rather it is in a *mixed state* with probability 50% for finding spin-up and 50% for spin-down. A mixed state cannot be written in terms of a wave function (or ket state), but is specified by a *density matrix*. We will take up this topic further in Sec. 2.5.

The second example of an entangled state is the spatial wave function

$$\Phi(\mathbf{r}_1, \mathbf{r}_2) = C\left(e^{i\mathbf{k}\cdot\mathbf{r}_1} e^{i\mathbf{k}'\cdot\mathbf{r}_2} + e^{i\mathbf{k}'\cdot\mathbf{r}_1} e^{i\mathbf{k}\cdot\mathbf{r}_2}\right), \tag{1.33}$$

Again, this wave function cannot be written as a product of the form $\Phi(\mathbf{r}_1, \mathbf{r}_2) = \psi_\alpha(\mathbf{r}_1)\psi_\beta(\mathbf{r}_2)$. Another way of writing (1.33) is

$$|\Phi\rangle = C\left(|\mathbf{k}\rangle|\mathbf{k}'\rangle + |\mathbf{k}'\rangle|\mathbf{k}\rangle\right). \tag{1.34}$$

The analogy with the triplet spin-state (1.32) is clear. Normalization of the wave function is determined by the coefficient C, but it has no bearing on the entanglement of the two particles.

---

[10] There are two more triplet states, $|\uparrow\rangle|\uparrow\rangle$ and $|\downarrow\rangle|\downarrow\rangle$, that are not entangled; these states are called triplets because there are three states that have a certain property in common, as discussed in Sec. 4.7.

1.3.4 THE POSTULATES OF QUANTUM MECHANICS

Quantum mechanics can be formulated in terms of a few postulates, i.e., theoretical principles based on experimental observations. The goal of this section is to introduce such principles. We shall state the postulates, first using Dirac notation, and then restate them in terms of wave functions.

Postulates

1. At each instant of time, t, the state of a physical system is represented by a vector (sometimes called a ket) $|\psi(t)\rangle$ in the vector space of states.
2. Every observable attribute of a physical system is described by an operator that acts on the ket that describes the system.
3. The only possible result of the measurement of an observable $\mathcal{A}$ is one of the eigenvalues of the operator $\hat{A}$ representing the observable. An observable must be represented by a Hermitian operator.

    Since measurement results are real numbers, the eigenvalues of operators corresponding to observables are real. The operator representing an observable is often called an observable operator, or observable for short. All eigenvalues of Hermitian operators are real, and their eigenvectors are orthogonal (and can be normalized), $\langle \phi_i | \phi_j \rangle = \delta_{ij}$.
4. When a measurement of an observable $\mathcal{A}$ is made on a generic state $|\psi\rangle$, the probability of obtaining an eigenvalue a_i is given by $|\langle \phi_i | \psi \rangle|^2$, where $|\phi_i\rangle$ is the eigenstate of the observable operator $\hat{A}$ with eigenvalue a_i.

    The complex number $\langle \phi_i | \psi \rangle$ is known as the "probability amplitude" to measure $\mathcal{A}_i$ as the value for $\mathcal{A}$ in the state $|\psi\rangle$.
5. Immediately after the measurement of an observable $\mathcal{A}$ that has yielded a value a_i, the state of the system is the normalized eigenstate $|\phi_i\rangle$.

    With the system initially in state $|\psi\rangle$, measurement of an observable $\mathcal{A}$ collapses the wave function. If the result of the measurement is a_i, the wave function collapses to state $|\phi_i\rangle$.

    "Collapse of the wave function" is the most controversial postulate. A viewpoint on this controversy will be presented in Secs. 2.4 and 2.5.4.

    There are several other formulations of the measurement postulates [the measurement operator formulation, POVM (positive operator-valued measure) representation, etc.] but the present projective measurement formulation is by far the most common.
6. The time evolution of the state of a quantum system is specified by the state vector $|\psi(t)\rangle = \hat{\mathcal{U}}(t,t_0)|\psi(t_0)\rangle$, where the operator $\hat{\mathcal{U}}$ is unitary ($\hat{\mathcal{U}}\hat{\mathcal{U}}^\dagger = 1$), and therefore preserves the normalization of the associated ket, and is called the *evolution operator*:

$$|\psi(t)\rangle = \hat{\mathcal{U}}(t,t_0)|\psi(t_0)\rangle. \tag{1.35}$$

For a system with a time independent Hamiltonian $\hat{H}$, $\hat{\mathcal{U}}(t,t_0) = \exp(-i\hat{H}(t-t_0)/\hbar)$. In general (i.e., even for time-dependent Hamiltonians)

$$i\hbar \frac{\partial \hat{\mathcal{U}}(t,t_0)}{\partial t} = \hat{H}(t)\hat{\mathcal{U}}(t,t_0). \tag{1.36}$$

This is equivalent to saying that $|\psi(t)\rangle$ satisfies the Schrödinger equation, $i\hbar \frac{\partial}{\partial t}|\psi(t)\rangle = \hat{H}|\psi(t)\rangle$.

7. The state space of a composite system is the tensor product of the state spaces of the constituent systems:

$$|\psi\rangle_{N\text{-particle}} = \sum_{\alpha\beta...\zeta} C_{\alpha\beta...\zeta} |\alpha\rangle |\beta\rangle ... |\zeta\rangle. \tag{1.37}$$

The RHS contains the tensor product of N single-particle states, and the coefficients C are complex amplitudes such that $|\psi\rangle_{N\text{-particle}}$ is normalized. Hence, the quantum mechanical vector space required for describing many particles

1.3 A First Taste of Quantum Mechanics

is huge. Consider the size of the space for N spin 1/2 particles, where the size of the space for each particle is 2. Each particle can be in an arbitrary superposition of two basis states, $|\uparrow\rangle$ and $|\downarrow\rangle$, so the N-particle state can be in an arbitrary superposition of 2^N basis states. For $N = 10$ there are $2^{10} = 1024$ basis states, but for 100 particles, $2^N \approx 1.27 \times 10^{30}$ (and this without considering the positional degrees of freedom of the particles). The size of the vector space grows exponentially with the number of particles!

Wave function form of the postulates

For the sake of emphasis, and to reinforce the relation between the vector space and wave function notations of quantum mechanics, we restate the postulates in wave function notation.

1. At each instant of time, t, the state of a physical system is represented by a wave function that can be written in coordinate space in the form $\psi(x,t) (\equiv \langle x | \psi(t) \rangle)$, where x represents all the coordinates of the particles in the system.
2. Every observable attribute of a physical system is described by an operator that acts on the wave functions that describe the system.
   For example, the momentum operator can be represented by the operator, $(\hbar/i)\nabla_x$, and the position operator by x.
3. The only possible result of the measurement of an observable $\mathcal{A}$ is one of the eigenvalues of the operator $\hat{A}$ representing the observable.
   Observable operators are Hermitian. All eigenvalues of Hermitian operators are real, and their eigenfunctions are orthogonal:
   $$\langle \phi_i | \phi_j \rangle = \int dx\, \phi_i^*(x) \phi_j(x) = \delta_{ij}.$$
4. When a measurement of an observable $\mathcal{A}$ is made on a generic wave function $\psi(x)$, the probability of obtaining an eigenvalue a_i is given by $|\langle \phi_i | \psi \rangle|^2 = \left| \int dx\, \phi_i^*(x) \psi(x) \right|^2$, where ϕ_i is the eigenfunction of the observable operator $\hat{A}$ with eigenvalue a_i.
   The complex number $\langle \phi_i | \psi \rangle$, is called the "probability amplitude" to measure a_i in a measurement of $\mathcal{A}$ in the state ψ.
5. Immediately after measurement of an observable $\mathcal{A}$ has yielded a value a_i, the system is in the state represented by the normalized eigenfunction ϕ_i.
   With the system initially in the state represented by the wave function ψ, measurement of an observable $\mathcal{A}$ collapses the wave function. If the result of the measurement is a_i, the wave function collapses to the normalized wave function ϕ_i.
6. The time evolution of the state of a quantum system is specified by the wave function $\psi(t) = \mathcal{U}(t,t_0) \psi(t_0)$, for some unitary operator $\mathcal{U}$ called the evolution operator, which preserves the normalization of the associated wave function. For a time independent Hamiltonian, $\mathcal{U}(t,t_0) = \exp(-iH(t-t_0)/\hbar)$. In general (i.e., for time-dependent Hamiltonians), $i\hbar \frac{\partial \mathcal{U}(t,t_0)}{\partial t} = H(t) \mathcal{U}(t,t_0)$.
   This is equivalent to saying that $\psi(t)$ satisfies the Schrödinger equation, $i\hbar \frac{\partial}{\partial t} \psi(t) = \hat{H} \psi(t)$.
7. The wave function of a composite system is the sum of products of the wave functions of the constituent systems:
   $$\psi(x_1, \ldots, x_n)_{\text{N-particle}} = \sum_{\alpha\beta\ldots\zeta} C_{\alpha\beta\ldots\zeta} \phi_\alpha(x_1) \phi_\beta(x_2) \ldots \phi_\zeta(x_N).$$

1.3.5 TIME-DEPENDENT AND -INDEPENDENT SCHRÖDINGER EQUATIONS

There is a significant difference in the way a state is propagated in time in quantum mechanics and in classical mechanics. In order to describe the propagation in time of a quantum mechanical state, and also in order to better understand the nature of a stationary quantum mechanical state (i.e., a quantum state that has a trivial time dependence corresponding to a change of phase of the state with time), it helps to understand how a classical state is propagated in time. The quantum propagation in time is described mathematically using the Hamiltonian, i.e., the "energy" operator. The concept of a Hamiltonian comes from classical mechanics. The reader who has not encountered the concept of a Hamiltonian in classical mechanics might want to have a quick look at Secs. 16.1–16.7, linked to the book web page, before continuing to read this chapter. This will build familiarity with the Hamiltonian, and will allow a better understanding of just how different quantum and classical mechanics are.

In quantum mechanics, propagation in time of the wave function of the system, let us call it $\Psi(t)$ here, is given by a wave equation, called the Schrödinger equation, which contains the Hamiltonian, $\hat{H}$ of the system. The Hamiltonian is constructed by taking the classical Hamiltonian function $H(\mathbf{p},\mathbf{r},t)$, where $\mathbf{p}$ and $\mathbf{r}$ are the momenta and coordinates of the system (see Sec. 16.2), and turning it into an operator H, which acts on the wave function $\Psi(t)$. The Schrödinger equation is given by

$$i\hbar \frac{\partial}{\partial t}\Psi(t) = H\Psi(t). \quad (1.38)$$

[or, equivalently, $i\hbar \frac{\partial}{\partial t}|\Psi(t)\rangle = \hat{H}|\Psi(t)\rangle$]. For a particle of mass m in a potential $V(\mathbf{r},t)$, the Hamiltonian is $H(t) = \frac{\mathbf{p}^2}{2m} + V(\mathbf{r},t)$ [or equivalently, $\hat{H}(t) = \frac{\hat{\mathbf{p}}^2}{2m} + \hat{V}(t)$]. We shall see in Sec. 1.3.6 that in classical mechanics, as well as in quantum mechanics, the momentum is the generator of space translations (and the Hamiltonian is the generator of time translations). This property determines that the quantum mechanical momentum operator is given by $\mathbf{p} = (\frac{\hbar}{i}\nabla)$. (This result is further supported by the analogy of the resulting wave equation for matter wave, the time-dependent Schrödinger equation, to the wave equation for light. The latter is second order in time, whereas the former is first order in time, yet the nature of plane waves for free particles and plane waves for light in vacuum is strikingly similar, as we shall soon see, and the nature of superposition of solutions for the two wave equations is completely analogous.) The time-dependent Schrödinger equation takes the form[11]

$$i\hbar \frac{\partial \Psi}{\partial t} = -\frac{\hbar^2}{2m}\nabla^2 \Psi + V(\mathbf{r},t)\Psi. \quad (1.39)$$

Let us first consider the case of a time-independent potential. Then, it is possible to write the wave function as the product of a spatial function and a time-dependent function $\Psi(\mathbf{r},t) = \psi(\mathbf{r})\phi(t)$, as can be shown by the method of *separation of variables*. Substituting this product form into the time-dependent Schrödinger equation, dividing the resulting equation by the product $\psi(\mathbf{r})\phi(t)$, and moving the parts of the resulting equation that depend on $\mathbf{r}$ and t to opposite sides of the equation results in the following expression:

$$\frac{-\frac{\hbar^2}{2m}\nabla^2 \psi(\mathbf{r}) + V(\mathbf{r})\psi(\mathbf{r})}{\psi(\mathbf{r})} = i\hbar \frac{\partial \phi(t)/\partial t}{\phi(t)}. \quad (1.40)$$

Hence, we can conclude that both sides of the resulting equation must be equal to a constant independent of $\mathbf{r}$ and t, since a function that depends only on $\mathbf{r}$ cannot equal a function that depends only on t unless both functions are constants. Since the units of both sides of the equation are energy, we will call the constant E. We now have the two equations obtained by setting the LHS of Eq. (1.40) equal to E, and the RHS of (1.40) equal to E. We can integrate the equation for ϕ to obtain

$$\phi(t) = \phi_0 \exp(-iEt/\hbar). \quad (1.41)$$

The spatial wave function $\psi(\mathbf{r})$ satisfies the equation,

$$-\frac{\hbar^2}{2m}\nabla^2 \psi + V(\mathbf{r})\psi = E\psi, \quad (1.42)$$

which is called the *time-independent Schrödinger equation*.

---

[11] To write the Schrödinger equation in curvilinear coordinates, see Sec. C.4 in Appendix C, and in particular see Eq. (C.53) to express ∇^2 in curvilinear coordinates.

1.3 A First Taste of Quantum Mechanics

For a vanishing potential, $V(\mathbf{r}) = 0$, it is easy to see by direct substitution that the plane wave, $\psi(\mathbf{r}) = C_\mathbf{p} e^{i\mathbf{p}\cdot\mathbf{r}/\hbar}$, where $C_\mathbf{p}$ is some normalization constant (which may or may not depend on $\mathbf{p}$), is a solution with energy $E = p^2/2m$. The full wave function is then given by the product form

$$\Psi(\mathbf{r},t) = C_\mathbf{p} e^{i(\mathbf{p}\cdot\mathbf{r}/\hbar - \frac{p^2}{2m}t/\hbar)}. \tag{1.43}$$

In analogy, the plane wave solution to the wave equation for light, Eq. (1.2), is

$$\mathbf{E}(\mathbf{r},t) = \mathbf{E}_\mathbf{k} e^{i(\mathbf{k}\cdot\mathbf{r} - \omega t)}, \tag{1.44}$$

where $\mathbf{E}(\mathbf{r},t)$ is the spatially and temporally dependent electric field, $\mathbf{E}_\mathbf{k}$ is a constant vector, and (1.44) is a solution provided $\omega = kc$. The energy of the 'photon' corresponding to this field is $E = \hbar\omega = \hbar kc$, and its momentum is $\mathbf{p} = \hbar\mathbf{k}$.

In the previous paragraphs, we explicitly considered the case of a system consisting of one particle, but we could have considered any number of particles, as long as the Hamiltonian is not explicitly time-dependent, and a similar result would have been obtained. For example, the N-particle Hamiltonian for particles in an external potential $U(\mathbf{r})$ interacting with each other through an interaction potential $V(\mathbf{r}_i - \mathbf{r}_j)$ is

$$H = \sum_{i=1}^{N} \left[\frac{\mathbf{p}_i^2}{2m} + U(\mathbf{r}_i) \right] + \frac{1}{2} \sum_{i \neq j=1}^{N} V(\mathbf{r}_i - \mathbf{r}_j). \tag{1.45}$$

The separation of variables method can again be used to write the wave function as the product of spatial and temporal functions, $\Psi(\{\mathbf{r}_i\},t) = \psi(\{\mathbf{r}_i\})\phi(t)$. The time-independent Schrödinger equation for the spatial wave function becomes

$$\boxed{H\psi_\alpha(\mathbf{r}_1, \mathbf{r}_2, \ldots, \mathbf{r}_N) = E_\alpha \psi_\alpha(\mathbf{r}_1, \mathbf{r}_2, \ldots, \mathbf{r}_N).} \tag{1.46}$$

The eigenvalues E_α and eigenfunctions $\psi_\alpha(\mathbf{r}_1, \mathbf{r}_2, \ldots, \mathbf{r}_N)$ of this equation determine the time-dependent wave functions, $\Psi_\alpha(\mathbf{r}_1, \mathbf{r}_2, \ldots, \mathbf{r}_N, t) = \psi_\alpha(\mathbf{r}_1, \mathbf{r}_2, \ldots, \mathbf{r}_N) e^{-iE_\alpha t/\hbar}$, which are products of a function of $\{\mathbf{r}_i\}$ and a function of t. Linear combinations of these solutions are also a solution of the time-dependent Schrödinger equation (1.38); depending upon initial conditions, a linear combination of solutions may be required. That is, if at time $t = 0$, the initial state is of the form $\Psi(\mathbf{r}_1, \mathbf{r}_2, \ldots, \mathbf{r}_N, 0) = \sum_\alpha c_\alpha \psi_\alpha(\mathbf{r}_1, \mathbf{r}_2, \ldots, \mathbf{r}_N)$, then the time-dependent solution that satisfies this initial condition is

$$\Psi(\mathbf{r}_1, \mathbf{r}_2, \ldots, \mathbf{r}_N, t) = \sum_\alpha c_\alpha \psi_\alpha(\mathbf{r}_1, \mathbf{r}_2, \ldots, \mathbf{r}_N) e^{-iE_\alpha t/\hbar}. \tag{1.47}$$

Problem 1.3

(a) Use the separation of variables method to show that a solution $\psi(x,y)$ to the time-independent Schrödinger equation $[H_x(x) + H_y(y)]\psi(x,y) = E\psi(x,y)$ can be written as $\psi(x,y) = \varphi(x)\phi(y)$ with $\varphi(x)$ and $\phi(y)$ satisfying the equations, $H_x(x)\varphi(x) = \varepsilon\varphi(x)$ and $H_y(x)\phi(x) = \epsilon\phi(x)$, if $\varepsilon + \epsilon = E$.

(b) Generalize (a) to the case of a multidimensional Hamiltonian that is additive,
$H(x_1, x_2, \ldots, x_n) = H_1(x_1) + \cdots + H_n(x_n)$.

Answer: (b) $H\psi(x_1, x_2, \ldots, x_n) = E\psi(x_1, x_2, \ldots, x_n)$, $\psi(x_1, \ldots, x_n) = \varphi_1(x_1) \ldots \varphi_n(x_n)$, $E = \varepsilon_1 + \cdots + \varepsilon_n$, and $H_i(x_i)\varphi_i(x_i) = \varepsilon_i\varphi_i(x_i)$.

1.3.6 MOMENTUM, ENERGY, AND ANGULAR MOMENTUM

In classical mechanics, the energy of a system is specified by the Hamiltonian (see Secs. 16.2 and 16.4), and therefore it should come as no surprise that the energy operator in quantum mechanics is the Hamiltonian. Moreover, in classical mechanics, the Hamiltonian is the generator of time-translations, as it is in quantum mechanics (see below). Quite generally, the space–time symmetries of displacements (i.e., translations), rotations and time-translations can be used to derive the quantum mechanical operators for momentum, angular momentum, and energy, just as they are used in classical mechanics to derive these classical quantities. These operators can also be obtained using wave–particle duality arguments; the latter method is perhaps more intuitive, but less appealing.

For systems that are homogeneous, the state of the system must be invariant under displacement, i.e., under translation of the system in space (or translation of the coordinate system used to describe the system). Similarly, systems that are spherically symmetric are invariant under rotations. Furthermore, even if the system is not spatially homogeneous or spherically symmetric, translating or rotating the system (or the coordinate system used to describe the system) does not change it in any significant way as long as the system is not in an external potential. The same is true of any Galilean transformation (i.e., translations, rotations, boosts of the system by a uniform velocity, or time-translations) applied to the system (or of the coordinate system), as we shall see below. For each such space–time transformation there is a corresponding transformation of observables and of state vectors, both classically and quantum mechanically.

Transformation operators in quantum mechanics must be unitary (or anti-unitary, as is the case with the time-inversion transformation, but we shall not consider this case at present), $U^{-1} = U^\dagger$. This requirement arises because a transformation U which changes states, $|\psi\rangle \to |\psi'\rangle = U|\psi\rangle$, must keep inner products unchanged, $\langle\phi|\psi\rangle \to \langle\phi'|\psi'\rangle = (\langle\phi|U^\dagger)U|\psi\rangle = \langle\phi|\psi\rangle$, hence $U^\dagger U = 1$. To determine how operators are affected by transformations, consider the transformation of the state vector $\mathcal{O}|\psi\rangle$: $\mathcal{O}|\psi\rangle \to U(\mathcal{O}|\psi\rangle) = U\mathcal{O}U^\dagger U|\psi\rangle$, where we have inserted unity in the form $U^\dagger U$ to obtain the last equality. Since this argument holds for any state vector $|\psi\rangle$, we conclude an opertor is transformed as follows:

$$\boxed{\mathcal{O} \to \mathcal{O}' = U\mathcal{O}U^\dagger.} \tag{1.48}$$

It is useful to incorporate the displacement parameter of the transformation within the symbol for the transformation operator, e.g., if we translate the system by the vector $\mathbf{R}$ in coordinate space, the operator that represents this translation in Hilbert space will be denoted as $U(\mathbf{R})$. The translation of the coordinate of a particle located at $\mathbf{r}$ in configuration (coordinate) space by the vector $\mathbf{R}$ will be denoted by $\mathcal{T}(\mathbf{R})$, i.e., $\mathcal{T}(\mathbf{R})\mathbf{r} = \mathbf{r} + \mathbf{R}$. Note that translation of the coordinate system by $\mathbf{R}$ means that the particle coordinate goes to $\mathbf{r} - \mathbf{R}$, so translating the particle and translating the coordinate system are very different operations. Note also the distinction between the operator that translates a coordinate of a particle, $\mathcal{T}(\mathbf{R})$, and the operator that translates a state vector in Hilbert space, $U(\mathbf{R})$.

If we first displace the system by $\mathbf{R}$, and then carry out another displacement by $\mathbf{R}'$, the resulting transformation operator is given by the product $U(\mathbf{R}')U(\mathbf{R})$. But this resulting transformation is equivalent to a transformation by a displacement $\mathbf{R}' + \mathbf{R}$. Hence, we must have (at least up to a phase),

$$\boxed{U(\mathbf{R}' + \mathbf{R}) = U(\mathbf{R}')U(\mathbf{R}).} \tag{1.49}$$

The group of displacements is *abelian*, i.e., $U(\mathbf{R}')U(\mathbf{R}) = U(\mathbf{R})U(\mathbf{R}')$. We mention parenthetically that rotational transformations are not necessarily abelian (see below), but the group property that the product of two rotations is another rotation does hold, i.e., the product of two rotational transformations of a state vector corresponds to another rotational transformation.

A unitary operator U can be written in terms of a Hermitian operator K as follows: $U(\mathbf{R}) = e^{iK(\mathbf{R})}$. In other words, if $U(\mathbf{R})$ is unitary, $K(\mathbf{R})$ is Hermitian (and vice versa). It is easy to see that $UU^\dagger = e^{iK}e^{-iK^\dagger} = e^{iK}e^{-iK} = 1$, and $U^\dagger U = e^{-iK^\dagger}e^{iK} = e^{-iK}e^{iK} = 1$. Moreover, from Eq. (1.49) we find that $K(\mathbf{R}' + \mathbf{R}) = K(\mathbf{R}) + K(\mathbf{R}')$.

The inner product of the transformed state $|\psi'\rangle = U(\mathbf{R})|\psi\rangle$ with $|\mathbf{r}\rangle$ yields the wave function $\langle\mathbf{r}|\psi'\rangle = \psi'(\mathbf{r}) = \langle\mathbf{r}|U(\mathbf{R})|\psi\rangle$. Let us try to understand what this wave function is. First, we note that it follows from the definition of the displacement operator $U(\mathbf{R})$ that $U(\mathbf{R})|\mathbf{r}\rangle = |\mathbf{r} + \mathbf{R}\rangle$, and by taking the Hermitian conjugate of this equation, $\langle\mathbf{r} + \mathbf{R}| =$

1.3 A First Taste of Quantum Mechanics

$\langle \mathbf{r}|U^\dagger(\mathbf{R})$. But $U^\dagger(\mathbf{R})$ is the inverse of $U(\mathbf{R})$ (by the unitarity property), as is $U(-\mathbf{R})$, hence $U^\dagger(\mathbf{R}) = U(-\mathbf{R})$. We therefore find that $\langle \mathbf{r}+\mathbf{R}| = \langle \mathbf{r}|U(-\mathbf{R})$ and $\langle \mathbf{r}-\mathbf{R}| = \langle \mathbf{r}|U(\mathbf{R})$. Hence, $\langle \mathbf{r}|\psi\rangle \to \langle \mathbf{r}|\psi'\rangle = \langle \mathbf{r}|U(\mathbf{R})|\psi\rangle = \langle \mathbf{r}-\mathbf{R}|\psi\rangle$, and therefore

$$\psi(\mathbf{r}) \to \psi'(\mathbf{r}) \equiv \langle \mathbf{r}|\psi'\rangle = \langle \mathbf{r}|U(\mathbf{R})|\psi\rangle = U(\mathbf{R})\psi(\mathbf{r}) = \psi(\mathbf{r}-\mathbf{R}) \,. \tag{1.50}$$

Hence, $\psi(\mathbf{r}-\mathbf{R}) = U(\mathbf{R})\psi(\mathbf{r})$, from which we can easily surmise,

$$\psi(\mathbf{r}+\mathbf{R}) = \langle \mathbf{r}|U^{-1}(\mathbf{R})|\psi\rangle = U^{-1}(\mathbf{R})\psi(\mathbf{r}) = U(-\mathbf{R})\psi(\mathbf{r}) \,. \tag{1.51}$$

An explicit construction of $U(\mathbf{R})$ will now be presented.

Consider a system consisting of a single particle in 3D with wave function $\psi(\mathbf{r})$, and apply an infinitesimal transformation of coordinates so that $\mathbf{r} \to \mathbf{r} + \delta\mathbf{r}$. Let us expand the wave function $\psi(\mathbf{r}+\delta\mathbf{r})$ about $\mathbf{r}$:

$$\psi(\mathbf{r}+\delta\mathbf{r}) = \psi(\mathbf{r}) + \nabla\psi \cdot \delta\mathbf{r} + \cdots = [1 + \delta\mathbf{r}\cdot\nabla + \cdots]\psi(\mathbf{r}). \tag{1.52}$$

Hence, the coordinate transformation, $\mathbf{r} \to \mathbf{r}+\delta\mathbf{r}$, leads to application of the operator $[1+\delta\mathbf{r}\cdot\nabla+\cdots]$, which, according to the previous paragraph, is transformation operator $U(-\delta\mathbf{r})$, so $U(-\delta\mathbf{r}) = [1 + i\,\delta\mathbf{r}\cdot(\frac{\hbar}{i}\nabla)/\hbar + \cdots]$. Note that in the second term we arbitrarily multiplied and divided by $i\hbar$ for convenience in the discussion that follows. The operator $U(-\delta\mathbf{r})$ is the transformation operator for translation of the wave function by $-\delta\mathbf{r}$, not by $\delta\mathbf{r}$, as explained in the previous paragraph; hence $U(\delta\mathbf{r}) = \{1 + i[-\delta\mathbf{r}\cdot(\frac{\hbar}{i}\nabla)/\hbar] + \cdots\}$. This infinitesimal displacement operator can be written in terms of the infinitesimal Hermitian operator $\delta K = [-\delta\mathbf{r}\cdot(\frac{\hbar}{i}\nabla)/\hbar]$ as follows: $U(\delta\mathbf{r}) = e^{i\delta K} = e^{i[-\delta\mathbf{r}\cdot(\frac{\hbar}{i}\nabla)/\hbar]} = [1 - i\,\delta\mathbf{r}\cdot(\frac{\hbar}{i}\nabla)/\hbar + \cdots]$. Thus, the operator for translation by any coordinate $\mathbf{R}$ is

$$U(\mathbf{R}) = e^{-i\mathbf{R}\cdot(\frac{\hbar}{i}\nabla)/\hbar}. \tag{1.53}$$

To conclude, we have seen that

$$\boxed{\psi(\mathbf{r}+\mathbf{R}) = U(-\mathbf{R})\psi(\mathbf{r}) = e^{i\mathbf{R}\cdot(\frac{\hbar}{i}\nabla)/\hbar}\psi(\mathbf{r}).} \tag{1.54}$$

We say that the operator $(\frac{\hbar}{i}\nabla)$ is the *generator* of the translations, and the operator $e^{i\mathbf{R}\cdot(\frac{\hbar}{i}\nabla)/\hbar}$ translates the wave function by $\mathbf{R}$. The operator $(\frac{\hbar}{i}\nabla)$ is the momentum operator, as we shall see below.

Generators of Galilean Transformations

The direct connection of the generator of translations and the momentum can be understood in the broad context of generators of Galilean transformations.[12] Quite generally, space–time symmetries include symmetry under transformations comprised of rotations, translations, and transformations between uniformly moving frames of reference [21–24]. The latter are Galilean (or relativistically, Lorentz) transformations that boost the velocities of one coordinate system relative to another (or boost the velocity of a particle within a certain reference frame). If the velocity is small compared to the speed of light, Lorentz transformations reduce to Galilean transformations. The set of all such nonrelativistic transformations (including translations, rotations, velocity boosts, and time-translations) are the elements of a group called the Galilei group. Under a general Galilean transformation, the coordinate $\mathbf{r}$ and the velocity $\dot{\mathbf{r}}$ transform as follows:

$$\mathbf{r} \to \mathbf{r}' = \mathcal{R}\mathbf{r} - \mathbf{v}t - \mathbf{R}, \quad \dot{\mathbf{r}} \to \dot{\mathbf{r}}' = \mathcal{R}\dot{\mathbf{r}} - \mathbf{v}. \tag{1.55}$$

Here $\mathbf{v}$ is a velocity boost vector, $\mathbf{R}$ is a displacement vector, and $\mathcal{R}$ is a 3×3 rotation matrix.

---

[12] Galilean transformations transform between the coordinates of two reference frames that differ by constant relative motion within the constructs of Newtonian physics [21, 22]. Commonly, the definition of Galilean transformations is broadened to include not only velocity boosts, but also translations, rotations, and time-translations. In the context of relativistic mechanics, Galilean transformations are replaced by Lorentz transformations [21–24].

In classical mechanics, the generator of translations is the momentum **p**. Hence, by association with classical mechanics, the operator $\frac{\hbar}{i}\nabla$ that was derived above to be the generator of translations in quantum mechanics must be the momentum operator $\hat{\mathbf{p}}$ (see the discussion of plane waves in the next section to confirm this connection). The transformation operator $U(\mathbf{R})$ that represents the translation of the particle by a vector $\mathbf{R}$ (or, if you like, translation of the coordinate system by a vector $-\mathbf{R}$) transforms the wave function $\psi(\mathbf{r})$ to $\psi(\mathbf{r} - \mathbf{R}) = U(\mathbf{R})\psi(\mathbf{r})$ and transforms the operator $\hat{\mathbf{r}}$ to $\hat{\mathbf{r}} - \mathbf{R}$ is given by $U(\mathbf{R}) = e^{-i\hat{\mathbf{p}}\cdot\mathbf{R}/\hbar}$, i.e.,

$$\psi(\mathbf{r} - \mathbf{R}) = e^{-i\hat{\mathbf{p}}\cdot\mathbf{R}/\hbar}\psi(\mathbf{r}), \quad e^{-i\hat{\mathbf{p}}\cdot\mathbf{R}/\hbar}\hat{\mathbf{r}}\,e^{i\hat{\mathbf{p}}\cdot\mathbf{R}/\hbar} = \hat{\mathbf{r}} - \mathbf{R}. \tag{1.56}$$

Moreover, in classical mechanics, the generator of rotations is the angular momentum, **J**, the generator of boosts in velocity is $\mathbf{G} \equiv m\mathbf{r}$, and the generator of time-translations is the Hamiltonian, H [9, 24].

In order to cement (or at least reinforce) the connection between the operator $\frac{\hbar}{i}\nabla$, which is the generator of translations, and the momentum operator, $\hat{\mathbf{p}}$, let us apply $\frac{\hbar}{i}\nabla$ to a plane wave state. We do so in the next subsection, and then return to Galilean transformations immediately thereafter.

Plane Waves

Plane waves are eigenfunctions of the momentum operator $\frac{\hbar}{i}\nabla$:

$$\frac{\hbar}{i}\nabla\psi(\mathbf{r}) = \mathbf{p}\,\psi(\mathbf{r}). \tag{1.57}$$

The eigenvalue of $\frac{\hbar}{i}\nabla$ appearing on the RHS of (1.57), is denoted **p**; we can denote the eigenfunction as $\psi_{\mathbf{p}}(\mathbf{r})$. The solution to Eq. (1.57) is

$$\psi_{\mathbf{p}}(\mathbf{r}) = C\,e^{i\mathbf{p}\cdot\mathbf{r}/\hbar}, \tag{1.58}$$

where C is a constant, as can be verified by direct substitution into Eq. (1.57).

Since $\frac{\hbar}{i}\nabla$ is the momentum operator $\hat{\mathbf{p}}$, the kinetic energy operator for a particle of mass m is $\frac{\hat{\mathbf{p}}^2}{2m} = -\frac{\hbar^2}{2m}\nabla^2$. The plane wave function (1.58) is also an eigenfunction of the kinetic energy operator, with eigenvalue $E = \mathbf{p}^2/2m$. Thus, $E = \mathbf{p}^2/2m$ is the eigenvalue of the free particle Hamiltonian $\hat{H} = \hat{\mathbf{p}}^2/2m$, and the plane wave (1.58) is the eigenfunction of the free particle Hamiltonian. The wave vector $\mathbf{k} \equiv \mathbf{p}/\hbar$ is related to the energy through the quadratic *dispersion relation* $E = \hbar^2 k^2/(2m)$. Clearly this is a very different dispersion relation than obtained for photons in Eq. (1.4).

Translation of a plane wave introduces an additional phase multiplying the wave function:

$$\psi_{\mathbf{p}}(\mathbf{r} + \mathbf{R}) = U(-\mathbf{R})\,\psi_{\mathbf{p}}(\mathbf{r}) = e^{i\mathbf{p}\cdot\mathbf{R}/\hbar}\,\psi_{\mathbf{p}}(\mathbf{r}). \tag{1.59}$$

The phase of the phase factor on the RHS of (1.59) is the scalar product of the momentum eigenvalue **p** and the displacement **R**.

The plane wave state (1.58) is not normalizable in the usual sense. In order to determine what kind of normalization of plane waves is possible, it is useful to introduce the Dirac delta function. We shall do this immediately after the next subsection.

As we shall see, plane waves are very useful in quantum mechanics; they are often used as basis functions with which to expand wave functions.

Generators of Galilean Transformations Continued

Let us consider an infinitely small rotation, $\delta\boldsymbol{\varphi}$ by the angle $\delta\varphi$ about a rotation axis $\hat{\boldsymbol{\varphi}}$. The change in the coordinate vector $\delta\mathbf{r}$ resulting from such a rotation is given by

$$\delta\mathbf{r} = \delta\boldsymbol{\varphi} \times \mathbf{r}. \tag{1.60}$$

1.3 A First Taste of Quantum Mechanics

The change in the wave function due to such an infinitesimal rotation is given by

$$\psi(\mathbf{r} + \delta\mathbf{r}) = \psi(\mathbf{r}) + \delta\mathbf{r} \cdot \nabla\psi = \psi(\mathbf{r}) + \delta\boldsymbol{\varphi} \times \mathbf{r} \cdot \nabla\psi$$
$$= \psi(\mathbf{r}) + i\frac{\delta\boldsymbol{\varphi}}{\hbar} \cdot \left(\mathbf{r} \times \left(\frac{\hbar}{i}\right)\nabla\right)\psi = \left(1 + i\frac{\delta\boldsymbol{\varphi}}{\hbar} \cdot \mathbf{L}\right)\psi. \quad (1.61)$$

This infinitesimal rotation, when repeated many times, yields a finite rotation as specified below. Thus, the operator that generates the unitary rotational transformation is the angular momentum operator $\mathbf{L} \equiv \mathbf{r} \times \mathbf{p} = \mathbf{r} \times (-i\hbar)\nabla$. We therefore conclude that

$$\boxed{\psi(\mathcal{R}_{\boldsymbol{\varphi}}\mathbf{r}) = U(-\boldsymbol{\varphi})\psi(\mathbf{r}) = e^{i\boldsymbol{\varphi}\cdot\mathbf{L}/\hbar}\psi(\mathbf{r}).} \quad (1.62)$$

The unitary operator for rotation by $\boldsymbol{\varphi}$ is

$$U(\boldsymbol{\varphi}) = e^{-i\boldsymbol{\varphi}\cdot\mathbf{L}/\hbar}, \quad (1.63)$$

and its power series expansion is $e^{-i\boldsymbol{\varphi}\cdot\mathbf{L}/\hbar} = 1 - i\boldsymbol{\varphi}\cdot\mathbf{L}/\hbar + \cdots$, where the first two terms on the RHS are sufficient for very small $\boldsymbol{\varphi}$. Here $\mathcal{R}_{\boldsymbol{\varphi}}\mathbf{r}$ is the rotation of the coordinate $\mathbf{r}$ by the angle φ about a rotation axis $\hat{\boldsymbol{\varphi}}$, and $U(\boldsymbol{\varphi})$ is the transformation of the state vector in Hilbert space due to the rotation. Note again the inverse relation that exists between the effects of coordinate transformations on state vectors and transformations on coordinates. We will see in Sec. 3.3 that the 3×3 rotation matrix $\mathcal{R}_{\boldsymbol{\varphi}} = e^{-i\boldsymbol{\varphi}\cdot\mathbf{L}/\hbar}$, where $\mathbf{L}$ is the 3×3 representation of the angular momentum operator.

We should mention that rotational transformations about *different* rotational axes do not commute, e.g., $\mathcal{R}_{\varphi_x}\mathcal{R}_{\varphi_y} \neq \mathcal{R}_{\varphi_y}\mathcal{R}_{\varphi_x}$. Moreover, different components of the angular momentum operator also do not commute, e.g., $L_xL_y \neq L_yL_x$. Hence, the unitary transformation operators that correspond to rotations about different axes also do not commute, e.g., $U_x(\phi_x)U_y(\phi_y) \neq U_y(\phi_y)U_x(\phi_x)$, i.e., the order of rotations about different axes is important.

We will study rotational transformations at length in Sec. 3.3. The discussion here is only a preamble to the subject of rotations and angular momentum.

Discussion of unitary transformations that boost the velocity of a particle by a certain velocity vector, and that accelerate a particle will be delayed until Chapter 2, Sec. 2.9.4, after we have developed some more expertise.

It is easy to determine the generator of time-translations by expanding the wave function $\psi(t + \delta t)$ in a Taylor series about time t:

$$\psi(t + \delta t) = \psi(t) + \left(\frac{\partial}{\partial t}\psi\right)_t \delta t + \cdots \quad (1.64)$$

One of the postulates of quantum mechanics, see Sec. 1.3.4, which is essentially equivalent to the time-dependent Schrödinger equation, is that the time evolution of a state vector is given by

$$\psi(t) = \mathcal{U}(t, t_0)\psi(t_0), \quad (1.65)$$

where $\mathcal{U}$ is called the *evolution operator*, and

$$\boxed{i\hbar\frac{\partial \mathcal{U}(t, t_0)}{\partial t} = H(t)\mathcal{U}(t, t_0).} \quad (1.66)$$

Here $H(t)$ is the Hamiltonian, which may or may not be dependent on time. Hence, using (1.65) (or using the time-dependent Schrödinger equation), $\psi(t + \delta t) = \mathcal{U}(t + \delta t, t)\psi(t) = [1 - i\frac{\delta t H(t)}{\hbar} + \cdots]\psi(t)$. Thus, we see that the generator of time-translations is the Hamiltonian. It is easy to verify by direct substitution into (1.66) that for a time-independent Hamiltonian, H,

$$\mathcal{U}(t, t_0) = e^{-iH(t-t_0)/\hbar}. \quad (1.67)$$

Clearly, the operator $\mathcal{U}(t, 0)$ is the evolution operator from time $t = 0$ to time t. The evolution operator satisfies the condition, $\mathcal{U}(t_1 + t_2, 0) = \mathcal{U}(t_1 + t_2, t_1)\mathcal{U}(t_1, 0)$, regardless of whether the Hamiltonian is time-dependent or not. For a time-independent Hamiltonian, $\mathcal{U}(t_2, t_1)$ depends only upon the time-difference $t_2 - t_1$. Moreover, $\mathcal{U}(t, t_0)$ is unitary, and $\mathcal{U}(t_0, t)\mathcal{U}(t, t_0) = 1$.

Note that, the relation that exists between the effects of coordinate transformations on state vectors and transformations on coordinates, Eqs (1.50) and (1.51), does not exist with regard to time transformations. Time appears only as a parameter in the wave function, and this is different from the way that space variables are treated in nonrelativistic quantum mechanics.

1.3.7 DIRAC DELTA FUNCTIONS

The Dirac *delta function* is defined such that it vanishes everywhere except at one point and there it is infinite, and its integral equals unity. It can be defined as the limit of a normalized gaussian function as follows:

$$\delta(x) \equiv \lim_{\sigma \to 0} \frac{1}{\sqrt{2\pi\sigma^2}} e^{-\frac{x^2}{2\sigma^2}}. \tag{1.68}$$

Mathematicians would say that the Dirac delta function is not a well-defined function, but rather a *generalized function*. The delta function $\delta(x - x_0)$ is simply a shifted delta function that vanishes everywhere except at $x = x_0$; clearly, $\delta(x - x_0) = \lim_{\sigma \to 0} \frac{1}{\sqrt{2\pi\sigma^2}} e^{-\frac{(x-x_0)^2}{2\sigma^2}}$. By definition of the delta function, the following integral, whose integrand contains $\delta(x - x_0)$, depends only on the integrand at the point x_0:

$$\int_{-\infty}^{\infty} dx \, g(x) \, \delta(x - x_0) = g(x_0). \tag{1.69}$$

If $g(x) = 1$, the integral is unity; if $g(x) = x^3$, the integral equals x_0^3, etc.

The eigenvalue equation for the position operator, $\hat{x}\psi(x) = x_0\psi(x)$, where x_0 is the real eigenvalue, has the formal solution,

$$\psi(x) = \delta(x - x_0), \tag{1.70}$$

i.e., $\psi(x)$ is the delta function centered at x_0.

Problem 1.4

(a) Show that $\lim_{\epsilon \to 0^+} \frac{1}{\pi} \frac{\epsilon}{x^2 + \epsilon^2} = \delta(x)$.

(b) Show that $\lim_{\epsilon \to 0^+} \frac{1}{2\epsilon} e^{-|x|/\epsilon} = \delta(x)$.

Answer: (a) For $x \neq 0$, $\lim_{\epsilon \to 0} \frac{\epsilon}{x^2 + \epsilon^2} = 0$. For any value of ϵ, $\frac{1}{\pi} \int dx \frac{\epsilon}{x^2 + \epsilon^2} = \text{sgn}(\epsilon)$. (b) $\frac{1}{2\epsilon} \int_{-\infty}^{\infty} dx \, e^{-|x|/\epsilon} = 1$.

Integrals of the form $\int_{-\infty}^{\infty} dx \, g(x) \delta(f(x))$ can be evaluated easily by making a change of variables. Letting $y = f(x)$, so $dy = (df/dx) \, dx$ and $x = f^{-1}(y)$, where f^{-1} is the inverse of the function f, we find,

$$\int_{-\infty}^{\infty} dx \, g(x) \, \delta(f(x)) = \sum_i \left| \frac{df}{dx} \right|_{x=x_i}^{-1} g(f^{-1}(x_i)), \tag{1.71}$$

where x_i are the roots of the equation $f(x) = 0$. For example, $\int_{-\infty}^{\infty} dx \, g(x) \delta(ax) = |a|^{-1} g(0)$. Another example involves the following integral, with $a > 0$:

$$\int_{-\infty}^{\infty} dx \, g(x) \, \delta(ax^2 - y_0) = \frac{1}{2\sqrt{ay_0}} g(\sqrt{y_0/a}) + \frac{1}{2\sqrt{ay_0}} g(-\sqrt{y_0/a}).$$

1.3 A First Taste of Quantum Mechanics

The derivative of a delta function, $\delta'(x)$, i.e., $\frac{d}{dx}\delta(x)$, is another generalized function related to the delta function. Its behavior can be understood by considering $\int_{-\infty}^{\infty} dx\, g(x)\delta'(x)$, and integrating by parts to find

$$\int_{-\infty}^{\infty} dx\, g(x)\, \delta'(x) = -\int_{-\infty}^{\infty} dx\, g'(x)\, \delta(x) = -g'(0). \tag{1.72}$$

Another related function is the step function, often called the *Heaviside step function*, $\Theta(x) = \int_{-\infty}^{x} dx'\, \delta(x')$; $\Theta(x) = 0$ for $x < 0$ and $\Theta(x) = 1$ for $x \geq 0$.

The 3D delta function, $\delta(\mathbf{r}) \equiv \delta(x)\delta(y)\delta(z)$, is sometimes denoted by $\delta^{(3)}(\mathbf{r})$. It has the following properties:

$$\int d\mathbf{r}\, g(\mathbf{r})\, \delta(a\mathbf{r}) = |a|^{-3} g(0), \tag{1.73}$$

$$\int d\mathbf{r}\, g(\mathbf{r})\, \delta(a\mathbf{r} - \mathbf{y}_0) = |a|^{-3} g(\mathbf{y}_0/a). \tag{1.74}$$

The 2D delta function can be used in two-dimensional integrals, e.g., it can be used to specify a specific solid angle, as in $\delta^{(2)}(\Omega_\mathbf{p} - \Omega_{\mathbf{p}'})$, see below.

An important equation involving delta functions is

$$\int d\mathbf{r}\, e^{i(\mathbf{k}-\mathbf{k}')\cdot\mathbf{r}} = (2\pi)^3 \delta(\mathbf{k} - \mathbf{k}'). \tag{1.75}$$

[Typically, we shall not explicitly write out three-dimensional integrals but only use a single integral sign, just as we have in (1.75).] To show that this equation is correct, multiply both sides of Eq. (1.75) by an arbitrary function $\tilde{f}(\mathbf{k}')$ and integrate over $\mathbf{k}'$ to obtain

$$\int d\mathbf{k}'\, \tilde{f}(\mathbf{k}') \int d\mathbf{r}\, e^{i(\mathbf{k}-\mathbf{k}')\cdot\mathbf{r}} = (2\pi)^3 \tilde{f}(\mathbf{k}). \tag{1.76}$$

The integral on the left hand side (LHS) of (1.76) can be computed by first doing the integral over $\mathbf{k}'$ and noting that it yields the Fourier transform of the function $\tilde{f}$, i.e., (1.76) becomes, after dividing the resulting equation by $(2\pi)^3$:

$$(2\pi)^{-3/2} \int d\mathbf{r}\, f(\mathbf{r})\, e^{i\mathbf{k}\cdot\mathbf{r}} = \tilde{f}(\mathbf{k}), \tag{1.77}$$

where f is the inverse Fourier transform of $\tilde{f}$.[13] The quantity on the LHS of (1.77) is, by definition, the Fourier transform of f, denoted by $\tilde{f}$. We have thereby verified that (1.75) is valid.

Plane waves cannot be normalized to unit norm, but they can be "momentum-normalized" or "energy-normalized" or "flux-normalized". Since,

$$\langle \psi_{\mathbf{p}'} | \psi_{\mathbf{p}} \rangle = C_{\mathbf{p}'}^* C_{\mathbf{p}} \int d\mathbf{r}\, e^{i(\mathbf{p}-\mathbf{p}')\cdot\mathbf{r}/\hbar} = C_{\mathbf{p}'}^* C_{\mathbf{p}} (2\pi\hbar)^3 \delta(\mathbf{p} - \mathbf{p}'), \tag{1.78}$$

the delta function normalized plane wave state is taken to have normalization coefficient $C_\mathbf{p} = (2\pi\hbar)^{-3/2}$, so $\langle \psi_{\mathbf{p}'} | \psi_{\mathbf{p}} \rangle = \delta(\mathbf{p} - \mathbf{p}')$ where

$$\boxed{\psi_\mathbf{p}(\mathbf{r}) = \langle \mathbf{r} | \mathbf{p} \rangle = (2\pi\hbar)^{-3/2}\, e^{i\mathbf{p}\cdot\mathbf{r}/\hbar}.} \tag{1.79}$$

---

[13] The Fourier transform of a function $f(x)$ in one dimension is defined as follows: $\tilde{f}(k) = (2\pi)^{-1/2} \int_{-\infty}^{\infty} dx\, f(x) e^{ikx}$. The inverse transform is defined as $f(x) = (2\pi)^{-1/2} \int_{-\infty}^{\infty} dk\, \tilde{f}(k) e^{-ikx}$. The three-dimensional Fourier transform of the function $f(\mathbf{r})$ is defined as, $\tilde{f}(\mathbf{k}) = (2\pi)^{-3/2} \int_{-\infty}^{\infty} d\mathbf{r}\, f(\mathbf{r}) e^{i\mathbf{k}\cdot\mathbf{r}}$, and the inverse transform is $f(\mathbf{r}) = (2\pi)^{-3/2} \int_{-\infty}^{\infty} d\mathbf{k}\, \tilde{f}(\mathbf{k}) e^{-i\mathbf{k}\cdot\mathbf{r}}$. See Appendix D and Refs. [25, 26].

Often, the variable commonly used is not the momentum **p**, but rather the wavenumber **k** = **p**/$\hbar$. Then the plane wave in d-dimensions takes the form $\psi_\mathbf{k}(\mathbf{r}) = (2\pi)^{-d/2} e^{i\mathbf{k}\cdot\mathbf{r}}$.

Let us explicitly consider plane waves in 1D. Normalization is then carried out in one dimension so that $\langle \psi_p | \psi_{p'} \rangle = \delta(p-p')$. For an energy-normalized plane wave, $\langle \psi_E | \psi_{E'} \rangle = \delta(E-E')$ [note that $\delta(\frac{p^2}{2m} - \frac{p'^2}{2m}) = \frac{m}{p}\delta(p-p')$]. Hence, the momentum and energy normalized 1D plane waves are given by

$$\psi_p(x) = \left(\frac{1}{2\pi\hbar}\right)^{1/2} e^{ipx/\hbar}, \tag{1.80}$$

$$\psi_E(x) = \left(\frac{m}{p}\right)^{1/2} \left(\frac{1}{2\pi\hbar}\right)^{1/2} e^{ipx/\hbar}, \tag{1.81}$$

respectively. These solutions can be written in terms of the wave number $k = p/\hbar$. The wave function $\psi_k(x)$ is normalized so that $\langle \psi_k | \psi_{k'} \rangle = \delta(k-k')$ and the properly normalized wave function is $\psi_k(x) = (2\pi)^{-1/2} e^{ikx}$. Thus, the energy-normalized 1D plane waves are flux normalized [this will be used in the discussion of the probability flux vector in the paragraph containing Eq. (1.99)].

To energy normalize a 3D plane wave, note that

$$\delta(\mathbf{p} - \mathbf{p}') = \frac{\delta(p - p')}{p^2}\delta(\Omega_\mathbf{p} - \Omega_{\mathbf{p}'}), \tag{1.82}$$

since $\int d\mathbf{p} f(\mathbf{p})\delta(\mathbf{p}-\mathbf{p}') = \int dp\, p^2 \int d\Omega_\mathbf{p} f(\mathbf{p})\delta(\mathbf{p}-\mathbf{p}') = f(\mathbf{p}')$. The factor $1/p^2$ on the RHS of (1.82) serves to cancel the p^2 in the volume element $d\mathbf{p} = p^2 dp d\Omega_\mathbf{p}$. Since $\delta(\frac{p^2}{2m} - \frac{p'^2}{2m}) = \frac{m}{p}\delta(p-p')$, we see that $\delta(\mathbf{p}-\mathbf{p}') = \frac{p}{m}\frac{1}{p^2}\delta(\frac{p^2}{2m} - \frac{p'^2}{2m})\delta(\Omega_\mathbf{p} - \Omega_{\mathbf{p}'})$. Hence, we conclude that $C_{E,\Omega_\mathbf{p}} = (mp)^{1/2}(2\pi\hbar)^{-3/2}$, i.e.,

$$\psi_{E,\Omega_\mathbf{p}}(\mathbf{r}) = (mp)^{1/2}(2\pi\hbar)^{-3/2} e^{i\mathbf{p}\cdot\mathbf{r}/\hbar}, \tag{1.83}$$

and, $\langle \psi_{E',\Omega_{\mathbf{p}'}} | \psi_{E,\Omega_\mathbf{p}} \rangle = \delta(E-E')\delta^{(2)}(\Omega_\mathbf{p} - \Omega_{\mathbf{p}'}) = (m/p)\delta(\mathbf{p}-\mathbf{p}')$.

1.3.8 POSITION AND MOMENTUM STATES, |x⟩ AND |p⟩

The state vector $|\mathbf{p}\rangle$ in Hilbert space for a particle with momentum **p**, is said to have *position representation* $\psi_\mathbf{p}(\mathbf{r}) = \langle \mathbf{r}|\mathbf{p}\rangle = (2\pi\hbar)^{-3/2} e^{i\mathbf{p}\cdot\mathbf{r}/\hbar}$ (in 1D, $\psi_p(x) = \langle x|p\rangle = (2\pi\hbar)^{-1/2} e^{ipx/\hbar}$). The state vector in Hilbert space for a particle located at coordinate x is $|x\rangle$ (and at coordinate **r** is $|\mathbf{r}\rangle$). Since the set of states $\{|x\rangle\}$ are complete, $\int dx\, |x\rangle\langle x| = \mathbf{1}$, we can use these states as a basis. Furthermore, the set of states $\{|p\rangle\}$ is also complete, so they too can be used as a basis.

Let us ask, what is $\langle \mathbf{r}|\mathbf{x}\rangle$? To answer this question, we insert the unity operator, in the form $\mathbf{1} = \int d\mathbf{p}\, |\mathbf{p}\rangle\langle\mathbf{p}|$, between the bra and the ket in $\langle \mathbf{r}|\mathbf{x}\rangle$,

$$\langle \mathbf{r}|\mathbf{x}\rangle = \int d\mathbf{p}\, \langle \mathbf{r}|\mathbf{p}\rangle \langle \mathbf{p}|\mathbf{x}\rangle, \tag{1.84}$$

and use (1.79) to obtain

$$\langle \mathbf{r}|\mathbf{x}\rangle = (2\pi\hbar)^{-3} \int d\mathbf{p}\, e^{i\mathbf{p}\cdot(\mathbf{r}-\mathbf{x})/\hbar} = \delta(\mathbf{r}-\mathbf{x}). \tag{1.85}$$

Compare this with Eq. (1.70), the 1D case. We conclude that $|\mathbf{x}\rangle$ is an eigenstate of the operator $\hat{\mathbf{r}}$ with eigenvalue **x**, $\hat{\mathbf{r}}|\mathbf{x}\rangle = \mathbf{x}|\mathbf{x}\rangle$, and $\langle \mathbf{r}|\mathbf{x}\rangle$ is the eigenfunction of the operator **r** with eigenvalue **x**, $\mathbf{r}\langle\mathbf{r}|\mathbf{x}\rangle = \mathbf{x}\langle\mathbf{r}|\mathbf{x}\rangle$, and moreover $\langle\mathbf{r}|\mathbf{x}\rangle$ vanishes unless **r** = **x**.

Similarly, by inserting the identity operator written in the form $\mathbf{1} = \int d\mathbf{r}\, |\mathbf{r}\rangle\langle\mathbf{r}|$ into the middle of $\langle \mathbf{p}'|\mathbf{p}\rangle$, it is easy to show that

$$\langle \mathbf{p}'|\mathbf{p}\rangle = \delta(\mathbf{p}-\mathbf{p}'). \tag{1.86}$$

1.3 A First Taste of Quantum Mechanics

The ket $|\mathbf{p}\rangle$ is the eigenstate of the operator $\hat{\mathbf{p}}$ with eigenvalue $\mathbf{p}$, $\hat{\mathbf{p}}|\mathbf{p}\rangle = \mathbf{p}|\mathbf{p}\rangle$, and the braket $\langle \mathbf{p}'|\mathbf{p}\rangle$ vanishes unless $\mathbf{p}' = \mathbf{p}$.

Problem 1.5

(a) Show that $\langle x|\hat{x}|x'\rangle = x\delta(x-x')$.
(b) Show that $\langle p|\hat{p}|p'\rangle = p\,\delta(p-p')$.
(c) Show that $\langle p|\hat{x}|x\rangle = (2\pi\hbar)^{-1/2}\, x\, e^{-ipx/\hbar}$.

1.3.9 EHRENFEST'S THEOREM

The time-dependent Schrödinger equation can be used to calculate time derivatives of expectation values of dynamical variables, $\langle \mathcal{O}\rangle(t) \equiv \langle \Psi(\mathbf{r},t)|\mathcal{O}|\Psi(\mathbf{r},t)\rangle = \int d\mathbf{r}\, \Psi^*(\mathbf{r},t)\mathcal{O}\Psi(\mathbf{r},t)$. The time-derivative, $\frac{d}{dt}\langle \mathcal{O}\rangle$, can be directly related to the time derivatives of dynamical variables in classical mechanics (see Chapter 16, linked to the book web page). For example, let us consider a single particle in an external potential, so $H = \frac{\mathbf{p}^2}{2m} + V(\mathbf{r})$. The expectation values of the position and momentum are $\langle \mathbf{r}\rangle \equiv \int d\mathbf{r}\, \Psi^*(\mathbf{r},t)\mathbf{r}\Psi(\mathbf{r},t)$ and $\langle \mathbf{p}\rangle \equiv \int d\mathbf{r}\, \Psi^*(\mathbf{r},t)\mathbf{p}\Psi(\mathbf{r},t)$. We shall calculate the time derivatives of these quantities, noting that only the wave functions in the expectation values vary with time. Using the time-dependent Schrödinger equation we can express the time derivatives of the expectation values as follows:

$$\frac{d}{dt}\langle \mathbf{r}\rangle = \frac{1}{i\hbar}\langle \mathbf{r}H - H\mathbf{r}\rangle, \tag{1.87a}$$

$$\frac{d}{dt}\langle \mathbf{p}\rangle = \frac{1}{i\hbar}\langle \mathbf{p}H - H\mathbf{p}\rangle. \tag{1.87b}$$

The expectation values on the LHS of (1.87) contain quantities called commutators. The *commutator* of operators $\hat{A}$ and $\hat{B}$ is defined as follows:[14]

$$\boxed{[\hat{A},\hat{B}] \equiv \left(\hat{A}\hat{B} - \hat{B}\hat{A}\right).} \tag{1.88}$$

The RHSs of Eqs (1.87) involve the commutators $[\mathbf{r},H]$ and $[\mathbf{p},H]$, respectively. These commutators do not vanish when the Hamiltonian is given by $\frac{\mathbf{p}^2}{2m} + V(\mathbf{r})$, as we shall soon see.

Problem 1.6

(a) Derive Eqs (1.87) using the time-dependent Schrödinger equation (1.38) and its complex conjugate, $-i\hbar \frac{\partial}{\partial t}\Psi^*(t) = H\Psi^*(t)$.
(b) Show that for any time-independent operator $\mathcal{O}$, $\frac{d}{dt}\langle \mathcal{O}\rangle = \frac{1}{i\hbar}\langle [\mathcal{O},H]\rangle$.

More generally, for a time-independent operator, $\mathcal{O}$, $\langle \mathcal{O}\rangle = \frac{1}{i\hbar}\langle [\mathcal{O},H]\rangle$, whereas for a time-dependent operator, $\mathcal{O}(t)$,

$$\boxed{\frac{d}{dt}\langle \mathcal{O}(t)\rangle = \frac{1}{i\hbar}\langle [\mathcal{O}(t),H]\rangle + \left\langle \frac{\partial \mathcal{O}(t)}{\partial t}\right\rangle.} \tag{1.89}$$

---

[14] It is also common to define the *anticommutator* of $\hat{A}$ and $\hat{B}$ as $\{\hat{A},\hat{B}\} \equiv \left(\hat{A}\hat{B} + \hat{B}\hat{A}\right)$.

To calculate the effect of an operator that is a commutator, $[\hat{A}, \hat{B}]$, it is easiest to apply the commutator to a wave function to determine what it does. For example, let us calculate $[p_x, x]\phi(\mathbf{r})$:

$$[p_x, x]\phi(\mathbf{r}) = p_x(x\phi(\mathbf{r})) - xp_x\phi(\mathbf{r}) = \frac{\hbar}{i}\phi(\mathbf{r}). \tag{1.90}$$

Hence $[p_x, x] = \hbar/i$, since (1.90) is true for any wave function $\phi(\mathbf{r})$. Moreover, it is clear that this generalizes to $[p_i, r_j] = \frac{\hbar}{i}\delta_{ij}$.

As another example, let us calculate $[\mathbf{p}, H]\phi(\mathbf{r})$:

$$[\mathbf{p}, H]\phi(\mathbf{r}) = \left\{\left(\mathbf{p}\frac{\mathbf{p}^2}{2m} - \frac{\mathbf{p}^2}{2m}\mathbf{p}\right) + (\mathbf{p}V(\mathbf{r}) - V(\mathbf{r})\mathbf{p})\right\}\phi(\mathbf{r}). \tag{1.91}$$

The first term in the square parenthesis on the RHS of Eq. (1.91) vanishes, since the order of the operators can be interchanged there, but the second term does not vanish since $\mathbf{p}(V(\mathbf{r})\phi(\mathbf{r})) = (\hbar/i)[(\boldsymbol{\nabla}V)\phi + V\boldsymbol{\nabla}\phi]$, and therefore the second term on the RHS of Eq. (1.91) equals $\frac{\hbar}{i}(\boldsymbol{\nabla}V)\phi$. Hence, since this holds for any function ϕ, we conclude that $[\mathbf{p}, H] = (\hbar/i)(\boldsymbol{\nabla}V)$. Now, let us calculate $[\mathbf{r}, H]\phi(\mathbf{r})$:

$$[\mathbf{r}, H]\phi(\mathbf{r}) = \left\{\left(\mathbf{r}\frac{\mathbf{p}^2}{2m} - \frac{\mathbf{p}^2}{2m}\mathbf{r}\right) + (\mathbf{r}V(\mathbf{r}) - V(\mathbf{r})\mathbf{r})\right\}\phi(\mathbf{r}). \tag{1.92}$$

The second term in the curly parenthesis on the RHS of Eq. (1.92) vanishes, and the first term can be easily evaluated to be $(\mathbf{p}/m)\phi$. Since this relation holds for any function ϕ, we conclude that $[\mathbf{r}, H] = \mathbf{p}/m$.

We can now substitute the commutators that we have evaluated back into (1.87) to finally obtain:

$$\frac{d}{dt}\langle\mathbf{r}\rangle = \frac{\langle\mathbf{p}\rangle}{m}, \tag{1.93a}$$

$$\frac{d}{dt}\langle\mathbf{p}\rangle = -\langle\boldsymbol{\nabla}V\rangle. \tag{1.93b}$$

Equations (1.93) are called *Ehrenfest's theorem* after the physicist and mathematician Paul Ehrenfest; they are very similar to the classical equations of motion of Hamilton (see Sec. 16.2 which is available on the book web page), to the extent that the quantity $-\langle\boldsymbol{\nabla}V\rangle$ appearing on the RHS of (1.93b) is interpreted as the average force on the particle. The general form of Ehrenfest's theorem can be stated as follows: the time derivative of the expectation value of a quantum mechanical operator that is time-independent, i.e., does not explicitly depend on time, is equal to $\frac{1}{i\hbar}$ times the expectation value of the commutator of that operator with the Hamiltonian of the system (see part (d) of Problem 1.7).

Problem 1.7

(a) Complete the algebra to derive Eq. (1.90).
(b) Explicitly show that $[\mathbf{r}, H] = \mathbf{p}/m$ for $H = \frac{\mathbf{p}^2}{2m} + V(\mathbf{r})$.
(c) Redo the algebra to derive (1.87) to obtain the time derivative of the expectation value of an operator $\hat{A}(t)$ that is time dependent, i.e., calculate $\frac{d}{dt}\langle\hat{A}(t)\rangle$, to obtain the general form of Ehrenfest's theorem.

Problem 1.8

Prove the following commutator identities:

(a) $[\hat{A}, \hat{B}] = -[\hat{B}, \hat{A}]$.
(b) $[\hat{A}_1 + \hat{A}_2, \hat{B}] = [\hat{A}_1, \hat{B}] + [\hat{A}_2, \hat{B}]$.

1.3 A First Taste of Quantum Mechanics

(c) $[\hat{A}_1\hat{A}_2, \hat{B}] = \hat{A}_1[\hat{A}_2, \hat{B}] + [\hat{A}_1, \hat{B}]\hat{A}_2$.

(d) Prove the anticommutator identities:
 (1) $\{\hat{A}, \hat{B}\} = \{\hat{B}, \hat{A}\}$,
 (2) $\{\hat{A}_1\hat{A}_2, \hat{B}\} = \hat{A}_1\{\hat{A}_2, \hat{B}\} + [\hat{A}_1, \hat{B}]\hat{A}_2$,
 (3) $\{\hat{A}_1\hat{A}_2, \hat{B}\} = \hat{A}_1\{\hat{A}_2, \hat{B}\} + \{\hat{A}_1, \hat{B}\}\hat{A}_2$.

(e) Prove the identity $[\hat{A}, \hat{B}\hat{C}] = \{\hat{A}, \hat{B}\}\hat{C} - \hat{B}\{\hat{A}, \hat{C}\}$.

Problem 1.9

Prove the following commutator relations:

(a) $[(\hat{p}_x)^2, \hat{x}] = \frac{2\hbar}{i}\hat{p}_x$.

(b) $[\hat{p}_x, (\hat{x})^2] = \frac{2\hbar}{i}\hat{x}$.

(c) $[(\hat{p}_x)^2, (\hat{x})^2] = \frac{2\hbar}{i}(\hat{p}_x\hat{x} + \hat{x}\hat{p}_x) = \frac{2\hbar}{i}(2\hat{x}\hat{p}_x + \frac{\hbar}{i})$.

(d) $[\mathbf{r} \times \mathbf{p}, \mathbf{r}^2] = 0$.

1.3.10 ONE-DIMENSIONAL WAVE EQUATIONS

If the motion of a particle depends on only a single coordinate, say x, the 1D time-independent Schrödinger equation is

$$-\frac{\hbar^2}{2m}\frac{d^2\psi}{dx^2} + V(x)\psi = E\psi. \tag{1.94}$$

For the 1D Schrödinger equation, it is useful to define the local wave vector $K(x)$ by the relation $K^2(x) \equiv \frac{2m}{\hbar^2}(E - V(x))$, since then Eq. (1.94) can be put in the compact and simple form $\frac{d^2\psi}{dx^2} = -K^2(x)\psi$. This equation is a 1D *Helmholtz equation*.

Note that, if the potential for a particle in 3D is given by a sum of the form, $V(\mathbf{r}) = V_1(x) + V_2(y) + V_3(z)$, the time-independent Schrödinger equation separates so that the wave function is given by a product of wave functions, $\psi(\mathbf{r}) = \psi_1(x)\psi_2(y)\psi_3(z)$, each of which satisfies a 1D equation, and the total energy is given by $E = E_1 + E_2 + E_3$.

Problem 1.10

Prove the statement above by writing the 3D time-independent Schrödinger equation and use the separation of variables method (divide the equation by $\psi_1(x)\psi_2(y)\psi_3(z)$, then separate the functions of the variables x, y and z in the resulting equation).

Bound state solutions of a 1D potential are nondegenerate (only one eigenfunction has a given eigenenergy). To prove this, assume the opposite; suppose $\psi_1(x)$ and $\psi_2(x)$ are two solutions with the same energy. Then $\frac{d^2\psi_1}{dx^2}/\psi_1 = \frac{d^2\psi_2}{dx^2}/\psi_2 = -K^2(x)$, hence $\frac{d^2\psi_1}{dx^2}\psi_2 - \frac{d^2\psi_2}{dx^2}\psi_1 = 0$. Integrating, we find $\frac{d\psi_1}{dx}\psi_2 - \frac{d\psi_2}{dx}\psi_1 =$ constant (independent of x), and since, for bound states $\psi_1(\infty) = \psi_2(\infty) = 0$, we find the constant equals zero and hence $\frac{d\psi_1}{dx}\psi_1 = \frac{d\psi_2}{dx}\psi_2$, which can be integrated to obtain $\psi_1 = C\psi_2$, where C is some constant. Thus, the two solutions are identical. (Note that this result is not applicable to 1D on a ring of length L with periodic boundary conditions, since the bound state wave function need not vanish at infinity.)

For a potential that tends to finite limits as $x \to \pm\infty$, it is easy to determine the asymptotic form of the wave function. For energy $E < V(-\infty)$ and $E < V(\infty)$, only isolated eigenenergies are possible, and the wave function must asymptotically go to zero exponentially as $x \to \pm\infty$. The energy eigenvalues must, in general, be numerically determined by integrating the Schrödinger equation from the left and from the right, and matching the logarithmic derivative of the wave functions, $\frac{d(\ln\psi)}{dx} = \frac{d\psi}{dx}/\psi(x)$ at some intermediate value of x, say at $x = x_m$. The isolated energies where this matching is possible are the bound state energies.

Quite generally, the asymptotic form of the wave function for $E < V(\infty)$, is given by $\psi(x) \xrightarrow[x\to\infty]{} c_\infty e^{-\kappa_\infty x}$, where $\kappa_\infty \equiv \sqrt{|E - V(\infty)|}/\hbar$, and, depending upon whether $x \to -\infty$ is open,

$$\psi(x) \xrightarrow[x\to-\infty]{} \begin{cases} c_{-\infty} e^{\kappa_{-\infty} x} \text{ for } E < V(-\infty), \\ c_{-\infty} \cos(K_{-\infty} x + \delta_{-\infty}) \text{ for } E > V(-\infty), \end{cases} \quad (1.95)$$

where $\kappa_{-\infty} \equiv \sqrt{|E - V(-\infty)|}/\hbar$ and $K_{-\infty} \equiv \sqrt{E - V(-\infty)}/\hbar$ and $\delta_{-\infty}$ is a constant phase that is determined by the boundary conditions.

If $E > V(\infty)$, $\psi(x) \xrightarrow[x\to\infty]{} c_\infty \cos(K_\infty x + \delta_\infty)$, and Eq. (1.95) still applies as $x \to -\infty$.

The momentum and energy normalization of the 1D continuum wave functions have already discussed, see Eqs (1.80) and (1.81). Bound state wave functions are normalized to unity, $\langle\psi|\psi\rangle \equiv \int dx\,|\psi(x)|^2 = 1$.

1.3.11 PARTICLE-IN-A-BOX AND PIECEWISE-CONSTANT POTENTIALS

The solutions to the time-independent Schrödinger equation for a piecewise-constant potential are known analytically. It is instructive to work out the solutions for such potentials, and we shall do so for the cases shown in Fig. 1.14. Matching of the wave function at the discontinuities of the potential, and assigning the correct asymptotic forms of the wave function are the two constraints that need to be implemented.

The general solution to the equation, $\frac{d^2\psi}{dx^2} = -K^2\psi$, for $K^2 \equiv 2m(E - V_0)/\hbar^2 > 0$ is, $\psi(x) = Ae^{iKx} + Be^{-iKx}$, and for $K^2 \equiv 2m(E - V_0)/\hbar^2 < 0$ is, $\psi(x) = A'e^{\kappa x} + B'e^{-\kappa x}$, where $\kappa^2 \equiv 2m(V_0 - E)/\hbar^2 > 0$. The constant coefficients, A and B, or A' and B', are arbitrary and must be determined by matching the boundary and continuity conditions for the wave function and its derivative, as explained below.

First consider the infinite square well shown in Fig. 1.14(a). Since the potential is infinite outside the well, the wave function must vanish outside $0 \le x \le L$. Inside the box, $k^2 = 2mE/\hbar^2 > 0$, and since the wave function must vanish at the origin, we have $\psi(x) = A\sin(kx)$. Moreover, since $\psi(L) = 0$, we conclude that only values of $k = n\pi/L$, where $n = 1, 2, \ldots$ are possible. Note that only nonnegative integers are taken, since the negative integers only change the sign of the wave function, and therefore to not produce a different solution. We shall normalize the wave functions $\psi(x) = A\sin(n\pi x/L)$

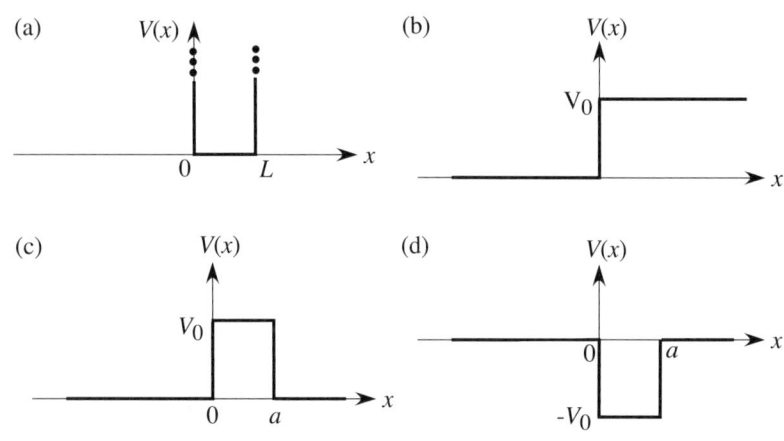

FIG 1.14 Piecewise-constant potentials. (a) Infinite square well. (b) Square step potential. (c) Square barrier potential. (d) Finite depth square well.

1.3 A First Taste of Quantum Mechanics

so that $\langle\psi|\psi\rangle = \int_0^L dx\,|\psi(x)|^2 = 1$, hence $A = \sqrt{2/L}$. Thus the bound state solutions are given by

$$\boxed{\psi_n(x) = \sqrt{\frac{2}{L}}\sin\left(\frac{n\pi x}{L}\right).} \tag{1.96}$$

and the energy eigenvalues are

$$\boxed{E_n = \frac{\hbar^2 k_n^2}{2m} = \frac{\hbar^2 \pi^2 n^2}{2mL^2}.} \tag{1.97}$$

Problem 1.11

Consider the 1D time-independent Schrödinger equation with zero potential in the region $0 \le x \le L$, with periodic boundary conditions, $\psi(0) = \psi(L)$ and $\psi'(0) = \psi'(L)$. This corresponds to wrapping the line $0 \le x \le L$ into a ring so that the points $x=0$ and $x=L$ correspond to the same the point.

(a) Verify that $\psi(x) = A\,e^{ikx} + B\,e^{-ikx}$ satisfies the Schrödinger equation.
(b) Show that either A or B should be zero.
(b) Find the values of k that satisfy the boundary conditions.
(c) Find the amplitude A that normalizes the wave function.
(d) Determine the probability of finding the particle between x and $x + dx$.
(e) Calculate the expectation values $\langle x^2 \rangle$ and $\langle \frac{p^2}{2m} \rangle$.

Answers: (b) $k = 2\pi n/L$. (c) $A = 1/\sqrt{L}$. (d) $P(x)dx = dx/L$. (e) $\langle x^2 \rangle = L^2/3$, $\langle \frac{p^2}{2m} \rangle = \frac{2n^2\pi^2}{mL^2}$.

The solutions are plotted in Fig. 1.15 for the lowest three energy eigenvalues. The number of nodes in the wave function $\psi_n(x)$ equals $n - 1$. The more nodes, the higher the energy because the kinetic energy (i.e., the second derivative in the Schrödinger equation) increases as n^2. The lowest energy solution ($n=1$) has finite energy $\frac{\hbar^2\pi^2}{2mL^2}$. Clearly, the discrete nature of the allowed energies and the minimum possible energy are very different from the classical behavior of a particle in a box (infinite square well) potential. The solutions with odd (even) n are symmetric (antisymmetric) with respect to inversion about the center of the well, $x - L/2 \to -(x - L/2)$. This symmetry upon flipping in the sign of the spatial coordinate is called *parity*, and results because the potential is symmetric under the inversion transformation. We defer the consideration of parity symmetry [see the discussion near Eq. (1.109) and Sec. 2.9.2].

Problem 1.12

For a particle of mass m in a hard-wall box, $0 \le x \le L$:

(a) Determine the expectation value $\langle \psi_n|x|\psi_n \rangle$.
(b) Determine the expectation values $\langle \psi_n|p|\psi_n \rangle$ and $\langle \psi_n|p^2|\psi_n \rangle$.
(c) Find an analytic expression for the expectation value $\langle \psi_n|p^2xp^2|\psi_n \rangle$.
(d) For a particle in the ground state, find the probability of finding in the region $0 \le x \le L/3$.

Answers: (a) $\langle \psi_n|x|\psi_n \rangle = L/2$. (b) $\langle \psi_n|p|\psi_n \rangle = 0$, $\langle \psi_n|p^2|\psi_n \rangle = \frac{\hbar^2\pi^2 n^2}{L^2}$.
(c) $\left(\frac{\hbar^2\pi^2 n^2}{L^2}\right)^2 \frac{L}{2}$. (d) $P = \frac{2}{L}\int_0^{L/3} dx\,\sin^2 \pi x/L \approx 0.1955$.

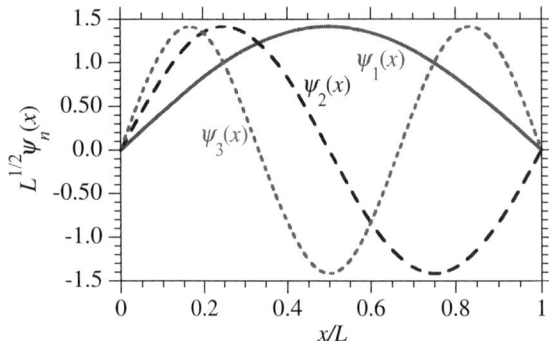

FIG 1.15 The three lowest energy eigenfunctions for a particle in a box (infinite square well) versus x.

Next we consider scattering off the step potential shown in Fig. 1.14(b). The nature of a scattering problem is very different from that of a bound state problem. Here the total energy E is given and we seek reflection and transmission coefficients for the scattering. We consider the scattering with incident wave from the left which has a reflected component and a transmitted component if $E > V_0$. You will be asked to consider the case where the incident wave comes from the right in a problem below. We denote the amplitudes of the reflected and transmitted waves by r and t. Aside from an overall multiplicative constant which has no physical bearing (see discussion of the probability current density below), the wave function takes the form

$$\psi(x) = \begin{cases} e^{ikx} + r e^{-ikx} & \text{for } x < 0, \\ t e^{iKx} & \text{for } x > 0, \end{cases} \quad (1.98)$$

where $K^2 = 2m(E - V_0)/\hbar^2$. Despite the fact that the potential is discontinuous at $x=0$, the wave function and its first derivative must be continuous [since the density and flux (see next paragraph) must be continuous]. Setting $\psi(0^-) = \psi(0^+)$, we find that r and t satisfy the equation $1 + r = t$, and setting $\frac{d}{dx}\psi(0^-) = \frac{d}{dx}\psi(0^+)$, we find $1 - r = \frac{K}{k}t$; hence $t = \frac{2}{1+K/k}$, $r = \frac{1-K/k}{1+K/k}$. Figure 1.16a plots the real and imaginary parts of the wave function for $V_0/E = 3/4$, so $K = k/2$, and the transmission and reflection amplitudes are $t = 4/3$ and $r = 1/3$, respectively.

Problem 1.13

Consider the scattering off the step potential in Fig. 1.14(b) with incident wave from the right for $E > V_0$. Take the wave function to be of the form

$$\psi(x) = \begin{cases} t' e^{-ikx} & \text{for } x < 0, \\ e^{-iKx} + r' e^{iKx} & \text{for } x > 0, \end{cases}$$

(a) Determine the reflection and transmission amplitudes r' and t'.
(b) Show that $t' \to 0$ as $E \to V_0$ from above, and note how counterintuitive this is.
Answers: (a) $r' = \frac{K-k}{K+k}$, $t' = \frac{2K}{K+k}$. (b) $K \to 0$ as $E \to V_0$, hence $t' \to 0$.

In order to better understand the nature of the reflection and transmission in this case, where the asymptotic momentum is different as $x \to \pm\infty$, we need to develop the concept of the *probability flux vector*, sometimes also called the *probability current density*. We shall do so here in arbitrary dimension, since the arguments here are dimension-independent. Consider a specific region of coordinate space, V. Let us calculate the volume integral (in 3D or in 1D or arbitrary dimension d), $\int_V d\mathbf{r}\, |\psi(\mathbf{r},t)|^2$, which is the probability of finding the particle in the region. The rate of change of this probability is given by

$$\frac{\partial}{\partial t} \int_V d\mathbf{r}\, |\psi(\mathbf{r},t)|^2 = \int_V d\mathbf{r} \left(\frac{\partial \psi^*}{\partial t}\psi + \psi^* \frac{\partial \psi}{\partial t} \right) = \frac{i\hbar}{2m} \int_V d\mathbf{r}\, (\psi^* \nabla^2 \psi - \nabla^2 \psi^* \psi)$$

$$= \frac{i\hbar}{2m} \int_V d\mathbf{r}\, \nabla \cdot (\psi^* \nabla \psi - (\nabla \psi^*)\psi),$$

1.3 A First Taste of Quantum Mechanics

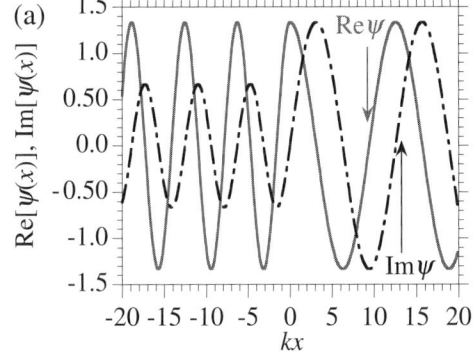

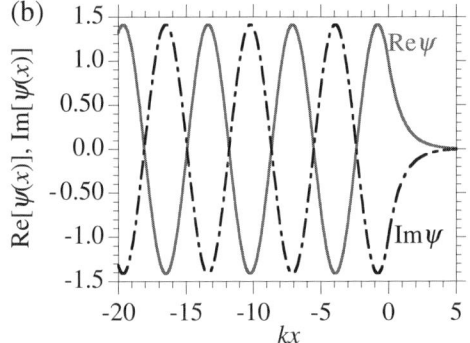

FIG 1.16 Real and imaginary parts of the wave function for the step potential shown in Fig. 1.14(b) with the step at $x = 0$. (a) RHS is open [ψ in Eq. (1.98)] and $K/k = 1/2$, and (b) RHS is closed [ψ of Eq. (1.103)] and $\kappa/k = 1$.

where we used the divergence theorem, Eq. (C.22), in the last equality. Since the region V is arbitrary, we conclude that

$$\boxed{\frac{\partial n(\mathbf{r},t)}{\partial t} + \nabla \cdot \mathbf{J}(\mathbf{r},t) = 0,} \quad (1.99)$$

where $n(\mathbf{r},t) = |\psi(\mathbf{r},t)|^2$ is the *probability density* and $\mathbf{J}$ is the *probability flux vector* (sometimes called the *probability current density*),

$$\boxed{\begin{aligned}\mathbf{J}(\mathbf{r},t) &= \mathrm{Re}\,[\psi^*(\mathbf{r},t)\frac{\mathbf{p}}{m}\psi(\mathbf{r},t)] \\ &= \frac{\hbar}{2mi}\{\psi^*(\mathbf{r},t)\nabla\psi(\mathbf{r},t) - [\nabla\psi^*(\mathbf{r},t)]\psi(\mathbf{r},t)\}.\end{aligned}} \quad (1.100)$$

Equation (1.99) is the *continuity equation*, $\partial n(\mathbf{r},t)/\partial t + \nabla \cdot \mathbf{J}(\mathbf{r},t) = 0$, which expresses the conservation of probability (and/or particle number). Integrating (1.99) over a volume V, and applying the divergence theorem (see Appendix C.3) to change the volume integral of the divergence of a vector into the surface integral of the vector gives,

$$\frac{\partial}{\partial t}\int_V d\mathbf{r}\,n(\mathbf{r},t) + \oint_S d\mathbf{S}\cdot\mathbf{J}(\mathbf{r},t) = 0. \quad (1.101)$$

The rate of change of probability for the particle to be within the volume V plus the net outward flux through the surface S surrounding the volume, $\oint_S d\mathbf{S}\cdot\mathbf{J}(\mathbf{r},t)$, vanishes. The probability current density can be rewritten as $\mathbf{J}(\mathbf{r},t) = \mathrm{Re}(\psi^*\hat{\mathbf{v}}\psi)$ with the velocity operator $\hat{\mathbf{v}} \equiv \hat{\mathbf{p}}/m = (\hbar/i)\nabla/m$. This is probably the easiest way to remember the probability current density.

Problem 1.14

(a) For $\psi(\mathbf{r},t) = \mathcal{N}e^{i(\mathbf{k}\cdot\mathbf{r}-\omega t)}$, calculate $\mathbf{J}(\mathbf{r},t)$ and $\langle\psi(t)|\frac{\hat{\mathbf{p}}}{m}|\psi(t)\rangle$.

(b) For the wave packet, $\psi(x,t) = \frac{1}{[2\pi\sigma^2(t)]^{1/4}}e^{-\frac{(x-\hbar\kappa t/m)^2}{4\sigma^2(t)}+i\kappa x+i\frac{p_0^2}{2m}\frac{t}{\hbar}}$, calculate $J_x(x,t)$ and $\langle\psi(t)|\frac{\hat{p}_x}{m}|\psi(t)\rangle$.

Answers: (a) $\mathbf{J}(\mathbf{r},t) = \frac{\mathcal{N}^2\hbar\mathbf{k}}{m}$, $\langle\psi(t)|\frac{\hat{\mathbf{p}}}{m}|\psi(t)\rangle = \frac{\mathcal{N}^2\hbar\mathbf{k}V}{m}$.

(b) $J_x(x,t) = \frac{\hbar\kappa}{m}\frac{1}{[2\pi\sigma^2(t)]^{1/2}}e^{-\frac{(x-\hbar\kappa t/m)^2}{2\sigma^2(t)}}$, $\langle\psi(t)|\frac{\hat{p}_x}{m}|\psi(t)\rangle = \hbar\kappa/m$.

When applied to the scattering off the step potential in Fig. 1.14b, conservation of flux, $\oint d\mathbf{S}\cdot\mathbf{J}=0$, gives the relation, $K|t|^2 + k|r|^2 = k$, i.e., $\frac{(1-K/k)^2+4K/k}{(1+K/k)^2} = 1$. The probability current density in the incident wave is $k \times 1$, in the reflected wave is $k|r|^2$, and in the transmitted wave is $K|t|^2$. We define the transmission coefficient $\mathcal{T}$ of the particle as the ratio of the probability current density in the transmitted wave to that in the incident wave,

$$\mathcal{T} = \frac{K|t|^2}{k} = \frac{K}{k}|t|^2. \quad (1.102)$$

Similarly we can define the reflection coefficient $\mathcal{R}$ as the ratio of the density in the reflected wave to that in the incident wave. Hence, $\mathcal{R} = 1 - \mathcal{T} = 1 - \frac{K}{k}|t|^2$. For the step potential in Fig. 1.14(b) with $V_0/E = 3/4$, so $K = k/2$, the transmission coefficient is $\mathcal{T} = 8/9$ and the reflection coefficient $\mathcal{R} = 1/9$.

For the step potential of Fig. 1.14(b), but now for $E < V_0$, i.e., considering the case where the region $x > 0$ of the step potential is classically forbidden, the wave function incident from the left must have a reflection probability $|r|^2$ equal to unity, and the wave function in the classically forbidden region must decay away as x increases. The form of the wave function is therefore given by

$$\psi(x) = \begin{cases} e^{ikx} + r e^{-ikx} & \text{for } x < 0, \\ t e^{-\kappa x} & \text{for } x > 0, \end{cases} \quad (1.103)$$

where $\kappa^2 = 2m(V_0 - E)/\hbar^2$. Equating the wave function on the two sides of $x = 0$, we find $1 + r = t$, and equating the derivative on the two sides, we find $1 - r = (i\kappa/k)t$; hence $t = \frac{2}{1+i\kappa/k}$ and $r = \frac{1-i\kappa/k}{1+i\kappa/k}$. Figure 1.16a plots the real and imaginary parts of the wave function for $V_0/E = 6/3$, i.e., $\kappa/k = 1$. The wave function decays very quickly in the classically forbidden region; it falls to $1/e$ of its magnitude after a distance $x = \kappa^{-1}$. The reflection probability is, of course, unity $|r|^2 = 1$ since all particles are eventually reflected by the barrier (i.e., all flux is reflected, despite the finite probability of finding particles in the forbidden region near the barrier edge).

Let us now turn to the barrier potential in Fig. 1.14(c). We shall first consider the case when the energy is less than the potential barrier height, $E < V_0$, and flux is incident from the left. The wave function with the right boundary conditions is

$$\psi(x) = \begin{cases} e^{ikx} + r e^{-ikx} & \text{for } x < 0, \\ A e^{\kappa x} + B e^{-\kappa x} & \text{for } 0 \leq x \leq a, \\ t e^{ikx} & \text{for } x > a. \end{cases} \quad (1.104)$$

Matching wave function and derivative at $x = 0$, we obtain the equations, $1 + r = A + B$, and $1 - r = \frac{\kappa}{ik}(A - B)$, and at $x = a$ we obtain, $Ae^{\kappa a} + Be^{-\kappa a} = te^{ika}$, and $\frac{\kappa}{ik}(Ae^{\kappa a} - Be^{-\kappa a}) = te^{ika}$. These four linear equations can be solved for the four unknowns r, A, B, and t. We explicitly present the formulas only for the transmission and reflection amplitudes:

$$t = \frac{4ie^{-ika}k\kappa}{(-\kappa^2 + k^2 + 2ik\kappa)e^{\kappa a} + (\kappa^2 - k^2 + 2ik\kappa)e^{-\kappa a}}, \quad (1.105)$$

$$r = \frac{(k^2 + \kappa^2)(e^{\kappa a} - e^{-\kappa a})}{(-\kappa^2 + k^2 + 2ik\kappa)e^{\kappa a} + (\kappa^2 - k^2 + 2ik\kappa)e^{-\kappa a}}. \quad (1.106)$$

The transmission and reflection amplitudes satisfy the condition, $|t|^2 + |r|^2 = 1$. Figure 1.17(a) shows the wave function for $\kappa/k = 1/4$ ($E/V_0 \approx 0.9411$) and $a = 3$. The reflection and transmission coefficients, $|t|^2$ and $|r|^2$, are plotted versus energy (up to the barrier height) in Fig. 1.17(b). For small energies, the transmission is exponentially small. The ratio of the transmission to reflection amplitudes is given by $t/r = \frac{4i\kappa k}{(k^2+\kappa^2)(e^{\kappa a}-e^{-\kappa a})}e^{-ika}$, so if the factor e^{-ika} is taken out of the transmission amplitude and is instead taken to multiply the plane wave factor e^{ikx} in (1.104), the reflection and transmission amplitudes are $\pi/2$ out of phase. It should be mentioned for future reference that the phase of the transmission and reflection amplitudes can be determined (i.e., can be measured) if interference experiments are carried out. Figure 1.17(c) shows the wave functions for barrier penetration of a wide barrier and a narrow barrier.

For $E > V_0$ (not shown in Fig. 1.17), the form for the wave function in the region $0 \leq x \leq a$ is given by $\psi(x) = A e^{iKx} + B e^{-iKx}$.

1.3 A First Taste of Quantum Mechanics

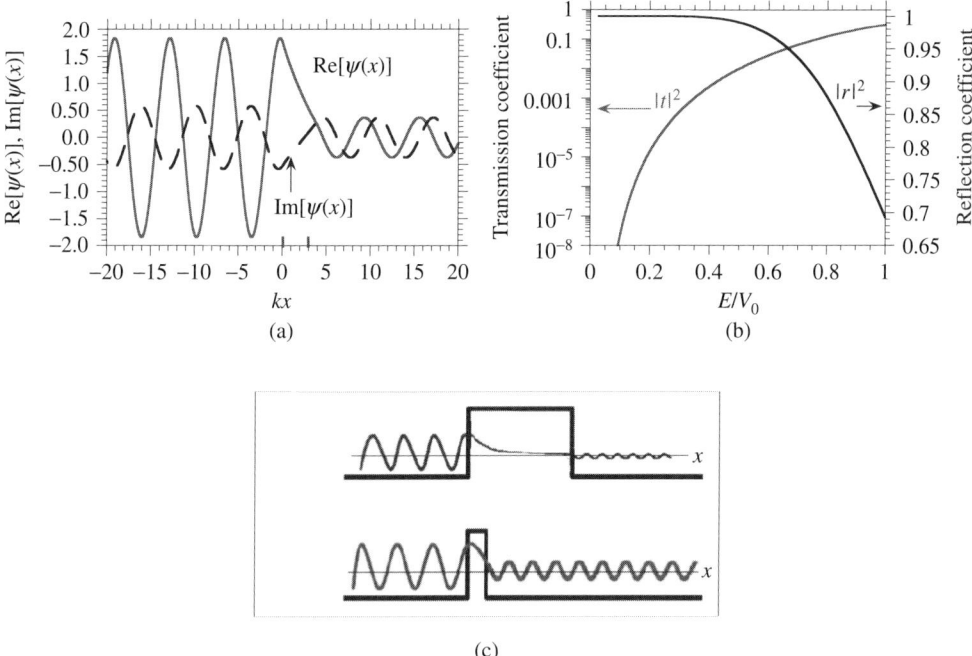

FIG 1.17 (a) Real and imaginary parts of the wave function, $\psi(x)$ versus kx, for the barrier potential in Eq. (1.104) with $\kappa/k = 1/4$ ($E/V_0 \approx 0.9411$) and $a = 3$. (b) Transmission and reflection coefficients $|t|^2$ and $|r|^2$ versus E/V_0. (c) Real part of the wave function $\psi(x)$ versus coordinate x for barrier penetration of a wide and a narrow barrier (the imaginary part behaves similarly).

Problem 1.15

Calculate the reflection and transmission coefficients for the barrier potential in Fig. 1.14(c) for $E > V_0$.
Answer: $|t|^2 = \frac{4k^2K^2}{4k^2K^2+(k^2-K^2)^2\sin^2(Ka)}$, $|r|^2 = 1 - |t|^2$. The transmission $|t|^2$ versus energy can show oscillations, due to the sine function term in the denominator of the expression for $|t|^2$, similar to the oscillations in Fig. 1.19.

Bound States in a Potential Well

The finite well potential in Fig. 1.14(d) possesses at least one bound state, no matter how small the well depth, and as the well depth increases, more bound states develop. Let us search for the bound state energy E that is negative. The wave function outside the potential well must decay exponentially as $x \to -\infty$, and as $x \to +\infty$, so we take the form of the wave function to be:

$$\psi(x) = \begin{cases} Ae^{\kappa x} & \text{for } x < 0, \\ Be^{iKx} + Ce^{-iKx} & \text{for } 0 \le x \le a, \\ De^{-\kappa x} & \text{for } x > a. \end{cases} \quad (1.107)$$

where $\kappa = \sqrt{2m|E|/\hbar^2}$ and $K = \sqrt{2m(V_0 - |E|)/\hbar^2}$. Matching the wave function and its derivative at $x = 0$ yields the equations, $A = B + C$, $\frac{\kappa}{K}A = i(B - C)$, and matching the wave function and its derivative at $x = a$ yields the equations,

$Be^{iKa} + Ce^{-iKa} = De^{-\kappa a}$, $iK(Be^{iKa} - Ce^{-iKa}) = -\kappa De^{-\kappa a}$. These equations can be written as a matrix equation,

$$\mathbf{M}\begin{pmatrix} A \\ B \\ C \\ D \end{pmatrix} = \begin{pmatrix} 0 \\ 0 \\ 0 \\ 0 \end{pmatrix}, \tag{1.108}$$

where the matrix $\mathbf{M}$ is a 4×4 matrix, whose determinant must be zero for there to be a bound state solution. Setting the determinant of the matrix $\mathbf{M}$ equal to zero yields an equation for the bound state energies.

In order to simplify the calculation of the bound state energies and eigenstates, let us make use of parity symmetry (symmetry under inversion of coordinates) to classify the eigenfunctions of the potential shown in Fig. 1.14(d). To do so, let us shift the potential by $-a/2$ so that it is symmetric under inversion of the coordinates, $x \to -x$, i.e., $V(-x) = V(x)$. Then the wave functions must have a definite symmetry under this transformation of coordinates, as we shall now show. Applying the inversion transformation $\mathcal{P}$ (i.e., the parity operator) to the Schrödinger equation, $H\psi = E\psi$, to obtain $\mathcal{P}H\psi = E\mathcal{P}\psi$, inserting unity in the form $\mathcal{P}^{-1}\mathcal{P}$ between H and ψ on the right hand side of the equation, $\mathcal{P}H\mathcal{P}^{-1}\mathcal{P}\psi = E\mathcal{P}\psi$, and noting that $\mathcal{P}H\mathcal{P}^{-1} = H$ because the potential is symmetric, we find that both $\psi(x)$ and $\mathcal{P}\psi(x) \equiv \psi(-x)$ satisfy the same wave equation. Unless the wave function is degenerate (i.e., there are more than one solution of the wave equation with the same energy), this implies that $\psi(-x) = \varepsilon\psi(x)$. Applying the parity operator again, we find that $\varepsilon^2 = 1$, hence, $\varepsilon = \pm 1$. Thus, all nondegenerate eigenfunctions are either even or odd under parity inversion. These wave functions said to have *even parity* or *odd parity*. In our case, the even solutions are of the form

$$\psi(x) = \begin{cases} Ae^{\kappa x} & \text{for } x < -a/2, \\ B\cos(Kx) & \text{for } -a/2 \leq x \leq a/2, \\ Ae^{-\kappa x} & \text{for } x > a/2. \end{cases} \tag{1.109}$$

Matching the even wave function and its derivative yields, $Ae^{-\kappa a/2} = B\cos(Ka/2)$, $\kappa Ae^{-\kappa a/2} = KB\sin(Ka/2)$, hence

$$\tan(Ka/2) = \kappa/K. \tag{1.110}$$

The odd solutions are of the form

$$\psi(x) = \begin{cases} Ae^{\kappa x} & \text{for } x < -a/2, \\ B\sin(Kx) & \text{for } -a/2 \leq x \leq a/2, \\ -Ae^{-\kappa x} & \text{for } x > a/2. \end{cases} \tag{1.111}$$

Matching the odd wave function and its derivative yields, $Ae^{-\kappa a/2} = -B\sin(Ka/2)$, $\kappa Ae^{-\kappa a/2} = KB\cos(Ka/2)$, hence

$$\cot(Ka/2) = -\kappa/K. \tag{1.112}$$

No odd bound state exists unless $V_0 a^2 > \pi^2\hbar^2/(8m)$, one bound state exists if $\pi^2\hbar^2/(8m) < V_0 a^2 \leq 9\pi^2\hbar^2/(8m)$, etc.

The transcendental equations (1.110) and (1.112) must be solved numerically to obtain the eigenvalues for the even and odd solutions respectively. Figure 1.18 shows the graphical solution of Eq. (1.110) for the even bound states. The dashed curve shows κ/K versus E/V_0, and the solid curves plot $\tan(Ka/2)$

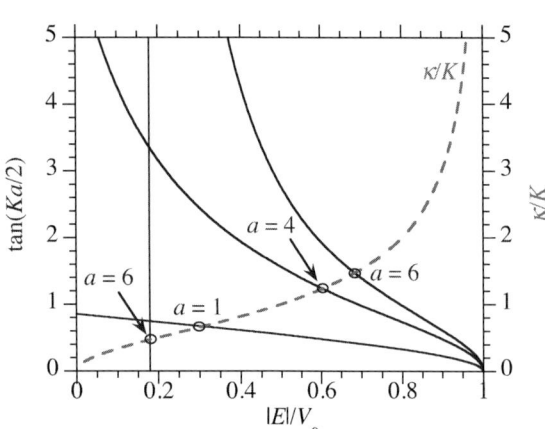

FIG 1.18 Graphical solution of Eq. (1.110) for the even parity bound state energies of a square well potential for three different values of a: $a = 1$, $a = 4$, and $a = 6$. The function $\tan(Ka/2)$ is drawn in various shades of blue for different a, and κ/K is drawn as a red dot-dashed curve. Solutions are indicated by open circles.

1.3 A First Taste of Quantum Mechanics

versus E/V_0 for various values of a. The intersection point(s) of the dashed curve and the solid curve, indicated by open circles, specify the bound state energy (energies). We used units where $V_0 = m = \hbar = 1$, and we solved Eq. (1.110) graphically for three values of a: $a = 1$, $a = 4$, and $a = 6$. For $a = 1$ and $a = 4$ there is only one bound state of even parity, but for $a = 6$, two bound state energies are obtained. As a is increased still further, more and more even bound states will be obtained. A similar graphical solution can be implemented for (1.112) to find the odd bound state energies in the square well potential.

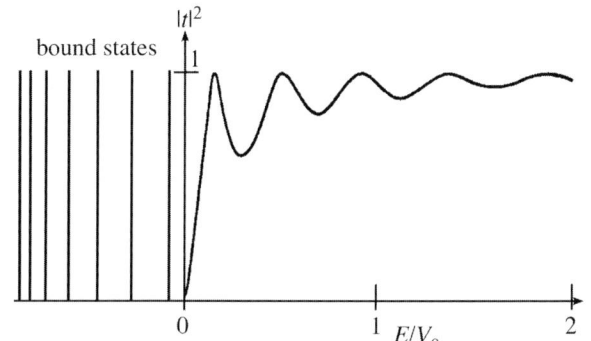

FIG 1.19 Bound state energies, and transmission $|t|^2$ versus $E/|V_0|$ for a square well with $\sqrt{2m|V_0|/\hbar^2}\, a = 3\pi$. The resonances in the transmission for $E > 0$ are clearly seen.

Clearly, as $V_0 \to \infty$, the solutions go to the even and odd solutions of the infinite square well, and the tunneling of the wave function into the classically forbidden regions becomes exponentially small.

The bound state energies for the square well potential in Fig. 1.14(d) and the transmission $|t|^2$ versus $E/|V_0|$, with $\sqrt{2m|V_0|/\hbar^2}\, a = 3\pi$ are shown in Figure 1.19. The peaks in the transmission are resonances (see Sec. 12.6 in the Scattering Chapter for a complete discussion of resonances, but basically, the idea is that the particle gets stuck in the well for a long time at these energies), which appear at the energies of the bound state levels of an infinite square well of the same width.

In 1D (and also 2D, see Sec. 12.9.1) an attractive potential always has a bound state, regardless of how small the well depth V_0 or width a [see Fig. 1.14(d)]. To show this, consider a potential with well depth that satisfies $V_0 \ll \hbar^2/(ma^2)$. The physical interpretation of this condition is that the well depth is much smaller than the kinetic energy the particle would have if it were totally confined in the well. Let us move the origin of the coordinate system to someplace near the middle of the well. We hypothesize that the magnitude of the bound state energy, $|E|$, is much smaller than V_0; this will be confirmed by our result. Hence, we neglect E on the RHS of the Schrödinger equation, $d^2\psi/dx^2 = \frac{2m}{\hbar^2}[V(x) - E]\psi$, within the well. Integrating from $-a/2$ to $a/2$ we find,

$$\left.\frac{d\psi}{dx}\right|_{-a/2}^{a/2} = \frac{2m}{\hbar^2} \int_{-\infty}^{\infty} dx\, V(x)\psi(x), \tag{1.113}$$

where we have extended the integral from $-\infty$ to ∞ but the integrand vanishes beyond $\pm a/2$ since the potential vanishes there. Without loss of generality, we can take the wave function to be unity within the well, and to be of the form $\psi(x) = e^{\pm \kappa x}$ to the right and left of the well, where $\hbar^2\kappa^2/(2m) = |E|$. Substituting into (1.113) we find, $-2\kappa = \frac{2m}{\hbar^2} \int_{-\infty}^{\infty} dx\, V(x)$, hence

$$|E| = \frac{m}{2\hbar^2}\left[\int_{-\infty}^{\infty} dx\, V(x)\right]^2. \tag{1.114}$$

In accordance with our hypothesis, the bound state energy is small, in fact, second-order small in the well-depth and well-width.

2D and 3D Wells

The solution of the Schrödinger equation for piecewise-constant potentials in 2D (or 3D) can often also be obtained analytically. Solutions to the Schrödinger equation in a potential that separates, e.g., $V(x,y) = V_1(x) + V_2(y)$, can be formed as products of the wave functions for the 1D potentials. For example, for the 2D infinite box potential, the solutions are given by $\psi_{n_x,n_y}(x,y) = \sqrt{\frac{4}{L_xL_y}} \sin(\frac{n_x\pi x}{L_x}) \sin(\frac{n_y\pi y}{L_y})$, and their energies are $E_{n_x,n_y} = \hbar^2(k_{n_x}^2 + k_{n_y}^2)/(2m) = \frac{\hbar^2\pi^2}{2m}\left(\frac{n_x^2}{L_x^2} + \frac{n_y^2}{L_y^2}\right)$. For a 3D infinite box potential, the solutions are given by

$$\psi_{n_x,n_y,n_z}(x,y,z) = \sqrt{\frac{8}{L_xL_yL_z}} \sin\left(\frac{n_x\pi x}{L_x}\right) \sin\left(\frac{n_y\pi y}{L_y}\right) \sin\left(\frac{n_z\pi y}{L_z}\right), \tag{1.115}$$

and the energy eigenvalues are

$$\boxed{E_{n_x,n_y,n_z} = \frac{\hbar^2(k_{n_x}^2 + k_{n_y}^2 + k_{n_z}^2)}{2m} = \frac{\hbar^2\pi^2}{2m}\left(\frac{n_x^2}{L_x^2} + \frac{n_y^2}{L_y^2} + \frac{n_z^2}{L_z^2}\right).} \tag{1.116}$$

For $L_x = L_y = L_z \equiv L$, the lowest energy corresponds to $n_x = n_y = n_z = 1$, i.e., $E_{1,1,1} = 3\frac{\hbar^2\pi^2}{2mL^2}$ (note that the n_i cannot equal zero, for otherwise the wave function vanishes). The first excited state corresponds to $n_y = n_z = 1$ and $n_x = 2$ and the permutations of these quantum numbers, i.e., $E_{2,1,1} = E_{1,2,1} = E_{1,1,2} = 6\frac{\hbar^2\pi^2}{2mL^2}$. Hence, the first excited state is triply *degenerate*. The second excited state is also triply degenerate, $E_{2,2,1} = E_{2,1,2} = E_{1,2,2} = 9\frac{\hbar^2\pi^2}{2mL^2}$.

A *quantum well* potential is one that confines particles, originally free to move in three dimensions, to two dimensions so they are free to move only in a planar region, a *quantum wire* potential is one that confines particles to move only in a line and confines them in the two orthogonal directions, and a *quantum dot* potential is one that confines particles in all three directions so they are localized near a certain point. Semiconductor nanostructures that confines the motion of conduction band electrons, valence band holes (missing electrons), or excitons (pairs of conduction band electrons and valence band holes) in one, two or three dimensions can be readily fabricated. The Schrödinger equation for electrons can be easily solved for such structures. For example, we shall do so for quantum wells and wires in Sec. 9.1.1, and for spherical quantum dot structures in Sec. 3.2.2.

Problem 1.16

(a) Calculate $\langle \psi_{n_x,n_y,n_z}|x|\psi_{n_x,n_y,n_z}\rangle$, $\langle \psi_{n_x,n_y,n_z}|p_x|\psi_{n_x,n_y,n_z}\rangle$, $\langle \psi_{n_x,n_y,n_z}|p_x^2|\psi_{n_x,n_y,n_z}\rangle$.
(b) Calculate $\langle \psi_{n_x,n_y,n_z}|\mathbf{p}^2|\psi_{n_x,n_y,n_z}\rangle$.
(c) Calculate $\langle \psi_{n_x,n_y,n_z}|x^2|\psi_{n_x,n_y,n_z}\rangle$.
(d) For $L_x = L_y = L_z \equiv L$, find the energy and degeneracy of the third excited state.

Answer: (a) $\langle \psi_{n_x,n_y,n_z}|x|\psi_{n_x,n_y,n_z}\rangle = L_x/2$, $\langle \psi_{n_x,n_y,n_z}|p_x|\psi_{n_x,n_y,n_z}\rangle = 0$. (b) $2mE_{n_x,n_y,n_z} = \hbar^2\pi^2\left(\frac{n_x^2}{L_x^2} + \frac{n_y^2}{L_y^2} + \frac{n_z^2}{L_z^2}\right)$. (c) $\frac{(2n_x^2\pi^2-3)L_x^2}{6n_x^2\pi^2}$. (d) $E_{3,1,1} = E_{1,3,1} = E_{1,1,3} = 11\frac{\hbar^2\pi^2}{2mL^2}$.

Tunneling Through a Double Barrier: Resonances

Let us now consider the potential in Fig. 1.20(a). This potential with $L_b \to \infty$ is similar to the potential in Fig. 1.14(d), which has bound states. Therefore, it should come as no surprise that the transmission $|t|^2$ has peaks at energies where the potential with $L_b \to \infty$ has bound states [see Fig. 1.20(b)]. These peaks are called *resonances*; they are the quasi-bound states that are vestiges of the bound states of the potential with $L_b \to \infty$. The nature of these resonances is somewhat different from those we encountered in Fig. 1.19; see Sec. 12.6 for a full discussion of resonances.

1.3 A First Taste of Quantum Mechanics

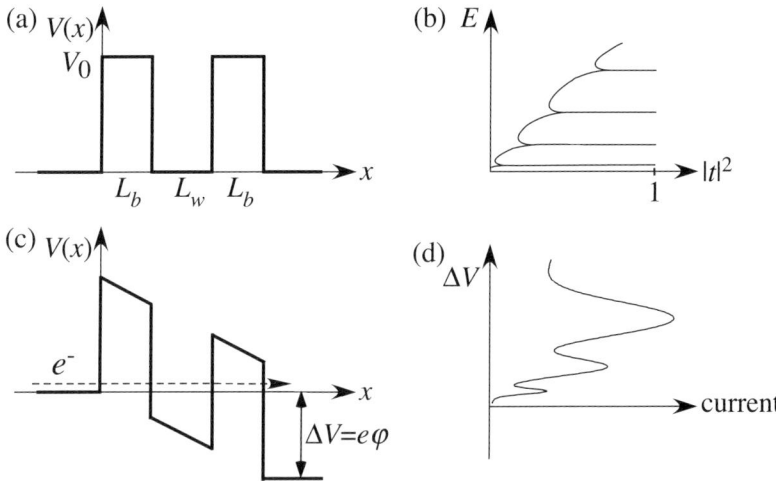

FIG 1.20 Tunneling through barriers. (a) One-dimensional double barrier potential. (b) Transmission through the barriers versus energy. (c) The potential with a bias voltage φ applied. (d) Schematic illustration of the current versus bias potential ΔV.

The potential in Fig. 1.20(a) is a simple model of a solid-state device composed of a multilayer structure formed by a stacking of two semiconductor single crystal thin films of different composition or doping. The double barrier resonance structure can be made of GaAs/GaAlAs layers in which an undoped GaAs is sandwiched between two potential barriers of GaAlAs. The device is then connected to leads, and a bias potential $\Delta V = e\varphi$, where ΔV has units of energy and φ of electric potential (volts), can be imposed across the device so that a current I flows. Figure 1.20(c) shows the potential versus position. Experimentally one can measure the current versus voltage curve, $I(\varphi)$ (from which the differential conductance $g = dI/d\varphi$ can be determined). The current is due to electrons to the left of the double barrier that move to the right provided the electrons are within a range of energies ΔV. The question of how the measured current can be calculated from the solution of the Schrödinger equation will be addressed in future chapters, but here we note that when the bias voltage is small, the conductance is directly proportional to the transmission coefficient. Figure 1.20(d) schematically depicts the current expected through the device as a function of ΔV. The quasi-bound states in the well affect the tunneling, and peaks in the current versus voltage result. The occurrence of peaks in the $I(\varphi)$ curve indicate that the differential conductance changes sign, being positive to the left of the peak and negative to the right. This property is a key element in the quantum electronics device known as the *resonant tunneling diode*.

Metal–Vacuum and Semiconductor–Vacuum Interfaces

True metal–vacuum, semiconductor–metal or semiconductor–vacuum interfaces are not discontinuous. The potential energy of the electrons change continuously over an interval whose dimensions are of the order of the interatomic distances in the metal or semiconductor. The potential energy near the surface can be written approximately as $V(x) = -V_0 \left(1 + e^{x/a}\right)^{-1}$, which approximates to the previously used discontinuous potential as $a \to 0$. Figure 1.21 plots $V(x)$ versus x. The solution of the Schrödinger equation, $-\frac{\hbar^2}{2m}\frac{d^2\psi}{dx^2} - V_0 \left(1 + e^{x/a}\right)^{-1} \psi = E\psi$, can be written in terms of a hypergeometric function [see Eq. (B.33) in Appendix B] and the reflection probability can also be obtained analytically. We shall not pause to do so here.

Electrons in metals can essentially be regarded as free particles. Consider a thin metal film grown on a semiconductor surface. The free electrons in the metal film cannot escape the metal into air because of the work function and cannot escape into the semiconductor if the band edge of the semiconductor is lower in energy valence band in the metal, i.e., if the valence band in the metal is in the energy gap of the semiconductor. Figure 1.22 schematically shows the arrangement, the electron energy distributions in the metal and semiconductor and the potentials that the electrons experience in the vacuum, metal, and semiconductor. The electrons are confined perpendicular to the surface to be within the metal. This leads to a set of standing waves, allowing only specific wave vectors in the perpendicular direction (no such restriction exists in the parallel direction and electrons are free to propagate with any wave vectors in the parallel directions). The electronic states form energy bands, as shown in Fig. 1.22, one for each wave vector allowed by the confinement. The energy of the bands depends on film thickness, and therefore thickness-dependent properties result. For example, the work function can show an oscillatory thickness dependence. The oscillations occur because the quantum well states shift to lower energies as the film thickness increases. At regular thickness intervals, new states become populated as the state energies decrease with increasing film thickness. But film thickness is not a continuous variable; it varies in steps of the thickness of a layer of metal atoms. If the period of the oscillations is incommensurate with the layer thickness, a beat period appears in transport properties.

The idea of energy eigenvalues of a periodic potential, such as a sinusoidal potential or a series of square well potentials, that form energy bands, as shown in Fig. 1.23 will be discussed at length in Chapter 9. Any periodic potential, whether the potential is in 1D, 2D, or 3D, will have eigenvalues that consist of bands, with band gaps between the bands, as shown schematically in Fig. 1.23.

1.3.12 THE DELTA FUNCTION POTENTIAL

Another simple model potential whose properties can be solved analytically is the *delta function potential*, $V(x) = \tilde{V}_0 \delta(x)$. The potential strength parameter $\tilde{V}_0$, which has units of energy times length, can be either positive (repulsive potential) or negative (attractive potential). Away from the origin, the solutions to the Schrödinger equation,

$$-\frac{\hbar^2}{2m}\frac{d^2\psi}{dx^2} + \tilde{V}_0 \delta(x)\psi = E\psi, \quad (1.117)$$

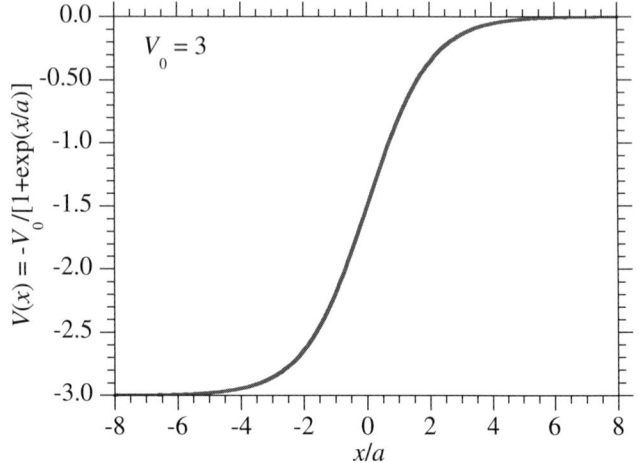

FIG 1.21 Interface potential $V(x) = -V_0(1 + e^{x/a})^{-1}$.

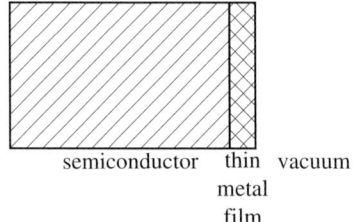

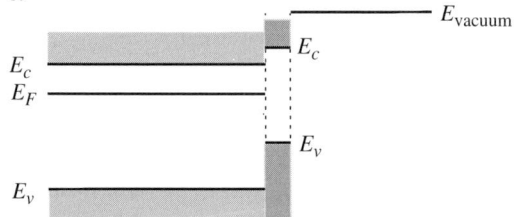

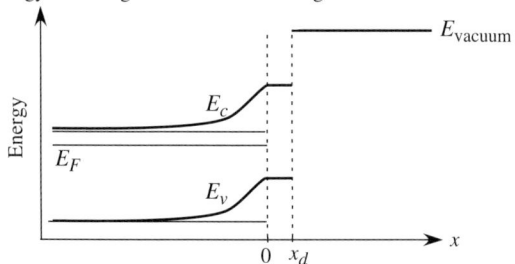

FIG 1.22 Thin metal-film semiconductor interface.

1.3 A First Taste of Quantum Mechanics

are plane waves, $e^{\pm ikx}$ with $k^2 = 2mE/\hbar^2$ if $E > 0$, and $e^{\pm \kappa x}$ with $\kappa^2 = 2m|E|/\hbar^2$ if $E < 0$ (for the case when $\tilde{V}_0 < 0$). At the origin, one can match the wave function and its derivative to the right of the origin and to the left as follows:

$$\psi(0^-) = \psi(0^+) \equiv \psi(0),$$
$$\psi'(0^-) - \psi'(0^+) = \frac{2m\tilde{V}_0}{\hbar^2}\psi(0). \quad (1.118)$$

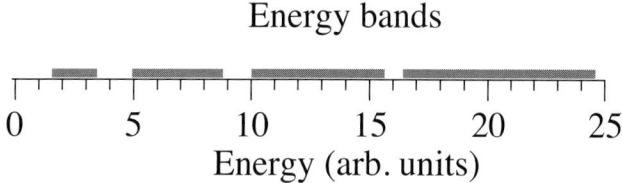

FIG 1.23 Lowest lying energy bands of a one-dimensional potential.

The latter equation follows by integrating the Schrödinger equation over a small region containing the origin, i.e., $\int_{-\epsilon}^{\epsilon} dx [-\frac{\hbar^2}{2m}\frac{d^2\psi}{dx^2} + \tilde{V}_0 \delta(x)\psi - E\psi = 0] \xrightarrow[\epsilon \to 0]{} -\frac{\hbar^2}{2m}[\psi'(0^-) - \psi'(0^+)] + \tilde{V}_0 \psi(0) = 0$. We shall consider two problems with delta function potentials; a scattering wave impinging on a delta function potential from the left, and a bound state problem in an *attractive* delta function potential.

For the scattering problem, let us take the wave function to be of the form $\psi(x) = e^{ikx} + re^{-ikx}$ for $x < 0$, and $\psi(x) = te^{ikx}$ for $x > 0$. Matching gives $1 + r = t$, and $ik(1-r) = t(ik + \frac{2m\tilde{V}_0}{\hbar^2})$. Hence, $t = \frac{1}{1-im\tilde{V}_0/k}$ and $r = \frac{im\tilde{V}_0/k}{1-im\tilde{V}_0/k}$.

For the 1D bound state in an attractive delta function potential, the wave function takes the form $\psi(x) = A e^{\kappa x}$ for $x < 0$, and $\psi(x) = B e^{-\kappa x}$ for $x > 0$. Matching gives $A = B$ and $E = -\frac{m\tilde{V}_0^2}{2\hbar^2}$. Only one bound state exists for the attractive delta function potential in 1D, with energy E proportional to the square of the strength of the attractive potential.

1.3.13 WAVE PACKETS

We have determined the continuum wave functions for piecewise-constant potentials in Sec. 1.3.11. These wave functions, $\psi_E(x)$, are eigenstates of the Hamiltonian. But how are these wave functions related to classical motion of a particle in these potentials? In order to obtain classical-like motion of a particle, we must start from a wave function of a form that is perceptibly different from zero only in a very small region of space in order to mimic the characteristic of a classical particle that is localized about a given point in space at a given time. Such a wave function is called a *wave packet*. We can make such wave packets by taking superpositions of the energy eigenstates that we calculated above.

For example, suppose that at time $t=0$ our wave function takes the form $\psi(x, t=0) = \mathcal{N} \exp(-\frac{(x-x_0)^2}{4\sigma^2})e^{ikx}$, where x_0 is a point very far to the left of a barrier potential centered near the origin. This initial wave packet is localized (centered) around the point x_0 and has central momentum $p = \hbar k$, with a spread of momenta around this central momentum. The initial state $\psi(x,0)$ can be expanded in energy eigenstates (having the right asymptotic behavior, e.g., coming from the left), $\psi_E(x)$, in the form $\psi(x,0) = \int dE\, b(E)\psi_E(x)$, and the amplitudes $b(E)$ can be ascertained by projection of this initial condition onto $\psi_E(x)$ to be given by $b(E) = \int dx\, \psi_E^*(x)\psi(x,0)$. The time-dependent wave packet is then given by $\psi(x,t) = \int dE\, b(E)\psi_E(x)e^{-iEt/\hbar}$. This type of wave function expansion in terms of a superposition of eigenstates of the Hamiltonian is very general. Given any initial wave packet at time $t=0$, $\psi(x,0) = \psi_0(x)$, the solution of the time-dependent Schrödinger equation having this initial condition can be obtained as a superposition of eigenstates of the Hamiltonian, $\phi_m(x)$, in the form, $\psi(x,t) = \sum_m b_m \phi_m(x)e^{-iE_m t/\hbar}$, by finding the amplitudes b_m such that at time $t=0$, $\psi(x,t=0) = \psi_0(x)$. If the spectrum contains both a discrete and continuous region,

$$\psi(x,t) = \sum_m b_m \phi_m(x)e^{-iE_m t/\hbar} + \int dE\, b(E)\, \phi_E(x)e^{-iEt/\hbar}. \quad (1.119)$$

Quite generally, this method can be used to form time-dependent wave packets in terms of basis states that are eigenstates of the Hamiltonian (see Sec. 6.6, and particularly see Fig. 6.20) for an illustration of how a wave packet propagates as a function of time.

Problem 1.17

Consider the 1D Gaussian wave packet, $\psi_0(x) = \mathcal{N} \exp\left(-\frac{x^2}{4\sigma^2}\right) e^{i\kappa x}$ moving in free space so $H = p^2/2m$.

(a) Calculate $\langle x \rangle_{t=0}$, $(\Delta x)_{t=0}$, $\langle p \rangle_{t=0}$, $(\Delta p)_{t=0}$.
(b) Determine the Fourier transform of the wave packet, $\psi_0(k)$.
(c) Find an analytic expression for time-dependent wave packet, $\psi(x,t) = \frac{1}{\sqrt{2\pi}} \int_{-\infty}^{\infty} dk \, \psi_0(k) e^{-i\omega(k)t}$, with

$\psi(x, t=0) = \psi_0(x)$ and $\omega(k) = \frac{\hbar k^2}{2m}$.

(d) Calculate $\langle x \rangle_t$, $(\Delta x)_t$, $\langle p \rangle_t$, $(\Delta p)_t$.

Answers: (b) $\psi_0(k) = \mathcal{N}\sqrt{2}\sigma \, e^{-\sigma^2(\kappa-k)^2}$.

(c) $\psi(x,t) = \mathcal{N} \frac{[2\pi\sigma^2]^{1/4}}{[2\pi\sigma^2(t)]^{1/4}} e^{-\frac{(x-\hbar\kappa t/m)^2}{4\sigma^2(t)} + i\kappa(x-\hbar\kappa t/m) + i\phi(t)}$,

where $\sigma^2(t) = \sigma^2 + i\frac{\hbar t}{2m}$ and $\phi(t) = \frac{\hbar^2\kappa^2}{2m}\frac{t}{\hbar}$.

(d) $\langle x \rangle_t = \hbar\kappa t/m$, $(\Delta x)_t = \sigma\sqrt{1 + \frac{\hbar^2 t^2}{4m^2\sigma^4}}$, $\langle p \rangle_t = \hbar\kappa$, $(\Delta p)_t = \frac{\hbar}{2\sigma}$.

1.3.14 THE LINEAR POTENTIAL AND QUANTUM TUNNELING

Quantum tunneling can occur through regions where the potential energy is larger than the total energy E. We have already seen an example of quantum tunneling through a piecewise-constant barrier potential in Sec. 1.3.11, where we learned that the general solution of the Schrödinger equation for piecewise-constant potential $V_0 > E$, $-\frac{\hbar^2}{2m}\frac{d^2\psi}{dx^2} + V_0\psi = E\psi$, is $\psi(x) = a\exp(\kappa x) + b\exp(-\kappa x)$ where $\kappa \equiv \sqrt{2m(V_0-E)/\hbar^2} > 0$.

Quantum tunneling through a region where the potential has a constant slope can be described in terms of well-known functions called *Airy functions* (see [27]), i.e., the solution of the Schrödinger equation for a linear potential, $V(x) = \mathcal{F}x$, is given in terms of Airy functions. A linear potential is experienced by a particle of charge q in a constant electric field, $\mathcal{E}$, where the constant $\mathcal{F} = -q\mathcal{E}$ and the force F is $F = -dV/dx = -\mathcal{F} = q\mathcal{E}$. The Schrödinger equation takes the form

$$\frac{d^2\psi}{dx^2} - \frac{2m}{\hbar^2}(\mathcal{F}x - E)\psi = 0. \tag{1.120}$$

By making the transformation $z = ax + b$ with the coefficients $a = (2m\mathcal{F}/\hbar^2)^{1/3}$ and $b = -(2m\mathcal{F}/\hbar^2)^{1/3}E/\mathcal{F}$, Eq. (1.120) takes the form

$$\frac{d^2\psi}{dz^2} = z\psi(z). \tag{1.121}$$

The general solution of this equation is given in terms of Airy functions [see Eq. (B.10)]: $\psi(z) = c\,\text{Ai}(z) + d\,\text{Bi}(z)$. Figure 1.24 plots the Airy functions. The function $\text{Ai}(z)$ decays exponentially in the classically forbidden region, $z > 0$, whereas the unction $\text{Bi}(z)$ exponentially increases in this region. Both functions oscillate in the classically allowed region (where $E > V$), $z < 0$. The asymptotic forms of these functions as $z \to \pm\infty$ is given in Eqs (B.12–B.15). The magnitude of the oscillation in the classically allowed region decreases as $|z|^{-1/4}$, and the functions $\text{Ai}(z)$ and $\text{Bi}(z)$ are out of phase in this region. In the classically forbidden region $\text{Ai}(z)$ decays exponentially as $\text{Ai}(z) \sim e^{-\int_0^z dz\, z^{1/2}} = e^{-\frac{2}{3}z^{3/2}}$.

The important experimental technique of scanning tunneling microscopy (STM) is used to view surfaces of conducting materials at nanometer resolution uses tunneling of electrons from the surface of the material to the STM tip. Figure 1.25 shows the potential energy that an electron feels near the surface when a attractive potential is applied. This is a simplified

1.3 A First Taste of Quantum Mechanics

1D view of the potential wherein we neglect the other dimensions orthogonal to the tip to closest-contact-point on the surface. The potential experienced by electrons upon tunneling out of the conducting material is due to the applied potential $-e\mathcal{E}z$, where $\mathcal{E}$ is the electric field strength, and to the attractive image potential, $V_{\text{image}}(z)$, that results from the effective polarization of the conductor due to repulsion of electrons in the conductor near the tunneling electron, $V_{\text{image}}(z) = -\frac{e^2}{4z}$. Hence, the total potential that an electron in the region $z > 0$ experiences, $V(z) = -e\mathcal{E}z - \frac{e^2}{4z}$, has a barrier that electrons with energy equal to the Fermi energy E_F (the energy of the highest occupied state in a conductor) must overcome in order to tunnel to the tip, as shown in Fig. 1.25. We shall see in Sec. 7.2.1 that, to a good approximation, the tunneling probability is proportional to $\left| \int_0^{z_{\text{tp}}} dz\, e^{-\sqrt{\frac{2m}{\hbar^2}[E-V(z)]}} \right|^2$, where in our case the turning point (defined as the coordinate position z_{tp} at which the potential $V(z)$ equals the energy E), $z_{\text{tp}} = \frac{W}{e\mathcal{E}}$, and the energy $E = E_F$. The quantity appearing in the exponential of this expression is the local momentum $p(z) = \sqrt{2m[E - V(z)]}$ divided by $\hbar$; so, to a good approximation, the amplitude for tunneling is given by the exponential of $-\int dz\, p(z)/\hbar$, where the integral is over the classically forbidden region of coordinates.

1.3.15 THE HARMONIC OSCILLATOR

Let us consider the quantum mechanics of a particle in a harmonic oscillator potential $V(x) = \frac{1}{2}kx^2$, where k is the spring force constant that is related to the frequency of the motion (see Sec. 16.1, linked to the book web page) by the relation $\omega = \sqrt{k/m}$, i.e., $V(x) = \frac{m\omega^2}{2}x^2$. The time-independent Schrödinger equation is

$$\left(-\frac{\hbar^2}{2m}\frac{d^2}{dx^2} + \frac{1}{2}m\omega^2 x^2 \right) \psi = E\psi. \quad (1.122)$$

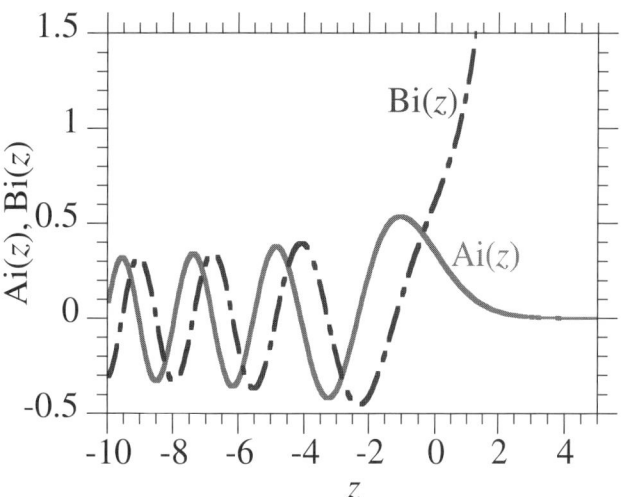

FIG 1.24 Airy functions Ai(z) and Bi(z) versus z.

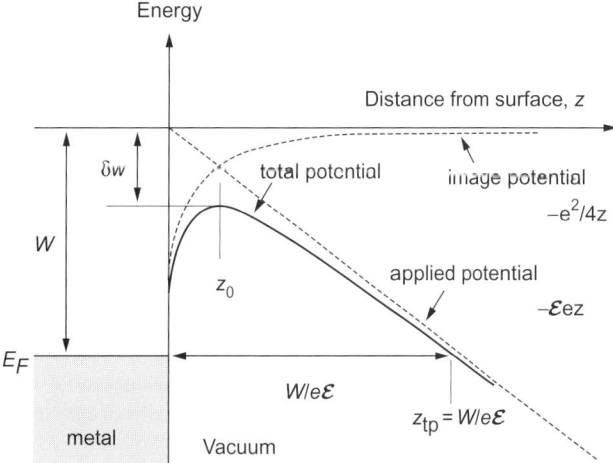

FIG 1.25 Electron tunneling from a metal surface in the presence of an external electric field $\mathcal{E}$ that yields an applied potential $-e\mathcal{E}z$ experienced by the electrons that tunnel into the vacuum region. See text for details.

It is convenient to define new dimensionless variables; a dimensionless length variable $y \equiv x/l_{\text{ho}}$, a dimensionless energy variable $\mathcal{E} \equiv E/E_{\text{ho}}$ and a dimensionless momentum variable $p_y \equiv p_x/p_{\text{ho}}$ and use these variables in considering the quantum harmonic oscillator problem. Here, l_{ho}, E_{ho}, and p_{ho} are the "natural" length, energy, and momentum values that are determined by the harmonic oscillator potential in the system of units we are using. We shall employ the same dimensionless variable approach for re-writing the quantum problem for any type of power law potential of the form, $V(x) = C_j x^j$, whether the power j can be a positive or negative integer (Coulomb potential, $j = -1$, van der Waals potential $j = -6$, dipole–dipole potential, $j = -3$). This is a valuable dimensional analysis technique that should be learned once-and-for-all, so we will go through the technique in detail. We need to find the appropriate variables l_{ho}, E_{ho}, and p_{ho}. To do so, let us first multiply Eq. (1.122) by $(-2m/\hbar^2)$ to obtain $\left(\frac{d^2}{dx^2} - \frac{m^2\omega^2}{\hbar^2}x^2 + \frac{2m}{\hbar^2}E\right)\psi = 0$. Since the first term in the parenthesis has units of $1/l^2$, each of the terms in the parenthesis must have the same dimension. By equating the units of the first and second terms, we find that $\frac{1}{[x^2]} = \frac{m^2\omega^2}{\hbar^2}[x^2]$, where $[\cdot]$ means "units of $\cdot$". Hence, we conclude that $[x] \equiv l_{ho} = \sqrt{\frac{\hbar}{m\omega}}$. By equating the units of the second and third terms in the parenthesis, $\frac{m^2\omega^2}{\hbar^2}\frac{\hbar}{m\omega} = \frac{2m}{\hbar^2}[E]$, we find $[E] \equiv E_{ho} = \hbar\omega$. The units of momentum are those of $\hbar/[x]$, since $p = \frac{\hbar}{i}\frac{\partial}{\partial x}$, so $[p] \equiv p_{ho} = \hbar/[x] = \sqrt{\hbar m\omega}$. We have thus determined the parameters l_{ho}, p_{ho}, and E_{ho}:

$$\boxed{l_{ho} = \sqrt{\frac{\hbar}{m\omega}}, \quad p_{ho} = \sqrt{\hbar m\omega}, \quad E_{ho} = \hbar\omega.} \quad (1.123)$$

The dimensionless units of length, momentum, and energy are therefore x/l_{ho}, p/p_{ho} and E/E_{ho}.

We can rewrite Eq. (1.122) using the dimensionless variables $y = x/l_{ho}$ and $\mathcal{E} \equiv E/E_{ho}$ by substituting $x = y l_{ho}$ and $E = \mathcal{E}\hbar\omega$ to obtain,

$$\left(-\frac{1}{2}\frac{d^2}{dy^2}\psi(y) + \frac{1}{2}y^2\right)\psi(y) = \mathcal{E}\psi(y), \quad (1.124)$$

or, upon rearranging,

$$\frac{d^2}{dy^2}\psi(y) + \left(2\mathcal{E} - y^2\right)\psi(y) = 0. \quad (1.125)$$

This is the time-independent Schrödinger equation for the harmonic oscillator in dimensionless units.

It is particularly easy to solve Eq. (1.125) for large y. Then, we can neglect $2\mathcal{E}$ in the second term on the LHS of the equation, obtaining $\left(\frac{d^2}{dy^2} - y^2\right)\psi(y) = 0$. If we substitute $\psi(y) = \exp(-y^2/2)$, we can see that this is a solution, since 1 can be neglected in comparison with y^2 [note that $\psi(y) = \exp(y^2/2)$ is also a solution, but the wave function must remain finite for all x (since the probability is proportional to the absolute square of the wave function), so we rule this wave function out]. Now, given the asymptotic form we have just found, let us try a solution of the form $\psi(y) = H(y)\exp(-y^2/2)$ and find $H(y)$. Upon substituting this form into Eq. (1.125), the equation we obtain for $H(y)$ is

$$\frac{d^2H}{dy^2} - 2y\frac{dH}{dy} + (\beta - 1)H = 0, \quad (1.126)$$

where we have defined $\beta \equiv 2\mathcal{E}$. Let us find $H(y)$ by assuming it can be written as a power series. Since the coefficient of the term with the highest derivative is unity, it follows from the theory of differential equations that the solution of Eq. (1.126) can have no singularities for finite y, and a power series solution has an infinite radius of convergence. Upon substituting the power series $H(y) = \sum_{j=0}^{\infty} a_j y^j$ into (1.126) and equating the coefficient of each power of y to zero, we find the recursion formula $a_{j+2} = \frac{2j+1-\beta}{(j+2)(j+1)}a_j$. Hence, the series terminates if $\beta = 2n + 1$ for some integer n. Moreover, if the series does not terminate, the function $H(y)$ diverges as e^{y^2} as $y \to \pm\infty$. We therefore conclude that the energy eigenvalues are given by $E_n = (n + 1/2)\hbar\omega$ with n an integer, and the functions $H_n(y)$ are polynomials of order n. These polynomials are called *Hermite polynomials*, and the differential equation (1.126), with $\beta = 2n + 1$ is called the Hermite

1.3 A First Taste of Quantum Mechanics

differential equation. The properties of these polynomials are well known [27]. Their *generating function* is given by

$$\exp(-t^2 + 2ty) = \sum_{n=0}^{\infty} H_n(y) \frac{t^n}{n!}. \tag{1.127}$$

The *recursion relation* for the Hermite polynomials are

$$2yH_n(y) = H_{n+1}(y) + 2nH_{n-1}(y), \tag{1.128}$$

and

$$\frac{d}{dy}H_n(y) = 2nH_{n-1}(y). \tag{1.129}$$

The lowest few Hermite polynomials are: $H_0(y) = 1$, $H_1(y) = 2y$, $H_2(y) = 4y^2 - 2$, $H_3(y) = 3y^3 - 12y$. One can readily normalize the wave functions, $\psi_n(x) = N_n H_n\left(\frac{x}{l_{ho}}\right) \exp\left(-\frac{x^2}{2l_{ho}^2}\right)$, i.e., find the normalization coefficients N_n such that $\int_{-\infty}^{\infty} dx\, |\psi(x)|^2 = 1$: $N_n = (\pi^{1/2} 2^n n!)^{-1/2}$. Upon converting back to dimensional variables, the normalized harmonic oscillator eigenfunctions (which have dimension $(\text{length})^{-1/2}$) are

$$\boxed{\psi_n(x) = \left(\frac{m\omega}{\pi\hbar}\right)^{1/4} \frac{1}{2^{n/2}\sqrt{n!}} H_n\left(\frac{x}{l_{ho}}\right) \exp\left[-\left(\frac{x^2}{2l_{ho}^2}\right)\right],} \tag{1.130}$$

and the eigenenergies are

$$\boxed{E_n = (n + 1/2)\hbar\omega.} \tag{1.131}$$

Figure 1.26 shows the six lowest eigenstate wave functions $\psi_n(x)$ of the harmonic oscillator potential as a function of x, superimposed over the harmonic potential with the x axes shifted to coincide with the energy $E_n = (n + 1/2)\hbar\omega$. The probability for finding the particle as a function of x is $P_n(x) = |\psi_n(x)|^2$. The eigenfunctions $\psi_n(x)$ are not only normalized, but they are orthonormal, $\langle \psi_n | \psi_{n'} \rangle = \delta_{n,n'}$, because, as we shall learn in Chapter 2, eigenfunctions of Hermitian operators are orthogonal. The solutions with even (odd) n are symmetric (antisymmetric) with respect to inversion about the center of the well, $x \rightarrow -x$. This parity symmetry upon flipping in the sign of the spatial coordinate results because the potential is symmetric under the inversion transformation (see Sec. 2.9.2).

Problem 1.18

(a) Using the generating function (1.127), show that $H_n(y) = \frac{\partial^n}{\partial t^n} \exp(-t^2 + 2ty)\Big|_{t=0}$.
(b) Confirm that $H_1(y)$, $H_2(y)$, and $H_3(y)$ satisfy (a).
(c) Use the recursion relation (1.128) to determine $H_4(y)$ given $H_2(y)$ and $H_3(y)$ (see above).
(d) Confirm that your result in (c) satisfies (1.129).

The harmonic oscillator potential has only a discrete spectrum; no continuum exists. The bound-state energy eigenvalues are equally spaced, $\Delta E_n \equiv E_{n+1} - E_n = \hbar\omega$, and the lowest energy eigenvalue is nonzero, $E_0 = \hbar\omega/2$. E_0 is called the *zero-point energy* of the oscillator. The lowest energy wave function ψ_0 has no nodes, the first excited state wave function ψ_1 has one node, the second excited state has two nodes, etc.

The spatial probability distribution of the ground state wave function is

$$P(x) = |\psi_0(x)|^2 = \left(\frac{1}{\pi l_{ho}^2}\right)^{1/2} \exp\left[-\left(\frac{x^2}{l_{ho}^2}\right)\right]. \tag{1.132}$$

A normal Gaussian probability distribution is given by $P(x) = \frac{1}{\sqrt{2\pi\sigma^2}} e^{-\frac{x^2}{2\sigma^2}}$ where σ is the width of the distribution. Comparing with (1.132), we find that the spatial width σ of the density for the ground state harmonic oscillator wave function is such that

$$\sigma^2 = \frac{l_{ho}^2}{2} = \frac{\hbar}{2m\omega}. \qquad (1.133)$$

In other words, if we write the ground state harmonic oscillator wave function in terms of the width of the density, σ, we obtain

$$\psi_{n=0}(x) = \left(\frac{1}{2\pi\sigma^2}\right)^{1/4} \exp\left(-\frac{x^2}{4\sigma^2}\right). \qquad (1.134)$$

Wave functions allow us to calculate expectation values of dynamical variables, $\mathcal{O}$, by calculating the expectation values (i.e., the diagonal matrix elements),

$$\langle \psi_n | \mathcal{O} | \psi_n \rangle = \int dx\, \psi_n^*(x) \mathcal{O} \psi_n(x). \qquad (1.135)$$

Moreover, the recursion relations (1.128) and (1.129) can be used to obtain analytic expressions for expectation values of the position and momentum operators, and their powers. As we shall see in later chapters, transitions from level n' to level n will involve matrix elements of various dynamical variables $\mathcal{O}$ of the form $\langle \psi_n | \mathcal{O} | \psi_{n'} \rangle = \int dx\, \psi_n^*(x) \mathcal{O} \psi_{n'}(x)$. In Problem 1.19 you are asked to calculate various expectation values and matrix elements of powers of position and momentum operators.

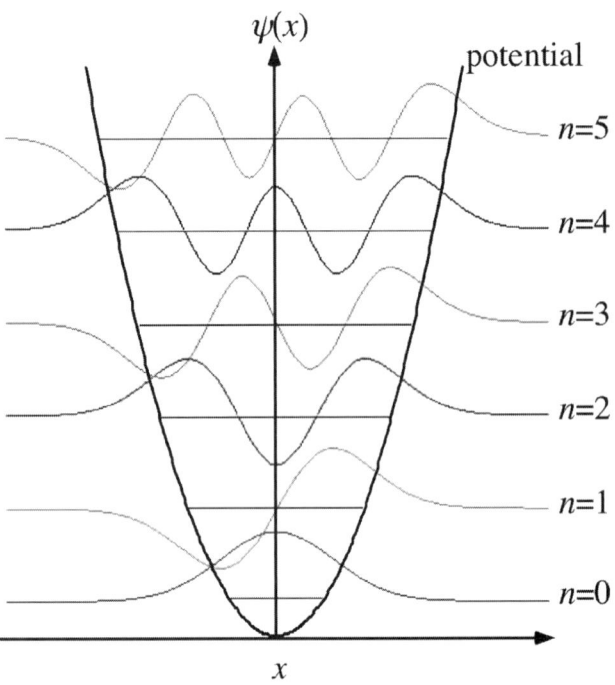

FIG 1.26 The lowest six eigenstate wave functions for the harmonic oscillator potential versus position x.

We shall return to the harmonic oscillator problem in Sec. 2.7.2 where we solve it in an entirely different fashion, using *raising and lowering operators*. This elegant and powerful matrix mechanics method is at the heart of quantum mechanical treatments applied to solve numerous quantum mechanical problems. However, we need to develop additional tools before presenting the matrix mechanics method, and that is done in Chapter 2.

Problem 1.19

(a) Calculate $\langle \psi_n | x | \psi_n \rangle$, $\langle \psi_n | p | \psi_n \rangle$, $\langle \psi_n | x^2 | \psi_n \rangle$, $\langle \psi_n | p^2 | \psi_n \rangle$.
(b) Calculate $\langle \psi_n | V(x) | \psi_n \rangle$ and $\langle \psi_n | \frac{p^2}{2m} | \psi_n \rangle$.
(c) Calculate $\langle \psi_n | x | \psi_{n'} \rangle$ and $\langle \psi_n | p | \psi_{n'} \rangle$.
(d) Calculate $\sum_{n'} \langle \psi_n | x | \psi_{n'} \rangle \langle \psi_{n'} | x | \psi_n \rangle$ and show that it equals $\langle \psi_n | x^2 | \psi_n \rangle$.
(e) Show that $\sum_{n'} (\langle \psi_n | x | \psi_{n'} \rangle \langle \psi_{n'} | p | \psi_j \rangle - \langle \psi_n | p | \psi_{n'} \rangle \langle \psi_{n'} | x | \psi_j \rangle) = \frac{\hbar}{i} \delta_{nj}$.

Answers: (a) $\langle x \rangle = 0$, $\langle p \rangle = 0$, for reasons associated with parity symmetry. The expectation value $\langle \psi_n | x^2 | \psi_n \rangle = \frac{\hbar}{m\omega}(n + 1/2)$, can be obtained using Eq. (1.128) by applying one power of x in $x^2 = x \cdot x$ to both $|\psi_n\rangle$

1.3 A First Taste of Quantum Mechanics

and $\langle \psi_n |$ in the matrix element, and $\langle \psi_n | p^2 | \psi_n \rangle = m\hbar\omega(n+1/2)$ can be obtained by applying Eq. (1.129) in the same way.

(b) $\langle \psi_n | V(x) | \psi_n \rangle = \langle \psi_n | \frac{p^2}{2m} | \psi_n \rangle = \frac{1}{2}\hbar\omega(n+1/2)$.

(c) Using the recursion relation (1.128) and the orthonormality properties of the wave functions, we find, $\langle \psi_n | x | \psi_{n'} \rangle = \sqrt{\frac{\hbar}{2m\omega}} \left(\sqrt{n}\, \delta_{n,n'+1} + \sqrt{n+1}\, \delta_{n,n'-1} \right)$. Using the recursion relation (1.129) and orthonormality, we find $\langle \psi_n | p | \psi_{n'} \rangle = m\omega \sqrt{\frac{\hbar}{2m\omega}} (-i) \left(\sqrt{n}\, \delta_{n,n'+1} - \sqrt{n+1}\, \delta_{n,n'-1} \right)$.

The Formalism of Quantum Mechanics

2

The first systematic formulation of the propositions and mathematical structure of quantum mechanics was set out by Paul A. M. Dirac in his book entitled *The Principles of Quantum Mechanics* [5] in 1930, just 4 years after the Schrödinger equation was developed. This was quickly followed in 1932 by John von Neumann's axiomatic formulation of the foundations of quantum mechanics [28]. Quantum mechanics is couched in the language of linear vector spaces and in probability theory. We shall assume that the reader is familiar with the material in the initial discussion of the structure of quantum mechanics presented in Sec. 1.3, in the linear algebra Appendix A, and with probability theory.

This chapter begins in Sec. 2.1 with a short reminder of Dirac notation for vector spaces with an inner product, and the completeness and orthogonality conditions in Hilbert space. Position and momentum representations are the focus of Sec. 2.1.1. Basis-state expansion methods are treated in Sec. 2.1.2. The properties of Hermitian and anti-Hermitian operators and compatible operators are considered in Sec. 2.2, and the uncertainty principle is explained in Sec. 2.3. A discussion of measurements in quantum mechanics is presented in Sec. 2.4 (this topic is revisited in Sec. 2.5.4 using density matrix language). The density matrix formulation of quantum mechanics is considered in Sec. 2.5. The Wigner representation is discussed in Sec. 2.6, and Schrödinger and Heisenberg representations are formulated in Sec. 2.7. The correspondence principle and the classical limit of quantum mechanics are considered in Sec. 2.8, and finally Sec. 2.9 takes up the topic of symmetry and conservation laws in quantum mechanics.

2.1 HILBERT SPACE AND DIRAC NOTATION

Quantum states are represented by vectors in a Hilbert space $\mathcal{H}$, as defined in Appendix A. Dirac notation simplifies the mathematical language required for handling manipulations in $\mathcal{H}$. In Dirac notation, a quantum state ψ is represented by a *ket* vector $|\psi\rangle$ in the Hilbert space $\mathcal{H}$. A Hilbert space is an inner product space that is complete and separable (in quantum mechanics, an infinite number of states of a system can exist, and then the notion of separability is needed). For any two vectors $|\psi\rangle$ and $|\chi\rangle$, the inner product $\langle\chi|\psi\rangle$ is a complex number that specifies their overlap. The length of a vector $|\psi\rangle$ is given by $\sqrt{\langle\psi|\psi\rangle}$. A dual space of vectors can be defined; a vector $\langle\psi|$ in the dual space is called a *bra*. If this paragraph is not clear to you, you should read (or reread) Appendix A before continuing.

In quantum mechanics, we assume that a complete basis of kets $\{|\phi_n\rangle\}$ exists. This is a general property of Hilbert spaces. Hence, any ket can be written as a superposition (linear combination) of basis kets [see Eq. (2.2)]. The basis vectors can be taken to be orthogonal (they are often taken to be eigenstates of a Hermitian operator, which are orthogonal). Completeness and orthogonality (orthonormality) can be written as [see Eq. (A.32) in Appendix A]

$$\left[\sum_n |\phi_n\rangle\langle\phi_n|\right] = \mathbf{1}, \qquad (2.1a)$$

$$\langle\phi_i|\phi_j\rangle = \delta_{ij}, \qquad (2.1b)$$

where $\mathbf{1}$ is the unit operator in $\mathcal{H}$. If $\mathcal{H}$ is infinite dimensional, the completeness relation should be understood as a limit (see Appendix A). Any ket $|\psi\rangle$ can be expanded in terms of basis vectors,

$$|\psi\rangle = \left[\sum_n |\phi_n\rangle\langle\phi_n|\right]|\psi\rangle = \sum_n |\phi_n\rangle\langle\phi_n|\psi\rangle. \qquad (2.2)$$

A dual space vector $\langle\psi|$ is called a *bra* and "lives" in the space dual to the space of ket vectors. A bra $\langle\psi|$ can be expanded as

$$\langle\psi| = \langle\psi|\left[\sum_n |\phi_n\rangle\langle\phi_n|\right] = \sum_n \langle\psi|\phi_n\rangle\langle\phi_n|, \tag{2.3}$$

where (2.3) can be obtained by taking the Hermitian conjugate of (2.2). The corresponding wave functions in position space (see Sec. 1.3.8) can be written as

$$\langle x|\psi\rangle = \langle x|\left[\sum_n |\phi_n\rangle\langle\phi_n|\right]|\psi\rangle = \sum_n \langle x|\phi_n\rangle\langle\phi_n|\psi\rangle = \sum_n c_n\phi_n(x), \tag{2.4}$$

$$\langle\psi|x\rangle = \langle\psi|\left[\sum_n |\phi_n\rangle\langle\phi_n|\right]|x\rangle = \sum_n \langle\psi|\phi_n\rangle\langle\phi_n|x\rangle = \sum_n c_n^*\phi_n^*(x), \tag{2.5}$$

where $c_n \equiv \langle\phi_n|\psi\rangle$. Equation (2.5) can be obtained by complex conjugation of (2.4) since $\langle\psi|x\rangle = \langle x|\psi\rangle^* = \psi^*(x)$. Operators can be written in terms of basis vectors as matrices

$$\hat{A} = \left[\sum_n |\phi_n\rangle\langle\phi_n|\right]\hat{A}\left[\sum_m |\phi_m\rangle\langle\phi_m|\right] = \sum_{nm} |\phi_n\rangle A_{nm}\langle\phi_m|, \tag{2.6}$$

where $A_{nm} = \langle\phi_m|\hat{A}|\phi_n\rangle$ is the matrix representing the operator $\hat{A}$ in the basis $\{|\phi_n\rangle\}$. Arbitrary matrix elements of operators can be calculated as follows:

$$\langle\chi|\hat{A}|\psi\rangle = \langle\chi|\left[\sum_n |\phi_n\rangle\langle\phi_n|\right]\hat{A}\left[\sum_m |\phi_m\rangle\langle\phi_m|\right]|\psi\rangle$$
$$= \langle\chi|\phi_n\rangle A_{nm}\langle\phi_m|\psi\rangle \equiv \chi_n A_{nm}\phi_m. \tag{2.7}$$

In the last line of (2.7), we used Einstein notation, and therefore, the sums over m and n are implied by the repeated indices. Using Einstein notation, (2.2) can be written as $|\psi\rangle = c_n|\phi_n\rangle$, with $c_n = \langle\phi_n|\psi\rangle$, and similarly, (2.3) as $\langle\psi| = \langle\phi_n|c_n^*$.

In order to elucidate Eqs (2.1) through (2.3), let us consider an example of a basis set for a two-level system, e.g., a spin 1/2 particle. A set of basis vectors that span the Hilbert space are

$$|\phi_1\rangle = |\uparrow\rangle = \begin{pmatrix}1\\0\end{pmatrix}, \quad |\phi_2\rangle = |\downarrow\rangle = \begin{pmatrix}0\\1\end{pmatrix}. \tag{2.8}$$

Hence,

$$|\phi_1\rangle\langle\phi_1| = \begin{pmatrix}1\\0\end{pmatrix}(1\ 0) = \begin{pmatrix}1&0\\0&0\end{pmatrix}, \tag{2.9}$$

$$|\phi_2\rangle\langle\phi_2| = \begin{pmatrix}0\\1\end{pmatrix}(0\ 1) = \begin{pmatrix}0&0\\0&1\end{pmatrix}, \tag{2.10}$$

and (2.1) becomes

$$\left[\sum_n |\phi_n\rangle\langle\phi_n|\right] = |\phi_1\rangle\langle\phi_1| + |\phi_2\rangle\langle\phi_2| = \begin{pmatrix}1&0\\0&1\end{pmatrix}, \tag{2.11a}$$

$$\langle\phi_1|\phi_1\rangle = (1\ 0)\begin{pmatrix}1\\0\end{pmatrix} = 1, \quad \langle\phi_1|\phi_2\rangle = (1\ 0)\begin{pmatrix}0\\1\end{pmatrix} = 0, \quad \text{etc.} \tag{2.11b}$$

2.1 Hilbert Space and Dirac Notation

Equation (2.2) becomes

$$|\psi\rangle = \sum_n c_n |\phi_n\rangle = c_1 \begin{pmatrix} 1 \\ 0 \end{pmatrix} + c_2 \begin{pmatrix} 0 \\ 1 \end{pmatrix} = \begin{pmatrix} c_1 \\ c_2 \end{pmatrix}, \qquad (2.12)$$

where $c_j = \langle \phi_j | \psi \rangle$, and similarly, (2.3) takes the form

$$\langle \psi | = \sum_n c_n^* \langle \phi_n | = c_1^* (1\ 0) + c_2^* (0\ 1) = (c_1^*\ c_2^*). \qquad (2.13)$$

A three-level system can be described similarly but with three-dimensional vectors, $|\psi\rangle^\dagger = (c_1^*\ c_2^*\ c_3^*)$ (similar to the algebra of three-dimensional coordinate vectors but where the vector components are complex, i.e., $\mathbf{C}^3$ rather than $\mathbf{R}^3$). The harmonic oscillator is another example; it is an infinite-dimensional vector space that is discrete (see Secs. 1.3.15 and 2.7.2).

Problem 2.1

(a) Write the arbitrary operator $\hat{A}$ as a matrix for a two-level system using Eq. (2.6).
(b) Write $(\hat{A}|\psi\rangle)^\dagger$ in matrix notation for a two-level system.

Answer: (a) $\hat{A} = \begin{pmatrix} A_{11} & A_{12} \\ A_{21} & A_{22} \end{pmatrix}$. (b) $(\hat{A}|\psi\rangle)^\dagger = (c_1^*\ c_2^*) \begin{pmatrix} A_{11}^* & A_{21}^* \\ A_{12}^* & A_{22}^* \end{pmatrix}$.

Problem 2.2

Write Eqs (2.1) through (2.3) using a harmonic oscillator basis, retaining only the N lowest energy basis states.
Hint: $|\phi_j\rangle^\dagger = (0\ 0\ \ldots\ 0\ 1\ 0\ 0\ \ldots 0)$, where the 1 is in the jth column.

A word of caution is in order with respect to the lack of consistency in the literature, as well as in this book, in using the hat notation for operators. Often, the hat is not explicitly used despite the fact that one is considering an operator; it is simply implied.

2.1.1 POSITION AND MOMENTUM REPRESENTATIONS

The position space wave function, $\psi(\mathbf{r})$, corresponding to a state $|\psi\rangle$ is given by $\psi(\mathbf{r}) = \langle \mathbf{r} | \psi \rangle$; it is called the *position representation* of state $|\psi\rangle$ (see Sec. 1.3.8).

The same state can be represented in the *momentum representation* by the momentum-space wave function, $\tilde{\psi}(\mathbf{p}) = \langle \mathbf{p} | \psi \rangle$ (note that often the tilde is not written). By inserting a complete set of states $\{|\mathbf{r}\rangle\}$ between the bra $\langle \mathbf{p}|$ and the ket $|\psi\rangle$, i.e., by inserting the unity operator, $\mathbf{1} = \int d\mathbf{r} |\mathbf{r}\rangle \langle \mathbf{r}|$, we can write $\tilde{\psi}(\mathbf{p})$ as follows:

$$\tilde{\psi}(\mathbf{p}) \equiv \langle \mathbf{p} | \psi \rangle = \int d\mathbf{r} \langle \mathbf{p} | \mathbf{r} \rangle \langle \mathbf{r} | \psi \rangle. \qquad (2.14)$$

Using Eq. (1.79) to evaluate $\langle \mathbf{p} | \mathbf{r} \rangle$, $\langle \mathbf{p} | \mathbf{r} \rangle = \langle \mathbf{r} | \mathbf{p} \rangle^* = (2\pi\hbar)^{-3/2} e^{-i\mathbf{p}\cdot\mathbf{r}/\hbar}$, we find

$$\boxed{\tilde{\psi}(\mathbf{p}) = (2\pi\hbar)^{-3/2} \int d\mathbf{r}\, e^{-i\mathbf{p}\cdot\mathbf{r}/\hbar}\, \psi(\mathbf{r}).} \qquad (2.15)$$

Thus, the momentum representation of the wave function, $\tilde{\psi}(\mathbf{p})$, is the Fourier transform (see Appendix D) of the position representation wave function, $\psi(\mathbf{r})$. Often, one simply denotes $\tilde{\psi}(\mathbf{p})$ by the symbol $\psi(\mathbf{p})$, despite the danger involved in this notation. In 1D, Eq. (2.15) is given by $\tilde{\psi}(p) \equiv \langle p | \psi \rangle = (2\pi\hbar)^{-1/2} \int dx\, e^{-ipx/\hbar}\, \psi(x)$.

In the position representation, the quantity $\langle \mathbf{r} | \hat{\mathbf{r}} | \psi \rangle$ is simply equal to $\mathbf{r}\psi(\mathbf{r})$. In momentum space,

$$\langle \mathbf{p} | \hat{\mathbf{r}} | \psi \rangle = \int d\mathbf{r} \langle \mathbf{p} | \mathbf{r} \rangle \langle \mathbf{r} | \hat{\mathbf{r}} | \psi \rangle = (2\pi\hbar)^{-3/2} \int d\mathbf{r}\, e^{-i\mathbf{p}\cdot\mathbf{r}/\hbar}\, \mathbf{r}\, \psi(\mathbf{r}) = i\hbar \nabla_{\mathbf{p}} \tilde{\psi}(\mathbf{p}). \quad (2.16)$$

Hence, the operator $\hat{\mathbf{r}}$ in momentum space is $i\hbar \nabla_{\mathbf{p}}$, and the operator $g(\hat{\mathbf{r}})$ in momentum space is $g(i\hbar \nabla_{\mathbf{p}})$.

In the momentum representation, the quantity $\langle \mathbf{p} | \hat{\mathbf{p}} | \psi \rangle$ is simply equal to $\mathbf{p}\tilde{\psi}(\mathbf{p})$. In position space,

$$\langle \mathbf{r} | \hat{\mathbf{p}} | \psi \rangle = \int d\mathbf{p} \langle \mathbf{r} | \mathbf{p} \rangle \langle \mathbf{p} | \hat{\mathbf{p}} | \psi \rangle = (2\pi\hbar)^{-3/2} \int d\mathbf{r}\, e^{-i\mathbf{p}\cdot\mathbf{r}/\hbar}\, \mathbf{p}\, \tilde{\psi}(\mathbf{p}) = \frac{\hbar}{i} \nabla_{\mathbf{r}} \psi(\mathbf{r}). \quad (2.17)$$

That is, the operator $\hat{\mathbf{p}}$ in position space is $\frac{\hbar}{i}\nabla_{\mathbf{r}}$, as we already know from Chapter 1 (see Sec. 1.3.6). Note that $\mathbf{p}\psi(\mathbf{r})$ can be written as $\langle \mathbf{r} | \hat{\mathbf{p}} | \psi \rangle$. Clearly, the operator $f(\hat{\mathbf{p}})$ in position space is $f\left(\frac{\hbar}{i}\nabla_{\mathbf{r}}\right)$.

2.1.2 BASIS-STATE EXPANSIONS

An important method for solving quantum problems, e.g., obtaining eigenvalues and eigenvectors of an operator (e.g., the Hamiltonian), is to expand the operator and the eigenvectors in terms of basis states $\{|\phi_j\rangle\}$. This maps the original problem onto a matrix eigenvalue–eigenvector problem. The matrix obtained for Hermitian operators by expanding it in a basis is Hermitian. Therefore, its eigenvalues are real and its eigenvectors can be made orthonormal (see Sec. A.2 in the Appendix). It is <u>crucial</u> to understand this method. The only approximation made using this method is that, in practice, it is required to truncate the basis to a finite number of basis states. Numerically, it is easy to follow the convergence of the eigenvalues to make sure that enough basis states have been taken.

Given the time-independent Schrödinger equation, $H|\psi\rangle = E|\psi\rangle$, one expands the state $|\psi\rangle$ in a set of orthonormal basis states $\{|\phi_j\rangle\}$, i.e., $|\psi\rangle = \sum_j c_j |\phi_j\rangle$, where $c_j = \langle \phi_j | \psi \rangle$, and orthonormality means $\langle \phi_j | \phi_i \rangle = \delta_{ji}$. Inserting the completeness relation, $\sum_j |\phi_j\rangle \langle \phi_j| = 1$, into the Schrödinger equation we obtain

$$H \left(\sum_j |\phi_j\rangle \langle \phi_j| \right) |\psi\rangle = E|\psi\rangle. \quad (2.18)$$

Applying the bra $\langle \phi_i |$ from the left, we find the matrix eigenvalue equation,

$$\sum_j H_{ij} c_j = E\, c_i, \quad (2.19)$$

where $H_{ij} = \langle \phi_i | H | \phi_j \rangle$. We now truncate the number of basis states to N states, so the Hamiltonian matrix $\{H_{ij}\}$ is of size $N \times N$ (the only approximation made in this method). Convergence as a function of the number of basis states can and should be checked by increasing the number of basis states taken and monitoring the convergence of the eigenvalues. The eigenvalue E can be obtained by solving the determinantal equation, $|H_{ji} - E\delta_{ji}| = 0$, thereby obtaining N eigenvalues E_k, $k = 1, \ldots, N$. Once the kth eigenvalue is known, the vector of amplitudes for the kth eigenvector $|\psi_k\rangle$ is obtained by solving the linear set of equations, $\sum_j H_{ij} c_j^{(k)} = E_k\, c_i^{(k)}$, or, using a simplified notation,

$$\boxed{\sum_{j=1}^{N} H_{ij} c_{jk} = E_k\, c_{ik}.} \quad (2.20)$$

The second index on the amplitudes c_{ik} (which was written above as a superscript) is often written as a subscript [as in (2.20)], and specifies that these are the amplitudes of the kth eigenvector $|\psi_k\rangle$. The kth eigenvector $|\psi_k\rangle$ can be written as

2.1 Hilbert Space and Dirac Notation

$|\psi_k\rangle = \sum_j c_{jk}|\phi_j\rangle$. Schematically, the eigenvalue–eigenvector problem of Eq. (2.20) is of the form

$$\begin{pmatrix} \ddots & & \\ & \mathbf{H} & \\ & & \ddots \end{pmatrix} \begin{pmatrix} \\ c \\ \end{pmatrix}_k = E_k \begin{pmatrix} \\ c \\ \end{pmatrix}_k,$$

where the Hermitian Hamiltonian matrix is expressed in the basis $\{|\phi_j\rangle\}$, i.e., $\mathbf{H} = \{H_{ij}\}$. One obtains the eigenvectors by solving the linear equations for the amplitudes c_{jk} of all k. The methods of linear algebra for determining the eigenvalues E_k by solving the determinantal equation $|\mathbf{H} - E\mathbf{1}| = 0$ and then solving the linear equations for the eigenvectors c_{jk} are discussed in detail in Sec. A.2 of the Appendix.

A judicious choice of basis states can often reduce the number of basis states needed in the calculation. If the Hamiltonian H is close to a zero-order Hamiltonian, H_0, with known eigenstates $\{|\phi_j\rangle\}$, i.e., if $H = H_0 + V$, with V "small", then the eigenstates $\{|\phi_j\rangle\}$ can be a good choice of basis states.

Problem 2.3

(a) Why is the relation $\sum_j c_{jk}^* c_{jk'} = \delta_{k,k'}$ true for two distinct eigenvalues E_k and $E_{k'}$, but not necessarily true for two degenerate eigenvalues?

(b) Review the Gram–Schmidt orthogonalization scheme (Sec. A.1.1 of Appendix A) for orthogonalization that can be used to make all eigenvectors of a Hermitian matrix orthogonal. That is, the eigenvectors for a degenerate eigenvalue can be diagonalized using Gram–Schmidt scheme.

Answer: (a) See the proof in Sec. A.2.3 of Appendix A).

Note that the basis-set expansion method turns quantum mechanical calculations into matrix calculations. This method was introduced by Werner Heisenberg and Pascual Jordan. The matrix representation of quantum mechanics is referred to as *Heisenberg matrix mechanics*.

Basis-set expansion methods can also be applied to calculate the dynamics of quantum systems. Suppose we have a system with a time-independent Hamiltonian H and the system starts off in a state that is not an eigenstate of the Hamiltonian. We want to determine how the system evolves as a function of time, $|\Psi(t)\rangle$. The preferred basis set to use for this problem is the set of eigenstates of the Hamiltonian, $\{|\psi_j\rangle\}$. Since this set is complete, we can expand the initial state in terms of this set, and since the set is orthonormal, it is simple to calculate the b_j amplitudes of the initial state,

$$|\Psi(0)\rangle = \sum_j b_j |\psi_j\rangle, \quad b_j = \langle \psi_j | \Psi(0) \rangle. \tag{2.21}$$

Furthermore, since we know the time dependence of the energy eigenstates, the time dependence of a superposition of energy eigenstates is also simple,

$$|\Psi(t)\rangle = \sum_j b_j e^{-iE_j t/\hbar} |\psi_j\rangle. \tag{2.22}$$

We will return to basis-state expansion methods to solve some problems where the Hamiltonian is time-dependent in Secs. 6.6, 6.7, and 7.1.1.

2.2 HERMITIAN AND ANTI-HERMITIAN OPERATORS

The importance of Hermitian operators as operators that can be measured, i.e., operators that can represent physical observables, has already been noted in Sec. 1.3.2, and has been reinforced as part of the postulates of quantum mechanics. Properties of Hermitian operators are reviewed in Appendix A; recall that there we defined the Hermitian conjugate of an operator $\hat{O}$, $\hat{O}^\dagger$, via the relation between the matrix elements of $\hat{O}$ in Eq. (A.36). A Hermitian operator $\hat{\mathcal{H}}$ was defined so that $\hat{\mathcal{H}} = \hat{\mathcal{H}}^\dagger$. Moreover, we defined an anti-Hermitian operator such that it satisfies the relation $\hat{\mathcal{A}} = -\hat{\mathcal{A}}^\dagger$. Matrix representations of Hermitian and anti-Hermitian operators yield Hermitian and anti-Hermitian matrices. The properties of the eigenvalues and eigenvectors of Hermitian and anti-Hermitian matrices were discussed in Sec. A.2.3 of Appendix A.

In some cases, care must be exercised in forming a Hermitian operator from a classical observable. For example, if the classical Hamiltonian of a system contains a term of the form $\mathbf{p}\cdot\mathbf{r}$, this term is not a Hermitian quantum operator. The appropriate Hermitian quantum operator is $(\mathbf{p}\cdot\mathbf{r}+\mathbf{r}\cdot\mathbf{p})/2$, i.e., one must symmetrize the operator. For example, consider the Hamiltonian for a charged particle in an electromagnetic field, $H(\mathbf{r},\mathbf{p},t) = \frac{1}{2m}\left(\mathbf{p}-\frac{q}{c}\mathbf{A}(\mathbf{r},t)\right)^2 + q\varphi(\mathbf{r},t)$. In writing $(\mathbf{p}-(q/c)\mathbf{A}(\mathbf{r},t))^2$ as a quantum operator, one must properly symmetrize; the Hermitian quantum operator is given by $\mathbf{p}^2 - (q/c)[\mathbf{p}\cdot\mathbf{A}(\mathbf{r},t)+\mathbf{A}(\mathbf{r},t)\cdot\mathbf{p}]/2 + q^2\mathbf{A}^2(\mathbf{r},t)/c^2$.

2.2.1 COMPATIBLE OPERATORS AND DEGENERACY

Two observables (i.e., Hermitian operators) are *compatible* if they share the same eigenfunctions (but they do not necessarily have the same eigenvalues). Consequently, two compatible observables can be simultaneously precisely measured (see the Uncertainty Principle in the next section). Two compatible operators $\hat{Q}$ and $\hat{R}$ can be put into the form:

$$\hat{Q} = \sum_n |\phi_n\rangle q_n \langle\phi_n|, \tag{2.23}$$

$$\hat{R} = \sum_n |\phi_n\rangle r_n \langle\phi_n|.$$

Compatible operators commute with one another. This is trivial to show using Eq. (2.23) and the orthonormality condition for eigenvectors of Hermitian operators:

$$\hat{Q}\hat{R} = \sum_n |\phi_n\rangle q_n \langle\phi_n| \sum_m |\phi_m\rangle r_m \langle\phi_m| = \sum_n |\phi_n\rangle q_n r_n \langle\phi_n| = \hat{R}\hat{Q}. \tag{2.24}$$

For example, the momentum operator $\hat{\mathbf{p}}$ and the kinetic energy operator $\hat{T} = \hat{\mathbf{p}}^2/2m$ are compatible, and the postition operator $\hat{\mathbf{r}}$ and the potential energy operator $\hat{V}(\mathbf{r})$ are compatible.

The converse is also true: any two commuting Hermitian operators are compatible, as we shall now show. Consider two Hermitian operators (or matrices) $\hat{Q}$ and $\hat{R}$ and diagonalize them, so $\hat{Q}$ takes the form,

$$\langle q,v|\hat{Q}|q',v'\rangle = q\delta_{qq'}\delta_{vv'}. \tag{2.25}$$

The index v accounts for the possibility of *degeneracy* of the eigenvalues, where degeneracy means that two or more eigenvectors of the operator have the same eigenvalues. Now, suppose the Hermitian operator (matrix) $\hat{R}$ commutes with $\hat{Q}$. Taking matrix elements of $[\hat{Q},\hat{R}] = 0$, we find:

$$\langle q,v|\hat{Q}\hat{R}-\hat{R}\hat{Q}|q',v'\rangle = (q-q')\langle q,v|\hat{R}|q',v'\rangle = 0. \tag{2.26}$$

For $q \neq q'$, we conclude that the off diagonal in "q" elements of $\hat{R}$ vanish, i.e., $\langle q,v|\hat{R}|q',v'\rangle = \hat{R}^{(q)}_{vv'}\delta_{qq'}$. We can now diagonalize the vv' blocks of $\hat{R}$ since this does not affect the rest of the matrix. Hence, the operators (matrices) $\hat{Q}$ and $\hat{R}$ are now compatible.

The statement, two compatible operators have simultaneous eigenvectors, can be generalized to the case when we have more than two mutually compatible operators,

$$[R_1,R_2] = [R_1,R_3] = [R_2,R_3] = \ldots = 0. \tag{2.27}$$

Assume that we have found a maximal set of commuting observables, i.e., we cannot add any more observables to our list of commuting observables appearing in (2.27). The eigenvalues of individual operators $\{R_i\}$ may have degeneracies, but if we specify a combination of eigenvalues, $(r_{n_1,1}, r_{n_2,2}, r_{n_3,3}, \ldots)$, then the corresponding simultaneous eigenvector of $R_1, R_2, R_3, \ldots$ is uniquely specified. We can use a collective index $K \equiv (r_{n_1,1}, r_{n_2,2}, r_{n_3,3}, \ldots)$ to identify the eigenvector. The orthogonality relations for $|K\rangle = |r_{n_1,1}, r_{n_2,2}, \ldots\rangle$ are

$$\langle K|K'\rangle = \delta_{K,K'} = \delta_{r_{n_1,1}, r_{n'_1,1}} \delta_{r_{n_2,2}, r_{n'_2,2}} \cdots, \tag{2.28}$$

and the completeness relation is

$$\sum_K |K\rangle \langle K| = \mathbf{1}. \tag{2.29}$$

2.3 THE UNCERTAINTY PRINCIPLE

Diffraction of light and particle waves cannot be reconciled with the idea that light or particles move in paths (well-defined trajectories). Nevertheless, a measurement of the coordinates of a particle can always be performed with any desired accuracy, at least in principle. In quantum mechanics, a (somewhat) localized particle can be described by a wave function $\psi(\mathbf{r}, t)$ (in Sec. 1.3.13, we used the term wave packet) that is a superposition (sum) over energies and wave vectors as follows:

$$\psi(\mathbf{r}, t) = \int d\mathbf{k} \int dE \, \tilde{\psi}(\mathbf{k}, E) \, e^{i(\mathbf{k}\cdot\mathbf{r} - Et/\hbar)}. \tag{2.30}$$

The amplitudes $\tilde{\psi}(\mathbf{k}, E)$ can be viewed as the Fourier transform of $\psi(\mathbf{r}, t)$ (see Appendix D). If one attempts to measure the location of the particle at a given time, the resulting measurement can yield a result anywhere within the extent of the wave function. It is, of course, more likely to find the particle in regions where the amplitude $\psi(\mathbf{r}, t)$ is large since $|\psi(\mathbf{r}, t)|^2$ is the probability density function for finding the particle at position $\mathbf{r}$ at time t. However, there is no theoretical limitation in making the probability density as narrow as possible, so the position of the particle at a given time can be very well specified. However, if the wave packet $\psi(\mathbf{r}, t)$ is very well localized, then the amplitude $\tilde{\psi}(\mathbf{k}, E)$ will be delocalized in momentum space (note that the momentum is simply related to the wave vector, $\mathbf{p} = \hbar \mathbf{k}$) so the product in the uncertainty of the position and the momentum cannot be smaller than a certain value.

There is another way to view this uncertainty in position and momentum. If two observables are represented by commuting operators, then one can measure the physical observables with simultaneously arbitrary accuracy. However, if the operators do not commute, as is the case for x and p_x (or y and p_y), then a simultaneous measurement will not be exactly repeatable. There will be a spread in the measurement results, such that the product of the standard deviations will exceed a minimum value. The minimum of the product of the standard deviations depends on the observables; more specifically on the commutator of the observables. This is one way to state the *Heisenberg Uncertainty Principle*. We shall prove the uncertainty principle in this section.

An uncertainty principle also exists for energy and time. We may wish to measure the energy E emitted during the time interval Δt corresponding to an atomic process, e.g., the energy of an excited state that decays radiatively with a certain lifetime. The minimum uncertainty in the electromagnetic wave energy, E, is then related to the time interval Δt and is given by

$$\Delta E \, \Delta t = \hbar \Delta \omega \, \Delta t \geq \hbar/2.$$

This uncertainty is due to Fourier expansion [see Eq. (D.26) of Appendix D] of the wave function and is formally similar to the uncertainty in position and momentum due to the Fourier expansion in Eq. (2.30). Note that t is not an operator in nonrelativistic quantum mechanics, so there is no commutator relation between E and t. Hence, at least at first sight, it is different from the *Uncertainty Principle for non-commuting operators*, such as x and p, but in fact the two are related [e.g., see Problem 2.4(d)]. We shall now take up the latter subject.

The Heisenberg's Uncertainty Principle states:

$$(\Delta A)^2 (\Delta B)^2 \geq \left| \frac{\langle [A, B] \rangle}{2} \right|^2. \tag{2.31}$$

Here, expectation values are calculated with some arbitrary state $|\psi\rangle$. To prove this inequality, we will need some mathematical machinery. One of the required ingredients is the *Cauchy–Schwartz Inequality* (sometimes called the Schwartz Inequality),

$$\langle \phi | \phi \rangle \langle \xi | \xi \rangle \geq |\langle \phi | \xi \rangle|^2. \tag{2.32}$$

The proof of the Schwartz Inequality is given in Appendix A [see Eq. (A.9)]. Another ingredient is the following lemma.

Lemma: Let $\alpha \equiv A - \langle A \rangle$ and $\beta \equiv B - \langle B \rangle$. Then $[\alpha, \beta] = [A, B]$. The proof of the Lemma is simple:

$$[\alpha, \beta] = (A - \langle A \rangle)(B - \langle B \rangle) - (B - \langle B \rangle)(A - \langle A \rangle)$$

$$[\alpha, \beta] = AB - \langle B \rangle A - \langle A \rangle B + \langle A \rangle \langle B \rangle$$

$$- BA + \langle B \rangle A + \langle A \rangle B - \langle A \rangle \langle B \rangle$$

$$[\alpha, \beta] = AB - BA = [A, B].$$

Now, with the definitions of α and β, we have that

$$(\Delta A)^2 = \langle (A - \langle A \rangle)^2 \rangle = \langle \alpha^2 \rangle, \quad (\Delta B)^2 = \langle (B - \langle B \rangle)^2 \rangle = \langle \beta^2 \rangle.$$

Hence,

$$(\Delta A)^2 (\Delta B)^2 = \langle \alpha^2 \rangle \langle \beta^2 \rangle.$$

That is, $(\Delta A)^2 (\Delta B)^2 = \langle \psi | \alpha^2 | \psi \rangle \langle \psi | \beta^2 | \psi \rangle$. Let us define the following kets,

$$|j\rangle \equiv \alpha |\psi\rangle \quad \text{and} \quad |k\rangle \equiv \beta |\psi\rangle.$$

With these definitions, $(\Delta A)^2 (\Delta B)^2 = (\langle \psi | \alpha)(\alpha | \psi \rangle)(\langle \psi | \beta)(\beta | \psi \rangle)$ can be written as $(\Delta A)^2 (\Delta B)^2 = \langle j | j \rangle \langle k | k \rangle \geq |\langle j | k \rangle|^2$, where the Cauchy–Schwartz inequality has been used [see Eq. (A.9) in Appendix A]. Thus,

$$(\Delta A)^2 (\Delta B)^2 \geq |\langle j | k \rangle|^2$$

i.e., $(\Delta A)^2 (\Delta B)^2 \geq (\text{Re}(\langle j | k \rangle))^2 + (\text{Im}(\langle j | k \rangle))^2.$

Hence, clearly,

$$(\Delta A)^2 (\Delta B)^2 \geq (\text{Im}(\langle j | k \rangle))^2 = \left| \frac{\langle j | k \rangle - \langle k | j \rangle}{2i} \right|^2$$

$$(\Delta A)^2 (\Delta B)^2 \geq \left| \frac{\langle \psi | \alpha \beta | \psi \rangle - \langle \psi | \beta \alpha | \psi \rangle}{2i} \right|^2$$

$$(\Delta A)^2 (\Delta B)^2 \geq \left| \frac{\langle [\alpha, \beta] \rangle}{2} \right|^2$$

$$(\Delta A)^2 (\Delta B)^2 \geq \left| \frac{\langle [A, B] \rangle}{2} \right|^2.$$

This completes the proof of the Uncertainty Principle.

2.4 The Measurement Problem

Applying the Uncertainty Principle to the operators x and p_x, we find [see Eq. (1.90) for the commutation relation]

$$(\Delta x)^2 (\Delta p_x)^2 \geq \left|\frac{\langle [x, p_x]\rangle}{2}\right|^2 = (\hbar/2)^2. \tag{2.33}$$

No matter what state the system is in, the product of Δx and Δp must be larger than or equal to $\hbar/2$. In particular, we can calculate $\Delta x\, \Delta p_x$ for the ground state 1D harmonic oscillator wave function $\psi(x) = \frac{1}{(2\pi\sigma^2)^{1/4}} \exp(-\frac{x^2}{4\sigma^2})$. Then, it is easy to show that $\langle x\rangle = 0$, $(\Delta x)^2 = \sigma^2$, $\langle p\rangle = 0$, and $(\Delta p)^2 = \frac{\hbar^2}{4\sigma^2}$, so the Gaussian wave function is a minimum uncertainty wave function, i.e., the uncertainty is the minimum allowed by the Uncertainty Principle.

Problem 2.4

Consider the 1D Gaussian wave packet, $\psi_0(x) = \mathcal{N}\exp\left(-\frac{x^2}{4\sigma^2}\right)e^{i\kappa x}$.

(a) Determine the Fourier transform of the wave packet, $\tilde{\psi}_0(k)$.
(b) Find $\psi(x,t)$ given the initial wave packet $\psi_0(x)$ and Hamiltonian $H = \frac{p_x^2}{2m}$.
(c) Calculate $\langle x\rangle_t$, $(\Delta x)_t$, $\langle p\rangle_t$, and $(\Delta p)_t$ for the wave packet in part (d).
(d) Determine the product $(\Delta x)_t (\Delta p)_t$.

Answers: Note that this problem is basically a repeat of Problem 1.17.

(a) $\tilde{\psi}_0(k) = \mathcal{N}\sqrt{2}\sigma\, e^{-\sigma^2(\kappa-k)^2}$.

(b) $\psi(x,t) = \mathcal{N}\frac{[2\pi\sigma^2]^{1/4}}{[2\pi\sigma^2(t)]^{1/4}} e^{-\frac{(x-\hbar\kappa t/m)^2}{4\sigma^2(t)} + i\kappa(x-\hbar\kappa t/m) + i\phi(t)}$

where $\sigma^2(t) = \sigma^2 + i\frac{\hbar t}{2m}$ and $\phi(t) = \frac{\hbar^2\kappa^2}{2m}\frac{t}{\hbar}$.

(c) $\langle x\rangle_t = \hbar\kappa t/m$, $(\Delta x)_t = \sigma\sqrt{1 + \frac{\hbar^2 t^2}{4m^2\sigma^4}}$, $\langle p\rangle_t = \hbar\kappa$, $(\Delta p)_t = \frac{\hbar}{2\sigma}$.

(d) $(\Delta x)_t (\Delta p)_t = \frac{\hbar}{2}\sqrt{1 + \frac{\hbar^2 t^2}{4m^2\sigma^4}}$.

Problem 2.5

For the free-particle wave packet $\psi(x,t)$ you found in Problem 2.4(b), show that $\Delta E \Delta t > \hbar$, where here we define $\Delta E \equiv \langle \frac{p^2}{2m}\rangle - \frac{\langle p\rangle^2}{2m}$, and Δt is the time required for most of the wave packet to pass a fixed point x_0.

2.4 THE MEASUREMENT PROBLEM

There appears to be a discrepancy associated with the relationship of the time evolution postulate and the measurement postulates. In quantum mechanics, a pure state of a physical system is completely described by the wave function ψ (or the state vector $|\psi\rangle$). $|\psi\rangle$ yields information about the system by specifying the probabilities of the results of measurements made on the system by a measurement apparatus. The time evolution postulate specifies the deterministic change of the state of an isolated system with Hamiltonian $H(t)$ according to the time-dependent Schrödinger equation,

$$i\hbar\frac{\partial |\psi(t)\rangle}{\partial t} = H(t)|\psi(t)\rangle. \tag{2.34}$$

Consider an isolated system consisting of the subsystem to be measured plus the measurement apparatus. The measurement postulates appear to specify a fundamentally different type of evolution for the state function of the subsystem that is to be measured. The subsystem evolves *discontinuously* by measurement into one of the eigenstates $\phi_1, \phi_2, \ldots$ of the operator $\hat{O}$ that is being measured, so that the initial state ψ is changed into one of these eigenstates, say $|\phi_i\rangle$, with probability $|\langle\phi_i|\psi\rangle|^2$. Are these two types of evolution consistent? Specifically, during the measurement process, can the state of the *whole* system (subsystem and measurement apparatus) be described by deterministic evolution? If it can, why is it necessary to invoke discontinuous evolution of the subsystem? If it cannot, then there are isolated systems that do not obey the time evolution postulate. There have been many attempts to derive the measurement postulates from the other postulates. To some extent, "the measurement problem" in quantum mechanics is still open and is an active field of research. Nevertheless, there has never been a discrepancy between quantum mechanical results calculated using the measurement postulates and experimental results.

Let us consider a simple example, or schematic illustration, of a measurement, viewed in a totally quantum mechanical setting, where both the system being measured and the measurement apparatus are treated quantum mechanically. We are given a two-level system (e.g., a spin 1/2 particle) that starts out in state $|\psi_0\rangle = a_0|\uparrow\rangle + b_0|\downarrow\rangle$ and an experimental apparatus that starts out in a definite "ready to measure" pure state $|A\rangle_0 \equiv |A_0\rangle$. The meaning of pure state will be elaborated below; for now, note only that we have assumed that the apparatus is initially in a specific quantum state that can be described by a state vector (or, if you like, a wave function). The fundamental postulates of quantum mechanics specify that the combined system-apparatus state evolves by unitary evolution. By measurement of the spin 1/2 system we mean that, after the interaction of the spin with the apparatus, we are able to determine which spin state the system is in by examining the apparatus. The combined system-apparatus starts out in the initial state, $|\Psi_0\rangle = |\psi_0\rangle|A_0\rangle = (a_0|\uparrow\rangle + b_0|\downarrow\rangle)|A_0\rangle$. By virtue of the unitary evolution postulate and the necessity that the state of the two-level system be determined by examining the apparatus, the combined system must have evolved into the state,

$$|\Psi\rangle = (a|\uparrow\rangle|A_\uparrow\rangle + b|\downarrow\rangle|A_\downarrow\rangle), \quad (2.35)$$

where $|A_\uparrow\rangle$ and $|A_\downarrow\rangle$ are apparatus states that allow us to determine the state of the spin 1/2 system. If we find the apparatus in state $|A_\uparrow\rangle$, we know that the spin 1/2 system is in state $|\uparrow\rangle$, and if we find the apparatus in state $|A_\downarrow\rangle$, we know that the spin 1/2 system is in state $|\downarrow\rangle$. But the probability of finding the apparatus in state $|A_\uparrow\rangle$ on any given realization of the experiment is $|a|^2$, and the probability of finding the apparatus in state $|A_\downarrow\rangle$ on any given realization of the experiment is $|b|^2$. This is the quantum view of a measurement, wherein both the system being measured and the measurement apparatus are treated as quantum systems. Note that the measurement problem remains; it has just been pushed one step backward. Given (2.35) as the state of the total system after the measurement, the necessity for the collapse (measurement) postulate still exists.

One of the factors that we have not considered in the example of the previous paragraph is that macroscopic systems (such as most measurement apparatuses) are never isolated from their environments. Hence, they do not evolve according to the Schrödinger equation, which is applicable only to a closed system, but suffer from the natural loss of "quantum coherence" which can leak into the environment. A lot more will be said about the effects of decoherence in what follows, e.g., in Sec. 2.5.4, we shall see how the measurement problem can be resolved (or at least ameliorated) by considering decoherence via interaction with an environment. This will require us to use a density matrix formulation of quantum states, a topic we take up now.

2.5 MIXED STATES: DENSITY MATRIX FORMULATION

The *density matrix* ρ of a pure state $|\Psi\rangle$ is an operator defined by $\rho \equiv |\Psi\rangle\langle\Psi|$. For example, given a two-level spin system in state $|\Psi\rangle = (a|\uparrow\rangle + b|\downarrow\rangle)$, the density matrix is given by

$$\begin{aligned}\rho = |\Psi\rangle\langle\Psi| &= (a|\uparrow\rangle + b|\downarrow\rangle)(a^*\langle\uparrow| + b^*\langle\downarrow|) \\ &= (aa^*|\uparrow\rangle\langle\uparrow| + ab^*|\uparrow\rangle\langle\downarrow| + ba^*|\downarrow\rangle\langle\uparrow| + bb^*|\downarrow\rangle\langle\downarrow|) \\ &= \left(a\begin{pmatrix}1\\0\end{pmatrix} + b\begin{pmatrix}0\\1\end{pmatrix}\right)(a^*(1\ 0) + b^*(0\ 1)) \\ &= \begin{pmatrix}aa^* & ab^* \\ ba^* & bb^*\end{pmatrix}. \end{aligned} \quad (2.36)$$

2.5 Mixed States: Density Matrix Formulation

In the first line of Eq. (2.36), we used the definition of the density matrix as the outer product of a ket vector and a bra vector (see Sec. A.2.1). In the second line, we simply opened the parenthesis. In the third line, we introduced the row vector notation (i.e., the representation) in (1.14) for the kets $|\uparrow\rangle$ and $|\downarrow\rangle$ and took their Hermitian conjugates to form the bra vectors that are represented by row vectors, to form the representation of the first line. In the last line, we multiplied the row and column vectors to obtain 2×2 matrices and collected these to form one 2×2 matrix. The diagonal elements of the density matrix are the probabilities of the up and down spin states, $P_\uparrow = aa^*$ and $P_\downarrow = bb^*$, and the off diagonal elements, $\rho_{\uparrow,\downarrow} = ab^*$ and $\rho_{\downarrow,\uparrow} = ba^*$, are called the *coherences*. They depend on the phase of the amplitudes a and b. We shall have much to say about the physical interpretation and significance of the coherences in the chapters that follow.

The density matrix of a system in a pure state $|\Psi(t)\rangle$ at time t is $\rho(t) = |\Psi(t)\rangle\langle\Psi(t)|$. The density matrix evolves according to the *Liouville-von Neumann equation*,

$$i\hbar \frac{\partial}{\partial t}\rho(t) = [H(t), \rho(t)], \tag{2.37}$$

as easily verified by taking the time derivative of $|\Psi(t)\rangle\langle\Psi(t)|$. Clearly, from its definition, $\rho(t) = \mathcal{U}(t,0)\rho(0)\mathcal{U}^\dagger(t,0)$, where $\mathcal{U}(t,0)$ is the evolution operator of Eq. (1.35).

A quantum system in an *ensemble of pure states*, $|\phi_i\rangle$, with probabilities $p_i > 0$, is represented by the density matrix

$$\rho = \sum_i p_i |\phi_i\rangle\langle\phi_i|, \quad \sum_i p_i = 1. \tag{2.38}$$

The density matrix is sometimes called the *statistical operator*, particularly in the context of statistical mechanics; sometimes it is called the *density operator*.

The density matrix has trace unity, $\text{Tr}\,\rho = 1$, since $\text{Tr}\,\rho = \sum_i p_i \text{Tr}\,|\phi_i\rangle\langle\phi_i| = \sum_i p_i$, and the sum of the probabilities must equal unity. The density matrix must be Hermitian, $\rho^\dagger = \rho$. Moreover, since the quantities p_i are probabilities, $p_i \geq 0$. Hence, the expectation value of the density matrix for any arbitrary state $|\varphi\rangle$ is greater or equal to zero, i.e., the density matrix is a *positive operator*. This is easy to show as follows:

$$\langle\varphi|\rho|\varphi\rangle = \sum_i p_i \langle\varphi|\phi_i\rangle\langle\phi_i|\varphi\rangle = \sum_i p_i |\langle\varphi|\phi_i\rangle|^2 \geq 0.$$

Positivity implies that the eigenvalues of the density matrix must be nonnegative. Any Hermitian positive operator having trace unity can be written in the form $\sum_i \lambda_i |\phi_i\rangle\langle\phi_i|$, where λ_i are *real nonnegative eigenvalues* with $\sum_i \lambda_i = 1$ and $\{|\phi_i\rangle\}$ are orthonormal vectors. This is equivalent to saying that any Hermitian operator (matrix) can be diagonalized, and furthermore, if positive, the eigenvalues are nonnegative.

Problem 2.6

(a) Prove that the eigenvalues of a density matrix must be nonnegative.
(b) Prove that the sum of the eigenvalues of the density matrix equal unity.
(c) Prove that the diagonal elements of a density matrix are nonnegative.

Problem 2.7

Consider a pure state, $|\phi_\vartheta\rangle = \frac{1}{\sqrt{2}}\left(|\uparrow\rangle + e^{i\vartheta}|\downarrow\rangle\right)$.

(a) Write the density matrix $\rho_\vartheta = |\phi_\vartheta\rangle\langle\phi_\vartheta|$ as a 2×2 matrix.
(b) Suppose ϑ is uniformly distributed over the interval $[0, 2\pi]$. Write the mixed state density matrix $\bar{\rho} \equiv \frac{1}{2\pi}\int_0^{2\pi} d\vartheta\,\rho_\vartheta$ as a 2×2 matrix.
(c) Find $\bar{\rho} \equiv \int d\vartheta\,P(\vartheta)\rho_\vartheta$ if $P(\vartheta) = \frac{1}{\sqrt{4\pi\gamma}}e^{-\frac{\vartheta^2}{4\gamma}}$ and $|\phi_\vartheta\rangle = \left(a|\uparrow\rangle + b\,e^{i\vartheta}|\downarrow\rangle\right)$.

Answers: (a) $\rho_\vartheta = \frac{1}{2}\begin{pmatrix} 1 & e^{-i\vartheta} \\ e^{i\vartheta} & 1 \end{pmatrix}$. (b) $\bar{\rho} = \frac{1}{2}\begin{pmatrix} 1 & 0 \\ 0 & 1 \end{pmatrix}$.
(c) $\bar{\rho} = \begin{pmatrix} |a|^2 & ab^* e^{-\gamma} \\ a^* b e^{-\gamma} & |b|^2 \end{pmatrix}$.

Any mixed state at time $t=0$, $\rho(0)$, can be written in the form of (2.38), and evolves in time as

$$\rho(t) = \mathcal{U}(t,0)\rho(0)\mathcal{U}^\dagger(t,0) = \sum_i p_i \mathcal{U}(t,0)|\phi_i\rangle\langle\phi_i|\mathcal{U}^\dagger(t,0), \tag{2.39}$$

where $\mathcal{U}(t,0)$ is the evolution operator of the system. The density matrix in (2.39) satisfies Eq. (2.37).

If we perform a measurement corresponding to an observable $\mathcal{A}$ on the system with density matrix ρ, the expectation value obtained is given by

$$\langle \mathcal{A} \rangle = \mathrm{Tr}\,\mathcal{A}\rho. \tag{2.40}$$

If the system is in state (2.38), we obtain $\langle \mathcal{A} \rangle = \sum_i p_i \langle\phi_i|\mathcal{A}|\phi_i\rangle$. Moreover, if $|\phi_i\rangle$ are eigenstates of the operator $\mathcal{A}$ with eigenvalues $\mathcal{A}_i$, we find $\langle \mathcal{A} \rangle = \sum_i p_i \mathcal{A}_i$. Hence, the probability of obtaining a measured value $\mathcal{A}_i$ is p_i.

A density matrix description of a subsystem is necessary when incomplete information regarding the whole system results due to averaging over degrees of freedom not explicitly taken into account in the description of the subsystem. That part of the whole system that will be fully described is called "the subsystem", and that part associated with the degrees of freedom that will not be explicitly treated is called "the bath" or "the environment." This breakup of the whole system into subsystem and bath is similar to that used in statistical mechanics. Averaging over the degrees of freedom not explicitly taken into account results in a mixed state of the subsystem even if the whole system is in a pure state; the subsystem can no longer be described in terms of a wave function. Such *mixed states* of the subsystem must be described in terms of a *density matrix* [see Eq. (2.45)] and cannot be described by a pure state. The equation of motion for the density matrix is called the *Liouville-von Neumann equation* or the *Bloch density matrix equation* because of its similarity to the Bloch equation for spin systems. The density matrix formalism was developed independently by Lev Landau and John von Neumann. We now take up this topic.

Reduction schemes for eliminating the degrees of freedom of the bath have been extensively studied (for details see Sec. 17.3 on the book web page). For example, in the context of the interaction of light with matter, optical Bloch equations for the density matrix describing the matter have become a standard method to determine the dynamics of a system described by a finite number of states (e.g., the two-level system) that undergoes interactions with a bath composed of many degrees of freedom. In the context of the optical Bloch equations, the modes of the radiation field participating in spontaneous emission from the subsystem levels are adiabatically eliminated (i.e., are reduced out of the problem). Elimination of the bath degrees of freedom leads to a density matrix to describe the ground- and excited-state populations (diagonal elements of the density matrix) and coherences (off-diagonal elements of the density matrix) of the subsystem.

Readers familiar with statistical mechanics will recognize the density matrix of a (sub-)system in equilibrium with a thermal reservoir at temperature T,

$$\rho_T = Z^{-1}\sum_i e^{-\beta E_i}|\psi_i\rangle\langle\psi_i|, \quad Z = \sum_j e^{-\beta E_j}, \tag{2.41}$$

where $\beta = (k_B T)^{-1}$, $|\psi_i\rangle$ and E_i are the ith energy eigenstate and eigenenergy and Z is the *partition function*, $Z = \sum_j e^{-\beta E_j} = \mathrm{Tr}\,e^{-\beta H}$ where H is the Hamiltonian. In position representation, $\rho_T(\mathbf{r},\mathbf{r}') = Z^{-1}\sum_i e^{-\beta E_i}\langle\mathbf{r}|\psi_i\rangle\langle\psi_i|\mathbf{r}'\rangle$, and the partition function can be written as $Z = \mathrm{Tr}\,e^{-\beta H} = \int d\mathbf{r}\,\langle\mathbf{r}|\rho_T|\mathbf{r}\rangle = \int d\mathbf{r}\,\rho_T(\mathbf{r},\mathbf{r})$. Hence, the thermal density matrix can be written as

$$\boxed{\rho_T = \frac{e^{-\beta H}}{Z}, \quad Z = \mathrm{Tr}\,e^{-\beta H},} \tag{2.42}$$

2.5 Mixed States: Density Matrix Formulation

and the thermal average of an operator $\mathcal{A}$ is

$$\langle \mathcal{A} \rangle_T = \text{Tr}\, \mathcal{A} \rho_T = \frac{\text{Tr}\, \mathcal{A}\, e^{-\beta H}}{\text{Tr}\, e^{-\beta H}} = \frac{\sum_i e^{-\beta E_i} \langle \psi_i | \mathcal{A} | \psi_i \rangle}{\sum_j e^{-\beta E_j}}. \tag{2.43}$$

A useful observable operator is the *projector* (or *projection operator*)[1] onto a state, say $|\phi_j\rangle$, $\mathcal{P}_{\phi_j} = |\phi_j\rangle \langle \phi_j|$. For a mixed state that can be written as $\rho = \sum_i p_i |\phi_i\rangle \langle \phi_i|$, the probability of finding the system in state $|\phi_j\rangle$ is $\langle \mathcal{P}_{\phi_j} \rangle = \text{Tr}\, \mathcal{P}_{\phi_j} \rho = \sum_i p_i \langle \phi_i | \mathcal{P}_{\phi_j} | \phi_i \rangle = p_j$, i.e., the expectation value of $\mathcal{P}_{\phi_j}$ is just the probability of finding the system in state $|\phi_j\rangle$. The probability of finding the system in an arbitrary state $|\vartheta\rangle$ is given by

$$\langle \mathcal{P}_\vartheta \rangle = \sum_i p_i \langle \phi_i | \mathcal{P}_\vartheta | \phi_i \rangle = \sum_i p_i\, |\langle \phi_i | \vartheta \rangle|^2. \tag{2.44}$$

The reduced density matrix for a subsystem S of a physical system composed of subsystems S and B (the bath) is defined by the trace over the bath degrees of freedom, i.e., by the *partial trace*,

$$\rho_S = \text{Tr}_B\, \rho_{SB}. \tag{2.45}$$

Here, Tr_B indicates a trace over degrees of freedom of the bath B, and ρ_{SB} is the density matrix of the whole system. Even if the whole system SB is in a pure state that can be described by a wave function, subsystem S (or subsystem B) cannot in general be described by a pure state if subsystems S and B interact with each other, i.e., are entangled (see Sec. 1.3.3). A mixed state representation in terms of the density matrix for subsystem S, ρ_S, is necessary.

We now give three examples of density matrices. Consider first the singlet state that was introduced in Sec. 1.3.3, $|\text{singlet}\rangle = \frac{1}{\sqrt{2}} (|\uparrow\rangle_1 |\downarrow\rangle_2 - |\downarrow\rangle_1 |\uparrow\rangle_2)$, where particles 1 and 2 are spatially separated. Suppose we are not interested in or cannot measure particle 2, perhaps because it has undergone some decoherence (it interacted with other degrees of freedom that are not under control), but we are intensely interested in particle 1. The state of particle 1 is obtained by taking the trace over particle 2 (in this case, we can call it the bath):

$$\rho_1 = \text{Tr}_2 \{|\text{singlet}\rangle \langle \text{singlet}|\}$$

$$= \frac{1}{2} \sum_\alpha {}_2\langle \alpha | \; [\; (|\uparrow\rangle_1 |\downarrow\rangle_2 - |\downarrow\rangle_1 |\uparrow\rangle_2)\, ({}_1\langle \uparrow |\, {}_2\langle \downarrow | - {}_1\langle \downarrow |\, {}_2\langle \uparrow |)\;]\; |\alpha\rangle_2$$

$$= \frac{1}{2} \left(|\uparrow\rangle \langle \uparrow | + |\downarrow\rangle \langle \downarrow | \right) = \frac{1}{2} \begin{pmatrix} 1 & 0 \\ 0 & 1 \end{pmatrix}, \tag{2.46}$$

where the last equality is in the representation of the density matrix in the basis given by (1.14). The probability of finding particle 1 in state spin-up is 50%, as is the probability of finding it in state spin-down, but particle 1 is not in a coherent superposition of spin-up and spin-down; it is in a mixed state with no discernible amplitudes of spin-up and spin-down (certainly no discernible phase to the amplitudes).

The second example involves the two-particle spatial wave function in volume V,

$$\Psi(\mathbf{r}_1, \mathbf{r}_2) = \frac{1}{\sqrt{2}(\sqrt{V})^2} \left(e^{i\mathbf{k} \cdot \mathbf{r}_1} e^{i\mathbf{k}' \cdot \mathbf{r}_2} + e^{i\mathbf{k}' \cdot \mathbf{r}_1} e^{i\mathbf{k} \cdot \mathbf{r}_2} \right). \tag{2.47}$$

Upon forming the density matrix for the system consisting of the two particles, and tracing over particle 2, we find

$$\rho_1 = \text{Tr}_2 \{|\Psi\rangle \langle \Psi|\} = \frac{1}{V} \frac{1}{2} \left(|\mathbf{k}\rangle \langle \mathbf{k}| + |\mathbf{k}'\rangle \langle \mathbf{k}'| \right). \tag{2.48}$$

Hence, the system is in an "incoherent superposition" of the two momentum states with 50% probability for being in each state.

---

[1] A projector, or projection operator, $\mathcal{P}$, is an operator that is Hermitian and *idempotent*, $\mathcal{P}^2 = \mathcal{P}$. The eigenvalues of such operators are either 0 or 1.

The next example involves the excitation of a ground-state hydrogen atom by a short pulse of unidirectional high energy electrons with a probability distribution function $P(E)$ for electron kinetic energy E (the shorter the pulse duration, the wider the distribution in energy). We chose the basis set for the hydrogen atom to be the energy and angular momentum eigenstates, denoted by $\phi_j \equiv \psi_{nlm}(\mathbf{r})$ (see Sec. 3.2.6). As a result of inelastic electron scattering by the hydrogen atom, the hydrogen atom can be left in an excited state. The angular distribution of the scattered electrons after collision depends on the final excited state $\psi_{nlm}(\mathbf{r})$. The amplitude at time t of state $\psi_{nlm}(\mathbf{r})$ produced by scattering an electron of initial kinetic energy E that is scattered into a solid angle Ω is denoted as $c_{E,\Omega,nlm}(t)$ and the hydrogenic state created by the electron scattering after the collision is

$$\Psi_{E,\Omega}(\mathbf{r},t) = \sum_{nlm} c_{E,\Omega,nlm}(t) \psi_{nlm}(\mathbf{r}). \tag{2.49}$$

The amplitude $c_{E,\Omega,nlm}(t)$ for scattering an electron of energy E so that it is scattered into a solid angle Ω can be calculated using quantum scattering methods. If the scattering angle of the scattered electrons is not measured, i.e., is averaged over, the state created by the scattering electrons is not a pure state; one must integrate over initial scattering energies and over all final scattering angles to obtain the representation of the (mixed) state obtained for the hydrogen atom if the angular distribution of the scattered electron is not measured:

$$\rho(\mathbf{r},\mathbf{r}',t) = \int dE\, P(E) \int d\Omega\, \Psi_{E,\Omega}(\mathbf{r},t) \Psi^*_{E,\Omega}(\mathbf{r}',t). \tag{2.50}$$

The expectation value of an operator O within the mixed state is given by

$$\langle O(t) \rangle = \int dE\, P(E) \int d\Omega\, \langle \Psi_{E,\Omega}(t) | O | \Psi_{E,\Omega}(t) \rangle = \int dE\, P(E) \int d\Omega \sum_{n'l'm'} \sum_{nlm} c^*_{E,\Omega,n'l'm'}(t)\, c_{E,\Omega,nlm}(t)\, O_{n'l'm',nlm}, \tag{2.51}$$

where

$$O_{n'l'm',nlm} = \int d\mathbf{r}' \int d\mathbf{r}\, \psi^*_{n'l'm'}(\mathbf{r}') O(\mathbf{r}',\mathbf{r}) \psi_{nlm}(\mathbf{r}). \tag{2.52}$$

The expectation value in Eq. (2.51) can be written as

$$\langle O(t) \rangle = \int d\mathbf{r}' \int d\mathbf{r}\, O(\mathbf{r}',\mathbf{r}) \rho(\mathbf{r},\mathbf{r}',t) = \mathrm{Tr}\,[O\,\rho(t)], \tag{2.53}$$

where $\rho(t)$ is the time-dependent density matrix of the hydrogen atom. In coordinate representation, it is given by

$$\rho(\mathbf{r},\mathbf{r}',t) = \int dE\, P(E) \int d\Omega\, \Psi_{E,\Omega}(\mathbf{r},t) \Psi^*_{E,\Omega}(\mathbf{r}',t) = \int dE\, P(E) \int d\Omega \sum_{n'l'm'} \sum_{nlm} c_{E,\Omega,n'l'm'}(t)\, c^*_{E,\Omega,nlm}(t)\, \psi_{n'l'm'}(\mathbf{r}) \psi^*_{nlm}(\mathbf{r}'). \tag{2.54}$$

It is useful to define the density matrix elements, $\rho_{n'l'm',nlm}(t)$, such that

$$\boxed{\rho(\mathbf{r},\mathbf{r}',t) = \sum_{n'l'm'} \sum_{nlm} \rho_{n'l'm',nlm}(t)\, \psi_{n'l'm'}(\mathbf{r}) \psi^*_{nlm}(\mathbf{r}').} \tag{2.55}$$

By comparing Eqs (2.54) and (2.55), we find that the density matrix elements $\rho_{n'l'm',nlm}(t)$ are given by

$$\rho_{n'l'm',nlm}(t) = \int dE\, P(E) \int d\Omega\, c_{E,\Omega,n'l'm'}(t)\, c^*_{E,\Omega,nlm}(t). \tag{2.56}$$

2.5 Mixed States: Density Matrix Formulation

It is important to understand that the state $\Phi(\mathbf{r},t) = \int dE\, P(E) \int d\Omega\, \Psi_{E,\Omega}(\mathbf{r},t)$ does not properly describe the state of the hydrogen atom after the collision with the electrons when the electrons have been scattered into various scattering angles, and these angles have not been observed. Taking the expectation value of an operator O in such a state, i.e., $\langle O(t) \rangle$, does not properly determine the expectation value of the operator O at time t. In the description of the hydrogen atom used here, the free-electron scattering angle and energy have been "traced over," i.e., averaged over, and the state of the hydrogen atom is a mixed state described by a density matrix.

Problem 2.8

(a) Generalize the measurement postulates so that they apply to a system in a mixed state.
(b) Verify that the generalization reduces to the standard statement of the postulates when the system is in a pure state, $\rho = |\Psi\rangle\langle\Psi|$.
(c) Generalize the seventh postulate to a system in a mixed state.

Answer: (a) The first measurement postulate is generalized to the following:
When a measurement of an observable $\mathcal{A}$ is made on a mixed state described by a density matrix ρ, the probability of obtaining an eigenvalue a_i is given by $\text{Tr}\,[|\phi_i\rangle\langle\phi_i|\rho] = \langle\phi_i|\rho|\phi_i\rangle$, where $|\phi_i\rangle$ is the eigenvector of the observable operator $\hat{\mathcal{A}}$ with eigenvalue a_i. The second measurement postulate is generalized (trivially) to the following:
Immediately after measurement of an observable $\mathcal{A}$ has yielded a value a_i, the system is in the pure state represented by the normalized eigenfunction $|\phi_i\rangle$, i.e., the pure-state density matrix $|\phi_i\rangle\langle\phi_i|$.
(c) The state space of a composite system is the tensor product of the state spaces of the constituent systems:
$\rho_{N\text{-particle}} = \rho_A \otimes \rho_B \otimes \ldots \otimes \rho_N$. Examples will be discussed in Secs. 6.1 and 6.2.

Problem 2.9

(a) Determine the probability of finding the system in state $|\mathbf{r}\rangle$ if the state of the system is given by the density matrix ρ. Hint: use the projection operator $\hat{P}_\mathbf{r} = |\mathbf{r}\rangle\langle\mathbf{r}|$.
(b) Determine the probability of finding the system in state $|\mathbf{p}\rangle$ if the state of the system is given by the density matrix ρ.
(c) Calculate the expectation value of the projection operator $\mathcal{P}_\varphi = |\varphi\rangle\langle\varphi|$, using the position space representation.

Answer: (a) $P(\mathbf{r}) = \text{Tr}(\rho \hat{P}_\mathbf{r}) = \sum_i \langle\phi_i|\mathbf{r}\rangle\langle\mathbf{r}|\rho|\phi_i\rangle = \langle\mathbf{r}|\rho|\mathbf{r}\rangle$, where the last step is perhaps easiest to understand by using a position basis, $|\phi_i\rangle \to |\mathbf{r}'\rangle$. (b) $P(\mathbf{p}) = \langle\mathbf{p}|\rho|\mathbf{p}\rangle$. (c) $\langle\mathcal{P}_\varphi\rangle = \text{Tr}\,\rho\mathcal{P}_\varphi = \int d\mathbf{r}d\mathbf{r}'\, \varphi(\mathbf{r})\rho(\mathbf{r},\mathbf{r}')\varphi^*(\mathbf{r}')$.

2.5.1 MANY-PARTICLE SYSTEMS: CORRELATION FUNCTIONS

Let us now consider a system composed of many particles. For a pure state of an N-particle system, the coordinate-space density matrix can be written as

$$\rho(x_1,\ldots,x_N,x_1',\ldots,x_N') = \langle x_1,\ldots,x_N|\hat{\rho}|x_1',\ldots,x_N'\rangle = \psi(x_1,\ldots,x_N)\psi^*(x_1',\ldots,x_N'), \tag{2.57}$$

where we use the usual abbreviation $x \equiv \mathbf{r}, m_s$ for space and spin coordinates. If no spin coordinates are required, we just neglect m_s (i.e., no spin degrees of freedom are necessary). The expectation value of a general N-particle operator $\hat{O}$ is given by

$$\langle \hat{O} \rangle = \text{Tr}\,\hat{O}\rho = \int dx_1 \int dx_2 \ldots \int dx_N\, \hat{O}\,\rho(x_1,x_2,\ldots,x_N,x_1,x_2,\ldots,x_N)$$
$$= \int dx_1 \int dx_2 \ldots \int dx_N\, \psi^*(x_1,x_2,\ldots,x_N)\hat{O}\psi(x_1,x_2,\ldots,x_N). \tag{2.58}$$

More generally, for a mixed state of N-particles,

$$\rho(x_1,\ldots,x_N,x_1',\ldots,x_N') = \langle x_1,\ldots,x_N|\hat{\rho}|x_1',\ldots,x_N'\rangle \quad (2.59)$$
$$= \sum_i p_i \psi_i(x_1,\ldots,x_N)\psi_i^*(x_1',\ldots,x_N').$$

Even in the more general case of a mixed state, the expectation value of a general N-particle operator $\hat{O}$ is still given by the first line of Eq. (2.58). The operators that are typically dealt with in quantum mechanics are one- or two-particle operators, and these operators can be calculated from *reduced density matrices*. The reduced single-particle density matrix is defined as

$$\rho(x_1,x_1') = N \int dx_2 \ldots \int dx_N \rho(x_1,x_2,\ldots,x_N,x_1',x_2,\ldots,x_N), \quad (2.60)$$

where the factor N normalizes the reduced single-particle density matrix such that $\int dx_1 \rho(x_1,x_1) = N$ (rather than unity). Similarly, the reduced two-particle density matrix is defined as

$$\rho_2(x_1,x_2,x_1',x_2') = \frac{N(N-1)}{2} \int dx_3 \ldots \int dx_N \rho(x_1,x_2,x_3,\ldots,x_N,x_1',x_2',x_3,\ldots,x_N), \quad (2.61)$$

where $N(N-1)/2$ is a convenient normalization. If the density matrix is a function of time, so are the reduced density matrices. Moreover, unequal-time density matrices can be defined to generalize the equal-time density matrices, e.g.,
$$\rho(x_1,t_1,\ldots,x_N,t_N,x_1',t_1',\ldots,x_N',t_N') = \sum_i p_i \psi_{i,j_1}(x_1,t_1)\ldots\psi_{i,j_N}(x_N,t_N)\psi_{i,j_1}^*(x_1',t_1')\ldots\psi_{i,j_N}^*(x_N',t_N').$$
The *first-order spatial coherence*, often called the *first-order correlation function*, is defined by

$$g^{(1)}(x,x') \equiv \frac{\rho(x,x')}{\sqrt{n(x)n(x')}}, \quad (2.62)$$

where $\rho(x,x')$ is the single-particle density matrix and the density $n(x)$ is given by the diagonal element of $\rho(x,x')$, i.e., $n(x) \equiv \rho(x,x)$ is a measure of the visibility of interference between parts of the atomic gas coming from x and x'. Furthermore, temporal coherence can be defined in terms of the *first-order temporal correlation function*,

$$g^{(1)}(x,t,x,t') \equiv \frac{\rho(x,t,x,t')}{\sqrt{n(x,t)n(x,t')}}. \quad (2.63)$$

The *pair correlation function*, $g^{(2)}(x_1,x_2)$, often called the *second-order correlation function*, is defined in terms of the diagonal element of $\rho_2(x_1,x_2,x_1',x_2')$ as

$$g^{(2)}(x_1,x_2) \equiv \frac{\rho_2(x_1,x_2,x_1,x_2)}{n(x_1)n(x_2)}, \quad (2.64)$$

where, clearly, $n(x_i) \equiv \rho(x_i,x_i)$. For very large $|x_1 - x_2|$,

$$g^{(2)}(x_1,x_2) \xrightarrow[|x_1-x_2|\to\infty]{} 1. \quad (2.65)$$

Temporal pair correlation functions can also be defined in a similar fashion. The concepts of first-order correlation function and pair correlation function are very useful in studies of many-body systems. We shall have more to say about these quantities in Chapter 14 on many-body theory. There, we shall come back to Eqs (2.62) and (2.64), analyze them, and rewrite them in second-quantized language.

2.5 Mixed States: Density Matrix Formulation

2.5.2 PURITY AND VON NEUMANN ENTROPY

How can one decide whether a given density matrix represents a pure state or a mixed state? One measure, denoted the *purity* of a state, is given by

$$\mathbb{P}(\rho) \equiv \operatorname{Tr} \rho^2. \quad (2.66)$$

If the purity is unity, i.e., $\mathbb{P}(\rho) = 1$, the state is a pure state, and if $\mathbb{P}(\rho) < 1$, the state is a mixed state. Since a density matrix is a positive Hermitian operator, it can always be put into the form $\rho = \sum_i p_i |\phi_i\rangle\langle\phi_i|$, where p_i are real nonnegative eigenvalues with $\sum_i p_i = 1$ and $\{|\phi_i\rangle\}$ are orthonormal vectors. Hence, $\operatorname{Tr} \rho^2 = \operatorname{Tr} \left(\sum_i p_i |\phi_i\rangle\langle\phi_i|\right)^2 = \operatorname{Tr} \left(\sum_i p_i^2 |\phi_i\rangle\langle\phi_i|\right)^2 = \sum_i p_i^2$.

If the sum is over only one state, the state is pure and the purity is unity, whereas if more than one state is present in the sum, the state is mixed and the purity is less than unity. If $\operatorname{Tr} \rho^2 = \operatorname{Tr} \rho$, the eigenvalues of ρ are 0 and 1. Since $\operatorname{Tr} \rho = 1$, the sum of these eigenvalues is 1. If the purity of the density matrix is unity, there is only a single eigenvector $|\varphi\rangle$ that satisfies $\rho|\varphi\rangle = |\varphi\rangle$ and we can write ρ as $\rho = |\varphi\rangle\langle\varphi|$. Thus, when diagonalized, the density matrix of a pure state takes the form

$$\rho = \begin{pmatrix} 0 & & & & & 0 \\ & \ddots & & & & \\ & & 0 & & & \\ & & & 1 & & \\ & & & & 0 & \\ & & & & & \ddots \\ 0 & & & & & 0 \end{pmatrix}.$$

Another measure of whether a state is pure is the *von Neumann entropy* of ρ defined as

$$S(\rho) \equiv -\operatorname{Tr} \rho \log \rho. \quad (2.67)$$

For the incoherent density matrix, $\rho = \sum_i p_i |i\rangle\langle i|$, we find $S(\rho) = -\sum_i p_i \log p_i$. For a pure state, only one p_i is non zero and it equals unity, so $S(\rho) = 0$; for a mixed state $S(\rho) > 0$. Note that the von Neumann entropy is unitless (it does not have units of thermodynamic entropy). The logarithm is typically evaluated in base 2 if one is considering spin 1/2 systems, so, e.g., if $p_1 = p_2 = 1/2$, $S(\rho) = \log_2 2 = 1$. Figure 2.1 plots the von Neumann entropy of a spin 1/2 density matrix, $S(\rho) = -[p \log_2 p - (1-p) \log_2(1-p)]$ versus p. The maximum entropy occurs for $p = (1-p) = 1/2$, where $S = 1$, and the entropy vanishes for $p = 0$ and $p = 1$.

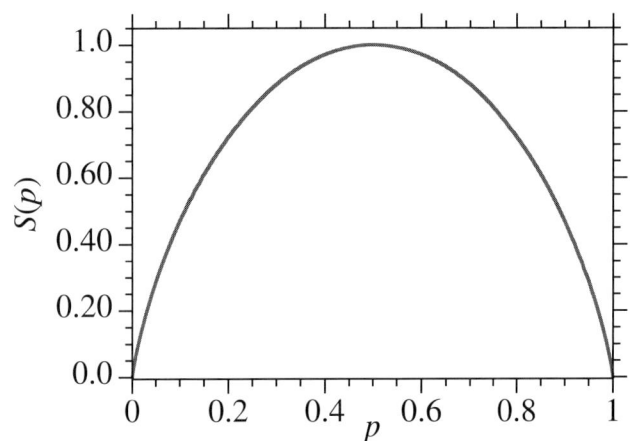

FIG 2.1 The von Neumann entropy of a spin 1/2 incoherent density matrix, $S(\rho) = -[p \log_2 p + (1-p) \log_2(1-p)]$ versus p.

2.5.3 DISTANCE BETWEEN STATES

There are several ways to define the distance between two quantum states. One way is the *trace distance*:

$$D(\rho_1, \rho_2) \equiv \frac{1}{2} \operatorname{Tr} |\rho_1 - \rho_2|, \quad (2.68)$$

where, for an arbitrary matrix M, $|M|$ is defined as the positive square root of $M^\dagger M$, $|M| \equiv \sqrt{M^\dagger M}$.[2] The trace distance $D(\rho_1, \rho_2)$ is a measure of the distance between density matrices, satisfying $0 \leq D \leq 1$, and it yields a measure of the

---

[2] $M^\dagger M$ is Hermitian, so it has a "full" set of eigenvalues λ_i and $\operatorname{Tr} |M| = \sum_i \sqrt{\lambda_i}$.

physical distinguishability of quantum states. A unitary transformation of the density matrices, $\rho_i \to U\rho_i U^\dagger$, does not change their trace distance. As an example of the calculation of the trace distance, let us take a two-level density matrix case with $\rho_1 = \rho = 1 + \begin{pmatrix} z & x-iy \\ x+iy & -z \end{pmatrix}$ and

$$\rho_2 = \rho + \delta\rho = \rho + \begin{pmatrix} \delta z & \delta x - i\delta y \\ \delta x + i\delta y & -\delta z \end{pmatrix}, \tag{2.69}$$

where this form is required so that ρ_2 remains a density matrix. Then, $D(\rho_1, \rho_2) = (1/2)\mathrm{Tr}\sqrt{\delta\rho^\dagger \delta\rho} = (1/2)\sqrt{\delta x^2 + \delta y^2 + \delta z^2}\,\mathrm{Tr}\,1 = \sqrt{\delta x^2 + \delta y^2 + \delta z^2}$.

Another distance measure between two quantum states is called the *fidelity* and is defined as

$$F(\rho_1, \rho_2) \equiv \mathrm{Tr}\sqrt{\rho_1^{1/2} \rho_2 \rho_1^{1/2}}. \tag{2.70}$$

This measure is also invariant under unitary transformations. For the above example, it does not produce a simple answer; nevertheless, this measure has many of the nice features of the trace distance. As an example, suppose we take ρ_1 to be a pure state, $\rho_1 = |\psi\rangle\langle\psi|$, and $\rho_2 = \rho$. Then,

$$F(\rho_1, \rho_2) = \sqrt{\langle\psi|\rho|\psi\rangle}. \tag{2.71}$$

Hence, the fidelity is the square root of the overlap of the pure state and the density matrix.

Problem 2.10

Prove Eq. (2.71) by considering the matrix $\rho_1^{1/2}$ and showing that it equals ρ_1 for the case of $\rho_1 = |\psi\rangle\langle\psi|$.

2.5.4 THE MEASUREMENT PROBLEM REVISITED

Let us now revisit the measurement problem discussed in Sec. 2.4, armed with the density matrix formalism. We again consider a two-level system in contact with a measurement apparatus, but now examine the effects of coupling of the measurement apparatus to the environment. Macroscopic systems, even if they are small, are never isolated from their environments, so if the measurement apparatus is a macroscopic system, we need to consider the effects of its coupling to the environment. In the treatment here, we will closely follow Zurek [29].

A density matrix corresponding to the pure state in Eq. (2.35), $|\Psi\rangle = (a|\uparrow\rangle|A_\uparrow\rangle + b|\downarrow\rangle|A_\downarrow\rangle)$, can be used to describe the probability distribution over the alternative outcomes. The outcomes of the measurement can be made independent of one another by taking the pure-state density matrix, $\rho_{\mathrm{pure}} = |\Psi\rangle\langle\Psi|$,

$$\rho_{\mathrm{pure}} = (a|\uparrow\rangle|A_\uparrow\rangle + b|\downarrow\rangle|A_\downarrow\rangle)(a^*\langle\uparrow|\langle A_\uparrow| + b^*\langle\downarrow|\langle A_\downarrow|), \tag{2.72}$$

and canceling the off-diagonal terms that determine the quantum correlations (i.e., the entanglement), so that a reduced density matrix with only classical correlations emerges, we obtain:

$$\rho_{\mathrm{dec}} = |a|^2 |\uparrow\rangle\langle\uparrow| \, |A_\uparrow\rangle\langle A_\uparrow| + |b|^2 |\downarrow\rangle\langle\downarrow| \, |A_\downarrow\rangle\langle A_\downarrow|. \tag{2.73}$$

The decohered density matrix ρ_{dec} is easier to interpret as a description of a measurement than ρ_{pure}, although both contain identical probabilities for finding spin-up or spin-down, because ρ_{dec} does not have any quantum correlation between the system and the apparatus; only classical type correlation is present. The density matrix (2.73) emerges from (2.72) if we add an environmental variable to the description of the measurement process and trace over the environment as follows.

We consider a quantum two-level system, the apparatus and the environment, where the environment is also a quantum system. The combined (two-level-system)-apparatus environment starts out in the initial state, $|\Psi_0\rangle \equiv |\psi_0\rangle|A_0\rangle|E_0\rangle =$

$(a_0|\uparrow\rangle + b_0|\downarrow\rangle) |A_0\rangle |E_0\rangle$, where $|E_0\rangle$ is the initial state of the environment and $|A_0\rangle$ is the initial state of the measurement apparatus. We now parallel the treatment in Sec. 2.4. By virtue of the interaction of the two-level system and the apparatus, and the unitary evolution postulate, Eq. (1.35), the combined (two-level-system)-apparatus subsystem evolves into the pure state $(a|\uparrow\rangle|A_\uparrow\rangle + b|\downarrow\rangle|A_\downarrow\rangle)$, so that

$$|\Psi_0\rangle \Rightarrow |\Psi\rangle = (a|\uparrow\rangle|A_\uparrow\rangle + b|\downarrow\rangle|A_\downarrow\rangle) |E_0\rangle. \tag{2.74}$$

Following the first step of the measurement process, i.e., the evolution (2.74) that establishes a correlation between the two-level system and the apparatus, or concurrent with it, the apparatus and the environment similarly interact and become correlated:

$$|\Psi\rangle \Rightarrow |\Psi'\rangle = a|\uparrow\rangle|A_\uparrow\rangle|E_\uparrow\rangle + b|\downarrow\rangle|A_\downarrow\rangle|E_\downarrow\rangle. \tag{2.75}$$

The apparatus-environment interaction has extended the correlation to the environment. When the state of the environment $|E_\uparrow\rangle$, corresponding to the state $|A_\uparrow\rangle$ of the apparatus, and $|E_\downarrow\rangle$, corresponding to the state $|A_\downarrow\rangle$ of the apparatus are orthogonal, $\langle E_i|E_j\rangle = \delta_{ij}$, the density matrix for the combined system-apparatus subsystem can be obtained by tracing over the information in the uncontrolled and unknown degrees of freedom of the environment:

$$\text{Tr}_E(|\Psi'\rangle\langle\Psi'| = |a|^2 |\uparrow\rangle\langle\uparrow| |A_\uparrow\rangle\langle A_\uparrow| + |b|^2 |\downarrow\rangle\langle\downarrow| |A_\downarrow\rangle\langle A_\downarrow| = \rho_{\text{dec}}. \tag{2.76}$$

In contrast to the treatment in Sec. 2.4 using (2.35), here, using the density matrix in Eq. (2.76), the collapse postulate is not necessary to describe the results of the experiment that measures the state of the two-level system. Invoking decoherence of the apparatus due to its interaction with the environment, we have finessed the need for the collapse measurement postulate.

For completeness, we mention that it is possible to do other types of measurements besides the projective measurements discussed in the postulates. There are *weak measurements*, *demolition measurements*, *POVM measurements* (Positive Operator-Valued Measure), and *von Neumann measurements*.

2.6 THE WIGNER REPRESENTATION

The Wigner representation of the density matrix, $W(\mathbf{p}, \mathbf{r}, t)$, provides information about the state of the system in phase space, $\{(\mathbf{p}, \mathbf{r})\}$. It allows both a position and a momentum view of the state of a system, in contradistinction to the wave function representation in position space, $\psi(\mathbf{r})$, which provides information about the position distribution, or the wave function representation in momentum space, $\psi(\mathbf{p})$, which yields the momentum distribution. The Wigner function for a system with N degrees of freedom is defined in terms of the density matrix, $\rho(t)$, as[3]

$$W(\mathbf{p}, \mathbf{r}, t) = (2\pi\hbar)^{-N} \int d\mathbf{u}\, e^{i\mathbf{p}\cdot\mathbf{u}/\hbar} \rho\left(\mathbf{r} - \frac{\mathbf{u}}{2}, \mathbf{r} + \frac{\mathbf{u}}{2}, t\right), \tag{2.77}$$

where the integral is over N coordinate dimensions, and all the vectors are N-dimensional (for n particles in 3D, $N = 3n$). If the system is in a pure state, $\hat{\rho} = |\psi\rangle\langle\psi|$, (2.77) takes the form

$$W(\mathbf{p}, \mathbf{r}, t) = (2\pi\hbar)^{-N} \int d\mathbf{u}\, e^{i\mathbf{p}\cdot\mathbf{u}/\hbar} \psi\left(\mathbf{r} - \frac{\mathbf{u}}{2}, t\right) \psi^*\left(\mathbf{r} + \frac{\mathbf{u}}{2}, t\right). \tag{2.78}$$

---

[3] Eq. (2.77) can be written, $W(\mathbf{p}, \mathbf{r}, t) = (2\pi\hbar)^{-N} \int d\mathbf{u}\, e^{i\mathbf{p}\cdot\mathbf{u}/\hbar} \langle \mathbf{r} - \frac{\mathbf{u}}{2}|\rho(t)|\mathbf{r} + \frac{\mathbf{u}}{2}\rangle$.

Defining the coordinate vectors $\mathbf{R} = \mathbf{r} + \mathbf{u}/2$ and $\mathbf{R}' = \mathbf{r} - \mathbf{u}/2$, Eq. (2.77) can be written as

$$W\left(\mathbf{p}, \frac{\mathbf{R}+\mathbf{R}'}{2}, t\right) = (2\pi\hbar)^{-N} \int d(\mathbf{R}-\mathbf{R}')\, e^{i\mathbf{p}\cdot(\mathbf{R}-\mathbf{R}')/\hbar} \rho(\mathbf{R}', \mathbf{R}, t), \tag{2.79}$$

where $\rho(\mathbf{R}', \mathbf{R}, t)$ in the integrand should be expressed as a function of $(\mathbf{R}+\mathbf{R}')/2$ and $\mathbf{R} - \mathbf{R}'$. $W(\mathbf{p}, \frac{\mathbf{R}+\mathbf{R}'}{2}, t)$ is a function of the momentum variable $\mathbf{p}$, the Fourier transform variable of the relative coordinate $(\mathbf{R} - \mathbf{R}')$, and the "center of mass coordinate" $\mathbf{R}_{CM} \equiv (\mathbf{R} + \mathbf{R}')/2$. The inverse transform required to obtain the density matrix $\rho(\mathbf{R}', \mathbf{R}, t)$ in terms of the Wigner function is

$$\rho(\mathbf{R}', \mathbf{R}, t) = \int d\mathbf{p}\, e^{-i\mathbf{p}\cdot(\mathbf{R}-\mathbf{R}')/\hbar} W\left(\mathbf{p}, \frac{\mathbf{R}+\mathbf{R}'}{2}, t\right). \tag{2.80}$$

Problem 2.11

Show that for a pure state, (a) $W(\mathbf{p}, \mathbf{r}, t)$ is real. (b) $|W(\mathbf{p}, \mathbf{r}, t)| \leq (\pi\hbar)^{-N}$, by defining the normalized wave functions $\phi_1(\mathbf{u}, t) \equiv 2^{-N/2} e^{-i\mathbf{p}\cdot\mathbf{u}/\hbar} \psi(\mathbf{r} - \frac{\mathbf{u}}{2}, t)$ and $\phi_2(\mathbf{u}, t) \equiv 2^{-N/2} \psi(\mathbf{r} + \frac{\mathbf{u}}{2}, t)$, writing the Wigner function as $W(\mathbf{p}, \mathbf{r}, t) = (\pi\hbar)^{-N} \int d\mathbf{u}\, \phi_1^*(\mathbf{u}, t) \phi_2(\mathbf{u}, t) = (\pi\hbar)^{-N} \langle \phi_1 | \phi_2 \rangle$, and using the Cauchy–Schwartz inequality, Eq. (A.9), $|\langle \phi_1 | \phi_2 \rangle|^2 \leq \langle \phi_1 | \phi_1 \rangle \langle \phi_2 | \phi_2 \rangle$.

Integrating the Wigner function $W(\mathbf{p}, \mathbf{r}, t)$ over the momentum $\mathbf{p}$, and making use of the equation $\int d\mathbf{p}\, e^{-i\mathbf{p}\cdot\mathbf{u}/\hbar} = (2\pi\hbar)^N \delta(\mathbf{u})$, yields the probability distribution for finding the system at coordinate $\mathbf{r}$,

$$P(\mathbf{r}, t) = \int d\mathbf{p}\, W(\mathbf{p}, \mathbf{r}, t) = \rho(\mathbf{r}, \mathbf{r}, t). \tag{2.81}$$

Integrating (2.77) over coordinates $\mathbf{r}$ yields, after twice inserting a complete set of momentum states, $\int d\mathbf{p} |\mathbf{p}\rangle\langle\mathbf{p}| = 1$, into $\langle \mathbf{r} - \frac{\mathbf{u}}{2} | \rho(t) | \mathbf{r} + \frac{\mathbf{u}}{2} \rangle$, the probability distribution function for finding the system at momentum $\mathbf{p}$,

$$P(\mathbf{p}, t) = \int d\mathbf{r}\, W(\mathbf{p}, \mathbf{r}, t) = \rho(\mathbf{p}, \mathbf{p}, t). \tag{2.82}$$

Integrating $W(\mathbf{p}, \mathbf{r}, t)$ over both momenta and coordinates yields unity,

$$\int d\mathbf{p} \int d\mathbf{r}\, W(\mathbf{p}, \mathbf{r}, t) = 1. \tag{2.83}$$

Note from (2.83) that W has dimensions of $[d\mathbf{p}\, d\mathbf{r}]^{-1}$; this accounts for the units of the factor of $(2\pi\hbar)^{-N}$ on the RHS of the definition of the Wigner function (2.78). More explicitly, ψ in the integrand of (2.78) has units $[\mathbf{r}]^{-1/2}$, so $\psi^*\psi$ has units $[\mathbf{r}]^{-1}$ (as is clear from the equation, $\int d\mathbf{r}\, \psi^*\psi = 1$). The factor $(2\pi\hbar)^{-N}$ in (2.78) guarantees that the units of $\int d\mathbf{p} \int d\mathbf{r}\, W(\mathbf{p}, \mathbf{r}, t)$ are dimensionless and it equals unity. Unfortunately, the Wigner function is not a probability distribution function (despite the fact that $P(\mathbf{r}, t)$ and $P(\mathbf{p}, t)$ are); it can be negative in regions of phase space, as shown below.[4] It is sometimes called a *quasi-probability distribution*.

---

[4] A quantity known as the *Husimi distribution*, a Gaussian smoothing of the Wigner function, is defined in a manner that guarantees it to be nonnegative, and therefore, it can have a probability interpretation. But the Husimi distribution does not satisfy relations (2.81) or (2.82). We shall not discuss the Husimi distribution here; the interested reader is referred to Ballentine [9].

2.6 The Wigner Representation

As an example, consider the calculation of the Wigner representation of a pure-state Gaussian wave function, $\psi(x) = \frac{1}{(2\pi\sigma^2)^{1/4}} e^{-\frac{x^2}{4\sigma^2}}$. We need to evaluate $\rho(\frac{x-y}{2}, \frac{x+y}{2}) = \psi(\frac{x-y}{2})\psi^*(\frac{x+y}{2})$:

$$\psi\left(\frac{x-y}{2}\right)\psi^*\left(\frac{x+y}{2}\right) = \frac{1}{(2\pi\sigma^2)^{1/2}} e^{\frac{-(x-y)^2}{16\sigma^2}} e^{\frac{-(x+y)^2}{16\sigma^2}}.$$

The Fourier transform of this quantity is the Wigner function,

$$W(p,x) = (2\pi\hbar)^{-1} \int dy\, e^{ipy/\hbar}\, \psi\left(\frac{x-y}{2}\right)\psi^*\left(\frac{x+y}{2}\right).$$

Note that the Fourier transform of a Gaussian is a Gaussian. You are asked to carry out this Fourier transform in Problem 2.12(a). Note also that in this case $W(p,x)$ is nonnegative.

Problem 2.12

(a) Complete the calculation of the Wigner representation for the ground-state harmonic oscillator wave function in 1D.

(b) Calculate the Wigner function for the wave function $\psi(x) = (2\pi\sigma^2)^{-1/4} e^{\frac{-[x-x_0(t)]^2}{4\sigma^2}} e^{ip_0 x/\hbar}$.

Answer: (a) $W(p,x) = (\pi\hbar)^{-1} e^{\frac{-p^2}{2\sigma_p^2}} e^{\frac{-x^2}{2\sigma^2}}$, where $\sigma_p = \hbar/2\sigma$.

(b) $W(p,x) = (\pi\hbar)^{-1} e^{\frac{-(p-p_0)^2}{2\sigma_p^2}} e^{\frac{-[x-x_0(t)]^2}{2\sigma^2}}$.

Problem 2.13

(a) Calculate the Wigner function for the two-particle entangled wave function $\psi(x_1, x_2) = \int dp\, e^{i(x_1-x_2-x_0)p} = 2\pi\delta(x_1-x_2-x_0)$.

(b) Calculate the Wigner function $W(p_1, p_2, x_1, x_2)$ for $\psi(x_1, x_2) = \frac{1}{\sqrt{2\pi\sigma_p^2}} \int dp\, e^{-\frac{p^2}{2\sigma_p^2}} e^{i(x_1-x_2-x_0)p}$.

Answer: (a) $W(p_1, p_2, x_1, x_2) = (2\pi\hbar)^{-2} \int du_1 du_2\, e^{i(p_1 u_1 + p_2 u_2)}\, \psi(x_1 - u_1/2, x_2 - u_2/2)\, \psi^*(x_1 + u_1/2, x_2 + u_2/2)$. Now substitute the wave function to obtain $W(p_1, p_2, x_1, x_2) = (\hbar)^{-2} \int du_1 du_2\, e^{i(p_1 u_1 + p_2 u_2)}\, \delta(x_1 - u_1/2 - x_2 + u_2/2 - x_0)\, \delta(x_1 + u_1/2 - x_2 - u_2/2 - x_0)$. The first delta function implies $u_1 = 2(x_1 - x_2 - x_0) + u_2$. Substituting this value of u_1 into the integral yields $W(p_1, p_2, x_1, x_2) = \hbar^{-2} \int du_2\, e^{i[p_1(2(x_1-x_2-x_0)+u_2)+p_2 u_2]} \times \frac{1}{2}\delta(x_1 - x_2 - x_0) = \pi\hbar^{-2} e^{2ip_1(x_1-x_2-x_0)}\, \delta(p_1+p_2)\, \delta(x_1-x_2-x_0)$. Alternatively, one could use the wave function $\phi(x_1, x_2) = e^{i(x_1-x_2-x_0)p}$, and finally do the integrals $\int dp\, \phi(x_1, x_2)$. This is the way to proceed in part (b).

The Wigner representation of the density matrix for the pure state given by the sum of two Gaussians separated by a distance $2c$,

$$\psi(x) = \sqrt{\frac{1}{2(1+e^{-\frac{c^2}{4\sigma^2}})}} \frac{1}{(2\pi\sigma^2)^{1/4}} \left[e^{\frac{-(x-c)^2}{4\sigma^2}} + e^{\frac{-(x+c)^2}{4\sigma^2}} \right], \qquad (2.84)$$

yields a Wigner function that is negative in some regions of phase space,

$$W(p,x) = \frac{1}{\left(1+e^{-\frac{c^2}{2\sigma^2}}\right)} e^{\frac{-(p)^2}{2\sigma_p^2}} \left[e^{\frac{-(x-c)^2}{2\sigma^2}} + e^{\frac{-(x+c)^2}{2\sigma^2}} + 2 e^{\frac{-x^2}{2\sigma^2}} \cos(2cp) \right], \qquad (2.85)$$

where $\sigma_p = \hbar/2\sigma$. No matter how far apart the two gaussians are, the Wigner function continues to be negative near $(p = \pm \frac{\pi}{2c}, x = 0)$. Figure 2.2 plots the Wigner function in (2.85) in phase space.

In analogy with the definition (2.77) of the Wigner representation for the density matrix, the *Wigner representation* for *any* operator $\mathcal{A}$, i.e., the *Wigner transform* of the operator whose position representation is $\mathcal{A}(\mathbf{r}, \mathbf{r}', t)$, is defined as

$$\boxed{\mathcal{A}_w(\mathbf{p}, \mathbf{r}, t) \equiv \int d\mathbf{u} \, e^{i\mathbf{p}\cdot\mathbf{u}/\hbar} \mathcal{A}\left(\mathbf{r} - \frac{\mathbf{u}}{2}, \mathbf{r} + \frac{\mathbf{u}}{2}, t\right).} \qquad (2.86)$$

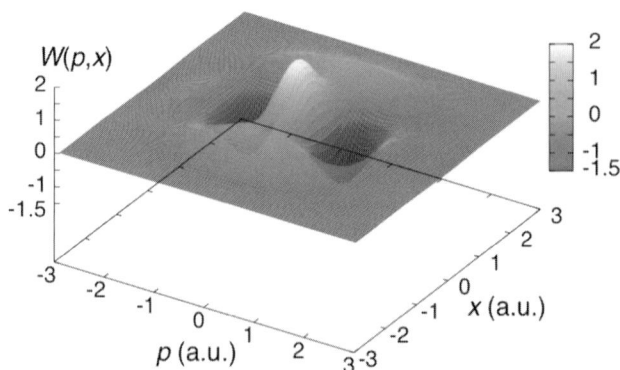

FIG 2.2 Wigner function $W(p,x)$ of the sum of two Gaussians, Eq. (2.85), with $2c = 3$ Bohr and $\sigma = 0.5$ Bohr.

This is almost identical to the definition of the Wigner function in (2.77). The only difference is the factor of $(2\pi\hbar)^{-N}$; this difference is necessary so that the units of the Wigner representation of operators come out reasonably. For example, if we consider the operator $V(\mathbf{r}, \mathbf{r}') = \langle \mathbf{r}|V|\mathbf{r}'\rangle = V(\mathbf{r})\delta(\mathbf{r} - \mathbf{r}')$, we have that $V_w(\mathbf{p}, \mathbf{r}) = V(\mathbf{r})$. As another example, consider the Wigner representation of the kinetic energy operator T, $\langle \mathbf{p}|T|\mathbf{p}'\rangle = \frac{p^2}{2m}\delta(\mathbf{p} - \mathbf{p}')$. In position space,

$$\langle \mathbf{r}|T|\mathbf{r}'\rangle = \int d\mathbf{p} \int d\mathbf{p}' \, \langle \mathbf{r}|\mathbf{p}\rangle \left[\frac{p^2}{2m}\delta(\mathbf{p}-\mathbf{p}')\right]\langle \mathbf{p}'|\mathbf{r}'\rangle = \frac{1}{(2\pi)^3}\int d\mathbf{p}' \, \frac{p'^2}{2m} e^{i\mathbf{p}'\cdot(\mathbf{r}-\mathbf{r}')/\hbar}.$$

Applying this expression to $\langle \mathbf{r} - \frac{\mathbf{u}}{2}|T|\mathbf{r} + \frac{\mathbf{u}}{2}\rangle$ and substituting into (2.86), we find $T_w(\mathbf{p}, \mathbf{r}) = \frac{p^2}{2m}$. Thus, the Wigner representation of operators having a simple momentum representation is also simple. Note that the Wigner representation of an operator can be written [by twice inserting a complete set of momentum states, $\int d\mathbf{p} \, |\mathbf{p}\rangle\langle \mathbf{p}| = 1$, into $\langle \mathbf{r} - \frac{\mathbf{u}}{2}|\mathcal{A}(t)|\mathbf{r} + \frac{\mathbf{u}}{2}\rangle$ that appears on the RHS of (2.86)] as the Fourier transform of the momentum space representation of the operator as follows:

$$\mathcal{A}_w(\mathbf{p}, \mathbf{r}, t) = (2\pi\hbar)^{-N} \int d\mathbf{p}' \, e^{-i\mathbf{p}'\cdot\mathbf{r}/\hbar} \mathcal{A}\left(\mathbf{p} - \frac{\mathbf{p}'}{2}, \mathbf{p} + \frac{\mathbf{p}'}{2}, t\right), \qquad (2.87)$$

Note that the Wigner representation of the density matrix is the Wigner function up to a constant,

$$\rho_w(\mathbf{p}, \mathbf{r}, t) = (2\pi\hbar)^N W(\mathbf{p}, \mathbf{r}, t). \qquad (2.88)$$

The average of a dynamical variable $\mathcal{A}(t)$ in the state specified by the density matrix ρ is given by $\langle \mathcal{A}\rangle = \text{Tr}(\rho\mathcal{A})$. We can express this average in terms of the Wigner function $W(\mathbf{p}, \mathbf{r}, t)$. To do so, let us first represent the trace of $\rho\mathcal{A}$ in position representation:

$$\langle \mathcal{A}(t)\rangle = \text{Tr}[\rho(t)\mathcal{A}(t)] = \int d\mathbf{r} \int d\mathbf{r}' \, \rho(\mathbf{r}', \mathbf{r})\mathcal{A}(\mathbf{r}, \mathbf{r}', t). \qquad (2.89)$$

We now express $\rho(\mathbf{r}', \mathbf{r}, t)$ on the RHS of (2.89) using Eq. (2.80):

$$\langle \mathcal{A}(t)\rangle = \int d\mathbf{r} \int d\mathbf{r}' \int d\mathbf{p} \, e^{-i\mathbf{p}\cdot(\mathbf{r}-\mathbf{r}')/\hbar} W\left(\mathbf{p}, \frac{\mathbf{r}+\mathbf{r}'}{2}, t\right) \mathcal{A}(\mathbf{r}, \mathbf{r}', t). \qquad (2.90)$$

2.6 The Wigner Representation

Rewriting this equation in terms of the variables $\mathbf{R} = \frac{\mathbf{r}+\mathbf{r}'}{2}$ and $\mathbf{u} = \mathbf{r} - \mathbf{r}'$ and renaming the integration variable $\mathbf{R}$ to be $\mathbf{r}$, we find

$$\langle \mathcal{A}(t) \rangle = \int d\mathbf{r} \int d\mathbf{p}\, W(\mathbf{p},\mathbf{r},t) \int d\mathbf{u}\, e^{-i\mathbf{p}\cdot\mathbf{u}/\hbar} \mathcal{A}\left(\mathbf{r}+\frac{\mathbf{u}}{2}, \mathbf{r}-\frac{\mathbf{u}}{2}, t\right). \tag{2.91}$$

Recalling the definition of the Wigner representation of an operator, Eq. (2.86), and using (2.88), we can write this expectation value of the operator $\mathcal{A}(t)$ as

$$\langle \mathcal{A}(t) \rangle = (2\pi\hbar)^{-N} \int d\mathbf{r} \int d\mathbf{p}\, \rho_w(\mathbf{p},\mathbf{r},t)\, \mathcal{A}_w(\mathbf{p},\mathbf{r},t). \tag{2.92}$$

By comparing (2.89) and (2.92), we conclude that,

$$\mathrm{Tr}\,(AB) = (2\pi\hbar)^{-N} \int d\mathbf{r} \int d\mathbf{p}\, \mathcal{A}_w(\mathbf{p},\mathbf{r},t)\, \mathcal{B}_w(\mathbf{p},\mathbf{r},t). \tag{2.93}$$

The product of operators in the Wigner representation can be expressed as

$$[\hat{A}\hat{B}]_w(\mathbf{p},\mathbf{r}) = [\hat{A}]_w(\mathbf{p},\mathbf{r}) \exp\left\{\frac{i\hbar}{2}\Lambda\right\} [\hat{B}]_w(\mathbf{p},\mathbf{r}), \tag{2.94}$$

where the differential operator Λ is

$$\Lambda = \overleftarrow{\nabla}_\mathbf{p} \cdot \overrightarrow{\nabla}_\mathbf{r} - \overleftarrow{\nabla}_\mathbf{r} \cdot \overrightarrow{\nabla}_\mathbf{p}, \tag{2.95}$$

as you will show in Problem 2.14, and the gradient operators act to the left or to the right as indicated. The symmetrized and antisymmetrized products are therefore given by

$$\frac{1}{2}[\hat{A}\hat{B} + \hat{B}\hat{A}]_w(\mathbf{p},\mathbf{r}) = [\hat{A}]_w(\mathbf{p},\mathbf{r}) \cos\left\{\frac{\hbar}{2}\Lambda\right\} [\hat{B}]_w(\mathbf{p},\mathbf{r}), \tag{2.96}$$

$$\frac{1}{2}[\hat{A}\hat{B} - \hat{B}\hat{A}]_w(\mathbf{p},\mathbf{r}) = i[\hat{A}]_w(\mathbf{p},\mathbf{r}) \sin\left\{\frac{\hbar}{2}\Lambda\right\} [\hat{B}]_w(\mathbf{p},\mathbf{r}). \tag{2.97}$$

We shall have occasion to use these results in studying the dynamics of the Wigner function and in deriving semiclassical expressions for the Wigner function in the next subsection.

Problem 2.14

(a) Carry out the algebra leading to (2.87) by twice inserting a complete set of momentum states, $\int d\mathbf{p}\, |\mathbf{p}\rangle\langle\mathbf{p}| = 1$.
(b) Prove Eq. (2.94) by showing that $\int d\mathbf{u}\, e^{i\mathbf{p}\cdot\mathbf{u}/\hbar} \langle \mathbf{r} - \frac{\mathbf{u}}{2} | \hat{A}\hat{B} | \mathbf{r} + \frac{\mathbf{u}}{2} \rangle$
$= \int d\mathbf{u}\, e^{i\mathbf{p}\cdot\mathbf{u}/\hbar} \langle \mathbf{r} - \frac{\mathbf{u}}{2} | \hat{A} | \mathbf{r} + \frac{\mathbf{u}}{2} \rangle \exp\left\{\frac{i\hbar}{2}\Lambda\right\} \int d\mathbf{u}'\, e^{i\mathbf{p}\cdot\mathbf{u}'/\hbar} \langle \mathbf{r} - \frac{\mathbf{u}'}{2} | \hat{B} | \mathbf{r} + \frac{\mathbf{u}'}{2} \rangle$, by inserting a complete set of states on the LHS between $\hat{A}$ and $\hat{B}$.

In order to find the dynamical equation for the Wigner function $W(\mathbf{p},\mathbf{r},t)$ [or $\rho_w(\mathbf{p},\mathbf{r},t)$], let us take the Wigner transform (2.86) of the Liouville-von Neumann equation for the density matrix, $i\hbar\frac{\partial}{\partial t}\rho(t) = [H(t), \rho(t)]$. When the Hamiltonian is given by $H(\mathbf{p},\mathbf{r},t) = \frac{\mathbf{p}^2}{2m} + V(\mathbf{r},t)$, for either one particle in 3D or for an N-dimensional system, we obtain $i\hbar\frac{\partial}{\partial t}\rho_w(\mathbf{p},\mathbf{r},t) = (T\rho - \rho T)_w(\mathbf{p},\mathbf{r}) + (V\rho - \rho V)_w$. In Problem 2.15, you will show that

$$(T\rho - \rho T)_w(\mathbf{p},\mathbf{r}) = \frac{\mathbf{p}}{m} \cdot \frac{\hbar}{i} \nabla_\mathbf{r} \rho_w(\mathbf{p},\mathbf{r},t). \tag{2.98}$$

The Wigner transform of $(V\rho - \rho V)$ does not yield a simple expression, so we shall simply leave it in symbolic form and write the *quantum Liouville equation* as:

$$\boxed{\frac{\partial}{\partial t}\rho_w(\mathbf{p},\mathbf{r},t) + \frac{\mathbf{p}}{m}\cdot\nabla_\mathbf{r}\rho_w(\mathbf{p},\mathbf{r},t) = -\frac{i}{\hbar}(V\rho - \rho V)_w(\mathbf{p},\mathbf{r},t).} \tag{2.99}$$

The quantity $(V\rho - \rho V)_w$ appearing on the RHS of (2.99) can be expanded in a power series in position representation:

$$(V\rho - \rho V)_w(\mathbf{p},\mathbf{r},t) = \int d\mathbf{u}\, e^{i\mathbf{p}\cdot\mathbf{u}/\hbar}\rho\left(\mathbf{r}-\frac{\mathbf{u}}{2},\mathbf{r}+\frac{\mathbf{u}}{2},t\right)\left[V\left(\mathbf{r}-\frac{\mathbf{u}}{2}\right) - V\left(\mathbf{r}+\frac{\mathbf{u}}{2}\right)\right]$$

$$= \int d\mathbf{u}\, e^{i\mathbf{p}\cdot\mathbf{u}/\hbar}\rho\left(\mathbf{r}-\frac{\mathbf{u}}{2},\mathbf{r}+\frac{\mathbf{u}}{2},t\right)\left[-2\sum_{\text{odd } j}\frac{(\mathbf{u}/2)^j}{j!}(\nabla_\mathbf{r})^j V\right].$$

Carrying out the integral over **u** yields,

$$(V\rho - \rho V)_w(\mathbf{p},\mathbf{r},t) = \left[-2\sum_{\text{odd } j}\frac{\left(\frac{\hbar}{2i}\nabla_\mathbf{p}\right)^j}{j!}\rho_w(\mathbf{p},\mathbf{r},t)\cdot(\nabla_\mathbf{r})^j V(\mathbf{r})\right]. \tag{2.100}$$

The RHS is an odd power series in $\hbar$, with lowest order term $i\hbar\nabla_\mathbf{p}\rho_w(\mathbf{p},\mathbf{r},t)\cdot\nabla_\mathbf{r} V(\mathbf{r})$. Substituting only this lowest order term into (2.99) yields,

$$\boxed{\frac{\partial}{\partial t}\rho_w(\mathbf{p},\mathbf{r},t) + \frac{\mathbf{p}}{m}\cdot\nabla_\mathbf{r}\rho_w(\mathbf{p},\mathbf{r},t) + \mathbf{f}\cdot\nabla_\mathbf{p}\rho_w(\mathbf{p},\mathbf{r},t) = 0,} \tag{2.101}$$

where the force is $\mathbf{f} = -\nabla_\mathbf{r} V(\mathbf{r})$. This equation is called the *Liouville equation* [see Eq. (16.33) on the book web page]. It is also called the *Vlasov equation* or the *collisionless Boltzmann equation* in fluid mechanics and plasma physics. Note that $\hbar$ does not appear in this lowest order equation; Eq. (2.101) exactly describes how a classical gas moves in phase space. The higher order terms in (2.100) give rise to quantum corrections of the equation of motion of ρ_w.

Problem 2.15

Obtain (2.98) starting from
$(T\rho - \rho T)_w(\mathbf{p},\mathbf{r}) = -\frac{\hbar^2}{2m}\int d\mathbf{u}\, e^{i\mathbf{p}\cdot\mathbf{u}/\hbar}[(\nabla_\mathbf{r}^2\psi(\mathbf{r}-\frac{\mathbf{u}}{2}))\psi^*(\mathbf{r}+\frac{\mathbf{u}}{2}) - \psi(\mathbf{r}-\frac{\mathbf{u}}{2})(\nabla_\mathbf{r}^2\psi^*(\mathbf{r}+\frac{\mathbf{u}}{2}))]$ by replacing $\nabla_\mathbf{r}^2$ with $\nabla_\mathbf{u}^2$ and integrating once by parts.

Problem 2.16

(a) Explicitly evaluate the first two terms in (2.100) for 1D.
(b) Evaluate $-\frac{i}{\hbar}(V\rho - \rho V)_w(p,x,t)$ for the 1D harmonic oscillator potential $V(x) = \frac{m\omega^2}{2}x^2$ for which only the $j=1$ term in Eq. (2.100) contributes and obtain the equation of motion for $\rho_w(p,x,t)$ [or $W(p,x,t)$]. Note that $\hbar$ does not appear in the resulting equation of motion for the Wigner function for the harmonic oscillator.
(c) Verify that the solution to equation in part (b) is

$$\rho_w(p,x,t) = f(p^2 - m^2\omega^2 x^2, -\omega^{-1}\ln(m\omega x + p) - t).$$

(d) Find $\rho_w(p,x,t)$, given the initial condition,

$$\rho_w(p,x,0) = \frac{1}{(2\pi\sigma_p^2\, 2\pi\sigma_x^2)^{1/2}}\exp\left[-\frac{(p-p_0)^2}{2\sigma_p^2} - \frac{x^2}{2\sigma_x^2}\right]. \tag{2.102}$$

2.7 Schrödinger and Heisenberg Representations

Answer: (a) $i\hbar \left[\partial_p \rho_w(p,x,t) \partial_x V(x) - \frac{\hbar^2}{24} \partial_p^3 \rho_w(p,x,t) \partial_x^3 V(x) \right]$.
(b) $\partial_t \rho_w(p,x,t) + (p/m) \partial_x \rho_w(p,x,t) = m\omega^2 x \, \partial_p \rho_w(p,x,t)$.
(d) Write the initial density matrix in terms of the variables $y = p^2 - m^2\omega^2 x^2$ and $z = -\omega^{-1} \ln(m\omega x + p)$ by solving for $p(y,z)$ and $x(y,z)$, i.e., substitute $p = \frac{1}{2} e^{\omega z}(e^{-2\omega z} + y)$, $x = \frac{1}{2m\omega} e^{\omega z}(e^{-2\omega z} - y)$, into $\rho_w(p,x,0)$. Then, substitute $z \to z - t$ into $\rho(y,z)$ to obtain $\rho_w(y,z,t)$. Finally, reexpress the resulting expression for $\rho_w(y,z,t)$ in terms of p and x by letting $y = p^2 - m^2\omega^2 x^2$ and $z = -\omega^{-1}\ln(m\omega x + p)$.

Problem 2.17

(a) Verify that the solution to $\partial_t \rho_w(p,x,t) + \frac{p}{m} \partial_x \rho_w(p,x,t) = 0$ is given by any function of the form $\rho_w(p,x,t) = f[p, x - (p/m)t]$.
(b) Generalize part (a) to the 3D case, $\frac{\partial}{\partial t}\rho_w(\mathbf{p},\mathbf{r},t) + \frac{\mathbf{p}}{m} \cdot \nabla_\mathbf{r} \rho_w(\mathbf{p},\mathbf{r},t)$.
(c) Find $\rho_w(p,x,t)$, given the initial condition (2.102), and plot for two times.

Answer: (b) $\rho_w(\mathbf{p},\mathbf{r},t) = f[\mathbf{p}, \mathbf{r} - (\mathbf{p}/m)t]$. (c) $\rho_w(p,x,t) = \frac{1}{(2\pi\sigma_p^2 \, 2\pi\sigma_x^2)^{1/2}} \exp\left[-\frac{(p-p_0)^2}{2\sigma_p^2} - \frac{[x-(p/m)t]^2}{2\sigma_x^2}\right]$. See Fig. 2.3.

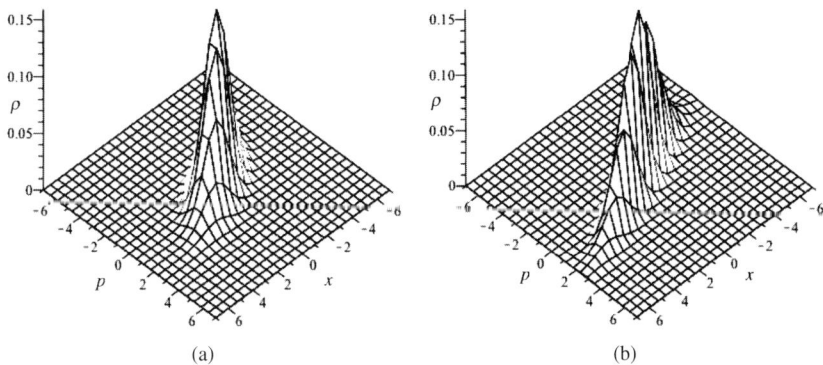

FIG 2.3 Spreading in phase space of the solution to the Vlasov equation due to diffusion, $\rho_w(p,x,t) = (2\pi\sigma_p\sigma_x)^{-1} e^{-\frac{(p)^2}{2\sigma_p^2}} e^{-\frac{[x-(p/m)t]^2}{2\sigma_x^2}}$, with $\sigma_x = \sigma_p = 1$. (a) $t = 1$ and (b) $t = 2$ (the wave packet is Gaussian and symmetric at $t = 0$).

2.7 SCHRÖDINGER AND HEISENBERG REPRESENTATIONS

In quantum mechanics, there are several approaches for treating the time dependence of states (pure or mixed) and operators. In the approach we have been using up to now, time dependence is carried by the state, i.e., in $|\psi(t)\rangle$ for a pure state and in the density matrix $\rho(t)$ for a mixed (or pure) state. This approach is called the *Schrödinger representation*. In it, a pure state evolves as the solution of the time-dependent Schrödinger equation,

$$|\psi_S(t)\rangle = \mathcal{U}(t,0)|\psi_S(0)\rangle, \tag{2.103}$$

where the evolution operator for a time-independent Hamiltonian is $\mathcal{U}(t,0) = e^{-iHt/\hbar}$ or for a time-dependent Hamiltonian, the evolution operator $\mathcal{U}(t,0)$ is the formal solution of the equation,

$$i\hbar \frac{\partial}{\partial t} \mathcal{U}(t,0) = H(t)\mathcal{U}(t,0), \quad U(0,0) = 1. \tag{2.104}$$

The density matrix evolves according to Eq. (2.39), i.e., $\rho_S(t) = \mathcal{U}(t,0)\rho_S(0)\mathcal{U}^\dagger(t,0)$. $\rho_S(t)$ satisfies the equation of motion, $\frac{\partial}{\partial t}\rho_S(t) = \frac{-i}{\hbar}[H(t), \rho_S(t)]$. In the Schrödinger representation, the state of the system evolves in time, i.e., the wave function ψ_S and the density matrix ρ_S depend explicitly on time, and operators are usually time independent [unless they depend explicitly on time]. When computing the quantum average at time t of a time-independent operator $\mathcal{A}_S$, as in (2.40), the time dependence enters through the states, i.e., $\langle A \rangle_t = \text{Tr}\, \mathcal{A}_S \rho_S(t)$, which for a pure state reduces to $\langle A \rangle_t = \langle \psi_S(t) | \mathcal{A}_S | \psi_S(t) \rangle$.[5]

In the *Heisenberg representation*, the time dependence is removed from the state and put onto the operators. The wave function in the Heisenberg representation is given by

$$|\psi_H(t)\rangle \equiv \mathcal{U}^{-1}(t,0)|\psi_S(t)\rangle = |\psi_S(0)\rangle, \tag{2.105}$$

and is therefore time independent. The density matrix in the Heisenberg representation is also time independent. To determine how to write an operator in this representation, consider a state $\mathcal{A}_S(t)|\psi_S(t)\rangle$ in the Schrödinger representation and transform it to the Heisenberg representation:

$$\mathcal{U}^{-1}(\mathcal{A}_S(t)|\psi_S(t)\rangle) = \mathcal{U}^{-1}\mathcal{A}_S(t)\mathcal{U}\mathcal{U}^{-1}|\psi_S(t)\rangle = \mathcal{A}_H(t)|\psi_H(t)\rangle, \tag{2.106}$$

hence,

$$\mathcal{A}_H(t) = \mathcal{U}^{-1}(t,0)\mathcal{A}_S(t)\mathcal{U}(t,0). \tag{2.107}$$

We inserted unity in the form $\mathcal{U}\mathcal{U}^{-1}$ in the middle equation of (2.106), and in the last equation we used the definition of $|\psi_H(t)\rangle$ and thereby identified $\mathcal{A}_H(t)$ as the quantity shown in (2.107). The time derivative of $\mathcal{A}_H(t)$ can be calculated as follows: $\frac{\partial \mathcal{A}_H(t)}{\partial t} = \mathcal{U}^{-1}\frac{\partial \mathcal{A}_S(t)}{\partial t}\mathcal{U} + \frac{i}{\hbar}\mathcal{U}^{-1}H(t)\mathcal{A}_S(t)\mathcal{U} - \frac{i}{\hbar}\mathcal{U}^{-1}\mathcal{A}_S(t)H(t)\mathcal{U}$, i.e., after inserting unity in the form $\mathcal{U}\mathcal{U}^{-1}$ into the last two terms of the RHS of this equation,

$$\frac{\partial \mathcal{A}_H(t)}{\partial t} = \mathcal{U}^{-1}\frac{\partial \mathcal{A}_S(t)}{\partial t}\mathcal{U} + \frac{i}{\hbar}[H, \mathcal{A}_H(t)]. \tag{2.108}$$

This equation is called the *Heisenberg equation of motion* for the operator $\mathcal{A}_H(t)$. For an operator that is not explicitly time dependent, $\frac{\partial \mathcal{A}_H(t)}{\partial t} = \frac{i}{\hbar}[H, \mathcal{A}_H(t)]$. As is clear from (2.105), the time derivative of the wave function in the Heisenberg representation vanishes, $\frac{\partial}{\partial t}|\psi_H(t)\rangle = 0$, i.e., $|\psi_H\rangle$ is time independent. The density matrix in the Heisenberg representation, $\rho_H = \mathcal{U}^{-1}(t,0)\rho_S(t)\mathcal{U}(t,0) = \rho_S(0)$, is also time independent, i.e., $\dot{\rho}_H = 0$. Expectation values are given by $\langle A \rangle_t = \text{Tr}\, \mathcal{A}_H(t)\rho_H$.

Using the Heisenberg picture, the equations of motion for the position and momentum operators are $\frac{\partial \mathbf{r}_H}{\partial t} = \frac{i}{\hbar}[H, \mathbf{r}_H]$ and $\frac{\partial \mathbf{p}_H}{\partial t} = \frac{i}{\hbar}[H, \mathbf{p}_H]$. Often the subscript H is not explicitly indicated. For the Hamiltonian $H = \frac{\mathbf{p}^2}{2m} + V(\mathbf{r})$, we find the Heisenberg equations of motion, $\frac{\partial \mathbf{r}}{\partial t} = \mathbf{p}/m$, and $\frac{\partial \mathbf{p}}{\partial t} = -\frac{\partial V(\mathbf{r})}{\partial \mathbf{r}}$; taking averages of these equations in the state $|\psi_H\rangle$, we obtain Ehrenfest's theorem, Eq. (1.87). These equations appear to be the classical equations of motion; the expectation value of the first equation is $\frac{\partial \langle \mathbf{r} \rangle}{\partial t} = \langle \mathbf{p} \rangle / m$, which is exactly the classical equation of motion. The expectation value of the second equation is $\frac{\partial \langle \mathbf{p} \rangle}{\partial t} = -\langle \frac{\partial V(\mathbf{r})}{\partial \mathbf{r}} \rangle$, whereas the classical equation of motion would be $\frac{\partial \langle \mathbf{p} \rangle}{\partial t} = -\frac{\partial V(\langle \mathbf{r} \rangle)}{\partial \langle \mathbf{r} \rangle}$. Expanding the RHS of the former equation about $\langle \mathbf{r} \rangle$ we find,

$$\frac{\partial \langle \mathbf{p} \rangle}{\partial t} = -\frac{\partial V(\langle \mathbf{r} \rangle)}{\partial \langle \mathbf{r} \rangle} - \frac{1}{2}\sum_{i,j}\frac{\partial^2 V(\langle \mathbf{r} \rangle)}{\partial \langle \mathbf{r}_i \rangle \partial \langle \mathbf{r}_j \rangle}\langle (x_i - \langle x_i \rangle)(x_j - \langle x_j \rangle)\rangle + \ldots . \tag{2.109}$$

In Sec. 7.2.2, we shall analyze the difference between this equation and the corresponding classical equation.

---

[5] An operator in the Schrödinger representation can be a function of time, so in general, we should write $\mathcal{A}_S(t)$.

2.7 Schrödinger and Heisenberg Representations

In either the Schrödinger or Heisenberg representation, the time derivative of the expectation value of an operator is given by

$$\frac{d}{dt}\langle \mathcal{A}\rangle_t = \left\langle \left(\frac{\partial \mathcal{A}}{\partial t} + \frac{1}{i\hbar}[\mathcal{A}, H]\right)\right\rangle_t. \quad (2.110)$$

Specifically, in the Schrödinger representation, for a pure state,

$$\frac{d}{dt}\langle \mathcal{A}\rangle_t = \int d\bar{x}\, \psi_S^*(\bar{x}, t)\left(\frac{\partial \mathcal{A}_S}{\partial t} + \frac{1}{i\hbar}[\mathcal{A}_S, H_S]\right)\psi_S(\bar{x}, t),$$

and for a mixed state, $\frac{d}{dt}\langle \mathcal{A}\rangle_t = \text{Tr}\left(\frac{\partial \mathcal{A}_S}{\partial t} + \frac{1}{i\hbar}[\mathcal{A}_S, H_S]\right)\rho_S(t)$. The Heisenberg representation expressions are also simple to derive.

It is of interest to determine whether the analog of the virial theorem of Sec. 16.6.2 (see the book web page) is true in quantum mechanics, i.e., whether

$$\frac{d}{dt}\langle \mathbf{r}\cdot\mathbf{p}\rangle = \left\langle \frac{\mathbf{p}^2}{2m}\right\rangle - \langle \mathbf{r}\cdot\nabla V\rangle. \quad (2.111)$$

By using (2.110), we see that $\frac{d}{dt}\langle \mathbf{r}\cdot\mathbf{p}\rangle = \langle[\mathbf{r}\cdot\mathbf{p}, H]\rangle$, for $H = \frac{\mathbf{p}^2}{2m} + V(\mathbf{r})$, so for this type of Hamiltonian, the quantum virial theorem is valid. Hence, for this type of Hamiltonian, whenever $\frac{d}{dt}\langle \mathbf{r}\cdot\mathbf{p}\rangle = 0$, $\left\langle \frac{\mathbf{p}^2}{2m}\right\rangle = \langle \mathbf{r}\cdot\nabla V\rangle$. For the Harmonic oscillator Hamiltonians of Sec. 1.3.15, this is the case; hence Problem 1.19 could be easily solved using the virial theorem.

Problem 2.18

(a) Show that $[\mathbf{r}\cdot\mathbf{p}, H] = 0$ yields $\frac{\mathbf{p}^2}{2m} - \mathbf{r}\cdot\nabla V = 0$.

(b) Show that $\frac{d}{dt}\langle \mathbf{r}\cdot\mathbf{p}\rangle = \langle V\rangle/2$ for a Coulomb potential.

Problem 2.19

Consider the evolution operator $\mathcal{U}(t, t_0)$ and transform it using a time-dependent unitary operator $S(t)$, $\mathcal{U}(t, t_0) \to \mathcal{U}'(t, t_0) = S(t)\mathcal{U}(t, t_0)S^\dagger(t)$. Using the definition of the evolution operator, $\mathcal{U}'(t + dt, t_0) = (1 - iH'(t)dt/\hbar)\mathcal{U}'(t, t_0)$. Using this equation, show that

$$H'(t) = i\hbar\frac{\partial \mathcal{U}'}{\partial t}(\mathcal{U}')^\dagger = SHS^\dagger + i\hbar\frac{\partial S}{\partial t}S^\dagger. \quad (2.112)$$

2.7.1 INTERACTION REPRESENTATION

An intermediate representation that takes out the time dependence of part of the Hamiltonian from the state and puts it into the operators is called the *interaction representation*. Breaking the full Hamiltonian of a system into two parts, $H(t) = H_0 + H_1(t)$, the state vector in the interaction representation is defined by

$$|\psi_I(t)\rangle \equiv \mathcal{U}_0^{-1}(t, 0)|\psi_S(t)\rangle, \quad (2.113)$$

where

$$\mathcal{U}_0(t, 0) = \exp(-iH_0 t/\hbar), \qquad \frac{\partial \mathcal{U}_0(t, 0)}{\partial t} = \frac{-iH_0}{\hbar}\mathcal{U}(t, 0). \quad (2.114)$$

The time derivative of the state in the interaction representation is

$$\frac{\partial |\psi_I(t)\rangle}{\partial t} = \frac{-iH_{1,I}(t)}{\hbar} |\psi_I(t)\rangle, \qquad (2.115)$$

where

$$H_{1,I}(t) = \mathcal{U}_0^{-1}(t,0) H_1(t) \mathcal{U}_0(t,0) = e^{iH_0 t/\hbar} H_1(t) e^{-iH_0 t/\hbar}. \qquad (2.116)$$

Problem 2.20

Do the algebra leading to Eq. (2.115).

Answer: $\frac{\partial |\psi_I(t)\rangle}{\partial t} = \frac{\partial \mathcal{U}_0^{-1}(t,0)}{\partial t} |\psi_S(t)\rangle + \mathcal{U}_0^{-1}(t,0) \frac{\partial |\psi_S(t)\rangle}{\partial t}$
$= \mathcal{U}_0^{-1}(t,0) \left(\frac{iH_0}{\hbar} + \frac{-i(H_0+H_1)}{\hbar} \right) |\psi_S(t)\rangle = \frac{-i}{\hbar} e^{iH_0 t/\hbar} H_1(t) e^{-iH_0 t/\hbar} |\psi_I(t)\rangle.$

In order to determine how to write an operator in this representation, let us consider a state given by $\mathcal{A}_S(t)|\psi_S(t)\rangle$ in the Schrödinger representation and transform it to the interaction representation:

$$\mathcal{U}_0^{-1} \mathcal{A}_S(t) |\psi_S(t)\rangle = \mathcal{U}_0^{-1} \mathcal{A}_S(t) \mathcal{U}_0 \mathcal{U}_0^{-1} |\psi_S(t)\rangle = \mathcal{A}_I(t) |\psi_I(t)\rangle,$$

hence,

$$\mathcal{A}_I(t) = \mathcal{U}_0^{-1} \mathcal{A}_S(t) \mathcal{U}_0 = e^{iH_0 t/\hbar} \mathcal{A}_S(t) e^{-iH_0 t/\hbar}. \qquad (2.117)$$

The time derivative of $\mathcal{A}_I(t)$ can be calculated as follows:

$$\frac{\partial \mathcal{A}_I(t)}{\partial t} = \mathcal{U}_0^{-1} \frac{\partial \mathcal{A}_S(t)}{\partial t} \mathcal{U} + \frac{i}{\hbar} \mathcal{U}_0^{-1} H_0 \mathcal{A}_S(t) \mathcal{U}_0 - \frac{i}{\hbar} \mathcal{U}_0^{-1} H_0 \mathcal{A}_S(t) \mathcal{U}_0.$$

After inserting unity in the form $\mathcal{U}_0 \mathcal{U}_0^{-1}$ into the last two terms of the RHS of this equation we obtain

$$\frac{\partial \mathcal{A}_I(t)}{\partial t} = \mathcal{U}_0^{-1} \frac{\partial \mathcal{A}_S(t)}{\partial t} \mathcal{U}_0 + \frac{i}{\hbar} [H_0, \mathcal{A}_I(t)]. \qquad (2.118)$$

In the interaction representation, the density matrix evolves according to the equation,

$$\frac{\partial}{\partial t} \rho_I(t) = \frac{-i}{\hbar} [H_{1,I}(t), \rho_I(t)]. \qquad (2.119)$$

Expectation values are obtained as follows: $\langle \mathcal{A} \rangle_t = \text{Tr} \, \mathcal{A}_I(t) \rho_I(t)$.

We delay the presentation of examples that use the interaction representation to future chapters.

2.7.2 HARMONIC OSCILLATOR RAISING–LOWERING OPERATORS

Using basis-set expansion, we can represent operators as matrices. For the harmonic oscillator problem, the eigenvalues and eigenvectors of the Hamiltonian can be obtained analytically, so this yields an alternative method for solving the quantum harmonic oscillator problem to that of Sec. (1.3.15). We shall develop this approach here.

In Sec. (1.3.15) we introduced harmonic oscillator units and obtained a dimensionless Hamiltonian written in terms of a dimensionless coordinate y and a dimensionless momentum p:

$$H = \frac{1}{2} \left(p^2 + y^2 \right). \qquad (2.120)$$

2.7 Schrödinger and Heisenberg Representations

Here the dimensionless momentum, $p = \frac{1}{i}\frac{d}{dy}$, and the dimensionless coordinate, y, satisfy the commutation relation $[y,p] = i$, as can easily be checked by applying $[y,p]$ to an arbitrary function $f(y)$.[6] We are now going to define dimensionless raising and lowering operators for the harmonic oscillator problem. The *lowering (destruction)* and *raising (creation)* operators are defined as follows:

$$a = \frac{1}{\sqrt{2}}(y + ip), \quad a^\dagger = \frac{1}{\sqrt{2}}(y - ip). \quad (2.121)$$

Note that the lowering and raising operators are non-Hermitian. Note also that we can invert the Eq. (2.121) to find $y = (2)^{-1/2}(a + a^\dagger)$ and $p = -i(2)^{-1/2}(a - a^\dagger)$. Let us motivate the definition of these operators. If p and y were not operators, but c-numbers, we could write the Hamiltonian (2.120) as $H = (p^2 + y^2)/2 = (y - ip)(y + ip)/2 = (a^\dagger)(a)$, thus "factoring" the Hamiltonian. But since y and p do not commute, we shall see that H will not quite equal $a^\dagger a$. Nevertheless, the introduction of these operators considerably simplifies the search for the eigenvalues and eigenvectors of the Hamiltonian. It is easy to see from the commutation relation $[y,p] = i$, by direct substitution of the expressions of y and p in terms of a and $a^\dagger$, that the lowering and raising operators a and $a^\dagger$ must satisfy the commutation relations,

$$[a, a^\dagger] = 1, \quad [a, a] = 0, \quad [a^\dagger, a^\dagger] = 0. \quad (2.122)$$

In terms of these operators, the dimensionless Hamiltonian can be written as $H = \frac{1}{2}(a^\dagger a + a a^\dagger)$, or, equivalently,

$$H = \left(a^\dagger a + \frac{1}{2}\right). \quad (2.123)$$

Hence, the eigenvalues and eigenstates of H are determined by those of the operator $N \equiv a^\dagger a$. To calculate them, we shall use the operator equations,

$$[N, a] = -a, \quad [N, a^\dagger] = a^\dagger, \quad (2.124)$$

which can be easily proved using the operator identity, $[AB, C] = A[B, C] + [A, C]B$, along with the commutation relations (2.122). Let us denote an eigenstate of N having eigenvalue ν by $|\nu\rangle$, i.e., $N|\nu\rangle = \nu|\nu\rangle$. Applying (2.124a) to $|\nu\rangle$, we obtain the equation, $Na|\nu\rangle = (\nu - 1)a|\nu\rangle$. Hence, $a|\nu\rangle$ is an eigenstate of the operator N with eigenvalue $\nu - 1$. Applying (2.124b) to $|\nu\rangle$, we obtain $Na^\dagger|\nu\rangle = (\nu + 1)a^\dagger|\nu\rangle$. Hence, $a^\dagger|\nu\rangle$ is an eigenstate with eigenvalue $\nu + 1$. This is why a and $a^\dagger$ are called lowering and raising operators.

Problem 2.21

(a) Rewrite Eq. (2.121) in terms of the dimensional momentum and position variables p_x and x, i.e., in terms of operators p_x and x for which the dimensional Hamiltonian is $H = \frac{1}{2}\left(\frac{p_x^2}{m} + m\omega^2 x^2\right)$.

Answer:

$$a = \sqrt{\frac{1}{2}}\sqrt{\frac{m\omega}{\hbar}}\left(x + i\frac{1}{m\omega}p_x\right), \quad (2.125a)$$

$$a^\dagger = \sqrt{\frac{1}{2}}\sqrt{\frac{m\omega}{\hbar}}\left(x - i\frac{1}{m\omega}p_x\right). \quad (2.125b)$$

(b) Using the commutation relation $[x, p_x] = i\hbar$, show that the dimensionless raising and lowering operators in Eqs (2.125a) and (2.125b) satisfy the commutation relations (2.122). To do so, write x and p_x in terms of a and $a^\dagger$ by inverting (2.125a) and (2.125b) and substitute these expressions into the commutator.

---

[6] Explicitly, $[y,p]f(y) = (yp - py)f(y) = -i[y\frac{df}{dy} - \frac{d}{dy}(yf(y))] = if(y)$.

To determine v, we consider the square of the length of the state $a|v\rangle$, $(\langle v|a^\dagger\rangle(a|v\rangle)) = \langle v|a^\dagger a|v\rangle = \langle v|N|v\rangle = v\langle v|v\rangle$. This quantity must be nonnegative, hence $v \geq 0$. By applying a repeatedly to $|v\rangle$, we obtain eigenvectors with eigenvalues $v, v-1, v-2, \ldots$. But this conflicts with the condition that these numbers must be nonnegative, unless the sequence terminates with the value $v = 0$. For $v = 0$, $a|v\rangle = 0$, i.e., $a|v\rangle$ is the zero vector, and further applications of the lowering operator again give the zero vector. Hence, the eigenvalues of the operator N are integers; the lowest integer is $v = 0$ and the eigenvector corresponding to this eigenvalue is denoted $|0\rangle$. The next lowest eigenvalue is $v = 1$ and the eigenvector corresponding to this eigenvalue is denoted $|1\rangle$. Thus, from now on, instead of using v to denote the eigenvalue and $|v\rangle$ the eigenvector of the operator N, we shall use n and $|n\rangle$, $N|n\rangle = n|n\rangle$.

We have already seen that $a^\dagger|n\rangle$ is proportional to $|n+1\rangle$, i.e., $a^\dagger|n\rangle = C_n|n+1\rangle$. To calculate C_n, note that $|C_n|^2 = (\langle n|a)(a^\dagger|n\rangle) = \langle n|aa^\dagger|n\rangle = (n+1)\langle n+1|n+1\rangle = (n+1)$, i.e., $|C_n| = \sqrt{n+1}$. Arbitrarily choosing the phase of C_n so that C_n is real, we obtain,

$$a^\dagger|n\rangle = \sqrt{n+1}|n+1\rangle. \tag{2.126}$$

Similarly, considering $a|n\rangle$, we find

$$a|n\rangle = \sqrt{n}|n-1\rangle, \tag{2.127}$$

except if $n = 0$ in (2.127), then the equation yields $a|0\rangle = 0$. Iterating (2.126) yields,

$$|n\rangle = \frac{(a^\dagger)^n}{\sqrt{n!}}|0\rangle. \tag{2.128}$$

Hence, the eigenvalue/eigenvector equation for the dimensionless Hamiltonian is, $H|n\rangle = (n+1/2)|n\rangle$, or, more explicitly,

$$\left(a^\dagger a + \frac{1}{2}\right)|n\rangle = \left(n + \frac{1}{2}\right)|n\rangle. \tag{2.129}$$

The eigenstates $|n\rangle$ are called *number states*, or sometimes *Fock states* after Vladimir A. Fock.

Taking the inner product of Eqs (2.126), (2.127), and (2.129) with $|n'\rangle$, we find the expressions for the matrix elements of $a^\dagger$, a, and $a^\dagger a$ in the number representation are as follows:

$$\langle n'|a^\dagger|n\rangle = \sqrt{n+1}\,\delta_{n',n+1}, \tag{2.130a}$$

$$\langle n'|a|n\rangle = \sqrt{n}\,\delta_{n',n-1}. \tag{2.130b}$$

$$\langle n'|a^\dagger a|n\rangle = n\,\delta_{n',n}. \tag{2.130c}$$

The operator $\hat{N} \equiv \hat{a}^\dagger \hat{a}$ is called the *number operator* and is often denoted by the symbol $\hat{n}$. The matrices representing $\hat{a}$ and $\hat{n}$ in the number representation are given explicitly by

$$\hat{a} = \begin{pmatrix} 0 & \sqrt{1} & 0 & 0 & \cdots \\ 0 & 0 & \sqrt{2} & 0 & \cdots \\ 0 & 0 & 0 & \sqrt{3} & \cdots \\ 0 & 0 & 0 & 0 & \cdots \\ \vdots & \vdots & \vdots & \vdots & \ddots \end{pmatrix}, \quad \hat{n} = \begin{pmatrix} 0 & 0 & 0 & 0 & \cdots \\ 0 & 1 & 0 & 0 & \cdots \\ 0 & 0 & 2 & 0 & \cdots \\ 0 & 0 & 0 & 3 & \cdots \\ \vdots & \vdots & \vdots & \vdots & \ddots \end{pmatrix}. \tag{2.131}$$

The matrix representing $\hat{a}^\dagger$ in the number representation is simply the transpose of $\hat{a}$ in (2.131) since $\hat{a}$ is real. The Hamiltonian matrix representing the operator $\hat{H} = (\hat{n} + 1/2)$ is clearly diagonal (the half means half times the unit matrix $\hat{1}$), and the eigenvalues are simply the diagonal elements. Putting back the dimensions, $\hat{H} = \hbar\omega(\hat{n} + 1/2)$.

2.7 Schrödinger and Heisenberg Representations

Problem 2.22

(a) Determine the matrices for x and p_x in the number representation.
(b) Determine the matrices for x^2, p_x^2, and the harmonic oscillator Hamiltonian, $H = \frac{1}{2}\left(\frac{p_x^2}{m} + m\omega^2 x^2\right)$, in the number representation.
(c) Calculate $\langle n|x|n\rangle$, $\langle n|p|n\rangle$, $(\Delta x)^2$, and $(\Delta p)^2$.

Answers: (a) $x = \sqrt{\frac{\hbar}{2m\omega}}(a + a^\dagger)$ and $p = -im\omega\sqrt{\frac{\hbar}{2m\omega}}(a - a^\dagger)$, so

$$x = \sqrt{\frac{\hbar}{2m\omega}}\begin{pmatrix} 0 & \sqrt{1} & 0 & 0 & \cdots \\ \sqrt{1} & 0 & \sqrt{2} & 0 & \cdots \\ 0 & \sqrt{2} & 0 & \sqrt{3} & \cdots \\ 0 & 0 & \sqrt{3} & 0 & \cdots \\ \vdots & \vdots & \vdots & \vdots & \ddots \end{pmatrix}, \qquad (2.132)$$

$$p = -im\omega\sqrt{\frac{\hbar}{2m\omega}}\begin{pmatrix} 0 & \sqrt{1} & 0 & 0 & \cdots \\ -\sqrt{1} & 0 & \sqrt{2} & 0 & \cdots \\ 0 & -\sqrt{2} & 0 & \sqrt{3} & \cdots \\ 0 & 0 & -\sqrt{3} & 0 & \cdots \\ \vdots & \vdots & \vdots & \vdots & \ddots \end{pmatrix}. \qquad (2.133)$$

(c) $\langle n|x|n\rangle = 0$, $\langle n|p|n\rangle = 0$, $(\Delta x)^2 = \frac{\hbar}{m\omega}(n + 1/2)$, and $(\Delta p)^2 = m\hbar\omega(n + 1/2)$.

The ground-state wave function in coordinate space, $\psi_0(x)$, can be obtained by representing the lowering operator in the expression, $\langle x|a|0\rangle = a\psi_0(x) = 0$ (which you will better understand after doing Problem 2.23), using Eq. (2.125a), which yields the differential equation,

$$\left[\frac{\partial}{\partial x} + \frac{m\omega}{\hbar}x\right]\psi_0(x) = 0. \qquad (2.134)$$

The solution to (2.134) is

$$\psi_0(x) = A_0 \exp\left(-\frac{m\omega x^2}{2\hbar}\right), \qquad (2.135)$$

and normalizing this wave function yields $A_0 = \left(\frac{m\omega}{\pi\hbar}\right)^{1/4}$. The excited-state wave functions in coordinate space, $\psi_n(x)$, can be obtained as follows:

$$\psi_n(x) = \langle x|n\rangle = \langle x|\frac{(a^\dagger)^n}{\sqrt{n!}}|0\rangle = \frac{\left[\sqrt{\frac{m\omega}{2\hbar}}\left(x - \frac{\hbar}{m\omega}\frac{\partial}{\partial x}\right)\right]^n}{\sqrt{n!}}\langle x|0\rangle$$

$$= \frac{\left(\frac{m\omega}{\pi\hbar}\right)^{1/4}}{2^{n/2}\sqrt{n!}}\left[\sqrt{\frac{m\omega}{\hbar}}\left(x - \frac{\hbar}{m\omega}\frac{\partial}{\partial x}\right)\right]^n \exp\left(-\frac{m\omega x^2}{2\hbar}\right). \qquad (2.136)$$

In the Schrödinger representation, states evolve in time; application of the evolution operator to the harmonic oscillator Fock state $|n\rangle$ gives

$$\mathcal{U}(t,0)|n\rangle = e^{-i(n+1/2)\omega t}|n\rangle. \qquad (2.137)$$

So, as we have just seen, in position representation, $\psi_n(x) = \langle x|n\rangle$, and the time-dependent wave function is given by $\psi_n(x,t) = \langle x|\mathcal{U}(t,0)|n\rangle$.

Problem 2.23

Prove that $\langle x|a|0\rangle = a\psi_0(x)$ by carrying out the following procedure.

(a) Insert a complete set of states, $\int dy\, |y\rangle\langle y|$, between a and $|0\rangle$ on the LHS of the equation.
(b) Use the definition of a in Eq. (2.125a) in your result from (a).
(c) Insert a complete set of momentum eigenstates, $\int dp\, |p\rangle\langle p|$, into the matrix element $\langle x|p|y\rangle$ to the right of the momentum operator in order to evaluate this matrix element.

Let us now write the equations of motion for the lowering and raising operators in the Heisenberg representation. These equations of motion will be solved analytically. In the Heisenberg representation, the operators a and $a^\dagger$ become $a_H(t) = e^{iHt/\hbar} a e^{-iHt/\hbar}$ and $a_H^\dagger(t) = e^{iHt/\hbar} a^\dagger e^{-iHt/\hbar}$.

$$\frac{\partial a_H(t)}{\partial t} = \frac{i}{\hbar}[H_H, a_H(t)] = i\omega[a_H^\dagger(t) a_H(t), a_H(t)] = -i\omega a_H(t), \tag{2.138a}$$

$$\frac{\partial a_H^\dagger(t)}{\partial t} = \frac{i}{\hbar}[H_H, a_H^\dagger(t)] = i\omega[a_H^\dagger(t) a_H(t), a_H^\dagger(t)] = i\omega a_H^\dagger(t). \tag{2.138b}$$

The solutions to these equations are

$$a_H(t) = a e^{-i\omega t}, \tag{2.139a}$$

$$a_H^\dagger(t) = a^\dagger e^{i\omega t}. \tag{2.139b}$$

The Hamiltonian in the Heisenberg representation is not time dependent because of the cancellation of $e^{-i\omega t}$ and $e^{i\omega t}$ in the product of the raising and lowering operators, $H_H = \hbar\omega[a_H^\dagger(t)a_H(t) + 1/2] = \hbar\omega(a^\dagger a + 1/2)$.

In summary, we defined number states (i.e., Fock states), $|n\rangle$, in (2.128). Application of the raising and lowering operators to these states is detailed in (2.126) and (2.127). The Fock states can be used as a basis since they are orthonormal and span the set of all states. They are one of the most useful basis sets for dealing with quantum many-body processes, as detailed in Chapter 14.

This is the second time we considered the harmonic oscillator problem. The first treatment in Sec. 1.3.15 was carried out using the Schrödinger wave function method; the treatment here used the Heisenberg matrix mechanics method wherein operators are represented by matrices and state vectors in Hilbert space are represented by row vectors. The first method involves solving the Schrödinger equation as a differential equation, and the second method involves diagonalizing the Hamiltonian matrix and thereby obtaining its eigenvalues and eigenvectors. In the particular case of the harmonic oscillator, we were able to carry out the diagonalization to obtain the eigenvalues and eigenvectors analytically. In general, once we obtain the Hamiltonian matrix, numerical solution to obtain the eigenvalues and eigenvectors is the only option, and is carried out upon truncating the Hamiltonian matrix to a finite matrix.

2.7.3 COHERENT STATES AND SQUEEZED STATES

The concept of a coherent state was first introduced into quantum optics by Roy J. Glauber in 1963. One could argue that the development of coherent states marks the birth of the field of *quantum optics*. Coherent states are a convenient tool for describing a variety of phenomena, including the radiation emitted by a laser, improved measurement characteristics of certain devices via *squeezing* so as to "beat" the uncertainty principle, and use in correlation spectroscopies, just to mention a few applications.

2.7 Schrödinger and Heisenberg Representations

A coherent state $|\alpha\rangle$ is defined as an eigenstate of the non-Hermitian annihilation (lowering) operator a,

$$\boxed{a|\alpha\rangle = \alpha|\alpha\rangle.} \qquad (2.140)$$

Upon expanding the coherent state $|\alpha\rangle$ in a complete set of Fock (i.e., number) states and using Eqs (2.128) and (2.137), we find:

$$|\alpha\rangle = \left(\sum_{n=0}^{\infty} |n\rangle\langle n|\right)|\alpha\rangle = \sum_{n=0}^{\infty} |n\rangle\left\langle 0\left|\frac{a^n}{\sqrt{n!}}\right|\alpha\right\rangle. \qquad (2.141)$$

Applying the operator $\frac{a^n}{\sqrt{n!}}$ to $|\alpha\rangle$ on the RHS of (2.141), we find

$$|\alpha\rangle = \langle 0|\alpha\rangle \sum_{n=0}^{\infty} \frac{\alpha^n}{\sqrt{n!}} |n\rangle. \qquad (2.142)$$

The matrix element $\langle 0|\alpha\rangle$ appearing on the RHS of Eq. (2.142) can be determined by imposing the normalization condition

$$\langle \alpha|\alpha\rangle = \left(\sum_{m=0}^{\infty}\langle 0|\frac{(a^\dagger)^m}{\sqrt{m!}}|\alpha\rangle\langle m|\right)\left(\sum_{n=0}^{\infty}|n\rangle\langle 0|\frac{a^n}{\sqrt{n!}}|\alpha\rangle\right) = 1, \qquad (2.143)$$

to find $|\langle 0|\alpha\rangle|^2 = \left(\sum_{n=0}^{\infty}|\alpha|^{2n}/n!\right)^{-1} = \exp(-|\alpha|^2)$. Taking $\langle 0|\alpha\rangle$ to be real, we can write Eq. (2.142) as

$$\boxed{|\alpha\rangle = e^{-|\alpha|^2/2} \sum_{n=0}^{\infty} \frac{\alpha^n}{\sqrt{n!}} |n\rangle.} \qquad (2.144)$$

It is easy to see from (2.144) that the probability of finding the number eigenstate $|n\rangle$ in the coherent state $|\alpha\rangle$ is given by

$$P_n(\alpha) = |\langle n|\alpha\rangle|^2 = e^{-|\alpha|^2} \frac{|\alpha|^{2n}}{n!}. \qquad (2.145)$$

Hence, the probability of finding n 'photons' in the coherent state is distributed in a Poisson probability distribution function with mean $|\alpha|^2$.

The time evolution of a coherent state $|\alpha\rangle$ in the Schrödinger representation is obtained by applying the evolution operator to the coherent state $|\alpha\rangle$ to obtain

$$\mathcal{U}(t,0)|\alpha\rangle = e^{-|\alpha|^2/2 - i\omega t/2} \sum_{n=0}^{\infty} \frac{(\alpha e^{-i\omega t})^n}{\sqrt{n!}} |n\rangle. \qquad (2.146)$$

Problem 2.24

Verify the following expectation values for the coherent state $|\alpha\rangle$.

(a) $\langle \hat{n}\rangle_\alpha \equiv \langle \alpha|a^\dagger a|\alpha\rangle = |\alpha|^2 = \sum_{n=0}^{\infty} nP_n$.

Hint: One way to prove the last part of this relation is by inserting the identity $\left(\sum_{n=0}^{\infty} |n\rangle\langle n|\right) = 1$ into $\langle \alpha|a^\dagger a|\alpha\rangle$.

(b) $\langle \hat{n}^2\rangle_\alpha \equiv \langle \alpha|\hat{n}^2|\alpha\rangle = \sum_{n=0}^{\infty} n^2 P_n = |\alpha|^2(|\alpha|^2 + 1)$.

(c) $\langle (\Delta \hat{n})^2\rangle_\alpha \equiv \langle \alpha|\hat{n}^2|\alpha\rangle - \langle \alpha|\hat{n}|\alpha\rangle^2 = |\alpha|^2$, hence $\langle (\Delta \hat{n})^2\rangle_\alpha = \langle \hat{n}\rangle_\alpha$.

(d) Prove that $\hat{n}|\alpha\rangle = e^{-|\alpha|^2/2} \sum_{n=0}^{\infty} n\frac{\alpha^n}{\sqrt{n!}} |n\rangle$.

(e) Use part (d) to prove Eq. (2.146). **Hint:** Use Eq. (2.137).

Problem 2.25

Show that $\langle \alpha | \alpha' \rangle = \exp(-|\alpha|^2/2 + \alpha'\alpha^* - |\alpha'|^2/2)$ and that $|\langle \alpha | \alpha' \rangle|^2 = \exp(-|\alpha - \alpha'|^2)$.

Problem 2.26

(a) Write the equation $\langle x|a|\alpha\rangle = \alpha \langle x|\alpha\rangle$ as $\left[\sqrt{\frac{1}{2}\sqrt{\frac{m\omega}{\hbar}}}(x + \frac{\hbar}{m\omega}\frac{\partial}{\partial x}) - \alpha\right]\psi_\alpha(x) = 0$, and solve this equation for the coherent state wave function, $\psi_\alpha(x) = \left(\frac{m\omega}{\pi\hbar}\right)^{1/4} e^{-\frac{m\omega}{2\hbar}(x - \sqrt{\frac{2\hbar}{m\omega}}\operatorname{Re}\alpha)^2 + i\sqrt{\frac{2\hbar}{m\omega}}\operatorname{Im}\alpha\, x}$.

(b) Show that for ψ_α, $\langle x \rangle = \sqrt{\frac{2\hbar}{m\omega}}\operatorname{Re}\alpha$ and $\langle p \rangle = \sqrt{2m\hbar\omega}\operatorname{Im}\alpha$.

The coherent states are complete (they can be used as a basis),

$$\frac{1}{\pi}\int d^2\alpha\, |\alpha\rangle\langle\alpha| = 1. \tag{2.147}$$

Here, the integral is over both real and imaginary parts of α, i.e., $d^2\alpha \equiv d\alpha_R d\alpha_I$ so $\int d^2\alpha \equiv \int_{-\infty}^{\infty} d\alpha_R \int_{-\infty}^{\infty} d\alpha_I$. Note that coherent states are not as simple to use as a basis as Fock states; they are overcomplete, as is clear from the nonorthogonality you derived in Problem 2.25, so

$$|\alpha\rangle = \frac{1}{\pi}\int d^2\alpha' |\alpha'\rangle\langle\alpha'|\alpha\rangle = \frac{1}{\pi}\int d\alpha' |\alpha'\rangle \exp(-|\alpha|^2/2 + \alpha'^*\alpha - |\alpha'|^2/2). \tag{2.148}$$

Thus, any pure state $|\psi\rangle$ can be expanded in terms of coherent states, $|\psi\rangle = \frac{1}{\pi}\int d^2\alpha |\alpha\rangle\langle\alpha|\psi\rangle$, and any density matrix can be expanded as

$$\rho = \frac{1}{\pi}\int d^2\alpha\, P(\alpha, \alpha^*)|\alpha\rangle\langle\alpha|, \tag{2.149}$$

where $P(\alpha, \alpha^*)$ is real and normalized, $\int d^2\alpha\, P(\alpha, \alpha^*) = 1$.

Equation (2.144) can be written entirely in terms of raising operators that operate on the vacuum state by noting that $|n\rangle = a^{\dagger n}|0\rangle/\sqrt{n!}$:

$$|\alpha\rangle = e^{-|\alpha|^2/2}\sum_{n=0}^{\infty}\frac{(\alpha a^\dagger)^n}{n!}|0\rangle = e^{-|\alpha|^2/2}e^{\alpha a^\dagger}|0\rangle. \tag{2.150}$$

It is often convenient to rewrite (2.150) in terms of the displacement operator

$$D(\alpha) \equiv \exp(\alpha a^\dagger - \alpha^* a), \tag{2.151}$$

by noting that, $a|0\rangle = 0$, hence,

$$e^{-|\alpha|^2/2}e^{\alpha a^\dagger}|0\rangle = \exp(-|\alpha|^2/2)\exp(\alpha a^\dagger)\exp(-\alpha^* a)|0\rangle, \tag{2.152}$$

and using the following identity for operators:

$$e^{(A+B)} = e^A e^B e^{-[A,B]/2} = e^B e^A e^{[A,B]/2} \quad \text{if } [A,[A,B]] = [B,[A,B]] = 0. \tag{2.153}$$

This latter result is known as the *Baker–Hausdorff theorem* (or the *Campbell–Baker–Hausdorff theorem*). Using (2.153) and (2.152), (2.150) becomes

$$|\alpha\rangle = D(\alpha)|0\rangle. \tag{2.154}$$

2.7 Schrödinger and Heisenberg Representations

The operator $D(\alpha)$ is called a *displacement operator* because it displaces the vacuum state $|0\rangle$ to the coherent state $|\alpha\rangle$ (see Fig. 2.4).

Problem 2.27

(a) For $A = \alpha a^\dagger$ and $B = -\alpha^* a$, show that $[A, B] = |\alpha|^2$.
(b) Prove $e^{\alpha a^\dagger - \alpha^* a} = e^{-|\alpha|^2/2} e^{\alpha a^\dagger} e^{-\alpha^* a}$ using the Baker–Hausdorff theorem.

Problem 2.28

Let A and B be two operators that do not necessarily commute. Expand the function $F(x) = \exp(xA) B \exp(-xA)$ in a series in x about $x = 0$.

(a) Show that $F'(0) = [A, B]$, $F''(0) = [A, [A, B]]$, and $F^{(n)}(0) = [A, \ldots, [A, B]]$ with n commutators appearing in this expression.
(b) Substitute into the Taylor series to show that $F(x) = B + x[A, B] + \frac{x^2}{2!}[A, [A, B]] + \ldots$. This relation is called the operator expansion theorem.
(c) Show that $[\exp(xA) B \exp(-xA)]^n = \exp(xA) B^n \exp(-xA)$.
(d) Show that $\exp(xA) F(B) \exp(-xA) = F(\exp(xA) B \exp(-xA))$.
(e) Generalize (d) to show that for any operator G that has an inverse, $G F(B) G^{-1} = F(G B G^{-1})$.

Problem 2.29

(a) Using the operator expansion theorem you derived in the previous problem, show that
$\exp(xN) a \exp(-xN) = a \exp(-x)$ and $\exp(xN) a^\dagger \exp(-xN) = a^\dagger \exp(x)$.
(b) Show that $\exp(\alpha a^\dagger) F(a, a^\dagger) \exp(-\alpha a^\dagger) = F(a - \alpha, a^\dagger)$.
(c) Show that $\exp(-\alpha^* a) F(a, a^\dagger) \exp(\alpha^* a) = F(a, a^\dagger - \alpha^*)$.
(d) Show that $e^{\alpha a^\dagger - \alpha^* a} F(a, a^\dagger) e^{-\alpha a^\dagger + \alpha * a} = F(a - \alpha, a^\dagger - \alpha^*)$.

Squeezed States and the Uncertainty Principle

Let us consider two noncommuting Hermitian conjugate operators, X_1 and X_2 (e.g., x and p). The commutator of these operators can be written as the product of i and another Hermitian operator that we call X_3,

$$[X_1, X_2] = iX_3. \tag{2.155}$$

The Heisenberg uncertainty relation shows that the product of the uncertainties of two operators X_1 and X_2 satisfies the inequality,

$$\Delta X_1 \, \Delta X_2 \geq \frac{1}{2}|\langle X_3 \rangle|, \tag{2.156}$$

where $\Delta X \equiv \sqrt{\langle X^2 \rangle - \langle X \rangle^2}$, and the expectation values are calculated with a given state $|\psi\rangle$. A state $|\psi\rangle$ is called a *minimum uncertainty state* if $\Delta X_1 \, \Delta X_2 = \frac{1}{2}|\langle X_3 \rangle|$, and it is called *squeezed* if the variance of one of its observable operators, say ΔX_1, satisfies

$$(\Delta X_1)^2 < \frac{1}{2}|\langle X_3 \rangle|. \tag{2.157}$$

Moreover, if, in addition,

$$\Delta X_1 \, \Delta X_2 = \frac{1}{2}|\langle X_3 \rangle|, \tag{2.158}$$

the state $|\psi\rangle$ is called an *ideal squeezed state*. The quantum fluctuations of a squeezed state in one observable, say X_1, are reduced below $|\langle X_3\rangle|/2$ at the expense of the fluctuations in the other observable.

Consider the Hermitian operators $X_1 \equiv \frac{1}{2}(a+a^\dagger)$ and $X_2 \equiv \frac{1}{2i}(a-a^\dagger)$. The commutator of these operators is given by $[X_1, X_2] = i\frac{1}{2}$, i.e., $X_3 = \frac{1}{2}$ is a constant. The Heisenberg uncertainty principle takes the form

$$\Delta X_1 \, \Delta X_2 \geq \frac{1}{2}. \tag{2.159}$$

Now let us consider the expectation values of X_1 and X_2, and the variances ΔX_1^2 and ΔX_2^2 for the coherent state $|\alpha\rangle = D(\alpha)|0\rangle$ defined in (2.154):

$$\langle \alpha | a | \alpha \rangle = (\langle X_1 \rangle_\alpha + i\langle X_2 \rangle_\alpha) = \langle 0 | D^\dagger(\alpha) a D(\alpha) | 0 \rangle = \alpha, \tag{2.160}$$

i.e., $\langle X_1 \rangle_\alpha = \operatorname{Re}\alpha$ and $\langle X_2 \rangle_\alpha = \operatorname{Im}\alpha$. Furthermore,

$$\langle \alpha | \Delta X_1^2 | \alpha \rangle = \langle 0 | \Delta X_1^2 | 0 \rangle = \frac{1}{4}, \tag{2.161}$$

$$\langle \alpha | \Delta X_2^2 | \alpha \rangle = \langle 0 | \Delta X_2^2 | 0 \rangle = \frac{1}{4}. \tag{2.162}$$

Hence, the coherent states $|\alpha\rangle$ (including $|0\rangle$) are indeed minimum uncertainty states. Figure 2.4(a) shows the coherent vacuum state $|0\rangle$ and the displaced coherent state $|\alpha\rangle$; these wave packets are plotted in the X_1-X_2 plane.

Squeezed states offer the possibility of beating the quantum uncertainty limit in measurements. Such states can be generated using a nonlinear phase-dependent interaction, as first observed by R. E. Slusher in an atomic sodium gas experiments in 1985. Note that neither the Fock state, $|n\rangle$, nor the coherent state, $|\alpha\rangle$, are squeezed states. For the coherent state $\Delta X_1 = \Delta X_2 = 1/2$, whereas for the Fock state $\Delta X_1 = \Delta X_2 = (2n+1)/2$. A squeezed state can be obtained from a coherent state by applying the *squeezing operator* to it:

$$S(\xi) \equiv e^{\xi^* a^2/2 - \xi a^{\dagger 2}/2}. \tag{2.163}$$

This squeezing operator $S(\xi)$ can be applied to either the vacuum state $|0\rangle$ (which is a trivial coherent state) or the coherent state $|\alpha\rangle$, i.e., $|\xi\rangle \equiv S(\xi)|0\rangle$ and $|\alpha, \xi\rangle \equiv S(\xi)|\alpha\rangle$. These "quadrature" squeezed states are shown in Fig. 2.4(b). Quantum states can be number squeezed or phase squeezed, as well as squeezed in q or p (position or momentum). The latter are shown in Fig. 2.4(b) and the former in Fig. 2.4(c). Number squeezed states can be described as simply a rotation of $D(\alpha)S(|\xi|)|0\rangle$ by angle $\theta = \arctan(\operatorname{Im}(\alpha)/\operatorname{Re}(\alpha))$ and phase squeezed states are obtained by an additional rotation of the number squeezed states by $\pi/2$. More information about squeezed states can be found in Ref. [18], Chapter 9, and Ref. [30]. These references discuss how squeezing of light beams can be implemented using nonlinear optics.

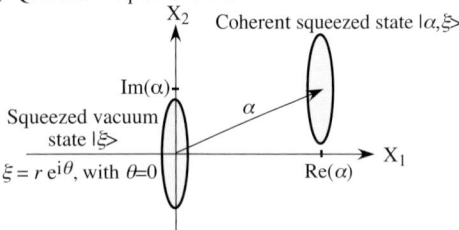

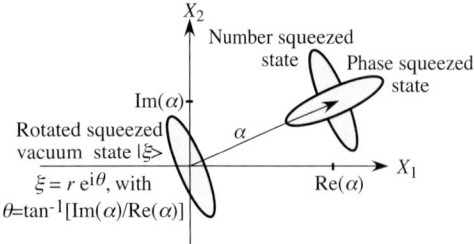

FIG 2.4 Coherent states and squeezed states in phase space. (a) Coherent state $|0\rangle$ and displaced coherent state $|\alpha\rangle$, (b) quadrature squeezed (i.e., position and momentum squeezed) states, and (c) number and phase squeezed states. *Source: Band, Light and Matter, Fig. 9.9, p. 548*

2.8 The Correspondence Principle and the Classical Limit

Other types of coherent states and squeezed states, not associated with those of the boson creation and annihilation operators of the linear harmonic oscillator, can be constructed. The best known example of such states is those associated with the angular momentum operator **J** (see Sec. 6.1.9). Other examples of squeezed states are described in Refs. [31–33].

> **Problem 2.30**
> Using the operator $X_1 \equiv x = \sqrt{\frac{\hbar}{2m\omega}}(a + a^\dagger)$, and $X_2 = p = -im\omega\sqrt{\frac{\hbar}{2m\omega}}(a - a^\dagger)$, so that $X_3 = -i[X_1, X_2] = -\hbar$, show that $\langle \alpha|a|\alpha \rangle = \langle 0|D^\dagger(\alpha)aD(\alpha)|0 \rangle = \sqrt{\frac{\hbar}{2m\omega}}(\langle X_1 \rangle_\alpha + i\sqrt{\frac{1}{m\omega}}\langle X_2 \rangle_\alpha) = \alpha$ and $\langle \alpha|\Delta X_1^2|\alpha \rangle = \langle 0|\Delta X_1^2|0 \rangle = \frac{\hbar^2}{2}$.

2.8 THE CORRESPONDENCE PRINCIPLE AND THE CLASSICAL LIMIT

Although the rules of quantum mechanics are highly successful in describing microscopic objects such as atoms, molecules, and even condensed phase systems such as crystals and metals, experiments show that macroscopic systems such as bicycles, spinning tops, soccer balls, etc., can be accurately described by classical mechanics. Nevertheless, it seems reasonable that the ultimate laws of physics must be independent of the size of the physical objects being described. Bohr's *correspondence principle* was motivated by this belief, and it explains how highly excited objects behave classically even though the underlying laws of nature are quantum mechanical. The correspondence principle, first invoked by Niels Bohr in 1923, states that the behavior of quantum mechanical systems reduce to classical physics in the limit of large quantum numbers. We should note that Bohr's formulation of the correspondence principle is not correct in general, and counterexamples have been found (e.g., highly excited rotational-vibrational states of diatomic molecules close to the dissociation threshold).

A demonstration of how large quantum numbers can give rise to classical behavior is provided by the one-dimensional quantum harmonic oscillator. The quantum mechanical probability distribution function for finding a particle at position x if it is in the nth eigenstate of the harmonic oscillator is given by $|\psi_n(x)|^2$. The classical probability distribution function $P_{cl}(x)$ for finding a particle at position x can be found as follows. The classical trajectory of the particle is given by $x(t) = a \sin(\omega t + \theta)$ where the amplitude a is related to the particle energy E by $E = m\omega^2 a^2/2$. Now, if the phase angle θ is randomly distributed, i.e., $P(\theta) = (2\pi)^{-1}$, we can calculate $P_{cl}(x)$ by noting that $P_{cl}(x)\,dx = P_{cl}(\theta)\,d\theta = (2\pi)^{-1}\,d\theta$. Since $dx = a\cos(\omega t + \theta)d\theta = a[1 - \sin^2(\omega t + \theta)]^{1/2}d\theta = [a^2 - x^2]^{1/2}d\theta$, we find

$$P_{cl}(x) = \frac{1}{\pi\sqrt{a^2 - x^2}}. \quad (2.164)$$

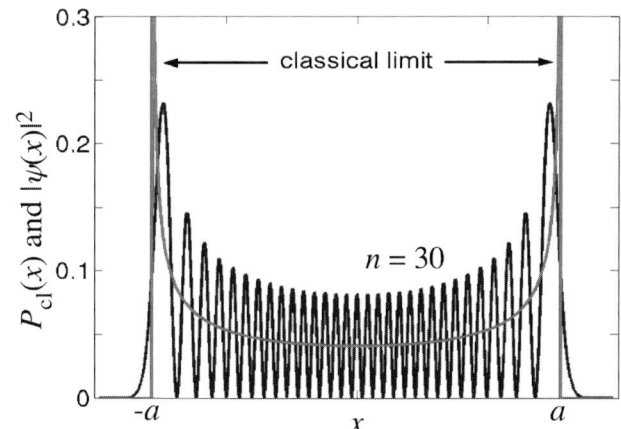

FIG 2.5 The classical probability, $P_{cl}(x) = [\pi(a^2 - x^2)]^{-1/2}$, of finding the particle at position x in a harmonic potential, and the probability of finding a particle at position x in the nth eigenstate, $P_n(x) = |\psi_n(x)|^2$, for $n = 30$. Here, a is the classical turning point, i.e., $E = m\omega^2 a^2/2 = \hbar\omega(n + 1/2)$.

Figure 2.5 plots $P_{cl}(x)$ and $|\psi_n(x)|^2$ for $n = 30$. Clearly, the two curves are similar. Moreover, the higher n, the better the correspondence. Furthermore, if we take a superposition of several wave functions with quantum numbers n around a central large quantum number (a wave packet), the correspondence becomes even better because the fast oscillations seen in Fig. 2.5 average out.

Bohr argued that classical physics does not emerge from quantum physics in the same way that non relativistic classical mechanics emerge as an approximation of relativistic mechanics at small velocities. Rather, classical physics exists independently of quantum theory and cannot be derived from it. Moreover, the Bohr correspondence principle falls short of describing how systems that are not very highly excited are well described by classical physics. So, how does classical physics arise out of quantum physics? We will have a lot more to say about this subject in Chapter 17, linked to the book web page.

2.9 SYMMETRY AND CONSERVATION LAWS IN QUANTUM MECHANICS

Symmetry plays an important role in quantum mechanics. It simplifies the solution of quantum problems, such as the hydrogen atom, it allows for the recognition of system properties, such as conserved quantities and invariances, and it allows for the classification of molecules, crystals, and elementary particles. It even allows for the unification of the fundamental forces, e.g., electromagnetism, electroweak, and the strong force.

In quantum mechanics, a symmetry operator $\mathcal{O}$ operates on states $\{|\psi\rangle\}$ in Hilbert space. Symmetries can be discrete, like the spatial inversion symmetry, or continuous, like the rotations around an axis. An operator $\mathcal{O}$ that commutes with the Hamiltonian, $H\mathcal{O} = H\mathcal{O}$, is conserved since the Heisenberg representation (2.108) [note that we are considering an operator that is not explicitly time dependent] gives

$$\frac{\partial \mathcal{O}}{\partial t} = \frac{i}{\hbar}[H, \mathcal{O}] = 0. \tag{2.165}$$

Hence, the Hamiltonian and the conserved operator can be simultaneously diagonalized, and then the eigenfunctions of the Hamiltonian are also eigenfunctions of $\mathcal{O}$. More explicitly, using $H\mathcal{O} = H\mathcal{O}$, we find that if ψ_n is an eigenfunction of H belonging to eigenvalue E_n, then $(\mathcal{O}\psi_n)$ is also an eigenfunction of H belonging to the same eigenvalue (see Sec. 2.2.1 on compatible operators):

$$H(\mathcal{O}\psi_n) = \mathcal{O}(H\psi_n) = E_n(\mathcal{O}\psi_n). \tag{2.166}$$

We have already pointed out some consequences of invariance in Sec. 1.3.6. Invariance of the Hamiltonian under displacements necessitates that the momentum is a constant of motion. Invariance under time displacements implies that the Hamiltonian is conserved, i.e., the energy of the system is conserved. Invariance under rotations requires that angular momentum is a constant of motion (this topic will be considered at length in Chapter 3). An important theorem first formulated by Emmy Noether shows that continuous symmetries are related to conservation laws.

In Sec. 2.9.1, we consider symmetry under exchange of particles; in Sec. 2.9.2, we discuss invariance under inversion, which leads to conservation of parity; and in Sec. 2.9.3, we consider the symmetry known as time-reversal invariance. In Sec. 3.6, we shall treat the consequences of symmetry on matrix elements of dynamical operators.

2.9.1 EXCHANGE SYMMETRY

In Sec. 8.2, we shall extensively treat invariance under exchange of identical particles, which requires symmetrization of the wave function of identical bosonic particles (particles with integer spin), and the antisymmetrization of the wave function of identical fermionic particles (particles with half-integer spin). The latter leads to the *Pauli exclusion principle*. The Pauli exclusion principle, formulated by Wolfgang Pauli in 1925, states that a many-electron wave function must be antisymmetric with respect to the interchange of any two electrons, and this ensures that only one electron can occupy a given quantum state (and similarly for any many-fermion wave function). The treatment of fermion wave functions in their antisymmetric form in terms of what is now called Slater determinants was developed by Paul A. M. Dirac in a famous 1926 paper, where he writes:

> *An antisymmetrical eigenfunction vanishes identically when two of the electrons are in the same orbit. This means that in the solution of the problem with antisymmetrical eigenfunctions there can be no stationary states with two or more electrons*

2.9 Symmetry and Conservation Laws in Quantum Mechanics

in the same orbit, which is just Pauli's exclusion principle. The solution with symmetrical eigenfunctions, on the other hand, allows any number of [particles] to be in the same orbit, so that this solution cannot be the correct one for the problem of electrons in an atom.

The same symmetry occurs for all fermionic particles,

$$\Psi(x_1, x_2, \ldots, x_i, \ldots, x_j, \ldots, x_N) = -\Psi(x_1, x_2, \ldots, x_j, \ldots, x_i, \ldots, x_N), \quad (2.167)$$

where $x_i = (\mathbf{r}_i \sigma_i)$ is the position and the spin projection of particle i. The statistical properties of systems composed of fermions at low temperatures was studied by Enrico Fermi in 1926; it is known as *Fermi–Dirac statistics*. The statistical properties of bosonic systems was developed by S. Bose and A. Einstein in 1924–1925 and is called *Bose–Einstein statistics*.

The Pauli exclusion principle is one of the most important principles in physics because it underpins many of the characteristic properties of matter, from the large-scale stability of matter to the existence of the periodic table of the elements. Figure 2.6 highlights some of the consequences of particle exchange symmetry. Identical boson wave functions must be symmetric under the interchange of any two particles. Hence, identical bosons can occupy the same quantum state. Identical fermion wave functions must be antisymmetric under interchange of particles, and therefore, fermions cannot occupy the same state. The lowest energy state that can be formed with five fermions (say, five spin-up electrons) is shown on the RHS of Fig. 2.6. One puts the fermions into the lowest level available, and once this level is occupied, no additional identical fermions can be added to it. One "builds up" a multiparticle fermionic ground-state wave function by adding particles to consecutively higher levels. This is called the *aufbau* (building up, in German) principle for constructing multiparticle fermion ground-state wave functions. A complete discussion of identical particles will be presented in Chapter 8.

Summarizing the symmetrization postulate for identical particles: for identical bosonic particle states, the wave function must be symmetric with respect to interchange of any two of them [leaves the wave function unchanged as in Eq. (2.167) but without the minus sign], and for identical fermionic particle states, the wave function must be antisymmetric with respect to all the particles so that interchange of any two of them changes the sign of the wave function [as in Eq. (2.167)]. How this symmetrization is implemented will be discussed in Sec. 8.2.

Symmetry with respect to particle exchange

Bosons

$\psi(\mathbf{x}_1, \mathbf{x}_2) \rightarrow +\psi(\mathbf{x}_2, \mathbf{x}_1)$

Multiple state occupation possible

S. Bose, 1924
A. Einstein, 1924-5

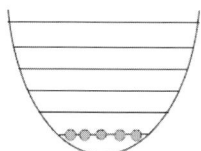

Fermions

$\psi(\mathbf{x}_1, \mathbf{x}_2) \rightarrow -\psi(\mathbf{x}_2, \mathbf{x}_1)$

Pauli Exclusion Principle

W. Pauli, 1925
E. Fermi, 1926
P. A. M. Dirac, 1926

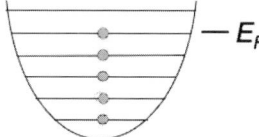

FIG 2.6 The zero-temperature occupation of single-particle states for bosons and fermions due to symmetry under particle interchange. The wave function for identical bosons must be symmetric under the interchange of any two particles, whereas for identical fermions, it must be antisymmetric. Therefore, bosons can occupy the same quantum state, whereas fermions cannot.

2.9.2 INVERSION SYMMETRY

Symmetry with respect to displacements and rotations yields the concepts of momentum conservation and angular momentum conservation. Similarly, invariance under space inversion, i.e., the simultaneous change of sign of all the spatial coordinates of the particles of a system, yields the concept of *parity conservation*. Unlike displacements and rotations, parity is a discrete symmetry. Classically, invariance under inversion does not lead to a conservation law, in contradistinction to the way that invariance under translations leads to conservation of momentum and invariance under rotations leads to conservation of angular momentum. But quantum mechanically if the inversion (or parity) operator, $\mathcal{P}$, which inverts the system through the origin,

$$\mathcal{P}\psi(\mathbf{r}) = \psi(-\mathbf{r}), \tag{2.168}$$

corresponds to a symmetry operation (commutes with the Hamiltonian),

$$i\hbar \frac{\partial \mathcal{P}}{\partial t} = [\mathcal{P}, H] = 0. \tag{2.169}$$

We then say that parity is conserved.

The transformation of the following vector operators under inversion apply:

$$\mathbf{r} \rightarrow -\mathbf{r}, \quad \mathbf{p} \rightarrow -\mathbf{p}, \quad \mathbf{J} \rightarrow \mathbf{J}, \tag{2.170}$$

i.e., $\mathcal{P}\mathbf{r}\mathcal{P}^{-1} = -\mathbf{r}$, $\mathcal{P}\mathbf{p}\mathcal{P}^{-1} = -\mathbf{p}$ and $\mathcal{P}\mathbf{J}\mathcal{P}^{-1} = \mathbf{J}$, where $\mathbf{J}$ is the angular momentum of the system [here $\mathbf{J}$ could be the total angular momentum given by the sum of the orbital angular momentum $\mathbf{L}$ and the spin angular momentum $\mathbf{S}$ (the internal angular momentum of particles), $\mathbf{J} = \mathbf{L} + \mathbf{S}$, or it could be just the orbital angular momentum, or just the spin angular momentum. These concepts will be explained in detail in Sec. 3.1]. The transformation law of the angular momentum in (2.170) results since both $\mathbf{r}$ and $\mathbf{p}$ in $\mathbf{L} = \mathbf{r} \times \mathbf{p}$ change sign upon applying an inversion transformation, so $\mathbf{L}$ remains unchanged, and $\mathbf{S}$ and $\mathbf{J}$ must transform like $\mathbf{L}$.

The eigenvalues of the parity operator are easy to find. Consider the eigensystem equation, $\mathcal{P}\psi(\mathbf{r}) = \varepsilon_p \psi(\mathbf{r})$, where ε_p is the eigenvalue of the parity operator, and again apply the parity operator to obtain $\mathcal{P}^2\psi(\mathbf{r}) = \varepsilon_p^2 \psi(\mathbf{r})$. Since $\mathcal{P}^2 = 1$, we conclude that $\varepsilon_p^2 = 1$, hence,

$$\varepsilon_p = \pm 1. \tag{2.171}$$

Eigenfunctions with eigenvalue $+1$ are said to be *even* under parity and eigenfunctions with eigenvalue -1, *odd*.

The 1D and the 3D harmonic oscillator Hamiltonians, the hydrogen atom Hamiltonian, and any central-field Hamiltonian, etc., commute with the inversion operator. Hence, the eigenstates of these Hamiltonians are simultaneously eigenstates of parity and are either odd or even under inversion. But a Hamiltonian such as $\frac{p^2}{2m} + Kx^3$ does not commute with the inversion operator $\mathcal{P}$, and the eigenstates need not have definite parity. In solid-state physics, there are many crystal structures that are not parity invariant.

The weak interaction of elementary particles, the interaction that is responsible for β-decay, is not invariant under inversion, i.e., does not commute with $\mathcal{P}$. In weak-interaction decay processes, final states that are superpositions of opposite parity states can be created. As first predicted by T. D. Lee and C. N. Yang in 1956, the angular distribution of decay products depends on pseudoscalars, such as $\mathbf{S} \cdot \mathbf{p}$, where $\mathbf{S}$ is spin operator for the decaying particle, and are odd under inversion. Lee and Yang won the Nobel Prize in 1957 for their work on parity violation in weak-interaction processes.

2.9.3 TIME-REVERSAL SYMMETRY

The *time-reversal* transformation sends $t \rightarrow -t$, hence, it reverses the velocity of particles but does not affect their positions. Classically, if $\mathbf{r}(t)$ is a solution to $m\frac{d^2}{dt^2}\mathbf{r}(t) = -\nabla V(\mathbf{r})$, then $\mathbf{r}(-t)$ is also a solution. Note that a dissipation

2.9 Symmetry and Conservation Laws in Quantum Mechanics

term such as $-\gamma \frac{d}{dt}\mathbf{r}$ on the RHS of the equation of motion would make this untrue since this term changes sign under the transformation. Moreover, a magnetic force term of the form $(q/c)\frac{d\mathbf{r}}{dt} \times \mathbf{B}$ would also violate this symmetry, unless $\mathbf{B}$ were reversed, as it would be if the particles that create this field had their velocity reversed. The universe itself does not show symmetry under *time reversal*, Today we know that the dynamical laws of nature may break time-reversal symmetry; the weak force is known not to be time-reversal symmetric and kaon decay is an example of such time-reversal symmetry breaking.

Time reversal (another name for this transformation is motion reversal) transforms dynamical variables as follows:

$$\mathbf{r} \to \mathbf{r}, \quad \mathbf{p} \to -\mathbf{p}, \quad \mathbf{J} \to -\mathbf{J}. \tag{2.172}$$

i.e., $\mathcal{T}\mathbf{r}\mathcal{T}^{-1} = \mathbf{r}$, etc., where we have denoted the time-reversal operator by $\mathcal{T}$. If the time-reversal operator commutes with the Hamiltonian of the system, it is a constant of the motion. By considering the Heisenberg equation of motion for the time-reversal operator, $i\hbar \frac{\partial \psi}{\partial t} = H\psi$, for a Hamiltonian that commutes with $\mathcal{T}$, it appears that changing t to $-t$ is equivalent to complex conjugation of the equation. Hence, perhaps we can use the operator of complex conjugation as the time-reversal operator. Let us apply this *guess* for $\mathcal{T}$ to the Heisenberg equation of motion:

$$\mathcal{T}\left(i\hbar\frac{\partial}{\partial t}\right)\mathcal{T}^{-1}\mathcal{T}\psi = \mathcal{T}H\mathcal{T}^{-1}\mathcal{T}\psi, \tag{2.173}$$

where we have inserted unity in the form $\mathcal{T}^{-1}\mathcal{T}$ into the LHS and RHS of this equation. For a Hamiltonian that commutes with $\mathcal{T}$, we have

$$-i\hbar\frac{\partial}{\partial t}(\mathcal{T}\psi) = H(\mathcal{T}\psi), \tag{2.174}$$

since $\mathcal{T}\left(i\hbar\frac{\partial}{\partial t}\right)\mathcal{T}^{-1} = -\left(i\hbar\frac{\partial}{\partial t}\right)$. If $(\mathcal{T}\psi) = \psi^*$, Eq. (2.174) is automatically satisfied since it is simply the complex conjugate of the original time-dependent Schrödinger equation, so perhaps the time-reversal operator is just the complex conjugation operator. We shall see in Sec. 4.4 that this is valid only when no half-integer spin degrees of freedom are present.

The complex-conjugation operator, often denoted $\mathcal{K}$, is not a linear operator, but rather an *antilinear operator* [see Appendix A, Eqs (A.47) and (A.48), and the text associated with these equations for the properties of antilinear operators]. Not only is $\mathcal{K}$ antilinear, it is antiunitary [see Eq. (A.48) and the surrounding text]. At this point, please read the paragraph in Appendix A on antilinear operators, and then return here.

As we shall see, the time-reversal operator is of the form of a product of a unitary operator and the complex conjugate operator, $\mathcal{U}\mathcal{K}$, where the unitary operator $\mathcal{U}$ is the unit operator for the spinless case. Applying such a transformation operator to a state vector $|\psi\rangle$, we get $|\tilde{\psi}\rangle \equiv \mathcal{U}\mathcal{K}|\psi\rangle$. Applying the same transformation operator, $\mathcal{U}\mathcal{K}$, to another such state $|\xi\rangle$ to obtain the state $|\tilde{\xi}\rangle$, and forming $\langle\tilde{\xi}|\tilde{\psi}\rangle$, we find

$$\langle\tilde{\xi}|\tilde{\psi}\rangle = \langle\xi|\psi\rangle^*. \tag{2.175}$$

If a system is time-reversal invariant and ψ is a stationary wave function of the system, the time-reversed wave function, $\mathcal{T}\psi$, describes a state with the same energy since $H(\mathcal{T}\psi) = \mathcal{T}H\psi = E(\mathcal{T}\psi)$. There are two possibilities: (1) ψ and $\mathcal{T}\psi$ are proportional to one another and describe the same state or (2) ψ and $\mathcal{T}\psi$ are linearly independent and describe two degenerate states. In the former case, $\mathcal{T}\psi = \tau\psi$, where τ is the eigenvalue of the time-reversal operator, and further application of $\mathcal{T}$ to $\mathcal{T}\psi$ yields $\mathcal{T}^2\psi = \mathcal{T}(\tau\psi) = \tau^*\mathcal{T}\psi = |\tau|^2\psi$. Clearly, for this nondegenerate and integer spin case, $\mathcal{T}^2 = \mathbf{1}$, since reversing the time twice does nothing, hence $\tau = e^{i\vartheta}$, where ϑ is a constant angle. Moreover, energy eigenfunctions $\psi_E(\mathbf{r})$ that are nondegenerate are real (or, more generally, a real function times a phase factor independent of position or momentum), as we can see by noting that $\psi_E(\mathbf{r})$ and $\psi_E^*(\mathbf{r})$ represent the same state, hence, they must be equal, up to a phase factor, $\psi_E^*(\mathbf{r}) = e^{i\vartheta_E}\psi_E(\mathbf{r})$. Thus, the wave function for a nondegenerate state is real (up to a constant phase factor of magnitude unity).

Consider the plane wave states $\psi_\mathbf{p}(\mathbf{r}) = \langle \mathbf{r}|\mathbf{p}\rangle = (2\pi)^{-3/2} e^{i\mathbf{p}\cdot\mathbf{r}/\hbar}$. These states are degenerate and are therefore case (2) states. Applying the time-reversal operator, $\mathcal{T}\psi_\mathbf{p}(\mathbf{r}) = \psi_\mathbf{p}^*(\mathbf{r}) = (2\pi)^{-3/2} e^{-i\mathbf{p}\cdot\mathbf{r}/\hbar} = \psi_{-\mathbf{p}}(\mathbf{r})$. Note also that in momentum space, $\mathcal{T}\psi(\mathbf{p}) = \psi^*(-\mathbf{p})$ since

$$\mathcal{T}|\psi\rangle = \mathcal{T}\int d\mathbf{p}\, |\mathbf{p}\rangle\langle \mathbf{p}|\psi\rangle = \int d\mathbf{p'}\, |-\mathbf{p'}\rangle\langle \mathbf{p'}|\psi\rangle^* = \int d\mathbf{p}\, |\mathbf{p}\rangle\langle -\mathbf{p}|\psi\rangle^*. \tag{2.176}$$

This result can be easily seen to yield $\mathcal{T}|\mathbf{p}\rangle = |-\mathbf{p}\rangle$.

The nondegenerate case [case (2)] is not possible for particles with half-integer-spin angular momentum. We shall explicitly consider the effect of time-reversal to these cases, e.g., for a spin $1/2$ particle, in Sec. 4.4. We conclude the discussion of time-reversal symmetry in this section by emphasizing that, quite generally, $\mathcal{T} = \mathcal{U}\mathcal{K}$; for spin-zero particles, $\mathcal{U} = 1$. For particles with nonzero spin S, the wave function is a spinor with $2S+1$ components and $\mathcal{U}$ is a $(2S+1) \times (2S+1)$ matrix. We will determine this matrix in Sec. 4.4.

2.9.4 ADDITIONAL GENERATORS OF GALILEAN TRANSFORMATIONS

This section can be skipped on a first reading.

In Sec. 1.3.6, we discussed the generators for Galilean transformations of translations, rotations, and time translations. Here, we return to consider boost and acceleration transformations.

The unitary transformation of a state that corresponds to a boost of the velocity $\mathbf{V} \equiv \frac{d\mathbf{r}}{dt} = \dot{\mathbf{r}}$ of a system by velocity $\mathbf{v}$ is $U_\mathbf{v} = e^{im\mathbf{r}\cdot\mathbf{v}/\hbar}$. I.e., the generator for velocity boosts is the quantum operator $\mathbf{Q} = m\mathbf{r}$ (see Sec. 1.3.6). The operator $\mathbf{Q}$ generates a displacement of the velocity in the sense [see Eq. (1.48)],

$$e^{im\mathbf{r}\cdot\mathbf{v}/\hbar}\, \mathbf{V}\, e^{-im\mathbf{r}\cdot\mathbf{v}/\hbar} = \mathbf{V} - \mathbf{v}, \tag{2.177}$$

or in momentum space,

$$e^{im\mathbf{r}\cdot\mathbf{v}/\hbar}\, \mathbf{p}\, e^{-im\mathbf{r}\cdot\mathbf{v}/\hbar} = \mathbf{p} - m\mathbf{v}, \tag{2.178}$$

in a fashion similar to $\mathbf{p}$ generating a displacement in coordinate space,[7]

$$e^{-i\mathbf{p}\cdot\mathbf{d}/\hbar}\, \mathbf{r}\, e^{i\mathbf{p}\cdot\mathbf{d}/\hbar} = \mathbf{r} - \mathbf{d}. \tag{2.179}$$

Note that the boost generator Q commutes with the operator $\hat{\mathbf{r}}$, hence, so does the boost $U_\mathbf{v}$,

$$e^{im\mathbf{r}\cdot\mathbf{v}/\hbar}\, \mathbf{r}\, e^{-im\mathbf{r}\cdot\mathbf{v}/\hbar} = \mathbf{r}. \tag{2.180}$$

Note further that the boost operator changes the kinetic energy as follows,

$$e^{im\mathbf{r}\cdot\mathbf{v}/\hbar}\, \frac{\mathbf{p}^2}{2m}\, e^{-im\mathbf{r}\cdot\mathbf{v}/\hbar} = \frac{(\mathbf{p} - m\mathbf{v})^2}{2m}, \tag{2.181}$$

In order to better understand how a boost transformation works, let us apply $U_\mathbf{v} \equiv e^{im\mathbf{r}\cdot\mathbf{v}/\hbar}$ to the plane wave state $\psi_\mathbf{p}(\mathbf{r}, t) = C\, e^{i\mathbf{p}\cdot\mathbf{r}/\hbar - iEt/\hbar}$:

$$U_\mathbf{v}\psi_\mathbf{p}(\mathbf{r},t) = e^{im\mathbf{r}\cdot\mathbf{v}/\hbar}\, C\, e^{i\mathbf{p}\cdot\mathbf{r}/\hbar - iEt/\hbar} = C\, e^{i(\mathbf{p}+m\mathbf{v})\cdot\mathbf{r}/\hbar - iEt/\hbar}. \tag{2.182}$$

We conclude that the boost transformation operator $U_\mathbf{v} \equiv e^{im\mathbf{r}\cdot\mathbf{v}/\hbar}$ is a unitary operator that changes the momentum of the state of the system from $\mathbf{p}$ to $\mathbf{p} + m\mathbf{v}$ (not $\mathbf{p} - m\mathbf{v}$). Since the kinetic energy of a plane wave $e^{i(\mathbf{p}+m\mathbf{v})\cdot\mathbf{r}/\hbar}$ is

---

[7] Eqs (2.178) and (2.179) can be obtained by expanding the exponential operators $e^{-i\mathbf{p}\cdot\mathbf{d}/\hbar}$ and $e^{i\mathbf{p}\cdot\mathbf{d}/\hbar}$ in a power series and using the commutation relations $[p_i, x_j] = (\hbar/i)\delta_{ij}$., e.g., $e^{-i\mathbf{p}\cdot\mathbf{d}/\hbar}\, \mathbf{r}\, e^{i\mathbf{p}\cdot\mathbf{d}/\hbar} = \mathbf{r} - i[\mathbf{p}, \mathbf{r}]\cdot\mathbf{d}/\hbar = \mathbf{r} - \mathbf{d}$.

2.9 Symmetry and Conservation Laws in Quantum Mechanics

$E' = \frac{(\mathbf{p}+m\mathbf{v})^2}{2m} = E + \mathbf{p}\cdot\mathbf{v} + m\mathbf{v}^2/2$, we can write (2.182) as

$$e^{im\mathbf{r}\cdot\mathbf{v}/\hbar}\,\psi_{\mathbf{p}}(\mathbf{r},t) = e^{i(\mathbf{p}\cdot\mathbf{v}+m\mathbf{v}^2/2)t/\hbar}\,\psi_{\mathbf{p}+m\mathbf{v}}(\mathbf{r},t). \tag{2.183}$$

We have used the fact that $\psi_{\mathbf{p}+m\mathbf{v}}(\mathbf{r},t) = C\,e^{i(\mathbf{p}+m\mathbf{v})\cdot\mathbf{r}/\hbar - iE't/\hbar}$ to obtain the final equality in (2.183). Similarly, $U_{-\mathbf{v}} = e^{-im\mathbf{r}\cdot\mathbf{v}}$ applied to $\psi_{\mathbf{p}}(\mathbf{r},t)$ gives $e^{i(-\mathbf{p}\cdot\mathbf{v}+m\mathbf{v}^2/2)t/\hbar}\,\psi_{\mathbf{p}-m\mathbf{v}}(\mathbf{r},t)$. These boosts affect the momentum (and velocity) rather than the position.

The unitary transformation that boosts the velocity $\mathbf{V} \equiv \frac{d\mathbf{r}}{dt} = \dot{\mathbf{r}}$ of a state by velocity $\mathbf{v}$ and translates it by $\mathbf{v}t$ is

$$U_{\mathbf{G}} = e^{i(m\mathbf{r}-\mathbf{p}t)\cdot\mathbf{v}/\hbar}. \tag{2.184}$$

The generator of this type of velocity boost is the operator $\mathbf{G} \equiv m\mathbf{r} - \mathbf{p}t$. The operator $\mathbf{G}$ generates a displacement in velocity space in the sense,

$$e^{im\mathbf{G}\cdot\mathbf{v}/\hbar}\,\mathbf{V}\,e^{-i\mathbf{G}\cdot\mathbf{v}/\hbar} = \mathbf{V} - \mathbf{v}, \tag{2.185}$$

or in momentum space,

$$e^{i\mathbf{G}\cdot\mathbf{v}/\hbar}\,\mathbf{p}\,e^{-i\mathbf{G}\cdot\mathbf{v}/\hbar} = \mathbf{p} - m\mathbf{v}, \tag{2.186}$$

and generates a displacement in coordinate space,

$$e^{i\mathbf{G}\cdot\mathbf{v}/\hbar}\,\mathbf{r}\,e^{-i\mathbf{G}\cdot\mathbf{v}/\hbar} = \mathbf{r} - \mathbf{v}t. \tag{2.187}$$

Note that $U_{\mathbf{G}} \neq e^{im\mathbf{r}\cdot\mathbf{v}/\hbar}\,e^{-i\mathbf{p}t\cdot\mathbf{v}/\hbar}$ since $\mathbf{r}$ and $\mathbf{p}$ do not commute. To transform the wave function $\psi(\mathbf{r},t)$ into a frame moving with velocity $\mathbf{v}$ and displaced by the vector $\mathbf{v}t$, one applies $U_{\mathbf{G}}$ to the wave function.

If, in a coordinate system undergoing rotational acceleration, such as the coordinate axes rotating with the Earth, the position, velocity, and acceleration of a particle is given by $\mathbf{r}$, $\mathbf{v} = \dot{\mathbf{r}}$, and $\mathbf{a} = \dot{\mathbf{v}} = \ddot{\mathbf{r}}$, the velocity and acceleration of the particle in the *space-fixed* (inertial) coordinate system is $\mathbf{v}_{sf} = \mathbf{v} + \boldsymbol{\Omega}\times\mathbf{r}$ and $\mathbf{a}_{sf} = \mathbf{a} + \boldsymbol{\Omega}\times\mathbf{v}_{sf} = \mathbf{a} + 2\boldsymbol{\Omega}\times\mathbf{v} + \boldsymbol{\Omega}\times(\boldsymbol{\Omega}\times\mathbf{r})$. Here, $\boldsymbol{\Omega}$ is the angular velocity of the rotating coordinate system. There are, of course, consequences if we want to describe the particle in the rotating coordinate system, e.g., the force in the accelerating coordinate system is given by $\mathbf{F} = \mathbf{F}_{sf} - 2m\boldsymbol{\Omega}\times\mathbf{v} - m\boldsymbol{\Omega}\times(\boldsymbol{\Omega}\times\mathbf{r})$, where $\mathbf{F}_{sf}$ is the force in the space-fixed coordinate system, the last term in this equation is the *centrifugal force*, and the next to last term is the *Coriolis force*. In quantum mechanics, the Hamiltonian in a rotating frame of reference also includes extra terms, Coriolis and centrifugal terms. The transformation operator to the rotating frame is given by $U_{\mathbf{G}}$ in (2.184) with a velocity $\mathbf{v}_{\text{eff}}$ determined as follows. The velocity in the inertial frame is $\dot{\mathbf{r}}_{sf} = \dot{\mathbf{r}} + \boldsymbol{\Omega}\times\mathbf{r}$, where $\boldsymbol{\Omega}$ is the angular velocity, and $\boldsymbol{\Omega}\times\mathbf{r}$ is the effective velocity of the rotating coordinate system, $\mathbf{v}_{\text{eff}} \equiv \boldsymbol{\Omega}\times\mathbf{r}$. Hence, using (2.184), we find $U_{\mathbf{G}}(\mathbf{v}_{\text{eff}}) = e^{i\mathbf{G}\cdot\mathbf{v}_{\text{eff}}/\hbar} = e^{i(m\mathbf{r}-\mathbf{p}t)\cdot(\boldsymbol{\Omega}\times\mathbf{r})/\hbar} = e^{-i\boldsymbol{\Omega}\cdot(\mathbf{r}\times\mathbf{p})t/\hbar}$; and the rotating frame,

$$\psi'(\mathbf{r},t) = U_{\mathbf{G}}\psi(\mathbf{r},t) = e^{-i\boldsymbol{\Omega}\cdot\mathbf{L}t/\hbar}\psi(\mathbf{r},t). \tag{2.188}$$

The (time-dependent unitary transformation) operator $U_{\mathbf{G}} = e^{-i\boldsymbol{\Omega}\cdot\mathbf{L}t/\hbar}$ applied to the wave function gives the new wave function in the frame undergoing rotational acceleration. As we shall learn in Sec. (2.7) [see Problem 2.19, Eq. (2.112)], the Hamiltonian of the system in the rotating frame is given by

$$H'(\mathbf{r},t) = U_{\mathbf{G}}HU_{\mathbf{G}}^{\dagger} + i\hbar\frac{\partial U_{\mathbf{G}}}{\partial t}U_{\mathbf{G}}^{\dagger} = U_{\mathbf{G}}HU_{\mathbf{G}}^{\dagger} + \boldsymbol{\Omega}\cdot\mathbf{L}. \tag{2.189}$$

Coriolis and centrifugal terms will in general be present in this Hamiltonian.

In the problem below, you will introduce transformations of the form $\mathbf{r} \to \mathbf{r}' = \mathbf{r} + \boldsymbol{\xi}(t)$ into the Schrödinger equation for general $\boldsymbol{\xi}(t)$ and determine the solution in the transformed frame. Then you will compare your result with the application of a unitary transformation of the wave function.

Problem 2.31

(a) Consider the Schrödinger equation $i\hbar \frac{\partial \psi}{\partial t} = -\frac{\hbar^2}{2m}\nabla^2 \psi + V(\mathbf{r},t)\psi$. Make the following transformation:

$$\mathbf{r} \to \mathbf{r}' = \mathbf{r} + \boldsymbol{\xi}(t), \qquad t \to t' = t.$$

Consider explicitly the following three cases:

$$\mathbf{r} \to \mathbf{r}' = \mathbf{r} + \mathbf{R}, \quad \mathbf{r} \to \mathbf{r}' = \mathbf{r} + \mathbf{v}t, \quad \mathbf{r} \to \mathbf{r}' = \mathbf{r} + \mathbf{a}t^2/2,$$

i.e., take the cases $\boldsymbol{\xi}(t) = \mathbf{R}$, $\boldsymbol{\xi}(t) = \mathbf{v}t$, and $\boldsymbol{\xi}(t) = \mathbf{a}t^2/2$. Rewrite the Schrödinger equation using the variables $\mathbf{r}'$ and t'. After this transformation, you can change t' back to t. **Hint:** Note that $\nabla = \nabla'$ and $\frac{\partial}{\partial t} = \dot{\boldsymbol{\xi}} \cdot \nabla' + \frac{\partial}{\partial t'}$, and substitute these expression into the Schrödinger equation.

(b) Write the wave function as $\psi(\mathbf{r},t) = u(\mathbf{r}',t)\, e^{i[f(\mathbf{r}',t)]}$, and choose

$$f(\mathbf{r}',t) = (m/\hbar)\left(-\dot{\boldsymbol{\xi}} \cdot \mathbf{r}' + \frac{1}{2}\int^{t}(\dot{\boldsymbol{\xi}}(t''))^2 dt''\right).$$

Find the equation satisfied by $u(\mathbf{r}',t)$.

(c) Show that $U_G \psi$, with $\mathbf{v}(t) = \dot{\boldsymbol{\xi}}(t)$, is consistent with your results in (a) and (b).

Angular Momentum and Spherical Symmetry

3

Rotational symmetry plays an important role in many physical systems, such as atoms, and more generally, spherically symmetic systems (see Sec. 16.6). Just as linear momentum is conserved in homogeneous systems, angular momentum is conserved in isotropic systems. In classical mechanics, one finds that a conserved quantity exists if a system is unchanged upon rotation of the system in space; the conserved quantity is called the *angular momentum*, **L**. For a system of particles, the total angular momentum is given by $\mathbf{L} \equiv \sum_\alpha \mathbf{r}_\alpha \times \mathbf{p}_\alpha$, where α is a summation index over the particles in the system [see Eq. (16.23) in Sec. 16.4]. Angular momentum is vital in treating such systems. But even in systems that are not isotropic, it is often useful to expand the state of the system in a basis of states composed of eigenstates of angular momentum. In this chapter, we study angular momentum in quantum mechanics.

We begin by introducing quantum mechanical angular momentum operators in Sec. 3.1. In Sec. 3.2, we discuss a number of spherically symmetric systems, including the spherical quantum dot, the 3D harmonic oscillator, the Morse Oscillator, the van der Waals and Lennard-Jones potentials, and the hydrogen atom. Then in Sec. 3.3, we establish the connection between rotations and angular momentum. Section 3.4 shows how to add angular momentum in quantum mechanics, Sec. 3.5 introduces tensor operators as well as vector and spinor fields, and finally Sec. 3.6 shows how symmetry considerations can be employed to evaluate matrix elements of dynamical variables. The topic of angular momentum for spin 1/2 systems is left largely to Chapter 4, and the dynamics of such systems is taken up in the first few sections of Chapter 6.

We note parenthetically that the mathematical study of symmetry is embodied in the theory of groups, and the rotation group is an important subtopic of group theory. We will refer to some group theory concepts in this chapter and others. Appendix E summarizes topics in group theory that are useful for quantum mechanics, including the rotation groups (O_3 and $SU(2)$), the permutation group, which will be used in discussing identical particles, and the point and space groups, which are useful in treating crystals, solid-state physics, and diatomic and polyatomic molecules. The reader is advised to look over this appendix to gain familiarity with the subject of group theory.

3.1 ANGULAR MOMENTUM IN QUANTUM MECHANICS

We have already mentioned angular momentum and its quantization in Sec. 1.1.5. Moreover, in Sec. 1.3.6, we showed that angular momentum is the generator of rotations. Specifically, Eq. (1.63) specifies that the unitary rotation operator for a single particle is expressible in terms of its orbital angular momentum operator,

$$\mathbf{L} = \mathbf{r} \times \mathbf{p} = \mathbf{r} \times (-i\hbar)\nabla. \tag{3.1}$$

The cartesian components of the orbital angular momentum operator are

$$L_x = yp_z - zp_y, \quad L_y = zp_x - xp_z, \quad L_z = xp_y - yp_x, \tag{3.2}$$

which, when written in quantum mechanics as differential operators, take the form

$$L_x = -i\hbar \left[y\frac{\partial}{\partial z} - z\frac{\partial}{\partial y} \right], \tag{3.3a}$$

$$L_y = -i\hbar \left[z\frac{\partial}{\partial x} - x\frac{\partial}{\partial z} \right], \tag{3.3b}$$

$$L_z = -i\hbar \left[x\frac{\partial}{\partial y} - y\frac{\partial}{\partial x} \right]. \tag{3.3c}$$

The square of the angular momentum operator is

$$L^2 = \mathbf{L} \cdot \mathbf{L} = L_x^2 + L_y^2 + L_z^2. \tag{3.4}$$

The components of the angular momentum operator do not commute with one another, e.g., $[L_x, L_y] = i\hbar L_z$ (which is sometimes written using the notation $[L_1, L_2] = i\hbar L_3$). The commutation relations for orbital angular momentum operators can be written succinctly using the Einstein summation convention as

$$[L_i, L_j] = i\hbar\, \varepsilon_{ijk}\, L_k, \tag{3.5}$$

where ε_{ijk} is the Levi-Civita symbol (also called the permutation symbol; see Appendix C), which has the properties $\varepsilon_{123} = 1$, an even number of permutations of the indices 123 in ε_{123} also yields 1, an odd number of permutations of the subscripts 123 yields -1, e.g., $\varepsilon_{132} = -1$, and if any of the subscripts ijk are equal, then $\varepsilon_{ijk} = 0$. The square of the angular momentum operator commutes with each of the components,

$$[L^2, L_i] = 0. \tag{3.6}$$

Problem 3.1

(a) Using the commutation relations of the position and momentum operators and the properties of commutators derived in Problem 1.8, show that $[L_x, L_y] = i\hbar L_z$.
(b) Show that $[L_i, L_j] = i\hbar \varepsilon_{ijk} L_k$.
(c) Show that $[L^2, L_i] = 0$.
(d) Show that the operator $\mathbf{r} \times \mathbf{p}$ is Hermitian if $\mathbf{r}$ and $\mathbf{p}$ are Hermitian.

To accommodate spin (i.e., internal) angular momentum as well as orbital angular momentum, we define the total angular momentum operator $\mathbf{J}$. Depending upon the system under study, the angular momentum may be just orbital in nature, or just spin, or the sum of both orbital and spin. From the general invariance of isotropic systems, we know that the generator of rotations for systems of particles with spin is the total angular momentum. We shall take up the study of spin angular momentum in the next chapter. We posit that angular momentum is additive, e.g., the total angular momentum in a system having both orbital ($\mathbf{L}$) and spin ($\mathbf{S}$) angular momenta is the sum of the two, $\mathbf{J} = \mathbf{L} + \mathbf{S}$,[1] and that the components of $\mathbf{J}$ also obey the commutation relations

$$[J_i, J_j] = i\hbar\, \varepsilon_{ijk}\, J_k. \tag{3.7}$$

If the operator $\mathbf{J}$ is the sum of an orbital and a spin angular momentum, it acts in the Hilbert space $\mathcal{H} = \mathcal{H}_{\text{space}} \otimes \mathcal{H}_{\text{spin}}$, which is a direct product of spatial and spin spaces. The elements of $\mathcal{H}$ are written as a sum of terms of the form $|\Psi\rangle \equiv |\psi\rangle \otimes |\chi\rangle$, where $|\psi\rangle$ is a space ket and $|\chi\rangle$ is a spinor ket (often the $\otimes$ will be omitted). The action of $\mathbf{J}$ on $|\Psi\rangle$ implies that $\mathbf{L}$ acts on $|\psi\rangle$ and $\mathbf{S}$ acts on $|\chi\rangle$. We shall often use the symbol $\mathbf{J}$ to denote an arbitrary angular momentum operator since many statements regarding the properties of angular momentum operators are valid for $\mathbf{L}$ (orbital), $\mathbf{S}$ (spin), and $\mathbf{J}$ (total) angular momenta.

---

[1] A more detailed account of angular momentum addition is presented in Sec. 3.4; see Eq. (3.148).

3.1 Angular Momentum in Quantum Mechanics

Angular momentum is measurable, hence $\mathbf{J}$ is Hermitian, $\mathbf{J}^\dagger = \mathbf{J}$, and the eigenvalues of J_i, $i = x, y, z$, are real. From the commutation relations (3.7), it follows that the square of the angular momentum operator, $J^2 = \mathbf{J} \cdot \mathbf{J}$, commutes with each of the components,

$$[J^2, J_i] = 0, \tag{3.8}$$

just like in the orbital angular momentum case. Hence, there exists a complete set of common eigenvectors of J^2 and any one component of $\mathbf{J}$. The component J_z is usually chosen to have common eigenvectors with J^2, and hence, the two can be simultaneously diagonalized or, equivalently, they share a common set of eigenstates denoted as $|\beta m\rangle$ such that

$$J^2 |\beta, m\rangle = \hbar^2 \beta |\beta, m\rangle, \tag{3.9a}$$

$$J_z |\beta, m\rangle = \hbar m |\beta, m\rangle. \tag{3.9b}$$

We shall discuss below an elegant technique enabling us to construct the states $|\beta, m\rangle$ as linear combinations of states $|\psi\rangle \otimes |\chi\rangle \in \mathcal{H}$. The quantum number m is often called the magnetic (or, particularly in the case of orbital angular momentum, azimuthal) quantum number.

By taking the inner product of (3.9a) with $|\beta, m\rangle$ and using the relation $J^2 = J_x^2 + J_y^2 + J_z^2$, we find that $\langle \beta, m | J^2 | \beta, m \rangle = \langle \beta, m | J_x^2 | \beta, m \rangle + \langle \beta, m | J_y^2 | \beta, m \rangle + \langle \beta, m | J_z^2 | \beta, m \rangle$. Given that $\langle \beta, m | J_x^2 | \beta, m \rangle = (\langle \beta, m | J_x^\dagger)(J_x | \beta, m \rangle) \geq 0$ since the inner product of a vector with itself cannot be negative, and similarly for J_y^2, we conclude that $\beta \geq m^2$, i.e., for a given β, there is a maximum $(+\sqrt{\beta})$ and minimum $(-\sqrt{\beta})$ value of m that is possible. To obtain the eigenvalues β and m, it is convenient to introduce the raising and lowering angular momentum operators. We do so in the next subsection.

Problem 3.2

Show that an operator $\mathcal{O}$ that commutes with two components of the angular momentum operator also commutes with the third component.

Hint: Consider the commutator $[\mathcal{O}, [J_i, J_j]]$.

Problem 3.3

Calculate the commutator $[J_x^2, J_z]$.

Answer: $[J_x^2, J_z] = J_x[J_x, J_z] + [J_x, J_z]J_x = -i(J_xJ_y + J_yJ_x)$.

3.1.1 ANGULAR MOMENTUM RAISING AND LOWERING OPERATORS

The angular momentum raising and lowering operators are defined as follows:

$$J_- = J_x - iJ_y, \quad J_+ = J_x + iJ_y. \tag{3.10}$$

These relations can be inverted:

$$J_x = \frac{J_+ + J_-}{2}, \quad J_y = \frac{J_+ - J_-}{2i}. \tag{3.11}$$

Note that $J_\pm$ are not Hermitian; rather, $J_\pm^\dagger = J_\mp$. We can derive the following commutation relations using the commutation relations (3.7):

$$[J_z, J_+] = \hbar J_+, \quad [J_z, J_-] = -\hbar J_-, \quad [J_+, J_-] = 2\hbar J_z. \tag{3.12}$$

Using (3.11) to express J_x and J_y in terms of J_+ and J_- and the commutation relations (3.12), it is easy to derive the relations

$$J^2 = \frac{J_+ J_- + J_- J_+}{2} + J_z^2, \tag{3.13a}$$

$$J^2 = J_+ J_- + J_z^2 - \hbar J_z, \tag{3.13b}$$

$$J^2 = J_- J_+ + J_z^2 + \hbar J_z. \tag{3.13c}$$

Applying (3.12a) to $|\beta, m\rangle$ and rearranging the resulting equation, we find

$$J_z J_+ |\beta, m\rangle = \hbar(m+1) J_+ |\beta, m\rangle. \tag{3.14}$$

Hence, $J_+|\beta, m\rangle$ is an eigenvector of J_z with eigenvalue $m + 1$, $J_+|\beta, m\rangle = c_+|\beta, m+1\rangle$; J_+ raised the eigenvalue of the ket $|\beta, m\rangle$ by one. This is why it is called a raising operator. Similarly, applying (3.12b) to $|\beta, m\rangle$ yields,

$$J_z J_- |\beta, m\rangle = \hbar(m-1) J_- |\beta, m\rangle. \tag{3.15}$$

Hence, $J_-|\beta, m\rangle$ is an eigenvector of J_z with eigenvalue $m - 1$, $J_-|\beta, m\rangle = c_-|\beta, m-1\rangle$. This is why J_- is called a lowering operator.

For a given β, there is a maximum value of $m \equiv j$ for which application of J_+ onto $|\beta, j\rangle$ yields the zero vector,

$$J_+|\beta, j\rangle = 0, \tag{3.16}$$

because there is no state $c_+|\beta, j+1\rangle$ with $(j+1) > \sqrt{\beta}$, as we have shown at the end of the previous subsection. Applying J_- to (3.16) and using (3.13c), we find

$$J_- J_+ |\beta, j\rangle = (J^2 - J_z^2 - \hbar J_z)|\beta, j\rangle = 0. \tag{3.17}$$

We conclude that

$$\beta = j(j+1). \tag{3.18}$$

Similarly, there is a minimum value of $m \equiv \iota$ for which application of J_- onto $|\beta, j\rangle$ yields the zero vector,

$$J_-|\beta, \iota\rangle = 0, \tag{3.19}$$

because there is no state $c_-|\beta, \iota - 1\rangle$ with $(\iota - 1) < -\sqrt{\beta}$. Applying J_+ to (3.19) and using (3.13), we find

$$J_+ J_- |\beta, \iota\rangle = (J^2 - J_z^2 + \hbar J_z)|\beta, \iota\rangle = 0. \tag{3.20}$$

We conclude that $\beta + \iota(-\iota + 1) = 0$, which together with (3.18) gives $\iota = -j$. The range of m is therefore given by $-j \leq m \leq j$. Because the raising operator J_+ increases the m (magnetic or azimuthal) quantum number in units of $1(\hbar)$, J_+ applied consecutively to $|j, -j\rangle$ an integer number of times will lead to the state $|j, j\rangle$, we conclude that $2j$ must be an integer, i.e., j is either an integer or a half integer. From now on, we denote the normalized angular momentum eigenstates by $|j, m\rangle$, where $J^2|j, m\rangle = \hbar^2 j(j+1)|j, m\rangle$ and $J_z|j, m\rangle = \hbar m|j, m\rangle$. The lowest value of j possible is $j = 0$, and for this value of j, only $m = 0$ is possible. For $j = 1/2$, m can take on the values $m = -1/2$ and $1/2$. For $j = 1$, m can take on the values $m = -1, 0, 1$. For $j = 3/2$, m can take on the values $m = -3/2, -1/2, 1/2, 3/2$, etc.

We can obtain the amplitude c_- defined immediately after Eq. (3.15) by the relation, $J_-|\beta, m\rangle = c_-|\beta, m-1\rangle$, by multiplying the equations

$$J_-|j, m\rangle = c_-|j, m-1\rangle$$

$$\langle j, m|J_+ = c_-^* \langle j, m-1|$$

3.1 Angular Momentum in Quantum Mechanics

by one another to obtain $\langle j,m|J_+J_-|j,m\rangle = c_-^* c_- = |c_-|^2$. Using $J_+J_- = (J^2 - J_z^2 + \hbar J_z)$, we find that $|c_-|^2 = \langle j,m|[j(j+1)\hbar^2 - \hbar^2 m^2 + \hbar^2 m]|j,m\rangle$, i.e., $|c_-|^2 = [j(j+1) - m^2 + m]\hbar^2$, or

$$|c_-|^2 = (j+m)(j-m+1)\hbar^2. \tag{3.21}$$

Hence, we find that

$$J_-|j,m\rangle = \sqrt{(j+m)(j-m+1)}\,\hbar\,|j,m-1\rangle. \tag{3.22}$$

Similarly, we can derive

$$J_+|j,m\rangle = \sqrt{(j-m)(j+m+1)}\,\hbar\,|j,m+1\rangle. \tag{3.23}$$

Note that Eqs (3.22) and (3.23) can also be written as follows:

$$J_-|j,m\rangle = \sqrt{j(j+1) - m(m-1)}\,\hbar\,|j,m-1\rangle, \tag{3.24a}$$

$$J_+|j,m\rangle = \sqrt{j(j+1) - m(m+1)}\,\hbar\,|j,m+1\rangle, \tag{3.24b}$$

Matrix elements of the angular momentum operators are now easy to compute:

$$\langle j',m'|J_-|j,m\rangle = \hbar\sqrt{(j+m)(j-m+1)}\,\delta_{j',j}\delta_{m',m-1}, \tag{3.25a}$$

$$\langle j',m'|J_+|j,m\rangle = \hbar\sqrt{(j-m)(j+m+1)}\,\delta_{j',j}\delta_{m',m+1}, \tag{3.25b}$$

$$\langle j',m'|J_z|j,m\rangle = \hbar\,m\,\delta_{j',j}\delta_{m',m}, \tag{3.25c}$$

$$\langle j',m'|J^2|j,m\rangle = \hbar^2\,j(j+1)\,\delta_{j',j}\delta_{m',m}. \tag{3.25d}$$

Figure 3.1 shows the $(J_+/\hbar)$ matrices for $j = 0, 1/2, 1, 3/2$. The $J_-/\hbar$ matrices are obtained by taking the Hermitian transpose of these matrices (but since the matrices are real, taking the transpose is sufficient). The matrices for J_x, J_y and J_z are given explicitly, in the basis-set representation $|j,m\rangle$ for $j = 1$, by:

$$J_x = \frac{\hbar}{\sqrt{2}}\begin{pmatrix} 0 & 1 & 0 \\ 1 & 0 & 1 \\ 0 & 1 & 0 \end{pmatrix}, \quad J_y = \frac{\hbar}{\sqrt{2}}\begin{pmatrix} 0 & -i & 0 \\ i & 0 & -i \\ 0 & i & 0 \end{pmatrix},$$

$$J_z = \hbar\begin{pmatrix} 1 & 0 & 0 \\ 0 & 0 & 0 \\ 0 & 0 & -1 \end{pmatrix}. \tag{3.26}$$

The explicit representation of the basis states for $j = 1$ is

$$\begin{pmatrix} 1 \\ 0 \\ 0 \end{pmatrix} \equiv |1,1\rangle, \quad \begin{pmatrix} 0 \\ 1 \\ 0 \end{pmatrix} \equiv |1,0\rangle, \quad \begin{pmatrix} 0 \\ 0 \\ 1 \end{pmatrix} \equiv |1,-1\rangle, \tag{3.27}$$

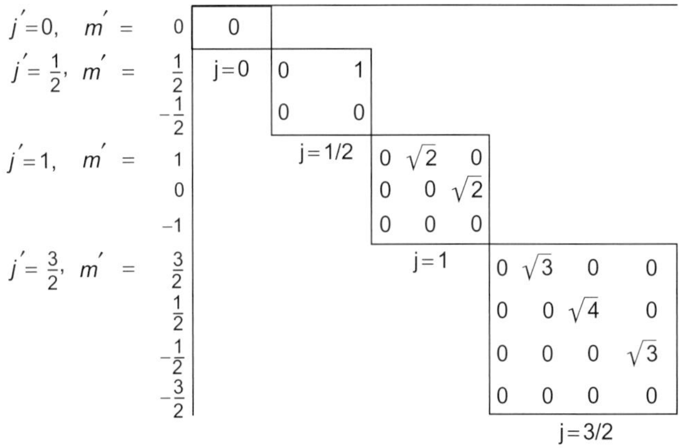

FIG 3.1 The matrices $(J_+/\hbar)$ in the basis-set representation $|j,m\rangle$ for $j = 0, 1/2, 1, 3/2$. The matrices $(J_-/\hbar)$ are obtained by taking the transpose. The matrices for J_x and J_y are obtained by adding $(J_+/\hbar)$ and $(J_-/\hbar)$ as per Eq. (3.11).

and an arbitrary linear combination of the basis states is given by

$$\begin{pmatrix} a \\ b \\ c \end{pmatrix} = a \begin{pmatrix} 1 \\ 0 \\ 0 \end{pmatrix} + b \begin{pmatrix} 0 \\ 1 \\ 0 \end{pmatrix} + c \begin{pmatrix} 0 \\ 0 \\ 1 \end{pmatrix} \equiv a|1,1\rangle + b|1,0\rangle + c|1,-1\rangle. \tag{3.28}$$

Problem 3.4

Using the matrices in Eq. (3.26) for $J = 1$, show that

$$J_x^2 + J_y^2 + J_z^2 = 2\hbar^2 \begin{pmatrix} 1 & 0 & 0 \\ 0 & 1 & 0 \\ 0 & 0 & 1 \end{pmatrix}.$$

3.1.2 ELECTRON SPIN: $J = 1/2$

Systems with zero orbital angular momentum, $\mathbf{L} = 0$, can still have an internal angular momentum. It is customary to call an internal angular momentum a *spin angular momentum*, or simply *spin*, and denote it by the symbol $\mathbf{S}$. Hence, when no orbital angular momentum is present, $\mathbf{J} = \mathbf{S}$. The best known example is that of electron spin, with $j = 1/2$. The basis states $|j, m\rangle = |\frac{1}{2}, \pm\frac{1}{2}\rangle$ for spin 1/2 can be taken to be

$$|\uparrow\rangle \equiv \begin{pmatrix} 1 \\ 0 \end{pmatrix} \equiv \left|\frac{1}{2}, \frac{1}{2}\right\rangle, \quad |\downarrow\rangle \equiv \begin{pmatrix} 0 \\ 1 \end{pmatrix} \equiv \left|\frac{1}{2}, -\frac{1}{2}\right\rangle, \tag{3.29}$$

and an arbitrary linear combination $|\chi\rangle$ of the basis states is given by

$$|\chi\rangle = \begin{pmatrix} a \\ b \end{pmatrix} = a \begin{pmatrix} 1 \\ 0 \end{pmatrix} + b \begin{pmatrix} 0 \\ 1 \end{pmatrix} \equiv a\left|\frac{1}{2}, \frac{1}{2}\right\rangle + b\left|\frac{1}{2}, -\frac{1}{2}\right\rangle. \tag{3.30}$$

Such states are called *spinors*. It is useful to define the Pauli spin operators for $j = 1/2$, $\hat{\sigma}_i$, where $i = x, y, z$, as follows: $\hat{\mathbf{S}} \equiv \frac{\hbar}{2}\hat{\boldsymbol{\sigma}}$. Equation (3.25) can be used to obtain explicit expressions for the matrices σ representing the operators $\hat{\sigma}$. These 2×2 Pauli spin matrices operate on spinors. The standard form for the 2×2 Pauli spin matrices is

$$\sigma_x = \begin{pmatrix} 0 & 1 \\ 1 & 0 \end{pmatrix}, \quad \sigma_y = \begin{pmatrix} 0 & -i \\ i & 0 \end{pmatrix}, \quad \sigma_z = \begin{pmatrix} 1 & 0 \\ 0 & -1 \end{pmatrix}. \tag{3.31}$$

The raising and lowering spin operators are defined as $S_+ \equiv S_x + iS_y = \hbar\sigma_+$ and $S_- \equiv S_x - iS_y = \hbar\sigma_-$, but the operator $S_z = (\hbar/2)\sigma_z$, hence

$$S_+ = \hbar\sigma_+ = \hbar \begin{pmatrix} 0 & 1 \\ 0 & 0 \end{pmatrix}, \quad S_- = \hbar\sigma_- = \hbar \begin{pmatrix} 0 & 0 \\ 1 & 0 \end{pmatrix}, \quad S_z = \frac{\hbar}{2}\sigma_z = \frac{\hbar}{2}\begin{pmatrix} 1 & 0 \\ 0 & -1 \end{pmatrix}. \tag{3.32}$$

Clearly, $\sigma_+ \binom{0}{1} = \binom{1}{0}$ and $\sigma_- \binom{1}{0} = \binom{0}{1}$, so σ_+ and σ_- are the spin raising and lowering operators.

We will also use the matrices σ_x, σ_y, and σ_z in discussing quantum gates since qubits, which are two-level quantum systems, can be represented in the form (3.28), and therefore, transformations of qubits can be written in terms of the Pauli spin matrices (see Sec. 5.2.3). In Secs. 3.3.2 and 4.2.2, we will discuss rotations of spins and thereby complete the discussion of transformations of two-level systems started in Sec. 5.2.3. Moreover, Chapter 4 is all about spin 1/2 particles, and the first part of Chapter 6 discusses spin dynamics. Furthermore, any two-level system can be described in terms of spin; in fact the language of treating the statics and dynamics of any two-level quantum system is the language of spin, as will be explained in Sec. 6.1.

3.1 Angular Momentum in Quantum Mechanics

Problem 3.5

(a) Find the eigenvalues of the J_x matrix for angular momentum of 1/2 and 1.
(b) Find the normalized eigenvector with the highest eigenvalue of J_x.

Answer: For $j = 1/2$, the eigenvalues are $\hbar/2$ and $-\hbar/2$, and the normalized eigenvectors are: $|\chi_{+1/2}\rangle = \frac{1}{\sqrt{2}}\begin{pmatrix}1\\1\end{pmatrix}$
and $|\chi_{-1/2}\rangle = \frac{1}{\sqrt{2}}\begin{pmatrix}1\\-1\end{pmatrix}$. For $j = 1$, the eigenvalues are $\hbar$, 0, and $-\hbar$; the eigenvector for $\hbar$ is $\begin{pmatrix}1/2\\1/\sqrt{2}\\1/2\end{pmatrix}$.

3.1.3 ANGULAR MOMENTUM IN SPHERICAL COORDINATES

The components of the angular momentum operator $\mathbf{L}$ defined in Eq. (3.3a) can be written as differential operators in spherical coordinates using Eqs. (16.45) [$r = \sqrt{x^2+y^2+z^2}$, $\phi = \arctan(y/x)$, $\theta = \arccos(z/r)$, see file linked to the book web page] to express the cartesian coordinates appearing on the RHS of Eqs. (16.45) in terms of spherical coordinates. Using the expression

$$\frac{\partial}{\partial x_i} = \frac{\partial r}{\partial x_i}\frac{\partial}{\partial r} + \frac{\partial \theta}{\partial x_i}\frac{\partial}{\partial \theta} + \frac{\partial \phi}{\partial x_i}\frac{\partial}{\partial \phi}, \tag{3.33}$$

Problem 3.6

(a) Using the matrices in (3.31), show that $S_x^2 + S_y^2 + S_z^2 = \frac{3}{4}\hbar^2\begin{pmatrix}1&0\\0&1\end{pmatrix}$, i.e., $S^2 = (1/2)(1/2+1)\hbar^2 \mathbf{1}$.
(b) Find the eigenvalues and eigenvectors of the J_y matrix for angular momentum of 1/2.

Answer: (b) For $j = 1/2$, the eigenvalues are $\hbar/2$ and $-\hbar/2$ and the normalized eigenvectors are:
$|\chi_{+1/2}\rangle = \frac{1}{\sqrt{2}}\begin{pmatrix}1\\-i\end{pmatrix}$ and $|\chi_{-1/2}\rangle = \frac{1}{\sqrt{2}}\begin{pmatrix}1\\i\end{pmatrix}$.

---

we obtain the derivatives with respect to the cartesian coordinates appearing on the RHS of (3.2), where Eqs. (16.45) can be used to obtain expressions for the derivatives of the spherical coordinates with respect to the cartesian coordinates in Eq. (3.33). Carrying out the algebra, we obtain,

$$L_z = -i\hbar\frac{\partial}{\partial \phi}, \tag{3.34}$$

$$L_x = i\hbar\left(\sin\phi\frac{\partial}{\partial \theta} + \frac{\cos\phi}{\tan\theta}\frac{\partial}{\partial \phi}\right), \tag{3.35}$$

$$L_y = i\hbar\left(-\cos\phi\frac{\partial}{\partial \theta} + \frac{\sin\phi}{\tan\theta}\frac{\partial}{\partial \phi}\right). \tag{3.36}$$

Using these expressions, we obtain the expression for the operators $\mathbf{L} = \mathbf{e}_x L_x + \mathbf{e}_y L_y + \mathbf{e}_z L_z$, where $\mathbf{e}_i$ is the unit vector along the ith axis, and L^2, in terms of spherical coordinates:

$$L^2 = L_x^2 + L_y^2 + L_z^2 = -\hbar^2\left[\frac{1}{\sin\theta}\frac{\partial}{\partial \theta}\left(\sin\theta\frac{\partial}{\partial \theta}\right) + \frac{1}{\sin^2\theta}\frac{\partial^2}{\partial \phi^2}\right]. \tag{3.37}$$

We would like to find the simultaneous eigenfunctions, $Y_{lm}(\theta,\phi)$, of the commuting operators L^2 and L_z,

$$L^2 Y_{lm}(\theta,\phi) = \hbar^2 l(l+1) Y_{lm}(\theta,\phi), \quad (3.38a)$$

$$L_z Y_{lm}(\theta,\phi) = \hbar m Y_{lm}(\theta,\phi), \quad (3.38b)$$

that are properly normalized and complete:

$$\int_{-1}^{1} d\cos\theta \int_{0}^{2\pi} d\phi\, Y^*_{l'm'}(\theta,\phi) Y_{lm}(\theta,\phi) = \delta_{l',l}\delta_{m',m}, \quad (3.39a)$$

$$\sum_{l=0}^{\infty} \sum_{m=-l}^{l} Y^*_{lm}(\theta',\phi') Y_{lm}(\theta,\phi) = \delta(\cos\theta - \cos\theta')\delta(\phi' - \phi). \quad (3.39b)$$

We do so in the next subsection. The normalized eigenfunctions $Y_{lm}(\theta,\phi)$ are called *spherical harmonics* because they are the angular part of the solution to Laplace's equation, $\nabla^2 f(\mathbf{r}) = 0$, in spherical coordinates (the solutions of Laplace's equation are called harmonic functions).

3.1.4 SPHERICAL HARMONICS

We shall now discuss some basic feature of the spherical harmonics. Let us begin by considering an eigenfunction of L_z only, $L_z \Phi_\alpha(\phi) = \hbar\alpha \Phi_\alpha(\phi)$. Upon using the spherical coordinate representation of the operator L_z, we obtain $-i\hbar \frac{\partial}{\partial \phi} \Phi_\alpha(\phi) = \hbar\alpha \Phi_\alpha(\phi)$. We have not yet taken the magnetic (azimuthal) quantum number is an integer; we shall rederive this condition from the continuity of the wave function below. The eigenfunction solution can be easily obtained by solving the differential equation, and we find that $\Phi_\alpha(\phi) = Ae^{i\alpha\phi}$. Normalization of Φ_α takes the form, $\int_0^{2\pi} d\phi |\Phi_\alpha(\phi)|^2 = 2\pi|A|^2 = 1$, so we take $A = \frac{1}{\sqrt{2\pi}}$ and $\Phi_\alpha(\phi) = \frac{1}{\sqrt{2\pi}} e^{i\alpha\phi}$. Periodicity of the function gives the condition, $\Phi_\alpha(\phi + 2\pi) = \Phi_\alpha(\phi)$, which implies $\alpha = m = 0, \pm 1, \pm 2, \pm 3 \ldots$. Hence, the angular momentum about the z-axis is quantized in units of $\hbar$, and the possible results of a measurement of L_z are $\hbar m$. Summarizing, we have found that

$$L_z \Phi_m(\phi) = \hbar m \Phi_m(\phi) \quad (3.40)$$

and

$$\Phi_m(\phi) = \frac{1}{\sqrt{2\pi}} e^{im\phi}. \quad (3.41)$$

Orthogonality of the eigenstates of L_z is expressed in the form

$$\int_0^{2\pi} d\phi\, \Phi_m^*(\phi) \Phi_n(\phi) = \delta_{mn}, \quad (3.42)$$

and completeness of the eigenstates requires that we can expand any function in these eigenstates, i.e., $\psi(\phi) = \sum_m a_m \Phi_m(\phi)$, where the amplitudes a_m are given by $a_m = \int_0^{2\pi} \Phi_m^*(\phi) \psi(\phi) d\phi$.

Problem 3.7

(a) For a 2D geometry in polar coordinates, ρ, ϕ, where $x = \rho \cos\phi$ and $y = \rho \sin\phi$, show that $\nabla^2 = \left(\frac{\partial^2}{\partial \rho^2} + \frac{1}{\rho} \frac{\partial}{\partial \rho} \right) + \frac{1}{\rho^2} \frac{\partial^2 \Phi}{\partial \theta^2}$.

3.1 Angular Momentum in Quantum Mechanics

(b) Show that the solution to the Schrödinger equation in this geometry can be written as $\psi(\rho,\phi) = \sum_{m=-\infty}^{\infty} R_m(\rho)\Phi_m(\phi)$, where $\Phi_m(\phi)$ are given in (3.41).

(c) Write the Schrödinger equation for $R_n(\rho)$ for an arbitrary potential $V(\rho)$.

Answer: (c) $\left[\dfrac{d^2}{d\rho^2} + \dfrac{1}{\rho}\dfrac{d}{d\rho} - \dfrac{m^2}{\rho^2} + \dfrac{2mE}{\hbar^2} - \dfrac{2m}{\hbar^2}V(\rho)\right]R_m(\rho) = 0$.

We now return to the eigenvalues/vectors for L^2, $L^2 Y_{lm}(\theta,\phi) = \hbar^2 l(l+1) Y_{lm}(\theta,\phi)$, where $L^2 = -\hbar^2\left[\dfrac{1}{\sin\theta}\dfrac{\partial}{\partial\theta}\left(\sin\theta\dfrac{\partial}{\partial\theta}\right) + \dfrac{1}{\sin^2\theta}\dfrac{\partial^2}{\partial\phi^2}\right]$, and seek a solution of the form $Y_{lm}(\theta,\phi) = \Theta_{lm}(\theta)\Phi_m(\phi) = \Theta_{lm}(\theta)\dfrac{1}{\sqrt{2\pi}}e^{im\phi}$. Upon substituting this form, we find that Θ_{lm} satisfies,

$$\left[\frac{1}{\sin\theta}\frac{\partial}{\partial\theta}\left(\sin\theta\frac{\partial}{\partial\theta}\right) - \frac{1}{\sin^2\theta}m^2 + l(l+1)\right]\Theta_{lm}(\theta) = 0. \tag{3.43}$$

By defining the variable $\mu = \cos\theta$ and writing (3.43) in terms of μ, we obtain the differential equation known as the *Legendre equation*, whose solution is a well-known special function [27]:

$$\frac{d}{d\mu}\left[(1-\mu^2)\frac{d\Theta_{lm}(\mu)}{d\mu}\right] + \left[l(l+1) - \frac{m^2}{(1-\mu^2)}\right]\Theta_{lm}(\mu) = 0. \tag{3.44}$$

Here $l = 0, 1, 2, \ldots$ and $l \geq |m|$. For $m = 0$, $\Theta_{l,m=0}(\mu) \equiv P_l(\mu)$, where P_l is called the *Legendre polynomial* of order l; the lowest few Legendre polynomials are as follows:

$$P_0(\mu) = 1, \quad P_1(\mu) = \mu, \quad P_2(\mu) = \frac{1}{2}(3\mu^2 - 1).$$

The *generating function* for the Legendre polynomials is

$$g(\mu, t) \equiv (1 - 2\mu t + t^2)^{-1/2} = \sum_{l=0}^{\infty} P_l(\mu) t^l, \tag{3.45}$$

so clearly $P_l(\mu) = (l!)^{-1} d^l g(\mu,t)/dt^l|_{t=0}$. The *Rodrigues formula*, which allows for the calculation of the Legendre polynomials via differentiation, is given by

$$P_l(\mu) = \frac{1}{2^l l!}\left(\frac{d}{d\mu}\right)^l (\mu^2 - 1)^l, \tag{3.46}$$

and the recurrence relation, which can be used to obtain higher order Legendre polynomials from lower ones, is given by

$$P_{l+1}(\mu) = \frac{1}{l+1}\left[(2l+1)\mu P_l(\mu) - l P_{l-1}(\mu)\right]. \tag{3.47}$$

Problem 3.8

(a) Use (3.47) to find $P_3(\mu)$, given P_1 and P_2.
(b) Use the generating function (3.45) to obtain $P_2(\mu)$. **Hint:** Differentiate twice with respect to t and then set $t = 0$.

The functions Θ_{lm} for $m \neq 0$ are proportional to the *associated Legendre polynomials*, denoted by P_l^m, $\Theta_{lm}(\mu) \equiv N_{lm} P_l^m(\mu)$, where N_{lm} is the normalization constant, to be discussed shortly. The associated Legendre polynomials with $m > 0$ can be generated from the Legendre polynomials as follows:

$$P_l^m(\mu) = (1 - \mu^2)^{|m|/2} \left(\frac{d}{d\mu}\right)^{|m|} P_l(\mu) \quad \text{for } m > 0. \tag{3.48}$$

Substituting (3.48) into the LSH of (3.45) yields the Rodrigues formula for P_l^m for all m, $-l \leq m \leq l$:

$$P_l^m(\mu) = \frac{1}{2^l l!} (1 - \mu^2)^{m/2} \left(\frac{d}{d\mu}\right)^{l+m} (\mu^2 - 1)^l. \tag{3.49}$$

The Legendre polynomials take the following values at $\mu = \pm 1$:

$$P_l(1) = 1, \quad P_l(-1) = (-1)^l, \quad P_l^m(1) = P_l^m(-1) = 0 \quad \text{if } m \neq 0. \tag{3.50}$$

Recurrence relations and the generating function for the P_l^m can be developed using (3.49). The associated Legendre polynomials obey the orthogonality relations,

$$\int_{-1}^{1} d\mu \, P_{l'}^m(\mu) P_l^m(\mu) = \frac{(l+m)!}{(l-m)!} \frac{2}{2l+1} \delta_{l,l'}. \tag{3.51}$$

Hence, properly normalized spherical harmonics are obtained with normalization constant, $N_{lm} = (-1)^{m+|m|} \left[\frac{(l-m)!}{(l+m)!} \frac{2l+1}{2}\right]^{1/2}$:

$$\boxed{Y_{lm}(\theta, \phi) = (-1)^{m+|m|} \left[\frac{(2l+1)}{4\pi} \frac{(l-m)!}{(l+m)!}\right]^{1/2} P_l^m(\cos\theta) e^{im\phi}.} \tag{3.52}$$

These functions are orthonormal, $\int_0^{2\pi} d\phi \int_{-1}^{1} d\cos\theta \, Y_{l'm'}^*(\theta, \phi) Y_{lm}(\theta, \phi) = \delta_{l,l'} \delta_{m,m'}$ [see (3.39a)]. The orthonormality can be expressed in terms of an integral over solid angles, $d\Omega = \sin\theta \, d\theta \, d\phi$, as

$$\int d\Omega \, Y_{l'm'}^*(\theta, \phi) Y_{lm}(\theta, \phi) = \delta_{l',l} \delta_{m',m}. \tag{3.53}$$

Completeness of these functions, (3.39b), means that any angular function $f(\theta, \phi)$ can be expanded in terms of the spherical harmonics,

$$f(\theta, \phi) = \sum_{l=0}^{\infty} \sum_{m=-l}^{l} a_{lm} Y_{lm}(\theta, \phi), \tag{3.54}$$

where

$$a_{lm} = \int d\Omega \, Y_{lm}^*(\theta, \phi) f(\theta, \phi). \tag{3.55}$$

The lowest few spherical harmonics are given by:

$$Y_{00}(\theta, \phi) = \frac{1}{\sqrt{4\pi}} \quad Y_{10}(\theta, \phi) = \sqrt{\frac{3}{4\pi}} \cos\theta = \sqrt{\frac{3}{4\pi}} \frac{z}{r}$$

$$Y_{11}(\theta, \phi) = -\sqrt{\frac{3}{8\pi}} \sin\theta \exp(i\phi) = -\sqrt{\frac{3}{8\pi}} \frac{(x+iy)}{r}$$

$$Y_{20}(\theta, \phi) = \sqrt{\frac{5}{16\pi}} \left(3\cos^2\theta - 1\right) = \sqrt{\frac{5}{16\pi}} \frac{(3z^2 - r^2)}{r^2}$$

3.1 Angular Momentum in Quantum Mechanics

$$Y_{21}(\theta,\phi) = -\sqrt{\frac{15}{8\pi}} \sin\theta \cos\theta \exp(i\phi) = -\sqrt{\frac{15}{8\pi}} \frac{(xz+iyz)}{r^2}$$

$$Y_{22}(\theta,\phi) = \sqrt{\frac{15}{32\pi}} \sin^2\theta \exp(2i\phi) = \sqrt{\frac{15}{32\pi}} \frac{(x^2-y^2+2ixy)}{r^2}$$

Clearly, $r^l Y_{lm}$ is a homogeneous function in x, y, z of order l [i.e., $f(tx,ty,tz) = t^l f(tx,ty,tz)$]. From (3.52), we see that

$$Y_{l,-m}(\theta,\phi) = (-1)^m Y^*_{lm}(\theta,\phi), \quad (3.56)$$

which can be used to obtain Y_{2-1} and Y_{2-2} from Y_{21} and Y_{22}, respectively, etc. Moreover, for $m=0$ and $m=l$,

$$Y_{l0}(\theta,\phi) = \sqrt{\frac{(2l+1)}{4\pi}} P_l^0(\cos\theta) \quad (3.57)$$

and

$$Y_{ll}(\theta,\phi) = (-1)^l \left[\frac{(2l+1)(2l)!}{4\pi \, 2^{2l}(l!)^2}\right] (\sin\theta)^l e^{il\phi}. \quad (3.58)$$

From (3.50), we see that

$$Y_{lm}(0,0) = \sqrt{\frac{(2l+1)}{4\pi}} \delta_{m,0}. \quad (3.59)$$

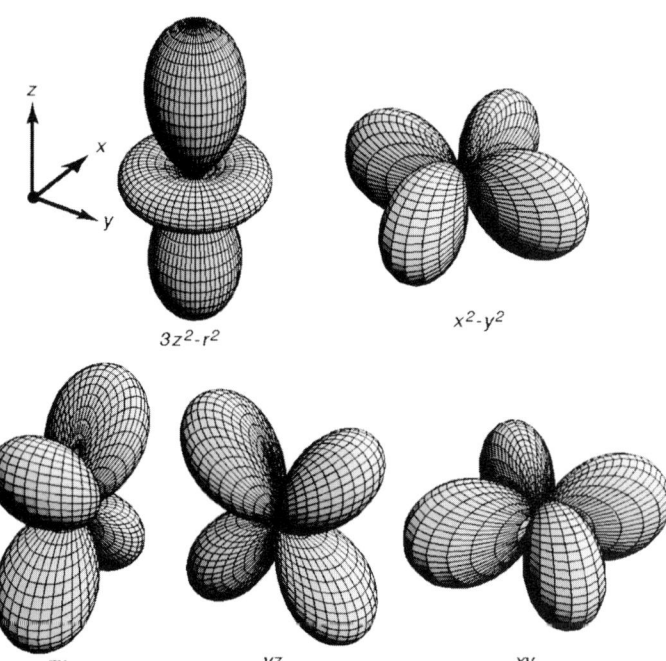

FIG 3.2 Real linear combinations of the $l=2$ spherical harmonics.

Figure 3.2 shows real linear combinations of $l=2$ spherical harmonics. In such visualizations of the spherical harmonics, the distance from origin corresponds to magnitude (modulus) of the plotted harmonic. A nice visualization of the spherical harmonics is available on the web:

http://www.vis.uni-stuttgart.de/~kraus/LiveGraphics3D/java_script/SphericalHarmonics.html

---

Problem 3.9

Consider the wave function $\psi(\phi) = \frac{1}{3\sqrt{\pi}}[1 + 2i\cos(3\phi)]$.

(a) Normalize the wave function $\psi(\phi)$.
(b) Find the amplitudes a_m in the expansion $\psi(\phi) = \sum_{m=-\infty}^{\infty} a_m \Phi_m(\phi)$, where $\Phi_m(\phi)$ are given in Eq. (3.41).
(c) If one measures L_z, what are the possible results and their corresponding probabilities.
(d) Calculate the expectation value $\langle L_z \rangle$ within the state ψ.

Answers: (a) Normalized wave function $\psi(\phi) = \sqrt{\frac{1}{6\pi}}[1 + 2i\cos(3\phi)]$. (b) $a_0 = \sqrt{2}/3$, $a_{\pm 3} = i\sqrt{2}/3$. (c) 0, $\pm 3\hbar$ with $P_m = |a_m|^2$. (d) $\langle L_z \rangle = 0$.

3.2 SPHERICALLY SYMMETRIC SYSTEMS

We have already seen that the Hamiltonian for two particles interacting via a potential can be reduced to a Hamiltonian involving the center-of-mass motion and the relative motion. If the potential is spherically symmetric, the solution of the relative motion problem is best carried out in spherical coordinates. The relative kinetic energy operator, $T = \frac{\mathbf{p}_r^2}{2\mu} = -\frac{\hbar^2}{2\mu}\nabla^2$, where μ is the reduced mass, is given in spherical coordinates in terms of the Laplacian,

$$\nabla^2 = \frac{1}{r}\frac{\partial^2}{\partial r^2}r - \frac{L^2}{\hbar^2 r^2}. \tag{3.60}$$

So, the kinetic energy operator in spherical coordinates can be written as

$$T = \frac{1}{2\mu}\left(p_r^2 + \frac{L^2}{r^2}\right) = \frac{-\hbar^2}{2\mu}\left[\frac{1}{r}\frac{\partial^2}{\partial r^2}r - \frac{L^2}{\hbar^2 r^2}\right], \tag{3.61}$$

where the *radial momentum operator*, p_r, is

$$p_r = \frac{\hbar}{i}\frac{1}{r}\frac{\partial}{\partial r}r \tag{3.62}$$

and $p_r^2 = -\hbar^2 \left(\frac{1}{r}\frac{\partial}{\partial r}r\right)\left(\frac{1}{r}\frac{\partial}{\partial r}r\right) = -\hbar^2 \frac{1}{r}\frac{\partial^2}{\partial r^2}r$.

Problem 3.10

Prove that the Laplacian operator, ∇^2, commutes with the angular momentum operators, L_i.
Hint: Use Eqs (3.34) through (3.37) and (3.60).

The time-independent Schrödinger equation for the relative motion in a spherically symmetric potential becomes

$$\left[\frac{1}{r}\frac{\partial^2}{\partial r^2}r - \frac{L^2}{\hbar^2 r^2} + \frac{2\mu}{\hbar^2}(E - V(r))\right]\psi(\mathbf{r}) = 0. \tag{3.63}$$

The wave function $\psi(\mathbf{r})$ can be expanded in spherical harmonic functions,

$$\psi(\mathbf{r}) = \sum_{l,m} a_{lm} R_l(r) Y_{lm}(\theta, \phi), \tag{3.64}$$

and upon substituting into (3.63), we find that different values of l and m, i.e., different partial waves, do not couple with each other, and the *radial wave function* $R_l(r)$ satisfies the equation,

$$\left[\frac{1}{r}\frac{\partial^2}{\partial r^2}r - \frac{l(l+1)}{r^2} + \frac{2\mu}{\hbar^2}(E - V(r))\right]R_l(r) = 0. \tag{3.65}$$

Since the radial wave function equation, (3.65), called the *radial Schrödinger equation*, does not contain the magnetic (azimuthal) quantum number m, $R_l(r)$ does not depend on m. Substituting $R_l(r) = \frac{f_l(r)}{r}$ into Eq. (3.65) yields

$$\left[\frac{\partial^2}{\partial r^2} - \frac{l(l+1)}{r^2} + \frac{2\mu}{\hbar^2}(E - V(r))\right]f_l(r) = 0. \tag{3.66}$$

The wave function $\psi(\mathbf{r})$ must be finite everywhere, including the origin, therefore $R_l(r)$ must be finite. Hence, $f_l(r)$ must vanish at $r = 0$ at least as fast as r. Note that Eq. (3.66) is of the form of a 1D Schrödinger equation,

3.2 Spherically Symmetric Systems

$\left[\frac{\partial^2}{\partial r^2} + \frac{2\mu}{\hbar^2}(E - U_l(r))\right] f_l(r) = 0$, with an effective radial potential energy given by

$$U_l(r) = V(r) + \frac{\hbar^2 l(l+1)}{2\mu r^2}. \tag{3.67}$$

The second term on the RHS of (3.67) is called the *centrifugal energy*. This 1D equation must be solved with the boundary condition $f_l(0) = 0$.

Let us consider the limiting form of $f_l(r)$ as $r \to 0$ and as $r \to \infty$. For $r \to 0$, we assume that $\lim_{r \to 0} V(r) r^2 = 0$, so near the origin, Eq. (3.66) is well approximated by $\left[\frac{\partial^2}{\partial r^2} - \frac{l(l+1)}{r^2}\right] f_l(r) = 0$, the other terms (the potential term and the term proportional to $k^2 f_l$) being small in comparison with the ones retained. The solution to this differential equation that vanishes at $r = 0$ is

$$f_l(r) = c_l r^{l+1} \text{ as } r \to 0. \tag{3.68}$$

The probability of finding the particle between r and $r + dr$ is proportional to $r^2 |R_l(r)|^2 dr = |f_l(r)|^2 dr$, so near the origin, the probability is proportional to r^{2l+2}, which becomes smaller the larger l.

The asymptotic form of $f_l(r)$ as $r \to \infty$ depends upon whether we are dealing with a bound state or a scattering state. Let us assume that the potential $V(r) \to 0$ faster than r^{-2} as $r \to \infty$. For a bound state in the lth partial wave with energy $E < 0$, the asymptotic solution to (3.66) is

$$f_l(r) = c_l e^{-\sqrt{\frac{2\mu|E|}{\hbar^2}} r} \text{ as } r \to \infty. \tag{3.69}$$

Hence, the wave function decays exponentially at large r. The coefficient c_l is determined by normalization,

$$\int_0^\infty dr\, r^2 |R_l(r)|^2 = \int_0^\infty dr\, |f_l(r)|^2 = 1.$$

For a continuum state with energy $E > 0$, the asymptotic form of Eq. (3.66) becomes the free-particle radial Schrödinger equation,

$$\frac{d^2}{dr^2} f_l - \frac{l(l+1)}{r^2} f_l + k^2 f_l = 0, \tag{3.70}$$

where $k^2 = \frac{2\mu E}{\hbar^2}$. The solutions to Eq. (3.70) are *Riccati-Bessel functions*, $\hat{j}_l(kr)$ and $\hat{n}_l(kr)$ [see Appendix B, Eqs (B.25) through (B.30)], so asymptotically,

$$f_l(r) \xrightarrow[r \to \infty]{} a_l \hat{j}_l(kr) + b_l \hat{n}_l(kr). \tag{3.71}$$

The Riccati-Bessel functions are given in terms of the *spherical Bessel functions* $j_l(z)$ and $n_l(z)$ as follows: $\hat{j}_l(z) = z j_l(z)$ and $\hat{n}_l(z) = z n_l(z)$ [27]. Figure 3.3 shows the regular and irregular Riccati-Bessel functions versus z for $l = 0, 1, 2$. The spherical-Bessel functions can in turn be written in terms of the Bessel functions $J_l(z)$ and $N_l(z)$ as follows: $j_l(z) = \sqrt{\frac{\pi}{2z}} J_{l+1/2}(z), n_l(z) = \sqrt{\frac{\pi}{2z}} N_{l+1/2}(z)$, so $\hat{j}_l(z) = \sqrt{\frac{\pi z}{2}} J_{l+1/2}(z), \hat{n}_l(z) = \sqrt{\frac{\pi z}{2}} N_{l+1/2}(z)$ (note that in the literature, sometimes the Neumann functions n_l and N_l are denoted by y_l and Y_l, respectively). At small r,

$$\hat{j}_l(kr) \to \frac{(kr)^{l+1}}{(2l+1)!!} \text{ as } r \to 0, \tag{3.72a}$$

$$\hat{n}_l(kr) \to -\frac{(2l-1)!!}{(kr)^l} \text{ as } r \to 0, \tag{3.72b}$$

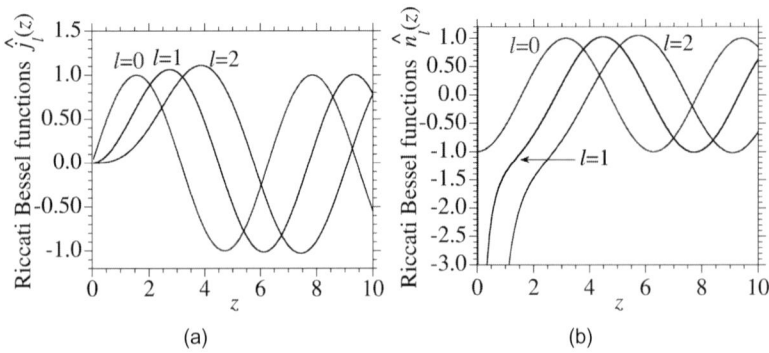

FIG 3.3 (a) Regular, $\hat{j}_l(z)$, and (b) irregular, $\hat{n}_l(z)$, Riccati-Bessel functions versus z for $l = 0, 1, 2$.

and the asymptotic forms of the Riccati-Bessel functions at large r are

$$\hat{j}_l(z) \xrightarrow[r \to \infty]{} \sin(z - l\pi/2),$$
(3.73a)

$$\hat{n}_l(z) \xrightarrow[r \to \infty]{} -\cos(z - l\pi/2).$$
(3.73b)

Hence, the asymptotic form of $f_l(r)$ for scattering states ($E > 0$) at large r takes the form

$$f_l(r) \xrightarrow[r \to \infty]{} A_l \sin[kr - l\pi/2 + \delta_l(k)].$$
(3.74)

The *scattering phase shifts* $\delta_l(k)$ are important quantities and characterize the scattering. We will have a lot more to say about them in Chapter 12.

Summarizing, the wave function for the relative motion in a spherically symmetric potential is given by

$$\psi(\mathbf{r}) = \sum_{l=0}^{\infty} \sum_{m=-l}^{m} a_{lm} \frac{f_l(r)}{r} Y_{lm}(\theta, \phi),$$
(3.75)

where $f_l(0) = 0$. The amplitudes a_{lm} depend on the initial conditions. For a bound state of a given angular momentum, only one of these amplitudes is nonzero. For a continuum state of only one partial wave (e.g., $l = m = 0$), only one of these amplitudes is nonzero, but for an initial condition corresponding to a plane wave, many amplitudes a_{lm} must in general be included. As far as the center-of-mass motion, if there is no external potential present, then the center-of-mass wave function, $\Psi_{\mathbf{P}}(\mathbf{R})$, is a plane wave with center-of-mass momentum $\mathbf{P}$ and an expansion of the form (3.75) can be made for $\Psi_{\mathbf{P}}(\mathbf{R})$ as discussed in the next subsection.

Problem 3.11

Determine the probability for measuring specific values of L^2 and L_z given the following orbitals:

(a) $\psi(\mathbf{r}) = R(r)[\cos\theta]^2$,
(b) $\psi(\mathbf{r}) = R(r)[\sin\theta]^5 \cos(5\phi)$ [Hint: see Eq. (3.58)],
(c) $\psi(\mathbf{r}) = R(r) \sin\theta \sin\phi$,
(d) $\psi(\mathbf{r}) = g(r)(x + y + z)$,
(e) $\psi(\mathbf{r}) = g(r)(x^2 + y^2 - 2z^2)$,
(f) $\psi(\mathbf{r}) = R(r) \cos\theta \sin\phi$. [This is tricky. Hint: see Eq. (3.55). Just determine the lowest few L^2.]

Answers: (a) Note that $[\cos\theta]^2 = \sqrt{\frac{16\pi}{45}} Y_{20} + \sqrt{\frac{4}{3}} Y_{00}$, so $P_{0,0} = 5/9, P_{2,0} = 4/9$.
(b) $P_{5,5} = 1$. (c) $P_{1,1} = 1$. (d) $P_{1,1} = 1/3, P_{1,0} = 1/3, P_{1,-1} = 1/3$. (e) $P_{2,0} = 1$. (f) Even powers of L^2, and $L_z = \pm 1$ are populated.

Problem 3.12

A wave function $\psi(\mathbf{r})$ is an eigenfunction of eigenstate of L^2 and L_z with eigenvalues $\hbar^2 l(l+1)$ and $\hbar m$. Prove the following expectation value expressions: $\langle L_x \rangle = \langle L_y \rangle = 0$, $\langle L_x^2 \rangle = \langle L_y^2 \rangle = \frac{\hbar^2}{2}[l(l+1) - m^2]$.

3.2 Spherically Symmetric Systems

3.2.1 ANGULAR MOMENTUM DECOMPOSITION OF PLANE WAVES

A momentum normalized plane wave state with momentum $\mathbf{p}$ is given by $\psi_\mathbf{p}(\mathbf{r}) = (2\pi\hbar)^{-3/2} \exp(i\mathbf{k} \cdot \mathbf{r})$ [see Eq. (1.79)], where $\mathbf{k} = \mathbf{p}/\hbar$ is the wave vector, and the normalization condition is $\langle \psi_{\mathbf{p}'} | \psi_\mathbf{p} \rangle = \delta(\mathbf{p}' - \mathbf{p})$. The energy normalized plane wave state is given by $\psi_{E,\Omega_\mathbf{p}}(\mathbf{r}) = \left(\frac{m}{p}\right)^{1/2} (2\pi\hbar)^{-3/2} \exp(i\mathbf{k} \cdot \mathbf{r})$ [see Eq. (1.83)]. The plane wave could be for the center-of-mass coordinate, in which case we would use the notation $\mathbf{R}$ for the center-of-mass coordinate, $\mathbf{P}$ for the center-of-mass momentum, and $\Psi_\mathbf{P}(\mathbf{R})$ for the momentum normalized plane wave state. If the potential for the relative motion vanishes, the plane wave could be for the relative motion. In either case, it might be useful to decompose the plane wave into angular momentum partial waves. The decomposition of a plane wave into spherical waves involves finding the amplitudes a_{lm} in (3.75) so that

$$\exp(i\mathbf{k} \cdot \mathbf{r}) = \sum_{l=0}^{\infty} \sum_{m=-l}^{m} a_{lm} j_l(kr) Y_{lm}(\theta, \phi). \quad (3.76)$$

Here, the radial wave functions are the spherical-Bessel functions $j_l(kr)$. We know that both the plane wave $\exp(i\mathbf{k} \cdot \mathbf{r})$ and $j_l(kr) Y_{lm}(\theta, \phi)$ are solutions of the Schrödinger equation for free particles, and both are also regular at the origin, so it remains only to find the amplitudes a_{lm}; we shall see that they are given by

$$a_{lm} = 4\pi i^l Y_{lm}^*(\theta_\mathbf{k}, \phi_\mathbf{k}), \quad (3.77)$$

where $\theta_\mathbf{k}$ and $\phi_\mathbf{k}$ are the polar angles of the wave vector $\mathbf{k}$. Let us first consider the case where the wave vector is in the z direction and (3.76) takes on the somewhat simpler form [see Eqs (3.59) and (3.57)],

$$\boxed{\exp(ikz) = \sum_{l=0}^{\infty} i^l (2l+1) j_l(kr) P_l(\cos\theta).} \quad (3.78)$$

Equation (3.78) can be proven by multiplying the equation, $\exp(ikr\cos\theta) = \sum_{l'} a_{l'} j_{l'}(kr) P_{l'}(\cos\theta)$ by $P_l(\cos\theta)$, integrating over θ, and using the orthogonality of the Legendre polynomials, Eq. (3.51), to obtain:

$$\int_{-1}^{1} d\mu\, e^{ikr\mu} P_l(\mu) = \frac{2a_l}{2l+1} j_l(kr). \quad (3.79)$$

Now, let us integrate the LHS of (3.79) by parts to obtain

$$\left.\frac{P_l(\mu) e^{ikr\mu}}{ikr}\right|_{-1}^{1} - \frac{1}{ikr} \int d\mu\, e^{ikr\mu} P_l'(\mu) = \frac{[e^{ikr} - (-1)^l e^{-ikr}]}{ikr} - \mathcal{O}\left(\frac{1}{r^2}\right), \quad (3.80)$$

where we have used (3.50) to evaluate $P_l(\pm 1)$, and we have indicated that the second term on the RHS of (3.80) goes to zero as $\frac{1}{r^2}$ at large r. The factor in square parenthesis on the RHS of (3.80) can be written as $[e^{ikr} - (-1)^l e^{-ikr}] = e^{il\pi/2}[e^{ikr-il\pi/2} - e^{-ikr+il\pi/2}] = 2i^{l+1} \sin(kr - l\pi/2)$. Using the asymptotic form of the Riccati-Bessel function, $j_l(kr) \to \frac{\sin(kr-l\pi/2)}{kr}$ [see Eq. (3.73)], we finally obtain the following expression from (3.79):

$$\frac{2i^l \sin(kr - l\pi/2)}{kr} = \frac{2a_l}{2l+1} \frac{\sin(kr - l\pi/2)}{kr}. \quad (3.81)$$

That is, we have shown that $a_l = i^l(2l+1)$, as stated in Eq. (3.78). Hence, using (3.57), (3.78) can be written as $\exp(ikz) = \sum_{l=0}^{\infty} i^l \sqrt{4\pi(2l+1)} j_l(kr) Y_{l0}(\theta, \phi)$. Moreover, for arbitrary vector $\mathbf{k}$, Eq. (3.78) can be immediately generalized to $\exp(i\mathbf{k} \cdot \mathbf{r}) = \sum_{l=0}^{\infty} i^l(2l+1) j_l(kr) P_l(\mathbf{n_k} \cdot \mathbf{n})$, where $\mathbf{n_k}$ is the unit vector in the direction of $\mathbf{k}$ and $\mathbf{n}$ is the unit vector in the direction of **r**. Using the *addition theorem for spherical harmonics* [25],

$$P_l(\mathbf{n_k} \cdot \mathbf{n}) = \frac{4\pi}{2l+1} \sum_{m=-l}^{m} Y^*_{lm}(\theta_{\mathbf{k}}, \phi_{\mathbf{k}}) Y_{lm}(\theta, \phi), \quad (3.82)$$

we can obtain (3.76) and (3.77) from (3.78). Incidentally, a special case of the addition theorem for the spherical harmonics obtained for $l=1$ is

$$\cos(\mathbf{n_k} \cdot \mathbf{n}) = \cos\theta_{\mathbf{k}} \cos\theta + \sin\theta_{\mathbf{k}} \sin\theta \cos(\phi_{\mathbf{k}} - \phi). \quad (3.83)$$

Summarizing this subsection, we have found that plane waves can be expanded in terms of spherical harmonics as follows:

$$\boxed{\exp(i\mathbf{k} \cdot \mathbf{r}) = 4\pi \sum_{l,m} i^l \frac{\hat{j}_l(kr)}{kr} Y^*_{lm}(\theta_{\mathbf{k}}, \phi_{\mathbf{k}}) Y_{lm}(\theta, \phi),} \quad (3.84)$$

where the Riccati-Bessel function $\hat{j}_l(kr)$ satisfies Eq. (3.70).

3.2.2 SPHERICAL QUANTUM DOT

Let us consider bound states in a spherical quantum dot. Initially, let us take the potential outside the dot to be infinitely high, so the boundary condition at the dot radius, r_d, is $\psi(r_d) = 0$. The radial Schrödinger equation is (3.66), and the potential $V(r) = 0$ inside the spherical box, hence the regular solution that vanishes at $r=0$ and at $r=r_d$ is $f_l(r) = \hat{j}_l(kr)$, where the only allowed values of k are such that kr_d equals one of the nodes $z_{n,l}$ of the Riccati-Bessel function of order l, $\hat{j}_l(z_{n,l}) = 0$, i.e., $k_{n,l} = z_{n,l}/r_d$ (except for $k=0$). For $l=0$, $\hat{j}_{l=0}(z) = \sin(z)$, so $z_{n,0} = n\pi$ and $k_{n,0} = n\pi/r_d$. Figure 3.4 plots the lowest energy eigenvalues $E_{n,l}$ for the lowest few values of n and $l=0$ and $l=1$ versus r_d.

If the potential V is finite for $r > r_d$, then the Schrödinger equation (3.66) in this region becomes $\frac{d^2}{dr^2}f_l - \frac{l(l+1)}{r^2}f_l - \kappa^2 f_l = 0$, where $\kappa^2 = \frac{2\mu(V-E)}{\hbar^2} > 0$. The solution to this equation that decays with increasing r is given in terms of the Riccati-Bessel function $\hat{h}_l^{(+)}(z) \equiv z h_l^{(+)}(z)$ [see Appendix B, Eq. (B.32)], with complex argument $z = i\kappa r$. Asymptotically, $\hat{h}_l^{(+)}(i\kappa r) \xrightarrow[r \to \infty]{} e^{-\kappa r - il\pi/2}$, i.e., it decays exponentially at large r. This function must be matched onto the solution $\hat{j}_l(kr)$ for $r < r_d$ by setting the wave functions and their derivatives equal at $r=r_d$. Matching $(df_l/dr)/f_l$ at $r=r_d$,

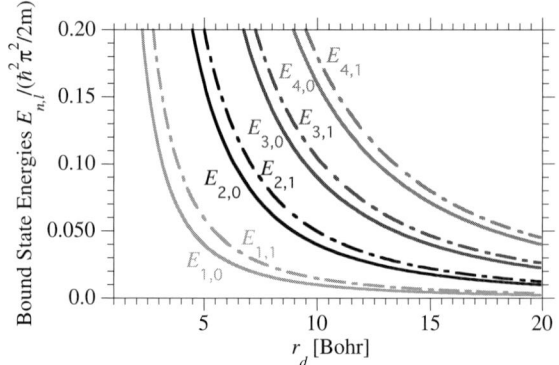

FIG 3.4 Bound-state energies $|E_{n,l}|$ of a spherical quantum dot for $l=0$ and $l=1$ versus the quantum dot radius r_d.

$$\left.\frac{d\hat{j}_l(kr)/dr}{\hat{j}_l(kr)}\right|_{r=r_d} = \left.\frac{d\hat{h}_l^{(+)}(i\kappa r)/dr}{\hat{h}_l^{(+)}(i\kappa r)}\right|_{r=r_d}, \quad (3.85)$$

yields an equation for the bound-state energies, E_l of the lth partial wave bound states of the finite depth spherical quantum dot.

3.2 Spherically Symmetric Systems

Unlike in 1D and 2D, where there is at least one bound state for any attractive potential, here there is a threshold below which no bound state exists. To clarify this point, we consider s-wave bound states. Equation (3.85) takes the form $k \cot kr_d = -\kappa$. Multiplying by r_d, and defining dimensionless parameters $q \equiv kr_d$ and $y \equiv \sqrt{2\mu V/\hbar^2}r_d$, the matching equation reads $q \cot q = -\sqrt{y^2 - q^2}$. For $y = \pi/2$ this equation has solution $q = \pi/2$, corresponding to binding energy $E_{1,0} = \hbar^2\pi^2/(8\mu r_d^2)$. However, for $y < \pi/2$, no solution exists. In terms of the original parameters, if $8\mu V r_d^2/(\pi^2\hbar^2) < 1$, the potential is too weak to support a bound state. Note that the number of bound states depends solely on y.

Problem 3.13

(a) Find the minimal value of $y = y_2$ for which there is a second s-wave bound state, and the binding energy of this state for given r_d and μ.

(b) Plot the function $q \cot q = -\sqrt{y_2^2 - q^2}$ as function of q and find the binding energy of the first bound state for $y = y_2$ for given r_d and μ.

Answer: (a) $y_2 = 3\pi/2$. $E_{2,0} = 9\hbar^2\pi^2/(8\mu r_d^2)$.
(b) $q_1 = 2.73$, $\Rightarrow E_{1,0} = \hbar^2(2.73)^2/(2\mu r_d^2)$.

The spectrum of emitted radiation via spontaneous emission by electrons in the quantum dot is composed of transition frequencies obtained when states $|nlm\rangle$ decay radiatively to lower lying states $|n'l'm'\rangle$. Transitions can occur if $l' = l \pm 1$ or $l' = l$ and $m' = m \pm 1$ or $m' = m$. These *selection rules* for optically allowed transitions are similar to those for hydrogenic optical transitions. Optically allowed selection rules will be discussed in Sec. 3.6.1.

3.2.3 THE 3D HARMONIC OSCILLATOR

The solution of the 3D harmonic oscillator with potential $V(r) = \frac{m\omega^2}{2}r^2$ can be obtained as a product wave function $\Psi_{n_x,n_y,n_z} = \psi_{n_x}(x)\psi_{n_y}(y)\psi_{n_z}(z)$ with energy eigenvalues $E_{n_x,n_y,n_z} = \hbar\omega(n_x + n_y + n_z + 3/2) = \hbar\omega(n + 3/2)$, where $n = n_x + n_y + n_z$. The problem can also be solved in spherical coordinates. For arbitrary partial wave l, the wave function takes the form $\psi(\mathbf{r}) = \frac{f_{n,l}(r)}{r}Y_{lm}(\theta,\phi)$; $f_{n,l}$ then satisfies the equation $\frac{d^2}{dr^2}f_{n,l} + \frac{2\mu}{\hbar^2}\left(E_{n,l} - \frac{l(l+1)\hbar^2}{2\mu r^2} - \frac{m\omega^2}{2}r^2\right)f_{n,l} = 0$. The degree of degeneracy of the nth level, $d(n)$, is equal to the number of ways in which n can be divided into the sum of nonnegative integers, $d(n) = (n+1)(n+2)/2$. The effective radial potential is given by $U_l(r) = \frac{l(l+1)\hbar^2}{2mr^2} + \frac{m\omega^2}{2}r^2$. Asymptotically,

$$f_{n,l}(r) \to r^{l+1} \text{ as } r \to 0, \quad f_{n,l}(r) \xrightarrow{r \to \infty} \exp\left(-\frac{r^2}{2\hbar/(m\omega)}\right). \tag{3.86}$$

Based upon our knowledge of the 1D solutions, we can seek solutions of $f_{n,l}$ in the form $f_{n,l}(r) = \xi_{n,l}(\rho)\exp\left(-\frac{\rho^2}{2}\right)$, where $\rho = r/l_{ho}$ is the dimensionless radius, and $l_{ho} = \sqrt{\frac{\hbar}{m\omega}}$ is the harmonic oscillator length. $\xi_{n,l}(\rho)$ can be expressed in terms of the generalized Laguerre polynomials which are specialized forms of the confluent hypergeometric functions [27] [see also Appendix B, (Eq. B.36)]:

$$f_{n,l}(r) = N_{kl}\rho^{l+1}\exp\left(-\frac{\rho^2}{2}\right)L_k^{l+1/2}(\rho), \quad k = (n-l)/2,$$

where N_{kl} is a normalizaion constant. The degeneracy d_n of energy level $E_n = \hbar\omega(n+3/2)$ can be determined by noting that for a given n, if we choose a particular n_x, then $n_y + n_z = n - n_x$. There are $n - n_x + 1$ possible two-tuples $\{n_y, n_z\}$, n_y can take on the values 0 to $n - n_x$, and for each n_y the value of n_z is fixed. Hence $d_n = \sum_{n_x=0}^{n}(n - n_x + 1)$. You are asked to calculate the degeneracy in Problem 3.14.

Problem 3.14

(a) Determine the degeneracies of the energy eigenvalues of a particle in a spherically symmetric harmonic oscillator potential.

(b) Show that the lowest energy eigenfunction corresponds to an $l = 0$ function, and the second level to a set of $l = 1$ functions.

Answer: (a) The degeneracy of the nth level with energy $E_n = \hbar\omega(n + 3/2)$ is the number of ways n can be divided into the sum of three positive integral or zero numbers. From the discussion in the text, $d_n = \sum_{n_x=0}^{n}(n - n_x + 1) = n(\sum_{n_x=0}^{n} 1) - \sum_{n_x=0}^{n} n_x + 1 = (n+1)(n+2)/2$. The energy $E_n = \hbar\omega(n + 3/2)$ can also be written as $E_n = \hbar\omega(2k + l + 3/2)$, where $2k = n - l$ is the the number of zeros of the radial wave function $f_{n,l}$, where k can take on values $k = 0, 1, 2, \ldots$. We can also calculate d_n using the spherically symmetric formulation, but we shall not do so here.

3.2.4 THE MORSE OSCILLATOR

An important model potential that has the property of being finite asymptotically at large r, and having a harmonic oscillator dependence near its minimum, is the Morse oscillator potential,

$$V(r) = D\{1 - \exp[-b(r - r_e)]\}^2 - D. \tag{3.87}$$

Here D is the well depth, i.e., the dissociation energy of a diatomic molecule having this potential, r_e is the equilibrium internuclear distance of the diatomic molecule, and b is a parameter that we will shortly relate to the vibrational frequency of the potential. The potential can be expanded near the minimum of the potential at $r = r_e$, where $V(r_e) = -D$, and the lowest order terms are $V(r) = Db^2(r - r_e)^2 - D$. Hence, the harmonic frequency for motion near the minimum, ω, is related to the coefficients in the Morse potential via the relation $Db^2 = \frac{\mu\omega^2}{2}$, i.e., $b = \sqrt{\frac{\mu\omega^2}{2D}}$. Figure 3.5 shows the Morse potential and its harmonic approximation near the minimum of the potential as a function of the radial coordinate r.

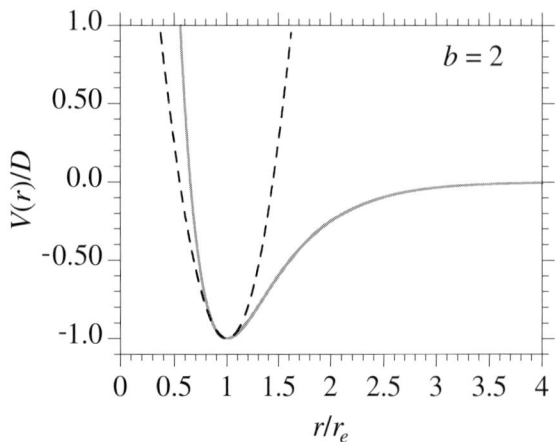

FIG 3.5 Morse potential and its harmonic approximation near the minimum of the potential as a function of the relative radial coordinate r. *Source:* Band, Light and Matter, Fig. C.1, p. 606

For arbitrary partial wave l, the wave function takes the form $\psi(\mathbf{r}) = \frac{f_l(r)}{r} Y_{lm}(\theta, \phi)$; f_l then satisfies the equation $\frac{d^2 f_l}{dr^2} + \frac{2\mu}{\hbar^2}\left(E - \frac{l(l+1)\hbar^2}{2\mu r^2} - V(r)\right) f_l = 0$. The effective radial potential is again given by $U_l(r) = V(r) + \frac{l(l+1)\hbar^2}{2\mu r^2}$. The bound-state wave functions behave asymptotically as

$$f_l(r) \to r^{l+1} \text{ as } r \to 0, \quad f_l(r) \xrightarrow[r \to \infty]{} e^{-\sqrt{\frac{2\mu|E|}{\hbar^2}}r}. \tag{3.88}$$

Analytic solutions for the bound states the Morse potential exist for $l = 0$. The bound-state energies are

$$E_{n,l=0} = \hbar\omega\left[(n + 1/2) - x_e(n + 1/2)^2\right] - D, \tag{3.89}$$

where $x_e \equiv \frac{\hbar\omega}{4D}$ is the dimensionless *anharmonicity constant*; the highest bound state for $l = 0$ corresponds to the integer $n_{\max} = [a - 1/2]$, where $a \equiv x_e^{-1}$. For small anharmonicity, $x_e \ll 1$, the lowest eigenvalues are harmonic-oscillator-like, but as n increases, the spacing between

3.2 Spherically Symmetric Systems

eigenvalues decreases. The eigenfunctions for $l=0$ are given by

$$f_{n,l=0}(r) = Nz^{a-n}e^{-z/2+br/2}L_n^{2a-2n-1}(z), \qquad (3.90)$$

where N is a normalization constant, $z \equiv 2ae^{-b(r-r_e)}$, and $L_n^\alpha(z)$ is the generalized Laguerre polynomial [27], which is the solution to the differential equation, $x\frac{d^2y}{dz^2} + (\alpha+1-z)\frac{dy}{dz} + ny = 0$. We shall not pause to analyze the properties of these eigenfunctions.

3.2.5 VAN DER WAALS AND LENNARD-JONES POTENTIALS

The long-range part of the ground-state potential between two atoms in closed-shell configurations is known as the *van der Waals potential*, named after the Dutch scientist Johannes van der Waals; its dependence on internuclear distance r is $V(r) = -\frac{C_6}{r^6}$ with coefficient $C_6 > 0$. It arises from an induced dipolar interaction between the ground-state atoms (dipole-induced-dipole interaction), and it can be calculated via second-order perturbation theory (see Sec. 7.3.1). The range of internuclear distances over which this interaction potential is valid is much larger than the internuclear distance at which the electrons of each atom overlap. The wave function satisfying the time-independent Schrödinger equation for the relative motion between the atoms is expanded as $\psi(\mathbf{r}) = \frac{f_l(r)}{r}Y_{lm}(\theta,\phi)$, and $f_l(r)$ then satisfies the equation

$$\frac{d^2}{dr^2}f_l + \left(\frac{2\mu E}{\hbar^2} - \frac{l(l+1)}{r^2} + \frac{2\mu C_6/\hbar^2}{r^6}\right)f_l = 0. \qquad (3.91)$$

It is convenient to define van der Waals (VdW) units as follows. The same method used to determine the units of length, energy, and momentum as was used in Sec. 1.3.15 to determine harmonic oscillator units can be used here. The resulting VdW unit of length is

$$l_{\text{vdw}} \equiv (2\mu C_6/\hbar^2)^{1/4}, \qquad (3.92)$$

and the dimensionless coordinate is defined as $y = r/l_{\text{vdw}}$. The VdW unit of energy is $E_{\text{vdw}} \equiv \hbar^2/(2\mu l_{\text{vdw}}^2)$ and the dimensionless energy is defined as $\mathcal{E} \equiv E/E_{\text{vdw}}$. The VdW unit of momentum is taken as $\hbar/l_{\text{vdw}}$. Using these units, Eq. (3.91) becomes

$$\frac{d^2}{dy^2}f_l + \left(\mathcal{E} - \frac{l(l+1)}{y^2} + \frac{1}{y^6}\right)f_l = 0. \qquad (3.93)$$

The analytic properties of f_l can be analyzed (and have been), but we shall not pause to do so.

The Lennard-Jones potential, $V_{LJ}(r) = \frac{C_{12}}{r^{12}} - \frac{C_6}{r^6}$, named after the British mathematician and physicist John Lennard-Jones, is often used as an approximate model for the isotropic part of a total (repulsion plus attraction) diatomic molecular potential as a function of internuclear distance; it is often called the 6–12 potential. The long-range attractive part of the Lennard-Jones potential is simply the VdW potential; the repulsive part of this potential does not accurately represent the potential in the inner internuclear coordinate region of real molecules – it is only a convenient representation for a strongly repulsive potential. The minimum of this potential is at $r_{\min} = (2C_{12}/C_6)^{1/6}$, where $V(r_{\min}) = -\frac{(C_6)^2}{4C_{12}}$. Figure 3.6 shows the effective potentials $U_l(r) = \frac{l(l+1)\hbar^2}{2\mu r^2} +$

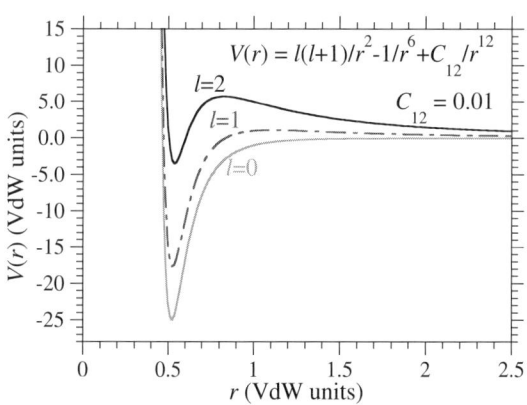

FIG 3.6 Lennard-Jones effective potentials for $l = 0, 1, 2$ versus the relative coordinate r.

$\frac{C_{12}}{r^{12}} - \frac{C_6}{r^6}$ for $l = 0, 1, 2$ versus r in van der Waals units with $C_{12} = 0.01$, so the minimum of the potential for $l = 0$ is potential for $l = 0$ is $r_{\min} = (2C_{12}/C_6)^{1/6} = (0.02)^{1/6} \approx 0.521$ VdW units.

Other potentials of the form $V(r) = -C_n/r^n$ are useful for describing interactions between atoms or between ions and atoms. Dipole–dipole interactions yield the case $n = 3$, and ion interactions with an induced dipole gives $n = 4$. The length and energy scales for $V(r) = -C_n/r^n$ are

$$l_n \equiv (2\mu C_n/\hbar^2)^{1/(n-2)}, \quad E_n \equiv \frac{\hbar^2}{2\mu l_n^2}. \tag{3.94}$$

3.2.6 THE HYDROGEN ATOM

The center-of-mass Hamiltonian for a hydrogenic atom, with a nucleus of charge Ze and mass m_N, is $H = -\frac{\hbar^2}{2\mu}\nabla^2 - \frac{Ze^2}{[4\pi\varepsilon_0]r}$. The reduced mass $\mu = \frac{m_e m_N}{m_e + m_N}$ is approximately equal to the electron mass, m_e (this is an excellent approximation since even for hydrogen $m_N = m_p \approx 1836 m_e$). The factor $[4\pi\varepsilon_0]$ in the Coulomb potential $V(\mathbf{r}) = -\frac{Ze^2}{[4\pi\varepsilon_0]r}$ is required in SI units, but this factor should be replaced by unity in Gaussian units. The time-independent Schrödinger equation is $\nabla^2 \psi(\mathbf{r}) + \frac{2m_e}{\hbar^2}\left(E + \frac{Ze^2}{[4\pi\varepsilon_0]r}\right)\psi(\mathbf{r}) = 0$. Substituting the wave function written as $\psi(\mathbf{r}) = R_l(r)Y_{lm}(\theta,\phi)$, where $R_l(r) = \frac{f_l(r)}{r}$, into the Schrödinger equation and using Gaussian units (setting $[4\pi\varepsilon_0] = 1$), we obtain the following equation for f_l:

$$\frac{d^2}{dr^2}f_l + \left(\frac{2m_e E}{\hbar^2} - \frac{l(l+1)}{r^2} + \frac{2m_e Ze^2}{\hbar^2}\frac{1}{r}\right)f_l = 0. \tag{3.95}$$

Before proceeding, we introduce atomic units (a.u.) since it is easier to analyze hydrogenic atom properties using atomic units. We also take this opportunity to review SI and Gaussian units.

SI, Gaussian, and Atomic Units

The International System of Units (SI) is an MKS (meter kilogram second) system in which the fundamental unit of current is called the Ampère (hence this system is also known as the MKSA system of units). An Ampère [A] is a unit of current flow defined as the current, when flowing in each of two infinitely long, parallel wires of negligible cross-sectional area separated by a distance of 1 m in vacuum, causes a transverse force per unit length of 2×10^{-7} newton/m to act between the wires. The secondary unit of charge is known as a Coulomb; one Coulomb (abbreviated C) is equal to the current obtained when one Ampère of current flows for one second. The charge of an electron is $1.602176462 \times 10^{-19}$ C. The volt is a derived unit in terms of force per unit length per unit charge.

A system of units that is perhaps more suitable for microscopic problems involving electric and magnetic phenomena is the Gaussian system of units. In this system of units, the unit of charge is called the statcoulomb; it is sometimes called an electrostatic unit (or simply esu). One statcoulomb equals $1/(3 \times 10^9)$ C. Therefore, the charge of an electron is $4.803204197 \times 10^{-10}$ statcoulombs (esu). The statvolt is the unit of voltage, i.e., force per unit length per unit charge. One statvolt equals 300 volts. Current is measured in units of statampères; a current of one statampère means the flow of one statcoulomb. The conversion from SI (MKSA) to Gaussian units is summarized in Table 3.1.

A convenient system of units for atomic and molecular physics is the atomic system of units (a.u.). In this system, the unit of charge equals the absolute value of the electron charge, $e = 4.8032 \times 10^{-10}$ statcoulomb (esu) ($= 1.602 \times 10^{-19}$ C), the unit of mass is the electron mass $m_e = 9.109 \times 10^{-28}$ g, and the unit of angular momentum is $\hbar = 1.054 \times 10^{-27}$ erg s. Hence in atomic atomic units (a.u.), $|e| = 1$, $m_e = 1$, and $\hbar = 1$. An important dimensionless parameter in atomic, molecular, and optical physics is the *fine structure constant*, α:

$$\boxed{\alpha = \frac{e^2}{\hbar c} = 7.297352533 \times 10^{-3} \approx 1/137 \quad \left\{\alpha = \frac{e^2}{[4\pi\varepsilon_0]\hbar c} \text{ in SI units}\right\}.} \tag{3.96}$$

3.2 Spherically Symmetric Systems

Table 3.1 *Conversion Table from SI (MKSA) to Gaussian units. All factors of 3 (apart from exponents) should be replaced by 2.99792458, arising from the numerical value of the velocity of light. (See Jackson,* Classical Electrodynamics *[13].)*

Physical quantity (symbol)	SI	Gaussian equivalent
Length (l)	1 m	100 cm
Mass (m)	1 kg	1000 g
Time (t)	1 s	1 s
Frequency (ν)	1 Hz	1 Hz
Force (F)	1 newton	10^5 dyne
Work [energy] (W [U])	1 joule	10^7 erg
Power (P)	1 watt	10^7 erg s^{-1}
Charge (q)	1 Coulomb (C)	3×10^9 statcoulomb (esu)
Charge density (ρ)	C m^{-3}	3×10^3 statcoul cm^{-3} (esu cm^{-3})
Current (I)	1 ampere (amp)	3×10^9 statampere (statamp)
Electric field (**E**)	1 volt m^{-1}	$(1/3) \times 10^{-4}$ statvolt cm^{-1}
Potential (V)	1 volt	1/300 statvolt
Resistance (R)	ohm	$(1/9) \times 10^{-11}$ s cm^{-1}
Capacitance (C)	1 farad (C/V)	9×10^{11} cm
Magnetic induction (**B**)	1 tesla	10^4 gauss
Magnetic field (**H**)	1 ampere m^{-1}	$4\pi \times 10^{-3}$ oersted
Intensity (I)	1 W m^{-2}	10^3 erg s^{-1} cm^{-2}

This is the expansion constant employed in Quantum Electrodynamics, and since it is small, Quantum Electrodynamics expansions converge quickly. In a.u., the speed of light in vacuum is approximately 137, i.e., $c \approx 137 \, [v_0]$, where $v_0 = e^2/\hbar$ is the *Bohr velocity*, as is evident by setting e and $\hbar$ equal to unity in Eq. (3.96). Hence, the atomic unit of velocity is the speed of light divided by about 137. Moreover, the atomic unit of energy (the Hartree) is given by $\alpha^2 m_e c^2$, i.e., the product of the electron rest mass energy and the square of the fine structure constant.

Let us derive the atomic unit of length, also called the *Bohr radius* or simply the *Bohr*, a_0, the atomic unit of energy, also called the *Hartree*, E_0, and the atomic unit of momentum, p_0, using Eq. (3.95). The units of the differential operator $\frac{d^2}{dr^2}$ appearing in the first term on the LHS of (3.95) is $1/a_0^2$. The term containing $\frac{2m_e Z e^2}{\hbar^2} \frac{1}{r}$ must have the same units. Hence, equating $1/a_0^2$ and $\frac{m_e e^2}{\hbar^2} \frac{1}{a_0}$ (the charge Z is not included in determining the atomic unit of length – it has no units and for hydrogen $Z=1$, neither is the factor 2), we obtain

$$a_0 \equiv \frac{\hbar^2}{m_e e^2}. \tag{3.97}$$

Now, equating the term $\frac{2m_eE}{\hbar^2}$ with the term $\frac{2m_eZe^2}{\hbar^2}\frac{1}{r}$, we obtain

$$E_0 \equiv \frac{e^2}{a_0} = \frac{m_e e^4}{\hbar^2} = \alpha^2 mc^2. \tag{3.98}$$

Since the units of momentum are those of $\frac{\hbar}{i}\frac{d}{dr}$, we conclude that the atomic unit of momentum is

$$p_0 \equiv \frac{\hbar}{a_0} = \frac{m_e e^2}{\hbar}. \tag{3.99}$$

Problem 3.15

(a) For the Hamiltonian $H = -\frac{\hbar^2}{2\mu}\nabla^2 - C_3/r^3$ with a "charge dipole" $-C_3/r^3$ molecular potential, find the appropriate units of length, momentum, and energy in terms of μ and C_3 when C_3 is given in units of Hartree $\times a_0^3$. Express your answer in atomic units.

(b) For the sodium molecule, the long-range molecular Hamiltonian is $H = -\frac{\hbar^2}{2m\mu}\nabla^2 - C_6/r^6$, with $\mu = m_{Na}/2 = 11.5$ atomic mass units (amu) and $C_6 = 1556$ a.u. Find l_{vdw} and E_{vdw} in a.u. Note: 1 amu $= 1822.889\, m_e$.

Answers: (a) $l_3 = 2(\mu/m_e)C_3$, $p_3 = [2(\mu/m_e)C_3]^{-1}$, $E_3 = \frac{1}{8(\mu/m_e)^3 C_3^2}$.

(b) $l_{vdw} = (2\mu C_6/\hbar^2)^{1/4} = 90.0\, a_0$, $E_{vdw} = \hbar^2/(2\mu l_{vdw}^2) = 2.92 \times 10^{-9}$ Hartree.

A useful relation exists involving the fine structure constant and the three lengths: the Bohr radius, a_0, the *Compton wavelength* divided by 2π (the de Broglie wavelength of an electron "moving at the speed of light"), $\lambda_c = \frac{\hbar}{m_e c} \approx 3.862 \times 10^{-11}$ cm, and the *classical electron radius*, $r_e = \frac{e^2}{m_e c^2} \approx 2.818 \times 10^{-13}$ cm. The classical electron radius r_e is defined such that, if the electron charge was contained in a sphere of this radius, the electrical energy would be mc^2, i.e., $mc^2 = e^2/r_e$. The Compton wavelength and the classical electron radius will be useful quantities in discussing electron-photon scattering (see Sec. 7.4.3). The relation is:

$$a_0 = \frac{\lambda_c}{\alpha} = \frac{r_e}{\alpha^2}, \tag{3.100}$$

or, $\alpha^2 a_0 = \alpha \lambda_c = r_e$. Thus, $a_0 = 1$ Bohr $\approx 5.292 \times 10^{-9}$ cm, $\lambda_c \approx 1/137\,[a_0] = 3.862 \times 10^{-11}$ cm, and $r_e \approx (1/137)^2\,[a_0] \approx 2.818 \times 10^{-13}$ cm.

The atomic unit of electrical potential ($\phi = q/r$) is e/a_0, the atomic unit of energy ($V = q\phi = q^2/r$) is e^2/a_0, and the atomic unit of electric field ($E = qr/r^3$) is e/a_0^2. In Gaussian units (and therefore atomic units), the units of magnetic induction B and magnetic field H are equal to that of electric field E. Hence, the atomic unit of B and H is e/a_0^2. The magnetic field in the rest frame of an electron due to a proton a distance 1 Bohr away and moving with velocity $\mathbf{v}$ ($\mathbf{H} = \frac{\mathbf{v}}{c} \times e\mathbf{r}/r^3$), with $\mathbf{v}$ orthogonal to $\mathbf{r}$, is given by $|B| = (v/c)(e/a_0^2)$, which is approximately 1/137 in atomic units of magnetic field if v is the Bohr velocity (since $v_0/c \approx 1/137$). Numerically, one atomic unit of intensity is about 8.825×10^{17} W/cm^2.

The units of charge, mass, length, velocity, momentum, time, energy, electrical potential (energy/charge), frequency, force, electric field, magnetic induction, magnetic field, intensity, and magnetic moment in atomic units are summarized in Table 3.2.

Summarizing, the atomic units of length, energy, and momentum are:

Length unit = Bohr, $a_0 \equiv \frac{1}{\alpha}\frac{\hbar}{m_e c} = \frac{[4\pi\varepsilon_0]\hbar^2}{e^2 m_e} = 5.292 \times 10^{-11}$ m

Energy unit = Hartree, $E_0 \equiv \alpha^2 m_e c^2 = \frac{e^2}{a_0} = 4.360 \times 10^{-18}$ J $= 27.21$ eV

Momentum unit, $p_0 \equiv \hbar/a_0 = \frac{m_e e^2}{\hbar} = 1.993 \times 10^{-24}$ Kg m/s

3.2 Spherically Symmetric Systems

Table 3.2 *Atomic units (a.u.).*

Unit [Name]	Symbol and Gaussian formula	Numerical value [Gaussian (SI)]
Charge	e	$4.803204197 \times 10^{-10}$ statcoulomb (1.602176×10^{-19} C)
Mass	m_e	$9.10938188 \times 10^{-28}$ g
Angular momentum	$\hbar$	$1.0545716 \times 10^{-27}$ erg s
Length [Bohr]	$a_0 = \frac{\hbar^2}{m_e e^2}$	$5.291772083 \times 10^{-9}$ cm
Velocity [Bohr velocity]	$v_0 = e^2/\hbar = \alpha c$	$(1/137.036)c = 2.1877 \times 10^8$ cm/s
Momentum [Bohr^{-1}]	$p_0 = m_e e^2/\hbar$	1.9926×10^{-19} g cm/s
Time	$a_0/v_0 = \hbar^3/m_e e^4$	2.4189×10^{-17} s
Energy [Hartree]	$e^2/a_0 = m_e e^4/\hbar^2$	4.3590×10^{-11} erg
Electrical potential	$e/a_0 = m_e e^3/\hbar^2$	0.09076 statvolt (27.210 V)
Frequency	$v_0/a_0 = e^2/(a_0\hbar) = m_e e^4/\hbar^3$	4.1341×10^{16} s^{-1}
Force	e^2/a_0^2	8.2377×10^{-3} dynes
Electric field (**E**)	$e/a_0^2 = m_e^2 e^5/\hbar^4$	1.71510×10^7 statvolt/cm (5.1453×10^{11} V/m)
Magnetic induction (**B**)	$(e/a_0^2) = m_e^2 e^5/\hbar^4$	1.71510×10^7 gauss (1715.1 T)
Magnetic field (**H**)	$(e/a_0^2) = m_e^2 e^5/\hbar^4$	1.71510×10^7 oersted (1.3648×10^9 ampere m^{-1})
Intensity (I)	ce^2/a_0^4	8.825×10^{24} erg s^{-1} cm^{-2} (8.825×10^{17} W/cm^2 = 8.825×10^{21} W/m^2)
Electric dipole moment (p)	ea_0	2.5415×10^{-18} statcoulomb cm (8.4784×10^{-30} C m) = 2.5415 Debye
Magnetic moment (μ_0) Bohr magneton	$\mu_0 = e\hbar/(2m_e c)$	0.9274×10^{-20} erg/gauss (1.40 MHz/G)

The radial Schrödinger equation with a Coulomb potential for f_l is transformed to the following form in atomic units (setting $e = \hbar = m_e = 1$):

$$-\frac{\hbar^2}{2m_e}\frac{d^2 f_l}{dr^2} + \left[\frac{l(l+1)\hbar^2}{2m_e r^2} - \frac{Ze^2}{r} - E\right] f_l = 0 \rightarrow -\frac{1}{2}\frac{d^2 f_l}{dr^2} + \left[\frac{l(l+1)}{2r^2} - \frac{Z}{r} - \mathcal{E}\right] f_l = 0.$$

On the RHS of the latter equation, $\mathcal{E}$ is in atomic units, i.e., $\mathcal{E} = E/E_0$, and r is dimensionless, i.e., is taken to be in atomic units.

The Coulomb Radial Wave Function

The radial wave function $f_l(r)$ in a.u. satisfies the equation

$$-\frac{1}{2}\frac{d^2 f_l}{dr^2} + \left[\frac{l(l+1)}{2r^2} - \frac{Z}{r}\right]f_l = \mathcal{E}f_l. \quad (3.101)$$

Asymptotically, bound-state wave functions behave as follows:

$$f_l(r) \to r^{l+1} \quad \text{as } r \to 0, \quad (3.102)$$

$$f_l(r) \xrightarrow[r \to \infty]{} e^{-\sqrt{2|\mathcal{E}|}\, r}. \quad (3.103)$$

Given this behavior, we take the ansatz form, $f_l(r) = r^{l+1} w_l(r) e^{-\sqrt{2|\mathcal{E}|}\, r}$, where $w_l(r)$ is to be determined. The differential equation for $w_l(r)$ obtained by substituting this form into (3.101) is

$$r\frac{d^2 w_l}{dr^2} + \left[2(l+1) - 2\sqrt{2|\mathcal{E}|}\, r\right]\frac{dw_l}{dr} + 2\sqrt{2|\mathcal{E}|}\,(Z - (l+1))\, w_l = 0. \quad (3.104)$$

Let us assume that the function $w_l(r)$ can be expanded in a power series, $w_l(r) = \sum_{k=0}^{\infty} a_k r^k$. Substituting this into Eq. (3.104), we get

$$\sum_{k=0}^{\infty}\left[(k(k+1) + 2(k+1)(l+1))a_{k+1} + (2Z - 2\sqrt{2|\mathcal{E}|}(k+l+1))a_k\right] r^k = 0.$$

Setting the coefficient of r^k equal to zero, we obtain the recurrence relation for the series,

$$\frac{a_{k+1}}{a_k} = \frac{2[\sqrt{2|\mathcal{E}|}(k+l+1) - Z]}{k(k+1) + 2(k+1)(l+1)}. \quad (3.105)$$

In order for the series to remain finite at large r, it must terminate at some value of k. Termination is only possible if the numerator on the RHS of (3.105) vanishes for some value of k, say $k_{\max}$, $Z = \sqrt{2|\mathcal{E}|}(k_{\max} + l + 1)$. Defining the integer $n \equiv k_{\max} + l + 1$, the condition for termination of the series [so the numerator on the RHS of (3.105) vanishes] becomes

$$\mathcal{E}_n = -\frac{Z^2}{2n^2}. \quad (3.106)$$

The integers n and l must be such that $k_{\max} \geq 0$, i.e., $l \leq n-1$. We conclude that the hydrogenic energy eigenvalues depend only on n, which is called the *principal quantum number*, not on l or m, and are given by

$$\boxed{E_{n,l,m} = -\frac{Z^2}{2n^2}\frac{e^2}{a_0} = -\frac{Z^2}{2n^2}\frac{m_e e^4}{\hbar^2}.} \quad (3.107)$$

(To be more exact, m_e should be replaced by the reduced mass). Equation (3.107) is known as the *Balmer formula*.[2] We shall return to the spectrum of hydrogenic atoms shortly, after describing the wave functions.

The radial wave function depends on the *quantum numbers n and l* and is given, in a.u., by $R_{nl} = f_{nl}/r = r^l\, e^{-Zr/n} w_{nl}(r)$, where $w_{nl}(r) = \sum_{k=0}^{\infty} a_k r^k$, with (3.105) determining all the a_k, except for the first value, a_0, which will be determined by normalization of the wave function. The function $w_{nl}(r)$, with r is in a.u., can be identified as a confluent hypergeometric

---

[2] For comparison, the hydrogenic energy eigenvalues of the relativistic Dirac equation are $E_{n,j} = m_e c^2 \left\{1 + \frac{Z^2\alpha^2}{[n-j-1/2+[(j+1/2)^2-Z^2\alpha^2]^{1/2}]^2}\right\}$, where j is the total electronic angular momentum. To order α^4, $E_{n,j} \approx m_e c^2 \left\{1 - \frac{Z^2\alpha^2}{2n^2} - \frac{Z^4\alpha^4}{2n^4}[n/(j+1/2) - 3/4] + \ldots\right\}$. The first term is the rest mass energy, the second term is the Balmer formula, and the last term incorporates the spin–orbit splitting that depends on j (see Sec. 4.5).

3.2 Spherically Symmetric Systems

function, $w_{nl}(r) = {}_1F_1(-n+l+1, 2(l+1), 2Zr/n)$ [see Appendix B, Eq. (B.36)], which is in fact an associated Laguerre polynomial, $L_{n+l}^{2l+1}(2Zr/n)$. The full wave function is given by

$$\psi_{nlm}(\mathbf{r}) = R_{nl}(r)\,Y_{lm}(\theta,\phi), \quad R_{nl}(r) = (Z/a_0)^{3/2} N_{nl} F_{nl}\left(\frac{2Zr}{na_0}\right), \tag{3.108}$$

$$N_{nl} = \sqrt{\frac{2^3}{n^3}\frac{(n-l-1)!}{2n(n+l)!}}, \quad F_{nl}(x) = x^l e^{-x/2} L_{n-l-1}^{2l+1}(x). \tag{3.109}$$

Since the spherical harmonics are normalized, normalization of the full wave function requires

$$\int_0^\infty dr\, r^2 R_{nl}^2(r) = 1. \tag{3.110}$$

The first few radial wave functions are (note that r is not in Bohr, i.e., we explicitly include a_0):

$$R_{10}(r) = 2\left(\frac{Z}{a_0}\right)^{3/2} \exp(-Zr/a_0)$$

$$R_{21}(r) = \frac{1}{\sqrt{3}}\left(\frac{Z}{2a_0}\right)^{3/2}\left(\frac{Zr}{a_0}\right)\exp\left(\frac{-Zr}{2a_0}\right)$$

$$R_{20}(r) = 2\left(\frac{Z}{2a_0}\right)^{3/2}\left(1-\frac{Zr}{2a_0}\right)\exp\left(\frac{-Zr}{2a_0}\right)$$

$$R_{32}(r) = \frac{4}{27\sqrt{10}}\left(\frac{Z}{3a_0}\right)^{3/2}\left(\frac{Zr}{a_0}\right)^2 \exp\left(\frac{-Zr}{3a_0}\right)$$

$$R_{31}(r) = \frac{4\sqrt{2}}{9}\left(\frac{Z}{3a_0}\right)^{3/2}\left(1-\frac{Zr}{6a_0}\right)\left(\frac{Zr}{a_0}\right)\exp\left(\frac{-Zr}{3a_0}\right)$$

$$R_{30}(r) = 2\left(\frac{Z}{3a_0}\right)^{3/2}\left(1-\frac{2Zr}{3a_0}+\frac{2Z^2r^2}{27a_0^2}\right)\exp\left(\frac{-Zr}{3a_0}\right)$$

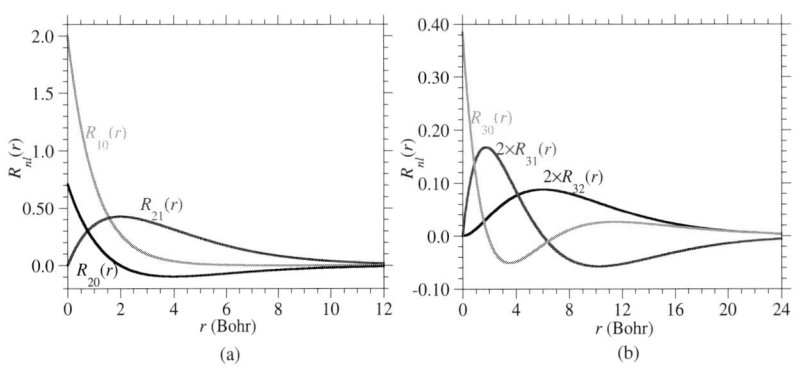

FIG 3.7 Hydrogen radial wave functions $R_{nl}(r)$ versus radial coordinate r.

Figure 3.7 plots the lowest radial wave functions versus r.

The total probability density at point $\mathbf{r}$ is given by $\mathcal{P}(\mathbf{r}) = |\psi_{nlm}(\mathbf{r})|^2 = R_{nl}^2(r)\,|Y_{lm}(\theta,\phi)|^2$, and the radial probability density, i.e., the probability to find the electron in a spherical shell between r and $r+dr$ is

$$P(r)dr = \int_{\theta,\phi} d\Omega\,\mathcal{P}(\mathbf{r})\,r^2\,dr = r^2 R_{nl}^2(r)dr.$$

(3.111)

The expectation values of r^p, $\langle \psi_{nlm}|r^p|\psi_{nlm}\rangle = \int_0^\infty dr\, r^{p+2} R_{nl}^2 = \int_0^\infty dr\, r^p f_{nl}^2$, for p equal to a positive or negative integer can be evaluated analytically, e.g.,

$$\langle n\,l'\,m'|r|n\,l\,m\rangle = \delta_{l',l}\,\delta_{m',m}\,n^2\frac{a_0}{Z}\left(1+\frac{1}{2}\left[1-\frac{l(l+1)}{n^2}\right]\right), \tag{3.112a}$$

$$\langle n\,l\,m|r^2|n\,l\,m\rangle = n^4\left(\frac{a_0}{Z}\right)^2\left(1+\frac{3}{2}\left[1-\frac{l(l+1)-1/3}{n^2}\right]\right), \tag{3.112b}$$

$$\langle n\,l\,m|r^{-1}|n\,l\,m\rangle = \frac{Z}{n^2 a_0}, \tag{3.112c}$$

$$\langle n\,l\,m|r^{-2}|n\,l\,m\rangle = \frac{Z^2}{n^3(l+1/2)a_0^2}, \tag{3.112d}$$

$$\langle n\,l\,m|r^{-3}|n\,l\,m\rangle = \frac{Z^3}{n^3(l+1)(l+1/2)l a_0^3}. \tag{3.112e}$$

Problem 3.16

(a) Without looking back at the chapter or the table above, write the atomic units of length, momentum, and energy in terms of the constants $\hbar, m_e, e$ (and if you like, c).
(b) Determine the matrix elements $\langle n'\,l'\,m'|H|n\,l\,m\rangle$ where H is the hydrogenic Hamiltonian.
(c) What is the degeneracy of the levels with a given value of the principal quantum number n.
(d) Calculate the matrix elements $\langle n'\,l'\,m'|L^2|n\,l\,m\rangle$, $\langle n'\,l'\,m'|L^4|n\,l\,m\rangle$, and $\langle n'\,l'\,m'|L_z|n\,l\,m\rangle$.

Problem 3.17

The normalized wave function for the ground-state hydrogen-like atom (H, He$^+$, Li^{++}, etc.) with nuclear charge Ze is $\psi(\mathbf{r}) = A e^{-\beta r}$. Without looking back at the chapter:

(a) Find A in terms of β.
(b) Find β in terms of the fundamental constants $e, m_e, \hbar$, and Z.
(c) Find the energy E in terms of $e, m_e, \hbar$, and Z.
(d) Find the expectation value of r, $\langle\psi|r|\psi\rangle$, given $\int_0^\infty dr\, r^3 e^{-2\beta r} = \frac{3}{8\beta^4}$.
(e) Find the value of r with maximum probability of finding the electron.

Answers: (a) $A = \sqrt{\frac{\beta^3}{\pi}}$. (b) $\beta = \frac{Ze^2 m_e}{\hbar^2} = \frac{Z}{a_0}$. (c) $E = E = -\frac{Z^2}{2}mc^2\left(\frac{e^2}{\hbar c}\right)^2 = -\frac{Z^2}{2}\frac{m_e e^4}{\hbar^2}$. (d) $\langle\psi|r|\psi\rangle = \frac{3a_0}{2Z}$.
(e) $\frac{d}{dr}(r^2|\psi(\mathbf{r})|^2) = 0$, yields $r_m = \frac{a_0}{Z}$.

The spectrum of the hydrogen atom includes positive energy eigenvalues, where $\mathcal{E}$ is continuous and extends from zero to infinity. These eigenvalues are infinitely degenerate; to each value of $\mathcal{E}$ there corresponds an infinite number of states with different l from 0 to ∞, and for any value of l, $m = -l, \ldots, l$. The radial eigenfunctions of the continuous spectrum take the form $R_{kl}(r) = C_{kl}(kr)^l e^{ikr}{}_1F_1\left(\frac{i}{k}+l+1, 2(l+1), 2ikr\right)$, where F is the confluent hypergeometric function and $\mathcal{E} = k^2/2$ (in a.u.). Asymptotically, at large r,

$$R_{kl}(r) \to \frac{1}{r}\sin\left(kr + \frac{\log 2kr}{k} - l\pi/2 + \sigma_l\right),$$

3.2 Spherically Symmetric Systems

where the *Coulomb phase shift*, $\sigma_l = \arg \Gamma(l + 1 - \frac{i}{k})$, and Γ is the gamma function [27]. We treat scattering from a Coulomb potential in Sec. 12.5.5.

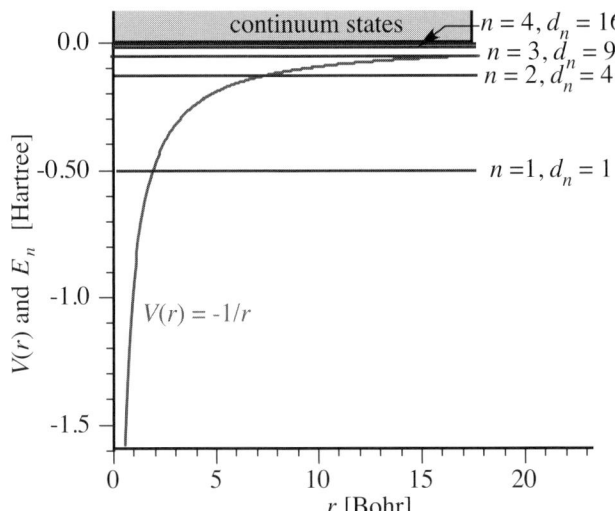

FIG 3.8 Hydrogen energy levels plotted on the same scale as the Coulomb potential. The degeneracy d_n of the bound-state levels is specified.

The Hydrogen Atom Spectrum

We derived the bound-state spectrum of hydrogenic atoms in Eq. (3.107), $E_n = -\frac{Z^2}{2n^2}\frac{e^2}{a_0}$. For the hydrogen atom, $Z = 1$, these discrete bound-state energies go up in energy from the lowest level corresponding to $n = 1$, for which $\mathcal{E}_1 = -\frac{1}{2}$. The first excited level, $n = 2$, has $\mathcal{E}_2 = -\frac{1}{2 \times 4}$, $n = 3$ has $\mathcal{E}_3 = -\frac{1}{2 \times 9}$, etc., all the way up in energy as $n \to \infty$ to zero energy, as shown in Fig. 3.8. States in the continuum exist for $E > 0$, but we shall not discuss them at present (see Sec. 12.5.5). For a given n, there are n values of l, $l = 0, \ldots, n-1$. For each l, there are $2l + 1$ values of m, $m = -l, -l+1, \ldots, l-1, l$, i.e., the degeneracy of the l states is $d_l = 2l+1$. The degree of degeneracy d of a hydrogenic bound-state level n is $d_n = n^2$, without consideration of spin degrees of freedom, which doubles this number (see Chapter 4). The energy levels E_n versus l and the degeneracy d_l of the various l levels are shown in Fig. 3.9. Note that the degeneracy of the various l levels belonging to the same n is special to hydrogenic atoms. This special degeneracy can be traced back to the presence of an additional conserved quantity in the Coulomb potential, the Runge–Lenz vector [see Eq. (16.79), linked to the book web page]. Orbitals of other atoms, e.g., He, Li, Na, etc., do possess this additional degeneracy.

Excited states can decay to lower states via emission of photons. The photon frequency emitted in the decay of state nlm to state $n'l'm'$, if it can occur (there are selection rules that restrict the transitions that can occur radiatively; see Sec. 3.6.1) is equal to $\nu_{nn'} = (E_n - E_{n'})/h$, which depends only on the principal quantum numbers n and n',

$$\nu_{nn'} \left(= \frac{c}{\lambda_{nn'}}\right) = \frac{E_n - E_{n'}}{h} = -\frac{Z^2}{2}\left(\frac{1}{n^2} - \frac{1}{n'^2}\right)\frac{m_e e^4}{\hbar^3}.$$

(3.113)

The transitions to $n' = 1$ are called *Lyman transitions*. These transitions are all at very high energy; the lowest of the Lyman transitions, $n = 2 \to n = 1$, is at 3/8 Hartree = 10.2 eV. Transitions to $n' = 2$ are called *Balmer transitions*. These are at much lower energies; Fig. 1.1 shows these transitions. Energy levels and transition energies of atoms and atomic ions are conveniently represented in *Grotrian diagrams* of the form shown in Fig. 3.10.

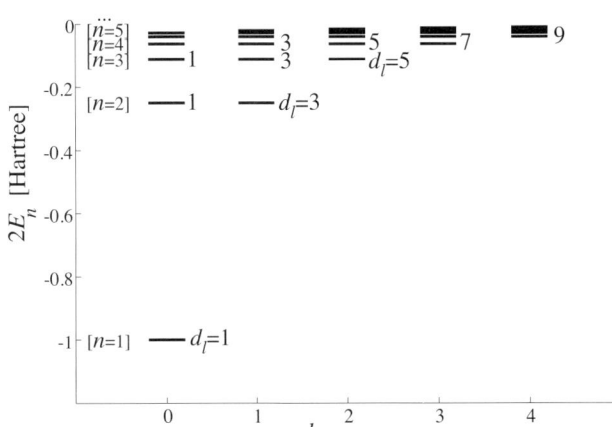

FIG 3.9 Hydrogen energy levels, E_n, and the degeneracies $d_l = 2l + 1$ of the various orbital angular momentum l levels. The spin degeneracy is 2 (spin-up and spin-down states are degenerate), but it has not been indicated in the figure.

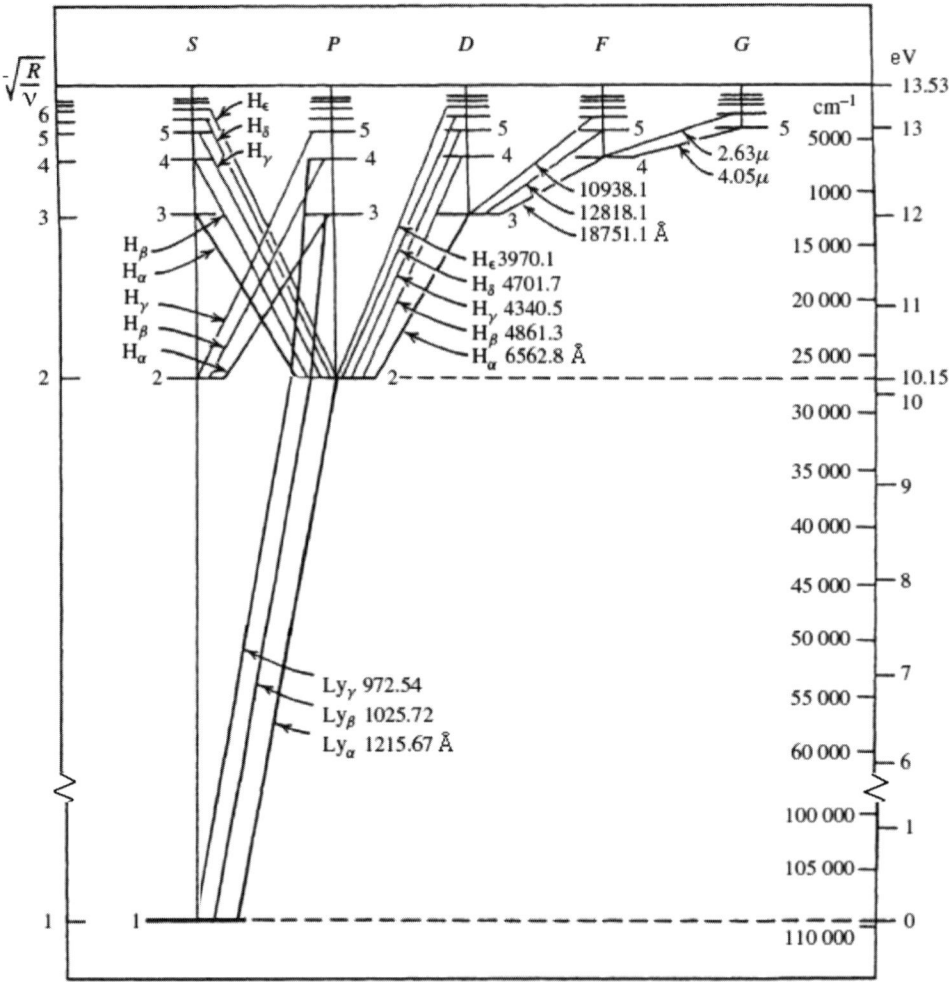

FIG 3.10 Grotrian diagram for hydrogen. The Lyman transitions are denoted by Ly$_\alpha$, where α are greek letters in ascending order of energy, the Balmer transitions are denoted H$_\alpha$, etc. The ordinate scales are in eV, cm^{-1}, and $-\sqrt{R/\nu}$, where $R = (1/2)\frac{m_e e^4}{c\hbar^3} = 109677.6$ cm^{-1} is the *Rydberg constant* (half a Hartree) and ν is the transition frequency in units of cm^{-1}. The values whose units are not explicitly shown are wavelengths in Angstroms (10^{-10} m). Reproduced from Grotrian [35].

3.3 ROTATIONS AND ANGULAR MOMENTUM

We saw in Chapter 1 that translation operators are generated by the momentum operator **p**. For a plane wave, $\psi(\mathbf{r} - \mathbf{a}) = e^{i\mathbf{p}\cdot(\mathbf{r}-\mathbf{a})/\hbar} = e^{-i\frac{\mathbf{p}\cdot\mathbf{a}}{\hbar}} e^{i\frac{\mathbf{p}\cdot\mathbf{r}}{\hbar}}$, and for an arbitrary wave function ψ [see Eq. (1.54)], $\psi(\mathbf{r} - \mathbf{a}) = e^{-i\frac{\mathbf{p}\cdot\mathbf{a}}{\hbar}} \psi(\mathbf{r})$. We also saw that rotation of a wave function by an angle φ about an axis $\hat{\boldsymbol{\varphi}}$ is generated by the angular momentum operator **J** [see Eq. (1.62)]. An arbitrary rotation of the wave function of a system is given by

$$\psi(\mathfrak{R}_{-\boldsymbol{\varphi}}\mathbf{r}) = U(\boldsymbol{\varphi})\psi(\mathbf{r}) = e^{-i\boldsymbol{\varphi}\cdot\mathbf{J}/\hbar}\psi(\mathbf{r}). \tag{3.114}$$

3.3 Rotations and Angular Momentum

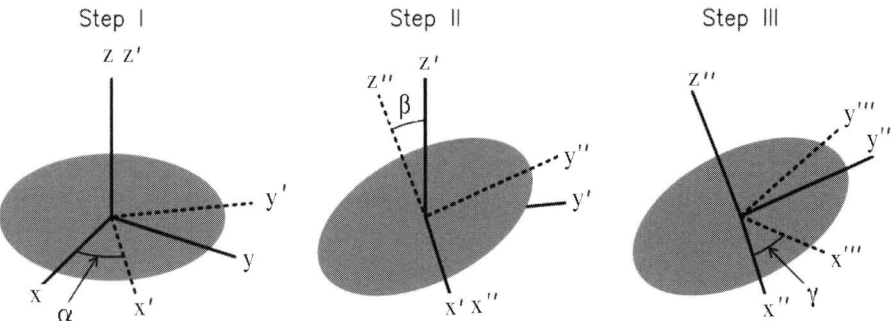

FIG 3.11 Euler angles as conventionally defined in classical mechanics. In quantum mechanics, step II corresponds to a rotation by angle β about the y'-axis, rather than the x'-axis, as explained in the text.

Here $\mathfrak{R}_{-\varphi}\mathbf{r}$ is the rotation of the coordinate $\mathbf{r}$ by the angle $-\varphi$ about the rotation axis $\hat{\boldsymbol{\varphi}}$, $\mathbf{J}$ is the total angular momentum operator, and $U(\boldsymbol{\varphi})$ is the rotation operator that rotates the state by the angle φ about $\hat{\boldsymbol{\varphi}}$. Rotating the wave function $\psi(\mathbf{r})$ by $\boldsymbol{\varphi}$ is equivalent to rotating the coordinate system by $-\boldsymbol{\varphi}$.

A particular example of a rotation operator that rotates a wave function is the operator that rotates about the $\hat{\mathbf{z}}$-axis by an angle φ, $U(\hat{\mathbf{z}}, \varphi) = e^{-i\frac{\varphi J_z}{\hbar}}$; this operator is diagonal in the standard representation where J_z is diagonal. Note that the rotation operator that generates a rotation of a state (a wave function) by an angle φ about an axis $\hat{\mathbf{n}}$ is a function of both φ and the unit vector $\hat{\mathbf{n}}$: $U(\hat{\mathbf{n}}, \varphi) = e^{-i\varphi\hat{\mathbf{n}}\cdot\mathbf{J}/\hbar}$. A unit vector is specified by two angles, so the rotation requires three parameters (angles) to completely define the rotation. A common parameterization of rotations is via the three *Euler angles*. This is the subject of the next subsection.

> **Problem 3.18**
>
> Prove that ∇^2 is invariant under rotations, i.e., under orthogonal transformations of the coordinate system.
>
> **Answer:** There are many ways to prove this. Perhaps the easiest is to note that $\nabla^2 = \frac{-2m}{\hbar^2}\mathbf{p}\cdot\mathbf{p}$, and the scalar product $\mathbf{p}\cdot\mathbf{p}$ is rotationally invariant. Another method is to express ∇^2 in spherical coordinates using (3.60), $\nabla^2 = \frac{1}{r}\frac{\partial^2}{\partial r^2} r - \frac{\mathbf{L}^2}{\hbar^2 r^2}$, and note that each of the terms on the RHS of this equation is rotationally invariant, i.e, each commutes with the angular momentum operators (see Problem 3.10), and are therefore rotationally invariant.

3.3.1 EULER ANGLES α, β, γ AND THE ROTATION MATRIX

Leonhard Euler defined a set of three angles to describe the orientation of a rigid body in a 3D space. A rigid body can be subjected to a sequence of three rotations described in terms of the Euler angles, α, β, γ, to orient the object in any desired way. For the moment, let us consider a coordinate system $0xyz$. The coordinate system can be oriented in any desired orientation, in three steps, as follows. In step I of the sequence, rotate the coordinate axes around the z-axis by an angle α. The resulting coordinate axes are called x', y', and z' as shown in Fig. 3.11. Then, in step II, rotate the coordinate system about the new x-axis, x', by an angle β. The new axes are called x'', y'', and z''. Finally, in step III, rotate the system about z'' by an angle γ. The resulting coordinate system, $0x'''y'''z'''$, is oriented in a completely arbitrary way, depending on the specific Euler angles α, β, γ, used in carrying out the sequence. In quantum mechanics, we use a set of three steps that differs from that shown in Fig. 3.11, only in that the middle rotation is around the y'-axis by angle β, rather than the x'-axis. This convention is used in quantum mechanics because rotation operators corresponding to rotations around the y-axis are real, as opposed to rotation operators corresponding to rotations around the x-axis, which are in general complex, so rotation about the y-axis is simpler to handle (see next section). Hence, the operator $\mathcal{R}_{\alpha\beta\gamma}$ that rotates a state (or wave function) according to this sequence is given by

$$\mathcal{R}_{\alpha\beta\gamma} = \mathcal{R}_\gamma \mathcal{R}_\beta \mathcal{R}_\alpha = e^{-i\mathbf{J}\cdot(\gamma\hat{\mathbf{z}}'')/\hbar} e^{-i\mathbf{J}\cdot(\beta\hat{\mathbf{y}}')/\hbar} e^{-i\mathbf{J}\cdot(\alpha\hat{\mathbf{z}})/\hbar}. \tag{3.115}$$

The expression in (3.115) for the rotation operator is inconvenient since each of the three rotations is performed about an axis belonging to a different coordinate system. It is possible to transform all the operators on the RHS of (3.115) to a common coordinate system. Applying transformation (1.48) to $\mathcal{R}_\beta$, we find $\mathcal{R}_\beta = U(\hat{\mathbf{z}}, \alpha) U(\hat{\mathbf{y}}, \beta) U(\hat{\mathbf{z}}, \alpha)^\dagger$. Similarly, $\mathcal{R}_\gamma$ can be written as

$$\mathcal{R}_\gamma = \mathcal{R}_\beta U(\hat{\mathbf{z}}, \alpha) U(\hat{\mathbf{z}}, \gamma) U(\hat{\mathbf{z}}, \alpha)^\dagger \mathcal{R}_\beta^\dagger.$$

Substituting these expressions into (3.115), we finally obtain, after some algebra, the following expression for $\mathcal{R}$:

$$\boxed{\mathcal{R}_{\alpha\beta\gamma} = e^{-i\alpha J_z/\hbar} e^{-i\beta J_y/\hbar} e^{-i\gamma J_z/\hbar}.} \tag{3.116}$$

Thus, the rotation operator $\mathcal{R}_{\alpha\beta\gamma}$ can be obtained by first rotating about the original z-axis by an angle γ, then rotating about the original y-axis by angle β, and finally rotating about the original z-axis by angle α.

3.3.2 ROTATION AND D FUNCTIONS

Matrix elements of this rotation operator involving the angular momentum eigenvectors take the form,

$$\langle j' m' | \mathcal{R}_{\alpha\beta\gamma} | j m \rangle = e^{-im'\alpha} d^j_{m'm}(\beta) e^{-im\gamma} \delta_{j',j}, \tag{3.117}$$

i.e., they are diagonal in j, as expected, where

$$d^j_{m'm}(\beta) = \langle j m' | e^{-i\beta J_y/\hbar} | j m \rangle. \tag{3.118}$$

The <u>real</u> functions $d^j_{m'm}(\beta)$ are called reduced rotation matrices, or simply d functions. It is convenient to define the symbol $D^{(j)}_{m',m}(\alpha\,\beta\,\gamma) \equiv \langle j m' | \mathcal{R}_{\alpha\beta\gamma} | j m \rangle$. This symbol is often called the *rotation function*, or rotation matrix. From the definition of the rotation function as a matrix element, it is easy to show that

$$\left(D^{(j)}_{m',m}(\alpha\,\beta\,\gamma)\right)^* = D^{(j)}_{m,m'}(-\gamma\,-\beta\,-\alpha). \tag{3.119}$$

Furthermore, since the product of two rotations is also a rotation, it is clear that the product, $\mathcal{R}_{\alpha'\beta'\gamma'}\mathcal{R}_{\alpha\beta\gamma}$ is also an rotation operator, $\mathcal{R}_{\alpha''\beta''\gamma''}$, obtained by composition of rotations, hence,

$$\sum_\mu D^{(j)}_{m,\mu}(\alpha'\,\beta'\,\gamma') D^{(j)}_{\mu,m'}(\alpha\,\beta\,\gamma) = D^{(j)}_{m,m'}(\alpha''\,\beta''\,\gamma''). \tag{3.120}$$

Problem 3.19

Prove Eqs (3.119) and (3.120) using the definition of the rotation function.

Rotation functions can be used to rotate angular momentum eigenstates as follows:

$$\boxed{\mathcal{R}_{\alpha\beta\gamma} | j m \rangle = \sum_{j',m'} | j' m' \rangle \langle j' m' | \mathcal{R}_{\alpha\beta\gamma} | j m \rangle = \sum_{m'} | j m' \rangle D^{(j)}_{m',m}(\alpha\,\beta\,\gamma).} \tag{3.121}$$

3.3 Rotations and Angular Momentum

Thus, rotation of an angular momentum eigenvector is reduced to matrix multiplication by a rotation matrix. For example, rotating a spherical harmonic wave function using Eq. (3.121) and noting that $Y_{lm}(\theta, \phi) = \langle \theta, \phi | lm \rangle$, we obtain,

$$\mathcal{R}_{\alpha\beta\gamma} Y_{lm}(\theta, \phi) = \sum_{m'} Y_{lm'}(\theta, \phi) D^{(j)}_{m',m}(\alpha\,\beta\,\gamma). \tag{3.122}$$

Moreover, we know from (3.114) that $\mathcal{R}_{\alpha\beta\gamma} Y_{lm}(\theta, \phi) = Y_{lm}[(\Re(\alpha\,\beta\,\gamma))^{-1}(\theta, \phi)]$. Taking $\theta = 0$ and $\phi = 0$ and using (3.59), we find

$$Y_{lm}[(\Re(\alpha\,\beta\,\gamma))^{-1}(0,0)] = \sqrt{\frac{(2l+1)}{4\pi}} D^{(j)}_{0,m}(\alpha\,\beta\,\gamma). \tag{3.123}$$

Hence, inverting the rotation in (3.123), we obtain

$$Y_{lm}[\Re(\alpha\,\beta\,\gamma)(0,0)] = \sqrt{\frac{(2l+1)}{4\pi}} D^{(j)}_{0,m}(-\gamma\,-\beta\,-\alpha),$$

which is equivalent to $Y_{lm}(\beta, \alpha) = \sqrt{\frac{(2l+1)}{4\pi}} D^{(j)}_{0,m}(-\gamma\,-\beta\,-\alpha)$. Complex conjugating this equation and using (3.119), we finally obtain

$$D^{(j)}_{m,0}(\alpha\,\beta\,\gamma) = \sqrt{\frac{4\pi}{(2l+1)}} Y^*_{lm}(\beta, \alpha). \tag{3.124}$$

The d functions were first evaluated by Eugene P. Wigner by means of group theory (the rotation functions are sometimes called Wigner functions for this reason); Wigner obtained the expression:

$$d^j_{m'm}(\beta) = \sqrt{(j+m)!(j-m)!(j+m')!(j-m')!} \sum_k \frac{(-1)^k (\cos\frac{\beta}{2})^{2j+m-m'-2k} (\sin\frac{\beta}{2})^{m'-m+2k}}{(j-m'-k)!(j+m-k)!(k+m'-m)!k!}. \tag{3.125}$$

These real functions satisfy the following properties:

$$d^j_{m'm}(\beta) = (-1)^{m-m'} d^j_{mm'}(\beta), \tag{3.126a}$$

$$d^j_{-m'-m}(\beta) = d^j_{mm'}(\beta) = d^j_{m'm}(-\beta), \tag{3.126b}$$

$$d^j_{m'm}(\pi - \beta) = (-1)^{j-m'} d^j_{m'm}(\beta). \tag{3.126c}$$

For any j,

$$d^j_{m'm}(2\pi) = (-1)^{2j} d^j_{m'm}(0) = (-1)^{2j} \delta_{m',m}. \tag{3.127}$$

Aspects of group theory that are useful in quantum mechanics are reviewed in Chapter E of the Appendix. One of the important results of group theory is the group orthogonality theorem. The group orthogonality theorem as applied to the rotation functions reads,

$$\int d(\cos\beta)\, d\alpha\, d\gamma \left(D^{(j)}_{\mu,m}(\alpha\beta\gamma)\right)^* D^{(j')}_{\mu',m'}(\alpha\beta\gamma) = \frac{8\pi^2}{2j+1} \delta_{j',j} \delta_{\mu',\mu} \delta_{m',m}. \tag{3.128}$$

Integration is over the Euler angles; the integral over β is similar to the integral over the angle θ in spherical coordinates, $\int_{-1}^{1} d(\cos\beta)\ldots$, and the integral over α and γ is from zero to 2π. Eq. (3.128) shows that the d functions satisfy the orthogonality relation,

$$\int_{-1}^{1} d(\cos\beta) d^{j}_{m'm}(\beta) d^{j'}_{m'm}(\beta) = \frac{2}{2j+1}\delta_{jj'}. \quad (3.129)$$

Explicit expressions for the case $j=1/2$ and $j=1$ can be easily obtained using either Eq. (3.129) or the properties of the angular momentum operators. For $j=1/2$, the basis states are given by $\left|\frac{1}{2}\frac{1}{2}2\frac{1}{2}\right\rangle = \begin{pmatrix}1\\0\end{pmatrix}, \left|\frac{1}{2}-\frac{1}{2}\right\rangle = \begin{pmatrix}0\\1\end{pmatrix}$, as specified in Eq. (3.27). The d functions are given by matrix elements of the rotation operator $e^{-i\vartheta n_y \cdot s}$. For the $j=1/2$ case, The formula $e^{-i(\vartheta/2)\mathbf{n}_y\cdot\boldsymbol{\sigma}} = [\cos(\vartheta/2)\mathbf{1} - i\sin(\vartheta/2)\sigma_y]$ can be derived by expanding the exponential in a power series and noting that $\sigma_y\sigma_y = 1$ [see Eq. (4.6)], hence $\sigma_y^m = 1$ for even m and $\sigma_y^m = \sigma_y$ for odd m. An arbitrary rotation operator about the unit vector $\mathbf{n}$ by an angle ϑ for spin $1/2$ states can be written as follows:

$$e^{-i\vartheta \mathbf{n}\cdot\mathbf{S}/\hbar} = e^{-i\frac{\vartheta}{2}\mathbf{n}\cdot\boldsymbol{\sigma}} = \cos(\vartheta/2)\mathbf{1} - i\sin(\vartheta/2)\,\mathbf{n}\cdot\boldsymbol{\sigma}. \quad (3.130)$$

Note that a rotation of a spin state by an angle of 2π about any axis multiplies the spin state by (-1). Writing an arbitrary unit vector as $\mathbf{n} = (\sin\theta\cos\phi, \sin\theta\sin\phi, \cos\theta)$, the 2×2 matrix $\mathbf{n}\cdot\boldsymbol{\sigma}$ appearing on the RHS of (3.130) takes the form

$$\mathbf{n}\cdot\boldsymbol{\sigma} = \begin{pmatrix} \cos\theta & e^{-i\phi}\sin\theta \\ e^{i\phi}\sin\theta & -\cos\theta \end{pmatrix}, \quad (3.131)$$

and the general rotation operator is given by

$$e^{-i\vartheta\mathbf{n}\cdot\mathbf{S}/\hbar} = \cos(\vartheta/2)\begin{pmatrix}1&0\\0&1\end{pmatrix} - i\sin(\vartheta/2)\begin{pmatrix}\cos\theta & e^{-i\phi}\sin\theta \\ e^{i\phi}\sin\theta & -\cos\theta\end{pmatrix}. \quad (3.132)$$

In particular, for $\mathbf{n} = \hat{\mathbf{y}}$, i.e., $\theta = \pi/2$ and $\phi = \pi/2$, we obtain the d function

$$d^{(\frac{1}{2})}(\vartheta) = \begin{pmatrix} \cos(\vartheta/2) & -\sin(\vartheta/2) \\ \sin(\vartheta/2) & \cos(\vartheta/2) \end{pmatrix}. \quad (3.133)$$

(Sometimes the superscript of the d functions are put in parenthesis.) For $j=1$, the basis states are the three component vectors,

$$|1,1\rangle = \begin{pmatrix}1\\0\\0\end{pmatrix}, \quad |1,0\rangle = \begin{pmatrix}0\\1\\0\end{pmatrix}, \quad |1,-1\rangle = \begin{pmatrix}0\\0\\1\end{pmatrix}. \quad (3.134)$$

For $\mathbf{n} = (\sin\theta\cos\phi, \sin\theta\sin\phi, \cos\theta)$, the 3×3 matrix $\mathbf{n}\cdot\mathbf{L}/\hbar$ that appears in the exponent of the rotation matrices is given by

$$\mathbf{n}\cdot\mathbf{L}/\hbar = \begin{pmatrix} \cos\theta & \frac{1}{\sqrt{2}}e^{-i\phi}\sin\theta & 0 \\ \frac{1}{\sqrt{2}}e^{i\phi}\sin\theta & 0 & \frac{1}{\sqrt{2}}e^{-i\phi}\sin\theta \\ 0 & \frac{1}{\sqrt{2}}e^{i\phi}\sin\theta & \cos\theta \end{pmatrix}. \quad (3.135)$$

Explicit expressions for the rotation functions for $j=1$ can be obtained by exponentiating $-i\vartheta\mathbf{n}\cdot\mathbf{L}/\hbar$. For rotation about the y-axis by angle ϑ,

$$d^{(1)}(\vartheta) = \begin{pmatrix} \cos^2(\vartheta/2) & -\frac{\sin\vartheta}{\sqrt{2}} & \sin^2(\vartheta/2) \\ \frac{\sin\vartheta}{\sqrt{2}} & \cos\vartheta & -\frac{\sin\vartheta}{\sqrt{2}} \\ \sin^2(\vartheta/2) & \frac{\sin\vartheta}{\sqrt{2}} & \cos^2(\vartheta/2) \end{pmatrix}. \quad (3.136)$$

For arbitrary rotations of $j=1$ states, use Eq. (3.117).

3.3 Rotations and Angular Momentum

Problem 3.20

(a) Calculate $\sum_{m'} m' |d^{(j)}_{m'm}(\beta)|^2$.
(b) Calculate $\sum_{m'} m'^2 |d^{(j)}_{m'm}(\beta)|^2$.

Hint: Take the expectation value of the operators J_z and J_z^2 with the state in Eq. (3.121) with $\alpha = \gamma = 0$.

Answers: (a) $\sum_{m'} m' |d^{(j)}_{m'm}(\beta)|^2 = m \cos \beta$, (b) $\sum_{m'} m'^2 |d^{(j)}_{m'm}(\beta)|^2 = \frac{1}{2}[j(j+1)\sin^2 \beta + m^2(3\cos^2 \beta - 1)]$.

Problem 3.21

(a) Show that it is impossible to rotate the state $|1,1\rangle$ into $|1,0\rangle$. Hint: Apply $d^{(1)}$ in (3.136) to $|1,1\rangle$.
(b) Show that the expectation value of the angular momentum operators in state $|1,0\rangle$ vanish, i.e., $\langle 1,0|J_x|1,0\rangle = \langle 1,0|J_y|1,0\rangle = \langle 1,0|J_z|1,0\rangle = 0$.
(c) Show that $\langle 1,0|\mathcal{R}^\dagger_{\alpha\beta\gamma} J_x \mathcal{R}_{\alpha\beta\gamma}|1,0\rangle = \langle 1,0|\mathcal{R}^\dagger_{\alpha\beta\gamma} J_y \mathcal{R}_{\alpha\beta\gamma}|1,0\rangle = \langle 1,0|\mathcal{R}^\dagger_{\alpha\beta\gamma} J_z \mathcal{R}_{\alpha\beta\gamma}|1,0\rangle = 0$.

3.3.3 RIGID-ROTOR EIGENFUNCTIONS

In this subsection, we shall show that the rotation functions $D^{(j)}_{m,m'}(\alpha \beta \gamma)$ are eigenfunctions of the angular momentum of a rigid body, where the angles α, β, and γ determine the orientation of the principal axes $x'\,y'\,z'$ fixed in the body relative to space fixed axes $x\,y\,z$. In fact, the rotation functions are eigenfunctions of $\mathbf{J}^2$, $J_z \equiv \mathbf{J} \cdot \hat{\mathbf{z}}$, and $J_{z'} \equiv \mathbf{J} \cdot \hat{\mathbf{z}}'$, with eigenvalues $j(j+1)\hbar^2$, $m\hbar$, and $m'\hbar$, respectively. Think of the rigid body as a spinning top, but the top may or may not have cylindrical symmetry. If it has an axis of symmetry and the z' body-fixed axis is oriented in this direction, then $\mathbf{J}^2$, J_z, as well as $J_{z'}$ are constants of the motion. If it does not, then $\mathbf{J}^2$ and J_z are still constants of the motion, but not $J_{z'}$.

The rotational Hamiltonian for a rigid body is

$$H = \frac{J_{x'}^2}{2I_{x'}} + \frac{J_{y'}^2}{2I_{y'}} + \frac{J_{z'}^2}{2I_{z'}}, \qquad (3.137)$$

where $J_{x'} = \mathbf{J} \cdot \hat{\mathbf{x}}'$, and similarly for $J_{y'}$ and $J_{z'}$, and the quantities $I_{x'}$, $I_{y'}$, and $I_{z'}$ are the principal moments of inertia of the body along the principal axes [21, 22].[3] The Hamiltonian (3.137) is called the *rigid-rotor Hamiltonian*. The angular momentum components $J_{i'}$ in the coordinate system rotating with the rigid body do not have the usual commutation relations of the angular momentum components along space fixed coordinates because the axes change dynamically. Let us evaluate the commutator, $[J_{i'}, J_{j'}] = [\mathbf{J} \cdot \hat{\mathbf{x}}'_i, \mathbf{J} \cdot \hat{\mathbf{x}}'_j]$, using the commutator identity, $[AB,CD] = A[B,C]D + AC[B,D] + [A,C]DB + C[A,D]B$. The position vectors $\mathbf{x}'_i$ commute with each other but do not commute with the angular momentum operators. Indeed, we have the commutation relation $[J_{i'}, x_{j'}] = i\hbar\varepsilon_{i'j'k'} x_{k'}$. Had we been considering fixed unit vectors, this commutation relation would not be relevant, but with the unit vectors along the body-fixed axes, we find, after using the above commutator identity and doing a bit of algebra, that

$$[J_{i'}, J_{j'}] = -i\hbar \varepsilon_{i'j'k'} J_{k'}, \qquad (3.138)$$

where an "extra" minus sign is present on the RHS of (3.138) relative to (3.7). This extra minus sign changes the sign of some matrix elements.

---
[3] The moment of inertia tensor is defined as $I_{ij} = \int d\mathbf{r}\, \rho(\mathbf{r})(r^2 \delta_{ij} - x_i x_j)$, where $\rho(\mathbf{r})$ is the mass density. The principal moments of inertia are obtained by diagonalizing this tensor.

Let us consider the rigid-rotor Hamiltonian for some limiting cases. For the case when all three principal moments of inertia of the body are equal, $I_{x'} = I_{y'} = I_{z'} \equiv I$, the Hamiltonian corresponds to that of a *spherical top* and can be written as

$$H = \frac{\mathbf{J}^2}{2I}. \tag{3.139}$$

Since $\mathbf{J}^2 = J_{x'}^2 + J_{y'}^2 + J_{z'}^2 = J_x^2 + J_y^2 + J_z^2$, the eigenvalues are $E_J = \frac{\hbar^2 J(J+1)}{2I}$. The degeneracy of these eigenvalues is $(2J+1)^2$ since there are $2J+1$ values of J_z and $2J+1$ values of $J_{z'}$ with the same value of J. Figure 11.8(a) in Sec. 11.4 shows the spectrum of the spherical top Hamiltonian, while 11.8(b) shows the absorption spectrum for a thermally populated molecule.

For the case where only two of the moments of inertia are the same, $I_{x'} = I_{y'} \neq I_{z'}$, we have the *symmetric top* Hamiltonian,

$$H = \frac{\mathbf{J}^2}{2I_{x'}} + \left(\frac{1}{2I_{z'}} - \frac{1}{2I_{x'}}\right) J_{z'}^2. \tag{3.140}$$

The eigenvalues of this Hamiltonian are

$$E_{J,m'} = \frac{\hbar^2}{2}\left[\frac{J(J+1)}{I_{x'}} + \left(\frac{1}{I_{z'}} - \frac{1}{I_{x'}}\right)(m')^2\right], \tag{3.141}$$

where the eigenvalues of the operator $J_{z'}$ are $\hbar m'$. The degeneracy with respect to values of m' of the spherical top is now partly removed, but values of m' differing only in sign have the same energy. These degenerate states have opposite directions of the angular momentum relative to the axis of the top. Thus, the energy levels of a symmetrical top are doubly degenerate if $m' \neq 0$. Moreover, the $2J+1$ degeneracy associated with J_z still exist.

The eigenfunctions of the symmetric top are the common eigenfunctions of the operators $\mathbf{J}^2$, J_z, and $J_{z'}$. Using Eq. (3.121), we can write the wave function of the state of the symmetric top described in terms of fixed coordinates x, y, z, $|\psi_{j,m}\rangle$, in terms of the wave functions of states described in terms of the axes x', y', z' fixed in the symmetric top, i.e., attached to the rigid body, $|\phi_{jm'}\rangle$, $|\psi_{j,m}\rangle = \sum_{m'} |\phi_{jm'}\rangle D_{m',m}^{(j)}(\alpha\,\beta\,\gamma)$. The Euler angle dependence of the wave function is given by the rotation functions on the RHS of this equation. Moreover, if we want a wave function with well-defined angular momentum component m' along z', $|\Psi_{j,m,m'}\rangle$, then only that specific term in the sum on the RHS of this equation is required:

$$|\Psi_{j,m,m'}\rangle = |\phi_{jm'}\rangle D_{m',m}^{(j)}(\alpha\,\beta\,\gamma). \tag{3.142}$$

Normalizing the wave functions such that $\int\int d(\cos\beta)\,d\alpha\,d\gamma\,|\Psi_{j,m,m'}|^2 = 1$, determines the normalization coefficient, and we obtain

$$|\Psi_{j,m,m'}\rangle = e^{i\varphi}\sqrt{\frac{8\pi^2}{2j+1}}\,D_{m',m}^{(j)}(\alpha\,\beta\,\gamma), \tag{3.143}$$

where the phase angle φ can be chosen arbitrarily.

When all the principal moments of inertia are different, $I_{x'} \neq I_{y'} \neq I_{z'}$, i.e., the *asymmetric top* case, the eigenvalues and eigenfunctions cannot be obtained analytically; numerical solutions of the secular equations for the eigenenergies and linear equations for the eigenfunctions must be obtained. The eigenstates now do not have definite values of the quantum number m', hence, we seek eigenstates given by linear combinations of the form

$$|\psi_{j,m}\rangle = \sum_{m'} c_{m'} |\Psi_{j,m,m'}\rangle. \tag{3.144}$$

Substituting into the Schrödinger equation, $H|\psi_{j,m}\rangle = E_j|\psi_{j,m}\rangle$, and taking inner products with the states $|\Psi_{j,m,m'}\rangle$ yields

$$\sum_{\mu'} \left(\langle \Psi_{j,m,m'}|H|\Psi_{j,m,\mu'}\rangle - E_j \delta_{m',\mu'} \right) c_{\mu'} = 0. \quad (3.145)$$

The roots of the secular equation obtained from (3.145) are the energy eigenvalues of the asymmetric top, and then (3.145) can be used to determine the amplitudes $\{c_{\mu'}\}$ that diagonalize the Hamiltonian.

The above discussion holds for either integer or half-integer angular momenta j, however, if no spin degrees of freedom exist, only integer angular momentum can result. For molecules with unpaired electrons, or for deformed nuclei with an odd number of nucleons, the angular momentum may be half integer.

3.4 ADDITION (COUPLING) OF ANGULAR MOMENTA

The states of two particles having angular momenta j_1 and j_2 can also be classified in terms of the resulting angular momentum obtained upon adding the angular momenta of the particles, as will be shown here. The product state

$$|j_1 j_2 m_1 m_2\rangle \equiv |j_1 m_1\rangle \otimes |j_2 m_2\rangle, \quad (3.146)$$

is an eigenvector of $\mathbf{J}^2_{(1)}, J_{z(1)}, \mathbf{J}^2_{(2)}, J_{z(2)}$ with eigenvalues $\hbar^2 j_1(j_1+1), \hbar m_1, \hbar^2 j_2(j_2+1), \hbar m_2$, but generally, it is not an eigenstate of the total angular momentum squared, $\mathbf{J}^2 = \mathbf{J}_{(1)} \otimes \mathbf{1}_{(2)} + \mathbf{1}_{(1)} \otimes \mathbf{J}_{(2)}$.[4] Our objective is to construct the eigenfunctions of $\mathbf{J}^2$ and J_z using proper linear combinations of states $|j_1 j_2 m_1 m_2\rangle$ defined in Eq. (3.146). Let us denote the eigenfunctions of $\mathbf{J}^2$ and J_z composed of the two-particle states by $|j_1 j_2 J M\rangle$. The following constraints apply to the quantum numbers J and M:

$$|j_1 - j_2| \leq J \leq |j_1 + j_2|, \quad (3.147a)$$

$$M = m_1 + m_2, \quad (3.147b)$$

where $|M| \leq J$. Thus, the quantum numbers J and j_1, j_2 are related by a *triangle inequality*, and the sum of the z-components of the angular momentum, $m_1 + m_2$, is conserved. For example, if $j_1 = 2$ and $j_2 = 3$, the total angular momentum can take on any of the following values: $2 + 3 = 1, 2, 3, 4, 5$. Furthermore, the z-component of total angular momentum is fully determined by the sum of m_1 and m_2.

Two simple examples clarify the addition of angular momentum. The two angular momenta $\mathbf{J}_{(1)}$ and $\mathbf{J}_{(2)}$ may correspond to the orbital and spin angular momenta of a particular particle,

$$\mathbf{J} = \mathbf{L} + \mathbf{S} = \mathbf{L} \otimes \mathbf{1} + \mathbf{1} \otimes \mathbf{S}, \quad (3.148)$$

or they may correspond to the spin of two different particles,

$$\mathbf{J} = \mathbf{S}_1 + \mathbf{S}_2 = \mathbf{S}_1 \otimes \mathbf{1} + \mathbf{1} \otimes \mathbf{S}_2. \quad (3.149)$$

In the former case, the following commutation relations obtain $[J^2, L^2] = 0$, $[J^2, S^2] = 0$, $[L^2, S^2] = 0$, $[J^2, J_i] = 0$, $[L^2, L_i] = 0$, $[S^2, S_i] = 0$; in the latter case we have $[J^2, S_1^2] = 0$, $[J^2, S_2^2] = 0$, $[J^2, S_2^2] = 0$, $[S_1^2, S_2^2] = 0$, $[S_1^2, S_{i,1}] = 0$, $[S_2^2, S_{i,2}] = 0$.

We shall make extensive use of angular momentum coupling when discussing the states of multielectron atoms and molecules, but also when discussing spin–orbit interactions and hyperfine interactions in atoms and molecules.

---

[4] Each term in the sum is a tensor product of operators acting in the appropriate Hilbert spaces of particle 1 (left factor) and 2 (right factor). Similarly, for the components, e.g., $J_z = J_{z(1)} \otimes \mathbf{1}_{(2)} + \mathbf{1}_{(1)} \otimes J_{z(2)}$

3.4.1 CLEBSCH–GORDAN COEFFICIENTS AND 3j SYMBOLS

The set of all product states $|j_1 j_2 m_1 m_2\rangle \equiv |j_1 m_1\rangle |j_2 m_2\rangle$ (see Eq. (3.146)) forms a complete orthogonal basis in the product Hilbert space $\mathcal{H} = \mathcal{H}_1 \otimes \mathcal{H}_2$ of particles 1 and 2. So is the set of all total angular momentum vectors $|j_1 j_2 J M\rangle$. Hence, there is a unitary transformation relating the two bases, (actually orthogonal since it is real):

$$|j_1 j_2 J M\rangle = \sum_{m_1 m_2} \langle J M | j_1 j_2 m_1 m_2 \rangle |j_1 m_1\rangle |j_2 m_2\rangle, \qquad (3.150)$$

$$|j_1 j_2 m_1 m_2\rangle = \sum_{JM} \langle j_1 j_2 m_1 m_2 | JM \rangle |j_1 j_2 J M\rangle. \qquad (3.151)$$

The transformation coefficients, $\langle J M | j_1 j_2 m_1 m_2 \rangle$, are called *Clebsch–Gordan coefficients* or vector coupling coefficients. The meaning of (3.150) is that the state $|j_1 j_2 J M\rangle$ (which is an eigenfunction of $\mathbf{J}^2$ and J_z) is composed of a specific linear combination of states $|j_1 m_1\rangle |j_2 m_2\rangle$, where the coefficients of the linear combination are the Clebsch–Gordan (CG) coefficients. These coefficients will be determined below. It will be convenient to call the quantum numbers of the total angular momentum and its z-component, J and M, by the symbols j_3 and m_3 respectively. We will often shorten the notation by not explicitly indicating the j_1 and j_2 quantum numbers in the CG coefficients and writing only $\langle j_3 m_3 | j_1 j_2 m_1 m_2 \rangle$. Eq. (3.150) then becomes, $|j_1 j_2 j_3 m_3\rangle = \sum_{m_1} \langle j_3 m_3 | j_1 j_2 m_1 m_2 \rangle |j_1 m_1\rangle |j_2 m_2\rangle$. The CG coefficients are constructed so as to be real,

$$\langle j_1 j_2 m_1 m_2 | j_3 m_3 \rangle = \langle j_3 m_3 | j_1 j_2 m_1 m_2 \rangle. \qquad (3.152)$$

The orthonormality of the eigenfunctions $|j_1 m_1\rangle |j_2 m_2\rangle$ and $|j_1 j_2 J M\rangle$ leads to the orthogonality relations for the coefficients

$$\sum_{m_1 m_2} \langle j_3 m_3 | j_1 j_2 m_1 m_2 \rangle \langle j_1 j_2 m_1 m_2 | j'_3 m'_3 \rangle = \delta_{j_3, j'_3} \delta_{m_3, m'_3}, \qquad (3.153a)$$

$$\sum_{j_3 m_3} \langle j_1 j_2 m_1 m_2 | j_3 m_3 \rangle \langle j_3 m_3 | j_1 j_2 m'_1 m'_2 \rangle = \delta_{m_1, m'_1} \delta_{m_2, m'_2}. \qquad (3.153b)$$

Problem 3.22

Calculate all the CG coefficients $\langle \frac{1}{2} m_1 \frac{1}{2} m_2 | \frac{1}{2} \frac{1}{2} J M \rangle$.

Answer: The singlet and three triplet states are given by,

$$|\tfrac{1}{2}\tfrac{1}{2} 0 0\rangle = \tfrac{1}{\sqrt{2}}(|\tfrac{1}{2}\tfrac{1}{2}\tfrac{1}{2}\tfrac{-1}{2}\rangle - |\tfrac{1}{2}\tfrac{1}{2}\tfrac{-1}{2}\tfrac{1}{2}\rangle)$$

$$|\tfrac{1}{2}\tfrac{1}{2} 1 0\rangle = \tfrac{1}{\sqrt{2}}(|\tfrac{1}{2}\tfrac{1}{2}\tfrac{1}{2}\tfrac{-1}{2}\rangle + |\tfrac{1}{2}\tfrac{1}{2}\tfrac{-1}{2}\tfrac{1}{2}\rangle), \quad |\tfrac{1}{2}\tfrac{1}{2} 1 1\rangle = |\tfrac{1}{2}\tfrac{1}{2}\tfrac{1}{2}\tfrac{1}{2}\rangle, \quad |\tfrac{1}{2}\tfrac{1}{2} 1 \bar{1}\rangle = |\tfrac{1}{2}\tfrac{1}{2}\tfrac{-1}{2}\tfrac{-1}{2}\rangle.$$

The coefficients are obtained by applying the appropriate bra states $\langle \tfrac{1}{2}\tfrac{1}{2} m_1 m_2|$ on the left and using the orthogonality of these states. The nonzero values, using the shorthand notation, $\langle \tfrac{1}{2} m_1 \tfrac{1}{2} m_2 | JM \rangle$, are:

$\langle \tfrac{1}{2}\tfrac{1}{2}\tfrac{1}{2}\tfrac{1}{2} | 1 1\rangle = \langle \tfrac{1}{2} -\tfrac{1}{2}\tfrac{1}{2} -\tfrac{1}{2} | 1 -1\rangle = 1$, $\langle \tfrac{1}{2}\tfrac{1}{2}\tfrac{1}{2} -\tfrac{1}{2} | 1 0\rangle = 1/\sqrt{2}$, $\langle \tfrac{1}{2} -\tfrac{1}{2}\tfrac{1}{2}\tfrac{1}{2} | 1 0\rangle = 1/\sqrt{2}$,
$\langle \tfrac{1}{2}\tfrac{1}{2}\tfrac{1}{2} -\tfrac{1}{2} | 0 0\rangle = 1/\sqrt{2}$, $\langle \tfrac{1}{2} -\tfrac{1}{2}\tfrac{1}{2}\tfrac{1}{2} | 0 0\rangle = -1/\sqrt{2}$.

Problem 3.23

Determine all possible angular and spin wave functions of a p electron in an atom having definite quantum numbers J, M.

3.4 Addition (Coupling) of Angular Momenta

Answer: For a p electron in an atom, we can have $J = 1/2, 3/2$ and $M = -J, -J+1, \ldots, J$. There are six possible wave functions,

$$\langle \hat{\mathbf{r}} | l s J M \rangle = \sum_{m_l=-1}^{1} \sum_{m_s=-\frac{1}{2}}^{\frac{1}{2}} \langle 1 m_l \tfrac{1}{2} m_s | l s J M \rangle Y_{l m_l}(\hat{\mathbf{r}}) |m_s\rangle,$$

where $|m_s\rangle = |\pm 1/2\rangle$ are the spinors $|\uparrow\rangle$ and $|\downarrow\rangle$. After explicit computation of the C-G coefficients:

$\langle \hat{\mathbf{r}} | 1 \tfrac{1}{2} \tfrac{3}{2} \tfrac{3}{2} \rangle = Y_{11}(\hat{\mathbf{r}}) |\uparrow\rangle$, $\langle \hat{\mathbf{r}} | 1 \tfrac{1}{2} \tfrac{3}{2} -\tfrac{3}{2} \rangle = Y_{1-1}(\hat{\mathbf{r}}) |\downarrow\rangle$,

$\langle \hat{\mathbf{r}} | 1 \tfrac{1}{2} \tfrac{3}{2} \tfrac{1}{2} \rangle = \sqrt{\tfrac{2}{3}} Y_{10}(\hat{\mathbf{r}}) |\uparrow\rangle + \sqrt{\tfrac{1}{3}} Y_{11}(\hat{\mathbf{r}}) |\downarrow\rangle$,

$\langle \hat{\mathbf{r}} | 1 \tfrac{1}{2} \tfrac{3}{2} -\tfrac{1}{2} \rangle = \sqrt{\tfrac{2}{3}} Y_{10}(\hat{\mathbf{r}}) |\downarrow\rangle + \sqrt{\tfrac{1}{3}} Y_{1-1}(\hat{\mathbf{r}}) |\uparrow\rangle$

$\langle \hat{\mathbf{r}} | 1 \tfrac{1}{2} \tfrac{1}{2} \tfrac{1}{2} \rangle = -\sqrt{\tfrac{1}{3}} Y_{10}(\hat{\mathbf{r}}) |\uparrow\rangle + \sqrt{\tfrac{2}{3}} Y_{11}(\hat{\mathbf{r}}) |\downarrow\rangle$,

$\langle \hat{\mathbf{r}} | 1 \tfrac{1}{2} \tfrac{1}{2} -\tfrac{1}{2} \rangle = -\sqrt{\tfrac{2}{3}} Y_{11}(\hat{\mathbf{r}}) |\uparrow\rangle + \sqrt{\tfrac{1}{3}} Y_{10}(\hat{\mathbf{r}}) |\downarrow\rangle$.

Problem 3.24

Express the state $|\tfrac{1}{2} \tfrac{1}{2}\rangle \otimes |10\rangle$ in terms of the total angular momentum states $|\tfrac{3}{2} \tfrac{1}{2}\rangle$ and $|\tfrac{1}{2} \tfrac{1}{2}\rangle$.

Answer: This is the inverse operation of angular momentum addition.

$|\tfrac{1}{2} \tfrac{1}{2}\rangle \otimes |10\rangle = \sqrt{\tfrac{2}{3}} |\tfrac{3}{2} \tfrac{1}{2}\rangle + \sqrt{\tfrac{1}{3}} |\tfrac{1}{2} \tfrac{1}{2}\rangle$.

The CG coefficients satisfy Eq. (3.147), i.e., $|j_1 - j_2| \leq j_3 \leq j_1 + j_2$ and $m_3 = m_1 + m_2$. Moreover, $j_1 + j_2 + j_3$ must be an integer.

One can calculate the values of the CG coefficients by successive application of the lowering operator to Eq. (3.150) with $m_1 = j_1$, $m_2 = j_2$, $j_3 = j_1 + j_2$, and $m_3 = j_3$. With these values, we have

$$\langle J = (j_1 + j_2)\, M = (j_1 + j_2) | j_1 j_2 \, m_1 = j_1 \, m_2 = j_2 \rangle = +1.$$

Applying the lowering operator $J_- = J_{-(1)} + J_{-(2)}$ to Eq. (3.150) and using Eq. (3.24) yields an equation with m's lowered by one; the coefficients in this equation are the CG coefficients. Successively applying the lowering operator to the resultant equations gives the CG coefficients $\langle J = (j_1 + j_2)\, M = m_3 | j_1 j_2 \, m_1 m_2 \rangle$ for all values of M. One must then find the CG coefficients for $J = (j_1 + j_2 - 1)$ and $M = J$; this can be done by noting that $|J = (j_1 + j_2 - 1)\, M = (j_1 + j_2 - 1)\rangle$ must be orthogonal orthogonal to $|J = (j_1 + j_2)\, M = (j_1 + j_2 - 1)\rangle$. In writing out the orthogonality relation, one uses the phase convention that $\langle JJ | j_1 j_1 j_2 \, J - j_1 \rangle > 0$ to determine the phase of the coefficients appearing in $|J = (j_1 + j_2)\, M = (j_1 + j_2 - 1)\rangle$. One then lowers the m quantum numbers by applying $J_- = J_{-(1)} + J_{-(2)}$ to successively obtain all the CG coefficients for the various values of M for this value of J, and one continues this procedure of orthogonalizing and successively applying the lowering operator, until all the CG coefficients for a given j_1 and j_2 are determined. Giulio Racah used this technique to obtain the following general formula:

$$\langle j_3\, m_3 | j_1 j_2 \, m_1 m_2 \rangle = \sqrt{\frac{(2j_3 + 1)\,(j_3 + j_1 - j_2)!\,(j_3 - j_1 + j_2)!\,(j_1 + j_2 - j_3)!}{(j_1 + j_2 + j_3 + 1)!}}$$

$$\sqrt{\frac{(j_3 - m_3)!\,(j_3 + m_3)!}{(j_1 - m_1)!(j_1 + m_1)!(j_2 - m_2)!(j_2 + m_2)!}}$$

$$\sum_k \frac{(-1)^{k+j_2+m_2}(j_3 + j_2 - m_1 - k)!\,(j_1 - m_1 + k)!}{(j_3 - j_1 + j_2 - k)!\,(j_3 + m_3 - k)!\,k!\,(k + j_1 - j_2 - m_3)!}. \tag{3.154}$$

The sum over k runs over all integers that do not lead to negative factorials. Computer codes to calculate the CG coefficients can be easily obtained.

A more symmetrical coefficient directly related to the CG coefficients, called the *Wigner 3j symbol*, or simply 3j symbol for short, is given by

$$\begin{pmatrix} j_1 & j_2 & j_3 \\ m_1 & m_2 & -m_3 \end{pmatrix} = \frac{(-1)^{j_1-j_2+m_3}}{\sqrt{2j_3+1}} \langle j_3 \, m_3 | j_1 \, j_2 \, m_1 \, m_2 \rangle. \tag{3.155}$$

By using definition (3.155) of the 3j symbol, Eq. (3.150) can be written as

$$|j_1 j_2 j_3 m_3\rangle = (-1)^{j_1-j_2+m_3}(2j_3+1)^{1/2} \sum_{m_1 m_2} \begin{pmatrix} j_1 & j_2 & j_3 \\ m_1 & m_2 & -m_3 \end{pmatrix} |j_1 j_2 m_1 m_2\rangle, \tag{3.156}$$

where $m_3 = m_1 + m_2$, $|j_1 - j_2| \le j_3 \le j_1 + j_2$, and the sum over magnetic quantum numbers involves $m_i = -j_i, \ldots, j_i$. Note that we have used the shorthand notation $|j_1 j_2 m_1 m_2\rangle \equiv |j_1 m_1\rangle |j_2 m_2\rangle$.

The 3j symbol vanishes unless $m_1 + m_2 - m_3 = 0$, (j_1, j_2, j_3) form the sides of a triangle [i.e., satisfy (3.152a)], and $j_1 + j_2 + j_3$ is an integer. Furthermore, the 3j symbols satisfy the permutation symmetry,

$$\begin{pmatrix} j_1 & j_2 & j_3 \\ m_1 & m_2 & m_3 \end{pmatrix} = (-1)^{j_1+j_2+j_3} \begin{pmatrix} j_2 & j_1 & j_3 \\ m_2 & m_1 & m_3 \end{pmatrix}, \tag{3.157}$$

i.e., they change sign when two columns are interchanged if $j_1 + j_2 + j_3$ is odd, and they are invariant under cyclic permutation of the columns. Moreover, the following symmetry exists when the magnetic quantum numbers change sign:

$$\begin{pmatrix} j_1 & j_2 & j_3 \\ -m_1 & -m_2 & -m_3 \end{pmatrix} = (-1)^{j_1+j_2+j_3} \begin{pmatrix} j_2 & j_1 & j_3 \\ m_2 & m_1 & m_3 \end{pmatrix}. \tag{3.158}$$

A special case of the 3j symbols that is worthy of note is

$$\begin{pmatrix} j_1 & j_2 & 0 \\ m_1 & m_2 & 0 \end{pmatrix} = \frac{(-1)^{j_1-m_1}}{\sqrt{2j_1+1}} \delta_{j_1, j_2} \delta_{m_1, -m_2}, \tag{3.159}$$

which corresponds to the CG relation,

$$\langle 0\,0 | j_1 \, j_2 \, m_1 \, m_2 \rangle = \frac{(-1)^{j_1-m_1}}{\sqrt{2j_1+1}} \delta_{j_1, j_2} \delta_{m_1, -m_2}. \tag{3.160}$$

When $j_1 = j_2$, this reduces to the simple relation,

$$\begin{pmatrix} j & j & 0 \\ m & -m & 0 \end{pmatrix} = \frac{(-1)^{j-m}}{\sqrt{2j+1}}$$

which corresponds to the CG relation, $\langle 0\,0 | j\,j\,m\,-m \rangle = (-1)^{j-m}/(2j+1)$.

Tables of CG coefficients can be found at http://www.en.wikipedia.org/wiki/Table_of_Clebsch-Gordan_coefficients. A web CG calculator can be found at the URL, http://personal.ph.surrey.ac.uk/~phs3ps/cgjava.html. A web 3j calculator can be found at http://plasma-gate.weizmann.ac.il/369j.html.

Problem 3.25

For two particles, one of angular momentum 1 and the other of angular momentum 1/2 having z-projection of their total angular momentum equal to $M = 1/2$, what are the states of total angular momentum J, $|J, M = 1/2\rangle$ that can be obtained and what are amplitudes of the states $|1, m_l\rangle$ and $|1/2, m_s\rangle$ needed to make these states?

3.4 Addition (Coupling) of Angular Momenta

Answer: $|3/2, 1/2\rangle = \sqrt{\frac{1}{3}}|1,1\rangle|1/2,-1/2\rangle + \sqrt{\frac{2}{3}}|1,0\rangle|1/2,1/2\rangle$,
$|1/2, 1/2\rangle = \sqrt{\frac{2}{3}}|1,1\rangle|1/2,-1/2\rangle - \sqrt{\frac{1}{3}}|1,0\rangle|1/2,1/2\rangle$.

Problem 3.26

(a) Show that $\langle j_1 j_2 m_1 m_2 | j_3 m_3 \rangle = (-1)^{j_1+j_2-j_3} \langle j_2 j_1 m_2 m_1 | j_3 m_3 \rangle$.

(b) Show $\langle j_1 j_2 m_1 m_2 | j_3 m_3 \rangle = \sqrt{\frac{2j_3+1}{2j_2+1}} (-1)^{j_1-m_1} \langle j_1 j_3 m_1 - m_3 | j_2 - m_2 \rangle$.

(c) Show $\langle j_1 j_2 m_1 m_2 | j_3 m_3 \rangle = \sqrt{\frac{2j_3+1}{2j_1+1}} (-1)^{j_2+m_2} \langle j_3 j_2 - m_3 m_2 | j_1 - m_1 \rangle$.

3.4.2 CLEBSCH–GORDAN SERIES

Clebsch–Gordan coefficients can be used to obtain simple expressions for products of the rotation functions that were introduced in Sec. 3.3.2. Upon application of the rotation operator $\mathcal{R}_{\alpha\beta\gamma}$ defined in (3.116) on Eq. (3.150) and using (3.121), we obtain

$$\sum_{M'} |JM'\rangle D^{(J)}_{M',M}(\alpha\beta\gamma) = \sum_{m_1 m_2} \langle JM | j_1 j_2 m_1 m_2 \rangle \sum_{m'_1} |j_1 m'_1\rangle D^{(j_1)}_{m'_1,m_1}(\alpha\beta\gamma) \sum_{m'_2} |j_2 m'_2\rangle D^{(j_2)}_{m'_2,m_2}(\alpha\beta\gamma). \qquad (3.161)$$

Again using (3.150) to express $|JM'\rangle$ on the LHS of (3.161) as a sum over states $|j_1 m'_1\rangle|j_2 m'_2\rangle$, we find

$$D^{(J)}_{M',M}(\alpha\beta\gamma) = \sum_{m_1 m_2 m'_1 m'_2} \langle JM' | j_1 j_2 m'_1 m'_2 \rangle \langle JM | j_1 j_2 m_1 m_2 \rangle D^{(j_1)}_{m'_1,m_1}(\alpha\beta\gamma) D^{(j_2)}_{m'_2,m_2}(\alpha\beta\gamma). \qquad (3.162)$$

Multiplying by CG coefficients and summing, this equation can be reexpressed as

$$D^{(j_1)}_{m'_1,m_1}(\alpha\beta\gamma) D^{(j_2)}_{m'_2,m_2}(\alpha\beta\gamma) = \sum_{JMM'} \langle j_1 j_2 m'_1 m'_2 | JM' \rangle \langle j_1 j_2 m_1 m_2 | JM \rangle D^{(J)}_{M',M}(\alpha\beta\gamma). \qquad (3.163)$$

This equation, known as the *Clebsch–Gordan series*, can be expressed schematically as

$$D^{(j_1)} \otimes D^{(j_2)} = D^{(j_1+j_2)} \oplus D^{(j_1+j_2-1)} \oplus \ldots \oplus D^{(|j_1-j_2|)}. \qquad (3.164)$$

By multiplying Eq. (3.163) by $(D^{(J)}_{m'_3,m_3})^*$, integrating over and using (3.128), we find

$$\int d(\cos\beta)\, d\alpha\, d\gamma\, (D^{(j_3)}_{m'_3,m_3})^* D^{(j_1)}_{m'_1,m_1}(\alpha\beta\gamma) D^{(j_2)}_{m'_2,m_2}(\alpha\beta\gamma) = \frac{8\pi^2}{2j_3+1} \langle j_1 j_2 m'_1 m'_2 | j_3 m'_3 \rangle \langle j_1 j_2 m_1 m_2 | j_3 m_3 \rangle. \qquad (3.165)$$

Equation (3.163) can be used to find the product of two spherical harmonics by using (3.124) to express the rotation functions in terms of spherical harmonics:

$$Y^*_{l_1 m_1}(\beta,\alpha) Y^*_{l_2 m_2}(\beta,\alpha) = \sum_{LM} \sqrt{\frac{(2l_1+1)(2l_2+1)}{4\pi(2L+1)}} \langle l_1 l_2 m_1 m_2 | LM \rangle \langle l_1 l_2 00 | L0 \rangle Y^*_{LM}(\beta,\alpha). \qquad (3.166)$$

If we multiply (3.166) by $Y^*_{l_3 m_3}$ and integrate, we obtain the following integral, which will be useful to determine matrix elements of operators involving angular momentum:

$$\int d\Omega\, Y^*_{l_3 m_3}(\beta,\alpha) Y_{l_1 m_1}(\beta,\alpha) Y_{l_2 m_2}(\beta,\alpha) = \sqrt{\frac{(2l_1+1)(2l_2+1)}{4\pi(2l_3+1)}} \langle l_1 l_2 m_1 m_2 | l_3 m_3 \rangle \langle l_1 l_2 00 | l_3 0 \rangle. \qquad (3.167)$$

Problem 3.27

Write Eqs (3.162), (3.163), (3.165), (3.166), and (3.167) in terms of 3j symbols rather the CG coefficients.

3.5 TENSOR OPERATORS

A coordinate vector $\mathbf{r} = (x, y, z) \equiv (x_1, x_2, x_3)$ transforms under rotation of the coordinate system according to the transformation rule, $x_i \to x'_i = \sum_j \mathfrak{R}_{ij} x_j$. An arbitrary vector in 3D, $\mathbf{V}$, transforms the same way. A second-rank tensor, T_{ij}, transforms like the product of two coordinate vectors, $x_i x_j$, i.e.,

$$T'_{ij} = \sum_{kl} \mathfrak{R}_{ik} \mathfrak{R}_{jl} T_{kl}. \tag{3.168}$$

Similarly for higher rank tensors,

$$T'_{ij\ldots k} = \sum_{lm\ldots n} \mathfrak{R}_{il} \mathfrak{R}_{jm} \ldots \mathfrak{R}_{kn} T_{lm\ldots n}. \tag{3.169}$$

A spherical vector operator (e.g., the position operator or the angular momentum operator), transforms under rotation as $\hat{x}_q \to \hat{x}'_q = \mathcal{R}_{\alpha\beta\gamma} \hat{x}_q \mathcal{R}^{-1}_{\alpha\beta\gamma} = \sum_{q'} \hat{x}_{q'} D^{(1)}_{q',q}(\alpha\,\beta\,\gamma)$, where on the RHS the subscripts are taken to be equal to ± 1 and 0, not x, y, z. This concept will now be generalized. A spherical tensor operator of rank k is defined to be one that transforms as

$$\boxed{\mathcal{R}_{\alpha\beta\gamma} \hat{T}^{(k)}_q \mathcal{R}^{-1}_{\alpha\beta\gamma} = \sum_{q'} \hat{T}^{(k)}_{q'} D^{(k)}_{q',q}(\alpha\,\beta\,\gamma).} \tag{3.170}$$

The rank k is a nonnegative integer and q and q' are magnetic quantum numbers, so $-k \le q, q' \le k$. For simplicity of notation, we drop the hat for operators. Spherical tensors can be formed from products of vectors (see below); this specifies the relationship between spherical tensors $T^{(k)}_q$ and Cartesian tensors $T_{ij\ldots k}$, as explained below. From the definition of the rotation function, this can be cast in the form,

$$\mathcal{R} T^{(k)}_q \mathcal{R}^{-1} = \sum_{q'} T^{(k)}_{q'} \langle k\,q' | \mathcal{R} | k\,q \rangle. \tag{3.171}$$

For infinitesimal rotation, $\mathcal{R} = e^{-i\delta\boldsymbol{\varphi}\cdot\mathbf{J}/\hbar} \approx 1 - i\delta\boldsymbol{\varphi}/\hbar \cdot \mathbf{J}/\hbar$ and (3.171) become

$$\frac{-i\delta\boldsymbol{\varphi}}{\hbar} \cdot [\mathbf{J}, T^{(k)}_q] = \frac{-i\delta\boldsymbol{\varphi}}{\hbar} \cdot \sum_{q'} T^{(k)}_{q'} \langle k\,q' | \mathbf{J} | k\,q \rangle, \tag{3.172}$$

which must be true for any infinitesimal rotation. Hence, any tensor operator $T^{(k)}_q$ has the following commutation properties with the angular momentum:

$$[J_+, T^{(k)}_q] = \sqrt{k(k+1) - q(q-1)}\,\hbar\,T^{(k)}_{q+1}, \tag{3.173a}$$

$$[J_-, T^{(k)}_q] = \sqrt{k(k+1) - q(q+1)}\,\hbar\,T^{(k)}_{q-1}, \tag{3.173b}$$

$$[J_z, T^{(k)}_q] = q\,\hbar\,T^{(k)}_q. \tag{3.173c}$$

Spherical tensors $T^{(k)}_q$ having the transformation properties specified above are called *irreducible tensor operators* because they transform according to a given well-specified irreducible representation of the angular momentum. k is

3.5 Tensor Operators

called the rank of the irreducible representation and q labels its components. The term irreducible comes from the theory of group representations, a topic discussed in Sec. E.5.1 of Appendix E. Coordinate operators (**r**) and angular momentum operators (**L**) transform as $k = 1$ tensors (i.e., as vectors or pseudovectors, respectively). The nine components of the tensor operator T_{ij} of a general second-rank-tensor, written out as a column vector of dimension 9, transforms with a 9×9 transformation matrix, i.e., as discussed in Appendix E, the set of these matrices forms a representation of the rotation group of dimension 9. These tensors can be *reduced* into ones that transform as scalars ($k=0$), vectors ($k=1$), and $k=2$ irreducible tensor operators. These irreducible tensors are labeled $T_0^{(0)}$, $\{T_q^{(1)}\}$ with $q = 1, 0, -1$ and $\{T_q^{(2)}\}$ with $q = 2, 1, 0, -1, -2$. More explicitly, we can expand $U_i V_j$ as follows:

$$U_i V_j = \frac{\mathbf{U} \cdot \mathbf{V}}{3} \delta_{ij} + \frac{U_i V_j - U_j V_i}{2} + \left(\frac{U_i V_j + U_j V_i}{2} - \frac{\mathbf{U} \cdot \mathbf{V}}{3} \delta_{ij} \right). \quad (3.174)$$

The first term transforms as a scalar under rotations, the second as a vector (it corresponds to the components of the axial vector $\mathbf{U} \times \mathbf{V}$), and the last term is a symmetric traceless tensor that transforms as $\{T_q^{(2)}\}$ with $q = 2, 1, 0, -1, -2$. Similarly, third and higher rank tensors can be broken down into their irreducible representations, e.g., the 27 components of the tensor $U_i V_j W_k$ give rise to one irreducible tensor of rank 3 (7 components), two of rank 2 (5 components each), three vectors (rank 1 tensors having 3 components each) and one scalar (rank 0) [the scalar is $(\mathbf{U} \times \mathbf{V}) \cdot \mathbf{W}$, the three vectors are $(\mathbf{U} \cdot \mathbf{V}) \mathbf{W}$, $(\mathbf{V} \cdot \mathbf{W}) \mathbf{U}$, $(\mathbf{W} \cdot \mathbf{U}) \mathbf{V}$, and the three rank 2 components $\varepsilon_{ikl} U_j V_k W_j$, $\varepsilon_{ikl} V_j W_k U_j$, $\varepsilon_{ikl} U_j W_k V_j$. We have already encountered examples of irreducible tensors; the spherical harmonics Y_{lm} are tensors of rank l. They transform as in (3.171) and satisfy the commutation relations (3.173).

The irreducible vector components of rank unity satisfying Eqs (3.171) and (3.173) are given in term of the Cartesian components of the vector by[5]

$$\boxed{V_{+1} \equiv -\frac{1}{\sqrt{2}}(V_x + iV_y), \quad V_0 \equiv V_z, \quad V_{-1} \equiv \frac{1}{\sqrt{2}}(V_x - iV_y).} \quad (3.175)$$

The scalar product of a vector with itself is given in terms of the irreducible vector components by $\mathbf{V} \cdot \mathbf{V} = -V_{+1} V_{-1} - V_{-1} V_{+1} + V_0 V_0$, and more generally, the scalar product of two vectors is given by

$$\mathbf{U} \cdot \mathbf{V} = -U_{+1} V_{-1} - U_{-1} V_{+1} + U_0 V_0. \quad (3.176)$$

The product of two irreducible tensors is not irreducible, yet it is easy to form an irreducible tensor from the product of two irreducible tensors:

$$\boxed{T_q^{(k)}(k_1, k_2) = \sum_{q_1, q_2} \langle k\, q | k_1\, k_2\, q_1\, q_2 \rangle R_{q_1}^{(k_1)} S_{q_2}^{(k_2)}.} \quad (3.177)$$

Here k can take on the values from $|k_1 - k_2|$ to $k_1 + k_2$ and $q = q_1 + q_2$. For example, if $k_1 = k_2 \equiv k$, the resulting zero-rank tensor is

$$T_0^{(0)}(k,k) = \frac{(-1)^k}{\sqrt{2k+1}} \sum_{q=-k}^{k} (-1)^{-q} R_q^{(k)} S_{-q}^{(k)}.$$

For $k = 1$, this is nothing but the inner product of the two vectors.

---

[5] Note that $J_+ \equiv J_x + iJ_y$ and $J_- \equiv J_x - iJ_y$, defined in (3.10), are <u>not</u> defined according to (3.175), but rather $J_+ = -\sqrt{2} J_{+1}$ and $J_- = \sqrt{2} J_{-1}$ (i.e., $J_{+1} = -\frac{1}{\sqrt{2}} J_+ = -\frac{1}{\sqrt{2}}(J_x + iJ_y)$ and $J_{-1} = \frac{1}{\sqrt{2}} J_- = \frac{1}{\sqrt{2}}(J_x - iJ_y)$). This similarity of notation can lead to some confusion. Note also that the rank-one tensors in (3.175) could be labeled $V_q^{(1)}$ with $q = +1, 0, -1$, according to the notation used above.

Problem 3.28

Prove that the factors in the definitions for V_{+1} and V_{-1} in (3.175) are required so that the commutation relations in Eq. (3.173) are satisfied, given $V_0 \equiv V_z$.

Problem 3.29

The vector $\mathbf{r}$ expressed in terms of spherical coordinates is $(x, y, z) = r(\sin\theta\cos\phi, \sin\theta\sin\phi, \cos\theta)$. Write $\mathbf{r}$ as a spherical tensor.

Answer: The spherical notation for a vector $\mathbf{v}$ as spherical tensor $T_q^{(k)}$ is
$v_{+1}^{(1)} = -\frac{1}{\sqrt{2}}(v_x + iv_y)$, $v_{-1}^{(1)} = \frac{1}{\sqrt{2}}(v_x - iv_y)$, $v_0^{(1)} = v_z$. Therefore, for $\mathbf{r}$,
$r_{+1}^{(1)} = -\frac{r}{\sqrt{2}}\sin\theta e^{i\phi} = \sqrt{\frac{4\pi}{3}} r Y_{11}(\hat{\mathbf{r}})$, $r_{-1}^{(1)} = \frac{r}{\sqrt{2}}\sin\theta e^{-i\phi} = \sqrt{\frac{4\pi}{3}} r Y_{1-1}(\hat{\mathbf{r}})$, $r_0^{(1)} = r\cos\theta = \sqrt{\frac{4\pi}{3}} r Y_{10}(\hat{\mathbf{r}})$.

Problem 3.30

(a) Find $T_0^{(2)}(k_1 = 1, k_2 = 1)$ formed from the rank-one irreducible tensors $R^{(1)}$ and $S^{(1)}$.
(b) Show that the inner product of two vectors can be expressed in terms of irreducible representations,
$\mathbf{R} \cdot \mathbf{S} \propto T_0^{(0)}(1, 1) = \sum_{q_1, q_2} \langle 0\,0|1, 1, q_1\,q_2\rangle R_{q_1}^{(1)} S_{q_2}^{(1)}$.

Answer: (b) $T_0^{(0)} = \sum_{q=-1}^{1}\langle 1\,q\,1\,-q|0\,0\rangle R_q^{(1)} S_{-q}^{(1)} = -\frac{1}{\sqrt{3}} \sum_{q=-1}^{1}(-1)^q R_q^{(1)} S_{-q}^{(1)} = -\frac{1}{\sqrt{3}} \mathbf{R} \cdot \mathbf{S}$.

3.5.1 IRREDUCIBLE REPRESENTATIONS OF THE DENSITY MATRIX

As an example of the use of irreducible representations, let us consider the density operator (i.e., the density matrix) ρ and expand it in terms of irreducible tensors. The density operator ρ is formed as the sum of products of a ket and a bra, as evident from Eq. (2.38). We can use kets and bras that are state vectors that transform as a given irreducible representation, i.e., we can use kets $|\alpha JM\rangle$ and bras $\langle \alpha' J'M'|$, where α and α' indicate a set of other quantum numbers that characterize the states. To form the density matrix, we need to take the outer product of the ket and bra to get an operator, $|\alpha JM\rangle\langle \alpha' J'M'|$, but we want the resulting operator to transform as an irreducible representation, similar to what we did in (3.177), where we multiplied $R_{q_1}^{(k_1)}$ by $S_{q_2}^{(k_2)}$ and then formed an irreducible representation from the products. Here, the only caveat is that the ket and the bra are in different vector spaces; the ket is in the state-vector Hilbert space and the bra is in the dual space. This will introduce a slight variation in the formation of the irreducible representation when compared to (3.177). The density matrix can be written as,

$$\rho = \sum_{\alpha JM}\sum_{\alpha' J'M'} |\alpha JM\rangle\langle \alpha JM|\rho|\alpha' J'M'\rangle\langle \alpha' J'M'| = \sum_{\alpha,\alpha',J,J'} \sum_{k=|J-J'|}^{J+J'} \sum_{q=-k}^{k} \rho_q^{(k)}(\alpha,\alpha',J,J')\, T_q^{(k)}(\alpha,\alpha',J,J'), \quad (3.178)$$

where in the second equality we defined the irreducible tensor basis functions

$$T_q^{(k)}(\alpha,\alpha',J,J') \equiv \sum_{M,M'} (-1)^{J'-M'}\langle k\,q|J\,J'\,M\,-M'\rangle\, |\alpha JM\rangle\langle \alpha' J'M'|, \quad (3.179)$$

3.5 Tensor Operators

and the coefficients $\rho_q^{(k)}(\alpha, \alpha', J, J') = \text{Tr}[\rho T_q^{(k)}(\alpha, \alpha', J, J')]$, so

$$\rho_q^{(k)}(\alpha, \alpha', J, J') = \sum_{M,M'} (-1)^{J'-M'} \langle k q | J J' M - M' \rangle \, \rho_{\alpha J M, \alpha' J' M'}, \quad (3.180)$$

where we have used the notation $\rho_{\alpha J M, \alpha' J' M'} = \langle \alpha J M | \rho | \alpha' J' M' \rangle$. In the Clebsch–Gordan coefficient, we took $M' \to -M'$ and added the factor $(-1)^{-M'}$ because the bra transforms as the complex conjugate of the ket [see Eq. (3.56)] and the factor $(-1)^{J'}$ is added by convention. Note that the zero-rank component is proportional to the trace of the density matrix, $\rho_0^{(0)}(\alpha, \alpha, J, J') = \delta_{JJ'} (2J+1)^{-1/2} \sum_M \rho_{\alpha J M, \alpha J M}$. It is clear that $\rho_q^{(k)}$ can be rotated with a rotation matrix $D_{q',q}^{(k)}$, as in (3.170).

Problem 3.31

(a) Substitute (3.179) and (3.180) into (3.178) to demonstrate that the last equality in (3.178) is correct.
(b) Show that $\text{Tr}[(T_{q'}^{(k')}(\alpha', \alpha, J', J))^\dagger T_q^{(k)}(\alpha, \alpha', J, J')] = \delta_{k'k} \delta_{q'q}$.
(c) Prove the following for matrix elements of irreducible tensor basis functions:
$\langle \alpha J M | T_q^{(k)}(\alpha, \alpha', J, J') | \alpha' J' M' \rangle = (-1)^{J'-M'} \langle k q | J J' M - M' \rangle$.

Figure 3.12 shows the density matrix components of an $n=2$ hydrogenic atomic state without including spin degrees of freedom. The four diagonal elements are the probabilities for occupation of states, $P_{nlm} = \rho_{nlm,nlm}$, but the off-diagonal elements provide additional information about the state of the $n=2$ manifold. For the special case of coherent (i.e., pure) state of hydrogen [see the discussion in the paragraph containing Eq. (2.49)], $\Psi(\mathbf{r}) = \sum_{n,l,m} c_{nlm} \psi_{nlm}(\mathbf{r})$, $P_{nlm} = \rho_{nlm,nlm} = |c_{nlm}|^2$. The off-diagonal elements $\rho_{nlm,nl'm'} = c_{nlm} c^*_{nl'm'}$ give information about the *alignment* and *orientation* of the state as well as *multipole moments* and time derivatives of multipole moments of the state [36, 37]. But the multipole moments are a means of representing the density matrix even for a mixed state. The components of the density matrix can be represented in terms of irreducible tensor components, as in Fig. 3.13, instead of being represented as in Fig. 3.12. The orientation of the state is given by the rank-1 tensors $\rho^{(1)}$ and the alignment by the rank-2 tensors $\rho^{(2)}$. For example, the orientation along the z-axis of the p state of hydrogen atom $n=2$ manifold is given by

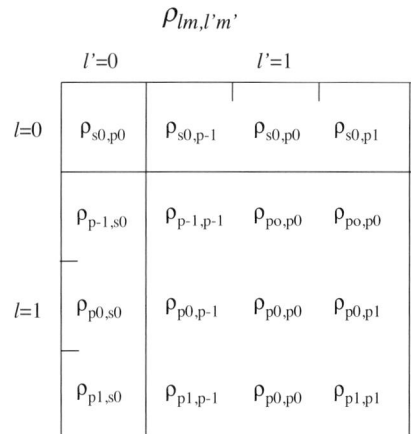

FIG 3.12 Components of the density matrix for an $n=2$ atomic state.

$$\langle L_z \rangle \propto \rho_0^{(1)}(n, n, 1, 1) \propto (\rho_{np1,np1} - \rho_{np-1,np-1}), \quad (3.181)$$

and the alignment of this manifold along the z-axis is given by

$$\langle 3L_z^2 - \mathbf{L}^2 \rangle \propto \rho_0^{(2)}(n, n, 1, 1) \propto (\rho_{np1,np1} - 2\rho_{np01,np0} + \rho_{np-1,np-1}). \quad (3.182)$$

The dipole moment along the z-axis of the $n=2$ manifold is given by the s–p coherence $\rho_{ns0,np0}$, i.e.,

$$\langle z \rangle \propto \text{Re}\, \rho_0^{(1)}(n, n, 0, 1) \propto \text{Re}\, \rho_{ns0,np0}. \quad (3.183)$$

whereas the time derivative dipole moment along the z-axis of the $n=2$ manifold is given by

$$\langle \dot{z} \rangle \propto \text{Im} \, \rho_0^{(1)}(n,n,0,1) \propto \text{Im} \, \rho_{ns0,np0}. \quad (3.184)$$

The expectation values in (3.181 through 3.184) are all independent properties of the $n=2$ atomic state manifold. Moreover, multipole components with different q are also independent. Clearly, higher n manifolds have additional nonvanishing multipole moments of $\mathbf{L}$, $\mathbf{r}$, and $\dot{\mathbf{r}}$.

Of course, for the case of a pure state, $\Psi(\mathbf{r})$, these multipole moment components are determined by the complex amplitudes c_{nlm} for $n=2$. For example, dipole moment along the z-axis of the $n=2$ manifold is given by the s–p coherence $\rho_{ns0,np0} = c_{ns0} c_{np0}^*$ with $n=2$.

The symmetry properties of the density matrix can help determine the optical characteristics of a sample. For example, the absorption and refractive index of a state of matter with a density matrix that is symmetric around the direction of propagation of a light beam will not depend on the polarization of the light beam (as is the case if the wave vector is propagating along the z-axis and only $q=0$ components are present). The absorption and refractive index of a state of matter with a density matrix that contains only a zero-rank tensor will be optically isotropic.

$l=0$	$l=1$	$l=2$	$l=3$
$\rho^{(0)}$	$\rho^{(1)}$	$\rho^{(2)}$	$\rho^{(3)}$
$\rho^{(1)}$	$\rho^{(0)}\rho^{(1)}\rho^{(2)}$	$\rho^{(1)}\rho^{(2)}\rho^{(3)}$	$\rho^{(2)}\rho^{(3)}\rho^{(4)}$
$\rho^{(2)}$	$\rho^{(1)}\rho^{(2)}\rho^{(3)}$	$\rho^{(0)}\rho^{(1)}\rho^{(2)}\rho^{(3)}\rho^{(4)}$	$\rho^{(1)}\rho^{(2)}\rho^{(3)}\rho^{(4)}\rho^{(5)}$
$\rho^{(3)}$	$\rho^{(2)}\rho^{(3)}\rho^{(4)}$	$\rho^{(1)}\rho^{(2)}\rho^{(3)}\rho^{(4)}\rho^{(5)}$	$\rho^{(0)}\rho^{(1)}\rho^{(2)}\rho^{(3)}\rho^{(4)}\rho^{(5)}\rho^{(6)}$

FIG 3.13 Irreducible representations $\rho_q^{(k)}$ of the density matrix for an $n=3$ atomic state (an $n=2$ state can be obtained by eliminating the $l=2$ and $l'=2$ rows and and columns). The various q components are not explicitly shown.

3.5.2 VECTOR FIELDS

Before introducing vector spherical harmonics and multipole expansions of a vector field (a set of vectors that depend on spatial or momentum variables), let us consider how a vector field $\mathbf{V}(\mathbf{r})$ transforms under rotation. To do so, we expand the vector field in unit vectors along a set of space fixed axes, $\mathbf{e}_i$, $i=1,2,3$,

$$\mathbf{V}(\mathbf{r}) = \sum_i V_i(\mathbf{r}) \mathbf{e}_i. \quad (3.185)$$

Rotating the vector field by an angle φ using the rotation operator $\Re = e^{-i\boldsymbol{\varphi}\cdot\mathbf{J}/\hbar} = e^{-i\boldsymbol{\varphi}\cdot\mathbf{L}} e^{-i\boldsymbol{\varphi}\cdot\mathbf{S}}$, we find

$$\mathbf{V}'(\mathbf{r}) = e^{-i\boldsymbol{\varphi}\cdot\mathbf{J}/\hbar} \sum_i V_i(\mathbf{r}) \mathbf{e}_i = \sum_i \left(e^{-i\boldsymbol{\varphi}\cdot\mathbf{L}} V_i(\mathbf{r})\right) \left(e^{-i\boldsymbol{\varphi}\cdot\mathbf{S}} \mathbf{e}_i\right), \quad (3.186)$$

where $e^{-i\boldsymbol{\varphi}\cdot\mathbf{L}} V_i(\mathbf{r}) = V_i(\Re^{-1}\mathbf{r})$. The two contributions upon rotating a vector field are (1) the variation of the field components at different field points, induced by the differential operator $\boldsymbol{\varphi} \cdot \mathbf{L}$, and (2) the change generated by $\boldsymbol{\varphi} \cdot \mathbf{S}$ due to the reorientation of the vector field components when the field is rotated. The operator $\mathbf{S}$ is a 3×3 matrix for the spin-one representation of the rotation group for the vector field. The spin part of the rotation can be represented as a cross product, $(\boldsymbol{\varphi} \cdot \mathbf{S})\mathbf{e}_i = \boldsymbol{\varphi} \times \mathbf{e}_i$. Taking the angle of rotation to be small, i.e., using $\boldsymbol{\varphi} = \delta\boldsymbol{\varphi}$, (3.186) becomes

$$\mathbf{V}'(\mathbf{r}) \approx \mathbf{V}(\mathbf{r}) - i \sum_i \left[(\delta\boldsymbol{\varphi} \cdot \mathbf{L}) V_i(\mathbf{r})\mathbf{e}_i + (\delta\boldsymbol{\varphi} \times \mathbf{e}_i) V_i(\mathbf{r}) \right]. \quad (3.187)$$

3.5 Tensor Operators

In order to proceed, it is convenient to define the unit vectors $\mathbf{e}_+$, $\mathbf{e}_-$, and $\mathbf{e}_0$ as:

$$\mathbf{e}_{+1} = -\frac{1}{\sqrt{2}}(\mathbf{e}_x + i\mathbf{e}_y), \quad \mathbf{e}_0 = \mathbf{e}_z, \quad \mathbf{e}_{-1} = \frac{1}{\sqrt{2}}(\mathbf{e}_x - i\mathbf{e}_y). \tag{3.188}$$

For comparison, see (3.174). We can now define the vector spherical harmonics using the unit vectors defined in (3.188)

$$\mathbf{Y}_M^{(L),l,1} = \sum_{m,q} \langle L M | l\, 1\, m\, q \rangle\, Y_{lm}\, \mathbf{e}_q. \tag{3.189}$$

Clearly, we must have $L = l \pm 1$ or $L = l$. The set of $2L+1$ fields $\mathbf{Y}_M^{(L),l,1}$ with $M = -L, \ldots, L, L$, transform under rotation amongst themselves as components of a components of a tensor of rank L, i.e., they form a set that transforms as an irreducible representation under rotations. They form a complete set of states for expanding the angular dependence of vector fields and satisfy the orthogonality conditions

$$\int d\Omega\, (\mathbf{Y}_{M'}^{(L'),l',1}(\theta,\phi))^* \, \mathbf{Y}_M^{(L),l,1}(\theta,\phi) = \delta_{l',l}\delta_{L',L}\delta_{M',M}. \tag{3.190}$$

The vector spherical harmonics are useful in considering the solutions to the vector wave equation, $(\nabla^2 + k^2)\mathbf{A} = 0$, known also as the vector Helmholtz equation, where $\mathbf{A}$ can be either the vector potential, the electric field or the magnetic field. We shall consider such solutions shortly.

Problem 3.32

(a) Show that $S_+ \mathbf{e}_0 = \sqrt{2}\mathbf{e}_{+1}$.
(b) Show that $S_{-1}\mathbf{e}_0 = -\sqrt{2}\mathbf{e}_-$.
(c) Prove that $J_z \mathbf{r} = (L_z + S_z)(x\mathbf{e}_x + y\mathbf{e}_y + z\mathbf{e}_z) = 0$.
(d) Show that $\mathbf{r}$ can be written as $\mathbf{r} = \sqrt{\frac{4\pi}{3}}\, r \sum_m Y_{1\,-m}\mathbf{e}_m = -\sqrt{4\pi}\, r\, \mathbf{Y}_0^{(0),1,1}(\theta,\phi)$.

3.5.3 SPINOR FIELDS

This subsection can be skipped until after reading Sec. 4.2.3.

The spinor spherical harmonics are the analogs of the vector spherical harmonics defined in (3.189). As we have already seen and as we shall again take up in the Chapter 4 [see Eq. (4.12)], any spinor field can be expanded in terms of the spin basis functions $|\frac{1}{2}, m_s = \frac{1}{2}\rangle = \begin{pmatrix}1\\0\end{pmatrix}$ and $|\frac{1}{2}, m_s = -\frac{1}{2}\rangle = \begin{pmatrix}0\\1\end{pmatrix}$ by multiplying the spinor basis function by amplitudes that are functions of $\mathbf{r}$ to obtain the following ket in spin space:[6]

$$\langle \mathbf{r} | u \rangle = \sum_{m_s=-\frac{1}{2}}^{\frac{1}{2}} \psi_{m_s}(\mathbf{r}) \left|\tfrac{1}{2}, m_s\right\rangle = \psi_{1/2}(\mathbf{r})\begin{pmatrix}1\\0\end{pmatrix} + \psi_{-1/2}(\mathbf{r})\begin{pmatrix}0\\1\end{pmatrix}. \tag{3.191}$$

We can form an irreducible set of spinor spherical-harmonic fields as follows:

$$|\chi_{m_j}^{(j),l,1/2}\rangle = \sum_{m,m_s} \langle j\, m_j | l\, \tfrac{1}{2}\, m\, m_s \rangle\, |l m\rangle |\chi_{m_s}\rangle. \tag{3.192}$$

---

[6] With this notation, the ket is denoted $|u\rangle = \sum_{m_s} |\psi_{m_s}\rangle|\frac{1}{2}, m_s\rangle$.

For a given l, j can take the values $j = l \pm 1/2$. By evaluating the Clebsh–Gordan coefficients, we find

$$\langle \theta, \phi | \chi_{m_j}^{(j=l\pm 1), l, 1/2} \rangle = \frac{1}{\sqrt{2l+1}} \begin{pmatrix} \pm \sqrt{l \pm m_j + \frac{1}{2}} \, Y_{l, m_j - 1/2}(\theta, \phi) \\ \sqrt{l \mp m_j + \frac{1}{2}} \, Y_{l, m_j + 1/2} Y_{l, m_j - 1/2}(\theta, \phi) \end{pmatrix}. \tag{3.193}$$

The set of $2j+1$ spinor fields $\chi_{m_j}^{(j, l, 1/2)}$, with $m_j = -j, \ldots, j$, transform under rotation amongst themselves as components of a tensor rank j. The irreducible set of spinor fields form a complete set of states for expanding the angular dependence of spinor fields,

$$\langle \mathbf{r} | u \rangle = \sum_{j, m_j, l} a_{j, m_j, l}(r) \chi_{m_j}^{(j), l, 1/2}(\theta, \phi). \tag{3.194}$$

Problem 3.33

(a) Determine the orthogonality properties of the spinor fields defined in (3.192).
(b) Expand the spinor plane wave function, $e^{i\mathbf{k} \cdot \mathbf{r}} | \chi_{m_s} \rangle$ in the irreducible set of spinor fields.
(c) Use (3.84) for the plane wave and consider the quantity

$$\mathcal{Y}_{M, m_s}^{(J)l}(\hat{\mathbf{r}}, \hat{\mathbf{k}}) = \left[\sum_{m' m'_s} \langle JM | l \tfrac{1}{2} m' m'_s \rangle Y_{lm'}(\hat{\mathbf{r}}) | \chi_{m'_s} \rangle \right] \left[\sum_m \langle JM | l \tfrac{1}{2} m m_s \rangle Y_{lm}^*(\hat{\mathbf{k}}) \right], \tag{3.195}$$

where the left bracket on the RHS of (3.195) is the spinor $|\chi_M^{(J), l, 1/2}(\hat{\mathbf{r}})\rangle$, which is an eigenfunction of J^2 and J_z. Noting that the sum over J, M of the product of the Clebshs in (3.195) gives a $\delta_{mm'} \delta_{m_s m'_s}$ section rule, we find

$$e^{i\mathbf{k} \cdot \mathbf{r}} \chi_{m_s} = 4\pi \sum_{lJM} i^l \frac{\hat{j}_l(kr)}{kr} \mathcal{Y}_{M, m_s}^{(J)l}(\hat{\mathbf{r}}, \hat{\mathbf{k}}). \tag{3.196}$$

Answer: (a) $\langle \chi_{m'_j}^{(j'), l', 1/2} | \chi_{m_j}^{(j), l, 1/2} \rangle = \int d\Omega \, (\chi_{m'_j}^{(j'), l', 1/2}(\Omega))^* \chi_{m_j}^{(j), l, 1/2}(\Omega) = \delta_{l' l} \delta_{j' j} \delta_{m'_j m_j}$.

3.5.4 MULTIPOLE EXPANSIONS

We have already encountered a multipole expansion when we expanded the plane wave in spherical harmonics [see Eq. (3.84)] and multipole expansions of the density matrix [see Eq. (3.183)]. Another expansion in terms of multipole moments that you may be familiar with from electricity and magnetism courses is the expansion of the Coulomb potential in spherical harmonics,

$$\frac{1}{|\mathbf{r} - \mathbf{r}'|} = \sum_{l, m} \frac{4\pi}{2l+1} \frac{r_<^l}{r_>^{l+1}} Y_{lm}^*(\theta', \phi') Y_{lm}(\theta, \phi), \tag{3.197}$$

where $r_< = \min(r, r')$, and $r_> = \max(r, r')$. Using (3.197), the potential due to a charge distribution, $V(\mathbf{r}) = \int d\mathbf{r}' \frac{\rho(\mathbf{r}')}{|\mathbf{r} - \mathbf{r}'|}$, can be written (for sufficiently large r) as

$$\boxed{V(\mathbf{r}) = \sum_{l, m} \sqrt{\frac{4\pi}{2l+1}} Q_{lm}^* \frac{1}{r^{l+1}} Y_{lm}(\theta, \phi),} \tag{3.198}$$

3.6 Symmetry Considerations

where the *tensor multipole moments* of the charge distribution are given by

$$Q_{lm} = \sqrt{\frac{4\pi}{2l+1}} \int d\mathbf{r}' \, \rho(\mathbf{r}') \, r'^l \, Y_{lm}(\theta', \phi'). \tag{3.199}$$

Clearly, this can be generalized to the case of a discrete charge distribution, $\rho(\mathbf{r}') = \sum_i q_i \delta(\mathbf{r}' - \mathbf{r}_i)$, wherein the multipole moments Q_{lm} can be easily calculated by substituting this into the RHS of (3.199). $Q_{00} = \int d\mathbf{r}' \, \rho(\mathbf{r}') = q$, which is the charge of the system, the Q_{1m} terms are the components of the electric dipole moment [e.g., [e.g., $Q_{10} = \int d\mathbf{r}' \, \rho(\mathbf{r}') z' = p_z$, where $\mathbf{p}$ is the electric dipole moment vector], the Q_{2m} terms are the quadrupole moment components [e.g., $Q_{20} = \frac{1}{2} \int d\mathbf{r}' \, \rho(\mathbf{r}') (3z'^2 - r'^2)$], etc. Away from the region encompassing the charges, the potential $V(\mathbf{r})$ in (3.198) satisfies the equation, $\nabla^2 V = 0$.

The Coulomb energy of a charge distribution is given by $U = \sum_{i,j>i} \frac{q_i q_j}{|\mathbf{r}_i - \mathbf{r}_j|} = \frac{1}{2} \sum_{i,j \neq i} \frac{q_i q_j}{|\mathbf{r}_i - \mathbf{r}_j|}$. For a continuous charge distribution,

$$U = \frac{1}{2} \iint d\mathbf{r}' d\mathbf{r} \frac{\rho(\mathbf{r}) \rho(\mathbf{r}')}{|\mathbf{r}' - \mathbf{r}|} = \frac{1}{2} \int d\mathbf{r} \rho(\mathbf{r}) V(\mathbf{r}). \tag{3.200}$$

But if we have two different charge distributions located far apart from one another, the Coulomb interaction energy of these two distributions is given by

$$U = \iint d\mathbf{r}' d\mathbf{r} \frac{\rho_a(\mathbf{r}) \rho_b(\mathbf{r}')}{|\mathbf{r}' - \mathbf{r}|} = \int d\mathbf{r} \rho_a(\mathbf{r}) V_b(\mathbf{r}), \tag{3.201}$$

where $V_b(\mathbf{r})$ is the potential at position $\mathbf{r}$ due to the charge distribution ρ_b. Expanding (3.201) in terms of multipole moments and the coordinate $\mathbf{r}$ from one-charge distribution to the other, we obtain:

$$U(\mathbf{r}) = \frac{q_a q_b}{r} - \left(\frac{3(\mathbf{p}_a \cdot \mathbf{r})(\mathbf{p}_b \cdot \mathbf{r}) - (\mathbf{p}_a \cdot \mathbf{p}_b) r^2}{r^5} \right) + \sum_{l \geq 2, m} \frac{Q_{a,lm} Q_{b,lm}}{r^l}. \tag{3.202}$$

A generalization of the multipole expansion of the plane wave, Eq. (3.84), which satisfies the Helmholtz equation, $(\nabla^2 + k^2) \psi_\mathbf{k}(\mathbf{r}) = 0$, is the multipole expansion of the vector Helmholtz equation, $(\nabla^2 + k^2) \mathbf{A} = 0$. Plane waves are expanded in terms of functions of the form $\phi_{lm} = 4\pi i^l \frac{\hat{j}_l(kr)}{kr} Y_{lm}(\theta, \phi)$; the solutions of the vector wave equation fall into the categories of longitudinal, transverse electric and transverse magnetic solutions, which can be expanded in terms of the following forms respectively:

$$\mathbf{A}_{lm} = \nabla \phi_{lm}, \tag{3.203a}$$

$$\mathbf{A}^e_{lm} = \nabla \times \mathbf{L} \phi_{lm}, \tag{3.203b}$$

$$\mathbf{A}^m_{lm} = \mathbf{L} \phi_{lm}. \tag{3.203c}$$

3.6 SYMMETRY CONSIDERATIONS

The matrix elements of a physical quantity $\mathcal{O}$,

$$\langle \psi_{\beta,j} | \mathcal{O} | \psi_{\alpha,i} \rangle = \int d\mathbf{r} \psi^*_{\beta,j}(\mathbf{r}) \mathcal{O} \psi_{\alpha,i}(\mathbf{r}), \tag{3.204}$$

where the subscripts α and β distinguish different energy levels, and i and j refer to different states belonging to the same degenerate level, which may vanish due to symmetry considerations. Rotating the states (or rotating the coordinates) can help determine whether the matrix elements (3.204) vanish or not due to rotational symmetry. Moreover, other transformations, such as inversion of the coordinates or time-reversal, can also help determine whether the matrix elements vanish for symmetry reasons. The mathematical language most helpful in this assessment is group theory (Appendix E), more specifically, the irreducible representations of symmetry groups, which we will take up in Sec. E.5.1. However, we have developed enough mathematical tools to begin discussion of the consequences of symmetry in the determination of whether matrix elements vanish. We shall now investigate what inversion, time-reversal symmetry, and rotational symmetry can tell us about matrix elements of operators $\mathcal{O}$ that transform in specific ways.

3.6.1 SELECTION RULES

Selection rules determine whether a given transition is forbidden, on the basis of the symmetry of the initial and final states and of the operator responsible for the transition. Selection rules can result due to various symmetry considerations, including rotational invariance, inversion symmetry invariance (parity), time-reversal invariance, exchange symmetry, etc.

Eugene Wigner was the first to employ group-theoretical considerations to interpret selection rules in spectroscopy (1926–1927), starting first with atomic spectroscopy, but the approach was quickly extended to molecular and nuclear spectroscopy. He invoked the transformation properties of energy eigenstates of the system with respect to operations that leave the system as a whole invariant, such as space rotations, inversions, reflections, time-reversal, and permutations of the electrons.

In what follows, we first consider parity and time-reversal selection rules and then introduce the Wigner–Eckart theorem for rotational invariance considerations. A general method of obtaining selection rules for arbitrary symmetry is discussed in Sec. 11.6.

3.6.2 INVERSION SYMMETRY

As discussed in Sec. 2.9.2, the following vector operators transform under space inversion as follows:

$$\mathbf{r} \to -\mathbf{r}, \quad \mathbf{p} \to -\mathbf{p}, \quad \mathbf{J} \to \mathbf{J}. \tag{3.205}$$

What does the inversion operator do to the wave function of the hydrogen atom, $\psi_{nlm}(\mathbf{r}) = R_{nl}(r)Y_{lm}(\theta,\phi)$? Transforming $\mathbf{r} \to -\mathbf{r}$ corresponds to $r \to r$, $\theta \to \pi - \theta$ (i.e., $\cos\theta \to -\cos\theta$), and $\phi \to \phi + \pi$. Therefore, using (3.49), we find $P_l^m(\cos\theta) \to (-1)^{l+m} P_l^m(\cos\theta)$, and $e^{im\phi} \to (-1)^m e^{im\phi}$, hence, we conclude that

$$\mathcal{P}\psi_{nlm}(\mathbf{r}) = (-1)^l \psi_{nlm}(\mathbf{r}), \tag{3.206}$$

i.e., the parity ε_{nlm} of the hydrogenic wave function ψ_{nlm} depends only upon the angular momentum quantum number l, $\varepsilon_{nlm} = (-1)^l$.

Suppose ϕ_i and ϕ_f are parity eigenstates, i.e., $\mathcal{P}\phi_i = \varepsilon_i \phi_i$, $\mathcal{P}\phi_f = \varepsilon_f \phi_f$. The matrix element $\langle \phi_f | \mathcal{O} | \phi_i \rangle$ of an operator $\mathcal{O}$ which commutes with the inversion operator vanishes if $\varepsilon_f = -\varepsilon_i$. Similarly, the matrix element of an odd operator, i.e., one which anticommutes with the inversion operator, $\mathcal{PO} = -\mathcal{OP}$, vanishes unless $\varepsilon_f = \varepsilon_i$. The proof is simple:

$$\langle \phi_f | \mathcal{O} | \phi_i \rangle = \langle \phi_f | \mathcal{P}^{-1} \mathcal{P} \mathcal{O} \mathcal{P}^{-1} \mathcal{P} | \phi_i \rangle = \varepsilon_f \varepsilon_i \langle \phi_f | \mathcal{P} \mathcal{O} \mathcal{P}^{-1} | \phi_i \rangle. \tag{3.207}$$

Hence, if $\mathcal{O}$ commutes with $\mathcal{P}$, the matrix element vanishes if $\varepsilon_f \varepsilon_i = -1$, and if it anticommutes, it vanishes unless $\varepsilon_f \varepsilon_i = 1$.

For example, the operators $\mathbf{r}$ and $\mathbf{p}$ each anticommutes with the inversion operator, therefore the initial and final states, ϕ_i and ϕ_f, in the matrix elements of this operator must have opposite parity if the matrix element is nonvanishing. We shall see in Sec. 7.4.2 that this implies that the initial and final states in electric dipole transitions must have opposite parity. Another example involves the operator $(\mathbf{p} \cdot \mathbf{A})(\mathbf{k} \cdot \mathbf{r})$, for electric quadrupole transitions, where $\mathbf{A}$ is the constant part of the vector potential for light (see Sec. 7.4.2). This operator commutes with the inversion operator so the initial and final states in the matrix elements of this operator must have the same parity if the matrix element is nonvanishing.

3.6 Symmetry Considerations

3.6.3 TIME-REVERSAL SYMMETRY

The *time-reversal* transformation sends $t \to -t$, thereby reversing the velocity of particles but not affecting their positions. It transforms dynamical variables as follows:

$$\mathbf{r} \to \mathbf{r}, \quad \mathbf{p} \to -\mathbf{p}, \quad \mathbf{J} \to -\mathbf{J}. \tag{3.208}$$

What does the time-reversal operator do to the wave function of the hydrogen atom, $\psi_{nlm}(\mathbf{r}) = R_{nl}(r) Y_{lm}(\theta, \phi)$? For a wave function of a spinless particle, the time-reversal operator corresponds to the complex conjugation operator, $\mathcal{K}$ (see Sec. 2.9.3), so

$$\mathcal{K}\psi_{nlm}(\mathbf{r}) = \psi_{nlm}^*(\mathbf{r}) = (-1)^m \psi_{nl-m}(\mathbf{r}), \tag{3.209}$$

i.e., the wave functions ψ_{nlm} are not in general eigenfunctions of the time-reversal operator. Since analysis of time reversal symmetry for systems having particles with spin is different from that for spinless particles, we postpone the discussion of time reversal symmetry to Sec. 4.4, after gaining more familiarity with spinors. Here we simply quote some results pertaining to restrictions dictated by time reversal invariance on matrix elements of operators. In analyzing the time reversal properties of matrix elements of an operator $\mathcal{O}$ (not necessarily Hermitian), some of the arguments used in the analysis of space inversion need to be modified because the time reversal operator $\mathcal{T}$ is antilinear and its action on bras from the right is ill defined (Dirac notation was designed for linear operators, not antilinear ones). In order to avoid this, it is useful to adopt a rule that antilinear operators act only on ket states (from the left) and not on bra states (from the right). Thus, if A is an antilinear operator, we interpret its matrix elements $\langle \beta | A | \alpha \rangle$ as $\langle \beta | (A | \alpha \rangle)$ and not as $(\langle \beta | A) | \alpha \rangle$. For a given ket $|\alpha\rangle$ we define $|\tilde{\alpha}\rangle \equiv \mathcal{T}|\alpha\rangle$. Once the ket $|\tilde{\alpha}\rangle$ is defined, we can define its dual vector $\langle \tilde{\alpha} |$. In Sec. 4.4, it will be shown that: (1) If the Hamiltonian H is invariant under time reversal and a given eigenket $|\psi_n\rangle$ of H is non degenerate, then the corresponding energy eigenfunction $\psi_n(\mathbf{x}) = \langle \mathbf{x} | \psi_n \rangle$ can be chosen to be real. (2) The matrix element of $\mathcal{O}$ between any two ket states $|\phi_i\rangle$ and $|\phi_f\rangle$ obeys the identity,

$$\langle \phi_f | \mathcal{O} | \phi_i \rangle = \langle \tilde{\phi}_i | \mathcal{T} \mathcal{O}^\dagger \mathcal{T}^{-1} | \tilde{\phi}_f \rangle. \tag{3.210}$$

As a special case, suppose that $\mathcal{O}$ is hermitian and has a definite parity $\tau_\mathcal{O} = \pm 1$ under time reversal, $\mathcal{T}\mathcal{O}^\dagger \mathcal{T}^{-1} = \tau_\mathcal{O} \mathcal{O}$. Assume further, that $|\phi_i\rangle$ and $|\phi_f\rangle$ have definite parities $\tau_i, \tau_f = \pm 1$ under time reversal, $|\tilde{\phi}_i\rangle = \tau_i |\phi_i\rangle$ and $|\tilde{\phi}_f\rangle = \tau_i |\phi_f\rangle$. Then,

$$\langle \phi_f | \mathcal{O} | \phi_i \rangle = \tau_i \tau_f \tau_\mathcal{O} \langle \phi_i | \mathcal{O}^\dagger | \phi_f \rangle = \tau_i \tau_f \tau_\mathcal{O} \langle \phi_f | \mathcal{O} | \phi_i \rangle^* . \tag{3.211}$$

Thus, if the matrix element is real (e.g., if $|\phi_f\rangle = |\phi_i\rangle$), it will vanish unless $\tau_i \tau_f \tau_\mathcal{O} = 1$. In general, nothing can be said about matrix elements with states that are not eigenfunctions of the time-reversal operator.

3.6.4 WIGNER–ECKART THEOREM

The evaluation of matrix elements of tensor operators is greatly simplified by means of the results derived in Sec. 3.5. These methods allow us to determine angular momentum selection rules for matrix elements, as summarized in what is known as the *Wigner–Eckart theorem*, which fully determines the angular part of the matrix elements.

The matrix elements of a tensor operator between initial and final states with definite angular momentum have a simple geometrical dependence on the magnetic quantum numbers. Let $T_q^{(k)}$ be a tensor operator of rank k and magnetic quantum number q, and consider the matrix element $\langle \alpha' j' m' | T_q^{(k)} | \alpha j m \rangle$, where α and α' are possible additional quantum numbers describing the states. $T_q^{(k)} | \alpha j m \rangle$ transforms under rotations as $D^{(k)} \otimes D^{(j)}$. The irreducible components of $T_q^{(k)} | \alpha j m \rangle$ are formed by taking linear combinations that form a specific rank K object having magnetic quantum number Q,

$$|\beta K Q\rangle = \sum_{q,m} \langle K Q | k j q m \rangle T_q^{(k)} | \alpha j m \rangle. \tag{3.212}$$

The inverse of this relation is

$$T_q^{(k)}|\alpha j m\rangle = \sum_{K,Q}\langle K Q|k j q m\rangle |\beta K Q\rangle. \tag{3.213}$$

Taking the inner product with $|\alpha' j' m'\rangle$, we obtain

$$\langle \alpha' j' m'|T_q^{(k)}|\alpha j m\rangle = \sum_{K,Q}\langle \alpha' j' m'|\beta K Q\rangle \langle K Q|k j q m\rangle. \tag{3.214}$$

The matrix element $\langle \alpha' j' m'|\beta K Q\rangle$ vanishes unless $j' = K$ and $m' = Q$, i.e., $\langle \alpha' j' m'|\beta K Q\rangle = \text{const}\,\delta_{j'K}\delta_{m'Q}$, and the constant is given the name $(-1)^{2k}\langle \alpha' j'||T^{(k)}||\alpha' j\rangle$, so we obtain

$$\boxed{\langle \alpha' j' m'|T_q^{(k)}|\alpha j m\rangle = (-1)^{2k}\langle \alpha' j'||T^{(k)}||\alpha j\rangle \langle j' m'|k j q m\rangle.} \tag{3.215}$$

Note that the constant $\langle \alpha' j'||T^{(k)}||\alpha j\rangle$, called the *reduced matrix element*, is independent of m, m' and q. Equation (3.215) is called the Wigner–Eckart Theorem. It can be written in terms of $3j$ symbols using the definition in Eq. (3.155):

$$\langle \alpha' j' m'|T_q^{(k)}|\alpha j m\rangle = \frac{(-1)^{j'+k+q}}{\sqrt{2j'+1}}\langle \alpha' j'||T^{(k)}||\alpha j\rangle \begin{pmatrix} j & k & j' \\ m & q & -m' \end{pmatrix}. \tag{3.216}$$

In summary, the angular momentum selection rules that are embodied in the Wignerd–Eckart theorem are:

$$\boxed{\Delta j = j - j' = 0, \pm 1, \ldots \pm k, \quad j = j' \neq 0, \quad q = m' - m.} \tag{3.217}$$

The first equation is a consequence of the triangle inequality for (j, j', k), and the middle equation indicates that, although j can equal j', this is not possible if $j = 0$ (since $k \neq 0$). For the special case of dipole-allowed electromagnetic radiation (see Sec. 7.4.2) selection rules, $k = 1$, i.e., the tensor corresponds to a rank-one tensor (a vector), and $\Delta j = 0, \pm 1$. For electric quadrupole radiation, $k = 2$ and $\Delta j = 0, \pm 1, \pm 2$.

Problem 3.34

Prove that a system with the magnitude of the total angular momentum **J** equal to 0 or 1/2 cannot have an electric quadrupole moment.

Let us consider some examples. When we compare Eq. (3.215) with the matrix elements of the spherical harmonics, Eq. (3.167), i.e.,

$$\int d\Omega\, Y_{lm}^*(\Omega) Y_{LM}(\Omega) Y_{lm}(\Omega) = \sqrt{\frac{(2l+1)(2L+1)}{4\pi(2l'+1)}}\langle l L m M|l' m'\rangle \langle l L 0 0|l' 0\rangle$$

we obtain the expression,

$$\langle l'||Y^{(k)}||l\rangle = \sqrt{\frac{(2l+1)(2L+1)}{4\pi(2l'+1)}}\langle l L 0 0|l' 0\rangle, \tag{3.218}$$

3.6 Symmetry Considerations

for the spherical harmonics reduced matrix element. As another example, consider matrix elements of the angular momentum operator. By comparing

$$\langle \alpha' j' m' | J_z | \alpha' j m \rangle = \hbar m \, \delta_{m',m} \delta_{j',j} \delta_{\alpha',\alpha} \tag{3.219}$$

with (3.215), we find the angular momentum reduced matrix element to be

$$\langle \alpha' j' || J^{(1)} || \alpha' j \rangle = \hbar \sqrt{j(j+1)} \delta_{j',j} \delta_{\alpha',\alpha}. \tag{3.220}$$

A simple relation for matrix elements with $j = j'$ of rank-one tensors, called the *projection theorem*, can be derived using the Wigner–Eckart theorem:

$$\boxed{\langle \alpha' j m' | V_q | \alpha j m \rangle = \frac{\langle \alpha' j m' | \mathbf{J} \cdot \mathbf{V} | \alpha j m \rangle}{\hbar^2 j(j+1)} \langle \alpha' j m' | J_q | \alpha j m \rangle.} \tag{3.221}$$

The angular momentum matrix element appearing on the RHS of (3.221) is determined using (3.220) to be $\langle \alpha' j m' | J_q | \alpha j m \rangle = \hbar \sqrt{j(j+1)} \langle j' m' | k j q m \rangle$. Note that the quantity $\frac{\mathbf{J}(\mathbf{J} \cdot \mathbf{V})}{\hbar^2 j(j+1)}$ appearing on the RHS of (3.221) is the component of the vector $\mathbf{V}$ along the unit vector in the direction of the angular momentum vector. The proof of the projection theorem goes as follows. Let us evaluate $\langle \alpha' j m' | \mathbf{J} \cdot \mathbf{V} | \alpha j m \rangle = \sum_q (-1)^q \langle \alpha' j m' | J_{-q} V_q | \alpha j m \rangle$ by inserting a complete set of states between J_{-q} and V_q:

$$\langle \alpha' j m' | \mathbf{J} \cdot \mathbf{V} | \alpha j m \rangle = \sum_{q \alpha'' j'' m''} (-1)^q \langle \alpha' j m' | J_{-q} | \alpha'' j'' m'' \rangle \langle \alpha'' j'' m'' | V_q | \alpha j m \rangle. \tag{3.222}$$

The matrix element of J_{-q} is diagonal in α and j, and the matrix element of V_q can be evaluated using the Wigner–Eckart theorem to obtain:

$$\langle \alpha' j m' | \mathbf{J} \cdot \mathbf{V} | \alpha j m \rangle = \sum_{q m''} (-1)^q \langle j m' | J_{-q} | j m'' \rangle \langle j m'' | 1 j q m \rangle \langle \alpha' j || V^{(1)} || \alpha j \rangle. \tag{3.223}$$

By setting $\mathbf{V} = \mathbf{J}$ in (3.223), the coefficient $\sum_{q m''} (-1)^q \langle j m' | J_{-q} | j m'' \rangle \langle j m'' | 1 j q m \rangle$ on the RHS can be evaluated and we find

$$\langle \alpha' j m' | \mathbf{J} \cdot \mathbf{V} | \alpha j m \rangle = \hbar \sqrt{j(j+1)} \langle \alpha' j || V^{(1)} || \alpha j \rangle. \tag{3.224}$$

By dividing the Wigner–Eckart equalities

$$\langle \alpha' j' m' | V_q | \alpha j m \rangle = \langle j' m' | 1 j q m \rangle \langle \alpha' j' || V^{(1)} || \alpha j \rangle,$$

$$\langle \alpha' j m' | J_q | \alpha j m \rangle = \langle j m' | 1 j q m \rangle \langle \alpha' j || J^{(1)} || \alpha j \rangle,$$

one by the other, and using (3.224) and (3.220), we finally obtain the projection theorem result, Eq. (3.221). We will use of the projection theorem in evaluating magnetic dipole moment g-factors in Chapter 4.

We shall delay the specification of the selection rules for spontaneous emission and for absorption for atoms to Sec. 7.4.2 and for molecules to Sec. 11.6.

Problem 3.35

Calculate the reduced matrix element $\langle 1/2 || \sigma^{(1)} || 1/2 \rangle$.

Answer: Use $\langle 1/2, 1/2 | \sigma_z | 1/2, 1/2 \rangle = 1$, and note that in the language of spherical tensor operators $\langle 1/2, 1/2 | \sigma_z | 1/2, 1/2 \rangle = \langle 1/2, 1/2 | \sigma_0^{(1)} | 1/2, 1/2 \rangle = \langle \frac{1}{2} 1 \frac{1}{2} 0 | \frac{1}{2} \frac{1}{2} \rangle \langle 1/2 || \sigma^{(1)} || 1/2 \rangle$. Since the CG coefficient, $\langle \frac{1}{2} 1 \frac{1}{2} 0 | \frac{1}{2} \frac{1}{2} \rangle = \frac{1}{\sqrt{3}}$, we obtain, $\langle 1/2 || \sigma^{(1)} || 1/2 \rangle = \sqrt{3}$.

> **Problem 3.36**
>
> Calculate the reduced matrix element $\langle l_3||Y^{(l_2)}||l_1\rangle$ by letting $m_1 = m_2 = m_3 = 0$ in (3.167) and comparing with (3.215).
>
> **Answer:** In the notation of tensor operators, the LHS of (3.167) is written as $\langle l_3 m_3|Y_{l_2 m_2}|l_1 m_1\rangle$. Let $m_1 = m_2 = m_3 = 0$ on the RHS of (3.167) to obtain
>
> $\langle l_3 0|Y_{l_2 0}|l_1 0\rangle = \sqrt{\frac{(2l_1+1)(2l_2+1)}{4\pi(2l_3+1)}} \langle l_1 l_2 00|l_3 0\rangle^2 = \langle l_1 l_2 00|l_3 0\rangle \langle l_3||Y^{(l_2)}||l_1\rangle$.
>
> Therefore, $\langle l_3||Y^{(l_2)}||l_1\rangle = \sqrt{\frac{(2l_1+1)(2l_2+1)}{4\pi(2l_3+1)}} \langle l_1 l_2 00|l_3 0\rangle$.

3.6.5 $6j$ AND HIGHER COEFFICIENTS

In coupling three angular momentum states, we may use an uncoupled representation

$$|j_1\, m_1\rangle|j_2\, m_2\rangle|j_3\, m_3\rangle, \tag{3.225}$$

or one in which the vectors couple to a resultant total angular momentum J and z-component M, i.e., an eigenstate of $\mathbf{J}^2 = (\mathbf{j}_1 + \mathbf{j}_2 + \mathbf{j}_2)^2$ and $J_z = j_{1z} + j_{2z} + j_{3z}$. These states are not unique, and a further quantum number is required. There are three possibilities: we may couple $\mathbf{j}_1$ and $\mathbf{j}_2$ to form $\mathbf{J}_{12}$, then add $\mathbf{j}_3$ vectorially to give $\mathbf{J}$. That is, first we couple $\mathbf{j}_1$ and $\mathbf{j}_2$,

$$|(j_1 j_2) J_{12} M_{12}\rangle = \sum_{m_1,m_2} |j_1\, m_1\rangle|j_2\, m_2\rangle \langle J_{12} M_{12}|j_1\, m_1 j_2\, m_2\rangle, \tag{3.226}$$

and then couple $\mathbf{J}_{12}$ and $\mathbf{j}_3$,

$$|((j_1 j_2) J_{12} j_3) JM\rangle = \sum_{M_{12}=-J_{12}}^{J_{12}} \sum_{m_3=-j_3}^{j_3} |(j_1 j_2) J_{12} M_{12}\rangle|j_3 m_3\rangle \langle J_{12} M_{12} j_3 m_3|JM\rangle. \tag{3.227}$$

There are two remaining alternatives, of which we explicitly present the one coupling $\mathbf{j}_2$ and $\mathbf{j}_3$ to form $\mathbf{J}_{23}$, then add $\mathbf{j}_1$ vectorially to give $\mathbf{J}$, i.e.,

$$|(j_1,(j_2 j_3)J_{23})JM\rangle = \sum_{m_1=-j_1}^{j_1} \sum_{M_{23}=-J_{23}}^{J_{23}} |j_1 m_1\rangle|(j_2 j_3)J_{23} M_{23}\rangle \langle j_1 m_1 J_{23} M_{23}|JM\rangle, \tag{3.228}$$

and then

$$|((j_1 j_2)J_{12} j_3)JM\rangle = \sum_{J_{23}} |(j_1,(j_2 j_3)J_{23})JM\rangle \langle (j_1,(j_2 j_3)J_{23})J|((j_1 j_2)J_{12} j_3)J\rangle. \tag{3.229}$$

The transformation coefficient $\langle (j_1,(j_2 j_3)J_{23})J|((j_1 j_2)J_{12} j_3)J\rangle$ is a scalar and independent of M. Racah defined the W function,

$$\langle (j_1,(j_2 j_3)J_{23})J|((j_1 j_2)J_{12} j_3)J\rangle = \sqrt{(2J_{12}+1)(J_{23}+1)}\, W(j_1 j_2 J j_3; J_{12} J_{23}), \tag{3.230}$$

whose normalization was chosen to simplify its symmetry properties. Wigner defined the $6j$ symbol

$$\begin{Bmatrix} j_1 & j_2 & J_{12} \\ j_3 & J & J_{23} \end{Bmatrix} = (-1)^{j_1+j_2+j_3+J} W(j_1 j_2 J j_3; J_{12} J_{23}), \tag{3.231}$$

3.6 Symmetry Considerations

which has somewhat higher symmetry than the W function; it is invariant under interchange of any two columns and also interchange of the upper and lower arguments in each of any two columns.

Web-based 6j calculators can be found at the URLs,

http://plasma-gate.weizmann.ac.il/369j.html
http://www-stone.ch.cam.ac.uk/wigner.shtml.

The number of possible coupling modes rapidly increases with the number of angular momentum vectors coupled. Transformations between the various modes involve more complex coefficients, e.g., the coupling of four angular momenta involves two resultant pairs of angular momenta, and the 9j symbol is defined in this connection. The interested reader is referred to Refs. [38, 39].

As an example of the use of 6j coefficients, consider the matrix element of a tensor operator of orbital angular momentum of a system, $T^{(k)}$, in which orbital and spin angular momentum, L and S, are coupled to obtain the total angular momentum J in both the initial and final states. The reduced matrix element of the tensor operator $T^{(k)}$ can be worked out in terms of a product of a 6j coefficient and the orbital angular momentum reduced matrix element of $T^{(k)}$:

$$\langle LSJ||T^{(k)}||L'S'J'\rangle = \delta_{SS'} D_{\text{couple}} \langle L||T^{(k)}||L'\rangle, \tag{3.232}$$

where the line strength coefficient D_{couple} is

$$D_{\text{couple}} = (-1)^{L+S+J'+k} (2J+1)^{1/2}(2J'+1)^{1/2} \begin{Bmatrix} L & S & J \\ J' & k & L' \end{Bmatrix}. \tag{3.233}$$

Spin 4

The history of spin was briefly outlined in Sec. 1.1.5. Recall that in 1922, the Stern–Gerlach experiment showed that silver atoms are separated into two components upon traversing an inhomogeneous magnetic field. In 1925, Goudsmit and Uhlenbeck proposed that electrons have a spin of 1/2 (in units of $\hbar$). About a year later, Pauli introduced a "two-valued quantum degree of freedom" (i.e., spinors, see below). Pauli was skeptical about the assignment of a spin 1/2 to electrons until he learned of Llewellyn Thomas's work (1926) that resolved a factor of two discrepancy between experimental spin–orbit splitting results and Goudsmit and Uhlenbeck's calculations. The discrepancy resulted because account was not taken of the noninertial (accelerating) reference frame of an electron as it circles the nucleus. The factor of 1/2 that resolved the discrepancy is called the *Thomas precession factor* (see Sec. 4.5). By 1927, it was pretty much accepted that the electron has a spin of 1/2. In 1928, Paul Dirac developed the relativistic theory of the electron; the *Dirac equation* (discussed in Sec. 13.6.3) correctly describes electron spin. The Thomas precession factor appears naturally in the Dirac equation, as does the spin–orbit interaction, the Zeeman splitting in the presence of a magnetic field, and many additional effects arising due to electron spin. It also paved the way to the understanding of the positron, the electron antiparticle having electric charge $+e$ and spin 1/2. Spin angular momentum exists for many elementary particles. In 1932, James Chadwick discovered the neutron; it too has spin 1/2. Otto Stern measured the spin of the proton in 1933 to be $\hbar/2$, and I. I. Rabi measured it more accurately in 1934. The neutrino, the elusive particle that was postulated by Pauli in 1930 to account for energy conservation in beta decay (and first detected in 1956), also has spin 1/2.[1]

In this chapter, we study electron (and nuclear) spin. Section 4.1 (re-)introduces spin angular momentum operators and Sec. 4.2 defines spinors and spin-orbitals. Sec. 4.3 treats a charged particle with spin in a magnetic field, and Sec. 4.4 deals with the time-reversal properties of spinors. Spin-orbit and hyperfine interactions are discussed in Secs. 4.5 and 4.6. In Sec. 4.7, we discuss the spin–spin interactions in singlet and triplet states. Finally, in Sec. 4.8, we consider magnetic resonance. The material in this chapter will be heavily used in our discussion of quantum information in Chapters 5, 6 that treat many spin-related topics, including the Bloch sphere picture of spins, the density matrix description of spin degrees of freedom, and the use of the spin formalism to treat arbitrary two-level systems (spin is the prototype two level system).

We note in passing that electron spin can be used in nanotechnology devices (although to-date, few do); this new nanotechnology concept is called *spintronics*. In spintronic devices, one generates a current of spin-polarized electrons, and one incorporates device elements that are sensitive to the spin polarization of the electrons, and perhaps also elements that change the current of electrons depending on the spin state. For more on spintronics, see Sec. 9.7.

4.1 SPIN ANGULAR MOMENTUM

The commutation relations for angular momentum operators are [see Eq. (3.5)] $[\hat{J}_i, \hat{J}_j] = i\hbar \varepsilon_{ijk} \hat{J}_k$, where ε_{ijk} is the Levi-Civita symbol (each of the indices i, j, k can take on the value 1, 2, 3, or, if you like x, y, z) and Einstein's summation convention is used. Electron spin angular momentum operators satisfy the same commutation relations,

$$[\hat{S}_i, \hat{S}_j] = i\hbar \varepsilon_{ijk} \hat{S}_k. \qquad (4.1)$$

---

[1] Initially, it was thought to be massless, but we now know it has finite but extremely small mass.

Just as for an eigenstate of angular momentum, $\hat{\mathbf{J}}^2 = \hbar^2 j(j+1)$, an eigenstate of spin with $s = 1/2$ has $\hat{\mathbf{S}}^2 = \hbar^2 s(s+1)$, i.e.,[2]

$$\hat{\mathbf{S}}^2 = \hat{S}_x^2 + \hat{S}_y^2 + \hat{S}_z^2 = (3/4)\hbar^2 \hat{\mathbf{1}}. \tag{4.2}$$

In analogy with the general angular momentum case, see Eq. (3.10), we can define the spin raising and lowering operators,

$$\hat{S}_+ = \hat{S}_x + i\hat{S}_y, \quad \hat{S}_- = \hat{S}_x - i\hat{S}_y. \tag{4.3}$$

It is convenient to define dimensionless operators that have commutation relations similar to angular momentum operators. For spin 1/2 particles, such operators are the *Pauli spin operators*, $\hat{\sigma}_i$, defined such that $\hat{S}_i = \hbar \hat{\sigma}_i / 2$. These operators obey the following commutation relations:

$$[\hat{\sigma}_i, \hat{\sigma}_j] = 2i\varepsilon_{ijk}\, \hat{\sigma}_k. \tag{4.4}$$

In analogy with (4.3), the Pauli spin raising and lowering operators, $\hat{\sigma}_+$ and $\hat{\sigma}_-$, are defined such that $\hat{S}_+ = \hbar \hat{\sigma}_+$ and $\hat{S}_- = \hbar \hat{\sigma}_-$, i.e.,

$$\hat{\sigma}_+ \equiv \frac{\hat{\sigma}_x + i\hat{\sigma}_y}{2}, \quad \hat{\sigma}_- \equiv \frac{\hat{\sigma}_x - i\hat{\sigma}_y}{2}. \tag{4.5}$$

Problem 4.1

(a) Show that the Pauli spin operators obey the equation

$$\hat{\sigma}_i \hat{\sigma}_j = \delta_{ij} + i\varepsilon_{ijk}\, \hat{\sigma}_k. \tag{4.6}$$

(b) Using the formula you proved in (a), prove that

$$(\mathbf{a} \cdot \hat{\boldsymbol{\sigma}})(\mathbf{b} \cdot \hat{\boldsymbol{\sigma}}) = \mathbf{a} \cdot \mathbf{b} + i(\mathbf{a} \times \mathbf{b}) \cdot \hat{\boldsymbol{\sigma}}. \tag{4.7}$$

(c) Show that $\hat{\sigma}_+$ and $\hat{\sigma}_-$ have the following anticommuter: $\{\hat{\sigma}_+, \hat{\sigma}_-\} = 1$.
(d) Show that $[\hat{\sigma}_+, \hat{\sigma}_-] = \sigma_z$. Note that the unit operator $\hat{\mathbf{1}}$ multiplying δ_{ij} and $\mathbf{a} \cdot \mathbf{b}$ on the RHSs of (4.6) and (4.7) respectively has not be explicitly written.

4.2 SPINORS

A *spinor* is a two-dimensional vector, $\begin{pmatrix} a \\ b \end{pmatrix}$, with complex components a and b. Spinors were first applied in physics by Wolfgang Pauli; the term spinor was coined by Paul Ehrenfest. The properties of spinors will be presented in this section.

A natural basis for the two component spinors is given by the vectors $\begin{pmatrix} 1 \\ 0 \end{pmatrix}$ and $\begin{pmatrix} 0 \\ 1 \end{pmatrix}$, so the general spinor $\begin{pmatrix} a \\ b \end{pmatrix}$ can be represented as $\begin{pmatrix} a \\ b \end{pmatrix} = a\begin{pmatrix} 1 \\ 0 \end{pmatrix} + b\begin{pmatrix} 0 \\ 1 \end{pmatrix}$. By convention, $\begin{pmatrix} 1 \\ 0 \end{pmatrix}$ and $\begin{pmatrix} 0 \\ 1 \end{pmatrix}$ are denoted in any of the following possible ways:

$$|\uparrow\rangle \equiv |\alpha\rangle \equiv \chi_\uparrow \equiv \begin{pmatrix} 1 \\ 0 \end{pmatrix}, \quad |\downarrow\rangle \equiv |\beta\rangle \equiv \chi_\downarrow \equiv \begin{pmatrix} 0 \\ 1 \end{pmatrix}. \tag{4.8}$$

We shall see shortly that the spin operator $\hat{S}_z$ has the 2×2 matrix representation, $\frac{\hbar}{2}\begin{pmatrix} 1 & 0 \\ 0 & -1 \end{pmatrix}$, which has eigenvalues $\pm\hbar/2$. The eigenvector of $\hat{S}_z$ with eigenvalue $\hbar/2$ is the spin-up spinor $|\uparrow\rangle = \begin{pmatrix} 1 \\ 0 \end{pmatrix}$, and the eigenvector with eigenvalue $-\hbar/2$ is the spin-down spinor $|\downarrow\rangle = \begin{pmatrix} 0 \\ 1 \end{pmatrix}$.[3] Clearly, these eigenvectors of the Hermitian operator $\hat{S}_z$ are orthogonal, and are of unit

---

[2] The unit operator on the RHS of (4.2) is sometimes not explicitly written but is implied.
[3] The z-axis is often assigned arbitrarily, but if a static magnetic field is present, the z-axis is assigned to be along the field.

4.2 Spinors

length, as is easy to explicitly check. The orthogonormality relations of these spinors are as follows:

$$\langle \alpha | \alpha \rangle = 1, \quad \langle \beta | \beta \rangle = 1, \quad \langle \beta | \alpha \rangle = 0, \quad \langle \alpha | \beta \rangle = 0. \tag{4.9}$$

The first and third equations in (4.9), written in the form of two-component vectors, take the form

$$\langle \alpha | \alpha \rangle = \begin{pmatrix} 1 & 0 \end{pmatrix} \begin{pmatrix} 1 \\ 0 \end{pmatrix} = 1, \quad \langle \beta | \alpha \rangle = \begin{pmatrix} 0 & 1 \end{pmatrix} \begin{pmatrix} 1 \\ 0 \end{pmatrix} = 0. \tag{4.10}$$

It is important to note that a pure state $|\chi\rangle$ of any two-level system [e.g., a quantum bit (qubit), see Sec. 5.2] can be represented as a spinor by a superposition of basis functions,

$$|0\rangle \equiv |\uparrow\rangle \equiv |\alpha\rangle = \begin{pmatrix} 1 \\ 0 \end{pmatrix}, \quad |1\rangle \equiv |\downarrow\rangle \equiv |\beta\rangle = \begin{pmatrix} 0 \\ 1 \end{pmatrix}, \tag{4.11}$$

i.e.,

$$|\chi\rangle = a|\alpha\rangle + b|\beta\rangle = a\begin{pmatrix} 1 \\ 0 \end{pmatrix} + b\begin{pmatrix} 0 \\ 1 \end{pmatrix} = \begin{pmatrix} a \\ b \end{pmatrix}. \tag{4.12}$$

The inner product of two arbitrary spinors $|\chi_1\rangle$ and $|\chi_2\rangle$ can be written as

$$\langle \chi_2 | \chi_1 \rangle = \begin{pmatrix} a_2^* & b_2^* \end{pmatrix} \begin{pmatrix} a_1 \\ b_1 \end{pmatrix} = a_2^* a_1 + b_2^* b_1. \tag{4.13}$$

Hence, in order to normalize the spinor $|\chi\rangle = \begin{pmatrix} a \\ b \end{pmatrix}$ whose length squared is $\langle \chi | \chi \rangle = a^*a + b^*b$, we can divide the spinor by its length,

$$|\chi\rangle \Rightarrow \frac{|\chi\rangle}{\langle \chi | \chi \rangle^{1/2}} = \frac{1}{(|a|^2 + |b|^2)^{1/2}} \begin{pmatrix} a \\ b \end{pmatrix}.$$

Note that a and b may be functions of a coordinate variable **r** and/or of time t. We shall make use of this possibility below in constructing spin-orbitals. Most often one uses amplitudes a and b that are such that $|a|^2 + |b|^2 = 1$, i.e., spinors are taken to have unit length, $\langle \chi | \chi \rangle = 1$.

Spinor matrix elements of operators can be written in the form

$$\langle \chi_2 | \hat{A} | \chi_1 \rangle = \begin{pmatrix} a_2^* & b_2^* \end{pmatrix} \begin{pmatrix} A_{11} & A_{12} \\ A_{21} & A_{22} \end{pmatrix} \begin{pmatrix} a_1 \\ b_1 \end{pmatrix}, \tag{4.14}$$

where the operator $\hat{A}$ has been represented as a 2×2 matrix A acting in spin space. (see Sec. 4.2.1)

Problem 4.2

An electron is in the spin state $|\chi\rangle = \frac{1}{3}|\uparrow\rangle - \frac{2\sqrt{2}}{3}|\downarrow\rangle$.

(a) Determine the probabilities to measure the electron in states $|\uparrow\rangle$ and $|\downarrow\rangle$.
(b) Calculate the expectation value $\langle \chi | \hat{S}_z | \chi \rangle$ using $\hat{S}_z = \frac{\hbar}{2}\begin{pmatrix} 1 & 0 \\ 0 & -1 \end{pmatrix}$.

4.2.1 PAULI MATRICES

Since the spin of an electron can be represented as a spinor with two components, the spin operators $\hat{S}_i$ (and the Pauli spin operators $\hat{\sigma}_i$) can be represented as 2×2 matrices (see Sec. 3.1.2). Moreover, since the spin matrices must satisfy the same commutation relations as the spin operators and since spin has the units of angular momentum, we can write the 2×2 spin matrices S_i in terms of 2×2 dimensionless matrices σ_i, called *Pauli spin matrices*, as $S_i \equiv (\hbar/2)\sigma_i$, where σ_i satisfy the commutation relations (4.4). The following set of Pauli spin matrices satisfy the commutation relations for the

Pauli spin operators (4.4) [see (3.31)]:

$$\sigma_x = \begin{pmatrix} 0 & 1 \\ 1 & 0 \end{pmatrix}, \quad \sigma_y = \begin{pmatrix} 0 & -i \\ i & 0 \end{pmatrix}, \quad \sigma_z = \begin{pmatrix} 1 & 0 \\ 0 & -1 \end{pmatrix}. \tag{4.15}$$

Problem 4.3

(a) Calculate the 2×2 matrices σ_x^2, σ_y^2, and σ_z^2.
(b) Demonstrate that the 2×2 matrices, $S_x = \hbar\sigma_x/2$, $S_y = \hbar\sigma_y/2$, and $S_z = \hbar\sigma_z/2$ with the Pauli matrices given in (4.15), satisfy the commutation relations (4.1).
(c) Show that the Pauli spin matrices σ_x, σ_y, and σ_z of (4.15) satisfy (4.4).
(d) Calculate the eigenvectors and eigenvalues of the matrices S_x S_y, and S_z.
(e) The Pauli spin matrices are not unique in the sense that other sets of 2×2 matrices satisfy the commutation relations (4.4). Show that a cyclic permutation of the Pauli matrices, $\tilde{\sigma}_x = \sigma_y$, $\tilde{\sigma}_y = \sigma_z$, $\tilde{\sigma}_z = \sigma_x$, also satisfies these commutation relations.
(f) Show that $\text{Tr}\,\sigma_i\sigma_j = 2\delta_{ij}$.

Answer: (a) $\sigma_x^2 = \sigma_y^2 = \sigma_z^2 = \begin{pmatrix} 1 & 0 \\ 0 & 1 \end{pmatrix}$. (d) $\pm \hbar/2$.

Problem 4.4

(a) Write out the 2×2 representations of the operators $S_+ = S_x + iS_y$ and $S_- = S_x - iS_y$.
(b) Show that S_+, when applied to the state $|\alpha\rangle$ yields the zero state and when applied to the state $|\beta\rangle$ yields $\hbar|\alpha\rangle$.
(c) Show that $\sigma_+ \equiv \frac{\sigma_x + i\sigma_y}{2} = \begin{pmatrix} 0 & 1 \\ 0 & 0 \end{pmatrix}$ and $\sigma_- \equiv \frac{\sigma_x - i\sigma_y}{2} = \begin{pmatrix} 0 & 0 \\ 1 & 0 \end{pmatrix}$.
(d) Show that $\sigma_+^2 = \sigma_-^2 = \mathbf{0}$.
(e) Show that $\sigma_+\sigma_- = (1+\sigma_z)/2 = \begin{pmatrix} 1 & 0 \\ 0 & 0 \end{pmatrix}$, $\sigma_-\sigma_+ = (1-\sigma_z)/2 = \begin{pmatrix} 0 & 0 \\ 0 & 1 \end{pmatrix}$.

Answer: (a) $S_+ = \hbar \begin{pmatrix} 0 & 1 \\ 0 & 0 \end{pmatrix}$, $S_- = \hbar \begin{pmatrix} 0 & 0 \\ 1 & 0 \end{pmatrix}$.

Any Hermitian 2×2 matrix A can be expanded using Pauli matrices,

$$A = c_0 \mathbf{1} + c_j \sigma_j, \tag{4.16}$$

where Einstein summation convention has been employed and the coefficients c_0, c_1, c_2, c_3 (or c_0, c_x, c_y, c_z) are real, and the unit 2×2 matrix is $\mathbf{1} = \begin{pmatrix} 1 & 0 \\ 0 & 1 \end{pmatrix}$. Moreover, any non-Hermitian 2×2 matrix can be expanded in this way with complex coefficients.

Problem 4.5

(a) Using Eq. (4.16), show that $\mathbf{c} = \frac{1}{2}\text{Tr}[\boldsymbol{\sigma} A]$ and $c_0 = \frac{1}{2}\text{Tr}[A]$.

4.2 Spinors

The spin operator in the direction of the unit vector $\mathbf{n}$ is $\mathbf{n} \cdot \mathbf{S} = \frac{\hbar}{2}\mathbf{n} \cdot \boldsymbol{\sigma}$. Since, in spherical coordinates, $\mathbf{n} = (\sin\theta\cos\phi, \sin\theta\sin\phi, \cos\theta)$, the 2×2 matrix $\mathbf{n} \cdot \mathbf{S}$ is given by

$$\mathbf{n} \cdot \mathbf{S} = \frac{\hbar}{2}\mathbf{n} \cdot \boldsymbol{\sigma} = \frac{\hbar}{2}\begin{pmatrix} \cos\theta & e^{-i\phi}\sin\theta \\ e^{i\phi}\sin\theta & -\cos\theta \end{pmatrix}. \tag{4.17}$$

Problem 4.6

(a) Calculate $\langle\uparrow|\mathbf{n}\cdot\boldsymbol{\sigma}|\uparrow\rangle$ and $\langle\downarrow|\mathbf{n}\cdot\boldsymbol{\sigma}|\uparrow\rangle$.

(b) Show that the eigenvalues of the 2×2 matrix in (4.17) are $\pm\hbar/2$ and calculate the eigenvector for eigenvalue $-\hbar/2$.

(c) Determine the probability that a measurement of the polarization along the $\mathbf{n}$-axis will yield spin-down given the state $|\uparrow\rangle$ polarized along the z-axis.

Hint: Calculate the projection of $|\uparrow\rangle$ on the eigenvector with eigenvalue $-\hbar/2$ that you calculated in (b), i.e., calculate $|\langle\beta_{\mathbf{n}}|\uparrow\rangle|^2$. **Answer:** $\sin^2(\theta/2)$.

Problem 4.7

(a) Find the eigenvalues and normalized eigenvectors of the matrix $S_x + S_z = \frac{\hbar}{2}\begin{pmatrix} 1 & 1 \\ 1 & -1 \end{pmatrix}$.

(b) Find the spherical angles θ and ϕ of $\mathbf{n} = (\sin\theta\cos\phi, \sin\theta\sin\phi, \cos\theta)$ so that the matrix $(S_x + S_z)/\sqrt{2}$ can be written as $\mathbf{n}\cdot\mathbf{S}$ and find the eigenvalues and normalized eigenvectors of $\mathbf{n}\cdot\mathbf{S}$.

Answers: (a) $E_1 = \hbar/2 \times \sqrt{2}$, $\xi_1 = \frac{1}{[(\sqrt{2}-1)^2+1]^{1/2}}\begin{pmatrix} 1 \\ \sqrt{2}-1 \end{pmatrix}$.

$E_{-1} = -\hbar/2 \times \sqrt{2}$, $\xi_{-1} = \frac{1}{[(\sqrt{2}+1)^2+1]^{1/2}}\begin{pmatrix} 1 \\ -(\sqrt{2}+1) \end{pmatrix}$.

(b) $\theta = \pi/4$ and $\phi = 0$. The eigenvectors are identical to those in (a), and the eigenvalues are multiplied by $1/\sqrt{2}$.

The eigenvector of $\mathbf{n}\cdot\boldsymbol{\sigma}$ with eigenvalue 1 is $\begin{pmatrix} e^{-i\phi}\sin\theta \\ 1-\cos\theta \end{pmatrix}$ and with eigenvalue -1 is $\begin{pmatrix} -e^{-i\phi}\sin\theta \\ 1+\cos\theta \end{pmatrix}$. Normalizing the eigenvectors, we obtain up to an arbitrary phase factor (e.g., $e^{i\phi/2}$; see Problem 4.9),

$$|\alpha_{\mathbf{n}}\rangle = \begin{pmatrix} e^{-i\phi/2}\cos(\theta/2) \\ e^{i\phi/2}\sin(\theta/2) \end{pmatrix}, \quad |\beta_{\mathbf{n}}\rangle = \begin{pmatrix} -e^{-i\phi/2}\sin(\theta/2) \\ e^{i\phi/2}\cos(\theta/2) \end{pmatrix}, \tag{4.18}$$

with eigenvalues 1 and -1 respectively. Any pure state of a two-level system, $|\chi\rangle = a|\alpha\rangle + b|\beta\rangle = \begin{pmatrix} a \\ b \end{pmatrix}$, can be written as the eigenvector $|\alpha_{\mathbf{n}}\rangle$ of some operator of the form $\mathbf{n}\cdot\mathbf{S}$ (or $\mathbf{n}\cdot\boldsymbol{\sigma}$), with the appropriate unit vector $\mathbf{n}$, having eigenvalue $+\hbar/2$ (or $+1$). From (4.18), it is clear that the angles θ and ϕ can be chosen so that $a = e^{-i\phi/2}\cos(\theta/2)$ and $b = e^{i\phi/2}\sin(\theta/2)$.

Problem 4.8

(a) Use the trigonometric identities $\cos^2(\theta/2) = (1+\cos\theta)/2$ and $\sin^2(\theta/2) = (1-\cos\theta)/2$ to obtain the normalized eigenstates in (4.18).

(b) Given the spin state $|\alpha_{\mathbf{n}}\rangle$, if the z-component of the spin is measured, what possible values can result in the measurement?

(c) What are the probabilities of obtaining the possible measured values?
(d) What is the expectation value of the operator S_z in state $|\alpha_\mathbf{n}\rangle$?
(e) If one measures $\mathbf{n} \cdot \mathbf{S}$, what are the possible measurement outcomes and their probabilities?

Answer: (b) $\hbar/2$, $-\hbar/2$. (c) $P_+ = \cos^2(\theta/2)$, $P_- = \sin^2(\theta/2)$. (d) $(\hbar/2)[\cos^2(\theta/2) - \sin^2(\theta/2)]$. (e) $\hbar/2$, $P=1$.

Problem 4.9

Write the spinor $|\tilde{\alpha}_\mathbf{n}\rangle = \begin{pmatrix} e^{-i\phi}\cos(\theta/2) \\ \sin(\theta/2) \end{pmatrix}$ in terms of the components of the unit vector $\mathbf{n} = (\sin\theta\cos\phi, \sin\theta\sin\phi, \cos\theta) = (x, y, z)$.

Answer: $|\tilde{\alpha}_\mathbf{n}\rangle = \begin{pmatrix} (\cos\phi - i\sin\phi)\sqrt{(1+\cos\theta)/2} \\ \sqrt{(1-\cos\theta)/2} \end{pmatrix} = 2^{-1/2}\begin{pmatrix} \frac{x-iy}{\sqrt{1-z}} \\ \sqrt{1-z} \end{pmatrix}$

Let us consider a system consisting of two spin 1/2 particles in a two-particle state $|\Psi\rangle$. It is sometimes relevant to find the expectation value of a correlated measurement of the spin along one direction for one of the particles and the spin along another direction for the other particle,

$$\langle\Psi|\,(\mathbf{n}_1 \cdot \boldsymbol{\sigma}_1)(\mathbf{n}_2 \cdot \boldsymbol{\sigma}_2)\,|\Psi\rangle. \tag{4.19}$$

For example, consider the singlet state, $|\Psi\rangle = \frac{1}{\sqrt{2}}(|\uparrow\rangle|\downarrow\rangle - |\downarrow\rangle|\uparrow\rangle)$. If we invoke the rotational invariance of the singlet state, then without loss of generality, we can choose $\mathbf{n}_1$ to be along the z-axis, and obtain for the singlet state,

$$\langle\Psi|(\mathbf{n}_1 \cdot \boldsymbol{\sigma}_1)(\mathbf{n}_2 \cdot \boldsymbol{\sigma}_2)|\Psi\rangle = \frac{1}{2}[\langle\downarrow_2|\mathbf{n}_2 \cdot \boldsymbol{\sigma}_2|\downarrow_2\rangle - \langle\uparrow_2|\mathbf{n}_2 \cdot \boldsymbol{\sigma}_2|\uparrow_2\rangle] = -\mathbf{n}_1 \cdot \mathbf{n}_2. \tag{4.20}$$

In obtaining the last equality in (4.20), we used the fact that $\langle\downarrow_2|\mathbf{n}_2 \cdot \boldsymbol{\sigma}_2|\downarrow_2\rangle = -\mathbf{n}_2 \cdot \hat{\mathbf{z}}$, $\langle\uparrow_2|\mathbf{n}_2 \cdot \boldsymbol{\sigma}_2|\uparrow_2\rangle = \mathbf{n}_2 \cdot \hat{\mathbf{z}}$, and that $\hat{\mathbf{z}} \equiv \mathbf{n}_1$. This result will be used in describing correlated measurements of the spin of two spin 1/2 particles in our discussion of Bell inequalities in Sec. 5.8.

4.2.2 ROTATION OF SPINORS

As already discussed in Sec. 3.3.1, we can rotate the state $|\uparrow\rangle$ about the y-axis by an angle ϑ using the rotation operator $e^{-i\vartheta\mathbf{n}_y \cdot \frac{\boldsymbol{\sigma}}{2}}$ as follows:

$$|\vartheta\rangle = e^{-i\vartheta\mathbf{n}_y \cdot \frac{\boldsymbol{\sigma}}{2}}|\uparrow\rangle = [\cos(\vartheta/2)\mathbf{1} - i\sin(\vartheta/2)\sigma_y]\begin{pmatrix}1\\0\end{pmatrix} = \begin{pmatrix}\cos(\vartheta/2)\\\sin(\vartheta/2)\end{pmatrix}. \tag{4.21}$$

The formula $e^{-i\frac{\vartheta}{2}\mathbf{n}_y \cdot \boldsymbol{\sigma}} = [\cos(\vartheta/2)\mathbf{1} - i\sin(\vartheta/2)\sigma_y]$ can be derived by expanding the exponential in a power series and noting that $\sigma_y\sigma_y = \mathbf{1}$ [see Eq. (4.6)], hence, $\sigma_y^m = \mathbf{1}$ for even m and $\sigma_y^m = \sigma_y$ for odd m. Eq. (4.21) can be easily generalized; the rotation operator for spin 1/2 states about an arbitrary unit vector $\mathbf{n}$ by an arbitrary angle ϑ can be written as follows:

$$\boxed{U_\mathbf{n}(\vartheta) \equiv e^{-i\vartheta\mathbf{n}\cdot\mathbf{S}/\hbar} = e^{-i\frac{\vartheta}{2}\mathbf{n}\cdot\boldsymbol{\sigma}} = \cos(\vartheta/2)\mathbf{1} - i\sin(\vartheta/2)\mathbf{n}\cdot\boldsymbol{\sigma}.} \tag{4.22}$$

Note that a rotation of a spin state by an angle of 2π about any axis multiplies the spin state by (-1).

Problem 4.10

(a) Prove that $e^{i\vartheta\mathbf{n}\cdot\boldsymbol{\sigma}} = \cos\vartheta\,\mathbf{1} + i\sin\vartheta\,\mathbf{n}\cdot\boldsymbol{\sigma}$.

(b) Prove that $e^{a+\mathbf{b}\cdot\boldsymbol{\sigma}} = e^a\left[\cosh(|\mathbf{b}|)\mathbf{1} + \sinh(|\mathbf{b}|)\frac{\mathbf{b}\cdot\boldsymbol{\sigma}}{|\mathbf{b}|}\right]$.

4.2 Spinors

The projector $\mathcal{P}_\vartheta = |\vartheta\rangle\langle\vartheta|$ onto the state $|\vartheta\rangle$ defined in (4.21) can be determined using half-angle trigonometric identities:

$$\mathcal{P}_\vartheta = \frac{1}{2}\begin{pmatrix} (1+\cos\vartheta) & \sin\vartheta \\ \sin\vartheta & (1-\cos\vartheta) \end{pmatrix} = \frac{1}{2}(1 + \cos\vartheta\,\sigma_z + \sin\vartheta\,\sigma_x). \quad (4.23)$$

This result generalizes easily; for a spin state corresponding to the spin pointing up along the direction $\hat{\mathbf{u}}$, the projector onto this state is given by

$$\mathcal{P}_{\hat{\mathbf{u}}} = |\alpha_{\hat{\mathbf{u}}}\rangle\langle\alpha_{\hat{\mathbf{u}}}| = \frac{1}{2}\left(1 + \hat{\mathbf{u}}\cdot\boldsymbol{\sigma}\right). \quad (4.24)$$

We can calculate probability of measuring spin-up along the polarization direction $\hat{\mathbf{u}}$ if the system is in an arbitrary spin state $|\psi\rangle$ as follows: $\langle\psi|\mathcal{P}_{\hat{\mathbf{u}}}|\psi\rangle = \frac{1}{2}\langle\psi|\left(1+\hat{\mathbf{u}}\cdot\boldsymbol{\sigma}\right)|\psi\rangle = \frac{1}{2}\left(1 + \langle\psi|\hat{\mathbf{u}}\cdot\boldsymbol{\sigma}|\psi\rangle\right)$.

Let us now consider systems consisting of two spin 1/2 particles in a two-particle pure state Ψ. We can calculate the joint probability of having the first particle in the up-state $|\alpha_{\hat{\mathbf{n}}_1}\rangle$ along polarization direction $\hat{\mathbf{n}}_1$ and the second particle in the up-state $|\alpha_{\hat{\mathbf{n}}_2}\rangle$ along polarization direction $\hat{\mathbf{n}}_2$ by evaluating the expectation value $\langle\Psi|\mathcal{P}_{\mathbf{n}_1}\mathcal{P}_{\mathbf{n}_2}|\Psi\rangle$ as follows:

$$\langle\Psi|\mathcal{P}_{\mathbf{n}_1}\mathcal{P}_{\mathbf{n}_2}|\Psi\rangle = \frac{1}{4}[\langle\Psi|(\mathbf{n}_1\cdot\boldsymbol{\sigma}_1)(\mathbf{n}_2\cdot\boldsymbol{\sigma}_2)|\Psi\rangle + \langle\Psi|(\mathbf{n}_1\cdot\boldsymbol{\sigma}_1)|\Psi\rangle$$
$$+ \langle\Psi|(\mathbf{n}_2\cdot\boldsymbol{\sigma}_2)|\Psi\rangle + 1]. \quad (4.25)$$

In particular, for the spin-singlet state, $|\Psi\rangle = \frac{1}{\sqrt{2}}(|\uparrow\rangle|\downarrow\rangle - |\downarrow\rangle|\uparrow\rangle)$, we can use Eq. (4.20) to obtain

$$\langle\Psi|\mathcal{P}_{\mathbf{n}_1}\mathcal{P}_{\mathbf{n}_2}|\Psi\rangle = \frac{1}{4}\left[1 - \cos(\theta_{\mathbf{n}_1\cdot\mathbf{n}_2})\right] = \frac{1}{2}\sin^2(\theta_{\mathbf{n}_1\cdot\mathbf{n}_2}/2), \quad (4.26)$$

where we have made use of a trigonometric half-angle formula to obtain the latter equality, and defined the angle $\theta_{\mathbf{n}_1\cdot\mathbf{n}_2}$ whose cosine is $\mathbf{n}_1\cdot\mathbf{n}_2$.

4.2.3 SPIN-ORBITALS

A spin-orbital $u(\mathbf{x})$ is a spinor whose components depend upon the coordinate of the particle,

$$u(\mathbf{x}) = \begin{pmatrix} \varphi_1(\mathbf{r}) \\ \varphi_2(\mathbf{r}) \end{pmatrix}. \quad (4.27)$$

Often (but certainly not always) a spin-orbital can be written in the form of a product of an orbital (a wave function that is a function of coordinates) $\phi(\mathbf{r})$ and a spinor χ (for convenience, we have not written the spinor using Dirac notation, $|\chi\rangle$), $u(\mathbf{x}) = \phi(\mathbf{r})\chi$. This is the case in (4.27) when $\varphi_1(\mathbf{r}) = a\phi(\mathbf{r})$ and $\varphi_2(\mathbf{r}) = b\phi(\mathbf{r})$. In any case, a spin-orbital $u(\mathbf{x})$ is a spinor, i.e., a two-component wave function. We use the notation that the variable $\mathbf{x}$ denotes not only the position of the particle $\mathbf{r}$ but also the spin degree of freedom. As an example, consider the following spin-orbital in a central potential that has well-defined principal quantum number n, orbital angular momentum l, and magnetic (azimuthal) quantum number m_l, as well as a well-defined projection of spin, denoted by the spin magnetic quantum number m_s:

$$u_{nlm_l m_s}(\mathbf{x}) = R_{nl}(r)Y_{lm_l}(\theta,\phi)\,\chi_{m_s}. \quad (4.28)$$

This spin-orbital is a product of an orbital and a two-component vector (i.e., a spinor). Spin-orbitals will be used to treat electronic structure (see Chapter 10) and *spin-orbit coupling* (see Sec. 4.5), which is an interaction occurring in atoms that splits some atomic spectral lines, and which can be described in terms of the Hamiltonian, $H_{so} = \xi(r)\mathbf{L}\cdot\mathbf{S}$, where $\xi(r)$ depends only on the magnitude of $\mathbf{r}$. H_{so} must be added to the Hamiltonian $H = T + V(r)$ to obtain the Hamiltonian for an electron in a central potential. We shall soon show that $[H_{so}, L^2] = [H_{so}, J^2] = [H_{so}, J_z] = 0$, hence the eigenfunctions of $\mathbf{L}\cdot\mathbf{S}$, u_{nljm_j}, can be constructed as linear combinations of the spinors $u_{nlm_lm_s}$ using the laws of angular momentum addition,

$$u_{nljm_j} = \sum_{m_l m_s} \langle lm_l sm_s | jm_j \rangle u_{nlm_l m_s}. \tag{4.29}$$

Thus, while the spinor $u_{nlm_l m_s}$ can be written as a product of space and spin factors, u_{nljm_j} has space and spin parts entangled [as in (4.27)]. This is a specific example of the fact that spatial and spin degrees of freedom of a single particle can be entangled. The spinor representing such entanglement is of the form of a sum of products of spatial and spin functions. The Schrödinger equation for the spin-orbital wave function u_{nljm_j} is

$$\left[-\frac{1}{2}\nabla_{\mathbf{r}}^2 + V(r) + \xi(r)\mathbf{L}\cdot\mathbf{S} \right] u_{nljm_j} = E_{nlj}\, u_{nljm_j}. \tag{4.30}$$

The energy eigenvalue E_{nlj} depends on a quantum number j representing the sum of the spin and orbital angular momentum of the electron (see Sec. 4.5), and depends on m_j if an external magnetic field is present. Spin-orbitals are often eigenfunctions of Hermitian operators, and then an orthonormalality condition would apply,

$$\langle u_\mu | u_\lambda \rangle = \int dx\, u_\mu^*(x) u_\lambda(x) = \delta_{\mu\lambda}, \tag{4.31}$$

e.g., for spinors that can be written as a product state, $\langle u_{n'l'm'_l m'_s} | u_{nlm_l m_s} \rangle = \int d\mathbf{r}\, \phi^*_{n'l'm'_l}(\mathbf{r}) \phi_{nlm_l}(\mathbf{r}) \langle \chi_{m'_s} | \chi_{m_s} \rangle = \delta_{n'n}\delta_{l'l}\delta_{m'_l m_l}\delta_{m'_s m_s}$, but this holds even for the case of entangled spinors, as in the spinors of Problem 4.11(b). Spinor spherical harmonics can be used as a basis in which to expand spin-orbitals [you may now go back and read Sec. 3.5.3].

Problem 4.11

(a) Find two one-electron eigenstates of the operator $L_z + S_z$ for an electron in an $L=1$ orbital with eigenvalue $\hbar/2$.
(b) Determine the linear combinations of the states in (a), which are eigenstates of $J^2 = (\mathbf{L}+\mathbf{S})^2$.
(c) What eigenstate of the operator $L_x + S_x$ for an electron in a $L=1$ orbital has eigenvalue $3/2$.

Answers: (a) $|1,1\rangle |\tfrac{1}{2},-\tfrac{1}{2}\rangle$, $|1,0\rangle |\tfrac{1}{2},\tfrac{1}{2}\rangle$. (b) $|\tfrac{3}{2},\tfrac{1}{2}\rangle = \sqrt{\tfrac{1}{3}}|1,1\rangle |\tfrac{1}{2},-\tfrac{1}{2}\rangle + \sqrt{\tfrac{2}{3}}|1,0\rangle |\tfrac{1}{2},\tfrac{1}{2}\rangle$, $|\tfrac{1}{2},\tfrac{1}{2}\rangle = \sqrt{\tfrac{2}{3}}|1,1\rangle |\tfrac{1}{2},\tfrac{-1}{2}\rangle - \sqrt{\tfrac{1}{3}}|1,0\rangle |\tfrac{1}{2},\tfrac{1}{2}\rangle$. (c) See Problem 3.5 in Sec. 3.1.2.

4.3 ELECTRON IN A MAGNETIC FIELD

The magnetic energy U of a magnetic moment $\boldsymbol{\mu}$ in a magnetic field $\mathbf{H}$ is given by the expression $U = -\boldsymbol{\mu}\cdot\mathbf{H}$ (see Eq. (1.11)). This magnetic energy is called the *Zeeman energy*, after Pieter Zeeman who won the Nobel prize in 1902 for his studies of magnetism (along with Hendrik A. Lorentz), and the Hamiltonian is called the Zeeman Hamiltonian,

$$\boxed{H = -\boldsymbol{\mu}\cdot\mathbf{H}.} \tag{4.32}$$

The magnetic moment of an electron is the sum of its spin magnetic moment and its orbital magnetic moment, $\boldsymbol{\mu}_{\text{el}} = \boldsymbol{\mu}_s + \boldsymbol{\mu}_l$. The spin magnetic moment of an electron, $\boldsymbol{\mu}_s$, is given in terms of its spin $\mathbf{S}$ by the formula

$$\boxed{\boldsymbol{\mu}_s = -g\,\mu_B\,\frac{\mathbf{S}}{\hbar},} \tag{4.33}$$

where the quantity μ_B is called the *Bohr magneton*, $\mu_B \equiv \frac{e\hbar}{2m_e c}$. The Bohr magneton is the only factor on the RHS of (4.33) that has units; it has units of energy per magnetic field. In SI units, $\mu_B = \frac{e\hbar}{2m_e}$ and is numerically equal to 927.400 899(37)

4.3 Electron in a Magnetic Field

$\times 10^{-26}$ JT^{-1}. The factor g is called the *Landé g-factor* or simply the *g-factor* of the electron (it is sometimes also denoted by g_s where the subscript s is for spin); it is dimensionless and is almost exactly equal to 2, $g = 2.0023193043737(82)$.[4] As noted in the footnote, sometimes the electron g-factor is taken to be negative, i.e., the minus sign in $\boldsymbol{\mu}_s = -g\mu_B \mathbf{S}/\hbar$ is incorporated into the g-factor. We do not do so here. The deviation of g from 2 is due to quantum electrodynamic corrections (note the precision to which g is known) and is called the anomalous magnetic moment. One convenient set of units in which to remember the magnitude of the Bohr magneton is 1.4 MHz/Gauss (more precisely, 1.39962 MHz/Gauss). The magnitude of the electron spin magnetic moment $\boldsymbol{\mu}_s = -g\mu_B \mathbf{S}/\hbar$ is 1.4 MHz/G since $g \approx 2$ and $S/\hbar = 1/2$; hence, the Zeeman energy of an electron with its spin aligned with a 1 Gauss magnetic field (a magnetic field of 1 Tesla is 10,000 Gauss) is $+1.4$ MHz.

The orbital magnetic moment of an electron, $\boldsymbol{\mu}_l$, is given in terms of its orbital angular momentum $\mathbf{L}$ by

$$\boldsymbol{\mu}_l = -\mu_B \frac{\mathbf{L}}{\hbar}. \quad (4.34)$$

Notice that the orbital g-factor is unity [and therefore has not been explicitly inserted into the RHS of (4.34)]. The spin magnetic moment of the electron is in the opposite direction to the spin, and the orbital magnetic moment is opposite in direction to the orbital angular momentum.

The Heisenberg equation of motion for electron spin is

$$\frac{\partial \mathbf{S}}{\partial t} = \frac{i}{\hbar}[H, \mathbf{S}] = \frac{ig\mu_B}{\hbar}[\mathbf{S} \cdot \mathbf{H}, \mathbf{S}]. \quad (4.35)$$

Writing (4.35) in terms of components, $\frac{\partial S_i}{\partial t} = -g\mu_B \varepsilon_{ijk} S_k H_j$, and noting that the quantity on the RHS is a cross product, we find

$$\boxed{\frac{\partial \mathbf{S}}{\partial t} = -g\mu_B \mathbf{S} \times \mathbf{H}.} \quad (4.36)$$

This equation is called the *Bloch equation* for the electron spin. Making use of the definition of the electron magnetic moment, $\boldsymbol{\mu}_s = -g\mu_B \mathbf{S}/\hbar$, Eq. (4.36) can also be written as an equation of motion for the electron magnetic moment:

$$\frac{\partial \boldsymbol{\mu}_s}{\partial t} = -g\mu_B \boldsymbol{\mu}_s \times \mathbf{H}. \quad (4.37)$$

Thus, both the spin of the electron and the magnetic moment of the electron precess around an external magnetic field. The frequency of precession, called the *Larmor frequency*, is given by $\omega_0 \equiv g\mu_B|\mathbf{H}|$.

For nuclei, the nuclear magnetic moment and nuclear spin are related by

$$\boxed{\boldsymbol{\mu}_N = g_N \mu_N \mathbf{I}/\hbar,} \quad (4.38)$$

where the nuclear magneton is defined as $\mu_N = \frac{e\hbar}{2M_p c}$ (in SI units $\mu_N = \frac{e\hbar}{2M_p}$) and $\mathbf{I}$ denotes the nuclear spin angular momentum (it is confusing that the symbols for the nuclear magnetic moment and the nuclear magneton are so similar, but this is standard notation). Note that the proton mass M_p appears in the expression for the nuclear magneton; nuclear magnetic moments are therefore roughly 1000 times smaller than the electron magnetic moment. The quantity g_N is the nucleon g-factor; nucleon g-factors are of order of magnitude unity [e.g., the proton g-factor is $g_p = 5.585\,694\,713(46)$ and the neutron g-factor is $g_n = -3.82608545(90)$]. The Heisenberg equation of motion for a nuclear magnetic moment, which is derived in a fashion analogous to (4.37), is

$$\boxed{\frac{\partial \boldsymbol{\mu}_N}{\partial t} = g_N \mu_N \boldsymbol{\mu}_N \times \mathbf{H}.} \quad (4.39)$$

---

[4] Often (4.33) is written as $\boldsymbol{\mu}_s = g_e \mu_B \mathbf{S}/\hbar$ where the electron g-factor $g_e = -g$ is negative, to conform with (4.38) which is used for all other particles besides electrons. In any case, what is important to note is that the electron magnetic moment vector is opposite to the spin of the electron.

Equation (4.39) is called the Bloch equation for the nuclear magnetic moment. Clearly, the nuclear magnetic moment also precesses around an external magnetic field, as illustrated in Fig. 4.1. The Larmor precession frequency for nuclei is given by $\omega_0 = g_N \mu_N H$.

The *magnetization* $\mathbf{M}$ of a system containing magnetic moments is the product of the density of the magnetic moments, n, and the expectation value of the magnetic moment vector, $\mathbf{M} \equiv n \langle \boldsymbol{\mu} \rangle$, where $\langle \boldsymbol{\mu} \rangle$ is the average magnetic moment of the atoms in the system. For example, in a sample having nuclei of density n, the nuclear magnetization is $\mathbf{M}_N = n g_N \mu_N \langle \mathbf{I} \rangle / \hbar$, whereas, for electrons in an s-state ($l=0$), the electronic magnetization is $\mathbf{M}_{el} = -n g_s \mu_B \langle \mathbf{S} \rangle / \hbar$ where n is now the density of electrons. Note that if either the density n or the expectation value $\langle \boldsymbol{\mu} \rangle$ has a spatial variation, so does the magnetization vector, $\mathbf{M}(\mathbf{r}, t)$. Using (4.36) or (4.39), we find that the Bloch equations can be written in terms of the magnetization vector as

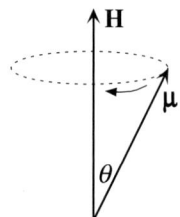

FIG 4.1 Precession of a magnetic moment $\boldsymbol{\mu}$ around a magnetic field $\mathbf{H}$.

$$\frac{\partial \mathbf{M}}{\partial t} = g \mu \mathbf{M} \times \mathbf{H}, \quad (4.40)$$

where the sign on the RHS of (4.40) must be changed for the electronic magnetization. Hence, the magnetization precesses around the external magnetic field.

Figure 4.2 shows the Zeeman energy of a spin 1/2 particle in the presence of a static magnetic field. For an electron, the lower state corresponds to $m_s = -1/2$ and the upper state to $m_s = +1/2$,

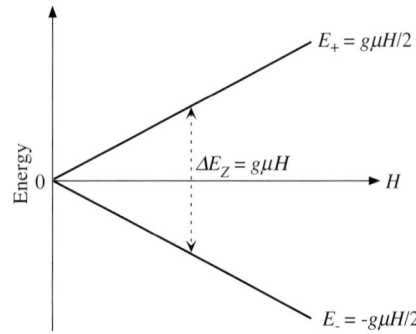

FIG 4.2 Zeeman energy levels of a spin 1/2 particle in a static magnetic field $\mathbf{H}$. The Zeeman energy splitting, ΔE_Z, depends on the product of magnetic field strength and the magnetic moment.

whereas for a proton, the lower state has $m_I = +1/2$ and the upper state $m_I = -1/2$. The energy splitting, $\Delta E_Z = g \mu H$, for the electron is over 1000 times larger than for a proton or for other nuclei since the magnetic moment of an electron is so much larger.

Problem 4.12

For a spin 1/2 particle in a magnetic field $\mathbf{H} = \frac{H_0}{\sqrt{2}}(1, 0, 1)$, the Hamiltonian is

$$H = -\boldsymbol{\mu} \cdot \mathbf{H} = -\frac{\mu_B H_0}{\sqrt{2}}(\sigma_x + \sigma_z) = -\frac{\mu_B H_0}{\sqrt{2}} \begin{pmatrix} 1 & 1 \\ 1 & -1 \end{pmatrix}.$$

(a) Calculate the eigenvalues and eigenvectors of H.

Answer: $E_\pm = \pm \mu_B H_0$. The unnormalized eigenvectors are $|\chi_+\rangle = \begin{pmatrix} -\sqrt{2}+1 \\ 1 \end{pmatrix}, |\chi_-\rangle = \begin{pmatrix} \sqrt{2}+1 \\ 1 \end{pmatrix}$.

(b) If the particle is in state $|\downarrow\rangle$ (spin-down along z), what are the probabilities of measuring each of the energy eigenvalues?

Hint: Square the projection of the initial state $\begin{pmatrix} 0 \\ 1 \end{pmatrix}$ onto the <u>normalized</u> eigenvectors to get the probabilities. **Answer:** $P_+ = \frac{2-\sqrt{2}}{4}, P_- = \frac{2+\sqrt{2}}{4}$.

4.3 Electron in a Magnetic Field

4.3.1 CHARGED PARTICLE IN A MAGNETIC FIELD: ORBITAL EFFECTS

Many magnetic field effects on atoms and molecules are associated with spin, but orbital effects are also present. In general, these effects need to be considered together. In this section we study magnetic field effects on atoms. Magnetic field effects in condensed matter physics and in low-dimensional systems will be discussed in future chapters.

The Hamiltonian for a charged particle having mass m and charge q in the presence of an electromagnetic field was determined in Chapter 16 (see the book web page), starting from a Lagrangian formulation [see Eq. (16.95)]:

$$H = \frac{1}{2m}\left(\mathbf{p} - \frac{q}{c}\mathbf{A}\right)^2 + V. \quad (4.41)$$

Here $\mathbf{p} = -i\hbar\nabla$ is the momentum operator and $V = q\varphi$ is the potential energy experienced by the particle (e.g., in the case of an electron in an atom, V is the Coulomb potential), i.e., φ is the *scalar potential* and $\mathbf{A}$ is the vector potential. Within the classical theory, all measurable quantities do not explicitly depend on the vector potential, but rather on the magnetic field. This is not the case within the quantum theory, as we shall see in the discussion of the Aharonov–Bohm effect in Sec. 9.5.2. In quantum mechanics, the vector potential takes on a "life of its own." The first term in the Hamiltonian operator, Eq. (4.41), can be expanded so that the Hamiltonian contains a linear and a quadratic term in the vector potential:

$$H = \frac{1}{2m}\left\{\mathbf{p}^2 - \frac{q}{c}(\mathbf{p}\cdot\mathbf{A} + \mathbf{A}\cdot\mathbf{p}) + \left(\frac{q}{c}\mathbf{A}\right)^2\right\} + V, \quad (4.42)$$

where the vector potential $\mathbf{A}(\mathbf{r},t)$ is such that its curl yields the magnetic field, $\mathbf{H} = \nabla\times\mathbf{A}$ (see the discussion in Sec. 9.5 and related material in Refs. [23, 34]). If we now add the spin degree of freedom of the particle by including the Zeeman Hamiltonian (4.32), Eq. (4.42) becomes

$$H = \frac{1}{2m}\left\{\mathbf{p}^2 - \frac{q}{c}(\mathbf{p}\cdot\mathbf{A} + \mathbf{A}\cdot\mathbf{p}) + \left(\frac{q}{c}\mathbf{A}\right)^2\right\} + V - g\mu_B\frac{\mathbf{S}}{\hbar}\cdot\mathbf{H}. \quad (4.43)$$

The Hamiltonian (4.43) can also be obtained from the lowest order nonrelativistic reduction of the relativistic Dirac equation for an electron (see Sec. 13.6.3) and yields,

$$i\hbar\frac{\partial\psi}{\partial t} = \left[\frac{(\mathbf{p} - (-e/c)\mathbf{A}(\mathbf{r},t))^2}{2m} + V(\mathbf{r}) - \boldsymbol{\mu}_s\cdot\mathbf{H}(\mathbf{r},t)\right]\psi(\mathbf{r},t). \quad (4.44)$$

It is of interest to write the Hamiltonian in (4.44) for the case of a constant, spatially homogeneous magnetic field $\mathbf{H}$. Then, the vector potential can be taken to be

$$\mathbf{A} = \frac{1}{2}\mathbf{H}\times\mathbf{r}, \quad (4.45)$$

as can easily be verified by taking the curl of $\mathbf{A}$. Substituting Eq. (4.45) into (4.43), we obtain, after some algebra (see Problem 4.15),

$$H = \left[\frac{\mathbf{p}^2}{2m} + V(\mathbf{r}) - (\boldsymbol{\mu}_l + \boldsymbol{\mu}_s)\cdot\mathbf{H} + \frac{e^2}{8mc^2}(\mathbf{H}\times\mathbf{r})^2\right]. \quad (4.46)$$

The third term in the square brackets on the RHS of Eq. (4.46) contains the interaction of the orbital magnetic moment with the magnetic field, where $\boldsymbol{\mu}_l = -\mu_B\mathbf{L}/\hbar$ is the *orbital magnetic moment* and μ_B is the Bohr magneton [see Eq. (4.34)], as well as the interaction of the spin magnetic moment with the magnetic field, $\boldsymbol{\mu}_s\cdot\mathbf{H}$. The last term in the square brackets is quadratic in the magnetic field strength and is called the *diamagnetic* term since it gives rise to *diamagnetism* (see Sec. 9.5.9).

For future reference, we mention that it is often convenient to write the Zeeman energy of an atom in a state $|\alpha LSJM_J\rangle$. This is particularly true when spin–orbit interactions are included. The expectation value in state $|\alpha LSJM_J\rangle$ of the Zeeman

Hamiltonian, $H_{\text{Zeeman}} = -(\boldsymbol{\mu}_l + \boldsymbol{\mu}_s) \cdot \mathbf{H}$, as:

$$\langle H_{\text{Zeeman}} \rangle_{\alpha LSJM_J} = -\langle (\boldsymbol{\mu}_l + \boldsymbol{\mu}_s) \rangle_{\alpha LSJM_J} \cdot \mathbf{H} = g_{LSJ} \mu_B \langle \mathbf{J} \rangle_{LSJM_J} \cdot \mathbf{H}. \quad (4.47)$$

g_{LSJ} is called the Landé g-factor; it can be determined by noting that

$$\boldsymbol{\mu}_l + \boldsymbol{\mu}_s = -\frac{\mu_B}{\hbar}(\mathbf{L} + g_s \mathbf{S}) = -\frac{\mu_B}{\hbar}[\mathbf{J} + (g_s - 1)\mathbf{S}], \quad (4.48)$$

and evaluating the matrix element $\langle \mathbf{S} \rangle_{LSJM_J}$ by showing that it is proportional to $\langle \mathbf{J} \rangle_{LSJM_J}$, as in Sec. 3.6.4.

Problem 4.13

Show that if the vector potential satisfies the "Coulomb gauge condition," $\nabla \cdot \mathbf{A} = 0$, then $[\mathbf{p}, \mathbf{A}] = 0$ and therefore, $\mathbf{p} \cdot \mathbf{A} + \mathbf{A} \cdot \mathbf{p} = 2\mathbf{A} \cdot \mathbf{p}$. Check if this always holds for a uniform field $\mathbf{H}_0$, which can be written in terms of the vector potential $\mathbf{A}(\mathbf{r}) = \mathbf{H}_0 \times \mathbf{r}/2$.

Problem 4.14

(a) What are the energies of a spin-up and spin-down electron in a magnetic field of strength $H = 0.1$ Tesla. Give your answer in GHz, using the fact that the Bohr magneton $\mu_B = \frac{e\hbar}{2m_e} = 13.9962$ GHz/T.

(b) What are the energies of a spin-up and spin-down proton in a magnetic field of strength $H = 0.1$ Tesla. Use the fact that the nuclear magneton $\mu_N = \frac{e\hbar}{2M_p} = 7.62$ MHz/T, and the proton g-factor is $g_p = 5.585$.

Answers: (a) $E_{m_s} = g\mu_B H m_s$, so $E_{1/2} = 1.39962$ GHz, $E_{-1/2} = -1.39962$ GHz. (b) $E_{m_I} = -g_p \mu_N m_s H$, so $E_{1/2} = -2.126$ MHz, $E_{-1/2} = 2.126$ MHz.

Problem 4.15

Carry out the algebra to go from (4.43) to (4.46) as follows.

(a) Expand $(\mathbf{p} - (-e/c)\mathbf{A}(\mathbf{r}, t))^2$, making sure to keep the order of ($\mathbf{p}$ and $\mathbf{r}$) right, since they do not commute.
(b) Substitute for $\mathbf{A}$ using Eq. (4.45).
(c) Rewrite the expressions such as $\mathbf{p} \cdot (\mathbf{H} \times \mathbf{r})$ as $\mathbf{H} \cdot (\mathbf{r} \times \mathbf{p})$, by making use of the Levi-Civita symbol.
(d) Note that $\mathbf{L} = \mathbf{r} \times \mathbf{p}$, to finally obtain Eq. (4.46).

Problem 4.16

Use the relation $\sigma_i \sigma_j = \delta_{ij} + i\varepsilon_{ijk} \sigma_k$ to show that

$$\frac{[\boldsymbol{\sigma} \cdot (\mathbf{p} - (q/c)\mathbf{A})][\boldsymbol{\sigma} \cdot (\mathbf{p} - (q/c)\mathbf{A})]}{2m} = \frac{(\mathbf{p} - (q/c)\mathbf{A})^2}{2m} - \frac{q\hbar}{2mc} \boldsymbol{\sigma} \cdot \mathbf{H}.$$

We shall have more to say about orbital and spin magnetic field effects in Sec. 9.5, where we treat paramagnetic and diamagnetic effects in atoms and solids, as well as transport properties of electrons and holes in the presence of magnetic fields, but for the time being, we return to the topic of orbital and spin effects in the simplest atom, hydrogen, before going on to describe spin–orbit and hyperfine effects in atoms.

4.3 Electron in a Magnetic Field

Hydrogen Atom in a Magnetic Field: Chaos

Upon application of a uniform magnetic field to a hydrogen atom, the nature of part of the spectrum changes dramatically from that of the spectrum without such a field. The Hamiltonian for the hydrogen atom in a magnetic field shows classical chaos (see Sec. 16.13 linked to the book web page), and the quantum spectrum reflects the nature of the chaotic classical behavior. Taking the vector potential to be given by (4.45) and assuming that the nuclear mass is infinite (so as to finesse any questions regarding the center of mass motion of the atom taken does not separate out) the Hamiltonian becomes

$$H = \frac{\mathbf{p}^2}{2m_e} - \frac{Ze^2}{r} + \frac{\mu_B}{\hbar}(\mathbf{L} + g_s \mathbf{S}) \cdot \mathbf{H} + \frac{e^2}{8mc^2}(\mathbf{H} \times \mathbf{r})^2. \quad (4.49)$$

With the magnetic field taken to be along the z-direction, we obtain

$$H = \frac{\mathbf{p}^2}{2m_e} - \frac{Ze^2}{r} + \frac{\gamma}{2}(L_z + 2S_z) + \frac{m_e \gamma^2}{8}(x^2 + y^2), \quad (4.50)$$

where $\gamma = eH/(m_e c)$ and $Z = 1$ for a hydrogen nucleus. Clearly, there is cylindrical symmetry around the z-axis, so we can write the wave function, $\Psi(\mathbf{r}) = \psi(\rho, z)e^{im\theta}\chi_{m_s}$, and the Hamiltonian H in cylindrical coordinates, $\rho = \sqrt{x^2 + y^2}$, z and $\theta = \tan^{-1}(y/x)$ [see Eq. (16.60)], where

$$H(p_\rho, p_z, \rho, z) = \frac{p_\rho^2 + p_z^2 + \frac{\hbar^2 m^2}{\rho^2}}{2m_e} - \frac{Ze^2}{\sqrt{\rho^2 + z^2}} + \frac{\hbar \gamma}{2}(m + 2m_s) + \frac{m_e \gamma^2}{8}\rho^2. \quad (4.51)$$

Here the constants of the motion are $L_z = \hbar m$ and $S_z = \hbar m_s$. In atomic units, the dimensionless Hamiltonian takes the form

$$H = -\frac{1}{2}\left(\frac{1}{\tilde{\rho}}\frac{\partial}{\partial \tilde{\rho}}\left(\tilde{\rho}\frac{\partial}{\partial \tilde{\rho}}\right) + \frac{\partial^2}{\partial \tilde{z}^2} + \frac{m^2}{\tilde{\rho}^2}\right) - \frac{Z}{\sqrt{\tilde{\rho}^2 + \tilde{z}^2}} + \frac{\beta}{2}(m + 2m_s) + \frac{\beta^2}{8}\tilde{\rho}^2, \quad (4.52)$$

where the dimensionless constant $\beta \equiv \frac{\hbar^3 \gamma}{m_e e^4} = \frac{\hbar^3}{m_e^2 c e^3}H$ is proportional to the magnetic field strength. The only exact constants of the motion for the Hamiltonian in (4.52) are the energy, the orbital and spin angular momenta m and m_s, and the parity. The Hamiltonian is nonintegrable [for given values of m and m_s, the Hamiltonian in (4.51) is two dimensional and there is only one constant of the motion, i.e., the energy – see Sec. 16.12.1 linked to the book web page] and the classical trajectories are free to explore the entire phase space on the energy shell. For low-lying energies or low field strengths (or both), the dynamics is regular. However, for sufficiently high energy or field strengths (or both), the classical dynamics is chaotic, i.e., above a certain value of β (magnetic field strength), the phase space is classically chaotic for a given region of Rydberg states (see Fig. 4.3). The quantum problem was studied by Delande and Gay in Ref. [40]. They answered the question, how does classical chaos manifest itself in the quantum spectrum of eigenstates, and found that a well-defined signature of chaos can be obtained from the quantum eigenstates by the resulting

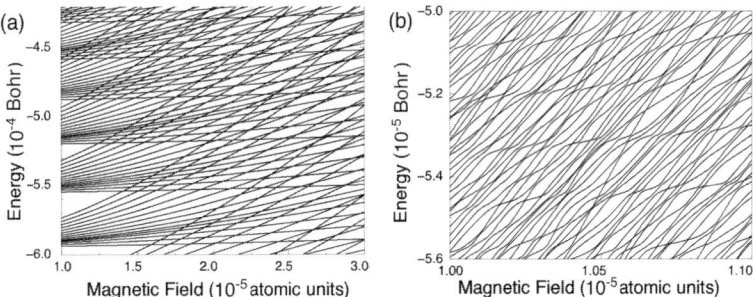

FIG 4.3 Calculated energy levels of a hydrogen atom versus magnetic field for Rydberg states of the $L_z = 0$, even parity series. (a) Low-energy region where the energy levels (quasi-)cross. The quantum eigenstates can be categorized by good quantum numbers. (b) High-energy region [note factor of 1/10 in energy scale relative to (a)] where the classical dynamics is chaotic, the good quantum numbers are lost and the energy levels strongly repel each other. The strong fluctuations in the energy levels are characteristic of a chaotic behavior. (*Figure provided by Dominique Delande*)

fluctuations in the energy levels. "Upon increasing the energy, the calculated statistics of the eigenvalues as a function of the magnetic field evolve from Poisson to Gaussian orthogonal ensemble according to the regular or chaotic character of the classical motion" (we shall discuss this in Chapter 13).[5] They numerically calculated a large number of energy levels in the regular and the chaotic regimes and found that the energy levels repel each other, i.e., these are avoided crossings. Figure 4.3 shows the energy levels of hydrogen as a function of the magnetic field strength. At low magnetic field, Fig. 4.3(a), there are level (quasi-)crossings and it is easy to follow the eigenstates as a function of field strength. At higher magnetic field strenghs, Fig. 4.3(b), the size of the avoided crossings increase and individual states progressively loose their identities, i.e., the good quantum numbers are destroyed.

4.4 TIME-REVERSAL PROPERTIES OF SPINORS

We considered time-reversal for zero-spin systems in Sec. 3.6.3 where we showed that the time-reversal operator is an antilinear operator (more specifically, an antiunitary operator), the properties of which are discussed in Appendix A. We now generalize that treatment by determining the time-reversal operator for particles with spin.

We have seen in Sec. 2.9.3, Eq. (2.172), that angular momentum transforms under time-reversal as $\mathcal{T}\mathbf{J}\mathcal{T}^{-1} = -\mathbf{J}$, hence the spin angular momentum must as well:

$$\mathcal{T}\mathbf{S}\mathcal{T}^{-1} = -\mathbf{S}. \tag{4.53}$$

In the standard representation, S_x is a real matrix, S_y is an imaginary matrix, and S_z is real [see Eq. (4.15)] for spin 1/2 and Eq. (3.26) for spin 1]. Hence $\mathcal{K}S_x\mathcal{K} = S_x$, $\mathcal{K}S_y\mathcal{K} = -S_y$, and $\mathcal{K}S_z\mathcal{K} = S_z$, where $\mathcal{K}$ is the antilinear complex conjugation operator satisfying $\mathcal{K}^{-1} = \mathcal{K}$. Clearly, this is not the transformation required for time reversal.

To satisfy Eq. (4.53), we write $\mathcal{T} = \mathcal{U}\mathcal{K}$, where $\mathcal{U}$ is a linear operator and we require

$$\mathcal{U}S_x\mathcal{U}^{-1} = -S_x, \quad \mathcal{U}S_y\mathcal{U}^{-1} = S_y, \quad \mathcal{U}S_z\mathcal{U}^{-1} = -S_z. \tag{4.54}$$

Moreover, since the correct transformation of spatial variables is produced by the complex conjugation operator, $\mathcal{U}$ must satisfy

$$\mathcal{U}\mathbf{r}\mathcal{U}^{-1} = \mathbf{r}, \quad \mathcal{U}\mathbf{p}\mathcal{U}^{-1} = \mathbf{p}. \tag{4.55}$$

Hence, $\mathcal{U}$ affects only spin variables. Furthermore, it is clear from (4.54) that $\mathcal{U}$ corresponds to a rotation about the y-axis by 180 degrees,

$$\mathcal{U} = e^{-i\pi S_y/\hbar}, \tag{4.56}$$

and therefore, $\mathcal{U}$ and $\mathcal{T}$ are represented by $(2S+1) \times (2S+1)$ matrices, with

$$\boxed{\mathcal{T} = e^{-i\pi S_y/\hbar}\mathcal{K}.} \tag{4.57}$$

Equation (4.57) is valid for spin 1/2 as well as any finite-spin system. Hence, Eq. (4.57) specifies the time-reversal operator, and it has the following transformation properties:

$$\boxed{\mathcal{T}\mathbf{r}\mathcal{T}^{-1} = \mathbf{r}, \quad \mathcal{T}\mathbf{p}\mathcal{T}^{-1} = -\mathbf{p}, \quad \mathcal{T}\mathbf{L}\mathcal{T}^{-1} = -\mathbf{L}, \quad \mathcal{T}\mathbf{S}\mathcal{T}^{-1} = -\mathbf{S}.} \tag{4.58}$$

Problem 4.17

Prove that $\mathcal{U}$ is unitary if $\mathcal{T} = \mathcal{U}\mathcal{K}$ is antiunitary and $\mathcal{K}^{-1} = \mathcal{K}$.

---

[5] Although the Hamiltonian does not commute with $\mathcal{T}$, there is another antiunitary operator that does commute with the Hamiltonian. That is the reason that the level spacing distribution is GOE and not GUE.

4.4 Time-Reversal Properties of Spinors

Problem 4.18

Prove that one can write the time-reversal operator for a spin 1/2 particle as $\mathcal{T} = -i\sigma_y \mathcal{K}$.
Answer: From Eq. (4.22), $U_{\mathbf{n}_y}(\pi) = \cos(\pi/2)\mathbf{1} - i\sin(\pi/2)\sigma_y = -i\sigma_y$.

Problem 4.19

Apply the time-reversal operator $\mathcal{T}$ to the spinor state $\psi_{nlm}(\mathbf{r})|\chi\rangle$, thereby generalizing (3.209).
Answer: $\mathcal{T}\psi_{nlm}(\mathbf{r})|\chi\rangle = \psi^*_{nlm}(\mathbf{r})\left[-i\sigma_y \begin{pmatrix} a^* \\ b^* \end{pmatrix}\right] = (-1)^m \psi_{nl-m}(\mathbf{r})\begin{pmatrix} -b^* \\ a^* \end{pmatrix}$.

If the Hamiltonian H and $\mathcal{T}$ commute, then the eigenstates $|\psi_n\rangle$ of H and $\mathcal{T}|\psi_n\rangle$ have exactly the same eigenvalues since applying $\mathcal{T}$ to both sides of the equation $H|\psi_n\rangle = E_n|\psi_n\rangle$ yields $H(\mathcal{T}|\psi_n\rangle) = E_n(\mathcal{T}|\psi_n\rangle)$. But are the states $|\psi_n\rangle$ and $\mathcal{T}|\psi_n\rangle$ in fact one and the same state? If the answer is yes, these two states can differ by at most a phase factor $e^{i\varphi}$. Then,

$$\mathcal{T}^2|\psi_n\rangle = \mathcal{T}(\mathcal{T}|\psi_n\rangle) = \mathcal{T}(e^{i\varphi}|\psi_n\rangle) = e^{-i\varphi}\mathcal{T}|\psi_n\rangle = e^{-i\varphi}(e^{i\varphi}|\psi_n\rangle) = |\psi_n\rangle.$$

This result is impossible for half-integer angular momentum particles, because then $\mathcal{T}^2|\psi_n\rangle = -|\psi_n\rangle$. Hence, for half-integer angular momentum particles, $|\psi_n\rangle$ and $\mathcal{T}|\psi_n\rangle$ are distinct states that are degenerate. This degeneracy is called *Kramers degeneracy*, named after Henrik Kramers. Thus, *any system with an odd number of electrons is at least two-fold degenerate*.

However, the presence of an external magnetic field lifts the degeneracy, and the Hamiltonian then no longer commutes with $\mathcal{T}$, i.e., time reversal is violated and $\mathcal{T}H\mathcal{T}^{-1} \neq H$. Specifically, the Zeeman term $-\boldsymbol{\mu} \cdot \mathbf{H}$ changes sign under time-reversal, since $\boldsymbol{\mu}$ is proportional to $\mathbf{S}$. Moreover, the vector potential terms $\mathbf{p} \cdot \mathbf{A} + \mathbf{A} \cdot \mathbf{p}$ change sign under time-reversal, since the direction of the momentum $\mathbf{p}$ is reversed. Hence, the Hamiltonian in Eq. (4.46) is not invariant under time-reversal. Reversing the directions of $\boldsymbol{\mu}$ and $\mathbf{p}$ has the same effect as reversing the direction of magnetic field $\mathbf{H}$, leaving $\boldsymbol{\mu}$ and $\mathbf{p}$ unchanged, so the relation

$$\mathcal{T}H(\mathbf{H})\mathcal{T}^{-1} = H(-\mathbf{H}), \tag{4.59}$$

is formally correct. However, it should be stressed that the time reversal operator acts on the dynamical variables $\boldsymbol{\mu}$ and $\mathbf{p}$ but not on the external fields.

Problem 4.20

Prove that the degree of degeneracy must be even if the Hamiltonian of a system is invariant under time reversal and $\mathcal{T}^2|\psi\rangle = -|\psi\rangle$.
Answer: Consider an eigenstate $|\psi_n\rangle$, so $H|\psi_n\rangle = E_n|\psi_n\rangle$. The state $\mathcal{T}|\psi_n\rangle$ is distinct from $|\psi_n\rangle$ but is degenerate with it. The state $\mathcal{T}^2|\psi_n\rangle = -|\psi_n\rangle$ also has eigenvalue E_n, but it is the essentially the same as state $|\psi_n\rangle$ (not distinct). The state $\mathcal{T}^3|\psi_n\rangle = -\mathcal{T}|\psi_n\rangle$ also has eigenvalue E_n but is essentially the same as $\mathcal{T}|\psi_n\rangle$, state $\mathcal{T}^4|\psi_n\rangle = |\psi_n\rangle$, etc., for higher powers of $\mathcal{T}$. So, there are essentially only two distinct degenerate states arising from any eigenvector via time-reversal invariance.

Time Reversal Invariance and Matrix Elements of Operators

Let us return to the analysis of the restriction posed on matrix elements of operators that are invariant under time reversal that we began in Sec. 3.6.3. There we indicated that the operation of an antilinear operator on bras from the right is ill defined, therefore, T should be applied only on kets. Once a ket $|\tilde{\alpha}\rangle = T|\alpha\rangle$ is constructed, the corresponding bra $\langle\tilde{\alpha}|$ is uniquely defined. The technique for doing this takes into account the fact that an application of $\mathcal{K}$ on a real basis state $|n\rangle$ leaves it intact. Expanding $|\alpha\rangle$ in a complete real basis $\{|n\rangle\}$ we can write, $|\tilde{\alpha}\rangle = T|\alpha\rangle = \mathcal{U}\mathcal{K}|\alpha\rangle = \mathcal{U}\mathcal{K}\sum_n |n\rangle\langle n|\alpha\rangle = \sum_n \langle n|\alpha\rangle^* \mathcal{U}\mathcal{K}|n\rangle = \sum_n \langle n|\alpha\rangle^* \mathcal{U}|n\rangle = \sum_n \langle \alpha|n\rangle \mathcal{U}|n\rangle$. The corresponding bra for $|\tilde{\alpha}\rangle$ can now be obtained easily, but we will need the bra for the ket $|\tilde{\beta}\rangle$,

$$|\tilde{\beta}\rangle = \sum_n \langle\beta|n\rangle \mathcal{U}|n\rangle \quad \Rightarrow \quad \langle\tilde{\beta}| = \sum_n \langle n|\beta\rangle \langle n|\mathcal{U}^\dagger . \tag{4.60}$$

The inner product $\langle\tilde{\beta}|\tilde{\alpha}\rangle$ is expressible in terms of the untilded kets $|\alpha\rangle$ and $|\beta\rangle$,

$$\langle\tilde{\beta}|\tilde{\alpha}\rangle = \sum_{mn} \langle m|\beta\rangle \langle m|\mathcal{U}^\dagger \mathcal{U}|n\rangle \langle \alpha|n\rangle = \sum_n \langle\alpha|n\rangle\langle n|\beta\rangle = \langle\alpha|\beta\rangle = \langle\beta|\alpha\rangle^* . \tag{4.61}$$

Our next task is to relate the matrix elements of a given linear operator $\mathcal{O}$ (not necessarily hermitian) between barred and unbarred states, avoiding an action of antilinear operators on bra states. The central identity proved below is,

$$\langle\beta|\mathcal{O}|\alpha\rangle = \langle\tilde{\alpha}|T\mathcal{O}^\dagger T^{-1}|\tilde{\beta}\rangle . \tag{4.62}$$

The proof uses the result (4.61). Defining

$$|\gamma\rangle \equiv \mathcal{O}^\dagger|\beta\rangle \quad \Leftrightarrow \quad \langle\gamma| = \langle\beta|\mathcal{O}, \tag{4.63}$$

we have,

$$\langle\beta|\mathcal{O}|\alpha\rangle = \langle\gamma|\alpha\rangle = \langle\tilde{\alpha}|\tilde{\gamma}\rangle = \langle\tilde{\alpha}|T\mathcal{O}^\dagger|\beta\rangle = \langle\tilde{\alpha}|T\mathcal{O}^\dagger T^{-1}T|\beta\rangle = \langle\tilde{\alpha}|T\mathcal{O}^\dagger T^{-1}|\tilde{\beta}\rangle . \tag{4.64}$$

This completes the proof. The difference between the action of unitary and anti-unitary discrete operations is now clear by comparing with the example of applying the space inversion (parity) transformation $\mathcal{P}$. Denoting $\mathcal{P}|\alpha\rangle = |\tilde{\alpha}\rangle$, we have (action on bra states from the right is now permitted),

$$\langle\beta|\mathcal{O}|\alpha\rangle = \langle\beta|\mathcal{P}^{-1}\mathcal{P}\mathcal{O}\mathcal{P}^{-1}\mathcal{P}|\alpha\rangle = \langle\tilde{\alpha}|\mathcal{P}\mathcal{O}\mathcal{P}^{-1}|\tilde{\beta}\rangle . \tag{4.65}$$

The significant difference is that T replaces the role of bra and kets. This is physically understood as follows: Matrix elements are usually calculated between initial and final states, but since T reverses the time, it swaps the role of initial and final states.

Time Reversal Invariance and Reality of Eigenfunctions

Consider a system whose Hamiltonian is invariant under time reversal, $THT^{-1}=H$ (no external magnetic field is present), and let $|\psi_n\rangle$ be a *non-degenerate eigenstate of H*. The corresponding configuration space wave eigenfunction, $\langle\mathbf{r}|\psi_n\rangle = \psi_n(\mathbf{r})$ is real. Before presenting the proof we note that in this case there is no Kramers degeneracy. This situation occurs not only for spinless particles, but also in electronic systems with an even number of electrons. In this case $\mathbf{r}$ stands for the many-particle space coordinates (the spin content is encoded in the ket $|\psi_n\rangle$). The proof of the statement goes as follows. Applying T to $H|\psi_n\rangle = E_n|\psi_n\rangle$, and using the time-reversal invariance of H,

$$TH(T^{-1}T)|\psi_n\rangle = HT|\psi_n\rangle = TE_n|\psi_n\rangle = E_n T|\psi_n\rangle . \tag{4.66}$$

Thus, $|\psi_n\rangle$ and $T|\psi_n\rangle$ are two states with the same energy E_n, and since E_n is non-degenerate, $|\psi_n\rangle$ and $T|\psi_n\rangle$ must be the same state, up to a phase factor. The corresponding wave functions are, $\langle\mathbf{r}|\psi_n\rangle$ and $\langle\mathbf{r}|T|\psi_n\rangle = \langle\mathbf{r}|\tilde{\psi}_n\rangle = \langle\tilde{\mathbf{r}}|\tilde{\psi}_n\rangle = \langle\mathbf{r}|\psi_n\rangle^*$. The second equality is due to the fact that $|\tilde{\mathbf{r}}\rangle = |\mathbf{r}\rangle$ while the third equality results from (4.61). We have shown that $\psi_n(\mathbf{r}) = \psi_n^*(\mathbf{r})$, i.e., the non-degenerate eigenfunctions can be chosen to be real.

Problem 4.21

(a) Check the validity of the reality theorem for a free spinless particle of mass m moving on a ring of length L subject to the the Hamiltonian $H = -\frac{\hbar^2}{2m}\frac{d^2}{dx^2}$ whose wave function satisfies periodic boundary conditions, $\psi(x+L) = \psi(x)$.

(b) Consider a similar system as in (a) for a spin 1/2 particle. Check the validity of the Kramers theorem.

(c) Assume now that there are two electrons on the ring, with the Hamiltonian $H = -\frac{\hbar^2}{2m}\left[\frac{d^2}{dx_1^2} + \frac{d^2}{dx_2^2}\right] + J\mathbf{S}_1 \cdot \mathbf{S}_2$, for $J > 0$, where the two electron wave function is required to satisfy periodic boundary conditions, $\psi(x_1+L, x_2) = \psi(x_1, x_2+L) = \psi(x_1, x_2)$. Determine the ground-state wave function and check its degeneracy. Explain your finding within the Karmers theorem.

Answer: (a) Denote $k_n = \frac{2\pi n}{L}, n = 0, 1, 2, \ldots$. The eigenvalues are $E_n = \frac{\hbar^2 k_n^2}{2m}$ and the eigenfunctions are, $\psi_n^\pm(x) = \frac{1}{\sqrt{L}}e^{\pm ik_n x}$. For $n = 0$ the eigenvalue $E_0 = 0$ is non degenerate and the wave function $\psi_0(x) = \frac{1}{\sqrt{L}}$ is real. For $n > 0$ the wave functions $\psi_n^\pm(x)$ belong to the same eigenvalue E_n. The wave functions are not real but this does not contradict the above theorem because E_n is degenerate.

(b) The wave functions are $\psi_n^\pm(x, \sigma) = \psi_n^\pm(x)\chi_\sigma$ where $\psi_n^\pm(x)$ are defined in (a). The eigenvalue E_0 is now doubly degenerate, while eigenvalues E_n with $n > 0$ are fourfold degenerate. Recall that Kramers theorem states that each level is *at least* two-fold degenerate.

(c) The wave functions is a product of space part $\psi(x_1, x_2)$ and spin part χ_{SM} where $\mathbf{S} = \mathbf{S}_1 + \mathbf{S}_2$ and $M = \frac{1}{\hbar}[S_{1z} + S_{2z}]$. In the ground-state the space part is symmetric and the spin part is a spin singlet (that is antisymmetric). Thus, the total wave function and total energy is, $\Psi(x_1, x_2, S, M) = \frac{1}{L}\chi_{S=M=0}$, $E_0 = -\frac{3J\hbar^2}{4}$. The ground-state is not degenerate but this does not contradict Kramer's theorem because the number of electrons is even.

4.5 SPIN–ORBIT INTERACTION IN ATOMS

The absorption and emission spectra of hydrogen atoms and alkali-metal atoms reveal spectral lines that are closely spaced pairs of lines, called doublets. This splitting is called fine-structure splitting. It is due to spin-orbit interaction in the excited states of the atoms between the electronic spin and the electronic angular momentum of the single unpaired electron in the highest occupied orbital.

Before we model the spin-orbit interaction, we present some facts about the spin-orbit splitting of hydrogen and alkali atoms. Let us first consider the 17.2 cm^{-1} splitting of the 589 nm line of Na resulting from the $3p \to 3s$ optical transition, that arises due to the energy difference of the $3p\,^2P_{3/2}$ and $3p\,^2P_{1/2}$ excited states.[6] Alkali spectra consist of three distinct series, the principal series, the sharp series and the diffuse series. The strongest lines are those in the principal series arising from $np\,^2P_{1/2}$ and $np\,^2P_{3/2}$ transitions to the ground $^2S_{1/2}$ state. The splittings in this series diminish with increasing n toward the $np \to 3s$ series limit as $n \to \infty$ in all the alkali. The splitting between the $np\,^2P_{1/2}$ and $np\,^2P_{3/2}$ states of Na is 5.6, 2.5 and 1.3 cm^{-1} for $n = 4, 5$ and 6 respectively. The spin-orbit splitting increases with increasing nuclear charge in the alkalis. The weaker sharp and diffuse series are also readily observed. The transitions from $ns\,^2S_{1/2}$ states to the first excited ^2P state form the sharp series and the transition from $nd\,^2D_{5/2,3/2}$ states to the first excited ^2P state form the diffuse series. These series converge to a common limit as $n \to \infty$. The splitting between

---

[6] We are using standard atomic term symbol notation here. The $3p$ indicates an atomic orbital with $n = 3$ and $l = 1$, and in the term symbol $^{2S+1}L_J$, the superscript 2 indicates the total spin degeneracy, so $2S + 1 = 2$ corresponds to a doublet state, the P indicates the total electronic orbital momentum, $L = 1$, and the subscript indicates the total electronic angular momentum J, which can be 3/2 or 1/2 here. See Sec. 10.3 for more on this notation.

the $3d\,^2D_{5/2}$ and $3d\,^2D_{3/2}$ states of Na is only 0.1 cm^{-1}. The diffuse series have an additional close satellite line to the red of the doublet, i.e., the red component of the doublet is actually two lines, thereby causing their diffuse appearance due to the splitting of the $nd\,^2D_{5/2}$ and $nd\,^2D_{3/2}$ states. Figure 4.4 schematically shows the reason for the diffuse nature of the spectral lines originating from $^2D \to {}^2P$ transitions [in Fig. 4.4(b)], and for comparison, simple doublet principal transitions $^2P \to {}^2S$ [in Fig. 4.4(a)].

The Hamiltonian for the coupling of the electron spin with an external magnetic field is $H = -\boldsymbol{\mu}_s \cdot \mathbf{H}$ [see Eq. (4.46)]. In the rest frame of an electron in orbit around a nucleus, an additional magnetic field is present due to the electric field $\mathbf{E}$ due to the charge of the proton in its rest frame (i.e., $\mathbf{E} = -\nabla V(\mathbf{r})$, where the electrical potential of the nucleus is $V(\mathbf{r}) = Ze/r$, so $\mathbf{E} = Ze\mathbf{r}/r^3$). This electric

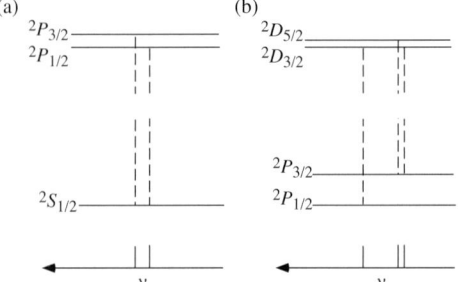

FIG 4.4 Comparison of the principal and diffuse series transitions and spectra. (a) Simple doublet in the $^2P \to {}^2S$ principal transitions. (b) Compound doublet in the $^2D \to {}^2P$ diffuse transitions. Reproduced with permission from Wiley from Ref. [18], Fig. 6.2, p. 322.

field is Lorentz transformed into a magnetic field in the electron's rest frame (Ref. [24], Chapter 24). The magnetic field $\mathbf{H}_{\text{add}}$ in the moving frame of the electron is given by

$$\mathbf{H}_{\text{add}} = -\frac{\mathbf{v}}{c} \times \mathbf{E} = -\frac{\mathbf{p}}{mc} \times \mathbf{E}. \tag{4.67}$$

That is, the magnetic field that results in the electron's rest frame from the transformed static electric field is given by Eq. (4.67). Using this magnetic field in the expression $H = -\boldsymbol{\mu}_s \cdot \mathbf{H}$, we obtain the spin–orbit interaction Hamiltonian *except of a factor of* 1/2. The correct Hamiltonian is obtained by substituting the magnetic field in Eq. (4.67) into the expression

$$H_{\text{so}} = -\boldsymbol{\mu}_s \cdot \mathbf{H}_{\text{add}}/2. \tag{4.68}$$

The reason for the additional factor of 1/2 was first explained by L. H. Thomas in 1926, and therefore, the precession of an electron spin in the magnetic field caused by the moving nucleus is called Thomas precession. The factor of 1/2 is due to the fact that the rest frame of the electron is not an inertial frame, hence, the Hamiltonian needs to be corrected to account for the fact that the electron is in an accelerating frame. An electric field with a component perpendicular to the electron velocity causes an additional acceleration of the electron perpendicular to its instantaneous velocity, leading to a curved electron trajectory so the electron moves in a rotating frame of reference, and this provides additional electron precession, so the net precession is half the naive result (see Ref. [34] for a classical derivation of the Thomas precession factor of 1/2). In any case, the correct Hamiltonian is also obtained directly from the Dirac equation. After substituting (4.67) into (4.68) and using $\mathbf{E}(\mathbf{r}) = Ze\mathbf{r}/r^3$, we obtain:

$$H_{\text{so}} = \boldsymbol{\mu}_s \cdot \frac{\mathbf{p}}{mc} \times (Ze\mathbf{r}/r^3)/2 = \frac{Ze^2}{2m_e^2 c^2 r^3} \frac{\hbar}{2} \boldsymbol{\sigma} \cdot \mathbf{L} \tag{4.69}$$

where we used $\boldsymbol{\mu}_s = -g_s \mu_B (\mathbf{S}/\hbar) \approx -\frac{e\hbar}{m_e c} \frac{\boldsymbol{\sigma}}{2}$ and $\mathbf{r} \times \mathbf{p} = \mathbf{L}$. This spin–orbit Hamiltonian is often written in the form

$$\boxed{H_{\text{so}} = \xi(r) \frac{\mathbf{L} \cdot \mathbf{S}}{\hbar^2}, \quad \langle H_{\text{so}} \rangle_{nlms} = A \frac{\mathbf{L} \cdot \mathbf{S}}{\hbar^2},} \tag{4.70}$$

where $\xi(r) = \frac{Ze^2 \hbar^2}{2m_e^2 c^2 r^3}$ is the spin–orbit coefficient of the electron (it has units of energy), and $A \equiv \langle \xi(r) \rangle_{nlm}$ is called the *spin–orbit coupling constant*. It is easy to evaluate $\mathbf{L} \cdot \mathbf{S}$ by noting that the square of the total electronic angular momentum, $\mathbf{J} = \mathbf{L} + \mathbf{S}$, contains the term $\mathbf{L} \cdot \mathbf{S}$ (see Problem 4.23). [This technique for evaluating a scalar product of angular momentum vectors, such as $\mathbf{L} \cdot \mathbf{S}$, will be employed again when we evaluate the hyperfine splitting in the next

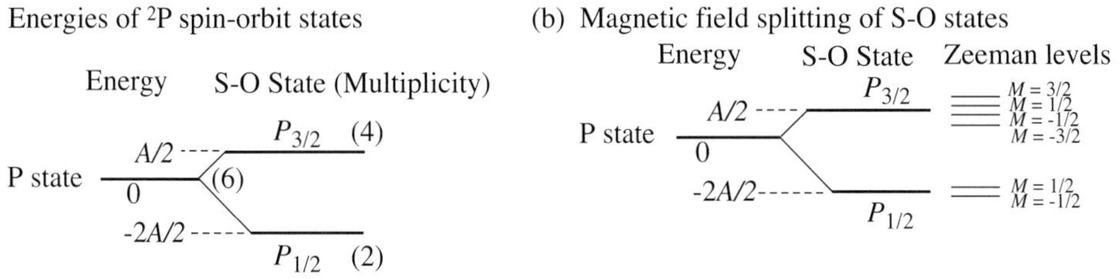

4.5 Spin–Orbit Interaction in Atoms

(a) Energies of ^{2}P spin-orbit states

```
              Energy    S-O State (Multiplicity)
                        _____ P_{3/2}  (4)
              A/2 ----/
   P state     0      (6)
              -2A/2 --\_____ 
                            P_{1/2}  (2)
```

(b) Magnetic field splitting of S-O states

```
              Energy    S-O State    Zeeman levels
                        _____ P_{3/2}  ≡ M = 3/2, 1/2, -1/2, -3/2
              A/2 ----/
   P state     0       
              -2A/2 --\_____ P_{1/2}  = M = 1/2, -1/2
```

FIG 4.5 (a) Spin–orbit splitting of a ^{2}P state into ^{2}P$_{3/2}$ and ^{2}P$_{1/2}$ states. (b) Zeeman splitting of the spin–orbit states.

section.] In order to find the *spin–orbit splitting* of two states, the *difference* of their spin–orbit energies must be taken, as described in the next paragraph.

The magnitude of the spin–orbit coupling, $\langle \xi(r) \rangle_{nlm}$, can be ascertained by noting that $\langle r^{-3} \rangle_{nlm} = Z^3 a_0^{-3}/[n^3(l+1)(l+1/2)l]$ for hydrogenic states [see Eq. (3.112e)], where the Bohr radius $a_0 = \hbar^2/(me^2)$, hence,

$$A = \langle \xi(r) \rangle_{nlm} = \frac{Z^4 \alpha^2 (mc^2 \alpha^2)}{n^3(l+1)(l+1/2)l}. \tag{4.71}$$

$(mc^2\alpha^2)$ is the atomic unit of energy (27.21 eV) and $\alpha^2 \approx (1/137)^2$, so

$$A \approx \frac{Z^4}{n^3(l+1)(l+1/2)l} \text{ meV}.$$

Hence, the spin–orbit energy of level $|n, j, l, s\rangle$ is[7]

$$\boxed{\langle H_{so} \rangle_{njls} = \frac{A}{2}[j(j+1) - l(l+1) - s(s+1)].} \tag{4.72}$$

The splitting of ^{2}P states into ^{2}P$_{3/2}$ and ^{2}P$_{1/2}$ spin–orbit states is given by

$$\langle H_{so} \rangle_{^2P_{3/2}} - \langle H_{so} \rangle_{^2P_{1/2}} = \frac{A}{2}\left\{\left[\frac{3}{2}\left(\frac{3}{2}+1\right) - 1(1+1) - \frac{1}{2}\left(\frac{1}{2}+1\right)\right]\right.$$

$$\left. - \left[\frac{1}{2}\left(\frac{1}{2}+1\right) - 1(1+1) - \frac{1}{2}\left(\frac{1}{2}+1\right)\right]\right\} = \frac{A}{2}\{[1] - [-2]\} = \frac{3}{2}A. \tag{4.73}$$

Note that the average of the spin–orbit interaction over *all* spin-multiplet levels vanishes (see Fig. 4.5).

In the presence of an external magnetic field, the spin–orbit states split into $2j+1$ equally spaced levels with spacing proportional to the magnetic field strength due to the Zeeman Hamiltonian (4.32). The details of the splitting are presented in Sec. 7.3.1. Figure 4.5(a) shows the splitting of a ^{2}P due to spin–orbit interaction, and Fig. 4.5(b) shows the Zeeman splitting of these states.

Problem 4.22

Starting from the Zeeman Hamiltonian $H_{so} = -\boldsymbol{\mu}_s \cdot \mathbf{H}$, derive $H_{so} = \frac{g_s \mu_B}{2 m_e c} \mathbf{E} \cdot (\mathbf{p} \times \boldsymbol{\sigma})$ for a moving electron in an electric field.

---

[7] It is useful to compare this result with the energies in the footnote after Eq. (3.107).

Hint: The effective **H** field felt by a moving electron is $\mathbf{H} = -\mathbf{v} \times \mathbf{E}/c$ [or if we symmetrize, $\mathbf{H} = \frac{1}{2mc}(\mathbf{E} \times \mathbf{p} - \mathbf{p} \times \mathbf{E})$]. We should point out that a term, $\hbar \nabla \cdot \mathbf{E}$, should be added to $\mathbf{E} \cdot (\mathbf{p} \times \boldsymbol{\sigma})$ since it is of the same order of magnitude. However, it does not involve $\boldsymbol{\sigma}$. This additional term is called the Darwin term; it originates from a higher order term in the nonrelativistic reduction of the Dirac equation than those specified in (4.43).

Problem 4.23

(a) Show that $\mathbf{L} \cdot \mathbf{S} = [\mathbf{J}^2 - \mathbf{L}^2 - \mathbf{S}^2]/2$, so $\langle \mathbf{L} \cdot \mathbf{S} \rangle_{njls} = \hbar^2[j(j+1) - l(l+1) - s(s+1)]/2$, where the total angular momentum quantum number j is such that $\mathbf{J}^2 = \hbar^2 j(j+1)$. **Hint:** $\mathbf{J}^2 = (\mathbf{L} + \mathbf{S})^2$.

(b) Estimate the spin–orbit splitting of the 3p $^2P_{3/2}$ and 3p $^2P_{1/2}$ states of Na. Compare your result with the experimental value of 17.2 cm^{-1} and consider possible reasons for the discrepancy.[a] Note that 1 eV $\approx$ 8065 cm^{-1}.

Problem 4.24

(a) Consider the ^{87}Rb excited state ... $4s^2 3d^{10} 4p^6 5p$ ^2P. (Only the single unpaired electron in the highest occupied 5p orbital contributes to the angular momentum.) What are the possible values of the total electronic angular momentum, J.

(b) Given the fine-structure splitting Hamiltonian, $H_{so} = A\mathbf{L} \cdot \mathbf{S}/\hbar^2$, with $A = 4\frac{2}{3}$ THz for ^{87}Rb, calculate the spin–orbit energies of the J states you specified in part (a).

Answers: (a) $J = 1/2, 3/2$.
(b) $E_{so}(j) = \frac{A}{2}[J(J+1) - S(S+1) - L(L+1)] = \frac{A}{2}[J(J+1) - 3/4 - 2]$. So, $E_{so}(j=1/2) = -A = -4\frac{2}{3}$ THz, $E(j=3/2) = A/2 = 2\frac{1}{3}$ THz.

We have analyzed the spin–orbit coupling in hydrogen and in alkali atoms which have one electron in an unfilled shell but delay consideration of spin–orbit coupling of multielectron atoms to Sec. 10.9. In multielectron systems, each electron can interact with the magnetic fields generated by other moving charged particles. Hence, in Sec. 10.9, sums of spin–orbit Hamiltonians, each of the form given in Eq. (4.70), will contribute to the spin–orbit energy of a multielectron atom. Spin–orbit coupling plays important roles also in other systems, including molecular systems, nuclei, and solid-state systems. In nuclei, spin–orbit coupling is responsible for the shell structure of nuclei where the excitation energies of medium and heavy nuclei display a beautiful pattern of magic numbers. Spin-orbit coupling in solid-state physics leads to a number of important effects including mixing of valence and conduction bands and heavy holes in semiconductors, the Anderson transition in two dimensions, topological insulators, weak anti-localization and spin relaxation, etc. These topics will be discussed in Chapters 9 and 13.

4.6 HYPERFINE INTERACTION

Examination of the absorption spectra of the alkali atoms under high resolution shows further structure (additional splittings) of spin–orbit lines due to the interaction of the spin of the outer electron with the nuclear spin. Only if the nuclear spin is nonzero is there a splitting. It is much smaller than the fine-structure splitting of states with nonvanishing **L** due to spin–orbit interactions. The interaction of the spin of electrons with the spin of the nucleus of the atom is called the *hyperfine interaction* and the resultant splitting is called *hyperfine splitting*. For example, let us consider ^{23}Na. The

---

[a] The reader is invited to read Sec. 71 of Ref. [2] entitled "Wave functions of the outer electrons near the nucleus".

4.6 Hyperfine Interaction

nuclear spin of ^{23}Na is $I = 3/2$, so the total angular momentum, F, of the 3s ^{2}S$_{1/2}$ state can be $F = |J+I|, \ldots, |J-I|$, i.e., $F = 1$ or $F = 2$. The 3s ^{2}S$_{1/2}$ $F = 1$ and $F = 2$ states are split by 1712 MHz. As another example, the 21-cm (1420 MHz) hydrogen line that is used so often to learn about the structure of the galaxy and of the universe is due to the hyperfine splitting of the 1s ^{2}S$_{1/2}$ $F = 0$ and $F = 1$ states.

Consider the interaction of two magnetic moments. A magnetic moment generates a magnetic field. The vector potential $\mathbf{A}(\mathbf{r})$ at space point $\mathbf{r}$ due to a magnetic moment $\boldsymbol{\mu}$ located at $\mathbf{r}=0$ is

$$\mathbf{A}(\mathbf{r}) = \left[\frac{\mu_0}{4\pi}\right] \boldsymbol{\mu} \times (-\nabla r^{-1}) = \left[\frac{\mu_0}{4\pi}\right] \boldsymbol{\mu} \times \mathbf{r}/r^3, \tag{4.74}$$

so the resulting magnetic field $\mathbf{H}(\mathbf{r}) = \nabla \times \mathbf{A}(\mathbf{r})$ is [23, 34]

$$\mathbf{H}(\mathbf{r}) = \left[\frac{\mu_0}{4\pi}\right] \left\{ \boldsymbol{\mu} \frac{8\pi}{3} \delta(\mathbf{r}) + \frac{3\mathbf{r}(\mathbf{r} \cdot \boldsymbol{\mu}) - r^2 \boldsymbol{\mu}}{r^5} \right\}. \tag{4.75}$$

The delta function term on the RHS of (4.75) arises because $\nabla^2 r^{-1} = -4\pi\delta(\mathbf{r})$, which gives rise to the *Fermi contact* hyperfine interaction (see below). The factor $\left[\frac{\mu_0}{4\pi}\right]$, where $\mu_0 = 4\pi \times 10^{-7}$ N A^{-2} is the permeability of free space, present in SI units but absent in Gaussian units [the square parenthesis is a reminder of its necessity in SI units]. The magnetic Hamiltonian of two interacting magnetic moments, $\boldsymbol{\mu}_1$ and $\boldsymbol{\mu}_2$, that are not at the same position is given by $H_{\text{mag}} = -\boldsymbol{\mu}_2 \cdot \mathbf{H}_1$, so

$$\boxed{H_{\text{mag}} = -\left[\frac{\mu_0}{4\pi}\right] \left(\frac{3(\boldsymbol{\mu}_1 \cdot \mathbf{r})(\boldsymbol{\mu}_2 \cdot \mathbf{r}) - (\boldsymbol{\mu}_1 \cdot \boldsymbol{\mu}_2) r^2}{r^5} \right).} \tag{4.76}$$

Figure 4.6 shows four different configurations of dipole moments that are the same distance away from one another. The lowest energy configuration is in (a), the next lowest is (b), then comes (c), and the highest is (d).

Problem 4.25

Given two spin-1/2 particles in states $\left|\frac{1}{2}, \frac{1}{2}\right\rangle$, calculate
$\mathbf{S}_1 \cdot \mathbf{S}_2 \left[\left|\frac{1}{2}, \frac{1}{2}\right\rangle \left|\frac{1}{2}, \frac{1}{2}\right\rangle \right]$. Note that $\mathbf{S}_T = \mathbf{S}_1 + \mathbf{S}_2$, so $\mathbf{S}_T^2 = \mathbf{S}_1^2 + \mathbf{S}_2^2 + 2\mathbf{S}_1 \cdot \mathbf{S}_2$.

Answer: $\mathbf{S}_1 \cdot \mathbf{S}_2 \left[\left|\frac{1}{2}, \frac{1}{2}\right\rangle \left|\frac{1}{2}, \frac{1}{2}\right\rangle \right] = \hbar^2/2 \left[\left|\frac{1}{2}, \frac{1}{2}\right\rangle \left|\frac{1}{2}, \frac{1}{2}\right\rangle \right]$.

Problem 4.26

Calculate the energies of the four dipole moment configurations of $\boldsymbol{\mu}_1$ and $\boldsymbol{\mu}_2$ shown in Fig. 4.6, with $|\boldsymbol{\mu}_1| = |\boldsymbol{\mu}_2| \equiv \mu$ and relative coordinate vector $\mathbf{r}$ of the same length in all configurations.

Problem 4.27

What is the form of the Hamiltonian in (4.76) if the magnetic moments are located at $\mathbf{r}_1$ and $\mathbf{r}_2$?
Answer: Let $\mathbf{r} \to \mathbf{r}_2 - \mathbf{r}_1$ in (4.76).

FIG 4.6 Four configurations of two dipole moments, in order of increasing energy.

In considering the *hyperfine interaction Hamiltonian* $H_{hf} = -\boldsymbol{\mu}_{el} \cdot \mathbf{H}_N$, we recall that electrons in $l=0$ states have finite probability of being located at the nucleus, so the delta function term in the Hamiltonian (4.75) contributes. Moreover, for $l=0$ states, the expectation value of the $1/r^3$ terms vanishes. For aribtrary l, the hyperfine Hamiltonian $H_{hf} = -\boldsymbol{\mu}_s \cdot \mathbf{H}_N$ coupling the electron magnetic moment with the magnetic field generated by the nuclear spin is

$$H_{hf}(\mathbf{r}) = \left[\frac{\mu_0}{4\pi}\right] \frac{(2\mu_B)(g_N \mu_N)}{\hbar^2} \left\{ \frac{8\pi}{3} (\mathbf{S} \cdot \mathbf{I}) \delta(\mathbf{r}) + \left[\frac{3(\mathbf{S} \cdot \mathbf{r})(\mathbf{I} \cdot \mathbf{r}) - (\mathbf{S} \cdot \mathbf{I}) r^2}{r^5}\right] \right\}. \qquad (4.77)$$

The delta function term on the RHS of Eq. (4.77) is called the *Fermi contact term*, named after Enrico Fermi who first used it to calculate hyperfine splittings in 1930. Its expectation value is nonvanishing only for s-wave ($l=0$) electronic states since only these states have finite density at the origin. The order of magnitude of the expectation value of the *dipole–dipole* $1/r^3$ hyperfine interaction term for states with $l \neq 0$ is

$$\frac{m_e}{M_p} Z^3 \alpha^2 \frac{e^2}{a_0} = \frac{m_e}{M_p} Z^3 \alpha^4 m_e c^2.$$

This is smaller than the expectation value of the fine-structure interaction (which also is proportional to $1/r^3$) by a factor of $m_e/(ZM_p)$, i.e., by at least a factor of 1000. The contact term yields the following expression for the hyperfine energy of s-wave ($l=0$) electronic states:

$$\langle H_{hf} \rangle_{n, l=0, s, j=s, f} = \left[\frac{\mu_0}{4\pi}\right] \frac{(2\mu_B)(g_N \mu_N)}{\hbar^2} \frac{8\pi}{3} (\mathbf{S} \cdot \mathbf{I}) |\psi_{n, l=0, m=0}(0)|^2. \qquad (4.78)$$

In Problem 4.28, you will calculate $\mathbf{S} \cdot \mathbf{I}$ and find an expression for $|\psi_{n,l=0,m=0}(0)|^2$ for hydrogenic atoms. Then you can calculate $\langle H_{hf} \rangle_{n,l=0,s,j=s,f}$ and obtain a general expression for the splitting between f states with $l=0$.

Problem 4.28

(a) Calculate $\langle \mathbf{S} \cdot \mathbf{I} \rangle$. Hint: Define $\mathbf{F} = \mathbf{S} + \mathbf{I}$, and square both sides of this equation.
(b) Determine an expression for $|\psi_{n,l=0,m=0}(0)|^2$ for hydrogenic atoms.
(c) Write out the Hamiltonian that couples the electron orbital magnetic moment with the magnetic field generated by the nuclear spin, $-\boldsymbol{\mu}_l \cdot \mathbf{H}_N$, in terms of $\mathbf{I}$ and $\mathbf{L}$.
(d) Why is there no contact term in the Hamiltonian of (c).

Answers: (a) $\langle \mathbf{S} \cdot \mathbf{I} \rangle = \hbar^2 [F(F+1) - S(S+1) - I(I+1)]/2$.
(b) Using (3.108) and (3.109), $\psi_{n,0,0}(0) = (Z/a_0)^{3/2} \sqrt{\frac{22}{n^5}} L_{n-1}^1(0) \frac{1}{\sqrt{4\pi}}$, where $L_{n-1}^1(0) = n$, hence $|\psi_{n,l=0,m=0}(0)|^2 = Z^3/(\pi a_0^3 n^3)$.
(c) $H_{hf}(\mathbf{r}) = \left[\frac{\mu_0}{4\pi}\right] \frac{(\mu_B)(g_N \mu_N)}{\hbar^2} \left[\frac{3(\mathbf{L} \cdot \mathbf{r})(\mathbf{I} \cdot \mathbf{r}) - (\mathbf{L} \cdot \mathbf{I}) r^2}{r^5}\right]$.

Problem 4.29

(a) The nuclear spin of the ^{87}Rb atom is $I=3/2$. What are the possible values of the total atomic angular momentum, F, of the ... $4s^2 3d^{10} 4p^6 5s$ ^2S ground state.
(b) Find energies of the ^{87}Rb ground-state hyperfine levels given the hyperfine Hamiltonian $H_{hf} = B \frac{\mathbf{S} \cdot \mathbf{I}}{\hbar^2}$, where $B = 3.4$ GHz.

Answers: (a) $F = 1, 2$.
(b) $E_{hf}(F) = \frac{B}{2}[F(F+1) - S(S+1) - I(I+1)] = \frac{B}{2}[F(F+1) - 3/4 - 15/4]$. So, $E_{hf}(F=1) = -1.25B = -4.25$ GHz, $E_{hf}(F=2) = 0.75B = 2.55$ GHz.

4.6 Hyperfine Interaction

Figure 4.8 shows the fine and hyperfine structure of ^{87}Rb. The hyperfine splitting of the ground-state and the excited-state fine and hyperfine splitting are not to scale. The fine-structure splitting of the excited state is much bigger than the ground-state hyperfine splitting, which in turn is much bigger than the excited-state hyperfine splittings.

Upon applying Eq. (4.78) to hydrogen, one finds that the splitting of the $F=1$ and $F=0$ hydrogen levels of the ground state is 1420 MHz (5.87×10^{-6} eV) corresponding to an emission wavelength of 21 cm. See Fig. 4.7 for a description of the $F=1$ and $F=0$ states participating in this transition. This line has been extensively used in astrophysical studies to determine relative velocities of different regions within our Milky Way galaxy, as well as to determine the relative velocities of other galaxies from ours, via the Doppler shift of the line. The hydrogen maser (Microwave Amplification by Stimulated Emission of Radiation) operating on this transition is also used to check the variability of the lunar and earth secular acceleration using satellite data. For additional discussion of hyperfine interactions, see, e.g., Bethe and Salpeter [4].

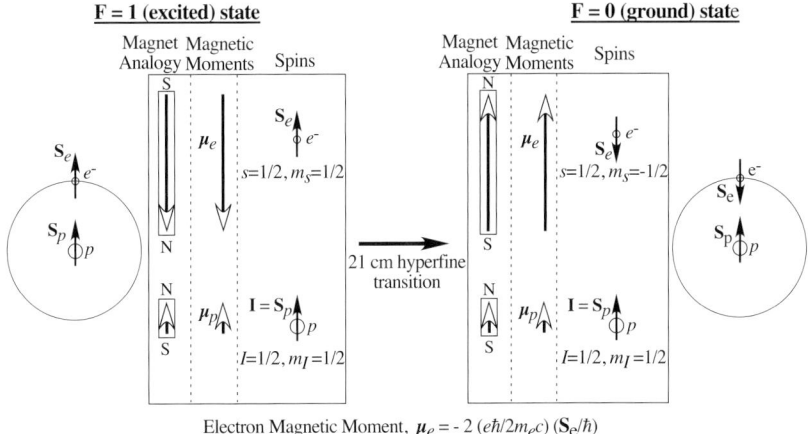

FIG 4.7 Hyperfine $F=1$ and $F=0$ states of the $1s\,^2S_{1/2}$ ground electronic state of hydrogen and a magnet analogy to explain why the $F=0$ is the ground state. The spins and magnetic moments of the electron and proton are shown schematically in both $F=1$ and $F=0$ states. The magnetic moments are not drawn to scale; the magnetic moment of the proton is smaller than that of the electron by a factor of about 1000. The $F=1$ state depicted here is the stretched state with $M_F=1$. If no external magnetic field is present, the three $F=1$ states (with $M_F=1,0,-1$) are degenerate in energy. The $F=0$ state is lower in energy than the $F=1$ states by 1420 MHz. Emission from the $F=1 \to F=0$ yields a photon with 21-cm wavelength. Reproduced with permission from Wiley from Ref. [18], Fig. 1.2, p. 13.

4.6.1 ELECTRIC QUADRUPOLE HYPERFINE INTERACTION

Nuclei with nuclear spin greater than 1/2 can have an *electric quadrupole moment*, i.e., $Q_{2m} = \int d\mathbf{r}\, \rho(\mathbf{r}) r^2 Y_{2m}$, $m = -2,\ldots,2$, where ρ is the charge distribution of the nucleus [see Eq. (3.199)], in addition to having magnetic dipole moment. Only nuclei (or particles) with $I > 1/2$ possess quadrupole moments (see Problem 3.34). The quadrupole moment of a nucleus interacts with the quadrupole moment of the electronic charge distribution [see Eq. (3.198)] and this results in an additional contribution to the atomic energy. Such couplings are also called hyperfine interactions. If the total electron spin angular momentum vanishes, the main hyperfine splitting of $S=0$ electronic states arises from the quadrupole interaction of the nucleus with the electrons. Tables of the electric quadrupole moments of selected nuclei can be found

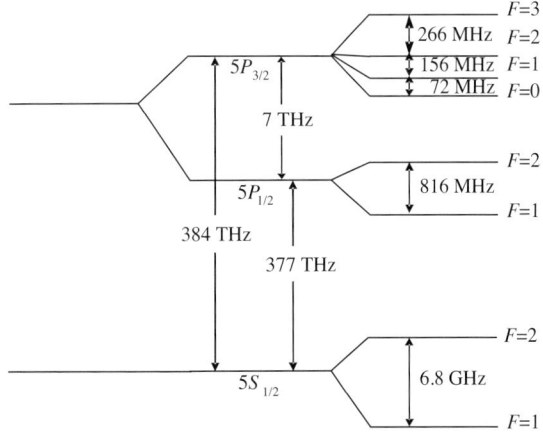

FIG 4.8 Fine and hyperfine structure of ^{87}Rb, which has a nuclear spin of $I=3/2$. The energy level spacings are not to scale.

4.6.2 ZEEMAN SPLITTING OF HYPERFINE STATES

In the presence of an external magnetic field **H**, the hyperfine components with $F = 1$ split into three Zeeman sublevels with $M_F = 0, \pm 1$. The $F = 0$ state is not split, but as we shall see, its energy changes at sufficiently large magnetic field strengths, i.e., at field strengths where the Zeeman energy of the electronic spin moment in the external magnetic field, $-\boldsymbol{\mu}_s \cdot \mathbf{H}$, is comparable to or larger than the interaction energy between nuclear magnetic moment and electron magnetic moment. At very large magnetic fields, when $-\boldsymbol{\mu}_s \cdot \mathbf{H}$ is much larger than the hyperfine energy, the states with $m_s = \pm 1/2$ having Zeeman energy $-g_s \mu_0 H m_s$ are split by the hyperfine coupling. This uncoupling of angular momentum by the magnetic field at large magnetic field strengths is called the *Paschen-Back regime*, or the strong field regime (see the large field regime in Figs 4.9 and 4.10). For large magnetic fields, the Hamiltonian for $l = 0$ states,

$$H = \left[\frac{\mu_0}{4\pi}\right] \frac{(2\mu_B)(g_N \mu_N)}{\hbar^2} \frac{8\pi}{3} |\psi_{n,l=0,m=0}(0)|^2 (\mathbf{S} \cdot \mathbf{I})$$
$$- (\boldsymbol{\mu}_s + \boldsymbol{\mu}_N) \cdot \mathbf{H}, \quad (4.79)$$

can be diagonalized in the $|s, m_s\rangle |I, m_I\rangle$ basis to obtain the eigenenergies. The eigenenergies as a function of magnetic field strength are shown for the hydrogen atom in Fig. 4.9 and for ^{23}Na in Fig. 4.10. For hydrogen, $I = 1/2$, so ground electronic state contains $F = 0$ and $F = 1$ states, whereas for ^{23}Na, $I = 3/2$ and the ground state contains $F = 1$ and $F = 2$ states.

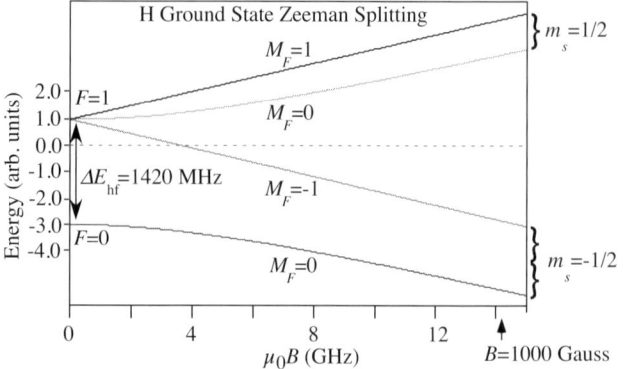

FIG 4.9 Zeeman splitting of the ground state of hydrogen as a function of magnetic induction field strength.

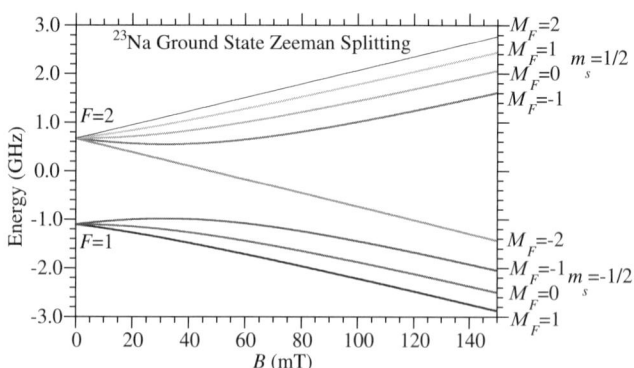

FIG 4.10 Zeeman splitting of the $^2S_{1/2}$ ground state of ^{23}Na versus magnetic induction field strength. Reproduced with permission from Wiley, from Band, *Light and Matter*, Fig. 6.5, p. 336.

Problem 4.30

(a) Given your results in Problem 4.14, what is the Zeeman energy of a ground-state hydrogen atom in the state $^2S_{1/2} |F = 1, M_F = 1\rangle$ state in a magnetic field of 0.1 Tesla (for the purpose of this question, neglect the all internal energies and only consider the Zeeman energy).

(b) Set up and solve the hyperfine plus Zeeman eigenvalue problem for the $F = 0, 1$ hydrogen ground electronic states in the $|F, M_F\rangle$ basis (which can be formed using the $|s, m_s\rangle |I, m_I\rangle$ states) by constructing the 4×4 Hamiltonian matrix for the sum of the Zeeman and hyperfine interactions and calculate the eigenvalues as a function of magnetic field shown in Fig. 4.9.

Answers: (a) $E_{F=1,M_F=1} = \langle \Psi_{F=1,M_F=1}| - (\boldsymbol{\mu}_s + \boldsymbol{\mu}_N) \cdot \mathbf{H}|\Psi_{F=1,M_F=1}\rangle = 1.39962$ GHz $+ (-2.126)$ MHz ≈ 1.3975 GHz. (b) Letting $\mu_S = g_e\mu_B = -g\mu_B$, and writing the Hamiltonian $\mathcal{H} = (J/\hbar^2)(\mathbf{S} \cdot \mathbf{I}) - B(\mu_S S_z + \mu_I I_z)$ in the $|F,M_F\rangle = |1,1\rangle, |1,-1\rangle, |1,0\rangle, |0,0\rangle$ basis,

$$\mathcal{H} = \begin{pmatrix} J/4 - (\mu_S + \mu_I)B & 0 & 0 & 0 \\ 0 & J/4 + (\mu_S + \mu_I)B & 0 & 0 \\ 0 & 0 & J/4 & (\mu_S - \mu_I)B \\ 0 & 0 & (\mu_S - \mu_I)B & -3J/4 \end{pmatrix}.$$

Now diagonalize the 2×2 matrix.

4.7 SPIN-DIPOLAR INTERACTIONS

The interaction of two magnetic dipole moments, $\boldsymbol{\mu}_1$ and $\boldsymbol{\mu}_2$, at $\mathbf{r}_1$ and $\mathbf{r}_2$ is given by $-\boldsymbol{\mu}_2 \cdot \mathbf{H}_1(\mathbf{r})$, where $\mathbf{r} = \mathbf{r}_2 - \mathbf{r}_1$, and the magnetic field $\mathbf{H}_1(\mathbf{r})$ is given by Eq. (4.75). Hence, using (4.76), the spin-dipolar interaction potential is given by

$$V_{SS}(r) = -\left[\frac{\mu_0}{4\pi}\right]\frac{(2\mu_B)^2}{\hbar^2}\frac{3(\mathbf{S}_1 \cdot \mathbf{r})(\mathbf{S}_2 \cdot \mathbf{r}) - \mathbf{S}_1 \cdot \mathbf{S}_2 r^2}{r^5}. \tag{4.80}$$

Note that we have not added the Fermi contact term [the first term on the RHS of Eq. (4.77)] here, but considered only the long range part of the potential (if the particles are electrons, or are nuclei, the repulsive Coulomb potential will insure that their wave function vanishes for zero distance between them).

It is of interest to calculate the spin-dipolar interaction energy of two spin 1/2 particles that are in a singlet and triplet spin state given by

$$|S=0, M_S = 0\rangle = \frac{1}{\sqrt{2}}(|\uparrow\rangle|\downarrow\rangle - |\downarrow\rangle|\uparrow\rangle) \tag{4.81}$$

and

$$|S=1, M_S = 0\rangle = \frac{1}{\sqrt{2}}(|\uparrow\rangle|\downarrow\rangle + |\downarrow\rangle|\uparrow\rangle). \tag{4.82}$$

We can consider these particles to be electrons, but they can just as well be two spin 1/2 nuclei, or two spin 1/2 atoms. We need to evaluate the matrix elements of this potential. To do so, we will use the expression $\mathbf{S}_1 \cdot \mathbf{S}_2 = [\mathbf{S}^2 - (\mathbf{S}_1^2 + \mathbf{S}_2^2)]/2 = [\mathbf{S}^2 - 3\hbar^2/2]/2$. Hence,

$$\mathbf{S}_1 \cdot \mathbf{S}_2 (|\uparrow\rangle|\downarrow\rangle - |\downarrow\rangle|\uparrow\rangle) = -\frac{3\hbar^2}{4}(|\uparrow\rangle|\downarrow\rangle - |\downarrow\rangle|\uparrow\rangle),$$

$$\mathbf{S}_1 \cdot \mathbf{S}_2 (|\uparrow\rangle|\downarrow\rangle + |\downarrow\rangle|\uparrow\rangle) = \frac{\hbar^2}{4}(|\uparrow\rangle|\downarrow\rangle + |\downarrow\rangle|\uparrow\rangle). \tag{4.83}$$

For simplicity, let us take the $\hat{z}$-axis along the distance from one particle to the other. Then,

$$(\mathbf{S}_1 \cdot \mathbf{r})(\mathbf{S}_2 \cdot \mathbf{r})(|\uparrow\rangle|\downarrow\rangle \pm |\downarrow\rangle|\uparrow\rangle) = -\frac{\hbar^2}{4}r^2(|\uparrow\rangle|\downarrow\rangle \pm |\downarrow\rangle|\uparrow\rangle). \tag{4.84}$$

The matrix elements can now be easily evaluated. We find:

$$\langle S=0, M_S=0|V_{SS}(r)|S=0, M_S=0\rangle = 0, \tag{4.85}$$

$$\langle S=1, M_S=0|V_{SS}(r)|S=1, M_S=0\rangle = \left[\frac{\mu_0}{4\pi}\right]\frac{(2\mu_B)^2}{r^3}. \tag{4.86}$$

Thus, the spin-dipolar interaction energy of the singlet state vanishes and that of the triplet state $|S=1, M_S = 0\rangle$ is repulsive and falls off as r^{-3}.

Problem 4.31

(a) Calculate the spin-dipolar interaction energy for the triplet state $|S=1, M_S = 1\rangle \equiv |\uparrow\rangle|\uparrow\rangle$, where the $\hat{z}$-axis is taken along $\hat{\mathbf{r}}$.
(b) Explain the sign of the energy of the triplet state in part (a) and the triplet state $|S=1, M_S = 0\rangle$ in Eq. (4.86).
(c) Find all the eigenvalues and eigenvectors of $\mathbf{S}_1 \cdot \mathbf{S}_2$.

Answers: (a) $-\left[\frac{\mu_0}{4\pi}\right](2\mu_B)^2 \frac{1}{2r^3}$. (b) $|\uparrow\rangle|\uparrow\rangle$ is the lowest energy state, and $|S=1, M_S = 0\rangle$ is an excited state. (c) Eigenvalues are $\hbar^2[S(S+1) - 3/2]/2$. See (4.83) for eigenvectors.

Problem 4.32

(a) Show that the operator $P_{12} \equiv \frac{1}{2} + \frac{2}{\hbar^2}\mathbf{S}_1 \cdot \mathbf{S}_2 = \frac{1}{2} + \frac{1}{2}\boldsymbol{\sigma}_1 \cdot \boldsymbol{\sigma}_2$ exchanges (permutes) the spins of any spin wave function involving the two spins 1 and 2, i.e., it is the permutation operator for spins.
(b) Show that the operator $\mathcal{S} = \frac{1}{2}(1 + P_{12}) = \frac{3}{4} + \frac{1}{\hbar^2}\mathbf{S}_1 \cdot \mathbf{S}_2 = \frac{1}{4}(3 + \boldsymbol{\sigma}_1 \cdot \boldsymbol{\sigma}_2)$ is a projection operator that projects any two-electron spin state onto symmetric ("triplet") states, in the sense that if $\mathcal{S}$ is applied to an antisymmetric state such as $|\psi_1\rangle|\psi_2\rangle - |\psi_2\rangle|\psi_1\rangle$, it yields zero, and if applied to a symmetric state $|\psi_1\rangle|\psi_2\rangle + |\psi_2\rangle|\psi_1\rangle$, it yields the same state. We can therefore call the projection operator $\mathcal{S}$ by the name P_{triplet}.
(c) Show that $\mathcal{A} \equiv \frac{1}{2}(1 - P_{12}) = \frac{1}{4} - \frac{1}{\hbar^2}\mathbf{S}_1 \cdot \mathbf{S}_2 = \frac{1}{4}(1 - \boldsymbol{\sigma}_1 \cdot \boldsymbol{\sigma}_2)$ is a projector onto antisymmetric ("singlet") states, i.e., $\mathcal{A}|\psi_1\rangle|\psi_2\rangle = \frac{1}{2}(|\psi_1\rangle|\psi_2\rangle - |\psi_2\rangle|\psi_1\rangle)$. We can therefore call the projection operator $\mathcal{A}$ by the name P_{singlet}.
(d) Show that $P_{\text{singlet}} + P_{\text{triplet}} = 1$.

Comment: In Chapter 8, we shall use the symmetrization and antisymmetrization operators for two two-level systems, $\mathcal{S} = \frac{1}{2}(1 + P_{12})$ and $\mathcal{A} = \frac{1}{2}(1 - P_{12})$, respectively, where $\mathcal{S} + \mathcal{A} = 1$, to properly symmetrize boson and fermion wave functions.

Problem 4.33

(a) Express the projection operator $P_{\uparrow\uparrow} = |\uparrow\uparrow\rangle\langle\uparrow\uparrow|$ that projects onto the triplet state $|\uparrow\rangle|\uparrow\rangle$ in terms of Pauli spin matrices.
(b) Do the same for the triplet state $|\downarrow\rangle|\downarrow\rangle$.

Answers: $P_{\uparrow\uparrow} = \frac{1}{4}(1 + \sigma_z)_1(1 + \sigma_z)_2$, $P_{\downarrow\downarrow} = \frac{1}{4}(1 - \sigma_z)_1(1 - \sigma_z)_2$.

The spin-dipolar Hamiltonian for a many-body spin system is given by the sum over all pairs of particles of the Hamiltonian (4.80):

$$\mathcal{H} = -\frac{1}{2}\left[\frac{\mu_0}{4\pi}\right]\frac{(2\mu_B)^2}{\hbar^2} \sum_{i,j \neq i} \frac{3(\mathbf{S}_i \cdot \mathbf{r}_{ij})(\mathbf{S}_j \cdot \mathbf{r}_{ij}) - \mathbf{S}_i \cdot \mathbf{S}_j r_{ij}^2}{r_{ij}^5}. \tag{4.87}$$

Heisenberg Spin Hamiltonian

In the dipolar Hamiltonian (4.87), the energy depends on the relative orientation of the magnetic moments $\boldsymbol{\mu}_i, \boldsymbol{\mu}_j$ and the radius vectors $\mathbf{r}_{ij}$ joining them. Consider atoms in a crystal lattice. When a pair of atoms are located at lattice points $\mathbf{r}_i$ and $\mathbf{r}_j$, we can form a scalar interaction of the form $J(\mathbf{r}_{ij})\mathbf{S}_i \cdot \mathbf{S}_j$, where $\mathbf{S}_i$ is the spin operator for the atom located at point

$\mathbf{r}_i$ and $\mathbf{r}_{ij} = \mathbf{r}_i - \mathbf{r}_j$. The physical origin of this interaction will be discussed in Chapter 9. It is *not* related to the dipolar interaction, but, rather, to a combination of Coulomb repulsion between electrons and the Pauli exclusion principle. The coefficient $J_{ij}(\mathbf{r}_{ij})$ is called the *exchange coupling* or *exchange integral*. After integration over electron coordinates, the resulting *exchange coupling* coefficients drop off rapidly (exponentially, not as $1/r_{ij}^3$) with the distance $\mathbf{r}_{ij}$ between atoms, so often only nearest neighbor terms are kept. If $J_{ij} > 0$, the interaction is called ferromagnetic, whereas if $J_{ij} < 0$, it is called antiferromagnetic. A ferromagnetic interaction tends to align spins, whereas an antiferromagnetic interaction tends to antialign them. For a lattice of magnetic atoms subject to a constant magnetic field, the Hamiltonian has the form,

$$\mathcal{H} = -\frac{1}{2} \sum_{i,j \neq i} J_{ij}(\mathbf{r}_{ij}) \mathbf{S}_i \cdot \mathbf{S}_j - g\mu_B \sum_i \mathbf{H} \cdot \mathbf{S}_i. \tag{4.88}$$

This is referred to as the *Heisenberg spin Hamiltonian*, which is often taken as the starting point for studying magnetic phenomena in condensed matter physics. Moreover, often an anisotropic interaction is used to model ferromagnets, with the z-component term $S_{z,i}S_{z,i+1}$ having different strength, $J_{i,i+1}^z$, than the x and y-components; these are called *anisotropic Heisenberg spin Hamiltonians*. More on the Heisenberg spin Hamiltonian is discussed in Chapter 9.

A simpler related model that can be easily solved in 1D is the so-called *Ising model*, named after Ernst Ising, uses nearest neighbor terms in (4.88) and replaces the operators $\mathbf{S}_i \cdot \mathbf{S}_j$ by <u>numbers</u> $\mathcal{S}_i \mathcal{S}_j$, where $\mathcal{S}_i$ and $\mathcal{S}_j$ are taken to be either $+1$ or -1, to obtain an expression for the energy

$$\mathcal{E} = -J \sum_{\langle i,j \rangle} \mathcal{S}_i \mathcal{S}_j. \tag{4.89}$$

Here $\langle i,j \rangle$ indicates all nearest neighbor pairs. Such models have been extensively studied; the 1D model was solved by Ising in 1925 and the 2D square-lattice version was solved by Lars Onsager in 1944. We shall not pause to consider such problems here.

More general spin Hamiltonians that describe spin degrees of freedom in molecules and in solids can contain a large number of terms, representing the Zeeman interaction of the magnetic moments of the electrons and the nuclei with an external field, fine-structure level splitting due to indirect effects of the crystal field in solids, hyperfine structure due to the presence of nuclear magnetic dipole, and electric quadruple moments in the central ion or ligand ions in a solid [41].

Problem 4.34

(a) Show that (4.88) with $\mathbf{H}=0$ can be written using (3.176) as

$$\mathcal{H} = \frac{1}{2} \sum_{i,j \neq i} J_{ij}(\mathbf{r}_{ij}) \left(S_{+1,i} S_{-1,j} + S_{-1,i} S_{+1,j} - S_{0,i} S_{0,j} \right). \tag{4.90}$$

(b) Rewrite (4.90) in terms of S_+ and S_- (recall their distinction from S_{+1} and S_{-1}).

Answer: $\mathcal{H} = -\dfrac{1}{2} \sum_{i,j \neq i} J_{ij}(\mathbf{r}_{ij}) \left[\dfrac{1}{2}(S_{+,i} S_{-,j} + S_{-,i} S_{+,j}) + S_{0,i} S_{0,j} \right]. \tag{4.91}$

4.8 INTRODUCTION TO MAGNETIC RESONANCE

Magnetic resonance phenomena involve the absorption or emission of electromagnetic radiation by electrons or atomic nuclei in response to the application of magnetic fields. Magnetic resonance phenomena include nuclear magnetic resonance (NMR) as well as electron spin resonance (ESR), which is sometimes also called electron paramagnetic resonance (EPR). Electron magnetic resonance was first observed by Y. K. Zavoisky (sometimes spelled Zavoysky) in experiments on salts of the iron group of elements. Electron magnetic resonance (EMR) occurs only in elements with unfilled electronic shells (i.e., unpaired electron spin states). NMR was invented in 1946 by Edward Purcell and Felix Bloch and their colleagues. It has many important applications such as molecular structure determination, dynamical studies both in the liquid and solid state, and magnetic resonance imaging.

At the simplest level, magnetic resonance phenomena can be understood in terms of a simple model with spins that interact with an effective static magnetic field comprised of an external magnetic field together with the internal magnetic field due to the presence of other nearby spins, and a radio-frequency (rf) electromagnetic field which can cause transitions between states with different magnetic quantum numbers. In the presence of a static magnetic field, a single spin 1/2 system is split into two energies, as shown in Fig. 4.2 which shows the Zeeman energy splitting, $\Delta E_Z = g_N \mu_N H_0$, as a function of magnetic field strength. In nuclear magnetic resonance experiments the external static and the rf fields are used to control the spin degrees of freedom of the nuclei in atoms, molecules and condensed phase materials, whereas in ESR experiments, electron spins are controlled. The Zeeman Hamiltonian for a single spin in the presence of a magnetic field is $H = -\mu \cdot \mathbf{H}(t)$, where $\mathbf{H}(t)$ is the time-dependent magnetic field due to a static field along the $\hat{\mathbf{z}}$-axis and a rf field along the $\hat{\mathbf{x}}$-axis, $\mathbf{H}(t) = H_0 \hat{\mathbf{z}} + H_1 \cos(\omega t) \hat{\mathbf{x}}$, but sometimes (particularly in NMR imaging of humans) a circularly polarized rf field of the form $\mathbf{H}(t) = H_0 \hat{\mathbf{z}} + H_1 \{\cos(\omega t) \hat{\mathbf{x}} + \sin(\omega t) \hat{\mathbf{y}}\}$ is used. The couplings between the spins in a material should also be incoporrated into the model of magnetic resonance phenomena, for these couplings can give rise to shifts, splittings, broadening and decoherence of the transitions that are studied. The spin-dipolar coupling Hamiltonian given by (4.87) could be used for this purpose. But the simplest approach is simply to add a contribution to the magnetic field felt by a single spin due to the other spins that are nearby. In liquids with low viscosity, spin-dipolar interactions are rapidly averaged to zero because the directions $\hat{\mathbf{r}}_{ij}$ change rapidly as a function of time and the average of the spin-dipolar Hamiltonian goes to zero as the averaging becomes rapid compared to the dipolar coupling energy scale. Nevertheless, the fluctuations due to these interactions broaden the observed rf transitions and, if the directions $\hat{\mathbf{r}}_{ij}$ don't strictly average to zero, slightly shift them as well. After excitation by an rf pulse, the spins, in turn, produce an rf signal, i.e., a time-dependent magnetization of the sample produces an rf field, which can be measured and Fourier transformed to obtain the frequency spectrum of the sample. The peaks in the spectrum, their splittings and their widths are signatures of the spin states and their environment.

The NMR spectrum of the nuclei with spin 1/2 that are commonly studied in NMR experiments, i.e., the spin 1/2 nuclei ^{1}H, ^{13}C, ^{14}N, ^{19}F, ^{31}P, in a constant static homogeneous magnetic field of 1 T is proportional to the g-factor of the nuclei, g_N, and shown in Fig. 4.11. These spectral peaks are nominally at frequencies corresponding to the Zeeman splitting ΔE_Z of these nuclei divided by Planck's constant, $\nu = \Delta E_Z/h$. The actual location of the peak frequencies of different NMR resonance signals, $\nu_{\text{expt}} = \Delta E_Z/h = g_N \mu_N H_{\text{local}}/h$, depend on the local static magnetic field at the location of the nuclei, H_{local}, which are the sum of the external magnetic field strength H_0 and the internal induced magnetic field at the position of the nucleus in question resulting from electron motion and the electron magnetic moments, $H_{\text{local}} = H_0 + H_{\text{induced}}$, i.e., the resonance frequencies ν_{expt} depend on the environment of the nuclei. The spectrum is also influenced by quadrupole field strengths at the location of the nucleus. Although the resonance signals of, say protons, at specific locations in different molecules are distinct and well separated, an unambiguous frequency often cannot be directly assigned. Therefore, one often adds a standard compound to the sample under study, and this compound acts as a well-defined reference signal. This added reference sample should not interfere with the resonances observed for the molecules being studied (i.e., it should be an inert material).

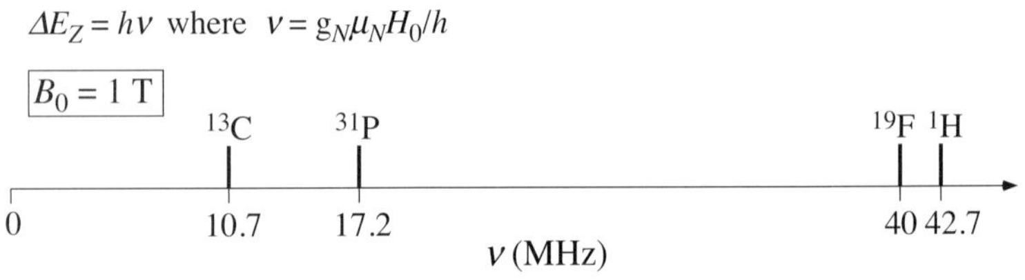

FIG 4.11 Zeeman energy splittings, $\Delta E_Z = g_N \mu_N H_0$, of the spin 1/2 nuclei ^{1}H, ^{13}C, ^{14}N, ^{19}F, ^{31}P in a static magnetic induction field of $B_0 = 1$ T (the splittings are linear with B_0, so simply multiply the splitting shown here by the value of B_0 in Tesla to obtain the splittings for arbitrary B_0). The Landé g-factors of these nuclei are $g_P = 5.585$, $g_{[^{19}F]} = 5.2546$, $g_{[^{31}P]} = 2.2610$, and $g_{[^{13}C]} = 1.4044$ respectively.

4.8 Introduction to Magnetic Resonance

4.8.1 THE ROTATING-WAVE APPROXIMATION

Let us consider the Zeeman Hamiltonian (4.32) for an NMR experiment with a static and an rf magnetic field, i.e., $H = -\boldsymbol{\mu} \cdot (H_0\hat{\mathbf{z}} + H_1 \cos(\omega t)\hat{\mathbf{x}})$, or, more specifically, for a spin 1/2 nucleus, $H = -g\mu \frac{\sigma}{2} \cdot (H_0\hat{\mathbf{z}} + H_1 \cos(\omega t)\hat{\mathbf{x}})$. In matrix form, this becomes

$$H(t) = \frac{\hbar}{2} \begin{pmatrix} \Delta & 2\Omega \cos(\omega t) \\ 2\Omega \cos(\omega t) & -\Delta \end{pmatrix}, \tag{4.92}$$

where

$$\Delta = \frac{g\mu H_0}{\hbar}, \quad \Omega = \frac{g\mu H_1}{2\hbar}. \tag{4.93}$$

We can break the Hamiltonian up into a zero-order term and a first-order term in Ω, where the first order term originates from the rf field,

$$H(t) = H_0 + H_1(t) = \frac{\hbar}{2} \left[\begin{pmatrix} \Delta & 0 \\ 0 & -\Delta \end{pmatrix} + \begin{pmatrix} 0 & 2\Omega \cos(\omega t) \\ 2\Omega \cos(\omega t) & 0 \end{pmatrix} \right]. \tag{4.94}$$

The first-order term in the interaction representation (see Sec. 2.7.1), is

$$\mathcal{H}_{1,I}(t) = e^{iH_0 t/\hbar} \mathcal{H}_{1,S}(t) e^{-iH_0 t/\hbar}, \tag{4.95}$$

while the zeroth-order term in the interaction representation is simply $\mathcal{H}_{0,I}(t) = H_0$. After some algebra, we find,

$$\mathcal{H}_{1,I}(t) = \frac{\hbar\Omega}{2} \begin{pmatrix} 0 & 2\cos(\omega t) e^{-i\Delta t} \\ 2\cos(\omega t) e^{i\Delta t} & 0 \end{pmatrix}, \tag{4.96}$$

and using $\cos(\omega t) = (e^{i\omega t} + e^{-i\omega t})/2$, we can approximate (4.96) as

$$\mathcal{H}_{1,I,\text{RWA}}(t) \approx \frac{\hbar\Omega}{2} \begin{pmatrix} 0 & e^{i(\omega-\Delta)t} \\ e^{-i(\omega-\Delta)t} & 0 \end{pmatrix}, \tag{4.97}$$

where we have dropped the quickly oscillating terms $e^{\pm i(\omega+\Delta)t}$. This approximation is called the rotating-wave approximation (RWA), and it is an excellent approximation when $\omega \approx \Delta$, since then the non-RWA terms oscillate with frequency $\omega + \Delta$, and therefore average out. If we take $\omega = \Delta$ in (4.97), the Hamiltonian reduces to $(\hbar\Omega/2)\sigma_x$. If, at time $t=0$, we start in the lowest energy state (see Fig. 4.2), the probability of being in the excited state (say state b) oscillates in time, $P_b(t) = |\sin(\Omega t/2)|^2$, i.e., the rf field transfers population from one level to the other in a periodic fashion. In Sec. 6.1.1, we will calculate the population transfer when $\omega \neq \Delta$.

We can also calculate the magnetization, defined as the product of the density times the expectation value of the magnetic moment, $\mathbf{M} = n\langle\boldsymbol{\mu}\rangle$, by solving the time-dependent Schrödinger equation for the wave function $\psi(t)$ in the interaction representation, within the RWA and then forming $\langle\psi(t)|\boldsymbol{\mu}|\psi(t)\rangle$. The solution to the Schrödinger equation is often carried out using time-dependent perturbation theory, see Sec. 7.3.3. We shall not pause to do so here; instead we determine the magnetization $\mathbf{M}(t)$ in a different way in Sec. 4.8.2. The time-dependent magnetization turns out to satisfy the Bloch equation,

$$\frac{\partial \mathbf{M}}{\partial t} = \gamma \mathbf{M} \times \mathbf{H}(t). \tag{4.98}$$

where we have defined $\gamma \equiv g_N \mu_N$. Before dealing with the dynamics of the magnetization, we take up the subject of how the magnetization decays with time due to interaction of the spins with other degrees of freedom, so as to come into steady state at long timescales.

4.8.2 SPIN RELAXATION AND THE BLOCH EQUATION

Introducing a resonant or near resonant rf pulse disturbs the spin system from thermal equilibrium. In due course, equilibrium is restored by a process known as *spin relaxation* whereby an exchange of energy occurs between the spin system and the surrounding thermal reservoir. The equilibrium, without the presence of the rf field, is characterized by a magnetic polarization with magnetization $\mathbf{M}_0 = M_0\hat{\mathbf{z}}$ directed along the longitudinal magnetic field, $\mathbf{H}_0 = H_0\hat{\mathbf{z}}$. The process of restoration to this equilibrium magnetization is called *longitudinal relaxation*. Phenomenologically, the equilibration can be described as follows:

$$\frac{dM_z}{dt} = -\frac{1}{T_1}(M_z - M_0), \tag{4.99}$$

whose solution is, $M_z(t) = M_z(0)\, e^{-t/T_1} + M_0\,[1 - e^{-t/T_1}]$, where T_1 is known as the *longitudinal relaxation time* and $M_z(0)$ is the initial value of the z-component of the magnetization. One might think that the lifetime of the transverse magnetization, M_x and M_y, is characterized by the same relaxation time. But the relaxation time for spins to come into thermal equilibrium among themselves, T_2, known as the *spin–spin relaxation time* is often significantly shorter than T_1, and in general, $T_2 \leq T_1$. The phenomenological equations for the transverse relaxation are

$$\frac{dM_x}{dt} = -\frac{1}{T_2}M_x, \quad \frac{dM_y}{dt} = -\frac{1}{T_2}M_y, \tag{4.100}$$

which have solutions $M_x(t) = M_x(0)\, e^{-\frac{t}{T_2}}$, $M_y(t) = M_y(0)\, e^{-\frac{t}{T_2}}$. Combining Eqs (4.99) and (4.100) with the *Bloch equations* [see (4.40)] we obtain,

$$\dot{M}_x = \gamma(\mathbf{M}\times\mathbf{H})_x - \frac{M_x}{T_2}, \quad \dot{M}_y = \gamma(\mathbf{M}\times\mathbf{H})_y - \frac{M_y}{T_2}, \quad \dot{M}_z = \gamma(\mathbf{M}\times\mathbf{H})_z - \frac{(M_z - M_0)}{T_1}. \tag{4.101}$$

Calculations using the Bloch equations can be carried out using either a circularly polarized rf field in the x-y plane (in addition to a dc field in the z-direction), $\mathbf{H}(t) = H_0\hat{\mathbf{z}} + H_1\{\cos(\omega t)\hat{\mathbf{x}} + \sin(\omega t)\hat{\mathbf{y}}\}$, or a linearly polarized rf field along the x-axis, $\mathbf{H}(t) = H_0\hat{\mathbf{z}} + H_1\{\cos(\omega t)\hat{\mathbf{x}}$. For simplicity, we assume that the rf field strength is small, $\gamma H_1 \ll T_2^{-1}$. The rf magnetization is then linear in H_1. For the linearly polarized rf field, one must make a rotating wave approximation to solve for the rf magnetization in closed form. For the circularly polarized rf field, one must go into a reference frame rotating with the field. In both cases, one obtains the following expressions for the rf magnetization $\mathbf{M}_1$ (in the circularly polarized case these are components in the rotating reference frome) (see Abragam, *Principles of Nuclear Magnetism* [42]):

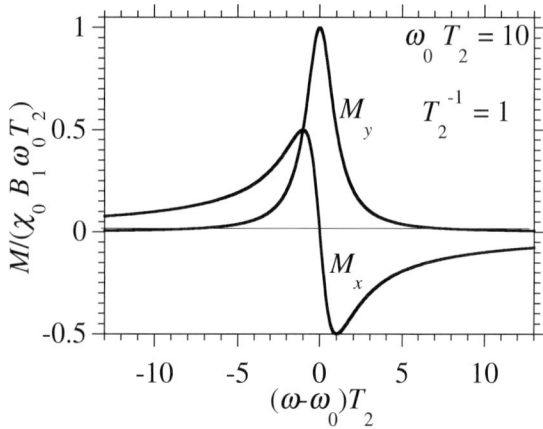

FIG 4.12 M_x and M_y versus frequency.

$$M_{1x}(\omega) = \chi_0 \frac{(\omega_0 T_2)(\omega_0 - \omega)T_2}{1 + (\omega - \omega_0)^2 T_2^2} H_1,$$

$$M_{1y}(\omega) = \chi_0 \frac{\omega_0 T_2}{1 + (\omega - \omega_0)^2 T_2^2} H_1. \tag{4.102}$$

We have defined the parameters $\omega_0 \equiv \gamma H_0$, $\omega_1 \equiv \gamma H_1$, and we take $M_0 \equiv \chi_0 H_0$. Figure 4.12 shows $M_{1x}(\omega)$ and $M_{1y}(\omega)$ plotted versus $(\omega - \omega_0)T_2$. The Lorentzian form of M_{1y} is familiar from near-resonance absorption phenomena (see Fig. 7.7) and the dispersion lineshape of M_{1x} is familiar from the frequency dependence of the refractive index. The full width at half maximum in frequency space is $\Delta\omega = T_2^{-1}$. The rf magnetic susceptibility can be defined as $\chi_{xx} = M_{1x}/H_1$ (and $\chi_{yx} = M_{1y}/H_1$ in the circularly polarized case). The rate of work done on the spins by the rf field, $dE/dt = -\mathbf{M}(t)\cdot\dot{\mathbf{H}}(t)$, depends on the relative phase of the

4.8 Introduction to Magnetic Resonance

magnetization $\mathbf{M}(t)$ and the rf field $\mathbf{H}_1(t)$. If the magnetic moments responsible for the magnetization flip from the low energy state along the static magnetic field to the higher energy state in the opposite direction, energy is transferred from the field to the spins (absorption). In the opposite case, the work is negative and energy is transferred from the spins to the field (induced emission).

> **Problem 4.35**
> Do the algebra leading to the RWA solution of Eq. (4.101) for a linearly polarized rf field.

4.8.3 NUCLEAR SPIN HAMILTONIAN

The magnetic moments of electrons produce a magnetic field that affects nuclear Zeeman energies. When an external magnetic field $\mathbf{H}_0$ is applied to a medium, the actual magnetic field felt by a nucleus is therefore not just the external magnetic field $\mathbf{H}_0$ but the field produced by neighboring magnetic moments, including electron and nuclear magnetic moments. We now consider the nuclear spin Hamiltonian and then consider the chemical shift experienced by nuclei in materials.

An atomic nucleus in its ground state with odd atomic number $A = Z + N$ composed of Z protons and N neutrons has a nonzero spin $\mathbf{I}$, and the corresponding nuclear magnetic moment is $\boldsymbol{\mu}_N = g_N \mu_N \mathbf{I}$. The nuclear g-factor is a property of the nucleus and can be either positive or negative. The corresponding Zeeman energy is $E_Z = -\hbar g_N \mu_N I_z H$, where H is the local magnetic field strength, and the essence of NMR spectroscopy involves measuring the energy difference $\Delta E = \hbar \omega_0 = \hbar g_N \mu_N H$ for $I = 1/2$. Furthermore, nuclei with $I > 1/2$ interact with electric fields via their quadrupole moments. The internal magnetic fields which contribute to the local magnetic field at the location of a nuclei arises from the magnetic moments of nearby electrons and nuclei. Imagine a lattice occupied by identical nuclei located at sites $\mathbf{R}_j$. The nuclear spin Hamiltonian is formally written as,

$$H_N = -g_N \mu_N \sum_j \left[\underbrace{H_0 I_{jz}}_{\text{Zeeman}} + \underbrace{H_1 [I_{jx} \cos \omega t + I_{jy} \sin \omega t]}_{\text{rf field}} \right] + \underbrace{H_0 \sum_j \mathbf{I}_j \sigma_j \mathbf{H}_0}_{\text{chemical shifts}}$$

$$+ \underbrace{\sum_{j<k} \mathbf{I}_j \mathbb{J}_{jk} \mathbf{I}_k}_{\text{scalar couplings}} + \underbrace{\sum_{j<k} \mathbf{I}_j \mathbb{D}_{jk} \mathbf{I}_k}_{\text{dipolar interactions}} + \underbrace{\sum_j \mathbf{I}_j \mathbb{Q}_{jk} \mathbf{I}_j}_{\text{Quadrupolar interactions}} + \underbrace{\sum_j \mathbf{I}_j A_j \mathbf{S}}_{\text{Knight shifts}}. \quad (4.103)$$

In addition to the nuclear Zeeman term and the interaction of the nuclear spins with the rf field (whose polarization was taken to rotate), the Hamiltonian H_N contains:

- The *chemical shift* arising from the interaction of the nuclear spin with the magnetic fields due to the electrons in the surrounding molecular orbitals. Here σ_j is the *shielding tensor* (see next section).
- The scalar coupling, namely, the indirect interaction of the magnetic moments of neighboring nuclear spins with one another through the electrons in the bonds between them.
- The dipolar coupling due to the direct interaction of the magnetic moments of neighboring nuclear spins.
- The quadrupolar interaction of a nuclear spin with $I > 1/2$ with the surrounding electric fields.
- The Knight/paramagnetic shift is similar to the chemical shift but is due to the conduction electrons in metals or unpaired electrons in radicals.

4.8.4 CHEMICAL SHIFTS

In molecules or in condensed matter (liquids or solids), nuclei are surrounded by electron clouds and neighboring nuclei that produce an internal magnetic field at the position of the nuclei, in addition to the applied external fields. These internal fields are characteristic of the local electron environment. The principal influence of the surrounding electrons is to produce a magnetic shielding of the applied magnetic field. The effect of the internal magnetic field is to produce what is known as chemical shifts in the NMR signals. In other words, the internal field results in a Larmor precession frequency, $\omega_0 = g_N \mu_N H$, that is slightly shifted, because the local field H is no longer just the external magnetic field, but contains contributions from the chemical environment of the nucleus. Hence, chemical shifts depend strongly on the environment and are typically of order of a few parts per million (ppm) in ^{1}H NMR, but several hundred ppm in ^{13}C NMR and ^{31}P NMR. The chemical shift in ppm is given by $\delta = 10^6 \times (\mu_{measured} - \mu_{reference})/\mu_{reference}$. The distribution of proton and carbon chemical shifts associated with different functional groups is shown in Figs. 4.13(a) and (b) respectively. The ranges of the chemical shifts are only approximate and may not encompass all compounds of a given class. For the proton shifts, the ranges specified for OH and NH protons are wider than those for most CH protons. This is due to hydrogen bonding variations at different sample concentrations. Figure 4.14 shows the NMR spectrum of the ethanol molecule, CH$_3$CH$_2$OH, which shows three chemically distinct hydrogen atom sites. Note the "intensity ratios" of 3:2:1 (the relative areas under the spectral peaks) corresponding to the number of protons in the group yielding the spectral peak. The additional splittings are due to spin–spin interactions with neighboring groups of protons (e.g., the methylene peak is split into a quartet by strong spin–spin interaction with the three protons of the methyl group). The proton in the OH group does not split the signals from the other groups (presumably because it is labile and hops around rapidly between ethanol molecules). In general, if a nuclear spin is coupled to n equivalent spin 1/2 particles, its peak is split into an $(n+1)$ multiplet with peak intensities within the multiplets proportional to the binomial coefficients $\binom{n}{r}$, $r = 0, 1, \ldots, n$. For example, the methylene protons can be exposed to the magnetic fields arising from the three methyl group protons that could have spin configuration $|\uparrow\uparrow\uparrow\rangle$, or the three configurations with two up-spins and one down-spin, or the three configurations with two down-spins and one up-spin, or $|\downarrow\downarrow\downarrow\rangle$. Hence, the splitting into four peaks with ratios 1:3:3:1.

Problem 4.36

Explain why it is possible to predict the proton NMR spectrum without considering the coupling of the two CH$_2$ protons with each other.

Answer: The coupling between the two CH$_2$ protons may be ignored because they are magnetically equivalent (i.e., they are the same isotopic species, there is a molecular symmetry operation that exchanges the two protons, and they have identical spin-dipolar couplings with all the other spins in the molecule). Hence, they both get shifted in exactly the same way as a result of their interaction, and no splitting results. The three CH$_3$ protons are also magnetically equivalent.

In solids, the environment of nuclei is rotationally anisotropic. Hence, shielding effects of the electron cloud around the nuclei have a tensorial character, reflecting the possibility that the field applied in one direction, say z, can result in an induced field along some other axis, say x. Thus, the Zeeman spin Hamiltonian should be generalized to $H_Z = -g_N \mu_N I_j \mathcal{O}_{jk} H_k$, where the tensor $\mathcal{O}_{jk}$ depends on the symmetry of environment of the nucleus in the solid.

4.8.5 FOURIER TRANSFORM NMR

NMR signals are often weak and in many cases are not substantially larger than the noise generated by the NMR spectrometer. An improvement in the ratio of the NMR signal to noise can be obtained by signal averaging over many experimental measurements, taking advantage of the fact that the noise contribution is random and therefore averages out. With n repetitions of the experiment, the signal will increase by n, whereas the noise will increase only by $\sqrt{n}$, i.e., the signal to noise will increase as $n^{1/2}$. However, this repetition of the experiment costs time.

4.8 Introduction to Magnetic Resonance

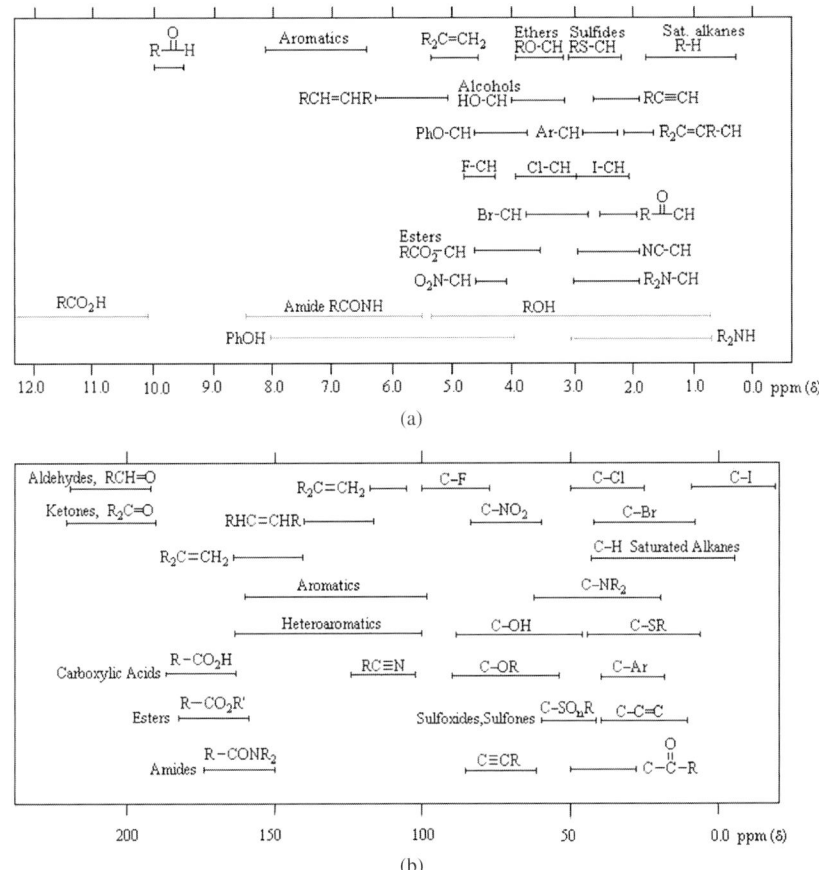

FIG 4.13 (a) Proton chemical shift ranges and (b) carbon chemical shift ranges for samples in CDCl$_3$ solution. The δ scale is relative to tetramethylsilane [Si(CH$_3$)$_4$] at $\delta = 0$. (*Reproduced with permission from Prof. William Reusch from* `http://www2.chemistry.msu.edu/faculty/reusch/VirtTxtJml/Spectrpy/nmr/nmr1.htm`

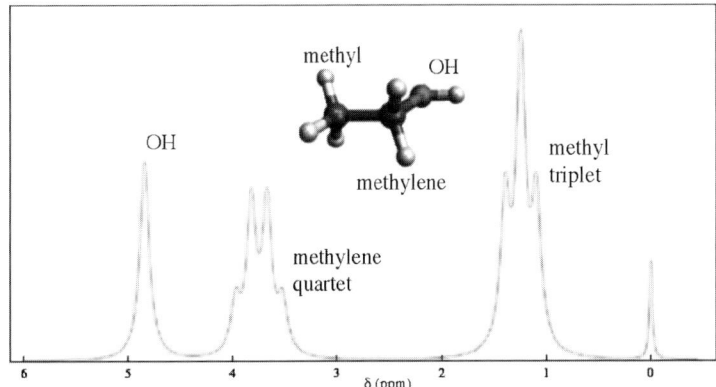

FIG 4.14 NMR spectrum of ethanol. The parameter δ is the fractional deviation of the chemical shift measured in parts per million from that of tetramethylsilane. (*Reproduced with permission from Prof. S. M. Blinder from* `http://demonstrations.wolfram.com/NuclearMagneticResonanceSpectrumOfEthanol`)

A cw NMR experiment requiring 1 Hz resolution over a 1000 Hz spectral width can require 1000 s to complete the measurement of the spectrum. However, such an experiment can be completed in 1 s with pulsed NMR. In pulsed NMR, short pulses of rf radiation with central frequency in the middle of the desired NMR spectral range are used, i.e., short pulses of temporal width τ_p and carrier frequency ω in the center of the desired range are applied to the sample. A frequency bandwidth $\Delta\omega$ roughly equal to the inverse of the pulse duration, $\Delta\omega \approx \tau_p^{-1}$, is thereby obtained. Applying such a pulse to a set of nuclear spins simultaneously excites all the NMR transitions within the bandwidth $\Delta\omega$ centered about the central frequency ω. Detailed treatment of Fourier transform NMR and such topics as magnetic resonance imaging (1D, 2D, and 3D Fourier imaging), solid-state NMR, and spin-echo techniques are outside the scope of this book. The interested reader is referred to Refs. [42].

Quantum Information   5

What is information? We understand this concept intuitively, but it is difficult to define rigorously. We could simply define information as a message received and understood or as a collection of data from which conclusions may be drawn. Information has an important role in physics, like energy and momentum, but until recently, its significance in physics was only alluded to through the concept of entropy.

Information Theory deals with the quantification of information and the methods for its efficient encoding for transmission and its detection and error correction due to transmission and reception problems. The new fields of *Quantum Information* and *Quantum Computation* (which can be considered a subfield of quantum information) are still in their infancy, and many outstanding questions remain. The central idea is to replace the classical bit, which can take on one of the two values, 0 or 1, by a quantum two-level system, described in terms of two orthonormal state vectors, $|0\rangle$ and $|1\rangle$, which span a two-dimensional Hilbert space containing all possible linear combinations, $a|0\rangle + b|1\rangle$, with $a, b \in \mathbb{C}$ and $|a|^2 + |b|^2 = 1$. A few of the many fundamental questions related to quantum information are: How does nature allow or prevent information from being expressed, manipulated, and secured? How does quantum mechanics affect information processing? For example, how do the uncertainty principle and the measurement collapse postulate affect quantum information processing? For quantum computing, what types of problems would a quantum computer be most useful at solving (in comparison with a classical computer)?

The study of information and computation using physics began with the analysis of the thermodynamic costs of elementary information processing by Landauer, Keyes, Bennett, and others during the 1960s and 1970s [43]. In the early 1980s, Benioff [44] and Feynman [45] started considering whether computation can be carried out on the scale of quantum physics (i.e., atomic length and energy scales), or, equivalently, whether systems that behave quantum mechanically can be used as information processing systems. Today, quantum information and quantum computation are thriving fields. The promise of fast algorithms within quantum computation, together with the practical implementation of concepts in quantum information and quantum cryptography, has stimulated widespread interest [46, 47]. The field of quantum information holds promise for a number of important applications. A short list includes *quantum computing*, protecting data from duplication (*counterfeit protection*), transmitting data from one party to another so that the data cannot be read by a third party (*key distribution/cryptography*), transmitting data from one party to another so that the receiver can be assured that the data was not corrupted during passage through the channel (*authentication*), transmitting data from one party to a second party so that a third party can later confirm that the second party did not alter it and that it was produced by the first party (*digital signature*), dividing data among n parties so that no $n - 1$ of them can reconstruct data, but all n working together can (*secret sharing*), and *quantum money*. To be more specific, we mention two applications of quantum information. The security of today's most common cryptographic systems used in banking systems, web browsers, etc. relies on the difficulty of factoring large numbers. The Shor quantum factorization algorithm [48] running on a quantum computer would compromise these systems because it turns factorization from a "hard" problem to an "easy" one. Another important application of a quantum computer is to simulate quantum systems [45]. Computational resources required for simulating quantum systems on classical computers grow exponentially with their size. Simulating a quantum system using another quantum system (or a quantum computer) should alleviate the necessity of using inefficient classically based simulation.

The topics presented in this chapter are not ordinarily treated in quantum mechanics courses and textbooks. However, the evergrowing importance of quantum information, its popularity, its breadth of concepts, and the growing research activity in this area during the past decade require exposing quantum mechanics students to this field.

This chapter starts with a brief review of classical information theory in Sec. 5.1, including the basic notions of a Turing machine, Shannon entropy, computational gates, classical cryptography, and computational complexity (readers well versed in the classical information theory can move directly to Sec. 5.2). Section 5.2 discusses the fundamental concepts of quantum bits (qubits), entanglement, Bell states, GHZ states, Schmidt decomposition, and mixed quantum states. This is followed by discussions of quantum gates and quantum circuits, which are then employed to treat such topics as dense coding, data compression, quantum teleportation, and quantum cryptography. Section 5.3 focuses on quantum computing. Several quantum computing procedures are analyzed, including the Deutsch, Deutsch–Jozsa, Grover, quantum Fourier transform, and Shor factorization algorithms. This section concludes with a discussion of quantum simulations. Decoherence is a major problem for quantum information processing and quantum computing, and this is the subject of Sec. 5.4. As with classical information processing and classical computing, error correction is an important issue for quantum information and computing; quantum error correction is considered in Sec. 5.5. Experimental systems that are promising candidates for realizing quantum information and computing protocols are discussed in Sec. 5.6. Finally, Sec. 5.7 introduces the EPR paradox, and Sec. 5.8 considers Bell inequalities.

5.1 CLASSICAL COMPUTATION AND CLASSICAL INFORMATION

The English mathematician, logician, cryptanalyst, and computer scientist, Alan Turing (1912–1954) was one of the founders of computer science. He built on the work of Charles Babbage (1791–1871), who conceived of most of the essential elements of a modern computer. Turing improved Babbage's computational engine to obtain what is now known as the *universal Turing machine*. Turing's machine was not meant to be a practical computing device but rather a concept of how computation can be performed; it clarified exactly what a calculating machine might be capable of [49]. The Turing machine is conceptually rich enough to address sophisticated mathematical questions, yet, is sufficiently simple to be subject to detailed analysis.

Alonzo Church's formulation of a computer, intertwined with Turing's, form a basis for the formal theory of computation known as the *Church–Turing principle*. It states that a Turing machine provides a precise definition of an algorithm, or methodological procedure, by which every function that would naturally be regarded as computable can be computed. A universal Turing machine consists of a string of data in the form of bits (strings of 0s and 1s), often called a "tape," that can be moved back and forth over an active element known as the read/write "head" that also possesses a property known as "state," and a "program" that is a set of instructions. The tape is the computer's storage medium, which can be used for input and output. The head can write either 0 or 1 on the tape at the position right underneath it. A set of instructions is supplied before the calculation begins, controlling how the head should modify the active tape data and move it. At each step of the calculation, the machine may modify the active tape data below the head to be 0 or 1, change the state of the head to be one of the states $q_1, \ldots, q_m$, and then move the tape one unit to the left or right. The states $q_1, \ldots, q_m$ of the head include two special states, q_s and q_h, called the starting state and the halting state. The head starts off in state q_s, and the calculation ends if the head enters state q_h. The Turing machine is the paradigm for investigating computability and many other concepts in computer science.

In 1948, Claude Elwood Shannon introduced a basic concept in digital communication, as used in computers, optical and magnetic storage media, and telecommunication systems, thereby laying the foundation for information theory. Building on Turing's ideas for a model computer that uses data in the form of 0s and 1s, he proposed to convert any kind of input data (pictures, sounds, text, etc.) into a string of bits (0 and 1) that could be sent along a wire. The amount of transmitted information is quantified in terms of the amount of disorder (i.e., entropy) contained in the data. Optimal communication of data is achieved by removing all redundancy (i.e., reaching the limit wherein the data stream is a completely random string of 0s and 1s). In addition to revolutionizing the field of communications, his information theory had a major effect on such diverse fields as genetics, computer science, code breaking, and neuroscience.

5.1.1 INFORMATION AND ENTROPY

It is important to quantify the amount of information in a message, for a variety of reasons. This can be carried out using Shannon's information theory. Before doing so, let us describe the format of messages that are to be sent. Consider a string (or sequence) of n symbols $(x_1, x_2, \ldots, x_n)$ that needs to be transferred from a sender to a receiver. These symbols could be composed from the English alphabet or the ASCII (American Standard Code for Information Interchange) character set, or it could be composed of bits that can take on the values 0 and 1. The length of the message, i.e., the number n

5.1 Classical Computation and Classical Information

of symbols in the sequence, may take any finite value (two extremes would be a two-letter word or the Encyclopedia Britannica), but we will assume that n is very large. Of course, each symbol may appear many times in a message. The number of *different* possible symbols, m, is a property of the alphabet. For example, $m = 31$ for ASCII and $m = 2$ for bits. The probability of appearance of different symbols in a given message is usually not the same. In a message written in English, the number of letters e is usually larger than the number of letters z. When the message is very large, we may speak of the probability of appearance of each symbol. Thus, in an English message, the probabilities of the appearance of the different letters are $p_a, p_b, \ldots$, whereas in a message composed of bits, we have just p_0 and p_1. The collection of different symbols $(x_1, x_2, \ldots, x_m)$, together with their probabilities $(p_1, p_2, \ldots, p_m)$, defines a *classical information source*. For example, if we assume that the information source contains bits 0 and 1 with probabilities $p_0 = 0.6$ and $p_1 = 0.4$, a string of 1000 bits is generated by sampling randomly from the source. Of the 1000 bits, about 600 bits should be 0s and about 400 should be 1s. The occurrence of a given symbol x_i in a string $(x_1, x_2, \ldots, x_n)$ with probability $p(x_i)$ in a message is referred to as an *event*. Thus, the appearance of each symbol 1 in the string of 1000 bits introduced above is an event that occurs with probability 0.4.

Once a message is generated, it can be transferred from sender to receiver, and it is often desired to compress the message. Data compression is a fundamental problem in information theory. Is it possible to transfer an n symbol message in a form that employs less than n symbols? For example, there is a great deal of redundancy in English that can be exploited for the purpose of compression; using shorter bit strings to represent high-probability symbols can cut down the length of a message. This problem was solved by Shannon [50], who quantified the amount of information in a message as the minimum communication resources needed to convey the message. This minimum is given by the *Shannon entropy*, which will be discussed later on in this section. Before formulating Shannon's analysis in Sec. 5.1.3, we need to clarify the notion of information content.

Information Content

There is a certain amount of information obtained by observing the occurrence of an event having probability p. The information, denoted as $i(p)$, should be defined in terms of the probability p. There are several desired properties (or axioms) to be satisfied by $i(p)$, listed below, together with their motivations.

1. $i(p) \geq 0 \Rightarrow$ Information is a nonnegative quantity.
2. $i(p = 1) = 0 \Rightarrow$ If an event has probability 1, there is no information gained in its occurrence.
3. For two *independent events*, e_1 and e_2, appearing with probability p_1 and p_2, respectively, $i(p_1, p_2) = i(p_1) + i(p_2) \Rightarrow$ the information gained from observing two independent events is the sum of the two pieces of information.
4. $i(p)$ is a continuous and monotonic function of p ($0 < p \leq 1$).

Problem 5.1

Show that (1) $i(p^2) = 2i(p)$ and (2) $i(p^k) = ki(p)$.

From these properties, it is clear that we can take

$$i(p) = -\log_b p \quad \text{(choosing the basis } b \text{ is a matter of convenience).} \tag{5.1}$$

Information Source and Random Variables

From the point of view of information storage and information transfer, all messages of length n containing symbols from the same source [specified by its m distinct symbols and m probabilities (x_i, p_i), $i = 1, 2, \ldots, m$] are equivalent; they have the same Shannon entropy (to be defined below). Thus, from an information storage and information transfer point of view, it is immaterial whether a message of length n is composed of well-formulated English text or a sequence generated by random sampling of ASCII symbols with the same probabilities $p_a, p_b, \ldots$ as they appear in a meaningful English text.

By using this equivalence, a *discrete information source* is modeled as a set of *independent identically distributed (i.i.d.) random variables* $X_1, X_2, \ldots, X_m$. A random variable X_i can be sampled to yield a symbol x_i with probability $p(x_i)$. In other words, the symbol x_i is sampled with probability $p(x_i) = p_i$. It is assumed that sampling of a symbol is independent of the previously sampled symbols. This is not always true; think of the combination of the letters "th" appearing in English. Nevertheless, we shall make this assumption. Generalizations to treat the case of correlations between symbols have been developed, but we shall not consider them here. We then have a probability distribution that characterizes the discrete source, $P(X_1, X_2, \ldots, X_m) \equiv P(X)$,

$$P(X) = \{p_1, p_2, \ldots, p_m\}, \quad p(x_i | \{x_{j<i}\}) = p(x_i) = p_i, \quad \sum_{i=1}^{m} p_i = 1. \tag{5.2}$$

The first equality gives the probability of the symbols, the second equality is a statement of independence of the probabilities, and the third equality is a condition required for probabilities. A string of n symbols $(x_1, x_2, x_3, \ldots, x_n)$ (usually with $n \gg m$) generated by (sampled from) the source forms a message.

5.1.2 SHANNON ENTROPY

Suppose we have a discrete information source that uses a set of m distinct symbols $(x_1, x_2, \ldots, x_m)$ from which we sample a string of symbols of length $n \gg m$ with probabilities $(p_1, p_2, \ldots, p_m)$. The corresponding information content is $[i(p_1), i(p_2), \ldots, i(p_m)]$. What is the average amount of information we get from each symbol in the string? If we observe the symbol x_i, we will obtain $i(p_i) = -\log p_i$ information. In a long string of n observations, we should expect to see $\approx np_i$ occurrences of x_i. Hence, for n independent observations, we expect that, on the average, we will get the total information,

$$\langle I_{\text{total}} \rangle = -\sum_{i=1}^{m} (np_i) \log p_i. \tag{5.3}$$

Thus, the average information obtained per observed symbol is

$$\langle I \rangle = \frac{1}{n} \langle I_{\text{total}} \rangle = -\sum_{i=1}^{m} p_i \log p_i. \tag{5.4}$$

A useful way to think about the average information is in terms of an expectation value. Let us denote the set of information contents belonging to a discrete source X as $I(X) = \{i(p_1), i(p_2), \ldots, i(p_m)\}$ and refer to it as the information content of the source. The expectation value $E[I(X)]$ is the *Shannon entropy*,[1]

$$\boxed{H[P(X)] = E[I(X)] = -\sum_{i=1}^{m} p_i \log p_i, \quad \sum_{i=1}^{m} p_i = 1.} \tag{5.5}$$

Shannon formulated his theory for a source of two symbols ($m = 2$), $x_1 = 0$ and $x_2 = 1$ with probabilities $p(x_1) = p$ and $p(x_2) = 1 - p$, respectively. In this case, the Shannon entropy is

$$H[p, 1-p] = -[p \log_2 p + (1-p) \log_2 (1-p)]. \tag{5.6}$$

The Shannon entropy H versus p is plotted in Fig. 2.1 (where it is denoted by the symbol $\mathcal{S}$).

---

[1] Note the similarity of Shannon entropy [Eq. (5.5)] and von Neumann entropy of a density matrix, $\mathcal{S}(\rho) = -\text{Tr} \rho \log_2 \rho$ defined in Sec. 2.5.2. If we take a density matrix of the form $\rho = \sum_i |\phi_i\rangle p_i \langle \phi_i|$, the entropy is given by $\mathcal{S} = -\sum_i p_i \log_2 p_i$, which is clearly related to Eq. (5.5).

5.1 Classical Computation and Classical Information

> **Problem 5.2**
>
> (a) Prove that $H(p, 1-p) \leq 1$ and that equality holds only for $p = 1/2$.
> (b) Prove that $H(p, 1-p)$ is a concave function of p.
>
> **Hint:** Show that $H'' \leq 0$.

By using the entropy concept, Shannon was able to obtain important results regarding information storage and transfer and laid the foundations for contemporary information, coding, and communication theory. He developed a general model for communication systems and a set of theoretical tools for analyzing such systems. The basic model consists of three parts: a sender (or source), a channel, and a receiver. It also includes encoding and decoding elements and noise within the channel. The problems of data compression and noisy channels (see Sec. 5.1.3 below) are then addressed. In particular, it was shown that information contained in a string of n symbols (bits in this case) can be compressed to nR bits and still be completely recovered. Here, $0 \leq H(X) \leq R \leq 1$ (recall that $H \leq 1$ as proved in Problem 5.2). In this sense, we speak of *reliable compression scheme*, and R is referred to as *the compression rate*. This quantity will be employed in Sec. 5.1.3, where we formulate Shannon noiseless channel encoding theorem.

Although Shannon demonstrated the possibility of information compression, explicit methods for doing so were only developed much later. Numerous methods have been developed [51] (Chapter 20), including the *Huffman coding* method, which is an entropy coding that is commonly used in final stages of compression, the *arithmetic coding* method, which is a variable-length entropy encoding used in lossless data compression that encodes a message into a single number f, $0 \leq f < 1$, and the *Lempel–Ziv algorithm*, which is a variable-to-fixed length coding method that is used in the "compress" utility in Unix operating systems and in GIF image format files.

Properties of Shannon Entropy

Some properties of Shannon entropy for the general case (not necessarily for a source of bits) are listed below.

1. The Shannon entropy $H(X)$ is a continuous function of p_i. If all p_i are equal, $p_i = \frac{1}{m}$, then H is maximal.
2. The Gibbs inequality:

$$H(X) \leq \log m, \text{ with equality iff } \quad p_i = \frac{1}{m}, \; i = 1, 2, \ldots, m. \tag{5.7}$$

3. H is monotonically increasing with m:

$$H_m(p_1, p_2, \ldots, p_m) \leq H_m\left(\frac{1}{m}, \frac{1}{m}, \ldots, \frac{1}{m}\right) \leq H_{m+1}\left(\frac{1}{m+1}, \frac{1}{m+1}, \ldots, \frac{1}{m+1}\right). \tag{5.8}$$

4. The "closeness" of two probability distributions, $P(X)$ and $Q(X)$, of the same random variable X can be measured by their *relative entropy* defined by

$$H[P(X) \,||\, Q(X)] = \sum_{i=1}^{m} p(x_i) \log_2 \frac{p(x_i)}{q(x_i)} = -H[P(X)] - \sum_{i=1}^{m} p(x_i) \log_2 q(x_i). \tag{5.9}$$

5. The relative entropy is nonnegative, $H[P(X) \,||\, Q(X)] \geq 0$, with equality only for $P(X) = Q(X)$. This follows from the inequality $\ln x = \log_2 x / \ln 2 \leq x - 1$ for all positive x, with equality if and only if $x = 1$. Rearranging the last inequality yields $-\log_2 x \geq (x-1)/\ln 2$, and applying this inequality gives

$$H[P(X) \,||\, Q(X)] = -\sum_{i=1}^{m} p(x_i) \log_2 \frac{q(x_i)}{p(x_i)} \geq \frac{1}{\ln 2} \sum_{i=1}^{m} p(x_i)\left[1 - \frac{q(x_i)}{p(x_i)}\right] = 0. \tag{5.10}$$

Equality results only if $p(x_i) = q(x_i)$ for all i.

6. For two different random variables, X and Y, the information content associated with their joint distribution $P(X, Y)$ is given by the *joint entropy*,

$$H[P(X, Y)] = -\sum_{i=1}^{m}\sum_{j=1}^{n} p(x_i, y_j) \log_2 p(x_i, y_j). \tag{5.11}$$

The joint entropy measures the information content in the pair (X, Y).

7. The *conditional entropy* of X knowing Y is defined by

$$H[P(X|Y)] = H[P(X, Y)] - H[Q(Y)] = -\sum_{i=1}^{m}\sum_{j=1}^{n} p(x_i, y_j) \log_2 \frac{p(x_i, y_j)}{q(y_j)}. \tag{5.12}$$

The conditional entropy is a measure of the uncertainty of X, given that we know the value of Y.

8. The *mutual information content* of X and Y measures how much information X and Y have in common:

$$H[P(X) : Q(Y)] = H[P(X)] + H[Q(Y)] - H[P(X, Y)]. \tag{5.13}$$

Problem 5.3

Prove that $H(X) \leq \log m$ and that equality holds if $p_i = 1/m$ (note that this includes an answer to Problem 5.2 as a special case).

Guidance: Show that $H(X) - \log m = \sum_{i=1}^{m} p_i \log \frac{1}{np_i} \leq 0$.

Problem 5.4

Prove the following relationships.

(a) $H[P(X) : Q(Y)] = H[P(X)] - H[P(X|Y)]$.
(b) $H[P(X) : Q(Y)] = H[Q(Y) : P(X)]$.
(c) $H[P(Y|X)] \geq 0$.
(d) $H[P(X) : Q(Y)] \leq H[Q(Y)]$.
(e) $H[P(X) : Q(Y)] = H[Q(Y)]$ if $Y = f(X)$.
(f) $H[P(Y|X)] \leq H[Q(Y)]$ and therefore $H[P(X) : Q(Y)] \geq 0$.

5.1.3 DATA COMPRESSION

Now, we are in a position to discuss data compression. What is the minimum number of bits needed to store or send a given piece of information? This is a fundamental question in information theory. Data compression, sometimes also called *source coding* in computer science, is the encoding of information using fewer bits than in an unencoded representation of the information. Many computer users employ the *zip* file format to compress files; also used are Huffman coding, arithmetic coding, and the Lempel–Ziv algorithm, which are mentioned in Sec. 5.1.1. Data compression of classical information relies on the Shannon quantification of the amount of information in a message [50] and Shannon entropy, which specifies the minimum number of bits needed to convey a message. It is summarized in Shannon's noiseless channel coding theorem, formulated below. With the advent of quantum information theory, data compression also becomes relevant for quantum information. The analogous analysis for quantum information is encoded in Schumacher's noiseless channel coding theorem (see Sec. 5.2.6).

5.1 Classical Computation and Classical Information

Data Compression in Classical Information Theory

Let us consider a string of n bits generated by (sampled from) a classical information source. We want to compress it into a string of nR bits ($0 < R < 1$) and send it to a receiver for decompression, so that we fully retrieve the initial information. In the following, we assume an information source composed of two symbols (bits) with the random variable $X = (X_1, X_2)$ determining a bimodal distribution $p(0) = p$ and $p(1) = 1 - p$. The idea of classical data compression is to distinguish between n bit sequences that are highly probable (referred to as *typical sequences*) and n bit sequences that are not likely (referred to as *atypical sequences*). For large n, an n bit sequence has $\approx pn$ 0 bits and $\approx (1-p)n$ 1 bits, and with the i.i.d. assumption on the source, the probability for the occurrence of this sequence is

$$p(x_1, x_2, \ldots, x_n) = \prod_{i=1}^{n} p(x_i) = p^{np}(1-p)^{n(1-p)} \approx 2^{-nH(X)}, \tag{5.14}$$

where $H(X)$ given in Eq. (5.6) is the Shannon entropy for this bimodal distribution. According to Shannon's theorem (formulated below), $H(X)$ is the lower bound on the compression rate R. Hence, for every number R such that $0 \leq H(X) \leq R \leq 1$, information encoded in an n bit sequence can be reliably compressed into a sequence containing $nR \leq n$ bits and then sent to a receiver for decompression and full recovery. Since $H(X)$ is smaller the farther p is from $1/2$, the compression becomes more and more effective as p is farther from $1/2$. The fact that we need only nR bits to achieve reliable data compression shows that of 2^n possible n bit sequences, there are at most 2^{nH} typical n bit sequences (appearing with high probability) and the rest are atypical.

Example: Suppose we have a message encoded in a string of 10^6 bits that need to be communicated. We want to reduce the number of bits in a message, $n \to nR$. If $p = 1/2$, the two possible outcomes, 0 and 1, occur with equal probability, and each outcome requires one bit of information to transmit. To send the entire sequence, we will require one million bits. Now, suppose the distribution is non-uniform, $p \neq (1-p)$, and for specificity, suppose $p(1) = 1/1000$. In a string of 10^6 bits, there will be about 10^3 1s. Rather than transmitting the results $\{x_n\}$ of each bit, we can just transmit the numbers 1s; the rest of the numbers are 0's. Each 1 has a position in the sequence: a number between 1 and 10^6. Specifying a single position requires about 20 bits, hence transmitting 10^3 20 bit numbers exhausts all information content using only around 2×10^4 bits. This already gives $R = 2 \times 10^4/10^6 = 0.02$.

Further improvement is achieved by noting that instead of encoding the absolute positions of the 1s, we can just specify the distance to the next 1, which takes fewer bits. On the average, the distance between two 1s will be around 10^3 positions (only rarely will the distance exceed 4×10^3 positions. Numbers in the range 1 to 4,000 can be encoded in 12 bits. Hence, a sequence of one million bits containing about 1,000 1s can be transmitted in just 12,000 bits, on average. This already gives $R = 1.2 \times 10^4/10^6 = 0.012$. Shannon theorem asserts that a lower bound for R is $R \geq H(X) = -0.001 \log_2(0.001) - 0.999 \log(0.999) = 0.0114$.

Problem 5.5

Work out the same procedure for $p = 1/200$ and compare your result for R with the Shannon bound.

The methods and techniques for designing an efficient compression algorithm will not be presented here. We end this section by presenting Shannon's theorem without proof.

Shannon's Noiseless Channel Coding Theorem. Given a classical information source characterized by m i.i.d. random variables $X = \{X_i\}$ whose entropy is $H(X)$. There exists a reliable compression scheme of rate $R > H(X)$ for the information source, whereas for $R < H(X)$, any compression scheme will not be reliable.

Error Correction of Classical Information. In classical information processing, data is transferred as a sequence of bits, but given a data stream that is to be sent, the data is often broken up into bytes; 1 byte = 8 bits. To alleviate errors, data is sometimes repeated or resent more than once. Error correction and detection are of tremendous importance in maintaining data integrity across noisy channels. An often-used form of error correction involves setting a parity bit. The number of 1 bits in a block of data is counted and then a "parity bit" after the data block is set or cleared if the number of 1 bits is odd or even. The parity bit can be used to detect an error in the transmitted data, and if detected, the data can be re-sent. Error correction is also a central issue in quantum information processing (see Sec. 5.5).

5.1.4 CLASSICAL COMPUTERS AND GATES

Classical computers operate on bits of 0s and 1s. A (classical) *gate* is an operator acting on a given ordered sequence of k bits $(b_1, b_2, \ldots, b_k)$ (the input), which results in an ordered sequence of l bits $(\beta_1, \beta_2, \ldots \beta_l)$ (the output), with $k \geq l$,

$$G(b_1, b_2, \ldots, b_k) = \beta_1, \beta_2, \ldots \beta_l, \quad b_i = 0, 1, \quad \beta_i = 0, 1. \tag{5.15}$$

The simplest gate is the NOT gate. It is a single-bit operation ($k = l = 1$) negating the input bit, i.e., $0_{in} \to 1_{out}$ and $1_{in} \to 0_{out}$ (see Problem 5.17 and Table 5.1). There are several gates with $k = 2$ and $l = 1$. The AND gate with $k = 2$ and $l = 1$ has 1_{out} only if both the inputs to the gate are 1_{in}; if neither input or only one input to the gate is 1_{in}, a 0_{out} output results. An OR gate has 1_{out} if one or both input bits are 1_{in}; if neither input is 1_{in}, a 0_{out} output results. The NAND gate has output 0_{out} only if both the inputs to the gate are 1_{in}; if one or both inputs are 0_{in}, a 1_{out} output results. Therefore, it is called a NAND gate, which stands for (NOT × AND). Clearly, it is equivalent to using an AND and then a NOT gate. A NOR gate has output 1_{out} only if both the inputs to the gate are 0_{in}. Tables 5.1–5.5 show the truth table for these gates.

Table 5.1 *NOT gate.*

Input	Output
0	1
1	0

Table 5.2 *AND gate.*

Input$_1$	Input$_2$	Output
0	0	0
0	1	0
1	0	0
1	1	1

Table 5.3 *OR gate.*

Input$_1$	Input$_2$	Output
0	0	0
0	1	1
1	0	1
1	1	1

Table 5.4 *NAND gate.*

Input$_1$	Input$_2$	Output
0	0	1
0	1	1
1	0	1
1	1	0

Table 5.5 *NOR gate.*

Input$_1$	Input$_2$	Output
0	0	1
0	1	0
1	0	0
1	1	0

5.1 Classical Computation and Classical Information

Problem 5.6

Show that the gates described above can be written in terms of the notation of Eq. (5.15) and denoting addition and multiplication modulo 2 by $\oplus$ and $\otimes$, as follows:
NOT gate: $G(b) = b \oplus 1$
AND gate: $G(b_1, b_2) = b_1 \otimes b_2$
OR gate: $G(b_1, b_2) = b_1 \oplus b_2 \oplus b_1 \otimes b_2$
NAND gate: $G(b_1, b_2) = b_1 \otimes b_2 \oplus 1$
NOR gate: $G(b_1, b_2) = b_1 \oplus b_2 \oplus b_1 \otimes b_2$.

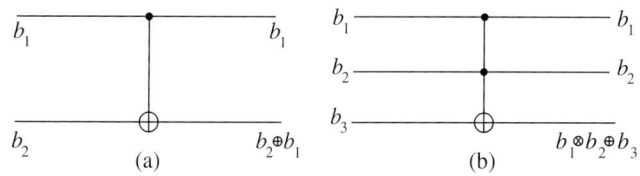

FIG 5.1 (a) Schematic circuit for a CNOT gate. The control bit x controls the output of bit y. (b) The Toffoli gate, where the first two bits control the output of the third bit.

The controlled-NOT or CNOT (sometimes written as C-NOT or as XOR) gate flips the second bit if and only if the first bit is 1_{in}. It is an example of a classical gate with $k = l = 2$. Figure 5.1(a) shows a schematic circuit for a CNOT gate; the control bit x is not affected, whereas the y bit undergoes NOT provided the x bit is turned "on." Table 5.6 shows the CNOT gate. Formally, it is written as,

$$\text{CNOT gate: } G(b_1, b_2) = b_1, b_1 \oplus b_2. \quad (5.16)$$

Within the group of gates with $k = l = 3$, an important reversible multiple input–output logic gate is the Toffoli gate, which is sometimes called the "controlled-controlled-not" gate, introduced by Tommaso Toffoli in 1980:

$$\text{Toffoli gate: } G(b_1, b_2, b_3) = b_1, b_2, b_1 \otimes b_2 \oplus b_3. \quad (5.17)$$

If the first two bits are set to 1_{in}, this gate flips the third bit [see Fig. 5.1(b)]. Table 5.7 shows the input and the output of a Toffoli gate. Any reversible function can be computed on a classical computer as a concatenation of the Toffoli gate on different inputs. For example, the logical AND gate on b_1 and b_2 can be obtained if we input $b_3 = 0$; the last bit will contain $b_1 \otimes b_2 \oplus 0 = \text{AND}(b_1, b_2)$. To implement the NOT gate on the third bit, set the first two bits to be 1.

The gates discussed above are implemented in classical computers and classical information processors. Over the past 50 years, there has been amazing growth in the power of computers using silicon-based integrated circuits. The rate of this progress is summarized in "Moore's law," attributed to Intel co-founder Gordon Moore, which states that the power of computational devices has roughly doubled every 18 months or increased 10-fold every 5 years. Unfortunately, this growth rate may not continue for much longer, as this increase in computing power requires a corresponding decrease in the size of the transistors on the chip, and this shrinking process cannot continue indefinitely since the transistors will eventually reach the atomic size scale. It has been estimated that this limit will be reached by about 2012, and further progress will require a conceptually different approach.

Table 5.6 *CNOT gate.*

Input$_1$	Input$_2$	Output$_1$	Output$_2$
0	0	0	0
0	1	0	1
1	0	1	1
1	1	1	0

Table 5.7 *Toffoli gate.*

Input$_1$	Input$_2$	Input$_3$	Output$_1$	Output$_2$	Output$_3$
0	0	0	0	0	0
0	0	1	0	0	1
0	1	0	0	1	0
0	1	1	0	1	1
1	0	0	1	0	0
1	0	1	1	0	1
1	1	0	1	1	1
1	1	1	1	1	0

5.1.5 CLASSICAL CRYPTOGRAPHY

Cryptography (in Greek, "crypto" means secret and "graphy" means writing) is the study of secret writing. The art and science of cryptography is concerned with developing algorithms to conceal the context of a message from all except the intended recipient(s) and/or verification of correctness of a message received by the recipient (i.e., authentication of the message and the sender). Cryptography has a 4000-year history. The ancient Egyptians enciphered some of their hieroglyphic writing, Julius Caesar used a replacement algorithm that is now named after him to encode messages (see below), certain words in the Bible are enciphered, the Enigma Rotor machine was a class of cipher machines that was heavily used to encipher and decipher messages during World War II, and so on. A relevant modern example of a secret communication you may wish to carry out is giving your credit card number to a merchant over the internet to make a purchase, hopefully without any malevolent third party intercepting your credit card number. The way this is done is to use a cryptographic protocol. Nowadays, cryptography is the basis of many technological solutions to communications security problems arising in modern computer and information storage/retrieval networks. *Cryptanalysis* or *codebreaking* are the principles and methods of transforming an unintelligible message enciphered by the sender back into an intelligible message *without* prior knowledge of the encoding algorithm used for concealment.

The sender typically starts off with *plaintext*, i.e., the original intelligible message that he wants to send to the recipients in such a fashion that others will not be able to decipher the message should it fall into their hands. Therefore, the sender creates *ciphertext*, which is the transformed message that is unintelligible, using a *cipher*, i.e., an algorithm for transforming the original plaintext into ciphertext by transposition, substitution methods, etc. In doing so, the sender uses a *key*, which is a critical part of the algorithm of the cipher, that is known only to the sender and receiver(s). The sender is said to *encipher* or *encode* the plaintext into ciphertext using a cipher and a key. The receiver must *decipher* or *decode* the ciphertext back into plaintext using a cipher and a key if he has to understand the message. If P denotes the plaintext, C the ciphertext, and let the symbol $\mathcal{T}(K)$ denote the family of invertible transformations that transform the plaintext into ciphertext, i.e., given the *cryptographic system*, which depends on the key K, the processes of encryption and decryption can be represented symbolically as:

$$C = \mathcal{T}(K)P, \quad P = \mathcal{T}^{-1}(K)C. \tag{5.18}$$

The cryptographic system (the family of transforms $\mathcal{T}$) can be public or private, but the key K is secret. A private-key or secret-key encryption algorithm is one where the sender and the recipient share a common but secret key K.

As an example of a cryptographic algorithm, let us consider the Caesar cipher that was referred to above. This is a truly simple cipher. The algorithm is: replace each letter of a message by a letter at a fixed distance away from it in the alphabet. For example, if we replace each letter by the sixth higher letter in the alphabet, we map the alphabet as follows:

ABCDEFGHIJKLMNOPQRSTUVWXYZ
GHIJKLMNOPQRSTUVWXYZABCDEF

The encryption transformation is

$$\mathcal{T}(K) : i \rightarrow i + K (\text{mod } 26),$$

5.1 Classical Computation and Classical Information

and the decryption algorithm is

$$\mathcal{T}^{-1}(K) : i \to i - K \pmod{26},$$

where in our special case, $K = 7$. Thus,

"I CAME I SAW I CONQUERED"

is transformed into

"P JHTL P ZHC P JVUXBLYLK."

To add complexity, one could also transpose the letters of the ciphertext, i.e., write the letters backward as

"KLYLBXUVJ P CHZ P LTHJ P".

Cryptanalysis of the Caesar cipher can be carried out by trying the replacement of each letter by another letter. This leads to $26! = 4.032914 \times 10^{26}$ possibilities. Alternatively, if we know that the algorithm is a Caesar cipher, we need to try only 26 possibilities.

One of the important tools in cryptanalysis is *letter frequency analysis*. Languages have a distribution of the frequency of occurrence of letters that are not uniform. For example, the letter "e" in English is the most commonly occurring letter. One can set up a table of frequency of occurrence of letters, of double letters (e.g., th, he, in, er) and triple letters, and use these tables to analyze sufficiently long ciphertext. To do this, one compares the frequency of occurrence of single letters, double letters, etc., that occur in the ciphertext with that of the language one believes the message is in.

The one known truly unbreakable cipher was devised in 1918 by Gilbert Vernam and Joseph Mauborgne and is called either a *one-time pad* or a *Vernam cipher*. The original design and the modern version of one-time pads are based on the binary alphabet. The plaintext is converted into a sequence of 0s and 1s, using some publicly known rule, e.g., the ASCII binary-equivalent representation. The key is another sequence of 0s and 1s of the same length or longer than the plaintext. Each bit of the plaintext is combined with the respective bit of the key, according to the rules of addition in base 2, i.e., $0 + 0 = 0, 0 + 1 = 1, 1 + 0 = 1$, and $1 + 1 = 0$. Since the key is a random sequence of 0s and 1s, the resulting ciphertext is also random and completely scrambled unless one knows the key. The plaintext can be recovered by adding the cryptogram and the key in base 2 and reusing the publicly known rule.

In the context of the alphabetic coding of

"I CAME I SAW I CONQUERED,"

one would need a one-time pad of at least 19 letters, chosen at random, to be added to the 19 letters of the plaintext (modulo 26) to encrypt the message. This same one-time pad could then be used to "subtract" from the ciphertext to obtain the original message.

One-time pads suffer from a serious practical limitation known as the *key distribution problem*. Potential users have to agree secretly and in advance on the key, i.e., the long, random sequence of 0s and 1s used to decipher the message. Once this is done, the key can be used for enciphering and deciphering, even if the resulting cryptograms are publicly transmitted. However, the key must be established between the sender and the receiver by means of a secure channel (whatever that means—see below). Users that are far apart, to guarantee perfect security, have to already be in possession of the cryptographic key, equal in size to all the messages they might later wish to send. Moreover, even if a "secure" channel is available, this security can never be guaranteed. In principle, any *classical* private channel can be monitored without the sender or receiver knowing that the eavesdropping has taken place.

So far, we have been considering *private key cryptosystems*, i.e., the two parties that wish to communicate share a private key that only they know. A second type of cryptosystem is the so-called *public key cryptosystem*. Public key cryptography does not rely on sharing a secret key in advance. Instead, the receiver of the message(s) simply publishes a public key, which is made available to the general public. The sender uses this public key to encrypt a message and sends it to the receiver. A third party cannot use the public key to decrypt the message because the encryption transformation is chosen, so that it is extremely difficult to invert, given only knowledge of the public key. To make inversion easy, the receiver has a secret key matched to the public key, which together enable him to easily perform the decryption. The secret key is known only to the receiver. Public key cryptosystems solve the key distribution problem by making it unnecessary for a shared private key to be distributed. An example of a public key cryptosystem is the *RSA algorithm* developed by Ronald Rivest, Adi Shamir, and Leonard Adleman in 1997. The decryption stage of RSA is closely related to factorization (see http://en.wikipedia.org/wiki/RSA). The security of RSA arises from the fact that with today's factoring algorithms running on a classical computer, factoring is hard (see Sec. 5.1.6). However, Shor's factorization algorithm (running on a quantum computer) would make RSA virtually useless. This application of quantum computers for breaking of public key cryptographic systems has made quantum computation a very attractive subject.

5.1.6 COMPUTATIONAL COMPLEXITY

Computers are designed and used to solve many distinct kinds of problems using a large variety of calculation techniques; some of these problems are computationally easy, whereas some are exceptionally hard. Computational problems can be classified according to the difficulty of their solutions. In computer science, the notion of *difficulty* is rigorously defined within the theory of computational complexity, which describes the scalability of algorithms that solve computational problems. Specifically, as the size of the input to an algorithm increases, at what rates do the computing resources (run time and memory requirements) grow.

An example of a computational problem that is (thought to be) computationally difficult is the *factoring* (or factorization) problem: given an (odd) integer, determine its prime factors.[2] The factorization problem cannot be solved efficiently by any known classical computing algorithm. The computational effort grows exponentially with the size of the integer to be factored. Yet, within the theory of computational complexity, it has not been proved to be exponentially hard. When Shor published his algorithm for factorization using a quantum computer in 1997 [48] (see Sec. 5.3.4), the field of quantum computing was set ablaze. The hope that the nature of quantum computational complexity would be different (better) than that of classical computational complexity was kindled. It is unclear whether this hope will come to fruition.

In the theory of computational complexity, problems and algorithms are divided into complexity classes. Let us consider *decision problems* that take some string as an input and require either YES or NO as an output. If there is an algorithm that runs on a Turing machine, which is able to produce the correct answer for any input string of length L in at most $c L^k$ steps, where k and c are some constants independent of the input string, then the problem is said to be solvable in *polynomial time* and is in the class called *deterministic polynomial time* (P). The problems in this class can be solved by a deterministic Turing machine in polynomial time. The class *non-deterministic polynomial time* (NP) consists of all those decision problems whose solutions can be *verified* in polynomial time, i.e., problems in this class with the answer YES (NO) have simple and fast proofs that the answer is indeed YES (NO). Yet, there is no efficient way to determine a solution to such problems. In an equivalent but alternative definition, NP is the set of decision problems solvable in polynomial time by a non-deterministic Turing machine.[3] One of the most important open questions of complexity theory is whether the complexity class P is the same as NP, or whether it is only a subset of NP, as generally believed. If the answer to this equation is affirmative, then NP problems can also be computed in polynomial time. Finally, the class *NP-complete* (NPC) includes the most difficult problems in NP. Class NPC is the smallest subclass of NP that could remain outside P. A decision problem is in NPC if (1) it is in NP and (2) it is NP-hard, i.e., every other problem in NP is reducible to it (reduction is a transformation of one problem into another problem). Then it is said to be complete for NP. In 1970s, Stephen A. Cook, Richard Karp, and Leonid Levin proved that *if* an efficient algorithm for any NPC problem was found, it could be adapted to solve all other NP problems, i.e., that all problems in NP would actually be in the class P. For example, if we had a polynomial time algorithm for solving a problem in class NPC, we could solve all NP problems in

---

[2] The fundamental theorem of arithmetic states that every positive integer greater than 1 has a unique prime factorization.

[3] The term non-deterministic refers to a non-deterministic Turing machine. A deterministic Turing machine performs a given calculation with certainty, but an element of randomness is present in a non-deterministic (or probabilistic) Turing machine. The machine can execute several operations with a given probability for each. The need to modify the original Turing machine emerged when it was realized that numerous decision problems cannot be solved with certainty in polynomial time but can be answered with high probability using the notion of randomized algorithms. This requires a simple modification of the strong Church–Turing principle, so that it states that any algorithmic process can be simulated efficiently using a probabilistic Turing machine.

polynomial time. An example of an NPC problem is the "traveling salesman problem": Given a number of cities located on points in a map, what is the shortest round-trip route that visits each city exactly once and then returns to the starting city? This question has immediate relevance to many optimization problems encountered in reality, but unfortunately, no known polynomial time solution exists for this. As of today, an efficient algorithm for solving an NPC problem has not been found, and many computer scientists believe that P ≠ NP. Thus, only one hope remains for solving NP problems in polynomial time and that is to broaden the scope of what we mean by a computer.

5.2 QUANTUM INFORMATION

Quantum information deals with information stored and manipulated subject to the laws of quantum mechanics. The fundamental building block of quantum information is the two-level system called a quantum bit or *qubit*. In principle, it is possible to use three-level quantum systems (sometimes called *qutrits*, see Sec. 6.2), or larger multilevel quantum systems, instead of qubits, but most quantum information studies are carried out with qubits. In this section, we first introduce the notion of qubit, discuss the fundamental concept of qubit entanglement and then introduce the operations on qubits in terms of unitary transformations referred to as *quantum gates*. Finally, the concept of *quantum circuits* is introduced.

5.2.1 QUBITS

The elementary information unit in classical computers is the bit, which can take a value of either 0 or 1. The quantum analog of a bit is a two-state system called a quantum bit or *qubit*. A two-state quantum system is described in terms of a two-dimensional Hilbert space $\mathcal{H}^{(2)}$, for which we can define an orthonormal basis consisting of two vectors, denoted by $|0\rangle$ and $|1\rangle$ in Dirac notation and identified in two-component spinor notation as,

$$|0\rangle \equiv \begin{pmatrix} 1 \\ 0 \end{pmatrix}, \quad |1\rangle \equiv \begin{pmatrix} 0 \\ 1 \end{pmatrix}. \tag{5.19}$$

An arbitrary qubit is any unit vector in $\mathcal{H}^{(2)}$,

$$|\chi\rangle = a|0\rangle + b|1\rangle \equiv \begin{pmatrix} a \\ b \end{pmatrix}, \tag{5.20}$$

where the complex numbers a and b are such that $|a|^2 + |b|^2 = 1$. A possible physical realization of a qubit as two-state system is a spin 1/2 particle, wherein state $|0\rangle$ would be the up-state $|\uparrow\rangle$ and state $|1\rangle$ would be the down-state $|\downarrow\rangle$. Analysis of two-level systems in quantum information often uses the same nomenclature as that of spin 1/2 particles. However, note that numerous other physical realizations of two-state systems have been considered, including ultracold atoms, ions, nuclei with spin 1/2 (in the context of NMR quantum computers), quantum dot systems, superconducting quantum interference devices, etc.

Problem 5.7

(a) Using the fact that $\langle 0|0\rangle = \langle 1|1\rangle = 1$ and $\langle 0|1\rangle = \langle 1|0\rangle = 0$, show that the commutator
$[\,|0\rangle\langle 1|, |1\rangle\langle 0|\,] = |0\rangle\langle 0| - |1\rangle\langle 1|$.

(b) Write the equation in (a) as a 2×2 matrix equation using the representation (5.19).

A quantum computer, or a quantum information processor, requires a large number of two-state systems. When the system consists of n qubits, the corresponding Hilbert space is the tensor product $\mathcal{H} = \bigotimes_{i=1}^{n} \mathcal{H}^{(2)}(i)$, and a basis in $\mathcal{H}$ consists of 2^n vectors. In the present framework, qubits are distinguishable, i.e., they need not be symmetrized multi-qubit states (but *could* be symmetrized), since we know which qubit is which. In the notation used in quantum information, a convenient basis, sometimes called *the computational basis* is

$$\left.\begin{array}{l} |0\rangle \otimes |0\rangle \otimes \ldots \otimes |0\rangle \otimes |0\rangle \\ |0\rangle \otimes |0\rangle \otimes \ldots \otimes |0\rangle \otimes |1\rangle \\ |0\rangle \otimes |0\rangle \otimes \ldots \otimes |1\rangle \otimes |0\rangle \\ \quad \vdots \\ |0\rangle \otimes |1\rangle \otimes \ldots \otimes |1\rangle \otimes |1\rangle \\ |1\rangle \otimes |0\rangle \otimes \ldots \otimes |0\rangle \otimes |0\rangle \\ \quad \vdots \\ |1\rangle \otimes |1\rangle \otimes \ldots \otimes |0\rangle \otimes |1\rangle \\ |1\rangle \otimes |1\rangle \otimes \ldots \otimes |1\rangle \otimes |0\rangle \\ |1\rangle \otimes |1\rangle \otimes \ldots \otimes |1\rangle \otimes |1\rangle \end{array}\right\} 2^n \text{ basis states.} \quad (5.21)$$

The ordering of these basis states from top to bottom in Eq. (5.21) is similar to the consecutive integers in binary arithmetic. Each basis state $|j_1 j_2 \ldots j_{n-1} j_n\rangle$, $(j_i = 0, 1)$ in Eq. (5.21) can be compactly represented by a ket $|j\rangle$ ($j = 0, 1, \ldots, 2^n - 1$) whose binary expansion is

$$|j\rangle = |j_1 2^{n-1} + j_2 2^{n-2} + \cdots + j_{n-1} 2^1 + j_n 2^0\rangle \equiv |j_1 j_2 \ldots j_{n-1} j_n\rangle. \quad (5.22)$$

In many quantum operations on qubits (e.g., in the quantum Fourier transform), the number j appears in arithmetic expressions, and therefore, it is more convenient to use its decimal expansion. Nevertheless, since no confusion should arise, we denote any n-qubit basis state in Eq. (5.21) either as $|j\rangle$, where $0 \leq j \leq 2^n - 1$ is a decimal number, or as $|j_1 j_2 \ldots j_{n-1} j_n\rangle$, where $j_i = 0, 1$, taking into account the correspondence between them as specified in Eq. (5.22).

We can "run" algorithms using initial states, $|\Psi_{\text{in}}\rangle \in \mathcal{H}$, represented by tensor products of superposition of single qubit states: $|\Psi_{\text{in}}\rangle = \bigotimes_{i=1}^{n} |\chi_i\rangle$, where $|\chi_i\rangle$ is a superposition state of qubit i, $|\chi_i\rangle = (a_i|0_i\rangle + b_i|1_i\rangle)$. If we denote the basis states in Eq. (5.21) by $|j_1, j_2, \ldots, j_n\rangle$, an arbitrary (initial) state can be written in three equivalent forms,

$$|\Psi_{\text{in}}\rangle = \sum_{j_1, \ldots, j_n = 0}^{1} \alpha_{j_1, j_2, \ldots, j_n} |j_1, \ldots, j_n\rangle = \sum_{j=0}^{2^n - 1} \alpha_j |j\rangle = \bigotimes_{i=1}^{n} (a_i|0\rangle + b_i|1\rangle). \quad (5.23)$$

The 2^n amplitudes $\alpha_{j_1, j_2, \ldots, j_n}$ can be determined in terms of the amplitudes $\{a_i\}$ and $\{b_i\}$ by equating the coefficients of the basis states. The product form in Eq. (5.23) establishes the reason of why quantum computing has the power of parallelism. By operating on initial superposition of states $(a_i|0\rangle + b_i|1\rangle)$, algorithms run all initial starting combinations of 0s and 1s simultaneously. An example of an n-qubit unentangled state is the state having a superposition of equal 2^n coefficients $\alpha_{j_1, j_2, \ldots, j_n}$ in Eq. (5.23): $\frac{1}{\sqrt{2^n}} \sum_{j_1, j_2, \ldots, j_n = 0}^{1} |j_1, j_2, \ldots, j_n\rangle$. This state is unentangled because it can be written as a product of the form $\bigotimes_{i=1}^{n} (|0\rangle + |1\rangle)$. An example of an n-qubit entangled state is $|0\rangle^{\otimes n} + |1\rangle^{\otimes n}$ (see Sec. 5.2.2).

Problem 5.8

(a) Using the representation $|0\rangle = \begin{pmatrix} 1 \\ 0 \end{pmatrix}$ and $|1\rangle = \begin{pmatrix} 0 \\ 1 \end{pmatrix}$, and the basis set convention (5.21), show that the two-qubit basis is given by

$$|0\rangle \otimes |0\rangle = \begin{pmatrix} 1 \\ 0 \\ 0 \\ 0 \end{pmatrix}, \quad |0\rangle \otimes |1\rangle = \begin{pmatrix} 0 \\ 1 \\ 0 \\ 0 \end{pmatrix}, \quad |1\rangle \otimes |0\rangle = \begin{pmatrix} 0 \\ 0 \\ 1 \\ 0 \end{pmatrix}, \quad |1\rangle \otimes |1\rangle = \begin{pmatrix} 0 \\ 0 \\ 0 \\ 1 \end{pmatrix}. \quad (5.24)$$

We will often use the notation $|00\rangle$, $|01\rangle$, $|10\rangle$, and $|11\rangle$ for these basis states.

(b) Write the three-qubit basis states as eight-dimension unit vectors.

5.2 Quantum Information

5.2.2 QUANTUM ENTANGLEMENT AND BELL STATES

Quantum entanglement for two-particle systems in a pure state was introduced in Sec. 1.3.3. It will be useful to redefine it, so that the definition also applies to *multiqubit states* of the form (5.23), as a property of states in a product Hilbert space. Moreover, it will be useful to redefine it for mixed states as well and for arbitrary n-level systems. First, consider a product of two spaces $\mathcal{H} = \mathcal{H}_1 \otimes \mathcal{H}_2$. A pure state $|\Psi\rangle \in \mathcal{H}$ is said to be entangled if it cannot be written as a single tensor product of states $|\psi_1\rangle \in \mathcal{H}_1$ and $|\psi_2\rangle \in \mathcal{H}_2$, i.e., $|\Psi\rangle \neq |\psi_1\rangle \otimes |\psi_2\rangle$. Rather, it must be written as a *sum* of such products, for example, $|\Psi\rangle = a|\psi_1\rangle \otimes |\psi_2\rangle + b|\phi_1\rangle \otimes |\phi_2\rangle$.

Quantum entanglement occurring in multipartite qubit states is a powerful computation and information resource. For example, entanglement allows for dense coding of quantum information (see Sec. 5.2.5) and teleportation of quantum information (see Sec. 5.2.7). One of the central questions in quantum information is how to quantify entanglement for arbitrary n qubit states. For two-qubit pure states, there is a set of four *maximally entangled* states called *Bell states* denoted as $|BS\rangle$ and defined as follows:

$$|\Phi^+\rangle = \frac{1}{\sqrt{2}}[|00\rangle + |11\rangle], \quad |\Phi^-\rangle = \frac{1}{\sqrt{2}}[|00\rangle - |11\rangle], \quad \text{(5.25a)}$$

$$|\Psi^+\rangle = \frac{1}{\sqrt{2}}[|01\rangle + |10\rangle], \quad |\Psi^-\rangle = \frac{1}{\sqrt{2}}[|01\rangle - |10\rangle]. \quad \text{(5.25b)}$$

Clearly, Bell states cannot be written as a tensor product of single qubit states. In Sec. 5.2.3, we will present methods for explicitly constructing Bell states.

The Einstein Podolsky Rosen (EPR) paradox [20], Bell inequalities [52–55], and many quantum information and quantum computation algorithms use the Bell states extensively. Bell states and EPR concepts play an important role in testing fundamental concepts in quantum mechanics. EPR and Bell inequalities will be discussed in Secs 5.7 and 5.8.

An important property of Bell states $|BS\rangle$ (and entangled pure states in general) is: When a pure state density matrix $\rho_{BS} = |BS\rangle\langle BS|$ is constructed from a Bell state, and is partially traced over (say) the second qubit, $\text{Tr}_2 [|BS\rangle\langle BS|] = \frac{1}{2}[|0\rangle\langle 0| + |1\rangle\langle 1|]$, the result is an incoherent state (mixed density matrix). In the case of Bell states, the first qubit, left after tracing out the second qubit, has 50% probability to be in either state $|0\rangle$ or $|1\rangle$.

The entangled pair of qubits in a Bell state could be spin 1/2 particles, where the state $|\downarrow\rangle$ can represent the qubit $|0\rangle$ and $|\uparrow\rangle$ can represent the qubit $|1\rangle$, or alternatively, the qubits could be photons, where a horizontal photon $|\leftrightarrow\rangle$ can represent the qubit $|0\rangle$ and a vertical photon $|\updownarrow\rangle$ can represent the qubit $|1\rangle$ as shown in Fig. 5.17 in Sec. 5.6.5. A method of entangling photons using degenerate down-conversion (degenerate difference frequency generation) will be described in Sec. 5.6.5. All one-qubit unitary transformations can be implemented using beam splitters and phase shifters, which will be described in Sec. 5.6.5. Moreover, beam splitters and phase shifters can be used to turn photons that are entangled in one of the Bell states into one of the other Bell states. So, photons are viable qubits that can be used for quantum information processing. The only problem is that photons do not interact (at least not in vacuum), so it might appear that two-qubit controlled gates (see below) cannot be engineered for photons, and these are essential for universal quantum computation (a universal set of gates are a set of gates such that any function can be computed using these gates). It was shown that *non-deterministic* two-qubit controlled gates for photons can be engineered [56], as described in Sec. 5.6.5.

An example of an entangled n-qubit state can be constructed using n photons in n different spatial modes (think of photon modes as being specified by the wavevector $\mathbf{k}$, such that photons in different modes correspond to photons with different wavevectors). Since photons are bosons, their state vector must be symmetrized (see Chapter 8). Such symmetrized states can be constructed by applying the symmetrization operator, $\mathcal{S}$,[4] to the state $|\mathbf{k}_1\rangle_1 |\mathbf{k}_2\rangle_2 \ldots |\mathbf{k}_n\rangle_n$:

$$|\Psi_{\mathcal{S}}\rangle \equiv \sqrt{n!}\,\mathcal{S}\,|\mathbf{k}_1\rangle_1 |\mathbf{k}_2\rangle_2 \ldots |\mathbf{k}_n\rangle_n = \frac{1}{\sqrt{n!}}\Big[|\mathbf{k}_1\rangle_1 |\mathbf{k}_2\rangle_2 \ldots |\mathbf{k}_n\rangle_n + |\mathbf{k}_2\rangle_1 |\mathbf{k}_1\rangle_2 \ldots |\mathbf{k}_n\rangle_n$$
$$+ \cdots + |\mathbf{k}_n\rangle_1 |\mathbf{k}_{n-1}\rangle_2 \ldots |\mathbf{k}_1\rangle_n\Big]. \quad \text{(5.26)}$$

---
[4] The symmetrization operator, $\mathcal{S}$, [see Sec. 8.1, Eq. (8.7)], is the sum of all possible permutations of the n-qubits divided by $n!$. The additional factor of $\sqrt{n!}$ in Eq. (5.26) is for normalization, as detailed in Sec. 8.6.

There are $n!$ terms in Eq. (5.26) corresponding to the $n!$ permutations of the photons. One of the challenges of all-optical quantum computing is how to engineer such states, but, as mentioned earlier, another challenge is how to construct two-qubit gates for photons, since they do not interact in vacuum.

Problem 5.9

(a) Write the Bell states (5.25a) as four-component vectors using the notation of Problem 5.8.
(b) Write the density matrix for $|\Phi^-\rangle$ and $|\Psi^-\rangle$ as 4×4 matrices.

Answer:

$$|\Phi^+\rangle = \frac{1}{\sqrt{2}}\begin{pmatrix}1\\0\\0\\1\end{pmatrix}, \quad |\Phi^-\rangle = \frac{1}{\sqrt{2}}\begin{pmatrix}1\\0\\0\\-1\end{pmatrix}, \quad |\Psi^+\rangle = \frac{1}{\sqrt{2}}\begin{pmatrix}0\\1\\1\\0\end{pmatrix}, \quad |\Psi^-\rangle = \frac{1}{\sqrt{2}}\begin{pmatrix}0\\1\\-1\\0\end{pmatrix}. \quad (5.27)$$

$$\rho_{\Phi^-} = \frac{1}{2}\begin{pmatrix}1 & 0 & 0 & -1\\0 & 0 & 0 & 0\\0 & 0 & 0 & 0\\-1 & 0 & 0 & 1\end{pmatrix}, \quad \rho_{\Psi^-} = \frac{1}{2}\begin{pmatrix}0 & 0 & 0 & 0\\0 & 1 & -1 & 0\\0 & -1 & 1 & 0\\0 & 0 & 0 & 0\end{pmatrix}. \quad (5.28)$$

Entanglement Entropy

Let us try to quantify the degree of entanglement in a bipartite state, a topic that was briefly discussed in Sec. 2.5.2. We introduce a quantity called *entanglement entropy* that serves as a measure for the degree of entanglement for a bipartite pure state. Let $|\Psi\rangle \in \mathcal{H}_1 \otimes \mathcal{H}_2$ be an entangled state (e.g., $|\Psi\rangle = |BS\rangle$) and let $\rho_{12} = |\Psi\rangle\langle\Psi|$ be the pure state density matrix constructed from $|\Psi\rangle$. The reduced mixed density matrices for the two subsystems is obtained by partial tracing: $\rho_1 = \text{Tr}_2 \rho_{12}$, $\rho_2 = \text{Tr}_1 \rho_{12}$. The entanglement entropy of the bipartite pure state ρ_{12} is defined as,

$$\mathcal{E}(\rho_{12}) \equiv \mathcal{S}(\rho_1) = \mathcal{S}(\rho_2), \quad (5.29)$$

where $\mathcal{S}(\rho) = -\text{Tr}\,\rho \log_2 \rho$ is the von Neumann entropy of ρ. The proof that $\mathcal{S}(\rho_1) = \mathcal{S}(\rho_2)$ will be given below after discussing the Schmidt decomposition.

Problem 5.10

(a) Show that the entanglement entropy of $|BS\rangle\langle BS|$ equals 1.
(b) Consider the entangled state $|\Psi\rangle = 0.8|00\rangle + 0.6|11\rangle$. Show that the entanglement entropy $\mathcal{E}(\rho = |\Psi\rangle\langle\Psi|) = 0.699722$.
(c) Show that the result $\mathcal{E}(|BS\rangle\langle BS|) > \mathcal{E}(\rho = |\Psi\rangle\langle\Psi|)$ is not accidental, namely, $|BS\rangle\langle BS|$ has the maximal entanglement entropy $\mathcal{E}(|BS\rangle\langle BS|)$ among all pure states constructed from entangled states.

Greenberger–Horne–Zeilinger States

Entangled states can be composed of more than two qubits, e.g., the three-qubit state known as the *Greenberger–Horne–Zeilinger* (GHZ) state [52],

$$|\text{GHZ}\rangle = \frac{1}{\sqrt{2}}(|000\rangle + |111\rangle), \quad (5.30)$$

5.2 Quantum Information

or the n-qubit GHZ state,

$$|GHZ\rangle = \frac{|0\rangle^{\otimes n} + |1\rangle^{\otimes n}}{\sqrt{2}}. \tag{5.31}$$

These states are not a simple tensor product of the states of each qubit but rather a superposition that describes quantum correlations (entanglement) between the qubits. The entanglement between the n-qubits means that the state of the system cannot be specified in terms of the state of each of the n qubits individually (the state is not a product state).[5] Note that this sort of state is what we called a Schrödinger cat state at the end of Sec. 1.3.1. There we noted that these kinds of states tend to decohere quickly by virtue of interactions with an outside environment.

Although there is no standard measure for multi-qubit state entanglement (particularly for mixed states), the GHZ states are believed to be maximally entangled. Note that the trace over the third qubit (denoted by Tr_3) in a three-qubit GHZ state yields a maximally mixed two-qubit state:

$$\text{Tr}_3\left[(|000\rangle + |111\rangle)(\langle 000| + \langle 111|)\right] = \frac{1}{2}\left[|00\rangle\langle 00| + |11\rangle\langle 11|\right].$$

Entanglement of qubits will be discussed from the perspective of correlation between the particles in Sec. 6.1.8, where we consider two-qubit entanglement, three-qubit entanglement, mutli-qubit entanglement, and also qutrit entanglement (Sec. 6.2) as well as "continuous-variable entanglement" (Sec. 6.5).

Schmidt Decomposition

An important and useful result concerning pure states of a bipartite system emerges from the *Schmidt decomposition theorem* of linear algebra (see below). We have in mind two Hilbert spaces $\mathcal{H}_A$ and $\mathcal{H}_B$ of respective dimensions n_A and n_B and their tensor product space $\mathcal{H}_{AB} = \mathcal{H}_A \otimes \mathcal{H}_B$. The theorem states that for pure states $|\psi\rangle \in \mathcal{H}_{AB}$, there exist two orthonormal sets of states $\{|\xi_{i,A}\rangle \in \mathcal{H}_A\}$ and $\{|\varphi_{i,B}\rangle \in \mathcal{H}_B\}$ with $i = 1, 2, \ldots, \mathcal{N}_S$, where $\mathcal{N}_S = \min(n_A, n_B)$, such that the state $|\psi\rangle$ can be written as a sum,

$$\boxed{|\psi\rangle = \sum_{i=1}^{\mathcal{N}_S} \lambda_i |\xi_{i,A}\rangle|\varphi_{i,B}\rangle, \quad 0 \leq \lambda_i \leq 1, \quad \sum_{i=1}^{\mathcal{N}_S} \lambda_i^2 = 1.} \tag{5.32}$$

The numbers λ_i are called *Schmidt coefficients*, and their number $\mathcal{N}_S$ is called the *Schmidt number*. The basis functions $|\xi_{i,A}\rangle$ and $|\varphi_{i,B}\rangle$ are called *Schmidt basis functions*. Note that we can write Eq. (5.32) as $|\psi\rangle = \sum_i \sqrt{p_i}\,|\xi_{i,A}\rangle|\varphi_{i,B}\rangle$, where p_i are the occupation probabilities for the states $|\xi_{i,A}\rangle$ and $|\varphi_{i,A}\rangle$, i.e., $\lambda_i = \sqrt{p_i}$.

The proof of Eq. (5.32) uses the singular value decomposition theorem [57]. Consider the expansion of the pure state $|\psi\rangle \in \mathcal{H}_A \otimes \mathcal{H}_B$ in terms of bases $\{|j\rangle \in \mathcal{H}_A$ and $\{|k\rangle \in \mathcal{H}_B$, $|\psi\rangle = \sum_{j,k} a_{jk}|j\rangle|k\rangle$. If $\dim \mathcal{H}_A \neq \dim \mathcal{H}_B$, then the matrix $a = \{a_{jk}\}$ is not a square matrix and therefore cannot be diagonalized. The theorem states that the matrix a can be written as a product $a = u d v^\dagger$, where u and v are unitary matrices and d is a diagonal matrix with nonnegative elements d_{ii}; these nonzero diagonal elements of d are called the *singular values* of the matrix a. Thus,

$$|\psi\rangle = \sum_{j,k} a_{jk}|j\rangle|k\rangle = \sum_{i=1}^{\mathcal{N}_S} \sum_{j,k} u_{ji} d_{ii} v^\dagger_{ki}|j\rangle|k\rangle. \tag{5.33}$$

Letting $|\xi_{i,A}\rangle = \sum_j u_{ji}|j\rangle$, $|\varphi_{i,B}\rangle = \sum_k v^\dagger_{ki}|k\rangle$, and $\lambda_i = d_{ii}$, yields Eq. (5.32). The singular value decomposition of a matrix a is pictorially shown in Fig. 5.2.

---

[5] The state in Eq. (5.31) is sometimes called a *noon* state, since it can be written as $|\text{noon}\rangle = (|n\rangle_0|0\rangle_1 + |0\rangle_0|n\rangle_1)/\sqrt{2}$, which is a superposition of n particles in state 0 and zero particles in state 1, and vise versa. A more general form of Eq. (5.31) that contains a phase angle is $|GHZ\rangle = (|0\rangle^{\otimes n} + e^{i\theta}|1\rangle^{\otimes n})/\sqrt{2}$.

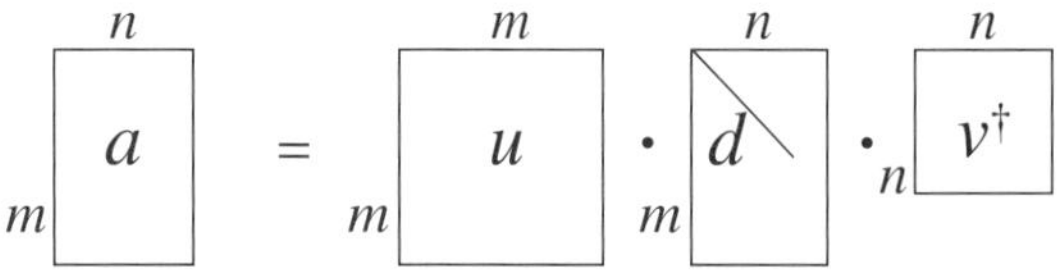

FIG 5.2 Singular value decomposition of an $m \times n$ matrix a into $a = u d v^\dagger$. The bullets imply matrix multiplication. u is an $m \times m$ matrix, d is an $m \times n$ matrix, and v is an $n \times n$ matrix.

Problem 5.11

(a) Show that $\{|\xi_{i,A}\rangle\}$ and $\{|\varphi_{i,B}\rangle\}$ are orthonormal sets of vectors using the unitarity of u and v.
(b) Under what condition on the amplitudes a_{jk}, is the condition $\sum_i \lambda_i^2 = 1$ satisfied?

Problem 5.12

If ABC is a composite system composed of three subsystems, A, B, and C, are there pure states $|\psi\rangle$ of ABC that cannot be written in the form $|\psi\rangle = \sum_i \lambda_i |\xi_{i,A}\rangle |\varphi_{i,B}\rangle |\chi_{i,C}\rangle$, where the λ_i are nonnegative, $\sum_i \lambda_i^2 = 1$ and the basis states are orthonormal? That is, can we always write $|\psi\rangle_{ABC} = \sum_i \sqrt{p_i} |\xi_{i,A}\rangle |\varphi_{i,B}\rangle |\chi_{i,C}\rangle$ with orthogonal basis functions?

Answer: Schmidt decomposition of tripartite pure states is not possible in general. If it was possible, the partial trace of the density matrix ρ_{ABC} over A and B would have the same "spectrum" of probabilities as would the trace of ρ_{ABC} over C (this is called a strong "equal spectrum" condition of the partial traces). Now, consider the state $\frac{1}{2}(|00\rangle_{AB} + |11\rangle_{AB})|0\rangle_C$, and show that this "equal spectrum" condition is not satisfied.

Schmidt Decomposition and Entanglement Entropy: The Schmidt decomposition theorem can be used to determine the entanglement entropy of an entangled pure state [defined in Eq. (5.29)],

$$\rho_{AB} = |\psi\rangle\langle\psi| = \sum_{i=1}^{N_S} \lambda_i |\xi_{i,A}\rangle |\varphi_{i,B}\rangle \sum_{j=1}^{N_S} \lambda_j \langle\xi_{j,A}|\langle\varphi_{j,B}|. \qquad (5.34)$$

Taking the partial trace of ρ_{AB} over B yields,

$$\rho_A = \text{Tr}_B[\rho_{AB}] = \sum_{i=1}^{N_S} \lambda_i^2 |\xi_{i,A}\rangle\langle\xi_{i,A}|, \qquad (5.35)$$

which is a mixed state of subsystem A. Similarly, taking partial trace of ρ_{AB} over A yields,

$$\rho_B = \text{Tr}_A[\rho_{AB}] = \sum_{i=1}^{N_S} \lambda_i^2 |\varphi_{i,B}\rangle\langle\varphi_{i,B}|, \qquad (5.36)$$

which is a mixed state of subsystem B. We conclude that there is spectral equality of ρ_A and ρ_B, i.e., since $\{|\xi_{i,A}\rangle \in \mathcal{H}_A\}$ and $\{|\varphi_{i,B}\rangle \in \mathcal{H}_B\}$ are orthonormal sets, then $\{\lambda_i^2\}$, $(i = 1, 2, \ldots N_S)$ are the nonzero eigenvalues of both ρ_A acting in $\mathcal{H}_A$ and ρ_B acting in $\mathcal{H}_B$. Thus,

If $|\psi\rangle \in \mathcal{H}_A \otimes \mathcal{H}_B$ and $\rho_{AB} = |\psi\rangle\langle\psi|$ is *the pure density operator constructed from* $|\psi\rangle$, then $\rho_A = \text{Tr}_B[\rho_{AB}]$ and $\rho_B = \text{Tr}_A[\rho_{AB}]$ have the same spectrum, $\{\lambda_i^2\}$, $(i = 1, 2, \ldots N_S)$.

Since λ_i^2 is the probability for observing the states $|\xi_{iA}\rangle\langle\xi_{iA}|$ and $|\varphi_{iB}\rangle\langle\varphi_{iB}|$,

$$S_A = S_B = -\sum_{i=1}^{N_S} \lambda_i^2 \log_2 \lambda_i^2, \qquad (5.37)$$

5.2 Quantum Information

i.e., the von Neumann entropies of the subsystems of a pure state are equal. These range between 0 (for a pure state, $\mathcal{N}_S = 1$) and $\log_2 \mathcal{N}_S$. Unless only one of the λ_i is nonzero, i.e., the Schmidt number is unity, the purity of the subsystems is not unity, and their von Neumann entropy are non-zero. Clearly, this *"entanglement entropy"* is not the entropy we consider in statistical mechanics and thermodynamics; here, we start with a system in a pure state, divide the system into two subsystems, and find that the entropy of the whole system is not the sum of the entropies of the subsystems; this von Neumann entropy is not extensive; hence, it is not always identical to the thermodynamic entropy.

Problem 5.13

(a) Take the partial trace over subsystem B of the pure state density matrix $\rho = |\psi\rangle\langle\psi|$ with $|\psi\rangle = \sum_{j,k} a_{jk} |j\rangle|k\rangle$ and show that $\rho_A = \text{Tr}_B \, \rho = aa^\dagger$.

(b) Find ρ_A for the Bell state $|\Phi^+\rangle$.

(c) Calculate the entropy $\mathcal{S}_A$.

Answers: (a) $\rho_A = \text{Tr}_B \, \rho = \sum_k a_{jk} a^*_{kj}$. (b) For the Bell state $|\Phi^+\rangle$, $a = \begin{pmatrix} \frac{1}{\sqrt{2}} & 0 \\ 0 & \frac{1}{\sqrt{2}} \end{pmatrix}$, so $\rho_A = aa^\dagger = \begin{pmatrix} \frac{1}{2} & 0 \\ 0 & \frac{1}{2} \end{pmatrix}$.

(c) $\mathcal{S}_A = 1$.

The Schmidt coefficients, λ_i, can be used to distinguish separable pure states from entangled pure states. A separable pure state is characterized by a vector of Schmidt coefficients with only one nonvanishing entry, whereas the Schmidt vector of an entangled state has at least two nonvanishing components. A pure state has maximal entanglement entropy if its Schmidt coefficient vector is

$$\{\lambda_i\} = \{1/\sqrt{\mathcal{N}_S}, \ldots, 1/\sqrt{\mathcal{N}_S}\}, \tag{5.38}$$

where $\mathcal{N}_S$ is the Schmidt number. The Schmidt coefficients relate the degree of entanglement of pure bipartite states to the von Neumann entropy of the corresponding reduced density matrices of the two subsystems composing the system—a pure reduced density matrix with vanishing von Neumann entropy corresponds to a separable state, whereas a maximally entangled state leads to the maximum von Neumann entropy density matrix. It is easy to verify that Eq. (5.38) maximizes the entropy of the subsystems, $\mathcal{S}(\{\lambda_i\}) = -\sum_{i=1}^{\mathcal{N}_S} \lambda_i^2 \log_2 \lambda_i^2$, by maximizing the von Neumann entropy $\mathcal{S}(\{\lambda_i\})$ with respect to the λ_i under the density matrix trace constraint condition $\mathcal{G}(\{\lambda_i^2\}) \equiv \sum_{i=1}^{\mathcal{N}_S} \lambda_i^2 = 1$ by forming the function $\mathcal{F}(\{\lambda_i\}) = \mathcal{S} + \gamma \mathcal{G}$, where γ is a Lagrange multiplier, setting $\frac{\partial \mathcal{F}}{\partial \lambda_i} = 0$ for all i, and determining γ, such that the constraint condition is satisfied [25].

Mixed-State Entanglement

Not only pure states but also mixed states of a quantum system consisting of two or more subsystems can be entangled. Consider two quantum systems A and B with corresponding density matrices ρ_A and ρ_B, and let ρ_{AB} be the density matrix of the composite system. Is $\rho_{AB} = \rho_A \otimes \rho_B$? The answer is affirmative only if there is no correlation between the two systems. (Correlation between systems will be discussed in Sec. 6.1.8.) The nature of the correlations between systems is related to the notion of *separability*. Two systems A and B are said to be separable if the density matrix ρ_{AB} can be written as a convex sum of product states,

$$\rho_{AB} = \sum_{i=1}^M p_i \, \rho_A^i \otimes \rho_B^i, \quad p_i > 0, \quad \sum_{i=1}^M p_i = 1, \tag{5.39}$$

where ρ_A^i and ρ_B^i ($i = 1, 2, \ldots, M$) are proper density matrices (hermiticity, unit trace, and semipositivity). A mixed state was defined by Werner [58] to be entangled if it is *inseparable*, i.e., if it *cannot* be written in the form (5.39).

Unfortunately, this definition is not constructive. Since it is defined through a negative statement, it is difficult, in general, to decide whether a given mixed state is separable or entangled.

Extending these definitions, a mixed state of an N-partite system is *separable* if it can be written as a convex sum of product states,

$$\rho = \sum_{i=1}^{M} p_i \, \rho_1^i \otimes \rho_2^i \ldots \otimes \rho_n^i, \quad p_i > 0, \quad \sum_{i=1}^{M} p_i = 1, \tag{5.40}$$

otherwise, it is entangled. A quantitative measure of mixed state entanglement in multi-partite systems has proven to be difficult to devise.

A simple test to determine whether a mixed state ρ of a bipartite system is entangled was suggested by Asher Peres [59, 60]. He showed that a necessary condition for separability is that the matrix obtained by *partial transposition* of ρ has only nonnegative eigenvalues. Partial transposition is defined as follows. Given the density matrix $\rho_{m\mu,n\nu}$, where Latin indices refer to the first subsystem and Greek indices to the second one, the transformation

$$\rho_{m\mu,n\nu} \Longrightarrow \sigma_{m\mu,n\nu} \equiv \rho_{n\mu,m\nu} \tag{5.41}$$

is the partial transpose of ρ with respect to the first subsystem.[6] Performing the partial transposition of Eq. (5.39) and recalling that ρ_A^i and ρ_B^i are (have the properties of) density matrices, it follows that none of the eigenvalues of the transposed density matrix is negative. This is a necessary condition for Eq. (5.39) to hold. It turns out that the necessary condition is also a sufficient one for two-qubit (and qubit–qutrit) systems [60]. This criterion is often called the *Peres–Horodecki entanglement condition*.

Further discussion of qubit (and qutrit, i.e., three-level system) entanglement is detailed in Sec. 6.1.8, where we discuss both classical and quantum correlations in terms of the correlation functions.

Problem 5.14

Consider the maximally mixed two-qubit density matrix state in the basis $\{|00\rangle, |01\rangle, |10\rangle, |11\rangle\}$: $\rho_{AB} = \frac{1}{4} \mathbf{1}_{4\times 4}$.

(a) Calculate the entropy S_{AB}.
(b) Calculate ρ_A and ρ_B.
(c) What is the von Neumann entropy $S_A = S_B$.
(d) Is ρ_{AB} separable?

Hint: Use the Peres criterion.

Answers: (a) $S_{AB} = -4 \left(\frac{1}{4} \log_2 \frac{1}{4} \right) = 2.$ (b) $\rho_A = \rho_B = \frac{1}{2} \begin{pmatrix} 1 & 0 \\ 0 & 1 \end{pmatrix}.$

(c) $S_A = S_B = -2 \left(\frac{1}{2} \log_2 \frac{1}{2} \right) = 1.$ (d) The four eigenvalues are $\frac{1}{4}$. Hence, ρ_{AB} is separable.

Problem 5.15

The density matrix corresponding to the two-qubit pure state $\frac{1}{\sqrt{2}}(|00\rangle + |11\rangle)$ (i.e., $\frac{1}{\sqrt{2}}(|\uparrow\uparrow\rangle + |\downarrow\downarrow\rangle)$) can be written as

$$\rho_{AB} = \frac{1}{2} \begin{pmatrix} 1 & 0 & 0 & 1 \\ 0 & 0 & 0 & 0 \\ 0 & 0 & 0 & 0 \\ 1 & 0 & 0 & 1 \end{pmatrix}.$$

---

[6] Similarly, we could transpose the indices for the second subsystem, ρ^{T_B}: $\rho_{m\mu,n\nu} \Longrightarrow \rho_{m\nu,n\mu}$.

5.2 Quantum Information

(a) Calculate the entropy S_{AB}.
(b) Calculate ρ_A and ρ_B.
(c) What is the von Neumann entropy $S_A = S_B$.
(d) Is ρ_{AB} separable?

Answers: (a) $S_{AB} = 0$. (b) $\rho_A = \rho_B = \frac{1}{2}\begin{pmatrix} 1 & 0 \\ 0 & 1 \end{pmatrix}$. (c) $S_A = S_B = -2\left(\frac{1}{2}\log_2 \frac{1}{2}\right) = 1$.
(d) The four eigenvalues of the matrix σ are $\frac{1}{3}$ ($\times 3$) and $-1/2$. Hence, it is entangled not separable.

Problem 5.16

Consider the density matrix state

$$\rho_{AB} = \frac{1}{8}\begin{pmatrix} 1 & 0 & 0 & 0 \\ 0 & 3 & 3 & 0 \\ 0 & 3 & 3 & 0 \\ 0 & 0 & 0 & 1 \end{pmatrix}. \quad (5.42)$$

(a) Calculate the entropy S_{AB}. Answer: $\frac{3}{4}(\log_2(3/2) - 1)$.
(b) Calculate ρ_A and ρ_B.
(c) What is the von Neumann entropy $S_A = S_B$.
(d) Is ρ_{AB} separable?

Distillation and Dilution of Entanglement

Entanglement can be distilled from quantum systems that are not fully entangled to make more fully entangled systems. *Entanglement distillation* is the extraction of pairs of qubits in Bell states (e.g., the singlet state, $|\Psi^-\rangle = \frac{1}{\sqrt{2}}[|01\rangle - |10\rangle]$, or any other Bell state) from some large number of copies of an inseparable state, by means of local quantum operations and classical communication (LOCC). The procedure can be described as follows: two observers each has n qubits coming from (partially) entangled pairs prepared in a given state ρ, which is partially entangled (i.e, not maximally entangled). Each observer can perform local operations on his or her n qubits and exchange classical information with the other observer. They can obtain a pair of entangled qubits in singlet-state form, or nearly singlet form (the rest of the qubits being discarded), or better yet, they can obtain several entangled qubits in nearly singlet-state form. If they manage to do this, they have distilled pure entanglement from the mixed state qubits. It has been shown that *any* inseparable two-qubit states can be distilled to a singlet-state form with enough copies of the qubit pairs [61]. Given n qubit pairs, if m copies of a Bell state can be obtained with high fidelity [see Eq. (2.70)], the $n : m$ ratio can be defined as the distillable entanglement of the initial qubit pair.

Entanglement dilution is the reverse process: Using local operations and classical communication (LOCC), a large number of copies of a Bell state, say the singlet state, can be converted into a large number of qubits in a specific bipartite inseparable pure state $|\psi\rangle$ or into a specific bipartite inseparable density matrix state ρ, again with high fidelity.

5.2.3 QUANTUM GATES

In analogy with the introduction of classical gates, quantum gates can be introduced. Quantum gates are unitary operators in the appropriate multi-qubit Hilbert space. There are several important differences between classical and quantum gates that should be emphasized. (1) Unlike some of the classical gates (such as AND, OR, NAND, and NOR), the number of input and output components are identical ($k = l$ in the notation used for classical gates). Moreover, the gates are represented by unitary matrices, and therefore, each gate has an inverse. (2) Quantum gates depend on continuous parameters, so that in principle, there is a continuously infinite number of them. Fortunately, as will be shown below, there are only a few important types of gates that are needed to perform all the required qubit manipulations. Moreover, the gates operating in multi-qubit spaces can be written as a direct product of single- and two-qubit gates. Therefore, it is sufficient to consider only these two classes of quantum gates (however, we also discuss several three-qubit gates below).

Single-Qubit Gates

A *single-qubit gate* is any unitary transformation on a single qubit. In general, a 2×2 unitary matrix depends on four continuous parameters. There are four complex matrix elements (eight real numbers) and four conditions implied by unitarity. To define the most common single-qubit gates, recall the definition of the unit 2×2 matrix and the three Pauli matrices,

$$\mathbf{1} = \begin{pmatrix} 1 & 0 \\ 0 & 1 \end{pmatrix}, \quad \sigma_x = \begin{pmatrix} 0 & 1 \\ 1 & 0 \end{pmatrix}, \quad \sigma_y = \begin{pmatrix} 0 & -i \\ i & 0 \end{pmatrix}, \quad \sigma_z = \begin{pmatrix} 1 & 0 \\ 0 & -1 \end{pmatrix}. \tag{5.43}$$

For an arbitrary qubit $|\chi\rangle = a|0\rangle + b|1\rangle$ ($|a|^2 + |b|^2 = 1$), represented as a two-component spinor, a *single-qubit gate* is a unitary transformation that is represented by a unitary matrix U ($UU^\dagger = U^\dagger U = \mathbf{1}$) that operates on the qubit:

$$|\chi\rangle = \begin{pmatrix} a \\ b \end{pmatrix} \to U|\chi\rangle = \begin{pmatrix} U_{11} & U_{12} \\ U_{21} & U_{22} \end{pmatrix} \begin{pmatrix} a \\ b \end{pmatrix}. \tag{5.44}$$

Often used single-qubit gates are the identity gate, the NOT gate, and the Y gate,

$$U_1 \equiv \mathbf{1}, \quad U_x \equiv \sigma_x, \quad Y \equiv -i\sigma_y. \tag{5.45}$$

Problem 5.17

(a) For a unit vector $\hat{\mathbf{u}}$, use the identity $U_{\hat{\mathbf{u}}}(\beta) \equiv e^{-i\beta \hat{\mathbf{u}} \cdot \boldsymbol{\sigma}/2} = \cos\left(\frac{\beta}{2}\right) \mathbf{1} - i\sin\left(\frac{\beta}{2}\right) \hat{\mathbf{u}} \cdot \boldsymbol{\sigma}$ [see Eq. (4.22)] and show that for $\hat{\mathbf{u}} = (1,0,0)$, U_x is given by $-i$ times a 180° rotation about the x axis.

(b) Show that for $\hat{\mathbf{u}} = (0,1,0)$, $U_y(\beta) = e^{-i\beta\sigma_y/2} = \begin{pmatrix} \cos\frac{\beta}{2} & -\sin\frac{\beta}{2} \\ \sin\frac{\beta}{2} & \cos\frac{\beta}{2} \end{pmatrix}$.

We have already seen in Sec. 4.2 that $U_{\hat{\mathbf{u}}}(\beta)$ performs a spinor (single-qubit) rotation by an arbitrary angle β around the $\hat{\mathbf{u}}$ axis. Note that $U_{\hat{\mathbf{u}}}(\beta = 2\pi) = -\mathbf{1}$ not $\mathbf{1}$. This is a general property of particles with fractional spin and was discussed in Sec. 4.2.

Another useful single-qubit operation is the phase gate, including the π-phase gate as a special case,

$$P(\phi) = \begin{pmatrix} 1 & 0 \\ 0 & e^{i\phi} \end{pmatrix}, \quad P(\pi) = \sigma_z. \tag{5.46}$$

The operation of phase gates on basis states $|n\rangle$, $n = 0, 1$ can be compactly expressed as $P(\phi)|n\rangle = e^{in\phi}|n\rangle$. Finally, perhaps one of the most useful single-qubit gates in quantum information and computation applications is the Hadamard gate H,

$$H = \sigma_x U_y(\pi/2) = \begin{pmatrix} 0 & 1 \\ 1 & 0 \end{pmatrix} \frac{1}{\sqrt{2}} \begin{pmatrix} 1 & -1 \\ 1 & 1 \end{pmatrix} = \frac{1}{\sqrt{2}} \begin{pmatrix} 1 & 1 \\ 1 & -1 \end{pmatrix}. \tag{5.47}$$

Since H is a unitary, real symmetric matrix, it satisfies $HH^\dagger = H^2 = \mathbf{1}$.

5.2 Quantum Information

> **Problem 5.18**
>
> (a) Show that the Hadamard gate is a unitary matrix.
> (b) Show that $HH = 1$.
> (c) Show that H is equivalent, up to a phase, to a 180° rotation about the unit vector $\hat{u} = (\hat{x} + \hat{y})/\sqrt{2}$.
>
> **Hint:** Use Problem 5.6 and $\sigma \cdot \hat{u} = (\sigma_x + \sigma_y)/\sqrt{2}$.

In Sec. 6.1.2, we develop a geometric picture of single-qubit gates. Here, a pure state of a two-level system (i.e., $|\chi\rangle = a|0\rangle + b|1\rangle$ with complex amplitudes a and b such that $|a|^2 + |b|^2 = 1$) can be described in terms of a unit vector $\hat{n} = (\sin\beta\cos\alpha, \sin\beta\sin\alpha, \cos\beta)$, which lies on a unit sphere, called the Bloch sphere (or, particularly for photon qubits, the Poincaré sphere). The polar and azimuthal angles of $\hat{n}$ are related to the coefficients a and b by $a = \cos\frac{\beta}{2} e^{-i\alpha/2}$ and $b = \sin\frac{\beta}{2} e^{i\alpha/2}$, and $\hat{n} = \langle\sigma\rangle$. Single-qubit unitary operators transform one unit vector into another. It is often useful to think in terms of this geometric picture.

Two-Qubit Gates

Two-qubit gates are unitary operations that simultaneously affect two-qubits. Together with single-qubit gates, they can generate all possible unitary operations in n-qubit space. Another important property of two (and higher) qubit gates is that some of them are *controlled gates*. Controlled two-qubit gates affect one qubit depending on the state of the other qubit.

> **Problem 5.19**
>
> Verify that the standard representation of the two-qubit operators $\sigma_z \otimes \sigma_x$ and $\sigma_x \otimes \sigma_z$ are given by the 4×4 matrices,
>
> $$\sigma_z \otimes \sigma_x = \begin{pmatrix} 0 & 1 & 0 & 0 \\ 1 & 0 & 0 & 0 \\ 0 & 0 & 0 & -1 \\ 0 & 0 & -1 & 0 \end{pmatrix}, \quad \sigma_x \otimes \sigma_z = \begin{pmatrix} 0 & 0 & 1 & 0 \\ 0 & 0 & 0 & -1 \\ 1 & 0 & 0 & 0 \\ 0 & -1 & 0 & 0 \end{pmatrix}. \quad (5.48)$$

The *controlled-phase gate* (or controlled-z) CP(ϕ) is defined by[7]

$$\mathrm{CP}(\phi)|m,n\rangle = \exp(imn\phi)|m,n\rangle,$$

$$\mathrm{CP}(\phi)\begin{pmatrix} a \\ b \\ c \\ d \end{pmatrix} = \begin{pmatrix} 1 & 0 & 0 & 0 \\ 0 & 1 & 0 & 0 \\ 0 & 0 & 1 & 0 \\ 0 & 0 & 0 & e^{i\phi} \end{pmatrix} \begin{pmatrix} a \\ b \\ c \\ d \end{pmatrix} = \begin{pmatrix} a \\ b \\ c \\ e^{i\phi} d \end{pmatrix}, \quad (5.49)$$

The action **1** on the first qubit, or the phase gate on the second qubit, is controlled by whether the first qubit is in the state $|0\rangle$ or $|1\rangle$.

The CNOT gate is another two-qubit gate that can be implemented experimentally by applying a $U_y(\pi/4)$ transformation (see Problem 5.6) on qubit 2, a controlled two-qubit π phase gate, CP(π), and then a $U_y(-\pi/4)$ transformation

---

[7] The gate can be defined either by its action on the two-qubit basis states $|m,n\rangle$ ($m,n = 0,1$) or as a 4×4 matrix acting on the four-dimensional vector of coefficients in the expansion of a general qubit state $a|00\rangle + b|01\rangle + c|10\rangle + d|11\rangle$ in the vector notation of Problem 5.7. In Eqs (5.49), we use both, and in Eq. (5.50), we use the latter.

on qubit 2, i.e., CNOT = $[U_y(-\pi/4)]_2 \, CP(\pi)[U_y(\pi/4)]_2$. The result of implementing this gate is as follows: qubit 1 is unchanged and qubit 2 is unchanged if qubit 1 was in state $|0\rangle$, but qubit 2 is flipped if qubit 1 was in state $|1\rangle$. The action of the CNOT gate on two qubits can be represented as,

$$\mathrm{CNOT}|m,n\rangle = |m, n \oplus m\rangle,$$

$$\mathrm{CNOT}\begin{pmatrix}a\\b\\c\\d\end{pmatrix} = \begin{pmatrix}1&0&0&0\\0&1&0&0\\0&0&0&1\\0&0&1&0\end{pmatrix}\begin{pmatrix}a\\b\\c\\d\end{pmatrix} = \begin{pmatrix}a\\b\\d\\c\end{pmatrix}. \tag{5.50}$$

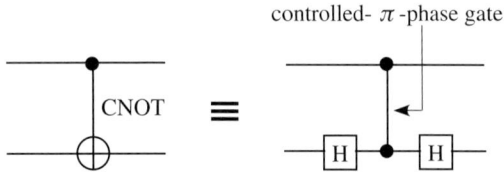

FIG 5.3 A controlled-not gate, graphically symbolized by the quantum circuit drawn on the left, can be constructed using two one-qubit Hadamard gates and the central controlled-π-phase gate [see Eq. (5.49)], as shown by the quantum circuit on the right.

The CNOT gate can also be implemented using a three-gate circuit (see discussion in Sec 5.2.9), as shown in Fig. 5.3. The two boxes containing the letter H represent one-qubit Hadamard gates, and the central gate is a two-qubit controlled-π phase-shift gate. This gate performs the transformation $CP(\pi)|m,n\rangle = \exp(imn\pi)|m,n\rangle$.

The SWAP gate swaps two qubits, $\mathrm{SWAP}|m,n\rangle = |n,m\rangle$, i.e.,

$$\mathrm{SWAP}\begin{pmatrix}a\\b\\c\\d\end{pmatrix} = \begin{pmatrix}1&0&0&0\\0&0&1&0\\0&1&0&0\\0&0&0&1\end{pmatrix}\begin{pmatrix}a\\b\\c\\d\end{pmatrix} = \begin{pmatrix}a\\c\\b\\d\end{pmatrix}. \tag{5.51}$$

A two-qubit ϕ phase gate that entangles two qubits by multiplying each basis state by a phase θ_{nm}, such that $\theta_{00} + \theta_{11} - \theta_{01} - \theta_{10} \equiv \phi \pmod{2\pi}$ is given by

$$P(\phi)|m,n\rangle = \exp(i\theta_{mn})|m,n\rangle, \quad e^{i(\theta_{00}+\theta_{11}-\theta_{01}-\theta_{10})} = e^{i\phi}. \tag{5.52}$$

This family of gates includes the control-phase gate, $CP(\phi)$, Eq. (5.49) as a special case. The $P(\phi)$ gates can be combined with unitary single-qubit Rabi shifts, $\exp(i\alpha_m)\exp(i\beta_n)$, to create the $CP(\phi)$ gate, where $\theta_{mn}+\alpha_m+\beta_n = mn\phi \pmod{2\pi}$, while $\theta_{00} + \theta_{11} - \theta_{01} - \theta_{10}$ is invariant under these shifts. The phase ϕ has an intrinsic physical feature in that it parameterizes uniquely the entanglement power of the gate and thus is intimately connected with the coupling strength of the two qubits during evolution. The control phase gate with phase $\phi = \pi$ can further combine with single-qubit operators to form the control-not gate, Eq. (5.50). Each of these two-qubit gates can be combined with a set of generators for single-qubit gates to form a universal set of gates for quantum computation (universal in the sense that any computation can be carried out using a universal set of gates). In practice, a physical system may evolve more naturally to a gate in $P(\pi)$ other than CNOT or $CP(\pi)$. Therefore, in designing a gate, it is better to aim less restrictively for any one of the equivalent $P(\phi)$ gates.

An important subset of the two-qubit gates are those that can be written as $|1\rangle\langle 1| \otimes \mathbf{1} + |0\rangle\langle 0| \otimes \mathbf{U}$, where $\mathbf{1}$ is the single-qubit identity operation and $\mathbf{U}$ is some other single-qubit gate. Such a two-qubit gate is called a *controlled-U gate*. The action $\mathbf{1}$ or $\mathbf{U}$ on the second qubit is controlled by whether the first qubit is in the state $|0\rangle$ or $|1\rangle$. The controlled-U gate corresponding to a rotation of the second qubit around the x-axis by an angle β can be represented in a matrix form as

$$\mathrm{Controlled-}U_{\hat{x}}(\beta) = \begin{pmatrix}1&0&0&0\\0&1&0&0\\0&0&\cos\frac{\beta}{2}&-\sin\frac{\beta}{2}\\0&0&\sin\frac{\beta}{2}&\cos\frac{\beta}{2}\end{pmatrix}. \tag{5.53}$$

5.2 Quantum Information

Three-Qubit Gates

Three-qubit gates are represented by 8×8 unitary matrices U operating in an eight-dimensional Hilbert space (a general state in this space can be written as $|\psi\rangle = \sum_{j=1}^{8} a_j |j-1\rangle$, where $j = 1$ has $|000\rangle, \ldots, j = 8$ has $|111\rangle$). Two important three-qubit gates are:

- The *Toffoli gate*, where the nonzero elements are $U_{i,i} = 1$ ($i = 1, 2, 3, 4, 5, 6$) and $U_{7,8} = U_{8,7} = 1$. This is a control-flip gate. Operating on a basis state $|m, n, l\rangle$, it flips the third qubit only if the first two qubits are in a state $|1, 1\rangle$, that is, $U|m, n, l\rangle = |m, n, l \oplus nm\rangle$.
- The *Fredkin gate*, where the nonzero elements are $U_{i,i} = 1$ ($i = 1, 2, 3, 4, 5, 8$) and $U_{6,7} = U_{7,6} = 1$. This is a control-swap gate. Operating on a basis state $|m, n, l\rangle$, it swaps the first two qubits only if the third qubit is $|1\rangle$, that is, $U|m, n, l\rangle = |m \oplus l \oplus mn, n \oplus l \oplus nm\rangle$.

Quantum Gates to Make Bell States

The Bell states (5.25a) can be created using CNOT and the rotation gates $[U_y(\theta)]_1$ as follows:

$$|\Phi^+\rangle = \text{CNOT}\,[U_y(\pi/2)]_1\,|00\rangle, \quad |\Phi^-\rangle = \text{CNOT}\,[U_y(-\pi/2)]_1\,|00\rangle, \tag{5.54a}$$

$$|\Psi^+\rangle = \text{CNOT}\,[U_y(\pi/2)]_1\,|01\rangle, \quad |\Psi^-\rangle = \text{CNOT}\,[U_y(-\pi/2)]_1\,|01\rangle. \tag{5.54b}$$

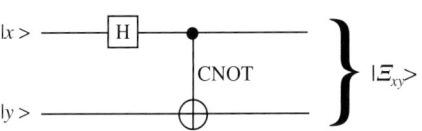

In Eqs (5.54a), the $[U_y(\theta)]_1$ rotation is applied to the first qubit, and then the CNOT gate is applied. Another way of making the Bell states is to apply the product gate CNOT H_1 to the input states $|x, y\rangle$, as illustrated in Fig. 5.4. The result is,

FIG 5.4 Quantum circuit to create the Bell states, $|\Phi^+\rangle \equiv |\Xi_{00}\rangle$, $|\Psi^+\rangle \equiv |\Xi_{01}\rangle$, $|\Phi^-\rangle \equiv |\Xi_{10}\rangle$, and $|\Psi^-\rangle \equiv |\Xi_{11}\rangle$.

$$\text{CNOT}\,H_1\,|x, y\rangle = \frac{1}{\sqrt{2}}[|0, y\rangle + (-1)^x |1, \bar{y}\rangle] = |\Xi_{xy}\rangle. \tag{5.55}$$

where $\bar{y}$ is the negation of y. The four Bell states are obtained as special cases,

$$|\Phi^+\rangle = |\Xi_{00}\rangle, \quad |\Psi^+\rangle = |\Xi_{01}\rangle, \quad |\Phi^-\rangle = |\Xi_{10}\rangle, \quad |\Psi^-\rangle = |\Xi_{11}\rangle. \tag{5.56}$$

Universal Quantum Gates

The number n of qubits required for quantum algorithms can be large. For example, in the Shor factorization algorithm for factoring $55 = 11 \times 5$ uses $n = 26$ qubits (Hilbert space dimension $N = 2^{26}$). Clearly, operating with quantum gates (matrices) of this size is not practical. Hence, it is necessary to simulate any desired n-qubit operation by a product of simple gates. Mathematically, it is possible to express any unitary matrix U as a product of unitary $N \times N$ matrices composed of blocks of 2×2 matrices U^{ij}. However, this result is of little practical use, since the number of such matrices required in the construction is sometimes huge. Moreover, the 2×2 matrices U^{ij} depend on continuous parameters, and any inaccuracy in their numerical value, no matter how small, leads to large error if they are used repeatedly. A possible way out of this problem is to give up the quest for an exact representation and to be content with an *approximate representation* of n-qubit operations (as clarified below), without significantly adversely affecting the results of the computation. Once this approximation procedure is accepted, the problem of inaccurate continuous parameters can be circumvented by the use of *discrete gates* (think of the Hadamard gate as an example). Such a construction, if it exists, is rather appealing, and the few discrete gates required for its implementation are referred to as a *universal set*. More specifically,

- A set of discrete gates is said to be universal for quantum computing if it can approximate any n-qubit unitary operation to an arbitrary precision using the operators in the set within a quantum circuit.

Note that for classical computation, the concept of universality is much simpler. There is one such gate, the Toffoli gate, Eq. (5.17), which is the building block for all other gates.

The universal set of gates for quantum computing is surprisingly small. It has been shown that an arbitrary unitary operation on n qubits can be approximated using a few single-qubit gates and one two-qubit CNOT gate [62]. The single-qubit gates include the Hadamard and two phase gates, $CP(\pi/2)$ and $CP(\pi/4)$ (the latter is also referred to as $\pi/8$ gate for historical reasons). The universal set listed above is by no means unique, and other sets of universal gates exist as well. For example, controlled-phase gates and single-qubit gates form a universal set.

The proof of the universality property will not be presented here. Instead, we survey the main steps of its construction. In the first step, one shows that the Hadamard and two phase gates, $CP(\pi/2)$ and $CP(\pi/4)$, can be used to approximate any other single-qubit gate to arbitrary accuracy. Second, single-qubit and CNOT gates can be used to construct an arbitrary two-qubit gate. The third step combines these gates for pairs of qubits, making it possible to simulate an arbitrary unitary operation on n qubits. However, this is not sufficient because the second and third steps require the use of single-qubit gates with continuous parameters, and this is vulnerable to errors. In the crucial fourth step, a *discrete* set of operators is identified, in terms of which an arbitrary n-qubit gate can be approximated. This is the required universal set. The two remaining tasks are two-fold. First, the construction of an arbitrary n-qubit gate from the universal set using error-correcting codes, making it immune against errors. Second, the notion of *approximating an n-qubit operation* should be clarified.

A set of quantum gates, $U_1, U_2, \ldots, U_p$, approximates a desired unitary transformation U with precision ϵ if $||U - U_1 \ldots U_p|| \leq \epsilon$, where $||\cdot||$ denotes the operator norm, i.e., the largest singular value. We can define a set of elementary gates as universal if every unitary operator operating on a fixed number of qubits can be approximated to precision ϵ using on the order of $\log \epsilon^{-1}$ elementary gates. Clearly, some unitary operators take more one- and two-qubit gates to approximate than others.

5.2.4 NO-CLONING THEOREM

The no-cloning theorem emphasizes the inaccessibility of quantum information. It shows that one cannot make a backup of quantum information in the form of qubits whose states are not explicitly known. That is, quantum mechanics guarantees that one cannot accurately copy unknown qubits.

Theorem: An unknown quantum state cannot be cloned.

Proof Assume the opposite. Let U be a two-qubit gate that clones the qubit $|\gamma\rangle$, i.e.,

$$U(|\gamma\rangle|0\rangle) = |\gamma\rangle|\gamma\rangle \ \forall \ \gamma, \tag{5.57}$$

where the operator U does not depend on the unknown state γ. Consider explicitly the state $|\gamma\rangle = \frac{1}{\sqrt{2}}(|\alpha\rangle + |\beta\rangle)$ that we would like to clone. Then

$$U(|\gamma\rangle|0\rangle) = \frac{1}{\sqrt{2}} [U(|\alpha\rangle|0\rangle) + U(|\beta\rangle|0\rangle)]$$
$$= \frac{1}{\sqrt{2}} [(|\alpha\rangle|\alpha\rangle) + (|\beta\rangle|\beta\rangle)] \neq |\gamma\rangle|\gamma\rangle. \tag{5.58}$$

We have a contradiction. Therefore, there is no unitary operator that is capable of cloning (i.e., making a copy) of an arbitrary qubit.

Problem 5.20

(a) The controlled-not gate CNOT_{12} maps $|a\rangle \otimes |b\rangle \longmapsto |a\rangle \otimes |a \oplus b\rangle$ for $a, b \in \{0, 1\}$. Show that CNOT_{12} maps $|a\rangle \otimes |0\rangle \longmapsto |a\rangle \otimes |a\rangle$.

(b) Show that no unitary 4×4 matrix U exists, so that

$$U\left[\begin{pmatrix} a \\ b \end{pmatrix} \otimes \begin{pmatrix} 1 \\ 0 \end{pmatrix}\right] = \begin{pmatrix} a \\ b \end{pmatrix} \otimes \begin{pmatrix} a \\ b \end{pmatrix}, \tag{5.59}$$

for $|a|^2 + |b|^2 = 1$ and $a, b \neq 0$. That is, $ab = 0$ for Eq. (5.59) to be true.

Hint: (a) Take $b = 0$. (b) Use superposition to write the RHS of Eq. (5.59) as the sum of four terms. Now, on the LHS of Eq. (5.59), write the vector that U operates on as $\begin{pmatrix} a \\ 0 \end{pmatrix} \otimes \begin{pmatrix} 1 \\ 0 \end{pmatrix} + \begin{pmatrix} 0 \\ b \end{pmatrix} \otimes \begin{pmatrix} 1 \\ 0 \end{pmatrix}$. U operating on this vector must give $\begin{pmatrix} a \\ 0 \end{pmatrix} \otimes \begin{pmatrix} a \\ 0 \end{pmatrix} + \begin{pmatrix} 0 \\ b \end{pmatrix} \otimes \begin{pmatrix} 0 \\ b \end{pmatrix}$. Explain why this is so. Then compare this result with the RHS to show that the result follows only if $ab = 0$.

5.2.5 DENSE CODING

Dense coding is a procedure by which two classical bits can be transmitted by sending a single qubit. Consider two parties, Alice and Bob, who are far apart. Alice has two classical bits of information that she wants to transmit to Bob. Can she achieve her goal by sending Bob only one qubit? The answer is affirmative, as demonstrated by Charles Bennett and Stephen Wiesner in 1992 [63]. Alice and Bob initially share two entangled qubits, say in the Bell state $|\Phi^+\rangle = \frac{1}{\sqrt{2}}[|00\rangle + |11\rangle]$. Alice has the first qubit, whereas Bob has the second. They can apply a single-qubit gate to their qubit, if they are in possession of both qubits, they can apply a two-qubit gate, and they can measure the qubits that are in their possession. We first specify what Alice needs to do if she wants to send the classical bits $(0,0)$, (01), $(1,0)$, and (11) and then specify what Bob needs to do. Suppose Alice wants to send Bob the two classical bits $(0,0)$, then she simply sends her qubit as it is to Bob and the two qubits will be in entangled form $|\Phi^+\rangle$. To send "01," she applies the gate σ_z of Eq. (5.43) to her qubit and sends, leading to the entangled state $|\Psi^+\rangle$. To send "10," Alice applies the NOT gate σ_x and sends, leading to $|\Phi^-\rangle$, and to send "11," she applies $i\sigma_y$ and sends, leading to $|\Psi^-\rangle$. Thus, the classical bits Alice wants to send and the corresponding states that Bob has are [see Eq. (5.55)],

$$00: |\Phi^+\rangle = \frac{1}{\sqrt{2}}[|00\rangle + |11\rangle] = |\Xi_{00}\rangle, \quad 01: |\Psi^+\rangle = \frac{1}{\sqrt{2}}[|01\rangle + |10\rangle] = |\Xi_{01}\rangle, \tag{5.60a}$$

$$10: |\Phi^-\rangle = \frac{1}{\sqrt{2}}[|00\rangle - |11\rangle] = |\Xi_{10}\rangle, \quad 11: |\Psi^-\rangle = \frac{1}{\sqrt{2}}[|01\rangle - |10\rangle] = |\Xi_{11}\rangle. \tag{5.60b}$$

Using Eqs (5.55) and (5.56), Bob can apply the *inverse transformation* to get

$$|x, y\rangle = [\text{CNOT } H_1]^{-1} |\Xi_{xy}\rangle \tag{5.61}$$

and measure the two qubits in the computational basis. The result of this projective measurement lets him determine the numbers x, y, exactly the classical bits Alice wanted to transmit. Thus, entanglement is an information resource; Alice sent only one qubit and effectively transmitted two classical bits of information.

Problem 5.21

Show that after applying a CNOT gate and then applying a Hadamard gate to qubit 1 Bob will have the following states, depending on which of the states in Eq. (5.60a) he has $|00\rangle$, $|01\rangle$, $|10\rangle$, and $-i|11\rangle$.

5.2.6 DATA COMPRESSION OF QUANTUM INFORMATION

A quantum analog of data compression was developed by Jozsa and Schumacher in 1994 and Schumacher in 1995 [64, 65] and is called the *Schumacher quantum noiseless channel coding theorem*.[8] The idea behind data compression for quantum information is to collect a large number, $n \gg 1$, of systems with density matrix ρ and encode the joint state of these systems into some smaller system [64, 65]. The smaller system is transmitted down a channel, and at the receiving end, the joint state is decoded into n systems of the same type as the original system. The final density matrix of each of the n received states is ρ', and the process is considered successful if ρ' is sufficiently close to ρ. A measure of the similarity between two density matrices ρ and σ is the fidelity f defined as [See Eq. (2.70)]:

$$f(\rho, \sigma) \equiv \text{Tr}\sqrt{\rho^{\frac{1}{2}} \sigma \rho^{\frac{1}{2}}}. \tag{5.62}$$

When ρ and σ are both pure states, the fidelity is simply the overlap of the pure states.

The von Neumann entropy $S(\rho) = -\text{Tr}\,\rho \log_2 \rho$, where ρ is the density operator describing an ensemble of states of the quantum system, can be used to quantify the amount of information in the quantum state. This is to be compared with the classical Shannon entropy for the classical case. The density matrix can be written as $\rho = \sum_x p(x)|x\rangle\langle x|$, where the states $|x\rangle$ are an orthogonal set and $p(x)$ are the eigenvalues of the density matrix.

The goal of quantum data compression is to find the smallest transmitted system, which permits fidelity $f = 1 - \epsilon$ for $\epsilon \ll 1$. Let us consider for simplicity two-level systems; the total state of n two-level systems is represented by a vector in a Hilbert space of dimension $2n$. However, if the von Neumann entropy $S(\rho) < 1$, then it is almost certain in the limit of large n that, in any given realization, the state vector actually falls in a typical subspace of Hilbert space. Jozsa and Schumacher showed that the dimension of the typical subspace is $2^{nS(\rho)}$. Hence, only $nS(\rho)$ qubits are required to represent the quantum information faithfully, i.e., the logarithm of the dimensionality of Hilbert space is a useful measure of quantum information. Furthermore, the encoding and decoding operation is blind: it does not depend on knowledge of the exact states being transmitted. This is a powerful general result. No assumptions about the exact nature of the quantum states is made. For example, the quantum states need not be orthogonal. If the states to be transmitted were mutually orthogonal, the whole problem would reduce to one of classical information.

These ideas form the basis of Schumacher's noiseless channel coding theorem, the quantum analogue of Shannon's theorem formulated for the classical case. To proceed, we need to clarify the concept of a quantum information source. Although there is no unique definition, the central role is played by entanglement and leads to the following definition: A quantum i.i.d. source is described as a pair $(\mathcal{H}, \rho)$, where $\mathcal{H}$ is a Hilbert space and $\rho = \sum_x p(x)|x\rangle\langle x|$ as defined above, is a mixed state density matrix on that space. Such a mixed state ρ can be obtained by partial tracing over a larger system whose density matrix is pure; the mixed state ρ results becuase of the entanglement between $\mathcal{H}$ and the rest of the system that is traced out. In classical information theory, a classical information source generates a sequence $(x_1, x_2, \ldots, x_n)$ that belongs to the set of d^n different sequences, and encodes classical information that can be compressed, transmitted and decompressed. Analogously, a quantum information source can randomly generate n mixed states. The state $\rho_1 \otimes \rho_2 \ldots \otimes \rho_n \in \mathcal{H}^{\otimes n}$, whose dimension is d^n, encodes quantum information that can be compressed, transmitted and decompressed. Hence, if the dimension of $\mathcal{H}$ is d (e.g., the dimension of $\mathcal{H}$ is 2 for qubits, 3 for qutrits and d for "qudits"), n qubits of information can be contained in a space $\mathcal{H}^{\otimes n}$ of dimension $2^{n \log_2 d}$ space.

The compression procedure, $\mathcal{C}$, similar to the procedure used for classical bits (see Sec. 5.1.2), transforms these quantum mixed states into states residing in a 2^{nR} dimensional (compressed) space, whereas the subsequent decompression, $\mathcal{D}$, transforms them back to the original space. Just as in the classical case, the procedure can be viewed as transforming strings of $n \log_2 d$ qubits into strings of nR qubits, where R is the rate of the compression scheme. If the corresponding fidelity is arbitrarily close to one for large n, the procedure is said to be reliable. The goal is to achieve reliable compression with smaller and smaller rate R, and Schumacher's theorem states that the lower bound of R that keeps the compression reliable is the von-Neumann entropy. Explicitly,

---

[8] The issue of data compression is directly related to the issue of the resources needed to store or transmit the state of a quantum system via the system's density matrix ρ.

5.2 Quantum Information

Schumacher's noiseless channel coding theorem: For a given quantum i.i.d. source, $\{\mathcal{H}, \rho\}$, a compression scheme of rate R is reliable if $R > S(\rho)$ and not reliable if $R < S(\rho)$.

5.2.7 QUANTUM TELEPORTATION

Quantum teleportation allows two parties that are far apart to exchange *unknown qubits* among them even in the absence of quantum communication channels between them. Teleportation serves as an ingredient in several computation and communication tasks. Suppose Alice and Bob share a pair of entangled qubits, say in a Bell state $|\Phi^+\rangle = \frac{1}{\sqrt{2}}[|00\rangle + |11\rangle]$, where the left part is Alice's and the right one is Bob's. Alice possesses a third qubit $|\phi\rangle = a|0\rangle + b|1\rangle$, which she wants to send to Bob, but she does not know a and b and there is no quantum communication channel between them. If Alice knew $|\phi\rangle$, she could send Bob a classical message using dense coding, but communicating an unknown qubit using classical communication channel is harder. A solution to this problem was suggested by Charles Bennett [66]. As in dense coding, quantum entanglement is used as an information resource. The required steps (see Fig. 5.5) are as follows:

1. Alice takes her unknown qubit $|\phi\rangle$ that together with state $|\Phi^+\rangle$ gives the initial state of the three qubits:

$$|\Psi_0\rangle = |\phi\rangle \otimes |\Phi^+\rangle = \frac{1}{\sqrt{2}}[a(|000\rangle + |011\rangle) + b(|100\rangle + |111\rangle)], \quad (5.63)$$

   where in each term, the first two states are Alice's and the third is Bob's.

2. Alice applies a CNOT gate on her two qubits (see the CNOT operation in Fig. 5.5). This can be carried out term by term using Eq. (5.50). The result is

$$|\Psi_1\rangle = \text{CNOT}_A |\Psi_0\rangle = \frac{1}{\sqrt{2}}[a(|000\rangle + |011\rangle) + b(|110\rangle + |101\rangle)]. \quad (5.64)$$

3. Alice applies the Hadamard gate on her unknown qubit (i.e., on each left basis state in $|\Psi_1\rangle$). This yields $|\Psi_2\rangle = H_1 |\Psi_1\rangle$, where

$$|\Psi_2\rangle = \frac{1}{2}[a(|0\rangle + |1\rangle)(|00\rangle + |11\rangle) + b(|0\rangle - |1\rangle)(|10\rangle + |01\rangle)]$$
$$= \frac{1}{2}[|00\rangle(a|0\rangle + b|1\rangle) + |01\rangle(a|1\rangle + b|0\rangle) + |10\rangle(a|0\rangle - b|1\rangle) + |11\rangle(a|1\rangle - b|0\rangle)]. \quad (5.65)$$

4. Alice now measures the state of her two qubits, and this collapses the state onto one of four different possibilities, yielding two classical bits. If Alice measures state $|00\rangle$, the qubit Bob has is in the state $|\phi\rangle$; if she measures $|01\rangle$, Bob's qubit is in state $a|1\rangle + b|0\rangle$; if she measures $|10\rangle$, Bob's qubit is in state $a|0\rangle + b|1\rangle$; and, if she measures $|11\rangle$, Bob's qubit is in state $a|1\rangle - b|0\rangle$. Alice now sends the outcome information as *two classical bits* to Bob. Bob's qubit is in a state that depends on Alice's measurement according to the list:

$$00 \longmapsto [a|0\rangle + b|1\rangle], \quad 01 \longmapsto [a|1\rangle + b|0\rangle],$$
$$10 \longmapsto [a|0\rangle - b|1\rangle], \quad 11 \longmapsto [a|1\rangle - b|0\rangle]. \quad (5.66)$$

5. To obtain the initial unknown qubit $|\phi\rangle = a|0\rangle + b|1\rangle$, Bob needs to apply the appropriate single-qubit unitary gate: He does nothing if Alice's measurement yielded 00; if Alice measured 01, Bob applies the gate X; if she measured 10, Bob applies Z; and if she measured 11, Bob first applies X and then Z. The operations applied by Bob can be represented as $Z^{m_1} X^{m_2}$, where m_1 and m_2 are the classical bits that result from Alice's measurements.

Note that quantum teleportation does not allow communication of information faster than the classical channel used by Alice to convey the information about the results of her measurements to Bob.

5.2.8 QUANTUM CRYPTOGRAPHY

Quantum mechanics holds the promise for secure communications because a measurement of an unknown quantum system changes its state. Hence, accurate copying of an unknown qubit is impossible, as we have already seen from the no-cloning theorem in Sec. 5.2.4. Moreover, the changes caused by eavesdropping on, i.e., measurement of, quantum information can be detected. Thus, the laws of physics assure security of quantum information. To carry out secure quantum communication at the speed of light, very weak laser pulses (i.e., basically, single photon pulses, which rarely have more than one photon per pulse, see Sec. 5.6.5) can be used. These photons can be transmitted through air or fibers and can be detected with special receivers (see Sec. 5.6.5). Secure quantum communications can be implemented using quantum "key" distribution. Many existing systems use a quantum cryptography protocol known as Bennett–Brassard 1984, or BB84, which generates the secure quantum key that can be used to encode messages sent between two parties, Alice and Bob, to securely exchange information. By using single-photon sources to encode information, and single photon detectors, Alice and Bob can detect attempts at eavesdropping. The distribution of secret quantum keys to encrypt and decrypt messages insures the security of the information from eavesdroppers. Alice and Bob can send and receive photons in four different orientations to represent $|0\rangle$ and $|1\rangle$ qubits. Each photon is sent in one of two polarization modes, either vertical or horizontal orientations of the electric field, $\leftrightarrow$ or $\updownarrow$, or $\pm$ 45° orientations (see Fig. 5.6). In each polarization mode, one orientation represents $|0\rangle$, and the other represents $|1\rangle$. Alice randomly chooses both polarization modes of the photon and an orientation for each photon. Bob places polarizers before his photon detectors, so that the polarization mode of the photons can be measured. He randomly chooses between the two polarization modes when he tries to detect a photon. If he chooses the same mode that Alice used for a particular photon, then Bob always measures the correct orientation, and hence, its true bit value. But if he chooses a different polarization mode, he may get the wrong bit value for that photon. To determine a shared key from a stream of photons, Alice uses a conventional communications channel to tell Bob which mode she used for each photon, without revealing its bit value. Bob tells Alice which photons he measured using the correct polarization mode, without sharing their values. Then they both discard the other bits, i.e., the one Bob measured with the wrong polarization mode. The correct measurements constitute the secret key that Alice and Bob now share. Figure 5.6 shows a schematic illustration of the BB84 protocol.

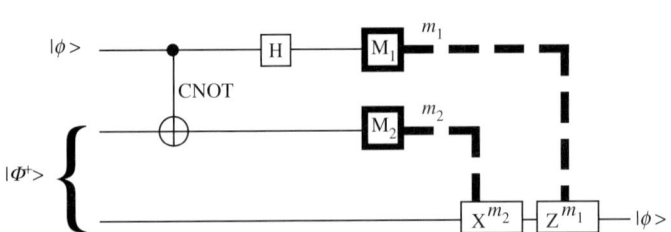

FIG 5.5 Teleportation of a qubit. Alice possesses two qubits, the unknown qubit $|\phi\rangle$ and one of an entangled pair of qubits (the second is in Bob's possession). Qubits are denoted by single horizontal lines. Alice performs a CNOT operation using the unknown qubit and her half of the entangled qubits, then performs a Hadamard gate on the unknown qubit (it has not changed as a result of the CNOT), and then measures the qubits in her possession to obtain the classical bits m_1 and m_2. Transfer of classical information from Alice to Bob is denoted by the heavy dashed lines. Bob performs unitary operations on his qubit based on the results sent to him by Alice.

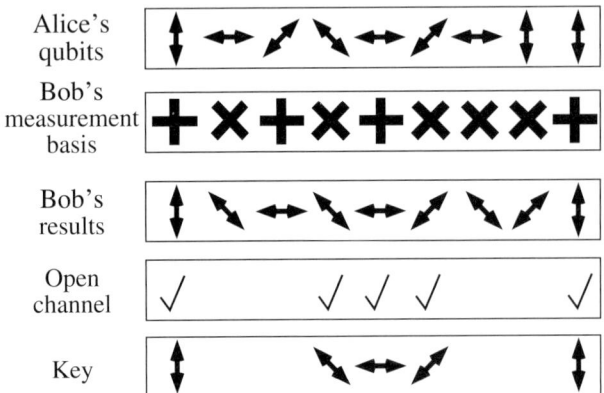

FIG 5.6 Schematic illustration of the Bennett–Brassard 1984 quantum cryptography protocol. Alice sends $|0\rangle$ and $|1\rangle$ qubits encoded in photon polarization in two different polarization modes. Bob randomly chooses a polarization mode for his measurement basis. Bob and Alice communicate over an open classical channel and thereby determine a secret key.

5.2 Quantum Information

The BB84 protocol insures that the message cannot be intercepted without the sender and the receiver finding out about it. If an eavesdropper, Eve, tries to eavesdrop on the message being sent, she will not be able to determine the polarization of a photon without altering or destroying it. If the photon is destroyed, it will not reach Bob and will not contribute to the key. If Eve sends a replacement photon, she may send the wrong bit value of the photon, because she used the wrong polarization mode to intercept the original photon. Hence, for photons detected with the wrong polarization mode, Eve may introduce errors into the key. If Alice and Bob detect an unusual number of errors in the key, they will be alerted to the presence of an eavesdropper.

If the idealized version of the BB84 protocol is secure, the real version may not be. Charles Bennett (the first "B" in BB84) recalled that the very first quantum-cryptographic system built used a high-voltage power supply to switch the polarization of the photons. "The power supply hummed differently depending on whether the voltage was being applied. If you listened, you could hear it." This quote underlines the dangers in assuming that even systems "assured to be secure by the laws of physics" are infallible.

BB84 can be generalized, so that other states and bases can be used. For example, Bennett suggested a simplified protocol in 1992, called B92, in which Alice, instead of the four pairwise orthogonal states, employs only two non-orthogonal states of the four states used in BB84. We will not present this protocol, but we note that in both protocols, the impossibility of distinguishing between nonorthogonal states without destroying the qubits lies at the heart of the protocol. It is impossible for an eavesdropper to distinguish between Alice's states without disrupting the correlation between the qubits that Alice and Bob finally keep.

5.2.9 QUANTUM CIRCUITS

We have already encountered the notion of quantum circuits in Figs 5.3 and 5.4. Here, they are more systematically introduced. A quantum circuit is a graphical representation of a series of operations by quantum gates and measurements on n-qubits. Alternatively, they can be viewed as diagrams describing the process of quantum computation. A quantum circuit represents a well-defined quantum algorithm that is written as an algebraic expression where the initial state appears on the right and the operations (quantum gates and measurements) operate one after the other from right to left. Graphically, however, the initial state appears on the left and the order of operations is from left to right.

The skeleton of a quantum circuit is a set of parallel lines (called wires) drawn horizontally. Each wire represents a qubit (or sometimes a collection of qubits), and the graph describes schematically the evolution of the n-qubit system with time. Wires cannot merge or bifurcate since this would imply irreversible logical gates.

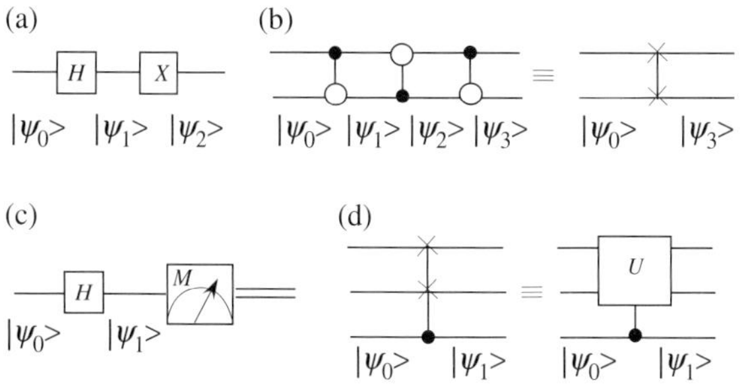

Single qubit gates are represented as a square on top of the wire at the appropriate position in the circuit, whereas two-qubit (controlled) gates are represented by vertical lines connecting the pertinent qubits. For example, in a two-qubit CNOT gate, the controlling qubit is denoted by a full circle, while the controlled qubit is denoted by an empty circle. A general n-qubit gate that is specified *ad hoc* for a given algorithm is denoted as U and represented by a square with n qubits entering it from the left and leaving it from the right. It may be controlled by other qubits. A measurement operation is denoted by M and is represented by a square containing a measurement apparatus drawing. Some examples of simple quantum circuits are shown in Fig. 5.7. In Fig. 5.7(a), the quantum circuit contains a single qubit. If the input qubit is $|\psi_0\rangle = |1\rangle$, then $|\psi_1\rangle = \frac{1}{\sqrt{2}}(|0\rangle - |1\rangle)$ and $|\psi_2\rangle = \frac{1}{\sqrt{2}}(-|0\rangle + |1\rangle)$.

FIG 5.7 Simple quantum circuits. (a) Single qubit Hadamard and X gates operating on the input state $|\psi_0\rangle$. The corresponding quantum algorithm is, $|\psi_1\rangle = H|\psi_0\rangle$ and $|\psi_2\rangle = X|\psi_1\rangle = XH|\psi_0\rangle$. (b) The circuit describes the operation of three staggered CNOT gates, which is equivalent to the SWAP gate. (c) A circuit containing a measurement operation, M, followed by two lines, which indicate that the result is one of several possible states. (d) Fredkin three-qubit gate is a special case of a circuit containing a controlled U gate.

In Fig. 5.7(b), the quantum circuit contains three staggered CNOT gates, representing graphically the algorithm for generating a SWAP gate. In Fig. 5.7(c), the quantum circuit contains a measurement element, after which the quantum system collapses into one component of a sum over basis states with the appropriate amplitude. That is the reason for the appearance of two adjacent lines. Finally, in Fig. 5.7(d), the quantum circuit represented the Fredkin gate as an example of the use of a controlled U gate.

Problem 5.22

(a) Determine $|\psi_2\rangle$ in circuit (a) if $|\psi_0\rangle = a|0\rangle + b|1\rangle$.
(b) Determine $|\psi_2\rangle$ in circuit (b) if $|\psi_0\rangle = a|0,1\rangle + b|1,0\rangle$.
(c) Determine the output in quantum circuit (c) if $|\psi_0\rangle = a|0\rangle + b|1\rangle$.
(d) Determine $|\psi_1\rangle$ in circuit (d) if $|\psi_0\rangle = |1,0,1\rangle$.

Answers: (a) $\frac{1}{\sqrt{2}}[(a-b)|0\rangle + (a+b)|1\rangle]$. (b) $a|1,1\rangle + b|01\rangle$. (c) Either $\frac{1}{\sqrt{2}}(a+b)|0\rangle$ or $\frac{1}{\sqrt{2}}(a-b)|1\rangle$. (d) $|0,1,1\rangle$.

5.2.10 QUANTUM COMPUTING DESPITE MEASUREMENT

The difficult part of developing quantum algorithms for computing is extracting the information that is contained in the qubits after the quantum gates are applied, because measurement of the qubits destroys the qubits by projecting them. We shall see in Sec. 5.3 how, despite this difficulty, quantum computation is still possible.

In the hope of alleviating this difficulty, one might consider using what is called a *weak measurement* [67], which is a type of quantum measurement where the measured system is very weakly coupled to the measuring device. After the measurement, the measuring device points to a "weak value" indicating the state of the measured system, but the system is not disturbed by the measurement. This may seem to contradict some basic concepts of quantum theory, since any quantum system should, on measurement of an observable, end up in the eigenstate corresponding to the detected eigenvalue, but in fact this does not contradict any quantum principles.

There are other types of measurement that can be considered. For example, a *nondemolition measurement* (or nondestructive measurement) [68] is a measurement that preserves the integrity of the system and the value of the measured observable, thereby allowing the system to be measured repeatedly (this kind of measurement corresponds to the kind of projective measurement envisioned in the quantum postulates). Moreover, there are measurements that demolish the system entirely (such as the measurement of photons by photodetectors) and therefore do not correspond to projective measurements. Furthermore, there are more general forms of measurement known as POVM (Positive Operator-Valued Measure) and involve projection of a quantum state on a set of nonorthogonal states. In this chapter, we shall consider only the projective type of measurement (also called von Neumann measurement) discussed in the quantum postulates in Sec. 1.3.4.

5.3 QUANTUM COMPUTING ALGORITHMS

There are three known categories of quantum computing algorithms that seem to provide advantage over known classical algorithms: (1) Algorithms based on quantum versions of the Fourier transform (e.g., the Deutsch and Deutsch–Jozsa algorithms, Shor's factoring algorithm, and the discrete logarithm algorithm). (2) Quantum search algorithms (e.g., the Grover search algorithm). (3) Quantum simulation algorithms, whereby a quantum computer simulates a quantum system. We now briefly describe each of these classes of algorithms.

We do not yet know for sure whether quantum computers can be used to efficiently (in polynomial time) solve problems in NP. A negative result in this direction exists. One approach to solve problems in NP on a quantum computer is to try to use quantum parallelism to search through all the possible solutions to the problem in parallel. It is known that this approach cannot work for some problems in NP, but this does not rule out the possibility that other quantum algorithms

5.3 Quantum Computing Algorithms

will work. Quantum computers can be used to solve some problems (e.g., factoring), which *may* be in NP but not in P, but this is an open issue since the complexity class of the factorization problem is not yet known.

It is of interest to develop a theory of quantum computational complexity and relate it to classical computational complexity theory. Unfortunately, little has been done in this respect. This is still a wide open research area.

Before considering specific quantum computing algorithms, let us try to understand the basics of how a computation can be performed using qubits and quantum gates. Suppose we want to compute the function $f : i_1 i_2 \ldots i_n \longmapsto f(i_1, \ldots, i_n)$, from n to n bits. We want an n-qubit system to evolve according to the time evolution operator $\hat{U}_f$:

$$|i_1, i_2, \ldots, i_n\rangle \longmapsto \hat{U}_f |i_1, i_2, \ldots, i_n\rangle \equiv |f(i_1, \ldots, i_n)\rangle. \tag{5.67}$$

We must find the Hamiltonian $\hat{H}_f$ that generates this evolution, i.e., determine the Hamiltonian whose evolution operator $\hat{U}_f$ solves the equation,

$$i\hbar \frac{\partial \hat{U}_f(t)}{\partial t} = \hat{H}_f(t) \hat{U}_f(t). \tag{5.68}$$

Applying the unitary operator that computes f, Eq. (5.67), we obtain

$$\frac{1}{\sqrt{2^n}} \sum_{i_1, i_2, \ldots, i_n=0}^{1} |i_1, i_2, \ldots, i_n\rangle \longmapsto \frac{1}{\sqrt{2^n}} \sum_{i_1, i_2, \ldots, i_n=0}^{1} |f(i_1, i_2, \ldots, i_n)\rangle, \tag{5.69}$$

by virtue of the linearity of quantum mechanics. Thus, applying the operator $\hat{U}_f$ once computes f simultaneously on all 2^n possible inputs. Herein lies the tremendous power of parallelism. For convenience, we can employ the enumeration scheme (5.22) and use the notation $x \equiv i_1, i_2, \ldots, i_n$, so Eq. (5.69) reads,

$$\frac{1}{\sqrt{2^n}} \sum_{x=0}^{2^n-1} |x\rangle \longmapsto \frac{1}{\sqrt{2^n}} \sum_{x=0}^{2^n-1} |f(x)\rangle. \tag{5.70}$$

But, how do we extract the information about $f(x)$ out of the quantum mechanical system? To extract quantum information, one has to measure the system, but the measurement process causes collapse of the wave function, i.e., the state of the system is projected to only one of the many possible states available, and most of the information that has been computed is lost. To gain advantage of quantum parallelism, one needs to combine it with another quantum characteristic: interference. We must arrange the cancellation occurring due to interference in such a way that only the computations we are interested in remain, and the rest cancel out. The combination of parallelism and interference gives quantum computation its power. Let us now see how this works out in the examples of quantum computation presented below.

5.3.1 DEUTSCH AND DEUTSCH–JOZSA ALGORITHMS

The Deutsch algorithm [62] is a paradigm for "massive quantum parallelism." It can be stated in its most simple form as follows: given a function $f(x) : \{0, 1\} \to \{0, 1\}$, construct a quantum circuit to calculate $f(0) \oplus f(1)$. Note that both the domain and the target of f are one bit (not qubit). The Deutsch–Jozsa algorithm generalizes it to a function $f(x_1, x_2, \ldots, x_n) \to \{0, 1\}$ whose domain is an n bit space, while its target is a single bit space. The algorithm is designed to answer the question whether f is constant or balanced (i.e., it assumes the value 0 on half of its 2^n arguments and 1 on the other half).

The U_f Gate

In our discussion of quantum circuits (see Sec. 5.2.9), we introduced a general purpose operator U, which is defined *ad hoc* based on the problem to be solved. In a problem requiring the evaluation of a function $f : \{x_1, x_2, \ldots, x_n\} \to \{0, 1\}$, this operator is referred to as *the U_f gate*. It is an n qubit gate that transforms a state $|x, y\rangle$ in $n + 1$ qubit space into a state

$|x, y \oplus f(x)\rangle$ in the same space. Here, $|x\rangle$, with $x = 0, 1, \ldots, 2^n - 1$, enumerates all 2^n n-qubit states and $|y\rangle$ with $y = 0, 1$ is a single qubit state. The n-qubit state $|x\rangle$ is called the data register, and the right qubit is called the target register. Thus,

$$|x, y\rangle \longmapsto U_f |x, y\rangle = |x, y \oplus f(x)\rangle. \tag{5.71}$$

In the Deutsch algorithm, $n = 1$ and $f : \{0, 1\} \to \{0, 1\}$. If the data register is in state $1/\sqrt{2}[|0\rangle + |1\rangle]$ and the target register is initially in state $|0\rangle$, then the resulting two-qubit state is $\frac{1}{\sqrt{2}}[|0, f(0)\rangle + |1, f(1)\rangle]$. The U_f gate evaluates $f(x)$ for two values of x simultaneously. The two terms contain information about both $f(0)$ and $f(1)$. This is a manifestation of quantum parallelism.

In the Deutsch–Jozsa algorithm, $f : \{x_1, x_2, \ldots, x_n\} \to \{0, 1\}$ and U_f acts on a n-qubit state. The procedure for getting a sum over states $|x, f(x)\rangle$ is achieved by two steps. First, applying Hadamard transform on the n data qubits starting from the initial state $|\psi_0\rangle = |0\rangle$ yields,

$$H^{\otimes n}|0, 0, \ldots, 0\rangle = \left(\frac{1}{\sqrt{2}}\right)^n [|0\rangle + |1\rangle]^n = \frac{1}{\sqrt{2^n}} \sum_{i_n=0}^{1} \cdots \sum_{i_1=0}^{1} |i_1, \ldots, i_n\rangle$$

$$= \frac{1}{\sqrt{2^n}} \sum_{x=0}^{2^n-1} |x\rangle \equiv |\psi_{\text{WH}}\rangle. \tag{5.72}$$

The Walsh–Hadamard state $|\psi_{\text{WH}}\rangle$ is an equally weighted coherent superposition of all 2^n distinct basis states [see (5.21)]. Second, parallel evaluation of a function $f(x)$ with an n bit input x and a 1 bit output, $f(x)$, can be performed by applying the n-qubit gate U_f, which for a given $n + 1$-qubit state $|x, 0\rangle$ results in,

$$|x\rangle|0\rangle \longmapsto U_f |x\rangle|0\rangle = |x\rangle|f(x)\rangle. \tag{5.73}$$

When applied to the state in Eq. (5.72), U_f yields the following result:

$$U_f \left[H^{\otimes n}|0, 0, \ldots, 0\rangle|0\rangle\right] = \frac{1}{\sqrt{2^n}} \sum_{x=0}^{2^n-1} |x\rangle|f(x)\rangle. \tag{5.74}$$

Thus, "Massive quantum parallelism" enables all values of the function f to be evaluated simultaneously.

Problem 5.23

(a) Calculate the two-qubit Walsh–Hadamard transform transformation $H^{\otimes 2}$ applied to $|0\rangle \otimes |0\rangle$.
(b) Calculate the two-qubit Walsh–Hadamard transform transformation $H^{\otimes 2}$ applied to $|1\rangle \otimes |1\rangle$.

Answers: (a) $H^{\otimes 2}|0\rangle \otimes |0\rangle = H|0\rangle \otimes H|0\rangle = \frac{1}{2}(|0\rangle + |1\rangle)(|0\rangle + |1\rangle)$.
(b) $H^{\otimes 2}|1\rangle \otimes |1\rangle = H|1\rangle \otimes H|1\rangle = \frac{1}{2}(|0\rangle - |1\rangle)(|0\rangle - |1\rangle)$.

Unfortunately, this form of "massive quantum parallelism" is not useful unless we can extract an information about $f(x)$ for all x. In the single-qubit example, measurement of the state gives *either* $|0, f(0)\rangle$ *or* $|1, f(1)\rangle$, *not both*. In the n-qubit example, measurement of $\sum_{x=0}^{2^n-1} |x\rangle|f(x)\rangle$ yields the pair $(x, f(x))$ for only one value of x. How do we extract information about more than one value of $f(x)$ from superposition states? This question is answered by the two algorithms detailed below, which involve a simple modification of the gate described above, together with the use of interference.

5.3 Quantum Computing Algorithms

The Deutsch Algorithm

In the single-qubit example, we follow the algorithm shown in Fig. 5.8, and let the output register start off in state $|1\rangle$, so that the initial two-qubit state is $|0\rangle|1\rangle$. Applying the Hadamard transformation to both qubits, $H_1 H_2 |0\rangle|1\rangle = \left(\frac{1}{\sqrt{2}}\right)^2 [|0\rangle + |1\rangle][|0\rangle - |1\rangle]$, and applying the gate U_f gives

$$U_f [H_1 H_2 |0\rangle|1\rangle] = \begin{cases} \pm\left(\frac{1}{\sqrt{2}}\right)^2 [|0\rangle + |1\rangle][|0\rangle - |1\rangle] & \text{if } f(0) = f(1) \\ \pm\left(\frac{1}{\sqrt{2}}\right)^2 [|0\rangle - |1\rangle][|0\rangle - |1\rangle] & \text{if } f(0) \neq f(1) \end{cases}. \quad (5.75)$$

In deriving Eq. (5.75), we used the fact that $U_f |x\rangle[|0\rangle - |1\rangle] = (-1)^{f(x)} |x\rangle[|0\rangle - |1\rangle]$. Applying a final Hadamard transformation to the first qubit yields

$$H_1 U_f [H_1 H_2 |0\rangle|1\rangle] = \begin{cases} \pm\frac{1}{\sqrt{2}} |0\rangle [|0\rangle - |1\rangle] & \text{if } f(0) = f(1) \\ \pm\frac{1}{\sqrt{2}} |1\rangle [|0\rangle - |1\rangle] & \text{if } f(0) \neq f(1) \end{cases}. \quad (5.76)$$

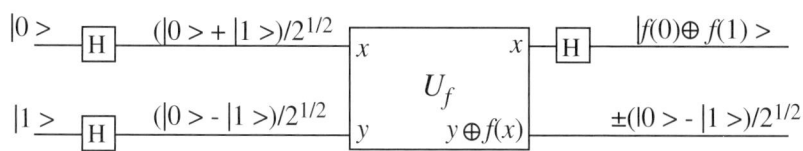

FIG 5.8 Schematic of the Deutsch algorithm.

Suppose the aim of the calculation is to differentiate between the following two alternative types of functions: the *constant* function, $f(x)$ is the same for all values of x, and the *balanced* function, $f(x) = 1$ for half the values of x and $f(x) = 0$ for the other half. Since $f(0) \oplus f(1) = 0$ if $f(0) = f(1)$ and $f(0) \oplus f(1) = 1$ if f is balanced, Eq. (5.76) can be written as

$$H_1 U_f [H_1 H_2 |0\rangle|1\rangle] = \pm\frac{1}{\sqrt{2}} |f(0) \oplus f(1)\rangle [|0\rangle - |1\rangle]. \quad (5.77)$$

By measuring the first qubit, we determine $f(0) \oplus f(1)$. This is a property of f, which is global; it depends on the values of the function at all values of x.

Deutsch–Jozsa Algorithm

The Deutsch–Jozsa algorithm generalizes the Deutsch algorithm for a function $f(x_1, x_1, \ldots, x_n) \to \{0, 1\}$ defined on n-bit states and has a one bit target space. Here, the specific challenge is to construct an efficient algorithm to decide whether $f(x)$ is constant, say $f(x) = 1$, or balanced. A balanced function from n bit domain to one bit target assumes the value 0 on 2^{n-1} arguments from its domain and 1 on the other 2^{n-1} arguments.

We start with n data qubits plus one target qubit in state, $|0, 0, \ldots, 0\rangle \otimes |1\rangle$. After applying the Walsh–Hadamard transform to the data qubits and obtaining Eq. (5.72), a Hadamard transformation H is applied to the target qubit $|1\rangle$ to obtain the state

$$|\psi_1\rangle = \frac{1}{\sqrt{2^n}} \sum_{x=0}^{2^n-1} |x\rangle \frac{1}{\sqrt{2}} [|0\rangle - |1\rangle]. \quad (5.78)$$

Then the operator U_f is applied to obtain

$$|\psi_2\rangle = \frac{1}{\sqrt{2^n}} \sum_{x=0}^{2^n-1} (-1)^{f(x)} |x\rangle \frac{1}{\sqrt{2}} [|0\rangle - |1\rangle]. \quad (5.79)$$

The data qubits now have the function values $f(x)$ stored in the amplitudes of the superposition state. Another Walsh–Hadamard transformation $H^{\otimes n}$ is applied to the data qubits, yielding the state

$$|\psi_3\rangle = \sum_{z,x=0}^{2^n-1} \frac{(-1)^{x \cdot z + f(x)}}{2^n} |x\rangle \frac{[|0\rangle - |1\rangle]}{\sqrt{2}}, \qquad (5.80)$$

where $x \cdot z$ is the modulo 2 bitwise inner product of x and z. [To understand Eq. (5.80), note that for $x = 0$ or $x = 1$, $H|x\rangle = 2^{-1/2} \sum_z (-1)^{xz} |z\rangle$; hence, $H^{\otimes n}|x_1,\ldots,x_n\rangle = 2^{-n/2} \sum_{z_1,\ldots,z_n} (-1)^{x_1 z_1 + \ldots + x_n z_n} |z_1,\ldots,z_n\rangle$]. Now, the data register is measured. It is surprising that the measured amplitude for the state $|0,\ldots,0\rangle$ is $(-1)^{f(x)}/2^n$. Therefore, for the case where $f(x) = $ constant, the amplitude for $|0,\ldots,0\rangle$ is $+1$ or -1, depending on the constant value $f(x)$ takes. Since $|\psi_3\rangle$ is of unit length, all the other amplitudes must be zero, and an observation will yield 0 for all qubits in the query register. If instead $f(x)$ is balanced, then the positive and negative contributions to the amplitude for $|0,\ldots,0\rangle$ cancel, leaving an amplitude of zero, and a measurement must yield a result other than 0 on at least one qubit in the date register. So, if all zeros are measured, the function is constant, and otherwise the function is balanced.

We shall see that the Deutsch–Jozsa algorithm actually makes use of the quantum Fourier transform algorithm (as do the discrete logarithm algorithm and the Shor factorization algorithm) via the Walsh–Hadamard transformation $H^{\otimes n}$. This is one of three classes of known quantum algorithms that provide advantages over classical algorithms, where the other two classes are quantum search algorithms and quantum simulation algorithms. Let us now consider quantum search algorithms.

5.3.2 THE GROVER SEARCH ALGORITHM

The quantum search algorithm was discovered by Lov Grover in 1996 [69]. It solves the following problem: given a database with N stored elements, with no prior knowledge about the structure of the information, find the element satisfying a given property. This problem requires approximately $N/2$ operations (on the average) with classical algorithms, but the quantum search algorithm requires approximately $\sqrt{N}$ operations, i.e., the quantum algorithm offers a quadratic speedup, as compared with the exponential speedup of the quantum Fourier transform (FT) algorithm. Moreover, Grover's algorithm is probabilistic in the sense that it gives the correct answer with high probability, but not certainty. Furthermore, on completing the algorithm, the data base is destroyed via the measurement carried out at the end of the algorithm.

Grover's algorithm is an inverting algorithm. Given a function $y = f(x)$ that can be evaluated on a quantum computer, this algorithm allows the determination of x when y is given. It is related to search of a database, since a function can be determined that produces a particular value of y if x matches a desired entry in the database.

Suppose we want to search a space of N elements, say, to find the names of persons from their list of phone numbers. To do so, we construct an *index x* corresponding to those elements, which is an integer in the range $[0, N-1]$, as shown in Fig. 5.9. Let us assume that $N = 2^n$, so the index can be stored in n (classical) bits. Furthermore, let us assume that the search problem has M solutions, with $0 \leq M \leq N$. Below, we assume, for simplicity, that only one phone number can meet the search criteria.

The solution to the search problem can be developed in terms of a function $f(x)$ with input x, which is an integer in the range $[0, N-1]$ (the index x defined in the previous paragraph), with $f(x) = 1$ if x is a solution to the search problem, and $f(x) = 0$

index (x)	name	phone number
0	A	d_0
1	B	d_1
2	C	d_1
3	D	d_2
4	D	d_3
5	F	d_4
.	.	.
.	.	.
.	.	.
N-1	Z	d_{N-1}

FIG 5.9 Schematic of a telephone directory to be searched in the Grover algorithm. The names are alphabetically arranged, but the phone numbers are randomly assigned.

5.3 Quantum Computing Algorithms

if x is not a solution. We define a unitary operator U_f (sometimes called an *oracle*), which acts on the states $|x\rangle|q\rangle$, where $|x\rangle$ are n-qubit states and $|q\rangle$ is an additional qubit (sometimes called the *oracle qubit*), such that

$$|x\rangle|q\rangle \longmapsto U_f|x\rangle|q\rangle = |x\rangle|q \oplus f(x)\rangle. \tag{5.81}$$

The oracle qubit $|q\rangle$ is flipped if $f(x) = 1$ and is unchanged otherwise. (If more than one solution to the search problem exists, more than one oracle qubit is required.) To determine whether x is a solution to our search problem, one can prepare $|x\rangle|0\rangle$, apply the oracle U_f, and check to see whether the oracle qubit has been flipped to $|1\rangle$. In the actual Grover algorithm, we apply U_f with the oracle qubit initially in the state $\frac{|0\rangle - |1\rangle}{\sqrt{2}}$, in a fashion similar to the Deutsch–Jozsa algorithm (i.e., we apply the Hadamard transformation to the oracle qubit in the initial state $|1\rangle$, thereby obtaining $|q\rangle = \frac{|0\rangle - |1\rangle}{\sqrt{2}}$). If x is not a solution to the search problem, applying U_f to the state $|x\rangle \frac{|0\rangle - |1\rangle}{\sqrt{2}}$ does not change the state of the $n+1$-qubit system, but if x is a solution, it gives the state $-|x\rangle \frac{|0\rangle - |1\rangle}{\sqrt{2}}$, i.e.,

$$|x\rangle \frac{|0\rangle - |1\rangle}{\sqrt{2}} \longmapsto U_f|x\rangle|q\rangle = (-1)^{f(x)}|x\rangle \frac{|0\rangle - |1\rangle}{\sqrt{2}}. \tag{5.82}$$

The Grover algorithm starts by considering $n = \log_2 N$ qubits in the state $|0\rangle$ and applying a Walsh–Hadamard transform to them, as in Eq. (5.72):

$$H^{\otimes n}|0, 0, \ldots, 0\rangle = \frac{1}{\sqrt{2^n}} \sum_{x=0}^{2^n-1} |x\rangle \equiv |\psi_{\text{WH}}\rangle. \tag{5.83}$$

In addition to the n qubits, a set of work qubits called the oracle qubits are employed, where the initial state of each oracle qubit is $|1\rangle$, and a Hadamard transformation is applied to each oracle qubit. For simplicity, we have assumed that only one phone number can meet the search criteria; then, only one oracle qubit is required. Then, the oracle operator U_f will be applied to $|\psi_{\text{WH}}\rangle \left(\frac{|0\rangle - |1\rangle}{\sqrt{2}}\right) = \left(\frac{1}{\sqrt{2^n}} \sum_{x=0}^{2^n-1} |x\rangle\right) \left(\frac{|0\rangle - |1\rangle}{\sqrt{2}}\right)$ to obtain

$$U_f \left(\frac{1}{\sqrt{2^n}} \sum_{x=0}^{2^n-1} |x\rangle \frac{|0\rangle - |1\rangle}{\sqrt{2}} \right) = \sum_{x=0}^{2^n-1} (-1)^{f(x)} |x\rangle \frac{|0\rangle - |1\rangle}{\sqrt{2}}. \tag{5.84}$$

A schematic representation of the Grover algorithm for this case is shown in Fig. 5.10. After the preparation of $|\psi_{\text{WH}}\rangle \left(\frac{|0\rangle - |1\rangle}{\sqrt{2}}\right) \equiv \left(\frac{1}{\sqrt{2^n}} \sum_{x=0}^{2^n-1} |x\rangle\right) \left(\frac{|0\rangle - |1\rangle}{\sqrt{2}}\right)$, the *Grover operator* G is applied on the order of $\sqrt{N} = 2^{n/2}$ times, where the Grover operator G is defined as the product of the oracle operator U_f and the "inversion about the average operator," D, i.e., $G = DU_f$. The structure of the operator D, called the *inversion about the average operator*, is detailed below. Then, the state of the n-qubit register is measured. The output, i.e., the sequence of 0s and 1s that are measured, is a binary representation of index x of the desired element, with probability close to unity.

The "inversion about the average operator," D, is defined as follows:

$$D \equiv H^{\otimes n} \mathcal{O}_{\text{phase}} H^{\otimes n} = H^{\otimes n}(2|0\rangle\langle 0| - 1)H^{\otimes n} = 2|\psi_{\text{WH}}\rangle\langle \psi_{\text{WH}}| - 1. \tag{5.85}$$

It operates only on the n-qubit register. The conditional phase-shift operator appearing in D, $\mathcal{O}_{\text{phase}} = (2|0\rangle\langle 0| - 1)$ is sandwiched between two Walsh–Hadamard transforms. When the operator D is applied to the arbitrary n-qubit superposition state $\sum_{x=0}^{2^n-1} c_x |x\rangle$, it yields

$$\sum_{x=0}^{2^n-1} c_x |x\rangle \longmapsto D \sum_{x=0}^{2^n-1} c_x |x\rangle = \sum_{x=0}^{2^n-1} (2\langle c \rangle - c_x)|x\rangle, \tag{5.86}$$

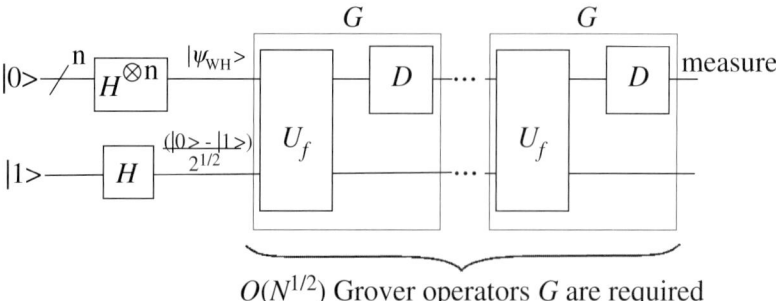

FIG 5.10 The quantum search algorithm. The oracle operator U_f employs a work qubit, called the oracle qubit, for its implementation. The Grover operator $G = DU_f$ is the product of the oracle operator U_f and the "inversion about the average operator" $D = H^{\otimes n}\mathcal{O}_{\text{phase}}H^{\otimes n}$ (see text), which operates only on the n-qubit register. The measurement performed at the end of the quantum search algorithm involves only the n-qubit register.

where $\langle c \rangle = N^{-1}\sum_{x=0}^{2^n-1} c_x$. In words, the D operator maps each amplitude c_x into $2\langle c \rangle - c_x$. We shall see that this means that the amplitude c_s of the state $|x_s\rangle$ representing the solution is increased by $\mathcal{O}(1/\sqrt{N})$ on each application of the Grover operator G.

The Grover operator can be regarded as a rotation in the two-dimensional space spanned by the solution vector $|x_s\rangle$ to the search problem and the space orthogonal to it, $\sum_{x \neq x_s} |x\rangle$ (if more than one solution exists, we need solution vectors, but here we assume there is only one). An arbitrary n-qubit superposition state $|\psi\rangle = \sum_{x=0}^{2^n-1} c_x |x\rangle$ can be expressed as the superposition

$$|\psi\rangle = c_s|x_s\rangle + \sum_{x \neq x_s} c_x|x\rangle \equiv c_s|x_s\rangle + c_\perp|x_\perp\rangle. \tag{5.87}$$

The application of G on $|\psi\rangle$ can be understood by noting that the oracle U_f performs a reflection about the vector $|x_\perp\rangle$, which is perpendicular to $|x_s\rangle$ in the plane defined by $|x_s\rangle$ and $|x_\perp\rangle$, i.e., $U_f(c_\perp|x_\perp\rangle + c_s|x_s\rangle) = c_\perp|x_\perp\rangle - c_s|x_s\rangle$. Moreover, D performs a reflection about the vector $|\psi_{\text{WH}}\rangle$ in the plane defined by $|x_s\rangle$ and $|x_\perp\rangle$. The product of two reflections is a rotation, so $G = DU_f = e^{i\sigma_y\theta}$. For large N, the rotation angle θ satisfies $\cos\theta = \sqrt{\frac{N-1}{N}}$, and therefore, $\theta \approx \sin\theta = \frac{1}{\sqrt{N}}$. After $\mathcal{O}(\sqrt{N})$ rotations, with high probability, the measurement yields an item satisfying $f(i) = 1$. In other words, repeated application of G rotates the state vector close to $|x_s\rangle$. Thus, measurement in the computational basis produces the integer x_s in base 2 with high probability, i.e., it produces a solution to the search problem.

Let us summarize the components used in the Grover algorithm. The Walsh–Hadamard operator applied at the beginning of the algorithm on the state $|0\rangle$ corresponds to making a Fourier transformation (see Sec. 5.3.3). This generates the uniform vector $|\psi_{\text{WH}}\rangle$. The Grover algorithm operates in the Hilbert space of n qubits plus the oracle qubit(s). It applies the Grover operator G, which is given by the product of the oracle operator U_f, and the "inversion about the average operator" D. [Recall that G uses the function $f : \{0,1\}^n \mapsto \{0,1\}$, where $f(i) = 0$ if the ith item does not satisfy the search criteria, and $f(i) = 1$ if it does.] In general, there are n_c items that satisfy the search criteria, such that $f(i) = 1$ a total of n_c times. We have assumed in our discussion that $n_c = 1$, but the algorithm can be generalized to the case $n_c > 1$. In this case, we need only one oracle qubit. After every iteration G, the amplitude of the solution increases by $\mathcal{O}(1/\sqrt{N})$. Therefore, after $\mathcal{O}(\sqrt{N})$ iterations, the probability to find the data register in state solution state approaches unity.

Example: Consider the case $n = 3$, i.e., $N = 8$. Then $i = 0, 1, \ldots, 7$ correspond to $|0,0,0\rangle, |0,0,1\rangle, \ldots, |1,1,1\rangle$. The starting state is,

$$|\Psi\rangle = |\psi\rangle \otimes \left[\frac{1}{\sqrt{2}}(|0\rangle - |1\rangle)\right] = \frac{1}{\sqrt{8}}\sum_{i=0}^{7} |i\rangle \otimes \left[\frac{1}{\sqrt{2}}(|0\rangle - |1\rangle)\right], \tag{5.88}$$

5.3 Quantum Computing Algorithms

where $|\psi\rangle = |\psi_{WH}\rangle$. Since the right qubit, $\frac{1}{\sqrt{2}}(|0\rangle - |1\rangle)$, is not altered during manipulations, it will be omitted in the analysis below. Suppose we are searching for the element with $|i=5\rangle = |1,0,1\rangle$. The Grover algorithm then consists of a series of operations (the quantum circuit denoted by C_G) operating on $|\psi\rangle$ whose purpose is to increase the overlap $\langle C_G \psi | 1,0,1 \rangle = \langle C_G \psi | 5 \rangle$ until it is close to unity. Controlling the overlap is facilitated by considering two orthogonal three-qubit states, $|5\rangle = |1,0,1\rangle$ and $|\phi\rangle \perp |5\rangle$, defined as,

$$|\phi\rangle \equiv \frac{1}{\sqrt{7}} \sum_{i=0}^{7} (1 - \delta_{i,5})|i\rangle, \quad (5.89)$$

and expressing $|\psi\rangle$ as a linear combination of $|\phi\rangle$ and $|5\rangle$,

$$|\psi\rangle = \frac{\sqrt{7}}{\sqrt{8}}|\phi\rangle + \frac{1}{\sqrt{8}}|5\rangle. \quad (5.90)$$

The overlap $\langle \psi | 5 \rangle = 1/\sqrt{8}$ is small, and the task of the Grover algorithm is to improve it. To construct the grover gate G, we consider the function $f(i) = -\delta_{i,5}$ ($i = 0, 1, \ldots, 7$) and design the special gate U_f, such that $U_f |i\rangle = (1 - 2\delta_{i,5})|i\rangle$. The Grover gate is then given by $G = DU_f = 2(|\psi\rangle\langle\psi| - 1)U_f$. Geometrically, it is a composition of reflection with respect to $|\phi\rangle$ (in terms of U_f) followed by a reflection with respect to $|\psi\rangle$ (in terms of D). The geometrical interpretation of the Grover algorithm is displayed in Fig. 5.11. Denoting by $\theta/2$ the angle between $|\phi\rangle$ and $|\psi\rangle$, we have $\theta = 2\arccos\sqrt{7/8} = \arccos(3/4)$. In the first step, U_f is applied, yielding

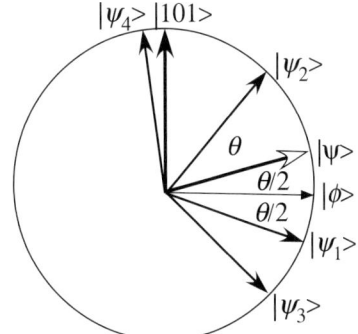

FIG 5.11 Geometric interpretation of the Grover search algorithm. The initial state $|\psi\rangle$, marked by white arrow and defined in Eq. (5.88), is decomposed into a linear combination of orthogonal states $|\phi\rangle$, Eq. (5.89), and $|5\rangle \equiv |101\rangle$. Initially, its overlap with $|5\rangle$ is small, and the aim of Grover's algorithm is to increase the overlap within a few applications of $G = DU_f$. Application of U_f on $|\psi\rangle$ reflects it around $|\phi\rangle$ leading to $|\psi_1\rangle$. Subsequent application of D reflects $|\psi_1\rangle$ around $|\psi\rangle$, leading to $|\psi_2\rangle$ with improved overlap with $|5\rangle$. Another application of U_f and D leads to $|\psi_3\rangle$ and then $|\psi_4\rangle$, and the probability of measuring $|5\rangle$ in $|\psi_4\rangle$ is already close to 95%.

$$|\psi_1\rangle = U_f |\psi\rangle = |\psi\rangle - \frac{1}{\sqrt{2}}|5\rangle = \frac{\sqrt{7}}{\sqrt{8}}|\phi\rangle - \frac{1}{\sqrt{8}}|5\rangle, \quad (5.91)$$

and then D is applied on $|\psi_1\rangle$, leading to

$$|\psi_2\rangle = (2|\psi\rangle\langle\psi| - I)|\psi_1\rangle = \frac{1}{2}|\psi\rangle + \frac{1}{\sqrt{2}}|5\rangle = \sqrt{\frac{7}{32}}|\phi\rangle + \frac{5}{\sqrt{32}}|5\rangle. \quad (5.92)$$

The overlap $\langle \psi_2 | 5 \rangle = 1/\sqrt{2}$ is indeed better. The next (and last) application of G is as follows:

$$U_f |\psi_2\rangle = |\psi_3\rangle = \frac{1}{2}|\psi\rangle - \frac{3}{\sqrt{8}}|5\rangle = \sqrt{\frac{7}{32}}|\phi\rangle - \frac{5}{\sqrt{32}}|5\rangle, \quad (5.93)$$

$$D|\psi_3\rangle = |\psi_4\rangle = (2|\psi\rangle\langle\psi| - I)|\psi_3\rangle = \sqrt{\frac{7}{128}}|\phi\rangle - \frac{11}{\sqrt{128}}|5\rangle. \quad (5.94)$$

The overlap $\langle \psi_4 | 5 \rangle = 11/\sqrt{128}$, and hence, the probability to find $|5\rangle = |101\rangle$ in $|\psi_4\rangle$ is $\left|11/\sqrt{128}\right|^2 = 0.945$.

5.3.3 QUANTUM FOURIER TRANSFORM

The quantum Fourier transform is the only known tool in quantum computation that yields an exponential advantage over classical computational methods. Therefore, we shall consider it in detail.

The (classical) Fourier transform (FT) is an important operation used in science, engineering, and economics (see Appendix D). The discrete FT transforms a set of (in general complex) numbers $x_0, x_1, \ldots, x_{N-1}$ to another set $y_0, y_1, \ldots, y_{N-1}$, as follows:

$$y_k = \frac{1}{\sqrt{N}} \sum_{j=0}^{N-1} e^{2\pi i jk/N} x_j. \tag{5.95}$$

This transformation is norm preserving, $\sum_k |y_k|^2 = \sum_j |x_j|^2$. For computational convenience, the dimension N is usually taken to be a power of 2, i.e., $N = 2^n$. The classical fast Fourier transform (FFT) algorithm requires $N \log N = n 2^n$ computational steps. The *quantum Fourier transform* algorithm, invented by Peter Shor, offers an exponential speedup; it can be evaluated in polynomial time on a quantum computer, i.e., $\mathcal{O}(n^2)$ gates are required for an n-qubit FT.

The n-qubit basis states, $|0\rangle, \ldots, |N-1\rangle$, where $N = 2^n$, can be represented using the notation $|j\rangle$, where $0 \leq j \leq 2^n - 1$ [in Eq. (5.70), we used the notation $|x\rangle$]. Recall that the binary representation of j is given by $j = j_1 2^{n-1} + j_2 2^{n-2} + \ldots + j_{n-1} 2^1 + j_n 2^0 \equiv j_1 j_2 \ldots j_{n-1} j_n$ [see Eq. (5.22)]. The quantum FT operator, U_{QFT}, is defined by its effect on the basis state $|k\rangle$ by

$$|k\rangle \longmapsto U_{QFT}|k\rangle = \frac{1}{\sqrt{N}} \sum_{j=0}^{N-1} e^{2\pi i jk/N} |j\rangle. \tag{5.96}$$

When this transformation is applied to an arbitrary n-qubit state $\sum_{j=0}^{N-1} x_j |j\rangle$, we obtain

$$\sum_{j=0}^{N-1} x_j |j\rangle \longmapsto U_{QFT} \sum_{j=0}^{N-1} x_j |j\rangle = \sum_{k=0}^{N-1} y_k |k\rangle, \tag{5.97}$$

where

$$y_k = \frac{1}{\sqrt{N}} \sum_{j=0}^{N-1} e^{2\pi i jk/N} x_j. \tag{5.98}$$

The transformation U_{QFT} is unitary, as we shall shortly see, and therefore, it can be implemented on a quantum computer. Another equivalent way of writing the transformation (5.96) uses the notation of binary fractions,

$$[0.j_l j_{l+1} \ldots j_{n-1} j_n] \equiv j_l/2^1 + j_{l+1}/2^2 + \ldots + j_{n-1}/2^{n-l} + j_n/2^{n-l+1}, \tag{5.99}$$

where $0.j \in [0, 1)$. Explicitly, it reads,

$$U_{QFT}|k\rangle = \frac{1}{2^{n/2}} (|0\rangle + e^{2\pi i [0.k_n]} |1\rangle)(|0\rangle + e^{2\pi i [0.k_{n-1} k_n]} |1\rangle) \ldots (|0\rangle + e^{2\pi i [0.k_1 \ldots k_n]} |1\rangle). \tag{5.100}$$

This representation allows us to construct an efficient quantum circuit computing the FT, and a proof that the quantum FT is unitary. The equivalence of Eqs (5.96) and (5.100) can be demonstrated by the following algebra:

$$U_{QFT}|k\rangle = \frac{1}{2^{n/2}} \sum_{j=0}^{2^n-1} e^{2\pi i jk 2^{-n/2}} |j\rangle$$

$$= \frac{1}{2^{n/2}} \sum_{j_1=0}^{1} \ldots \sum_{j_n=1}^{1} e^{2\pi i k(\sum_{l=1}^{n} j_l 2^{-l})} |j_1 \ldots j_{n-1} j_n\rangle$$

5.3 Quantum Computing Algorithms

$$= \frac{1}{2^{n/2}} \prod_{l=1}^{n} \sum_{j_l=0}^{1} e^{2\pi i k j_l 2^{-l}} |j_l\rangle = \frac{1}{2^{n/2}} \prod_{l=1}^{n} \left(|0\rangle + e^{2\pi i k 2^{-l}}|1\rangle\right)$$

$$= \frac{1}{2^{n/2}} \left(|0\rangle + e^{2\pi i [0.k_n]}|1\rangle\right) \left(|0\rangle + e^{2\pi i [0.k_{n-1}k_n]}|1\rangle\right) \cdots \left(|0\rangle + e^{2\pi i [0.k_1\ldots k_n]}|1\rangle\right). \quad (5.101)$$

This product representation of the unitary transformation can be implemented by the circuit shown in Fig. 5.12. In Fig. 5.12, the swap of the qubits $\{|\varphi_m\rangle\}$ to get the order of the qubits as in Eq. (5.100) is not shown (it needs to be added). By explicit construction, this transformation is unitary, since each gate it is composed of is unitary.

Let us consider the components of this transformation. The controlled-phase gates add a phase factor to the coefficient of the basis state $|1\rangle$; e.g., $R_j \equiv \begin{pmatrix} 1 & 0 \\ 0 & e^{2\pi i 2^{-j}} \end{pmatrix}$ adds the phase $e^{2\pi i 2^{-j}}$. The Hadamard gate applied to the qubit $|k_1\rangle$ gives

$$|k_1\rangle \longmapsto H|k_1\rangle = \frac{|0\rangle + e^{2\pi i [0.k_1]}|1\rangle}{\sqrt{2}}, \quad (5.102)$$

since $e^{2\pi i [0.k_1]} = -1$ when $k_1 = 1$ and $e^{2\pi i [0.k_1]} = +1$ when $k_1 = 0$; similarly, a Hadamard gate applied to the qubit $|k_m\rangle$ gives

$$|k_m\rangle \longmapsto H|k_m\rangle = \frac{|0\rangle + e^{2\pi i [0.k_m]}|1\rangle}{\sqrt{2}}. \quad (5.103)$$

Let us follow the first qubit as the gates are applied to it in succession. The Hadamard gate applied to the first qubit in Fig. 5.12 gives Eq. (5.102), which when followed by application of R_2 gives

$$\frac{|0\rangle + e^{2\pi i [0.k_1 k_2]}|1\rangle}{\sqrt{2}}.$$

Applying R_3, R_4, through R_n yields

$$|\varphi_n\rangle = \frac{|0\rangle + e^{2\pi i [0.k_1 k_2 \ldots k_n]}|1\rangle}{\sqrt{2}}.$$

Similarly for the other qubits in the figure. In Fig. 5.12, we have used the notation, $|\varphi_j\rangle \equiv (|0\rangle + e^{2\pi i [0.k_{n-j+1}k_{n-j+2}\ldots k_n]} |1\rangle)/\sqrt{2}$, to denote the output qubits corresponding to the various input qubits.

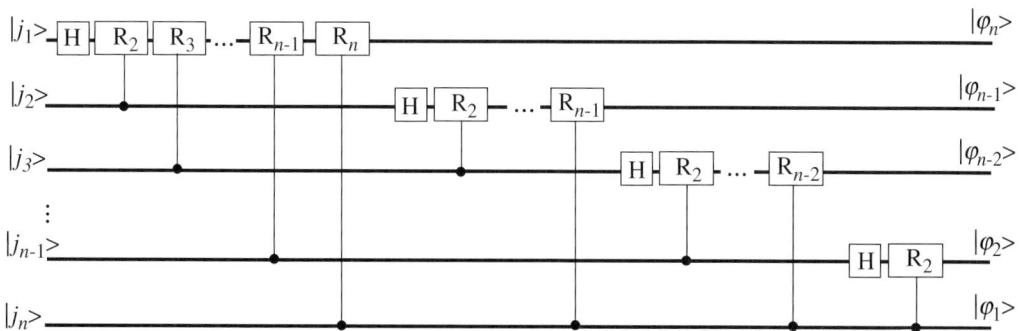

FIG 5.12 Circuit diagram for the quantum Fourier transform based on the product representation of the transformation given in Eq. (5.100). The final swap of the qubits at the very end of the circuit is not shown.

To obtain Eq. (5.100), the qubits shown in Fig. 5.12 must be swapped. The SWAP gate [see Eq. (5.50)] can be implemented by three CNOT gates, as shown in Fig. 5.13. Swapping $|\varphi_i\rangle$ and $|\varphi_j\rangle$ is implemented as follows:

$$|\varphi_i\rangle|\varphi_j\rangle \longmapsto \mathrm{CNOT}_{12}|\varphi_i\rangle|\varphi_j\rangle = |\varphi_i\rangle|\varphi_i \oplus \varphi_j\rangle$$

$$|\varphi_i\rangle|\varphi_i \oplus \varphi_j\rangle \stackrel{\mathrm{CNOT}_{21}}{\longmapsto} |\varphi_i \oplus (\varphi_i \oplus \varphi_j)\rangle|\varphi_i \oplus \varphi_j\rangle = |\varphi_j\rangle|\varphi_i \oplus \varphi_j\rangle$$

$$|\varphi_j\rangle|\varphi_i \oplus \varphi_j\rangle \stackrel{\mathrm{CNOT}_{12}}{\longmapsto} |\varphi_j\rangle|(\varphi_i \oplus \varphi_j) \oplus \varphi_j\rangle = |\varphi_j\rangle|\varphi_i\rangle. \quad (5.104)$$

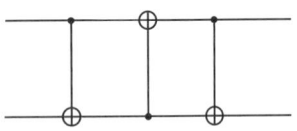

FIG 5.13 Circuit diagram for the swap of two qubits, Eq. (5.104).

The total number of gates used in the quantum FT is as follows: The first qubit is subjected to n gates in Fig. 5.12. The second qubit to $n-1$, and so on, so the total number of gates in Fig. 5.12 is $n + (n-1) + \ldots + 1 = n(n+1)/2$. Moreover, $n/2$ swaps are required at the end, each of which required three CNOT gates, making for a total of $n^2/2 + 2n$ gates. Hence, the order of the algorithm is $\mathcal{O}(n^2)$. This is an exponential speedup relative to the $\mathcal{O}(n\,2^n)$ operations needed for the FFT on a classical computer.

Given the exponential speedup of the quantum FT, one might think that it can be used to efficiently solve a great variety of classical problems. However, to date, only a limited number of problems have been shown to be amenable to the quantum FT algorithm, including the Shor prime factorization algorithm, period finding, and discrete logarithm determination.

One problem associated with the use of the quantum FT is that it is difficult to accurately prepare the input state of a quantum register in an arbitrary state (5.97), $|\psi\rangle_{\mathrm{in}} = \sum_{j=0}^{N-1} x_j |j\rangle$, which is typically an entangled state of N qubits. Another problem is measuring the amplitudes y_k of the output state (5.97), unless the input represents a periodic function, for then only one or just a few output states $|k\rangle$ have appreciable probabilities $|y_k|^2$. Until these problems are overcome, the quantum FT will not be used to speedup the calculation of FT problems that are so important for engineering, science, and data processing problems.

5.3.4 SHOR FACTORIZATION ALGORITHM

Shor's algorithm focuses on the factorization of a large integer into its prime factors. A *prime number* is an integer greater than 1, which has only itself and 1 as factors, e.g., 2, 3, 5, 7, 11. The *fundamental theorem of arithmetic* states that if N is an integer greater than 1, it can be represented in exactly one way, apart from rearrangement, as a product of one or more primes, i.e., it can be factorized in the form

$$N = p_1^{N_1} p_2^{N_2} \ldots p_n^{N_n}, \quad (5.105)$$

where $p_1, \ldots, p_n$ are distinct prime numbers (divisibility by 2, 3, and 5 is easy to determine, so these primes are naturally excluded) and $N_1, \ldots, N_n$ are positive integers. The *factorization problem* can be stated as follows: Given a positive composite integer N, find its factorization according Eq. (5.105). A subproblem of the factoring problem is as follows: Given a composite integer N, find an integer m, between 1 and $N/2$, that divides N. This is what Shor's algorithm does. It is relatively easy to factor small integers, e.g., $693 = 3^2 \times 7 \times 11$. However, no efficient algorithm is known for factorization on a classical computer. But Shor's quantum factorization algorithm *is* computationally efficient on a quantum computer.

Shor's algorithm is composed of two parts: (1) Reduction of the factoring problem to the order-finding problem, as discussed below, which can be done on a classical computer. (2) The quantum algorithm to solve the order-finding problem.

Shor's algorithm: preliminaries

Before discussing Shor's factorization algorithm, it is necessary to develop some mathematical background in *number theory* and *modular arithmetic*. In the analysis below, all symbols $a, b, N, r, x, k, \ldots$ are assumed integers unless otherwise specified.

5.3 Quantum Computing Algorithms

1. **Divisibility and greatest common divisor:** We say that b *divides* a if $a/b = k$ (an integer). This is written as $b \mid a$. If b does not divide a, we write it as $b \nmid a$. The *greatest common divisor* of two integers a and b is denoted as $\gcd(a,b)$ and is defined as the largest integer that divides both a and b, e.g., $\gcd(9,21) = 3$ and $\gcd(5,9) = 1$. In the latter case, we say that a and b do not have a common factor. A useful theorem states that if $a = kb + r$, $r > 0$, then $\gcd(a,b) = \gcd(b,r)$. Determining $\gcd(a,b)$ can be accomplished in polynomial time in L using *Euclid's algorithm* on a classical computer, where L is the size in bits of $\text{Max}(a,b)$.

2. **Euclid's algorithm to find $\gcd(a,b)$**
   (a) Make sure that $a > b$. First, divide a by b to get
   $a = k_1 b + r_1$. By the theorem stated above, $\gcd(a,b) = \gcd(b,r_1)$.
   (b) Divide b by r_1 to get
   $b = k_2 r_1 + r_2, \Rightarrow \gcd(a,b) = \gcd(b,r_1) = \gcd(r_1,r_2)$.
   (c) Divide r_1 by r_2 to get
   $r_1 = k_3 r_2 + r_3, \Rightarrow \gcd(a,b) = \gcd(b,r_1) = \gcd(r_1,r_2) = \gcd(r_2,r_3)$.
   (d) ... divide r_m by r_{m+1} to get
   $r_m = k_{m-2} r_{m+1} + r_{m+2}, \Rightarrow \gcd(a,b) = \ldots = \gcd(r_{m+1},r_{m+2})$.
   (e) When $r_m = k_{m+2} r_{m+1} + 0$, the chain is complete:
   $\gcd(a,b) = \cdots = \gcd(r_m, r_{m+1}) = r_{m+1}$.
   Example: Find $\gcd(6825, 1430)$.
   $6825 = 4 \times 1430 + 1105 \Rightarrow k_1 = 4, r_1 = 1105$. $1430 = 1 \times 1105 + 325 \Rightarrow k_2 = 1, r_2 = 325$.
   $1105 = 3 \times 325 + 130 \Rightarrow k_3 = 3, r_3 = 130$.
   $325 = 2 \times 130 + 65 \Rightarrow k_4 = 2, r_4 = 65$.
   $130 = 2 \times 65 + 0, \Rightarrow k_5 = 2, r_5 = 0$.
   $\gcd(6825, 1430) = \ldots = \gcd(130, 65) = 65$.

3. **Modular equality:** We say that a *equals b modulo N* if N divides $a - b$, e.g., a can be written as $a = kN + b$, where k and b are integers and $0 < b < N$. This is often written as,

$$a = b[N] \Leftrightarrow N \mid (a-b) \Leftrightarrow a = kN + b \tag{5.106}$$

4. **Computing high powers modulo N:** Application of the Shor algorithm can require computation of high powers modulo N. This requires polynomial time on a classical computer. For example, suppose we take $N = 87$ and $x = 3$ and want to check $7^{19}[87]$. First, we write 19 in a binary basis, $19 = 16 + 2 + 1$. Then, $7^{19}[87] = 7^{16}[87]\, 7^2[87]\, 7^1[87]$. Each power can be written as successive powers of 2, and all powers and multiplications are done modulo 87.

5. **The order of x with respect to N and the order finding problem:** Suppose $1 < x < N-1$ and r is the smallest integer, such that $x^r = 1[N]$. r is then called the *order of x with respect to N*.

Problem 5.24

Find the order of 4 with respect to 9 and of 5 with respect to 12.
Answer: 3 ($4^3 = 64 = 1[9]$) and 4 ($5^4 = 625 = 1[12]$).

Finding the order r is referred to as the *order-finding problem*. Solution of the order problem is at the heart of Shor's quantum algorithm.

6. **Relation of the order to the factorization problem:** The case $r = 2$ is of special importance, due to the following theorem:

Theorem: *Let N be a composite number as in Eq. (5.105), which is L bits long and let $1 < x < N - 1$ a solution of $x^2 = 1[N]$. Then at least $\gcd(x-1, N)$ and $\gcd(x+1, N)$ is a non-trivial factor of N. (According to step 1, these can be computed with $O(L^3)$ steps.)*

Proof:

$$x^2 = 1[N] \Rightarrow N \mid (x+1)(x-1), \tag{5.107}$$

i.e., N has common factor either $(x+1)$ or $(x-1)$. Take N odd, then x is also odd, i.e., $1 < x-1 < x+1 < N$. Hence, for the number x satisfying the above condition, the factor $(x+1$ and/or $x-1)$ is not trivial. If r is even, then $y \equiv x^{r/2}$ satisfies $y^2 = 1[N]$, and we have $r = 2$.

7. **Continued fraction (CF):** The concept of CF has many applications in arithmetic. Here, it is used in answering the following question: Given a rational number k/q and an integer $N \ll q$, what is the closest rational number to k/q whose denominator $r < N$? This number is denoted as $R_{kq}(N)$. The procedure for finding $R_{kq}(N)$ follows a couple of guidelines. (1) Starting from k/q, every rational number $f < 1$ is written as an inverse of its inverse, $f = 1/f^{-1}$, so that $f^{-1} > 1$. (2) Every rational number $g > 1$ is written as a sum of its integer and fractional parts, $g = [g] + \{g\}$, so that $\{g\} < 1$. The n's CF approximation to k/q is obtained when $\{g\} < 1$ at the n's step is replaced by 0. The desired approximation obtains when the denominator of the next approximation exceeds N. When $\{g\} = 0$, the CF series converges to k/q.

As an example, consider the case $k = 4915$, $q = 8192$, $N = 55$.

$$\frac{k}{q} = \frac{1}{\frac{8192}{4915}} = \frac{1}{1 + \frac{3277}{4915}} \approx \frac{1}{1+0} = 1.$$

$$\frac{3277}{4915} = \frac{1}{\frac{4915}{3277}} = \frac{1}{1 + \frac{1638}{3277}} \Rightarrow \frac{k}{q} = \frac{1}{1 + \frac{3277}{4915}} = \frac{1}{1 + \frac{1}{1+\frac{1638}{3277}}} \approx \frac{1}{1 + \frac{1}{1+0}} = \frac{1}{2}.$$

$$\frac{1638}{3277} = \frac{1}{\frac{3277}{1638}} = \frac{1}{2 + \frac{1}{1638}} \Rightarrow \frac{k}{q} = \frac{1}{1 + \frac{1}{1+\frac{1638}{3277}}} = \frac{1}{1 + \frac{1}{1+\frac{1}{2+\frac{1}{1638}}}} \approx \frac{1}{1 + \frac{1}{1+\frac{1}{2+0}}} = \frac{3}{5}.$$

(with very small error).

Since, in the next approximation, the denominator becomes larger than $N = 55$, we conclude that the answer to the above question is $R_{4915,8192}(55) = \frac{3}{5}$. Actually, since $\frac{4915}{8192} = 0.59997559$, this approximation could be guessed anyhow.

Shor's Algorithm: Step by Step

The task is to find a nontrivial divisor of an odd integer N of length L in bits.

1. **Choosing a random number:** The process starts by randomly choosing an odd number $1 < x < N$ and checking that $x \nmid N$, e.g., $\gcd(x, N) = 1$ (if $x \mid N$, the factoring problem is partially solved). Using the Euclid's algorithm, this check requires $O(L^3)$ steps on a classical computer. It leaves us with the quantum computational problem of finding an (even) order r of x with respect to N. In our example, with $N = 55$, we choose $x = 13$.
2. **Determining the qubit space:** The appropriate number of qubits to be used naturally depends on N. Consider a number $q = 2^\ell$ satisfying

$$2N^2 < q = 2^\ell < 3N^2. \tag{5.108}$$

Then, we need to work with 2ℓ qubits, and the dimension of the corresponding Hilbert space is $n = 2^{2\ell}$.

Example: For $N = 55$, we need to find ℓ, such that $2 \times (55)^2 = 6050 < 2^\ell < 9075 = 3 \times (55)^2$. This is satisfied by $\ell = 13$ as $2^{13} = 8192$. The number of qubits in this case is then $2\ell = 2 \times 13 = 26$.

This Hilbert space is a tensor product of two spaces, $\mathcal{H} = \mathcal{H}_a \otimes \mathcal{H}_b$, each containing ℓ qubits and therefore have dimension 2^ℓ. A state in $\mathcal{H}$ can be written as

$$|a, b\rangle = |a\rangle \otimes |b\rangle = |a_1, a_2, \ldots a_\ell\rangle \otimes |b_1, b_2, \ldots, b_\ell\rangle, \tag{5.109}$$

5.3 Quantum Computing Algorithms

$$a = \sum_{i=1}^{\ell} a_i 2^{\ell-i}, \; b = \sum_{i=1}^{\ell} b_i 2^{\ell-i}, \; 0 \le a, b \le q-1,$$

$$a_i = 0, 1, \; b_i = 0, 1, \; i = 1, 2, \ldots, \ell.$$

In the notation above, $|a, b\rangle$ are referred to as *first and second registers*. The numbers a, b are determined according to our state enumeration defined in Eq. 5.21. Having defined the basis states in $\mathcal{H}$, we write the general states as appropriate linear combinations,

$$|\Psi\rangle = \sum_{a=0}^{q-1} \sum_{b=0}^{q-1} c_{ab} |a, b\rangle, \quad \sum_{a=0}^{q-1} \sum_{b=0}^{q-1} |c_{ab}|^2 = 1. \tag{5.110}$$

3. **Walsh–Hadamard transformation on the first register:** Now comes the actual quantum mechanical operations implemented as gates applied to the qubits. The initial state is chosen to be $|\Psi_0\rangle = |a = 0, b = 0\rangle$. The first operation is to apply the Walsh–Hadamard transform $\bigotimes_{i=1}^{\ell} H$ on the first register, where $H = \frac{1}{\sqrt{2}} \begin{pmatrix} 1 & 1 \\ 1 & -1 \end{pmatrix}$. By using Eq. (5.72), we see that the state obtained is

$$|\Psi_1\rangle = [\bigotimes_{i=1}^{\ell} H|0\rangle] \otimes |0\rangle = \frac{1}{\sqrt{q}} \sum_{a=0}^{q-1} |a, 0\rangle. \tag{5.111}$$

In our example, $N = 55$, $\ell = 13$, $q = 2^\ell = 8192$, and the state after the first step is

$$|\Psi_1\rangle = \frac{1}{\sqrt{2^{13}}} \left(|0, 0\rangle + |1, 0\rangle + \ldots + |2^{13} - 1, 0\rangle \right). \tag{5.112}$$

4. **Applying U_f:** The operator U_f defined in Eq. (5.71) is now applied on $|\Psi_1\rangle$. The function $f(a)$ in the present case is $x^a[N]$, so $U_f |a, 0\rangle = |a, x^a[N]\rangle$. Note that $1 < x^a[N] < N \ll q = 2^\ell$. Hence, the second register contains states with numbers smaller than N. After applying U_f, the state of the system becomes

$$|\Psi_2\rangle = \frac{1}{\sqrt{q}} \sum_{a=0}^{q-1} |a, x^a[N]\rangle. \tag{5.113}$$

In our example (see above for computing high powers modulo N),

$$|\Psi_2\rangle = \frac{1}{\sqrt{2^{13}}} [|0, 13^0[55]\rangle + |1, 13^1[55]\rangle + |2, 13^2[55]\rangle + |3, 13^3[55]\rangle + \ldots$$
$$+ |9, 13^9[55]\rangle + \ldots |2^{13} - 1, 13^{2^{13}-1}[55]\rangle]$$
$$= \frac{1}{\sqrt{2^{13}}} [|0, 1\rangle + |1, 13\rangle + |2, 4\rangle + \cdots + |9, 28\rangle + \ldots |2^{13} - 1, 2\rangle]. \tag{5.114}$$

5. **Measuring the second register only:** The interpretation from the point of view of quantum measurement theory is as follows: On measuring the second register, the state $|\Psi_2\rangle$ of Eq. 5.113 will collapse in such a way that the second register will be one of the states $|a, x^a[N]\rangle$, say

$$x^{a_0}[N] \equiv c, \; 0 \le c < N - 1. \tag{5.115}$$

After this collapse, there will still remain a sum on a, which runs on all the as in the first register for which $x^a[N] = x^{a_0}[N] = c$. The set of as satisfying this condition is denoted by A. To identify all the elements of A, recall the order $1 < r < N - 1$, defined as the smallest integer, such that $x^r[N] = 1$. Of course this implies periodicity in r, such that

$x^{mr} = 1[N]$ for any integer m, i.e.,

$$A = \{a_0, a_0 + r, a_0 + 2r, \ldots a_0 + (M-1)r\}, \text{ such that} \tag{5.116}$$
$$x^{a_0} = c, \ x^{mr} = 1[N], \ m = 1, 2, \ldots (M-1), \ a_0 + (M-1)r \leq q - 1.$$

The number M is determined by the conditions that $0 \leq a \leq q - 1$, i.e., $(M-1)r \lesssim (q-1)$. Clearly, $M \approx \frac{q}{r} \gg 1$. After step 3, the state of the system collapses to:

$$|\Psi_3\rangle = \frac{1}{\sqrt{M}} \sum_{a \in A} |a, c\rangle = \frac{1}{\sqrt{M}} \sum_{m=0}^{M-1} |a_0 + mr, c\rangle. \tag{5.117}$$

In this expression, we know a_0 and c, but of course we do not know r nor M. In our example, we assume that the result of measuring the second register is $c = 28$, so that $a_0 = 9$ and

$$|\Psi_3\rangle = \frac{1}{\sqrt{M}} \sum_{m=0}^{M-1} |9 + mr, 28\rangle. \tag{5.118}$$

6. **Applying QFT U_q to the first register:** For a given basis state in the sum, the operation reads,

$$U_q|a_0 + mr\rangle = \frac{1}{\sqrt{q}} \sum_{k=0}^{q-1} e^{2\pi i \frac{k(a_0 + mr)}{q}} |k\rangle \equiv \frac{1}{\sqrt{q}} \sum_{k=0}^{q-1} e^{2\pi i \frac{ka_0}{q}} e^{2\pi i \frac{kmr}{q}} |k\rangle \equiv \frac{1}{\sqrt{q}} \sum_{k=0}^{q-1} e^{2\pi i \frac{ka_0}{q}} \eta(k)^m |k\rangle, \ \eta(k) = e^{2\pi i \frac{kr}{q}}. \tag{5.119}$$

Consequently, the state of the system now becomes,

$$U_q|\Psi_3\rangle = |\Psi_4\rangle = \frac{1}{\sqrt{qM}} \sum_{k=0}^{q-1} e^{2\pi i \frac{ka_0}{q}} \left(\sum_{m=0}^{M-1} \eta(k)^m \right) |k\rangle \otimes |c\rangle$$
$$= \frac{1}{\sqrt{qM}} \sum_{k=0}^{q-1} e^{2\pi i \frac{ka_0}{q}} \frac{1 - \eta(k)^M}{1 - \eta(k)} |k\rangle \otimes |c\rangle \tag{5.120}$$

in our example, $|\Psi_4\rangle = \sum_{k=0}^{2^{13}-1} \frac{e^{2\pi i \frac{9k}{2^{13}}}}{\sqrt{2^{13}M}} \frac{1 - \eta(k)^M}{1 - \eta(k)} |k\rangle \otimes |28\rangle.$

7. **Measuring the first register:** The probability $P(k)$ of observing a component of $|\Psi_4\rangle$ corresponding to a specific k is given by,

$$P(k) = \frac{1}{qM} \left| \frac{1 - \eta(k)^M}{1 - \eta(k)} \right|^2. \tag{5.121}$$

Note that $P(k)$ does not contain information about c, and its information on x (the random number chosen in step 1) is hidden (see below). For large M, the probability $P(k)$ is very sharply peaked at M/q when $\eta(k) \to 1$. Therefore, the result of measurement will give a value of k very close to one of the peaks (to which the state will collapse) with very high probability. From the definition in Eq. (5.119), such a peak occurs when $kr/q = l$ (an integer). Although k and q are known, this Diophantic equation does not completely identify r because, although $r = m\ell$ with integers m and ℓ such that $k\ell/q = l/m$ is also an integer, we cannot conclude that $\ell = r$. For example, if $k = 10, q = 5$, and $r = 6$, we find $k \times 6/q = l = 12 \to r = 6$ but also $k \times 2/q = l = 4 \to r = 2$ or $k \times 3/q = l = 6 \to r = 3$.

5.3 Quantum Computing Algorithms

8. **Identifying r from the peaks of $P(k)$:** To identify r, we should use the CF method. If r is a multiple of some smaller integer, say $r = m\ell$ (m, ℓ integers) and $k\ell/q = I/m$ is an integer, then the CF method can be used to identify ℓ. To get r from ℓ requires further test. Starting with ℓ, note that although k/q does not, in general, fall exactly on values $I/(m\ell)$, it may fall very close to it. So the question is, what is the closest rational number to $k/q < 1$ whose denominator does not exceed N? This is exactly the question that is solved by the CF procedure; therefore, ℓ can be evaluated. Now we can find r by examining which power $x^{m\ell} = 1[N]$, and this implies $r = m\ell$. If r is even the factorization problem is thereby solved. If r is odd, one need to choose another x and repeat the procedure until an even r is found. In our example, a plot of $P(k)$ versus k looks like in Fig. 5.14.

Inspection of the locations of peaks indicates (even without employing CF) that all of them occur at points very close to $\frac{k}{q} = \frac{I}{5m}$, where I and m are integers. This implies that $\ell = 5$ and it remains to check which m satisfies $x^{m\ell} = 13^{5m} = 1[55]$. The result is $m = 4$, and therefore $r = 20$. Since r is even, $y = 13^{r/2}[55] = 13^{10}[55] = 34$, and it satisfies $y^2 = 1156 = 21 \times 55 + 1 = 1[55]$. Consequently, $y+1$ and/or $y-1$ have a common factor with 55, which can be found in polynomial time. In our case, both of them do, that is, $\gcd(y+1, N) = \gcd(35, 55) = 5$, $\gcd(y-1, N) = \gcd(33, 55) = 11$.

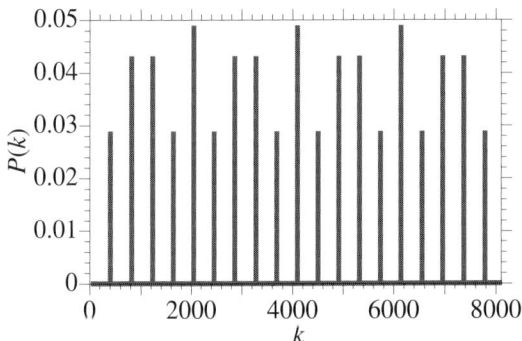

FIG 5.14 $P(k)$ defined in Eq. 5.121 for $N = 55$, $\ell = 13$, $q = 2^{13}$, $x = 13$, and $r = 20$. Alice (or Bob) do not know r and has to elucidate it from the location of the peaks, as explained in step 8.

The speedup of the Shor factorization algorithm is very impressive and is due to the quantum FT algorithm. However, as already mentioned, to-date, only a limited number of problems have been found that amenable to the quantum FT algorithm. All of them belong to the general class known as the *(Abelian) hidden subgroup problem*, which includes order-finding, prime factorization, period-finding, and discrete logarithm determination.

5.3.5 QUANTUM SIMULATION

Classical computers are inefficient in simulating quantum dynamics of compound systems because the number of degrees of freedom is extremely large as the number of subsystems increases. Consider a quantum system S, composed of subsystems $A, B, \ldots$. The corresponding Hilbert space is $\mathcal{H}_S = \mathcal{H}_A \otimes \mathcal{H}_B \otimes \ldots$. These Hilbert spaces are assumed to be finite dimensional, $\dim \mathcal{H}_A = N_A$, etc. The number of states that span $\mathcal{H}_S$ is $N_S = N_A \times N_B \times \ldots$. The Hamiltonian for the system is time independent and given by $H = \sum_j H_j + V$, where H_j is the Hamiltonian of subsystem j and V includes the interactions between the subsystems. Suppose, for simplicity, that the initial state $\psi(t_0)$ of the system is not entangled $\psi(0) = \psi_A(0)\psi_B(0)\ldots$. Then, the formal solution of the Schrödinger equation, $i\hbar \frac{\partial \psi}{\partial t} = H\psi(t)$ is $\psi(t) = \mathcal{U}(t,0)\psi(0) = e^{-iHt/\hbar}\psi(0)$. The exact determination of $e^{-iHt/\hbar}$ is, in many cases, not possible, and the numerical calculation is often impractical for a large Hamiltonian matrix of size $N_S \times N_S$. In some cases, calculation of the evolution can be carried out (see Sec. 7.1) by choosing a basis of size N_S of time-independent functions $\{\phi_n\}$ and diagonalizing the Hamiltonian matrix $H_{nm} \equiv H_{S,nm} = \langle \phi_n | H_S | \phi_m \rangle$, thereby obtaining its eigenvalues $\{\lambda\}$ and eigenstates $\{|\lambda\rangle\}$. Then one has,

$$\mathcal{U}(t,0) = e^{-iH_S t/\hbar} = \sum_{\lambda=1}^{N_S} |\lambda\rangle e^{-i\lambda t/\hbar} \langle\lambda|. \tag{5.122}$$

Another possibility is to expand the wave function as $\psi(t) = \sum_{n=1}^{N_S} c_n(t)\phi_n$ and solve the set of differential equations for the amplitudes $c_n(t)$,

$$i\hbar \frac{dc_n}{dt} = \sum_{m=1}^{N_S} H_{nm} c_m, \quad c_n(0) = \langle \phi_n | \psi(0) \rangle, \tag{5.123}$$

For either method, even for a small number of systems, e.g., $n_S = 10$, and say 20 states per system, the matrix size becomes $N_S = 20^{10}$; hence, the calculation of the dynamics even for this relatively small system is prohibitive.

With a scalable quantum computer, we need only $j \approx \log_2 N_S$ qubits ($2^j \approx N_S$, i.e., the size of the qubit register of the quantum computer should be comparable with that of the system. The qubit register of the computer starts off in state $|0_1 0_2 \ldots 0_j\rangle$. Then the register is changed to the state $|\psi_0\rangle$ corresponding to the initial state that is to be simulated. If the initial state is not entangled, but rather a simple product state, this step requires only a small number of single-qubit gates. Now, a sequence of gates satisfying $\mathcal{U}(t, 0) = \ldots U_2 U_1$ needs to be designed and applied on $\psi(0)$ to obtain $|\psi(t)\rangle$. Finally, the quantum register whose state corresponds to that of the final state $|\psi(t)\rangle$ of the simulated system must be measured. One problem regarding the measurement occurs if the final state $|\psi(t)\rangle$ is not an eigenstate of the quantum register, i.e., if $|\psi(t)\rangle$ is not an eigenstate of the time-independent Hamiltonian H_S of the system at the final time and the measurement is of the energy of the system, many repetitions of this procedure may have to be performed to attain a reliable probability distribution for the final state. Provided the number of required repetitions does not grow exponentially with system size, but only polynomially, the algorithm will still be efficient.

5.4 DECOHERENCE

Quantum information is extremely fragile, due to interactions between the system and its environment. These interactions cause the system to lose its quantum nature, a process called *decoherence*. Decoherence is the result of entanglement between a quantum system and its (unobserved) environment, and it results in degradation of the purity of a quantum state; it can result in the loss of purity of a one-body state, of two-body entanglement, or of many-body entanglement. Decoherence is often blamed for the fact that the world is mostly classical despite the fact that quantum theory provides all the governing principles. Larger bodies lose coherence more quickly than small ones. This is the essential ingredient in producing nearly instantaneous decay of entanglement between two large bodies. Preservation of coherence is crucial for quantum information processing and quantum computing, so control of decoherence is an important issue for quantum information. In the context of quantum information and quantum computation, we can think of decoherence as noise processes in quantum processing. The noise may arise in a quantum gate, within a measurement or in the preparation of the initial qubits used.

The coherence time of a quantum mechanical system, τ_{coh}, is defined as the time for which the system remains quantum mechanically coherent. This should be compared with the time to perform elementary unitary transformations involving one and two qubits, τ_{op}. The ratio, $\tau_{op} : \tau_{coh}$, is an important dimensionless number that characterizes a quantum information processing system; the smaller the ratio, the better.

In addition to its importance for quantum computing and quantum information, decoherence is also relevant to three important problems: (1) the "quantum measurement" problem (see Sec. 2.5.4), (2) the explanation of the emergence of the classical world from quantum mechanics, and (3) the arrows of time problem, i.e., why does time never go backward? In the context of quantum computing and information aspects of decoherence, the relevant questions are: How does decoherence evolve in time? How does decoherence affect N-qubit entanglement? How does it depend on N? How can decoherence be controlled and/or tested experimentally? We are going to delay our study of these questions and of decoherence and dissipation in general, until Chapter 17, to be linked to the book web page, where we take up some of these topics in detail.

5.5 QUANTUM ERROR CORRECTION

Since decoherence is unavoidable and leads to errors in quantum computations, it is absolutely essential to develop a method for fault tolerant quantum computation. The problem of noise occurring in classical computation was considered by von Neumann [70] in the 1950s. Using the concept of *redundant information*, he showed how to compute when the elements of computation are faulty. Each bit of information is redundantly encoded in more than one bit (say, in three bits), and one checks to see whether the bits are the same. If not, the "incorrect" bit is discarded. In the context of quantum information and computation, the analogous procedure is referred to as *quantum error correction*, which is a necessary ingredient in any quantum computation scheme.

5.5 Quantum Error Correction

> **Problem 5.25**
>
> (a) Given a bit error probability of p, find the expression for the probability P for obtaining a bit error on encoding a bit in three bits.
> (b) Calculate the probability P for $p = 1.0 \times 10^{-4}$.
>
> **Answer:** (a) The probability that two bits are faulty is given, using the binomial probability distribution, by $P_2 = 3p^2(1-p)$, and the probability of three faulty bits is $P_3 = p^3$. Hence, $P = P_2 + P_3 = 3p^2 - 2p^3$.

There are several reasons why fault-tolerant quantum computation is more important and more complicated than for classical computation.

1. **Myriad of possible errors:** An error in prescribing a qubit state $|\psi\rangle = a|0\rangle + b|1\rangle$ can take on a continuum of possibilities since, unlike for classical bits, a continuum variation of the amplitudes a and b is possible. The two extremes are either a very small change, $a \to a[1+O(\varepsilon)]$, which accumulates after $[1/\varepsilon]$ operations, or a *phase error* such as $|0\rangle \to -|0\rangle$.
2. **Qubit collapse after measurement:** We cannot simply compare two qubits by projective measurement without destroying them.
3. **No-cloning constraints:** Encoding the same qubit more than once (in order to build a quantum error correction code) is more problematic than the analogous classical procedure on bits, since the no-cloning theorem restricts our ability to copy an unknown qubit. Hence, protecting quantum information by coding qubits redundantly is problematic.
4. **Faulty gate operations:** Errors can result also from faulty quantum gates. The reliability of a quantum gate can be quantified as follows: Suppose that an ideal gate described by the operator $\hat{U}$ transforms the state $|\psi\rangle$ to $|\psi_f\rangle = \hat{U}|\psi\rangle$, while an imperfect gate $\hat{U}'$ yields $|\psi'_f\rangle = \hat{U}'|\psi\rangle$. The *fidelity* $\mathcal{F}(\hat{U})$ of the quantum gate $\hat{U}$ can be defined in terms of the overlap

$$\mathcal{F}_{\psi,\hat{U}'}(\hat{U}) = |\langle \psi_f | \psi'_f \rangle|^2 = |\langle \psi | \hat{U}^\dagger \hat{U}' | \psi \rangle|^2. \tag{5.124}$$

The minimum over $|\psi\rangle$ and over $\hat{U}'$ of $\mathcal{F}_{\psi,\hat{U}'}$ is one way to define gate fidelity:

$$\mathcal{F}(\hat{U}) \equiv \min_{\psi,\hat{U}'} \mathcal{F}_{\psi,\hat{U}'}(\hat{U}). \tag{5.125}$$

An alternative way of defining the fidelity is to use $F_{\hat{U}'}(\hat{U}) = \text{Tr}\sqrt{\hat{U}'^{1/2}\hat{U}\hat{U}'^{1/2}}$ [see Eq. (5.62)] and then take $F(\hat{U}) = \min_{\hat{U}'} F_{\hat{U}'}(\hat{U})$. In any case, the gate failure rate p is given in terms of the fidelity by $p = 1 - F$.

5. **Interacting qubits:** Qubits interact with each other through two-qubit gates; hence, errors can propagate through the application of two-qubit gates, from one qubit to another.

Despite the problems mentioned above, algorithms for quantum error correction have been developed:

- Peter Shor's nine-qubit-code [71] encodes one logical qubit in nine qubits. It corrects for arbitrary errors in a single qubit.
- Andrew Steane's algorithm [72] does the same with seven instead of nine qubits.
- Stabilizer codes were developed to correct errors in several qubits systems. The simplest error-correcting codes correct single-qubit errors but fail when two or more errors occur in the encoding block. By using stabilizer codes, one can, at least in principle, correct any desired number of errors per block of qubits.
- Alexei Kitaev developed ideas for implementing fault-tolerant computing using topological methods [73] wherein information stored in the topology of a system will be robust against noise.
- A *threshold theorem* exists showing that if quantum error correction codes are concatenated, and the error rate of individual quantum gates is below a certain threshold, it is possible to perform resilient quantum computation, and therefore, decoherence and imprecision are no longer insurmountable obstacles to realizing a quantum computation.

Let us consider an error in a single-qubit state $|\psi\rangle = a|0\rangle + b|1\rangle$. Quantum error correction codes to correct an error in this single qubit employ *syndrome measurements*. A multi-qubit measurement that does not disturb the quantum information in the encoded state $|\psi\rangle$ but retrieves information about the error if this qubit is employed. A syndrome measurement can determine whether the qubit in state $|\psi\rangle$ has been corrupted. The measurement tells us not only whether the physical qubit was affected but also which of several possible ways it was affected. It is assumed that the error is either a bit flip, $|0\rangle \leftrightarrow |1\rangle$, i.e., $a|0\rangle + b|1\rangle \to b|0\rangle + a|1\rangle$, or a sign flip of the amplitude of $|1\rangle$, i.e., $a|0\rangle + b|1\rangle \to a|0\rangle - b|1\rangle$, or both, i.e., $a|0\rangle + b|1\rangle \to b|0\rangle - a|1\rangle$. These correspond to applying the operators σ_x, σ_z or $\sigma_x\sigma_z = -i\sigma_y$, respectively. The reason for this assumption is that the measurement of the syndrome has the projective effect of a quantum measurement, so even if the error due to the noise was arbitrary, it can be expressed as a superposition of basis operations σ_x, σ_y and σ_z. We can represent "no error" by the operator 1 applied to $|\psi\rangle$. Therefore, a general single qubit error can be represented by the operator

$$\mathcal{E}_{\text{error}} = \sqrt{1 - p_x - p_z - p_{xz}}\,1 + p_x^{1/2}\sigma_x + p_z^{1/2}\sigma_z + p_{xz}^{1/2}\sigma_x\sigma_z, \tag{5.126}$$

where p_i is the probability for error i. Error correcting codes are required for each of the error types described above.

The "general ideas" for designing these codes are as follows [47]:

1. Encode the state of a qubit in an entangled state of several qubits, e.g., adding two or more additional qubits and entangling the qubit in state $|\psi\rangle$ with the additional qubits. This entangled multi-qubit state belongs to a cleverly chosen code subspace of the multi-qubit Hilbert space, such that an arbitrary single-qubit error on any qubit in the subspace takes its state to an orthogonal subspace uniquely associated with that particular qubit and the error type. Hence, the error-correcting code leads to a state orthogonal to the original uncorrupted state for every possible error.
2. Perform multi-qubit measurements on the subspace, which can distinguish between the uncorrupted state and all other states resulting from any single-qubit error. Such measurements do not disclose the encoded data, but reveal the error syndrome that identifies the type and location of the error.
3. Knowing the error syndrome, perform the error correction by applying the appropriate transformation to the corrupted qubit.

Shor's Quantum Error Correction Algorithm

To see why the classical redundancy correction algorithm is inadequate for quantum error correction, suppose a single qubit $a|0\rangle + b|1\rangle$ is encoded as,

$$a|0\rangle + b|1\rangle \to a|000\rangle + b|111\rangle. \tag{5.127}$$

Consider first an error caused by, say, a flip of the first qubit,

$$a|000\rangle + b|111\rangle \xrightarrow{\text{error}} a|100\rangle + b|011\rangle. \tag{5.128}$$

Can one perform a measurement and decide which qubit is flipped? If the first qubit is measured, the result is either 0 or 1, which is the same as for the nondamaged state. However, quantum mechanics enables measurement on two-qubit states as well. Thus, for a three-qubit state, $|nm\ell\rangle$ measurement of observables such as $m \oplus \ell$ or $n \oplus \ell$ are admissible (recall that $n \oplus \ell = (n+\ell)\,[2]$). For the undamaged state, Eq. (5.127) gives 0 and 0, but for the damaged state, Eq. (5.128) gives 0 and 1. A simple check indicates that this combination is possible only if the first qubit is flipped, and we are then instructed to flip it back. Thus, the error caused by qubit flip can in principle be corrected by use of a superposition as in Eq. (5.127), and the possibility to perform collective measurements on two-qubits at once. However, the procedure in Eq. (5.127) does not yet provide any protection against phase errors, for if any one of the three qubits undergoes a phase error, then the encoded state transforms into $a|000\rangle - b|111\rangle$ and the quantum information is damaged. Moreover, the probability of phase errors now becomes three times larger after using the code of Eq. (5.127) due to qubit redundancy. To overcome this problem without deleteriously affecting qubit flip error correction, the redundancy in phase is superimposed. Explicitly, the single qubit $a|0\rangle + b|1\rangle$ is now encoded as,

$$a|0\rangle + b|1\rangle \rightarrow a|\mathbf{0}\rangle + b|\mathbf{1}\rangle, \tag{5.129}$$

$$|\mathbf{0}\rangle = \frac{1}{\sqrt{8}}[(|000\rangle + |111\rangle)(|000\rangle + |111\rangle)(|000\rangle + |111\rangle)],$$

$$|\mathbf{1}\rangle = \frac{1}{\sqrt{8}}[(|000\rangle - |111\rangle)(|000\rangle - |111\rangle)(|000\rangle - |111\rangle)]. \tag{5.130}$$

Both $|\mathbf{0}\rangle$ and $|\mathbf{1}\rangle$ are a tensor product of three factors $|\psi_1\rangle \otimes |\psi_2\rangle \otimes |\psi_3\rangle$ each containing three qubits and prepared in the same quantum state, with triple bit redundancy. A single bit flip can then be corrected in any factor by the method discussed above.

A phase change error might occur turning

$$|000\rangle \pm |111\rangle \xrightarrow[\text{error}]{} |000\rangle \mp |111\rangle, \quad \text{in one of the factors.} \tag{5.131}$$

The relative phase of the damaged factor is distinct from the phases of the other two. By the "majority rule" discussed above, we can identify it. Identification is executed not by measuring the relative phase in each factor (because such measurement would disturb the information contained in a and b), but rather by comparing the phases of pairs of factors. Technically, we need to measure a six-qubit observable $\mathcal{O}$ that flips qubits 1 through 6. Since two bit flips leaves a state intact, we have $\mathcal{O}^2 = 1$, so it has eigenvalues ± 1. A pair of factors $|\psi_i\rangle \otimes |\psi_j\rangle$ with the same sign is an eigenstate of $\mathcal{O}$ with eigenvalue $+1$, whereas a pair of factors $|\psi_i\rangle \otimes |\psi_j\rangle$ with opposite sign is an eigenstate of $\mathcal{O}$ with eigenvalue -1. By measuring $\mathcal{O}$ for a second pair of factors, we can determine which one has a different sign than the other two. Finally, we need to apply a unitary phase transformation to one of the qubits in the damaged pair and reverse the sign. The error is thereby corrected.

For further details on quantum error correction, the interested reader is referred to Refs [46, 47].

5.6 EXPERIMENTAL IMPLEMENTATIONS

There are a number of schemes that are being explored for implementing quantum information processing and quantum computing. The physical systems that have been proposed as quantum logic gates include the following:

- Ion traps [74].
- Neutral atoms in optical lattices [75, 76].
- High-Q optical cavities [77].
- Nuclear magnetic resonance (NMR) [78].
- All-optical Quantum computing [56].
- Solid-state qubits (semiconductor quantum-dot [79] and Josephson-junction devices [80]).

Any experimental implementation of quantum information processing must meet the DiVincenzo requirements for a quantum information system [81]:

I. A scalable physical system with well characterized qubits.
II. The ability to initialize the state of the qubits to a simple well-specified state.
III. Long decoherence times, much longer than the gate times.
IV. A universal set of quantum gates (single qubit rotations, two-qubit gate such as CNOT or controlled phase).
V. A qubit-specific read-out (i.e., measurement) capability.
  Let us now briefly review these quantum information processing schemes.

5.6.1 ION TRAPS

Ions can be easily confined and suspended in free space using electromagnetic fields. The qubits can be stored in the internal states of each ion, and lasers can be applied to induce single-qubit operations. Two-qubit gates can be implemented via the collective quantized motion of the ions in the trap; these collective interactions are strong since the ions interact through the Coulomb force. The first proposed scheme for a quantum computer, which considered a trapped ion system, was suggested by Ignacio Cirac and Peter Zoller in 1995 [74]. They showed how a two-qubit controlled-NOT quantum gate could be implemented in an ion trap system. In the same year, a controlled-NOT gate was experimentally realized. The fundamental operations of a quantum computer have been demonstrated experimentally with high fidelity in trapped ion systems and entanglement of up to eight qubits has been demonstrated. The trapped ion quantum computer system is one of the most promising architectures for a scalable, universal quantum information processor.

Experimentally, one places ions in a linear Paul trap, which consists of four parallel conducting rods that serve as electrodes. The distance from the surface of each electrode to the common axis is comparable to the radius of the electrodes, which is typically about 1 mm. An oscillating rf potential $V_0 \cos(\Omega t)$ is applied to two opposite electrodes, while the other pair of electrodes are grounded. For sufficiently high rf frequency Ω, a charged ion feels a time-averaged potential $m\omega_\perp^2/2$, where m is the mass of the ion and $\omega_\perp$ is the frequency associated with the transverse motion of the ion in the plane perpendicular to the common axis; typically, $\omega_\perp$ is in the MHz range. Thus, a chain of ions, cooled to low temperature, $T < 1$ mK, is strongly trapped by the effective potential in the transverse direction and confined to the lowest transverse state and trapped in the axial direction along the common axis by static potential end caps.

In an rf trap, the dynamics of the trapped ions in the axial direction, taken as the z-axis, can be described to a reasonable level of approximation in terms of normal modes of an ionic 1D crystal, since the transverse frequencies ω_x, ω_y are very high compared with the axial frequency ω_z, and the motion in these directions is frozen out (only the lowest mode in x and y is populated). The effective 1D Hamiltonian is

$$H_{\text{eff}} = \sum_i \left[\frac{p_i^2}{2m} + \frac{m}{2}\left(\omega_z^2 z_i^2 + \cos(\Omega t) v_z^2 z_i^2\right) \right] + \frac{1}{2} \sum_{i,j} \frac{e^2}{|z_i - z_j|}, \quad (5.132)$$

where p_i is the z component of the ith particle momentum and v_z is the modulation frequency. Letting $z_i(t) = z_i^{(0)} + q_i(t)$, we can determine the equilibrium position of the ith ion, $z_i^{(0)}$, by solving the equation, $(\partial V_{\text{eff}}/\partial z_i) = 0$, where $V_{\text{eff}} = \frac{m\omega_z^2}{2}\sum_i z_i^2 + \frac{1}{2}\sum_{i,j}\frac{e^2}{|z_i - z_j|}$. Expanding the effective Hamiltonian about the equilibrium position and retaining up to quadratic terms in the coordinates $q_i(t)$ yields a quadratic Hamiltonian whose normal modes can be determined (see Sec. 16.9 attached to the book web page). The normal mode with the lowest frequency is the center-of-mass mode with all of the ions rigidly oscillating together. The first excited mode is a breathing mode, and all other modes have higher frequencies.

Laser cooling of the ions by means of interaction with near-resonant laser light is used to decrease the ion temperature to less than 1 mK and gets the ions into the lowest vibrational state of the trap. Laser beams in opposite directions are applied, and by means of the Doppler effect, ions moving toward the laser propagation direction will preferentially absorb light from this beam and thus decrease their velocity. After many cycles of absorption and re-emission of light, the ions are cooled to a temperature $T_D = \hbar\gamma/(2k_B)$, known as the *Doppler cooling temperature*, where γ is the inverse lifetime of the excited electronic state used in the absorption process (see Sec. 6.1.6).

One-qubit gates are easy to implement with near-resonant lasers that excite the ionic two-level system, as long as the ions are spatially separated by more than a wavelength of the laser light, so that they can be individually addressed, and as long as the ion oscillation distance in the trap is small compared to the wavelength. This latter condition is called the *Lamb-Dicke criterion* and is quantified by the Lamb–Dicke parameter, $kl_{\text{ho}} \equiv \frac{2\pi}{\lambda}\sqrt{\frac{\hbar}{m\omega}}$, where λ is the laser wavelength and l_{ho} is the harmonic oscillator length for ions in the trap. The Lamb–Dicke criterion is $kl_{\text{ho}} \ll 1$; this condition is required so that the motional state of the ions remains largely unchanged on optical excitation. One-qubit gates are implemented with lasers that excite the qubits under two conditions: First, the inequality $d > \lambda$ should be satisfied between the average separation distance d between the ions and the wavelength of the laser light λ, so that atoms can be independently addressed by light beams. Second, the Lamb–Dicke criterion requires that the ion oscillation distance in the trap be much smaller than λ, i.e., $kl_{\text{ho}} \equiv \frac{2\pi}{\lambda}\sqrt{\frac{\hbar}{m\omega}} \ll 1$.

Two-qubit gates that are fast and have high fidelity are less trivial to implement for ions. The Cirac–Zoller CNOT gate is one solution [74]. It flips the state of a target qubit, $|\downarrow_2\rangle \leftrightarrow |\uparrow_2\rangle$, only when the control qubit is, say, in state $|\downarrow_1\rangle$. It

5.6 Experimental Implementations

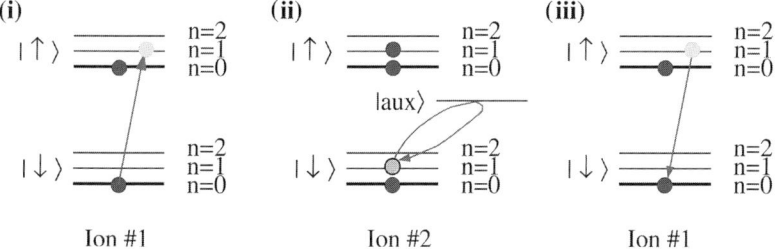

FIG 5.15 Cirac–Zoller CNOT gate. A phase gate is constructed by performing the sequence illustrated in this figure. (a) π-pulse on the first blue sideband on the first ion to map the internal state to the collective vibrational state. (b) 2π-pulse between the $|\downarrow, n=1\rangle$ state and an auxiliary state $|aux\rangle$ on the second ion, resulting in a π phase shift on the state $|\downarrow, n=1\rangle$. (c) a π-pulse on the first blue sideband on the first ion to map the vibrational state back to the internal state. A controlled-NOT gate can be constructed from a phase gate with a $\pi/2$-pulse on the second qubit before and after the phase gate as described in the text. Reproduced from P. J. Lee et al. Phase Control of Trapped Ion Quantum Gates, J. Opt. B: Quantum Semiclass. Opt. **7**, S371 (2005).

can be implemented by cooling the breathing motional mode ν of the two ions to the $|0_\nu\rangle$ ground state and performing the three steps:

(1) Apply a $\pi/2$-pulse on the target qubit with associated phase ϕ.
(2) Apply a π phase gate on the two ions (see Fig. 5.15).
(3) Apply a $-\pi/2$-pulse on the target qubit with phase ϕ.

Steps 1 and 3 are achieved by focusing radiation on the target ion only and applying on-resonance $\pi/2$ pulse (that is on for a specific period of time long enough to put the system in a coherent 50% superposition of ion number two, see Sec. 6.1.1). Step 1 results in the evolution

$$\alpha|\uparrow_2\rangle + \beta|\downarrow_2\rangle \to \frac{(\alpha + e^{-i\phi}\beta)}{\sqrt{2}}|\uparrow_2\rangle + \frac{(\beta - e^{i\phi}\alpha)}{\sqrt{2}}|\downarrow_2\rangle. \quad (5.133)$$

Step 3 is identical to Step 1 except the phase is shifted by π, i.e., Step 3 is Step 1 reversed. Step 2, involves what is known as a π-pulse, which completely moves the population between two states. Step 2 can be implemented as shown in Fig. 5.15 [74]. Other schemes of entangling trapped ion qubits are reviewed in Ref. [82], including the two-qubit gate proposed by Sorensen and Molmer that uses laser radiation tuned close to the motional sidebands and involving destructive interference to eliminate the dependence of rates and revolution frequencies on vibrational quantum numbers.

5.6.2 NEUTRAL ATOMS IN OPTICAL LATTICES

An optical lattice is a periodic optical structure created by standing wave laser beams that give rise to the optical potential felt by the atoms in which cooled atoms are trapped (see Fig. 5.16). Ideally, one atom is trapped in each lattice site. The first proposals for using neutral atoms in optical lattices for quantum computing appeared in 1999 [75, 76]. Neutral atoms in their electronic ground state couple extremely weakly to the environment. However, neutral atoms also couple very weakly to each other, making two-qubit gates difficult to implement, but interatomic couplings can be created on demand by induced electric dipole–dipole interactions or ground state collisions. The ability to turn on interactions only when needed is highly advantageous because it reduces coupling to the environment and the spread of errors during computation. The weak atomic interactions also make it relatively straightforward to trap and cool neutral atoms in large numbers, with favorable implications for scaling to many qubits and perhaps parallel processing. While both Refs [75, 76] suggest controlled collisions between atoms trapped on different lattice sites for conditional logic operations, single-qubit gates are implemented by addressing each atom with a laser beam to change the electronic state of the atoms. A two-qubit gate is implemented by bringing the atoms together and letting their wave functions overlap. During this overlap a phase that depends on the two-qubits state, $|m, n\rangle \equiv |m\rangle \otimes |n\rangle$, is accumulated because of the atom–atom (molecular) interaction. Two-qubit phase gates based on this scheme were demonstrated in Ref. [83]. Since the atom–atom interaction is weak the atoms need to interact for a relatively long time to build up the phase due to their interaction. To achieve efficient computation and to be faster than decoherence processes, it is desirable to generate gates that operate as fast as possible. The slow two-qubit collisional gate is one of the two serious obstacles to quantum computing with neutral atoms in optical lattices, the other is decoherence. Optimal control theory has been used to determine the time dependence of potential to bring the atoms together, so they interact and generate two-qubit collisional gates, which operate as fast as possible while maintaining high fidelity of the gate [84].

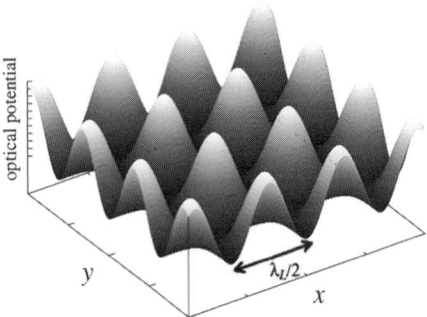

FIG 5.16 The spatial interference pattern of lasers creates an optical lattice potential that confines atoms in a region smaller than the optical wavelength, λ_L. The potential is proportional to the laser intensity and inversely proportional to the laser detuning from resonance of the atomic transition.

The loading of an optical lattice with one atom per site using a Bose–Einstein condensate (BEC) source has been proposed and studied. These loading methods provide a means for initializing a coherent qubit system that can then be used to perform elementary quantum logic operations. Hence, a system of atoms in an optical lattice holds out the prospect for parallelism and scalability, making them near-ideal systems for testing quantum computing methodologies.

One-qubit gates for atoms in an optical lattice are easy to implement with lasers, as long as the Lamb–Dicke criterion is met, but making fast two-qubit gates with high fidelity is more problematic. The basic idea of realizing controlled-phase gates with a two-particle system in an optical lattice potential is: The external potential initially localizes the particles far enough apart so that they may be considered independent. It is then varied in time, so that wave function of the atoms in different lattice sites overlap causing correlations due to atom–atom interaction. The external potential is finally restored to its initial shape, so that the two atoms no longer interact and are found in a new correlated state,

$$\text{CP}(\phi)|m, n\rangle = \exp(imn\phi) |m, n\rangle, \tag{5.134}$$

with an acquired controlled phase ϕ given by the integral over time of the state-dependent energy difference of the qubits that results from their interaction during their overlap. Optimization of such two-qubit gates has been studied in Ref. [84].

5.6.3 CAVITY BASED QUANTUM COMPUTING

A number of cavity quantum electrodynamics (QED) based schemes for implementing quantum information processing and communication have been suggested. In cavity QED one is able to prepare quantum states for the electromagnetic field stored in a resonant cavity. Moreover, one can entangle the field with atoms having a transition frequency close to resonance with the cavity frequency of the traverse mode of the cavity. Cavities in the microwave and the optical domain have been used. In cavity QED, the electromagnetic energy $\frac{1}{4\pi}E^2 \mathcal{V}$ stored in the volume $\mathcal{V}$ of the cavity mode is taken equal to half a photon energy, $\hbar\omega/2$, i.e., $E^2 = \frac{\hbar\omega}{8\pi\mathcal{V}}$. Then, the atom–cavity coupling energy $\hbar g = \mathbf{d} \cdot \mathbf{E}$ between the atomic dipole moment $\mathbf{d}$ and the field can also be large, in the sense that g is large compared with the rates at which energy is dissipated out of the system by spontaneous emission from the atom and the rate at which energy leaks out of the cavity. Quantum computing with atoms in cavities requires a precisely tunable atom–cavity coupling g. Cavity QED quantum computing schemes are scalable to a relatively large number of qubits represented by single atoms placed in an optical cavity. The photon mode of the cavity plays the role of the quantum data bus, which conveys information between the qubits. The reader interested in more details on cavity-based quantum computing is referred to Ref. [77].

5.6 Experimental Implementations

5.6.4 NUCLEAR MAGNETIC RESONANCE SYSTEMS

Nuclei having nonvanishing spin can have quantum information encoded into their spin degrees of freedom. Direct manipulation and detection of nuclear spin states is well developed and is the basis for *nuclear magnetic resonance* (NMR) spectroscopy and imaging techniques (see Sec. 4.8), wherein a static magnetic field and radiofrequency electromagnetic waves are used to control the nuclear spin degrees of freedom of atoms and molecules. NMR spectroscopy has a wealth of diverse coherent manipulations of spin dynamics and can therefore be considered for quantum information processing applications.

NMR studies began in 1946, when the groups of Felix Bloch and Edward Purcell first observed the magnetic induction of nuclear spins, and this field has rapidly blossomed and led to many important applications, such as molecular structure determination, dynamics studies in both the liquid and the solid states, and magnetic resonance imaging used today for medical diagnostics. NMR techniques are typically used to control and observe a macroscopic number of nuclei. Some commonly used spin 1/2 nuclei used for NMR include ^{1}H, ^{13}C, ^{14}N, ^{19}F, and ^{31}P (NMR of higher spin nuclei can also be carried out). When these nuclei are put into a static magnetic field, the Zeeman splitting of the levels, ΔE_Z is proportional to the product of the nuclear magnetic moment and the magnetic field strength, $\Delta E_Z = g_N \mu_N B$, where g_N is the nuclear g-factor and $\mu_N = \frac{e\hbar}{2M_p c}$ is the nuclear magneton. Figure 4.2 shows the Zeeman energies and their splittings ΔE_Z as a function of the static magnetic field strength (see the discussion in Sec. 4.3 of Zeeman splitting, and the discussion of NMR in Sec. 4.8). When the radiofrequency ω is such that the rf photon energy $\hbar \omega$ equals the Zeeman spitting ΔE_Z, transitions between the levels are induced.

Ideas and instrumentation from NMR spectroscopy have been used in quantum information processing experiments [78]. Two problems exist regarding the use of NMR for quantum computation. (1) The nuclear magnetic moment is small, hence, a large number ($\gtrsim 10^4$) of nuclear spins must be present to produce a measurable signal. Usually, the output of an NMR measurement is an average over all the spins in a macroscopic volume element. How does one do quantum computation with an ensemble of spins? Can the average output of an ensemble of quantum spins be useful for quantum computation? (2) It would seem that a system used for NMR computation would have to be prepared in a pure state not in a high entropy mixed state as most NMR experimental systems used today. Moreover, there is an additional issue of the coherence time of nuclear spins as it affects quantum computation algorithms. An early connection of NMR to coherent information processing is the spin echo effect discovered in 1950 by E. Hahn who demonstrated that inhomogeneous interactions could be refocused to the extent that the phase of the nuclear spins retain information about the local field. Refocusing makes use of the fact that rf pulses that produce $180°$ rotation about the x-axis, $R_x(\pi) = e^{-i\pi\sigma_x/2}$, can be used to reverse the time evolution, so that different spins initially pointing in the same direction that defocus due to different local fields are refocused by using a π pulse.

NMR systems are robust quantum systems with coherences that can be precisely manipulated, and decoherence rates that can be slow on the time scale of the qubit interactions. Single-qubit gates are implemented by applying ac magnetic fields and two-qubit gates by spin–exchange interaction between the nuclear spins. A large number of molecules in a liquid state can serve as many quantum registers operating in parallel. The measurement involves detection of the average magnetization of the whole sample. Liquid-state NMR schemes have been realized with a small number of qubits ($\lesssim 10$) but are not scalable because of the difficulty of robust initialization in a thermal ensemble and because of difficulties in resolving the NMR frequencies of individual qubits and the measurement signal when the number of qubits becomes large. Scalable NMR schemes have been proposed for solid-state systems with dopants that serve as qubits implanted at a regular separation from each other in a semiconductor material.

One-qubit gates can be implemented directly. For example, a simple NOT gate, which interconverts $|0\rangle$ and $|1\rangle$, can be implemented as a rotation about axes in the xy-plane that can be achieved using rf pulses, while rotations about the z-axis can be accomplished by using periods of free precession under the Zeeman Hamiltonian, $H = -\boldsymbol{\mu}_N \cdot \mathbf{B} = -g_N \mu_N B S_z/\hbar$ [see Sec. 4.3, Eq. (4.32)]. Two-qubit gates can be implemented using the CNOT gate and one-qubit gates. The CNOT gate can be implemented using a three-gate circuit, as shown in Fig. 5.3, where the two-qubit controlled-π-phase gate performs the transformation $|1\rangle|1\rangle \xrightarrow{\pi} -|1\rangle|1\rangle$, while leaving all other states unchanged [see (5.49) and substitute $\phi = \pi$]. These gates can be easily implemented in two spin systems, allowing quantum computers with two qubits to be easily constructed. However, for larger spin systems, the process becomes more complicated. It is not possible to simply use pulse sequences designed for two spin systems, as it is necessary to consider the evolution of all the additional spins in the system. But, it is possible to refocus the evolution of these spins under their chemical shift and angular momentum coupling interactions using spin-echo techniques. The simplest method is to nest spin echoes within one another, so that all the undesirable interactions are removed, but this simple approach requires an exponentially large number of refocusing pulses. This problem can be overcome by using efficient refocusing sequences.

5.6.5 ALL-OPTICAL QUANTUM COMPUTERS

Photons have two orthogonal polarization states that can be easily manipulated and controlled and are therefore natural candidates for representing qubits. They propagate quickly over long distances in optical single-mode fibers, almost without undergoing absorption and decoherence. The polarization of a photon can be put into arbitrary linear combinations of the two orthogonal polarization states, as shown in Fig. 5.17. Moreover, the quantum information stored in a photon tends to stay there, i.e., decoherence of the polarization state of a photon is often not a very serious problem. A unitary transformation that effects the photon polarization is easy to affect using waveplates [18] so one-qubit gates are easy to implement. The problem is that photons do not interact with one another, at least not in vacuum, and therefore, it is difficult to make a two-qubit gate for photons. Effective interaction does occur in media, due to the Kerr effect, i.e., the intensity dependence of the refractive index of materials, but this effect is very small for a few photons; normally, only when one uses intense laser sources is the effect significant. Attempts at achieving a large enough Kerr nonlinearity using electromagnetically induced transparency in atomic media, and other schemes involving a Kerr nonlinearity have been explored. An alternative technique to induce an effective interaction between photons is via projective measurements with photodetectors using a protocol in which probabilistic two-photon gates are implemented using teleportation with high probability [56]. The trick is "to prepare an appropriate entangled state suitable for teleportation with the desired gate already applied" before using it for the teleportation protocol. The problem then becomes that of preparing the entangled state (non-deterministically) and implementing the requisite measurement in the protocol [56]. We shall not give details of the protocol (the interested reader is referred to Ref. [56]).

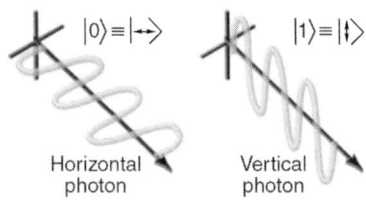

FIG 5.17 Photon qubits. A horizontal photon $|\leftrightarrow\rangle$ can represent the qubit $|0\rangle$, and a vertical photon $|\updownarrow\rangle$ can represent the qubit $|1\rangle$.

The key ingredients in all-optical quantum computers include the following: (1) A source of single photons that are produced on demand (this is also required for other quantum information applications such as quantum cryptography via the BB84 algorithm and quantum teleportation of qubits). (2) A means of performing polarization rotation and phase-shifting of single photons. This can be accomplished with birefringent waveplates and polarization beam splitters. (3) A two-photon conditional gate. Reference [56] proposed an optical CNOT gate by teleportation. (4) A measurement scheme capable of detecting the polarization states of a single photon. One possibility for the latter is superconducting single-photon detectors, which demonstrate a combination of the picosecond response time and high quantum efficiency, such as those based on ultrathin, submicron-width NbN structures.

5.6.6 SOLID-STATE QUBITS

To create a solid-state qubit, one needs a two-level, solid-state quantum system. Efforts to make solid-state qubits have been focused on superconductors and semiconductors. The advantages of solid-state systems for mass-produced devices is clear, but initially it was not obvious that solid-state qubit systems could be created with sufficiently long coherent storage times for the qubits, fast qubit manipulation, single qubit measurement capability, and scalable methods for entangling spatially separated matter-based qubits. As we shall see, progress with these goals has been substantial.

5.6 Experimental Implementations

The superconducting approach uses macroscopic size devices. Superconductors have an advantage in being able to maintain coherence because electrons in a superconductor condense into *Cooper pairs* (spin-singlet paired electrons of opposite momentum that are bound together by an arbitrarily small attractive interaction mediated by phonon coupling) are very stable at low temperature. They form a superfluid and are able to move through the superconductor without interactions, since it takes an amount of energy, known as the *energy gap*, to break up a Cooper pair. This gap is an order of magnitude greater than the typical energy available. Hence, qubit states can be transformed via unitary transformations without breaking up the Cooper pairs and jeopardizing the quantum coherence of the system [80].

Single-electron states of semiconductor quantum dots have also been explored as coherent qubit systems for quantum information uses. [79] Two different ways to achieve a semiconductor qubit using electrons have been studied. (1) In a magnetic field, the Zeeman splitting of the electron spin within a quantum dot results in a two-level system. An electron spin confined within a semiconductor nanostructure can be relatively stable against decoherence and is easily manipulated electrically and/or optically. (2) The spatial wave function of a single-electron state in a double-well potential is also a two-level system. The double-well potential can be formed by two adjacent quantum dots coupled, for instance, by quantum mechanical tunneling between the two localized states. Tunneling between the adjacent dots lifts the degeneracy and allows for mixed and entangled states.

Superconducting Qubits

A key component in most superconducting qubits is a *Josephson junction*, which consists of a thin layer (a few nanometers) of insulating aluminum oxide, sandwiched between two superconducting layers of aluminum, which is superconducting when cooled below temperatures of 1.2 K. Cooper pairs can tunnel through the insulating layer and couple the superconducting wave functions on either side of the barrier. Most circuits for superconducting qubits contain Josephson junctions. For example, a charge qubit consists of a small volume of superconductor, known as a Cooper-pair box [80], that is connected to a weak Josephson junction and driven with a gate voltage, V_g, through a capacitor. If $V_g = e/C_g$, where C_g is the capacitance of the gate, then the states in which there are zero extra Cooper pairs, $|0\rangle$, and $|1\rangle$, an extra Cooper pair in the box, have the same energy. However, quantum tunneling through the Josephson junction results in two new quantum states, one is a symmetric superposition, $|0\rangle + |1\rangle$, whereas the other is an antisymmetric superposition, $|0\rangle - |1\rangle$. These new quantum states differ in energy, and these superposition states form the basis of the "charge qubit." Interaction between charge qubits is via their charge. A two-qubit CNOT gate using a pair of superconducting quantum bits has recently been demonstrated.

Quantum Dot Qubits

The first quantum dot option was studied theoretically by Loss and DiVincenzo [85]. Two-qubit gates in such systems have been discussed in Ref. [86], which considered a structure composed of a 10-nm thick layer of AlGaAs sandwiched between two thicker layers of GaAs. The lower substrate of GaAs is *n*-doped and provides free electrons that accumulate at the upper interface between the AlGaAs and GaAs, forming a 2D electron gas. An array of metallic contacts (i.e., gates), is lithographically imprinted on the top of the upper GaAs layer about 100 nm above the 2D electron gas. Electrons can be confined in these quantum dot regions. Voltages applied to the gates restrict the movement of the electrons in the *x-y* plane. Quantum control involving initialization, spin rotation, and detection, can be obtained using electrically controlled radiofrequency pulses, but this method is rather slow for the construction of quantum computing circuits operating at useful clock speeds. Optical manipulation of electron spin allows faster operation, e.g., by first initializing the spin state using optical pumping, then rotating the spin by a laser pulse through stimulated rapid adiabatic passage (STIRAP) [see Sec. 6.4], and finally by detecting the spin state using optical pumping and photon detection.

The spin state of an individual electron represents a qubit with basis states $|0\rangle \equiv |\uparrow\rangle$ and $|1\rangle \equiv |\downarrow\rangle$, where the *z*-axis is taken normal to the surface of the structure. By applying a static magnetic field $B_\perp(x)$ perpendicular to the surface having a large gradient along the *x*-axis, one induces a Zeeman shift of the spin-up and spin-down components of the electrons, $E_{j,\pm} = \pm 2\mu B_\perp(x_j)$, where x_j is the *x*-position of *j*th dot.

In the second option mentioned in Sec. 5.6.6, potential wells for the electrons can be induced and controlled by gate-electrodes that determine potential barriers between neighboring potential minima. By manipulating the voltages applied to the metal gates, a 2D electron gas under the gates can be depleted and then doped, so that each quantum dot contains a single electron occupying the lowest energy level. Tunneling between the adjacent dots lifts the degeneracy and allows for mixed and entangled states of the two split states. Hence, the two split levels can be used as a two-level system. Single-qubit operations correspond to transferring population from one level to another. A two-qubit swap operation can be achieved by applying a pulsed inter-dot gate voltage, so the exchange constant in the Heisenberg Hamiltonian becomes time dependent, $H_{\text{SWAP}}(t) = J(t)\,\boldsymbol{\sigma}_A \cdot \boldsymbol{\sigma}_B$, where the Pauli matrices operate in the space of the two-level system for each quantum dot. The unitary operation is then given by $U_{\text{SWAP}} = e^{-i \int_0^T dt\, H_{\text{SWAP}}(t)}$ and the duration of the pulse is chosen, so that the integral in time over time of J(t) gives π (mod 2π).

Two-qubit gates for solid-state charge-qubit architectures involving spins in quantum dots have been proposed, but we shall not discuss these here. Another solid-state system suggested for implementing a quantum computer are nitrogen-vacancy (NV) diamond centers [87]. The NV center is a point defect in the diamond lattice consisting of a nearest-neighbor pair of a nitrogen atom, which substitutes for a carbon atom, and a lattice vacancy. The $^3A_2$ ground state of the NV center consists of two unpaired electrons in a triplet state. It is possible to selectively address transitions between the $m_s = 0$ and the $m_s = 1$ states of the triplet in the presence of an external magnetic field, so that each NV center behaves like a two-level subspace of the spin triplet. Schemes for implementing two-qubit gates in this system have been suggested.

5.7 THE EPR PARADOX

The Einstein, Podolsky, and Rosen (EPR) paradox [20], set out in 1935, challenged the interpretation of quantum mechanics as a correct and complete theory. The EPR paradox is related to quantum information theory, the subject of this chapter, because it involves entanglement, which is an information resource. Hence, we present a short discussion of the EPR paradox, and Bell inequalities, which quantify the EPR paradox, in this chapter.

The EPR paradox is often presented via a variation of the original argument, due to Bohm and Aharonov [88], in terms of experiments performed on two correlated two-level systems (e.g., two spin polarized electrons or two polarized photons).[9] Consider the state for the two particles in the entangled singlet state $|\Psi^-\rangle$ that can taken in the form,

$$|\Psi^-\rangle = \frac{1}{\sqrt{2}}(|+_A\rangle_1|-_A\rangle_2 - |-_A\rangle_1|+_A\rangle_2), \qquad (5.135)$$

where $|\pm_A\rangle_1$ are orthonormal eigenstates for operator $\mathcal{A}_1 = \mathbf{n}_A \cdot \boldsymbol{\sigma}_1$ of particle 1 with eigenvalues ± 1, and $|\pm_A\rangle_2$ are orthonormal eigenstates for operator $\mathcal{A}_2 = \mathbf{n}_A \cdot \boldsymbol{\sigma}_2$ of particle 2 with eigenvalues ± 1. The unit vector $\mathbf{n}_A$ can point in any direction. For example, the singlet state can be written in terms of the basis states that are eigenstates of the operator $\mathcal{A}_1 = \sigma_{1,z}$ and $\mathcal{A}_2 = \sigma_{2,z}$ (i.e., $\mathbf{n}_A = \hat{z}$) with eigenvalues ± 1, i.e.,

$$|\Psi^-\rangle \equiv \frac{1}{\sqrt{2}}(|\uparrow\rangle_1|\downarrow\rangle_2 - |\downarrow\rangle_1|\uparrow\rangle_2). \qquad (5.136)$$

Suppose we measure $\mathcal{A}_1$ for particle 1. If the result of the measurement is the eigenvalue $+1$, particle 2 must be in state $|\downarrow\rangle_2$. Hence, we would predict with certainty that after the measurement of particle 1 in eigenstate $|\uparrow\rangle_1$ of $\mathcal{A}_1$, 2 is in eigenstate $|\downarrow\rangle_2$ of $\mathcal{A}_2$ with unit probability.

Problem 5.26

Denote by $|\rightarrow\rangle_i$ and $|\leftarrow\rangle_i$ ($i = 1, 2$) eigenfunctions of $\sigma_{i,x}$ with eigenvalues $+1$ and -1. Show that $|\Psi^-\rangle$ can be written as

---

[9] In the photon case, we would consider positive and negative helicity polarized photons, or two linearly polarized photons with polarization states $|\uparrow\rangle$ and $|\rightarrow\rangle$. A photon is an important example of a qubit since it can propagate at the speed of light and therefore is ideal for quantum information systems requiring propagation from one site to another.

5.8 Bell's Inequalities

$$|\Psi^-\rangle = |\Psi_x^-\rangle \equiv \frac{1}{\sqrt{2}}(|\rightarrow\rangle_1|\leftarrow\rangle_2 - |\leftarrow\rangle_1|\rightarrow\rangle_2). \tag{5.137}$$

Hint: Express $|\uparrow\rangle$ and $|\downarrow\rangle$ as linear combinations of $|\rightarrow\rangle$ and $|\leftarrow\rangle$ for each particle.

Furthermore, we can rewrite Eq. (5.136) in terms of eigenstates of different operators, $\mathcal{B}_1 = \mathbf{n}_B \cdot \boldsymbol{\sigma}_1$ and $\mathcal{B}_2 = \mathbf{n}_B \cdot \boldsymbol{\sigma}_2$, where $\mathbf{n}_B$ is some other unit vector not equal to $\mathbf{n}_A$ (see Problem 5.26), because the state is rotationally invariant. This is convenient if we carry out the measurement of the operator $\mathcal{B}_1 = \mathbf{n}_B \cdot \boldsymbol{\sigma}_1$ on particle 1 and $\mathcal{B}_2 = \mathbf{n}_B \cdot \boldsymbol{\sigma}_2$ on particle 2. For example, as shown in Problem 5.26, we could take $\mathcal{A}_1 \equiv \sigma_{1,z}$, $\mathcal{A}_2 \equiv \sigma_{2,z}$ or alternatively, $\mathcal{B}_1 \equiv \sigma_{1,x}$, $\mathcal{B}_2 \equiv \sigma_{2,x}$. If the result of the measurement of $\mathcal{B}_1$ is eigenvalue ± 1, then particle 2 must be in eigenstate $|\mp\rangle_2$ of $\mathcal{B}_2$. EPR asked the question, what happens when the two operators on particle 2, $\mathcal{A}_2$ and $\mathcal{B}_2$, do not commute? Then, depending on what measurement is performed on particle 1 ($\mathcal{A}_1$ or $\mathcal{B}_1$), we predict with certainty the values of physical quantities represented by noncommuting operators ($\mathcal{A}_2$ or $\mathcal{B}_2$) without in any way interacting with this particle, i.e., without in any way disturbing the system. This led EPR to claim, "if without in any way disturbing a system, we can predict with certainty (i.e., with probability equal to unity), the value of a physical quantity, then there exists an element of physical reality corresponding to that quantity." Since the wave function contains a complete description of the two-particle system, it seems that the EPR argument establishes that it is possible to assign reality to two different states, the eigenstates of the operator $\mathcal{A}_2 = \sigma_{2,z}$ and the eigenstates of the operator $\mathcal{B}_2 = \sigma_{2,x}$. But, according to quantum mechanics, two physical quantities represented by operators that do not commute *cannot have simultaneous reality*. Hence, we are forced to abandon one of the following two assertions: (1) The wave function is a complete description of the system, or (2) the states of spatially separated systems are independent of each other. Therefore, EPR concluded that the quantum mechanical description of physical reality is incomplete. The EPR argument that shows that quantum mechanics cannot be a complete theory makes use of two assumptions, locality and realism. Locality prohibits any influences between events in space like separated regions. Realism means that all measurement outcomes depend on pre-existing properties of objects that are independent of the measurement procedure.

The EPR argument regarding the meaning of the expression "without in any way disturbing the system" was criticized by Niels Bohr. While he agreed that there is no question of a mechanical disturbance of the system, and that "our freedom of handling the measuring instruments is characteristic of the very idea of experiment ... we have a completely free choice whether we want to determine the one or the other of these quantities ...," Bohr argued that "there is essentially the question of an influence on the very conditions which define the possible types of predictions regarding the future behavior of the system." That is, the choice of experiment performed on the first system determines the predictions that can be made for the experiments performed on the second system. Yet, no experiment performed on the second system, without knowledge of this choice, could reveal the occurrence of a disturbance to the second system, thereby determining the choice of the first experiment.

If we accept the conclusion of EPR, a more complete physical theory than quantum mechanics, with additional (perhaps inaccessible) variables might exist. John S. Bell showed that, such a theory, possessing locality and realism, leads to predictions that are incompatible with quantum mechanics. Experiments have shown that local realism theories (and the "hidden variable" theories developed with this concept in mind) are wrong and that quantum mechanics is right! We shall now take up Bell's arguments.

5.8 BELL'S INEQUALITIES

Bell's theorem [53–55], originally formulated in 1964, shifted the discussion from the abstract realm of local realism and hidden-variable theories applied to entangled states, to the realm where stringent predictions, in the form of algebraic expressions obtained upon using these theories, could be tested experimentally. These kinds of expressions, now called *Bell inequalities*, are violated by quantum mechanics. Therefore, an appropriate experiment should be able to resolve the issue. Such an experiment was suggested by Bell. Specifically, he showed that there is an upper limit to the correlation of distant events if one assumes the validity of local realism and that this upper limit is violated by quantum mechanics.

Bell's formalism allowed quantitative questions to be asked and answered experimentally. Quantum mechanics could thereby be tested in a way that it never was before.

This section contains three subsections. First, we heuristically derive the Bell inequalities, relate them to the arguments of EPR and show that they are violated by quantum mechanics. Second, we analyze the Bell inequalities based on the notion of the hidden variable scenario. Finally, we present the Bell inequalities in a more universal context.

5.8.1 BELL'S INEQUALITIES AND THE EPR PARADOX

Bell's idea was to study correlations arising when measuring projections of two spins along several directions. When these measurements are carried on an entangled state, such as the spin-singlet state $|\Psi^-\rangle = (|\uparrow\rangle|\downarrow\rangle - |\downarrow\rangle|\uparrow\rangle)/\sqrt{2}$, and local realism is assumed to be valid, he derived an inequality between the correlations, that can be tested experimentally. Let $A_i(\alpha)$ denote a measurement designed to determine the spin projection of particle i ($i = 1, 2$) along the direction given by the unit vector $\mathbf{n}_\alpha$, with possible measurement outcomes $a_i(\alpha) = \pm 1$. Here, α denotes the two spherical angles that determine the direction of the unit vector $\mathbf{n}_\alpha$. $A_i(\alpha)$ can be envisioned as arising from the measurement of a spin component in a Stern–Gerlach experiment. For convenience, the spin projection is measured in units of $\hbar/2$; hence, the values ± 1. We stress that $A_i(\alpha)$ is not necessarily a measurement operator in the orthodox quantum mechanical sense, since we do not necessarily use an orthodox quantum mechanical theory, but some theory that supersedes quantum mechanics such as envisioned by EPR. Later on, when we treat the measurement within quantum mechanics, we will use the operator, $\hat{A}_i(\alpha) = \hat{\sigma}_i \cdot \mathbf{n}_\alpha$.

The possible outcomes of measuring $A_i(\alpha)$ in a spin-singlet state $|\Psi^-\rangle$ are

$$a_i(\alpha) = \pm 1, \tag{5.138}$$

independent of α, as is clear from Problem 5.26. We consider four possible directions, $\mathbf{n}_\alpha$, $\mathbf{n}_\beta$, $\mathbf{n}_\gamma$, and $\mathbf{n}_\delta$, and inspect the quantity

$$Q(\alpha, \beta, \gamma, \delta) \equiv a_1(\alpha)[a_2(\gamma) - a_2(\delta)] + a_1(\beta)[a_2(\gamma) + a_2(\delta)]. \tag{5.139}$$

Is it possible to determine the allowed values of Q that can be obtained experimentally by simply using Eq. (5.138)? According to the EPR local realism concept, the answer is affirmative, since the results of two measurements pertaining to particles 1 and 2, such as $a_1(\alpha)$ and $a_2(\gamma)$, have *simultaneous* definite values. In other words, according to the EPR argument, the measurement of $A_1(\alpha)$ associated with particle 1 does not affect the results of the measurement of $A_2(\gamma)$ associated with particle 2, and vice versa. Thus, the local realism hypothesis implies the validity of the following equality,

$$|Q| = |a_1(\alpha)[a_2(\gamma) - a_2(\delta)] + a_1(\beta)[a_2(\gamma) + a_2(\delta)]| = 2, \tag{5.140}$$

because one of the expressions in the square brackets vanishes, while the other equals ± 2. Now consider the quantity $\overline{Q}$ obtained by averaging Q over many measurements performed on identical systems. In each measurement, Q can take on the values ± 2, hence $|\overline{Q}| \leq 2$. The following variant of Bell's inequality, referred to as the *CHSH inequality*, named after Clauser, Horne, Shimony, and Holt [89] is obtained on averaging the RHS of Eq. (5.139),[10]

$$\boxed{\left|\overline{a_1(\alpha)a_2(\gamma)} - \overline{a_1(\alpha)a_2(\delta)} + \overline{a_1(\beta)a_2(\gamma)} + \overline{a_1(\beta)a_2(\delta)}\right| \leq 2.} \tag{5.141}$$

Is Eq. (5.141) compatible with quantum mechanics? If quantum mechanics is complete and correct, the measurement of $A_i(\alpha)$ corresponds to the measurement of the operator $\hat{A}_i(\alpha) = \hat{\sigma}_i \cdot \mathbf{n}_\alpha$, and the average value of the correlation, say $\hat{A}_1(\alpha)\hat{A}_2(\gamma)$ when the system is in the spin-singlet state corresponds to [see Eq. (4.20)]

---

[10] Inequality (5.141) is relevant not just for quantum mechanical systems. For example, it can be used for analyzing an experiment where two people separated by a large distance are tossing identical coins. In such an experiment, this inequality should be satisfied, because there is no influence whatsoever between the two measurement outcomes. But in quantum mechanics, this inequality does not hold because quantum mechanics is not local.

5.8 Bell's Inequalities

$$\overline{a_1(\alpha)a_2(\gamma)} \Rightarrow \left\langle \Psi^- \left| \hat{A}_1(\alpha)\hat{A}_2(\gamma) \right| \Psi^- \right\rangle = -\mathbf{n}_\alpha \cdot \mathbf{n}_\gamma. \quad (5.142)$$

In a pure state, such as $|\Psi^-\rangle$, the expectation value of the correlation has the same operational meaning as the average $\overline{a_1(\alpha)a_2(\gamma)}$ discussed above. Therefore, the compatibility of quantum mechanics with Bell's inequality (5.141) can be determined by substituting the expectation value of the quantum correlation on the LHS of Eq. (5.141) instead of the barred averages. By making a judicious choice of the unit vectors $\mathbf{n}_\alpha$, $\mathbf{n}_\beta$, $\mathbf{n}_\gamma$, and $\mathbf{n}_\delta$, it is possible to check if the inequality (5.141) is satisfied by computing the four correlations using Eq. (5.142). This is easily done if we let all four unit vectors lie in a plane, e.g., by using the polar angles $\beta = 0$, $\alpha = 90^0$, $\delta = 135^0$ and $\gamma = 225^0$, and letting $\phi = 0$. According to Eq. (5.142), quantum mechanics gives $\langle a_1(\alpha)a_2(\gamma)\rangle = \langle a_1(\beta)a_2(\gamma)\rangle = \langle a_1(\beta)a_2(\delta)\rangle = 2^{-1/2}$, whereas $\langle a_1(\alpha)a_2(\delta)\rangle = -2^{-1/2}$. Hence,

$$|\langle a_1(\alpha)a_2(\gamma)\rangle - \langle a_1(\alpha)a_2(\delta)\rangle| + |\langle a_1(\beta)a_2(\gamma)\rangle + \langle a_1(\beta)a_2(\delta)\rangle| = 2\sqrt{2} > 2. \quad (5.143)$$

Experiments verify Eq. (5.143); they are incompatible with Eq. (5.141) [90]. In other words, quantum mechanics violates the Bell inequality that was derived as a consequence of the assumptions used by EPR. To summarize,

(1) In 1935, EPR argued, based on the analysis of correlations of entangled states, that quantum theory is incomplete, i.e., something is missing in the existing theory. They assumed "local realism," which means that a measurement performed on a particle cannot affect the result of a measurement performed on another particle located in a space-like separated region.
(2) In 1964, John Bell showed that, if local realism is true, then the inequality Eq. (5.141) should be satisfied, and, moreover, that the laws of quantum mechanics as applied to a pair of entangled quantum particles lead to its violation. Experiment must determine which theory is valid. Experimental tests have verified that the theoretical prediction based on quantum mechanics is correct.
(3) Consequently, local realism cannot be a correct picture of the physical world. In an entangled state, it is not possible to perform measurement on a single particle without affecting measurements of the other.

5.8.2 BELL'S ANALYSIS USING HIDDEN VARIABLES

Bell argued that if the "complete" theory that EPR were hoping for exists, it should incorporate an additional parameter(s), whose knowledge would render measurements independent for two spatially separated objects, i.e., there must be a real parameter λ, which determines completely the results of correlated measurements made on entangled states such as the singlet-state $|\Psi^-\rangle$. This parameter is absent from the quantum mechanical description (as we know it). The parameter λ is known as a hidden-variable since it is not directly accessible to experiment. The entangled state depends upon λ, hence, the results $a_i(\lambda, \theta)$ of measuring the operator $A_i(\theta)$ depends on λ as well.

The key quantities in Bell's analysis are spin correlations, which require the averaging over products such as $a_1(\lambda, \alpha)a_2(\lambda, \beta)$. The locality property postulated by EPR is encoded in the requirement that the measured results $a_1(\lambda, \alpha)$ and $a_2(\lambda, \gamma)$ are such that a_1 depends only on α and not on γ, and similarly, a_2 depends only on γ and not on α. Let us assume that the hidden variable λ is statistically distributed according to a distribution function $\rho(\lambda)$ such that

$$\rho(\lambda) \geq 0, \quad \int d\lambda \, \rho(\lambda) = 1. \quad (5.144)$$

One might have thought that it would never be possible to disprove a hidden variable theory of this kind without getting access to the hidden variables. However, Bell came up with a way of testing hidden variable theories by looking at the correlations between the results of measurements of entangled states.

For simplicity, let us assume that all unit vectors used in the test lie in a given plane, so that the direction of the unit vectors are determined by their polar angles. Consider now an experiment in which the operators $A_1(0)$ and $A_2(\theta)$ are measured and the products $a_1(\lambda, 0)a_2(\lambda, \theta)$ are averaged on a large number of pairs. Under the assumption of local realism, these measurements do not affect each other. According to the hidden-variable prescription, λ is not directly accessible, but the correlation between the measurements is given by,

$$R(0,\theta) \equiv \overline{a_1(0)a_2(\theta)} = \int d\lambda\, \rho(\lambda) a_1(\lambda, 0) a_2(\lambda, \theta). \tag{5.145}$$

Now consider a second set of measurements in which the first detector is not changed but the second one is set to measure $A_2(\phi)$. These measurements yield a second correlation $R(\phi)$. From the two sets of measurements we can determine the difference,

$$R(0,\theta) - R(0,\phi) = \int d\lambda\, \rho(\lambda) a_1(\lambda, 0)[a_2(\lambda, \theta) - a_2(\lambda, \phi)]. \tag{5.146}$$

Note that for given angles and given λ the spins of particles 1 and 2 are perfectly anti-correlated,

$$a_2(\lambda, \theta) = -a_1(\lambda, \theta), \quad a_2(\lambda, \phi) = -a_1(\lambda, \phi). \tag{5.147}$$

Also, recall from Eq. (5.138) that $a_i(\lambda, \theta) = \pm 1$, so that $a_i(\lambda, \theta)^2 = 1$. Substitution of Eq. (5.147) into Eq. (5.146) leads to,

$$R(0,\theta) - R(0,\phi) = -\int d\lambda\, \rho(\lambda) a_1(\lambda, 0)[a_1(\lambda, \theta) - a_1(\lambda, \phi)]$$

$$= -\int d\lambda\, \rho(\lambda) a_1(\lambda, 0) a_1(\lambda, \theta)[1 - a_1(\lambda, \theta) a_1(\lambda, \phi)]. \tag{5.148}$$

Taking the absolute value and using the triangle inequality [see (A.8) in Appendix A] we obtain the inequality,

$$|R(0,\theta) - R(0,\phi)| \le \int d\lambda\, \rho(\lambda)[1 - a_1(\lambda, \theta) a_1(\lambda, \phi)]$$

$$\le 1 + \int d\lambda\, \rho(\lambda)[1 + a_1(\lambda, \theta) a_2(\lambda, \phi)]. \tag{5.149}$$

The last integral on the RHS of Eq. (5.149) is the two-particle spin correlation function between measurements of particle spins along directions θ and ϕ, which must be dependent only on the difference $\theta - \phi$, hence, this term is just $R(\theta - \phi)$. Thus, we arrive at the variant of Bell's inequality for the hidden-variable theory,

$$\boxed{|R(0,\theta) - R(0,\phi)| \le 1 + R(0, \theta - \phi).}$$

(5.150)

Armed with this prediction, which is based on the hidden-variable assumption, we can now check if quantum mechanics, which, by Eq. (5.142) predicts $R(0,\theta) = -\cos\theta$, which obeys the above inequality. Let us apply Eq. (5.150) to the case when the Stern–Gerlach polarization axes are as shown in Fig. 5.18(a). Then, $(\mathbf{n}_1 \cdot \mathbf{n}_2) = \cos(\theta)$, $(\mathbf{n}_3 \cdot \mathbf{n}_1) = \cos(2\theta)$ and $(\mathbf{n}_3 \cdot \mathbf{n}_2) = \cos(\theta)$, i.e., $\phi = 2\theta$, and inequality (5.150) reads,

$$|\cos 2\theta - \cos\theta| \le 1 - \cos\theta. \tag{5.151}$$

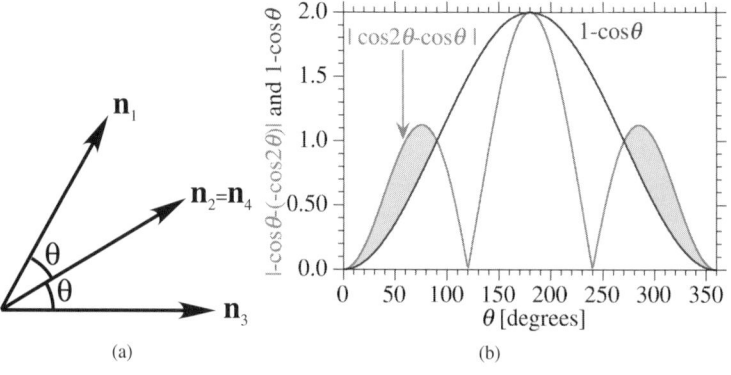

FIG 5.18 (a) Polarization vectors $\mathbf{n}_1$, $\mathbf{n}_2$, and $\mathbf{n}_3$ for Bell's inequality. (b) Plot of $|R(\mathbf{n}_1, \mathbf{n}_2) - R(\mathbf{n}_1, \mathbf{n}_3)|$ and $1 + R(\mathbf{n}_2, \mathbf{n}_3)$ versus θ.

5.8 Bell's Inequalities

Plotting $|\cos 2\theta - \cos \theta|$ and $1 - \cos \theta$ versus θ in Fig. 5.18(b), it is clear that the regions $0 < \theta < 90$ and $270 < \theta < 360$ degrees are outside the regime of validity of the Bell inequality (5.150).

We conclude that quantum mechanics is not consistent with a local deterministic hidden variables theory. A similar procedure can be used to obtain Bell inequalities for the other maximally entangled Bell states in Eq. (5.25a).

Relation of the Bell and CHSH Inequalities

We have considered two similar forms of the Bell inequalities in Eq. (5.141) (also referred to as the CHSH inequality) and in Eq. (5.150). Let us determine whether there is a difference in the initial assumptions leading to these inequalities.

To clarify this point, recall the main ingredients from which the analysis starts. In the standard EPR setup, a source emits oppositely directed particles (e.g., electrons) in the spin-entangled singlet state. After the particles are far enough from each other (a distance $\ell > c\Delta t$, where Δt is the time of the measurement after emission) two experimenters 1 and 2 randomly choose spatial axes $\hat{\mathbf{n}}_1$ and $\hat{\mathbf{n}}_2$, respectively, and measure the spins projections of the particles along these axes using their respective devices. The possible outcomes of these measurements are $a_1(\hat{\mathbf{n}}_1) = \pm 1$ and $a_2(\hat{\mathbf{n}}_2) = \pm 1$.

Consider first the inequality Eq. (5.150). To derive it, it is imagined that experimenter 2 performs also a measurement along a third axis $\hat{\mathbf{n}}_3$ with possible results $a_2(\hat{\mathbf{n}}_3) = \pm 1$. This second experiment by experimenter 2 is not really performed, and we will comment about this notion at the end of this section. The inequality involves the correlations $R(\hat{\mathbf{n}}, \hat{\mathbf{n}}') = \overline{a_1(\hat{\mathbf{n}})a_2(\hat{\mathbf{n}}')}$ and for arbitrary three directions $\hat{\mathbf{n}}_1, \hat{\mathbf{n}}_2, \hat{\mathbf{n}}_3$ it reads,

$$|R(\hat{\mathbf{n}}_1, \hat{\mathbf{n}}_2) - R(\hat{\mathbf{n}}_1, \hat{\mathbf{n}}_3)| \leq 1 + R(\hat{\mathbf{n}}_2, \hat{\mathbf{n}}_3). \quad (5.152)$$

This inequality was derived on the basis of a hidden variable theory with a parameter λ. Is there a need to attribute a special interpretation of λ as something that takes us away from the central axioms of quantum mechanics? For example, perhaps λ could be thought of as a quantum number of the singlet state, therefore nothing is "hidden" at this stage. The two assumptions that led to the derivation of the inequality are determinism and locality. Determinism means that the outcomes are determined since there exist functions $a_1(\lambda, \hat{\mathbf{n}}_1, \hat{\mathbf{n}}_2) = \pm 1$ and $a_2(\lambda, \hat{\mathbf{n}}_1, \hat{\mathbf{n}}_2) = \pm 1$ that yield the outcomes recorded by the apparatus of experimenters 1 and 2 once the parameter λ is specified. Classical mechanics is deterministic, but in quantum mechanics, we can only specify the probability $P(a_i|\lambda, \hat{\mathbf{n}}_1, \hat{\mathbf{n}}_2)$ that a certain measurement outcome will occur. Locality means that each outcome recorded by experimenter i is determined without being influenced by the setup of the other experimenter (e.g., axis $\hat{\mathbf{n}}_j$) since it is far away. Hence, the functions $a_1(\lambda, \hat{\mathbf{n}}_1, \hat{\mathbf{n}}_2)$ and $a_2(\lambda, \hat{\mathbf{n}}_1, \hat{\mathbf{n}}_2)$ should have the form, $a_1(\lambda, \hat{\mathbf{n}}_1) = \pm 1$, $a_2(\lambda, \hat{\mathbf{n}}_2) = \pm 1$. But quantum mechanics is not a local theory. The experimental violation of Eq. (5.152) implies that quantum mechanics is not a deterministic and local theory. Hence, the parameter λ cannot simply be the quantum numbers of a quantum state. It is then necessary to endow λ with the characteristics of a hidden variable with a probability distribution $\rho(\lambda)$, in terms of which the correlation is defined as

$$R(\hat{\mathbf{n}}_1, \hat{\mathbf{n}}_2) = \int d\lambda \rho(\lambda) a_1(\lambda, \hat{\mathbf{n}}_1) a_2(\lambda, \hat{\mathbf{n}}_2, \lambda). \quad (5.153)$$

This leads us directly to the Bell inequality (5.152).

Next, consider the derivation of the CHSH inequality (5.141). In this case, one does not assume that the outcomes of measurements are fixed by the functions $a_1(\lambda, \hat{\mathbf{n}}_1)$ and $a_2(\lambda, \hat{\mathbf{n}}_2)$. Rather, the first assumption is that there are probabilities for arriving at various possible outcomes. Thus, instead of the assumption pertaining to determinism (realized by the functions $a_1(\lambda, \hat{\mathbf{n}}_1)$ and $a_2(\lambda, \hat{\mathbf{n}}_2)$), one assumes that there are two probability functions $P(a_1|\lambda, \hat{\mathbf{n}}_1)$ and $P(a_2|\lambda, \hat{\mathbf{n}}_2)$. $P(a_i|\lambda, \hat{\mathbf{n}}_i)$ is the probability that experimenter i observes the result $a_i(\lambda, \hat{\mathbf{n}}_i)$. On the other hand, locality is assumed here, as it is assumed in deriving inequality (5.153). The probability depends only on factors that are locally accessible to the pertinent experimenter. The probability for different outcomes a_1 depends only on $\hat{\mathbf{n}}_1$ and not on the distant axis setting $\hat{\mathbf{n}}_2$ or outcome a_2. Thus, the basic theory to be tested is assumed to be local but not necessarily deterministic. The expectation value of the product of two measurement results is,

$$R(\hat{\mathbf{n}}_1, \hat{\mathbf{n}}_2) = \int d\lambda \rho(\lambda) \bar{a}_1(\lambda, \hat{\mathbf{n}}_1) \bar{a}_2(\lambda, \hat{\mathbf{n}}_2), \quad (5.154)$$

where

$$\bar{a}_i(\lambda, \hat{\mathbf{n}}_i) = \sum_{a_i=\pm 1} a_i P(a_i|\lambda, \hat{\mathbf{n}}_i) = P(a_i = 1|\lambda, \hat{\mathbf{n}}_i) - P(a_i = -1|\lambda, \hat{\mathbf{n}}_i), \quad (5.155)$$

is the predicted average value of the measurement in an experiment performed by i along an axis $\hat{\mathbf{n}}_i$ and with particles in the state λ. The derivation of the inequality employs the correlations involving the results of two measurements by each experimenter, $i = 1, 2$, along $\hat{\mathbf{n}}_{ia}$ and $\hat{\mathbf{n}}_{ib}$. Based on the probabilistic picture specified above, the manipulations involve some rearrangements and taking absolute values, leading to the CHSH inequality

$$|R(\hat{\mathbf{n}}_{1a}, \hat{\mathbf{n}}_{2a}) - R(\hat{\mathbf{n}}_{1a}, \hat{\mathbf{n}}_{2b})| + |R(\hat{\mathbf{n}}_{1b}, \hat{\mathbf{n}}_{2b}) + R(\hat{\mathbf{n}}_{1b}, \hat{\mathbf{n}}_{2a})| \leq 2, \quad (5.156)$$

In summary, the derivation of Eq. (5.152) is based on the assumption that any experimental result is either ± 1, while in deriving Eq. (5.156) the assumption is that there is a certain probability that the result will be ± 1 (this is tacitly assumed in deriving Eq. (5.141) through the averaging procedure). However, this 'difference' is just semantics, because we can regard the classical probability as arising from a hidden random variable, and that is implied by introducing the distribution $\rho(\lambda)$ and the expression (5.153) for the correlations.

Finally, let us comment about the introduction of imagined but unperformed experiments. Dealing with correlations involving more than two settings leads us to a conceptual problem because the same experimenter cannot perform simultaneous measurements along different axes. Literally, the apparatus cannot be simultaneously aligned along both directions. In the context of Bell's inequality, it requires us to think of specific products $a_1(\lambda, \hat{\mathbf{n}}_{1a})a_1(\lambda, \hat{\mathbf{n}}_{1b})$ of outcomes of experiments one of them actually performed and the other not. From this point of view, the fact that quantum mechanics refutes the pertinent inequalities is attributed to the fact that *unperformed experiments have no results* as was pointed out by Peres [91].

5.8.3 GENERAL ASPECTS OF BELL'S INEQUALITIES

Although the Bell inequalities are usually presented in terms of measurements on entangled pairs of quantum particles (e.g., the spin-singlet state of spin 1/2 particles, or entangled photon states), they can be presented in a way that has nothing to do with quantum physics, or even physics. Perhaps the simplest statement of Bell's inequalities goes as follows [92]. For any collection of objects that have three different properties, A, B and C, the number of objects that have property B but not C, and the number of objects that have property A and not C, obey the following inequality:

$$\mathcal{N}(A, \overline{B}) + \mathcal{N}(B, \overline{C}) \geq \mathcal{N}(A, \overline{C}), \quad (5.157)$$

where the overline indicates the logical NOT operation. Let us call this relationship Bell's inequality. Note that the properties need not be independent; there can be correlations between the parameters. For example, in a group of N people, we could have the properties: $A \equiv$ MALE, $B \equiv$ BLUE-EYES, and $C \equiv$ BLONDE-HAIR. Clearly, the latter two properties are correlated.

Proof: First note that $\mathcal{N}(A, \overline{B}, C) + \mathcal{N}(\overline{A}, B, \overline{C}) \geq 0$. Now, add $\mathcal{N}(A, \overline{B}, \overline{C}) + \mathcal{N}(A, B, \overline{C})$ to both sides of this expression to obtain

$$\mathcal{N}(A, \overline{B}, C) + \mathcal{N}(A, \overline{B}, \overline{C}) + \mathcal{N}(\overline{A}, B, \overline{C}) + \mathcal{N}(A, B, \overline{C}) \geq \mathcal{N}(A, \overline{B}, \overline{C}) + \mathcal{N}(A, B, \overline{C}), \quad (5.158)$$

which can be reduced to Eq. (5.157), $\mathcal{N}(A, \overline{B}) + \mathcal{N}(B, \overline{C}) \geq \mathcal{N}(A, \overline{C})$, since, on the RHS, either B or $\overline{B}$ must be true, and on the LHS, C or $\overline{C}$ must be true in the first two terms, and A or $\overline{A}$ must be true in the last two terms. This completes the proof.

The proof made use of the assumption that the properties exist whether they are measured or not, i.e., all measurement outcomes depend on pre-existing attributes of objects that are independent of the measurement (realism). For example, when we collected the terms $\mathcal{N}(A, \overline{B}, \overline{C}) + \mathcal{N}(A, B, \overline{C})$ to get $\mathcal{N}(A, \overline{C})$, we assumed that either $\overline{B}$ or B is true for every member, whether they are measured or not. The inequality has nothing to do with quantum mechanics, or physics for that

5.8 Bell's Inequalities

matter. Note that we could have stated Bell's inequality, Eq. (5.157), in terms of probabilities rather than numbers:

$$P(A, \overline{B}) + P(B, \overline{C}) \geq P(A, \overline{C}), \quad (5.159)$$

where $P(x, y)$ is the joint probability of x and y.[11]

Let us try to apply Bell's inequalities to a single electron spin. For example we can consider a beam of electrons and take the three parameters of Bell's inequality to be as follows:

$$A : |\uparrow\rangle \text{ along } z\text{-axis}, \quad B : |\uparrow\rangle \text{ along } \hat{u} = \frac{1}{\sqrt{2}}(\hat{x} + \hat{z}), \quad C : |\uparrow\rangle \text{ along } x\text{-axis}. \quad (5.160)$$

Bell's inequality Eq. (5.159), written in terms of spins, reads

$$P(|\uparrow\rangle_z, |\downarrow\rangle_u) + P(|\uparrow\rangle_u, |\downarrow\rangle_x) \geq P(|\uparrow\rangle_z, |\downarrow\rangle_x). \quad (5.161)$$

Note that $\overline{|\uparrow\rangle_{\hat{v}}} = |\downarrow\rangle_{\hat{v}} = |\uparrow\rangle_{-\hat{v}}$, for any direction $\hat{v}$. To measure the number of electrons (or the probability of finding electrons) that are spin-up along $\hat{z}$, and spin-up along $-\hat{u}$, $P(A, \overline{B})$, we would have to set up a Stern-Gerlach apparatus to measure the spin along $\hat{z}$, and set up a Stern–Gerlach apparatus to measure the spin along $\hat{u}$. But in quantum mechanics, if we measure the spin along $\hat{z}$, the spin state is destroyed, so we cannot do the measurement of the spin along $\hat{z}$ with the spin in the same state as it had initially. Hence, Bell's inequality cannot be applied directly to a single spin.

Now apply Eq. (5.159) to the spin-singlet; the inequality (5.159) reads,

$$P(|\uparrow\rangle_{\mathbf{n}_1}, |\downarrow\rangle_{\mathbf{n}_2}) + P(|\uparrow\rangle_{\mathbf{n}_2}, |\downarrow\rangle_{\mathbf{n}_3}) \geq P(|\uparrow\rangle_{\mathbf{n}_1}, |\downarrow\rangle_{\mathbf{n}_3}). \quad (5.162)$$

Clearly, this is equivalent to

$$P(|\uparrow\rangle_{\mathbf{n}_1}, |\uparrow\rangle_{-\mathbf{n}_2}) + P(|\uparrow\rangle_{\mathbf{n}_2}, |\uparrow\rangle_{-\mathbf{n}_3}) \geq P(|\uparrow\rangle_{\mathbf{n}_1}, |\uparrow\rangle_{-\mathbf{n}_3}), \quad (5.163)$$

where we have flipped both the spin and the polarization direction. We can now use Eq. (4.26), which gives the quantum mechanical expression for the probability of finding the first spin 1/2 particle polarized along $\mathbf{n}_1$ and the second spin 1/2 particle polarized along $\mathbf{n}_2$, $P(|\uparrow\rangle_{\mathbf{n}_1}, |\uparrow\rangle_{\mathbf{n}_2}) = [1 - (\mathbf{n}_1 \cdot \mathbf{n}_2)]/4$. Substituting into Eq. (5.163) yields:

$$1 - (\mathbf{n}_1 \cdot \mathbf{n}_2) - (\mathbf{n}_2 \cdot \mathbf{n}_3) \geq -(\mathbf{n}_1 \cdot \mathbf{n}_3). \quad (5.164)$$

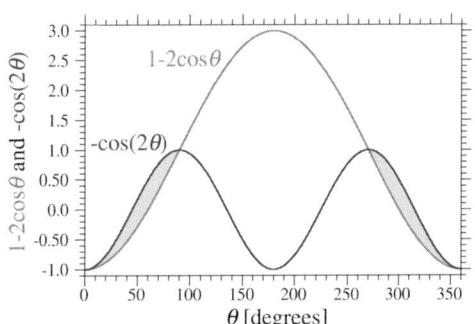

Figure 5.19 plots the LHS and RHS of inequality 5.164) for the spin-singlet state and the polarizations of the Stern-Gerlach apparatus set as shown in Fig. 5.18(a). The quantum results are outside the regime of validity of the Bell inequality (5.163) in the regions $0 < \theta < 90$ and $270 < \theta < 360$ degrees.

Bell's inequality (5.159) (as well as the standard correlation formalism of Bell's inequality) can also be applied to measurements of the polarization of correlated light photons, in a manner similar to measurements of the spin of spin 1/2 particles. A measurement of Bell inequalities was first carried out on pairs of polarized photons by Alain Aspect

FIG 5.19 $P(|\uparrow\rangle_{\mathbf{n}_1}, |\uparrow\rangle_{-\mathbf{n}_2}) + P(|\uparrow\rangle_{\mathbf{n}_2}, |\uparrow\rangle_{-\mathbf{n}_3}) = 1 - 2\cos\theta$ and $P(|\uparrow\rangle_{\mathbf{n}_1}, |\uparrow\rangle_{-\mathbf{n}_3}) = -\cos(2\theta)$ versus θ for the arrangement of polarizations $\mathbf{n}_1$, $\mathbf{n}_2$, and $\mathbf{n}_3$.

---

[11] In the context of probability theory, sometimes the joint probability $P(x, y)$ is written as $P(x \cap y)$, and sometimes the joint probability is written in terms of a conditional probability, where the conditional probability of x given y, is denoted by the symbol $P(x|y)$, and is defined by $P(x|y) \equiv P(x, y)/P(y)$.

et al. [90]. Photons emitted by excited calcium atoms in a radiative cascade from a $J = 0$ state to a $J = 1$ state, emitting a photon of wavelength $\lambda_1 = 422.7$ nm, and then from this state to another $J = 0$ state, emitting a photon of wavelength $\lambda_2 = 551.3$ nm. Each photon that passes through a collimator, impinges on an acousto-optical switch, which pseudo-randomly directs the photon toward one of two polarization analyzers. Each analyzer measures the linear polarization along one direction, and the detector outputs are checked for coincidences to find correlations between them. The results were in agreement with quantum mechanics, yet violated Bell's inequality. Moreover, the general form of Eq. (5.159) is applicable to many other types of physical systems.

Bell's inequalities are a direct consequence of probability theory. They specify conditions that a set of joint probability distributions must satisfy, assuming local realism. But quantum mechanics argues against the simultaneous exact measurement of two quantities whose operators do not commute. This modifies the probabilities for some correlation measurements. The Bell inequalities provide a test that can be compared with quantum mechanics and with experiment. If hidden variable theories of the type used by Bell in developing his inequalities were an accurate representation of nature, observed joint probability distributions would have to obey the inequalities. But quantum mechanics predicts that the Bell inequalities are violated under certain conditions, so hidden variable theories are not compatible with quantum mechanics. Experiments have shown that the predictions based on the quantum mechanics are correct and those based on hidden-variable theories are incorrect i.e., quantum mechanics is vindicated by these experiments.

Quantum Dynamics and Correlations

6

In this chapter we study the dynamics of quantum systems and the correlations of multipartite (many-particle) systems. We begin by considering the quantum dynamics of two-level systems in Sec. 6.1. Quantum dynamics can be calculated by applying unitary transformations to state vectors in Hilbert space (or to wave functions). Some physical systems, however, are best modeled as quantum sub-systems that are coupled to other degrees of freedom (a bath). Then, if the degrees of freedom of the bath are traced over, the dynamics of the sub-system is not unitary, as explained in Sec. 6.1.3. We introduce this kind of dynamics in this chapter, but delay a thorough treatment of this important topic to Chapter 17 which will be linked to the book web page. Two important applications of the dynamics of two level systems is to model the absorption and dispersion of light by a medium, modeled as two-level atoms in Sec. 6.1.5, and the Doppler cooling and optical trapping of atoms, discussed in Secs. 6.1.6 and 6.1.7. The correlation of two-level systems (entanglement for pure states, and either classical correlation or entanglement for mixed states) is considered in Sec. 6.1.8. Section 6.1.9 introduces the N-two-level system Bloch sphere, and Sec. 6.1.10 considers Ramsey fringe spectroscopy. We then consider three-level systems in Sec. 6.2, first treating a single three-level system and the process of stimulated Raman adiabatic passage (STIRAP), and then treating two or more correlated three-level systems in Sec. 6.2.1. Section 6.3 discusses the classification of correlation and entanglement (unfortunately, entanglement of mixed states is only partially understood, even for two-level systems). In Sec. 6.4, we consider three-level system dynamics. After a brief introduction to continuous-variable systems (the $n \to \infty$ limit of n level systems) in Sec. 6.5, we treat wave packet dynamics in an external potential in Sec. 6.6. In Sec. 6.7, we consider the dynamics of systems with time dependent Hamiltonians. Section 6.8 briefly considers optimal control theory for quantum systems, i.e., how to control the parameters of a Hamiltonian to obtain a desired final state of the system.

6.1 TWO-LEVEL SYSTEMS

Many kinds of systems can be modeled as a two-level system. We have already extensively discussed the spin 1/2 system, which is certainly a prime example, in the previous chapter. The polarization degrees of freedom of photons correspond to a two-level system, with right- and left-circular polarized photons being the analogy to spin-up and spin-down. The rotation of the polarization of a photon in a medium that is exposed to an applied magnetic field whose direction is along the photon propagation direction is called the *Faraday effect* and corresponds to the rotation of a spin 1/2 particle. Another two-level system is an atom, where we restrict our interest to two levels with energies, say, E_a and E_b. We can study the dynamics of these two levels coupled by an electromagnetic field that is almost in resonance with the transition between these two levels, and this will also correspond to the rotation of a spin 1/2 particle. If the photon energy is $\hbar\omega$, then the detuning from resonance is given by $\Delta = \omega - (E_b - E_a)/\hbar$. If the photon energy is less than the energy difference, then the detuning is negative, and if it is greater, then the detuning is positive. The radiation field couples the two levels with a coupling strength $\hbar\Omega/2$. Another important example of a two-level system is the two-site system shown in Fig. 6.1 (the potential wells could be harmonic-like or any other shape), where in this case, the energy difference between the two levels is $\hbar\delta = E_b - E_a$ and the coupling, say, due to tunneling between the two sites, is $\hbar\Omega/2$. The dynamics of a spin 1/2 particle, the polarization of a photon, the two-level system in a nearly resonant electromagnetic field, and the two-site systems in Fig. 6.1 can all be treated in same manner, and we shall take up this subject now.

The wave function in all four cases can be written as $\Psi(t) = \psi_a(t)\phi_a + \psi_b(t)\phi_b = \begin{pmatrix} \psi_b(t) \\ \psi_a(t) \end{pmatrix}$, where ϕ_i, $i = a, b$ are the internal state basis functions $[\phi_b \equiv |b\rangle = \begin{pmatrix} 1 \\ 0 \end{pmatrix}, \phi_a \equiv |a\rangle = \begin{pmatrix} 0 \\ 1 \end{pmatrix}]$ and $\psi_i(t)$, $i = a, b$ are the time-dependent amplitudes of the internal states. The Hamiltonian for these systems (for the fourth case, approximately) is given by

$$H = \hbar \begin{pmatrix} \delta & \Omega/2 \\ \Omega/2 & 0 \end{pmatrix}, \quad (6.1)$$

Twice the coupling matrix element $H_{ab} = H_{ba}$ between the levels a and b is called the Rabi frequency, Ω. The Hamiltonian in (6.1) is sometimes written as

$$H = \frac{\hbar \delta}{2} \mathbf{1} + \frac{\hbar}{2} \begin{pmatrix} \delta & \Omega \\ \Omega & -\delta \end{pmatrix}. \quad (6.2)$$

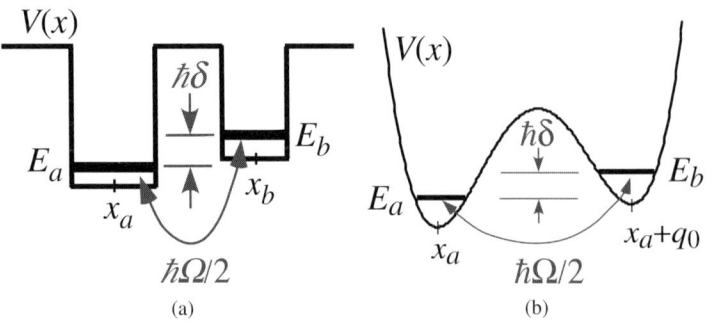

FIG 6.1 Two-site systems with energy difference of the two levels equal to $\hbar\delta$ and coupling $\hbar\Omega/2$. (a) Square well system and (b) parabolic wells.

The first term in Eq. (6.2) introduces a shift $\hbar\delta/2$ of both eigen-energies $E_{1,2} = \frac{\hbar}{2}\left(\delta \pm \sqrt{\delta^2 + \Omega^2}\right)$ and will just add a phase factor $e^{it\delta/2}$ to the time-dependent wave function components when calculating dynamics. Therefore, we can consider the Hamiltonian without the first term,

$$H = \frac{\hbar}{2} \begin{pmatrix} \delta & \Omega \\ \Omega & -\delta \end{pmatrix}, \quad (6.3)$$

whose energy eigenvalues are

$$E_{1,2} = \pm \frac{\hbar}{2} \sqrt{\delta^2 + \Omega^2}. \quad (6.4)$$

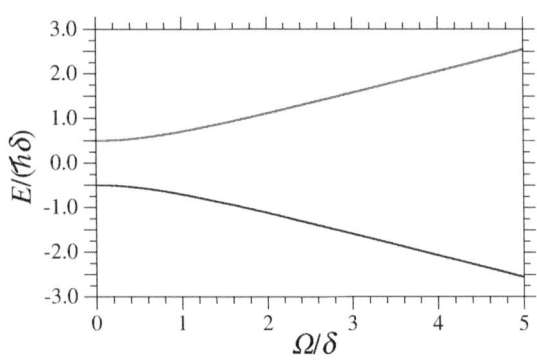

FIG 6.2 Energy eigenvalues in Eq. (6.4) as a function of Ω/δ.

Figure 6.2 plots these energy eigenvalues as a function of Ω (in units of $1/\delta$).

The Hamiltonian in (6.1) is appropriate for all two-level systems; but in the context of the two-level spin system, it is often used as an approximation to treat the interaction of the spin with both an external static magnetic field and a radio-frequency field. In this section, we shall study the dynamics of such a system. In general, the detuning and Rabi frequency can be time dependent, i.e., $\delta(t)$ and $\Omega(t)$, but for now, we take them to be constant in time.

The most general Hamiltonian for a two-level system (spin 1/2 particle) is [save for a term proportional to the 2×2 unit matrix, see Eq. (6.2)] $H = \mathbf{\Omega} \cdot \mathbf{S} = (\Omega_x S_x + \Omega_y S_y + \Omega_z S_z)$, where the coefficients Ω_x, Ω_y, and Ω_z must be real for H to be Hermetian. Written in terms of Pauli spin matrices,

$$H = \frac{\hbar}{2}\mathbf{\Omega} \cdot \boldsymbol{\sigma} = \frac{\hbar}{2}(\Omega_x \sigma_x + \Omega_y \sigma_y + \Omega_z \sigma_z) = \frac{\hbar}{2}\begin{pmatrix} \Omega_z & \Omega_x - i\Omega_y \\ \Omega_x + i\Omega_y & -\Omega_z \end{pmatrix}. \quad (6.5)$$

6.1 Two-Level Systems

The Hamiltonian in Eq. (6.3) corresponds to using $\mathbf{\Omega} = (\Omega, 0, \delta)$ in Eq. (6.5), i.e., the precession axis $\hat{\mathbf{\Omega}}$ is tilted relative to the z-axis at an angle $\theta = \arctan(\Omega/\delta)$, and Eq. (6.3) can be written as

$$H = \frac{\hbar}{2}(\Omega\,\sigma_x + \delta\,\sigma_z). \tag{6.6}$$

In ket-bra notation, Eq. (6.3) can be written as

$$H = \frac{\hbar\delta}{2}(|b\rangle\langle b| - |a\rangle\langle a|) + \frac{\hbar\Omega}{2}(|b\rangle\langle a| + |a\rangle\langle b|), \tag{6.7}$$

where the kets $|a\rangle = \binom{0}{1}$ and $|b\rangle = \binom{1}{0}$ correspond to the ground and excited states of the two-level system, and the most general Hamiltonian (6.5) takes the form

$$H = \frac{\hbar\Omega_z}{2}(|b\rangle\langle b| - |a\rangle\langle a|) + \frac{\hbar\Omega_x}{2}(|b\rangle\langle a| + |a\rangle\langle b|) + i\frac{\hbar\Omega_y}{2}(|b\rangle\langle a| - |a\rangle\langle b|). \tag{6.8}$$

Problem 6.1

(a) Write the Hamiltonian $H = \hbar \begin{pmatrix} \delta & \Omega/2 \\ \Omega/2 & 0 \end{pmatrix}$ in terms of the Pauli matrices and the unit matrix.

(b) Calculate the energy eigenvalues and eigenvectors of the Hamiltonian.

Answer: (a) $H = \frac{\hbar\delta}{2}\mathbf{1} + \hbar\left[\frac{\delta}{2}\sigma_z + \frac{\Omega}{2}\sigma_x\right]$. (b) $E_\pm = \frac{\delta}{2} \pm \frac{1}{2}\sqrt{\delta^2 + \Omega^2}$. Unnormalized eigenvectors:

$$|v_+\rangle = \begin{pmatrix} -\frac{\Omega}{(\delta - \sqrt{\delta^2+\Omega^2})} \\ 1 \end{pmatrix}, \quad |v_-\rangle = \begin{pmatrix} -\frac{\Omega}{(\delta + \sqrt{\delta^2+\Omega^2})} \\ 1 \end{pmatrix}.$$

6.1.1 TWO-LEVEL DYNAMICS (SPIN DYNAMICS)

The dynamics of a two-level system is governed by the Schrödinger equation, $i\hbar\frac{\partial |\psi(t)\rangle}{\partial t} = H|\psi(t)\rangle$, and if the Hamiltonian is time independent, i.e., $\mathbf{\Omega}$ is constant in time, then the solution $|\psi(t)\rangle$ can be written in terms of the evolution operator

$$\mathcal{U}(t,0) = e^{-iHt/\hbar} = e^{-i(\mathbf{\Omega}\cdot\mathbf{S})t/\hbar} = e^{-i(\mathbf{\Omega}\cdot\boldsymbol{\sigma})t/2}, \tag{6.9}$$

and the time-dependent spinor is given by $|\psi(t)\rangle = \mathcal{U}(t,0)|\psi(0)\rangle$. Equation (6.9) can be viewed as a rotation operator; hence, the spinor precesses about the unit vector $\hat{\mathbf{\Omega}}$ at a rate $|\mathbf{\Omega}|$. The evolution operator $\mathcal{U}(t,0) = e^{-iHt/\hbar}$ for the Hamiltonian in Eq. (6.3) [or, equivalently, Eq. (6.6)] is

$$\mathcal{U}(t,0) = \begin{pmatrix} \cos(\frac{\Omega_g t}{2}) + \frac{i\delta}{\Omega_g}\sin(\frac{\Omega_g t}{2}) & \frac{i\Omega}{\Omega_g}\sin(\frac{\Omega_g t}{2}) \\ \frac{i\Omega}{\Omega_g}\sin(\frac{\Omega_g t}{2}) & \cos(\frac{\Omega_g t}{2}) - \frac{i\delta}{\Omega_g}\sin(\frac{\Omega_g t}{2}) \end{pmatrix}. \tag{6.10}$$

Here we defined the *generalized Rabi frequency*, $\Omega_g \equiv \sqrt{|\Omega|^2 + \delta^2}$, which is the energy difference between the energy eigenvalues in (6.4) (in units of $\hbar$). At zero detuning, $\delta = 0$, Eq. (6.10) becomes

$$\mathcal{U}(t) = \begin{pmatrix} \cos(\frac{\Omega t}{2}) & i\sin(\frac{\Omega t}{2}) \\ i\sin(\frac{\Omega t}{2}) & \cos(\frac{\Omega t}{2}) \end{pmatrix}. \tag{6.11}$$

Problem 6.2

Derive Eq. (6.10) starting from $\mathcal{U}(t,0) = e^{-i(\delta\sigma_z + \Omega\sigma_x)t/2}$, by using Eq. (4.22), $e^{-i\frac{\vartheta(t)}{2}\mathbf{n}\cdot\boldsymbol{\sigma}} = \cos[\vartheta(t)/2]\mathbf{1} - i\sin[\vartheta(t)/2]\mathbf{n}\cdot\boldsymbol{\sigma}$, and identifying the angle $\vartheta(t)$ and the unit vector $\mathbf{n}$.

Answer: $\vartheta(t) = \Omega_g t$ and $\mathbf{n} = (\Omega/\Omega_g, 0, \delta/\Omega_g)$.

Let the spinor start at $t = 0$ in state $|\psi(0)\rangle \equiv \binom{\psi_b(0)}{\psi_a(0)} = \binom{0}{1}$, so the initial spin-vector points along the $-z$-axis (the spin-down vector), and it precesses as a function of time about the vector $\hat{\boldsymbol{\Omega}}$ at a rate $|\boldsymbol{\Omega}| = \sqrt{\delta^2 + \Omega^2} = \Omega_g$. This picture of a vector precessing around $\boldsymbol{\Omega}$ will be clarified in what follows. Applying the evolution operator in Eq. (6.10), we find the following time dependence of the spinor wave function $|\psi(t)\rangle = \binom{\psi_b(t)}{\psi_a(t)}$:

$$|\psi(t)\rangle = \mathcal{U}(t)|\psi(0)\rangle = \mathcal{U}(t)\binom{0}{1} = \begin{pmatrix} i\frac{\Omega}{\Omega_g}\sin(\frac{\Omega_g t}{2}) \\ \cos(\frac{\Omega_g t}{2}) - i\frac{\delta}{\Omega_g}\sin(\frac{\Omega_g t}{2}) \end{pmatrix}. \quad (6.12)$$

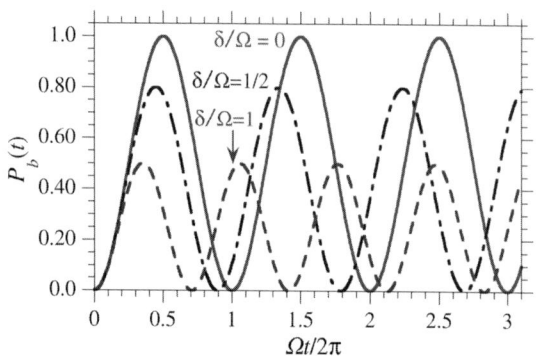

FIG 6.3 Rabi oscillations. Probability of being in state b, $P_b(t) = |\frac{\Omega}{\Omega_g}\sin(\frac{\Omega_g t}{2})|^2$, versus t for several values of δ.

The amplitude of the excited state (top component) starts off at $t = 0$ as zero, $\psi_b(0) = 0$, and oscillates in time with angular frequency $\Omega_g/2$ and has magnitude Ω/Ω_g, i.e., the maximum probability of being in the excited state (top) component is $|\Omega/\Omega_g|^2$. Figure 6.3 shows the *Rabi oscillations* (named after I. I. Rabi) in the probability $P_b(t) = |\frac{\Omega}{\Omega_g}\sin(\frac{\Omega_g t}{2})|^2$ for several values of δ; $P_b(t)$ oscillates as a function of t with peak magnitude $|\Omega/\Omega_g|^2$. The ground state probability $P_a(t) = 1 - P_b(t)$ (amplitude squared of the bottom component) starts out as unity at $t = 0$ and oscillates, returning to unity at times $t = 2\pi m/\Omega_g$, where m is an integer.

Note that the ground state has been represented in several equivalent ways: (1) the basis state $|\phi_a\rangle \equiv |a\rangle$, (2) the two-component vector $\binom{0}{1}$, and (3) the "spin" state representation $|\downarrow\rangle$.

We have already noted [see Eq. (4.18)] that any spinor can be represented as $|\psi\rangle = |\mathbf{n}\rangle$, where the unit vector $\mathbf{n} = (\sin\theta\cos\phi, \sin\theta\sin\phi, \cos\theta)$ lies on the unit sphere. If the components of the spinor change with time, $|\psi(t)\rangle = |\mathbf{n}(t)\rangle$, then the position of the unit vector on the surface of the unit sphere changes with time. Up to a phase factor [e.g., $e^{-i\phi/2}$, see Eq. (4.18), see also Eq. (6.12)], the spin state pointing along unit vector $\mathbf{n}$ can be written as

$$|\mathbf{n}\rangle \equiv |\alpha_{\mathbf{n}}\rangle = [\cos(\theta/2)|\uparrow\rangle + e^{i\phi}\sin(\theta/2)|\downarrow\rangle] = \begin{pmatrix} \cos(\theta/2) \\ e^{i\phi}\sin(\theta/2) \end{pmatrix}, \quad (6.13)$$

Any two-level pure state can be written in this form (we just need the appropriate unit vector $\mathbf{n}$). For example, the ground state $|\downarrow\rangle$ corresponds to the state $|\alpha_{\mathbf{n}}\rangle$ with polar angle $\theta = \pi$, and for definiteness, azimuthal angle $\phi = 0$. The state $|\alpha_{\mathbf{n}}\rangle$ in Eq. (6.13) is often designated as the coherent state $|\theta, \phi\rangle$. In Sec 6.1.2, the ground state will be represented by the point at the bottom of the Bloch sphere, the vector $\mathbf{n}$ will be called the *Bloch vector*, and the state $|\mathbf{n}\rangle \equiv |\theta, \phi\rangle \equiv |\alpha_{\mathbf{n}}\rangle$ will be represented as a point at solid angle (θ, ϕ) on the Bloch sphere. Two-level dynamics means that the Bloch (unit) vector $\mathbf{n}$ on the Bloch sphere varies with time on the Bloch sphere, i.e., $|\mathbf{n}(t)\rangle \equiv |\theta(t), \phi(t)\rangle = |\alpha_{\mathbf{n}(t)}\rangle$.

Problem 6.3

Express $|\psi(t)\rangle$ in Eq. (6.12) in the form given by Eq. (6.13).

Answer: $\theta = 2\arccos[(\Omega/\Omega_g)\sin(\Omega_g t/2)]$, $\phi = -\arcsin\left[\frac{\cos(\Omega_g t/2)}{\sqrt{1-(\Omega/\Omega_g)^2\sin^2(\Omega_g t/2)}}\right]$.

6.1.2 THE BLOCH SPHERE PICTURE

To better understand two-level system dynamics, it is useful to consider the expectation value of σ, $\mathbf{n}(t) = \langle\boldsymbol{\sigma}\rangle_t = \text{Tr}[\boldsymbol{\sigma}\rho(t)]$, where $\rho(t)$ is the time-dependent density matrix, $\rho(t) = |\psi(t)\rangle\langle\psi(t)|$ [the density matrix of pure and mixed

6.1 Two-Level Systems

states is defined in Sec. 2.5]. Note that to obtain the magnetization **M** of a collection of spins (atoms or molecules with a net spin), we must multiply $\mathbf{n}(t)$ by the Bohr magneton, $\mu_B = \frac{e\hbar}{2m_e c}$, the g-factor of the atom or molecule, and the density $\mathcal{N}$ of the spins, i.e., the magnetization is defined by $\mathbf{M}(t) \equiv \mathcal{N}\langle\boldsymbol{\mu}\rangle_t = (\mathcal{N}g\mu_B/2)\langle\boldsymbol{\sigma}\rangle_t = (\mathcal{N}g\mu_B/2)\mathbf{n}(t)$. In the context of the two-level system, it is common to call the "dimensionless magnetization" vector **n** the *Bloch vector*, and to denote its x, y, and z components by the symbols u, v, and w. The quantity w is called the *population inversion*, i.e., the difference of the populations of states a and b. The Bloch vector is given in terms of the wave function components by the relations

$$u(t) \equiv n_x(t) = \mathrm{Tr}[\sigma_x \rho(t)] = \rho_{ba}(t) + \rho_{ab}(t) = 2\,\mathrm{Re}[\psi_b^*(t)\psi_a(t)], \quad (6.14a)$$

$$v(t) \equiv n_y(t) = \mathrm{Tr}[\sigma_y \rho(t)] = -i(\rho_{ba}(t) - \rho_{ab}(t)) = 2\,\mathrm{Im}[\psi_b^*(t)\psi_a(t)], \quad (6.14b)$$

$$w(t) \equiv n_z(t) = \mathrm{Tr}[\sigma_z \rho(t)] = \rho_{bb}(t) - \rho_{aa}(t) = |\psi_b(t)|^2 - |\psi_a(t)|^2. \quad (6.14c)$$

Note that when $\psi_a(t) = 1$, $\psi_b(t) = 0$, the population inversion $w(t) = -1$.

The u and v components of the Bloch vector are called *coherences* because when the two-level system is in a coherent state, u and v are in general nonzero (in the case of two levels coupled by a radiation field, they are proportional to the real and imaginary parts of the transition dipole moment). For a pure state, the Bloch vector is just another way of parameterizing the amplitudes $\psi_a(t)$ and $\psi_b(t)$; these two complex numbers are restricted by the condition $|\psi_a(t)|^2 + |\psi_b(t)|^2 = 1$; so, only three independent real numbers (u, v, w) are necessary to determine the amplitudes. Figure 6.4 plots the precession of the Bloch vector around $\boldsymbol{\Omega}$, for $\delta = 0$ and $\delta \neq 0$, thereby illustrating the comments in the previous paragraph about **n** precessing about $\boldsymbol{\Omega}$. The Bloch vector $\mathbf{n}(t) = (u(t), v(t), w(t))$ moves on the *Bloch sphere*, $\mathbf{n}^2(t) = u^2(t) + v^2(t) + w^2(t) = 1$.

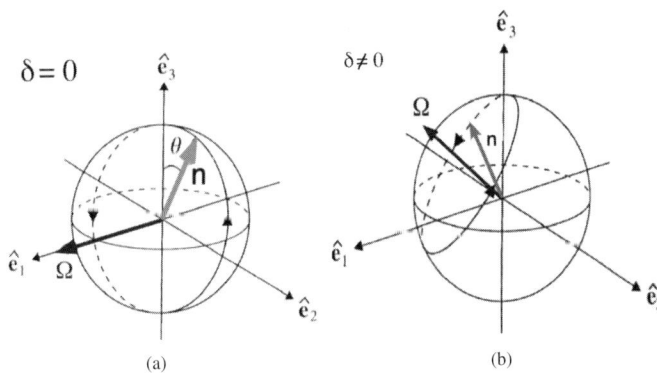

FIG 6.4 Precession of the Bloch vector **n** around the vector $\boldsymbol{\Omega}$ for (a) $\delta = 0$, and (b) $\delta \neq 0$. After Scully and Zubairy [30]. Reproduced from M.O. Scully and M.S. Zubairy, Quantum Optics, (Cambridge Univ. Press, 1997).

As discussed earlier, the vector $\boldsymbol{\Omega} = (\Omega, 0, \delta)$ may be a function of time, i.e., the detuning and Rabi frequency can be time dependent, $\delta = \delta(t)$ and $\Omega = \Omega(t)$.[1] If controllable, the time dependence of $\boldsymbol{\Omega}(t)$ can be used to move the Bloch vector **n** to any desired position on the Bloch sphere.

Instantaneously, the rate of change of the Bloch vector will be given by $\boldsymbol{\Omega}(t) \times \mathbf{n}(t)$ (see below). Using the properties of the trace of a matrix, we find $\mathbf{n}(t) \equiv \mathrm{Tr}[\boldsymbol{\sigma}\,\rho(t)] = \mathrm{Tr}[\boldsymbol{\sigma}(t)\rho(0)] = \langle\boldsymbol{\sigma}\rangle_t$, where $\boldsymbol{\sigma}(t) = \mathcal{U}(t)\boldsymbol{\sigma}\mathcal{U}(t)^{-1}$. Furthermore,

$$\rho(t) = \mathcal{U}(t)\rho(0)\mathcal{U}(t)^{-1} = |\mathbf{n}(t)\rangle\langle\mathbf{n}(t)|, \quad (6.15)$$

and using (4.16) and the properties of the Pauli σ matrices, we find

$$\boxed{\rho(t) = \frac{1}{2}[1 + \mathbf{n}(t) \cdot \boldsymbol{\sigma}].} \quad (6.16)$$

ρ is a 2×2 Hermitian matrix with trace equal to unity; therefore, $3\ (= 2 \times 2 - 1)$ independent real parameters are required to parameterize it. These parameters can be the three expectation values $n_i \equiv \langle\sigma_i\rangle$. One way to understand Eq. (6.16) is to represent the state vector as $|\psi(t)\rangle \equiv |\mathbf{n}(t)\rangle = \cos(\theta(t)/2)|\uparrow\rangle + e^{i\phi(t)}\sin(\theta(t)/2)|\downarrow\rangle$, form $|\psi(t)\rangle\langle\psi(t)|$,

---

[1] The Rabi frequency vector $\boldsymbol{\Omega}(t)$ can represent the effects of a pulsed electromagnetic field on a two-level system: the Rabi frequency Ω turns on and off as the field turns on and off, and the frequency detuning δ can change with time. Moreover, the polarization of the field may vary with time, so all three components of $\boldsymbol{\Omega}(t)$ may be nonzero.

represent the ket-bra basis states in terms of the σ matrices, and use $\mathbf{n}(t) = (\sin\theta(t)\cos\phi(t), \sin\theta(t)\sin\phi(t), \cos\theta(t))$. Since $\rho(t) = |\mathbf{n}(t)\rangle\langle\mathbf{n}(t)| = \frac{1}{2}(1 + \mathbf{n}(t)\cdot\boldsymbol{\sigma})$, we can represent $\rho(t)$ as the 2×2 matrix

$$\rho(t) = \begin{pmatrix} \cos^2\frac{\theta(t)}{2} & \frac{1}{2}e^{-i\phi(t)}\sin\theta(t) \\ \frac{1}{2}e^{i\phi(t)}\sin\theta(t) & \sin^2\frac{\theta(t)}{2} \end{pmatrix}. \tag{6.17}$$

Problem 6.4

Show that Eq. (6.17) is equivalent to using Eqs (6.16) and (4.17).

Figure 6.5 summarizes the three representations of a two-level system, the spin-vector picture, the Bloch sphere picture, and the density matrix picture wherein the density matrix can be represented in terms of Pauli spin matrices.[2]

Problem 6.5

(a) Plot the states $\frac{1}{\sqrt{2}}(|\downarrow\rangle + |\uparrow\rangle)$, $\frac{1}{\sqrt{2}}(|\downarrow\rangle - |\uparrow\rangle)$, $\frac{1}{\sqrt{2}}(|\downarrow\rangle + i|\uparrow\rangle)$, and $\frac{1}{\sqrt{2}}(|\downarrow\rangle - i|\uparrow\rangle)$ on the Bloch sphere.

(b) Plot the density matrix state $\rho = \frac{1}{2}(|\downarrow\rangle\langle\downarrow| + |\uparrow\rangle\langle\uparrow|)$ in the Bloch sphere.

(c) Plot the density matrix obtained by averaging $\begin{pmatrix} \cos^2\frac{\theta}{2} & \frac{1}{2}e^{-i\phi}\sin\theta \\ \frac{1}{2}e^{i\phi}\sin\theta & \sin^2\frac{\theta}{2} \end{pmatrix}$ over ϕ with a uniform probability distribution in the interval $[0, 2\pi]$.

Answers: (a) See Fig. 6.5. (b) The point at the origin. (c) The point at $(x, y, z) = (0, 0, \cos^2\frac{\theta(t)}{2} - \sin^2\frac{\theta(t)}{2})$.

Representations of a two-level spin system

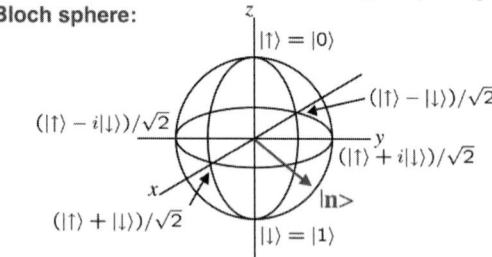

FIG 6.5 Three representations of a two-level system. (a) The spinor $|\mathbf{n}\rangle$ in terms of the spin-vector $\mathbf{n} \equiv \langle\boldsymbol{\sigma}\rangle = (\sin\theta\cos\phi, \sin\theta\sin\phi, \cos\theta)$, (b) the Bloch sphere, $n_x^2 + n_y^2 + n_z^2 = 1$, and the vector $\mathbf{n}$ pointing from the origin to a point on the sphere, and (c) the density matrix, $\rho = |\mathbf{n}\rangle\langle\mathbf{n}|$, written in terms of the spin-vector $\mathbf{n}$ and the Pauli σ matrices, $\rho = \frac{1}{2}(1 + \mathbf{n}\cdot\boldsymbol{\sigma})$. The Bloch vector $\mathbf{n}$ is the expectation value of the spin.

---

[2] If a dissipative process is present, e.g., decoherence of the transition dipole components u and v due to collisions with other atoms or spontaneous emission from the excited state to levels outside the two-level system, the trajectory of the Bloch vector as a function of time will enter the interior of the Bloch sphere. We shall study the effects of decoherence in Sec. 6.1.3 and Chapter 17, linked to the book web page.

6.1 Two-Level Systems

The time-dependent Schrödinger equation, $i\hbar \frac{\partial |\psi(t)\rangle}{\partial t} = H|\psi(t)\rangle$, with Hamiltonian (6.5) can be rewritten in terms of Bloch vector $\mathbf{n}(t)$ as

$$\boxed{\frac{\partial \mathbf{n}(t)}{\partial t} = \mathbf{\Omega} \times \mathbf{n}(t),} \quad (6.18)$$

In the Bloch sphere picture, the *Bloch equation* (6.18) determines the dynamics and specifies that the Bloch vector precesses about $\mathbf{\Omega}$. You will verify Eq. (6.18) in Problem 6.6(a) using the Schrödinger equation. Another way of obtaining Eq. (6.18) is to note that the density matrix satisfies the Liouville–von Neumann equation, $i\hbar \frac{\partial}{\partial t} \rho = [H, \rho]$. Substituting the expression $\rho(t) = \frac{1}{2}(1 + \mathbf{n}(t) \cdot \boldsymbol{\sigma})$ obtained in Eq. (6.16) for ρ and the Hamiltonian of Eq. (6.5) into the Liouville–von Neumann equation, multiplying both sides of the resulting equation by $\boldsymbol{\sigma}$, and taking the trace, we obtain

$$i\hbar \frac{\partial \mathbf{n}(t)}{\partial t} = \frac{1}{2} \text{Tr} \, \boldsymbol{\sigma} \left[\frac{\hbar}{2} \mathbf{\Omega} \cdot \boldsymbol{\sigma}, (1 + \mathbf{n} \cdot \boldsymbol{\sigma}) \right]. \quad (6.19)$$

Using the Pauli spin matrix identities including Eq. (4.7), we obtain Eq. (6.18).

Problem 6.6

(a) Check that the Bloch equation is equivalent to the time-dependent Schrödinger equation by differentiating $|\psi(t)\rangle\langle\psi(t)|$ with respect to time and using the Schrödinger equation to derive the Bloch equation.
(b) Carry out the algebra leading from Eq. (6.19) to Eq. (6.18).
(c) Write the three components of the Bloch equation in terms of $(u(t), v(t), w(t))$ and $(\Omega_x, \Omega_y, \Omega_z)$.

Problem 6.7

For the general spin Hamiltonian, $H = h_0 + \frac{\hbar}{2} \mathbf{\Omega} \cdot \boldsymbol{\sigma}$, similar to Eq. (6.5), show that the equation of motion for the density matrix is

$$\frac{\partial \rho}{\partial t} = -\frac{i}{\hbar}[H, \rho] = \boldsymbol{\sigma} \cdot (\mathbf{\Omega} \times \mathbf{n}(t)), \quad (6.20)$$

where $\mathbf{n}(t) = \text{Tr} \, \boldsymbol{\sigma} \rho(t)$. Note that Eq. (6.20) is equivalent to Eqs (6.16) and (6.18).

Problem 6.8

Find the trace distance $D(\rho_1, \rho_2)$ [see Eq. (2.68)] between the density matrices $\rho_1 = \frac{1}{2}[1 + \mathbf{n}_1 \cdot \boldsymbol{\sigma}]$ and $\rho_2 = \frac{1}{2}[1 + \mathbf{n}_2 \cdot \boldsymbol{\sigma}]$.
Answer: $D(\rho_1, \rho_2) = \frac{1}{2} \text{Tr} \, |(\mathbf{n}_1 - \mathbf{n}_2) \cdot \boldsymbol{\sigma}/2| = |\mathbf{n}_1 - \mathbf{n}_2|/2$. Hence, the trace distance is half the distance of the Bloch vectors on the Bloch sphere.

The "Bloch sphere" representation was originally invented to describe the two-level system consisting of the polarization state of light by the French mathematician Jules Henri Poincaré in the second half of the 19th century. Hence, for light polarization, this picture is called the *Poincaré sphere*. It provides a convenient representation of the polarization state of a photon (or a light ray) and is equivalent to the Bloch sphere description of a two-level system described earlier.

Adopting a basis set $|R\rangle$ and $|L\rangle$, representing right- and left-circular polarized photons, respectively, a photon of any polarization can be represented, within an overall phase by the superposition

$$|\theta,\phi\rangle = \cos\frac{\theta}{2}|R\rangle + e^{i\phi}\sin\frac{\theta}{2}|L\rangle, \tag{6.21}$$

where the angles θ and ϕ define the point on the surface of the unit sphere (the Poincaré sphere in this case) whose south and north poles represent the states $|L\rangle$ and $|R\rangle$, in exact analogy with the Bloch sphere $|\downarrow\rangle$ and $|\uparrow\rangle$ states Equation (6.21) for photon wave functions is a direct analogy to Eq. (6.13) for spin wave functions. Points in the northern (southern) hemisphere represent right (left) elliptical polarizations, and the points on the equator represent all possible linear polarizations. The orthogonal horizontal and vertical linear polarizations are given by $|H\rangle = \frac{1}{\sqrt{2}}(|R\rangle - |L\rangle)$ and $|V\rangle = \frac{1}{\sqrt{2}}(|R\rangle + |L\rangle)$, respectively, and they appear at diametrically opposite points on the equator. An incoherent polarization state is represented by a point within the Poincaré sphere. The most general state of polarization can be written as

$$\rho = \frac{1}{2}(S_0\mathbf{1} + \mathbf{S}\cdot\boldsymbol{\sigma}), \tag{6.22}$$

where the four parameters $S_0, S_x(\equiv S_1), S_y(\equiv S_2)$, and $S_z(\equiv S_3)$ are called *Stokes parameters*. They can be expressed in terms of expectation values of products of the electric field strength components perpendicular to the wave vector and the relative phase of the electric field components [18]. For a pure photon state, $S_0 = 1$. If the polarization state of the photon changes as a function of time, then the Stokes parameters change with time.

6.1.3 COUPLING TO A BATH: DECOHERENCE

The fundamental problem that limits the quantum coherence of a spin (or of a quantum two-level system, or more generally, any quantum system) is its unavoidable interaction with the environment. For example, an electron spin in a semiconductor interacts with nuclear spins via hyperfine interaction, with impurity spins, etc. During these interactions, the phase relations between the various states of the system are quickly lost. Hence, the spin can no longer be represented as a coherent superposition of two quantum states $|\uparrow\rangle$ and $|\downarrow\rangle$. This process is called *decoherence*. If the state of the two-level spin system is represented by the Bloch vector $\mathbf{n}(t) \equiv \langle\boldsymbol{\sigma}\rangle = \text{Tr}[\boldsymbol{\sigma}\rho(t)] = (u(t), v(t), w(t))$, then, due to decoherence, the Bloch vector will move off the Bloch sphere and into its interior, even if the conservation of probability condition, $|\psi_a|^2 + |\psi_b|^2 = 1$ is still satisfied (e.g., the center of the Bloch sphere corresponds to the density matrix $\rho = 1/2$, so $|\psi_a|^2 + |\psi_b|^2 = 1/2 + 1/2 = 1$).

To see how decoherence happens, we can consider the dynamics of a spin described by the Hamiltonian H_S (S refers to system), which is coupled to a bath of external degrees of freedom described by the Hamiltonian H_B via the interaction Hamiltonian H_{SB}. The total Hamiltonian is $H = H_S + H_{SB} + H_B$. Let us assume that the spin system has been prepared in a pure state $|\xi(0)\rangle$, and that the bath is initially in the state $|\phi(0)\rangle$, so that at $t=0$, there are no correlations between the system and the bath, and the state of the composite system at $t=0$ is separable, $|\Psi(0)\rangle = |\xi(0)\rangle|\phi(0)\rangle$. Due to system-bath interaction, the spin system and the bath develop quantum correlations, and the composite wave function $|\Psi(t)\rangle$ will no longer be separable. Since we are not interested in the detailed description of the bath, we describe the properties of the spin system by the reduced density matrix $\rho_S(t) = \text{Tr}_B|\Psi(t)\rangle\langle\Psi(t)|$, where Tr_B denotes trace over the bath degrees of freedom. The general density-matrix description that results, after elimination of the bath degrees of freedom, will be presented in Chapter 17, linked to the book web page. In the simplest Markov approximation limit that results, the u and v components of the Bloch vector decay into the interior of the Bloch sphere with a decay time called T_2 and w decays to -1 with a decay time T_1 as follows:

$$\dot{u} = -\frac{u}{T_2}, \quad \dot{v} = -\frac{v}{T_2}, \quad \dot{w} = -\frac{w+1}{T_1}. \tag{6.23}$$

The first two equations account for the decoherence loss and the last equation for population relaxation, which results in excited state (up-spin) decay to the ground state (down-spin) because of environment couplings. The parameters T_1 and

6.1 Two-Level Systems

T_2 are called the *relaxation time* and *decoherence time*, respectively, and are sometimes referred to as *longitudinal* and *transverse* relaxation times. The solution to Eq. (6.23) is

$$u(t) = u(0)\, e^{-\frac{t}{T_2}}, \quad v(t) = v(0)\, e^{-\frac{t}{T_2}}, \quad w(t) = -1 + (w(0) + 1)\, e^{-\frac{t}{T_1}}. \quad (6.24)$$

In terms of the density matrix, ρ, Eq. (6.23) takes the form

$$\dot{\rho}_{ab} = -\frac{\rho_{ab}}{T_2}, \quad \dot{\rho}_{ba} = -\frac{\rho_{ba}}{T_2}, \quad \dot{\rho}_{bb} = -\frac{\rho_{bb}}{T_2}, \quad \dot{\rho}_{aa} = \frac{\rho_{bb}}{T_1}. \quad (6.25)$$

The full density matrix equations, including the decay terms in Eq. (6.25) and the Hamiltonian dynamics terms that arise from H_S [see Eq. (2.37)], which we will call H (without the subscript, for ease of notation), are, therefore, given by

$$\boxed{\frac{d}{dt}\rho(t) = \frac{-i}{\hbar}[H(t), \rho(t)] - \Gamma \rho(t),} \quad (6.26)$$

which, written in components, is

$$\frac{d}{dt}\rho_{ij}(t) = \frac{-i}{\hbar}\sum_k [H_{ik}(t)\rho_{kj} - \rho_{ik}H_{kj}] - \sum_{kl} \Gamma_{ijkl}\, \rho_{kl}. \quad (6.27)$$

The "superoperator" Γ_{ijkl} is given in terms of the decay times by

$$\Gamma_{ijkl} = \begin{cases} -T_1^{-1} & \text{for } ijkl = 2222 \\ T_1^{-1} & \text{for } ijkl = 1122 \\ -T_2^{-1} & \text{for } ijkl = 1212 \\ -T_2^{-1} & \text{for } ijkl = 2121 \end{cases}. \quad (6.28)$$

If spontaneous emission is the *only* decay mechanism, $T_2 = 2T_1$. Sometimes, the inverse decay times, i.e., the decay rates, $\Gamma \equiv T_2^{-1}$ and $\gamma \equiv T_1^{-1}$, are specified rather than the decay times.

For a two-level system with Hamiltonian (6.6) [or (6.7)], the density matrix equations (6.26) are[3]

$$\dot{\rho}_{ba} = -\frac{i}{2}\Omega(\rho_{bb} - \rho_{aa}) + [i(-\delta) - \Gamma]\rho_{ba}, \quad (6.29a)$$

$$\dot{\rho}_{bb} - \dot{\rho}_{aa} = i\Omega(\rho_{ba}^* - \rho_{ba}) - \gamma[(\rho_{bb} - \rho_{aa}) + 1]. \quad (6.29b)$$

In (u, v, w) notation, these equations take the form:

$$\boxed{\frac{du}{dt} = (-\delta)v - \Gamma u,} \quad (6.30a)$$

$$\boxed{\frac{dv}{dt} = -(-\delta)u + \Omega w - \Gamma v,} \quad (6.30b)$$

$$\boxed{\frac{dw}{dt} = -\Omega v - \gamma(w + 1).} \quad (6.30c)$$

---

[3] The reason for using $(-\delta)$ in Eqs (6.29a) and (6.30a) will become apparent in Sec. 6.1.4, where $\delta \to -\Delta$ in the Hamiltonian.

This version of the Bloch equations slightly differs from Eq. (6.18), as it now includes decay terms. They also describe the optical coupling of a two-level system to a radiation field (see Sec. 6.1.4). In terms of the Bloch vector **n**, they can be written as

$$\frac{d\mathbf{n}(t)}{dt} = \mathbf{\Omega} \times \mathbf{n}(t) - \mathbf{\Gamma}\,\mathbf{n}(t), \tag{6.31}$$

where $\mathbf{\Gamma}$ is the decay tensor. For time-dependent fields, the Rabi frequency can be time dependent, $\mathbf{\Omega}(t)$.

In the Bloch sphere picture, the γ decay term drives the Bloch vector to $n_z(=w) = -1$, i.e., the population decays to the ground state, and the Γ decay terms drive the Bloch vector to the z-axis, i.e., it causes decay of $n_x(=u)$ and $n_y(=v)$ to zero.

The phenomenological decay operators $\mathbf{\Gamma}$ appear in a term for the change of the density matrix that is proportional to the present value of the density matrix and not on the past (or future) values of the density matrix. Processes for which the evolution of the system is based on the present (and not past) state of the system are called *Markov processes*. Here, the Markov process involves a first-order differential evolution equation in time for the density matrix. Not every such equation will produce positive probabilities or a positive semi-definite density operator. The conditions on the decay terms for this positivity condition is that they are of Lindblad form [93] (see also Sec. 17.3.1 linked to the book web page):

$$\frac{d}{dt}\rho = \frac{i}{\hbar}[\rho, H] + \sum_k \Gamma_k (2 O_k \rho O_k^\dagger - \{O_k^\dagger O_k, \rho\}). \tag{6.32}$$

The second term on the right-hand side of Eq. (6.32) has the Markovian Lindblad form [93] and gives rise to dissipation effects of the "bath" on the density matrix. Such terms may be used to depict decay due to spontaneous emission, atom-surface interactions, motional effects, black-body radiation, and other environmental effects. The Lindblad operators O_k are determined from the nature of the system-bath coupling, and the coefficients Γ_k are the corresponding coupling parameters. The Markovian Lindblad form of quantum master equations is the analog of the Fokker–Planck equation for the dynamics of the open systems and has been used to model a variety of systems, from the harmonic oscillator to the two-level system.

For a two-level system, the Lindblad operator σ_z affects dephasing (decay) of the coherence without affecting the population of the ground or excited states. Spontaneous emission is described using the raising operator σ_+ as the Lindblad operator:

$$\frac{d}{dt}\rho = \frac{i}{\hbar}[\rho, H] + \Gamma(2\sigma_+\rho\sigma_- - \sigma_+\sigma_-\rho - \rho\sigma_+\sigma_-). \tag{6.33}$$

This equation is equivalent to Eq. (6.30a) with $\Gamma = \gamma/2$.

6.1.4 PERIODICALLY DRIVEN TWO-LEVEL SYSTEM

The periodically driven two-level system is a paradigm that can be used for better understanding of many fundamental phenomena in physics. The dynamics of superconducting Josephson devices, the two-level quantum dot system driven by near-resonant (or far off-resonant) radiation, and the two-level atom in a single-frequency electromagnetic field are but three examples. The periodically driven two-level system has been widely studied using a variety of methods, including the rotating-wave approximation, the time-averaging method, and perturbation theory. The Hamiltonian can be written using a coupling between the two levels which is sinusoidally varying, $V(t) = \hbar\Omega\sin(\omega t)$, so the Hamiltonian (6.6) becomes

$$H(t) = \hbar\Omega\sin(\omega t)(|b\rangle\langle a| + |a\rangle\langle b|) + \frac{\hbar\delta}{2}(|b\rangle\langle b| - |a\rangle\langle a|) = \hbar\begin{pmatrix} \frac{\delta}{2} & \Omega\sin(\omega t) \\ \Omega\sin(\omega t) & -\frac{\delta}{2} \end{pmatrix}. \tag{6.34}$$

Figure 6.6 shows two different extreme coupling cases, the nearly resonant case $\omega \approx \delta$ and the multiphoton resonant case $n\omega \approx \delta$. Writing the solution of the time-dependent Schrödinger equation as $|\psi(t)\rangle = \psi_a(t)|a\rangle + \psi_b(t)|b\rangle = \begin{pmatrix}\psi_b(t)\\\psi_a(t)\end{pmatrix}$, the

6.1 Two-Level Systems

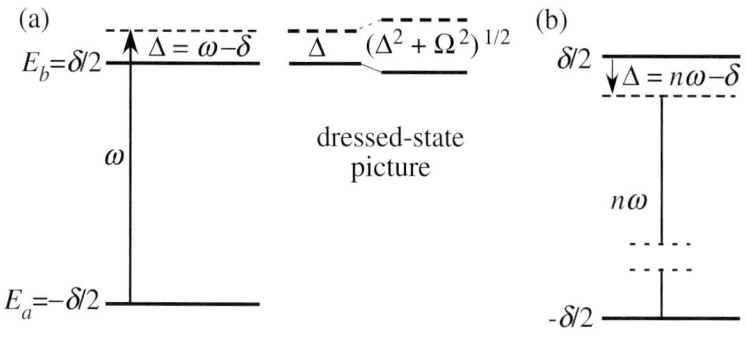

FIG 6.6 Two-level system with a periodically driven coupling with frequency ω. (a) ω almost in resonance with the energy difference δ between the levels (the light is blue-detuned here, i.e., $\Delta > 0$). The right panel of (a) shows the dressed-state picture that results from the rotating-wave approximation (see the Rotating-Wave Approximation section). (b) Multiphoton case with $n\omega \approx \delta$ with $n \gg 1$ ($\Delta < 0$ here).

differential equations for the amplitudes are

$$i\frac{d\psi_b}{dt} = \frac{\delta}{2}\psi_b + \Omega \sin(\omega t)\psi_a, \tag{6.35a}$$

$$i\frac{d\psi_a}{dt} = -\frac{\delta}{2}\psi_a + \Omega \sin(\omega t)\psi_b. \tag{6.35b}$$

A solution to Eqs (6.35) can be obtained by performing a rotation about the y-axis by an angle of $\pi/2$, $|\psi(t)\rangle \to |\phi(t)\rangle = \binom{\phi_b(t)}{\phi_a(t)} = e^{-i(\pi/4)\sigma_y}|\psi(t)\rangle$, and

$$H(t) \to \mathcal{H}(t) = e^{-i(\pi/4)\sigma_y}H(t)e^{i(\pi/4)\sigma_y} = \frac{\hbar\delta}{2}\sigma_x + \hbar\Omega \sin(\omega t)\sigma_z. \tag{6.36}$$

With this transformation, the equations for the wave functions separate

$$i\frac{d\phi_b}{dt} = \frac{\delta}{2}\phi_a + \Omega \sin(\omega t)\phi_b, \tag{6.37a}$$

$$i\frac{d\phi_a}{dt} = \frac{\delta}{2}\phi_b - \Omega \sin(\omega t)\phi_a. \tag{6.37b}$$

Differentiating the second equation and substituting the first into the resultant equation yields

$$\frac{d^2\phi_a}{dt^2} + \{-i\Omega\omega \cos(\omega t) + \frac{\delta^2}{4} + [\Omega \sin(\omega t)]^2\}\phi_a = 0. \tag{6.38}$$

Similarly,

$$\frac{d^2\phi_b}{dt^2} + \{i\Omega\omega \cos(\omega t) + \frac{\delta^2}{4} + [\Omega \sin(\omega t)]^2\}\phi_b = 0. \tag{6.39}$$

The solutions to Eqs (6.38) and (6.39) are given in terms of the *Heun confluent functions*. Focusing attention on ϕ_b, making transformations for both independent and dependent variables [94],

$$z(t) = \sin^2(\omega t/2), \quad \phi_b(z) = e^{\frac{2i\Omega}{\omega}z}\phi(z), \tag{6.40}$$

using $\frac{d}{dt} = \dot{z}\frac{d}{dz}$ and $\frac{d^2}{dt^2} = \ddot{z}\frac{d}{dz} + \dot{z}^2\frac{d^2}{dz^2}$, we obtain after some algebra,

$$\phi_{zz} + \left(\frac{4i\Omega}{\omega} + \frac{1/2}{z} + \frac{1/2}{z-1}\right)\phi_z + \left[\frac{\frac{4i\Omega}{\omega}z - \left(\frac{2i\Omega}{\omega} + \frac{\delta^2}{4\omega^2}\right)}{z(z-1)}\right]\phi = 0. \tag{6.41}$$

This equation can be written in the form

$$\frac{d^2\phi}{dz^2} + \left(\alpha + \frac{\beta+1}{z} + \frac{\gamma+1}{z-1}\right)\frac{d\phi}{dz} + \frac{qz+p}{z(z-1)}\phi = 0. \tag{6.42}$$

where $\alpha = \frac{4i\Omega}{\omega}$, $\beta+1 = \gamma+1 = \frac{1}{2}$, $q = \frac{4i\Omega}{\omega} \equiv \zeta + \alpha(\beta+\gamma+2)/2$, and $p = -\left(\frac{2i\Omega}{\omega} + \frac{\delta^2}{4\omega^2}\right) \equiv \eta + \beta/2 + (\gamma - \alpha(\beta+1)/2$. Equation (6.42) is called the *Heun confluent equation*; it has two singular regular points at $z = 0$ and $z = 1$ and an irregular singularity at $z = \infty$, and it has two linearly independent solutions, $\phi_1(z) = HC(\alpha, \beta, \gamma, \zeta, \eta, z)$ and $\phi_2(z) = z^{-\beta}HC(\alpha, -\beta, \gamma, \zeta, \eta, z)$, and $HC(\cdot)$ are the Heun confluent functions [94]. ϕ_b is a superposition of these solutions, $\phi_b = c_1\phi_1 + c_2\phi_2$. The coefficients c_1 and c_2 are determined by the initial values of $\phi_b(0)$ and $\dot{\phi}_b(0)$.

Problem 6.9

Carry out the algebra leading to Eq. (6.41).

Rotating-Wave Approximation

The "standard" treatment of the periodically driven two-level system is to use the *rotating-wave approximation* (RWA) (see Sec. 4.8.1). The idea is to perform a time-dependent transformation of $\binom{\psi_a}{\psi_b}$ in Eq. (6.35) that is equivalent to working in a "rotating frame." In this frame, one distinguishes two terms; the first, an *antiresonant* or *counter rotating* term, strongly oscillates (rotates) in time, whereas the second term slowly oscillates in time. Within the RWA, the former term is neglected since its effects nearly "average out" over time. One is then left with an equation for the latter, which is governed by a time-independent Hamiltonian and can be solved by methods as discussed in the beginning of this chapter. One starts with Eq. (6.35) and performs the unitary transformation,

$$\psi_a(t) = e^{-i\delta_a t}\varphi_a(t), \quad \psi_b(t) = e^{-i\delta_b t}\varphi_b(t), \tag{6.43}$$

and uses $\sin(\omega t) = \frac{e^{i\omega t} - e^{-i\omega t}}{2i}$ to obtain[4]

$$i\frac{d\varphi_b}{dt} = \left(\frac{\delta}{2} - \delta_b\right)\varphi_b + \Omega\left(\frac{e^{i\omega t} - e^{-i\omega t}}{2i}\right)e^{i(\delta_b - \delta_a)t}\varphi_a, \tag{6.44a}$$

$$i\frac{d\varphi_a}{dt} = \left(-\frac{\delta}{2} - \delta_a\right)\varphi_a + \Omega\left(\frac{e^{i\omega t} - e^{-i\omega t}}{2i}\right)e^{-i(\delta_b - \delta_a)t}\varphi_b. \tag{6.44b}$$

If we now choose δ_a and δ_b such that $\omega = \delta_b - \delta_a$, then one of the terms proportional to Ω in each of the above equations no longer rotates (e.g., $e^{i(-\omega+\delta_b-\delta_a)t} = 1$, and the other term rotates basically twice as quickly as before, so it can be neglected. If we arbitrarily set $\delta_a = -\delta/2$ and take $\omega = \Delta + \delta$ (see Fig. 6.6(a), where we arbitrarily chose $\Delta > 0$), the *detuning* Δ is given by $\Delta = \omega - \frac{E_b - E_a}{\hbar} = \omega - \delta$, and we obtain

$$i\frac{d}{dt}\binom{\varphi_b}{\varphi_a} = \begin{pmatrix} -\Delta & i\frac{\Omega}{2} \\ -i\frac{\Omega}{2} & 0 \end{pmatrix}\binom{\varphi_b}{\varphi_a}. \tag{6.45}$$

---

[4] Recall that a unitary transformation $\psi = \mathcal{U}(t)\varphi(t)$ in $i\hbar\frac{\partial \psi(t)}{\partial t} = H\psi(t)$ yields $i\hbar\frac{\partial \varphi(t)}{\partial t} = [\mathcal{U}^\dagger(t)H\mathcal{U}(t) - i\hbar\mathcal{U}^\dagger(t)\dot{\mathcal{U}}(t)]\varphi(t)$.

6.1 Two-Level Systems

Applying a further transformation, $\varphi_a \to -i\varphi_a$ turns the complex Hermitian (time independent) Hamiltonian matrix on the RHS of Eq. (6.45) into a real symmetric Hamiltonian, and we find

$$i\frac{d}{dt}\begin{pmatrix}\varphi_b \\ \varphi_a\end{pmatrix} = \begin{pmatrix}-\Delta & \frac{\Omega}{2} \\ \frac{\Omega}{2} & 0\end{pmatrix}\begin{pmatrix}\varphi_b \\ \varphi_a\end{pmatrix}. \tag{6.46}$$

Note that the energy eigenvalues of the Hamiltonian in Eq. (6.46) are the same as those in Eq. (6.45), $E_\pm = (-\Delta \pm \sqrt{\Delta^2 + \Omega^2})/2$. This is the same Hamiltonian as in Eq. (6.1) except that $\delta \to -\Delta$ [see Fig. 6.6(a) to discern the reason].

In what follows, we need to generalize the treatment above by taking the interaction between the two levels in the form, $V(t) = \hbar\Omega_0 \cos(\omega t + \phi_0) = \hbar\Omega_0 \frac{e^{i\omega t + i\phi_0} + e^{-i\omega t - i\phi_0}}{2}$, where ϕ_0 is a phase angle (which may have some temporal and spatial dependence – see below). Carrying out the procedure used earlier for this case yields

$$i\frac{d\varphi_b}{dt} = \left(\frac{\delta}{2} - \dot\delta_b\right)\varphi_b + \Omega_0\left(\frac{e^{i\omega t + i\phi_0} + e^{-i\omega t - i\phi_0}}{2}\right)e^{i(\delta_b - \delta_a)t}\varphi_a, \tag{6.47a}$$

$$i\frac{d\varphi_a}{dt} = \left(-\frac{\delta}{2} - \dot\delta_a\right)\varphi_a + \Omega_0\left(\frac{e^{i\omega t + i\phi_0} + e^{-i\omega t - i\phi_0}}{2}\right)e^{-i(\delta_b - \delta_a)t}\varphi_b. \tag{6.47b}$$

and on choosing δ_a and δ_b as earlier, we obtain

$$i\frac{d}{dt}\begin{pmatrix}\varphi_b \\ \varphi_a\end{pmatrix} = \begin{pmatrix}-\Delta & \frac{\Omega^*}{2} \\ \frac{\Omega}{2} & 0\end{pmatrix}\begin{pmatrix}\varphi_b \\ \varphi_a\end{pmatrix}, \tag{6.48}$$

where $\Omega \equiv \Omega_0 e^{i\phi_0}$.

The pure state density matrix $\rho(t) = |\psi(t)\rangle\langle\psi(t)|$ in the RWA, and its equation of motion, are now derived. For this purpose, it is convenient to define $\sigma_{ba}(t)$, $\sigma_{ab}(t)$, $\sigma_{bb}(t)$ and $\sigma_{bb}(t)$ through the relations,

$$\rho_{ba}(t) = \sigma_{ba}(t)e^{-i[\omega t + \phi_0]}, \quad \rho_{ab}(t) = \sigma_{ab}(t)e^{i[\omega t + \phi_0]} = \rho_{ba}^*(t) = \sigma_{ba}^*(t)e^{i[\omega t + \phi_0]}, \tag{6.49}$$

$$\rho_{bb}(t) - \rho_{aa}(t) = \sigma_{bb}(t) - \sigma_{aa}(t). \tag{6.50}$$

The RWA density matrix satisfies equations similar to Eqs (6.29a) and (6.30a):

$$\boxed{\dot\sigma_{ba} = \frac{i}{2}\Omega_0(\sigma_{bb} - \sigma_{aa}) + [i(\Delta + \dot\phi_0) - \Gamma]\sigma_{ba},} \tag{6.51a}$$

$$\boxed{\dot\sigma_{bb} - \dot\sigma_{aa} = i\Omega_0(\sigma_{ba} - \sigma_{ab}^*) - \gamma[(\sigma_{bb} - \sigma_{aa}) + 1],} \tag{6.51b}$$

In the (u, v, w) notation, where $u = \sigma_{ba} + \sigma_{ba}^* = 2\,\text{Re}\,\sigma_{ba}$, $v = -i(\sigma_{ba} - \sigma_{ba}^*) = 2\,\text{Im}\,\sigma_{ba}$, $w = \sigma_{bb} - \sigma_{aa}$,

$$\boxed{\frac{du}{dt} = -(\Delta + \dot\phi_0)v - \Gamma u,} \tag{6.52a}$$

$$\boxed{\frac{dv}{dt} = (\Delta + \dot\phi_0)u + \Omega_0 w - \Gamma v,} \tag{6.52b}$$

$$\boxed{\frac{dw}{dt} = -\Omega_0 v - \gamma(w + 1).} \tag{6.52c}$$

In the context of the coupling of an atom (modeled as a two-level system) to light, Eqs (6.51a) and (6.52) are called the *optical Bloch equations*.

Sometimes the Rabi frequency Ω can be written as the product of the transition dipole moment matrix element between the ground and excited states, μ_{ba}, and the slowly varying envelope of the electric field, A [see Eq. (7.88)], $\hbar\Omega = 2\mu_{ba}A/\hbar$ ($=\mu_{ba}E_0$). The slowly varying envelope, A, and therefore Ω (or E_0), may be a function of space and time (the light field may be a pulse – see the next few subsections).

Problem 6.10

Write Eq. (6.52) in vector-matrix notation.

Answer: $\frac{d}{dt}\begin{pmatrix} u \\ v \\ w \end{pmatrix} = \begin{pmatrix} -\Gamma & -(\Delta+\dot{\phi}_0) & 0 \\ (\Delta+\dot{\phi}_0) & -\Gamma & \Omega_0 \\ 0 & -\Omega_0 & -\gamma \end{pmatrix}\begin{pmatrix} u \\ v \\ w \end{pmatrix} - \begin{pmatrix} 0 \\ 0 \\ \gamma \end{pmatrix}.$

Adiabatic Limit

We can easily obtain solutions to the dynamics of a two-level system in a pulse of light with a time-dependent Rabi frequency, $\Omega(t)$, when the temporal duration of the light pulse, τ_p, is very much larger than the decay times T_1 and T_2 using an adiabatic approximation. Setting all three optical Bloch equations into steady state (all time derivatives in Eq. (6.30a) are set equal to zero), we find[5]

$$u = \frac{\Omega\Delta/\Gamma^2}{1+\Delta^2/\Gamma^2+\Omega^2/(\gamma\Gamma)}, \tag{6.53a}$$

$$v = (-1)\frac{\Omega/\Gamma}{1+\Delta^2/\Gamma^2+\Omega^2/(\gamma\Gamma)}, \tag{6.53b}$$

$$w = (-1)\frac{1+\Delta^2/(\gamma\Gamma)}{1+\Delta^2/\Gamma^2+\Omega^2/(\gamma\Gamma)}. \tag{6.53c}$$

The excited state population, ρ_{bb} is, therefore, given by

$$\rho_{bb} = \frac{1}{2}\frac{\Omega^2/(\gamma\Gamma)}{1+\Delta^2/\Gamma^2+\Omega^2/(\gamma\Gamma)} = \frac{1}{2}\frac{\Omega^2\Gamma/\gamma}{\Delta^2+\Gamma^2+\Omega^2\Gamma/\gamma}. \tag{6.54}$$

The quantity $s \equiv \Omega^2/(\gamma\Gamma)$ is often called the *saturation parameter*.

For large Rabi frequency, *power broadening* of the transition occurs. The excited state population ρ_{bb} in Eq. (6.54) can be written in terms of an effective transition width Γ_{eff}, as $\rho_{bb} = \frac{1}{2}\frac{\Omega^2\Gamma/\gamma}{\Delta^2+\Gamma_{\text{eff}}^2}$, where

$$\Gamma_{\text{eff}}^2 = \Gamma^2 + \Omega^2\frac{\Gamma}{\gamma} = \Gamma^2\left(1+\frac{\Omega^2}{\gamma\Gamma}\right) = \Gamma^2(1+s). \tag{6.55}$$

For extremely large Rabi frequency, i.e., for very large saturation parameter, $s \gg 1$, and small detuning, $\Delta^2/\Gamma^2 \ll 1$, the system is in a superposition of ground and excited states, i.e., $w \approx 0$ and $\rho_{aa} \approx \rho_{bb} \approx 1/2$.

For small Rabi frequency, $\Omega^2/(\gamma\Gamma) \ll 1$, the system is mostly in the ground state, i.e., the steady-state population difference is given by $w \approx -1$, but this is a very crude approximation. A better approximation obtains if $T_2 \ll T_1, \tau_p$, since we can set the coherences into steady state (set the time derivatives of u and v equal to zero in the Bloch equations) to obtain

$$u = -\frac{\Delta}{\Gamma}v, \quad v = \frac{\Omega}{\Gamma(1+\Delta^2/(\gamma\Gamma))}w. \tag{6.56}$$

Substituting the expression for v into the differential equation (6.52c) for w, we find

$$\frac{dw}{dt} = -\frac{\Omega}{\Gamma(1+\Delta^2/(\gamma\Gamma))}w - \gamma(w+1). \tag{6.57}$$

---

[5] To generalize to the case with finite $\dot{\phi}_0$ discussed above, let $\Omega \to \Omega_0$, $\Delta \to \Delta + \dot{\phi}_0$.

6.1 Two-Level Systems

This equation, which is in the form of an ordinary driven population rate equation, $\frac{dw}{dt} = -aw + b$, has analytic solution $w(t) = e^{-at}\left[w(0) + (b/a)\left(e^{at} - 1\right)\right]$. Given the initial population inversion $w(0)$, the adiabatic solution to the Bloch equation is completely determined by setting dw/dt to zero. The steady-state ground state population is $\rho_{aa} = 1 - \rho_{bb}$. In Sec. 7.4.4, we calculate the excited state population versus time using perturbation theory. The result is plotted in Fig. 7.7.

6.1.5 ATOMS IN AN ELECTROMAGNETIC FIELD: DISPERSION AND ABSORPTION

There are many physical phenomena that can be treated using the formalism described above, e.g., absorption and dispersion of atoms in an atomic gas, atom cooling and trapping by light, inhomogeneous broadening of optical transitions, self-induced transparency, spin echoes, and optical spin echoes (see Ref. [18], Chapter 9). In the next few subsections we shall consider absorption and dispersion of light by atoms in an atomic gas, and Doppler cooling of atoms and trapping of atoms, using the optical Bloch equations when an electromagnetic field, with electric vector $\mathbf{E}(\mathbf{r}, t)$, is applied to a gas of atoms.

We begin by considering a dilute gas of atoms irradiated by light. Modeling starts by assuming that the light is in near resonance with two levels of the atom, say, the ground electronic state of the atom and a specific excited electronic state. The interaction energy of the light and an atom (a two-level system) is, $U = -\mathbf{p} \cdot \mathbf{E}$, or, if the dipole moment in the atom is induced by the field $\mathbf{E}$, $U = -\frac{1}{2}\mathbf{p} \cdot \mathbf{E}$), where the atomic polarization is $\mathbf{p} = \text{Tr}(\rho\boldsymbol{\mu})$, where $\boldsymbol{\mu}$ is the transition dipole operator of the atomic transition (which can be expressed in terms of u, v, w of Eq. (6.14a). We first consider atoms interacting with a single-frequency electromagnetic field of frequency ω, i.e., $\mathbf{E}(t) = \text{Re}\left[\mathbf{e}_0 E_0 e^{-i\omega t - i\phi_0}\right]$, where $\mathbf{e}_0$ is the unit vector in the direction of the electric field and ϕ_0 is the phase angle of the field. Then the transition dipole moment induced by a field also oscillates with frequency ω. We shall take the transition dipole matrix element μ_{ba} to be real, $\mu_{ba} = \mu_{ab}$. Since the transition dipole moment is induced by the field, it is natural to take the transition dipole moment along the direction of the field, i.e., $\mathbf{e}_\mu = \mathbf{e}_0$. The optical potential felt by the atoms in the gas due to the light-matter interaction is

$$U_{\text{opt}}(\mathbf{r}, t) = -\frac{1}{2}\text{Tr}(\rho\boldsymbol{\mu}) \cdot \mathbf{E}(\mathbf{r}, t) = \frac{1}{2}(\rho_{ba} + \rho_{ab})\mu_{ba}\,\mathbf{e}_0 \cdot \mathbf{E}(\mathbf{r}, t) = \frac{1}{2} \times$$
$$[\sigma_{ba}(t)e^{-i[\omega t + \phi_0]} + \sigma_{ab}(t)e^{i[\omega t + \phi_0]}]\mu_{ba}\,\mathbf{e}_0 \cdot [\mathbf{A}(\mathbf{r}, t)e^{-i[\omega t + \phi_0]} + \mathbf{A}^*(\mathbf{r}, t)e^{i[\omega t + \phi_0]}], \quad (6.58)$$

where we have written the electromagnetic field in terms of the slowly varying envelope $\mathbf{A}$ as $\mathbf{E}(\mathbf{r}, t) = \mathbf{e}_0[A(\mathbf{r}, t)e^{-i[\omega t + \phi_0]} + A^*(\mathbf{r}, t)e^{i[\omega t + \phi_0]}]$. The optical force is given by Ehrenfest's theorem and involves calculating the expectation value of the operator $\dot{\hat{\mathbf{p}}} = -\nabla \hat{U} = \nabla(\hat{\boldsymbol{\mu}} \cdot \mathbf{E}(\mathbf{r}, t))$. Hence, the optical force is the expectation value of the optical force operator,

$$\vec{F}_{\text{opt}}(\mathbf{r}, t) = \frac{1}{2}\text{Tr}(\rho\hat{\boldsymbol{\mu}}) \cdot \vec{\nabla}\mathbf{E}(\mathbf{r}, t), \quad (6.59)$$

where we used the arrow vector notation as distinguished from the boldfaced vectors to clarify that the force is a vector as a result of the gradient. The trace picks out the off-diagonal elements of the density operator, i.e.,

$$\vec{F}_{\text{opt}}(\mathbf{r}, t) = \frac{1}{2}(\rho_{ba} + \rho_{ab})\mu_{ba}\,\vec{\nabla}\left[\mathbf{e}_0 \cdot \mathbf{E}(\mathbf{r}, t)\right]. \quad (6.60)$$

We shall complete the calculation of the optical force on an atom due to a light field in Sec. 6.1.6 after a short discussion of absorption and the refractive index of light.

Problem 6.11

(a) Obtain an explicit expression for the solution to Eq. (6.57).
(b) Show that the $t \to \infty$ limit of the solution is equivalent to Eq. (6.53c).

Refractive Index and Absorption

Light propagating in a medium is affected by the polarization of the medium, as discussed in Ref. [18], Sec. 3.3.3 (see also the discussion below). To lowest order in the electric field strength, the polarization vector $\mathbf{P}$ is proportional to the electric field of the light, $\mathbf{P} = \chi \mathbf{E}$, where χ is the susceptibility and is related to the complex dielectric constant [see Eq. (7.221a)]. For a dilute gas of atoms of density $\mathcal{N}$, the polarization is given by the product of the density and the transition dipole moment of the atoms $\mathbf{p}$, $\mathbf{P} = \mathcal{N}\mathbf{p} = \mathcal{N}\mu_{ba}(u+iv)\mathbf{e}_0$. More explicitly, for light of frequency ω, $\chi(\omega)$ gives the polarization of the medium when light with frequency ω propagates in the medium, $\mathbf{P}(\omega) = \mathcal{N}\mathbf{p}(\omega) = \chi(\omega)\mathbf{E}(\omega)$, and the displacement vector is given (in Gaussian units) by $\mathbf{D}(\omega) = \varepsilon(\omega)\mathbf{E}(\omega) = \mathbf{E}(\omega) + 4\pi\mathbf{P}(\omega) = [1 + 4\pi\chi(\omega)]\mathbf{E}(\omega) = \{E(\omega) + 4\pi\mathcal{N}\mu_{ba}[u(\omega) + iv(\omega)]\}\mathbf{e}_0$.[6] $\chi(\omega)$ is called the *optical susceptibility* at frequency ω and is the response of the medium to the light. In summary, for a medium composed of atoms whose optical properties are well approximated by two levels (the ground state and the excited state produced by the optical excitation), the real and imaginary parts of $\chi(\omega)$ [which are proportional to the *refractive index* $n(\omega)$ and the *absorption coefficient* $\alpha(\omega)$] are proportional to the real and imaginary parts of the atomic transition dipole moment $\mathbf{p}$, which are given by u and v, respectively. Thus, the refractive index $n(\omega)$ and absorption coefficient $\alpha(\omega)$ are related to the real and imaginary parts of the susceptibility $\chi(\omega)$. We shall now explicitly make the connection of the susceptibility χ (which is proportional to the dipole moment $\mathbf{p}$) and the absorption and dispersion of a medium.

The absorption and dispersion of a medium can be understood using the Faraday and Ampère equations:

$$\nabla \times \mathbf{E} = -\frac{1}{c}\frac{\partial \mathbf{B}}{\partial t}, \quad \nabla \times \mathbf{H} = \frac{4\pi}{c}\mathbf{J} + \frac{1}{c}\frac{\partial \mathbf{D}}{\partial t}. \tag{6.61}$$

We shall assume that the material is not a conductor, and therefore, no current flows in the medium, $\mathbf{J} = 0$. Now, we substitute the constitutive equations $\mathbf{D} = \varepsilon\mathbf{E} = \mathbf{E} + 4\pi\mathbf{P}$ and $\mathbf{H} = \mu^{-1}\mathbf{B}$ into these equations to eliminate the fields $\mathbf{D}$ and $\mathbf{B}$ in favor of $\mathbf{E}$ and $\mathbf{H}$ to obtain the equations $\nabla \times \mathbf{E} = -\frac{\mu}{c}\frac{\partial \mathbf{H}}{\partial t}$ and $\nabla \times \mathbf{H} = \frac{\varepsilon}{c}\frac{\partial \mathbf{E}}{\partial t}$. By taking the curl of the first equation and substituting into the time derivative of the second equation, and making use of the identity $\nabla \times \nabla \times \mathbf{E} = -\nabla^2\mathbf{E} + \nabla(\nabla \cdot \mathbf{E})$, we see that $\mathbf{E}$ satisfies the *wave equation*,

$$\nabla^2\mathbf{E} - \nabla(\nabla \cdot \mathbf{E}) - \frac{\mu\varepsilon}{c^2}\frac{\partial^2\mathbf{E}}{\partial t^2} = 0. \tag{6.62}$$

If no free charges are present $\nabla \cdot \mathbf{E} = \varepsilon^{-1}\nabla \cdot \mathbf{D} = 0$ (Gauss's law) and the middle term in Eq. (6.62) vanishes.

Problem 6.12

(a) Show that Eq. (6.62) can be written as $\nabla^2\mathbf{E} - \nabla(\nabla \cdot \mathbf{E}) - \frac{\mu}{c^2}\frac{\partial^2\mathbf{E}}{\partial t^2} = 4\pi\frac{\mu}{c^2}\frac{\partial^2\mathbf{P}}{\partial t^2}$. Hence, the second derivative with respect to time of the polarization is a source for the electromagnetic field.

(b) Rederive the wave equation, retaining the current in the Ampère equation and substitute the constitutive equation $\mathbf{J} = \sigma\mathbf{E}$. By comparing with the wave equation obtained with $\sigma = 0$, show that the imaginary part of the polarization can be related to the conductivity σ.

For a single frequency field, $\mathbf{E}(\mathbf{r},t) = \mathrm{Re}\,[\mathbf{e}_0 E_0(\mathbf{r})e^{i(\mathbf{k}\cdot\mathbf{r}-\omega t)}]$, taking the direction of propagation as the z-axis, so $e^{i\mathbf{k}\cdot\mathbf{r}} = e^{i(\omega n(\omega)/c)z}e^{-\alpha z/2}$, substituting into the wave equation (6.62), taking $\varepsilon(\omega) = 1 + 4\pi\chi(\omega)$ and the permeability $\mu = 1$ for a nonmagnetic medium, we find

$$\mathbf{k} = \hat{\mathbf{z}}\frac{\omega(\varepsilon)^{1/2}}{c} = \hat{\mathbf{z}}[n(\omega)\omega/c - i\alpha(\omega)/2], \tag{6.63}$$

where $n(\omega) = \mathrm{Re}\,[1 + 4\pi\chi(\omega)]^{1/2} = \mathrm{Re}\,\{1 + 4\pi\mathcal{N}\mu_{ba}[u(\omega) + iv(\omega)]/E_0\}^{1/2} \approx 1 + 2\pi\mathcal{N}\mu_{ba}u(\omega)/E_0$ is the refractive index and $\alpha(\omega) = 2\,\mathrm{Im}\,k = (2\omega/c)\,\mathrm{Im}\,[1 + 4\pi\chi(\omega)]^{1/2} \approx (4\pi\omega/c)\mathcal{N}\mu_{ba}v(\omega)/E_0$ is the absorption coefficient. The

---

[6] u and v are proportional to $E(\omega)$ to lowest order in field strength, but higher-order terms are present and give rise to saturation effects.

6.1 Two-Level Systems

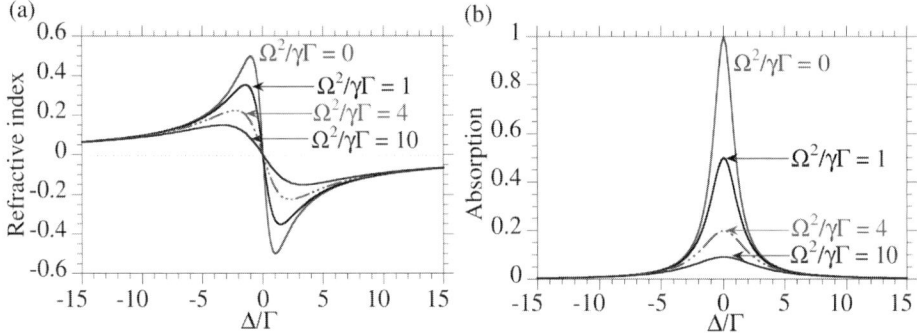

FIG 6.7 (a) Refractive index n [$\propto u(\omega)/E_0$] and (b) absorption α [$\propto v(\omega)/E_0$] versus detuning, Δ/Γ, for several values of the saturation parameter $s \equiv \Omega^2/(\gamma\Gamma)$. Here, the scales for the refractive index and the absorption are arbitrary.

intensity, $I = \frac{c}{8\pi}|\mathbf{E}|^2$ in Gaussian units ($I = |\mathbf{E}|^2/2$ in SI units) decays exponentially with propagation distance, $e^{-\alpha z}$, and the phase velocity of light is $v_{\text{ph}} = c/n = c/\text{Re}(\sqrt{\varepsilon})$. Figure 6.7 shows the refractive index $n(\omega)$ [$\propto u(\omega)/E_0$] and the absorption $\alpha(\omega)$ [$\propto v(\omega)/E_0$] versus ω/Δ for various laser intensities [$\propto \Omega^2$]. The distinctive resonance dispersion curve of the index of refraction is shown in part (a), and the Lorentzian lineshape of the absorption coefficient is shown in part (b). As Ω (which is proportional to the electric field strength) increases, the dispersion and absorption become *power broadened*, and the transition becomes saturated (the ground state population is depleted, as discussed earlier).

6.1.6 DOPPLER COOLING OF ATOMS

The idea of Doppler cooling of atoms was first conceived of in 1975 by T. W. Hänsch and A. L. Schawlow and, independently, by D. Wineland and H. Dehmelt. Doppler cooling of atoms involves the exchange of momentum, energy, and entropy between the light field and the atoms. It often involves the use of three mutually orthogonal pairs of counter-propagating laser beams with frequency red-detuned from an atomic transition.

Suppose an atom is moving with a velocity $+v$ along the x-axis in the presence of two laser beams with frequency ω that counter propagate along the x-axis. The momentum of the photons in these beams are $\hbar k$ and $-\hbar k$, respectively, and their energy is $\hbar\omega$, where $\hbar\omega < \hbar\delta$ (the beams are redshifted from resonance). In the reference frame of the atom, the photons in beam from the left are redshifted to frequency $\omega - kv$, whereas the photons in the beam from the right are blueshifted to frequency $\omega + kv$. The latter photons are preferentially absorbed, since they are closer to resonance. The excited atom now has slightly lower velocity since it absorbed a right-moving photon with momentum $-\hbar k$ ($mv_f = mv - \hbar k > 0$). The excited atom decays, emitting a photon in a dipole radiation pattern [see Eq. (7.108)]. The process of preferential absorption of photons from the beam from the right continues until the atom is substantially slowed. In contrast, an atom moving to the left with velocity $-v$ will preferentially absorb right-moving photons, hence it will also slow. Thus, a cloud of atoms will cool. Now, let us describe the process quantitatively.

Let us first consider a gas of two-level atoms in the presence of an electromagnetic field $\mathbf{E}(\mathbf{r}, t)$ of frequency ω,

$$\mathbf{E}(\mathbf{r}, t) = \mathbf{e}_0 E_0(\mathbf{r}) \cos[\omega t + \phi_0(\mathbf{r})] \equiv \mathbf{e}_0 E_0(\mathbf{r}) \cos\varphi(\mathbf{r}, t), \quad (6.64)$$

where $\mathbf{e}_0$ is the polarization of the field. The phase ϕ_0 takes the form[7]

$$\phi_0(\mathbf{r}) = -\mathbf{k} \cdot \mathbf{r} + \Phi_0(\mathbf{r}). \quad (6.65)$$

---

[7] For a plane wave the additional phase $\Phi_0(\mathbf{r})$ vanishes, but for a Gaussian beam focused at the origin, $\Phi_0(\mathbf{r}) = \tan^{-1}(z/z_0) + k(x^2 + y^2)/(2z^2[1 + (z_0/z)^2]^2)$, where z_0 is the *Rayleigh range* and the arctan factor is called the *Guoy phase* [18].

The Hamiltonian of an atom in the gas consists of kinetic energy term and an internal energy term in the two-level space, $\hat{H}_A = p^2/(2m)\mathbf{1} + (\hbar\delta/2)\sigma_z$, and the atom-field interaction operator, $\hat{U}(\mathbf{r}, t)$, is given in the dipole approximation by

$$\hat{U}(\mathbf{r}, t) = -\frac{1}{2}\hat{\boldsymbol{\mu}} \cdot \mathbf{E}(\mathbf{r}, t) = -\frac{1}{2}(\hat{\boldsymbol{\mu}} \cdot \mathbf{e}_0)\, E_0(\mathbf{r}) \cos(\omega t + \phi(\mathbf{r})). \tag{6.66}$$

The transition dipole moment operator is $\hat{\boldsymbol{\mu}} = \boldsymbol{\mu}_{ba}(|b\rangle\langle a| + |a\rangle\langle b|) = \boldsymbol{\mu}_{ba}\sigma_x$, the dipole moment matrix element is taken to be real, $\boldsymbol{\mu}_{ba} = \boldsymbol{\mu}_{ab} = \mathbf{e}_{ba}\mu_{ba}$, and the unit vector $\mathbf{e}_{ba}$ can be taken equal to $\mathbf{e}_0$ since the dipole moment is induced. The effective potential experienced by an atom in the presence of this field at position $\mathbf{r}$ and time t is $U(\mathbf{r}, t) = \mathrm{Tr}\,\hat{U}(\mathbf{r}, t) = -\frac{1}{2}\mathbf{p} \cdot \mathbf{E}(\mathbf{r}, t)$, where the induced transition dipole moment of the atom is

$$\mathbf{p} = \mathrm{Tr}(\rho\boldsymbol{\mu}) = \rho_{ba}\boldsymbol{\mu}_{ab} + \rho_{ab}\boldsymbol{\mu}_{ba} = \boldsymbol{\mu}_{ba}\{\sigma_{ba}\,e^{-i[\omega t+\phi_0(\mathbf{r})]} + \sigma_{ab}\,e^{i[\omega t+\phi_0(\mathbf{r})]}\}. \tag{6.67}$$

To find the average force exerted on an atom by an external electromagnetic field, we use the Heisenberg equations of motion for the external degrees of freedom and take their trace and their time average,

$$\left\langle\!\!\left\langle\frac{d\mathbf{r}}{dt}\right\rangle\!\!\right\rangle_t = \left\langle\!\!\left\langle\frac{\mathbf{p}_{\mathrm{at}}}{m}\right\rangle\!\!\right\rangle_t, \quad \left\langle\!\!\left\langle\frac{d\mathbf{p}_{\mathrm{at}}}{dt}\right\rangle\!\!\right\rangle_t = \mathbf{F} = -\langle\!\langle\nabla_{\mathbf{r}} U\rangle\!\rangle_t, \tag{6.68}$$

where we introduced the double brackets $\langle\!\langle\bullet\rangle\!\rangle_t$ to indicate both quantum (inner bracket) and time (outer bracket) averages, since we are dealing with quickly oscillating electromagnetic fields. In (6.68), $\mathbf{p}_{\mathrm{at}}$ is the atomic momentum (to be distinguished from $\mathbf{p}$, which in (6.67) is taken to be the transition dipole moment). We first take the quantum average:

$$\mathbf{F} = \left\langle\!\!\left\langle -\nabla\left(-\frac{1}{2}\boldsymbol{\mu}_{ba} \cdot \mathbf{e}_0\,(|b\rangle\langle a| + |a\rangle\langle b|)\,E_0(\mathbf{r})\cos\varphi(\mathbf{r}, t)\right)\right\rangle\!\!\right\rangle_t$$

$$= \left\langle\!\!\left\langle \frac{1}{2}\boldsymbol{\mu}_{ba} \cdot \mathbf{e}_0\,(|b\rangle\langle a| + |a\rangle\langle b|)\,\{\nabla(E_0)\cos\varphi + E_0\sin\varphi\,\nabla\phi_0\}\right\rangle\!\!\right\rangle_t$$

$$= \left\langle \frac{1}{2}\boldsymbol{\mu}_{ba} \cdot \mathbf{e}_0\,\mathrm{Tr}(\rho(|b\rangle\langle a| + |a\rangle\langle b|))\,\{(\nabla E_0)\cos\varphi + E_0\sin\varphi\,\nabla\phi_0\}\right\rangle_t$$

$$= \frac{\boldsymbol{\mu}_{ba}\cdot\mathbf{e}_0}{2}\,\langle(\rho_{ba} + \rho_{ab})\,\{\nabla E_0(\mathbf{r})\cos\varphi(\mathbf{r}, t) + E_0(\mathbf{r})\sin\varphi(\mathbf{r}, t)\,\nabla\phi_0\}\rangle_t. \tag{6.69}$$

We then express the off-diagonal density matrix elements in terms of the Bloch vector components u and v using Eq. (6.49), write $\cos\varphi(\mathbf{r}, t)$ and $\sin\varphi(\mathbf{r}, t)$ as complex exponentials and then take the time average to obtain

$$\boxed{\mathbf{F} = \frac{1}{2}\boldsymbol{\mu}_{ba} \cdot \mathbf{e}_0\,[u\,\nabla E_0(\mathbf{r}) + v\,E_0(\mathbf{r})\,\nabla\phi_0].} \tag{6.70}$$

The first term on the RHS is the *far-off resonance force* (or *dipole force*) and the last term is the *friction force* (or *dissipative force*). For a plane wave field, the gradient of Φ_0 in $\nabla\phi_0$ on the RHS of Eq. (6.70) vanishes, so $\nabla\phi_0 = -\mathbf{k}$, and $\nabla E_0(\mathbf{r}) = 0$, so the far-off resonance force vanishes; only the friction force remains. We can write Eq. (6.70) in terms of the Rabi frequency, $\hbar\Omega = \mu_{ba}E_0$:

$$\mathbf{F} = \frac{\hbar}{2}\,[u\,\nabla\Omega(\mathbf{r}) + v\,\Omega(\mathbf{r})\,\nabla\phi_0]. \tag{6.71}$$

Now, consider an atom in the field of two laser beams, one propagating to the right, $\mathbf{E}_R(\mathbf{r}, t) = \mathbf{e}_0 E_0(\mathbf{r})\cos\varphi_R(\mathbf{r}, t) = \mathbf{e}_0 E_0(\mathbf{r})\cos[\omega t+\phi_{0,R}(\mathbf{r})]$ with $\phi_{0,R}(\mathbf{r}) = -\mathbf{k}\cdot\mathbf{r}+\Phi_{0,R}(\mathbf{r})$, and one propagating to the left, $\mathbf{E}_L(\mathbf{r}, t) = \mathbf{e}_0 E_0(\mathbf{r})\cos\varphi_L(\mathbf{r}, t) = \mathbf{e}_0 E_0(\mathbf{r})\cos[\omega t+\phi_{0,L}(\mathbf{r})]$ with $\phi_{0,L}(\mathbf{r}) = \mathbf{k}\cdot\mathbf{r}+\Phi_{0,L}(\mathbf{r})$. Note that for plane waves, $\Phi_{0,R}(\mathbf{r}) = \Phi_{0,L}(\mathbf{r}) = 0$, $\dot\phi_{0,R} = -\mathbf{k}\cdot\mathbf{v}$, and $\dot\phi_{0,L} = \mathbf{k}\cdot\mathbf{v}$, where $\mathbf{v}$ is the velocity of the atom. The terms with $\dot\phi_0$ account for the Doppler shift of the photons in the rest frame of the atoms. Substituting $v_{R,L} = (-1)\dfrac{\Omega/\Gamma}{1+(\Delta\mp\mathbf{k}\cdot\mathbf{v})^2/\Gamma^2+\Omega^2/(\gamma\Gamma)}$ into the RHS of Eq. (6.70), noting that

6.1 Two-Level Systems

$\nabla \phi_0 = \mp \mathbf{k}$ for the R and L beams, and using the definition of the saturation parameter, $s \equiv \Omega^2/(\gamma\Gamma)$, we find

$$\mathbf{F}_{R,L} = \mp \frac{1}{2} \frac{\gamma s}{1 + s + (\Delta \mp \mathbf{k} \cdot \mathbf{v})^2/\Gamma^2} \hbar \mathbf{k}. \tag{6.72}$$

We can interpret this result as originating from the product of three factors: (1) the rate of spontaneous emission γ, (2) the excited state population fraction $\rho_{bb} = \frac{1}{2} \frac{\Omega^2/(\gamma\Gamma)}{1+\Delta^2/\Gamma^2+\Omega^2/(\gamma\Gamma)}$, and (3) the photon momentum $\mp\hbar\mathbf{k}$ provided to the atom on absorbing a photon. The force on the atom is the sum of the forces due to the two laser beams. Expanding the expressions for the forces due to the R and L beams in the small parameter $\mathbf{k} \cdot \mathbf{v}/\Delta$ and adding contributions from the two counter-propagating beams, we find that the net force is

$$\mathbf{F}_{\text{friction}} = \mathbf{F}_R + \mathbf{F}_L = \frac{2\gamma \hbar k^2 \Delta}{\Gamma^2} \frac{s}{(1+s+(\Delta/\Gamma)^2)^2} \mathbf{v} = \beta \mathbf{v}, \tag{6.73}$$

where

$$\beta = \frac{2\gamma \hbar k^2 s \Delta}{\Gamma^2 [1 + s + (\Delta/\Gamma)^2]^2}. \tag{6.74}$$

Note that the force is *opposite* in direction to the velocity for red-detuned laser light, i.e., for $\Delta < 0$, $\beta < 0$. We can now calculate the rate of change of the kinetic energy due to the cooling process when $\beta < 0$:

$$\frac{dE}{dt} = \frac{d}{dt}\left(\frac{mv^2}{2}\right) = mv\frac{dv}{dt} = \beta v^2 = \frac{2\beta}{m} E. \tag{6.75}$$

In addition to cooling due to photon absorption from the laser beams, heating due to spontaneous emission from the excited state of the gas atoms occurs because of the momentum imparted to an atom on photon emission. Cooling and heating eventually balance each other, and the system will reach equilibrium. The rate of heating can be estimated by averaging over spontaneous emission events whose direction is stochastic. Let us count photon absorption and emission events. During the short interval of time δt, an atom can absorb δN_R photons from the R beam and δN_L from the L beam. The total number of absorptions events (that are separated by spontaneous emissions to the ground state) is equal to $\delta N = \delta N_R + \delta N_L$. Within the time interval δt, the change of momentum due to emission and absorption events is equal to $\delta p = \hbar k (\delta N_R - \delta N_L) + \sum_j \hbar k_j$, where the first two terms are due to absorption for the laser beams and the last term corresponds to the momentum kicks caused by spontaneous emission. From this expression, we can obtain the mean square value of the momentum change,

$$\overline{(\delta p)^2} = \overline{(\hbar k(\delta N_R - \delta N_L))^2} + \sum_{jj'} \hbar^2 k_j k'_j, \tag{6.76}$$

which can be used to estimate the average change of the atomic kinetic energy. Note that the subsequent spontaneous emission events are independent; hence, the average of the sum of k_j's is zero and in the last sum only diagonal terms $j = j'$ contribute. Additionally, we assume that the average k_j^2 is of order k^2, to obtain $\overline{\delta N}\hbar^2 k^2$ for the last term on the RHS of Eq. (6.76). To calculate the contribution from the first term on the RHS of Eq. (6.76), we assume $\overline{\delta N_R} = \overline{\delta N_L}$, and that both N_R and N_L satisfy the Poisson distribution. In this case, $\overline{(\delta N_{R,L})^2} = \overline{(\delta N_{R,L})}^2 + \overline{(\delta N_{R,L})}$. After some algebra, we obtain $\overline{\delta N}\hbar^2 k^2$; hence, the rate of change of the total energy due to heating is given by

$$\frac{dE_{\text{heat}}}{dt} = \frac{1}{2m}\frac{\overline{(\delta p)^2}}{\delta t} = \frac{1}{2m} 2\frac{\overline{\delta N}}{\delta t}\hbar^2 k^2. \tag{6.77}$$

The expression (6.76) is quadratic in $N_R - N_L$. Since δN is due to absorption from the two counter-propagating beams, it equals $2R$, where $R = \gamma \rho_{bb}$. We now have all the ingredients in hand to calculate the final temperature reached by the Doppler cooling process. We have to set only the rate of heating equal to the rate of cooling to determine the temperature to which the gas cools. When these rates are equal to each other,

$$\frac{dE_{\text{total}}}{dt} = -\frac{2\beta}{m}E + \frac{2}{m}\frac{\Gamma s}{2}\frac{\hbar^2 k^2}{(1+s+(\Delta/\Gamma)^2)} = 0 \tag{6.78}$$

and cooling stops. This equation determines the resulting value of the kinetic energy E, which is $E = k_B T/2$ by the equipartition theorem. The resulting temperature is called the *Doppler temperature*, T_D. For small s,

$$k_B T_D = \frac{\hbar\Gamma}{2}\left(\frac{\Gamma}{\Delta} + \frac{\Delta}{\Gamma}\right). \tag{6.79}$$

Minimizing this temperature by taking best detuning Δ, i.e., setting the derivative $dT_D/d\Delta = 0$, yields

$$\Delta = -\gamma/2 \quad \text{and} \quad k_B T_D = \hbar\gamma/2. \tag{6.80}$$

For sodium, $T_D = 240\,\mu\text{K}$, while for cesium, $T_D = 130\,\mu\text{K}$.

When the concept of atom cooling was being experimentally tested in the lab, the temperature obtained turned out to be *lower* than the Doppler cooling temperature T_D. Also, the frequency dependence found was different from Eq. (6.79). The reason for the discrepancy is the multilevel nature (the spin-orbit and hyperfine splittings) of the alkali atoms that were cooled. The mechanism leading to temperatures lower than the Doppler cooling temperature is called "Sisyphus cooling" or "polarization gradient cooling." An extensive account of polarization gradient cooling, and cooling and trapping of atoms in general, is given in the review articles by the 1997 Nobel prize laureates Chu, Cohen-Tannoudji, and Phillips [95].

6.1.7 OPTICAL TRAPPING OF ATOMS

Trapping of atoms with the dipole force was first suggested by V. S. Letokhov in 1968. In 1970, Arthur Ashkin first reported the detection of the far-off resonance force on micrometer-sized particles. Later, Ashkin and colleagues from Bell Labs reported the first observation of *optical tweezers*, a tightly focused beam of light capable of holding microscopic particles stable in three dimensions. In 1986, Steven Chu used the far-off resonance force to trap neutral atoms.

We have seen in Sec. 6.1.6 that the optical force exerted on an atom is composed of two components, the *far-off resonance force* or *dipole force* and the *friction force* or *dissipative force*, as detailed in Eq. (6.70). The dissipative force gives rise to *radiation pressure* and is used for laser cooling, but for optical trapping, the far-off resonance force is the important component. On substituting the expression for u from Eq. (6.53a) into Eq. (6.70), and using $\hbar\Omega = \mu_{ba}E_0$, we obtain

$$\mathbf{F}_{\text{dip}} = \frac{\hbar\Delta/\Gamma^2}{1+\Delta^2/\Gamma^2+\Omega^2/(\gamma\Gamma)}\nabla\Omega^2(\mathbf{r}). \tag{6.81}$$

The far-off resonance force is the time-averaged force arising from the interaction of the transition dipole induced by the oscillating electric field of the light with the gradient of the electric field amplitude, as specified in Eq. (6.70). Focusing the light beam affects the field gradient, and detuning the optical frequency relative to the atomic transition controls the sign of this force on the atom. If the light is tuned below resonance ($\Delta < 0$), the atom is attracted to the center of the light beam, whereas "blue-detuned" light ($\Delta > 0$) repels it.

A number of configurations have been used to trap atoms using far-off resonance traps (FORTs) that rely on the dipole optical force [18, 95], including the laser beam configuration with two confocal counter-propagating red-detuned beams with orthogonal polarizations is shown in Fig. 6.8. *Magneto-optical traps* (MOTs) that produce cold-trapped neutral atoms use both laser cooling and magneto-optical trapping. Moreover, configurations have been developed that both trap and cool atoms with a form of Doppler cooling referred to as *optical molasses*, since the dissipative optical force resembles the viscous drag on a body moving through molasses. Optical molasses can cool neutral atoms to temperatures colder than a MOT. In optical molasses, cold atoms accumulate in a region where three orthogonal pairs

6.1 Two-Level Systems

of laser beams intersect. With the correct choice of polarizations for the laser beams, this configuration can trap atoms by the addition of a magnetic field gradient created by two coils with currents in opposite directions that produce a quadrupole magnetic field. A typical sodium MOT can cool atoms down to 300 μK, whereas optical molasses can cool the atoms down to an order of magnitude cooler. For more information about cooling and trapping of atoms, see Ref. [95].

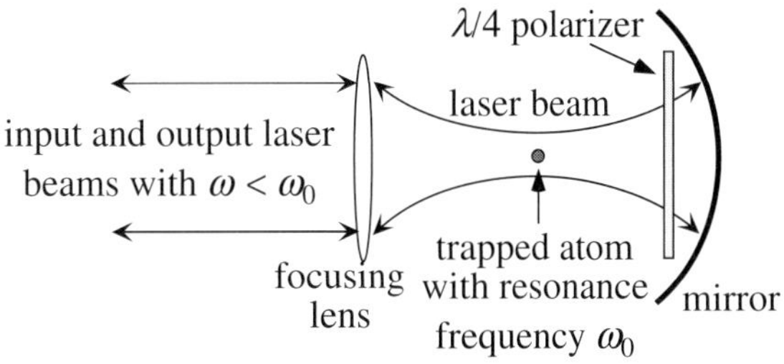

6.1.8 TWO OR MORE CORRELATED "SPINS"

FIG 6.8 Far-off resonance laser trap for atoms, with two confocal counter-propagating red-detuned laser beams having orthogonal polarizations.

Let us consider two two-level systems (or a bipartite qubit system, such as the polarization state of two photons or two spin 1/2 particles) and call them A and B. For two uncorrelated spins (or two uncorrelated photons), we can write their density matrix in the product form $\rho_{AB} = \rho_A \rho_B$ (i.e., the tensor product of the density matrices, $\rho_A \otimes \rho_B$). For two correlated spins, the density matrix must include another term that accounts for the correlation of the spins. It can be written in the form

$$\rho_{AB} = \frac{1}{4}\left[(1 + \mathbf{n}_A \cdot \boldsymbol{\sigma}_A)(1 + \mathbf{n}_B \cdot \boldsymbol{\sigma}_B) + \boldsymbol{\sigma}_A \cdot \mathbf{C}_{AB} \cdot \boldsymbol{\sigma}_B\right], \tag{6.82}$$

where the 3×3 tensor $\mathbf{C}_{AB}$ specifies the correlations between the two spins,

$$C_{ij,AB} = \langle \sigma_{i,A}\sigma_{j,B}\rangle - \langle \sigma_{i,A}\rangle\langle \sigma_{j,B}\rangle = \langle \sigma_{i,A}\sigma_{j,B}\rangle - n_{i,A}\,n_{j,B}. \tag{6.83}$$

ρ_{AB} can be represented as a 4×4 Hermitian matrix with trace unity, so 15 (= 4×4 - 1) parameters are required to parameterize it. The three components of $\mathbf{n}_A$, the three components of $\mathbf{n}_B$, and the nine components C_{ij}, where we no longer explicitly show the subscripts A and B, are sufficient for this purpose. All the parameters appearing in Eq. (6.82) can be experimentally determined by a combination of single-particle measurements for $\mathbf{n}_A$ and $\mathbf{n}_B$ and Bell measurements for the correlation matrix $\mathbf{C}$.

For example, let us determine the density matrix ρ_{AB} for the spin singlet state, (4.81). The expectation value of the projection of the first spin A and the second spin B on any axis must vanish, $\langle \boldsymbol{\sigma}_A \rangle = \langle \boldsymbol{\sigma}_B \rangle = 0$, so $\mathbf{n}_A = \mathbf{n}_B = 0$. Moreover, the expectation value $\langle \sigma_{i,A}\sigma_{j,B}\rangle = -\delta_{ij}$ (the spins are oppositely polarized), so

$$\rho_{\text{singlet}} = \frac{1}{4}(1_A 1_B - \boldsymbol{\sigma}_A \cdot \boldsymbol{\sigma}_B) = \frac{1}{2}\begin{pmatrix} 0 & 0 & 0 & 0 \\ 0 & 1 & -1 & 0 \\ 0 & -1 & 1 & 0 \\ 0 & 0 & 0 & 0 \end{pmatrix}. \tag{6.84}$$

This density matrix is also a projection operator onto the singlet state [note that the eigenvalues of the 4×4 matrix $\boldsymbol{\sigma}_A \cdot \boldsymbol{\sigma}_B$ are -3 and 1 (which is triply degenerate), so the eigenvalues of ρ_{singlet} are 1 and 0 (triply degenerate)]. In Problem 6.14, you will work out the density matrix of the Bell state $|\Psi^+\rangle = \frac{1}{\sqrt{2}}[|\uparrow\downarrow\rangle + |\downarrow\uparrow\rangle]$, which is one of the three components of the triplet state, and the other two Bell states as well. The density matrix

$$\rho_{\Psi^+} = \frac{1}{4}(1_A 1_B + \boldsymbol{\sigma}_A \cdot \boldsymbol{\sigma}_B - 2\sigma_{z,A}\sigma_{z,B}) = \frac{1}{2}\begin{pmatrix} 0 & 0 & 0 & 0 \\ 0 & 1 & 1 & 0 \\ 0 & 1 & 1 & 0 \\ 0 & 0 & 0 & 0 \end{pmatrix} \tag{6.85}$$

is also a projection operator, as are the other density matrices you will derive in Problem 6.14.

Problem 6.13

Using the density matrix in Eq. (6.82), show that $\langle \sigma_{i,A} \sigma_{j,B} \rangle = C_{ij} + \langle \sigma_{i,A} \rangle \langle \sigma_{i,A} \rangle$, thereby verifying Eq. (6.83).
Hint: Recall that $\langle \mathcal{O} \rangle = \operatorname{Tr} \rho \mathcal{O}$.

Problem 6.14

(a) Determine the density matrix ρ_{Ψ^+} for the $M = 0$ triplet (4.82), $|\Psi^+\rangle = \frac{1}{\sqrt{2}}[|\uparrow\downarrow\rangle + |\downarrow\uparrow\rangle]$ and show that it projects onto $|\Psi^+\rangle$.
(b) Determine the density matrix for the "Bell states," $|\Phi^+\rangle = \frac{1}{\sqrt{2}}[|\uparrow\uparrow\rangle + |\downarrow\downarrow\rangle]$, $|\Phi^-\rangle = \frac{1}{\sqrt{2}}[|\uparrow\uparrow\rangle - |\downarrow\downarrow\rangle]$.

Answers: (a) $\rho_{\Psi^+} = \frac{1}{4}(1_A 1_B + \sigma_{x,A} \sigma_{x,B} + \sigma_{y,A} \sigma_{y,B} - \sigma_{z,A} \sigma_{z,B})$.
(b) $\rho_{\Phi^+} = \frac{1}{4}(1_A 1_B + \sigma_{x,A} \sigma_{x,B} - \sigma_{y,A} \sigma_{y,B} + \sigma_{z,A} \sigma_{z,B})$, $\rho_{\Phi^-} = \frac{1}{4}(1_A 1_B - \sigma_{x,A} \sigma_{x,B} + \sigma_{y,A} \sigma_{y,B} + \sigma_{z,A} \sigma_{z,B})$.

Problem 6.15

For the entangled state $|\psi\rangle = [2 \cosh(2\theta)]^{-1/2}[e^{-\theta}|\uparrow\uparrow\rangle + e^{\theta}|\downarrow\downarrow\rangle]$:
(a) Determine the expectation values $\langle \boldsymbol{\sigma}_A \rangle$, $\langle \boldsymbol{\sigma}_B \rangle$.
(b) Determine the variances $\langle (\Delta \sigma_{i,A})^2 \rangle$, $\langle (\Delta \sigma_{i,B})^2 \rangle$ for $i = x, y, z$.
(c) Determine $\langle \mathbf{J} \rangle$, and $\langle (\Delta J_i)^2 \rangle$ for $i = x, y, z$, where $\mathbf{J} = \frac{\hbar}{2}(\boldsymbol{\sigma}_A + \boldsymbol{\sigma}_B)$.
(d) Show that the state is a minimum uncertainty state in the sense that $\langle \Delta J_x \rangle \langle \Delta J_y \rangle = \frac{\hbar}{2} |\langle J_z \rangle|$.
(e) Determine the density matrix. Hint: Note that $J_i = \frac{\hbar}{2}(\sigma_{i,A} + \sigma_{i,B})$ and squaring yields an expression for $\sigma_{i,A} \sigma_{i,B}$.

Answers: (a) $\langle \sigma_x \rangle = \langle \sigma_y \rangle = 0$, $\langle \sigma_z \rangle = -\tanh(2\theta)$ for both A and B.
(b) $\langle (\Delta \sigma_i)^2 \rangle = 1$ for all i and A and B.
(c) $\langle J_x \rangle = \langle J_y \rangle = 0$, $\langle J_z \rangle = -\hbar \tanh(2\theta)$, $\langle (\Delta J_x)^2 \rangle = \frac{\hbar^2}{2}[1 + \operatorname{sech}(2\theta)]$, $\langle (\Delta J_y)^2 \rangle = \frac{\hbar^2}{2}[1 - \operatorname{sech}(2\theta)]$,
$\langle (\Delta J_z)^2 \rangle = \hbar^2 \operatorname{sech}^2(2\theta)$.
(e) $\rho_\psi = \frac{1}{4}\{[1_A - \tanh(2\theta)\sigma_{z,A}][1_B - \tanh(2\theta)\sigma_{z,B}] + \operatorname{sech}(2\theta)(\sigma_{x,A} \sigma_{x,B} - \sigma_{y,A} \sigma_{y,B}) + \operatorname{sech}^2(2\theta)\sigma_{z,A} \sigma_{z,B}\}$.

Let us consider the classically correlated two-qubit states of the form $\rho_{AB} = \sum_i p_i \rho_i^A \rho_i^B$, with $p_i > 0$ and $\sum_i p_i = 1$, first introduced in Sec. 5.2.2 [see Eq. (5.39)]. When two terms are present, $i = 1, 2$, we have $\rho_{AB} = p_1 \rho_1^A \rho_1^B + p_2 \rho_2^A \rho_2^B$, and using Eq. (6.16) for the single-qubit density matrices,

$$\rho_{AB} = \frac{1}{4} \left[p_1 (1 + \mathbf{n}_{A,1} \cdot \boldsymbol{\sigma}_A)(1 + \mathbf{n}_{B,1} \cdot \boldsymbol{\sigma}_B) + p_2 (1 + \mathbf{n}_{A,2} \cdot \boldsymbol{\sigma}_A)(1 + \mathbf{n}_{B,2} \cdot \boldsymbol{\sigma}_B) \right]. \tag{6.86}$$

After some algebra, we find that the coefficients $\mathbf{n}_A$, $\mathbf{n}_B$, and $C_{ij,AB}$ in Eq. (6.82) are given by, $\mathbf{n}_A = p_1 \mathbf{n}_{A,1} + p_2 \mathbf{n}_{A,2}$, $\mathbf{n}_B = p_1 \mathbf{n}_{B,1} + p_2 \mathbf{n}_{B,2}$, and $C_{ij,AB} = p_1 n_{i,A,1}[n_{j,B,1} - (p_1 n_{j,B,1} + p_2 n_{j,B,2})] + p_2 n_{i,A,2}[n_{j,B,2} - (p_1 n_{j,B,1} + p_2 n_{j,B,2})] + p_2 n_{i,A,2}$, for the classically correlated two-qubit state example of Eq. (6.86). The fact that the C matrix is nonvanishing proves that these states are correlated.

Three correlated spins can be described by the density matrix

$$\rho_{ABC} = \frac{1}{8}[(1 + \mathbf{n}_A \cdot \boldsymbol{\sigma}_A)(1 + \mathbf{n}_B \cdot \boldsymbol{\sigma}_B)(1 + \mathbf{n}_C \cdot \boldsymbol{\sigma}_C) + \boldsymbol{\sigma}_A \cdot \mathbf{C}_{AB} \cdot \boldsymbol{\sigma}_B$$
$$+ \boldsymbol{\sigma}_A \cdot \mathbf{C}_{AC} \cdot \boldsymbol{\sigma}_C + \boldsymbol{\sigma}_B \cdot \mathbf{C}_{BC} \cdot \boldsymbol{\sigma}_C + \sum_{ijk} \sigma_{i,A} \sigma_{j,B} \sigma_{k,C} D_{ijk}], \tag{6.87}$$

6.1 Two-Level Systems

where $\mathbf{C}_{AB}$, $\mathbf{C}_{AC}$, and $\mathbf{C}_{BC}$ are the two-particle correlation matrices and the tensor that specifies the three-spin correlations is

$$D_{ijk} = \langle \sigma_{i,A}\sigma_{j,B}\sigma_{k,C}\rangle - \langle \sigma_{i,A}\rangle\langle \sigma_{j,B}\rangle\langle \sigma_{k,C}\rangle. \tag{6.88}$$

The density matrices of four-particle and higher states can be constructed similarly, but with increased complexity.

Problem 6.16

(a) Determine ρ for the GHZ-like state $\frac{1}{\sqrt{2}}[|\uparrow\uparrow\uparrow\rangle - |\downarrow\downarrow\downarrow\rangle]$.

(a) Determine ρ_{GHZ} for the GHZ state $\frac{1}{\sqrt{2}}[|\uparrow\uparrow\uparrow\rangle + |\downarrow\downarrow\downarrow\rangle]$.

Answer:

(a) $\rho = \frac{1}{8}(1_A 1_B 1_C + \sigma_{z,A}\sigma_{z,B} + \sigma_{z,A}\sigma_{z,C} + \sigma_{z,B}\sigma_{z,C} + \sigma_{x,A}\sigma_{x,B}\sigma_{x,C} + \sigma_{x,A}\sigma_{y,B}\sigma_{y,C} + \sigma_{y,A}\sigma_{x,B}\sigma_{y,C} + \sigma_{y,A}\sigma_{y,B}\sigma_{x,C})$.

(b) $\rho_{GHZ} = \frac{1}{8}(1_A 1_B 1_C + \sigma_{z,A}\sigma_{z,B} + \sigma_{z,A}\sigma_{z,C} + \sigma_{z,B}\sigma_{z,C} + \sigma_{x,A}\sigma_{x,B}\sigma_{x,C} - \sigma_{x,A}\sigma_{y,B}\sigma_{y,C} - \sigma_{y,A}\sigma_{x,B}\sigma_{y,C} - \sigma_{y,A}\sigma_{y,B}\sigma_{x,C})$.

Problem 6.17

(a) Given parameters $\mathbf{n}_A$, $\mathbf{n}_B$, and $\mathbf{C}_{ij}$ for the bipartite qubit density matrix ρ_{AB} in Eq. (6.82), how does one determine whether the density matrix corresponds to a pure state?

(b) If $C_{ij} = 0$ for all i,j, what are the conditions on $\mathbf{n}_A$ and $\mathbf{n}_B$ such that ρ_{AB} is a pure state?

(c) If $\mathbf{n}_A = \mathbf{n}_B = 0$, what are the conditions on $\mathbf{C}_{AB}$, so that the density matrix corresponds to a pure state?

Answers: (a) For the density matrix to be a pure state, the condition $\rho_{AB}^2 - \rho_{AB} = 0$ must be satisfied.
(b) Using Eq. (4.6), we find $n_A^2 = n_B^2 = 1$. I.e., each qubit is in a pure state.
(c) Hint: Solve the set of equations $\rho_{AB}^2 - \rho_{AB} = 0$ for the parameters $\mathbf{C}_{AB}$ after using Eq. (4.6). For the case of a diagonal $\mathbf{C}_{AB}$ matrix, one obtains the following results:

$\{C_{xx} = 1, C_{yy} = 1, C_{zz} = -1\}, \{C_{xx} = -1, C_{yy} = 1, C_{zz} = 1\}, \{C_{xx} = 1, C_{yy} = -1, C_{zz} = 1\}, \{C_{xx} = -1, C_{yy} = -1, C_{zz} = -1\}$

These are the C coefficients for the four Bell states, and the pure states referred to in this problem are the Bell states.

Problem 6.18

(a) Determine the parameters in Eq. (6.82) for the density matrix $\rho_{AB} = \frac{1}{2}[|\uparrow\uparrow\rangle\langle\uparrow\uparrow| + |\downarrow\downarrow\rangle\langle\downarrow\downarrow|]$.

(b) Write the density matrix in the basis of states $|\uparrow\uparrow\rangle$, $|\downarrow\uparrow\rangle$, $|\uparrow\downarrow\rangle$, and $|\downarrow\downarrow\rangle$. [Notice the order in which the basis functions are written; the first spin varies first. Compare this with (5.21) and with (5.50).]

(c) Determine the parameters in Eq. (6.82) for the density matrix
$\rho_{AB} = [2\cosh(2\theta)]^{-1/2}[e^{-\theta}|\uparrow\uparrow\rangle\langle\uparrow\uparrow| + e^{\theta}|\downarrow\downarrow\rangle\langle\downarrow\downarrow|]$.

Answer: (a) $\mathbf{n}_A = \mathbf{n}_B = 0$, $C_{ij} = 0$ for all i,j, except $C_{zz} = 1$.

(b)
$$\rho_{AB} = \frac{1}{4}\left[\begin{pmatrix}1 & 0 & 0 & 0\\ 0 & 1 & 0 & 0\\ 0 & 0 & 1 & 0\\ 0 & 0 & 0 & 1\end{pmatrix} + \begin{pmatrix}1 & 0 & 0 & 0\\ 0 & -1 & 0 & 0\\ 0 & 0 & -1 & 0\\ 0 & 0 & 0 & 1\end{pmatrix}\right]$$

(c) All the correlation coefficients vanish, except for $C_{zz} = \text{sech}(2x)^2$.

> **Problem 6.19**
>
> (a) Determine the characteristics of the density matrix $\rho_{AB} = \frac{1}{3}[|\uparrow\uparrow\rangle\langle\uparrow\uparrow| + |\rightarrow\rightarrow\rangle\langle\rightarrow\rightarrow| + |\odot\odot\rangle\langle\odot\odot|]$.
>
> (b) Determine the characteristics of $\rho_{AB} = \frac{1}{6}[|\uparrow\uparrow\rangle\langle\uparrow\uparrow| + |\downarrow\downarrow\rangle\langle\downarrow\downarrow| + |\rightarrow\rightarrow\rangle\langle\rightarrow\rightarrow| + |\leftarrow\leftarrow\rangle\langle\leftarrow\leftarrow| + |\odot\odot\rangle\langle\odot\odot| + |\otimes\otimes\rangle\langle\otimes\otimes|]$.
>
> **Answer:** (a) $\mathbf{n}_A = \mathbf{n}_B = (1/3, 1/3, 1/3)^t$, and the correlation matrix is
>
> $$C = \frac{1}{9}\begin{pmatrix} 2 & -1 & -1 \\ -1 & 2 & -1 \\ -1 & -1 & 2 \end{pmatrix}. \tag{6.89}$$
>
> (b) $\mathbf{n}_A = \mathbf{n}_B = \mathbf{0}$ and $C = 0$.

An identical description of the density matrix for the polarization degrees of freedom of two photons can be made using Eq. (6.82). For example, the equivalent of the correlated "singlet state" (6.84) corresponds to the unpolarized two-photon state $\frac{1}{\sqrt{2}}(|R\rangle|L\rangle - |L\rangle|R\rangle)$. Similarly, the density matrix for the polarization degrees of freedom of three-photons can be described using Eq. (6.87).

When the two-particle, three-particle, etc., two-level systems undergo dynamics, the parameters $\mathbf{n}_A$, $\mathbf{n}_B$, C_{ij}, etc. become time dependent. Knowing the Hamiltonian (or the Liouville operator) that defines the dynamics, the time dependence of these parameters can be determined by solving the Liouville–von Neumann density matrix equation.

6.1.9 THE N-TWO-LEVEL SYSTEM BLOCH SPHERE

Let us now consider N-two-level systems. One possible choice of basis states for such a system is the set of direct product states $|m_1, m_2, \ldots, m_N\rangle \equiv \prod_{i=1}^{N} |m_i\rangle$, where $m_i = -1/2$ corresponds to the ground state a of Sec. 6.1 (or qubit state $|0\rangle$) and $m_i = +1/2$ corresponds to the excited state b (or qubit state $|1\rangle$) of the ith qubit.[8] Another representation of such states is in terms of the collective angular momentum operators

$$\mathbf{J} = \sum_{i=1}^{N} \mathbf{S}_i, \quad J_z = \sum_{i=1}^{N} S_{zi}, \tag{6.90}$$

where $\mathbf{S}_i$ and S_{zi} are the vector spin operator and its z-projection for particle i. The basis set of states $|J, M\rangle$ are eigenstates of $\mathbf{J}^2$ and J_z,

$$\mathbf{J}^2|J, M\rangle = \hbar^2 J(J+1)|J, M\rangle, \quad J_z|J, M\rangle = \hbar M|J, M\rangle, \tag{6.91}$$

and they are linear combinations of the product states $\prod_{i=1}^{N} |m_i\rangle$ (for the remainder of this section, we shall set $\hbar = 1$ and $J = N/2$). Yet another set of states that can be used as a basis set for the system are the generalized coherent states of the $SU(2)$ Lie algebra [31, 96], which are parameterized by the two polar angles θ and ϕ corresponding to rotations of the fully stretched atomic state $|J, -J\rangle$, first about the y-axis by an angle θ and then about the z-axis by an angle ϕ. The $SU(2)$ coherent state $|\theta, \phi\rangle$ is given by[9]

$$|\theta, \phi\rangle \equiv \exp(-i\phi J_z) \exp(-i\theta J_y) |J, -J\rangle = \exp(\alpha J_+ - \alpha^* J_-) |J, -J\rangle, \tag{6.92}$$

---

[8] Note that here we consider N two-level systems where each can be in the same spin state. Clearly, they are not identical fermions (unless they are all in different spatial modes which are not being explicitly represented here).

[9] This is a superposition state of N spins. It should not be confused with the state of a single spin $|\theta, \phi\rangle$ defined in Eq. (6.13). The state in Eq. (6.92) involves the product of N spin 1/2 particles in state (6.13).

6.1 Two-Level Systems

with $\alpha = (\theta/2)\exp(-i\phi)$. Hence, the state of the N two-level system is represented by a Bloch vector that lies on an N two-level Bloch sphere. The definition (6.92) can be used to obtain the following expansion of $|\theta,\phi\rangle$ in terms of the $|J,M\rangle$ states (see Problem 6.20):

$$|\theta,\phi\rangle = \left[1+\tan^2\left(\frac{\theta}{2}\right)\right]^{-J} \sum_{M=-J}^{J} \left[\tan\left(\frac{\theta}{2}\right)e^{-i\phi}\right]^{J+M} \binom{2J}{J+M}^{1/2} |J,M\rangle. \quad (6.93)$$

Figure 6.9 plots the probabilities $P_M(\theta) = \frac{\tan^{2(J+M)}(\theta/2)}{[1+\tan^2(\theta/2)]^{2J}}\binom{2J}{J+M}$ of state $|J,M\rangle$ in the expansion (6.93) of the coherent state $|\theta,\phi\rangle$ versus M and θ for $J=10$. These probabilities are independent of ϕ.

The expectation values of the angular momentum operators for the coherent states $|\theta,\phi\rangle$ can be calculated using Eq. (6.92) or (6.93):

$$\langle\theta,\phi|J_x|\theta,\phi\rangle = J\sin\theta\cos\phi, \quad \langle J_y\rangle = J\sin\theta\sin\phi, \quad \langle J_z\rangle = J\cos\theta. \quad (6.94)$$

Hence, the Bloch vector, $\langle\mathbf{J}\rangle/J$, is restricted to the Bloch sphere of unit radius, since $\langle J_x\rangle^2 + \langle J_y\rangle^2 + \langle J_y\rangle^2 = J^2$. Moreover, the standard deviation of the angular momentum operators for the coherent states are given by

$$\Delta J_x = \frac{J}{2}\left(1-\sin^2\theta\cos^2\phi\right), \quad \Delta J_y = \frac{J}{2}\left(1-\sin^2\theta\sin^2\phi\right), \quad \Delta J_z = \frac{J}{2}\sin^2\theta, \quad (6.95)$$

where the variances are defined as $(\Delta J_i)^2 \equiv \langle\theta,\phi|J_i^2|\theta,\phi\rangle - \langle\theta,\phi|J_i|\theta,\phi\rangle^2$. The variance of the total angular momentum is given by $|\Delta\mathbf{J}|^2 = \langle\mathbf{J}^2\rangle - \langle\mathbf{J}\rangle^2 = J(J+1) - (J)^2 = J$. The Heisenberg uncertainty relations of the angular momentum operators for the coherent states give, for example,

$$\Delta J_x \Delta J_y \geq \frac{1}{2}|\langle J_z\rangle|. \quad (6.96)$$

For any given pure state of an N two-level system, $|\Phi\rangle$, one can plot the value of the probability $P(\theta,\phi) = |\langle\theta,\psi|\Phi\rangle|^2$ in the θ-ϕ plane or on the Bloch sphere. The Bloch sphere representation of the N two-level system is the generalization of the Bloch sphere representation for one two-level system to N such levels. The generalized Bloch sphere is defined such that the Bloch vector for the state $|J,-J\rangle$ points to the south pole and $|J,J\rangle$ points to the north pole. The points located on the equator ($\theta = \pi/2$) have equal probability of finding $N/2$ of the two-level systems in the ground state and $N/2$ in the excited state. The state $|\theta,\phi\rangle$ is represented as a wave packet on the Bloch sphere, the center of the wave packet at the angles θ, ϕ with $0 \geq \theta \geq 2\pi$ and $0 \geq \phi \geq \pi$ (the "physics" spherical coordinate convention).

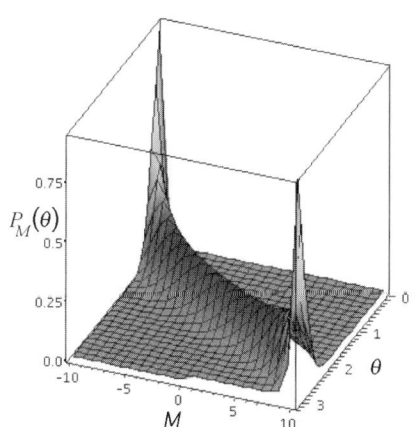

FIG 6.9 The probability $P_M(\theta)$ versus M and θ of obtaining $|J,M\rangle$ given the coherent state $|\theta,\phi\rangle$, for $J=10$ [see Eq. (6.93)].

Figure 6.10 shows the expectation values of $\langle\mathbf{J}\rangle/J$ for the coherent states, as well as the standard deviations ΔJ_x and ΔJ_y for 10 such states. Coherent states for which inequality (6.96) is an equality are referred to as *intelligent states* or *ideal coherent states*. From Eqs (6.95), we obtain that SU(2) intelligent states are obtained for $\phi = 0, \pi/2, \pi, 3\pi/2$, and arbitrary θ, as depicted by the dashed curves in Fig. 6.10. A subset of the intelligent states is the minimum uncertainty state with $\phi = 0, \pi/2, \pi, 3\pi/2$, and $\theta = \pi/2$, for which the RHS of Eq. (6.96) is minimized, with $\Delta J_x \Delta J_y = 0$ (depicted by the solid blue lines in Fig. 6.10 – note, however, that these states have nonzero standard deviation ΔJ_z, but this is not properly depicted in the figure). Although the states with $\theta = 0$ and arbitrary ϕ (large yellow disks) are also intelligent, their values of the product $\Delta J_x \Delta J_y = J/2$ are in fact maximal and are larger than $\Delta J_x \Delta J_y = J/4$ obtained for the nonintelligent states depicted by the smaller cyan disks.

In the Bloch sphere picture [for both the case of one two-level system (as in Sec. 6.1.2) and the N-two-level system (as in Fig. 6.10)], the width of states, $\Delta J_x/J$, $\Delta J_y/J$, and $\Delta J_z/J$ can be shown on the Bloch sphere, as well as the central

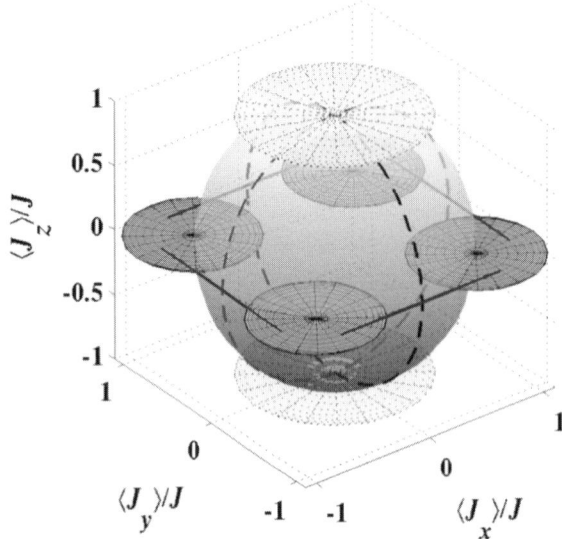

FIG 6.10 Bloch sphere (shaded shell) and coherent states of $SU(2)$. Dashed black curves mark intelligent coherent states. Ellipses depict ΔJ_x and ΔJ_y standard deviations for 10 coherent states (standard deviations ΔJ_z are not depicted): the $|J,-J\rangle$ and $|J,J\rangle$ states (large discs), the four nonintelligent states (smaller discs) and the four squeezed, minimum uncertainty states (solid lines). Reproduced from Ref. [33]. (For the color version of this figure, please refer the color plate section at the end of book.)

position $\langle \mathbf{J} \rangle / J$. See, for example, Fig. 6.14, where the widths of the states, as well as their position on the Bloch sphere are shown. Figure 6.11(a) shows another view of the state with $\langle \mathbf{J} \rangle$ along the direction $\hat{y}$, (b) shows a state which has 50% of its N atoms in state up and in state down, with no definite phase between them, i.e., $|\uparrow\rangle^{N/2}|\downarrow\rangle^{N/2}$, and (c) shows a squeezed number state, with a wave packet appearing on the Bloch sphere which is an ellipse rather than a circle.

To summarize, the pure states of an N two-level system can be represented as wave packets on a Bloch sphere. One can plot the value of the probability $P(\theta,\phi) = |\langle \theta,\phi|\Phi\rangle|^2$ on the Bloch sphere. As long as the state $|\Phi(t)\rangle$ is time dependent, this probability can evolve as a function of time. If the system starts off in state $|J,-J\rangle$, with $J = N/2$ (all the atoms in the ground state), the dynamics can be represented by a generalized Bloch vector that moves on a generalized Bloch sphere, starting from the south pole at the initial time (see Fig. 6.14). Section 6.1.10 presents such dynamics.

In general, for an N two-level system, there can be Bloch spheres with different values of J, where $0 \le J \le N/2$ for even N and J, where $1/2 \le J \le N/2$ for odd N. However, for an initial state $|J,-J\rangle$ that evolves as a pure state via a conservative Hamiltonian (e.g., the Bose-Hubbard Hamiltonian discussed in Sec. 18.5.3 to be linked to the book web page), only one Bloch sphere exists, since J can take on only the value $N/2$, and remains so if $\mathbf{J}^2$ is conserved, and the state of the system can be followed as a function of time on the Bloch sphere. Mixed states (density matrices) for an N two-level system will in general lie within the generalized Bloch sphere.

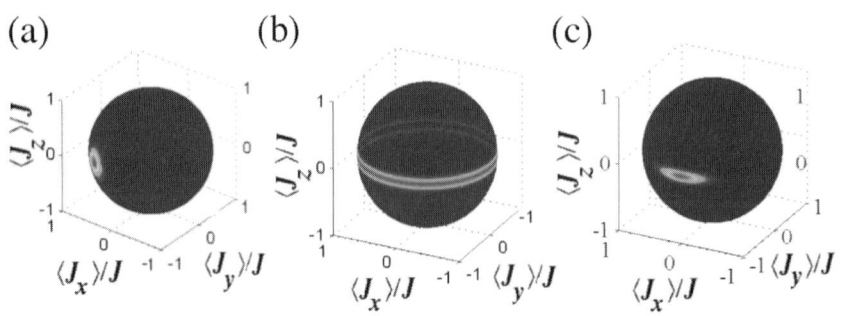

FIG 6.11 Bloch sphere representation for (a) a coherent state, (b) a fragmented, i.e., $|\uparrow\rangle^{N/2}|\downarrow\rangle^{N/2}$, and (c) a squeezed state. (For the color version of this figure, please refer the color plate section at the end of book.)

Problem 6.20

Derive Eq. (6.93) for $|\theta,\phi\rangle$, starting from the single-spin coherent state (6.13), $|\mathbf{n}\rangle = \cos(\theta/2)|\uparrow\rangle + e^{i\phi}\sin(\theta/2)|\downarrow\rangle$.

(a) Use the binomial theorem to express $|\theta,\phi\rangle = |\mathbf{n}\rangle^N$ in the form of a sum $\sum_n c_n |\uparrow\rangle^n |\uparrow\rangle^{N-n}$. I.e., determine c_n.

(b) Use the fact that $\cos^n(\theta/2) \sin^{N-n}(\theta/2) = \cos^N(\theta/2) \tan^{N-n}(\theta/2)$ to rewrite the result obtained in (a).
(c) Use the fact that $\cos^N(\theta/2) = (1 + \tan^2(\theta/2))^{-N}$ to write
$$|\theta, \phi\rangle = (1 + \tan^2(\theta/2))^{-N} \sum_{n=0}^{N} \tan^{N-n}(\theta/2) e^{i(N-n)\phi} \binom{N}{n} |\uparrow\rangle^n |\downarrow\rangle^{N-n}.$$
(c) Let $N = 2J$ and $n = J + M$, in the result derived in (c).
(d) Now use the fact that the properly normalized and symmetrized state with n spin-up and $N - n$ spin-down particles is given by $\binom{2J}{J+M}^{-1/2} |J, M\rangle$ as implied in Eq. (8.43).

6.1.10 RAMSEY FRINGE SPECTROSCOPY

Let us now study what happens to a system composed of two-level atoms subjected to two temporally separated pulses of radiation of central frequency ω close to the resonance frequency of the two levels, $\omega_0 = (E_b - E_a)/\hbar$. This technique was developed by Norman Ramsey (in 1950), who received the Nobel prize in 1989 for the invention of separated oscillatory field spectroscopy, which is sometimes called *Ramsey fringe spectroscopy* or *Ramsey double-resonance spectroscopy*. In this method, a long time-period between the application of two nearly resonant coherent fields makes the Ramsey resonance very narrow and thus suitable for high-performance atomic clocks[10] and precision measurements. The method has since become a widely used technique for determining resonance frequencies to high precision. For example, in ^{133}Cs fountain clock experiments that use this method [97], the observed linewidth is less than 1 Hz out of 9,192,631,770 Hz (this microwave hyperfine transition, $|F = 4, M_F = 0\rangle \leftrightarrow |F = 3, M_F = 0\rangle$, defines the second). As of 2005, the frequency uncertainty $\Delta \nu / \nu$ is 5×10^{-16}.

Let us first consider a single atom. For a two-level atom in an intense short near-resonant pulse with central frequency ω, the wave function can be written as, $|\psi(t)\rangle = a_g(t) \exp(-i(E_a/\hbar + \omega)t)|a\rangle + a_b(t) \exp(-iE_bt/\hbar)|b\rangle$. The Hamiltonian in the interaction representation and rotating-wave approximation takes the form given in Eq. (6.3), where $\Omega = 2\mu A/\hbar$ is the Rabi frequency, A is the slowly varying envelope of the electric field [see Eq. (7.88)], μ is the transition dipole moment, and $\Delta = (E_b - E_a - \hbar\omega_0)/\hbar$ is the detuning from resonance of the laser frequency ω. The solution of the optical Bloch equations for the two-level atom, and real envelope A, is given in terms of the unitary evolution operator for the two-level system by Eq. (6.10). In the Ramsey method, the system, initially in the ground state $|a\rangle$, is subjected to two pulses separated by a delay time T,

$$\Omega(t) = \begin{cases} \Omega & \text{if } 0 \leq t \leq \tau_p, \\ 0 & \text{if } \tau_p < t < T + \tau_p, \\ \Omega & \text{if } T + \tau_p \leq t \leq T + 2\tau_p, \end{cases} \quad (6.97)$$

with $\Omega \tau_p = \pi/2$ and $T \gg \tau_p$. From expression (6.10) for the evolution operator, it is clear that the effect of the first pulse is to evolve the initial ground state $|a\rangle$ into the superposition state $(|a\rangle + i|b\rangle)/\sqrt{2}$, i.e., the initial ground state is rotated by angle $\pi/2$ about the x-axis. In a one-particle Bloch-sphere picture with $u = \text{Re}(a_a^* a_b)$, $v = \text{Im}(a_a^* a_b)$, and $w = (|a_b|^2 - |a_a|^2)/2$, the Bloch vector (u, v, w) is projected by the first pulse into the uv plane. During the delay time between pulses, the system carries out phase oscillations, corresponding to rotation of the Bloch vector in the uv plane with frequency Δ. Finally, the second pulse rotates the vector again by an angle of $\Omega_g \tau_p$ about the u-axis. Measuring the final projection of the Bloch vector on the w-axis as a function of T for fixed detuning Δ, one obtains fringes of fixed amplitude Ω/Ω_g and frequency Δ. Alternatively, fixing T and measuring $w(t > T + 2\tau_p)$ as a function of the detuning Δ results in a power-broadened fringe pattern of amplitude Ω/Ω_g and frequency $2\pi/T$. The resulting probability to be in the excited state is given by

$$P_e = \frac{4\Omega^2}{\Omega_g^2} \sin^2\left(\frac{1}{2}\Omega_g \tau_p\right) \left(-\cos\left(\frac{1}{2}\Omega_g \tau_p\right) \cos\left(\frac{1}{2}\Delta T\right) + \frac{\Delta}{\Omega_g} \sin\left(\frac{1}{2}\Omega_g \tau_p\right) \sin\left(\frac{1}{2}\Delta T\right)\right)^2. \quad (6.98)$$

---

[10] An atomic clock is a clock that uses an electronic transition frequency of atoms (with a long-lifetime excited state) as a frequency standard for timekeeping. Atomic clocks are the most accurate time and frequency standards known. They are used to control the frequencies of television and radio broadcasts, the Global Positioning System, scientific instruments, etc.

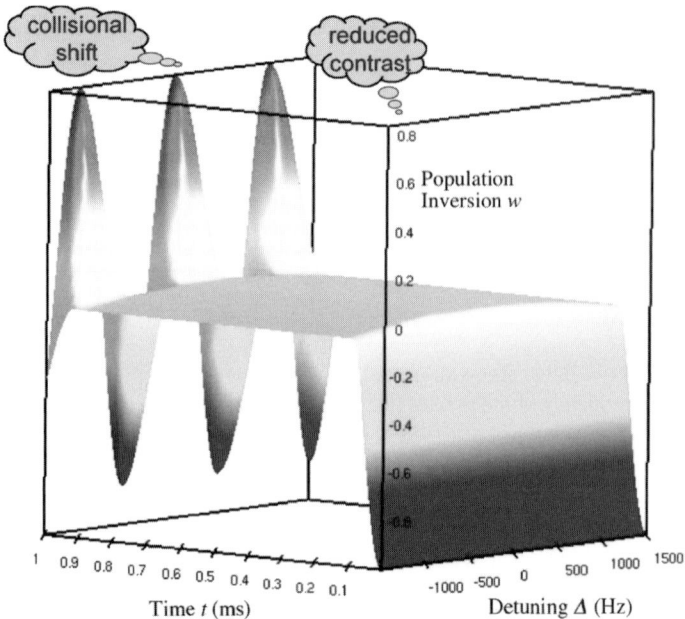

FIG 6.12 Population inversion w versus time t and detuning Δ for a Ramsey separated-field experiment. Atom–atom interactions reduce the contrast in the Ramsey fringes and shift the resonance frequency.

Figure 6.12 shows the population inversion $w = 2P_e - 1$ versus time and detuning Δ using a Ramsey separated field method for one atom in an optical lattice site. The final time corresponds to the time at which the Ramsey clock signal is measured as a function of detuning Δ, i.e., either the population of the ground or the excited state is measured as a function of Δ. Note that the excited state population at the final time is unity at zero detuning and that the population inversion oscillates as a function of detuning. Figure 6.13 shows the Ramsey fringes obtained in a cold Cs fountain clock, where cooled atoms are exposed to a vertical laser beam to toss a cloud of atoms upward in a "fountain"-like action, and then the lasers are turned off. The cloud of atoms travel upwards about a meter high through a microwave-filled cavity. Under the influence of gravity, the cloud of atoms then falls back down through the microwave cavity.

It is easy to generalize this treatment to a time-dependent Rabi frequency $\Omega(t) = 2\mu A(t)/\hbar$ due to a pulse of light which turns on and off with a finite rate, and having pulse

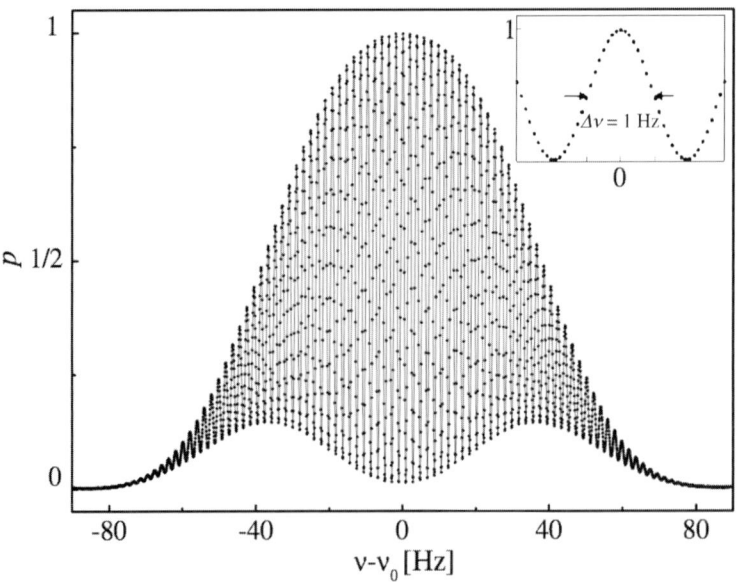

FIG 6.13 Ramsey fringes of a Cs fountain clock. The transition probability p, which is the ratio of the population of the $|F = 4, M_F = 0\rangle$ state divided by the sum of the $|F = 3, M_F = 0\rangle$ and $|F = 4, M_F = 0\rangle$ populations, plotted versus detuning. The fringe width is 1 Hz, as can be seen from the inset. Reproduced with permission from P. Lemonde et al. [97] "Cold-Atom Clocks on Earth and in Space," in Frequency Measurement and Control, Springer-Verlag, Berlin, 2001, Figure 3.

6.2 Three-Level Systems

area, $\int_0^{\tau_p} \Omega(t')dt'$. For a $\pi/2$ pulse, $\int_0^{\tau_p} \Omega(t')dt' = \pi/2$, where τ_p is the pulse duration. For the Ramsey pulse sequence, $\int_0^{\tau_p} \Omega(t')dt' = \pi/2$, $\int_{\tau_p}^{T+\tau_p} \Omega(t')dt' = 0$, and $\int_{T+\tau_p}^{T+2\tau_p} \Omega(t')dt' = \pi/2$.

In a gas containing many atoms, there can be a shift of the resonance frequency and a reduced contrast of the Ramsey oscillations as a result of atom–atom interactions, as shown in Fig. 6.12. Nevertheless, measuring the final electronic state of many atoms allows for better measurement accuracy. This can be understood as follows. The stability of an atomic clock is determined by the following parameters: (1) The width of the atomic transition frequency $\Delta\nu$ for a single atom. In Ramsey fringe spectroscopy, $\Delta\nu$ is the spectral width of a single fringe, which is equal to the inverse separation time between the pulses $1/T$. (2) The number of atoms being measured, N. (3) The number of measurement cycles which is given by the total measurement time τ of the experiment relative to the fountain cycle duration T_c. The clock uncertainty is quantified by the *Allan standard deviation* $\sigma(\tau)$, which has the following dependence on the above parameters:

$$\sigma(\tau) = \frac{\Delta\nu}{\nu_0}\left(\frac{T_c/\tau}{2\pi^2 N}\right)^{1/2}, \qquad (6.99)$$

where $\nu_0 = \omega_0/(2\pi)$ is the transition frequency. Hence, the clock uncertainty decreases as one over the square root of the number of atoms measured and square root of the number of fountain cycles used.

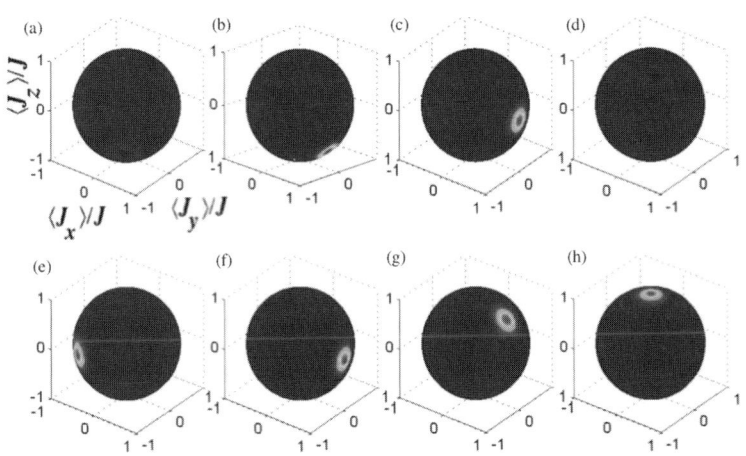

FIG 6.14 N-atom Bloch sphere representation of a Ramsey double-resonance experiment for $T\Delta = 2\pi$ and $N = 100$. The wave packet at (a) the south pole at the initial time, (b) halfway through the first $\pi/2$ pulse, (c) end of first $\pi/2$ pulse, (d)–(f) 1/3, 2/3, and 3/3 of the way through the "free-induction" (i.e., time-delay) cycle, (g) halfway through second $\pi/2$ pulse, and (h) final state at the north pole. (For the color version of this figure, please refer the color plate section at the end of book.)

The population dynamics of the N atoms can be followed on the N-atom SU(2) Bloch sphere, as shown in Fig. 6.14, where we plot the wave packet at several times during the Ramsey double-resonance experiment for the near-resonance case, $T\Delta = 2\pi$. Starting from the initial state $|J, -J\rangle$, the first $\pi/2$ pulse moves the initial wave packet to the y-axis, which then rotates in the x-y plane until it returns to the y-axis, and the final $\pi/2$ pulse moves the pulse to the final state, which is the state $|J, J\rangle$ for the detuning used. The wave packet is shown via its coloring and its widths along the three axes, $\Delta J_x/J$, $\Delta J_y/J$, and $\Delta J_z/J$, where $J = N/2$, are clearly discernible. The widths, ΔJ_i, are proportional to $\sqrt{J} = \sqrt{N/2}$, so $\Delta J_i/J \propto N^{-1/2}$. If the time delay T was such that $T\Delta = \pi$, the final pulse would wind up at south pole, i.e., the population would return to the ground state; as a function of T, $\langle J_z \rangle$ varies periodically from $+1$ to -1. If the initial state were squeezed in J_z, the clock could be made more accurate [98]. Atom–atom interaction shifts the observed clock frequency and reduces the contrast of the fringes as a function of the detuning Δ. Spontaneous emission of the excited state also reduces fringe contrast (and brings the Bloch sphere wave packet into the sphere) and widens the resonance, but does not shift the resonance frequency.

6.2 THREE-LEVEL SYSTEMS

The next level of complexity up after the two-level system is the three-level system. The three levels may correspond to the spin components of a particle with $S = 1$, bound states in a three-site potential, as shown in Fig. 6.15, or they could simply be the three lowest levels of an atom.

We begin our discussion of three-level systems by considering three-level system density matrices. The density matrix for a single *qutrit* (quantum three-level system) can be represented using the eight traceless Hermitian *Gell-Mann matrices*, λ_i,

$$\lambda_1 = \begin{pmatrix} 0 & 1 & 0 \\ 1 & 0 & 0 \\ 0 & 0 & 0 \end{pmatrix}, \quad \lambda_2 = \begin{pmatrix} 0 & -i & 0 \\ i & 0 & 0 \\ 0 & 0 & 0 \end{pmatrix}, \quad \lambda_3 = \begin{pmatrix} 1 & 0 & 0 \\ 0 & -1 & 0 \\ 0 & 0 & 0 \end{pmatrix},$$

$$\lambda_4 = \begin{pmatrix} 0 & 0 & 1 \\ 0 & 0 & 0 \\ 1 & 0 & 0 \end{pmatrix}, \quad \lambda_5 = \begin{pmatrix} 0 & 0 & -i \\ 0 & 0 & 0 \\ i & 0 & 0 \end{pmatrix}, \quad \lambda_6 = \begin{pmatrix} 0 & 0 & 0 \\ 0 & 0 & 1 \\ 0 & 1 & 0 \end{pmatrix}, \quad (6.100)$$

$$\lambda_7 = \begin{pmatrix} 0 & 0 & 0 \\ 0 & 0 & -i \\ 0 & i & 0 \end{pmatrix}, \quad \lambda_8 = \frac{1}{\sqrt{3}} \begin{pmatrix} 1 & 0 & 0 \\ 0 & 1 & 0 \\ 0 & 0 & -2 \end{pmatrix},$$

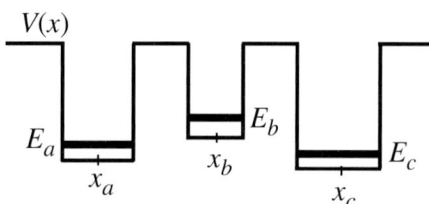

FIG 6.15 Three-square well potentials that specify a three-level system. Coupling of the levels result due to tunneling from one well into another.

which act on kets $|\psi\rangle = (b_x, b_y, b_z)^t$ that correspond to the representation of a three-level system. These matrices are directly related to the SU(3) generators T_i, $T_i \equiv \lambda_i/2$ that satisfy the commutation relations, $[T_i, T_j] = i \sum_{k=1}^{8} f_{ijk} T_k$, with parameters f_{ijk} called *structure constants* and have the property, $\text{Tr}\, \lambda_i \lambda_j = 2\delta_{ij}$ [99, 100]. The density matrix for a three-level system can be written in terms of the Gell-Mann matrices as follows:

$$\rho = \frac{1}{3}\left(1 + \frac{3}{2}\langle\lambda_i\rangle\lambda_i\right), \quad (6.101)$$

where Einstein summation notation is used. ρ is a 3×3 Hermitian matrix with unit trace and, therefore, requires eight (3 × 3 -1) independent real parameters to parameterize it. These parameters can be taken to be the eight expectation values $\langle\lambda_i\rangle$.

Problem 6.21

(a) Find the structure constants f_{ijk} such that $[\lambda_i, \lambda_j] = i\sum_{k=1}^{8} f_{ijk}\lambda_k$.
(b) Explicitly demonstrate that $\text{Tr}\, \lambda_i \lambda_j = 2\delta_{ij}$.

Answer: $f_{123} = 1, f_{147} = f_{246} = f_{257} = f_{345} = -f_{156} = -f_{367} = 1/2, f_{458} = f_{678} = \sqrt{3}/2$.

6.2.1 TWO OR MORE THREE-LEVEL CORRELATED SYSTEMS

We can easily generalize to two or more three-level systems. A three-level bipartite density matrix can be parameterized in the form

$$\rho_{AB} = \frac{1}{9}\left[\left(1 + \frac{3}{2}\langle\lambda_{i,A}\rangle\lambda_{i,A}\right)\left(1 + \frac{3}{2}\langle\lambda_{i,B}\rangle\lambda_{i,B}\right) + \lambda_{i,A}C_{ij}\lambda_{j,B}\right], \quad (6.102)$$

where the Einstein summation convention is used and

$$C_{ij} = \frac{9}{4}\left(\langle\lambda_{i,A}\lambda_{j,B}\rangle - \langle\lambda_{i,A}\rangle\langle\lambda_{j,B}\rangle\right) \quad (6.103)$$

6.2 Three-Level Systems

is the correlation coefficient between the operators $\lambda_{i,A}$ and $\lambda_{j,B}$. The two-particle density matrix ρ_{AB} can be represented as a 9×9 Hermitian matrix with trace unity, so 80 ($9 \times 9 - 1$) parameters are required to parameterize it. The eight components of $\langle \lambda_{i,A} \rangle$, eight components of $\langle \lambda_{i,B} \rangle$, and 64 components of C_{ij} are sufficient for this purpose. The three-level three-particle density matrix can be parameterized similarly, in a fashion analogous to the three-particle qubit case:

$$\rho_{ABC} = \frac{1}{27}[\prod_I \left(1 + \frac{3}{2}\sum_i \langle \lambda_{i,I}\rangle \lambda_{i,I}\right) + \sum_{I,J}\sum_{i,j} \lambda_{i,I} C_{ij,IJ}\lambda_{j,J}$$

$$+ \sum_{I,J,K}\sum_{i,j,k} \lambda_{i,A}\lambda_{j,B}\lambda_{k,C} D_{ijk,IJK}], \tag{6.104}$$

and

$$D_{ijk,IJK} = \frac{27}{8}(\langle \lambda_{i,I}\lambda_{j,J}\lambda_{k,K}\rangle - \langle \lambda_{i,I}\rangle\langle \lambda_{j,J}\rangle\langle \lambda_{k,K}\rangle). \tag{6.105}$$

There are alternative parameterizations of the density matrix of three-level (and more generally, n-level) systems. For example, a one-particle three-level system can be parameterized in terms of its dipole- and quadrupole-moment tensor operator components, $T_q^{(1)}$ and $T_q^{(2)}$ (see Sec. 3.5), can be used in the expansion of the density matrix [36, 37], instead of the operators λ_i; the expectation values $\rho_q^{(k)}$ [see Eqs (3.178)–(3.180) in Sec. 3.5] can be used instead of the eight expectation values $\langle \lambda_i\rangle$, $i = 1,\ldots,8$. Since a dipole-moment $T_q^{(1)}$ has three components and a quadrupole $T_q^{(2)}$ has five components, the dipole and quadrupole tensor operators are sufficient. For a two-particle three-level system, the 64 components of C_{ij} can be replaced by the 64 expectation values of the irreducible tensor operators that can be formed using two particles each having $J = 1$ [36, 37].

To gain experience with qutrits, consider the two-qutrit pure states

$$|\psi^{(1)}\rangle = \frac{1}{\sqrt{2}}(|+1\rangle_A|-1\rangle_B + |-1\rangle_A|+1\rangle_B),$$

$$|\psi^{(2)}\rangle = \frac{1}{\sqrt{2}}(|+1\rangle_A|+1\rangle_B + |-1\rangle_A|-1\rangle_B). \tag{6.106}$$

Note that the single-particle states $|+1\rangle$, $|0\rangle$, and $|-1\rangle$ with projection of angular momentum $+1, 0,$ and -1 along the z-axis are given in terms of the cartesian basis states $|v_1\rangle \equiv (1,0,0)^t$, $|v_2\rangle \equiv (0,1,0)^t$, and $|v_3\rangle \equiv (0,0,1)^t$ as follows: $|+1\rangle = \frac{1}{\sqrt{2}}(1,i,0)^t$, $|-1\rangle = \frac{1}{\sqrt{2}}(1,-i,0)^t$, and $|0\rangle = (0,0,1)^t$. The states in Eq. (6.106) look somewhat like the Bell states $|\Psi^+\rangle = \frac{1}{\sqrt{2}}[|\uparrow\downarrow\rangle + |\downarrow\uparrow\rangle]$ and $|\Phi^+\rangle = \frac{1}{\sqrt{2}}[|\uparrow\uparrow\rangle + |\downarrow\downarrow\rangle]$, respectively. However, the density matrix in the form of Eq. (6.102) for these pure states have nonvanishing parameters $\langle \lambda_{8,A}\rangle = \langle \lambda_{8,B}\rangle = 1/\sqrt{3}$. The C_{ij} matrices for these states are given by

$$C_{ij}^{(1)} = \frac{9}{4}\begin{pmatrix} 1 & 0 & 0 & 0 & 0 & 0 & 0 & 0 \\ 0 & -1 & 0 & 0 & 0 & 0 & 0 & 0 \\ 0 & 0 & 1 & 0 & 0 & 0 & 0 & 0 \\ 0 & 0 & 0 & 0 & 0 & 0 & 0 & 0 \\ 0 & 0 & 0 & 0 & 0 & 0 & 0 & 0 \\ 0 & 0 & 0 & 0 & 0 & 0 & 0 & 0 \\ 0 & 0 & 0 & 0 & 0 & 0 & 0 & 0 \\ 0 & 0 & 0 & 0 & 0 & 0 & 0 & 0 \end{pmatrix}, \quad C_{ij}^{(2)} = \frac{9}{4}\begin{pmatrix} -1 & 0 & 0 & 0 & 0 & 0 & 0 & 0 \\ 0 & 1 & 0 & 0 & 0 & 0 & 0 & 0 \\ 0 & 0 & 1 & 0 & 0 & 0 & 0 & 0 \\ 0 & 0 & 0 & 0 & 0 & 0 & 0 & 0 \\ 0 & 0 & 0 & 0 & 0 & 0 & 0 & 0 \\ 0 & 0 & 0 & 0 & 0 & 0 & 0 & 0 \\ 0 & 0 & 0 & 0 & 0 & 0 & 0 & 0 \\ 0 & 0 & 0 & 0 & 0 & 0 & 0 & 0 \end{pmatrix}. \tag{6.107}$$

These states are not maximally entangled. Maximally entangled two-qutrit states can take the form $\frac{1}{\sqrt{3}}(\pm|v_1\rangle|v_1\rangle \pm |v_2\rangle|v_2\rangle \pm |v_3\rangle|v_3\rangle)$. These states have $\langle \lambda_i\rangle = 0$ for all i and diagonal correlation matrices C_{ij}.

6.3 CLASSIFICATION OF CORRELATION AND ENTANGLEMENT

In this section, we categorize classically correlated and quantum-correlated states (see the Werner definition in Sec.5.2.2) in terms of the correlation matrix **C** [see Eq. (6.82) for the two qubit case, or Eq. (6.102) for the two qutrit case]. We shall do so in terms of the number of nonzero singular values [57, 279–281], $\{d_i\}$, of the correlation matrix **C** (recall the singular value decomposition of a matrix $C = U d V^\dagger$, where U and V are orthogonal and d is diagonal). Entanglement is harder to characterize; the entanglement of bipartite qubit systems can be characterized using the Peres–Horodecki criterion [59, 60] (see Sec. 5.2.2). It would be useful to reformulate this criterion in terms of the parameters used in forming the density matrix and interpret the result physically, but this is not easy to do.

The correlation matrix **C** quantifies the correlation of bipartite states. A bipartite correlation measure for an n-level and m-level system is based on the $(n^2 - 1) \times (m^2 - 1)$ correlation matrix **C** as follows:

$$\mathcal{E}_C \equiv \frac{n_<^2}{4(n_<^2 - 1)} \operatorname{Tr} \mathbf{C}\mathbf{C}^T = \frac{n_<^2}{4(n_<^2 - 1)} \sum_{i,j} C_{ij} C_{ji}^T. \quad (6.108)$$

Here, $n_< = \min(n, m)$. $\mathcal{E}_C = \frac{n_<^2}{4(n_<^2-1)} \operatorname{Tr}(\rho_{AB} - \rho_A \rho_B)^2$ is a nonnegative real number. If **C** is a normal matrix [57, 279–281], $\operatorname{Tr} \mathbf{C}\mathbf{C}^T$ equals the sum of the squares of its eigenvalues. But **C** need not be normal, and if not, the trace equals the sum of the squares of $\{d_i\}$. $\mathcal{E}_C$ is basis independent, hence any rotation in Hilbert space leaves it unchanged. The normalization factor $n_<^2/[4(n_<^2 - 1)]$ in Eq. (6.108) ensures that the maximum possible value of $\mathcal{E}_C$ is unity. $\mathcal{E}_C$ measures both classical and quantum correlation.

Let us explicitly consider two qubits. Two-qubit classically correlated states take the form

$$\rho^{CC} = \frac{1}{4} \sum_{k \geq 2} p_k (1 + \mathbf{n}_{A,k} \cdot \boldsymbol{\sigma}_A)(1 + \mathbf{n}_{B,k} \cdot \boldsymbol{\sigma}_B), \quad (6.109)$$

with $\sum_k p_k = 1$ and $p_k > 0$. This density matrix can be written in the form of Eq. (6.82) with Bloch vectors

$$\mathbf{n}_A = \sum_k p_k \mathbf{n}_{A,k}, \quad \mathbf{n}_B = \sum_k p_k \mathbf{n}_{B,k}, \quad (6.110)$$

and correlation matrix

$$C_{ij} = \sum_k p_k n_{i,A,k} \left[n_{j,B,k} - \sum_l p_l n_{j,B,l} \right]. \quad (6.111)$$

For example, for classically correlated mixed states of the form

$$\rho^{CC} = (2 \operatorname{sech}^2(2\theta))^{-1/2} (e^{-\theta} |\downarrow\uparrow\rangle \langle\downarrow\uparrow| + e^{\theta} |\uparrow\downarrow\rangle \langle\uparrow\downarrow|),$$

we find that all the correlation coefficients vanish, except for $C_{zz} = -\operatorname{sech}^2(2\theta)$, the density matrix in representation (6.82) is $\rho^{CC} = \frac{1}{4}\left(1_A 1_B - \operatorname{sech}^2(2\theta) \sigma_{z,A} \sigma_{z,B}\right)$, and the classical-correlation measure is $\mathcal{E}_C^{CC} = \frac{1}{3} \operatorname{sech}^4(2\theta)$.

For pure two-qubit states, the number of nonzero singular values (NSVs) of **C** is zero for unentangled states (**C** vanishes) and three for entangled states. For classically correlated states with two terms in the sum [see Eq. (6.109)], only one NSV occurs, two NSVs occur for three terms, three NSVs occur for four or more terms, and for entangled (i.e., quantum-correlated) mixed states, there are three NSVs. These cases are summarized in Fig. 6.16. If the number of NSVs is less than or equal to two, the state is not entangled; only states with three NSVs can be entangled, but further tests to determine if they are entangled are required. Entangled mixed states can be differentiated from classically correlated states with three NSVs by applying the Peres–Horodecki partial transposition condition [59] to the density matrices with three NSVs [which corresponds to changing the sign of $n_{y,B}$ and the matrix elements C_{iy}^{AB} that multiply $\sigma_{y,B}$ in Eq. (6.82) and determining whether the resulting ρ is still a genuine density matrix – if it is, the state is classically

6.3 Classification of Correlation and Entanglement

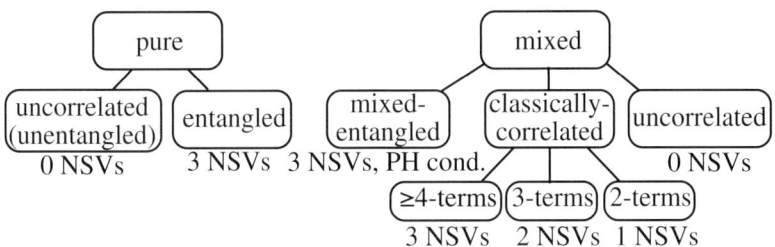

FIG 6.16 Classification of two-qubit states. Categories can be experimentally distinguished by measuring $\mathbf{n}_A$, $\mathbf{n}_B$, and using Bell measurements [52] to determine the $\mathbf{C}$ matrix.

correlated, i.e., unentangled but classically correlated]. The *only* categories that cannot be distinguished without use of the Peres–Horodecki condition are the mixed entangled and the classically correlated states with three NSVs.

An eye-opening example involves the Werner two-qubit density matrix composed of a sum of the singlet state $|\Psi^-\rangle$ and the maximally mixed state $\mathbf{1}/4$, $\rho^W_{\Psi^-} = p|\Psi^-\rangle\langle\Psi^-| + \frac{1-p}{4}\mathbf{1}$, or, the more general Werner (GW) two-qubit density matrix,

$$\rho^{GW}_{\psi^-} = p|\psi^-\rangle\langle\psi^-| + \frac{1-p}{4}\mathbf{1}, \tag{6.112}$$

where $|\psi^-\rangle = (2\cosh(2\theta))^{-1/2}(e^{-\theta}|\uparrow\downarrow\rangle - e^{\theta}|\downarrow\uparrow\rangle)$. ρ^{GW} reduces to ρ^W for $\theta = 0$. For $\rho^{GW}_{\psi^-}$,

$$\mathbf{n}_A = -\mathbf{n}_B = -p\tanh(2\theta)\,\hat{\mathbf{z}} \tag{6.113}$$

and

$$\mathbf{C}^{GW}_{\psi^-} = -p\begin{pmatrix} \mathrm{sech}(2\theta) & 0 & 0 \\ 0 & \mathrm{sech}(2\theta) & 0 \\ 0 & 0 & 1-p+p\,\mathrm{sech}^2(2\theta) \end{pmatrix}. \tag{6.114}$$

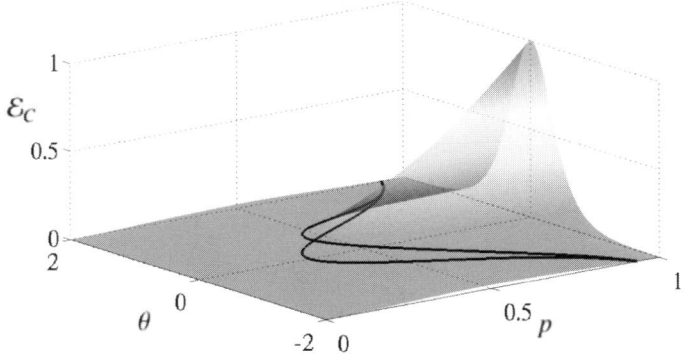

FIG 6.17 $\mathcal{E}_C(p,\theta)$ versus p and θ for the generalized Werner density matrix ρ^{GW} and the Peres–Horodecki entanglement criterion limit, $p[1 + 2\,\mathrm{sech}(2\theta)] = 1$, drawn on the p-θ plane and projected onto the $\mathcal{E}_C$ surface. The states within the region $p[1 + 2\,\mathrm{sech}(2\theta)] > 1$ are entangled.

The Peres–Horodecki entanglement criterion [59] shows that this state is entangled if $p[(1 + 2\,\mathrm{sech}(2\theta)] \geq 1$. Figure 6.17 plots the Peres–Horodecki criterion limit and the correlation measure, $\mathcal{E}_C(p,\theta) = p^2[1 - p + (2+p)\,\mathrm{sech}^2(2\theta)]$, for the generalized Werner state. Note that the Peres–Horodecki criterion is not obtainable from $\mathbf{C}$ alone but can be obtained using the invariant parameters $\xi \equiv \sum_i e_i - \frac{\mathbf{n}_A \cdot \mathbf{C} \cdot \mathbf{n}_B}{\mathbf{n}_A \cdot \mathbf{n}_B}$ and $\mathbf{n}_A \cdot \mathbf{n}_B$. More explicitly, $p[1 + 2\,\mathrm{sech}(2\theta)] = -\xi + \sqrt{\xi^2/4 - \mathbf{n}_A \cdot \mathbf{n}_B}$, so the Peres–Horodecki condition reads

$$-\frac{\xi}{2} + \frac{-\xi + \sqrt{\xi^2 - 4\,\mathbf{n}_A \cdot \mathbf{n}_B}}{2} \geq 1, \tag{6.115}$$

Unfortunately, the physical significance of the Peres–Horodecki entanglement criterion is not yet clear.

Criteria for determining the entanglement of two qutrits do not exist. Moreover, multipartite (≥ 3 particles) entanglement is also not well understood. Given the importance of entanglement as an information resource, these are extremely important unsolved problems.

Problem 6.22

(a) Determine $\mathbf{n}_A$, $\mathbf{n}_B$, and $\mathbf{C}$ for the generalized Werner state $\rho_{\phi^-}^{GW} = p|\phi^-\rangle\langle\phi^-| + \frac{1-p}{4}\mathbf{1}$ where
$|\phi^-\rangle = (2\cosh(2\theta))^{-1/2}(e^{-\theta}|\uparrow\uparrow\rangle - e^{\theta}|\downarrow\downarrow\rangle)$.

(b) Determine $\mathbf{n}_A$, $\mathbf{n}_B$, and $\mathbf{C}$ for the generalized Werner states $\rho_{\psi^\pm}^{GW} = p|\psi^\pm\rangle\langle\psi^\pm| + \frac{1-p}{4}\mathbf{1}$ where
$|\psi^\pm\rangle = (2\cosh(2\theta))^{-1/2}(e^{-\theta}|\uparrow\downarrow\rangle \pm e^{\theta}|\downarrow\uparrow\rangle)$.

6.3.1 ENTANGLEMENT WITNESS OPERATORS

Equation (5.39) defined the notion of a separable state, and through it, an entangled state was defined by Werner as one that is not separable. The importance of distinguishing between entangled and unentangled states is evident, given the central role played by entanglement in quantum information science, and in elucidating fundamental aspects of quantum mechanics in general. It would be useful to construct operators designed to distinguish between separable and entangled states; such operators are referred to as *entanglement witness operators*. It might seem natural to base the search for entanglement witness operators on the criterion of whether states violate the CHSH version of the Bell inequality, Eq. (5.156) (hereafter referred to as the CHSH inequality). Separable states do not violate these inequalities, and it was previously believed that all entangled (nonseparable) states violate the CHSH inequality, but this is in fact not the case, as was shown by Werner [58] who gave a counter example.

Although it can be proven that an entanglement witness operator always exists, its actual identification and construction is not always easy. Particularly since the intuitive assumption that all entangled states violate a Bell inequality is incorrect and one cannot simply use the CHSH inequality as a bona fide criterion. A useful tool in the quest for constructing entanglement witness operators is related to the concept of partial transpose, defined in Eq. (5.41). Separable states have positive semidefinite partial transpose; hence, all states that do not have positive semidefinite partial transpose are not separable. In other words, positive semidefinite partial transpose of a density operator is a necessary condition for separability; for two qubits or a qubit and qutrit, this condition is also sufficient.

A thorough analysis of entanglement witness operators goes beyond the scope of this book; hence, we shall only define them here. For a bipartite system with Hilbert space $\mathcal{H} = \mathcal{H}_A \otimes \mathcal{H}_B$, we denote by $\mathcal{S}$ the set of all density operators $\{\rho_S\}$ that admit a separable form, (5.39), and by $\mathcal{E}$, the set of nonseparable density operators that are entangled by Werner's criterion, assuming $\mathcal{E} \neq \emptyset$. The definition of an entanglement witness operator $\mathcal{W}$ is

A Hermitian operator $\mathcal{W}$ acting in $\mathcal{H}$ is an entanglement witness if:

(1) $\text{Tr}[\mathcal{W}\rho_S] \geq 0 \ \forall \rho_S \in \mathcal{S}$.
(2) There exists at least one state $\rho_\mathcal{E} \in \mathcal{E}$ for which $\text{Tr}[\mathcal{W}\rho_\mathcal{E}] < 0$.

If there exists a separable density operator $\rho_S \in \mathcal{S}$ for which condition (1) is an equality, $\text{Tr}[\mathcal{W}\rho_S] = 0$, then $\mathcal{W}$ is said to be an *optimal entanglement witness*, $\mathcal{W}_{\text{opt}}$.

Note that all states $\{\rho_S\}$ have non-negative trace with $\mathcal{W}$, $\text{Tr}[\mathcal{W}\rho_S] \geq 0$, but not necessarily all states $\{\rho_\mathcal{E}\}$ have negative trace with $\mathcal{W}$. In other words, $\text{Tr}[\mathcal{W}\rho] \geq 0$ is a necessary condition for separability, whereas $\text{Tr}[\mathcal{W}\rho] < 0$ is a sufficient condition for nonseparabilty (entanglement). We say that an entangled density operator $\rho_\mathcal{E}$ for which $\text{Tr}[\mathcal{W}\rho_\mathcal{E}] < 0$ is "detected by the entanglement witness operator" $\mathcal{W}$. The existence of an entanglement witness operator is guaranteed by a theorem stating that, if $\rho \in \mathcal{E}$, then there exists an entanglement witness $\mathcal{W}$ such that $\text{Tr}[\mathcal{W}\rho_S] \geq 0$ for all $\rho_S \in \mathcal{S}$, and $\text{Tr}[\mathcal{W}\rho] < 0$.

In addition to entanglement witness operators, one can also define *CHSH witness operators*, which are designed to distinguish between states that satisfy the CHSH inequality (including also entangled states) and those who do not. A (hermitian) operator $\mathcal{U}$ in this class satisfies $\text{Tr}[\mathcal{U}\rho_{\text{CHSH}}] \geq 0$ for all states ρ_{CHSH} that obey the CHSH inequality. It has

6.4 THREE-LEVEL SYSTEM DYNAMICS

been shown [101] that a CHSH witness operator is obtained by shifting an optimal entanglement witness operator by a mere constant, $\mathcal{U} = \mathcal{W}_{\text{opt}} + \gamma \mathbf{I}$, where $\gamma > 0$ is confined to a finite interval on the positive real axis.

6.4 THREE-LEVEL SYSTEM DYNAMICS

Many quantum processes involve three levels. Quantum optical processes of this type include stimulated Raman scattering and two-photon absorption [Sec. 7.4.3], stimulated Raman adiabatic passage (STIRAP) [Sec. 7.8.2], electromagnetically induced transparency, coherent population trapping, slow light, and lasing without inversion; all these processes fall into the category of the interaction of a three-level system with two laser fields. We shall set up the formalism to treat these phenomena here. We first consider a Λ-system configuration with the ground state 1 and first excited state 3 being optically coupled to a highly excited state 2 [see Fig. 6.18(a)]. A V-system, with lower state, 2, coupled to two upper states, 1 and 3, is shown in Fig. 6.18(b), and a two-photon absorption system, wherein level 1 is the ground state, level 2 is the first excited state, and level 3 is the second excited state reachable from level 2 by absorption of a photon (of a different frequency), is depicted in Fig. 6.18(c).

Just like in the treatment of a two-level system, we can write the time-dependent wave function in the form

$$\Psi(\mathbf{r},t) = a_1(t)\phi_1(\mathbf{r}) + a_2(t)\phi_2(\mathbf{r}) + a_3(t)\phi_3(\mathbf{r}), \qquad (6.116)$$

where $a_i(t)$ is the time-dependent amplitude for state i. For the level structure in Fig. 6.18(a) or its dressed state counterpart 6.19(a), the Hamiltonian can be written as

$$H = \hbar \begin{pmatrix} 0 & \Omega^*(t)/2 & 0 \\ \Omega(t)/2 & \Delta & \Omega'(t)/2 \\ 0 & \Omega'^*(t)/2 & \delta \end{pmatrix}, \qquad (6.117)$$

where we used the rotating-wave approximation to eliminate quickly oscillating terms, and we allowed for the possibility that the Rabi frequencies are complex and time dependent. Let us also allow state 2 to decay with rate γ by adding a phenomenological decay term to the rate equation for a_2. Then, the time dependent Schrödinger equation for the amplitudes $a_i(t)$ is

$$i\hbar \frac{d}{dt} \begin{pmatrix} a_1 \\ a_2 \\ a_3 \end{pmatrix} = \hbar \begin{pmatrix} 0 & \Omega^*(t)/2 & 0 \\ \Omega(t)/2 & \Delta & \Omega'(t)/2 \\ 0 & \Omega'^*(t)/2 & \delta \end{pmatrix} \begin{pmatrix} a_1 \\ a_2 \\ a_3 \end{pmatrix} - \frac{i\gamma}{2} \begin{pmatrix} 0 \\ a_2 \\ 0 \end{pmatrix}. \qquad (6.118)$$

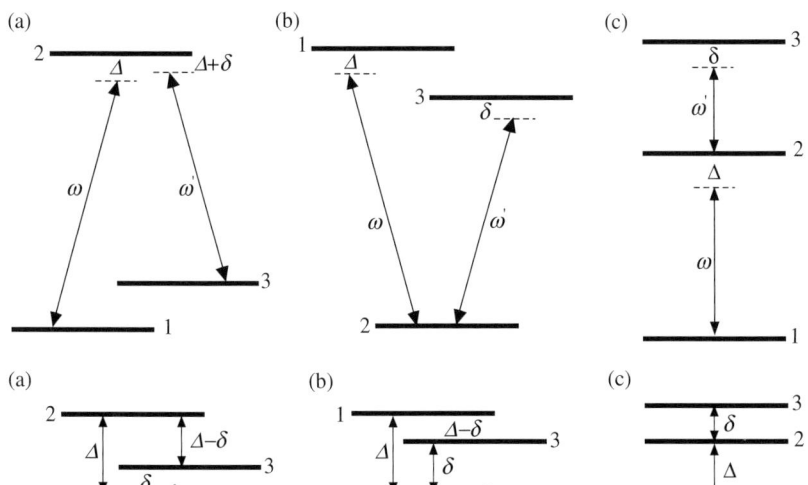

FIG 6.18 Optically coupled three level systems. (a) Λ-system, (b) V-system, (c) ladder-system (two-photon absorption or emission). Reproduced with permission from Band, Light and Matter, Wiley, Figure 9.7, p. 537.

FIG 6.19 Dressed three-level systems. (a) Λ system, (b) V system, (c) ladder (two-photon absorption or emission) system.

The rate of change of a_2 will eventually come into steady state if the rates of change of the Rabi frequencies are much slower than the decay rate γ. The steady state value of a_2 can then be solved by setting $da_2/dt = 0$:

$$a_2 = -\left(\frac{\Omega(t)}{2(\Delta - i\gamma)}a_1(t) + \frac{\Omega'(t)}{2(\Delta - i\gamma)}a_3(t)\right). \tag{6.119}$$

Substituting a_2 from Eq. (6.119) into the differential equations for a_1 and a_3, we, thereby, adiabatically eliminate a_2 and obtain the equations of motion for a_1 and a_3:

$$i\hbar\frac{d}{dt}\begin{pmatrix}a_1\\a_3\end{pmatrix} = \hbar\begin{pmatrix}\frac{|\Omega(t)|^2}{4(\Delta-i\gamma)} & \frac{\Omega^*(t)\Omega'(t)}{4(\Delta-i\gamma)}\\ \frac{\Omega(t)\Omega'^*(t)}{4(\Delta-i\gamma)} & \frac{|\Omega(t)|^2}{4(\Delta-i\gamma)}+\gamma\end{pmatrix}\begin{pmatrix}a_1\\a_3\end{pmatrix}. \tag{6.120}$$

The Hamiltonian on the RHS of Eq. (6.120) is not Hermitian. This is the price we pay for not adequately treating decay in (6.118); a density matrix treatment is necessary for a more correct treatment. Taking the limit as $\gamma \ll \Delta$, we obtain the effective two-level Hamiltonian

$$H = \hbar\begin{pmatrix}\frac{|\Omega(t)|^2}{4\Delta} & \frac{\Omega^*(t)\Omega'(t)}{4\Delta}\\ \frac{\Omega(t)\Omega'^*(t)}{4\Delta} & \frac{|\Omega'(t)|^2}{4\Delta}+\gamma\end{pmatrix}. \tag{6.121}$$

The effective Rabi frequency coupling levels 1 and 3 is proportional to the product of the Rabi frequencies and the inverse of the detuning, $\Omega_{\text{eff}} = \frac{\Omega^*(t)\Omega'(t)}{4\Delta}$.

Problem 6.23

(a) Carry out the algebra to show that Eq. (6.119) is obtained when γ is large.
(b) Carry out the algebra leading to Eqs (6.120) and (6.121).

The Hamiltonian corresponding to the two-photon absorption-level structure in Fig. 6.18(c), or its dressed state counterpart in Fig. 6.19(c), is

$$H = \hbar\begin{pmatrix}0 & \Omega^*(t)/2 & 0\\ \Omega(t)/2 & \Delta & \Omega'^*(t)/2\\ 0 & \Omega'(t)/2 & \Delta+\delta\end{pmatrix}. \tag{6.122}$$

Note the differences relative to Eq. (6.117). Both transitions, $1 \to 2$ and $2 \to 3$, specify absorption processes, and the energy difference between 1 and 3 is $\Delta + \delta$. The Hamiltonian corresponding to the V-level structure in Fig. 6.18(b) or its dressed state counterpart 6.19(b) is

$$H = \hbar\begin{pmatrix}0 & \Omega(t)/2 & 0\\ \Omega^*(t)/2 & \Delta & \Omega'^*(t)/2\\ 0 & \Omega'(t)/2 & \delta\end{pmatrix}. \tag{6.123}$$

There are many phenomena at the forefront in research in quantum optics that involve the dynamics of three-level systems. These phenomena are best described by a density matrix description of three-level systems, just as two-level phenomena, e.g., self-induced transparency, Ramsey separated-field spectroscopy, etc., are best described by a density matrix formalism. The three-level density matrix $\rho(t)$ satisfies the Liouville–von Neumann density matrix equation, $\frac{d}{dt}\rho(t) = \frac{-i}{\hbar}[H(t), \rho(t)] - \Gamma\rho(t)$, where the decay matrix for the three-level system can be easily formed in a fashion similar to the two-level case. We will not pause to develop all the interesting and important topics mentioned above using the density matrix method. The interested reader is referred to the density matrix formalism description in Refs. [18, 30, 102, 103].

6.5 CONTINUOUS-VARIABLE SYSTEMS

Previously in this chapter, we studied the correlations of two or more two-level systems, and two or more three level systems. We can continue to study correlations of higher level systems in roughly the same way, but correlations between continuous variable systems are quite different from those of N-level systems. In this section, we consider correlations in continuous variable systems.

A simple continuous-variable 1D one-particle system is characterized by the 1D density matrix $\rho(x, x')$, whose Fourier transform is the Wigner function,

$$W(p, x) = (2\pi\hbar)^{-1} \int du \, e^{ipu/\hbar} \rho(x - u/2, x + u/2). \tag{6.124}$$

The Wigner function [hence also the density matrix $\rho(x, x')$] can be specified fully in terms of the characteristic function $\varphi(\tau_1, \tau_2)$
defined by

$$\varphi(\tau_1, \tau_2) \equiv \langle e^{i(\tau_1 x + \tau_2 p)} \rangle = \int \int dx \, dp \, W(p, x) \, e^{i(\tau_1 x + \tau_2 p)}. \tag{6.125}$$

The expectation value of products of position and momentum operators can be determined using the characteristic function as follows:

$$\langle \ldots x^{n_3} p^{n_2} x^{n_1} \rangle = (-i)^{n_1 + n_2 + n_3 + \ldots} \ldots \frac{\partial^{n_3}}{\partial \tau_1^{n_3}} \frac{\partial^{n_2}}{\partial \tau_2^{n_2}} \frac{\partial^{n_1}}{\partial \tau_1^{n_1}} \varphi(\tau_1, \tau_2) \bigg|_{\tau_1 = 0, \tau_2 = 0}. \tag{6.126}$$

Furthermore, the characteristic function can be obtained if all the above expectation values are known,

$$\varphi(\tau_1, \tau_2) = \sum_{n_1, n_2, n_3, \ldots} \frac{i^{n_1} i^{n_2} i^{n_3} \ldots}{n_1! n_2! n_3! \ldots} \ldots \tau_1^{n_3} \tau_2^{n_2} \tau_1^{n_1} \langle \ldots x^{n_3} p^{n_2} x^{n_1} \rangle. \tag{6.127}$$

The above formalism is straightforwardly extended to N-particle states that are functions of continuous variables, although the expressions become more complex with increasing N. Two-particle states are characterized by the density matrix $\rho(x_1, x_2, x_1', x_2')$, whose Fourier transform is the Wigner function $W(p_1, p_2, x_1, x_2)$, in direct analogy with Eq. (6.124). The characteristic function $\varphi(\tau_{11}, \tau_{12}, \tau_{21}, \tau_{22})$ can be defined by

$$\varphi(\tau_{11}, \tau_{12}, \tau_{21}, \tau_{22}) \equiv \langle e^{i(\tau_{11} x_1 + \tau_{12} x_2 + \tau_{21} p_1 + \tau_{22} p_2)} \rangle$$

$$= \int dx_1 \, dx_2 \, dp_1 \, dp_2 \, W(p_1, p_2, x_1, x_2) \, e^{i(\tau_{11} x_1 + \tau_{12} x_2 + \tau_{21} p_1 + \tau_{22} p_2)}. \tag{6.128}$$

It can be expanded in terms of mean values as follows:

$$\varphi(\tau_{11}, \tau_{12}, \tau_{21}, \tau_{22}) = \sum_{n_1, n_2, n_3, n_4, \ldots} \frac{i^{n_1} i^{n_2} i^{n_3} i^{n_3} i^{n_4} \ldots}{n_1! n_2! n_3! n_4! \ldots} \ldots \tau_{22}^{n_4} \tau_{21}^{n_3} \tau_{12}^{n_2} \tau_{11}^{n_1} \langle \ldots p_2^{n_4} p_1^{n_3} x_2^{n_2} x_1^{n_1} \rangle. \tag{6.129}$$

For example, the correlation function of the position of the two particles is given in terms of the characteristic function $\varphi(\tau_{11}, \tau_{12}, \tau_{21}, \tau_{22})$ as

$$\langle x_2^{n_2} x_1^{n_1} \rangle = (-i)^{n_1 + n_2} \frac{\partial^{n_2}}{\partial \tau_{12}^{n_2}} \frac{\partial^{n_1}}{\partial \tau_{11}^{n_1}} \varphi(\tau_{11}, \tau_{12}, \tau_{21}, \tau_{22}) \bigg|_{\tau_{11} = 0, \tau_{12} = 0, \tau_{21} = 0, \tau_{12} = 0}. \tag{6.130}$$

The quantities in (6.130) are the generalizations of the correlation matrix $C_{ij,AB}$ in (6.83) for two qubits, and the quantities in (6.126) are the generalizations of the vectors $\mathbf{n}_A$ and $\mathbf{n}_B$ that appear in the density matrix parameterization (6.82) for two qubits.

Now, let us consider two particles in three dimensions. If no external force acts on the two particles (or if the external potential is harmonic and isotropic), the Hamiltonian can be written in terms of a sum of the Hamiltonians for the center-of-mass and relative variables, $H(\mathbf{p}_1, \mathbf{p}_2, \mathbf{r}_1, \mathbf{r}_2) = H(\mathbf{P}, \mathbf{X}) + H(\mathbf{p}, \mathbf{x})$. Then, if there is no correlation between the center-of-mass and relative variables, the density matrix can be written in product form, $\rho(\mathbf{r}_1, \mathbf{r}_2, \mathbf{r}'_1, \mathbf{r}'_2) = \rho(\mathbf{x}, \mathbf{x}') \mathcal{R}(\mathbf{X}, \mathbf{X}')$, where $\rho(\mathbf{x}, \mathbf{x}')$ and $\mathcal{R}(\mathbf{X}, \mathbf{X}')$ are the density matrices for the relative and center-of-mass degrees of freedom. Then, each of the density matrices $\rho(\mathbf{x}, \mathbf{x}')$ and $\mathcal{R}(\mathbf{X}, \mathbf{X}')$ can be expanded in terms of irreducible representations of the rotation group, as described in Sec. 3.5.1, using the irreducible tensor basis functions $T_q^{(k)}(\alpha, \alpha', J, J')$ defined in (3.179). This type of simple factorizable representation is not possible for a two-particle system subjected to an external force, such as two electrons bound to a nucleus, or even two particles in a box potential; in this case, the center-of-mass degrees of freedom cannot be separated out from the relative motion degree of freedom.

We have already seen that for a general bipartite system (even one having continuous variables), it is useful to define the two-body correlated piece of the density matrix, $\rho_{\text{corr}} \equiv \rho - \rho_A \rho_B$, which, together with the quantities, $\rho_A \equiv \text{Tr}_B \, \rho$, $\rho_B \equiv \text{Tr}_A \, \rho$, can be used to write the density matrix as

$$\rho = \rho_A \rho_B + \rho_{\text{corr}}. \tag{6.131}$$

The two-body correlated part of the density matrix, ρ_{corr}, does not contribute to expectation values of any quantity involving only one of the two particles. It contributes *only* to the correlation between the two particles, whereas the term $\rho_A \rho_B$ does not contribute to correlation. Note that the breakup of the density matrix in Eq. (6.131) is similar to the decomposition in Eqs (6.82) and (6.102). But, how do we parametrize ρ_{corr} in a way that will allow easy extraction of the physics of the correlation between the particles when continuous variables are involved? Such a parameterization should generalize the bipartite parameterization of qubits and qutrits in Eqs (6.82) and (6.102), respectively, and adapt it to the continuous variable case. Since expectation values of the irreducible operators can be used to parameterize ρ_A and ρ_B, we expect that the expectation value of products of irreducible tensor basis functions for particles A and B might be a useful parameterization for ρ_{corr}. Hence, the correlated part of the density matrix, ρ_{corr}, appearing in Eq. (6.131), can be expanded in the products of tensors for each of the particles, $T_{q_A}^{(k_A)}(\alpha_A, \alpha'_A, J_A, J'_A) T_{q_B}^{(k_B)}(\alpha_B, \alpha'_B, J_B, J'_B)$ is the generalization of the correlated term on the RHS of Eq. (6.82) which contains the product of the Pauli matrices, $\sigma_A \sigma_B$. The generalization of the parameterization of ρ_{corr} corresponding to $C_{ij,AB} = \langle \sigma_{i,A} \sigma_{j,B} \rangle - \langle \sigma_{i,A} \rangle \langle \sigma_{i,B} \rangle$ in Eq. (6.82) to the continuous variable system is given by the variance of the tensor products:

$$\langle T_{q_A}^{(k_A)}(J_A, J'_A) T_{q_B}^{(k_B)}(J_B, J'_B) \rangle - \langle T_{q_A}^{(k_A)}(J_A, J'_A) \rangle \langle T_{q_B}^{(k_B)}(J_B, J'_B) \rangle. \tag{6.132}$$

We can denote such parameters by $\rho_{q_A, q_B}^{(k_A, k_B)}(J_A, J'_A, J_B, J'_B)$, where, for the sake of simplicity, we have not explicitly indicated the dependence of the parameters on the quantum numbers α, α' for the two particles. Thus, the correlated density matrix in Eq. (6.131) can be expanded in the form,

$$\rho_{\text{corr}} = \sum_{k_A, k_B, q_A, q_B} \rho_{q_A, q_B; \text{corr}}^{(k_A, k_B)}(J_A, J'_A, J_B, J'_B) T_{q_A}^{(k_A)}(J_A, J'_A) T_{q_B}^{(k_B)}(J_B, J'_B). \tag{6.133}$$

The terms containing $\rho^{(k_A, k_B)}$ with $k_A = k_B = 1$ represent dipole–dipole bipartite correlations. High-order multipole terms may also be present. Moreover, we could couple k_A and k_B to make a total K and q_A and q_B to make a total Q to classify the correlation in terms of total angular momentum.

As an example, consider highly excited states of the helium atom wherein both electrons are in Rydberg states (high principal quantum number states). Such doubly excited Rydberg states with the two electrons on opposite sides of the nucleus (and, therefore, are maximally separated) have a dipole moment component for electron A and an oppositely directed dipole moment for electron B, and therefore, nonzero $\rho^{(k_A=1, k_B=1)}$ terms will be present in the density matrix expansion. Moreover, since the electrons are correlated, we also expect nonvanishing $\rho_{\text{corr}}^{(k_A=1, k_B=1)}$ components. In discussing the doubly excited Rydberg states, we have not accounted for the necessity to antisymmetrize the electronic wave function. The procedure for doing so is discussed in Sec. 8.5.

6.6 WAVE PACKET DYNAMICS

A wave packet is a localized wave form, i.e., a function $\psi(\mathbf{r},t)$, which is localized in space and propagates in time. The magnitude squared of the wave function, $|\psi(\mathbf{r},t)|^2$, is the probability distribution function for finding the particle at position $\mathbf{r}$ at time t. Clearly, if the wave function is localized in position, there is a minimum momentum dispersion that is possible due to the uncertainty principle. Moreover, if the wave function is tightly localized, and the external potential does not change rapidly over the localized region, the wave function is likely to spread in space due to dispersion (i.e., due to the effects of the kinetic energy operator operating on a wave function composed of many momentum components).

To better understand the nature of wave packet dynamics, we propagate the normalized 1D Gaussian wave packet with initial state given by

$$\psi(x,0) = \frac{1}{(2\pi\sigma^2)^{1/4}} e^{-\frac{(x-x_0)^2}{4\sigma^2} + ip_0(x-x_0)/\hbar}. \tag{6.134}$$

This wave packet is centered around the point, $x = x_0$, and has central momentum p_0. We want to obtain $\psi(x,t)$ for the case where the potential vanishes, $V(x) = 0$. Later, we shall also consider wave packet dynamics for the case of a time-dependent harmonic potential whose minimum $\mathbf{r}_0(t)$ varies with time and whose frequencies can also vary with time (these quantities may be constant as a special case),

$$V(\mathbf{r},t) = \frac{m}{2} \sum_{i=x,y,z} \omega_i(t)^2 [x_i - x_{i0}(t)]^2 \equiv V_0(\mathbf{r} - \mathbf{r}_0(t), t). \tag{6.135}$$

The dynamics are governed by the Schrödinger equation, $i\hbar \frac{\partial \psi}{\partial t} = \left[-\frac{\hbar^2 \nabla^2}{2m} + V(\mathbf{r},t)\right]\psi(\mathbf{r},t)$. We could consider other potentials, but these cases will give a sufficiently clear picture of the nature of wave packet dynamics.

For the $V(x) = 0$ 1D case, we can expand the initial wave function in Eq. (6.134) in plane waves as follows: $\psi(x,0) = \int dp\, c(p) e^{ipx/\hbar}$. Taking the Fourier transform of this expression, i.e., multiplying this equation by $e^{-ipx/\hbar}$ and integrating over space, we find

$$(2\pi)\,c(p) = \int dx\, e^{-ipx/\hbar} \psi(x,0) = \sqrt{4\pi\sigma^2}\, e^{-\sigma^2(p-p_0)^2/\hbar^2 - i(p-p_0)x_0/\hbar}. \tag{6.136}$$

For $V(x) = 0$, the eigenstates are plane waves; hence, the time dependence is obtained by multiplying the p component of the wave function by $e^{i\frac{p^2 t}{2m\hbar}}$, i.e., $\psi(x,t) = \int dp\, c(p) e^{i\frac{p^2 t}{2m\hbar}} e^{ipx/\hbar}$. Substituting Eq. (6.136) into this expression and integrating over p, we find

$$\psi(x,t) = \frac{1}{[2\pi\sigma^2(t)]^{1/4}} e^{-\frac{(x-x_0-p_0 t/m)^2}{4\sigma^2(t)} + ip_0 x/\hbar + i\phi(t)}, \tag{6.137}$$

where the time-dependent (complex) width parameter of the Gaussian is

$$\sigma^2(t) = \sigma^2 \left(1 + i\frac{\hbar t}{m\sigma^2}\right) \tag{6.138}$$

and the time-dependent phase of the wave function, $\phi(t)$, is given by

$$\phi(t) = \frac{p_0^2}{2m} \frac{t}{\hbar}. \tag{6.139}$$

The time-dependent spatial width, $|\sigma(t)|$, increases with time (see below), as expected, due to dispersion. Figure 6.15 illustrates the spreading of a moving wave packet. Given the time-dependent wave function (6.137), we can calculate the average position, $\langle x \rangle_t = x_0 + p_0 t/m$; the average momentum, $\langle p \rangle_t = p_0$; the variance of the position as a function of time,

$$(\Delta x)^2(t) = \sigma^2 \left(1 + \frac{\hbar^2 t^2}{m^2 \sigma^4}\right); \tag{6.140}$$

and the variance of the momentum as a function of time, $(\Delta p)^2 = \frac{\hbar^2}{4\sigma^2}$. Hence, the product of the standard deviations obeys the uncertainty principle,

$$\Delta p \, \Delta x = \frac{\hbar}{2}\left(1 + \frac{\hbar^2 t^2}{m^2 \sigma^4}\right)^{1/2} \geq \frac{\hbar}{2}. \qquad (6.141)$$

Initially, the uncertainty is the minimum possible, $(\Delta p \, \Delta x)_{t=0} = \frac{\hbar}{2}$, but as time increases, the uncertainty $\Delta p \, \Delta x$ increases. For large t, the width Δx, and therefore the product $\Delta p \, \Delta x$, increases linearly with t. This spreading of the wave function with time does not correspond to what is observed for classical particles (a baseball does not grow in size with time).

Problem 6.24

(a) Show that Eq. (6.137) is the solution to the Schrödinger equation for $V(x) = 0$.
(b) Derive the variance $(\Delta x)^2$ in Eq. (6.140) from the time-dependent (complex) width parameter $\sigma^2(t)$ in Eq. (6.138).

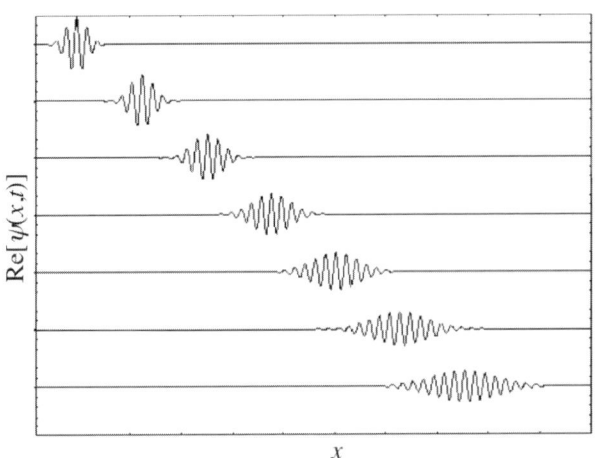

FIG 6.20 Schematic of $\text{Re}[\psi(x,t)]$ versus x at several times, increasing from top to bottom, illustrating the spreading of the wave packet as it propagates in free space.

A similar procedure can be carried out for the harmonic potential, $V(x) = m\omega^2 x^2/2$. The wave function at time $t = 0$ is expanded in harmonic oscillator eigenstates, $\psi(x,0) = \sum_n c_n \psi_n(x)$, and the amplitudes c_n are calculated by computing the inner products,

$$c_n = \int dx \, \psi_n(x) \psi(x,0). \qquad (6.142)$$

On substituting these amplitudes back into the equation

$$\psi(x,t) = \sum_n c_n e^{-i(n+1/2)\omega t} \psi_n(x), \qquad (6.143)$$

we obtain the temporal dependence of the wave function. Let us now consider the more general case of an initial wave function for the harmonic oscillator potential (6.135). The solution of the Schrödinger equation, given a potential of the form (6.135), can be written as

$$\psi(\mathbf{r},t) = \psi_0(\mathbf{r} - \mathbf{R}(t), t) \exp\{i[\mathbf{P}(t) \cdot \mathbf{r}/\hbar - \phi(t)]\}, \qquad (6.144)$$

where $\psi_0(\mathbf{r},t)$ satisfies the time-dependent Schrödinger equation with $\mathbf{r}_0(t) = 0$, and the ith component of the position vector of the center of the wave packet, $\mathbf{R}(t)$, satisfies the equation of motion

$$\ddot{R}_i + (\omega_i(t))^2 (R_i - x_{i0}(t)) = 0. \qquad (6.145)$$

The time-dependent momentum in Eq. (6.144) is given by $\mathbf{P}(t) = m\dot{\mathbf{R}}(t)$, and the time-dependent phase $\phi(t) = \sum_i \phi_i(t)$ is

$$\phi(t) = \frac{m}{2\hbar} \sum_i \int_0^t \left\{\omega_i^2(t')[R_i^2(t') - x_{i0}^2(t')] + [\dot{R}_i(t')]^2\right\} dt'. \qquad (6.146)$$

6.7 Time-Dependent Hamiltonians

The proof is straightforward. Assume that $\psi_0(\mathbf{r},t)$ satisfies Eq. (6.144) with $\mathbf{r}_0 = 0$, then substitute the solution (6.143) into Eq. (6.144) for arbitrarily varying $\mathbf{r}_0(t)$ to obtain

$$-m\ddot{\mathbf{R}} \cdot \mathbf{r} + \hbar\dot{\phi} = \frac{m}{2}\dot{\mathbf{R}}^2 + \frac{1}{2}m\sum_i \omega_i^2 \left[(x_i - x_{i0}(t))^2 - (x_i - R_i(t))^2\right]\psi_0. \tag{6.147}$$

Note that the center of the wave packet, $\mathbf{R}(t)$, does not in general adiabatically follow the center of the potential, $\mathbf{r}_0(t)$, and that the phase factor $\exp\{i[\mathbf{P}(t) \cdot \mathbf{r}/\hbar - \phi(t)]\}$ in Eq. (6.144) affects the coherence properties of the wave packet, which are given in terms of the coherence function $C(\boldsymbol{\rho}, \tau; t) = \int d\mathbf{r}\, \psi^*(\mathbf{r}+\boldsymbol{\rho}, t+\tau)\psi(\mathbf{r},t)$. The spatial width of the wave packet is generally a function of time, as in the case for $V(x) = 0$.

Problem 6.25

(a) For the 1D potential $V(x) = m\omega^2 x^2/2$, substitute the ansatz, $\psi(x,t) = N e^{-\alpha(t)(x-x_t)^2 + ip_t(x-x_t)/\hbar + \gamma_t}$, where x_t, p_t, and γ_t are functions of t, into the Schrödinger equation. By equating powers of $x - x_t$ in the exponent, find the differential equations for $\alpha(t)$, x_t, p_t, and γ_t.
(b) Solve the differential equations for x_t, p_t.
(c) Refer the study by E. J. Heller [104] that completes the solution by solving the differential equations for $\alpha(t)$ and γ_t.

Answers: (a) $\dot{\alpha}(t) = -2\alpha^2(t)/m - m\omega^2$, $\dot{x}(t) = p(t)/m$, $\dot{p}(t) = -m\omega^2 x(t)$, $\dot{\gamma}(t) = i\hbar\alpha(t)/m + p(t)\dot{x}(t) - E$.
(b) $x(t) = x_0 \sin(\omega t + \phi)$, and $p(t) = m\omega x_0 \cos(\omega t + \phi)$.

For an arbitrary time-independent Hamiltonian, if the wave function at time $t = 0$ is expanded in eigenstates $\phi_n(\mathbf{r})$, $\psi(\mathbf{r},0) = \sum_n c_n \phi_n(\mathbf{r})$, and the amplitudes c_n are calculated by computing the inner products,

$$c_n = \int d\mathbf{r}\, \phi_n^*(\mathbf{r})\psi(\mathbf{r},0), \tag{6.148}$$

the time-dependent wave packet is given by

$$\psi(\mathbf{r},t) = \sum_n c_n e^{-iE_n t/\hbar} \phi_n(\mathbf{r}), \tag{6.149}$$

where E_n is the nth eigenvalue.

6.7 TIME-DEPENDENT HAMILTONIANS

The dynamics of systems that evolve with a Hamiltonian that explicitly depends on time, $\hat{H}(t)$, are more difficult to treat than the dynamics of nonstationary states in a time-independent Hamiltonian. The energy is not conserved in such systems and there are no stationary states. A special case of experimental relevance is the study of systems in which the Hamiltonian $\hat{H}(t) = \hat{H}_0 + \hat{V}(t)$ can be decomposed into a sum of a time-independent part $\hat{H}_0$ and an additional small time-dependent perturbation $\hat{V}(t)$ that induces transitions between eigenstates of $\hat{H}_0$. It will be the subject of Sec. 7.3.3 where we shall apply a method for solving linear differential equations based on variation of constants, as originally suggested by Dirac in 1926.

Let us consider a time-dependent perturbation acting for a finite time, i.e., suppose that the perturbation $\hat{V}(t)$ acts only during a finite interval of time or that it diminishes rapidly as $t \to \pm\infty$. Let the system be in the jth stationary state ϕ_j of H_0 with energy E_j before the perturbation is switched on at time $t = 0$. The time-dependent wave function for $t > 0$ is expanded as

$$\psi(t) = \sum_k a_k(t) e^{-iE_k t/\hbar} \phi_k. \quad (6.150)$$

Substituting Eq. (6.150) into the time-dependent Schrödinger equation and projecting onto state k, we obtain,

$$i\hbar \dot{a}_k(t) = e^{iE_k t/\hbar} \sum_i \langle \phi_k | \hat{V}(t) | \phi_i \rangle a_i(t) e^{-iE_i t/\hbar}. \quad (6.151)$$

For small $\hat{V}(t)$, we can expand the amplitudes in powers of the perturbation, $a_k = \delta_{kj} + a_k^{(1)} + a_k^{(2)} + \ldots$, where $|\phi_j\rangle$ is the state before the interaction is turned on, and $a_k^{(1)}$ is first order in $\hat{V}(t)$, etc. Substituting $a_k^{(0)} = \delta_{kj}$ into the RHS of Eq. (6.151), we find to first order

$$a_j(\infty) = 1 - \frac{i}{\hbar} \int_0^\infty dt\, V_{jj}(t) \quad (6.152a)$$

and

$$a_k^{(1)}(\infty) = -\frac{i}{\hbar} \int_0^\infty dt\, e^{i(E_k - E_i)t/\hbar} V_{kj}(t). \quad (6.152b)$$

To first order in the interaction, the quantities $|a_k^{(1)}(\infty)|^2$ give the probability of finding the system in state k after the perturbation has ceased. Second- and high-order perturbation corrections will be discussed in Sec. 7.3.3. In particular, we will consider the case when $V_{kj}(t)$ is periodic in time, an example of which is the interaction of matter with an electromagnetic field, which will be treated in Sec. 7.4.

6.8 QUANTUM OPTIMAL CONTROL THEORY

Let us consider control of a time-dependent perturbation in order to obtain a desired condition (objective) at the end of the perturbation. Optimal control theory involves mathematical optimization methods for deriving control policies for the perturbation so as to obtain a specified objective after the perturbation is over. These methods are an extension of the calculus of variations and have been advanced in the latter half of the twentieth century by Lev Pontryagin, Richard Bellman and collaborators and many others. In quantum optimal control theory, i.e., in the application of optimal control theory to quantum mechanical systems, a system is driven during some time period by one or more external fields whose temporal dependence is optimized to obtain a certain objective (e.g., to maximize the probability to be found in a prescribed final state at the final time). Using the time-dependent Schrödinger equation, we can propagate a wave function from some initial time to some final time. The Hamiltonian could contain external fields, such as the electromagnetic field coming from a laser or some other time-dependent parameter. The time-dependent parameters are determined so as to optimize a given objective at the final time. Several books review the field [105–107]. Here, we briefly introduce this subject.

Consider a quantum mechanical system governed by the Hamiltonian $\hat{H}(\epsilon(t))$ that depends parametrically on a "control field" $\epsilon(t)$. For example, the control field could be the slowly varying envelop of an external electromagnetic field applied to the system. This field drives the system from a specified initial state $|\psi_0\rangle$ to a final state $|\psi(T)\rangle$. We wish to find the optimal control field $\epsilon_{\text{opt}}(t)$ such that it maximizes a given objective, which is expressed as a functional, $J[\psi]$ of the final state $\psi(T)$. We constrain the search so that the control field can only take on physical values, e.g., the electric field cannot acquire uncontrolled large values. Hence, we prescribe the maximization of the functional

$$J[\psi, \xi, \epsilon] = J_1[\psi] + J_2[\epsilon] + J_3[\psi, \xi, \epsilon], \quad (6.153)$$

6.8 Quantum Optimal Control Theory

where $J_1[\psi]$ is the objective functional that is an expectation value of some definite-positive operator $\hat{O}$ at the final time

$$J_1[\psi] = \langle \psi(T) | \hat{O} | \psi(T) \rangle, \tag{6.154}$$

$J_2[\epsilon]$ is a control function associated with the field (see below), $\xi(t)$ is a Lagrange multiplier (see below), and $J_3[\psi, \xi, \epsilon]$ is a control functional associated with the Schrödinger equation of motion (see below). For example, the objective functional could involve a projection onto a specific state $|\phi\rangle$; then the operator $\hat{O} = |\phi\rangle\langle\phi|$ is the projection operator onto this state, or, if you prefer $\hat{O} = |\phi(T)\rangle\langle\phi(T)|$. Note that we might, or might not, want to take the limit as $T \to \infty$ for some applications. The functional $J_2[\epsilon]$ constrains the values of the control field $\epsilon(t)$. For example, to maximize the objective $J_1[\psi]$ while minimizing the fluence,[11] we can use

$$J_2[\psi] = -\alpha \int_0^T dt\, |\epsilon(t)|^2, \tag{6.155}$$

where the constant α is called *penalty*. The constraint may be of the form of the total fluence equal to a given value, e.g., $E = \int_0^T dt\, |\epsilon(t)|^2$. Then, the parameter α will be chosen (later) to ensure this particular fluence. The functional $J_3[\psi, \xi, \epsilon]$ should ensure that the Schrödinger equation is satisfied throughout the dynamics

$$J_3[\psi, \xi, \epsilon] = -2\,\text{Im} \int_0^T dt\, \left\langle \xi(t) \left| i\hbar \frac{\partial}{\partial t} - \hat{H}[\epsilon(t)] \right| \psi(t) \right\rangle. \tag{6.156}$$

The auxiliary state $|\xi(t)\rangle$ plays the role of a time-dependent Lagrange multiplier. To maximize the functional $J[\psi, \xi, \epsilon]$, we must find the solution to the corresponding Euler–Lagrange equations for $\psi(t)$, $\epsilon(t)$, and $\xi(t)$, i.e., $\frac{\delta J}{\delta \psi} = 0$, $\frac{\delta J}{\delta \epsilon} = 0$, and $\frac{\delta J}{\delta \xi} = 0$. Moreover, we need to take the variation of $J[\psi, \xi, \epsilon]$ with respect to $\psi(T)$ and $\xi(0)$ and set these to zero, $\frac{\delta J}{\delta \psi(T)} = 0$ and $\frac{\delta J}{\delta \xi(0)} = 0$. In carrying out the variations, there are two points that should be noted. The quantities $\epsilon(t)$ and $\epsilon^*(t)$ enter independently into the functional in Eq. (6.153), so the variation with respect to these variables should be done independently, $\frac{\delta J}{\delta \epsilon} = 0$, $\frac{\delta J}{\delta \epsilon^*} = 0$. The term $-2\,\text{Im} \int_0^T dt\, \langle \xi(t) | i\hbar \frac{\partial}{\partial t} | \psi(t) \rangle$ that appears in Eq. (6.153) can be integrated by parts to yield

$$-2\,\text{Im} \int_0^T dt\, \langle \xi(t) | i\hbar \frac{\partial}{\partial t} | \psi(t) \rangle = -\hbar\,\text{Re}\left[\langle \xi | \psi \rangle \big|_0^T + \int_0^T dt\, \left(\langle \xi(t) | \dot{\psi}(t) \rangle + \langle \dot{\xi}(t) | \psi(t) \rangle \right) \right]. \tag{6.157}$$

Carrying out the functional derivatives of $J[\psi, \xi, \epsilon]$ leads to the *quantum optimal control theory* equations:

$$\left(i\hbar \frac{\partial}{\partial t} - \hat{H}[\epsilon(t)] \right) |\psi(t)\rangle = 0, \tag{6.158a}$$

$$\left(i\hbar \frac{\partial}{\partial t} - \hat{H}^\dagger[\epsilon(t)] \right) |\xi(t)\rangle = 0, \tag{6.158b}$$

$$|\psi(0)\rangle = |\psi_0\rangle, \tag{6.158c}$$

$$|\xi(T)\rangle = \hat{O}|\psi(T)\rangle, \tag{6.158d}$$

$$\alpha\,\epsilon(t) = \text{Im}\left\langle \xi(t) \left| \frac{\partial \hat{H}}{\partial \epsilon} \right| \psi(t) \right\rangle. \tag{6.158e}$$

---

[11] The number of photons per unit section incident on the system.

The control equations (6.158a) are coupled, and a self-consistent solution satisfying all the equations must be obtained. This requires an iterative scheme wherein the input control function ϵ is modified to produce an output control function, which is then used as input for the following iteration, etc. The simplest control scheme is straight iteration: given a trial control function $\epsilon^{(k)}$, where k is the iteration index, the output functional $F[\epsilon^{(k)}]$ is constructed using the following steps:

1. Propagate from $|\psi(0)\rangle = |\psi_0\rangle$ to $|\Psi(T)\rangle$ with $\epsilon^{(k)}(t)$.
2. Propagate backwards from $|\chi(T)\rangle = \hat{O}|\psi(T)\rangle$ to $|\chi(0)\rangle$ with $\epsilon^{(k)}(t)$. During the evolution, calculate the output field $F[\epsilon^{(k)}]$ given by

$$\alpha F[\epsilon^{(k)}](t) = \text{Im}\langle\chi(t)|\frac{\partial \hat{H}}{\partial \epsilon}|\Psi(t)\rangle. \tag{6.159}$$

3. Define $\epsilon^{(k+1)}(t) = F[\epsilon^{(k)}(t)]$ and repeat from step 1 until convergence is reached, i.e., until $F[\epsilon](t) = \epsilon(t)$ up to a given error criterion.

Unfortunately, the straightforward iteration approach does not always converge. One possibility is to set $\epsilon^{(k+1)} = \epsilon^{(k)} + \gamma F[\epsilon^{(k)}]$, where the parameter γ may be determined by performing a line-search optimization, such that $\epsilon^{(k+1)}$ produces the maximal objective J. Another approach is the monotonically convergent algorithm introduced in Ref. [108]. Given the trial control function $\epsilon^{(k)}$, the output $F[\epsilon^{(k)}]$ is constructed using the following steps:

1. Propagate from $|\psi(0)\rangle = |\psi_0\rangle$ to $|\psi(T)\rangle$ with $\epsilon^{(k)}(t)$.
2. Propagate backwards from $|\chi(T)\rangle = \hat{O}|\psi(T)\rangle$ to $|\chi(0)\rangle$ using control function

$$\alpha\tilde{\epsilon}(t) \equiv \text{Im}\langle\chi(t)|\hat{V}|\psi(t)\rangle, \tag{6.160}$$

where $\hat{V} \equiv \frac{\partial \hat{H}}{\partial \epsilon}$. The integrand on the RHS of Eq. (6.160) is formed using $|\chi(t)\rangle$ from the present iteration and the previously obtained $|\psi(t)\rangle$.

3. Propagate forward from $|\psi'(0)\rangle = |\psi_0\rangle$ to $|\psi'(T)\rangle$ using the output field $F[\epsilon^{(k)}](t)$, where

$$\alpha F[\epsilon^{(k)}](t) = \text{Im}\langle\chi(t)|\hat{V}|\psi'(t)\rangle. \tag{6.161}$$

Then, define $\epsilon^{(k+1)}(t) \equiv F[\epsilon^{(k)}](t)$ and proceed to the next iteration. Note that one does not need to explicitly carry out step 1 again, since it repeats step 3 in the previous iteration. Convergence is obtained once the iteration yields a fixed point $F[\epsilon](t) = \epsilon(t)$ up to a given error criterion.

Many modifications and extensions to the quantum optical control techniques discussed here are possible. For example, multiple objectives can be considered, time-dependent targets (i.e., objectives) can be treated, spectral and fluence constraints of the external fields can be added, etc.

A computer code to carry out the optimal control theory calculations is available on the web as a part of the octopus project (whose primary aim is time-dependent density functional calculations) at the URL:
http://www.tddft.org/programs/octopus/wiki/index.php/Tutorial:Basic_QOCT

Approximation Methods 7

A host of approximation methods for solving the time-independent and -dependent Schrödinger equations are developed in this chapter. The rich physics revealed in the application of these methods ensures that this chapter is not just about approximation methods. Section 7.1 presents the powerful basis set expansion method for solving time-dependent problems (the application to time-independent problems is presented in Sec. 2.1.2). Semiclassical approximation methods are presented in Sec. 7.2, including the WKB approximation in Sec. 7.2.1. Section 7.3 presents perturbation theory, starting with time-independent nondegenerate perturbation theory in Sec. 7.3.1, with application to systems exposed to external magnetic and external electric fields, and then degenerate perturbation theory in Sec. 7.3.2. Time-dependent perturbation theory is discussed in Sec. 7.3.3. An important application of time-dependent perturbation theory for the dynamics of charged particles in electromagnetic fields is presented in Sec. 7.4. Absorption and spontaneous emission are discussed in Sec. 7.4.1, electric dipole and multipole expansion methods for treating the dynamics of atoms and molecules in a radiation field are discussed in Sec. 7.4.2, and Rayleigh, Raman, and Brillouin two-photon transitions are discussed in Sec. 7.4.3. Section 7.5 presents exponential and nonexponential (power law) decay of excited states. The variational method is presented in Sec. 7.6. The sudden approximation is presented in Sec. 7.7, and the adiabatic approximation is presented in Sec. 7.8. Three examples of the adiabatic approximation are explicitly worked out: chirped pulse adiabatic passage in Sec. 7.8.1, stimulated Raman adiabatic passage (STIRAP) in Sec. 7.8.2, and the Landau–Zener problem in Sec. 7.8.3. In Sec. 7.8.4, we discuss the concept of generalized displacements and forces; in Sec. 7.8.5, we introduce Berry phases; and finally, in Sec. 7.9.4, we discuss linear response theory, the Kubo formalism, and the fluctuation-dissipation theorem.

7.1 BASIS-STATE EXPANSIONS

One of the most important approximation methods uses a basis set to expand the wave function of a quantum system, and truncates the basis set at a finite (but sufficiently large) number of basis functions. The truncation step is the approximation. This method leads to a matrix representation of the system. It was developed in Sec. 2.1.2 for the time-independent Schrödinger equation. Now, we shall apply it to the time-dependent problems.

7.1.1 TIME-DEPENDENT BASIS SET EXPANSIONS

As discussed in Sec. 2.1.2, given an initial state $|\Psi(0)\rangle$ at time $t = 0$, which is not an eigenstate of the time-independent Hamiltonian of the system, H, a basis set expansion method can be used to evolve the state of the system. If we use a basis set composed of eigenstates of H, $\{|\psi_k\rangle\}$, with eigenvalues $\{E_k\}$, the evolution is particularly simple,

$$|\Psi(t)\rangle = \sum_k b_k\, e^{-iE_k t/\hbar}|\psi_k\rangle, \qquad (7.1)$$

where the coefficients b_k are time independent, and the initial state is given by $|\Psi(0)\rangle = \sum_k b_k|\psi_k\rangle$. On projecting with $\langle\psi_j|$, we find the coefficient b_j: thus $b_k = \langle\psi_k|\Psi(0)\rangle$, and $|\Psi(t)\rangle = \sum_k \langle\psi_k|\Psi(0)\rangle\, e^{-iE_k t/\hbar}|\psi_k\rangle$.

In some cases, it is preferable to use other basis sets to carry out the expansion, e.g., when our interest is focused on the momentum properties of the dynamics, we might want to use momentum eigenstates instead of energy eigenstates. For a complete orthonormal set of basis states $\{|\phi_j\rangle\}$ that are *not* eigenstates of H, an arbitrary initial state can be expanded as $|\Psi(0)\rangle = \sum_j a_j|\phi_j\rangle$. Moreover, expanding the time-dependent wave function $|\Psi(t)\rangle$ in terms of the eigenstates $\{|\psi_j\rangle\}$ of the Hamiltonian, which in turn are expanded in terms of the noneigenstate basis states, $\{|\phi_j\rangle\}$, $|\psi_k\rangle = \sum_j c_{jk}|\phi_j\rangle$, we find

$$|\Psi(t)\rangle = \sum_k b_k\, e^{-iE_k t/\hbar}|\psi_k\rangle = \sum_k b_k\, e^{-iE_k t/\hbar} \sum_j c_{jk}|\phi_j\rangle. \qquad (7.2)$$

We used the fact that $|\psi_j\rangle$ are eigenstates of the Hamiltonian to get the time dependence, and the expansion coefficients c_{jk} of the eigenstates in terms of the basis states $\{|\phi_j\rangle\}$. The amplitudes b_k of the eigenstates $|\psi_k\rangle$ in the expansion (7.2) can be determined by setting $t = 0$ in Eq. (7.2) and equating the result with the initial condition $|\Psi(0)\rangle$. If $|\Psi(0)\rangle = \sum_j a_j |\phi_j\rangle$, then we obtain, from Eq. (7.2) at $t = 0$, $\sum_j c_{jk} b_k = a_j$, or, inverting the matrix c,

$$b_k = \sum_{j'} (c^{-1})_{kj'} a_{j'}. \tag{7.3}$$

Hence, the time-dependent wave function satisfying the initial condition is

$$|\Psi(t)\rangle = \sum_k \sum_{j'} (c^{-1})_{kj'} a_{j'} e^{-iE_k t/\hbar} |\psi_k\rangle = \sum_{kj'j} (c^{-1})_{kj'} a_{j'} e^{-iE_k t/\hbar} c_{jk} |\phi_j\rangle. \tag{7.4}$$

The initial wave packet $\Psi(\mathbf{r}, 0)$ develops in time as

$$\Psi(\mathbf{r}, t) = \sum_{kj'j} (c^{-1})_{kj'} a_{j'} e^{-iE_k t/\hbar} c_{jk} \phi_j(\mathbf{r}). \tag{7.5}$$

Problem 7.1

(a) Show that $\Psi(\mathbf{r}, t)$ of Eq. (7.5) equals $\sum_j a_j \phi_j(\mathbf{r})$ at $t = 0$.
(b) Show that Eq. (7.5) reduces to $|\Psi(t)\rangle = \sum_k a_k e^{-iE_k t/\hbar} |\phi_k\rangle$ if the $\{|\phi_j\rangle\}$ are eigenstates of H.
(c) Given the initial wave function $\Psi(\mathbf{r}, 0) = \langle \mathbf{r} | \Psi(0) \rangle$, show that $b_k = \sum_{j'} (c^{-1})_{kj'} \int \phi_{j'}^*(\mathbf{r}) \Psi(\mathbf{r}, 0)$.

Answer: (a) $\sum_k c_{jk} (c^{-1})_{kj'} = \delta_{jj'}$. (b) Then, $c_{jk} = \delta_{jk}$ and the result follows. (c) The result follows by taking the overlap of Eq. (7.2) at $t = 0$ with $\langle \phi_{j'} |$ and inverting the c matrix.

7.2 SEMICLASSICAL APPROXIMATIONS

The predictions of quantum mechanics must smoothly approach those of classical mechanics when applied on macroscopic systems. This statement is at the heart of the *correspondence principle* (see Sec. 2.8). In this section, we explore several semiclassical approximations wherein we take the limit $\hbar \to 0$. This procedure can be carried out by expanding the relevant expressions in powers of $\hbar$ and identifying the terms proportional to $\hbar^0$ with the classical ones. When small powers of $\hbar$ in this expansion are still retained, the formalism is referred to as *semiclassical quantum mechanics*.

7.2.1 THE WKB APPROXIMATION

The WKB approximation, named after Gregor Wentzel, Hendrik Anthony Kramers, and Leon Brillouin (it is sometimes called the JWKB approximation, where the "J" stands for Harold Jeffreys), is a technique for obtaining approximate solutions to the time-independent Schrödinger equation in one dimension (and for radially symmetric problems in 3D). It can be used to calculate bound-state energies and quantum tunneling rates through potential barriers.

Consider a particle of energy E moving through a region where the potential $V(x)$ is (nearly) constant. In classically allowed regions, where $E > V(x)$, the wave function is approximately of the form $a e^{ikx} + b e^{-ikx}$, where $k = \sqrt{2m(E - V)}/\hbar$. In classically forbidden regions, $E < V(x)$, and the wave function is of the form $a e^{Kx} + b e^{-Kx}$, where $K = \sqrt{2m(V - E)}/\hbar$. A potential $V(x)$ is *slowly varying* over distance scales of the local wavelength $\lambda(x)$ if

$$\lambda(x) \frac{dV/dx}{V(x)} \ll 1. \tag{7.6}$$

7.2 Semiclassical Approximations

The local wavelength in classically allowed regions is $\lambda(x) = 2\pi/\kappa(x)$, where $\kappa(x) = k(x) = \sqrt{2m(E - V(x))}/\hbar$ in classically allowed regions, whereas $\kappa(x) = K(x) = \sqrt{2m(V(x) - E)}/\hbar$ in classically forbidden regions. As we shall see, the local wave function in classically allowed regions will take the form, $\psi(x) = a(x) e^{i \int^x k(z)\, dz} + b(x) e^{-i \int^x k(z)\, dz}$, and in classically forbidden regions, $\psi(x) = a(x) e^{-\int^x K(z)\, dz} + b(x) e^{\int^x K(z)\, dz}$, where we shall specify the coefficients $a(x)$ and $b(x)$ shortly.

To derive the WKB approximation, consider the following form, with $A(x)$ taken to be real, for the wave function in the classically allowed regions:

$$\psi(x) = A(x)\, e^{i\phi(x)}. \tag{7.7}$$

This is the same form that we used to discuss Hamilton–Jacobi theory in Sec. 16.2, linked to the book web page, of the classical mechanics chapter linked to the web page of the book. The first derivative of the wave function is $\psi'(x) = [A'(x) + iA\phi']\, e^{i\phi(x)}$, and the second derivative is $\psi''(x) = [A''(x) + 2iA'\phi' + iA\phi'' - A(\phi')^2]\, e^{i\phi(x)}$. Inserting the second derivative into the Schrödinger equation yields

$$A''(x) + 2iA'\phi' + iA\phi'' - A(\phi')^2 = -k^2(x) A, \tag{7.8}$$

which is equivalent to the following two real equations:

$$\boxed{(A^2 \phi')' = 0,} \tag{7.9a}$$

$$\boxed{A'' = A\left[(\phi')^2 - k^2(x)\right].} \tag{7.9b}$$

The solution to Eq. (7.9a) is

$$A(x) = \frac{C}{\sqrt{\phi'(x)}}. \tag{7.10}$$

An approximate solution to Eq. (7.9b) can be obtained by assuming $|A''(x)| \ll |k^2(x) A(x)|$ (this is the key WKB approximation), yielding

$$\frac{d\phi}{dx} = \pm k(x). \tag{7.11}$$

In terms of the local de Broglie wavelength, the condition $|A''(x)| \ll |k^2(x) A(x)|$ for validity of the WKB approximation can be written as,

$$\frac{1}{2\pi}\left|\frac{d\lambda(x)}{dx}\right| \ll 1, \Rightarrow \lambda(x)\left|\frac{dV/dx}{E - V(x)}\right| \ll 1. \tag{7.12}$$

Thus, the WKB wave function, in classically allowed regions that are not too close to turning points, takes the form, $\psi(x) = \frac{C}{\sqrt{k(x)}} e^{\pm i \int^x dz\, k(z)}$. The solution has been conveniently written in terms of the indefinite integral of the local wavevector. The integration constant can be absorbed into C, which thereby becomes complex. The general form of the wave function in classically allowed regions is

$$\psi(x) = \frac{C_+}{\sqrt{k(x)}} e^{i \int^x dz\, k(z)} + \frac{C_-}{\sqrt{k(x)}} e^{-i \int^x dz\, k(z)}, \tag{7.13}$$

which can be written as

$$\psi(x) = \frac{D}{\sqrt{k(x)}} \cos\left(\int^x dz\, k(z) + \varphi\right). \tag{7.14}$$

In classically forbidden regions, $E < V(x)$, $k = iK$, $K(x) = \sqrt{2m(V(x) - E)}/\hbar$; so, the WKB form of the wave function becomes

$$\psi(x) = \frac{B_+}{\sqrt{K(x)}} e^{\int^x dz\, K(z)} + \frac{B_-}{\sqrt{K(x)}} e^{-\int^x dz\, K(z)}. \tag{7.15}$$

The boundary conditions on the wave function determine the amplitudes B_+ and B_-. If the forbidden region is to the right of the rightmost turning point (e.g., x_t in Fig. 7.1), then the wave function should vanish as $x \to \infty$, hence $B_+ = 0$. Similarly, if the forbidden region is to the left of the leftmost turning point (e.g., x'_t in Fig. 7.1) then the wave function should vanish as $x \to -\infty$, hence $B_- = 0$. If the classically forbidden region is located between two finite turning points $x_t < x'_t$ [e.g., tunneling through a repulsive barrier $V(x)$, so that $E < V(x)$ for $x_t \leq x \leq x'_t$], the boundary conditions are determined by matching the wave function (7.17) computed at x_t, and x'_t with the wave functions in the allowed regions to the right of x'_t and to the left of x_t.

To better understand the relationship of the WKB approximation with the semiclassical limit, in Problem 7.2 you are asked to derive the WKB approximation by substituting $\psi(x) = e^{if(x)/\hbar}$ into the time-independent Schrödinger equation, $-\frac{\hbar^2}{2m}\frac{d^2\psi}{dx^2} + V(x)\psi = E\psi$ and then expand $f(x)$ in powers of $\hbar$. On collecting terms of the resulting equation in powers of $\hbar$ and truncating the expansion of $f(x)$ after the second term, Eq. (7.13) could be recovered. So, the WKB approximation can be viewed as a power series expansion in powers of $\hbar$ that is truncated after the second term.

Problem 7.2

Derive the WKB approximation using the form, $\psi(x) = e^{if(x)/\hbar}$, substituting this form into the Schrödinger equation, expanding the function f in powers of $\hbar$, $f(x) = f_0(x) + \hbar f_1(x) + \hbar^2 f_2(x) + \ldots$, and collecting terms in the resulting equation in powers of $\hbar$.

Hint: Note that $\ln(-z) = \ln(z) + i n\pi$.

Unfortunately, a simple generalization of the WKB approximation to 2D, 3D, or higher dimensions does not exist. The local momentum $p(\mathbf{r})$ [or local wavevector $k(\mathbf{r}) = p(\mathbf{r})/\hbar$] is trivially generalized to higher dimension, but there is a problem to uniquely define the local vector momentum $\mathbf{p}(\mathbf{r})$. The nature of the problem is illustrated effectively by considering a degenerate eigenvalue. The eigenfunctions corresponding to the degenerate eigenvalue can have completely different nodal structure, i.e., the direction of the local flux [see Eq. (1.100)] can be very different for each of these states, even though the local momentum is the same.

WKB Connection Formulas

At a turning point, x_t, $E = V(x_t)$, the local momentum vanishes and the local wavelength $\lambda(x_t) \to \infty$. Thus, condition (7.6) is not satisfied near a turning point. Therefore, the WKB approximation is invalid near turning points. However, near a turning point, the potential can be expanded as

$$V(x) = E + V'(x_t)(x - x_t) + \cdots, \tag{7.16}$$

and higher order terms can be dropped. The Schrödinger equation becomes, $\psi''(x) = k^2(x)\psi(x) \approx \frac{2m V'(x_t)}{\hbar^2}(x - x_t)\psi(x)$. On transforming to the independent variable $z = [2mV'(x_t)/\hbar^2]^{1/3}(x - x_t)$, we obtain $\frac{d^2\psi}{dz^2} = z\psi$, whose solution is given in terms of Airy functions [see Appendix B, Eq. (B.11) and Sec. 1.3.14]:

$$\psi(x) = a\,\mathrm{Ai}[z(x)] + b\,\mathrm{Bi}[z(x)]. \tag{7.17}$$

The asymptotic expansion of the RHS of Eq. (7.17) is given by Appendix B, Eqs (B.12)–(B.15).

7.2 Semiclassical Approximations

Consider the case where $V(x) > E$ for $x > x_t$ (see Fig. 7.1). For x sufficiently far from the turning point, $\psi(x) = \frac{B_-}{\sqrt{K(x)}} e^{-\int^x dy\, K(y)}$ for $x > x_t$, whereas the wave function takes the form of Eq. (7.13) for $x < x_t$. Close to a turning point, a good approximation for the wave function is

$$\psi(x) = a\, \text{Ai}[z(x)]. \tag{7.18}$$

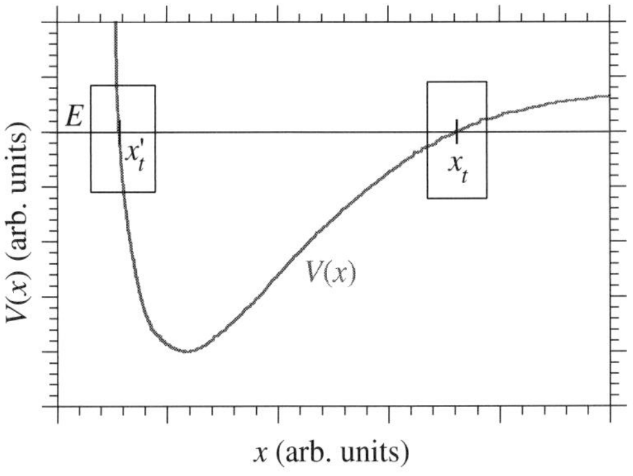

To determine the coefficients C_+ and C_- for the wave function in the region $x < x_t$ in terms of B_-, we must follow the variation in the wave function from positive $x - x_t$ to negative. To make this connection, note that $\int_{x_t}^x dy\, k(y) = \frac{2}{3\hbar}[2mV'(x_t)]^{1/2}(x - x_t)^{3/2}$, and similarly for $\int_{x_t}^x dy\, K(y)$. Hence, the forms of the wave functions in Eqs (7.13) and (7.15) are similar to the asymptotic forms of the Airy function in Eqs (B.12)–(B.13). Using these relations, we find that for this case, $a = 2\pi^{1/2}B_-$, $b = 0$, and $B = 2B_-$, and in Eq. (7.14), $\varphi = -\pi/4$. Thus, for $x < x_t$,

$$\psi(x) = \frac{2D}{\sqrt{k(x)}} \sin\left[\int_x^{x_t} dy\, k(y) + \pi/4\right]. \tag{7.19}$$

FIG 7.1 Classical turning points. For x within one of the boxes, the Airy function form, Eq. (7.18), is a reasonable approximation for the wave function.

Bound states can be studied using the WKB approximation for both the right and left turning points (see Fig. 7.1). By matching the wave function near x_t', one finds that the wave function in the region $x > x_t'$ is

$$\psi(x) = \frac{2D}{\sqrt{k(x)}} \sin\left[\int_{x_t'}^x dy\, k(y) - \pi/4\right]. \tag{7.20}$$

By equating the argument of the sine functions in Eqs (7.19) and (7.20), we find the bound-state condition

$$\int_{x_t'(E)}^{x_t(E)} dy\, p(y) = \left(n + \frac{1}{2}\right)\pi\hbar, \tag{7.21}$$

for $n = 0, 1, 2, \ldots$. This 1D semiclassical quantization is very similar to the original Bohr quantization condition developed for bound state energy levels in 1913, as discussed in Sec. 1.1.2; it was also introduced by A. Sommerfeld and W. Wilson in 1915. Only energies E at which this condition is satisfied can be bound state energies. The WKB approximation was developed in 1926, and it served to cement the connection between the Bohr quantization condition and the quantization resulting for bound states as obtained from the full Schrödinger equation.

Similar WKB connection formulas as those derived here for potential wells, as illustrated in Fig. 7.1, can be obtained for potential barriers (see Sec. 12.7.2).

Problem 7.3

Estimate the transmission of a particle of energy E incident from the left on a potential, $V(x) = 0$ for $x < 0$, $V(x) = U_0 - Fx$ for $x > 0$, when the condition $U_0 \gg E$ is satisfied and the slope F is not too large.

Answer: $T \approx \exp\left[-\frac{2\sqrt{2m}}{\hbar} \int_0^{\frac{U_0-E}{F}} dx\, (U_0 - Fx)^{1/2}\right] = \exp\left[-\frac{4\sqrt{2m}}{3\hbar F}\left(U_0^{3/2} - E^{3/2}\right)\right]$. For the details of the matching conditions on the WKB wave functions, see discussion of (Eqs. 12.429)–(12.438).

> **Problem 7.4**
>
> For the harmonic potential, $V(x) = m\omega^2 x^2/2$, show that the bound state energies can be obtained from the condition $\hbar \int_{-x_t}^{x_t} dx\, k(x) = (n + 1/2)\pi\hbar$. This is the *Bohr–Sommerfeld quantization condition* of the "old" quantum theory.
>
> **Answer:** The integral of the local momentum, $\hbar \frac{2m}{\hbar^2} \int_{-x_t}^{x_t} dx \sqrt{E - m\omega^2 x^2/2}$, can be evaluated analytically. Define the variable $z^2 = m\omega^2 x^2/2E$, and note that the turning points are given by $x_t^2 = 2E/\omega^2$. The integral $\int dt \sqrt{1-t^2} = (t\sqrt{1-t^2} + \sin^{-1} t)/2$, hence $\int_{-1}^{1} dt \sqrt{1-t^2} = \pi/2$. We thereby obtain $E_n = \hbar\omega(n + 1/2)$.

WKB for Radial Problems

The WKB approximation cannot be applied directly to the radial Schrödinger equation, $\left[\frac{d^2}{dr^2} - \frac{l(l+1)}{r^2} + \frac{2\mu}{\hbar^2}(E - V(r)) \right] f_l(r) = 0$. This is because the singularity of the effective potential $U_l(r) = V(r) + \frac{\hbar^2 l(l+1)}{2\mu r^2}$ at the origin may not be sufficiently well separated from the turning point x'_t for the behavior of the wave function at this turning point to be treated correctly. This is hinted at by the fact that the WKB approximation does not reproduce the $f_l(r) \sim (kr)^{l+1}$ behavior of the wave function as $r \to 0$. To fix this problem, the region $0 \le r < \infty$ can be mapped to $-\infty < x < \infty$ by the transformation $r = e^x$. With this change of variables,

$$f_l(r) = e^{x/2} \phi_l(x), \tag{7.22}$$

the singularity is moved to $x = -\infty$, and the Schrödinger equation becomes

$$\left[\frac{d^2}{dx^2} - \kappa_l^2(x) \right] \phi_l(x) = 0, \tag{7.23}$$

where $\kappa_l^2(x) = e^{2x}\left[k^2 - \frac{2\mu}{\hbar^2} V(e^x)\right] - (l + 1/2)^2$. Therefore, using Eq. (7.20), $f_l(r) = e^{2x} \frac{2D}{\sqrt{\kappa_l(x)}} \sin\left[\int_{x'_t}^{x} dz\, \kappa_l(z) - \pi/4\right]$, and on transforming back to r, the WKB approximation for the radial wave function in the region $r > r'_t$ becomes,

$$f_l(r) = \frac{2D}{\sqrt{k_l(r)}} \sin\left[\int_{r'_t}^{r} dr'\, k_l(r') - \pi/4\right], \tag{7.24}$$

where

$$k_l^2(r) = k^2 - \frac{2\mu}{\hbar^2} V(r) - \frac{(l+1/2)^2}{r^2}. \tag{7.25}$$

Therefore, the mapping introduced above replaces $l(l+1)$ by $(l+1/2)^2$ in the effective potential appearing in the WKB approximation for the radial wave function. The application of the WKB method to scattering problems will be discussed further in Sec. 12.7.2.

> **Problem 7.5**
>
> Estimate the transmission of an alpha particle of energy E and zero angular momentum in a nucleus of the radius r_0 with potential $V(r) = -U_0$ for $r < r_0$ and $V(r) = \kappa/r$ for $r > r_0$.
>
> **Answer:** $T \approx \exp\left[-\frac{2\sqrt{2m}}{\hbar} \int_{r_0}^{\kappa/E} dr\, (\kappa/r - E)^{1/2}\right]$. On integrating, $T \approx \exp\left\{-\frac{2\kappa\sqrt{2m/E}}{\hbar}\left[\arccos(\sqrt{Er_0/\kappa}) - \sqrt{Er_0/\kappa}\right.\right.$ $\left.\left.(1 - Er_0/\kappa)\right]\right\}$. For the details of the matching conditions on the WKB wave functions, see discussion of Eqs. (12.429)–(12.438).

7.2.2 SEMICLASSICAL TREATMENT OF DYNAMICS

In the previous few sections, we developed a semiclassical approximation to 1D tunneling and bound state problems. Semiclassical approximations to dynamical problems can also be developed. To this end, in Sec. 2.7, beginning with the Ehrenfest Theorem, Eq. (1.93), we expanded the RHS of the equation $\frac{d}{dt}\langle \mathbf{p} \rangle = -\langle \boldsymbol{\nabla} V \rangle_t \equiv \langle \mathbf{F}(\mathbf{r}) \rangle_t$ about the classical position $\langle \mathbf{r} \rangle_t$, thereby obtaining the expansion (2.109), whose lowest order term in $\hbar$, $\langle \mathbf{F}(\mathbf{r}) \rangle_t \approx \mathbf{F}(\langle \mathbf{r} \rangle_t)$, is Newton's classical equation of motion. The higher order terms give quantum corrections. The lowest order quantum correction term is proportional to $\langle (\Delta \mathbf{r})^2 \rangle_t$. Therefore, it is related to the spatial width of the quantum distribution.

At the end of Sec. 2.6 we discussed the Wigner function, and showed that the lowest order of the equation of motion for the Wigner function becomes the classical Liouville equation, sometimes called the collisionless Boltzmann equation [see Eq. (2.101)]. More explicitly, the Wigner function $W(\mathbf{p}, \mathbf{r}, t)$, which can be used for both pure and mixed states, can be expanded in powers of $\hbar$ to obtain a semiclassical expansion of the density matrix $\rho(\mathbf{r}', \mathbf{r}, t)$ by applying the expansion to the Fourier transform, $(2\pi\hbar)^{-N} \int d\mathbf{p}\, e^{i\mathbf{p} \cdot (\mathbf{r}' + \mathbf{r})/\hbar} W(\mathbf{p}, \mathbf{r}, t)$. The technical details of this procedure will not be developed here (see Ref. [109]).

7.2.3 SEMICLASSICAL HAMILTON–JACOBI EXPANSION

The classical limit of the time-dependent Schrödinger equation can be obtained by letting, $\psi(\mathbf{r}, t) = A(\mathbf{r}, t) e^{iS(\mathbf{r}, t)/\hbar}$, where both A and S are taken to be real. Substitution into the Schrödinger equation and separation of real and imaginary parts of the resulting equation yields,

$$\frac{\partial A}{\partial t} + \frac{A}{2m}\nabla^2 S + \frac{1}{m}\boldsymbol{\nabla} A \cdot \boldsymbol{\nabla} S = 0, \tag{7.26a}$$

$$\frac{\partial S}{\partial t} + \frac{1}{2m}(\boldsymbol{\nabla} S)^2 + V = \frac{\hbar^2}{2m}\frac{\nabla^2 A}{A}. \tag{7.26b}$$

These equations are also derived in the classical mechanics chapter, Sec. 16.12 [see Eqs. (16.180)], which is available on the web page for the book. Equation (7.26b) is the Hamilton–Jacobi equation if the term on the RHS is set to 0, i.e., if $\hbar \to 0$. These equations were interpreted in terms of the probability density $P(\mathbf{r}, t) = |\Psi|^2 = [A(\mathbf{r}, t)]^2$ and the probability flux vector $\mathbf{J}(\mathbf{r}, t) = \frac{P\boldsymbol{\nabla} S}{m}$. We then defined the velocity field, $\mathbf{v}(\mathbf{r}, t) \equiv \mathbf{J}/P = \boldsymbol{\nabla} S/m$, and took the gradient of Eq. (7.26b) to obtain

$$\frac{\partial \mathbf{v}}{\partial t} + (\mathbf{v} \cdot \boldsymbol{\nabla})\mathbf{v} = -\frac{1}{m}\boldsymbol{\nabla}(V + V_Q), \tag{7.27}$$

where the "quantum potential" $V_Q(\mathbf{r}, t) \equiv -\frac{\hbar^2}{2m}\frac{\nabla^2 A}{A}$ vanishes as $\hbar \to 0$. In this limit, Eq. (7.27) yields *Euler's equation* for hydrodynamic fluid flow, which is equivalent to Newton's law of motion.

Another semiclassical method important for high-energy scattering applications is the Eikonal Approximation, which will be discussed in Sec. 12.7.4.

7.3 PERTURBATION THEORY

Exact solutions to the Schrödinger equation are known only for a very small number of problems. However, numerical solutions can be obtained if the number of dimensions of the problem is small. Another approach that can be used for many types of quantum problems involves breaking the problem into a zero-order problem whose solution is known and a perturbation whose effect is small. Time-independent and -dependent perturbation theory methods will be presented below.

7.3.1 NONDEGENERATE PERTURBATION THEORY

In this subsection, we present nondegenerate perturbation theory for a time-independent Hamiltonian H, where the full Hamiltonian is broken into a zero-order Hamiltonian $H^{(0)}$ whose eigenvalues and eigenvectors, $E_n^{(0)}$ and $\psi_n^{(0)}$, are known, and a perturbation Hamiltonian $H^{(1)}$, $H = H^{(0)} + H^{(1)}$. This type of perturbation theory is sometimes called *Rayleigh–Schrödinger perturbation theory*.

For example, for a one-dimensional Hamiltonian with kinetic and potential energies, T and $V(x)$, respectively, the Hamiltonian $H = T + V$ can be broken into a zero-order Hamiltonian, $H^{(0)} = T + V^{(0)}$, where T is the kinetic energy and $V^{(0)}$ is a potential for which one knows the eigenvalues and eigenvectors of $H^{(0)}$, and a perturbation Hamiltonian, $H^{(1)} = V - V^{(0)}$. The Schrödinger equation for $H^{(0)}$ is,

$$H^{(0)}\psi_n^{(0)} = E_n^{(0)}\psi_n^{(0)}. \quad (7.28)$$

For a harmonic potential, $V^{(0)} = (m\omega^2/2)x^2$, $E_n^{(0)} = (n+1/2)\hbar\omega$, and the eigenvectors $\psi_n^{(0)}$ are Gaussian functions times Hermite polynomials for the harmonic oscillator [see Appendix Eq. (1.130)].

An elegant method of applying perturbation theory is to multiply the perturbation Hamiltonian by a perturbation parameter, λ, whose value will be put to unity at the end of the calculation. Taking the full Hamiltonian as $H = H^{(0)} + \lambda H^{(1)}$ and writing out the full wave function and the full energy eigenvalue for the nth state as a sum of terms with increasing powers of the perturbation parameter:

$$\psi_n = [\psi_n^{(0)} + \lambda \psi_n^{(1)} + \lambda^2 \psi_n^{(2)} + \cdots], \quad (7.29a)$$

$$E_n = [E_n^{(0)} + \lambda E_n^{(1)} + \lambda^2 E_n^{(2)} + \cdots]. \quad (7.29b)$$

The full Schrödinger equation for the nth order state can be written by substituting these forms into the full Schrödinger equation, $H\psi_n = E_n\psi_n$,

$$\boxed{[H^{(0)} + \lambda H^{(1)}][\psi_n^{(0)} + \lambda \psi_n^{(1)} + \cdots] = [E_n^{(0)} + \lambda E_n^{(1)} + \lambda^2 E_n^{(2)} + \cdots][\psi_n^{(0)} + \lambda \psi_n^{(1)} + \cdots].} \quad (7.30)$$

Equating powers of λ on both sides of the equation, we find

$$H^{(0)}\psi_n^{(1)} + H^{(1)}\psi_n^{(0)} = E_n^{(0)}\psi_n^{(1)} + E_n^{(1)}\psi_n^{(0)}, \quad (7.31a)$$

$$H^{(0)}\psi_n^{(2)} + H^{(1)}\psi_n^{(1)} = E_n^{(0)}\psi_n^{(2)} + E_n^{(1)}\psi_n^{(1)} + E_n^{(2)}\psi_n^{(0)}, \quad (7.31b)$$

$$H^{(0)}\psi_n^{(j+1)} + H^{(1)}\psi_n^{(j)} = E_n^{(0)}\psi_n^{(j+1)} + E_n^{(1)}\psi_n^{(j)} + \cdots + E_n^{(j+1)}\psi_n^{(0)}, \quad (7.31c)$$

where the last equation is for arbitrary j. We now expand $\psi_n^{(1)}$ in terms of the complete set of zero-order eigenfunctions, except for the zero-order function $\psi_n^{(0)}$ which is explicitly excluded from the sum since it is already included in the first term on the RHS of Eq. (7.29a),

$$\psi_n^{(1)} = \sum_{k \neq n} c_{nk}^{(1)} \psi_k^{(0)}, \quad (7.32)$$

or, more generally, for any order perturbation,

$$\psi_n^{(j)} = \sum_{k \neq n} c_{nk}^{(j)} \psi_k^{(0)}. \quad (7.33)$$

Taking the inner product of Eq. (7.31a) with $\psi_n^{(0)}$ yields the first-order energy correction:

$$\boxed{E_n^{(1)} = \langle \psi_n^{(0)} | H^{(1)} | \psi_n^{(0)} \rangle,} \quad (7.34)$$

7.3 Perturbation Theory

where we have used the orthogonality of the zero-order eigenstates. Taking the inner product of Eq. (7.31a) with $\psi_k^{(0)}$, we find the coefficients $c_{nk}^{(1)}$,

$$c_{nk}^{(1)} = \frac{\langle \psi_k^{(0)} | H^{(1)} | \psi_n^{(0)} \rangle}{E_n^{(0)} - E_k^{(0)}}. \tag{7.35}$$

Substituting these coefficients into Eq. (7.33), we obtain

$$\boxed{\psi_n^{(1)} = \sum_{k \neq n} \frac{\langle \psi_k^{(0)} | H^{(1)} | \psi_n^{(0)} \rangle}{E_n^{(0)} - E_k^{(0)}} \psi_k^{(0)}.} \tag{7.36}$$

Taking the inner product of Eq. (7.31b) with $\psi^{(0)}$, we find

$$\boxed{E_n^{(2)} = \sum_{k \neq n} c_{nk}^{(1)} \langle \psi_n^{(0)} | H^{(1)} | \psi_k^{(0)} \rangle = \sum_{k \neq n} \frac{\left| \langle \psi_k^{(0)} | H^{(1)} | \psi_n^{(0)} \rangle \right|^2}{E_n^{(0)} - E_k^{(0)}}.} \tag{7.37}$$

For $n = 0$, i.e., the ground state, the second-order correction is *always negative*; the lowest energy state is always further lowered in energy by the second-order correction to the energy. Taking the inner product of Eq. (7.31b) with $\psi_k^{(0)}$, we find the second-order coefficient,

$$c_{nk}^{(2)} = \frac{1}{E_n^{(0)} - E_k^{(0)}} \sum_{j \neq n} \frac{\langle \psi_k^{(0)} | H^{(1)} | \psi_j^{(0)} \rangle \langle \psi_j^{(0)} | H^{(1)} | \psi_n^{(0)} \rangle - \langle \psi_j^{(0)} | H^{(1)} | \psi_n^{(0)} \rangle E_n^{(1)}}{E_n^{(0)} - E_j^{(0)}}, \tag{7.38}$$

which can be substituted into Eq. (7.33) to obtain the wave function to second order. This procedure can be continued to obtain higher order (e.g., third order) perturbation results.

Perturbative Magnetic Field Effects

As we have already seen in Sec. 4.3, an external magnetic field splits atomic levels and thereby removes the degeneracy of nonzero total angular momentum states. This is called the Zeeman effect. Consider the splitting of atomic levels having definite values of the quantum numbers L, S, and J, by a weak magnetic field, assuming that L–S coupling is a good approximation. The spin–orbit splitting of the atomic levels $|L, S, J, M_J\rangle$ is given by (4.72). The Zeeman Hamiltonian (4.32) for these levels, which is linear in the magnetic field, is

$$H_{\text{Zeeman}} = -\boldsymbol{\mu} \cdot \mathbf{H} = \frac{\mu_B}{\hbar}(\mathbf{L} + g\mathbf{S}) \cdot \mathbf{H}. \tag{7.39}$$

In the first-order perturbation theory approximation, the splitting is determined by the mean values of the perturbation in the unperturbed states. Taking the direction of the magnetic field as the z axis, the energy perturbations of the spin–orbit states due to the presence of a magnetic field is given by

$$\Delta E_{J,L,S,M_J} = \frac{\mu_B H}{\hbar} \langle (L_z + gS_z) \rangle = \frac{\mu_B H}{\hbar} \langle L, S, J, M_J | [J_z + (g-1)S_z] | L, S, J, M_J \rangle. \tag{7.40}$$

The expectation value $\langle J_z \rangle$ equals $\hbar M_J$. The expectation value $\langle S_z \rangle$ can be found using the projection theorem, Eq. (3.215):

$$\langle S_z \rangle = \hbar M_J \left(\frac{\mathbf{S} \cdot \mathbf{J}}{\mathbf{J}^2} \right) = \hbar M_J \frac{J(J+1) - L(L+1) + S(S+1)}{2J(J+1)}.$$

Hence, the expectation value of Eq. (4.32) in Eq. (7.40) for the state $|J, L, S, M_J\rangle$ yields

$$\Delta E_{L,S,J,M_J} = \mu_B M_J H \left[1 + (g-1)\frac{J(J+1) - L(L+1) + S(S+1)}{2J(J+1)} \right]. \tag{7.41}$$

Thus, each of the spin–orbit states is split into $2J+1$ equally spaced levels, with spacing proportional to H (and continue). $g_{J,L,S} \equiv \left[1 + (g-1)\frac{J(J+1) - L(L+1) + S(S+1)}{2J(J+1)} \right]$ is called the Landé g-factor, and the linear Zeeman splitting of the fine structure states in magnetic field can be written as

$$\Delta E_{\text{Zeeman}} = \mu_B g_{J,L,S} H M_J. \tag{7.42}$$

Perturbative Electric Field Effects

The presence of a static electric field can modify the orbits of electrons, so that they become elliptical with the major axis of the ellipse oriented in the direction of the field. Suppose an atom or molecule is in a state with permanent dipole moment $\mathbf{d}$. In a static electric field $\mathbf{E}$, the matter-field interaction energy is $\mathcal{E} = -\mathbf{d} \cdot \mathbf{E}$. Thus, the resulting change of energy of the dipole due to the field is linear with electric field strength. We can write the term of the Hamiltonian describing the interaction of the atom with the external electric as $\sum_i q_i \mathbf{r}_i \cdot \mathbf{E}$, where the sum is over all charged particles in the atom. This interaction term is called the Stark Hamiltonian, which is named after Johannes Stark (1874–1957), who won the Nobel Prize in 1919 for his work on the splitting of spectral lines in the presence of electric fields in 1913. Stark splitting of spectral lines is used as a diagnostic in laboratory and astrophysical plasmas.

For atoms, the linear shift in energy with electric field strength can only occur if there is a degeneracy of the energy levels having different angular momentum, such as occurs in excited hydrogen states (recall that the diagonal matrix elements $\langle nlm|z|nlm \rangle = 0$ and that $\langle l,m| \cos\theta |l \pm 1, m\rangle = \frac{l^2 - m^2}{(4l^2 - 1)^{1/2}}$). Hence, the lowest order nonvanishing perturbation is second-order (see Problem 7.6).

For atoms (or molecules) that do not have a permanent dipole moment, the electric field will induce a dipole moment, and this *induced dipole moment*, $\mathbf{d}_{\text{in}}$, interacts with the field, so that the interaction energy $\mathcal{E} = -(1/2)\mathbf{d}_{\text{in}} \cdot \mathbf{E}$. We can write this energy by expanding the induced dipole moment as a power series in the electric field in the form,

$$\mathcal{E} = -\frac{1}{2}\mathbf{d}_{\text{in}} \cdot \mathbf{E} \approx -\frac{1}{2} E_i \alpha_{ij} E_j + \cdots, \tag{7.43}$$

where $d_{\text{in},i} = \alpha_{ij} E_j + \cdots$. Therefore, the splitting of the energy levels is proportional to the square of the static electric field, and the second-order perturbation correction to the energy of a state is given by

$$\Delta \mathcal{E}^{(2)} = -\frac{1}{2} E_i \alpha_{ij} E_j, \tag{7.44}$$

where the *polarizability tensor* α_{ij} is a rank two symmetric tensor. Taking the z-axis in the direction of the field, we obtain

$$\Delta \mathcal{E}^{(2)} = -\frac{1}{2}\alpha_{zz} E^2. \tag{7.45}$$

The polarizibility tensor α_{ij} for an atom or molecule in state $|n\rangle$ is

$$\alpha_{ij}^{(n)} = -2 \sum_{m \neq n} \frac{\langle n|d_i|m\rangle \langle m|d_j|n\rangle}{E_n - E_m}. \tag{7.46}$$

7.3 Perturbation Theory

The polarizability depends on its unperturbed state, $|n\rangle$, and on its quantum number M_J. The lowest-order induced dipole moment is linear in the field,

$$d_i^{(n)} = \frac{\partial \Delta \mathcal{E}^{(2)}}{\partial E_i} = \sum_j \alpha_{ij}^{(n)} E_j, \tag{7.47}$$

and the energy in Eq. (7.44) can be written as

$$\Delta \mathcal{E}^{(2)} = -\frac{1}{2} \mathbf{d}^{(n)} \cdot \mathbf{E}. \tag{7.48}$$

Problem 7.6

The spherical top Hamiltonian is $H = \frac{\mathbf{L}^2}{2I}$ [see Eq. (3.139)]. The eigenvalues and eigenstates are $E_{l,m} = \frac{\hbar^2}{2I}l(l+1)$ and $Y_{lm}(\theta, \phi)$, respectively. These states are $2l+1$-fold degenerate. In the presence of an electric field $\mathbf{E}$ along the z-axis, an additional Hamiltonian, $H^{(1)} = -\mathbf{d} \cdot \mathbf{E} = -dE\cos\theta$, results. Calculate the perturbed energy levels.

Answer: The nonzero matrix elements, $\langle l', m'|H^{(1)}|l, m\rangle$, only have $m' = m$ and $l' = l \pm 1$, and $\langle l, m|\cos\theta|l \pm 1, m\rangle = \frac{l^2 - m^2}{(4l^2-1)^{1/2}}$. The second-order energy correction is $E_{l,m}^{(2)} = \frac{2I(Ed)^2}{\hbar^2}\sum_{l'}\frac{\langle l,m|\cos\theta|l',m\rangle}{l(l+1)-l'(l'+1)} = \frac{2I(Ed)^2}{\hbar^2 l(l+1)}\frac{l(l+1)-3m^2}{2(2l-1)(2l+3)}$. States with m and $-m$ remain degenerate. Also note that $E_{0,0}^{(2)} < 0$.

Problem 7.7

Calculate the ground-state polarizability of the hydrogen atom.

Answer: Due to the spherical symmetry of the $1s$ state, ψ_0, the polarizability is a scalar, $\alpha_{ij} = \alpha \delta_{ij}$. Using Eq. (7.46) with $d_z = ez$, we find, $\alpha = -2e^2 \sum_k \frac{\langle 0|z|k\rangle\langle k|z|0\rangle}{E_0 - E_k}$. It is convenient to define an operator $\hat{b}$, such that $\hat{z} = (m/\hbar)\dot{\hat{b}} = \frac{im}{\hbar^2}[\hat{H}, \hat{b}]$, so $\langle k|\hat{z}|0\rangle = \frac{-im}{\hbar^2}(E_0 - E_k)\langle k|\hat{b}|0\rangle$, and therefore,

$$\alpha = \frac{2ime^2}{\hbar^2}\langle 0|\hat{z}\hat{b}|0\rangle. \tag{7.49}$$

$\hat{z}\psi_0 = (m/\hbar)\dot{\hat{b}}\psi_0 = \frac{im}{\hbar}[\hat{H}\hat{b} - \hat{b}\hat{H}]\psi_0$, and by defining $\hat{b}\psi_0(\mathbf{r}) = b(\mathbf{r})\psi_0(\mathbf{r}) = f(r)\cos\theta\,\psi_0(\mathbf{r})$, we can determine the differential equation satisfied by $f(r)$: $f''/2 + f'/r - f/r^2 + (\psi_0'/\psi_0)f' = ir$. The solution is $f(r) = -ira_0(a_0 + r/2)$, which can be verified by direct substitution. Using Eq. (7.49) we find that $\alpha = \frac{2i}{a_0}\langle 0|rf\cos^2\theta|0\rangle = \frac{2i}{3a_0}\langle 0|rf|0\rangle = 9a_0^3/2$.

7.3.2 DEGENERATE PERTURBATION THEORY

Let us now consider the case when the zero-order Hamiltonian $H^{(0)}$ has degenerate eigenvalues and the perturbation $H^{(1)}$ can remove the degeneracy. The approach in the previous subsection cannot be used directly for this case since the difference of the zero-order energies appearing in the denominators of the expressions developed there vanishes. A degenerate perturbation theory must be specially developed. Suppose the nth eigenvalue has degeneracy $d(n)$, i.e., there are $d(n)$ eigenvectors with eigenvalue $E_n^{(0)}$:

$$H^{(0)}\psi_{nj}^{(0)} = E_n^{(0)}\psi_{nj}^{(0)}, \quad j = 1, \ldots, d(n). \tag{7.50}$$

We again write $H = H^{(0)} + H^{(1)}$ and multiply the perturbation Hamiltonian by a parameter, λ, whose value will be put to unity at the end of the calculation: $H = H^{(0)} + \lambda H^{(1)}$. We can write the perturbed wave functions and the energy eigenvalues for the nth state as a sum of terms with increasing powers of the perturbation parameter using an index $\gamma = 1, \ldots, d(n)$, which specifies the perturbed eigenvectors and eigenvalues,

$$\psi_n^\gamma = \left[\sum_{j=1}^{d(n)} a_j^\gamma \psi_{nj}^{(0)} \right] + \lambda \psi_n^{\gamma(1)} + \lambda^2 \psi_n^{\gamma(2)} + \cdots, \tag{7.51a}$$

$$E_n^\gamma = E_n^{(0)} + \lambda E_n^{\gamma(1)} + \lambda^2 E_n^{\gamma(2)} + \cdots. \tag{7.51b}$$

In what follows, we take the first-order term in the wave function, $\psi_n^{\gamma(1)}$, to have contributions only from states *outside* the set $\{\psi_{nj}^{(0)}\}$ of zero-order state with eigenvalue $E_n^{(0)}$, i.e.,

$$\psi_n^{\gamma(1)} = \sum_{m \neq n} \sum_{l=1}^{d(m)} c_{n;ml}^{\gamma(1)} \psi_{ml}^{\gamma(0)}. \tag{7.52}$$

Inserting Eqs (7.51) and (7.52) into the time-independent Schrödinger equation, $H\psi_n^\gamma = E_n^\gamma \psi_n^\gamma$, and taking the terms that are first order in λ, we obtain

$$H^{(0)} \psi_n^{\gamma(1)} + H^{(1)} \left[\sum_{j=1}^{d(n)} a_j^\gamma \psi_{nj}^{(0)} \right] = E_n^{(0)} \psi_n^{\gamma(1)} + E_n^{\gamma(1)} \left[\sum_{j=1}^{d(n)} a_j^\gamma \psi_{nj}^{(0)} \right]. \tag{7.53}$$

Taking the inner product with $\psi_{nj}^{(0)}$, we obtain

$$\boxed{\sum_{j=1}^{d(n)} \langle \psi_{nk}^{(0)} | H^{(1)} | \psi_{nj}^{(0)} \rangle a_j^\gamma = E_n^{\gamma(1)} a_k^\gamma.} \tag{7.54}$$

This equation is in the standard matrix eigenvalue form, and the $d(n)$ eigenvalues $E_n^{\gamma(1)}$, $\gamma = 1, \ldots, d(n)$, are obtained from the determinantal equation:

$$\boxed{\left| \langle \psi_{nk}^{(0)} | H^{(1)} | \psi_{nj}^{(0)} \rangle - E_n^{\gamma(1)} \delta_{kj} \right| = 0.} \tag{7.55}$$

The eigenvector $\{a^\gamma\}$ corresponding to the eigenvalue $E_n^{\gamma(1)}$ can be computed from the linear equations (7.53) on substituting the eigenvalue into Eq. (7.54).

To obtain the coefficients $\{c_{n;ml}^{\gamma(1)}\}$, substitute the amplitudes $\{a_j^\gamma\}$ and the eigenvalue $E_n^{\gamma(1)}$ into Eq. (7.53) and take the inner product with $\psi_{ml}^{\gamma(0)}$,

$$c_{n;ml}^{\gamma(1)} = \sum_{j=1}^{d(n)} a_j^\gamma \frac{\langle \psi_{ml}^{(0)} | H^{(1)} | \psi_{nj}^{(0)} \rangle}{E_n^{(0)} - E_m^{(0)}}. \tag{7.56}$$

The expansion of the wave function to first order is given by

$$\psi_n^\gamma = \left[\sum_{j=1}^{d(n)} a_j^\gamma \psi_{nj}^{(0)} \right] + \sum_{m \neq n} \sum_{l=1}^{d(m)} c_{n;ml}^{\gamma(1)} \psi_{ml}^{\gamma(0)}, \tag{7.57}$$

with the coefficients $\{a_j^\gamma\}$ and $\{c_{n;ml}^{\gamma(1)}\}$ obtained as described above. With this result, we can obtain the second-order correction to the energy, $E_n^{\gamma(2)}$.

7.3 Perturbation Theory

Problem 7.8

A 2D hard-wall square well with sides of length L has energy eigenvalues $E_{n_x,n_y} = \frac{\hbar^2\pi^2}{2mL^2}(n_x^2 + n_y^2)$. The first excited state of the well is degenerate, $E_{1,2} = E_{2,1} = \frac{5\hbar^2\pi^2}{2mL^2} \equiv \varepsilon$. What are the perturbed energy eigenvalues if $H^{(1)} = -Kxy$?

Answer: Taking $-L/2 < x, y < L/2$, the wave functions are $\psi_{1,2} = (2/L)\cos\frac{\pi x}{L}\sin\frac{\pi y}{L}$ and $\psi_{2,1} = (2/L)\sin\frac{\pi x}{L}\cos\frac{2\pi y}{L}$. The perturbation matrix elements are $\Omega \equiv \langle 1,2|H^{(1)}|2,1\rangle = -K(2/L)^2\int_{-L/2}^{L/2}dx\,x\cos\frac{2\pi x}{L}\sin\frac{\pi x}{L}\int_{-L/2}^{L/2}dy\,y\sin\frac{\pi y}{L}\cos\frac{2\pi y}{L} = -\frac{16}{9\pi^2}KL^2 = \langle 2,1|H^{(1)}|1,2\rangle$. The Hamiltonian matrix is given by $H = \begin{pmatrix} \varepsilon & -\Omega \\ -\Omega & \varepsilon \end{pmatrix}$, and the eigenenergies are $E = \varepsilon \pm \Omega$.

Problem 7.9

Consider the $n = 2$ manifold of the hydrogen atom in the presence of an electric field, so the perturbation Hamiltonian is $H^{(1)} = eEz$. Calculate the perturbed energy eigenvalues and eigenvectors.

Hint: The only nonvanishing perturbation matrix element is $\langle\psi_{200}|H^{(1)}|\psi_{210}\rangle = -3eEa_0$.

7.3.3 TIME-DEPENDENT PERTURBATION THEORY

Consider a Hamiltonian containing a time-dependent potential, so that the full Hamiltonian can be written as, $H = H_0 + V(t)$, where $H_0 = T + V_0$ is time independent. The time-dependent wave function $\Psi(\mathbf{r}, t)$ satisfies the time-dependent Schrödinger equation $i\hbar\frac{\partial}{\partial t}\Psi(\mathbf{r}, t) = [H_0 + V(t)]\Psi(\mathbf{r}, t)$ and can be expanded in terms of a complete set of zero-order orthonormal eigenfunctions $\psi_j^{(0)}$ satisfying the eigenvalue equations,

$$H_0\psi_j^{(0)} = E_j\psi_j^{(0)}, \tag{7.58}$$

in the form

$$\Psi(\mathbf{r}, t) = \sum_j c_j(t)\psi_j^{(0)}(\mathbf{r})e^{-iE_jt/\hbar}. \tag{7.59}$$

The amplitudes $c_j(t)$ are time-dependent expansion coefficients, which can be computed by substituting Eq. (7.59) into the time-dependent Schrödinger equation. Taking the inner product with $\psi_j^{(0)}e^{-iE_jt/\hbar}$, we obtain a set of differential equations for the amplitudes c_j:

$$i\hbar\frac{d}{dt}c_j(t) = \sum_k c_k(t)\langle\psi_j^{(0)}|V(t)|\psi_k^{(0)}\rangle e^{i(E_j-E_k)t/\hbar}. \tag{7.60}$$

At the initial time $t = t_0$ when $V(t)$ is turned on, only the initial state i is populated, hence, $c_j(t_0) = \delta_{ji}$. We can now approximate $c_k(t)$ on the right hand side of Eq. (7.58) by the zero-order approximation $c_k(t) = \delta_{ki}$ to obtain the first-order approximation for $c_j^{(1)}(t)$:

$$\boxed{c_j^{(1)}(t) = \frac{1}{i\hbar}\int_{t_0}^t dt'\,\langle\psi_j^{(0)}|V(t')|\psi_i^{(0)}\rangle e^{i(E_j-E_i)t'/\hbar}.} \tag{7.61}$$

A second-order approximation is obtained by substituting this expression into the RHS of Eq. (7.60) for $c_j(t)$ to obtain

$$c_j^{(2)}(t) = \frac{1}{(i\hbar)^2} \sum_k \int_{t_0}^{t} dt'' \int_{t_0}^{t''} dt' \, \langle \psi_j^{(0)}|V(t'')|\psi_k^{(0)}\rangle \langle \psi_k^{(0)}|V(t')|\psi_i^{(0)}\rangle e^{i[(E_j-E_k)t''+(E_k-E_i)t']/\hbar}. \quad (7.62)$$

This procedure can be repeated by direct substitution of $c_j^{(k)}(t)$ into the RHS of Eq. (7.60) to obtain an expression for $c_j^{(k+1)}(t)$, and so on. The time-dependent expansion coefficient $c_j(t)$ starting from state i is given by $c_{ji}(t) = c_{ji}^{(0)}(t) + c_{ji}^{(1)}(t) + c_{ji}^{(2)}(t) + \cdots$.

A suggestive form in which to couch perturbation theory can be obtained in terms of the interaction representation form of the perturbation, $V_I(t) = e^{iH_0 t/\hbar} V(t) e^{-iH_0 t/\hbar}$ (see Sec. 2.7.1). The matrix element of this operator is $\langle \psi_j^{(0)}|V_I(t)|\psi_i^{(0)}\rangle = \langle \psi_j^{(0)}|V(t)|\psi_i^{(0)}\rangle e^{i(E_j-E_i)t/\hbar}$, which is the integrand of the RHS of Eq. (7.61). Using this notation, we have

$$c_j^{(1)}(t) = \frac{-i}{\hbar} \int_{t_0}^{t} dt' \, \langle \psi_j^{(0)}|V_I(t')|\psi_i^{(0)}\rangle, \quad (7.63)$$

$$c_j^{(2)}(t) = \left(\frac{-i}{\hbar}\right)^2 \sum_k \int_{t_0}^{t} dt'' \int_{t_0}^{t''} dt' \, \langle \psi_j^{(0)}|V_I(t'')|\psi_k^{(0)}\rangle \langle \psi_k^{(0)}|V_I(t')|\psi_i^{(0)}\rangle. \quad (7.64)$$

In the second-order term, the integration over t'' and t' is evaluated as shown in Fig. 7.2(a). Instead, we could evaluate the integral over the same area of the upper triangular region as shown in Fig. 7.2(b). Hence, we can write the integral in Eq. (7.64) as

$$\int_{t_0}^{t} dt'' \int_{t_0}^{t''} dt' \, (V_I(t''))_{jk}(V_I(t'))_{ki} = \frac{1}{2}\left[\int_{t_0}^{t} dt'' \int_{t_0}^{t''} dt' \, (V_I(t''))_{jk}(V_I(t'))_{ki} + \int_{t_0}^{t} dt' \int_{t_0}^{t'} dt'' \, (V_I(t''))_{jk}(V_I(t'))_{ki}\right]. \quad (7.65)$$

If we interchange the names of the variables t' and t'' in the second term on the RHS of Eq. (7.65), we can write the whole RHS as

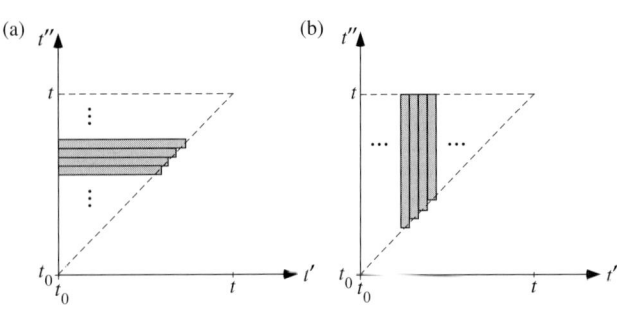

FIG 7.2 Integration over t' and t'' in the integral on the RHS of Eq. (7.62) is as in (a). The integral can also be carried out as in (7.65) as schematically indicated in (b).

$$\frac{1}{2}\int_{t_0}^{t} dt'' \int_{t_0}^{t''} dt' \, T\left[(V_I(t''))_{jk}(V_I(t'))_{ki}\right], \quad (7.66)$$

where the *time-ordered product* is defined by

$$T[V_I(t'')V_I(t')] \equiv \begin{cases} V_I(t'')V_I(t') & \text{if } t'' > t' \\ V_I(t')V_I(t'') & \text{if } t' > t'' \end{cases}. \quad (7.67)$$

With this time-ordered notation, we can write the transition amplitude as an infinite sum to all orders in perturbation theory in the form

$$c_j(t) = \langle \psi_j^{(0)}|U_I(t,t_0)|\psi_i^{(0)}\rangle, \quad (7.68)$$

7.3 Perturbation Theory

where

$$U_I(t, t_0) = 1 + \sum_{n=1}^{\infty} \frac{1}{n!} \left(\frac{-i}{\hbar}\right)^n \int_{t_0}^{t} \ldots \int_{t_0}^{t} dt_1 \ldots dt_n \, T\left[V_I(t_1) \ldots V_I(t_n)\right]. \quad (7.69)$$

This can be written compactly as $U_I(t, t_0) = T \exp\left[\frac{-i}{\hbar} \int_{t_0}^{t} dt' \, V_I(t')\right]$.

Piecewise-Constant Perturbation

As a first application of time-dependent perturbation theory, let us take the case of piecewise-constant perturbation: $V(t) = 0$ for $t < 0$ and $V(t) = V(\mathbf{r})$ for $t \geq 0$. Starting in state i, we have $c_j(0) = c_j^{(0)}(t) = \delta_{ji}$, so

$$c_j^{(1)}(t) = \frac{V_{ji}}{i\hbar} \int_0^t dt' \, e^{i(E_j - E_i)t'/\hbar} = \frac{V_{ji}}{E_j - E_i}\left(1 - e^{i(E_j - E_i)t/\hbar}\right), \quad (7.70)$$

where $V_{ji}(t') = \langle \psi_j^{(0)} | V(t') | \psi_i^{(0)} \rangle$, and

$$|c_j^{(1)}(t)|^2 = \frac{4|V_{ji}|^2}{|E_j - E_i|^2} \sin^2\left(\frac{(E_j - E_i)t}{2\hbar}\right). \quad (7.71)$$

The dependence of the RHS of Eq. (7.71), which can be written as $\frac{|V_{ji}|^2}{\hbar^2} 4 \left(\frac{\sin(\omega t/2)}{\omega}\right)^2$, with $\omega \equiv (E_j - E_i)/\hbar$, becomes sharply peaked around $\omega = 0$ as time t increases. This is illustrated in Fig. 7.3, where we plot $4\left(\frac{\sin(\omega t/2)}{\omega}\right)^2$ versus ω or several values of t. For transitions to states having energy equal to the initial energy, $E_j = E_i$, Eq. (7.71) becomes

$$|c_j^{(1)}(t)|^2 = \frac{|V_{ji}|^2 t^2}{\hbar^2}, \quad (7.72)$$

using $\lim_{x \to 0} \sin(x)/x = 1$.

Suppose there is a group of eigenstates $\{\psi_j\}$ of H_0, with energies E_j near E_i that are very close to each other. We want to find the transition rate at long time from the initial state i into any state belonging to this group of final states. For convenience, we assume that their energies form a continuous energy spectrum with density $\rho(E)$ in the vicinity of E_i. To find the transition rate $w_{\{j\},i}$ (transition probability $P_{\{j\},i}$ per unit time), we need to sum the amplitudes $|c_{ij}(t)|^2$ over these final states. The sum is evaluated as follows:

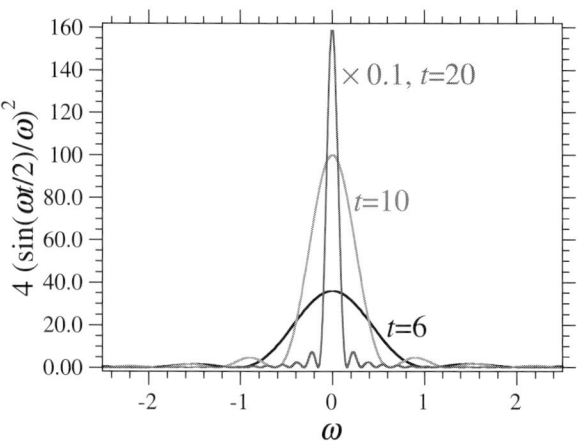

FIG 7.3 Plot of $4\left(\frac{\sin(\omega t/2)}{\omega}\right)^2$ versus ω for $t = 6$, $t = 10$, and $t = 20$. The curve for $t = 20$ has been multiplied by a factor of 0.1 so as to fit with the scale used.

$$\sum_j |c_j^{(1)}(t)|^2 = \int dE_j \, \rho(E_j) \frac{|V_{ji}|^2}{\hbar^2} 4 \left(\frac{\sin\left[(E_j - E_i)t/2\hbar\right]}{(E_j - E_i)/\hbar}\right)^2, \quad (7.73)$$

At large times, we can take the limit,

$$\lim_{t \to \infty} 4 \left(\frac{\sin(\omega t/2)}{\omega} \right)^2 = 2\pi t \, \delta(\omega), \tag{7.74}$$

which, when substituted into Eq. (7.73), yields

$$\lim_{t \to \infty} P_{\{j\},i} = t \frac{2\pi}{\hbar^2} \int dE_j \, \rho(E_j) \, |V_{ji}|^2 \delta((E_j - E_i)/\hbar). \tag{7.75}$$

Thus, the transition probability per unit time at large times, $w_{\{j\},i} \equiv \frac{dP_{j,i}}{dt}$, i.e., the rate of change of the probability of the transition from state i to states of energy E_j, is given by the t-independent expression,

$$\boxed{w_{\{j\},i} = \frac{2\pi}{\hbar} \, \overline{|V_{ji}|^2} \, \rho(E_i).} \tag{7.76}$$

where the bar indicates the average over the matrix elements squared with final states at energy E_j. Equation (7.76) is an extremely important and useful result; it is called the *Fermi Golden Rule* for transition rates, $w_{j,i} \equiv \frac{dP_{j,i}}{dt}$. This result was first obtained by Gregor Wentzel but was named the "Golden Rule" by Enrico Fermi; today, it is called the *Fermi Golden Rule* and is one of the most important results of perturbation theory.

Second-order perturbation theory for a perturbation $\Theta(t)V(\mathbf{r})$ turned on at $t = 0$ is determined using Eq. (7.62):

$$c_j^{(2)}(t) = \frac{1}{(i\hbar)^2} \sum_k \frac{V_{jk} V_{ki}}{i(E_k - E_i)/\hbar} \int_0^t dt'' \, e^{i(E_j - E_k)t''/\hbar} [e^{i(E_k - E_i)t''/\hbar} - 1]. \tag{7.77}$$

The first term in the integrand on the RHS of Eq. (7.77), $e^{i(E_j - E_i)t/\hbar}$, has the same time dependence as the integrand in the first-order expression (7.70), and the same arguments as those made above apply. The second term oscillates rapidly as $t \to \infty$, and therefore, its contribution is "averaged out." Hence, the sum of the first- and second-order rates is

$$w_{\{j\},i} = \frac{2\pi}{\hbar} \, \overline{\left| V_{ji} + \sum_k \frac{V_{jk} V_{ki}}{E_i - E_k} \right|^2} \, \rho(E_i). \tag{7.78}$$

The intermediate states $|k\rangle$ in the second-order terms need not have the same energy as the initial state (and therefore, the final state), but the system does not remain in these states, i.e., the virtual transitions to these intermediate states are not energy conserving, and the system only passes through these states. In many applications, the symmetry demands that one or the other of the first and second order terms vanish, so only one of these terms contributes to the transition rate.

Harmonic Perturbation

An important application of time-dependent perturbation theory is to treat time-dependent potentials of the form

$$V(t) = V_\omega e^{-i\omega t} + V_\omega^\dagger e^{i\omega t}, \tag{7.79}$$

where V_ω may depend on $\mathbf{r}$ or $\mathbf{p}$ (or other degrees of freedom of the system); for simplicity of notation, we drop the subscript on V_ω in what follows. For example, this is the form of the potential for matter interacting with a single-mode

7.3 Perturbation Theory

electromagnetic field, and gives rise to absorption and emission of light. Substituting Eq. (7.79) into the first-order expression (7.61), we obtain

$$c_j^{(1)}(t) = \frac{1}{i\hbar}\int_0^t dt' \left[V_{ji}\, e^{i(-\omega+\omega_{ji})t'} + (V^\dagger)_{ji}\, e^{i(\omega+\omega_{ji})t'} \right]$$

$$= \frac{-1}{\hbar}\left[V_{ji}\frac{e^{i(-\omega+\omega_{ji})t}-1}{-\omega+\omega_{ji}} + (V^\dagger)_{ji}\frac{e^{i(\omega+\omega_{ji})t}-1}{\omega+\omega_{ji}} \right], \tag{7.80}$$

where $\omega_{ji} \equiv (E_j - E_i)/\hbar$. This result is similar to that of the constant potential case. Taking only the first term on the RHS (dropping the second term) gives

$$P_j^{(1)}(t) = \frac{1}{\hbar^2}|V_{ji}|^2\, 4 \left(\frac{\sin(\omega_{ji}-\omega)t/2}{(\omega_{ji}-\omega)} \right)^2. \tag{7.81}$$

Had we dropped the first term and kept only the second, a similar result with $(\omega_{ji}-\omega) \to (\omega_{ji}+\omega)$ would have been obtained. It is not necessary to keep both terms, as we shall see shortly. Note that at very small times t,

$$P_j^{(1)}(t) \sim \frac{|V_{ji}|^2}{\hbar^2}t^2. \tag{7.82}$$

We shall use this result in Sec. 7.4.4.

When we keep only the first term on the RHS of Eq. (7.80), and take the limit $t \to \infty$, $|c_j^{(1)}(\infty)|^2$ is significant only when $E_j = E_i + \hbar\omega$ (recall Fig. 7.3). Using Eq. (7.74) yields,

$$P_j^{(1)}(\infty) = \frac{2\pi}{\hbar}t\,|V_{ji}|^2\,\delta(E_j - E_i - \hbar\omega). \tag{7.83}$$

For the second term on the RHS of Eq. (7.80), the probability is significant only for $E_j = E_i - \hbar\omega$. Hence, the first term is *on resonance* for absorption of a photon of frequency ω if the denominator of the first term vanishes, whereas the second term is *on resonance* for stimulated emission when the denominator of the second term vanishes. Therefore, the transition rate into a group of levels $\{j\}$, $w_{\{j\},i} = \lim_{t\to\infty}[P_j^{(1)}(t)/t]$, is given by

$$w_{\{j\},i} = \frac{2\pi}{\hbar}\begin{cases} |V_{ji}|^2\,\rho(E_j)\,\delta(E_j - E_i + \hbar\omega) & \text{(stimulated emission)} \\ |(V^\dagger)_{ji}|^2\,\rho(E_j)\,\delta(E_j - E_i - \hbar\omega) & \text{(absorption)}. \end{cases} \tag{7.84}$$

where the top row is for stimulated emission of radiation and the bottom row is for absorption.

The second-order perturbation theory amplitude arising from the first term on the RHS of Eq. (7.79) [for photon absorption (see Sec. 7.4)] is

$$c_j^{(2)}(t) = \frac{1}{(i\hbar)^2}\sum_k \frac{V_{jk}V_{ki}}{i(\omega_{ki}-\omega)} \int_0^t dt''\, e^{i(\omega_{jk}-\omega)t''}[e^{i(\omega_{ki}-\omega)t''} - 1]. \tag{7.85}$$

The two-photon emission amplitude and the two-photon Raman scattering are similar in form. We shall use these results in treating Rayleigh and Raman scattering, and two-photon absorption and emission processes, in Sec. 7.4.3.

Problem 7.10

(a) Carry out the calculation of the integral over time [see Eq. (7.85)] in the two-photon absorption rate.
(b) Calculate the similar integrals that appear in the rate for two-photon emission and Raman scattering.

Rotating-Wave Approximation

In the *rotating-wave approximation* (RWA), one retains only one of the oscillating terms of the potential (7.79). For example, in photon absorption between two levels of an atom or a molecule, say from state i to state j, we keep the first term of Eq. (7.79) with $\omega \approx \omega_{ji}$, whereas for emission, we keep the second term. We used the RWA in Eq. (7.91) when we retained only the first term on the RHS of Eq. (7.81) and in Sec. 6.1.4. The RWA is applicable even for pulses of light with finite bandwidth, as described in the next subsection. The RWA is often made in spin resonance problems with a rotating magnetic field; this application was the origin of the name RWA.

Nearly Harmonic Perturbation

Let us now consider an electromagnetic pulse with finite bandwidth centered around frequency ω and take an interaction potential of the form

$$V(t) = E(t)\mathcal{O}, \tag{7.86}$$

where $E(t)$ is the electric field and $\mathcal{O}$ is the operator that couples the electric field to the atom or molecule. In the dipole approximation, $\mathcal{O}$ is proportional to the transition dipole operator (see Sec. 7.4.2). The pulse is turned on at early times and turns off at late times, so we take the time t_0 in Eqs (7.61) and (7.64) to be $-\infty$. The first-order perturbation theory expression now becomes

$$c_j^{(1)}(t) = \frac{1}{i\hbar}\int_{-\infty}^{t} dt'\, E(t')\mathcal{O}_{ji} e^{i\omega_{ji}t'}. \tag{7.87}$$

Let us define the Fourier expansion of the slowly varying envelope (SVE) as follows[1]:

$$E(t') = \frac{1}{\sqrt{2\pi}} \int_{-\infty}^{\infty} d\omega'\, A(\omega + \omega') e^{-i(\omega+\omega')t'}, \tag{7.88}$$

where the function $A(\cdot)$ is peaked at ω [$A(\omega + \omega')$ falls off as $|\omega'|$ increases]'. Since the field $E(t')$ must be real, in the limiting case of a single-frequency field, $E(t') = A(\omega)e^{-i\omega t'} + A(-\omega)e^{i\omega t'}$, with $A(-\omega) = A^*(\omega)$. For a (real) electric field of finite bandwidth, we must have $A(-\omega - \omega') = A^*(\omega + \omega')$. Substituting Eq. (7.88) into Eq. (7.87), and taking the limit as $t \to \infty$ in Eq. (7.87), we find, on carrying out the integral over t',

$$c_j^{(1)}(\infty) = \frac{\sqrt{2\pi}}{i\hbar} A(\omega_{ji})\mathcal{O}_{ji}. \tag{7.89}$$

Writing $A(\omega_{ji})$ in terms of a magnitude and phase, $A(\omega_{ji}) = |A(\omega_{ji})|e^{i\Phi(\omega_{ji})}$, we find that the probability for the transition from state i to state j is

$$P_j(\infty) = \frac{2\pi}{\hbar^2} |A(\omega_{ji})|^2 |\mathcal{O}_{ji}|^2. \tag{7.90}$$

---

[1] The definition of the SVE $A(t)$ is given by the relation, $E(t) = A(t)e^{-i\omega t}$.

7.4 Dynamics in an Electromagnetic Field

Thus, the probability of being in state j is proportional to the square of the amplitude of the Fourier component of the optical field at resonance, and the phase $\Phi(\omega_{ji})$ does not affect the excitation probability.

The second-order perturbation theory expression for the two-photon absorption amplitude $c_j^{(2)}(\infty)$ obtained for an incident pulse of light can be determined using Eq. (7.62), with the first term on the RHS of (7.86):

$$c_j^{(2)}(\infty) = \frac{1}{(i\hbar)^2} \sum_k \int_{-\infty}^{\infty} dt'' \int_{-\infty}^{t''} dt' \, \mathcal{O}_{jk} E(t'') \mathcal{O}_{ki} E(t') e^{i[(\omega_{jk}-\omega)t''+(\omega_{ki}-\omega)t']}. \tag{7.91}$$

Substituting the Fourier transform expression in Eq. (7.88) yields

$$c_j^{(2)}(\infty) = \frac{1}{(i\hbar)^2 2\pi} \sum_k \mathcal{O}_{jk} \mathcal{O}_{ki} \int_{-\infty}^{\infty} d\omega' \int_{-\infty}^{\infty} d\omega'' A(\omega+\omega') A(\omega+\omega'') \times \int_{-\infty}^{\infty} dt'' \int_{-\infty}^{t''} dt' \, e^{i[(\omega_{jk}-\omega-\omega'')t''+(\omega_{ki}-\omega-\omega')t']}. \tag{7.92}$$

The integral over t' equals $\frac{e^{i[(\omega_{jk}-\omega-\omega'')+\omega_{ki}-\omega-\omega')]t''}}{i(\omega_{jk}-\omega-\omega'')}$, the integral over t'' yields $2\pi\delta(\omega_{ji}-2\omega-\omega'-\omega'')$, and the integral over ω'', which contains the delta function, can be evaluated analytically; hence, the two-photon transition amplitude is

$$c_j^{(2)}(\infty) = \frac{-i}{(i\hbar)^2} \sum_k \mathcal{O}_{jk} \mathcal{O}_{ki} \int_{-\infty}^{\infty} d\omega \frac{A(\omega) A(\omega_{ji}-\omega)}{\omega - \omega_{ki}}. \tag{7.93}$$

Here, $\omega + \omega'$ was renamed as ω, so the integral $\int_{-\infty}^{\infty} d\omega' \frac{A(\omega+\omega')A(\omega_{ji}-\omega-\omega')}{\omega+\omega'-\omega_{ki}} \rightarrow \int_{-\infty}^{\infty} d\omega \frac{A(\omega)A(\omega_{ji}-\omega)}{\omega-\omega_{ki}}$. The transition probability can be controlled by manipulating the spectral phase function of the pulse using pulse shaping techniques to create constructive interference in the integral on the RHS of Eq. (7.93) [110]. Similar considerations apply for stimulated Raman scattering (see below), wherein one photon with frequency ω is absorbed from the radiation field and one photon with frequency ω' is emitted.

7.4 DYNAMICS IN AN ELECTROMAGNETIC FIELD

We now apply perturbation theory to atoms and molecules in an electromagnetic field. The Hamiltonian of a particle of charge q and mass m interacting with electromagnetic radiation was given in Eq. (16.95) of Chapter 16, linked to the book web page, as

$$H(\mathbf{r},\mathbf{p},t) = \frac{1}{2m}\left(\mathbf{p} - \frac{q}{c}\mathbf{A}(\mathbf{r},t)\right)^2 + q\varphi(\mathbf{r},t). \tag{7.94}$$

For an electron (charge $q = -e$), we must add the Hamiltonian for the interaction of the spin with the magnetic field, i.e., the Zeeman Hamiltonian, $-\boldsymbol{\mu} \cdot \mathbf{H} = g\frac{e\hbar}{2mc}\frac{\boldsymbol{\sigma}}{2} \cdot \mathbf{H}$. It is convenient to use the Coulomb gauge (see Sec. 16.7, linked to the book web page), $\nabla \cdot \mathbf{A} = 0$, for the vector potential to eliminate one of the cross terms arising from the first term on the RHS of Eq. (7.94) [$\mathbf{p} \cdot \mathbf{A} + \mathbf{A} \cdot \mathbf{p} = 2\mathbf{A} \cdot \mathbf{p}$], so the Hamiltonian becomes

$$H(\mathbf{r},\mathbf{p},t) = \frac{1}{2m}\mathbf{p}^2 + (-e)\varphi(\mathbf{r},t) + \frac{e}{mc}\mathbf{A}(\mathbf{r},t) \cdot \mathbf{p} + \frac{e^2}{2mc^2}\mathbf{A}^2(\mathbf{r},t) + g\frac{e\hbar}{2mc}\frac{\boldsymbol{\sigma}}{2} \cdot \mathbf{H}. \tag{7.95}$$

Here, $\mathbf{H} = \nabla \times \mathbf{A}$, and in the Coulomb gauge, the scalar field appearing in Eq. (7.95) is just the instantaneous Coulomb potential due to the charge density ρ, $\varphi(\mathbf{r},t) = \int d\mathbf{r}' \frac{\rho(\mathbf{r}',t)}{|\mathbf{r}-\mathbf{r}'|}$ (see Sec. 16.7 in the classical mechanics chapter linked to the book web page).

The radiation field energy flux density, S, also called the *intensity* of the field, equals the energy density u times the speed of light c, where $u = \frac{1}{8\pi}\overline{(E^2 + B^2)}$, and the average is over a small time interval including at least several cycles of the radiation field [18, 34]. Hence,

$$S = cu = \frac{c}{8\pi}\left(\left|-\frac{1}{c}\frac{\partial \mathbf{A}}{\partial t}\right|^2 + |\nabla \times \mathbf{A}|^2\right). \tag{7.96}$$

Let us consider a single-mode field, i.e., a monochromatic field with only one wavevector $\mathbf{k}$ and polarization $\hat{\boldsymbol{\varepsilon}}$, $\mathbf{A}(\mathbf{r},t) = A_0 \hat{\boldsymbol{\varepsilon}} \cos(\mathbf{k}\cdot\mathbf{r} - \omega t)$. The vector potential satisfies the wave equation,

$$\left(\nabla^2 - \frac{1}{c^2}\frac{\partial^2}{\partial t^2}\right)\mathbf{A} = 0, \tag{7.97}$$

so, $|\mathbf{k}| = \omega/c$. The unit vector $\hat{\boldsymbol{\varepsilon}}$ specifies the polarization direction; it is orthogonal to the wavevector, $\mathbf{k}\cdot\hat{\boldsymbol{\varepsilon}} = 0$, in order for the Coulomb gauge condition to be satisfied. Hence, using Eq. (7.96), the energy flux density (i.e., the intensity) is

$$S = \frac{1}{2\pi}\frac{\omega^2}{c}|A_0|^2. \tag{7.98}$$

The intensity is often given in units of W/cm² (despite this being in mixed units) or W/m² in SI or in units of erg/(s cm²) in Gaussian units.

With the single-mode form of the vector potential, the light-matter interaction term $\mathbf{A}(\mathbf{r},t)\cdot\mathbf{p}$ in the Hamiltonian of Eq. (7.95), using $\cos(\mathbf{k}\cdot\mathbf{r} - \omega t) = \frac{1}{2}\left[e^{i(\mathbf{k}\cdot\mathbf{r}-\omega t)} + e^{-i(\mathbf{k}\cdot\mathbf{r}-\omega t)}\right]$, we obtain

$$V(\mathbf{r},\mathbf{p},t) = -\frac{q}{2mc}\left[A_0 e^{i(\mathbf{k}\cdot\mathbf{r}-\omega t)} + A_0^* e^{-i(\mathbf{k}\cdot\mathbf{r}-\omega t)}\right]\hat{\boldsymbol{\varepsilon}}\cdot\mathbf{p}. \tag{7.99}$$

The $A_0 e^{i(\mathbf{k}\cdot\mathbf{r}-\omega t)}$ term gives rise to photon absorption using Eq. (7.84) with

$$(V^\dagger)_{ji} = -\frac{qA_0}{2mc}\left(e^{i\mathbf{k}\cdot\mathbf{r}}\hat{\boldsymbol{\varepsilon}}\cdot\mathbf{p}\right)_{ji}, \tag{7.100}$$

whereas $A_0^* e^{-i(\mathbf{k}\cdot\mathbf{r}-\omega t)}$ gives rise to emission using Eq. (7.84) with

$$V_{ji} = -\frac{qA_0^*}{2mc}\left(e^{-i\mathbf{k}\cdot\mathbf{r}}\hat{\boldsymbol{\varepsilon}}\cdot\mathbf{p}\right)_{ji}. \tag{7.101}$$

The absorption and emission arising from the Zeeman term in Eq. (7.95) are discussed in Sec. 7.4.2. It gives rise to higher order effects (magnetic dipole interactions).

The absorption cross-section, σ_{abs}, is defined as the energy per unit time absorbed by the atom divided by the energy flux of the radiation field, S. The energy per unit time absorbed is given by $\hbar\omega\, w_{ji}$, where the rate w_{ji} is given by

$$w_{ji} = \frac{2\pi q^2 |A_0|^2}{\hbar 4m^2 c^2}\left|\left(e^{i\mathbf{k}\cdot\mathbf{r}}\hat{\boldsymbol{\varepsilon}}\cdot\mathbf{p}\right)_{ji}\right|^2 \delta(E_j - E_i - \hbar\omega). \tag{7.102}$$

7.4 Dynamics in an Electromagnetic Field

Hence, on setting $q = -e$, the absorption cross-section is

$$\boxed{\sigma_{\text{abs}} = \frac{\hbar\omega \, w_{ji}}{S} = \frac{4\pi^2 \hbar \, e^2}{m^2 \omega \, \hbar c} \left| \left(e^{i\mathbf{k}\cdot\mathbf{r}} \hat{\boldsymbol{\varepsilon}} \cdot \mathbf{p} \right)_{ji} \right|^2 \delta(E_j - E_i - \hbar\omega).} \qquad (7.103)$$

Here, the dimensionless factor $\alpha \equiv \frac{e^2}{\hbar c}$ is the fine-structure constant. The delta function has units of energy^{-1}, $\frac{\hbar}{m\omega}$ has the units of length^2 (remember the harmonic oscillator length in Sec. 1.3.15), and $\mathbf{p}^2$ from the matrix element, divided by m from the leftmost fraction, has units of energy. Hence, the cross-section has units of length^2, consistent with its definition.

In 1927, Paul Dirac showed how to quantize the free electromagnetic field by associating each mode of the radiation field with a quantized harmonic oscillator. Subsequently, the general theory of matter interacting with a radiation fields was developed, and it was further generalized into quantum electrodynamics upon quantizing the radiation field (see Sec. 14.1.7). Sin-Itiro Tomonaga, Julian Schwinger and Richard Feynman were jointly awarded with a Nobel prize in physics in 1965 for their work on quantum electrodynamics.

7.4.1 SPONTANEOUS AND STIMULATED EMISSION OF RADIATION

We shall apply the Fermi Golden Rule to calculate the rate for spontaneous and stimulated emission per unit solid angle from an excited state atom or molecule, $dw_{ji}/d\Omega$. To do so, we will need the density of allowed photon states per unit solid angle $d\Omega$ and per unit energy interval $d(\hbar\omega)$ per unit volume, which can be written as

$$d\rho(\omega) = \frac{V}{(2\pi)^3} \frac{k^2 dk d\Omega}{d(\hbar\omega)} = \frac{V\omega^2 d\Omega}{(2\pi)^3 \hbar c^3}, \qquad (7.104)$$

where $k = \omega/c$ and $d\mathbf{k} = k^2 dk d\Omega$. This is basically the same calculation as in Sec. 9.1, where we calculate the density of states per unit energy of a free electron [see Eq. (9.19) – note that the symbol Ω has a different meaning there]. Note also that we have yet to account for the two orthogonal polarization modes of the photon with momentum $\mathbf{k}$ having polarization vectors $\hat{\boldsymbol{\varepsilon}}^{(1)}$ and $\hat{\boldsymbol{\varepsilon}}^{(2)}$, respectively (see Fig. 7.4). Using Eqs (7.84) and (7.101), we find the differential rate

$$\frac{dw_{ji}}{d\Omega} = \frac{2\pi e^2}{\hbar m^2 c^2} \int d(\hbar\omega) \left| A_0^* \left(e^{-i\mathbf{k}\cdot\mathbf{r}} \hat{\boldsymbol{\varepsilon}}^{(\alpha)*} \cdot \mathbf{p} \right)_{ji} \right|^2 \frac{V\omega^2}{(2\pi)^3 \hbar c^3} \delta(E_j - E_i - \hbar\omega). \qquad (7.105)$$

Integrating this expression over energy and solid angle, summing over the two polarization states of the radiation, making the approximation $e^{-i\mathbf{k}\cdot\mathbf{r}} \approx 1$, and setting $\mathbf{p} = im\omega_{ji}\mathbf{r}$ (i.e., making the *electric dipole approximation* – see next section), we obtain

$$w_{ji} = \frac{e^2 \omega_{ji}^4 V}{(2\pi)^2 \hbar^2 c^5} \sum_\alpha \int d\Omega \, |A_0|^2 \left| \hat{\boldsymbol{\varepsilon}}^{(\alpha)*} \cdot \langle \psi_j | \mathbf{r} | \psi_i \rangle \right|^2. \qquad (7.106)$$

The integral over solid angle and sum over the two polarization states can be carried out on noting the orientation of the vector $\mathbf{r}_{ji}$ relative to the triad $\hat{\mathbf{k}}$ and the orientation of the polarization unit vectors $\hat{\boldsymbol{\varepsilon}}^{(1)}$ and $\hat{\boldsymbol{\varepsilon}}^{(2)}$ (the electric vector is perpendicular to $\hat{\mathbf{k}}$, as shown in Fig. 7.4), and using the relations,

$$\hat{\boldsymbol{\varepsilon}}^{(\alpha)} \cdot \mathbf{r}_{ji} = |\mathbf{r}_{ji}| \cos \Theta^{(\alpha)} = \begin{cases} |\mathbf{r}_{ji}| \sin\theta \cos\phi & \text{for } \alpha = 1, \\ |\mathbf{r}_{ji}| \sin\theta \sin\phi & \text{for } \alpha = 2. \end{cases} \qquad (7.107)$$

The sum over α and the integral over solid angle, using

$$\int_0^{2\pi} d\phi \int d\cos\theta \, [\sin^2\theta (\cos^2\phi + \sin^2\phi)] = 2\pi \int_{-1}^1 dx \, (1 - x^2) = \frac{8\pi}{3}, \qquad (7.108)$$

yields $w_{ji} = \frac{2e^2\omega_{ji}^4 V}{3\pi\hbar^2 c^5}|A_0|^2|\mathbf{r}_{ji}|^2$. The term in the square brackets in the integrand of Eq. (7.108) gives the dipole radiation angular distribution pattern for unpolarized light.

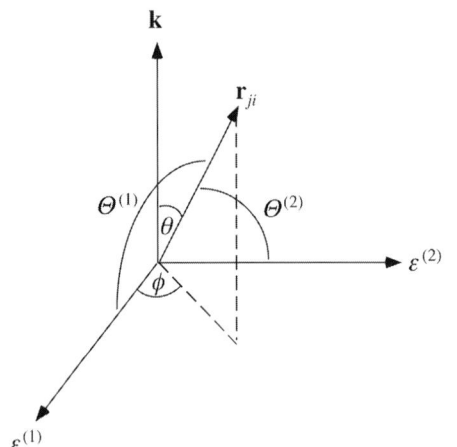

FIG 7.4 Orientation of $\mathbf{r}_{ji}$ relative to $\hat{\mathbf{k}}, \hat{\boldsymbol{\varepsilon}}^{(1)}$ and $\hat{\boldsymbol{\varepsilon}}^{(2)}$ used for the calculation of $\hat{\boldsymbol{\varepsilon}}^{(\alpha)} \cdot \mathbf{r}_{ji}$ for $\alpha = 1$ and $\alpha = 2$. *Source:* Band, Light and Matter, Fig. 5.1, p. 307

To obtain the value of $|A_0|^2$ for spontaneous and stimulated emissions, we draw on the second quantized form of the vector potential operator obtained in Sec. 14.1.7:

$$\mathbf{A}(\mathbf{r},t) = \sum_{\mathbf{k}'}\sqrt{\frac{2\pi c^2\hbar}{V\omega'}}\sum_{\alpha'=1}^{2}\left(a_{\mathbf{k}',\alpha'}\hat{\boldsymbol{\varepsilon}}^{(\alpha')}e^{i(\mathbf{k}'\cdot\mathbf{r}-\omega't)} + h.c.\right). \quad (7.109)$$

Here, $a_{\mathbf{k}',\alpha'}$ and its Hermitian conjugate $a_{\mathbf{k}',\alpha'}^\dagger$ are the photon lowering and raising operators for mode $\{\mathbf{k}',\alpha'\}$. When no photons are initially present in mode $\{\mathbf{k}',\alpha'\}$, the raising operator creates one photon, $\langle 1_{\mathbf{k}',\alpha'}|a_{\mathbf{k}',\alpha'}^\dagger|0_{\mathbf{k}',\alpha'}\rangle = 1$, so $|A_0|^2 = \frac{2\pi c^2\hbar}{V\omega}$. Hence, we obtain the final result

$$\boxed{w_{ji} = \frac{4}{3}\frac{e^2}{\hbar c}\frac{\omega_{ji}^3}{c^2}|\langle\psi_j|\mathbf{r}|\psi_i\rangle|^2.} \quad (7.110)$$

If we multiply this rate by the photon energy $\hbar\omega_{ji}$, we obtain a formula for the rate of change of energy, which is the generalization of the famous classical *Larmor energy loss formula*.

When $N_{\mathbf{k},\alpha}$ photons are in mode $\{\mathbf{k}',\alpha'\}$, the raising operator creates one additional photon, $\langle(N_{\mathbf{k},\alpha}+1)|a_{\mathbf{k},\alpha}^\dagger|N_{\mathbf{k},\alpha}\rangle = \sqrt{(N_{\mathbf{k},\alpha}+1)}$, and the rate (7.110) is multiplied by a factor of $(N_{\mathbf{k},\alpha}+1)$, i.e., we obtain *stimulated emission*.

Problem 7.11

(a) Carry out the integration over angles leading to Eq. (7.110).

(b) The *oscillator strength* for a transition $i \to j$ is defined as $f_{ji} \equiv \frac{2m\omega_{ji}}{\hbar}|\langle\psi_j|\mathbf{r}|\psi_i\rangle|^2$. Express the decay rate w_{ji} in terms of f_{ji}.

(c) Explain why the oscillator strength is a dimensionless quantity.

(d) Prove the Thomas-Reiche-Kuhn sum rule, $\sum_j f_{ji} = 1$ (for an atom with Z electrons, $\sum_j f_{ji} = Z$).
Hint: Note that $\sum_j|\langle\psi_j|\mathbf{r}|\psi_i\rangle|^2 = \sum_j\langle\psi_i|\mathbf{r}|\psi_j\rangle \cdot \langle\psi_j|\mathbf{r}|\psi_i\rangle$, use completeness and consider the commutator $[\mathbf{r},[\mathbf{r},H]]$ to show that $[x_k,[x_l,H]] = \frac{\hbar^2}{m}\delta_{kl}$.

(e) Show that the Thomas-Reiche-Kuhn sum rule can be written as $\frac{2m}{\hbar^2}\sum_j(E_j-E_i)|\langle\psi_j|\hat{\mathbf{n}}\cdot\mathbf{r}|\psi_i\rangle|^2 = Z$, where the direction $\hat{\mathbf{n}}$ is arbitrary.

Answers: (b) $w_{ji} = \frac{2e^2 f_{ji}\omega_{ji}^2}{3mc^3}$. (c) Noting that $f_{ji} \equiv \frac{3mc^3 w_{ji}}{2e^2\omega_{ji}^2}$ and that mc^2 has units of energy, e^2/c has units of energy × time, ω_{ji}^2 has units of time^{-2}, and w_{ji} has units of time^{-1}, we find that f_{ji} is indeed dimensionless.

7.4.2 ELECTRIC DIPOLE AND MULTIPOLE RADIATION

The matrix element in Eq. (7.103) for absorption processes contains a plane wave factor, which can be expanded as follows:

$$(\mathcal{O}) \equiv \left(e^{i\mathbf{k}\cdot\mathbf{r}}\hat{\boldsymbol{\varepsilon}}\cdot\mathbf{p}\right) = \left[1 + i\mathbf{k}\cdot\mathbf{r} - \frac{1}{2}(\mathbf{k}\cdot\mathbf{r})^2 + \cdots\right]\hat{\boldsymbol{\varepsilon}}\cdot\mathbf{p}. \quad (7.111)$$

7.4 Dynamics in an Electromagnetic Field

A similar expansion can be performed for the *emission* matrix element appearing in Eq. (7.105). The wavevector has length $|\mathbf{k}| = \frac{\omega}{c} = \frac{2\pi}{\lambda}$, so $\mathbf{k} \cdot \mathbf{r} \approx \frac{2\pi R_{\text{system}}}{\lambda}$. The inequality $\mathbf{k} \cdot \mathbf{r} \ll 1$ is valid from the microwave through the X-ray region of the spectrum for atomic and molecular systems since R_{system} is of order nanometers or less and $\lambda \gg R_{\text{system}}$. Therefore, the lowest order term of the matrix element is $(\hat{\boldsymbol{\varepsilon}} \cdot \mathbf{p})_{ji} = \hat{\boldsymbol{\varepsilon}} \cdot \langle \psi_j^{(0)} | \mathbf{p} | \psi_i^{(0)} \rangle$. Taking the expectation value of the expression $\mathbf{p} = \frac{m}{i\hbar}[\mathbf{r}, H_0]$, we find

$$\langle \psi_j^{(0)} | \mathbf{p} | \psi_i^{(0)} \rangle = \frac{m}{i\hbar} \langle \psi_j^{(0)} | [\mathbf{r}, H_0] | \psi_i^{(0)} \rangle = im\omega_{ji} \langle \psi_j^{(0)} | \mathbf{r} | \psi_i^{(0)} \rangle. \tag{7.112}$$

Therefore, the lowest order term of Eq. (7.111) is

$$\mathcal{O}_{E1} = im\omega \,\hat{\boldsymbol{\varepsilon}} \cdot \mathbf{r} \tag{7.113}$$

and is called the *electric dipole approximation*. The transition involving this lowest order term of Eq. (7.111) is called an *electric dipole* (E1) transition, because the quantity $q\mathbf{r}$ is the electric dipole moment operator. In this approximation,

$$\sigma_{\text{abs}} = 4\pi^2 \alpha \,\omega_{ji} \left| \hat{\boldsymbol{\varepsilon}} \cdot \langle \psi_j^{(0)} | \mathbf{r} | \psi_i^{(0)} \rangle \right|^2 \delta(\omega_{ji} - \omega). \tag{7.114}$$

When $\langle \psi_j^{(0)} | \mathbf{r} | \psi_i^{(0)} \rangle = 0$, it is necessary to consider the next order term in Eq. (7.111), $i(\mathbf{k} \cdot \mathbf{r})(\hat{\boldsymbol{\varepsilon}} \cdot \mathbf{p})$. It is convenient to rewrite this term as

$$\frac{i}{2} \left\{ [\mathbf{k} \cdot (\mathbf{rp} + \mathbf{pr}) \cdot \hat{\boldsymbol{\varepsilon}}] + [(\mathbf{k} \cdot \mathbf{r})(\hat{\boldsymbol{\varepsilon}} \cdot \mathbf{p}) - (\hat{\boldsymbol{\varepsilon}} \cdot \mathbf{p})(\mathbf{k} \cdot \mathbf{r})] \right\}. \tag{7.115}$$

An *electric quadrupole* (E2) radiative transition results from the first term involving the symmetric dyadic, $(\mathbf{rp} + \mathbf{pr})$, whereas *magnetic dipole* (M1) radiative transitions result from the second term. Using vector identities [see (C.18)], the second term of Eq. (7.115) can be written as,

$$(\mathbf{k} \cdot \mathbf{r})(\hat{\boldsymbol{\varepsilon}} \cdot \mathbf{p}) - (\hat{\boldsymbol{\varepsilon}} \cdot \mathbf{p})(\mathbf{k} \cdot \mathbf{r}) = (\mathbf{k} \times \hat{\boldsymbol{\varepsilon}}) \cdot (\mathbf{r} \times \mathbf{p}), \tag{7.116}$$

where $i(\mathbf{k} \times \hat{\boldsymbol{\varepsilon}})$ is the leading term in the expansion of a magnetic field $\mathbf{H}$ and $(\mathbf{r} \times \mathbf{p}) = \mathbf{L} = \boldsymbol{\mu}_L/\mu_B$. Hence, the name magnetic dipole transition. The E2 term can be rewritten, using $(\mathbf{rp} + \mathbf{pr}) = \frac{im}{\hbar}[H_0, \mathbf{rr}]$, where $\mathbf{rr}$ is the dyatic product of two position operators, as

$$\frac{i}{2} \mathbf{k} \cdot (\mathbf{rp} + \mathbf{pr}) \cdot \hat{\boldsymbol{\varepsilon}} = \frac{m\omega}{2} \mathbf{k} \cdot \mathbf{rr} \cdot \hat{\boldsymbol{\varepsilon}}, \tag{7.117}$$

and because $\mathbf{k} \cdot \hat{\boldsymbol{\varepsilon}} = 0$ (the photon is transverse), we can add $-\frac{1}{3}|\mathbf{r}|^2 \delta_{ij}$ to obtain the electric quadrupole operator

$$\mathcal{O}_{E2} = \frac{m\omega}{2} \mathbf{k} \cdot \left(\mathbf{rr} - \frac{1}{3} |\mathbf{r}|^2 \mathbf{1} \right) \cdot \hat{\boldsymbol{\varepsilon}}. \tag{7.118}$$

For the M1 transition, we need to add the spin magnetic moment interaction, $-\boldsymbol{\mu} \cdot \mathbf{H}/2 \propto (i/2)\mu_B \hbar \boldsymbol{\sigma} \cdot (\mathbf{k} \times \hat{\boldsymbol{\varepsilon}}) = (i/2)2\mu_B \mathbf{S} \cdot (\mathbf{k} \times \hat{\boldsymbol{\varepsilon}})$, to the magnetic dipole term to obtain the full magnetic dipole moment operator:

$$\mathcal{O}_{M1} = \frac{i}{2}(\mathbf{k} \times \hat{\boldsymbol{\varepsilon}}) \cdot (\mathbf{L} + 2\mathbf{S}) = \frac{i}{2}(\mathbf{k} \times \hat{\boldsymbol{\varepsilon}}) \cdot (\mathbf{J} + \mathbf{S}). \tag{7.119}$$

The hyperfine transition in atomic hydrogen from $F = 1$ to $F = 0$ at 1420 MHz is an M1 transition. The rate for such a spontaneous emission transition is 2.9×10^{15} s^{-1}, since the radiation frequency is small (and magnetic dipole transitions have smaller probabilities than electric dipole transitions when $\mathbf{k} \cdot \mathbf{r} \ll 1$), so the lifetime of the $F = 1$ state is about 10 million years. Despite the long lifetime, this is an extremely important transition for astrophysical observations.

The operators $\mathcal{O}_{E1}$, $\mathcal{O}_{E2}$, or $\mathcal{O}_{M1}$ can be used instead of $\left(e^{-i\mathbf{k}\cdot\mathbf{r}}\hat{\boldsymbol{\varepsilon}} \cdot \mathbf{p}\right)$ inside the matrix element of Eq. (7.103) to obtain E1, E2, or M1 cross-sections, and inside (7.102) for transition rates. Lifetimes of excited atomic states that decay via E1 spontaneous emission in the visible frequency range are typically of the order of 10^{-8} s, whereas M1 and E2 transitions are longer by roughly a factor of $(kr_{atom})^{-2} = [\lambda/(2\pi r_{atom})]^2$, i.e., the decay rates are smaller by a factor of $(kr_{atom})^2$. This is clear from the additional factor of $i\mathbf{k} \cdot \mathbf{r}$ in Eq. (7.111) for E2, instead of the factor 1 for E1 and from the factor $(\mathbf{k} \times \hat{\mathbf{e}}) \cdot (\mathbf{L})$ in Eq. (7.119) whose magnitude is $|\mathbf{k}||\mathbf{r} \times \mathbf{p}|$ and is therefore larger than the factor $|\mathbf{p}|$ in Eq. (7.111) by (kr_{atom}). For the 1420 MHz (21 cm) hyperfine transition in hydrogen, the factor in the lifetime is $[\lambda/(2\pi r_{atom})]^2 = 21/[2\pi \times 0.52 \times 10^{-8}]^2 \approx 4.1 \times 10^{17}$.

Selection rules for spontaneous emission, stimulated emission, and absorption of atoms for the various multipole moments are now presented. Using the parity selection rules of Sec. 3.6.2 and the Wigner–Eckart theorem, Eq. (3.215) [i.e., using the inequality (3.147) on the Clebsch–Gordan coefficient in Eq. 3.215], we obtain:

E1: $\Delta J = 0, \pm 1$ except $0 \not\leftrightarrow 0$. $\Delta L = 0, \pm 1$ except $0 \not\leftrightarrow 0$. $\Delta M_L = 0, \pm 1$. $\Delta S = 0$. $\Delta M_S = 0$. Parity change.
M1: $\Delta J = 0, \pm 1$ except $0 \not\leftrightarrow 0$. $\Delta M_J = 0, \pm 1$. No parity change.
E2: $\Delta J = 0, \pm 1, \pm 2$ except $0 \not\leftrightarrow 0, 1$, and $1/2 \not\leftrightarrow 1/2$. $\Delta L = 0, \pm 1, \pm 2$ except $0 \not\leftrightarrow 0, 1$. $\Delta M_L = 0, \pm 1, \pm 2$. $\Delta S = 0$. $\Delta M_S = 0$. No parity change.

Moreover, the equality condition on the magnetic quantum numbers in Eq. (3.147) yields the condition, $m_f = q + m_i$, where q is the magnetic quantum number of the multipole operator in the transition and m_i and m_f are the initial and final magnetic quantum numbers, respectively.

Higher order transitions, e.g., M2 and E3, have been observed. Their intensities $\left(\text{i.e., their oscillator strengths, } f_{ji} \equiv \frac{3mc^3 w_{ji}}{2e^2 \omega_{ji}^2}\right)$ are small.

Problem 7.12

Determine the angular distributions for the two polarizations of light from an M1 transition that involves a change in spin from (a) $s = 1, m_s = 1$ to $s = 0$ and (b) $s = 1, m_s = 0$ to $s = 0$.

Hint: M1 transitions involve $\mathcal{O}_{M1} = \frac{i}{2}(\mathbf{k} \times \hat{\mathbf{e}}) \cdot (\mathbf{L} + 2\mathbf{S})$, hence the matrix elements required are of the form $\langle 1, m_s | \mathcal{O}_{M1} | 0, 0 \rangle \propto (\mathbf{k} \times \hat{\mathbf{e}}) \cdot \langle 1, m_s | \mathbf{S} | 0, 0 \rangle$.

7.4.3 THOMSON, RAYLEIGH, RAMAN, AND BRILLOUIN TRANSITIONS

Rayleigh scattering is the elastic scattering of light by particles much smaller than the wavelength of the light, and is named after Lord Rayleigh (John William Strutt) who first described the process around 1870. Rayleigh scattering can be represented in the form of the elastic scattering process $A + \hbar\omega \to A' + \hbar\omega$, where A and A' represent initial and final states of the particle, respectively, that have exactly the same energy if the emitted photon frequency equals the incident photon frequency. Both energy and momentum are conserved in the scattering process. Since the scattered photon need not be in the same direction as the incident photon, a small amount of momentum transfer $\hbar \Delta \mathbf{k}$ is imparted to the particle (hence the prime on A'). Therefore, the kinetic energy transferred to the particle, $(\hbar \Delta \mathbf{k})^2/2M$, where M is the mass of the scattering particle, must be removed from the energy of the scattered photon relative to the energy of the initial photon, but this is usually a negligible fraction of the incident photon energy $\hbar\omega$, so we have not put a prime on the scattered photon frequency. If light is scattered off a free charged particle, the process is called Compton scattering, named after Arthur H. Compton (the low-frequency limit, $\hbar\omega \ll Mc^2$, is called Thomson scattering, which is named after J.J. Thomson). For a high-energy photon, with energy much larger than the binding energy of an atom or a molecule, one can neglect the binding energy of the electron in the atom or molecule and consider the scattering of the photon off the atom as Thomson scattering.

Inelastic light scattering can be represented in the form $A + \hbar\omega \to B + \hbar\omega'$, where conservation of energy dictates $E_A + \hbar\omega = E_B + \hbar\omega'$. When particles are excited via photons to an excited electronic-vibrational-rotational state, the inelastic

7.4 Dynamics in an Electromagnetic Field

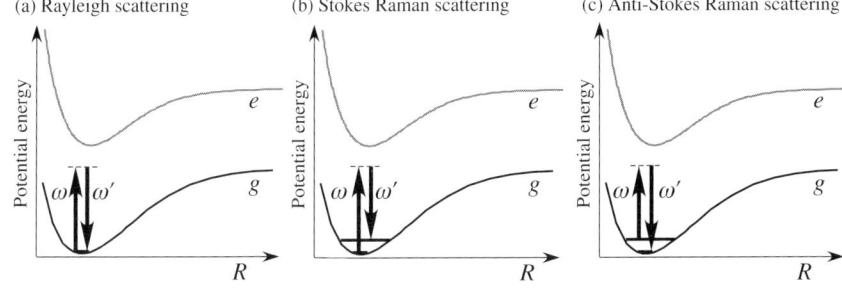

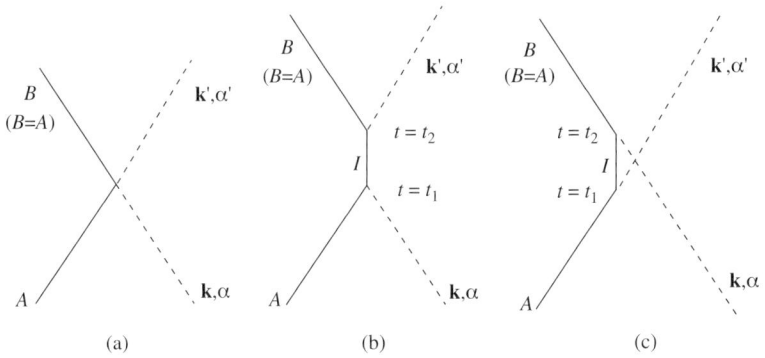

FIG 7.5 Rayleigh and Raman scattering from a diatomic molecule.

FIG 7.6 Feynman diagrams for light scattering. Time increases upward. The first (seagull) graph contributes only for elastic (Rayleigh) scattering, while the second and third graphs contribute to both Rayleigh and Raman scattering. *Source:* Band, Light and Matter, Fig. 5.4, p. 315

light scattering process is called *Raman scattering*, which is named after C.V. Raman who first observed this effect in liquids in 1928; it was first predicted by A. Smekal in 1923. If $\omega > \omega'$, i.e., $E_A < E_B$, the Raman transition is called a *Stokes transition*, whereas if $\omega < \omega'$, i.e., the scattered light is to the blue of the incident light and the matter is de-excited, the transition is known as an *anti-Stokes transition*. Rayleigh and Raman scattering processes occur in atoms, molecules, and even condensed phase materials. Figure 7.5 depicts Rayleigh and Raman transitions in a diatomic molecule. If the incident photon is in resonance with the transition to the excited electronic state, or very close to resonance, the process is called a *resonant Raman transition*. Selection rules for Rayleigh and Raman processes in diatomic molecules will be discussed in Sec. 11.6. You will develop the selection rules for atoms in Problem 7.13.

When excitation (de-excitation) of the matter involves production (destruction) of a sound wave (rather than an electronic or vibrational or rotational excitation), the process is known as Brillouin scattering. A Brillouin scattering process can also be viewed as a sound wave modulating the optical dielectric constant and resulting in an exchange of energy between the electromagnetic wave and the acoustical wave.

Compton, Rayleigh, Raman, and Brillouin scattering processes involve the destruction of a photon and the creation of another photon. Therefore, the vector potential appears in the transition amplitude of these processes to second order. Since the light-matter interaction Hamiltonian (7.95) contains a linear term $\mathbf{A} \cdot \mathbf{p}$ and a quadratic term $\mathbf{A}(\mathbf{r}, t) \cdot \mathbf{A}(\mathbf{r}, t)$, these processes can be obtained via first-order perturbation theory in $\mathbf{A}^2(\mathbf{r}, t)$ and second-order perturbation theory in $\mathbf{A} \cdot \mathbf{p}$.

The two amplitudes, $c^{(1)}_{B,\mathbf{k}'\alpha'}(t)$ and $c^{(2)}_{B,\mathbf{k}'\alpha'}(t)$, given in Eqs (7.80) and (7.85), respectively, must be added together to give rise to Rayleigh and Raman scattering amplitude $M_{BA} = c^{(1)}_{B,\mathbf{k}'\alpha'}(t) + c^{(2)}_{B,\mathbf{k}'\alpha'}(t)$. The sum of these amplitudes is proportional to

$$M_{BA} = \delta_{BA} \hat{\boldsymbol{\varepsilon}} \cdot \hat{\boldsymbol{\varepsilon}}' - \frac{1}{m} \sum_j \left(\frac{(\mathbf{p}_{Bj} \cdot \hat{\boldsymbol{\varepsilon}}')(\mathbf{p}_{jA} \cdot \hat{\boldsymbol{\varepsilon}})}{E_j - E_A - \hbar\omega} + \frac{(\mathbf{p}_{Bj} \cdot \hat{\boldsymbol{\varepsilon}})(\mathbf{p}_{jA} \cdot \hat{\boldsymbol{\varepsilon}}')}{E_j - E_A - \hbar\omega'} \right). \tag{7.120}$$

They are represented graphically by the space-time diagrams shown in Fig. 7.6. In these diagrams, known as Feynman diagrams, named after Richard Feynman, who first used them to represent perturbation theory amplitudes for quantum electrodynamics. Feynman diagrams constitute an elegant and compact method to represent every term in perturbation theory by a digram composed of lines and vertices. In these diagrams, the time axis is vertical. A solid line represents a state of an electron in the atom or molecule (i.e., an electron in a given state, say A or B) and a dashed line represents a photon. Except for the initial and final lines (those at the very bottom and very top of the diagram), the interior lines represent particle propagators and the vertices represent interaction strengths. Momentum and energy conservation are automatically satisfied. Using a list of a few rules, every diagram can be easily calculated. An intersection point between matter and photon lines, called a vertex, represents an interaction between matter and light. The first (seagull) graph in Fig. 7.6 represents $c^{(1)}_{B,\mathbf{k}'\alpha'}(t)$ with $B = A$ and shows the instantaneous interaction of the two photons $\mathbf{k}\alpha$ and $\mathbf{k}'\alpha'$ with the matter. In the second graph, representing one of the terms in the amplitude $c^{(2)}_{B,\mathbf{k}'\alpha'}(t)$, state A absorbs photon $\mathbf{k}\alpha$ at time $t = t_1$ and turns into state I, and then state I emits photon $\mathbf{k}'\alpha'$ and turns into state B at time $t = t_2$. In the third graph, state A first emits photon $\mathbf{k}'\alpha'$ and turns into state I at $t = t_1$, and then state I absorbs photon $\mathbf{k}\alpha$ at time $t = t_2$. The details regarding the amplitudes $c^{(1)}_{B,\mathbf{k}'\alpha'}(t)$ and $c^{(2)}_{B,\mathbf{k}'\alpha'}(t)$ are determined using the first- and second-order expressions, Eqs (7.80) and (7.85) [or for the nearly harmonic approximation, Eqs (7.89) and (7.93)]. The transition probability for scattering into $d\Omega$ is

$$\frac{dw}{d\Omega} = \int dE \frac{|c^{(1)}_{B,\mathbf{k}'\alpha'}(t) + c^{(2)}_{B,\mathbf{k}'\alpha'}(t)|^2}{t} \rho_{E,d\Omega} = \frac{2\pi}{\hbar} r_e^2 \left(\frac{c^2\hbar}{2V\sqrt{\omega\omega'}}\right)^2 \frac{V}{(2\pi)^3} \frac{\omega'^2}{\hbar c^3} d\Omega |M_{BA}|^2, \quad (7.121)$$

where $r_e = \frac{e^2}{m_e c^2}$ is the classical electron radius. The differential cross-section is obtained by dividing $dw/d\Omega$ by the incident flux density, c/V:

$$\frac{d\sigma}{d\Omega} = r_e^2 \frac{\omega'}{\omega} |M_{BA}|^2. \quad (7.122)$$

For further details on Rayleigh, Raman, and Compton scattering cross-sections, consult the end of Chapter 5 of Band's *Light and Matter* [18] and Chapter 2 of Sakurai [111]. The formalism of Feynman diagrams in many-body theory is further elaborated in Chapter 18 which is linked to the book web page.

Here we review some of the conclusions regarding Rayleigh and Raman cross-sections. The unpolarized (sum over final polarizations and average over initial polarization) Rayleigh (i.e., elastic) scattering cross-section is proportional to the incident photon frequency to the fourth power, ω^4, to the induced dipole moment squared of the scattering particle, and to $(1 + \cos^2 \theta)$, where θ is the angle between the incident and the scattered photon wavevector. The total (integrated differential) Rayleigh scattering cross-section is

$$\sigma_{\text{Rayleigh}} = \frac{8\pi}{3} r_e^2 \frac{\omega^4}{\omega_0^4}, \quad (7.123)$$

where ω_0 is the "typical" optical absorption frequency of the scattering particle (and $\omega \ll \omega_0$). The ω^4 dependence is the reason the sky is blue and the sun is red at sunset; high-energy photons scatter more strongly than low-energy photons.

Thomson scattering corresponds to the opposite limit, in which the incident photon energy is much larger than the atomic binding energy. Hence, one can ignore all but the seagull graph in Fig. 7.6, and the differential cross-section for scattering of light by a free electron (for photon energies much smaller than the rest mass of the electron) is given by

$$\frac{d\sigma_{\text{Thomson}}}{d\Omega} = r_e^2 |\hat{\boldsymbol{\varepsilon}} \cdot \hat{\boldsymbol{\varepsilon}}'|^2, \quad (7.124)$$

and the unpolarized differential Thompson scattering cross-section is

$$\frac{d\sigma_{\text{Thomson}}}{d\Omega} = \frac{r_e^2}{2}(1 + \cos^2 \theta). \quad (7.125)$$

7.4 Dynamics in an Electromagnetic Field

The total cross-section, obtained by integrating over all scattering angles, is $\sigma_{\text{Thomson}} = \frac{8\pi}{3} r_e^2$, which equals 6.65×10^{-25} cm^2. This result can be compared with the total Rayleigh scattering cross-section, which is much smaller since $\omega \ll \omega_0$.

Problem 7.13

(a) Determine the selection rules for Raman transitions in atoms. Hint: Raman amplitudes involve a product of two-dipole transition amplitudes.

(b) Determine the selection rules for atomic Rayleigh transitions.

7.4.4 DECAY WIDTH

The rates of spontaneous emission calculated in the previous sections have not accounted for the finite lifetime of the excited states. We shall now consider how this modifies the first-order perturbation expressions for absorption and emission rates. The probability of finding an state j excited at time $t = 0$ decreases with time as e^{-t/τ_j} due to its finite lifetime τ_j; hence, its probability amplitude $c_j(t)$ decreases as $e^{-t/2\tau_j}$ (see next section). Furthermore, the probability of finding a state j that is excited via absorption from an initial state i also decreases with time due to its finite lifetime τ_j, as we shall now calculate. Moreover, the linewidth of the excitation is affected by the decay of the excited state. Equation (7.60) can be modified to include a decay term due to the finite lifetime of the state:

$$i\hbar \frac{d}{dt} c_j(t) = \sum_k c_k(t) \langle \psi_j^{(0)} | V(t) | \psi_k^{(0)} \rangle e^{i(E_j - E_k)t/\hbar} - \frac{i\hbar}{2\tau_j} c_j(t). \quad (7.126)$$

We can now substitute $c_k(t) = \delta_{ki}$ into the RHS, and use $V(t)$ in the form (7.79), to solve for $c_j^{(1)}(t)$ using the fact that the solution to the differential equation $\frac{dy}{dt} = b(t) - ay$ is $y(t) = e^{-at}\left[y(0) + \int_0^t dt'\, e^{at'} b(t')\right]$, where the initial condition we need is $y(0) = 0$:

$$c_j^{(1)}(t) = \frac{-1}{\hbar}\left[V_{ij} \frac{e^{i(-\omega+\omega_{ij}+i/2\tau_j)t}-1}{-\omega+\omega_{ji}+i/2\tau_j} + (V^\dagger)_{ij} \frac{e^{i(\omega+\omega_{ji}+i/2\tau_j)t}-1}{\omega+\omega_{ij}+i/2\tau_j}\right], \quad (7.127)$$

Therefore, the probability for populating state j as $t \to \infty$ is *not* proportional to $\delta(E_j - E_i - \hbar\omega)$ for absorption, as found in Eq. (7.83), but rather is given by

$$P_j^{(1)}(\infty) = |c_j^{(1)}(\infty)|^2 = \frac{|V_{ij}|^2}{[\hbar\omega - (E_j - E_i)]^2 + \left(\frac{\hbar}{2\tau_j}\right)^2}. \quad (7.128)$$

This frequency dependence of the transition probability has a *Lorentz lineshape*. The probability of finding the system in state j due to absorption from state i using continuous wave (cw) radiation at a given frequency ω is given in first-order perturbation theory by a Lorentzian distribution about line center $E_j - E_i = \hbar\omega_0$ with decay width given by $\hbar$ times half the decay rate $\gamma_j = \tau_j^{-1}$. Figure 7.7 plots $P_j^{(1)}(t)/P_j^{(1)}(\infty)$ as a function of time, where at large times,

$$P_j^{(1)}(\infty) = \frac{|V_{ij}|^2/\hbar^2}{\Delta^2 + (\gamma_j/2)^2} = \frac{|V_{ij}|^2}{(\hbar\gamma/2)^2} \frac{(\gamma/2)^2}{\Delta^2 + (\gamma_j/2)^2}, \quad (7.129)$$

and $\Delta \equiv \hbar\omega - (E_j - E_i) = \hbar(\omega - \omega_0)$ is the detuning from resonance. The insert shows the Lorentzian profile of the probability $P_j^{(1)}(\infty)$ plotted versus detuning Δ. The probability of excitation is largest at line center, $\Delta = 0$, and falls off in a Lorentzian fashion with detuning Δ.

7.4.5 DOPPLER SHIFT

The expression (7.128) for the transition probability for scattering a photon of frequency ω is appropriate in the rest frame of the atom or molecule. If the atom is moving with velocity $\mathbf{v}$ relative to the photon source (i.e., the laboratory frame), the frequency of the photon in the rest frame of the atom or molecule, ω', and the angle θ' in the rest frame, where θ is the angle between the momentum of the photon and the velocity of the atom in the lab frame, are given by

$$\omega' = \gamma_v(\omega - \mathbf{k} \cdot \mathbf{v}) \approx \omega - \mathbf{k} \cdot \mathbf{v} = \omega(1 - \beta \cos\theta), \tag{7.130a}$$

$$\tan\theta' = \frac{\sin\theta}{\gamma_v(\cos\theta - \beta)}. \tag{7.130b}$$

Here, γ_v is the *Lorentz factor*, $\gamma_v = (1 - \beta^2)^{-1/2}$, and $\beta = v/c$ is the ratio of the speed v to the speed of light. For nonrelativistic velocities, $v \ll c$, we can substitute $\omega' = \omega - \mathbf{k} \cdot \mathbf{v}$ instead of ω into the denominator of the RHS of Eq. (7.128) to obtain

$$P_j^{(1)}(\infty) = \frac{|V_{ij}|^2}{(\omega - \omega_0 - \mathbf{k} \cdot \mathbf{v})^2 + (\gamma_j/2)^2}.$$

Doppler shifts give rise to many phenomena, including broadening of spectral lines due to the distribution of velocities of atoms and molecules in a gas, and the spectral shift of light emitted from high-velocity bodies. The expansion of the universe is known from the redshift of spectral lines from stars and galaxies very far away. This spectral shift has been found to be (nearly) proportional to the distance of the object from us; the farther an object from us, the greater the red-shift due to the greater recessional velocity of object. The (nearly) linear relation between distance and recessional velocity is called *Hubble's Law* and is one of the three pillars of big-bang cosmology.

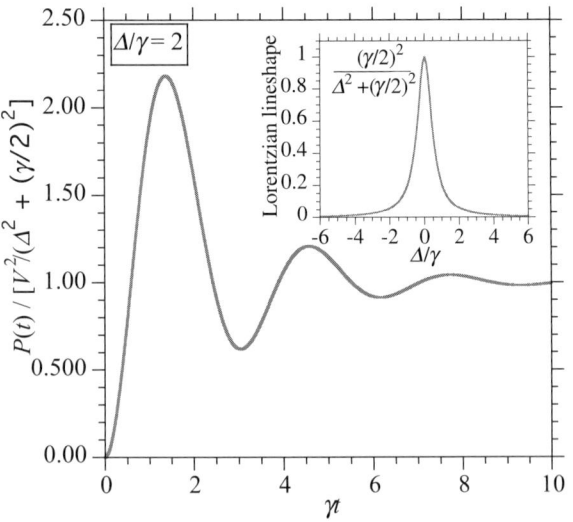

FIG 7.7 $P^{(1)}(t)$ [divided by $P^{(1)}(\infty)$] versus γt for $\Delta = 2\gamma$. The inset shows the Lorentzian profile of $P^{(1)}(\infty) = \frac{4|V_{ij}|^2}{\gamma^2} \frac{(\gamma/2)^2}{\Delta^2 + (\gamma/2)^2}$ versus detuning Δ. The Lorentzian lineshape function, $L(\omega) \equiv \frac{(\gamma/2)^2}{\Delta^2 + (\gamma/2)^2}$, appears in Eq. (7.129).

Problem 7.14

Develop an expression that is first order in $\mathbf{v}$ for the Doppler shifted probability $P_j^{(1)}(\infty)$.

Problem 7.15

The Lorentz transformation of the four-component wavevector $(k_0, \mathbf{k}) = (\omega/c, \mathbf{k})$ from the rest frame at which the photon with this wavevector was emitted to a new frame moving with velocity $\mathbf{v}$ relative to the rest frame is [34]:

$$k_0' = \gamma_v(k_0 - \mathbf{k} \cdot \mathbf{v}/c), \quad k_\parallel' = \gamma_v(k_\parallel - k_0 v/c), \quad \mathbf{k}_\perp' = \mathbf{k}_\perp'.$$

Show that Eqs (7.130) follow from these relations.

7.5 EXPONENTIAL AND NONEXPONENTIAL DECAY

Given a system in state i that decays with a constant rate $w(=\gamma)$, the probability of finding the system in the same state at time $t+dt$ can be obtained by noting that the probability for not decaying in time period dt is $(1-w\,dt)$; so, if the probability of finding the system in state i at time t is $P_i(t)$, we have $P_i(t+dt)=P_i(t)(1-w\,dt)$, i.e., $dP_i[=P_i(t+dt)-P_i(t)]=-P_i(t)\,w\,dt$. The solution of this equation is the exponential decay law

$$P_i(t)=P_i(t_0)e^{-w(t-t_0)}. \quad (7.131)$$

This treatment assumes that there are no populated states that can decay into state i and that the rate w (i.e., the decay time $\tau \equiv 1/w$) is time independent.

The *quantum Zeno effect*, predicted in 1977 by Misra and Sudarshan [112], is a dramatic demonstration of the collapse of the wave function, which produces the "watched pot never boils" effect and shows that the exponential decay law does not always apply. Consider an atom in an excited state, which is subjected to repeated measurements. Each observation collapses the wave function, resetting the clock and delaying the expected transition to the lower state. It works as follows. Let us take the initial state to be the excited state, $\Psi(t=0)=\psi_2$, having a natural lifetime τ for transition to the ground state ψ_1. For times significantly less than τ, the probability of a transition is $P_{2\to 1}(t)=t/\tau$. If we make a measurement after a time t, the probability that the system is still in the upper state is $P_2(t)=1-t/\tau$. Suppose we find it to be in the upper state. The wave function thus collapses back to ψ_2. If we make a second measurement, say at time $2t$, the probability that the system is still in the upper state is $P_2(t)=(1-t/\tau)^2 \approx 1-2t/\tau$, which is the same as it would have been had we never made the measurement at t. However, for extremely short times, we see from Eqs (7.73) and (7.81) that the probability of a transition is not proportional to t, but rather to t^2; more specifically, $P_{2\to 1}(t)=\frac{|V|^2}{\hbar^2}t^2$ [see Eq. (7.82)]. Hence, the probability that the system is in the upper state after two measurements is $P_2(t)=(1-\frac{|V|^2}{\hbar^2}t^2)^2 \approx 1-2\frac{|V|^2}{\hbar^2}t^2$. If we would never have made the first measurement, the probability would have been $1-\frac{|V|^2}{\hbar^2}(2t)^2 \approx 1-4\frac{|V|^2}{\hbar^2}t^2$. Thus, measurement of the system after time t decreased the net probability of finding the system in the lower state. If we examine the system at n time periods, $T, 2T, 3T, \ldots, nT$, the probability that the system is in the upper state after n measurements is

$$P_2(nT)=\left[1-\frac{|V|^2}{\hbar^2}\left(\frac{T}{n}\right)^2\right]^n \approx 1-\frac{|V|^2}{n\hbar^2}T^2. \quad (7.132)$$

In the $n \to \infty$ limit, this probability equals unity. The system never decays!

Note that modifying the density of states $\rho(E_j)$ that appears in Eq. (7.73) can modify the Zeno effect, and, in fact, it can turn into an *anti-Zeno effect* wherein the decay of a system perturbed by measurements is faster than the decay of the system when it is left unperturbed. The density of states $\rho(E_j)$ can be affected by modifying the cavity (i.e., the resonator cavity) for the electromagnetic field that interacts with the atom or molecule, or, more generally, by modifying the "bath" to which the system is coupled.

7.6 THE VARIATIONAL METHOD

The variational method is one way of finding approximations to the ground state or an excited state of a system. Given a trial function Ψ_t for a stationary state, where the subscript t stands for "trial," the expectation value $E[\Psi_t]$ of the Hamiltonian operator is given by $E[\Psi_t]=\frac{\langle\Psi_t|H|\Psi_t\rangle}{\langle\Psi_t|\Psi_t\rangle}$. We seek the ground-state energy (and perhaps wave function) corresponding to the Hamiltonian H. Typically, we want the trial function to be normalized, $\langle\Psi_t|\Psi_t\rangle=1$, so, the denominator equals unity and is therefore not required. The energy $E[\Psi_t]=\langle\Psi_t|H|\Psi_t\rangle$ is a *functional* of Ψ_t since its value depends on the form of a function, rather than a single variable (see Secs. 16.10 and 16.10.1, linked to the book web page, for an introduction to functionals and functional derivatives). Suppose we vary Ψ_t by an arbitrarily small amount, e.g., by changing the parameters on which it depends, like the volume of the system or the external fields applied to the system. We indicate the general variation of the wave function by writing

$$\Psi_t \to \Psi_t + \delta\Psi_t. \tag{7.133}$$

We still want to insure that the new function is normalized, i.e.,

$$\delta\langle\Psi_t|\Psi_t\rangle = \langle\delta\Psi_t|\Psi_t\rangle + \langle\Psi_t|\delta\Psi_t\rangle = 0. \tag{7.134}$$

$\langle\Psi_t|\delta\Psi_t\rangle$ is the complex conjugate of $\langle\delta\Psi_t|\Psi_t\rangle$; so, $\langle\Psi_t|\delta\Psi_t\rangle$ must be complex.

To find the minimum of the expectation value $E[\Psi_t]$ under the constraint $\langle\Psi_t|\Psi_t\rangle = 1$, we form the functional

$$\mathcal{L} = \langle\Psi_t|H|\Psi_t\rangle - \lambda(\langle\Psi_t|\Psi_t\rangle - 1), \tag{7.135}$$

where λ, the *Lagrange multiplier*, serves to insure the normalization condition [25]. We then vary the wave function, such that $\delta\mathcal{L} = 0$:

$$\delta\mathcal{L} = \langle\delta\Psi_t|H - \lambda|\Psi_t\rangle + \text{c.c.} = 0. \tag{7.136}$$

Carrying out the variation independently for the bra and ket, i.e., taking the variation of both $\langle\Psi_t|$ and $|\Psi_t\rangle$, and noting that Eq. (7.136) must be valid for arbitrary variations, we find

$$H|\Psi_t\rangle = \lambda|\Psi_t\rangle. \tag{7.137}$$

It is clear from this equation that the Lagrange multiplier λ is in fact the expectation value of the Hamiltonian, $\lambda = E$.

The variational method that yields a numerical procedure for finding energy eigenvalues and eigenfunctions on expanding the trial wave function in a set of basis functions $\{\phi_i\}$ that are not necessarily orthogonal, $\Psi_t = \sum_i c_i \phi_i$, with arbitrary coefficients $\{c_i\}$. We want to obtain the ground-state energy and wave function, i.e., we want to find the coefficients c_i that minimize the energy. To do so, we form the function, $\mathcal{L} = \sum_{ij} c_j^* c_i \langle\phi_j|H|\phi_i\rangle - E\left(\sum_{ij} c_j^* c_i \langle\phi_j|\phi_i\rangle - 1\right)$, where E is the Lagrange multiplier, which will turn out to be the energy. Taking the variation, we find

$$\sum_j \delta c_j^* \sum_i \left(H_{ji} - ES_{ji}\right) c_i + \text{c.c.} = 0, \tag{7.138}$$

where $H_{ji} = \langle\phi_j|H|\phi_i\rangle$ and $S_{ji} = \langle\phi_j|\phi_i\rangle$. Since the δc_j are arbitrary, the coefficients c_i and the energy E must satisfy the generalized eigenvalue equations,

$$\boxed{\sum_i \left(H_{ji} - ES_{ji}\right) c_i = 0.} \tag{7.139}$$

The energy eigenvalues E are solutions of the secular equation $|H_{ji} - ES_{ji}| = 0$, and the coefficients c_i can be calculated as a linear system of equations once the energy eigenvalues are determined. If the basis functions are orthogonal, $S_{ji} = \delta_{ji}$, this problem reduces to the ordinary eigenvalue problem. Equation (7.139) with $S_{ji} \neq \delta_{ji}$ is called a *generalized eigenvalue problem*. Solving Eq. (7.139) yields not only the ground state but also the excited state. The eigenvectors corresponding to different eigenvalues are orthogonal as a consequence of the hermiticity of H and S.

Problem 7.16

(a) Set up the variational approximation for the first excited state trial wave function Φ_t, which is orthogonal to the ground-state wave function Ψ_t in terms of an orthogonal set of basis functions $[\langle\phi_j|\phi_i\rangle = \delta_{ji}]$, i.e., $\Phi_t = \sum_i d_i \phi_i$ and $\Psi_t = \sum_i c_i \phi_i$, where the coefficients $\{c_i\}$ have been determined already from the ground-state variational

7.7 The Sudden Approximation

problem, by forming the function

$$\mathcal{L} = \langle \Phi_t|H|\Phi_t\rangle - \lambda_1(\langle \Phi_t|\Phi_t\rangle - 1) - \lambda_2\langle \Phi_t|\Psi_t\rangle - \lambda_3\langle \Psi_t|\Phi_t\rangle,$$

and expressing this function in terms of the coefficients $\{c_i\}$ and $\{d_i\}$.

(b) Take the variation with respect to the coefficients $\{d_i\}$ to obtain a set of equations for the excited state energy and the coefficients $\{d_i\}$.

(c) Show that the equations derived in (b) are satisfied by the first excited state energy obtained using Eq. (7.139).

We can obtain approximations to the eigenvalues of H by choosing a trial function Ψ_t that depends on a given set of parameters, and by varying these parameters to find the stationary points of the functional $E[\Psi_t] = \frac{\langle \Psi_t|H|\Psi_t\rangle}{\langle \Psi_t|\Psi_t\rangle}$, but, in general, these stationary points are neither maxima nor minima, but only inflection points or saddle points in the space of variational parameters. Nevertheless, the following theorem, known as the *variational theorem*, ensures that we can obtain an upper bound to the lowest eigenvalue of a Hermitian operator. If $H = H^\dagger$ and E_0 is the lowest eigenvalue of H, then for any Ψ_t,

$$E_0 \leq \frac{\langle \Psi_t|H|\Psi_t\rangle}{\langle \Psi_t|\Psi_t\rangle}. \tag{7.140}$$

Let us use the eigenvector expansion of $|\Psi_t\rangle$ to prove this theorem, $|\Psi_t\rangle = \sum_i |\psi_i\rangle\langle\psi_i|\Psi_t\rangle$, where $H|\psi_i\rangle = E_i|\psi_i\rangle$. Using orthonormality and completeness of the eigenvectors, we find

$$\langle \Psi_t|H|\Psi_t\rangle = \sum_{i,j}\langle\Psi_t|\psi_i\rangle\langle\psi_i|H|\psi_j\rangle\langle\psi_j|\Psi_t\rangle = \sum_i \langle\Psi_t|\psi_i\rangle E_i \langle\psi_i|\Psi_t\rangle$$

$$\geq E_0 \sum_i \langle\Psi_t|\psi_i\rangle\langle\psi_i|\Psi_t\rangle = E_0\langle\Psi_t|\Psi_t\rangle. \tag{7.141}$$

Thus, we have proved the theorem.

If we increase the size of a finite incomplete basis set by adding one additional (linearly independent) basis function, then the eigenvalues of the Hamiltonian in the larger basis interleave the eigenvalues in the previous basis, in such a way that (1) the smallest eigenvalue in the larger basis is smaller or equal to that for the smaller basis and (2) the first excited eigenvalue of the larger basis Hamiltonian lies between the first and the second eigenvalues of the smaller basis Hamiltonian matrix, etc. This is called the *interleaving theorem*.

Problem 7.17

Carry out a variational calculation for the ground-state energy of a bound electron in the screened Coulomb potential, $V(r) = e^{-\kappa r}e^2/r$ using a trial wave function of the form $\Psi_t(r) = (\alpha^3/\pi)^{1/2}e^{-\alpha r}$.

(a) Calculate the expectation value of the Hamiltonian $\langle\Psi_t|H|\Psi_t\rangle$.

(b) Determine the equation that yields the optimum value for the variational parameter α for any value of κ.

(c) Verify that for $\kappa = 0$, the optimum is $\alpha = e^2 m/\hbar^2$.

Answer: (a) $\langle\Psi_t|H|\Psi_t\rangle = \frac{\hbar^2\alpha^2}{2m} - \frac{4e^2\alpha^3}{(\kappa+2\alpha)^2}$. (b) $\frac{d\langle\Psi_t|H|\Psi_t\rangle}{d\alpha} = 0$.

7.7 THE SUDDEN APPROXIMATION

Suppose that the Hamiltonian of a system changes suddenly from $H(t) = H_0$ for $t \leq t_0$ to $H(t) = H_1$ for $t \geq t_1$, where the time interval $T = t_1 - t_0$ is small. In the limit as $T \to 0$, the dynamical state of the system remains unchanged, i.e.,

$$\lim_{T \to 0} \mathcal{U}(t_1, t_0) = 1. \tag{7.142}$$

Hence, the state of a system whose Hamiltonian changes rapidly from one to another remains in the same state, $\psi(t_1) = \psi(t_0)$. The condition for the validity of the sudden approximation is,

$$\Delta \bar{H} T \ll \hbar, \tag{7.143}$$

where $(\Delta \bar{H})^2 \equiv \langle 0|\bar{H}^2|0\rangle - \langle 0|\bar{H}|0\rangle^2$, $|0\rangle$ is the initial state of the system at time t_0 and $\bar{H} = \frac{1}{T}\int_{t_0}^{t_1} dt\, H(t)$ is the time average of the Hamiltonian over the time interval $[t_0, t_1]$. Equation (7.143) is simply the time-energy uncertainty relation.

As an example of the sudden approximation, consider the 1D delta function potential, $V(x) = \tilde{V}_0 \delta(x)$, with $\tilde{V}_0 < 0$, discussed in Sec. 1.3.12. Suppose the state of the system is initially in the ground state, with energy $E_g = -\frac{m\tilde{V}_0^2}{2\hbar^2}$ and wave function $\psi_g(x) = A\, e^{-\kappa|x|}$, where $\kappa = \frac{m^2\tilde{V}_0^2}{2\hbar^4}$, and normalization requires $A = \sqrt{\kappa}$. At time $t = 0$, the strength of the potential is suddenly (in a time much less than the characteristic time $\tau = \hbar/E_g = \frac{2\hbar^3}{m\tilde{V}_0^2}$) changed from $\tilde{V}_0$ to $\tilde{U}_0$, which also is less than 0. Immediately after the change of the potential, we are assured by the sudden approximation that the wave function remains in the ground state, $\psi_g(x)$. But the eigenstates of the new Hamiltonian are given by

$$\psi_0(x) = \sqrt{\mathcal{K}}\, e^{-\mathcal{K}|x|},$$

$$\psi_{k,\pm}(x) = \begin{cases} e^{\pm ikx} + \frac{im\tilde{U}_0/k}{1 - im\tilde{U}_0/k} e^{\mp ikx} & \text{for } (\pm x < 0), \\ \frac{1}{1 - im\tilde{U}_0/k} e^{\pm ikx} & \text{for } (\pm x > 0), \end{cases} \tag{7.144}$$

where $\mathcal{K} = \frac{m^2\tilde{U}_0^2}{2\hbar^4}$ and $k = \sqrt{2mE/\hbar^2}$. For $t > 0$ the wave function for the system can be expanded in terms of these eigenstates,

$$\Psi(x,t) = c_0 \psi_0(x)\, e^{-iE_0 t/\hbar} + \int_0^\infty dk \left[c_+(k) \psi_{k,+}(x)\, e^{-iEt/\hbar} + c_-(k) \psi_{k,-}(x)\, e^{-iEt/\hbar} \right], \tag{7.145}$$

At $t = 0$, $\psi_g(x) = \Psi(x, 0)$, so

$$\sqrt{\kappa}\, e^{-\kappa|x|} = c_0 \psi_0(x) + \int_0^\infty dk \left[c_+(k) \psi_{k,+}(x)\, e^{-iEt/\hbar} + c_-(k) \psi_{k,-}(x) \right]. \tag{7.146}$$

We can now find the amplitudes c_0, $c_+(k)$, and $c_-(k)$ by noting that the states ψ_0, $\psi_{k,+}$ and $\psi_{k,-}$ are orthogonal.

Problem 7.18

Find the probability for being in the new ground state ψ_0 immediately after the change of the strength of the delta function potential.

Answer: $P_0 = |c_0|^2 = \left| \int_{-\infty}^\infty dx\, \psi_0^*(x) \psi_g(x) \right|^2 = \frac{4\tilde{U}_0^2 \tilde{V}_0^2}{(\tilde{U}_0^2 + \tilde{V}_0^2)^2}$. Clearly $P_0 = 1$ if $\tilde{U}_0 = \tilde{V}_0$.

Problem 7.19

(a) The definition of the projector onto states other than the initial state $|0\rangle$ is $Q_0 \equiv 1 - |0\rangle\langle 0|$ (provided $|0\rangle$ has unit norm). Show that the probability of finding the system in other than the initial state is given by

$$P_{Q_0} = \langle 0|\mathcal{U}^\dagger(t_1, t_0)\, Q_0\, \mathcal{U}(t_1, t_0)|0\rangle.$$

(b) Show that Eq. (7.143) follows from the condition $P_{Q_0} \ll 1$.

Hint: Expand $\mathcal{U}(t_1, t_0)$ in powers of the Hamiltonian.

An example of a Hamiltonian that switches quickly from H_0 to H_1 is $H(t) = \frac{1-\tanh[(t-\tau)/T]}{2} H_0 + \frac{1+\tanh[(t-\tau)/T]}{2} H_1$. As $T \to 0$, the switch at time $t = \tau$ from H_0 to H_1 is sudden. The state immediately after the switch will be equal to the state immediately preceding it.

7.8 THE ADIABATIC APPROXIMATION

In many experiments, it is often possible to *slowly*, i.e., adiabatically, change an external parameter of a system, be it the strength of an externally applied field or the volume occupied by the system, etc. The first adiabatic-passage population transfer experiments were carried out by M. Loy in 1974 [113] utilizing a fixed frequency laser field and a slowly ramped dc electric field, which generated a Stark shift that modified the transition frequency in a molecule as a function of time, i.e., a linearly varying Stark shift was applied, so that the transition frequency was swept through the laser frequency as a function of time. The population of the ground state was thereby completely transferred to an excited state. Applications of adiabatic passage to move population from state to state in atoms and molecules have become increasingly successful as control of the pulse duration, pulse shape, and frequency chirp of laser pulses has improved.

The control of experimental results using slowly varying external parameters is achieved using ideas based on the *Adiabatic Theorem*, which we now introduce. Consider the dynamics of a system based on the time-dependent Schrödinger equation, $i\hbar\, d\psi/dt = H(t)\psi$, with a time-dependent Hamiltonian. When $H(t)$ varies slowly, an eigenstate $u_n(-\infty)$ of the initial Hamiltonian, $H(-\infty)$, remains in the same eigenstate $u_n(t)$ of the instantaneous Hamiltonian $H(t)$, where the instantaneous eigenstate $u_n(t)$ is such that

$$H(t)u_n(t) = E_n(t)u_n(t). \tag{7.147}$$

The theorem tells us [roughly – there is a caveat – see Eq. (7.153)] that if the initial state $\psi(-\infty) = \psi_{\text{in}}$, is an eigenstate $u_n(-\infty)$, then as long as the Hamiltonian varies slowly, the system will evolve to a wave function proportional to $u_n(\infty)$ as $t \to \infty$. If the initial state is a superposition, $\psi_{\text{in}} = \psi(-\infty) = \sum_n c_n(-\infty)u_n(-\infty)$, it evolves to

$$\psi(\infty) = \sum_n c_n(\infty)u_n(\infty) \exp\left(\frac{-i}{\hbar} \int_{-\infty}^{\infty} dt'\, E_n(t')\right). \tag{7.148}$$

To prove the Adiabatic theorem, expand the time-dependent wave function in the eigenstates of the time-dependent Hamiltonian,

$$\psi(t) = \sum_n c_n(t) u_n(t) \exp\left(\frac{-i}{\hbar} \int_{-\infty}^{t} dt'\, E_n(t')\right). \tag{7.149}$$

Substitute this expansion into the time-dependent Schrödinger equation and take the inner product with $u_i(t)$ to obtain an equation of motion for the time-dependent coefficients $c_n(t)$:

$$\frac{dc_i(t)}{dt} = -\sum_n c_n(t) \left\langle u_i(t) \left| \frac{du_n(t)}{dt} \right.\right\rangle \exp\left(\frac{-i}{\hbar} \int_{-\infty}^{t} dt'\, [E_n(t') - E_i(t')]\right). \tag{7.150}$$

An expression for $\langle u_i(t)|\frac{du_n(t)}{dt}\rangle$ can be obtained by differentiating Eq. (7.147) with respect to time and taking matrix elements with $u_i(t)$:

$$\left\langle u_i(t) \left| \frac{du_n(t)}{dt} \right.\right\rangle = \frac{\langle u_i(t)|\frac{dH(t)}{dt}|u_n(t)\rangle}{E_n(t) - E_i(t)} \quad \text{for } i \neq n. \tag{7.151}$$

Hence, for $i \neq n$,

$$\frac{dc_i(t)}{dt} = -\sum_n c_n(t) \frac{\langle u_i(t) | \frac{dH(t)}{dt} | u_n(t) \rangle}{E_n(t) - E_i(t)} e^{-i/\hbar \int_{-\infty}^{t} dt' [E_n(t') - E_i(t')]}. \qquad (7.152)$$

If the Hamiltonian varies sufficiently slowly, in the sense that

$$\int_{-\infty}^{t} dt' \left| \frac{\langle u_i(t') | \frac{dH(t')}{dt'} | u_n(t') \rangle}{[E_n(t') - E_i(t')]} \right| \ll 1 \text{ for } n \neq i, \qquad (7.153)$$

[see Problem 7.21(c) for the treatment of $n \neq i$, and Problem 7.21(a-b) for $n = i$], then $\frac{dc_i(t)}{dt} \approx 0$, i.e., the amplitudes can evolve, so that $c_i(t) \approx c_i(-\infty)$. Hence, the wave function at time t is given in the adiabatic approximation by

$$\boxed{\psi(t) \approx \sum_n c_n(-\infty) u_n(t) e^{-i \int_{-\infty}^{t} dt' E_n(t')}} \qquad (7.154)$$

Problem 7.20

Prove Eq. (7.152).

Problem 7.21

(a) Show that $\langle u_i(t) | \frac{du_i(t)}{dt} \rangle$ is imaginary by considering the time derivative of the equation $\langle u_i(t) | u_i(t) \rangle = 1$.
(b) Define $\alpha_i(t)$ by the relation $\langle u_i(t) | \frac{du_i(t)}{dt} \rangle = i\alpha_i(t)$. Now define a new eigenfunction $|u_i'(t)\rangle = |u_i(t)\rangle e^{i\gamma_i(t)}$ and calculate $\langle u_i'(t) | \frac{du_i'(t)}{dt} \rangle$. For what choice of $\gamma_i(t)$ is $\langle u_i'(t) | \frac{(du_i'(t))}{dt} \rangle = 0$?
(c) Substitute $c_n(t) = \delta_{n,0}$ into Eq. (7.152) and solve for $c_i(t)$ for $i \neq 0$, assuming that all quantities appearing in the integrand are constant in time.
(d) Prove Eq. (7.154) by considering the term with $n = i$ in Eq. (7.152).

7.8.1 CHIRPED PULSE ADIABATIC PASSAGE

A beautiful and often used adiabatic passage technique can be realized by changing the frequency of a laser with time, so that it sweeps through the frequency of a transition. Consider a two-level system with energy levels E_1 and E_2 and an allowed optical transition dipole moment μ_{21}. We apply an electromagnetic field of the form $E(t) = A(t)e^{-i\omega_0 t} + \text{c.c.}$, with central frequency equal to $\omega_0 \approx (E_2 - E_1)/\hbar = \omega_{21}$. Here, $A(t)$ is the slowly varying envelope (SVE) of the field, and we take the pulse to be a Gaussian function of time with width σ, centered at time τ_p, and chirped:

$$A(t) = E_0 e^{[\frac{i}{2} \frac{d\omega}{dt}(t-\tau_p)^2 - \frac{(t-\tau_p)^2}{2\sigma^2}]}. \qquad (7.155)$$

Here, $\omega(t) = \omega_0 + \frac{d\omega}{dt}(t - \tau_p)$ is the instantaneous frequency at time t of the chirped pulse. Such a chirped pulse is depicted in Fig. 7.8(a). To explain the nature of chirped pulse absorption processes with this SVE, let us write the Rabi frequency $\Omega(t) = 2\mu_{21}A(t)/\hbar$ as a complex time-dependent quantity,

$$\Omega(t) = \Omega_0 e^{[\frac{i}{2} \frac{d\omega}{dt}(t-\tau_p)^2 - \frac{(t-\tau_p)^2}{2\sigma^2}]}. \qquad (7.156)$$

7.8 The Adiabatic Approximation

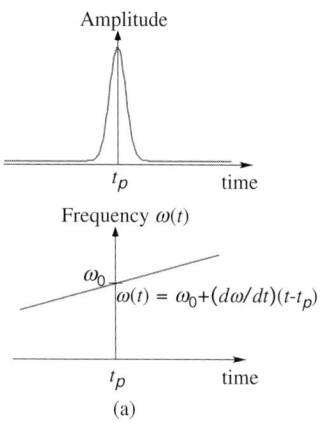

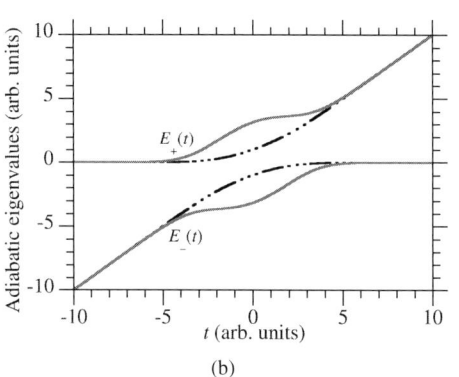

FIG 7.8 (a) A linearly chirped Gaussian pulse, $\Omega_0 e^{[\frac{i}{2}\frac{d\omega}{dt}(t-\tau_p)^2 - \frac{(t-\tau_p)^2}{2\sigma^2}]}$. (b) Adiabatic eigenvalues $E_\pm(t)$ of the two-level Hamiltonian (7.159) with $\Delta = 0$, $t_p = 0$, for medium (dot-dashed curves) and large Rabi frequency strength Ω_0.

The dynamics are conveniently described in terms of the Hamiltonian (6.2):

$$H(t) = \hbar \begin{pmatrix} 0 & \Omega^*/2 \\ \Omega/2 & \Delta \end{pmatrix}, \quad (7.157)$$

where the complex conjugate of Ω in the upper-right matrix element in Eq. (7.157) is required because of the complex nature of the Rabi frequency, and where $\Delta = \omega - \omega_{21}$ is the detuning. After making the transformation $\psi(t) \equiv (\psi_1(t), \psi_2(t)) \rightarrow \Psi(t) \equiv (\Psi_1(t), \Psi_2(t))$ given by

$$\begin{pmatrix} \Psi_1(t) \\ \Psi_2(t) \end{pmatrix} = \begin{pmatrix} \psi_1(t) \exp[i\frac{d\omega}{dt}(t-t_p)^2/2] \\ \psi_2(t) \end{pmatrix}, \quad (7.158)$$

the two-level chirped Hamiltonian for the new state vector $(\psi_1(t), \psi_2(t))$ becomes

$$H(t) = \hbar \begin{pmatrix} \frac{d\omega}{dt}(t-t_p) & |\Omega|/2 \\ |\Omega|/2 & \Delta \end{pmatrix}. \quad (7.159)$$

This Hamiltonian can be diagonalized to obtain the instantaneous eigenvalues and eigenvectors of the time-dependent Hamiltonian. If we assume that the temporal duration of the pulse is small compared to the spontaneous decay time of the excited state, the Hamiltonian dynamics can be modeled using the instantaneous eigenvalues and eigenvectors. Here, the criterion for validity of the adiabatic approximation is

$$\frac{d}{dt}\arctan\left(\frac{\Omega(t)}{\Delta + \frac{d\omega}{dt}(t-t_p)}\right) \ll \sqrt{\left[\Delta + \frac{d\omega}{dt}(t-t_p)\right]^2 + \Omega^2(t)}, \quad (7.160)$$

Then, the adiabatic theorem ensures that a state that is initially an eigenvector of the Hamiltonian remains an eigenvector of the time-dependent Hamiltonian throughout the course of the temporal evolution, provided the conditions of the adiabatic theorem are met. Figure 7.8(b) plots the eigenvalues of the Hamiltonian (7.159) for a positively linearly chirped pulse (i.e., $d\omega/dt$ is time independent and positive) with $t_p = 0$, $|\Omega(t)| = \Omega_0 \exp(-t^2/2\sigma^2)$ and $\Delta = 0$ for medium and large strength Ω_0. If the chirp rate is small enough and the magnitude of the Rabi frequency is sufficiently large to ensure the validity of the adiabatic approximation, the system remains in the adiabatic eigenstate that it started on.

Problem 7.22

(a) Calculate the eigenvalues of the Hamiltonian (7.159).

Answer: $E_\pm = \frac{\hbar}{2}\left(\Delta + \frac{d\omega}{dt}(t-t_p) \pm \sqrt{\left[\Delta - \frac{d\omega}{dt}(t-t_p)\right]^2 + \Omega^2}\right).$

(b) Schematically draw the eigenvalues as a function of time for a positive chirp and for weak Rabi frequency $|\Omega(t)|$ with Gaussian temporal dependence centered at the time t^*, where $\frac{d\omega}{dt}(t^* - t_p) = \Delta$.
Hint: First draw the limit of zero Rabi frequency.

7.8.2 STIMULATED RAMAN ADIABATIC PASSAGE

Consider a three-level Λ-system as shown in Fig. 6.18(a), and an electromagnetic field composed of two light pulses, that turn on and off, and have central frequencies, ω' and ω. Pulse sequences exist that can completely move the population from the ground level to the terminal level without ever populating the intermediate level. This can be accomplished if (1) the central pump frequency is not very far from resonance with the $1 \leftrightarrow 2$ transition, (2) the Stokes frequency is not far from resonance with the $2 \leftrightarrow 3$ transition, (3) the indirect $1 \leftrightarrow 3$ transition is on-resonance, and (4) the peak pump and Stokes Rabi frequencies $\Omega_p(t)$ and $\Omega_S(t)$ are large and temporally delayed relative to one another (as we shall now describe). The process we discuss here is called STIRAP (*stimulated rapid adiabatic passage*). The STIRAP method can be used to completely move population from level 1 to level 3 without ever populating level 2.

In the limit, when the pulses are sufficiently long that the adiabatic approximation is valid, we know from the Adiabatic Theorem that an eigenstate $u_n(t)$ of the initial Hamiltonian, $H(-\infty)$, remains in the same eigenstate $u_n(t)$ of the instantaneous Hamiltonian with eigenvalue $E_n(t)$: $H(t) u_n(t) = E_n(t) u_n(t)$. For the Hamiltonian in Eq. (6.117) with real pump and Stokes Rabi frequencies $\Omega_p(t)$ and $\Omega_S(t)$ and with $\delta = 0$ (i.e., the indirect $1 \leftrightarrow 3$ transition is on-resonance), the analytic expressions for the eigenvalues/eigenvectors are tractable; the three eigenvalues are

$$E_0(t) = 0, \quad E_\pm(t) = \frac{\hbar}{2}\left(\Delta \pm \sqrt{\Delta^2 + \Omega_p^2(t) + \Omega_S^2(t)}\right), \quad (7.161)$$

for any t. The eigenvector $u_0(t)$ is given by

$$u_0(t) = \begin{pmatrix} \cos\Theta(t) \\ 0 \\ -\sin\Theta(t) \end{pmatrix} = \begin{pmatrix} \frac{\Omega_S(t)}{\sqrt{\Omega_p^2(t)+\Omega_S^2(t)}} \\ 0 \\ \frac{-\Omega_p(t)}{\sqrt{\Omega_p^2(t)+\Omega_S^2(t)}} \end{pmatrix}, \quad (7.162)$$

i.e., $\tan\Theta = \Omega_p(t)/\Omega_S(t)$. For a pulse sequence with Stokes pulse preceding the pump pulse (the so-called counterintuitive pulse ordering), as shown in Fig. 7.9, the eigenvector $u_0(t)$ with eigenvalue $E_0(t)$ starts at $t = -\infty$ in the ground state and develops into the terminal state at $t = \infty$, without ever containing any population in the intermediate state. Figure 7.9 shows the probabilities $P_1(t) = \cos^2\Theta(t)$, $P_2(t) = 0$, and $P_3(t) = \sin^2\Theta(t)$ versus time, for the Stokes and pump pulses shown in the figure. Clearly, the population adiabatically and fully transfers from the ground state to the terminal state. As long as the Stokes and pump Rabi frequencies are sufficiently large, and the process is sufficiently adiabatic, the dynamics are very well approximated by the adiabatic dynamics described by Eqs (7.161) and (7.162).

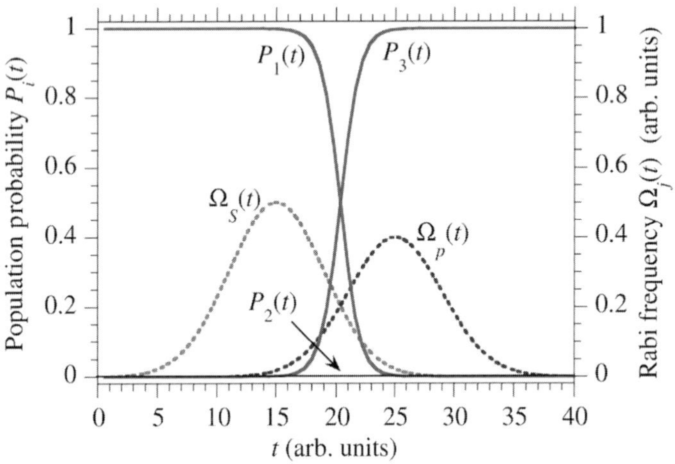

FIG 7.9 Probabilities $P_1(t)$, $P_2(t) = 0$, and $P_3(t)$ versus time, for the counterintuitive Stokes and pump pulse sequence having Rabi frequencies $\Omega_S(t)$ and $\Omega_p(t)$, respectively, sufficiently intense pulses, and $\delta = 0$. *Source:* Band, Light and Matter, Fig. 9.8, p. 541

7.8 The Adiabatic Approximation

> **Problem 7.23**
>
> What kind of Stokes and pump pulse sequence is necessary to adiabatically move population from the "terminal" state 3 to the "initial" state 1 in Fig. 7.9?
>
> **Answer:** The pump should precede the Stokes pulse, since then Θ would start off at $\pi/2$ and would evolve to zero.

7.8.3 THE LANDAU–ZENER PROBLEM

A paradigm time-dependent quantum mechanical problem related to adiabaticity is manifest in what is called the *Landau–Zener transition*. This refers to a problem first solved separately by Lev Landau, Clarence Zener, and Ernst Stueckelberg in 1932, wherein a two-level system whose energy difference varies linearly in time with a rate of change α is coupled by a constant coupling matrix element V. The time-dependent 2×2 Hamiltonian matrix can be written as

$$H(t) = \begin{pmatrix} \epsilon_1 & V \\ V & \epsilon_2 + \alpha t \end{pmatrix}. \tag{7.163}$$

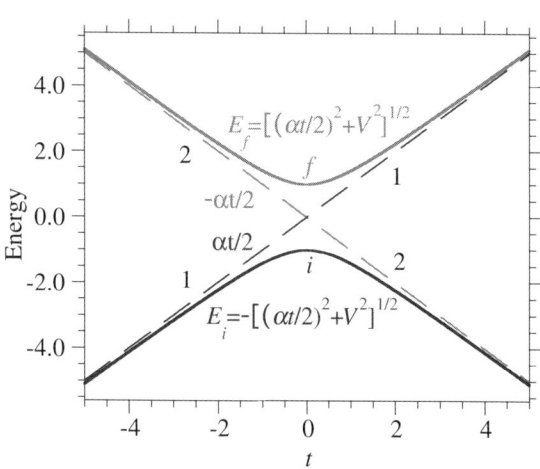

FIG 7.10 Diagonal Hamiltonian matrix elements (dashed lines) in Eq. (7.165) and eigenvalues of the Hamiltonian (solid curves), Eq. (7.166), with $\alpha = 2$ and $V = 1$.

[Note the similarity to the chirped pulse Hamiltonian (7.159) in Sec. 7.8.1]. We need to find the transition probability P_{LZ} of finding the system in state 2 as $t \to \infty$, if it was in the state 1 as $t \to -\infty$ (see Fig. 7.10). For infinitely slow rate of energy change, $\alpha \to 0$, the adiabatic theorem ensures that the system stays on the initial *adiabatic* state i. (Note that states 1 and 2 are called *diabatic* and the states i and f are called *adiabatic* – see Fig. 7.10.) For finite α, the transition probability P_{LZ} exponentially depends on the inverse rate of change of the energy difference:

$$\boxed{P_{LZ} = 1 - \exp\left(-\frac{\pi V^2}{2\hbar\alpha}\right).} \tag{7.164}$$

At slow rates α, the probability is close to unity, and at fast rates, it vanishes exponentially (see Fig. 7.12). Let us now derive Eq. (7.164), noting that this expression is not analytic in α as $\alpha \to 0$; therefore, a perturbation expansion in α is not valid near $\alpha = 0$. To proceed, apply a gauge transformation (i.e., a phase transformation) and transform Eq. (7.163) into the following more symmetric form that facilitates the algebra:

$$H(t) = -\frac{\alpha t}{2}\sigma_z + V\sigma_x = \begin{pmatrix} -\alpha t/2 & V \\ V & \alpha t/2 \end{pmatrix}. \tag{7.165}$$

The relative slope between the two channel potentials is still α, and the transition probability (7.164) applies also to Eq. (7.165). Figure 7.10 shows the diagonal elements of the Hamiltonian in Eq. (7.165) as well as the energy eigenvalues as a function of time.

The instantaneous eigenvalues of the Hamiltonian matrix in Eq. (7.165) are the solutions of the secular equation, $\det[\mathbf{H}(t) - E\mathbf{1}] = (\alpha t/2 - E)(-\alpha t/2 - E) - V^2 = 0$. The two roots, which we label i and f, are

$$E_{i,f}(t) = \pm\sqrt{(\alpha t/2)^2 + V^2}. \tag{7.166}$$

These roots are equal when the square root in Eq. (7.166) vanishes, i.e., when $t_0 = \pm 2iV/\alpha$, and then $E_{i,f}(t_0) = 0$. Near t_0, the time dependence of the eigenvalues are $E_{i,f}(t) = \pm \alpha\sqrt{t_0}\sqrt{t-t_0}$; hence, $E_j(t)$ where $j = i, f$ have a branch point at $t = t_0$. For real times, the difference between the eigenvalues is minimum for $t = 0$, where $E_{i,f}(0) = \pm V$. The eigenvalue–eigenvector equations are:

$$\begin{pmatrix} -\alpha t/2 & V \\ V & \alpha t/2 \end{pmatrix} \begin{pmatrix} u_{1j} \\ u_{2j} \end{pmatrix} = E_j(t) \begin{pmatrix} u_{1j} \\ u_{2j} \end{pmatrix}. \quad (7.167)$$

Equations (7.167) are *identical* at time $t = t_0$ for $j = i$ and f, where $E_i(t_0) = E_f(t_0) = 0$; hence, only one eigenfunction exists at $t = t_0$. At $t = t_0$, $u_{2i}(t_0) = u_{2f}(t_0) = 0$, and near $t = t_0$, $u_{2i}(t) = u_{2f}(t) \propto \sqrt{t - t_0}$.

Problem 7.24

(a) Determine the normalized eigenvectors (u_{1i}, u_{2i}) and (u_{1f}, u_{2f}) for the Landau–Zener Hamiltonian (7.165).
(b) Determine the RHS of Eq. (7.152) to obtain an explicit expression for Eq. (7.168).

We can use Eq. (7.152) for the time rate of change of the transition amplitudes to derive the transition probability for the Landau–Zener problem:

$$\frac{dc_f(t)}{dt} = -c_i(t) \frac{\alpha}{2} \frac{u_{1f}^*(t)u_{1i}(t) - u_{2f}^*(t)u_{2i}(t)}{\sqrt{(\alpha t/2)^2 + V^2}} e^{-2i/\hbar \int_{-\infty}^{t} dt' \sqrt{(\alpha t'/2)^2 + V^2}}. \quad (7.168)$$

Since $c_f(-\infty) = 0$,

$$c_f(\infty) = \int_{-\infty}^{\infty} dt \frac{dc_f(t)}{dt}. \quad (7.169)$$

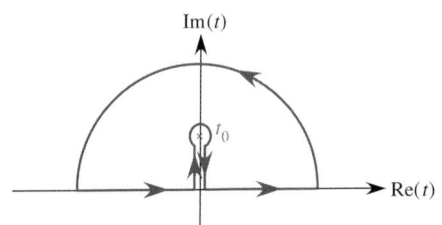

FIG 7.11 Contour of integration for calculating the Landau–Zener probability using the integral in Eq. (7.169). The singularity lies off the real axis.

Note that here c_f is the amplitude of the second *adiabatic* state and c_i is the amplitude of the first *adiabatic* state; these are not the same as states 1 and 2 in Eqs (7.163) and (7.165) (see Fig. 7.10). $c_f(\infty)$ is evaluated by contour integration choosing the contour shown in Fig. 7.11. The amplitude $c_f(\infty)$ can also be calculated by evaluating the integral on the RHS of Eq. (7.169) by the steepest descent method (stationary phase approximation) [25],

$$\int dt\, g(t) e^{iF(t)} \approx g(t_0)\sqrt{\frac{2\pi}{|F''(t_0)|}} e^{iF(t_0)}, \quad (7.170)$$

where t_0 is the time at which $F'(t) = 0$. In our case,

$$F(t) = (-2/\hbar) \int_{-\infty}^{t} dt' \sqrt{(\alpha t'/2)^2 + V^2},$$

and $t_0 = 2iV/\alpha$. Hence,

$$|c_f(\infty)| \approx e^{\text{Im}[iF(t_0)]},$$

$\text{Im}[iF(t_0)] = -\frac{\pi V^2}{4\hbar\alpha}$ (see Problem 7.25) and,

$$|c_f(\infty)|^2 = K e^{-\frac{\pi V^2}{2\hbar\alpha}}, \quad (7.171)$$

7.8 The Adiabatic Approximation

where the prefactor K cannot be accurately determined by the steepest descent method. We conclude that, up to a multiplicative constant, K, which will turn out to be equal to be unity, the Landau–Zener transition probability is $P_{LZ} = 1 - |c_f(\infty)|^2$, as given in Eq. (7.164). The Landau–Zener transition probability is plotted as a function of $\hbar\alpha/V^2$ in Fig. 7.12.

Problem 7.25

(a) Calculate the indefinite integral $\int dt \sqrt{a^2 - t^2}$ and show that it equals $\frac{1}{2}\left(t\sqrt{a^2 - t^2} + a^2 \arctan\left[\frac{t}{\sqrt{a^2 - t^2}}\right]\right)$.

(b) From (a), show that $\int_0^a dt \sqrt{a^2 - t^2} = \frac{\pi a^2}{4}$.

(c) Show that (b) is relevant for the calculation of the integral, $F(t_0) = (-2/\hbar)\int_0^{t_0} dt' \sqrt{(\alpha t'/2)^2 + V^2}$, where $t_0 = 2iV/\alpha$.

(d) Use Eq. (7.170) to determine $c_f(\infty)$, not just its absolute value.

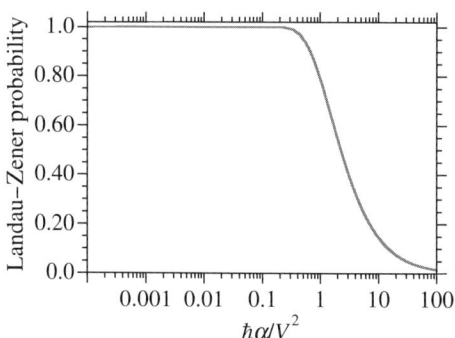

FIG 7.12 Landau–Zener probability, $P_{LZ} = 1 - \exp\left(-\frac{\pi V^2}{2\hbar\alpha}\right)$, versus $\hbar\alpha/V^2$.

Another way of deriving the Landau–Zener result is to directly solve the time-dependent Schrödinger equation

$$i\hbar \frac{\partial}{\partial t}\begin{pmatrix}\psi_1(t)\\ \psi_2(t)\end{pmatrix} = \begin{pmatrix}-\alpha t/2 & V\\ V & \alpha t/2\end{pmatrix}\begin{pmatrix}\psi_1(t)\\ \psi_2(t)\end{pmatrix}. \quad (7.172)$$

A dimensional analysis of Eq. (7.172), similar to that of Sec. 1.3.15 for the harmonic oscillator problem, shows that the natural unit of time for Eq. (7.172) is $\sqrt{\hbar/\alpha}$, so we define a transformation to dimensionless time τ and dimensionless coupling strength, λ,

$$t \to \tau = \sqrt{\frac{\hbar}{\alpha}}\,t, \qquad \lambda = \frac{V}{\sqrt{\hbar\alpha}}. \quad (7.173)$$

Equation (7.172), in these units, is

$$i\frac{\partial}{\partial \tau}\begin{pmatrix}\psi_1(\tau)\\ \psi_2(\tau)\end{pmatrix} = \begin{pmatrix}-\tau/2 & \lambda\\ \lambda & \tau/2\end{pmatrix}\begin{pmatrix}\psi_1(\tau)\\ \psi_2(\tau)\end{pmatrix}, \quad (7.174)$$

Initially, as $\tau \to -\infty$, only one component (the lower level corresponding to energy $\frac{\alpha}{2}t$) is occupied,

$$\psi(\tau) = \begin{pmatrix}0\\ \psi_2(\tau)\end{pmatrix}, \quad (7.175)$$

where

$$\psi_2(\tau) = e^{i\phi(\tau)}, \quad \text{as } \tau \to -\infty, \quad (7.176)$$

and $\phi(\tau)$ is a real function of τ, so $|\psi_2(\tau)|^2 \to 1$ as $\tau \to -\infty$. The solution as $\tau \to \infty$ can be written as,

$$\psi(\tau) = \begin{pmatrix}A_1\,e^{if_1(\tau)}\\ A_2\,e^{if_2(\tau)}\end{pmatrix}, \quad \text{as } \tau \to +\infty, \quad (7.177)$$

$$|A_1|^2 + |A_2|^2 = 1, \quad f_1(\tau), f_2(\tau) \in \mathbb{R}, \quad (7.178)$$

where the symmetry of the problem dictates that

$$-f_1(\tau) = f_2(\tau) \equiv f(\tau), \quad (7.179)$$

as can be seen by comparing the two Eqs (7.172) at large τ where the coupling V can be dropped. Our objective is to find the probability $|A_1|^2$ of finding the system in state 1 as $\tau \to \infty$. The leading-order term appearing in $f(\tau)$ and $\phi(\tau)$ can be determined by solving Eqs (7.174) after the coupling V is dropped at large $|\tau|$, since coupling between the levels is negligible when the diagonal elements of the matrix in Eq. (7.174) are wildly disparate. Thereby, we find

$$f(\tau) \approx \frac{1}{4}\tau^2, \quad \phi(\tau) \approx -\frac{1}{4}\tau^2, \quad |\tau| \to \infty. \tag{7.180}$$

Differentiating the bottom equation in Eq. (7.174) with respect to time, substituting the top differential equation into the resulting equation, and reusing the bottom equation, we obtain

$$\ddot{\psi}_2(\tau) + \left(\frac{1}{4}\tau^2 - \rho\right)\psi_2(\tau) = 0, \tag{7.181}$$

where the complex parameter $\rho \equiv -\frac{1}{2}(\lambda^2 + i)$, and the initial condition for $\psi_2(\tau)$ is

$$\psi_2(\tau \to -\infty) \to e^{i\phi(\tau)}. \tag{7.182}$$

Equation (7.181) is analyzed in Ref. [114], page 118 [see Eq. (12)]. The solutions are expressed in terms of *parabolic cylinder functions*, $D_{-i\frac{\lambda^2}{2}}(\pm e^{i\frac{\pi}{4}}\tau)$ [see Eq. (13) of Ref. [114]]. The choice of sign is dictated by the initial condition (7.182). As $\tau \to -\infty$, we use Eq. (1) on page 122 of Ref. [114]. The argument z of $D_{-i\frac{\lambda^2}{2}}(z)$ must satisfy $-\frac{3}{4}\pi < \arg z < \frac{3}{4}\pi$. This is clearly satisfied with the minus sign, because for $\tau < 0$, $\arg[-e^{i\frac{\pi}{4}}\tau] = \frac{\pi}{4}$. The solution of Eq. (7.181) with the initial condition (7.182) is

$$\psi_2(\tau) = e^{-\pi\frac{\lambda^2}{8}} D_{-i\frac{\lambda^2}{2}}(-e^{i\frac{\pi}{4}}\tau), \tag{7.183}$$

$$\psi_2(\tau) \to e^{-i(\frac{1}{4}\tau^2 + \frac{\lambda^2}{2}\log|\tau|)}, \quad \tau \to -\infty. \tag{7.184}$$

For $\tau > 0$, the level 1 corresponds to the upper level, while level 2 corresponds to the lower level. The asymptotic form of $e^{-\pi\frac{\lambda^2}{8}} D_{-i\frac{\lambda^2}{2}}(-e^{i\frac{\pi}{4}}\tau)$ as $\tau \to +\infty$ contains two terms, one with $e^{-i\frac{\tau^2}{4}}$ and the other with $e^{i\frac{\tau^2}{4}}$. Note that as $\tau \to +\infty$, the argument $z = -e^{i\frac{\pi}{4}}\tau$ satisfies $-\frac{5}{4}\pi < \arg z = -\frac{3}{4}\pi < -\frac{\pi}{4}$. Hence, we can use Eq. (3) on page 123 of Ref. [114] and find,

$$e^{-\pi\frac{\lambda^2}{8}} D_{-i\frac{\lambda^2}{2}}(-e^{i\frac{\pi}{4}}\tau) \to e^{-\frac{\pi\lambda^2}{4}} e^{-i(\frac{1}{4}\tau^2 + \frac{\lambda^2}{2}\log|\tau|)} + \frac{\sqrt{\pi}}{\lambda\Gamma(-i\frac{\lambda^2}{2})} e^{-\frac{\pi\lambda^2}{4} + i\frac{\pi}{4}} e^{i(\frac{1}{4}\tau^2 - \frac{\lambda^2}{2}\log|\tau|)}, \quad \tau \to +\infty. \tag{7.185}$$

Therefore, the probability $|A_1|^2$ of transition to level 1 and the probability $|A_2|^2$ for staying at level 2 are

$$|A_1|^2 = |e^{-\frac{\pi\lambda^2}{4}}|^2 = e^{-\pi\frac{\lambda^2}{2}} = e^{-\frac{\pi V^2}{2\hbar\alpha}}, \tag{7.186}$$

$$|A_2|^2 = \left|\frac{\sqrt{\pi}}{\lambda\Gamma(-i\frac{\lambda^2}{2})} e^{-\frac{\pi\lambda^2}{4} + i\frac{\pi}{4}}\right|^2 = 1 - |A_1|^2 = 1 - e^{-\frac{\pi V^2}{2\hbar\alpha}}. \tag{7.187}$$

This shows that the prefactor K multiplying the exponential in Eq. (7.171) is unity, i.e., $|A_1|^2 = 1 \times e^{-\frac{\pi V^2}{2\hbar\alpha}}$.

Let us summarize the significance of the Landau–Zener treatment. A process is adiabatic if a parameter in the Hamiltonian is varied slowly with respect to an "internal" time-scale. In the Landau–Zener problem, there are two time scales that can be formed from the quantities in the 2×2 Hamiltonian matrix (7.165): the inverse of the coupling rate, $\hbar/V$,

7.8 The Adiabatic Approximation

where V is the off-diagonal coupling, and V/α, where α is the sweep rate in units of energy per unit time. The ratio of these two time scales, $(V/\alpha)/(\hbar/V)$, is the LZ parameter (up to a factor on the order of unity) in Eq. (7.164). When this dimensionless parameter gets large (i.e., α gets small), the Landau–Zener probability goes to unity.

Systems having two discrete energy levels that cross in time and that are coupled via an interaction potential have a finite transition probability that depends on how fast the levels approach each other. The phenomena of this sort are often referred to as Landau–Zener *tunneling*.

7.8.4 GENERALIZED DISPLACEMENTS AND FORCES

The adiabatic approximation can be used to approximate the response of a physical system to slowly varying external perturbations, such as external fields that are turned on slowly or a slowly changing volume. This problem is formulated using the concept of generalized forces and generalized displacements. Generically, the perturbations are described in terms of slowly varying time-dependent parameters $\{x_i(t)\}$ that describe how the Hamiltonian of the system varies with time,

$$H = H(q, p; x_1(t), x_2(t), \ldots). \tag{7.188}$$

The components of the vector $\mathbf{x}(t) = (x_1(t), x_2(t), \ldots, x_N(t))$ are called *generalized displacements*, and the components of the vector $\dot{\mathbf{x}}(t)$ are called *generalized velocities*. For simplicity, let us initially assume that the set of generalized displacements is finite. The extension to the case of a continuous set will be made in Sec. 7.9. The response of the system to these perturbations is encoded in a set of *generalized forces*, which are the operators defined as the negative of the derivatives of the Hamiltonian with respect to the generalized displacements $\{x_i\}$,

$$\boxed{\mathcal{F}_i \equiv -\frac{\partial H}{\partial x_i}.} \tag{7.189}$$

The *generalized force vector* is given by $\mathcal{F} = (\mathcal{F}_1, \mathcal{F}_2, \ldots, \mathcal{F}_N)$. Note that if x_i has units of length (or volume, electric field, or magnetic field), $\mathcal{F}_i$ has units of force (or pressure, polarization, or magnetization).

The generalized forces determine the power absorbed or released by the system due to time variation of the generalized displacements. To show this, note that $E(t) = \langle H(t) \rangle = \text{Tr}[H(t)\rho]$ is the instantaneous energy of the system governed by the time-dependent Hamiltonian $H(t)$ and the density matrix $\rho(t)$. If the system is in a pure state $|\Psi(t)\rangle$, then $E(t) = \langle \Psi(t)|H|\Psi(t)\rangle$, and the time rate of change of energy, $dE(t)/dt = \langle d\Psi(t)/dt|H|\Psi(t)\rangle + \langle \Psi(t)|dH/dt|\Psi(t)\rangle + \langle \Psi(t)|H|d\Psi(t)/dt\rangle$ can be shown, using the Hellman–Feynman theorem (see Sec. 11.3.3) to equal $dE(t)/dt = \langle \Psi(t)|dH/dt|\Psi(t)\rangle$,

$$\boxed{\frac{dE}{dt} = \left\langle \frac{dH}{dt} \right\rangle = -\sum_i \langle \mathcal{F}_i \rangle \dot{x}_i = -\langle \mathcal{F} \rangle \cdot \dot{\mathbf{x}}.} \tag{7.190}$$

In the more general case of a mixed state, the time rate of change of an expectation value of any operator, $\langle \mathcal{O} \rangle$, is given by,

$$\frac{d\langle \mathcal{O} \rangle}{dt} = \text{Tr}\left(\frac{\partial \mathcal{O}}{\partial t}\rho(t) + \mathcal{O}\frac{\partial \rho(t)}{\partial t}\right) = \text{Tr}\left(\frac{\partial \mathcal{O}}{\partial t}\rho(t) - \frac{i}{\hbar}\mathcal{O}[H, \rho(t)]\right)$$
$$= \text{Tr}\left(\frac{\partial \mathcal{O}}{\partial t}\rho(t) + \frac{i}{\hbar}[H, \mathcal{O}]\rho(t)\right) = \left\langle \frac{\partial \mathcal{O}}{\partial t} \right\rangle + \frac{i}{\hbar}\langle [H, \mathcal{O}] \rangle, \tag{7.191}$$

where the Liouville equation, $\dot{\rho} = -(i/\hbar)[H, \rho]$, has been used. Hence, for $\mathcal{O}(t) = H(t)$, we recover Eq. (7.190).

The concept of generalized force is essential for linear response theory as discussed below. Before taking up this topic, let us first consider the dependence of the instantaneous wave function on the generalized displacement vector $\mathbf{x}(t)$.

Generalized Forces to First Order in $\dot{\mathbf{x}}(t)$

An expression for the expectation value of the generalized forces exerted on a system, $\langle \mathcal{F}_k(t) \rangle$, on experiencing time-varying generalized displacement, $x_j(t)$, can be derived to first order in $\dot{\mathbf{x}}(t)$ as follows. Consider the amplitudes of the adiabatic basis states whose dynamics are given by Eq. (7.202),

$$\frac{dc_n}{dt} = -\frac{i}{\hbar}(E_n - \mathbf{A}_{nn} \cdot \dot{\mathbf{x}})\, c_n - \frac{i}{\hbar} \sum_{m \neq n} W_{nm}\, c_m, \tag{7.192}$$

where the perturbation matrix W_{nm} is defined by

$$W_{nm}(t) \equiv -\mathbf{A}_{nm}(t) \cdot \dot{\mathbf{x}}(t), \tag{7.193}$$

for $n \neq m$. We can think of the system as starting out at time $t = -\infty$, and, if needed, slowly turning on the perturbation. Without the W_{nm} terms, the solution to Eq. (7.192) is

$$c_n(t) = \exp\left\{ \frac{i}{\hbar} \left[\int_{-\infty}^{t} dt'\, E_n(t') - \hbar \int_{\mathbf{x}(-\infty)}^{\mathbf{x}(t)} d\mathbf{x}' \cdot \mathbf{A}_{nn}(\mathbf{x}') \right] \right\} c_n(-\infty),$$

hence,

$$|\psi(t)\rangle = \exp\left\{ \frac{-i}{\hbar} \left[\int_{-\infty}^{t} dt'\, E_n(t') - \int_{\mathbf{x}(-\infty)}^{\mathbf{x}(t)} d\mathbf{x}' \cdot \mathbf{A}_{nn}(\mathbf{x}') \right] \right\} |n(\mathbf{x}(t))\rangle. \tag{7.194}$$

To first order in $\dot{\mathbf{x}}(t)$, the solution is

$$|\psi(t)\rangle = e^{-i\Phi_n(t)}|n(\mathbf{x}(t))\rangle + \sum_{m \neq n} \frac{\int_{-\infty}^{t} dt'\, W_{mn}(t') e^{-i[\Phi_n(t') - \Phi_m(t')]}}{E_n(\mathbf{x}(t)) - E_m(\mathbf{x}(t))} e^{-i\Phi_m(t)} |m(\mathbf{x}(t))\rangle, \tag{7.195}$$

where $\Phi_n(t) = \hbar[\int_{-\infty}^{t} dt'\, E_n(t')]$ is the phase in Eq. (7.194). Note that the procedure used to obtain the wave function $|\psi(t)\rangle$ in Eq. (7.195) is a generalization of the adiabatic procedure in Sec. 7.8 (and in Sec. 7.8.5). We can now write the expectation value of the generalized force, $\mathcal{F}_k = -\frac{\partial H}{\partial x_k}$, $\langle \psi(t)|\mathcal{F}_k|\psi(t)\rangle$, as a sum of a zeroth-, first-, and second-order contribution in $\dot{\mathbf{x}}$. Substituting the zero-order wave function (7.194) into $\langle \psi(t)|\mathcal{F}_k|\psi(t)\rangle$ yields a conservative force,

$$\langle \mathcal{F}_k \rangle = \langle \psi(t)|\mathcal{F}_k|\psi(t)\rangle \approx -\left\langle n \left| \frac{\partial \mathcal{H}}{\partial x_k} \right| n \right\rangle = -\frac{\partial}{\partial x_k} \langle n|\mathcal{H}|n\rangle, \tag{7.196}$$

where we used the Hellman–Feynman theorem (see Sec. 11.3.3) to obtain the rightmost equality. The first order in $\dot{\mathbf{x}}$ contribution can be obtained by substituting Eq. (7.195) into $\langle \psi(t)|\mathcal{F}_k|\psi(t)\rangle$ to yield an expression that is first order in $\dot{x}_j$. This procedure is a bit complicated, so we do not pursue it here.

7.8.5 BERRY PHASE

We have seen in Sec. 7.8 how the state of a quantum system evolves under adiabatic changes. It turns out that, apart from the change given in Eq. (7.154), there is an additional phase factor that depends on the curve traced out by $\mathbf{x}(t)$ in the N-dimensional parameter space of the generalized displacements, provided $N > 1$. It was originally suggested by Michael V. Berry in 1984 when he considered Hamiltonians of the form (7.148), which contain time-dependent parameters [115].

To derive the Berry phase, it is convenient to re-express the adiabatic formalism developed in Sec. 7.8 in terms of rates of change of the generalized displacements in the Hamiltonian, $\dot{\mathbf{x}}(t)$. We use the fact that the rates of change of the

7.8 The Adiabatic Approximation

amplitudes of the adiabatic basis functions in Eq. (7.152) are given in terms of the matrix elements $\left\langle u_i(t) \left| \frac{dH(t)}{dt} \right| u_n(t) \right\rangle$, which can be expressed in terms of the generalized forces:

$$\left\langle u_i(t) \left| \frac{dH(t)}{dt} \right| u_n(t) \right\rangle = \sum_k \left\langle u_i \left| \frac{\partial H}{\partial x_k} \right| u_n \right\rangle \dot{x}_k = -\sum_k \langle u_i | \mathcal{F}_k | u_n \rangle \dot{x}_k. \tag{7.197}$$

The dynamics of the time-dependent Schrödinger equation, $i\hbar \frac{d|\psi(t)\rangle}{dt} = H(\mathbf{x}(t))|\psi\rangle$, are analyzed in terms of the adiabatic eigenstates $|u_n(t)\rangle \equiv |n(\mathbf{x}(t))\rangle$, which satisfy the time-independent Schrödinger equation,

$$H(\mathbf{x}(t))|n(\mathbf{x}(t))\rangle = E_n(\mathbf{x}(t))|n(\mathbf{x}(t))\rangle. \tag{7.198}$$

If the system is prepared in state $|n(\mathbf{x}(0))\rangle$ at $t = 0$, and the Hamiltonian varies adiabatically, the state of the system at time t is

$$|\psi(t)\rangle = \exp\left\{\frac{-i}{\hbar} \int_0^t dt'\, E_n(\mathbf{x}(t'))\right\} e^{i\gamma_n(t)} |n(\mathbf{x}(t))\rangle, \tag{7.199}$$

where $\gamma_n(t)$ is a *geometrical phase*, which satisfies the condition

$$\dot{\gamma}_n(t) = i\langle n(\mathbf{x}(t))|\nabla n(\mathbf{x}(t))\rangle \cdot \dot{\mathbf{x}}(t). \tag{7.200}$$

Moreover, for a system that is initially in a superposition of eigenstates and evolves adiabatically, the state at time t according to the adiabatic theorem is

$$|\psi(t)\rangle = \sum_n c_n(0) \exp\left\{\frac{-i}{\hbar} \int_0^t dt'\, E_n(\mathbf{x}(t'))\right\} e^{i\gamma_n(t)} |n(\mathbf{x}(t))\rangle. \tag{7.201}$$

Equation (7.199) can be derived by substituting the adiabatic expansion of the wave function (7.201) into the time-dependent Schrödinger equation to obtain

$$\frac{dc_n(t)}{dt} = -\frac{i}{\hbar} E_n(t) c_m + \frac{i}{\hbar} \sum_m \mathbf{A}_{nm}(t) \cdot \dot{\mathbf{x}}(t)\, c_m(t), \tag{7.202}$$

where we have defined the "vector potential" matrix elements,

$$\boxed{\mathbf{A}_{mn} \equiv \frac{\hbar}{i} \langle m(\mathbf{x})|\nabla n(\mathbf{x})\rangle.} \tag{7.203}$$

Equation (7.202) generalizes Eq. (7.152) in the sense that it explicitly involves $\dot{\mathbf{x}}(t)$. Integrating Eq. (7.202) with respect to time yields Eq. (7.199), with the geometric phase factor given by Eq. (7.200), as can be understood on noting that only the term with $m = n$ on the RHS of Eq. (7.202) contributes if $\dot{\mathbf{x}}(t)$ is small. We know that this is the case due to the adiabatic theorem. Hence, Eq. (7.202) becomes, $\frac{dc_n(t)}{dt} = -\frac{i}{\hbar}[E_n(t) - \mathbf{A}_{nn}(t) \cdot \dot{\mathbf{x}}(t)] c_n(t)$, and the solution is $c_n(t) = \exp\left\{\frac{-i}{\hbar} \int_0^t dt'\, E_n(\mathbf{x}(t'))\right\} e^{i\gamma_n(t)}$, where

$$\gamma_n(t) = i\int_0^t dt'\, \langle n(\mathbf{x}(t))|\nabla n(\mathbf{x}(t))\rangle \cdot \dot{\mathbf{x}}(t) = -\hbar^{-1} \int_{\mathbf{x}(0)}^{\mathbf{x}(t)} d\mathbf{x}' \cdot \mathbf{A}_{nn}(\mathbf{x}'), \tag{7.204}$$

as can be checked by direct substitution.

The matrix elements of the vector potential $\mathbf{A}_{mn}$ can be expressed in terms of the matrix elements of the generalized force $\mathcal{F}_{k,mn}$. To show this, differentiate the equation $\langle m(\mathbf{x})|H|n(\mathbf{x})\rangle = E_n(\mathbf{x})\delta_{mn}$ with respect to x_k for $m \neq n$, to obtain $\frac{\partial}{\partial x_k}\langle m(\mathbf{x})|H(\mathbf{x})|n(\mathbf{x})\rangle = 0$, hence,

$$A_{k,mn} = \frac{\hbar}{i}\frac{\langle m(\mathbf{x})|\frac{\partial H}{\partial x_k}|n(\mathbf{x})\rangle}{E_m - E_n} = -\frac{\hbar}{i}\frac{\mathcal{F}_{k,mn}}{E_m - E_n}. \tag{7.205}$$

Problem 7.26

Prove that $\langle n(\mathbf{x})|\nabla n(\mathbf{x})\rangle$ is imaginary, and therefore, $\gamma_n(t)$ is real provided $\gamma_n(0)$ is.

Hint: Take the gradient of the equation $\langle n(\mathbf{x})|n(\mathbf{x})\rangle = 1$.

If the displacement parameters $\mathbf{x}(t)$ trace out a closed curve C in the N-dimensional generalized displacement parameter space, so that at time $t = T$, $\mathbf{x}(T) = \mathbf{x}(0)$, the *Berry phase*, $\gamma_n \equiv \gamma_n(T)$, can be expressed in terms of a contour integral over Eq. (7.200),

$$\boxed{\gamma_n = i\oint_C d\mathbf{x}\cdot\langle n(\mathbf{x})|\nabla n(\mathbf{x})\rangle = -\hbar^{-1}\oint_C d\mathbf{x}\cdot\mathbf{A}_{nn}(\mathbf{x}).} \tag{7.206}$$

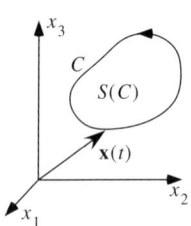

FIG 7.13 The dynamics of the external parameters **x**-space. For the adiabatic change of one turn along the closed loop C, the geometrical phase γ_n obtained for a state $|n(\mathbf{x}(0))\rangle$ for a loop in parameter space is given by the surface integral (7.210) of the vector $\mathbf{B}_n(\mathbf{x})$ in Eq. (7.211) penetrating through the surface $S(C)$ surrounded by loop C.

See Fig. 7.13 for a schematic of the motion of the parameter vector in a closed loop in a 3D parameter space. We know from Problem 7.26 that the geometric phase (or Berry phase) γ_n is real. The definition (7.206) makes sense only if γ_n is gauge invariant, i.e., if the gauge transformation,

$$|n(\mathbf{x})\rangle \to e^{i\Lambda_n(\mathbf{x})}|n(\mathbf{x})\rangle, \tag{7.207}$$

does not affect physical behavior for arbitrary (well behaved) real functions $\Lambda_n(\mathbf{x})$. Indeed, the gauge transformation (7.207) on the wave function induces a gauge transformation on the vector potential,

$$\mathbf{A}_{nn} \to \mathbf{A}_{nn} + \nabla\Lambda_n(\mathbf{x}), \tag{7.208}$$

and since $\oint_C \nabla\Lambda_n(\mathbf{x})\cdot d\mathbf{x} = 0$, the definition of the Berry phase is gauge invariant.

Let us now develop a connection between the geometric phase and the generalized forces. If the generalized displacement parameter space is 3 dimensional, the line integral on the RHS of Eq. (7.206), using the Stokes Theorem, can be expressed as a surface integral,

$$\gamma_n = i\int_S d\mathbf{S}\cdot\nabla\times\langle n(\mathbf{x})|\nabla n(\mathbf{x})\rangle = i\int_S d\mathbf{S}\cdot\langle\nabla n(\mathbf{x})|\times|\nabla n(\mathbf{x})\rangle$$

$$= i\int_S d\mathbf{S}\cdot\sum_{m\neq n}\langle\nabla n(\mathbf{x})|m(\mathbf{x})\rangle\times\langle m(\mathbf{x})|\nabla n(\mathbf{x})\rangle. \tag{7.209}$$

7.8 The Adiabatic Approximation

We have inserted a complete set of states and excluded the term with $m = n$ since $\langle n(\mathbf{x})|\nabla n(\mathbf{x})\rangle$ is purely imaginary, but γ_n must be real. Using Eq. (7.205), we can rewrite Eq. (7.209) as

$$\gamma_n = -i \int_S d\mathbf{S} \cdot \sum_{m \neq n} \frac{\mathbf{A}_{nm} \times \mathbf{A}_{mn}}{(E_m - E_n)^2}. \tag{7.210}$$

Since we know that γ_n is real, the integral must be imaginary, i.e., Eq. (7.210) can be written as $\gamma_n = \int_S d\mathbf{S} \cdot \mathbf{B}_n(\mathbf{x})$, where

$$\mathbf{B}_n(\mathbf{x}) = \text{Im} \sum_{m \neq n} \frac{\mathbf{A}_{nm}(\mathbf{x}) \times \mathbf{A}_{mn}(\mathbf{x})}{[E_m(\mathbf{x}) - E_n(\mathbf{x})]^2}. \tag{7.211}$$

Equations (7.206) and (7.209) can be recast into the vector form,

$$\mathbf{B}_n(\mathbf{x}) = \nabla \times \mathbf{A}_{nn}(\mathbf{x}), \quad \gamma_n = \int_S d\mathbf{S} \cdot \mathbf{B}_n(\mathbf{x}). \tag{7.212}$$

The vector $\mathbf{B}_n(\mathbf{x})$ is the curl of the vector potential $\mathbf{A}_{nn}(\mathbf{x})$. Hence, it resembles a magnetic field vector.

A system that does not return to its original state when transported around a closed loop in parameter space (therefore, the wave function for the system has a nontrivial Berry phase) is said to be *nonholonomic* (the term holonomy is derived from the greek word holos, meaning whole, and nomos, meaning law or rule). There are many classical analogs to the Berry phase. For example, if you take your arm, held with your hand held tight against your leg, then move your arm by 90 degrees so that it is straight out in front of you, then move it out to the side so that it is sticking straight out from your body, and then move it back down to your side, you will note that your hand is at a 90 degree angle from where it was in the beginning (try it!). This 90 degree angle is a geometric phase.

Example 1: Berry Phase in a Two-Level System
Berry phase in a two-level system can be realized when an electron is subject to a time-dependent magnetic field,

$$\mathbf{B}(t) = B\left(\sin\theta(t)\cos\phi(t), \sin\theta(t)\sin\phi(t), \cos\theta(t)\right),$$

where the generalized displacements $\theta(t)$ and $\phi(t)$ are periodic functions of time with the same period T. For definiteness, take $\mathbf{B}(0) = \mathbf{B}(T) = B(0,0,1)$. Denoting the 2D generalized displacement vector by $\mathbf{x}(t) = (\theta(t), \phi(t))$, the adiabatic Hamiltonian [see also Eq. (6.5)] is $H(\mathbf{x}) = -(g\mu_B/\hbar)\mathbf{B}(\mathbf{x}) \cdot \mathbf{S}$. At time t, the two instantaneous eigenvectors of $H(\mathbf{x})$ are, up to a phase [see Eq. (4.18)],

$$|n_+(\mathbf{x})\rangle = \begin{pmatrix} e^{-i\phi}\cos(\theta/2) \\ \sin(\theta/2) \end{pmatrix}, \quad |n_-(\mathbf{x})\rangle = \begin{pmatrix} -e^{-i\phi}\sin(\theta/2) \\ \cos(\theta/2) \end{pmatrix}, \tag{7.213}$$

with corresponding eigenvalues $\pm g\mu_B B/2$. At $(\theta = 0, \phi = 0)$, $|n_+(0,0)\rangle = \binom{1}{0}$ and $|n_-(0,0)\rangle = \binom{0}{1}$. The above states can be multiplied by arbitrary phases that may depend on the generalized displacements [see Eqs (4.18) and (7.207)]. Following definition (7.203), the corresponding vector potentials are

$$\mathbf{A}_{++} = -i\hbar\langle n_+(\mathbf{x})|\nabla n_+(\mathbf{x})\rangle = (0, -\hbar\sin^2(\theta/2)), \tag{7.214}$$

$$\mathbf{A}_{--} = -i\hbar\langle n_-(\mathbf{x})|\nabla n_-(\mathbf{x})\rangle = (0, -\hbar\cos^2(\theta/2)), \tag{7.215}$$

where $\nabla = (\frac{\partial}{\partial\theta}, \frac{\partial}{\partial\phi})$. Let us start from $|n_+(0,0)\rangle$ and change the direction of the field adiabatically in three steps: (1) from the z-axis $(0,0,1)$ to the direction $(\sin\theta, 0, \cos\theta)$ [this would be the x-axis if $\theta = \pi/2$], (2) then to the direction $(0, \sin\theta, \cos\theta)$ [this would be the y-axis if $\theta = \pi/2$], and (3) finally back to the z-axis $(0,0,1)$. If this variation is very slow, the electron spin direction follows the magnetic field direction and the electron wave function acquires a Berry phase. Since only the ϕ component of $\mathbf{A}_{++}$ is nonzero, the Berry phase will be accumulated only during the second step, and the Berry phase at the end of the magnetic field sweep is

$$\gamma_+ = -\hbar^{-1} \oint_C \mathbf{A}_{++} \cdot d\mathbf{s} = \sin^2(\theta/2) \int_0^{\frac{\pi}{2}} d\phi = \frac{\pi}{2} \sin^2(\theta/2). \tag{7.216}$$

Similarly,

$$\gamma_- = -\hbar^{-1} \oint_C \mathbf{A}_{--} \cdot d\mathbf{s} = \cos^2(\theta/2) \int_0^{\frac{\pi}{2}} d\phi = \frac{\pi}{2} \cos^2(\theta/2). \tag{7.217}$$

The Berry phase is half the solid angle subtended by the closed curve. For example, if $\theta = \pi/2$, the Berry phases are $\gamma_+ = \gamma_- = \pi/4$, and the solid angle corresponds to the area within two meridians and a quarter of the equator, which is 1/8 of the solid angle of a sphere, that is, $\pi/2$.

Now consider starting with an initial superposition state $2^{-1/2}(|n_+(0,0)\rangle + |n_-(0,0)\rangle)$. Carry out a closed circuit as explained above. After completing the cycle, we obtain a state $\psi(T)$ in which each component acquires its own Berry phase. Suppose we now measure σ_x. The state $\psi(T)$ is then partially projected onto a state $2^{-1/2}(|n_+(0,0)\rangle + |n_-(0,0)\rangle)$, which is an eigenstate of the operator σ_x with eigenvalue 1. The probability of measurement is given by

$$P = \frac{1}{2}|e^{i(E_+T/\hbar+\gamma_+)} + e^{i(E_-T/\hbar+\gamma_-)}|^2 = \frac{1}{2}|e^{i(g\mu_B BT/\hbar+\gamma_+-\gamma_-)} + 1|^2, \tag{7.218}$$

where $E_\pm = \pm g\mu_B B/2$, $\gamma_+ = \cos^2(\theta/2)\pi/2$ and $\gamma_- = \sin^2(\theta/2)\pi/2$. Hence, the probability depends on the Berry phase difference $\gamma_+ - \gamma_-$.

Example 2: A Conducting Ring Near a Magnetic Moment

Another experimentally feasible system implementing Berry phase physics involving a two-level system consists of an electron confined to a nanoscopic conducting ring, subject to a magnetic field produced by a magnetic atom [such as a dysprosium (Dy) atom] placed above the ring. The magnetic field lines meet the ring along a polar angle θ and form the shape of a crown, as illustrated in Fig. 7.14.

The Hamiltonian, the eigenspinors, and the vector potentials are the same as above; the only difference is that here the ring serves as a waveguide for the electron motion. Along this ring, θ is fixed and ϕ varies from 0 to 2π. The Berry phase

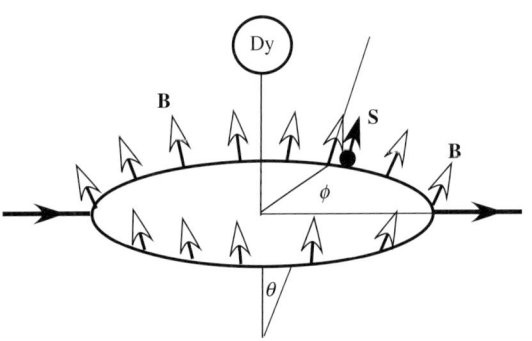

FIG 7.14 Electron on a metallic ring subject to a crown-shaped magnetic field produced by a magnetic atom (Dy, e.g., has one of the largest atomic magnetic moments) placed above the ring. If the electron moves slowly along the ring, it aligns its spin with the magnetic field lines. When the electron completes one cycle, its wave function gains a Berry phase $\gamma = \pi(1 - \cos\theta)$. The ring is part of an electric circuit with left and right leads (marked by black arrows). Electrons entering the ring from the left lead are split, and the corresponding amplitudes of the spinor wave functions in both arms of the ring interfere before leaving to the right lead. The conductance is sensitive to the phase difference of these spinor wave functions, i.e., a non-Abelian Berry phase affects the conductance.

7.9 Linear Response Theory

corresponding to $n_+(\mathbf{x})$ is

$$\gamma_+ = \int_0^{2\pi} \sin^2(\theta/2)\, d\phi = \pi[1 - \cos\theta]. \tag{7.219}$$

Problem 7.27

Verify the gauge invariance of the Berry phase.

(a) Using Eq. (7.207), apply the gauge transformation $|n_+(\mathbf{x})\rangle \to e^{i\phi/2}|n_+(\mathbf{x})\rangle$ that yields the spinors (4.18), and show that the new vector potential is $\mathbf{A}_{++} = \frac{\hbar}{2}\left(-\sin^2\frac{\theta}{2}, \cos^2\frac{\theta}{2}\right)$.

(b) Use the definition (7.206) to calculate the new Berry phase, and show that the result is the same as in Eq. (7.219).

(c) Following Eq. (7.212), show that the "magnetic field," also called the *Berry curvature*, is $\mathbf{B}_+(\mathbf{x}) = \partial_\theta A_{+\phi} - \partial_\phi A_{+\theta} = \frac{1}{2}\sin\theta$. Prove its gauge invariance.

(d) Recalculate γ_+ using the second part of Eq. (7.212). The result must coincide with Eq. (7.219) (modulo 2π).

Berry's original proposal for measuring the geometrical phase involved an interference experiment between two polarized beams of particles with spin that were split and then recombined after one of them went through a nontrivial closed circuit in which the magnetic field varied but returned to its original position. The projective measurement experiment suggested in the first example is different; it determines the Berry phase *difference* between two states, $\gamma_+ - \gamma_-$ and not the Berry phase of a given state. But in the second example, interference is exploited to determine the Berry phase of a given state as explained in the caption of Fig. 7.14.

7.9 LINEAR RESPONSE THEORY

When a physical system in equilibrium is subject to an external perturbation, it is driven out of equilibrium. Generically, the observables can depend nonlinearly on the external perturbation, but if the perturbation is weak, then, to good approximation, one need to only keep the first-order (linear) term in the external perturbation, and the coefficients in front of the first power of the external perturbation can be calculated in equilibrium. This is the essence of linear response theory. For example, in Ohm's law relating the direct current and the voltage, $V = IR$, the constant R is determined by the system in equilibrium at zero current.

The general structure of linear response theory is formulated as follows: Suppose there are weak external perturbations $\{x_j(\mathbf{r}', t')\}$ at space-time point $(\mathbf{r}', t')$ acting on a system initially in equilibrium. These perturbations generalize the displacements $\{x_j(t)\}$ of Eq. (7.188) in that they may depend on position. How do such perturbations affect the expectation values of a set of observables O_i (which could be the generalized forces $\mathcal{F}_i \equiv -\delta H/\delta x_i$)? The expectation values $\langle O_i(\mathbf{r}, t)\rangle$ can be expressed as,

$$\langle O_i(\mathbf{r}, t)\rangle = \int d\mathbf{r}'\, dt'\, \chi_{ij}(\mathbf{r}, \mathbf{r}', t - t') x_j(\mathbf{r}', t'), \tag{7.220}$$

where the matrix χ_{ij} of response coefficients are a property of the system in equilibrium [see Eqs (7.221) for examples]. The aim of linear response theory is to calculate these *response functions*, which are also referred to as *response kernels* or *susceptibilities*.

Our first task is to define the susceptibilities and determine their properties, whereas our second task is to find a procedure to compute them. An expression for susceptibility in terms of the system parameters is referred to as a *Kubo formula*. Obtaining an exact Kubo formula requires the full solution of the equilibrium problem. Often, a system is composed of many interacting particles, and a solution is, in general, not available. But approximations can be made. Here, we present the main ingredients of linear response and the Kubo formalism. We will return to this topic in Chapter 14 to present some sophisticated approximation methods.

7.9.1 SUSCEPTIBILITIES

As examples of linear response relations, consider the constitutive relations of electromagnetic theory: the polarization **P** of a medium due to the presence of an external electric field, **E**, the magnetization of a medium, **M**, due to the presence of an external magnetic field, **H**, and the current **J** induced in a medium by the presence of an external electric field. These relations are written as

$$P_i(\mathbf{r},t) = \int d\mathbf{r}' \int_{-\infty}^{\infty} dt' \, \chi_{ij}(\mathbf{r},\mathbf{r}',t-t') E_j(\mathbf{r}',t'), \qquad (7.221a)$$

$$M_i(\mathbf{r},t) = \int d\mathbf{r}' \int_{-\infty}^{\infty} dt' \, \chi_{ij}^M(\mathbf{r},\mathbf{r}',t-t') H_j(\mathbf{r}',t'), \qquad (7.221b)$$

$$J_i(\mathbf{r},t) = \int d\mathbf{r}' \int_{-\infty}^{\infty} dt' \, \sigma_{ij}(\mathbf{r},\mathbf{r}',t-t') E_j(\mathbf{r}',t'). \qquad (7.221c)$$

Here, $i,j = x,y,z$ are Cartesian indices. In an anisotropic media, the response to a vector field perturbation is itself a vector, not necessarily parallel to the perturbation vector, and the susceptibilities are tensors. The *electric susceptibility tensor* χ_{ij} is related to the *dielectric tensor* ϵ_{ij}, $\epsilon_{ij} = \delta_{ij} + 4\pi \chi_{ij}$ (in SI units, $\epsilon_{ij} = \epsilon_0 \delta_{ij} + \chi_{ij}$). The *magnetic susceptibility tensor* χ_{ij}^M is related to the magnetic permeability tensor μ_{ij} by the relation $\mu_{ij} = \delta_{ij} + 4\pi \chi_{ij}^M$ (in SI units, $\mu_{ij} = \mu_0 \delta_{ij} + \chi_{ij}^M$), and σ_{ij} is the conductivity tensor, which is related to the inverse of the resistance R, which appears in Ohm's law, $V = IR$. Similar equations can be derived for any system in which external parameters change with time [117], e.g., the thermal conductivity coefficient relates the heat flow to the temperature gradient across the medium, the Young modulus relates the stress to the strain applied to the medium.

Generically, linear response coefficients are nonlocal in space and time. If the external perturbation acts at point $\mathbf{r}'$ at time t', it affects the response of the system at another point $\mathbf{r}$ at a later time $t > t'$. The restriction $t > t'$ is due to *causality*: No response is possible from events occurring in the future, i.e., the response is retarded. Susceptibilities must then include a factor $\Theta(t - t')$, where $\Theta(\tau)$ is the *step function* [$\Theta(\tau) = 0$ for $\tau < 0$ and $\Theta(\tau) = 1$ for $\tau > 0$]:

$$\chi(t-t') = \Theta(t-t') \tilde{\chi}(t-t'). \qquad (7.222)$$

Causality ensures that $\chi_{kj}(\tau) = 0$ for $\tau < 0$. Thus, susceptibilities are retarded functions. It has already been stressed that the susceptibilities are to be calculated in equilibrium, where the system is invariant under time translation. Hence, they depend only on the time difference $t - t'$.

Since the susceptibilities contain the step function $\Theta(t - t')$, the time integrals in Eq. (7.221) are taken over the interval $-\infty < t' \le t$. But the temporal duration over which the system remembers an earlier external perturbation, called the characteristic *relaxation time*, also effectively limits the integration region. The spatial integrals over $\mathbf{r}'$ in Eq. (7.221) are over the whole system volume, but, for a given $\mathbf{r}$, the spatial extent over which significant contributions occur are typically over the range of the characteristic *mean free path*.

Translation-Invariant Media

For systems that are not translation invariant, the susceptibilities depend separately on $\mathbf{r}$ and $\mathbf{r}'$, while for translationally invariant systems, the response coefficients depend on space only through the variable $(\mathbf{r} - \mathbf{r}')$. Hence, if the system is translation invariant and stationary, the spatial and temporal dependence of the susceptibilities is given by

7.9 Linear Response Theory

$\chi(\mathbf{r} - \mathbf{r}', t - t')$. Then, both space and time integrals in Eqs (7.221) are convolutions, and the Fourier transforms [i.e., $F(\mathbf{q}, \omega) = \frac{1}{(2\pi)^2} \int d\mathbf{r} \, dt \, e^{-i(\mathbf{q}\cdot\mathbf{r}-\omega t)} F(\mathbf{r}, t)$] of these equations become

$$P_i(\mathbf{q}, \omega) = \chi_{ij}(\mathbf{q}, \omega) E_j(\mathbf{q}, \omega), \tag{7.223a}$$

$$M_i(\mathbf{q}, \omega) = \chi_{ij}^M(\mathbf{q}, \omega) H_j(\mathbf{q}, \omega), \tag{7.223b}$$

$$J_i(\mathbf{q}, \omega) = \sigma_{ij}(\mathbf{q}, \omega) E_j(\mathbf{q}, \omega). \tag{7.223c}$$

For generalized displacements that are vector fields, the susceptibilities are 3×3 matrices, but in general, there may be any number N of perturbations [generalized dispacements, see Eq. (7.188)], and the susceptibilities $\boldsymbol{\chi}(\mathbf{q}, \omega) = \{\chi_{ij}(\mathbf{q}, \omega)\}$ are then $N \times N$ matrices.

Kramers–Kronig Relations

The susceptibilities $\chi_{ij}(\mathbf{q}, \omega)$ are generally complex functions of $\mathbf{q}$ and ω. Causality implies an important relation between the real and the imaginary parts of $\chi_{ij}(\mathbf{q}, \omega)$, known as the *Kramers–Kronig relation*. To derive this relation, we need to expose two important properties of $\chi_{ij}(\mathbf{q}, \omega)$ when considered as a complex function of the *complex variable* ω. Since $\mathbf{q}$ and the indices i, j are fixed, they will not be explicitly specified in the discussion below.

(1) $\chi(\omega)$ (and in fact any Fourier component of a retarded function) is an analytic function of ω in the upper half plane $\text{Im}(\omega) > 0$. To prove this, take $t < 0$ and note that

$$0 = \chi(t < 0) = \frac{1}{\sqrt{2\pi}} \int d\omega \, e^{-i\omega t} \chi(\omega), \quad (t < 0). \tag{7.224}$$

For $t < 0$, $\chi(t)$ can also be written as an integral over the contour displayed in Fig. 7.15(a) consisting of part of the real axis, $-R \leq \text{Re}(\omega) \leq R$, and an upper semicircle of radius R, letting $R \to \infty$. $\chi(t)$ will vanish (as it should for $t < 0$) only if $\chi(\omega)$ is analytic in the upper half plane, which proves statement 1.

(2) If $\chi(\omega)$ is analytic in the upper half plane, $\text{Im}(\omega) > 0$, and $\chi(\omega) \to 0$ as $|\omega| \to \infty$, then

$$\chi(\omega) = \frac{1}{i\pi} \mathcal{P} \int_{-\infty}^{\infty} d\omega' \, \frac{\chi(\omega')}{\omega' - \omega}, \tag{7.225}$$

where $\mathcal{P}$ denotes the Cauchy principal part of the integral. To prove it, consider the integral

$$I(\omega) \equiv \int_C d\omega' \, \frac{\chi(\omega')}{\omega' - \omega}, \tag{7.226}$$

where C is the contour shown in Fig. 7.15(b), consisting of two parts of the real axis, $-R \leq \text{Re}(\omega') \leq \omega - r$ and $\omega + r \leq \text{Re}(\omega') \leq R$, an upper semicircle of radius R, and a small upper semicircle of radius r centered at ω; and let

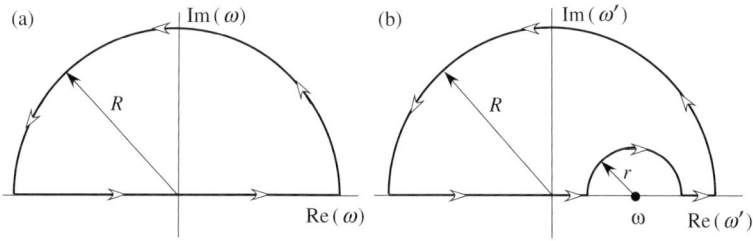

FIG 7.15 (a) Integration contour for Eq. (7.224). (b) Integration contour for Eq. (7.226).

$R \to \infty$ and $r \to 0$. The integrand is an analytic function of ω' inside C, and therefore, $\lim_{R \to \infty} I(\omega) = 0$. Moreover, the only contribution to the integral comes from the real axis. As $r \to 0$, we have, by definition,

$$\left[\int_{-\infty}^{\omega-r} + \int_{\omega+r}^{\infty} \right] d\omega' \, \frac{\chi(\omega')}{\omega' - \omega} = \mathcal{P} \int_{-\infty}^{\infty} d\omega' \, \frac{\chi(\omega')}{\omega' - \omega}, \tag{7.227}$$

while the integral over the small semicircle $\omega' - \omega = re^{i\theta}$ yields

$$\lim_{r \to 0} i \int_{\pi}^{0} d\theta \, \chi(\omega + re^{i\theta}) = -i\pi \chi(\omega). \tag{7.228}$$

Equating the sum of the LHSs of Eqs (7.227) and (7.228) to 0, and dividing by $i\pi$, we obtain Eq. (7.225). Finally, taking real and imaginary parts of Eq. (7.225),

$$\text{Re}[\chi(\omega)] = \frac{1}{\pi} \mathcal{P} \int_{-\infty}^{\infty} d\omega' \, \frac{\text{Im}[\chi(\omega')]}{\omega' - \omega}, \tag{7.229a}$$

$$\text{Im}[\chi(\omega)] = \frac{-1}{\pi} \mathcal{P} \int_{-\infty}^{\infty} d\omega' \, \frac{\text{Re}[\chi(\omega')]}{\omega' - \omega}. \tag{7.229b}$$

These equations are called the *Kramers–Kronig relations*. They show that the real part of the susceptibility at frequency ω is determined by an integral over all frequencies of the imaginary part of the susceptibility, and similarly for $\text{Im}[\chi(\omega)]$.

Problem 7.28

Use the Kramers–Kronig relations to show that the general linear response equation (7.220) can be written as,

$$\langle O_i(\mathbf{r}, t) \rangle = 2i \int d\mathbf{r}' \int_{-\infty}^{t} dt' \sum_j \bar{\chi}_{ij}(\mathbf{r}, \mathbf{r}', t') x_j(\mathbf{r}', t'), \tag{7.230}$$

where $\bar{\chi}_{ij}(t') = \frac{1}{\sqrt{2\pi}} \int d\omega \, e^{i\omega t'} \text{Im} \chi_{ij}(\omega)$.

7.9.2 KUBO FORMULAS

Kubo formulas are explicit expressions for susceptibilities within the linear response formalism. These expressions involve *correlation functions* of operators, e.g., $\langle A(\mathbf{r}, t) B(\mathbf{r}', t') \rangle$,[2] where the operators are expressed in the interaction representation. The susceptibilities are given in retarded form [see Eq. (7.241)], where the RHS of the equation involves $\Theta(t - t') \langle A(\mathbf{r}, t) B(\mathbf{r}', t') \rangle$, so the susceptibilities obey a causality condition. The Kubo formula, sometimes called the *Green–Kubo formula*, was first derived independently by Melville S. Green and by Ryogo Kubo in the early 1950s [116]. The derivation of the Kubo formula is presented below, and then it will be applied to obtain the response functions for conductivity and conductance.

---

[2] Note that these correlation functions are not directly related to the single-particle or pair correlation functions, $g^{(1)}(x, x')$ and $g^{(2)}(x_1, x_2)$, respectively, defined in Sec. 2.5.1.

7.9 Linear Response Theory

Before presenting the derivation, we need to standardize some notation, because we will be concerned with operators given in different representations (Schrödinger, Heisenberg, and Interaction), keeping in mind that the perturbations will be time dependent, even in the Schrödinger representation. Operators in the Heisenberg and interaction representations will always carry the subscript H and I, respectively, whereas operators in the Schrödinger representation will carry the subscript S (if they appear in an equation that contains operators written in other representations). $\rho(t)$, or $\rho_S(t)$, is used to denote the density matrix in the Schrödinger representation with the time-dependent perturbation switched on, whereas ρ_0 denotes the (time-independent) density matrix in the Schrödinger representation at thermal equilibrium. We shall write the expectation value for an operator O as $\langle O(t) \rangle = \text{Tr}[\rho_S(t) O_S(t)] = \text{Tr}[\rho_I(t) O_I(t)] = \text{Tr}[\rho_H(t) O_H(t)]$. The notation $\langle O_S \rangle_0 = \text{Tr}[\rho_0 O_S]$ (or with the subscript H or I) will be used for averaging operators in thermal equilibrium.

Consider a quantum system in thermal equilibrium at temperature T, described by a time-independent Hamiltonian H_0, with eigenstates $|n\rangle$ and corresponding eigenvalues E_n. Beginning at time $t = 0$,[3] the system is subject to a time-dependent perturbation, $H_1(t)$. The Hamiltonian is then[4]

$$H(t) = H_0 + H_1(t), \quad H_1(t < 0) = 0. \tag{7.231}$$

For $t < 0$, the density operator and the quantum thermodynamic average of an operator O are given by [$\beta = (k_B T)^{-1}$],

$$\rho_0 = \frac{e^{-\beta H_0}}{Z_0}, \quad \langle O \rangle_0 = \text{Tr}[\rho_0 O], \quad Z_0 = \text{Tr}\, e^{-\beta H_0} = \sum_n e^{-\beta E_n}, \tag{7.232}$$

where Z_0 is the *partition function* for the system with Hamiltonian H_0. If we know the exact density operator $\rho(t)$, we can compute $\langle O(t) \rangle \equiv \text{Tr}[\rho(t) O]$. If $\rho(t)$ is not known to all orders in H_1, then within linear response, it can be expanded in powers of H_1 up to first order. We shall do so within the interaction representation formalism. An operator $X_S(t)$ in the Schrödinger representation has the following form in the interaction representation (denoted by a subscript I, see Sec. 2.7.1), $X_I(t) = e^{iH_0 t/\hbar} X_S(t) e^{-iH_0 t/\hbar}$ (we shall use this definition with $X_S = \rho_S$ and H_1, thereby defining ρ_I and $H_{1,I}$). The Liouville equation for $\rho_I(t)$ is then $\partial \rho_I(t)/\partial t = -\frac{i}{\hbar}[H_{1,I}(t), \rho_I]$, with initial condition $\rho_I(0) = \rho_S(0) = \rho_0$. The solution to first order in $H_{1,I}$ is

$$\rho_I(t) \approx \rho_S(0) - \frac{i}{\hbar} \int_0^t dt' [H_{1,I}(t'), \rho_S(0)]. \tag{7.233}$$

Now, let us apply the operator $O_I(t)$ on both sides and take the trace, noting that $\text{Tr}[O_I \rho_I] = \text{Tr}[O_S \rho_S]$, and $\text{Tr}(A[B, C]) = \text{Tr}(C[A, B])$. This yields,

$$\text{Tr}(\rho(t) O) \approx \text{Tr}(\rho_0 O) - \frac{i}{\hbar} \int_0^t dt'\, \text{Tr}\left(\rho_0 [O_I(t), H_{1,I}(t')]\right). \tag{7.234}$$

The integrand on the RHS of Eq. (7.234) contains a correlation function. Because the time t at which the operator O is observed is always larger that the time t' where the perturbation is applied, the correlation is retarded,

$$C^R_{O,H_1}(t, t') \equiv -\frac{i}{\hbar} \Theta(t - t') \langle [O_I(t), H_{1,I}(t')] \rangle_0. \tag{7.235}$$

The subscript 0 on $\langle O \rangle_0$ indicates trace with ρ_0. At $T = 0$, $\langle O \rangle_0 \to \langle GS|O|GS \rangle$ where $|GS\rangle$ is the ground state of H_0. If the perturbation is slowly turned on from $t = -\infty$, the lower limit of the integrand in Eq. (7.234) should be $-\infty$. In performing the time integral in Eq. (7.234), it is useful to multiply the correlation by a decay factor $e^{-\eta|t-t'|}$ to

---

[3] The choice of beginning at $t = 0$ is but one possibility. Often, we want to begin at $t = -\infty$ and slowly turn on the perturbation (see below).
[4] Alternatively, one can slowly turn on the perturbation $H_1(t)$ from time $t = -\infty$, as in the next section.

guarantee convergence and assure that the integrand decay at large $|t - t'|$ and then take the limit as $\eta \to 0^+$ at the end of calculations. Thus, we finally have

$$\langle O(t)\rangle = \langle O\rangle_0 + \lim_{\eta \to 0^+} \int_{0 \text{ or} -\infty}^{\infty} dt'\, C^R_{O,H_1}(t,t')e^{-\eta|t-t'|}, \tag{7.236}$$

which is one version of the Kubo formula.

The reason for using the notation $C^R_{O,H_1}(t,t')$ and not $\chi(t-t')$ in Eqs (7.235) and (7.236) is that when $[H_1(t), H_1(t')] \neq 0$, the correlation depends on t and t' separately and not just on $t - t'$.

Response to a Time-Dependent Generalized Displacement

We now discuss the somewhat simpler and physically more realistic case where the perturbation $H_1(t)$ is caused by a time-dependent generalized displacement $x(t)$ that is coupled to an operator that is time independent in the Schrödinger representation. The generalized displacement $x(t)$ is not an operator, rather, it is a c-number that is coupled to some operator $\mathcal{F}$ of the system, so that $H_1(t) = -x(t)\mathcal{F}$. For example, $x(t)$ can be an external vector potential and $\mathcal{F}$ the current operator. Or $x(t) = g\mu_B H_z(t)$ is proportional to an external time-dependent magnetic field and $\mathcal{F}$ is an electron spin operator. Because $x(t)$ is a c-number and $\mathcal{F}$ is time independent, $[H_1(t), H_1(t')] = 0$; hence, the necessity for time ordering is avoided.

By response, one means the change of an ensemble-averaged physical observable $\langle O(t)\rangle$ to the external perturbation $H_1(t)$. Of special interest is $O = \mathcal{F}$. The applicability of linear response theory is restricted to the regime where, to a good approximation, $\langle O(t)\rangle$ changes linearly with the force. Hence, we assume that $x(t)$ is sufficiently weak to ensure that the response is linear. We assume that the system, in the absence of the perturbation, is in a stationary state described by the density matrix $\rho_0 = e^{-\beta H_0}/Z_0$ with *partition function* $Z_0 = \text{Tr}[e^{-\beta H_0}]$.

The Hamiltonian takes the simple form,[5]

$$H = H_0 - x(t)\mathcal{F} = H_0 + H_1(t). \tag{7.237}$$

In the Schrödinger representation, $\mathcal{F}$ is a constant operator at all times and $x(t)$ is a small c-number function depending on time with $\lim_{t \to \pm\infty} x(t) = 0$. Note that in the language of generalized displacements and forces, $x(t)$ is the displacement and $\mathcal{F}$ is the force. After $H_1(t)$ is switched on (say, slowly, from $t = -\infty$), the density matrix becomes time dependent, $\rho_0 \to \rho(t)$, and similarly, for any operator O, $\langle O\rangle_0 = \text{Tr}[\rho_0 O] \to \langle O(t)\rangle \equiv \text{Tr}[\rho(t)O]$.

The linear response hypothesis is

$$\langle O(t)\rangle = \langle O\rangle_0 + \int_{-\infty}^{t} dt'\, \chi(t-t')x(t'). \tag{7.238}$$

To obtain an expression for χ, we replace $H_{1,I}(t')$ in Eq. (7.234) by $-x(t)\mathcal{F}_I(t')$ and equate the result with Eq. (7.238) to get,

$$\chi(t-t') = \frac{i}{\hbar}\Theta(t-t')\langle[O_I(t), \mathcal{F}_I(t')]\rangle_0, \tag{7.239}$$

Note the difference compared with Eq. (7.235); the dependence on $t - t'$ is due to $[H_1(t), H_1(t')] = 0$, whereas the upper limit on the integral is due to causality, $\chi(t-t') = 0$ for $t' > t$. This guarantees the analyticity of the Fourier component of the susceptibility $\chi(\omega)$ in the upper half ω plane and enables the utilization of the Kramers–Kronig relations (7.229).

---

[5] This can be easily generalized to $H = H_0 - \sum_k x_k(t)\mathcal{F}_k$, see below.

7.9 Linear Response Theory

It is useful to generalize this result for the Hamiltonian $H = H_0 - \sum_k x_k(t)\mathcal{F}_k$ and for system operators (generalized forces) $O_k = \mathcal{F}_k$ [see Eq. (7.189)]. Direct application of the above analysis yields,

$$\langle \mathcal{F}_k(t) \rangle = \sum_j \int_{-\infty}^{\infty} \chi_{kj}(t-t')\, x_j(t')\, dt', \qquad (7.240)$$

where the susceptibility χ_{kj} is given by

$$\chi_{kj}(t-t') = \frac{i}{\hbar}\Theta(t-t')\langle [\mathcal{F}_{k,I}(t), \mathcal{F}_{j,I}(t')] \rangle_0. \qquad (7.241)$$

Generalized Conductance Matrix

One of the frequently measured response functions is the generalized conductance. Conductance is not limited to electrical conductance; its significance depends on the nature of the generalized displacements and forces. The occurrence of time-translation invariance enables a simple derivation of the generalized conductance in terms of the Fourier components of the susceptibility, which also enables its separation into dissipative and nondissipative parts. We shall now develop expressions for the generalized conductance.

The Fourier transform of the generalized force $\langle \mathcal{F}_k(t) \rangle$ in Eq. (7.240) can be evaluated by noting that the Fourier transform of a convolution is a product [making use of the Faltung (folding) theorem for Fourier transforms],

$$\langle \mathcal{F}_k(\omega) \rangle = \sum_j \chi_{kj}(\omega) x_j(\omega), \qquad (7.242)$$

where $\chi_{kj}(\omega)$ is the Fourier transform of the susceptibility $\chi_{kj}(\tau)$,

$$\chi_{kj}(\omega) = \frac{i}{\hbar}\frac{1}{\sqrt{2\pi}}\int_0^{\infty} dt\, e^{i\omega t} \langle [\mathcal{F}_{k,I}(t), \mathcal{F}_{j,I}(0)] \rangle_0. \qquad (7.243)$$

Let us express the generalized force in terms of the Fourier transform of the velocities $v_j(t)$,

$$v_j(\omega) \equiv \frac{1}{\sqrt{2\pi}}\int_{-\infty}^{\infty} dt\, e^{-i\omega t} \dot{x}_j(t). \qquad (7.244)$$

Problem 7.29

(a) Show that $v_j(\omega) = i\omega x_j(\omega)$.

Guidance: Write the Fourier expansion of $x_j(t)$, apply $\frac{d}{dt}$, and then take the inverse transform.

Using the result of Problem 7.29, Eq. (7.242) can be written as,

$$\langle \mathcal{F}_k(\omega) \rangle = -\frac{i}{\omega}\sum_j \chi_{kj}(\omega) v_j(\omega). \qquad (7.245)$$

The *generalized conductance matrix* $G_{kj}(\omega)$ is defined as the real part of the coefficient of $v_j(\omega)$ in Eq. (7.245), i.e.,

$$G_{kj}(\omega) = \frac{\text{Im}[\chi_{kj}(\omega)]}{\omega} = \frac{1}{\omega} \int_0^\infty d\tau \, \chi_{kj}(\tau) \sin(\omega\tau). \tag{7.246}$$

Problem 7.30

Show that the DC limit of the generalized conductance matrix is readily obtained from its definition in Eq. (7.246),

$$G_{kj}(\omega = 0) = \lim_{\omega \to 0} \frac{\text{Im}[\chi_{kj}(\omega)]}{\omega} = \int_0^\infty d\tau \, \tau \chi_{kj}(\tau). \tag{7.247}$$

Note that Eqs (7.242) and (7.245) are virtually identical. By taking their inverse Fourier transform, both equations yield Eq. (7.240), thereby expressing the generalized force in terms of the generalized displacements $\mathbf{x}(t)$ *not* the generalized velocities.

The DC generalized conductance $G_{kj}(\omega = 0)$ can be written as the sum of a symmetric matrix, $\eta_{kj} = \eta_{jk}$, and an antisymmetric matrix, $B_{kj} = -B_{jk}$:

$$G_{kj}(\omega = 0) \equiv \eta_{kj} + B_{kj}. \tag{7.248}$$

The nonconservative (dissipative) part of the conductance matrix $G_{kj}(\omega = 0)$, which is also called the *dissipation coefficient* or *friction coefficient*, is given by the symmetric matrix η_{kj}. To see it, recall that the dissipation is the rate at which energy is lost by the system [see Eq. (7.190)] is given (in the adiabatic limit) by power,

$$P(t) \equiv \dot{E}(t) = -\sum_j \langle \mathcal{F}_j(t) \rangle v_j(t), \tag{7.249}$$

Taking Fourier transform and the DC limit $\omega \to 0$, employing Eqs (7.245) and (7.247), we find,

$$P(\omega \to 0) = \sum_{kj} G_{kj}(0) v_k(0) v_j(0) = \sum_{kj} \eta_{kj}(0) v_k(0) v_j(0), \tag{7.250}$$

since the contribution of B_{ij} vanishes by (anti)symmetry. Thus, only the symmetric matrix, η_{kj}, contributes to dissipation.

The Kubo Formula for Electrical Conductivity

Let us apply the Kubo formalism to electrical conductivity. Consider a system of N electrons in a solid of volume $\mathcal{V}$. To employ Eq. (7.236), we need to specify the observable operator O and the perturbation H_1. The external electric field is distinct from the local electric field acting on the electrons, because there are internal fields that are generated when the external field is applied. Therefore, when we speak of "the electric field" $\mathbf{E}(\mathbf{r}, t)$, it will mean the resultant field acting on the electrons.

The system is assumed to respect translation invariance, so that the conductivity depends on $\mathbf{r} - \mathbf{r}'$. This is not true on the atomic scale, but it is justified in solids when the current at $\mathbf{r}$ is averaged over many unit cells surrounding $\mathbf{r}$. At the end, we will present the expression for the conductivity tensor for systems without translation invariance. The electromagnetic field to which the system responds is assumed to have a single frequency and wavevector,

$$\mathbf{E}(\mathbf{r}, t) = \mathbf{E}(\mathbf{q}, \omega) \, e^{i(\mathbf{q} \cdot \mathbf{r} - \omega t)} + \text{c.c.}. \tag{7.251}$$

7.9 Linear Response Theory

Consider only the $e^{i(\mathbf{q}\cdot\mathbf{r}-\omega t)}$ term, which can be written in terms of a vector potential $\mathbf{A}(\mathbf{r},t)$, $\mathbf{E}(\mathbf{r},t) = -\frac{1}{c}\frac{\partial \mathbf{A}(\mathbf{r},t)}{\partial t}$,

$$\mathbf{A}(\mathbf{r},t) = \mathbf{A}(\mathbf{q},\omega)\, e^{i(\mathbf{q}\cdot\mathbf{r}-\omega t)} = \frac{ic}{\omega} \mathbf{E}(\mathbf{q},\omega) e^{i(\mathbf{q}\cdot\mathbf{r}-\omega t)}. \tag{7.252}$$

This simplifying assumption is justified within linear response formalism. With these assumptions, the Fourier transform of Eq. (7.221c) is

$$J_\alpha(\mathbf{q},\omega) = \sum_\beta \sigma_{\alpha\beta}(\mathbf{q},\omega) E_\beta(\mathbf{q},\omega), \tag{7.253}$$

where the conductivity is generically complex, $\sigma_{\alpha\beta} = \text{Re}[\sigma_{\alpha\beta}] + i\,\text{Im}[\sigma_{\alpha\beta}]$. In the *DC limit*, $\mathbf{q} \to 0$ and $\omega \to 0$, the conductivity is real.

The Hamiltonian of the system is,

$$H = H_0 + H_1(t) = \left[\sum_i \frac{\mathbf{p}_i^2}{2m} + U\right] + \frac{e}{2mc}\sum_i [\mathbf{p}_i \cdot \mathbf{A}(\mathbf{r}_i,t) + \mathbf{A}(\mathbf{r}_i,t)\cdot \mathbf{p}_i + \frac{e}{c}\mathbf{A}^2(\mathbf{r}_i,t)], \tag{7.254}$$

where $\mathbf{p}_i$ is the momentum operator of particle i, H_0 is the many-body Hamiltonian of the electrons in the crystal in the absence of the electric field, and $H_1(t)$ is the perturbation due to the field, with $H_1(t<0)=0$. The potential U includes all other interactions. Within the linear response formalism, the $\mathbf{A}^2$ terms are often neglected, since A is small (but see the diamagnetic current term below). The current density *operator* $\mathbf{J}(\mathbf{r},t)$ is expressed in terms of the velocity operators,

$$\mathbf{v}_i = \frac{d\mathbf{r}_i}{dt} = -i[\mathbf{r}_i,H] = \frac{1}{m}[\mathbf{p}_i + \frac{e}{c}\mathbf{A}(\mathbf{r}_i,t)], \tag{7.255}$$

as follows:

$$\mathbf{J}(\mathbf{r},t) = \frac{-e}{2\mathcal{V}} \sum_i [\delta(\mathbf{r}-\mathbf{r}_i)\mathbf{v}_i + \mathbf{v}_i \delta(\mathbf{r}-\mathbf{r}_i)]$$

$$= \frac{-e}{2m\mathcal{V}}\sum_i [\delta(\mathbf{r}-\mathbf{r}_i)\mathbf{p}_i + \mathbf{p}_i \delta(\mathbf{r}-\mathbf{r}_i)] - \frac{e^2}{mc\mathcal{V}}\sum_i \delta(\mathbf{r}-\mathbf{r}_i)\mathbf{A}(\mathbf{r}_i,t) \equiv \mathbf{j}(\mathbf{r},t) + i\frac{ne^2}{m\omega}\mathbf{E}(\mathbf{r},t), \tag{7.256}$$

where the $\mathbf{j}$ is from the momentum operator terms, and

$$n(\mathbf{r}) = \sum_i \delta(\mathbf{r}-\mathbf{r}_i) \tag{7.257}$$

is the local density of carriers, and the second term on the RHS of Eq. (7.256), the *diamagnetic term*, is obtained by employing Eq. (7.252). This diamagnetic current flows with no deformation of the state of the system. The object is to express $\langle \mathbf{J}(\mathbf{r},t)\rangle$ in terms of the field $\mathbf{E}(\mathbf{r},t)$ within linear response, but since the second term on the RHS of Eq. (7.256) is not an operator, our task of calculating $\langle \mathbf{J}(\mathbf{r},t)\rangle$ is reduced to calculating $\langle \mathbf{j}(\mathbf{r},t)\rangle$.

If we work in the *Coulomb gauge*, $\nabla \cdot \mathbf{A}(\mathbf{r},t) = \mathbf{q} \cdot \mathbf{A}(\mathbf{q},\omega) = 0$, and neglect quadratic terms in $\mathbf{A}$, the interaction Hamiltonian $H_1(\mathbf{q},t')$ can be written as,

$$H_1(\mathbf{q},t') = -\frac{1}{c}\int d\mathbf{r}\, \mathbf{j}(\mathbf{r},t')\cdot \mathbf{A}(\mathbf{r},t') = \frac{i}{\omega}e^{-i\omega t'} \mathbf{E}(\mathbf{q},\omega)\cdot \mathbf{j}(\mathbf{q},t'), \tag{7.258}$$

where

$$\mathbf{j}(\mathbf{q},t') = \int d\mathbf{r}\, e^{i\mathbf{q}\cdot\mathbf{r}} \mathbf{j}(\mathbf{r},t'), \quad \mathbf{j}(-\mathbf{q},t') = \mathbf{j}^\dagger(\mathbf{q},t'). \tag{7.259}$$

The two operators that should appear in the retarded correlation (7.235) are then identified as $\mathbf{j}_I(\mathbf{r},t)$, and $H_{1,I}(\mathbf{q},t')$. From Eq. (7.234), we get

$$\langle \mathbf{J}(\mathbf{r},t)\rangle = \langle \mathbf{j}(\mathbf{r},t)\rangle_0 - \frac{i}{\hbar}\int_0^t dt'\, \langle [\mathbf{j}_I(\mathbf{r},t), H_{1,I}(\mathbf{q},t')]\rangle_0 + i\frac{ne^2}{m\omega}\mathbf{E}(\mathbf{r},t). \tag{7.260}$$

The first term on the RHS of Eq. (7.260) vanishes because there is no current in the absence of an electric field. Employing Eq. (7.258), we compute the correlation component for a given cartesian component $\alpha = x, y, z$, and get

$$[j_{\alpha,I}(\mathbf{r},t), H_{1,I}(\mathbf{q},t')] = \frac{ie^{-i\mathbf{q}\cdot\mathbf{r}}}{\omega}\sum_\beta [j_{\alpha,I}(\mathbf{r},t), j_{\beta,I}(\mathbf{q},t')] E_\beta(\mathbf{r},t) e^{i\omega(t-t')}. \tag{7.261}$$

Thus, the observable current is expressed in terms of a coefficient multiplying the electric field. This coefficient is identified with the electric conductivity. To smear out atomic fluctuations, the RHS of Eq. (7.261) is averaged over $\mathbf{r}$ and divided by $\mathcal{V}$, employing Eq. (7.259). Taking advantage of time translation invariance to express the conductivity tensor in $(\mathbf{q}, \omega)$ space, finally gives,

$$\sigma_{\alpha\beta}(\mathbf{q},\omega) = \frac{1}{\omega\mathcal{V}}\int_0^\infty dt\, e^{i(\omega+i\eta)t}\left\langle\left[j_{\alpha,I}^\dagger(\mathbf{q},t), j_{\beta,I}(\mathbf{q},0)\right]\right\rangle_0 + \lim_{\eta\to 0^+} i\frac{ne^2}{m\omega}\delta_{\alpha\beta}, \tag{7.262}$$

where the limit $\eta \to 0^+$ must be taken.

The DC Limit

The second term on the RHS of Eq. (7.262) is the diamagnetic contribution, since it originates from the A^2 term in the Hamiltonian. This term diverges as $\omega \to 0$. Physically, however, we know that the conductivity should stay finite and must be real at $\omega = 0$. This means that in the DC limit, the diamagnetic term should be cancelled with the imaginary part of the first term (sometimes referred to as the *paramagnetic contribution*). In the uniform case $\mathbf{q} = 0$, this can be shown, based on the following definitions and equalities,

$$\mathbf{d} \equiv -e\sum_i \mathbf{r}_i, \quad \mathbf{J} = -e\sum_i \mathbf{v}_i = \dot{\mathbf{d}} = \frac{i}{\hbar}[H, \mathbf{d}], \quad [\mathbf{d}, \mathbf{J}] = i\frac{Ne^2\hbar}{m}. \tag{7.263}$$

Now, noting that $\mathbf{d} = \mathbf{d}^\dagger$, we re-write Eq. (7.262) as,

$$\sigma_{\alpha\beta}(\omega) - i\frac{ne^2}{m\omega}\delta_{\alpha\beta} = \frac{1}{\hbar\omega\mathcal{V}}\int_0^\infty dt\, e^{i(\omega+i\eta)t}\frac{d}{dt}\langle[d_{\alpha,I}(t), j_{\beta,I}(0)]\rangle_0$$

$$= -\frac{1}{\hbar\omega\mathcal{V}}\langle[d_{\alpha,I}(0), J_{\beta,I}(0)]\rangle_0 - \frac{i}{\hbar\mathcal{V}}\int_0^\infty dt\, e^{i(\omega+i\eta)t}\langle[d_{\alpha,I}(t), j_{\beta,I}(0)]\rangle_0. \tag{7.264}$$

Employing Eq. (7.263), the second term on the LHS and the first term on the RHS of Eq. (7.264) cancel each other, and we are then left with

$$\sigma_{\alpha\beta}(\omega) = -\frac{i}{\hbar\mathcal{V}}\int_0^\infty dt\, e^{i(\omega+i\eta)t}\langle[d_{\alpha,I}(t), j_{\beta,I}(0)]\rangle_0, \tag{7.265}$$

which is finite at $\omega = 0$.

7.9 Linear Response Theory

Spectral Representation

If, in Eq. (7.262) for $\mathbf{q} = 0$, we insert a complete set of eigenstates $|n\rangle$ of H_0 with energies E_n between any pair of operators $\rho_0, \hat{j}_{\alpha,I}^\dagger(t)$ and $j_\beta(0)$ the time dependence is lifted to the exponents $e^{\pm i E_n t/\hbar}$ and can be integrated (here the role of η is crucial). The result is

$$\sigma_{\alpha\beta}(\omega) = \frac{i}{\omega V} \sum_{m,n} \frac{e^{-\beta E_n}}{Z_0} \frac{\langle n|j_\alpha|m\rangle \langle m|j_\beta|n\rangle}{\hbar(\omega + i\eta) - (E_m - E_n)} + i\frac{ne^2}{m\omega}\delta_{\alpha\beta}. \tag{7.266}$$

Problem 7.31

(a) Using the spectral representation (7.266) show that

$$\text{Re}\,\sigma_{xx}(\omega) = \frac{\pi}{\omega V} \sum_{m,n} \frac{e^{-\beta E_n}}{Z_0} |\langle n|j_x|m\rangle|^2 \delta(\hbar\omega - E_m + E_n). \tag{7.267}$$

(b) Discuss the relation between the above result and the expression for the power absorbed by the system using the Fermi Golden Rule.

Kubo Formula for the Conductivity of Nonuniform Systems

For completeness, we give the expression of the conductivity tensor $\sigma_{\alpha\beta}(\mathbf{r}, \mathbf{r}', t - t')$ for systems without translation invariance, as appears in Eq. (7.221),

$$\sigma_{\alpha\beta}(\mathbf{r}, \mathbf{r}'; t - t') = -i\Theta(t - t')\left\{\frac{1}{\hbar}\langle[j_{\alpha,I}(\mathbf{r},t), j_{\beta,I}(\mathbf{r}',t')]\rangle_0 + \delta_{\alpha\beta}\frac{e^2}{m}n(\mathbf{r})\delta(\mathbf{r} - \mathbf{r}')\right\}. \tag{7.268}$$

The first term (multiplied by $\hbar$),

$$\Pi_{\alpha\beta}^R(\mathbf{r}, \mathbf{r}'; t - t') \equiv -i\Theta(t - t')\langle[j_{\alpha,I}(\mathbf{r},t), j_{\beta,I}(\mathbf{r}',t')]\rangle_0, \tag{7.269}$$

is the lowest order contribution to the *retarded current–current correlation*. The Fourier transform of $\sigma_{\alpha\beta}(\mathbf{r}, \mathbf{r}'; t - t')$ is

$$\sigma_{\alpha\beta}(\mathbf{r}, \mathbf{r}', \omega) = \frac{i}{\omega}\Pi_{\alpha\beta}^R(\mathbf{r}, \mathbf{r}'; \omega) + i\delta_{\alpha\beta}\frac{e^2}{m\omega}n(\mathbf{r})\delta(\mathbf{r} - \mathbf{r}'), \tag{7.270}$$

where,

$$\Pi_{\alpha\beta}^R(\mathbf{r}, \mathbf{r}'; \omega) = \frac{-i}{\sqrt{2\pi}}\int_0^\infty dt\, e^{i\omega t}\Pi_{\alpha\beta}^R(\mathbf{r}, \mathbf{r}'; t - 0), \tag{7.271}$$

Therefore, the real part of $\sigma_{\alpha\beta}(\mathbf{r}, \mathbf{r}', \omega)$ is

$$\text{Re}\left[\sigma_{\alpha\beta}(\mathbf{r}, \mathbf{r}', \omega)\right] = -\frac{1}{\hbar\omega}\text{Im}\left[\Pi_{\alpha\beta}^R(\mathbf{r}, \mathbf{r}'; \omega)\right]. \tag{7.272}$$

Its DC limit, $\omega \to 0$, may be obtained using Eq. (7.247):

$$\lim_{\omega \to 0}\text{Re}\left[\sigma_{\alpha\beta}(\mathbf{r}, \mathbf{r}', \omega)\right] = \int_0^\infty dt\, t\, \Pi_{\alpha\beta}^R(\mathbf{r}, \mathbf{r}'; t). \tag{7.273}$$

Kubo Formula for Conductivity Using Generalized Forces

To derive the conductivity within the framework of generalized forces, we consider each component $A_\alpha(\mathbf{r}, t)$ as a set of infinite adiabatic parameters, $x_{\mathbf{r}}(t) = A_x(\mathbf{r}, t)$. The generalized force is the current operator $\mathbf{j}(\mathbf{r}, t)$, derived from the Hamiltonian $H_1(t) = \int d\mathbf{r}\, \mathbf{j}(\mathbf{r}, t) \cdot \mathbf{A}(\mathbf{r}, t)$, Eq. (7.256) by performing the functional derivative, $\mathbf{j}(\mathbf{r}, t) = -\delta H_1(t)/\delta \mathbf{A}(\mathbf{r}, t)$. According to Eq. (7.241), the corresponding susceptibiity is given by the first term on the RHS of Eq. (7.268).

The Kubo Formula for Conductance

The conductivity σ is the coefficient appearing before the electric field $\mathbf{E}$ in the expression for the current density $\mathbf{J}$; it is a property of the material. The conductance G is the coefficient appearing before the potential difference V in the expression for the current I; it is a property of the specific sample. For a homogeneous material, where σ is constant, the conductance of a sample of length L and cross-section area A is related to the conductivity σ as,

$$G = \sigma \frac{A}{L}. \tag{7.274}$$

For inhomogeneous samples, this relation is inadequate. Every quantum system has a typical temperature-dependent *phase breaking length* ℓ_φ beyond which quantum coherence is lost. In the last few decades, it has become possible to fabricate and study the electronic properties of mesoscopic systems whose length L is shorter than ℓ_φ at low temperatures, i.e., quantum coherence is maintained throughout the entire system. Thus, expression (7.274) is not valid for *mesoscopic conductors* at low temperatures (see Chapter 13). It is then necessary to use the Kubo formula for conductance rather than for conductivity. Consider, for example, a sample in the form of a slab stretched along the x direction with a potential difference V between its two ends. Denote by A the area of the cross-section of the slab a distance x from the left end. Then, the current along x through the slab at time t is given by

$$\langle I(t)\rangle = \int d\mathbf{r}_\perp \langle J(\mathbf{r}_\perp, x, t)\rangle = \int d\mathbf{r}_\perp d\mathbf{r}'_\perp dx' dt' \sigma(\mathbf{r}_\perp, x, \mathbf{r}'_\perp, x', t - t') E(\mathbf{r}'_\perp, x', t'), \tag{7.275}$$

where $\mathbf{r}_\perp$ are the coordinates on the transverse cross-section area, and for isotropic material, $\sigma = \sigma_{xx}$ and $E = E_x$. Employing Eq. (7.268) and focusing on the real part, we have,

$$\langle I(t)\rangle = -\frac{i}{\hbar} \int d\mathbf{r}_\perp d\mathbf{r}'_\perp dx' dt' \Theta(t - t') \langle [J_{x,I}(\mathbf{r}_\perp, x, t), J_{x,I}(\mathbf{r}'_\perp, x', t')]\rangle_0 E(\mathbf{r}'_\perp, x', t'). \tag{7.276}$$

The integral over $d\mathbf{r}'_\perp dx'$ can be performed on a surface of constant E_x whose rightmost point is x'. Therefore, the transverse integrations can be performed separately, which leads to a current–current correlation independent of x or x'. Since $\int dx' E_x(x', t') = V(t')$,

$$\langle I(t)\rangle = -\frac{i}{\hbar} \int dt' \Theta(t - t') \langle [I_I(t), I_I(t')]\rangle_0 V(t'). \tag{7.277}$$

With the help of Eq. (7.273), the linear response expression for the DC conductance becomes,

$$G = \frac{1}{\hbar} \mathrm{Im} \int_0^\infty dt\, t\, \langle [I_I(t), I_I(0)]\rangle_0. \tag{7.278}$$

7.9.3 ONSAGER RECIPROCAL RELATIONS

Symmetry relations exist between the generalized susceptibilities $\chi_{ij}(\mathbf{q}, \omega)$ due to basic physical principles, including the Kramers–Kronig relations. From the defining equation (7.240), it follows that if the generalized displacements $x_j(t)$

7.9 Linear Response Theory

are real and the generalized forces $\mathcal{F}_j(t)$ are real, so too is $\chi_{ij}(t)$. Then, from the definition of the Fourier transform $\chi_{ij}(\omega) = \frac{1}{\sqrt{2\pi}} \int_0^\infty dt e^{i\omega t} \chi_{ij}(t)$, where the dependence on **q** is suppressed, we find

$$\chi_{ij}(-\omega) = \chi_{ij}^*(\omega). \tag{7.279}$$

Writing the defining equation (7.240) for real monochromatic perturbations (and their velocities),

$$x_j(t) = \frac{1}{2}[a_j e^{i\omega t} + a_j^* e^{-i\omega t}], \quad \dot{x}_j(t) = \frac{i\omega}{2}[a_j e^{i\omega t} - a_j^* e^{-i\omega t}], \tag{7.280}$$

the corresponding response is,

$$\langle \mathcal{F}_j(t) \rangle = \frac{1}{2} \sum_j [\chi_{ij}(\omega) e^{i\omega t} a_j + \chi_{ij}(-\omega) e^{-i\omega t} a_j^*], \tag{7.281}$$

which is real, following Eq. (7.279). We can go further with this simple case and substitute the above expressions for $\langle \mathcal{F}_j(t) \rangle$ and $\dot{x}_j(t)$ into Eq. (7.249) to calculate the rate of energy dissipation. After averaging over a period $2\pi/\omega$, the oscillatory terms vanish, and the averaged dissipation rate is,

$$\overline{P} = \frac{i\omega}{4} \sum_{ij} [\chi_{ij}^*(\omega) - \chi_{ji}(\omega)] a_j a_j^*. \tag{7.282}$$

Now, recall the expression for $\overline{P}$ using the Fermi golden rule for Harmonic perturbation and compare it with Eq. (7.282). Assume that the system is in some particular stationary state $|n\rangle$ of the unperturbed Hamiltonian H_0 with energy E_n. Then, the rate of energy absorption by the system (averaged over one period) as a response to the harmonic perturbations $x_j(t)$ in Eq. (7.280) is,

$$\overline{P} = \frac{\pi\omega}{2\hbar} \sum_{ij} a_i a_j^* \sum_m \{[\mathcal{F}_i]_{mn}[\mathcal{F}_j]_{nm} \delta(\omega + \omega_{nm}) - [\mathcal{F}_i]_{nm}[\mathcal{F}_j]_{mn} \delta(\omega + \omega_{mn})\}, \tag{7.283}$$

where $\omega_{nm} = (E_n - E_m)/\hbar$ and the operator $\mathcal{F}_j$ is written in the Schrödinger representation. Comparing the two expressions for $\overline{P}$, we find

$$\chi_{ij}^*(\omega) - \chi_{ji}(\omega) = -\frac{2\pi i}{\hbar} \sum_m \{[\mathcal{F}_i]_{mn}[\mathcal{F}_j]_{nm} \delta(\omega + \omega_{nm}) - [\mathcal{F}_i]_{nm}[\mathcal{F}_j]_{mn} \delta(\omega + \omega_{mn})\}. \tag{7.284}$$

These results enable the derivation of symmetry relations among the generalized susceptibilities $\chi_{ij}(\omega)$. Assume that no external magnetic field is present and the system is time reversal invariant. Then it is possible to find a representation such that the matrices $[\mathcal{F}_i]_{mn}$ are real and symmetric. Therefore, the RHS of Eq. (7.284) is symmetric under $i \leftrightarrow j$ and so must be its LHS, so that

$$\chi_{ij}^*(\omega) - \chi_{ji}(\omega) = \chi_{ji}^*(\omega) - \chi_{ij}(\omega) \Rightarrow \chi_{ij}(\omega) + \chi_{ij}^*(\omega) = \chi_{ji}(\omega) + \chi_{ji}^*(\omega), \tag{7.285}$$

i.e., Re $\chi_{ij}(\omega) =$ Re $\chi_{ji}(\omega)$. Since the real and imaginary parts of $\chi_{ij}(\omega)$ are related by the Kramers–Kronig relation, the imaginary part is also symmetric. Hence, we find,

$$\boxed{\chi_{ij}(\omega) = \chi_{ji}(\omega), \text{ (for time-reversal invariant systems).}} \tag{7.286}$$

When the system is subject to an external magnetic field **H** (not as a generalized displacement), time reversal invariance is broken, and the wave function is not real. Rather, they satisfy the relation $\psi(\mathbf{H}) = \psi^*(-\mathbf{H})$. Therefore, the matrix elements of the generalized forces satisfy $[\mathcal{F}_i]_{nm}(\mathbf{H}) = [\mathcal{F}_i]_{mn}(-\mathbf{H})$. Hence, the RHS of Eq. (7.284) will not change under simultaneous operations $i \leftrightarrow j$ and $\mathbf{H} \to -\mathbf{H}$. Therefore, instead of Eq. (7.285) we will now have,

$$\chi_{ij}^*(\omega, \mathbf{H}) - \chi_{ji}(\omega, \mathbf{H}) = \chi_{ji}^*(\omega, -\mathbf{H}) - \chi_{ij}(\omega, -\mathbf{H}). \tag{7.287}$$

An additional symmetry relation for $\chi_{ij}(\omega, \mathbf{H})$ is obtained if one regards the Kramers–Kronig relations (7.229) as a linear integral operator (kernel) denoted by $i\hat{K}$ (see Problem 7.32),

$$\chi_{ij}(\omega, \mathbf{H}) = i \int d\omega' \hat{K}(\omega, \omega') \chi_{ij}(\omega', \mathbf{H}). \tag{7.288}$$

Problem 7.32

Write the kernel $\hat{K}(\omega, \omega')$ as a 2×2 matrix operating on the real and imaginary parts of χ_{ij} and check that it is Hermitian.

Answer: The susceptibility in Eq. (7.288) is a two-component vector $\chi_{ij}(\omega, \mathbf{H}) = \begin{pmatrix} \mathrm{Re}\chi_{ij}(\omega,\mathbf{H}) \\ i\mathrm{Im}\chi_{ij}(\omega,\mathbf{H}) \end{pmatrix}$. Equation (7.288) is equivalent to Eqs (7.229), where the 2×2 matrix $\hat{K}$ is $\hat{K} = \begin{pmatrix} 0 & \frac{-1}{\pi}\mathcal{P}\frac{1}{\omega'-\omega} \\ \frac{-1}{\pi}\mathcal{P}\frac{1}{\omega'-\omega} & 0 \end{pmatrix}$, which is Hermitian.

The Hermitian conjugate of Eq. (7.288) is $\chi_{ij}^*(\omega, \mathbf{H}) = -i \int d\omega' \hat{K}(\omega, \omega') \chi_{ij}^*(\omega', \mathbf{H})$. Adding Eq. (7.288) and its Hermitian conjugate leads to the relation,

$$\chi_{ij}^*(\omega, \mathbf{H}) + \chi_{ji}(\omega, \mathbf{H}) = -i\hat{K}[\chi_{ij}^*(\omega, \mathbf{H}) - \chi_{ji}(\omega, \mathbf{H})], \tag{7.289}$$

hence any symmetry of the sum is also a symmetry of the difference, and hence, of χ_{ij} itself. Since the sum is symmetric under simultaneous operations $i \leftrightarrow j$ and $\mathbf{H} \to -\mathbf{H}$, for systems in an external magnetic field:

$$\boxed{\chi_{ij}(\omega, \mathbf{H}) = \chi_{ji}(\omega, -\mathbf{H}).} \tag{7.290}$$

Finally, one may encounter a correlation $\langle \mathcal{F}_i \mathcal{F}_j \rangle$ such that at least one of the generalized forces is odd under time reversal. If both of them are odd, then the relations (7.286) and (7.290) remain intact, but if only one of the generalized forces is odd under time reversal, say $\mathcal{F}_i$ then $[\mathcal{F}_i]_{nm} = -[\mathcal{F}_i]_{mn}$, the RHS of Eq. (7.284) changes sign when $i \leftrightarrow j$. Therefore, the corresponding relations are,

$$\chi_{ij}(\omega, \mathbf{H}) = -\chi_{ji}(\omega, -\mathbf{H}), \quad (\mathcal{F}_i \text{ or } \mathcal{F}_j \text{ is odd under time-reversal}), \tag{7.291}$$

including the case $\mathbf{H} = 0$.

7.9.4 FLUCTUATION–DISSIPATION THEOREM

The *fluctuation–dissipation theorem* was first formulated by Harry Nyquist in 1928, and later proven by Herbert Callen and Theodore Welton in 1951 [118]. It relates the irreversible dissipation occurring in a system which is affected by an external force driving it out of equilibrium, and the fluctuations of the system in thermal equilibrium [see Eq. (7.303)]. Our understanding of the relation between fluctuations of a system and its response to an external perturbation dates back to Einstein's 1905 famous work on Brownian motion, which was continued by John Johnson, Harry Nyquist, Lars Onsager, and then by Callen and Welton. These works lead up to Ryogo Kubo's description of linear response theory.

We now formulate the fluctuation dissipation theorem and derive it. To simplify notation, we assume that there is a single external perturbation; the generalization to multiple generalized displacements is trivial. Moreover, only the time dependence of operators will be specified (not their position, $\mathbf{r}$). The same symbol will be used for a function of time and its Fourier transform, [e.g., $\chi(t)$ and $\chi(\omega)$]. Recall our convention that $\mathcal{F}_S(t)$ or $\mathcal{F}(t)$ refer to the operator in

7.9 Linear Response Theory

the Schrödinger picture (at time $t = 0$, we simply use $\mathcal{F}$), $\mathcal{F}_H(t)$ is its representation in the Heisenberg picture, while $\mathcal{F}_I(t)$ is its representation in the interaction picture. Averages performed on an operator O when the system is in thermal equilibrium are denoted as $\langle O \rangle_0 = \text{Tr}[\rho_0 O]$. They are quantum thermodynamic averages using a thermal density matrix. Averages out of equilibrium are denoted as $\langle O(t) \rangle = \text{Tr}[\rho_S(t) O]$, where $\rho_S(t) = \rho(t)$ is the time-dependent density matrix in the Schrödinger representation.

The basic ingredients for discussing the fluctuation-dissipation theorem are now in place. Section 7.9.2 covered the coupling of a generalized displacement [sometimes denoted by $x(t)$, but below denoted by $f(t)$] of a given system to the generalized force $\mathcal{F}$; the Hamiltonian for the system was given in Eq. (7.237). The generalized force $\mathcal{F}$ is a Hermitian operator. In Eq. (7.237), H_0 can be viewed as the Hamiltonian of the system in equilibrium. Equations (7.240) and (7.241) showed that $\langle \mathcal{F}(t) \rangle$ is a linear functional of $f(t)$ where the coefficient $\chi(t-t')$ is given by the retarded correlation function of $\mathcal{F}(t)$ [see Eq. (7.241)]. The Kubo formula,

$$\langle \mathcal{F}(t) \rangle = \langle \mathcal{F} \rangle_0 + \int_{-\infty}^{\infty} dt' \, \chi(t-t') f(t'), \qquad (7.292)$$

$$\chi(t-t') = \frac{i}{\hbar} \Theta(t-t') \langle [\mathcal{F}_I(t), \mathcal{F}_I(t')] \rangle_0 = \frac{1}{\sqrt{2\pi}} \int_{-\infty}^{\infty} d\omega \, e^{-i\omega(t-t')} \chi(\omega) \qquad (7.293)$$

will be used in developing the fluctuation-dissipation theorem. For simplicity of notation, we temporarily set $\hbar = 1$ and restore it in Eq. (7.303). When the system with Hamiltonian H_0 is in equilibrium at temperature T, the quantity $\mathcal{F} - \langle \mathcal{F} \rangle$ will fluctuate with variance $\langle (\mathcal{F} - \langle \mathcal{F} \rangle)^2 \rangle$ characterized by a power spectrum $S(\omega)$ [see Eq. (7.294)]. This power spectrum is the Fourier component of the two-point correlation of $\mathcal{F}$ at different times, $S(t-t')$, i.e.,

$$S(t-t') = \langle \mathcal{F}_I(t) \mathcal{F}_I(t') \rangle_0 = \frac{1}{\sqrt{2\pi}} \int_{-\infty}^{\infty} d\omega \, e^{-i\omega(t-t')} S(\omega). \qquad (7.294)$$

Because $H_1(t) = -f(t)\mathcal{F}$, where $f(t)$ is a c-number and, in the Schrödinger representation, $\mathcal{F}$ is time independent, $S(\tau) = S(-\tau)$. The difference between this correlation and $\chi(t-t')$ of Eq. (7.293) is that $S(t-t')$ is not retarded and the operators do not appear in a commutator. The fluctuation-dissipation theorem relates the imaginary part of $\chi(\omega)$ that characterizes the irreversible approach of a system to its equilibrium via dissipation, to $S(\omega)$, the power spectrum of the fluctuations of $\mathcal{F}_I$ around its equilibrium value.

Derivation of the relation between $S(\omega)$ and $\text{Im}\chi(\omega)$ uses the spectral decomposition of $S(t)$ and $\chi(t)$. The Fourier transform of $S(t)$ is the inverse of Eq. (7.294),

$$S(\omega) = \frac{1}{\sqrt{2\pi}} \int_{-\infty}^{\infty} dt \, e^{i\omega t} \langle \mathcal{F}_I(t) \mathcal{F}_I(0) \rangle_0. \qquad (7.295)$$

The quantum thermodynamic averages are carried out with respect to the equilibrium density matrix. Let $\{|\phi_\alpha\rangle\}$ denote the complete set of eigenfunctions of the time-independent Hamiltonian H_0, with corresponding energies $\{\varepsilon_\alpha\}$. By definition,

$$\langle \mathcal{F}_I(t) \mathcal{F}_I(0) \rangle_0 = \sum_\alpha \langle \phi_\alpha | \mathcal{F}_I(t) \mathcal{F}_I(0) | \phi_\alpha \rangle e^{-\beta(\varepsilon_\alpha - F)}, \qquad (7.296)$$

where the free energy F is related to the equilibrium partition function Z by

$$F = -k_B T \ln Z = -k_B T \ln \sum_\alpha e^{\beta \varepsilon_\alpha}. \qquad (7.297)$$

Note that $\mathcal{F}_I(0) = \mathcal{F}_S \equiv \mathcal{F}$. Inserting expansion (7.296) into (7.295), introducing the identity operator $\sum_\gamma |\phi_\gamma\rangle\langle\phi_\gamma|$ between $\mathcal{F}_I(t)$ and $\mathcal{F}_I(0) = \mathcal{F}$, and performing the time integration and setting $\hbar = 1$, we find

$$S(\omega) = \sqrt{2\pi} \sum_{\alpha\gamma} e^{-\beta(\varepsilon_\alpha - F)} |\langle\phi_\alpha|\mathcal{F}(0)|\phi_\gamma\rangle|^2 \, \delta(\varepsilon_\gamma - \varepsilon_\alpha - \omega). \tag{7.298}$$

Because $S(\tau) = S(-\tau)$, $S(\omega)$ is real. The temporal correlation function is,

$$S(t - t') = \sum_{\alpha\gamma} e^{-\beta(\varepsilon_\alpha - F)} |\langle\phi_\alpha|\mathcal{F}(0)|\phi_\gamma\rangle|^2 \, e^{-i(\varepsilon_\gamma - \varepsilon_\alpha)(t - t')}. \tag{7.299}$$

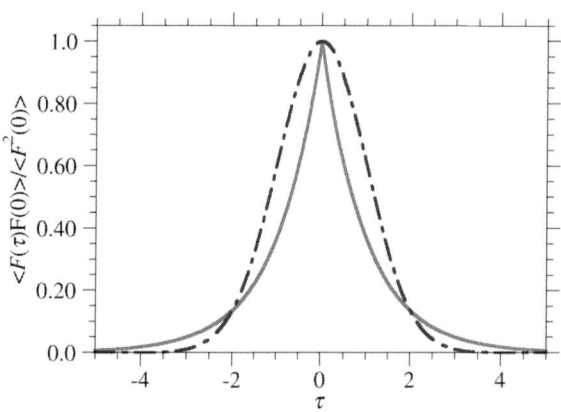

FIG 7.16 Two typical correlation functions $\frac{\langle\mathcal{F}_I(\tau)\mathcal{F}\rangle_0}{\langle\mathcal{F}^2\rangle_0}$ versus τ: Gaussian function (dot-dashed curve) and exponential function $e^{-a|\tau|}$ (solid curve).

Within the linear response formalism, S is independent of the generalized displacement $f(t)$. Figure 7.16 shows typical behavior of $C(\tau) \equiv \frac{\langle\mathcal{F}_I(\tau)\mathcal{F}\rangle_0}{\langle\mathcal{F}^2\rangle_0} = \frac{S(\tau)}{S(0)}$ versus τ. It typically falls off rapidly (exponentially or as a Gaussian) as a function of τ.

Next, we apply the same procedure to $\chi(\omega)$,

$$\chi(\omega) = \frac{i}{\sqrt{2\pi}} \int_{-\infty}^{\infty} dt\, e^{i\omega t} \Theta(t) \langle [\mathcal{F}_I(t), \mathcal{F}_I(0)] \rangle_0, \tag{7.300}$$

to obtain

$$\chi(\omega) = -\sqrt{2\pi}\, e^{\beta F} \sum_{\gamma\alpha} (e^{-\beta\varepsilon_\gamma} - e^{-\beta\varepsilon_\alpha}) \frac{|\langle\phi_\alpha|\mathcal{F}|\phi_\gamma\rangle|^2}{\varepsilon_\gamma - \varepsilon_\alpha + \omega}. \tag{7.301}$$

Using the Kramers–Kronig relation (7.229b), we obtain

$$\operatorname{Im}\chi(\omega) = \frac{\sqrt{2\pi}}{2}(1 - e^{-\beta\omega}) \sum_{\alpha\gamma} e^{-\beta(\varepsilon_\alpha - F)} |\langle\phi_\alpha|\mathcal{F}|\phi_\gamma\rangle|^2 \delta(\varepsilon_\alpha - \varepsilon_\gamma - \omega). \tag{7.302}$$

From Eq. (7.298), the RHS of Eq. (7.302) is proportional to $S(\omega)$. Restoring $\hbar$ units, we obtain the fluctuation-dissipation theorem,

$$\boxed{S(\omega) = \frac{2\hbar}{1 - e^{-\beta\hbar\omega}} \operatorname{Im}\chi(\omega).} \tag{7.303}$$

Thus, employing spectral decomposition, we established a direct link between the power spectrum $S(\omega)$ and the dissipative part of the response function $\chi(\omega)$.

Example 1:
Consider a free particle in a 1D system of length L with periodic boundary conditions subjected to a perturbation $-f(t)\mathcal{F}$ where the generalized displacement $f(t)$ is coupled to the generalized force, $\mathcal{F} \equiv V(x) = 2V_0 \cos k_1 x$, where $k_1 = \frac{2\pi}{L}$. The Hamiltonian is,

$$H = -\frac{\hbar^2}{2m}\frac{d^2}{dx^2} - f(t)2V_0 \cos k_1 x \equiv H_0 - f(t)V(x), \tag{7.304}$$

where $f(t)$ is an arbitrary dimensionless function. The eigenfunctions of H_0 are normalized plane waves,

$$\langle x|k\rangle = \frac{1}{\sqrt{L}} e^{ikx}, \quad \varepsilon_k = \frac{\hbar^2 k^2}{2m}, \quad k = n_k k_1, \quad n_k = 0, \pm 1, \pm 2, \ldots. \tag{7.305}$$

7.9 Linear Response Theory

Note that

$$\langle k'|V|k\rangle = \frac{V_0}{L}(\delta_{k',k+k_1} + \delta_{k',k-k_1}). \tag{7.306}$$

Substitution of these expressions into Eq. (7.299) yields,

$$S(t) = \frac{V_0 e^{\beta F}}{L^2} \sum_{kk'} e^{-\beta \varepsilon_k}(\delta_{k',k+k_1} + \delta_{k',k-k_1}) e^{\frac{i}{\hbar}(\varepsilon_k - \varepsilon_{k'})t}. \tag{7.307}$$

To exploit the delta function constraints, define the velocity $v_1 \equiv \hbar k_1/m$ and note that $(\varepsilon_k - \varepsilon_{k \pm k_1}) = \mp \frac{\hbar}{2} v_1 (k - k_1)$. Thus, we are left with a single sum on k,

$$S(t) = \frac{V_0 e^{\beta F - i v_1 k_1 t/2}}{L^2} \sum_k e^{-\beta \varepsilon_k}(e^{i v_1 k t/2} + e^{-i v_1 k t/2}). \tag{7.308}$$

If we define $\alpha^2 \equiv \beta \frac{\hbar^2}{2m} = \beta \varepsilon_k/k^2$ and employ the approximation $\frac{1}{L}\sum_k \to \frac{1}{2\pi}\int dk$, we obtain a Gaussian correlation function,

$$S(t) = \sqrt{\frac{V_0^2}{\pi \alpha^2 L^2}} e^{\beta F - i v_1 k_1 t/2 - \frac{(v_1 t)^2}{16 \alpha^2}}. \tag{7.309}$$

The Fourier transform of the correlation function (see Problem 7.33) gives the loss via the fluctuation-dissipation theorem, Eq. (7.303).

Problem 7.33

Show that the power spectrum is given by,

$$S(\omega) = \frac{V_0 e^{\beta F}}{L} \sqrt{\frac{8}{\pi v_1^2}} \exp\left[-\alpha^2 \left(\frac{\omega}{v_1} - \frac{k_1}{2}\right)^2\right], \tag{7.310}$$

where by definition of the partition function is $Z = e^{-\beta F} = \sum_k e^{-\beta \varepsilon_k}$, and an integral approximation of the sum gives, $e^{\beta F} = \sqrt{\frac{\beta \hbar^2}{2m\pi L^2}}$.

Example 2:
Consider again a free particle in 1D with external potential $V = ap/\hbar = -ia\frac{d}{dx}$, where the units of the constant a is such that V has dimension of energy and $f(t)$ is dimensionless. The Hamiltonian is

$$H = -\frac{\hbar^2}{2m}\frac{d^2}{dx^2} + f(t) ia \frac{d}{dx}. \tag{7.311}$$

In Problem 7.34 you will complete the analysis.

Problem 7.34

Following an analysis similar to that of Example 1, show that the power spectrum is,

$$S(\omega) = S_0 \delta(\omega), \quad S_0 = \frac{1}{\sqrt{2\pi} L^2} e^{\beta F} \sum_k e^{-\beta \varepsilon_k} a^2 k^2 = \frac{2ma^2}{\sqrt{2\pi} L^2 \hbar^2 \beta}. \tag{7.312}$$

Answer: The matrix element of V between plane waves is, $\langle k|V|k'\rangle = \frac{ka}{L}\delta_{kk'}$. Using Eq. (7.299),

$$S(t) = \frac{1}{L^2} e^{\beta F} \sum_k e^{-\beta \varepsilon_k} a^2 k^2 \tag{7.313}$$

is independent of t. Taking the Fourier transform, we get the first result of Eq. (7.312). The second equality for S_0 is obtained by noting that $\sum_k k^2 e^{-\beta \varepsilon_k} = -\frac{2m}{\hbar^2} \frac{\partial Z(\beta)}{\partial \beta}$, where $Z(\beta) = L\sqrt{\frac{2m\pi}{\beta \hbar^2}}$.

Historic Applications of the Fluctuation-Dissipation Theorem

The fluctuation-dissipation theorem makes the connection between the spontaneous fluctuations in the system in equilibrium, as described by the non-retarded correlation function $S(\omega)$, and the response of the system to external perturbations, as determined by the susceptibility.

Historically, three applications of the fluctuation-dissipation theorem have been especially important (all of these applications were developed before the proof of the theorem by Callen and Welton was published):

1. Brownian motion. A body moving with velocity $\dot{x}$ that experiences a friction force $\mathcal{F} = -\eta \dot{x}$ also experiences a fluctuating force that is related to the friction coefficient η by the relation, $\int_{-\infty}^{\infty} d\tau \, \langle [\mathcal{F}_I(\tau), \mathcal{F}_I(0)]\rangle_0 = 2k_B T \eta$, i.e., $S(\omega) = S_0 \delta(\omega)$.
2. Dissipative Harmonic Oscillator. Consider the fluctuations of a particle in a harmonic oscillator potential of frequency ω in thermal equilibrium at a temperature T. Just as in Brownian motion, the integral over the velocity–velocity correlation function gives the diffusion coefficient for the Brownian motion of the particle, $\int_{-\infty}^{\infty} d\tau \, \langle [v_I(\tau), v_I(0)]\rangle_0 = D$. The velocity–velocity correlation for a free particle can be calculated by using the fluctuation-dissipation theorem and taking the limit as $\omega \to 0$. When written in the form, $D = \mu k_B T$, where μ is the mobility, defined as the average drift velocity divided by the applied external field (see Sec. 9.2.1), this relation is called the Einstein–Smoluchowski relation.
3. Fluctuating Currents. These fluctuations are called thermal or *Johnson–Nyquist noise*. The word noise indicates fluctuations of a given measurable quantity about its mean (it may be related not only to thermal fluctuations but also to shot noise — see below). Thermal noise occurs at frequencies $\omega \leq k_B T/\hbar$. A closed electrical circuit containing a resistor, whose conductance is $G = 1/R$, where R is the resistance, experiences fluctuations of the current $I(t)$, $\int_{-\infty}^{\infty} d\tau \, \langle [I_I(\tau), I_I(0)]\rangle_0 = 2Gk_B T$, where T is the temperature. The spectral density in equilibrium ($V = 0$) is $S = 4k_B T G$, independent of frequency. The term *white noise* means that $S(\omega)$ is independent of frequency. (Another possibility is to observe voltage fluctuations $\langle V^2 \rangle = 4k_B T R \Delta \omega$, where $\Delta \omega$ is the bandwidth over which the voltage is measured.) Electrical noise can also be described in terms of current fluctuations $\langle I_I(\omega) I_I(0)\rangle_0$ in a conductor subject to a voltage difference V. For $V \neq 0$, the noise increases and becomes frequency dependent. Experimentally, the noise power is directly measurable by the quantity of heat collected in a cold reservoir.

Another important fluctuation phenomenon, referred to as *shot noise*, is important at very low temperatures and when the current is very small. In contrast to thermal (Johnson–Nyquist) noise, resulting from the thermal excitations of electrons, shot noise is due to the discreteness of the electrical charge (the semiclassical view is that the individual electron wave packets arrive at the detector at quasi-random[6] times). We will say more about shot noise in Chapter 13. At low temperature, fluctuations are still related to dissipation via the fluctuation-dissipation theorem, i.e., shot noise is properly included in the theorem.

In Chapter 17, linked to the book web page, we shall return to the study of fluctuations and dissipation in system–bath interactions.

---

[6] Quasi because electrons repel each other, so electron arrival times have a whole in their arrival time distribution. See Hanbury Brown–Twiss effect for fermionic particles.

Identical Particles

8

In classical mechanics, we can keep track of individual particles, even if they are identical. Consider two identical particles undergoing a collision as shown in Fig. 8.1. Knowing their initial positions and momenta at time $t = 0$, their motion as a function of time can be followed. Hence, we can think of the particles as being numbered, as in Fig. 8.1, and following their trajectories, we can keep track of which particle is which. But in quantum mechanics, particle trajectories cannot be followed; this would involve determining the position of each particle at each instant of time (which disturbs the particles). Moreover, the uncertainty principle does not allow determining both the position and momentum with infinite precision. Thus, while the two situations shown in Fig. 8.1 can be distinguished classically, they cannot be distinguished quantum mechanically, even in principle. Quantum mechanically, identical particles entirely lose their individual identity and are completely indistinguishable. This principle of indistinguishability of identical particles plays a fundamental role in quantum theory and is called *exchange symmetry* (see Sec. 2.9.1). The name describes the symmetry of the wave function on exchange of identical particles. Before discussing exchange symmetry in detail, we develop the mathematics of permutation symmetry, using permutation group concepts (see the brief review of group theory and symmetry in Appendix E).

8.1 PERMUTATION SYMMETRY

An arbitrary permutation, i.e., rearrangement, of N positive integer numbers, or N objects (particles), is represented as

$$P = \begin{pmatrix} 1 & 2 & \dots & N \\ \alpha(1) & \alpha(2) & \dots & \alpha(N) \end{pmatrix}, \qquad (8.1)$$

where $\alpha(i)$ is an integer $\leq N$ and the function α is 1-to-1 and onto [no duplication of numbers in the second row on the RHS of (8.1) and all the positive integers up to and including N are present]. The symbol on the RHS of Eq. (8.1) should be read as follows: replace the number 1 by the number (or object) $\alpha(1)$, 2 by $\alpha(2)$, ..., and N by $\alpha(N)$. The permutations of N objects form a group having $N!$ elements, denoted by S_N, since multiplication of permutations can be defined (see below) and all the group axioms (associativity, identity, and inverse) are met. Arbitrary permutations can be decomposed into products of *two-cycles*, $P_{ij} = \binom{i \, j}{j \, i}$; this *two-cycle* is a permutation that replaces i with j and j with i. Two-cycles are also called *transpositions*. In quantum mechanics, we can define the *permutation operator* P_{ij} to be the operator that affects the replacement of particle i with particle j and j with i (P_{ij} permutes particles i and j).

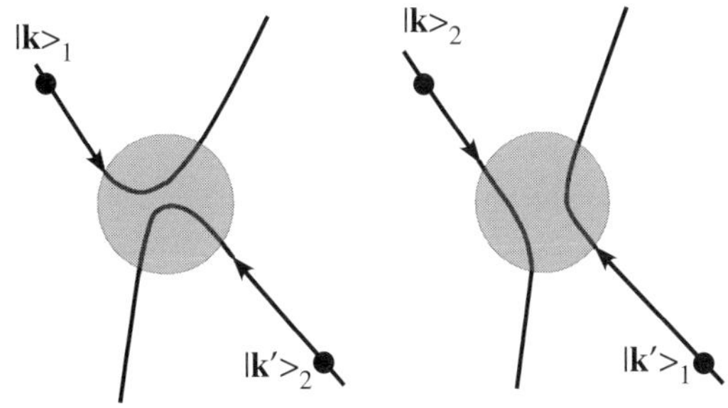

Two permutations can be multiplied and the result is also a permutation; if P_1 maps $n \to \alpha(n)$ and P_2 maps $n \to \beta(n)$, then $P_2 P_1$ maps $n \to \gamma(n) = \beta[\alpha(n)]$. For example, suppose

$$P_1 = \begin{pmatrix} 1 & 2 & 3 & 4 & 5 \\ 2 & 4 & 3 & 5 & 1 \end{pmatrix}, \quad P_2 = \begin{pmatrix} 1 & 2 & 3 & 4 & 5 \\ 5 & 4 & 1 & 2 & 3 \end{pmatrix}, \qquad (8.2)$$

then the product $P_2 P_1$ is the permutation

$$P_2 P_1 = \begin{pmatrix} 1 & 2 & 3 & 4 & 5 \\ 4 & 2 & 1 & 3 & 5 \end{pmatrix}. \qquad (8.3)$$

Note that first P_1 is applied, then P_2. Figure 8.2 shows how the product permutation is obtained.

FIG 8.1 Scattering of two identical particles 1 and 2 which can be distinguished in classical mechanics but cannot be distinguished in quantum mechanics. The shaded region is where the particles interact strongly.

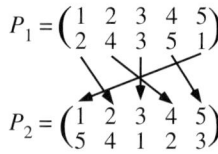

$$P_1 = \begin{pmatrix} 1 & 2 & 3 & 4 & 5 \\ 2 & 4 & 3 & 5 & 1 \end{pmatrix}$$

$$P_2 = \begin{pmatrix} 1 & 2 & 3 & 4 & 5 \\ 5 & 4 & 1 & 2 & 3 \end{pmatrix}$$

FIG 8.2 The product, $P_2 P_1$, of the permutations in (8.2) is obtained by following the arrows, e.g., $1 \to 2 \to 4$, $2 \to 4 \to 2$, etc., to obtain (8.3).

A permutation is *even* if it can be written as a product of an even number of two-cycles, and *odd* otherwise. An example of an even permutation is

$$P = \begin{pmatrix} 1 & 2 & 3 & 4 \\ 2 & 1 & 4 & 3 \end{pmatrix} = (12)(34), \tag{8.4}$$

since it can be written as an even number of two-cycles. Note that the two-cycle (34) on the right is applied first and (12) afterward. An example of an odd permutation is

$$P = \begin{pmatrix} 1 & 2 & 3 & 4 & 5 \\ 2 & 3 & 4 & 1 & 5 \end{pmatrix} = (14)(13)(12). \tag{8.5}$$

The one-cycle (5) is often left out for concise notation. The even permutations of N objects is a proper subgroup of the permutation group S_N.

A *cyclic permutation* $P = (12 \ldots N)$ is a permutation of the form

$$P = (12 \ldots N) \equiv \begin{pmatrix} 1 & 2 & \ldots & N-1 & N \\ 2 & 3 & \ldots & N & 1 \end{pmatrix}. \tag{8.6}$$

A cyclic permutation can be written as a product of two-cycles in the form, $(a_1 a_2 \ldots a_N) = (a_1 a_N)(a_1 a_{N-1}) \ldots (a_1 a_2)$. A three-cycle is a cyclic permutation of three numbers, e.g., (123), or $(6, 9, 13) = \begin{pmatrix} 6 & 9 & 13 \\ 9 & 13 & 6 \end{pmatrix}$. A permutation is said to be in disjoint cycle form if it is written so that the various pairs of cycles which define it have no number in common.

Problem 8.1

(a) Show that $P_{12}[= (12)]$ and $P_{23}[= (23)]$ do not commute, i.e., the permutation group is not abelian.
(b) Demonstrate that $P_{12}^{-1} = P_{12}$.
(c) Show that $(123)^{-1} = \begin{pmatrix} 1 & 2 & 3 \\ 3 & 1 & 2 \end{pmatrix}$.

Problem 8.2

Show that the product $P_1 P_2$ of the permutations in (8.2) is $P_1 P_2 = \begin{pmatrix} 1 & 2 & 3 & 4 & 5 \\ 1 & 5 & 2 & 4 & 3 \end{pmatrix}$.

The parity of a permutation P, δ_P, is defined to be $+1$ for even and -1 for odd permutations. We can write $\delta_P = (-1)^P$ with understanding that P in the exponent is even (odd) if the permutation P is even (odd).

A permutation P of the N arguments of a function $f(\mathbf{r}_1, \mathbf{r}_2, \ldots, \mathbf{r}_N)$ is defined such that $Pf(\mathbf{r}_1, \mathbf{r}_2, \ldots, \mathbf{r}_N) = f(\mathbf{r}_{\alpha(1)}, \mathbf{r}_{\alpha(2)}, \ldots, \mathbf{r}_{\alpha(N)})$. For example, consider a function $f(\mathbf{r}_1, \mathbf{r}_2, \ldots, \mathbf{r}_N)$ of N coordinate vectors that is a symmetric function of its arguments. If we apply *any* permutation P to this function, then $Pf(\mathbf{r}_1, \mathbf{r}_2, \ldots, \mathbf{r}_N) = (+1)f(\mathbf{r}_1, \mathbf{r}_2, \ldots, \mathbf{r}_N)$. If $f(\mathbf{r}_1, \mathbf{r}_2, \ldots, \mathbf{r}_N)$ is antisymmetric, and we apply a permutation P to it, then $Pf(\mathbf{r}_1, \mathbf{r}_2, \ldots, \mathbf{r}_N) = (-1)f(\mathbf{r}_1, \mathbf{r}_2, \ldots, \mathbf{r}_N)$ if P is odd (i.e., the eigenvalue corresponding to an antisymmetric function is -1 if P is odd), whereas, $Pf(\mathbf{r}_1, \mathbf{r}_2, \ldots, \mathbf{r}_N) = (+1)f(\mathbf{r}_1, \mathbf{r}_2, \ldots, \mathbf{r}_N)$ if P is even.

The *symmetrization operator* for N particles, $\mathcal{S}$, is given by

$$\boxed{\mathcal{S} \equiv \frac{1}{N!} \sum_P P,} \tag{8.7}$$

8.1 Permutation Symmetry

and the *antisymmetrization operator* for N particles, $\mathcal{A}$, is

$$\mathcal{A} \equiv \frac{1}{N!} \sum_P \delta_P P. \qquad (8.8)$$

Clearly, any nonzero constant times these operators also does the job. Application of $\mathcal{S}$ to any function $f(\mathbf{r}_1, \mathbf{r}_2, \ldots, \mathbf{r}_N)$ yields a symmetric function (or the zero function, if the initial function is antisymmetric), and application of $\mathcal{A}$ yields an antisymmetric function (or the zero function if the initial function is symmetric).

Problem 8.3

(a) Determine the symmetrization and antisymmetrization operators for the spin degrees of freedom of two spin 1/2 particles in terms of the permutation operator P_{12} (see Problem 4.33).
(b) Show that these operators can be written as $\mathcal{S} = \frac{3}{4}\mathbf{1} + \frac{1}{4}\boldsymbol{\sigma}_1 \cdot \boldsymbol{\sigma}_2$ and $\mathcal{A} = \frac{1}{4}(1 - \boldsymbol{\sigma}_1 \cdot \boldsymbol{\sigma}_2)$.
(c) Write these operators as 4×4 matrices using the basis (5.24).
(d) Write P_{12} as a 4×4 matrices using the basis (5.24).
(e) Explicitly write $\mathcal{S}$ and $\mathcal{A}$ for 3 particles in terms of permutation operators.

Answers: (a) $\mathcal{S} = \frac{1}{2}(1 + P_{12}), \mathcal{A} = \frac{1}{2}(1 - P_{12}).$

(c) $\mathcal{S} = \frac{1}{2}\begin{pmatrix} 2 & 0 & 0 & 0 \\ 0 & 1 & 1 & 0 \\ 0 & 1 & 1 & 0 \\ 0 & 0 & 0 & 2 \end{pmatrix}, \mathcal{A} = \frac{1}{2}\begin{pmatrix} 0 & 0 & 0 & 0 \\ 0 & 1 & -1 & 0 \\ 0 & -1 & 1 & 0 \\ 0 & 0 & 0 & 0 \end{pmatrix}.$ (d) $P_{12} = \begin{pmatrix} 1 & 0 & 0 & 0 \\ 0 & 0 & 1 & 0 \\ 0 & 1 & 0 & 0 \\ 0 & 0 & 0 & 1 \end{pmatrix}.$

(e) Hint: There are six terms in the sums.

Problem 8.4

For any N, show that $\mathcal{S}^2 = \mathcal{S}$, $\mathcal{A}^2 = \mathcal{A}$ and $\mathcal{S}\mathcal{A} = \mathcal{A}\mathcal{S} = 0$.
Hint: First show that for any permutation P, the following holds: $P\mathcal{S} = \mathcal{S}P = \mathcal{S}$, $P\mathcal{A} = \mathcal{A}P = \delta_P \mathcal{A}$.

8.1.1 THE SYMMETRIC GROUP S_N

The symmetric group S_N, sometimes called the *permutation group* (but this term is often restricted to subgroups of the symmetric group), provides the mathematical language necessary for treating identical particles. The elements of the group S_N are the permutations of N objects, i.e., the permutation operators we discussed above. There are $N!$ elements in the group S_N, so the order of the group is $N!$. The unit element, I, of S_N is

$$I = \begin{pmatrix} 1 & 2 & 3 & 4 & 5 \\ 1 & 2 & 3 & 4 & 5 \end{pmatrix}, \qquad (8.9)$$

and the inverse of the element in (8.1) is

$$\begin{pmatrix} \alpha(1) & \alpha(2) & \ldots & \alpha(N) \\ 1 & 2 & \ldots & N \end{pmatrix}. \qquad (8.10)$$

Note that we typically rearrange the columns of this form so that the numbers in the upper row are ordered consecutively from 1 to N.

The following important results are known for the symmetric group. Every group G of order N is isomorphic to a subgroup of S_N; this result is known as Cayley's theorem. The set of all even permutations of N objects forms a subgroup of S_N known as the alternating group, denoted as A_N; it has order $N!/2$. The conjugate classes of S_N correspond to the cycle structures of permutations. Two elements of S_N are conjugate if and only if they consist of the same number of disjoint cycles of the same lengths; e.g., in S_5, (123)(45) and (145)(23) are conjugate, but these elements are not conjugate to (12)(35).

8.1.2 YOUNG TABLEAUX

A convenient bookkeeping technique for imposing permutation symmetry known as Young tableaux was developed by the British mathematician Alfred Young in 1901. It provides the means to describe the representations of the symmetric group and the general linear group $GL(N)$. Here, we use Young tableaux to characterize the exchange symmetry of particles with (and without) spin.

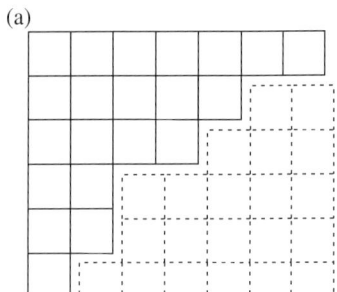
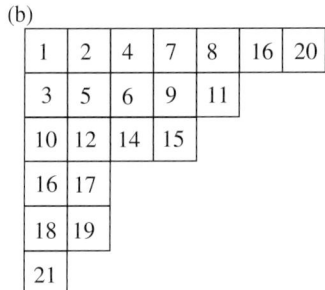

FIG 8.3 A Young tableau for a function $f(\mathbf{r}_1, \mathbf{r}_2, \ldots, \mathbf{r}_{21})$. There are 21 spatial variables, hence 21 boxes. (a) A particular partition for 21 variables. The partition with dashed lines is complementary to the solid-lined partition. (b) The right hand side shows a particular Young tableau.

Let us consider a function $f(\mathbf{r}_1, \mathbf{r}_2, \ldots, \mathbf{r}_N)$ of N spatial variables and determine its symmetry properties with respect to the variables. The N variables $\mathbf{r}_1, \mathbf{r}_2, \ldots, \mathbf{r}_N$, i.e., the suffixes 1, 2, 3, ..., N, can be divided into $K \leq N$ sets with n_i elements in set i, $i = 1, 2, \ldots, K$, with $n_1 + n_2 + \cdots n_K = N$. This division can be portrayed by a *Young tableau* (plural tableaux), as shown in Fig. 8.3 for $N = 21$, in which each of the numbers $n_1, n_2, \ldots$ is represented by a row of n_i cells with $\sum_i n_i = N$, i.e., in Fig. 8.3, $\sum_i n_i = 21$. The lines are placed in order of decreasing length, so the diagram, called a *partition*, contains not only successive horizontal rows with nonincreasing length but also vertical columns of non-increasing length from left to right, as in Fig. 8.3(a). The *complementary partition* is drawn in dashed lines. The various partitions are labeled by a *partition number* λ. Now, one of the integers 1, 2, 3, ..., N is to be placed into each square of the partition, such that numbers increase from left to right in each row and increase from top to bottom in each column, as in Fig. 8.3(b). This construction is called a Young tableau [2].

One now symmetrizes the function $f(\mathbf{r}_1, \mathbf{r}_2, \ldots, \mathbf{r}_N)$ with respect to the variables in each row. Hence, antisymmetrization with respect to a pair of variables in the same row gives zero identically. Having chosen one variable from each row, we can, without loss of generality, regard them as being in the first cells in each row, i.e., after symmetrization, the order of the variables among the cells in each row is immaterial. Let us now antisymmetrize with respect to these variables. Then, deleting the first column, we antisymmetrize with respect to variables chosen one from each row in the reduced diagram; these variables can again be deleted by deleting this row. Continuing this process, we finally have the function first symmetrized with respect to the variables in each row and then antisymmetrized with respect to the variables in each column. However, after antisymmetrization, the resulting function is in general not symmetric with respect to the variables in each row of the tableau. The symmetry is preserved only with respect to the variables in the cells of the first row which project beyond the other rows. Having distributed the N variables in various ways among the rows of a Young tableau, the distribution among the cells in each row being immaterial, except for the criteria of increasing numbers in the rows and columns, we thus obtain a series of functions, which are transformed linearly into one another when the variables are permuted in any manner. Not all these functions are linearly independent—the number of independent functions is in general less than the number of possible distributions of the variables among the rows of the diagram.

8.2 Exchange Symmetry

To illustrate Young tableaux techniques for a simple example, consider the spin states of a two-electron system $\chi(\sigma_1, \sigma_2)$. Spin Young tableaux are composed of cells corresponding to spin states of an electron represented by boxes, $\boxed{1}$ for spin-up and $\boxed{2}$ for spin-down. The two-electron spin functions $\chi(\sigma_1, \sigma_2)$, i.e., $N = 2$, can have variables σ_1 and σ_2 that are either $\uparrow$, represented by a cell containing the number 1, or $\downarrow$, represented by a cell containing the number 2. The three symmetric states corresponding to the three possible orientations of the spin-triplet state and the antisymmetric state corresponding to the spin singlet are shown in Fig. 8.4.

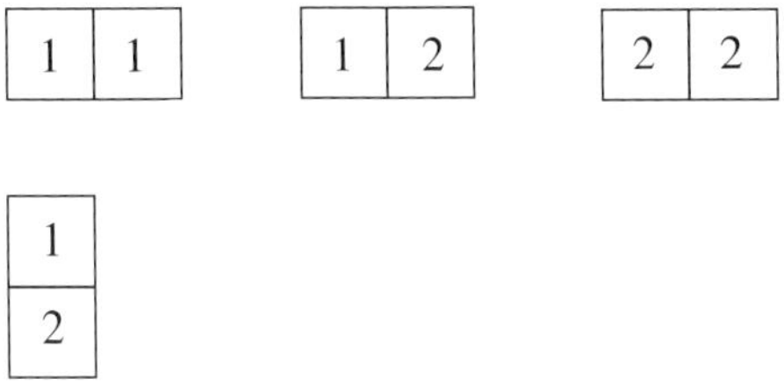

Now consider the spin wave functions of N electrons, $\chi(\sigma_1, \sigma_2, \ldots, \sigma_N)$. Their kinds of symmetry with respect to permutations of the particles are given by the same Young diagrams as we considered for two spins. The spin variables $\sigma_1, \sigma_2, \ldots, \sigma_N$ can take only the two values corresponding to cells in the Young tableau containing only 1 and 2. Since a function antisymmetric with respect to any two variables vanishes when these variables take the same value, the Young tableau for the spin functions can contain columns of only one or two cells. Hence, only one or two rows can exist in these tableau; a higher number of rows is impossible for spin 1/2 Young diagrams. Figure 8.5 shows three of the possible six spin Young tableau for 10 electrons. The tableaux for the $S = 1, 2, 3$ and 4 states with 10 electrons can be easily obtained from the $S = 0$ tableaux by consecutively moving boxes from the second row to the first or from the $S = 5$ tableau by consecutively moving boxes from the first row to the second.

Young tableau can be used to determine irreducible representations of the symmetric group (see the group theory Appendix E for a discussion of irreducible representations). The dimension of the irreducible representation corresponding to a given partition is equal to the number of different Young tableaux that can be obtained from the partition (the number of ways of filling in numbers into the partition). We shall discuss the use of Young tableaux for symmetrizing the wave function of identical particles that are written as the product of a spatial function and a spin function in Sec. 8.6. Young tableaux can also be used to construct representations of the general linear group, but we shall not discuss this topic.

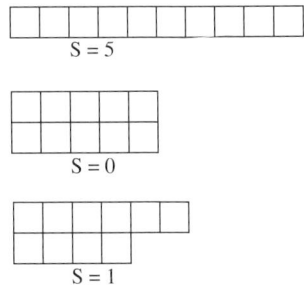

FIG 8.4 Young tableau for two-electron spin functions. The tableaux in the first row correspond to $|\uparrow\rangle|\uparrow\rangle + |\uparrow\rangle|\uparrow\rangle, |\uparrow\rangle|\downarrow\rangle + |\downarrow\rangle|\uparrow\rangle$ and $|\downarrow\rangle|\downarrow\rangle + |\downarrow\rangle|\downarrow\rangle$, respectively. The second row tableau corresponds to $|\uparrow\rangle|\downarrow\rangle - |\downarrow\rangle|\uparrow\rangle$.

FIG 8.5 Young partitions for spin functions of 10 electrons. The top tableaux is totally symmetric in the 10 electrons and corresponds to $S = 5$. The middle tableaux corresponds to $S = 0$, and the bottom tableaux corresponds to $S = 1$. $S = 1, 2, 3, 4$ and 5 tableaux can be obtained from the $S = 0$ tableau by consecutively moving boxes from the second row to the first.

8.2 EXCHANGE SYMMETRY

Consider a wave function of N identical particles. As discussed in Sec. 2.9.1, exchanging any two of them must yield a wave function which is identical to the initial wave function, up to a phase factor,

$$P_{ij}\Psi(\mathbf{x}_1, \ldots, \mathbf{x}_i, \ldots, \mathbf{x}_j, \ldots, \mathbf{x}_N) \equiv \Psi(\mathbf{x}_1, \ldots, \mathbf{x}_j, \ldots, \mathbf{x}_i, \ldots, \mathbf{x}_N)$$
$$= e^{i\alpha} \Psi(\mathbf{x}_1, \ldots, \mathbf{x}_i, \ldots, \mathbf{x}_j, \ldots, \mathbf{x}_N). \quad (8.11)$$

Applying the permutation operator again yields the identity operator, i.e., $P_{ij}^2 = 1$, so $e^{2i\alpha} = 1$, and the eigenvalues of P_{ij} are ± 1. The term exchange symmetry is appropriate because, as we shall see below, no observable physical quantity should change after exchanging two identical particles.

Problem 8.5

Consider two identical particles, each of which has one-particle states represented in coordinate representation by the orthogonal wave functions $\psi_\alpha(\mathbf{r})$ and $\psi_\beta(\mathbf{r})$. Define the symmetric and antisymmetric and wave functions of the system, $\Psi^{(S)}(\mathbf{r}_1, \mathbf{r}_2) = \psi_\alpha(\mathbf{r}_1)\psi_\beta(\mathbf{r}_2) + \psi_\beta(\mathbf{r}_1)\psi_\alpha(\mathbf{r}_2)$ and $\Psi^{(A)}(\mathbf{r}_1, \mathbf{r}_2) = \psi_\alpha(\mathbf{r}_1)\psi_\beta(\mathbf{r}_2) - \psi_\beta(\mathbf{r}_1)\psi_\alpha(\mathbf{r}_2)$. Let us suppose that there is no symmetrizafion rule for identical particles so the system would in general have the following wave function: $\Psi(\mathbf{r}_1, \mathbf{r}_2) = a\Psi^{(S)} + b\Psi^{(A)}$, with $|a|^2 + |b|^2 = 1$. Calculate the probability of finding a particle at $\mathbf{r}_1$ and another one at $\mathbf{r}_2$.

Answer: $P(\mathbf{r}_1, \mathbf{r}_2) = |\Psi(\mathbf{r}_1, \mathbf{r}_1)|^2 + |\Psi(\mathbf{r}_2, \mathbf{r}_1)|^2 = 2|a|^2|\Psi^{(S)}(\mathbf{r}_1, \mathbf{r}_2)|^2 + |b|^2|\Psi^{(A)}(\mathbf{r}_1, \mathbf{r}_2)|^2$. $P(\mathbf{r}_1, \mathbf{r}_2)$ depends on $|a|^2$ and $|b|^2$. We shall see below that a and b can only be such that $|a|^2 = 1, |b|^2 = 0$, or $|a|^2 = 0, |b|^2 = 1$.

8.2.1 SYMMETRIZATION POSTULATE

All elementary particles have an intrinsic angular momentum called spin. There are two types of particles, with different permutation properties, depending on spin: Bosons and Fermions. The application of the permutation operator on elementary particles depends on the spin of the particles. This dependence is summarized in the symmetrization postulate.

Symmetrization Postulate:

- For integer spin particles (bosons) $[S = (0, 1, 2, \ldots)\hbar]$, the exchange of any two bosons leaves the wave function unchanged:

$$P_{ij}\Psi(\mathbf{x}_1, \ldots, \mathbf{x}_i, \ldots, \mathbf{x}_j, \ldots, \mathbf{x}_N) = +\Psi(\mathbf{x}_1, \ldots, \mathbf{x}_i, \ldots, \mathbf{x}_j, \ldots, \mathbf{x}_N). \tag{8.12}$$

- For half-integer spin particles (fermions) $[S = (1/2, 3/2, \ldots)\hbar]$, the exchange of any two fermions changes the sign of the wave function:

$$P_{ij}\Psi(\mathbf{x}_1, \ldots, \mathbf{x}_i, \ldots, \mathbf{x}_j, \ldots, \mathbf{x}_N) = -\Psi(\mathbf{x}_1, \ldots, \mathbf{x}_i, \ldots, \mathbf{x}_j, \ldots, \mathbf{x}_N). \tag{8.13}$$

For the fermion case, consider, for example, an N electron state wave function, $\psi(\mathbf{x}_1, \mathbf{x}_2, \ldots, \mathbf{x}_N)$. According to the symmetrization postulate, the wave function must be antisymmetrized. Therefore, the Young partition for the spatial wave function multiplying the spin wave function must be *complementary* to the spin partition if the total wave function is to be antisymmetric. This is related to the *Pauli exclusion principle* which states that only one electron can occupy a given electron state; otherwise, the antisymmetrized wave function would vanish.

For the boson case, an N boson wave function, $\psi(\mathbf{x}_1, \mathbf{x}_2, \ldots, \mathbf{x}_N)$, must be symmetrized. For example, for N hydrogen atoms or N ^{7}Li atoms, the N atom wave function must be symmetrized with respect to all the particles. Since the total wave function must be symmetric for bosons, the tableaux corresponding to the spatial and spin parts of the wave function must be identical, i.e., the Young partitions for the spatial and spin parts of the wave function must be identical.

Any two-particle wave function can be written as a sum of products of single-particle states, and the corresponding kets can be written as $|\psi\rangle = \sum_{\alpha\beta} C_{\alpha\beta} |\alpha\rangle_1 |\beta\rangle_2$ where $|\alpha\rangle_1$ and $|\beta\rangle_2$ are single-particle states. *By definition of the two-cycle,* P_{12} operating on the state $|\alpha\rangle_1 |\beta\rangle_2$, $P_{12}|\alpha\rangle_1|\beta\rangle_2 = |\alpha\rangle_2|\beta\rangle_1 = |\beta\rangle_1|\alpha\rangle_2$, hence $P_{12}|\psi\rangle = \sum_{\alpha\beta} C_{\alpha\beta} |\beta\rangle_1 |\alpha\rangle_2$. Therefore,

$$P_{12}\psi(\mathbf{x}_1, \mathbf{x}_2) = \langle \mathbf{x}_1, \mathbf{x}_2 | P_{12} | \psi \rangle = \langle \mathbf{x}_1, \mathbf{x}_2 | \sum_{\alpha\beta} C_{\alpha\beta} |\beta\rangle_1 |\alpha\rangle_2 = \sum_{\alpha\beta} C_{\alpha\beta} \langle \mathbf{x}_2 | \alpha \rangle \langle \mathbf{x}_1 | \beta \rangle. \tag{8.14}$$

Note that we have just shown that

$$\boxed{P_{12}\psi(\mathbf{x}_1, \mathbf{x}_2) = \psi(\mathbf{x}_2, \mathbf{x}_1) \text{ where } [x \text{ means } \mathbf{r} \text{ and spin}].} \tag{8.15}$$

8.2 Exchange Symmetry

Exchanging two particles in a wave function is equivalent to changing the coordinates of the particles (i.e., exchanging their positions and spin). From Eqs. (8.12) and (8.13), we know that for bosons, $\psi(\mathbf{x}_2, \mathbf{x}_1) = \psi(\mathbf{x}_1, \mathbf{x}_2)$, whereas for fermions, $\psi(\mathbf{x}_2, \mathbf{x}_1) = -\psi(\mathbf{x}_1, \mathbf{x}_2)$.

The Hamiltonian of a set of indistinguishable particles must be unchanged by permutations of the particles. Moreover, the principle of indistinguishability of identical particles implies that dynamical states that differ only by a permutation of identical particles cannot be distinguished. Hence, for any two particles, e.g., 1 and 2,

$$\langle \psi | B | \psi \rangle = \left\langle \psi \left| P_{12}^{\dagger} B P_{12} \right| \psi \right\rangle, \tag{8.16}$$

for any $|\psi\rangle$ and dynamical operator B. Therefore, $B = P_{12}^{\dagger} B P_{12}$, so $P_{12} B = B P_{12}$. All physical observables of a many-identical-particle system must be invariant under permutations, i.e., commute with permutation operators.

For eigenstates of the total angular momentum of two identical particles with $j_1 = j_2 = j$, the symmetry of the states under exchange of the two particles is determined by the property of the Clebsch–Gordan coefficients, $\langle jjmm'|JM\rangle = (-1)^{J-2j}\langle jjm'm|JM\rangle$ (see Problem 3.26). The eigenstate of the total angular momentum is

$$|jj\, JM\rangle = \sum_{mm'} |jm\rangle_1 |jm'\rangle_2 \langle jjm'm|JM\rangle$$

$$= \frac{1}{2} \sum_{mm'} \left[|jm\rangle_1 |jm'\rangle_2 + (-1)^{J-2j} |jm'\rangle_1 |jm\rangle_2 \right] \langle jjmm'|JM\rangle. \tag{8.17}$$

Upon interchanging particles 1 and 2, state $|jj\, JM\rangle$ is multiplied by $(-1)^{J-2j}$. Hence, if $J - 2j$ is odd (even), the state is antisymmetric (symmetric). Thus, both the symmetric boson states and the antisymmetric fermion states will have even J only, odd J states have the wrong symmetry in both cases.

The interchange of two identical spinless particles is equivalent to inversion of their relative coordinate, $\mathbf{r} = \mathbf{r}_2 - \mathbf{r}_1 \to -\mathbf{r}$. If the total orbital angular momentum l of the two particle system is a good quantum number, the properties of the spherical harmonics $Y_{lm}(\theta, \phi)$ imply that the result of inversion $\theta \to \pi - \theta$ and $\phi \to \phi + \pi$ is to multiply the spatial wave function of the two particles by $(-1)^l$. Hence, the orbital angular momentum of a system of two spinless identical particles interacting via central potential must be even.

If we interchange two identical particles with spin $s = 1/2$, say, electrons, then the total wave function of the system must be antisymmetric. For two particles, the total wave function can be written as a product of a spatial and spin function. Hence, if the coordinate wave function is symmetric, the spin function must be antisymmetric, and vice versa.

Problem 8.6

For a system of two identical particles, each of which can be in one of n quantum states, show that there are $n(n+1)/2$ symmetric, and $n(n-1)/2$ antisymmetric states.

Answer: For n single-particle states, there are n symmetric states with both particles in the same state, and $n(n-1)/2$ symmetric states with the particles in different states, so the total number of symmetric states is $n(n+1)/2$. The number of antisymmetric states is $n(n-1)/2$, because states with both particles in the same state are precluded.

Problem 8.7

For two particles having spin J, show that the ratio of symmetric to antisymmetric spin states is $(J+1)/J$.

Answer: For particles with spin J, there are $n = 2J+1$ single-particle states with different M_J. Hence, the number of symmetric states is $n_{\text{sym}} = (J+1)(2J+1)$ and the number of antisymmetric states is $n_{\text{antisym}} = J(2J+1)$. The ratio is therefore $(J+1)/J$.

Problem 8.8

Consider the wave function of a system of two identical spinless particles that is an eigenfunction of the relative orbital angular momentum of the two particles, **L**. Show that the quantum number l must be even.

Hint: Let $\Psi(\mathbf{r}_1, \mathbf{r}_2) = \Phi(\frac{\mathbf{r}_1+\mathbf{r}_2}{2})\psi_l(\mathbf{r}_2 - \mathbf{r}_1)$. Here, inversion is equivalent to permutation. Inversion multiplies $\psi_l(\mathbf{r}_2 - \mathbf{r}_1)$ by $(-1)^l$.

Problem 8.9

Explain the effects of exchange symmetry in determining the energy of the Fermi contact term for singlet and triplet states of identical fermionic particles that have a magnetic moment.

Answer: For the Fermi contact term to be nonzero, the two particles must be at the same position in space. For a singlet state, the spin wave function is antisymmetric so the spatial part of the wave function must be symmetric, and the particles can be at the same position. For a triplet state, the spin part of the wave function is symmetric, so the spatial part of the wave function must be antisymmetric, and therefore, the two particles cannot be at the same position. Hence, the Fermi contact term vanishes.

8.3 PERMANENTS AND SLATER DETERMINANTS

As discussed in Appendix A.2.2, a *determinant* of a square $N \times N$ matrix B is a scalar, denoted by $\det(B)$ or $|B|$, defined by $\det(B) = \sum_P (-1)^P \prod_{i=1}^N b_{i,P(i)}$, where $P(i) = \alpha(i)$ as per Eq. (8.1). It is often written in the form:

$$\det(B) = \begin{vmatrix} b_{11} & b_{12} & b_{13} & \cdots & b_{1N} \\ b_{21} & & & & b_{2N} \\ b_{31} & & \ddots & & b_{3N} \\ \vdots & & & & \vdots \\ b_{N1} & b_{N2} & b_{N3} & \cdots & b_{NN} \end{vmatrix}. \tag{8.18}$$

A *permanent* of a square $N \times N$ matrix B is a scalar, denoted by $\text{perm}(B)$ defined by $\text{perm}(B) = \sum_P \prod_{i=1}^N b_{i,P(i)}$, i.e., it is a determinant without the minus signs. We shall use permanents when we deal with bosons.

Slater determinants, named after John Slater, are simply determinants composed of spin orbitals that are used for wave functions of multi-fermion systems such as atoms or molecules. Simple examples of Slater determinants for two-electron systems, which involve 2×2 determinants, will be used to construct two-particle atomic states in the next section, see Eqs. (8.24)–(8.27). For N identical fermions in a product wave function form, $\psi(\mathbf{x}_1, \ldots, \mathbf{x}_N) = u_\alpha(\mathbf{x}_1)u_\beta(\mathbf{x}_2)\ldots u_\nu(\mathbf{x}_N)$, where the single-particle wave functions for the particles are given by spin orbitals, $u_\beta(\mathbf{x}_i)$, the antisymmetrized wave function is

$$\boxed{\Psi(\mathbf{x}_1, \ldots, \mathbf{x}_N) = \frac{1}{\sqrt{N!}} \sum_P (-1)^P P u_\alpha(\mathbf{x}_1) u_\beta(\mathbf{x}_2) \ldots u_\nu(\mathbf{x}_N) \equiv \sqrt{N!}\,\mathcal{A}\,\psi.} \tag{8.19}$$

For spin orbitals $u_\beta(\mathbf{x}_i) = \phi_\beta(\mathbf{r}_i)\chi_i$, the wave function can be written in terms of a Slater determinant,

$$\Psi(1, 2, \ldots, N) = \frac{1}{\sqrt{N!}} \begin{vmatrix} \phi_1(1)\alpha_1 & \phi_1(1)\beta_1 & \cdots & \phi_n(1)\beta_1 \\ \phi_1(2)\alpha_2 & \phi_1(2)\beta_2 & & \\ \vdots & \vdots & & \\ \phi_1(N)\alpha_N & \phi_1(N)\beta_N & \cdots & \phi_n(N)\beta_N \end{vmatrix}. \tag{8.20}$$

8.4 Simple Two- and Three-Electron States

For many-electron states with spin-paired wave functions (singlets), the wave function can be written in terms of Slater determinants with an equal number of spin-up and spin-down spinors.

An often utilized shorthand notation uses bars to indicate the spin-down state β, with a lack of a bar indicating the spin-up state α. With this notation, Eq. (8.20) becomes

$$\Psi(1,2,\ldots,N) = \frac{1}{\sqrt{N!}} \begin{vmatrix} \phi_1(1) & \overline{\phi_1}(1) & \cdots & \overline{\phi_n}(1) \\ \phi_1(2) & \overline{\phi_1}(2) & & \\ \vdots & \vdots & & \\ \phi_1(N) & \overline{\phi_1}(N) & \cdots & \overline{\phi_n}(N) \end{vmatrix}. \quad (8.21)$$

An even more concise notation is to show the diagonal terms only, e.g., for the beryllium atom ground state wave function, one can write $\Psi(1,2,3,4) = \frac{1}{\sqrt{4!}} |1s(1)\,\overline{1s}(2)\,2s(3)\,\overline{2s}(4)|$. Sometimes one simply writes $\Psi = |1s\,\overline{1s}\,2s\,\overline{2s}|$, where the normalization and the coordinate indices are implied.

Quite generally, electronic wave functions take the form

$$\Psi(\mathbf{x}_1, \mathbf{x}_2, \ldots, \mathbf{x}_N) = \frac{1}{\sqrt{N!}} \sum_P (-1)^P P\, u(\mathbf{x}_1, \mathbf{x}_2, \ldots, \mathbf{x}_N), \quad (8.22)$$

where $u(\mathbf{x}_1, \mathbf{x}_2, \ldots, \mathbf{x}_N)$ is a multicomponent spinor for the N electrons, and the wave function in Eq. (8.22) has been antisymmetrized by applying $\sqrt{N!}\,\mathcal{A}$ to $u(\mathbf{x}_1, \mathbf{x}_2, \ldots, \mathbf{x}_N)$.

If the wave function for N identical bosons is constructed from single-paticle spin orbitals $u_\beta(\mathbf{x}_i)$, then the wave function is given by

$$\Psi(\mathbf{x}_1, \mathbf{x}_2, \ldots, \mathbf{x}_N) = \frac{1}{\sqrt{N!}} \sum_P P\, u_\alpha(\mathbf{x}_1) u_\beta(\mathbf{x}_2) \ldots u_\nu(\mathbf{x}_N) \equiv \sqrt{N!}\, \mathcal{S}\, \psi. \quad (8.23)$$

Problem 8.10

Two noninteracting identical particles occupy the two energy levels, E_0 and E_n, in a one-dimensional quadratic potential, $V(x) = m\omega^2(x-x_e)^2/2$. Calculate $\langle x_1 x_2 \rangle$ for bosonic and fermionic particles.
Answer: $\langle x_1 x_2 \rangle = \frac{1}{2}(\langle x_1 \rangle_0 \langle x_2 \rangle_n + \langle x_1 \rangle_n \langle x_2 \rangle_0 \pm \langle x_1 \rangle_{0n} \langle x_2 \rangle_{n0} \pm \langle x_1 \rangle_{n0} \langle x_2 \rangle_{0n} =$
$\langle x \rangle_0 \langle x \rangle_n \pm |\langle x \rangle_{0n}|^2$. Now, $\langle x \rangle_0 = \langle x \rangle_n = x_e$ but $\langle x \rangle_{0n} = x_e$ only if n is odd. Otherwise, you need to calculate the matrix element.

8.4 SIMPLE TWO- AND THREE-ELECTRON STATES

The simplest examples of multielectron systems are two-electron systems. For the two-electron wave function corresponding, e.g., to the ground state of the helium atom, one takes the spin-singlet with both electrons in the lowest orbital and the wave function takes the form:

$$\Psi_s(\mathbf{x}_1, \mathbf{x}_2) = \frac{1}{\sqrt{2!}} \begin{vmatrix} u_\alpha(\mathbf{x}_1) & u_\beta(\mathbf{x}_1) \\ u_\alpha(\mathbf{x}_2) & u_\beta(\mathbf{x}_2) \end{vmatrix} = \frac{1}{\sqrt{2}} \begin{vmatrix} \phi_{100}(\mathbf{r}_1)\chi_\uparrow(1) & \phi_{100}(\mathbf{r}_1)\chi_\downarrow(1) \\ \phi_{100}(\mathbf{r}_2)\chi_\uparrow(2) & \phi_{100}(\mathbf{r}_2)\chi_\downarrow(2) \end{vmatrix}$$

$$= \phi_{100}(\mathbf{r}_1)\phi_{100}(\mathbf{r}_2) \frac{1}{\sqrt{2}} \left[\chi_\uparrow(1)\chi_\downarrow(2) - \chi_\downarrow(1)\chi_\uparrow(2) \right]. \quad (8.24)$$

In terms of spin-up-down notation, $\Psi_s(\mathbf{x}_1, \mathbf{x}_2) = \phi_{100}(\mathbf{r}_1)\phi_{100}(\mathbf{r}_2) \frac{1}{\sqrt{2}} [|\uparrow\downarrow\rangle - |\downarrow\uparrow\rangle]$. Two-electron triplet states must have the spatial part of the wave function antisymmetric, since the spin part is symmetric. Hence, the spatial orbitals must be different; otherwise, the spatial wave function will vanish. The triplet with one electron in the 1s and the other electron in the 2s orbital, and both spins pointing up, is given by the following wave function:

$$\Psi_t(\mathbf{x}_1, \mathbf{x}_2) = \frac{1}{\sqrt{2!}} \begin{vmatrix} u_\alpha(\mathbf{x}_1) & u_\beta(\mathbf{x}_1) \\ u_\alpha(\mathbf{x}_2) & u_\beta(\mathbf{x}_2) \end{vmatrix} = \frac{1}{\sqrt{2}} \begin{vmatrix} \phi_{100}(\mathbf{r}_1)\chi_\uparrow(1) & \phi_{200}(\mathbf{r}_1)\chi_\uparrow(1) \\ \phi_{100}(\mathbf{r}_2)\chi_\uparrow(2) & \phi_{200}(\mathbf{r}_2)\chi_\uparrow(2) \end{vmatrix}$$

$$= \frac{1}{\sqrt{2}} [\phi_{100}(r_1)\phi_{200}(r_2) - \phi_{200}(r_1)\phi_{100}(r_2)] \chi_\uparrow(1)\chi_\uparrow(2), \qquad (8.25)$$

i.e., $\Psi_{t,M_s=1}(\mathbf{x}_1,\mathbf{x}_2) = \frac{1}{\sqrt{2}}[\phi_{100}(r_1)\phi_{200}(r_2) - \phi_{200}(r_1)\phi_{100}(r_2)][|\uparrow\uparrow\rangle]$. This state has unit projection of the spin on the z axis. Similarly, the triplet state with spin projection $M_s = -1$ is,

$$\Psi_{t,M_s=-1}(\mathbf{x}_1, \mathbf{x}_2) = \frac{1}{\sqrt{2}} [\phi_{100}(r_1)\phi_{200}(r_2) - \phi_{200}(r_1)\phi_{100}(r_2)][|\downarrow\downarrow\rangle], \qquad (8.26)$$

and the triplet state with a spin projection $M_s = 0$ is

$$\Psi_t(\mathbf{x}_1, \mathbf{x}_2) = \frac{1}{\sqrt{2}} \begin{vmatrix} \phi_{100}(\mathbf{r}_1) & \phi_{200}(\mathbf{r}_1) \\ \phi_{100}(\mathbf{r}_2) & \phi_{100}(\mathbf{r}_2) \end{vmatrix} \times \frac{1}{\sqrt{2}} [\chi_\uparrow(1)\chi_\downarrow(2) + \chi_\downarrow(1)\chi_\uparrow(2)], \qquad (8.27)$$

i.e., $\Psi_{t,M_s=0}(\mathbf{x}_1,\mathbf{x}_2) = [\phi_{100}(r_1)\phi_{200}(r_2) - \phi_{200}(r_1)\phi_{100}(r_2)][|\uparrow\downarrow\rangle + |\downarrow\uparrow\rangle]/2$.

The tableaux corresponding to the spin wave functions in Eqs. (8.24)–(8.27) are those in Fig. 8.4 going from left to right and then down. Note that the partition for the spatial wave function multiplying the spin wave function in these equations is the complementary partition to that of the spin wave function, if the total wave function is to be antisymmetric.

On the other hand, if the total wave function is to be symmetric, as it must for bosons, the tableaux corresponding to the spatial and spin parts of the wave function must be identical.

As another example, consider the wave function for the (three-electron) lithium atom ground state

$$\Psi(1, 2, 3) = \frac{1}{\sqrt{6}} [1s(1)\alpha_1 \ 1s(2)\beta_2 \ 2s(3)\alpha_3 - 1s(1)\alpha_1 \ 2s(2)\alpha_2 \ 1s(3)\beta_3 - 1s(1)\beta_1 \ 1s(2)\alpha_2 \ 2s(3)\alpha_3$$
$$+ 1s(1)\beta_1 \ 2s(2)\alpha_2 \ 1s(3)\alpha_3 + 2s(1)\alpha_1 \ 1s(2)\alpha_2 \ 1s(3)\beta_3 - 2s(1)\alpha_1 \ 1s(2)\beta_2 \ 1s(3)\alpha_3]. \qquad (8.28)$$

This wave function can be written as a Slater determinant, and we developed shorthand notation for such a wave function,

$$\Psi(1, 2, 3) = \frac{1}{\sqrt{3!}} \begin{vmatrix} 1s(1) & \overline{1s}(1) & 2s(1) \\ 1s(2) & \overline{1s}(2) & 2s(2) \\ 1s(3) & \overline{1s}(3) & 2s(3) \end{vmatrix}. \qquad (8.29)$$

An even more compact notation is $\Psi = |1s \ \overline{1s} \ 2s|$.

Problem 8.11

For the Slater determinants $\Psi_1 = |u_i u_j|$ and $\Psi_2 = |u_k u_l|$ composed of orthonormal spin orbitals u_i, u_j, u_k, u_k, show that $\langle \Psi_2 | \Psi_1 \rangle = \delta_{ki}\delta_{lj} - \delta_{li}\delta_{kj}$.

Problem 8.12

(a) Form all possible partitions, and all possible Slater determinants for the spin orbitals of lithium involving orbitals ϕ_{100}, ϕ_{200}.

(b) Write the ground state wave function for beryllium in terms of a Slater determinant.

8.5 Exchange Symmetry for Two Two-Level Systems

Answer: (b) $|\psi_{100}|\uparrow\rangle \ \psi_{100}|\downarrow\rangle \ \psi_{200}|\uparrow\rangle \ \psi_{200}|\downarrow\rangle| = \frac{1}{\sqrt{4!}} \times$

$$\begin{vmatrix} \psi_{100}(\mathbf{r}_1)|\uparrow\rangle_1 & \psi_{100}(\mathbf{r}_1)|\downarrow\rangle_1 & \psi_{200}(\mathbf{r}_1)|\uparrow\rangle_1 & \psi_{200}(\mathbf{r}_1)|\downarrow\rangle_1 \\ \psi_{100}(\mathbf{r}_2)|\uparrow\rangle_2 & \psi_{100}(\mathbf{r}_2)|\downarrow\rangle_2 & \psi_{200}(\mathbf{r}_2)|\uparrow\rangle_2 & \psi_{200}(\mathbf{r}_2)|\downarrow\rangle_2 \\ \psi_{100}(\mathbf{r}_3)|\uparrow\rangle_3 & \psi_{100}(\mathbf{r}_3)|\downarrow\rangle_3 & \psi_{200}(\mathbf{r}_3)|\uparrow\rangle_3 & \psi_{200}(\mathbf{r}_3)|\downarrow\rangle_3 \\ \psi_{100}(\mathbf{r}_4)|\uparrow\rangle_4 & \psi_{100}(\mathbf{r}_4)|\downarrow\rangle_1 & \psi_{200}(\mathbf{r}_4)|\uparrow\rangle_4 & \psi_{200}(\mathbf{r}_4)|\downarrow\rangle_4 \end{vmatrix}.$$

Problem 8.13

Write the Slater determinant of the following excited state of helium $(1s0)(2p-1)$ $^3P_2$, $M_J = 0$ state, where the meaning of $(1s0)$ and $(2p-1)$ are as follows: $(1s0) \equiv \psi_{n=1,l=0,m_l=0}(\mathbf{r})$ and $(2p-1) \equiv \psi_{n=2,l=1,m_l=-1}(\mathbf{r})$, and $M_j = 0$ means that projection of the total electronic angular momentum in the $\hat{z}$ direction is 0.

Answer: $\Psi(\mathbf{x}_1, \mathbf{x}_2) = |\psi_{100}|\uparrow\rangle \ \psi_{21-1}|\uparrow\rangle| = \frac{1}{\sqrt{2}} \begin{vmatrix} \psi_{100}(\mathbf{r}_1)|\uparrow\rangle_1 & \psi_{21-1}(\mathbf{r}_1)|\uparrow\rangle_1 \\ \psi_{100}(\mathbf{r}_2)|\uparrow\rangle_2 & \psi_{21-1}(\mathbf{r}_2)|\uparrow\rangle_2 \end{vmatrix}.$

8.5 EXCHANGE SYMMETRY FOR TWO TWO-LEVEL SYSTEMS

The general form for a density matrix for two two-level systems (e.g., the spin degrees of freedom of two electrons) was given in Sec. 6.1.8, Eq. (6.82), as $\rho_{AB} = \frac{1}{4}[(1 + \mathbf{n}_A \cdot \boldsymbol{\sigma}_A)(1 + \mathbf{n}_B \cdot \boldsymbol{\sigma}_B) + \boldsymbol{\sigma}_A \cdot \mathbf{C}_{AB} \cdot \boldsymbol{\sigma}_B]$. For bosonic and fermionic systems (e.g., two identical bosonic or fermionic atoms which are treated as two-level atoms), the density matrix must be properly symmetrized by applying the symmetrization operator $\mathcal{S}$ or antisymmetrization operator $\mathcal{A}$, respectively, e.g., $\rho_{AB}^{\text{sym}} = \mathcal{S}\rho_{AB}\mathcal{S}$. For two particles, $\mathcal{S} = \frac{1}{2}(1 + P_{AB})$ and $\mathcal{A} = \frac{1}{2}(1 - P_{AB})$. Therefore, for two two-level atoms of bosonic and fermionic character, the density matrices are

$$\rho_{AB}^{\text{sym}} = \frac{1}{4}(1 + P_{AB})\rho_{AB}(1 + P_{AB}), \tag{8.30}$$

$$\rho_{AB}^{\text{anti}} = \frac{1}{4}(1 - P_{AB})\rho_{AB}(1 - P_{AB}). \tag{8.31}$$

In Problem 4.32, you showed that $\mathcal{S} = \frac{3}{4} + \frac{1}{4}\boldsymbol{\sigma}_A \cdot \boldsymbol{\sigma}_B$ is a projection operator which projects onto the triplet state manifold, i.e., it symmetrizes spin-states, and $\mathcal{A} = \frac{1}{4} - \frac{1}{4}\boldsymbol{\sigma}_A \cdot \boldsymbol{\sigma}_B$ is a projection operator which projects onto the singlet state manifold, i.e., it antisymmetrizes spin states. Hence, for two spins,

$$\rho_{AB}^{\text{sym}} = \left(\frac{3}{4} + \frac{1}{4}\boldsymbol{\sigma}_A \cdot \boldsymbol{\sigma}_B\right)\rho_{AB}\left(\frac{3}{4} + \frac{1}{4}\boldsymbol{\sigma}_A \cdot \boldsymbol{\sigma}_B\right), \tag{8.32}$$

$$\rho_{AB}^{\text{anti}} = \left(\frac{1}{4} - \frac{1}{4}\boldsymbol{\sigma}_A \cdot \boldsymbol{\sigma}_B\right)\rho_{AB}\left(\frac{1}{4} - \frac{1}{4}\boldsymbol{\sigma}_A \cdot \boldsymbol{\sigma}_B\right). \tag{8.33}$$

Since $\mathcal{S} + \mathcal{A} = 1$, we can write an unsymmetrized density matrix ρ_{AB} as

$$\rho_{AB} = (\mathcal{S} + \mathcal{A})\rho_{AB}(\mathcal{S} + \mathcal{A}) = \rho_{AB}^{\text{sym}} + \rho_{AB}^{\text{anti}} + \mathcal{S}\rho_{AB}\mathcal{A} + \mathcal{A}\rho_{AB}\mathcal{S}. \tag{8.34}$$

If the state must be antisymmetric, $\rho_{AB} = \rho_{AB}^{\text{anti}} = \mathcal{A}\rho_{AB}\mathcal{A}$, then the state must be pure singlet, $\rho_{AB}^{\text{anti}} = \frac{1}{4}(1 - \boldsymbol{\sigma}_A \cdot \boldsymbol{\sigma}_B)$; it cannot be a mixed state nor can it contain any triplet component. Moreover, if one takes an arbitrary nonsymmetric density matrix, ρ_{AB}, and antisymmetrizes it, $\rho_{AB} \to \mathcal{A}\rho_{AB}\mathcal{A}$, the resultant density matrix will be a number, say α times the singlet density matrix,

$$\rho_{AB} \to \mathcal{A}\rho_{AB}\mathcal{A} = \alpha \frac{1}{4}(1 - \boldsymbol{\sigma}_A \cdot \boldsymbol{\sigma}_B), \tag{8.35}$$

and $\alpha = \operatorname{Tr} \rho_{AB}\mathcal{A} = \operatorname{Tr} \mathcal{A}\rho_{AB}\mathcal{A}$.

> **Problem 8.14**
>
> Express the term ρ_{AB}^{anti} on the RHS of (8.34) in terms of $\mathbf{n}_A$, $\mathbf{n}_B$ and $\mathbf{C}_{AB}$.
> **Answer:** $\rho_{AB}^{\text{anti}} = \alpha \frac{1}{4}(1 - \boldsymbol{\sigma}_A \cdot \boldsymbol{\sigma}_B)$, where $\alpha = \frac{1}{4}\left[1 - (\mathbf{n}_A \cdot \mathbf{n}_B + \sum_i C_{ii})\right]$. Note that $\alpha = \operatorname{Tr} \rho_{AB}^{\text{anti}} = \operatorname{Tr} \mathcal{A}\rho_{AB}\mathcal{A}$.

Suppose spatial degrees of freedom need to be included for the description of the two two-level atoms, in addition to the internal degrees of freedom discussed above. For two particles, the wave function and density matrix can always be written as a product of an internal (i.e., spin) part and an external (i.e., space) part. A symmetric density matrix (8.32) for the internal (e.g., spin) degrees of freedom must be multiplied by a symmetric [antisymmetric] density matrix for the spatial degrees of freedom ($\mathbf{r}_A$, $\mathbf{r}_B$) for bosons [fermions]; and an antisymmetric density matrix (8.33) must be multiplied by an antisymmetric [symmetric] density matrix for the spatial degrees of freedom for bosons [fermions], so that the full density matrix has the right exchange symmetry. Exchange symmetry has drastic consequences regarding the scattering of two identical two-level atoms, as we shall see in Sec. 12.5.6. In particular, for bosons, only odd partial waves (spatially antisymmetric states) are possible when the internal state of the two two-level systems is antisymmetric, ρ_{AB}^{anti}, and only even partial waves (spatially symmetric states) are possible with ρ_{AB}^{sym}, and vise versa for fermions. Moreover, at ultra-low collision energies, all partial waves but s-waves are frozen out because of the centrifugal barrier for higher partial waves, so only the s-wave can collide, and this further restricts the nature of collisions of two identical two-level atoms.

8.6 MANY-PARTICLE EXCHANGE SYMMETRY

If we write the total wave function for a multiparticle system as a sum of products of a spatial function and a spin function, there arises the question of what Young tableau to use for the spin function given a Young tableau for the coordinate function (and vise versa), so that the total wave function is antisymmetric under particle interchange for half-integer particles and symmetric for integer spin particles. For bosons, the symmetry of the spin and the coordinate functions must be given by the same Young tableau, and the complete wave function Ψ is the product of the two. For fermions, the symmetry of the spin and coordinate functions must be given by complimentary tableaus, e.g., as shown in Fig. 8.3(a). The full wave function Ψ can always be written as a sum of products of space and spin function.

For identical fermions, we typically do not consider total wave functions given by the product of a spatial and a spin function. Rather, we write the wave function as an antisymmetrized product of spin-orbitals, $u_\beta(\mathbf{x}_i)$, and then the full wave function is given by Eq. (8.19), where we used the product form $\psi = u_\alpha(\mathbf{x}_1)u_\beta(\mathbf{x}_2)\ldots u_\nu(\mathbf{x}_N)$. More generally, normalized fermionic wave functions take the form

$$\Psi(\mathbf{x}_1, \mathbf{x}_2, \ldots, \mathbf{x}_N) = \frac{1}{\sqrt{N!}} \sum_P (-1)^P P u(\mathbf{x}_1, \mathbf{x}_2, \ldots, \mathbf{x}_N), \tag{8.36}$$

where $u(\mathbf{x}_1, \mathbf{x}_2, \ldots, \mathbf{x}_N)$ is a wave function for the N identical fermions, and the wave function in Eq. (8.36) has been antisymmetrized and then multiplied by $\sqrt{N!}$ to normalize the wave function.

A multicomponent spinor can always be written in terms of single-particle spinors as

$$|u\rangle = \sum_{\alpha\beta\gamma\ldots} C_{\alpha\beta\gamma\ldots}|u_\alpha\rangle|u_\beta\rangle|u_\gamma\rangle\ldots, \tag{8.37}$$

8.6 Many-Particle Exchange Symmetry

so

$$u(\mathbf{x}_1, \mathbf{x}_2, \ldots, \mathbf{x}_N) = \langle \mathbf{x}_1 \ldots \mathbf{x}_N | u \rangle = \sum_{\alpha\beta\gamma\ldots} C_{\alpha\beta\gamma\ldots} \langle \mathbf{x}_1 | u_\alpha \rangle \langle \mathbf{x}_2 | u_\beta \rangle \ldots \quad (8.38)$$

Eq. (8.38) can be substituted into (8.36) and (8.22) to obtain the wave functions for fermionic or bosonic states that are sufficiently general to include the effects of interaction of the fermions or the bosons. For example, on substitution of (8.37) into Eq. (8.36) for fermionic states, we obtain

$$\Psi(\mathbf{x}_1, \mathbf{x}_2, \ldots, \mathbf{x}_N) = \frac{1}{\sqrt{N!}} \sum_P (-1)^P P \sum_{\alpha\beta\gamma\ldots} C_{\alpha\beta\gamma\ldots} \langle \mathbf{x}_1 | u_\alpha \rangle \langle \mathbf{x}_2 | u_\beta \rangle \ldots \quad (8.39)$$

For identical bosons, populating <u>distinct</u> orthonormal spin-orbitals the wave function is given by

$$\Psi(\mathbf{x}_1, \mathbf{x}_2, \ldots, \mathbf{x}_N) = \frac{1}{\sqrt{N!}} \sum_P P u_\alpha(\mathbf{x}_1) u_\beta(\mathbf{x}_2) \ldots u_\nu(\mathbf{x}_N) \equiv \sqrt{N!} \, \mathcal{S} \, \psi_{12\ldots N}. \quad (8.40)$$

The wave function for identical bosonic particles in a multicomponent wave function takes the form

$$\Psi(\mathbf{x}_1, \mathbf{x}_2, \ldots, \mathbf{x}_N) = \frac{1}{\sqrt{N!}} \sum_P P u(\mathbf{x}_1, \mathbf{x}_2, \ldots, \mathbf{x}_N), \quad (8.41)$$

where $u(\mathbf{x}_1, \mathbf{x}_2, \ldots, \mathbf{x}_N)$ is a wave function for the N identical bosons, and the wave function in Eq. (8.41) has been symmetrized by applying the symmetrization operator $\mathcal{S}$ (8.7) to $u(\mathbf{x}_1, \mathbf{x}_2, \ldots, \mathbf{x}_N)$ and then multiplied by $\sqrt{N!}$ to normalize the wave function. However, (8.40) or (8.41) is not necessarily properly normalized, as we shall now see.

Let us explicitly consider a system of N identical bosons, and let $p_1, p_2, \ldots, p_N$ be the labels of the occupied single-particle states, where some of these labels may be the same. The wave function $\Psi(\mathbf{x}_1, \mathbf{x}_2, \ldots, \mathbf{x}_N)$ is given by a sum of products of the form $\psi_{p_1}(\mathbf{x}_1)\psi_{p_2}(\mathbf{x}_2)\ldots\psi_{p_N}(\mathbf{x}_N)$. We shall take $\psi_{p_1}, \psi_{p_2}, \ldots$ to be normalized, orthogonal wave functions. For example, for two bosons in different single-particle states,

$$\Psi(\mathbf{x}_1, \mathbf{x}_2) = \frac{1}{\sqrt{2}} [\psi_{p_1}(\mathbf{x}_1)\psi_{p_2}(\mathbf{x}_2) + \psi_{p_1}(\mathbf{x}_2)\psi_{p_2}(\mathbf{x}_1)]. \quad (8.42)$$

In the general case of N bosons,

$$\boxed{\Psi(\mathbf{x}_1, \mathbf{x}_2, \ldots, \mathbf{x}_N) = \left(\frac{N_1! N_2! \ldots}{N!}\right)^{1/2} \sum \psi_{p_1}(\mathbf{x}_1)\psi_{p_2}(\mathbf{x}_2)\ldots\psi_{p_N}(\mathbf{x}_N),} \quad (8.43)$$

where the sum is taken over all possible permutations of the coordinates (or the labels $p_1, p_2, \ldots, p_N$), and the factor $\left(\frac{N_1! N_2! \ldots}{N!}\right)^{1/2}$ is for normalization, where the numbers N_i determine how many of these suffixes have the same value p_i, such that $\sum_i N_i = N$. In calculating $\langle \Psi | \Psi \rangle$, all terms vanish except the squared modulus of each term in the sum on the RHS of (8.43). There are $\frac{N!}{N_1! N_2! \ldots}$ terms, so (8.43) is a properly normalized wave function. Note that if the wave functions $\{\psi_{p_j}\}$ are not orthogonal, the normalization must be computed differently.

Problem 8.15

Consider three identical spin 1 particles.

(a) Assume the space part of the wave function is symmetric under particle exchange, and let $|m_1 m_2 m_3\rangle$ with $m_i = 0, 1, -1 \equiv 0, 1, \bar{1}$ denote a spin basis state of the three spin 1 particles. Construct the normalized spin states for each of the following three cases: (I) $m_i = 1$, for $i = 1, 2, 3$. (II) $m_i = m_j = 1$, and $m_k = 0$. (III) $m_i \neq m_j \neq m_k \neq m_i$.

(b) Now take the spatial part of the wave function to be antisymmetric.

Answer: (a) For a symmetric spatial wave function, the spin wave function, $|\Xi\rangle$, must also be symmetric. Hence, $|\Xi_I\rangle = |111\rangle$, $|\Xi_{II}\rangle = \frac{1}{\sqrt{3}}(|110\rangle + |101\rangle + |011\rangle)$, and $|\Xi_{III}\rangle = \frac{1}{\sqrt{6}}(|1\bar{1}0\rangle + |10\bar{1}\rangle + |01\bar{1}\rangle + |0\bar{1}1\rangle + |\bar{1}10\rangle + |\bar{1}01\rangle)$. $|\Xi_I\rangle$ has $S = 3$, and $|\Xi_{II}\rangle$ and $|\Xi_{III}\rangle$ are linear combinations of $S = 1$ and $S = 3$.

(b) The spin function must now be totally antisymmetric. This is not possible for cases I and II (the corresponding Young tableau consists of a three box column with non-decreasing entries going downwards). For case III, the antisymmetric spin state is given by $|\Xi_{III}\rangle = \frac{1}{\sqrt{6}}(|10\bar{1}\rangle - |1\bar{1}0\rangle + |0\bar{1}1\rangle - |01\bar{1}\rangle + |\bar{1}10\rangle - |\bar{1}01\rangle)$, with total spin $S = 0$.

Problem 8.16

Consider two electronic eigenstates $|g\rangle$ and $|e\rangle$ of an atom, and suppose we have two atoms, one in the superposition state $|\psi_1\rangle = \alpha|g\rangle + \beta|e\rangle$, and the other in the superposition state $|\psi_2\rangle = \gamma|g\rangle + \delta|e\rangle$.

(a) For identical bosonic atoms with a spatial wave function which is symmetric, the electronic wave function of the two atoms is in state $|\Psi_S\rangle = \frac{1}{\sqrt{2}}[|\psi_1\rangle|\psi_2\rangle + |\psi_2\rangle|\psi_1\rangle]$; find $\langle\Psi_S|\Psi_S\rangle$.

(b) For identical fermionic atoms with a spatial wave function which is symmetric, what is the electronic wave function of the two atoms $|\Psi_f\rangle$? Now find $\langle\Psi_f|\Psi_f\rangle$.

(c) If the electronic energy of state $|g\rangle$ is $E_g = 0$, and the energy of state $|e\rangle$ is $E_e = 1$ eV, what are the expectation values of the electronic states in (a) and (b)?

Answer: (a) $\langle\Psi_S|\Psi_S\rangle = 1 + |\alpha\gamma^2 + \beta\delta^*|^2$, (b) $|\Psi_f\rangle = |\Psi_A\rangle = \frac{1}{\sqrt{2}}[|\psi_1\rangle|\psi_2\rangle - |\psi_2\rangle|\psi_1\rangle]$. $\langle\Psi_A|\Psi_A\rangle = 1 - |\alpha\gamma^2 + \beta\delta^*|^2$. (c) $E = \frac{|\beta|^2 + |\delta|^2}{\langle\Psi_S|\Psi_S\rangle}$ and $E = \frac{|\beta|^2 + |\delta|^2}{\langle\Psi_A|\Psi_A\rangle}$.

Problem 8.17

(a) How many energy levels are there for N spin $1/2$ particles, due to spin degeneracy, with a particular value of spin S, when no magnetic field is present and spin–spin interactions are not considered?

(b) What is the total number of energy levels (summed over all spin S).

Answer: (a) There are $f(M_s) = \frac{N!}{(N/2+M_s)!(N/2-M_s)!}$ ways of making a state with $M_s = \sum_i m_{s,i}$. For a given S, there are $2S + 1$ possible values of M_s, $M_s = S, S-1, \ldots, -(S-1), -S$. Therefore, for a given S, there are $n(S) = f(S) - f(S+1)$ different energy levels with a given value of S. Why? Because there are $f(S)$ energy levels with spin greater or equal to S, but $f(S+1)$ of them belong to higher values of S. (b) The total number of different energy levels is $n = \sum_S n(S) = f(S_{\min})$. Thus, for even N, $n = f(0) = \frac{N!}{[(N/2)!]^2}$, and $n = f(1/2) = \frac{N!}{(N/2+1/2)!(N/2-1/2)!}$ for odd N.

In the next Chapter, we shall see that, for a system in equilibrium at finite temperature T, the distribution of occupation of the energy eigenstates depends on whether the particles are bosons or fermions. The distribution function for fermions is called the *Fermi-Dirac distribution,* and for bosons, it is called the *Bose–Einstein distribution.* These distributions are radically different at low temperature. In Chapter 14, we shall develop many-body theory in terms of creation and annihilation operators that have exchange symmetry built into their definition; this formalism keeps track of the proper exchange symmetry of the state of the many-body system.

Electronic Properties of Solids 9

This chapter is an introduction to the electronic properties of solids. Solids are materials with a high degree of rigidity; they do not flow like liquids or gases. They can be crystalline (atoms or molecules arranged in a regular array) or amorphous (e.g., glasses, rubbers, and solid polymers). Solids can be roughly classified into metals, semiconductors, and insulators depending on their ability to transfer charge and energy (heat). The nature of electron dynamics in these materials determines their electrical and thermal properties. Metals are most often hard, shiny, and malleable and have good electrical and thermal conductivity at room temperature. Examples include iron, gold, silver, copper, aluminum, and alloys such as brass and steel. Exceptions include liquid metals, such as mercury- and gallium-based alloys, that are liquid at room temperature. Semiconductors are characterized by transport properties that are tunable with temperature near room temperature. They conduct poorly at low temperature, but their conductance improves greatly at high temperatures. The conductivity of semiconductors can be significantly enhanced by doping impurities that add electrons to the conduction bands or deplete electrons from the valence band. Commonly used semiconductors are silicon, germanium, selenium, gallium-arsenide, and silicon-carbide. Insulators are characterized by extremely high resistivity. They conduct electricity and heat very poorly. A few examples are plastics, styrofoam, paper, rubber, and glass.

We begin this chapter by considering the properties of the free electron gas in three spatial dimensions (3D) as well as in 2D and 1D in Sec. 9.1. We then present classical and semiclassical theories of electrical and thermal conductivity in Sec. 9.2, still within the framework of the free electron gas. We then discuss some basic concepts in solid-state physics, including crystal structure in Sec. 9.3, electrons in a periodic potential in Sec. 9.4, magnetic field effects on electrons in solids in Sec. 9.5, semiconductors in Sec. 9.6. Section 9.7 is devoted to the rapidly growing field of spintronics, and Sec. 9.8.1 discusses elementary excitations in solids. Finally a short description of insulators is presented in Sec. 9.9. In Chapters 10 and 15, we take up the task of actually calculating electronic structure; the former deals with a mean-field method called the *Hartree-Fock approximation* and its generalization called configuration interaction, while the latter treats density functional theory. These methods can also be applied to calculate the electronic structure of solids.

9.1 THE FREE ELECTRON GAS

Valence electrons in solids, and in particular, in metals, experience the potential of neighboring atoms and not just the potential of a single atom. In metals, these electrons leave their nascent atoms and become itinerant throughout the entire material. They form what is called an *electron gas*. Transport properties of solids (electronic, and to some degree, thermal) are determined, to a large extent, by the properties of the electron gas. If the interaction between electrons is taken into account, this is a generic *many-body problem* (see Chapter 14). However, often a *single-particle approximation*, where the electrons feel only an effective mean-field potential, offers an adequate description. The question why an interacting fermionic system can be described within a single-particle picture will be answered in Chapter 14. In the crudest approximation, electrons in metals are assumed to be free. Studying the *free electron gas* serves as an important benchmark and reference point for more realistic treatments, especially, for electrons in a periodic potential.

This section introduces the basic techniques for treating a system of free electrons, using the Schrödinger equation for free particles (electrons) that obey Fermi statistics. Transport properties of solids (particularly metals) are largely dependent on the density of electrons that are free to move through the entire solid. Within the framework of the free electron gas, we consider a large number N of noninteracting itinerant electrons that move in a solid without the presence of an external potential. The electrons are confined within a macroscopic volume $\mathcal{V} = L^d$ that is a cube of length L, where $d = 1, 2, 3$ is the relevant space dimension. The *average density* $n \equiv N/\mathcal{V}$ remains constant as $N, \mathcal{V} \to \infty$. This limiting procedure is called the *thermodynamic limit*.

Let us first consider a free electron gas in three-dimensional (3D) space. The Hamiltonian $H = \sum_{i=1}^{N} \mathbf{p}_i^2/2m$ contains only the kinetic energy, where m is the electron mass and $\mathbf{p}$ is the momentum operator. In an *infinite volume* ($L = \infty$), plane waves form a complete set of free electron wave functions satisfying the time-independent Schrödinger equation, $H\psi = E\psi$,

$$\psi_{\mathbf{k}}(\mathbf{r}) = (2\pi)^{-3/2} e^{i\mathbf{k}\cdot\mathbf{r}}, \quad E_{\mathbf{k}} = \frac{\hbar^2 k^2}{2m} = \frac{\hbar^2}{2m}\left(k_x^2 + k_y^2 + k_z^2\right), \tag{9.1}$$

where $-\infty < k_x, k_y, k_z < \infty$.

When L is finite, the boundary conditions on the wave function must be specified. For *periodic boundary conditions*, $\psi(\mathbf{r} + \hat{\mathbf{u}}L) = \psi(\mathbf{r})$, where $\hat{\mathbf{u}} = \hat{\mathbf{x}}, \hat{\mathbf{y}}, \hat{\mathbf{z}}$, the components of $\mathbf{k}$ must be quantized in units of $2\pi/L$,

$$\mathbf{k}_{n_x,n_y,n_z} = \frac{2\pi}{L}(n_x, n_y, n_z), \quad n_i = 0, \pm 1, \pm 2, \ldots, \quad i = x, y, z. \tag{9.2}$$

The normalized wave functions and eigenenergies are

$$\psi_{n_x,n_y,n_z}(\mathbf{r}) = \frac{1}{L^{3/2}} e^{i\mathbf{k}_{n_x,n_y,n_z}\cdot\mathbf{r}}, \tag{9.3}$$

$$E_{n_x,n_y,n_z} = \frac{\hbar^2 k_{n_x,n_y,n_z}^2}{2m} \equiv \mathcal{E}_0 \left(n_x^2 + n_y^2 + n_z^2\right), \tag{9.4}$$

where $\mathcal{E}_0 = \frac{2\hbar^2\pi^2}{mL^2}$. This relation is similar to Eq. (9.1) except that now $\mathbf{k}$ is quantized. The single-particle ground state has energy $E_{0,0,0} = 0$, and the excited energies are a sum of three squares of integers times $\mathcal{E}_0$. Except for the ground state, all energies are degenerate.

Problem 9.1

(a) How many different states are there in Eq. (9.1) if $0 \leq n_x, n_y, n_z \leq 7$?
(b) How many states have the same energy, E_{234}?
(c) How many states have the same energy, E_{017}?

Answers: (a) $N = 336$. (b) All the six energies E_{i_2,i_3,i_4} are degenerate, where (i_2, i_3, i_4) are permutations of (234).
(c) **Hint:** Besides the six energies E_{i_0,i_1,i_7}, there are other integers n_x, n_y, n_z, such that $n_x^2 + n_y^2 + n_z^2 = 0^2 + 1^2 + 7^2 = 50$.

Hard-wall boundary conditions correspond to an electron moving freely in a cube $0 \leq x, y, z \leq L$ with an infinite potential outside it. The wave function must vanish on the cube's faces, i.e., $\psi(\mathbf{0}) = \psi(\hat{\mathbf{u}}L) = 0$. Hence, the normalized wave functions and their energies are

$$\psi_{n_x,n_y,n_z}(\mathbf{r}) = \left(\frac{2}{L}\right)^{3/2} \sin(k_x x)\sin(k_y y)\sin(k_z z)$$

$$= \left(\frac{2}{L}\right)^{3/2} \sin\left(\frac{n_x \pi x}{L}\right) \sin\left(\frac{n_y \pi y}{L}\right) \sin\left(\frac{n_z \pi z}{L}\right), \tag{9.5}$$

$$E_{n_x,n_y,n_z} = \frac{\hbar^2 k^2}{2m} \equiv \mathcal{E}_0\left(n_x^2 + n_y^2 + n_z^2\right), \tag{9.6}$$

where $\mathcal{E}_0 \equiv \frac{\hbar^2 \pi^2}{2mL^2}$. The quantized wavevectors are $\mathbf{k} = \frac{\pi}{L}(n_x, n_y, n_z)$ with $n_x, n_y, n_z = 1, 2, \ldots$, and the single-particle ground-state energy is $3\mathcal{E}_0$. Note that n_i cannot be equal to zero, for otherwise the wave function vanishes.

The many-electron wave function can be written as a product (or sum of products) of single-particle states. According to the *Pauli exclusion principle*, at most one electron of a given spin can occupy each of these single-particle states such as (9.3) or (9.5). Ordering the energies $E_{n_x n_y n_z}$ as $E_0 \leq E_1 \ldots \leq E_N \leq E_{N+1} \ldots$, with spin and orbital degeneracy included,

9.1 The Free Electron Gas

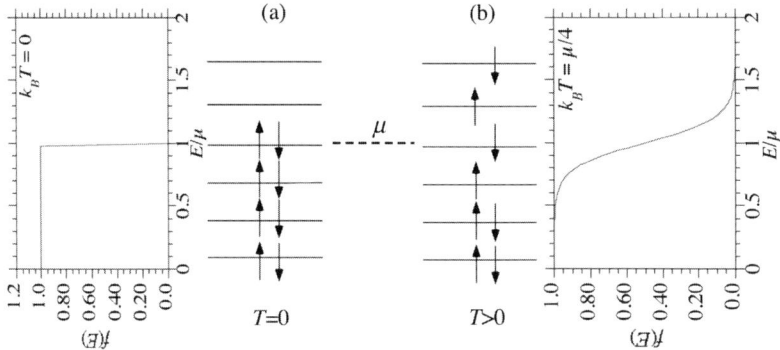

FIG 9.1 Level filling of electrons. (a) At zero temperature, the single-particle levels are filled from bottom up. Each level contains an electron with spin state $|\uparrow\rangle$ and an electron with spin state $|\downarrow\rangle$. The levels are filled until all N electrons are used up ($N = 8$ in this example). The highest occupied level determines the Fermi energy (dashed line), denoted here by μ because, at $T = 0$, the chemical potential coincides with the Fermi energy. (b) At finite T, each level is occupied with electrons according to the Fermi distribution (displayed on the right), see Sec. 9.1.2.

the lowest N-electron wave function, $\Psi_0(\mathbf{r}_1, \ldots, \mathbf{r}_N)$, is constructed as a product of N single-particle states (9.5) having the N lowest energies (but the necessity to antisymmetrize with respect to interchange of electrons requires that we build a Slater determinant of these single-particle states, so a sum of products is actually obtained – as discussed in detail in Sec. 8.3). This corresponds to successively "filling up" the spatial states, starting with the lowest energy state, putting two electrons in each state, one electron with spin-up and the other with spin-down, until no more electrons are available. An illustration of this *Aufbau principle* (in German, Aufbau means "building-up") is illustrated in Fig. 9.1(a). The energy of the highest-energy filled state defines the *Fermi energy*, $E_F \equiv E_N$, which is named after Enrico Fermi who developed the statistical laws for fermions known as *Fermi statistics*. This procedure is valid at zero temperature, $T = 0$. When $T > 0$, the probability that a quantum state with energy E_n ($0 \leq n < \infty$) is occupied is given by $P_n = e^{-\beta E_n} / \sum_j e^{-\beta E_j}$, where $\beta = (k_B T)^{-1}$.

To find a relation between the total number of electrons N and the Fermi energy E_F, we need to answer the following question: For a given single-particle energy E, what is the number $N(E)$ of single-particle states whose energies $E_{n_x, n_y, n_z} \leq E$? We first answer the question within the hard-wall boundary condition scheme. It is evident from Eq. (9.6) that the number of spatial quantum states $|\psi_{n_x, n_y, n_z}\rangle$ equals the number of 3-tuples with positive integers $n_x, n_y, n_z > 0$, such that $n_x^2 + n_y^2 + n_z^2 < E/E_0$. The required number $N(E)$ of states, including spin degeneracy, is twice this number. Figure 9.2 shows that these integer points occupy $1/8$ of the volume of a sphere of radius $r = \sqrt{E/E_0}$, i.e., $\frac{N(E)}{2} = \frac{1}{8} \frac{4\pi}{3} r^3$, where the factor $1/2$ is due to spin degeneracy. Thus,

$$N(E) = \frac{\pi}{3} r^3 = \frac{\pi}{3} \left(\frac{E}{E_0}\right)^{3/2} = \frac{V}{3\pi^2} \left[\frac{2mE}{\hbar^2}\right]^{3/2}, \tag{9.7}$$

where we have used E_0 as defined above and recall that $V = L^3$. Thus, the density of electrons that fill energy levels up to the Fermi energy is given by

$$n_e = N(E_F)/V = \frac{(2m)^{3/2}}{3\hbar^3 \pi^2} E_F^{3/2}. \tag{9.8}$$

We also define the *Fermi momentum* k_F in terms of the Fermi energy and the *Fermi velocity* as

$$k_F \equiv \sqrt{\frac{2mE_F}{\hbar^2}}, \quad v_F \equiv \frac{\hbar k_F}{m}. \tag{9.9}$$

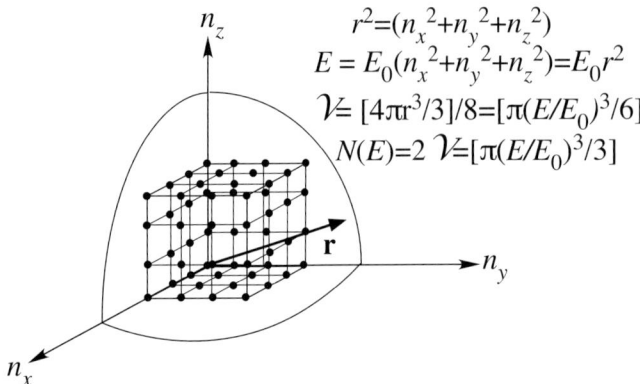

FIG 9.2 Constant energy surface in quantum number space (n_x, n_y, n_z) for hard-wall boundary conditions. A grid of (n_x, n_y, n_z) is shown. The number of spatial states in the volume of a section of the sphere of radius r equals $\mathcal{V} = [4\pi r^3/3]/8$, and the total number of states (including spin) is $N(E) = 2\mathcal{V} = [\pi r^3/3] = \frac{\pi}{3}\left(\frac{E}{E_0}\right)^{3/2}$.

Combining Eqs (9.8) and (9.9), we arrive at the useful relation,

$$n_e = \frac{k_F^3}{3\pi^2}. \tag{9.10}$$

It is convenient to introduce a quantity called the *Wigner–Seitz radius*, r_s, which is defined as the radius of a sphere around a single conduction electron in physical space, such that the total volume of the system divided by the total number of electrons is the volume of the Wigner–Seitz sphere, i.e., $V/N = \frac{4\pi r_s^3}{3}$, so

$$r_s \equiv \left(\frac{3}{4\pi n_e}\right)^{\frac{1}{3}}. \tag{9.11}$$

For metals, the ratio $r_s : a_0$, where a_0 is the Bohr radius, ranges between 1.8 (Be) and 5.6 (Cs). Practical expressions can easily be derived for k_F, v_F, and E_F in terms of r_s and a_0, which simplify their numerical estimates:

$$k_F = \left(\frac{9\pi}{4}\right)^{\frac{1}{3}}\frac{1}{r_s} = \frac{1.92}{r_s} = 3.63\frac{a_0}{r_s}\,\text{Å}^{-1}, \tag{9.12a}$$

$$v_F = \frac{\hbar k_F}{m} = 4.20 \times 10^8 \frac{a_0}{r_s}\,\text{cm/s}, \tag{9.12b}$$

$$E_F = \frac{\hbar^2 k_F^2}{2m} = \frac{e^2}{2a_0}(k_F a_0)^2 = 50.1\left(\frac{a_0}{r_s}\right)^2\,\text{eV}. \tag{9.12c}$$

An important quantity in electron gas systems is the *single-particle density of states* at energy E, $D(E)$. It is defined as the number of states per unit energy at energy E. Assuming that the system size, L, is large, the energy can be regarded as a continuous variable. Then, the number of states between E and $E + dE$ is $dN(E) = D(E)dE$. Differentiating Eq. (9.7) with respect to E, we find

$$D(E) \equiv \frac{dN(E)}{dE} = \frac{V}{2\pi^2}\left(\frac{2m}{\hbar^2}\right)^{3/2} E^{1/2}. \tag{9.13}$$

Clearly, $N(E)$ is obtained from $D(E)$ by integration,

$$N(E) = \int_{E_0}^{E} dE'\, D(E'), \tag{9.14}$$

where E_0 is the lowest energy in the spectrum (any spectrum is bounded from below).

Because $D(E)$ is proportional to the volume of the system, it is useful to define the *number of states per unit energy per unit volume* of the free electron gas,

$$\boxed{g(E) \equiv \frac{D(E)}{V} = \frac{1}{2\pi^2}\left(\frac{2m}{\hbar^2}\right)^{3/2} E^{1/2}.} \tag{9.15}$$

9.1 The Free Electron Gas

It is instructive to derive the above expressions from somewhat different conceptual point of view. Quite generally, the number of states with energy less than or equal to E, $N(E)$,[1] can be calculated by integrating the density of states in phase space under the condition that the energy is $\leq E$, i.e., for one particle,

$$N(E) = \frac{1}{h^3} \int\int d\mathbf{r}\, d\mathbf{p} = \frac{1}{(2\pi)^3} \int\int d\mathbf{r}\, d\mathbf{k}, \tag{9.16}$$

and for N noninteracting particles, the many-particle density of states is

$$N(E) = \frac{1}{h^{3N}} \int\int d^N\mathbf{r}\, d^N\mathbf{p} = \frac{1}{(2\pi)^{3N}} \int\int d^N\mathbf{r}\, d^N\mathbf{k}. \tag{9.17}$$

In Eq. (9.16), as well as in Eq. (9.17), the integral is evaluated in that region of phase space for which the energy is less than or equal to E. The single-particle density of states for a free particle, $D(E)$, can also be determined in the continuum limit using Eq. (9.1), $E = \frac{\hbar^2 k^2}{2m}$, by calculating the number of states in phase space with energy $\leq E$ as follows:

$$N(E) = \frac{1}{(2\pi\hbar)^3} V \int d\mathbf{p} = \frac{V}{(2\pi\hbar)^3}\hbar^3 \int d\mathbf{k} = \frac{V}{(2\pi)^3} \frac{4\pi}{3} k^3. \tag{9.18}$$

Here, we used spherical symmetry to write the volume element in k-space as $d\mathbf{k} = 4\pi k^2 dk$. Differentiating this expression for the number of states in phase space with respect to energy, $D(E) = \frac{dN(E)}{dE}$, we obtain the following result.

$$D(E) = \frac{V}{(2\pi)^3} 4\pi k^2 \frac{dk}{dE} = \frac{V}{2\pi^2} k^2 \frac{dk}{dE}. \tag{9.19}$$

On substituting $\frac{dk}{dE} = \left(\frac{dE}{dk}\right)^{-1} = \frac{m}{\hbar^2 k}$ and expressing k as a function of E, i.e., $k = \sqrt{2mE}/\hbar$, we obtain the 3D density of states, $D_{3D}(E) = \frac{V}{4\pi^2}\left(\frac{2m}{\hbar^2}\right)^{3/2} E^{1/2}$. (Alternatively, it is easy to see that $N(E) = \frac{V}{(2\pi)^3} \frac{4\pi}{3}(2mE/\hbar^2)^{3/2}$, and this expression can be differentiated with respect to E.) For electrons, the RHS of Eq. (9.19) is multiplied by 2, because electrons have two spin states, the 3D electron density of states is then,

$$\boxed{D_{3D}(E) = \frac{V}{2\pi^2}\left(\frac{2m}{\hbar^2}\right)^{3/2} E^{1/2}.} \tag{9.20}$$

Clearly, we have re-derived Eq. (9.13).

Problem 9.2

Consider N free particles of mass m in a volume V. Calculate the *many-particle density of states*.

Answer: Let us denote the many-particle density of states with the same symbol, $N(E) = \frac{1}{(2\pi\hbar)^{3N}} V^N \int d^{3N}p$, where the momentum integral is over a $3N$-sphere in momentum space whose radius squared is $p^2 = 2mE$. The integral can be evaluated as follows: $\int d^{3N}p = S_n \int_0^p dp'\, p'^{n-1} = \frac{S_n}{n} p^n$, where $n = 3N$ and S_n is the hypersurface area in dimension n. As shown in http://mathworld.wolfram.com/Hypersphere.html, $S_n = \frac{2\pi^{n/2}}{\Gamma(n/2)}$. The well-known cases are $S_2 = 2\pi$ and $S_3 = 4\pi$. Writing $\int d^{3N}p = \frac{S_n}{n} p^n = \frac{S_{3N}}{3N}(2mE)^{3N/2}$, we obtain $N(E) = \frac{S_{3N} V^N}{3N(2\pi\hbar)^{3N}}(2mE)^{3N/2}$. Hence, the 3D many-particle density of states, $D_{3D}(E)$, is given by

---

[1] In statistical mechanics, it is common to denote the number of states with energy less than or equal to E by the symbol $\Phi(E) [\equiv N(E)]$ and to denote the density of states at energy E by the symbol $\Omega(E) \equiv \frac{d\Phi(E)}{dE} = \frac{dN(E)}{dE} = D(E)$ (see Reif [119]). The entropy S of a system in equilibrium is given by $S = k_B \ln \Omega$; this equation is inscribed on Ludwig Boltzmann's headstone.

$$D(E) = \frac{dN(E)}{dE} = \frac{S_{3N}(2m)^{3N/2}\mathcal{V}^N}{2(2\pi\hbar)^{3N}}E^{3N/2-1}.$$

This result does not take into account the internal (spin) states of the particles. Clearly, the many-particle density of states per unit energy increases very rapidly with energy for large N.

Problem 9.3

Determine the density of states $D(E)$ using the periodic boundary condition case of Eq. (9.2)–(9.4). Show that Eq. (9.20) applies.

Parenthetically, we remark that much of the discussion above applies to photon states. If we consider the electromagnetic field in a volume $\mathcal{V}$, the modes of the electromagnetic field are very similar to the modes (i.e., levels) of the electrons specified in Eq. (9.5) [and their energies are given by $E_{n_x,n_y,n_z} = \hbar|\mathbf{k}|c = \tilde{E}_0(n_x^2 + n_y^2 + n_z^2)^{1/2}$, which is not so different from Eq. (9.6)]. The difference is that photons are bosons and therefore obey Bose–Einstein statistics, as opposed to the Fermi–Dirac statistics obeyed by electrons. Every quantum state can accommodate any number of photons, and no aufbau principle exists for photons. At zero temperature, all bosons occupy the ground state.

9.1.1 DENSITY OF STATES IN 2D AND 1D SYSTEMS

Let us apply the same kind of reasoning as used in the previous section to obtain the single-particle density of states of a thin 2D layer as shown in the inset of Fig. 9.4(b). The eigenstates and eigenenergies are similar to Eqs (9.6)–(9.7), except that now $L_z \ll L_x = L_y \equiv L$, so

$$\psi_{n_x,n_y,n_z}(\mathbf{r}) = \left(\frac{2}{L}\right)\left(\frac{2}{L_z}\right)^{1/2}\sin\left(\frac{n_x\pi x}{L}\right)\sin\left(\frac{n_y\pi y}{L}\right)\sin\left(\frac{n_z\pi z}{L_z}\right), \quad (9.21)$$

and

$$E_{n_x,n_y,n_z} = \frac{\hbar^2\pi^2}{2mL^2}\left(n_x^2 + n_y^2\right) + \frac{\hbar^2\pi^2}{2mL_z^2}n_z^2 \equiv E_{n_xn_y} + E_{n_z}. \quad (9.22)$$

The energy spacings of standing waves along the z axis are much larger than those along the x and y axes, i.e.,

$$\Delta_{E_z} \equiv E_{n_z+1} - E_{n_z} = \frac{\hbar^2\pi^2}{2mL_z^2}(2n_z + 1) \gg E_{n_x+1,n_y} - E_{n_x,n_y}. \quad (9.23)$$

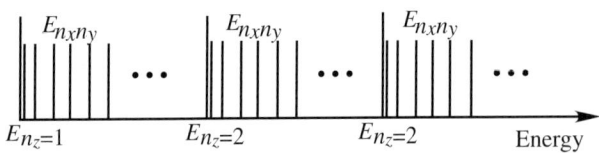

FIG 9.3 Single-particle energy levels of the free electron gas in a box of sides L_x, L_y and L_z, with $L_z \ll L_x, L_y$.

Therefore, it is justified to compute $D_{2D}(E)$ in a given energy interval $E_{n_z} \leq E \leq E_{n_z+1}$, as shown in Fig. 9.3, and the result is independent of n_z.

We may then focus on the two-dimensional (2D) quadrant of positive integers (n_x, n_y) and count the number of states within a quarter of a circle of radius $r = \sqrt{E/E_0}$. The number of single-particle states in this area (spin degeneracy included) is $N(E) = \frac{\pi r^2}{2} = \pi\left(\frac{E}{2E_0}\right) = \frac{mL^2}{\pi\hbar^2}E$. Differentiating

9.1 The Free Electron Gas

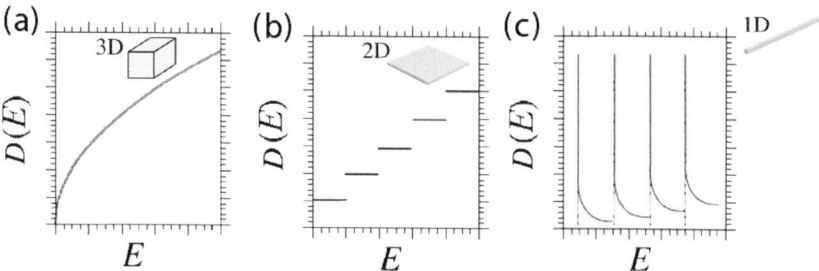

FIG 9.4 Density of states in 3D, 2D, and 1D structures. (a) The $D(E) \propto E^{1/2}$ dependence of Eq. (9.13) for 3D structures. (b) In 2D (see inset), the density of states is constant, whereas (c) in 1D structures (see inset), $D(E) \propto E^{-1/2}$. In (b) and (c), higher transverse modes open up as the energy is increased.

with respect to E to obtain the density of states, we find

$$D_{2D}(E) = \frac{dN(E)}{dE} = \frac{mL^2}{\pi \hbar^2}. \quad (9.24)$$

Figure 9.4(b) shows the 2D density of states. In this interval, $D_{2D}(E)$ is independent of energy.

Problem 9.4

Calculate the number of levels $E_{n_x n_y}$ between E_{n_z} and E_{n_z+1}.

Answer: The length of the energy interval is Δ_{E_z} as defined in Eq. (9.23) and $D_{2D}(E) = \frac{mL^2}{\pi \hbar^2}$ independent of E. The number of levels then equals $\Delta_{E_z} D_{2D}(E)$.

The same reasoning can be applied to a quasi-1D system. In recent years, it has become possible to fabricate quasi-one-dimensional systems, which maintain quantum coherence along a micron-sized length scale. Such a *quantum wire* is schematically illustrated in the inset of Fig. 9.4(c). The eigenstates and eigenenergies are again given by Eqs (9.21) and (9.22), but, taking the wire along x, we now have $L_x \gg L_y = L_z \equiv L$. The threshold energies are then $E_{n_y n_z} = \frac{\hbar^2 \pi^2 (n_y^2 + n_z^2)}{2mL^2}$, and their spacings are much larger than $E_{n_x+1} - E_{n_x}$, where $E_{n_x} = \frac{p_x^2}{2m} = \frac{\hbar^2 \pi^2}{2mL_x^2} n_x^2$. The 1D density of states then refers to the density of levels E_{n_x}. The integrated density of states and the density of states are given by,

$$N(E) = 2 \times \frac{1}{2\pi \hbar} L_x \Delta p_x = \frac{\sqrt{2mL_x}}{\hbar \pi} \sqrt{E}, \quad (9.25)$$

$$D_{1D}(E) = \frac{dN}{dE} = \frac{\sqrt{2m}}{2\hbar \pi} L_x E^{-1/2}, \quad (9.26)$$

i.e., $D_{1D}(E) \propto E^{-1/2}$. The density of states diverges near the energies where additional transverse modes open, i.e., at energies $\frac{5\hbar^2 \pi^2}{2mL^2}, \frac{8\hbar^2 \pi^2}{2mL^2}$, etc., as shown in Fig. 9.4(c). However, the integrated density of states remains well-behaved.

Problem 9.5

(a) Consider N free particles of mass m in a 2D area A and calculate the many-particle density of states.
(b) Consider the 1D case and calculate the density of states.

Answer: (a) $N(E) = \frac{1}{(2\pi\hbar)^{2N}} A^N \int d^{2N}p$, where the integral is evaluated as $\int d^{3N}p = S_n \int_0^p dp' \, p'^{n-1} = \frac{S_n}{n} p^n$, with $n = 2N$. Hence, $N(E) = \frac{S_{2N} A^N}{2N(2\pi\hbar)^{2N}} (2mE)^N$, and $D_{2D}(E) = \frac{dN(E)}{dE} \propto E^{N-1}$.

9.1.2 FERMI–DIRAC DISTRIBUTION

At zero temperature, all states with energy E below the Fermi energy E_F are populated, whereas states with energy $E > E_F$ are empty. At finite temperature, $T > 0$, the occupation distribution is rounded off. Level occupation of electrons is determined by the Fermi–Dirac distribution, $f(E, T)$, which was first introduced by Enrico Fermi in 1926. When the system is found in thermal equilibrium at a given temperature T, $f(E, T)$ is defined as the probability that a quantum state at energy E is occupied. We now briefly discuss the Fermi–Dirac distribution; a detailed discussion can be found in study by Reif [119] and Landau and Lifshitz [120].

At zero temperature, $f(E, T = 0) = 1 - \Theta(E - E_F)$, where $\Theta(x)$ is the step function [$\Theta(x) = 0$ for $x < 0$ and $\Theta(x) = 1$ for $x \geq 0$], as shown in Fig. 9.5. At finite temperature, $T > 0$, the Fermi–Dirac function gives the average number of fermions in a state with energy E :[2]

$$f(E, T) = \left[e^{\frac{E-\mu}{k_B T}} + 1 \right]^{-1}. \quad (9.27)$$

Here, μ is the *chemical potential*. The filling of energy levels according to the Fermi statistics at finite temperature is schematically illustrated in Fig. 9.1(b) (right panel). The chemical potential, μ, depends on temperature and should be determined in a consistent manner, such that the number of particles in the system equals N (the relation between μ and N is discussed below). Note that when $T = 0$, Eq. (9.27) reduces to

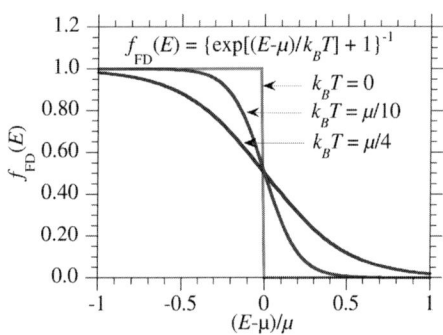

FIG 9.5 Fermi–Dirac distribution versus energy for $k_B T = 0$, $\mu/10$ and $\mu/4$. *Source: Band, Light and Matter, Fig. 6.34, p. 396*

$f(E, T = 0) = 1 - \Theta(E - \mu)$, and

$$\mu(T = 0) = E_F. \quad (9.28)$$

The dependence of the chemical potential on temperature is weak in metals and semiconductors, and in many cases, the deviation of μ from E_F can be neglected. At any temperature, $f(E = \mu, T) = 1/2$. This can serve as a definition of the chemical potential. When $E \gg \mu$, the Fermi–Dirac distribution tends to the Boltzmann distribution, that is, $f(E \gg \mu, T) \to \exp[-(E - \mu)/k_B T]$. Figure 9.5 plots the Fermi–Dirac distribution versus energy for $k_B T = 0$, $\mu/10$, and $\mu/4$. One should compare the Fermi–Dirac distribution presented in this section with the Planck blackbody radiation law [see Sec. 1.1.3, Eq. (1.5)] and the Bose–Einstein distribution for integer spin particles,

$$f_{BE}(E, T) = \left[e^{\frac{E-\mu}{k_B T}} - 1 \right]^{-1}, \quad (9.29)$$

which is the thermal distribution for Bose–Einstein particles, and can be used to derive the Planck blackbody radiation law (see Landau and Lifshitz [120]). Note that no chemical potential is necessary for photon distribution, i.e., $\mu = 0$, because there is no restriction on the total number of photons. The distribution in Eq. (9.29) for $\mu = 0$ is shown in Fig. 9.6.

---

[2] Note that $f(E, T)$ is not a probability distribution function. It is average electron occupation of states at temperature T.

9.1 The Free Electron Gas

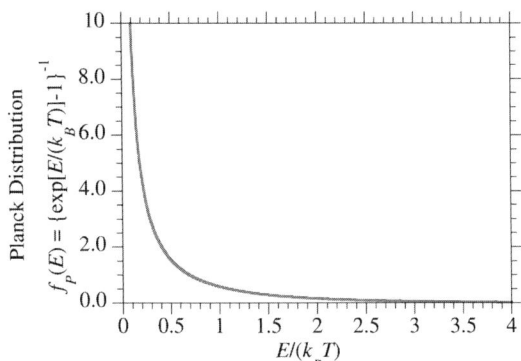

FIG 9.6 Planck distribution given by $f_P(E,T) = \left[e^{\frac{E}{k_B T}} - 1\right]^{-1}$.

Now, let us consider the relation between the chemical potential μ and the number of particles N. This relation touches upon an important concept in quantum statistical mechanics. Often an electron gas in thermal equilibrium is set in contact with a very large electron reservoir at temperature T, so that exchange of particles takes place between the system and the reservoir. Therefore, the number of electrons in the system is not fixed; it might fluctuate with time. However, the average number of electrons obtained by integrating $N(E)$ with weight $f(E,T)$ over the entire energy spectrum between some lowest energy E_0 and ∞ (the spectrum is bounded from below) is fixed by the chemical potential μ. In terms of statistical mechanics, a system of particles interacting with large reservoir at temperature T such that its number of particles fluctuates around average number is modeled by a *grand canonical ensemble*. The actual procedure is straightforward. Assume for simplicity that $E_0 = 0$, and let $\mathcal{P}(E,T)dE$ denote the probability of electrons to occupy states in the energy interval between E and $E + dE$. The distribution $\mathcal{P}(E,T)$ is given by the product of the density of states $D(E)$ [which, for free electrons in 3D, is given by Eq. (9.20)] and the average occupation of levels is given by the Fermi–Dirac distribution, $f(E,T)$, i.e.,

$$\mathcal{P}(E,T) = D(E)f(E,T). \tag{9.30}$$

The chemical potential is formally defined, so that the integral over energy of $\mathcal{P}(E,T)$ gives the number N of electrons,

$$N = \frac{\mathcal{V}}{2\pi^2}\left(\frac{2m}{\hbar^2}\right)^{3/2} \int_0^\infty dE\, \mathcal{P}(E,T) = \frac{\mathcal{V}}{2\pi^2}\left(\frac{2m}{\hbar^2}\right)^{3/2} \int_0^\infty dE\, \frac{E^{1/2}}{e^{\frac{E-\mu}{k_B T}} + 1}$$

$$\xrightarrow{T \to 0} \int_0^{\mu(T=0)} dE\, D(E) = \frac{\mathcal{V}}{(2\pi)^2}\left(\frac{2m}{\hbar^2}\right)^{3/2}\frac{2}{3}[\mu(T=0)]^{3/2}. \tag{9.31}$$

Hence,

$$\mu(T=0) = E_F = \frac{\hbar^2}{2m}\left(\frac{3\pi^2 N}{\mathcal{V}}\right)^{2/3} = \frac{\hbar^2}{2m}\left(3\pi^2 n_e\right)^{2/3}. \tag{9.32}$$

At finite temperature, the integral in the first part of Eq. (9.31) cannot be carried out in closed form, and a number of techniques can be used to approximate the integral, depending on the temperature.

Problem 9.6

(a) Work out the algebra leading to Eq. (9.31).
(b) Plot the occupation density $\mathcal{P}(E) = D(E)f(E,T)$ versus E for nearly free electrons and $k_B T = \mu/10$.
(c) Calculate the average energy of free electrons $\langle E \rangle = \int dE\, E\mathcal{P}(E,T)$, and the electronic-specific heat, $C_{el} = \frac{\partial \langle E \rangle}{\partial T} = \int dE\, E \frac{\partial \mathcal{P}(E,T)}{\partial T}$.

Answer: (c) $\langle E \rangle \approx Nk_B T$, $C_{el} \approx \pi^2 Nk_B \frac{k_B T}{2\mu}$. Because $\frac{k_B T}{2\mu} = \frac{n_e \mathcal{V}}{mv_F^2}$, the heat capacity per volume is $c_{el} = C_{el}/\mathcal{V} \approx \pi^2 k_B^2 T \frac{n_e}{mv_F^2}$.

The integral defined in Eq. (9.31) is a special case of an observable $O(E)$ (such as $N(E)$) that needs to be integrated with weight given by $f(E, T)$. We now describe a technique to efficiently calculate such integrals. It is more convenient to define the observable $O(E)$ over the entire interval $-\infty < E < \infty$ with the understanding that $O(E < E_0) = 0$. Therefore, the quantity to be calculated is $I(\mu) \equiv \int_{-\infty}^{\infty} dE\, O(E) f(E)$. The structure of $f(E)$ is such that $-f'(E)$ is sharply peaked around $E = \mu$. This suggests that integrating by parts could simplify the calculation. Denoting by $F(E)$ the integral of $O(E)$, i.e., $O(E) = dF(E)/dE$, we obtain

$$I(\mu) = \int_{-\infty}^{\infty} dE\, F(E) \left[-f'(E)\right]. \tag{9.33}$$

Problem 9.7

(a) Show that $f'(E)$ is an even function of $E - \mu$.
(b) Show that $f'(E)$ goes rapidly to zero as $E \to \pm\infty$, and therefore, the limits of the integral in Eq. (9.33) are justified.

The next step is to expand $F(E)$ near $E = \mu$. Defining $F^{(n)}(\mu) \equiv [d^n F(E)/dE^n]_{E=\mu}$ and separating even and odd derivative orders yield,

$$F(E) = \sum_{n=0}^{\infty} \left\{ \frac{1}{(2n)!} F^{(2n)}(\mu)(E-\mu)^{2n} + \frac{1}{(2n+1)!} F^{(2n+1)}(\mu)(E-\mu)^{(2n+1)} \right\}. \tag{9.34}$$

The first term in this expansion is $F(\mu) = \int_{-\infty}^{\mu} dE\, O(E)$; moreover, $F^{(2n)}(\mu) = O^{(2n-1)}(\mu)$. Inserting this expansion into Eq. (9.33), noting that odd power contributions vanish, and using the fact that $\int_{-\infty}^{\infty} dE\,[-f'(E)] = 1$, we find

$$I(\mu) = \int_{-\infty}^{\mu} dE\, O(E) + \sum_{n=1}^{\infty} \frac{1}{(2n)!} \int_{-\infty}^{\infty} dE (E-\mu)^{2n} \left[-f'(E)\right] O^{(2n-1)}(\mu). \tag{9.35}$$

The integral over energy can be carried out by defining x, such that $(E-\mu) \equiv x k_B T$ and defining the sequence of numbers,

$$a_n = \frac{1}{(2n)!} \int_{-\infty}^{\infty} dx\, x^{2n} \frac{e^x}{(e^x+1)^2}. \tag{9.36}$$

Then, Eq. (9.35) becomes the *Sommerfeld expansion*,

$$I(\mu) = \int_{-\infty}^{\mu} dE\, O(E) + \sum_{n=1}^{\infty} a_n [k_B T]^{2n} O^{(2n-1)}(\mu). \tag{9.37}$$

The coefficients a_n can be written as $a_n = (1 - 2^{-2(n-1)})\zeta(2n)$, where $\zeta(z) = \sum_{k=1}^{\infty} 1/k^z$ is the celebrated *Riemann zeta function*, which can be defined for complex number z. For $z = 2n$, the factorized form $\zeta(2n) \equiv \frac{1}{2} \frac{(2\pi)^{2n}}{(2n)!} H_n$ relates the zeta function with the *Bernoulli's numbers* H_n. The first few Bernoulli's numbers are $H_{n=1,2,\ldots,5} = 1/6, 1/30, 1/42, 1/30, 1/66$.

When this procedure is carried out for the calculation of the average energy and the specific heat per unit volume, keeping only the H_1 terms, one obtains a somewhat better approximation than the one derived in Problem 9.6,

$$u \equiv \frac{\langle E \rangle}{V} \approx u_0 + \frac{\pi^2}{6}(k_B T)^2 D(E_F), \qquad (9.38)$$

where u_0 is the ground-state energy per unit volume. The electron gas specific heat at constant volume is then,

$$\boxed{c_{el} = \left(\frac{\partial u}{\partial T}\right)_V = \frac{\pi^2}{3} k_B^2 T D(E_F).} \qquad (9.39)$$

Using Eq. 9.20 for the free electron gas density of states $D(E)$ yields,

$$c_{el} = \frac{\pi^2}{2} \frac{k_B T}{E_F} n_e k_B. \qquad (9.40)$$

It is instructive to compare Eq. (9.39) for c_{el} with the classical expression for the specific heat of an ideal gas, $c_{cl} = \frac{3}{2} n_e k_B$. The ratio,

$$\frac{c_{cl}}{c_{el}} = \frac{\pi^2}{3} \frac{k_B T}{E_F} \approx 10^{-2} \quad \text{at room temperature}, \qquad (9.41)$$

is proportional to temperature, but it is rather small even at room temperature. This is the reason for the absence of any noticeable contribution of the electrons to the specific heat of metals at room temperature. Qualitatively, it can be understood by inspecting the behavior of $f(E, T)$ with temperature as shown in Fig. 9.5. The increment in the electron energy with increased temperature occurs because electrons below E_F instead fill empty levels with energy above E_F. The density of these electrons is approximately $D(E_F)$, and their gain in energy is about $k_B T$, hence the relation (9.39).

Problem 9.8

Determine whether the chemical potential μ for electrons increases or decreases with temperature.

Hint: Consider Eq. (9.31) as an implicit relation $F(\mu, T) = 0$ between μ and T and use the implicit function theorem to formally compute $\frac{d\mu}{dT} = -\frac{\partial F}{\partial T} / \frac{\partial F}{\partial \mu}$.

The results obtained for the free electron gas will be employed in Secs 9.2 and 9.6 for the derivation of transport and thermodynamic properties of metals and semiconductors.

9.2 ELEMENTARY THEORIES OF CONDUCTIVITY

In this section, we discuss elementary theories of electrical (and, to some extent, thermal) conductivity of metals. The electrical conductivity of a solid is determined by its ability to sustain charge transport in a specified direction under the application of an external electric field. In metals, valence electrons tend to leave their parent atoms and become itinerant. The atoms then become positive ions, which undergo small oscillations around their fixed positions. The itinerant electrons are free to roam through the entire medium. Electron transport is affected by the electric field of these ions, as well as electron–electron interaction and other externally applied fields (e.g., external electric and magnetic fields). Here, we focus on conducting properties near thermal equilibrium, where the external fields cause only small perturbations and Ohm's law yields a linear relation between the current density and the applied electrical field. For specificity, we consider electron transport, although, as we shall see below, there are charge carriers other than electrons. In semiconductors, in addition to electrons, holes (the absence of electrons) can also participate in charge transport (see Sec. 9.6). On the other hand, in metals, charge is carried solely by electrons. For simplicity, the analysis here is carried out assuming that there is a single type of charge carrier (such as electrons in metals). Within the present elementary approach, extension to the case where there are two (or more) types of charge carriers (such as electrons and holes in semiconductors) is straightforward.

The two elementary models of electron transport under the influence of an external field in metals are the classical *Drude model* and the *Sommerfeld theory*, which improves upon the former by employing the basic principles of Fermi statistics implied by the Pauli exclusion principle. Improving these simple models for the conductivity of a solid requires solving the Schrödinger equation for interacting electrons and static ions. This is a formidable task that is facilitated by Ohm's law, which constitutes a general framework referred to as *linear response* theory. More detailed framework uses the *fluctuation-dissipation theorem*, which yields an expression for the conductivity in terms of a correlation function of the system in thermal equilibrium, as discussed in Secs 7.9.2 and 7.9.4.

Although most solids encountered in nature are nonmetallic, metals play a central role in solid-state physics due (partially) to their high conductivity. As described below, the high electrical conductivity of metals is due to the overlap in energy of the conduction band and the valence band. In metals, there is often only one band of relevance, and the Fermi energy is within this band (see Fig. 9.26). The electrical conductivity, σ, of a good metal may be as high as 10^{10} (Ohm cm)$^{-1}$ ($\sigma = 9 \times 10^{21}$ s^{-1} in Gaussian units) at a temperature of 1 K. In contrast, the conductivity of a representative insulator may be as low as 10^{-22} (Ohm cm)$^{-1}$ ($\sigma = 9 \times 10^{-11}$ s^{-1} in Gaussian units). This is an enormous diapason of 21 orders of magnitude in the conductivity. (A discussion of Gaussian units, and the conversion from SI units to Gaussian units, is presented in Sec. 3.2.6.)

Many phenomena related to electronic and heat transport in clean metals can be understood within a mean-field theory wherein each conduction electron is assumed to be affected by the average potential created by all the other electrons and the external potentials. This translates into a practical independent-particle calculation scheme, which often serves as a good approximation. The *Landau–Fermi liquid theory* [121] explains why, in many cases, a system of interacting fermions can be analyzed within the independent particle formalism. In this theory, charge carriers are not the bare electrons, but rather, *dressed electrons*, in the sense that they carry with them traces of the collective degrees of freedom. These dressed electrons are referred to as quasi-particles. Although they behave as free charged particles with spin 1/2 and charge $-e$, their mass is usually different from the electron mass. We shall continue to call these quasi-particles electrons, but m_e is replaced by an *effective mass* m_e^*. It should be noted that sometimes a system of interacting fermions cannot be treated solely within Landau–Fermi liquid theory and requires the use of a bona fide many-body theory. When Landau–Fermi liquid theory is inadequate, the underlying physics is entirely different compared with that of Fermi liquid systems.

The Drude and Sommerfeld models for charge (and heat) transport in metals assume not only an independent-particle picture but also that electrons that are nearly free. The classical *Drude theory* is based on the application of Newtonian mechanics to particles moving in a dissipative media characterized by the mean time between collisions, τ. Despite its apparent qualitative success, it is marred by its inadequate account of the *Wiedemann–Franz law* (see below). This flaw is alleviated by the *Sommerfeld Theory of Metals*, which modifies the Drude theory to account for the nature of electron statistics that affects the occupation of quantum states.

9.2.1 DRUDE THEORY OF CONDUCTIVITY

Generically, a conducting system is not isotropic or homogeneous, and the electric field inside the system might depend on space and time, i.e., $\mathbf{E} = \mathbf{E}(\mathbf{r}, t)$. However, for the sake of simplicity, let us assume that the system, e.g., a macroscopic piece of solid, is isotropic and homogeneous and that the applied electric field is uniform and time independent. Application of a static electric field across the material results in a motion of charge carriers, i.e., a *current density* $\mathbf{J}(\mathbf{r})$, so that $\mathbf{J}(\mathbf{r}) \cdot d\mathbf{S}$ is the amount of charge per unit time crossing an area element $d\mathbf{S}$ centered at $\mathbf{r}$. When $\mathbf{E}$ is uniform (as assumed here), $\mathbf{J}$ is also uniform.

> **Problem 9.9**
>
> Check that the unit of the electron current density given by the expression $\mathbf{J}(\mathbf{r}, t) = (-e)n_e(\mathbf{r}, t)\mathbf{v}(\mathbf{r}, t)$ is charge per unit time per unit area. Here, $\mathbf{v}$ is the electron velocity and n_e is the electron density.

9.2 Elementary Theories of Conductivity

The conductivity σ is a quantity that characterizes the response of a system to an external electric field $\mathbf{E}$. In the simplest approximation, known as Ohm's law, the current density $\mathbf{J}$ is linearly proportional to the applied static electric field: $\mathbf{J} = \sigma \mathbf{E}$. Application of an electric field results in a force $\mathbf{F} = -e\mathbf{E}$ acting on the electrons. But experimentally, this leads to a stationary current $\mathbf{J}$. Hence, on the average, electrons are not simply accelerated; their equations of motion must contain an effective damping force in addition to the force due to the electric field. This damping is characterized by a collision time τ, which is a central quantity in the analysis of electrical transport; τ is the average time between collisions that the electrons experience in the material. The conductivity σ is proportional to τ. Treating transport properties of electrons in metals within Newtonian mechanics with conservative and damping forces was proposed by Paul Drude in 1900. It can also be used to obtain the conductivity of ions in solution.

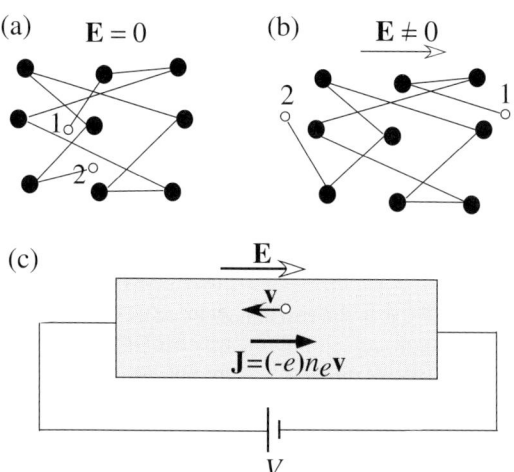

FIG 9.7 Schematic illustration of the Drude model. Charge carriers (taken as valence electrons), represented by empty circles, move randomly in a crystal composed of localized cations, represented by black circles, and execute thermal motion around their equilibrium position. Occasionally, the carrier electrons collide with the ions (and very occasionally with each other). Only ions participating in a collision are shown. (a) In the absence of an electric field, an electron initially at a point $\mathbf{r}_1$ at time t_1 moves randomly and collides with the lattice ions. After a time, $t_2 - t_1$, it is at a point $\mathbf{r}_2$. Without an external electric field being applied, the position of $\mathbf{r}_2$ with respect to $\mathbf{r}_1$ is isotropic; no net motion of the electrons occurs. (b) In the presence of an electric field $\mathbf{E}$, electrons are acted upon by a constant force $\mathbf{F} = -e\mathbf{E}$ and inelastically collide with ions and impurities, so that their motion is damped. There is a global motion of the electrons directed opposite to the field. If τ is the average time between collisions, then the average velocity is $\bar{\mathbf{v}} = -e\mathbf{E}\tau/m$. (c) Current is directed along the field, $\mathbf{J} = (-e)n_e\bar{\mathbf{v}} = (e^2 n_e \tau/m)\mathbf{E}$, where n_e is the density of free electrons.

In the Drude model, schematically depicted in Fig. 9.7, electrons under the influence of an external electric field move in the solid and occasionally collide with the ions comprising the solid and impurities and also with each other. These collisions lead to energy dissipation. The dynamics of charge carriers is then governed by Newton's equations of motion for a charged particle in an electric field within a dissipative environment. The statements made above for electrons also hold for other carriers of charge $q \neq -e$, such as holes of charge $+e$, that are required to describe conduction in semiconductors (see below). In a metal, the conductivity σ is solely due to the motion of electrons, but their effective mass m_e^* might be different from m_e (see Sec. 9.4.6).

The derivation of the Drude expression for electrical conductivity in a constant electric field is straightforward. The starting point is Newton's equation of motion for a particle of mass m and charge q (in a metal, $m = m_e^*$ and $q = -e$, but here we will keep the notation general). If only the electric field $\mathbf{E}$ is present (no friction force is considered), the equation of motion is simply $m\ddot{\mathbf{r}} = q\mathbf{E}$. This equation must now be modified by including a friction force to describe electrical resistance. Friction ensures that the charged particles do not continuously accelerate in the presence of external field. The friction may be due to collisions with other particles, e.g., impurities, or with phonons (vibrational degrees of freedom of the atoms comprising the material). The friction force can be modeled as being proportional to the particle velocity and in a direction opposite to the velocity. Thus,

$$m\dot{\mathbf{v}} = q\mathbf{E} - \frac{m}{\tau}\mathbf{v}. \tag{9.42}$$

The friction force is inversely proportional to the average time τ elapsed between two successive collisions of the electrons (for dimensional reasons, a parameter with dimensions of time must appear in the denominator of the friction force). This central quantity depends on the metal purity and temperature. For high-purity copper at liquid helium temperature, $\tau \approx 10^{-9}$ s. Without an external field, the friction term in Eq. (9.42) ensures exponential decay of electron velocity with time. In the presence of an external field, the time-averaged velocity is constant. Consequently, taking the average over sufficiently long times in Eq. (9.42), the average of the acceleration appearing on the LHS of the equation vanishes, $\bar{\dot{v}}=0$, due to the steady-state condition, and we obtain an expression for the averaged velocity of the electron,

$$\bar{v} = \frac{q\tau}{m}E. \qquad (9.43)$$

Hence, the *mobility* (defined as the average velocity per unit electric field strength), is $\mu = |\bar{v}|/|E| = e\tau/m$, and the current per unit area, **J**, of charged particles is

$$J = nq\bar{v} = \frac{nq^2\tau}{m}E \equiv \sigma E, \qquad (9.44)$$

where n is the density of the itinerant charged particles. Note that, by dimensional analysis, **J** has units of statcoulomb s^{-1} cm^{-2} (or in SI units, C s^{-1} cm^{-2}), and the units of the RHS of Eq. (9.44) are in accordance with this. Hence, the *electrical conductivity*, σ, is expressed in s^{-1}, or in SI units (Ohm cm)$^{-1}$. If there is more than one type of charge carrier, $J = \sum_\alpha n_\alpha q_\alpha \bar{v}_\alpha$, where $\bar{v}_\alpha = \frac{q_\alpha \tau_\alpha}{m_\alpha} E$. We have just derived Ohm's law relating the current density with the electric field (or with the potential difference, as demonstrated below) and obtained an expression for the electrical conductivity. For itinerant electrons of charge $-e$, density n_e, and effective mass m_e^*, the Drude expression for the conductivity is

$$\sigma_e = \frac{n_e e^2 \tau}{m_e^*}. \qquad (9.45)$$

Equation (9.44) can be written in a more standard form, $I = V/R$, where R is the resistance, by noting that the (uniform and time independent) electric field E is given by the potential across the system divided by the length of the system, $E = V/L$, and that the current I is the product of the current per unit area times the area, $I = JA$. Substituting these relations in Eq. (9.43) and multiplying by the area A, we find $I = (\sigma A/L)V = V/R$. Hence, the conductance is $\Sigma = (A/L)\sigma$, and its inverse, the resistance, is

$$R = \frac{L}{A}\frac{1}{\sigma} = \frac{L}{A}\frac{m_e^*}{n_e e^2 \tau} \equiv \frac{L}{A}\rho, \qquad (9.46)$$

where we have defined the resistivity ρ, which is the inverse of the conductivity, $\rho = 1/\sigma$. The resistance R is the product of the resistivity ρ and $\frac{L}{A}$, just as the conductance, $\Sigma = R^{-1}$ is given by $\Sigma = A\sigma/L$. The resistivity and conductivity are determined by the properties of the material and temperature. The resistance and conductance are also related to the geometry of the conducting material (length and cross-sectional area).

Problem 9.10

Analytically solve Eq. (9.42) with initial condition $v(0) = v_0$ and show that at times $t \gg \tau$, $v(t) = \frac{q\tau}{m}E$.

Hint: The general solution to $\dot{y} = b - y/\tau$ is $y(t) = e^{-t/\tau}[y_0 + b\int_0^t dt'\, e^{t'/\tau}]$.

The Drude derivation of the conductivity (9.45) does not make any assumptions regarding the collision time τ. Because the resistivity ρ is measurable, the Drude expression can be used to determine τ: $\tau = \frac{m_e^*}{\rho n_e e^2}$. For numerics, it is useful to recall the Wigner–Seitz radius r_s defined by Eq. (9.11). Denoting the resistivity measured in units of μ-Ohm cm by the symbol $\tilde{\rho}$, the Drude relation yields the following estimate for collision time:

$$\tau = \frac{m_e^*}{\rho n_e e^2} = \frac{4\pi m_e^*}{3e^2}\frac{1}{\rho}r_s^3 = \left(\frac{0.22}{\tilde{\rho}}\right)\left(\frac{r_s}{a_0}\right)^3 \times 10^{-14} \text{ s}. \qquad (9.47)$$

In metals, r_s/a_0 ranges between 2 and 6.

9.2 Elementary Theories of Conductivity

The Drude formula can be tested to determine whether the average distance between collisions, the *mean free path* $\ell = \bar{v}\tau$, is commensurate with inter-atomic distances in the solid. Estimating $\bar{v}$ using the statistical equipartition of energy, $m_e^* \bar{v}^2/2 = 3k_B T/2$, one has $1\,\text{Å} < \ell < 10\,\text{Å}$ at room temperature, which is a typical distance between atoms in a metal. This seems to corroborate the Drude idea that collisions occur mainly between electrons and ions executing thermal motion around their equilibrium position. However, these estimates are too crude and naïve. For one thing, the actual values of ℓ are at least an order of magnitude larger at room temperature and three orders of magnitude larger at very low temperatures. Moreover, the actual electron speed (the Fermi velocity v_F, unlike $\bar{v}$ in Drude's theory) is nearly temperature independent. Briefly, although the Drude expression is physically intuitive, the main problems are, the collision time τ is not known and the velocity used is not appropriate. Indeed, when v_F is used instead of $\bar{v}$ and the value of τ is known or determined independently, the Drude expression can serve as a good reference point for more realistic calculations. As an example, consider the case of copper. The mean free path of a conduction electron is defined as $\ell = v_F \tau$, where v_F is the Fermi velocity. For pure copper at liquid helium temperatures, 4 K, $\tau = 2 \times 10^{-9}$ s, whereas at room temperature, 300 K, $\tau = 2 \times 10^{-14}$ s. The Fermi velocity of copper is $v_F = 1.57 \times 10^{-6}$ cm, so $\ell(4\,\text{K}) = 0.3$ cm and $\ell(300\,\text{K}) = 3\,\mu\text{m}$. The conductivity of copper can be calculated by using the electron density $n_e = 8.5 \times 10^{22}$ cm^{-3}, $m_e^* = m_e = 9.1 \times 10^{-28}$ g, $e = 4.8 \times 10^{-10}$ statcoulomb, and Eq. (9.45); we find $\sigma(300\,\text{K}) = 5 \times 10^{17}$ s^{-1}.

Because Drude theory does not tell us anything about the value of τ, it is desirable to test it against an experimentally measurable quantity, which does not depend on τ. As far as electrical transport is concerned, there are at least two such quantities. The first concerns the alternating current (AC) conductivity of a metal, which will be discussed below. As a byproduct, we also calculate the propagation of electromagnetic radiation in metals and the formation of charge density oscillations within the Drude model. A second test involves analyzing the Hall coefficient, which is discussed in Sec. 9.5.6 in relation to the *Hall effect*. In addition to testing the model within the realm of charge transport, there are also τ-independent observables related to heat transfer, such as the Wiedemann–Franz ratio and the thermopower, which will be analyzed below within the Drude model.

The Drude model is solely based on classical concepts. But, as we shall soon see, quantum mechanics plays a crucial role in transport phenomena, and the Drude model must be modified to account for quantum considerations required to correctly estimate the collision time τ. As we shall see in this chapter, other dramatic manifestations of quantum mechanics in electronic transport abound. One of the most dramatic is *superconductivity*. For some materials (e.g., aluminum, lead, mercury, niobium, vanadium), as temperature is lowered below a certain (material dependent) *critical temperature* T_c, the resistance decreases sharply to zero [$\rho < 10^{-23}$ Ohm cm, which is 14 orders of magnitude below the resistivity of pure copper at 4 K, $\rho_{\text{Cu}}(4\,\text{K}) = 10^{-9}$ Ohm cm], and the material is said to become superconducting. Superconductivity was discovered in 1911 by H. Kamerlingh-Onnes in mercury at temperatures below the critical temperature $T_c = 4.2$ K. It will be discussed in Chapter 14 (in the supplementary material on the book web page).

AC Electrical Conductivity of a Metal

Now consider the response of a metal to a uniform time-dependent monochromatic electrical field within the Drude model. For convenience, the electric field at frequency ω assumes a complex form,

$$\mathbf{E}(t) = \mathbf{E}(\omega) e^{-i\omega t}, \tag{9.48}$$

with the understanding that the real part of this quantity is taken at the end of the calculation. Equation (9.42) can be solved in a steady state by setting $\mathbf{v}(t) = \mathbf{v}(\omega) e^{-i\omega t}$ and substituting this form, and Eq. (9.48), into (9.42) to obtain

$$\mathbf{v}(\omega) = \frac{q\tau/m}{1 - i\omega\tau} \mathbf{E}(\omega). \tag{9.49}$$

The resulting (complex) current, $\mathbf{J}(t) = \mathbf{J}(\omega) e^{-i\omega t}$, is proportional to the electric field, and the proportionality constant is the AC conductivity $\sigma(\omega)$,

$$\mathbf{J}(\omega) = nq\mathbf{v}(\omega) = \sigma(\omega)\mathbf{E}(\omega), \tag{9.50}$$

with

$$\sigma(\omega) = \frac{\sigma_0}{1 - i\omega\tau}, \quad \sigma_0 = \frac{nq^2\tau}{m}. \tag{9.51}$$

The measurable current is given by Re[$\mathbf{J}(t)$]. Equation (9.51) and the expression for the *complex AC conductivity* $\sigma(\omega)$ play important roles in the analysis of propagation of electromagnetic radiation in matter (magnetic field effects are usually much smaller than those of the electric field). Although, in many cases, the electric field is not uniform and also depends on space, $\mathbf{E}(\mathbf{r}, t)$, when the wavelength λ of the electric field is much larger than the *mean free path*, $\ell \equiv v\tau$, it is reasonable to relate the Fourier components of the local current density and the electric field as follows:

$$\mathbf{J}(\mathbf{r}, \omega) = \sigma(\omega)\,\mathbf{E}(\mathbf{r}, \omega). \tag{9.52}$$

This can generally be justified when visible light, with wavelength 400 nm $< \lambda <$ 700 nm, or shorter wavelength light, propagates in a metal. Equation (9.52) is called the *Ohm's law constitutive equation*.

Propagation of Electromagnetic Waves in Metals

Equation (9.52), together with the Faraday and Ampére equations,

$$\nabla \times \mathbf{E} = -\frac{1}{c}\frac{\partial \mathbf{H}}{\partial t}, \quad \nabla \times \mathbf{H} = \frac{4\pi}{c}\mathbf{J} + \frac{1}{c}\frac{\partial \mathbf{E}}{\partial t}, \tag{9.53}$$

can be used to derive a wave equation for electromagnetic radiation in *plasmas* (a plasma is a conducting fluid, and an electron gas certainly qualifies). Taking the time dependence of the electric field to be $\mathbf{E}(\mathbf{r}, t) = \mathbf{E}(\mathbf{r})\,e^{-i\omega t}$, and using Eq. (9.52), we obtain the Helmholtz equation (see Problem 9.11),

$$\nabla^2 \mathbf{E} = -\frac{\omega^2}{c^2}\left(1 + \frac{4\pi i\sigma(\omega)}{\omega}\right)\mathbf{E} \equiv -\frac{\omega^2}{c^2}\epsilon(\omega)\mathbf{E}. \tag{9.54}$$

Here, $\epsilon(\omega)$ is the complex dielectric function. By using the Drude expression (9.51), and taking the high-frequency limit $\omega\tau \gg 1$, we find

$$\epsilon(\omega) = 1 - \frac{\omega_p^2}{\omega^2}, \quad \omega_p \equiv \sqrt{\frac{4\pi n_e e^2}{m_e^*}}, \tag{9.55}$$

where ω_p is the *plasma frequency*. For frequencies $\omega > \omega_p$, the propagation of the field within the metal is not attenuated. In the high-frequency limit, the dielectric constant $\epsilon(\omega)$ is real and negative (positive) for $\omega < \omega_p$ ($\omega > \omega_p$), and the corresponding solution of Eq. (9.54) decays (propagates). The plasma frequency does not depend on τ (but the high-frequency limit does). To obtain an estimate on the value of $\omega_p\tau$ for electrons in metals, we can express τ in terms of the resistivity through Eq. (9.47),

$$\omega_p\tau = 160\left(\frac{r_s}{a_0}\right)^{\frac{3}{2}}\left[\frac{\mu\Omega\text{-cm}}{\tilde{\rho}}\right], \tag{9.56}$$

where $\tilde{\rho}$ denotes the resistivity in units of μ-Ohm cm [see Eq. (9.47)]. For alkali metals, the frequency above which transparency is established is of the order of 8×10^{15} Hz. The wavelength λ_p of the electromagnetic field below which the metal is transparent is of the order of 300 nm.

Problem 9.11

(a) Derive the wave equation by taking the curl of Faraday's equation and using Ampére's equation on the RHS of the resulting equation.

(b) Now derive the Helmholtz equation using $\mathbf{E}(\mathbf{r}, t) = \mathbf{E}(\mathbf{r})\,e^{-i\omega t}$ and Eq. (9.52).

9.2.2 THERMAL CONDUCTIVITY OF METALS

Electrons in a metal can transport thermal energy from one side of the sample to the other. The *thermal current density*, $\mathbf{J}^q(\mathbf{r}, t)$, at a point $\mathbf{r}$ and time t is defined such that $\mathbf{J}^q(\mathbf{r}, t) \cdot d\mathbf{A}$ is the energy per unit time, which at time t crosses a small area $d\mathbf{A}$ centered at $\mathbf{r}$. The dimension of $\mathbf{J}^q$ is energy/(area − time). Whereas charge current results from application of an external electric field, i.e., a potential gradient (see Fig. 9.7), a thermal current results by applying a temperature gradient across the material. To produce a thermal current in a metal, an external temperature gradient can be maintained across the metal by heating one end and/or cooling the other end. For a modest temperature gradient ∇T, the *Fourier law of heat conduction* states that the thermal current is given by the linear response equation,

$$\mathbf{J}^q = -\kappa \nabla T, \tag{9.57}$$

where κ is the *thermal conductivity coefficient*; it has dimension $[\kappa]$ = energy area^{-1} time^{-1} K^{-1}. The minus sign on the RHS of Eq. (9.57) ensures that thermal current is directed from hot to cold for $\kappa > 0$.

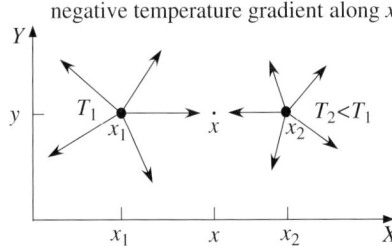

FIG 9.8 Schematic illustration of heat transfer within the Drude model. Electrons emerge from collisions with atoms 1 and 2 in all directions at energies $\varepsilon[T(x_1)] > \varepsilon[T(x_2)]$ (indicated by arrow length). This inequality leads to a thermal current from left to right. The thermal current is calculated in the text at point x, $x_1 < x < x_2$.

Metals conduct heat better than insulators because the dominant contribution to thermal current is carried by electrons, and electrons are free to move through metals. The *Wiedemann–Franz law* states that the ratio κ/σ of thermal to electrical conductivities is directly proportional to temperature with a universal (material independent) proportionality constant. This suggests that the Drude model can be used to study heat transfer in metals in the same way as it is applied for charge transfer. The mechanism of heat conduction is elucidated with the help of Fig. 9.8, which, for simplicity, is drawn for a two-dimensional sample. The calculation focuses attention on the x component of the thermal current. A temperature gradient is maintained across the sample from left to right. The energy ε of electrons at any point in the sample solely depends on the temperature and, hence, on the x coordinate, i.e., $\varepsilon(x) = \varepsilon[T(x)]$. The average electron velocity $v(x)$ should also depend on x, but for moderate temperature gradients, the deviation of $v(x)$ from the average thermal equilibrium velocity v is extremely small. Electrons that have their last collision with ions 1 and 2 located at points x_1 and x_2, respectively, emerge after the collision in either left or right direction with equal probability and reach the point $x_1 < x < x_2$. Our aim is to calculate the thermal current J_x^q at x. By definition, $x = x_1 + v\tau = x_2 - v\tau$. The contribution of electrons emerging rightward (leftward) from x_1 (x_2) to the thermal current at x is given by the product of their density $n/2$, their velocity v ($-v$), and their energy $\varepsilon[T(x_1)]$ ($\varepsilon[T(x_2)]$). Combining the two opposite contributions,

$$J_x^q = \frac{n}{2} v \{\varepsilon[T(x_1)] - \varepsilon[T(x_2)]\} = \frac{n}{2} v \{\varepsilon[T(x - v\tau)] - \varepsilon[T(x + v\tau)]\}. \tag{9.58}$$

Expanding $\varepsilon[T(x \pm v\tau)]$ around x up to first order yields

$$\mathbf{J}^q \approx -n v^2 \tau \frac{d\varepsilon}{dT} \frac{dT}{dx} \hat{\mathbf{x}}. \tag{9.59}$$

This relation can be easily extended to the case of a 3D sample. Instead of the factor v^2, applicable in one dimension, one uses the 3D averaged velocity squared, $\langle v^2 \rangle/3$. Moreover, recalling that $\frac{d\varepsilon}{dT} = c_{\mathrm{el}}$, the electron contribution to the specific heat, we arrive at the expression

$$\mathbf{J}^q = -\frac{1}{3} c_{\mathrm{el}} \tau \langle v^2 \rangle \nabla T, \tag{9.60}$$

which, using Eq. (9.57), gives the thermal conductivity,

$$\kappa = \frac{1}{3} c_{el} \tau \langle v^2 \rangle = \frac{1}{3} c_{el} \ell v, \qquad (9.61)$$

where $\ell = v\tau$ is the mean free path. This heuristic argument is discussed in many textbooks, e.g., Ref. [122]. A more careful analysis to be carried out in Sec. 9.4.9 confirms that Eq. (9.59) is sound, provided its constituents are microscopically well defined. Using expressions (9.45) and (9.61) for the electrical and thermal conductivities in the Drude model, the ratio between thermal and electrical conductivities is obtained as,

$$\frac{\kappa}{\sigma} = \frac{m \langle v^2 \rangle c_{el}}{3ne^2}, \qquad (9.62)$$

which is independent on the collision time τ. Assuming the electrons behave as a classical ideal gas, $m\langle v^2 \rangle /2 = 3k_B T/2$ and $c_{el} = c_{cl} = 3nk_B/2$, can be inserted into the RHS of (9.62), to obtain the Drude version of the *Wiedemann-Franz law*,

$$\frac{\kappa}{\sigma T} = \frac{3}{2} \left(\frac{k_B}{e} \right)^2 = 1.24 \times 10^{-13} \left(\frac{\text{erg}}{\text{statcoul K}} \right)^2. \qquad (9.63)$$

This result is about half of that found experimentally. This might be regarded as a great success of the Drude model, but in fact, this is rather misleading. At room temperature, the electronic contribution to the specific heat is about two orders of magnitude smaller than $3nk_B/2$, whereas the electronic velocity squared is about two orders of magnitude larger than $3k_B T/m$. We shall see below that the application of classical ideal gas theory to the problem of itinerant electrons in metals is not justified. In the derivation of the Wiedemann–Franz law, these two errors for the values of the specific heat and electron velocity squared compensate each other. But there is at least one experimental quantity, which can be calculated within the Drude model for which such a cancelation does not occur, as we now discuss.

The Thermo-Electric (Seebeck) Effect

If a temperature gradient is maintained in a long finite wire, then an electric field directed opposite to the temperature gradient develops, as schematically illustrated in Fig. 9.9. This is called the *thermo-electric* or *Seebeck effect*.

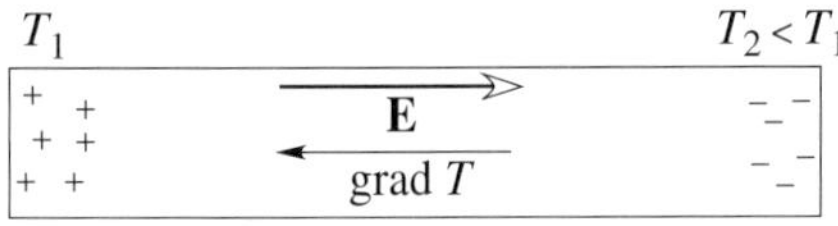

In steady state, excess electrons accumulate at the cold end of the wire and an excess of positive charge accumulates on the other side of the wire. No additional global charge transport occurs within the wire. The velocity $\mathbf{v}_T$ of the electrons due to the temperature gradient must be compensated by the velocity $\mathbf{v}_E$ of electrons due to the resulting electric field $\mathbf{E}$, such that $\mathbf{v}_T + \mathbf{v}_E = 0$. The relation between $\mathbf{E}$ and ∇T is given in linear response by,

$$\mathbf{E} = Q \nabla T, \qquad (9.64)$$

where $Q(<0)$ is referred to as the *thermopower*.

In analyzing this effect, the assumption of weak dependence of the velocity on the position of the electron is retained (as in the calculation of κ discussed in relation to Fig. 9.8). Thus, $|\mathbf{v}_1| > |\mathbf{v}_2|$ in that figure, as discussed above. In a long 1D wire, the average electronic velocity at a point x due to the temperature gradient is

FIG 9.9 The Seebeck effect. In a long finite metallic wire whose left and right ends are kept at temperatures $T_1 > T_2$ (the temperature gradient is directed leftward), a transient thermal current is generated as given by Eq. (9.60). When steady state is reached, an excess negative (positive) charge is accumulated on the right (left). Because no additional charge transport occurs, an electric field opposite to the temperature gradient results.

$$v_T = \frac{1}{2}[v_1 - v_2] = \frac{1}{2}[v(x - v\tau) - v(x + v\tau)] = -v\tau \frac{dv}{dx} = -\tau \frac{d}{dx}\left[\frac{1}{2}\langle v^2 \rangle\right]. \qquad (9.65)$$

9.2 Elementary Theories of Conductivity

This expression can be easily generalized to a three-dimensional sample, with the result,

$$\mathbf{v}_T = -\frac{\tau}{6}\frac{d\langle v^2\rangle}{dT}\nabla T. \quad (9.66)$$

The velocity $\mathbf{v}_E$ is proportional to the electric field $\mathbf{E}$ as in Eq. (9.43), although here the electric field is generated inside the system due to the temperature gradient,

$$\mathbf{v}_E = -\frac{e\tau}{m}\mathbf{E}. \quad (9.67)$$

The requirement, $\mathbf{v}_T + \mathbf{v}_E = 0$, enables expressing $\mathbf{E}$ in terms of ∇T, and by using $(nm/2)d\langle v^2\rangle/dT = c_{el}$, we derive the following expression for the thermopower,

$$Q = -\frac{c_{el}}{3ne}. \quad (9.68)$$

When the (classical) ideal gas expression $c_{el} = c_{cl} = 3nk_B/2$ is used, we find, $Q = -\frac{k_B}{2e} = -0.43 \times 10^{-4}$ volt K^{-1}. This estimate is two orders of magnitude larger than the experimentally observed value. The reason is that Q is proportional to the specific heat; as already explained, the classical expression for the electronic specific heat overestimates the electronic specific heat by two orders of magnitude. The reason why the Drude model estimate of Q is two order of magnitudes larger than experiment, while the Drude model estimate of the Wiedemann-Franz law, Eq. (9.63), almost agree, is that in the Wiedemman-Franz law, the ratio κ/σ is proportional to $c_{el}\langle v^2\rangle$, and the two orders of magnitude overestimate of c_{el} is compensated for by the two order of magnitude underestimate of $\langle v^2\rangle$. In the analysis of Q, only the overestimated specific heat appears whereas the underestimated $\langle v^2\rangle$ does not appear as a factor. Hence, quantum mechanics must be used in analysis of electrical and thermal conductivities.

9.2.3 SOMMERFELD THEORY OF TRANSPORT IN METALS

The inadequacy of the Drude model for describing transport properties of electrons in metals stems from the fact that quantum aspects of the electron occupation of states are completely ignored. The Sommerfeld theory takes account of the Fermi statistics of the electrons and thereby corrects the Drude theory. It significantly affects the estimates of the electron velocity. The classical Maxwell–Boltzmann velocity distribution function [119, 120], $f_{MB}(\mathbf{v})d\mathbf{v}$, gives the probability that the particle velocity lies between $\mathbf{v}$ and $\mathbf{v} + d\mathbf{v}$:

$$f_{MB}(\mathbf{v}) = n\left(\frac{m}{2\pi k_B T}\right)^{\frac{3}{2}} e^{-\frac{mv^2}{2k_B T}}. \quad (9.69)$$

Hence, the density of particles with velocity $\mathbf{v}$ is $n_\mathbf{v} = nf_{MB}(\mathbf{v})$, where $n = N/\mathcal{V}$ is the particle density. This distribution leads to an average velocity $\langle v\rangle = \sqrt{3k_B T/m}$ and specific heat $c_{el} = 3nk_B/2$. The use of these quantities in the Drude theory is inadequate, as discussed earlier. To correct this, Sommerfeld used the conceptual framework of the Drude model [including the basic relation (9.45) for the conductivity], but used the Fermi–Dirac velocity distribution (instead of the Maxwell–Boltzmann distribution) to calculate the average velocity and the specific heat for a free electron gas, as discussed in Sec. 9.1. To find the velocity distribution of electrons in a metal as deduced from Fermi statistics, consider a small volume $d\mathbf{k}$ in $\mathbf{k}$-space, such that the variation of the Fermi–Dirac function, Eq. (9.27), within $d\mathbf{k}$ is negligible. As we have seen in Sec. 9.1, the number of single-electron states in this $\mathbf{k}$-space volume element is $2 \times \frac{\mathcal{V}}{(2\pi)^3}d\mathbf{k}$, including spin degeneracy. The probability that a level $\mathbf{k}$ is occupied is $f(E_\mathbf{k})$, where $f(E)$ is the Fermi function, Eq. (9.27), and for a free electron gas, $E_\mathbf{k} = \frac{\hbar^2 k^2}{2m}$. Therefore, the density (number per volume in physical space) of electrons in this $\mathbf{k}$-space volume element is

$$dn_\mathbf{k} = \frac{n}{4\pi^3}f(E_\mathbf{k})\,d\mathbf{k}. \quad (9.70)$$

Because we are interested in the velocity distribution, we explicitly set $E_\mathbf{k} = \hbar^2 k^2/(2m)$ and note that $d\mathbf{k} = \frac{\partial \mathbf{k}}{\partial \mathbf{v}} d\mathbf{v} = \frac{m^3}{\hbar^3} d\mathbf{v}$. Hence, the probability distribution for the electron velocity is

$$f(\mathbf{v}) = \frac{m^3}{4\pi^3 \hbar^3} \frac{1}{\exp\left(\frac{mv^2/2 - \mu}{k_B T}\right) + 1}, \tag{9.71}$$

rather than the classical Maxwell–Boltzmann distribution (9.69). In Sommerfeld theory, the expressions for the electrical conductivity σ and the thermal conductivity κ remain the same as those derived using Drude theory, but the average velocity and specific heat are not those derived from the classical ideal gas theory, $\langle v \rangle = \sqrt{3 k_B T / m}$, and $c_{el} = c_{cl} = \frac{3}{2} n k_B$, but those derived from the quantum theory of the free electron gas as discussed in Eq. (9.40),

$$v \equiv \langle v \rangle = \int v f(\mathbf{v}) d\mathbf{v} = \sqrt{\frac{2 E_F}{m}}, \quad c_{el} = \frac{\pi^2}{2} \frac{k_B T}{E_F} n k_B. \tag{9.72}$$

Comparing Sommerfeld Theory with Experiment

The results derived from Eq. (9.72) can now be compared with experiment once the collision time τ, which is an input parameter into both the Drude and the Sommerfeld theory, is ascertained. The mean free path is given by $\ell = v_F \tau$, and by using Eq. (9.47) for τ and Eq. (9.12b) for v_F, we find,

$$\ell = 92 \left(\frac{r_s}{a_0}\right)^2 \frac{1}{\tilde{\rho}} \quad [\text{Å}]. \tag{9.73}$$

With the estimate, $\tilde{\rho} \approx 10$ at room temperature, and $r_s/a_0 \approx 3$, the mean free path is about 100 Å.

Next, consider the Wiedemann–Franz ratio; with $\sigma = \frac{ne^2 \tau}{m}$ and $\kappa = v^2 \tau c_{el}/3$, as in Drude theory, but with $v_F = \sqrt{2 E_F/m}$ and c_{el} given by Eq. (9.40), the ratio is

$$\frac{\kappa}{\sigma T} = \frac{\pi^2 k_B^2}{3 e^2} = 2.44 \times 10^{-8} \text{ W} \Omega \text{ K}^{-2}. \tag{9.74}$$

The number on the RHS is referred to as the *Lorentz number*. It is in good agreement with experiment (recall that the Drude's result is also close to experiment, but that is misleading due to over-estimation of the specific heat and under-estimation of the average velocity).

Finally, consider the thermopower Q and make use of Eq. (9.40) for the specific heat in the Sommerfeld theory to obtain

$$Q = -\frac{\pi^2 k_B}{6 e}\left(\frac{k_B T}{E_F}\right) = -1.42 \times 10^{-4} \left(\frac{k_B T}{E_F}\right) \text{ V K}^{-1}. \tag{9.75}$$

This result compares well with experiment, being smaller than the Drude value, Eq. (9.68), by a factor $k_B T/E_F$, which is about 10^{-2} at room temperature.

9.3 CRYSTAL STRUCTURE

A crystal is a solid containing a large number of atoms, molecules, or ions that form a periodic structure, i.e., the constituents, which we refer to as atoms, are packed in a regularly ordered, repeating pattern. Apart from their small oscillations, that are ignored for the time being, the position of the atoms is fixed in space. Crystal structures can be generated starting from a group of atoms, called the *basis*, or the *motif* by crystallographers, and regularly ordering this group in a periodically repeating pattern in 3D, 2D, or 1D structures, as shown in Fig. 9.10 for a 2D crystal structure. In this figure, the basis (motif) consists of two atoms, schematically represented by a circle and a square, and it repeats in a periodic

9.3 Crystal Structure

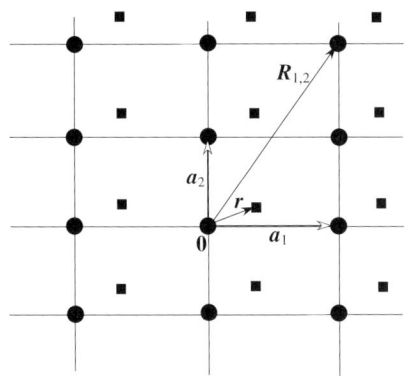

FIG 9.10 A two-dimensional crystal constructed from a rectangular Bravais lattice (the solid circles arranged in a rectangular pattern) and a basis consisting of two different atoms (represented as solid circles and squares). The two primitive lattice vectors, $\mathbf{a}_1$ and $\mathbf{a}_2$, define the primitive rectangular unit cell. Bravais lattice vectors take the form $\mathbf{R}_{n_1,n_2} = n_1\mathbf{a}_1 + n_2\mathbf{a}_2$, e.g., $\mathbf{R}_{1,2} = \mathbf{a}_1 + 2\mathbf{a}_2$. The position vectors of the two atoms forming the basis with reference to the Bravais lattice point are $\mathbf{0}$ and $\mathbf{r} = x\mathbf{a}_1 + y\mathbf{a}_2$ with $|x|, |y| \leq 1$.

pattern. We are most familiar with crystals such as diamonds, rock salts, quartz, and gemstones, with clearly visible planar facets and sharp edges, but a myriad of solid-state materials, including metals, have a crystal structure.

The purpose of this section is to introduce the basic geometrical tools required for the analysis of crystal structures. A central concept in this respect is that of a *lattice*, an infinite periodic arrangement of identical groups of points. Each group may contain several points, e.g., in Fig. 9.10, each group contains two points. If it contains more than a single point, the lattice has a basis (or a motif). Although the concepts of crystal and lattice are sometimes used interchangeably, we find it useful to reserve the former for the physical system of atoms and molecules having a finite (albeit large) extent and the latter for the geometrical structure of periodically arranged points, which is regarded as an infinite array. Lattices are defined in three, two, and one space dimensions.

9.3.1 BRAVAIS LATTICES AND CRYSTAL SYSTEMS

The theory of space lattices was introduced by the French physicist Auguste Bravais in the middle of the 19th century. A 3D (simple) *Bravais lattice* is constructed in terms of a triad of noncoplanar vectors $\mathbf{a}_1$, $\mathbf{a}_2$, and $\mathbf{a}_3$, which are called as *primitive lattice vectors*. It consists of the collection (an infinite set) of points defined by position vectors $\{\mathbf{R}_{n_1 n_2 n_3}\}$, including a point $\mathbf{R}_{000} = \mathbf{0}$ (arbitrarily chosen to be used as an origin),

$$\mathbf{R}_{n_1,n_2,n_3} \equiv n_1\mathbf{a}_1 + n_2\mathbf{a}_2 + n_3\mathbf{a}_3 \quad (n_1, n_2, n_3 \in \text{integers}). \tag{9.76}$$

The pattern of points in a Bravais lattice looks identical when viewed from any two points $\mathbf{r}$ and $\mathbf{r}'$, not necessarily belonging to the lattice, if and only if the difference vector $\mathbf{r} - \mathbf{r}'$ is expressible as a vector $\mathbf{R}_{n_1,n_2,n_3}$ as defined in Eq. (9.76). Sometimes, the subscripts n_1, n_2, and n_3 will be omitted when no confusion might arise.

The vectors $\mathbf{R}$ given in Eq. (9.76) define the *Bravais lattice points*. From this definition, it is clear that the difference vectors $\mathbf{R} - \mathbf{R}'$, referred to as *Bravais lattice vectors*, is also expressible as in Eq. (9.76). As a periodic structure, a 3D Bravais lattice has three *lattice periods* $\mathbf{a}_1$, $\mathbf{a}_2$, and $\mathbf{a}_3$ in the sense that the lattice looks the same from a point $\mathbf{r}$ and a point $\mathbf{r} + \mathbf{a}_i$, or a point $\mathbf{r} + \sum_i n_i \mathbf{a}_i$ for any integers n_i. Any experimentally observable quantity $O(\mathbf{r})$ depending on space point $\mathbf{r}$ is a periodic function $O(\mathbf{r} + \sum_i n_i \mathbf{a}_i) = O(\mathbf{r})$, where $i = 1, 2, 3$. This implies $O(\mathbf{r} + \mathbf{R}) = O(\mathbf{r})$ for any lattice vector $\mathbf{R}$. The lengths $|\mathbf{a}_1|$, $|\mathbf{a}_2|$, and $|\mathbf{a}_3|$ are called *lattice constants*. The lattice constants and the three angles defined by $\cos\theta_i = \hat{\mathbf{a}}_j \cdot \hat{\mathbf{a}}_k$, $i \neq j \neq k$, constitute the six *lattice parameters*. Recall that a grouping of atoms that repeats itself in a crystal is referred to as a basis (motif). We may augment a simple Bravais lattice into a *Bravais lattice with a basis* by attaching to each point $\mathbf{R}$, a finite number J of points (i.e., a motif) whose position vectors $\mathbf{r}_j$, $j = 1, 2, \ldots, J$ are defined with respect to the lattice point $\mathbf{R}$, and expressed as linear combinations of the same three basis vectors $\mathbf{a}_1$, $\mathbf{a}_2$, and $\mathbf{a}_3$, albeit with noninteger real coefficients (see Fig. 9.10),

$$\mathbf{r}_j = x_j\mathbf{a}_1 + y_j\mathbf{a}_2 + z_j\mathbf{a}_3, \quad |x_j|, |y_j|, |z_j| \leq 1, \quad j = 1, 2, \ldots, J. \tag{9.77}$$

A *unit cell* is a volume element $\mathcal{V}$ in a lattice with the following properties: (1) It contains at least one Bravais lattice point $\mathbf{R}$; (2) the entire space can be filled by such closely packed identical volume elements $\{\mathcal{V}_i\}$, $i = 1, 2, \ldots$. Two unit cell volume elements $\mathcal{V}_i$ and $\mathcal{V}_j$ are related to each other by simple translation. It is useful to add a third

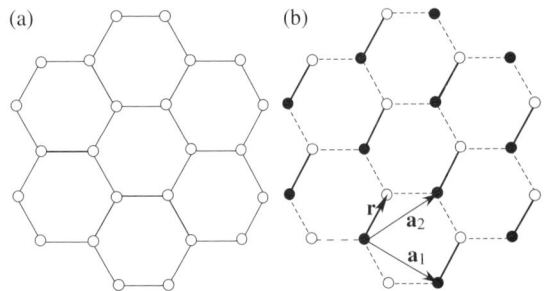

FIG 9.11 (a) A 2D honeycomb lattice. You will answer the question whether it is a simple Bravais lattice in Problem 9.12(a). (b) A 2D honeycomb lattice can be regarded as a triangular Bravais lattice with a basis (motif). Its relation to a Bravais lattice with a basis is a part of Problem 9.12(b). The Bravais lattice is composed of the black circles, and in each lattice point, there are two atoms, black and white connected by a vector **r**. The dashed lines connecting black and white atoms serve to underline the honeycomb structure. In 2004, a single layer of graphite was fabricated, wherein the carbon atoms (empty circles) were arranged on the vertices of a 2D honeycomb lattice. This material is called *graphene*, and its properties are discussed in Chapter 13.

property: (3) A unit cell has the point group symmetry of the lattice. For example, it may be constructed as a parallelepiped with volume $\mathcal{V} = (m_1 \mathbf{a}_1 \times m_2 \mathbf{a}_2) \cdot m_3 \mathbf{a}_3$, with integer m_i.

A *primitive unit cell* is a unit cell which contains just a single Bravais lattice point **R**, and, by definition, also the atoms in the motif attached to it. An example of a primitive unit cell is a parallelepiped defined by taking $m_1 = m_2 = m_3 = 1$ in the example above. Its volume is

$$\mathcal{V}_{cell} = |(\mathbf{a}_1 \times \mathbf{a}_2) \cdot \mathbf{a}_3|. \tag{9.78}$$

The concept of a primitive unit cell is also meaningful for 2D and 1D lattice structures. In 2D, a primitive unit cell is an aerial element containing just one Bravais lattice point. From these definitions, it is evident that a primitive unit cell is the minimal object satisfying properties 1 and 2. If it satisfies property 3 as well, it can be regarded as the "atom" of the lattice. Because the lattice volume is tiled by unit cells, the volume of all unit cells is the same (independent of their shape), and as given by Eq. (9.78).

Problem 9.12

(a) With the help of Fig. 9.11(a), check whether a 2D honeycomb lattice is a simple Bravais lattice. Answer: No.
(b) With the help of Fig. 9.11(b) show that the 2D honeycomb lattice can be regarded as a Bravais lattice with a basis (motif) containing two atoms in a primitive unit cell. Write down the two components of $\mathbf{a}_1, \mathbf{a}_2$, and **r** in a Cartesian system (feel free to choose the origin and axes at your convenience).

Nearest Neighbors and Coordination Number

For a given Bravais lattice point **R**, the lattice points closest to it are called *nearest neighbor points* and their number, p is the *coordination number*. More precisely, we say that the lattice point $\mathbf{R}_m$ ($m = 1, 2, \ldots, p$) is a nearest neighbor of **R** if the unit cells containing **R** and $\mathbf{R}_m$ touch each other in at least one point.

Problem 9.13

Find the coordination number for the simple cubic, body-centered, and face-centered lattices.

Answer: 6, 8, and 12.

Classification of Lattices

In 1850, Bravais demonstrated that there are 14 different types of Bravais lattices in three dimensions (they will be enumerated below). In two dimensions, there are just five different types of Bravais lattices: oblique, rectangular, centered rectangular, hexagonal, and square. The five 2D Bravais lattices are shown in Figure 9.12.

Because of the periodic structure of the lattice, symmetry operations exist that leave it invariant. These operations can be divided into *translations, rotations, inversions, and reflections*. Mathematically, these symmetry operations form

9.3 Crystal Structure

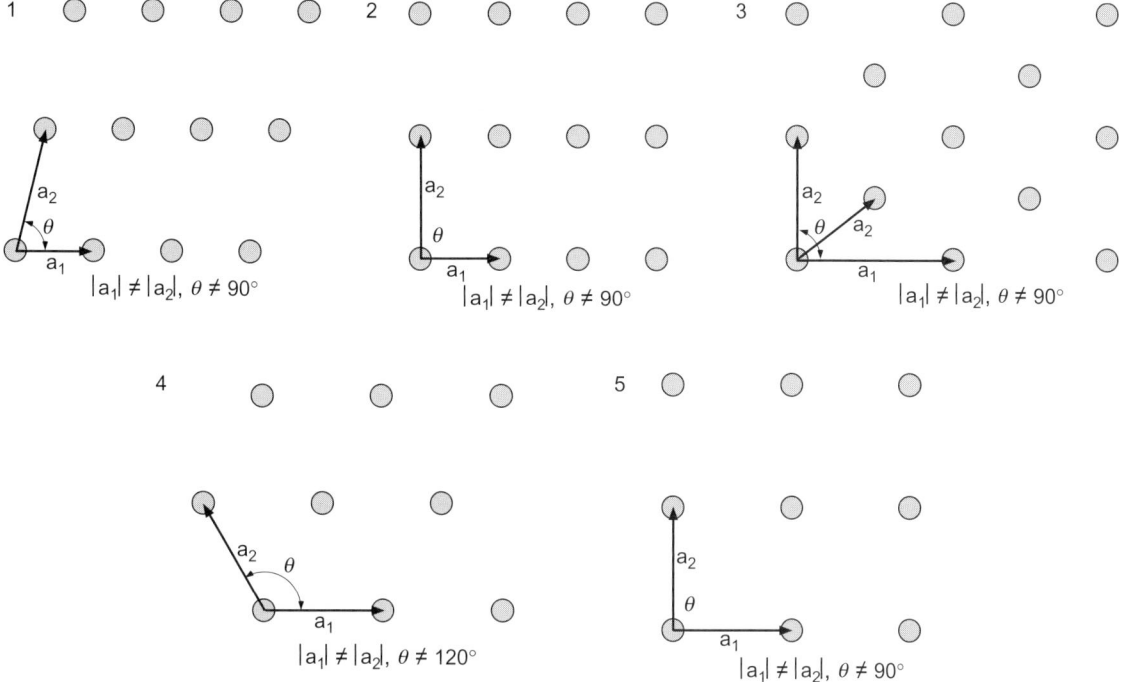

FIG 9.12 The five two-dimensional Bravais lattices in 2D: oblique, rectangular, centered rectangular, hexagonal, and square. For the centered rectangular Bravais lattice, the vectors a_1 and a_2' (not a_2) form a basis.

a finite group (for each given lattice structure), referred to as a *space group* whose properties can be analyzed in terms of (discrete) group theory. For a brief review of the group theory aspects of crystals, i.e., point groups and space groups, see Appendix E.1 and Refs [123, 124].

A *point group* is a group of symmetry operations that leave a point fixed. In what follows, we restrict ourselves to consider only discrete point groups. The elements of a point group are constructed from combinations of eight elementary symmetry elements, as discussed in Appendix E.1 and are defined as follows: E (identity element), C_2 (rotation by 180°), C_3 (120° rotation), C_4 (90° rotation), C_6 (60° rotation), i (inversion), σ_v (reflection in a plane passing through an axis of symmetry), and σ_h (reflection in a plane perpendicular to an axis of symmetry). These symmetry elements form 32 possible point groups, and they correspond to 32 crystal classes. The lattice classes are divided into seven lattice systems (cubic, hexagonal, trigonal, tetragonal, orthorhombic, monoclinic, and triclinic) as follows:

1. Cubic – three mutually perpendicular equal axes. The cubic system is often called isometric.
2. Hexagonal – four axes, a_1, a_2, a_3, and c, where a_1, a_2, and a_3 lie in a plane and are of equal length and intersect at angle of 120°, and the c is perpendicular to the plane. (a_1, a_2, a_3 are not linearly independent).
3. Trigonal – four axes, a_1, a_2, a_3, and c, where a_1, a_2, and a_3 lie in a plane and are of equal length and intersect at angles of 60°, and the c is perpendicular to the plane. This system is sometimes included with the hexagonal system. It is often called rhombohedral. (a_1, a_2, a_3 are not linearly independent).
4. Tetragonal – three mutually perpendicular axes, two of which are of equal length.
5. Orthorhombic – three mutually perpendicular unequal length axes.
6. Monoclinic – three unequal length axes, two intersecting at acute or obtuse angles and the third (b) axis perpendicular to the other two axes.
7. Triclinic – three unequal length axes that intersect at acute or obtuse angles.

Table 9.1 *Crystal systems and point groups.*

Crystal system (No. lattices) [lattice type]	Crystal point group (Schönflies)	Crystal point group (international)
Cubic (3) [P, I, F]	T	23
	O	432
	T_h	m3
	T_d	$\bar{4}$3m
	O_h	m3m
Hexagonal (1) [P]	C_6	6
	D_6	622
	C_{6v}	6mm
	C_{6h}	6/m
	D_{6h}	$\bar{6}$m2
	C_{3h}	$\bar{6}$
	D_{3h}	$\bar{6}$m2
Trigonal [rhombohedral] (1) [R]	C_3	3
	D_3	32
	C_{3i}	$\bar{3}$
	C_{3v}	3m
	D_{3d}	$\bar{6}$m
Tetragonal (2) [P, I]	C_4	4
	D_4	422
	S_4	$\bar{4}$
	C_{4h}	4/m
	C_{4v}	4mm
	D_{2d}	$\bar{4}$2m
	D_{4h}	4/mmm
Orthorhombic (4) [P, I, C, F]	D_2	222
	C_{2v}	mm2
	D_{2h}	mmm
Monoclinic (2) [P, C]	C_2	2
	C_s	m
	C_{2h}	2/m
Triclinic (1) [P]	C_1	1
	C_i	$\bar{1}$

The 14 Bravais lattices in 3D (which are also enumerated in the first column of Table 9.1) are shown in Fig. 9.13. They can be constructed following the seven geometric classifications, i.e., the seven crystal systems. For example, the cubic system has three types of crystal space lattices: the simple cubic (given the symbol P for primitive), the body-centered cubic (given the symbol I and sometimes represented as BCC), and the face-centered cubic (given the symbol F and sometimes represented as FCC). The simple cubic lattice has an atom at the corners of a cube; the body-centered cubic has, in addition, an atom at the center of the cube; and the face-centered cubic has atoms at the center of each face of the cube.

The composition of two identical Bravais lattices shifted with respect to each other is, in general, not a simple Bravais lattice. Rather, it is a Bravais lattice with two points per unit cell. An important example is the diamond structure obtained from two FCC lattices shifted by 1/4 of the diagonal along the diagonal. Figure 9.14(a) shows an FCC lattice, and Fig. 9.14(b) shows two shifted FCC lattices forming the diamond structure. A familiar representative is zinc-sulfide (ZnS), where the zinc atoms occupy the sites of one FCC and the sulfur atoms occupy the sites of the second FCC.

9.3 Crystal Structure

> **Problem 9.14**
>
> Express all the FCC lattice sites in Fig. 9.14(a) in terms of the basis vectors $\mathbf{a}_1, \mathbf{a}_2$, and $\mathbf{a}_3$ connecting the origin 0 with the sites marked 1, 2, and 3.

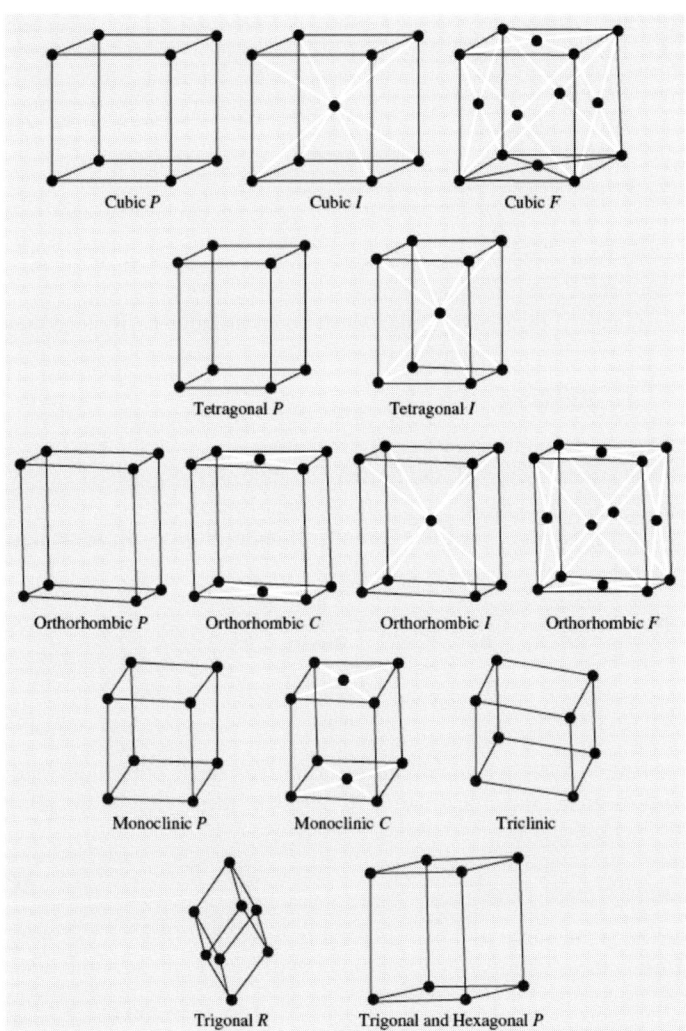

FIG 9.13 The 14 conventional unit cells of the space Bravais lattices.

Space Groups

Table 9.1 specifies the 32 groups belonging to the various crystal systems, using both Schönflies and International classification symbols, and the 14 Bravais lattices (enumerated in the first column of the table). Recall that P designates primitive, I designates body-centered, C designates centered, F designates face-centered and R designates rhombohedral. In the cubic system, P is called simple cubic, I is body-centered cubic, and F is face-centered cubic. In the international classification, the notation n/m indicates an n-fold symmetry axis and a mirror reflection (the m stands for mirror), and the notation $\bar{n}$ indicates an n-fold symmetry axis and an inversion symmetry. In the Schönflies classification, standard point group notation is used for the groups (see Appendix E.1).

The unit cells of space lattices are shown in Fig. 9.13. A *lattice translation operation T* (or *crystal translation*) is defined by displacement of the crystal by a lattice vector $\mathbf{R}_{n_1,n_2,n_3}$. A lattice translation operation affects all points but leaves the crystal invariant. Note that the unit cells shown in Fig. 9.13 are not the minimum volume unit cells (also called primitive unit cells – there is a density of one lattice point per primitive cell). The 32 point groups can combine with the *group of lattice translation symmetries* to form 230 possible combinations called *space groups*.

9.3.2 THE RECIPROCAL LATTICE

Any local physically measurable quantity in a crystal, such as the electron density $n(\mathbf{r})$, for example, is invariant under lattice symmetry operations due to crystal periodicity. Invariance under translations implies periodicity,

$$n(\mathbf{r} + \mathbf{R}) = n(\mathbf{r}), \tag{9.79}$$

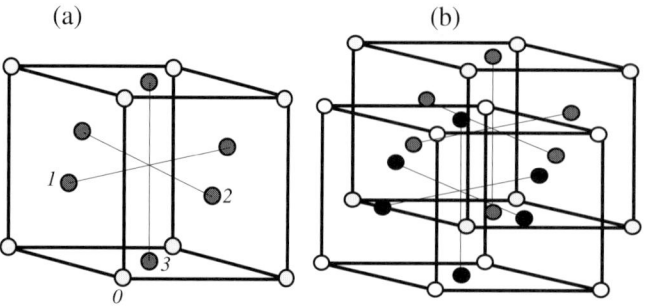

FIG 9.14 (a) Face-centered cubic (FCC) Bravais lattice. A suitable basis consist of the three vectors $\mathbf{a}_1$, $\mathbf{a}_2$, and $\mathbf{a}_3$ connecting the origin 0 with sites marked 1, 2, and 3. (b) Two shifted FCC lattices form a Bravais lattice with two atoms in the unit cell. When atoms of one element occupy the site of the first FCC lattice (shaded circles) and atoms of a second element occupy the sites of the second lattice (black and white circles), we obtain a *Zinc Blende structure*.

for any Bravais lattice vector **R**. Therefore, it can be expanded as a *discrete Fourier series*,

$$n(\mathbf{r}) = \sum_{\mathbf{G}} n_{\mathbf{G}} e^{i\mathbf{G}\cdot\mathbf{r}}, \tag{9.80}$$

where $n_{\mathbf{G}}$ are the Fourier components of the electron density. Invariance of the density on translation by **R**, according to Eq. (9.79), requires that

$$\mathbf{G}\cdot\mathbf{R} = 2\pi m, \tag{9.81}$$

where m is an integer. Equation (9.80), with the auxiliary condition (9.81), is the Fourier series for the electron density, which has the lattice periodicity. For the Bravais lattice vectors **R**, the (infinite discrete) set of points **G** satisfying Eq. (9.81) defines the *reciprocal lattice* with respect to the *direct lattice*, i.e., the original Bravais lattice. A *reciprocal lattice basis* is a triad of vectors $\mathbf{b}_i$, $i = 1, 2, 3$, defined in terms of the three (primitive) lattice vectors $\mathbf{a}_1$, $\mathbf{a}_2$, and $\mathbf{a}_3$ of the direct lattice,

$$\mathbf{b}_1 = 2\pi \frac{\mathbf{a}_2 \times \mathbf{a}_3}{\mathbf{a}_1 \cdot \mathbf{a}_2 \times \mathbf{a}_3}, \quad \mathbf{b}_2 = 2\pi \frac{\mathbf{a}_3 \times \mathbf{a}_1}{\mathbf{a}_1 \cdot \mathbf{a}_2 \times \mathbf{a}_3}, \quad \mathbf{b}_3 = 2\pi \frac{\mathbf{a}_1 \times \mathbf{a}_2}{\mathbf{a}_1 \cdot \mathbf{a}_2 \times \mathbf{a}_3}. \tag{9.82}$$

It is easy to verify the orthogonality relation,

$$\mathbf{b}_i \cdot \mathbf{a}_j = 2\pi \delta_{ij}. \tag{9.83}$$

The vectors $\mathbf{b}_i$ are the axes vectors of the reciprocal lattice. As $\mathbf{a}_1$, $\mathbf{a}_2$, and $\mathbf{a}_3$ are the primitive vectors of the direct lattice, $\mathbf{b}_1$, $\mathbf{b}_2$, and $\mathbf{b}_3$ are referred to as the primitive vectors of the reciprocal lattice. In terms of these vectors, the allowed vectors **G** satisfying Eq. (9.81) are expressed as,

$$\mathbf{G} = \mathbf{G}_{m_1,m_2,m_3} = m_1\mathbf{b}_1 + m_2\mathbf{b}_2 + m_3\mathbf{b}_3, \tag{9.84}$$

where m_1, m_2, and m_3 are integers whose geometrical significance will be explained in Eq. (9.87). The vectors **q** of the reciprocal lattice are the *reciprocal lattice vectors*. Only these vectors enter the Fourier expansion of the periodic function $n(\mathbf{r})$. Therefore, Eq. (9.80) can be explicitly written as

$$n(\mathbf{r}) = \sum_{\mathbf{G}} n_{\mathbf{G}} e^{i\mathbf{G}\cdot\mathbf{r}}, \tag{9.85}$$

Following the definition (9.84), it is evident that *a reciprocal lattice of a Bravais lattice is a Bravais lattice*. Therefore, all the definitions specified for the direct lattice apply to the reciprocal lattice as well. In particular, we can define a primitive unit cell in the reciprocal lattice whose volume is

$$V(\text{reciprocal lattice primitive unit cell}) = \mathbf{b}_1 \cdot (\mathbf{b}_2 \times \mathbf{b}_3) = \frac{(2\pi)^3}{\mathbf{a}_1 \cdot (\mathbf{a}_2 \times \mathbf{a}_3)}. \tag{9.86}$$

9.3 Crystal Structure

> **Problem 9.15**
> (a) A direct Bravais lattice is defined in terms of the basis vectors $\mathbf{a}_1 = (1, 2, 3)a_0$, $\mathbf{a}_2 = (3, 1, 2)a_0$, and $\mathbf{a}_3 = (2, 3, 1)a_0$. Find the direct lattice parameters and the basis vectors of the reciprocal lattice.
> (b) Find the relation among $\mathbf{m} = (m_1, m_2, m_3)$, $\mathbf{n} = (n_1, n_2, n_3)$, and the integer m in Eq. (9.81).
>
> **Answers:** (a) $|a_1| = |a_2| = |a_3| = \sqrt{14}a_0$, $\cos\theta_1 = \cos\theta_2 = \cos\theta_3 = \frac{11}{14}$. (b) $\mathbf{m} \cdot \mathbf{n} = m$.

Wigner–Seitz Cell and Brillouin Zone

We stressed that the choice of the (primitive) unit cell is not unique, as long as it satisfies properties 1 and 2 and preferably property 3 listed above. A convenient choice is the *Wigner–Seitz cell*, named after Eugene Wigner and Frederick Seitz, that is defined for both the direct and the reciprocal lattice as follows:

Direct lattice: A Wigner–Seitz cell is defined about a Bravais lattice point $\mathbf{R}$ as the points $\mathbf{r}$ in space that are closer to that lattice point than to any of the other lattice points $\mathbf{R}' \ne \mathbf{R}$, i.e., $|\mathbf{r} - \mathbf{R}| \le |\mathbf{r} - \mathbf{R}'|$ [see Fig. 9.16(b)]. Properties 1 and 2 listed after Eq. (9.77) assure that the entire space can be tiled by Wigner–Seitz cells obtained by lattice translation of a single cell. Moreover, it satisfies point 3, i.e., it maintains the full (point) symmetry of the lattice. To construct it geometrically, consider a given Bravais lattice point $\mathbf{R}$ and draw the vectors $\mathbf{R}_i - \mathbf{R}$ connecting it to all its p nearest neighbor lattice points (recall that p is the coordination number). At the midpoint $(\mathbf{R}_i + \mathbf{R})/2$ of each vector, draw a plane normal to it.

> **Problem 9.16**
> For primitive basis vectors $\mathbf{a}_i = x_i\hat{\mathbf{x}} + y_i\hat{\mathbf{y}} + z_i\hat{\mathbf{z}}$, find the equation of a plane bisecting the vector connecting $\mathbf{R} = 0$ and $\mathbf{R}_1 = \mathbf{a}_1$.
>
> **Answer:** Let $\mathbf{r} = x\hat{\mathbf{x}} + y\hat{\mathbf{y}} + z\hat{\mathbf{z}}$ be a point on the plane. Then it should satisfy the relation $(\mathbf{r} - \mathbf{a}_1/2) \cdot \mathbf{a}_1 = 0$. These planes intersect each other and define a polyhedron, which is the Wigner–Seitz primitive cell (see Fig. 9.15).

Reciprocal lattice: A similar construction in a reciprocal space is carried out around a reciprocal lattice point $\mathbf{G}$, thereby defining the *Wigner–Seitz cell in a reciprocal space*. It contains the points $\mathbf{k}$ that are closest to $\mathbf{G}$ than to any other reciprocal lattice point $\mathbf{G}' \ne \mathbf{G}$. It will be shown below that the energy $\varepsilon(\mathbf{k})$ of an electron in a crystal is a continuous and periodic function of $\mathbf{k}$, so that the corresponding energies are closest to $\varepsilon(\mathbf{G})$. The Wigner–Seitz cell in a reciprocal space is called the *first Brillouin zone*, denoted as BZ_1. As an example, consider the diamond crystal structure, which also corresponds to the Si and Ge structures. There are 4 nearest neighbors and 12 next nearest neighbors for each atom. The unit cube contains eight atoms with the bonds between atoms being covalent bonds in a tetrahedral arrangement. The diamond structure is composed of two face-centered cubic lattices that are displaced from each other by a translation

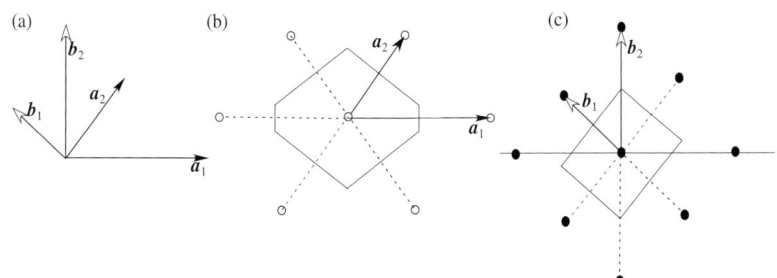

FIG 9.15 (a) Primitive vectors $\mathbf{a}_1$ and $\mathbf{a}_2$ in the direct lattice and $\mathbf{b}_1$ and $\mathbf{b}_2$ in the reciprocal lattice. (b) Wigner–Seitz unit cell in the direct lattice. (c) Brillouin zone in the reciprocal lattice.

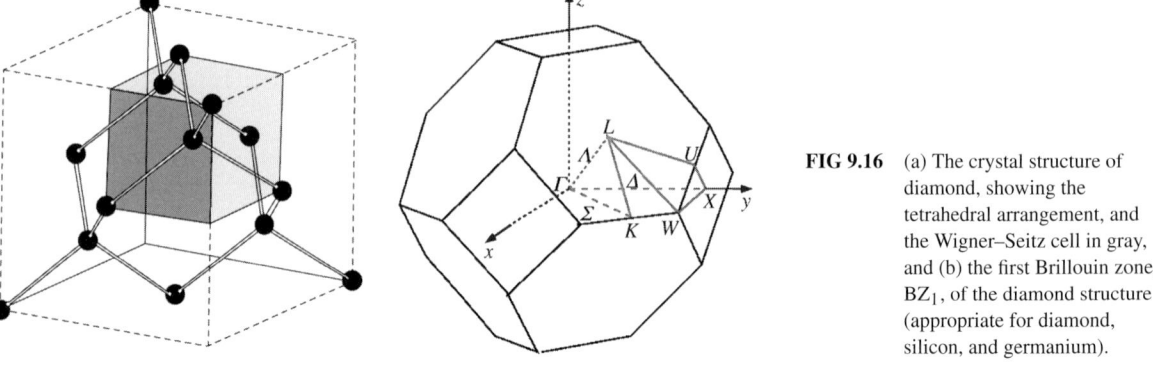

FIG 9.16 (a) The crystal structure of diamond, showing the tetrahedral arrangement, and the Wigner–Seitz cell in gray, and (b) the first Brillouin zone, BZ_1, of the diamond structure (appropriate for diamond, silicon, and germanium).

vector along the body diagonal by a length equal to one-quarter of the body diagonal. The unit cell in a direct space is shown in Fig. 9.16(a), whereas the Wigner–Seitz cell in a reciprocal space (i.e., the first Brillouin zone, BZ_1) is shown in Fig. 9.16(b). Anticipating the need to study electron energy levels $\varepsilon(\mathbf{k})$, $\mathbf{k} \in BZ_1$, it is useful to define a set of lines joining certain points in the Brillouin zone, along which the $\mathbf{k}$ vector varies and the energy is traced. These points are defined as follows: the Γ point $\mathbf{k} = 0$, the K points at the center of the lines, L and X points at the center of faces of the Brillouin zone boundary, and W points at the vertices of the Brillouin zone boundary. For example, the points in a reciprocal space in Fig. 9.16(b) show the Brillouin zone of Si; they will be used in Fig. 9.66, which presents the energy eigenvalues of the electrons in silicon as a function of wavevector.

Lattice Planes and Miller Indices

We briefly introduce some important geometrical concepts for analyzing electron and crystal properties in solids. Let us consider a given Bravais lattice and its reciprocal lattice. A *lattice plane* is a plane containing at least three noncollinear points of the direct lattice.

Problem 9.17

Show that the lattice points on a lattice plane form a two-dimensional Bravais lattice.

There are an infinite set of equally spaced parallel lattice planes; these include all the lattice points in the lattice. This set is called a *family of lattice planes*. To distinguish between different families, we look for an algorithm to characterize (enumerate) a given family of planes. The first method is to choose a unit cell in the lattice and inspect a plane belonging to the family, which cuts the axes along the basis vectors $\mathbf{a}_1$, $\mathbf{a}_2$, and $\mathbf{a}_3$ at distances $\frac{a_1}{h}$, $\frac{a_2}{k}$, and $\frac{a_3}{l}$, respectively, from the origin, where (h, k, l) are rational numbers. The smallest triple with a given ratio $h : k : l$ is referred to as the Miller indices characterizing the family of planes. For example, in Fig. 9.17(a), the plane cuts $\mathbf{a}_1$ and $\mathbf{a}_3$ in the middle and $\mathbf{a}_2$ is cut at its end, so that $(h, k, l) = (2, 1, 2)$. In Fig. 9.17(b), the plane cuts the axes at $a_1, a_2/2$, and ∞, so that $(h, k, l) = (1, 2, 0)$. Finally, in Fig. 9.17(c), it cuts the axes at $a_1, -a_2$, and $a_3/2$, so that $(h, k, l) = (1, -1, 2) \equiv (1, \bar{1}, 2)$. The distance between two adjacent planes belonging to the family (h, k, l) is denoted as d_{hkl}.

The second method is more elegant; it is based on a theorem relating families of (direct) lattice planes to vectors in the reciprocal lattice.

Theorem: For any family of lattice planes with distance d, there is a reciprocal lattice vector perpendicular to it, whose length is $\frac{2\pi}{d}$. Conversely, for every reciprocal lattice vector $\mathbf{G}$ of (shortest) length $\frac{2\pi}{d}$ there is a family of planes in the direct lattice perpendicular to $\mathbf{G}$ whose distance is d.

9.3 Crystal Structure

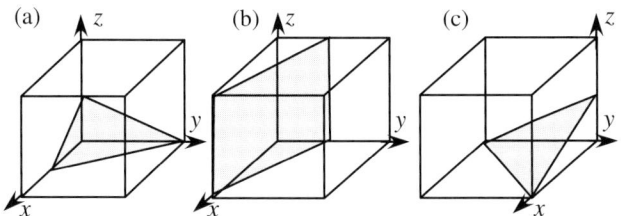

FIG 9.17 (a) A plane belonging to a family of lattice planes in a cubic lattice with Miller indices $(h,k,l) = (2,1,2)$. (b) A plane belonging to another family has Miller indices $(h,k,l) = (1,2,0)$. (c) A plane belonging to a third family has Miller indices, $(h,k,l) = (1,\bar{1},2)$.

The proof of this theorem will not be given here. It is based on definition (9.82) of the reciprocal lattice, which implies $e^{i\mathbf{G}\cdot\mathbf{R}} = 1$, as well as on the fact that a plane wave $e^{i\mathbf{k}\cdot\mathbf{r}}$ has the same value for all lattice points $\mathbf{r}$ belonging to a family of planes perpendicular to the vector $\mathbf{k}$ whose distance d from one another is an integer number of wavelengths. Once this theorem is established, one can characterize a family of lattice planes by specifying the shortest length vector $\mathbf{G}$ in a reciprocal space perpendicular to this family of planes. This vector can be written as a linear combination basis vectors,

$$\mathbf{G} = \mathbf{G}_{hkl} = h\mathbf{b}_1 + k\mathbf{b}_2 + l\mathbf{b}_3, \quad (9.87)$$

where (h,k,l) are integers, by definition of the reciprocal lattice. They are the *Miller indices* as defined in the first method.

X-ray Scattering from a Crystal

Crystal structures are experimentally determined by diffraction of electrons, electromagnetic waves, or neutrons. The physical basis for diffraction experiments is the interference effects due to phase differences between reflected or transmitted waves, which are elastically scattered from the atoms in the crystal. Scattering of X-rays from a crystal and analyzing the reflected intensity is a common procedure for determining crystal structures. The wavelength of the radiation should be comparable with the interatomic distances, which are of the order of a few Å, i.e., for electromagnetic radiation, the characteristic wavelength is in the X-ray region of the spectrum. Analysis of X-ray scattering from a periodic solid enables the elucidation of the position of the atoms forming the crystal. This procedure was pioneered by Max von Laue (Physics Nobel Prize in 1914) and W. H. Bragg and W. L. Bragg (Physics Nobel Prize in 1915) and is called *Bragg scattering*.

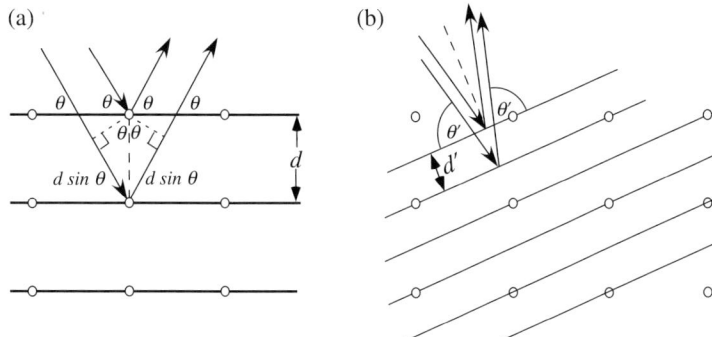

FIG 9.18 (a) An X-ray beam is scattered at an angle θ from a series of horizontal crystal planes at distance d from one another. The optical path length difference between the reflected beams scattered from two adjacent lattice planes is $2d\sin\theta$. The two beams interfere constructively if this difference is an integer multiple of λ, the X-ray wavelength. (b) The same beam is also scattered from a different family of crystal planes with different scattering angle θ' and different spacing d' between two adjacent planes.

In Bragg scattering formalism, an X-ray impinging on a solid crystal is reflected from a series of equally spaced lattice planes as indicated in Fig. 9.18. Consider an incoming X-ray beam impinging at an angle θ on a series of lattice planes with distance d between two adjacent planes. The specular reflection of two adjacent lattice planes interfere coherently if their optical path difference is a multiple of λ, the X-ray wavelength. That is,

$$d\sin\theta = n\lambda, \quad n = \text{positive integer.} \quad (9.88)$$

This is known as the *Bragg condition* for reflection of order n. It is possible to obtain information on lattice planes by tracing the directions along which the reflected waves emerge with high intensity. For this purpose, it is preferable to have a monochromatic X-ray source.

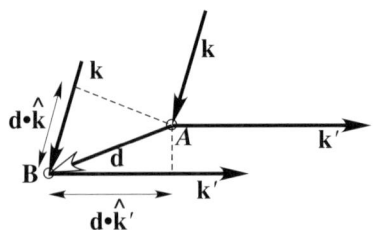

FIG 9.19 von Laue analysis of scattering from a crystal. An X-ray beam impinges and scattered from the atoms composing the crystal. Two X-ray beams with wavevector directed along $\hat{\mathbf{k}}$ are reflected from two atoms located at points **A** and **B** and scatter into wavevector directed along $\hat{\mathbf{k}}'$. The optical path difference between the two rays is $\mathbf{d} \cdot \hat{\mathbf{k}} - \mathbf{d} \cdot \hat{\mathbf{k}}' = \mathbf{d} \cdot (\hat{\mathbf{k}} - \hat{\mathbf{k}}')$, where $\mathbf{d} = \mathbf{B} - \mathbf{A}$. Constructive interference results if this path difference equals $n\lambda$, where n is an integer.

A somewhat difference approach is suggested by von Laue, based on analysis of directions in the reciprocal lattice. It does not require the inspection of lattice planes or a specular reflection. Consider, as in Fig. 9.19, two monochromatic X-rays of wavevector $\mathbf{k} = \frac{2\pi}{\lambda}\hat{\mathbf{k}}$, impinging on a crystal in a direction $\hat{\mathbf{k}}$ and reflected from two atoms located at **A** and **B** in the direct lattice, along another direction $\hat{\mathbf{k}}'$. The vector connecting the two points is $\mathbf{d} = \mathbf{B} - \mathbf{A}$. Constructive interference between the two reflected rays occurs if the optical path difference between them is an integer multiple of λ, that is,

$$\mathbf{d} \cdot (\hat{\mathbf{k}} - \hat{\mathbf{k}}') = n\lambda, \quad n = \text{positive integer}, \tag{9.89}$$

which is the *von Laue condition* for reflection at order n. Because **A** and **B** are Bravais lattice points, then $\mathbf{d} = \mathbf{B} - \mathbf{A}$ is a Bravais lattice vector, and the von Laue condition (9.89) (after multiplication by $2\pi/\lambda$) becomes

$$\boxed{\mathbf{d} \cdot (\mathbf{k} - \mathbf{k}') = 2\pi n,} \tag{9.90}$$

which, on comparison with Eq. (9.81), shows that constructive interference occurs along directions where the *momentum transfer* $\mathbf{k} - \mathbf{k}' = \mathbf{G}$ is a reciprocal lattice vector. The above analysis applies not only to X-rays but also to electrons or neutrons with de Broglie wavelength $\lambda = h/p$ of a few angstroms.

Problem 9.18

For elastic scattering, $|\mathbf{k}'| = |\mathbf{k}|$. Find the angles $\theta_\mathbf{G}$ (with respect to $\mathbf{k}$) where there is a constructive interference.

Answer: $\cos\theta_\mathbf{G} = 1 - \frac{G^2}{2k^2}$.

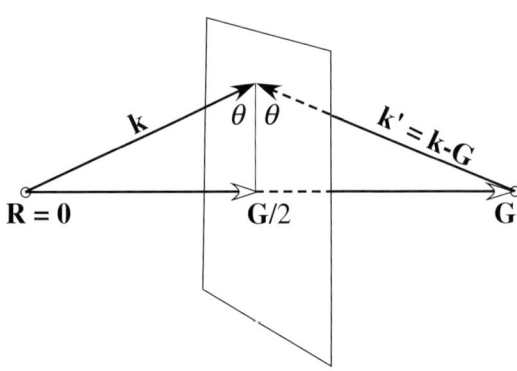

FIG 9.20 Geometrical interpretation of the von Laue analysis of scattering from a crystal. An X-ray of wavevector $\mathbf{k}$ is reflected coherently in the direction $\mathbf{k}' = \mathbf{k} - \mathbf{G}$ if and only if the tip of the vector $\mathbf{k}$ lies on a Bragg plane. See analysis related to Eq. 9.91.

The von Laue elastic scattering condition implies

$$|\mathbf{k}| = |\mathbf{k}'| = |\mathbf{G} - \mathbf{k}|. \tag{9.91}$$

This has a useful geometrical interpretation: Squaring both sides of Eq. (9.91) yields

$$\mathbf{G} \cdot \left(\mathbf{k} - \frac{1}{2}\mathbf{G}\right) = 0. \tag{9.92}$$

Thus, the points $\mathbf{k} - \mathbf{G}/2$ lie on a plane, which bisects the vector $\mathbf{G}$ (such a plane is called *a Bragg plane*). Consequently, *an X-ray of wavevector $\mathbf{k}$ is reflected coherently in the direction $\mathbf{k} - \mathbf{G}$ if and only if the tip of $\mathbf{k}$ lies on a Bragg plane*. This is schematically shown in Fig. 9.20.

9.3 Crystal Structure

Equivalence of Bragg and von Laue Interference Conditions

To demonstrate the equivalence of the two approaches, let us first derive the Bragg condition assuming the von Laue condition is satisfied. An X-ray of initial wavenumber $\mathbf{k}$ is reflected, such that the wavevector of the reflected beam is $\mathbf{k}' = \mathbf{k} - \mathbf{G}$, where $\mathbf{G}$ is a reciprocal lattice vector. For elastic scattering $|\mathbf{k}| = |\mathbf{k}'|$, that is, the two vectors have the same inclination angle θ with respect to the Bragg plane (see Fig. 9.20). Thus, the Bragg plane appearing in the von Laue formulation is a representative of a family of lattice planes in the direct lattice, which are perpendicular to $\mathbf{G}$, and the angle θ appearing in the von Laue description is identical with the angle θ of scattering from a family of planes in the direct lattice as defined in Bragg formulation. Moreover, from the properties of the reciprocal lattice, we know that $\mathbf{G} = m\mathbf{G}_0$, where $\mathbf{G}_0$ is the shortest reciprocal lattice vector in the direction of $\mathbf{G}$ and m is a positive integer. According to the theorem presented above, the length of $\mathbf{G}_0$ is $\frac{2\pi}{d}$, where d is the distance between the family of direct lattice planes. Combining this result and inspecting Fig. 9.20, we may write

$$G = \frac{2m\pi}{d} = 2k \sin \theta. \tag{9.93}$$

Finally, writing $k = \frac{2\pi}{\lambda}$ immediately implies the Bragg condition for constructive interference, $2d \sin \theta = m\lambda$, as in Eq. (9.88). Consequently, the von Laue condition $\mathbf{k} - \mathbf{k}' = \mathbf{G}$ is identical with the Bragg condition $2d \sin \theta = m\lambda$ with,

$$\cos \theta = \hat{\mathbf{k}} \cdot \hat{\mathbf{k}}', \quad \frac{2\pi}{\lambda} = |\mathbf{k}| = |\mathbf{k}'|, \quad \frac{2m\pi}{d} = |\mathbf{G}|.$$

Experimental analysis of X-ray scattering is more conveniently carried out within the Laue formulation (based on reciprocal lattice vector analysis) than within the Bragg formulation (based on resolution of the direct lattice into families of lattice planes). The pattern of X-ray reflections consists of isolated points (Bragg peaks) corresponding to the allowed scattering angles determined by the von Laue (or Bragg) conditions. It gives valuable information on the structure of the reciprocal lattice, because once we know $\mathbf{k}$ and $\mathbf{k}'$, we know $\mathbf{k} - \mathbf{k}'$, which must be a vector in the reciprocal lattice. Once the structure of the reciprocal lattice is known, the determination of the direct lattice can be made.

X-ray Scattering from a Mono-atomic Crystal with a Motif

Consider a Bravais lattice with a motif. The von Laue condition (9.89) assumes that the rays scattered from each atom in a Barvais lattice interfere coherently. Hence, a set of sharp peaks in the reflected intensity results from the constructive interference of reflected beams from atoms in various crystal planes. Analysis of the modified intensity yields valuable information on the structure of the unit cell of the crystal. Let us consider for simplicity a lattice with a motif consisting of n identical atoms located at points $\mathbf{r}_j$, $j = 1, 2, \ldots J$. The phase difference between rays scattered from atoms located at points $\mathbf{r}_i$ and $\mathbf{r}_j$ is seen from Fig. 9.19 is $(\mathbf{k} - \mathbf{k}') \cdot (\mathbf{r}_i - \mathbf{r}_j) = \mathbf{G} \cdot (\mathbf{r}_i - \mathbf{r}_j)$. The amplitude of the reflected beam is then proportional to the *geometrical structure factor*,

$$S_\mathbf{G} = \sum_{j=1}^{J} e^{i\mathbf{G} \cdot \mathbf{r}_j}, \tag{9.94}$$

and the intensity at the Bragg peak is proportional to $|S_\mathbf{G}|^2$. However, because there are other factors determining the intensity at the Bragg peaks, the main use of the above property is in cases when $S_\mathbf{G} = 0$. Consider for example a body-centered cubic lattice (b.c.c). It can be viewed as a simple cubic lattice with a motif containing two identical atoms, one at $\mathbf{0} = (0, 0, 0)$ and the other at $\mathbf{r} = \frac{a}{2}(\hat{\mathbf{x}} + \hat{\mathbf{y}} + \hat{\mathbf{z}})$, where a is the (simple cubic) lattice constant. Therefore,

$$S_\mathbf{G} = e^{i\mathbf{G} \cdot \mathbf{0}} + e^{i\mathbf{G} \cdot \frac{a}{2}(\hat{\mathbf{x}} + \hat{\mathbf{y}} + \hat{\mathbf{z}})} = 1 + e^{i\pi(h+k+l)}, \tag{9.95}$$

using Eqs (9.87) and (9.83). Thus, in a BCC lattice, the scattering amplitude from Bragg planes with odd sum of $h + k + l$ of Miller indices vanishes.

9.3.3 QUASICRYSTALS

The orthodox definition of a crystal requires that its structure is perfectly periodic and displays well-defined translational and point symmetries as discussed above. The definition of a crystal has been recently broadened to include structures that are not perfectly periodic. Periodicity is a special case of a *long-range order* wherein a structure repeats itself periodically. The X-ray diffraction pattern from a perfectly periodic crystal consists of discrete isolated points (Bragg peaks) in momentum space in accordance with the von Laue condition $\mathbf{k} - \mathbf{k}' = \mathbf{G}$, where $\mathbf{G}$ is a reciprocal lattice vector. We now know that nonperiodic crystal structures having X-ray diffraction patterns consisting of discrete points exist. Structures having this property are referred to as *quasicrystals*. A quasicrystal is an extension of the notion of a crystal to structures that are not perfectly periodic yet have a long-range order. Quasicrystals can be defined as condensed phase structures that are both ordered and nonperiodic. They form patterns that fill space in 2D or 3D but lack translational symmetry. Nevertheless, they produce a Bragg diffraction pattern, but unlike crystals that have a simple repeating structure in reciprocal space, quasicrystals have a more complex structure, as described below.

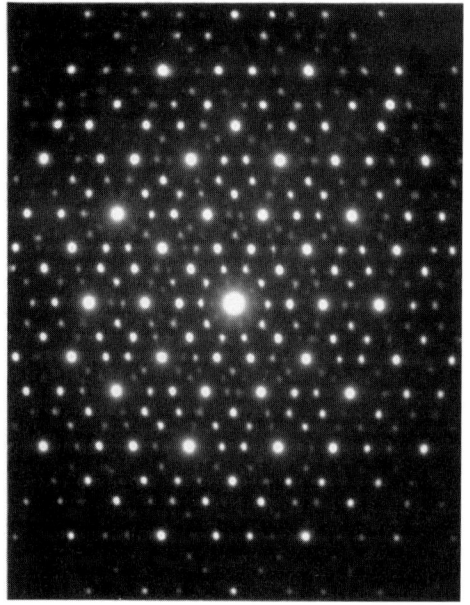

FIG 9.21 Diffraction diagram of a quasicrystal exhibiting 5-fold or 10-fold rotational symmetry.

The first experimental measurements of such structures were made by Dan Shechtman in 1984 [126] (the Nobel prize in physics was awarded to him in 2011). He obtained an electron diffraction pattern of an Al-Mn alloy with sharp reflection peaks and 10-fold symmetry, similar to the diffraction pattern shown in Fig. 9.21. Mathematical considerations of quasicrystals preceded their experimental observation (e.g., Johannes Kepler, 1571–1630, noted the existence of quasicrystals). 2D quasicrystal structures were formulated in 1974 and 1976 by Roger Penrose using a set of several types of tiles that nonperiodically tiled the plane. Several other nonperiodic tilings of the plane were independently suggested by Robert Ammann. Figure 9.22(a) shows the 5-fold symmetric Penrose tiling known as the rhombus tiling (P3), and Fig. 9.22(b) shows the 8-fold symmetric Ammann–Beenker tiling.

Diffraction and Quasiperiodicity

Intuitively, we may distinguish the structure of a solid material to be anywhere between perfectly periodic and completely amorphous. In this section, we consider crystal structures that are not perfectly periodic but, in some sense, are not very far off. To give a more quantitative understanding of this concept, we recall that an X-ray diffraction pattern of a periodic crystal has a discrete structure. However, periodicity is not a requisite for a long-range order. A system is said to be quasiperiodic if it is not perfectly periodic but it has a discrete X-ray diffraction pattern (hence, it has a long-range order). As in the case of periodic crystals, the X-ray pattern tells us directly about wavevector space. Specifically, consider a system of atoms located at points $\mathbf{R}$ occupying a volume $\mathcal{V}$, with N atoms in this volume. The density of points is $\rho(\mathbf{r}) = \sum_{\mathbf{R} \in \mathcal{V}} \delta(\mathbf{r} - \mathbf{R})$, and the intensity of radiation is proportional to $|S_{\mathbf{k}}|^2$, where

$$S_{\mathbf{k}} = \lim_{N \to \infty} \frac{1}{N} \int d\mathbf{r}\, \rho(\mathbf{r})\, e^{i\mathbf{k}\cdot\mathbf{r}} = \lim_{N \to \infty} \frac{1}{N} \sum_{\mathbf{R}} e^{i\mathbf{k}\cdot\mathbf{R}}, \qquad (9.96)$$

9.3 Crystal Structure

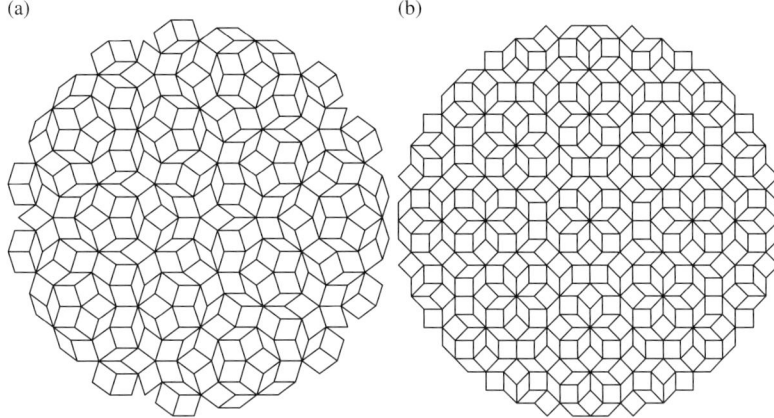

FIG 9.22 (a) The 5-fold symmetric nonperiodic rhombic Penrose tiling (P3). (b) The 8-fold symmetric nonperiodic Ammann–Beenker tiling.

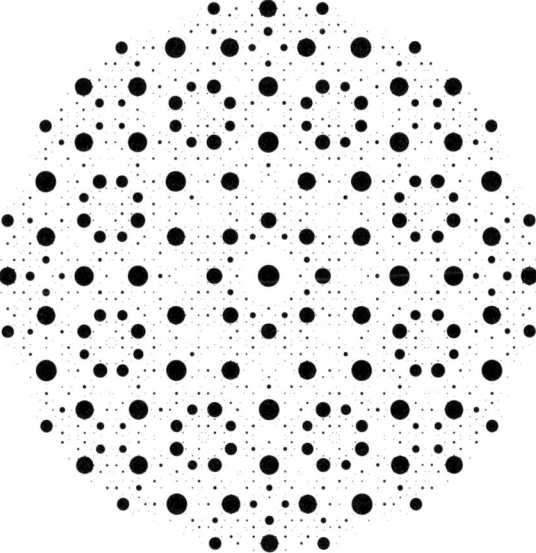

FIG 9.23 Diffraction pattern from the Ammann–Beenker tiling of Fig. 9.22(b). The area of each disc is proportional to the intensity of the peak, with cutoff set at 0.1% of the central intensity.

is obtained by Fourier transforming the density. For example, the diffraction pattern of the Ammann–Beenker tiling with scatterers of equal strength on all points is shown in Fig. 9.23. The calculation of $S_\mathbf{k}$ for general nonperiodic structures is often nontrivial.

Incommensurate Chain of Atoms: Aperiodic structures such as Penrose or Amman-Beenker tilings have an X-ray diffraction pattern with sharp peaks. Let us now consider a very simple system that displays order but its X-ray diffraction pattern is dense. For a 1D periodic structure composed of lattice points (atoms) equally spaced on a line having density given by $\rho_1(x) = \sum_n \delta(x - na)$, the Fourier transform of this structure is $S_{1k} = \sum_m \delta(k - 2\pi m/a)$. A structure with points located in two series of different repetition length, i.e., $\rho(x) = \sum_{n_1,n_2} \delta(x - n_1 a) + \delta(x - n_2 \alpha a)$, has a long-range order, but is not periodic if $0 < \alpha < 1$ is an irrational number. Then, there is no coincidence in space between the two periodic components of the whole structure. Its Fourier transform is

$$S_k = \sum_{m_1, m_2} \delta \left[k - \frac{2\pi}{a}(m_1 + m_2/\alpha) \right]. \tag{9.97}$$

The density of points in reciprocal space is higher than for the regular periodic structure; it is "dense" for irrational numbers α.

Modified Definition of a Crystal

The International Union of Crystallography redefined the term crystal in 1996 to mean "any solid having an essentially discrete diffraction diagram," thus shifting the essential attribute of crystals from position to momentum space. With this new definition, the incommensurate crystals (i.e., the quasicrstals or aperiodic crystals) fall into the definition of a crystal.

Bulk quasicrystals tend to be rather brittle, and this limits the applications that are possible for quasicrystals. Coatings made of thin layers of quasicrystals can still be hard, and quasicrystals have been used as materials for surface coatings, e.g., quasicrystalline coatings of frying pans. These coatings typically have a low-friction coefficient and do not conduct heat well, yet they are thermally stable. Another application of quasicrystals is as a reversible storage medium for hydrogen, which can be imbedded into the quasicrystal. Other possible novel applications may exist in linear and nonlinear optics.

9.4 ELECTRONS IN A PERIODIC POTENTIAL

In this section, we shall study the physics of electrons subject to periodic potentials. Metals are composed of atoms (or molecules) ordered periodically, whose valence electrons form an electron gas. Each electron experiences the sum of the atomic (ionic) potentials, and because the atoms form a crystal, the potential is periodic. The problem of determining the eigenvalues and eigenfunctions of electrons in a periodic potential is solved within Bloch theory, the central topic in this section. Strictly speaking, the physics of electrons in a solid is a many-body problem, as electrons interact with each other. However, most of the important experimentally relevant features such as the band structure of the spectrum can be deduced from an *independent electron model* that uses a *mean-field approximation*, where electrons move independently in an effective potential, $V(\mathbf{r})$. It was not until Lev Landau developed *Fermi liquid theory* around 1956 that the simple mean-field picture we shall paint below was put on firm ground.

The potential $V(\mathbf{r})$ experienced by the electrons is periodic with period determined by the crystal structure of the solid,

$$V(\mathbf{r} + \mathbf{R}) = V(\mathbf{r}), \tag{9.98}$$

where $\mathbf{R}$ is a Bravais lattice vector as defined in Eq. (9.76). The energy spectrum of a Hamiltonian with a periodic potential is composed of allowed energy bands separated by energy gaps (i.e., *band gaps*), as will be described in the following paragraphs. For "good" metals, the main physical properties (such as the band structure, transport properties, specific heat, electric, and magnetic field response) are well described in terms of the (nearly) free electron gas model with weak perturbation caused by the periodic potential. This was evident even at the very beginning of the quantum era. One of the pioneers who developed the theory of electrons moving in periodic potentials, Felix Bloch, reminisced as follows: "When I started to think about it, I felt that the problem was to explain how the electrons could sneak by all the ions in a metal...By straight Fourier analysis I found to my delight that the wave differed from the plane wave of free electrons only by a periodic modulation."

9.4.1 FROM ATOMIC ORBITS TO BAND STRUCTURE

The energy levels of a single atom form a discrete set of energies and a continuum of levels above the ionization energy. Within a solid material, atoms are located close to one another and cannot be treated independently; the energy levels are modified due to interaction. Consider a periodic arrangement of N identical atoms on a straight line

9.4 Electrons in a Periodic Potential

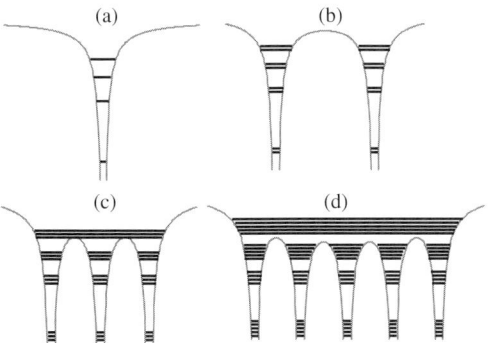

FIG 9.24 Splitting of atomic energies due to interaction between $N = 1, 2, 3, 4$ equidistant atoms arranged along a straight line (schematic). The potential seen by an electron (solid curves) and the energy levels are drawn as straight lines. (a) $N = 1$. Atomic levels in an atomic potential. (b–d) The number of split states equals the number of atoms that interact, and as the number increases, bands of allowed states, separated by band gaps, are formed.

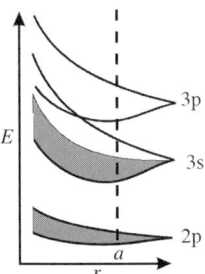

FIG 9.25 Band structure of atoms as a function of distance between atoms r. $r = a$ is the specific distance between atoms in the crystal. The asymptotic atomic levels are labeled to be appropriate for sodium. (Adapted from Fig. 1.1 of Pankove [127].)

with a distance a between two adjacent atoms. A qualitative picture of the energy levels can be deduced as follows. For $N = 2$ (two identical atoms), the energy levels vary as the atoms are brought together, i.e., they form energy eigenstates of a molecular potential, which depends on the internuclear distance a. At any given finite, but large, internuclear distance between two atoms, the energy levels of the atoms are split by the interaction of the atoms. As the number N of atoms along the line increases, the number of split levels increases, as depicted in Fig. 9.24.

The levels, whose number increases with N, are arranged in well-defined separate energy bands, such that the typical energy difference, δ, between two adjacent levels belonging to the same band is rather small ($\delta \propto N^{-2/d}$, where d is the space dimension), whereas the typical energy gap E_g between two adjacent bands saturates (hence, it is much larger than δ). In the limit $N \to \infty$, $\delta \to 0$, and the energy levels in each group form a continuous domain, referred to as an *energy band*. This scenario holds for two- and three-dimensional crystals as well, although the dependence of potential energy on position (as in Fig. 9.24) is much richer. Because we focus on macroscopic samples in solid-state physics, we can assert that the energy levels form continuous bands. The width of a given band (difference between highest and lowest levels in the same band) is called the *bandwidth*. The energy difference E_g between two adjacent bands is called the *band gap*. Figure 9.25 schematically illustrates the *band structure* formation resulting from this 1D arrangement as a function of the distance between atoms, r, where the value of $r = a$ corresponds to the true period of the crystal. In 2D and 3D, details of the band pattern depends on the crystal structure. The deep atomic levels (e.g., the $1s$, $2s$ and $2p$ levels in solid sodium) hardly overlap, and therefore, the resulting bands are very narrow (i.e., for the $2p$ levels, the $2 \times 6 \times N$ levels are packed into a very narrow energy range). When the spacing between two adjacent atoms is comparable with the radius of the electrons in a given atomic orbit, the levels split into a band of levels. For sodium metal, the $3s$ levels split into a wide band. At absolute zero temperature, all electronic levels are filled up to the Fermi energy, E_F. Recall that for metals at zero temperature, E_F is the energy below which all the levels are occupied, while above it, all levels are empty. For semiconductors, the definition of the Fermi level is somewhat modified (see below).

9.4.2 BAND STRUCTURE AND ELECTRON TRANSPORT

Figure 9.26 schematically shows the band structure of metals, semimetals, semiconductors, and insulators, and clarifies the classification of crystalline solids into these categories. Let us first consider semiconductors (e.g., Si) and band insulators (see Sec. 9.9 for a classification of insulators). Both have a completely filled band and an empty band above it, separated by an energy gap (also referred to as a *band gap*) of width E_g. The electrons in the filled band are the valence atomic electrons, which are no longer localized around their parent atoms. This filled band is called *the valence band*. Electrons in a filled band do not yield a net current because this requires a net momentum, but in a filled band, all momentum states are occupied. This situation occurs in band insulators and semiconductors at zero temperature. The band above the valence band is called the *conduction band*. In metals, it partially occupied, and in semiconudctors at finite temperature, it is slightly occupied. Electrons that occupy it are able to freely move if an electric field is applied to the material.

When the valence band is full and the conduction band is empty, as in the case for insulators and semiconductors at zero temperature, no current can flow in the material. The only way to achieve electron motion is to excite them into the conduction band. This requires an excitation energy to overcome the gap E_g, which, depending on the size of E_g, may be supplied by the ambient temperature provided $k_B T \simeq E_g$. The difference between semiconductors and insulators is the size of the energy gap. When the energy gap is wide

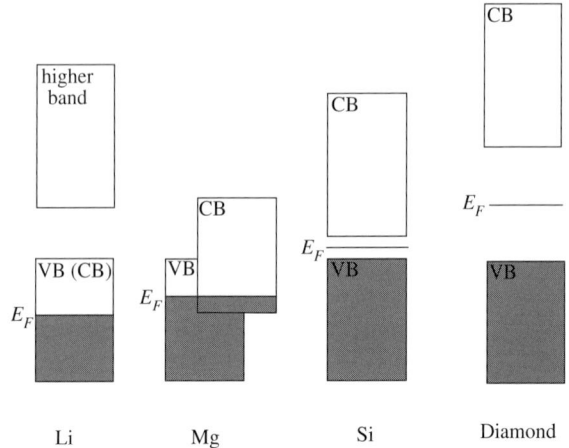

FIG 9.26 Schematic energy band diagram showing valence and conduction bands (VB and CB) for metals (e.g., Na and Li), semimetals (e.g., Mg), semiconductors (e.g., Si, Ge, GaAs), and insulators (such as diamond, quartz, and sodium chloride). The horizontal coordinate represents a momentum quantum number **k** on which the energy depends. In some cases, the maximum of the valence band and the minimum of the conduction band occur at the same **k**, and in other cases, they are shifted (as for semimetals indicated in the figure, but this is possible also for semiconductors, e.g. Si). *Source*: Band, Light and Matter, Fig. 6.32, p. 394

(e.g., 1.5 eV for diamond in Fig. 9.26), electrons do not have any chance to achieve the required excitation energy to jump from the valence band to the conduction band at room temperature (where $k_B T \approx 1/40$ eV), and no current is possible. The material is then referred to as *band insulator*. If the gap is much narrower, electron excitation is possible at room temperature. In this case, current can exist and the material is referred to as semiconductor (direct or indirect according to whether the maximum of the valence band and the minimum of the conduction band occur at the same **k**).

When the upper band is partially filled, the system can conduct even at zero temperature. This is the case for metals, where the valence band is not full, and hence, it can be regarded also as the conduction band. In sodium, for example, the valence band consists of $3s$ atomic electrons, one electron from each atom. If there are N atoms, the band has $2N$ levels; hence, it is half filled. Consequently, sodium is a good conductor because electrons at the Fermi level have empty states available to them that can be occupied under the influence of an externally applied electric field. Finally, there are situations where valence and conduction bands slightly overlap and their extremum points occur at different **k** (see Fig. 9.26, a second panel from the left that depicts Mg). In that case, the system is referred to as *semimetal*.

9.4.3 PERIODIC POTENTIAL AND BAND FORMATION

Within the independent-particle model, let us for the moment assume that the free motion of electrons is perturbed only weakly by the periodic potential of the crystal. At least in the case of metals, this assumption is justified and it explains why the electronic spectrum of metals is given by bands and band gaps. As will be shown (see Sec. 1.1.4), energy gaps develop in part of the otherwise free particle spectrum, $E_\mathbf{k} = \hbar^2 k^2 / (2m)$. Because of the periodic potential of the metal, an electron is scattered from a state of initial wavenumber **k** into a state of final wavenumber **k**′,

9.4 Electrons in a Periodic Potential

such that $\mathbf{k} - \mathbf{k}' = \mathbf{G}/2$, where $\mathbf{G}$ is a vector of the reciprocal lattice. For simplicity, let us consider the problem of a linear (1D) solid with lattice constant a. The low-energy portions of the band structure can be understood as follows. Nearly, free electrons with wavenumber close to the Bragg points $k = \pm n\pi/a$ undergo Bragg scattering off the periodic lattice potential. For diffraction of a wave of wavevector k, the Bragg scattering condition $(k + G)^2 = k^2$ becomes $k = \pm G/2 = \pm n\pi/a$, where $G = 2n\pi/a$ is a reciprocal lattice vector and n is an integer. Bragg scattering significantly affects the spectrum E_k near these k points, because it leads to gap opening as a result of level repulsion (see below). The most important reflections (leading to the first energy gap) occur at $k = \pm \pi/a$. In this simple 1D crystal, the first Brillouin zone, BZ_1 is the region $-\pi/a \leq k \leq \pi/a$. Other energy gaps occur at $|n| > 1$. As we shall see below, the wave functions at $k = \pm \pi/a$ are not traveling waves $\exp(\pm i\pi x/a)$, but rather they are superpositions of these waves. Figure 9.27 shows the free particle spectrum versus wavevector (light dashed curve) and the nearly free particle spectrum for a weak periodic potential (heavy solid curve) where a gap develops at the reciprocal lattice vectors.

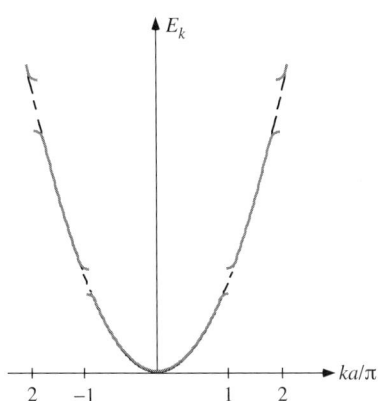

FIG 9.27 Energy bands and band gaps. The free dispersion relation is shown as a light dashed curve, and the heavy solid curve shows the allowed bands.

As discussed above, the energy gaps in the spectrum (or the density of states) determine whether a solid is a band insulator, semiconductor, or conductor. The energy bands $E_\mathbf{k}$ (sometimes written $E(\mathbf{k})$) versus wavevector $\mathbf{k}$ are typically plotted in reciprocal lattice space in the restricted region of the first Brillouin zone, e.g., in 1D, BZ_1, $-\pi/a \leq k \leq \pi/a$, where the higher energy bands are folded back into this region of k-space in the first Brillouin zone. The energy gap E_g is associated with the first Bragg reflection at $k = \pm \pi/a$, as discussed in the previous paragraph. The Fermi energy of a metal lies within a band, whereas the Fermi energy of insulators and semiconductors lies in the band gap. The gap of an insulator is very large (typically on the order of 5 eV) and much smaller for semiconductors (typically less than an eV).

To quantify the above statements, one can seek the solution of the Schrödinger equation,

$$-\frac{\hbar^2}{2m}\nabla^2 \psi_\mathbf{k}(\mathbf{r}) + V(\mathbf{r})\psi_\mathbf{k}(\mathbf{r}) = E_\mathbf{k}\psi_\mathbf{k}(\mathbf{r}), \qquad (9.99)$$

with a periodic potential, $V(\mathbf{r} + \mathbf{R}) = V(\mathbf{r})$. We will see below that the most general solution of Eq. (9.99) has a *Bloch function* form $\psi_\mathbf{k}(\mathbf{r}) = e^{i\mathbf{k}\cdot\mathbf{r}} u_\mathbf{k}(\mathbf{r})$ with $u_\mathbf{k}(\mathbf{r} + \mathbf{R}) = u_\mathbf{k}(\mathbf{r})$. The meaning of the wavevector subscript $\mathbf{k}$ on the wave function and energy eigenvalues and the question of boundary conditions on the wave function are clarified below. There are very few potentials for which Eq. (9.99) has an analytic solution. The *Kronig–Penney potential*, shown in Fig. 9.28, is one such potential (see Problem 9.19), and the periodic sinusoidal potential, whose solution will be given in terms of Mathieu functions in Sec. 9.4.6, is another.

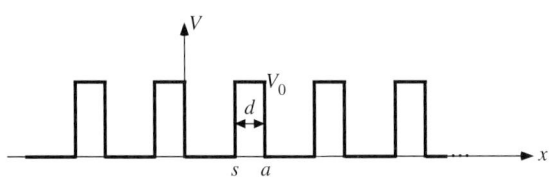

FIG 9.28 The 1D periodic Kronig–Penney potential.

Problem 9.19

Develop an analytical solution to the band structure of the 1D Kronig–Penney potential (see Fig. 9.28) by matching the wave function and its derivative at the points where the potential is discontinuous as follows:

(a) Write the wave function in the region where $V = 0$ in $0 \leq x \leq s$ as $\psi_k(x) = A_k e^{iKx} + B_k e^{-iKx}$, and in the region where $V = V_0 < E_k$ in $a \leq x \leq a$ as $\psi_k(x) = C_k e^{\beta x} + B_k e^{-\beta x}$, where $\beta = [2m(V - E_k)/\hbar^2]^{1/2}$, and match the wave function and its first derivative at $x = s$.

(b) Write the wave function in the form $\psi_k(x) = e^{ikx} u_k(x)$, where $u_k(x)$ is periodic, $u_k(x + a) = u_k(x)$ to obtain the wave function in other regions.

(c) Apply periodicity over N regions, $u_k(-Na/2) = u_k(Na/2)$ to determine the remaining coefficient.

Answers: (a) $U(x) = \sum_{n=-\infty}^{\infty} V(x - X_n)$, $V(x - X_n) = V_0 \Theta(x - X_n + \frac{s}{2})\Theta(X_n + \frac{s}{2} - x)$. (b) $\psi_I = Ae^{iqx} + Be^{-iqx}$, $\psi_{II} = Ce^{ipx} + Be^{-ipx}$, where $p = \sqrt{2mE}/\hbar$, and $q = \sqrt{2m(E - V_0)}/\hbar$. (c) $A + B = C + D$, $q(A - B) = p(C - D)$. (d) The Bloch condition is $\psi(x + a) = e^{ika}\psi(x)$ where k is the crystal momentum to be determined. Matching the function, we find, $\psi_I(-d) = e^{ika}\psi_{II}(s) \Leftrightarrow Ae^{-iqs} + Be^{iqs} = e^{ika}(Ce^{ipd} + De^{-ipd})$, and matching the derivative, $\psi_I'(-d) = e^{ika}\psi_{II}'(s) \Leftrightarrow q(Ae^{-iqs} - Be^{iqs}) = e^{ika}p(Ce^{ipd} - De^{-ipd})$.

9.4.4 BLOCH WAVE FUNCTIONS AND ENERGY BANDS

The eigenfunctions of the single-particle Schrödinger equation in a periodic potential are the Bloch wave functions briefly mentioned after Eq. (9.99). They can be viewed as the analogs of electron wave functions in free space (plane waves) for a periodic potential. Bloch wave functions are especially useful for elucidating the physics of metals and semiconductors where valence electrons are able to leave their parent atoms and feel a periodic potential. Bloch wave functions are used for calculating numerous physical properties of metals and semiconductors. The fact that the spectrum of a particle in a periodic potential consists of energy bands and gaps results naturally within the Bloch theory.

Bloch Theorem

In 1928, Felix Bloch proved what is now known as the *Bloch theorem*[3]: The eigenfunctions of the Schrödinger equation with a periodic potential, Eq. (9.99), can be written in the form of a plane wave, $e^{i\mathbf{k}\cdot\mathbf{r}}$, times a function $u_{n\mathbf{k}}(\mathbf{r})$ having the periodicity of the potential:

$$\psi_{n\mathbf{k}}(\mathbf{r}) = e^{i\mathbf{k}\cdot\mathbf{r}} u_{n\mathbf{k}}(\mathbf{r}), \quad \text{where } u_{n\mathbf{k}}(\mathbf{r} + \mathbf{R}) = u_{n\mathbf{k}}(\mathbf{r}). \tag{9.100}$$

The corresponding energy is $E_{n\mathbf{k}}$. The index n labels the different bands but will be omitted when no confusion arises, and for infinite systems, $\mathbf{k}$ is a continuous wave vector. From the periodicity of $u_\mathbf{k}$, we see that the full wave function has the property

$$\psi_\mathbf{k}(\mathbf{r} + \mathbf{R}) = e^{i\mathbf{k}\cdot\mathbf{R}} \psi_\mathbf{k}(\mathbf{r}), \tag{9.101}$$

a property shared by plane waves. Note that the Bloch functions defined above are not localized, just as the plane waves for free electrons are not localized. In analogy with free electrons, we can define the *group velocity* of an electron whose energy is $E_{n\mathbf{k}}$ as

$$\mathbf{v}_{n\mathbf{k}} = \frac{1}{\hbar}\nabla_\mathbf{k} E_{n\mathbf{k}}. \tag{9.102}$$

---

[3] Gaston Floquet developed a similar theorem in 1883 in connection with solutions to linear differential equations.

9.4 Electrons in a Periodic Potential

This relation plays an important role in the study of the response of Bloch electrons to external fields (see Sec. 9.4.9).

To prove Bloch's theorem, consider the unitary operators for translation by Bravais lattice vectors, $\mathcal{U}(\mathbf{R})$, defined by their action on a state $\psi_\mathbf{k}(\mathbf{r})$,

$$\mathcal{U}(\mathbf{R})\psi_\mathbf{k}(\mathbf{r}) = \psi_\mathbf{k}(\mathbf{r}+\mathbf{R}). \tag{9.103}$$

Note that we are using a slightly different notation than in Eq. (1.51); here, we use $\mathcal{U}(\mathbf{R}) = U^{-1}(\mathbf{R})$. The single electron Hamiltonian $H = \frac{\mathbf{p}^2}{2m} + V(\mathbf{r})$ commutes with $\mathcal{U}(\mathbf{R})$,

$$\mathcal{U}(\mathbf{R})H = H\mathcal{U}(\mathbf{R}), \tag{9.104}$$

because $\mathcal{U}(\mathbf{R})V(\mathbf{r})\mathcal{U}^\dagger(\mathbf{R}) = V(\mathbf{r}+\mathbf{R}) = V(\mathbf{r})$ and $\mathcal{U}(\mathbf{R})\frac{\mathbf{p}^2}{2m}\mathcal{U}^\dagger(\mathbf{R}) = \frac{\mathbf{p}^2}{2m}$, hence,

$$\mathcal{U}(\mathbf{R})H\psi_\mathbf{k}(\mathbf{r}) = H\mathcal{U}(\mathbf{R})\psi_\mathbf{k}(\mathbf{r}). \tag{9.105}$$

Moreover, the translation operators commute,

$$\mathcal{U}(\mathbf{R})\mathcal{U}(\mathbf{R}') = \mathcal{U}(\mathbf{R}')\mathcal{U}(\mathbf{R}) = \mathcal{U}(\mathbf{R}+\mathbf{R}'), \tag{9.106}$$

and therefore, $\mathcal{U}(\mathbf{R})\mathcal{U}(\mathbf{R}')\psi(\mathbf{r}) = \psi_\mathbf{k}(\mathbf{r}+\mathbf{R}+\mathbf{R}')$. Equations (9.104) and (9.106) show that the Hamiltonian and the translation operators for all Bravais lattice vectors form a set of commuting operators. Therefore, the eigenstates of H can also be chosen to be simultaneous eigenstates of all $\mathcal{U}(\mathbf{R})$, i.e.,

$$H\psi_\mathbf{k}(\mathbf{r}) = E_\mathbf{k}\psi_\mathbf{k}(\mathbf{r}), \tag{9.107a}$$
$$\mathcal{U}(\mathbf{R})\psi_\mathbf{k}(\mathbf{r}) = c_\mathbf{k}(\mathbf{R})\psi_\mathbf{k}(\mathbf{r}). \tag{9.107b}$$

Because $\mathcal{U}(\mathbf{R})$ is unitary, $|c_\mathbf{k}(\mathbf{R})|^2 = 1$. Furthermore, from Eq. (9.106), we see that

$$c_\mathbf{k}(\mathbf{R})c_\mathbf{k}(\mathbf{R}') = c_\mathbf{k}(\mathbf{R}+\mathbf{R}'); \tag{9.108}$$

hence, we can always write $c_\mathbf{k}(\mathbf{R})$, with $\mathbf{R} = \mathbf{R}_{n_1,n_2,n_3} = n_1\mathbf{a}_1 + n_2\mathbf{a}_2 + n_3\mathbf{a}_3$ being a Bravais lattice vector, as a plane wave

$$c_\mathbf{k}(\mathbf{R}) = e^{i\mathbf{k}\cdot\mathbf{R}}. \tag{9.109}$$

Equations (9.107b), (9.103), and (9.109) prove the Bloch theorem in the form of Eq. (9.101). Note that this proof goes through even if interactions between electrons are present. Hence, the Bloch theorem is very general. Note that, although $u_\mathbf{k}(\mathbf{r})$ is periodic with period $\mathbf{R}$, $\psi_\mathbf{k}(\mathbf{r})$ is not, because $c_\mathbf{k}(\mathbf{R})$ is in general not equal to 1.

Born–von Karman Periodic Boundary Conditions

Let us now consider the boundary conditions for the wave functions and the quantization of the wavevector $\mathbf{k}$. The Bloch functions as defined above are not normalizable. As in many problems in condensed matter physics, it is useful to confine the system in a large box and impose boundary conditions. In Sec. 9.1, we considered either periodic or hard-wall boundary conditions for free electrons in a box whose faces lie on Cartesian planes. Here, instead of a rectangular box, it is natural to consider a parallelepiped determined by the vectors $\mathbf{L}_1$, $\mathbf{L}_2$, and $\mathbf{L}_3$, which are oriented along the primitive lattice vectors, such that

$$\mathbf{L}_1 = N_1\mathbf{a}_1, \quad \mathbf{L}_2 = N_2\mathbf{a}_2, \quad \mathbf{L}_3 = N_3\mathbf{a}_3, \tag{9.110}$$

where N_1, N_2, and N_3 are positive integers. The number N of sites in the crystal is then $N = N_1 N_2 N_3$, and the volume of the box is

$$\mathcal{V}_\text{box} = \mathbf{L}_1 \cdot \mathbf{L}_2 \times \mathbf{L}_3 = N_1 N_2 N_3 (\mathbf{a}_1 \cdot \mathbf{a}_2 \times \mathbf{a}_3) = N\mathcal{V}_\text{cell}. \tag{9.111}$$

We can let the wave functions satisfy *Born–von Karman periodic boundary conditions*,

$$\psi_\mathbf{k}(\mathbf{r}+\mathbf{L}_1) = \psi_\mathbf{k}(\mathbf{r}+\mathbf{L}_2) = \psi_\mathbf{k}(\mathbf{r}+\mathbf{L}_3) = \psi_\mathbf{k}(\mathbf{r}). \tag{9.112}$$

To insure these boundary conditions, the wavevectors are written in terms of the reciprocal lattice vectors, Eq. (9.82), which are quantized as follows:

$$\boxed{\mathbf{k}_{v_1,v_2,v_3} = v_1 \mathbf{b}_1 + v_2 \mathbf{b}_2 + v_3 \mathbf{b}_3, \quad \text{with } \mathbf{b}_i \cdot \mathbf{a}_j = 2\pi \delta_{ij}.} \tag{9.113}$$

Here, $v_1 = m_1/N_1$, $v_2 = m_2/N_2$, and $v_3 = m_3/N_3$, with positive integers m_1, m_2, and m_3 ranging up to N_1, N_2, and N_3. Because $0 \leq v_1, v_2, v_3 \leq 1$, the vector $\mathbf{k}$ is confined within the primitive unit cell of the reciprocal lattice. It is customary to let the integers m_1, m_2, and m_3 run over negative integers, such that the wavevectors $\mathbf{k}$ are contained in the (first) Brillouin zone, e.g., when the Brillouin zone is a parallelepiped, $-N_1/2 \leq m_1 \leq N_1/2$, $-N_2/2 \leq m_2 \leq N_2/2$, $-N_3/2 \leq m_3 \leq N_3/2$, for N_1, N_2, and N_3 also. Because the volume of the primitive cell in the reciprocal lattice and the Brillouin zone are equal, we arrive at an important result:

The number N of allowed quantum wavenumbers $\mathbf{k}$ in a primitive cell of the reciprocal lattice (or in the first Brillouin zone) is equal to the number of primitive unit cells in the direct lattice.

In term of these quantized values of $\mathbf{k}$, the eigenvalues of the translation operators $\mathcal{U}(\mathbf{R})$ defined in Eq. (9.107b) become,

$$c_{\mathbf{k}}(\mathbf{R}) = (e^{2\pi i v_1})^{n_1} (e^{2\pi i v_2})^{n_2} (e^{2\pi i v_3})^{n_3}$$
$$= (e^{2\pi i n_1 m_1/N_1})(e^{2\pi i n_2 m_2/N_2})(e^{2\pi i n_3 m_3/N_3}). \tag{9.114}$$

Once the Bloch wave functions are defined within a finite box, Eq. (9.111), they can be normalized to unity. Because $u_{\mathbf{k}}(\mathbf{r})$ is periodic, the normalization of $\psi_{\mathbf{k}}(\mathbf{r})$ over $\mathcal{V}_{\text{box}}$ is translated into that of $u_{\mathbf{k}}(\mathbf{r})$ over $\mathcal{V}_{\text{cell}}$. If the Bloch function is expressed as in Eq. (9.100), then

$$1 = \int_{\mathcal{V}_{\text{box}}} d\mathbf{r}\, |\psi_{\mathbf{k}}(\mathbf{r})|^2 = \int_{\mathcal{V}_{\text{box}}} d\mathbf{r}\, |u_{\mathbf{k}}(\mathbf{r})|^2 = N \int_{\mathcal{V}_{\text{cell}}} d\mathbf{r}\, |u_{\mathbf{k}}(\mathbf{r})|^2. \tag{9.115}$$

On the other hand, if we adopt the normalization as per the free electron gas with periodic boundary conditions, as in Eq. (9.3),

$$\psi_{\mathbf{k}}(\mathbf{r}) = \frac{e^{i\mathbf{k}\cdot\mathbf{r}}}{\sqrt{\mathcal{V}_{\text{box}}}} u_{\mathbf{k}}(\mathbf{r}), \tag{9.116}$$

then

$$\frac{1}{\mathcal{V}_{\text{cell}}} \int_{\mathcal{V}_{\text{cell}}} d\mathbf{r}\, |u_{\mathbf{k}}(\mathbf{r})|^2 = 1. \tag{9.117}$$

The Bloch functions defined according to Eq. (9.116) are orthonormal,

$$\int_{\mathcal{V}_{\text{box}}} d\mathbf{r}\, \psi_{\mathbf{k}}(\mathbf{r})^* \psi_{\mathbf{q}}(\mathbf{r}) = \frac{1}{\mathcal{V}_{\text{box}}} \int_{\mathcal{V}_{\text{box}}} d\mathbf{r}\, e^{i(\mathbf{q}-\mathbf{k})\cdot\mathbf{r}} u_{\mathbf{k}}(\mathbf{r})^* u_{\mathbf{q}}(\mathbf{r}) = \frac{1}{N\mathcal{V}_{\text{cell}}} \sum_{\mathbf{R}} e^{i(\mathbf{q}-\mathbf{k})\cdot\mathbf{R}}$$
$$\times \int_{\mathcal{V}_{\text{cell}}} d\mathbf{r}\, e^{i(\mathbf{q}-\mathbf{k})\cdot\mathbf{r}} u_{\mathbf{k}}(\mathbf{r})^* u_{\mathbf{q}}(\mathbf{r}) = \frac{\delta_{\mathbf{kq}}}{\mathcal{V}_{\text{cell}}} \int_{\mathcal{V}_{\text{cell}}} d\mathbf{r}\, |u_{\mathbf{k}}(\mathbf{r})|^2 = \delta_{\mathbf{kq}}. \tag{9.118}$$

We took $\mathbf{r} \to \mathbf{R} + \mathbf{r}$ in the third equality, where $\mathbf{r}$ is in the unit cell.

9.4 Electrons in a Periodic Potential

The Vector $\hbar\mathbf{k}$ as Crystal Momentum

The wavevector $\mathbf{k}$ appearing in the Bloch wave function (9.100) defines the *crystal momentum* or *quasi-momentum*, $\hbar\mathbf{k}$, for electrons in a periodic crystal. The latter is related to, but not equal to, the electron momentum, as can be seen by calculating the expectation value of the momentum operator with the Bloch wave function,

$$\langle\psi_\mathbf{k}|\frac{\hbar}{i}\nabla|\psi_\mathbf{k}\rangle = \hbar\mathbf{k}\langle\psi_\mathbf{k}|\psi_\mathbf{k}\rangle + \langle u_\mathbf{k}|\frac{\hbar}{i}\nabla|u_\mathbf{k}\rangle = \hbar\mathbf{k} + \langle u_\mathbf{k}|\frac{\hbar}{i}\nabla|u_\mathbf{k}\rangle, \quad (9.119)$$

because $\langle\psi_\mathbf{k}|\psi_\mathbf{k}\rangle = 1$ by Eq. (9.118). The second term on the RHS of this equation can be evaluated by expanding the periodic function $u_\mathbf{k}$ in plane waves with wavevectors in the first Brillouin zone, BZ_1,

$$u_\mathbf{k}(\mathbf{r}) = \sum_{\mathbf{q}\in BZ} d_\mathbf{q}(\mathbf{k})e^{i\mathbf{q}\cdot\mathbf{r}}. \quad (9.120)$$

Substituting into the RHS of Eq. (9.119) yields

$$\langle\psi_\mathbf{k}|\frac{\hbar}{i}\nabla|\psi_\mathbf{k}\rangle = \hbar\mathbf{k} + \sum_{\mathbf{q}\in BZ} \hbar\mathbf{q}|d_\mathbf{q}(\mathbf{k})|^2. \quad (9.121)$$

The relation between the momentum $\mathbf{p}$ (i.e., the expectation of the momentum operator $\hat{\mathbf{p}}$) and crystal momentum $\hbar\mathbf{k}$ is

$$\langle\psi_\mathbf{k}|\hat{\mathbf{p}}|\psi_\mathbf{k}\rangle = \hbar\mathbf{k} + \sum_{\mathbf{q}\in BZ} \hbar\mathbf{q}|d_\mathbf{q}(\mathbf{k})|^2. \quad (9.122)$$

Problem 9.20

Use Eq. (9.118) to prove that the amplitudes $d_\mathbf{q}$ are normalized, such that $\sum_{\mathbf{q}\in BZ}|d_\mathbf{q}(\mathbf{k})|^2 = 1$.

The *crystal momentum* $\hbar\mathbf{k}$ enters the selection rules for transitions that govern collision processes in crystals. If an electron in state $\psi_{n\mathbf{k}}(\mathbf{r})$ absorbs a phonon (i.e., a quantized vibrational mode of the lattice) of wavevector $\mathbf{k}_{ph}$, due to a vibrationally inelastic collision, the selection rule for the crystal momentum is $\mathbf{k} + \mathbf{k}_{ph} = \mathbf{k}' + \mathbf{q}$. Here, $\mathbf{k}'$ is the final crystal momentum and $\mathbf{q}$ is one of the reciprocal lattice vectors, i.e., the electron is scattered from a state with crystal momentum $\hbar\mathbf{k}$ to a state with crystal momentum $\hbar(\mathbf{k} + \mathbf{k}_{ph} - \mathbf{q})$.

Finally, it is worth pointing out that in many cases, when the system size becomes very large, summation over wavevector $\mathbf{q}$ can be replaced by integration over a continuous variable $\mathbf{q}$:

$$\frac{1}{V}\sum_\mathbf{q} f(\mathbf{q}) \rightarrow \frac{1}{(2\pi)^3}\int d\mathbf{q}\, f(\mathbf{q}). \quad (9.123)$$

Problem 9.21

Obtain the Schrödinger equation for $u_\mathbf{k}(\mathbf{r})$ starting from Eq. (9.99).

Answer: $\left[-\frac{\hbar^2}{2m}\nabla^2 + \frac{\hbar}{m}\mathbf{k}\cdot\mathbf{p} + V(\mathbf{r})\right]u_\mathbf{k}(\mathbf{r}) = \left(E_\mathbf{k} - \frac{\hbar^2 k^2}{2m}\right)u_\mathbf{k}(\mathbf{r})$. This will be the starting point for $\mathbf{k}\cdot\mathbf{p}$ perturbation theory in Sec. 9.6.7.

9.4.5 SCHRÖDINGER EQUATION IN RECIPROCAL LATTICE SPACE

As indicated in Sec. 9.3.2, any periodic function with the lattice period $\mathbf{R}$ can be expanded in Fourier series in terms of the reciprocal lattice vectors $\mathbf{q}$ defined in Eq. (9.84), (e.g., see expansion (9.85) of the density). This type of expansion can be used to determine the solutions of the Schrödinger equation (9.99). First, let us expand the periodic potential,

$$V(\mathbf{r}) = \sum_{\mathbf{q}} V_{\mathbf{q}} e^{i\mathbf{q}\cdot\mathbf{r}}, \quad V_{\mathbf{q}} = V_{-\mathbf{q}}^*, \tag{9.124}$$

where the latter equality insures that $V(\mathbf{r})$ is real. Then, using Eq. (9.120), expand the wave function $\psi_{\mathbf{k}}(\mathbf{r})$ as,

$$\psi_{\mathbf{k}}(\mathbf{r}) = e^{i\mathbf{k}\cdot\mathbf{r}} u_{\mathbf{k}}(\mathbf{r}) = e^{i\mathbf{k}\cdot\mathbf{r}} \sum_{\mathbf{q}} d_{\mathbf{q}}(\mathbf{k}) e^{i\mathbf{q}\cdot\mathbf{r}}. \tag{9.125}$$

Here, the vector $\mathbf{k}$ is quantized, as in Eq. (9.113), with N different wavevectors $\mathbf{k}$. Although, in principle, an infinite number of coefficients $d_{\mathbf{q}}(\mathbf{k})$ are required, in practice, a small finite number $M_{\mathbf{k}}$ is often sufficient. Hence, the original problem has been recast into a set of $M_{\mathbf{k}}$ linear equations. Indeed, substituting Eqs (9.124) and (9.125) into the Schrödinger equation, and using the orthogonality of the plane waves, yields a set of algebraic equations for the unknown coefficients $d_{\mathbf{q}}(\mathbf{k})$,

$$\boxed{\left(\frac{\hbar^2(\mathbf{k}+\mathbf{q})^2}{2m} - E_{\mathbf{k}}\right) d_{\mathbf{q}}(\mathbf{k}) + \sum_{\mathbf{q}'} V_{\mathbf{q}'} d_{\mathbf{q}-\mathbf{q}'}(\mathbf{k}) = 0.} \tag{9.126}$$

For any fixed $\mathbf{k}$, Eq. (9.126) is a standard eigenvalue–eigenvector equation, which can be compactly written as

$$H(\mathbf{k})\,\mathbf{d}(\mathbf{k}) = E_{\mathbf{k}}\,\mathbf{d}(\mathbf{k}), \tag{9.127a}$$

$$[H(\mathbf{k})]_{\mathbf{q}\mathbf{q}'} = \left[\frac{\hbar^2(\mathbf{k}+\mathbf{q})^2}{2m}\right]\delta_{\mathbf{q}\mathbf{q}'} + V_{\mathbf{q}-\mathbf{q}'}. \tag{9.127b}$$

The vectors $\mathbf{q}$ in reciprocal lattice space can be ordered and numbered, so that, in writing the matrix $[H(\mathbf{k})]_{\mathbf{q}\mathbf{q}'}$, the subscripts $\mathbf{q}\mathbf{q}'$ refer to the corresponding numbers. The eigenvalues $E_{\mathbf{k}}$ can be obtained by solving the secular equation,

$$\det[H(\mathbf{k}) - E_{\mathbf{k}}\mathbf{1}] = 0, \tag{9.128}$$

where $\mathbf{1}$ is the unit matrix of dimension $M_{\mathbf{k}}$. For any $\mathbf{k}$, there are $M_{\mathbf{k}}$ equations, whose solution yields the eigenvalues $E_{n\mathbf{k}}$ with $n = 1, 2, \ldots M_{\mathbf{k}}$, and then the eigenvectors $\mathbf{d}_n(\mathbf{k})$ can be determined by solving a set of linear equations. The quantum number n is the *band index*, which together with the wavevector $\mathbf{k}$ characterizes the wave functions $\psi_{n\mathbf{k}}(\mathbf{r}) = e^{i\mathbf{k}\cdot\mathbf{r}} u_{n\mathbf{k}}(\mathbf{r})$ and eigenvalues $E_{n\mathbf{k}}$ for a Schrödinger problem in a periodic potential. The eigenvalues (and eigenfunctions) are periodic in $\mathbf{k}$ with periods of the reciprocal lattice vectors, $\mathbf{q}$,

$$\boxed{E_{n(\mathbf{k}+\mathbf{q})} = E_{n\mathbf{k}}, \quad \psi_{n(\mathbf{k}+\mathbf{q})}(\mathbf{r}) = \psi_{n\mathbf{k}}(\mathbf{r}).} \tag{9.129}$$

Hence, it is possible (but not necessary) to restrict the crystal momentum $\mathbf{k}$ within BZ_1.

As an example, consider a 1D lattice of period a, and a primitive reciprocal lattice vector $G = 2\pi/a$. The reciprocal lattice vectors are numbered as qG, $q = 0, \pm 1, \ldots$. Let us further assume that the expansion (9.124) of the potential has only two components, $V_1 = V_{-1} = V$ and $V_{|q|>1} = 0$, so the potential in coordinate space is $V(x) = V(e^{iGx} + e^{-iGx}) = 2V\cos(Gx)$. The Hamiltonian (9.127b) has the form,

$$[H(k)]_{qq'} = \frac{\hbar^2}{2m}(k+qG)^2 \delta_{qq'} + V(\delta_{q,q'+1} + \delta_{q,q'-1}).$$

9.4 Electrons in a Periodic Potential

With integers $-2 \leq q \leq 2$, the Hamiltonian then takes the form

$$\begin{pmatrix} \frac{\hbar^2(k-2G)^2}{2m} & V & 0 & 0 & 0 \\ V & \frac{\hbar^2(k-G)^2}{2m} & V & 0 & 0 \\ 0 & V & \frac{\hbar^2 k^2}{2m} & V & 0 \\ 0 & 0 & V & \frac{\hbar^2(k+G)^2}{2m} & V \\ 0 & 0 & 0 & V & \frac{\hbar^2(k+2G)^2}{2m} \end{pmatrix}. \quad (9.130)$$

Although the matrix (9.127b) is generically large, it is often sufficient to consider only a small q range of integers. For illustration, we consider the 5×5 matrix shown in Eq. (9.130). It has five eigenvalues E_{nk}, $n = 1, 2, 3, 4, 5$, thereby defining five different energy bands. Inclusion of additional rows and columns in Eq. (9.130) will yield additional roots and will modify the energies of the five roots E_{nk}. After defining the Mathieu functions, we shall analyze in more detail the case of a 2×2 matrix, which is qualitatively adequate when **k** is close to the Bragg points **G**/2.

9.4.6 SINUSOIDAL POTENTIAL: MATHIEU FUNCTIONS

It is instructive to study a relatively simple 1D model with a periodic potential. The Schrödinger equation for a particle of mass m in a sinusoidal potential $V(z) = V_0 \sin^2(k_L z) = (V_0/2)[1 - \cos(2k_L z)]$ can be written as $\{p^2/2m + (V_0/2)[1 - \cos(2k_L z)]\}\psi(z) = E\psi(z)$. Using the dimensionless variable $x = k_L z$, this equation can be rewritten as

$$\frac{d^2\psi(x)}{dx^2} + [a - 2q\cos 2x]\psi(x) = 0, \quad (9.131)$$

where

$$a \equiv (2E - V_0)/2E_R, \quad q = -V_0/4E_R, \quad E_R \equiv \frac{\hbar^2 k_L^2}{2m}. \quad (9.132)$$

Here, a and q are dimensionless energy and potential strength, respectively. Equation (9.131) is the dimensionless *Mathieu equation*, first introduced by Émile Mathieu in 1868 for analyzing the motion of elliptical membranes. The solutions are called *Mathieu functions*; they are presented in Abramowitz and Stegun [27]. Mathieu equations with solutions satisfying periodic as well as hard-wall boundary conditions are of relevance for a number physical applications. In the context of solid-state physics, only the former ones are relevant. For periodic boundary conditions, the solutions can be written in a Bloch function form. Even (cosine-like) solutions of Eq. (9.131) are denoted by the symbol $C(a,q,x)$ and odd solutions by $S(a,q,x)$. For $q=0$, we have $C(a,0,x) = \cos(\sqrt{a}x)$ and $S(a,0,x) = \sin(\sqrt{a}x)$. For nonzero q, the Mathieu functions C and S are periodic in x only for certain characteristic eigenvalues denoted as $a_n(q)$ and $b_n(q)$, where n is a positive integer. The even and odd Mathieu functions with characteristic values $a_n(q)$ and $b_n(q)$ are often denoted as $ce_n(x,a)$ and $se_n(x,b)$ and are known as the elliptic cosine and elliptic sine functions, respectively. Note that the Bloch theorem does not require $\psi_k(x)$ to be periodic; only $u_k(x)$ need be periodic. Existence of periodic solutions is a specific feature of the Mathieu equation that is a special case of the Bloch equation.

Now let us use the formalism developed above to obtain the solution of Eq. (9.131). The potential $\cos(2x)$ has a (dimensionless) period $x_p = \pi$, and the reciprocal lattice vector is $b = 2\pi/x_p = 2$. Therefore, the Brillouin zone is $-1 \leq k \leq 1$. The potential (9.124) is,

$$\cos 2x = V_1 e^{2ix} + V_{-1} e^{-2ix}, \quad V_1 = V_{-1} = \frac{1}{2}.$$

The solutions of Eq. (9.131) can be represented as Bloch wave functions, $\psi_k(x)$, using the expansion (9.125)

$$\psi_k(x) = e^{ikx} u_k(x) = e^{ikx} \sum_{q=-\infty}^{\infty} d_q(k) e^{2iqx}, \quad \left(-\frac{\pi}{2} \leq x \leq \frac{\pi}{2}\right), \quad (-1 \leq k \leq 1). \quad (9.133)$$

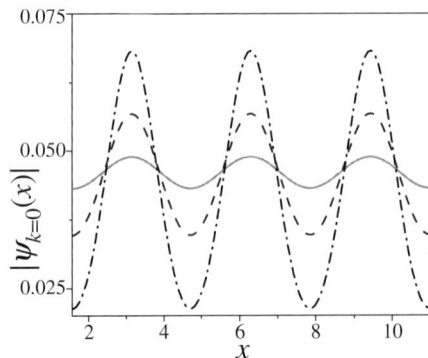

FIG 9.29 $|\psi_{k=0}(x)|$ of Eq. (9.131) with periodic boundary conditions for $V_0 = 0.5\,E_R$ (solid curve), $V_0 = 2\,E_R$ (dashed curve), and $V_0 = 5\,E_R$ (dot-dashed curve). The wave functions are periodic and increasingly localized at the center of each site as V_0 increases.

Equation (9.126) then reads

$$[(k+q)^2 - \varepsilon(k)]d_q(k) + \frac{1}{2}[d_{q+1}(k) + d_{q-1}(k)] = 0. \tag{9.134}$$

This is an eigenvalue equation for $\varepsilon_n(k)$ with eigenvectors $\mathbf{d}_n(k)$, where n is the band index. The energy is $E_n(k) = \frac{\hbar^2 k_L^2}{2m}\varepsilon_n(k)$. The restriction $k \in [-1, 1]$ as a continuous variable assumes an infinite lattice. For a finite lattice including $2N$ periods, the Born–von Karman condition $\psi_k(x) = \psi_k(x+2Na) = \psi_k(x+2N\pi)$ implies that k is quantized with $k_j = -\frac{j}{N}$, $-N \le j \le N$. A solution of Eq. (9.131) for $k = 0$ is shown in Fig. 9.29.

An important application of sinusoidal potentials is in the description of atoms in a standing wave light field, because the intensity of a standing-wave light field is sinusoidal; hence, Eq. (9.131) applies. For such a light field with period $\lambda_L/2 = \pi/k_L$, atoms feel a periodic optical potential with strength V_0 proportional to the intensity of the field. The parameter E_R is then called the recoil energy, because this is the energy imparted to the atom on absorption of a photon of wavevector k_L.

The Wave Function Near a Band-Edge

Pursuing our analysis of the 1D case, the deviation of the energy E_{nk} from the free electron energy $E_k = \frac{\hbar^2 k^2}{2m}$ is most notable when the wavevector k is at the Brillouin zone boundary, $k = G/2$. Gap formation occurs when the diagonal elements $(k+qG)^2$ of the Hamiltonian are degenerate and the potential V acts as a perturbation leading to *avoided level-crossing*. Therefore, it is sufficient to concentrate on the 2×2 matrix obtained from Eq. (9.130) by selecting $q = 0, -1$. The restricted eigenvalue equation for $k = G/2$ then reads,

$$\begin{pmatrix} \frac{\hbar^2 G^2}{8m} - E & V \\ V & \frac{\hbar^2 G^2}{8m} - E \end{pmatrix} \begin{pmatrix} d_0(k) \\ d_{-1}(k) \end{pmatrix} = 0. \tag{9.135}$$

The two eigenvalues are given by

$$E_{n=\pm,k=\frac{G}{2}} = \frac{\hbar^2 G^2}{8m} \pm V. \tag{9.136}$$

Thus, the periodic potential leads to avoided crossing. The levels that crossed at $k = G/2$ in the absence periodic potential now have an avoided crossing. This creates a band gap, $E_g = 2V$, at the Brillouin zone boundary (the Bragg plane in 3D), as illustrated in Fig. 9.30.

The corresponding eigenvectors are

$$\mathbf{d}_+ = \frac{1}{\sqrt{2}}\begin{pmatrix}1 \\ -1\end{pmatrix}, \quad \mathbf{d}_- = \frac{1}{\sqrt{2}}\begin{pmatrix}1 \\ 1\end{pmatrix}, \tag{9.137}$$

and the wave functions at the zone boundary, following Eq. (9.125), are standing waves,

$$\psi_{n=\pm, k=G/2}(x) = \frac{1}{\sqrt{2}} e^{iGx/2}(e^{-iGx} \pm 1) = \frac{1}{\sqrt{2}}(e^{-iGx/2} \pm e^{iGx/2}). \tag{9.138}$$

9.4 Electrons in a Periodic Potential

When k is in the vicinity of $G/2$, the dispersion relation determined by the 2×2 matrix extracted from Eq. (9.130) is given by

$$E_{n=\pm,k} = \frac{1}{2}\left(\frac{\hbar^2 k^2}{8m} + \frac{\hbar^2 (k-G/2)^2}{8m}\right) \pm \left[\left(\frac{\hbar^2 k^2}{8m} - \frac{\hbar^2 (k-G/2)^2}{8m}\right)^2 + V^2\right]^{1/2}. \quad (9.139)$$

Besides opening a gap $E_g = 2V$ in the vicinity of $k = G/2$, a weak periodic potential has a sizable effect on the density of states at the zone boundary, where the gradient of $E(k)$ goes to zero. To obtain the density of states, i.e., the number of states in the energy ranges between E and $E + dE$, we first determine the element of volume in wavevector space, $d\mathbf{k}$, associated with a differential energy dE and then integrate over the reciprocal lattice points, such that the energy is between E and $E + dE$,

$$D(E) = \frac{dN(E)}{dE} = \frac{dN(E)}{dk}\frac{dk}{dE} = 2\frac{V}{(2\pi\hbar)^3}\oint_S \frac{dS}{|\nabla_{\mathbf{k}} E(\mathbf{k})|}. \quad (9.140)$$

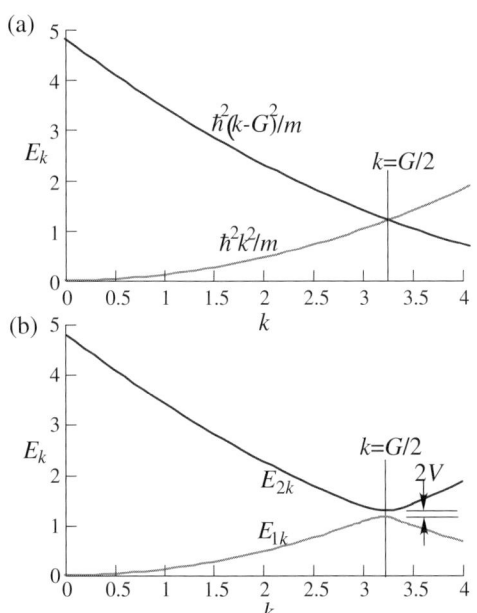

FIG 9.30 Gap opening that results in two bands E_{1k} and E_{2k}. (a) The two free-particle like curves $\frac{\hbar^2 k^2}{2m}$ and $\frac{\hbar^2(k-G)^2}{2m}$ cross each other at $k = G/2$ where they are degenerate. (b) Due to the periodic potential, the levels are repelled and crossing is avoided. A gap $E_G = 2V$ occurs at $k = G/2$, separating the lower energy (valence) band E_{1k} from the upper energy (conduction) band E_{2k}.

Because $E_{\mathbf{k}}$ [which we will sometimes write as $E(\mathbf{k})$] is periodic, its gradient must vanish at specific points, leading to a singularity structure of the density of states. These singularities are referred to as *Van Hove singularities*. Away from the Bragg point $k = \pm G/2$, the wave functions are plane waves with $E_{\mathbf{k}} = \frac{\hbar^2 k^2}{2m}$. This gives the free electron density of states, $D(E) = \frac{V}{2\pi^2}(\frac{2m}{\hbar^2})^{3/2} E^{1/2}$, as can be easily verified by noting that $|\nabla_{\mathbf{k}} E(\mathbf{k})| = \hbar^2 |k|/m$ and $\int dS = 4\pi k^2$.

Effective Mass

The example detailed above for a 1D crystal is illustrative of what happens in higher dimensions, but in two- and three-dimensional crystals, the electronic band structure is more complicated. For each band n, the energies $E_n(\mathbf{k})$ are functions of three variables k_x, k_y, and k_z, and the functional form may be rather complex. However, for describing transport properties, it is sufficient to focus only on the valence and conduction bands denoted here as $E_v(\mathbf{k})$ and $E_c(\mathbf{k})$, respectively, as illustrated in Fig. 9.26. Suppose the local maximum of the valence band $E_v(\mathbf{k})$ occurs at some wavenumber $\mathbf{k}_v$. The conduction band energy $E_c(\mathbf{k})$ might have a minimum at some other wavenumber $\mathbf{k}_c$. In the following, we analyze the energy of an electron as a function of wavevector near the band extrema, i.e., the minima of the conduction band and the maxima of the valence band. For notational convenience, the subscripts x, y, and z are replaced by 1, 2, and 3 and the extremum E_0 which occurs at some wavevector $\mathbf{k}_0$ (either $\mathbf{k}_v$ for the valence band or $\mathbf{k}_c$ for the conduction band) is shifted to $E_0 = 0$, so, the Taylor expansion of $E(\mathbf{k})$

around $\mathbf{k}_0$ starts with the quadratic term,

$$E(\mathbf{k}) = \sum_{ij=1}^{3} a_{ij}(k-k_0)_i(k-k_0)_j + \ldots. \tag{9.141}$$

The coefficients a_{ij} have the same dimension as $\frac{\hbar^2}{2m}$, so it is useful to define an *effective mass tensor*, $M^* = \{m_{ij}^*\}$, as

$$\boxed{[(M^*)^{-1}]_{ij} \equiv \frac{1}{\hbar^2}\left(\frac{\partial^2 E_{\mathbf{k}}}{\partial k_i \partial k_j}\right)_{\mathbf{k}=\mathbf{k}_0},} \tag{9.142}$$

so that

$$E(\mathbf{k}) = \frac{\hbar^2}{2}\sum_{ij} q_i [(M^*)^{-1}]_{ij} q_j + \ldots, \tag{9.143}$$

where $\mathbf{q} \equiv \mathbf{k}-\mathbf{k}_0$. It is clear from the definition (9.142) that the mass tensor M is symmetric. For the valence band, $E_v(\mathbf{k}_0)$ is a local maximum; therefore, the matrix M is negative definite there, and the bilinear form (9.143) is negative, whereas for the conduction band, $E_c(\mathbf{k}_0)$ is a local minimum, M is positive definite, and the bilinear form (9.143) is positive. Although the choice of cartesian coordinate frame is dictated by the crystal geometry, it is important to note that as far as the analysis of the mass tensor is concerned, there is always a cartesian coordinate system in which the mass tensor M is diagonal, $M_{ij} = m_i \delta_{ij}$. In this *principal axes coordinate system*, the energy of an electron as a function of wavevector near the band extremum can be written as

$$E(\mathbf{k}) = \sum_{i=1}^{3} \frac{\hbar^2 q_i^2}{2m_i}, \tag{9.144}$$

where $m_i < 0$ for the valence band and $m_i > 0$ for the conduction band. The quantities m_i have the dimension of mass, and hence, they are referred to as effective masses. When the anisotropy is not large, it is useful to define the effective mass near a band extremum as,

$$m^* = [\det M^*]^{\frac{1}{3}} = [m_1 m_2 m_3]^{\frac{1}{3}}. \tag{9.145}$$

The following points should be noted:

- We show below that the effective mass enters the dynamics of electrons in periodic crystals, especially their response to external fields. Therefore, the definition (9.142) is not just a formal analogy with the free electron formalism.
- The absolute value and sign of m^* are determined by the *curvature* of the energy surface $E(\mathbf{k})$ at $\mathbf{k}_0$, and might be very different from the free electron mass m_e.
- The effective electron mass m^* is *negative* near the top of the valence band. This has deep physical implications, related to the concept of *holes*, to be explained below. On the other hand, the effective electron mass m^* near the top of the conduction band is positive.
- In metals, the Fermi energy is typically higher than the bottom of the conduction band. Electron transport occurs in the vicinity of the Fermi energy, hence, it is useful to carry out the Taylor expansion of $E(\mathbf{k})$ around the Fermi momentum, $\mathbf{k}_F$, i.e., a vector whose tip is on the fermi surface, and this expansion typically contains a linear term $[\nabla_\mathbf{k} E(\mathbf{k})]_{\mathbf{k}_F} \cdot (\mathbf{k}-\mathbf{k}_F)$. Therefore, the effective mass for electrons at the Fermi energy depends on the location of the vector $\mathbf{k}_F$ on the Fermi surface. It is still defined as in (9.142), but with $\mathbf{k}_0$ replaced by $\mathbf{k}_F$.
- The effective mass of an electron in semiconductors is determined at the bottom of the conduction band, and the effective mass of the hole is determined at the top of the valence band (see below).

9.4 Electrons in a Periodic Potential

9.4.7 TIGHT-BINDING MODEL

Section 9.4.5 focused on Bloch wave functions, which can be regarded as analogs of the free electron states. Given the similarity of Bloch wave functions and plane waves, it is clear that they are especially relevant when the valence electrons are not tightly bound to their parent atoms. Hence, one might expect Bloch functions to be adequate for describing the properties of electrons in ideal metals, but less effective for the description of electronic properties in transition metals or semiconductors where the nature of atomic orbitals should be accounted for in detail. In this section, we discuss one scheme for addressing this issue, the *tight-binding model*, while in the next section, we study a second approach, *Wannier functions*.

The tight-binding model is an independent electron (single-particle) model, wherein the description of electronic states starts from the limit of isolated atomic orbitals; hence, the name tight binding. It is based on the intuitive picture that an electron is mainly affected by its parent atom and the effect of all other atoms can be taken as a perturbation. Let us denote the position of an electron by $\mathbf{r}$ and assume that its parent atom is at $\mathbf{R} = 0$. The atomic potential of the parent atom is $V_0(\mathbf{r})$, whereas the potential of all other atoms is denoted by $U(\mathbf{r})$, such that the sum, $V_0(\mathbf{r}) + U(\mathbf{r}) = V(\mathbf{r})$, is the periodic potential used to derive the Bloch theorem. The Hamiltonian H for the electron in the crystal, which stresses the decomposition of the full potential into V_0 and U, reads,

$$H = H_0 + U(\mathbf{r}) = -\frac{\hbar^2}{2m}\nabla^2 + V(\mathbf{r}), \tag{9.146a}$$

$$H_0 = -\frac{\hbar^2}{2m}\nabla^2 + V_0(\mathbf{r}), \quad U(\mathbf{r}) \equiv \sum_{\mathbf{R}\neq 0} V_0(\mathbf{r}-\mathbf{R}), \quad V(\mathbf{r}) = V_0(\mathbf{r}) + U(\mathbf{r}). \tag{9.146b}$$

The eigenfunctions of H_0 satisfy the atomic Schrödinger equation, $H_0\phi_\lambda(\mathbf{r}) = E_\lambda \phi_\lambda(\mathbf{r})$, where λ represents the full set of the atomic quantum numbers and $\phi_\lambda(\mathbf{r})$ is the corresponding atomic wave function. In the simple case of a central potential, $\lambda = (\bar{n}lm)$, where $\bar{n}$ is the principal quantum number and lm are the orbital angular momentum and its projection. An arbitrary bound state localized around the atom at the origin $\mathbf{R} = 0$ can be expanded as

$$\varphi(\mathbf{r}) = \sum_\lambda a_\lambda \phi_\lambda(\mathbf{r}), \tag{9.147}$$

where the coefficients a_λ must properly encode the effect of $U(\mathbf{r})$. We shall see below that, in addition to the atomic quantum numbers λ, the coefficients a_λ depend on the crystal wavenumber $\mathbf{k}$ and the band index n that will be dropped to simplify the notation. Note that the continuum states of the atom are excluded from the sum. A wave function that satisfies the Bloch condition (9.101) (which is *not* a localized bound state) can be constructed by forming a linear combination of such localized atomic states,

$$\psi_\mathbf{k}(\mathbf{r}) = \sum_\mathbf{R} e^{i\mathbf{k}\cdot\mathbf{R}} \varphi(\mathbf{r}-\mathbf{R}), \tag{9.148}$$

composed of a sum of strongly localized atomic wave functions multiplied by the phase factors $e^{i\mathbf{k}\cdot\mathbf{R}}$. The requirement that the wave function $\psi_\mathbf{k}(\mathbf{r})$ satisfy the Schrödinger equation,

$$H\psi_\mathbf{k}(\mathbf{r}) = [H_0 + U(\mathbf{r})]\psi_\mathbf{k}(\mathbf{r}) = E_\mathbf{k}\psi_\mathbf{k}(\mathbf{r}), \tag{9.149}$$

determines the coefficients a_λ in expansion (9.147). Indeed, multiplying Eq. (9.149) on the left by $\phi_\mu^*(\mathbf{r})$, using the expansions (9.147) and (9.148) and integrating over $\mathbf{r}$ yields the condition,

$$\sum_\lambda [A(\mathbf{k})]_{\mu\lambda}\, a_\lambda = E_\mathbf{k} \sum_\lambda [B(\mathbf{k})]_{\mu\lambda}\, a_\lambda. \tag{9.150}$$

where

$$[A(\mathbf{k})]_{\mu\lambda} = \sum_{\mathbf{R}} e^{i\mathbf{k}\cdot\mathbf{R}} \int d\mathbf{r}\, \phi_\mu^*(\mathbf{r}) H \phi_\lambda(\mathbf{r}-\mathbf{R}), \qquad (9.151)$$

$$[B(\mathbf{k})]_{\mu\lambda} = \sum_{\mathbf{R}} e^{i\mathbf{k}\cdot\mathbf{R}} \int d\mathbf{r}\, \phi_\mu^*(\mathbf{r}) \phi_\lambda(\mathbf{r}-\mathbf{R}). \qquad (9.152)$$

Equation (9.150) is a generalized eigenvalue problem with eigenvalue $E_\mathbf{k}$ for the matrix $A(\mathbf{k})$ (with matrix $B(\mathbf{k})$ on the RHS in place of the identity).[4] If we define a new matrix $C(\mathbf{k})$,

$$[C(\mathbf{k})]_{\mu\lambda} \equiv \sum_{\mathbf{R}} e^{i\mathbf{k}\cdot\mathbf{R}} \int d\mathbf{r}\, \phi_\mu^*(\mathbf{r}) U(\mathbf{r}) \phi_\lambda(\mathbf{r}-\mathbf{R}). \qquad (9.153)$$

Equation (9.150) then takes the form,

$$\sum_\lambda [C(\mathbf{k})]_{\mu\lambda}\, a_\lambda = (E_\mathbf{k} - E_\mu) \sum_\lambda [B(\mathbf{k})]_{\mu\lambda}\, a_\lambda, \qquad (9.154)$$

The "eigenvalue" $(E_\mathbf{k} - E_\mu)$ of Eq. (9.154) is the shift of the band energy $E_\mathbf{k}$ with respect to the original atomic energy E_μ. The size of this eigenvalue problem is determined by the number of eigenstates of the atomic system retained in expansion (9.148). In the simple case of a central potential such that $\lambda = (\bar{n}lm)$ one usually indicates the quantum numbers $\bar{n}l$ pertaining to an open atomic shell. There are $d = 2l+1$-degenerate levels, so one solves a d-dimensional matrix problem for each $\mathbf{k}$.

The only case where one can legitimately neglect the interactions with all other levels is for atomic s orbitals. In this approximation, the matrix equation (9.154) is of dimension 1×1, and $a_\mu = 1$ with all the other a amplitudes vanishing. Then, $C(\mathbf{k})_{\mu\mu} = (E_\mathbf{k} - E_\mu) B(\mathbf{k})_{\mu\mu}$, so

$$E_\mathbf{k} = E_\mu + \frac{C(\mathbf{k})_{\mu\mu}}{B(\mathbf{k})_{\mu\mu}}. \qquad (9.155)$$

It is convenient to separate contributions in the infinite sums over Bravais lattice vectors in the definitions of the $B(\mathbf{k})$ and $C(\mathbf{k})$ matrices into the contributions determined by the distances of the Bravais lattice points $\mathbf{R}$ from the point $\mathbf{R}=0$. The first contribution is from $\mathbf{R}=0$, the second from the nearest neighbor contributions, then the contribution of next nearest neighbors, etc. The $\mathbf{R}=0$ contribution to $B(\mathbf{k})$ is unity, and its contribution to $C(\mathbf{k}) = \int d\mathbf{r}\, U(\mathbf{r}) |\phi_\mu(\mathbf{r})|^2$ is negative, reflecting the attraction the other nuclei produce on the electron. Denoting the contributions to the integrals in $B(\mathbf{k})$ and $C(\mathbf{k})$ from $\mathbf{R} \neq 0$ by

$$\beta_\mu(\mathbf{R}) = \int d\mathbf{r}\, \phi_\mu^*(\mathbf{r}) \phi_\mu(\mathbf{r}-\mathbf{R}), \qquad (9.156)$$

$$\gamma_\mu(\mathbf{R}) = \int d\mathbf{r}\, \phi_\mu^*(\mathbf{r}) U(\mathbf{r}) \phi_\mu(\mathbf{r}-\mathbf{R}), \qquad (9.157)$$

respectively, noting that symmetry dictates $\beta_\mu(-\mathbf{R}) = \beta_\mu(\mathbf{R})$ and $\gamma_\mu(-\mathbf{R}) = \gamma_\mu(\mathbf{R})$ and that in all Bravais lattice sums both $\mathbf{R}$ and $-\mathbf{R}$ are present, we can rearrange the solution for $E_\mathbf{k}$ in Eq. (9.155) for the energy bands originating from atomic s orbitals as follows:

$$E_\mathbf{k} = E_\mu + \frac{\int d\mathbf{r}\, U(\mathbf{r}) |\phi_\mu(\mathbf{r})|^2 + \sum_{\mathbf{R}\neq 0} \cos(\mathbf{k}\cdot\mathbf{R}) \gamma_\mu(\mathbf{R})}{1 + \sum_{\mathbf{R}\neq 0} \cos(\mathbf{k}\cdot\mathbf{R}) \beta_\mu(\mathbf{R})}. \qquad (9.158)$$

Both $\beta_\mu(\mathbf{R})$ and $\gamma_\mu(\mathbf{R})$ become exponentially small for large $\mathbf{R}$ due to the localized character of the atomic wave functions ϕ_μ. Hence, the sums over $\mathbf{R}$ in Eq. (9.158) can be truncated at some large R.

---

[4] The eigenvalue depends on the band index n, but we have dropped it to simplify the notation.

9.4 Electrons in a Periodic Potential

Problem 9.22

(a) Prove that $\psi_{\mathbf{k}}(\mathbf{r})$ defined in Eq. (9.148) satisfies the Bloch condition, $\psi_{\mathbf{k}}(\mathbf{r}+\mathbf{R}) = e^{i\mathbf{k}\cdot\mathbf{R}}\psi_{\mathbf{k}}(\mathbf{r})$ for a Bravais lattice vector $\mathbf{R}$.

(b) Show that Eq. (9.148) can be inverted by multiplying by $e^{-i\mathbf{k}\cdot\mathbf{R}'}$ and integrating over $\mathbf{k}$:

$$\varphi(\mathbf{r}-\mathbf{R}) = \int d\mathbf{k}\, e^{-i\mathbf{k}\cdot\mathbf{R}}\psi_{\mathbf{k}}(\mathbf{r}). \tag{9.159}$$

(c) Check whether functions $\psi_{\mathbf{k}}(\mathbf{r})$ and $\psi_{\mathbf{k}'}(\mathbf{r})$ defined in Eq. (9.148), but with $\mathbf{k} \neq \mathbf{k}'$, are orthogonal. **Answer:** No.

9.4.8 WANNIER FUNCTIONS

Wannier functions for a given band are defined in terms of the Bloch functions $\psi_{\mathbf{k}}$ of that band by

$$w(\mathbf{r},\mathbf{R}) \equiv \frac{1}{\sqrt{\mathcal{V}_{BZ}}}\int d\mathbf{k}\, e^{-i\mathbf{k}\cdot\mathbf{R}}\psi_{\mathbf{k}}(\mathbf{r}) = \frac{1}{\sqrt{\mathcal{V}_{BZ}}}\int d\mathbf{k}\, e^{-i\mathbf{k}\cdot(\mathbf{R}-\mathbf{r})} u_{\mathbf{k}}(\mathbf{r}), \tag{9.160}$$

where the $\mathbf{k}$ integration is carried over BZ_1, $\mathbf{R}$ is a Bravais lattice point, and $\mathcal{V}_{BZ}$ is the volume of BZ_1. These functions were introduced by Gregory H. Wannier in 1937 and have been extensively used as an alternative representation to Bloch wave functions. Wannier functions provide a local, atomic-orbital-like description of electronic wave functions. We shall display below the connection with tight-binding wave functions. Conventionally, Wannier functions are determined from the Bloch functions using the Wannier transformation (9.160).

Let us compare the Wannier function (9.160) with an arbitrary bound state localized around the atom at the origin $\mathbf{R} = 0$ as defined in Eq. (9.147). To do so, we multiply Eq. (9.160) by $e^{i\mathbf{k}'\cdot\mathbf{R}}$ and sum over $\mathbf{R}$, as in Eq. (9.148), and thereby obtain

$$\psi_{\mathbf{k}}(\mathbf{r}) = \sqrt{\mathcal{V}_{BZ}}\sum_{\mathbf{R}} e^{i\mathbf{k}\cdot\mathbf{R}} w(\mathbf{r},\mathbf{R}). \tag{9.161}$$

Equation (9.161) is equivalent to Eq. (9.148) if $w(\mathbf{r},\mathbf{R})$ is of the form $w(\mathbf{r}-\mathbf{R})$, i.e., if w is a function only of $\mathbf{r}-\mathbf{R}$. But this is indeed the case, as can be seen from Eq. (9.160) recalling that $u_{\mathbf{k}}(\mathbf{r})$ is periodic on the direct lattice due to the Bloch theorem. Thus, if both $\mathbf{r}$ and $\mathbf{R}$ are shifted by a given Bravais lattice vector, w is unchanged. Hence, Eq. (9.160) implies

$$w(\mathbf{r},\mathbf{R}) = w(\mathbf{r}-\mathbf{R}). \tag{9.162}$$

It is instructive to compare Eq. (9.162) with Eq. (9.159). The difference between the tight-binding functions and the Wannier functions is that there is no approximation in the Wannier function definition, whereas the tight-binding wave function is taken to be strictly localized on an atomic site and has no amplitude off this site, whereas the Wannier function $w(\mathbf{r}-\mathbf{R})$ can have amplitude off site $\mathbf{R}$.

Another important property of Wannier functions is their orthogonality. For a given band, the Wannier functions for different Bravais lattice points are orthogonal,

$$\int d\mathbf{r}\, w(\mathbf{r}-\mathbf{R})w(\mathbf{r}-\mathbf{R}') = \delta_{\mathbf{R},\mathbf{R}'}. \tag{9.163}$$

The proof is based on the fact that the transformation (9.160) is unitary and that the Bloch functions $\psi_{n\mathbf{k}}(\mathbf{r})$ are orthonormal on the unit cell, as shown in Eq. (9.118). Assuming a finite lattice with N sites with Born–von Karman periodic conditions, as in Eq. (9.112), we find that the integral on the LHS of Eq. (9.163) is

$$\frac{1}{N}\sum_{\mathbf{k},\mathbf{k}'} e^{i(\mathbf{k}\cdot\mathbf{R}-\mathbf{k}'\cdot\mathbf{R}')}\int d\mathbf{r}\,\psi_{\mathbf{k}}(\mathbf{r})^*\psi_{\mathbf{k}'}(\mathbf{r}) = \frac{1}{N}\sum_{\mathbf{k},\mathbf{k}'} e^{i(\mathbf{k}\cdot\mathbf{R}-\mathbf{k}'\cdot\mathbf{R}')}\delta_{\mathbf{k},\mathbf{k}'} = \delta_{\mathbf{R},\mathbf{R}'}.$$

The orthogonality property often turns the Wannier functions to be of greater use than atomic orbitals centered on different lattice sites that were used in the tight-binding model, because the latter are, generically, not orthogonal.

9.4.9 ELECTRIC FIELD EFFECTS

The response of a solid to a (possibly time-dependent) external electric field $\mathbf{E}(\mathbf{r},t)$ determines whether it is a metal or an insulator. On applying an electric field across a metal, a current density $\mathbf{J}(\mathbf{r},t)$ is generated. Finding the precise relation between $\mathbf{J}$ and $\mathbf{E}$ is a central challenge in the physics of metals. On the other hand, applying an electric field across an insulator does not generate current, but rather a polarization density is generated within the material. In this section, we discuss the response of metals and insulators to an external electric field. The response of Bloch electrons to an external magnetic field, $\mathbf{H}(\mathbf{r},t)$, is discussed in Sec. 9.5.

The Hamiltonian for an electron in a periodic potential $V(\mathbf{r})$ in the presence of an applied electric field $\mathbf{E}(\mathbf{r},t) = -\nabla \varphi(\mathbf{r},t)$ is

$$H = p^2/2m + V(\mathbf{r}) - e\varphi(\mathbf{r},t). \tag{9.164}$$

The full potential experienced by the electron, $V(\mathbf{r}) - e\varphi(\mathbf{r})$, is no longer periodic (and is time dependent), so Bloch's theorem no longer applies. The response of the system to a *weak* external electric field is most easily taken into account perturbatively, as described in Sec. 7.3. If the field is not very weak, perturbation theory is inadequate.

A successful yet simple approach for analyzing the response of an electron in a metal to an external electric (and/or magnetic) field is based on a semiclassical treatment. The semiclassical formalism is especially appropriate when the fields $\mathbf{E}(\mathbf{r},t)$ (or $\mathbf{H}(\mathbf{r},t)$) are smooth and slowly varying. Within the semiclassical picture, the position $\mathbf{r}$ and quasi-momentum $\hbar \mathbf{k}$ of an electron in a crystal are treated as classical variables that evolve in time according to Newton's laws. The quantum mechanical content is provided by the band energy $E_{n\mathbf{k}}$, which is assumed to be given in terms of the band index n and the wavevector $\mathbf{k}$. This assumption has profound consequences on the ensuing electron dynamics because $E_{n\mathbf{k}}$ is a periodic function of $\mathbf{k}$ with period $\mathbf{G}$ (a reciprocal lattice vector). Thus, despite the fact that the wavevector $\mathbf{k}(t)$ evolves in time, it is limited to lie within a primitive cell of the reciprocal lattice (usually in BZ_1). We will also assume that the band index n is constant in time (no band crossing), so it will not be specified below. This assumption can be justified [122] if

$$e\,|\mathbf{E}(\mathbf{r},t)|\,a \ll \frac{E_g^2}{E_F}, \tag{9.165a}$$

$$\hbar \omega_c \ll \frac{E_g^2}{E_F}, \tag{9.165b}$$

where a is a typical length of the order of the lattice constant, $\omega_c = \frac{eH}{mc}$ ($\omega_c = \frac{eH}{m}$ in SI units) is the cyclotron frequency, and E_g and E_F are the gap and Fermi energies, respectively. The LHS of Eq. (9.165a) is the electrical energy imparted to the system when the electron travels a distance a. Violation of either condition causes a transition between the bands, referred to as *electric breakdown* or *magnetic breakdown*. We assume that these inequalities are maintained and hence focus our analysis within a given band n. Within the semiclassical approach, the evolution equations for $\mathbf{r}$ and $\mathbf{k}$ are [recall Eq. (9.102)]

$$\frac{d\mathbf{r}}{dt} = \mathbf{v_k} = \frac{1}{\hbar}\nabla_{\mathbf{k}}E_{\mathbf{k}}, \tag{9.166a}$$

$$\hbar \frac{d\mathbf{k}}{dt} = -e[\mathbf{E}(\mathbf{r},t) + \frac{1}{c}\mathbf{v_k} \times \mathbf{H}(\mathbf{r},t)]. \tag{9.166b}$$

In this section, we concentrate on the response to an external electric field and set $\mathbf{H}(\mathbf{r},t) = 0$ in Eq. (9.166b). In Sec. 9.5.6, we shall analyze the response to an external magnetic field and the combined response to both fields.

9.4 Electrons in a Periodic Potential

Equation (9.166b) is consistent with the fact that the force on an electron is $\mathbf{F} = -e\mathbf{E}$ and is related to the rate of change of the electron energy $E_\mathbf{k}$ as,

$$\frac{dE_\mathbf{k}}{dt} = \mathbf{F} \cdot \frac{d\mathbf{r}}{dt} = -e\mathbf{E}(\mathbf{r},t) \cdot \mathbf{v_k}. \tag{9.167}$$

On the other hand,

$$\frac{dE_\mathbf{k}}{dt} = \nabla_\mathbf{k} E_\mathbf{k} \cdot \frac{d\mathbf{k}}{dt} = \hbar \mathbf{v_k} \cdot \frac{d\mathbf{k}}{dt}, \tag{9.168}$$

where we have used Eq. (9.102) to express $\nabla_\mathbf{k} E_\mathbf{k}$ in terms of $\mathbf{v_k}$. By comparing Eqs (9.167) and (9.168), we see that

$$\mathbf{v_k} \cdot [\hbar \frac{d\mathbf{k}}{dt} + e\mathbf{E}(\mathbf{r},t)] = 0. \tag{9.169}$$

Eqs (9.166a) and (9.166b) imply Eq. (9.169), but the converse is not necessarily true, because any vector perpendicular to $\mathbf{v_k}$, which is added to the square brackets, does not affect Eq. (9.169). Let us now draw some consequences from the semiclassical approach encoded in Eqs (9.165) and (9.169).

1. Equation (9.166a) shows that the velocity of the electron is equal to the group velocity of the wave packet, which seems plausible. Equation (9.166b) is somewhat less transparent and implies that in an external electric field, the time rate of change of the crystal momentum vector is equal to the applied *external* force. There is also an internal force due to the periodic potential $V(\mathbf{r})$, which does not appear in Eq. (9.166b). The reason is that this latter force is accounted for through the energy function $E_\mathbf{k}$. This again underscores the fact that $\hbar \frac{d\mathbf{k}}{dt}$ is not the electron's physical momentum, because its time variation does not equal the total force.

2. In a uniform and constant electric field $\mathbf{E}$, Eq. (9.166b) is easily solved:

$$\mathbf{k}(t) = \mathbf{k}(0) - \frac{1}{\hbar} e\mathbf{E}\, t. \tag{9.170}$$

Hence,

$$\mathbf{v_k}(t) = \mathbf{v}_{[\mathbf{k}(0) - e\mathbf{E}t/\hbar]}, \tag{9.171}$$

and because $\mathbf{v_k}$ is a periodic function of $\mathbf{k}$ with period $\mathbf{G}$ (a reciprocal lattice vector), it is a bounded function. Therefore, $\mathbf{v_k}(t)$ must be an oscillating function of time. This can be easily checked in the special case that $\mathbf{E} \parallel \mathbf{G}$, where $\mathbf{v_k}(t)$ is explicitly a periodic function of time, as we can see by the following argument. In Fig. 9.31, the energy $E_\mathbf{k}$ and the group velocity $v_\mathbf{k}$ are plotted versus $\mathbf{k}$ in one direction, but according to Eq. (9.170), this reflects the time dependence of these two quantities, with the understanding that when the zone boundary is reached, the graph is shifted back to the other zone boundary. In Problem 9.23, you will determine the period.

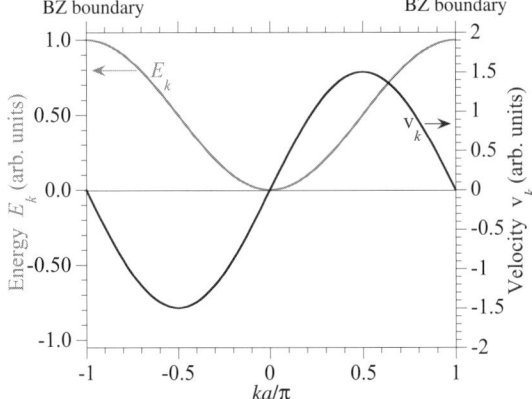

FIG 9.31 $E_\mathbf{k}$ and $v_\mathbf{k}$ for a 1D lattice of period a [see discussion following Eqs (9.170) and (9.171)].

Consider an electron with $\mathbf{k}(0) = 0$ and apply an electric field along a reciprocal lattice vector $\mathbf{G}$ so that from Eq. (9.170), $\mathbf{k} \propto \mathbf{G}t$. It is accelerated, for small t, with constant acceleration, but eventually v_k reaches a maximum, after which the electron decelerates, even though the external force is trying to accelerate it. At the zone boundary, the velocity vanishes and then changes sign, and so on. This somewhat counterintuitive picture of an oscillatory current as a response to a constant and uniform electric field, referred to as *Bloch oscillations*, is due to the periodic potential whose effects are encoded here through the energy $E_\mathbf{k}$. Bloch oscillations are difficult to observe in periodic solid-state electron systems due to electron–phonon collisions. They have been observed in ultracold atomic gases in the presence of a periodic optical potential created by retro-reflected single-frequency laser beams, which can create a 2D or 3D egg-crate potential that the atoms experience.

3. The current produced in response to an external electric field depends crucially on whether the band is completely filled or partially filled. The contribution of a given state to the current is given by the product of electron density and electron velocity for that state. The velocity is given by Eq. (9.166a), and calculation of the electron density employs the semiclassical concept of *number of states in phase space* that we have already encountered in Eq. (9.18). Accordingly, the number of electron states in the volume element $d\mathbf{r}d\mathbf{p}$ near a phase space point $(\mathbf{r},\mathbf{p})$ is

$$dN(\mathbf{r},\mathbf{p}) = 2\frac{d\mathbf{r}\,d\mathbf{p}}{h^3} = \frac{d\mathbf{r}\,d\mathbf{k}}{4\pi^3}, \tag{9.172}$$

where the factor 2 is due to spin degeneracy. To obtain the electron density, we need to divide by the volume element $d\mathbf{r}$ and multiply by the occupation probability (the Fermi function) $f(E_\mathbf{k})$. In a filled band, all the energies are well below the chemical potential μ, so the Fermi function can be well approximated by $f(E_\mathbf{k}) = \theta(E_F - E_\mathbf{k})$. Thus, the density of electrons with momentum between $\mathbf{k}$ and $\mathbf{k} + d\mathbf{k}$ is $\frac{d\mathbf{k}}{4\pi^3}$ for all points $\mathbf{k}$ in BZ_1. In other words, the phase space density of electrons in a filled band is $\frac{1}{4\pi^3\hbar^3}$. But, is this statement valid for all times as $\mathbf{k}(t)$ evolves in time according to the semiclassical equations (9.166a) and (9.166a)? The answer is affirmative, due to *Liouville's theorem* (see Sec. 16.4.2, linked to the book web page) which shows that if a set of points $\{\mathbf{r}(t),\mathbf{p}(t)\}$ occupy a certain volume Ω_t in phase space, then at a later time, $t' > t$, the set of points $\{\mathbf{r}(t'),\mathbf{p}(t')\}$ occupy the same volume, $\Omega_{t'} = \Omega_t$ (see Fig. 9.32).

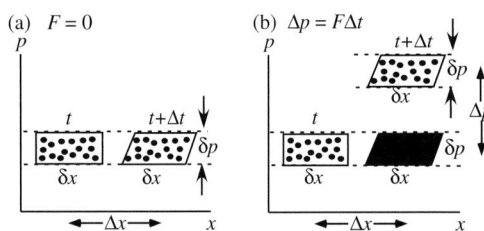

FIG 9.32 Illustration of the Liouville theorem in one dimension, where phase space consists of points $(x(t), p(t))$. (a) In the absence of an external force, a patch of rectangular area $\Omega_t = \delta x \delta p$ contains N points $(x_i(t), p_i(t))$, $i = 1, 2, \ldots, N$. At time $t' = t + \Delta t$, each point moves rightward a distance $\Delta x_i = v_i \Delta t$, where $v_i = p_i/m$ is the constant velocity of particle i. Points at higher p_i move faster, and the N points now occupy a parallelogram having the same area $\Omega_{t'} = \Omega_t$. (b) If there is a constant force, it exerts an impulse that shifts the parallelogram upward by adding a linear momentum $\Delta p = F \Delta t$, leaving the area of the parallelogram unchanged.

We are now in a position to calculate the charge current density $\mathbf{J}_C$ of a filled band at time t in response to a uniform static electric field:

$$\mathbf{J}_C = -\frac{e}{4\pi^3}\int_{\mathbf{k}\in BZ} d\mathbf{k}\,\mathbf{v}_\mathbf{k} = -\frac{e}{4\hbar\pi^3}\int_{\mathbf{k}\in BZ} d\mathbf{k}\,\nabla_\mathbf{k} E_\mathbf{k}. \tag{9.173}$$

This current vanishes identically because the integral of the gradient of a periodic function ($E_\mathbf{k}$) over the Brillouin zone (or any other primitive cell) is identically zero! That is, filled bands do not contribute to the charge current density, and non-zero conduction occurs only in materials with partially filled bands. A similar result is valid for the *energy current density*,

$$\mathbf{J}_E = \frac{1}{4\pi^3}\int_{\mathbf{k}\in BZ} d\mathbf{k}\,E_\mathbf{k}\mathbf{v}_\mathbf{k} = \frac{1}{8\hbar\pi^3}\int_{\mathbf{k}\in BZ} d\mathbf{k}\,\nabla_\mathbf{k}[E_\mathbf{k}]^2 = 0. \tag{9.174}$$

We have seen, in connection with Eq. (9.113), that the number of levels in each band is twice the number of primitive cells in the crystal (the factor 2 is due to spin degeneracy). Hence, all bands can be full or empty only if the number of electrons in a primitive cell is even. This is a necessary (but not sufficient) condition for a material to be an electrical (and thermal) insulator.

9.4 Electrons in a Periodic Potential

4. The situation is entirely different for partially filled bands. Moreover, there is a profound difference between an almost empty band [see Fig. 9.33(a)], where the relevant energies $E_{\mathbf{k}}$ are close to the band minimum, and an almost filled band [Fig. 9.33(b)], where the relevant energies close to E_F are close to the band maximum. The latter case brings us to the important concept of *holes* (see below). In either case, the charge current density $\mathbf{J}_C$ (below, we will often omit the subscript C) at zero temperature is given by an integral over momentum, as in Eq. (9.173), albeit only over occupied states not the entire Brillouin zone,

$$\mathbf{J} = -\frac{e}{4\hbar\pi^3}\int_{E_{\mathbf{k}}<E_F} d\mathbf{k}\,\nabla_{\mathbf{k}}E_{\mathbf{k}} = \frac{e}{4\hbar\pi^3}\int_{E_{\mathbf{k}}>E_F} d\mathbf{k}\,\nabla_{\mathbf{k}}E_{\mathbf{k}}, \qquad (9.175)$$

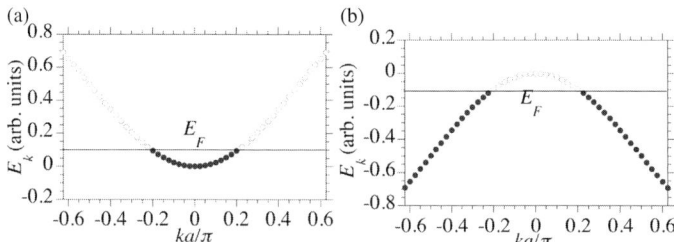

FIG 9.33 (a) An almost empty band. At $T = 0$, the electrons occupy the few levels above the band minimum, and the charge current is calculated according to the first term on the RHS of Eq. (9.175). (b) An almost filled band. At $T = 0$, the few levels above the Fermi energy and below the band maximum are empty, and the current is calculated according to the second term on the RHS of Eq. (9.175). These empty levels can be viewed as "occupied" by holes, which can be regarded as particles with positive charge $+e$.

which, generically, is different from zero. For the case of an almost empty band, it is natural to integrate over the occupied states in Eq. (9.175) (the first term on the RHS). However, it is more convenient for an almost filled band to use the fact that the integral (9.173) over the entire Brillouin zone vanishes, and therefore, the integral in Eq. (9.175) can be performed over the *unoccupied* states (the last term on the RHS), *with a + sign in front of the integral*. This looks like we have "particles" with charge $+e$ occupying levels $E_{\mathbf{k}} > E_F$, whose group velocity $\mathbf{v}_{\mathbf{k}} = \hbar^{-1}\nabla_{\mathbf{k}}E_{\mathbf{k}}$ is calculated as if these levels were occupied by electrons (although these levels are empty). These "particles" are referred to as *holes*. We can analyze the response of holes to a static uniform electric field by recalling the equation of motion (9.166b), which pertains to electrons. If this equation is used for the empty levels at the top of a valence band [Fig. 9.33(b)], the mass tensor is negative definite as we have seen in discussion near Eq. (9.143). To simplify the discussion, assume that the maximum of the band occurs at a point $\mathbf{k}_0$ with cubic symmetry, so that in the principal axis system, $[M]_{ij} = -m^*\delta_{ij}$ with $m^* > 0$. Hence, $E_{\mathbf{k}} = E_{\mathbf{k}_0} - \frac{\hbar^2(\mathbf{k}-\mathbf{k}_0)^2}{2m^*}$ leading to the following expressions for velocity and acceleration of an electron level near the top of the valence band [the empty levels in Fig. 9.33(b)],

$$\mathbf{v}_{\mathbf{k}} = \frac{1}{\hbar}\nabla E_{\mathbf{k}} = -\frac{\hbar}{m^*}(\mathbf{k}-\mathbf{k}_0), \qquad (9.176)$$

so

$$\mathbf{a}_{\mathbf{k}} = \dot{\mathbf{v}}_{\mathbf{k}} = -\frac{\hbar}{m^*}\dot{\mathbf{k}} = -\frac{1}{m^*}(-e\mathbf{E}) = -\frac{1}{m^*}\mathbf{F}. \qquad (9.177)$$

The acceleration is *opposite* to the applied external force, $\mathbf{F} = -e\mathbf{E}$. Once again, the reason for this somewhat counter intuitive scenario is that the periodic potential, which gives rise to the band energy $E_{\mathbf{k}}$, is not included as an external force. If we now view this *electron-empty* level $E_{\mathbf{k}}$ as if it is occupied by a fictitious particle of charge $+e$ and mass $m^* > 0$, we restore the intuitive picture with $\mathbf{a}_{\mathbf{k}} = e\mathbf{E}/m^* = \mathbf{F}/m^*$; the acceleration has the same direction of the external force.

In summary, a hole is a (fictitious) particle with charge $+e$ and spin $1/2$ whose semiclassical phase space coordinates $\mathbf{r}(t), \mathbf{k}(t)$ obey the semiclassical equations (9.166a) and (9.166b), but its energy $E_{\mathbf{k}}$ is always larger than E_F, so that the corresponding level is unoccupied. The mass tensor of a hole is positive definite,

$$[(M^*)^{-1}]_{ij} = -\frac{1}{\hbar^2}\left(\frac{\partial^2 E_{\mathbf{k}}}{\partial k_i \partial k_j}\right)_{\mathbf{k}=\mathbf{k}_0=\text{ band maximum}}. \tag{9.178}$$

The hole concept, which is most useful in an almost filled band, may look like just a formal definition, but it has profound physical consequences in solid-state physics, as we shall soon see.

Finally, if the mass tensor for electrons is calculated for an almost empty band using Eq. (9.142), and that for holes is calculated for an almost filled band as in Eq. (9.178), the Newton equation in an external electric field can be written as,

$$M^* \mathbf{a_k} = \pm e\mathbf{E}, \tag{9.179}$$

with $+$ for holes and $-$ for electrons, and M^* is the mass tensor defined in Eq. (9.142).

Problem 9.23

Show that if $\mathbf{E} \parallel \mathbf{G}$, the group velocity is periodic in time with period $T = \frac{\hbar}{\alpha e}$, where $\alpha = E/G$.

Hole conduction is as important as electron conduction in semiconductors, because holes can significantly contribute to electric conductance. Consider, for example, the case of a vacancy near the top of an otherwise filled energy band. The missing electron would have had energy $E(\mathbf{k})$ and momentum $\mathbf{k}$ (in units of $\hbar$). We can think of the hole as having energy $-E(\mathbf{k})$ and momentum $-\mathbf{k}$, i.e., $\mathbf{k}_h = -\mathbf{k}_e$ and $E_h(\mathbf{k}_h) = -E_e(\mathbf{k}_e)$. Moreover, because $\mathbf{k}_h = -\mathbf{k}_e$, we find $E_h(\mathbf{k}_h) = -E_e(\mathbf{k}_e) = -E_e(-\mathbf{k}_h)$. The motion of an electron in the presence of an applied electric field $\mathbf{E}$ is given by $d(\hbar \mathbf{k}_e)/dt = -e\mathbf{E}$ (because the charge of an electron is $-e$). A hole with momentum $\mathbf{k}_h = -\mathbf{k}_e$ moves according to the equation of motion, $d(\hbar \mathbf{k}_h)/dt = \mathbf{F}_h$. Because the hole has a positive charge (it is the absence of a negatively charged particle), we find the equation of motion

$$\frac{d(\hbar \mathbf{k}_h)}{dt} = \mathbf{F}_h = +e[\mathbf{E} + (\mathbf{v}_h/c) \times \mathbf{H}], \tag{9.180}$$

where we have also included the Lorentz force due to the presence of a magnetic field. The hole velocity, $\mathbf{v}_h$, must be the same as the velocity of the electron (as demonstrated below), hence, $\mathbf{v}_h = \mathbf{v}_e$. However, because $\mathbf{k}_h = -\mathbf{k}_e$, we conclude that the mass of the hole must be the negative of the electron mass, $m_h = -m_e^*$.

The velocity of an electron is given by $\mathbf{v}_e = \nabla_{\mathbf{k}} E(\mathbf{k})/\hbar$. If only this electron is missing from the band, the net current $\mathbf{J}$ is equal to that of an unpaired electron with momentum $-\mathbf{k}$; hence, $\mathbf{J} = (-e)\mathbf{v}_e(-\mathbf{k}) = (-e)(-\mathbf{v}_e(\mathbf{k})) = e\mathbf{v}_e(\mathbf{k})$. To be consistent with using a positive charge for the hole, the velocity of the hole must equal the velocity of the electron, $\mathbf{v}_h(\mathbf{k}) = \mathbf{v}_e(\mathbf{k})$, as described above. Thus, $\mathbf{v}_h(\mathbf{k}) = \hbar^{-1} \nabla_{\mathbf{k}} E(\mathbf{k})$, where $E(\mathbf{k})$ denotes the energy of the electron in the state with momentum $\hbar \mathbf{k}$. Hence, we conclude that $\mathbf{v}_h(\mathbf{k}) = \hbar^{-1} \nabla_{\mathbf{k}} E_e(\mathbf{k}_e) = \hbar^{-1} \nabla_{\mathbf{k}} E_h(\mathbf{k}_h)$. In summary,

$$\mathbf{v}_h(\mathbf{k}) = \hbar^{-1} \nabla_{\mathbf{k}} E_h(\mathbf{k}_h) = \mathbf{v}_e(\mathbf{k}). \tag{9.181}$$

9.5 MAGNETIC FIELD EFFECTS

In this section, we analyze the response of electrons or holes to an external magnetic field. Magnetic fields of strengths from zero to a few tens of tesla can be experimentally generated in the laboratory. Probing materials using an external magnetic field is an essential tool for elucidating their electronic properties. A magnetic field affects both orbital and spin degrees of freedom. It serves as one of the most efficient ways to identify the structure of the Fermi surface of a metal

9.5 Magnetic Field Effects

or a semiconductor. In addition to probing properties of materials, application of a magnetic field leads to numerous new and sometimes spectacular phenomena.

Due to the central role played by magnetic fields in studying electronic properties in solids, the present analysis is somewhat long. Following a discussion of the Schrödinger equation in a magnetic field, we introduce the Aharonov–Bohm and Aharonov–Casher effects. Then we introduce the Hall effect in its classical formulation, followed by a semi-classical analysis of a system consisting of an electron in a periodic potential subject to an external magnetic field. This discussion naturally leads to the study of the de Haas van Alphen and Shubnikov-de Haas effects. Then we consider the quantum Hall effect, Landau levels and the 2D electron gas in a magnetic field, followed by the integer and fractional quantum Hall effects. Finally, we discuss paramagnetism and diamagnetism and close the section by introducing the notion of magnetic order, which is responsible for ferromagnetism and antiferromagnetism. Most of the analysis in this section remains within the single-particle formalism.

9.5.1 ELECTRON IN A MAGNETIC FIELD

Classically, a magnetic field $\mathbf{H}(\mathbf{r},t)$ affects the motion of a charged particle through the Lorentz force, $\mathbf{F} = \frac{q}{c}\mathbf{v} \times \mathbf{H}(\mathbf{r},t)$, where q is the charge of the particle, $\mathbf{v}$ is its velocity, and $\mathbf{H}(\mathbf{r},t)$ is the magnetic field at the spacetime point $(\mathbf{r},t)$ of the particle (the factor of $1/c$ is required in Gaussian units but should be omitted in SI units). We have seen in Sec. 4.3.1 that the magnetic field can be expressed as the curl of a *vector potential*, $\mathbf{A}(\mathbf{r},t)$, so that $\mathbf{H}$ satisfies the Maxwell equation, $\nabla \cdot \mathbf{H} = 0$, i.e., $\mathbf{H} = \nabla \times \mathbf{A}$. Within the classical theory, all measurable quantities depend only on the magnetic field, and the difference between two vector potentials producing the same magnetic field is physically unobservable. In quantum theory, the significance of the vector potential is upgraded, as we shall see.

The time-dependent Schrödinger equation for the wave function of a charged particle with spin in a magnetic field is [see Eq. (4.43)],

$$i\hbar \frac{\partial \psi}{\partial t} = \mathcal{H}\psi = \left[\frac{1}{2m}(\mathbf{p} - \frac{q}{c}\mathbf{A})^2 + V - \boldsymbol{\mu}_s \cdot \mathbf{H} \right]\psi, \quad (9.182)$$

where $\mathcal{H}$ is the Hamiltonian of a charged particle in a magnetic field $\mathbf{H}$ and a potential $V(\mathbf{r})$ and $\boldsymbol{\mu}_s = -g\mu_B \mathbf{S}/\hbar$ is the magnetic moment of the electron, where the g-factor is taken to be roughly equal to 2. The periodic potential V is related to the electrostatic potential φ simply by $V = q\varphi$. Equation (9.182) has several properties that distinguishes it from the Schrödinger equation in the absence of an external magnetic field. Obviously, the Hamiltonian $\mathcal{H}$ is not time-reversal invariant, because time-reversal reverses the sign of the momentum, $\mathbf{p} \to -\mathbf{p}$, and the magnetic moment, $\boldsymbol{\mu}_s \to -\boldsymbol{\mu}_s$, leading to a different Hamiltonian.

Gauge Freedom

Equation (9.182) is unchanged if the vector potential $\mathbf{A}$, the scalar potential φ, and the wave function ψ are modified as follows:

$$\mathbf{A} \to \mathbf{A} + \nabla \chi, \quad \varphi \to \varphi - \frac{1}{c}\frac{\partial \chi}{\partial t}, \quad \psi \to e^{i\frac{q}{\hbar c}\chi}\psi. \quad (9.183)$$

Here, $\chi(\mathbf{r},t)$ is a single valued and differentiable function of space and time but otherwise arbitrary. Equation (9.183) is a manifestation of *gauge invariance*, which plays a fundamental role in the theory of fields.

Problem 9.24

(a) Show that the two vector potentials, $(A_x, A_y, A_z) = (0, -Bx, 0)$ and $(A_x, A_y, A_z) = (By/2, -Bx/2, 0)$, lead to the same uniform magnetic field $\mathbf{H} = B\hat{\mathbf{z}}$. The first choice is referred to as *the Landau gauge*, whereas the second one is *the symmetric gauge*.

(b) Determine the function $\chi(\mathbf{r})$, which relates these vector potentials according to Eq. (9.183).

Hint: Because the vector potentials is a linear function of the coordinates, Eq. (9.183) suggests that $\chi(\mathbf{r})$ is bilinear in x, y. Try the form $\chi(\mathbf{r}) = \gamma xy$ and find the constant γ by substitution.
(c) Prove that the electric and magnetic fields do not change and that the Schrödinger equation (9.182) does not change on making the transformation (9.184).

The velocity operator: According to the Heisenberg equations of motion, the velocity operator for a charged particle in a magnetic field is

$$\mathbf{v} = \dot{\mathbf{r}} = \frac{1}{i\hbar}[\mathbf{r}, \mathcal{H}] = \frac{1}{m}\left(\mathbf{p} - \frac{q}{c}\mathbf{A}\right) \equiv \frac{\mathbf{\Pi}}{m}, \tag{9.184}$$

where $\mathbf{\Pi} = \mathbf{p} - \frac{q}{c}\mathbf{A}$ is the *covariant momentum*. The vector potential enters the kinetic energy part of the Hamiltonian $T = \frac{m}{2}\mathbf{v}^2 = \frac{\mathbf{\Pi}^2}{2m}$. The different components of the velocity operator do not commute among themselves (see Problem 9.25). Furthermore, the equation of motion (9.184) for the momentum is

$$\dot{\mathbf{\Pi}} = \frac{1}{i\hbar}[\mathbf{\Pi}, \mathcal{H}] = \frac{1}{i\hbar}[\mathbf{\Pi}, V(\mathbf{r})] = \frac{1}{i\hbar}[\mathbf{p}, V(\mathbf{r})]. \tag{9.185}$$

The velocity operator $\mathbf{v} = \mathbf{\Pi}/m$ determines the (matter) current density, which in the presence of magnetic field reads,

$$\mathbf{J}(\mathbf{r}, t) = \text{Re}\left[\psi^* \mathbf{v} \psi\right] = \text{Re}\left[\frac{1}{m}\psi^{\dagger}\left(\mathbf{p} - \frac{q}{c}\mathbf{A}\right)\psi\right], \tag{9.186}$$

which now contains an electromagnetic part, $-\frac{q}{c}\mathbf{A}|\psi|^2$. We will see below that there are circumstances where the latter is the only contribution to the current.

Problem 9.25

Verify the commutation relations,

$$[v_i, v_j] = i\varepsilon_{ijk}\frac{q\hbar}{m^2 c}H_k. \tag{9.187}$$

The consequences of Eqs (9.184), (9.185), and (9.187) to electron trajectories in a periodic potential will be discussed in Sec. 9.5.6.

9.5.2 AHARONOV–BOHM EFFECT

In classical electrodynamics, vector and scalar potentials $(\mathbf{A}, \varphi)$ are introduced as auxiliary fields designed in such a way that the magnetic and electric fields are derived from them:

$$\mathbf{H} = \nabla \times \mathbf{A}, \tag{9.188}$$

$$\mathbf{E} = -\nabla\varphi - \frac{1}{c}\frac{\partial \mathbf{A}}{\partial t}. \tag{9.189}$$

Equation (9.188) insures that the Maxwell equation, $\nabla \cdot \mathbf{H} = 0$, is satisfied. Likewise, Eq. (9.189) insures that, in the absence of a magnetic field, $\mathbf{A} = 0$, $\nabla \times \mathbf{E} = 0$. In classical electrodynamics all measurable quantities depend solely on $\mathbf{H}$, and the vector potential $\mathbf{A}$ itself has no measurable effects; the vector potential is simply a convenient mathematical tool for describing dynamics of charged particles and it has no physical consequences. In quantum mechanics, the situation is different. Yakir Aharonov and David Bohm [129] predicted in 1959 that there are situations in which a vector potential can have a tangible physical effect. In quantum mechanics, a charged particle is directly affected by the vector potential $\mathbf{A}$, even when the magnetic field $\mathbf{H} = \nabla \times \mathbf{A}$ vanishes over the whole region in which the particle

9.5 Magnetic Field Effects

is confined. This is known as the *Aharonov–Bohm effect*. As shown below, it originates from quantum interference of charged particle wave packets moving in a doubly connected region, even when the magnetic field vanishes over the region where the particles can propagate.

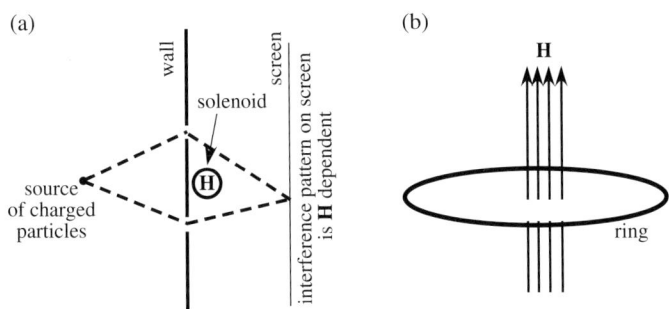

FIG 9.34 (a) Schematic of an Aharonov–Bohm interference experiment. (b) A current carrying ring threaded by a magnetic field. No magnetic field acts on the electrons in the ring, but the current through the ring nevertheless depends on the magnetic flux, $\Phi = \int d\mathbf{S} \cdot \mathbf{H}$, threading the ring.

The Aharonov–Bohm effect can be manifested in many different physical systems. The first experiment that confirmed the effect used electron interferometry. Consider a charged particle source and a double slit diffraction apparatus, as shown in Fig. 9.34(a). A long (virtually infinite) solenoid carrying a stationary current I is placed between the two slits, perpendicular to the (x, y) plane of the figure. A constant magnetic field $H\hat{\mathbf{z}}$ is confined inside the solenoid. $H = 0$ outside the solenoid but obviously $\mathbf{A} \neq 0$. The charged particles (e.g., electrons) from the source cannot reach the solenoid. When the current I is varied, the strength of the magnetic field inside the solenoid is changed. Experimentally, the Aharonov–Bohm effect is manifested as it is found that the intensity of electrons reaching the screen on the right oscillates as function of the current. In other words, the interference pattern that results at the screen depends on the magnetic flux through the cylinder, whereas the intensity of the magnetic field outside the solenoid is zero.

In a second setup, the Aharonov–Bohm effect is manifested for electrons moving on a metallic ring of radius R and circumference $L = 2\pi R$, as shown in Fig. 9.34(b). The ring is threaded by a solenoid of radius $r_0 < R$, so that a constant magnetic field $\mathbf{H}$ results along the axis perpendicular to the plane of the ring. The magnetic flux is given by [23, 34]

$$\Phi = \int d\mathbf{S} \cdot \mathbf{H} = \oint d\mathbf{x} \cdot \mathbf{A}, \quad (9.190)$$

where $d\mathbf{S}$ is an element of an area bounded by the ring and $d\mathbf{x}$ is a line element along the ring. The last equality in Eq. (9.190) is due to Stokes' theorem. Because the magnetic field is zero on the ring, we can choose a gauge so that the vector potential is constant on the ring,

$$\mathbf{A} = \frac{\Phi}{2\pi R}\hat{\mathbf{x}}, \quad (9.191)$$

where $0 \leq x \leq L \equiv 2\pi R$ is the coordinate and $\hat{x}$ is a unit vector along the ring. It is easily verified that on the ring, where $\mathbf{A} = (\Phi/L)\hat{\mathbf{x}}$, the magnetic field vanishes, $\mathbf{H} = \nabla \times \mathbf{A} = 0$. This simple system exhibits the Aharonov–Bohm effect. The geometry of a ring enclosing a flux tube is topologically special in that the domain on which particles move (the ring) is doubly connected in the sense that it is not possible to shrink the closed loop around the solenoid into a point. As shown below, in this system, the Aharonov–Bohm effect is manifested through the oscillatory dependence of the magnetization of the ring as function of the magnetic field produced by the solenoid.

Problem 9.26

Find a vector potential in all space for an infinite solenoid of radius r_0 with a uniform magnetic field H generated by a current in the solenoid.

Answer: Let $\Phi = \pi r_0^2 H$. $\mathbf{A} = \begin{cases} \frac{\Phi}{2\pi r}\hat{\mathbf{x}}, & (r > r_0), \\ \frac{\Phi r}{2\pi r_0^2}\hat{\mathbf{x}}, & (r \leq r_0). \end{cases}$

Electrons on a Ring: Persistent Currents

The Aharonov–Bohm effect plays an important role in *mesoscopic systems* [131, 132] (see Sec. 13.2.1) where, at low temperature, quantum coherence can be maintained throughout the entire system. Impressive progress in experimental techniques have enabled fabrication of micron-sized conducting rings. When such rings are subject to magnetic fields, the main effect is due to the magnetic flux Φ threading the ring (the magnetic field acting on the particles in the ring vanishes). Consequently, the spin of the electrons does not play a role. Therefore, electrons can be treated as spinless fermions of charge $-e$ and mass m. The time-independent Schrödinger equation for electrons moving on the ring is

$$\frac{1}{2m}\left(p + \frac{e}{c}\frac{\Phi}{L}\right)^2 \psi = E\psi, \qquad (9.192)$$

where $p = -i\hbar\frac{d}{dx}$ is the momentum operator and $[p, H] = 0$ because Φ is independent of x. The periodic solutions are plane waves, $\psi(x) = L^{-1/2} e^{ikx}$. Because the wave function must be single valued, the wavenumbers are quantized, $k = k_n = 2n\pi/L$, with $n = 0, \pm 1, \pm 2 \ldots$. Substituting in Eq. (9.192), one obtains the energy eigenvalues E_n,

$$E_n(\Phi) = \frac{1}{2m}\left(\hbar k_n + \frac{e\Phi}{cL}\right)^2 = \frac{\hbar^2}{2mR^2}\left(n + \frac{\Phi}{\Phi_0}\right)^2, \qquad (9.193)$$

where Φ_0 is a unit of quantum flux called a "fluxon,"

$$\boxed{\Phi_0 \equiv \frac{hc}{e} = 4.1414 \times 10^{-7} \quad \text{Gauss-cm}^2.} \qquad (9.194)$$

Before analyzing the spectrum, it is useful to explore the gauge invariance that arises in this problem. Following the discussion near Eq. (9.183), the physics encoded in the vector potential and wave function $(\mathbf{A}, \psi)$, is the same as that in $(\mathbf{A} + \nabla\chi(x), e^{\frac{ie}{\hbar c}\chi(x)}\psi)$, where $\chi(x)$ is a single-valued differentiable function of x. It is tempting to use this gauge freedom by choosing

$$\eta(x) \equiv e^{i\chi(x)} = e^{-\frac{ie}{\hbar c}\int^x \mathbf{A}\cdot d\mathbf{x}}, \qquad (9.195)$$

to arrive at the false conclusion that the vector potential can be eliminated. However, the gauge transformation $\psi(x) \to \phi(x) = \eta(x)\psi(x)$ is generically not a bona fide gauge transformation, because it is path dependent (see Fig. 9.35). Arriving at the point x from an initial point x_0 in the clockwise direction results a phase factor $\eta_+(x)$, whereas for the anticlockwise direction results a phase factor $\eta_-(x)$. It is easy to see that $\eta_+(x)/\eta_-(x) = e^{-2\pi i\Phi/\Phi_0}$. Hence,

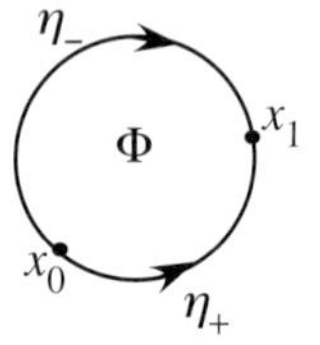

FIG 9.35 Gauge transformations defined in Eq. (9.195) are path dependent. Performing the integral between x_0 and x counterclockwise (clockwise) yields phase factors $\eta_+(x)$ ($\eta_-(x)$).

- If the magnetic flux Φ through the ring is not an integer multiple of the quantum flux Φ_0, the vector potential cannot be eliminated by a gauge transformation. The quantity $\eta(x)$ is then referred to as a *nonintegrable phase factor*.
- On the other hand, if $\Phi = m\Phi_0$ (m integer), the vector potential can be eliminated, and the physics with $m \neq 0$ is the same as that for $m = 0$.

This manifestation of gauge invariance is a special case of a more general theorem, which states that no physical experiment performed outside the solenoid can distinguish between the cases Φ and $\Phi + m\Phi_0$.

9.5 Magnetic Field Effects

Problem 9.27

(a) Write the Schrödinger equation for the gauge transformed wave function $\phi(x) = e^{\frac{ie}{\hbar c}\frac{\Phi}{L}x}\psi(x) = e^{-\frac{ie}{\hbar c}\int^x \mathbf{A}\cdot d\mathbf{x}}\psi(x)$ is,

$$\frac{\mathbf{p}^2}{2m}\phi(x) = E\phi(x). \qquad (9.196)$$

(b) Find the solutions to Eq. (9.196) and calculate the energy eigenvalues.

Hint: The physical wave function $\psi(x)$ (not $\phi(x)$) must be single valued. The solution of Eq. (9.196) is $\phi(x) = L^{-1/2}e^{ikx}$ and the energy is $E = \frac{\hbar^2 k^2}{2m}$. $\psi(x)$ is single valued if and only if $k = \frac{2n\pi}{L} + \frac{\Phi}{\Phi_0 L}$.

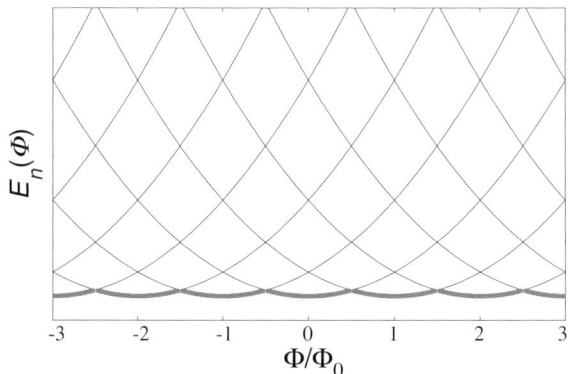

FIG 9.36 Energy eigenvalues $E_n(\Phi)$ versus magnetic flux Φ/Φ_0. Level crossing at $\Phi = m\Phi_0$ and $\Phi = (m+1/2)\Phi_0$ is a property of the clean system. Any amount of disorder or imperfection leads to avoided crossing. For any fixed value of the flux, the levels should be ordered from low to high. The ground-state energy of a single electron is shown as the thick line. With this ordering, the spectrum is a periodic function of the flux with period Φ_0.

Figure 9.36 displays the energy eigenvalues $E_n(\Phi)$ versus flux Φ in units of Φ_0 for a number of n values. It is composed of shifted parabolas $E_n(\Phi)$ with $E_n(\Phi = n\Phi_0) = 0$. The dependence of the energy spectrum on the magnetic flux is the manifestation of the Aharonov–Bohm effect for electrons on a ring. It is easily seen that the spectrum is symmetric with respect to reflection about the lines $\Phi = m\Phi_0$ and $\Phi = (m+1/2)\Phi_0$. When level crossing occurs at these points, the levels should be continued as if this crossing is avoided (e.g., the ground-state energy $\varepsilon_0(\Phi)$ is indicated by the thick curve). This procedure introduces the physical energies $\varepsilon_0(\Phi) \leq \varepsilon_1(\Phi) \leq \ldots$. With this convention, the spectrum has several remarkable properties:

- The energy is a periodic function of the flux with period $\Phi_0 \equiv hc/e$, i.e., $\varepsilon_n(\Phi + m\Phi_0) = \varepsilon_n(\Phi)$. Note that the bare energies, Eq. (9.193), are not periodic in Φ.
- The energy ε_n is a symmetric function of the flux, $\varepsilon_n(\Phi) = \varepsilon_n(-\Phi)$, as expected for spinless particles. This symmetry remains also if there is scattering and electron–electron interaction in the ring. On the other hand, the symmetry with respect to reflection through the lines $(m+1/2)\Phi_0$ is less robust; it is a result of a special geometry and lack of scattering.

Let us now assume that the ring is coupled to an electron reservoir at Fermi energy E_F. Then, at zero temperature, the ground-state energy is given by the sum of single-particle energies of levels below E_F,

$$E_{gs} = \sum_{\varepsilon_n < E_F} \varepsilon_n, \qquad (9.197)$$

which following our analysis of the spectrum, satisfies the conditions,

$$E_{gs}(\Phi) = E_{gs}(-\Phi) = E_{gs}(\Phi + m\Phi_0). \qquad (9.198)$$

One of the important consequences of the Aharonov–Bohm effect in the ring geometry is that there can exist a non-zero steady-state current in the ring [132]. As we shall see below, the existence of the current is related also to the fact that in a magnetic field, time-reversal symmetry is violated and the Hamiltonian is complex. This current is not the usual conduction current where electrons in a metal at the Fermi energy exchange energy with an external electric field. Rather, all the electrons in the levels up to the Fermi energy contribute to the current, and the current is an equilibrium property of the system. There is no dissipation associated with this current. The current operator can be written as (see Problem 9.28 below)

$$\hat{I} = -c\frac{\partial H}{\partial \Phi} = \frac{-e}{L}\left[\frac{1}{m}(p + \frac{e\Phi}{cL})\right] = \frac{-e}{L}\hat{v}, \quad (9.199)$$

where $\hat{v}$ is the velocity operator. The observable current is obtained by taking the expectation value of $\hat{I}$. In Problem 9.28, you will show that the slope $dE_n/d\Phi$ in Fig. 9.36 determines the contribution of ψ_n to the total current.

Problem 9.28

Prove that the current of an electron in the nth level, $I_n \equiv \langle \psi_n | \hat{I} | \psi_n \rangle$ is given by $I_n = -c\frac{dE_n}{d\Phi}$, and therefore, $I_n = -\frac{eh}{mL^2}(n + \frac{\Phi}{\Phi_0})$.

Hint: If $\psi_n(x)$ is a normalized solution of $H\psi_n = E_n\psi_n$, then $E_n = \langle \psi_n | H | \psi_n \rangle$. Hence, $\frac{dE_n}{d\Phi} = \langle \frac{\partial \psi_n}{\partial \Phi} | H | \psi_n \rangle + \langle \psi_n | \frac{\partial H}{\partial \Phi} | \psi_n \rangle + \langle \psi_n | H | \frac{\partial \psi_n}{\partial \Phi} \rangle$. Prove that only the second term on the RHS of this equation is nonvanishing using the fact that $\langle \psi_n | \psi_n \rangle = 1$. See the proof of the Hellman–Feynman theorem in Sec. 11.3.3, which is in complete analogy with the proof required here.

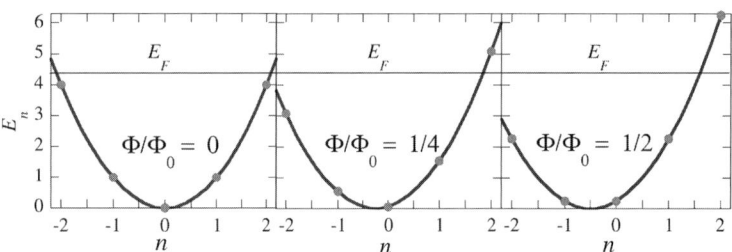

FIG 9.37 Energy eigenvalue $E_n(\Phi)$ of Eq. (9.193) versus n (plotted here as a continuous variable) for magnetic fluxes $\Phi = (0.0, 0.25, 0.5)\Phi_0$. The current at zero temperature is contributed from discrete states (integer values of n) below the Fermi energy and is proportional for each level to the slope $\frac{dE_n}{dn}$. For $\Phi = 0$ and $\Phi = 0.5\Phi_0$, the slopes cancel each other and the total current vanishes. For $\Phi = 0.25\Phi_0$, the slopes do not cancel and $I \neq 0$.

To analyze the total current contributed by all levels below the Fermi energy, it is useful to plot E_n, given by Eq. 9.193, as a function of n for fixed flux Φ. Note that $\frac{dE_n}{d\Phi} = \frac{1}{\Phi_0}\frac{dE_n}{dn}$. This is shown in Fig. 9.37, where E_n is plotted as a function of n for $\Phi/\Phi_0 = 0, 0.25$, and 0.5. The total current at zero temperature is [see Eq. (9.197)]

$$I = \sum_{E_n < E_F} I_n = -c\frac{dE_{gs}}{d\Phi}. \quad (9.200)$$

As a result of the properties of the spectrum listed above, it is clear that the current is a periodic function of Φ with period Φ_0 and is an *antisymmetric* function of the flux,

$$I(\Phi) = -I(-\Phi), \quad I(\Phi + m\Phi_0) = I(\Phi). \quad (9.201)$$

Let us check what happens if the ring is not "clean," i.e., a disordered potential energy is present. In this case, the electrons can be scattered from numerous fixed scattering centers. Such scattering centers are related to imperfections and impurities. Experimentally they are often unavoidable. In many cases, the positions and strengths of these impurities are random; this kind of potential is called *quenched disorder*. The main consequences of disorder in the ring are:

9.5 Magnetic Field Effects

1. Level crossings shown in Fig. 9.36 are avoided, and the dependence of energy levels on the flux is smooth.
2. The periodicity and symmetry relations of the energy and current, Eqs (9.198) and (9.201), remain unchanged. The Fourier expansion of the current is

$$I = \sum_{k=1}^{\infty} I_k \sin(2\pi k\Phi/\Phi_0). \tag{9.202}$$

   In some experiments, the current is found to have an effective period of $\Phi_0/2$ because the dominant Fourier components have $k = 2N$, where N is an integer.
3. We will see in Sec. 9.9 that in the presence of static disorder, the wave functions in one-dimension decay exponentially at large distance with some characteristic length scale ξ, which depends on energy and strength of disorder. ξ is referred to as the *localization length*, and the phenomenon is called *Anderson localization*. Thus, if $L > \xi$, we expect the current to decay exponentially with ring size.
4. Surprisingly, for $L < \xi$, the current persists.

Despite its simplicity, this model of a 1D ring threaded by a magnetic flux reveals numerous fundamental concepts in quantum mechanics, including the Aharonov–Bohm effect, quantum coherence, gauge invariance, persistent currents in mesoscopic systems, disorder effects, Anderson localization, and topological effects.

Aharonov–Casher Effect

Another related effect occurs when a particle with non-zero magnetic moment $\boldsymbol{\mu} = g\mu_B \mathbf{S}/\hbar$ moves in a doubly connected region (e.g., a ring) threaded by a perpendicular line of charge density λ (note that we discuss this effect here despite the fact that there need not be a magnetic field present because it is strongly related to the Aharonov–Bohm effect from the point of view of gauge transformations). The particle need not even be charged; it can be a neutron or a neutral atom with spin. The setup is similar to that in Fig. 9.34 except that the particle has spin, and instead of a solenoid with magnetic field $\mathbf{H} = H\hat{\mathbf{z}}$ threading the ring, here there is a line of charge. A comparison of the ring geometries of the Aharonov–Bohm and Aharonov–Casher effects is shown in Fig. 9.38.

In the two-slit experimental setup with a line of charge instead of a magnetic field shown in Fig. 9.34, the intensity of particles at a given point on the screen will oscillate as a function of λ, even if the particles are not charged. This effect is called the *Aharonov–Casher effect* [138], and it is due to spin–orbit interaction. A heuristic explanation of the effect ensues from the following argument: A particle subject to an electric field $\mathbf{E}$ feels, in its rest frame, an effective magnetic field $\mathbf{H}_{\text{eff}} = (\mathbf{v}/c) \times \mathbf{E} = \frac{1}{mc}\mathbf{p} \times \mathbf{E}$, where $\mathbf{v}$ is the particle's velocity and $\mathbf{p}$ is its momentum. This results in an effective Zeeman Hamiltonian, $H_Z \equiv -\boldsymbol{\mu} \cdot \mathbf{H}_{\text{eff}} = g(\mu_B/mc)(\mathbf{S}/\hbar) \cdot \mathbf{p} \times \mathbf{E}$, where $g \approx 2$. This operator must be properly put into Hermitian form (see Sec. 2.2), i.e., $H_Z = g(\mu_B/4mc)(\boldsymbol{\sigma} \cdot \mathbf{p} \times \mathbf{E} - \boldsymbol{\sigma} \cdot \mathbf{E} \times \mathbf{p})$. For an electron subject to an electric field $\mathbf{E}$, the spin–orbit term of the Hamiltonian is written as,

$$H_{\text{so}} = \alpha(\mathbf{p} \cdot \boldsymbol{\sigma} \times \mathbf{E} + \boldsymbol{\sigma} \times \mathbf{E} \cdot \mathbf{p}), \tag{9.203}$$

where $\alpha = \frac{g\mu_B}{4mc}$ specifies the spin-orbit interaction strength. In contrast with the Hamiltonian $\mathcal{H}$ defined in Eq. (9.182), the Hamiltonian H_{so} is invariant under time reversal that reverses the signs of both $\mathbf{p}$ and $\boldsymbol{\sigma}$.

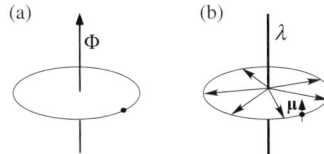

FIG 9.38 An illustration of the Aharonov–Bohm and Aharonov–Casher effects. (a) Electron moving on a metallic ring threaded by a magnetic flux Φ. When the electron encircles the ring, its wave function gains a phase $e^{2\pi i\phi}$, where $\phi = \Phi/\Phi_0$. The spin of the electron does not play a role. (b) Electron with magnetic moment $\boldsymbol{\mu}$ moving on a metallic ring threaded by a charged wire of longitudinal charge density λ. This results in a radial electric field $\mathbf{E} = \frac{\lambda}{L}\hat{\mathbf{r}}$ (black arrows), where R is the radius of the ring. When the electron encircles the ring, its wave function (a two-component spinor) is multiplied by a unitary matrix $e^{2\pi i\gamma\sigma_z}$, where $\gamma = \frac{g\mu_B\lambda}{2hc}$.

The Hamiltonian H_{so} is a part of the *Pauli Hamiltonian*, which arises in the context of working out the nonrelativistic limit of the Dirac equation by expanding it in the Foldy–Wouthuysen scheme [10]. The Hamiltonian H_{so} Eq. (9.203) is responsible for a myriad of phenomena in solid-state physics. It is valid even when the orbital angular momentum L is not a good quantum number.

Problem 9.29

Show that in the case of a central potential, $V(\mathbf{r}) = V(r)$, where the electric field is $\mathbf{E} = -\frac{1}{-e}\nabla V = \frac{\mathbf{r}}{er}\frac{dV}{dr}$ and the spin–orbit Hamiltonian can be written as $H_{so} = f(r)\mathbf{L}\cdot\mathbf{S}$.

Hint: Use the identity $\mathbf{L} = \mathbf{r} \times \mathbf{p}$. Answer: $H_{so} = \frac{e\hbar}{4m^2c^2}\frac{1}{r}\frac{dV}{dr}\mathbf{L}\cdot\mathbf{S}$.

The Aharonov–Casher effect can be understood by considering the ring geometry illustrated in Fig. 9.38(b). An electron moving on a ring of radius R experiences a static electric field $\mathbf{E} = \frac{\lambda}{L}\hat{\mathbf{r}}$ produced by a charged wire with constant charge per unit length λ, stretched along the z axis passing through the ring's center (here, $L = 2\pi R$). It is convenient to write the Hamiltonian as

$$\mathcal{H} = \frac{\mathbf{p}^2}{2m} + H_{so} = \frac{1}{2m}\left(\mathbf{p} + \frac{g\mu_B}{2c}\boldsymbol{\sigma}\times\mathbf{E}\right)^2 - \frac{(g\mu_B\mathbf{E})^2}{8mc^2}. \quad (9.204)$$

In the ring geometry, the magnitude of the electric field is constant and it points radially outward, $\mathbf{E} = (\lambda/L)\hat{\mathbf{r}}$. The last term on the RHS of Eq. (9.204) is constant, $\bar{E} = \frac{(g\mu_B\mathbf{E})^2}{8mc^2}$. The Schrödinger equation on the ring involves only the angular direction along the ring, and the angular component of $\boldsymbol{\sigma}\times\mathbf{E}$ results from the z component of $\boldsymbol{\sigma}$:

$$\frac{1}{2m}\left(p + \frac{e}{c}\frac{\Phi_{AC}}{L}\sigma_z\right)^2\psi = (E + \bar{E})\psi, \quad (9.205)$$

where the quantity $\Phi_{AC} \equiv \frac{g\mu_B\lambda}{2e}$ has units of magnetic flux.

Let us consider the Aharonov–Bohm and Aharonov–Casher effects from the gauge invariance point of view. The similarity of Eqs (9.192) and (9.205) is striking. In Eq. (9.192), Φ/L is the vector potential on the ring responsible for the Aharonov–Bohm effect, whereas in Eq. (9.205), $\Phi_{AC}\sigma_z/L$ can be interpreted as a 2×2 matrix vector potential on the ring responsible for the Aharonov–Casher effect. This analogy is not perfect because in the latter case, the energy eigenvalue is shifted by $\bar{E}(\Phi_{AC})$, hence strictly speaking, $E(\Phi_{AC})$ is *not* periodic in Φ_{AC}. However, the ratio between $\bar{E} = \left[\frac{e}{c}\frac{\Phi_{AC}}{L}\right]^2$ and the kinetic energy at p_F can be shown to be extremely small; so, let us assume for the moment that $\bar{E}$ can be safely neglected. Then, Φ/L and $\Phi_{AC}\sigma_z/L$ appear on the same footing referred to as $U(1)$ and $SU(2)$ vector potentials, respectively. The reason for this group theoretical nomenclature is that the phase factor $\eta(x)$ [Eq. (9.195)] is an element of the $U(1)$ group represented here by complex numbers on the unit circle, whereas the matrix,

$$\eta_{AC}(x) = e^{-i\frac{e}{\hbar c}\int^x dx \frac{\Phi_{AC}}{L}\sigma_z}, \quad (9.206)$$

is an element of the $SU(2)$ group represented here by complex 2×2 unitary matrices with determinant equal to unity. In general, $SU(2)$ gauge transformations involve different Pauli matrices, such as,

$$e^{-i\frac{e}{\hbar c}\int^x dx \frac{\Phi_{AC}}{L}\sigma_i}, \quad (i = x, y, z),$$

and two matrices encoding $SU(2)$ gauge transformations need not commute. In other words, the $U(1)$ gauge group is Abelian, whereas the $SU(2)$ gauge group is not. Gauge theories play a central role in many fields of modern physics. Non-Abelian gauge theories were introduced into physics in a seminal 1954 paper by Yang and Mills.

9.5 Magnetic Field Effects

The energy spectrum can be understood in analogy with Eqs (9.192) and (9.193); the wave functions for spin up electrons ($|+\rangle \equiv |\uparrow\rangle$) and spin down electrons ($|-\rangle \equiv |\downarrow\rangle$) are $\psi_n^{(\pm)}(x) = L^{-1} e^{ik_n x} |\pm\rangle$, where $k_n = \frac{2\pi n}{L}$ with $n = 0, \pm 1, \pm 2, \ldots$, and the energy eigenvalues are

$$E_n^{(\pm)} = \frac{\hbar^2}{2mR^2}(n \pm \gamma)^2 - \bar{E} = \frac{\hbar^2}{2mR^2}(n^2 \pm 2n\gamma), \qquad (9.207)$$

where $\gamma = \Phi_{AC}/\Phi_0 = \frac{g\mu_B \lambda}{2hc} = \frac{ge\lambda}{2\pi mc^2}$ and $\bar{E} = \frac{(g\mu_B E)^2}{8mc^2} = \frac{\hbar^2 \gamma^2}{2mR^2}$.

9.5.3 THE HALL EFFECT AND MAGNETORESISTANCE

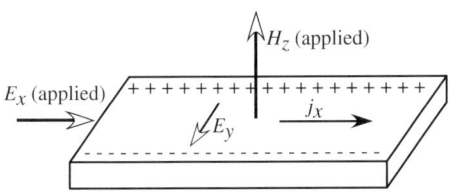

FIG 9.39 A schematic illustration of the classical Hall effect.

In 1879, Edwin H. Hall studied the influence of a magnetic field on a current carrying conductor. Consider the geometry illustrated in Fig. 9.39. Application of an electric field E_x along the x-direction generates a current j_x. Moreover, on application of a perpendicular magnetic field H_z along the z direction, a transverse field E_y develops along the y direction. This phenomenon is called the *Hall effect*.

The two-dimensional sample geometry plays an important role in the physics of the Hall effect. In an ideal two-dimensional geometry, where the thickness of that Hall bar in Fig. 9.39 shrinks to zero, Ohm's law, $R = \rho L/A$, relating the resistance of a sample of length L and cross sectional area A to the resistivity ρ of the material, shows that R and ρ have the same physical dimensions, because the "area" A has the dimension of length. This means that the resistance of a square sample and the resistivity are numerically identical. Hence, one may speak of resistance and resistivity (conductance and conductivity) of a square sample interchangeably.

Classical Analysis of the Hall Effect

The analysis of the Hall effect based on classical electrodynamics requires that we take into account that the system displayed in Fig. 9.39 is not isotropic. Ohm's law $\mathbf{J} = \sigma \mathbf{E}$, or, equivalently, $\mathbf{E} = \rho \mathbf{J}$, relating current density $\mathbf{J} = (J_x, J_y)$ and the electric field $\mathbf{E} = (E_x, E_y)$ requires that the resistivity ρ and the conductivity σ be 2×2 matrices,

$$\rho = \begin{pmatrix} \rho_{xx} & \rho_{xy} \\ \rho_{yx} & \rho_{yy} \end{pmatrix}, \quad \sigma = \rho^{-1} = \begin{pmatrix} \sigma_{xx} & \sigma_{xy} \\ \sigma_{yx} & \sigma_{yy} \end{pmatrix}, \qquad (9.208)$$

which are called the *resistivity tensor* and *conductivity tensor*, respectively.

The longitudinal resistivity in the presence of a magnetic field, ρ_{xx}, is referred to as *magnetoresistance*, whereas the transverse resistivity ρ_{xy} is called the *Hall resistance*. The Hall resistance is related to the *Hall coefficient* R_H,

$$R_H = \frac{\rho_{xy}}{H_z}. \qquad (9.209)$$

Let us determine ρ_{xx} and ρ_{xy} using a classical analysis. An application of a constant field E_x leading to a stationary longitudinal current density J_x implies that there is a dissipation. In other words, the total force affecting the rate of change of momentum **p** of the charge carrier includes a friction (or a damping) term characterized by a collision time τ,

$$\frac{d\mathbf{p}}{dt} = q\left(\mathbf{E} + \frac{1}{mc}\mathbf{p}\times\mathbf{H}\right) - \frac{\mathbf{p}}{\tau}. \tag{9.210}$$

Here, $q = -e$ for electrons and $q = e$ for holes. In steady state, $d\mathbf{p}/dt = 0$, so the average momentum is expressible in terms of the external fields, in a fashion similar to that in the Drude approximation of Sec. 9.2.1. Recalling that $\mathbf{J} = q_c n_c \mathbf{v} = q_c n_c \mathbf{p}/m$, where n_c is the *carrier density*, and using the expression for the Drude conductivity, $\sigma_0 \equiv \frac{n_c q_c^2 \tau}{m}$, and the expression $\omega_c = \frac{q_c H}{mc}$ for the cyclotron frequency, Eq. (9.210) with $d\mathbf{p}/dt = 0$ yields,

$$\sigma_0 E_x = \omega_c \tau J_y + J_x, \tag{9.211a}$$
$$\sigma_0 E_y = -\omega_c \tau J_x + J_y. \tag{9.211b}$$

These equations can be written in matrix form, $\mathbf{E} = \rho \mathbf{J}$ or $\mathbf{J} = \sigma \mathbf{E}$:

$$\begin{pmatrix} E_x \\ E_y \end{pmatrix} = \frac{1}{\sigma_0}\begin{pmatrix} 1 & \omega_c\tau \\ -\omega_c\tau & 1 \end{pmatrix}\begin{pmatrix} J_x \\ J_y \end{pmatrix} \equiv \begin{pmatrix} \rho_{xx} & \rho_{xy} \\ \rho_{yx} & \rho_{yy} \end{pmatrix}\begin{pmatrix} J_x \\ J_y \end{pmatrix}, \tag{9.212a}$$

$$\begin{pmatrix} J_x \\ J_y \end{pmatrix} = \frac{\sigma_0}{1+(\omega_c\tau)^2}\begin{pmatrix} 1 & -\omega_c\tau \\ \omega_c\tau & 1 \end{pmatrix}\begin{pmatrix} E_x \\ E_y \end{pmatrix} \equiv \begin{pmatrix} \sigma_{xx} & \sigma_{xy} \\ \sigma_{yx} & \sigma_{yy} \end{pmatrix}\begin{pmatrix} E_x \\ E_y \end{pmatrix}. \tag{9.212b}$$

One then immediately finds the magnetoresistance ρ_{xx}, the Hall resistance ρ_{xy}, and the Hall coefficient R_H as,

$$\rho_{xx} = \frac{1}{\sigma_0}, \quad R_H = \frac{\rho_{xy}}{H_z} = \frac{\frac{H_z}{n_c q_c c}}{H_z} = \frac{1}{n_c q_c c}. \tag{9.213}$$

According to Eq. (9.213), $R_H < 0$ for electrons and $R_H > 0$ for holes. Because R_H is directly measurable, the ability to identify the sign of the charge carriers through the Hall coefficient is a real godsend.

The above results are general. Further important information is obtained in the Hall bar geometry of Fig. 9.39 where no transverse current is possible, $J_y = 0$. Using $J_y = 0$ in Eqs (9.211), the transverse field (also termed *Hall field*) is obtained as $E_y = -\omega_c \tau E_x$. Hence, the current $\mathbf{J} = J_x \hat{\mathbf{x}} + J_y \hat{\mathbf{y}} = J_x \hat{\mathbf{x}}$ is not parallel to the field $\mathbf{E} = E_x \hat{\mathbf{x}} + E_y \hat{\mathbf{y}} = E_x \hat{\mathbf{x}} - \omega_c \tau E_x \hat{\mathbf{y}}$. The angle θ_H between $\mathbf{J}$ and $\mathbf{E}$,

$$\cos\theta_H = \frac{\mathbf{J}\cdot\mathbf{E}}{|\mathbf{J}||\mathbf{E}|} = -\frac{1}{\sqrt{1+(\omega_c\tau)^2}}, \tag{9.214}$$

is referred to as the *Hall angle*. In the Hall bar geometry analyzed above, $\tan\theta_H = -\omega_c\tau$. But the geometry where $J_y = 0$ is an idealization. To measure the *Hall voltage*, $V_H = V_y$, one has to set up a voltmeter through which a current can pass, so generically, $J_y \neq 0$. In bulk materials there is, in general, a non-zero *Hall current* perpendicular to the electric field. This current is nondissipative, because the motion of the charge carriers is perpendicular to the applied force. In this more general case, the expression for the Hall angle is less simple.

Equation (9.213) shows that the magnetoresistance ρ_{xx} does not change due to the presence of a magnetic field. However, this result is obtained within a simplified picture, where the electrons are treated classically. At low temperatures, when quantum effects are important, the magnetoresistance (and the Hall resistance) depend on the magnetic field in a fascinating way (see Sec. 9.5.8).

A glance at Eqs (9.212a) and (9.212b) reveals a pattern of symmetries obeyed by the Hall resistance ρ_{xy} and the magnetoresistance ρ_{xx}:

$$\rho_{xx}(H) = \rho_{xx}(-H), \quad \rho_{xy}(H) = -\rho_{xy}(-H), \quad \rho_{xy}(H) = -\rho_{yx}(H). \tag{9.215}$$

This is a particular case of an *Onsager relation* where reciprocal relations among transport coefficients are derived by incorporating microscopic reversibility into a statistical mechanical treatment of irreversible linear processes [139].

9.5 Magnetic Field Effects

9.5.4 LANDAU QUANTIZATION

Now, let us consider the effect of a static uniform magnetic field on the energy levels of a free electron. The field is taken to be along the z-axis, $\mathbf{H} = H\hat{\mathbf{z}} = \nabla \times \mathbf{A}$, where $\mathbf{A}$ is a vector potential. We already stressed the freedom in the choice of the vector potential due to gauge invariance. The two conventional choices of the vector potential $\mathbf{A}$ are the *Landau gauge* and the *symmetric gauge*:

$$\mathbf{A} = -Hx\hat{\mathbf{y}} \quad \text{(Landau gauge)}, \tag{9.216a}$$

$$\mathbf{A} = \frac{1}{2}\mathbf{r} \times \mathbf{H} = \frac{1}{2}(Hy\hat{\mathbf{x}} - Hx\hat{\mathbf{y}}) \quad \text{(symmetric gauge)}. \tag{9.216b}$$

Because, in either case, $\mathbf{A}$ lies in the (x, y) plane, the Schrödinger equation is

$$\left[\frac{1}{2m}\left[\left(\mathbf{p}_\perp + \frac{e}{c}\mathbf{A}\right)^2 + p_z^2 \right] - \boldsymbol{\mu}_s \cdot \mathbf{H} \right] \Psi(\mathbf{r}) = E\Psi(\mathbf{r}). \tag{9.217}$$

Here, $\mathbf{p}_\perp = p_x\hat{\mathbf{x}} + p_y\hat{\mathbf{y}}$ is the component of the momentum operator in the (x, y) plane, $\boldsymbol{\mu}_s = g_e\mu_B\mathbf{S}/\hbar$ is the spin magnetic moment operator of the electron, $g_e = -g \approx -2$, and $\mathbf{S}$ is the electron spin operator. The objective is to find the two-component spinor eigenfunctions $\Psi(\mathbf{r})$ and the eigenvalues E. The commutation relations of the Hamiltonian $\mathcal{H}$,

$$[p_z, \mathcal{H}] = [S_z, \mathcal{H}] = 0, \tag{9.218}$$

show that p_z, S_z and $\mathcal{H}$ can be simultaneously diagonalized, hence the wave function and energy can be written as,

$$\Psi(\mathbf{r}) = e^{ik_z z}\psi(x, y)\chi_{m_s}, \tag{9.219a}$$

$$E = \varepsilon + \frac{\hbar^2 k_z^2}{2m} + g_e\mu_B m_s H. \tag{9.219b}$$

Here, k_z is the wavenumber along the magnetic field (the eigenvalue of p_z is $\hbar k_z$) and χ_{m_s} is a spin eigenfunction of S_z with eigenvalue $m_s = \pm 1/2$. The 2D Schrödinger equation for $\psi(x, y)$ is

$$\mathcal{H}_{2D}\psi(x, y) = \frac{1}{2m}\left(\mathbf{p}_\perp + \frac{e}{c}\mathbf{A}\right)^2 \psi(x, y) = \varepsilon\,\psi(x, y), \tag{9.220}$$

whose detailed solution depends on the choice of gauge, as we shall see now.

Solution in the Landau gauge

Using the Landau gauge, Eq. (9.216a), the 2D Hamiltonian is

$$\mathcal{H}_{2D} = \frac{1}{2m}\left[p_x^2 + \left(p_y + \frac{eHx}{c}\right)^2 \right]. \tag{9.221}$$

Translation invariance exists along the y direction, because $[p_y, \mathcal{H}_{2D}] = 0$. Hence, the solution takes the general form,

$$\psi(x, y) = Ce^{iky}\phi(x), \tag{9.222}$$

where C is a normalization constant. Substituting Eq. (9.222) into the Schrödinger equation (9.220) with Hamiltonian (9.221), we find

$$\left[\frac{p_x^2}{2m} + \frac{1}{2m}\left(\hbar k + \frac{eHx}{c}\right)^2 \right]\phi(x) = \varepsilon\,\phi(x). \tag{9.223}$$

Equation (9.223) can be rewritten as a 1D harmonic oscillator problem for $\phi(x)$,

$$\left[\frac{p_x^2}{2m} + \frac{m}{2}\omega_c^2(x - x_k)^2\right]\phi(x) = \varepsilon\,\phi(x), \tag{9.224}$$

where the *cyclotron frequency* ω_c is $\omega_c \equiv \frac{e|H|}{mc}$ and the *guiding center* parameter is $x_k = -\frac{c\hbar k}{e|H|}$. The harmonic oscillator length for this problem is called the *magnetic length* and is given by

$$l_H \equiv \sqrt{\frac{\hbar}{m\omega_c}} = \sqrt{\frac{\hbar c}{e|H|}}, \tag{9.225}$$

so the guiding center x_k can be written in terms of l_H as follows:

$$x_k \equiv -\frac{c\hbar k}{e|H|} = -k l_H^2. \tag{9.226}$$

The *Landau wave functions* and the *Landau energies* are then given by,

$$\psi_{nk}(x,y) = A e^{iky}\phi_{n,k}(x) = A\frac{e^{iky}}{\sqrt{2^n n!\pi^{1/2}l_H}} e^{-\frac{(x-x_k)^2}{2l_H^2}} H_n\!\left(\frac{x-x_k}{l_H}\right), \tag{9.227a}$$

$$\varepsilon_n = (n + 1/2)\hbar\omega_c, \quad n = 0, 1, 2, \ldots, \tag{9.227b}$$

where $H_n(z)$ is the Hermite polynomial and A is a normalization constant for the plane wave function [see below after Eq. 9.229]. The resulting energy levels E_{n,m_s,k_z} are given in Eq. 9.219b with ε_n given in Eq. (9.227b). The Landau energies ε_n are plotted in Fig. 9.40 as a function of H. Note that for an electron, $g\mu_B H \approx \hbar\omega_c$, so

$$E_{nm_s k_z} \approx (n + m_s + 1/2)\hbar\omega_c + \frac{\hbar^2 k_z^2}{2m}. \tag{9.228}$$

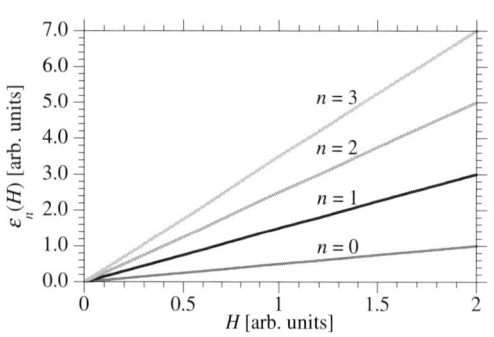

FIG 9.40 The lowest four Landau levels $\varepsilon_n = (n + 1/2)\hbar e H/(mc)$ plotted as a function of H. In a system with fixed Fermi energy E_F, the line $\varepsilon = E_F$ crosses the nth Landau level at the magnetic field $H_n = mcE_F/[\hbar e(n + 1/2)]$.

The definition of the magnetic length contains $\hbar$; hence, its interpretation must be quantum mechanical in nature, just as the interpretation of the harmonic oscillator length. The classical motion of a particle with energy ε in the plane perpendicular to the magnetic field corresponds to a circle of radius $R_c = \sqrt{2m\varepsilon}/(e|H|)$, the *cyclotron radius*. But quantum mechanically, particle trajectories are not sharply defined; due to the uncertainty relation, the standard deviation of the particle's position is proportional to l_H, as you will show in Problem 9.30. However, regarding k as $p_y/\hbar$, the guiding center parameter $x_k = \frac{p_y c}{eH}$ does not contain $\hbar$ and is, therefore, expected to have a classical interpretation. Inspection of the Landau wave function in Eq. (9.227a) shows that the exponent of the Gaussian factor is maximal at $x = x_k$. Hence, the guiding center x_k is interpreted as the center of the circular motion of the electron.

Problem 9.30

Calculate the standard deviation $\Delta x \equiv \sqrt{\langle x^2 \rangle - \langle x \rangle^2}$ in the lowest Landau level function $\psi_{0,x_k}(x,y)$.

9.5 Magnetic Field Effects

Boundary Conditions, Degeneracy, and Quantized Wavenumbers

The Landau energy ε_n does not depend on k, therefore, if the system is infinite, and the wavenumbers k_z and k are continuous variables with $-\infty < k_z, k < \infty$, then ε_n and E_{n,m_s,k_z} are infinitely degenerate. However, it is more realistic, to confine the motion to a box, $0 \leq x \leq L_x$, $0 \leq y \leq L_y$, and $0 \leq z \leq L_z$, and to impose appropriate boundary conditions whenever possible. We thereby obtain a discrete spectrum with finite degeneracy. As far as the motion along the magnetic field direction and the y direction is concerned, periodic boundary conditions can be imposed, $\Psi(x, y + L_y, z) = \Psi(x, y, z + L_z) = \Psi(x, y, z)$, leading to quantization of the wavenumbers,

$$k_z = \frac{2\pi n_z}{L_z}, \quad \text{with } n_z = 0, \pm 1, \pm 2 \ldots, \tag{9.229}$$

$$k_y = k = \frac{2\pi n_y}{L_y}, \quad \text{with } n_y = 0, 1, 2, \ldots, N_L. \tag{9.230}$$

Accordingly, the normalization prefactor A in Eq. (9.227a) equals $1/\sqrt{L_y}$. Thus, the degeneracy of ε_n is given by the number N_L, which will now be determined. In principle, no boundary conditions along x are required because the wave function decays as $x \to \pm\infty$. However, if the system is confined along x between $x = 0$ and $x = L_x$, it is necessary to confine the guiding centers accordingly, i.e., $0 \leq x_k = k l_H^2 \leq L_x$. Employing Eq. (9.230), the maximum allowed value of x_k is $x_k(\max) \equiv (2\pi N_L)/L_y \, l_H^2 = L_x$. Thus, we arrive at the important result,

$$\boxed{N_L = \frac{L_x L_y}{2\pi l_H^2} = \frac{HL_x L_y}{\frac{hc}{e}} = \frac{\Phi}{\Phi_0}} \quad \text{(Landau level degeneracy)}. \tag{9.231}$$

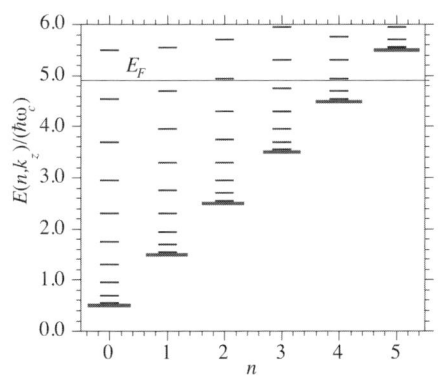

FIG 9.41 The energies $E(n, k_z) = (n + 1/2)\hbar\omega_c + 2\hbar^2\pi^2 n_z^2/(mL_z^2)$ versus n for fixed value of the magnetic field H. The longer bottom line in each column is the 2D Landau energy ε_n, and the shorter lines above indicate the quantized longitudinal energies $\hbar^2 k_z^2/(2m) = 4\pi^2\hbar^2 n_z^2/(2mL_z^2)$. The degeneracy of each level is N_L. The energy spacing between the energies ε_n (the shorter lines) and the degeneracy N_L increases linearly with H.

Hence, the degeneracy N_L of any Landau level ε_n (spin degeneracy not included) is given by the ratio between the magnetic flux through the planar area, $\Phi = BL_xL_y$, and the quantum unit of flux Φ_0. The quantized energies $E(n, k_z)$ (not including Zeeman splitting) are displayed as a function of n in Fig. 9.41.

The density of states at energy E can be calculated as follows: For a given column n, and for $E > \varepsilon_n(H) = (n + 1/2)\hbar\omega_c$, the available energy for the 1D longitudinal motion along the z axis is $E - \varepsilon_n$. Using Eq. 9.26 and recalling that each Landau level is N_L-fold degenerate, the corresponding 1D density of states is

$$D_n(E, H) = N_L \frac{\sqrt{2m}}{h} L_z \frac{1}{\sqrt{E - \varepsilon_n(H)}}. \tag{9.232}$$

When we sum up all the levels in the columns, i.e., sum over $\varepsilon_n < E$, we find

$$D(E, H) = \sum_{\varepsilon_n < E} D_n(E, H) = \frac{\sqrt{2m} e H L_x L_y L_z}{h^2 c} \sum_{\varepsilon_n < E} \frac{1}{\sqrt{E - \varepsilon_n(H)}}. \tag{9.233}$$

Problem 9.31

(a) Determine the number of Landau levels (longer lines in Fig. 9.41) below the Fermi energy E_F.

Answer: $N_F(E_F) = \left[\dfrac{E_F - \frac{1}{2}\hbar\omega_c}{\hbar\omega_c}\right]$, where the square brackets denotes integer value.

(b) Calculate the number $Z(E_F)$ of states with energies $E(n, k_z) \leq E_F$.

Hint: Use the result of (a) and integrate the density of states (9.233).

Answer: $Z(E_F) = \dfrac{\sqrt{2m}eHL_xL_yL_z}{h^2 c} \sum_{n=0}^{N_F-1} \sqrt{E_F - \varepsilon_n}$.

9.5.5 2D ELECTRON GAS IN A PERPENDICULAR MAGNETIC FIELD

In a clean two-dimensional system subject to a perpendicular magnetic field, the energy of electrons is restricted to the Landau levels. There are many phenomena related to the dependence of the Fermi energy and the ground-state energy on the magnetic field, e.g., the de Haas van Alphen effect and the quantum Hall effect (see below). In a system where the Fermi energy is fixed but the magnetic field varies, the number of particles cannot be conserved i.e., the system is not closed, but rather is attached to a particle reservoir, which can exchange particles with the system. In statistical mechanics, a collection of such systems is referred to as a *grand canonical ensemble*, wherein the number of particles is not fixed but the average number of particles is well specified [119, 120]. On the other hand, if the system is closed and the particle number is fixed, the Fermi energy depends on the external magnetic field. A collection of such systems is referred to as a *canonical ensemble*. In both cases, the variation of physical quantities with the magnetic field is somewhat unexpected. The discussion below will concentrate on the orbital effects of the magnetic field (the spin physics is encoded in the Zeeman energy).

The density of states of free electrons in 2D in the presence of a perpendicular magnetic field is

$$D_{2D}(E, H) = \sum_n \delta(E - \varepsilon_n), \qquad (9.234)$$

where the Landau energy ε_n is defined in Eq. (9.227b). It should be stressed that realistic systems are not perfectly clean. In most cases, there is some potential scattering off imperfections or impurities. This kind of disorder leads to broadening of the delta functions, and the density of states takes the form of a sum of broadened peaks centered at ε_n. For weak enough disordered potential of strengths much less than $\hbar\omega_c$, the peaks are well separated, and their distribution is close to Gaussian.

System with Fixed Fermi Energy

It is instructive to compare the number N_L of degenerate states in a given Landau level with the number of electron states in an energy interval of width $\Delta E = \hbar\omega_c$ in two dimensions, in the absence of a magnetic field. Consider a planar rectangle of area L_xL_y and employ Eq. (9.24) for the density of states in 2D to find that the latter equals

$$D_{2D}\hbar\omega_c = \dfrac{m\omega_c}{\hbar}\dfrac{L_xL_y}{\pi} = N_L. \qquad (9.235)$$

If we divide the positive energy axis into segments of width $\hbar\omega_c$ centered at the Landau energies $\varepsilon_n = (n + 1/2)\hbar\omega_c$, it is clear that in each such segment, the number of levels is N_L, whether or not there is a magnetic field present. Consequently, when the Fermi energy is an integer multiple, say m, of $\hbar\omega_c$, the number of states below the Fermi energy is mN_L with and without the presence of a magnetic field. The difference is that, in the presence of magnetic field, every group of N_L states is restricted to the corresponding Landau level energy, whereas in the absence of magnetic field, they are distributed uniformly within the energy interval. Because the 2D density of states D_{2D} is independent of energy, the total energy of

9.5 Magnetic Field Effects

the filled states is the same, with and without magnetic field. The situation is different when E_F is a half-integer multiple of $\hbar\omega_c$, i.e., if E_F coincides with a Landau level, because in this case, the electrons subject to the magnetic field occupy this upper Landau level. For fixed Fermi energy (i.e., in the grand canonical ensemble), this analysis is depicted in Fig. 9.42.

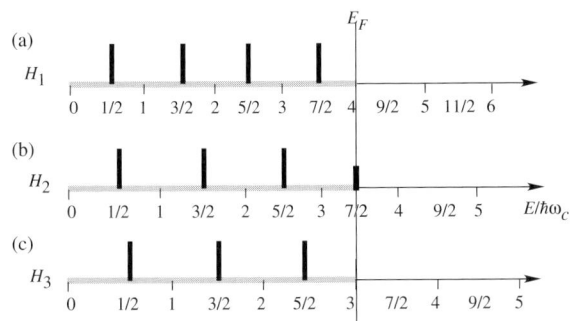

FIG 9.42 Occupation of electron states in a two-dimensional sample in the presence of a perpendicular magnetic field, given a fermi energy E_F. The grey line indicates the occupation of electron states in the absence of the magnetic field. (a) The magnetic field, H_1, is such that $4\hbar\omega_c = E_F$. Four Landau levels $\varepsilon_n = (n + 1/2)\hbar\omega_c$ with $n = 0, 1, 2$, and 3 are completely filled so that the number of electrons is $4N_L$. In the absence of a magnetic field, the number of filled electron states in each interval of width $\hbar\omega_c$ of the grey line is also equal to N_L, hence, the number of electrons below E_F is also equal to $4N_L$. In 2D, and in the absence of a magnetic field, the electrons are equally distributed, ($\rho(E) = $ constant). The ground-state energy in both cases is the same. (b) The field H_2 ($H_2 > H_1$) is such that the fourth Landau level $\varepsilon_3 = \frac{7}{2}\hbar\omega_c$ coincides with the Fermi energy. This upper Landau level is half-filled (shorter vertical thick line) and the electron energy ε_3 is higher than the energies in the interval $3\hbar\omega_c \leq \varepsilon_k \leq E_F$ on the grey line. The ground-state energy in the presence of the magnetic field is higher than that in its absence. (c) $H_3 > H_2$ and $3\hbar\omega_c = E_F$, and the situation is similar to that of (a).

The above analysis enables the calculation of the ground-state energy as a function of magnetic field strength. At zero temperature, all states below E_F are filled with electrons. We have already calculated the number $N_F = \left[\frac{E_F - \hbar\omega_c/2}{\hbar\omega_c}\right]$ of (filled) Landau levels below E_F in Problem 9.31. Suppose we take N_F electrons and put one in each level ε_n. Their energies sum to

$$\frac{1}{2}\hbar\omega_c[1 + 3 + \ldots + (2N_F - 1)] = \frac{1}{2}\hbar\omega_c N_F^2.$$

After multiplying by N_L, the ground-state energy is obtained,

$$E_{gs} = \frac{1}{2}\hbar\omega_c N_L N_F^2, \tag{9.236}$$

which is plotted as a function of H and $1/H$ in Fig. 9.43. In both cases, the energy undergoes sharp oscillations associated with fields H_n, where the Fermi energy coincides with the Landau level energy ε_n. The reason for plotting E_{gs} as a function of $1/H$ is that in this case, the period of oscillations is almost constant. This remarkable finding can be heuristically explained using semiclassical analysis. In classical electrodynamics, an electron restricted to the x–y plane and subject to a magnetic field $\mathbf{H} = H\hat{\mathbf{z}}$ is affected by the *Lorentz force* $\mathbf{F} = \frac{e}{c}\mathbf{v} \times \mathbf{H}$, where $\mathbf{v}$ is the electron velocity vector in the plane. Classically, the electron moves in a circle with radius $R = v/\omega_c$. In the semiclassical approach, we first compare the kinetic energy of the electron with the quantum mechanical Landau level energy, $mv_n^2/2 = \varepsilon_n = (n+1/2)\hbar\frac{eH}{mc}$, and then define a 2D wavenumber $\mathbf{k}_{\perp n}$ through the relation $\mathbf{v}_n = \frac{\hbar \mathbf{k}_{\perp n}}{m}$. An electron with wavenumber $\mathbf{k}_{\perp n}$ executes a circular motion in k-space as well, encircling an area,

$$S_n(H) = \pi k_{\perp n}^2 = \frac{2\pi e}{\hbar c}H(n+1/2). \tag{9.237}$$

For fixed H, the area of orbits in k-space is quantized, and the area between successive orbits is

$$\Delta S \equiv S_n - S_{n-1} = \frac{2\pi e}{\hbar c}H. \tag{9.238}$$

Now, let us assume that for some value H_n, the Fermi energy E_F coincides with the Landau energy $\varepsilon_n = \hbar\frac{eH_n}{mc}(n+1/2)$. Accordingly, $k_{\perp n} = k_F$ is just the Fermi wavenumber, which is independent of n for fixed E_F. Therefore, if we change H from H_n to H_{n+1}, the areas must satisfy

$$S_n(H_n) = S_{n+1}(H_{n+1}) = S = \pi k_F^2, \tag{9.239}$$

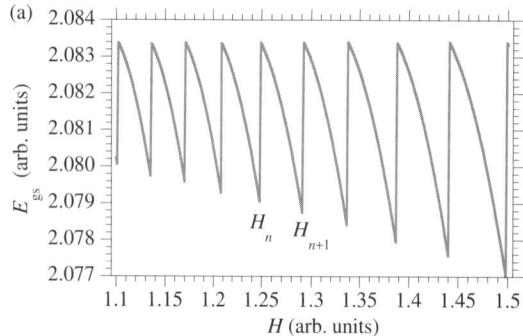

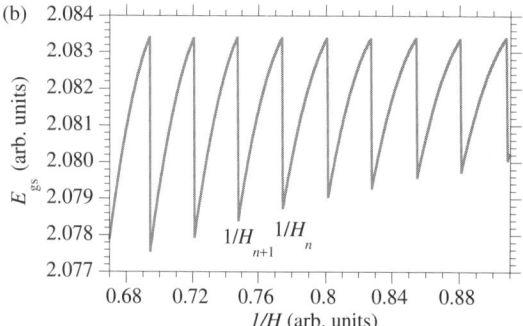

FIG 9.43 Ground-state energy of a 2D closed system of free electrons in a perpendicular magnetic field with fixed Fermi energy (grand canonical ensemble), as given by Eq. (9.236). (a) Ground-state energy, E_{gs}, versus H and (b) E_{gs} versus $1/H$. In the latter case, $H_n^{-1} - H_{n+1}^{-1} = \frac{2\pi e}{\hbar c S}$, where S is the area in k-space of the 2D Fermi surface. Between successive jumps, S is constant, as shown in Eq. (9.325). Strictly speaking, due to a slowly varying envelope, E_{gs} is not periodic in $1/H$. However, the period of oscillations is constant.

which, employing Eq. 9.237, implies a $1/H$ period of

$$\boxed{\left(\frac{1}{H_n} - \frac{1}{H_{n+1}}\right) = \frac{2\pi e}{\hbar c S}.} \qquad (9.240)$$

Thus, in 2D, the period of oscillations of E_{gs} plotted versus $1/H$ is related to an area enclosed by the Fermi "surface" (curve). The analysis leading to Eqs (9.237)–(9.240) can be regarded as a special case of Onsager analysis [140] (see below), which is valid for noncircular orbits as well, with the replacement of the factor $1/2$ by a number $0 < \gamma < 1$ depending on the details of the system under study. The result of these calculations will be useful in the analysis of the de Haas–van Alphen effect later on.

System with Fixed Number of Particles

For the sake of completeness, we also carry out the analysis of a closed 2D system with fixed number N of electrons (i.e., a canonical ensemble). The details of this task are left as a problem. An important quantity in this case is the ratio between the electron number N and the degeneracy of the Landau level (spin included)

$$\nu \equiv \frac{N}{2N_L}, \qquad (9.241)$$

which is referred to as the *filling factor*. Because, by Eq. (9.231), $N_L \propto H$, one has $\nu \propto H^{-1}$. The dependence of such experimental quantities as the longitudinal and Hall resistance on the filling factor constitute one of the most fascinating effects in contemporary solid-state physics, i.e., the quantum Hall effect.

Problem 9.32

For a closed 2D system with fixed number of electrons, N, find the dependence of the Fermi energy E_F on the magnetic field H.

Hint: First, show that the number of filled Landau levels is $N_F = [\frac{N-1}{N_L(H)}] + 1$. Then show that $E_F = (N_F + 1/2)\hbar\omega_c$. The dependence of the Fermi energy on the magnetic field is displayed in Fig. 9.44.

Problem 9.33

Find the ground-state energy of the system studied in Problem 9.32 as a function of the magnetic field.

Hint: The energy of N_F-filled Landau levels (you have calculated N_F in Problem 9.32) is given in Eq. 9.236. The number of electrons in the highest unfilled Landau level is $N - N_F N_L$. The energy of these extra electrons should be

9.5 Magnetic Field Effects

added to that of the N_F filled levels. The result is $E_{gs} = (\hbar\omega_c/2)[N(2N_F + 1) - N_F N_L(N_F + 1)]$. The energy behavior of E_{gs} as a function of H and as function of $1/H$ is similar to that shown in Fig. 9.43. With some modifications, the analysis detailed in connection with the previous case of fixed Fermi energy applies also here. In particular, the constant period of oscillations of E_{gs} as a function of $1/H$ reflects the fact that the number of particles is fixed. More precisely, the periodicity condition (9.240) is the same, but the value of S is now given by $S = \frac{2\pi^2 N}{L_x L_y}$.

Softening Boundary Conditions Along x

So far, the boundary conditions of the wave function (9.219a) along x have not been specified. In Eq. (9.227a), we were content to limiting the guiding centers x_k to stay within the range $[0, L_x]$. We asserted that the Landau wave functions $\psi_{nk}(x, y)$ of Eq. (9.227a) fall off as Gaussians when $|x - x_k| \gg l_H$. The fact that there are periodic boundary conditions along y and z and Gaussian falloff along x is disturbing not only from an esthetic point of view but also physically, because this results in an artificial edge effect. Thus, for large values of k, $x_k \sim L_x$, and the wave function that is not small near the edge, $x = L_x$ is abruptly cut-off. It is not possible to find a solution of the Schrödinger equation that is periodic in *both* x and y. A possible way to partially remove this flaw is taken from Bloch theory. For convenience, we explain this procedure for the lowest Landau level functions $\psi_{0,k}(x, y) = \psi_k(x, y)$. Consider the wave function,

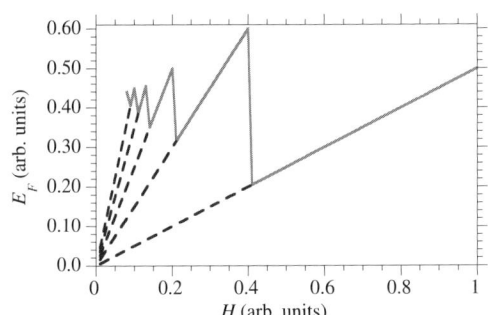

FIG 9.44 Fermi energy E_F (zigzag curve) of a 2D electron gas at zero temperature with fixed number of particles versus magnetic field H. The dashed lines show the energies $\varepsilon_n(H)$ encountered in Fig. 9.40. For strong magnetic fields (above $H = 0.41$ in the figure), all the electrons occupy the lowest Landau level. As the magnetic field is decreased, the height and the degeneracy N_L of the Landau levels decrease, and E_F decreases until, eventually, there are not enough states on the lowest Landau level, and electrons have to occupy the second Landau level, and so on.

$$\psi_k(x, y) = \frac{1}{\sqrt{L_y \pi^{1/2} l_H}} \sum_{m=-\infty}^{\infty} e^{i(k + \frac{mL_x}{l_H^2})y} e^{-\frac{(x - x_k - mL_x)^2}{2l_H^2}}.$$

(9.242)

This function is defined on the whole (x, y) plane and satisfies the conditions,

$$\psi_k(x, y + L_y) = \psi_k(x, y), \quad \psi_k(x + L_x, y) = e^{iL_x y/l_H^2} \psi(x, y).$$

(9.243)

The function so defined is a solution of the Schrödinger equation belonging to the lowest Landau level that is periodic in y and is "almost" periodic in x, because it gains a y-dependent phase factor after adding L_x to x.

Problem 9.34

Prove that $\psi_k(x, y)$ defined by Eq. (9.242) satisfies the boundary conditions (9.243).

Answer: If we replace $y \to y + L_y$, the first exponent is multiplied by a factor $e^{ikL_y} e^{i\frac{mL_x L_y}{l_H^2}}$. The first factor is unity due to the quantization condition on k, Eq. (9.230). The second factor is unity following the definition (9.231) of the Landau level degeneracy N_L. If we replace $x \to x + L_x$, the term inside the brackets in the second exponent on the RHS of Eq. (9.242) becomes $(x - x_k - (m - 1)L_x)$. Changing the summation variable to $m = n + 1$ induces an extra factor $e^{iL_x y/l_H^2}$.

Solution in the Symmetric Gauge

Let us now return to Eq. (9.217) with **A** given in the symmetric gauge, Eq. (9.216b). Once again we concentrate on Eq. (9.220), which now reads,

$$\frac{1}{2m}\left(\mathbf{p}_\perp + \frac{e}{2c}\mathbf{r}\times\mathbf{H}\right)^2 \psi(x,y) = E\psi(x,y). \tag{9.244}$$

Problem 9.35

(a) For a static uniform magnetic field, $\mathbf{H}=H\hat{\mathbf{z}}$, prove that $\frac{1}{2}(\mathbf{p}\cdot\mathbf{r}\times\mathbf{H}+\mathbf{r}\times\mathbf{H}\cdot\mathbf{p}) = \mathbf{r}\times\mathbf{p}\cdot\mathbf{H} \equiv -HL_z$.

(b) Prove that $[L_z, H_{2D}] = 0$, where $L_z = -i\hbar(x\frac{\partial}{\partial y} - y\frac{\partial}{\partial x})$.

By employing rotational invariance, we can work in cylindrical coordinates ρ, θ, and z, wherein the full 3D Schrödinger equation (including spin variables) is

$$\left\{\frac{-\hbar^2}{2m}\left[\frac{1}{\rho}\frac{\partial}{\partial\rho}\left(\rho\frac{\partial}{\partial\rho}\right) + \frac{\partial^2}{\partial z^2} + \frac{1}{\rho^2}\frac{\partial^2}{\partial\theta^2}\right] - (\boldsymbol{\mu}_l + \boldsymbol{\mu}_s)\cdot\mathbf{H} + \frac{e^2H^2}{2mc^2}\rho^2\right\}\Psi(\rho,\theta,z) = E\Psi(\rho,\theta,z), \tag{9.245}$$

where $\boldsymbol{\mu}_l = \mu_B\mathbf{L}/\hbar$ and $\boldsymbol{\mu}_s = \mu_B\mathbf{S}/\hbar$. The planar part is a 2D harmonic oscillator Hamiltonian with force constant $\frac{e^2B^2}{mc^2}$ [compare with Eq. (9.224)]. Separation of variables is possible in the form

$$\Psi(\rho,\theta,z) = R(\rho)\frac{1}{\sqrt{2\pi}}e^{im_l\theta}e^{ik_z z}\chi_{m_s}, \tag{9.246}$$

where χ_{m_s} is the spin function. Substituting into Eq. (9.245) yields

$$\left\{\frac{-\hbar^2}{2m}\left[\frac{1}{\rho}\frac{\partial}{\partial\rho}\left(\rho\frac{\partial}{\partial\rho}\right) + k_z^2 + \frac{m^2}{\rho^2}\right] + (m_l + gm_s)\mu_B H + \frac{m}{2}\omega_c^2\rho^2\right\}R(\rho) = ER(\rho), \tag{9.247}$$

where $m_l = 0, \pm 1, \ldots$ is the integer eigenvalue of L_z and $m_s = \pm 1/2$ is the spin projection quantum number (in the presence of magnetic field, the quantization axis is chosen along the magnetic field direction). The solution to Eq. (9.247) is given by the radial function,

$$R(\rho) = Ce^{-\frac{\rho^2}{4l_H^2}}\rho^{|m_l|}L_n^{|m_l|+1}\left(\frac{\rho^2}{2l_H^2}\right), \tag{9.248}$$

where $L_n^{|m_l|+1}(x)$ is a generalized Laguerre polynomial, C is a normalization constant, and the eigenenergies,

$$E_{nm_lm_s} = \hbar\omega_c\left[n + \frac{1}{2}(|m_l| - m_l + 2m_s + 1)\right] + \frac{\hbar^2 k_z^2}{2m},$$

are the same as those derived within the Landau gauge, see Eq. (9.228). This is to be expected because energies (in fact, any measurable quantity) should be independent of gauge. In the present analysis using the symmetric gauge, the degeneracy pattern is as follows: for positive m_l, the eigenenergies are independent of m_l, whereas for negative m_l, they fall on levels with higher n.

9.5 Magnetic Field Effects

Complex Analysis

An elegant approach for studying the physics of electrons in two-dimensional systems subject to a strong perpendicular magnetic field employs complex analysis within the symmetric gauge. It uses the powerful concept of analytic functions and proves to be useful for treating an interacting electronic system in a strong magnetic field, which is at the origin of the fractional quantum Hall effect (see below). Here, we introduce the basic definitions and some simple relations.

We begin by defining the *covariant momentum operators*, which are proportional to the velocity operators,

$$(\Pi_x, \Pi_y) \equiv \left(p_x + \frac{\hbar y}{2l_H^2}, p_y - \frac{\hbar x}{2l_H^2}\right). \tag{9.249}$$

They obey the commutation relation,

$$[\Pi_x, \Pi_y] = \frac{i\hbar}{l_H^2}. \tag{9.250}$$

The Hamiltonian can be written in terms of these operators as

$$H_{2D} = \frac{1}{2m}\left[(\Pi_x - i\Pi_y)(\Pi_x + i\Pi_y) - \frac{\hbar^2}{2l_H^2}\right]. \tag{9.251}$$

Next, we introduce the complex variables,

$$z = x + iy, \quad z^* = x - iy, \quad |z| = \sqrt{x^2 + y^2} = \rho. \tag{9.252}$$

According to Eq. (9.246), the lowest Landau level function with $n = m_l = 0$ can be written as

$$R_0(x, y) = Ce^{-\frac{zz^*}{4l_H^2}}. \tag{9.253}$$

We now define derivatives with respect to z and z^*,

$$\frac{\partial}{\partial z} \equiv \frac{1}{2}\left(\frac{\partial}{\partial x} - i\frac{\partial}{\partial y}\right), \quad \frac{\partial}{\partial z^*} \equiv \frac{1}{2}\left(\frac{\partial}{\partial x} + i\frac{\partial}{\partial y}\right). \tag{9.254}$$

The crucial point is that z and z^* can be considered as independent variables, although one is the complex conjugate of the other. Thus, for any analytic function $f(z)$, one has $\frac{\partial}{\partial z^*}f(z^*) = \frac{\partial}{\partial z^*}f(z) = 0$. Clearly, the following commutation relations are satisfied:

$$\left[\frac{\partial}{\partial z}, z\right] = \left[\frac{\partial}{\partial z^*}, z^*\right] = 1. \tag{9.255}$$

Now, we would like to write H_{2D} Eq. (9.242) in terms of $\frac{\partial}{\partial z}$ and $\frac{\partial}{\partial z^*}$. For that purpose, we introduce two differential operators,

$$D_1 \equiv 2\frac{\partial}{\partial z^*} + \frac{z}{2l_H^2}, \quad D_2 \equiv 2\frac{\partial}{\partial z} - \frac{z^*}{2l_H^2}, \tag{9.256}$$

which satisfy the following relations:

$$[D_1, z] = 0, \quad [D_2, D_1] = \frac{2}{l_H^2}. \tag{9.257}$$

Employing expression (9.251) for H_{2D}, it is not difficult to deduce that

$$H_{2D} = -\frac{\hbar^2}{2m}D_2D_1 + \frac{1}{2}\hbar\omega_c = \frac{\hbar\omega_c}{2}\left(-4\frac{\partial^2}{\partial z \partial z^*} - z\frac{\partial}{\partial z} + z^*\frac{\partial}{\partial z^*} + \frac{zz^*}{4}\right). \tag{9.258}$$

Problem 9.36

(a) Prove the second equality in Eq. (9.258).
(b) Verify that $[H_{2D}, D_2] = \hbar\omega_c D_2$.

It is reassuring to note that $D_1 R_0(x, y) = 0$ [see Eq. (9.221)], and hence, $H_{2D} R_0(x, y) = \hbar\omega_c R_0(x, y)/2$, so that $R_0(x, y)$ belongs to the lowest Landau level as constructed. In fact, from Eq. (9.257), $D_1[z^{|m_l|} R_0(x, y)] = 0$ for any integer m_l, so that all these functions belong to the lowest Landau level, and according to Eq. (9.246), the integer m_l is the orbital angular momentum, because $z^{|m_l|} = \rho^{|m_l|} e^{im_l\theta}$. Because any linear combination of such functions has the same eigenvalue, we see that a general function in the lowest Landau level can be written as

$$\phi_0(x, y) = e^{-\frac{zz^*}{4l_H^2}} \sum_{m_l=0}^{\infty} a_{m_l} z^{m_l} = e^{-\frac{zz^*}{4l_H^2}} f(z), \tag{9.259}$$

where the coefficients are such that the radius of convergence of the power series is infinite. $f(z)$ is analytic in the whole complex z plane, i.e., it is an *entire function*. Finally, from the second relation in Eq. (9.258), we see that D_2 has the following property:

$$H_{2D}\psi(x, y) = \left(n + \frac{1}{2}\right)\hbar\omega_c \psi(x, y) \rightarrow H_{2D}[D_2\psi(x, y)] = \left(n + 1 + \frac{1}{2}\right)\hbar\omega_c [D_2\psi(x, y)], \tag{9.260}$$

i.e., D_2 is a *raising operator* similar to J_+ familiar from angular momentum algebra. This immediately suggests that higher Landau level eigenfunctions are obtained from the lower ones by repeated application of D_2,

$$\psi_{n,m_l}(x, y) = D_2^n[z^{m_l} e^{-\frac{zz^*}{4l_H^2}}], \quad E_n = \left(n + \frac{1}{2}\right)\hbar\omega_c. \tag{9.261}$$

9.5.6 ELECTRON SUBJECT TO PERIODIC POTENTIAL AND MAGNETIC FIELD

Application of an external magnetic field to crystals leads to many fascinating effects. Both the orbital motion and the spin of electrons are affected. As far as the orbital motion is concerned, there is a remarkable difference between the action of a magnetic field on a free electron gas and on electrons in a crystal. In the first part of this section, we analyze the semiclassical motion of crystal electrons in a uniform magnetic field. In the second part, we discuss other aspects such as the Shubnikov–de Hass effect and de Haas–van Alphen effect.

Within the semiclassical framework, the semiclassical equations in the presence of a uniform and static magnetic field (again assuming no band crossing) are as follows:

$$\dot{\mathbf{r}} = \mathbf{v_k} = \frac{1}{\hbar}\nabla E_\mathbf{k}, \tag{9.262a}$$

$$\hbar\dot{\mathbf{k}} = -\frac{e}{c}\mathbf{v_k} \times \mathbf{H}. \tag{9.262b}$$

Problem 9.37

(a) Prove that the component of $\mathbf{k}$ in the direction of the field is conserved.

Hint: Assume for convenience that $\mathbf{H} = H\hat{\mathbf{z}}$ (not necessarily a symmetry axis of the crystal) and decompose $\mathbf{k} = k_z\hat{\mathbf{z}} + \mathbf{k}_\perp$. Then scalar multiply Eq. (9.262b) by $k_z\hat{\mathbf{z}}$.

(b) Prove that the energy $E_\mathbf{k}$ is conserved.

Hint: Use the chain rule to show that $\dot{E}_\mathbf{k} = \hbar\mathbf{v_k} \cdot \dot{\mathbf{k}}$ and use Eq. (9.262b) to show that this scalar product vanishes.

9.5 Magnetic Field Effects

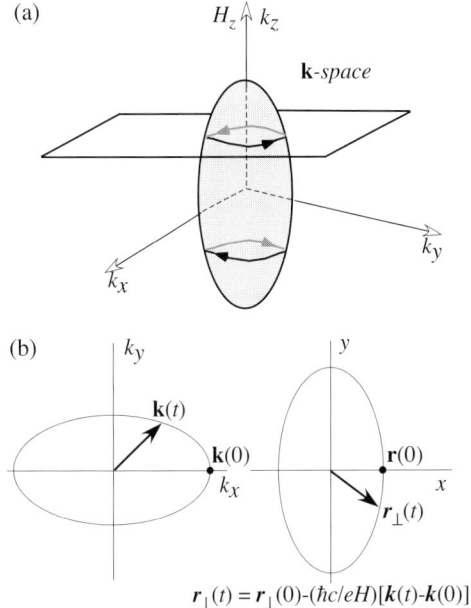

FIG 9.45 (a) Electron (upper) closed orbit and hole (lower) closed orbit obtained through the intersection of a plane perpendicular to the magnetic field with a surface of constant energy [see discussion in the text following Eqs (9.262a) and (9.262b)]. (b) Orbit in **k**-space of the vector $[\mathbf{k}(t) - \mathbf{k}(0)]$ (left panel) and in **k**-space of the vector $[\mathbf{r}_\perp(t) - \mathbf{r}_\perp(0)]$ (right panel). See discussion in the text after Eqs (9.263a) and (9.263b). Note that the prefactor in front of $[\mathbf{k}(t) - \mathbf{k}(0)]$ in Eq. (9.263b) is $-l_H^2$, where l_H is the magnetic length defined in Eq. (9.225).

The two conservation laws derived in Problem 9.37 imply that the electron trajectory in **k**-space is obtained by an intersection of the surface of constant energy and a plane perpendicular to the magnetic field. The sense is opposite to the planar component of the vector product $\nabla E_\mathbf{k} \times \mathbf{H}$. This means that closed orbits in **k**-space surrounding *valleys* [electron orbits according to Fig. 9.33(a)] have opposite sense than those surrounding *hills* [hole orbits according to Fig. 9.33(b)]. This result is illustrated in Fig. 9.45(a).

For free electrons, the surface of constant energy is a sphere (or ellipsoid), and the motion in **r**- and **k**-spaces is along a circle perpendicular to the direction of the magnetic field. It should be stressed that in realistic situations, where the surfaces of constant energy are geometrically less simple and when the magnetic field does not coincide with a symmetry axis, the motion in **k**-space is more complicated and the orbit need not be closed. In that case, it is more transparent to display the motion in a repeated Brillouin zone picture.

Although in **k**-space the motion is planar, in configuration space, the curve $\mathbf{r}(t)$ is not necessarily planar. However, its projection $\mathbf{r}_\perp(t)$ on a plane perpendicular to **H** can be easily traced [122]. Taking the cross product of both sides of Eq. (9.262b) with $\hat{\mathbf{H}} = \hat{\mathbf{z}}$ and using the vector identity, $\hat{\mathbf{b}} \times \mathbf{a} \times \mathbf{b} = \mathbf{a} - \hat{\mathbf{b}}(\hat{\mathbf{b}} \cdot \mathbf{a}) = \mathbf{a}_\perp$, we obtain

$$\hat{\mathbf{H}} \times \hbar \dot{\mathbf{k}} = -\frac{eH}{c}\mathbf{r}_\perp(t), \tag{9.263a}$$

$$\mathbf{r}_\perp(t) = \mathbf{r}_\perp(0) - \frac{\hbar c}{eH}\hat{\mathbf{H}} \times [\mathbf{k}(t) - \mathbf{k}(0)]. \tag{9.263b}$$

Thus, the trajectory $[\mathbf{r}_\perp(t) - \mathbf{r}_\perp(0)]$ in **r**-space is traced out by $-\frac{eH}{\hbar c}[\mathbf{k}_\perp(t) - \mathbf{k}_\perp(0)]$, and for an electron, this motion corresponds to a clockwise rotation around $\hat{\mathbf{H}}$, as illustrated in Fig. 9.45(b).

In metals, the mobile electrons are those with energy near E_F. Therefore, experiments designed to extract information on closed orbits are useful for determining the properties of the Fermi surface. Such an experiment was carried out by de Haas and van Alphen in 1930 (see below). Its analysis requires knowledge of some geometrical properties of closed orbits in **k**-space. The time difference it takes the electron to move on the closed orbit between two points $\mathbf{k}_1$ and $\mathbf{k}_2$ is

$$t_2 - t_1 = \int_{\mathbf{k}_1}^{\mathbf{k}_2} \frac{dk}{|\dot{\mathbf{k}}|} = \frac{\hbar^2 c}{eH} \int_{\mathbf{k}_1}^{\mathbf{k}_2} \frac{dk}{|\frac{\partial E}{\partial \mathbf{k}_\perp}|}, \tag{9.264}$$

using Eq. (9.166b) for $\dot{\mathbf{k}}$. Here, dk is a line element connecting $\mathbf{k}_1$ and $\mathbf{k}_2$ along an orbit in **k**-space. The denominator of the integrand in the second expression on the RHS is the component of $\nabla_\mathbf{k} E$ on the plane perpendicular to **H**. In terms of finite differences, $|\frac{\Delta E}{\Delta \mathbf{k}_\perp}|$ is just the ratio between the energy difference of two closed orbits at $E_{\mathbf{k}_\perp}$ and $E_{\mathbf{k}_\perp} + \Delta E$ and

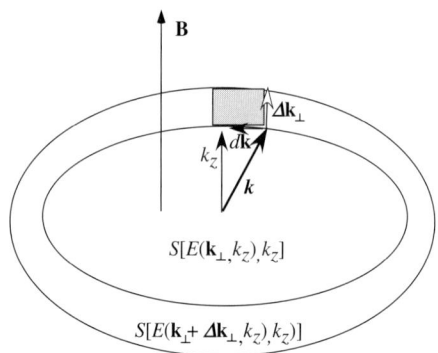

FIG 9.46 Two closed orbits in k-space in a plane perpendicular to the magnetic field, for fixed k_z. The inner orbit corresponds to energy $E(\mathbf{k}_\perp, k_z)$, and the outer orbit corresponds to energy $E(\mathbf{k}_\perp + \Delta \mathbf{k}_\perp, k_z)$.

the corresponding increment $|\Delta \mathbf{k}_\perp|$ of $\mathbf{k}_\perp$. Moreover, $k|\Delta \mathbf{k}_\perp|$ is an area element of the annulus between these two orbits. When integrated between $\mathbf{k}_1$ and $\mathbf{k}_2$, it gives the area ΔS_{12} of the strip. Thus, if $T = t_2 - t_1$ is the period of the orbital motion and $S[E(\mathbf{k}_\perp, k_z), k_z)]$ is the k-space area encircled by an orbit perpendicular to $\mathbf{H}$ at fixed k_z, then (see Fig. 9.46),

$$T = \frac{\hbar^2 c}{eH} \frac{\partial S[E(\mathbf{k}_\perp, k_z), k_z)]}{\partial E}. \tag{9.265}$$

The notation $S[E(\mathbf{k}_\perp, k_z), k_z)]$ is somewhat awkward, but it stresses the dependence of E on k_z and of the area S on k_z due to the position in k-space of the plane perpendicular to $\mathbf{H}$. Whenever possible we will simply use E instead of $E(\mathbf{k}_\perp, k_z)$.

Equation (9.265), when combined with the time–energy uncertainty relation, leads to an important result. A closed orbit in 2D corresponds to a bound state at energy $\varepsilon_n(\mathbf{k}_\perp, k_z)$, so that for a fixed k_z, the energies of the three dimensional systems are quantized, $E_n = \varepsilon_n(\mathbf{k}_\perp, k_z) + \varepsilon(k_z)$ (for free electrons, $\varepsilon_n = (n + 1/2)\hbar\omega_c$ and $\varepsilon(k_z) = \frac{\hbar^2 k_z^2}{2m}$). The relevant uncertainty relation reads,

$$\Delta E_n = (E_{n+1} - E_n) = \frac{h}{T} = \left[\frac{\partial S(E, k_z)}{\partial E}\right]^{-1} \frac{4\pi^2 eH}{hc}, \tag{9.266}$$

where the last equality is due to Eq. (9.265). Noting that

$$\frac{\partial S(E, k_z)}{\partial E} \approx \frac{S(E_{n+1}, k_z) - S(E_n, k_z)}{E_{n+1} - E_n}, \tag{9.267}$$

we find

$$\boxed{\Delta S_n(E_n, k_z) = S(E_{n+1}, k_z) - S(E_n, k_z) = \frac{4\pi^2 eH}{hc} = 4\pi^2 \frac{H}{\Phi_0}.} \tag{9.268}$$

Equation 9.268 is the analog of Eq. 9.238, which was previously derived for a simpler case of free electrons in two dimensions. Another way of looking at this result is to notice from Eq. (9.264) and the subsequent analysis that one can write $E_n dt = \frac{\hbar^2 c}{eH} dS_n$. Integrating both sides and employing the Bohr–Sommerfeld quantization condition, we obtain

$$\int_0^T E_n dt = (n + \gamma) h = \frac{\hbar^2 c}{eH} S_n, \tag{9.269}$$

where $0 \leq \gamma < 1$ does not depend on n (it is related to a quantity called the *Maslov index*, which is well known in semiclassical analysis but will not be elaborated upon here). Thus, we arrive at the expression

$$S(E_n, k_z) = (n + \gamma) \Delta S_n(E_n, k_z), \tag{9.270}$$

which is originally due to Onsager. If ΔS_n is substituted from Eq. (9.268), we obtain

$$S(E_n, k_z) = (n + \gamma) 4\pi^2 \frac{H}{\Phi_0}, \tag{9.271}$$

which is the analog of Eq. (9.237).

9.5 Magnetic Field Effects

> **Problem 9.38**
>
> Use the relation between closed orbits in **r**- and **k**-spaces (Fig. 9.45) to show that if the flux through an electron orbit in **r** is quantized as $n\Phi_0$, then an approximate form of Eq. (9.270) is obtained where $\gamma = 0$.
>
> **Hint:** Let A_n denote the area in **r**-space, such that $HA_n = n\Phi_0$. Use the analysis leading to Fig. 9.45 to show that $A_n = S_n l_H^4$, where l_H is the magnetic length defined in Eq. (9.225).

Semiclassical Motion in Electric and Magnetic Fields

Let us now inspect the semiclassical motion when, in addition, there is a static uniform electric field **E**. Taking the cross product of Eq. (9.166b) with $\hat{\mathbf{H}}$ as before, now results in a modification of Eq. (9.263b):

$$\mathbf{r}_\perp(t) = \mathbf{r}_\perp(0) - \frac{\hbar c}{eH}\hat{\mathbf{H}} \times [\mathbf{k}(t) - \mathbf{k}(0)] + \mathbf{u}t, \tag{9.272a}$$

$$\mathbf{u} \equiv \frac{cE}{H}\hat{\mathbf{E}} \times \hat{\mathbf{H}}. \tag{9.272b}$$

In the special case, $\mathbf{E} \perp \mathbf{H}$, we can go a bit further by defining a modified "band" energy $\bar{E}_\mathbf{k} \equiv E_\mathbf{k} - \hbar\mathbf{k}\cdot\mathbf{u}$, such that Eq. (9.262b) becomes

$$\hbar\dot{\mathbf{k}} = -\frac{e}{\hbar c}\nabla \bar{E}_\mathbf{k} \times \mathbf{H}. \tag{9.273}$$

Consequently, the motion in **k**-space is executed along a curve obtained by an intersection of a plane perpendicular to **H** with a surface of constant $\bar{E}_\mathbf{k}$. Although $\bar{E}_\mathbf{k}$ is not periodic in **k**, it is a conserved quantity.

The above analysis of semiclassical dynamics in perpendicular electric and magnetic fields enables the investigation of fundamental physical phenomena of electrons in a periodic potential when they are exposed to an external magnetic field, to which we now turn.

Hall Effect in 3D Periodic Systems

The analysis of the Hall effect in terms of classical mechanics is too naïve, because it completely ignores quantum mechanical effects. Let us then briefly examine the Hall coefficient and magnetoresistance for electrons in periodic crystals within the semiclassical picture developed above. The phenomena is mainly relevant at high magnetic fields (above 1 T $= 10^4$ gauss). Under this condition, it follows from Eq. (9.272b) that $\bar{E}_\mathbf{k}$ is very close to $E_\mathbf{k}$.

> **Problem 9.39**
>
> Estimate the energy $\hbar|\mathbf{k}||\mathbf{u}|$ for $H = 1$ T assuming an electric field **E**, such that $eEa_0 = 10^{-6}$ Ry, where a_0 is the Bohr radius.
>
> **Hint:** Use the estimate $k \approx 1/a_0$.
> **Answer:** $\approx 10^{-6}$ Ry, which justifies the approximation $\bar{E}_\mathbf{k} \approx E_\mathbf{k}$.

The analysis itself is rather subtle and will not be detailed here (see Ref. [122]). The expression for the Hall coefficient and the magnetoresistance depends crucially on whether all occupied (or all unoccupied) orbits are closed or whether some of them are open. In the first case (all occupied or all unoccupied orbits are closed), it is reasonable to assume that at high magnetic fields and for clean samples, the period T for completing an orbit is rather short (in the free-electron language, it means $\omega_c\tau \gg 1$, and experimentally, a value of 100 is indeed achievable). The component of the electron velocity on a plane perpendicular to the magnetic field can be deduced from the semiclassical equations (9.272a), (9.272b), and (9.273),

$$\mathbf{v}_\perp = -\frac{\hbar c}{eH}\hat{\mathbf{H}} \times \frac{1}{\tau}[\mathbf{k}(0) - \mathbf{k}(-\tau)] + \mathbf{u}, \qquad (9.274)$$

where $\mathbf{u}$ is defined in Eq. (9.272b). Note that the inverse square of the magnetic length $l_H = \sqrt{\frac{\hbar c}{eH}}$ appears on the RHS of Eq. (9.274). Because $[\mathbf{k}(0) - \mathbf{k}(-\tau)]$ is bounded for closed orbits, and,

$$\lim_{\frac{\tau}{T} \to \infty} \mathbf{v}_\perp = \mathbf{u} = \frac{cE}{H}\hat{\mathbf{E}} \times \hat{\mathbf{H}}, \qquad (9.275)$$

the correction due to the first term in Eq. (9.274) is smaller than $\mathbf{u}$ by a factor $(\omega_c \tau)^{-1}$. The contribution of this term to the conductivity tensor, $\bar{\sigma}_{e,h}$ (e for electrons and h for holes), decays as H^{-2} at a large magnetic field. The expression for the perpendicular component of the current contributed from closed (electron or hole) orbits with densities n_e or n_h then reads,

$$\mathbf{J}_\perp = -\frac{n_e ec}{H}\mathbf{E} \times \hat{\mathbf{H}} + \bar{\sigma}_e \mathbf{E} \quad \text{(closed electron orbits)}, \qquad (9.276a)$$

$$\mathbf{J}_\perp = \frac{n_h ec}{H}\mathbf{E} \times \hat{\mathbf{H}} + \bar{\sigma}_h \mathbf{E} \quad \text{(closed hole orbits)}, \qquad (9.276b)$$

such that $H^2 \bar{\sigma}_e$ and $H^2 \bar{\sigma}_h$ saturate as $H \to \infty$. Therefore, the current at high magnetic field is dominated by the contribution of $\mathbf{u}$, which is perpendicular to the electric field, and hence, it is nondissipative. Intuitively, the strong magnetic field corroborates the possibility of electrons to absorb energy from the external electric field. The fact that there is a current component perpendicular to the electric field is expressed in terms of a nonzero Hall coefficient $R_H = \rho_{xy}/H$, which in the high field limit reads,

$$R_H(\text{electrons}) = -\frac{1}{n_e ec} \quad \text{(closed electron orbits)}, \qquad (9.277a)$$

$$R_H(\text{holes}) = \frac{1}{n_h ec} \quad \text{(closed hole orbits)}. \qquad (9.277b)$$

The situation is different if there are open orbits at energy $E = E_F$ (closed and open orbits are shown in Fig. 9.50 of Sec. 9.5.7). In this case, the magnetic field, no matter how strong, cannot prevent the current from having a component along the electric field. More precisely, if an open orbit at some point $\mathbf{r}$ in real space is directed along $\hat{\mathbf{n}}(\mathbf{r})$, then whenever $\hat{\mathbf{n}}(\mathbf{r}) \cdot \mathbf{E} \neq 0$, there is a current component along $\hat{\mathbf{n}}(\mathbf{r})$. Instead of Eqs (9.276a) and (9.276b), we now have (dropping the $\mathbf{r}$ dependence),

$$\mathbf{J} = \sigma_1 \hat{\mathbf{n}}(\hat{\mathbf{n}} \cdot \mathbf{E}) + \boldsymbol{\sigma}_2 \cdot \mathbf{E}, \qquad (9.278)$$

such that $\boldsymbol{\sigma}_2$ vanishes as H^{-2} in the high field limit as do $\bar{\sigma}_e$ and $\bar{\sigma}_h$ for closed orbits, see second terms on the RHS of Eqs (9.276a) and (9.276b). Note that in expression (9.278), $\boldsymbol{\sigma}_2$ is a tensor, hence the bold notation and the dot operation denoting a conductivity matrix acting on a field vector. On the other hand, σ_1 is a scalar satisfying $\sigma_1 \to$ constant as $H \to \infty$. Expression (9.278) implies that (1) the expression for the Hall coefficient deviates from its simple form specified in Eqs (9.277a) and (9.277b), and (2) the geometrical structure of the first term on the RHS of Eq. (9.278) shows that the high field magnetoresistance is not bounded as $H \to \infty$ [122]. To show this, consider again the Hall bar geometry of Fig. 9.39, but assume now that the direction $\hat{\mathbf{n}}$ of an open orbit in coordinate space is *not* along the current. In the high field limit, this can happen only if $(\hat{\mathbf{n}} \cdot \mathbf{E}) \to 0$ as $H \to \infty$. Therefore, we may decompose the electric field as in Fig. 9.47,

$$\mathbf{E} = E_1 \hat{\mathbf{n}} + E_2 \hat{\mathbf{n}} \times \hat{\mathbf{H}}. \qquad (9.279)$$

9.5 Magnetic Field Effects

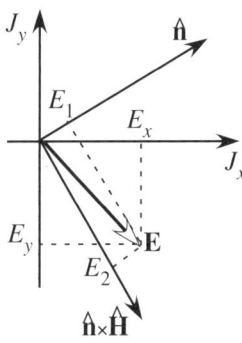

FIG 9.47 Fields and current in the Hall bar geometry of Fig. 9.39 [see Eqs (9.278) and (9.279)]. In the high field limit, $E_1 \to 0$ and $\mathbf{E} \perp \hat{\mathbf{n}}$.

In the high field limit, E_1 vanishes and E_2 saturates. The magnetoresistance is, by definition,

$$\rho_{xx} = \frac{\mathbf{E} \cdot \hat{\mathbf{J}}}{J} \to \frac{E_2}{J} \hat{\mathbf{n}} \times \hat{\mathbf{H}} \cdot \hat{\mathbf{J}}. \quad (9.280)$$

To find E_2, we insert the expression for $\mathbf{E}$ from Eq. (9.279) into the expression for $\mathbf{J}$ in Eq. (9.278) and get, in the high field limit,

$$\mathbf{J} = \sigma_1 \hat{\mathbf{n}} E_1 + \sigma_2 \cdot \hat{\mathbf{n}} \times \hat{\mathbf{H}} E_2. \quad (9.281)$$

Scalar multiplication by $\hat{\mathbf{n}} \times \hat{\mathbf{H}}$ yields,

$$\hat{\mathbf{n}} \times \hat{\mathbf{H}} \cdot \mathbf{J} = \hat{\mathbf{n}} \times \hat{\mathbf{H}} \cdot \sigma_2 \cdot \hat{\mathbf{n}} \times \hat{\mathbf{H}} E_2. \quad (9.282)$$

Therefore, we have

$$\frac{E_2}{J} = \frac{\hat{\mathbf{n}} \times \hat{\mathbf{H}} \cdot \hat{\mathbf{J}}}{\hat{\mathbf{n}} \times \hat{\mathbf{H}} \cdot \sigma_2 \cdot \hat{\mathbf{n}} \times \hat{\mathbf{H}}}.$$

Inserting this expression for E_2/J into expression (9.280) for ρ_{xx} finally yields,

$$\rho_{xx} = \frac{(\hat{\mathbf{n}} \times \hat{\mathbf{H}} \cdot \hat{\mathbf{J}})^2}{\hat{\mathbf{n}} \times \hat{\mathbf{H}} \cdot \sigma_2 \cdot \hat{\mathbf{n}} \times \hat{\mathbf{H}}}, \quad (9.283)$$

which is not bounded as $H \to \infty$ because $\sigma_2 \to 0$, as noted after Eq. (9.278).

Thus, unlike in free electron gas where the magnetoresistance ρ_{xx} is constant and field independent as indicated in Eq. (9.213), in the open orbit scenario, ρ_{xx} is field dependent and divergent as $H \to \infty$. This unbounded magnetoresistance for strong magnetic fields is associated with the occurrence of Fermi surface with open orbits.

9.5.7 DE HAAS–VAN ALPHEN AND SHUBNIKOV–DE HAAS EFFECTS

We have seen above that Landau quantization of electron energies in an applied magnetic field leads to an oscillatory behavior of numerous physical quantities versus magnetic field. Figure 9.43 showed that the ground-state energy of a 2D system of (free) electrons in a perpendicular magnetic field with fixed Fermi energy displays periodic oscillations as a function of $1/H$ with period $\frac{2\pi e}{\hbar c S}$, where S is the area in k-space encircled by a closed planar electron orbit moving on the Fermi surface. Following the Onsager analysis that results in Eq. (9.270), it becomes possible to relate numerous experimental quantities to the structure of the Fermi surface. This is a godsend, because in metals, the geometry of the Fermi surface is directly related to transport coefficients. Here, we discuss two important quantities whose oscillatory dependence on magnetic field enables the study of the shape of the Fermi surface: the magnetic susceptibility χ and the conductivity σ. The corresponding physical phenomena are known as the *de Haas–van Alphen* and the *Shubnikov–de Haas* effects. In both experiments, the oscillations of $\chi(H)$ and $\sigma(H)$ as a function of $1/H$ exhibit a remarkable robustness.

de Haas–van Alphen Effect

A powerful technique for studying the geometry of the Fermi surface in metals is the *de Haas–van Alphen effect*, first observed by W. J. de Haas and P. M. van Alphen in 1930 [142]. It involves the oscillatory variation of numerous physical quantities as the magnetic field H varies. When plotted as a function of $1/H$, the period of oscillations is nearly constant. The de Haas–van Alphen effect can be observed at strong fields (perhaps exceeding several tesla) and at low temperatures, especially in pure metals, where the quantization of the electron orbits is not ruined by collisions with impurities. The de Haas–van Alphen effect was originally explained by Lev Landau in 1930 [143] as resulting from the quantization of closed electronic orbits in a magnetic field. The usefulness of the de Haas–van Alphen effect as a powerful tool for studying the geometry of the Fermi surface for Bloch electrons was pointed out by Lars Onsager in 1952 [140], who derived Eq. (9.270).

Consider for example the magnetization and magnetic susceptibility of a sample, defined as,

$$\mathbf{M}(\mathbf{H}) \equiv -n \nabla_{\mathbf{H}} E_{\text{gs}}(\mathbf{H}), \quad \chi_{ij}(\mathbf{H}) \equiv \frac{\partial M_i(\mathbf{H})}{\partial H_j}, \qquad (9.284)$$

where E_{gs} is the ground-state energy and n is the density of the particles possessing a magnetic moment (usually electron and holes). In the original experiment, the magnitude of the magnetization divided by the magnetic field was plotted against the field strength. A graphical illustration of the original experimental results is shown in Fig. 9.48.

A simple and elegant experimental technique is based on the *torque method*. A metallic crystal is suspended by a stiff wire and the torque exerted on the wire is measured as a function of the applied external magnetic field. Figure 9.49 schematically shows what is expected by measurement of the torque that is due to an applied magnetic field on a silver crystal.

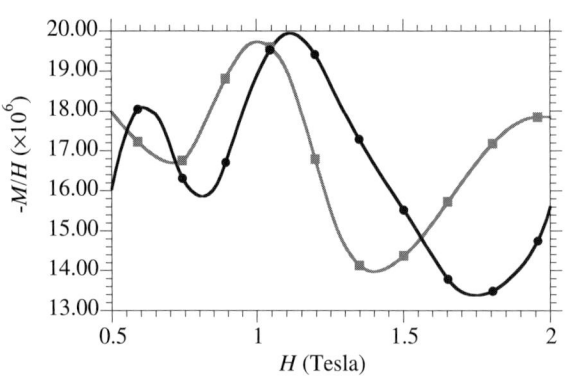

FIG 9.48 Graphical reproduction of the results of the original de Haas–van Alphen experiment [142]. The absolute value of the magnetization per gram divided by the magnetic field strength is plotted against the field strength at temperature $T = 14.2 K$ for two orientations of a bismuth crystal. The units on the ordinate are erg Gauss^{-2} gm^{-1} cm^{-3}.

For the simplified case of free electrons in two dimensions, the explanation of the de Haas–van Alphen effect is based on the analysis, which led to Fig. 9.43(b), i.e., the ground-state energy oscillates as a function of $1/H$ with period,

$$\Delta \left(\frac{1}{H} \right) = \frac{2\pi e}{\hbar c S} \equiv \frac{1}{F}, \qquad (9.285)$$

where $F = \frac{\hbar c S}{2\pi e}$ is the de Haas–van Alphen "frequency." Similar oscillatory behavior is expected also for the magnetic susceptibility χ. It is necessary to analyze the more realistic situation pertaining to Bloch electrons in three space dimensions. We have seen in Sec. 9.5.6 that in the presence of a uniform magnetic field $\mathbf{H} = H\hat{\mathbf{z}}$, Bloch electrons move on k-space orbits obtained through the intersection of planes perpendicular to the magnetic field $\mathbf{H}$ with surfaces of constant energy (actually the Fermi surface, see Fig. 9.45). Once the energy is fixed at E_F, the only parameter determines the orbit and its area is k_z (this is the case whether z is a crystal symmetry axis). We have also shown in Eq. 9.271 that the corresponding areas $S(E_F, k_z)$ are quantized, so that only those k_{zn} for which the quantization condition (9.271) (with $E_n \to E_F$) is satisfied are allowed. Each such permissible orbit with its own encircled area produces its own pattern of periodic oscillations like that in Fig. 9.43. The corresponding periods are of course different, and the combined effect can be smeared out. However, it can be shown that the main contribution comes

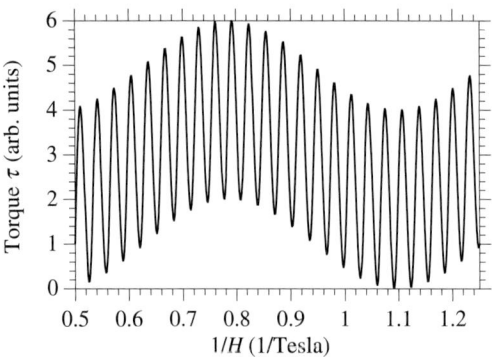

FIG 9.49 The torque τ on a suspended chunk of silver as a function of the inverse magnetic field.

9.5 Magnetic Field Effects

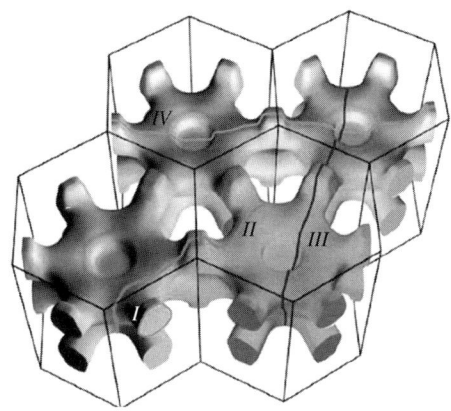

FIG 9.50 An illustration of orbits on the Fermi surface of UPt$_3$. Curve I is a closed extremal orbit encircling a minimal area; curves II and III are closed extremal orbits that extend over more than one Brillouin zone; and curve IV is an open orbit. Reproduced with permission from Ref. [144]. (For the color version of this figure, please refer to the color plate section at the end of book.)

from the extremal orbits satisfying,

$$\frac{\partial S(E_F, k_z)}{\partial k_z} = 0. \quad (9.286)$$

An example of closed extremal and open orbits on the Fermi surface of UPt$_3$ is given in Fig. 9.50.

The reason for the dominant contribution of extremal orbits to the de Haas–van Alphen effect is, first, that the density of the allowed points $\{k_z\}$ is proportional to $[\partial S(E_F, k_z)/\partial k_z]^{-1}$, yielding a large density of orbits with almost identical areas near extremal k_z points. Second, the density of states $D(E_F, H)$, Eq. (9.233), is expected to display a square root singularity at values of H, such that the quantization condition,

$$S(E_F, k_z) = (n + \gamma) 4\pi^2 \frac{H}{\Phi_0}, \quad (9.287)$$

is satisfied by an extremal orbit encircling an area $S(E_F, k_z)$ [122]. Thus, in a narrow sense, the de Haas–van Alphen effect reveals the structure of extremal (closed) orbits on the Fermi surface.

The de Haas–van Alphen effect displays a high degree of robustness. Several factors, such as Landau level broadening, field inhomogeneity, too large frequency, and electron–electron interaction, might combine to attenuate it. The requirement for low enough frequency and high enough magnetic field is reasonable, because the phase of oscillations $2\pi F/H$ must not vary by much more than 2π for a closed orbit of a large area S since, otherwise, the interference will mask the amplitude of oscillations. Fortunately, many metals have extremal orbits encircling small areas of the Fermi surface (see Fig. 9.50 orbit *I* for example), so that according to Eq. (9.285), it implies a small frequency, F. As far as electron–electron interaction is concerned, it was explicitly demonstrated that the de Haas–van Alphen effect is virtually insensitive to electron–electron interaction [145].

Shubnikov–de Haas Effect

Another magnetic phenomenon, related to the de Haas–van Alphen effect, is the oscillatory behavior of the *magneto-conductivity* with external magnetic field. This is called the *Shubnikov–de Haas effect* [146]. Whereas the de Haas–van Alphen effect concerns the magnetic susceptibility, which is an *equilibrium property* [and hence, it gets contributions from all energy levels below the Fermi energy, as evident from Eq. (9.284)], the Shubnikov–de Haas effect is intended to probe a transport property, which occurs due to electrons at or near the Fermi surface. For a system with fixed Fermi

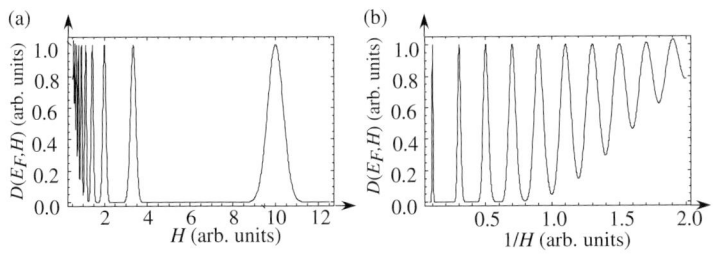

FIG 9.51 Density of states of electrons in two dimensions for fixed Fermi energy. The Landau levels are assumed to be broadened due to scattering by impurities. (a) $D_2(E_F, H)$, (b) $D_2(E_F, 1/H)$.

energy E_F, the magnetoconductivity displays maxima when the Landau level ε_n crosses the Fermi energy as the magnetic field is varied and displays minima when the Fermi energy is located between two Landau levels. According to the definition of the filling factor ν (see Eq. (9.241)), the maxima occur at integer ν and the minima occur at half-integer ν.

The explanation of the Shubnikov–de Haas effect is based on our previous analysis in Secs 9.5.4 and 9.5.6. From Eq. (9.234), we see that the density of states has a maximum value when $E = \varepsilon_n$ and minimum when $E = (\varepsilon_n + \varepsilon_{n+1})/2$. Following the discussion near Fig. 9.40, we concluded that E_F crosses the Landau level ε_n at a magnetic field

$$H_n = \frac{mcE_F}{\hbar e(n+1/2)}. \tag{9.288}$$

Because the conductivity is proportional to the density of states at the Fermi energy, it has a maximum at H_n. In three dimensions with periodic potential, this simple expression should be replaced by

$$H_n = \frac{\Phi_0 S(E_n, k_z)}{4\pi^2(n+\gamma)}, \tag{9.289}$$

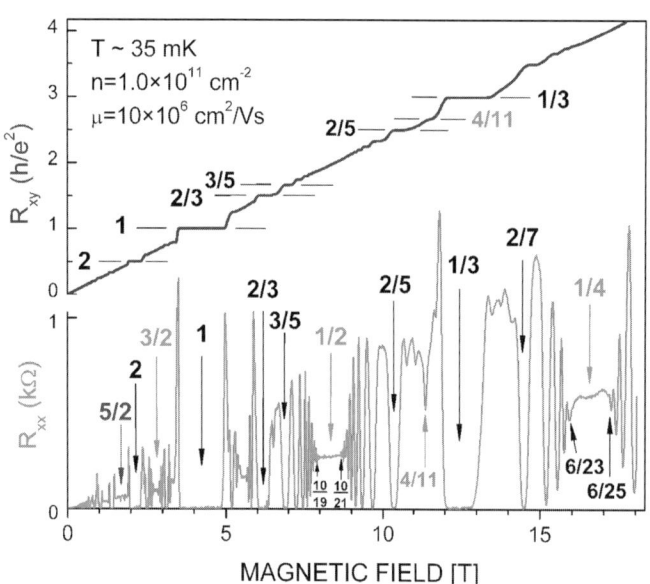

FIG 9.52 Longitudinal resistance $R_{xx} = \rho_{xx}$ and Hall resistance $R_H = R_{xy} = \rho_{xy}$ as function of the magnetic field between 0 and 20 T. For field below 2 T, the behavior of R_{xx} is characterized by the Shubnikov–de Haas oscillations (see the lower H part of the upper panel in Fig. 9.51, while R_H follows the linear behavior (9.209). As H increases, the regions $H \in \Delta E_n$ for which $R_{xx} = 0$ and $R_H = \frac{h}{ne^2}$ broaden and the plateaux of R_H displaying the IQHE for $n = 2, 1$ are clearly visible (for $n = 1$, $H = 4$ T). For $H > 5$ T, one enters the FQHE domain, and numerous plateaux at $R_H = \frac{h}{\nu e^2}$ with $\nu = q/p$ with p odd are visible. Courtesy of Dr. Wei Pan.

as a consequence of Eq. (9.271). Because $H_n - H_{n+1}$ decays as $1/n^2$, the distance between maxima is very small for weak magnetic fields (see Fig. 9.40) and increases at higher magnetic fields. On the other hand, when plotted as a function of $1/H$, the magnetoconductivity display periodic oscillations with maxima at $1/H_n$ and oscillation period,

$$\Delta\left(\frac{1}{H}\right) = \frac{\hbar e}{mcE_F}. \tag{9.290}$$

Thus, the Fermi energy can be extracted either from the period (9.290) or by plotting $n = \frac{mcE_F}{\hbar eH_n} - 1/2$ as a function of $1/H_n$ where one obtains a straight line whose slope yields E_F.

9.5.8 THE QUANTUM HALL EFFECT

The quantum Hall effect (QHE) is a remarkable phenomenon observed in 2D electron systems (usually quantum wells in semiconductors) that are subjected to a strong perpendicular magnetic field at low temperatures. It is a spectacular demonstration of the quantum nature of electron dynamics. Its manifestation, as shown in Fig. 9.52, consists of quantized values of the Hall conductance σ_{xy} in units of e^2/h, which has dimension of inverse ohms and is the quantum unit of conductance. Depending on whether these quantized values are integers or some rational numbers with odd denominator, the phenomena are divided into the *integer quantum Hall effect* (IQHE) or the *fractional Hall effect* (FQHE). The IQHE was first experimentally observed by von Klitzing et al. in 1980 [147]. The experimental setup requires a six-probe Hall bar, as

9.5 Magnetic Field Effects

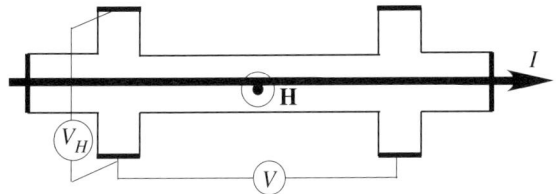

FIG 9.53 Experimental setup for detecting the quantum Hall effect.

shown in Fig. 9.53. The quantization of σ_{xy} is found to be extremely accurate (up to eight digits) that it serves as an effective precise method for determining the value of the fine-structure constant, $\alpha = \frac{e^2}{\hbar c}$. In the IQHE, the Hall conductance is found to be quantized and assumes only integer values of e^2/h (the quantum unit of conductance), i.e., $\sigma_{xy} = \nu e^2/h$, where $\nu = 1, 2, \ldots$ is the filling factor defined in Eq. (9.241). In very strong magnetic fields (usually above 10 Tesla), and in very clean systems, the FQHE is observed, wherein the Hall conductance goes through a series of rational multiples of e^2/h, i.e., $\sigma_{xy} = \nu e^2/h$ with $\nu = 2/7, 1/3, 2/5, 3/5$, etc. In both the IQHE and the FQHE, the values of ν correspond to the filling fractions defined in Eq. (9.241).

The Integer Quantum Hall Effect and Landau Levels

The semiclassical theoretical analysis of the Hall effect discussed in Sec. 9.5.6 for 3D systems with a periodic potential requires significant modification to treat the IQHE physics occurring in 2D systems, where the periodic potential does not play a dominant role. In the bulk picture, we have in mind a very large two-dimensional system, where the spectrum consists of a series of Landau levels with identical spacings, $\Delta = \hbar\omega_c$, between them. Because the IQHE is observed in systems that are not perfectly clean, the Landau levels are broadened as explained after Eq. (9.234). The position of the Fermi energy determines the value of both the longitudinal conductance σ_{xx} and the Hall conductance σ_{xy}. One of the conclusions from Fig. 9.52 is that the magnetoresistance ρ_{xx} and Hall resistance ρ_{xy} of 2D (electron or hole) systems at strong magnetic fields should be considered simultaneously. From Eq. (9.209), we see that when the Hall resistance ρ_{xy} is plotted as a function of the magnetic field H, one obtains a straight line whose slope is the Hall coefficient R_H. Let us now consider Eq. (9.209) at a magnetic field H for which there are exactly n-filled Landau levels. If the system is clean, and the density of states is given by Eq. (9.234), then the field H can be anywhere between $(mcE_F)/[\hbar e(n + 3/2)]$ and $(mcE_F)/[\hbar e(n + 1/2)]$ [see also Fig. 9.40 and Eq. (9.288)]. However, because the Landau levels are broadened and the density of states is as shown in Fig. 9.51, it is clear that this condition can be fulfilled only when the magnetic field H is located at points where the density of states vanishes. In any case, we see from Fig. 9.51, that for a strong enough magnetic field, there is an interval ΔE_n, where H can vary without affecting the number n of filled Landau levels. For example, in the notation of Fig. 9.51 (upper panel), $4 < \Delta_1 < 8.5$ and $2.5 < \Delta_2 < 3$.

The total number of carriers (ignoring spin degeneracy for the moment) is then given by $N_c = nN_L$, where N_L is the Landau level degeneracy defined in Eq. (9.231). If we substitute the value of N_L from Eq. (9.231) in Eq. (9.209), we immediately find the value of ρ_{xy}. When $H \in \Delta E_n$, the density of states vanishes, and there are no charge carriers at the Fermi energy. Hence, the *magnetoconductances* σ_{xx}, σ_{yy} vanish. The conductance and resistance matrices, σ and ρ, are the *matrix inverse* of each other. Hence, the *magnetoresistances* ρ_{xx}, ρ_{yy} vanish as well, while $\rho_{xy} = 1/\sigma_{xy}$. These results that are valid only when $H \in \Delta E_n$ mark the integer quantum Hall effect (IQHE), because they are valid when the filling factor [see Eq. (9.241)], $\nu = N_c/N_L = n$ is an integer. The IQHE is expressed in a compact form in the following set of relations:

$$\boxed{\rho_{xy} = \frac{h}{ne^2}, \quad \sigma_{xy} = n\frac{e^2}{h}, \quad \rho_{xx} = \rho_{yy} = 0, \quad \sigma_{xx} = \sigma_{yy} = 0.} \quad (9.291)$$

When spin degeneracy is included, a factor 2 multiplies the conductivity and divides the resistivity. The above results are remarkable for several reasons.

1. In two dimensions, the resistance and the resistivity are identical, both of them are measured in Ohms. The quantity h/e^2 then has the dimensions of Ohms and serves as the *quantum unit of resistance* (about 25 KΩ). Similarly, conductance and conductivity have identical units, and the quantity e^2/h is the *quantum unit of conductance*.

2. The tensor structure of the conductivity and resistivity implies that when $H \in \Delta E_n$, the longitudinal components of the resistivity and conductivity vanish together. This can occur only if the charge carriers are localized in this region. Localization is a key point to interpret IQHE.
3. Whereas ρ_{xx} is determined by the properties of the charge carriers on the Fermi energy, hence, it is regarded as a transport property, the Hall resistance is determined by the number of filled Landau levels below the Fermi energy. In other words, it is an equilibrium property, like the magnetization or the specific heat.

The Integer Hall Effect: Hall Transition

So far we have focused on Eq. (9.291), where the longitudinal resistances vanish and the Hall resistance (or conductance) is quantized. We now consider the deviation from this strict quantization within the IQHE (single-particle) formalism. For clean systems, the conditions specified in Eq. (9.291) occur when

$$(mcE_F)/[\hbar e(n + 3/2)] < H < (mcE_F)/[\hbar e(n + 1/2))], \qquad (9.292)$$

whereas for impure systems, they occur for $H \in \Delta E_n$. Expressed in terms of Fermi energy, the relations (9.291) are valid for $E_{n-1} < \varepsilon_F < E_n$. The obvious question is what happens when the Fermi energy is varied continuously between two such intervals, for example, $\varepsilon \approx E_n$. In this case, the Hall conductance σ_{xy} need not be quantized, and σ_{xx} does not vanish. These features are clearly seen in Fig. 9.52. The Hall resistance between plateaux (top curve) is not quantized, and the magnetoresistance between plateaux (bottom curve) is peaked. This scenario is referred to as the *Hall transition*. For amorphous (impure) systems, the Hall transition occurs exactly when $\varepsilon_F = E_n$, because when $\varepsilon_F \neq E_n$, there is either a gap in the density of states or no gap but the states are localized. In this case, the Hall transition is in fact an Anderson transition between an Anderson insulator valid for $\varepsilon_F \neq E_n$ and a conducting state for $\varepsilon_F = E_n$. The physics of the Anderson transition is discussed in Sec. 9.9.

Integer Hall Effect: Bulk and Edge Pictures

For understanding the physics of the Hall effect, it is essential to distinguish between the notions of bulk systems and systems with edges. A bulk system without edges is an idealization. Yet, it serves as an important tool for elucidating the integer quantum Hall effect. In particular, for a disordered system, it is used to elucidate the Hall transition, and its relation to the Anderson metal–insulator transition. Theoretically, a bulk system is analyzed by either assuming it to be infinite or by adopting periodic boundary conditions in all directions [however, see discussion following Eq. (9.243)]. A beautiful example for the analysis of the Hall transition in a bulk system with random potential has been developed by J. Chalker and P. Coddington based on percolation theory. Assuming the disordered potential to be smooth, the dynamics of electrons can be characterized by *skipping orbits*: The guiding centers [see discussion after Eq. (9.230)] move on equipotential surfaces in closed orbits around hills or inside valleys of the potential. Electron orbits at different energies constitute a structure similar to equal height lines in a topographic map. At low Fermi energy, all orbits are closed, and there is no orbit that crosses the entire system. The states are then localized, and the longitudinal conductance vanishes. When the Fermi energy increases there are larger and larger orbits until there is at least one infinite orbit. This situation corresponds to the Hall transition when the Fermi energy crosses the Landau level and entails $\sigma_{xx} > 0$. This intuitive picture is then mapped on a network model of random (quantum) resistors that proves to be extremely useful in the study of the integer Hall effect.

Experimentally, however, every system has edges, and the solution of the Schrödinger equation for a system with edges reveals the occurrence of states that carry current along the edges, i.e., *edge states*. Edge states also occur in classical electromagnetism when the Newton equations are solved for two-dimensional electron inside a finite strip (see Fig. 9.54).

Edge states in a Hall bar are schematically illustrated in Fig. 9.55, where we see that the edge states are collected in

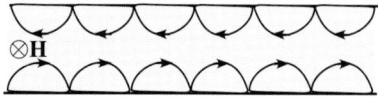

FIG 9.54 Classical electron orbits in a finite clean strip subject to a perpendicular magnetic field.

9.5 Magnetic Field Effects

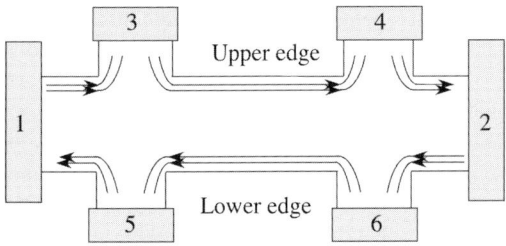

FIG 9.55 Edge-state propagation in a Hall bar subject to a strong perpendicular magnetic field. When a negative bias is applied in contact 1, extra electrons are injected into edge states propagating on the upper edge from left to right (see text for a detailed analysis).

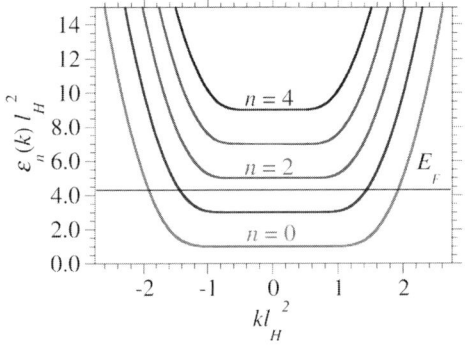

FIG 9.56 Electron energies $\varepsilon_n(k)$ (bent Landau levels) obtained by the solution of Eq. (9.223) with hard wall boundary conditions at $x = \pm a \, (= \pm 1$ here) and periodic boundary conditions along y. The dimensionless quantities $\varepsilon_n(k)l_H^2$ are plotted versus kl_H^2. When $kl_H^2 = \pm a$, the electron guiding center lies on the boundary of the system, whereas states with $|kl_H^2| \ll a$ lie well inside the sample and consists of degenerate Landau levels $\varepsilon_n l_H^2 = 2n + 1$. States with $|kl_H^2| \simeq a$ are edge states whose energies lie within the gap between two Landau levels. States with $|kl_H^2| \gg a$ should be discarded because the corresponding guiding centers lie outside the boundaries.

the transverse ports 3–4 and 5–6. This suggests that the Hall conductance σ_{xy} is intimately related with the number of edge states. To substantiate this scenario, let us inspect the electron spectrum in a two-dimensional system subject to a perpendicular magnetic field within the geometry of a cylinder. The solution of the Schrödinger equation is worked out in Sec. 12.9.4 [see Eq. (12.812a)]. Figure 9.56, a modified version of Fig. 9.56, shows the electron energies $\varepsilon_n(k)$ plotted versus kl_H^2. Due to the existence of edges, the Landau levels bend upward when the guiding centers located at kl_H^2 are close to the edge. Each time when the Landau energy crosses the Fermi energy from below, the Hall transition takes place and σ_{xx} peaks. When the Fermi energy lies between the n and $n + 1$ Landau levels, it crosses $n + 1$ edge states on each side (in the figure, the Fermi energy is between levels $n = 1$ and $n = 2$). The direction of the current on one edge is opposite to that on the other edge, because it is proportional to the velocity $v = \frac{\partial \varepsilon_k}{\partial k}$. This picture derived for a clean system is somewhat smeared by the presence of disorder, but the edge-state physics remains valid because edge-state electrons stick to their direction, as shown in Fig. 9.54. For that reason, edge states are said to be *chiral*. The only way to reverse the direction of the motion of electrons in a given edge state is by diverting them to the other side of the sample. This requires a very strong scattering, which is usually absent in materials such as Si or GaAs.

In summary, we briefly highlight the essentials of the bulk and edge pictures:

- **Bulk picture:** (a) When $E_n < \varepsilon_F < E_{n+1}$, there is a gap in the density of states (or the states are localized). Then, $\sigma_{xy} = ne^2/h$ counts the number of Landau levels below ε_F and $\sigma_{xx} = 0$. (b) When $\varepsilon_F = E_n$, $(n-1)e^2/h < \sigma_{xy} < ne^2/h$ and σ_{xx} has a peak.

- **Edge state picture:** (a) When $E_n < \varepsilon_F < E_{n+1}$, there are n edge states on each edge of the system that close the gap between two Landau levels. The Hall conductance is proportional to the number of edge states at the Fermi energy and $\sigma_{xx} = 0$. (b) When $\varepsilon_F = E_n$, the number of edge states is not defined, and states on the Fermi energy can carry current. Hence, σ_{xx} is peaked and σ_{xy} is not quantized.

Integer Hall Effect: Laughlin Description of Quantized σ_{xy}

A beautiful analysis due to Robert Laughlin explains the quantization of the Hall conductance σ_{xy} and its relation to the number of edge states [148]. Consider, as in Fig. 9.57, an electron on a cylinder of radius R, whose axis is along $\hat{\mathbf{x}}$, subject

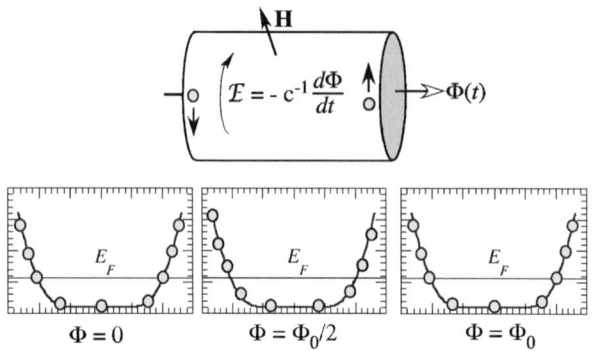

FIG 9.57 Laughlin derivation of the quantization of σ_{xy}. Upper panel: Electrons on a cylinder with radial magnetic field and slowly varying time-dependent flux that generates a tangential (nonconserving) electric field. There are two (chiral) edge states, one on each side moving in opposite directions. A slow variation of the flux from 0 to Φ_0 transfers one electron per edge state from one side to the other. Lower panel: The electron spectrum on the lowest Landau level as the flux changes $\Phi = 0 \to \Phi_0/2 \to \Phi_0$.

to a radial magnetic field. In addition, a slowly varying time-dependent magnetic flux $\Phi(t)$ is applied along the axis of the cylinder, generating an electromotive force $\mathcal{E} = -c^{-1}d\Phi/dt$. The Schrödinger equation (12.812a) derived in Sec. 12.9.4 is slightly modified as follows: (1) The Landau gauge is chosen as $A_y = Hx$, and the wave function $e^{iky}\psi_k(x)$ is periodic in y. The radius of the cylinder R is used as a unit of length, so that all quantities are dimensionless, $x \to x/R$ and the energy $\varepsilon = 2mER^2/\hbar^2$, where E is the physical energy. Thus, we have the eigenvalue problem,

$$\left[-\frac{d^2}{dx^2} + \left(k - \phi(t) - \frac{R^2}{\ell_H^2} x \right)^2 \right] \psi_k(x) = \varepsilon_k \psi_k(x),$$

$$\psi_k(-a/2R) = \psi_k(a/2R) = 0, \qquad (9.293)$$

where $\phi(t) = \Phi(t)/\Phi_0$. With this choice of units, $k = 0, \pm 1, \pm 2, \ldots \pm K$, such that $2K + 1 = N_L = 4\pi a R H/\Phi_0$ is the number of states (not all of them are degenerate) on the Landau level. Therefore, changing $\phi(t)$ adiabatically from 0 to 1 moves all the energies ε_k slowly along the dispersion curve $\varepsilon(k)$ as in Fig. 9.57, displaying a single-edge state per edge on the lowest Landau level. At the end of this procedure, the eigenvalues must be identical to the original ones because changing ϕ by one unit does not affect the spectrum. The net result is that a single electron is transferred from one edge to the other. The Hall current is related to the Hall conductance as

$$\frac{dQ}{dt} = -\frac{1}{c}\sigma_{xy}\frac{d\Phi}{dt} \Rightarrow \sigma_{xy} = -c\frac{dQ}{d\Phi}. \qquad (9.294)$$

If there are n edge states, then $dQ = -ne$, and by construction, $d\Phi = \Phi_0 = hc/e$, therefore, $\sigma_{xy} = ne^2/h$.

The Fractional Quantum Hall Effect

When the external magnetic field becomes very strong (typically $H > 10$ Tesla), the degeneracy N_L of the Landau levels can exceed the total number of charge carriers. Consequently, the filling factor $\nu < 1$. In 1982, experiments revealed the quantization of the Hall conductance for a series of rational values of the filling factor, $\nu = \frac{q}{p}$, where q and p are integers and p is odd [149]. The occurrence of rational number quantized conductivity,

$$\sigma_{xy} = \nu \frac{e^2}{h}, \quad \nu = \frac{q}{p}, \qquad (9.295)$$

is referred to as the *fractional quantum Hall effect* (FQHE). Unlike the IQHE case, where localization is an essential ingredient that causes ρ_{xx} and σ_{xx} to vanish when the Fermi energy is away from the Landau levels, the FQHE is observed only in very clean samples where the effect of disorder is negligible. The "rational" quantization of σ_{xy} and the vanishing of ρ_{xx} and σ_{xx} are solely due to the presence of a gap (and not due to localized states). Soon, it became clear that electron–electron interaction plays a crucial role in the underlying physics. Thus, the FQHE is a many-body problem and the gap occurs in the many-body sense.

Many-body theory will be studied in Chapter 14, but its application to the FQHE is limited because there is no small parameter enabling an effective use of perturbation theory. Fortunately, in some cases, it is possible to use variational

9.5 Magnetic Field Effects

wave functions. This approach will be briefly outlined below. The many-electron Hamiltonian in the symmetric gauge can be written as,

$$\mathcal{H} = \frac{1}{2m}\sum_{i=1}^{N}\left(\mathbf{p}_i + \frac{e}{2c}\mathbf{H}\times\mathbf{r}_i\right)^2 + \sum_{i<j=1}^{N}\frac{e^2}{|\mathbf{r}_i - \mathbf{r}_j|}. \quad (9.296)$$

If the field is very strong, then in the absence of the Coulomb interaction between electrons, the wave function is a Slater determinant of single-particle states belonging to the lowest Landau level. When Coulomb interaction is switched on, it is still reasonable to assume that for strong magnetic fields that prevent Landau level mixing, the wave function can be written as a combination of Slater determinants built from single-particle states belonging to the lowest Landau level. This construction is rather cumbersome because a Slater determinant basis is inefficient for diagonalizing the Coulomb interaction. The question is how to construct an appropriate N-electron wave function on the lowest Landau level taking into account the Pauli principle and the Coulomb interaction and avoid the Slater determinant formalism. It is evident that the interaction cannot be treated as a perturbation. Because the many-body problem defined by the Hamiltonian in Eq. (9.296) cannot be solved exactly, a possible way to approach a solution is by using variational wave functions. Using the complex analysis that led to Eq. (9.259), Laughlin suggested a variational ground-state wave function for interacting electrons projected on the lowest Landau level, for the case $\nu = 1/p$ [150], which has the form,

$$\Psi(\mathbf{r}_1,\mathbf{r}_2,\ldots\mathbf{r}_N) = \Psi(z_1,z_2,\ldots z_N) = \prod_{i<j=1}^{N}(z_i - z_j)^p\, e^{-\sum_{k=1}^{N}|z_k|^2/4l_H^2}. \quad (9.297)$$

Here, $z_i = x_i + iy_i$, where (x_i, y_i) are the coordinates of the ith electron in the plane perpendicular to the magnetic field (see Sec. 9.5.5). Numerical calculations carried out for few electrons systems indicate that the Laughlin wave function has a substantial overlap with the exact wave function. Thus, the expectation value of the Hamiltonian (9.296) with the Laughlin wave function yields an energy that is close to the true ground state E_0. Because p is an odd integer, the wave function is antisymmetric under particle exchange. It should be noted that for high magnetic fields, the system will be spin polarized, and the antisymmetry properties must be manifested in the spatial part of the wave function. It can be shown that the Landau level filling fraction for the Laughlin state is indeed given by $1/p$ [150]. Due to the Gaussian factor on the RHS of Eq. (9.297), the wave function occupies a relatively small region in a $2N$ Euclidean space, so it is sometimes referred to as a liquid drop. Moreover, the ground state is gapped, in the sense that the first excited state lies a finite energy above the ground-state energy E_0. In this sense, the "liquid" is incompressible. Thus, in both the IQHE and the FQHE cases, the magnetic field is such that the ground state is gapped. In the IQHE, the gap is due to Landau level splitting, and in the FQHE, it is due to many-body effects. Although electron–electron interactions play a key role (see Sec. 13.7.4), the exact form of these interactions is not so important in the ground state.

Problem 9.40

Assume there are N electrons in the system described by the Laughlin wave function defined above. Denote by $\mathcal{H}_{2D}(i)$ the kinetic energy operator defined in Eq. (9.258), and let $T_N \equiv \sum_{i=1}^{N}\mathcal{H}_{2D}(i)$ be the kinetic energy operator for the N electron system. Show that $T_N\Psi(z_1,z_2,\ldots z_N) = \frac{N}{2}\hbar\omega_c$.

Answer: From Eqs (9.246) and (9.248), in the symmetric gauge, we know that $z_i^n e^{-\frac{|z_i|^2}{4l_H^2}}$ belongs to the lowest Landau level for particle i, where $n = 0, 1, 2, \ldots$. When the product appearing in the Laughlin wave function is opened, every term is a certain product of the lowest Landau level single-electron states, that is an eigenstate of the kinetic energy operator with eigenvalue $\frac{N}{2}\hbar\omega_c$.

9.5.9 PARAMAGNETISM AND DIAMAGNETISM

In the first part of this section, we consider the magnetization resulting from bound electrons, assuming that the magnetic interaction between the atoms is weak. The magnetization is then given by the product of the atomic magnetic moment and the density of the atoms in the material. In the second part, we consider the response of a metal (basically, a free electron gas) to a weak magnetic field. The treatment employed here is based on the independent electron approach (although electron–electron interaction is not completely ignored). The response of these systems to a weak magnetic field is referred to as *paramagnetism* if the induced moments are parallel to the magnetic field and as diamagnetism if the induced moments are antiparallel to the magnetic field. In both cases, the magnetic moment vanishes in the absence of an external magnetic field. There are materials that are magnetic even in the absence of an external magnetic field. These spectacular magnetic properties are called *ferromagnetism* (occurring in Fe, Co, Ni, Gd, Dy, etc.) and *antiferromagnetism* (occurring in FeO, MnO, NiO, CoO, etc.), and they cannot be explained without taking into account the interaction between electrons. The theory of *quantum magnetism* for these and other related phenomena is briefly discussed below in Sec. 9.5.10.

Magnetization and Magnetic Susceptibility

Consider an insulator composed of N identical atoms occupying a volume $\mathcal{V}$, so the density of atoms is $n = N/\mathcal{V}$. Our aim is to analyze the response of such a system to a constant and uniform weak magnetic field $\mathbf{H}$. Because the response is contributed from the individual atoms, this requires the assumption that the external magnetic field $\mathbf{H}$ is the same as the magnetic field that acts on the atoms inside the material. Strictly speaking, this is not always so, but for many substances, it is an adequate assumption. The quantities that need to be calculated in this context is the *magnetization vector* $\mathbf{M}(\mathbf{H})$, which can be defined as the average magnetic moment per unit volume, and the *susceptibility tensor* $\chi_{ij}(\mathbf{H}) \equiv \partial \mathbf{M}_i / \partial \mathbf{H}_j$. Denoting the ground-state energy of the N atom system as $\mathcal{E}_0(\mathbf{H})$, the magnetization vector and the susceptibility tensor at zero temperature are defined as,

$$\mathbf{M}_i(\mathbf{H}) = -\frac{1}{\mathcal{V}} \frac{\partial \mathcal{E}_0(\mathbf{H})}{\partial H_i}, \quad \chi_{ij}(\mathbf{H}) = \frac{\partial \mathbf{M}_i(\mathbf{H})}{\partial H_j}, \qquad (9.298)$$

where i and j are Cartesian indices. At finite temperature, the ground-state energy in Eq. (9.298) should be replaced by the free energy, $\mathcal{F}(\mathbf{H}, T)$, which is defined as,

$$e^{-\beta \mathcal{F}(\mathbf{H},T)} = \mathcal{Z}(\mathbf{H}, T) \equiv \sum_n e^{-\beta \mathcal{E}_n(\mathbf{H})}, \qquad (9.299)$$

where $\beta = 1/(k_B T)$ and $\mathcal{Z}(\mathbf{H}, T)$ is the *partition function*. To make the discussion simple, the system is assumed to be isotropic and uniform, so that $\mathbf{M}$ is parallel to $\mathbf{H} = H\hat{\mathbf{z}}$. Then, $M(H) = M_z(H)$ and $\chi(H)$ are scalars,

$$M(H) = -\frac{1}{\mathcal{V}} \frac{\partial \mathcal{E}_0(H)}{\partial H}, \quad \chi(H) = \frac{\partial M(H)}{\partial H}. \qquad (9.300)$$

The response is said to be *paramagnetic* if $\chi > 0$, while for $\chi < 0$, the response is said to be *diamagnetic*. Paramagnetism and diamagnetism are equilibrium properties, and are generally rather weak (compared with ferromagnetism, see Sec. 9.5.10).

Problem 9.41

Show that in the limit $T \to 0$, $\mathcal{F}(\mathbf{H}, T) \to \mathcal{E}_0(\mathbf{H})$.

The task of calculating the magnetization and susceptibility of insulators is facilitated by the fact that in many cases, the magnetic interaction between different atoms can be neglected, and a reasonable approximation is then to compute

9.5 Magnetic Field Effects

the magnetic response of the entire solid by multiplying the response of a single atom by the atomic density n. This brings us to the problem of calculating the response of a single atom. Denoting the magnetic field–dependent discrete energy levels of a *single atom* by $E_n(H)$, the magnetization and susceptibility at zero temperature are given by

$$M(H) = -n\frac{\partial E_0(H)}{\partial H}, \quad \chi(H) = -n\frac{\partial^2 E_0(H)}{\partial H^2}. \tag{9.301}$$

At finite temperature, the atomic ground-state energy $E_{gs}(H)$ should be replaced by the free energy $F(H,T)$, defined by

$$e^{-\beta F(H,T)} = Z(H,T) \equiv \sum_n e^{-\beta E_n(H)}, \tag{9.302}$$

where $Z(H,T)$ is the single-atom partition function. The problem of determining the magnetic response of the insulator at zero temperature then reduces to that of finding the dependence of the atomic ground-state energy $E_0(H)$ on the magnetic field.

Atomic Magnetization

Consider a single atom with $\mathcal{N}$ electrons in a static uniform magnetic field. In the absence of electron–electron interaction, the Hamiltonian is

$$H = \sum_{i=1}^{\mathcal{N}} \left[\frac{1}{2m}\left[\mathbf{p}_i + \frac{e}{c}\mathbf{A}(\mathbf{r}_i)\right]^2 + V(r_i) + g_0\frac{\mu_B}{\hbar}\mathbf{S}_i \cdot \mathbf{H} \right], \tag{9.303}$$

where $\mathbf{p}_i$, $\mathbf{r}_i$, and $\mathbf{S}_i$ are momentum, position, and spin operators, respectively, for electron i, ($i = 1,2,\ldots\mathcal{N}$), $V(r_i)$ is the potential energy, $\mathbf{A}(\mathbf{r}) = (1/2)(\mathbf{r} \times \mathbf{H})$ is the vector potential, $g_0 \approx 2$, and $\mu_B = \frac{e\hbar}{2mc}$ is the Bohr magneton. The Schrödinger equation $H\psi = E\psi$ for finite $\mathbf{H}$ cannot be solved analytically, even for the hydrogen atom (see Sec. 4.3.1). However, if the magnetic field is sufficiently weak, one can use perturbation theory (see Sec. 7.3) to calculate the magnetic field corrections $\Delta E_n(H)$ to the low-lying atomic energies. To this end, the kinetic energy term in Eq. (9.303) combined with the Zeeman energy is rewritten as,

$$\sum_{i=1}^{\mathcal{N}}\left[\frac{1}{2m}\left[\mathbf{p}_i+\frac{e}{c}\mathbf{A}(\mathbf{r}_i)\right]^2 + g_0\frac{\mu_B}{\hbar}\mathbf{S}_i \cdot \mathbf{H}\right] = \sum_{i=1}^{\mathcal{N}}\left[p_i^2 + \frac{e^2 B^2}{8mc^2}\left(x_i^2+y_i^2\right)\right] + \frac{\mu_B}{\hbar}[\mathbf{L}+g_0\mathbf{S}] \cdot \mathbf{H}, \tag{9.304}$$

where $\mathbf{L} \equiv \sum_{i=1}^{\mathcal{N}} \mathbf{r}_i \times \mathbf{p}_i = \sum \mathbf{L}_i$ is the total orbital angular momentum, $\mathbf{S} \equiv \sum_{i=1}^{\mathcal{N}} \mathbf{S}_i$ is the total spin, and $\mathbf{J} = \mathbf{L} + \mathbf{S}$ is total angular momentum.

> **Problem 9.42**
>
> Derive Eq. (9.304).

The magnetic field–dependent part of the Hamiltonian, to be considered as a perturbation, is

$$\mathcal{H}_I = \frac{\mu_B}{\hbar}(\mathbf{L}+g_0\mathbf{S}) \cdot \mathbf{H} + \sum_{i=1}^{\mathcal{N}} \frac{e^2 H^2}{8mc^2}(x_i^2+y_i^2). \tag{9.305}$$

Noting that the magnetic susceptibility (9.301) involves second derivative with respect to H, the perturbation expansion is formulated below up to order H^2. Let us denote the $\mathcal{N}$ electron atomic state with energy E_n by the ket $|n\rangle$. In the absence of the magnetic field, the unperturbed energies are ε_n, $n = 0, 1, \ldots$, where $\varepsilon_0 \leq \varepsilon_1 \leq \varepsilon_2 \ldots$. The levels are perturbed on application of an external uniform magnetic field $\mathbf{H}$, $\varepsilon_n \to E_n(H) = \varepsilon_n + \Delta E_n(H)$. If the levels ε_n are nondegenerate, the perturbation correction $\Delta E_n(H)$ is given by,

$$\Delta E_n(H) = \frac{\mu_B}{\hbar}\mathbf{H} \cdot \langle n|\mathbf{L} + g_0\mathbf{S}|n\rangle + \sum_{m \neq n} \frac{|\frac{\mu_B}{\hbar}\mathbf{H} \cdot \langle n|(\mathbf{L} + g_0\mathbf{S})|m\rangle|^2}{\varepsilon_n - \varepsilon_m}$$

$$+ \frac{e^2 H^2}{8mc^2}\left\langle n\left|\sum_{i=1}^{\mathcal{N}}(x_i^2 + y_i^2)\right|n\right\rangle. \tag{9.306}$$

To calculate the magnetization, we need to compute $-\frac{\partial \Delta E_n(H)}{\partial H}$ and substitute it in Eq. (9.301). The result crucially depends on the orbital and spin angular momentum dependence of the atomic wave functions. When the atomic excitation energies are much larger than $k_B T$, we can assume $T = 0$ and simply use Eq. (9.301) with the ground state $|n = 0\rangle$ in Eq. (9.306). Several scenarios are now possible, which are briefly discussed below.

1. Closed Shell Atoms
The atomic ground state $|0\rangle$ for atoms with closed shells is characterized by the following properties:

$$\mathbf{J}|0\rangle = \mathbf{L}|0\rangle = \mathbf{S}|0\rangle = 0. \tag{9.307}$$

Generically, the ground state is nondegenerate; hence, Eq. (9.306) is applicable. Moreover, the first two terms on RHS of Eq. (9.306) vanish as a consequence of Eqs (9.307), so one is left solely with the orbital contribution to the susceptibility,

$$\chi = -n\frac{e^2}{4mc^2}\left\langle 0\left|\sum_{i=1}^{\mathcal{N}}(x_i^2 + y_i^2)\right|0\right\rangle = -n\frac{e^2}{6mc^2}\left\langle 0\left|\sum_{i=1}^{\mathcal{N}}r_i^2\right|0\right\rangle < 0, \tag{9.308}$$

where we have used the fact that electronic wave functions of atoms with closed shells are spherically symmetric, and this implies $\langle 0|x_i^2|0\rangle = \langle 0|y_i^2|0\rangle = \frac{1}{3}\langle 0|r_i^2|0\rangle$. The susceptibility is hence negative, and therefore, the response is diamagnetic. This contribution of the orbital motion to the magnetic response is referred to as *Larmor diamagnetism*. Many, although not most, elements are diamagnetic, including the noble atoms and solids composed of noble atoms. Quantitatively, the resulting magnetization is extremely small compared with other types of magnetic response.

Problem 9.43

Compute the Larmor susceptibility of the He atom by taking the orbital part of the ground state wave function of He as a symmetrized product of hydrogentic $1s$ wave functions with $Z = 2$, $|0\rangle = |\phi_0(r_1)\phi_0(r_2)\rangle$ and calculating the expectation value $\langle 0|r_i^2|0\rangle$.

---

2. Atoms with One Partially Filled Shell
In the absence of strong spin–orbit interaction, an atomic orbital is characterized by the single-electron quantum numbers n and l and the energy ε_{nl} is $2(2l + 1)$ fold degenerate. This number, denoted as N_{max}, is the maximum number of electrons that can be accommodated in a given atomic shell. For example, transition metal atoms have an open d shell ($l = 2$, $N_{max} = 10$) and rare earth atoms have an open f shell ($l = 3$, $N_{max} = 14$). For small spin–orbit coupling, the *Russel–Saunders Coupling* scheme takes $\mathbf{L}$ and $\mathbf{S}$ to be good quantum numbers, and the atomic ground state $|0\rangle$ is then an approximate eigenstate of S^2, S_z, L^2, L_z, J^2, and J_z with respective eigenvalues $S(S + 1)$, M_S, $L(L + 1)$, M_L, $J(J + 1)$, and M_J. Therefore, the degeneracy of the ground state is $2J + 1$. Hund's rules determine the values of J, L, and S in a ground-state atom with a partially filled atomic shell once the number of electrons $N \leq N_{max}$ in this shell is given (see Sec. 10.9).

9.5 Magnetic Field Effects

Problem 9.44

Show that if $\mathbf{J}|0\rangle = 0$, then $\langle n|\mathbf{L} + g_0\mathbf{S}|n\rangle = 0$.

Hint: Use the Wigner–Eckhart theorem (which implies that the expectation value of a vector operator in an eigenstate of J^2 and J_z is proportional to the expectation value of $\mathbf{J}$).

Let us now return to Eq. (9.306) and calculate the atomic susceptibility for cases of partially filled shells, starting with the special case $N = 2l$ and $J = 0$, where the ground state is not degenerate. Based on the solution of the above problem, the first term on the RHS of Eq. (9.306) does not contribute to the susceptibility, and we then have

$$\chi = n\left[2\mu_B^2 \sum_{m \neq 0} \frac{|\langle n|(L_z + g_0 S_z)|m\rangle|^2}{\varepsilon_m - \varepsilon_0} - \frac{e^2}{4mc^2}\langle 0|\sum_{i=1}^{N}(x_i^2 + y_i^2)|0\rangle\right]. \tag{9.309}$$

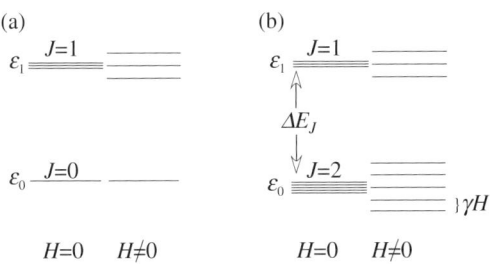

FIG 9.58 First and second energy levels in an atom with unfilled shell in the absence and the presence of an external magnetic field. (a) Nondegenerate ground state with $J = 0$ and an excited state with $J = 1$. The magnetic response is calculated within nondegenerate perturbation theory according to Eq. (9.309). (b) Five fold degenerate ground state with $J = 2$ and an excited state with $J = 1$. The free energy is calculated within degenerate perturbation theory (pertaining to the five degenerate levels of the ground state) according to Eq. (9.312) from which the magnetization is calculated in Eq. (9.313).

Both terms within the square brackets are positive. The response due to the first term is paramagnetic. This *Van Vleck paramagnetism* competes with the Larmor diamagnetism due to the second term, which has already been discussed above. The present analysis is valid when the energies of the excited levels with $J \neq 0$ are much larger than the Zeeman energy separating two levels in an excited state with $J \neq 0$ as illustrated in Fig. 9.58(a). This assumption can be satisfied for a weak enough magnetic field. Note that in both cases where $J|0\rangle = 0$, the perturbation expansion for the energy starts at order H^2.

Finally, we discuss the situation $\mathbf{J}|0\rangle \neq 0$ when there is an open shell with $N \neq 2l$. In this case, the ground-state N electron energy E_J is $2J + 1$ fold degenerate (all states $|LSJM\rangle$ belonging to a given J multiplet have the same energy E_J), and perturbation theory as in Eq. (9.306) is inapplicable. In principle, degenerate perturbation theory is required, but here the perturbation $(\mu_B/\hbar)\mathbf{H} \cdot \langle n|\mathbf{L} + g_0\mathbf{S}|n\rangle$ does not couple different states but just removes the degeneracy by unequally shifting the original levels, as illustrated in Fig. 9.58(b).

Problem 9.45

(a) Use the results of the Wigner–Eckhart theorem to prove the following relation:

$$\langle LSJM|\mathbf{L} + g_0\mathbf{S}|LSJM'\rangle = g(LSJ)\langle LSJM|\mathbf{J}|LSJM'\rangle\delta_{MM'}. \tag{9.310}$$

(b) Find the value of the *Landé factor* $g(LSJ)$.

Hint: First prove that if the equality (9.310) holds, then $\langle LSJM|(\mathbf{L} + g_0\mathbf{S}) \cdot \mathbf{J}|LSJM'\rangle = g(LSJ)J(J+1)\hbar^2\delta_{MM'}$. Then, calculate $\langle LSJM|\mathbf{L}\cdot\mathbf{J}|LSJM'\rangle$ and $\langle LSJM|\mathbf{S}\cdot\mathbf{J}|LSJM'\rangle$ using the technique developed in Problem 4.23(a). Taking the electron g-factor to be $g_0 \approx 2$, the result is $g(LSJ) = \left[\frac{S(S+1)-L(L+1)}{2J(J+1)}\right] + \frac{3}{2}$.

As a consequence of Eq. (9.310), $\mathbf{L} + g_0\mathbf{S} = g(LSJ)\mathbf{J}$. Hence, as far as a response to magnetic field is concerned, the atom in its ground state can be considered as a point particle with magnetic moment,

$$\boldsymbol{\mu} = -g(LSJ)\frac{\mu_B}{\hbar}\mathbf{J}. \tag{9.311}$$

Note that within the present approach, the contribution of higher levels (above the ground state) is neglected. This is justified only if the energy of the excited electronic states is much greater than $k_B T$. The Zeeman energy splitting between adjacent energy levels in the ground state is equal to $\gamma H \equiv g(LSJ)\mu_B H$. Denoting the excitation energy to the next J multiplet by ΔE_J [see Fig. 9.58(b)], we assume that $\gamma H < k_B T \ll \Delta E_J$, which generally can be satisfied for weak enough magnetic fields. Then the magnetization $M(H)$ and the susceptibility $\chi(H)$ are derived from the free energy,

$$F = -k_B T \ln\left[\sum_{M_J=-J}^{J} e^{-\beta \gamma H M_J}\right]. \tag{9.312}$$

Denoting $x \equiv \beta \gamma J H$, we find[5]

$$M(H) = -n\frac{\partial F}{\partial H} = -n\gamma \frac{\sum_{M_J=-J}^{J} M_J e^{-\beta \gamma H M_J}}{\sum_{M_J=-J}^{J} e^{-\beta \gamma H M_J}} = n\gamma J B_J(x), \tag{9.313}$$

where

$$B_J(x) \equiv \frac{2J+1}{2J}\coth\frac{2J+1}{2J}x - \frac{1}{2J}\coth\frac{1}{2J}x \tag{9.314}$$

is the *Brillouin function*. In the limit $\gamma H \ll k_B T$, which is often easily satisfied, the magnetic susceptibility is approximately given by

$$\chi(H) = \frac{\partial M}{\partial H} \approx n\frac{(g\mu_B)^2}{3}\frac{J(J+1)}{k_B T}, \quad (\gamma H \ll k_B T). \tag{9.315}$$

This result is the *Curie law* of paramagnetism. Although, strictly speaking, it is not a law but an approximation (it retains only the first term of a power series in H), it is valid in a myriad of physical systems. Figure 9.59 plots $M(H)$ of Eq. (9.313) versus x for $J = 1/2, 1$, and $3/2$, and the dashed curves plot the Brillouin functions (B_1 falls right on M for $J = 1$).

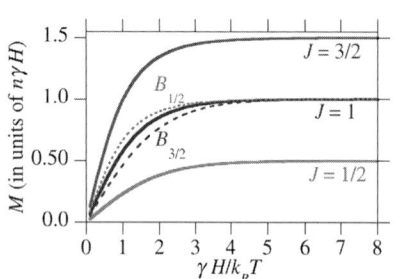

FIG 9.59 Magnetization $M(H)$ versus $\gamma H/k_B T$ for $J = 1/2, 1$, and $3/2$. The dashed curves show the Brillouin functions $B_J(x)$.

Problem 9.46

(a) Use the method described in the footnote to calculate the sum $\mathcal{S}(-J, J) \equiv \sum_{i=-J}^{J} z^i$.

(b) Derive Eq. (9.313) by letting $z = e^x$ and using the result obtained in (a).

---

[5] The sum in Eq. (9.312) is easy to analytically evaluate by noting that the sum, $\mathcal{S}(1, J) \equiv \sum_{i=1}^{J} z^i = z + \cdots + z^J$ and $z\mathcal{S}(1, J) = z\sum_{i=1}^{J} z^i = z^2 + \cdots + z^{J+1}$. Subtracting these two equations yields $\mathcal{S}(1, J) = \sum_{i=1}^{J} z^i = (1 - z^{J+1})/(1 - z)$. You will evaluate the sum defined as $\mathcal{S}(-J, J)$ in Problem 9.46 to obtain an analytic formula for Eq. (9.312).

9.5 Magnetic Field Effects

It is worth noting that the basic assumption of uncorrelated atoms on which Eqs (9.301) are based is not always justified. The main source of deviation is due to the reduction of the atom's symmetry due to the crystal field, which an atom residing within the solid feels. This results in a *crystal field splitting*, which tends to reduce the symmetry of the single atom from a spherical to a lower symmetry. Once spherical symmetry is lost, the orbital angular momentum is not a good quantum number anymore. This is referred to as *angular momentum quenching*. The Curie law (9.315) can still be used, albeit with S instead of J (as if $L = 0$).

Pauli Paramagnetism and Landau Diamagnetism

In metals, we are mainly concerned with the contribution of conduction electrons (not bound electrons) to the magnetization of the metal. Within the single-particle picture, the magnetic response can be calculated rather straightforwardly. It is composed of two contributions, one due to the electron spin and the other due to the orbital degrees of freedom.

Electron Spin Contribution: Pauli Paramagnetism

The contribution of the electron spin to the susceptibility of a metal is referred to as *Pauli paramagnetism*. Let n denote the density of conduction electrons and $n_\pm$ denote the density of conduction electrons with spin parallel $(+)$ or antiparallel $(-)$ to the magnetic field. Clearly,

$$n = n_+ + n_- = \int dE\, g(E) f(E), \tag{9.316}$$

where $g(E) = D(E)/\mathcal{V}$ denotes the density of states per unit energy [see Eqs (9.13) and (9.15), where these quantities are defined for the free electron gas], and $f(E)$ is the Fermi distribution [Eq. (9.27)]. When $n_+ \neq n_-$, there is a nonzero magnetization given by,

$$M(H) = -\mu_B (n_+ - n_-). \tag{9.317}$$

To relate the densities $n_\pm$ to the applied magnetic field, we consider the system at zero temperature. We already noted that for metals, $k_B T \ll E_F$, so that, to a good approximation, calculations involving density of electrons near the Fermi energy can be carried out at $T = 0$. In the absence of a magnetic field at zero temperature, the states are filled up to the Fermi energy E_F. In the presence of a magnetic field, the magnetic energy of the electrons with $m_s = 1/2$ (spin aligned with the magnetic field H and therefore, the magnetic moment aligned against the field) is $\mu_B H$ and those with $m_s = -1/2$ is $-\mu_B H$. Therefore, the electrons redistribute themselves in a manner shown in Fig. 9.60. If $g_\pm(E)$ denotes the densities of states (number of states per unit energy per unit volume), then clearly

$$n_\pm = \int dE\, g_\pm(E) f(E). \tag{9.318}$$

A glance at Fig. 9.60 indicates that the magnetic field shifts the energies of spin-up (spin-down) electrons by $\pm \mu_B H$, so that $g_\pm(E) = g(E \mp \mu_B H)$. Because $\mu_B H \ll E_F$, we can expand these densities near E to first order and obtain the density difference,

$$n_+ - n_- = \int dE\, [g(E - \mu_B H) - g(E + \mu_B H)] f(E) \approx -\mu_B H \int dE\, g'(E) f(E)$$

$$= -\mu_B H \int dE\, g(E)[-f'(E)] \approx -g(E_F) \mu_B H, \tag{9.319}$$

where we used integration by parts and for $k_B T \ll E_F$, $-f'(E) \approx \delta(E - E_F)$. Using (9.317), we find the magnetization and the Pauli susceptibility,

$$\boxed{M(H) = -\mu_B H(n_+ - n_-) = \mu_B^2 H g(E_F), \quad \chi_{\text{Pauli}} = \frac{\partial M}{\partial H} = \mu_B^2 g(E_F) > 0.} \tag{9.320}$$

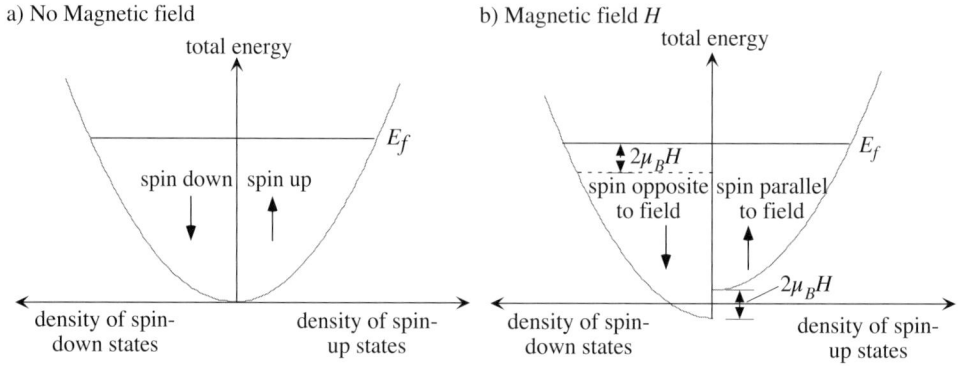

FIG 9.60 Electron density of states $g(E)$ (horizontal axis) is plotted as a function of energy (vertical axis) for the two spin orientations. (a) At zero magnetic field, $g(E) = g_+(E) = g_-(E) = \frac{1}{2\pi^2}\left(\frac{2m}{\hbar^2}\right)^{3/2} E^{1/2}$. (b) At finite magnetic field, the curves are shifted $\pm \mu_B H$, leading to Pauli paramagnetism in a metal as explained in the text.

Landau Diamagnetism

The discussion above focused on the paramagnetic response of a free electron gas resulting from the interaction of the electron spin with the magnetic field. Let us now consider the interaction of the orbital motion of the electrons with the magnetic field. We show that it leads to a diamagnetic response. For realistic systems, such calculations should be carried out with Bloch electronic wave functions but this turns out to be complicated; nevertheless, the calculation presented here, based on Ref. [151] for the free electron gas, is quite instructive. For an electron in a three-dimensional volume $V = L_x L_y L_z$ subject to a constant magnetic field $H\hat{z}$, the energies $E_{n k_z}$ and their degeneracy N_L (see Eq. (9.231)) are as follows:

$$E_{n k_z} = \left(n + \frac{1}{2}\right)\hbar\omega_c + \frac{\hbar^2 k_z^2}{2m} \equiv \varepsilon_n + \frac{\hbar^2 k_z^2}{2m}, \quad N_L = \frac{m\omega_c}{\hbar}\frac{L_x L_y}{\pi}, \tag{9.321}$$

where we recall that $\omega_c = \frac{eH}{mc}$ is the cyclotron frequency and $\varepsilon_n = (n + 1/2)\hbar\omega_c$ are the Landau energies. The first task is to calculate the number of states $\mathcal{N}(E)$ below a given energy E. For a given Landau energy ε_n, the energy available for the 1D motion along the magnetic field direction is $E_z \equiv E - \varepsilon_n$ (see Fig. 9.41). The corresponding 1D density of states is given in Eq. (9.26). When integrated to give the total number of states, we see that each level ε_n with degeneracy N_L contributes $\frac{L_z\sqrt{2m}}{\pi\hbar}\sqrt{E - \varepsilon_n}$ states below E due to the longitudinal motion along z. Therefore,

$$\mathcal{N}(E) = \frac{L_z\sqrt{2m}}{\pi\hbar} N_L \sum_{\varepsilon_n < E} \sqrt{E - \varepsilon_n} = \frac{2\sqrt{2m}VeH}{h^2 c}\sum_{\varepsilon_n < E} \sqrt{E - (2n+1)\mu_B H}. \tag{9.322}$$

With this expression in hand, we can now calculate the free energy at finite temperature T and chemical potential μ. Assuming there are N electrons in the system, the following statistical mechanics expression is obtained,

$$F = N\mu - 2k_B T \int_{\varepsilon_0}^{\infty} dE \frac{d\mathcal{N}(E)}{dE} \ln(1 + e^{\frac{\mu-E}{k_B T}}) = N\mu - 2k_B T \int_{\varepsilon_0}^{\infty} dE\, \mathcal{N}(E) f(E), \tag{9.323}$$

where $f(E)$ is the Fermi distribution, and the second expression is obtained via integration by parts, using $\mathcal{N}(\varepsilon_0) = 0$. Because $f(E)$ decays exponentially for $E \gg \mu$ the integrals converge. Precise evaluation of the integral requires power

9.5 Magnetic Field Effects

expansion of $f(E)$ and the use of Poisson's summation formula. This technical part will not be detailed here. The result, assuming the constraint that the chemical potential $\mu \gg k_B T$ and $\mu_B H$ is

$$F = F_0 + \frac{m^{\frac{3}{2}}(\mu_B H)^2 \mathcal{V}\sqrt{\mu}}{3\pi^2 \hbar^3 \sqrt{2}}, \quad (9.324)$$

where F_0 is the free energy in the absence of magnetic field. Accordingly, the susceptibility for Landau diamagnetism is

$$\boxed{\chi_{\text{Landau}} = -\frac{1}{\mathcal{V}}\frac{\partial^2 F}{\partial H^2} = -\frac{e^2}{12\pi^2 \hbar c^2}\sqrt{\frac{2\mu}{m}} \xrightarrow[T \to 0]{} -\frac{e^2 k_F}{12\pi^2 mc^2} = -\frac{1}{3}\chi_{\text{Pauli}}.} \quad (9.325)$$

As $T \to 0$, $\mu \to E_F = \frac{\hbar^2 k_F^2}{2m}$.

9.5.10 MAGNETIC ORDER

A regular pattern in alignment of the atomic magnetic moments in a material is called magnetic order. The most familiar example is that of a permanent magnet, wherein the magnetic moments are aligned. (They may have been aligned under the application of an external magnetic field, but they remain aligned even after the external field is switched off.) These materials are either *ferromagnetic* or *ferrimagnetic*. Roughly speaking, in ferromagnets, the magnetic moments are aligned in a parallel orientation, whereas in ferrimagnets, more than one type of atomic magnetic moment is present and the magnetic moments of the atoms on different sublattices are generally aligned in opposite directions. The magnetization of ferromagnetic and ferrimagnetic materials is much larger than in paramagnetic or diamagnetic materials. Magnetic order also exists in materials that do not form permanent magnets and are referred to as *antiferromagnets*. Heuristically, the magnetic moments in these materials are aligned in a regular pattern with neighboring spins pointing in opposite directions. See Fig. 9.61 for examples of ferromagnetic, antiferromagnetic, and ferrimagnetic structures. Magnetically ordered materials lose their magnetic order above a certain material-dependent critical temperature T_c. Above the critical temperature, the magnetic moments are randomly ordered and the materials are usually paramagnetic.

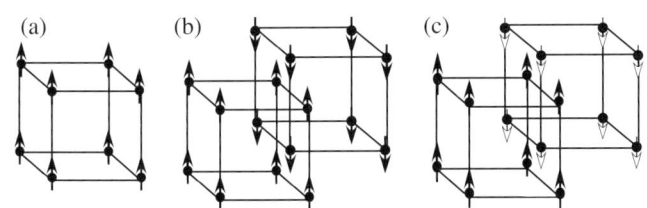

FIG 9.61 Ferromagnetic, antiferrromagnetic, and ferrimagnetic structures. (a) The magnetic moments of ions on the sites of a simple cubic lattice pointing in the same direction. (b) Antiferromagnetic arrangement of magnetic moments of identical ions on the sites of two interpenetrating simple cubic lattices that point in opposite directions. (c) Ferrimagnetic arrangement of the magnetic moments of different ions on the sites of two interpenetrating simple cubic lattices that point in opposite directions but their magnitudes are not equal.

The origin of the magnetic interactions that cause magnetic order is not fully understood, despite the enormous effort that has been invested. Qualitatively, the picture is that magnetism in solids is a many-body phenomenon that requires the existence of localized ions and itinerant electrons. The analysis of the energy levels of the hydrogen molecule (two ions and two electrons) given below shows that the magnetic interactions are due to the familiar Coulomb interaction between electrons combined with the Pauli exclusion principle. We will encounter spin-dependent *exchange interactions* in the discussion of Hund's rules, applicable for partially filled atomic shells (see Sec. 10.9), and we will encounter it again in the discussion of the energy difference between the ground state singlet and triplet levels of the hydrogen molecule (see Sec. 10.10.2). This type of interaction is much stronger (three to five orders of magnitude) than dipole–dipole interactions or spin–orbit interactions, which are too weak to account for ferromagnetism, ferrimagnetism, or antiferromagnetism. In the next subsection, we discuss a toy model showing how the spin-dependent interaction arises in a structure similar to the hydrogen molecule.

This system contains only two electrons (and two protons), compared with an Avogadro number of electrons in a typical bar magnet. Nevertheless, it sheds some light on the origin of and mechanism for the magnetic interactions in solids.

Problem 9.47

Consider the dipolar interaction between two electrons having magnetic moments $\mu_i = -g(\mu_B/2)\sigma_i$, located at points $\mathbf{r}_i$ with $i = 1, 2$, and separated a distance $|\mathbf{r}| = |\mathbf{r}_1 - \mathbf{r}_2|$, $U = \frac{1}{r^3}[\mu_1 \cdot \mu_2 - 3(\mu_1 \cdot \hat{\mathbf{r}})(\mu_2 \cdot \hat{\mathbf{r}})]$. Estimate the magnitude of this interaction in eV for a distance $r \approx 2$ Å.

Answer: $\left(\frac{e\hbar}{mc}\right)^2 r^{-3} = \left(\frac{4.8 \times 10^{-10} \cdot 1.06 \times 10^{-27}}{9.11 \times 10^{-28} \cdot 3.0 \times 10^{10}}\right)^2 (2.0 \times 10^{-8})^{-3}$ erg $\cdot\, 6.24 \times 10^{11}$ eV/erg $= 2.7 \times 10^{-5}$ eV.

Exchange Interactions: Heisenberg Spin Hamiltonian

Before considering a macroscopic solid-state metallic system, let us first consider a simple system where magnetic interaction might arise and study the energy levels of a two-electron system, focusing on both the space and spin degrees of freedom of the electrons. A more detailed study of the hydrogen molecule will be presented in Sec. 10.10.2. A good approximation for the ground states of the hydrogen molecule for the singlet and triplet states is given by the *Valence Bond* (or *Heitler–London*) approximation [see Eqs (10.94), (10.102), and (10.104)],

$$\Psi_s(x_1, x_2) = \frac{1}{\sqrt{2}} [\phi_a(\mathbf{r}_1)\phi_b(\mathbf{r}_2) + \phi_b(\mathbf{r}_1)\phi_a(\mathbf{r}_2)] \frac{1}{\sqrt{2}} [|\uparrow\downarrow\rangle - |\downarrow\uparrow\rangle],$$

$$\Psi_{t,m}(x_1, x_2) = \frac{1}{\sqrt{2}} [\phi_a(\mathbf{r}_1)\phi_b(\mathbf{r}_2) - \phi_b(\mathbf{r}_1)\phi_a(\mathbf{r}_2)] |t, m\rangle, \tag{9.326}$$

where ϕ_a and ϕ_b are atomic orbitals centered on atoms a and b, respectively, $m = -1, 0, 1$, and

$$|t, 1\rangle = |\uparrow\uparrow\rangle, \quad |t, 0\rangle = \frac{1}{\sqrt{2}} [|\uparrow\downarrow\rangle - |\downarrow\uparrow\rangle], \quad |t, -1\rangle = |\downarrow\downarrow\rangle. \tag{9.327}$$

Writing the two-electron wave functions as a product of space and spin parts, $\Psi(x_1, x_2) = \psi(\mathbf{r}_1, \mathbf{r}_2)|\chi\rangle$, it is clear that the spatial parts have the following symmetry under electron exchange: $\psi_s(\mathbf{r}_2, \mathbf{r}_1) = \psi_s(\mathbf{r}_1, \mathbf{r}_2)$ and $\psi_{t,m}(\mathbf{r}_2, \mathbf{r}_1) = -\psi_{t,m}(\mathbf{r}_1, \mathbf{r}_2)$, i.e., the spatial parts of the singlet and triplet are symmetric and antisymmetric, respectively, under electron interchange. Restricting ourselves to the four-dimensional subspace spanned by $|\Psi_s\rangle$ and $|\Psi_{t,m}\rangle$ and carrying out the calculation of the eigenvalues of the molecular Hamiltonian as detailed in Sec. 10.10.2, we find that the singlet-state eigenvalue is lower in energy than the triplet-state eigenvalues (which are triply degenerate) over the whole range of internuclear distances, R, but converge asymptotically as $R \to \infty$, as shown in Fig. 10.7. For fixed R,

$$\mathcal{H}|\Psi_s\rangle = E_s|\Psi_s\rangle, \quad \mathcal{H}|\Psi_{t,m}\rangle = E_t|\Psi_{t,m}\rangle, \tag{9.328}$$

and using the projection operators onto the singlet and triplet states given in Sec. 8.5, we write the 4×4 Hamiltonian as

$$\mathcal{H} = \frac{1}{4}(E_s + E_t)\mathbf{1} - \frac{E_s - E_t}{\hbar^2} \mathbf{S}_1 \cdot \mathbf{S}_2. \tag{9.329}$$

The Hamiltonian in Eq. (9.329) is composed of a constant term $(E_s(R) + E_t(R))/4$ and a spin-dependent part, $\mathcal{H} = -J(R)\mathbf{S}_1 \cdot \mathbf{S}_2$ of the form specified in Eq. (4.90), that is the basic ingredient of the *Heisenberg spin Hamiltonian*. Here, the coefficient $J(R)$, called the *exchange coupling* or *exchange integral*, is given by $J(R) = [E_s(R) - E_t(R)]/\hbar^2$. The ground-state singlet and excited triplet potential curves, $E_s(R)$ and $E_t(R)$, are plotted versus R in Fig. 10.7, and, as is clear from the figure, the energy difference $\hbar^2 J$ is four to five orders of magnitude larger than the estimate of the spin-dipolar coupling energy that you calculated in Problem 9.47.

9.5 Magnetic Field Effects

The above model considers only two atoms, whereas magnetically ordered materials contain a huge number of atoms. The generalization to a system of $N \gg 1$ atoms is based on intuitive reasoning. For atoms located on a Bravais lattice at points $\{\mathbf{R}_i\}$, an exchange interaction $J(\mathbf{R}_i - \mathbf{R}_j)$ exists between spins $\mathbf{S}_i$ and $\mathbf{S}_j$. In the presence of an external magnetic field $\mathbf{H} = H\hat{\mathbf{z}}$, the *Heisenberg spin Hamiltonian* is [see Eq. (4.88)]

$$\mathcal{H} = -\frac{1}{2}\sum_{i,j\neq i} J(\mathbf{R}_{ij})\,\mathbf{S}_i \cdot \mathbf{S}_j - g\mu_B \sum_i \mathbf{H}\cdot\mathbf{S}_i. \tag{9.330}$$

For a single type of magnetic atom, only one exchange interaction is present. In the simplest version of the Heisenberg Hamiltonian, exchange interaction is taken only between nearest neighbors, i.e.,

$$J(\mathbf{R}_{ij}) = \begin{cases} J & \text{(if } i \text{ and } j \text{ are nearest neighbor sites).} \\ 0 & \text{(otherwise).} \end{cases} \tag{9.331}$$

When $J > 0$ ($J < 0$), neighboring spins prefer to align parallel (antiparallel) to each other and the coupling is said to be ferromagnetic (antiferromagnetic). Because exchange interactions decay exponentially with $\mathbf{R}_{ij}$, the assumption of nearest neighbor exchange is reasonable. Variations of the Heinseberg Hamiltonian include the *anisotropic Heisenberg model* with exchange interaction $J_x S_{ix} S_{jx} + J_y S_{iy} S_{jy} + J_z S_{iz} S_{jz}$ between nearest neighbor sites, e.g., the *XXZ model* has $J_x = J_y \neq J_z$. The Heisenberg model is used to study systems of arbitrary spins, not just spin 1/2 electrons, i.e., localized magnetic ions with total angular momentum $S \geq 1/2$. The physics resulting from the Heisenberg spin Hamiltonian is very rich, although it is not adequate for describing all phenomena of quantum magnetism. The Hamiltonian (9.330) appears to be deceptively simple, but its analysis requires the full machinery of the many-body problem. Even when the ground state and the lowest lying states are known, it is still difficult to extract measurable magnetic properties as a function of temperature and external magnetic field. A systematic approach to determine the magnetization and the spin–spin correlations is not yet in hand. In Sec. 9.8.3, we shall discuss the approach for a one-dimensional arrangement of spins.

When the exchange coupling is ferromagnetic ($J > 0$), the low-energy spectrum is known in all three dimensions ($d = 1, 2, 3$) and the corresponding eigenstates are referred to as *spin waves*. Knowing the spin wave spectrum enables an approximate evaluation of the magnetization $M(T)$ as $T \to 0$, with the result [122],

$$M(T) \approx M(0)\left[1 - \frac{\mathcal{V}(k_B T)^{\frac{3}{2}}}{NS}\int \frac{d\mathbf{q}}{(2\pi)^d}\frac{1}{e^{(SJ/2)\sum_{\mathbf{R}}(\mathbf{q}\cdot\mathbf{R})^2} - 1}\right]. \tag{9.332}$$

Written as $M(T) \approx M(0)[1 - (T/T_0)^{3/2}]$, this is known as the *Bloch $T^{3/2}$ law*.

Examples of Magnetically Ordered Structures

In nonmagnetically ordered materials, and in the absence of an external magnetic field, the atomic magnetic moments point in random directions and the magnetization vanishes identically at any temperature. However, materials with strongly interacting magnetic moments become magnetically ordered below a critical temperature, T_C. This temperature is referred to as the Curie temperature T_C for ferromagnets and ferrimagnets, and the Néel temperature T_N for antiferromagnets. Below T_C, the magnetization of ferromagnets and ferrimagnets does not vanish even in the absence of an external magnetic field; this is referred to as *spontaneous magnetization*. The ground state of a perfect ferromagnet has all magnetic moments pointing along the same direction, and the magnetization equals $M_s(T=0) \equiv ng\mu_B S/\hbar$, where S is the angular momentum quantum number of the magnetic atom. In the ground state of ferrimagnet, the directions of neighboring magnetic moments point in opposite directions, but they are not equal in magnitude, as in $MnO\cdot Fe_2O_3$ and $Y_3Fe_5O_{12}$, for example. The relatively new rare earth magnets have exceptionally large magnetizations (e.g., the neodymium magnet $Nd_2Fe_{14}B$, which is sometimes denoted NdFeB). They contain lanthanide elements that have large magnetic moments due to well-localized f-orbitals.

Table 9.2 *Table of some ferromagnets and antiferromagnets.*

Material	T_C (K)	$M_{s(T=0)}$ (Gauss)	Material	T_N (K)
Fe	1043	1752	MnO	122
Co	1388	1446	FeO	198
Ni	627	510	CoO	291
Gd	293	1980	NiO	600
Dy	85	3000	VS	1040
NdFeB	580	16000		

In antiferromagnets, magnetic order prevails below T_N, but the magnetization vanishes. Intuitively, we can think of the ground state of an antiferromagnet as an ordered arrangement of magnetic moments, such that the directions of neighboring moments are equal in magnitude but point in opposite directions. However, this description is too naïve; the structure of the antiferromagnetic ground state is more involved. Antiferromagnetism plays a role in giant magnetoresistance (see Sec. 9.7).

A quantitative measure of magnetic order that applies for all three classes of materials is the *spin–spin correlation function*,

$$C(\mathbf{R}_i - \mathbf{R}_j) \equiv \langle \mathbf{S}(\mathbf{R}_i) \cdot \mathbf{S}(\mathbf{R}_j) \rangle - \langle \mathbf{S}(\mathbf{R}_i) \rangle \cdot \langle \mathbf{S}(\mathbf{R}_j) \rangle, \qquad (9.333)$$

where the brackets refer to quantum thermodynamic averaging. If $C(\mathbf{R})$ tends to a constant as $R \to \infty$, we say that the system displays *long range magnetic order*. Sometimes $C(\mathbf{R}) \sim R^{-\alpha}$ decays as a power in 2D magnetic materials (see below). A representative functional behavior in isotropic media is given by,

$$C(\mathbf{R}) = \frac{A}{R^{d-2+\eta}} e^{-\frac{R}{\xi(T)}}, \qquad (9.334)$$

where $\xi(T)$ is the correlation length that diverges as a power, $\xi(T) \sim (T - T_c)^{-\nu}$. Here ν and η are two critical exponents characterizing the transition and d is the space dimension. Above the critical temperature, the correlation function (9.333) decays exponentially.

Examples of ferromagnetic, antiferromagnetic, and ferrimagnetic structures are shown in Fig. 9.61(a–c). Representative ferromagnetic and antiferromagnetic materials are tabulated in Table 9.2.

Thermodynamic Behavior Near T_C and T_N

Determining the magnetic properties of magnetically ordered materials near their critical temperature is very hard. Here, we simply mention some magnetic properties of three-dimensional (infinite) systems, including the temperature dependence of the magnetization near T_c, which we relate to the concept of phase transitions.

In ferromagnets, the magnitude of the magnetization $M(T)$ tends to 0 as a power law when T approaches T_C from below, while the susceptibility $\chi(T)$ and the specific heat diverge as a power law when T approaches T_C from above,

$$M(T) \sim (T_C - T)^\beta, \quad \chi(T) \sim (T - T_C)^{-\gamma}, \quad c(T) \sim (T - T_C)^{-\alpha}. \qquad (9.335)$$

The *critical exponents* β, γ, and α are *universal*; their numerical values, $\beta \approx 0.35$, $\gamma \approx 1.35$, and $\alpha \approx 0.1$ are material independent. The material dependence enters through the Curie temperature T_C and the appropriate prefactors.

Considered as function of T, the magnetization is continuous but its derivative is singular at T_C. Indeed, something unusual happens at T_C; the material is not magnetically ordered for $T > T_C$ and goes through a phase transition at T_C; so, magnetic order exists for $T < T_C$. We are familiar with singular behavior of materials as the temperature crosses a

9.5 Magnetic Field Effects

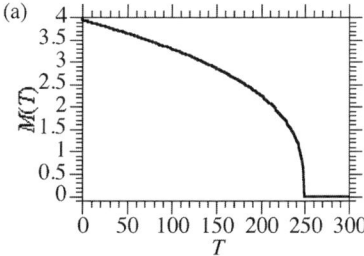
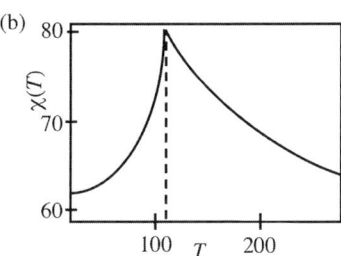

FIG 9.62 (a) Magnetization $M(T)$ (arbitrary units) of a ferromagnetic material versus temperature for a Curie temperature, $T_C = 250$ K (schematic). (b) Susceptibility $\chi(T) = \partial M/\partial H$ (arbitrary units) versus temperature of an antiferromagnetic material for a Néel temperature of $T_N = 110$ K (schematic).

critical value, e.g., the melting or the boiling temperature. In these cases, some quantities (e.g., density and specific heat) are *discontinuous* functions of temperature. The material changes its phase from solid to liquid or from liquid to vapor. In this case, we say that there is a *first-order phase transition*.

In contrast, magnetic materials remain solids at T_C or T_N, but some quantities, such as the magnetization (or the inverse susceptibility), vanish as a power law $(T - T_C)^p$, with universal critical exponents. This is a signature of a *second-order thermodynamic phase transition*. The system's magnetization above T_C vanishes at zero external magnetic field, because the system is isotropic and no direction is preferred. Below T_C, the system starts to be ordered and the magnetization is finite. Because the value of $M(T)$ reflects the degree of order for ferromagnets, it is referred to as an *order parameter*. Figure 9.62(a) displays a typical behavior of the magnetization $M(T)$ for a ferromagnet with $T_C = 250$ K. In antiferromagnets, the magnetization vanishes identically, but the magnetic susceptibility can be measured by a low-energy neutron scattering. Low-energy neutrons are an appropriate probe because their magnetic moments are coupled to the electronic spins. This results in peaks in the intensity of scattered neutrons that can be resolved from the ubiquitous Bragg peaks because they are extremely sensitive to temperature and an external magnetic field. Above T_N, the susceptibility follows a Curie–Weiss law (9.336),

$$\chi(T) = \frac{C}{T + \Theta}, \qquad (9.336)$$

where C and Θ are parameters that can be determined from experiment. As T_N is crossed from above, $\chi(T)$ has a kink and then it decreases, as depicted in Fig. 9.62(b), which displays typical behavior of the susceptibility $\chi(T)$ for an antiferromagnet.

From this analysis, we surmise that what happens at T_C or T_N is not a structural change, such as melting or boiling, but rather a loss of isotropy. An isotropic system above T_C is no longer isotropic below T_C as ferromagnetic order develops, because its magnetic moments are aligned along a preferred direction. In other words, rotational invariance, which prevails above T_C, is broken below T_C. Breaking a continuous symmetry, such as a rotational symmetry, at a critical temperature is a signature of a thermodynamic (temperature driven) second-order phase transition, as occurs in magnetic materials in 3D.

In lower dimensions, $d = 1, 2$, the situation is different. An important result known as the *Mermin–Wagner theorem* [122] states that continuous symmetries (e.g., rotational symmetry) cannot be spontaneously broken at a finite temperature in systems with dimension $d \leq 2$. This theorem is relevant for understanding the result obtained in Problem 9.48. In $d = 2$, the magnetization vanishes at any finite temperature (also below T_C), but the spin–spin correlation function (9.333) decays as a power (compared with exponential decay for $T > T_C$). This type of phase transition, where the correlation $C(R)$ decays as a power but there is no breaking of a continuous symmetry, is referred to as a *Kosterlitz–Thouless–Berezinsky* transition [131]. For $d = 1$, there is no phase transition whatsoever. The ground state of the ferromagnetic Heisenberg Hamiltonian ($J > 0$) is perfectly magnetized, so that at $T = 0$, $M(T)$ is finite, but magnetic order is absent at a finite temperature, $M(T) = 0$. The ground state of the antiferromagnetic ($J < 0$) Heisenberg Hamiltonian at zero magnetic field is known, and $C(R)$ indeed tends to a constant as $R \to \infty$, but once $T > 0$, $C(R)$ decays exponentially.

Problem 9.48

By expanding the exponent appearing on the RHS of Eq. (9.332) for small q, show that for $d = 1, 2$, the **q** integration diverges at small q.

Domains and Hysteresis: Dipolar Interactions

According to the analysis developed so far, a ferromagnetic material well below T_C should have all its moments aligned along the same direction. However, an iron bar at room temperature is not magnetized, although T_C for iron is about 1000 K. It can be magnetized by subjecting it to an external magnetic field. The reason for this discrepancy is related to the magnetic dipolar interactions between spins, which were so far neglected because they are much weaker than the exchange interaction (see Problem 9.47). The expression of the exchange constant $J(R)$ derived in Sec. 10.10.2 shows that it decays exponentially with R because it is proportional to the overlap between atomic orbitals on different sites separated by R. Hence, dipolar interactions are expected to be relevant if the atoms are well separated, because they fall off as R^{-3}.

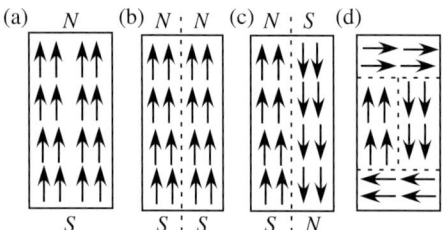

FIG 9.63 (a) A bar magnet with all magnetic moments aligned and pointing from south to north. (b) The magnet is cut into two adjacent narrower parallel bar magnets. Parallel spins have negative exchange energy (since $J > 0$) and positive dipolar energy (see Problem 4.26). When the latter is dominant (occurring for large enough separation), the configuration is unstable. (c) A more stable configuration than in (a) or (b) is achieved when one of the magnets is reversed. The spin configuration has two oppositely oriented domains. (d) Further stability is achieved when the spin configuration form kind of vortex with four domains.

To illustrate this argument for a ferromagnetic material, let us consider a bar magnet with all moments pointing from south to north [see Fig. 9.63(a)]. Cutting it in the middle, from south to north [see Fig. 9.63(a)], we obtain a pair of two identical and parallel bar magnets, which are in contact with each other along the cut. Experience teaches us that this configuration is unstable, and a configuration consisting of two magnets with *antiparallel configuration* has lower energy. Thus, a construction of *ferromagnetic domains* with magnetic moments pointing in different directions can form. In our example, a lower energy state can be achieved if the magnetic moments of the domains are arranged in a head-to-tail configuration. This is schematically illustrated in Fig. 9.63. An important element in this construction is the structure of the boundary between two domains (referred to as *domain wall* or *Bloch wall*). The physics of formation of domains is not well understood. It is possible to show that an abrupt change of orientation between two domains is energy costly, and a smooth change is preferred.

Domain formation affects the behavior of the magnetization $M(H, T)$ as a function of the external field H at $T < T_C$. The curve of $M(H, T)$ versus H for $T < T_C$ shown in Fig. 9.64 is referred to as the *hysteresis curve*. The magnetization saturates at points s and s' at the saturation magnetization, $M_s = ng\mu_B S/\hbar$, where S is the angular momentum quantum number of the magnetic atom, when all the magnetic moments are aligned. The magnetization at the point r (and r') in the figure, at which the external magnetic

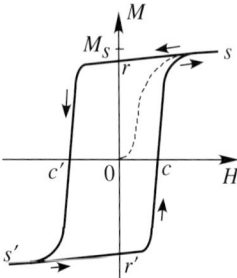

FIG 9.64 Hysteresis in the magnetization M versus magnetic field H of a ferromagnet.

field vanishes, is called the *remnant magnetization* M_r, and the magnetic field at points c and c' at which the magnetization vanishes is called the *coercive field* H_c.

Clearly, the magnetization depends on the history of the ferromagnetic material. Ferromagnetic hysteresis occurs because ferromagnets are composed of small *ferromagnetic domains*. Within each domain, the local magnetization is saturated at $M_s(T)$ for temperature T, but, in general, the direction of magnetization in different domains are not parallel. The increase in the magnetization on applying or increasing the strength of an external magnetic field occurs by (a) modification of the volume of the domains, so that the volume of favorably oriented domains grows and that of unfavorably oriented domains decreases, and (b) in somewhat stronger magnetic fields, the direction of magnetization of domains reorient with increasing magnetic field. Hence, magnetization of ferromagnetic materials below the Curie temperature involves the orientation of macroscopic domains by external magnetic fields.

9.6 SEMICONDUCTORS

A semiconductor is a solid, which, from the band structure point of view, could be classified as an insulator, but its conductivity, when measured at finite temperature, is nonzero and sometimes reaches that of poor metals. Recall that a simple metal is characterized by a partially filled (valence conduction) band, whereas band insulators are characterized by a filled valence band and an empty conduction band. The central quantity that distinguishes semiconductors from band insulators is the size of the band gap, E_g, between the top of the valence band and the bottom of the conduction band. At zero temperature, a material with a finite gap cannot carry direct current. But, if $E_g \sim k_B T$, there is a finite probability, proportional to the Boltzmann factor $\exp[-E_g/(k_B T)]$, that electrons will be excited and occupy the lowest levels in the conduction band, leaving unoccupied states (holes) near the top of the valence band. The thermally excited electrons occupy the conduction band and, therefore, can move in response to an electric field, so the material has a nonzero conductivity. Similarly, hole conductivity can be realized in the valence band. Such a material is referred to as an *intrinsic semiconductor*. There are other mechanisms for partially populating the conduction band and/or partially vacating the valence band, thereby leading to nonzero conductivity. The most commonly used method in this context is *doping*, i.e., inserting a small concentration of foreign atoms which either release electrons into the conduction band (donor atoms) or attract and bind electrons from the valence band (acceptor atoms), thereby generating holes. A doped semiconductor is called an *extrinsic semiconductor*. Because the energy required to excite an electron from the donor level to the conduction band is much smaller than the band gap, the conductivity of an extrinsic semiconductor is non-negligible even if $k_B T < E_g$. Another method of exciting electrons from the valence to the conduction band is realized by excitations with light of an appropriate wavelength. The conductivity of an extrinsic semiconductor can be raised by increasing the temperature, by doping the material with impurity atoms, or by illuminating it with light (photoconductivity). Because the thermal excitation probability is proportional to $\exp[-\Delta E/(k_B T)]$, where ΔE is the required excitation energy, the sensitivity of conductivity to variation of temperature in semiconductors is rather dramatic compared with that of metals with conductivity $\sigma = ne^2\tau/m$, where the dependence of the carrier lifetime τ on temperature is, in general, governed by a weak power law. Moreover, although τ decreases with increasing temperature, the conductivity of semiconductors increases with temperature. Semiconductors are characterized by very large thermopower, about two orders of magnitude larger than that of metals (see below). Unlike metals, the current in semiconductors can be carried by holes and electrons. Measurement of the Hall coefficient can determine the dependence on the density of carriers and their charge (electrons or holes). These (and numerous other) phenomena are consistent with the band theory developed in Sec. 9.3.

Semiconductors are the source of a myriad of electronic devices. Commonly used semiconductor materials in electronic devices are silicon (Si), germanium (Ge), gallium arsenide (GaAs), gallium aluminum arsenide (Ga$_{1-x}$Al$_x$As), cuprous oxide (Cu$_2$O), selenium (Se), indium antimonide (InSb), and silicon carbide (SiC). Electronic devices based on semiconductors include transistors, rectifiers, modulators, detectors, and photocells. GaAs and Ga$_{1-x}$Al$_x$As are often used in photonic devices, including light-emitting diodes (LEDs) and semiconductor lasers. The first transistors and integrated electronic circuits were not made from Si but from Ge. The reasons are that Ge is easier to purify and it has a higher

Table 9.3 *Band gap energy, E_g, between valence and conduction bands for semi-conductors and insulators (from Kittel, Introduction to Solid State Physics, 7th Ed. (1996), Table 1, p. 201 [125]).*

Material (type of gap)	E_g (eV) (300 K) (λ_g (μm))	E_g (eV) (0 K)
Si (i)	1.14 (1.08)	1.17
Ge (i)	0.67 (1.85)	0.67
InAs (d)	0.35 (3.54)	0.43
InSb (d)	0.18 (6.89)	0.24
InP (d)	1.25 (0.99)	1.25
GaAs (d)	1.43 (0.87)	1.52
GaP (i)	2.26 (0.55)	2.32
GaSb (d)	0.78 (1.59)	0.81
Te (d)	0.35 (0.71)	0.33
PbS (d)	0.34–0.37 (3.64–3.35)	0.286
PbSe (d)	0.27 (4.6)	0.165
ZnO (d)	3.2 (0.39)	3.436
ZnS (d)	3.6 (0.34)	3.91
Diamond (i)	5.33 (0.23)	5.33

(i) = indirect gap, (d) = direct gap. (Direct and indirect gaps are defined in the next section.)

mobility than Si for both electrons and holes. However, the bandgap in Ge is only 0.67 eV, compared with 1.1 eV for Si. Therefore, the performance of Ge transistors degrades more quickly with temperature. In 1950s, methods of preparing high-purity Si were developed. Combined with the development of a method for producing a glassy oxide layer on the silicon surface during growth, which prevents impurity contamination, silicon quickly became the material of choice for the semiconductor industry.

Table 9.3 gives the values of the band gap between valence and conduction bands for a number of semiconductors and insulators at 300 K and 0 K. The slight dependence of E_g on temperature is due to the thermal expansion of the crystal, which changes the periodicity of the crystal, and the nonadiabatic effects that lattice vibrations have on the band structure.

9.6.1 SEMICONDUCTOR BAND STRUCTURE

Materials for which the conduction band minimum and the valence band maximum occur at the same wavevector **k** are called *direct gap* semiconductors; otherwise, the extremum points are shifted by a vector **q** and the material is called an *indirect gap* semiconductor. A schematic illustration of direct and indirect gap structures is shown in Fig. 9.65.

A transition involving the excitation of an electron from the top of the valence band to the bottom of the conduction band in an indirect gap semiconductor requires change of the electron momentum by **q**. This shift may result from absorption (or emission) of a phonon, i.e., a sound wave excitation, so as to conserve momentum. However, such an

9.6 Semiconductors

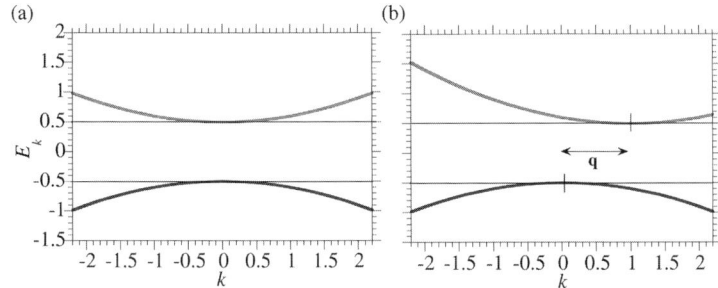

FIG 9.65 Schematic illustration of band structure for (a) direct and (b) indirect gap semiconductors.

excitation is typically much less efficient than in a direct gap material, which does not require phonon absorption or emission for an electron transition. Therefore, transitions involving photoexcitation of an electron or photoemission by electron–hole recombination are much more likely in a direct bandgap material, such as GaAs, InP, and InGaAs, than an indirect bandgap material, such as Si or Ge.

The electronic properties of semiconductors are largely determined by the excited electrons in the conduction bands and the holes left behind in the valence band. Consider the situation depicted in Fig. 9.33. Following our discussion of the effective mass in Sec. 9.4.6 and adapting the principal axes coordinate system [see Eq. (9.144)], we write the electron and hole energies as,

$$E_e(\mathbf{k}) = E_c + \sum_{i=1}^{3} \frac{\hbar^2 k_i^2}{2m_i^{(e)}} \quad \text{(for electrons)}, \tag{9.337a}$$

$$E_h(\mathbf{k}) = E_v - \sum_{i=1}^{3} \frac{\hbar^2 (k_i - \mathbf{q})^2}{2m_i^{(h)}} \quad \text{(for holes)}, \tag{9.337b}$$

where E_c is the bottom of the conduction band, E_v is the top of the valence band, and the electron mass ($m_i^{(e)}$) and hole mass ($m_i^{(h)}$) might be different. In this quadratic dispersion relation approximation, the constant energy surfaces near $\mathbf{k} = 0$ are ellipsoids in $\mathbf{k}$-space centered at the extrema (the $\mathbf{k}$-space coordinates of the top and bottom of the bands). For example, silicon has a diamond structure; hence, its reciprocal lattice is face centered, and its Brillouin zone is a truncated octahedron. The conduction band has six minima in the $(1,0,0), (\bar{1},0,0), \ldots, (0,0,1), (0,0,\bar{1})$ directions (see discussion on Miller indices in Sec. 9.3.2). By symmetry, all six ellipsoids are generated by revolution.

Figure 9.66 shows the band structure for silicon (see Fig. 9.16(b) for the corresponding Brillouin zone and important points in $\mathbf{k}$-space). Note that the top of the valence band, at the point labeled Γ, *is not* directly below the bottom of the conduction band, labeled X, i.e., silicon is an indirect gap material.

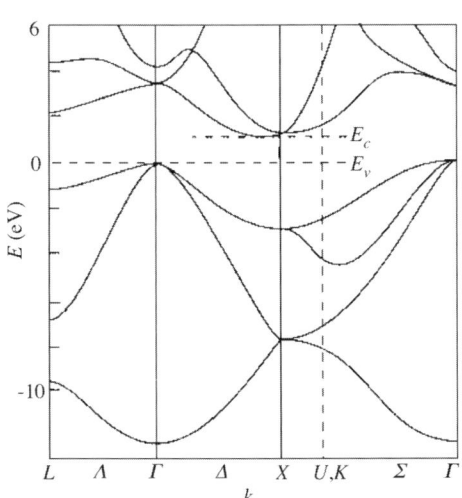

FIG 9.66 Energy bands versus wavevector k in silicon. The points in reciprocal lattice space are labeled as in Fig. 9.16(b).

Germanium has the same lattice structure the conduction band minima occur at the Brillouin zone boundaries (so they are half ellipsoids). The minima are located at the eight points $(1,1,1), (\bar{1},1,1), \ldots, (\bar{1},\bar{1},\bar{1})$. Every two opposite half ellipsoids can be thought of as a full ellipsoid by translating one by a reciprocal lattice vector toward the other. The symmetry of the lattice implies that all four are ellipsoids of revolution.

9.6.2 CHARGE CARRIER DENSITY

Conduction in semiconductors is due to transport of both electrons and holes. We can calculate the transport properties of a semiconductor in thermal equilibrium as a function of the density of electrons and holes. The chemical potential μ is located within the band gap, and for an extrinsic semiconductor, its value depends on the degree of doping. It should be determined self-consistently, in the sense that fixing μ yields the densities, which in turn determine μ, and so on. For temperatures $T \ll E_g/k_B$, we will assume the following inequalities,

$$E_c - \mu \simeq \mu - E_v \gg k_B T, \tag{9.338}$$

and check for self-consistency. These relations are easily satisfied at room temperature, $T = 300$ K $\approx 1/40$ eV, and for typical gap energies of order $E_g \simeq 0.2$ eV ≈ 2400 K, thereby enabling a reasonable range of values for μ. Recall that the electron density of states $g_e(E)$ is Eq. (9.15), so the density of electrons at energies $E > E_c$ is given by,

$$n_e = \int_{E_c}^{\infty} dE\, g_e(E) f(E,T) = \int_{E_c}^{\infty} dE\, \frac{g_e(E)}{e^{\frac{E-\mu}{k_B T}} + 1}. \tag{9.339}$$

Similarly, the density of holes at energies $E < E_v$ is given by,

$$n_h = \int_{-\infty}^{E_v} dE\, g_h(E)[1 - f(E,T)] = \int_{-\infty}^{E_v} dE\, \frac{g_h(E)}{e^{\frac{\mu-E}{k_B T}} + 1}. \tag{9.340}$$

With the approximation (9.338), we obtain the following estimates:

$$f(E,T) \approx e^{-\frac{E-\mu}{k_B T}}, \quad 1 - f(E,T) \approx e^{-\frac{\mu-E}{k_B T}}. \tag{9.341}$$

Substituting in Eqs (9.339) and (9.340), we find

$$\boxed{n_e = \left[\int_{E_c}^{\infty} dE\, g_e(E) e^{-\frac{E-E_c}{k_B T}}\right] e^{-\frac{E_c-\mu}{k_B T}} \equiv N_e e^{-\frac{E_c-\mu}{k_B T}},} \tag{9.342}$$

$$\boxed{n_h = \left[\int_{-\infty}^{E_v} dE\, g_h(E) e^{-\frac{E_v-E}{k_B T}}\right] e^{-\frac{\mu-E_v}{k_B T}} \equiv N_h e^{-\frac{\mu-E_v}{k_B T}}.} \tag{9.343}$$

Due to the exponentially decaying factors in the integrand for n_e and n_h, the range of integration is a few $k_B T$ from the band extrema. In this range, the quadratic approximations (9.337) are adequate, and therefore,

$$g_e(E) = \frac{(m_e^*)^{\frac{3}{2}}}{\hbar^3 \pi^2}\sqrt{2(E - E_c)}, \quad g_h(E) = \frac{(m_h^*)^{\frac{3}{2}}}{\hbar^3 \pi^2}\sqrt{2(E_v - E)}, \tag{9.344}$$

where $(m^*)^3 = m_1 m_2 m_3$ is an effective mass expressing the geometric mean of the masses in Eqs (9.337). The integrals in the formulas for N_e and N_h can now be calculated analytically to obtain [122],

$$N_e = \left(\frac{m_e^*}{m}\right)^{\frac{3}{2}} \left(\frac{T(\text{K})}{300}\right)^{\frac{3}{2}} \times 2.5 \times 10^{19}\ \text{cm}^{-3}, \tag{9.345}$$

9.6 Semiconductors

$$N_h = \left(\frac{m_h^*}{m}\right)^{\frac{3}{2}} \left(\frac{T(\text{K})}{300}\right)^{\frac{3}{2}} \times 2.5 \times 10^{19} \text{ cm}^{-3}. \tag{9.346}$$

Note that although both n_e and n_h depend on the chemical potential μ (which is not known *a priori*), their product is independent of μ, i.e.,

$$\boxed{n_e n_h = N_e N_h \, e^{-\frac{E_v - E_c}{k_B T}} = N_e N_h \, e^{-\frac{E_g}{k_B T}}.} \tag{9.347}$$

This is referred to as the *law of mass action*. Note that this result holds for both intrinsic and doped semiconductors. For an intrinsic semiconductor, $n_e = n_h$, and

$$n_e = n_h = \sqrt{N_e N_h} \, e^{-\frac{E_g}{2k_B T}} = \left(\frac{m_e^* m_h^*}{m^2}\right)^{\frac{3}{4}} \left(\frac{T(\text{K})}{300}\right)^{\frac{3}{2}} e^{-\frac{E_g}{2k_B T}} \times 2.5 \times 10^{19} \text{ cm}^{-3}. \tag{9.348}$$

Defining the chemical potential μ_i for the intrinsic case, such that in Eqs (9.342) and (9.343), $n_i \equiv n_e(\mu_i) = n_h(\mu_i)$, as given in Eq. (9.348), we obtain

$$\mu_i = E_v + \frac{E_g}{2} + \frac{3k_B T}{4} \ln(m_h^*/m_e^*). \tag{9.349}$$

Hence, the intrinsic carrier concentration of electrons and holes, n_i, decreases exponentially with $E_g/2k_B T$. If $m_h^* = m_e^*$ (or if $T \to 0$), $\mu_i = E_g/2$, and in general, the chemical potential depends on the mass ratio m_h^*/m_e^*. In the extrinsic case, it also depends on the degree of doping. Typical intrinsic concentrations for Si, GaAs, and Ge are $n_i = 1.38 \times 10^{10}$, 9.00×10^8, and 2.33×10^{13} respectively.

In semiconductors, the current is carried by both electrons and holes. The mobility of the electrons $\tilde{\mu}_e$ and holes $\tilde{\mu}_h$ (the tilde is to distinguish the mobility from the chemical potential) is given by

$$\tilde{\mu}_e = \frac{e\tau_e}{m_e^*}, \quad \tilde{\mu}_h = \frac{e\tau_h}{m_h^*}, \tag{9.350}$$

and the conductivity is

$$\sigma = (n_e e \, \tilde{\mu}_e + n_h e \, \tilde{\mu}_h) = (n_e e^2 \tau_e / m_e^* + n_h e^2 \tau_h / m_h^*). \tag{9.351}$$

Experimentally measured room temperature mobilities and electron and hole masses are shown in Table 9.4.

Problem 9.49

The gap energy for Si is 1.206 eV. Calculate the concentrations of free carriers in intrinsic silicon at $T = 350$ K and the relative change $\Delta n/n$ if the temperature changes by 0.2%.

Answer: For an intrinsic semiconductor, $n_i = n_e = n_h$. Hence, we can use Eq. (9.348) and Table 9.4 for the effective masses to get $n_i = 4.86 \times 10^{11}$ cm^{-3}. Taking logarithm of Eq. (9.348), we find $\frac{dn_i}{n_i} = \left(\frac{3}{2} + \frac{E_g}{2k_B T}\right) \frac{dT}{T}$.

In GaAs (and similar materials), the value taken for the hole mass m_h^* includes the contribution from two valence bands, one of which has a heavy hole mass m_{hh}^* and another one has a light hole mass m_{lh}^*, and $(m_h^*)^{3/2} \equiv (m_{hh}^*)^{3/2} + (m_{lh}^*)^{3/2}$. For example, in GaAs, $m_{hh}^* = 0.68 m_e$, $m_{lh}^* = 0.12 m_e$, and hence $m_h^* = 0.47 m_e$.

Table 9.4 *Electron and hole masses and experimentally measured room temperature mobilities.*

Material	m_e^*/m_e	m_{lh}^*/m_e	μ_e (cm^2/Vs)	μ_h (cm^2/Vs)
Si	0.92 ($\parallel$) 0.19 ($\perp$)	0.54 ($\parallel$) 0.15 ($\perp$)	1300	430
Ge	1.59 ($\parallel$) 0.08 ($\perp$)	0.33 ($\parallel$) 0.043 ($\perp$)	4500	3500
InAs	0.026	0.025	77,000	750
InSb	0.015	0.021	33,000	460
InP	0.073	0.078	4600	150
GaAs	0.07	0.12	8800	400
GaSb	0.047	0.06	4000	1400
PbS	0.25	0.25	550	600
Diamond	0.2	0.25	1800	1200

Problem 9.50

(a) Carry out the algebra leading to Eqs (9.345), (9.346), and (9.347).
(b) Develop the kinetic argument proving that the product $n_e n_h$ is constant at a given temperature by developing rate equations for dn_e/dt and dn_h/dt due to $e + h$ recombination and generation by blackbody radiation. Equate the rates to zero at equilibrium to show that this yields $n_e n_h = \text{const}(T)$.
(c) Show that for an intrinsic semiconductor, $n_e = n_i e^{\beta(\mu - \mu_i)}$ and $n_h = n_i e^{-\beta(\mu - \mu_i)}$.

Answer: (b) In steady state, $dn_e/dt = A(T) - B(T)n_e n_h = 0$ and $dn_h/dt = A(T) - B(T)n_e n_h = 0$, where $A(T)$ is the rate at which photons generate electron–hole pairs and $B(T)$ is the recombination rate coefficient. Hence, $n_e n_h = A(T)/B(T)$.

9.6.3 EXTRINSIC SEMICONDUCTORS

In this section, we concentrate on extrinsic semiconductors, which can conduct current even in the presence of large gap. The probability for thermal excitations is proportional to the Boltzmann factor $\exp[-E_g/(k_B T)]$, so when $E_g \gg k_B T$, the material cannot carry direct current and remains an insulator. However, for numerous materials, it is possible to replace a certain fraction of the original atoms by foreign atoms with different number of valence electrons. This procedure is referred to as *doping*, and the foreign atoms are termed *impurities*. An insulating material that has been doped is referred to as an *extrinsic semiconductor*. If the impurity atom provides one or more additional electrons than the replaced atom, the impurity is called a *donor*, whereas if it provides fewer electrons than the atom it replaced, it is called an *acceptor*. For example, if an As atom replaces a Ge atom in a germanium crystal, it acts as a donor; similarly, a Si atom replacing a Ga atom in a GaAs crystal is a donor (see the ionization energies and electron affinities in Fig. 10.3).

9.6 Semiconductors

If the impurity atom does not occupy a lattice site but is lodged in an interstitial position, its outer electrons can participate in conduction, so the interstitial atom is a donor. An atom that is missing from a lattice site robs the crystal of electrons that can participate in conduction, and therefore, the vacancy is an acceptor.

Donor and Acceptor Impurities

Figure 9.67 shows the energy levels of a donor and an acceptor atom in a semiconductor. The donor or *n-type impurity* [*n*-type because it contributes negatively charged particles (electrons) to the conduction band] is neutral at $T = 0$, but at finite temperature, the electrons that were bound to it at $T = 0$ can escape and enter the conduction band, while the donor remains positively ionized. Similarly, an electron from the valence band can bind to an impurity that is an acceptor atom, leaving a hole in the valence band. An acceptor atom is called a *p-type impurity*. The acceptor atom is then negatively ionized.

The effects of impurities on conduction in an otherwise insulating material with $E_g \gg k_B T$ depends on the energy levels of the impurities. Specifically, let us consider an As impurity (valence 5) in a group IV material such as Ge (valence 4). The As impurity will then be a donor. After releasing one electron, it becomes very similar to its Ge neighbors except that it now becomes an ion with charge $+e$, which tends to attract electrons. In the ideal case, the released electron is nearly free to move around the entire crystal, and thus it can contribute to the current. A crucial question is whether the ionization energy E_i is much smaller than E_g. The ionization energy of an As atom is 9.81 eV, which is larger than a typical gap energy (0.5–2.0 eV). However, it modifies the ionization energy in a dramatic way. First, the electric field of the As$^+$ ion is reduced by a factor of the inverse of the Ge dielectric constant, $\varepsilon \approx 16$. (A classical electrostatic argument in a quantum mechanical system can be justified since the extent of the electron wave function in the As atom is about two orders of magnitude larger than the lattice constant.) Second, the electron wave function is a superposition of low-energy conduction band states modified by the presence of the As$^+$ ion. It has an effective mass m^*, which is about $0.1m$, where m is the free electron mass.

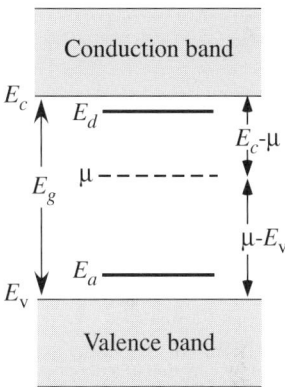

FIG 9.67 Energy levels of a donor impurity, E_d, and an acceptor impurity, E_a, in an extrinsic semiconductor for which $E_g \gg k_B T$. The chemical potential μ is close to the middle of the gap $\mu \approx E_v + \frac{E_g}{2}$, so that $E_c - \mu \gg k_B T$ and $\mu - E_v \gg k_B T$.

Problem 9.51

An electron of mass m^* is moving in the central field produced by a positive point charge $+e$. The static dielectric constant is ϵ. Express its Bohr radius r_0 in terms of the hydrogen atom Bohr radius a_0 and its ionization energy E_i in terms of the Rydberg constant, Ry $= 13.6$ eV.
Answers: $r_0 = \varepsilon \frac{m}{m^*} a_0$, $E_i = \frac{m^*}{m \varepsilon^2}$ Ry.

By using the results of Problem 9.51 with $m^*/m = 0.1$ and $\varepsilon = 20$, we get $r_0 = 200 a_0$ and $E_i = 0.0034$ eV, which is indeed much smaller than E_g. Consequently, the effect of donor doping is to introduce electronic levels at energies E_d

just below the conduction band such that $0 < E_c - E_d \ll E_g$, as indicated in Fig. 9.67. Similarly, the effect of acceptor doping is to introduce hole levels at energies E_a just above the valence band such that $0 < E_a - E_v \ll E_g$.

Impurities also contribute to the carrier densities, and the density of electrons in the conduction band need not be equal to that of holes in the valence band, i.e., $n_e - n_h \equiv \Delta n \neq 0$ in general. Recall, however, that the law of mass action, Eq. (9.347), holds also for extrinsic semiconductors, hence $n_e n_h = n_i^2$, where n_i is the common value of n_e and n_h for the intrinsic semiconductor. From the two relations, $n_e - n_h \equiv \Delta n$ and $n_e n_h = n_i^2$, we find

$$n_e = \frac{1}{2}\left(\sqrt{(\Delta n)^2 + 4n_i^2} + \Delta n\right), \quad n_h = \frac{1}{2}\left(\sqrt{(\Delta n)^2 + 4n_i^2} - \Delta n\right). \tag{9.352}$$

The donor impurity contribution is significant ($\Delta n = n_e - n_h \gg n_i$) when $n_e \gg n_i$, with $n_h = n_i^2/n_e \ll n_i$, and the acceptor contribution is significant ($n_h - n_e \gg n_i$) when $n_e \gg n_i$, with $n_e = n_i^2/n_h \ll n_i$. From the solution of Problem 9.50(c), we obtain

$$\frac{\Delta n}{n_i} = 2\sinh\beta(\mu - \mu_i), \tag{9.353}$$

which highlights the importance of doping (adding impurity levels) in modifying the concentration of charge carriers. In deriving these results, we have assumed inequalities (9.338), which must hold for both μ_i and μ.

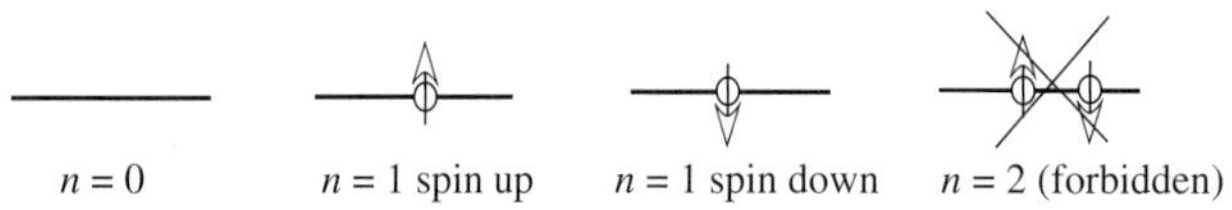

To estimate Δn and μ, we need the average number $\langle n \rangle$ of electrons occupying a given impurity level whose energy is E_d at temperature T and chemical potential μ. Let us assume thermal equilibrium and use the partition function in Eq. (9.299). Because the concentration of impurities is low, each impurity can be treated separately. A single donor impurity can be considered as an open system attached to the rest of the lattice, which serves as a particle bath at the equilibrium temperature T. Assuming the level has a single orbital and that putting two electrons on the same level costs too much (Coulomb) energy, the number of electrons (spin projection indicated) in that level can take values $n = 0, 1\uparrow$, or $1\downarrow$ (see Fig. 9.68). The Boltzmann weight for having n electrons in the level is $w_n = e^{-\beta(E_d - \mu n)}$, independent of spin; hence,

FIG 9.68 Possible occupation of a donor level used in Eq. (9.354). The level can accommodate no electrons and single electron with either spin projection. Although the Pauli principle allows two electron occupation with opposite spin projections, this is excluded here due to high energy cost.

$$\langle n \rangle = \frac{\sum_{n=0,1,1} n w_n}{\sum_{n=0,1,1} w_n} = \frac{2e^{-\beta(E_d - \mu)}}{1 + 2e^{-\beta(E_d - \mu)}}. \tag{9.354}$$

Finally, if the total number of donor impurities per unit volume is N_d, then the average number of electrons per unit volume occupying all the donor levels is

$$n_d = \frac{N_d}{1 + \frac{1}{2}e^{\beta(E_d - \mu)}}. \tag{9.355}$$

Similar considerations lead to the analogous formula for hole concentration in the acceptor levels,

$$n_a = \frac{N_a}{1 + \frac{1}{2}e^{\beta(\mu - E_a)}}. \tag{9.356}$$

9.6 Semiconductors

The density of ionized donor and acceptor impurities at energies E_d and E_a just above the top of the valence band is,

$$n_d^+ = N_d - n_d = \frac{N_d}{1 + \frac{1}{2} e^{\beta(\mu - E_d)}}, \tag{9.357}$$

$$n_a^- = N_a - n_a = \frac{N_a}{1 + \frac{1}{2} e^{\beta(E_a - \mu)}}. \tag{9.358}$$

In the limit, $E_d - \mu$, $\mu - E_a \gg k_B T$, to be discussed later, $n_d, n_a \to 0$ and $n_d^+, n_a^- \to N_d, N_a$.

Extrinsic Semiconductor Carrier Densities at Thermal Equilibrium

Recall that the density of electrons in the conduction band of an *intrinsic* semiconductor is equal to the density of holes in the valence band, $n_e = n_h$. Now, we consider the doping of a semiconductor with N_d and N_a donor and acceptor impurities per unit volume. How do the electrons and holes redistribute at temperature $T \neq 0$? The densities of electrons in the conduction band and in the donor levels are denoted as n_e and n_d, while those of holes in the valence band and the acceptor levels are denoted as n_h and n_a. At $T = 0$, electrons populate the lowest levels, consistent with the Pauli principle; the valence band and all acceptor levels are filled, $N_d - N_a$ donor levels are filled, and the conduction band is empty. Suppose $N_d > N_a$ and $T > 0$, the electron density n_c in the conduction band is nonzero and the donor electron density n_d is modified from its $T = 0$ value $N_d - N_a$. This gain $n_e + n_d - (N_d - N_a) > 0$ is obtained by forming holes in the acceptor levels and in the valence band with densities n_a and n_h. Thus,

$$n_e + n_d = N_d - N_a + n_h + n_a. \tag{9.359}$$

The temperature dependence of n_e and n_h is given in Eqs (9.342) and (9.343) and in Problem 9.50, and the temperature dependence of n_d and n_a are given in Eqs (9.355) and (9.356). This lets us estimate the chemical potential μ and hence the carrier densities. Following the assumptions in the caption of Fig. 9.67 [E_d (E_a) just slightly below (above) E_c (E_v)] and generalizing them for donor and acceptor impurities,

$$E_d - \mu \gg k_B T, \quad \mu - E_a \gg k_B T. \tag{9.360}$$

Hence, the exponents in the denominators in Eqs (9.355) and (9.356) are very large as well as $n_d \ll N_d$ and $n_a \ll N_a$. The ionization of the donor and acceptor levels is then virtually complete. Thus,

$$n_e - n_h \equiv \Delta n = N_d - N_a. \tag{9.361}$$

By using Eqs (9.352) and (9.353), we obtain,

$$n_e = \frac{1}{2}\left(\sqrt{(N_d - N_a)^2 + 4n_i^2} + \frac{1}{2}(N_d - N_a)\right), \tag{9.362}$$

$$n_h = \frac{1}{2}\left(\sqrt{(N_d - N_a)^2 + 4n_i^2} - \frac{1}{2}(N_d - N_a)\right), \tag{9.363}$$

$$N_d - N_a = 2n_i \sinh \beta(\mu - \mu_i). \tag{9.364}$$

As long as the inequalities (9.360) are satisfied, which is the case if μ is not very far from μ_i [defined in Eq. (9.349)], Eqs (9.362), (9.363), and (9.364) are valid. This covers a large range of concentrations, from the intrinsic regime $|N_a - N_d| \ll n_i$ to deep into the extrinsic regime $|N_a - N_d| \gg n_i$. In the intrinsic regime, an expansion of the square root in Eqs (9.363) yields

$$n_e \approx n_i + \frac{1}{2}(N_d - N_a), \quad n_h \approx n_i - \frac{1}{2}(N_d - N_a). \tag{9.365}$$

In the extrinsic regime, $n_i \ll |N_d - N_a|$,

$$n_e = N_d - N_a, \quad n_h = \frac{n_i^2}{N_d - N_a}, \quad (N_d > N_a),$$

$$n_e = \frac{n_i^2}{N_a - N_d}, \quad n_h = N_a - N_d, \quad (N_a > N_d). \tag{9.366}$$

Hence, in the extrinsic semiconductor case, most of the charge carriers occupy one of the corresponding bands (electrons in the conduction band if $N_d \gg N_a$, or holes in the valence band if $N_a \gg N_d$), whereas the carrier density in the other band is very low.

If donors are present with density n_d and no acceptors are present, then the concentration of electrons at finite T increases with increasing n_d by virtue of donor ionization. However, the product $n_e n_h$ remains as determined in Eq. (9.347), so the density of holes decreases below its intrinsic value. The charge density must be equal to the density of ionized donors, n_d^+, because each conduction electron results from the ionization of a donor atom, $n_e = n_d^+$, and the intrinsic hole and electron densities are suppressed. In Problem 9.52, you are asked to calculate the equilibrium electron density in the conduction band.

Problem 9.52

Donors are doped into silicon at a concentration of $N_d = 5 \times 10^{15}/\text{cm}^3$. Calculate the concentration of electrons and holes at $T = 300$ K, and the relative change of concentrations following a temperature rise by 1 K.

Answer: At 300 K, the impurities are fully ionized and the concentration is much higher than $n_i = 4.86 \times 10^{11}$ (as calculated in Problem 9.49). Therefore, the majority (electron) concentration coincides with that of the impurities, $n_e = N_D = 5 \times 10^{15}/\text{cm}^3$. From the law of mass action, $n_e n_h = n_i^2$, and we have $n_h = n_i^2/n_e = 4.72 \times 10^7/\text{cm}^3$. The relative changes of the law of mass action yields $\frac{dn_h}{n_h} = 2\frac{dn_i}{n_i} - \frac{dn_e}{n_e} \approx 2\frac{dn_i}{n_i} = \left(3 + \frac{E_G}{k_B T}\right)\frac{dT}{T} = 0.166$. This approximation is justified because the excess electron concentration due to thermal electron-hole excitations is much smaller than the concentration itself. The second equality is proved in Problem 9.49. Because the changes are caused by thermal electron-hole excitations, charge neutrality implies $dn_e = dn_h$.

9.6.4 INHOMOGENEOUS SEMICONDUCTORS: *p-n* JUNCTIONS

Our discussion so far has been focused on homogeneous semiconductors in which the concentration of impurities is uniform. Most of the important applications and technological achievements of semiconductor physics, and the myriad of semiconductor devices, involve *inhomogeneous semiconductors* wherein the concentrations of donor and acceptor impurities vary in space and are meticulously tuned as a function of position. The principles governing the dynamics of electrons and holes are generally based on the semiclassical model introduced in Sec. 9.5.6. The basic challenge in this context is to find out how the densities of charge carriers and the corresponding currents are distributed within the material and how they respond to an external field. One of the simplest structures involving inhomogeneous semiconductors is a quasi-one dimensional crystal (e.g., in a narrow slab) with longitudinal coordinate $-\infty < x < \infty$, where the densities of donors and acceptors vary oppositely and monotonically between zero on one side and some saturation values on the other side as x varies between $-\infty$ and $+\infty$. A structure with majority of donors on one side and majority of acceptors on the other side is called a *p-n* junction. Thus, *p*-type and *n*-type doped semiconductors in contact form a *p-n* junction. It has numerous important device applications, including rectifiers, photovoltaic detectors, voltage regulators (Zener diodes), tuners, light-emitting diodes (LEDs), laser diodes, photodetectors, and solar cells. Figure 9.69(a) shows a schematic diagram of a *p-n* junction. An example of a smooth profile is

$$N_d(x) = \frac{N_d}{2}\left(1 + \tanh\frac{x}{a}\right), \quad N_a(x) = \frac{N_a}{2}\left(1 - \tanh\frac{x}{b}\right),$$

9.6 Semiconductors

where N_d and N_a are the donor and acceptor saturation values and a and b are tunable length scales. However, modeling the inhomogeneity by an abrupt junction,

$$N_d(x) = N_d\Theta(x), \quad N_a(x) = N_a\Theta(-x), \qquad (9.367)$$

simplifies the calculations.

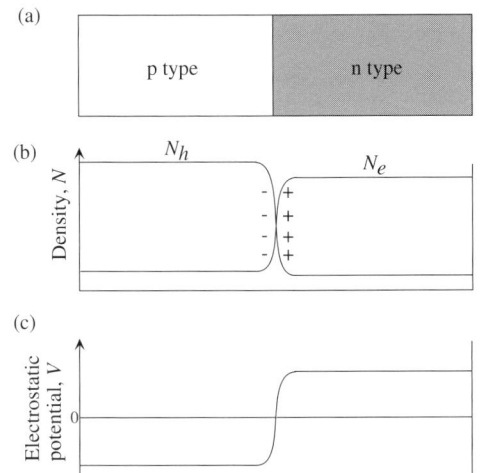

FIG 9.69 (a) Schematic diagram of a *p-n* junction, (b) the density of holes and electrons and the positively and negatively charged layers, and (c) the potential across the bilayer.
Source: Band, Light and Matter, Fig. 6.38, p. 402

Qualitative description of a *p-n* junction

Let us briefly explain the electrostatics and electrodynamics of a *p-n* junction. On the left side of the junction, most of the acceptor atoms (*p*-type impurities) become negatively charged ions; free holes of the same concentration are present so that electrical neutrality is maintained. On the right side, most of the n-type donor atoms become positively charged ions; free electrons in the conduction band of the same concentration are present. There is a small amount of minority carriers in both p-type and n-type doped materials. The holes on the *p* side near the interface diffuse into the *n* side and the electrons on the *n* side diffuse into the *p* side. Hence, an excess of negatively charged ionized acceptor atoms are left behind on the *p* side, and an excess of positively charged ionized donor atoms are left behind on the *n* side. The region near the interface contains only fixed charges: positive ions on the *n*-side and negative ions on the *p*-side. As a result, a narrow region on both sides of the junction is depleted of mobile charge carriers. This region is called the *depletion layer*. The double layer of charge shown in Fig. 9.69(b) gives rise to an electric field $\mathbf{E}(\mathbf{r})$ pointing from *p* to *n*, which inhibits further diffusion. The electrostatic potential $\varphi(\mathbf{r})$ varies across the bilayer as shown on the bottom of Fig. 9.69(c) [recall that the hole (electron) potential energy is $e\varphi(\mathbf{r})$ ($-e\varphi(\mathbf{r})$)]. Thus, there is a barrier in the potential energy for electrons to cross the junction from right to left (from n-type to p-type material) and a barrier for the holes to cross the junction from left to right (from *p*-type to *n*-type). The density of electrons and holes in thermal equilibrium are given in terms of $\varphi(\mathbf{r})$ by

$$n_e(\mathbf{r}) = C_e e^{\beta e\varphi(\mathbf{r})}, \quad n_h(\mathbf{r}) = C_h e^{-\beta e\varphi(\mathbf{r})}, \qquad (9.368)$$

where C_e and C_h are constants. Recombination of electrons and holes occurs within the junction, and the current I_r resulting from the flow of electrons and holes into the junction to feed the recombination is balanced by a small current flow, I_g, of generated electrons and holes, which diffuse across the junction, i.e., $I_r = -I_g$.

When an external voltage is applied to the junction, the flow of majority carriers is altered. For a forward-biased junction, the voltage V applied to the *p*-region is positive, so that an electric field is produced in a direction opposite to that of the built-in field. The presence of the external bias voltage causes a departure from equilibrium and misalignment of the Fermi levels in the *p*- and *n*-regions, as well as in the depletion layer. When a positive voltage is applied to the *p*-type region and a negative voltage to the *n*-type region, a large current flows, but if the voltage is reversed (this situation is called reversed bias), a very small current flows. The net effect of the forward bias is a reduction in the height of the potential energy barrier by an amount eV. For forward-biased voltage, the current increases because the barrier is lowered for both electrons and holes. The majority carrier current increases exponentially with the voltage as $\exp(eV/k_BT)$, so the net current also increases exponentially with V. The forward-biased current is given by $I(V) = I_g[\exp(-eV/k_BT) - 1]$ for forward bias. For reversed voltage, the electrons cannot surmount the barrier, since the reversed voltage augments the potential barrier between the *p* and *n* regions. The reverse-biased current is given by $I(V) = I_g[\exp(-eV/k_BT) - 1]$ for backward (reversed) bias. Figure 9.70 shows the current–voltage characteristics of a *p-n* junction. If an alternating

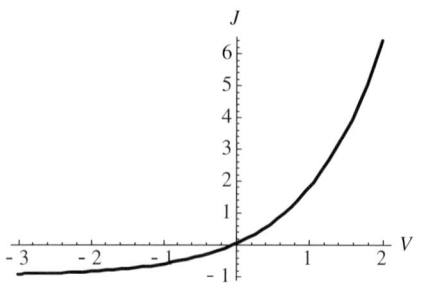

FIG 9.70 Current versus voltage in a *p-n* junction [see Eq. (9.382)].

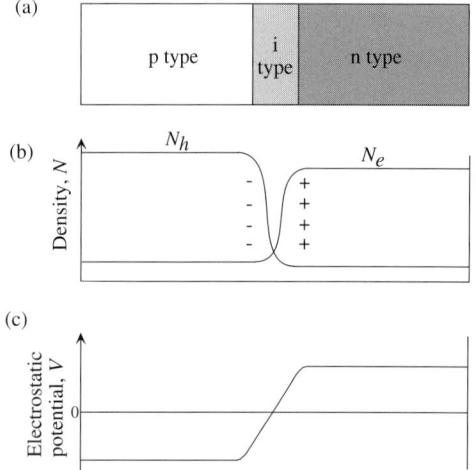

FIG 9.71 (a) Schematic diagram of a *p-i-n* junction, (b) the density of holes and electrons and the positively and negatively charged layers, and (c) the potential across the bilayer. *Source:* Band, Light and Matter, Fig. 6.40, p. 404

voltage is applied, then the current will flow mostly in one direction, so the current is rectified. Later we present a quantitative analysis of the *p-n* junction and analyze the origin of the current I_g. We will show that I_g has contributions from both electrons and holes.

Photovoltaic effect

Consider irradiation of an unbiased *p-n* junction with light of frequency sufficient to create electron-hole pairs (photon energy $\geq$ gap energy). The additional charge carriers generated by this process diffuse into the junction and an electric field gradient develops across the junction. The separation of the charge carriers produces a forward voltage across the barrier (the electric field of the photo-induced carriers is opposite to the field of the junction). Thus, the light drives a current that delivers electrical power to an external circuit. This effect is called the *photovoltaic effect*. It can be used to detect light or to produce energy as in solar cells.

***p-i-n* junction**

Another type of junction that finds applications is a *p-i-n* junction, which is made by inserting a layer of intrinsic semiconductor material between the *p*-type and *n*-type materials. The region over which the electric field now varies includes the whole width of the *i* region. Figure 9.71 shows a diagram of a *p-i-n* junction, the density of holes and electrons, the positively and negatively charged layers, and the potential across the bilayer. This kind of junction has small capacitance due to the large width of the junction, and consequently, the response of the junction is fast. The *p-i-n* junction is often used in semiconductor photodiodes.

Analysis of p-n Junctions in Equilibrium

In this section, a more quantitative analysis of the principles governing the physics of *p-n* junctions is discussed. For simplicity, we will continue to assume the abrupt junction profile as in Eq. (9.367). When there is neither an applied external field nor current, the junction is in equilibrium and all quantities assume their equilibrium values at the given temperature T. The fact that $N_a(x)$ and $N_d(x)$ vary in space implies that the carrier densities $n_e(x)$ and $n_h(x)$ depend on space as well. This generates an electrostatic potential profile $\varphi(x)$ as shown in Fig. 9.69(c). $n_e(x)$ and $n_h(x)$ are nearly uniform except within the depletion layer (whose thickness is about 10-1000 nm). Within the depletion layer, the charge density is much smaller than away from it.

The semiclassical formalism in Sec. 9.5.6 is applicable when the variation of the potential $\varphi(x)$ is small: the electrostatic energy change over a lattice constant should be smaller than the band gap E_g. This condition is satisfied in most cases of interest, except perhaps within the depletion layer.

Our task is to deduce the profile of carrier densities $n_e(x)$ and $n_h(x)$ and the potential profile $\varphi(x)$ given the doping profiles $N_d(x)$ and $N_a(x)$. In a semiclassical picture, the dynamics of electrons in a given band n with energy dispersion $E_n(\mathbf{k})$ subject to an electrostatic potential $\varphi(x)$ can be described classically using the Hamiltonian $H_{\text{class}} = E_n(\mathbf{k}) - e\varphi(x)$.

9.6 Semiconductors

We can generalize Eqs (9.342) and (9.343) for the equilibrium carrier densities,

$$n_e(x) = N_e(T)\, e^{-\beta[E_c - \mu - e\varphi(x)]}, \quad n_h(x) = N_h(T)\, e^{-\beta[\mu - E_v + e\varphi(x)]}, \quad (9.369)$$

where the pre-factors $N_e(T)$ and $N_h(T)$ are defined in Eqs (9.345) and (9.346).

Problem 9.53

(a) Assume that for $x \to \infty$, all the donor impurities are ionized, and for $x \to -\infty$, all the acceptor impurities are anti-ionized (this follows from the analysis leading to expressions (9.366). Use Eq. (9.369) to obtain the relations,

$$N_d = N_e e^{-\beta(E_c - e\varphi(\infty) - \mu)}, \quad N_a = N_h e^{-\beta(\mu - E_v + e\varphi(-\infty))}. \quad (9.370)$$

(b) Show that the electrostatic energy difference across the junction is,

$$e[\varphi(\infty) - \varphi(-\infty)] \equiv e\Delta\varphi = E_g + k_B T \ln \frac{N_d N_a}{N_e N_h}, \quad (9.371)$$

where $E_g = E_c - E_v$ is the gap energy.

(c) Use the results of (a) and (b) and the fact that $e[\varphi(\infty) - \varphi(-\infty)] \approx E_g \gg k_B T$ to show that

$$n_e(x) = N_d e^{-\beta e[\varphi(\infty) - \varphi(x)]}, \quad n_h(x) = N_a e^{-\beta e[\varphi(x) - \varphi(-\infty)]}. \quad (9.372)$$

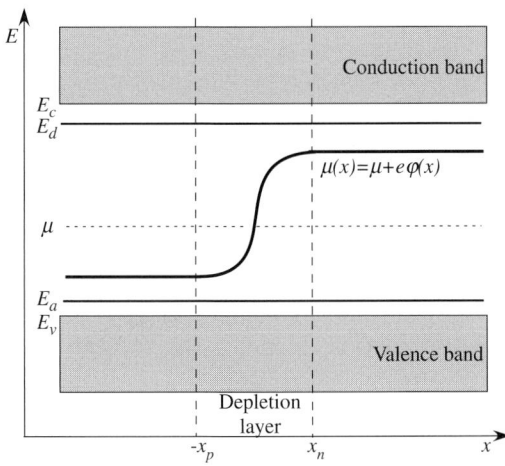

FIG 9.72 Energy parameters of a p-n junction versus position. The function $\mu(x) = \mu + e\varphi(x)$ serves as a local chemical potential such that relations (9.369) for the densities in the inhomogeneous case are obtained from Eqs (9.342) and (9.343) (valid for the equilibrium carrier densities in the homogeneous case) by simply replacing μ by $\mu(x)$.

Note that the potential $\varphi(x)$ is not yet known, since it is related to the charge densities $n_e(x)$ and $n_h(x)$ through the Poisson equation. Therefore, it must be determined self-consistently. The potential $\varphi(x)$ is monotonic and varies appreciably only within the depletion layer, as shown in Fig. 9.69. Figure 9.72 schematically illustrates the energy parameters as a function of position of a p-n junction. Following Eq. (9.369), it is reasonable to consider the quantity $\mu(x) \equiv \mu + e\varphi(x)$ as a local effective chemical potential. Of course, once the system is in thermal equilibrium, its chemical potential μ is constant.

To use the Poisson equation,

$$\frac{d^2\varphi(x)}{dx^2} = -\frac{4\pi\rho(x)}{\varepsilon}, \quad (9.373)$$

to determine $\varphi(x)$, we need to know the charge density $\rho(x)$, which has contributions from ionized impurities and carriers in the band. The impurities are fully effective in the sense that donor impurities are ionized and left as negatively charged ions, whereas acceptor impurities are positively charged ions. This is accurate far away from the depletion layer where on the left $N_a - N_d \gg n_i$ and on the right $N_d - N_a \gg n_i$, and the analysis leading to Eq. (9.366) applies. This remains true over the entire sample. Inspecting Eqs (9.355) and (9.356), we see that when the chemical potential is such that $\mu - E_a \gg k_B T$ $(E_d - \mu \gg k_B T)$, and the impurity densities n_a (n_d)

decrease (or, equivalently, the ionization degree increases) as shown in Fig. 9.72. At the far left, where the acceptor impurities are ionized, the inequality $\mu - E_a \gg k_B T$ is sharpened. Similarly, at the far right, where the donor impurities are already ionized, the inequality $E_d - \mu \gg k_B T$ is sharpened. Collecting these results, we obtain the charge density,

$$\rho(x) = e[N_d(x) - N_a(x) - n_e(x) + n_h(x)], \qquad (9.374)$$

where $N_d(x)$ and $N_a(x)$ are given by Eq. (9.367) with prefactors N_d and N_a as in Eqs (9.345) and (9.346). Combining this result with Eq. (9.372) and noting from Fig. 9.72 that in the depletion layer $e|\varphi(x) - \varphi(\pm\infty)| \gg k_B T$, the following pattern emerges: outside the depletion layer, $n_e = N_d$ on the n side and $n_h = N_a$ on the p side, and hence, the charge density $\rho(x)$ virtually vanishes. Inside the depletion layer, $n_{e,h}(x)$ fall quickly beyond $x_{n,p}$ and $N_{d,a}(x) = N_{d,a}$ as indicated in Eq. (9.367). To complete the picture, we need to take into account thermal excitations of electrons from the valence band in the n region and possible excitation of electrons into the conduction band in the p region. In both cases, these events are rare but of importance. In the first case, there are a few holes generated on the n side, where the majority of charge carriers are electrons. In the second case, there are a few electrons generated on the p side, where the majority of charge carriers are holes. On each side of the depletion layer, we then have *majority and minority charge carriers*. These results are illustrated in Fig. 9.73.

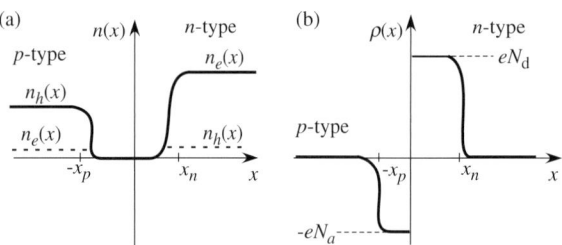

FIG 9.73 (a) Carrier densities of electrons ($n_e(x)$) and holes ($n_h(x)$) across a *p-n* junction in equilibrium. The dashed line indicate a very small amount of minority charge carriers. (b) Charge density $\rho(x)$ across a *p-n* junction in equilibrium, as given by Eq. (9.374).

When the expression (9.374) for $\rho(x)$ is inserted into the LHS of the Poisson equation (9.373) and expressions (9.370) are used for the carrier densities, we obtain a nonlinear differential equation for $\varphi(x)$. To gain more insight about the variation of densities across a junction, let us adopt a crude approximation inspired by the potential profile in Fig. 9.72. (A physical justification for this approximation is detailed in Ref. [122].) The limiting points $x_n > 0 > -x_p$ of the depletion layer are defined by the property that $\varphi(x > x_n) = \varphi(\infty)$ and $\varphi(x < -x_p) = \varphi(-\infty)$. Following the result of Eq. (9.372) derived in Problem 9.53, one has $n_e(x > x_n) = N_d$ and $n_h(x < -x_p) = N_a$. On the other hand, deep inside the depletion layer, $-x_p < x < x_n$, $|\varphi(x) - \varphi(\pm\infty)| \gg k_B T$, so according to Eq. (9.372), $n_e \ll N_d$ and $n_h \ll N_a$, implying $\rho(x) \approx e[N_d(x) - N_a(x)]$. The shape of the potential curve $\mu(x)$ in Fig. 9.72 for $x \geq -x_p$ is parabolic to lowest order in $(x + x_p)$. Similarly, for $x \leq x_n$, $\mu(x)$ is an upside-down parabola to lowest order in $(x - x_n)$. Combined with the Poisson equation, we then obtain the following approximation for $\varphi(x)$:

$$\varphi(x) = \varphi(-\infty)\Theta[-(x+x_p)] + \left\{\varphi(-\infty) + \left[\frac{2\pi e N_a}{\varepsilon}\right](x+x_p)^2\right\}\Theta(-x)\Theta(x+x_p)$$

$$+ \left\{\varphi(\infty) - \left[\frac{2\pi e N_d}{\varepsilon}\right](x-x_n)^2\right\}\Theta(x)\Theta(x_n-x) + \varphi(\infty)\Theta(x-x_n). \qquad (9.375)$$

Problem 9.54

(a) Show that $\varphi(x)$ and $\varphi'(x)$ are continuous at $x = -x_p$ and $x = x_n$.
(b) Check that continuity of $\varphi'(x)$ at $x = 0$ implies $N_d x_n = N_a x_p$.
(c) Check that continuity of $\varphi(x)$ at $x = 0$ implies $\left[\frac{2\pi e}{\varepsilon}\right](N_a x_p^2 + N_d x_n^2) = \Delta\varphi = \varphi(\infty) - \varphi(-\infty)$.
(d) Calculate the electric field $E_x = -\frac{d\varphi(x)}{dx}$ and plot it against x.

Answer: Draw straight segments joining the points $E_x(-x_p) = 0$, $E_x(0) = -\frac{4\pi e N_a}{\varepsilon}$, $E_x(x_n) = 0$.

9.6 Semiconductors

From the solution of Problem 9.54, estimates of the positions of the edges of the depletion layer are

$$x_n = \sqrt{\frac{qN_a}{N_d(N_a+N_d)}}, \quad x_p = \sqrt{\frac{qN_d}{N_a(N_a+N_d)}}, \quad q \equiv \frac{\varepsilon \Delta\varphi}{2\pi e}, \tag{9.376}$$

where q has dimensions of inverse length. Typical numerical values are $\varepsilon\Delta\varphi = 1$ V, $N_{d,a} = 10^{16}\text{cm}^{-3}$, and $x_{n,p} = 10^3$ Å. The field in the depletion layer is about $\Delta\varphi/(x_n+x_p)$, which is of the order of 10^6 V/m.

p-n Junction in a Static Field: Rectification

Let us now consider an external uniform and static electric field applied along the x-axis. Recall that in the absence of an applied field, the thickness of the depletion layer is of the order of 10^3 Å. In this layer, following Fig. 9.73, the carrier densities $n_e(x)$ and $n_h(x)$ drop to zero. Consequently, the electric resistance of the depletion layer is very high (the conductance must be proportional to the density of carriers as is noted from the Drude formula). Applying a potential difference V across the semiconductor (e.g., by using the junction as an element within a circuit), the main part of the potential drop will occur over the depletion layer, and therefore, the modified potential drop across the layer is approximately

$$\Delta\phi = \Delta\varphi - V. \tag{9.377}$$

Inserting this potential drop into Eq. (9.376) leads to a modified value of $q \to \bar{q} = \frac{\varepsilon(\Delta\varphi-V)}{2\pi e}$, and $x_{n,p}$ becomes smaller (larger) than their value at $V=0$ for $V>0$ ($V<0$). Between $-x_p$ and x_n, the local potential is $\varphi(x) - V(x)$, where $V(x)$ is the linear potential due to the externally applied constant field. The effect of applying a potential drop is shown in Fig. 9.74.

One can view the conduction and valence bands as effectively bending due to the internal potential profile $\varphi(x)$ and the externally applied bias $V(x)$. The effective conduction and valence bands for electrons bend with x according to

$$E_c(x) = E_c - e(\varphi(x) + V(x)), \quad E_v(x) = E_v - e(\varphi(x) + V(x)), \quad \text{(for electrons)}. \tag{9.378}$$

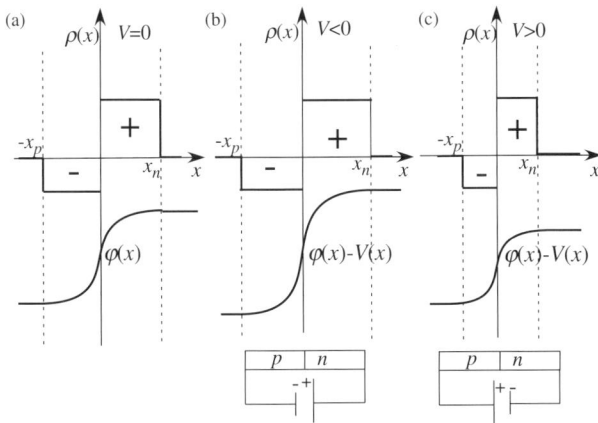

FIG 9.74 A *p-n* junction in equilibrium which is subject to a static potential drop V. (a) $V=0$. (b) For $V<0$, the depletion layer is extended, and the potential drop across the junction, $\Delta\varphi - V$ is increased compared to the potential drop $\Delta\varphi$ when $V=0$. (c) For $V>0$, the depletion layer is contracted, and the potential drop across the junction, $\Delta\varphi - V$ is reduced compared to the potential drop $\Delta\varphi$ when $V=0$. The charge density $\rho(x)$ is modified accordingly.

$E_c(x)$ is the energy of an electron wave packet formed from levels near the conduction band and localized at x, and similarly $E_v(x)$ for the valence band. For holes, replace $-e$ by $+e$ in Eqs 9.378 and the energy curves have opposite derivatives. *Band bending* is schematically depicted in Fig. 9.75.

Now consider the current in a *p-n* junction that is subject to a potential bias V, as depicted in Fig. 9.74(b) and (c) or, equivalently, Fig. 9.75(b) and (c). There are two types of electron current and two types of hole current. As shown in Fig. 9.76(a), there are very few electrons (full circles) that are generated on the *p* side; these electrons are referred to as *minority electron carriers*. Once a minority electron is generated, it immediately starts to move, as indicated by the black arrow numbered 2, toward the *n* side, where it "belongs." This process yields an *electron generation current*, J_e^g, of magnitude independent of the bias V. Second, an electron on the *n* side, where electrons are *majority electron carriers*, can be

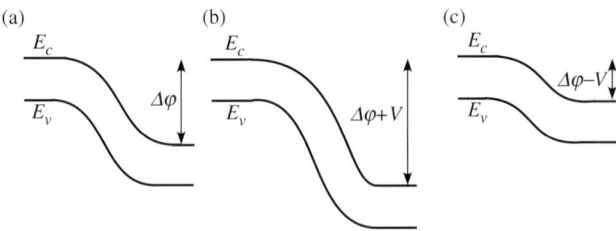

FIG 9.75 Schematic electron energy band diagram. The cases (a), (b), and (c) correspond to (a), (b), and (c), respectively, in Fig. 9.74.

thermally excited above the potential barrier $\Delta\phi$ [see Eq. (9.377)] and pass to the p side of the depletion layer (black arrow number 1). Once on the p side, this electron could combine with a hole, a process called *pair annihilation*, but in any case, it contributes to the current through the depletion layer. The resulting *electron diffusion current*, J_h^d, also called a *recombination current*, is very sensitive to V, because the probability for thermal activation above the potential barrier is exponentially small,

$$J_e^d(V) \propto e^{-\beta e \Delta\phi} = e^{-\beta e(\Delta\varphi - V)}. \tag{9.379}$$

For $V = 0$, the two currents must be equal in magnitude because J_e^g is independent of V, and there can be no electron current for $V = 0$. Thus,

$$J_e^d(0) = J_e^g \propto e^{-\beta e \Delta\varphi}. \tag{9.380}$$

Combining Eqs (9.379) and (9.380), we get the total electron current

$$J_e = J_e^d - J_e^g = J_e^g(e^{\beta eV} - 1). \tag{9.381}$$

An analogous analysis is valid for holes, and the reader can analyze the hole currents using Fig. 9.76(b) as a guide. Although the *matter currents* of electrons and holes have opposite signs, the *charge currents* have the same directions. The total charge current due to both electron and hole contributions is then given by

$$\boxed{J = (J_e^g + J_h^g)(e^{\beta eV} - 1).} \tag{9.382}$$

It is highly asymmetric as a function of V, growing exponentially for $V > 0$ and approaching a small negative constant at $V < 0$ (see Fig. 9.70). Thus, the p-n junction can function as a rectifier, effectively transmitting charge in one direction.

p-n Junction Out of Equilibrium

The above analysis does not account for the nonequilibrium situation which prevails once $V \neq 0$. In particular, the carrier densities $n_e(x)$ and $n_h(x)$ are no longer given by the Maxwellian distribution (9.369). Let us set up equations for the electron and hole currents and concentrations, which

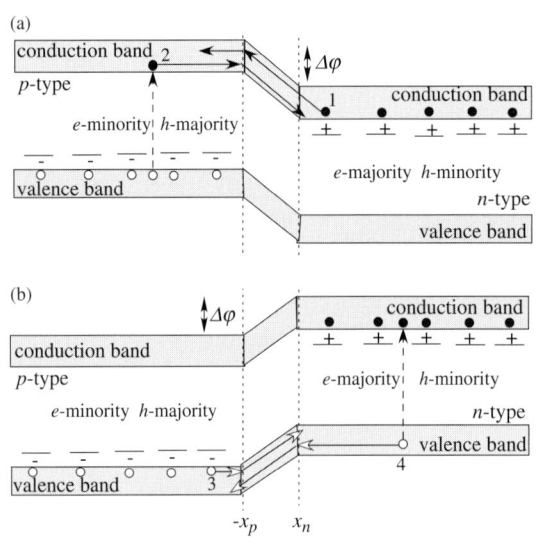

FIG 9.76 Current generation in a p-n junction. (a) Electron currents: Arrow 1 is the electron diffusion current, arrow 2 is the electron generation current, (b) Hole currents: Arrow 3 is the hole diffusion current, and arrow 4 is the hole generation current.

are applicable for $V \neq 0$. The fact that there is a concentration gradient [$n_e(x)$ and $n_h(x)$ depend on x] leads to a current. Denoting the electron and hole current densities along the x direction by J_e and J_h, we have

$$\boxed{J_e = en_e\mu_e E_x + eD_e \frac{dn_e}{dx}, \quad J_h = en_h\mu_h E_x - eD_h \frac{dn_h}{dx},} \tag{9.383}$$

9.6 Semiconductors

where the electron and hole mobilities are defined in Eq. (9.350). The first terms on the RHSs (proportional to the electric field) are called *drift currents*. The constants $D_e, D_h > 0$ are the electron and hole *diffusion constants*.

Problem 9.55

Show that to have vanishing current at equilibrium ($V = 0$), the mobilities and the diffusion constants must be related by

$$\mu_e = \beta e D_e, \quad \mu_h = \beta e D_h. \tag{9.384}$$

Guidance: Use Eqs (9.369) for the concentrations and $E_x = -\frac{d\varphi(x)}{dx}$. Equations (9.384) are referred to as the *Einstein relations*.

Equations (9.383) are useful in determining the ratio between minority and majority carriers on both sides of the depletion layer, e.g., $n_e(-x_p)/n_e(x_n)$ and $n_h(x_n)/n_h(-x_p)$. Consider first the case of no bias, $V = 0$, so the current vanishes, $J_e = J_h = 0$ in Eqs (9.383). Using the Einstein relations (9.384) and relating the field to the potential gradient, $E_x = -d\varphi(x)/dx$, one immediately finds

$$\frac{1}{n_e}\frac{dn_e}{dx} - \beta e \frac{d\varphi(x)}{dx} = 0, \quad \frac{1}{n_h}\frac{dn_h}{dx} + \beta e \frac{d\varphi(x)}{dx} = 0. \tag{9.385}$$

The first equation is now integrated between $-x_p$ and x_n while the second equation is integrated between x_n and $-x_p$, leading to,

$$\frac{n_e(-x_p)}{n_e(x_n)} = \frac{n_h(x_n)}{n_h(-x_p)} = e^{-\beta e \Delta \varphi}. \tag{9.386}$$

Problem 9.56

Use the law of mass action and the fact that the impurities are fully ionized for $x \geq x_n$ or $x \leq -x_p$, to show that $n_h(x_n) = n_i^2/N_d$ and $n_e(-x_p) = n_i^2/N_a$.

It is reasonable to adopt an approximation implying that this relation holds also out of equilibrium with the total potential $\Delta \phi = \Delta \varphi - V$ replacing the potential difference in equilibrium $\Delta \varphi$. Then, the inverse of the relations (9.386) read

$$\frac{n_e(x_n)}{n_e(-x_p)} = \frac{n_h(-x_p)}{n_h(x_n)} = e^{\beta e (\Delta \varphi - V)}. \tag{9.387}$$

Next, assuming that the densities of majority carriers at the interfaces are almost equal to their equilibrium values, we get,

$$n_e(x_n) \simeq n_e^{(0)}, \quad n_h(-x_p) \simeq n_h^{(0)}. \tag{9.388}$$

By using Eqs (9.386), we arrive at the relations for the minority carrier densities at the edge points *out of equilibrium* in terms of their values at equilibrium,

$$\boxed{n_e(-x_p) = n_e^{(0)}(-x_p) e^{\beta eV}, \quad n_h(x_n) = n_h^{(0)}(x_n) e^{\beta eV}.} \tag{9.389}$$

Continuity Equations: The electron and hole currents satisfy separate continuity equations. When the semiconductor is out of equilibrium, there are sources of charge that affect these continuity equations. If the number of charge carriers were conserved, the standard continuity equations would apply,

$$\frac{\partial J_e}{\partial x} + (-e)\frac{\partial n_e}{\partial t} = 0, \quad \frac{\partial J_h}{\partial x} + e\frac{\partial n_h}{\partial t} = 0, \tag{9.390}$$

However, out of equilibrium, the densities of charge carriers are not conserved, and the equations must be generalized. Denoting by $n_e^{(0)}, n_h^{(0)}$ the carrier concentrations at equilibrium, given by Eqs (9.369), the excess densities are $n_e(x,t) - n_e^{(0)}$ and $n_h(x,t) - n_h^{(0)}$. They are caused either by generation (e.g., an electron in the valence band is thermally excited, thereby adding an electron and a hole) or by recombination [where an electron from the conduction band dives into an empty location (a hole) in the valence band, thereby removing one particle of each kind]. These processes restore equilibrium; when $n_e > n_e^{(0)}$ and $n_h > n_h^{(0)}$ recombination is favored, while when $n_e < n_e^{(0)}$ and $n_h < n_h^{(0)}$ generation is more dominant. These observations suggest a simple model according to which the rate of change of carrier densities due to generation and recombination is proportional (with a negative coefficient) to the excess density,

$$\left[\frac{dn_e}{dt}\right]_{g-d} = -\frac{1}{\tau_e}(n_e - n_e^{(0)}), \quad \left[\frac{dn_h}{dt}\right]_{g-d} = -\frac{1}{\tau_h}(n_h - n_h^{(0)}), \tag{9.391}$$

where g–d is a shorthand notation for generation-diffusion and τ_e, τ_h are referred to as the *electron and hole lifetimes*. They quantify the decay rate of the excess densities toward zero if equilibrium is restored. The lifetimes are typically of order, $\tau_e, \tau_h \approx 10^{-3}$ s, which is much longer than collision times determining mobility and conductivity ($\tau^{(\text{coll})} \simeq 10^{-12}$ s). Collecting these results, we can now write the continuity equations as,

$$\boxed{\frac{\partial J_e}{\partial x} - e\frac{\partial n_e}{\partial t} = \frac{e}{\tau_e}(n_e - n_e^{(0)}), \quad \frac{\partial J_h}{\partial x} + e\frac{\partial n_h}{\partial t} = -\frac{e}{\tau_h}(n_h - n_h^{(0)}).} \tag{9.392}$$

To complete the analysis, recall that the electric field E_x appearing in Eq. (9.383) is related to the charge density $\rho(x,t)$, Eq. (9.374) through the Maxwell equation,

$$\frac{\partial E_x}{\partial x} = \frac{4\pi}{\varepsilon}\rho, \tag{9.393}$$

where ε is the dielectric constant of the medium. Equations (9.383), (9.392), and (9.393) are sufficient for determining the currents and carrier densities in a biased (out of equilibrium) p-n junction.

The Schockley Equation

To proceed, we make two simplifying assumptions. (1) The junction is abrupt at $x = 0$ with depletion layer between $-x_p$ on the p side and x_n on the n side, as shown in Figs 9.74 and 9.76. The (homogeneous and neutral) p and n regions are located at $-\infty < x \leq -x_p$ and $x_n \leq x < \infty$, respectively. In these neutral regions, the potential is constant and hence $E_x \simeq 0$. (2) The minority carrier densities (see Fig. 9.76) are small compared with the majority carrier densities. This allows neglect of the drift currents [the terms proportional to E_x in Eqs (9.383)] in the neutral regions near the junction. As a result,

$$J_e = eD_e\frac{\partial n_e}{\partial x}, \quad J_h = -eD_h\frac{\partial n_h}{\partial x}. \tag{9.394}$$

Substituting these expressions into the continuity equations (9.392), we find,

$$\frac{\partial n_e}{\partial t} = D_e\frac{\partial^2 n_e}{\partial x^2} - \frac{1}{\tau_e}(n_e - n_e^{(0)}), \quad \frac{\partial n_h}{\partial t} = D_h\frac{\partial^2 n_h}{\partial x^2} - \frac{1}{\tau_h}(n_h - n_h^{(0)}). \tag{9.395}$$

Equations (9.395) are particularly useful in the corresponding minority regions [$x < -x_p$ for $n_e(x,t)$ and $x > x_n$ for $n_h(x,t)$]. Although the p-n junction biased by a static potential V is out of equilibrium, still it is in a steady state in the

9.6 Semiconductors

sense that the minority carrier densities are constant in time,

$$\left[\frac{\partial n_e(x,t)}{\partial t}\right]_{x<-x_p} = \left[\frac{\partial n_h(x,t)}{\partial t}\right]_{x>x_n} = 0. \quad (9.396)$$

Equations (9.395) then reduce to

$$\frac{\partial^2 n_e}{\partial x^2} = \frac{1}{D_e \tau_e}[n_e - n_e^{(0)}(-x_p)], \quad \frac{\partial^2 n_h}{\partial x^2} = \frac{1}{D_h \tau_h}[n_h - n_h^{(0)}(x_n)]. \quad (9.397)$$

The solutions,

$$n_e(x < -x_p) = A_e e^{\frac{x}{L_e}} + n_e^{(0)}(-x_p), \quad n_h(x > x_n) = A_h e^{-\frac{x}{L_h}} + n_h^{(0)}(x_n), \quad (9.398)$$

contain two constants of integration, A_e and A_h, as well as the *electron and hole diffusion lengths*,

$$L_e = \sqrt{D_e \tau_e}, \quad L_h = \sqrt{D_h \tau_h}, \quad (9.399)$$

which measure the distances required for the densities to attain their equilibrium value.

Problem 9.57

Find the constants A_e and A_h in terms of the values of the solutions at the borders of the depletion layers, $n_e(-x_p)$ and $n_h(x_n)$.

Answer: $A_e = [n_e(-x_p) - n_e^{(0)}(-x_p)]e^{-\frac{x_p}{L_e}}$ and $A_h = [n_h(x_n) - n_h^{(0)}(x_n)]e^{\frac{x_n}{L_h}}$.

In the next step, we use Eq. (9.394) for the minority diffusion currents and express the densities on the LHS in terms of their edge values $n_e(-x_p)$ and $n_h(x_n)$ [substitute A_e and A_h from the solution of Problem 9.57 in Eq. (9.398)]. The result is

$$J_e(-x_p) = e\frac{D_e}{L_e}[n_e(-x_p) - n_e^{(0)}(-x_p)], \quad J_h(x_n) = e\frac{D_h}{L_h}[n_h(x_n) - n_h^{(0)}(x_n)]. \quad (9.400)$$

Substitute Eqs (9.389) to obtain closed-form expressions for the electron and hole currents in their respective minority sides,

$$J_e(-x_p) = e\frac{D_e}{L_e}n_e^{(0)}(-x_p)[e^{\beta eV} - 1], \quad J_h(x_n) = e\frac{D_h}{L_h}n_h^{(0)}(x_n)[e^{\beta eV} - 1]. \quad (9.401)$$

Both hole and electron currents are constant in the depletion layer between $-x_p$ and x_n, i.e., $J_e(-x_p) = J_e(x_n)$. We now sum the electron and hole currents at x_n to get the total current through the p-n junction,

$$\boxed{J = J_e(x_n) + J_h(x_n) = e\left(\frac{D_e}{L_e}n_e^{(0)}(-x_p) + \frac{D_h}{L_h}n_h^{(0)}(x_n)\right)[e^{\beta eV} - 1].} \quad (9.402)$$

Equation (9.402) is the *Schockley diode formula* for the current through an ideal p-n junction in terms of the applied bias and the junction parameters.

p-n Junction: Summary

The *p-n* junction is one of the most important semiconductor structures and is a spectacular example of quantum mechanics at work in a macroscopic system at room temperature. *p-n* junctions can be easily fabricated in semiconductors (such as Si) in which one can produce both *p* and *n* regions. The *p-n* junction opens the door for a myriad of applied semiconductor devices. In our discussion of the *p-n* junction, we addressed the following points:

1. The existence of both electrons and holes in a semiconductor leads to the generation of drift and diffusion currents for electron and hole currents.
2. Analysis of the energy levels in the *p-n* junction shows that there is a potential difference between the two sides of the depletion layer, which affects the spacial profiles of the carrier densities. This potential difference can be modulated by adding an external bias, thereby affecting the diffusion currents.
3. There is a small minority density of carriers, electrons on the *p* side, and holes on the *n* side of the junction, leading to small drift currents (processes 2 and 4 in Fig. 9.76), which are independent of the applied bias.
4. The effects of the applied potential on the drift and diffusion currents results in highly asymmetric current–voltage characteristic for positive and negative bias. This asymmetry is the hallmark of the *p-n* junction, and it is this property that makes it so useful.

9.6.5 EXCITONS

When an electron in a semiconductor or an insulator is excited (e.g., by light) from the valence band to the conduction band, it leaves a hole in the valence band. As we have seen, the hole behaves in many respects as a positively charge particle. An electron and a hole attract each other, and this can lead to a bound state, referred to as an *exciton*. In other words, an exciton is a Coulomb-correlated electron-hole pair that exists within a solid. As a zeroth order approximation, excitons may be regarded as a variant of the hydrogen atom, but the presence of the crystal environment modifies this simple picture. Excitons can be regarded as elementary excitations of a condensed matter system that can transport energy but does not transport net electric charge.

The interaction between the electron and the hole, which is responsible for exciton binding, is different from the interaction between two charges of opposite sign in free space. However, in a crude approximation, one may regard the exciton as a two-body problem where the two oppositely charged particles interact via an attractive Coulomb potential. The spatial extent of the exciton wave function depends on the strength of the attractive Coulomb interaction, which, in turn, depends on the degree of screening within the crystal. If the Coulomb interaction is not effectively screened and its energy exceeds (in magnitude) the kinetic energy of the electrons, then the exciton is formed within a single-crystal cell. This is the case in alkali-halide crystals and in many aromatic molecule crystals. This kind of bound electron-hole pair is not completely localized, since it can move from cell to cell and also correlates with pairs in other cells. Such a system is referred to as a *Frenkel exciton*. The binding energy of a Frenkel exciton, defined as the energy of its ionization leading to a non-correlated electron-hole pair, is of order 100-300 meV, and its spatial extent is of the order of the lattice constant. On the other hand, in many semiconductors, the Coulomb interaction is effectively screened. Then, the extent of the exciton wave function is a few tens of lattice constants and its binding energy is small (a few millielectron volt). This is referred to as a *Wannier–Mott* exciton. Figure 9.77 qualitatively illustrates Frenkel and Wannier–Mott excitons.

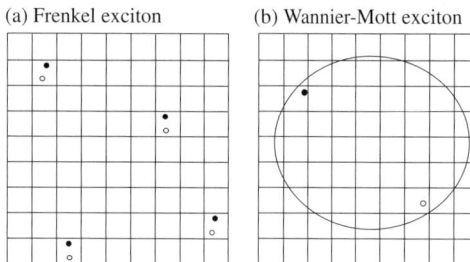

FIG 9.77 Schematic illustration of the spatial extent of excitons in a crystal. (a) Frenkel excitons occur in molecular crystals, where the Coulomb interaction is not effectively screened. Electrons (filled black circles) and holes (empty circles) form a Coulomb correlated pair whose extent is of the order of a lattice cell. Frenkel excitons can move in the crystal and also correlate with each other. (b) Wannier–Mott excitons occur in semiconductors, where the Coulomb interaction is effectively screened. Wannier–Mott excitons extend over many lattice cells.

We again stress that considering an exciton as a two-body problem is a crude approximation; an exciton is a many-body phenomenon that involves other electrons and crystal atoms. Excitons can be viewed as fictitious particles, referred

9.6 Semiconductors

to as *quasi-particles*, which result from low energy many-body *elementary excitations*. This topic will be introduced in Sec. 9.8. Other types of excitons such as *charge transfer excitons* will not be discussed.

Wannier–Mott Excitons

Consider a direct bandgap semiconductor where the valence and conduction band dispersions are parabolic, with minimum and maximum at $\mathbf{k} = 0$. The electron and hole energies are written as

$$E_c(\mathbf{k}) = \frac{\hbar^2 k^2}{2m_e^*}, \quad E_v(\mathbf{k}) = -E_g - \frac{\hbar^2 k^2}{2m_h^*}. \quad (9.403)$$

where m_e^* and m_h^* are the electron and hole effective masses and the reference energy is the bottom of the conduction band, $E_c(0) = 0$.

Problem 9.58

(a) Calculate the energy for one electron excitation from the valence band at momentum $\mathbf{K} - \mathbf{k}$ to the conduction band at momentum $\mathbf{k}$.

Answer: $E(\mathbf{k}, \mathbf{K}) \equiv E_c(\mathbf{k}) - E_v(\mathbf{K} - \mathbf{k}) = \varepsilon(\mathbf{K}) + \frac{\hbar^2 q^2}{2m^*}$ with

$$\varepsilon(\mathbf{K}) \equiv E_g + \frac{\hbar^2 K^2}{2M}, \quad M = m_e^* + m_h^*, \quad m^* = \frac{m_e^* m_h^*}{M}, \quad \mathbf{q} \equiv \mathbf{k} - \frac{m_e^*}{M}\mathbf{K}.$$

(b) Show that $\mathbf{q}$ is the relative momentum of the electron-hole pair.

Although this problem explicitly involves only an electron and a hole, the many-body aspects, such as Coulomb potentials from all the nuclei and the other electrons, are implicitly incorporated into the periodic potential and in the self-consistent calculation of the energy bands. If the electron-hole system is regarded as an exciton moving in a medium of dielectric constant ϵ, the energy levels can be calculated using the Coulomb potential, $-e^2/(\epsilon r)$, where $r = |\mathbf{r}_e - \mathbf{r}_h|$. The Schrödinger equation in the center of mass frame is that of a hydrogenic atom, for which the discrete energies are,

$$\varepsilon_{nlm} = -\frac{m^* e^4}{2\epsilon^2 \hbar^2} \frac{1}{n^2}, \quad (n = 1, 2, \ldots).$$

In addition, a continuum part of the spectrum is composed of energies above the bound states, $\varepsilon_{qlm} \approx \frac{\hbar^2 q^2}{2m^*}$. Denoting the internal energy of the electron-hole pair by ε_{eh} (either ε_{nlm} or ε_{qlm}), the excitation energy without interaction $E(\mathbf{k}, \mathbf{K})$ is calculated in Problem 9.58. The total energy including interaction for the bound exciton states becomes

$$E_{\mathbf{K}nlm} = \varepsilon(\mathbf{K}) + \varepsilon_{nlm}. \quad (9.404)$$

The exciton spectrum is schematically illustrated in Fig. 9.78. It can be observed in optical absorption experiments (see discussion later in Sec. 9.6.9). The wavevector of the photons is small compared with the reciprocal lattice vector, and therefore, the excited states near $\mathbf{K} = 0$ give the dominant contribution to the absorption process. The assumptions used in the above analysis are the effective mass approximation and the constant dielectric model of *Coulomb screening*. The physical

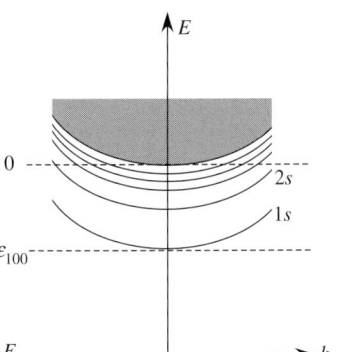

FIG 9.78 Spectrum of an electron-hole pair in an insulator as a function of the center of mass momentum K.

Table 9.5 *Several materials which accommodate Wannier–Mott Excitons and their relevant physical parameters.*

Semiconductor	E_g(eV)	m^*/m_e	E_B(eV)	a_B (Å)
PbTe	0.17	0.024	0.01	17,000
InSb	0.237	0.014	0.5	860
Ge	0.89	0.038	1.4	360
GaAs	1.52	0.066	4.1	150
InP	1.42	0.078	5.0	140
CdTe	1.61	0.089	10.6	80
ZnSe	2.82	0.13	20.4	60

origin of screening is the accumulation of opposite charges in the vicinity of the point charge. The quantitative analysis of screening, which is a many-body effect, will not be discussed here; the reader is referred to Chapter 16 in Ashcroft and Mermin [122]. The Wannier–Mott limit is justified, if

$$a_B \equiv \frac{\epsilon \hbar^2}{m^* e^2} \gg a_0, \tag{9.405}$$

i.e., if the exciton Bohr radius is much larger than the Bohr radius. It is well satisfied in narrow-gap semiconductors, where $\epsilon \gg 1$ and $m^* \ll m_e$. In GaAs, $\epsilon = 12.85$, $m^* = 0.067 m_e$, so $a_{0,\text{eff}} \approx 100$ Å, and $E_{n=1} \approx -5.5$ meV. A short list of semiconductors for which Wannier–Mott excitons that have been experimentally established is given in Table 9.5.

Frenkel Excitons

The approximations made in the analysis of the Wannier–Mott exciton are not justified in crystals with small dielectric constant and large effective masses m_e^* and m_h^*, such as *molecular crystals*, where the exciton radii are smaller than the lattice spacing. Then, the Coulomb interaction is not effectively screened and $a_B \simeq a_0$. The low n, $l = 0$ wave functions for the relative electron-hole motion is well localized, and the electron–hole pair effectively belong to a single lattice cell. This kind of exciton is called a *Frenkel exciton*. However, this notion of localized object is misleading, because any ion can be excited and the excitation energy can be transferred due to the strong coupling between the outer shells of the ions. In other words, a Frenkel exciton can propagate through the crystal while the ions themselves do not propagate. In addition to the relative motion, there is also translational motion of the center of mass, $\mathbf{R} = (m_e^* \mathbf{r}_e + m_h^* \mathbf{r}_h)/(m_e^* + m_h^*)$, of the electron-hole pair with wavevector $\mathbf{K}$. Thus, one can think of the Frenkel exciton as a well-localized molecular excitation, which is a superposition of states in which a single atom is excited while all other atoms stay in their ground state, and which propagates with wavevector $\mathbf{K}$ (with factor $e^{i\mathbf{K} \cdot \mathbf{R}_n}$), as shown in Fig. 9.79.

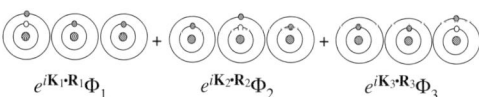

FIG 9.79 Illustration of a Frenkel exciton in a molecular "lattice" composed of $N = 3$ hydrogen-like molecules located at lattice points $\mathbf{R}_1, \mathbf{R}_2, \mathbf{R}_3$. A state Φ_n is such that $N - 1 = 2$ molecules are in a $1s$ state and one molecule at $\mathbf{R}_n$ is in a $2p$ state. The Frenkel exciton is a quantum mechanical superposition of N such states with coefficients $e^{i\mathbf{K} \cdot \mathbf{R}_n}$.

We now derive an expression for the unperturbed Frenkel exciton wave function. We first neglect the Coulomb interaction between electrons on different molecules and then add it as a perturbation. Let Φ_n denote a many-body state corresponding to the configuration in which the excited electron belongs to molecule located at $\mathbf{R}_n$. In molecular crystals, electrons are strongly localized about the molecules and are best described in terms of Wannier functions (see Sec. 9.4.8). Let us follow the description in Fig. 9.79, assuming a

9.6 Semiconductors

single electron for each molecule and neglecting the spin degrees of freedom. Although the molecular energy levels are affected by the crystal field, they will be labeled using the atomic nomenclature $1s, 2p, \ldots$. We denote by $w_\alpha(\mathbf{r}_i - \mathbf{R}_i)$, the Wannier function for an electron in level α ($\alpha = 1s, 2p$) localized near molecule i ($i = 1, 2, \ldots, N$). As an example, referring to the left configuration of Fig. 9.79 (corresponding to Φ_1), the Wannier function of the electron on the excited molecule 1 (on the left) is $w_{2p}(\mathbf{r}_1 - \mathbf{R}_1)$, while on molecules 2,3, the Wannier functions are $w_{1s}(\mathbf{r}_i - \mathbf{R}_i)$ ($i = 2, 3$). The function $\Phi_1(\mathbf{r}_1, \mathbf{r}_2, \mathbf{r}_3)$ must be antisymmetric in the coordinates of all three electrons. After constructing the N antisymmetric functions Φ_n (N is the number of lattice points), we obtain the unperturbed Frenkel exciton wave function,

$$\Psi_\mathbf{K} = \frac{1}{\sqrt{N}} \sum_n e^{i\mathbf{K} \cdot \mathbf{R}_n} \Phi_n, \tag{9.406}$$

which is antisymmetric.

Problem 9.59

Use the three Wannier functions constructed for the system in Fig. 9.79 to construct a function $\Phi_1(\mathbf{r}_1, \mathbf{r}_2, \mathbf{r}_3)$ of the three-electron system which is antisymmetric under the exchange of any two position vectors $\mathbf{r}_i \leftrightarrow \mathbf{r}_j$.

Answer:

$$\Phi_1 = \frac{1}{\sqrt{6}} \begin{vmatrix} w_{2p}(\mathbf{r}_1 - \mathbf{R}_1) & w_{1s}(\mathbf{r}_1 - \mathbf{R}_2) & w_{1s}(\mathbf{r}_1 - \mathbf{R}_3) \\ w_{2p}(\mathbf{r}_2 - \mathbf{R}_1) & w_{1s}(\mathbf{r}_2 - \mathbf{R}_2) & w_{1s}(\mathbf{r}_2 - \mathbf{R}_3) \\ w_{2p}(\mathbf{r}_3 - \mathbf{R}_1) & w_{1s}(\mathbf{r}_3 - \mathbf{R}_2) & w_{1s}(\mathbf{r}_3 - \mathbf{R}_3) \end{vmatrix}.$$

Subscript 1 in Φ_1 indicates that the excited electron belongs to site 1. Now write a Slater determinant function Φ_n for an arbitrary number N of atoms, with an excited electron on molecule n.

Next, we calculate the exciton energy. The unperturbed excitation energy $\varepsilon = E_{2p} - E_{1s}$ is assumed to be known. Thus, when the molecules (or atoms) are far apart, $\Psi_\mathbf{K}$ is an exact wave function describing exciton propagation at energy ε. The contribution to the exciton energy of the electron–electron interaction and the kinetic energy transfer from site to site can be calculated to first order using standard perturbation theory,

$$\Delta_C \equiv \left\langle \Psi_\mathbf{K} \left| \frac{1}{2} \sum_{m \neq n} \frac{e^2}{|\mathbf{r}_m - \mathbf{r}_n|} \right| \Psi_\mathbf{K} \right\rangle. \tag{9.407}$$

This expression can be decomposed into a double sum depending on $\mathbf{R}_n$ and $\mathbf{R}_m$, using the orthogonality of the Wannier functions. For example, the contributions to energy transfer are given by the following matrix elements:

$$U_{mn}(\mathbf{R}_n, \mathbf{R}_m) \equiv \int d\mathbf{r}_m d\mathbf{r}_n \, w_\alpha^*(\mathbf{r}_n - \mathbf{R}_n) w_\beta^*(\mathbf{r}_m - \mathbf{R}_m) \frac{e^2}{|\mathbf{r}_m - \mathbf{r}_n|} \times w_\beta(\mathbf{r}_n - \mathbf{R}_n) w_\alpha(\mathbf{r}_m - \mathbf{R}_m), \tag{9.408}$$

where $\alpha = 1p$ and $\beta = 1s$. The fact that matrix elements of the sum of two-body potentials of N-body wave functions [as in Eq. (9.407)] can be written in terms of four single-particle wave functions [as in Eq. (9.408)], because the Slater determinant is a sum of products of single-particle wave functions. For large separation distance $|\mathbf{R}_{nm}| \equiv |\mathbf{R}_n - \mathbf{R}_m|$, the denominator $|\mathbf{r}_n - \mathbf{r}_m|$ is treated by multipole expansion,

$$U_{mn}(\mathbf{R}_n, \mathbf{R}_m) \approx D(\mathbf{R}_{nm}) = \frac{1}{R_{nm}^3} \left[d^2 - 3 \frac{(\mathbf{d} \cdot \mathbf{R}_{nm})^2}{R_{nm}^2} \right], \tag{9.409}$$

where **d** is the intra-atomic (or molecular) electric dipole matrix element,

$$\mathbf{d} = -e \int d\mathbf{r}\, w_\alpha^*(\mathbf{r})\, \mathbf{r}\, w_\beta(\mathbf{r}). \tag{9.410}$$

Performing the sum over n and m, the first order in e^2 exciton energy is

$$E_\mathbf{K} = \varepsilon + \sum_{n \neq 0} e^{-i\mathbf{K}\cdot\mathbf{R}_n} D(\mathbf{R}_n) = \varepsilon + D(\mathbf{K}), \tag{9.411}$$

where $D(\mathbf{K})$ is the Fourier transform of $D(\mathbf{R})$. When $Ka_0 \ll 1$, the sum can be approximated by an integral, with the result (up to an additive constant)

$$D(\mathbf{K}) = -\frac{4\pi n}{3}\left(d^2 - 3\frac{(\mathbf{d}\cdot\mathbf{K})^2}{K^2}\right), \tag{9.412}$$

where n is the density of atoms or molecules. The case $\mathbf{K} \parallel \mathbf{d}$ ($\mathbf{K} \perp \mathbf{d}$) is referred to as a longitudinal (transverse) Frenkel exciton.

Problem 9.60

Show that if the atom is not excited, its electric dipole matrix element is zero.

Hint: Start from Eq. (9.410) with $\alpha = \beta$ and use the fact that the ground state of the atom is invariant under the parity operation $\mathbf{r} \to -\mathbf{r}$.

9.6.6 SPIN–ORBIT COUPLING IN SOLIDS

We already discussed spin–orbit coupling in Sec. 4.5, where we considered the fine structure of atoms. The spin–orbit coupling was derived by Lorentz by transforming the electric field of the atomic nucleus into the rest frame of the electron and accounting for the interaction of the magnetic field with the magnetic moment of the electron. Spin–orbit coupling is manifested in the splitting of atomic spectral lines from the radiative decay of excited states that are not s-states. We also encountered the spin–orbit potential in the Aharonov–Casher effect in Sec. 9.5.2. The analysis in the present section closely follows that of Ref. [152].

Spin–orbit coupling is important in many semiconductors. Consider an electron with effective mass m^* in a solid, subject to a potential $V(\mathbf{r})$. The potential induces an electric field $\mathbf{E} = \nabla V/e$, which, in turn, generates spin–orbit coupling. The spin–orbit Hamiltonian is [see Eq. (9.203)]

$$H_{\text{so}} = \lambda[\nabla V \cdot (\mathbf{p}\times\boldsymbol{\sigma}) + (\mathbf{p}\times\boldsymbol{\sigma})\cdot\nabla V], \tag{9.413}$$

where $\mathbf{p} = -i\hbar\nabla$ is the momentum operator, $\mathbf{S} = \hbar\boldsymbol{\sigma}/2$ is the spin operator, and

$$\lambda = -\frac{\hbar}{8(m^*)^2 c^2} \tag{9.414}$$

is the spin-orbit strength (for $m^* = m_e$, $\hbar\lambda \approx -1.85\times 10^{-6}\,\text{Å}^2$). The magnitude of the spin-orbit coupling depends on the speed of the electrons and the strength of the electric field acting on it, as well as on its effective mass, m^*. Large spin-orbit interaction is obtained when the Bloch electrons move close to the nuclei with velocities that are close to relativistic, and where the effective mass of the electrons is small.

The potential experienced by an electron in a crystal includes the periodic crystal part $V_c(\mathbf{r})$ and an "external" potential $V_{\text{ext}}(\mathbf{r})$ due to imperfections and any other external fields, $V(\mathbf{r}) = V_c(\mathbf{r}) + V_{\text{ext}}(\mathbf{r})$. Both parts of the potential affect the

9.6 Semiconductors

spin-orbit coupling, since the electric field in H_{so} is the gradient of the full potential. The full single-particle Hamiltonian, including spin-orbit, is

$$H = -\frac{\hbar^2}{2m^*}\nabla^2 + V(\mathbf{r}) + H_{so}. \qquad (9.415)$$

It is useful to absorb all the terms having the symmetry of the crystal into one term, which includes the intrinsic (crystal-field related) spin-orbit potential,

$$H_c \equiv -\frac{\hbar^2}{2m^*}\nabla^2 + V_c(\mathbf{r}) + \lambda[\nabla V_c(\mathbf{r}) \cdot (\mathbf{p} \times \boldsymbol{\sigma}) + (\mathbf{p} \times \boldsymbol{\sigma}) \cdot \nabla V_c(\mathbf{r})]. \qquad (9.416)$$

This intrinsic Hamiltonian has band energies $E_n(\mathbf{k})$.

Spin-orbit coupling is invariant under time reversal. For electrons and holes, the Kramers theorem holds, i.e., for an odd number of electrons, each energy level is at least two-fold degenerate.

Let us consider electrons in a cubic direct gap semiconductor, where the energy has a minimum at the center of the Brillouin zone, $\mathbf{k} = 0$. Because the spin-orbit interaction is even under time reversal, each level is at least two-fold degenerate (Kramers theorem). At $\mathbf{k} = 0$, this is the only degeneracy. The two states corresponding to this degenerate level are referred to as a *Kramers doublet*. We can write an effective Hamiltonian for the system as,

$$H_{eff} = E_n(\mathbf{k}) + H_{int} + \lambda[\nabla V_{ext}(\mathbf{r}) \cdot (\mathbf{p} \times \boldsymbol{\sigma}) + (\mathbf{p} \times \boldsymbol{\sigma}) \cdot \nabla V_{ext}(\mathbf{r})], \qquad (9.417)$$

where the spin-orbit energy due to the periodic potential is

$$H_{int} \equiv \mu_B \mathbf{h}(\mathbf{k}) \cdot \boldsymbol{\sigma}, \qquad (9.418)$$

with $\mathbf{h}(\mathbf{k})$ being the *intrinsic spin-orbit field* that depends on the details of the crystal structure, $\mathbf{h}(\mathbf{k})$ has units of magnetic field since μ_B has units of energy/magnetic-field, and is $\mathbf{k}$-dependent. Note that the time-reversal operation takes $\boldsymbol{\sigma} \to -\boldsymbol{\sigma}$ and $\mathbf{k} \to -\mathbf{k}$. Hence $\mathbf{h}(\mathbf{k}) \to -\mathbf{h}(-\mathbf{k})$, so that the intrinsic Hamiltonian H_{int} is time-reversal invariant. In crystals with inversion symmetry, where $\mathbf{h}(\mathbf{k}) = \mathbf{h}(-\mathbf{k})$, the intrinsic spin-orbit field must vanish. Some examples will be given later after the effect of spin-orbit on the band energies $E_n(\mathbf{k})$ is clarified.

Problem 9.61

Determine the energies and wave functions for an electron moving in the x–y plane subject to a constant electric field $\mathbf{E} = \mathcal{E}\hat{\mathbf{z}}$.

Guidance: Start from the Schrödinger equation with H given by Eq. (9.415) and check that $[H, \mathbf{p}] = 0$. The Hamiltonian is translation invariant, so $\psi(x, y) = Ce^{i(kx+qy)}v(k, q)$, where C is a normalization constant and $v(k, q)$ is a two-component spinor. After substituting in the Schrödinger equation, you should get a 2×2 eigenvalue problem, $A(k, q)v(k, q) = Ev(k, q)$. The solution yields the eigenvalues $E_1(k, q)$ and $E_2(k, q)$ and the eigenvectors.

Let us estimate the strength of spin-orbit effects on the structure of the energy bands $E_n(\mathbf{k})$ of the semiconductor GaAs, whose crystal structure is zinc-blende, see Fig. 9.14(b). In the tight-binding picture, electrons near the band extrema are described by Wannier functions that are similar to molecular wave functions but are affected also by other molecules in the crystal. The electron wave functions at the top of the valence band are p-orbitals, because the corresponding molecular Wannier function is in a state of angular momentum $l = 1$. In the absence of spin-orbit coupling, there are $2(2l + 1)$ degenerate levels, but on including spin-orbit, then at the Γ point (the top of the valence band at $\mathbf{k} = 0$), the degeneracy is partially lifted and states with total angular momentum $j = 3/2$ have higher energy than states with $j = 1/2$. The energy difference is referred to as *spin-orbit gap* or *split-off gap* Δ_0. We shall see later that for $\mathbf{k} \neq 0$, there is further splitting. States with $j = 3/2$ and $j_z = \pm 3/2$ have larger effective mass than states with $j = 3/2$ and $j_z = \pm 1/2$, and hence, we may refer to the quasi-particles in these two groups as *heavy and light holes*, respectively. The energy of the lowest (spin-orbit) band with $j = j_z = 1/2$ is Δ_0 below the light- and heavy-hole bands. Finally, the conduction band states are s-wave orbitals, which are unaffected by spin-orbit coupling. At the Γ point ($\mathbf{k} = 0$), each group of states form a basis of an irreducible representation of the point group of zinc-blende structures, which is denoted by

$$\Gamma_6 = \left|\tfrac{1}{2}, \tfrac{1}{2}\right\rangle, \left|\tfrac{1}{2}, -\tfrac{1}{2}\right\rangle, \quad \Gamma_7 = \left|\tfrac{1}{2}, \tfrac{1}{2}\right\rangle, \left|\tfrac{1}{2}, -\tfrac{1}{2}\right\rangle, \quad \Gamma_8 = \left|\tfrac{3}{2}, \tfrac{1}{2}\right\rangle, \left|\tfrac{3}{2}, -\tfrac{1}{2}\right\rangle, \left|\tfrac{3}{2}, \tfrac{3}{2}\right\rangle, \left|\tfrac{3}{2}, -\tfrac{3}{2}\right\rangle. \tag{9.419}$$

Γ_6 refers to the conduction band while Γ_7 and Γ_8 refer to the valence band.

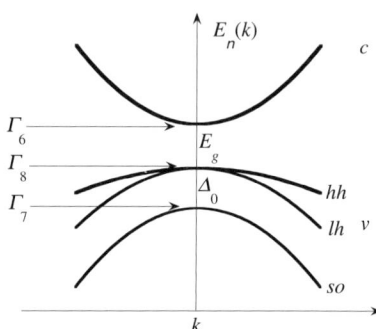

FIG 9.80 Schematic drawing of the energy bands in zinc-blende semiconductors near the Γ point. The conduction band originates from a doubly degenerate s orbital, whereas the valence band originates from a p orbital, and consists of three doubly degenerate bands, the heavy hole (hh) band with $(j, j_z) = (3/2, \pm 3/2)$, the light hole ($lh$) band with $(j, j_z) = (3/2, \pm 1/2)$, and the spin-orbit split-off (so) band with $(j, j_z) = (1/2, \pm 1/2)$, sometimes called the *lower heavy hole* band. The Γ symbols designate the angular momentum states $|jj_z\rangle$ of the electrons in the bands.

In Fig. 9.80, the band energies $E_n(\mathbf{k})$ are plotted as a function of the Bloch momentum $\mathbf{k}$ near the Γ point (along the direction to the K point). The Γ_8 valence band is four-fold degenerate at $\mathbf{k} = 0$ but split off into two doubly degenerate (heavy and light hole) bands for $\mathbf{k} \neq 0$ due to reduced symmetry.

9.6.7 k · p PERTURBATION THEORY

The behavior of the eigenvalues and eigenfunctions of the electronic Hamiltonian near the band extrema (the Γ points) in reciprocal lattice space is important for electronic conduction in semiconductors. One technique for calculating this behavior is $\mathbf{k} \cdot \mathbf{p}$ perturbation theory. The essence of the formalism is to treat the $\mathbf{k} \cdot \mathbf{p}$ term in the Hamiltonian as a small perturbation. The idea is to compute the Bloch wave functions $u_{n\mathbf{k}}(\mathbf{r})$ and energies $E_n(\mathbf{k})$ at small k, starting from $u_{n,\mathbf{k}=0}(\mathbf{r})$ and $E_n(\mathbf{k}=0)$. Problem 9.21 showed that for a spinless particle in the periodic potential $V(\mathbf{r})$, where the wave function is written in the Bloch form, $\psi_{n\mathbf{k}}(\mathbf{r}) = e^{i\mathbf{k} \cdot \mathbf{r}} u_{\mathbf{k}}(\mathbf{r})$, the equation for the Bloch function $u_{\mathbf{k}}(\mathbf{r})$ is

$$\left[-\frac{\hbar^2}{2m^*}\nabla^2 + \frac{\hbar}{m^*}\mathbf{k} \cdot \mathbf{p} + \frac{\hbar^2 k^2}{2m^*} + V(\mathbf{r})\right] u_{n\mathbf{k}}(\mathbf{r}) = E_n(\mathbf{k}) u_{n\mathbf{k}}(\mathbf{r}). \tag{9.420}$$

Problem 9.62

In the Schrödinger equation $H\psi(\mathbf{r}) = \varepsilon \psi(\mathbf{r})$ with the Hamiltonian (9.415), substitute $\psi_{\nu\mathbf{k}}(\mathbf{r}) = e^{i\mathbf{k} \cdot \mathbf{r}} v_{\nu\mathbf{k}}(\mathbf{r})$, where $\psi(\mathbf{r})$ and $v_{\nu\mathbf{k}}(\mathbf{r})$ are two component spinors. Show that the equation for the spinor $v_{\nu\mathbf{k}}(\mathbf{r})$ is

$$\left[-\frac{\hbar^2}{2m^*}\nabla^2 + \frac{\hbar}{m^*}\mathbf{k} \cdot \boldsymbol{\pi} + \frac{\hbar^2 k^2}{2m^*} + V(\mathbf{r}) + H_{so}\right] v_{\nu\mathbf{k}}(\mathbf{r}) = \varepsilon_\nu(\mathbf{k}) u_{\nu\mathbf{k}}(\mathbf{r}), \tag{9.421}$$

where $\boldsymbol{\pi} \equiv \mathbf{p} + 2m^*\lambda \boldsymbol{\sigma} \times \nabla V$.

Verify that the second term has the dimension of momentum.

Answer: $[\lambda] = L^2/\hbar$ [see after Eq. (9.414)]. $\nabla V = $ Energy/L. $(m^* L^2 E)/(\hbar L) = m^* L/T = m^* v = p$.

In the following, we use bra-ket notation: $u_{n\mathbf{k}}(\mathbf{r}) = \langle \mathbf{r} | n\mathbf{k} \rangle$ and $v_{\nu\mathbf{k}}(\mathbf{r}) = \langle \mathbf{r} | \nu\mathbf{k} \rangle$. For any $\mathbf{k}$, the kets $|n\mathbf{k}\rangle$ form a complete set of states. This is true in particular, for $\mathbf{k} = 0$, where we set $|n\rangle \equiv |n\mathbf{k} = 0\rangle$. The spinors $|n\sigma\rangle \equiv |n\rangle \otimes |\sigma\rangle$ form a complete set in spinor space. In the perturbation theory, we use these states as an unperturbed basis, and H_{so}

9.6 Semiconductors

is treated as a perturbation. Note that while $|n\sigma\rangle$ are eigenfunctions of σ_z, the kets $|\nu\mathbf{k}\rangle$ are not, since spin rotation invariance is violated in the presence of spin-orbit coupling. Therefore, the kets $|\nu\mathbf{k}\rangle$ are expanded as

$$|\nu\mathbf{k}\rangle = \sum_{n\sigma} c_{\nu n\sigma}(\mathbf{k})|n\sigma\rangle. \tag{9.422}$$

Substituting Eqs (9.422) into (9.421) and operating on the left by the bra $\langle m\sigma'|$, we obtain,

$$\sum_{n\sigma}\left[\left(E_n(0)+\frac{\hbar^2 k^2}{2m^*}\right)\delta_{mn}\delta_{\sigma'\sigma}+\frac{\hbar}{m^*}\mathbf{k}\cdot\mathbf{P}_{m\sigma';n\sigma}+D_{m\sigma';n\sigma}\right]c_{\nu n\sigma}(\mathbf{k})=\varepsilon_m(\mathbf{k})c_{\nu m\sigma'}(\mathbf{k}), \tag{9.423}$$

where

$$\mathbf{P}_{m\sigma';n\sigma}=\langle m\sigma'|\boldsymbol{\pi}|n\sigma\rangle, \quad D_{m\sigma';n\sigma}=\lambda\langle m\sigma'|\mathbf{p}\cdot\boldsymbol{\sigma}\times(\nabla V)|n\sigma\rangle. \tag{9.424}$$

The nondiagonal terms that couple different basis states $|n\sigma\rangle$ are the $\mathbf{k}\cdot\mathbf{P}$ term $\frac{\hbar}{m^*}\mathbf{k}\cdot\mathbf{P}_{m\sigma';n\sigma}$ and the spin-orbit term $D_{m\sigma';n\sigma}$. The latter is responsible for band splitting even at $\mathbf{k}=0$ where the $\mathbf{k}\cdot\mathbf{P}$ term vanishes.

The band structure near $\mathbf{k}=0$ for the zinc-blende structure, shown in Fig. 9.80, is due to the $\mathbf{k}\cdot\mathbf{P}$ and the spin-orbit terms. The calculations of the SO splitting must take into account the symmetry of the wave functions at the Γ point. The $\mathbf{k}\cdot\mathbf{p}$ perturbation theory results in a matrix eigenvalue problem, Eq. (9.423), and yields the band energies $\varepsilon_\nu(\mathbf{k})$ and the wave function coefficients $c_{\nu m\sigma'}(\mathbf{k})$ for the representations Γ_6, Γ_7, and Γ_8. We are interested only in a few adjacent bands, $\nu=1,2,\ldots,N$, and in their energies $\varepsilon_\nu(\mathbf{k})$ near $\mathbf{k}=0$ (the Γ point). In this case, the system of equations (9.423) is cut off at N and the other states $|n\sigma\rangle$ with $n>N$ can be included as an additional perturbation containing higher order terms in $\mathbf{k}$. This requires the *Schrieffer–Wolff transformation* [153, 154], which eliminates the high energy levels and renormalizes the low energy levels on which interest is focused.

The eigenvalue problem is then written as $\mathcal{H}\Psi=E\Psi$, where the effective Hamiltonian $\mathcal{H}$ is an $N\times N$ matrix acting in the N dimensional space of the bands, which have the point symmetry of the crystal. Focusing on Fig. 9.80, the total number of levels, Kramers degeneracy included, is 8. If all of them are kept, the 8×8 Hamiltonian matrix is known as the *Kane model*. Inclusion of more levels has also been considered, e.g., the 14×14 Kane model. However, in many cases, it is possible to restrict the discussion to fewer levels using the Schrieffer–Wolff technique. Consider the four-dimensional space ($N=4$) spanned by the light and heavy hole bands of the zinc-blende crystals (the space Γ_8 in Fig. 9.80). In terms of the angular momentum operator $\mathbf{J}$ with $j=3/2$, the effective 4×4 Hamiltonian is

$$\mathcal{H}=-\frac{\hbar^2}{2m^*}\left[\left(\gamma_1+\frac{5}{2}\gamma_2\right)k^2-2\gamma_3(\mathbf{J}\cdot\mathbf{k})^2+2(\gamma_3-\gamma_2)\left(J_x^2 k_x^2+J_y^2 k_y^2+J_z^2 k_z^2\right)\right]. \tag{9.425}$$

This is known as the *Luttinger (or Luttinger–Kohn) Hamiltonian*. The real and dimensionless Luttinger parameters, $\gamma_1,\gamma_2,\gamma_3$, have been determined for many specific semiconductor materials from experiment or *ab initio* calculations. The eigenvalues of $\mathcal{H}$ are still pairwise degenerate, as indicated by the two (instead of four) energy Γ_8 bands in Fig. 9.80. Setting the zero of energy at the top of the valence band, we find, after diagonalization,

$$E_\nu(\mathbf{k})=-\frac{\hbar^2}{2m^*}\left[\gamma_1 k^2\mp\sqrt{4\gamma_2^2 k^4+144(\gamma_3^2-\gamma_2^2)^2\left[(k_x k_y)^2+(k_x k_z)^2+(k_y k_z)^2\right]}\right]. \tag{9.426}$$

The heavy-hole band (light-hole band) corresponds to the minus (plus) sign in the square brackets on the RHS of Eq. (9.426). The wave functions away from the Γ point can be identified with the $|jj_z\rangle$ basis functions introduced in Eq. (9.419) only if $\mathbf{k}\cdot\hat{\mathbf{z}}=0$.

Intrinsic Spin-Orbit Coupling: Dresselhaus and Rashba Potentials

Now consider the field $\mathbf{h}(\mathbf{k})$ in Eq. (9.418). Just as the Luttinger Hamiltonian (9.425) is derived from the general equation (9.423) for the analysis of the four levels at the top of the valence band, one can derive an effective Hamiltonian for the two Kramers degenerate levels at the bottom of the conduction band, starting from the same equation (9.423). Inversion symmetry exists for n-doped 3D semiconductors with III–V and II–V groups, thereby simplifying matters. For the zinc blend materials such as GaAs and InAs, the 3D *Dresselhaus term* is, $V_D(3D) = \mathbf{h}_D(\mathbf{k}) \cdot \boldsymbol{\sigma}$, with

$$\mathbf{h}_D(\mathbf{k}) = \alpha_D[k_x(k_y^2 - k_z^2), k_y(k_z^2 - k_x^2), k_z(k_x^2 - k_y^2)] \quad \text{(in 3D)}, \tag{9.427}$$

where the value of α_D can be estimated from experimental data (≈ 20 eV Å^3). In two-dimensional systems confined along the [001] direction, the average $\langle \mathbf{h}_D(\mathbf{k}) \rangle_z$ should be taken, with $\langle k_z \rangle \approx 0$ and $\langle k_z^2 \rangle \approx (\pi/d)^2$. The value of π/d depends on the confinement potential $V(z)$, and for strong confinement $\pi/d \gg k_F$. For the 2D Dresselhaus term, $V_D^{(2D)} = \mathbf{h}_D(\mathbf{k}) \cdot \boldsymbol{\sigma}$, where the two-component vector $\mathbf{h}_D(\mathbf{k})$ is,

$$\mathbf{h}_D(\mathbf{k}) = \alpha_D[(k_y^2 k_x, -k_x^2 k_y) - \left(\frac{\pi}{d}\right)^2 (k_x, -k_y)] \quad \text{(in 2D)}. \tag{9.428}$$

In 2D systems, the occurrence of an asymmetric confining potential $V(z)$ leads to another kind of spin-orbit coupling term which, for the conduction band, is directly obtained from Eq. (9.417) after averaging the third term on its RHS. The resulting *Rashba term* is, $V_R^{(2D)} = \mathbf{h}_R(\mathbf{k}) \cdot \boldsymbol{\sigma}$, with

$$\mathbf{h}_R(\mathbf{k}) = \alpha_R(k_y, -k_x) = \alpha_R \hat{\mathbf{z}} \times \mathbf{k} \quad \text{(in 2D)}, \tag{9.429}$$

and approximately, $\alpha_R \approx 0.1$ eV Å, as estimated from experimental data.

Now consider the effect of confinement on 2D p-doped semiconductors. In the scheme shown in Fig. 9.80, confinement lifts the two Kramers degenerate heavy-hole levels with $(J, J_z) = (3/2, \pm 3/2)$ above the two Kramers degenerate light-hole levels with $(J, J_z) = (3/2, \pm 1/2)$. Keeping only the heavy-hole levels and applying the same perturbation procedure for eliminating the light-hole levels from the Luttinger Hamiltonian, we arrive at the Rashba term for holes,

$$V_{R-h}(2D) = i\alpha_h (k_-^3 \sigma_+ - k_+^3 \sigma_-) \quad \text{(in 2D)}, \tag{9.430}$$

where $k_\pm = k_x \pm i k_y$, $\sigma_\pm = \sigma_x \pm i\sigma_y$, and $\boldsymbol{\sigma}$ is the pseudospin operator in the two-level heavy hole system (while the real spin of the holes is $J = 3/2$). Note that, unlike for electrons, Eq. (9.429), the Rashba Hamiltonian for heavy holes is cubic in k.

Problem 9.63

Assume that the Rashba and Dresselhaus spin-orbit couplings in 2D systems are given by,

$$V_R^{(2D)} = \alpha(k_x \sigma_y - k_y \sigma_x), \quad V_D^{(2D)} = \beta(k_y \sigma_y - k_x \sigma_x). \tag{9.431}$$

Solve the Schrödinger equation $H\psi(x,y) = E\psi(x,y)$ with the Hamiltonian,

$$\mathcal{H} = \frac{\hbar^2 k^2}{2m} + V_R^{(2D)} + V_D^{(2D)}, \quad (\mathbf{k} = -i\nabla). \tag{9.432}$$

Guidelines: Assume $\psi(x,y) = e^{i(q_x x + q_y y)} u(q_x, q_y)$, where u is a two-component spinor. This leads to an algebraic eigenvalue problem with 2×2 matrix. The solution is, $u_\pm = \frac{1}{\sqrt{2}} \begin{pmatrix} 1 \\ \pm e^{-i\theta} \end{pmatrix}$, $\tan\theta = \frac{\alpha q_x + \beta q_y}{\alpha q_y + \beta q_x}$.

$E_\pm(\mathbf{q}) = \frac{\hbar^2 q^2}{2m} \pm \sqrt{(\alpha q_y + \beta q_x)^2 + (\alpha q_x + \beta q_y)^2}$.

9.6 Semiconductors

The Extrinsic Spin-Orbit Potential

The origin of the extrinsic spin-orbit potential is the electric fields due to impurities and to externally applied electric fields. Its effect on the Kramers degenerate conduction level is calculated by starting from the 8×8 Kane Hamiltonian of the electron and hole levels appearing in Fig. 9.80 and by using the elimination procedure discussed earlier, which leaves only two conduction band levels. The result is given by Eq. (9.413), with $\hbar\lambda \approx 5.3$ Å^2 for GaAs, and 120 Å^2 for InAs, which is about six orders of magnitude stronger than in vacuum. In some cases, it may even have the opposite sign. This enhancement is important for manipulating electron spins by external fields.

It is also possible to explain the effect of extrinsic fields on the hole bands. For example, for a 2D system, the extrinsic contribution to the $J = \pm 3/2$ valence band yields an additional term,

$$V_{\text{ext-holes}} = \lambda_h J_z (\mathbf{p} \times \nabla V_{\text{ext}})_z, \tag{9.433}$$

which should be added to the Luttinger Hamiltonian (9.425), where V_{ext} is the external potential due to imprefections and any other external fields and $\hbar\lambda_h \approx 22$ Å^2 for GaAs.

Spin-Orbit Coupling and Spin Transport Mechanisms

Spin-orbit coupling plays a crucial role in spintronics, which we discussed in Sec. 9.7. It focuses on spin transport and spin accumulation. Here, we briefly mention the role of spin orbit in spin transport mechanisms. Examples of intrinsic and extrinsic mechanisms are given.

Intrinsic Spin-Orbit Coupling and Spin Precession. The intrinsic spin-orbit coupling $\mathbf{h}(\mathbf{k})$ defined through Eq. (9.418) is equivalent to a local magnetic field related to the band structure. $\mathbf{h}(\mathbf{k})$ depends strongly on crystal symmetry and is different for electrons and holes. It also depends on space dimension $d = 2$ or $d = 3$. In a static situation, the particle magnetic moment tends to align along $\mathbf{h}(\mathbf{k})$. When an electric field $\mathbf{E} = E\hat{\mathbf{x}}$ is applied, the vector $\mathbf{k}$ evolves in time according to the semiclassical equations $\dot{\mathbf{k}} = e\mathbf{E}/\hbar$, and consequently, the effective field $\mathbf{h}(\mathbf{k})$ evolves according to $\dot{\mathbf{h}}(\mathbf{k}) = \nabla_{\mathbf{k}} \mathbf{h}(\mathbf{k}) \cdot e\mathbf{E}/\hbar$. For weak electric fields, the adiabatic approximation is justified. Then, the effective magnetic field $\mathbf{h}(\mathbf{k})$ varies slowly with time, and the electron has enough time to adjust its spin to lie along $\mathbf{h}(\mathbf{k})$, thereby undergoing spin precession.

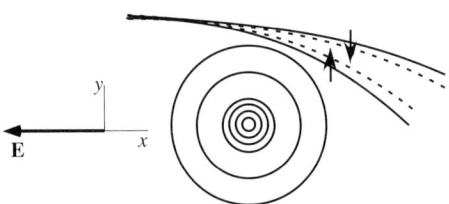

FIG 9.81 Skew-scattering and side-jump spin-orbit mechanisms of electron scattering from an attractive impurity potential of the form $V_{so} = \lambda \boldsymbol{\sigma} \cdot (\mathbf{k} \times \nabla V)$. An electric field $-E\hat{\mathbf{x}}$ drives electrons along $\hat{\mathbf{x}}$, where they encounter an attractive impurity. The classical trajectories are shown here. Skew scattering is manifested by different scattering angles for spin-up and spin-down electrons, whereas the side jump is manifested by a shift of the trajectories, which is radially inward for the down spin and outward for the up spin, as indicated by the dashed curves.

Extrinsic Spin Transport: Skew Scattering and Side Jump. The presence of extrinsic spin-orbit coupling leads to several spin transport mechanisms. They are manifested when an electron scatters from an impurity, whose potential $V(\mathbf{r})$ generates a spin-orbit force whose direction depends on the spin direction. For example, if the potential is central, $V(\mathbf{r}) = V(r)$, the spin-orbit potential has the standard form $\frac{\lambda}{r} \frac{dV}{dr} \mathbf{L} \cdot \mathbf{S}$ familiar from atomic physics. The cross-section for scattering from the impurity depends on spin orientation, a property known as *Mott skew scattering*. Skew scattering appears only for cubic orders of the potential or higher. Finally, there is another extrinsic mechanism known as the *side jump*, where the center of mass of a wave packet undergoes a discontinuous and finite sideways displacement due to scattering by a central potential in the presence of spin-orbit interaction. The direction of displacement with respect to the original trajectory depends on the spin direction. The skew scattering and the side jump mechanisms are schematically illustrated in Fig. 9.81.

9.6.8 SPIN HALL EFFECT

A remarkable phenomenon in semiconductors with strong spin-orbit coupling is the *spin Hall effect* (SHE), which was predicted by Dyakonov and Perel in 1971, but was observed experimentally only quite recently. In the SHE, a current flowing through a sample can lead to spin transport in a perpendicular direction, which leads to accumulation of spin polarization on the edges of a Hall-bar, in analogy with charge accumulation in the Hall effect. In contrast with the Hall effect, there is no magnetic field involved in the SHE. A schematic illustration stressing the difference between QHE and SHE is shown in Fig. 9.82.

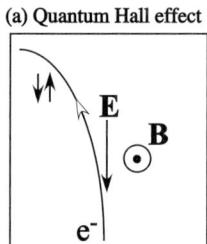
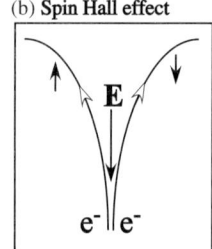

FIG 9.82 (a) The Hall effect. An in-plane electric field drives a current along the sample, and a perpendicular magnetic field leads to charge accumulation on the edges and to a transverse Hall voltage. (b) The spin Hall effect. An in-plane electric field drives a current along the sample, and the presence of strong spin-orbit coupling in the sample leads to accumulation of different spin polarizations on the edges. No Hall voltage is generated.

Depending on the origin of the spin-orbit potential, one distinguishes between *extrinsic* and *intrinsic* SHE. The extrinsic SHE is due to spin-orbit effects influencing spin-dependent scattering by static impurities. As discussed earlier, extrinsic spin-orbit coupling leads to spin transport mechanisms of skew scattering and side jump. The intrinsic SHE is due to bulk inversion asymmetry or structure asymmetry (strain) resulting from spin-orbit coupling terms in the single-particle carrier Hamiltonian, and it occurs even in the absence of scattering. This kind of spin transport mechanism is related to spin precession.

Definition of Spin Current

The definition of spin current is not as simple as that of charge current. The reasons are as follows: (1) the spin current cannot be easily observed experimentally and (2) unlike charge, spin is not conserved when the interactions depend on spin, as in the case with spin-orbit interaction. Consider a particle (e.g., an electron) whose spin operator is $\mathbf{S} = \hbar\boldsymbol{\sigma}/2$ and whose wave function $\psi(\mathbf{r})$ is a two-component spinor. The local charge density $\rho(\mathbf{r}) = e\psi^\dagger(\mathbf{r})\psi(\mathbf{r})$ and the charge current density $\mathbf{J}(\mathbf{r}) = e\text{Re}[\psi^\dagger(\mathbf{r})\mathbf{v}\psi(\mathbf{r})]$, where $\mathbf{v}$ is the velocity operator, are related by the continuity equation that results from charge conservation,

$$\nabla \cdot \mathbf{J} + \frac{\partial \rho}{\partial t} = 0. \quad (9.434)$$

In analogy with the charge density and the charge current density, the *local spin density* and the local spin current can be defined as

$$\mathbf{s}(\mathbf{r}) = \psi^\dagger(\mathbf{r})\mathbf{S}\psi(\mathbf{r}), \quad \mathbb{J}(\mathbf{r}) = \frac{1}{2}\text{Re}[\psi^\dagger(\mathbf{r})\{\mathbf{v},\boldsymbol{\sigma}\}\psi(\mathbf{r})]. \quad (9.435)$$

Note that we use lower case $\mathbf{s}(\mathbf{r})$ for spin density and upper case $\mathbf{S} = \hbar\boldsymbol{\sigma}/2$ for the spin operator. The spin density is a vector, whereas the spin current is a tensor having two Cartesian components, one for the direction of motion of the particles and one for polarization direction. The *spin current operator* is given by the anticommutator $\hat{\mathbb{J}}(\mathbf{r}) = \{\mathbf{v},\boldsymbol{\sigma}\}$. Note that the velocity operator $\mathbf{v}$ depends on spin (see Problem 9.64).

Problem 9.64

(a) Calculate the operator v_x for the 2D Rashba Hamiltonian, $H_R = \frac{\hbar^2 k^2}{2m} + \alpha_R(k_y\sigma_x - k_x\sigma_y)$, [see also Eq. (9.429)].
(b) Calculate the components of the spin current operator along x.

Answer: (a) $v_x = \frac{dx}{dt} = \frac{1}{i\hbar}[x, H_R] = -i\frac{\hbar k_x}{m} - \frac{\alpha_R}{i\hbar}[x, k_x\sigma_y] = -i\frac{\hbar k_x}{m} + \frac{\alpha_R}{\hbar}\sigma_y$.
(b) $\frac{1}{2}\{v_x,\sigma_x\} = -i\frac{\hbar k_x}{m}\sigma_x + i\frac{\alpha_R}{\hbar}\sigma_z$, $\frac{1}{2}\{v_x,\sigma_y\} = i\frac{\hbar k_x}{m}\sigma_y$, $\frac{1}{2}\{v_x,\sigma_z\} = i\frac{\hbar k_x}{m}\sigma_z - i\frac{\alpha_R}{\hbar}\sigma_x$.

9.6 Semiconductors

Equation (9.435) shows that spin is not conserved. The analog of the continuity equation for spin current and density is

$$\nabla \cdot \mathbb{J} + \frac{\partial \mathbf{s}(\mathbf{r})}{\partial t} = \psi^\dagger(\mathbf{r}) \frac{\partial \mathbf{S}}{\partial t} \psi(\mathbf{r}) = \psi^\dagger(\mathbf{r}) \frac{1}{i\hbar}[\mathbf{S}, H]\psi(\mathbf{r}) \equiv \mathbf{T}(\mathbf{r}), \qquad (9.436)$$

where $\mathbf{T}(\mathbf{r})$ is referred to as the *spin torque*. Thus, when $[\mathbf{S}, H] \ne 0$, the local spin current density $\mathbb{J}(\mathbf{r})$ is not conserved. In many realistic 3D systems, where inversion symmetry is present, the volume average of the torque vanishes, i.e., $\frac{1}{V} \int dV \mathbf{T}(\mathbf{r}) = 0$. This means that $\mathbf{T}(\mathbf{r})$ can be written as a divergence of a (tensor) function denoted as $\mathbb{P}(\mathbf{r})$ (a dipole torque density). Explicitly,

$$\mathbf{T}(\mathbf{r}) = -\nabla \cdot \mathbb{P}(\mathbf{r}). \qquad (9.437)$$

The quantity $\mathbb{J}(\mathbf{r}) + \mathbb{P}(\mathbf{r})$ then satisfies a continuity equation. Expressions for the spin Hall conductance for two important systems are presented later.

Spin Hall Conductance of Band Electrons in 2D

The relation of the spin Hall conductance to the electric field that drives the spin current is similar to that of the electric conductance with the electric field that drives the charge current. In the presence of spin-orbit interaction, an electric field along $\hat{\mathbf{x}}$ may lead to electrons moving along $\hat{\mathbf{y}}$ whose spin is polarized along $\hat{\mathbf{z}}$ (see Fig. 9.82). The current $\mathbb{J}_y^z$ is related to the field E_x by a component of the *spin Hall conductance* tensor, which has three components, two subscripts for the space directions and one superscript for the polarization direction,

$$\sigma_{xy}^z \equiv -\mathbb{J}_y^z / E_x. \qquad (9.438)$$

Once an expression for the spin current is established, and the eigenstates of the Hamiltonian are known, linear response theory relates the current to the electric field and yields an expression for the *spin conductance*. For example, consider a 2D system of electrons subject to Hamiltonian (9.432) in Problem 9.64. The energy eigenvalues are $E_\pm(\mathbf{q}) = \frac{\hbar^2 q^2}{2m} \pm \sqrt{(\alpha q_y + \beta q_x)^2 + (\alpha q_x + \beta q_y)^2}$ and eigenfunctions are $\langle x, y | \mathbf{q}\mu = \pm \rangle = e^{i(q_x x + q_y y)} u_\pm(q_x, q_y)$, where $u_\pm = \frac{1}{\sqrt{2}} \begin{pmatrix} 1 \\ \pm e^{-i\theta} \end{pmatrix}$, and $\tan \theta = \frac{\alpha q_x + \beta q_y}{\alpha q_y + \beta q_x}$. The linear response (Kubo) expression for the spin Hall conductance at finite frequency is,

$$\sigma_{xy}^z(\omega) = \frac{e}{\mathcal{V}(\omega + i\eta)} \int_0^\infty dt e^{i(\omega + i\eta)t} \sum_{\mathbf{q}\mu} f[E_\mu(\mathbf{q})] \langle \mathbf{q}\mu | [\hat{\mathbb{J}}_z^x(t), v_y(0)] | \mathbf{q}\mu \rangle, \qquad (9.439)$$

where $\mathcal{V}$ is the volume of the sample, $f(E)$ is the Fermi distribution, $\mu = \pm$, and the spin current and the velocity operators are taken in the Heisenberg representation. The DC conductance is obtained in the limit $\omega \to 0$, but this limiting procedure is far from simple. Equation (9.439) can be difficult to evaluate, but some simple cases can be easily calculated. For example, the Rashba spin-orbit Hamiltonian in 2D, $\mathcal{H} = \frac{\hbar^2 k^2}{2m} + \alpha(k_x \sigma_y - k_y \sigma_x)$, the spin Hall conductance is,

$$\sigma_{xy}^z(0) = -\sigma_{yx}^z(0) = \frac{e}{8\pi} \quad \text{(independent of } \alpha\text{)}. \qquad (9.440)$$

This result is somewhat useless because it does not take disorder, which is usually unavoidable, into account. Once it is included, the spin-orbit strength enters the spin conductance together with other energy scales such as the Fermi energy and the disorder strength. The latter is usually expressed as a ratio $\hbar/\tau$, where τ is the scattering rate,

$$\sigma_{xy}^z(0) = \frac{e}{8\pi} - \frac{e\hbar}{32\pi \tau \varepsilon_R} \arctan\left[\frac{4\varepsilon_R \tau}{\hbar}\left(1 + \frac{8\varepsilon_R \varepsilon_F \tau^2}{\hbar^2}\right)^{-1}\right], \qquad (9.441)$$

where the quantity $\varepsilon_R \equiv m\alpha_R^2/\hbar^2$ is called the *Rashba energy*.

Spin Hall Conductance of Holes in 3D

An important result concerns the spin Hall conductance of a 3D-hole system described by the Luttinger Hamiltonian equation (9.425). Application of the Kubo formula and inclusion of disorder of strength $\hbar/\tau$, the expression for the spin Hall conductance is determined by the Luttinger parameters γ_1, γ_2 and the wavenumbers of the heavy and light holes

$$k_{h/l} = \frac{k_F}{\sqrt{\gamma_1 \mp 2\gamma_2}}, \qquad (9.442)$$

where $k_F^2 = \frac{2mE_F}{\hbar^2}$ is the Fermi momentum,

$$\sigma_{xy}^z(0) = \frac{e}{4\pi^2} \frac{\gamma_1 + 2\gamma_2}{\gamma_2} \left[k_h - k_l - \int_{k_l}^{k_h} \frac{dk}{1 + \left(\frac{2\hbar\tau}{m}\gamma_2 k^2\right)^2} \right]. \qquad (9.443)$$

Note that Eq. (9.441) gives the conductivity of a two-dimensional system, whereas Eq. (9.443) describes conductivity of three-dimensional bulk, hence the different physical dimensions.

Anomalous Hall Effect

We discussed the Hall effect in metals (not to be confused with the quantum Hall effect occurring in semiconductors) in Sec. 9.5.3. The metal was assumed to be nonmagnetic, and the Hall effect is solely due to the magnetic induction **B** that is determined by the external magnetic field **H**. The Hall resistance ρ_H is given by $\rho_H = R_0 B$, where R_0 is the Hall coefficient defined in Eq. (9.277a). In Sec. 9.5.10, we discussed ferromagnetic materials, such as Fe, Co, Ni, Gd, and Dy, which posses a permanent magnetic moment **M**. In ferromagnetic materials, where spin-orbit coupling is intrinsic and strong, a non-zero anomalous Hall conductivity is observed. This is referred to as the *Anomalous Hall effect*. The anomalous Hall effect combines equilibrium polarization of a ferromagnet with spin-orbit interaction. This leads to electrical (charge) Hall currents transverse to an applied field.

The contribution to the Hall resistivity ρ_H due to the spontaneous magnetization can be calculated as follows. Assuming the external field **H**, the magnetic induction **B** and the magnetization **M** are parallel, we may express the Hall resistance in terms of their magnitudes. Then, ρ_H can usually be fitted by

$$\rho_H = R_0 B + 4\pi R_s M, \qquad (9.444)$$

The first term, $R_0 B$, represents the ordinary Hall contribution, which is inversely proportional to the density of carriers and has the same sign as their charge. It is a linear function of H, and in the Hall measurement geometry, $B = H$. The second term, $R_s M$, is referred to as the anomalous Hall term and is usually associated with the spin polarization of the conduction carriers and the spin-orbit interaction. The anomalous Hall term in Eq. (9.444) for ρ_H is proportional to the magnetization of the material. The mechanism leading to the anomalous Hall term (encoded in R_s) is still not completely understood. It is often assumed to have extrinsic origins involving processes such as skew scattering and side-jump processes (see discussion in connection with Fig. 9.81). Its magnitude depends on the concentration and scattering strength of impurities.

9.6.9 PHOTON INDUCED PROCESSES IN SEMICONDUCTORS

At the beginning of Sec. 9.6, we mentioned that excitation of electrons in semiconductors can be realized by excitation with light of an appropriate frequency. Let us consider such phenomena in more detail. Photoexcitation and photoemission cannot occur for photon energies well within the band gap in undoped semiconductors, except for excitonic states that exist near the bandgap edge (see later). As the photon energy approaches the bandgap energy, exciton absorption or exciton emission can become important (see Sec. 9.6.9). On increasing the photon energy even further beyond the bandgap energy, photoabsorption or photoemission is still possible, but the photoabsorption coefficient $\alpha(\nu)$ and the photoemission cross-section decrease in magnitude. Figure 9.83 depicts the photoabsorption and photoemission

9.6 Semiconductors

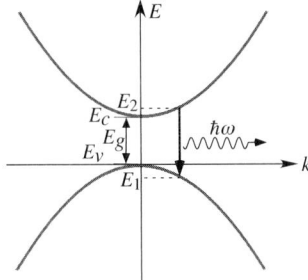

FIG 9.83 Photoemission in a direct bandgap semiconductor. The conduction band is populated due to thermal activation or electrical excitation. The photon energy $\hbar\omega$ in the absorption or emission process equals $E_2 - E_1$.

processes in a direct bandgap semiconductor. In direct bandgap materials, photoabsorption is time-reversed photoemission, but in an indirect bandgap material, photoabsorption and photoemission are very different, as illustrated in Fig. 9.84. In a photoabsorption event, an electron is created with momentum of the hole state that is created, but the electron quickly thermally relaxes back down to the electron states populated at finite temperature as a result of activation, as shown in Fig. 9.84(a). Table 9.6 presents wavelength ranges of strong absorption in a number of semiconductor materials.

Absorption and Emission Spectra

Transition rates for absorption and emission in semiconductors can be calculated using time-dependent first-order perturbation theory. In direct band gap semiconductors, the absorption and emission coefficients are given in terms of the electronic transition matrix elements from the valence band states to the conduction band states,

$$\langle \psi_{k,c} | H_{\text{int}} | \psi_{k,v} \rangle, \tag{9.445}$$

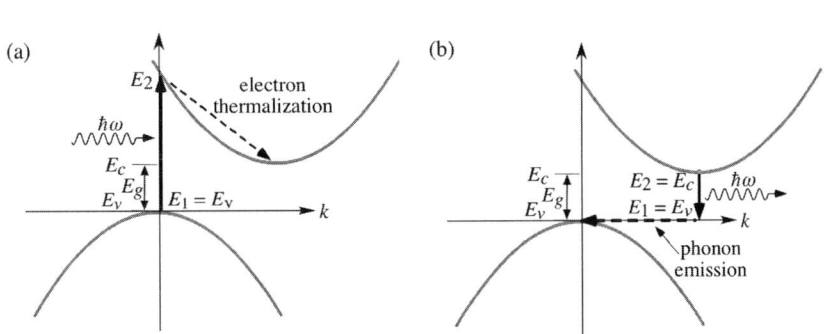

FIG 9.84 (a) Photo-absorption and (b) photo-emission in an indirect bandgap semiconductor. Electron thermalization occurs following photo-absorption to an energy $E_2 > E_c$. Phonon emission and photon emission occur simultaneously in the photo-emission process to conserve both energy and momentum. *Source:* Band, Light and Matter, Fig. 6.36, p. 399

Table 9.6 Strong absorption wavelength ranges of various semiconductor materials. These materials can be used as photon detectors in these wavelength ranges.

Detector material	Wavelength (μm)
Si	0.2–1.1
Ge	0.4–1.8
InAs	1.0–3.5
InGaAs	0.8–2.6
InSb (77 K) [0 K]	1.0–5.6 [1.0–5.2]
HgCdTe (77 K)	1.0–25.0

where $H_{\text{int}} = -\mathbf{E} \cdot \mathbf{d}$ is the electric dipole interaction Hamiltonian, $\psi_{k,c}$ and $\psi_{k,v}$ are the wave functions for an electron of wavevector k in the conduction and valence bands. The absorption and emission coefficients at frequency ω and temperature T are

$$\alpha_{cv}(\omega, T) \approx \int dk\, k^2 |\langle \psi_{k,c}|H_{\text{int}}|\psi_{k,v}\rangle|^2 f\left(-\frac{\hbar^2 k^2}{2m_v} - E_{vF}, T\right) \times \left[1 - f\left(E_g + \frac{\hbar^2 k^2}{2m_e^*} - E_F, T\right)\right]$$
$$\times \delta\left(E_g + \frac{\hbar^2 k^2}{2m_c} - \left(\hbar\omega - \frac{\hbar^2 k^2}{2m_v}\right)\right) \tag{9.446}$$

and

$$\epsilon_{vc}(\omega, T) \approx \int dk\, k^2 |\langle \psi_{k,c}|H_{\text{int}}|\psi_{k,v}\rangle|^2 f\left(E_g + \frac{\hbar^2 k^2}{2m_e^*} - E_F, T\right) \times \left[1 - f(-\frac{\hbar^2 k^2}{2m_v} - E_F, T)\right]$$
$$\times \delta\left(E_g + \frac{\hbar^2 k^2}{2m_c} + \hbar\omega + \frac{\hbar^2 k^2}{2m_v}\right), \tag{9.447}$$

respectively. The Fermi distribution $f(-\hbar^2 k^2/2m_v - E_{vF}, T)$ appearing on the RHS of the expression for the absorption coefficient gives the probability that the valence state with momentum $\hbar k$ is initially filled, and the function $[1 - f(E_g + k^2/2m_e^* - E_F, T)]$ gives the probability that the conduction state with momentum $\hbar k$ is initially vacant. In the expression for the emission coefficient, which is proportional to the emission cross-section $\sigma_{\text{em}}(\omega, T)$, the Fermi distribution $f(E_g + \hbar^2 k^2/2m_e^* - E_F, T)$ gives the probability that the valence state with momentum $\hbar k$ is filled and $[1 - f(-\hbar^2 k^2/2m_v - E_F, T)]$ gives the probability that the valence state with momentum $\hbar k$ is empty.

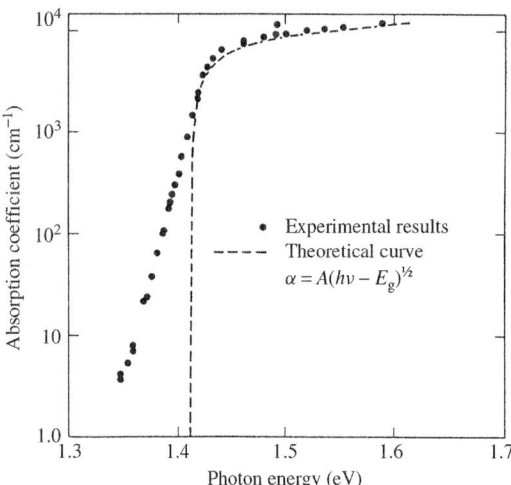

FIG 9.85 GaAs absorption coefficient versus frequency at room temperature. (Reproduced from Figure 3.10 in Pankove [58].) Reproduced with permission from J. I. Pankove, Optical Processes in Semiconductors, (Dover, NY, 1971).

The photon absorption coefficient $\alpha(\omega)$ in semiconductor materials becomes vanishingly small for frequencies smaller than the absorption edge frequency. Figure 9.85 shows the measured absorption coefficient of GaAs as a function of photon energy. Below approximately 1.3 eV, there is no measurable absorption because there are no states between the conduction band and the valence band, and therefore, the absorption from the valence band for frequencies less than 1.3 eV photon energy cannot proceed. Once the photon frequency is larger than the band gap frequency, absorption ensues.

Exciton Absorption and Emission

The hydrogen-like energy level diagram of excitonic states is schematically shown in Fig. 9.86. The Rydberg series converges to the bottom of the conduction band. Excited excitons can decay to lower energy ones, thereby giving up energy via emission of a low frequency photon of energy $h\nu = E_{n_i} - E_{n_f}$ or the electrons and the holes of an exciton can recombine, and then their emission is to the red of the band emission. The inverse process of photoabsorption to excitonic states can also occur. In these excitonic emission or absorption processes, the frequencies are to the red of the band emission and absorption. At high temperature, exciton absorption is often experimentally indistinguishable from temperature-smeared band-to-band absorption and is swamped by the tail of the absorption band. Only at very low temperature can these discrete

9.7 Spintronics

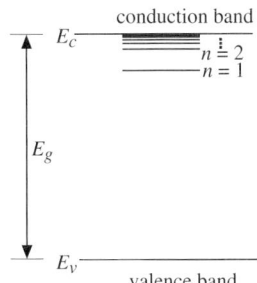

FIG 9.86 Energy levels of excitonic states (bound electron-hole pairs). *Source:* Band, Light and Matter, Fig. 6.43, p. 406

lines be observed. *Biexciton* levels, i.e., states composed of two bound excitons (two electrons and two holes) can sometimes also be observed at ultra-low temperatures.

If one applies an external electric field to a semiconductor, the absorption edge is shifted to lower frequency, and the field strongly influences exciton energy levels. The interaction of an electric field with an exciton is similar to the interaction of an electric field with a hydrogen atom. The reduction of the symmetry due to the electric field leads to splitting of degenerate levels and to field-dependent level shifts, i.e., there is a Stark shift of the energy levels. For strong enough fields, the exciton can dissociate (ionize) by a tunneling process.

For more detailed treatments of the spectroscopy and electronic properties of semiconductors, the reader is invited to consult, e.g., Ashcroft and Mermin, *Solid State Physics* [122], or Kittel, *Introduction to Solid State Physics* [125].

9.7 SPINTRONICS

Conventional electronic devices rely on transport and manipulation of electrical charge carriers – electrons in metals, and electrons and holes in semiconductors. *Spintronics* aims to manipulate the *spin* of charge carriers in solid-state materials (semiconductors, semiconductor heterostructures). Spintronics deals with the electric, magnetic, and optical properties of solids arising from the spin populations of charge carriers and their dynamics in and out of equilibrium. Important subtopics include spin injection, spin interactions, spin-orbit, hyperfine, and spin exchange couplings, spin relaxation and decoherence, magnetic long-range order in semiconductors, and spin-polarized current flow in low-dimensional semiconductor systems.

In metals, the use of electron spin in devices is already an important ingredient in computer hardware. In particular, giant magnetoresistance systems are used as hard disk read heads. In these devices, the phase associated with the spin wave function does not play a significant role.

To date, semiconductor spintronics devices are scarce, but the potential is large. Spintronics can play a role even in the absence of a magnetic field in nonmagnetic semiconductors, due to spin-orbit coupling (see Sec. 9.6.6). The fact that spin-orbit coupling in crystals is strongly enhanced by crystal field gradients and high electron velocities near the atomic nuclei is noteworthy. The dimensionless parameter of spin-orbit coupling in vacuum is of order 10^{-6} [the ratio of the electron kinetic energy $E(k) \approx 1$ eV and its rest energy $m_0 c^2 \approx 0.5$ MeV]. In semiconductors, the dimensionless parameter is Δ_{so}/E_{gap}, where Δ_{so} is the spin-orbit splitting of valence bands and E_{gap} is the semiconductor gap, both of order 1 eV.

In addition to the new physics revealed by semiconductor spintronics, there is a reason to believe that spintronic devices will be small, versatile, and robust. There may be promising applications for quantum information science for the following two reasons: (1) spin orientation of conduction electrons survives for a relatively long time (nanoseconds), while momentum states decohere after a few tens of femtoseconds, (2) information can be easily stored in spins. This information is carried by electrons and is read out at a terminal.

The three main requirements for spintronic devices are as follows: (1) efficient spin injection, (2) slow spin relaxation, and (3) accurate and reliable spin detection. Spintronics devices are highly energy efficient and generate less heat in operation than semiconductor devices. This property holds out the hope of higher integration levels in smaller devices, with low heat generation.

In this section, we discuss a few central concepts used in spintronics. We start by listing the main tools for manipulating spins, such as spin injection, F/N contacts and F/N/F junctions, magnetic tunneling junctions, using the concepts of spin relaxation and spin dephasing. Finally, a few spintronic devices will be considered, such as giant magnetoresistance devices, magnetic tunnel junctions, spin torque effect devices, and Datta–Das spin transistors. The discussion below is based on Refs. [156] and [157].

9.7.1 TOOLS FOR MANIPULATING SPINS

There are basically two ways to manipulate spins of electrons (and holes): magnetic and electric fields. The magnetic field acts through the Zeeman effect and is advantageous in materials where the effective mass is small and the Zeeman energy is high. The electric field controls the spin-orbit strength. In quantum wells with asymmetric confining potential restricting the charge carriers to move on the x–y plane, the spin–orbit part of the Hamiltonian is described by the Rashba term, defined in Eq. (9.431). The eigenvalues and eigenfunctions of the Rashba Hamiltonian in 2D are found in Problem 9.63. It is convenient to represent the Rashba term as a spin precession Hamiltonian in an effective magnetic field,

$$V_R^{(2D)} = \frac{1}{2} g\mu_B [\mathbf{H}(\mathbf{k}) \times \boldsymbol{\sigma}], \quad \mathbf{H}(\mathbf{k}) = \frac{2\alpha}{g\mu_B}(\mathbf{k} \times \hat{\mathbf{z}}), \tag{9.448}$$

where μ_B is the Bohr magneton (which is canceled out in $V_R^{(2D)}$). From this point of view, $V_R^{(2D)}$ describes spin precession in the effective field $\mathbf{H}(\mathbf{k})$.

Spin Injection

A key process in spintronics is injection of spin-polarized electrons from a ferromagnetic metal into a nonmagnetic conductor, usually a metal or a semiconductor. The devices are composed of interfaces (contacts) between ferromagnetic and nonmagnetic materials, denoted as F/N contacts. Spin-polarized electrons driven across an F/N contact result in a nonequilibrium spin accumulation in the nonmagnetic conductor, as is illustrated in Fig. 9.87.

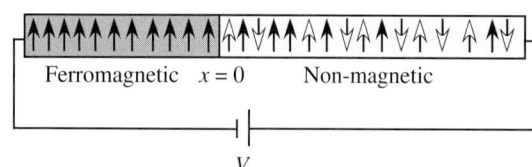

FIG 9.87 Spin injection in a ferromagnetic-nonmagnetic junction. Spin diffuses from the ferromagnetic into the nonmagnetic material, and the value of the magnetization $M(x,t)$ inside the nonmagnetic material is determined by the solution of the diffusion equations.

The microscopic theory of spin injection is rather involved. Injection into a metal and into a semiconductors are rather distinct. The physics describing transport across a ferromagnetic-semiconductor interface is not yet completely understood. It is useful to consider first a formalism based on a Boltzmann approach to charge and spin drift, which is common to both cases, and to discuss the more involved issues of spin injection into semiconductor material later. Spin drift and diffusion is, by itself, a very broad topic, and only its basics are discussed here.

An electron with a given spin orientation can undergo collisions with impurities, phonons, and other electrons, which make the spin decohere. A microscopic theory of spin relaxation should account for the physics at the F/N interface. As in the Drude model for transport of charge, spin relaxation inside a solid material is described in terms of a central quantity, τ, the average time between collisions. In Sec. 9.6.4, the diffusion of charge carriers (electrons and holes) in semiconductors is discussed. Similar considerations lead to analogous diffusion equations for particles with different spin orientations. In an environment characterized by spin-dependent interactions, the diffusion equations couple the densities $n_\uparrow$ and $n_\downarrow$ of spin-up and spin-down electrons. To obtain these diffusion equations, we assume that electrons carry out a random walk along the longitudinal direction x, subject to a small longitudinal electric field $\mathcal{E} = -\partial V/\partial x$. The probability p_+ of walking rightward is larger than the probability p_- of walking leftward (see Fig. 9.7). The new ingredient, as compared with the analysis of the Drude model, is that in each collision, there is a small probability w for electrons to flip their spin (if $w \simeq 10^{-3}$ to 10^{-6}, it requires a thousand to a million collisions to flip the spin of the electron, on average). The derivation of the diffusion equations for $n_\uparrow(x,t)$ and $n_\downarrow(x,t)$,

$$\frac{\partial n_\sigma}{\partial t} = D\frac{\partial^2 n_\sigma}{\partial x^2} - v_d \frac{\partial n_\sigma}{\partial x} - \frac{w}{\tau}(n_\sigma - n_{\bar\sigma}), \tag{9.449}$$

is elementary and will not be worked out here. Here, D is the diffusion constant, and $v_d = -\mu\mathcal{E}$ is the *drift velocity* given in terms of the mobility $\mu = e\tau/m$ [see discussion following Eq. (9.43)]. Subtracting one equation from the other results

9.7 Spintronics

in an equation for the spin density $s(x,t) \equiv n_\uparrow(x,t) - n_\downarrow(x,t)$,

$$\frac{\partial s}{\partial t} - \frac{\partial}{\partial x}\left(\mu \mathcal{E} s + D\frac{\partial s}{\partial x}\right) = -\frac{2ws}{\tau}. \tag{9.450}$$

Problem 9.65

By adding the two equations (9.449), show that the diffusion equation for the *charge density* $n(x,t) = n_\uparrow + n_\downarrow$ is,

$$\frac{\partial n}{\partial t} - \frac{\partial}{\partial x}\left(\mu \mathcal{E} n + D\frac{\partial n}{\partial x}\right) = 0. \tag{9.451}$$

The expression within the brackets on the RHS of Eq. (9.450) is the *spin particle current*,

$$J_s = -\left(\mu \mathcal{E} s + D\frac{\partial s}{\partial x}\right). \tag{9.452}$$

The *spin charge current* and the *spin conductivity* are

$$j_s = -eJ_s \equiv \sigma_s \mathcal{E} + eD\frac{\partial s}{\partial x}, \quad \sigma_s = e\mu s. \tag{9.453}$$

The *density spin polarization*, the *current spin polarization*, and the *conductivity spin polarization* are

$$P_n = \frac{n_\uparrow - n_\downarrow}{n_\uparrow + n_\downarrow} = \frac{s}{n}, \quad P_j = \frac{j_\uparrow - j_\downarrow}{j_\uparrow + j_\downarrow}, \quad P_\sigma = \frac{\sigma_s}{\sigma}. \tag{9.454}$$

Solution for $\mathcal{E} = 0$

Equation (9.450) can be solved in the absence of an electric field $\mathcal{E} = 0$. Assuming the contact between the ferromagnet and the nonmagnetic material is formed at $t = 0$, the initial condition is $s(x,0) = s_0 \delta(x)$. Equation (9.450) has the solution,

$$s(x,t) = \frac{s_0}{\sqrt{4\pi Dt}} e^{-\left(\frac{x^2}{4Dt} + \frac{2wt}{\tau}\right)}. \tag{9.455}$$

Thus, in the absence of electric field, the spin density decays in space as a Gaussian (the first term in the exponent) and relaxes in time through the second term. The width of the Gaussian $\sigma = \sqrt{D\tau/w} \equiv \sqrt{2}L_s$ at time $t = \tau/2w$ defines the *spin diffusion length*

$$L_s = \sqrt{D\tau/2w}, \tag{9.456}$$

which is the length of spin diffusion until relaxation. In metals, $L_s \simeq 0.1$ mm, whereas in semiconductors, $L_s \simeq 0.1\mu$.

Problem 9.66

Show that steady state solutions of the diffusion equation, (9.450), for $\mathcal{E} = 0$ and $\partial s/\partial t = 0$ are given by $s(x) = s_0 e^{-x/L_s}$.

The F/N Contact

Given an F/N contact integrated within a circuit, as shown in Fig. 9.87, the quantity of interest for spin injection is the current spin polarization $P_j(0)$ defined in Eq. (9.454). To determine it separately for the ferromagnetic and nonmagnetic regions, we need to specify the spin diffusion length L_s (9.456) and the conductivity spin polarization P_σ (9.454). These parameters will then carry a subscript F or N.

Following the values of L_s specified after Eq. (9.456), we see that the lengths of the ferromagnetic and nonmagnetic samples in Fig. 9.87 are much larger than the spin diffusion lengths, $L_F \gg L_{sF}$, $L_N \gg L_{sN}$. Denoting by R_F and R_N, the resistances of the ferromagnetic and nonmagnetic regions, the spin-current polarization was calculated to be [156],

$$P_j(0) = \frac{I_\uparrow - I_\downarrow}{I_\uparrow + I_\downarrow} = \frac{R_F P_{\sigma F}}{R_F + R_N}. \tag{9.457}$$

This expression assumes that the contact has no resistance at all. Otherwise, the resistance of the contact should be taken into account. If the nonmagnetic material is a metal, then $R_N \approx R_F$. On the other hand, if the nonmagnetic material is semiconductor, $R_N \gg R_F$ and $P_j(0) \approx \frac{R_F}{R_N} P_{\sigma F}$.

Semiconductor Spintronics

Although the use of spintronics in metals has found applications (such as giant magnetoresistance, see later), this is not the case for semiconductors. The hope behind the study of semiconductor spintronics is, first, to augment and improve existing electronic devices in semiconductor technology and, second, to exploit the semiconductor physics to introduce new directions of pure and applied research. By this, we mean the dominant role of spin-orbit coupling and the importance of heavy holes, as discussed in Sec. 9.6.6. Thus far, this goal has been elusive, but semiconductor spintronics has recently progressed significantly.

Semiconductor spintronics brings us into the fascinating field of *ferromagnetic semiconductors*. It requires the control of both the magnetic and the electric properties within the same material. The magnetic properties are expected to be sensitive to a bias voltage and an applied current, whereas the resistance is sensitive to the application of a magnetic field. If present at room temperature, it would be useful in numerous applications. Ferromagnetic semiconductors exist, albeit with low Curie temperatures (below 200 K), e.g., $Ga_{1-x}Mn_xAs$, Mn_xGe_{1-x}.

Spin Injection into Semiconductors

Because nearly all electronic components and devices rely on semiconductors, it is necessary to design an F/SC interface to incorporate spintronics technology in semiconductor devices. Unfortunately, maintaining spin polarization in semiconductors is quite difficult. The main reason is the lack of a good quality atomic interfaces between a ferromagnetic metal and a semiconductor. Poor interfaces can cause the electron spins to be randomized in the direction perpendicular to the interface. We just mentioned ferromagnetic semiconductors that are good candidates for solving the interface problem. However, magnetic semiconductors often work only at or below room temperature, and they are not strongly magnetic. In most experimental attempts to construct an F/SC contact, there is a large difference in conductance between the ferromagnetic metal and the semiconductor. Typical values are $R_{FM} \simeq 150\,\mu\Omega$ and $R_{SC} \simeq 20\,\Omega$. This is called *conductance mismatch*. One way to alleviate this problem is to insert a spin-dependent interface resistance that separates the ferromagnetic metal and the semiconductor. The current through the tunnel barrier depends on the density of states on the left and right of the tunnel barrier. Since the density of states in the ferromagnetic metal differs for spin-up and spin-down electrons, a tunnel barrier can provide a spin-dependent resistance. The tunnel barrier resistance must be matched with the semiconductor channel resistance to maximize magnetoresistance. Spin injection into silicon has recently been realized using this method.

Spin Dynamics

Our analysis of spin drift and spin injection based on the random walk picture and the diffusion equation was carried out in the absence of a magnetic field. Now suppose that the electrons undergoing the random walk are subjected to a uniform and constant magnetic field $\mathbf{H}_0$. Then, the RHS of Eq. (9.450) should be modified to include a precession term at frequency $\omega_0 = g\mu_B \mathbf{H}$, where $g \approx 2$ is the electron Landé g-factor. The analysis carried out in Sec. 6.1.3 shows that there are two relaxation times, τ_1 and τ_2. The longitudinal time T_1 for the spin parallel to the magnetic field is defined as $1/T_1 = \frac{2w}{\tau_1}$, and a second spin relaxation time T_2, for the spin perpendicular to the magnetic field, is defined as

9.7 Spintronics

$1/T_2 = \frac{2w}{\tau_2}$. At relatively weak magnetic fields, $H \leq 1$ Tesla, and in most conducting solids, the two times are roughly equal $T_1 \approx T_2 \equiv \tau_r$. Therefore, the spin diffusion equation (9.450) takes the form

$$\frac{\partial \mathbf{s}}{\partial t} = \mathbf{s} \times \boldsymbol{\omega}_0 + D\nabla^2 \mathbf{s} + \mu \mathcal{E} \nabla \mathbf{s} - \frac{\mathbf{s} - \mathbf{s}_0}{\tau_r}, \qquad (9.458)$$

where $\mathbf{s}_0$ is an equilibrium spin configuration that is reached due to spin relaxation (it can be shown to be very small in metals – and henceforth neglected here – but this is not the case in semiconductors, see later). The four terms on the RHS describe spin precession, spin diffusion, spin drift, and spin relaxation. Assuming charge neutrality, so that $\mathcal{E} = $ constant, Eq. (9.459) can be written as a continuity equation,

$$\frac{\partial \mathbf{s}}{\partial t} - \nabla [\mu \mathcal{E} \mathbf{s} - D \nabla \mathbf{s}] = \mathbf{s} \times \boldsymbol{\omega}_0 - \frac{\mathbf{s}}{\tau_r}, \qquad (9.459)$$

combining the generalized spin current (the expression in the square brackets) and the rate of change of spin density.

Equation (9.458) is the starting point for elucidating the dependence of the spin accumulation on the magnetic field applied perpendicular to the injected spin in the non-magnetic material, a phenomenon referred to as the *spintronic Hanle effect*.[6] The spin dynamics results from a combination of spin precession and drift motion in a diffusive media. As a consequence, the spin accumulation decreases with the field at weak magnetic fields and displays coherent decaying oscillations at strong fields. When the spin density is probed away from the spin injection points, it yields an averaged quantity, because different electrons have different transit times, and their spin precession angles differ from each other. At large transverse magnetic fields, the detected spin vanishes because this angle difference is comparable with the Larmor period.

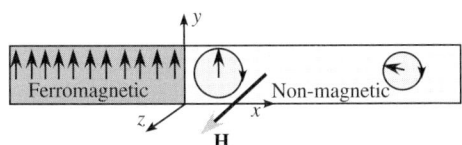

FIG 9.88 Geometry of the Hanle effect: In a ferromagnetic-nonmagnetic junction, a magnetic field is applied to the nonmagnetic component along the z direction. Following spin injection at initial spin density $s_y(0)$, the spin precesses around the magnetic field lines and the size of the spin density components at steady state, $s_x(x)$, $s_y(x)$, decay at large x.

To quantify this, consider the geometry of Fig. 9.88, where the propagation is along x and the polarization in the ferromagnetic material is along y. Imagine now that a magnetic field $\mathbf{H} = H\hat{\mathbf{z}}$ is applied at time $t = 0$ as in Fig. 9.88. The density of a spin injected at $x = 0$ at $t = 0$ is $\mathbf{s}(0,0) = (0, s_y(0), 0)$. The question to be addressed is, what is the *steady state* spin density $s_x(x)$ at $x > 0$ that is reached after some transient time? Qualitatively, the spin precesses around the magnetic field direction z (s_z is conserved) with relaxation taking place along the positive x direction, so $\lim_{x \to \infty} \mathbf{s}(x) = 0$. To analyze this picture quantitatively, put Eq. (9.458) into steady state with the constraint, $s_z \equiv 0$. Recalling the drift velocity $v_d = -\mu \mathcal{E}$, the steady-state limit of Eq. (9.458) is

$$0 = \dot{s}_x = \omega_0 s_y + D s_x'' - v_d s_x' - \frac{s_x}{\tau_r}, \qquad (9.460)$$

$$0 = \dot{s}_y = -\omega_0 s_x + D s_y'' - v_d s_y' - \frac{s_y}{\tau_r}, \qquad (9.461)$$

The asymptotic conditions (9.462) and the fact that the component s_z is not coupled to s_x and s_y suggest the simple solution $s_z = 0$, so we need only to determine the dynamics of s_x and s_y. The decay of spin at large distance from the

---

[6] The original Hanle effect, discovered in 1924, involves the radiation scattered resonantly by atoms in the presence of a weak magnetic field. The radiation undergoes changes in both its intensity and the direction of its polarization on changing the magnetic field strength. The essence of the effect involves the low frequency coherences between Zeeman magnetic sublevels of an atomic state and their destruction by the lifting of the sublevel degeneracy by the magnetic field.

junction and the relations between spin densities and spin currents, Eq. (9.453), together with the requirement of current conservation at the contact $x = 0$, imply the boundary conditions,

$$\mathbf{s}(\infty) = 0, \quad j_{sx} = e(D - v_d)s'_x(0) = 0, \quad j_{sy} = e(D - v_d)s'_y(0) = j_{s0}. \tag{9.462}$$

The solution of Eqs (9.460) and (9.461) with the boundary conditions (9.462) is straightforward [156] albeit lengthy. It is expressed in terms of the diffusion and drift lengths as well as a dimensionless parameter indicating the strength of drift over diffusion,

$$L_D = \sqrt{D\tau_r}, \quad L_d = v_d\tau_r, \quad \kappa = \frac{L_d}{2L_D}, \tag{9.463}$$

In terms of these quantities, we define the following dimensionless parameters,

$$\alpha_1 = \frac{1}{\sqrt{2}}\sqrt{1 + \kappa^2 + \sqrt{(1 + \kappa^2)^2 + (\omega_0\tau_r)^2}} - \kappa, \tag{9.464}$$

$$\alpha_2 = \frac{1}{\sqrt{2}}\sqrt{\sqrt{(1 + \kappa^2)^2 + (\omega_0\tau_r)^2} - (1 + \kappa^2)}. \tag{9.465}$$

The stationary spin densities are given in terms of these parameters by

$$s_x(x) = -\frac{j_{s0}L_D e^{-\alpha_1 x/L_D}}{eD[2(\kappa + \alpha_1)^2 + \alpha_2^2]}\left[(2\kappa + \alpha_1)\sin\frac{\alpha_2 x}{L_D} + \alpha_2\cos\frac{\alpha_2 x}{L_D}\right], \tag{9.466}$$

$$s_y(x) = -\frac{j_{s0}L_D e^{-\alpha_1 x/L_D}}{eD[2(\kappa + \alpha_1)^2 + \alpha_2^2]}\left[(2\kappa + \alpha_1)\sin\frac{\alpha_2 x}{L_D} - \alpha_2\cos\frac{\alpha_2 x}{L_D}\right]. \tag{9.467}$$

The occurrence of α_1 in the exponent indicates its role in the effective spin relaxation. The occurrence of α_2 in the oscillating function specifies its role in spin precession.

9.7.2 TUNNELING MAGNETORESISTANCE

When the electrical resistance of a material is affected by a magnetic filed, the measured quantity is referred to as *magnetoresistance*. The magnetic field acts on both the orbital and spin degrees of freedom. In an electrical circuit containing ferromagnetic materials, the role of spin degrees of freedom is dominant. Basic elements of such circuit are ferromagnetic-semiconductor-ferromagnetic or ferromagnetic-insulator-ferromagnetic junctions, collectively denoted as *magnetic tunnel junctions* (MTJs). Because the conductance of the material between the leads is poor, the relevant physical quantity is referred to as *tunneling magnetoresistance* (TMR). The TMR of a Fe/Ge/Co junction was first measured in 1975 and was shown to display sensitivity to the orientation of the mean magnetization in the Fe and Co leads. In particular, the magnetoresistance for antiparallel orientation is higher than with parallel orientations. This is schematically illustrated in Fig. 9.89.

Let us consider the density of states on the left and right ferromagnetic leads for spin-up and spin-down electrons and compare the parallel configuration with the antiparallel one, as illustrated in Fig. 9.90. Denoting by R_P and R_{AP}, the resistance of the junction for parallel and antiparallel orientations, respectively, the physical situation is characterized by the inequality $R_{AP} > R_P$. A dimensionless quantity expressing this is

$$\rho = \frac{R_{AP} - R_P}{R_P}. \tag{9.468}$$

9.7 Spintronics

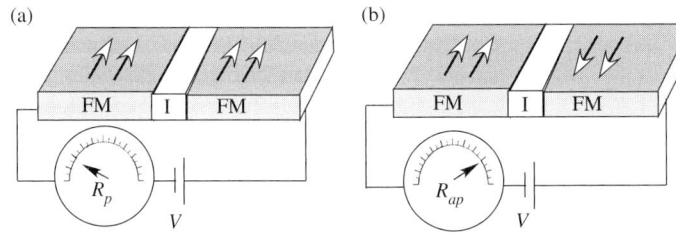

FIG 9.89 Illustration of TMR in an FM-I-FM magnetic tunnel junction stressing the occurrence of low magnetoresistance R_P when the magnetization vectors in the two leads are parallel (a) and higher magnetoresistance R_{AP} when the magnetization vectors in the two leads are antiparallel (b). A modified version of Fig. II.32 (upper part) in Ref. [157].

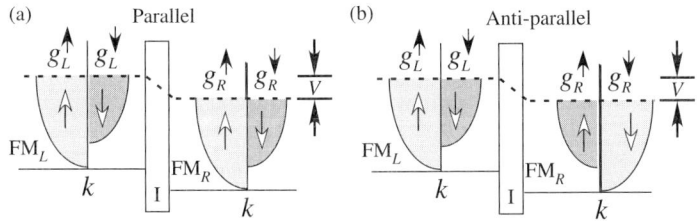

FIG 9.90 Density of state profiles of spin-up and spin-down electrons in an FM-I-FM magnetic tunnel junction under a potential bias V, for (a) parallel orientations and (b) antiparallel orientations corresponding, respectively, to panels (a) and (b) in Fig. 9.89. A small potential difference is maintained across the junction between the left (L) and right (R) ferromagnets, and the chemical potential is marked by the dashed line. The densities of states $g_{L,R}^{\uparrow,\downarrow}$ match in the parallel case and mismatch in the antiparallel case. This leads to the higher resistance in the latter case.

To calculate ρ, consider a microscopic model Hamiltonian for the system consisting of the left (L) and right (R) ferromagnets and the tunneling (T) between them. Such tunneling Hamiltonian is formally written as

$$H = H_L + H_R + H_T \equiv H_0 + H_T, \quad (9.469)$$

where H_L and H_R are the Hamiltonians of the left and right electrodes (ferromagnets) and govern the physics of the ferromagnetic materials, and H_T describes the tunneling process. Such a Hamiltonian is usually expressed in the formalism of second quantization (see Sec. 18.1.7 for details). In the absence of electron–electron interactions, H_L and H_R can be diagonalized. The corresponding basis states (kets) are written as $|L\mathbf{k}\sigma\rangle$, $|R\mathbf{k}\sigma\rangle$, where $\mathbf{k}$ is a crystal (Bloch) wavenumber of an electron and $\sigma = \uparrow, \downarrow$ is its spin projection. Thus, we have

$$H_L|L\mathbf{k}\sigma\rangle = \varepsilon_{L\mathbf{k}\sigma}|L\mathbf{k}\sigma\rangle,$$
$$H_R|R\mathbf{k}\sigma\rangle = \varepsilon_{R\mathbf{k}\sigma}|L\mathbf{k}\sigma\rangle, \quad (9.470)$$

where $\varepsilon_{L\mathbf{k}\sigma}$ and $\varepsilon_{R\mathbf{k}\sigma}$ are the single-particle energies. The tunneling rate due to H_T obtained using tunneling amplitudes

$$t_{\mathbf{kq}} \equiv \langle L\mathbf{k}|H_T|R\mathbf{q}\rangle, \quad (9.471)$$

which is the matrix element of the potential H_T responsible for tunneling between wave functions on the left and on the right sides, is usually small, since the wave functions decay exponentially beyond the interface, and their overlap is small. Thus, H_T is treated as a perturbation. The tunneling is assumed to be spin independent.

Now, assume that the junction is biased by a potential V. Within this framework, the tunneling rate at which electrons with spin σ are transferred from the left to the right electrode can be estimated by the Fermi golden rule,

$$W_{L\to R,\sigma}(V) = \frac{2\pi}{\hbar} \sum_{\mathbf{kq}} |t_{\mathbf{kq}}|^2 f(\varepsilon_{L\mathbf{k}\sigma})[1 - f(\varepsilon_{R\mathbf{q}\sigma})]\delta(\varepsilon_{L\mathbf{k}\sigma} - \varepsilon_{R\mathbf{q}\sigma} + eV). \quad (9.472)$$

By using a series of approximations, including the assumption that the tunneling amplitudes $t_{\mathbf{kq}} = t$ are momentum independent and replacing summation over $\mathbf{k}, \mathbf{q}$ by integration over energy

$$W_{L\to R,\sigma}(V) = \frac{2\pi|t|^2}{\hbar} \int_{-\infty}^{\infty} d\varepsilon\, g_L^\sigma(\varepsilon) g_R^\sigma(\varepsilon) f(\varepsilon)[1 - f(\varepsilon - eV)], \quad (9.473)$$

where $g_L^\sigma(\varepsilon)$ and $g_R^\sigma(\varepsilon)$ are the electron density of states in the leads defined in Fig. 9.90. The tunneling current is

$$I_\sigma(V) = e[W_{L\to R,\sigma}(V) - W_{R\to L,\sigma}(V)] = \frac{2\pi e|t|^2}{\hbar} \int_{-\infty}^{\infty} d\varepsilon\, g_L^\sigma(\varepsilon - eV) g_R^\sigma(\varepsilon)[f(\varepsilon - eV) - f(\varepsilon)]. \quad (9.474)$$

At low temperatures, $f(\varepsilon) \approx \Theta(\varepsilon_F - \varepsilon)$, and for small voltage difference, an expansion in V to first order yields the conductance for polarized electrons,

$$G_\sigma \approx 2\pi \frac{e^2}{\hbar} |t|^2 g_L^\sigma(\varepsilon_F) g_R^\sigma(\varepsilon_F). \tag{9.475}$$

The difference between parallel and antiparallel orientations enters through the density of states,

$$G_P = G_{P\uparrow} + G_{P\downarrow} = 2\pi \frac{e^2}{\hbar} |t|^2 \left[g_L^\uparrow(\varepsilon_F) g_R^\uparrow(\varepsilon_F) + g_L^\downarrow(\varepsilon_F) g_R^\downarrow(\varepsilon_F) \right], \tag{9.476}$$

$$G_{AP} = G_{AP\uparrow} + G_{AP\downarrow} = 2\pi \frac{e^2}{\hbar} |t|^2 \left[g_L^\uparrow(\varepsilon_F) g_R^\downarrow(\varepsilon_F) + g_L^\downarrow(\varepsilon_F) g_R^\uparrow(\varepsilon_F) \right]. \tag{9.477}$$

Finally, the quantity ρ defined in Eq. (9.468) is evaluated as,

$$\rho = \frac{R_{AP} - R_P}{R_P} = \frac{G_P - G_{AP}}{G_{AP}}. \tag{9.478}$$

Extension of this formulation to arbitrary orientation between the polarization directions in L and R is possible but is technically involved.

Spin Relaxation

Spin relaxation in spintronics can be determined using the analysis carried out in Sec. 6.1.3 by applying it to an ensemble of electrons (rather than a single electron). There are two relaxation processes: (1) relaxation of the spin component s_z along a constant magnetic field $H_0 \hat{z}$ toward its equilibrium value and (2) a loss of coherence of s_x and s_y in a time-dependent magnetic field in the x–y plane.

Before considering relaxation processes, let us recall the pure state case, starting from the equation of motion for the spin $\mathbf{S}$ of a particle of charge q in a magnetic field $\mathbf{H}$ and an electric field $\mathbf{E}$,

$$\frac{\partial \mathbf{S}}{\partial t} = \frac{q}{mc} \mathbf{S} \times \left[\frac{g}{2} \mathbf{H} - \left(\frac{g-1}{2} \right) \frac{\mathbf{v}}{c} \times \mathbf{E} + O\left(\frac{v}{c} \right)^2 \right]. \tag{9.479}$$

We consider the case where $\mathbf{E} = 0$ and a time-dependent magnetic field,

$$\mathbf{H}(t) = (H_1 \cos \omega t, -H_1 \sin \omega t, H_0). \tag{9.480}$$

is present. There are three frequencies involved, $\omega_i = g \mu_B H_i$, ($i = 0, 1$) and the planar field oscillation frequency ω. The Bloch equations for a single-particle spin are,

$$\frac{\partial \mathbf{S}}{\partial t} = g \mu_B \mathbf{S} \times \mathbf{H}. \tag{9.481}$$

If we consider a spin 1/2 particle with initial state $|\uparrow\rangle$, then the probability for the spin to be flipped at time t is given by

$$P_{\uparrow \to \downarrow}(t) = |\langle \downarrow (t) | \uparrow (0) \rangle|^2 = \frac{\omega_1^2}{2\Omega^2} (1 - \cos \Omega t), \tag{9.482}$$

where $\Omega = \sqrt{(\omega - \omega_0)^2 + \omega_1^2}$. $P_{\uparrow \to \downarrow}(t)$ displays a resonance at $\omega = \omega_0$. On resonance, $P_{\uparrow \to \downarrow}(t = \pi/\omega_1) = 1$ and the spin flips with certainty.

Let us now use the Bloch equations to analyze the spin dynamics of an ensemble of electrons. In addition to the magnetic field term, relaxation and decoherence (dephasing) terms are present. These two processes involve two different

9.7 Spintronics

time scales T_1 and T_2, respectively,

$$\frac{\partial S_x}{\partial t} = [\mathbf{S} \times \omega_0]_x - \frac{S_x}{T_2}, \quad \frac{\partial S_y}{\partial t} = [\mathbf{S} \times \omega_0]_y - \frac{S_y}{T_2}, \quad \frac{\partial S_z}{\partial t} = [\mathbf{S} \times \omega_0]_z - \frac{S_z - \bar{S}_z}{T_1}. \tag{9.483}$$

Here, $\bar{S}_z$ is the equilibrium value of S_z. From the analysis of Sec. 6.1.3, we recall that T_1 is the *spin relaxation time* and T_2 is the *spin dephasing time*. Physically, T_1 describes the relaxation of a nonequilibrium spin population, the diagonal elements of the spin density matrix, toward equilibrium. Thus, $1/T_1$ is the decay rate of S_z along the static field direction toward its equilibrium value $\bar{S}_z$. This decay is possible since, for $H_0 \neq 0$, the electrons exchange energy during the spin relaxation process, due to the magnetic energy difference between the initial and final equilibrium states. The time T_2 is the dephasing time of the spin component transverse to the static field $H_0\hat{z}$. Equivalently, $1/T_2$ is the decay rate of coherent spin oscillations or off-diagonal elements of the spin-density matrix (see Fig. 9.91).

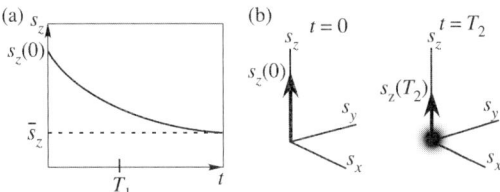

FIG 9.91 Qualitative interpretation of the Bloch equations (9.483). The initial state is $\mathbf{S}(0) = (0, 0, S_z(0))$. (a) Relaxation of $S_z(t)$ toward equilibrium $\bar{S}_z$ controlled by time T_1. (b) The fuzzy region indicates loss of coherence of S_x and S_y after time T_2.

In the Bloch equations (9.483), the relaxation and decoherence times T_1 and T_2 enter as phenomenological parameters. Calculating them within a microscopic theory requires, first, the identification of the relaxation and dephasing mechanisms. There are four main mechanisms that lead to spin relaxation of conduction electrons in semiconductors. First, the electron scatters from impurities or phonons. An electric field near impurities generates spin-orbit coupling. To take account of spin-orbit coupling, electron states are represented by spinors where, generically, both components are nonzero. Following scattering events, there is a finite probability to flip the spin. The spin flip resulting from electron-impurity spin-orbit coupling and its contribution to spin relaxation is referred to as the *Elliot–Yafet mechanism*, which is active in semiconductors with and without a center of inversion symmetry. The rate is small: an electron has to undergo about 10^5 scattering events before a spin flip occurs. Second, we have seen in Sec. 9.6.6 that in semiconductors without a center of symmetry, such as in the Zinc-blende group, the spin-orbit interaction manifests itself as an effective, momentum-dependent magnetic field. As an electron scatters by an impurity or a phonon, its momentum is altered, and hence, it feels a different spin-orbit induced effective magnetic field. Thus, the axis and frequency of precession vary randomly. Although, on average, the spin direction remains unchanged, its fluctuations increase as if the spin performs a random walk. This relaxation mechanism is referred to as the *Dyakonov–Perel mechanism*. Third, recall that if a semiconductor is doped with acceptors (that is, the semiconductor is p-doped), it has excess holes and there will be an exchange interaction between electrons and holes. An electron with a spin-up will exchange its spin with a hole with spin-down. This interaction preserves the total spin, but holes tend to relax their spins much faster than electrons, since the valence bands are susceptible to strong spin-orbit coupling. Thus, the Dyakonov–Perel mechanism relaxes the spin of the holes, and the result is an electron spin-flip, as in Fig. 9.92. This is the Bir–Aronov–Pikus mechanism. Fourth, in many semiconductors, such as GaAs, the atomic nuclei in their ground state have nonzero spin, and the hyperfine interaction leads to spin flips. The hyperfine interaction is most active when electrons are confined on impurities or in quantum dots (see Chapter 13). The extent of the electron wave function encompasses a region containing about 10^5 nuclear spins, and the hyperfine interaction leads to a spin-flip and, more significantly, spin dephasing (loss of phase coherence). Discussion of microscopic models that quantify these four mechanisms goes beyond the scope of this book.

9.7.3 SPINTRONIC DEVICES

The essence of spintronics is control and manipulation of electron spin. The hope is to construct new devices that are based on the magnetic nature of electrons. Instead of relying solely on the electrons charge to manipulate electron motion

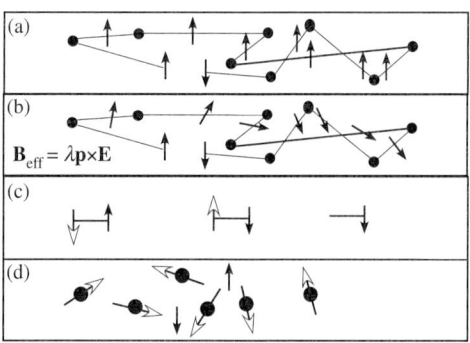

FIG 9.92 Four mechanisms of electron spin relaxation in semiconductors resulting from spin-flip processes. (a) The Elliot–Yafet mechanism: An electron undergoes many collisions with impurities and phonons, and after about 10^5 such scattering events, it has a finite probability to undergo a spin-flip. (b) Dyakonov–Perel mechanism: In semiconductors without center of symmetry, the spin-orbit coupling is equivalent to a momentum-dependent effective magnetic field. Following scattering by an impurity, the momentum is altered and the effective field changes with it. Precession axis and precession frequency change randomly. (c) Bir–Aronov–Pikus mechanism: Viewed from left to right, an electron exchanges spin with holes, and the hole's spin then relaxes very fast. (d) Spin can also be flipped due to hyperfine interaction with the atomic nuclei.

or to store information, spintronic devices also use electron spin. This is achieved by applying external magnetic fields and making use of spin-orbit coupling in the semiconductor. The advantage of spin-based devices is that they are nonvolatile, compared with charge-based devices. Small magnetic structures for applications in nonvolatile memory devices (i.e., memory that can retain stored information even when not powered) and magnetosensors have been developed, and many more are proposed. In the next sections, we briefly discuss a few spintronic devices.

Giant Magnetoresistance Devices

Giant magnetoresistance (GMR) is based on the physics of the magnetic tunnel junction discussed in relation to Figs 9.89 and 9.90, but the region between the two ferromagnetic leads is a good conductor that is a few nanometers thick. The idea is to use magnetization changes to increase the sensitivity of the electrical current in the electronic device. A ferromagnetic region acts as a *spin valve* in the sense that it enables the passage of electrons whose spin is parallel to the polarization vector of the ferromagnetic region while it blocks the transmission of electrons with opposite spin. Thus, in the configuration of Fig. 9.89, by applying an external magnetic field on the right electrode, it is possible to switch it from the parallel orientation as in Fig. 9.89(a) into the antiparallel orientation as in Fig. 9.89(b), and the resistance thereby increases significantly. This is the phenomenon of *giant magnetoresistance* (GMR) discovered in 1988. The quantity ρ defined in Eq. (9.478) that reaches up to 50% quantifies the system performance. GMR enabled an increase in the density of information stored in hard disks by about two orders of magnitude.

Magnetic Tunnel Junction (MTJ)

The same physics as discussed in connection with Figs 9.89 and 9.90 results in a better mechanism for a sensitive magnetic field sensor, based on transmission tunneling magnetoresistance (TMR). Like GMR, TMR in a multilayer junction filters one spin polarization over another, depending on the orientation of the left and right electrodes, which can be controlled by an external magnetic field. The two technologies differ in their filtering mechanism. As we have seen in Fig. 9.89, in an MTJ, the two magnetic layers are separated by a thin insulating layer. If a bias is maintained across the junction, the electron tunneling depends on the relative orientation of the two ferromagnetic plates. Recall from the discussion pertaining to Fig. 9.90 that the density of states differs for up and down spins, causing an intrinsic magnetization of the material. Electrons tunnel across the device depending on the availability of free states for the "right" spin direction. If the two magnetic layers are parallel, a majority of electrons in one will find enough states of similar orientation in the other, thereby reducing the resistance of the device. However, if the electrodes are antiparallel, electrons with both spin directions will encounter a closed valve in either of the two plates, resulting in a higher total resistance. The TMR effect is larger than the GMR effect by about one order of magnitude, making it a good candidate for magnetic sensors. Nowadays, they are used in hard drives.

Magnetic Spin Transistor

An ordinary transistor is based on an *n-p-n* junction, where a gate controls the voltage across the *p*-type semiconductor. Depending on the direction of the applied voltage, free electrons are either attracted toward the gate or repelled away

9.7 Spintronics

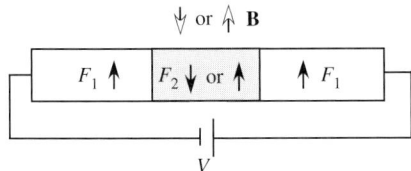

FIG 9.93 Magnetic spin transistor (schematic). An $F_1/F_2/F_1$ junction is integrated in an electric circuit. A magnetic field controls the direction of the magnetization vector in F_2 and thereby determines the resistance of the device.

from it. The voltage controls the flow of current between the two n-type semiconductors, allowing the transistor to occupy both "on" and "off" states. The problem with this transistor is its volatility. When power is shut off, the electrons in the p-type semiconductor diffuse throughout, destroying their previous "on" or "off" configuration. This is the reason why computers cannot be instantly turned on and off. In a magnetic transistor, shown in Fig. 9.93, magnetized ferromagnetic layers replace the role of n and p-type semiconductors. Much like in a spin-valve, substantial current can flow through parallel magnetized ferromagnetic layers. However, if in a three-layer structure, the middle layer is antiparallel to the two outside layers, the current flow would be quite restricted, resulting in a high overall resistance. If the two outside layers are pinned, and the middle layer is allowed to be switched by an external magnetic field, a magnetic transistor could be made, with on and off configurations depending on the orientation of the middle magnetized layer.

Spin Torque

Spintronic devices often require a contact F_1/F_2 between two ferromagnetic materials with different magnetization vectors $\mathbf{M}_1 \neq \mathbf{M}_2$. This is a domain wall configuration, as discussed in Sec. 9.5.10. In GMR or TMR structures, the relative orientation of magnetization affects the flow of the spin-polarized current. Therefore, controlling the relative orientations of $\mathbf{M}_1$ and $\mathbf{M}_2$ is crucial for the efficiency of the device. Suppose electrons flow from F_1 to F_2. After leaving F_1 they change their spin direction in F_2. In the absence of an external magnetic field, angular momentum conservation requires that $\mathbf{M}_2$ undergoes a small change as well. Therefore, we encounter a reverse effect wherein the flow of spin-polarized current between F_1 and F_2 transfers angular momentum from the carriers into the ferromagnet F_2 and alters the orientation of the magnetization $\mathbf{M}_2$. Changing $\mathbf{M}_2$ is equivalent to application of a torque on F_2. This phenomenon, *spin-transfer torque*, has been studied theoretically and experimentally. It can be used to switch magnetization direction without applying external magnetic fields (generating such magnetic fields using conventional electric currents is not always easy). Therefore, the possibility of switching the magnetization configuration in a multilayer system by directly using an electric current, rather than a magnetic field, is quite appealing.

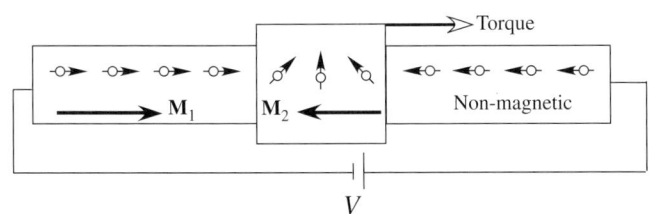

FIG 9.94 Spin torque effect (schematic). A current of spin polarized electrons can modify the direction of the moment of a ferromagnetic plate.

As an illustration, in Fig. 9.94, the junction (domain wall) $F_1/F_2/N$ is arranged such that the magnetization vectors are $\mathbf{M}_1 = M_1\hat{\mathbf{x}}$, $\mathbf{M}_2 = -M_2\hat{\mathbf{x}}$. Now, we consider sending a strongly polarized current in F_1 across the domain wall into the magnetic layer F_2 whose magnetization vector points in an opposite direction. At the F_2/N contact, the electrons propagate into N with their spin reversed, oriented along $-\hat{\mathbf{x}}$. Because the layer F_2 absorbed angular momentum, this process is followed by exerting a torque on the F_2 layer, that, in turn, affects the direction of $\mathbf{M}_2$. Thus, it is possible to change the direction of the moment by sending a spin-polarized electrical current through the magnet, instead of by applying a magnetic field.

Datta–Das Spin Transistor

In an electronic field effect transistor, electric charge is emitted from a source electrode and arrives at a drain electrode after being subjected to a gate potential that generates an electric field that controls the size of the channel through which the current flows. In this way, a small electric field can control large currents. The *Datta–Das spin transistor* is a device that can be regarded as the spin analog of the field effect transistor. It is a structure made from indium–aluminum–arsenide and indium–gallium–arsenide that enables 2D electron transport between the two ferromagnetic electrodes. The device, illustrated in Fig. 9.95, employs Bychkov-Rashba type spin-orbit coupling for its operation.

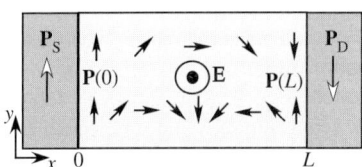

FIG 9.95 Datta–Das spin transistor. A two-dimensional semiconductor material (middle) is sandwiched between source and drain ferromagnetic leads with polarization vectors $\mathbf{P}_S = \uparrow$ and $\mathbf{P}_D = \downarrow$, respectively. The electrodes are held at small potential difference allowing charge current from source to drain. The spin-orbit interaction in the semiconductor is controlled by a gate inducing an in-plane electric field. Electrons moving from source to drain feel an effective in-plane magnetic field $\mathbf{h}(\mathbf{k}) = \alpha_R \hat{\mathbf{z}} \times \mathbf{k}$. Electrons injected from the source lead are initially polarized as $\mathbf{P}(0) = \mathbf{P}_S$. For weak spin-orbit coupling (upper channel), the precession is slow and the spin completes half cycle ending with polarization $\mathbf{P}(L) = \mathbf{P}_D$, and the conductance is maximal. For strong spin-orbit coupling (lower channel), the precession is faster and the spin completes full cycle ending with polarization $\mathbf{P}(L) = -\mathbf{P}_D$ and the resistance is maximal.

Suppose that the source and drain electrodes are polarized along polarization vectors $\mathbf{P}_S$ and $\mathbf{P}_D$. Electrons injected from the source (left ferromagnetic electrode) are polarized along $\mathbf{P}(0) = \mathbf{P}_S$. They reach the drain electrode with polarization $\mathbf{P}(L)$. If $\mathbf{P}(L) \neq \mathbf{P}_D$, the electron needs to align its magnetic moment along $\mathbf{P}_D$ and that costs an energy that is not available at very low temperature $K_B T < g\mu_B |\mathbf{P}_D|$. Hence, the electron is reflected back and the resistance of the device is enhanced. This kind of *spin valve* scenario is shown in Fig. 9.95. As in a field effect transistor, the top gate induces a perpendicular electric field $\mathbf{E} = E\hat{\mathbf{z}}$. According to Eq. (9.448), this induces a momentum-dependent effective magnetic field in the x–y plane, $\mathbf{h}(\mathbf{k}) = \alpha_R \hat{\mathbf{z}} \times \mathbf{k}$ [see Eq. (9.429)]. This effective field forces the electron spins to precess in the x–y plane.

Assuming ballistic transport (without scattering by impurities), the spins of the electrons precess with a fixed precession frequency proportional to the Rashba spin-orbit strength parameter α_R. Depending on the precession speed, the spin may precess at different angular speeds. In the configuration depicted in the upper part of Fig. 9.95, the electrons complete half a turn and smoothly enter the drain electrode, because they have their spin parallel to $\mathbf{P}_D$. The resistance of the device is then small. If the electrons precess faster and complete a full turn as in the lower part of Fig. 9.95, the electrons reach the drain, encounter an opposite polarization, and bounce back. This spin-valve scenario leads to enhanced resistance.

9.8 LOW-ENERGY EXCITATIONS

The topic of low energy excitations in solids has been very extensively studied (see Refs. [159, 160]). Here we briefly introduce it and give a few examples. We already pointed out in Sec. 9.6.5 that an exciton is an excited state of a many-electron system in a semiconductor or an insulator. This is an example of a low energy excitation that belongs to a collection of phenomena known as *elementary excitations in solids*. A closely related class of low energy excitations is known as *low energy collective excitations*. The difference between these two concepts is a bit blurry. In the 1960s, Nozières and Pines developed the theory for the elementary excitations of quasi-particles which transform into real particles for noninteracting Fermi systems, e.g., electrons and holes (and elecron-hole pairs). Collective excitations exist only due to interaction. These are modes describing the motion of a Fermi system as a whole: zero and first sound, plasmons, polaritons, etc. In Bose systems, the difference between elementary excitations and collective excitations is less obvious, because there are no phonons without elastic interaction, there are no magnons without exchange interaction, there is no second sound without bosonoic-particle interaction, and so on. Well defined Bose quasi-particles exist in cold-atom gases with atoms with integer angular momentum.

Collective excitations are distinct from single-particle excitations. As an example of the latter, consider a metal with a partially filled band at zero temperature. All states below the Fermi energy E_F are occupied, whereas all states above E_F are empty. An external field can excite an electron from its original state below E_F into a new energy state above E_F,

9.8 Low-Energy Excitations

and the empty level left below E_F is a hole. These *electron-hole excitations* are not collective excitations; they involve only a very few number of particles. Collective excitations can also be quantized and, as a result, behave like particles, in a fashion similar to photons, which are obtained by quantization of the electromagnetic field. Similar to photons, collective excitations have energy, momentum and a definite statistics (either bosonic or fermionic). For this reason, collective excitations are often referred to as *quasi-particles*.

As an example, consider electrons and ions in a solid, with the many-body Hamiltonian,

$$H = H_e + H_i + H_{ei}, \qquad (9.484)$$

where H_e is the electronic part, H_i is the ionic (vibrational excitation) part, and H_{ei} is the interaction Hamiltonian between electrons and ions. A full solution of the many-body problem described by the Hamiltonian (9.484) is out of reach. Consider the first two terms separately and then take into account the interaction between them. The low-energy collective excitations of H_e and H_i can sometimes be treated within the theory of small oscillations familiar from classical mechanics, and therefore, they can be regarded as a system of coupled quantum oscillators. After setting up the normal modes, the low-energy excited states are encoded by specifying the integer numbers n_j of quanta of mode j with frequency ω_j. These are the elementary excitations of the uncoupled systems. Examples are as follows: (1) The collective oscillations of H_i (the lattice vibrations of the ions in the solid), whose quantized modes are the phonons. Thesewill be discussed in Sec. 9.8.1. (2) Collective excitations of a system consisting of electrons in the background of the static ions in metals or *charge density waves*, whose quanta are called *plasmons*. These will be discussed in Sec. 9.8.2. (3) In a magnetic material, the deviation of spin direction from its ordered ground state configuration, which propagate from site to site, are *spin waves*. Their quanta are called *magnons*. These will be discussed in Sec. 9.8.3. The three examples of collective excitations mentioned above are schematically illustrated in Fig. 9.96. Moreover, in Sec. 9.8.4 we discuss the polaron, an excitation emerging when the electron and phonon systems are coupled, and in Sec. 9.8.5 we discuss the polariton, an excitation that arises due to strong coupling of electromagnetic waves with electric (or magnetic) dipole excitations of a crystal lattice.

FIG 9.96 Examples of collective excitations of many-particle systems (schematic): (a) Lattice vibrations (phonons), (b) Charge oscillations (plasmons), and (c) Spin waves (magnons).

9.8.1 PHONONS

In this section, we discuss the excitations of the crystal lattice; we denoted the crystal lattice Hamiltonian as H_i (the ionic Hamiltonian) in Eq. (9.484). Since our focus in this chapter is on electronic properties of solids and H_i does not involve electrons, the discussion of lattice vibrations will be brief. However, it should be stressed that the interaction between electrons and lattice vibrations is of utmost importance in solid state physics. For example, electron–phonon interaction plays a crucial role in the physics of superconductivity (see Sec. 18.9), and the interaction between electrons and phonons gives rise to the *polaron*, discussed later in Sec. 9.8.4.

Quantization of Lattice Vibrations

The Hamiltonian H_i contains the kinetic energy of the atoms (ions) and their mutual two-body interactions, and the theory of *small oscillations* is used to analyze the dynamics. In Sec. 9.8.4, we shall study the 1D case. Here, we list the main results for three dimensions. The basic assumption of the theory of small oscillations is that an atom located at a Bravais lattice point $\mathbf{R}_i$ can vibrate and therefore can be located at a nearby position, $\mathbf{r}_i = \mathbf{R}_i + \mathbf{u}_i$, where $|\mathbf{u}_i| \ll |\mathbf{R}_i - \mathbf{R}_j|$. One can expand the two-atom interaction $V(\mathbf{r}_i - \mathbf{r}_j)$ around $\mathbf{R}_i - \mathbf{R}_j$, keeping terms up to quadratic in $\mathbf{u}_i$. This leads to a description of lattice vibrations as a classical coupled oscillators, which can be written in terms of decoupled normal modes. The classical description can be quantized following the treatment of the harmonic oscillator in Sec. 2.7.2. Quantization is achieved by imposing the quantum commutation relation between position and momentum. The normal modes and their occupation are specified by the quantum numbers $n, \mathbf{k}, s$ where n is the oscillator level number, $\mathbf{k}$ is a wavenumber in reciprocal space, and $s = 1, 2, \ldots, S$ specifies the branch in each mode (see explanation below). The excitation energies of H_i are then given by

$$\varepsilon_{n\mathbf{k}s} = (n + 1/2)\hbar\omega_s(\mathbf{k}). \tag{9.485}$$

Similar to Bloch energies, the frequencies $\omega_s(\mathbf{k})$ are periodic functions of $\mathbf{k}$ with period $\mathbf{G}$ (a reciprocal lattice vector). These excitations of the lattice are referred to as *phonons*. Phonons behave as quantum mechanical particles with momentum $\hbar\mathbf{k}$ and energy $\hbar\omega_s(\mathbf{k})$. In analogy with the quantum harmonic oscillator in Sec. 2.7.2, we infer that phonons are bosons and obey Bose statistics.

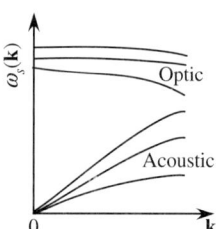

FIG 9.97 Examples of dispersion curves for 3D ionic crystal with two ions in the unit cell along a certain direction in reciprocal lattice space. There are three acoustic branches that are linear in small $\mathbf{k}$ and three optical branches that are flat at small $\mathbf{k}$.

A few words about the branch quantum number s are in order. In monoatomic crystals, $S = d$ (the space dimension) and the dispersion branches $\omega_s(\mathbf{k})$ are linear at small $\mathbf{k}$, with $\omega_s(0) = 0$. The corresponding excitations are referred to as *acoustic phonons*. In solids with more than one type of atom per unit cell, there are three additional branches for each additional atom. The dispersion laws for these additional branches satisfy $\omega_s(\mathbf{k} \to 0) \simeq c_s > 0$. These excitations are called *optical phonons* because in ionic crystals, such as sodium chloride, they can be excited by electromagnetic (usually infrared) radiation. An example of dispersion curves for a crystal with two atoms per unit cell is shown in Fig. 9.97.

Deviation from this somewhat ideal description of lattice vibrations as uncoupled oscillators arises when higher than quadratic terms are included in the expansion of the ion–ion potential energy, as required for highly excited states, leading to interaction between phonons. The derivation of the phonon Hamiltonian and the analysis of the electron–phonon interaction is carried out in Sec. 9.8.4 within the formalism of first quantization. It will be discussed further in Sec. 14.1.8 within the formalism of second quantization.

Phonon Contribution to the Specific Heat of Solids

In our study of the heat conductance of electrons in a solid and the Wiedemann–Franz law in Sec. 9.2.3, we calculated the electron contribution to the specific heat. Lattice vibrations also contribute to the specific heat of a solid. The calculation of this contribution requires knowledge of the free energy $F(T)$. It is obtained by averaging of the energy at finite temperature in terms of the partition function. A similar technique is used in the calculation of magnetization in Sec. 9.5.9 [see Eq. (9.299)] and in the electron population of impurities in semiconductors, as in Eq. (9.354). To be specific, we assume a gas of phonons at equilibrium temperature within a volume $\mathcal{V}$ and write down the partition function, and the averaged energy per unit volume,

$$Z = \sum_{n\mathbf{k}s} e^{-\beta \varepsilon_{n\mathbf{k}s}}, \quad F = -k_B T \ln Z, \quad u = -\frac{1}{\mathcal{V}}\frac{\partial F}{\partial \beta}. \tag{9.486}$$

Substitution of $\varepsilon_{n\mathbf{k}s}$ from Eq. (9.485) into the sum over n yields,

$$u = \frac{1}{\mathcal{V}} \sum_{\mathbf{k}s} \left[n_s(\mathbf{k}) + \frac{1}{2} \right] \hbar\omega_s(\mathbf{k}), \tag{9.487}$$

with bosonic occupation number,

$$n_s(\mathbf{k}) = \frac{1}{e^{\beta\hbar\omega_s(\mathbf{k})} - 1}. \tag{9.488}$$

9.8 Low-Energy Excitations

Problem 9.67
(a) Show that in the product, $\prod_{ks} \left(\sum_{n=0}^{\infty} e^{-\beta(2n+1)\hbar\omega_{ks}} \right)$, each energy ε_{nks} appears only once.
(b) Evaluate the sum in the brackets and show that $Z = \prod_{ks} n_s(\mathbf{k})$.
(c) Using these results, prove expression (9.487).

The 1/2 term on the RHS of Eq. (9.487) is present even at zero temperature; it is referred to as the *zero point energy*. The $n_s(\mathbf{k})$ term vanishes at zero temperature and determines the contribution of lattice vibrations to the specific heat per unit volume, $c_v = (\partial u / \partial T)_V$, whose evaluation requires knowledge of the frequencies $\omega_s(\mathbf{k})$. The classical temperature-independent expression $c_v^{cl} = 3(N/V)k_B$ due to Dulong and Petit is obtained in the high temperature limit.

We now develop rough estimates of c_v for high and low temperatures. As $\beta \to 0$ (high T), the exponent $\beta\hbar\omega_s(\mathbf{k}) \ll 1$, and a Taylor expansion is justified. The lowest order leads to c_v^{cl}, whereas higher order terms yield quantum corrections. When c_v^{cl} is subtracted and the expansion is terminated at the quadratic term, we find

$$c_v - c_v^{cl} = -\frac{\hbar^2 \beta}{12V} \sum_{ks} \omega_s^2(\mathbf{k}), \quad (\beta\hbar\omega_s(\mathbf{k}) \ll 1). \tag{9.489}$$

As $\beta \to \infty$ (low T), the exponent $\beta\hbar\omega_s(\mathbf{k})$ depends on the behavior of $\omega_s(\mathbf{k})$, and the contributions come only from vanishingly small frequencies. For lattices with a single atom per unit cell, $\lim_{|\mathbf{k}|\to 0} \omega_s(\mathbf{k}) = 0$, and these acoustic phonons contribute to the specific heat at low temperatures. For lattices with more than one atom per unit cell, there are also optical phonons for which $\lim_{|\mathbf{k}|\to 0} \omega_s(\mathbf{k}) > 0$. Hence, optical phonons do not contribute to c_v at very low temperatures. Keeping only the contribution of the acoustic phonons, a reasonable approximation is the linear spectrum of the form $\omega_s(\mathbf{k}) = a(\hat{\mathbf{k}})k$, which yields the approximate relation,

$$c_v \approx \frac{2\pi^2}{5} k_B \left(\frac{1}{\beta\hbar\langle a\rangle} \right)^3, \quad (T \to 0), \tag{9.490}$$

where $\langle a \rangle$ represents angular averaging of $a(\hat{\mathbf{k}})$.

To calculate the phonon contribution to the lattice specific heat at intermediate temperatures, we need to know the dispersion $\omega_s(\mathbf{k})$. Instead of specifying it for each given crystal structure, the main features can be elucidated using a crude approximation for $\omega_s(\mathbf{k})$. Two simple and useful dispersion formulations are due to Debye and to Einstein.

The Debye Model: In the Debye model, the dispersion is assumed to be linear in $\mathbf{k}$ and identical for all branches s, and the frequency is independent of the direction of the wavenumber:

$$\omega_s(\mathbf{k}) = ck \quad \text{(Debye model)}. \tag{9.491}$$

The constant c has the dimension of speed and is of the order of the speed of sound in solids, i.e., a few thousand meters per second. To calculate the specific heat, the summation over $\mathbf{k}$ in the second term on the RHS of Eq. (9.487) is approximated by an integral over volume in k-space within a sphere of radius k_D, the Debye wavenumber, which contains N $\mathbf{k}$ vectors. Because a point in k-space occupies a volume $(2\pi)^3/V$,

$$\frac{(2\pi)^3 N}{V} = \frac{4\pi}{3} k_D^3, \Rightarrow n = \frac{N}{V} = \frac{k_D^3}{6\pi^2} \quad \text{(phonon density)}. \tag{9.492}$$

Hence, the expression for c_v becomes

$$c_v = \frac{3\hbar c}{2\pi^2} \frac{\partial}{\partial T} \left[\int_0^{k_D} \frac{k^3 dk}{e^{\beta\hbar ck} - 1} \right]. \tag{9.493}$$

The result can be written using the *Debye frequency* ω_D and the *Debye temperature* Θ_D,

$$\omega_D \equiv ck_D, \quad k_B\Theta_D \equiv \hbar\omega_D. \tag{9.494}$$

The Debye temperature in solids is of the order of a few hundred Kelvin, e.g., $\Theta_D = 321K$ in NaCl. With these definitions, the Debye approximation for the specific heat yields

$$c_v = \frac{12\pi^4}{5}nk_B\left(\frac{T}{\Theta_D}\right)^3 \quad \text{(Debye expression for the specific heat)}. \tag{9.495}$$

Although the Debye approximation takes $\omega_s(0) = 0$, the contribution of optical phonons to the specific heat is tacitly included, because at $k \simeq k_D$, the acoustic phonon frequency flattens, as shown in Fig. 9.97. The linear approximation $\omega = ck$ for $k > k_D$ then adds to the contribution above that of the acoustic phonons. This extra contribution is attributed to the optical phonons.

The Einstein Model: In the Einstein model, the contribution of the acoustic phonons is identical to that of the Debye model, but the contribution of optical phonons is considered separately. Thus, in determining the contribution of the acoustic branches, the number of modes N used in determining the phonon density n as in Eq. (9.492) is not only equal to the total number of phonon modes but only to the number of acoustic modes. Because the acoustic modes are obtained from crystals with a single atom in the unit cell, N should be equal to the number of unit cells in the crystal. All the optical phonons are assumed to have the same frequency ω_E, independent of k, as becomes reasonable following Fig. 9.97. In this approximation, the contribution of the optical branches to the specific heat, denoted as c_v^{optical}, is simply evaluated. Each branch contributes $n\hbar\omega_E/(e^{\beta\hbar\omega_E} - 1)$ to the thermal energy. Assuming there are p optical branches, we obtain, after differentiating with respect to temperature,

$$c_v^{\text{optical}} = pnk_B\frac{(\beta\hbar\omega_E)^2 e^{\beta\hbar\omega_E}}{(e^{\beta\hbar\omega_E} - 1)^2}. \tag{9.496}$$

Comparing Electron and Phonon Contributions

We have written the contribution of the electrons to the specific heat using the Sommerfeld theory for electrons in metals in Eq. (9.72). The ratio between the electron contribution and the low temperature phonon contribution to the specific heat, Eq. (9.490),

$$\frac{c_v^{\text{electrons}}}{c_v^{\text{phonons}}} \approx \frac{5}{24\pi^2}Z\frac{\Theta_D^3}{T^2 T_F}, \tag{9.497}$$

where Z is the atom valence number (number of itinerant electrons that leave the atom and wander through the solid), and $T_F = E_F/k_B$ is the Fermi temperature. Thus, the electron contribution is dominant at low temperature, whereas the phonon contribution is dominant at high temperature. The two contributions are equal at

$$\bar{T} = \sqrt{\frac{5}{24\pi^2}Z\frac{\Theta_D^3}{T_F}}. \tag{9.498}$$

Since $T_F \approx 10\,\Theta_D$, $\bar{T}$ is of the order of several Kelvin. When $T < \bar{T}$, the specific heat of metals is linear in T, but for $T > \bar{T}$ a cubic dependence is obtained.

Phonon Density of States: van Hove Singularities

The occurrence of van Hove singularities discussed in Sec. 9.4.5 [see Eq. (9.140)] is also relevant here, this time in connection with the phonon spectrum. In calculating the contributions of phonons to the specific heat, we have replaced

9.8 Low-Energy Excitations

summation over **k** by integration. For a single mode and a single branch with dispersion $\omega(\mathbf{k})$, the prescription is

$$A \equiv \frac{1}{\mathcal{V}} \sum_{\mathbf{k}} f[\omega(\mathbf{k})] \to \frac{1}{(2\pi)^3} \int d\mathbf{k} f[\omega(\mathbf{k})]. \tag{9.499}$$

Because the value of **k** determines the frequency $\omega(\mathbf{k})$, it is useful to proceed further and integrate over frequency,

$$A = \int d\omega \, g(\omega) f(\omega), \quad g(\omega) = \frac{1}{(2\pi)^3} \int d\mathbf{k} \, \delta[\omega - \omega(\mathbf{k})]. \tag{9.500}$$

Here, $g(\omega)$ is the density of phonon states, i.e., the number of phonon states between ω and $\omega + d\omega$ per unit volume is $g(\omega)d\omega$. After imposing the delta function constraint, the density of states is expressed in terms of integral of the inverse velocity over a constant frequency surface $S(\omega)$ in **k**-space, determined by the constraint $\omega(\mathbf{k}) = \omega$, similar to expression (9.140),

$$g(\omega) = \frac{1}{(2\pi)^3} \int_{S(\omega)} \frac{dS}{|\nabla_\mathbf{k} \omega(\mathbf{k})|}. \tag{9.501}$$

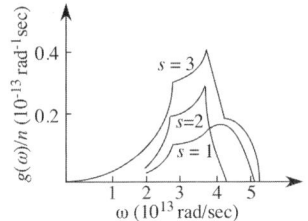

FIG 9.98 van Hove singularities appear in phonon density of states $g_s(\omega)$ (s=1,2,3) for the three acoustic branches obtained by neutron scattering on aluminum [Adapted from R. Stedman, L. Almquist and G. Nillson, Phys. Rev. **162**, 549 (1967)]. Permission obtained from The American Physical Society.

Because $\omega(\mathbf{k})$ is periodic in **k** [see remark after Eq. (9.485)], we expect $g(\omega)$ to exhibit van Hove singularities [as was pointed out regarding Eq. (9.140) for the density of states of Bloch electrons]. An illustration of van Hove singularities obtained by neutron scattering on aluminum is shown in Fig. 9.98.

9.8.2 PLASMONS

Plasmons are low energy collective excitations of an electron gas. Consider the electron gas in a metal, keeping in mind that the metal is electrically neutral. An external electric field acting on these electrons causes them to move and readjust. When the electrons are displaced, the positive charge left behind exerts an attractive force on the electrons, pulling them back to their original positions (see Fig 9.99). Once the external field is switched off, the restoring Coulomb interaction leads to charge density oscillations. These oscillations are collective, involving the entire electron gas. Since the physics of electrons in metals is governed by quantum mechanics, the charge oscillations are quantized. These quanta are called *plasmons*. The oscillation frequency, $\omega_p \equiv (4\pi e^2 n_e/m_e^*)^{1/2}$, where n_e is the electron density and m_e^* is its effective mass, is the *plasma frequency*. Typical plasmon frequencies are comparable with the frequencies of visible (or higher frequency) light. Plasmons have a finite lifetime and eventually decay by exciting single electrons.

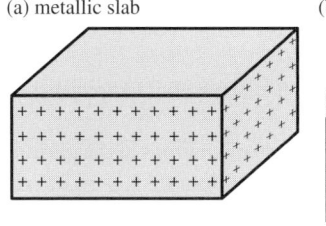
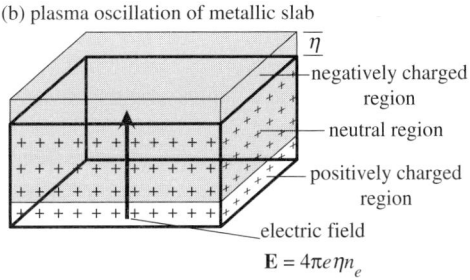

FIG 9.99 Plasmon excitation in a metallic slab. (a) Without plasma oscillation, no net charge density is present. (b) A plasma oscillation creates surface charge densities and an oscillating electric field (indicated by the thick arrow) results. *Source:* Band, Light and Matter, Fig. 4.16, p. 278

Plasmons can be excited by passing a high-energy electron through a metal or by reflecting an electron or a photon off a metal surface. The reflected or transmitted electron then shows an energy loss equal to integer multiples of the plasmon energy $\hbar\omega_p$. Plasmons play an important role in the optical properties of metals. Light of frequency $\omega < \omega_p$ is reflected because the electrons in the metal screen the optical electromagnetic field. Light of frequency $\omega > \omega_p$ is transmitted because electrons cannot respond fast enough to screen it. In many metals, the plasma frequency is in the ultraviolet region; hence, metals are often shiny (reflective) for light in the visible range.

To get a quantitative understanding of plasmons, suppose that the ions have a net positive charge ze and local density $n_i(\mathbf{r}, t)$. The electron density can be written as

$$n_e(\mathbf{r}, t) = z n_i(\mathbf{r}, t) + \delta n_e(\mathbf{r}, t), \tag{9.502}$$

where $|\delta n_e| \ll n_i$, because the local deviation from electrical neutrality is expected to be small. For simplicity, we assume that n_i is spatially uniform and time independent. The electric field that arises in a plasma is proportional to $\delta n_e(\mathbf{r}, t)$, and this field gives rise to a slow motion of electrons at speed $\mathbf{v}_e(\mathbf{r}, t)$. The continuity condition (which results from charge conservation), the equation of motion for an electron, and the Coulomb law Maxwell equation, take the following forms:

$$\frac{\partial \delta n_e(\mathbf{r}, t)}{\partial t} + \nabla \cdot [z n_i \mathbf{v}_e(\mathbf{r}, t)] = 0, \tag{9.503a}$$

$$m_e^* \frac{\partial \mathbf{v}_e(\mathbf{r}, t)}{\partial t} = -e\mathbf{E}, \tag{9.503b}$$

$$\nabla \cdot \mathbf{E}(\mathbf{r}, t) = -4\pi e \delta n_e(\mathbf{r}, t). \tag{9.503c}$$

In Eq. (9.503a), we approximated $n_e(\mathbf{r}, t)\mathbf{v}_e(\mathbf{r}, t)$ by $z n_i \mathbf{v}_e(\mathbf{r}, t)$. Taking the time derivative of Eq. (9.503a), and the divergence of Eq. (9.503b), as well as using (9.503c),

$$\frac{\partial^2 \delta n_e(\mathbf{r}, t)}{\partial t^2} + \frac{4\pi e^2 z n_i}{m_e^*} \delta n_e(\mathbf{r}, t) = 0. \tag{9.504}$$

This is a harmonic oscillator equation, with the plasma frequency, ω_p (defined above), approximated here as,

$$\omega_p \equiv \left(\frac{4\pi e^2 n_e}{m_e^*}\right)^{1/2} \approx \left(\frac{4\pi e^2 z n_i}{m_e^*}\right)^{1/2}. \tag{9.505}$$

Thus, the electron density $\delta n_e(\mathbf{r}, t)$ oscillates harmonically with frequency ω_p. Since there is no spatial dependence in this equation, this density displacement is uniform; the electron gas oscillates as a whole with respect to the positive background.

The behavior of the electric field propagating in the plasma can be deduced from the analysis before Eq. (9.55). Thus, the solution of Eq. (9.54) is

$$\mathbf{E}(\mathbf{r}, \omega) = \mathbf{E}_0 e^{i\kappa \hat{\mathbf{n}} \cdot \mathbf{r}}, \tag{9.506}$$

where $\hat{\mathbf{n}}$ is the direction of propagation and

$$\kappa = \frac{\omega}{c}\sqrt{\epsilon(\omega)}. \tag{9.507}$$

From Eq. (9.55), we see that the dielectric constant $\epsilon(\omega)$ is complex and the electric field decays as it propagates if $\omega < \omega_p$.

The above analysis shows that, in bulk metal, the electron gas can have collective longitudinal excitations at the plasma frequency ω_p. The electromagnetic field at frequency ω will decay if $\omega < \omega_p$ or propagate without decay if $\omega > \omega_p$.

Surface Plasmons: Coherent electron oscillations may exist also at the interface between two materials (e.g., a surface of a metal and air). They appear as charge density waves near the surface of the metal. Again, these oscillations are quantized, and the quanta are called *surface plasmons*.

9.8 Low-Energy Excitations

Surface plasmons have played a significant role in a variety of areas of fundamental and applied research, including (1) surface dynamics and surface-plasmon microscopy, (2) surface-plasmon resonance technology, (3) a wide range of photonic applications, and (4) surface-enhanced Raman Spectroscopy. Moreover, surface plasmons can be converted to an electrical signal via a photodetector in the metal plane or can be decoupled into freely propagating light by a defect or grating structure, thereby scattering the surface wave into free space.

The frequency of surface plasmons $\omega(\mathbf{k})$ depends on the wavevector of the exciting electromagnetic field (see later) but appears to be smaller than the plasma frequency in the bulk, ω_p. This means that the field emerging from these charge oscillations cannot penetrate deeply into the metal, but can develop outside the surface, as illustrated in Fig. 9.100. Surface plasmons can also form on nanoparticles or nanostructures, such as a small nanosphere or a hole drilled into a thin sheet of metal.

Instead of studying electron oscillations on the surface of a bulk metal, one may also consider electrons trapped in a thin metallic film on top of an insulator, as shown in Fig. 9.101.

Surface plasmons can be excited by the electromagnetic field of incident light, or by the passage of a pulsed charged particle beam. Strictly speaking, the quantum nature of plasmons requires that the exciting electromagnetic field be quantized to describe a beam of photons incident on the metal-air interface. The excited surface plasmons can

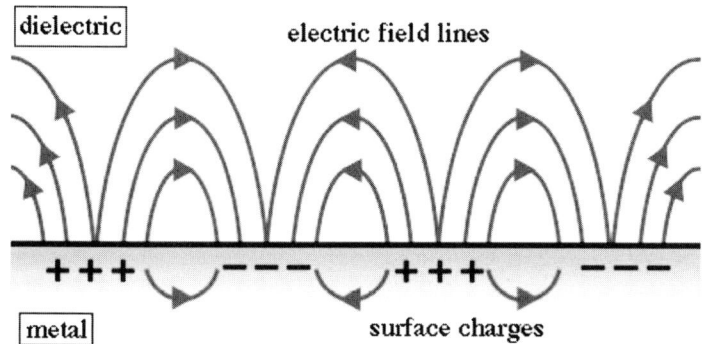

FIG 9.100 Electric field and charge distribution of a surface plasmon.

decay back into photons, as evidenced by the increased intensity of light radiated from the interface as the reflected intensity decreases. Analysis of the quantization of the electromagnetic field requires *second quantization* (see Chapter 14). Here, we briefly discuss the physics on the classical level.

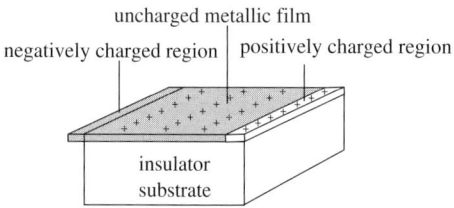

FIG 9.101 Surface plasmon in a thin-metal film on an insulator substrate.

Consider an interface between two media, 1 (dielectric), and 2 (metal), as shown schematically in Fig. 9.102. In the present discussion, it will be assumed that the dielectric constants are real. Assuming nonmagnetic materials ($\mu = 1$), writing Maxwell's equations for the electric field in each medium, and choosing a solution with $E_y = 0 = H_x$, the general solution for the electric field is

$$\mathbf{E}(z > 0) = E_0 e^{i(kx-\omega t) - \kappa_1(k,\omega)z}\left(1, 0, i\frac{k}{\kappa_1(k,\omega)}\right), \quad (9.508)$$

$$\mathbf{E}(z < 0) = E_0 e^{i(kx-\omega t) + \kappa_2(k,\omega)z}\left(1, 0, i\frac{k}{\kappa_2(k,\omega)}\right), \quad (9.509)$$

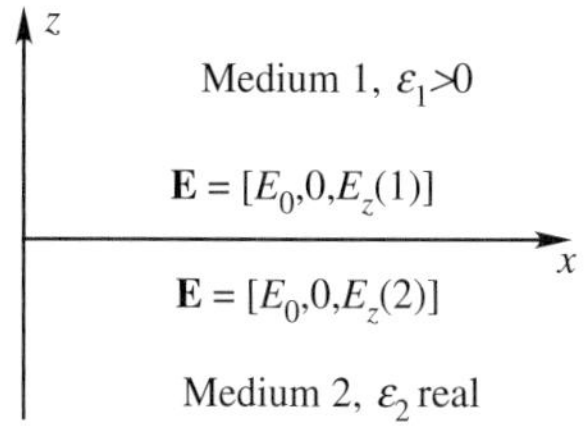

where

$$\kappa_{1,2}(k,\omega) = \left(k^2 - \varepsilon_{1,2}\frac{\omega^2}{c^2}\right)^{1/2}. \quad (9.510)$$

FIG 9.102 Interface between two dielectric media and the components of the electric field above and below the interface.

The analogous expression for the magnetic field **H** can be obtained from Maxwell's equations. There are three constraints that the solutions must satisfy. The tangential component of the electric field, E_x, and the perpendicular component of the magnetic field, H_z, should be continuous on traversing the interface, and the fields should vanish as $z \to \pm\infty$. The condition $E_x(z=0^+) = E_x(z=0^-)$ is automatically satisfied in the solutions (9.508) and (9.509). The continuity of H_z and the decay of the fields as $z \to \pm\infty$ are satisfied if

$$\varepsilon_1 \kappa_2(k,\omega) = -\varepsilon_2 \kappa_1(k,\omega), \tag{9.511}$$

$$\kappa_{1,2}(k,\omega) > 0. \tag{9.512}$$

The expression for ω derived from these equations is

$$\omega^2 = c^2 k^2 \left(\frac{1}{\varepsilon_1} + \frac{1}{\varepsilon_2} \right). \tag{9.513}$$

Note, however, that Eqs (9.511) and (9.512) can be satisfied only if $\varepsilon_2 < 0$ and Re $\varepsilon_1 > 0$. Moreover, to get a real-valued frequency, Eq. (9.513) can be satisfied only if $-\varepsilon_2 > \varepsilon_1$.

To obtain the dependence of ω on k, we need to know the dependence of ε_2 on ω. Assuming, for concreteness, that $\varepsilon_1 = 1$ (for air), and $\varepsilon_2 = 1 - \omega_p^2/\omega^2$ for the metal, we obtain the dispersion relation for the surface plasmons,

$$\omega^2(k) = \frac{1}{2}\omega_p^2 + c^2 k^2 - \left[\frac{1}{4}\omega_p^4 + c^4 k^4 \right]^{1/2}. \tag{9.514}$$

Thus, $\omega(k) \to \omega_p/\sqrt{2}$ as $k \to \infty$. It is readily verified that $\omega(k) < ck$. Therefore, the dispersion curves for the surface plasmons and the incident light do not intersect. Hence, it is not possible to match the frequency and wavevector of the surface plasmons to those of light in air. To circumvent this mismatch, the incident light is directed through a glass before reaching the metallic surface, thereby increasing the wavevector by the refractive index of the dielectric, n(dielectric) (see below in the discussion of resonance). These considerations are summarized in Fig. 9.103.

Surface Plasmon Resonance (SPR): When the frequencies of the electromagnetic field inside the dielectric and the surface plasmons on the metal surface match each other, the system exhibits a *surface plasmon resonance* (SPR). This occurs when the dispersion curves intersect (see Fig. 9.103). The interaction of the light and the electron charge distribution is resonant at a light frequency (wavelength) that depends strongly on the size, shape, composition, and environment of the surface structure. The fields associated with these modes are significantly enhanced.

Let us assume that we have an interface between two transparent dielectric media, such as glass and water. Light coming from the side of higher refractive index (glass) is partly reflected and partly refracted. Above the critical angle, there is no refraction, only total internal reflection. At

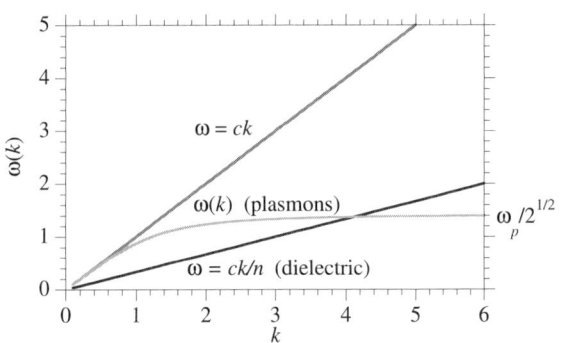

FIG 9.103 Dispersion curves for light in air, $\omega = ck$, light in a dielectric, $\omega = ck/n$ (dielectric) and for the surface plasmon $\omega(k)$ derived from Eq. (9.514). The surface plasmon dispersion curve intersects that for light in the dielectric at finite k. Note that $\omega(k)$ saturates at $\omega_p/\sqrt{2}$.

the same time, the electromagnetic field component penetrates a short distance into the medium with the lower refractive index (water), thereby creating evanescent waves. Now, we assume that the interface between the media is coated with a thin layer of metal (e.g., gold) and that the light is monochromatic and p-polarized (electric field polarized in the plane of incident k-vector and the surface normal). At a specific incident angle, the intensity of the reflected light is reduced,

9.8 Low-Energy Excitations

as there is a resonant energy transfer between evanescent waves and surface plasmons. The resonance conditions are influenced by the material adsorbed onto the thin metal film. An experimental setup taking into account the dependence of frequency of the incident wave on the glass refractive index and the incident angle (known as Kretschmann method) is shown in Fig. 9.104.

Plasmons in Nanoparticles: Plasma oscillations can be excited in metallic nanostructures as well. In this case, they are called *nanoparticle plasmons*. Nanoparticle plasmons are used as sensors in the sense that the characteristics of a given molecule or molecular cluster can be related to its plasmonic oscillations. The spatial dimension of the particles is such that light can penetrate the whole object and affect all the conduction electrons (see Fig. 9.105).

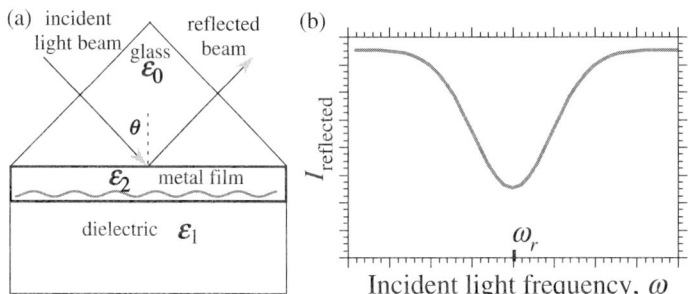

FIG 9.104 (a) Schematic illustration of Kretschmann's experimental setup for excitation of surface plasmons through a resonance. A thin metallic film is evaporated on glass. A light beam passing through the glass generates evanescent modes propagating inside the metal film. When the resonance condition is satisfied, surface plasmons (wavy curve) can be excited at the metal-dielectric interface. (b) Reflected intensity versus incident light frequency (schematic). The reflected intensity is normally high, except near the resonance frequency, ω_r, where the light energy is strongly absorbed by the surface plasmons.

The cloud of conduction electrons is displaced with respect to the positive ions, and the generated electric dipole exhibits a restoring force. Thus, we have a system that resembles an harmonic oscillator driven by light and damped by ohmic losses (generation of heat) or by decay of plasmons followed by an emission of photons. An important difference between surface plasmon resonance and nanoparticle plasmon resonance is that in the first case, sophisticated experimental techniques are required to match the wavevectors of light and the traveling electrons, whereas in the latter case, the light wavelength and the size of the nanoparticle are similar and the plasmons are effectively localized. Thus, nanoparticle plasmons can be localized irrespective of the light direction. The only condition that must be met is matching the wavevectors to achieve resonance with the nanoparticle plasmon oscillations. For applications, this is a significant advantage as compared with surface plasmons.

Plasmons are promising candidates for a number of devices. For example, they may propagate a long distance along the surface of a metal and, hence, may be used for information transfer on computer chips. They can generate high local electric fields at metal surfaces, and because the position and inten-

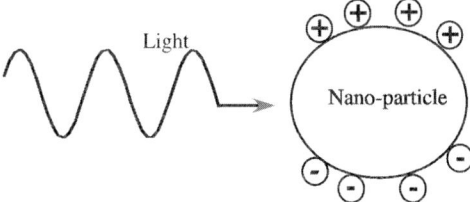

FIG 9.105 Light wave excites a plasmon resonance in a metallic nanoparticle.

sity of plasmon absorption and emission peaks are affected by molecular adsorption, plasmons can be used in molecular sensors. Furthermore, surface plasmons can confine light to very small dimensions, which might lead to optical trapping applications.

9.8.3 MAGNONS

When the atoms in a crystal have nonvanishing spins, their magnetic moments interact with each other. In Sec. 9.5.10, we stressed the magnetic exchange interactions involving itinerant electrons. The physics of localized spins interacting with each other is governed by a *Heisenberg spin Hamiltonian*, defined in Eq. (9.330). In this section, we discuss the Heisenberg spin Hamiltonian in 1D.

For ferromagnetic exchange, $J > 0$, the low lying (elementary) excitations are referred to as *magnons*. Consider a simple model with identical spin-1/2 atoms arranged in a 1D lattice and situated at points $x_n = na$, where a is the lattice constant (we shall take $a = 1$). The magnetic interaction between atoms is assumed to be nonzero only for nearest neighbor atoms. Denoting the spin operator of an atom located at x_n by $\mathbf{S}(n)$, interactions occur between $\mathbf{S}(n)$ and $\mathbf{S}(n \pm 1)$. In the Heisenberg model, the interaction between $\mathbf{S}(n)$ and $\mathbf{S}(n+1)$ has the form $J\mathbf{S}(n) \cdot \mathbf{S}(n+1)$. With these simplifications, the *Heisenberg spin Hamiltonian* [see Eq. (9.330)] takes the form

$$H = -J \sum_{n=1}^{N} \mathbf{S}(n) \cdot \mathbf{S}(n+1). \tag{9.515}$$

For convenience, we can take the periodic boundary condition, $\mathbf{S}(N+1) = \mathbf{S}(1)$. The sign of the exchange constant J determines the physical nature of the system. For $J > 0$, any two adjacent spins tend to align parallel to each other, while for $J < 0$, any two adjacent spins tend to align antiparallel to each other.[7] The Hamiltonian H acts on the 2^N dimensional product space of the spins. The evolution of the spin system by the unitary transformation $U = e^{-iHt/\hbar}$ is an example of a quantum gate.

The Heisenberg spin Hamiltonian describes an interacting many-body problem. We will discuss the spin physics of this many-body system in Chapter 14. Here we concentrate on the low energy spectrum for the case of ferromagnetic exchange $J > 0$, whose low energy excitations are spin waves or magnons [see Fig. 9.96(c)]. Before studying the low energy excitations, however, we need to identify the ground state of H for $J > 0$. We expect the ground state to be such that all spins are parallel,

$$|\Psi_0\rangle = |\uparrow_1\rangle \otimes |\uparrow_2\rangle \otimes \ldots \otimes |\uparrow_N\rangle \equiv |\uparrow_1 \uparrow_2 \ldots \uparrow_N\rangle. \tag{9.516}$$

This is indeed the case, as will be shown below. Note that

$$S_z(n)|\Psi_0\rangle = \frac{\hbar}{2}|\Psi_0\rangle, \quad \forall n. \tag{9.517}$$

In the absence of an external magnetic field, the Hamiltonian (9.516) has rotational symmetry, and the choice of quantization axis is arbitrary. Once this axis is chosen, any eigenstate with a given spin configuration is degenerate with the state obtained from it by reversing all spin directions. To prove that $|\Psi_0\rangle$ is the ground state, use the definition $S_\pm = \frac{1}{\sqrt{2}}(S_x \pm iS_y)$ and write the Hamiltonian as,

$$\mathcal{H} = -J \sum_{n=1}^{N} [S_+(n)S_-(n+1) + S_-(n)S_+(n+1) + S_z(n)S_z(n+1)]. \tag{9.518}$$

From the theory of angular momentum, recall that $|m\rangle$ is an eigenstate of S^2 and S_z, with eigenvalues $\hbar^2 s(s+1)$ and $\hbar m$, respectively, and

$$S_\pm |m\rangle = \hbar\sqrt{(s \mp m)(s \pm m + 1)}|m \pm 1\rangle. \tag{9.519}$$

Since we are considering the fully stretched state and $S_+|\uparrow\rangle = 0$, the first two terms on the RHS of Eq. (9.518) do not contribute. Consequently, we can consider only the last term $S_z(n)S_z(n+1)$ acting on $|\Psi_0\rangle$ and

$$\mathcal{H}|\Psi_0\rangle = -J \sum_{n=0}^{N} S_z(n)S_z(n+1)|\Psi_0\rangle = -JN\frac{\hbar^2}{4}|\Psi_0\rangle \equiv E_0|\Psi_0\rangle. \tag{9.520}$$

---

[7] Spin models are relevant for quantum information. A system of N spins is equivalent to the system of N qubits discussed in Chapter 5.

9.8 Low-Energy Excitations

Problem 9.68

Show that E_0 is doubly degenerate and find the state orthogonal to $|\Psi_0\rangle$ [defined in Eq. (9.516)] with energy E_0.
Answer: $|\bar{\Psi}_0\rangle = |\downarrow_1\rangle \otimes |\downarrow_2\rangle \otimes \ldots \otimes |\downarrow_N\rangle \equiv |\downarrow_1\downarrow_2 \ldots \downarrow_N\rangle$. Compared with Eq. (9.517), the eigenvalue of $S_z(n)$ is $S_z(n)|\bar{\Psi}_0\rangle = -\frac{\hbar}{2}|\bar{\Psi}_0\rangle \,\forall n$.

It is left to show that E_0 is indeed the ground state energy. Let us assume that there is another state $|\Phi_0\rangle$ with energy $\bar{E}_0$. Then, we use the fact that for Hermitian matrices the largest diagonal element is bounded by the largest eigenvalue and that the expectation value of $\mathbf{S}(n) \cdot \mathbf{S}(n+1)$ is smaller than S^2, to get

$$\bar{E}_0 = \langle \Phi_0|H|\Phi_0\rangle \geq -J\sum_{n=1}^{N}\text{Max}_\Psi\left[\langle \Psi|\mathbf{S}(n)\cdot\mathbf{S}(n+1)|\Psi\rangle\right] \geq -JNs^2 = E_0. \tag{9.521}$$

Having found the ground-state wave function and energy the next task is to find the low energy excited states. A low-energy excited state can be constructed as a linear combination of states obtained from the ground state $|\Psi_0\rangle$ by reversing the direction of one of the spins. Consider the state $|n\rangle$ obtained from $|\Psi_0\rangle$ by reversing a spin located on site n,

$$|n\rangle \equiv \frac{1}{\hbar}S_-(n)|\Psi_0\rangle = |\uparrow_1\uparrow_2\ldots\uparrow_{n-1}\downarrow_n\uparrow_{n+1}\ldots\uparrow_N\rangle. \tag{9.522}$$

$|n\rangle$ is not an eigenstate of H, but direct calculations show that the required combinations of states $|n\rangle$ yielding eigenstates of H are the one-magnon states or *spin waves*,

$$|k\rangle = \frac{1}{\sqrt{N}}\sum_{n=1}^{N}e^{ikn}|n\rangle, \tag{9.523}$$

where periodic boundary conditions require the relation $k = \frac{2\pi}{N}n_k$, with $n_k = 1, 2, \ldots, N$. The energy of these one-magnon states is

$$E_k = E_0 + \frac{2Js}{\hbar}\sin^2(k/2), \tag{9.524}$$

which is quadratic for small k. Problem 9.69 considers the eigenvalues and eigenstates of the Heisenberg spin Hamiltonian with a constant magnetic field.

Problem 9.69

Consider the Hamiltonian $\mathcal{H}(H) = \mathcal{H}(0) - \frac{g\mu_B H}{\hbar}\sum_{n=1}^{N}S_z(n)$, where $\mathcal{H}(0)$ is the zero-field Heisenberg Hamiltonian defined in Eq. (9.515).

(a) Show that $|\Psi_0\rangle$ defined in Eq. (9.516) is the unique ground state and that the ground-state energy is
$E_0(H) = E_0(0) - Ng\mu_B H/\hbar$.
(b) Show that the one-magnon states $|k\rangle$ in Eq. (9.523) are eigenstates of $\mathcal{H}$, with energies
$E_k(H) = E_k(0) + g\mu_B H/\hbar$.

Note that: (1) The system with ferromagnetic exchange $J > 0$ does not have a gap. When $N \to \infty$, the one-magnon spectrum starts at E_0. (2) The total spin of state $|k\rangle$ is

$$\left\langle k\left|\sum_{n=1}^{N}S_z(n)\right|k\right\rangle = Ns\hbar - \frac{\hbar}{2} = (N-1)\frac{\hbar}{2}. \tag{9.525}$$

(3) The probability for the reversed spin to be found at location n is $|\langle n|k\rangle|^2 = 1/N$. (4) In analogy with the discussion of phonons (because magnons are bosons), the mean number of magnons of wavenumber k at temperature T is

$$n(k) = \frac{1}{e^{\beta E_k} - 1}. \qquad (9.526)$$

Now let us consider the two-magnon states composed of two spins with reversed directions relative to the other spins in the ground state,

$$|k_1 k_2\rangle = \frac{1}{\sqrt{N(N-1)}} \sum_{n_1 < n_2} e^{ik_1 n_1 + k_2 n_2} S_-(n_1) S_-(n_2) |\Psi_0\rangle. \qquad (9.527)$$

The analysis of two (and higher) magnon states is more complicated than for the one-magnon states. The constraint $n_1 \neq n_2$ appears to be a serious obstacle and requires the construction of a set of relations from which the allowed values of k_1 and k_2 are derived together with the corresponding energies $E_{k_1 k_2}$. These are known as *Bethe ansatz equations*. The elucidation of the ground state energy and wave function, as well as one-magnon states in the antiferromagnetic case $J < 0$ is far more complicated than the ferromagnetic case. The ground state configuration for $J < 0$ is *not* equal to the classical one of alternating spin directions. We will not discuss the consequence of these results for the magnetic properties of the spin system. The interested reader is referred to Ref. [161].

9.8.4 POLARONS

The interaction between electrons and phonons, which has been discussed in Sec. 9.8.1, is responsible for numerous phenomena, including low-temperature superconductivity.[8] In this section, we discuss another important facet of electron–phonon interaction: the collective excitation called the *polaron*. For this purpose, we derive the electron–phonon interaction for a 1D system within the formalism of first quantization.

Heuristically, the electron–phonon interaction that gives rise to polarons can be described as follows: a local change in the electronic state in a crystal modifies the interaction between the atoms composing the crystal, and thereby affects the phonon spectrum. Self-consistently, a local change in the state of the lattice ions locally modifies the electronic states, and so on. This electron–phonon interaction is effective even at zero temperature. Another way to look at it is to start with an electron moving in an ionic crystal and polarizing the atoms surrounding it, causing them to oscillate. Figure 9.106 describes this qualitatively. The oscillating ions affect the motion of the electron. As we have seen, oscillating atoms in a crystal are described by phonons. The quantum state composed of an electron and a polarized phonon cloud is a collective excitation, i.e., a quasi-particle the polaron. It has its own peculiar characteristics, including effective mass,

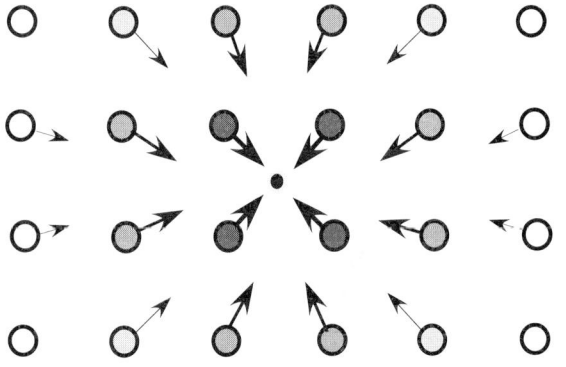

FIG 9.106 Schematic illustration of a polaron: An electron in an ionic crystal polarizes the atoms in its neighborhood and the deformed atomic configuration affects the electron state. The degree of polarization is indicated by the degree of shading, and the amplitude of oscillations is reflected by the thickness of the arrows. This many-body excitation consisting of a single electron surrounded by a cloud of phonons is a quasi-particle, with its own effective mass and momentum.

---

[8] A more thorough treatment of the electron–phonon interaction and of superconductivity requires second quantization, as discussed in Chapter 14.

9.8 Low-Energy Excitations

total momentum, spin, and energy. Other quantum numbers might be required to describe its internal state and its response to external fields. Polaron formation is a consequence of dynamic electron-lattice interaction, which is also responsible for scattering of charge carriers, phonon frequency renormalization, and screening of interaction between charge carriers in solids.

The *polaron problem* plays an important role in statistical mechanics and quantum field theory because it is one of the simplest examples of a nonrelativistic quantum particle interacting with a quantum field. It serves as a theoretical test bed for various tools to treat quantum field theories. The Feynman path integral formalism was originally applied to the polaron problem. Subsequently, it has become one of the important tools of statistical mechanics and quantum field theory.

The Polaron Hamiltonian

The derivation of electron interaction with lattice vibrations is based on the fact (used in Sec. 9.8.1) that the position of the ith atom (or ion) in the crystal is not fixed at lattice point $\mathbf{R}_i$. Rather, the atom undergoes small oscillations around $\mathbf{R}_i$. Its position is given by the vector $\mathbf{R}_i + \mathbf{u}_i$, where $|\mathbf{u}_i| \ll a$ (a is the lattice constant). The interaction potential between an electron located at space-point $\mathbf{r}$ and an ion located at $\mathbf{R}_i + \mathbf{u}_i$ is $V(\mathbf{r} - \mathbf{R}_i - \mathbf{u}_i)$. For example, in an ionic crystal with an unscreened Coulomb interaction between an electron and an ion of excess charge Ze, $V(\mathbf{x}) = -Ze^2/|\mathbf{x}|$. Expanding the interaction around $\mathbf{r} - \mathbf{R}_i$ to first order in $\mathbf{u}_i$ is

$$V(\mathbf{r} - \mathbf{R}_i - \mathbf{u}_i) \approx V(\mathbf{r} - \mathbf{R}_i) - \nabla V(\mathbf{r} - \mathbf{R}_i) \cdot \mathbf{u}_i. \tag{9.528}$$

When this interaction is summed over all lattice points we get two terms. The first $\sum_i V(\mathbf{r} - \mathbf{R}_i)$ is the periodic potential between the electron and the atoms when they are fixed at the lattice points $\mathbf{R}_i$. This term is treated within the familiar Bloch theory. The second term, $-\sum_i \nabla V(\mathbf{r} - \mathbf{R}_i) \cdot \mathbf{u}_i$, is the interaction of the electron with lattice vibrations. When the atoms in the lattice undergo small oscillations, the interaction energy of the electron and the lattice is modified. But what causes the oscillations of the atoms around the lattice points $\mathbf{R}_i$? These oscillations require energy, and at very low temperatures, the energy comes from the interaction with electrons. Therefore, an interaction of electrons with atoms leads to oscillations in the position of the atoms, which then affects the interaction, and so on. Having identified the term responsible for the electron–phonon interaction, we can now write the electron–phonon Hamiltonian within the general structure of the Hamiltonian in Eq. (9.484).

To simplify the derivation, we consider an electron moving in a 1D atomic crystal composed of N identical ions of mass M oscillating near lattice points x_n with small amplitudes, $|u_n| \ll a$, where $a = x_{n+1} - x_n$ is the lattice constant. The wave function of a single electron and the N ions is $\Psi(x, \{u_n\})$. We write the three terms on the RHS of Eq. (9.484) appropriate for the present system. First, the pure lattice phonon Hamiltonian consists of a collection of coupled Harmonic oscillators,

$$H_{\text{ph}} = -\frac{\hbar^2}{2M} \sum_n \frac{d^2}{du_n^2} + \frac{1}{2} \sum_{nm} D(x_n - x_m) u_n u_m, \tag{9.529}$$

where

$$D(x_n - x_m) = \left[\frac{d^2 U(x)}{dx^2}\right]_{x = x_n - x_m} \tag{9.530}$$

can be viewed as a real symmetric matrix obtained by expanding the atom–atom potential energy $U(x_n + u_n - x_m - u_m)$ to second order in the amplitudes (recall the theory of small oscillations). Second, the potential seen by the electron [the second term on the RHS of (9.528)] in this 1D model is

$$H_{\text{e-ph}} = -\lambda \sum_n \frac{d}{dx} V(x - x_n) u_n, \tag{9.531}$$

where λ is a dimensionless constant controlling the strength of the interaction. Finally, the electron Bloch Hamiltonian

$$H_e = -\frac{\hbar^2}{2m}\frac{d^2}{dx^2} + \sum_n V(x - x_n) \tag{9.532}$$

is responsible for the band structure of the electron spectrum. To proceed further, let us assume that the N atoms are arranged in a circular chain. The matrix D can be diagonalized, and its eigenvectors and eigenvalues are denoted as v_q and ω_q, respectively, where $q = \frac{2\pi}{N} n_q$, $n_q = 0, \pm 1, \ldots$. By using the Fourier transform relation

$$u_n = \frac{1}{\sqrt{N}} \sum_q v_q e^{iqx_n}, \tag{9.533}$$

the phonon Hamiltonian (9.529) can be written as a sum of oscillators,

$$H_{\text{ph}} = -\frac{\hbar^2}{2M} \sum_q \frac{d^2}{dv_q^2} + \frac{1}{2} \sum_q M\omega_q^2 v_q^2. \tag{9.534}$$

Next, we expand $V(x - x_n)$,

$$V(x - x_n) = \frac{1}{\sqrt{N}} \sum_q V(q) e^{iq(x-x_n)}, \quad V'(x - x_n) = \frac{i}{\sqrt{N}} \sum_q qV(q) e^{iq(x-x_n)}. \tag{9.535}$$

Inserting this expansion of V' and the expansion (9.533) of u_n into expression (9.531) for the electron–phonon Hamiltonian and using the orthogonality of the plane wave functions, we find

$$H_{\text{e-ph}} = -\lambda \frac{i}{\sqrt{N}} \sum_q qV(q) e^{iqx} v_q. \tag{9.536}$$

The Schrödinger equation for the electron phonon problem is then

$$\left[H_e + H_{\text{ph}} + H_{\text{e-ph}}\right] \Psi(x, \{v_q\}) = E\Psi(x, \{v_q\}), \tag{9.537}$$

where now $\Psi(x, \{v_q\})$ depends on the electron coordinate x and on the normal mode amplitudes v_q.

The Free Phonon Wave Function

In the absence of coupling between the electron and the atoms, we have

$$\Psi(x, \{v_q\}) = \psi_\alpha(x) \Phi_\beta(\{v_q\}), \tag{9.538}$$

i.e., a product of an electronic and atomic wave functions. The phonon wave function $\Phi_\beta(\{v_q\})$ describes the small oscillations of atoms near their lattice points $\{x_n\}$. Because the phonon Hamiltonian (9.534) is represented as a sum of independent oscillators, the free phonon wave function is written as a product,

$$\Phi_\beta(\{v_q\}) = \prod_q \eta_{m_q}(v_q), \tag{9.539}$$

where $\eta_{m_q}(v_q)$ is a harmonic oscillator wave function of order m_q in the variable v_q with energy $(m_q + \frac{1}{2})\hbar\omega_q$, so that $\beta = \{m_q\}$. The free phonon wave function (9.539) is compactly represented by a ket $|m_{q_1} m_{q_2} \ldots\rangle$. The ket $|0\rangle \equiv |0, 0, \ldots\rangle$ is said to be a *zero phonon state*, whereas the ket $|q\rangle \equiv |0, 0, \ldots, 1_q, 0, \ldots\rangle$ is a one-phonon state of energy $\hbar\omega_q$, and so on.

9.9 Insulators

> **Problem 9.70**
>
> Calculate the matrix element $\langle n_{q_1} n_{q_2} \ldots | v_p | m_{q_1} m_{q_2} \ldots \rangle$.
>
> **Answer:** $\sqrt{\frac{\hbar}{2M\omega_p}}[\sqrt{m_p}\delta_{n_p,m_p-1} + \sqrt{m_p+1}\delta_{n_p,m_p+1}]$.

Correction to the Electron Energy

Consider the case of weak coupling, $\lambda \ll 1$, where perturbation theory is justified. Let us consider a product state $|k\rangle \otimes |0\rangle$ of an electron in a Bloch state $\psi_k(x)$ with energy ε_k, and a zero phonon state with zero point energy E_0. We are interested in the perturbed energy $\varepsilon_k + \Delta_k$ due to the electron–phonon coupling. From the solution of Problem 9.70, it is clear that the first-order perturbation correction vanishes, and second-order perturbation theory couples the zero phonon state $|0\rangle$ to the single phonon states $|1_q\rangle$. The factor e^{iqx} in H_{e-ph} on the RHS of Eq. (9.536) requires that the electronic wave function $\psi_k(x)$ have nonzero matrix element with the wave function $\psi_{k-q}(x)$ having energy ε_{k-q}. The fact that H_{e-ph} connects the initial state $|k\rangle \otimes |0\rangle$ and the final state $|k-q\rangle \otimes |1_q\rangle$ means that momentum is conserved. Combining these results, we find

$$\Delta_k = -\frac{\lambda^2}{N}\sum_q \frac{\hbar}{2M\omega_q} \frac{[qV(q)]^2}{\varepsilon_k - \varepsilon_{k-q} - \hbar\omega_q}. \quad (9.540)$$

> **Problem 9.71**
>
> Use the result of Problem 9.70 and the formalism of second-order perturbation theory to prove Eq. (9.540).

If the electron energies are approximated by those of free electrons, $\varepsilon_k \approx \frac{\hbar^2 k^2}{2m}$, we see that for $k < \frac{q}{2} + \frac{m\omega_q}{\hbar q}$, the contribution to Δ_k is negative. In particular, for $k = 0$, $\varepsilon_k = 0$, the electron acquires a negative energy due to interaction with phonons; the polaron becomes weakly trapped.

9.8.5 POLARITONS

A polariton is a quasi-particle that arises due to strong coupling of electromagnetic waves with electric (or magnetic) dipole excitations of a crystal lattice, i.e., with any vibrational excitation that can couple to light. There are different types of polaritons, including phonon–polaritons, resulting from coupling of an infrared photon with an optic phonon, exciton–polaritons, resulting from coupling of visible light with an exciton, inter-subband-polaritons, resulting from coupling of an infrared photon with an inter-band vibrational excitation, and surface plasmon-polaritons, resulting from coupling of surface plasmons with photons. The dispersion relation for these excitations typically involves level repulsion (anticrossing) between the vibrational and optical mode frequencies involved. The phonons involved can be longitudinal acoustic or transverse acoustic phonons or longitudinal optical or transverse optical phonons (transverse modes oscillate with the atomic displacement perpendicular to the wavevector, whereas longitudinal modes have atomic displacement along the wavevector). For more on polaritons, see Refs. [18, 125, 251].

9.9 INSULATORS

An insulator is a material with vanishingly small DC electrical conductivity. This definition is valid assuming: (1) the applied electric field is weak and static, (2) the temperature is low, and (3) the volume of the system is large. Charge transport in solids involves moving electrons and holes, which are subject to Coulomb interaction among themselves.

Hence, conductivity of insulators is a many-body problem. Nevertheless, in numerous cases, the interaction between electrons can be treated within Landau–Fermi liquid theory, which results in an effective theory of noninteracting (quasi-) particles, see Sec. 18.5. On the other hand, there are several cases where the Landau–Fermi Liquid description is inadequate, and the many-body aspects resulting from the electron–electron and the electron-phonon interaction are manifest. When such a system cannot be described in terms of weakly interacting quasi-particles, it is then referred to as a *strongly correlated electron system*.

When electron correlations are insignificant, the occurrence of an insulating phase can be due to a finite gap in the spectrum of single-electron extended states. In metals, there is a band of single-particle states with wave functions that are spread over the whole sample, as described by Bloch theory, and the Fermi energy is within the band. In insulators, either the density of states vanishes at the Fermi energy or the density of states is still finite, but the wave functions of the single-particle states decay exponentially at large distance. This kind of band of localized states applies to amorphous materials in which the Bloch theory does not apply. When electron correlations are significant, the distinction between a conductor and an insulator is more subtle. In ordered (nonamorphous) systems, an insulator is characterized by a gap for charge excitations in the many-body spectrum. The physics of the insulating phase in strongly correlated systems with disorder is not yet fully understood.

Insulators are classified according to the mechanisms that cause the gap and the nature of this gap. It is useful to first classify insulators into two main groups and then divide each group into classes. The first group includes those materials for which the mechanism responsible for the insulating phase can be explained within the single-particle picture; this group includes *band insulators*, *Peierls insulators*, and *Anderson insulators*. Recently, it was noticed that in some band insulators, the Bloch wave functions $u_\mathbf{k}(\mathbf{r})$ in a filled band have interesting properties related to their topology in reciprocal space. One manifestation of these properties is that when the system has edges, there are current-carrying states along these edges, despite the fact that when the system does not have edges (is infinite) it is an insulator. Thus, band insulators are subdivided into ordinary band insulators and *topological insulators*. The second group includes materials whose insulating behavior is due to electron correlations; their nature can be understood in terms of cooperative many-electron phenomena. The main class in this group is *Mott insulators*, which can be further subdivided into *Mott–Hubbard insulators* and *Mott–Heisenberg insulators*.

9.9.1 DEFINING INSULATORS

At the beginning of Sec. 9.2, we discussed the Drude theory of electrical conductivity, which is a classical linear response treatment appropriate for weak external fields. We obtained Ohm's law, whose simplest form for a static uniform field and isotropic material is $\mathbf{J} = \sigma \mathbf{E}$. However, many materials are not isotropic. In addition, the applied electric field $\mathbf{E}(\mathbf{r}, t)$ need not be constant and/or uniform. Within the linear response framework, we can concentrate on a single Fourier component for the electric field $\mathbf{E}(\mathbf{r}, t)$ and the current density $\mathbf{J}(\mathbf{r}, t)$ at a space-time point $(\mathbf{r}, t)$,

$$\mathbf{E}(\mathbf{r}, t) = \mathbf{E}(\mathbf{q}, \omega) e^{i(\mathbf{q} \cdot \mathbf{r} - \omega t)}, \quad \mathbf{J}(\mathbf{r}, t) = \mathbf{J}(\mathbf{q}, \omega) e^{i(\mathbf{q} \cdot \mathbf{r} - \omega t)}. \tag{9.541}$$

In Sec. 7.9, we obtained a linear response expression relating the Cartesian components of the current and the electric field, which defines the *conductivity tensor*, $\sigma_{\alpha\beta}(\mathbf{q}, \omega)$,

$$J_\alpha(\mathbf{q}, \omega) = \sum_\beta \sigma_{\alpha\beta}(\mathbf{q}, \omega) E_\beta(\mathbf{q}, \omega). \tag{9.542}$$

The limit $\mathbf{q} \to 0$ and $\omega \to 0$ corresponds to a uniform and static field. The conductivity tensor $\sigma_{\alpha\beta}$ is generally complex and both its real and imaginary parts have physical significance. The calculation of $\sigma_{\alpha\beta}$ starting from a microscopic Hamiltonian is an important problem in solid-state physics. As shown in Sec. 7.9.2, $\sigma_{\alpha\beta}(\mathbf{q}, \omega)$ can be obtained using the Kubo formula [see Eq. (7.262)]. One of the subtle points in applying the Kubo formula is that, in Eq. (9.542), the electric field includes both the externally applied field and a field due to the presence of other charges in the solid. We are interested in determining the response of the system to the external field. The Kubo formula is often applied under

9.9 Insulators

the assumption that the system is infinite. Hence, the relation between $\sigma_{\alpha\beta}$ and the resistance measured in a finite system subject to a voltage difference should be carefully examined.

A solid is said to be a metal or an insulator depending on whether or not the conductivity tensor $\sigma_{\alpha\beta}(\mathbf{q}, \omega)$ vanishes in the limit $\mathbf{q} \to 0$, $\omega \to 0$ and $T \to 0$. The $T \to 0$ restriction ensures that the system remains coherent on macroscopic length scales. With these points in mind, restricting ourselves to weak external fields and zero temperature, the definition of an insulator is obtained by requiring that the static electrical conductivity vanishes within the following limiting scheme [164],

$$\lim_{T \to 0} \lim_{\mathbf{q} \to 0} \lim_{\omega \to 0} \mathrm{Re}[\sigma_{ij}(\mathbf{q}, \omega)] = 0. \tag{9.543}$$

The three limiting procedures do not always commute and the order as expressed in Eq. (9.543) is important.

A central question is whether it is possible to drive a metal-insulator transition at zero temperature by tuning external parameters such as pressure, magnetic field, or chemical potential. Transitions that are not driven by temperature are referred to as *quantum phase transitions* [162]. Quantum phase transitions are sharp in the sense that the conductivity (as a function of the driving parameter) vanishes either as a step or as a power smaller than unity. When $T \neq 0$, the sharp picture of quantum phase transitions is blurred.

9.9.2 CLASSIFICATION OF INSULATORS

Insulators are classified according to the mechanisms responsible for the occurrence of an insulating phase. As already mentioned, insulators are divided into two groups. The first group includes insulators for which the occurrence of the gap is due to electron–ion interactions, whereas the second contains insulators for which the occurrence of the gap is due to electron–electron interactions (or, more generally, electron correlations). The most notable representatives of the first group are

1. *Band insulators*. The insulating phase emerges due to the interaction of electrons with the periodic array of atoms in the crystal. The gap is understood using the Bloch theory.
2. *Topological insulators*. These are electronic band insulators in 2D or 3D systems with the following two properties:
    (1) *The occurrence of edge states*. Systems having boundaries (edges in 2D or surfaces in 3D systems) can have states that are localized at the boundaries. In a topological insulator that has definite geometrical boundaries, there is a finite number of conducting states with energies inside the gap that propagate along the boundaries of the sample. These edge states are protected against disorder and interactions. Depending on the symmetries of the physical system under time reversal, these states may also conduct spin, in addition to charge.
    (2) *Existence of a topological quantum number*. This property is more subtle and concerns bulk (infinite) systems without edges. The eigenfunctions in a filled band depend on a set of continuous quantum numbers [like the Bloch functions $\psi_\mathbf{k} = e^{i\mathbf{k} \cdot \mathbf{r}} u_\mathbf{k}(\mathbf{r})$ that depend on $\mathbf{k}$ in the periodic reciprocal lattice]. By changing these quantum numbers adiabatically, as with Berry phases (see Sec. 7.8.4), a set of nontrivial integer quantum numbers arise, which are independent of the details of the potential and are immune to small perturbations. Relations exist between the topological quantum numbers that occur in bulk systems (without boundaries) and the number of edge states that occur when the systems have boundaries [see property (1)].
    As an example of a topological insulator, consider a 2D electronic system with periodic potential $V(x, y)$ of period a subject to a perpendicular magnetic field H such that $Ha^2 = \frac{p}{q} \Phi_0$ [163]. Here, p and q are integers with no common divisors and $\Phi_0 = hc/e$ is the quantum flux unit. In the Landau gauge, Eq. (9.221) becomes

$$\mathcal{H}_{2D} = \frac{1}{2m}\left[p_x^2 + \left(p_y + \frac{hpx}{a^2 q}\right)^2\right]. \tag{9.544}$$

The corresponding eigenfunctions $\psi_\mathbf{k}(\mathbf{r}) = e^{i\mathbf{k} \cdot \mathbf{r}} u_\mathbf{k}(\mathbf{r})$ are still Bloch wave functions provided k_x, k_y are defined within a *magnetic Brillouin zone*, with unit vectors $(\frac{2\pi}{aq}\hat{\mathbf{x}}, \frac{2\pi}{a}\hat{\mathbf{y}})$. The spectrum has the pattern of bands separated by gaps, and when bands are either filled or empty, we have a band insulator. What is more subtle is that, in a sense described later, this is not a trivial band insulator, but, rather, a topological insulator. As was shown in Ref. [163], when the Fermi energy is in a gap, the quantity

$$n = \frac{i}{2\pi} \sum \int_{MBZ} d\mathbf{k} \left[\left\langle \frac{\partial u_\mathbf{k}}{\partial k_x} \bigg| \frac{\partial u_\mathbf{k}}{\partial k_y} \right\rangle - \left\langle \frac{\partial u_\mathbf{k}}{\partial k_y} \bigg| \frac{\partial u_\mathbf{k}}{\partial k_x} \right\rangle \right], \quad (9.545)$$

is an integer. Here, the integration is carried out on the magnetic Brillouin zone and the sum is taken over the fully occupied bands below the Fermi energy. The above expression, after being multiplied by the quantum unit of conductance e^2/h can be reduced to the Kubo formula for the calculation of the Hall conductance σ_{xy}. From Laughlin's arguments (see Figs. 9.56 and 9.56), the Hall conductance in units of e^2/h is equal to the number n of edge states crossed by the Fermi energy in a system with boundaries. Thus, when the Fermi energy is in the gap between two Landau levels, the integer quantum Hall system is a topological insulator, and the value of the topological number in the bulk system is equal to the number of edge states in the system that has boundaries. The term "topological" is justified, because this is a global property of the Bloch wave functions, independent of the details of the periodic potential. The integer n is referred as the *first Chern number*. Due to the presence of the magnetic field, time-reversal symmetry is broken, and the edge state carries charge, but not spin. In Chapter 13, we will encounter topological insulators in which time-reversal symmetry is conserved, and the edge state are capable of carrying spin. These systems, predicted in 2005, are referred to as *quantum spin Hall systems*.

3. *Peierls insulators*. The insulating phase is due to electron interaction with static lattice deformations.
4. *Anderson insulators*. The insulating phase is due to the presence of disorder. In these insulators, electron interaction with nonmagnetic impurities and other lattice imperfections lead to localized states.

In the second group of insulators, the formation of a gap is due to electron correlations, and is a bona fide many-electron problem. A representative of this class is the *Mott insulator*. Within a finer classification, one may further distinguish between *Mott–Heisenberg* and *Mott–Hubbard insulators*, depending on whether or not there is long range order of the local magnetic moments that are formed as a result of electron correlations.

Often, more than one mechanism can be present, complicating the problem of identifying the mechanism responsible for a metal-insulator transition.

A more detailed discussion of insulators can be found on the web page (see below), associated with this book.
`https://sites.google.com/site/thequantumbook/Insulators`

Electronic Structure of Multielectron Systems

10

The electronic many-body problem in the presence of nuclear point charges is central to much of physics, chemistry, and nanotechnology. We start this chapter by introducing the full nonrelativistic time-independent Schördinger equation for such problems in Sec. 10.1. We shall assume that the positions of nuclei are fixed and try to determine the wave function of the electrons, which is antisymmetric under exchange of electrons (as we have seen in Chapter 8, the Pauli exclusion principle requires that many-electron wave functions should be antisymmetric with respect to the interchange of any two electrons). Electronic structure of atoms, molecules, and solids is usually calculated in this fixed nuclei approximation, since the nuclei move so much slower than electrons (but corrections beyond the *Born-Oppenheimer approximation* are sometimes important and will be discussed in Sec. 11.3). In Sec. 10.2, we introduce Slater and Gaussian type orbitals, the two commonly used basis functions for calculating electronic structure. Section 10.3 discusses *term symbols* for atoms. A further simplifying approximation for calculating electronic structure is the *Hartree–Fock approximation*, which is a *mean-field* treatment, wherein each electron feels the mean field due to all the other electrons. First, we treat two-electron systems in Sec. 10.4 before considering multielectron systems in Secs 10.5 and 10.6, starting with the Hartree approximation and then antisymmetrizing to obtain the Hartree–Fock approximation. Koopmans' theorem is discussed in Sec. 10.7, trends in the atomic radii of atoms depending on their position in the periodic table are considered in Sec. 10.8, and the fine and hyperfine structures of multielectron atomic systems are then treated as perturbations in Sec. 10.9. *Hund's rules* for determining which term symbol has the lowest atomic energy are also treated in this section. Electronic structure of molecules is considered in Sec. 10.10. The Hückel Approximation is introduced in Sec. 10.10.3, and Sec. 10.11 briefly considers electronic structure of metals using this approximation. Section 10.12 considers more accurate (but more complicated) approaches to calculating electronic structure, e.g., adding additional Slater determinants to the wave function in a procedure called *configuration interaction*, concludes this chapter. Textbooks that extensively treat the electronic structure of atoms and molecules include Refs [165–168]. Additional topics related to molecular structure will be introduced in the next chapter. The electron many-body problem is considered further in Chapter 14 on many-body theory, and the most popular method today for calculating ground electronic state structure, density functional theory, is treated in Chapter 15.

10.1 THE MULTIELECTRON SYSTEM HAMILTONIAN

Let us begin by writing the Hamiltonian for $\mathcal{N}$ charged particles with mass M_i and charge $\mathcal{Z}_i e$ interacting via Coulomb potentials,

$$H = -\sum_{i=1}^{\mathcal{N}} \frac{\hbar^2}{2M_i} \nabla_i^2 + \sum_{i=1}^{\mathcal{N}-1} \sum_{j>i} \frac{\mathcal{Z}_i \mathcal{Z}_j e^2}{[4\pi\varepsilon_0] r_{ij}}. \tag{10.1}$$

The Coulomb potentials have been written in a form that is also appropriate in SI units by inserting the factor $1/[4\pi\varepsilon_0]$; this factor should be set to unity in Gaussian units. In this chapter, we shall use atomic units, so not only is this factor set to unity, but $m_e = \hbar = e = 1$. In atomic units, the expression for the potential energy of N electrons interacting via Coulomb potentials is

$$\sum_{i=1}^{N-1} \sum_{j>i} \frac{1}{r_{ij}} = \frac{1}{r_{12}} + \frac{1}{r_{13}} + \cdots + \frac{1}{r_{23}} + \frac{1}{r_{24}} + \cdots + \frac{1}{r_{34}} + \frac{1}{r_{35}} + \cdots + \frac{1}{r_{N-1\,N}}, \tag{10.2}$$

and the full Hamiltonian of N electrons and N_n nuclei is

$$H = -\sum_{\alpha}^{N_n} \frac{\nabla^2_{R_\alpha}}{2m_\alpha} - \frac{1}{2}\sum_{i=1}^{N} \nabla_i^2 + \sum_{\alpha}^{N_n-1}\sum_{\beta>\alpha}^{N_n} \frac{Z_\alpha Z_\beta}{R_{\alpha\beta}} - \sum_{i=1}^{N}\sum_{\alpha}^{N_n} \frac{Z_\alpha}{r_{i\alpha}} + \sum_{i}^{N-1}\sum_{j>i}^{N} \frac{1}{r_{ij}}, \quad (10.3)$$

where $m_\alpha = M_\alpha/m_e$. We used $\mathbf{R}_\alpha$ to denote the coordinate of the αth nucleus, $\mathbf{R}_{\alpha\beta} \equiv \mathbf{R}_\beta - \mathbf{R}_\alpha$ for the relative coordinate vector between nuclei β and α, and $\mathbf{r}_{i\alpha} \equiv \mathbf{r}_i - \mathbf{R}_\alpha$ for the coordinate vector between an electron i and a nucleus α. The full nonrelativistic time-independent Schördinger equation for this system is

$$H\Psi(\mathbf{r}_1,\ldots,\mathbf{r}_N,\mathbf{R}_1,\ldots,\mathbf{R}_{N_n}) = E\Psi(\mathbf{r}_1,\ldots,\mathbf{r}_N,\mathbf{R}_1,\ldots,\mathbf{R}_{N_n}). \quad (10.4)$$

We begin our consideration of this equation by assuming that the nuclei are pinned at specific positions, $\{\mathbf{R}_\alpha\}$, and this reduces our interest to the wave function $\Psi(\mathbf{r}_1,\ldots,\mathbf{r}_N;\mathbf{R}_1,\ldots,\mathbf{R}_{N_n})$ where the nuclear coordinates are taken to be fixed parameters.

10.2 SLATER AND GAUSSIAN TYPE ATOMIC ORBITALS

In performing quantum mechanical calculations on multielectron atoms, molecules and condensed-phase systems, one assumes that each electron is in an atomic or molecular orbital. Molecular orbitals are formed from atomic orbitals, i.e., molecular orbitals are taken as a coherent superposition of orbitals having centers on several nuclei. Atomic orbitals can be taken in the form of *Slater Type Orbitals* (STOs), which are hydrogen-like orbitals. A $1s$ STO is given by

$$\xi(\mathbf{r}-\mathbf{R}) = \left(\frac{\zeta^3}{\pi}\right)^{1/2} e^{-\zeta|\mathbf{r}-\mathbf{R}|}, \quad (10.5)$$

where $\mathbf{R}$ is the center of the orbital (e.g., $\mathbf{R} = 0$ for an atom centered at the origin), the factor in front of the exponential on the RHS of Eq. (10.5) is the normalization coefficient, and the coefficient ζ can be optimized (or chosen to be hydrogen-like in the simplest calculations). The STO with quantum numbers nlm take the form

$$\xi_{n_i l_i m_i}(\boldsymbol{\rho}) = N_{n_i l_i m_i}\, g_{n_i l_i}(\rho)\, e^{-\zeta_{n_i l_i m_i}|\rho|}\, Y_{l_i m_i}(\theta,\phi), \quad (10.6)$$

where $\boldsymbol{\rho} = \mathbf{r} - \mathbf{R}$, the angles θ and ϕ are the polar and azimuthal angles of $\boldsymbol{\rho}$, and $g_{n_i l_i}(\rho)$ is a polynomial of order $n-l-1$. The simplest optimization procedure for the STOs is to take $\zeta_{nlm} = \frac{Z_{\text{eff}}}{na_0}$ instead of the hydrogen-like value $\zeta_{nlm} = \frac{Z}{na_0}$, and minimize the energy with respect to Z_{eff}. Better approximations for atomic and molecular orbitals are obtained using a coherent superposition of STOs, $\phi_j = \sum_k c_{jk} \xi_k$, where the coefficients c_{jk} can be optimized.

The atomic or molecular or condensed phase wave function is given by an antisymmetrized product of orbitals. The goal of quantum chemistry calculations is to find the values of the coefficients c_i (and the coefficients $\zeta_{n_i l_i m_i}$, if they are allowed to vary), which minimize the energy of the system, e.g., via the variational method.

In molecules, STOs are difficult to calculate with because they are not easily orthogonalized to orbitals centered on other atoms. Therefore, calculations are often done with Gaussian type orbitals (GTOs) instead of STOs. A Gaussian $1s$ orbital is of the form

$$\xi_i(\mathbf{r}-\mathbf{R}) = \left(\frac{2\alpha_i}{\pi}\right)^{3/4} e^{-\alpha_i|\mathbf{r}-\mathbf{R}|^2}, \quad (10.7)$$

where α_i is the square inverse width of the ith orbital. Higher order Cartesian Gaussians take the form

$$\xi_{i;jkl}(\mathbf{r} - \mathbf{R}) = N_{i;jkl}(x - R_x)^j (y - R_y)^k (z - R_z)^l e^{-\alpha_i |\mathbf{r} - \mathbf{R}|^2}, \tag{10.8}$$

where $N_{i;jkl}$ is the normalization factor. The Gaussian orbitals are taken to be normalized, but if they are centered on different atoms, they will not be orthogonal. The overlap integrals

$$S_{12;ij} = \int d\mathbf{r} \xi_i^*(\mathbf{r} - \mathbf{R}_1) \xi_j(\mathbf{r} - \mathbf{R}_2), \tag{10.9}$$

depend on $\mathbf{R}_1 - \mathbf{R}_2$, and are relatively simple to calculate. One problem with the Gaussian orbitals is that they do not have the cusp-like structure of the wave functions right near the atomic nuclei, and therefore many Gaussian orbitals with many different coefficients α_i are required to reasonably approximate the cusp.

10.3 TERM SYMBOLS FOR ATOMS

Multielectron atomic states can be labeled by the values of the total orbital, total spin, and total angular momentum for all the electrons together, L, S, and J. The states are labeled as follows: $^{2S+1}L_J$. Here, the J quantum number is necessary when one includes spin-orbit coupling, and $2S + 1$ is the spin degeneracy; $2S + 1 = 1$ is called a singlet state, $2S + 1 = 2$ is called a doublet state, $2S + 1 = 3$ is called a triplet state, etc. For example, the excited states of helium corresponding to one electron in $1s$ and one electron in $2p$ can be written as $1s\,2p\,^3P_2$, $1s\,2p\,^3P_0$, and $1s\,2p\,^1P_1$. For a two-electron atom with one electron in a $2p$ state and the other in a $3p$ state, we can have $2p\,3p\,^1D_2$, $2p\,3p\,^1P_1$, $2p\,3p\,^1S_0$ singlet states and $2p\,3p\,^3D_3$, $2p\,3p\,^3D_2$, $2p\,3p\,^3D_1$, $2p\,3p\,^3P_2$, $2p\,3p\,^3P_1$, $2p\,3p\,^3P_0$, $2p\,3p\,^3S_1$ triplet states. In Sec. 10.9, we present the *Hund's rules* for determining which term symbol has the lowest energy. For example, the electronic configuration of the ground state of silicon and its term symbol is given by $1s^2\,2s^2\,2p^6\,3s^2\,3p^2\,^3P_0$. Sometimes, only the open shell electrons are specified; then this state would be designated by $\ldots 3p^2\,^3P_0$.

In building term symbols for atoms, if two or more electrons are in the same orbital, not all possible term symbols can be formed because of the necessity to antisymmetrize the atomic wave function, and restrictions arise due to the Pauli exclusion principle. For example, let us consider, a configuration of three equivalent p electrons, for the time being without taking spin-orbit into account. The projection of the orbital angular momentum for each electron, m_l, can take the values $m_l = 1, 0, -1$, and the projection of the spin takes the values $m_s = 1/2, -1/2$, so there are six possible levels $|m_l, m_s\rangle$ for each of the three electrons. The three electrons can be in any three of these levels, but no two electrons can be in the same level (i.e., spin-orbital). It is easy to explicitly list the resulting three-electron states $|M_L, M_S\rangle$, which take the following values: $|M_L, M_S\rangle = |2, 1/2\rangle, |1, 1/2\rangle, |0, 3/2\rangle, |0, 1/2\rangle$ (as well as those with negative M_L and M_S, which we do not write out explicitly). Counting the possible ^{2S+1}L terms that result for the three equivalent p electrons, we find that the only possibilities are one term of each of the types: ^2D, ^2P, ^4S. Thus, many of the possible terms that could be obtained without imposing exchange symmetry are absent.

Problem 10.1

Write the normalized Slater determinant of the $(1s0)^2(2p0)(2p1)\,^3D_3$, $M_J = 2$ excited state of beryllium, where $(1s0)$, $(2p1)$ and $(2p0)$ mean the following: $(1s0) \equiv \psi_{n=1,l=0,m_l=0}(\mathbf{r})$ and $(2p1) \equiv \psi_{n=2,l=1,m_l=1}(\mathbf{r})$, $(2p0) \equiv \psi_{n=2,l=1,m_l=0}(\mathbf{r})$.

Answer: We must have $M_S = 1$ and $M_L = 1$, so the two unpaired spins both have spin-up.

10.4 TWO-ELECTRON SYSTEMS

Let us first consider the two-electron systems. The helium atom is such a system, as is the hydrogen molecule, but other two-electron systems could also be considered, e.g., the lithium ion, Li$^+$, doubly ionized beryllium, etc. For simplicity, we begin with an atomic system, specifically helium, so only one nuclear center need be considered.

10.4.1 THE HELIUM ATOM

The electronic helium Hamiltonian is

$$H = -\frac{1}{2}\nabla^2_{\mathbf{r}_1} - \frac{1}{2}\nabla^2_{\mathbf{r}_2} - \frac{2}{r_1} - \frac{2}{r_2} + \frac{1}{r_{12}}. \tag{10.10}$$

We can write this Hamiltonian as

$$H = h(\mathbf{r}_1) + h(\mathbf{r}_2) + \frac{1}{r_{12}}, \tag{10.11}$$

where $h(\mathbf{r}_i) \equiv -\frac{1}{2}\nabla^2_{\mathbf{r}_i} - \frac{2}{r_i}$ is the Hamiltonian of electron i, and $1/r_{12}$ is the Coulomb electron–electron potential.

We could try to treat helium using one of the approximation methods of Chapter 7. For example, if we treat the electron–electron term as the first-order perturbation, $H^{(1)} = r_{12}^{-1}$, the first-order contribution to the energy is (see Sec. 7.3.1)

$$\Delta E = E^{(1)} = \int d\mathbf{r}_1 d\mathbf{r}_2\, \psi^{(0)*} H^{(1)} \psi^{(0)}, \tag{10.12}$$

and the zeroth-order spatial part of the two-electron wave function $\psi^{(0)}$ is the product function,

$$\psi^{(0)} = \phi_1(\mathbf{r}_1)\phi_2(\mathbf{r}_2) = \left(\frac{Z^{3/2}}{\sqrt{\pi}}\right)^2 e^{-Zr_1} e^{-Zr_2} = \frac{Z^3}{\pi} e^{-Z(r_1+r_2)}, \tag{10.13}$$

with $Z = 2$. The antisymmetry of the total two-electron wave function then requires the spin function to be the singlet state. Since the Coulomb interaction does not depend on spin, the spin function is not explicitly written here. However, given the strength of the Coulomb interaction, the perturbation theory approach for treating the electron–electron interaction is a poor approximation. Below, we develop a more elaborate and more accurate approach using the self-consistent mean-field (SCF) approach.

Alternatively, a variational approach (see Sec. 7.6) starts with an expression for the expectation value of the Hamiltonian

$$\langle E \rangle \approx E_{\text{trial}} = \frac{\langle \phi_t | H | \phi_t \rangle}{\langle \phi_t | \phi_t \rangle} \geq E_0, \tag{10.14}$$

with some trial wave function containing parameters that can be determined by minimizing the expectation value with respect to the parameters. We shall use a variational approach within the SCF approach developed below.

A mean-field approach for treating interacting electrons in atoms and molecules is outlined in what follows. The parameters that appear in the mean-field wave function will be variationally varied to obtain the lowest energy.

10.4.2 THE HARTREE METHOD: HELIUM

Douglas Hartree first developed the mean-field theory for electron systems in the 1920s, but did not consider that electron wave functions must be antisymmetric with respect to exchange. Vladimir Fock then extended the theory to include antisymmetrization of the wave function, thereby developing the *Hartree-Fock method*. Let us first present Hartree's method, wherein we "guess" initial values for the individual atomic orbitals, perhaps in the form of an initial set of coefficients in a linear combination of STOs, and then self-consistently determine the atomic orbitals.

The mean-field iterative procedure is designed whereby the electron orbitals $\phi_i(\mathbf{r}_i)$ in the wave function (10.13) for helium are improved in a series of steps. We first consider electron 1 and assume that its interaction with the second electron (or with electrons number $2, 3, 4, \ldots$ in multielectron systems) is with the averaged electron density of the second electron. The interaction potential with the second electron is taken to be

$$V_1^{\text{eff}}(\mathbf{r}_1) = \int \frac{|\phi_2^{\text{init}}(\mathbf{r}_2)|^2}{r_{12}} d\mathbf{r}_2 = \left\langle \phi_2^{\text{init}}(2) \left| \frac{1}{r_{12}} \right| \phi_2^{\text{init}}(2) \right\rangle, \tag{10.15}$$

10.4 Two-Electron Systems

where the subscript in $\phi_2(2)$ refers to the quantum numbers of the state and the argument $(2) = (\mathbf{r}_2)$ refers to the coordinate, and the effective Hamiltonian is

$$H_1^{\text{eff}} = -\frac{1}{2}\nabla_1^2 - \frac{2}{r_1} + V_1^{\text{eff}}. \tag{10.16}$$

so the effective Hamiltonian acting on the wave function of electron 1 is given by sum, $H_1^{\text{eff}} \equiv h(\mathbf{r}_1) + V_1^{\text{eff}}(\mathbf{r}_1)$. The effective Schrödinger equation for electron 1, $H_1^{\text{eff}} \phi_1 = \varepsilon_1 \phi_1$, must be solved to get a new estimate for the function ϕ_1^{new} and the new effective orbital energy $\varepsilon_1^{\text{new}}$. Similarly, for electron 2: $H_2^{\text{eff}} \phi_2 = \varepsilon_2 \phi_2$, where

$$H_2^{\text{eff}} = -\frac{1}{2}\nabla_2^2 - \frac{2}{r_2} + V_2^{\text{eff}}, \tag{10.17}$$

with

$$V_2^{\text{eff}}(\mathbf{r}_2) = \left\langle \phi_1^{\text{init}}(1) \left| \frac{1}{r_{12}} \right| \phi_1^{\text{init}}(1) \right\rangle. \tag{10.18}$$

The effective Schrödinger equation for electron 2 must also be solved to get a new estimate for the function, ϕ_2^{new} and the new effective orbital energy $\varepsilon_2^{\text{new}}$.

Note that we have used the initial guessed wave functions to compute effective Hamiltonians and then the resulting equations are solved to get new orbitals. These new orbitals are then used to compute new effective Hamiltonians, and then the new resulting equations are solved to get new orbitals, etc. This procedure is carried out again and again, until the new orbitals and energies predicted by solving the equations get closer and closer to the orbitals and energies used in the previous step. In the $n + 1$ step ($n = 1, 2, 3, \ldots$), the effective potential

$$V_i^{(n+1)}(\mathbf{r}_i) \equiv \int d\mathbf{r}_j \frac{\left|\phi_j^{(n)}(\mathbf{r}_j)\right|^2}{|\mathbf{r}_i - \mathbf{r}_j|} = \left\langle \phi_j^{(n)} \left| \frac{1}{|\mathbf{r}_i - \mathbf{r}_j|} \right| \phi_j^{(n)} \right\rangle_{\mathbf{r}_j}, \tag{10.19}$$

is substituted into the effective Schrödinger equation

$$\left[-\frac{1}{2}\nabla_i^2 - \frac{2}{r_i} + V_i^{(n+1)}(\mathbf{r}_i) \right] \phi_i^{(n+1)} = \varepsilon_i^{(n+1)} \phi_i^{(n+1)}, \tag{10.20}$$

and one solves for $\phi_i^{(n+1)}$ and $\varepsilon_i^{(n+1)}$. A criterion for determining when to stop needs to be developed; the criterion typically chosen involves the calculated energy. When the energy continues to decrease and finally saturates (stops changing), the calculation is deemed to have converged. The resulting output orbitals yield a self-consistent field (SCF), i.e., the effective potentials, in which each of the electrons moves.

The Hartree energy is obtained as an expectation of the *exact Hamiltonian* (10.11) using the converged orbitals,

$$E_H = \int d\mathbf{r}_1 d\mathbf{r}_2 \, \psi^* H \psi = \langle \phi_1(1)\phi_2(2) | h(\mathbf{r}_1) + h(\mathbf{r}_2) + \frac{1}{r_{12}} |\phi_1(1)\phi_2(2)\rangle, \tag{10.21}$$

where we assume that the orbitals $\phi_1(1)$ and $\phi_2(2)$ are the converged normalized orbitals obtained from the SCF calculation.

This expression can be written in terms of the converged eigenvalues ε_i of the self-consistent equations (10.20). Since $H_1^{\text{eff}} \phi_1 = \varepsilon_1 \phi_1$ where

$$H_1^{\text{eff}} = -\frac{1}{2}\nabla_1^2 - \frac{2}{r_1} + V_1^{\text{eff}} = -\frac{1}{2}\nabla_1^2 - \frac{2}{r_1} + \langle \phi_2(2)| \frac{1}{r_{12}} |\phi_2(2)\rangle, \tag{10.22}$$

we find that

$$\varepsilon_1 = \langle \phi_1 | H_1^{\text{eff}} |\phi_1\rangle = \langle \phi_1(1)| - \frac{1}{2}\nabla_1^2 - \frac{2}{r_1} |\phi_1(1)\rangle + \langle \phi_1(1)\phi_2(2)| \frac{1}{r_{12}} |\phi_1(1)\phi_2(2)\rangle.$$

Hence,

$$\varepsilon_1 = \langle \phi_1(1)| h(1) |\phi_1(1)\rangle + \langle \phi_1(1)\phi_2(2)| \frac{1}{r_{12}} |\phi_1(1)\phi_2(2)\rangle \equiv I_1 + J_{12}, \qquad (10.23)$$

where $I_1 \equiv \langle \phi_1(1)| h(1) |\phi_1(1)\rangle$ is the expectation value of the kinetic energy plus the Coulomb potential energy of the electron and nucleus, and the *Coulomb repulsion integral* is

$$J_{12} \equiv \langle \phi_1(1)\phi_2(2)| \frac{1}{r_{12}} |\phi_1(1)\phi_2(2)\rangle = \int\int d\mathbf{r}_1 d\mathbf{r}_2 \frac{[\phi_1^*\phi_1][\phi_2^*\phi_2]}{r_{12}}. \qquad (10.24)$$

Similarly for the second electron, $H_2^{\text{eff}}\phi_2 = \varepsilon_2\phi_2$, where

$$H_2^{\text{eff}} = -\frac{1}{2}\nabla_2^2 - \frac{2}{r_2} + V_2^{\text{eff}} = -\frac{1}{2}\nabla_2^2 - \frac{2}{r_2} + \langle \phi_1(1)| \frac{1}{r_{12}} |\phi_1(1)\rangle, \qquad (10.25)$$

and

$$\varepsilon_2 = \langle \phi_2(2)| h(2) |\phi_2(2)\rangle + \langle \phi_1(1)\phi_2(2)| \frac{1}{r_{12}} |\phi_1(1)\phi_2(2)\rangle \equiv I_2 + J_{12}. \qquad (10.26)$$

Using the notation $\varepsilon_1 = I_1 + J_{12}$ and $\varepsilon_2 = I_2 + J_{12}$, we see that

$$\varepsilon_1 + \varepsilon_2 = I_1 + I_2 + 2J_{12}, \qquad (10.27)$$

and $\langle H \rangle = E = I_1 + I_2 + J_{12}$. The Hartree energy can be rewritten as

$$E_H = (I_1 + J_{12}) + (I_2 + J_{12}) - J_{12}, \qquad (10.28)$$

so

$$E_H = \varepsilon_1 + \varepsilon_2 - J_{12}, \qquad (10.29)$$

where $\varepsilon_1 + \varepsilon_2 = (I_1 + J_{12}) + (I_2 + J_{12})$. Note that the Coulomb repulsion integral is always positive. It accounts for the repulsion of two electron charge densities, $\rho(1) = -e|\phi_1(1)|^2$ and $\rho(2) = -e|\phi_2(2)|^2$, that are separated by a distance, r_{12}, and integrated over the coordinates of the two electrons.

The (one-center) Coulomb integrals can be evaluated using the multipole expansion (3.197). For example, taking a $1s^2$ orbital configuration with STO (10.5), the integral is particularly simple because the wave functions are independent of angles. More generally, for orbitals of arbitrary angular momentum, the integration over angles yields zero except when the integrand is independent of angles. The radial integration should be divided into two parts, one with $r_1 > r_2$ and the other with $r_2 > r_1$ (see Problem 10.2).

Problem 10.2

(a) Calculate the Coulomb integral for helium with an STO.
(b) Calculate the kinetic plus Coulomb potential energy for helium.

Answers: (a) $J = \int\int d\mathbf{r}_1 d\mathbf{r}_2 \frac{[\phi_1^*\phi_1][\phi_2^*\phi_2]}{r_{12}}$. Expanding r_{12}^{-1} using (3.197),

$$J = \left(\frac{\zeta^3}{\pi}\right)^2 (4\pi)^2 \int_0^\infty dr_2 r_2^2 e^{-2\zeta r_2} \left[\int_0^{r_2} dr_1 \frac{r_1^2 e^{-2\zeta r_1}}{r_2} + \int_{r_2}^\infty dr_1 \frac{r_1^2 e^{-2\zeta r_1}}{r_1}\right]$$

$$= \left(\frac{\zeta^3}{\pi}\right)^2 (4\pi)^2 \times \frac{5}{2^7 \zeta^5} = \frac{5}{8}\zeta \, [e^2/a_0].$$

(b) $I_1 + I_2 = (\zeta^2 - 2Z\zeta) \, [e^2/a_0]$.

Problem 10.3

(a) With STO trial function (10.5) for the helium atom, find the best ground state energy using the variational theorem.

(b) Interpret ζ in terms of a shielding or screening constant, σ, where, in general, $\zeta_{nlm} = \frac{Z_{\text{eff}}}{na_0} \equiv \frac{Z - \sigma_{nlm}}{na_0}$.

Answer: (a) $\langle H \rangle = (\zeta^2 - 2Z\zeta + 5\zeta/8) \, [e^2/a_0]$, therefore $\frac{d}{d\zeta}\langle H \rangle = 2\zeta - 2Z + 5/8 = 0$, which yields $\zeta = Z - 5/16 = 27/16$.

10.5 HARTREE APPROXIMATION FOR MULTIELECTRON SYSTEMS

It is easy to generalize the Hartree formalism detailed above for helium to multielectron atoms. The Hamiltonian Eq. (10.11) generalizes to

$$H = \sum_{i=1}^{N} h(\mathbf{r}_i) + \sum_{i<j} \frac{1}{r_{ij}} \equiv H^{(0)} + H^{(1)}, \quad h(\mathbf{r}) = -\frac{1}{2}\nabla^2(\mathbf{r}) - \frac{Z}{r} \tag{10.30}$$

where Z is the nuclear charge (the system may be an ion so Z need not be equal to N), and the spatial part of the N-electron wave function is taken to be a product of single-particle orbitals,

$$\psi(\mathbf{r}_1, \mathbf{r}_2, \ldots, \mathbf{r}_N) = \prod_{i=1}^{N} \phi_i(\mathbf{r}_i) = \phi_1(\mathbf{r}_1) \phi_2(\mathbf{r}_2) \ldots \phi_N(\mathbf{r}_N). \tag{10.31}$$

As in the two-electron case, the Hartree method is an iterative procedure for constructing an approximation for the electron orbitals $\phi_i(\mathbf{r}_i)$ as limits of sequences $\phi_i^{(0)}(\mathbf{r}_i), \phi_i^{(1)}(\mathbf{r}_i), \ldots$. Within iteration step $n+1$, the orbital $\phi_i^{(n+1)}(\mathbf{r}_i)$ is the solution of a Schrödinger equation that contains an effective potential $V_i^{(n)}(\mathbf{r}_i)$ [see Eq. (10.15)] calculated as,

$$V_i^{(n)}(\mathbf{r}_i) = \sum_{j \neq i} \int d\mathbf{r}_j \frac{\left|\phi_j^{(n)}(\mathbf{r}_j)\right|^2}{|\mathbf{r}_i - \mathbf{r}_j|} = \sum_{j \neq i} \left\langle \phi_j^{(n)} \left| \frac{1}{|\mathbf{r}_i - \mathbf{r}_j|} \right| \phi_j^{(n)} \right\rangle. \tag{10.32}$$

In analogy with Eq. (10.16), we construct the mean-field Hamiltonian $H_i^{(n+1)}$ that determines the $n+1$ step orbital $\phi_i^{(n+1)}(\mathbf{r}_i)$ and the corresponding energy $\varepsilon_i^{(n+1)}$ through the Schrödinger equation,

$$H_i^{(n+1)} \phi_i^{(n+1)}(\mathbf{r}_i) = [h(\mathbf{r}_i) + V_i^{(n)}(\mathbf{r}_i)] \phi_i^{(n+1)}(\mathbf{r}_i) = \varepsilon_i^{(n+1)} \phi_i^{(n+1)}(\mathbf{r}_i). \tag{10.33}$$

After convergence, the effective Hamiltonian for particle 1 is $H_1^{\text{eff}} \phi_1 = \varepsilon_1 \phi_1$ where $H_1^{\text{eff}} = -\frac{1}{2}\nabla_1^2 - \frac{Z}{r_1} + V_1^{\text{eff}}$, and

$$V_1^{\text{eff}} = \sum_{j=2}^{N} \left\langle \phi_j(j) \left| \frac{1}{r_{1j}} \right| \phi_j(j) \right\rangle = \left\langle \phi_2 \left| \frac{1}{r_{12}} \right| \phi_2 \right\rangle + \cdots + \left\langle \phi_N \left| \frac{1}{r_{1N}} \right| \phi_N \right\rangle. \tag{10.34}$$

For particle i,

$$H_i^{\text{eff}} \phi_i = \varepsilon_i \phi_i, \tag{10.35a}$$

$$H_i^{\text{eff}} = -\frac{1}{2}\nabla_i^2 - \frac{Z}{r_i} + V_i^{\text{eff}}, \tag{10.35b}$$

$$V_i^{\text{eff}} = \sum_{j \neq i} \langle \phi_j | \frac{1}{r_{ij}} | \phi_j \rangle = \sum_{j \neq i} \int d\mathbf{r}_j \frac{|\phi_j(\mathbf{r}_j)|^2}{r_{ij}}. \tag{10.35c}$$

The Hartree approximation to the energy is

$$\boxed{E_H = \sum_{i=1}^{N} \varepsilon_i - \sum_{i=1}^{N-1} \sum_{j>i} J_{ij},} \tag{10.36}$$

that is, $E_H = \varepsilon_1 + \varepsilon_2 + \cdots - J_{12} - J_{13} - J_{14} - \cdots - J_{23} - J_{24} - \cdots$, where

$$J_{ij} = \langle \phi_i \phi_j | \frac{1}{r_{12}} | \phi_i \phi_j \rangle = \int d\mathbf{r}_1 \int d\mathbf{r}_2 \frac{|\phi_i(\mathbf{r}_1)|^2 |\phi_j(\mathbf{r}_2)|^2}{r_{12}}, \tag{10.37}$$

and

$$\varepsilon_i = \langle \phi_i | H_i^{\text{eff}} | \phi_i \rangle = \langle \phi_i | h(\mathbf{r}_i) | \phi_i \rangle + \sum_{j \neq i} \langle \phi_i(1)\phi_j(2) | \frac{1}{r_{12}} | \phi_i(1)\phi_j(2) \rangle \equiv I_i + \sum_{j \neq i} J_{ij}, \tag{10.38}$$

For $N = 2$, Eq. (10.36) reduces to $E = \varepsilon_1 + \varepsilon_2 - J_{12}$ [see (10.29)].

10.6 THE HARTREE–FOCK METHOD

In the previous chapter, we saw that the wave function of identical fermionic particles can be obtained from a product wave function by applying the antisymmetrization operator, $\mathcal{A} = \frac{1}{N!} \sum_P (-1)^P P$. More specifically, the normalized wave function for multielectron systems with N electrons in well-specified spin-orbitals are antisymmetrized by applying $\sqrt{N!}\,\mathcal{A}$ to the product of the orbitals [see Eq. (8.19)]:

$$\Psi(x_1, x_2, \ldots, x_N) = \frac{1}{\sqrt{N!}} \sum_P (-1)^P P u_\alpha(x_1) u_\beta(x_2) \ldots u_\nu(x_N). \tag{10.39}$$

In the Hartree formalism of the previous section, antisymmetrization of the wave function was not implemented, the wave function was taken to be a product, and the Hartree energy was given in Eq. (10.36). As we shall see below, properly antisymmetrizing the wave function, we obtain the Hartree–Fock energy expression

$$\boxed{E_{\text{HF}} = \sum_{i=1}^{N} I_i + \sum_{i=1}^{N} \sum_{j>i} (J_{ij} - K_{ij}),} \tag{10.40}$$

for singlet states (where all the electrons are spin-paired), where I_i is the expectation value of the kinetic energy plus the electron–nucleus Coulomb potential energy of the ith orbital,

$$\boxed{I_i = \langle \phi_i | -\frac{1}{2}\nabla_i^2 - \frac{Z}{r_i} | \phi_i \rangle,} \tag{10.41}$$

10.6 The Hartree–Fock Method

the *direct Coulomb repulsion integral* is

$$J_{ij} = \langle \phi_i \phi_j | \frac{1}{r_{12}} | \phi_i \phi_j \rangle = \int \int d\mathbf{r}_1 d\mathbf{r}_2 \frac{[\phi_i^*(\mathbf{r}_1)\phi_i(\mathbf{r}_1)][\phi_j^*(\mathbf{r}_2)\phi_j(\mathbf{r}_2)]}{r_{12}}, \quad (10.42)$$

and the new ingredient that emerges due to antisymmetrization is the *Coulomb exchange integral* is[1]

$$K_{ij} = \langle \phi_i \phi_j | \frac{1}{r_{12}} | \phi_j \phi_i \rangle = \int \int d\mathbf{r}_1 d\mathbf{r}_2 \frac{[\phi_i^*(\mathbf{r}_1)\phi_j(\mathbf{r}_1)][\phi_j^*(\mathbf{r}_2)\phi_i(\mathbf{r}_2)]}{r_{12}}. \quad (10.43)$$

The Coulomb exchange integral arises from the antisymmetry of the electron wave function with respect to electron exchange. It has no classical analog, just as the Pauli principle of antisymmetrization has no classical analog. Although the Coulomb integrals are always positive, the exchange integrals are generally positive (but can be negative). Therefore, the exchange integrals generally serve to lower the energy of an electronic system relative to the Hartree approximation. Moreover, the exchange integrals are responsible for molecular binding that allows molecules to be stable.

For closed shell spin-singlet states, Eq. (10.40) can be written as

$$E = \sum_{i=1}^{n_{\text{orb}}} 2I_i + \sum_{i=1}^{n_{\text{orb}}} \sum_{j=1}^{n_{\text{orb}}} (2J_{ij} - K_{ij}), \quad (10.44)$$

where n_{orb} is the number of orbitals that are populated in the closed shell state.

The numerical calculation of the direct and exchange Coulomb integrals, as well as the kinetic energy and overlap integrals between orbitals centered on different atoms in a molecule is easier to carry out when using Gaussian basis sets rather than STOs. There are a number of different techniques for carrying out such integrals. One method is called the *fast multipole method*, which is based upon multipole expansion of r_{12}^{-1} [see Eq. (3.197) and Problem 10.2 for $l = 0$ orbitals]. The reader interested in learning about the calculation of the direct and exchange Coulomb integrals for other than $l = 0$ orbitals can refer to Refs [166, 167].

We now derive the Hartree–Fock energy by properly antisymmetrizing the electronic wave function. To begin with, recall the decomposition (10.30) of the Hamiltonian into the sum of the single-particle Hamiltonians, $H^{(0)}$, and the sum of the electron–electron interaction terms, $H^{(1)}$. Note that both $H^{(0)}$ and $H^{(1)}$ commute with the antisymmetrization operator, $[H^{(0)}, \mathcal{A}] = [H^{(1)}, \mathcal{A}] = 0$. The energy is given by the expectation value of the sum of these Hamiltonians:

$$E_{\text{HF}}[\Psi] = \langle \Psi | H | \Psi \rangle = \langle \Psi | H^{(0)} | \Psi \rangle + \langle \Psi | H^{(1)} | \Psi \rangle. \quad (10.45)$$

Now, use the properly antisymmetrized wave function $|\Psi\rangle$, Eq. (10.39), written as $|\Psi\rangle = \sqrt{N!}\mathcal{A}|\Psi_H\rangle$, where the Hartree wave function $|\Psi_H\rangle$ is simply a product of spin-orbitals. With this notation,

$$\langle \Psi | H^{(0)} | \Psi \rangle = N! \langle \Psi_H | \mathcal{A} H^{(0)} \mathcal{A} | \Psi_H \rangle = N! \langle \Psi_H | H^{(0)} \mathcal{A}^2 | \Psi_H \rangle = N! \langle \Psi_H | H^{(0)} \mathcal{A} | \Psi_H \rangle$$

$$= \sum_{i=1}^{N} \sum_{P} (-1)^P \langle \Psi_H | h_i P | \Psi_H \rangle = \sum_{i=1}^{N} \langle \Psi_H | h_i | \Psi_H \rangle = \sum_{\lambda} \langle u_\lambda(x_i) | h_i | u_\lambda(x_i) \rangle.$$

We have used the fact that $\mathcal{A}$ commutes with $H^{(0)}$, $\mathcal{A}^2 = \mathcal{A}$, and that only one of the terms in $\langle \Psi_H | h_i P | \Psi_H \rangle$ is nonzero (the one where $P = 1$). Defining $I_\lambda \equiv \langle u_\lambda(x_i) | h_i | u_\lambda(x_i) \rangle$, we find

$$\langle \Psi | H^{(0)} | \Psi \rangle = \sum_\lambda I_\lambda. \quad (10.46)$$

---

[1] Sometimes the notation $[ii|jj]$ is used to denote the direct Coulomb repulsion integral J_{ij}, and $[ij|ji]$ for the Coulomb exchange integral K_{ij}.

Furthermore, $\mathcal{A}$ commutes with $H^{(1)}$, hence

$$\langle\Psi|H^{(1)}|\Psi\rangle = N!\langle\Psi_H|\mathcal{A}H^{(1)}\mathcal{A}|\Psi_H\rangle = N!\langle\Psi_H|H^{(1)}\mathcal{A}^2|\Psi_H\rangle = N!\langle\Psi_H|H^{(1)}\mathcal{A}|\Psi_H\rangle$$

$$= \sum_{i>j}^{N}\sum_{j=1}^{N}\sum_{P}(-1)^P\langle\Psi_H|\frac{1}{r_{ij}}P|\Psi_H\rangle = \sum_{i>j}^{N}\sum_{j=1}^{N}\langle\Psi_H|\frac{1}{r_{ij}}(1-P_{ij})|\Psi_H\rangle$$

$$= \sum_{(\lambda,\mu)}\left[\langle u_\lambda(x_i)u_\mu(x_j)|\frac{1}{r_{ij}}|u_\lambda(x_i)u_\mu(x_j)\rangle - \langle u_\lambda(x_i)u_\mu(x_j)|\frac{1}{r_{ij}}|u_\mu(x_i)u_\lambda(x_j)\rangle\right]$$

$$= \frac{1}{2}\sum_{\lambda\neq\mu}\sum_{\mu}\left[\langle u_\lambda(x_i)u_\mu(x_j)|\frac{1}{r_{ij}}|u_\lambda(x_i)u_\mu(x_j)\rangle - \langle u_\lambda(x_i)u_\mu(x_j)|\frac{1}{r_{ij}}|u_\mu(x_i)u_\lambda(x_j)\rangle\right],$$

where $\lambda,\mu = \alpha,\beta,\ldots,\nu$. Thus,

$$\langle\Psi|H^{(1)}|\Psi\rangle = \frac{1}{2}\sum_{\lambda\neq\mu}\sum_{\mu}[J_{\lambda\mu} - K_{\lambda\mu}]. \tag{10.47}$$

We can conclude that the expectation value of the Hamiltonian is

$$\boxed{E_{\mathrm{HF}}[\Psi] = \langle\Psi|H_1|\Psi\rangle + \langle\Psi|H_2|\Psi\rangle = \sum_\lambda I_\lambda + \frac{1}{2}\sum_\lambda\sum_\mu[J_{\lambda\mu} - K_{\lambda\mu}],} \tag{10.48}$$

where $K_{\lambda\mu} = 0$ unless the spin-orbitals u_λ and u_μ have the same spin state, i.e., $K_{\lambda\mu}$ is proportional to $\delta_{m_s^\lambda,m_s^\mu}$. Note that each pair of occupied spin-orbitals appears twice in the sum in the last term on the RHS of (10.48), hence the factor 1/2 in this term.

We now consider the iterative algorithm for calculating the spin-orbitals. Regarding the spin-orbitals $u_\lambda(x)$ as variational parameters, the energy $E[\Psi]$ is stationary with respect to variations of the spin-orbitals u_λ, ($\lambda = \alpha,\beta,\ldots,\nu$), where the spin-orbitals are required to be orthonormal. The normalization and orthogonalization are imposed as a constraint, by forming the function

$$\mathcal{L}[\{u_\lambda\}] = E[\{u_\lambda\}] - \sum_\lambda\sum_\mu\varepsilon_{\lambda\mu}\langle u_\mu|u_\lambda\rangle, \tag{10.49}$$

where the $\{\varepsilon_{\lambda\mu}\}$ serve as a set of Lagrange multipliers. This optimization can be simplified by diagonalizing the Lagrange multiplier matrix to obtain a diagonal set of Lagrange multipliers ε_λ, so that the variational equations take the form

$$\delta\left(E - \sum_\lambda\varepsilon_\lambda\langle u_\lambda|u_\lambda\rangle\right) = 0. \tag{10.50}$$

Using the techniques of functional derivative (see Sec. 16.10.1 linked to the book web page and Ref. [169]), the resulting Hartree–Fock equations are

$$\boxed{\begin{aligned}&\left(-\frac{1}{2}\nabla_{\mathbf{r}_i}^2 - \frac{Z}{r_i}\right)u_\lambda(\mathbf{r}_i) + \sum_\mu\left[\int d\mathbf{r}_j u_\mu^*(\mathbf{r}_j)\frac{1}{r_{ij}}u_\mu(\mathbf{r}_j)\right]u_\lambda(\mathbf{r}_i)\\ &- \sum_\mu\left[\delta_{m_s^\mu,m_s^\lambda}\int d\mathbf{r}_j u_\mu^*(\mathbf{r}_j)\frac{1}{r_{ij}}u_\lambda(\mathbf{r}_j)\right]u_\mu(\mathbf{r}_i) = \varepsilon_\lambda u_\lambda(\mathbf{r}_i),\end{aligned}} \tag{10.51}$$

for $\lambda,\mu = \alpha,\beta,\ldots,\nu$. Evidently, the Lagrange multipliers ε_λ are the orbital energies. The exchange term has the non-local structure, $\int d\mathbf{r}_j V(\mathbf{r}_i,\mathbf{r}_j)u_\lambda(\mathbf{r}_j)$. Since the exchange integral $K_{\lambda\mu}$ vanishes unless the spin-orbitals u_λ and u_μ have the same spin state, we can write (10.51) as

10.6 The Hartree–Fock Method

$$\left(-\frac{1}{2}\nabla^2_{\mathbf{r}_i} - \frac{Z}{r_i}\right) u_\lambda(\mathbf{r}_i) + \sum_\mu \left[\int d\mathbf{r}_j u^*_\mu(\mathbf{r}_j) \frac{1}{r_{ij}} u_\mu(\mathbf{r}_j)\right] u_\lambda(\mathbf{r}_i)$$

$$- \sum_\mu \left[\delta_{m^\mu_s, m^\lambda_s} \int d\mathbf{r}_j u^*_\mu(\mathbf{r}_j) \frac{1}{r_{ij}} u_\lambda(\mathbf{r}_j)\right] u_\mu(\mathbf{r}_i) = \varepsilon_\lambda u_\lambda(\mathbf{r}_i). \tag{10.52}$$

These equations, in the form,

$$\left\{\left(-\frac{1}{2}\nabla^2_{\mathbf{r}_i} - \frac{Z}{r_i}\right) + \sum_\mu \left[V^d_\mu(\mathbf{r}_i) - V^{ex}_\mu(\mathbf{r}_i)\right]\right\} u^{(k+1)}_\lambda(\mathbf{r}_i) = \varepsilon^{(k+1)}_\lambda u^{(k+1)}_\lambda(\mathbf{r}_i), \tag{10.53}$$

are solved recursively (see Fig. 10.1), just as we discussed for the Hartree approximation, until the resulting energies $\varepsilon^{(k+1)}_\lambda$ (and wave functions) no longer change with increasing k. The resulting orbitals satisfy the Hartree–Fock equations

$$\left\{\left(-\frac{1}{2}\nabla^2_{\mathbf{r}_i} - \frac{Z}{r_i}\right) + V^d(\mathbf{r}_i) - V^{ex}(\mathbf{r}_i)\right\} u_\lambda(\mathbf{r}_i) = \varepsilon_\lambda u_\lambda(\mathbf{r}_i), \tag{10.54}$$

where the direct potential $V^d(\mathbf{r}_i)$ and the <u>nonlocal</u> exchange potential $V^{ex}(\mathbf{r}_i)$ can be read off using Eq. (10.51). The notation

$$J_\mu(\mathbf{r}_i) \equiv V^d_\mu(\mathbf{r}_i) = \left[\int u^*_\mu(\mathbf{r}_j) \frac{1}{r_{ij}} u_\mu(\mathbf{r}_j) d\mathbf{r}_j\right], \tag{10.55}$$

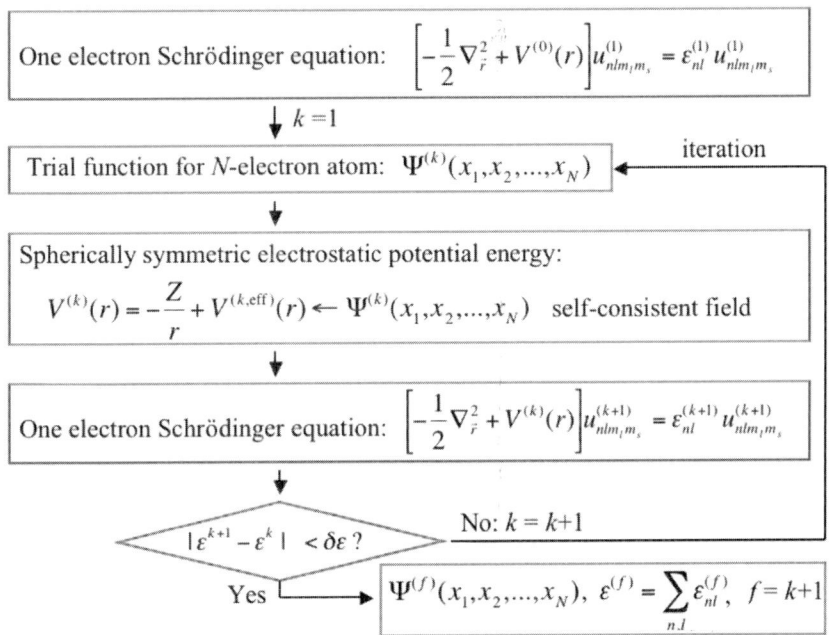

FIG 10.1 Schematic flow diagram of the Hartree–Fock method.

and

$$K_\mu(\mathbf{r}_i)u_\lambda(\mathbf{r}_i) \equiv V_\mu^{ex}(\mathbf{r}_i)u_\lambda(\mathbf{r}_i) = \left[\delta_{m_s^\mu,m_s^\lambda}\int u_\mu^*(\mathbf{r}_j)\frac{1}{r_{ij}}u_\lambda(\mathbf{r}_j)d\mathbf{r}_j\right]u_\mu(\mathbf{r}_i), \tag{10.56}$$

are sometimes used to denote the Coulomb and exchange integrals appearing in (10.51). Note that $K_\mu(\mathbf{r}_i) \equiv V_\mu^{ex}(\mathbf{r}_i)$ depends upon u_λ and is nonlocal; it is therefore an *integral operator*.[2] Note also that there is a big difference between $J_\mu(\mathbf{r}_i)$ and $J_{\lambda\mu}$, and $K_\mu(\mathbf{r}_i)$ and $K_{\lambda\mu}$ (the double index quantities being double integrals); hence, the similarity of notation is unfortunate (but it has become standard). Again, note that the exchange integral vanishes unless $m_s^\mu = m_s^\lambda$.

We have seen [Eq. (10.48)] that the Hartree–Fock expression for the energy can be written as a sum over occupied orbitals, which can be rewritten using the expression for the Hartree–Fock eigenenergies,

$$\boxed{\varepsilon_\lambda = I_\lambda + \sum_\mu (J_{\lambda\mu} - K_{\lambda\mu}),} \tag{10.57}$$

to obtain

$$\boxed{E = \sum_\lambda \varepsilon_\lambda - \frac{1}{2}\sum_{\lambda,\mu}(J_{\lambda\mu} - K_{\lambda\mu}).} \tag{10.58}$$

For closed shell spin-singlet states, Eq. (10.57) can be written as a sum over spatial orbitals,

$$E = \sum_{k=1}^{n_{orb}} 2\varepsilon_k - \sum_{k=1}^{n_{orb}}\sum_{l=1}^{n_{orb}}(2J_{kl} - K_{kl}), \tag{10.59}$$

where n_{orb} is the number of spatial orbitals populated in the closed shell state.

It is common to define the *Fock operator* f_i, which is the effective one-electron operator for the ith electron appearing on the LHS of Eq. (10.54). Its eigenfunctions are the Hartree–Fock orbitals $u_\lambda(\mathbf{r}_i)$ and its eigenvalues are the orbital energies ε_λ, i.e.,

$$f_i \equiv \left[\left(-\frac{1}{2}\nabla_{\mathbf{r}_i}^2 - \frac{Z}{r_i}\right) + V^d(r_i) - V^{ex}(r_i)\right], \tag{10.60}$$

so that

$$f_i u_\lambda(\mathbf{r}_i) = \varepsilon_\lambda u_\lambda(\mathbf{r}_i). \tag{10.61}$$

Moreover, one often defines the *Hartree–Fock potential*, or the "field" seen by the ith electron, to be $V_i^{HF} \equiv V^d(r_i) - V^{ex}(r_i)$. The solution of the Hartree–Fock eigenvalue problem (10.61) [equivalently, (10.54)] yields a set $\{u_\lambda\}$ of orthonormal Hartree–Fock spin-orbitals with orbital energies $\{\varepsilon_\lambda\}$. The N spin-orbitals with the lowest energies are occupied, and the Slater determinant formed from these orbitals is the Hartree–Fock ground state wave function. It is the best variational approximation to the ground state of the system. The *Hartree–Fock Hamiltonian*, $H_{HF} = \sum_{i=1}^N f_i$, has energy eigenvalue $\mathcal{E}_0 = \sum_{i=1}^N \varepsilon_i$ and eigenvalue given by the Slater determinant composed of the lowest $N = n_{orb}$ spin-orbitals. The total Hartree–Fock energy is given by (10.58).

---

[2] A better notation for $V_\mu^{ex}(\mathbf{r}_i)u_\lambda(\mathbf{r}_i)$ would be $\int d\mathbf{r}_j V_\mu^{ex}(\mathbf{r}_i, \mathbf{r}_j)u_\lambda(\mathbf{r}_j)$.

10.6 The Hartree–Fock Method

10.6.1 HARTREE–FOCK FOR HELIUM

It is instructive to go over the Hartree–Fock approximation for the simplest case of a two-electron system. The ground state of helium, which is a singlet spin state with each of the two electrons in a 1s spatial orbital, is denoted by $1s^2\,{}^1S_0$ and is given by

$$\Psi(x_1,x_2) = \frac{1}{\sqrt{2!}} \begin{vmatrix} u_\alpha(x_1) & u_\beta(x_1) \\ u_\alpha(x_2) & u_\beta(x_2) \end{vmatrix} = \frac{1}{\sqrt{2}} \begin{vmatrix} \phi_{100}(r_1)\chi_\uparrow(1) & \phi_{100}(r_1)\chi_\downarrow(1) \\ \phi_{100}(r_2)\chi_\uparrow(2) & \phi_{100}(r_2)\chi_\downarrow(2) \end{vmatrix}$$

$$= \phi_{100}(r_1)\phi_{100}(r_2)\frac{1}{\sqrt{2}}\left[\chi_\uparrow(1)\chi_\downarrow(2) - \chi_\uparrow(2)\chi_\downarrow(1)\right]$$

$$= \phi_{100}(r_1)\phi_{100}(r_2)\frac{1}{\sqrt{2}}\left[|\uparrow\downarrow\rangle - |\downarrow\uparrow\rangle\right]. \tag{10.62}$$

Using Eq. (10.59), the ground state energy of helium is

$$\langle E\rangle_{1s^2\,{}^1S_0} = 2\varepsilon_{1s} - (2J_{1s,1s} - K_{1s,1s}), \tag{10.63}$$

where the Hartree–Fock orbital energy ε_{1s}, (10.57), becomes

$$\varepsilon_{1s} = I_{1s} + (2J_{1s,1s} - K_{1s,1s}). \tag{10.64}$$

Hence, the ground state energy is

$$\langle E\rangle_{1s^2\,{}^1S_0} = I_{1s} + I_{1s} + 2J_{1s1s} - K_{1s1s} = 2I_{1s} + J_{1s1s}. \tag{10.65}$$

The exchange integral does not appear in the final result of (10.65) because $J_{1s1s} = K_{1s1s}$ and there is a partial cancellation. Using pure hydrogenic orbitals (i.e., $Z = 2$), $I_{nl} = -\frac{Z^2}{2n^2} \times 27.21$ eV, so for helium, $I_{1s} = -54.42$ eV, $J_{1s1s} = K_{1s1s} = (5/8)Z \times 27.21$ eV, and explicitly calculating the helium ground state energy, we find: $E_{He} = I_{1s} + I_{1s} + J_{1s1s} = -54.4 - 54.4 + 34.0$ eV $= -74.8$ eV. This result is in the ball park of the experimental value of -79.0 eV, but the error is 5%. Note that we have not varied the orbital exponents of the STOs (we could use $\zeta_{100} = \frac{Z_{\text{eff}}}{a_0}$ with the optimized value $Z_{\text{eff}} = 27/16 \approx 1.69$ rather than $Z = 2$, see Problem 10.3). If we use $\zeta = Z_{\text{eff}} = 27/16$, we obtain a total energy for He of $\langle H\rangle = (\zeta^2 - 2Z\zeta + 5\zeta/8)[e^2/a_0] = -77.48$ eV, thereby reducing the error to 2%. We could also use additional basis functions, etc., to further improve the results.

Hartree–Fock for Helium Excited States

As explained above, Hartree–Fock for the helium ground state yields the same results as a Hartree calculation, but this is not true for other states of helium. The first excited singlet state, $\Psi_{1s,2s,{}^1S_0} = \frac{1}{\sqrt{2}}[\phi_{100}(r_1)\phi_{200}(r_2) + \phi_{200}(r_1)\phi_{100}(r_2)]$ $\frac{1}{\sqrt{2}}[|\uparrow\downarrow\rangle - |\downarrow\uparrow\rangle]$, has energy

$$\langle E\rangle = \left\langle \frac{1}{\sqrt{2}}[1s(1)2s(2) + 2s(1)1s(2)] \middle| H \middle| \frac{1}{\sqrt{2}}[1s(1)2s(2) + 2s(1)1s(2)] \right\rangle, \tag{10.66}$$

which yields

$$\langle E\rangle_{1s,2s,{}^1S_0} = I_{1s} + I_{2s} + J_{1s2s} + K_{1s2s}. \tag{10.67}$$

Note that (10.66) dictates a plus sign here in front of K_{1s2s}. Here,

$$J_{1s2s} = \langle 1s(1)2s(2)|\frac{1}{r_{12}}|1s(1)2s(2)\rangle, \tag{10.68}$$

$$K_{1s2s} = \langle 1s(1)2s(2)|\frac{1}{r_{12}}|2s(1)1s(2)\rangle, \tag{10.69}$$

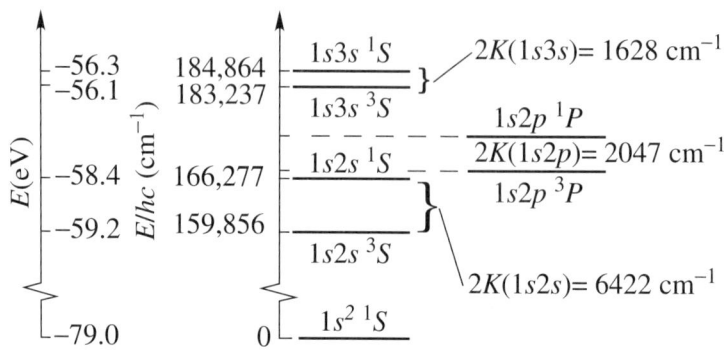

FIG 10.2 The lowest few states of helium. The experimental energies of the p states are $E(1s2p\,^3P) = 169{,}087$ cm^{-1} and $E(1s2p\,^1P) = 171{,}134$ cm^{-1}. Moreover I_{1s}, I_{2s}, J_{1s2p} and K_{1s2p} can be analytically calculated [4]. Note that the $1s2p\,^3P$ state is split by spin–orbit interaction, but this splitting is small on the scale shown here.

that is, $J_{1s2s} = \int\int \frac{1s(1)^2 2s(2)^2}{r_{12}} d\mathbf{r}_1 d\mathbf{r}_2$ and $K_{1s2s} = \int\int \frac{[1s(1)2s(1)]\cdot[1s(2)2s(2)]}{r_{12}} d\mathbf{r}_1 d\mathbf{r}_2$, and $I_{2s} = -13.6$ eV, $J_{1s2s} = 11.4$ eV, and $K_{1s2s} = 1.2$ eV. We could optimize the STOs to minimize the energy with respect to ζ_{100} and ζ_{200}; we shall not pause to work out the numerics of this optimization.

The lowest triplet helium state, $\frac{1}{\sqrt{2}}[\phi_{100}(r_1)\phi_{200}(r_2) - \phi_{200}(r_1)\phi_{100}(r_2)]|\uparrow\uparrow\rangle]$, can be treated in the same way, and it too can be pretty well described by wave functions of the form of a single Slater determinant. The expectation value of the Hamiltonian in this state is given by

$$\langle E \rangle_{3S} = \left\langle \frac{1}{\sqrt{2}}[1s(1)2s(2) - 2s(1)1s(2)] \middle| H \middle| \frac{1}{\sqrt{2}}[1s(1)2s(2) - 2s(1)1s(2)] \right\rangle, \quad (10.70)$$

and this yields

$$\langle E \rangle_{1s,2s,^3S_1} = I_{1s} + I_{2s} + J_{1s2s} - K_{1s2s}. \quad (10.71)$$

The first excited singlet is higher in energy than the lowest triplet state because K_{1s2s} is about 1.2 eV. Figure 10.2 shows the lowest few energy states of helium, where the energy differences are obtained from experiment. The experimental energy difference of the $1s^12s^1\,^1S$ and the $1s^12s^1\,^3S$ differs from the Hartree–Fock value of the exchange integral $K(1s2s)$ obtained from a STO calculation by about 30% (presumably due to correlation effects).

Problem 10.4

The lowest energy S terms of helium have the following measured energies relative to the ground state energy $E(1s^2\,^1S) = 0$: $E(1s^12s^1\,^1S) = 166{,}277$ cm^{-1} (20.615 eV), $E(1s^12s^1\,^3S) = 159{,}856$ cm^{-1} (19.819 eV), $E(1s^13s^1\,^1S) = 184{,}864$ cm^{-1} (22.919 eV); $E(1s^13s^1\,^3S) = 183{,}237$ cm^{-1} (22.718 eV). Obtain an estimate of exchange integral K in the $1s^12s^1$ and $1s^13s^1$ configurations based upon the experimental energies.
Answer: $K(1s^12s^1) = 3211$ and $K(1s^13s^1) = 814$ cm^{-1}.

10.7 KOOPMANS' THEOREM

Once an electronic structure calculation is completed, the results can be used to determine properties of the electronic system. For atoms, ionization energies, electron affinities, atomic radii, and other atomic characteristics can be determined.

10.7 Koopmans' Theorem

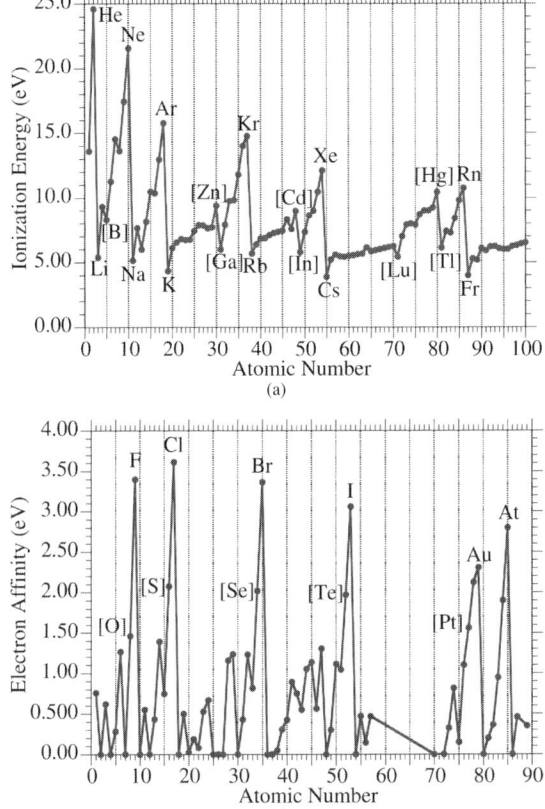

FIG 10.3 (a) Ionization energies of the elements. (b) Electron affinities of the elements. The negative ions of the elements with EA = 0.0 are not stable. *Source:* Band, Light and Matter, Fig. 6.6, p. 341

Although Hartree–Fock results are not particularly accurate, they do show general trends similar to those found using more accurate methods of calculation (e.g., using configuration interaction methods – see Sec. 10.12.1).

The *ionization energy* (IE) of an atom or molecule M is the minimum energy required to remove (to infinity) an electron, $M \to M^+ + e^-$, where M is neutral with N electrons and M^+ is its positive ion with $N-1$ electrons; it is defined by

$$\text{IE} \equiv E(M^+) - E(M) = E(N-1) - E(N), \quad (10.72)$$

where all energies on the RHS refer to ground state energies. Sometimes, the IE is called the *first ionization energy*. In general, the first ionization energy increases as we go from left to right across a row of the periodic table (there are, however, many exceptions), and decreases as we go down a column of the periodic table. Figure 10.3(a) shows the IEs of the elements.

The *electron affinity* of an atom or molecule is the energy given off when the neutral atom or molecule in the gas phase gains an extra electron to form a negatively charged ion. That is, the electron affinity (EA) of M, in the process, $M + e^- \to M^-$, where M^- is the negative ion of M, is defined as

$$\text{EA} \equiv E(M^-) - E(M) = E(N+1) - E(N). \quad (10.73)$$

A positive EA means that adding an electron to the atom is an exothermic process; e.g., the EA of chlorine is 3.61 eV/atom; it is the element that most strongly attracts extra electrons. In general, EAs become smaller as we go down a column of the periodic table. Alkali earth elements (Group IIA) and the noble gas atoms (Group VIIIA) do not form stable negative ions (their EAs are taken to be zero). Figure 10.3(b) shows the EAs of the elements. The *fundamental gap* of an atomic or molecular system is defined as $\Delta \equiv \text{IE} - \text{EA} = E(N-1) - E(N+1)$.

Koopmans' theorem, named after Tjalling Koopmans, is an approximation that results from the Hartree or Hartree–Fock approximation, in which the ionization energy of an atom or molecule is equal to the energy of the highest occupied molecular orbital (often abbreviated HOMO), and the electron affinity is the negative of the energy of the lowest unoccupied (i.e., virtual) molecular orbital (LUMO). Electron affinities calculated via Koopmans' theorem are usually quite poor. The derivation of Koopmans' theorem assumes that the electronic wave function of multielectron atom or molecule can be described as the product, or Slater determinant, of a set of one-electron orbitals, and assumes that upon the addition or subtraction of a single electron to, or from, the system, the mean field of the electrons does not change. In other words, there are two approximations in using Koopmans' theorem to estimate ionization energies, which limit the accuracy:

1. Differences in the "correlation energy" [defined as the exact energy of the atom or the molecule minus the mean field (Hartree–Fock) energy] of the electrons in the ion and neutral atom are ignored.
2. Electron "relaxation" of the remaining $N-1$ electrons is neglected.

Reiterating, Koopmans' theorem states: The ionization energy of an atom or molecule can be estimated as $IE = -\varepsilon_H$, which is minus the orbital energy of the HOMO. Moreover, the electron affinity is estimated as $EA = -\varepsilon_L$, which is minus the orbital energy of the lowest unoccupied orbital.

In the context of the (Hartree or) Hartree–Fock approximation for the ground state of helium, the ionization energy is given by Koopmans' theorem as $IE = -\varepsilon_1 = I_2 - E = E(M^+) - E(M)$, where

$$\varepsilon_1 = I_1 + J_{12} = (I_1 + I_2 + J_{12}) - I_2. \tag{10.74}$$

To obtain an accurate estimate of the IE, one should perform quantum mechanical calculations of the energies of both the neutral atom and the ion to get $E(M)$ and $E(M^+)$, from which the IE can be computed. Similarly, to get an accurate EA, one should separately calculate $E(M^-)$ and $E(M)$.

Problem 10.5

Calculate the first and second ionization energies of helium using Koopmans' theorem. (The experimental ionization energies are 24.58 eV and 54.40 eV.)

Answer: From Koopmans' theorem, $IE = -\varepsilon_1 = -(I_1 + J_{12}) = -(\zeta^2 - 2Z\zeta)/2 + 5\zeta/8)[e^2/a_0] \approx 24.38$, where we used $\zeta = 27/16$. The second IE is that of the helium ion, 54.4 eV.

10.8 ATOMIC RADII

Atomic radii can also be computed using electronic structure calculations. There are several ways to define the atomic radius of an atom; definitions used include: (1) $\langle r \rangle = \int d\mathbf{r}\, rn(\mathbf{r}) / \int d\mathbf{r}\, n(\mathbf{r})$, where $n(\mathbf{r})$ is the (charge) density calculated from the atomic orbitals, (2) $\bar{r} = \sqrt{\langle r^2 \rangle}$, and (3) the radius of maximum charge density. These alternative definitions can yield slightly different atomic radii. The general trends obtained are as follows:

- Atoms get bigger as one goes down the columns of the periodic table.
- Atoms get smaller as one goes across the rows of the periodic table.

Some tabulations of atomic radii use half the distance between two adjacent atoms in crystals of the elements. Then, the rare gas atomic radii are bigger than the halide atoms in the same row, but this is because the rare gas atoms do not form covalent bonds but rather form van der Waals bonds. The calculated atomic radii of a rare gas atom is considerably smaller than the corresponding halide in the same row.

Another important characteristic of atomic orbitals is their shielding or screening, as characterized by the constants σ_{nlm}, defined in terms of the coefficients ζ_{nlm} of the STO's in Eq. (10.6) by $\zeta_{nlm} = \frac{Z_{\text{eff}}}{na_0} \equiv \frac{Z - \sigma_{nlm}}{na_0}$. The screening constants for particular values of n and l, generally increase as a function of the atomic number Z, and the rate of increase depends strongly on n and l.

10.9 MULTIELECTRON FINE STRUCTURE: HUND'S RULES

Atomic states are categorized using *term symbols*, or Russell–Saunders term symbols as they are often called, $^{2S+1}L_J$, as discussed in Section 10.3. The general empirical rule for the lowest energy state of a given atomic electron configuration (e.g., 2p 3p) is called *Hund's rule*:

- The state with the largest S and the largest L (for this S) has the lowest energy.

For example, for the case of two electrons in a 2p 3p configuration, the ^3D state has the lowest energy. Hund's rule is based on making an electronic state with the electrons as far apart as possible; the largest S state has the most antisymmetric

10.9 Multielectron Fine Structure: Hund's Rules

coordinate part of the electronic wave function, and the largest L state has the electrons sitting farthest apart. There is an additional rule with regard to the lowest energy J state of the term found by the above Hund's rule: The lowest energy J state depends on whether the electron configuration of the term is more or less than a half-filled shell (see generalized Hund's rules below).

With many electrons orbiting around a nucleus, the total spin–orbit coupling operator can be written as sum of the spin–orbit interactions for each electron (in this section we do not use atomic units):

$$H_{so} = \sum_i \xi(r_i) \mathbf{l}_i \cdot \mathbf{s}_i / \hbar^2. \tag{10.75}$$

Here, the spin–orbit coupling constant for the ith electron is always positive and is given by $\xi(r_i) = -g_s \frac{e\hbar^2}{2m_e^2 c^2 r_i} \frac{dV(r_i)}{dr_i} \approx \frac{Ze^2\hbar^2}{m_e^2 c^2 r_i^3}$, since near the origin, $V(r_i) \approx Ze/r_i$. Using the Wigner–Eckart theorem of Sec. (3.6.4), the expectation value of H_{so} for a given $^{2S+1}L_J$ state gives

$$E_{so,J} = \langle \Psi(^{2S+1}L_J) | H_{so} | \Psi(^{2S+1}L_J) \rangle = \frac{A}{\hbar^2} \langle \Psi(^{2S+1}L_J) | \mathbf{L} \cdot \mathbf{S} | \Psi(^{2S+1}L_J) \rangle$$

$$= \frac{A}{2}[J(J+1) - S(S+1) - L(L+1)], \tag{10.76}$$

where the spin–orbit coupling constant of the atom, A, depends on L and S, but not J. The splitting between individual adjacent levels is

$$\Delta E_{so} = E_{so,J} - E_{so,J-1} = AJ. \tag{10.77}$$

This is known as *Landé's interval rule*. The spin–orbit coupling constant A can be either positive or negative. For atoms with electron configurations that are less than half filled, $A > 0$, and the multiplet level with the lowest energy is the one with the smallest possible J, whereas for atoms that have more than half-filled shells $A < 0$, and the level with the highest J has the lowest energy. The $A > 0$ case is said to be normal, and the $A < 0$ case is said to be inverted. Figure 4.5 shows the spin–orbit splitting of a ^2P state into $^2P_{3/2}$ and $^2P_{1/2}$ states for the $A > 0$ case.

The kind of coupling in Eqs (10.76) and (10.77) is called *L–S coupling* or *Russell–Saunders coupling*. Its predictions are in good agreement with observed spectra for many light atoms. For heavier atoms, another coupling scheme called *j–j coupling* provides better agreement with experiment. With larger nuclear charge, the spin–orbit interactions of individual electrons become as strong as spin–spin interactions or orbit–orbit interactions. Consequently, the spin and orbital angular momenta of individual electrons tend to couple to form individual electron angular momenta, i.e., $\mathbf{j}_1 = \mathbf{l}_1 + \mathbf{s}_1$, $\mathbf{j}_2 = \mathbf{l}_2 + \mathbf{s}_2$, etc., and the total electron angular momenta is $\mathbf{J} = \sum_i \mathbf{j}_i$. The only "good" quantum number in this case is J, where $|\mathbf{J}| = [J(J+1)]^{1/2}$. Levels no longer form narrowly spaced fine structure components of multiplets but are energetically mixed with levels of different l_i. Hence, the spectra of atoms with large Z are hard to understand because the spectrum is very crowded and not easy to assign.

The combination of Hund's and Landé's rules yields the following generalization of Hund's rule for the lowest energy multiplet corresponding to a given electron configuration:

> **Generalized Hund's Rule**: The largest S and the largest L (for this S), and the smallest (largest) value of J for less (more) than half-filled shells has the lowest energy.

This rule is called the *generalized Hund's rule*. Figure 10.4 illustrates Hund's rules for the case of an atom with one $4p$ and one $4d$ electron.

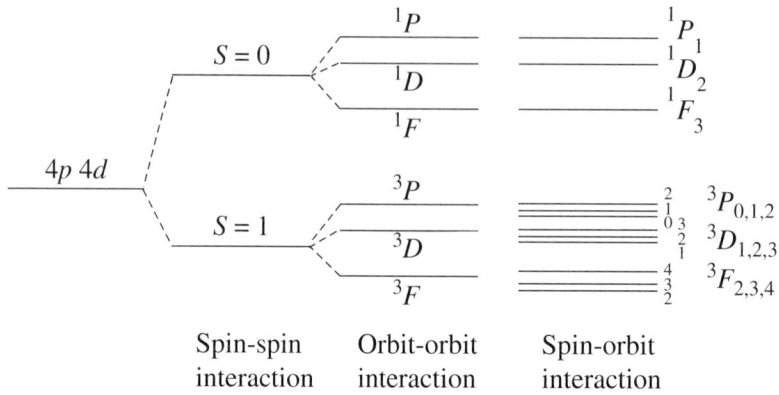

FIG 10.4 Hund's rules for a 4p 4d electron configuration. *Source:* Band, Light and Matter, Fig. 6.6, p. 347

There are additional terms such as spin-other-orbit coupling terms. We shall not discuss such effects at present. The interested reader is referred to Condon and Shortley [170].

Problem 10.6

(a) What are the possible atomic term symbols of the excited Scandium (^{45}Sc) configuration $1s^2 2s^2 2p^6 3s^2 3p^6 3d 4s 4p$? That is, specify the quantum numbers S, L, and J appearing in the term symbol $^{2S+1}L_J$.
(b) Using Hund's rules, determine the ground state atomic term symbol for the configuration in part (a).
(c) What is the electronic degeneracy of the ground state that you found in part (b)?
(d) The nuclear spin of ^{45}Sc is $I = 7/2$. What are the possible total angular momentum quantum numbers of the ground state?

Answer: (a) The total quantum numbers can take on the values $S = 3/2, 1/2$, $L = 3, 2, 1$. Hence, the atomic states that can be formed are as follows: $^4F_{9/2, 7/2, 5/2, 3/2}$, $^4D_{7/2, 5/2, 3/2, 1/2}$, $^4P_{5/2, 3/2, 1/2}$
$^2F_{7/2, 5/2}$, $^2D_{5/2, 3/2}$, $^2P_{3/2, 1/2}$.
(b) $^4F_{3/2}$.   (c) 4.   (d) $^4F_{3/2}$, $F = 5, 4, 3, 2$.

10.10 ELECTRONIC STRUCTURE OF MOLECULES

Hartree–Fock for molecules is similar to Hartree–Fock for atoms, since the Hamiltonian for molecules is similar to that for atoms. The only substantive difference regarding the electronic structure is that molecular symmetry should be built into the electronic wave function by composing molecular orbitals via a linear combination of atomic orbitals. Robert S. Mulliken and Friedrich Hund, building on the work of John Lennard-Jones, were primarily responsible for the early development of molecular orbital theory around 1927; Mulliken received the Nobel prize for his work in 1966. The linear combination of atomic orbitals approximation for molecular orbitals was first introduced by Sir John Lennard-Jones in 1929. In this approximation, molecular electronic wave functions are written as a Slater determinant with molecular orbitals composed of a linear combination of atomic spin-orbitals. The spatial part of the molecular spin-orbital is called a *molecular orbital* (MO). MOs are determined by a self-consistent field (SCF) calculation. We shall seek simple approximations for the MOs that enable qualitative understanding of chemical bonding. To illustrate how to build molecular

10.10 Electronic Structure of Molecules

orbitals, we take up the electronic structure of the simplest molecule, the hydrogen molecule, but only after treating the simplest molecular ion, H_2^+.

After the molecular electronic structure calculation is complete, the nucleon-nucleon Coulomb potential, $V_{NN} = \sum_{\alpha}^{N_n-1} \sum_{\beta>\alpha}^{N_n} \frac{Z_\alpha Z_\beta}{r_{\alpha\beta}}$, must be added to the electronic Hamiltonian, as must the nucleon kinetic energy; then the problem of nuclear motion can be dealt with, as discussed in the next chapter.

10.10.1 H_2^+: MOLECULAR ORBITALS

H_2^+ is a one-electron problem, and excellent methods exist for analyzing it [171, 172]. The electronic Hamiltonian is separable in confocal elliptic coordinates (i.e., prolate spheroidal coordinates) [18]. Analytic expressions for the electronic ground state molecular-ion energy can be obtained by using linear combinations of atomic orbitals centered on each of the protons that are written in terms of elliptic coordinates (see Sec. 6.3.5 of Ref. [18]) and developing expressions for the overlap integral and the matrix elements of the Hamiltonian. A simple approach to the problem uses a trial function for H_2^+ of the form

$$\phi = c_a \phi_{1s_a} + c_b \phi_{1s_b}, \tag{10.78}$$

can be used, where c_a and c_b are variational parameters, and the atomic orbitals are of the form

$$\phi_{1s_a} = \pi^{-1/2} \zeta^{3/2}(R) e^{-\zeta(R)r_a}, \qquad \phi_{1s_b} = \pi^{-1/2} \zeta^{3/2}(R) e^{-\zeta(R)r_b}. \tag{10.79}$$

Here, r_a and r_b are the distances of the electron to the two nuclei a and b, and $\zeta(R)$ is a parameter that may be optimized afterward to obtain the minimum energy. The MO (10.78) is a *linear combination of atomic orbitals* (LCAO), i.e., it is a LCAO-MO. Using (10.78) as a trial wave function, the variational equations (7.139) yield the secular equation

$$\begin{vmatrix} H_{aa} - ES_{aa} & H_{ab} - ES_{ab} \\ H_{ba} - ES_{ba} & H_{bb} - ES_{bb} \end{vmatrix} = 0, \tag{10.80}$$

where for a homonuclear diatomic molecule $H_{bb} = H_{aa}$, $S_{bb} = S_{aa} = 1$ and $S_{ba} = S_{ab}$. The determinantal equation (10.80) yields the two roots

$$E_1 = \frac{H_{aa} + H_{ab}}{1 + S_{ab}}, \qquad E_2 = \frac{H_{aa} - H_{ab}}{1 - S_{ab}}. \tag{10.81}$$

Substituting E_1 back into the linear equations that led to the determinantal equation (10.80) yields the (normalized) amplitude coefficients,

$$c_a = \frac{1}{\sqrt{2(1 + S_{ab})}}, \qquad c_b = \frac{1}{\sqrt{2(1 + S_{ab})}}. \tag{10.82}$$

For E_2, we obtain

$$c_a = \frac{1}{\sqrt{2(1 - S_{ab})}}, \qquad c_b = -\frac{1}{\sqrt{2(1 - S_{ab})}}. \tag{10.83}$$

Thus, the MO $\phi_1 \equiv \phi_{1s\sigma_g}$ with energy E_1 is symmetric with respect to reflection in a plane bisecting the internuclear axis, and the antisymmetric MO $\phi_2 \equiv \phi_{1s\sigma_u}$ with energy E_2 (i.e., the MO has a node in the plane bisecting the internuclear axis). The following expressions for the integrals $H_{aa}(R)$, $H_{ab}(R)$, and $S_{ab}(R)$ as a function of the internuclear coordinate

R can be determined using elliptical coordinates:

$$S_{ab}(R) = \left[1 + \zeta R + \frac{1}{3}(\zeta R)^2\right] e^{-\zeta R}, \quad (10.84)$$

$$H_{aa}(R) = H_{bb}(R) = \frac{e^2}{a_0}\left[\left(-\frac{1}{2} + \frac{a_0}{R}\right) - \frac{1}{\zeta R} + \left(1 + \frac{1}{\zeta R}\right)e^{-\zeta R}\right], \quad (10.85)$$

$$H_{ab}(R) = H_{ba}(R) = \frac{e^2}{a_0}\left[S_{ab}(R)\left(-\frac{1}{2} + \frac{a_0}{R}\right) - (1 + \zeta R)e^{-\zeta R}\right]. \quad (10.86)$$

The ground state energy of the hydrogen molecular ion is given by

$$E(R) = \frac{-e^2}{2a_0} + \frac{e^2}{R} + \frac{e^2}{a_0}\frac{e^{-\zeta R}[(\zeta R)^{-1} - \zeta R] - (\zeta R)^{-1}}{1 + S_{ab}(R)}. \quad (10.87)$$

The parameter $\zeta(R)$ can be optimized, i.e., the ground state energy can be minimized, by determining the best $\zeta(R)$ for each R.

The ground state MO with the energy (10.87) is denoted by

$$1s\,\sigma_g \equiv \frac{1}{\sqrt{2(1 + S_{ab})}}(\phi_{1s_a} + \phi_{1s_b}), \quad (10.88)$$

and the excited state wave function, having amplitudes (10.83), and therefore having a node in the plane bisecting the internuclear axis, is denoted by

$$1s\,\sigma_u \equiv \frac{1}{\sqrt{2(1 - S_{ab})}}(\phi_{1s_a} - \phi_{1s_b}). \quad (10.89)$$

The symbols g and u stand for gerade (even) and ungerade (odd) with respect to inversion about the center of mass of the molecule, and σ indicates that the projection of the electronic angular momentum about the internuclear axis is zero (and π in the MOs discussed in the next paragraph indicates projection equal to 1, see Sec. 11.2 for a more complete explanation of these symbols). The $1s\,\sigma_u$ MO is said to be antibonding, since it gives no buildup of charge between the two nuclei, and the $1s\,\sigma_g$ is bonding. The potential energies of the gerade and ungerade states are plotted in Fig. 10.5.

Excited H_2^+ MOs include: $2s\,\sigma_g$, $1s\,\sigma_u$, $2p_z\,\sigma_g \sim ((2p_z)_a - (2p_z)_b)$, $2p_z\,\sigma_u \sim ((2p_z)_a + (2p_z)_b)$, and the π MOs, $2p_x\,\pi_u \sim ((2p_x)_a + (2p_x)_b)$, $2p_x\,\pi_g \sim ((2p_x)_a - (2p_x)_b)$ (and the equivalent p_y MOs) as well as higher angular momentum MOs.

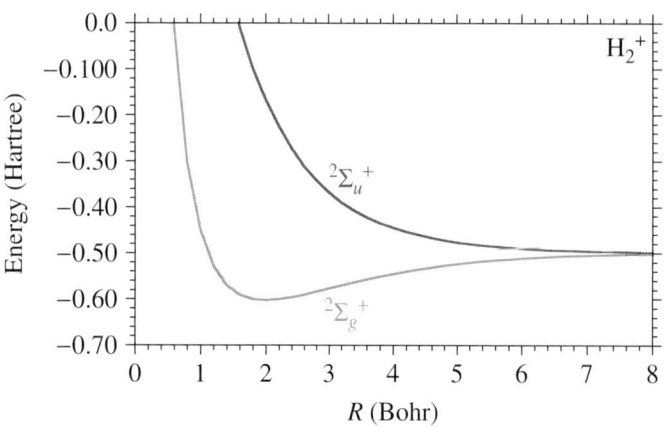

FIG 10.5 The two lowest H_2^+ potential energy curves are plotted versus R. *Source: Band, Light and Matter, Fig. 6.7, p. 351*

10.10 Electronic Structure of Molecules

Problem 10.7

Draw cross sections (as solid curves) and nodal surfaces (as dashed curves) for the $2p_z\,\sigma_g$, $2p_z\,\sigma_u$, $2p_x\,\pi_u$ and $2p_x\,\pi_g$ MOs. Compare your plots with those on the web page http://www.falstad.com/qmmo.

10.10.2 THE HYDROGEN MOLECULE

The Schrödinger equation for many-electron molecules (unlike H_2^+) cannot be solved exactly, so let us use the SCF treatment to approximate the electronic structure. The hydrogen molecule is the simplest molecule where electron interaction plays a role; hence, we treat it first. Denoting the nuclei by a and b, and the electrons by 1 and 2, as shown in Fig. 10.6, the Hamiltonian for H_2 is

$$H = -\frac{1}{2}\nabla^2_{\mathbf{r}_1} - \frac{1}{2}\nabla^2_{\mathbf{r}_2} - \frac{1}{r_{a1}} - \frac{1}{r_{b1}} - \frac{1}{r_{a2}} - \frac{1}{r_{b2}} + \frac{1}{r_{12}} + \frac{1}{R}. \tag{10.90}$$

There are two well-known SCF approaches to the hydrogen molecule electronic problem, the *molecular orbital method* and the *valence bond method*. Let us first consider the molecular orbital method.

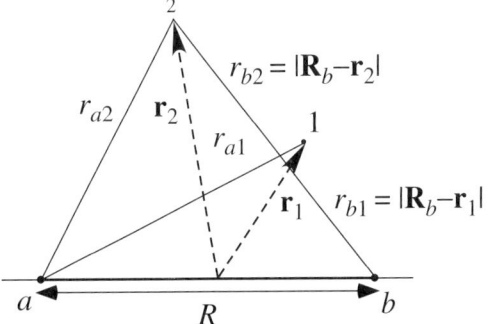

FIG 10.6 H_2 coordinates. The nuclei are labeled a and b, and the electrons are labeled 1 and 2.

Molecular Orbital Method

The ground state electronic configuration of H_2 is $(1s\,\sigma_g)^2$ and the Slater determinant for this configuration is

$$\Psi(1,2) = \frac{1}{\sqrt{2}}\begin{vmatrix} 1s\sigma_g(1)\chi_\uparrow(1) & 1s\sigma_g(1)\chi_\downarrow(1) \\ 1s\sigma_g(2)\chi_\uparrow(2) & 1s\sigma_g(2)\chi_\downarrow(2) \end{vmatrix}$$

$$= 1s\sigma_g(1)\,1s\sigma_g(2)\,\frac{[\chi_\uparrow(1)\chi_\downarrow(2) - \chi_\downarrow(1)\chi_\uparrow(2)]}{\sqrt{2}}. \tag{10.91}$$

The wave functions $1s\sigma_g$ are taken to be of the form (10.88). In order to calculate the matrix elements of the Hamiltonian, it is convenient to write the Hamiltonian in the form $H = H_1^0 + H_2^0 + \frac{1}{r_{12}}$, where H_1^0 and H_2^0 are the H_2^+ Hamiltonians for electrons 1 and 2. Hence,

$$\langle \Psi(1,2)|H|\Psi(1,2)\rangle = 2E(R) + \left\langle 1s\sigma_g(1)\,1s\sigma_g(2)\left|\frac{1}{r_{12}}\right|1s\sigma_g(1)\,1s\sigma_g(2)\right\rangle + \frac{1}{R}. \tag{10.92}$$

Here, $E(R)$ is the hydrogen molecular ion energy given by (10.87), and the Coulomb integral that appears on the RHS of (10.92) can be calculated by expanding r_{12}^{-1} in a multipole expansion [see (3.197)]. We shall not present the details of the evaluation of the integral here (see Slater [168]).

The ground state configuration of the hydrogen molecule is specified by the symbol $(1s\sigma_g)^2\,X^1\Sigma_g^+$, where X indicates the ground state, the superscript 1 indicates singlet, Σ indicates that the projection of the total electronic angular momentum about the diatomic axis is zero, and the subscript g indicates that the total two-electron wave function is even under inversion. The superscript + has to do with symmetry under reflection in a plane containing the diatomic axis (in point group notation, see Sec. 11.1.2, this reflection operator is called σ_v). A diatomic molecule is physically unchanged by reflection in a plane containing the diatomic axis. Hence, the wave function must remain unchanged or change sign under such a reflection operation. For Σ states, and only Σ states, the symmetry under this reflection is indicated in the state label by a superscript $+$ or $-$, for symmetric and antisymmetric configuration, respectively. Excited electronic configurations of H_2 include the lowest excited electronic state $(1s\sigma_g)(1s\sigma_u) \, b^3\Sigma_u^+$, which also correlates asymptotically to two hydrogen atoms in the ground state (see Fig. 10.7). The subscript u indicates that the total two-electron wave function is odd under inversion, and the symbol b will be explained in Chapter 11. Figure 10.7 shows the two potential energy curves correlating with two ground state hydrogen atoms versus internuclear distance. Higher lying excited states will be discussed in Chapter 11.

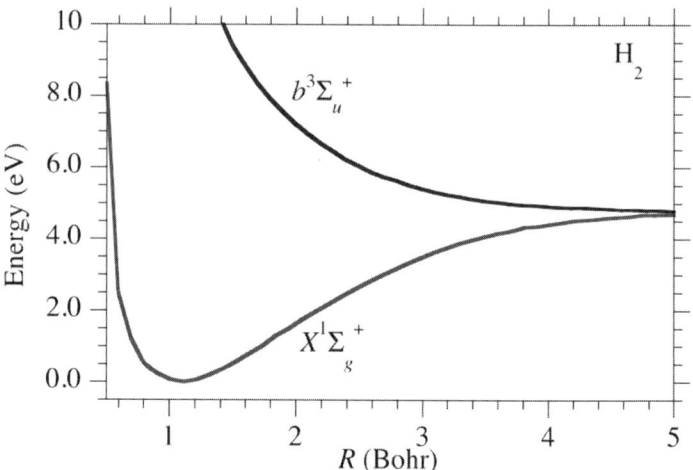

FIG 10.7 H_2 ground singlet and triplet potential curves, $(1s\sigma_g)^2 \, X^1\Sigma_g^+$ and $(1s\sigma_g)(1s\sigma_u) \, b^3\Sigma_u^+$ as a function of internuclear distance R. The singlet spin state has a spatially symmetric wave function under exchange of the two electrons, whereas the triplet spin state has a spatially antisymmetric wave function, as detailed in the next subsection.

Problem 10.8

Show that the molecular orbital approach leading to the wave function in Eq. (10.91) predicts dissociation into a mixture of atoms and ions by expanding the MOs in the product $1s\sigma_g(1) \, 1s\sigma_g(2)$ in terms of atomic orbitals. Hence, the wave function (10.91) does not yield a wave function that has the right asymptotic character at large R in which one electron is centered on one proton, and the other is centered on the other proton.

Answer: Using Eq. (10.88) to write $1s\sigma_g \sim (\phi_{1s_a} + \phi_{1s_b})$, we see that $(1s\sigma_g)^2 \, X^1\Sigma_g^+ \sim \phi_{1s_a}\phi_{1s_a} + \phi_{1s_b}\phi_{1s_b} + \phi_{1s_a}\phi_{1s_b} + \phi_{1s_b}\phi_{1s_a}$. The first two terms correspond asymptotically to the ionic structures H^-H^+ and H^+H^-, respectively, and only the last two terms correspond to two neutral atoms.

There are several ways of correcting the unphysical dissociation of the molecular state (10.91) discussed in Problem 10.8. One is to use the valence bond method discussed in the next section. Another is to include more basis states in the calculation, e.g., we can use trial wave functions with some $2p_0$ character mixed in (mixing in $2p_\pm$ functions would change the character of the wave function), i.e., we can use a MO of the form

$$\phi = [1s_a + c(2p_0)_a] + [1s_b - c(2p_0)_b], \quad (10.93)$$

where $(2p_0)_a = \frac{\beta^{5/2}}{4(2\pi)^{1/2}} r_a e^{-\beta r_a/2} \cos\theta_a \sim z_a e^{-\beta r_a/2}$. Mixing two or more atomic orbitals on the same atom is called *hybridization*, and the $[1s + c(2p_0)]$ function is called an *sp* hybrid atomic orbital. The wave function (10.93) is called a hybridized MO. Note that the hybridized MO has more charge on one side of the atom than on the other, i.e., the charge is polarized. Hybridization allows for the polarization of the $1s$ orbitals, and these types of atomic orbitals are important in molecule formation [because the charge density of the resulting MO swells in the region between the two protons

10.10 Electronic Structure of Molecules

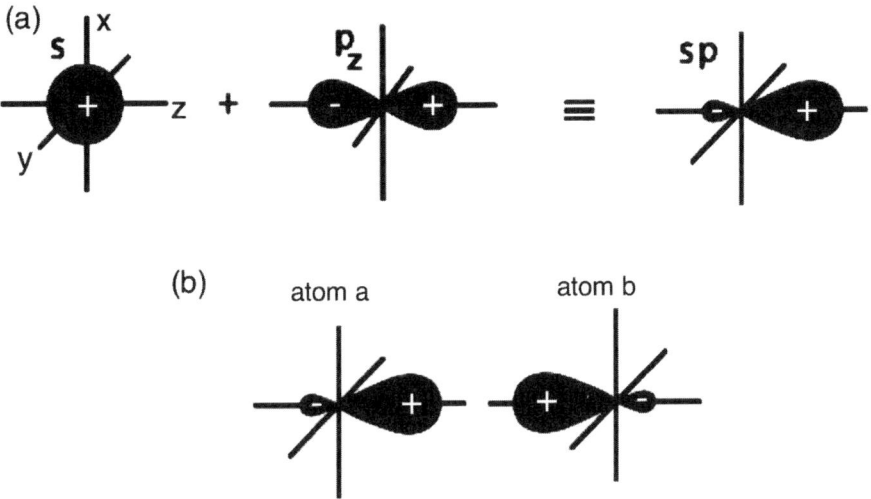

FIG 10.8 (a) *sp* hybrid orbital. (b) Molecular orbital made from a LCAOs, $\phi = [1s_a + c(2p_0)_a] + [1s_b - c(2p_0)_b]$.

of H_2, this requires the minus sign in (10.93), see Fig. 10.8]. One can then optimize the coefficient c to minimize the energy.

Valence Bond Method

Walter Heitler and Fritz London developed the *valence bond method* to treat the electronic structure of the hydrogen molecule in 1927. Their approach was extended by John C. Slater and Linus Pauling, and is today the most common way of treating chemical bonding in molecules; it is sometimes called the *valence bond* (VB) or *Heitler–London–Slater–Pauling method*. The valence bond method describes molecules as atomic cores (nuclei plus inner-shell electrons) and bonding valence electrons (clearly, in the case of H_2, both the electrons are valence electrons and the cores are the two protons). In the Heitler–London treatment of H_2, the wave function correlating asymptotically with two ground state hydrogen atoms is taken to be a superposition of the form

$$\Psi(1,2) = c_1\, \phi_{1s_a}(1)\phi_{1s_b}(2) + c_2\, \phi_{1s_b}(1)\phi_{1s_a}(2). \tag{10.94}$$

To determine the energy, we calculate $\langle \Psi(1,2)|H|\Psi(1,2)\rangle$, where

$$H = H_a(\mathbf{r}_1) + H_b(\mathbf{r}_2) - \frac{1}{r_{b1}} - \frac{1}{r_{a2}} + \frac{1}{r_{12}},$$

(the $1/R$ term that appears in (10.90) can be added after completing the calculation by simply adding it to the calculated energy eigenvalues). The variation with respect to the coefficients c_1, c_2 in the wave function (10.94) is carried out to obtain the lowest energy. This leads to the determinantal equation,

$$\begin{vmatrix} H_{11} - ES_{11} & H_{12} - ES_{12} \\ H_{21} - ES_{21} & H_{22} - ES_{22} \end{vmatrix} = 0, \tag{10.95}$$

where $H_{ij} = \langle \psi_i | H | \psi_j \rangle$, $S_{ij} = \langle \psi_i | \psi_j \rangle$, with $\psi_1 = \phi_{1s_a}(1)\phi_{1s_b}(2)$ and $\psi_2 = \phi_{1s_b}(1)\phi_{1s_a}(2)$. The diagonal matrix elements appearing in (10.95) are given by

$$H_{22} = H_{11} = 2\left\langle \phi_{1s_a} \left| -\frac{1}{2}\nabla_{\mathbf{r}}^2 - \frac{1}{r} \right| \phi_{1s_a} \right\rangle + \mathcal{J} = -1 + \mathcal{J}, \tag{10.96}$$

where $\mathcal{J}$ is a Coulomb integral [note difference from (10.42)] of the form

$$\mathcal{J} = \left\langle \phi_{1s_a}(1)\phi_{1s_b}(2) \left| -\frac{1}{r_{b1}} - \frac{1}{r_{a2}} + \frac{1}{r_{12}} \right| \phi_{1s_a}(1)\phi_{1s_b}(2) \right\rangle. \tag{10.97}$$

In (10.96), we made use of the fact that the hydrogen atom eigenenergy is $-1/2$ in atomic units (the simple Heitler–London treatment does not introduce an internuclear distance-dependent $\zeta(R)$ parameter, but this can be easily added). The off-diagonal elements appearing in (10.95) are given by

$$H_{21} = H_{12} = \left\langle \phi_{1s_a}(2)\phi_{1s_b}(1) \left| H_a(1) + H_b(2) - \frac{1}{r_{b1}} - \frac{1}{r_{a2}} + \frac{1}{r_{12}} \right| \phi_{1s_a}(1)\phi_{1s_b}(2) \right\rangle$$

$$= 2\left(-\frac{1}{2}S_{ab}S_{ab}\right) + \mathcal{K} = -S_{ab}^2 + \mathcal{K}, \tag{10.98}$$

where $\mathcal{K}$ is an "exchange integral" [note the difference from (10.43)],

$$\mathcal{K} = \left\langle \phi_{1s_b}(1)\phi_{1s_a}(2) \left| -\frac{1}{r_{b1}} - \frac{1}{r_{a2}} + \frac{1}{r_{12}} \right| \phi_{1s_a}(1)\phi_{1s_b}(2) \right\rangle, \tag{10.99}$$

and the overlap integrals are given by

$$S_{21} = S_{12} = S_{ab}^2, \tag{10.100}$$

and $S_{22} = S_{11} = 1$. The resulting determinantal equation is identical to that in Eq. (10.80), hence,

$$E_1 = \frac{H_{11} + H_{12}}{1 + S_{12}} = -1 + \frac{\mathcal{J} + \mathcal{K}}{1 + S_{ab}^2}, \quad E_2 = \frac{H_{11} - H_{12}}{1 - S_{12}} = -1 + \frac{\mathcal{J} - \mathcal{K}}{1 - S_{ab}^2}. \tag{10.101}$$

E_1 is the energy of the $^1\Sigma_g$ ground state, which has equal amplitudes $c_1 = c_2$ in the wave function Ψ in (10.94),

$$c_1 = \frac{1}{\sqrt{2(1 + S_{ab}^2)}}, \quad c_2 = \frac{1}{\sqrt{2(1 + S_{ab}^2)}}. \tag{10.102}$$

Note that the spin wave function for this state is

$$\frac{1}{\sqrt{2}}\left[\chi_\uparrow(1)\chi_\downarrow(2) - \chi_\downarrow(1)\chi_\uparrow(2)\right]. \tag{10.103}$$

The excited state with energy E_2 is the $^3\Sigma_u$ state. The triplet state could be in any one of the spin states $\frac{1}{\sqrt{2}}[\chi_\uparrow(1)\chi_\downarrow(2) + \chi_\downarrow(1)\chi_\uparrow(2)]$, $\chi_\uparrow(1)\chi_\uparrow(2)$ or $\chi_\downarrow(1)\chi_\downarrow(2)$; these three triplets are degenerate (in the absence of a magnetic field) and have the same spatial wave function. The amplitudes of the wave function Ψ in (10.94) are now given by

$$c_1 = -c_2 = \frac{1}{\sqrt{2(1 - S_{ab}^2)}}. \tag{10.104}$$

10.10 Electronic Structure of Molecules

We shall not pause to explicitly evaluate the integrals $\mathcal{J}$ and $\mathcal{K}$, but only note that the "exchange integral" $\mathcal{K}$ is negative, because of the inclusion of the terms $-\frac{1}{r_{b1}} - \frac{1}{r_{a2}}$ in (10.99) [which are not present in the Coulomb exchange integral K of (10.42)]. The "exchange integral" $\mathcal{K}$ is responsible for binding of H_2. Quite generally, the exchange coupling in molecular Hartree–Fock calculations is responsible for binding of molecules; without it, molecules would not be bound.

Better approximations for the atomic orbitals, e.g., taking $\phi_{1s_a} = (d_0(R) + d_1(R)z_a)e^{-\zeta(R)r_a}$, and similarly for ϕ_{1s_b}, will lower the energy obtained in a Heitler–London calculation.

Valence bond theory provides a more accurate picture of the electronic charge distribution when bonds are broken and formed during the course of a chemical reaction, as compared with the molecular orbital method. Specifically, valence bond theory correctly predicts the dissociation of homonuclear diatomic molecules into separate atoms, whereas the molecular orbital approach predicts dissociation into a mixture of atoms and ions, as can be seen by writing the hydrogen $(1s\sigma_g)^2 \, X^1\Sigma_g^+$ state out in terms of atomic orbitals with $1s\sigma_g \sim (\phi_{1s_a} + \phi_{1s_b})$ (see Problem 10.8). However, the molecular orbital method with configuration–interaction (see Sec. 10.12.1) can be designed to correct this problem, and thereby significantly improve the accuracy of the resulting energies.

10.10.3 THE HÜCKEL APPROXIMATION

A simple LCAO-MO method for the determination of energies of molecular orbitals of molecules with π electrons in conjugated hydrocarbon systems [covalently bonded atoms with alternating single and multiple (e.g. double) bonds], such as ethene, benzene, butadiene, cyclobutadiene, and conjugated polyenes was developed in 1930 by Erich Hückel and is called the Hückel molecular orbital approximation. It was later extended to conjugated molecules such as pyridine and pyrrole that contain atoms other than carbon. Figure 10.9 shows the linear polyene case and the cyclic case for $N = 6$ and $N = 8$ carbon atoms in a ring.

In the Hückel approximation, the π-electron Hamiltonian is approximated by a particularly simple form. In a minimal-basis-set calculation of a planar conjugated hydrocarbon, the only AOs of π symmetry are the carbon $2p\pi$ orbitals, whereby $2p\pi$ we mean the real $2p$ AOs that are perpendicular to the molecular plane. We can write $\psi_j = \sum_{k=1}^{n_C} c_{kj}\phi_k$, where ϕ_k is a $2p\pi$ AO on the kth carbon atom and n_C is the number of carbon atoms. The optimum values of the coefficients c_{kj} for the n_C lowest MOs ψ_j satisfy

$$\begin{pmatrix} H_{11} - ES_{11} & H_{12} - ES_{12} & H_{12} - ES_{12} & \cdots \\ H_{21} - ES_{21} & H_{22} - ES_{22} & H_{13} - ES_{13} & \cdots \\ \vdots & \vdots & \ddots & \vdots \end{pmatrix} \begin{pmatrix} c_1 \\ c_2 \\ \vdots \end{pmatrix} = 0, \quad (10.105)$$

The energy eigenvalues E_j are obtained by solving the determinantal equation $|\mathbf{H} - E\mathbf{S}| = 0$. For simplicity, the off-diagonal overlap integrals are neglected in the Hückel molecular orbital approximation, i.e., $S_{ij} = 0$ and $S_{ii} = 1$.

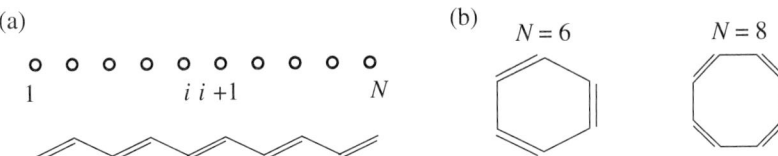

FIG 10.9 (a) Linear polyenes, with carbon atoms represented as open circles. Also represented are a series of alternating single/double bonds between the carbon atoms. (b) Cyclic hydrocarbons (with $N = 6$ and $N = 8$ carbon atoms) with conjugated bonds. The electronic structure of the π electron systems in these molecules can be approximated by the Hückel molecular orbital approximation.

Moreover, all nearest-neighbor Hamiltonian matrix elements are taken to be equal; they are denoted β, and usually $\beta < 0$. Furthermore, all nonnearest-neighbor Hamiltonian matrix elements are set to zero, and all diagonal Hamiltonian matrix elements are set equal; they are denoted α. The Hamiltonian thus becomes

$$H = \begin{pmatrix} \alpha & \beta & 0 & 0 & \cdots \\ \beta & \alpha & \beta & 0 & \cdots \\ 0 & \beta & \alpha & \beta & \cdots \\ 0 & 0 & \beta & \alpha & \cdots \\ \vdots & \vdots & & \ddots & \vdots \end{pmatrix} = \alpha \begin{pmatrix} 1 & \varepsilon & 0 & 0 & \cdots \\ \varepsilon & 1 & \varepsilon & 0 & \cdots \\ 0 & \varepsilon & 1 & \varepsilon & \cdots \\ 0 & 0 & \varepsilon & 1 & \cdots \\ \vdots & \vdots & & \ddots & \vdots \end{pmatrix}, \qquad (10.106)$$

where $\varepsilon \equiv \beta/\alpha$ (usually $\varepsilon < 0$). The eigenvalues of this Hamiltonian are

$$E_j = \alpha[1 + 2\varepsilon \cos(\pi j/(N+1))], \qquad (10.107)$$

$j = 1, \ldots, N$, and the (unnormalized) eigenvectors are $c_{k,j} = \sin(\pi k j/(N+1))$. The eigenvalue/eigenvector equations can be written as $\varepsilon(c_{k-1,j} + c_{k+1,j}) + c_{k,j} = E_j c_{k,j}$ (except for the first and last equation). Using simple trigonometric identities, the eigenvalues and eigenvectors can be seen to yield the equation, $\varepsilon(\sin[\pi(k-1)j/(N+1)] + \sin[\pi(k+1)j/(N+1)]) + \sin[\pi k j/(N+1)] = [1 + 2\varepsilon \cos(\pi j/(N+1)] \sin(\pi k j/(N+1))$. It is easy to verify this equality by noting the trigonometric identities, $\sin(x+y) = \sin(x)\cos(y) + \cos(x)\sin(y)$, hence $\sin(x-y) = \sin(x)\cos(y) - \cos(x)\sin(y)$; adding these two equations together, one obtains $\sin(x+y) + \sin(x-y) = 2\sin(x)\cos(y)$, which can be used to verify the above equation.

To illustrate the Hückel molecular orbital approximation, let us consider the straight-chain-conjugated polyene, e.g., $CH_2=CH-CH=CH-CH=CH_2$ with six carbon atoms called 1,3,5-hexatriene (or more generally, the molecule $CH_2=[CH-CH=]_n CH_2$ with $2n+2$ carbon atoms). The Hamiltonian is given by

$$H = \alpha \begin{pmatrix} 1 & \varepsilon & 0 & 0 & 0 & 0 \\ \varepsilon & 1 & \varepsilon & 0 & 0 & 0 \\ 0 & \varepsilon & 1 & \varepsilon & 0 & 0 \\ 0 & 0 & \varepsilon & 1 & \varepsilon & 0 \\ 0 & 0 & 0 & \varepsilon & 1 & \varepsilon \\ 0 & 0 & 0 & 0 & \varepsilon & 1 \end{pmatrix}. \qquad (10.108)$$

The eigenvalues in Eq. (10.107) for $N = 6$ are plotted as circles in Fig. 10.10.

Let us now consider a cyclic case, e.g., the benzene molecule, C_6H_6. The Hamiltonian matrix is now given by

$$H = \alpha \begin{pmatrix} 1 & \varepsilon & 0 & 0 & 0 & \varepsilon \\ \varepsilon & 1 & \varepsilon & 0 & 0 & 0 \\ 0 & \varepsilon & 1 & \varepsilon & 0 & 0 \\ 0 & 0 & \varepsilon & 1 & \varepsilon & 0 \\ 0 & 0 & 0 & \varepsilon & 1 & \varepsilon \\ \varepsilon & 0 & 0 & 0 & \varepsilon & 1 \end{pmatrix}. \qquad (10.109)$$

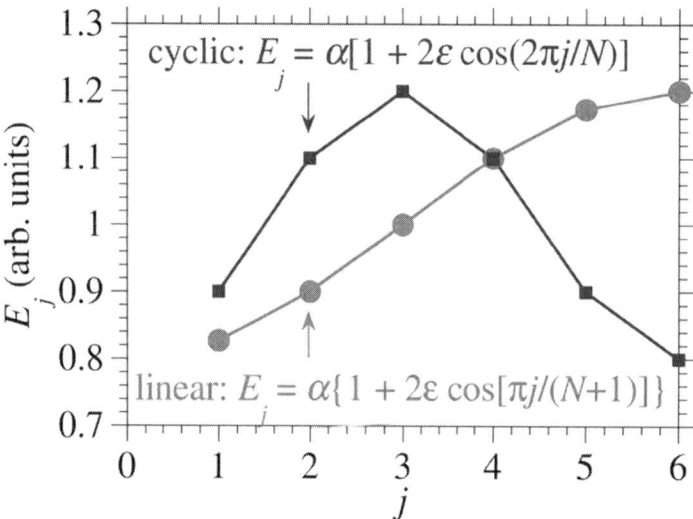

FIG 10.10 Hückel energy levels for a linear chain and for a cyclic molecule with $N = 6$. We have taken $\alpha = 1$ and $\varepsilon = -0.1$.

10.10 Electronic Structure of Molecules

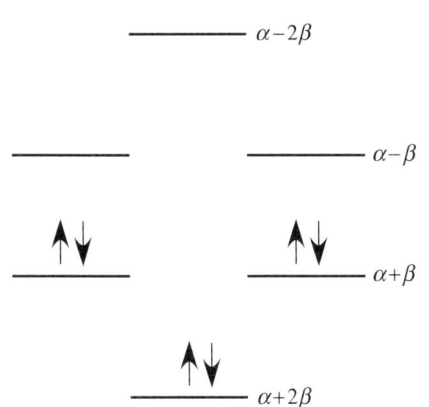

FIG 10.11 Hückel energy levels for benzene (identical to the squares in Fig. 10.10) and occupation of these levels by spin-up and spin-down electrons. The lowest three levels are occupied by spin-up and spin-down electrons. The parameter β (the nearest neighbor coupling matrix element) was taken to be negative (if it were positive, the labeling of the energy levels must be turned upside down). The states with energies $\alpha + 2\beta$ and $\alpha + \beta$ and π bonding orbitals. The unoccupied states correspond to π^* antibonding orbitals.

Note the additional nonvanishing coupling matrix elements in the upper-right and lower-left corners of the Hamiltonian matrix that result from the coupling of the first with the last atom when a linear chain is converted into a ring. The eigenvalues of the N atom cyclic Hamiltonian of the form (10.109) are

$$E_j = \alpha[1 + 2\varepsilon \cos(2\pi j/N)], \quad (10.110)$$

and the (unnormalized) eigenvectors are $c_{k,j} = \exp[i(2\pi kj/N)]$. Ring hydrocarbon $C_N H_N$. The eigenvalues of Eq. (10.110) are plotted as the squares in Fig. 10.10. The filling (i.e., the occupation) of the eigenvalues with spin-up and spin-down electrons is shown in Fig. 10.11; the six π electrons occupy the lowest three levels, which are π bonding orbitals. The states with energy $\alpha - \beta$ and $\alpha - 2\beta$ are antibonding π^* orbitals, and are unoccupied.

Problem 10.9

Use the Huckel approximation to find the molecular orbitals and their respective energies for the molecule butadiene (C_4H_6). Give your answers in terms of the α and β integrals.

Answer: To find the eigenvalues of the Hamiltonian matrix H in the equation

$$\begin{pmatrix} \alpha - E & \beta & 0 & 0 \\ \beta & \alpha - E & \beta & 0 \\ 0 & \beta & \alpha - E & \beta \\ 0 & 0 & \beta & \alpha - E \end{pmatrix} \begin{pmatrix} c_1 \\ c_2 \\ c_3 \\ c_4 \end{pmatrix} = 0,$$

it is easiest to divide H by $\alpha - E$ and let $x \equiv \frac{\beta}{\alpha - E}$. The determinantal equation, $1 - 3x^2 + x^4 = 0$, is a quadratic equation for x^2, which leads to $E = \alpha \pm \frac{1+\sqrt{5}}{2}\beta$ and $E = \alpha \pm \frac{1-\sqrt{5}}{2}\beta$. The eigenvectors corresponding to these eigenvalues can now be easily obtained.

Problem 10.10

(a) Determine the Hückel energy levels for 1,3,5,7-octatriene.
(b) Determine the amplitudes $\{c_{k,8}\}$ for the ground state energy E_8.

Problem 10.11

Given the molecule NC_2, which forms an isosceles triangle (see Fig. 10.12), determine the eigenenergies and eigenfunctions of the Hamiltonian matrix in the Hückel approximation with $H_{12} = \beta$, $H_{13} = H_{23} = 0.8\beta$.

1D Solid-State Hückel Approximation

The Hückel approximation can be used to calculate the states of a linear chain of atoms in the continuous limit, thereby solving the 1D solid-state band problem (see Sec. 9.4). From (10.107) with large N (so that $N + 1 \approx N$), the energy eigenvalues (in units of α) and the eigenfunctions, are given by

$$E_{k_j} = [1 + 2\varepsilon \cos(k_j a)], \quad \text{where} \quad k_j a = \frac{\pi j}{N}, \tag{10.111}$$

$$\psi_j(x_i) = \sqrt{\frac{2}{N}} \sin(k_j x_i), \quad \text{where} \quad x_i = ia. \tag{10.112}$$

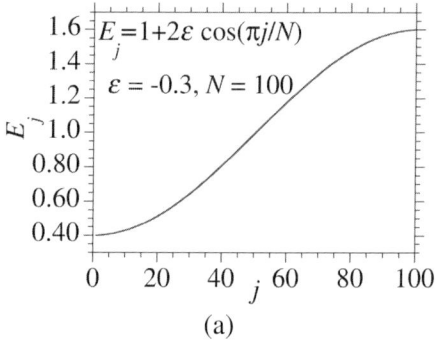

FIG 10.12 C_2N molecule.

We can take the continuous limit by letting k be a continuous variable,

$$E(k) = [1 + 2\varepsilon \cos(ka)], \quad \text{i.e.,} \quad k = \frac{1}{a} \arccos\left(\frac{E(k) - 1}{2\varepsilon}\right). \tag{10.113}$$

The range of the wave vector is, $0 \le ka \le \pi$, that the bandwidth of the energy band (in units of α) is $\Delta E = 4\varepsilon$. The 1D density of states [see Eq. (9.26)], $D_{1D}(E) = \frac{L}{2\pi} \frac{dk}{dE}$, where L is the length of the chain, is

$$D_{1D}(E) = \frac{L}{4\pi \varepsilon a} \left[1 - \left(\frac{E-1}{2\varepsilon}\right)^2\right]^{-1/2}, \tag{10.114}$$

since $d \arccos x/dx = -(1 - x^2)^{-1/2}$. Figure 10.13(b) shows the energy eigenvalues $E(k)$ and the density of states dk/dE versus wave vector k.

The Hückel approximation in solid-state physics is closely related with the tight binding approximation discussed in Chapter 9.

10.11 HARTREE–FOCK FOR METALS

The Hartree–Fock approximation can be used to obtain a many-electron wave function for metals as a Slater determinant of plane waves functions for the nearly free electrons. As a first approximation, one can use plane wave spin-orbitals, $\Psi = |u_{\lambda_1}(x_1) \ldots u_{\lambda_j}(x_j) \ldots u_{\lambda_N}(x_N)|$, with $u_{\lambda_j}(x_j) = \mathcal{V}^{-1/2} e^{i\mathbf{k}_{\lambda_j} \cdot \mathbf{r}_j} \chi_{\lambda_j}$, with each wave vector less than the Fermi momentum, $|\mathbf{k}_{\lambda_j}| \le k_F$, and take the positive charge distribution of the ions to be smeared out as a uniform distribution of positive charge with the same density as the electronic charge. The plane waves satisfy periodic boundary conditions, as explained in Sec. 9.1. Then, in the resulting Hartree–Fock equations (10.51), the potential of the ions will exactly cancel the Hartree potential. That is, the direct (Hartree) potential V_d in the Hartree–Fock potential $V^{HF} = V^d - V^{ex}$ of Eqs (10.60) and (10.61), is exactly canceled by the potential arising from the positively charged background. Hence, (10.60) becomes

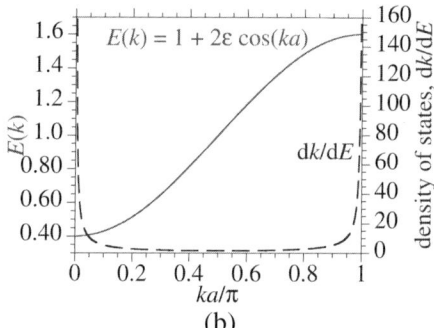

FIG 10.13 (a) Hückel energy levels for $N = 100$, $\varepsilon = -0.3$. (b) Energy eigenvalues in the continuous limit, $E(k)$ and the density of states, which is proportional to dk/dE versus the wave vector k. The full bandwidth is 4ε.

$$\left[-\frac{1}{2}\nabla^2_{\mathbf{r}_i} - V^{ex}(r_i)\right] u_\lambda(\mathbf{r}_i) = \varepsilon_\lambda u_\lambda(\mathbf{r}_i). \tag{10.115}$$

10.11 Hartree–Fock for Metals

The kinetic energy operator term is trivially calculated because the orbital is of plane wave form, and the nonlocal exchange potential is such that

$$\sum_{\mu} V_\mu^{ex}(\mathbf{r}_i) u_\lambda(\mathbf{r}_i) = \sum_{\mu} \left[\delta_{m_s^\mu, m_s^\lambda} \int u_\mu^*(\mathbf{r}_j) \frac{1}{r_{ij}} u_\lambda(\mathbf{r}_j) d\mathbf{r}_j \right] u_\mu(\mathbf{r}_i). \tag{10.116}$$

Substituting the plane wave for the orbitals, and noting that

$$\frac{1}{|\mathbf{r}_i - \mathbf{r}_j|} = \frac{4\pi}{V} \sum_{\mathbf{q}} \frac{e^{i\mathbf{q}\cdot(\mathbf{r}_i - \mathbf{r}_j)}}{q^2} \rightarrow \frac{4\pi}{(2\pi)^3} \int d\mathbf{q} \frac{e^{i\mathbf{q}\cdot(\mathbf{r}_i - \mathbf{r}_j)}}{q^2}, \tag{10.117}$$

we find

$$V^{ex}(\mathbf{r}_i) = \sum_{\mu} V_\mu^{ex}(\mathbf{r}_i) = \frac{4\pi}{(2\pi)^3} \int d\mathbf{q} \, e^{i\mathbf{q}\cdot\mathbf{r}_i} \int_{k' < k_F} \frac{d\mathbf{k}'}{(2\pi)^3} \frac{1}{|\mathbf{q} - \mathbf{k}'|^2}. \tag{10.118}$$

The sum over $\mathbf{q}$ in (10.117) does not include $\mathbf{q} = 0$, and the passage from summation to integration is discussed in Sec. 9.1. Taking the Fourier transformation of Eq. (10.115) with $u_\mathbf{k}(\mathbf{r}) = (2\pi)^{-\frac{3}{2}} e^{i\mathbf{k}\cdot\mathbf{r}} \chi$, the expression for the energies $\varepsilon_\mathbf{k}$ on the RHS of (10.115) becomes,

$$\varepsilon_\mathbf{k} = \frac{k^2}{2} - \frac{4\pi}{(2\pi)^3} \int_{k' < k_F} \frac{d\mathbf{k}'}{|\mathbf{k} - \mathbf{k}'|^2}. \tag{10.119}$$

The integral over the Fermi sphere can be carried out analytically by choosing $\mathbf{k}$ to lie along the z axis and using spherical coordinates (k', θ, ϕ) for the integration over $\mathbf{k}'$. Noting that the azimuthal angle ϕ does not appear in the integrand, one finds,

$$\frac{4\pi}{(2\pi)^3} \int_{k' < k_F} \frac{d\mathbf{k}'}{|\mathbf{k} - \mathbf{k}'|^2} = \frac{4\pi}{(2\pi)^3} 2\pi \int_0^{k_F} dk' \, k'^2 \int_{\theta=0}^{\pi} d\theta \frac{\sin\theta}{k^2 + k'^2 - 2kk'\cos\theta}, \tag{10.120}$$

Setting $\cos\theta = x$ and $\sin\theta \, d\theta = -dx$, the integral over θ becomes,

$$\int_{-1}^{1} dx \frac{k'^2}{k^2 + k'^2 - 2kk'x} = \frac{k'}{k} \log\left|\frac{k + k'}{k - k'}\right|. \tag{10.121}$$

The integral over k' can now be completed, and we thereby obtain, upon putting back $\hbar$ and e, the HF energies,

$$\boxed{\varepsilon_\mathbf{k} = \frac{\hbar^2 k^2}{2m} - \frac{2e^2}{\pi} k_F \left(\frac{1}{2} + \frac{k_F^2 - k^2}{4kk_F} \log\left|\frac{k + k_F}{k - k_F}\right| \right).} \tag{10.122}$$

The main features of these results are:

1. The exchange contribution is generally *negative* (the exchange integral is generally positive).
2. A somewhat disturbing result is that $d\varepsilon_k/dk$ diverges at k_F, i.e., the group velocity diverges. This is related to the long-range nature of the Coulomb potential, which, in k space leads to divergence of its Fourier component as $1/q^2$.

This is called an *infrared divergence* since it occurs at low q, i.e., low energy. In reality, the Coulomb interaction between electrons within a solid is screened, and becomes short ranged so the infrared divergence is shielded.

3. The total energy is obtained by summing the HF energies up to $k = k_F$. Taking spin degeneracy into account, the expression for the total energy of N electrons within the HF approximation is,

$$E_{\mathrm{HF}} = N\left[\frac{3}{5}E_F - \frac{3}{4\pi}e^2 k_F\right], \tag{10.123}$$

where Eqs (9.8) and (9.10) give E_F and k_F in terms of the electron density $n_e = N/\mathcal{V}$.

10.12 ELECTRON CORRELATION

This section can be skipped on a first reading.

The difference of the exact energy of an atom or molecule and the Hartree–Fock (mean field) energy is called the *correlation energy*. Several methods have been developed to calculate the correlation energy of atoms and molecules, at various levels of approximation. The first such method that we discuss is called configuration interaction.

10.12.1 CONFIGURATION INTERACTION

Higher order corrections to the wave function and energy can be obtained by mixing-in contributions from excited configurations, in a process called *configuration interaction* (CI) or *configuration mixing* (CM). A CI wave function takes the form,

$$\Psi = t_0 \Phi_0 + \sum_{ia} t_i^a \Phi_i^a + \sum_{ijab} t_{ij}^{ab} \Phi_{ij}^{ab} + \sum_{ijkabc} t_{ijk}^{abc} \Phi_{ijk}^{abc} + \cdots . \tag{10.124}$$

Here Φ_0 is the reference determinant for the state,

$$\Phi_0 = |u_1 \cdots u_n|, \tag{10.125}$$

where u_1 through u_n are the lowest occupied orbitals, and the determinantal wave functions Φ_i^a are given by a singly excited determinant, wherein one excites an occupied spin-orbital u_i to an unoccupied spin-orbital u_a:

$$\Phi_i^a = |u_1 \cdots u_{i-1} u_a u_{i+1} \cdots u_n|. \tag{10.126}$$

The doubly excited determinants, Φ_{ij}^{ab}, where $u_i \to u_a$ and $u_j \to u_b$ are

$$\Phi_{ij}^{ab} = |u_1 \cdots u_{i-1} u_a u_{i+1} \cdots u_{j-1} u_b u_{j+1} \cdots u_n|, \tag{10.127}$$

etc. One then minimizes the energy

$$E = \frac{\langle \Psi | \hat{H} | \Psi \rangle}{\langle \Psi | \Psi \rangle}, \tag{10.128}$$

with respect to the t coefficients appearing in Eq. (10.124).

Let us again take H$_2$ as an example, and consider a minimal basis LCAO-MO description, with each hydrogen atom having a $1s$ atomic orbital and molecular orbitals (MOs) are formed as a linear combination of atomic orbitals (LCAO). The symmetric combination leads to a bonding molecular orbital of gerade symmetry (symmetric with respect to inversion about the point centered between the nuclei) of form (10.88),

$$\phi_1 \equiv \phi_{1s\sigma_g} \equiv \frac{1}{\sqrt{2(1+S_{ab})}}\left(\phi_{1s_a} + \phi_{1s_b}\right), \tag{10.129}$$

10.12 Electron Correlation

and the antisymmetric combination to the ungerade MO of form (10.89),

$$\phi_2 \equiv \phi_{1s\sigma_u} \frac{1}{\sqrt{2(1-S_{ab})}} \left(\phi_{1s_a} - \phi_{1s_b}\right). \tag{10.130}$$

These MOs are orthonormal. Four spin-orbitals can be constructed using these two MOs, $u_1 \equiv u_{1s\sigma_g \uparrow}$, $u_2 \equiv u_{1s\sigma_g \downarrow}$, $u_3 \equiv u_{1s\sigma_u \uparrow}$, $u_4 \equiv u_{1s\sigma_u \downarrow}$. How many determinantal wave functions for hydrogen can be constructed using these spin-orbitals? Well, we can choose from four spin-orbitals for the first electron and then from the remaining three for the second orbital, but we need to divide by two because these two choices can be interchanged, so we find a total of $\frac{4!}{2!2!} = 6$ determinants. More generally, if we have $2K$ spin-orbitals, u_λ with $\lambda = 1, 2 \ldots, 2K$, and N electrons, we can form a total of $\binom{2K}{N} = \frac{(2K)!}{(2K-N)!N!}$ determinants. When all of these determinants are used in the CI calculation, it is called *full CI*. Even for a small system ($N \approx 10$) and a minimal basis set, the number of determinants that must be included in a full CI calculation is extremely large. Hence, in practice, one truncates the full CI expansion and uses only a small fraction of the possible determinants.

The discussion above regarding the number of required CI configurations should be modified by symmetry considerations, as is clear from the following. Within the space spanned by the minimal basis set for hydrogen, the CI wave functions Ψ are linear combinations of the six determinants discussed above. However, the Hartree–Fock ground state Φ_0 has two electrons in a gerade MO, hence is of g symmetry. The doubly excited determinant has two electrons in an ungerade orbital and hence is also of g symmetry. But the singly excited determinants have one electron in a gerade orbital and one in an ungerade orbital and are therefore of u symmetry. Therefore, the H_2 ground state minimal basis CI wave function, Ψ of (10.124), like the Hartree–Fock approximation to it, $\Phi_0 = |u_{1s\sigma_g \uparrow} \, u_{1s\sigma_g \downarrow}| = |u_1 u_2|$ of (10.125), is of g symmetry. Thus, only determinants of g symmetry can appear in the expansion (10.124) for the ground state,

$$\Psi = t_0 \Phi_0 + t_{12}^{34} \Phi_{12}^{34}, \tag{10.131}$$

where $\Phi_{12}^{34} \equiv |u_3 u_4|$. The coefficients t_0 and t_{12}^{34} and the exact energy are determined by diagonalizing the full CI matrix, which in this case is a 2×2 Hamiltonian matrix,

$$\mathbf{H} = \begin{pmatrix} \langle \Phi_0 | H | \Phi_0 \rangle & \langle \Phi_0 | H | \Phi_{12}^{34} \rangle \\ \langle \Phi_{12}^{34} | H | \Phi_0 \rangle & \langle \Phi_{12}^{34} | H | \Phi_{12}^{34} \rangle \end{pmatrix}, \tag{10.132}$$

where the operator H appearing in the matrix elements on the RHS of (10.132) is given in (10.90). The ground state is the eigenvector with the lowest eigenvalue. The evaluation of the matrix elements in (10.132) involve integrals we have already encountered, namely, I_i, the Coulomb integrals J_{ij} defined in (10.42), the exchange integrals K_{ij} defined in (10.43), but the off-diagonal elements involve integrals of the new form

$$\langle\langle ij|kl\rangle\rangle \equiv \langle u_i u_j | \frac{1}{r_{12}} | u_i u_j \rangle. \tag{10.133}$$

Problem 10.12

(a) Show that the full CI Hamiltonian matrix in the minimal H_2 basis is

$$\mathbf{H} = \begin{pmatrix} I_1 + I_2 + \langle\langle 12|12\rangle\rangle - \langle\langle 12|21\rangle\rangle & \langle\langle 12|34\rangle\rangle - \langle\langle 12|43\rangle\rangle \\ \langle\langle 34|12\rangle\rangle - \langle\langle 34|21\rangle\rangle & I_3 + I_4 + \langle\langle 34|34\rangle\rangle - \langle\langle 34|43\rangle\rangle \end{pmatrix}. \tag{10.134}$$

(b) By integrating out the spin variables, show that your result in (a) can be reduced to

$$\mathbf{H} = \begin{pmatrix} 2I_1 + \langle\langle \phi_1 \phi_1 | \phi_1 \phi_1 \rangle\rangle & \langle \phi_1 \phi_2 | \phi_2 \phi_1 \rangle \\ \langle \phi_2 \phi_1 | \phi_2 \phi_1 \rangle & 2I_2 + \langle\langle \phi_2 \phi_2 | \phi_2 \phi_2 \rangle\rangle \end{pmatrix}. \tag{10.135}$$

Problem 10.13

(a) How many spin-orbitals are there in the minimal basis set for benzene?
(b) How many electrons are there in a benzene molecule?
(c) Calculate the size of the full CI matrix if it would be formed from determinants.
(d) How many singly excited determinants are there?

Answers: (a) 72 spin-orbitals. (b) $N = 7 \times 6 = 42$. (c) $\binom{72}{42} = 164307576757973059488$. (d) $(2K - N)N = 1260$.

CI calculations suffer from a lack of *size consistency*. A method is size consistent if the quality of the results are independent of the size of the system, in the following sense. For a system containing two subsystems, a method is size consistent if the energy of system AB, computed when subsystems A and B are infinitely far apart, is equal to the sum of the energies of A and B when they are separately computed using the same method (so the energy of a dimer, with the two monomers very well separated, should equal twice the energy of the monomer). The Hartree Fock method satisfies the size consistency condition but CI does not, unless one is considering full CI which is size consistent.

The Frozen-Core Approximation

A full CI calculation is virtually impossible to implement except for small molecules and small basis sets. Hence, one typically resorts to a limited CI calculation. Often the *frozen-core approximation* is used, wherein excitations out of the inner-shell (i.e., the core) MOs of the molecule are not included. (Recall that Koopmans' theorem also makes use of a frozen orbital approximation.) The frozen core for light atoms (lithium to neon) consists of the $1s$ atomic orbital, whereas that for atoms sodium to argon consists of the atomic $1s$, $2s$, $2p_x$, $2p_y$, and $2p_z$ orbitals. For molecules, the frozen orbitals are inner-shell atomic orbitals of the atoms comprising the molecule. The frozen core orbitals are doubly occupied and all other molecular orbitals are orthogonal to the frozen core orbitals.

10.12.2 MOLLER–PLESSET MANY-BODY PERTURBATION THEORY

Configuration interaction calculations provide a systematic approach for going beyond the Hartree–Fock approximation using a single Slater determinant by including a superposition of determinants in the wave function that are successively singly excited, doubly excited, triply excited, etc., from a given reference configuration. Although CI is variational, it lacks size consistency (except for full CI). Perturbation theory provides an alternative systematic approach to finding the correlation energy. Although such calculations are size consistent, they are not variational in the sense that it does not, in general, give energies that are upper bounds to the exact energy. Moller–Plesset perturbation theory, which is a variant of Rayleigh–Schrödinger perturbation theory, can be made size consistent.

Rayleigh–Schrödinger perturbation theory was introduced in Sec. 7.3.1. The application of perturbation theory to a system composed of many interacting particles is called many-body perturbation theory (MBPT). A systematic treatment of MBPT will be presented in the many-body chapters that follow, within the formalism of second quantization. Here we introduce MBPT within first quantization. As an example of its application, we calculate the correlation energy for the ground state, by taking the zero-order Hamiltonian from the Fock operators f_i (10.60) of the HF-SCF method, and the difference of the full Hamiltonian and the zero-order Hamiltonian is taken to be the perturbation potential V. This method, which improves on the Hartree–Fock by adding electron correlation effects by means of perturbation theory, was developed in the early days of quantum mechanics (in 1934) by C. Moller and M. S. Plesset, and the procedure is therefore called Moller–Plesset perturbation theory. Applications of this method to molecular systems did not actually

10.12 Electron Correlation

begin until much later. In Moller–Plesset perturbation theory (MPPT), the zero-order Hamiltonian $H^{(0)}$ for the ground state is given by the sum of the one-electron Fock operators (10.60),

$$H_{\text{HF}} \equiv \sum_{i=1}^{N} f_i = \sum_{\lambda} \varepsilon_\lambda |u_\lambda\rangle\langle u_\lambda|. \tag{10.136}$$

where u_λ is the spin-orbital for the λth-filled spin-orbital [see Eq. (10.61)]. MPPT energy corrections are obtained with the perturbation

$$H^{(1)} \equiv H - H_{\text{HF}} - \langle \Phi_0|H - H_{\text{HF}}|\Phi_0\rangle, \tag{10.137}$$

where

$$H_{\text{HF}} \Phi_0 = \left(\sum_{\lambda} \varepsilon_\lambda\right) \Phi_0, \tag{10.138}$$

and

$$H^{(0)} = H_{\text{HF}} + \langle \Phi_0|H - H_{\text{HF}}|\Phi_0\rangle, \tag{10.139}$$

serves as the unperturbed zeroth-order operator. The HF ground state wave function Φ_0 is an eigenfunction of H_{HF} with eigenvalue E_0 given by the sum of the orbital energies of all the occupied spin-orbitals. Since the Slater determinant Φ_0 is an eigenfunction of H_{HF},

$$H^{(0)} \Phi_0 = \langle \Phi_0|H|\Phi_0\rangle \Phi_0. \tag{10.140}$$

The zeroth-order MP energy, $E_0^{\text{MP}} = \langle \Phi_0|H^{(0)}|\Phi_0\rangle = \langle \Phi_0|H|\Phi_0\rangle$ is therefore the Hartree–Fock energy, Eq. (10.58). The first-order MP energy is clearly zero

$$E_1^{\text{MP}} = \langle \Phi_0|H^{(1)}|\Phi_0\rangle = 0. \tag{10.141}$$

Thus, the lowest order MP correlation energy appears in second order. This result, called the *Moller–Plesset theorem*, states that the correlation potential does not contribute in first order to the exact electronic energy. An expression for the second-order MP energy, E_2^{MP}, can be computed from second-order Rayleigh–Schrödinger perturbation theory, Eq. (7.37), written in terms of doubly excited Slater determinants (the singly excited Slater determinants do not contribute due to the *Brillouin theorem*, which states that $\langle \Phi_0|H^{(0)}|\Phi_1\rangle = 0$). The result is:

$$E_2^{\text{MP}} = \sum_{i,j,a,b} \langle \phi_i(1)\phi_j(2)|r_{12}^{-1}|\phi_a(1)\phi_b(2)\rangle \times \frac{2\langle \phi_a(1)\phi_b(2)|r_{12}^{-1}|\phi_i(1)\phi_j(2)\rangle - \langle \phi_a(1)\phi_b(2)|r_{12}^{-1}|\phi_j(1)\phi_i(2)\rangle}{\varepsilon_i + \varepsilon_j - \varepsilon_a - \varepsilon_b}, \tag{10.142}$$

where i and j are occupied orbitals, and a and b are excited orbitals.

10.12.3 COUPLED CLUSTER METHOD

Another ab initio many-body method that, like MPPT, is size consistent but not variational, is the coupled cluster method (CC method). It was initially developed by Fritz Cöster and Hermann Kümmel in the 1950s for treating the many-body problem in nuclear physics, and was reformulated for electron correlation in atoms and molecules in the 1960s.

Let us define the operator C_1 which, when applied to the reference determinant Φ_0, yields a state containing *all* single excitations, Φ_i^a,

$$C_1 \Phi_0 = \sum_{i,a} c_a^i \Phi_i^a. \tag{10.143}$$

The amplitudes c_a^i are arbitrary at this point, and are to be determined. Similarly, we can define the operator C_2 which, when applied to the reference determinant Φ_0, yields a state containing all doubly excited determinants, Φ_{ij}^{ab},

$$C_2 \Phi_0 = \sum_{ij,ab} c_{ab}^{ij} \Phi_{ij}^{ab}. \tag{10.144}$$

Clearly, we can go on and define the operator C_3 producing all possible triple excitations, etc.

The CC method starts by considering the HF determinantal wave function $\Phi_0 = \det |u_1 \ldots u_N|$ and introduces the *cluster operator* C, which relates the exact electronic wave function Ψ to Φ_0 through the relation, $\Psi = e^C \Phi_0$. When e^C, called the *wave operator*, is applied to Φ_0, it creates a linear combination of Slater determinants that includes Φ_0 (the first term) and all its singly, doubly, ..., excited determinants. In other words, the cluster operator C operating on Φ_0 gives a linear combination of Slater determinants in which electrons from occupied spin-orbitals have been excited to virtual spin-orbitals. Specifically, C is the sum of the one-electron excitation operator C_1 defined in (10.143), the two-electron excitation operator C_2 defined in (10.144), ..., the N-electron excitation operator C_N:

$$C = C_1 + C_2 + \ldots + C_N. \tag{10.145}$$

In practice, this series is finite, because the number of occupied molecular orbitals is finite, as is the number of excitations to orbitals in the basis set. When the cluster operator C is assumed to be of the form $C = C_A + C_B$, as would be appropriate for infinitely separated systems A and B, then the wave operator $e^{(C_A+C_B)} = e^{C_A} e^{C_B}$, if C_A and C_B commute, and this ensures a desired form of the product of the wave function, $e^{(C_A+C_B)} \Phi_0 = e^{C_A} e^{C_B} \Phi_0$. If instead, a finite CI expansion is taken, $(C_A + C_B) \Phi_0$, one does not obtain a product but a sum, which is incorrect.

In order to simplify the task for finding the coefficients c appearing in Eqs (10.143) and (10.144), the expansion of C into individual excitation operators in (10.146) is terminated at the second or third level of excitation. There are several ways of writing the equations to determine the c coefficients, but the standard formalism results in a terminating set of equations which may be solved iteratively. We shall not present these methods here. The interested reader is referred to Refs [166, 167] for details.

A list of quantum chemistry and solid-state physics software programs for computing electronic structure, both open source and commercial, is available at http://en.wikipedia.org/wiki/Quantum_chemistry_computer_programs.

Molecules 11

The word molecule comes from the French word molécule, meaning "extremely minute particle." In this chapter, we again take up the study of molecules (molecular electronic structure calculations are discussed in Sec. 10.10). We first discuss the general use of symmetry to classify molecules in Sec. 11.1 (Appendix E presents some aspects of group theory, which are helpful in quantum mechanics). Section 11.1.1 discusses the formation of molecular orbitals (MOs) using symmetry and shows how group character tables expedite this task. Section 11.1.2 discusses character tables and the Mulliken symbols for molecules. The classification of the electronic states of diatomic molecules is presented in Sec. 11.2. Section 11.2.1 discusses the asymptotic form of diatomic molecular potentials, and Sec. 11.2.2 considers the electronic spin and angular momentum coupling in diatomic molecules, which is known as Hund's coupling. Section 11.3 introduces the Born–Oppenheimer approximation (BOA) that separates electronic and nuclear degrees of freedom of molecules (used in treating electronic structure, in Chapter 10), by making use of the fact that the nuclear degrees of freedom vary much slower than the electronic ones (therefore, it is a type of adiabatic approximation). We then discuss the Born–Oppenheimer basis representation, which removes the BOA but keeps the form of the wave function used in the BOA as a basis set to treat molecular problems. Section 11.4 treats the rotational and vibrational degrees of freedom of molecules. Section 11.5 develops the small oscillations theory for molecules and considers the symmetry of vibrational modes. Optical transitions in molecules are discussed in Sec. 11.6. Finally, Section 11.7 discusses the Franck–Condon principle for determining the changes that occur in nuclear degrees of freedom when molecules undergo optical transitions.

Molecules are most often studied experimentally using spectroscopic methods. Microwave spectroscopy uses photons with frequency in the range 10^9–10^{12} Hz (30 cm–0.03 cm) to probe the rotational degrees of freedom of molecules; infrared spectroscopy uses light with frequency 10^{12} to 4×10^{14} Hz (3 mm–780 nm) to probe (mostly) vibrational degrees of freedom; visible light with frequency 4×10^{14} to 8×10^{14} Hz (780–390 nm); and ultraviolet light with frequency 8×10^{14} to 4×10^{17} Hz (390–3 nm) are used to probe the electronic degrees of freedom of molecules. Various kinds of spectroscopies are used, including absorption, emission, and Raman scattering [including coherent anti-Stokes Raman spectroscopy (CARS)]. Moreover, nuclear magnetic resonance (NMR) is used to probe molecular structures with microwave radiation, and X-rays probe the crystal structure of molecules using light with frequency 10^{17}–10^{20} Hz (3–0.003 nm). Furthermore, photoacoustic (sometimes called acousto-optic) spectroscopy, the absorption of electromagnetic waves and the subsequent detection of the acoustic waves, a method first developed by Alexander Graham Bell in 1880, is a sensitive method used to study overtone absorption in molecules (e.g., $n = 0 \rightarrow n = 9$ of the CH stretch in benzene). There are many books that thoroughly cover molecular spectroscopy, e.g., Refs. [173–175].

11.1 MOLECULAR SYMMETRIES

Molecules are classified according to their symmetry. Their symmetry helps to understand chemical properties such as dipole moments and allowed spectroscopic transitions. Moreover, molecular symmetry is useful in characterizing molecular orbitals (MOs) and simplifies their calculation. As discussed in Sec. E.3.1 of Appendix E, molecules belong to one of the point groups C_1, C_s, C_i, C_n, C_{nv}, C_{nh}, S_{2n}, D_n, D_{nd}, D_{nh}, T, T_d, T_h, O, O_h, I, and I_h. The symmetry operations contained in each of these groups are listed there, and examples of group multiplication tables for the point groups are provided (see also Table 11.1). Only the integers $n = 1, 2, 3, 4$, and 6 are compatible with translational symmetry, and these give rise to the 32 point groups occurring in crystals. For free molecules, other values of n, such as $n = 5$, $n = 7$, etc., are possible. To determine what point group a given molecule belongs to, one can use the decision tree shown in Fig. 11.1. Knowing the geometry of the molecule, one simply works through the questions in the table, starting at the top to determine the point group for the molecule.

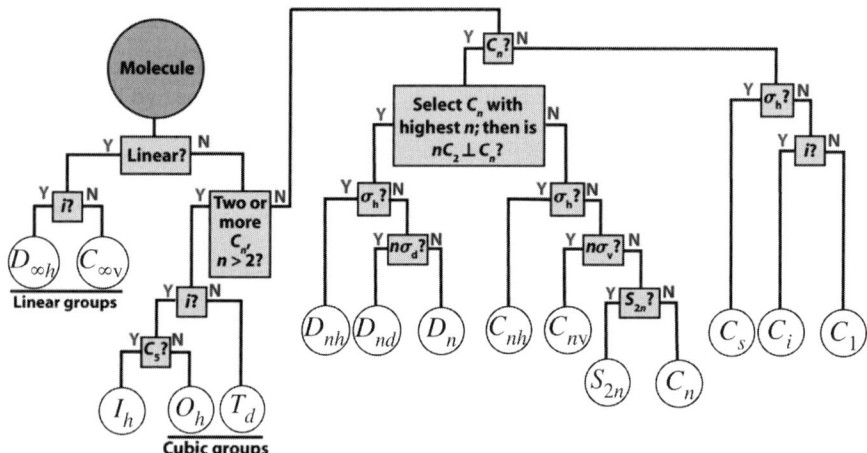

FIG 11.1 Point group decision tree. (After Shriver and Atkins, *Inorganic Chemistry*.) Permission obtained from Oxford University Press.

Problem 11.1

(a) Determine the point symmetry group of the molecule cis-CHFCHF (cis-1,2-difluoroethylene, see Fig. 11.2). Note that cis refers to the isomer in which substituents at opposite ends of a carbon-carbon double bond are on the same side of the bond (whereas trans means on the opposite side).
(b) What are the symmetry elements of the group?
(c) Build the group multiplication table.
(d) Construct the group representations of the group elements for the base functions: $1s$ (H_1), $1s$ (H_2), $2p_z$ (F_1), $2p_z$ (F_2), $2p_z$ (C_1), $2p_z$ (C_2).
Answer: (a) C_{2v}. (b) C_2 axis, another C_2 axis perpendicular to it, and two orthogonal planes of symmetry, i.e., σ_v and σ'_v. (c) See Table 11.1.

One of the applications of symmetry is analyzing molecular vibrations. The number and types of vibrational modes in the absorption, emission, and Raman spectra of a molecule can be determined based on just the point group of the molecule. Moreover, together with some basic knowledge of the potential energy surface of a molecule, its vibrational frequencies can be fully determined. We consider symmetry-adapted nuclear vibrational coordinates and develop small oscillations theory for molecules in Sec. 11.5.

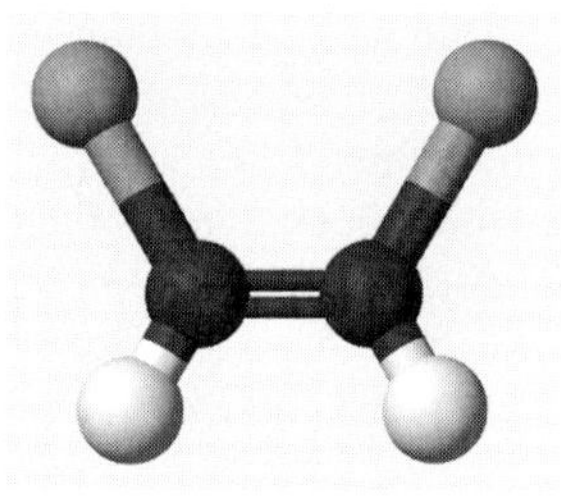

11.1.1 MOLECULAR ORBITALS AND GROUP THEORY

Many computational chemistry programs start by calculating MOs and electronic energies and can also determine vibrational and rotational frequencies

FIG 11.2 Ball-and-stick model of the molecule cis-1,2-difluoroethylene.

Table 11.1 *Group multiplication table for the group C_{2v}.*

C_{2v}	I	C_2	σ_v	σ'_v
I	I	C_2	σ_v	σ'_v
C_2	C_2	I	σ'_v	σ_v
σ_v	σ_v	σ'_v	I	C_2
σ'_v	σ'_v	σ_v	C_2	I

11.1 Molecular Symmetries

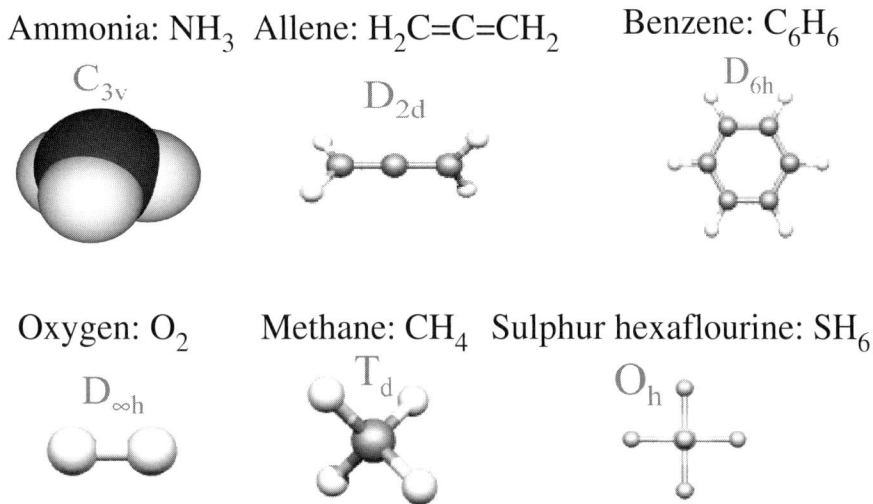

FIG 11.3 The point group symmetry of a number of molecules. See Sec. E.3.1 of Appendix E for a complete list of the point groups and representative molecules belonging to these groups.
(Adapted from http://www.staff.ncl.ac.uk/j.p.goss/symmetry/Molecules_pov.html)

of molecules of the equilibrium configuration of molecular states. The number of MOs used equals the number of the atomic orbitals included in the linear combination of atomic orbitals (LCAOs) expansion (see Sec. 10.10). If the molecule has symmetry, as is the case for the molecules shown in Fig. 11.3, there will be degenerate atomic orbitals, and they can be grouped in linear combinations called symmetry-adapted orbitals. The symmetry-adapted orbitals belong to the representations of the symmetry group of the molecule. The number of MOs that belong to a particular group representation equals the number of symmetry-adapted atomic orbitals. A symmetry operator, $\hat{A}$, applied to a degenerate electronic atomic wave function belonging to a given irreducible representation, say (α), $\phi_j^{(\alpha)}$, converts it to a linear combination of the atomic wave functions, $\hat{A}\phi_j^{(\alpha)} = \sum_k A_{jk}^{(\alpha)} \phi_k^{(\alpha)}$ (see Sec. E.5.4 of Appendix E). The effects of the symmetry operator $\hat{A}$ on the wave functions of an n-fold orbitally degenerate electronic orbital are specified by the n^2 amplitudes $A_{jk}^{(\alpha)}$, $j = 1, \ldots, n, k = 1, \ldots, n$. The number and the properties of irreducible representations of the point groups are fully characterized using the normal representation of the groups (see Sec. E.5.4). For example, for a molecule with D_{6h} symmetry (e.g., benzene, C_6H_6, see Fig. 11.3), the irreducible representations are A_{1g}, A_{2g}, B_{1g}, B_{2g}, E_{1g}, E_{2g}, A_{1u}, A_{2u}, B_{1u}, B_{2u}, E_{1u}, E_{2u}. The A_{1g} representation is the totally symmetric symmetry representation. Sections E.5.1–E.5.4 of Appendix E discuss how to determine the coefficients $A_{jk}^{(\alpha)}$ in general for any group and any irreducible representation.

The electronic ground state of the equilibrium configuration of many molecules belongs to the nondegenerate totally symmetric representation. Moreover, the electronic spins are usually all paired in the ground state, so the ground state is a singlet. For example, for benzene, the ground electronic state is $^1A_{1g}$. Note that the point group of the molecule in its equilibrium geometry can be very different from that at a nonequilibrium geometry. Furthermore, the equilibrium geometry of excited electronic states of the molecule is typically different from those of the ground state.

11.1.2 CHARACTER TABLES AND MULLIKEN SYMBOLS

In this section, we discuss character tables and the Mulliken symbols used in the tables (please read Appendix E, which discusses group theory and describes the point groups, if you are not familiar with group character tables). Character tables are the most often used device for summarizing the symmetry aspects of groups. The rows in a character table are labeled by the irreducible representation, and the columns are labeled by the classes. Tables E.3 and E.4 in Appendix E show the character tables for the groups C_{3v} and C_6. The following holds for character tables for the point groups:

- Isomorphic groups have the same representations and are often given together (isomorphic groups have a one-to-one correspondence between the elements of the groups which preserves the group multiplication properties – basically, isomorphic groups are the same group, but the names of the elements are different).
- The numbers in front of the symbols for the elements (actually, in front of the symbols for the classes) of the group in the top row of the table show the numbers of elements in the corresponding classes.
- The left-hand column shows the conventional names of the representations.
- Beside the symbols for the representations are placed the letters x, y, z; these show the representations by which the coordinates themselves are transformed. The z-axis is always taken along the principal axis of symmetry.
- Mulliken symbols are used to identify the irreducible representations as follows:

   A = singly degenerate state (i.e., a one-dimensional representation), symmetric with respect to rotation about the principal axis.

   B = singly degenerate state (i.e., a one-dimensional representation), antisymmetric with respect to rotation about the principal axis.

   E = doubly degenerate (i.e., a two-dimensional representation).

   T = triply degenerate representation. Numerical subscripts for Es and Ts follow certain rules that we will not elaborate on here.

   X_1 = subscript 1 on a symbol A or B indicates a symmetric representation upon C_2 rotation perpendicular to the principal axis, or, if such a C_2 is lacking, to a vertical plane of symmetry. Subscript 2 (as in X_2), indicates a change sign (antisymmety).

   $'$ = symmetric with respect to a horizontal symmetry plane σ_h

   $''$ = antisymmetric with respect to a horizontal symmetry plane σ_h

   X_g = subscript g (gerade, meaning symmetric in German) indicates symmetric under inversion in those groups containing a center of inversion symmetry element. X_u = subscript u (ungerade, antisymmetric) indicates antisymmetric under inversion.

In summary, the one-dimensional representations are denoted by the letters A, B, the two-dimensional ones by E, and the three-dimensional ones by T. The base functions of A representations are symmetric, and those of B representations are antisymmetric, with respect to rotations about a principal axis of the nth order. For A and B representations, a numerical subscript 1 (2) indicates symmetric (antisymmetric) behavior under a C_2 rotation perpendicular to the principal axis, or reflection in a vertical plane. The functions of different symmetry with respect to a reflection σ_h are distinguished by the number of primes (one or two). The suffixes g and u show the symmetry with respect to inversion.

Problem 11.2

Consider a triatomic molecule with three identical atoms on the vertices of an equilateral triangle. The symmetry of the molecule is D_{3h}. Find the symmetry-adapted MOs formed from a LCAOs formed from the s and the p atomic orbitals. Hint: Use the projection operators $P^{(\alpha)} \equiv \frac{f_\alpha}{g} \sum_A \left(\chi^{(\alpha)}\right)^* A$, where $\chi^{(\alpha)}$ is the character for irreducible representation α, as explained in Sec. E.5.4.

11.2 DIATOMIC ELECTRONIC STATES

The classification of diatomic electronic states is based on their asymptotic correlation to particular electronic states of the two atoms comprising the molecule as the distance between the atoms, R, tends to infinity. Let us consider a generic diatomic molecule AB, with atoms A and B in given electronic states, $(^{2S+1}L_J)_A$ and $(^{2S+1}L_J)_B$. The molecular states correlating asymptotically with atomic fragment states having projection of the total spin along the diatomic axis with

11.2 Diatomic Electronic States

values $S_Z = S_A + S_B, S_A + S_B - 1, \ldots, -S_A - S_B$, projection of the total orbital angular momentum along the diatomic axis with values $M_L = L_A + L_B, L_A + L_B - 1, \ldots, -L_A - L_B$, and projection of the total angular momentum along the diatomic axis with values $J_Z = J_A + J_B, J_A + J_B - 1, \ldots, -J_A - J_B$. So, L_Z S_Z, and J_Z denote the projection of the electronic orbital spin and total angular momentum along the diatomic axis, respectively. By convention, the absolute value of the projection of $\mathbf{L}/\hbar$ on the diatomic axis is labeled $\Lambda \equiv |L_Z|/\hbar$ (note that this quantum number is non-negative). The projection of $\mathbf{S}/\hbar$ is Σ ($\equiv S_Z/\hbar$) and $\Omega \equiv \Lambda + \Sigma$ (note that Σ and Ω can be negative). For a given Λ and S, Ω can take on the values $\Lambda + S, \Lambda + S - 1, \ldots, \Lambda - S$. The specific values of the Λ quantum number are denoted by the Greek capital letters $\Sigma, \Pi, \Delta, \ldots$, for 0, 1, 2, ..., respectively. That is, states with $\Lambda = 0$ are denoted as Σ states, states with $\Lambda = 1$ (i.e., $L_Z = \pm 1$) as Π states, states with $\Lambda = 2$ (i.e., $L_Z = \pm 2$) as Δ states, and so on. Molecular electronic states are referred to using the symbol $^{2S+1}\Lambda_\Omega$. The states with molecular term symbols $^{2S+1}\Lambda$ are split by spin-orbit splitting via an interaction energy $B\Lambda\Sigma$, where B is a constant, to the multiplets $^{2S+1}\Lambda_\Omega$. We shall elaborate on this splitting in Sec. 11.2.2.

For example, for atoms A and B, both asymptotically in ^2P states, for any value of J ($J = 1/2, 3/2$), the projection of the spin along the axis, Σ, can take the values -1, 0, and 1, the projection $\Lambda = 2, 1, 0$, and $\Omega = 3, 2, 1, 0, -1$, where $\Omega = 3$ clearly correlates with A and B both in $^2P_{3/2}$ states and $\Omega = 2$ correlates with at least one of the atoms in a $^2P_{3/2}$ state. These molecular states can be denoted as $^3\Delta_3, ^3\Delta_2, ^3\Delta_1, ^1\Delta_2, ^3\Pi_2, ^3\Pi_1, ^3\Pi_0, ^1\Pi_1, ^3\Sigma_1, ^1\Sigma_0$, and $^3\Sigma_{-1}$ where the superscript corresponds to the value of $2S+1$ and the subscript to the value of Ω. The total number of states can be calculated as follows: because the P atomic states are triply degenerate and the spin doublets are doubly degenerate, the total number of states equals $3 \times 3 \times 2 \times 2 = 36$. The same number of molecular states must exist, i.e., the number of molecular states $^{2S+1}\Lambda_\Omega$ must equal 36.

A diatomic molecule is physically unchanged by reflection in a plane containing the diatomic axis; this reflection operator is called σ_v (see Sec. 11.1.2). The wave function must remain unchanged or change sign under such a reflection operation. Therefore, electronic terms are given an extra superscript "+" or "−" to indicate the symmetry with respect to reflection in such a plane. For $\Lambda > 0$, one of the degenerate states with $L_Z = \pm\Lambda\hbar$ is "+" and the other is "−" and the superscript is omitted. However, Σ states, i.e., those with $\Lambda = 0$, are further distinguished according to whether they are even or odd under reflection in a plane containing the diatomic axis; these states are denoted as Σ^+ and Σ^-, respectively.

If atoms A and B are identical, i.e., the diatomic molecule is homonuclear, then an additional symmetry exists; that of inversion of the electronic coordinates through the center of mass of the nuclei. States that are gerade under inversion are labeled with a subscript g and states that are ungerade are labeled with a subscript u. For example, when the two atoms are identical, the $^3\Lambda$ and $^1\Lambda$ states are labeled as follows: $^3\Lambda_u, ^3\Lambda_g, ^1\Lambda_u$, and $^1\Lambda_g$. The number of states for identical ^2P atoms is 72, rather than the 36 for nonidentical ^2P atoms. However, if the two identical atoms are in the same electronic state, then the Pauli exclusion principle must be taken into consideration and the total number of possible states is again 36.

Molecular electronic states are composed of products of MOs (actually, of Slater determinants of MOs, see Chapter 8, and Secs. 10.10.1 and 10.10.2). MOs are named in a fashion similar to that used to describe the total electronic state of the molecule, but lower case Greek letters are used. MOs are labeled as follows: $nl\lambda_{g,u}$. Here, λ is the projection of the electron angular momentum along the molecular axis, and it is denoted by the Greek lowercase letters $\sigma, \pi, \delta, \ldots$, for 0, 1, 2, ..., respectively. The subscript g or u is present only for homonuclear diatomics, and the quantum numbers nl indicate the principal and angular momentum quantum numbers of the atomic orbitals comprising the MO. See Problem 10.7 for plots of the lowest σ and π orbitals of symmetry g and u. The product of the MOs determines the possible molecular symmetries. For example, the two-electron MO product $(1s\sigma)(2s\sigma)$ can yield $^1\Sigma$ or $^3\Sigma$ states, whereas $(2p\pi)^2$ can yield $^1\Sigma, ^3\Sigma, ^1\Delta$, and $^3\Delta$ states. If we also specify the possible values of Ω using the notation $^{2\Sigma+1}\Lambda_\Omega$, the states that can be obtained from the configuration $(2p\pi)^2$ are $^3\Sigma_{-1}, ^3\Sigma_1, ^1\Sigma_0, ^3\Delta_3, ^3\Delta_2, ^3\Delta_1$, and $^1\Delta_2$.

In Fig. 10.7, the potential energy curves of the hydrogen molecule correlating to two ground-state hydrogen atoms are plotted versus internuclear distance. Figure 11.4 shows the excited state hydrogen potential energy curves. The asymptotic atomic states to which the molecular curves correlate are indicated at the right side of the figure. The ground vibrational state of the various potentials is shown, as are the $v = 5$ vibrational states and some of the $v = 10$ states. An interesting feature is the double minimum E, F electronic state, which arises due to a curve crossing.

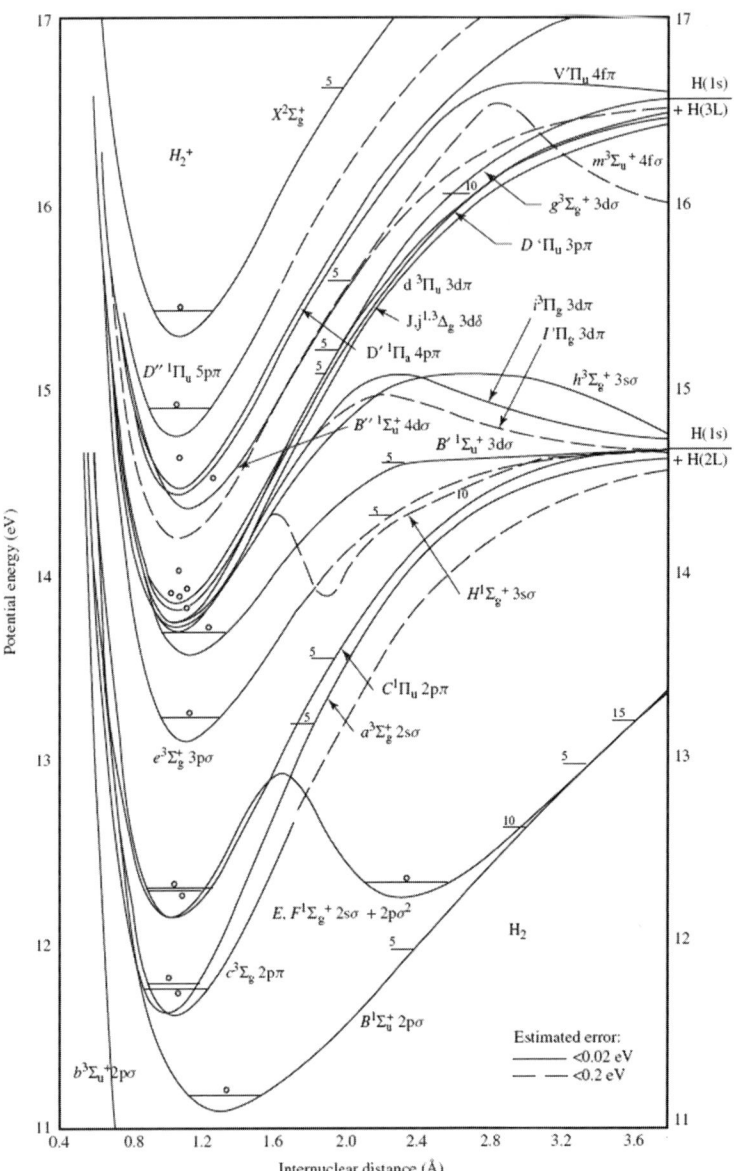

FIG 11.4 H$_2$ excited potential energy curves versus internuclear distance R. (Reproduced from Sharp [176], T. E. Sharp, Potential-energy curves for molecular hydrogen and its ions, Atomic Data 2, 119, 1971)

Problem 11.3

(a) Determine the possible term symbols for the diatomic phosphorous molecule, P$_2$, where the atoms are in the ^{4}S state.
(b) For the molecular states in part (a), what are the possible values of the absolute value of the projection of the total electronic angular momentum Ω on the axis.

Answers: (a) $^1\Sigma_g^+, ^3\Sigma_u^+, ^5\Sigma_g^+, ^7\Sigma_u^+$. (b) For $^1\Sigma_g^+$, $\Omega = 0$, for $^3\Sigma_u^+$, $|\Omega| = 0, 1$, for $^5\Sigma_g^+$, $|\Omega| = 0, 1, 2$, and for $^7\Sigma_u^+$, $|\Omega| = 0, 1, 2, 3$.

Problem 11.4

Determine the possible term symbols for the molecule HF where the atoms are in the ^2S and ^2P states.
Answer: $^{1,3}\Sigma^+, ^{1,3}\Pi$.

Problem 11.5

Determine the possible term symbols for the molecule O_2, where the atoms are in the ^3P state.
Answer: $2\,^1\Sigma_g^+, ^1\Sigma_u^-, ^1\Pi_g, ^1\Pi_u, ^1\Delta_g, ^3\Delta_g, ^5\Delta_g, 2\,^3\Sigma_u^+, ^3\Sigma_g^-, ^3\Pi_u, 2\,^5\Sigma_g^+, ^5\Sigma_u^-, ^5\Pi_g, ^5\Pi_u, ^5\Delta_g$. The numbers in front of the term symbols specify the number of terms of the type indicated, if the number is greater than unity. The ground state of O_2 is $^3\Sigma_g^-$.

Problem 11.6

(a) Given the electronic structure of Actinium, ^{89}Ac, ... $7s^2\, 6d$, what are the possible term symbols of the atom.
(b) What are the possible molecular term symbols for the diatomic Actinium molecule (parity symmetry need not be considered).
(c) Assuming that the total electronic angular momentum J of the atoms is the highest value it can take, what are the possible values of the projection of the total electronic angular momentum Ω on the diatomic axis of the molecule for each of the term symbols.

Answers: (a) $^{2S+1}L_J = {}^2D_{5/2}$ and $^2D_{3/2}$. (b) $^{2S+1}\Lambda_\Omega = {}^{1,3}\Lambda_\Omega$, where Λ can range over 4, 3, 2, 1, 0 and Ω can range over $\Lambda + \Sigma, \ldots, \Lambda - \Sigma$. (c) $\Omega = 5, 4, 3, 2, 1, 0, -1$.

Problem 11.7

Given the two-electron MO product $(2p\pi)(2d\delta)$ in a diatomic molecule, what are the quantum numbers $^{2S+1}\Lambda_\Omega$ of the molecular state that can be obtained.
Answer: $^3\Phi_4, ^3\Phi_3, ^3\Phi_2, ^1\Phi_3, ^3\Pi_2, ^3\Pi_1, ^3\Pi_0, ^1\Pi_1$.

11.2.1 INTERATOMIC POTENTIALS AT LARGE INTERNUCLEAR DISTANCES

The asymptotic form of the potential between two atoms that are very far apart can be determined using perturbation theory, regarding the two isolated atoms as the unperturbed system and the potential energy of their electrical interaction as the perturbation. The potential of two systems of charges at large relative distance R can be expanded in powers of R. The multipole expansion (3.198) of the potential energy due to the charge distribution of a system contains terms from the system charge, dipole moment, quadrupole moment, etc. For a neutral atom, the total charge is zero, so the expansion then begins with the dipole term, etc.

For two neutral atoms, the potential energy in Eq. (3.201) can be expanded in multipoles to obtain (3.202). The lowest-order term for two neutral atoms is dipole–dipole (R^{-3}), followed by dipole–quadrupole (R^{-4}), quadrupole–quadrupole and dipole–octupole terms (R^{-5}), and so on. The expectation value of the potential can now be used to get an expression for the first-order perturbation correction to the energy,

$$\Delta E^{(1)}(R) = \langle \psi_a \psi_b | U(R) | \psi_a \psi_b \rangle, \quad (11.1)$$

where $U(R)$ is given by (3.202). For the two atoms in S states, no interaction between the atoms is possible in first-order perturbation theory. Therefore, the interaction energy of the atoms is determined by second-order perturbation theory. Let us restrict ourselves to the dipole–dipole interaction term in Eq. (3.202), because it decreases least rapidly as R increases, i.e., as R^{-6}. Because the nondiagonal matrix elements of the dipole moment, e.g., $\langle \psi_{npm} | \mathbf{p} | \psi_{ns0} \rangle$, are in general different from zero, second-order perturbation theory yields a nonvanishing result which, being quadratic in U, is proportional to R^{-6}. The second-order perturbation energy for the lowest eigenvalue is always negative, and the atomic interaction energy is

$$\Delta E^{(2)}(R) = -\frac{C_6}{R^6}, \quad (11.2)$$

where the constant C_6 is positive. The attractive forces between S-state atoms at large distances are called *van der Waals forces*. If only one of the atoms is in an S-state, Eq. (11.2) still gives the interaction energy, because the first-order perturbation vanishes when the multipole moments of one of the atoms is zero. Now, the constant C_6 depends on the mutual orientation of the atoms.

However, if both atoms have nonzero orbital and nonzero total angular momenta, the situation is different. The expectation value of the dipole moment vanishes, because it has odd parity, but the expectation values of the quadrupole moment in states with $L \neq 0$ and $J \neq 0, 1/2$ are nonzero. The quadrupole–quadrupole term yields a finite first-order perturbation, and the interaction energy of the atoms goes as $\Delta E^{(1)}(R) = C_5/R^5$, where C_5 may be either positive or negative.

The interaction of two "identical" atoms that are in different electronic states must be treated differently. The state of two atoms at large internuclear distance has an additional degeneracy due to atom interchange, and degenerate first-order perturbation theory must be used. If the states of the two atoms have different parities, and angular momenta L differing by ± 1 or 0 (but not both zero), then the nondiagonal matrix elements of the transition dipole moment between these states do not vanish. First-order degenerate perturbation theory using the dipole–dipole interaction yields an interaction energy $U(R) = C_3/R^3$, where C_3 can have either sign. However, when the interaction of the atoms is averaged over all possible orientations, the first-order interaction energy vanishes and the second-order perturbation, Eq. (11.2), becomes the lowest-order interaction energy.

11.2.2 HUND'S COUPLING

In molecules, interactions between the electronic spin $\mathbf{S}$ and other angular momenta, such as the angular momentum arising from the rotation of the molecular axis, and the orbital electronic angular momentum, can split electronic states (just as spin-orbit interactions split electronic state energies in atoms). Coupling of electronic spin to the electronic orbital angular momentum (spin-orbit coupling) occurs in a fashion similar to atoms, yielding an effective interaction of the form $\mathcal{H}_{so} = A\mathbf{L} \cdot \mathbf{S}$. Here, A can be a function of the internuclear coordinate(s). The rotational Hamiltonian of a molecule is $\mathcal{H}_r = \frac{\mathbf{N}^2}{2\mu R^2}$, where $\mathbf{N}^2 = (\mathbf{J} - \mathbf{L} - \mathbf{S})^2$ is the nuclear angular momentum and $\mathbf{J}$ is the total angular momentum of the molecule (this ignores nuclear spin angular momenta). The rotation of the molecule as a whole, i.e., the orbital angular momentum of the nuclei $\mathbf{N}$, can also couple with the electronic spin, yielding a small (compared with $\mathcal{H}_{so}$) spin-rotation coupling of the form $\mathcal{H}_{sr} = \gamma(R)\mathbf{N}\cdot\mathbf{S}$, where $\gamma(R)$ is called the spin-rotation parameter and is a function of the internuclear distance R. Moreover, a spin-axis coupling exists, $\mathcal{H}_{sa} = \alpha \mathbf{n}\cdot\mathbf{S}$, where α is the spin-axis parameter and $\mathbf{n}$ is the unit vector in the direction of the diatomic axis, but this effect is very small. Furthermore, there is the coupling of the *nuclear* spin with $\mathbf{L}$, $\mathbf{S}$, and $\mathbf{N}$, but, as in the case of atoms, these interactions are much smaller, because the nuclear magnetic moment is much smaller than the electron magnetic moment. Nevertheless, the symmetry of the nuclear spin wave functions can be important for determining the spectra of homonuclear diatomic molecules.

11.2 Diatomic Electronic States

The complete Hamiltonian for a molecule in the absence of external fields is always diagonal in the total angular momentum quantum numbers J and M_J, i.e., in the absence of external fields, the molecular state is an eigenfunction of $\mathbf{J}^2$ and the space-fixed projection J_z with eigenvalues $J(J+1)\hbar^2$ and $M_J\hbar$, respectively. For a diatomic molecule, ignoring nuclear spin angular momenta,

$$\mathbf{J} = \mathbf{N} + \mathbf{L} + \mathbf{S} \equiv \mathbf{K} + \mathbf{S}. \tag{11.3}$$

The orbital angular momentum of the nuclei, $\mathbf{N}$, is perpendicular to the internuclear axis [note that is convenient for the discussion below to define the vector $\mathbf{K}$ as the sum of $\mathbf{N}$ and the electronic orbital angular momentum $\mathbf{L}$, as in Eq. (11.3)].

There are several energy scales that determine the relative strengths of the angular momentum coupling: (1) The electronic electrostatic interaction strength $|\Delta E|$ between two adjacent electronic levels with different values of Λ. (2) The spin-orbit coupling constant A appearing in the spin-orbit interaction energy $A\mathbf{L} \cdot \mathbf{S}/\hbar^2$. (3) The rotation constant $B = \hbar^2/(2\mu R_e^2)$, where R_e is the internuclear equilibrium distance, which determines the strength of the spin-rotation coupling. Friedrich Hund identified five limiting cases, labeled (a) through (e) [Hund's case (e) is of little importance – it does not occur in bound molecules], depending on the limiting values of the three energies, $|\Delta E|$, $|A|$, and B. The four Hund's cases are illustrated in Fig. 11.5

Hund's case (a)

Our discussion here will take $\hbar = 1$ (angular momentum variables will be unitless). In Hund's case (a), the electrostatic interactions are much larger than the spin-orbit interaction which in turn is much larger than the rotational energy, i.e., $|\Delta E| \gg |A| \gg B$. We first consider, the electron terms, neglecting rotation. $\Lambda \equiv |L_Z|$ and $\Sigma \equiv S_Z$ combine to give projection of the total angular momentum of the electrons about the axis of the molecule, $J_Z = L_Z + S_Z$; one can also define the quantum number $\Omega = \Lambda + \Sigma$, which can take on the values $\Omega = \Lambda + S$, $\Lambda + S - 1, \ldots, \Lambda - S$. For a given Λ, the terms with different Ω are split by the spin-orbit interaction $B\Lambda\Sigma$. For example, for $\Lambda = 1$, $S = 1/2$, the levels $^2\Pi_{1/2}$ and $^2\Pi_{3/2}$ are split; for the case of $\Lambda = 1$, $S = 1$, the levels $^3\Pi_0$, $^3\Pi_1$, and $^3\Pi_2$ are split, etc. With the molecular basis $|S, \Sigma, \pm\Lambda, J, \Omega, M_J\rangle$, the first-order perturbation of the spin-orbit coupling energy is given by the expectation value

$$E_{so} = \langle S, S_Z, L_Z, J, J_Z, M_J | A\mathbf{L} \cdot \mathbf{S} | S, S_Z, L_Z, J, J_Z, M_J \rangle = AL_Z S_Z, \tag{11.4}$$

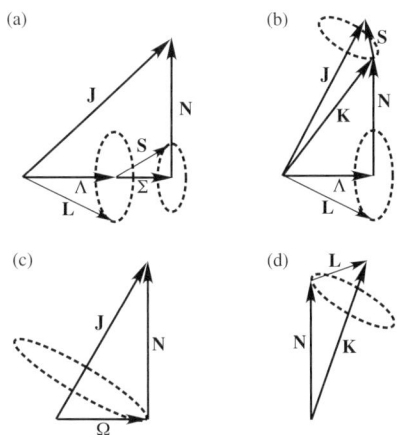

FIG 11.5 Vector diagram illustrating Hund's coupling cases (a), (b), (c), and (d). Case (a), $|\Delta E| \gg |A| \gg B$, is the most common case; one first couples Λ and Σ and treats rotational effects last. Case (b), $|\Delta E| \gg B \gg |A|$, is used for $\Lambda = 0$ molecules and light molecules. Case (c), $|A| \gg |\Delta E| \gg B$, is used for molecules with strong spin-orbit coupling (molecules with heavy atoms). Case (d), $B \gg |\Delta E| \gg |A|$, is used for $J \gg 1$ in light molecules.

where the latter equality follows because $\langle L_X \rangle = \langle L_Y \rangle = \langle S_X \rangle = \langle S_Y \rangle = 0$, so $E_{so} = A(R) M_L \Sigma$, where the spin-orbit parameter A depends on internuclear distance. It is more convenient to write this spin-orbit energy in terms of the quantum number Ω; defining $\mathcal{A}(R) \equiv A(R) L_Z$, and noting that $\Sigma = \Omega - \Lambda$, we can write $E_{so} = \mathcal{A}(R)(\Omega - \Lambda)$ and incorporate the term proportional to Λ in the internuclear potential $U(R)$. Similarly, the spin-axis coupling is proportional to Σ, $E_{sa} = \alpha \mathbf{n} \cdot \mathbf{S} = \tilde{\alpha}\Sigma$, and this term can be combined with E_{so} to give $E_{so} + E_{sa} = \tilde{\mathcal{A}}(R)\Omega$. Now let us add the rotational energy to $E_{so} + E_{sa}$. The rotational energy is given by the expectation value

$$E_r = \left\langle S, S_Z, L_Z, J, J_Z, M_J \left| \frac{\mathbf{N}^2}{2\mu R^2} \right| S, S_Z, L_Z, J, J_Z, M_J \right\rangle. \tag{11.5}$$

Because $\mathbf{N}^2 = (\mathbf{J} - \mathbf{L} - \mathbf{S})^2 = \mathbf{J}^2 - 2\mathbf{J} \cdot (\mathbf{L} + \mathbf{S}) + \mathbf{L}^2 + 2\mathbf{L} \cdot \mathbf{S} + \mathbf{S}^2$, $\langle \mathbf{S}^2 \rangle = S(S+1)$, and $\langle \mathbf{L} \cdot \mathbf{S} \rangle = L_Z S_Z$, the rotational energy can be written as, $E_r = B[J(J+1) - 2\langle \mathbf{J} \cdot (\mathbf{L} + \mathbf{S}) \rangle + L(L+1) + 2L_Z S_Z + S(S+1)]$, where the rotation constant

B depends on internuclear distance, $B(R)$. The term $\langle \mathbf{J} \cdot (\mathbf{L} + \mathbf{S}) \rangle$ can be reduced by noting that $\langle \mathbf{L} \rangle = \mathbf{n}\Lambda$ and $\langle \mathbf{S} \rangle = \mathbf{n}\Sigma$. Because $\langle \mathbf{J} \rangle \cdot \mathbf{n} = (\langle \mathbf{L} \rangle + \langle \mathbf{S} \rangle) \cdot \mathbf{n} = \Lambda + \Sigma = \Omega$, we find $\langle \mathbf{J} \cdot (\mathbf{L} + \mathbf{S}) \rangle = \Omega^2$. Again writing all the terms using Ω, we obtain the rotational energy in Hund's case (a) as $E_r = B(R)[J(J+1) - 2\Omega^2]$, which, when combined with $E_{so} + E_{sa}$, and with the internuclear potential, gives the effective potential energy formula,

$$U_J(R) = U(R) + \tilde{A}(R)\Omega + B(R)[J(J+1) - 2\Omega^2]. \tag{11.6}$$

At the internuclear equilibrium position of the molecule, $R = R_e$, the effective potential energy is $U_J(R_e) = U(R_e) + \tilde{A}(R_e)\Omega + B(R_e)[J(J+1) - 2\Omega^2]$.

Case (a) is the most common, except for $\Lambda = 0$ molecules, where case (b) mainly occurs (see next section), because the expectation value of the spin-orbit interaction for $\Lambda = 0$ vanishes.

Hund's case (b)

In Hund's case (b), the rotation of the molecule predominates over multiplet splitting (but electrostatic interactions are still larger than both), i.e., $|\Delta E| \gg B \gg |A|$. Therefore, we first consider the effect of rotation, neglecting the spin-orbit interaction, and then the spin-orbit is taken into account as a perturbation. Case (b) is sometimes found in very light molecules (such as H_2, CH, and OH) and $\Lambda = 0$ molecules, because the spin-orbit interaction is here comparatively weak, whereas the distances between the rotational levels are large because the moment of inertia $I = \langle \mu R^2 \rangle$ is small. Not only is the total angular momentum $\mathbf{J}$ conserved, but so is the sum of the orbital angular momentum of the electrons and the rotational angular momentum, $\mathbf{K} = \mathbf{N} + \mathbf{L}$. Hence, good basis states to use are simultaneous eigenfunctions of $\mathbf{S}^2$, $\mathbf{K}^2$, K_Z, $\mathbf{J}^2$, and J_z, denoted as $|S, K, K_Z, J, \Omega, M_J\rangle$. The rotational energy is given by $E_r = \langle S, K, K_Z, J, \Omega, M_J | \frac{\mathbf{N}^2}{2\mu R^2} | S, K, K_Z, J, \Omega, M_J\rangle$. A similar argument as used for case (a) shows that $E_r = B(R)K(K+1)$, i.e., $\mathbf{N} = \mathbf{K} - \mathbf{L}$, $\langle \mathbf{N}^2 \rangle = \langle \mathbf{K}^2 - 2\mathbf{K} \cdot \mathbf{L} + \mathbf{L}^2 \rangle$, and the expectation value of the last two terms in the $|S, K, K_Z, J, \Omega, M_J\rangle$ basis can be incorporated into the potential term $U(r)$. In addition, there is a small spin-axis coupling, $E_{sa} = \langle \mathcal{H}_{sa}\rangle = \alpha(R)\langle \mathbf{n} \cdot \mathbf{S}\rangle$ where $\alpha(R)$ is the spin-axis parameter. Because $\langle \mathbf{n} \cdot \mathbf{S}\rangle = c\langle \mathbf{K}\rangle \cdot \langle \mathbf{S}\rangle = c\langle \mathbf{J}^2 - \mathbf{K}^2 - \mathbf{S}^2\rangle/2$, and the constant of proportionality c in the relation $\langle \mathbf{n}\rangle = c\langle \mathbf{K}\rangle$ can be obtained by multiplying both sides by $\mathbf{K}$ noting that the eigenvalues of $\mathbf{n} \cdot \mathbf{K}$ and $\mathbf{K}^2$ are Λ and $K(K+1)$, respectively, we find that, $E_{sa} = \frac{\alpha(R)\Lambda}{2K(K+1)}(\mathbf{J}^2 - \mathbf{K}^2 - \mathbf{S}^2)$. Thus, the total effective potential energy is

$$U_K(R) = U(R) + B(R)K(K+1) + \frac{\alpha(R)\Lambda}{2K(K+1)}(\mathbf{J}^2 - \mathbf{K}^2 - \mathbf{S}^2). \tag{11.7}$$

Hund's case (c)

In this case, the spin-orbit coupling is so strong that Λ and Σ are no longer good quantum numbers, but $\Omega = \Lambda + \Sigma$ is a good quantum number, i.e., $|A| \gg |\Delta E| \gg B$. This is often the case with molecules containing rare-earth or other heavy elements. For example, in the molecule HgH, the spin-orbit coupling is close to 4000 cm^{-1} and the $X^2\Sigma$ and $A^2\Pi$ states are strongly mixed. An analysis of this case can be found in Ref. [177].

Hund's cases (d) and (e)

In Hund's case (d), the electrostatic interaction is so weak that the electronic wave function is not constrained to rotate with the molecule, i.e., $B \gg |\Delta E| \gg |A|$. This is the case for very high rotational levels, $J \gg 1$ in light molecules. In Hund's case (e), $|A| \gg B \gg |\Delta E|$, but this case is practically of no importance because it is not found in Nature.

Spin Uncoupling

Spin uncoupling, or decoupling, refers to the deviations from the idealized Hund's cases. This often happens as J increases so that a molecule falls within one of the Hund's cases for small J but a different Hund's case for large J. As J increases, intermediate coupling (i.e., spin uncoupling) results. For example, in a case (a) molecule, as J increases, $\mathbf{S}$ uncouples from the molecular axis and couples with $\mathbf{K}$ to go into case (b).

Λ-Doubling

The double degeneracy of the terms with $L_z = \pm\Lambda$ for $\Lambda \neq 0$ is only approximate and applies only so long as we neglect the effect of the rotation of the molecule on the electron state. When this interaction is taken into account, a term with $\Lambda \neq 0$ is split into two closely lying levels. This splitting is called Λ-doubling. Let us begin by considering singlet states, $S = 0$. The rotational energy is then given by $\langle \frac{\mathbf{N}^2}{2\mu R^2} \rangle = \langle \frac{1}{2\mu R^2}(\mathbf{K} - \mathbf{L})^2 \rangle$. The matrix elements of $\mathbf{K}^2$ and $\mathbf{L}^2$ are diagonal in Λ, so no splitting results, but off-diagonal matrix elements of $\mathbf{K} \cdot \mathbf{L}$ do not vanish, so this term gives rise to splitting. We will not pause to calculate the splittings here.

11.2.3 HYPERFINE INTERACTIONS IN DIATOMIC MOLECULES

The hyperfine structure in atomic spectra has been discussed in Sec. 4.6. There we showed that an *s*-wave electronic state interacts with the nuclear spin through the Fermi contact term, whereas orbitals with $l \neq 0$ interact via a $1/r^3$ dipole–dipole interaction. In the case of diatomic molecules, where the orbital angular momenta of electrons are not good quantum numbers (i.e., the potential possesses only axial symmetry and no classification of electronic states according to a total angular momentum quantum number can be made), the magnetic interaction of the nuclear moments with the electronic currents is more difficult to calculate.

Most molecules have zero total electronic spin angular momentum. When the nuclei have nuclear spin greater than 1/2, (only such nuclei possess quadrupole moments), the lowest-order hyperfine splitting of such molecules arises from the quadrupole interaction of the nuclei with the electrons. Denoting the quadrupole moment operators of a nucleus by the symbol $\hat{Q}_{ij}$ and averaging over the electron state for fixed nuclei, we obtain the interaction energy of a given nucleus with the electrons in a diatomic molecule,

$$V = \sum_{ij} \langle \hat{Q}_{ij} \rangle n_i n_j = \sum_{ij} \langle \hat{Q}_{ij} \rangle \left(n_i n_j - \frac{1}{3}\delta_{ij} \right), \tag{11.8}$$

where $\mathbf{n}$ is the unit vector along the diatomic axis and $\langle \hat{Q}_{ij} \rangle$ is the expectation value of the nuclear quadrupole tensor over the nuclear charge distribution. This quantity can be averaged over the rotation of the molecule to get an expression for the hyperfine energy splitting, but we shall not do so here.

If the spin of a nucleus is 1/2, the electric quadrupole moment vanishes, and then the lowest-order hyperfine splitting results from the direct magnetic interaction between the nuclear-dipole magnetic moment and the electron-dipole magnetic moments. Using Eq. (4.80) for the spin-dipolar interaction potential, the interaction of the nuclear spins with the electron spins in a molecule can be written as

$$V(r) = -\left[\frac{\mu_0}{4\pi}\right] \frac{(2\mu_B)(\mu_N)}{\hbar^2} \sum_\alpha g_{N_\alpha} \sum_i \frac{3(\mathbf{S}_\alpha \cdot \mathbf{r}_{\alpha i})(\mathbf{S}_i \cdot \mathbf{r}_{\alpha i}) - \mathbf{S}_\alpha \cdot \mathbf{S}_i r_{\alpha i}^2}{r_{\alpha i}^5}, \tag{11.9}$$

where i is for the electrons, α is for the nuclei, and $\mathbf{r}_{\alpha i}$ is the coordinate vector between the αth nucleus and the ith electron. To calculate the hyperfine splitting, this interaction potential must be averaged over the state of the molecule.

An additional hyperfine splitting of molecular levels results from the interaction of the nuclear magnetic moment with the rotation of the molecule. The rotating charges in the molecule create a magnetic field, and this magnetic field interacts with the nuclear magnetic moments. We shall not develop an expression for this interaction.

11.3 THE BORN-OPPENHEIMER APPROXIMATION

The BOA, named after Max Born and his then doctoral student, Robert Oppenheimer, involves a separation of electronic and nuclear motion based on the large difference between their timescales. In this sense, it is an adiabatic approximation. The adiabatic parameter is the ratio of the electron mass to the nuclear mass m_e/M_N (actually, to the power of 1/4). Because the nuclei move slowly compared with the electrons, the nuclear positions can be held fixed as the electronic wave function is calculated. In the BOA, we write the full wave function of the molecule as a product of an electronic wave function, ψ, and a nuclear wave function, χ:

$$\Psi(\mathbf{r}, \mathbf{R}) = \psi(\mathbf{r}, \mathbf{R})\chi(\mathbf{R}). \tag{11.10}$$

Here, $\mathbf{r}$ denotes all the electronic coordinates and $\mathbf{R}$ all the nuclear coordinates. For the time being, let us view the argument $\mathbf{R}$ of ψ on the RHS of Eq. (11.10) as a constant. If we are considering a given electronic state, say state j, the electronic wave function in question will be denoted $\psi_j(\mathbf{r}, \mathbf{R})$. Substituting the product form (11.10) into the full Schrödinger equation (10.4) with $H = T_N + T_{el} + V$, we get

$$(T_N + T_{el} + V)\,\psi_j(\mathbf{r}, \mathbf{R})\chi_j(\mathbf{R}) = E\psi_j(\mathbf{r}, \mathbf{R})\chi_j(\mathbf{R}). \tag{11.11}$$

If we now neglect the operation of the nuclear kinetic energy operator T_N on $\psi_j(\mathbf{r}, \mathbf{R})$ on the LHS of Eq. (11.11) and note that

$$(T_{el} + V)\psi_j(\mathbf{r}, \mathbf{R}) = \epsilon_j(\mathbf{R})\psi_j(\mathbf{r}, \mathbf{R}),$$

then on multiplying the resulting equation by $\psi_j^*(\mathbf{r}, \mathbf{R})$ and integrating over $\mathbf{r}$, we find the following equation for the nuclear wave function:

$$\left[T_N + \epsilon_j(\mathbf{R})\right] \chi_{j,\kappa}(\mathbf{R}) = E_{j,\kappa}\chi_{j,\kappa}(\mathbf{R}). \tag{11.12}$$

We have added a subscript κ to χ and E, where κ denotes all the quantum numbers (vibrational, rotational, and translational) of the nuclear motion. The electronic energy eigenvalues, $\epsilon_j(\mathbf{R})$, serve as nuclear potentials (and are often denoted as $V_j(R)$ in the context of nuclear motion problems), and the energy eigenvalues, $E_{j,\kappa}$, are the nuclear energy eigenvalues. The nuclear degrees of freedom include the molecule's center-of-mass translational coordinates, the vibrational degrees of freedom, and the rotational degrees of freedom of the molecule. The number of nuclear degrees of freedom equals $3N$ for a molecule with N nuclei; three center-of-mass degrees of freedom of the molecule, three rotational degrees of freedom, and $3N - 6$ vibrational degrees of freedom (for linear molecules, there are $3N - 5$ vibrational modes).

The BOA is a good approximation when an electronic potential energy surface is far from other potential energy surfaces. For diatomic molecules, the BOA fails near curve crossings of the diatomic potential energy curves and asymptotically where more than one potential curve tends to the same atomic term limit. For polyatomic molecules, the BOA fails at surface crossings that are often conical intersections, and asymptotically, when more than one potential energy surface correlates to given fragment states. Near the crossing or coalescence, nuclear derivative couplings (see Sec. 11.3.2) are responsible for the interaction of the surfaces. We now consider such potential energy intersections and the couplings near such crossings or pseudocrossings.

11.3.1 POTENTIAL ENERGY CROSSINGS AND PSEUDOCROSSINGS

In diatomic molecules, potential energy curves $V_j(R)$ for states of the same symmetry do not intersect but rather undergo avoided crossings. This is called the *no-crossing rule*, which can be understood as follows. Consider two potentials $V_1(R)$ and $V_2(R)$ that cross at $R = R_c$ and interact via a coupling interaction $V_{12}(R)$. The 2×2 potential matrix is

$$V(R) = \begin{pmatrix} V_1(R) & V_{12}(R) \\ V_{12}(R) & V_2(R) \end{pmatrix}. \tag{11.13}$$

The eigenvalues of this matrix are given by

$$V_\pm(R) = \frac{1}{2}\left(V_1(R) + V_2(R) \pm \sqrt{[V_1(R) - V_2(R)]^2 + 4V_{12}^2(R)}\right). \tag{11.14}$$

11.3 The Born-Oppenheimer Approximation

Figure 11.6 shows the two eigenvalues in the region where two (in this case linear) potentials, $V_1(R)$ and $V_2(R)$, interact via an interaction potential V_{12}. Only if the interaction potential $V_{12}(r)$ vanishes at the crossing point R_c of the two curves $V_1(R)$ and $V_2(R)$ (this rarely occurs) do the eigenvalues $V_-(R)$ and $V_+(R)$ intersect.

In polyatomic molecules, potential energy surfaces $V_j(\mathbf{R})$ of electronic states of the same symmetry can in general intersect, i.e., there is no noncrossing rule for polyatomics when $\mathbf{R}$ is multidimensional.

One important and common type of intersection of electronic states is the so-called *conical intersection*, depicted in Fig. 11.7. For two nuclear coordinates, say R_1 and R_2, the 2×2 potential matrix in (11.13), with $R \equiv (R_1, R_2)$, suppose $V_{12}(R) = V_{21}(R) = 0$ at some point R_c and $V_{11}(R_c) = V_{22}(R_c) \equiv V(R_c)$.

Let us expand the potentials $V_{ij}(R_1, R_2)$ about R_c and keep only linear terms,

$$V(R) = V(R_c)\mathbf{1} + \begin{pmatrix} \alpha R_1 + \beta R_2 & aR_1 + bR_2 \\ aR_1 + bR_2 & \gamma R_1 + \delta R_2 \end{pmatrix}. \quad (11.15)$$

Then, a conical intersection results for the diagonalized potential. We can write two conically intersecting potential energy surfaces in the form

$$V_\pm(R_1, R_2) = V_0 + c_1 R_1 + c_2 R_2 \pm \sqrt{d_1 R_1^2 + d_2 R_2^2 + d_3 R_1 R_2}. \quad (11.16)$$

An arbitrary translation and rotation of the coordinates will retain the conical form of the intersection.

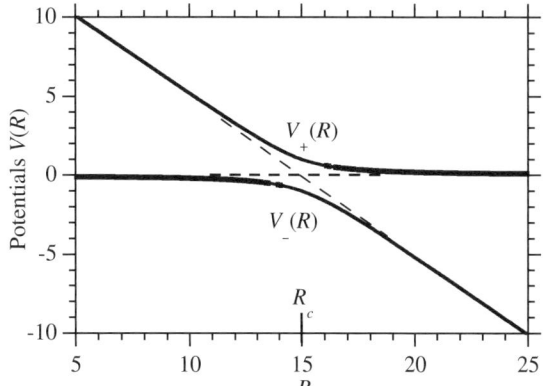

FIG 11.6 Two potentials $V_1(r)$ and $V_2(r)$ that cross and interact via a coupling potential $V_{12}(r)$ typically obey the *non-crossing rule*, unless, accidentally, $V_{12}(r)$ vanishes at the crossing point R_c.

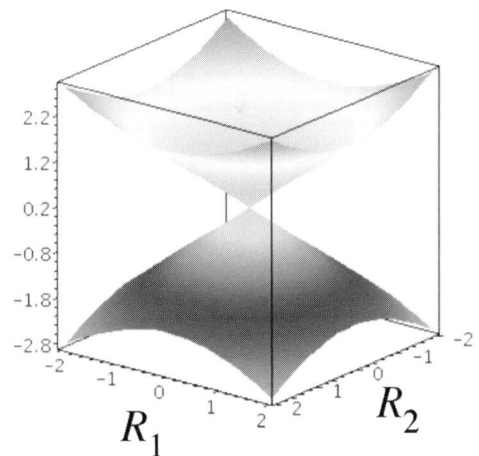

FIG 11.7 Conical intersection of two potential energy surfaces, $V_\pm(R_1, R_2) = \pm\sqrt{d_1 R_1^2 + d_2 R_2^2}$.

11.3.2 BORN–OPPENHEIMER NUCLEAR DERIVATIVE COUPLING

A general form for the wave function of a molecule that allows for coupling between different Born–Oppenheimer (BO) potentials due to the nuclear kinetic energy is given by

$$\Psi_{j,\kappa}(\mathbf{r}, \mathbf{R}) = \psi_j(\mathbf{r}, \mathbf{R})\chi_{j\kappa}(\mathbf{R}), \quad (11.17)$$

where we specify the electronic state by an index j and use the index κ to specify particular eigenvalues of the nuclear problem. Substituting this form into the full molecular Schrödinger equation yields

$$(T_N + T_{el} + V) \sum_{j,\kappa'} \psi_j(\mathbf{r}, \mathbf{R})\chi_{j\kappa'}(\mathbf{R}) = E_{i\kappa}\, \psi_i(\mathbf{r}, \mathbf{R})\chi_{i\kappa}(\mathbf{R}). \quad (11.18)$$

Multiplying Eq. (11.18) from the left by $\psi_i^*(\mathbf{r}, \mathbf{R})$, integrating over electronic coordinates $\mathbf{r}$, and neglecting vibrational-rotational coupling (taking into account only electronic coupling through the nuclear kinetic energy operator), we obtain,

$$E_{i\kappa}\chi_{i\kappa}(\mathbf{R}) = [T_N + \epsilon_i(\mathbf{R})]\chi_{i\kappa}(\mathbf{R}) + \sum_j \left(\langle \psi_i | T_N | \psi_j \rangle \chi_{j\kappa}(\mathbf{R}) + \sum_\alpha \frac{-\hbar^2}{2m_\alpha} \langle \psi_i | \nabla_{\mathbf{R}_\alpha} | \psi_j \rangle \cdot \nabla_{\mathbf{R}_\alpha}\chi_{j\kappa}(\mathbf{R}) \right). \quad (11.19)$$

The terms in the square parenthesis on the RHS of (11.19) are diagonal in the electronic state index, and the terms in the large parenthesis couple all electronic states ψ_j to state ψ_i. The first term has the nuclear kinetic energy operator operating only on the electronic wave functions to yield a matrix $B_{ij}(\mathbf{R}) \equiv \langle \psi_i | T_N | \psi_j \rangle$ that multiplies the ro-vibronic (i.e., rotational-vibrational) wave function $\chi_{j\kappa}(\mathbf{R})$. The second term has part of the nuclear kinetic energy operator operating on the electronic wave function and part on the ro-vibronic wave function $\chi_{j\kappa}(\mathbf{R})$ and yields terms that act like a vector potential in a magnetic field problem, $\mathbf{A}_{ij} \cdot \hat{\mathbf{p}} \equiv \sum_\alpha \frac{-\hbar^2}{2m_\alpha} \langle \psi_i | \nabla_{\mathbf{R}_\alpha} | \psi_j \rangle \cdot \nabla_{\mathbf{R}_\alpha}$. These are *nonabelian gauge terms* – an ordinary vector potential introduces an abelian gauge, but the indices ij on the "vector potential" $\mathbf{A}_{ij}$ yield nonabelian gauges. In general, both radial nuclear and angular nuclear kinetic energy terms can participate in the coupling. Sometimes, one type of term is more important than the other, but this depends on the specific system.

11.3.3 THE HELLMAN–FEYNMAN THEOREM

The Hellman–Feynman theorem, first derived by Hans Hellman in 1936 and independently discovered by Richard Feynman in 1939, states that, in the BOA, the forces on nuclei in molecules or solids are those that arise electrostatically, and the electric field that gives rise to the force is that obtained if the electron probability density were treated as a static distribution of negative electric charge.

The electronic energy is the minimum possible on the ground-state potential surface, $E_{el}(\mathbf{R}) = \min \langle \psi(\mathbf{r},\mathbf{R}) | H_{el}(\mathbf{r},\mathbf{R}) | \psi(\mathbf{r},\mathbf{R}) \rangle$, subject to the normalization condition $\langle \psi(\mathbf{R}) | \psi(\mathbf{R}) \rangle = 1$. The variational theorem states that an approximate wave function has an energy that is at or above the exact ground-state energy. For a general electron state $|\psi(\mathbf{r},\mathbf{R})\rangle$, the electronic energy depends on the state and on the atomic positions. To find the force on any particular atom, we can use the product rule to write

$$\frac{\partial E_{el}(\mathbf{R})}{\partial \mathbf{R}} = \left\langle \frac{\partial \psi(\mathbf{r},\mathbf{R})}{\partial \mathbf{R}} \middle| H_{el} \middle| \psi(\mathbf{r},\mathbf{R}) \right\rangle + \left\langle \psi(\mathbf{r},\mathbf{R}) \middle| \frac{\partial H_{el}}{\partial \mathbf{R}} \middle| \psi(\mathbf{r},\mathbf{R}) \right\rangle + \left\langle \psi(\mathbf{r},\mathbf{R}) \middle| H_{el} \middle| \frac{\partial \psi(\mathbf{r},\mathbf{R})}{\partial \mathbf{R}} \right\rangle. \quad (11.20)$$

Here, the matrix elements mean integration over electronic coordinates only (usually, we do not indicate the coordinates integrated over in matrix elements, but, for clarity, we make an exception here). Using the fact that $\langle \psi(\mathbf{R}) | H_{el} = \epsilon(\mathbf{R})\langle \psi(\mathbf{R})|$, $H_{el}|\psi(\mathbf{r},\mathbf{R})\rangle = \epsilon(\mathbf{R})|\psi(\mathbf{r},\mathbf{R})\rangle$, and $\frac{\partial}{\partial \mathbf{R}}\langle \psi(\mathbf{R})|\psi(\mathbf{r},\mathbf{R})\rangle = 0$, we find

$$\frac{\partial E_{el}(\mathbf{R})}{\partial \mathbf{R}} = \left\langle \psi(\mathbf{r},\mathbf{R}) \middle| \frac{\partial H_{el}}{\partial \mathbf{R}} \middle| \psi(\mathbf{r},\mathbf{R}) \right\rangle. \quad (11.21)$$

Because $\frac{\partial H_{el}}{\partial \mathbf{R}_\beta} = \frac{\partial}{\partial \mathbf{R}_\beta} \sum_{i,\alpha} \frac{-Z_\alpha e^2}{|\mathbf{r}_i - \mathbf{R}_\alpha|}$, the force on nucleus β is given by

$$\mathbf{F}_\beta = -\frac{\partial E_{el}(\mathbf{R})}{\partial \mathbf{R}_\beta} = Z_\beta e\, \mathbf{E}_{el}(\mathbf{R}_\beta), \quad (11.22)$$

where the electric field at nucleus β due to the electrons, $\mathbf{E}_{el}(\mathbf{R}_\beta)$, is

$$\mathbf{E}_{el}(\mathbf{R}_\beta) = \left\langle \psi(\mathbf{r},\mathbf{R}) \middle| \frac{\partial}{\partial \mathbf{R}_\beta} \sum_i \frac{-e}{|\mathbf{r}_i - \mathbf{R}_\beta|} \middle| \psi(\mathbf{r},\mathbf{R}) \right\rangle. \quad (11.23)$$

Thus, Eq. (11.22) shows us that the intermolecular forces may be calculated on the basis of straightforward classical electrostatics, where the electric field appearing on the RHS of (11.22) is given by the expectation value in (11.23).

11.4 ROTATIONAL AND VIBRATIONAL STRUCTURE

Let us first consider rotational motion of molecules. The rotational properties of a rigid body were discussed in Sec. 3.3.3. The Hamiltonian is

$$H = \frac{1}{2}\left(\frac{J_X^2}{I_X} + \frac{J_Y^2}{I_Y} + \frac{J_Z^2}{I_Z}\right) \tag{11.24}$$

where $J_X = \mathbf{J}\cdot\hat{\mathbf{X}}$ is the projection of the angular momentum on the body-fixed x-axis denoted X, and similarly for the other components; the quantities I_X, I_Y, and I_Z are the principal moments of inertia of the body along the principal axes [21, 22]. The angular momentum components J_X, J_Y and J_Z obey the "strange" commutation relations (3.138).

When all three principal moments of inertia of a molecule are equal, $I \equiv I_X = I_Y = I_Z$, the molecule is a *spherical top*. The energy eigenvalues are $E(J) = \frac{\hbar^2 J(J+1)}{2I} \equiv BJ(J+1)$, where B is called the *rotational constant* and the rotational eigenfunctions are spherical harmonics, $\langle\theta,\phi|\psi\rangle = Y_{JM}(\theta,\phi)$. Figure 11.8(a) shows the spectrum of the spherical top Hamiltonian, while Figure 11.8(b) shows the absorption spectrum, $\Delta E_{J+1,J} = 2B(J+1)$, for a molecular gas whose rotational levels are thermally populated.

When exactly two of the moments of inertia are the same, $I \equiv I_X = I_Y \neq I_Z$, the molecule is a *symmetric top*. Now, the fact that the angular momentum operators in (11.24) are body-fixed projections becomes important [see Eq. (3.138)]. The energy eigenvalues are given by

$$E(J,K) = \frac{\hbar^2 J(J+1)}{2I} + \frac{\hbar^2}{2}\left(\frac{1}{I_Z} - \frac{1}{I}\right)K^2 \equiv BJ(J+1) + (C-B)K^2, \tag{11.25}$$

where $K \equiv J_Z = -J,\ldots,J$, $B = \hbar^2/(2I)$, and $C = \hbar^2/(2I_Z)$. The rotational eigenfunctions are the rotation functions (see Sec. 3.3.2), $\psi_{J,M,K} = D^{(J)}_{M,K}(\phi\,\theta\,\xi)$, which should be properly normalized [see (3.143)]. The arguments are the Euler angles, which define the rotation of the axes of the symmetric top molecule with respect to the fixed axes.

An *asymmetric top* molecule is one with $I_X \neq I_Y \neq I_Z$. The eigenvalues and eigenvectors must be calculated by diagonalizing the Hamiltonian (11.24) in some basis, e.g., using the basis set expansion, $\psi_{J,M,\alpha} = \sum_K c_K^\alpha D^{(J)}_{M,K}(\phi\,\theta\,\xi)$.

Molecules are not truly rigid bodies. They vibrate as well as rotate. Moreover, there is coupling between the rotational and vibrational degrees of freedom. The moments of inertia depend on the positions of the atoms in the molecule, hence

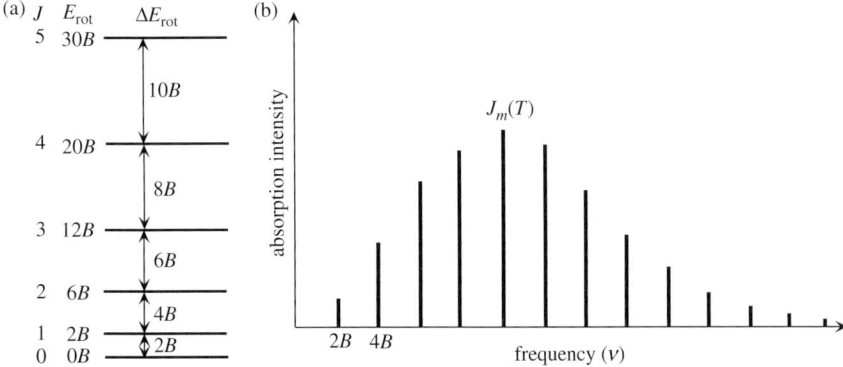

FIG 11.8 (a) Rotational energy levels of a spherical top diatomic molecule in the rigid rotor approximation and the dipole-allowed transitions between the levels. (b) Schematic drawing of the absorption spectrum at finite temperature. The absorption intensity is proportional to the probability for molecules to be in the absorbing state, $P(J,T) = g_J e^{-E(J)/k_B T}/Z$, where $Z \equiv \sum_J (2J+1) e^{-B_e J(J+1)/k_B T}/Z$ is the partition function and $g_J = (2J+1)$ is the degeneracy of level J, times the square of a transition dipole matrix element (see Sec. 11.6.1). The probability $P(J,T)$ has a maximum at $J_m(T)$.

molecular vibrations influence the moments of inertia, and rotation of the molecule can make it stretch, hence the rotations influence the vibrational motion. These effects couple the vibrational and rotational motion.

Without taking the vibrational-rotational coupling into effect, the vibrational-rotational Hamiltonian for a symmetric top molecule in a given electronic state can be written as

$$H = \sum_j \left(\frac{p_j^2}{2m_j} + m_j \omega_j^2 Q_j^2 + \kappa_j Q_j^3 + \ldots + Q_j^2 \sum_{k \neq j} \kappa_{jk} Q_k + \ldots \right) + B\mathbf{J}^2 + (C - B)K^2. \qquad (11.26)$$

Typically, the coefficients of the higher than quadratic terms in Q appearing in the large parenthesis are small and are often dropped. Then, the eigenvalues are $\hbar \sum_j [\omega_j(n_j + 1/2)] + BJ(J + 1) + (C - B)K^2$ and the vibrational-rotational eigenfunctions are given, to a good approximation, by

$$\Psi = \left(\frac{8\pi^2}{2J+1} \right)^{1/2} D_{M,K}^{(J)}(\phi, \theta, \xi) \prod_j \left(\frac{m_j \omega_j}{\pi \hbar} \right)^{1/4} \frac{1}{2^{n_j/2}\sqrt{n_j!}} H_n(q_j) e^{-\frac{q_j^2}{2}}, \qquad (11.27)$$

where $q_j = Q_j/\sqrt{\hbar/(m_j \omega_j)}$ is the dimensionless vibrational coordinate for the jth mode, where $j = 1, \ldots, 3N - 6$ for a general molecule with N atoms, and $D_{M,K}^{(J)}$ are the rotation functions defined in Sec. 3.3.2.

Imagine a molecule that is spinning quickly around some axis. Centrifugal and Coriolis forces will pull the atoms out to larger distances from the center. This will affect the molecular vibrations as well as the moments of inertia of the molecule relative to that of a nonrotating molecule. Similarly, imagine a vibrating molecule with large vibrational quantum number. Because the potential as a function of the vibrational coordinates, $V(\mathbf{R})$, is typically not symmetric around the equilibrium position $\mathbf{R}_e$, vibrations can modify the average position within the potential, and this in turn affects the moment of inertia of the molecule. One way to parameterize the energy eigenvalues of a molecule to account for the vibrational-rotational interaction is to allow the rotational constants to depend on the vibrational quantum numbers,

$$E_{n_j, J, K} = \sum_j \hbar \omega_j [(n_j + 1/2) - x_j (n_j + 1/2)^2] + BJ(J+1) + (C-B)K^2 - D[J(J+1)]^2 + \ldots \qquad (11.28)$$

where

$$B = B_e - \sum_j \alpha_j (n_j + 1/2),$$

$$(C - B) = (C_e - B_e) - \sum_j \beta_j (n_j + 1/2),$$

$$D = D_e - \sum_j \delta_j (n_j + 1/2).$$

Here, x_j is the *dimensionless anharmonicity constant* of mode j [see Eq. 3.89], the $D[J(J+1)]^2$ term results because the moments of inertia change with rotational quantum number due to the centrifugal and Coriolis forces, and the constants α_j, β_j, and δ_j specify the degree to which the *rotational constants* B, $C - B$, and D are affected by vibrational mode j.

As an example, let us consider the CO_2 molecule (see Fig. 16.6 in the Classical Mechanics Chapter linked to the book web page.) in the isotopic form $^{12}C^{16}O_2$. The ν_1 vibration is the symmetric stretch mode with frequency $\nu_1 = 1388$ cm^{-1} (to get the frequency in Hz, multiply by c). The bending mode frequency is $\nu_2 = 667$ cm^{-1}. The bending mode is doubly degenerate, with motion of the atoms in the plane of Fig. 16.6 and perpendicular to it comprising the two degenerate motions. The asymmetric stretch mode has frequency $\nu_3 = 2349$ cm^{-1}. The frequencies ω_j appearing in Eq. (11.28) are $\omega_j \equiv 2\pi \nu_j$. The constants appearing in Eq. (11.28) take the values $B_e = 0.39021$ cm^{-1}, $\alpha_1 = 0.00121$ cm^{-1}, $\alpha_2 = -0.00072$ cm^{-1}, $\alpha_3 = 0.00309$ cm^{-1}, and $D_e = 13.5 \times 10^{-8}$ cm^{-1}. The β_j are very small compared with D_e. Each rotational level J ($J \neq 0$) in Eq. (11.28) is degenerate. However, the interaction of the rotation and the degenerate bending vibration when the ν_2 mode is excited removes some of this degeneracy and gives rise to Λ-doubling via the $-\mathbf{J}\cdot\mathbf{L}/I$ term of the Hamiltonian that originates from $\frac{(\mathbf{J}-\mathbf{L})^2}{2I}$ [the latter term replaces $E = \frac{\hbar^2 J(J+1)}{2I} \equiv BJ(J+1)$ when vibrational angular momentum is present]. Here, $\mathbf{J}$ is the total angular momentum and $\mathbf{L}$ is the vibrational angular momentum of the degenerate bending vibration. The magnitude of the splitting is proportional to $J(J+1)(n_2+1)$ for $l = 1$. The $^{12}C^{16}O_2$ molecule has only even rotational levels because of Bose–Einstein statistics.

Electronic angular momentum can also couple to rotational angular momentum. For a diatomic molecule having nonzero projection of angular momentum Λ along the diatomic axis, this coupling can give rise to Λ-doubling. In this case, the rotational Hamiltonian for the molecule is $\frac{(\mathbf{J}-\mathbf{L})^2}{2I}$, where $\mathbf{J}$ is the total angular momentum and $\mathbf{L}$ is the electronic angular momentum. The double degeneracy of the terms with $\Lambda \neq 0$ is removed by the $-\mathbf{J}\cdot\mathbf{L}/I$ term in the expansion of the rotational Hamiltonian.

11.5 VIBRATIONAL MODES AND SYMMETRY

Symmetry-adapted nuclear coordinates of a molecule with a given symmetry can be created by applying symmetry-adapted projection operators to the set of internal nuclear coordinates. The projection operators are constructed with the aid of the character table of the molecular point group, as described in Sec. E.5.4. The projection operator for irreducible representation α is $P^{(\alpha)} \equiv \frac{f_\alpha}{g} \sum_A \left(\chi^{(\alpha)}\right)^* A$ [see Eq. (E.16) and also Eq. (E.17)]. Applying $P^{(\alpha)}$ to the row vector of dimension $3N$ composed of the position vectors of the N atoms in the molecule, $(x_1, y_1, z_1, \ldots, x_N, y_N, z_N)^\dagger$ (this vector is used to obtain the regular representation of the group when the group elements are applied to it) gives a vector with the transformation properties of the irreducible representation. This yields the symmetry of the normal modes but not the mode frequencies. To get the normal mode frequencies (as well as the modes themselves), one must calculate the molecular vibrations, e.g., by solving a small oscillations problem as described in Sec. 16.9 linked to the book web page. Using the notation developed there, the kinetic and potential energies, up to quadratic terms in the coordinate displacements, are

$$2T = \dot{x}_i m_{ij} \dot{x}_j, \quad 2V = x_i V_{ij} x_j, \tag{11.29}$$

where x_i and $\dot{x}_i$ are the small displacements and their time derivatives, m_{ij} is the equilibrium mass matrix, and V_{ij} is the potential matrix.

The method for finding the normal modes of molecules was pioneered by Edgar Bright Wilson. He denoted the inverse mass matrix $\mathbf{m}^{-1}$ by $\mathbf{G} \equiv \mathbf{m}^{-1}$ and the potential matrix $\mathbf{V}$ by $\mathbf{F} \equiv \mathbf{V}$, i.e., $2T = \dot{\mathbf{x}}\mathbf{G}^{-1}\dot{\mathbf{x}}$ and $2V = \mathbf{x}\mathbf{F}\mathbf{x}$. Hence, the method is often called the *GF matrix method*. One can introduce the *normal modes* $\mathbf{X}$ that are linear combinations of the small oscillations $\mathbf{x}$,

$$\mathbf{x} = \mathbf{L}\mathbf{X}, \tag{11.30}$$

such that the *transformation coefficients* $\mathbf{L}$ diagonalize the kinetic and potential energies,

$$2T = \dot{\mathbf{X}}^\dagger \mathbf{L}^\dagger \mathbf{G}^{-1} \mathbf{L} \dot{\mathbf{X}} = \dot{\mathbf{X}}^\dagger \mathbf{1} \dot{\mathbf{X}}, \quad 2V = \mathbf{X}^\dagger \mathbf{L}^\dagger \mathbf{F} \mathbf{L} \mathbf{X} = \mathbf{X}^\dagger \mathbf{\Lambda} \mathbf{X}. \tag{11.31}$$

Hence,

$$\mathbf{L}^\dagger \mathbf{G}^{-1} \mathbf{L} = \mathbf{1}, \quad \mathbf{L}^\dagger \mathbf{F} \mathbf{L} = \mathbf{\Lambda}, \tag{11.32}$$

where $\mathbf{\Lambda}$ is a diagonal matrix. Solving the first equation for $\mathbf{L}^\dagger$ and substituting in the second equation, then multiplying from the left by $\mathbf{L}$ yields,

$$\mathbf{G}\mathbf{F}\mathbf{L} = \mathbf{L}\mathbf{\Lambda}. \tag{11.33}$$

Equation (11.33) is an eigenvector/eigenvalue problem, and the solution requires solving for the eigenvalues $\mathbf{\Lambda}$ via the determinantal equation

$$|\mathbf{G}\mathbf{F} - \mathbf{\Lambda}| = 0, \tag{11.34}$$

which yields the square of mode frequencies, $\Lambda_{\alpha,\alpha} = \omega_\alpha^2$. One can then calculate the eigenvectors, i.e., the rows of the matrix **L**, which are the mode vibrations (they are already in the form of irreducible representations of the group of the molecule).

Problem 11.8

Consider the equilateral triatomic molecule in Problem 11.2.

(a) Find the decomposition of the nine-dimensional regular representation obtained from the Cartesian displacements of the atoms in irreducible representations of the group D_{3h}.
(b) Let the mass of each atom be m and the internuclear distance be l. Calculate the moment of inertia tensor and find the rotational energy eigenvalues.

Answer: (b) The center of mass is at $(l/2, l/(2\sqrt{3}), 0)$, while the vertices are at $(0, a/\sqrt{3}, 0)$, $(-a/2, -a/(2\sqrt{3}), 0)$, and $(a/2, -a/(2\sqrt{3}), 0)$. The moments of inertia are $I_1 = ml^2/2$, $I_2 = ml^2/2$, and $I_3 = ml^2$. Because two of the moments of inertia are equal, the energy eigenvalues are given by Eq. (11.25).

11.6 SELECTION RULES FOR OPTICAL TRANSITIONS

General selection rules for transition matrix elements $T_{fi} = \langle \psi_f | \mathcal{O} | \psi_i \rangle$, where $|\psi_i\rangle$ and $|\psi_f\rangle$ are the initial and final states of a molecular transition and $\mathcal{O}$ is the operator responsible for the transition, result from symmetry considerations. Taking the direct product of the representations for ψ_f, $\mathcal{O}$, and ψ_i, which we will assume to transform according to irreducible representation $\Gamma^{(\lambda)}$, $\Gamma^{(\mu)}$, and $\Gamma^{(\nu)}$, respectively, $\Gamma^{(\lambda)} \otimes \Gamma^{(\mu)} \otimes \Gamma^{(\nu)}$ yields the possible representations that can result in the matrix element

$$\Gamma^{(\lambda)} \otimes \Gamma^{(\mu)} \otimes \Gamma^{(\nu)} = \sum_\alpha a_\alpha \Gamma^{(\alpha)}. \tag{11.35}$$

That is, the matrix element can be expressed as a sum of irreducible representations, as can be shown using the group orthogonality theorem (see Sec. E.5.2 of Appendix E) by first applying it to $\Gamma^{(\mu)} \otimes \Gamma^{(\nu)}$ and then applying it to the product with $\Gamma^{(\lambda)}$. If the sum on the RHS of Eq. (11.35) does not contain the totally symmetric irreducible representation (e.g., A_1), the matrix element vanishes. For example, for the electric-dipole transitions for absorption or emission processes, the matrix elements of the electric-dipole moment $\mathbf{d}(Q)$ (which is in general a function of the nuclear coordinates Q of the molecule, and is determined by evaluating an electronic matrix element given by an integral over electronic coordinates) are

$$T_{fi} = -\mathbf{E} \cdot \langle \psi_f | \mathbf{d} | \psi_i \rangle = -\mathbf{E} \cdot \int dQ \, \psi_f(Q) \mathbf{d}(Q) \psi_i(Q). \tag{11.36}$$

For the electric-dipole process to occur, the integrand $\psi_f(Q)\mathbf{d}(Q)\psi_i(Q)$ in Eq. (11.36) must contain a component that transforms as the totally symmetric representation.

A proof that $\int dQ \, \phi_j^{(\alpha)} = 0$, when $\phi_j^{(\alpha)}$ is a basis function of irreducible representation α other than the totally symmetric representation proceeds as follows. The integral over all space is invariant with respect to any volume preserving transformation of the coordinate system, such as a symmetry transformation. Hence

$$\int dQ \, \phi_j^{(\alpha)} = \int dQ \, \hat{A} \phi_j^{(\alpha)} = \int dQ \sum_k A_{ki}^{(\alpha)} \phi_k^{(\alpha)} = \frac{1}{g} \int dQ \sum_k \sum_A A_{ki}^{(\alpha)} \phi_k^{(\alpha)}. \tag{11.37}$$

The third equality in (11.37) follows using (E.14) of Appendix E, and the fourth equality results by noting that the first three equalities are true for every symmetry operator $\hat{A}$, so if we sum over all symmetry operations and divide by the number of symmetry operations in the group, g, the fourth equality follows. Now, using the group orthogonality

11.6 Selection Rules for Optical Transitions

theorem (E.4) when applied to the case of β in Eq. (E.4) equals the totally symmetric representation, we find $\sum_A A_{ki}^{(\alpha)} = 0$ unless α equals the totally symmetric representation. The proof follows from the last equality in Eq. (11.37) and the previous sentence.

11.6.1 SELECTION RULES FOR DIATOMIC MOLECULES

We have already discussed selection rules for atoms in Sec. 3.6. For diatomic molecules, E1 selection rules yield the following properties for electronic transitions:

$\Delta J = 0, \pm 1$ except $J = 0 \not\leftrightarrow J = 0$ (for $\Omega = 0 \to \Omega = 0, \Delta J \neq 0$).

Parity change, $g \leftrightarrow u$.

$\Delta \Lambda = 0, \pm 1$ and $\Delta \Omega = 0, \pm 1$.

$\Sigma^{\pm} \leftrightarrow \Sigma^{\pm}$ and $\Sigma^{\pm} \not\leftrightarrow \Sigma^{\mp}$.

For weak spin-orbit coupling, $\Delta S = 0$ and $\Delta \Sigma = 0$.

For homonuclear diatomic molecule transitions the following selection rules must be satisfied: $g \leftrightarrow u, g \not\leftrightarrow g, u \not\leftrightarrow u$. Table 11.2 shows the E1 selection rules for diatomic molecules.

Rotational Transitions

The rotational absorption spectrum of a thermally populated spherical top diatomic molecule in the rigid rotor approximation is shown in Fig. 11.8(b). Figure 11.9(a) shows the spectrum of a spherical top molecule in a magnetic field. Also shown are the dipole-allowed transitions between the states. Figure 11.9(b) compares the spectrum of a spherical and a symmetric top molecule (in the absence of a magnetic field. The dipole-allowed transitions between the states of the symmetric top molecule are indicated.

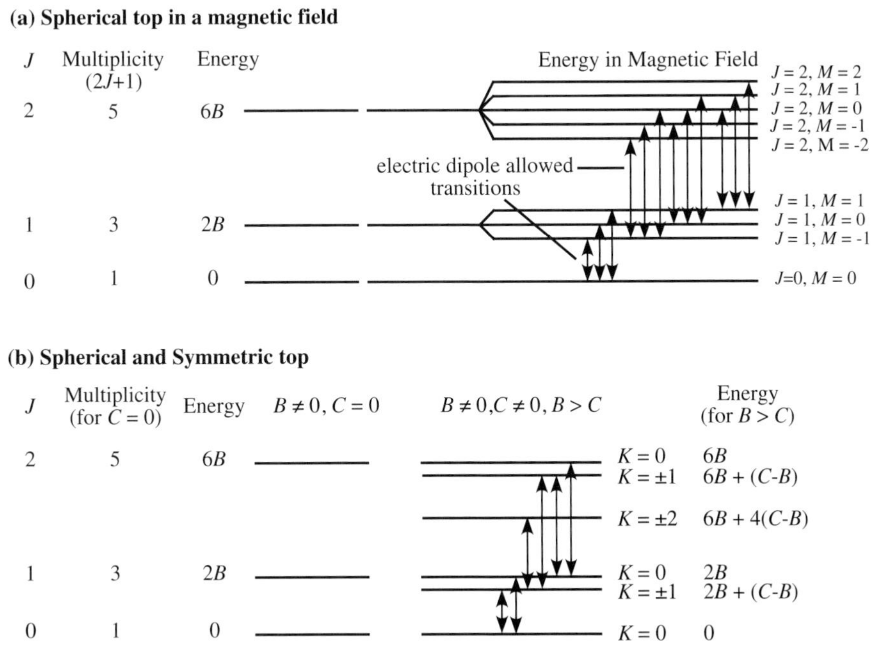

FIG 11.9 (a) Spectrum of a spherical top molecule in a magnetic field. $J = 0, 1, 2$ states are shown. (b) Spherical ($C = 0$) and symmetric top molecule (with $B > C$, i.e., an oblate molecule) rotational energy eigenvalues for $J = 0, 1, 2$. *Source*: Band, Light and Matter, Fig. 6.16, p. 372

Table 11.2 *Selection rules for E1 transitions in diatomics.*

Rule	Examples	Comments
$\Delta J = 0, \pm 1$, $J = 0 \not\leftrightarrow J = 0$	$J = 0 \leftrightarrow J = 1$ $J = 1 \leftrightarrow J = 1$ $J = 1 \leftrightarrow J = 2$	Total angular momentum selection rule.
$\Delta \Lambda = 0, \pm 1$	$\Sigma \leftrightarrow \Sigma, \Sigma \leftrightarrow \Pi, \Pi \leftrightarrow \Pi$ $\Sigma \not\leftrightarrow \Delta$	Orbital angular momentum projection on diatomic axis.
$\Delta S = 0$	$^1\Sigma \leftrightarrow {}^1\Pi_1, {}^3\Pi \leftrightarrow {}^3\Pi$ $^1\Sigma \not\leftrightarrow {}^3\Pi$	Slightly violated for heavy atom molecules.
$\Delta \Sigma = 0$	$^3\Pi_2 \leftrightarrow {}^3\Delta_3$ $^3\Pi_2 \not\leftrightarrow {}^1\Delta_2$	Slightly violated for heavy atom molecules.
$\Delta \Omega = 0, \pm 1$	$^3\Pi_2 \leftrightarrow {}^3\Delta_3$ $^3\Pi_1 \not\leftrightarrow {}^3\Delta_3$	Total angular momentum projection on diatomic axis.
$\Sigma^\pm \leftrightarrow \Sigma^\pm, \Sigma^\pm \not\leftrightarrow \Sigma^\mp$	$\Sigma^+ \leftrightarrow \Sigma^+, \Sigma^- \leftrightarrow \Sigma^-$ $\Sigma^+ \not\leftrightarrow \Sigma^-$	Relevant only for Σ states.
$g \leftrightarrow u, g \not\leftrightarrow g, u \not\leftrightarrow u$	$\Sigma_g^+ \leftrightarrow \Sigma_u^+$ $\Sigma_g^+ \not\leftrightarrow \Sigma_g^+, \Sigma_g^+ \not\leftrightarrow \Pi_g$	Relevant only for homonuclear diatomics.

Problem 11.9

Determine the selection rules on K for a symmetric top molecule given by the matrix element in the first-order light-matter interaction, $-\mathbf{E} \cdot \langle D_{M',K'}^{(J')} | \mathbf{d} | D_{M,K}^{(J)} \rangle$.

Hint: Take $\mathbf{E}$ to be along the z-axis and express $\mathbf{d}$ by rotating to the molecular frame.

Vibrational Transitions

Vibrational transitions occur through the transition matrix elements

$$V_{ji} = -\mathbf{E} \cdot \langle j | \mathbf{d}(Q) | i \rangle \approx -\mathbf{E} \cdot \left[\langle k | \mathbf{d} | i \rangle_{Q_e} + \nabla_{\mathbf{Q}} \langle k | \mathbf{d}(\mathbf{Q}) | i \rangle_{Q_e} \cdot (\mathbf{Q} - \mathbf{Q}_e) + \ldots \right] \quad (11.38)$$

As an example, Fig. 11.10 shows the infrared vibrational spectrum of the diatomic molecule DCl for $n = 0 \to n = 1$. The gap in the center of the band arises because the pure vibrational transition with $J = 0$ is forbidden for $\Sigma \leftrightarrow \Sigma$ electronic transitions. The parts of the band on the left- and right-hand side of the origin are the P branch (the lower energy branch with $J' = J'' - 1$) and the R branch (the higher energy branch with $J' = J'' + 1$), respectively. These lines are almost equally spaced since $(B_e' - B_e'')$ is small. Because B_e decreases with increasing v within a given electronic state, $(B_e' - B_e'') < 0$ and the quadratic term in the angular momentum is always negative. Therefore, a point can be reached for sufficiently large values of J'', where the spacing between successive lines decreases to zero in the R branch. Further increase in J'' yields decreased frequencies (see Fig. 11.12). This is called the band head. For vibrational-rotational spectra, the intensity at the band head is often weak. The individual rotation lines are doublets because two stable Cl isotopic forms, ^{35}Cl and ^{37}Cl, contribute to the spectrum. These isotopes have abundances 75.77% and 24.23%, respectively. The intensity of the spectral lines for each isotope is proportional to these abundances. The splittings of the lines in Fig. 11.10 arise because the vibrational frequency $\omega_v = \sqrt{k/\mu}$ is proportional to the inverse square root of the reduced mass of the molecule, μ, and furthermore, the rotational constant of the molecule, B_e, is inversely proportional to μ.

11.6 Selection Rules for Optical Transitions

Electronic Transitions

The electric-dipole moment operator is sandwiched between the initial state of the diatomic molecule $|\Psi_i\rangle = |\psi_i\rangle|\chi_{i,\text{vib}}\rangle|\chi_{i,\text{rot}}\rangle$ and the final state of the diatomic molecule $|\Psi_f\rangle = |\psi_f\rangle|\chi_{f,\text{vib}}\rangle|\chi_{f,\text{rot}}\rangle$ in $\langle\Psi_f|H_{\text{int}}|\Psi_i\rangle$, where the interaction operator is $H_{\text{int}} = -\mathbf{E}\cdot\mathbf{d}$ [see Eq. (5.63)]. The electric-dipole moment operator is the sum of an electronic part and a nuclear part, $d = d_{\text{el}} + d_{\text{nuc}}$, where d_{el} and d_{nuc} are given in terms of the coordinates of the electrons and the two nuclei relative to the center of mass of the molecule. The electric-dipole moment transition matrix element can be written as

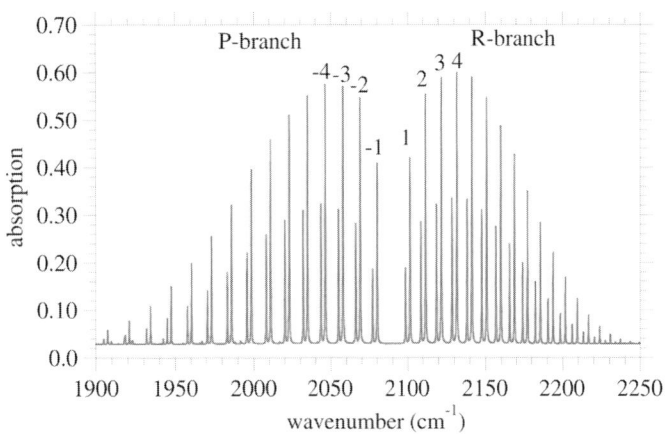

$$\langle\Psi_f|H_{\text{int}}|\Psi_i\rangle = -\mathbf{E}\cdot\langle\chi_{f,\text{rot}}|\langle\chi_{f,\text{vib}}|\left(\langle\psi_f|\mathbf{d}_{\text{el}}|\psi_i\rangle + \mathbf{d}_{\text{nuc}}\right)|\chi_{i,\text{vib}}\rangle|\chi_{f,\text{rot}}\rangle. \quad (11.39)$$

FIG 11.10 Infrared absorption spectrum of DCl for $n = 0 \to n = 1$. The gap in the center of the band arises because the pure vibrational transition with $\Delta J = 0$ is forbidden for $\Sigma \leftrightarrow \Sigma$ electronic transitions. The individual rotational lines are doublets because of the presence of two isotopic species $D^{35}Cl$ and $D^{37}Cl$. The former are more intense. *Source*: Band, Light and Matter, Fig. 6.10, p. 359

The electronic part of this matrix element can be evaluated by integrating over electronic coordinates to obtain the R-dependent transition dipole moment, $\mathbf{d}_{fi}(\mathbf{R}) = \langle\psi_f|\mathbf{d}_{\text{el}}|\psi_i\rangle + \mathbf{d}_{\text{nuc}}(\mathbf{R})\delta_{fi}$. The vibrational part of the matrix element can then be evaluated to obtain, $\mathbf{d}_{fi}(\hat{\mathbf{R}}) = \langle\chi_{f,\text{vib}}|\mathbf{d}_{fi}(\mathbf{R})|\chi_{i,\text{vib}}\rangle$. Then, the rotational matrix element can be evaluated. In calculating the electronic, vibrational, and rotational matrix elements, selection rules can be used to determine whether the matrix elements vanish due to symmetry. For example, the initial and final angular momentum of the molecule cannot differ by more than one unit of angular momentum because of selection rules based on symmetry. We shall develop the vibrational-rotational parts of the diatomic problem after considering the electronic part of the matrix elements.

For nonvanishing electric-dipole matrix elements, coupling the initial and final electronic states of a diatomic molecule, $\langle\psi_f|\mathbf{d}_{\text{el}}|\psi_i\rangle + \mathbf{d}_{\text{nuc}}(\mathbf{R})\delta_{fi}$, the following selection rule must be satisfied:

$$\Delta\Lambda = 0, \pm 1. \quad (11.40)$$

A $\Delta\Lambda = 0$ transition is called a *parallel transition*, and a $\Delta\Lambda = \pm 1$ transition is called a *perpendicular transition*. For $\Sigma \leftrightarrow \Sigma$ transitions, we already found that the following selection rules must be satisfied:

$$\Sigma^\pm \leftrightarrow \Sigma^\pm \text{ and } \Sigma^\pm \nleftrightarrow \Sigma^\mp.$$

The selection rules for homonuclear diatomic transitions are:

$$g \leftrightarrow u, \quad g \nleftrightarrow g, \quad u \nleftrightarrow u.$$

All these selection rules result from the vector form of the dipole moment operator. Table 11.2 summarizes the diatomic molecule selection rules for electric-dipole transitions.

Absorption and emission spectra of diatomic molecules in the visible and UV spectral regions typically involve transitions between different electronic, vibrational, and rotational states. For a given electronic transition, there is a series of bands for each (sufficiently intense) vibrational transition. Each vibrational band is composed of a series of rotational transitions. The most conspicuous feature of such spectra are band heads. When the projection of the electronic quantum number on the diatomic axis of the initial state, Λ, is nonzero, transitions with $\Delta J = 0$ (the Q branch transitions) as well as $\Delta J = \pm 1$ (the P and R branch transitions) can occur and an additional series of transitions with $\Delta J = 0$ is then observed. The photon energy emitted in a Q branch transition involving electronic, vibrational, and rotational transitions is

$$\hbar\omega = \Delta E_{el} + \hbar\omega_e(n' - n'')[1 - (n' + n'')x_e] + (B'_e - B''_e)J''(J'' + 1) \text{ for } J'' = J'. \tag{11.41}$$

Equation 11.41 applies for the initial and final vibrational-rotational energies of P and R transitions (Q-branches cannot occur for vibrational-rotational transitions within the same electronic state). The Q branch is usually narrow compared with the P and R branches since the terms proportional to $(B'_e + B''_e)$ in the spectra vanishes. In electronic transitions, B'_e and B''_e are in general significantly different since the equilibrium values R'_e and R''_e are in general different. Therefore, sharp band heads can appear because of the $(B'_e - B''_e)$ terms in Eq. (11.41). A band head can appear in either the P or the R branches since either B'_e and B''_e can be larger. Figure 11.11 shows the transition diagram for an electronic transition within the $n = 1 \to n = 0$ vibrational manifold that has P, Q, and R branches. The structure of an electronic band spectrum, with a band head in the P branch, is shown in Fig. 11.12. The upper part of the figure is known as a Fortrat diagram and shows the parabolic curves for the three branches. For further details on diatomic spectra, see Berry, Rice, and Ross, *Physical Chemistry* [178].

Rayleigh and Raman selection rules

Figure 7.5 depicts Rayleigh and Raman transitions in a diatomic molecule. The Feynmann diagrams for these two-photon light scattering processes are depicted in Fig. 7.6. The selection rules for these scattering processes in diatomic molecules are determined by the one-photon selection rules, because the second-order amplitude for these processes is a product of two one-photon matrix elements (from the initial to the intermediate state, and then from the intermediate to the final state of the Raman transition); the transition amplitude from state i to state j involves a sum of products of the form, $\sum_k V_{jk}V_{ki}/(E_k - E - i\varepsilon)$, where V_{jk} and V_{ki} are transition dipole matrix elements. We thereby obtain the selection rules: $\Delta J = 0, \pm 1, \pm 2$, but $\Delta J = 0$ is forbidden for a $J = 0$ state. The vibrational selection rules are easy to determine by expanding the dipole transition matrix elements $V_{ki} = \langle k|\mathbf{r}|i\rangle$ in terms of a power series in the internuclear coordinate R around the equilibrium position R_e,

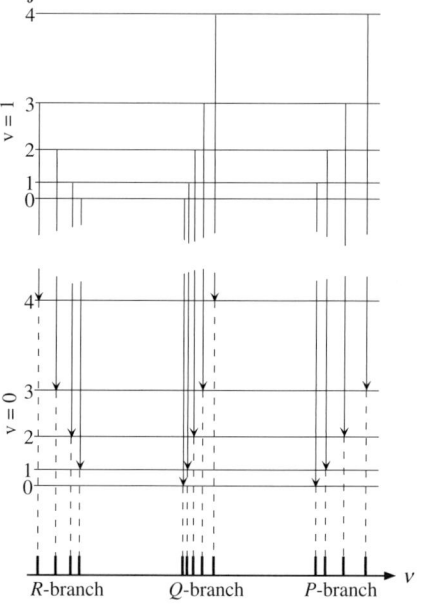

FIG 11.11 Electronic transitions for a diatomic molecule with P, Q, and R branches in the $n = 1 \to n = 0$ vibrational manifold.

$$V_{ki} = -\mathbf{E} \cdot \langle k|\mathbf{d}(R)|i\rangle \approx -\mathbf{E} \cdot \left[\langle k|\mathbf{d}|i\rangle_{R_e} + \frac{\partial}{\partial R} \langle k|\mathbf{d}(R)|i\rangle_{R_e} (R - R_e) + \ldots \right]. \tag{11.42}$$

For a harmonic oscillator potential, the selection rules are $\Delta n = 1$ for a Stokes transition and $\Delta n = -1$ for anti-Stokes transition; however, for nonharmonic potentials, this selection rule breaks down. Anti-Stokes transitions are typically less intense because of the lower Boltzmann population of the initial state.

11.7 THE FRANCK–CONDON PRINCIPLE

In general, no selection rules exist that restrict the vibrational quantum numbers for optically induced electronic transitions (the case of transitions between two harmonic potentials with the same minimum position is an exception and does have a $\Delta v = \pm 1$ selection rule). Nevertheless, a general principle, called the *Franck–Condon Principle*, named after

11.7 The Franck–Condon Principle

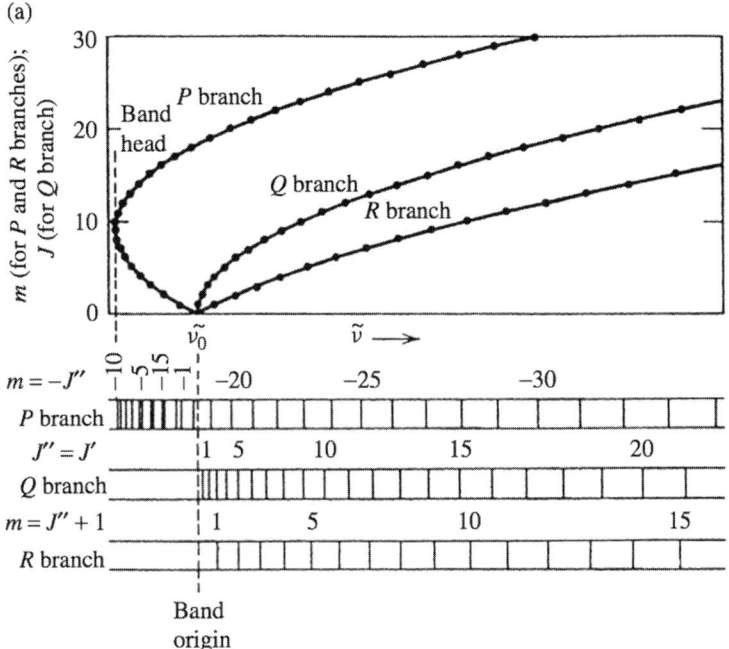

FIG 11.12 Structure of electronic band spectrum, with a band head in the P branch, the upper part of which is known as a Fortrat diagram showing the parabolic curves for the three branches. (Reproduced with permission from Fig. 7.10 of R. S. Berry et al [178].) R. S. Berry, S. A. Rice and J. Ross, Physical Chemistry, (Wiley, 1980).

James Franck and Edward Condon, determines what vibrational states will be excited on making an optically allowed electronic transition. An electron transition in a molecule takes place so rapidly in comparison with the motion of the nuclei that, immediately after the transition, *the nuclei still have nearly the same relative position and momenta as before the transition* (and this despite the fact that the position and momentum cannot be simultaneously determined exactly due to the Heisenberg uncertainty principle – hence the word nearly). The Franck–Condon principle can be developed into a tool for computing optical transition probabilities. To do so, let us consider the matrix element for a transition from an initial to a final state of a molecule via the light-matter interaction Hamiltonian V of (7.99) with initial and final electronic wave functions of the form (11.10), $\langle \Psi_f | V | \Psi_f \rangle_{\mathbf{r},\mathbf{R}} = \langle \psi_f \chi_{f\kappa'} | V | \psi_f \chi_{i\kappa} \rangle_{\mathbf{r},\mathbf{R}}$. The matrix element $\langle \Psi_f | V | \Psi_f \rangle_{\mathbf{r},\mathbf{R}}$ is then inserted in the Fermi Golden rule expression (7.76) to obtain the rate of the transition. It can be approximated by evaluating the electronic matrix element, $\langle \psi_f(\mathbf{r},\mathbf{R}) | V | \psi_f(\mathbf{r},\mathbf{R}) \rangle_{\mathbf{r}}$ at the *Condon points* $\mathbf{R}_c$ at which the vibronic wave functions have their largest overlap,

$$\boxed{\langle \psi_f \chi_{f\kappa'} | V | \psi_f \chi_{i\kappa} \rangle_{\mathbf{r},\mathbf{R}} \approx \langle \psi_f(\mathbf{r},\mathbf{R}_c) | V | \psi_f(\mathbf{r},\mathbf{R}_c) \rangle_{\mathbf{r}} \langle \chi_{f\kappa'} | \chi_{i\kappa} \rangle_{\mathbf{R}}.} \quad (11.43)$$

Here, the matrix element with the subscript **r** indicates integration over electronic coordinates **r**, and the matrix element with the subscript **R** indicates integration over vibrational-rotational coordinates **R**, and is called the *Franck–Condon integral*. This approximation makes use of the fact that the integrand over nuclear coordinates is highly peaked around the *Condon point*(s) $\mathbf{R}_C$, so the electronic matrix element, which is part of the integrand over nuclear coordinates, can be approximated by its value at the Condon points and can be taken out of the integral over nuclear coordinates. This approximation of pulling out the electronic matrix element from the nuclear matrix element is particularly appropriate if the matrix element is not strongly dependent on nuclear coordinates. The integrand of the remaining vibronic matrix element $\langle \chi_{f\kappa'} | \chi_{i\kappa} \rangle_{\mathbf{R}}$ that appears on the RHS of Eq. (11.43), which is called the *Franck–Condon factor*, is sharply peaked about the Condon points $\mathbf{R}_C$. That is, the turning points on the ground and excited electronic potential energy surfaces,

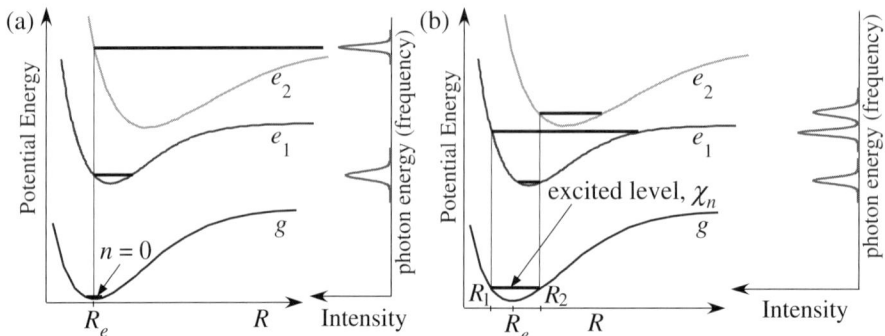

FIG 11.13 Schematic illustration of the Franck–Condon principle for a diatomic molecule. (a) Absorption from the ground vibrational state of the ground electronic potential, g, to the excited state potentials e_1 and e_2. (b) Absorption from an excited vibrational level (with vibrational wave function χ_n) of the ground electronic potential. The *Condon points* labeled R_1 and R_2 are the turning points of the excited vibrational state on the ground electronic potential and also the turning points on the excited electronic potential. *Source: Band, Light and Matter, Fig. 6.12, p. 362*

except for the ground vibrational level and very low vibrational excited levels, where the Condon points correspond to the minimum of the electronic potential energy surface, R_e. The energy $V_f(R_C)$ at which R_C intersects V_f is called the *Condon energy* E_C.

Figure 11.13 illustrates the Franck–Condon Principle for a diatomic molecule. Figure 11.13(a) shows the absorption from the ground vibrational level of the ground electronic potential. The peaks of the absorption intensity in the absorption spectrum (at the RHS of the figure) occur at photon energies corresponding to lengths of the vertical lines at R_e, where the length is measured from the ground vibrational level energy of V_g to the intersection with the first- and second-excited electronic states. Vibrational energy of the transition is marked by the heavy horizontal lines. The bound vibrational energy eigenvalues of the excited electronic states closest to this heavy horizontal line have the greatest probability of excitation. Closely lying vibrational states may also have significant probability for excitation, although less than the eigenvalue closest to the heavy horizontal line (the absorption spectrum may have discrete lines excited within the envelope shown in the spectrum). The energy eigenvalue on the second-excited state in Fig. 11.13(a) corresponds to a continuum level with energy just above the threshold for dissociation, and the spectrum in this case is continuous. Figure 11.13(b) shows the absorption from an excited vibrational level of the ground electronic state. In this case, the excited vibrational wave function $\chi_{i,v}(R)$ (not shown) has maxima near the classical turning points, labeled R_1 and R_2. The vertical lines from the turning points intersect the first- and second-excited electronic states at the energies corresponding to maxima of the absorption spectra. For the case in Fig. 11.13(b), there are two peaks in the absorption spectrum arising from the first-excited electronic state e_1.

11.7.1 BOUND-FREE MATRIX ELEMENTS

Consider transitions in a diatomic molecule from a bound state to a continuum state, e.g., photodissociation of a diatomic molecule. This is the case for the transition from g to e_2 shown in Fig. 11.13(a). But let us be more general and consider transitions from any bound state χ_{gn} on one potential V_g to a continuum state χ_{eE} on another potential V_e from which dissociation can occur (e.g., e_2). Taking only the vibrational part of the Franck–Condon matrix element on the RHS of Eq. (11.43) (i.e., for the moment neglecting the electronic matrix element that was pulled out of the nuclear matrix element), we can write the Franck–Condon factor as

$$f_n(E) \equiv \langle \chi_{eE} | \chi_{gn} \rangle = \langle E | \chi_n \rangle = \int dR\, \chi_E(R) \chi_n(R). \tag{11.44}$$

11.7 The Franck–Condon Principle

A good approximation for such matrix elements can be obtained by taking the continuum wave function $\chi_E(R)$ to be a delta function at the turning point of the excited state potential $V_e(R)$:

$$\chi_E(R) \approx |V'_e(R_t(E))|^{-1/2} \delta(R - R_t(E)). \quad (11.45)$$

The origin of the delta function at the turning point $R = R_t(E)$ can be understood as arising from the fact that the highest probability of finding system at a given position R is at the turning point, and the factor $|V'_e(R_t(E))|^{-1/2}$ can be understood as arising from the energy normalization of the continuum wave function $\chi_E(R)$. Substituting this expression in (11.44) yields

$$f_n(E) \approx |V'_e(R_t(E))|^{-1/2} \chi_n(R_t(E)). \quad (11.46)$$

This is the *reflection approximation* for the bound-continuum Franck–Condon matrix element. It can be described as follows: the Franck–Condon overlap integral $f_n(E)$ in Eq. (11.44) is proportional to the bound wave function χ_n of the monotonically decreasing function of energy, $w(E) \equiv R_t(E)$:

$$f_n(E) \approx \left| \frac{dw(E)}{dE} \right|^{1/2} \chi_n(w(E)), \quad \text{where } w(E) \equiv R_t(E). \quad (11.47)$$

We used the fact that at $R_t(E)$ is defined such that, $V_e(R_t(E)) = E$, and therefore, $V'_e(R_t(E)) = dE/dw$. Hence, the Franck–Condon factor $f_n(E)$ "reflects," one by one, all the zeroes, maxima, and minima of the bound wave function $\chi_n(R)$ as the energy E varies. The reflection approximation is valid even far away from the Condon energies E_C, where the probability for a transition is low. The Franck–Condon factor $f_n(E)$ can be inserted into the Fermi Golden rule, Eq. (7.76), to obtain the rate of photodissociation, which is proportional to $|f_n(E)|^2 \approx \left| \frac{dw(E)}{dE} \chi_n^2(w(E)) \right|$.

Problem 11.10

Given a ground potential curve and a nearly linear repulsive potential as shown in Fig. 11.14 explain the photodissociation spectrum obtained from the ground vibrational level and the $n = 3$ level, as shown in the figure, by relating it to the square of the vibrational wave function χ_n.

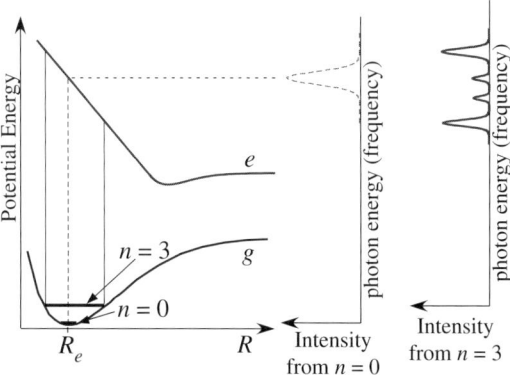

FIG 11.14 Schematic illustration of the reflection approximation for photodissociation of a diatomic molecule. Absorption intensity from the ground vibrational level (light gray-dashed curve) and the $n = 3$ level (solid black curve) of the ground electronic potential, g, to the nearly linear excited state potential e.

Scattering Theory   12

Scattering experiments play a crucial role in determining the structure of matter on the nanoscopic, microscopic, and mesoscopic scales, and help in elucidating the basic nature of interactions between colliding particles and their internal structure. For example, high-energy electron scattering from nuclei can be used to determine the charge distribution within nuclei. Compton scattering (photon–electron scattering) probes the fundamental nature of quantum electrodynamics, and scattering of light from water droplets gives rise to some of the most beautiful phenomena in nature, such as the rainbow and the glory. X-ray and electron scattering can reveal the structure of crystals and their symmetries. Quantum scattering theory, the subject of this chapter, employs many concepts that are unique to quantum mechanics, e.g., the wave nature of particles, the uncertainty principle, the quantization of spin and angular momentum. These are all central elements of quantum scattering theory with no classical analogs (classical concepts, such as the discrete symmetries of time reversal and parity, also play crucial roles in quantum scattering).

In a typical scattering experiment, a target is bombarded by a beam of projectile particles (e.g., electrons, nuclei, photons, atoms, molecules, or clusters), and target and projectile collide and scatter. Quantum scattering theory is a formalism to calculate (or predict) experimentally measurable quantities in such experiments (e.g., the angular distributions and internal-state distributions after the scattering—these quantities are usually expressed in terms of scattering *cross-sections*) and to extract information on the interaction between targets and projectiles from experimental scattering data.

The breadth of quantum scattering theory is enormous. There are many books devoted exclusively to scattering theory, including Refs [179–182]. The aim of this chapter is rather modest: to lay out the basic ingredients and concepts of the theory and to prepare the reader to work out concrete problems in scattering theory. We also include a few topics that are relevant to recent developments in nanoscience, such as scattering from disordered potentials, Feshbach and Fano resonances, scattering in low-dimensional systems, and Anderson localization.

Our analysis begins with a brief account of classical scattering theory in Sec. 12.1, and in Sec. 12.2, we present the main concepts of quantum scattering theory, starting from the time-dependent formalism (which may be skipped on first reading). Section 12.3 presents stationary scattering theory, where most practical aspects of scattering theory are developed. In Sec. 12.4, some formal scattering theory methods are introduced, including the Green's function method, which is also useful for numerous other branches of quantum theory, such as the many-body problem. Section 12.5 presents scattering from a spherically symmetric potential. Here, due to the conservation of angular momentum, the problem essentially reduces to a problem in one spatial (radial) dimension, and the partial wave expansion of the wave function and the phase shift analysis yield practical tools for calculating cross-sections. Section 12.6 details the important aspects of resonance scattering, including a discussion of Feshbach and Fano resonances. In Sec. 12.7, some approximation methods are introduced including variational methods, WKB, and the eikonal approximation. Scattering of particles with internal degrees of freedom is considered in Sec. 12.8 where the concept of channels is introduced and inelastic reactions are discussed. We conclude this chapter with Sec. 12.9, which includes an account of the theory of quantum scattering in low-dimensional systems (relevant for mesoscopic physics applications), scattering from a quasi-one-dimensional random potential, and Anderson localization.

12.1   CLASSICAL SCATTERING THEORY

Classical scattering is a branch of Newtonian mechanics [21–23]. In a classical scattering experiment, one can focus on the trajectory of a particle of mass m moving with an initial velocity $\mathbf{v}_0$ toward a scattering region in which the particle experiences a potential $V(\mathbf{r})$ as illustrated in Fig. 12.1. The center of potential is used as the origin of a fixed coordinate system. The trajectory of the particle starts at time t_0 at position $\mathbf{r}_0$ very far from the scattering center with particle energy $E = mv_0^2/2$. At very early time t slightly later than the initial time t_0, the trajectory is simply given by

$$\mathbf{r}(t) \xrightarrow[t,t_0 \to -\infty]{} \mathbf{r}_0 + \mathbf{v}_0(t - t_0). \tag{12.1}$$

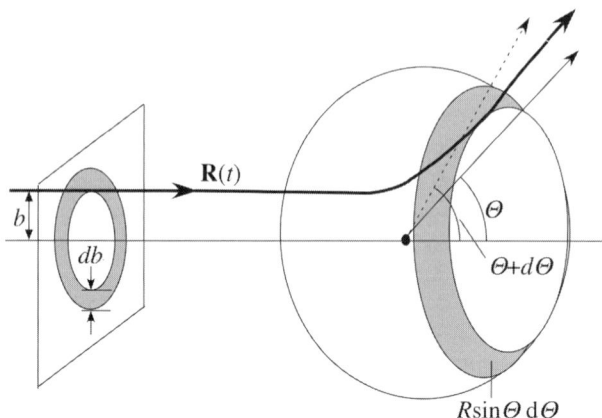

FIG 12.1 Scattering of a classical particle by a central (repulsive) potential. The particle moves toward the scattering region at an initial velocity $\mathbf{v}_0$ with an impact parameter b and when it enters the scattering zone, it is strongly repelled by the repulsive potential. At large distance R from the center, it reaches its final direction specified by the scattering angle Θ. Particles with impact parameter between b and $b+db$ scatter into a spherical annulus between Θ and $\Theta + d\Theta$ whose area is $2\pi R \sin \Theta d\Theta$.

Before proceeding with the classical analysis, note that this description must be altered when quantum scattering is considered, since the concept of particle trajectory where both position $\mathbf{r}$ and momentum $\mathbf{p}$ are sharply defined is not valid in quantum mechanics.

When the particle approaches the potential center, the trajectory is distorted. If the potential is repulsive, the distortion is "outward" as shown in Fig. 12.1, whereas if the potential is attractive, the trajectory is distorted inward. In this interaction region, the kinetic energy changes and part of the total energy E is transformed into a potential energy. When the particle recedes from the potential center, the force on it decreases and the trajectory becomes a straight line.

For an elastic scattering event, the particle kinetic energy after scattering is equal to its value before scattering. For simplicity, let us assume that the potential $V(\mathbf{r})$ is *spherically symmetric*, i.e., it is only a function of $|\mathbf{r}|$. Then, the angular momentum $\mathbf{L} = \mathbf{r} \times \mathbf{p}$ is conserved and the orbit of the particle remains in the plane perpendicular to $\mathbf{L}$, which includes the scattering center (the origin $\mathbf{r} = \mathbf{0}$). It is useful to define the *impact parameter*, $b \equiv |\mathbf{r}_0 \times \mathbf{v}_0|/v_0$, as the distance from the origin to the straight line directed along the asymptotic trajectory and parallel with the incident velocity $\mathbf{v}_0$ (see Fig. 12.1). The relation among b, $v_0 = \sqrt{2E/m}$, and L (the magnitude of $\mathbf{L}$) is

$$L = mv_0 b. \tag{12.2}$$

Problem 12.1

(a) Prove that angular momentum $\mathbf{L} = \mathbf{r} \times \mathbf{p}$ is conserved if V is a central potential.
   Hint: Calculate $\frac{d\mathbf{L}}{dt}$ and use Newton's second law, $\frac{d\mathbf{p}}{dt} = \mathbf{F}$, noting that for a central potential, $\mathbf{F}$ is along $\mathbf{r}$.
(b) Draw particle trajectories, as in Fig. 12.1, for an attractive potential.

The concept of a *differential cross-section* (and its integral over all scattering angles) is important in both classical and quantum scattering. In Sec. 12.3.1, we introduce these concepts within the formulation of quantum scattering. To define the differential cross-section for classical scattering, we consider a beam of particles that is incident on a scattering center located at the origin (see Fig. 12.1). In general, the angle at which a particle is scattered is a function of impact parameter b. The incident beam particle current density is $\mathbf{J}_{\text{in}} = \rho \mathbf{v}_0$, where ρ is the density of incoming particles and $\mathbf{v}_0$ is the initial velocity. The magnitude $J_{\text{in}} = \rho v_0$ of the current density is also referred to as the *incident beam intensity*. The number of particles per unit time crossing an annular area $2\pi b\, db$ perpendicular to $\mathbf{v}_0$ is given by $I_{\text{in}} = J_{\text{in}} 2\pi b\, db$. The same particles are scattered into a differential solid angle, $d\Omega = 2\pi \sin\Theta\, d\Theta$, where Θ is called the *scattering angle* (see Fig. 12.1). The number of particles scattered per unit time into $d\Omega$ is the scattered current I_{out}. Particle conservation dictates that I_{out} must equal I_{in}. Moreover, I_{out} must be proportional to $J_{\text{in}}\, d\Omega$, and the proportionality constant is defined as the *differential cross-section*, $\frac{d\sigma}{d\Omega}$, i.e., $I_{\text{out}} = \frac{d\sigma}{d\Omega} J_{\text{in}}\, d\Omega$. Equating $I_{\text{in}} = I_{\text{out}}$ yields

$$2\pi b\, db = \frac{d\sigma}{d\Omega}(2\pi \sin\Theta\, d\Theta). \tag{12.3}$$

12.1 Classical Scattering Theory

Thus, the differential cross-section $d\sigma/d\Omega$ is defined as

$$\frac{d\sigma}{d\Omega}d\Omega \equiv \frac{\text{Number of particles scattered into } d\Omega \text{ per unit time}}{\text{Incident beam intensity}} = \frac{I_{\text{out}}}{J_{\text{in}}} = \frac{\frac{d\sigma}{d\Omega}\rho v_0 \, d\Omega}{\rho v_0}.$$

It has the units of area, and from Eq. (12.3),

$$\boxed{\frac{d\sigma}{d\Omega} = \frac{b}{\sin\Theta}\left|\frac{db}{d\Theta}\right|}. \tag{12.4}$$

The absolute value is required in Eq. (12.4), since, for repulsive potentials, increasing in b decreases Θ.[1] In a central potential (see Sec. 16.6 linked to the book web page), the scattering cross-section does not depend on the azimuthal scattering angle ϕ (which is not explicitly shown in Fig. 12.1).

The challenge of classical scattering theory is to find the relation $b(\Theta, E)$ for a given scattering energy E and scattering angle Θ, since this function allows calculation of the cross-section. Deriving the function $b(\Theta, E)$ requires the solution of the classical equations of motion. The presence of symmetries leads to conservation laws and hence, the existence of integrals of the motion, as first formulated by Emmy Noether. The most familiar example of *Noether's theorem* is the invariance of the classical Hamiltonian under time translations leading to the conservation of energy. Another example is scattering from a central potential, which is invariant under space rotation, where the corresponding integral of the motion is the angular momentum.

The *total scattering cross-section* is the ratio of the number of particles scattered in all directions per unit time to the incoming flux. It is obtained by integrating the differential cross-section $d\sigma/d\Omega$ over all solid angles. For scattering off a central potential,

$$\sigma = \int d\Omega \frac{d\sigma}{d\Omega} = 2\pi \int_0^\pi d\Theta \sin\Theta \frac{d\sigma}{d\Omega} = 2\pi \int_0^\pi d\Theta\, b(\Theta)\left|\frac{db}{d\Theta}\right| = \pi b_{\max}^2, \tag{12.5}$$

where $b_{\max}$ is the largest impact parameter, which still depends on Θ. This implies that as long as b depends on Θ, the integrand does not vanish, and the total cross-section is simply given by the area obtained by projecting the scattering region on the yz plane. Thus, the scattering potential $V(r)$ does not vanish for $r < a$ but vanishes for $r > a$, hence,

$$\sigma = \pi a^2. \tag{12.6}$$

Therefore, the classical total scattering cross-section is infinite for any potential that does not vanish identically at large r no matter how fast it falls off. This somewhat counter intuitive outcome is remedied in the quantum scattering formulation introduced below. In particular, the total quantum mechanical cross-section remains finite for any potential that falls off faster than C/r^2.

For a gas of particles undergoing collisions, the scattering cross-section determines the rate of collisions, which given by $\Gamma = n\overline{|\mathbf{v}|}\sigma$, where n is the density of particles, and $\mathbf{v}$ is the average relative velocity of a colliding pair of particles. The mean time between collisions is $\tau = \Gamma^{-1}$.

These general considerations will now be used to solve a central problem in classical scattering, the scattering of two charged particles. This problem was first studied by Ernest Rutherford in 1911 in scattering experiments of α particles (charge $2e$) off a heavy nuclei (charge Ze). The derivation of the relation between the impact parameter b and the scattering angle Θ uses conservation laws to avoid the need to solve the differential equations arising from Newton's second law. The geometry of *Rutherford scattering*, i.e., the Coulomb scattering of two charged particles, is displayed in Fig. 12.2. For simplicity, let us consider a light particle of charge q_1 and mass m having an initial velocity $\mathbf{v}_0 = v_0\hat{\mathbf{x}}$ moving toward a heavy particle of charge q_2, initially at rest at the origin O in the laboratory frame. The conserved orbital angular

---

[1] In general, more than one impact parameter can contribute to scattering at a given scattering angle Θ, so a sum of such impact parameters is necessary on the RHS of Eq. (12.4).

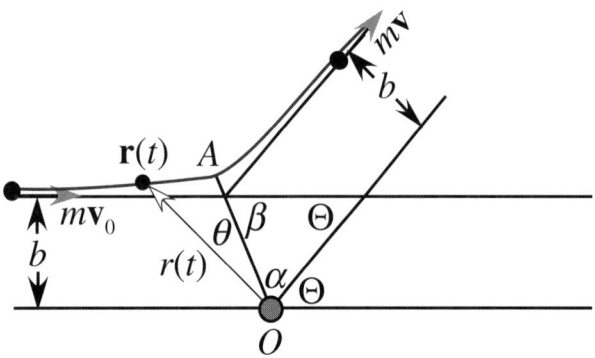

FIG 12.2 Geometry of Rutherford scattering.

momentum yields $\mathbf{L} = \mathbf{b} \times \mathbf{p}_0$, where $\mathbf{b} = b\hat{\mathbf{z}}$ and $\mathbf{p}_0 (= m\mathbf{v}_0)$ is the initial linear momentum. The particle moves in a plane perpendicular to $\mathbf{L}$ and is repelled by the Coulomb force, $\mathbf{F} = \frac{Kq_1q_2}{r^2}\hat{\mathbf{r}} \equiv C\frac{\hat{\mathbf{r}}}{r^2}$ [see Sec. 16.6 linked to the book web page—we use the notation discussed following Eq. (16.51)]. Its trajectory is given in terms of the function $r(\theta)$, where the angle θ is measured from the line OA. After reaching the point of closest approach A, it asymptotically recedes in a direction Θ with respect to its initial direction. Conservation of $|\mathbf{L}|$ implies,

$$mr^2\dot{\theta} = mv_0 b, \quad \Rightarrow \quad \dot{\theta} = \frac{v_0 b}{r^2}. \tag{12.7}$$

Let us compute the change of linear momentum along OA, which by symmetry bisects the angle $\pi - \Theta$, i.e., $\alpha = \frac{1}{2}(\pi - \Theta) = \beta$. From Fig. 12.2, elementary geometry, and noting that $\cos\beta = \cos(\pi/2 - \Theta/2) = \cos(\pi/2)\cos(\Theta/2) + \sin(\pi/2)\sin(\Theta/2) = \sin(\Theta/2)$, the projection of $\mathbf{p}_0$ along OA can be easily shown to equal $-mv_0 \sin(\Theta/2)$. Similarly, the projection of the final momentum $\mathbf{p} = m\mathbf{v}$ on OA is $mv_0 \sin(\Theta/2)$ ($v = v_0$ due to energy conservation). Noting that the projection of $\mathbf{F}$ on OA is $\frac{C}{r^2}\cos\theta$, the momentum change, $\Delta\mathbf{p} = \int dt\, \mathbf{F}$, projected on OA, is

$$2mv_0 \sin\frac{\Theta}{2} = \int_{\theta(-\infty)}^{\theta(\infty)} \frac{C}{r^2}\cos\theta \frac{dt}{d\theta}\, d\theta. \tag{12.8}$$

Inserting $dt/d\theta = \frac{r^2}{v_0 b}$ [see Eq. (12.7)] into the integrand of Eq. (12.8) and noting from Fig. 12.2 that $\theta(-\infty) = -\alpha = -\frac{1}{2}(\pi - \Theta) = -\theta(\infty)$, the integral over $d\theta$ from $-\alpha$ to α can be easily performed, leading to the relation,

$$2mv_0 \sin\frac{\Theta}{2} = \frac{2C}{bv_0}\cos\frac{\Theta}{2}, \quad \Rightarrow \quad b = \frac{Kq_1q_2}{mv_0}\cot\frac{\Theta}{2}. \tag{12.9}$$

A few comments are in order: (1) One can generalize this treatment to an arbitrary central force as $F(r) = \frac{C(r)}{r^2}$ and use Eq. (12.8) with $C(r)$ instead of C. However, this requires knowledge of the trajectory $r(\theta)$. (2) Although the analysis and Fig. 12.2 are for a repulsive force, the final expression is also valid for an attractive force with $Kq_1q_2 < 0$. Since by definition $b > 0$, the scattering angle must be negative for an attractive force, $-\pi \leq \Theta \leq 0$. (3) When $\Theta = 0$, the impact parameter diverges. This means that for any b, no matter how large, there will be scattering and the trajectory will deviate from its original direction. This reflects the fact that the Coulomb potential decays very slowly. (4) For a repulsive potential, $b = 0$ implies $\Theta = \pi$. Hence, a Coulomb repulsive potential is impenetrable when the scattering is head-on. (5) When the potential is attractive and $b \to 0$, one has $\Theta = \pi$ and the particle completes a full round trip encircling the scattering center.

Using Eq. 12.4, the differential cross-section for Coulomb scattering is

$$\boxed{\frac{d\sigma}{d\Omega} = \frac{b}{\sin\Theta}\left|\frac{db}{d\Theta}\right| = \left(\frac{Kq_1q_2}{4E}\right)^2 \frac{1}{\sin^4\frac{\Theta}{2}}.} \tag{12.10}$$

This is known as the *Rutherford scattering formula*. The differential cross-section is forward-peaked and diverges at $\Theta = 0$. From the discussion of the total cross-section and Eq. (12.5), it is clear that the integral over scattering angles diverges and the total cross-section is infinite. Remarkably, this classical result agrees with its quantum counterpart.

12.2 Quantum Scattering

> **Problem 12.2**
>
> (a) For scattering from an impenetrable sphere of radius, R, show that the relation between b and Θ is $b = R\cos(\Theta/2)$ and calculate the differential cross-section.
>
> (b) For a given impact parameter b, use the two conservation laws for energy $E = \frac{1}{2}mv_0^2$ and the magnitude of the angular momentum $L = m|\mathbf{v} \times \mathbf{r}|$ to compute the distance $r_A \equiv OA$ in Fig. 12.1 and the speed v_A of the particle at the point A for scattering from a repulsive Coulomb potential $V(r) = Kq_1q_2/r$.
>
> **Answers:** (a) Due to the specular reflection at the impinging point A, the line OA bisects the scattering angle Θ. Hence, the angle between the horizontal axis and the line $\vec{OA}$ is $\Theta/2$. Hence, $\frac{db}{d\Theta} = -\sin(\Theta/2)/2 \Rightarrow \frac{d\sigma}{d\Omega} = \frac{R^2 \sin\frac{\Theta}{2} \cos\frac{\Theta}{2}}{2\sin\Theta} = \frac{R^2}{4}$. (b) The initial value of L at $t \to -\infty$ is $L = L_0 = mv_0 b$. Noting that at this point the trajectory is perpendicular to OA, we find that $L = L_A = mv_A r_A$. The corresponding energies are $E_0 = mv_0^2/2$ and $E_A = mv_A^2/2$. The two equations $L_0 = L_A$ and $E_0 = E_A$ determine r_A and v_A, respectively.

12.2 QUANTUM SCATTERING

In the next few sections, we discuss the main ingredients of quantum scattering theory and introduce the central quantities of interest, such as the scattering operator (also called the S matrix), scattering amplitudes, and cross-sections.

In a scattering experiment, two subsystems collide. We may think of the collision of two elementary particles, such as electrons or nucleons, or the collision of an elementary particle and a composite subsystem (such as an atom, a molecule, or a nucleus) or the collision of two composite subsystems such as two atoms or an atom and a molecule. At some initial time t_i (eventually, we take $t_i \to -\infty$), the subsystems are far apart and noninteracting. This is also the case at a much later time t_f, after the collision (eventually, we take $t_f \to \infty$). The subsystems approach each other and interact during a short-time interval Δt sometime between t_i and t_f. During this time interval, the two subsystems can exchange energy, linear momentum, angular momentum, perhaps spin, etc. If the systems are composite, they might exchange particles as well or they may break up. Our primary focus of interest is the nature of the asymptotic states after the collision is completed, since these determine the scattering cross-sections.

12.2.1 TIME-DEPENDENT AND STATIONARY APPROACHES

There are two basic approaches to quantum scattering theory, the time-dependent approach and the stationary approach. The time-dependent formalism is conceptually closer to intuition; a scattering experiment is an event that evolves in time with initial, intermediate, and final stages. Although the time-dependent approach is an important part of the fundamental theory, the stationary approach is more practical as far as calculation of observables is concerned. Before going into details, it might be useful to give a brief description of these two approaches. In both formulations, one starts with the system's total Hamiltonian $H = H_0 + V$, where H_0 governs the dynamics of the noninteracting systems and V is responsible for scattering as it encodes the interaction between the two systems and should vanish when the systems are far apart before and after the collision. Both H_0 and V are time independent. We start with a brief description of these two concepts that must describe the same physics.

Time-dependent Approach

The time-dependent description is based on the time-dependent Schrödinger equations for the free (noninteracting) and the interacting system, which are written as,

$$i\hbar \frac{\partial |\Psi_0(t)\rangle}{\partial t} = H_0 |\Psi_0(t)\rangle, \qquad (12.11)$$

$$i\hbar \frac{\partial |\Psi(t)\rangle}{\partial t} = H |\Psi(t)\rangle. \qquad (12.12)$$

Here, $|\Psi_0(t)\rangle$ and $|\Psi(t)\rangle$ are ket states that are representable by wave functions once a representation is chosen, e.g., in configuration representation, $\langle \mathbf{r}|\Psi(t)\rangle = \Psi(\mathbf{r}, t)$ is the wave function. (We will occasionally refer to the kets as wave functions: this should be understood in the above sense.)

An experimentalist prepares the system at an initial time t_i (we will take $t_i \to -\infty$ in what follows) where the constituents are far apart and do not interact. The corresponding initial state, assumed to be pure, is denoted as $|\Psi_0^{in}(t)\rangle$. In the *absence* of interaction, this state would have propagated freely and evolves forward in time according to Eq. (12.11) all the way to any time t, namely, $|\Psi_0^{in}(t)\rangle = U_0(t, t_i)|\Psi_0^{in}(t_i)\rangle$, where $U_0(t, t_i) = e^{-iH_0(t-t_i)/\hbar}$. In particular, we are interested in time t close to 0 when the interaction is most effective. Within the same experiment, the experimentalist observes his system after the collision at time $t_f \to +\infty$ where, again, the constituents are far apart and do not interact. The corresponding final state is a solution of Eq. (12.11) denoted as $|\Psi_0^{out}(t)\rangle$. In what might seem as an academic exercise, we consider how such a state propagates *backward in time*, still in the absence of interaction. The corresponding formal expression for the state is $|\Psi_0^{out}(t)\rangle = U_0(t, t_f)|\Psi_0^{in}(t_f)\rangle$, where $t < t_f$.

What happens if there *is* interaction? The exact state $|\Psi(t)\rangle$ evolves under the application of the full evolution operator $U(t, t_i) = e^{-iH(t-t_i)/\hbar}$ from $t = t_i \to -\infty$ (where it coincides with $|\Psi_0^{in}(t_i)\rangle$). As $t \to +\infty$, it coincides with $|\Psi_0^{out}(t)\rangle$. Although the Hamiltonian is time independent, neither $|\Psi_0(t)\rangle$ nor $|\Psi(t)\rangle$ is required to be stationary. The central result of time-dependent scattering theory is that at any finite time t, there is a unitary time-independent *scattering operator S* mapping $|\Psi_0^{in}(t)\rangle$ onto $|\Psi_0^{out}(t)\rangle$. This mapping is formally written as

$$|\Psi_0^{out}(t)\rangle = S|\Psi_0^{in}(t)\rangle. \tag{12.13}$$

The interpretation of Eq. (12.13) is that the operator S takes the state $|\Psi_0^{in}(t)\rangle$, which has been prepared at time t_i and propagated forward in time by $U_0(t, t_i)$ to a finite time t, and transforms it onto a state $|\Psi_0^{out}(t_f)\rangle$, which is the state obtained at the time $t_f \to \infty$ and has been propagated backward in time by $U_0(t, t_f)$ to the same finite time t. The scattering operator S encodes all the relevant information required for interpretation of the scattering experiment. In Sec. 12.2.3, the construction of S will be discussed in detail.

Time-Independent Approach

In the time-independent approach to scattering theory, one employs the fact that the Hamiltonian describing the physical system does not depend explicitly on time, and the *scattering energy E* at which the experiment is conducted is determined by the experimentalist and is conserved throughout the scattering process. Consider, for example, the scattering of a beam of particles from a potential $V(\mathbf{r})$ that vanishes faster than r^{-2} as $r \to \infty$. The stationary wave function can be written as a product of a function of time and a function of space, $\Psi(\mathbf{r}, t) \equiv \langle \mathbf{r}|\Psi(t)\rangle = e^{-iEt/\hbar}\psi(\mathbf{r})$, where $\psi(\mathbf{r})$ is the solution of the time-independent Schrödinger equation, $H\psi(\mathbf{r}) = E\psi(\mathbf{r})$, with $H = \frac{\mathbf{p}^2}{2m} + V(\mathbf{r})$. When projectile particles are very far from the origin, $V = 0$, the solution of the time-independent Schrödinger equation, $\psi(\mathbf{r})$, encompasses both the incoming state representing the beam of particles approaching the potential center and the outgoing particles leaving it. The incoming state is a *plane wave*, $A\,e^{i\mathbf{k}\cdot\mathbf{r}}$, where A is a normalization constant taken hereafter to be $(2\pi)^{-3/2}$, and $E = \frac{\hbar^2 k^2}{2m}$. The outgoing solutions at very large distance $|\mathbf{r}|$ from the potential center are *outgoing spherical waves* $Af(\hat{\mathbf{r}})e^{ikr}/r$, where $f(\hat{\mathbf{r}})$ depends on the energy E and the direction of the vector from the origin to the receding particles. Combining the plane wave and outgoing spherical wave components, the wave function $\psi(\mathbf{r})$ is required to satisfy the following asymptotic boundary conditions,

$$\psi_{\mathbf{k}}^+(\mathbf{r}) \xrightarrow[r\to\infty]{} (2\pi)^{-\frac{3}{2}} \left[e^{i\mathbf{k}\cdot\mathbf{r}} + f(\hat{\mathbf{r}})\frac{e^{ikr}}{r} \right]. \tag{12.14}$$

The superscript $^+$ on ψ indicates that the spherical wave encoded in the second term on the RHS of Eq. (12.14) is *outgoing*. In order to develop scattering theory, we will also need to introduce *incoming* spherical waves, e^{-ikr}/r, although they do not have a direct experimental relevance. The corresponding wave function will be denoted by $\psi_{\mathbf{k}}^-(\mathbf{r})$. The *scattering amplitude* $f(\hat{\mathbf{r}})$ in Eq. (12.14) is an object of paramount interest in time-independent scattering theory since it is directly

12.2 Quantum Scattering

related to the scattering cross-section. We shall develop time-independent scattering theory below. It is more practical for obtaining explicit expressions for the measurable quantities in scattering experiments than the time-dependent formulation.

12.2.2 PREPARATION OF THE INITIAL STATE

In classical scattering, the particles that scatter have a definite position and linear momentum. This prescription should be modified in quantum mechanics since $\mathbf{r}$ and $\mathbf{p}$ cannot be simultaneously determined due to the uncertainty principle. Hence, some care should be exercised in analyzing the preparation of the initial state. For simplicity, we will first consider the preparation of a single-particle initial state appropriate for scattering from a fixed potential center. Later, we will extend the discussion to the situation where two particles collide with each other.

Before scattering, the particle is far removed from the scattering potential. Using Dirac notation, a quantum state of a free particle (possibly with internal degrees of freedom such an atom or a molecule) is described by a state vector $|\psi\rangle$ and its dynamics is governed by an appropriate Hamiltonian H_0 that does not contain the interaction potential since at large distance $V(\mathbf{r}) = 0$. The eigenstates of H_0, $|\eta_{\mathbf{k}n}\rangle \equiv |\mathbf{k}\rangle \otimes |n\rangle$, are tensor products of the ket $|\mathbf{k}\rangle$ ($\mathbf{k}$ specifying the wavevector of the particle) and a ket $|n\rangle$ encoding the quantum numbers of the internal wave function of the particle when it is not point particle (think of the scattering of a hydrogen atom from an external potential). Note that $|\mathbf{k}\rangle$ is an eigenstate of the linear momentum operator. Hence,

$$H_0|\eta_{\mathbf{k}n}\rangle = \varepsilon_{\mathbf{k}n}|\eta_{\mathbf{k}n}\rangle. \qquad (12.15)$$

As a simple example, consider a particle with no internal degrees of freedom except spin, i.e., an *elementary particle*. Then $|n\rangle \equiv |s\sigma\rangle$ is the particle's *spin state* (spinor), with spin s and z projection σ, such that

$$S^2|s\sigma\rangle = \hbar^2 s(s+1)|s\sigma\rangle, \quad S_z|s\sigma\rangle = \hbar\sigma|s\sigma\rangle, \qquad (12.16)$$

where S is the spin operator, S_z is its z projection, and the state $|\eta_{\mathbf{k}s\sigma}\rangle$ has the spatial representation,

$$\eta_{\mathbf{k}n}(\mathbf{r}) \equiv \langle \mathbf{r}|\eta_{\mathbf{k}s\sigma}\rangle = (2\pi)^{-\frac{3}{2}} e^{i\mathbf{k}\cdot\mathbf{r}}|s\sigma\rangle. \qquad (12.17)$$

Being a plane wave, the state $|\eta_{\mathbf{k}n}\rangle$ is neither localized nor square integrable. Experimental conditions require that particles should be confined within some macroscopic volume. Particle can indeed be confined by constructing a *wave packet* out of plane waves,

$$\gamma_{\mathbf{p}n}(\mathbf{r}) \equiv (2\pi)^{-\frac{3}{2}} \int d\mathbf{k}\, A(\mathbf{k}-\mathbf{p}) e^{i\mathbf{k}\cdot\mathbf{r}}|n\rangle = \int d\mathbf{k}\, A(\mathbf{k}-\mathbf{p})\,\eta_{\mathbf{k}n}(\mathbf{r}), \qquad (12.18)$$

where the function $A(\mathbf{k}-\mathbf{p})$ is the amplitude of the plane wave state $(2\pi)^{-3/2}e^{i\mathbf{k}\cdot\mathbf{r}}$ in the wave packet and $|n\rangle$ is the ket specifying the internal (spin) state. To check the extent to which the wave packet deviates from a plane wave, let us consider the Fourier transform of the amplitude $A(\mathbf{q})$,

$$a(\mathbf{r}) = (2\pi)^{-\frac{3}{2}} \int d\mathbf{q}\, A(\mathbf{q})\, e^{i\mathbf{q}\cdot\mathbf{r}}, \qquad (12.19)$$

in terms of which the wave packet $\gamma_{\mathbf{p}n}(\mathbf{r})$ is written as,

$$\gamma_{\mathbf{p}n}(\mathbf{r}) = a(\mathbf{r})\,\eta_{\mathbf{p}n}(\mathbf{r}). \qquad (12.20)$$

The use of $\mathbf{p}$ instead of $\mathbf{k}$ indicates that $\gamma_{\mathbf{p}n}(\mathbf{r})$ is not a momentum eigenstate. Rather, it is a product of a plane wave and an amplitude $a(\mathbf{r})$ that determines the extent of confinement. If the amplitude is square integrable, so is the wave packet. The normalization for the function $a(\mathbf{r})$ is

$$\int d\mathbf{r}\, |a(\mathbf{r})|^2 = 1, \quad \Rightarrow \quad \langle \gamma_{\mathbf{p}n}|\gamma_{\mathbf{p}'n'}\rangle = \delta_{\mathbf{p}\mathbf{p}'}\delta_{nn'}. \qquad (12.21)$$

Roughly speaking, $a(\mathbf{r})$ is smooth and nonzero within a volume of size L^3, which is the extent of the wave packet. Experimentally, L is macroscopic in the sense that it is much larger than any microscopic length scale such as the particle wavelength $\lambda = h/p$. Thus, Eq. (12.20) implies that on length scale smaller than L, the wave packet $\gamma_{\mathbf{p}n}(\mathbf{r})$ is virtually a plane wave. Another important property of wave packets is that they are not stationary. If at time $t = 0$ the wave function is given by $\gamma_{\mathbf{p}n}(\mathbf{r})$, its time evolution is determined by the *evolution operator* with the Hamiltonian H_0,

$$\gamma_{\mathbf{p}n}(\mathbf{r},t) = e^{-iH_0t/\hbar}\gamma_{\mathbf{p}n}(\mathbf{r}) \equiv U_0(t,0)\gamma_{\mathbf{p}n}(\mathbf{r}). \tag{12.22}$$

We can approximately evaluate $\gamma_{\mathbf{p}n}$ by employing Eq. (12.18) and expanding $\varepsilon_{\mathbf{k}n}$ in $\mathbf{k}$ around $\mathbf{p}$. In lowest order, the quadratic term is neglected, and the result is simple and transparent:

$$\gamma_{\mathbf{p}n}(\mathbf{r},t) = e^{-i\varepsilon_{\mathbf{p}n}t/\hbar}a(\mathbf{r} - \mathbf{v}t)\eta_{\mathbf{p}n}(\mathbf{r}). \tag{12.23}$$

Here, $\mathbf{v} = \mathbf{p}/m$ is the velocity of the wave packet. In this approximation, the wave packet does not spread or change shape, and moves with constant speed v. Strictly speaking, wave packets do spread, but for scattering problems, the approximation Eq. (12.23) can be satisfied to high accuracy for wave packets with large energy variance [179].

So far we have discussed the preparation of a single-incident particle, which is appropriate for problems involving scattering of a particle by a fixed potential. However, elementary scattering experiments typically involve two particles, and the notion of a fixed potential is an approximation where the mass of one of the scatterers is infinite. Therefore, the construction of two-particle wave packets is required. For two isolated particles denoted as 1 and 2 (assumed here to be nonidentical), the Hamiltonian is $H_0 = H_0^{(1)} + H_0^{(2)}$; its eigenstates are direct products of single-particle states [Eq. (12.15)],

$$|\Lambda_{\mathbf{k}_1 n_1 \mathbf{k}_2 n_2}\rangle = |\eta_{\mathbf{k}_1 n_1}^{(1)}\rangle \otimes |\eta_{\mathbf{k}_2 n_2}^{(2)}\rangle, \quad H_0|\Lambda_{\mathbf{k}_1 n_1 \mathbf{k}_2 n_2}\rangle = (\varepsilon_{\mathbf{k}_1 n_1} + \varepsilon_{\mathbf{k}_2 n_2})|\Lambda_{\mathbf{k}_1 n_1 \mathbf{k}_2 n_2}\rangle. \tag{12.24}$$

Since wave packets involve a convolution integral, as in Eq. (12.18), the two-particle wave packet remains a product,

$$\Gamma_{12}(\mathbf{r}_1\mathbf{r}_2) \equiv \langle \mathbf{r}_1\mathbf{r}_2|\Gamma_{\mathbf{p}_1 n_1 \mathbf{p}_2 n_2}\rangle = a_1(\mathbf{r}_1)a_2(\mathbf{r}_2)\,\eta_{\mathbf{p}_1 n_1}^{(1)}(\mathbf{r}_1)\eta_{\mathbf{p}_2 n_2}^{(2)}(\mathbf{r}_2). \tag{12.25}$$

Moreover, within the approximation discussed in obtaining Eq. (12.23), the evolution of $\Gamma_{12}(\mathbf{r}_1\mathbf{r}_2)$ in time is given by

$$\Gamma_{12}(\mathbf{r}_1\mathbf{r}_2,t) = a_1(\mathbf{r}_1 - \mathbf{v}_1 t)a_2(\mathbf{r}_2 - \mathbf{v}_2 t)\,e^{-\frac{i}{\hbar}(\varepsilon_{\mathbf{p}_1 n_1} + \varepsilon_{\mathbf{p}_2 n_2})t}\,\Gamma_{12}(\mathbf{r}_1\mathbf{r}_2). \tag{12.26}$$

12.2.3 TIME-DEPENDENT FORMULATION

In Sec. 12.2, we introduced the basic ingredients of time-dependent scattering theory. Now we elaborate on this subject. In scattering experiments, a system consisting of two colliding objects (e.g., elementary particles, atomic nuclei, atoms, or molecules) evolves in time in the following sense. In the remote past ($t \to -\infty$) and in the distant future ($t \to \infty$), the two objects are at a large distance from each other, whereas during a short-time period near $t = 0$, they interact. It then seems plausible to describe this dynamical behavior quantum mechanically within the time-dependent Schrödinger equation. This leads to the time-dependent formulation of scattering theory which is briefly explained in this section. We have already stressed that the time-independent approach is more convenient for the practical purpose of calculating physical observables resulting from the scattering, and hence, it constitutes the main body of this chapter. However, the time-dependent approach is important, first, because it gives us physical insight, and second, because several expressions that are used in the time-independent formalism are derived from the time-dependent one.

Möller Operators and the S Matrix

Consider two physical systems 1 and 2 approaching each other from a large distance, such that when the systems are far apart, they do not interact and are described by their separate respective Hamiltonian operators, H_1 and H_2. As the systems approach each other, the interaction between them, represented by an operator V, becomes significant. The dependence

12.2 Quantum Scattering

of V on the individual coordinates and spins of the systems 1 and 2 can be quite general, as long as it falls off sufficiently fast as a function of the distance r_{12} between the two systems. The time-independent Hamiltonian is then written as

$$H = H_1 + H_2 + V \equiv H_0 + V. \tag{12.27}$$

The ket $|\Psi(t)\rangle$ is the solution of the Schrödinger equation

$$i\hbar \frac{\partial}{\partial t} |\Psi(t)\rangle = H|\Psi(t)\rangle. \tag{12.28}$$

When $t \to \pm\infty$, the subsystems are far apart and do not interact. Therefore, there exist two solutions $|\Psi_0^{\text{in,out}}(t)\rangle$ of the free motion Schrödinger equations,

$$i\hbar \frac{\partial}{\partial t} |\Psi_0^{\text{in,out}}(t)\rangle = H_0 |\Psi_0^{\text{in,out}}(t)\rangle. \tag{12.29}$$

We stress that the functions $|\Psi_0^{\text{in,out}}(t)\rangle$ are defined for all $-\infty < t < \infty$ as solutions of Eqs. (12.29). To be more precise, let us fix t to have some finite value and consider states at very early time $t - T$ and very late time $t + T$, where $T \to \infty$. Then, we require that

$$\lim_{T \to \infty} ||\,|\Psi(t-T)\rangle - |\Psi_0^{\text{in}}(t-T)\rangle\,|| = 0, \quad \lim_{T \to \infty} ||\,|\Psi(t+T)\rangle - |\Psi_0^{\text{out}}(t+T)\rangle\,|| = 0. \tag{12.30}$$

Physically, the system is prepared in the distant past in state $|\Psi(t-T)\rangle = |\Psi_0^{\text{in}}(t-T)\rangle$ and evolves in the distant future into the state $|\Psi(t+T)\rangle = |\Psi_0^{\text{out}}(t+T)\rangle$. In terms of the evolution operators,

$$U(t_1, t_2) \equiv e^{-iH(t_1-t_2)/\hbar}, \quad U_0(t_1, t_2) \equiv e^{-iH_0(t_1-t_2)/\hbar}, \tag{12.31}$$

relations (12.30) imply the following evolution equations for $T \to \infty$:

$$|\Psi_0^{\text{in}}(t)\rangle = U_0(t, t-T)|\Psi_0^{\text{in}}(t-T)\rangle, \quad |\Psi_0^{\text{out}}(t)\rangle = U_0(t, t+T)|\Psi_0^{\text{out}}(t+T)\rangle,$$
$$|\Psi(t)\rangle = U(t, t-T)|\Psi_0^{\text{in}}(t-T)\rangle, \quad |\Psi(t)\rangle = U(t, t+T)|\Psi_0^{\text{out}}(t+T)\rangle. \tag{12.32}$$

Note that in the second equation in each row, the kets propagate backward in time. Using Eqs. (12.32), it is straightforward to construct an operator S, which, when operates on $|\Psi_0^{\text{in}}(t)\rangle$ yields $|\Psi_0^{\text{out}}(t)\rangle$. First, one expresses the same state $|\Psi(t)\rangle$ once in terms of $|\Psi_0^{\text{in}}(t)\rangle$ and once in terms of $|\Psi_0^{\text{out}}(t)\rangle$. The corresponding operators, Ω_+ and Ω_-, are referred to as *Möller operators*. Formally, the Möller operators are required to satisfy the following relations:

$$\Omega_+ |\Psi_0^{\text{in}}(t)\rangle = |\Psi(t)\rangle, \quad \Omega_- |\Psi_0^{\text{out}}(t)\rangle = |\Psi(t)\rangle. \tag{12.33}$$

Inspection of Eqs. (12.32) shows that application of a Möller operation is achievable in two steps, as shown schematically in Fig. 12.3 and explained in the caption. Consequently,

$$\Omega_+ \equiv \lim_{T \to \infty} e^{-iHT/\hbar} e^{iH_0 T/\hbar}, \quad \Omega_- \equiv \lim_{T \to \infty} e^{iHTt/\hbar} e^{-iH_0 T/\hbar}. \tag{12.34}$$

Problem 12.3

Explain the action of Ω_- in analogy with the explanation of the operation of Ω_+ detailed in the caption of Fig. 12.3.
Answer: Ω_- operates on $\Psi_0^{\text{out}}(t)$ in two steps. First, Ω_- carries it forward in time to $t + T$ with the free Hamiltonian H_0, and then it carries it backward in time to t with the full Hamiltonian H.

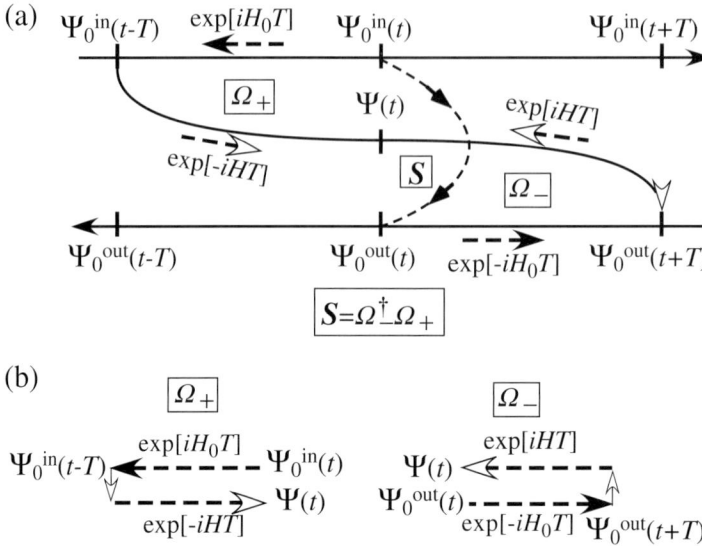

FIG 12.3 The action of Möller operators. For notational convenience, we take $\hbar = 1$. (a) Application of Ω_+ on $|\Psi_0^{in}(t)\rangle$ and Ω_- on $|\Psi_0^{out}(t)\rangle$ in this figure. Consider first the action of Ω_+ on $|\Psi_0^{in}(t)\rangle$. The state $|\Psi_0^{in}(t)\rangle$ is first propagated backward in time (dashed black arrow) to the distant past $t - T$ with $T > 0$ by the free evolution operator e^{iH_0T}. At that early times, it coincides with $|\Psi(t-T)\rangle$, which can be evolved forward to time t with the full evolution operator e^{-iHT} (dashed white arrow). The product of these two operations, where $T \to \infty$, yields the Möller operator Ω_+ defined in Eq. (12.34). A similar construction defines Ω_-. The S matrix, $S = \Omega_-^\dagger \Omega_+$, takes $|\Psi_0^{in}(t)\rangle$ to $|\Psi_0^{out}(t)\rangle$. (b) Summary of the action of the Möller operators.

The above limiting process is rather formal and suffers from the problem of taking the limit of a function that might contain oscillatory part. A standard way to treat this problem is to perform an averaging procedure that does not affect the nonoscillatory parts but eliminate the oscillatory parts. The mathematical operation is based on the definition

$$\lim_{t \to -\infty} f(t) = \lim_{\eta \to 0^+} \eta \int_{-\infty}^{0} d\tau\, e^{\eta \tau} f(\tau). \tag{12.35}$$

Within this formulation, the following integral is then an alternative definition of the Möller operators:

$$\Omega_\pm = \lim_{\eta \to 0^+} \mp i\eta \int_0^{\mp\infty} dt\, e^{iHt/\hbar} e^{-iH_0 t/\hbar} e^{\pm \eta t}. \tag{12.36}$$

The operation leading from $|\Psi_0^{in}(t)\rangle$ to $|\Psi_0^{out}(t)\rangle$ is obtained through substitution of the second equation in Eqs. (12.34) into the first and applying $\Omega_-^\dagger$, thereby obtaining

$$\boxed{|\Psi_0^{out}(t)\rangle = S|\Psi_0^{in}(t)\rangle, \quad S = \Omega_-^\dagger \Omega_+.} \tag{12.37}$$

Inspecting the definition of S in terms of Möller operators and the definition of the latter in terms of the corresponding evolution operators, it can be verified (under reasonable conditions) that S is unitary, i.e.,

$$SS^\dagger = S^\dagger S = \mathbf{1}, \tag{12.38}$$

12.2 Quantum Scattering

where $\mathbf{1}$ is the unit operator in the pertinent Hilbert space.[2] From a physical point of view, the unitarity of the S matrix reflects particle number conservation. It is useful to separate S into a unit matrix and a *transition operator*, T:

$$\boxed{S = \mathbf{1} - 2\pi i T.} \tag{12.39}$$

If no interaction has occurred, $S = \mathbf{1}$, i.e., $T = \mathbf{0}$. In terms of the transition operator T, the unitarity relation for the S matrix reads,

$$\mathbf{1} = S^\dagger S = (\mathbf{1} + 2\pi i T^\dagger)(\mathbf{1} - 2\pi i T) = \mathbf{1} - 2\pi i(T - T^\dagger) + 4\pi^2 T^\dagger T,$$

i.e.,

$$i(T^\dagger - T) = 2\pi T^\dagger T. \tag{12.40}$$

The relation between the transition operator T and measurable quantities (such as differential cross-sections) will be discussed in Sec. 12.4.

Scattering within the Interaction Representation

As we have seen above, the time-dependent approach to scattering theory is concerned with the evolution of the state vector $|\Psi(t)\rangle$ between very early and very late times. This evolution can also be analyzed within the *interaction representation* introduced in Sec. 2.7.1. This approach plays an important role in quantum field theory and many-body theory as discussed in Chapter 18, linked to the book web page. We shall briefly reformulate this approach here (with slightly different notation) in order to construct a perturbation series for $|\Psi(t)\rangle$. The starting point is the time-dependent Schrödinger equation

$$i\hbar \frac{\partial}{\partial t}|\Psi(t)\rangle = H|\Psi(t)\rangle = (H_0 + V)|\Psi(t)\rangle, \tag{12.41}$$

written for a general time-dependent state $|\Psi(t)\rangle$ of a quantum system governed by a time-independent Hamiltonian $H = H_0 + V$. It is further assumed that the eigenstates and eigenvalues of H_0 are known. The formal solution of Eq. (12.41) with the initial condition $|\Psi(t = t_0)\rangle = |\Psi_0\rangle$ is

$$|\Psi(t)\rangle = U(t, t_0)|\Psi_0\rangle \equiv e^{-iH(t-t_0)/\hbar}|\Psi_0\rangle, \tag{12.42}$$

and $U(t, t_0)$ is a unitary evolution operator that carries the state from time t_0 to a later time $t > t_0$. We shall now apply a unitary transformation using the operator $e^{iH_0 t/\hbar}$ and cast the Schrödinger equation in the interaction representation, using the following definitions that were introduced in Sec. 2.7.1:

$$|\Phi_I(t)\rangle \equiv e^{iH_0 t/\hbar}|\Psi(t)\rangle, \tag{12.43a}$$

$$V_I(t) \equiv e^{iH_0 t/\hbar} V e^{-iH_0 t/\hbar}, \tag{12.43b}$$

$$U_I(t, t_0) \equiv e^{iH_0 t/\hbar} U(t, t_0) e^{-iH_0 t_0/\hbar}. \tag{12.43c}$$

The state $|\Phi_I(t)\rangle$ satisfies the Schrödinger equation in the interaction picture,

$$i\hbar \frac{\partial}{\partial t}|\Phi_i(t)\rangle = V_I(t)|\Phi_I(t)\rangle, \tag{12.44}$$

which is equivalent to the original Schrödinger equation (12.41) since transformation [Eq. (12.43a)] is unitary. The formal solution of Eq. (12.44) is

$$|\Phi(t)\rangle = U_I(t, t_0)|\Phi(t_0)\rangle. \tag{12.45}$$

---

[2] The condition is $\text{Ran}\Omega^+ = \text{Ran}\Omega^-$, where $\text{Ran}A$ denotes the range of the operator A in the corresponding Hilbert space $\mathcal{H}$, i.e., the set $\{\langle A\hat{\mathbf{x}}|\hat{\mathbf{x}}\rangle \in \mathbb{C}\}$, where $\hat{\mathbf{x}} \in \mathcal{H}$ is a unit vector.

Problem 12.4

The Landau–Zener Hamiltonian of a two-level system, as discussed in Sec. 7.8.3, is given by $H_{LZ} = \varepsilon\sigma_x + \alpha t\sigma_z \equiv H_0 + V$. Use the identity $e^{i\beta \hat{n}\cdot\boldsymbol{\sigma}} = \cos\beta + i\hat{n}\cdot\boldsymbol{\sigma}\sin\beta$ (valid for any real β and a unit vector $\hat{n}$) to calculate $V_I(t)$ and write the time-dependent Schrödinger equation for H_{LZ} in the interaction representation.

Guidance: Following Eq. (12.43b) and denoting $\omega = \varepsilon/\hbar$, we have $V_I(t) = e^{i\omega\sigma_x t}\alpha t\sigma_z e^{-i\omega\sigma_x t} = (\cos\omega t - i\sigma_x\sin\omega t)\alpha t\sigma_z(\cos\omega t + i\sigma_x\sin\omega t)$. You need to simplify this expression by employing commutation relations of the Pauli matrices.

Problem 12.5

Verify the following relations using the above definitions:

$$U_I(t_3,t_2)U_I(t_2,t_1) = U_I(t_3,t_1), \quad U_I(t,t) = 1, \tag{12.46a}$$

$$U_I^\dagger(t,t_0) = U_I^{-1}(t,t_0) = U(t_0,t), \tag{12.46b}$$

$$i\hbar\frac{dU_I(t,t_0)}{dt} = V_I(t)U_I(t,t_0), \tag{12.46c}$$

$$U_I(t,t_0) = 1 - \frac{i}{\hbar}\int_{t_0}^t dt_1 V(t_1)U_I(t_1,t_0). \tag{12.46d}$$

Equation (12.46d) is suitable for perturbation series in $V_I(t)$, because it can be iterated. However, some care is required in this procedure because $V_I(t)$ is an operator, and generically,

$$[V_I(t), V_I(t')] \neq 0 \quad (t \neq t'). \tag{12.47}$$

If we define

$$X_n(t,t_0) \equiv \frac{1}{\hbar^n}\int_{t_0}^t dt_n V_I(t_n)\int_{t_0}^{t_n} dt_{n-1}V_I(t_{n-1})\ldots\int_{t_0}^{t_2} dt_1 V_I(t_1), \tag{12.48}$$

then the perturbation expansion for $U_I(t,t_0)$ reads,

$$U_I(t,t_0) = \sum_{n=0}^\infty \frac{(-i)^n}{n!} X_n(t,t_0). \tag{12.49}$$

A product of time-dependent operators such as $V_I(t_n)V_I(t_{n-1})\ldots V_I(t_1)$ in which the time arguments are ordered with $t_n > t_{n-1}\cdots > t_1$ is called a *time-ordered product*. Discussion of a method to evaluate the first few terms, X_n, in this expansion is presented in Chapter 7.

12.3 STATIONARY SCATTERING THEORY

We now introduce the basic concepts of time-independent (stationary) scattering theory and the definitions of important relations between the basic relevant quantities. The details of calculating important quantities such as phase shifts and cross-sections are presented in later sections.

For the sake of simplicity, it is assumed that the two colliding particles are point-like and structureless, i.e., they do not have internal degrees of freedom and therefore cannot be excited. Hence, only *elastic scattering* is possible, and

12.3 Stationary Scattering Theory

the initial and final kinetic energies are equal. Scattering of particles with internal degrees of freedom will be treated in Sec. 12.8. It is assumed that there is no external force affecting the particles, and they interact only through their mutual interaction potential, $V(\mathbf{r})$, where $\mathbf{r} = \mathbf{r}_1 - \mathbf{r}_2$ is the position vector joining the two particles. The interaction potential is assumed to fall off sufficiently fast at large $\mathbf{r}$ so that the particles are essentially free (noninteracting) at large $|\mathbf{r}|$. In Sec. 12.3.2, we shall show that this two-body scattering problem is equivalent to that of one-body scattering from a potential $V(\mathbf{r})$, i.e., potential scattering, as illustrated in Fig. 12.4. In this

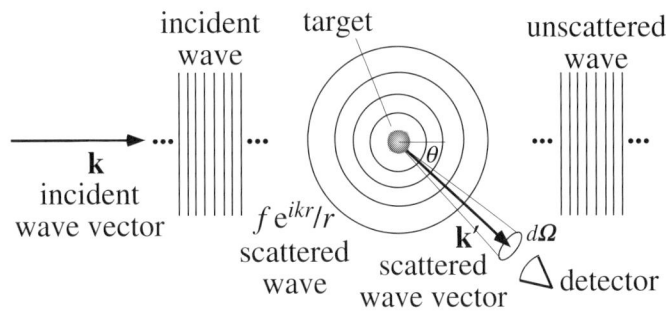

FIG 12.4 Illustration of potential scattering in quantum mechanics. Compare with the description of classical scattering in Fig. 12.1.

formulation, an incident particle represented as a wave packet impinges on the target with potential $V(\mathbf{r})$ centered at point $\mathbf{r} = 0$ (the scattering center). The initial direction of the wave packet is specified by its central wavevector, $\mathbf{k} = (0, 0, k)$, whose magnitude k determines the kinetic energy $E(k)$ of the incoming wave (i.e., the *scattering energy*). The wave packet is then scattered by the potential in all directions, and the scattered part of the wave function is an outgoing spherical wave. Generally, the scattering is not isotropic: At a point very far from the scattering center, the amplitude of the scattered wave moving along a direction $\mathbf{k}' = (k, \theta, \phi)$ is denoted as $f(\Omega)$, which depends on the solid angle $\Omega = (\theta, \phi)$ and the kinetic energy $E(k)$. The equality $|\mathbf{k}'| = |\mathbf{k}| = k$ implies that the scattering is elastic, and the kinetic energy after scattering is equal to the kinetic energy before scattering. The wavenumbers $\mathbf{k}$ and $\mathbf{k}'$ will appear as ubiquitously below. Their relation to the geometry of scattering for elastic scattering is summarized as follows:

$$\mathbf{k} = k\hat{\mathbf{z}}, \quad \mathbf{k}' = k\hat{\Omega} = k(\sin\theta\cos\phi, \sin\theta\sin\phi, \cos\theta). \tag{12.50}$$

Energy conservation for elastic scattering implies

$$k^2 = k'^2 = \frac{2mE}{\hbar^2}, \tag{12.51}$$

where E is the scattering energy. Equation (12.51) is also referred as the on-energy-shell condition.

A particle detector located at large distance from the scattering center collects particles propagating with final momentum $\mathbf{k}'$. Our interest is focused on the scattered spherical wave, or, more precisely, on the scattering amplitude $f(\Omega)$. A glance at Fig. 12.4 shows that at very small polar angle θ, a contribution to the scattering can result from interference between the spherical wave and the part of the incoming wave, which was not scattered. Experimentally, one is mostly interested in the scattered spherical wave. The problem of "disentangling" the incoming and scattered waves will be addressed below.

Although a potential scattering problem is fully encoded within the solution of a nonrelativistic Schrödinger equation for a single particle, it is by no means a simple problem. As discussed in connection with Eq. (12.14), the wave function $\psi(\mathbf{r})$ of the scattered particle should satisfy certain asymptotic boundary conditions, which need to be carefully specified and analyzed. The relevant physical information about the scattering process is derived from the wave function at large distance from the scattering center, where the potential $V(\mathbf{r})$ is absent. Also note that the wave function is not normalizable as is clear from the form of the asymptotic wave function in Eq. (12.14).

12.3.1 CROSS-SECTIONS

In order to further analyze the process of quantum mechanical scattering from a potential, let us recast Fig. 12.4 in somewhat different format, as shown in Fig. 12.5. Before scattering, the particles approach the scattering center from a large distance. All the particles are assumed to be identical and monoenergetic. They form a collimated beam whose

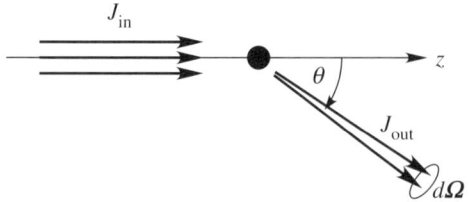

FIG 12.5 Schematic illustration of a scattering into differential solid angle $d\Omega$ at scattering angle θ. Incident flux J_{in} impinges on the target and outgoing flux J_{out} passes through the differential solid angle $d\Omega$. The incoming beam is highly collimated, and the amplitude of the incoming wave vanishes at the detector located far away from the scattering center in solid angle $d\Omega$.

width is much larger than the size of the scattering center (although the width is narrow on a macroscopic scale, see discussion in Sec. 12.2). If the wave packet amplitude $a(\mathbf{r})$ in Eq. (12.20) is assumed to be slowly varying, all the particles in the beam have the same initial wavenumber $\mathbf{k}_i = k\hat{\mathbf{z}}$, and hence, the initial state of each particle in the incoming beam can be represented by a plane wave, $(2\pi)^{-3/2}e^{ikz}$. It is further assumed that the density of particles in the beam is sufficiently low, so that interaction between particles can be ignored. However, the density cannot be too low to avoid poor statistics.

In the kinematics of scattering, an important role is played by particle currents as indicated in Fig. 12.5. The incoming beam current J_{in} is defined as the number of particles per unit time and per unit area that is perpendicular to the propagation direction of the beam. After scattering, those particles that move along a given radial direction specified by a wave vector $\mathbf{k}'$ are collected by a detector located at large distance r from the scattering center. More precisely, the detector is centered at a point whose polar coordinates are r, θ, ϕ, and covers a small area $r^2 d\Omega = r^2 \sin\theta d\theta d\phi$ on a sphere of radius r. The number of particles moving radially outward, crossing this differential area per unit time is denoted by $N(\theta, \phi)d\Omega = J_{out} r^2 d\Omega$, where J_{out} is the magnitude of the outgoing current density at the point (r, θ, ϕ). The ratio $N(\theta, \phi)/J_{in}$ defines the *differential cross-section*, a quantity of prime experimental interest:

$$\frac{d\sigma}{d\Omega} \equiv \frac{N(\theta, \phi)}{J_{in}}. \tag{12.52}$$

Note that $d\sigma/d\Omega$ has the dimension of area. If we rewrite Eq. (12.52) as $J_{in} d\sigma = N(\theta, \phi) d\Omega$, then the geometrical interpretation is evident: $J_{in} d\sigma$ is the number of particles crossing an area $d\sigma$ perpendicular to the initial beam per unit time and $N(\theta, \phi) d\Omega$ is the number of particles crossing the area $r^2 d\Omega$ per unit time.

The differential cross-section as defined in Eq. (12.52) is an experimentally measurable quantity. Its knowledge as a function of scattering angles and scattering energy provides important information on the scattering mechanism, the nature of the potential, and its symmetries. Another important quantity is the number of particles per unit time scattered in *all directions*. It is obtained by integrating the differential cross-section over all solid angles, and the relevant quantity is the *total cross-section* σ, the number of scattered particles crossing a sphere of radius R per unit time divided by the incoming flux, which has dimension of length squared:

$$\sigma = \int d\Omega \frac{d\sigma}{d\Omega}. \tag{12.53}$$

Incoming particles that do not undergo scattering are not counted in the total cross-section. Unlike the analogous quantity in classical scattering, the quantum mechanical total cross-section is finite if the potential decays faster than Cr^{-2} at large distance.

12.3.2 TWO-BODY COLLISIONS

In Section 12.3.1, we considered a single-particle scattering from a fixed potential $V(\mathbf{r})$. In most experiments, however, scattering involves two particles that collide. Examples are collision between two atoms, or two nuclei, or between an electron and an atom. In Secs. 16.6 (linked to the book web page) and 3.2, we show that the problem of collision between

12.3 Stationary Scattering Theory

two particles can be reduced to that of a single-particle scattering from a potential. More precisely, the relative motion of the two particles is mapped on the problem of potential scattering, while the motion of the center of mass of the two particles behaves as a free particle. In this subsection, we elaborate upon this analysis a bit further.

In cases where the colliding particles have speeds v such that $v/c \approx 1$, *relativistic kinematics* and *relativistic quantum mechanics* apply, rather than Newtonian kinematics and nonrelativistic quantum mechanics. Here, we assume that $v \ll c$, i.e., we consider only *nonrelativistic scattering*.

Consider nonrelativistic scattering of two spinless particles of masses m_1, m_2 and position coordinates $\mathbf{r}_1, \mathbf{r}_2$ interacting through a potential $V(\mathbf{r}_1, \mathbf{r}_2)$. The position vectors $\mathbf{r}_{1,2}$ are defined in the *laboratory frame*, which is a fixed frame of reference where experimental results (such as cross-sections) are collected. In this frame, the particles have velocities $\mathbf{v}_1, \mathbf{v}_2$ and respective linear momenta $\mathbf{p}_i = m_i \mathbf{v}_i$. It is assumed that apart from the potential $V(\mathbf{r}_1, \mathbf{r}_2)$, there are no other potentials in the problem (i.e., there are no external forces acting on the colliding particles).

If the two-body potential is translationally invariant, i.e., if it depends only on the relative position of the particles, $V(\mathbf{r}_1, \mathbf{r}_2) = V(\mathbf{r}_2 - \mathbf{r}_1)$, it is possible to completely separate the relative motion and the center of mass (CM) motion of the two-particle system, and the dynamics of the relative motion is equivalent to that of a single-particle scattering from the potential $V(\mathbf{r})$, where $\mathbf{r} = \mathbf{r}_2 - \mathbf{r}_1$. The CM motion is that of a free particle whose wave function is a plane wave. In this case, the scattering is most simply described in the *center of mass frame* in which the CM is at rest. As viewed from the CM frame, long before the collision takes place, the particles move head-on toward each other along a straight line, usually chosen to be the z axis in a Cartesian frame used to describe the pertinent kinematics. Long after the scattering, the particles move in opposite directions along a straight line whose direction is determined by the angles (θ, ϕ).

The procedure of transforming the two-body scattering problem (formulated in the laboratory frame) into a potential scattering problem (formulated in the CM frame) will now be explained. It should be stressed that once the scattering calculation is completed, it may be necessary to translate the results back into measurable quantities in the laboratory frame in order to compare it with the laboratory experimental results. This transformation is worked out below, after calculating the cross-section in the center of mass frame. The original, two-particle, Schrödinger equation written in the laboratory frame is

$$\left[\frac{\mathbf{p}_1^2}{2m_1} + \frac{\mathbf{p}_1^2}{2m_2} + V(\mathbf{r}_2 - \mathbf{r}_1) \right] \Psi(\mathbf{r}_1, \mathbf{r}_2) = E_{12} \Psi(\mathbf{r}_1, \mathbf{r}_2), \tag{12.54}$$

where $\mathbf{p}_i = -i\hbar \nabla_i, i = 1, 2$ are the momentum operators for the two particles in the laboratory frame, where the position vectors $\mathbf{r}_i$ are also defined. The *center of mass coordinate and momentum* $(\mathbf{R}, \mathbf{P})$, the *relative coordinate and momentum* $(\mathbf{r}, \mathbf{p})$, and the *total and reduced masses* (M, m) are

$$\mathbf{R} = \frac{m_1 \mathbf{r}_1 + m_2 \mathbf{r}_2}{m_1 + m_2}, \quad \mathbf{P} = \mathbf{p}_1 + \mathbf{p}_2,$$

$$\mathbf{r} = \mathbf{r}_2 - \mathbf{r}_1, \quad \mathbf{p} = \frac{m_1 \mathbf{p}_2 - m_2 \mathbf{p}_1}{m_1 + m_2},$$

$$M = m_1 + m_2, \quad m = \frac{m_1 m_2}{m_1 + m_2}. \tag{12.55}$$

In position representation, the quantum mechanical momentum operators are

$$\mathbf{P}\left[= (m_1 \mathbf{v}_1 + m_2 \mathbf{v}_2) \right] \Rightarrow \frac{\hbar}{i} \nabla_R, \quad \mathbf{p}\left[= m(\mathbf{v}_2 - \mathbf{v}_1) \right] \Rightarrow \frac{\hbar}{i} \nabla_r. \tag{12.56}$$

The Hamiltonian in Eq. (12.54) can be written in terms of these coordinates and momentum, and the Schrödinger equation is

$$\left[\frac{P^2}{2M} + \frac{p^2}{2m} + V(\mathbf{r}) \right] \Psi(\mathbf{R}, \mathbf{r}) = E_{12} \Psi(\mathbf{R}, \mathbf{r}). \tag{12.57}$$

This permits separation of variables, and the wave function can be taken in product form,

$$\Psi(\mathbf{R},\mathbf{r}) = F(\mathbf{R})\psi_{\mathbf{k}}^{+}(\mathbf{r}), \tag{12.58}$$

and the total energy is $E_{12} = E_{CM} + E$. The reason for the subscript $\mathbf{k}$ and superscript $+$ on the relative wave function $\psi_{\mathbf{k}}^{+}(\mathbf{r})$ will become evident shortly. It is useful to define scaled energies and potentials, which have units of inverse length squared,

$$\varepsilon_{CM} \equiv \frac{2M}{\hbar^2} E_{CM} \equiv K^2,$$

$$\varepsilon \equiv \frac{2m}{\hbar^2} E \equiv k^2,$$

$$v(\mathbf{r}) \equiv \frac{2m}{\hbar^2} V(\mathbf{r}). \tag{12.59}$$

The corresponding Schrödinger equations for the CM and relative degrees of freedom are

$$-\nabla_{\mathbf{R}}^{2} F_{\mathbf{K}}(\mathbf{R}) = K^2 F(\mathbf{R}), \quad [-\nabla_{\mathbf{r}}^{2} + v(\mathbf{r})]\psi_{\mathbf{k}}^{+}(\mathbf{r}) = k^2 \psi_{\mathbf{k}}^{+}(\mathbf{r}). \tag{12.60}$$

Clearly, the center of mass wave function is a plane wave, $F_{\mathbf{K}}(\mathbf{R}) = (2\pi)^{-3/2} e^{i\mathbf{K}\cdot\mathbf{R}}$, and the entire nontrivial physics of scattering is encoded in the second equation in Eq. (12.60). The problem of extracting physically measurable quantities from the Schrödinger equation for $\psi_{\mathbf{k}}(\mathbf{r})$ will be elaborated below.

Transforming to the Laboratory frame

Now, we change notation and use a prime to denote laboratory frame quantities and a subscript 0 to denote quantities before the collision takes place. Let us consider the elastic scattering of two particles, one of which is at rest in the laboratory frame. Figure 12.6 shows the particle velocities $\mathbf{v}_1$ and $\mathbf{v}_1'$ in the CM and laboratory frames, after the collision, for the case when particle 2 is at rest in the laboratory frame before the collision begins, i.e., $\mathbf{v}_{20}' = 0$. The CM frame moves with velocity $\mathbf{V} = \frac{m_1}{m_1+m_2}\mathbf{v}_{10}'$ relative to the laboratory frame, and the relative velocity of the two particles after the collision is $\mathbf{v}' = \mathbf{v}_2' - \mathbf{v}_1' = -\mathbf{v}_1' [= \mathbf{v} = \mathbf{v}_2 - \mathbf{v}_1]$. Simple geometrical manipulations (see Problem 12.6) lead to the transformation of the scattering angle in the center of mass frame to the laboratory frame,

$$\tan\theta_1' = \frac{m_2 \sin\theta}{m_1 + m_2 \cos\theta}, \quad \theta_2' = \frac{1}{2}(\pi - \theta). \tag{12.61}$$

For equal masses, $\theta_1' = \frac{1}{2}\theta$ and $\theta_2' = \frac{1}{2}(\pi - \theta)$. In the laboratory frame, the particle trajectories after the collision are also perpendicular to each other.

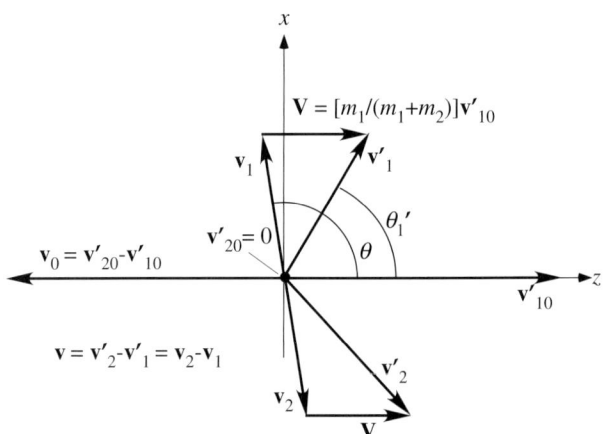

FIG 12.6 Relations between the particle velocities in the laboratory and CM frames, $(\mathbf{v}_1', \mathbf{v}_2')$ and $(\mathbf{v}_1, \mathbf{v}_2)$, where $\mathbf{V}$ is the CM velocity and $\mathbf{v}' = \mathbf{v}_2' - \mathbf{v}_1' = \mathbf{v} = \mathbf{v}_2 - \mathbf{v}_1$ is the relative velocity.

12.3 Stationary Scattering Theory

For elastic collisions, the velocities in the laboratory frame long after the collision has taken place are given by

$$\mathbf{v}'_1 = \frac{m_2}{m_1 + m_2}\mathbf{v} + \frac{m_1\mathbf{v}_1 + m_2\mathbf{v}_2}{m_1 + m_2}, \qquad (12.62a)$$

$$\mathbf{v}'_2 = -\frac{m_1}{m_1 + m_2}\mathbf{v} + \frac{m_1\mathbf{v}_1 + m_2\mathbf{v}_2}{m_1 + m_2}. \qquad (12.62b)$$

Multiplying these equations by m_1 and m_2, respectively, we find the momenta in the laboratory frame after the collision,

$$\mathbf{p}'_1 = m\mathbf{v} + \frac{m_1}{m_1 + m_2}(\mathbf{p}_1 + \mathbf{p}_2), \qquad (12.63a)$$

$$\mathbf{p}'_2 = -m\mathbf{v} + \frac{m_2}{m_1 + m_2}(\mathbf{p}_1 + \mathbf{p}_2), \qquad (12.63b)$$

Note that we can substitute $(\mathbf{p}'_1 + \mathbf{p}'_2) = \mathbf{p}'_{10}$ for $(\mathbf{p}_1 + \mathbf{p}_2)$ in the RHS of Eqs. (12.63). We can graphically illustrate these equations as follows. In Fig. 12.7, we draw a circle of radius $|\mathbf{p}|$ where the relative momentum $\mathbf{p}$ is along the line segment OC. The vectors $\mathbf{p}'_1$ and $\mathbf{p}'_2$ are the line segments AC and CB, respectively, and the CM momentum $\mathbf{P}$ is AB. Figure 12.7 can be easily used to show that $\theta'_2 = \frac{1}{2}(\pi - \theta)$ [see Problem 12.6(e)]. Moreover, using Fig. 12.6, it is easy to show that $\tan\theta'_1 = \frac{m_2 \sin\theta}{m_1 + m_2 \cos\theta}$ [see Problem 12.6(d)]. The sum of the angles θ'_1 and θ'_2 is the angle between the directions of motion of the particles after the collision in the laboratory frame. It is clear from Fig. 12.6 that $\theta'_1 + \theta'_2 > \pi/2$ for $m_1 < m_2$, $\theta'_1 + \theta'_2 < \pi/2$ for $m_1 > m_2$, and $\theta'_1 + \theta'_2 = \pi/2$ for $m_1 = m_2$. Note that the point C may be anywhere on the circle. No further information about the collision can be obtained from the laws of conservation of momentum and energy, i.e., the direction of the relative momentum vector $\mathbf{p}$ depends on the interaction of the particles and can lie anywhere on the circle.

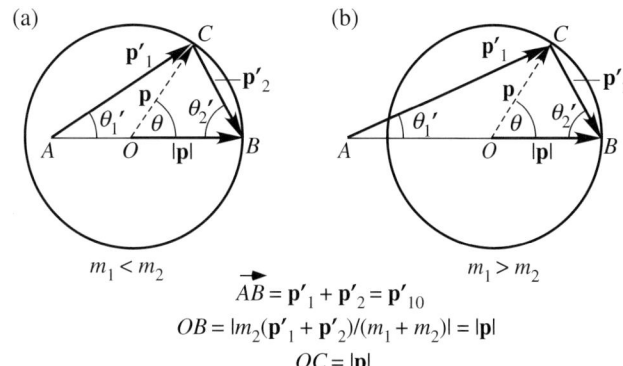

FIG 12.7 Relations between the particle momenta in the laboratory frame, $(\mathbf{p}'_1, \mathbf{p}'_2)$ and the relative and center of mass momenta $(\mathbf{p}, \mathbf{P})$ when the initial velocity of particle 2 in the laboratory frame is zero, $\mathbf{p}'_{20} = 0$. (a) $m_1 < m_2$ and (b) $m_1 > m_2$. For $m_1 = m_2$, point A lies on the circle of radius $\mathbf{p}$.

Problem 12.6

For two particles of mass m_1 and m_2, with particle 2 at rest in the laboratory frame and the initial velocity of particle 1 being $\mathbf{v}_{10}$,

(a) Show that $\mathbf{v}_1 = -\mathbf{V} + \mathbf{v}'_1$.
(b) Show that the CM velocity is given by $(m_1 + m_2)\mathbf{V} = m_1\mathbf{v}_{10}$.
(c) Using Fig. 12.6, show that $v_1\sin\theta = v'_1\sin\theta'_1$ and $v_1\cos\theta = -V + v'_1\cos\theta'_1$, where θ'_1 is the laboratory frame angle shown in Fig. 12.6.
(d) Using the results of (b) and (c), show that $\tan\theta'_1 = \frac{m_2\sin\theta}{m_1 + m_2\cos\theta}$.
(e) Using Fig. 12.7, which shows the momentum vectors $\mathbf{p}'_1, \mathbf{p}'_2, \mathbf{P} = \mathbf{p}'_1 + \mathbf{p}'_2$, and $\mathbf{p} = m_1\mathbf{p}'_2 - m_2\mathbf{p}'_1/(m_1 + m_2)$, and the angles θ'_1, θ'_2 and θ, prove that $\theta'_2 = \frac{1}{2}(\pi - \theta)$.
(f) Prove that $\frac{d\sigma}{d\Omega} = \frac{d\sigma}{d\Omega'}\frac{\sin\theta'_1}{\sin\theta}\left|\frac{d\theta'_1}{d\theta}\right|$.

> **Problem 12.7**
>
> Using Eqs. (12.62), determine the magnitudes of the velocities of the particles after the collision in the laboratory frame, $|\mathbf{v}'_1|$ and $|\mathbf{v}'_2|$, in terms of the particle masses, and the center of mass quantities $|\mathbf{v}|$ and θ. Hint: From Fig. 12.7, we see that $|\frac{m_2}{m_1+m_2}(\mathbf{p}'_1 + \mathbf{p}'_2)| = |m\mathbf{v}|$.
>
> **Answer:** $|\mathbf{v}'_1| = \frac{\sqrt{m_1^2+m_2^2+2m_1 m_2 \cos\theta}}{m_1+m_2} v$, $|\mathbf{v}'_2| = \frac{2m_1}{m_1+m_2} v \sin(\theta/2)$.

12.3.3 FROM WAVE FUNCTIONS TO CROSS-SECTIONS

Having demonstrated the equivalence of two-particle collision and a single-particle scattering from a potential, we now relate the solution of the Schrödinger equation for potential scattering [in its reduced form (12.60)],

$$h\psi(\mathbf{r}) \equiv [h_0 + v]\psi_{\mathbf{k}}^+(\mathbf{r}) \equiv [-\nabla_{\mathbf{r}}^2 + v(\mathbf{r})]\psi_{\mathbf{k}}^+(\mathbf{r}) = k^2 \psi_{\mathbf{k}}^+(\mathbf{r}), \quad (12.64)$$

to the scattering cross-section [Eq. (12.52)]. At large distance r from the origin, the wave function $\psi_{\mathbf{k}}^+(\mathbf{r})$ is written as a combination of a plane wave representing the incoming beam propagating in the $\mathbf{k} = k\hat{\mathbf{z}}$ direction and an outgoing spherical wave with amplitude $f(\Omega) = f(\theta, \phi)$, where (θ, ϕ) are the polar and azimuthal angles specifying the propagation direction of the outgoing particle. As already indicated in Eq. (12.14), the solution of Eq. (12.64) has the asymptotic form,

$$\psi_{\mathbf{k}}^+(\mathbf{r}) \xrightarrow[r\to\infty]{} (2\pi)^{-\frac{3}{2}} \left[e^{i\mathbf{k}\cdot\mathbf{r}} + f(\theta,\phi) \frac{e^{ikr}}{r} \right]. \quad (12.65)$$

The pre-factor $(2\pi)^{-3/2}$ can be replaced by any factor A as it does not affect the expression of the scattering cross-section when expressed as a ratio of outgoing flux and incoming current [see Eq. (12.52)]. The choice $A = (2\pi)^{-3/2}$ is adapted here because $(2\pi)^{-3/2} d\mathbf{k}$ is an appropriate volume element in momentum space. Note that $\psi_{\mathbf{k}}^+(\mathbf{r})$ is not normalized to unity, since a plane wave defined within an infinite volume cannot be normalized. This does not pose a problem since we are not interested in determining probabilities of finding particles at given locations but rather are interested in particle currents.

The subscript $\mathbf{k}$ indicates that the initial direction (before scattering) is along $\hat{\mathbf{k}}$ and that the scattering energy is $\varepsilon = k^2$. The superscript $+$ indicates an *outgoing spherical wave* e^{+ikr}/r. We may also consider solutions of the Schrödinger equation with *incoming spherical wave* asymptotic boundary conditions,

$$\psi_{\mathbf{k}}^-(\mathbf{r}) \xrightarrow[r\to\infty]{} (2\pi)^{-\frac{3}{2}} \left[e^{i\mathbf{k}\cdot\mathbf{r}} + f(\pi - \theta, \phi + \pi) \frac{e^{-ikr}}{r} \right]. \quad (12.66)$$

Although $\psi_{\mathbf{k}}^-(\mathbf{r})$ is a solution of the Schrödinger equation at $r \to \infty$, the incoming wave boundary conditions do not correspond to a physically realizable situation because they describe a wave going backward (in the sense of Fig. 12.4). This is indicated by the superscript $-$, which implies the minus sign in the exponent, and by the reversed direction of $\mathbf{k}'$ so $(\theta, \phi) \to (\pi - \theta, \phi + \pi)$. However, $\psi_{\mathbf{k}}^-(\mathbf{r})$ is theoretically important as it is employed within the Green's function formalism.

The central object that scattering theory aims to calculate is the scattering amplitude $f(\theta, \phi)$ in Eq. (12.65), which has the physical dimension of length. To demonstrate the relation of $f(\theta, \phi)$ to the differential cross-section, consider the local current density vector associated with the solution $\psi_{\mathbf{k}}^+(\mathbf{r})$ of Eq. (12.65),

$$\mathbf{J} = \text{Re}\left[\frac{\hbar}{im} \psi_{\mathbf{k}}^{+*}(\mathbf{r}) \nabla_{\mathbf{r}} \psi_{\mathbf{k}}^+(\mathbf{r}) \right]. \quad (12.67)$$

Inserting the asymptotic expression (12.65) for $\psi_{\mathbf{k}}^+(\mathbf{r})$ into the above expression gives the current at large distance long after the collision. It is evident from the resulting expression that $\mathbf{J}$ has contributions from the incoming plane wave, from the outgoing spherical wave, and from the interference term between the plane and the spherical waves. However, as stressed in Sec. 12.2.2 and illustrated in Fig. 12.5, the representation of the incoming wave as a plane wave is valid only for very small lateral width; the incident beam is highly collimated in the forward direction. Hence, at any angle $\theta > 0$, the contribution of the plane wave to the current tends to zero as $r \to \infty$ and one is then left with the two contributions:

$$\mathbf{J}^{(\text{in})} = \frac{\hbar k}{m}\hat{\mathbf{z}}, \tag{12.68a}$$

$$\mathbf{J}^{(\text{out})} = \frac{\hbar k}{m}\frac{|f(\theta,\phi)|^2}{r^2}\hat{\mathbf{r}}. \tag{12.68b}$$

The outgoing flux into the element of area $r^2 d\Omega$ is $J_r^{(\text{out})} r^2 d\Omega$. Therefore, according to Eq. (12.52), the differential cross-section in the center of mass frame is

$$\boxed{\frac{d\sigma}{d\Omega} = |f(\theta,\phi)|^2.} \tag{12.69}$$

This is a very useful result. Once the scattering wave function $\psi_{\mathbf{k}}^+(\mathbf{r})$ is known at large distance, the scattering amplitude $f(\theta,\phi)$ can be determined and the experimentally relevant quantity, $\frac{d\sigma}{d\Omega}$, can be evaluated. Note, however, that although the wave function is required in the asymptotic region, where $V(\mathbf{r}) = 0$, one typically obtains the scattering amplitude by solving the Schrödinger equation for the wave function (or sometimes $\psi^{-1}\frac{d\psi}{dr}$ or other related quantities) *everywhere* from very small r to very large r. Specific algorithms for calculating cross-sections by integrating the Schrödinger equation are presented below.

12.3.4 GREEN'S FUNCTION

One of the central issues in solving the Schrödinger equation for a scattering problem is how to incorporate the boundary conditions. One method for doing so is to employ the *Green's function* formalism, wherein the Schrödinger equation (12.64) and the boundary condition (12.65) are replaced by a single integral equation. A major advantage of this formalism is that it yields a direct relation between the potential $V(\mathbf{r})$ and the scattering amplitude $f(\theta,\phi)$. We now briefly introduce Green's functions for potential scattering and then use them to obtain an integral equation for the scattering wave function in configuration space. A more formal treatment is presented in Sec. 12.4.

In order to define the Green's function, consider the Hamiltonian operator $H = H_0 + V$ in which the first term H_0 is simple enough so that its eigenstates and eigenvalues are known, or at least can be calculated without much difficulty. In a standard potential scattering problem, H_0 is the kinetic energy operator, $H_0 = -\frac{\hbar^2}{2m}\nabla^2$. First, we introduce the *resolvent operators* of H and H_0, which are closely related to the corresponding Green's functions defined below. These are operators acting in the underlying Hilbert space and depend on a complex parameter z (having units of energy) and are defined as

$$G(z) \equiv (z - H)^{-1}, \tag{12.70}$$

$$G_0(z) \equiv (z - H_0)^{-1}. \tag{12.71}$$

Below we consider matrix elements of these operators between states defined in Hilbert space. Such matrix elements are functions of the complex variable z with certain analytic properties in the complex z plane. These can be used to define the analytic properties of the operators themselves. The analytic properties of $G(z)$ and $G_0(z)$ depend on the corresponding spectra of H and H_0. By definition, they are singular if z belongs to the spectrum of the Hamiltonian. For scattering problems with a potential V that falls off fast enough at large r, the spectrum of H consists of a continuous part extending along the positive real axis, $0 \le E < \infty$. It might also have a discrete spectrum on the negative part of the real axis (that is bounded from below). As for H_0, if it is taken to be the kinetic energy alone, its spectrum consists only of a continuous part extending along the positive part of the real axis, $0 \leq E < \infty$. Therefore, both $G(z)$ and $G_0(z)$, when considered as operator-valued functions of the complex parameter z, have a cut along the positive real axis.[3] The relevance of these operators to the actual scattering problem emerges when the complex variable z becomes real and positive, $z \to E > 0$, where E is the scattering energy. For example, let us rewrite the time-independent Schrödinger equation, $(H_0 + V)|\psi\rangle = E|\psi\rangle$, in a trivially different form,

$$(E - H_0)|\psi\rangle = V|\psi\rangle. \tag{12.72}$$

Now formally invert the operator $(E - H_0)$ and apply this inverse to both sides of Eq. (12.72). Recalling our statement on the analytic properties of the resolvents, the inverse of $(E - H_0)$ is ill defined when E belongs to the spectrum of H_0 (which usually consists of the positive real axis). To avoid this singularity, it is customary to add a small imaginary part to the energy $E \to E \pm i\eta$, $\eta > 0$, and let $\eta \to 0$ at the end of the calculations. As will be evident below, the sign in front of $i\eta$ is significant. This procedure defines two free Green's functions:

$$G_0^{\pm}(E) \equiv (E \pm i\eta - H_0)^{-1}. \tag{12.73}$$

The term "free" implies that they are associated with H_0, which describes free particles.

Applying G_0^+ to both sides of Eq. (12.72), we obtain $|\psi\rangle = G_0^+(E)V|\psi\rangle$. By definition, $|\psi\rangle$ satisfies the Schrödinger equation (12.72). However, this is not the solution that we are looking for, because it does not satisfy the asymptotic boundary conditions (12.65) as there is no incoming wave. This drawback is remedied by adding a plane wave solution $|\mathbf{k}\rangle$ of the free Schrödinger equation, $H_0|\mathbf{k}\rangle = E|\mathbf{k}\rangle = \frac{\hbar^2 k^2}{2m}|\mathbf{k}\rangle$ to the term $G_0^+(E)V|\psi\rangle$ on the RHS of the equation. This procedure does not affect the validity of the Schrödinger equation because $(E + i\eta - H_0)|\mathbf{k}\rangle = 0$. The state $|\psi_\mathbf{k}^+\rangle = |\mathbf{k}\rangle + G_0^+(E)V||\psi_\mathbf{k}^+\rangle$ is therefore a solution of the Schrödinger equation (12.72), which includes an incoming plane wave. It will be shown below that G_0^+ is associated with the outgoing spherical wave boundary conditions of $\psi_\mathbf{k}^+(\mathbf{r})$ (12.65), while G_0^- is associated with the incoming spherical wave boundary conditions of $\psi_\mathbf{k}^-(\mathbf{r})$ (12.66). Hence, we have used the notation $|\psi_\mathbf{k}^+\rangle$ for the former.

The result of these manipulations is an integral equation for $|\psi_\mathbf{k}\rangle$ with an inhomogeneous term given by $|\mathbf{k}\rangle$. Had we used G_0^- instead of G_0^+, we would have arrived at an equation for $|\psi_\mathbf{k}^-\rangle$, i.e., the wave function with incoming spherical wave boundary conditions [see Eq. (12.66)]. Hence, the equation for $|\psi_\mathbf{k}^{\pm}\rangle$ is written as

$$\boxed{|\psi_\mathbf{k}^{\pm}\rangle = |\mathbf{k}\rangle + G_0^{\pm}(E)V|\psi_\mathbf{k}^{\pm}\rangle.} \tag{12.74}$$

Equation (12.74) is the abstract form of the *Lippmann–Schwinger equation* in Hilbert space with outgoing and incoming wave boundary conditions, respectively. In order to obtain the Lippmann–Schwinger equation in position space, we take the inner product of Eq. (12.74) with $|\mathbf{r}\rangle$. On the LHS, we obtain the wave function $\psi_\mathbf{k}^{\pm}(\mathbf{r})$, and on the RHS, we can insert a complete set of states $\int d\mathbf{r}' |\mathbf{r}'\rangle \langle \mathbf{r}'|$ in between $G_0^{\pm}$ and V and a complete set $\int d\mathbf{r}'' |\mathbf{r}''\rangle \langle \mathbf{r}''|$ in between V and $|\psi_\mathbf{k}^{\pm}\rangle$, using the fact that V is local in position space, we find

$$\psi_\mathbf{k}^{\pm}(\mathbf{r}) = (2\pi)^{-\frac{3}{2}} e^{i\mathbf{k}\cdot\mathbf{r}} + \int d\mathbf{r}' \, G_0^{\pm}(\mathbf{r}, \mathbf{r}'; E) V(\mathbf{r}') \psi_\mathbf{k}^{\pm}(\mathbf{r}'). \tag{12.75}$$

The RHS of the integral equation (12.75) for $\psi_\mathbf{k}^{\pm}(\mathbf{r})$ contains $G_0^{\pm}(\mathbf{r}, \mathbf{r}'; E) = \langle \mathbf{r}|G_0^{\pm}(E)|\mathbf{r}'\rangle$, the configuration space representation of the free Green's function. The advantage of the integral equation formulation is that the boundary conditions are automatically incorporated into the equations. The integral on the RHS converges if the potential decays as $r \to \infty$ faster than r^{-1}. Scattering by a Coulomb potential, $V(r) = C/r$, requires separate discussion and will be presented later. Numerically, integral equations can be treated by matrix inversion methods as discussed below.

---

[3] A complex function $f(z)$ is said to have a cut along the positive real axis if for $0 < E < \infty$ and $\eta \to 0^+$, $f(E + i\eta) - f(E - i\eta) \neq 0$.

12.3 Stationary Scattering Theory

It is convenient to use the scaled form of the Schrödinger equation, Eq. (12.64), and define the scaled Green's function as,

$$g^{\pm}(k^2) = (k^2 \pm i\eta - h)^{-1}, \tag{12.76}$$

$$g_0^{\pm}(k^2) = (k^2 \pm i\eta - h_0)^{-1}, \tag{12.77}$$

i.e., $g^{\pm}(k^2) \equiv (2m/\hbar^2)G_0^{\pm}(E)$ (recall that $v(\mathbf{r}') \equiv (2m/\hbar^2)V(\mathbf{r}')$, $\varepsilon \equiv (2m/\hbar^2)E = k^2$, $h = (2m/\hbar^2)H$, etc.). The relation $(k^2 \pm i\eta - h_0)g_0^{\pm}(k^2) = 1$, which follows directly from the definition of $g_0^{\pm}(k^2)$, is represented in configuration space as,

$$[\nabla_r^2 + k^2 \pm i\eta]g_0^{\pm}(\mathbf{r}, \mathbf{r}'; k^2) = \delta(\mathbf{r} - \mathbf{r}'). \tag{12.78}$$

Note that $g_0^{\pm}(\mathbf{r}, \mathbf{r}'; k^2) = g_0^{\pm}(\mathbf{r} - \mathbf{r}'; k^2)$. Equation (12.75) then becomes

$$\psi_{\mathbf{k}}^{\pm}(\mathbf{r}) = (2\pi)^{-\frac{3}{2}} e^{i\mathbf{k}\cdot\mathbf{r}} + \int d\mathbf{r}'\, g_0^{\pm}(\mathbf{r} - \mathbf{r}'; k^2) v(\mathbf{r}') \psi_{\mathbf{k}}^{\pm}(\mathbf{r}'). \tag{12.79}$$

Problem 12.8

Consider the one-dimensional equation for the free Green's function, $[\frac{d^2}{dx^2} + k^2 \pm i\eta]g_0^{\pm}(x, x'; k^2) = \delta(x - x')$.

(a) Show that the solutions are $g_0^{\pm}(x, x') = -\frac{1}{2ik} e^{\pm ik|x-x'|}$.

(b) For $g_0^+(x, x')$, find the coefficient of e^{ikx} as $x \to \infty$ to first order in x'/x.

Answer: $e^{ik|x-x'|} \approx e^{ikx\left(1 - \frac{x'x}{x^2}\right)} = e^{ik(x-x')}$.

To relate $g_0^{\pm}$ with the corresponding spherical wave boundary conditions, consider a solution of Eq. (12.78) with either outgoing or incoming spherical wave boundary conditions as $r \to \infty$,

$$g_0^{\pm}(\mathbf{r} - \mathbf{r}'; k^2) = -\frac{e^{\pm ik|\mathbf{r} - \mathbf{r}'|}}{4\pi |\mathbf{r} - \mathbf{r}'|}. \tag{12.80}$$

To demonstrate the spherical wave boundary conditions, use the limiting form,

$$|\mathbf{r} - \mathbf{r}'| = r\sqrt{1 - \frac{2\mathbf{r}\cdot\mathbf{r}'}{r^2} + \frac{r'^2}{r^2}} \xrightarrow{r \to \infty} r\left[1 - \frac{\mathbf{r}\cdot\mathbf{r}'}{r^2} + O\left(\frac{1}{r^2}\right)\right]. \tag{12.81}$$

This immediately implies,

$$g_0^{\pm}(\mathbf{r} - \mathbf{r}'; k^2) \xrightarrow{r \to \infty} -\frac{e^{\pm ikr}}{4\pi r} e^{-ik\hat{\mathbf{r}}\cdot\mathbf{r}'} = -\frac{e^{\pm ikr}}{4\pi r} e^{-i\mathbf{k}'\cdot\mathbf{r}'}. \tag{12.82}$$

The last equality employs the fact that as $r \to \infty$, the direction of $\mathbf{r}$ approaches the final momentum $\mathbf{k}'$, so that

$$\lim_{r \to \infty} k\hat{\mathbf{r}} = \mathbf{k}'. \tag{12.83}$$

It is readily verified that $\psi_{\mathbf{k}}^+(\mathbf{r})$ satisfies the Schrödinger equation (12.64):

$$[\nabla_r^2 + k^2 + i\eta]\psi_{\mathbf{k}}^+(\mathbf{r}) = [\nabla_r^2 + k^2 + i\eta](2\pi)^{-\frac{3}{2}} e^{i\mathbf{k}\cdot\mathbf{r}} + \int d\mathbf{r}'\, \delta(\mathbf{r} - \mathbf{r}') v(\mathbf{r}') \psi_{\mathbf{k}}^+(\mathbf{r}') = v(\mathbf{r})\psi_{\mathbf{k}}^+(\mathbf{r}), \tag{12.84}$$

as the first term on the RHS vanishes. To check the boundary conditions of $\psi_{\mathbf{k}}(\mathbf{r})$ as $r \to \infty$, we can employ the asymptotic expansion of $g_0(\mathbf{r} - \mathbf{r}')$, Eq. (12.82), and extract the asymptotic form of the function $\psi_{\mathbf{k}}^+(\mathbf{r})$ in Eq. (12.79) as,

$$\psi_{\mathbf{k}}^+(\mathbf{r}) \xrightarrow[r\to\infty]{} (2\pi)^{-\frac{3}{2}} e^{i\mathbf{k}\cdot\mathbf{r}} - \frac{e^{ikr}}{4\pi r} \int d\mathbf{r}' \, e^{-i\mathbf{k}'\cdot\mathbf{r}'} v(\mathbf{r}') \psi_{\mathbf{k}}^+(\mathbf{r}') \tag{12.85}$$

$$\equiv (2\pi)^{-\frac{3}{2}} \left[e^{i\mathbf{k}\cdot\mathbf{r}} + f(\theta, \phi) \frac{e^{ikr}}{r} \right]. \tag{12.86}$$

where we have compared with Eq. (12.65) to show that the coefficient of the outgoing spherical wave e^{ikr}/r in the second term on the RHS of Eq. (12.85) is the scattering amplitude $f(\theta, \phi)$. Note that the application of ∇_r^2 to the RHS of Eq. (12.86) does not completely kill the second term (the outgoing spherical wave), but the nonzero part vanishes as $r \to \infty$.

Depending on the context, we shall use several notations for the arguments of the scattering amplitude, such as $f(\Omega)$, $f(\theta, \phi)$, $f(\mathbf{k}', \mathbf{k})$; all of these are equivalent. Sticking to the convention that the initial-state quantum numbers appear on the right, the order of the wavevectors is $\mathbf{k}', \mathbf{k}$. The notation $f(\mathbf{k}', \mathbf{k})$ is sometimes used when the variables are integrated over, and the wavenumbers might be *off the energy shell*, in the sense of Eq. (12.51). However, as far as the relation of $f(\mathbf{k}', \mathbf{k})$ to the scattering cross-section is concerned, for a given energy, the wavevectors $\mathbf{k}$ and $\mathbf{k}'$ must be constrained by the on-energy-shell condition (12.51). We, therefore, arrive at the key relation between the scattering amplitude $f(\Omega)$ and the scattering potential $V(\mathbf{r})$,

$$\boxed{\begin{aligned} f(\Omega) \equiv f(\mathbf{k}', \mathbf{k}) &= -\frac{(2\pi)^{\frac{3}{2}}}{4\pi} \int d\mathbf{r}' \, e^{-i\mathbf{k}'\cdot\mathbf{r}'} v(\mathbf{r}') \psi_{\mathbf{k}}^+(\mathbf{r}') \\ &= -\frac{(2\pi)^{\frac{3}{2}} m}{2\pi \hbar^2} \int d\mathbf{r}' \, e^{-i\mathbf{k}'\cdot\mathbf{r}'} V(\mathbf{r}') \psi_{\mathbf{k}}^+(\mathbf{r}') = -\frac{4\pi^2 m}{\hbar^2} \langle \mathbf{k}'|V|\psi_{\mathbf{k}}^+\rangle, \end{aligned}} \tag{12.87}$$

where the last equality is due to the plane-wave normalization $e^{-i\mathbf{k}'\cdot\mathbf{r}'} = (2\pi)^{\frac{3}{2}} \langle \mathbf{k}'|\mathbf{r}'\rangle$. Solving the integral equation (12.79) and inserting the solution in Eq. (12.87) constitutes a complete solution of the scattering problem. We shall see below that Eq. (12.87) can serve as a good starting point for numerous approximations. In the analysis leading to Eqs. (12.85) and (12.87), the simple expression (12.80) for the free Green's function $g_0^\pm(\mathbf{r} - \mathbf{r}')$ has been used. Clearly, if the full Green's function $g^\pm(\mathbf{r}, \mathbf{r}')$ can be computed, the scattering problem is essentially solved. The corresponding differential equation for $g^\pm(\mathbf{r}, \mathbf{r}')$ (which is in general not a function of only $\mathbf{r} - \mathbf{r}'$) is

$$[\nabla_r^2 + k^2 \pm i\eta - v(\mathbf{r})] g^\pm(\mathbf{r}, \mathbf{r}'; k^2) = \delta(\mathbf{r} - \mathbf{r}'). \tag{12.88}$$

Unfortunately, evaluation of the full Green's function in Eq. (12.88) is in general not possible. However, Eq. (12.88) can serve as a starting point for a number of approximation schemes. These will be discussed in Sec. 12.7. Examples are the *Eikonal approximation*, applicable for high energy and smooth potentials $v(\mathbf{r})$, and the *Born approximation*.

Green's Function and the Density of States

An important quantity in quantum scattering theory (and elsewhere in quantum mechanics) is the density of states $\rho(E)$. It is defined mainly for energies E belonging to the continuum part of the spectrum of H. Thus, $\rho(E)dE$ is defined as the number of states between E and $E+dE$. If there is a function of energy $A(E)$, which needs to be integrated over the energy spectrum, the volume element for this integration is naturally $\rho(E)dE$. This enables the derivation of an important relation between the Green's function and the density of states. Consider the full Green's function $G^+(E) \equiv (E+i\eta - H)^{-1}$. In its application to scattering theory, it is often represented in the configuration representation as the function $G^+(\mathbf{r}, \mathbf{r}'; E) = \langle \mathbf{r}|G^+(E)|\mathbf{r}'\rangle$. Now, let us assume that H has only a continuum spectrum and write G^+ in the energy representation. That

12.4 Aspects of Formal Scattering Theory

is, we consider a complete orthogonal set $\{|\lambda\rangle\}$ of eigenstates of H, such that $H|\lambda\rangle = \lambda|\lambda\rangle$. In this representation, the diagonal matrix element is

$$G^+(\lambda, \lambda; E) \equiv \langle \lambda | (E + i\eta - H)^{-1} | \lambda \rangle = \frac{1}{E + i\eta - \lambda}. \tag{12.89}$$

Multiplying both sides by $\rho(\lambda)d\lambda$ and integrating over the entire spectrum, the LHS becomes $\text{Tr}[G^+(E)]$, i.e.,

$$\text{Tr}[G^+(E)] = \int \frac{d\lambda \rho(\lambda)}{E + i\eta - \lambda}. \tag{12.90}$$

The imaginary part of the RHS of Eq. (12.90) is easily seen to be,

$$\text{Im Tr}[G^+(E)] = -\int d\lambda \rho(\lambda) \frac{\eta}{(E - \lambda)^2 + \eta^2}. \tag{12.91}$$

As $\eta \to 0$, we may use the following representation of the Dirac δ function,

$$\delta(x) = \lim_{\eta \to 0} \frac{1}{\pi} \frac{\eta}{x^2 + \eta^2}. \tag{12.92}$$

Thus, we arrive at the important relation,

$$\rho(E) = -\frac{1}{\pi} \text{Im Tr}[G^+(E)]. \tag{12.93}$$

Since the trace is representation independent, any representation can be used to evaluate it. This relation between density of states and the imaginary part of the trace of the Green's function can be used for any quantum mechanical system. It plays an important role in the study of the quantum many-body problem, to be discussed in Chapter 14, where the quantity $\text{Im}[\text{Tr } G]$ is referred to as the *spectral function*.

Bound States

Now, consider Eq. (12.89) in the situation where the spectrum of H also includes bound states consisting of a discrete set of points $\{E_n\}$ below the threshold E_0 of the continuous spectrum. Near an isolated point $E = E_n$, there is no need to add a small imaginary part, and

$$G(\lambda, \lambda, E_n) = \frac{1}{E_n - \lambda}. \tag{12.94}$$

Therefore, if the eigenvalues E_n are not degenerate, the bound state energies are simple poles of the Green's function.

12.4 ASPECTS OF FORMAL SCATTERING THEORY

The formulation of scattering theory within an abstract Hilbert space is referred to as *Formal Scattering Theory*. The motivation for pursuing this topic is its flexibility and its potential generalization. It is possible to study numerous different problems within the same formalism, e.g., the resolvent operators in Eqs. (12.70) and (12.71) are defined in the same way for many-particle systems, where H and H_0 are many-body Hamiltonians. This topic requires a certain degree of mathematical rigor but we will try to avoid it here as much as possible.

12.4.1 THE TRANSITION OPERATOR (THE T MATRIX)

State vectors and operators in the abstract Hilbert space spanned by the eigenstates of a given Hamiltonian H can be represented by vectors and matrices, respectively. The following state vectors and operators, and their representations, are of special importance in scattering theory:

$$|\mathbf{k}\rangle \rightarrow \langle\mathbf{r}|\mathbf{k}\rangle = (2\pi)^{-\frac{3}{2}} e^{i\mathbf{k}\cdot\mathbf{r}}, \tag{12.95a}$$

$$|\psi_{\mathbf{k}}^{+}\rangle \rightarrow \langle\mathbf{r}|\psi_{\mathbf{k}}^{+}\rangle = \psi_{\mathbf{k}}^{+}(\mathbf{r}), \tag{12.95b}$$

$$H_0 \rightarrow \langle\mathbf{r}|H_0|\mathbf{r}'\rangle = -\frac{\hbar^2}{2m}\delta(\mathbf{r}-\mathbf{r}')\nabla_{\mathbf{r}}^2, \tag{12.95c}$$

$$V \rightarrow \langle\mathbf{r}|V|\mathbf{r}'\rangle = \delta(\mathbf{r}-\mathbf{r}')V(\mathbf{r}), \tag{12.95d}$$

$$G_0 \rightarrow \langle\mathbf{r}|G_0|\mathbf{r}'\rangle = G_0(\mathbf{r}-\mathbf{r}') = -\frac{\hbar^2}{2m}\frac{e^{\pm ik|\mathbf{r}-\mathbf{r}'|}}{4\pi|\mathbf{r}-\mathbf{r}'|}, \tag{12.95e}$$

$$G_0 \rightarrow \langle\mathbf{k}'|G_0|\mathbf{k}\rangle = \delta(\mathbf{k}-\mathbf{k}')\frac{1}{E-\frac{\hbar^2 k^2}{2m}}. \tag{12.95f}$$

Plane wave states are normalized such that the completeness and orthogonality relations in the subspace $\mathcal{D}^+$ spanned by the positive energy eigenstates of the Hamiltonian are

$$\int d\mathbf{k}\, |\mathbf{k}\rangle\langle\mathbf{k}| = 1, \tag{12.96a}$$

$$\langle\mathbf{k}'|\mathbf{k}\rangle = \delta(\mathbf{k}-\mathbf{k}'). \tag{12.96b}$$

There are several starting points for developing formal scattering theory. We choose an approach related to the time-dependent scattering theory outlined in Sec. 12.2.1, as applied to the potential scattering Hamiltonian

$$H = \frac{p^2}{2m} + V(\mathbf{r}) \equiv H_0 + V. \tag{12.97}$$

The potential $V(\mathbf{r})$ is required to satisfy the condition,

$$r^2|V(\mathbf{r})| \xrightarrow[r\to\infty]{} 0. \tag{12.98}$$

Recall Eq. (12.33), $\Omega_{\pm}|\Psi^{\text{in,out}}(t)\rangle = |\Psi(t)\rangle$, with the Möller operators $\Omega_{\pm}$ defined in Eq. (12.34). For convenience, we take $t=0$ and write

$$|\Psi(0)\rangle = \Omega_{\pm}|\Psi^{\text{in,out}}(0)\rangle. \tag{12.99}$$

Now, the representation (12.36) for the Möller operators and the completeness relation of the momentum states $|\mathbf{k}\rangle$ on $\mathcal{D}^+$ with $H_0|\mathbf{k}\rangle = E_{\mathbf{k}}|\mathbf{k}\rangle = \frac{\hbar^2 k^2}{2m}|\mathbf{k}\rangle$ are used. First, we substitute Eq. (12.96a) between $\Omega_{\pm}$ and $|\Psi^{\text{in,out}}(0)\rangle$ in Eq. (12.99). Next, the time integral in Eq. (12.36) performed with H_0 is replaced by $E_{\mathbf{k}}$, to obtain the Möller operators in the energy representation,

$$\Omega_{\pm}(E_{\mathbf{k}}) \equiv \mp i\eta(E_k \pm i\eta - H)^{-1}, \tag{12.100}$$

Hence,

$$|\Psi(0)\rangle = \int d\mathbf{k}\, \Omega_{\pm}(E_{\mathbf{k}})|\mathbf{k}\rangle\langle\mathbf{k}|\Psi^{\text{in,out}}(0)\rangle, \tag{12.101}$$

12.4 Aspects of Formal Scattering Theory

where the limit $\eta \to 0^+$ should be taken. We keep the same notation for $\Omega_\pm(E_\mathbf{k})$ as for the abstract Möller operators. There is a direct relation between $\Omega_\pm(E)$ and the Green function $G^\pm(E) = (E \pm i\eta - H)^{-1}$. Based on the identity

$$(A - B)^{-1} = A^{-1} + A^{-1}B(A - B)^{-1}, \qquad (12.102)$$

which is also valid for operators that do not commute with each other. Using Eq. (12.100) for $\Omega_\pm$ and the identity (12.102) with $A = \pm i\eta$ and $B = (H - E_\mathbf{k})$, we get

$$\Omega_\pm(E_\mathbf{k}) \equiv 1 + G^\pm(E_\mathbf{k})(H - E_\mathbf{k}) = 1 + G^\pm(E_\mathbf{k})V. \qquad (12.103)$$

Inserting Eq. (12.103) into Eq. (12.101) we find

$$|\Psi(0)\rangle = \int d\mathbf{k}\, |\psi_\mathbf{k}^\pm\rangle \langle \mathbf{k}|\Psi^{\text{in,out}}(0)\rangle, \qquad (12.104)$$

where $|\psi_\mathbf{k}^\pm\rangle$ is an eigenstate of the Hamiltonian (12.97),

$$|\psi_\mathbf{k}^\pm\rangle = \Omega(E_k \pm i\eta)|\mathbf{k}\rangle \equiv |\mathbf{k}\rangle + (E_k \pm i\eta - H)^{-1}V|\mathbf{k}\rangle. \qquad (12.105)$$

Finally, again using the identity (12.102) for $G^\pm(E) = (E \pm i\eta - H_0 - V)^{-1}$, we can write the following integral equation for $|\psi_\mathbf{k}^\pm\rangle$:

$$|\psi_\mathbf{k}^\pm\rangle = |\mathbf{k}\rangle + (E_k \pm i\eta - H_0)^{-1}V|\psi_\mathbf{k}^\pm\rangle. \qquad (12.106)$$

This is just the Lippmann–Schwinger equation (12.74). Thus, we have established a direct relation between the abstract time-dependent formulation of scattering theory and the stationary formulation, which is more practical for calculations of experimental observables.

We have already stressed that, unlike the scattering state $|\psi_\mathbf{k}^+\rangle$, which has the transparent physical meaning of a plane wave and an outgoing spherical wave, the state $|\psi_\mathbf{k}^-\rangle$ represents a plane wave and an *incoming* spherical wave. It does not represent an experimentally realizable situation, yet the states $\{|\psi_\mathbf{k}^-\rangle\}$ form a complete set in the subspace of positive energy solutions of the Schrödinger equation, i.e., the subspace $\mathcal{D}^+$, and the linear operator transforming this set onto the set of states $\{|\psi_\mathbf{k}^+\rangle\}$ (i.e., the S matrix) is of utmost importance.

As has already been mentioned, it is useful to regard the energy variable as a complex number z and carry out all the operator manipulations avoiding at this stage the subtle question of how to approach the real axis. At the end, the complex variable z approaches the real (physical) energy axis either from above ($z = E + i\eta$) or from below ($z = E - i\eta$). For any operator $\mathcal{O}(z)$, we define

$$\mathcal{O}^\pm(E) \equiv \mathcal{O}(z \to E \pm i0), \qquad (12.107)$$

The most important operators used in formal scattering theory are

$$G_0(z) = (z - H_0)^{-1}, \qquad (12.108\text{a})$$
$$G(z) = (z - H)^{-1}, \qquad (12.108\text{b})$$
$$T(z) = V + VG(z)V = V + VG_0(z)T(z), \qquad (12.108\text{c})$$
$$\Omega(z) = 1 + G(z)V. \qquad (12.108\text{d})$$

These definitions imply some useful integral relations between the operators, which are called *operator Lippmann–Schwinger equations*,

$$G(z) = G_0(z) + G_0(z)VG(z) = G_0(z) + G_0(z)T(z)G_0(z), \qquad (12.109)$$
$$T(z) = V + VG_0(z)T(z), \qquad (12.110)$$
$$\Omega(z) = 1 + G_0(z)T(z). \qquad (12.111)$$

We have already defined the physical Green's function as a representation of the pertinent resolvent operators as $z \to E \pm i\eta$. Analogously, we define the T-matrix as the momentum space representation of $T(z \to E + i\eta)$:

$$T(\mathbf{k}', \mathbf{k}; E) \equiv \langle \mathbf{k}' | T(E + i\eta) | \mathbf{k} \rangle. \tag{12.112}$$

The connection between $T(\mathbf{k}', \mathbf{k}; E)$ and $S(\mathbf{k}', \mathbf{k}) \equiv \langle \mathbf{k}' | S | \mathbf{k} \rangle$ will be made in Sec. 12.4.2, where it will also be shown that the *on-shell* T-matrix, $T(\mathbf{k}', \mathbf{k}; E)$, with $k^2 = k'^2 = \varepsilon$, is directly related to the scattering amplitude $f(\theta, \phi)$. The Lippmann–Schwinger equations for the wave functions, $|\psi_\mathbf{k}^\pm\rangle$, based on the definitions (12.105) and Eqs. (12.108b)–(12.108d) and (12.109)–(12.111) are (see also Sec. 12.3.4)

$$|\psi_\mathbf{k}^\pm\rangle = |\mathbf{k}\rangle + G_0^\pm(E) V |\psi_\mathbf{k}^\pm\rangle \tag{12.113a}$$

$$= |\mathbf{k}\rangle + G^\pm(E) V |\mathbf{k}\rangle \tag{12.113b}$$

$$= |\mathbf{k}\rangle + G_0^\pm(E) T |\mathbf{k}\rangle. \tag{12.113c}$$

Equation (12.113a) is identical to expression (12.74). The asymptotic states $|\psi_\mathbf{k}^\pm\rangle$ are orthogonal,

$$\langle \psi_{\mathbf{k}'}^+ | \psi_\mathbf{k}^+ \rangle = \langle \mathbf{k}' | [1 + V G^-(E_{\mathbf{k}'})] | \psi_\mathbf{k}^+ \rangle$$

$$= \left\langle \mathbf{k}' \left| 1 + V \frac{1}{E_{\mathbf{k}'} - E_\mathbf{k} - i\epsilon} \right| \psi_\mathbf{k}^+ \right\rangle$$

$$= \left\langle \mathbf{k}' \left| 1 - \frac{1}{E_\mathbf{k} - H_0 + i\epsilon} V \right| \psi_\mathbf{k}^+ \right\rangle = \langle \mathbf{k}' | \mathbf{k} \rangle = \delta(\mathbf{k}' - \mathbf{k}). \tag{12.114}$$

The first equality is based on $\langle \psi_{\mathbf{k}'}^+ | = \langle \mathbf{k}' | [1 + V G^-(E_{\mathbf{k}'})]$. The second equality results from $(E_{\mathbf{k}'} - i\eta - H)^{-1} |\psi_\mathbf{k}^+\rangle = (E_{\mathbf{k}'} - i\eta - E_\mathbf{k})^{-1} |\psi_\mathbf{k}^+\rangle$, because $|\psi_\mathbf{k}^+\rangle$ is an eigenstate of H with an eigenvalue $E_\mathbf{k}$. The third equality results from the replacement of $E_\mathbf{k}$ by H_0 in the denominator, which is allowed when the operator acts leftward on the bra $\langle \mathbf{k} |$. Finally, the last equality is based on expressing $|\mathbf{k}\rangle$ in terms of $|\psi_\mathbf{k}^+\rangle$ using Eq. (12.114).

Each set of the scattering states $|\psi_\mathbf{k}^\pm\rangle$ is complete in the sense that it spans the subspace pertaining to the continuum spectrum of the Hamiltonian (12.97). Although most of our discussion focuses on $|\psi_\mathbf{k}^+\rangle$, we will encounter below the need to include $|\psi_\mathbf{k}^-\rangle$ in our calculations. The existence of two sets of eigenfunctions of H, which span the same subspace $\mathcal{D}^+$ is not surprising because each set is determined by different boundary conditions. Since boundary conditions are an integral part of the definition of H as a hermitian operator, the two sets correspond to different hermitian operators.

Note from Eq. (12.104) that $\psi_\mathbf{k}^\pm$ is the coefficient of $\langle \mathbf{k} | \Psi^{in, out}\rangle$ in the expression for $|\Psi(0)\rangle$. Hence, in the same way that in the time-dependent formulation the S matrix is an operator connecting $|\Psi^{out}\rangle$ and $|\Psi^{in}\rangle$ through Eq. (12.37), we expect to have an analogous operator in the time-independent formalism connecting $|\psi_\mathbf{k}^-\rangle$ with $|\psi_\mathbf{k}^+\rangle$ which is the corresponding S matrix. This will be demonstrated shortly.

An important expression involving the T matrix, the scattering potential, and the wave functions relate these somewhat abstract definitions to concrete physical observables. It is obtained by applying $\langle \mathbf{k}' |$ on Eq. (12.113b),

$$\boxed{T(\mathbf{k}', \mathbf{k}; E) = \langle \mathbf{k}' | V | \psi_\mathbf{k}^+ \rangle = \langle \mathbf{k}' | V [1 + G^+(E) V] | \mathbf{k} \rangle = \langle \mathbf{k}' | T(E + i\eta) | \mathbf{k} \rangle.} \tag{12.115}$$

Recalling the expression (12.87) for the scattering amplitude, using Eq. (12.115) and the plane wave normalization (12.95a), we can rewrite the scattering amplitude $f(\mathbf{k}', \mathbf{k})$ in Eq. (12.87) in terms of the T matrix element $\langle \mathbf{k}' | T | \mathbf{k} \rangle \equiv T(\mathbf{k}', \mathbf{k})$:

$$\boxed{f(\mathbf{k}', \mathbf{k}) = -\frac{4\pi^2 m}{\hbar^2} \langle \mathbf{k}' | T(E + i\eta) | \mathbf{k} \rangle = -\frac{4\pi^2 m}{\hbar^2} \langle \mathbf{k}' | V | \psi_\mathbf{k}^+ \rangle, \quad (k^2 = k'^2 = \varepsilon).} \tag{12.116}$$

To recover Eq. (12.87) from Eq. (12.116), we can insert $\int \mathbf{r} |\mathbf{r}\rangle\langle\mathbf{r}|$ between $\langle \mathbf{k} |$ and V and between V and $|\psi_\mathbf{k}^+\rangle$ and evaluate the integrals.

12.4 Aspects of Formal Scattering Theory

The dimension of the scattering amplitude is $[f] = L$ (length); hence, from Eq. (12.116), it is easily seen that $[\langle \mathbf{k}'|T|\mathbf{k}\rangle] = EL^3$, since $\langle \mathbf{k}'|T|\mathbf{k}\rangle = (2\pi)^{-3} \int d\mathbf{r}\, e^{-i(\mathbf{k}'-\mathbf{k})\cdot \mathbf{r}} T(\mathbf{r})$ and $T(\mathbf{r})$ has the dimension of energy, as does potential $V(\mathbf{r})$. Thus, the scattering amplitude can be expressed as a constant times the momentum space representation of the transition operator on the energy shell. Note that for a local potential, such as in Eq. (12.95d), one has

$$\langle \mathbf{k}'|V|\mathbf{k}\rangle = V(\mathbf{k}' - \mathbf{k}) = V(\mathbf{q}), \tag{12.117}$$

where $\mathbf{q} \equiv \mathbf{k}' - \mathbf{k}$ is the *momentum transfer* (actually wavenumber transfer).

Equation (12.116) is very useful and it shows that one is able to compute the scattering amplitude, and hence the differential cross-section, directly by solving the Lippmann–Schwinger integral equation (12.108c) in its momentum space representation,

$$T(\mathbf{k}', \mathbf{k}) = V(\mathbf{k}', \mathbf{k}) + \frac{2m}{\hbar^2} \int d\mathbf{k}''\, V(\mathbf{k}', \mathbf{k}'')[k^2 + i\eta - k''^2]^{-1} T(\mathbf{k}'', \mathbf{k}). \tag{12.118}$$

This is the Lippmann–Schwinger equation for the T-matrix in momentum space. Note that while $\mathbf{k}'$ and $\mathbf{k}$ are restricted to the energy shell, as in Eq. (12.51), the integration variable $\mathbf{k}''$ leaves the energy shell. The prefactor $\frac{2m}{\hbar^2}$ appears since the energy denominator of the Green's function is expressed in terms of squares of wavenumbers and not in terms of energy. When working in k space, it is convenient to scale $T(\mathbf{k}', \mathbf{k})$ and $V(\mathbf{k}', \mathbf{k})$ as follows:

$$T(\mathbf{k}', \mathbf{k}) = \frac{\hbar^2}{2m} t(\mathbf{k}', \mathbf{k}), \tag{12.119a}$$

$$V(\mathbf{k}', \mathbf{k}) = \frac{\hbar^2}{2m} v(\mathbf{k}', \mathbf{k}). \tag{12.119b}$$

The quantities $t(\mathbf{k}', \mathbf{k})$ and $v(\mathbf{k}', \mathbf{k})$ have the dimension of length, as does the scattering amplitude $f(\mathbf{k}', \mathbf{k})$. Equation (12.118), when rewritten in terms of $t(\mathbf{k}', \mathbf{k})$ and $v(\mathbf{k}', \mathbf{k})$, does not contain a prefactor:

$$t(\mathbf{k}', \mathbf{k}) = v(\mathbf{k}', \mathbf{k}) + \int d\mathbf{k}''\, v(\mathbf{k}', \mathbf{k}'')[k^2 + i\eta - k''^2]^{-1} t(\mathbf{k}'', \mathbf{k}). \tag{12.120}$$

The relation between the on-shell t matrix, $t(\mathbf{k}', \mathbf{k})$ (with $k^2 = k'^2 = \frac{2mE}{\hbar^2}$), and the scattering amplitude, $f(\mathbf{k}', \mathbf{k})$, follows from Eqs. (12.116) and (12.119a):

$$f(\mathbf{k}', \mathbf{k}) = -2\pi^2 t(\mathbf{k}', \mathbf{k}). \tag{12.121}$$

The integral equation (12.120) seems rather difficult to handle, since it involves three-dimensional integration. However, when symmetries are present, e.g., rotational symmetry, Eq. (12.120) can be replaced by a set of one-dimensional equations for the partial wave T matrices, as explained below.

Eq. (12.116), when combined with Eq. (12.110), suggests a natural scheme for approximating the scattering amplitude in ascending powers of the potential V, simply by iterating Eq. (12.110):

$$\boxed{T = V + VG_0T = V + VG_0V + VG_0VG_0V + \ldots .} \tag{12.122}$$

This result is referred to as the *Born series*. To lowest order, one simply replaces T by V, and we obtain the *Born approximation* for the T matrix. Making this replacement in (12.116), we arrive at the *Born approximation* for the scattering amplitude,

$$\boxed{f_B(\mathbf{k}', \mathbf{k}) = -\frac{4\pi^2 m}{\hbar^2} \langle \mathbf{k}'|V|\mathbf{k}\rangle = -2\pi^2 v(\mathbf{k}', \mathbf{k}).} \tag{12.123}$$

For example, consider the case where $V(\mathbf{r})$ is the Yukawa potential,

$$V(\mathbf{r}) = V_0 \frac{e^{-\mu r}}{\mu r}, \qquad (12.124)$$

which is used in models for interaction between nucleons and also in the Thomas–Fermi formalism for electron screening. This is a central potential, so the scattering amplitude depends only on the polar angle θ and not on the azimuthal angle ϕ. The momentum space representation gives

$$f_B(\theta) = -\frac{2mV_0}{\hbar^2 \mu} \frac{1}{q^2 + \mu^2} = -\frac{2mV_0}{\hbar^2 \mu} \frac{1}{2k^2 \sin^2 \frac{\theta}{2} + \mu^2}. \qquad (12.125)$$

Problem 12.9

Determine the total cross-section for scattering from the Yukawa potential within the Born approximation.

Solution: Denote by $c \equiv \frac{2mV_0}{\hbar^2 k^2}$ the dimensionless quantity expressing the ratio between the potential and kinetic energy and by $\alpha \equiv \frac{\mu}{k\sqrt{2}}$. Employing the relation $\sin^2 \frac{\theta}{2} = \frac{1 - \cos\theta}{2}$, we have

$$\frac{d\sigma}{d\Omega} = \frac{c^2}{2\mu^2} \frac{1}{(1 - \cos\theta + \alpha^2)^2}.$$

The total cross-section is obtained by integrating $\frac{d\sigma}{d\Omega}$ with the surface element $2\pi \sin\theta \, d\theta$. Letting $\cos\theta \equiv x$ we get

$$\sigma = \frac{2\pi c^2}{2\mu^2} \int_{-1}^{1} \frac{dx}{(1 - x + \alpha^2)^2} = \frac{2\pi c^2}{\mu^2 \alpha^2 (2 + \alpha^2)}.$$

Problem 12.10

Obtain the Born approximation for the scattering amplitude and the differential cross-section for a Coulomb potential by taking the limit $V_0 \to 0$, $\mu \to 0$, $\frac{V_0}{\mu} = q_1 q_2$.

Answer:

$$f_B(\theta) = -\frac{q_1 q_2}{2E \sin^2 \frac{\theta}{2}}, \quad \frac{d\sigma}{d\Omega} = \frac{q_1^2 q_2^2}{4E^2 \sin^4 \frac{\theta}{2}}. \qquad (12.126)$$

For the Coulomb potential, the expression obtained from the Born approximation coincides with the Rutherford formula, Eq. (12.10), and is the same as the exact quantum mechanical cross-section.

The Low Equation

Starting from (12.115) we can derive a nonlinear integral equation for $T(\mathbf{k}', \mathbf{k}; E)$ that proves to be useful. For that purpose, we employ the spectral representation of the resolvent operator $G^+(E) = G(E + i\eta)$ in terms of complete set of eigenkets $\{|\Psi_E\rangle\}$ of the full Hamiltonian H. The energies include the continuous spectrum $E_\mathbf{k}$ as well as bound state energies $-E_B < 0$. Thus, with $E^+ \equiv E + i\eta$,

$$G(E^+) = \sum_B \frac{|\Psi_{E_B}\rangle \langle \Psi_{E_B}|}{E + E_B} + \int d\mathbf{k}'' \frac{|\Psi_{E_\mathbf{k}''}\rangle \langle \Psi_{E_\mathbf{k}}|}{E^+ - E_{\mathbf{k}''}}. \qquad (12.127)$$

12.4 Aspects of Formal Scattering Theory

Inserting the spectral representation (12.127) for $G^+(E)$ into (12.115), we arrive at the *Low equation* for the T matrix,

$$T(\mathbf{k}',\mathbf{k};E^+) = V(\mathbf{k}',\mathbf{k}) + \sum_B \frac{\langle \mathbf{k}'|V|\Psi_{E_B}\rangle\langle\Psi_{E_B}|V|\mathbf{k}\rangle}{E^+ + E_B} + \int d\mathbf{k}'' \frac{T(\mathbf{k}',\mathbf{k}'';E^+)\left[T(\mathbf{k}'',\mathbf{k};E^+)\right]^*}{E^+ - E_{\mathbf{k}''}}. \tag{12.128}$$

12.4.2 THE S MATRIX AND MÖLLER OPERATORS

In Sec. 12.2.3, we introduced the S operator (S matrix) $S = \Omega_-^\dagger \Omega_+$, which was defined in terms of the Möller operators $\Omega_\pm$. It was also pointed out below Eq. (12.114) that the two sets of states $\{|\psi_\mathbf{k}^+\rangle\}$ and $\{|\psi_\mathbf{k}^-\rangle\}$ form two bases in $\mathcal{D}_+$ (the subspace of the Hilbert space corresponding to the positive spectrum). Here, we use the energy representation of the Möller operators $\Omega_\pm(E)$, Eq. (12.103), to form the transformation matrix from the basis $|\psi_\mathbf{k}^+\rangle$ to the basis $|\psi_\mathbf{k}^-\rangle$. This (unitary) transformation is the S matrix in the energy representation. In analogy with definition (12.37), the S operator in the energy representation is defined as,

$$S(E) = \Omega_-^\dagger(E)\Omega_+(E) = [1 + G^-(E)V]^\dagger[1 + G^+(E)V]. \tag{12.129}$$

Using the fact that the operators $\Omega_\pm(E)$ transform the states $|\mathbf{k}\rangle$ to the scattering states $|\psi_\mathbf{k}^\pm\rangle$ as in Eq. (12.113b), we find the momentum space representation of the S matrix to be

$$S(\mathbf{k}',\mathbf{k}) = \langle\mathbf{k}'|S|\mathbf{k}\rangle = \langle\psi_{\mathbf{k}'}^-|\psi_\mathbf{k}^+\rangle. \tag{12.130}$$

Thus, the matrix $S(\mathbf{k}',\mathbf{k})$ transforms the set of states $\{|\psi_{\mathbf{k}'}^-\rangle\}$ into $\{|\psi_\mathbf{k}^+\rangle\}$,

$$|\psi_\mathbf{k}^+\rangle = \int d\mathbf{k}' \, |\psi_{\mathbf{k}'}^-\rangle S(\mathbf{k}',\mathbf{k}). \tag{12.131}$$

The dimension of $S(\mathbf{k}',\mathbf{k})$ is L^3. We stress that the matrix element $S(\mathbf{k}',\mathbf{k})$ is limited to the representation of the S operator only in a subspace $\mathcal{D}^+$ of the full Hilbert space pertaining to the positive spectrum of the Hamiltonian H. In general, the Hamiltonian might also have bound state eigenvalues, which do not belong to the sub-space $\mathcal{D}^+$. The presence of bound states affects the low-energy scattering in a profound way, but this is effectively included in the present formalism even if bound states are not directly included in the basis of states forming the S matrix.

Employing the completeness (in $\mathcal{D}^+$) and orthogonality of the asymptotic states $|\psi_\mathbf{k}^\pm\rangle$, it is evident that the matrix $S(\mathbf{k}',\mathbf{k})$ is unitary [see also Eq. (12.38)],

$$\int d\mathbf{q}\, S^\dagger(\mathbf{k}',\mathbf{q})S(\mathbf{q},\mathbf{k}) = \int d\mathbf{q}\, S^*(\mathbf{q},\mathbf{k}')S(\mathbf{q},\mathbf{k}) = \delta(\mathbf{k}-\mathbf{k}'). \tag{12.132}$$

Since the S matrix relates incoming waves to outgoing waves, the unitarity of the S matrix imply flux conservation, i.e., the number of particles entering a volume bounded by a spherical shell around the scattering center per unit time is equal to the number of particles leaving this volume.

Let us return to the relation between the S matrix and the T matrix as defined in Eq. (12.115). Employing the definition (12.129) and the result (12.765) we can express the S matrix as

$$\begin{aligned}S(\mathbf{k}',\mathbf{k}) &= \langle\psi_{\mathbf{k}'}^-|\psi_\mathbf{k}^+\rangle = \langle\mathbf{k}'|\psi_\mathbf{k}^+\rangle + \langle\mathbf{k}'|V(E_{\mathbf{k}'} + i\eta - H)^{-1}|\psi_\mathbf{k}^+\rangle \\ &= \langle\mathbf{k}'|\psi_\mathbf{k}^+\rangle + (E_{\mathbf{k}'} + i\epsilon - E_\mathbf{k})^{-1}\langle\mathbf{k}'|V|\psi_\mathbf{k}^+\rangle.\end{aligned} \tag{12.133}$$

Employing Eq. (12.113b) for $|\psi_\mathbf{k}^+\rangle$ in $\langle\mathbf{k}'|\psi_\mathbf{k}^+\rangle$ appearing on the RHS of the last equality of (12.133) we find,

$$\begin{aligned}S(\mathbf{k}',\mathbf{k}) &= \langle\mathbf{k}'|\mathbf{k}\rangle + \langle\mathbf{k}'|(E_\mathbf{k}+i\eta-H_0)^{-1}V|\psi_\mathbf{k}^+\rangle + (E_{\mathbf{k}'}+i\eta-E_\mathbf{k})^{-1}\langle\mathbf{k}'|V|\psi_\mathbf{k}^+\rangle \\ &= \langle\mathbf{k}'|\mathbf{k}\rangle + [(E_{\mathbf{k}'}+i\eta-E_\mathbf{k})^{-1} + (E_\mathbf{k}+i\eta-E_{\mathbf{k}'})^{-1}]\langle\mathbf{k}'|V|\psi_\mathbf{k}^+\rangle.\end{aligned} \tag{12.134}$$

Digression: An important identity

Let $f(x)$ be a bounded smooth function on the interval $-X \leq x \leq X$ for some fixed $X > 0$. Consider the following integral:

$$\lim_{\eta \to 0^+} \int_{-X}^{X} dx \frac{f(x)}{x \pm i\eta} = \lim_{\eta \to 0^+} \left[\int_{-X}^{X} dx \frac{xf(x)}{x^2 + \eta^2} \mp i\eta \int_{-X}^{X} dx \frac{f(x)}{x^2 + \eta^2} \right]. \tag{12.135}$$

In the first term on the RHS, we can take the limit $\eta \to 0^+$ as long as $x \neq 0$; therefore, we can compute it formally as,

$$\lim_{\eta \to 0^+} \int_{-X}^{X} dx \frac{xf(x)}{x^2 + \eta^2} = \int_{-X}^{X} dx \frac{f(x) - f(0)}{x} + f(0) \int_{-X}^{X} \frac{dx}{x}.$$

The second term on the RHS vanishes while the first term is defined as *the principal value integral*, $\mathcal{P} \int \frac{f(x)}{x} \equiv \int_{-X}^{X} dx \frac{f(x) - f(0)}{x}$. The second term on the RHS of Eq. (12.135) is computed using the representation of the Dirac δ function as,

$$\delta(x) = \lim_{\eta \to 0^+} \frac{\eta}{x^2 + \eta^2}.$$

These results are summarized in the identity,

$$\boxed{\frac{1}{x \pm i\eta} = \mathcal{P}\left[\frac{1}{x}\right] \mp i\pi \delta(x).} \tag{12.136}$$

Combining Eq. (12.136) and Eq. (12.115), we finally obtain,

$$S(\mathbf{k}', \mathbf{k}) = \delta(\mathbf{k}' - \mathbf{k}) - 2\pi i \delta(E_{\mathbf{k}'} - E_{\mathbf{k}}) T(\mathbf{k}', \mathbf{k}). \tag{12.137}$$

This relation between the S matrix and the T matrix, together with the relation (12.116) between the T matrix and the scattering amplitude, leads to an important relation known as the *optical theorem*. First, note from Eq. (12.116) that $T(\mathbf{k}', \mathbf{k}) = -\frac{2\pi\hbar^2}{m} f(\mathbf{k}', \mathbf{k})$ and insert this into Eq. (12.137), so that the S matrix is given in terms of the scattering amplitude. Next, use this modified expression for the S matrix in the unitarity relation (12.132):

$$\int d\hat{\mathbf{k}}'' f^*(\mathbf{k}', \mathbf{k}'') f(\mathbf{k}'', \mathbf{k}) = \frac{4\pi}{2ik}[f(\mathbf{k}', \mathbf{k}) - f^*(\mathbf{k}, \mathbf{k}')]. \tag{12.138}$$

Recall that, as far as the scattering amplitude is concerned, both momenta are restricted on the energy shell $k^2 = k'^2 = k''^2 = 2mE/\hbar^2$ [this is guaranteed by the $\delta(E_{\mathbf{k}} - E_{\mathbf{k}'})$ term in Eq. (12.137)] and is the reason that the integration on the LHS of Eq. (12.138) is performed only on the angular part. On the other hand, note that when we consider the Lippmann–Schwinger equation (12.118) for the T matrix, the intermediate momentum $\mathbf{k}''$ is allowed to leave the energy shell. In the special case $\mathbf{k} = \mathbf{k}'$, the LHS of Eq. (12.138) is identified with the total cross-section, while in the RHS, we encounter the imaginary part of the forward scattering amplitude,

$$\boxed{\sigma = \int d\hat{\mathbf{k}}'' |f(\mathbf{k}, \mathbf{k}'')|^2 = \frac{4\pi}{k} \text{Im}[f(\mathbf{k}, \mathbf{k})].} \tag{12.139}$$

This result is known as the *optical theorem*. It relates the total cross-section to the imaginary part of the forward scattering amplitude. See also Eq. (12.170) for a discussion of the optical theorem.

12.5 CENTRAL POTENTIALS

A local interaction potential $V(\mathbf{r}_1 - \mathbf{r}_2)$ between two particles that depends only on $r = |\mathbf{r}| = |\mathbf{r}_1 - \mathbf{r}_2|$, i.e., $V(\mathbf{r}_1 - \mathbf{r}_2) = V(r)$, is referred to as a *spherically symmetric* or a *central potential* (see Sec. 3.2). It was shown in Sec. 12.3.2 that the two-body collision problem with interaction $V(\mathbf{r}_1 - \mathbf{r}_2)$ is reducible to that of single-particle scattering from a central potential $V(r)$. This section treats scattering of a particle of mass m and energy $E > 0$ from a central potential $V(r)$. The underlying Hamiltonian is

$$H = -\frac{\hbar^2}{2m}\nabla_\mathbf{r}^2 + V(r). \tag{12.140}$$

To simplify the discussion, we assume that the colliding particles are elementary, i.e., besides spin, they do not have internal structure.

12.5.1 CENTRAL POTENTIALS AND SPIN

The role of spin in connection with central potentials should be clarified. If the particles are spinless, their mutual interaction must be central. To show this, start from the general form of a two-body interaction $V(\mathbf{r}_1, \mathbf{r}_2)$. Invariance under translation dictates that it must be a function of $\mathbf{r} = \mathbf{r}_1 - \mathbf{r}_2$. Moreover, the interaction must be a scalar, and therefore, it can depend only on $|\mathbf{r}|$. In other words, the only scalar that can be constructed from the vector $\mathbf{r}$ is $r^2 = \mathbf{r} \cdot \mathbf{r}$. Thus, non-central potentials are relevant if the interaction depends also on spin. The corresponding spin operators, $\mathbf{s}_1$ and $\mathbf{s}_2$, can form scalars with $\mathbf{r}$ in such a way that the resulting interaction will not depend only on r. Such *non-central potentials* are intimately related to the particle spin and will be discussed in Sec. 12.8.8. If the two-body interaction of particles is spin-independent, the spin part of the two-particle wave function can be written in terms of the individual spin functions $\eta_{s_i\sigma_i}$, $i = 1, 2$, where s_i is the spin of particle i and σ_i is its projection along a given quantization axis, as

$$|s\sigma\rangle \equiv \eta_{s\sigma} = \sum_{\sigma_1\sigma_2}\langle s_1\sigma_1 s_2\sigma_2|s\sigma\rangle \eta_{s_1\sigma_1}\eta_{s_2\sigma_2}, \tag{12.141}$$

where $\langle s_1\sigma_1 s_2\sigma_2|s\sigma\rangle$ is the corresponding Clebsch–Gordan coefficient. Spin independence of the potential V is formally expressed as,

$$\langle s'\sigma'|V|s\sigma\rangle = \delta_{ss'}\delta_{\sigma\sigma'}V^s. \tag{12.142}$$

Hence, even if the scattered particles do have spin, the spin independence of the interaction encoded in the above equation implies that the corresponding spin wave functions are completely separated from the space coordinates and can be "factored out." All matrix representations of the relevant operators, e.g., the S matrix and the T matrix, are diagonal in the spin coordinates, as is the potential V in Eq. (12.142).

12.5.2 AXIAL SYMMETRY

The Hamiltonian (12.140) commutes with the orbital angular momentum operator $\mathbf{L}$. From Sec. 3.2, it is clear that the natural coordinate system for treating this kind of problem is the spherical coordinate system, r, θ, ϕ. However, the asymptotic boundary condition (12.66) destroys the spherical symmetry of the problem due to the presence of the plane wave. Still, the problem has cylindrical symmetry around the z axis, which is the direction of the incoming beam, $\mathbf{k} = (0, 0, k)$. The dependence of the wave function on the azimuthal angle ϕ enters only through a phase factor $e^{im\phi}$, but the scattering amplitude and the differential cross-section do not depend on ϕ at all. This symmetry is employed below in the *partial wave analysis* formalism.

For notational convenience, we continue to use scaled energy $k^2 = \frac{2mE}{\hbar^2}$ and potential $v(r) = \frac{2mV(r)}{\hbar^2}$, so Eqs. (12.64) and (12.65) adapted for scattering from central potential now read,

$$[-\nabla_\mathbf{r}^2 + v(r)]\psi_\mathbf{k}^+(\mathbf{r}) = k^2 \psi_\mathbf{k}^+(\mathbf{r}), \tag{12.143}$$

$$\psi_\mathbf{k}^+(\mathbf{r}) \xrightarrow[r\to\infty]{} (2\pi)^{-\frac{3}{2}} \left[e^{ikz} + f(\theta)\frac{e^{ikr}}{r} \right], \tag{12.144}$$

where, following Fig. 12.5,

$$\cos\theta = \hat{\mathbf{k}} \cdot \hat{\mathbf{k}}' = \lim_{r\to\infty} \hat{\mathbf{k}} \cdot \hat{\mathbf{r}}. \tag{12.145}$$

The differential and total cross-sections are given by,

$$\frac{d\sigma}{d\Omega} = |f(\theta)|^2, \tag{12.146a}$$

$$\sigma = 2\pi \int d\theta \, |f(\theta)|^2 \sin\theta. \tag{12.146b}$$

12.5.3 PARTIAL WAVE ANALYSIS

We shall now use the spherical symmetry of the potential in order to reduce the 3D Schrödinger equation (12.143) to a decoupled set of one-dimensional equations. The rotational symmetry is encoded by the relation $[H, \mathbf{L}] = 0$, where $\mathbf{L}$ is the angular momentum operator. Within the spherical coordinate system r, θ, ϕ, this symmetry enables the separation of variables. The key relation for this separation is the decomposition of the Laplacian as,

$$\nabla_\mathbf{r}^2 = \frac{\partial^2}{\partial r^2} + \frac{2}{r}\frac{\partial}{\partial r} - \frac{L^2}{\hbar^2 r^2}, \tag{12.147}$$

which is proved in Sec. 3.2, Eq. (3.60). Thus, Eq. (12.143) has basic solutions [see Eqs. (3.64) and (3.75)]

$$\psi_{lm}(r, \theta, \phi) = R_l(r) Y_{lm}(\theta, \phi), \tag{12.148}$$

which are eigenfunctions of L^2 and L_z,

$$L^2 \psi_{lm}(r, \theta, \phi) = \hbar^2 l(l+1) \, \psi_{lm}(r, \theta, \phi), \tag{12.149}$$

$$L_z \psi_{lm}(r, \theta, \phi) = \hbar m \, \psi_{lm}(r, \theta, \phi), \tag{12.150}$$

where $Y_{lm}(\theta, \phi)$ are the spherical harmonics. Inclusion of spin is achieved simply by multiplying $\psi_{lm}(r, \theta, \phi)$ by the spin function $\eta_{s\sigma}$ defined in Eq. (12.141). The two-particle wave function, including the spinor, is

$$\Psi_{lms\sigma}(r, \theta, \phi) = \psi_{lm}(r, \theta, \phi) \, \eta_{s\sigma}. \tag{12.151}$$

The angular and spin part can be combined together as a ket $|lms\sigma\rangle$ with configuration space representation given by,

$$\langle \hat{\mathbf{r}} | lms\sigma \rangle = Y_{lm}(\theta, \phi) \, \eta_{s\sigma}. \tag{12.152}$$

These are eigenfunctions of L^2, L_z, S^2 and S_z with corresponding eigenvalues $\hbar^2 l(l+1), m\hbar, \hbar^2 s(s+1)$, and $\hbar\sigma$. The functions $\langle \hat{\mathbf{r}} | lms\sigma \rangle$ can be regarded as geometric factors, in the sense that they reflect the symmetry of the problem, but they do not depend on the scattering potential. All the dynamics of the problem and its relation to the scattering potential $v(r)$ is then encoded in the radial wave functions $R_l(r)$, which play a central role in the present analysis.

Since the spin functions play no role here, the representation (12.152) is not actually used. The solution of the Schrödinger equation (12.143) is achieved by expanding the space part of the wave function $\psi_\mathbf{k}^+(\mathbf{r})$ in terms of the basic solutions $\psi_{lm}(r, \theta, \phi)$, such that the asymptotic condition (12.144) is satisfied. This procedure is referred to as *partial wave expansion*. As already mentioned, for central potentials, the wave function depends on ϕ only through a phase factor $e^{im\phi}$, and the scattering amplitude does not depend on ϕ. Hence, it is legitimate to take $m=0$ and expand the wave function in terms of angular functions $Y_{l0}(\theta\phi) = P_l(\cos\theta)$, i.e., the Legendre polynomials.

12.5 Central Potentials

Partial Wave Expansion of $f(\theta)$

The partial wave expansion of the scattering amplitude is carried out in several steps.

- In the first step, the partial wave expansion of the incoming plane wave is performed (see Sec. 3.2.1),

$$\frac{1}{(2\pi)^{\frac{3}{2}}} e^{ikz} = \frac{1}{(2\pi)^{\frac{3}{2}}} e^{ikr\cos\theta} = \frac{1}{(2\pi)^{\frac{3}{2}}} \sum_{l=0}^{\infty} (2l+1) i^l j_l(kr) P_l(\cos\theta). \tag{12.153}$$

- In the second step, the partial wave expansion of the scattering state $\psi_{\mathbf{k}}^{(+)}(r,\theta)$ is carried out,

$$\psi_{\mathbf{k}}^{(+)}(r,\theta) = \frac{1}{(2\pi)^{\frac{3}{2}}} \sum_{l=0}^{\infty} (2l+1) i^l R_l(r) P_l(\cos\theta), \tag{12.154}$$

In the absence of scattering ($v(\mathbf{r}) = 0$), $\psi_{\mathbf{k}}^{(+)}(r,\theta)$ reduces to the plane wave (12.153) and $R_l(r) \to j_l(kr)$.

- In the third step, the Schrödinger equation for the radial wave function $R_l(r)$ is derived. Substituting expansion (12.154) into the Schrödinger equation (12.143), using expression (12.147) for the Laplacian and the orthogonality of the Legendre polynomials, one arrives at the radial differential equation for $R_l(r)$,

$$\left[\frac{d^2}{dr^2} + \frac{2}{r}\frac{d}{dr} + k^2 - \frac{l(l+1)}{r^2} - v(r)\right] R_l(r) = 0. \tag{12.155}$$

Thus, as asserted above, the radial functions $R_l(r)$ contain all the dynamical information of the scattering problem.

In the absence of scattering, $v(r) = 0$, the basic solutions of Eq. (12.155) are the *spherical Bessel functions* and the *spherical Neumann functions* of order l, $j_l(kr)$ and $n_l(kr)$ [see Appendix B, Eqs. (B.28)]. While both $j_l(kr)$ and $n_l(kr)$ are real, $j_l(kr)$ is regular, whereas $n_l(kr)$ is singular at the origin $r = 0$.

Construction of linear combinations of $j_l(kr)$ and $n_l(kr)$, which at large distance r behave as outgoing and incoming spherical waves, is of special importance. These are the *spherical Hankel functions* $h_l^{\pm}(kr)$ defined as (see Appendix B),

$$h_l^{\pm}(kr) = j_l(kr) \pm i n_l(kr). \tag{12.156}$$

For large argument, $z = kr \to \infty$, the following asymptotic behavior will be useful,

$$j_l(z) \to \frac{\sin(z - \frac{l\pi}{2})}{z}, \quad n_l(z) \to -\frac{\cos(z - \frac{l\pi}{2})}{z}, \quad h_l^{\pm}(z) \to \mp i \frac{e^{\pm i(z - \frac{l\pi}{2})}}{z}. \tag{12.157}$$

It is sometimes useful to eliminate the first derivative term in Eq. (12.155) using the transformation

$$u_l(r) = kr R_l(r). \tag{12.158}$$

$u_l(r)$ satisfies the modified radial Schrödinger equation,

$$\left[\frac{d^2}{dr^2} + k^2 - \frac{l(l+1)}{r^2} - v(r)\right] u_l(r) = 0. \tag{12.159}$$

Since $R_l(r)$ must be regular at any point, including $r = 0$, we require that $u_l(r) \to 0$ as $r \to 0$ at least as $O(r)$. We will use either $R_l(r)$ or $u_l(r)$, depending upon which is more convenient. In the absence of scattering, the analogous solutions are the Riccati Bessel, Ricatti Neumann, and Ricatti Hankel functions defined as (see Eqs. (B.31) in Appendix B),

$$\hat{j}_l(z) = z j_l(z), \quad \hat{n}_l(z) = z n_l(z), \quad \hat{h}_l^{\pm}(z) = z h_l^{\pm}(z). \tag{12.160}$$

- Our task in the fourth step is to relate the solutions $R_l(r)$ of the radial Eq. (12.155) or (12.159) to the scattering amplitude $f(\theta)$) appearing in Eq. (12.143). To this end, we substitute in Eq. (12.144) the partial wave expansions of the incoming plane wave (12.153) and of the scattering state (12.154). Note that Eq. (12.144) is valid only at $r \to \infty$, and hence, we will need to use the asymptotic form of the radial functions $R_l(r)$. In this region, the potential $v(r)$ is negligibly small and can be omitted in Eq. (12.155). Equation (12.155) then reduces to a variant of the radial Bessel equation whose general solution is a combination of Bessel and Neumann functions ($j_l(kr)$ and $n_l(kr)$, respectively). The asymptotic form of either $R_l(r)$ or $u_l(r)$ is written as

$$R_l(r) \xrightarrow[r \to \infty]{} [\cos \delta_l(k) j_l(kr) - \sin \delta_l(k) n_l(kr)], \tag{12.161a}$$

$$u_l(r) \xrightarrow[r \to \infty]{} [\cos \delta_l(k) \hat{j}_l(kr) - \sin \delta_l(k) \hat{n}_l(kr)]. \tag{12.161b}$$

The coefficients are expressed in terms of trigonometric functions of an angle $\delta_l(k)$ referred to as *phase shift* for the angular momentum partial wave l. Knowledge of all phase shifts $\delta_l(k)$, $l = 0, 1, \ldots$ is equivalent to a complete solution of the scattering problem (see below). Equations (12.161a) and (12.161b) are valid at large r, where $v(r) \approx 0$. Taking r even larger, one can employ the asymptotic expressions for the Bessel and Neumann functions [see Eqs. (B.29a) in Appendix B],

$$j_l(kr) \xrightarrow[r \to \infty]{} \frac{\sin(kr - \frac{1}{2}l\pi)}{kr}, \quad n_l(kr) \xrightarrow[r \to \infty]{} -\frac{\cos(kr - \frac{1}{2}l\pi)}{kr}, \tag{12.162a}$$

$$[\cos \delta_l j_l(kr) - \sin \delta_l n_l(kr)] \to \frac{1}{kr} \sin\left(kr - \frac{1}{2}l\pi + \delta_l\right). \tag{12.162b}$$

Thus, at very large r, where the Bessel functions can be approximated by simple trigonometric functions, the radial functions $R_l(r)$ [or equivalently $u_l(r)$] have the asymptotic behavior,

$$R_l(r) \xrightarrow[r \to \infty]{} \frac{A_l}{kr} \sin\left(kr - \frac{1}{2}l\pi + \delta_l\right), \tag{12.163a}$$

$$u_l(r) \xrightarrow[r \to \infty]{} A_l \sin\left(kr - \frac{1}{2}l\pi + \delta_l\right), \tag{12.163b}$$

where the constant A_l will be determined using the fifth and sixth steps to be $A_l = e^{i\delta_l}$.
- In the fifth step, the asymptotic expansions (12.153) and (12.154) with $R_l(r)$ from (12.163a) are inserted in Eq. (12.144), which [after factoring out the common factor $(2\pi)^{-3/2}$ and taking r to be very large] is rewritten as,

$$\sum_{l=0}^{\infty} (2l+1) i^l A_l \frac{\sin(kr - \frac{1}{2}l\pi + \delta_l)}{kr} P_l(\cos\theta) = \sum_{l=0}^{\infty} (2l+1) i^l \frac{\sin(kr - \frac{1}{2}l\pi)}{kr} P_l(\cos\theta) + f(\theta) \frac{e^{ikr}}{r}. \tag{12.164}$$

- In the sixth step, the identity $\sin x = \frac{e^{ix} - e^{-ix}}{2i}$ is used and (12.164) is rewritten as a linear combination $ae^{ikr} + be^{-ikr} = 0$, which forces both constants a and b to vanish. The coefficient of e^{-ikr} does not involve $f(\theta)$ and can be made to vanish by choosing $A_l = e^{i\delta_l(k)}$. This leads to the relation,

$$\boxed{f(\theta) = \frac{1}{k} \sum_{l=0}^{\infty} (2l+1) e^{i\delta_l} \sin \delta_l P_l(\cos\theta).} \tag{12.165}$$

Equation (12.165) completes the procedure of partial wave analysis of the wave function for scattering by spherically symmetric potential. The relevant physics is encoded in the *phase shifts*, $\delta_l(k)$, and the main goal of scattering theory calculations is to compute the phases within some reasonable approximation. This task will be elaborated in the next section.

12.5 Central Potentials

Following the definition of the differential cross-section, Eq. (12.146a), it is readily found that

$$\frac{d\sigma}{d\Omega} = |f(\theta)|^2 = \frac{1}{k^2} \left| \sum_{l=0}^{\infty} (2l+1) e^{i\delta_l} \sin \delta_l P_l(\cos\theta) \right|^2. \tag{12.166}$$

When this expression is integrated over a spherical shell (at very large distance from the scattering center), one obtains the total cross-section (12.146b),

$$\sigma = \int d\Omega \frac{d\sigma}{d\Omega} = 2\pi \int_0^\pi d\theta \frac{d\sigma}{d\Omega} \sin\theta = \frac{4\pi}{k^2} \sum_{l=0}^{\infty} (2l+1) \sin^2 \delta_l. \tag{12.167}$$

It is useful at this point to compare this quantum mechanical result with the classical result (12.6). We have seen that the concept of total scattering cross-section in classical mechanics makes sense only if the range r_0 of the scattering potential is finite. For impact parameters b exceeding r_0, classical scattering does not take place. Translated into the quantum mechanical picture, this means that, roughly speaking, there is a maximal angular momentum $L_{max} \gtrsim kr_0$, such that the phase shift δ_l for $l > L_{max} \gtrsim kr_0$ is very small. Let us make the further assumption that for $l < L_{max}$, scattering is "maximal" in the sense that $\delta_l = \pi/2$ (this is the unitary bound of the partial wave cross-section imposed by flux conservation), hence,

$$\sigma \approx \frac{4\pi}{k^2} \sum_{l=0}^{L_{max}} (2l+1) \approx 4\pi r_0^2 \quad \text{(classical approximation).} \tag{12.168}$$

This means that for a potential of finite extent, the quantum mechanical total cross-section might exceed the classical total cross-section (i.e., the area of the scatterer) by a factor 4. It is a manifestation of the wave nature of particles and the unavoidable occurrence of diffraction.

Equations (12.166) and (12.167) show that the method of partial waves is easiest to carry out at low energy where $kr_0 \lesssim 10$, so that the summation is carried out over only a few partial waves. Indeed, accurate calculation of phase shifts δ_l involves substantial effort, since it requires the solution of the radial equation (12.155) separately for each partial wave l. Moreover, even after obtaining the phase shifts, carrying out the actual summation over partial waves is computationally expensive if the number of contributing partial waves is very large (if $kr_0 \gg 10$). An illuminating example of this difficulty can be taken from electromagnetic scattering, such as light scattered from water droplets. Then kr_0, the product of the wavenumber of visible light and the radius of the water droplet, is about 300, and the summation cannot be done in a straight-forward manner (for, among other reasons, it yields a rapidly oscillating function of k and θ). One has to employ more sophisticated methods of summation, e.g., the Sommerfeld–Watson transformation. Using this procedure, one finds some of the most fascinating phenomena in nature, such as the rainbow or the glory.

The optical theorem (12.139) relates the total cross-section (12.167) to the imaginary part of the forward ($\theta = 0$) scattering amplitude. Indeed, for $\theta = 0$, using $P_l(1) = 1$ in Eq. (12.165), one obtains the following expression for the forward scattering amplitude,

$$\boxed{f(0) = \frac{1}{k} \sum_{l=0}^{\infty} (2l+1) e^{i\delta_l} \sin \delta_l.} \tag{12.169}$$

The imaginary part of the RHS of the last equation multiplied by $\frac{4\pi}{k}$ is equal to the RHS of Eq. (12.167),

$$\sigma = \frac{4\pi}{k} \text{Im}[f(\theta = 0)]. \tag{12.170}$$

The fact that the total cross-section is related to the scattering amplitude in the forward direction can be simply understood as follows. The total cross-section is defined as the number of particles per unit time which are deflected from their initial direction $\mathbf{k} = k\hat{\mathbf{z}}$, divided by the incoming flux. This deflection of incoming particles is due to interference between the plane wave and the outgoing radial component of the wave function (12.144) at large r and at $\theta = 0$. When this interference scenario is presented in detail, the optical theorem holds more generally, e.g., in cases of scattering by non-central potentials, multichannel scattering, and scattering reactions (both these terms are defined later). The derivation of Eq. (12.170) and the relation (12.137) between the S and T matrices within formal scattering theory indicate that the optical theorem is just another manifestation of the unitarity of the S matrix, which, physically, expresses flux conservation.

Partial Wave S Matrix and Phase Shift

When l is a good quantum number, the scattering behavior is encoded into the radial wave function R_l, and more specifically, into the phase shift $\delta_l(k)$ (or equivalently, into the partial wave S matrix $S_l(k)$). To show this, it is useful to write the asymptotic expression (12.161a) for $R_l(r)$ as,

$$R_l(r) = h_l^-(kr) + S_l(k)h_l^+(kr). \tag{12.171}$$

The reason for using this form is as follows: Eq. (12.171) tells us that the scattering process for a given angular momentum quantum number l is composed of an incoming spherical wave approaching the origin $r = 0$ with unit amplitude and then reflected from the origin as an outgoing spherical wave with an amplitude S_l. Since no flux is lost in this process, the amplitude of the reflected wave should satisfy the unitarity condition $|S_l| = 1$. This amplitude is referred to as *the S matrix for partial wave l* (although for central potential scattering it is just a single number for a given l). It encodes all the relevant physics of the scattering for the lth partial wave. Comparing Eqs. (12.171) and (12.161a) and using the relations (12.157), one easily verifies that

$$S_l(k) = e^{2i\delta_l(k)}. \tag{12.172}$$

The numbers $S_l(k)$ can be used to describe partial wave expansions of scattering. The *partial wave scattering amplitude* $f_l(k)$ is defined as,

$$f_l(k) = \frac{1}{k}e^{i\delta_l(k)}\sin\delta_l(k) = \frac{S_l(k)-1}{2ik}, \tag{12.173}$$

where the latter equation follows since $e^{i\delta_l}\sin\delta_l = \frac{e^{2i\delta_l}-1}{2i}$. Hence, differential scattering cross-sections can be expressed in terms of S_l.

In a later section, we will study scattering between particles whose interaction is spin dependent, i.e., the potential is not spherically symmetric. In this case, the orbital angular momentum is not a good quantum number, and one encounters a bona fide S matrix with elements $S_{ll'}$ that are not diagonal in l. Nevertheless, the S matrix is unitary, $SS^\dagger = S^\dagger S = 1$, where 1 is the unit matrix in the appropriate space. The eigenvalues of the S matrix are then unimodular and can be written as $e^{2i\delta_n}$. The corresponding phases, δ_n, are referred to as *eigen-phase shifts*.

Let us now carry out the partial wave expansion of the T matrix. Recall that on the energy shell, $k^2 = k'^2 = \frac{2m}{\hbar^2}E$, and for central potentials, the t matrix $t(\mathbf{k}', \mathbf{k})$ is a function of the energy (or more conveniently of the magnitude of the momentum k) and the scattering angle θ, such that $\cos\theta = \hat{\mathbf{k}} \cdot \hat{\mathbf{k}}'$. Thus, on-shell we may write $t(\mathbf{k}', \mathbf{k}) = t(\cos\theta, k)$. The partial expansion reads,

$$t(\cos\theta, k) = \frac{1}{4\pi}\sum_{l=0}^{\infty}(2l+1)t_l(k)P_l(\cos\theta). \tag{12.174}$$

The inverse relation, obtained upon multiplying by a Legendre polynomial, integrating over angle and using the orthogonality of the Legendre functions, $\int_{-1}^{1}P_l(x)P_m(x)dx = \frac{2}{2l+1}\delta_{lm}$, is

$$t_l(k) = 2\pi\int_{-1}^{1}d(\cos\theta)\,P_l(\cos\theta)\,t(\cos\theta, k). \tag{12.175}$$

12.5 Central Potentials

From Eq. (12.121) relating $f(\mathbf{k}', \mathbf{k})$ to $t(\mathbf{k}', \mathbf{k})$ and from the partial wave expansion of $f(\mathbf{k}', \mathbf{k})$ in Eq. (12.165), the relation of $t_l(k)$ to the partial wave scattering amplitude $f_l(k)$, introduced in Eq. (12.173), is simply through a constant factor

$$t_l(k) = -\frac{2}{\pi} \frac{e^{i\delta_l} \sin \delta_l}{k} = -\frac{2}{\pi} f_l(k). \tag{12.176}$$

Combined with Eq. (12.173) we obtain,

$$S_l(k) = e^{2i\delta_l(k)} = 1 - i\pi k t_l(k) = 1 + 2ikf_l(k). \tag{12.177}$$

Partial Wave Lippmann–Schwinger Equation

The integral over momentum in the Lippmann–Schwinger Eq. (12.120) involves leaving the energy shell, i.e., $k''^2 \neq k^2 = \frac{2mE}{\hbar^2}$. Hence, if we want to expand Eq. (12.120) in partial waves, we must keep in mind that $k'' \neq k$ and write $t(\mathbf{k}'', \mathbf{k}) = t(\cos\theta, k'', k)$. Partial wave expansions of the t matrix and the potential (for arbitrary wavenumbers $\mathbf{q}', \mathbf{q}$ whether they are on-shell or not) reads,

$$t(\mathbf{q}', \mathbf{q}) = \frac{1}{4\pi} \sum_{l=0}^{\infty} (2l+1) t_l(q', q) P_l(\cos\theta), \tag{12.178}$$

$$v(|\mathbf{q} - \mathbf{q}'|) = v(\cos\theta, q, q') = \frac{1}{4\pi} \sum_{l=0}^{\infty} (2l+1) v_l(q, q') P_l(\cos\theta). \tag{12.179}$$

The inverse relations are

$$t_l(q', q) = 2\pi \int_0^\pi d\theta \, \sin\theta \, P_l(\cos\theta) t(\cos\theta; q', q), \tag{12.180}$$

$$v_l(q, q') = 2\pi \int_0^\pi d\theta \, \sin\theta \, P_l(\cos\theta) v(\cos\theta; q, q'). \tag{12.181}$$

Carrying out a partial wave expansion of the Lippmann–Schwinger equation (12.120), we obtain the following equation for the partial wave t matrix:

$$t_l(k', k) = v_l(k', k) + \int dk'' \, v_l(k', k'') \frac{k''^2}{k^2 + i\eta - k''^2} t_l(k'', k). \tag{12.182}$$

When we need the on-shell partial wave t matrix, we take $k' = k = \sqrt{\varepsilon}$, but k'' in the integral deviates from the on-shell constraint.

Problem 12.11

Find the partial wave potential $v_l(k', k)$ for the Yukawa potential, $V(\mathbf{r}) = V_0 \frac{e^{-\mu r}}{\mu r}$, introduced in Eq. (12.124). Note that its on-shell momentum space representation is given by the first term on the RHS of Eq. (12.125).

Answer: The off-shell momentum space representation is $v(\cos\theta, k, k') = -\frac{2mV_0}{\hbar^2 \mu} \frac{1}{k^2 + k'^2 - 2kk' \cos\theta}$. Using Eq. (12.181), we then obtain $v_l(k, k') = -\frac{2mV_0}{\hbar^2 \mu} \int d\theta \sin\theta \frac{P_l(\cos\theta)}{k^2 + k'^2 - 2kk' \cos\theta}$.

Relation between the Wave function and the T Matrix

The above analysis is now employed in order to write down an expression for the partial wave function $R_l(r)$ in terms of the partial wave T matrix. Let us start from the abstract equation (12.113c) and represent it in configuration space by applying the bra $\langle \mathbf{r}|$ on the left, replacing $G_0^+ T$ by $g_0^+ t$.

$$\psi_{\mathbf{k}}^+(\mathbf{r}) = \langle \mathbf{r}|\mathbf{k}\rangle + \langle \mathbf{r}|g_0^+ t|\mathbf{k}\rangle = (2\pi)^{-\frac{3}{2}} \left(e^{i\mathbf{k}\cdot\mathbf{r}} + (2\pi)^3 \int d\mathbf{r}' \, \langle \mathbf{r}|g_0^+|\mathbf{r}'\rangle \langle \mathbf{r}'|t|\mathbf{k}\rangle \right). \tag{12.183}$$

Note the mixed representation of the t matrix in the integrand. This is not an obstacle because using the configuration space representation (12.80) of the Green's function and taking the limit $r \to \infty$, we can follow the procedure leading to Eq. (12.85) and find,

$$\lim_{r \to \infty} \int d\mathbf{r}' \, \langle \mathbf{r}|g_0^+|\mathbf{r}'\rangle \langle \mathbf{r}'|t|\mathbf{k}\rangle = -\frac{e^{ikr}}{4\pi r} t(\mathbf{k}', \mathbf{k}), \tag{12.184}$$

because as $r \to \infty$, the direction of $\hat{\mathbf{r}}$ coincides with that of the final momentum $\hat{\mathbf{k}}'$. Hence,

$$\lim_{r \to \infty} \psi_{\mathbf{k}}^+(\mathbf{r}) = (2\pi)^{-\frac{3}{2}} \left(e^{i\mathbf{k}\cdot\mathbf{r}} - 2\pi^2 \frac{e^{ikr}}{r} t(\mathbf{k}', \mathbf{k}) \right) = (2\pi)^{-\frac{3}{2}} \left(e^{i\mathbf{k}\cdot\mathbf{r}} + \frac{e^{ikr}}{r} f(\theta) \right), \tag{12.185}$$

and the resulting expression,

$$f(\theta) = -2\pi^2 t(\mathbf{k}', \mathbf{k}), \tag{12.186}$$

is consistent with Eq. (12.121).

Now let us examine the partial wave content of Eq. (12.113c) or equivalently, Eq. (12.183), using the momentum space representation of the T matrix,

$$\psi_{\mathbf{k}}^+(\mathbf{r}) = (2\pi)^{-\frac{3}{2}} \left(e^{i\mathbf{k}\cdot\mathbf{r}} + \int d\mathbf{q} \, \frac{e^{i\mathbf{q}\cdot\mathbf{r}}}{k^2 + i\eta - q^2} t(\mathbf{q}, \mathbf{k}) \right). \tag{12.187}$$

We now expand all the three functions, $\psi_{\mathbf{k}}(\mathbf{r})$, $e^{i\mathbf{k}\cdot\mathbf{r}}$, and $\langle \mathbf{q}|t|\mathbf{k}\rangle$ in partial waves and use the completeness and orthogonality of the angular (Legendre) functions. This task has already been carried out in Eqs. (12.154), (12.153), and (12.175), and after its application, it yields the partial wave component of Eq. (12.183),

$$R_l(r) = j_l(kr) + \int dk' \, \frac{k'^2 j_l(k'r)}{k^2 + i\eta - k'^2} t_l(k', k). \tag{12.188}$$

Note that $t_l(k', k)$ is used here off-shell. At large distance, $r \to \infty$, the integral on the RHS of Eq. (12.188) can further be simplified. First, on changing variables from k' to $\varepsilon' = k'^2$, the integral over k' is computed using the identity (12.136). Second, the asymptotic form (12.162a) of the spherical-Bessel function $j_l(kr)$ is inserted. The asymptotic form of the integral is

$$\lim_{r \to \infty} \int dk' \, \frac{k'^2 j_l(k'r)}{k^2 + i\eta - k'^2} t_l(k', k) = -\pi \frac{e^{i(kr - l\pi/2)}}{r} t_l(k), \tag{12.189}$$

where $t_l(k) \equiv t_l(k, k)$ with the on-shell condition $k = k' = \sqrt{\varepsilon}$ [see Eq. (12.234)], hence,

$$R_l(r) \to j_l(kr) - \pi \frac{e^{i(kr - l\pi/2)}}{r} t_l(k). \tag{12.190}$$

Inserting this limit into the RHS of Eq. (12.188) and recalling the relation (12.176) between $t_l(k)$ and the phase shift, and expressing the spherical-Bessel function $j_l(kr)$ in terms of spherical Hankel functions $h_l^\pm$ leads us back to Eq. (12.171) relating the radial wave function directly to the S matrix and phase shift.

12.5 Central Potentials

Lippmann–Schwinger Equation for the Radial Function

Using the Green's function method for Eq. (12.155), we can obtain the Lippmann–Schwinger equation for the radial function $R_l(r)$. We seek the partial wave Green's function $g_l^+(r, r'; k)$ satisfying the inhomogeneous differential equation,

$$\left[\frac{d^2}{dr^2} + \frac{2}{r}\frac{d}{dr} + k^2 - \frac{l(l+1)}{r^2}\right] g_l^+(r, r'; k) = \delta(r - r'), \quad (12.191)$$

such that $g_l^+(r, r'; k)$ is regular at the origin and has the asymptotic form of an outgoing radial partial wave at large r,

$$g_l^+(r, r'; k) \xrightarrow[r \to \infty]{} a_l(kr') h_l^+(kr). \quad (12.192)$$

Here, $a_l(kr')$ is bounded and $h_l^+(kr)$ is the radial Hankel function of the first kind, Eq. (12.157). The solution of Eq. (12.191) with the boundary conditions (12.192) regular at the origin is

$$g_l^+(r, r'; k) = \theta(r' - r) j_l(kr) h_l^+(kr') + \theta(r - r') j_l(kr') h_l^+(kr). \quad (12.193)$$

Then a solution of Eq. (12.155) with the appropriate asymptotic conditions (12.161a) or equivalently (12.171) can be defined in terms of the *partial wave Lippmann–Schwinger equation for the wave function*,

$$R_l(r; k) = j_l(kr) + \int_0^\infty dr' \, g_l^+(r, r'; k) v(r') R_l(r'; k). \quad (12.194)$$

Problem 12.12

Prove the following equalities,

$$h_l^+(-kr) = (-1)^l [h_l^+(kr)]^*, \quad (12.195)$$

$$u_l(r; -k) = (-1)^{l+1} [u_l(r; k)]^*. \quad (12.196)$$

Solution: Following Eq. (10.1.16) in Ref. [27], $h_l^+(z) = i^{-l-1} z^{-1} e^{iz} \sum_{k=0}^l (n + \frac{1}{2}, k)(-2iz)^{-k}$, where $(l + \frac{1}{2}, k)$ are real functions of l and k. For $z = kr$ real, the equality (12.195) is obtained by replacing the sign of z in the above expression. For Eq. (12.196), note that $u_l(r; -k)$ satisfies the same equation as $u_l(r; k)$ and it is required to be a regular solution. Hence, $u_l(r; -k) = e^{i\alpha} u_l(r; k)$ for an arbitrary phase α. Following Eq. (12.163b) with $A = e^{i\delta_l(k)}$, we may write,

$$u_l(r; -k) \xrightarrow[r \to \infty]{} e^{i\delta_l(-k)} \sin[-kr - \tfrac{1}{2} l\pi + \delta_l(-k)] = e^{i\alpha} e^{i\delta_l(k)} \sin[kr - \tfrac{1}{2} l\pi + \delta_l(k)].$$

This equality is possible only if $\delta_l(-k) = -\delta_l(k)$ and $e^{i\alpha} = (-1)^{l+1}$.

Example 1: Scattering from an Impenetrable Sphere

As a first example consider the scattering from an impenetrable (i.e., hard) sphere, defined by,

$$v(r) = \begin{cases} \infty & (r < r_0) \\ 0 & (r > r_0) \end{cases}. \quad (12.197)$$

The wave function must vanish for $r \leq r_0$. Setting $R_l(r_0) = 0$ in Eq. (12.171) implies the relation,

$$S_l = e^{2i\delta_l} = -\frac{h_l^-(kr_0)}{h_l^+(kr_0)} \quad \text{(for hard sphere potential of radius } r_0\text{)}. \quad (12.198)$$

This relation is also equivalent to

$$\tan \delta_l = \frac{j_l(kr_0)}{n_l(kr_0)}. \tag{12.199}$$

In the low-energy limit $kr_0 \ll 1$, using the approximations of the Bessel and Neumann functions for small argument $z \ll 1$, one can go further and approximate,

$$\tan \delta_l \approx -\frac{(kr_0)^{2l+1}}{(2l+1)[(2l-1)!!]^2}. \tag{12.200}$$

This shows that at low energy ($kr_0 < 1$), $\tan \delta_l$ strongly falls off with l. For s-wave scattering ($l = 0$), the limit $k \to 0$ of the expressions for the differential and total cross-sections [Eqs. (12.166) and (12.167), respectively] exists due to the factor $\frac{1}{k^2}$, and one has,

$$\frac{d\sigma}{d\Omega} = r_0^2, \quad \sigma = 4\pi r_0^2 \tag{12.201}$$

in agreement with Eq. (12.168). In the high-energy limit, $kr_0 \gg 1$, we can use the result (12.199) in the partial wave expansions for either the differential cross-section (12.166) or the total cross-section (12.167), keeping in mind that, classically, the number of contributing partial waves is of the order of kr_0. Hence, the sum can be extended with $l \to \infty$, but it still turns out to be difficult to evaluate, especially for the differential cross-section where it includes rapidly oscillating terms. For the total cross-section, the summation reads,

$$\sigma = \frac{4\pi}{k^2} \sum_{l=0}^{\infty} \frac{(2l+1)j_l^2(kr_0)}{j_l^2(kr_0) + n_l^2(kr_0)}. \tag{12.202}$$

Using asymptotic expressions for the Bessel functions, it turns out that the main contribution to the sum comes from terms with $l < kr_0 + A(kr_0)^{\frac{1}{3}}$, where $A > 0$ is of order unity. Moreover, the sum can be approximated by $\frac{1}{2}(kr_0)^2[1 + O(kr_0)^{-\frac{2}{3}}]$. Thus, within this approximation, the total cross-section is,

$$\sigma = 2\pi r_0^2, \tag{12.203}$$

which is twice the classical result. Both the low-energy (12.201) and the high-energy (12.203) results are attributed to the wave nature of particles. However, the precise relation with the pertinent factors 4 and 2 is not completely established.

Example 2: Scattering from a Penetrable Sphere
A less trivial example is that of scattering from a *penetrable sphere* (a finite spherical "square well"),

$$v(r) = \begin{cases} -V_0 & (r < r_0) \\ 0 & (r > r_0) \end{cases}, \tag{12.204}$$

where $V_0 > 0$ is the depth of the potential well. Unlike the previous case of hard sphere, now the wave function $R_l(r)$ [which solves Eq. (12.155)] does not vanish at $r < r_0$ but should vanish at the origin for $l > 0$. It is then given by

$$R_l(r) = A j_l(\kappa r), \quad \text{for } r < r_0, \tag{12.205}$$

where

$$\kappa = \sqrt{k^2 + V_0} = k\sqrt{1 + \frac{V_0}{k^2}} \tag{12.206}$$

12.5 Central Potentials

is the modified wavenumber inside the potential well. The square root multiplying the wavenumber k in the second term on the RHS of Eq. (12.206) is in fact the refraction index,

$$n(k) = \frac{\kappa}{k} = \sqrt{1 + \frac{V_0}{k^2}}. \tag{12.207}$$

The matching procedure at $r = r_0$ is straightforward. Define

$$\gamma \equiv kr_0, \quad \zeta \equiv \kappa r_0, \tag{12.208}$$

and employ the following notation for logarithmic derivatives,

$$[j_l](\zeta) = \frac{1}{j_l(z)} \left(\frac{dj_l(z)}{dz}\right)_{z=\zeta}, \quad [h_l^\pm](\gamma) = \frac{1}{h_l^\pm(z)} \left(\frac{dh_l^\pm(z)}{dz}\right)_{z=\gamma}. \tag{12.209}$$

Then, one finds that

$$S_l(k) = \frac{h_l^-(\gamma)}{h_l^+(\gamma)} \frac{\kappa[j_l](\zeta) - k[h_l^-](\gamma)}{k[h_l^+(\gamma)] - \kappa[j_l](\zeta)}. \tag{12.210}$$

Problem 12.13

(a) Show that the phase shift for the penetrable sphere is

$$\tan \delta_l(k) = \frac{kj_l'(\gamma)j_l(\zeta) - \kappa j_l'(\zeta)j_l(\gamma)}{kn_l'(\gamma)j_l(\zeta) - \kappa j_l'(\zeta)n_l(\gamma)}$$

(b) Carry out the analysis for a repulsive square well potential.

Guidance: In the equations above, replace $V_0 \to -V_0 < 0$. Distinguish the two cases: (1) $k^2 - V_0 > 0$ and (2) $k^2 - V_0 < 0$. In case 1, $\kappa, n(k)$, and ζ are real, and all the equations remain the same. In case 2, $\kappa = i\sqrt{V_0 - k^2}$, $\zeta = \kappa r_0$ are pure imaginary and $n(k)$ is complex. Using the Rayleigh formulas in Abramowitz and Stegun [27], Eqs. (10.1.25) and (10.1.26) with $z \to ix$, you will find that

$$j_l(ix) = (-ix)^l \left(\frac{1}{x}\frac{d}{dx}\right)^l \frac{\sinh x}{x}, \quad n_l(ix) = i(-ix)^l \left(\frac{1}{x}\frac{d}{dx}\right)^l \frac{\cosh x}{x}. \tag{12.211}$$

Hence, $j_l(ix)$ is real (pure imaginary) for even (odd) l, while $n_l(ix)$ is real (pure imaginary) for odd (even) l. Show that S_l is unitary and that δ_l is real.

Although the method for evaluating S_l is straightforward, the calculation of the scattering amplitude (12.165) is more subtle. The usual expression,

$$f(\theta) = \frac{1}{2ik} \sum_{l=0}^{\infty} (2l+1) i^l (S_l - 1) P_l(\cos\theta). \tag{12.212}$$

might contain a large number of terms. As we have already stated, the number of terms with appreciable contribution to the partial wave summation is of order $\gamma + A\gamma^{\frac{1}{3}}$, $A \simeq 1$, and the convergence appears to be very slow. Even when the summation can be performed up to $\gamma \approx 300$, the resulting scattering amplitude exhibits strong fluctuations in θ and γ

(or equivalently the refractive index n). A method to overcome this problem has been suggested by Nussenzveig based on application of the *Watson transform*. It is especially adapted for a range of energies and refractive indices such that

$$\gamma^{\frac{1}{3}} \gg 1, \quad (n-1)\gamma^{\frac{1}{3}} \gg 1. \tag{12.213}$$

which includes the case of light scattering from water droplets. The basic idea is to rewrite Eq. (12.210) in terms of the cylindrical Bessel functions $H^{1,2}_{\lambda=l+\frac{1}{2}}(\gamma) = h_l^{\pm}(\gamma)$ and $J_{\lambda=l+\frac{1}{2}}(\zeta) = j_l(\zeta)$, and continue the expression,

$$S(\lambda, k) = \frac{H_\lambda^2(\gamma)\, \kappa[J_\lambda](\zeta) - k[H_\lambda^2](\gamma)}{H_\lambda^1(\gamma)\, k[H_\lambda^1(\gamma)] - \kappa[J_\lambda](\zeta)}, \tag{12.214}$$

into the complex λ plane. The Watson transform evaluates the partial wave summation (12.212) by a contour integral,

$$f(\theta, \gamma) = \frac{i}{2\gamma} \int_C d\lambda \, \frac{\lambda}{\cos \pi \lambda} P_{\lambda-\frac{1}{2}}(\cos\theta) e^{-i\pi\lambda}[1 - S(\lambda, \gamma)], \tag{12.215}$$

where C is the contour shown in Fig. 12.8. In principle, this contour integral can be evaluated using the residue theorem once the contour is deformed to include a large circle of radius R, such that the integrand vanishes on $|\lambda| = R$ as $R \to \infty$. As a function of the complex variable λ, the S matrix is a meromorphic function with simple poles, which are the roots of

$$k[H_\lambda^1(\gamma)] = \kappa[J_\lambda](\zeta). \tag{12.216}$$

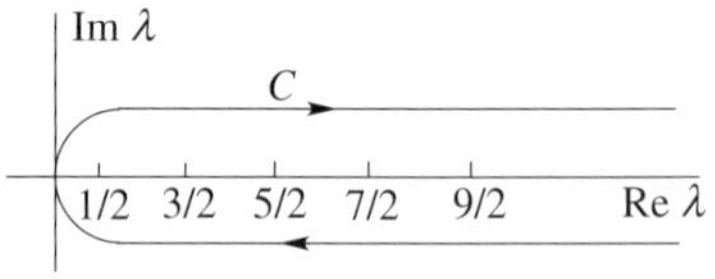

FIG 12.8 Integration contour C in the complex λ plane, for evaluating the scattering amplitude $f(\theta, \gamma)$ employing Eq. (12.215).

These are referred to as *Regge poles* and appear in numerous physical contexts. In particular, they play an important role in particle physics. Locating the Regge poles and performing the sum over residues require great effort and a high degree of mathematical sophistication, and we will not elaborate on this topic here.

Recipe for Numerical Evaluation of Phase Shifts

In the general case of an arbitrary central potential $v(r)$, an exact analytic expression for the phase shifts cannot be obtained, and a reliable method for evaluating the phase shifts δ_l is required. This can be achieved by using either numerical integration or approximation methods (see Sec. 12.7). The basic steps in a numerical procedure are as follows:

- Solve the differential Eq. 12.159 for $u_l(r)$ starting from the origin $r=0$ outward using the initial conditions $u_l(0) = 0, u_l(\Delta r) = C$, where Δr is an infinitesimal step and $C > 0$ is arbitrary.
- At a point r_0 beyond which $v(r)$ is negligible, evaluate the logarithmic derivative,

$$[u_l] \equiv \frac{1}{u_l(r_0)} \frac{du_l(r)}{dr}\bigg|_{r=r_0}. \tag{12.217}$$

- Comparing with the asymptotic form (12.161b), one finds,

$$\tan \delta_l = \frac{\hat{j}_l(kr_0)}{\hat{n}_l(kr_0)} \frac{k[\hat{j}_l] - [u_l]}{k[\hat{n}_l] - [u_l]}. \tag{12.218}$$

Instead of solving the differential Equation (12.159) for $u_l(r)$, one may solve the integral equation (12.194) for $R_l(r)$ and then compute $[u_l]$. Alternatively, one may try to solve the partial wave Lippmann–Schwinger integral equation (12.182) directly and employ (12.176) to evaluate the phase shift. In this case, care should be exercised in taking the limit $\eta \to 0$.

12.5 Central Potentials

Propagating Logarithmic Derivatives

Propagation of the wave function is often proved to be unstable. An alternative procedure is to write down an appropriate equation for another function whose propagation is more stable than that of the wave function itself. An example of such function is the logarithmic derivative of the wave function (or, in the present case, its inverse). Formally, the radial equation (12.159) for $u_l(r)$ can be rewritten as a first-order nonlinear differential equation for the inverse of its logarithmic derivative, $D_l \equiv u_l(r)/u'_l(r)$. The reason for this somewhat unusual choice is that since $u_l(0) = 0$, the logarithmic derivative is not defined at $r = 0$, while $D_l(0) = 0$. Simple manipulation yields,

$$\frac{dD_l(r)}{dr} = 1 - \left[\frac{l(l+1)}{r^2} + v(r) - k^2\right] D_l(r)^2, \quad D_l(0) = 0. \quad (12.219)$$

For $r \to \infty$, the asymptotic form of $D_l(r)$ is derived from (12.163b):

$$D_l(r) \xrightarrow[r \to \infty]{} \frac{1}{k} \cot\left(kr - \frac{1}{2}l\pi + \delta_l\right). \quad (12.220)$$

Integrating Eq. (12.219) from $r = 0$ outside and using Eq. (12.220) at very large r consists of a stable numerical procedure for calculating phase shifts.

Problem 12.14

Derive Eq. (12.219) from Eq. (12.159).

Answer: $D'_l = \frac{(u'_l)^2 - u_l u''_l}{(u'_l)^2} = 1 - \left[\frac{l(l+1)}{r^2} + v(r) - k^2\right] D_l(r)^2, \quad D_l(0) = 0.$

12.5.4 PHASE SHIFT ANALYSIS

In previous sections, we have established the relevance of phase shifts $\{\delta_l\}$ for the evaluation of cross-sections and suggested recipes for their evaluation. There are general properties of phase shifts, in particular, their dependence on energy and angular momentum which are important for the extraction of information on the scattering potential, that will be discussed in this subsection. For example, we elucidate the relation between the sign of the phase shift and the nature of the potential (repulsive or attractive). Another question is whether the behavior of phase shift at low energy $k \to 0$ can give us information on the existence of bound states at the pertinent angular momentum l (even though bound states occur at negative energy while phase shifts are defined only at positive energies). An important question (which will not be answered here due to its mathematical complexity) is related to the so-called *inverse scattering problem*: If one knows the phase shifts $\delta_l(k)$ for all k and l, is it possible to uniquely determine the scattering potential $v(r)$? The formalism described above allows us to draw some general results about the energy and angular momentum dependence of phase shifts and to answer some of these questions.

Phase shifts appear solely within trigonometric functions that can be transformed into expressions involving $\tan \delta_l$, hence they can typically only be determined up to multiples of π. For example, the asymptotic form of the radial wave function $u_l(r) \to e^{i\delta_l} \sin(kr - \frac{1}{2}l\pi - \delta_l)$ [see Eq. (12.163b)] remains unchanged (up to a sign) when $\delta_l \to \delta_l + n\pi$. Much of the discussion below is concerned with low-energy behavior of phase shifts and, as will be shown, $\lim_{k \to 0} \tan \delta_l(k) = 0$, so one might think of choosing $\delta_l(k = 0) = 0$. However, we shall see that it is preferable to adopt the constraint $\delta(k = \infty) = 0$, which implies that $\delta_l(0)$ is some integer multiple of π (including the case $\delta_l(0) = 0$).

Relation Between the Signs of $\delta_l(k)$ and $v(r)$

To derive a correlation between the signs of $\delta_l(k)$ and $v(r)$, we inspect Eq. (12.159) more closely. Since the analysis below employs two solutions of Eq. (12.159) at different energies, let us specify the corresponding energies (or wavenumbers) explicitly and write these solutions as $u_l(r; k_i)$ or $u_l(r; \varepsilon_i)$ with $i = 1, 2$. Our technique is based on analyzing the behavior of the logarithmic derivative,

$$[u_l(r)] \equiv \frac{u_l'(r)}{u_l(r)}, \qquad (12.221)$$

as function of energy $\varepsilon = k^2$. Dropping the angular momentum subscript l, we rewrite Eq. (12.159) in the form,

$$u'' + [k^2 - v_{\text{eff}}(r)]u = 0, \qquad (12.222)$$

where $v_{\text{eff}}(r) \equiv v(r) + l(l+1)/r^2$. Recall the required boundary conditions, $u(0) = 0$, $u(r) \to A \sin(kr - \frac{1}{2}l\pi + \delta_l(k))$. The functions $u(r; k)$ that solve Eq. (12.222) depend on the wavenumber k. Now, consider two solutions $u_1 = u(r; k_1)$ and $u_2 = u(r; k_2)$ corresponding to different wavevectors. Write Eq. (12.222) for each one of them, multiply by the other, and then subtract. The resulting equation is,

$$u_1 u_2'' - u_2 u_1'' = \frac{d}{dr}[u_1 u_2' - u_2 u_1'] \equiv \frac{d}{dr} W(u_1, u_2) = (\varepsilon_1 - \varepsilon_2) u_1 u_2. \qquad (12.223)$$

The quantity $W(u_1, u_2) = [u_1 u_2' - u_2 u_1']$ is often used for determining the independence of two solutions of a second-order differential equation; it is called the *Wronskian* of u_1 and u_2. The Wronskian of two linearly dependent solutions vanishes. We keep the same notation, but it should be stressed that in our case the two solutions correspond to different energies, and the quantity so defined is, strictly speaking, not a Wronskian. Integrating Eq. (12.223) between two points r_1, r_2 yields,

$$W(u_1, u_2)|_{r_1}^{r_2} = [u_1 u_2' - u_2 u_1']_{r_1}^{r_2} = (\varepsilon_1 - \varepsilon_2) \int_{r_1}^{r_2} dr\, u_1 u_2. \qquad (12.224)$$

Now we consider two solutions u_1, u_2, such that (1) they have the same logarithmic derivative at a fixed point, say d,

$$[u_1(d; \epsilon_1)] = [u_2(d; \epsilon_2)] \equiv q_d, \qquad (12.225)$$

and (2) the corresponding energies are infinitesimally close to each other, and hence the two solutions are close to each other in an appropriate norm,

$$\varepsilon_2 = \varepsilon_1 + \delta\epsilon$$
$$u_2 = u_1 + \delta u. \qquad (12.226)$$

Under these conditions, $W(u_1, u_2)|_{r=d} = 0$, and of course, $W(u_1, u_1) = 0$. Let us now study these quantities at another point $b \neq d$. Direct calculation yields,

$$W(u_1, u_1 + \delta u) = u_1^2 \delta\{u_1\}, \qquad (12.227)$$

where we use the notation, $\delta\{A\} \equiv \frac{\partial A}{\partial \varepsilon} \delta\varepsilon$. Collecting these results, we find,

$$\left.\frac{\partial [u_1(r; \varepsilon)]}{\partial \varepsilon}\right|_{r=b} = -\frac{1}{u_1^2(b; \varepsilon)} \int_d^b dr\, u_1^2(r; \varepsilon). \qquad (12.228)$$

Thus, we have established the following: If $u(r; \varepsilon)$ is a solution of Eq. (12.222) with $u(0; \varepsilon) = 0$ and $[u(d; \varepsilon)] \equiv \frac{1}{u(r;\varepsilon)} \frac{du(r;\varepsilon)}{dr}|_{r=d} = q_d$, then, as a function of energy ε, $[u(r; \varepsilon)]$ is a monotonic decreasing function above $r = d$. The

12.5 Central Potentials

precise functional form of $[u(r;\varepsilon)]$ depends on the scattering central potential $v(r)$. However, note that for $r > d$, where $v(r) = 0$, the relation (12.218) between the logarithmic derivative and the phase shift is independent of the potential.

We will now employ the above analysis in order to obtain a bound on the phase shift at large l and also to relate the sign of the phase shift and the sign of the (central) potential $v(r)$. To this end, let us check the deviation $k\eta_l$ of $[u_l(d;\varepsilon)]$ from the zero-potential case $v(r) = 0$ (restoring l dependence notation),

$$[u_l(d;\varepsilon)] = k\{[\hat{j}_l(kd)] + \eta_l\}. \tag{12.229}$$

Then, we rewrite Eq. (12.218) as,

$$\tan\delta_l = \frac{\eta_l^2 \hat{j}_l^2(kd)}{\eta_l^2 \hat{j}_l(kd)\hat{n}_l(kd) - 1}. \tag{12.230}$$

Expression (12.230) is exact but it makes sense only if, in Eq. (12.229), the inequality $|\eta_l| \ll |[\hat{j}_l(kd)]|$ holds. For $l \gg (kd)^2$, we can employ the asymptotic approximations for $j_l(kd)$ and $n_l(kd)$ in order to estimate the upper limit on $|\eta_l|$, $|\eta_l| \ll \frac{l}{kd}$ for which the use of expression (12.229) is useful. Within this constraint, we conclude from Eq. (12.230) the approximation,

$$\tan\delta_l \approx -\frac{\eta_l 2^{2l}(l!)^2(kd)^{2l+2}}{[(2l+1)!]^2}. \tag{12.231}$$

Equation (12.231) shows that $\tan\delta_l$ falls off rapidly at large l. It also shows that δ_l has the same sign as η_l. Note that η_l is the difference between logarithmic derivatives of the wave functions corresponding to $v(r) \ne 0$ and $v(r) = 0$. Roughly speaking, if the potential is repulsive, $v(r) > 0$, it pushes the wave function away from the origin, thereby causing it to have a larger logarithmic derivative at d compared with the $v = 0$ case. Conversely, an attractive potential pulls the wave function inside, and its logarithmic derivative at d is expected to be smaller than $[j_l(kd)]$. Thus, we arrive at the useful relations,

$$v(r) > 0 \Rightarrow \text{repulsive potential} \Rightarrow \eta_l > 0 \Rightarrow \delta_l < 0,$$
$$v(r) < 0 \Rightarrow \text{attractive potential} \Rightarrow \eta_l < 0 \Rightarrow \delta_l > 0. \tag{12.232}$$

Although the above analysis is based on the condition $l \gg kd$, it turns out to be valid even for small l.

Behavior of Phase Shifts at Low Energy

Let us now derive the low-energy behavior of the phase shifts $\delta_l(k)$ using their relation to the on-shell t matrix $t_l(k)$ and the S matrix through Eqs. (12.176) and (12.177). This extends the relations obtained for large l in Eqs. (12.230) and (12.231). The starting point is the off-energy shell partial wave t matrix $t_l(k,k';\varepsilon)$ as defined by the Lippmann–Schwinger equation (12.182). It can be considered as a momentum space representation $\langle k|\hat{t}_l(\varepsilon)|k'\rangle$ of an operator $\hat{t}_l(\varepsilon)$. This operator can be represented in configuration space $\langle r'|\hat{t}_l(\varepsilon)|r\rangle \equiv t_l(r',r;\varepsilon)$ by inserting two unit operators expressed in terms of integrals over a complete sets of radial states $\int dr'\, r'^2|r'\rangle\langle r'|$, $\int dr\, r^2|r\rangle\langle r|$, and noting that within this partial wave subspace, $\langle r|k\rangle = j_l(kr)$:

$$t_l(k',k;\varepsilon) = \langle k'|\hat{t}_l(\varepsilon)|k\rangle = \int_0^\infty dr'drr^2 r'^2 j_l(kr')j_l(kr)t_l(r',r;\varepsilon). \tag{12.233}$$

For small k, the Bessel function satisfies $j_l(kr) \approx (kr)^l$, and therefore, as $k \to 0$ on the energy shell,

$$t_l(k) = t_l(k,k) \to \left[\int_0^\infty drdr'\, t_l(r',r;\varepsilon \to 0)r'^{l+2}r^{l+2}\right]k^{2l} \equiv C_l(\varepsilon \to 0)k^{2l}, \tag{12.234}$$

where, in general, the limit $C_l(\varepsilon \to 0) \equiv C_l \neq 0$ exists. However, there is a subtle exception for $l = 0$ in case of zero-energy resonance (see below the discussion of the Levinson theorem). The convergence of the integral in the prefactor defining C_l requires that the potential $v(r)$ will fall off faster than r^{2l+2}. The corresponding expressions for the S matrix and the phase shift are derived immediately using Eqs. (12.176) and (12.177), i.e.,

$$S_l(k) \to 1 - i\pi C_l k^{2l+1}, \tag{12.235}$$

$$\tan \delta_l(k) \to -\frac{\pi}{2} C_l k^{2l+1}. \tag{12.236}$$

As a consequence, in the absence of zero-energy s-wave resonance, the phase shift for all partial waves behaves as $Ak^{2l+1} + n\pi$ for small k, where n is an integer. $A > 0$ and $n \geq 0$ for attractive potential, whereas $A < 0$ and $n = 0$ for repulsive potential. For $l = 0$, in the presence of zero energy resonance, n should be replaced by $(n + 1/2)$.

Effective Range Analysis

The discussion below is limited to the case of s-wave scattering, $l = 0$. In many cases, the low-energy phase shift $\delta_0(k)$ for $k \gtrsim 0$ is only weakly dependent on the detailed nature of the scattering potential $v(r)$. We have seen above in Eq. (12.236) that the s-wave phase shift $\delta_0(k)$ vanishes linearly (modulo $n\pi$) with the wavenumber k as $k \to 0$. Therefore, the s-wave scattering amplitude $f_0(k) = e^{i\delta} \sin \delta / k$ of (12.173) is expected to remain finite as $k \to 0$. The limit of the s-wave scattering amplitude $f_0(k)$ as $k \to 0$ defines an important quantity referred to as the *s-wave scattering length*, a_0, which has the dimension of length:

$$\lim_{k \to 0} f_0(k) = \lim_{k \to 0} \frac{e^{i\delta_0(k)} \sin \delta_0(k)}{k} \equiv -a_0, \tag{12.237a}$$

$$\delta_0(k) \simeq -ka_0 + n\pi, \tag{12.237b}$$

assuming no zero energy resonance (see discussion of Levinson theorem below), where the correction to Eq. (12.237b) will be shown to be of order $O(k^2)$. According to Eqs. (12.165) and (12.169), the scattering amplitude at zero energy, $f_{k=0}(\theta)$, which is of course isotropic, is equal to the s-wave scattering amplitude $f_0(k)$, hence,

$$f_{k=0}(\theta) = -a_0 \quad \text{(scattering amplitude at zero energy)}, \tag{12.238}$$

and the zero-energy total cross section is given by

$$\sigma = 4\pi a_0^2 \quad \text{(total cross-section at zero energy)}. \tag{12.239}$$

To substantiate Eq. (12.237a), the s-wave scattering amplitude can be rewritten as,

$$f_0(k) = \frac{e^{i\delta_0(k)} \sin \delta_0(k)}{k} = \frac{1}{k \cot \delta_0(k) - ik}. \tag{12.240}$$

Now recall Eq. (12.236), stating that as $k \to 0$, $\tan \delta_l(k) \simeq k^{2l+1}$. Therefore, the leading term in the expansion of $k \cot \delta_0(k)$ as a function of k is an even function of k.[4] As $k \to 0$, the term ik in the denominator on the RHS of Eq. (12.240) vanishes but the limit of $k \cot \delta_0(k)$ remains finite since $\tan \delta_0(k) \sim k$. This limit is denoted as $-1/a_0$, and hence, we get Eq. (12.237a). The scattering length a_0 for elastic scattering processes is real.

---

[4] Quite generally, the occurrence of only even powers of k in the expansion of a physical quantity $F(k)$ may also be related to the analyticity of this function in the energy variable $\varepsilon = k^2$. For a potential $v(r)$ that falls off sufficiently fast (i.e., it decays at least exponentially at large r), the analyticity close to the real ε axis is established, and hence, it can be expanded in power series of ε.

12.5 Central Potentials

Scattering Length and Range of the Potential
Comparing Eq. (12.239) with Eq. (12.201) for the total cross-section for scattering from an impenetrable sphere, one might get the false impression that the absolute value of the scattering length a_0 is nothing but the range r_0 of the potential. However, this is misleading. First, the two quantities have different physical content, in particular, the scattering length can be positive or negative. Second, there are cases where $|a_0| \gg r_0$. In the following two examples, this distinction is clarified.

Role of Scattering Length in Cold Atom Systems
In weakly interacting many-body systems, such as in a cold bosonic atomic gas, one often applies a *mean-field approximation* wherein the particle interacts with the mean-field of all other particles. At very low energies and for dilute gases, an excellent approximation for the interaction energy of an atom at space point $\mathbf{r}$ with the other atoms, within the mean-field approximation, is $U(\mathbf{r}) = \frac{4\pi \hbar a_0}{m} n(\mathbf{r})$, where $n(\mathbf{r})$ is the gas density at $\mathbf{r}$, m is the atomic mass, and a_0 is the s-wave scattering length of the bosonic atoms.

Efimov states
The quantum states of three identical bosonic atoms have been analyzed in terms of the two-body s-wave scattering length. When three identical bosons interact, an infinite series of excited three-body energy levels exist when a two-body state is exactly at the dissociation threshold. A corollary is that there exist bound states (called Efimov states) of three bosons even if the two-particle attraction is too weak to allow two bosons to form a pair. When $|a_0| \gg r_0$ (where r_0 is the range of the two-body potential), the number of three-body bound states increases as $N = \frac{1}{\pi} \log[|a_0|/r_0]$. These so-called Efimov states have recently been experimentally identified in cold atom systems.

Inclusion of $O(k^2)$ Corrections
As has been shown above, the first correction to Eqs. (12.237a) and (12.237b) is of order k^2. Our goal is to elucidate this correction and relate it to the scattering potential. The correction is shown to be positive, so we write it as,

$$\boxed{k \cot \delta_0(k) = -\frac{1}{a_0} + \frac{1}{2} r_e k^2 + O(k^4).} \tag{12.241}$$

This is the *effective range approximation*, with a_0 being the s-wave scattering length and r_e being the *effective range*. All potentials that produce the same scattering length and effective range are, in some sense, equivalent as far as low-energy scattering is concerned. The constants a_0 and r_e have very different physical content. The physical content of a_0 can be deduced from Eqs. (12.238) and (12.239), together with the analysis of the s-wave phase shift at low energy. If we use Eq. (12.161b) and modify the prefactor of the wave functions, such that

$$u_0(r) = C(k) \left(\sin kr \cot \delta_0 + \cos kr \right), \tag{12.242}$$

where $C(k)$ is finite at $k = 0$, then for very small k, we have,

$$\boxed{u_0(r) \to C(0) \left(1 - \frac{r}{a_0} \right).} \tag{12.243}$$

In addition to its relevance to the scattering cross-section and the low-energy wave function, the sign (and magnitude) of a has other important consequences. In the discussion of Levinson's theorem (see below), it is shown that $\delta(0) = n_b \pi$, where n_b is the number of s-wave bound states (except when there is a zero-energy resonance). Therefore, according to Eqs. (12.240) and (12.241), we can write

$$\text{sign}(a_0) = (-1)^{n_b}. \tag{12.244}$$

Thus, although the approximation (12.243) is rather crude, at least for $n_b = 1$, the occurrence of a single radial node is consistent with the general theorems relating bound states and radial nodes.

The scattering length plays an important role beyond two-body scattering, as is clear from the following two examples.

Let us now relate the effective range r_e appearing in Eq. (12.241) with the solution of the Schrödinger equation (12.222) for $u_0(r)$. For this purpose, we normalize $u_0(r)$ in Eq. (12.242), such that $C(k) = 1$ (this is always possible since the Schrödinger equation (12.222) is linear). Then, as $r \to \infty$,

$$u_0(r) \to \sin kr \cot \delta_0(k) + \cos kr. \tag{12.245}$$

The analysis detailed here is based on relations (12.223) and (12.224). First, they are used as before for $u_0(r)$ defined above. Second, they are applied on another radial function $v_0(r)$, which is simply the solution of the free radial s-wave Schrödinger equation, Eq. (12.222), albeit setting $v_{\text{eff}}(r) = 0$, which has the same form as $u_0(r)$ in Eq. (12.245) for all r:

$$\left(\frac{d^2}{dr^2} + k^2\right) v_0(r) = 0, \tag{12.246a}$$

$$v_0(r) = \sin kr \cot \delta_0(k) + \cos kr. \tag{12.246b}$$

$$v_0(0) = 1, \quad v_0'(0) = k \cot \delta_k(0). \tag{12.246c}$$

In the next step, we employ Eqs. (12.223) and (12.224) on two pairs of solutions (the energy of each solution will be indicated explicitly), $u_0(r; k)$, $u_0(r; q)$, on the one hand, and $v_0(r; k)$, $v_0(r; q)$, such that in each pair the energies ε_k and ε_q are different. Subtracting the ensuing relations one from the other and taking the upper limit of integration as $r \to \infty$ yields

$$W[v_0(r; q), v_0(r; k)]_{r=0} = (q^2 - k^2) \int_0^\infty dr [u_0(r; q) u_0(r; k) - v_0(r; q) v_0(r; k)], \tag{12.247}$$

where it is understood that the Wronskian satisfies

$$W[v_0(r; q), v_0(r; k)]_{r \to \infty} - W[u_0(r; q), u_0(r; k)]_{r \to \infty} \to 0. \tag{12.248}$$

Using Eq. (12.246c) to evaluate the Wronskian $W[v_0(r; q), v_0(r; k)]_{r=0}$, we then arrive at the useful relation,

$$k \cot \delta_0(k) - q \cot \delta_0(q) = (k^2 - q^2) \int_0^\infty dr [v_0(r; q) v_0(r; k) - u_0(r; q) u_0(r; k)]. \tag{12.249}$$

We now take the limit $q \to 0$ and recall that $q \cot \delta_0(q) \to -1/a_0$, leading to

$$k \cot \delta_0(k) = -\frac{1}{a_0} + \frac{1}{2} k^2 r_e + \dots, \tag{12.250}$$

where the expression we obtain for the effective range r_e is

$$r_e = 2 \int_0^\infty dr [v_0(r; 0) v_0(r; k) - u_0(r; 0) u_0(r; k)]. \tag{12.251}$$

12.5 Central Potentials

Note that Eq. (12.250) is identical to Eq. (12.241). Another useful relation that can be derived from Eq. (12.248) by letting $q \to k$ is as follows,

$$\frac{d[k \cot \delta_0(k)]}{dk^2} = \int_0^\infty dr [v_0^2(r;k) - u_0^2(r;k)]. \tag{12.252}$$

This relation is valid at any energy, since no energy limit has been employed in its derivation. To proceed further, let us assume that the central potential $V(r)$ is negligible for $r > r_0$. Then, the integral can be carried out up to r_0. Moreover, the integral over $v_0(r;k)^2$ is immediate following the definition of $v_0(k;r)$ in Eq. (12.246b). Using these facts and performing the derivative on the LHS of Eq. (12.252), we get an equality for $\frac{d\delta_0(k)}{dk}$, which is then modified into an inequality,

$$\frac{d\delta_0(k)}{dk} = \frac{1}{2k} \sin[2kr_0 + \delta_0(k)] + 2 \sin 2\delta_0(k) \int_0^{r_0} dr u_0^2(r;k) \geq -\left(r_0 + \frac{1}{2k}\right), \tag{12.253}$$

a relation originally derived by Wigner in 1955.

Jost Functions

Following our discussion of Green's functions and the resolvent operators in Sec. 12.3.4, it is evident that analytic continuation of functions of the energy E or of the momentum k from the real axis into the complex plane often serves as a powerful tool to investigate the properties of these functions. We have previously tacitly used this when we added a small imaginary part $\pm i\eta$ to the energy E as the argument of the Green's function, in order to control the asymptotic behavior of the wave function (outgoing spherical waves for $+i\eta$ and incoming spherical waves for $-i\eta$). In the analysis presented below, this concept is employed for partial wave amplitudes.

Consider Eq. (12.159) for $u_l(r)$ for $r > r_0$, such that $v_l(r > r_0) \approx 0$. Requiring $u_l(r)$ to be real, we write it as

$$u_l(r) = \frac{1}{2}[\mathcal{F}_l(k)\hat{h}_l^-(kr) + \mathcal{F}_l^*(k)\hat{h}_l^+(kr)], \tag{12.254}$$

where the Riccati–Hankel functions $\hat{h}_l^\pm(z)$ are defined in Eq. (12.160) [see also Eq. (12.157)]. The function $\mathcal{F}_l(k)$ is called the *Jost function*; it plays an important role in the study of the analytic properties of phase shifts. Using Eqs. (12.171) and (12.172), we immediately conclude that the phase of $\mathcal{F}_l^*(k)$ is just $\delta_l(k)$, so that

$$S_l(k) = e^{2i\delta_l(k)} = \frac{\mathcal{F}_l^*(k)}{\mathcal{F}_l(k)}. \tag{12.255}$$

Taking $r \to \infty$ in Eq. (12.254), one finds

$$\mathcal{F}_l(k) = 1 + \frac{i}{k} \int_0^\infty dr' \, \hat{h}_l^+(kr') v(r') u_l(r';k). \tag{12.256}$$

Using Eqs. (12.195), (12.196), and (12.256) and the fact that $u_l(r;k)$ is real, we find,

$$\mathcal{F}_l(-k) = \mathcal{F}_l^*(k) \quad \text{(for real } k\text{)}. \tag{12.257}$$

Hence,

$$S_l(k) = \frac{\mathcal{F}_l(-k)}{\mathcal{F}_l(k)} = e^{2i\delta_l(k)}, \tag{12.258}$$

from which we deduce that

$$\mathcal{F}_l(k) = e^{-i\delta_l(k)}, \qquad (12.259)$$

and also that (see Problem 12.12),

$$\delta_l(-k) = -\delta_l(k) \pmod{2n\pi}. \qquad (12.260)$$

Here, k can be negative as a special case of the more general situation, where k is allowed to be complex. As far as $u_l(r;k)$ is concerned, there is no problem in its continuation to complex k because the coefficient k^2 in Eq. (12.159) is analytic in k and the boundary conditions for real $u_l(r;k)$ are independent of k. Therefore, it is an entire function of k. In the case of the Jost function $\mathcal{F}_l(k)$, one notices from Eq. (12.256) that as $\operatorname{Im} k \to \infty$, $\hat{h}_l^+(kr) \simeq e^{-\operatorname{Im}[k]r}$, so that $\mathcal{F}_l(k)$ should be analytic in the region $\operatorname{Im} k > 0$, with

$$\lim_{\operatorname{Im}[k] \to \infty} \mathcal{F}_l(k) = 1. \qquad (12.261)$$

A more rigorous analysis shows that $\mathcal{F}_l(k)$ is continuous on the real axis $\operatorname{Im} k = 0$ and that if the potential decays at large r as $e^{-\alpha r}$ with some constant $\alpha > 0$, it is analytic also on part of the lower half k plane, $\operatorname{Im} k > -\alpha$.

Since $\mathcal{F}_l(k)$ appears in the denominator of the S matrix (12.258), our interest focuses on *zeros of the Jost function*. If, for some real $\kappa > 0$, the analytic continued function $\mathcal{F}_l(k)$ has a zero at $k = i\kappa$, $\mathcal{F}_l(k = i\kappa) = 0$ (this does not imply $\mathcal{F}_l^*(i\kappa) = 0$), then the wave function (12.254) at this point becomes,

$$u_l(r)|_{k=i\kappa} = \mathcal{F}_l^*(i\kappa)\hat{h}_l^+(i\kappa r), \qquad (12.262)$$

which decays exponentially as $r \to \infty$ and is therefore a bound state. Thus, a zero of $\mathcal{F}_l(k)$ on the positive imaginary k axis corresponds to a bound state at binding energy $-B = -\frac{\hbar^2 \kappa^2}{2m}$. Since bound-state energies are real, there are no zeros of $\mathcal{F}_l(k)$, which are not pure imaginary.

Problem 12.15

Prove that $\mathcal{F}_l(k)$ cannot vanish for real positive k.

Hint: Consider Eq. (12.254).

So far we discussed the zeros of the Jost function in the upper half k plane. The zeros of $\mathcal{F}_l(k)$ in the lower half plane, $\operatorname{Im} k < 0$, are of special importance as they are related to resonance scattering (see below). Consider a simple zero $\bar{k}$ in the lower half plane with large real part and small negative imaginary part,

$$\mathcal{F}_l(\bar{k} = k_1 - ik_2) = 0, \quad k_1 \gg k_2 > 0. \qquad (12.263)$$

For real k in the neighborhood of $\bar{k}$, we can expand

$$\mathcal{F}_l(k) \approx (k - \bar{k}) \left. \frac{d\mathcal{F}_l(k)}{dk} \right|_{k=\bar{k}}. \qquad (12.264)$$

Therefore, the phase shift is given by

$$\delta_l(k) = -\arg \left. \frac{d\mathcal{F}_l(k)}{dk} \right|_{k=\bar{k}} - \arctan \frac{k_2}{k - k_1} \equiv \phi_l - \arctan \frac{k_2}{k - k_1}. \qquad (12.265)$$

As k varies from $k < k_1$ to $k > k_1$, $\delta_l(k)$ varies from $\phi_l(k)$ to $\phi_l(k) + \pi$. This is the essence of *resonance phenomena*: A tendency of a physical system to oscillate at larger amplitude at some set of discrete energies. In the present context, the scattering amplitude varies substantially at a resonance energy compared with other energies. The zeros of the Jost function $\mathcal{F}_l(k)$ corresponding to bound states and resonances are schematically displayed in Fig. 12.9.

12.5 Central Potentials

Dispersion Relation for the Jost Function

In their study of propagation of electromagnetic radiation, Kronig and Kramers showed that the requirement that the speed of light propagating in media should be smaller than c, leads to a relation between the real and imaginary parts of the complex refraction index $n(\omega)$, where ω is the frequency of radiation. The dependence of n on ω is referred to as dispersion, and the above relation is called a *dispersion relation*. The Kronig–Kramers relations are based on the fact that the speed of propagation is less than c, which in a field-theoretic nomenclature is referred to as *microscopic causality*. The Kronig–Kramers relations play an important role in physical systems because causality implies that an analyticity condition is satisfied. We discussed these relations in Chapter 7 in connection with linear response theory.

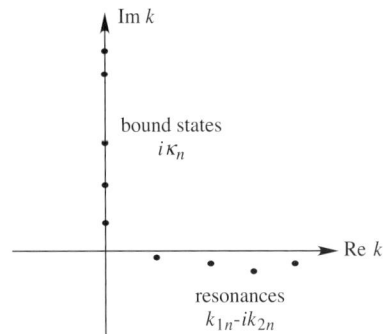

FIG 12.9 Schematic illustration of the zeros of the Jost function $\mathcal{F}_l(k)$ in the complex k plane. Poles at $k_n = i\kappa_n$, $n = 1, 2, 3$ with $\kappa_n > 0$ correspond to bound states, whereas poles at $k_n = k_{1n} - ik_{2n}$, $n = 1, 2, 3, 4$ and $k_{1n} \gg k_{2n} > 0$ correspond to resonances.

After defining the Jost function $\mathcal{F}_l(k)$ as a function of the complex wavenumber k, we derive dispersion relations for it and extract further information relating phase shifts to measurable quantities such as bound states and resonance energies. A detailed treatment of this fascinating topic goes beyond the scope of this book; only the essential concepts are introduced here.

Recall that $\mathcal{F}_l(k)$ is an analytic function of k in the upper half plane, including the real axis. Moreover, $\mathcal{F}_l(k) \to 1$ as $k \to \infty$ within this analyticity domain. Using Cauchy's Residue Theorem (see Appendix D, near Fig. D.3), we can write an integral relation, similar to the Kramers–Kronig relation obtained in Sec. 7.9, relating $\mathcal{F}_l(k)$ for any k such that $\operatorname{Im} k > 0$ in terms of $\mathcal{F}_l(k)$ with k on the real axis:

$$\mathcal{F}_l(k) - 1 = \frac{i}{2\pi} \int_{-\infty}^{\infty} dk' \frac{\mathcal{F}_l(k') - 1}{k' - k}. \tag{12.266}$$

As k approaches the real axis from above, this yields

$$\mathcal{F}_l(k) - 1 = \frac{i}{\pi} \mathcal{P} \int_{-\infty}^{\infty} dk' \frac{\mathcal{F}_l(k') - 1}{k' - k}, \tag{12.267}$$

where $\mathcal{P}$ indicates the principal part of the (otherwise singular) integral. Using the identity (12.136), we can rewrite Eq. (12.267) as

$$\mathcal{F}_l(k) = 1 - \frac{1}{\pi} \int_{-\infty}^{\infty} dk' \frac{\operatorname{Im} \mathcal{F}_l(k')}{k' - k + i\eta}. \tag{12.268}$$

Another equivalent form of (12.268) is

$$\operatorname{Re} \mathcal{F}_l(k) = 1 - \frac{\mathcal{P}}{\pi} \int_{-\infty}^{\infty} dk' \frac{\operatorname{Im} \mathcal{F}_l(k')}{k' - k}. \tag{12.269}$$

This expression defines a dispersion relation for the function $\mathcal{F}_l(k)$. It should be stressed that, since it is based on analyticity without any dynamical input, it cannot be regarded as an integral equation for $\mathcal{F}_l(k)$. A legitimate question that can be asked is whether the knowledge of phase shift $\delta_l(k)$ is sufficient for determining $\mathcal{F}_l(k)$. The answer is, in general, negative, since the phase shift $\delta_l(k)$ alone does not determine the location of bound states (as we have seen, these are zeros of $\mathcal{F}_l(k)$ at points $k_n = i\kappa_n$ with $\kappa_n > 0$, $n = 1, 2, \ldots n_l$). To determine $\mathcal{F}_l(k)$ from knowledge of $\delta_l(k)$ and k_n, consider the function

$$\bar{\mathcal{F}}_l(k) \equiv \left[\prod_{n=1}^{n_l} \frac{k + k_n}{k - k_n}\right] \mathcal{F}_l(k), \qquad (12.270)$$

which is analytic in the upper half plane and has no zeros there. Evidently, $\bar{\mathcal{F}}_l(k) \to 1$ as $k \to \infty$ within the analyticity domain. It is then justified to write a relation analogous to Eq. (12.267) for $\ln \bar{\mathcal{F}}_l(k)$,

$$\ln \bar{\mathcal{F}}_l(k) = i\frac{\mathcal{P}}{\pi} \int_{-\infty}^{\infty} dk' \frac{\ln \bar{\mathcal{F}}_l(k')}{k' - k}. \qquad (12.271)$$

We can now take the real parts of both sides and recall that $|\bar{\mathcal{F}}_l(k)| = |\mathcal{F}_l(k)|$, so

$$\ln |\bar{\mathcal{F}}_l(k)| = \ln |\mathcal{F}_l(k)| = -\frac{\mathcal{P}}{\pi} \int_{-\infty}^{\infty} \frac{\operatorname{Im} \ln \bar{\mathcal{F}}_l(k')}{k' - k} dk'. \qquad (12.272)$$

Recall from Eq. (12.259) that $\mathcal{F}_l(k) = |\mathcal{F}_l(k)| e^{-i\delta_l(k)}$, hence,

$$\operatorname{Im} \ln \bar{\mathcal{F}}_l(k) = -\delta_l(k) - i \sum_{n=1}^{n_l} [\ln(k + k_n) - \ln(k - k_n)]. \qquad (12.273)$$

Substitution of Eq. (12.273) into Eq. (12.272) yields an integral over a sum of logarithms, which can be carried out using contour integration. The result is

$$\ln |\mathcal{F}_l(k)| = \frac{\mathcal{P}}{\pi} \int_{-\infty}^{\infty} \frac{\delta_l(k')}{k' - k} dk' + \ln \prod_{n=1}^{n_l} \frac{k^2 + \kappa_n^2}{k^2}. \qquad (12.274)$$

We have already established that $\lim_{k \to \pm \infty} \delta_l(k) = 0$ since $\mathcal{F}_l(k) \to 1$ as a consequence of the boundary conditions on the Jost functions. Using $\ln |\mathcal{F}_l(k)| = \ln \mathcal{F}_l(k) + i\delta_l(k)$, we finally obtain

$$\ln \mathcal{F}_l(k) = \frac{1}{\pi} \int_{-\infty}^{\infty} dk' \frac{\delta_l(k')}{k' - k + i\eta} + \ln \prod_{n=1}^{n_l} \frac{k^2 + \kappa_n^2}{k^2}. \qquad (12.275)$$

Note that, unlike in Eq. (12.274), the integral is not a principal part integral. Exponentiating Eq. (12.275), we get an expression for $\mathcal{F}_l(k)$ in terms of the phase shift $\delta_l(k)$ and the bound-state energies $-\kappa_n^2$, as originally obtained by Jost and Kohn, and Newton:

$$\mathcal{F}_l(k) = \prod_{n=1}^{n_l} \left(1 + \frac{\kappa_n^2}{k^2}\right) e^{\frac{1}{\pi} \int_{-\infty}^{\infty} dk' \frac{\delta_l(k')}{k' - k + i\eta}}. \qquad (12.276)$$

Levinson Theorem

Scattering calculations involve the solution of the Schrödinger equation at a positive energy, $\varepsilon = k^2 = \frac{2m}{\hbar^2} E > 0$, and the evaluation of cross-sections at these energies. However, in our discussion of the Jost function, we saw that the radial Schrödinger equation (12.155) [or Eq. (12.159)] might have bound-state solutions $R_{nl}(r)$ at negative energies

12.5 Central Potentials

$\varepsilon_n = -\kappa_n^2 < 0$. Finding bound states requires finding the normalized eigenfunctions and the corresponding eigenvalues of the wave equation

$$\left[-\frac{d^2}{dr^2} - \frac{2}{r}\frac{d}{dr} + \frac{l(l+1)}{r^2} + v(r)\right] R_{nl}(r) = -\kappa_n^2 R_{nl}(r), \tag{12.277}$$

with $\int_0^\infty dr\, |R_{nl}(r)|^2 r^2 = 1$. Does knowledge of the phase shift $\delta_l(k)$ enable us to extract information pertinent to the bound states of Eq. (12.277), despite the fact that the phase shifts are obtained within a scattering problem at positive energies and the bound-state energies are obtained as solutions of the eigenvalue problem at negative energies?

To find out, we employ the dispersion relation (12.276). It proves useful to pass from the complex k variable to the complex energy variable $\varepsilon = k^2$. Taking $k = \sqrt{\varepsilon}$, it is evident that the upper half k plane is mapped on the first Riemann sheet in the complex energy plane with $0 \le \arg \varepsilon < 2\pi$ and the lower half of the k plane is mapped on the second Riemann sheet in the complex energy plane with $2\pi \le \arg \varepsilon < 4\pi$. In order to distinguish whether the energy is on the first (physical) or second sheet, it is therefore necessary to add a small imaginary part, $\pm i\eta$, to it. Now, consider the function

$$\mathcal{D}_l(\varepsilon) \equiv \mathcal{F}_l(\sqrt{\varepsilon}). \tag{12.278}$$

Following the dispersion relation (12.268) for $\mathcal{F}_l(k)$, the corresponding relation for $\mathcal{D}_l(\varepsilon)$ reads,

$$\mathcal{D}_l(\varepsilon) = 1 - \frac{1}{\pi} \int_0^\infty d\varepsilon' \, \frac{\mathcal{D}_l(\varepsilon')}{\varepsilon' - \varepsilon - i\eta}. \tag{12.279}$$

The function $\mathcal{D}_l(\varepsilon)$ is analytic in the entire complex energy plane except for a cut along the real axis between $\varepsilon = 0$ and $\varepsilon = \infty$. The quantities of physical interest related to $\mathcal{D}_l(\varepsilon)$ are obtained by letting the energy approach the real axis from above. From the relation between the functions $\mathcal{F}_l(k)$ and the phase shift $\delta_l(k)$, it is evident that the phase of $\mathcal{D}_l(\varepsilon)$ on the real axis is $-\delta_l(\varepsilon)$, i.e.,

$$\lim_{\eta \to 0^+} \arg \mathcal{D}_l(\varepsilon + i\eta) = -\delta(\varepsilon) \quad (\varepsilon \text{ real}). \tag{12.280}$$

Moreover, since $\mathcal{D}_l(\varepsilon - i\eta) = \mathcal{D}_l(\varepsilon + i\eta)^*$, we obtain

$$\lim_{\eta \to 0^+} \frac{\mathcal{D}_l(\varepsilon + i\eta)}{\mathcal{D}_l(\varepsilon - i\eta)} = e^{-2i\delta_l(\varepsilon)} = \frac{1}{S_l(\varepsilon)}, \tag{12.281}$$

where $S_l(\varepsilon) = S_l(k = \sqrt{\varepsilon})$ is the S matrix defined in Eq. (12.258). We now employ Eq. (12.276) for $\mathcal{F}_l(k) = \mathcal{D}_l(\varepsilon)$ and rewrite it in terms of energy ε recalling that the bound-state energies occur at $\varepsilon_n = -\kappa_n^2$:

$$\mathcal{D}_l(\varepsilon) = \prod_{n=1}^{n_l} \left(1 + \frac{\varepsilon_n}{\varepsilon}\right) e^{\frac{1}{\pi}\int_0^\infty d\varepsilon' \, \frac{\delta_l(\varepsilon')}{\varepsilon' - \varepsilon - i\eta}}. \tag{12.282}$$

We are now in a position to prove *Levinson's theorem*. Consider the (possibly complex) number defined by the contour integral,

$$I_l = -\frac{1}{2\pi i} \int_C dz \, \frac{d \ln \mathcal{D}_l(z)}{dz}, \tag{12.283}$$

where C is a closed contour in the complex plane shown in Fig. 12.10. It contains the ray from $(\eta, i\eta)$ to $(R, i\eta)$, followed by an almost complete circle between $(R, i\eta)$ and $(R, -i\eta)$ and a backward-oriented ray between $(R, -i\eta)$ and $(\eta, -i\eta)$ and then the contour is completed by a circle avoiding the origin between $(\eta, -i\eta)$ and $(\eta, i\eta)$. Since $\mathcal{D}_l(z)$ is an analytic function inside the contour, the only contribution to the integral comes from the zeros of $\mathcal{D}_l(z)$ at points $\varepsilon_n = -\kappa_n^2$. In order to include all these points, we take $R \to \infty$ and $\eta \to 0$. In this limit, $I_l = n_l$, the number of bound states for angular momentum l. The integral along the large circle vanishes, and we are left with the contributions from the small circle and those along the positive real axis. The contribution from the small circle around the origin requires some care. The threshold behavior of the Jost function is $\mathcal{F}_l(k) \approx k$ for $l = 0$ and $\mathcal{F}_l(k) \approx k^2$ for $l \neq 0$. The corresponding threshold behavior for $\mathcal{D}_l(z) \approx z^\gamma$ is $\gamma = 1/2$ for $l = 0$ and $\gamma = 1$ for $l > 0$. This contributes γ to the integral around the small circle, so

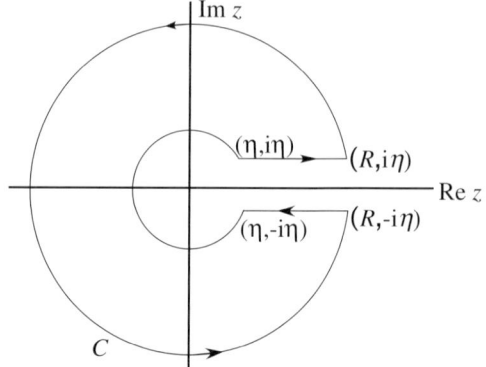

FIG 12.10 Integration contour for Eq. (12.283) in the z plane.

$$n_l = I_l = \gamma - \frac{1}{2\pi i}\int_0^\infty d\varepsilon \frac{d}{d\varepsilon} \ln \frac{\mathcal{D}_l(\varepsilon + i\eta)}{\mathcal{D}_l(\varepsilon - i\eta)}$$

$$= \gamma - \frac{1}{2\pi i}\int_0^\infty \frac{d \ln e^{-2i\delta_l(\varepsilon)}}{d\varepsilon} d\varepsilon = \gamma - \frac{1}{\pi}[\delta_l(\infty) - \delta_l(0)]. \quad (12.284)$$

Since we adopted the convention $\delta_l(\infty) = 0$, we arrive at Levinson's Theorem,

$$\boxed{\begin{aligned}\delta_0(0) &= \pi\left(n_0 + \frac{1}{2}\right) \quad \text{(if } \mathcal{D}_0(0) = 0\text{)} \\ \delta_l(0) &= \pi n_l \quad \text{(if } l > 0 \text{ or } \mathcal{D}_0(0) \neq 0\text{).}\end{aligned}} \quad (12.285)$$

The case $\mathcal{D}_0(0) = 0$ specified on the first line of Eq. (12.285) is referred to as a *zero-energy resonance*. In this case, the scattering length introduced in Eqs. (12.237a) and (12.241) is infinite. This is not just a case of academic interest. In recent experiments on systems of cold atoms, it was possible to control the scattering length using the mechanism of *Feshbach resonance* (discussed below) and tune it continuously through the point where it is infinite.

Ramsauer–Townsend Effect

We indicated in connection with Eq. (12.212) that the number of partial waves contributing to the scattering amplitude is of the order kr_0, where r_0 is the range of the potential. At very low energies, i.e., when $kr_0 \leq 1$, the main contribution to the total cross-section comes from $l = 0$, and the total cross-section at zero energy is $\sigma = \frac{4\pi}{k^2}\sin^2\delta_0 \approx 4\pi a_0^2$, where a_0 is the s-wave scattering length. In this energy regime, σ is very weakly dependent on energy. In the majority of cases, this weak dependence holds for a wide range of energies. There is, however, an interesting exception, which gives rise to the *Ramsauer–Townsend effect*, first observed in the elastic scattering of slow electrons off nobel gas atoms. The elastic scattering cross-section goes through a minimum value for electrons with a certain kinetic energy (about 0.7 eV for xenon gas). We shall now explain how this comes about, using qualitative arguments.

The question that needs to be addressed is how it is possible to have a wavenumber $0 < k_1 < r_0^{-1}$, such that $\sin\delta_0(k_1) = 0$ so the scattering cross-section vanishes. The cross-section should fall from $\sigma(k = 0) = 4\pi a^2$ to $\sigma(k = k_1) = (4\pi/k_1^2)\sin^2\delta_0(k_1) = 0$. The potential experienced by electron scattered off a noble gas atom is strongly attractive and

12.5 Central Potentials

supports many bound states. Moreover, the excitation energies of the atom ($\geq$ a few eV) are much higher than the kinetic energy of the scattered electrons (a fraction of an eV), so that the scattering is purely elastic. Levinson theorem's states that if the convention $\delta_0(\infty) = 0$ is adopted, then $\delta_0(0) = n_b \pi$, where n_b is the number of s-wave bound states supported by the atomic attractive potential. For atoms such as Xe or Kr, $n_b \gg 1$. Thus, between $k = 0$ and $k \to \infty$, $\delta_0(k)$ must go through $(n_b - 1)$ values k_n, $(n = 1, 2, \ldots, n_b - 1)$ for which $\delta_0(k_n) = (n_b - n)\pi$ and $\sin \delta_0(k_n) = 0$. For large n_b, it is likely that $k_1 < r_0^{-1}$ and the scattering is s-wave dominated. At these values of k_1, the elastic scattering cross-section vanishes.

12.5.5 SCATTERING FROM A COULOMB POTENTIAL

The scattering of two charged particles interacting via the Coulomb potential is an extremely important problem; it is relevant for many physical systems, including electron–electron and electron–atom scattering. We have seen in Sec. 12.3.2 that such a two-body problem is equivalent to scattering of a particle of mass m from a central potential,

$$V_{\text{Coul}}(r) = \frac{q_1 q_2}{r}. \tag{12.286}$$

However, it cannot be directly solved within the framework of scattering from central potential as discussed in Sec. 12.5 because the Coulomb central potential falls off too slowly. Indeed, the discussion of the asymptotic behavior associated with Eqs. (12.85) and (12.65) is based on the assumption that the potential $V(\mathbf{r})$ falls off faster than $1/r^2$ as $r \to \infty$. One of the consequences of the slow fall-off of the Coulomb potential is that the total cross-section σ diverges. This is easily verified if one tries to integrate the differential cross-section within the Born approximation [see Eq. (12.126)]. The reason is clear: every incident particle is scattered, even if its classical impact parameter is extremely large. Thus, the Coulomb potential requires special treatment. The technique we shall employ is to use parabolic coordinates in which the Schrödinger equation with a Coulomb potential is separable. Let x, y, z denote the Cartesian coordinates, and let $r = \sqrt{x^2 + y^2 + z^2}$ and $\tan \phi = \frac{y}{x}$ (note that r, z, ϕ are *not* cylindrical coordinates). The parabolic coordinates (sometimes called parabolic cylindrical coordinates) are defined as,

$$\xi = r + z, \quad \eta = r - z, \quad \phi = \arctan \frac{y}{x}. \tag{12.287}$$

The range of the parabolic coordinates are $0 \leq \xi \leq \infty$, $0 \leq \eta \leq \infty$ and $0 \leq \phi \leq 2\pi$. These coordinates are also convenient for describing problems in an electric field, such as the hydrogen atom in a constant electric field along the z axis. In these coordinates, the Schrödinger equation for the wave function $\psi(\xi, \eta, \phi)$ has the form,

$$-\left[\frac{4}{\xi + \eta} \frac{\partial}{\partial \xi}\left(\xi \frac{\partial}{\partial \xi}\right) + \frac{4}{\xi + \eta} \frac{\partial}{\partial \eta}\left(\eta \frac{\partial}{\partial \eta}\right) + \frac{1}{\xi \eta} \frac{\partial^2}{\partial \phi^2}\right]\psi + \frac{2q_1 q_2}{\eta + \xi}\psi = k^2 \psi, \tag{12.288}$$

where as before, $E = \frac{\hbar^2 k^2}{2m}$ is the scattering energy. In a central potential scattering problem, there is a symmetry with respect to the axis along the direction of the incoming beam, which is chosen to be the z axis. Therefore, we seek a solution separable in ξ, η, and ϕ,

$$\psi = \psi_1(\xi)\psi_2(\eta)e^{im\phi}. \tag{12.289}$$

Substituting into the Schrödinger equation (12.288), one arrives at a set of differential equations,

$$\frac{d}{d\xi}\left(\xi \frac{d\psi_1}{d\xi}\right) - \frac{1}{4}k^2 \xi \psi_1 - c_1 \psi_1 = 0, \tag{12.290}$$

$$\frac{d}{d\eta}\left(\eta \frac{d\psi_2}{d\eta}\right) - \frac{1}{4}k^2 \eta \psi_2 - c_2 \psi_2 = 0, \tag{12.291}$$

where the sum of the constants c_1 and c_2 yields

$$c_1 + c_2 = \frac{m q_1 q_2}{\hbar^2}. \tag{12.292}$$

> **Problem 12.16**
>
> Consider the hydrogen atom in a constant electric field $\mathbf{E} = E_0 \hat{\mathbf{z}}$. Formulate the Schrödinger equation in parabolic coordinates and show that it is separable in ξ and η.
> **Hint**: The corresponding potential is $U = -E_0 z = \frac{1}{2} E_0 (\xi - \eta)$. Equation (12.291) is obtained by multiplying Eq. (12.288) by $(\xi + \eta)$ and noting that the term $k^2 (\xi + \eta)$ does not hinder separability. When the electric potential $U = -E_0 z = E_0 (\xi - \eta)/2$ is multiplied by $(\xi + \eta)$, the resulting term $E_0 (\xi^2 - \eta^2)/2$ still permits separability.

Due to the slow fall-off of the Coulomb potential, the solutions ψ_1 and ψ_2 at large distance from the scattering center cannot be simply cast in the form of the usual scattering asymptotic boundary conditions specified in Eq. (12.65). We shall see below how these asymptotic conditions must be modified.

Written in parabolic coordinates, the plane wave and the outgoing radial wave $\frac{e^{ikr}}{r}$ take the form,

$$e^{ikz} = e^{i\frac{k}{2}(\xi - \eta)}, \quad e^{ikr} = e^{i\frac{k}{2}(\xi + \eta)}. \tag{12.293}$$

This suggests the following guess for $\psi_1(\xi)$,

$$\psi_1(\xi) = e^{\frac{1}{2} i k \xi}, \tag{12.294}$$

which indeed solves Eq. (12.290) with $c_1 = ik/2$. Equation (12.291) then becomes,

$$\frac{d}{d\eta}\left(\eta \frac{d\psi_2}{d\eta}\right) - \left(\frac{m q_1 q_2}{\hbar^2} - \frac{1}{2} ik\right) \psi_2 + \frac{1}{4} k^2 \eta \psi_2 = 0. \tag{12.295}$$

At this point, it is useful to introduce the dimensionless *Sommerfeld parameter*, $\beta \equiv (k a_C)^{-1}$, where $a_C = \hbar^2/(m q_1 q_2)$ is the "Coulomb unit of length,"

$$\beta \equiv \frac{m q_1 q_2}{\hbar^2 k}, \tag{12.296}$$

and express the solution for $\psi_2(\eta)$ as,

$$\psi_2(\eta) = e^{-\frac{i}{2} k \eta} f(\eta), \tag{12.297}$$

$$\eta \frac{d^2 f}{d\eta^2} + (1 - ik\eta) \frac{df}{d\eta} - \beta k f = 0, \tag{12.298}$$

whose solution is expressed in terms of the *confluent hypergeometric function* (see Morse and Feshbach [26], Chapter 5),

$$f(\eta) = A F(-i\beta, 1, ik\eta), \tag{12.299}$$

where A is a normalization constant assuring a unit velocity of the incoming wave (see below). Combining Eqs. (12.289), (12.290), (12.297), and (12.299), we have

$$\psi(\xi, \eta) = A e^{i\frac{k}{2}(\xi - \eta)} F(-i\eta, 1, ik\eta). \tag{12.300}$$

We are now ready to determine the large distance form of the wave function in terms of the behavior of the hypergeometric function for large argument. Specifying the constant A to be,

$$A = |\Gamma(1 - i\beta)| e^{-\beta \pi/2}, \tag{12.301}$$

12.5 Central Potentials

and denoting by $\gamma_\beta = \arg \Gamma(1 - i\beta)$, we obtain the following asymptotic form in spherical coordinates r, θ, ϕ with $\eta = r - z = r(1 - \cos\theta)$, $\xi = r + z = r(1 + \cos\theta)$,

$$\psi(r, z) \xrightarrow[r(1-\cos\theta)\to\infty]{} \left[e^{ik[z + \beta \log kr(1-\cos\theta) + \gamma_\beta]} \right] + \left[\frac{\beta}{kr(1 - \cos\theta)} \frac{e^{ik[r - \beta \log kr(1-\cos\theta) - \gamma_\beta]}}{r} \right]. \quad (12.302)$$

This estimate does not hold for the forward direction, $\theta = 0$; it is evident that expression (12.302) is distinct from the standard asymptotic form (12.65) in the sense that the denominator of the second term (the coefficient of the spherical wave) is singular in the forward direction $\theta = 0$. Note that there are logarithmic corrections to both the plane and the spherical waves. Within these constraints, the calculations of current and flux as in Eqs. (12.67), (12.68a) and (12.68b) proceed along the same lines. The scattering amplitude (defined up to a constant phase as the coefficient of the spherical wave) and the differential cross-section are then given by,

$$f_k(\theta) = -\frac{\beta}{k(1 - \cos\theta)} e^{-i\beta \log(1-\cos\theta)}, \quad (12.303)$$

$$\frac{d\sigma}{d\Omega} = |f_k(\theta)|^2 = \frac{\beta^2}{4k^2 \sin^4 \frac{\theta}{2}} = \frac{q_1^2 q_2^2}{4E^2 \sin^4 \frac{\theta}{2}}, \quad (12.304)$$

which is the same result (12.126) obtained within the Born approximation. It also coincides with the classical result, Eq. (12.10).[5] Coulomb scattering is an example of a quantum scattering problem that can be solved completely analytically, and as such it has enormous pedagogical value. We shall see below that quantum effects in Coulomb scattering emerge when the scattering occurs between two identical particles.

The Coulomb potential is often screened by particles of opposite charge that are attracted by the scattering center. For example, when electrons are scattered by atomic nuclei in solids, the Coulomb field of the atomic nuclei is screened, and the effective potential felt by the electrons can be represented by a Yukawa potential (12.124).

In the study of nuclear reactions between heavy nuclei, the electrons surrounding the nuclei do not play a significant role but do screen the Coulomb potential. The interaction between two such nuclei is the combination of a screened Coulomb interaction and a short-range nuclear force $V_N(\mathbf{r})$. Such scattering can be treated using partial wave analysis, but the analysis for the Coulomb potential must be modified relative to that for potentials that fall off faster than $1/r^2$.

Partial Wave Expansion in a Coulomb Potential

Let us consider Eq. (12.159) for the Coulomb potential,

$$\left[\frac{d^2}{dr^2} + k^2 - \frac{l(l+1)}{r^2} - \frac{2m}{\hbar^2} \frac{q_1 q_2}{r} \right] u_l(r) = 0. \quad (12.305)$$

The asymptotic form of $u_l(r)$, or $R_l(r) = \frac{u_l(r)}{kr}$, is no longer given by the relation (12.161a), since the Coulomb potential falls off slower than the centrifugal potential $\frac{l(l+1)}{r^2}$. The solution to Eq. (12.305) can be written in terms of hypergeometric functions as shown below. Unlike the solutions introduced in Eqs. (12.290) and (12.291), here the solution depends on the angular momentum number l. We define $\nu \equiv l + 1/2$ and introduce the dimensionless variable,

$$y \equiv 2ikr. \quad (12.306)$$

Using these definitions, Eq. (12.305) takes the form,

$$\left[\frac{d^2}{dy^2} - \frac{1}{4} + \frac{i\beta}{y} + \frac{\frac{1}{4} - \nu^2}{y^2} \right] u_l(y) = 0, \quad (12.307)$$

---

[5] We shall see in Sec. 12.5.6 that quantum signatures of Coulomb scattering are exposed when the scattering of two identical particles is analyzed.

[the dimensionless Sommerfeld parameter $\beta \equiv \frac{mq_1q_2}{\hbar^2 k}$ is defined in Eq. (12.296)]. Equation (12.307) is the Whittaker equation [see Eq. (13.1.31) in Abramowitz and Stegun [27]] with $\kappa \to$ in and $\mu \to v$. It has regular and singular solutions at $y = 0$ given, respectively, by $M_{in,v}(y)$ and $W_{in,v}(y)$, as specified in Abramowitz and Stegun's study [27], Eqs. (13.1.32) and (13.1.33):

$$M_{i\beta,v}(y) = e^{-\frac{y}{2}} y^{\frac{1}{2}+v} M\left(\frac{1}{2} + v - i\beta, 1 + 2v, y\right), \tag{12.308}$$

$$W_{i\beta,v}(y) = e^{-\frac{y}{2}} y^{\frac{1}{2}+v} U\left(\frac{1}{2} + v - i\beta, 1 + 2v, y\right). \tag{12.309}$$

Here, $M(a, b, y)$ and $U(a, b, y)$ are the regular and irregular confluent hypergeometric (Kummer) functions, given in Ref. [27], Eqs. (13.1.2) and (13.1.3). We will need both regular and irregular solutions to analyze the asymptotic states, just as in the case of a pure centrifugal potential, where we needed both $j_l(kr)$ (regular at $kr = 0$) and $n_l(kr)$ (singular at $kr = 0$), see Eq. (12.161a). The precise combination and normalization are somewhat a matter of convenience, as is the case in Eq. (12.161a). The standard form of the regular solution, including its normalization and dependence on l, is

$$F_l(r) = \sqrt{\frac{1}{2\pi}} \frac{\Gamma(l+1-i\beta)\Gamma(l+1+i\beta)}{\Gamma(2l+2)} e^{-i\pi/2(l+1-i\beta)} M_{i\beta,l+\frac{1}{2}}(2ikr), \tag{12.310}$$

while the irregular solution is,

$$G_l(r) = c_l W_{i\beta,l+\frac{1}{2}}(2ikr) + c_l^* W_{-i\beta,l+\frac{1}{2}}(-2ikr), \tag{12.311}$$

with $c_l = ie^{-i(\sigma_l - l\pi/2)} e^{\beta\pi/2}$ and σ_l is the Coulomb phase shift,

$$\sigma_l = \arg[\Gamma(l+1+i\beta)]. \tag{12.312}$$

The asymptotic approximations for $F_l(r)$ and $G_l(r)$ as $r \to \infty$ are

$$F_l(r) \xrightarrow[r\to\infty]{} \sqrt{\frac{2}{\pi}} \sin\left(kr - l\frac{\pi}{2} - \beta \log 2kr + \sigma_l\right), \tag{12.313}$$

$$G_l(r) \xrightarrow[r\to\infty]{} -\sqrt{\frac{2}{\pi}} \cos\left(kr - l\frac{\pi}{2} - \beta \log 2kr + \sigma_l\right). \tag{12.314}$$

The partial wave expansion developed above need not be used to obtain the scattering amplitude $f(\theta)$ through partial wave summation, since the latter is already known, given in Eq. (12.304). Rather, the solutions $F_l(r)$ and $G_l(r)$ can be used in problems of scattering from a combination of Coulomb and short-range potentials, where they replace the regular and irregular Ricatti–Bessel functions $\hat{j}_l(kr) = krj_l(kr)$ and $\hat{n}_l(kr) = krn_l(kr)$ as solutions of the Schrödinger equation in the asymptotic regime, where the short-range potential can be neglected. This problem is addressed below.

Scattering by Coulomb and Short-Range Potentials

In order to treat the problem of scattering from a Coulomb potential $v_c(r) = \frac{2m}{\hbar^2} \frac{q_1 q_2}{r}$ and a short-range potential $v(r)$ (of range d), it is useful to start with a screened Coulomb potential and then "remove" the screening by letting $R \to \infty$. For example, consider the case of abrupt screening

$$v_{sc}(r) = v_c(r) \Theta(R - r), \tag{12.315}$$

12.5 Central Potentials

where $\Theta(x)$ is the step function and the *screening radius* R is much greater than the range d of the short-range potential. Thus, we start with the radial equation of the form (12.159),

$$\left[\frac{d^2}{dr^2} + k^2 - \frac{l(l+1)}{r^2} - v(r) - v_{sc}(r)\right] u_l(r) = 0. \tag{12.316}$$

For $r < d$, both $v_c(r)$ and $v(r)$ are active, while for $d < r < R$, the short-range potential is negligible and $v_c(r)$ is active, together with the centrifugal potential. In this region, then, the solution can be written as a linear combination of Coulomb wave functions $F_l(r)$ and $G_l(r)$ defined in Eqs. (12.310) and (12.311). In analogy with Eq. (12.161a), we may then write, for $d < r < R$,

$$u_l(r) = A_l[\cos \tilde{\delta}_l(k) F_l(r) - \sin \tilde{\delta}_l(k) G_l(r)]. \tag{12.317}$$

Here, $\tilde{\delta}_l$ is the phase shift for the short-range potential $v(r)$ in the presence of the Coulomb potential $v_c(r)$, and A_l is a constant. It is clear that one cannot simply perform partial wave summation as in Eq. (12.165), because the incoming wave is not a simple plane wave. In other words, $\tilde{\delta}_l$ is not the phase shift δ_l. The latter is determined by the asymptotic form of $u_l(r)$ at $r \gg R$ (see below). Using the asymptotic forms of the Coulomb wave functions (12.313) and (12.314), we can approximate $u_l(r)$ (still in $d \ll r < R$) as,

$$u_l(r) \approx B_l \sin[\tilde{\delta}_l + \sigma_l + \gamma_l(r)], \tag{12.318}$$

where B_l is another constant, σ_l is the Coulomb phase shift (12.312), and

$$\gamma_l(r) \equiv kr - l\frac{\pi}{2} - \beta \log 2kr \tag{12.319}$$

is what left in the argument of the Coulomb wave functions at large distance in Eqs. (12.313) and (12.314).

Finally, for $r \gg R$, there is no interaction, and hence, u_l is the solution of the partial wave-free Schrödinger equation with the standard form [see Eq. (12.161b)],

$$u_l(r) = \sin\left(kr - l\frac{\pi}{2} + \delta_l\right), \tag{12.320}$$

where δ_l is the phase shift that should be inserted into the partial wave expansion of the scattering amplitude (12.165). To reiterate, the validity of the above expressions for the corresponding different domains in the radial coordinate r is as follows: Eq. (12.317) is valid for $r \geq d$ but not for $r \ll R$. Equation (12.318) is valid for $r \leq R$ but not for $r \gg d$, and Eq. (12.320) is valid for $r \gg R$. The mathematical procedure by which δ_l is evaluated as $R \to \infty$ is rather technical and will not be detailed here. At the end of this procedure, the partial wave scattering amplitude (12.173) can be written (up to an l independent phase) as,

$$f_l(k) = \frac{e^{2i\delta_l} - 1}{2ik} = \frac{e^{2i\sigma_l} - 1}{2ik} + \frac{e^{2i\sigma_l}(e^{2i\tilde{\delta}_l} - 1)}{2ik} + \tilde{f}_l(k), \tag{12.321}$$

such that the contribution of the residual term $\tilde{f}_l$ to the partial wave summation (12.165) is shown to be very small and can be neglected. The first term on the RHS of Eq. (12.321) is the partial wave Coulomb scattering amplitude. The second term is *Coulomb-modified* scattering amplitude of the short-range potential. We stress again that $\tilde{\delta}_l$ also depends on the Coulomb potential. It is not simply the phase shift obtained in the scattering from the short-range potential $v(r)$ alone. With this in mind, the scattering amplitude is a direct sum,

$$f(\theta) = f_C(\theta) + f_{SR}(\theta) \tag{12.322}$$

$$= \sum_{l=0}^{\infty} (2l+1) \left[\frac{e^{2i\sigma_l} - 1}{2ik} + \frac{e^{2i\sigma_l}(e^{2i\tilde{\delta}_l} - 1)}{2ik}\right] P_l(\cos\theta). \tag{12.323}$$

The differential cross-section derived from the above formalism is then,

$$\frac{d\sigma}{d\Omega} = |f(\theta)|^2 = |f_C(\theta) + f_{SR}(\theta)|^2, \tag{12.324}$$

which takes into account the interference between Coulomb and short-range potential scattering amplitudes. Expression (12.324) is not valid at the forward direction $\theta = 0$. Moreover, if the two scattered particles are identical, the corresponding statistics should be enforced (see below). From a practical point of view, the numerical procedure for calculating the differential cross-section is straightforward. First, one integrates the Schrödinger equation from $r = 0$ to large r, $r \gg d$, and by computing the logarithmic derivative of the wave function, one finds δ_l using Eq. (12.317). The Coulomb phase shift σ_l is given in Eq. (12.312). Thus, using Eqs. (12.322)–(12.324), one has all the input for evaluation of the scattering amplitudes and scattering cross-section.

12.5.6 SCATTERING OF TWO IDENTICAL PARTICLES

When the scattered particles are identical, exchange symmetry must be accounted for (see Chapter 8). The consequence to the calculation of experimentally observed quantities is the subject of this subsection.

For identical particles, the wave function should be properly symmetrized (for bosons) or antisymmetrized (for fermions). The wave function $\phi(\mathbf{r}_i, \eta_{s\sigma_i})$ of each of the two identical particles $i = 1,2$ with spin s is composed of a space part $\phi(\mathbf{r}_i)$, which is a function of the coordinate $\mathbf{r}_i$, and a spin part $\eta_{s\sigma_i}$ with spin projection $\sigma_i = -s, -s+1, \ldots, s$. It is assumed that the magnetic quantum numbers of the two particles are quantized along the same fixed axis. Using translation invariance, the space part of the wave function is a product of a plane wave for the center of mass motion and a function $\psi(\mathbf{r})$ for the relative coordinate $\mathbf{r} = \mathbf{r}_2 - \mathbf{r}_1$. The asymptotic form of the relative coordinate wave function is of the form (12.85). Since the symmetry of the wave function involves space and spin coordinates, it is necessary to include the spin part in the discussion of exchange symmetry, even though the two-body potential is spin independent. Before symmetrization, the two-particle scattering wave function in Eq. (12.79) is written as,

$$\psi_{\mathbf{k},\sigma_1,\sigma_2}(\mathbf{r}) = \psi_{\mathbf{k}}(\mathbf{r})\eta_{s\sigma_1}\eta_{s\sigma_2}. \tag{12.325}$$

We shall now treat the case of scattering by a spin-independent (central) potential. The case of spin-dependent potentials will be discussed in Sec. 12.8.8. The spin and space part of the wave function in Eq. (12.325) can be separately symmetrized because of the factorized form. Hence, the properly symmetrized or antisymmetrized wave function is

$$\Psi_{\mathbf{k}}(\mathbf{r};\sigma_1,\sigma_2) = \frac{[\psi_{\mathbf{k}}(\mathbf{r}) + \phi_r\psi_{\mathbf{k}}(-\mathbf{r})]}{\sqrt{2}} \frac{[\eta_{s\sigma_1}(1)\eta_{s\sigma_2}(2) + \phi_s\eta_{s\sigma_1}(2)\eta_{s\sigma_2}(1)]}{\sqrt{2}} \equiv \Psi_{\mathbf{k}}(\mathbf{r})F_{s,\sigma_1,\sigma_2}(1,2), \tag{12.326}$$

where $\Psi_{\mathbf{k}}(\mathbf{r})$ and F_{s,σ_1,σ_2} are the properly symmetrized two-particle space and spin functions. Note that the subscripts on the spin function F_{s,σ_1,σ_2} refer to the spin quantum numbers, whereas the arguments refer to particles, and $\phi_r = \pm 1$ and $\phi_s = \pm 1$ are the parity of the space and spin functions, respectively, under exchange. The symmetry of the two-particle state under exchange is given by,

$$p_{12} \equiv \phi_r\phi_s = 1 \text{ for bosons}, \quad p_{12} = -1 \text{ for fermions}. \tag{12.327}$$

For bosons ($p_{12} = +1$), there are two ways of obtaining $p_{12} = 1$ ($\phi_r = 1, \phi_s = 1$) or ($\phi_r = -1, \phi_s = -1$), and fermions ($p_{12} = -1$), one can have ($\phi_r = -1, \phi_s = 1$) or ($\phi_r = 1, \phi_s = -1$). Thus, there are two distinct functions $\Psi_{\mathbf{k}}(\mathbf{r};\sigma_1,\sigma_2)$ that can be formed in Eq. (12.326) for bosons and fermions.

As far as the scattering amplitude is concerned, it is the asymptotic form of the spatial part of the wave function that is important, hence the factor ϕ_r, the relative sign between $\psi_{\mathbf{k}}(\mathbf{r})$ and $\psi_{\mathbf{k}}(-\mathbf{r})$ in the RHS of Eq. (12.326), determines the effects of exchange on the scattering amplitude. Using Eq. (12.327) we may write $\phi_r = p_{12}\phi_s$. For bosons ($p_{12} = 1$), the relative sign is just $\phi_r = \phi_s$. For fermions ($p_{12} = -1$), the relative sign is $\phi_r = -\phi_s$. Generically, ϕ_s might be ± 1

12.5 Central Potentials

for fermions and bosons with nonzero spin. Let us inspect the space part of the wave function, recalling that in terms of spherical coordinates,

$$\mathbf{r} = (r, \theta, \phi), \quad -\mathbf{r} = (r, \pi - \theta, \phi + \pi). \tag{12.328}$$

Using the form suggested in Eq. (12.326), the space part of the scattering wave function with exchange symmetry taken into account has the asymptotic form,

$$\Psi_{\mathbf{k}}^{+}(\mathbf{r}) \xrightarrow[r \to \infty]{} \frac{1}{\sqrt{2}} \left\{ e^{i\mathbf{k} \cdot \mathbf{r}} + p_{12}\phi_s e^{-i\mathbf{k} \cdot \mathbf{r}} + \frac{e^{ikr}}{r} [f_k(\theta) + p_{12}\phi_s f_k(\pi - \theta)] \right\}, \tag{12.329}$$

where $f_k(\theta)$ is the scattering amplitude calculated without taking into account exchange symmetry, as in Eq. (12.165). In calculating the differential cross-section, we should pay attention to the following points: (1) The incoming current is unaffected by the symmetry operation. (2) The outgoing current is the sum of the currents of particles 1 and 2. This compensates the factor $1/2$ obtained by squaring the outgoing spherical wave part in Eq. (12.329). (3) The total spin $\mathbf{S} = \mathbf{s}_1 + \mathbf{s}_2$ is conserved if the potential is central, and the scattering occurs in a definite spin state. For central potentials, all values of $\mathbf{S}$ present in the initial wave function scatter identically. Therefore, it is not necessary to break up the wave function into the various total spin states present in the initial wave function (in contradistinction to the case when the potential is spin-dependent).

The different situations for which the analysis detailed above is relevant are given below.

(1) Scattering of two identical spinless bosons (e.g., two ^{4}He nuclei or two helium atoms). In this case, we have $p_{12} = 1$, $\phi_s = 1$ and the properly symmetrized scattering amplitude (denoted here as $F_k(\theta)$), and differential cross-section for *two identical spinless bosons* (or, more generally, two identical bosons with a symmetric spin function) is,

$$F_k(\theta) = f_k(\theta) + f_k(\pi - \theta), \tag{12.330}$$

$$\frac{d\sigma}{d\Omega} = |f_k(\theta) + f_k(\pi - \theta)|^2. \tag{12.331}$$

(2) Scattering of two identical bosons with nonzero spin such that both space and spin parts of the wave function are antisymmetric: $p_{12} = 1$, $\phi_s = -1$, $\phi_r = -1$. For two identical bosons with an antisymmetric spin function,

$$F_k(\theta) = f_k(\theta) - f_k(\pi - \theta), \tag{12.332}$$

$$\frac{d\sigma}{d\Omega} = |f_k(\theta) - f_k(\pi - \theta)|^2. \tag{12.333}$$

The recent progress in manipulating cold atoms makes the situation described in (1) and (2) experimentally accessible for atomic scattering.

The following two examples refer to identical fermions. For spin $1/2$ particles, the spin state is either triplet ($S = 1$ and $\phi_s = 1$) or singlet ($S = 0$ and $\phi_s = -1$.)

(3) *Two identical fermions with a symmetric spin function (triplet state for spin 1/2)*, e.g., two electrons or protons or ^{3}Li atoms. Then, $p_{12} = -1$, $\phi_s = 1$, and $\phi_r = -1$, and

$$F_k(\theta) = f_k(\theta) - f_k(\pi - \theta), \tag{12.334}$$

$$\frac{d\sigma}{d\Omega} = |f_k(\theta) - f_k(\pi - \theta)|^2. \tag{12.335}$$

(4) *Two identical fermions with antisymmetric spin function (singlet state for spin 1/2)*, $p_{12} = -1$, $\phi_s = -1$, $\phi_r = 1$,

$$F_k(\theta) = f_k(\theta) + f_k(\pi - \theta), \tag{12.336}$$

$$\frac{d\sigma}{d\Omega} = |f_k(\theta) + f_k(\pi - \theta)|^2. \tag{12.337}$$

In performing the partial wave expansion of the properly symmetrized amplitudes, $F_k(\theta)$, we note the following property of the Legendre polynomials,

$$P_l[\cos(\pi - \theta)] = P_l(-\cos\theta) = (-1)^l P_l(\cos\theta). \quad (12.338)$$

Therefore, we can expand as follows:

$$f_k(\theta) + f_k(\pi - \theta) = \frac{2}{k} \sum_{l \text{ even}} (2l+1) e^{i\delta_l(k)} \sin\delta_l(k), \quad (12.339)$$

$$f_k(\theta) - f_k(\pi - \theta) = \frac{2}{k} \sum_{l \text{ odd}} (2l+1) e^{i\delta_l(k)} \sin\delta_l(k). \quad (12.340)$$

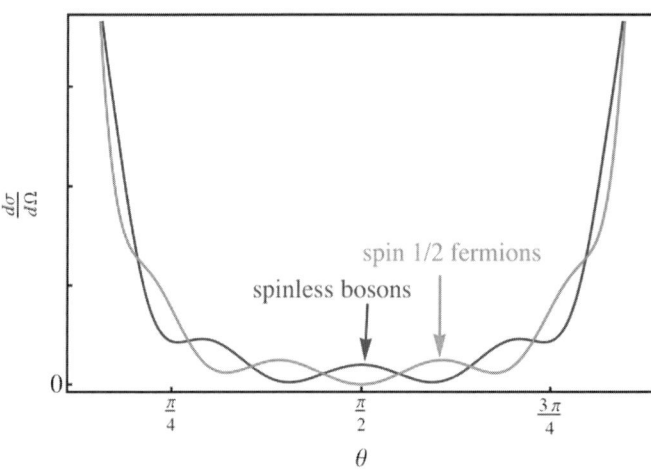

FIG 12.11 The properly symmetrized quantum differential cross-sections for two identical charged particles interacting via a pure Coulomb potential $V(r) = Z^2 e^2/r$ for two spinless bosons and two spin 1/2 fermions with parallel spins. The Sommerfeld parameter is taken to be $\beta = 5$.

This shows that for cases (2) and (3), the lowest partial wave is $l = 1$, i.e., p-wave scattering. For cold atoms at very low energy, this is rather crucial, because at such low energies, the centrifugal barrier prevents the atoms from approaching each other, and the cross-section is expected to be very small (barring a zero-energy resonance).

As an example of the effects of exchange symmetry, let us consider the case of Coulomb scattering of identical particles. Figure 12.11 plots the properly symmetrized Coulomb differential cross-section for two spinless bosons and two spin 1/2 fermions with parallel spins. The interference coming from the exchange symmetry is clearly seen. At $\theta = \pi/2$, the differential cross-section for spinless bosons shows constructive interference, whereas that for two spin 1/2 fermions with parallel spins shows destructive interference. No interference and no backward peak results for two distinct fermi particles ($\sigma_1 = 1/2$, $\sigma_2 = -1/2$).

12.6 RESONANCE SCATTERING

Low-energy scattering cross-sections sometimes display a series of peaks as a function of energy (e.g., in experiments of low-energy neutron scattering from atomic nuclei). If a peak occurs at zero energy, a weakly bound state at negative energy, $E = -B < 0$, just below threshold, may be present. If peaks occur at finite positive energy, they are often related to quasi-bound states, i.e., resonances. The relation between weakly bound states, resonances, and peaks in low-energy scattering cross-sections is the subject of this section.

12.6.1 INFLUENCE OF BOUND STATES

The dominant partial wave in low-energy scattering is $l = 0$ for spinless particles [see discussion near Eq. (12.200)] or $l = 1$ for identical fermions with a symmetric spin function. Consider s-wave scattering of spinless particles from a central potential $V(r)$ that supports a bound state at energy $E_B = -B$, as shown in Fig. 12.12. The figure shows an attractive square well potential (the well-depth $|V_0|$ of the potential can be much larger than the binding energy B of a bound

12.6 Resonance Scattering

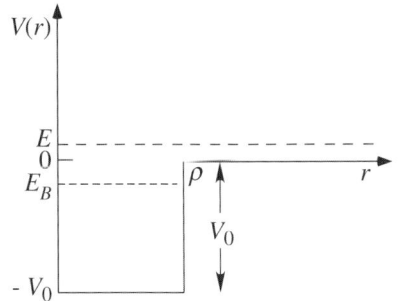

FIG 12.12 Low-energy scattering with energy E (upper dashed line) from a square well potential of depth V_0 that supports a weakly bound state at energy $E_B = -B < 0$ (lower dashed line).

state in this potential). Let us assume that a binding potential $V(r)$ falls off very quickly beyond some distance ρ, and

$$\rho \ll \frac{\hbar}{\sqrt{mB}}. \tag{12.341}$$

Then, at low scattering energy, $E < B$, the de Broglie wavelength of the scattered particle is larger than ρ. This enables us to easily estimate the low-energy scattering cross-section. We start from the radial Schrödinger equation for the $l = 0$ scattering wave function $u_0(r) = krR_0(r)$ [see Eq. (12.159)],

$$\left[\frac{d^2}{dr^2} + k^2 - v(r)\right] u_0(r) = 0. \tag{12.342}$$

From Eq. (12.162a), we know that $u_0(r)$ behaves asymptotically as,

$$u_0(r) \propto \sin[kr + \delta_0(k)], \tag{12.343}$$

where $\delta_0(k)$ is the s-wave phase shift. In addition to the scattering state $u_0(r)$, we will also need to consider the bound state wave function $u_B(r)$. Its behavior at $r > \rho$ where $v(r)$ is really small is,

$$u_B(r) \propto e^{-\kappa r}, \tag{12.344}$$

with $\kappa = \frac{\sqrt{2mB}}{\hbar}$. Taking the well-depth $|V_0| \gg B$, it is reasonable to assume that the two functions $u_B(r)$ and $u_0(r)$ are close to each other (i.e., their dependence on r is similar) inside the scattering region $r < \rho$. This enables us to relate the phase shift to the binding energy by comparing logarithmic derivatives at $r = \rho$,

$$\left[\frac{u'_B(r)}{u_B(r)}\right]_{r=\rho} = \left[\frac{u'_0(r)}{u_0(r)}\right]_{r=\rho}, \tag{12.345}$$

hence,

$$\tan[k\rho + \delta_0(k)] = -\sqrt{\frac{E}{B}}. \tag{12.346}$$

From (12.341), $k\rho \ll 1$, so $k\rho$ can be neglected in the argument of the tangent function to obtain,

$$\sin^2 \delta_0(k) = \frac{E}{E + B}. \tag{12.347}$$

Employing Eq. (12.202) for the total cross-section and restricting the sum to $l = 0$, we obtain,

$$\sigma_{l=0}(E) = \frac{4\pi}{k^2} \sin^2 \delta_0(k) = \frac{2\pi \hbar^2}{m} \frac{1}{E + B}, \tag{12.348}$$

which can be much larger than the classical cross-section $\sigma_{cl} = \pi \rho^2$.

12.6.2 RESONANCE CROSS-SECTIONS

When $l \neq 0$ and $v(r)$ is attractive, the effective potential, $v(r) + l(l+1)/r^2$, might have a "quasi-bound" state where resonances, sometimes called *shape resonances*, can be formed at positive energy, as shown in Fig. 12.13. Similar scenarios arise when the potential $v(r)$ contains a short-range strongly attractive part and a long-range repulsive part, such as the combination of nuclear forces and Coulomb repulsion between atomic nuclei. The term quasi-bound implies that these states do not have infinite lifetime; after some time, depending on height and width of the barrier (see below), tunneling out of the state occurs. A peak in the cross-section arising from a given partial wave l can arise when the phase shift $\delta_l(k)$ goes through $(n+1/2)\pi/2$; hence $\cot \delta_l(k)$ goes through zero as the scattering energy approaches the resonance energy E_r from below. In the vicinity of E_r,

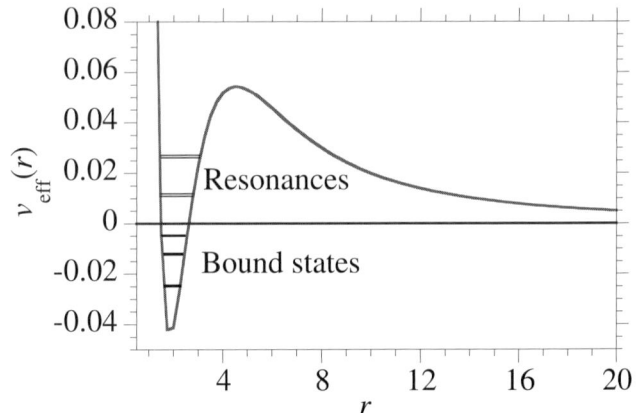

FIG 12.13 Effective potential $v_{\text{eff}}(r) = -v_0 e^{-\alpha r} + l(l+1)/r^2$ for $v_0 = 4$, $\alpha = 1$, $l = 1$, showing bound states at negative energies $E_B < 0$ and shape resonances at positive energies $E_r > 0$.

$$\cot \delta_l(k) \approx \frac{2}{\Gamma_l}(E - E_r) + O\left[(E - E_r)^2\right], \quad \left(E = \frac{\hbar^2 k^2}{2m}, \ \Gamma_l = -2\left[\frac{d \cot \delta_l}{dE}\right]^{-1}\right), \quad (12.349)$$

hence, $e^{-i\delta_l(E)} = \cos \delta_l(E) - i \sin \delta_l(E) \approx \frac{E - E_l + i\Gamma_l/2}{\sqrt{(E - E_r)^2 + (\Gamma_l/2)^2}}$. The contribution of this partial wave to the total cross-section in the vicinity of the resonance energy is then given by the *Breit–Wigner formula* (see Problem 12.17),

$$\sigma_l \approx \frac{4\pi}{k^2}(2l+1) \sin^2 \delta_l(k) = \frac{\pi}{k^2}(2l+1) \frac{\Gamma_l^2}{(E - E_r)^2 + \left(\frac{\Gamma_l}{2}\right)^2}. \quad (12.350)$$

Due to the factor k^2 in the denominator, the RHS of Eq. (12.350) is not strictly a Lorenzian, but it is quite reasonable to regard $\Gamma_l/2$ (we shall sometimes drop the subscript l) as the full width at half maximum. By application of the heuristic uncertainty relation between time and energy, we can interpret the time $\tau = 2\hbar/\Gamma$ as the lifetime of the resonance, beyond which it will tunnel outside the potential.

The effects of resonances on cross-sections can also be studied through the analytic properties of the Jost functions $\mathcal{F}_l(k)$ introduced in Eqs. (12.254) and (12.259). For this purpose, it is more convenient to consider $\mathcal{F}_l(k)$ and $\delta_l(k)$ as functions of energy E rather than momentum k. Therefore, we revisit the powerful tool of analytic continuation of the energy E from the real axis into the complex plane. To this end, recall the discussion related to Fig. 12.9 and perform the mapping $\varepsilon = k^2$, where both ε and k are considered as complex variables. For simplicity, we concentrate on a single bound state and a single resonance and drop the ubiquitous l dependence. A zero of $\mathcal{F}_l(k)$ at $k = i\kappa$ on the imaginary axis in the complex k plane is mapped onto bound states $-B < 0$ on the negative real axis in the ε plane. Similarly, a zero of $\mathcal{F}_l(k)$ at point $k_1 - ik_2$ slightly below the real axis in the complex k plane (whose argument is slightly smaller than 2π), $\bar{E} = E_R - \Gamma/2$, with $\Gamma > 0$, slightly below the real axis in the ε plane. Note that the argument of the resonance point is slightly below 4π. In the language of the theory of complex functions, a point on the complex plane whose argument is $2\pi < \theta \leq 4\pi$ is said to be located on the *second Riemman sheet*. Thus, for a bound state at energy $-B$, $\mathcal{F}(E = -B) = 0$, and for an isolated resonance, $\mathcal{F}(\bar{E} = E_R - i\frac{\Gamma}{2}) = 0$, where $E_R \gg \Gamma > 0$. The reason for the negative imaginary part $-i\Gamma/2$ is that if we consider $u(r)$ as a stationary state with complex "eigenvalue" $E = E_R - i\Gamma/2$, the time dependence of the wave function is given by $e^{-i\frac{Et}{\hbar}} = e^{-i\frac{E_R t}{\hbar}} e^{-\frac{\Gamma t}{2\hbar}}$, which decays exponentially with time, with

12.6 Resonance Scattering

decay time $\tau \equiv \hbar/\Gamma$. A zero of $\mathcal{F}(E)$ at $\bar{E} = E_R + i\frac{\Gamma}{2}$ would imply a state that evolves in time with a factor $e^{-i\bar{E}t/\hbar}$ that grows exponentially with time, which does not make sense.

Employing an expansion as in Eq. (12.265) and assuming (without loss of generality) that $\phi_l = 0$, we obtain (restoring the l dependence),

$$\mathcal{F}_l(E) \approx \left(E - E_l + i\frac{\Gamma_l}{2}\right). \tag{12.351}$$

Recall from Eqs. (12.254), (12.258), and (12.259) that at real positive energy $E > 0$, $\mathcal{F}_l(E) = e^{-i\delta_l(E)}$. We conclude that for real E and small $\Gamma > 0$, the S matrix is,

$$S_l(E) = e^{2i\delta_l(E)} = \frac{e^{i\delta_l(E)}}{e^{-i\delta_l(E)}} = \frac{E - E_l - i\frac{\Gamma_l}{2}}{E - E_l + i\frac{\Gamma_l}{2}}, \tag{12.352}$$

which is indeed unitary. From Eq. (12.352), one can evaluate $\sin^2 \delta_l(E)$ to obtain the contribution of the lth partial wave to the cross-section near resonance,

$$\boxed{\sigma_l(E \approx E_l) = \frac{4\pi}{k^2}(2l+1)\sin^2 \delta_l(E) = \frac{\pi}{k^2}(2l+1)\frac{\Gamma_l^2}{(E-E_l)^2 + \left(\frac{\Gamma_l}{2}\right)^2}.} \tag{12.353}$$

This is called the *Breit–Wigner resonance formula*. Within the approximations employed above, the contributions from several resonances add incoherently, leading to a series of (slightly distorted) Lorenzian peaks.

Problem 12.17

Use the identity $\sin^2 \delta = (1 - \cos 2\delta)/2 = (1 - \text{Re } e^{2i\delta})/2$ to prove Eq. (12.353) using Eq. (12.352).

12.6.3 FESHBACH RESONANCE

In Sec. 12.6.2, we analyzed the physics of shape resonances, which emerge in scattering from a potential with a quasi-bound state as depicted in Fig. 12.13. A different scenario of resonance scattering occurs when a scattering channel is coupled with another (closed) channel having a bound state, as shown in Fig. 12.14 (in the coupled system, the bound state is not truly bound, but rather, is quasi-bound and therefore has a finite lifetime, due to its interaction with the continuum states on the open channel). Such resonances called *Feshbach resonances* [183] are accessible in low-energy atom–atom collisions, nuclear collisions, and many other kinds of scattering events. In Fig. 12.14, two atoms with interaction potential V_C collide with low collision energy $E = \varepsilon_B - V_C(\infty)$. A different electronic potential energy V_B exists for the two-atom system, and this closed channel potential has a bound state at energy E_B.[6] Due to the coupling $V \equiv V_{CB}$ between the two electronic potentials, a resonance can appear in the scattering cross-section when ε_B on the open channel (C) is near the bound-state energy E_B on the closed channel (B) (B stands for bound, and V_B is often called the *closed channel potential*). In cold atom collisions, it is possible to experimentally control the energy E_B by tuning the strength of an external magnetic field, so that the whole potential energy curve V_B moves up or down in energy relative to the potential energy curve V_C. One can start with a magnetic field such that $\varepsilon_B < E_B$ and change the magnitude of the magnetic field so the potential V_B moves relative to V_C and E_B sweeps through the resonance, which occurs when $\varepsilon_B = E_B$; after the sweep, $\varepsilon_B > E_B$. When $\varepsilon_B \approx E_B$, the system exhibits a *Feshbach resonance*, which is distinct from the shape resonance

---

[6] If $\varepsilon_B < V_B(\infty)$, the potential V_B is energetically closed; hence, it is called the *closed channel potential*.

encountered above. At a Feshbach resonance, the scattering length becomes infinite; it changes sign as the energy passes through resonance, as shown in Fig. 12.15. When E_B is lowered below E, molecules can be trapped, forming true bound states. The formalism for analyzing Feshbach resonances was originally developed by Herman Feshbach in the context of nuclear reactions [183]. Other examples of Feshbach resonances occur in electron–atom and electron–ion scattering. The theory of Feshbach resonances will be analyzed in the first part of this section. Another formalism describing the coupling between a bound state and continuum states, leading to resonances, is known as *Fano resonance*. The corresponding physics will be analyzed in the second part of this section. Although the formalisms used by Feshbach and Fano are quite different, they both deal with very similar scattering phenomena.

Feshbach Resonance Scattering Formalism

Consider two systems C (for continuous) and B (for bound). System C (B) is subject to Hamiltonian $H_C = H_0 + V_C$ ($H_B = H_0 + V_B$) where the kinetic energy (or zeroth order Hamiltonian), H_0, is the same for both systems, while the potentials V_C and V_B are distinct (see Fig. 12.14). For convenience, we consider V_C and V_B to be spherically symmetric. The potential $V_C(r) \to 0$ as $r \to \infty$ faster than r^{-2}, and $V_B(r) \to V_B(\infty) > 0$ as $r \to \infty$. $V_B(r)$ supports a bound state at energy $E_B < V_B(\infty)$. It is useful to be able to shift V_B by adding to it a tunable constant.[7] Each system can be studied within the elementary procedure of solving the Schrödinger equation with scattering boundary condition for system C and bound-state boundary condition for system B. What happens when the two systems are coupled by a potential (denoted by $V \equiv V_{CB}$) as shown schematically in Fig. 12.14?

When systems B and C are coupled and the total energy E of the system is positive, there is only a scattering state with asymptotic conditions as in Eq. (12.85). Intuitively, however, the system may be trapped as a virtual bound state for a long time by V_B. Traditionally, one speaks of an open and closed channel (for more on scattering channels, see Sec. 12.8)[8] with corresponding

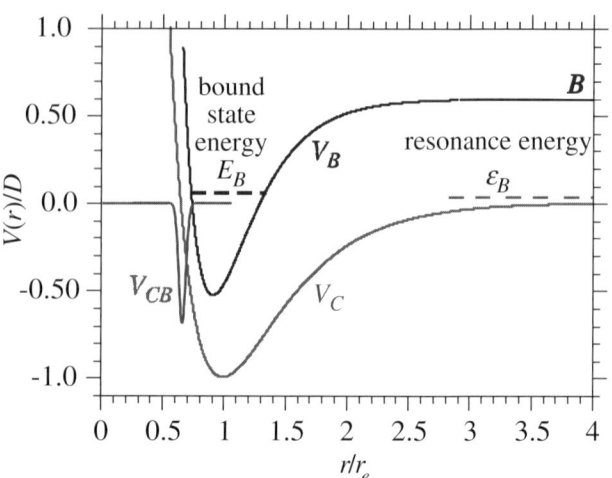

FIG 12.14 Feshbach resonance in the scattering of two atoms. Plotted are the two potentials (in units of the binding energy of potential V_C) versus internuclear separation r (in units of the equilibrium internuclear coordinate of the scattering potential r_e). The resonance energy E_r is similar to the energy of the bound-state energy E_B on potential V_B. The coupling potential $V \equiv V_{CB}$.

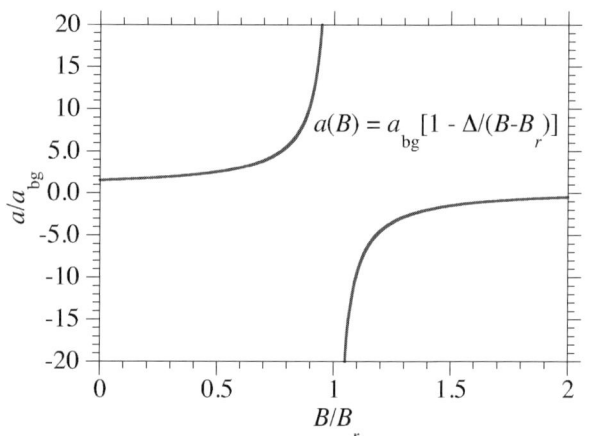

FIG 12.15 Modified scattering length versus external magnetic field strength near a Feshbach resonance.

---

[7] Experimentally, this can be accomplished, as discussed above, by introducing a magnetic field, if the atoms composing the diatomic molecule have a magnetic moment, and the magnetic moments of the atoms asymptotically obtained from channels B and C are different.
[8] From the strict definition of scattering channels, only one channel is present here.

12.6 Resonance Scattering

scattering and exponentially decaying bound state wave functions. The central question that will be answered below is how the coupling between the continuum states in the open channel and the bound state in the "closed channel" affects the scattering amplitude (or the scattering length, since we are concerned here with low-energy scattering).

The initial states (before coupling) are eigenstates of the corresponding Hamiltonians and satisfy the Schrödinger equations,

$$(H_0 + V_B)|\psi_B\rangle = E_B|\psi_B\rangle, \quad (H_0 + V_C)|\psi_{\mathbf{k}}^{+}\rangle = E_{\mathbf{k}}|\psi_{\mathbf{k}}^{+}\rangle, \tag{12.354}$$

where $|\psi_B\rangle$ is the bound-state wave function at energy $E_B < B$ and $|\psi_{\mathbf{k}}^{+}\rangle$ belongs to the set of continuous eigenstates of $H_0 + V_C$ with scattering energy $E_{\mathbf{k}} = \frac{\hbar^2 k^2}{2m} = E$. The kets $|\psi_B\rangle$ and $|\psi_{\mathbf{k}}^{+}\rangle$ satisfy orthogonality relations,

$$\langle \psi_B|\psi_B\rangle = 1, \quad \langle \psi_B|\psi_{\mathbf{k}}^{+}\rangle = 0, \quad \langle \psi_{\mathbf{k}}^{+}|\psi_{\mathbf{k}'}^{+}\rangle = \delta(\mathbf{k} - \mathbf{k}'), \tag{12.355}$$

and, in configuration space, they obey the asymptotic boundary conditions,

$$\langle \mathbf{r}|\psi_B\rangle \xrightarrow[r\to\infty]{} 0, \quad \langle \mathbf{r}|\psi_{\mathbf{k}}^{+}\rangle \xrightarrow[r\to\infty]{} (2\pi)^{-\frac{3}{2}}\left[e^{i\mathbf{k}\cdot\mathbf{r}} + f_0(\theta)\frac{e^{ikr}}{r}\right], \tag{12.356}$$

where θ is the scattering angle. The subscript 0 on the scattering amplitude indicates that we are considering a system where the coupling V is off. Denote by $\mathcal{B}$ the subspace spanned by the bound state $|\psi_B\rangle \in \mathcal{B}$ and by $\mathcal{C}$ the subspace of continuum states $|\psi_{\mathbf{k}}^{+}\rangle \in \mathcal{C}$. The corresponding projection operators (sometimes called *Feschbach projection operators*) are,

$$P_B = |\psi_B\rangle\langle\psi_B|, \quad P_C = 1 - P_B = \sum_{\mathbf{k}} P_{\mathbf{k}} = \sum_{\mathbf{k}} |\psi_{\mathbf{k}}^{+}\rangle\langle\psi_{\mathbf{k}}^{+}|. \tag{12.357}$$

Coupling between the two subspaces is induced by $V \equiv V_{CB}$, and the full Hamiltonian of the system is H. The system is completely specified as follows:

$$P_C H P_C = H_0 + V_C \equiv H_C, \quad P_B H P_B = H_0 + V_B \equiv H_B, \tag{12.358a}$$

$$P_C H P_B = V, \quad P_B H P_C = V^{\dagger}. \tag{12.358b}$$

The scattering state $|\Psi\rangle$ of the system corresponding to $E > 0$ is a solution of the stationary Schrödinger equation,

$$(H - E)|\Psi\rangle = 0. \tag{12.359}$$

With the definitions $|\Psi_C\rangle \equiv P_C|\Psi\rangle$, $|\Psi_B\rangle \equiv P_B|\Psi\rangle$, Eq. (12.359) can be recast as,

$$(H_C - E)|\Psi_C\rangle = -V|\Psi_B\rangle, \tag{12.360a}$$

$$(H_B - E)|\Psi_B\rangle = -V^{\dagger}|\Psi_C\rangle. \tag{12.360b}$$

Our goal is to solve these equations (at least approximately) at low but positive energy $E > 0$. The solution should be a scattering state that evolves from the unperturbed state $|\psi_{\mathbf{k}}^{+}\rangle$. There will be no bound state, despite the component $|\Psi_B\rangle$ in the full wave function. Our first task is to express both $|\Psi_B\rangle$ and $|\Psi_C\rangle$ in terms of $\psi_{\mathbf{k}}^{+}$ and use it for elucidating the influence of the coupling on the scattering amplitude. The formal solution of Eq. (12.360a) evolving from $|\psi_{\mathbf{k}}^{+}\rangle$ is given in terms of the Green's function of H_C with outgoing boundary conditions, $g_C^{+} = (E + i\eta - H_C)^{-1}$, as

$$|\Psi_C\rangle = |\psi_{\mathbf{k}}^{+}\rangle + g_C^{+} V|\Psi_B\rangle. \tag{12.361}$$

Substituting this expression into the RHS of Eq. (12.360b) and rearranging yields,

$$(E - H_B - V^{\dagger} g_C^{+} V)|\Psi_B\rangle = V^{\dagger}|\psi_{\mathbf{k}}^{+}\rangle. \tag{12.362}$$

We now formally invert the operator acting on $|\Psi_B\rangle$ and get an expression (without an inhomogeneous term) for $|\Psi_B\rangle$ in terms of $|\psi_{\mathbf{k}}^+\rangle$,

$$|\Psi_B\rangle = (E - H_B - V^\dagger g_C^+ V)^{-1} V^\dagger |\psi_{\mathbf{k}}^+\rangle. \tag{12.363}$$

Finally, this expression for $|\Psi_B\rangle$ is inserted into the RHS of Eq. (12.361) to obtain

$$|\Psi_C\rangle = [1 + g_C^+ V(E - H_B - V^\dagger g_C^+ V)^{-1} V^\dagger]|\psi_{\mathbf{k}}^+\rangle. \tag{12.364}$$

Using the configuration space representation of Eq. (12.364), the asymptotic behavior of the scattering state $\langle \mathbf{r}|\Psi_C\rangle$ can be analyzed, and the modification of the scattering amplitude (or the S matrix) is obtained. The physical content is encoded in the second term inside the square brackets on the RHS of Eq. (12.364), which is the correction to $|\psi_{\mathbf{k}}^+\rangle$ due to coupling with the bound state. The inverse operator appearing in the brackets acts in the subspace spanned by $|\Psi_B\rangle$. Inserting the projection $|\Psi_B\rangle\langle\Psi_B|$ into both sides we find,

$$g_C^+ V(E - H_B - V^\dagger g_C^+ V)^{-1} V^\dagger = g_C^+ V |\Psi_B\rangle (E - E_B - \langle\Psi_B|V^\dagger g_C^+ V|\Psi_B\rangle)^{-1} \langle\Psi_B|V^\dagger. \tag{12.365}$$

The quantity

$$\Sigma_B(E) \equiv \langle\Psi_B|V^\dagger g_C^+(E) V|\Psi_B\rangle \tag{12.366}$$

is referred to as the *self-energy* and plays an important role. The self-energy has both real and imaginary parts. The resonance energy E_r and width of the resonance Γ are defined as,

$$E_r \equiv E_B + \operatorname{Re} \Sigma_B, \quad \Gamma \equiv -2 \operatorname{Im} \Sigma_B. \tag{12.367}$$

When the resonance energy is close to the scattering energy, $E_r \approx E$, we have a *Feshbach resonance*. The width Γ of this resonance is determined by the imaginary part of the self-energy [see Eq. (12.367)]. In order to evaluate Γ, we use the spectral representation of the Green's function g_C^+ [employing the completeness relation (12.96a)] and substitute it into Eq. (12.366) to obtain,

$$\Gamma = -2 \operatorname{Im} \left[\int \frac{d\mathbf{k}}{(2\pi)^3} \frac{|\langle\Psi_B|V^\dagger|\psi_{\mathbf{k}}^+\rangle|^2}{E - E_{\mathbf{k}} + i\eta} \right]. \tag{12.368}$$

In this integration, the energy $E_{\mathbf{k}}$ should not be restricted to the energy shell. Employing the identity (12.136), (12.368) can be written as,

$$\Gamma \approx \frac{1}{4\pi^2} \int d\mathbf{k} \, |\langle\Psi_B|V^\dagger|\psi_{\mathbf{k}}^+\rangle|^2 \delta(E - E_{\mathbf{k}}). \tag{12.369}$$

At very low scattering energy E, the s-wave scattering dominates, hence, $|\psi_{\mathbf{k}}\rangle$ can be replaced by its s-wave component $R_0(r)$ [see the expansion (12.154)], which is isotropic. Integration over energy, using the delta function then yields,

$$\Gamma \approx \frac{m}{2\pi\hbar^2} k |\langle\Psi_B|V^\dagger|R_0\rangle|^2 \equiv \alpha k. \tag{12.370}$$

The normalization of the scattering states is such that the s-wave radial wave function R_0 is dimensionless, so that the dimension of $|\langle\Psi_B|V^\dagger|R_0\rangle|^2$ is equal to that of volume, ensuring that Γ has the dimension of energy. Equation (12.370), and the fact that $\Gamma \propto k$, will now be used in the calculation of the modified scattering length. When Eq. (12.370) is used in Eq. (12.365) and then the result is inserted in Eq. (12.364), we get

$$|\Psi_C\rangle = |R_0\rangle + g_C^+ V|\Psi_B\rangle \frac{1}{E - E_r + i\frac{\Gamma}{2}} \langle\Psi_B|V^\dagger|R_0\rangle. \tag{12.371}$$

We are now in a position to determine the asymptotic behavior of $\langle\mathbf{r}|\Psi_C\rangle$ starting from Eq. (12.371). The only task left is the calculation of $\langle\mathbf{r}|g_C^+ V|\Psi_B\rangle$. The coordinate representation of the s-wave Green's function $g_C^+(r, r')$ at large r can be

12.6 Resonance Scattering

expressed in terms of the regular and irregular s-wave solutions of the $l = 0$ Schrödinger equation,

$$g_C^+(r, r') = -\frac{m}{4\pi\hbar^2} e^{i\delta_0^{(0)}(E)} \frac{e^{ikr}}{kr} R_0(r'), \qquad (12.372)$$

where $\delta_0^{(0)}$ is the s-wave phase shift for scattering from V_C alone (without coupling to the bound-state system). At low energy, $\delta_0^{(0)} \to -k a_0^{(0)}$, where $a_0^{(0)}$ is the scattering length for scattering from V_C. Expression (12.372) is analogous to the second term on the RHS of Eq. 12.193, albeit with slightly different normalization of the asymptotic wave function. Due to this separable form, the exponential part can be pulled out of the matrix element $\langle r|g_C^+ V|\Psi_B\rangle$ and the remaining part is proportional to $\langle R_0|V|\Psi_B\rangle$. The product of this matrix element with the second one appearing on the RHS of Eq. (12.371) is proportional to Γ. Carrying out the algebra [184], we arrive at

$$\langle \mathbf{r}|g_C^+ V \Psi_B\rangle \langle \Psi_B|V^\dagger|R_0\rangle \to i\Gamma e^{2i\delta_0} \frac{e^{ikr}}{kr}. \qquad (12.373)$$

Finally, we go back to Eq. (12.371) and express the asymptotic form of $\langle \mathbf{r}|\Psi_C\rangle$ in terms of R_0 with coefficients that yield the modified S matrix. The asymptotic form of $R_0(r)$, according to Eq. (12.171), is

$$R_0(r) \to \frac{1}{kr}(e^{-ikr} - e^{2i\delta_0} e^{ikr}). \qquad (12.374)$$

Combining Eqs. (12.371), (12.373), and (12.374), we finally obtain

$$\langle \mathbf{r}|\Psi_C\rangle \to \frac{1}{kr}\left[e^{-ikr} - e^{ikr} e^{2i\delta_0^{(0)}}\left(1 - \frac{i\Gamma}{E - E_r + i\frac{\Gamma}{2}}\right)\right]. \qquad (12.375)$$

According to expression (12.171), the coefficient of the outgoing spherical wave is the partial-wave S matrix $S_l = e^{2i\delta_l}$ (here, we have $l = 0$). Hence,

$$S_0 = e^{2i\delta_0} = e^{2i\delta_0^{(0)}}\left[\frac{E - E_r - i\frac{\Gamma}{2}}{E - E_r + i\frac{\Gamma}{2}}\right] \equiv e^{2i(\delta_0^{(0)} + \theta_0)}, \quad \theta_0 = \arg\left(E - E_r - i\frac{\Gamma}{2}\right). \qquad (12.376)$$

When $E \to 0$, both $\delta_0^{(0)} \to 0$ and $\theta_0 \to 0$ (since $\Gamma = \alpha k \to 0$). The modified scattering length is then,

$$a_0 = -\lim_{E\to 0} \frac{1}{k}\tan(\delta_0^{(0)} + \theta_0) \approx -\lim_{E\to 0} \frac{1}{k}(\tan\delta_0^{(0)} + \tan\theta_0). \qquad (12.377)$$

Using $\tan\theta_0 = -\frac{\Gamma}{2(E-E_r)}$ and $\Gamma = \alpha k$, we find

$$a_0 = a_0^{(0)} + \frac{\alpha}{2E_r}. \qquad (12.378)$$

Tuning of the resonance energy by varying an external magnetic field (or an optical field) is an important experimental tool that can be used to control collisions. For example, in fermionic ultra-cold atom collisions, sweeping the strength of an external magnetic field through a Feshbach resonance can result in the formation of a large number of ultra-cold molecules composed of two fermionic atoms. The application of an external magnetic field enables control of the energy difference of the asymptotic molecular potential by the Zeeman energy of the atoms, $U_Z = -\boldsymbol{\mu} \cdot \mathbf{B}$. Since the magnetic dipole moments of the atoms to which the molecular potentials $V_C(r)$ and $V_B(r)$ asymptotically correlate are different, tuning the magnetic field strength tunes the asymptotic energy difference and therefore the resonance energy,

$E_r(B) = -\mu(B - B_r)$. Here, B_r is the magnetic field strength at which the resonance energy passes through zero. Substituting this expression into Eq. (12.378) yields

$$a(B) = a_{bg}\left(1 - \frac{\Delta}{B - B_r}\right), \qquad (12.379)$$

where we have denoted the background scattering length as $a_{bg} \equiv a_0^{(0)}$ and $\Delta = \alpha/(2a_0^{(0)}\mu)$. Figure 12.15 shows the scattering length dependence on the external magnetic field. The scattering length diverges at resonance $B = B_r$ and changes sign through resonance.

Feshbach Resonance Models

As an example of Feshbach resonance problem, consider the two-channel scattering with the 2×2 potential matrix $V(r) = \begin{pmatrix} V_0(r) & V_{01}(r) \\ V_{10}(r) & V_1(r) \end{pmatrix}$, where

$$\begin{aligned} V_0(r) &= -V_0[1 - \theta(r-\rho)], \\ V_1(r) &= (-V_1 + \varepsilon)[1 - \theta(r-\rho)] + \varepsilon\theta(r-\rho), \\ V_{01}(r) &= V_{10}(r) = U[1 - \theta(r-\rho)]. \end{aligned} \qquad (12.380)$$

Here, V_0 and V_1 are positive, $\varepsilon > 0$, $-V_0 + \varepsilon < 0$, and the scattering energy E is such that $\varepsilon > E > 0$ (see Fig. 12.16).

The Schrödinger equation for $l = 0$ for this problem is given by,

$$\left[-\frac{\hbar^2}{2\mu}\frac{d^2}{dr^2} + V_0(r) - E\right]u_0(r) = V_{01}(r)u_1(r),$$

$$\left[-\frac{\hbar^2}{2\mu}\frac{d^2}{dr^2} + V_1(r) - E\right]u_1(r) = V_{10}(r)u_0(r),$$

$$(12.381)$$

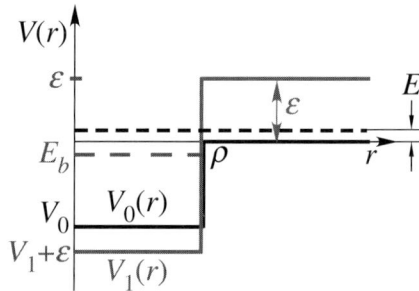

FIG. 12.16 A two-channel square well Feshbach resonance model. The open channel potential $V_0(r)$ is coupled to a closed channel potential $V_1(r)$ via a coupling potential $V_{01}(r) = V_{10}(r) = U[1 - \theta(r-\rho)]$, which is not shown. A Feshbach resonance occurs when the bound-state energy E_b on the closed channel potential $V_1(r)$ is adjusted via control of the difference between the asymptotic channel energies, ε, so that E_b (roughly) equals the scattering energy E (when the coupling strength U is small).

where μ is the reduced mass. For $r \geq \rho$, where $u_0(r)$ and $u_1(r)$ are uncoupled, we know that the form of the wave functions are $u_0(r) = B\sin(kr + \delta)$, and $u_1(r) = Ce^{-Kr}$, with $\hbar^2 k^2/(2\mu) = E$ and $\hbar^2 K^2/(2\mu) = \varepsilon - E$.

If $U = 0$ (no coupling), the closed channel wave function for $r < \rho$ is given by $u_1(r) = C\sin(\mathcal{K}r)$, where $\hbar^2\mathcal{K}^2/(2\mu) = E - V_1 - \varepsilon$, and for $r \geq \rho$, $u_1(r) = Ce^{-Kr}$, where $\hbar^2 K^2/(2\mu) = E - \varepsilon$. Matching the wave function and its derivative at $r = \rho$,

$$C\sin(\mathcal{K}\rho) = Ce^{-K\rho}, \quad C\mathcal{K}\cos(\mathcal{K}\rho) = -CKe^{-K\rho}, \qquad (12.382)$$

yields, upon dividing one equation by the other, the transcendental equation, $\tan(\mathcal{K}\rho) = -\frac{\mathcal{K}}{K}$. This is basically identical to Eq. (1.112) for the 1D square well odd bound states. No closed channel bound-state exists if $V_1\rho^2 < \pi^2\hbar^2/(8m)$, one bound state exists if $\pi^2\hbar^2/(8m) \leq V_1\rho^2 < 9\pi^2\hbar^2/(8m)$, etc. The open channel wave function for $U = 0$ is given by $u_0(r) = B\sin(k_0 r)$, where $\hbar^2 k_0^2/(2\mu) = E - V_0$, and for $r \geq \rho$, $u_0(r) = B\sin(kr + \delta_{bg})$, where the subscript bg stands for background.

12.6 Resonance Scattering

For $U \neq 0$ (coupling switched on), $u_0(r)$ and $u_1(r)$ are coupled for $r < \rho$, and we can define an orthogonal transformation $O = \begin{pmatrix} \cos\theta & \sin\theta \\ -\sin\theta & \cos\theta \end{pmatrix}$ to a new set of wave functions $u_P(r)$ and $u_Q(r)$, which are uncoupled, i.e.,

$$\begin{pmatrix} u_P(r) \\ u_Q(r) \end{pmatrix} = \begin{pmatrix} \cos\theta & \sin\theta \\ -\sin\theta & \cos\theta \end{pmatrix} \begin{pmatrix} u_0(r) \\ u_1(r) \end{pmatrix}, \tag{12.383}$$

where $O^t \begin{pmatrix} -V_0 & U \\ U & -V_1 + \varepsilon \end{pmatrix} O = \begin{pmatrix} v_P & 0 \\ 0 & v_Q \end{pmatrix}$ and

$$v_P = [(-V_0 - V_1 + \varepsilon) + \sqrt{(-V_0 + V_1 - \varepsilon)^2 + 4U^2}]/2,$$
$$v_Q = [(-V_0 - V_1 + \varepsilon) - \sqrt{(-V_0 + V_1 - \varepsilon)^2 + 4U^2}]/2. \tag{12.384}$$

The angle θ in Eq. (12.383) is such that $\tan 2\theta = 2V_{01}/(V_1 - V_2)$ for $r < \rho$. For $r \geq \rho$, the potential matrix V is already diagonal, so $\theta = 0$, $v_P = 0$ and $v_Q = \varepsilon$ (see Fig. 12.16), and $u_P = u_0$, $u_Q = u_1$. Although this model can be solved numerically, it is not quite simple enough to yield an analytic expression for the near threshold phase shift or the scattering length.

An even simpler two-channel scattering problem which can be solved analytically [185] is given by the 2×2 potential matrix $V(r) = \begin{pmatrix} V_o(r) & V_{01}(r) \\ V_{10}(r) & V_c(r) \end{pmatrix}$, where

$$V_o(r) = -V_o[1 - \theta(r - \rho)],$$
$$V_c(r) = -V_c[1 - \theta(r - \rho)] + V_\infty \theta(r - \rho),$$
$$V_{01}(r) = V_{10}(r) = \frac{2\pi\hbar^2 b}{\mu} \delta(\mathbf{r}) \frac{\partial}{\partial r} r.$$

Here, the open o and closed c channel potentials have well-depth V_o and V_c, respectively, which are positive, outside the well radius ρ, $V_\infty \gg 1$ (practically infinite, as shown in Fig. 12.17. The coupling potential $V_{01} = V_{10}$ is a regularized delta function (the regularization with $\frac{\partial}{\partial r} r$ is required since the wave function is singular at the origin) with strength proportional to the length b. E_b is the lowest bound-state energy in the closed channel relative to the asymptote of the open channel (see Fig. 12.16). This is a 3D spherically symmetric system, and the full 3D wave function can be expanded in the form, $\psi(\mathbf{r}) = \sum_l \frac{u_l(r)}{r} P_l(\theta)$. Only the $l = 0$ partial wave is relevant, since higher partial waves are frozen out at low scattering energy. The radial part of the Laplacian operator is equal to $\frac{1}{r} \frac{\partial^2}{\partial r^2} r$. The Schrödinger equations for the $l = 0$ radial wave functions, u_c and u_c, for energy E are given by,

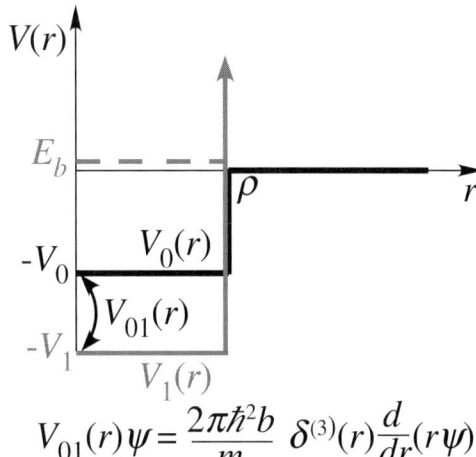

FIG 12.17 An example of a Feshbach resonance for a two-channel square well with a regularized delta function coupling. A Feshbach resonance occurs when the bound state on the closed channel potential $V_c \equiv V_1$, at energy E_b relative to the asymptote of the open channel potential, is adjusted, so that it corresponds to the scattering energy with which flux enters the open potential $V_o(r) \equiv V_0(r)$.

$$-\frac{\hbar^2}{2m}\left(\frac{u_o''(r)}{r} - 4\pi\delta(\mathbf{r})u_o(0)\right) + U_o(r)\frac{u_o(r)}{r} + V_{oc}(r)\frac{u_c(r)}{r} = E\frac{u_o(r)}{r}, \tag{12.385}$$

$$-\frac{\hbar^2}{2m}\left(\frac{u_c''(r)}{r} - 4\pi\delta(\mathbf{r})u_c(0)\right) + U_c(r)\frac{u_c(r)}{r} + V_{oc}^*(r)\frac{u_o(r)}{r} = E\frac{u_c(r)}{r}. \tag{12.386}$$

When the interaction potentials are finite everywhere, we must set $u_i(0) = 0$, i.e., for finite potentials, the wave function $\psi(\mathbf{r})$ should be finite, so $u_i(r) \to 0$ as $r \to 0$. However, for the delta function coupling potential V_{oc}, which is singular at the origin, we obtain

$$\frac{1}{r}\left(-\frac{\hbar^2}{2m}u_o''(r) + (U_o(r) - E)u_o(r)\right) + \frac{2\pi\hbar^2}{m}\left(u_o(0) + bu_c'(0)\right)\delta(\mathbf{r}) = 0, \qquad (12.387)$$

$$\frac{1}{r}\left(-\frac{\hbar^2}{2m}u_c''(r) + (U_c(r) - E)u_c(r)\right) + \frac{2\pi\hbar^2}{m}\left(u_c(0) + bu_o'(0)\right)\delta(\mathbf{r}) = 0, \qquad (12.388)$$

and the wave function can be singular (although probability distributions remain finite). To satisfy Schrödinger equation, we have to put

$$u_o(0) + bu_c'(0) = 0, \qquad u_c(0) + bu_o'(0) = 0. \qquad (12.389)$$

We shall not present the analytic solution, but only sketch the result. Figure 12.18 shows the scattering length a versus V_c, as calculated using Eqs. (9.386). The black dot indicates the value of the scattering length for the specific value of V_c to the left of the graph of a versus V_c, while the red curve shows the whole curve of a versus V_c. As long as the coupling parameter b is small, the resonance occurs when the bound state in the closed channel crosses threshold and is rather narrow. If b is large, this need not be the case.

When V_c is such that $E_b \lesssim 0$, a two-channel bound state can exist. We can write the wave functions for open and closed channels as

$$u_o(r) = \begin{cases} A_o \sin(q_o r - \eta_o) & \text{for } r < \rho, \\ B_0 e^{-\kappa\rho} & \text{for } r > \rho, \end{cases} \qquad (12.390)$$

$$u_c(r) = \begin{cases} A_c \sin(q_c r - \eta_c) & \text{for } r < \rho, \\ 0 & \text{for } r > \rho, \end{cases} \qquad (12.391)$$

where we now define $\hbar^2 q_o^2/(2\mu) \equiv V_o - \mathcal{E}_b$ for the open channel, and $\hbar^2 q_c^2/(2\mu) \equiv V_c - \mathcal{E}_b$ for the closed channel, with $|\mathcal{E}_b| > 0$ being the binding energy, i.e., the energy of the two-channel bound state equals $\mathcal{E}_b = -|\mathcal{E}_b|$, and $\hbar^2\kappa^2/(2\mu) \equiv |\mathcal{E}_b|$. In a fashion similar to the above procedure, we again find,

$$\tan\eta_o \tan\eta_c = b^2 q_o q_c. \qquad (12.392)$$

By considering the closed channel wave function at $r = \rho$, we obtain

$$\sin(q_c\rho + \eta_c) = 0, \implies \tan(q_c\rho) = -\tan\eta_c. \qquad (12.393)$$

The open channel boundary conditions at $r = \rho$ yields,

$$\kappa = -q_o/\tan(q_o\rho + \eta_o). \qquad (12.394)$$

12.6 Resonance Scattering

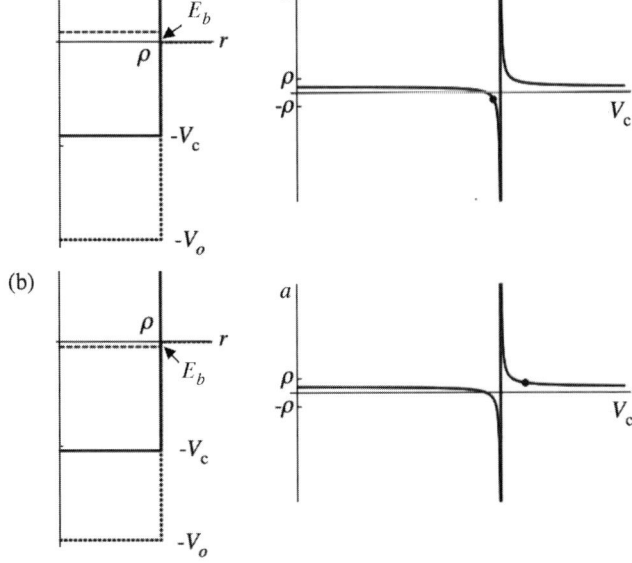

FIG 12.18 Scattering length a for the Feshbach resonance versus potential parameter V_c. The black dot indicates the value of the scattering length for the given V_c. (a) The bound state on the closed channel is above threshold on the open channel, and the scattering length is positive. (b) The bound state on the closed channel is below threshold on the open channel, and the scattering length is positive. A true two-channel bound state then exists, and if $a \gg \rho$, the two-channel bound-state energy is given by $\mathcal{E}_b \approx -\frac{\hbar^2}{2\mu a^2}$.

This equation can be re-expressed in terms of the parameters of the problem as follows. Noting that $\tan(q_o\rho + \eta_o) = [\tan(q_o\rho) + \tan\eta_o]/[1 - \tan(q_o\rho)\tan\eta_o]$, we finally obtain

$$\kappa = -q_o \frac{1 - \tan(q_o\rho)\tan\eta_o}{\tan(q_o\rho) + \tan\eta_o}. \quad (12.395)$$

The quantity $\tan\eta_o$ can be evaluated using Eqs. (12.392) and (12.393):

$$\tan\eta_o = -\frac{b^2 q_o q_c}{\tan(q_c\rho)}. \quad (12.396)$$

$$\kappa = -q_o \frac{\tan(q_c\rho) + b^2 q_o q_c \tan(q_o\rho)}{\tan(q_o\rho)\tan(q_c\rho) - b^2 q_o q_c}. \quad (12.397)$$

This is a transcendental equation for the bound-state energy. However, a simple approximation can be made to obtain an approximate formula for the bound-state energy when $a \gg \rho$. We see from Eq. (12.394) that, $a = \rho - \tan(q_o\rho + \eta_o)/q_o$. Substituting this expression for $\tan(q_o\rho+\eta_o)$ into Eq. (12.395), assuming that the two-channel bound-state energy is very close to threshold on the open channel, we obtain $\kappa \approx \frac{1}{a-\rho}$, hence,

$$\mathcal{E}_b \approx -\frac{\hbar^2}{2\mu(a-\rho)^2} \approx -\frac{\hbar^2}{2\mu a^2}. \quad (12.398)$$

12.6.4 FANO RESONANCE

The consequence of coupling between a discrete level (or a group of discrete levels) and an energetically close-by continuum of levels was developed by Ugo Fano in 1961 [186] in the context of studying the autoionization of atoms by scattered electrons. The experimental feature of the autoionization cross-section that is distinct from that of a Breit–Wigner resonance is the asymmetric lineshape of the cross-section (see Fig. 12.19). Consider an electron scattering from a He atom in its ground electronic state whose configuration is $(1s)^2 \, ^1S_0$. Electron scattering can excite the helium to the first state $(1s2s) \, ^3S_1$. A Breit–Wigner resonance peak appears in the excitation spectrum. But a different situation occurs when the helium is excited into a very high energy configuration, such as $(2s2p) \, ^1P_1$, which is located above the lowest ionization threshold. After excitation, such a state decays into $He^+ + e^-$ and the excitation spectrum displays a highly asymmetric lineshape, which vanishes (experimentally, nearly vanishes) on one side of the peak. This type of resonance feature is referred to as a Fano resonance.

The Fano and Feshbach resonance phenomena are similar. The analysis in both cases involves the mixing between discrete and continuum states that have energies close to each other. In the analysis of the Feshbach resonance discussed above, the focus was on the effect of the quasi-bound state on the scattering length (or the scattering cross-section), while in the Fano resonance case, the focus is on the autoionization of the bound state due to configuration mixing, which, again affects the autoionization cross-section. The Feshbach and Fano formalisms can be recast one into the other.

Recently, Fano resonances were observed in electronic quantum interferometers. In these devices, one path of the interferometer allows free passage of the electrons, whereas the other path contains a quantum dot that mimics an attractive potential well. Experiments show that when such an interferometer serves as a resistor, the corresponding conductance displays a Fano lineshape. This is attributed to interference between the continuum and the bound-state contributions to the conductance, which coherently interfere.

The analysis of Fano resonances starts from Eqs. (12.359) and (12.360) for the components of a scattering state $|\Psi\rangle$ at energy E. It is assumed that there is a single discrete state $|\varphi\rangle$ at energy E_B (before coupling is switched on), and a continuous set of states $\{\psi(E')\}$ with energies E'. It is useful here to adopt the energy representation for the states in the space C of continuous functions, using the notation $|\psi^+(E')\rangle$ instead of the scattering states $|\psi_{\mathbf{k}}^+\rangle$ used previously. For notational convenience, the superscript $+$ will be dropped. For reasons to become clear later on, we first assume that the energies E' form a discrete set of N levels with constant level separation d. The continuum limit will be taken at the appropriate stage.

The wave function can be written as,

$$|\Psi\rangle = \beta_E |\varphi\rangle + \sum_{E'} \gamma_E(E') |\psi(E')\rangle, \tag{12.399}$$

where β_E and $\gamma_E(E')$ are coefficients to be determined (their normalization will also concern us later on). Employing Eqs. (12.360) yields the Schrödinger equation for the coefficients β_E and $\gamma_E(E')$,

$$E_B \beta_E + \sum_{E'} V_{BE'}^* \gamma_E(E') = E\beta_E, \tag{12.400a}$$

$$V_{E'B} \beta_E + E' \gamma_E(E') = E\gamma_E(E') \Rightarrow \gamma_E(E') = \frac{V_{E'B} \beta_E}{E - E'}. \tag{12.400b}$$

Inserting the solution back into Eq. (12.400a) leads to

$$E_B \beta_E + \sum_{E'} \frac{|V_{E'B}|^2}{E - E'} \beta_E = E\beta_E. \tag{12.401}$$

After dividing Eq. (12.401) by β_E, one gets an implicit equation for the eigenvalues E_n ($n = 0, 1, \ldots, N, N+1$). This equation also appears in other branches of physics, and, as was originally shown by Rayleigh, $E_0 < E_1'$, $E_{N+1} > E_N'$, while $E_1' \leq E_n \leq E_N'$ ($n = 1, \ldots, N$). The question now is how to take the continuum limit. Technically, this is carried out by taking the limits $d \to 0$, $N \to \infty$, keeping Nd finite. Clearly, $\sum_{E'} \to \int dE' \rho(E')$, where $\rho(E')$ is the density of states $\psi(E')$. Handling the singularity in Eq. (12.401) depends on the boundary conditions. The standard procedure is to replace $E \to E + i\eta$ and take the limit $\eta \to 0^+$ using the formula [see Eq. (12.136)],

$$\frac{1}{x \pm i\eta} = \mathcal{P}\left[\frac{1}{x}\right] \mp i\pi \delta(x), \tag{12.402}$$

where $\mathcal{P}$ denotes the principal part of the integral. This implies that the solutions are complex. For time-reversal invariant systems, we seek real solutions. To achieve this goal, a procedure adapted by Fano is to take the continuum limit of the solution (12.400b) as,

$$\gamma_E(E') = V_{E'B} \beta_E \left[\mathcal{P} \frac{1}{E - E'} + z(E)\delta(E - E')\right], \tag{12.403}$$

containing an energy-dependent constant $z(E)$, which is to be determined self-consistently. First, let us obtain an expression for the phase shift within the Fano analysis, which will enable the determination of the scattering length, in a fashion similar to that in Eq. (12.378) in Feshbach theory. To this end, Eq. (12.403) for $\gamma_E(E')$ is inserted into the continuum version of Eq. (12.399) and then the r representation is taken on both sides, assuming s-wave scattering, $\langle r|\psi(E')\rangle \to A \sin[k(E')r + \delta_0(E')]$ (where for free particle, $k(E') = \sqrt{2mE'/\hbar^2}$). The contribution from $\langle r|\varphi\rangle$ decays as $r \to \infty$, and hence, one obtains

$$\langle r|\Psi\rangle = \Psi(r) \to \int dE' \rho(E') \gamma_E(E') A \sin[k(E')r + \delta_0(E')]. \tag{12.404}$$

12.6 Resonance Scattering

The integral can be estimated to give

$$\Psi(r) \to A' \sin[kr + \bar{\delta}_0(E)], \tag{12.405a}$$

$$\bar{\delta}_0(E) = \delta_0(E) - \tan^{-1}\frac{\pi}{z(E)} \equiv \delta_0(E) + \delta_r(E), \tag{12.405b}$$

where $\delta_r(E)$ is the *resonance phase shift*, since it undergoes a π jump at the *resonance energy* E_r, where $z(E_r) = 0$.

To determine $z(E)$, substitute Eq. (12.403) into Eq. (12.400a), which leads to

$$z(E) = \frac{E - E_B - F(E)}{\rho(E)|V_{E'B}|^2}, \tag{12.406a}$$

with

$$F(E) = \mathcal{P}\int dE' \frac{\rho(E')|V_{E'B}|^2}{E' - E}. \tag{12.406b}$$

Since the Fano resonance energy E_r is such that $z(E_r) = 0$,

$$E_r \equiv E_B + F(E_r). \tag{12.407}$$

Thus, $F(E)$ is the real part of the self-energy [see Eq. (12.366)], which, when added to the bare bound-state energy E_B gives the resonance anergy E_r.

The coefficients β_E and $\gamma_E(E')$ can be determined from the above equations, but the normalization conditions $\langle \Psi_E | \Psi_{\bar{E}} \rangle = \delta(E - \bar{E})$ should be applied with care. The result is as follows:

$$\beta_E = \frac{\sin \delta_r}{\pi \sqrt{\rho(E)}\, V_{EB}}, \quad \gamma_E(E') = \frac{V_{E'B} \sin \delta_r}{\pi \sqrt{\rho(E)}\, V_{EB}(E - E')} - \cos \delta_r\, \delta(E - E'). \tag{12.408}$$

It is instructive to examine $|\beta_E|^2$ as a function of E. Using Eq. (12.408), we find

$$|\beta_E|^2 = \frac{\rho(E)|V_{EB}|^2}{(E - E_r)^2 + \pi^2 \rho(E)^2 |V_{EB}|^4}, \tag{12.409}$$

which is a Lorenzian centered at E_r and half-width $\Gamma/2$, where

$$\Gamma = 2\pi |V_{EB}|^2 \rho(E). \tag{12.410}$$

Thus, instead of a bound state at energy E_B, we get a Lorenzian distribution of β_E centered at E_r with width Γ.

The fact that $z(E)$ is real and changes sign through resonance affects the transition rate between a given initial state $|i\rangle$ and the exact state $|\Psi\rangle$. If the transition is induced by an operator $\hat{T}$ (e.g., $\hat{T}$ can be a dipole operator), the transition matrix element can be decomposed, using Eqs. (12.399) and (12.408), as,

$$\langle \Psi | \hat{T} | i \rangle = \frac{\sin \delta_r}{\pi V_{EB}} \langle \Psi_B | \hat{T} | i \rangle - \cos \delta_r \langle \psi_E | \hat{T} | i \rangle, \tag{12.411a}$$

where

$$|\Psi_B\rangle = |\varphi\rangle + \mathcal{P}\int dE' \rho(E') \frac{V_{BE'}\psi(E')}{E - E'}. \tag{12.411b}$$

Consider first Eq. (12.411b), which indicates how the bound state $|\varphi\rangle$ is modified due to coupling with the continuum states. The difference between this expression and Eq. (12.363) is that in the latter case, it is assumed that Eq. (12.360b) has no solution in the absence of coupling, while here it is assumed that a bound-state solution exists even without coupling. Our main concern here, however, is Eq. (12.411a), which displays a Fano interference between transition into a (modified) bound state and transition into the unperturbed continuum. In a time-reversal invariant system for which all the matrix elements can be chosen to be real, there exists an energy E_0 defined by,

$$-\frac{E_0 - E_B - F(E_0)}{\pi \rho(E_0) |V_{E_0 B}|^2} = \frac{\pi V_{E_0 B} \langle \psi_{E_0} | \hat{T} | i \rangle}{\langle \Psi_B | \hat{T} | i \rangle}, \tag{12.412}$$

at which the transition amplitude $\langle \Psi | \hat{T} | i \rangle$ vanishes. There is destructive interference between the two contributions to the transition.

Although Eq. (12.411a) appears to involve several parameters, it can be recast in a more transparent and compact form. To this end, we define the following quantities:

$$\epsilon(E) \equiv -\cot \delta_r = \frac{E - E_r}{\frac{1}{2}\Gamma}, \quad q(E) \equiv \frac{\langle \Psi_B | \hat{T} | i \rangle}{\sqrt{\rho(E)} \pi V_{EB} \langle \psi_E | \hat{T} | i \rangle}. \tag{12.413}$$

Equation (12.412) guarantees that $\epsilon(E_0) + q(E_0) = 0$. In terms of ϵ and q, the Fano formula for the ratio of the corresponding transition rates reads,

$$\boxed{\left| \frac{\langle \Psi | \hat{T} | i \rangle}{\langle \psi_E | \hat{T} | i \rangle} \right|^2 = \frac{(q+\epsilon)^2}{1+\epsilon^2}.} \tag{12.414}$$

This so-called *Fano lineshape*, shown in Fig. 12.19, is the hallmark of the physics involving coupling between a bound state and continuum states. As a function of ϵ, the RHS of Eq. (12.414) displays a peak at $\epsilon = 1/q$ and vanishes at $\epsilon = -q$.

In the special case where the initial state $|i\rangle$ is a plane wave describing the relative motion of an electron and an atom in its ground state, and $\langle \mathbf{r} | \Psi \rangle$ is a scattering state with outgoing spherical wave asymptotic boundary conditions, the corresponding matrix element is nothing but the scattering amplitude. Thus, if $\hat{T}$ is a dipole operator, the matrix element $\langle \psi_E | \hat{T} | i \rangle$ is the scattering amplitude for the photoabsorption cross-section into the continuum, while $\langle \Psi | \hat{T} | i \rangle$ is the modified photoabsorption cross-section due to the presence of the bound state embedded in the continuum. The dependence of q on energy is much weaker than that of ϵ, and hence, q can be regarded as a constant parameter. For $|q| \gg |\epsilon|$, the line shape is Lorenzian, as for shape resonances. For small q, however, the transition cross-section is far from being Lorenzian, and the Fano zero at $\epsilon = -q$ and the Fano peak at $\epsilon = 1/q$ are close to each other.

Example: s-wave scattering

In the following example, we consider the coupling between continuous and discrete states in configuration space and point out some distinctions between Fano and Feshbach analyses.

In a three-dimensional system, we consider an s-wave scattering where system 1 describes a free particle [wave function $\psi(r)$] and system 2 describes a particle [wave function $\varphi(r)$] scattered from a

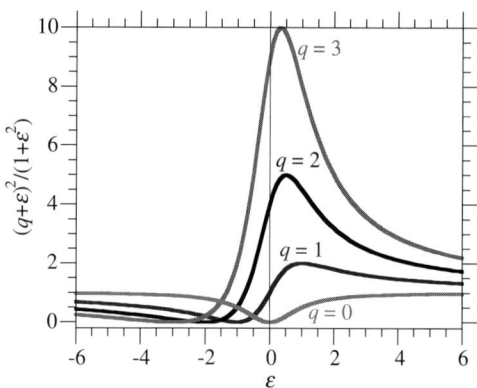

FIG 12.19 Fano lineshapes. The energy dependence of experimental resonance cross-sections can often be written in the form $\sigma(\epsilon) = \sigma_0 \frac{(q+\epsilon)^2}{1+\epsilon^2} + \sigma_b$, where σ_b is a background contribution to the cross-section and σ_0 is proportional to $|\langle \psi_E | \hat{T} | i \rangle|$. For negative q, invert the curves with respect to $\epsilon = 0$.

12.6 Resonance Scattering

spherical shell potential $-\lambda\delta(r-a)$. The two systems are coupled on the spherical shell due to a potential $-\alpha\delta(r-a)$. In the notation of (12.158) and (12.159), the coupled Schrödinger equations are

$$-\frac{d^2}{dr^2}\psi(r) - \alpha\delta(r-a)\varphi(r) = k^2\psi(r), \tag{12.415a}$$

$$-\frac{d^2}{dr^2}\varphi(r) - \lambda\delta(r-a)\varphi(r) - \alpha\delta(r-a)\psi(r) = q^2\varphi(r). \tag{12.415b}$$

The connection with Eqs. (12.400a) and (12.400b) are $\psi(E') \leftrightarrow \psi(r)$, $V_{E'B} = V_{BE'}^* \leftrightarrow \alpha\delta(r-a)$, $\frac{2mE}{\hbar^2} \leftrightarrow k^2$, and $\frac{2m(E-E_B)}{\hbar^2} \leftrightarrow q^2$ (not the Fano parameter). In the first equation, $k^2 > 0$, whereas in the second equation, there are two possibilities: If $q^2 > 0$, then we say that both systems 1 and 2 correspond to *open channels*. If, on the other hand, $q^2 < 0$, system 2 describes a bound state and is referred to as a *closed channel*. Due to Eq. (12.158), both $\psi(r)$ and $\varphi(r)$ vanish at $r = 0$. When $r \to \infty$, $\psi(r)$ is a combination of an incoming and an outgoing spherical wave, whereas $\varphi(r)$ is either an outgoing spherical wave ($q^2 > 0$) or an exponentially decaying function ($q^2 < 0$). Explicitly,

$$\psi(r) = \begin{cases} e^{-ikr} - S_{11}e^{ikr}, & r > a, \\ A(e^{ikr} - e^{-ikr}) & 0 < r < a. \end{cases} \qquad \varphi(r) = \begin{cases} -S_{21}e^{iqr}, & r > a, \\ C(e^{iqr} - e^{-iqr}) & 0 < r < a. \end{cases} \tag{12.416}$$

Here, S_{ji} denotes the corresponding S matrix element for final channel j and incoming channel i. The matching at $r = a$ requires continuity of the wave functions at $r = a$ and discontinuity of the derivatives as dictated by the Dirac delta function potential. When this is carried out, we obtain

$$S_{11} = \frac{\frac{q}{\sin qa} - \lambda e^{iqa} - \frac{\alpha^2}{k}e^{i(q-k)a}\sin ka}{D},$$

$$S_{21} = \frac{-2i\alpha \sin ka}{D},$$

$$D = \lambda e^{iqa} + \frac{\alpha^2}{k}e^{i(q+k)a}\sin ka - \frac{q}{\sin qa}. \tag{12.417}$$

When $q^2 > 0$, both channels are open and contain an outgoing spherical wave. Unitarity (flux conservation) implies $|S_{11}|^2 + |S_{21}|^2 = 1$. If $q^2 = -\kappa^2 < 0$,

$$S_{11} = \frac{\frac{\kappa}{\sinh \kappa a} - \lambda e^{-\kappa a} - \frac{\alpha^2}{k}e^{-(\kappa+ik)a}\sin ka}{\bar{D}},$$

$$S_{21} = \frac{-2i\alpha \sin ka}{\bar{D}},$$

$$\bar{D} = \lambda e^{-\kappa a} + \frac{\alpha^2}{k}e^{-(\kappa-ik)a}\sin ka - \frac{\kappa}{\sinh \kappa a}. \tag{12.418}$$

Here, only channel 1 is open, and S_{21} is, strictly speaking, not a generic S matrix element; thus, $|S_{11}|^2 = 1$. This is a slight distinction between the Feshbach and the Fano resonances; in the former case, channel 2 is always closed, due to the confining potential V_B in Eq. (12.354), see also Fig. 12.14. In this case, $S_{11} = e^{2i\delta}$, and at low energy, the relation of the phase shift to the scattering length is employed. It is technically trivial to remove V_B and "open" channel 2 but then $S_{11} \neq e^{2i\delta}$. By analogy with Eq. (12.376), the Fano resonance energy E_r and its width Γ should be read out from the pole of S_{11} at $E = E_r - i\Gamma/2$. The autoionization cross-section for s-wave scattering is $4\pi a^2$, where the scattering length a is given in Eq. (12.378). In addition to the autoionization cross-section, the Fano formalism yields also the probability $P_\varphi = |\langle\varphi|\psi\rangle|^2$ that, for $q^2 < 0$, φ is occupied.

Direct calculation yields,

$$P_\varphi = \frac{16\pi |AC|^2}{\varepsilon_B^2} \frac{\left[4\kappa^2 e^{\kappa a} - 2\kappa\lambda(1 - e^{-2\kappa a}) - 4\alpha^2 e^{-\kappa a}\sinh^2 \kappa a\right]^2}{[2\kappa + \lambda(1 - e^{-2\kappa a})]^2}, \qquad (12.419)$$

where the coefficients A, C appearing in Eq. (12.416) should be calculated by solving the matching equations in the same way that S_{11} and S_{21} were calculated. P_φ vanishes at some energy $\kappa_0^2 > 0$. This fixes the zero of the excitation spectrum line shape occurring at an energy that is shifted from E_B' by an energy κ_0^2.

Problem 12.18

(a) For $q^2 > 0$, prove the unitarity relation $|S_{11}|^2 + |S_{21}|^2 = 1$.
(b) For $q^2 < 0$, prove the unitarity relation $|S_{11}|^2 = 1$.
(c) Explain the difference between (a) and (b).

Solution: For (a), one cannot avoid brute-force calculations. For (b), one immediately notes that the numerator and the denominator in the expression for S_{11} are complex conjugates. In case (a), particle flux occurs in both channels, and the unitarity of the S matrix reflecting flux conservation should be computed for each channel separately. In case (b), only channel 1 is open, hence flux is present only in this channel.

Feshbach and Fano Resonances: Summary

(1) Feshbach and Fano resonances are very similar to each other. Both analyses resolve the subtle issues of treating the interaction of a discrete level embedded in, and interacting with, a continuum of states.
(2) Feshbach and Fano resonances are distinct from shape resonances that are characterized by a Breit–Wigner lineshape. The latter is a Lorenzian, symmetrically distributed around the peak, while the former is asymmetric and vanishes at a certain energy. However, it is sometimes difficult to distinguish between Feshbach or Fano resonances on the one hand and interference effects between Breit–Wigner resonances and a smooth background B, characterized by an amplitude such as, $F = B + \frac{\Gamma/2}{E - E_r + i\Gamma/2}$. $|F|^2$ can be parametrized by a Fano resonance lineshape with $q = 1/B$.
(3) The Feshbach formalism was originally developed for the study of nuclear reactions, but recently has played a central role in cold atom physics, where the atom–atom scattering length can be tuned via Feshbach resonances. In this framework, only one channel is open, and the expression (12.379) indicates that the scattering length passes through ∞ and changes sign at the resonance.
(4) The Fano formalism was originally developed to study autoionization processes, but recently it was shown to be relevant for mesoscopic physics. Examples from mesoscopic physics are related to electron interferometers, to be discussed in Sec. 13.2. The Fano expression (12.405b) for the modified phase shift can be easily reformulated as an expression for the scattering length, similar to Eq. (12.378). Thus, the low-energy scattering cross-section can be calculated in either theory.

12.7 APPROXIMATION METHODS

In Sec. 12.5.3, we presented a recipe for calculating phase shifts by numerical integration of the radial Schrodinger equation (12.159). This section focuses on approximation methods that can be applied to more general potential scattering problems, for central or non-central potentials. The first method to be discussed is the *Born approximation*, which consists of retaining only the first few terms in the Born series (12.122). The Born series is an operator relation, which is applicable also to problems involving scattering from non-central potentials. The second formalism is the *WKB approximation*,

12.7 Approximation Methods

which was introduced in Chapter 7. It is a semiclassical approximation, which is justified when the potential is slowly varying on the scale of the local particle wavelength. Then, *variational methods* of approximation are introduced, and we describe the *Kohn variational principle* for calculating phase shifts. The Kohn variational principle is also useful in finding the few lowest eigenvalues of a given Hamiltonian. Finally, the eikonal approximation is discussed. Similar to the WKB approximation, it is suitable for high-energy particle scattering from a smooth potential but does not require partial wave expansion, and, unlike the WKB approximation, it can directly be used in three-dimensional systems with non-central potentials. (The WKB approximation is basically a 1D semiclassical method, and a simple generalization to higher dimensions is not available.)

12.7.1 BORN APPROXIMATION

The Born approximation is based on successive iterations of the Lippmann–Schwinger equation in its various forms. These could be the abstract operator equations for the Green's function (12.109), for the T operator (12.110), the momentum representation of the partial wave Eq. (12.182) for the t matrix, the partial wave Lippmann–Schwinger Eq. (12.194) for the radial wave function, and other equations such as (12.118), (12.120), (12.122), and (12.85). A natural question concerns the rate of convergence of the Born series (12.122) for $z = E^+ = \frac{\hbar^2 k^2}{2m} + i\eta$,

$$T = V + V G_0^+ V + V G_0^+ V G_0^+ V + \ldots . \tag{12.420}$$

Determining the rate of convergence of the Born series is in general rather complicated. It is, however, illuminating to compare the ratio or the second and first term of the series. The Lippmann–Schwinger equation (12.120), to second order in $v = \frac{2m}{\hbar^2} V$, is

$$t^{(2)}(\mathbf{k}', \mathbf{k}) = v(\mathbf{k}', \mathbf{k}) + \int d\mathbf{k}''\, v(\mathbf{k}', \mathbf{k}'')[k'^2 + i\eta - k''^2]^{-1} v(\mathbf{k}'', \mathbf{k}). \tag{12.421}$$

The explicit expressions for the two terms on the RHS are

$$B_1(\mathbf{k}', \mathbf{k}) = v(\mathbf{k}' - \mathbf{k}) \equiv v(\mathbf{q}) = \int d\mathbf{r}\, e^{i\mathbf{q}\cdot\mathbf{r}} v(\mathbf{r}), \tag{12.422}$$

$$B_2(\mathbf{k}', \mathbf{k}) = \int d\mathbf{k}'' \frac{v(\mathbf{k}' - \mathbf{k}'')}{[k'^2 + i\eta - k''^2]} v(\mathbf{k}'' - \mathbf{k}). \tag{12.423}$$

Here, $\mathbf{q} = \mathbf{k} - \mathbf{k}'$ is the momentum transfer. Recall the on-energy shell condition $k^2 = k'^2 = \varepsilon$, so that $q^2 = 2\varepsilon(1 - \cos\theta)$. At high energy ε and at high momentum transfer squared q^2, it is reasonable to replace the denominator in the expression for B_2 simply by $1/\varepsilon$ and to approximate the ratio,

$$\frac{B_2(\mathbf{k}', \mathbf{k})}{B_1(\mathbf{k}', \mathbf{k})} \approx \frac{1}{\varepsilon} \frac{\langle \mathbf{k}'|v^2|\mathbf{k}\rangle}{\langle \mathbf{k}'|v|\mathbf{k}\rangle}. \tag{12.424}$$

For example, consider the Yukawa potential (12.124) whose momentum representation (12.125) is

$$v(\mathbf{q}) = \frac{v_0}{\mu} \frac{1}{q^2 + \mu^2}. \tag{12.425}$$

For $k = \sqrt{\varepsilon} \gg \mu$ and $q \gg \mu$, direct application of Eq. (12.424) yields

$$\frac{B_2(\mathbf{k}', \mathbf{k})}{B_1(\mathbf{k}', \mathbf{k})} \approx \frac{v_0}{2k\mu}. \tag{12.426}$$

Reliable estimates of higher order terms are rather difficult to carry out. For the Yukawa potential, one finds that the condition

$$\frac{2}{\pi} \ln\left(\frac{v_0}{2\mu^2}\right) < 1 \tag{12.427}$$

is sufficient for the convergence of the Born series (12.420) [179]. Note that this is close to the condition that the RHS of Eq. (12.426) is smaller than unity.

Often, the Born approximation works well even when it is not expected to do so. At the same time, there are examples where, counter to intuition, it does not work. For example, it gives poor results in the problem of scattering from a hard sphere even when the dimensionless parameter kr_0 satisfies the condition $kr_0 \ll 1$ leading to an s-wave scattering with a very small phase shift, $\delta_0 \approx -kr_0$ [see Eq. (12.200)].

12.7.2 WKB APPROXIMATION

The WKB approximation [187] for the solution of a 1D Schrödinger equation in the semiclassical approximation was introduced in Sec. 7.2.1, where the analysis focused on the calculation of bound-state energies based on the Bohr–Sommerfeld semiclassical theory of the quantization of energy levels. Here, we focus on its use for scattering problems. The technique using the connection formulae at the turning points is the same, and therefore, we will not repeat it here. In the study of transmission and reflection in generic 1D problems, the coordinate $-\infty < x < \infty$, but the WKB approximation is applicable also for the analysis of the radial equations obtained after partial wave expansion, where the one-dimensional coordinate is the distance between the scattered particle and the potential center with $0 \leq r < \infty$. The connection formulas developed in Sec. 7.2.1 are also applicable here with a slight modification. Starting from the radial equation (12.159), we will derive an expression for the phase shift $\delta_l(k)$. The WKB approximation is expected to be useful when the scattering potential varies slowly on the scale of the wavelength $\lambda = \frac{2\pi}{q(x)}$, where $q(x) = \frac{1}{\hbar}\sqrt{2m[E - V(x)]}$ is the local wavenumber (which includes the contribution from the potential),

$$\lambda(x) = \frac{h}{\sqrt{2m[E - V(x)]}} \ll 4\pi \frac{[E - V(x)]}{\left|\frac{dV(x)}{dx}\right|}. \qquad (12.428)$$

WKB Approximation: Transmission Through a Potential Barrier

Consider a one-dimensional potential $V(x)$ forming a smooth barrier as shown in Fig. 12.20. Defining $v(x) = 2mV(x)/\hbar^2$, the Schrödinger equation at energy $E = \hbar^2 k^2/(2m)$ is,

$$\left[\frac{d^2}{dx^2} + q^2(x)\right] \psi(x) = 0, \qquad (12.429)$$

with local wavenumber $q(x)$ defined as,

$$q^2(x) \equiv k^2 - v(x) = \varepsilon - v(x). \qquad (12.430)$$

At any given scattering energy $\varepsilon = k^2$, there are two turning points a and b defining two classically allowed regions $x < a$ and $x > b$ and a forbidden region $a < x < b$. In the region $a < x < b$, where $v(x) > \varepsilon$, the local wavenumber $q(x)$ becomes purely imaginary,

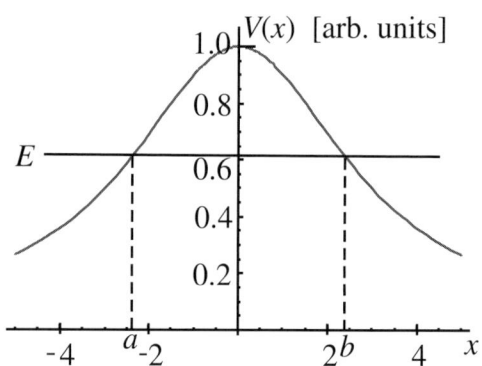

FIG 12.20 Transmission through a potential barrier $V(x)$ within the WKB approximation. The energy E determines the turning points $b > a$, with $V(a) = V(b) = E$.

$$q(x) = i\kappa(x) = i\sqrt{v(x) - \varepsilon}, \quad v(x) > \varepsilon, \quad \kappa(x) > 0. \qquad (12.431)$$

12.7 Approximation Methods

The corresponding wave functions in the WKB approximation are

$$\psi(x) = \frac{1}{\sqrt{q(x)}} \left[A_+ e^{i \int_a^x dy\, q(y)} + A_- e^{-i \int_a^x dy\, q(y)} \right], \quad (x < a), \tag{12.432}$$

$$\psi(x) = \frac{1}{\sqrt{\kappa(x)}} \left[C_+ e^{-\int_a^x dy\, \kappa(y)} + C_- e^{\int_a^x dy\, \kappa(y)} \right], \quad (a < x < b), \tag{12.433}$$

$$\psi(x) = \frac{1}{\sqrt{q(x)}} \left[B_+ e^{i \int_b^x dy\, q(y)} + B_- e^{-i \int_b^x dy\, q(y)} \right], \quad (x > b), \tag{12.434}$$

where $A_+, A_-, C_+, C_-, B_+,$ and B_- are constants. The connection formulae developed in Sec. 7.2.1 enable us to express the coefficients B_+, B_- as functions of the coefficients A_+, A_- in terms of the central quantity,

$$\tau \equiv e^{-\int_a^b dx\, \kappa(x)}. \tag{12.435}$$

Explicitly,

$$\begin{pmatrix} C_+ \\ C_- \end{pmatrix} = \frac{1}{2} \begin{bmatrix} \frac{\tau}{2} + \frac{2}{\tau} & i(\frac{2}{\tau} - \frac{\tau}{2}) \\ -i(\frac{2}{\tau} - \frac{\tau}{2}) & \frac{\tau}{2} + \frac{2}{\tau} \end{bmatrix} \begin{pmatrix} A_+ \\ A_- \end{pmatrix} \equiv M \begin{pmatrix} A_+ \\ A_- \end{pmatrix}. \tag{12.436}$$

Equation (12.436) defines the 2×2 matrix, M, transforming amplitudes from the left to the right side of the barrier. In the language of one-dimensional scattering theory, it is called the *transfer matrix*. Transfer matrices are encountered in many branches of physics, including statistical mechanics and field theory. From the theory of barrier penetration in Sec. 1.3.11, we know that when the conditions are chosen so that $A_+ = 1$ and $C_- = 0$, the *transmission amplitude* is given by C_+.

Problem 12.19

(a) For $A_+ = 1$ and $B_- = 0$ in Eq. (12.436), show that the transmission amplitude $t(\varepsilon)$ is given by,

$$t(\varepsilon) = B_+ = [M_{11}]^{-1} = \frac{2}{\frac{\tau}{2} + \frac{2}{\tau}}. \tag{12.437}$$

(b) Similarly, calculate the *reflection amplitude* $r(\varepsilon) = A_-$ in terms of the elements of M. (Hint: Show that $M_{21} = \frac{r(\varepsilon)}{t(\varepsilon)}$ and use your answer to part (a).)

Hence, the transmission coefficient in the WKB approximation is,

$$T(\varepsilon) = |t(\varepsilon)|^2 = \frac{4}{\left(\frac{\tau}{2} + \frac{2}{\tau}\right)^2} \approx \tau^2 = e^{-2 \int_a^b dx\, \kappa(x)}, \tag{12.438}$$

where the approximated estimates holds for $\tau \ll 1$.

We have indicated above that the WKB method is applicable also for 3D problems where the potential is spherically symmetric and the waves propagate along the radial coordinate r. As an example, let us consider a nuclear fission reaction corresponding to α particle decay. An alpha particle of charge $Z_1 = 2e$ in a heavy nucleus of charge $(Z_2 + 2)e$ is bound by the short-range attractive nuclear potential $V_N(r)$. In addition, a repulsive Coulomb potential $V(r) = Z_1 Z_2 e^2/r$ exists between the α particle and the Z_2 protons of the remaining nucleus. The combination of the nuclear potential and the

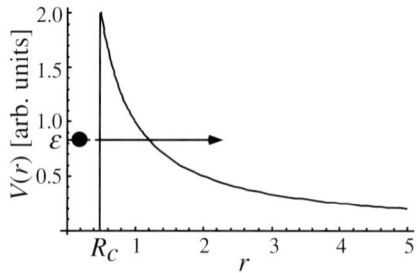

FIG 12.21 Schematic illustration of an α particle tunneling through the Coulomb barrier in nuclear fission. The barrier starts at the radius R_C, the range of the nuclear attraction. The transmission probability is approximately given by the Gamow factor (12.440).

repulsive Coulomb potential generates a radial potential, such that the α particle is quasi-bound at positive energy ε and can tunnel through the Coulomb potential (see Fig. 12.21). Let us calculate the transmission coefficient for this nuclear fission process.

Since the range R_C of the nuclear potential $V_N(r)$ is short (R is a few fermi, 1 fermi = 10^{-15} m), a reasonable approximation is to replace $V = V_N + V_C$ by the Coulomb potential tail,

$$V(r) = \theta(r - R_C) V_C(r). \tag{12.439}$$

The transmission coefficient in this case is obtained from Eq. (12.435) with $a = R_C$ and b is the right turning point where the local wavenumber vanishes, i.e., $\varepsilon = V_C(b)$. It can be approximated by the so-called *Gamow factor* for the fission probability,

$$T = e^{-\frac{2\pi \mu Z_1 Z_2 e^2}{\hbar^2 k}}, \tag{12.440}$$

where μ is the reduced mass of the alpha-nucleus system.

Problem 12.20

Calculate the transmission coefficient within the WKB approximation for a potential having the shape of an inverse parabola, $v(x) = v_0 - Cx^2$, with $v_0 > 0$ and $C > 0$, at energy $0 < \varepsilon < v_0$.

Hint: Find the two turning points $x_{1,2}$ by solving the equation $v(x) = \varepsilon$ and compute $|\tau|^2 = e^{-2\int_{x_1}^{x_2} dx \kappa(x)}$, where $\kappa(x) = \sqrt{v(x) - \varepsilon}$.

WKB Approximation for Phase Shifts

We now discuss the use of the WKB approximation for calculating phase shifts and estimate its effectiveness. Returning to Eq. (12.159), let us assume that the potential $v(r)$ falls off faster than $1/r$ as $r \to \infty$ and that its behavior at $r \to 0$ is less singular than r^{-2}. This excludes the Coulomb potential case, but it should be kept in mind that the phase shift is known exactly for the pure Coulomb potential, Eq. (12.312), while the problem of scattering by a Coulomb plus short-range potential can be cast in a form where the effective combined potential satisfies the above constraints as explained in Sec. 12.5.5. In the presence of a centrifugal potential $l(l+1)/r^2$, it is useful to define an effective potential

$$v_l(r) = v(r) + \frac{l(l+1)}{r^2}, \tag{12.441}$$

and a local wavenumber,

$$\kappa_l^2(r) = k^2 - v_l(r). \tag{12.442}$$

As far as potential scattering in three space dimensions is concerned, the important classical turning point r_t is defined as the largest r at which the local wavenumber $\kappa_l(r)$ vanishes. Indeed, for sufficiently large r, $\kappa_l^2(r) > 0$ and

$$\lim_{r \to \infty} \kappa_l^2(r) = k^2. \tag{12.443}$$

12.7 Approximation Methods

Thus, as r decreases from ∞, a turning point $r = r_t$ is reached, where $\kappa_l^2(r_t) = 0$, such that

$$\kappa_l^2(r) < 0, \text{ for } r < r_t, \quad \kappa_l^2(r) > 0, \text{ for } r > r_t. \tag{12.444}$$

This is valid for $l > 0$, but even for s-wave scattering, with $l = 0$, it occurs when the potential $v(r)$ has a weak singularity at $r = 0$, such as the case for the Yukawa potential. Regions for which $\kappa_l^2(r) < 0$ are classically forbidden, and regions for which $\kappa_l^2(r) > 0$ are classically allowed. In order to avoid cumbersome manipulations while still retaining the important points, it is assumed that $\kappa_l^2(r)$ is strictly monotonic (at least for some region around r_t),

$$\frac{d\kappa_l^2(r)}{dr} > 0. \tag{12.445}$$

There is a subtle distinction in the application of the WKB approximation for 1D and 3D problems regarding the analysis of the matching equations at the turning point. It is tempting to apply the procedure employed in 1D to the 3D case. However, this leads to some inaccuracy. The reason is that it is necessary to transform variables $r \to x$, such that $-\infty < x < \infty$. Let us transform both the radial coordinate r and the function $u_l(r)$ in Eq. 12.159 as follows:

$$r = \frac{1}{k} e^x, \quad 0 \leq r < \infty, \quad -\infty < x < \infty. \tag{12.446}$$

$$u_l(r) = e^{\frac{x}{2}} w(x). \tag{12.447}$$

Then, the Schrödinger equation (12.159) for $w(x)$ reads,

$$\left[\frac{d^2}{dx^2} + K^2(x) \right] w(x) = 0, \tag{12.448}$$

with variable momentum $K(x)$ defined by,

$$K^2(x) \equiv \left(1 - \frac{v(x)}{k^2} \right) e^{2x} - \left(l + \frac{1}{2} \right)^2. \tag{12.449}$$

Equation (12.448), which is exact, is now amenable to treatment using the WKB formalism as in the 1D problem analyzed in Sec. 7.2.1. Recall that for the scattering problem in 3D, we are interested in the wave function as $r \to \infty$. The rightmost turning point in position space, r_t, is mapped onto the rightmost turning point x_t at which $K^2(x_t)$ changes sign as,

$$K^2(x) < 0 \text{ for } x < x_t, \quad K^2(x) > 0 \text{ for } x > x_t. \tag{12.450}$$

Regions for which $K^2(x) < 0$ are classically forbidden, and regions for which $K^2(x) > 0$ are classically allowed. Thus, in the x variable, the physically relevant, classically allowed region, $K^2(x) > 0$ is located to the right of the rightmost turning point x_t, and the WKB solutions need to be calculated as $x \to \infty$. Taking these points into account, the analysis carried out above for the original one-dimensional problem can be carried out starting from Eq. (12.448) but here, unlike in the one-dimensional case, there is only one turning point x_t. Therefore, there is only one matching equation at the single turning point $= x_t$ and, moreover, the WKB solution to the left of x_t, where $K^2(x) < 0$, should decay at least exponentially as $x \to -\infty$. This implies that for $x < x_t$ only one Airy function, $A_i(x)$, is retained. We now use the analysis developed in Sec. (7.2.1) and the asymptotic form of the regular Airy function,

$$A_i(x) \to \pi^{-\frac{1}{2}} x^{-\frac{1}{4}} [c \sin(\zeta + \pi/4) + d \cos(\zeta + \pi/4)] \tag{12.451}$$

where c and d are constants and $\zeta = \frac{2}{3} x^{3/2}$. The WKB approximation for $w(x)$ at $x > x_t$ is

$$w(x) = \frac{C}{\sqrt{K(x)}} \sin \left[\int_{x_t}^{x} dx' \, K(x') + \frac{\pi}{4} \right], \tag{12.452}$$

where C is a constant, and the occurrence of $\frac{\pi}{4}$ is related to the discussion preceding Eq. (12.451). Getting back to r space, the local wavenumber is slightly modified from $\kappa_l(r)$ of Eq. (12.442) to,

$$q_l^2(r) = k^2 - v(r) - \frac{(l+\frac{1}{2})^2}{r^2}, \qquad (12.453)$$

$$q_l(r) \to k, \quad (\text{as } r \to \infty). \qquad (12.454)$$

The WKB approximation for $u_l(r)$ then reads,

$$u_l(r) = C\sqrt{\frac{k}{q_l(r)}} \sin\left[\int_{r_t}^{r} dr'\, q_l(r') + \frac{\pi}{4}\right]. \qquad (12.455)$$

Comparing with Eq. (12.163b) and recalling the limit (12.454), we find that $C = 1$ and

$$u_l(r) \xrightarrow[r \to \infty]{} \sin\left\{\int_{r_t}^{r} dr'\, [q_l(r') - k] + k(r - r_t) + \frac{\pi}{4}\right\}. \qquad (12.456)$$

Hence, according to Eq. (12.163b),

$$\delta_{l,\text{WKB}} = \int_{r_t}^{\infty} dr'\, [q_l(r') - k] - kr_t + (l + 1/2)\pi. \qquad (12.457)$$

12.7.3 THE VARIATIONAL PRINCIPLE

The variational principle is a useful tool for finding energy eigenvalues (see Sec. 7.6). Here, we use it to estimate scattering phase shifts [179]. The variational principle requires the minimization of a certain functional with constraints. For the eigenvalue problem, the functional to be minimized is $\langle \psi | H | \psi \rangle$ and the constraint is $\langle \psi | \psi \rangle = 1$. It is carried out by solving an unconstrained problem for a function ψ for which the functional

$$I[\psi] \equiv \langle \psi | H | \psi \rangle - \lambda(\langle \psi | \psi \rangle - 1) \qquad (12.458)$$

is a minimum with respect to all choices of ψ. The *Lagrange multiplier* λ [see Eq. (7.135)] is determined by the constraint $\langle \psi | \psi \rangle = 1$ and turns out to be the desired energy eigenvalue. The idea behind the use of the variation principle for calculation of phase shift is similar, but instead of the normalization constraint, the asymptotic form of the wave function is imposed. The starting point for the present discussion is Eq. (12.159) with slightly modified boundary conditions than Eq. (12.163b), obtained by dividing the wave function by the constant $\cos \delta_l$,

$$\left[\frac{d^2}{dr^2} + k^2 - \frac{l(l+1)}{r^2} - v(r)\right] u_l(r) = 0, \qquad (12.459)$$

$$u_l(r) \xrightarrow[r \to \infty]{} \sin\left(kr - \frac{1}{2}l\pi\right) + \tan \delta_l \cos\left(kr - \frac{1}{2}l\pi\right), \quad u_l(0) = 0. \qquad (12.460)$$

Denote by $w_l(r)$ a (real) variational wave function, which is well behaved for $0 \le r < \infty$ and satisfies similar constraints as $u_l(r)$ in Eq. (12.163b) albeit with a phase shift $\alpha_l(k) \ne \delta_l(k)$, since $w_l(r)$ is not the exact solution of Eq. (12.459), while $u_l(r)$ is. Thus, $w_l(0) = 0$, and

$$w_l(r) \xrightarrow[r \to \infty]{} \sin\left(kr - \frac{1}{2}l\pi\right) + \tan \alpha_l \cos\left(kr - \frac{1}{2}l\pi\right). \qquad (12.461)$$

12.7 Approximation Methods

The difference
$$d_l(r) \equiv w_l(r) - u_l(r), \tag{12.462}$$

clearly satisfies, $d_l(0) = 0$, and

$$d_l(r) \xrightarrow[r \to \infty]{} (\tan\alpha_l - \tan\delta_l) \cos\left(kr - \frac{1}{2}l\pi\right). \tag{12.463}$$

An estimate of the difference $d_l(r \to \infty)$ can be obtained using the functional

$$I[w_l] \equiv \int_0^\infty dr \, w_l(r) \left[\frac{d^2}{dr^2} + k^2 - v(r) - \frac{l(l+1)}{r^2}\right] w_l(r). \tag{12.464}$$

Indeed, suppose that $w_l(r)$ is wisely chosen to be close to $u_l(r)$ in some appropriate norm, and write

$$w_l(r) = u_l(r) + d_l(r), \quad ||d_l|| \ll ||u_l||, ||w_l||. \tag{12.465}$$

Inserting this decomposition in the integral on the RHS of Eq. (12.464) and recalling that $I[u_l] = 0$, the integral is evaluated to first order in d_l to yield,

$$I[w_l] = [u_l(r) d_l'(r) - d_l(r) u_l'(r)]_{r=0}^{r=\infty} = -k(\tan\alpha_l - \tan\delta_l). \tag{12.466}$$

The quantity $[I[w_l] + k \tan\delta_l]$ is stationary around $u_l(r)$ [to first order in $d_l(r)$], i.e.,

$$\delta \left[I[w_l] + k \tan\delta_l\right] = 0. \tag{12.467}$$

Since $I[u_l] = 0$ by the application of Eq. (12.459) in the definition (12.464), then, to order $||d_l||^2$, one arrives at the Kohn variational estimate of the phase shift,

$$\tan\delta_l \approx \tan\alpha_l + \frac{1}{k} I[w_l]. \tag{12.468}$$

The problem of finding an approximate function $w_l(r)$ is not discussed here. In principle, it can be approached by letting w_l depend on a few parameters x_i and then minimize the quantity

$$J_l \equiv [I[w_l] + k \tan\alpha_l], \tag{12.469}$$

with respect to the parameters x_i.

A similar procedure can also be applied to the scattering amplitude $f(\hat{\mathbf{k}}', \hat{\mathbf{k}})$ [see, e.g., Eq. (12.116), or Eqs. (12.143) and (12.144)]. Recall Eq. (12.143), which, for simplicity, is written here for a central potential $v(r)$,

$$\left[\nabla^2 + k^2 - v(r)\right] \psi_{\mathbf{k}}^+(\mathbf{r}) = 0, \tag{12.470}$$

with the boundary condition (12.144),

$$\lim_{r \to \infty} \psi_{\mathbf{k}}^+(\mathbf{r}) = e^{i\mathbf{k}\cdot\mathbf{r}} + f(\hat{\mathbf{k}} \cdot \hat{\mathbf{r}}) \frac{e^{ikr}}{r}, \tag{12.471}$$

where the factor $(2\pi)^{-3/2}$ is dropped for convenience and for central potential scattering, the scattering angle is $\theta = \hat{\mathbf{k}} \cdot \hat{\mathbf{r}}$ (at large distance, the unit vector $\hat{\mathbf{r}} = \hat{\mathbf{k}}'$). In analogy with Eq. (12.464), let $\phi_{\mathbf{k}}^+(\mathbf{r})$ be an approximate solution of Eq. (12.470) with asymptotic behavior similar to Eq. (12.471), albeit with an approximate scattering amplitude $\bar{f}(\theta)$,

$$\phi_{\mathbf{k}}^+(\mathbf{r}) \xrightarrow[r \to \infty]{} e^{i\mathbf{k}\cdot\mathbf{r}} + \bar{f}(\hat{\mathbf{k}} \cdot \hat{\mathbf{r}}) \frac{e^{ikr}}{r}. \tag{12.472}$$

The difference

$$D_{\mathbf{k}}(\mathbf{r}) \equiv \phi_{\mathbf{k}}^+(\mathbf{r}) - \psi_{\mathbf{k}}^+(\mathbf{r}) \qquad (12.473)$$

has the asymptotic behavior,

$$D_{\mathbf{k}}(\mathbf{r}) \xrightarrow[r \to \infty]{} \left[\bar{f}(\hat{\mathbf{k}} \cdot \hat{\mathbf{r}}) - f(\hat{\mathbf{k}} \cdot \hat{\mathbf{r}})\right] \frac{e^{ikr}}{r} \equiv \delta\left[f(\hat{\mathbf{k}} \cdot \hat{\mathbf{r}})\right] \frac{e^{ikr}}{r}. \qquad (12.474)$$

In analogy with the procedure detailed in connection with Eq. (12.464), the corresponding functional for the function $\phi_{\mathbf{k}}^+(\mathbf{r})$ is

$$I[\phi_{\mathbf{k}'}^+, \phi_{\mathbf{k}}^+] \equiv \int d\mathbf{r}\, \phi_{\mathbf{k}'}^+(\mathbf{r}) \left[\nabla^2 + k^2 - v(r)\right] \phi_{\mathbf{k}}^+(\mathbf{r}), \qquad (12.475)$$

while, due to Eq. (12.470), one has for the exact wave function $\psi_{\mathbf{k}}^+(\mathbf{r})$,

$$I[\psi_{\mathbf{k}'}^+, \psi_{\mathbf{k}}^+] = 0. \qquad (12.476)$$

Following the procedure detailed in Eqs. (12.465) and (12.466), we obtain, to first order in $D_{\mathbf{k}}(\mathbf{r})$

$$I[\phi_{\mathbf{k}'}^+, \phi_{\mathbf{k}}^+] = I[\psi_{\mathbf{k}'}^+ + D_{\mathbf{k}'}, \psi_{\mathbf{k}}^+ + D_{\mathbf{k}}]$$

$$= \int d\mathbf{r} \left(\psi_{\mathbf{k}'}^+(\mathbf{r}) \frac{\partial D_{\mathbf{k}}(\mathbf{r})}{\partial r} - \psi_{\mathbf{k}}^+(\mathbf{r}) \frac{\partial D_{\mathbf{k}'}(\mathbf{r})}{\partial r}\right). \qquad (12.477)$$

After some manipulations employing the asymptotic forms (12.471), (12.472), and (12.474), the result is

$$I[\phi_{\mathbf{k}'}^+, \phi_{\mathbf{k}}^+] = -4\pi\, \delta[f(-\hat{\mathbf{k}}' \cdot \hat{\mathbf{k}})], \qquad (12.478)$$

implying that

$$J \equiv I[\psi_{\mathbf{k}'}^+, \psi_{\mathbf{k}}^+] + 4\pi\, \delta[f(-\hat{\mathbf{k}}' \cdot \hat{\mathbf{k}})], \qquad (12.479)$$

is stationary with respect to small variation of $\psi_{\mathbf{k}'}^+$ and $\psi_{\mathbf{k}}^+$. This statement is the 3D analog of the relation (12.467). The 3D analog of the Kohn variational principle for the phase shift (12.468) is obtained in a similar way, namely, assuming $D_{\mathbf{k}}$ is small, one has an approximate relation for the scattering amplitude,

$$f(\hat{\mathbf{k}}' \cdot \hat{\mathbf{k}}) = \bar{f}(\hat{\mathbf{k}}' \cdot \hat{\mathbf{k}}) + \frac{1}{4\pi} I[\phi_{-\mathbf{k}'}^+, \phi_{\mathbf{k}}^+]. \qquad (12.480)$$

Note the use of the wave function with incoming spherical wave boundary conditions.

12.7.4 EIKONAL APPROXIMATION

The *eikonal approximation*[9] is good for treating the scattering of particles at high energy from a smooth potential [188, 189]. A detailed analysis can be found in Ref. [179]. The eikonal approximation is based on an analogy with classical scattering theory and uses the concept of impact parameter introduced in Sec. 12.1. It is also used in geometrical optics, where it was first developed.

The starting point is Eq. (12.64) for the wave function $\psi_{\mathbf{k}}^+(\mathbf{r})$ whose asymptotic form (12.65) defines the scattering amplitude $f(\theta, \phi)$:

$$[\nabla_{\mathbf{r}}^2 + k^2 + i\eta - v(\mathbf{r})]\psi_{\mathbf{k}}^+(\mathbf{r}) = 0. \qquad (12.481)$$

---

[9] Eikonal means image in Greek. In optics, the eikonal is the optical path length of a ray between object and image.

12.7 Approximation Methods

The conditions of high-energy scattering and smooth potential are quantitatively expressed as,

$$|v(r)| \ll k^2 \quad \text{for all } r, \tag{12.482}$$

$$\left|\frac{dv(r)}{dr}\right| \ll k\,|v(r)| \quad \text{for all } r, \tag{12.483}$$

which are almost identical with the conditions required for the validity of the WKB approximation. The first condition (12.482) implies the positivity of the local squared momentum in 3D,

$$k^2(\mathbf{r}) \equiv k^2 - v(\mathbf{r}) > 0. \tag{12.484}$$

From a classical point of view, the condition (12.482) implies that the flux of scattered particles is concentrated at small angles, so that the condition

$$\theta \ll 1 \tag{12.485}$$

can be safely employed.

The WKB analysis consists of writing the wave function in terms of modulus and phase, depending on the local momentum, but some care is required in applying it in 3D. The wave function $\psi_{\mathbf{k}}^+(\mathbf{r})$ is written as,

$$\psi_{\mathbf{k}}^+(\mathbf{r}) = A(\mathbf{r})\,e^{i\phi(\mathbf{r})}, \tag{12.486}$$

where $A(\mathbf{r})$ and $\phi(\mathbf{r})$ are real functions. The dimensionless phase $\phi(\mathbf{r})$ is referred to as the *action*. Inserting ansatz (12.486) into Eq. (12.481) results in

$$\left\{\nabla^2 A(\mathbf{r}) - A(\mathbf{r})[\nabla\phi(\mathbf{r})]^2 + k^2 - v(\mathbf{r})\right\} + i\left[\nabla A(\mathbf{r}) \cdot \nabla\phi(\mathbf{r}) + A(\mathbf{r})\nabla^2\phi(\mathbf{r})\right] = 0. \tag{12.487}$$

Since the real and imaginary parts on the LHS of Eq. (12.487) should vanish independently, we obtain two equations, for the amplitude $A(\mathbf{r})$ and the action $\phi(\mathbf{r})$. With the assumption $\left|\frac{\nabla^2 A(\mathbf{r})}{k^2(\mathbf{r})}\right| \ll 1$ [based on the estimates (12.483)], the real part of Eq. (12.487) leads to the following equation for the action,

$$[\nabla\phi(\mathbf{r})]^2 = k^2(\mathbf{r}). \tag{12.488}$$

Since the RHS of this equation is positive, by Eq. (12.484), it is formally possible to take the square root of both sides and obtain the action $\phi(\mathbf{r})$ as a line integral of the local momentum $k(\mathbf{r})$ along a certain curve joining some initial point $\mathbf{r}_0$ and the final point $\mathbf{r}$. In general, this integral is path dependent. The idea of the eikonal approximation is reminiscent of the classical theory of scattering where a particle starts its motion at $z = -\infty$ along a straight line parallel to the z axis at distance b from it (where b is the impact parameter). For simplicity, the method is explained here for the case of central potential, but extension to non-central potential is straightforward. Since, in this geometry, $\mathbf{b}$ is a vector lying in the plane perpendicular to $\hat{\mathbf{z}}$, the three-dimensional position vector $\mathbf{r}$ of the scattered particle is represented in cylindrical coordinates as $\mathbf{r} = (\mathbf{b}, z)$, and for central potential, we may write, $v(\mathbf{r}) = v(r) = v(\sqrt{b^2 + z^2})$. The eikonal representation of the solution of Eq. (12.488) is

$$\phi(\mathbf{r}) = \phi(\mathbf{b}, z) = kz + \int_{-\infty}^{z} dz' \left[\sqrt{k^2 - v(\sqrt{b^2 + z'^2})} - k\right]. \tag{12.489}$$

The constant of integration is chosen such that

$$\phi(\mathbf{b}, z) \to kz, \quad \text{as} \quad v\left(\sqrt{b^2 + z^2}\right) \to 0. \tag{12.490}$$

Employing the condition (12.482) and expanding the square root to first order in v/k, the action is approximated by,

$$\phi(\mathbf{b}, z) = kz - \frac{1}{2k} \int_{-\infty}^{z} dz' \, v\left(\sqrt{b^2 + z'^2}\right). \tag{12.491}$$

Within the eikonal approximation, the prefactor $A(\mathbf{r})$ in the expression (12.486) for the wave function is set equal to unity. This is consistent with the notion that the particle current is related to the phase of the wave function. Thus, Eq. (12.486) is rewritten as

$$\psi_{\mathbf{k}}^{+}(\mathbf{r}) = (2\pi)^{-\frac{3}{2}} e^{ikz} e^{-\frac{i}{2k} \int_{-\infty}^{z} dz' \, v\left(\sqrt{b^2 + z'^2}\right)}. \tag{12.492}$$

This expression can now be used in Eq. (12.87) for the scattering amplitude, yielding,

$$f_{\text{Eikonal}}(\mathbf{k}', \mathbf{k}) = -\frac{1}{4\pi} \int d\mathbf{r}' \, e^{-i\mathbf{k}' \cdot \mathbf{r}'} v(\mathbf{r}') e^{i\mathbf{k} \cdot \mathbf{r}'} e^{-\frac{i}{2k} \int_{-\infty}^{z'} dz'' \, v\left(\sqrt{x'^2 + y'^2 + z''^2}\right)}. \tag{12.493}$$

The replacement $kz' \to \mathbf{k} \cdot \mathbf{r}'$ is allowed since $\mathbf{k} = k\hat{\mathbf{z}}$. Without the last exponential factor in the integrand, the expression on the RHS of the above equation is identical with the Born approximation (12.123) for the scattering amplitude. The integral over $\mathbf{r}'$ can be carried out in cylindrical coordinates, $\mathbf{r}' = (b\cos\beta, b\sin\beta, z')$, so $d\mathbf{r}' = b\, db\, d\beta\, dz'$. Note that with this choice, $\mathbf{k} \cdot \mathbf{b} = 0$. Together with condition (12.485) this justifies the approximation,

$$(\mathbf{k} - \mathbf{k}') \cdot \mathbf{r}' = -kb\theta \cos\beta + O(\theta^2). \tag{12.494}$$

When this estimate is used to calculate the amplitude in Eq. (12.493), we find,

$$f_{\text{Eikonal}}(\mathbf{k}', \mathbf{k}) = \int_0^\infty b\, db \left[\int_0^{2\pi} e^{-ikb\theta \cos\beta} d\beta \right] \left(\int_{-\infty}^\infty v\left(\sqrt{b^2 + z'^2}\right) e^{-\frac{i}{2k} \int_{-\infty}^{z'} v\left(\sqrt{b^2 + z''^2}\right) dz''} dz' \right). \tag{12.495}$$

The integral in square brackets gives $2\pi J_0(kb\theta)$. To compute the remaining integral, the following change of variable is employed,

$$q(z) \equiv \int_{-\infty}^{z} dz' \, v\left(\sqrt{b^2 + z'^2}\right). \tag{12.496}$$

The integral in the second square brackets now reads,

$$\int dz \, \frac{dq(z)}{dz} e^{-\frac{i}{2k} q(z)} = 2ik \left[e^{-\frac{i}{2k} \int_{-\infty}^\infty dz \, v\left(\sqrt{b^2 + z^2}\right)} - 1 \right]. \tag{12.497}$$

Collecting these results, we finally obtain the scattering amplitude in the eikonal approximation,

$$\boxed{f_{\text{Eikonal}}(\mathbf{k}', \mathbf{k}) = -ik \int db\, b\, J_0(kb\theta) \left[e^{-\frac{i}{2k} \int_{-\infty}^\infty dz \, v\left(\sqrt{b^2 + z^2}\right)} - 1 \right].} \tag{12.498}$$

If the potential $v(r)$ has a finite range R, the domain of integration in b is $0 \leq b \leq R$, and when $b > R$, the expression within the square brackets vanishes.

12.8 PARTICLES WITH INTERNAL DEGREES OF FREEDOM

So far, our discussion of scattering has focused on collisions of point particles, where the only relevant dynamical coordinate is the relative position vector **r** between the two particles. Special attention was given to scattering from a spherically symmetric (central) potential, where partial wave analysis leads to a set of effective one-dimensional scattering problem, and different partial waves were not coupled. But scattering theory must provide tools for studying more general scattering scenarios, including cases where the colliding particles have internal degrees of freedom, e.g., spin.

12.8.1 SPIN

The simplest example of particles with internal degrees of freedom is particles with spin. We already considered spin degrees of freedom of colliding particles with spin-independent central potentials, but once a potential depends on spin, the spin states of particles before and after the collision need not be the same; the collision affects the internal degrees of freedom of the particle by changing its spin state.

12.8.2 COMPOSITE PARTICLES

In addition to the inclusion of spin as an internal degree of freedom, we also need to analyze collisions between objects having internal degrees of freedom, such as objects composed of several point particles. Different scenarios are encountered when at least one of the two colliding objects is a cluster of particles, which are bound together. Such colliding particles are referred to as *composite*. Examples are an atom composed of its nucleus and electrons, or a nucleus composed of constituent neutrons and protons, or a molecule composed of bound atoms. If the scattering energy is low (e.g., below the ionization energy of each atom), the only possible result of a collision can be to excite the internal states of the colliding particles at the expense of their relative kinetic energy, and the process is referred to as *inelastic scattering*. Low-energy scattering between two atoms, or two nuclei, or electron–atom scattering below the ionization threshold, belong to this class of collisions. If the scattering energy is above the breakup threshold, some constituents can be transferred among the colliding particles; moreover, disintegration (or breakup) might occur. In the latter case, the number of particles leaving the collision region might exceed two.

The following types of collisions between composite particles A and B (assumed to be initially at their ground state) are possible:

(1) *Elastic scattering*, $A + B \rightarrow A + B$, where the particles A and B keep their identities and their internal states before, during and after the collision.
(2) *Inelastic scattering*, $A + B \rightarrow A' + B'$, where the particles keep their identities but emerge from the collision in different internal states. Such internal states need not be an excited energy state, it might be just a different spin state (and in that case, the kinetic energy might still be conserved). Note that not both final particles need to be in different states.
(3) *Exchange (or rearrangement) reactions*, $A + B \rightarrow C + D$, where the colliding particles loose their identities as a result of the collision process but the number of particles after the collision remains two.
(4) *Breakup reactions*, $A + B \rightarrow C + D + F + \cdots$, where the final number of particles exceeds two.
(5) *Dissociation reactions*, $A \rightarrow B + C$, where the metastable state A breaks apart into two (or more) particles, and the inverse, *association reactions*, $B + C \rightarrow A$, wherein two particles, B and C, collide and form a metastable composite-particle state C. Examples are the molecular reactions, $N_2O_4 \leftrightarrow NO_2 + NO_2$.

In the following discussion, we limit ourselves to cases (1) and (2), wherein scattering occurs between two particles that keep their identities throughout the collision. Hence, the corresponding masses m_1 and m_2 are not changed during the collision and neither is the reduced mass $m = (m_1 m_2)/(m_1 + m_2)$.

12.8.3 CHANNELS

The internal degrees of freedom (such as spin and excitation energies) can change during the collision. Consequently, the eigenvalues of the S operator can change. Unlike the case studied for scattering of spinless point particles, where, according to Eq. (12.133), the S matrix, $S(\mathbf{k}', \mathbf{k}) = \langle \psi_{\mathbf{k}'}^-|\psi_{\mathbf{k}}^+\rangle$, depends only on relative wavenumbers before and after the collision through the asymptotic states $\psi_{\mathbf{k}}^{\pm}(\mathbf{r})$, here the asymptotic states also depend on the internal state quantum numbers as specified by the labels **a** for the pre-collision states and by **b** for the post-collision states (including $\mathbf{k}' \in \mathbf{b}$ and $\mathbf{k} \in \mathbf{a}$). Thus, instead of $|\psi_{\mathbf{k}}(\mathbf{r})\rangle$, we now have $|\psi_{\mathbf{a}}(\mathbf{r})\rangle$, and the S matrix elements are denoted as,

$$S_{\mathbf{ba}} = \langle \psi_{\mathbf{b}}^-|\psi_{\mathbf{a}}^+\rangle. \tag{12.499}$$

The extension of the formalism from scattering of point particles to scattering of composite particles is straightforward. Instead of a single Schrödinger equation describing the scattering of point particles, scattering theory of composite particles can be formulated in terms of a coupled set of Schrödinger or Lippmann–Schwinger equations. This formalism is referred to as *multichannel scattering theory*. A precise definition of the concept of channels is given below.

The definition of the S matrix in Eq. (12.499) makes sense only if the states $\{|\psi_{\mathbf{a}}^+\rangle\}$ and $\{|\psi_{\mathbf{b}}^-\rangle\}$ each form a complete set in the space $\mathcal{D}^+$ consisting of two composite particle states that are eigenstates of the Hamiltonian with positive total energy. An asymptotic state of two separate particles with assigned internal degrees of freedom is referred to as a *channel*. Channels are denoted in terms of kets $|\mathbf{a}\rangle, |\mathbf{b}\rangle \ldots$, where $\mathbf{a}, \mathbf{b}, \ldots$ are sets of quantum numbers, which completely characterize the corresponding states of the two composite particles. They contain relative and internal quantum numbers including spin. For example, if we consider the collision of two hydrogen atoms, an appropriate ket for a given channel is $|\mathbf{a}\rangle = |\mathbf{k}_a n_1 l_1 m_1 n_2 l_2 m_2 \ldots\rangle$, where $\mathbf{k}_a$ is the relative momentum and $\ldots$ refers to the additional internal quantum numbers specifying spin quantum numbers. The dependence of the relative momentum $\mathbf{k}_a$ on the internal quantum numbers a is evident: The possibility of energy transfer between relative motion and internal excitations implies that the corresponding relative wavevector depends on channel index a [see Eq. (12.56)]. The total scattering energy E is independent of channel index, i.e.,[10]

$$E = \frac{\hbar^2 k_a^2}{2m} + E_a = \frac{\hbar^2 k_b^2}{2m} + E_b. \tag{12.500}$$

The fact that scattering is inelastic implies that, generically,

$$k_a = \sqrt{\varepsilon - \varepsilon_a} \neq k_b = \sqrt{\varepsilon - \varepsilon_b}, \tag{12.501}$$

where,

$$\varepsilon = \frac{2mE}{\hbar^2}, \quad \varepsilon_{a,b} = \frac{2mE_{a,b}}{\hbar^2}. \tag{12.502}$$

It is sometimes useful to consider a more general relation $e(k)$ between momentum and kinetic energy instead of the quadratic one, $e(k) = \frac{\hbar^2 k_a^2}{2m}$. The density of states related to the function $e(k_a)$ is

$$n(e) = k_a^2 \frac{dk_a}{de(k_a)} = \frac{mk_a}{\hbar^2}, \tag{12.503}$$

where the last equality is true for the case $e(k) = \frac{\hbar^2 k_a^2}{2m}$. The dependence on the energy ε enters through Eq. (12.500).

Denoting the interaction between the two composite particles as V, we may write the full Hamiltonian as,

$$H = -\frac{\hbar^2}{2m}\nabla_{\mathbf{r}}^2 + H_1 + H_2 + V \equiv H_0 + V. \tag{12.504}$$

Here, m is the reduced mass and $\mathbf{r}$ is the vector connecting the two centers of mass of the two clusters. Channel kets are eigenstates of the Hamiltonian H_0 describing the relative motion and the internal dynamics of the noninteracting particles

---

[10] Equation (12.500) is the *on-shell* condition. As in the discussion of potential scattering, sometimes one needs to consider *off-shell* conditions. Then, Eq. (12.500) is not obeyed.

12.8 Particles with Internal Degrees of Freedom

governed by internal Hamiltonians H_1, H_2; hence, they form a complete set in $\mathcal{D}^+$. If the interaction V depends only on the vector $\mathbf{r}$ joining the centers of mass of the two particles, the problem basically reduces to that of structureless particles discussed before. A nontrivial situation occurs when V also depends on internal degrees of freedom (one important example is that V depends on spin). Once the potential V depends on kinematic variables other than the distance between the centers of masses of the two colliding particles, it cannot be regarded as a central potential. Hence, the present analysis includes the scattering from a non-central potential as a special case. Inspection of Eq. (12.501) raises the possibility that the (scaled) scattering energy is chosen such that $\varepsilon - \varepsilon_a > 0$, while $\varepsilon - \varepsilon_b < 0$. The corresponding wavenumbers are k_a and $k_b = i\kappa_b$, where κ_b is real and positive. The wave function in channel b decays as $r \to \infty$, and channel b is sometimes referred to as a *closed channel*. This process is an association reaction of the type $A + B \to C$.

Let $\mathbf{k}_a$ and E_a denote the relative momentum and the sum of the internal energies in a given stationary state of the system, $|\Phi_a\rangle$ denote the wave function constructed from the internal wave functions $|\Phi_{a_1}\rangle$ and $|\Phi_{a_2}\rangle$, and $\mathbf{a} = (\mathbf{k}_a a)$ denote the set of quantum numbers encoding the internal degrees of freedom a and the momentum $\mathbf{k}_a$ for the relative motion of particles 1 and 2. Including the spins and their projections in the quantum numbers $a_1(s_1 \nu_1)$, $a_2(s_2 \nu_2)$, and $a(s\nu)$, we have

$$|\Phi_a\rangle = \sum_{\nu_1 \nu_2} \langle s\nu | s_1 \nu_1 s_2 \nu_2\rangle |\Phi_{a_1}\rangle \otimes |\Phi_{a_2}\rangle, \tag{12.505a}$$

$$\mathbf{a} = (\mathbf{k}_a a), \tag{12.505b}$$

$$|\mathbf{a}\rangle = |\mathbf{k}_a\rangle \otimes |\Phi_a\rangle = |\mathbf{k}_a a\rangle, \tag{12.505c}$$

$$\langle \mathbf{r}|\mathbf{a}\rangle = (2\pi)^{-\frac{3}{2}} e^{i\mathbf{k}_a \cdot \mathbf{r}} |\Phi_a\rangle. \tag{12.505d}$$

Note the distinction in Eq. (12.505b) between $\mathbf{a}$ and a; the latter refers only to internal quantum numbers. Equation (12.505d) is an example of what is referred to as *partial representation*, where the dependence on $\mathbf{r}$ is specified explicitly while the dependence on internal degrees of freedom is specified in terms of abstract kets (and bras). With the above definitions, the following equations define the channel states,

$$(H_1 + H_2)|\Phi_a\rangle = E_a |\Phi_a\rangle, \tag{12.506a}$$

$$H_0|\mathbf{a}\rangle = E|\mathbf{a}\rangle = \left(\frac{\hbar^2 k_a^2}{2m} + E_a\right)|\mathbf{a}\rangle. \tag{12.506b}$$

Equation (12.506a) describes the internal states alone without the relative motion, while Eq. (12.506b) describes the internal states and the relative motion, but without the interaction between the objects.

12.8.4 ASYMPTOTIC STATES AND CROSS-SECTIONS

We are now in a position to define asymptotic states, scattering amplitudes, and cross-sections. The analysis is carried out in the center of mass frame of the two colliding particles. Transformation to the laboratory frame, if required, should be carried out as discussed in Sec. 12.3.2. Starting from the Hamiltonian (12.504), we write the Schrödinger equation as

$$H|\Psi_{\mathbf{a}}^+\rangle = E|\Psi_{\mathbf{a}}^+\rangle \tag{12.507}$$

for the wave function whose initial channel wave function is $|\mathbf{a}\rangle$ [see Eq. (12.505d)]. As $r \to \infty$, the function (12.507) is a combination of plane wave plus outgoing spherical waves for the relative motion multiplied by (possibly excited) internal states (12.505a). Adopting the procedure of (partial) representation as in Eq. (12.505d), the scattering wave function $\Psi_{\mathbf{a}}^+(\mathbf{r}) = \langle \mathbf{r}|\Psi_{\mathbf{a}}^+\rangle$ is required to have the following asymptotic behavior,

$$\Psi_{\mathbf{a}}^+(\mathbf{r}) \xrightarrow[r \to \infty]{} (2\pi)^{-\frac{3}{2}} \left[e^{i\mathbf{k}_a \cdot \mathbf{r}_a}|\Phi_a\rangle + \sum_b f_{\mathbf{ba}}(\Omega) \frac{e^{ik_b r}}{r} |\Phi_b\rangle\right]. \tag{12.508}$$

Here, Ω is the solid angle denoting the direction of the momentum $\mathbf{k}_b$ with respect to the initial direction $\mathbf{k}_a$. It specifies the direction of $\mathbf{r}$ as the two particles recede after the collision and $f_{\mathbf{ba}}(\Omega)$ is the amplitude for scattering from the initial channel $\mathbf{a}$ into an outgoing channel $\mathbf{b}$.[11] The differential cross-section from the initial channel $\mathbf{a}$ into an outgoing channel $\mathbf{b}$ is the ratio of the outgoing current into an area $r^2 d\Omega$ divided by the incoming flux. Since the magnitudes of k_a and k_b must not be the same, the appropriate formula is

$$\frac{d\sigma_{\mathbf{ba}}}{d\Omega} = \frac{k_b}{k_a}|f_{\mathbf{ba}}(\Omega)|^2. \tag{12.509}$$

Note that in case of elastic scattering, $b = a$, one must replace $\mathbf{k}_b$ with $\mathbf{k}'_a$ where in general $\mathbf{k}'_a \neq \mathbf{k}_a$, but $\mathbf{k}'^2_a = \mathbf{k}^2_a$, and the prefactor k_b/k_a is unity. The case $k_b \neq k_a$ will be considered in Sec. 12.8.9 where we discuss inelastic scattering. The total cross-section from an initial channel $\mathbf{a}$ is given by

$$\sigma_{\mathbf{a}} = \sum_{\mathbf{b}} \int d\Omega \, \frac{d\sigma_{\mathbf{ba}}}{d\Omega}. \tag{12.510}$$

The dynamics of the scattering process is determined by the interaction V, which, as stated above, depends on both the internal degrees of freedom and the relative position vector $\mathbf{r}$. If V is local (as will be assumed), then its *partial representation* [see definition following Eq. (12.505d)] can be written as

$$V(\mathbf{r}) = \sum_{ab} v_{ab}(\mathbf{r})|\Phi_a\rangle\langle\Phi_b|, \quad v_{ab}(\mathbf{r}) = \langle\Phi_a|V(\mathbf{r})|\Phi_b\rangle. \tag{12.511}$$

12.8.5 MÖLLER OPERATORS

Let us now develop the formal scattering theory, in the sense of Sec. 12.4, which is applicable for multichannel scattering. To this end we recall the time-dependent theory, where the evolution is governed by the Hamiltonian (12.504). The exact wave function $\Psi(t)$ is the solution of the time-dependent Schrödinger equation

$$i\hbar \frac{\partial}{\partial t}|\Psi(t)\rangle = H|\Psi(t)\rangle. \tag{12.512}$$

When $t \to \pm\infty$, the colliding particles are far apart and do not interact. Therefore, there exist two solutions $|\Phi^{\text{in}}_{\mathbf{a}}(t)\rangle$ and $|\Phi^{\text{out}}_{\mathbf{b}}(t)\rangle$ for the free-motion Schrödinger equations,[12]

$$i\hbar \frac{\partial}{\partial t}|\Phi^{\text{in}}_{\mathbf{a}}(t)\rangle = H_0|\Phi^{\text{in}}_{\mathbf{a}}(t)\rangle \quad t \to -\infty, \tag{12.513}$$

$$i\hbar \frac{\partial}{\partial t}|\Phi^{\text{out}}_{\mathbf{b}}(t)\rangle = H_0|\Phi^{\text{out}}_{\mathbf{b}}(t)\rangle \quad t \to \infty, \tag{12.514}$$

such that

$$\lim_{t \to -\infty} \| \, |\Psi(t)\rangle - |\Phi^{\text{in}}_{\mathbf{a}}(t)\rangle \, \| = 0,$$

$$\lim_{t \to \infty} \| \, |\Psi(t)\rangle - |\Phi^{\text{out}}_{\mathbf{b}}(t)\rangle \, \| = 0. \tag{12.515}$$

---

[11] The dependence of the scattering solid-angle Ω on the final channel index $\mathbf{b}$, i.e., $\Omega_{\mathbf{k}_b}$ is clear and will be dropped.
[12] The time-dependent states $|\Phi^{\text{in}}_{\mathbf{a}}(t)\rangle$ and $|\Phi^{\text{out}}_{\mathbf{b}}(t)\rangle$ also encode the relative motion and should not be confused with the stationary internal states $|\Phi_{a,b}\rangle$ defined through Eq. (12.506a) that encode only the internal degrees of freedom.

12.8 Particles with Internal Degrees of Freedom

In this time-dependent formalism, the direction of time also determines the time-dependent asymptotic states. Nevertheless, we retain the channel indices **a**, **b**; they will be used when the stationary states are introduced. The Möller operators,

$$\Omega_\pm = \lim_{t \to \mp\infty} e^{iHt/\hbar} e^{-iH_0 t/\hbar}, \tag{12.516}$$

carry the asymptotic states into the exact wave function at time t,

$$\Omega_+ |\Phi_{\bf a}^{\rm in}(t)\rangle = |\Psi(t)\rangle, \quad \Omega_- |\Phi_{\bf b}^{\rm out}(t)\rangle = |\Psi(t)\rangle. \tag{12.517}$$

These operations are illustrated in Fig. 12.22.

From Eq. (12.36), we find

$$\Omega_\pm = \lim_{\eta \to 0^+} \mp i\eta \int_0^{\mp\infty} dt\, e^{iHt/\hbar} e^{-iH_0 t/\hbar} e^{\pm\eta t}. \tag{12.518}$$

This equality is now inserted into Eq. (12.517). We insert unit operators in $\mathcal{D}^+$ expressed in terms of complete sets of channel wave functions $|{\bf a}\rangle$ and $|{\bf b}\rangle$ between Ω_+ and $|\Phi_{\bf a}^{\rm in}(t)\rangle$ and between Ω_- and $|\Phi_{\bf b}^{\rm in}(t)\rangle$. By definition, $H_0|{\bf a}\rangle = E_{\bf a}|{\bf a}\rangle$ and $H_0|{\bf b}\rangle = E_{\bf b}|{\bf b}\rangle$ [see Eq. (12.506b)]. Now, we formally perform the time integration and use the same manipulations as those leading from Eq. (12.36) to Eq. (12.100) and Eq. (12.104), and thereby get an expression for $|\Psi(t)\rangle$ in terms of $|\Phi_{\bf a}^{\rm in}(t)\rangle$ or $|\Phi_{\bf b}^{\rm out}(t)\rangle$,

$$|\Psi(t)\rangle = \int d{\bf a}\, e^{-iE_{\bf a}t/\hbar} |\psi_{\bf a}^+\rangle \langle {\bf a}|\Phi_{\bf a}^{\rm in}(0)\rangle, \tag{12.519}$$

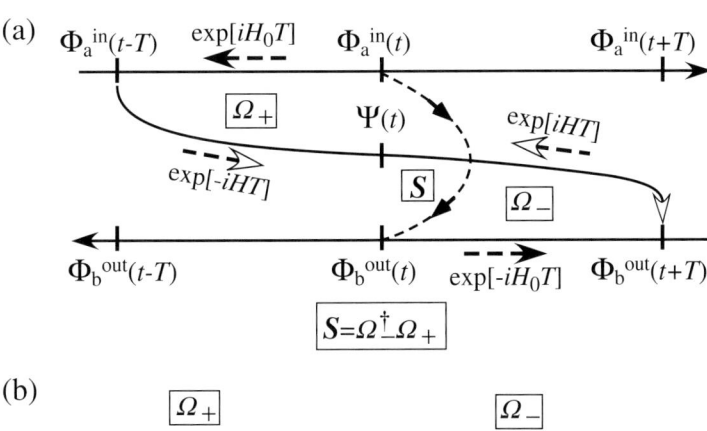

FIG 12.22 The action of Möller operators in the multichannel formalism. For notational convenience, we take $\hbar = 1$ in this figure. (a) Application of Ω_+ on $|\Phi_{\bf a}^{\rm in}(t)\rangle$ and Ω_- on on $|\Phi_{\bf b}^{\rm out}(t)\rangle$. The state $|\Phi_{\bf a}^{\rm in}(t)\rangle$ is first propagated backward in time (dashed black arrow) to the distant past $t - T$ with $T > 0$ by the free evolution operator $e^{iH_0 T}$. At that early time, it coincides with $|\Psi(t-T)\rangle$, which then evolves forward to the present time t by the evolution operator corresponding to the full Hamiltonian, e^{-iHT} (dashed white arrow). The product of these two operations as $T \to \infty$ yields the Möller operator Ω_+ defined in Eq. (12.517). A similar construction defines Ω_-. The S matrix, $S = \Omega_-^\dagger \Omega_+$, takes $|\Phi_{\bf a}^{\rm in}(t)\rangle$ to $|\Phi_{\bf b}^{\rm out}(t)\rangle$. (b) Summary of the action of the Möller operators.

$$|\Psi(t)\rangle = \int d\mathbf{b}\, e^{-iE_\mathbf{b}t/\hbar}|\psi_\mathbf{b}^-\rangle\langle\mathbf{b}|\Phi_\mathbf{b}^{\text{out}}(0)\rangle, \tag{12.520}$$

where $E_{\mathbf{a},\mathbf{b}} = \frac{\hbar^2 k_{a,b}^2}{2m} + E_{a,b}$.

If we work in energy instead of time space, we can use the energy shell condition (12.500) and replace $E_\mathbf{a} = E_\mathbf{b} = E$, i.e., the scattering energy [see Eq. (12.506b)]. In Eqs. (12.519) and (12.520), $|\psi_\mathbf{a}^+\rangle$ and $|\psi_\mathbf{b}^-\rangle$ are eigenstates of the Hamiltonian (12.504) belonging to the continuous spectrum,

$$|\psi_\mathbf{a}^+\rangle = \Omega_+(E)|\mathbf{a}\rangle \equiv |\mathbf{a}\rangle + (E + i\eta - H)^{-1}V|\mathbf{a}\rangle, \tag{12.521}$$

$$|\psi_\mathbf{b}^-\rangle = \Omega_-(E)|\mathbf{b}\rangle \equiv |\mathbf{b}\rangle + (E - i\eta - H)^{-1}V|\mathbf{b}\rangle. \tag{12.522}$$

We can now define the operators $G_0(z), G(z), \Omega(z)$, and $T(z)$ exactly as in Eqs. (12.108b)–(12.108d), this time with $H_0 = -\frac{\hbar^2}{2m}\nabla_\mathbf{r}^2 + H_1 + H_2$ instead of just $-\frac{\hbar^2}{2m}\nabla_\mathbf{r}^2$ as is the case of scattering of particles without internal degrees of freedom. Evidently, these new operators satisfy Eqs. (12.109)–(12.111) since these equations are quite general, irrespective of how the Hamiltonian is divided into its free part H_0 and the interaction part V. As before, the parameter z is an arbitrary complex number, but for describing the real physical situation, it should eventually approach the real axis, $z \to E^\pm = E \pm i\eta$, and the representation of operators expressing physical observables should be taken on the energy shell. Thus, in Eqs. (12.113a), (12.113b), (12.113c), (12.115), and (12.114), we can replace $\mathbf{k}$ and $\mathbf{k}'$ by $\mathbf{a}$ and $\mathbf{b}$, respectively. In particular, the *transition matrix* $T_\mathbf{ba}$ [the analog of $T(\mathbf{k}',\mathbf{k})$ in Eq. (12.115)] is given by,

$$T_\mathbf{ba} = \langle\mathbf{b}|T|\mathbf{a}\rangle = \langle\mathbf{b}|V|\psi_\mathbf{a}^+\rangle = \langle\mathbf{b}|V[1 + G^+(E)V]|\mathbf{a}\rangle. \tag{12.523}$$

This leads immediately to the analogous expression of Eq. (12.116) for the scattering amplitude. In other words, exactly as one arrives at Eq. (12.116), one now arrives at the fundamental relation,

$$f_\mathbf{ba}(\Omega) = -\frac{4m\pi^2}{\hbar^2}\langle\mathbf{b}|V|\psi_\mathbf{a}^+\rangle = -\frac{4m\pi^2}{\hbar^2}\langle\mathbf{b}|T|\mathbf{a}\rangle. \tag{12.524}$$

There is, however, at least one task remaining, namely, to show that the wave function $|\psi_\mathbf{a}^+\rangle$ has indeed the asymptotic form as in Eq. (12.508). Only then we can justify the analog of Eq. (12.116) for the multichannel case and suggest reliable algorithms for calculating experimental quantities. The analog of Eq. (12.113a) for $|\psi_\mathbf{a}^\pm\rangle$ is

$$|\psi_\mathbf{a}^\pm\rangle = |\mathbf{a}\rangle + G_0^\pm(E)V|\psi_\mathbf{a}^+\rangle, \tag{12.525}$$

$$|\psi_\mathbf{a}^\pm\rangle = |\mathbf{a}\rangle + G_0^\pm(E)T(E)|\mathbf{a}\rangle, \tag{12.526}$$

where $G_0^\pm(E) = (E \pm i\eta - H_0)^{-1} = (E \pm i\eta + \frac{\hbar^2}{2m}\nabla_\mathbf{r}^2 - H_1 - H_2)^{-1}$. In order to check if $\psi_\mathbf{a}^+(\mathbf{r}) = \langle\mathbf{r}|\psi_\mathbf{a}^+\rangle$ indeed has the asymptotic condition specified in Eq. (12.508), we need to construct the free Green's function with outgoing and incoming spherical wave asymptotic boundary conditions for the appropriate channels. This requires an extension of the analysis based on Eqs. (12.78) and (12.82) for the multichannel case, with $H_0 = -\frac{\hbar^2}{2m}\nabla_\mathbf{r}^2 + H_1 + H_2$. The desired Green's function is an operator in channel space as is evident from Eq. (12.525), where G_0 operates on a state in this space. Explicitly,

$$\left(-\frac{\hbar^2}{2m}\nabla_\mathbf{r}^2 + H_1 + H_2 - E\right)G_0^\pm(\mathbf{r},\mathbf{r}') = -\delta(\mathbf{r}-\mathbf{r}'), \tag{12.527}$$

$$G_0^\pm(\mathbf{r},\mathbf{r}') = \langle\mathbf{r}|G_0^\pm|\mathbf{r}'\rangle \to \sum_b \frac{e^{ik_b r}}{r} g(\mathbf{r}\cdot\mathbf{r}';k_b)|\Phi_b\rangle\langle\Phi_b|, \tag{12.528}$$

12.8 Particles with Internal Degrees of Freedom

where $k_\mathbf{b}^2 = \frac{2m}{\hbar^2}(E - \varepsilon_b)$, $\mathbf{k}_b \to k_b \hat{\mathbf{r}}$ as $r \to \infty$, and $g(x; k)$ is a bounded function. These conditions are realized by,

$$G_0^\pm(\mathbf{r}, \mathbf{r}') = -\sum_b \frac{e^{\pm i k_b |\mathbf{r}-\mathbf{r}'|}}{4\pi|\mathbf{r}-\mathbf{r}'|} |\Phi_b\rangle\langle\Phi_b|, \qquad (12.529)$$

$$G_0^\pm(\mathbf{r}, \mathbf{r}') \to -\sum_b \frac{e^{\pm i k_b r}}{r} e^{-i\mathbf{k}_b \cdot \mathbf{r}'} |\Phi_b\rangle\langle\Phi_b|, \qquad (12.530)$$

$$= -(2\pi)^{\frac{3}{2}} \sum_b \frac{e^{\pm i k_b r}}{r} |\Phi_b\rangle\langle\mathbf{b}|\mathbf{r}'\rangle, \qquad (12.531)$$

where the last equality results from Eq. (12.505d). The passage from Eq. (12.525) to Eq. (12.508) is now straightforward. First, one takes the partial representation of Eq. (12.525) and apply $\langle \mathbf{r}|$ on the left and "unity operator in relative motion space" $\int |\mathbf{r}'\rangle\langle\mathbf{r}'|d\mathbf{r}'$ between G_0 and V and between V and $|\psi_\mathbf{a}^+\rangle$. Then, one takes the limit $r \to \infty$ and employs Eq. (12.531). This leads directly to Eq. (12.508) with the scattering amplitude $f_{ab}(\Omega)$ given by Eq. (12.524).

12.8.6 THE MULTICHANNEL S MATRIX

The S operator (or S matrix) for scattering of particles with internal degrees of freedom is defined exactly as in Eq. (12.129),

$$S(E) = \Omega_-^\dagger(E)\Omega_+(E) = [1 + G^-(E)V]^\dagger[1 + G^+(E)V], \qquad (12.532)$$

except that in the present case V depends also on internal coordinates. Using the fact that the operators $\Omega_\pm(E)$ transform the states $|\mathbf{a}\rangle$ to $|\psi_\mathbf{a}^\pm\rangle$ as in Eqs. (12.521) and (12.522), we infer the representation of the S operator as a matrix in the Hilbert space spanned by states $|\mathbf{a}\rangle$, the so-called S matrix,

$$S_{\mathbf{ba}} = \langle \mathbf{b}|S|\mathbf{a}\rangle = \langle \psi_\mathbf{b}^-|\psi_\mathbf{a}^+\rangle. \qquad (12.533)$$

Let us again stress that once we speak of the matrix $S_{\mathbf{ba}}$, we limit ourselves to the representation of the S operator only in a subspace $\mathcal{D}^+$ of the full Hilbert space pertaining to the positive spectrum of the relative motion part of the Hamiltonian. Roughly speaking, bound states of the two particles are excluded from this subspace. In general, the Hamiltonian might also have bound states, which do not belong to the subspace $\mathcal{D}^+$. We have already seen that the presence of bound states affects the low-energy scattering in a profound way, but this can be included in the present formalism even if bound states are not directly presented in the basis of states forming the S matrix. The properties of the S matrix and its relation to the T matrix element $T_{\mathbf{ba}}$ defined in Eq. (12.523) can be worked out using methods similar to those used in Eqs. (12.133), (12.134), and (12.137), replacing $\mathbf{k}$ and $\mathbf{k}'$ by $\mathbf{a}$ and $\mathbf{b}$, respectively. First, let us indicate the orthogonality and normalization of the scattering states as in Eq. (12.114),

$$\langle\psi_\mathbf{b}^+|\psi_\mathbf{a}^+\rangle = \langle\mathbf{b}|[1 + VG^-(E_\mathbf{b})]|\psi_\mathbf{a}^+\rangle$$
$$= \langle\mathbf{b}|\left[1 + V\frac{1}{E_\mathbf{b} - E_\mathbf{a} - i\eta}\right]|\psi_\mathbf{a}^+\rangle = \langle\mathbf{b}|\left[1 - \frac{1}{E_\mathbf{a} - H_0 + i\eta}V\right]|\psi_\mathbf{a}^+\rangle$$
$$= \langle\mathbf{b}|\mathbf{a}\rangle = \delta_{\mathbf{ba}} = \delta(\mathbf{k}_b - \mathbf{k}_a)\delta_{ba}. \qquad (12.534)$$

Let us briefly explain how each step in the above equation is determined. The first equality is based on $\langle\psi_\mathbf{b}^+| = \langle\mathbf{b}|[1 + VG^-(E_\mathbf{b})]$ derived from Eq. (12.522). The second line results from noting that $(E_\mathbf{b} - i\eta - H)^{-1}|\psi_\mathbf{b}^+\rangle = (E_\mathbf{b} - i\eta - E_\mathbf{a})^{-1}|\psi_\mathbf{a}^+\rangle$, since $|\psi_\mathbf{a}^+\rangle$ is an eigenstate of H with an eigenvalue $E_\mathbf{a}$. The third equality results from the replacement of $E_\mathbf{b}$ by H_0 in the denominator, which is allowed when the operator acts leftward on the bra $\langle\mathbf{b}|$. Finally, the last equality is based on expressing $|\mathbf{a}\rangle$ in terms of $|\psi_\mathbf{a}^+\rangle$ using Eq. (12.521). Moreover, the orthogonality and normalization relation of the scattering states with incoming boundary conditions is $\langle\psi_\mathbf{b}^-|\psi_\mathbf{a}^-\rangle = \delta_{\mathbf{ba}}$. The overlap of incoming and outgoing scattering states defining the S matrix as in Eq. (12.533) has essentially been worked out in (12.133), Eqs. (12.134), and (12.137). With the appropriate modifications, we find,

$$S_{\mathbf{ba}} = \langle \psi_{\mathbf{b}}^-|\psi_{\mathbf{a}}^+\rangle = \langle \mathbf{b}|\mathbf{a}\rangle - 2\pi i \delta(E_{\mathbf{b}} - E_{\mathbf{a}})\langle \mathbf{b}|V|\psi_{\mathbf{a}}^+\rangle$$
$$= \delta(\mathbf{k}_b - \mathbf{k}_a)\delta_{ba} - 2\pi i \delta(E_{\mathbf{b}} - E_{\mathbf{a}})T_{\mathbf{ba}}(E_{\mathbf{a}} + i\eta), \quad (12.535)$$

where the T matrix element $T_{\mathbf{ba}}(E_{\mathbf{a}} + i\eta)$ is defined in Eq. (12.523) and the specification of the argument $E_{\mathbf{a}} + i\eta$ and the energy delta function underline the fact that this relation is valid on the energy shell.

The S matrix is unitary, as can be demonstrated from its definition (12.535). Indeed, using the orthogonality of the scattering states (12.534) and their completeness on the subspace $\mathcal{D}^+$ allow us to express $|\psi_{\mathbf{a}}^+\rangle$ in terms of $\{|\psi_{\mathbf{b}}^-\rangle\}$ and vice versa as,

$$|\psi_{\mathbf{a}}^+\rangle = \sum_{\mathbf{b}} |\psi_{\mathbf{b}}^-\rangle S_{\mathbf{ba}} = \sum_{\mathbf{b}} |\psi_{\mathbf{b}}^-\rangle \langle \psi_{\mathbf{b}}^-|\psi_{\mathbf{a}}^+\rangle \quad (12.536)$$

$$|\psi_{\mathbf{b}}^-\rangle = \sum_{\mathbf{a}} |\psi_{\mathbf{a}}^+\rangle [S^{-1}]_{\mathbf{ab}} = \sum_{\mathbf{a}} |\psi_{\mathbf{a}}^+\rangle \langle \psi_{\mathbf{a}}^+|\psi_{\mathbf{b}}^-\rangle. \quad (12.537)$$

Hence,

$$[S^{-1}]_{\mathbf{ab}} = [S_{\mathbf{ab}}]^*, \quad (12.538)$$

so that the matrix S is unitary,

$$\sum_{\mathbf{c}} S_{\mathbf{ac}} S_{\mathbf{bc}}^* = \delta_{\mathbf{ab}}. \quad (12.539)$$

Expressing the S matrix in terms of the T matrix using Eq. (12.535) and substituting it into the unitarity relation leads to the multichannel analog of the optical theorem. In the first step, one obtains the unitarity relation in terms of the T matrices as,

$$T_{\mathbf{ba}} - [T^\dagger]_{\mathbf{ba}} = -2\pi i \sum_{\mathbf{c}} \delta(E_{\mathbf{c}} - E_{\mathbf{a}})[T^\dagger]_{\mathbf{bc}} T_{\mathbf{ca}}. \quad (12.540)$$

Note that the on energy-shell delta function refers to the total energy, as in Eq. (12.500). As a special case, we take $\mathbf{a} = \mathbf{b}$ on both sides and find,

$$\mathrm{Im}[T_{\mathbf{aa}}] = -2\pi i \sum_{\mathbf{c}} \delta(E_{\mathbf{c}} - E_{\mathbf{a}})|T_{\mathbf{ca}}|^2. \quad (12.541)$$

In the second step, the sum over intermediate states is turned into an integral and the delta function in energy is used to reduce the number of integration variables,

$$\sum_{\mathbf{c}} \delta(E_{\mathbf{c}} - E_{\mathbf{a}}) \to \frac{m}{\hbar(2\pi)^3} \sum_{c} k_c \int d\Omega_{\mathbf{c}}. \quad (12.542)$$

In the third step, we use the relation (12.542) for calculating the RHS of Eq. (12.541) and also employ Eq. (12.524) to express the T matrix elements in terms of the scattering amplitude. Moreover, we recall Eq. (12.509), which enables us to put the differential cross-section from an initial state $\mathbf{a}$ to a final state $\mathbf{c}$ inside the integral (this requires putting $k_{\mathbf{a}}$ in the denominator, which is executed on both sides). Performing the integral yields the total cross-section between the same initial and final states $\mathbf{a}$ as in Eq. (12.510). Finally, we arrive at the *optical theorem for multichannel scattering*,

$$\boxed{\sigma_{\mathbf{a}} = \frac{4\pi}{k_{\mathbf{a}}} \mathrm{Im} f_{\mathbf{aa}}(\theta = 0).} \quad (12.543)$$

The argument $\theta = 0$ occurs since $\mathbf{a} = (\mathbf{k}_a, a)$ [see Eq. (12.505c)], which means that the final direction coincides with the initial direction.

12.8 Particles with Internal Degrees of Freedom

Symmetries of the S matrix

Here, we consider the invariance of the S matrix under several symmetry transformations. For this purpose, it is useful to separate the notations for the quantum numbers of a given channel **a** and single out (in addition to relative momentum $\mathbf{k}_a$) also the corresponding spins. The reason is that momenta and spins are the quantum numbers most affected by symmetry operations. Thus, if in a state $|\mathbf{a}\rangle$ the relative momentum between the two particles is $\mathbf{k}_a$, the spin projections of the two particles are σ_{1a}, σ_{2a} and all other quantum numbers (mainly the internal energy E_a) are collected under a (and similarly for channel b), we will sometime write,

$$S_{\mathbf{ba}} = \langle \mathbf{b}|S|\mathbf{a}\rangle = \langle \mathbf{k}_b \sigma_{1b}\sigma_{2b}b|S|\mathbf{k}_a\sigma_{1a}\sigma_{2a}a\rangle. \quad (12.544)$$

The on-shell condition is,

$$\varepsilon(\mathbf{k}_a) + \varepsilon_a = \varepsilon(\mathbf{k}_b) + \varepsilon_b, \quad (12.545)$$

where for free particles $\varepsilon(\mathbf{k}) = k^2$, but we may in principle consider other dependence of energy on momentum. These might be Bloch energies in crystal or linear dispersion energies $\varepsilon(\mathbf{k}) = v_F|\mathbf{k}|$ near the Dirac point in monolayer graphene (see Sec. 13.6). Alternatively, if the spins of the two particles are added to form a spin state $|sv\rangle$, the matrix representation of the S matrix is denoted as,

$$S_{\mathbf{ba}} = \langle \mathbf{b}|S|\mathbf{a}\rangle = \langle \mathbf{k}_b S_b v_b|S|\mathbf{k}_a S_a v_a\rangle. \quad (12.546)$$

The symmetries of the S matrix defined in Eq. (12.533) reflect the symmetries of the Hamiltonian H. If Π is a symmetry of H such that $\Pi H \Pi^{-1} = H$, then formally, this property holds also for S, $\Pi S \Pi^{-1} = S$. The symmetry Π might be a continuous one, such as translational invariance (whose generator is momentum operator $\mathbf{P}$), rotation (whose generator is the angular momentum operator $\mathbf{J}$ and combines space and spin rotations), or a discrete one, such as space inversion, also referred to as parity $\mathcal{P}$, space reflection, or time-reversal $\mathcal{T}$. The question that should be answered here is what are the constraints on the matrix elements $S_{\mathbf{ba}}$ that should be imposed due to the presence of a symmetry Π. Translation invariance and rotation symmetry have already been elaborated upon, and hence, the present discussion concerns the *discrete symmetries*, inversion, and time reversal in particular. If the symmetry operation Π is represented by a linear operator, the analysis is easily carried out using bra–ket notation,

$$S_{\mathbf{ba}} = \langle \mathbf{b}|S|\mathbf{a}\rangle = \langle \mathbf{b}\Pi^{-1}|\Pi S \Pi^{-1}|\Pi \mathbf{a}\rangle = \langle \Pi \mathbf{b}|S|\Pi \mathbf{a}\rangle$$
$$= S_{[\Pi \mathbf{b}][\Pi \mathbf{a}]}, \quad \text{(for a linear symmetry operation } \Pi\text{).} \quad (12.547)$$

Invariance Under Space Inversion

If Π corresponds to space inversion (parity), $\Pi = \mathcal{P}$, then, in the notation of Eqs. (12.546) and (12.547), the invariance of the Hamiltonian under space inversion implies,

$$S_{\mathbf{ba}} = S_{\mathbf{k}_b S_b v_b b; \mathbf{k}_a S_a v_a a} = S_{-\mathbf{k}_b S_b \mathcal{P} v_b \mathcal{P} b; -\mathbf{k}_a S_a \mathcal{P} v_a \mathcal{P} a}. \quad (12.548)$$

A special case of importance is when at least one of the scattered particles or both have spin, but do not have any additional internal structure. The S matrix elements are expressed as in Eq. (12.546). Since space inversion does not affect spin projection quantum number, the invariance under parity implies,

$$\langle \mathbf{k}_b S_b v_b|S|\mathbf{k}_a S_a v_a\rangle = \langle -\mathbf{k}_b S_b v_b|S|-\mathbf{k}_a S_a v_a\rangle. \quad (12.549)$$

This constraint is schematically depicted in Fig. 12.23, where the S matrix for scattering event Fig. 12.23(a) is related to that of scattering event Fig. 12.23(b).

Time-Reversal Invariance

Time reversal is represented by an antilinear and antiunitary operator $\mathcal{T}$. There are some pitfalls in trying to operate with antilinear operators on bras [9], and therefore, we will avoid this and stick to the rule that antilinear operators should operate only on kets and not on bras. Hence, the manipulations performed in Eq. (12.547) are not applicable here. In our convention of writing the S matrix elements $S_{\bf ba}$, the state appearing as a ket on the right, $|{\bf a}\rangle$, is the initial state and the state appearing on the left as a bra, $\langle{\bf b}|$, is the final state. The fact that time-reversal operation reverses the arrows of time (encoded by complex conjugation if no spin is involved) implies that the application of time reversal effectively replaces the roles of kets and bras. The required procedure for implementing time-reversal invariance starts from the representation of the S matrix in terms of asymptotic scattering states as in Eq. (12.533). Applying the time-reversal operators on these states, we obtain

$$S_{\bf ba} = \langle {\bf b}|S|{\bf a}\rangle = \langle \psi_{\bf b}^-|\psi_{\bf a}^+\rangle$$
$$= \langle \mathcal{T}\psi_{\bf b}^-|\mathcal{T}\psi_{\bf a}^+\rangle^* = \langle \psi_{\mathcal{T}{\bf a}}^-|\psi_{\mathcal{T}{\bf b}}^+\rangle$$
$$= \langle \mathcal{T}{\bf a}|S|\mathcal{T}{\bf b}\rangle = S_{\mathcal{T}{\bf a}\mathcal{T}{\bf b}}. \qquad (12.550)$$

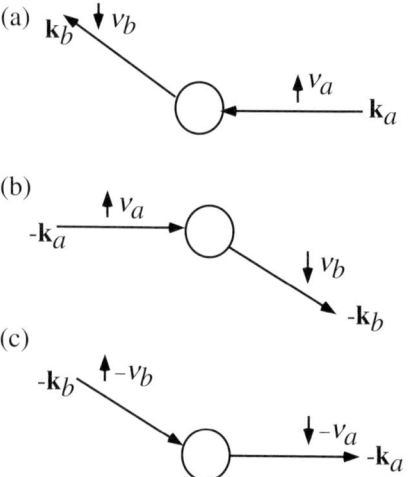

FIG 12.23 Scattering process for a light particle of spin $s = 1/2$ from a heavy spinless particle. Both particles are structureless. (a) A process described by S matrix $\langle {\bf k}_b, \downarrow |S|{\bf k}_a, \uparrow\rangle$. (b) Space inversion process described by an S matrix $\langle -{\bf k}_b, \downarrow |S| -{\bf k}_a, \uparrow\rangle$. (c) Time-reversed process described by an S matrix $(-1)\langle -{\bf k}_a, \downarrow |S| -{\bf k}_b, \uparrow\rangle$. Note that in our convention of writing the S matrix elements as $S_{\bf ba}$, the state appearing on the right, $|{\bf a}\rangle$, is the initial state before scattering. The direction of time follows the arrows. Thus, in (a) and (b) the initial state is $|\pm{\bf k}_a, \uparrow_a\rangle$ and the final state is $|\mp{\bf k}_b, \downarrow_b\rangle$, while in (c) the initial state is $|-{\bf k}_b, \downarrow\rangle$ and the final state is $-{\bf k}_a, \uparrow\rangle$. In other words, time reversal swaps the roles of initial and final states.

The equality (12.550) holds between amplitudes. The implication for cross-sections can be deduced with the help of Eq. (12.509),

$$k_a \left[\frac{d\sigma}{d\Omega}\right]_{\bf ba} = k_b \left[\frac{d\sigma}{d\Omega}\right]_{[\mathcal{T}{\bf a}][\mathcal{T}{\bf b}]}. \qquad (12.551)$$

As a special case, consider scattering of elementary (structureless) particles 1 and 2 with spins s_1, s_2, and corresponding projections m_1, m_2. The two-particle spin state of spin s and projection ν is

$$|s\nu\rangle = \sum_{m_1 m_2} \langle s_1 m_1 s_2 m_2 | s\nu\rangle |s_1 m_1\rangle \otimes |s_2 m_2\rangle. \qquad (12.552)$$

Then $|{\bf a}\rangle = |{\bf k}_a s_a \nu_a\rangle$ and $|{\bf b}\rangle = |{\bf k}_b s_b \nu_b\rangle$. Applying the time-reversal operator on a state $|s\nu\rangle$, we obtain

$$\mathcal{T}|s\nu\rangle = (-1)^{s+\nu}|s\bar{\nu}\rangle, \qquad (12.553)$$

where $\bar{\nu} = -\nu$. Employing this expression in Eq. (12.550), which displays the time-reversal invariance of the S matrix, we obtain the following relation,

$$\boxed{\langle {\bf k}_b s_b \nu_b |S|{\bf k}_a s_a \nu_a\rangle = (-1)^{s_b+\nu_b-s_a-\nu_a}\langle -{\bf k}_a s_b \bar{\nu}_a |S| -{\bf k}_b s_b \bar{\nu}_b\rangle.} \qquad (12.554)$$

The above constraint is schematically depicted in Fig. 12.23 displaying a scattering of a light spin 1/2 from a heavy spin 0 particle, so that $\nu = \pm 1/2$. Time-reversal invariance relates the processes (a) and (c) [by reversing the momenta, the spins and the time and multiplying by an appropriate phase].

12.8 Particles with Internal Degrees of Freedom

The Inverse Collision

From Eqs. (12.549) and (12.554), it is clear that there is no symmetry relation between $\langle \mathbf{k}_b s_b \nu_b | S | \mathbf{k}_a s_a \nu_a \rangle$ and $\langle \mathbf{k}_a s_a \nu_a | S | \mathbf{k}_b s_b \nu_b \rangle$. The former is referred to as the amplitude for the *direct collision*, while the latter is the amplitude for the *inverse collision*. An intriguing question is whether there is an equality between cross-sections of the direct and inverse collisions such as Eq. (12.551) albeit without the operation of $\mathcal{T}$. We will see below, when discussing the effect of spin–orbit coupling, that certain scattering amplitudes can be written as a sum of two functions, one is symmetric under $\mathbf{k}_a \leftrightarrow \mathbf{k}_b$ and the other is antisymmetric. The latter is referred to as skew scattering, which has the form $i\boldsymbol{\sigma} \cdot \mathbf{k}_a \times \mathbf{k}_b$.

Problem 12.21

Show that the term $i\boldsymbol{\sigma} \cdot \mathbf{k}_a \times \mathbf{k}_b$ is hermitian and invariant under parity and time-reversal transformations.

The presence of skew scattering then implies an asymmetry between direct and inverse collisions. On the other hand, a central potential does not produce skew scattering, and the symmetry is maintained. This is one facet of a general principle referred to as *detailed balance*. We may then write,

$$k_a \left[\frac{d\sigma}{d\Omega} \right]_{ba} \neq k_b \left[\frac{d\sigma}{d\Omega} \right]_{ab} \quad \text{(skew scattering present)}, \tag{12.555a}$$

$$k_a \left[\frac{d\sigma}{d\Omega} \right]_{ba} = k_b \left[\frac{d\sigma}{d\Omega} \right]_{ab} \quad \text{(skew scattering absent)}. \tag{12.555b}$$

12.8.7 SCATTERING FROM TWO POTENTIALS

A generalization of the Born approximation introduced in Sec. 12.7 can be applied to scattering problems involving two different potentials, e.g., scattering of two nuclei where in addition to the nuclear potential, the Coulomb potential also plays an important role. Consider the situation where the potential V is composed of two terms, $V \to V + U$, so Hamiltonian (12.504) reads,

$$H = -\frac{\hbar^2}{2m}\nabla_\mathbf{r}^2 + H_1 + H_2 + V + U \equiv H_0 + V + U. \tag{12.556}$$

According to Eq. (12.523), the required transition matrix element is,

$$T_{\mathbf{ba}} = \langle \mathbf{b} | V + U | \psi_\mathbf{a}^+ \rangle, \tag{12.557}$$

where $|\psi_\mathbf{a}^+\rangle$ is the solution of the Lippmann–Schwinger integral equation (12.521), albeit with the sum of the two potentials,

$$\begin{aligned}|\psi_\mathbf{a}^+\rangle &= |\mathbf{a}\rangle + (E + i\eta - H)^{-1}(V + U)|\mathbf{a}\rangle \\ &= |\mathbf{a}\rangle + (E + i\eta - H_0)^{-1}(V + U)|\psi_\mathbf{a}^+\rangle.\end{aligned} \tag{12.558}$$

So far, we have not used the solutions of equations such as Eq. (12.522) since they correspond to the unphysical situation of an incoming spherical wave following a scattering event. At this point, we will make use of these solutions for technical reasons, and consider a variant of Eq. (12.522) corresponding only to the second term of the potential, U. Explicitly, consider the state $|\phi_\mathbf{b}^-\rangle$ defined by

$$|\phi_\mathbf{b}^-\rangle = |\mathbf{b}\rangle + (E - i\eta - H_0)^{-1} U |\phi_\mathbf{b}^-\rangle. \tag{12.559}$$

The goal is to eliminate the state $|\mathbf{b}\rangle$ from Eq. (12.558) in favor of $|\phi_\mathbf{b}^-\rangle$ by using Eq. (12.559). Doing so, we find,

$$T_{\mathbf{ba}} = \langle \phi_\mathbf{b}^- | V | \psi_\mathbf{a}^+ \rangle + \langle \phi_\mathbf{b}^- | U | \mathbf{a} \rangle. \tag{12.560}$$

The second term on the RHS of Eq. (12.560) depends solely on the potential U. The first term represents a distortion of the expression in Eq. (12.523) where the state $\langle \mathbf{b} |$ is replaced by $\langle \phi_\mathbf{b}^- |$. In the second term, we re-express the bra $\langle \phi_\mathbf{b}^- |$ in terms of the free wave $\langle \mathbf{b} |$ using the bra version of Eq. (12.559),

$$\langle \phi_\mathbf{b}^- | = \langle \mathbf{b} | [1 - U(E + i\eta - H_0)]^{-1}, \tag{12.561}$$

where the sign change of the $i\eta$ term is due to the passage from ket to bra, and is crucial. Indeed, when this expression for $\langle \phi_\mathbf{b}^- |$ is substituted into the second term on the RHS of Eq. (12.560), the operator between $\langle \mathbf{b} |$ and $| \mathbf{a} \rangle$ is identified as the transition operator for the potential U alone, i.e.,

$$t_U \equiv [1 - U(E + i\eta - H_0)]^{-1} U = U + U G_0^+ t_U. \tag{12.562}$$

Hence, we arrive at the exact relation for the scattering from two potentials,

$$\boxed{T_{\mathbf{ba}} = \langle \phi_\mathbf{b}^- | V | \psi_\mathbf{a}^+ \rangle + \langle \mathbf{b} | t_U | \mathbf{a} \rangle.} \tag{12.563}$$

If the potential V is small and can be taken to first order, one can solve Eq. (12.558) without V and substitute the solution $|\phi_\mathbf{a}^+\rangle$ instead of $|\psi_\mathbf{a}^+\rangle$ in the first term on the RHS of Eq. (12.563), thus arriving at the *distorted wave Born approximation* (DWBA),

$$\boxed{T_{\mathbf{ba}}^{\text{DWBA}} = \langle \phi_\mathbf{b}^- | V | \phi_\mathbf{a}^+ \rangle + \langle \mathbf{b} | t_U | \mathbf{a} \rangle.} \tag{12.564}$$

12.8.8 SCATTERING OF PARTICLES WITH SPIN

In this section, we focus on the scattering between two particles with spins when the interaction potential V depends on spin. The partial wave expansion must be modified because the orbital angular momentum is no longer a good quantum number. Spin-dependent interactions can occur in several situations. The two most common situations are as follows. (1) The two particles have spins s_1 and s_2 and the interaction V depends on both spins (e.g., spin-dipolar coupling). (2) One of the particles has spin s, while the other does not have spin, and the interaction between the particles includes spin–orbit coupling. The analysis of scattering from a non-central potential is carried out below, first for nonidentical particles and then for identical particles with spin.

In the center of mass frame, scattering due to a spin-dependent potential can be described as follows. In the initial state, the two particles 1 and 2 have spins and spin projections $s_1 \sigma_1$ and $s_2 \sigma_2$, and momenta $\mathbf{k}$ and $-\mathbf{k}$. After the collision, they are in a final state, with spins and spin projections $s_1 \sigma_1'$ and $s_2 \sigma_2'$ and momenta $\mathbf{k}'$ and $-\mathbf{k}'$. The corresponding energies of relative motion are $\varepsilon(\mathbf{k})$ and $\varepsilon(\mathbf{k}')$. In the absence of a magnetic field, the internal energies do not depend on spin projections and can be taken to be zero. Therefore, the on-shell condition reduces simply to the equality,

$$\varepsilon(\mathbf{k}) = \varepsilon(\mathbf{k}') \equiv \varepsilon. \tag{12.565}$$

The Schrödinger equation (12.507) can be written in the center of mass frame (after multiplying by $\frac{2m}{\hbar^2}$) as,

$$[-\nabla_\mathbf{r}^2 + v(\mathbf{r})] \psi_\mathbf{k}^+(\mathbf{r}) = k^2 \psi^+(\mathbf{r}), \tag{12.566}$$

but the fact that the potential $v(\mathbf{r}) = (2m/\hbar^2) V(\mathbf{r})$ is non-central and depends on spin renders the partial wave expansion a bit more complicated than its central potential counterpart.

We have already seen in Chapter 4 [see also Eq. (12.552)] that for two-particle systems, it is convenient to employ angular momentum algebra to add the two spins and construct a two-particle spin function χ_{sv}, out of the two spin functions $|s_1 \sigma_1\rangle$ and $|s_2 \sigma_2\rangle$, with $\mathbf{s} = \mathbf{s}_1 + \mathbf{s}_2$ and $v = \sigma_1 + \sigma_2$. Explicitly,

$$\chi_{sv} = \sum_{\sigma_1 \sigma_2} \langle s_1 \sigma_1 s_2 \sigma_2 | s v \rangle |s_1 \sigma_1\rangle \otimes |s_2 \sigma_2\rangle. \tag{12.567}$$

12.8 Particles with Internal Degrees of Freedom

The kets $|a\rangle$ and $|b\rangle$ of the initial and final states defined in Eq. (12.505c) are

$$|a\rangle = |\mathbf{k} s \nu\rangle, \quad |b\rangle = |\mathbf{k}' s' \nu'\rangle, \tag{12.568}$$

where, generically, all the three quantum numbers might differ. The corresponding $\mathbf{r}$ space representation (12.505d) is a product of plane wave and spin function,

$$\langle \mathbf{r}|\mathbf{k} s \nu\rangle = (2\pi)^{-\frac{3}{2}} e^{i\mathbf{k}\cdot\mathbf{r}} \chi_{s\nu}, \quad \langle \mathbf{r}|\mathbf{k}' s' \nu'\rangle = (2\pi)^{-\frac{3}{2}} e^{i\mathbf{k}'\cdot\mathbf{r}} \chi_{s'\nu'}. \tag{12.569}$$

Following Eqs. (12.535) and using the notation of Eq. (12.544), the on-shell S matrix element is related to the T matrix and thereby to the scattering amplitude as,

$$\langle \hat{\mathbf{k}}' s' \nu'|S|\hat{\mathbf{k}} s \nu\rangle = \delta_{\hat{\mathbf{k}}\hat{\mathbf{k}}'} \delta_{ss'} \delta_{\nu\nu'} - 2\pi i n(\varepsilon) \langle \hat{\mathbf{k}}' s' \nu'|T(\varepsilon)|\hat{\mathbf{k}} s \nu\rangle, \tag{12.570}$$

where the density of states $n(\varepsilon)$ defined in Eq. (12.503) results from the energy conservation delta function. Note that the on-energy shell condition implies the dependence on energy and the initial and final directions. For a non-central potential, the scattering amplitude, generically, depends on $\cos\theta = \hat{\mathbf{k}}' \cdot \hat{\mathbf{k}}$ and the azimuthal angle ϕ. The unitarity of the on-shell S matrix is explictly expressed as,

$$\sum_{s''\nu''} \int d\Omega_{\hat{\mathbf{k}}''} \langle \hat{\mathbf{k}}' s' \nu'|S|\hat{\mathbf{k}}'' s'' \nu''\rangle \langle \hat{\mathbf{k}}'' s'' \nu''|S|\hat{\mathbf{k}} s \nu\rangle = \delta_{\hat{\mathbf{k}}\hat{\mathbf{k}}'} \delta_{ss'} \delta_{\nu\nu'}. \tag{12.571}$$

Employing the familiar relation (12.524) between the T matrix and scattering amplitude,

$$f_{\mathbf{k}}(\Omega_{\hat{\mathbf{k}}'}) = -\frac{4\pi^2 m}{\hbar^2} \langle \hat{\mathbf{k}}' s' \nu'|T(\varepsilon)|\hat{\mathbf{k}} s \nu\rangle, \tag{12.572}$$

$$\frac{d\sigma}{d\Omega_{\hat{\mathbf{k}}'}} = |f_{\mathbf{k}}(\Omega_{\hat{\mathbf{k}}'})|^2. \tag{12.573}$$

As in the case of scattering from a central potential, the calculation of the cross-section requires partial wave expansion and solution of the partial wave Schrödinger equation (or equivalently, the Lippmann–Schwinger equation). This is our next task.

Let us start from the abstract form of the Lippmann–Schwinger equation (12.526) for the scattering state $|\psi_a^+\rangle$ and represent it in configuration space by applying the bra $\langle\mathbf{r}|$ on the left. Recall that in the present case, $|a\rangle$ and $\langle\mathbf{r}|a\rangle$ are given by Eqs. (12.568) and (12.569), respectively. In carrying out this procedure, we encounter the kernel $\langle\mathbf{r}|G_0^+ T|a\rangle$, which is handled by introducing a unity operator $\int |a'\rangle\langle a'|da'$ between $\langle\mathbf{r}|$ and G_0^+. It is also useful to re-scale the T matrix and the non-central potential in their $(\mathbf{k} s \nu)$ representations as in Eqs. (12.119)

$$\langle \mathbf{k} s \nu|T|\mathbf{k}' s' \nu'\rangle = \frac{\hbar^2}{2m} \langle \mathbf{k} s \nu|t|\mathbf{k}' s' \nu'\rangle, \tag{12.574a}$$

$$\langle \mathbf{k} s \nu|V|\mathbf{k}' s' \nu'\rangle = \frac{\hbar^2}{2m} \langle \mathbf{k} s \nu|v|\mathbf{k}' s' \nu'\rangle. \tag{12.574b}$$

This yields the following equation [the analog of Eq. (12.187) for scattering by a non-central potential],

$$\psi_{\mathbf{k} s \nu}^+(\mathbf{r}) = (2\pi)^{-\frac{3}{2}} \left[e^{i\mathbf{k}\cdot\mathbf{r}} \chi_{s\nu} + \sum_{s'\nu'} \int d\mathbf{k}' \frac{e^{i\mathbf{k}'\cdot\mathbf{r}} \chi_{s'\nu'}}{\varepsilon(k) + i\eta - \varepsilon(k')} \langle \mathbf{k}' s' \nu'|t|\mathbf{k} s \nu\rangle \right]. \tag{12.575}$$

Following the analysis detailed in Eqs. (12.186) and (12.187), it is straightforward to write down the corresponding relations for the non-central potential case, and to arrive at an expression for the cross-section from the wave function, starting from Eq. (12.575). As $r \to \infty$, Eq. (12.575) takes the form,

$$\psi^+_{\mathbf{k}sv}(\mathbf{r}) \to (2\pi)^{-\frac{3}{2}} \left[e^{i\mathbf{k}\cdot\mathbf{r}} \chi_{sv} - 2\pi^2 \frac{e^{ikr}}{r} \sum_{s'v'} \langle \hat{\mathbf{r}} s'v' | t | \hat{\mathbf{k}} sv \rangle \chi_{s'v'} \right], \qquad (12.576)$$

which is the analog of Eq. (12.185). It has indeed the structure of (12.508) with $|a\rangle = |k\hat{\mathbf{k}}sv\rangle$ and $\langle b| = \langle k\hat{\mathbf{r}}s'v'|$ with the on-shell condition $k = \sqrt{\varepsilon}$ and

$$f_{\mathbf{k}}(\Omega_{\hat{\mathbf{k}}'}) = -2\pi^2 \langle \hat{\mathbf{r}} s'v' | t | \hat{\mathbf{k}} sv \rangle. \qquad (12.577)$$

Once again, recalling that, as $r \to \infty$, the direction of $\mathbf{r}$ coincides with that of the final momentum $\mathbf{k}'$, Eq. (12.577) is consistent with Eqs. (12.572) and (12.574a). Therefore, $-2\pi^2 \langle \hat{\mathbf{r}} s'v' | t | \hat{\mathbf{k}} sv \rangle$ is the scattering amplitude for scattering by non-central potential, and the differential cross-section from an initial state $|\mathbf{k}sv\rangle$ to a final state $\langle \mathbf{k}'s'v'|$ is,

$$\left[\frac{d\sigma}{d\Omega_{\hat{\mathbf{k}}'}} \right]_{k's'v';ksv} = |f_{\mathbf{k}}(\Omega_{\hat{\mathbf{k}}'})|^2 = 4\pi^4 |\langle \hat{\mathbf{r}} s'v' | t | \hat{\mathbf{k}} sv \rangle|^2. \qquad (12.578)$$

In many cases of practical interest, the incident beam is *unpolarized*. This means that if these particles have spins s_1, the beam contains a random mixture of spin orientations v_1 with equal weights $w_{v_1} = (2s_1 + 1)^{-1}$. Likewise, the collection of target particles having spin s_2 is a random mixture of spin orientations v_2 with equal weights $w_{v_2} = (2s_2 + 1)^{-1}$. Moreover, in many cases, the observer is not interested in the polarization of the outgoing particles. In this special situation, the differential cross-section is given by the incoherent sum,

$$\left\langle \frac{d\sigma}{d\Omega_{\hat{\mathbf{k}}'}} \right\rangle = w_{v_1} w_{v_2} \sum_{sv\,s'v'} \left[\frac{d\sigma}{d\Omega_{\hat{\mathbf{k}}'}} \right]_{k's'v';ksv}. \qquad (12.579)$$

This procedure is referred to as *averaging over initial states and summing over final states*.

In order to compute the scattering amplitude, it is necessary to expand all the three functions, $\psi_{\mathbf{k}sv}(\mathbf{r})$, $e^{i\mathbf{k}\cdot\mathbf{r}}\chi_{sv}$, and $\langle \mathbf{k}'s'v' | T | \mathbf{k}sv \rangle$ (or equivalently $\langle \hat{\mathbf{r}} s'v' | t | \hat{\mathbf{k}} sv \rangle$) in partial waves. Employing the completeness and orthogonality of the angular wave functions implies a relation involving the partial wave components, which is analogous to Eq. (12.188) derived for central potentials. Unlike the situation encountered in scattering from central potentials, however, the pertinent partial wave expansion basis is constructed from both spin and orbital functions.

First, we need to introduce some functions related to the addition of orbital and spin angular momentum. Consider first the *spinor spherical harmonics*.

$$\langle \hat{\mathbf{r}} | lsJM \rangle = \mathcal{Y}^{JM}_{ls}(\hat{\mathbf{r}}) = \sum_{mv} \langle lsmv | JM \rangle Y_{lm}(\hat{\mathbf{r}}) \chi_{sv}, \qquad (12.580)$$

which are eigenfunctions of L^2, S^2, J^2, and J_z with eigenvalues $l(l+1), s(s+1), J(J+1)$, and M, respectively. These may be regarded as the analogs of the spherical harmonics $Y_{lm}(\theta, \phi)$, properly modified for treating scattering of spin 1/2 particles. In the initial state $|\mathbf{k}sv\rangle$ (12.569), the spin projection v is specified, and therefore, its expansion in terms of spinor spherical harmonics should take this into account. To this end, we define the functions,

$$Q^{JM}_{l's'v';lsv}(\hat{\mathbf{r}}; \hat{\mathbf{k}}) = \sum_{m',m} \langle l'm's'v' | JM \rangle \langle lmsv | JM \rangle i^{l'} Y^{m'}_{l'}(\hat{\mathbf{r}}) Y^{m*}_{l}(\hat{\mathbf{k}}), \qquad (12.581)$$

$$X^{J}_{l's'v';lsv}(\hat{\mathbf{r}}; \hat{\mathbf{k}}) = \sum_{M} Q^{JM}_{l's'v';lsv}(\hat{\mathbf{r}}; \hat{\mathbf{k}}). \qquad (12.582)$$

12.8 Particles with Internal Degrees of Freedom

Problem 12.22

Prove the equality: $\sum_J X^J_{l's'v';lsv}(\hat{\mathbf{r}}; \hat{\mathbf{k}}) = \frac{2l+1}{4\pi} i^l P_l(\hat{\mathbf{r}} \cdot \hat{\mathbf{k}})$.

Hint: Use the addition theorem for spherical harmonics.

Problem 12.23

Show that $\sum_{v'} Q^J_{l's'v';lsv}(\hat{\mathbf{r}}; \hat{\mathbf{k}}) \chi_{s'v'}$ is an eigenfunction of J^2 and J_z.

Hint: Show that, after summation, you get a quantity proportional to $\langle \hat{\mathbf{r}} | lsJM \rangle$ defined in Eq. (12.580).

We are now in a position to expand the initial state, the S and T matrices and the wave function in terms of the functions defined above. Starting with expansion of the plane wave, let us first use the addition formula for spherical harmonics and rewrite Eq. (12.153) as

$$e^{ikz} = e^{ikr\cos\theta} = 4\pi \sum_{l=0}^{\infty} i^l j_l(kr) Y_l^{m*}(\hat{\mathbf{k}}) Y_l^m(\hat{\mathbf{r}}). \tag{12.583}$$

Using definitions (12.581) and (12.582), this can be written as

$$e^{i\mathbf{k}\cdot\mathbf{r}} \chi_{sv} = 4\pi \sum_{l,J} j_l(kr) X^J_{lsv;lsv}(\hat{\mathbf{r}}; \hat{\mathbf{k}}) \chi_{sv}. \tag{12.584}$$

Although this is certainly a more complicated representation of the same function, the corresponding angular functions will appear in the expansion of the S and T matrices, $\langle \hat{\mathbf{k}}'s'v'|S|\hat{\mathbf{k}}sv\rangle$ and $\langle \hat{\mathbf{k}}'s'v'|T|\hat{\mathbf{k}}sv\rangle$, to which we now turn. The basic technique is to introduce unit operators in terms of complete sets of states and employ orthogonality and conservation laws. First, we introduce the projection operator

$$P_1 \equiv \sum_{lms''v''} |lms''v''\rangle\langle lms''v''|, \tag{12.585}$$

between the ket $|\hat{\mathbf{k}}sv\rangle$ and the S operator, and use the relation,

$$\langle lms''v''|\hat{\mathbf{k}}sv\rangle = Y_{lm}(\hat{\mathbf{k}})\delta_{ss''}\delta_{vv''}. \tag{12.586}$$

Next, we introduce the projection operator

$$P_2 \equiv \sum_{JM} |lsJM\rangle\langle lsJM|, \tag{12.587}$$

between P_1 and the S operator. When a similar procedure is carried out between the bra $\langle \hat{\mathbf{k}}'s'v'|$ and the S operator, we encounter the matrix element,

$$\langle l's'J'M'|S(k)|lsJM\rangle = \delta_{JJ'}\delta_{MM'} S^J_{l's',ls}(k), \tag{12.588}$$

where the invariance of the S matrix under rotations has been used. Here, the energy (or the momentum k) is explicitly indicated, and the RHS of Eq. (12.588) is referred to as the $klsJ$ representation of the S matrix. It should be stressed that both P_1 and P_2 are unit operators in the appropriate spaces. After carrying out all manipulations, we get

$$\langle \hat{\mathbf{k}}'s'v'|S|\hat{\mathbf{k}}sv\rangle = \langle \hat{\mathbf{k}}'s'v'|P_1P_2SP_2P_1|\hat{\mathbf{k}}sv\rangle = \sum_{l'l,J} X^J_{l's'v';lsv}(\hat{\mathbf{k}}'; \hat{\mathbf{k}}) S^J_{l's',ls}(k). \tag{12.589}$$

A similar expression holds for the t matrix and the non-central potential v in momentum space. Note from Eq. (12.575) that the off-shell t matrix and potential might also be required. Thus, we have,

$$\langle \mathbf{k}'s'\nu'|t|\mathbf{k}s\nu\rangle = \sum_{l'l,J} X^J_{l's'\nu';ls\nu}(\hat{\mathbf{k}}';\hat{\mathbf{k}})t^J_{l's',ls}(k',k,\varepsilon), \tag{12.590}$$

$$\langle \mathbf{k}'s'\nu'|v|\mathbf{k}s\nu\rangle = \sum_{l'l,J} X^J_{l's'\nu';ls\nu}(\hat{\mathbf{k}}';\hat{\mathbf{k}})v^J_{l's',ls}(k',k), \tag{12.591}$$

where the on-shell condition here is simply $k' = k$, and $t^J_{l's',ls}(k',k) \equiv t^J_{l's',ls}(k)$ is the on-shell t matrix in the $klsj$ representation. Relation (12.570) between S and T (on energy shell) in the $\hat{\mathbf{k}}s\nu$ representation is translated with the help of the expansions (12.589) and (12.590) into an appropriate relation between S and T in the $klsj$ representation,

$$\boxed{S^J_{l's',ls}(k) = \delta_{l'l}\delta_{s's} - 2\pi i k \, t^J_{l's',ls}(k).} \tag{12.592}$$

Equations (12.589) and (12.590) are a consequence of the rotation invariance of the Hamiltonian. It is useful at this point to stress the consequences of unitarity and time-reversal invariance on $S^J_{l's',ls}(k)$. The unitarity relation is obtained from Eq. (12.571) expressed in the $\hat{\mathbf{k}}s\nu$ representation and the expansion (12.589),

$$\sum_{l''s''}[S^J_{l''s'',l's'}]^* S^J_{l''s'',ls} = \delta_{ll'}\delta_{ss'}. \tag{12.593}$$

The invariance of the S matrix under time reversal discussed previously in connection with Eq. (12.550) and Fig. 12.23 has consequences for the partial wave S matrix elements. For the sake of self-consistence, let us recall our earlier results within a slightly different reasoning. The time-reversal operator $\mathcal{T}$ applied on a scattering state $|\psi_\mathbf{a}^+\rangle$, such as defined in Eq. (12.521), can be formulated as follows,

$$\mathcal{T}|\psi_\mathbf{a}^+\rangle = |\psi_{\mathcal{T}\mathbf{a}}^-\rangle, \tag{12.594}$$

namely, it turn an outgoing scattering state into an incoming state whose quantum numbers are transformed accordingly. In particular, if $\mathbf{a} = (\mathbf{k}s\nu)$, the phase of the ket $|\mathbf{k}s\nu\rangle$ can be chosen, so that

$$\mathcal{T}|\mathbf{k}s\nu\rangle = (-1)^{s+\nu}|-\mathbf{k}s-\nu\rangle. \tag{12.595}$$

Therefore, the invariance of the S matrix under time reversal implies,

$$\langle \mathbf{k}'s'\nu'|S|\mathbf{k}s\nu\rangle = \langle \psi^-_{\mathbf{k}'s'\nu'}|\psi^+_{\mathbf{k}s\nu}\rangle = (-1)^{s'-s+\nu'-\nu}\langle \psi^+_{-\mathbf{k}s-\nu}|\psi^-_{-\mathbf{k}'s'-\nu'}\rangle$$
$$= (-1)^{s'-s+\nu'-\nu}\langle -\mathbf{k}s-\nu|S|-\mathbf{k}'s'-\nu'\rangle. \tag{12.596}$$

This relation is true also off-shell, but we will use it on-shell and expand both sides of the relation

$$\langle \hat{\mathbf{k}}'s'\nu'|S|\hat{\mathbf{k}}s\nu\rangle = (-1)^{s'-s+\nu'-\nu}\langle -\hat{\mathbf{k}}s-\nu|S|-\hat{\mathbf{k}}'s'-\nu'\rangle. \tag{12.597}$$

According to Eq. (12.589), we obtain the relation,

$$S^J_{l's',ls} = S^J_{ls,l's'}. \tag{12.598}$$

Thus, rotational invariance and time-reversal symmetry imply that in the $klsj$ representation, the S matrix (and the T matrix as well) is symmetric.

12.8 Particles with Internal Degrees of Freedom

Problem 12.24

Show that if there are n combinations for the pair (ls) (these are n channels for a given J), the matrix $S^J_{l's',ls}$ is determined by $\frac{1}{2}n(n+1)$ independent parameters.

Guidance: There are n^2 complex numbers, i.e., $2n^2$ real numbers forming the matrix elements of the S matrix. Subtract the number of symmetry relations (12.598) and the number of unitarity relations implied by the constraint $S^\dagger S = \mathbf{1}_{n\times n}$.

Problem 12.25

Show that for scattering of two spin 1/2 particles in the triplet state $s = 1$, the number of possible values of l' and l are $J+1$ and $J-1$ (i.e., $n=2$). Then, show that the S matrix can be diagonalized by a real 2×2 orthogonal matrix depending on a single parameter,

$$S = \begin{pmatrix} \cos\alpha & \sin\alpha \\ -\sin\alpha & \cos\alpha \end{pmatrix} \begin{pmatrix} e^{2i\delta_1} & 0 \\ 0 & e^{2i\delta_2} \end{pmatrix} \begin{pmatrix} \cos\alpha & -\sin\alpha \\ \sin\alpha & \cos\alpha \end{pmatrix}. \tag{12.599}$$

The two parameters δ_1 and δ_2 are called the *eigenphases* of S.

Solution: Since $s=1$, the spin function is symmetric, and the space part must be antisymmetric. Therefore, both l and l' must be odd. Moreover, l cannot differ from J by more than one unit since $\mathbf{J} = \mathbf{l} + \mathbf{s}$ and $s=1$. From the solution of the previous problem, we know that the number of independent real numbers determining S is 3. Two of them are the phase shifts and the third one is the angle α appearing in the matrix that diagonalizes S. A unitary 2×2 matrix that depends on a single parameter can be chosen to be real, that is, it is orthogonal. The group of 2×2 orthogonal matrices that depends on a single parameter are the rotations about a given axis, that has the form appearing in the question (the matrices commute with each other).

In analogy with the expansion of the initial state in Eq. (12.584), the expansion of the scattering state (12.575) in partial waves reads,

$$\psi_{\mathbf{k}s\nu}(\mathbf{r}) = 4\pi \sum_{l's'\nu'lJ} X^J_{l's'\nu';ls\nu}(\hat{\mathbf{r}};\hat{\mathbf{k}}) \chi_{s'\nu'} \psi^+_{l's',ls;Jk}(r). \tag{12.600}$$

Inserting the expansions (12.590) and (12.600) into Eq. (12.575) then leads to a relation between the wave function and the T matrix in the $klsj$ representation,

$$\psi^+_{l's',ls;Jk}(r) = \delta_{l'l}\delta_{s's}j_l(kr) + \int_0^\infty dk'\, k'^2 \frac{j_l(k'r)}{\varepsilon(k) + i\eta - \varepsilon(k')} t^J_{l's',ls}(k',k), \tag{12.601}$$

which is the analogous expression to Eq. (12.188) for scattering by a non-central potential. Using the same procedure as followed in Eq. (12.188), the expression relating the wave function to the partial wave scattering amplitude, together with its asymptotic limit, yields

$$\psi^+_{l's',ls;Jk}(r) = \delta_{l'l}\delta_{s's}j_l(kr) - \pi k \frac{e^{ikr}}{2kr} t^J_{l's',ls}(k) \to \frac{1}{kr}\left[-\delta_{l'l}\delta_{s's}e^{-ikr} + S^J_{l's',ls}(k)e^{ikr}\right]. \tag{12.602}$$

Recipe for Calculating Cross-sections

A procedure for calculating the partial wave functions $\psi^+_{l's',ls;Jk}(r)$ introduced in Eqs. (12.600)–(12.602) can be carried out starting from the Schrödinger equation (12.566) and expanding the wave function and the potential in partial waves.

The wave function expansion has already been executed in Eq. (12.600). Writing Eq. (12.566) in the form,

$$\left[\nabla_r^2 - v(\mathbf{r}) + k^2\right]\psi_\mathbf{k}(\mathbf{r}) = 0, \tag{12.603}$$

implies the following equality,

$$\int d\Omega_{\hat{\mathbf{k}}} d\Omega_{\hat{\mathbf{r}}} \sum_{v'} \chi^\dagger_{s'v'} X^{J*}_{l's'v';lsv}(\hat{\mathbf{r}};\hat{\mathbf{k}}) \left[\nabla_r^2 - v(\mathbf{r}) + k^2\right]\psi_\mathbf{k}(\mathbf{r}) = 0. \tag{12.604}$$

To complete the procedure, the wave function $\psi_\mathbf{k}(\mathbf{r})$ is now expanded as in (12.600), and the integration over solid angles is performed. These operations define the configuration space partial wave potential matrix in the Jls representation [in momentum space, it has already been introduced in Eq. (12.591)],

$$v^J_{l''s'';l's'}(r) = \frac{2s+1}{2J+1} \int d\Omega_{\hat{\mathbf{r}}} d\Omega_{\hat{\mathbf{k}}} \sum_{v''v'} \chi^\dagger_{s''v''} X^{J*}_{l''s''v'';lsv}(\hat{\mathbf{r}};\hat{\mathbf{k}}) v(\mathbf{r}) X^J_{l's'v';lsv}(\hat{\mathbf{r}};\hat{\mathbf{k}}) \chi_{s'v'}. \tag{12.605}$$

The dependence on v in X and in X^* cancels out. Except for the potential $v(\mathbf{r})$, all the functions appearing in Eq. (12.605) are of geometrical origin. Since the Hamiltonian $H_0 + v$ is hermitian, the time-reversal invariance implies that $v^J_{l's';ls}$ is real and symmetric,

$$v^J_{l's';ls} = \left[v^J_{l's';ls}\right]^* \tag{12.606}$$

$$v^J_{l's';ls} = v^J_{ls;l's'}. \tag{12.607}$$

It is now possible to formulate a recipe for the calculation of the partial wave S matrix defined through Eq. (12.589) and thereby the differential cross-section. The partial wave Schrödinger equation for the wave function (12.601) is,

$$\left[\frac{\partial^2}{\partial r^2} + \frac{2}{r}\frac{\partial}{\partial r} - \frac{l'(l'+1)}{r^2} + k^2\right]\psi^+_{l's',ls;Jk}(r) = \sum_{l''s''} v^J_{l's';l''s''}(r) \psi^+_{l''s'',ls;Jk}(r). \tag{12.608}$$

The implicit dependence on l is implied through the asymptotic boundary conditions (12.602). The procedure for obtaining the t matrix elements now goes as follows: First, in order to avoid the first-order derivative and set the boundary condition at $r = 0$, the wave function is written as,

$$\psi^+_{l's',ls;Jk}(r) = \frac{u_{l's',ls;Jk}(r)}{kr}, \tag{12.609}$$

such that

$$\left[\frac{\partial^2}{\partial r^2} - \frac{l'(l'+1)}{r^2} + k^2\right] u_{l's',ls;Jk}(r) = \sum_{l''s''} v^J_{l's';l''s''}(r) u_{l''s'',ls;Jk}(r). \tag{12.610}$$

Second, the set of coupled differential Eqs. (12.610) is integrated from $r = 0$ (where $u_{l's',ls;Jk}(0) = 0$) up to large distance R much beyond the range of the potential $v(\mathbf{r})$. Finally, the wave functions at R are matched with the asymptotic form as in Eq. (12.602),

$$u^+_{l's',ls;Jk}(R) = \delta_{l'l}\delta_{s's} kR j_l(kR) - \pi k \frac{e^{ikr}}{2} t^J_{l's',ls}(k) \to \left[-\delta_{l'l}\delta_{s's} e^{-ikR} + S^J_{l's',ls}(k) e^{ikR}\right]. \tag{12.611}$$

from which one extracts the S matrix or the t matrix elements. Using Eq. (12.590), the t matrix elements are obtained, from which the differential cross-section is evaluated according to Eq. (12.578).

12.8 Particles with Internal Degrees of Freedom

Scattering of Identical Particles with Spins

The above formulas must be modified if the two scattered particles are identical. The following analysis is similar to the one detailed in Sec. 12.5.6 near Eq. (12.329) with the necessary modifications arising due to the fact that the scattering potential is non-central. The starting point is the application of the exchange operator $\tilde{S}$ on the asymptotic wave function (12.576). The result is,

$$\tilde{S}\psi_{\mathbf{k}}^{+}(\mathbf{r}) \to \frac{1}{\sqrt{2}}\left[\left(e^{i\mathbf{k}\cdot\mathbf{r}} + p_{12}\phi_s e^{-i\mathbf{k}\cdot\mathbf{r}}\right)\chi_{sv}\right]$$

$$-\frac{1}{\sqrt{2}}\frac{e^{ikr}}{4\pi r}\left[\sum_{s'v'}(\langle\hat{\mathbf{r}}s'v'|t|\hat{\mathbf{k}}sv\rangle + p_{12}\phi_{s'}\langle-\hat{\mathbf{r}}s'v'|t|\hat{\mathbf{k}}sv\rangle)\chi_{s'v'}\right], \tag{12.612}$$

where $p_{12} = \pm 1$ for bosons and fermions, respectively, and $\phi_s \equiv (-1)^{2s_1-s}$. Accordingly, the properly symmetrized scattering amplitude from initial state $|\mathbf{k}sv\rangle$ to final state $\langle\mathbf{k}'s'v'|$ is the coefficient of the outgoing spherical wave e^{ikr}/r, i.e.,

$$f_{\mathbf{k}}(\Omega_{\hat{\mathbf{k}}'}) = -\frac{1}{4\pi\sqrt{2}}\left[\langle\hat{\mathbf{r}}s'v'|t|\hat{\mathbf{k}}sv\rangle + p_{12}\phi_{s'}\langle-\hat{\mathbf{r}}s'v'|t|\hat{\mathbf{k}}sv\rangle\right], \tag{12.613}$$

which is the analog of Eq. (12.577) when the two particles are identical. The cross-section is then calculated from $f_{\mathbf{k}}(\Omega_{\hat{\mathbf{k}}'})$ as in Eq (12.578) or (12.579).

Scattering of Spin $\frac{1}{2}$ and Spin 0 Particles

Consider the scattering of particle 1 with spin $s_1 = \frac{1}{2}$ and particle 2 with no spin ($s_2 = 0$). Examples of scattering of spin $\frac{1}{2}$ and spin 0 particles are neutron scattering from a spin 0 atomic nucleus and scattering of low-energy electrons by ground state ^{4}He atoms or other noble gas atoms. The total spin of the two-particle system is $s = s_1 = \frac{1}{2}$, and the dependence of the scattering potential on spin enters through the spin–orbit interaction (see below).

Using basic conservation laws of quantum mechanics, the dependence of the partial wave S and T matrices, defined within Eqs. (12.588)–(12.590) and (12.592), takes a simple form when compared with the structure for the case of two particles with spins. The conservation of spin $s = s' = \frac{1}{2}$ is of course obvious. In addition, if it is assumed that the potential $v(\mathbf{r})$ is even under parity, the parity of the initial and final states must be the same. The parity of the spherical harmonics $Y_l^m(\hat{\mathbf{r}})$ and the kets $|lsJM\rangle$ is $(-1)^l$. Therefore, parity conservation implies $(-1)^l = (-1)^{l'}$. The total angular momentum J is a good quantum number, and hence, for a given value of $J = 1/2, 3/2\ldots$, the orbital angular momentum can take only two values,

$$l_+ = J + \frac{1}{2} \quad l_- = J - \frac{1}{2}. \tag{12.614}$$

The S and T matrices and the potential matrix (12.605) are diagonal in l,

$$S^J_{l's';ls} = S^J_{l'\frac{1}{2};l\frac{1}{2}} = \delta_{l'l}S^J_l = \delta_{l'l}e^{2i\delta^J_l}, \tag{12.615}$$

$$t^J_{l's';ls} = t^J_{l'\frac{1}{2};l\frac{1}{2}} = \delta_{l'l}t^J_l, \tag{12.616}$$

$$v^J_{l's';ls} = v^J_{l'\frac{1}{2};l\frac{1}{2}} = \delta_{l'l}v^J_l. \tag{12.617}$$

The relation (12.592) between the partial wave S and T matrices takes the simple form,

$$S^J_l = 1 - 2\pi i k t^J_l, \tag{12.618}$$

$$t^J_l = -\frac{e^{i\delta^J_l}\sin\delta^J_l}{\pi k}. \tag{12.619}$$

The partial wave expansions (12.589)–(12.591) are also simplified, since in Eq. (12.582), the quantum numbers are restricted,

$$X^J_{l's'\nu';ls\nu}(\hat{\mathbf{r}};\hat{\mathbf{k}}) \to X^J_{l\frac{1}{2}\nu';l\frac{1}{2}\nu}(\hat{\mathbf{r}};\hat{\mathbf{k}}) = \sum_{m',m,M} \left\langle lm'\frac{1}{2}\nu'|JM\right\rangle\left\langle lm\frac{1}{2}\nu|JM\right\rangle i^l Y^{m'}_l(\hat{\mathbf{r}}) Y^{m*}_l(\hat{\mathbf{k}}), \qquad (12.620)$$

With the definitions (12.581) and (12.582), the expansion now reads,

$$\left\langle \hat{\mathbf{k}}'\frac{1}{2}\nu'|\mathcal{O}|\hat{\mathbf{k}}\frac{1}{2}\nu\right\rangle = \sum_{lmm'JM} Y^{m'}_l(\hat{\mathbf{k}}') Y^{m*}_l(\hat{\mathbf{k}}) \left\langle l\frac{1}{2}m'\nu'|JM\right\rangle\left\langle l\frac{1}{2}m\nu|JM\right\rangle \mathcal{O}^J_l, \qquad (12.621)$$

where $\mathcal{O} = S, t, \nu$.

Schrödinger Equation with Spin–Orbit Interaction

We focus on the problem of an electron (or other spin 1/2 particle) scattered from a central potential $V(\mathbf{r}) = V_c(r) = e\Phi(r)$ together with a spin–orbit potential,

$$V_{so}(\mathbf{r}) = -\frac{\hbar}{4m^2c^2}\frac{1}{r}\frac{dV_c}{dr}\boldsymbol{\sigma}\cdot\mathbf{L}, \qquad (12.622)$$

where $\mathbf{L} = \mathbf{r}\times\mathbf{p}$ is the orbital angular momentum operator. Defining $v_c(r) = \frac{2mV_c(r)}{\hbar^2}$, we can start our analysis from Eq. (12.603), where the presence of spin–orbit potential implies the following form of the non-central potential,

$$v(\mathbf{r}) = v_c(r) + v_{so}(r)\boldsymbol{\sigma}\cdot\mathbf{L}. \qquad (12.623)$$

The spin–orbit potential, with $v_{so}(r) = \frac{\lambda}{r}\frac{dv_c(r)}{dr}$, where $\lambda = \frac{e\hbar}{8m^2c^2}$, is non-central due to the occurrence of the spin operator. Equation (12.603) then reads,

$$\left[\nabla^2_{\mathbf{r}} + k^2 - v_c(r) - v_{so}(r)\boldsymbol{\sigma}\cdot\mathbf{L}\right]\psi^+_{\mathbf{k}\frac{1}{2}\nu}(\mathbf{r}) = 0. \qquad (12.624)$$

Problem 12.26

A particle of mass m and charge q is restricted to move on a sphere of radius R. It is acted upon by the Coulomb field generated by a charged particle of charge Q fixed at the center of the sphere. Find the energies and the wave functions for this system.

Guidance: On the sphere, $v_c(r) = Qq/R$ and $v_{so}(R)$ are constants. Therefore, the problem is reduced to that of finding the eigenvalues of $\mathbf{L}\cdot\mathbf{S}$. Due to spin–orbit interaction, $\mathbf{L}$ is not a good quantum number but $\mathbf{J} = \mathbf{L} + \mathbf{S}$. Use the identity $\mathbf{L}\cdot\mathbf{s} = [J(J+1) - L(L+1) - S(S+1)]/2$.

In principle, the formalism developed for the scattering of particles with spin can be repeated here, with the angular functions X^J restricted as in Eq. (12.620). It is useful, however, to elaborate on these steps and expose the simplifications resulting from the special form of the spin–orbit interaction, employing the conservation laws, rotation invariance, time-reversal invariance, and parity. The wave function is now expanded in partial waves as in Eq. (12.600), but due to the conservation laws, this expansion takes a simpler form. The partial wave components of the wave function depends only on klJ, which reduces to $kl_\pm$ for $J = l \pm \frac{1}{2}$, namely,

$$\psi^+_{l's',ls;Jk}(r) \to \psi^+_{klJ}(r) = \psi^+_{kl_\pm}(r) \quad \text{for } J = l \pm \tfrac{1}{2}, \qquad (12.625)$$

12.8 Particles with Internal Degrees of Freedom

and

$$\psi^+_{\mathbf{k}s\nu}(\mathbf{r}) = 4\pi \sum_{l\nu'J} X^J_{l\frac{1}{2}\nu';l\frac{1}{2}\nu}(\hat{\mathbf{r}};\hat{\mathbf{k}}) \chi_{\frac{1}{2}\nu'} \psi^+_{klJ}(r). \tag{12.626}$$

When the expansion (12.626) is substituted in Eq. (12.624), the following identities should be employed, which result from the definition (12.620),

$$\boldsymbol{\sigma} \cdot \mathbf{L} X^J_{l\frac{1}{2}\nu';l\frac{1}{2}\nu}(\hat{\mathbf{r}};\hat{\mathbf{k}}) \chi_{\frac{1}{2}\nu'} = l X^J_{l\frac{1}{2}\nu';l\frac{1}{2}\nu}(\hat{\mathbf{r}};\hat{\mathbf{k}}) \chi_{\frac{1}{2}\nu'} \quad (\text{for } J = l + \tfrac{1}{2}), \tag{12.627}$$

$$\boldsymbol{\sigma} \cdot \mathbf{L} X^J_{l\frac{1}{2}\nu';l\frac{1}{2}\nu}(\hat{\mathbf{r}};\hat{\mathbf{k}}) \chi_{\frac{1}{2}\nu'} = -(l+1) X^J_{l\frac{1}{2}\nu';l\frac{1}{2}\nu}(\hat{\mathbf{r}};\hat{\mathbf{k}}) \chi_{\frac{1}{2}\nu'} \quad (\text{for } J = l - \tfrac{1}{2}). \tag{12.628}$$

Employing Eqs. (12.627) and (12.628) and the orthogonality of spin and angular functions, the partial wave Schrödinger equation for the wave function $\psi^+_{klJ}(r)$ is obtained, which is the analog of Eq. (12.608). Using the notation of Eq. (12.625), $l_\pm = J \mp \tfrac{1}{2}$,

$$\left[\frac{d^2}{dr^2} + \frac{2}{r}\frac{d}{dr} - \frac{l(l+1)}{r^2} + k^2\right] \psi^+_{kl_\pm}(r) = v_{l_\pm}(r) \psi^+_{kl_\pm}(r), \tag{12.629}$$

where, following Eqs. (12.627) and (12.628), the potentials $v_{l_\pm}(r)$ are

$$v_{l_+}(r) = v_c(r) + l v_{so}(r), \tag{12.630}$$

$$v_{l_-}(r) = v_c(r) - (l+1) v_{so}(r), \tag{12.631}$$

which constitutes a remarkable simplification compared with the definition of partial wave potential in Eq. (12.605). In what follows, we use the same procedure recipe for calculating phase shift and cross-section as already discussed. Keeping in mind a fixed quantum number J, the first derivative is eliminated through the definition,

$$\psi^+_{kl_\pm}(r) = \frac{u_{kl_\pm}(r)}{kr}, \tag{12.632}$$

such that

$$\left[\frac{d^2}{dr^2} - \frac{l(l+1)}{r^2} + k^2\right] u_{kl_\pm}(r) = v_{l_\pm}(r) u_{kl_\pm}(r). \tag{12.633}$$

The differential Eqs. (12.633) are integrated from $r = 0$ (where $u_{kl_\pm}(0) = 0$) up to large distance R much beyond the range of the potential $v(\mathbf{r})$. The wave functions at R are then matched with the asymptotic form as in Eq. (12.602),

$$u_{kl_\pm}(kR) = kR j_{l_\pm}(kR) - \pi k \frac{e^{ikr}}{2} t^J_{l_\pm}(k) \to \left[-e^{-ikR} + S^J_{l_\pm}(k) e^{ikR}\right]. \tag{12.634}$$

from which one extracts the partial wave matrix S^J_l or t^J_l. Using Eq. (12.621), the t matrix elements $\langle \hat{\mathbf{k}}' \tfrac{1}{2}\nu' | t | \hat{\mathbf{k}} \tfrac{1}{2}\nu \rangle$ are obtained,

$$\left\langle \hat{\mathbf{k}}' \frac{1}{2}\nu' \middle| t \middle| \hat{\mathbf{k}} \frac{1}{2}\nu \right\rangle = \sum_{lmm'JM} Y_l^{m'}(\hat{\mathbf{k}}') Y_l^{m*}(\hat{\mathbf{k}}) \left\langle l\tfrac{1}{2}m'\nu' | JM \right\rangle \left\langle l\tfrac{1}{2}m\nu | JM \right\rangle t^J_l, \tag{12.635}$$

to which the scattering amplitude is simply related as in Eq. (12.577), and the differential cross-section is evaluated according to Eq. (12.578).

It is instructive, however, to exploit the consequences of rotation, parity, and time-reversal invariance in a somewhat different way. The t matrix elements on the LHS of Eq. (12.635) form a 2×2 matrix $\mathbf{t}$ in spin space. The 2×2 matrix $\mathbf{f}$ of scattering amplitudes $f_{\nu'\nu} = -2\pi^2 \langle \hat{\mathbf{k}}' \tfrac{1}{2}\nu' | t | \hat{\mathbf{k}} \tfrac{1}{2}\nu \rangle$ [see Eq. (12.577)] can be written in terms of the vector $\boldsymbol{\sigma}$ of Pauli matrices,

$$\mathbf{f} = a + i\boldsymbol{\sigma} \cdot \mathbf{b}, \tag{12.636}$$

where a is a scalar multiplying the unit 2×2 matrix and $\mathbf{b}$ is a vector (the prefactor i is introduced for convenience). Both a and $\mathbf{b}$ depend on the scattering angle θ (as well as on the energy) and should be calculated in terms of the partial amplitudes $t^J_{l_\pm}$. Rotational invariance dictates that the matrix $\mathbf{f}$ should be a scalar formed from the vectors $\hat{\mathbf{k}}$, $\hat{\mathbf{k}}'$, and $\boldsymbol{\sigma}$. This immediately implies that

$$a = h(\hat{\mathbf{k}} \cdot \hat{\mathbf{k}}') = h(\cos\theta), \tag{12.637}$$

$$i\boldsymbol{\sigma} \cdot \mathbf{b} = \boldsymbol{\sigma} \cdot \left[\hat{\mathbf{k}} \times \hat{\mathbf{k}}' + b_1(\cos\theta)\mathbf{k} + b_2(\cos\theta)\mathbf{k}'\right] g(\cos\theta). \tag{12.638}$$

Time-reversal invariance expressed in Eq. (12.554) or, equivalently, (12.597), implies that this expression must be invariant (up to a phase independent of the momenta) under sign change of spin and momenta and swapping initial and final momenta $\hat{\mathbf{k}} \leftrightarrow \hat{\mathbf{k}}'$, which requires $b_1 = b_2$. Moreover, invariance under space inversion (parity) expressed in Eq. (12.549) implies that $\mathbf{f}$ should be invariant under sign change of the momenta. This constraint eliminates the terms linear in momenta, namely, $b_1 = b_2 = 0$. Collecting these constraints, the most general form of the $\mathbf{f}$ matrix consistent with rotation, parity, and time-reversal invariance is of the form,

$$\mathbf{f} = h(\cos\theta) + i\boldsymbol{\sigma} \cdot \hat{\mathbf{k}} \times \hat{\mathbf{k}}' \, g(\cos\theta), \tag{12.639}$$

where it should be recalled that the scattering angle θ also enters in the vector product. Beyond the kinematics and symmetry constraints, the physical content of the scattering is encoded in the functions $h(\cos\theta)$ and $g(\cos\theta)$. These can formally be calculated in terms of the partial wave t matrix elements t^J_l [see discussion around Eq. (12.634)]. This task is carried out by comparing expression (12.639) with the sum on the RHS of Eq. (12.635) (multiplied by $-2\pi^2$). For each fixed ν, ν', the sum is divided into a first term containing only $\cos\theta$ and a second term with a factor $\sin\theta$ multiplying a function of $\cos\theta$. The result is

$$h(\cos\theta) = -\frac{1}{k} \sum_{l=0}^{\infty} \left[(l+1)t^J_{l_+} + l t^J_{l_-}\right] P_l(\cos\theta), \tag{12.640}$$

$$g(\cos\theta) = -\frac{1}{k} \sum_{l=1}^{\infty} \left[t^J_{l_+} - t^J_{l_-}\right] P_l'(\cos\theta), \tag{12.641}$$

where $P_l'(\cos\theta) = dP_l(\cos\theta)/d\cos\theta$. The total angular momentum J is determined by $J = l_+ - \frac{1}{2}$ or $J = l_- + \frac{1}{2}$. Equation (12.639) is convenient for analyzing the various contributions for the differential cross-section. First, it is evident from this expression that any dependence of the cross-section on spin direction cannot occur neither in the forward direction $\theta = 0$ nor in the backward direction $\theta = \pi$ (provided $g(\cos\theta)$ is regular at these points, which is normally the case). Second, in the scattering plane defined by the initial and final momentum unit vectors, the sign of the second term depends on the direction of the vector product $\hat{\mathbf{k}} \times \hat{\mathbf{k}}'$. This implies left-right asymmetry, referred to as *skew scattering*. In solid-state physics, it contributes to the *anomalous Hall effect*.

The differential cross-section for an unpolarized beam and unpolarized detector requires summation over final polarizations $\nu' = \pm 1/2$ and averaging over initial polarizations $\nu = \pm 1/2$,

$$\left\langle \frac{d\sigma}{d\Omega_{\hat{\mathbf{k}}'}} \right\rangle = \frac{1}{2}\text{Tr}(\mathbf{f}^\dagger \mathbf{f}) = |h(\cos\theta)|^2 + \sin^2\theta |g(\cos\theta)|^2. \tag{12.642}$$

In order to analyze the cross-section for polarized beam and detector, we need to define a quantization axis z. A natural choice is $\hat{\mathbf{z}}$ perpendicular to the scattering plane and points in the direction of $\hat{\mathbf{k}} \times \hat{\mathbf{k}}'$. In this case, $\boldsymbol{\sigma} \cdot \hat{\mathbf{k}} \times \hat{\mathbf{k}}' = \eta \sin\theta \sigma_z$ and

$$\mathbf{f} = h(\cos\theta) + i\eta \sin\theta \, g(\cos\theta) \, \sigma_z, \tag{12.643}$$

12.8 Particles with Internal Degrees of Freedom

where $\eta = \pm 1$ for scattering to the left or right, respectively. For particle polarized along $\pm \hat{z}$ (i.e., $\nu = \pm \frac{1}{2}$), Eq. (12.639) then reduces to,

$$f_{\nu'\nu} = \delta_{\nu'\nu} \left[h(\cos\theta) + i(-1)^{\frac{1}{2}-\nu} \eta \sin\theta \, g(\cos\theta) \right], \tag{12.644}$$

and there is no spin flip in this case. The cross-section for a polarized incoming beam is then given by,

$$\left[\frac{d\sigma}{d\Omega_{\hat{\mathbf{k}}'}} \right]_{\nu \leftarrow \nu} = |f_{\nu\nu}|^2 = |h(\cos\theta)|^2 + \sin^2\theta |g(\cos\theta)|^2 + 2\eta(-1)^{\frac{1}{2}-\nu} \mathrm{Im}[h(\cos\theta)g(\cos\theta)^*]. \tag{12.645}$$

As long as $\mathrm{Im}[h(\cos\theta)g(\cos\theta)^*] \neq 0$, the cross-section displays left-right asymmetry according to the sign of η.

Problem 12.27

If the initial beam is polarized along $\hat{\mathbf{x}}$ (i.e., $\nu_x = \frac{1}{2}$), find the amplitudes for direct and spin-flip scattering events into final states $\nu'_x = \pm \frac{1}{2}$.

Answer: Express the states polarized along $\pm \hat{\mathbf{x}}$ in terms of those along $\pm \hat{\mathbf{z}}$,

$$\left| \nu_x = \pm \frac{1}{2} \right\rangle = \frac{1}{\sqrt{2}} \left[\left| \nu_z = \frac{1}{2} \right\rangle \pm \left| \nu_z = -\frac{1}{2} \right\rangle \right], \tag{12.646}$$

and employ Eqs. (12.643) and (12.644) to compute $\langle \nu'_x = \pm \frac{1}{2} | \mathbf{f} | \nu_x = \frac{1}{2} \rangle$. This procedure yields

$$\left\langle \nu'_x = \frac{1}{2} \middle| \mathbf{f} \middle| \nu_x = \frac{1}{2} \right\rangle = \frac{1}{2} \left[f_{\frac{1}{2}\frac{1}{2}} + f_{-\frac{1}{2}-\frac{1}{2}} \right] = h(\cos\theta), \tag{12.647}$$

$$\left\langle \nu'_x = -\frac{1}{2} \middle| \mathbf{f} \middle| \nu_x = \frac{1}{2} \right\rangle = \frac{1}{2} \left[f_{\frac{1}{2}\frac{1}{2}} - f_{-\frac{1}{2}-\frac{1}{2}} \right] = i\eta \sin\theta \, g(\cos\theta). \tag{12.648}$$

In this case, there is no left-right asymmetry of the spin-flip cross-section.

12.8.9 INELASTIC SCATTERING AND SCATTERING REACTIONS

We now consider inelastic scattering of particles with internal degrees of freedom. The particles can be atoms, molecules, atomic nuclei, etc. In the laboratory frame, a standard experiment usually involves one particle, referred to as the projectile, that moves toward another particle at rest, referred to as the target. We distinguish between two cases: In the simpler case, internal degrees of freedom can be excited but the particles retain their identities. A more complicated case is when the target and projectile exchange particles, a process referred to as a *rearrangement collision*. An even more complicated situation occurs when at least one of the two colliding particles breaks up, a process referred to as *breakup reaction*. In case of breakup reactions, there are at least three bodies in the final state. Rearrangement and breakup processes will not be discussed here; their analysis requires new tools that are beyond our scope. Therefore, we restrict ourselves to cases (1) and (2), whereas cases (3) and (4) of the list in Sec. 12.8 are not considered here.

A simple example borrowed from the physics of nuclear reactions can help illustrate the above discussion. Consider the collision between two α particles. The α particle contains two protons and two neutrons. A successful model for describing low-energy states of light and medium nuclei is the so-called *nuclear shell model* that assigns a shell structure to the nuclei, in analogy with the shell structure of atomic electrons. According to this model, the α particle is a closed shell nucleus; hence, it is tightly bound. This is confirmed in experiments, showing that the binding energy of the α particle is about 28 Mev. Consider the collision with the projectile α particle having a kinetic energy of tens of Mev. As a result of the collision, the following final states can be obtained, depending on the value of the scattering energy used: (1) Two α particles that may or may not be in excited states. This is an example of an elastic or inelastic collision, where the latter is written as $\alpha + \alpha \to \alpha' + \alpha'$. (2) A proton and a neutron leave one of the α particles and become bound inside the other α particle. The particles leaving the scattering region are a ^{6}Li nucleus (with three protons and three neutrons) and a deuteron. This is an example of rearrangement reaction, written as $\alpha + \alpha \to {}^6\text{Li} + d$. (3) At least one of the α particles is broken up into several lighter particles. For example, if one of the α particles is broken up into a deuteron, a neutron, and a proton, the breakup reaction is written as $\alpha + \alpha \to \alpha' + d + n + p$. This process is possible only if the kinetic energy of the relative motion of the two particles is higher than the breakup energy of the α particle (the *breakup threshold*). There are many atom–atom, atom–molecule, and molecule–molecule collisions that can also serve as good examples of these types of reactions, but we shall not pause to present additional examples.

For inelastic scattering, the on-shell energy condition (12.500) is satisfied but the relative kinetic energies are not equal.

$$\varepsilon = \varepsilon(\mathbf{k}_a) + \varepsilon_a = \varepsilon(\mathbf{k}_b) + \varepsilon_b, \tag{12.649a}$$

$$\varepsilon(\mathbf{k}_a) \neq \varepsilon(\mathbf{k}_b). \tag{12.649b}$$

In this section, our goal is to modify the general framework introduced in Sec. 12.8.4 to inelastic scattering and to bring it into a workable calculation scheme.

The restriction to pure inelastic scattering means that the Hilbert space of positive energy states can be spanned by any basis composed of channel states $|\mathbf{a}\rangle$ as defined in Eq. (12.505c). Moreover, for any potential V that depends both on the internal coordinates and on the vector $\mathbf{r}$ between the centers of masses of the colliding particles, we can define the potential matrix $v_{ab}(\mathbf{r})$ as specified in Eqs. (12.511). Thus, the problem is formulated within a *multi-channel scattering theory* in the sense that the original Schrödinger equation (12.507) is replaced by a coupled set of equations. Let us expand $\Psi_\mathbf{a}^+(\mathbf{r})$ [whose asymptotic form is given in Eq. (12.508)] in terms of internal states as

$$\Psi_\mathbf{a}^+(\mathbf{r}) = \sum_b |\Phi_b\rangle \psi_{b\mathbf{a}}^+(\mathbf{r}). \tag{12.650}$$

Note that in this expansion, the internal states $|\Phi_b\rangle$ are abstract kets, while the coefficients $\psi_{b\mathbf{a}}^+(\mathbf{r})$ are c numbers (for a given value of the indices, they are numerical functions of $\mathbf{r}$, which will be determined below).

The functions $\psi_{b\mathbf{a}}^+(\mathbf{r})$ satisfy the set of coupled Schrödinger equations,

$$-\nabla^2 \psi_{b\mathbf{a}}^+(\mathbf{r}) + \sum_c v_{bc}(\mathbf{r}) \psi_{c\mathbf{a}}^+(\mathbf{r}) = \varepsilon \psi_{b\mathbf{a}}^+(\mathbf{r}). \tag{12.651}$$

With the asymptotic condition,

$$\psi_{b\mathbf{a}}^+(\mathbf{r}) \xrightarrow[r \to \infty]{} (2\pi)^{-\frac{3}{2}} \left[\delta_{ab} e^{i\mathbf{k}_a \cdot \mathbf{r}} + f_{b\mathbf{a}}(\Omega) \frac{e^{ik_b r}}{r} \right]. \tag{12.652}$$

Equations (12.651) and (12.652) define the multichannel scattering problem.

It is worth mentioning the situation where the scattered particles have both internal energy states and spin. This case is encountered in cold atom physics. The discussion of scattering of particles with spin in Sec. 12.8.8 is a special case of this more general situation, where the internal degrees of freedom include only the spin, and the scattering is elastic (if no external magnetic field is present). The channels of the S matrix are labeled by the spin and other internal degrees of freedom. In Eq. (12.568), the notation (12.505c) for channel states is modified to specify the spin degrees of freedom. Now, we modify it once more and define our channel states (kets) as

$$|\mathbf{a}\rangle = |\mathbf{k}_a s_a v_a a\rangle, \tag{12.653}$$

where a denotes all internal quantum numbers except spin, as in Eq. (12.544). The channel density of states n_a is given in Eq. (12.503). Note that when the on-shell condition (12.545) is imposed, it fixes the magnitude of the channel momenta $k_\mathbf{a}$ [as in Eq. (12.649a)], so that matrix elements of physical interest between kets (12.653) such as S and T matrices depend on $\hat{\mathbf{k}}_a$ and $\hat{\mathbf{k}}_b$ and on energy ε. The relation between the on-shell S and T matrices [analog of Eq. (12.570)] now reads,

$$\langle \hat{\mathbf{k}}_b s_b v_b b | S | \hat{\mathbf{k}}_a s_a v_a a \rangle = \delta_{\hat{\mathbf{k}}_b \hat{\mathbf{k}}_a} \delta_{s_b s_a} \delta_{v_b v_a} \delta_{ba} - 2\pi i \sqrt{n_b n_a} \langle \hat{\mathbf{k}}_b s_b v_b b | T | \hat{\mathbf{k}}_a s_a v_a a \rangle, \tag{12.654}$$

12.8 Particles with Internal Degrees of Freedom

and the unitarity of the on-shell S matrix is as in Eq. (12.571) with channel indices included (in the summation as well). The relation between the on-shell T matrix and the scattering amplitude [the analog of Eq. (12.572)] is

$$f_{ba}(\Omega) = -\frac{(2\pi)^2 m}{\hbar^2} \langle \hat{\mathbf{k}}_b s_b \nu_b b | T | \hat{\mathbf{k}}_a s_a \nu_a a \rangle, \tag{12.655}$$

and the differential cross-section is then given by

$$\frac{d\sigma}{d\Omega} = \frac{k_b}{k_a} |f_{ba}(\Omega)|^2 = \frac{(4\pi)^4 n_b n_a}{k_a^2} |\langle \hat{\mathbf{k}}_b s_b \nu_b b | T | \hat{\mathbf{k}}_a s_a \nu_a a \rangle|^2. \tag{12.656}$$

Partial Wave Analysis

The partial wave analyses for the S matrix, and the T matrix are performed as in Eqs. (12.589) and (12.590). Here, they are written explicitly since the channel momenta $\mathbf{k}_a$ and $\mathbf{k}_b$ might have different magnitudes (even on-shell):

$$\langle \hat{\mathbf{k}}_b s_b \nu_b b | S | \hat{\mathbf{k}}_a s_a \nu_a a \rangle = \sum_{l_b l_a, J} X^J_{l_b s_b \nu_b; l_a s_a \nu_a}(\hat{\mathbf{k}}'; \hat{\mathbf{k}}) S^J_{l_b s_b b, l_a s_a a}(\varepsilon), \tag{12.657}$$

$$\langle \mathbf{k}_b s_b \nu_b b | t | \mathbf{k}_a s_a \nu_a a \rangle = \sum_{l_b l_a, J} X^J_{l_b s_b \nu_b; l_a s_a \nu_a}(\hat{\mathbf{k}}'; \hat{\mathbf{k}}) t^J_{l_b s_b b, l_a s_a a}(k_b, k_a, \varepsilon). \tag{12.658}$$

The relation between partial wave S and t matrices can be deduced from Eq. (12.654),

$$S^J_{l_b s_b b, l_a s_a a} = \delta_{l_b s_b b, l_a s_a a} - 2\pi i \sqrt{n_b n_a} t^J_{l_b s_b b, l_a s_a a} = \delta_{l_b s_b b, l_a s_a a} - 2\pi i \sqrt{n_b n_a} \frac{2m}{\hbar^2} T^J_{l_b s_b b, l_a s_a a}, \tag{12.659}$$

which is the analog of Eq. (12.592). In Eq. (12.659), it is assumed for simplicity that the relative kinetic energy is k^2 as in Eq. (12.500). Once again, it should be stressed that generically, $k_a \neq k_b$, as is stated in Eq. (12.501). Unitarity and symmetry of the S matrix (a consequence of time-reversal invariance) are derived as in the case of elastic scattering (see the general discussion in Sec. 12.8.6). Explicitly, abbreviate $\alpha \equiv l_a s_a a$,

$$\sum_\gamma [S^J_{\gamma\beta}]^* [S^J_{\gamma\alpha}] = \delta_{\alpha\beta}, \tag{12.660}$$

$$S^J_{\alpha\beta} = S^J_{\beta\alpha}. \tag{12.661}$$

The next task is to perform partial wave analysis on the coupled set of Schrödinger Eq. (12.651). The partial wave potential is derived as in Eq. (12.605).

$$v^J_{l''s''a'';l's'a'}(r) = \frac{2s+1}{2J+1} \int d\Omega_{\hat{\mathbf{r}}} d\Omega_{\hat{\mathbf{k}}}$$

$$\left[\sum_{\nu''\nu'} \chi^\dagger_{s''\nu''} X^{J*}_{l''s''\nu'';lsv}(\hat{\mathbf{r}};\hat{\mathbf{k}}) v_{a''a'}(r) X^J_{l's'\nu';lsv}(\hat{\mathbf{r}};\hat{\mathbf{k}}) \chi_{s'\nu'} \right]. \tag{12.662}$$

The expansion of the wave function requires some modifications of the expansion (12.600), which will not be detailed here. The partial wave components of the wave function, $\psi^+_{l's'a';lsa;J}(r)$, satisfy the set of Schrödinger equations,

$$-\frac{d^2}{dr^2} \psi^+_{l's'a';lsa;J}(r) - \frac{2}{r}\frac{d}{dr}\psi^+_{l's'a';lsa;J}(r) + \frac{l'(l'+1)}{r^2}\psi^+_{l's'a';lsa;J}(r)$$

$$+ \sum_{l''s''a''} v^J_{l's'a';l''s''a''}(r) \psi^+_{l''s''a'';lsa;J}(r) = k^2_{a'} \psi^+_{l's'a';lsa;J}(r), \tag{12.663}$$

where the appearance of the channel momenta $k_a'^2$ on the RHS marks one of the differences from Eq. (12.608). In order to relate the wave function to the T matrix, we need to write down the Lippmann–Schwinger equation derived from Eq. (12.663). This requires the free Green's function satisfying

$$\left[-\frac{d^2}{dr^2} - \frac{2}{r}\frac{d}{dr} + \frac{l'(l'+1)}{r^2} - k_{a'}^2\right] g_{l'}^+(r, r'; k_a') = \delta(r - r'), \tag{12.664}$$

with outgoing boundary conditions, which is given in Eq. (12.193). Then one gets, in analogy with Eq. (12.194),

$$\psi_{l's'a';lsa;J}^+(r) = j_l(k_a'r)\delta_{l's'a',lsa} + \sum_{l''s''a''} \int d\bar{r}\, g_{l'}^+(r, \bar{r}; k_a') v_{l's'a';l''s''a''}^J(\bar{r}) \psi_{l''s''a'';lsa;J}^+(\bar{r}). \tag{12.665}$$

Using the formal relation $G_0^+ V|\psi^+\rangle = G_0^+ T|\phi\rangle$ within the Lippmann–Schwinger identities $|\psi^+\rangle = |\phi\rangle + G_0^+ V|\psi^+\rangle = |\phi\rangle + G_0^+ T|\phi\rangle$ [see Eqs. (12.113a) and (12.113c)], we can relate the wave function directly to the partial wave S matrix (or T matrix) introduced via Eq. (12.659),

$$\psi_{l's'a';lsa;J}^+(r) \rightarrow \delta_{l'l}\delta_{s's}\delta_{a'a}j_l(k_a'r) - \pi \frac{e^{ik_a'r - \frac{1}{2}l'\pi}}{r} t_{l's'a';lsa}^J(\varepsilon),$$

which is the analog of Eq. (12.190). Employing the relation (12.654), we finally obtain

$$\psi_{l's'a';lsa;J}^+(r) \rightarrow \frac{1}{2\sqrt{n_a n_{a'}}}\left[-\delta_{l'l}\delta_{s's}\delta_{a'a}\frac{e^{-ik_a r}}{r} + \frac{e^{ik_{a'}r}}{r} S_{l's'a';lsa}^J(\varepsilon)\right]. \tag{12.666}$$

In principle, it is possible to employ the above formalism for the calculation of cross-sections. Once the partial wave potentials are known from Eq. (12.662), the Schrödinger Eq. (12.663) or the Lippmann–Schwinger equations (12.665) should be solved for $\psi_{l's'a';lsa;J}^+(r)$ and the S matrix elements are deduced from the asymptotic behavior (12.666). The t matrix in the $\mathbf{k}sv$ representation is then constructed using Eq. (12.658) to which the scattering amplitude is related via Eq. (12.655).

Let us examine the low-energy behavior of the total cross-section from an initial state (channel) a to a final state a'. For that purpose, it is useful to express the total cross-section in terms of the partial wave T matrix elements. Expanding the RHS of Eq. (12.656) in partial waves and performing the angular integration over $d\Omega$ yield

$$\sigma_{a'a} = \pi \sum_J C_J \frac{n_{a'}n_a}{k_a^2} \sum_{l'l} |T_{l's'a';lsa}^J(\varepsilon)|^2, \tag{12.667}$$

where C_J is a spin-related factor. In many cases, the dominant contribution comes from a given set of quantum numbers J, l', l and summation is then avoided. This has an important consequence at low energy. As we have seen in 12.5.4, e.g., in Eq. (12.236), the low-energy behavior of the partial T matrix for *central potential* scattering is $t_l(\varepsilon) \simeq \epsilon^l = k^{2l}$. This can be extended for the case of non-central potential scattering (including the multichannel partial wave T matrix),

$$T_{l's'a';lsa}^J(k_{a'}, k_a, \varepsilon) \simeq \text{const} \times k_{a'}^{l'} k_a^l, \quad (\varepsilon = k_{a'}^2 + \varepsilon_{a'} = k_a^2 + \varepsilon_a), \tag{12.668}$$

hence,

$$\sigma_{a'a} \simeq \text{constant} \times n_{a'} n_a k_{a'}^{2l'} k_a^{2l-2}. \tag{12.669}$$

Inelastic Scattering: A Single-Channel Formalism

The multichannel scattering formalism developed above is designed to calculate all the scattering amplitudes $f_{ba}(\Omega_{\hat{b}})$ defined in Eq. (12.652), a task that often requires heavy calculational resources. In many cases, we are interested in the scattering amplitude $f_{aa}(\Omega)$, where the initial and final channels are identical, and the effect of coupling to other partial waves or other channels can be encoded in a phenomenological manner. This formalism is now presented.

12.8 Particles with Internal Degrees of Freedom

Partial wave expansion

The technique of eliminating inelastic channels is similar to the procedure employed to analyze Feshbach resonance. To clarify the procedure, consider Eq. (12.663) for the simple case, there is no spin and no internal degrees of freedom, and only two partial waves, l and l', are coupled. Substituting $u_l(r) = \frac{\psi_l(r)}{r}$, we obtain

$$\left[-\frac{d^2}{dr^2} + \frac{l(l+1)}{r^2} + v_{ll}(r)\right] u_l(r) + v_{ll'}(r) u_{l'}(r) = k^2 u_l(r),$$

$$\left[-\frac{d^2}{dr^2} + \frac{l'(l'+1)}{r^2} + v_{l'l'}(r)\right] u_{l'}(r) + v_{l'l}(r) u_l(r) = k^2 u_{l'}(r). \tag{12.670}$$

Now, formally express $u_{l'}(r)$ from the second equation in terms of $u_l(r)$ and substitute it into the first equation. This yields the following equation for $u_l(r)$:

$$\left[-\frac{d^2}{dr^2} + \frac{l(l+1)}{r^2} + v_{ll} + w_{ll;k^2}(r)\right] u_l(r) = k^2 u_l(r), \tag{12.671}$$

in which the coupling with $u_{l'}(r)$ is encoded within the potential,

$$w_{ll;k^2}(r) = v_{ll'}(r) \left[-\frac{d^2}{dr^2} + \frac{l'(l'+1)}{r^2} + v_{l'l'} - k^2 - i\eta_l\right]^{-1} v_{l'l}(r), \tag{12.672}$$

where $\eta_l \to 0^+$. Equation (12.673) is formally exact, but an evaluation of the potential $w_{ll;k^2}(r)$ is very difficult. Its dependence on energy, with the *retardation* boundary condition, causes it to be complex with positive imaginary part. Generalization of this formalism for the realistic case, where several partial waves are coupled, is straightforward. Equation (12.671) is a standard scattering problem for central potentials within a given partial wave l. In analogy with Eq. (12.171),

$$\psi_l(r) = \hat{h}_l^-(kr) + S_l(k)\hat{h}_l^+(kr), \tag{12.673}$$

where $\hat{h}_l^\pm(x)$ are the Ricatti Hankel functions with incoming $(-)$ and outgoing $(+)$ spherical wave boundary conditions. There is, however, an important difference between the present analysis and that of scattering from a spherically symmetric real potential, detailed in Sec. 12.5. Due to the coupling with other partial waves (or other channels), encoded by the (complex) potential $w_{ll;k^2}(r)$, the absolute value of the partial wave S matrix is less than unity.

$$S_l(\varepsilon) = \beta_l(\varepsilon) e^{2i\delta_l(\varepsilon)}, \quad 0 < \beta_l(\varepsilon) < 1, \tag{12.674}$$

where $\varepsilon = k^2$ and the phase shift $\delta_l(\varepsilon)$ is real. Flux is not conserved for each partial wave and $|S_l| < 1$, since part of it, $1 - |S_l|^2$, leaks into other partial waves. In the present formulation, this leakage is due to the fact that the scattering potential is complex with a positive imaginary part.

Problem 12.28

Reflection from an absorptive media. A particle moving in 1D, $-\infty < x < \infty$, is subject to a potential $v(x) = 0$ $(x \le 0)$, $v(x) = iv_0$ $(x > 0)$, and $v_0 > 0$. The wave function for $x \le 0$ is $\psi(x) = e^{ikx} + Re^{-ikx}$, where $k > 0$ is the wavenumber. Let us assume $\hbar^2/(2m) = 1$.

(a) Find $\psi(x)$ for $x > 0$, and the reflection amplitude R. Show that $|R| < 1$.
(b) What happens when $v_0 < 0$.

Answer: (a) $\psi(x) = Ae^{i\kappa x}$, $\kappa = \sqrt{k^2 - iv_0}$, $R = \frac{k - i\kappa}{k + i\kappa}$, $A = 1 + R$.

$\kappa = |\kappa| e^{-i\theta/2}$, $\tan \theta = \frac{v_0}{k^2} > 0$, $|R| = \left| \frac{k - i|\kappa|\cos\frac{\theta}{2} - |\kappa|\sin\frac{\theta}{2}}{k + i|\kappa|\cos\frac{\theta}{2} + |\kappa|\sin\frac{\theta}{2}} \right| < 1$.

(b) For $v_0 < 0$, $|R| > 1$ and the wave function on the right side diverges.

Instead of Eq. (12.165), the scattering amplitude is given by

$$f(\theta) = \frac{1}{2ik} \sum_{l=0}^{\infty} (2l+1)(S_l - 1) P_l(\cos\theta), \tag{12.675}$$

which reduces to Eq. (12.165) when $S_l = e^{2i\delta}$, but takes another form when $S_l = \beta_l e^{2i\delta}$. The differential and total *elastic* cross-sections are

$$\frac{d\sigma_e}{d\Omega} = \frac{1}{k^2} \left| \sum_{l=0}^{\infty} (2l+1)(S_l - 1) P_l(\cos\theta) \right|^2, \tag{12.676}$$

$$\sigma_e = \frac{\pi}{k^2} \sum_{l=0}^{\infty} (2l+1) |1 - S_l|^2. \tag{12.677}$$

Compare these with Eqs. (12.166) and (12.167). Let us also find the *inelastic cross-section*, σ_{in}. We already indicated that $1 - |S_l|^2$ is the fraction of flux that leaks from channel l, so the inelastic cross-section is

$$\sigma_{in} = \frac{\pi}{k^2} \sum_{l=0}^{\infty} (2l+1)\left(1 - |S_l|^2\right), \tag{12.678}$$

and the total cross-section is

$$\sigma_t = \sigma_e + \sigma_{in} = \frac{2\pi}{k^2} \sum_{l=0}^{\infty} (2l+1)(1 - \operatorname{Re} S_l). \tag{12.679}$$

Employing Eqs. (12.675) and (12.679), we arrive at the optical theorem (appropriate for inelastic scattering),

$$\sigma_t = \frac{4\pi}{k} \operatorname{Im} f(0). \tag{12.680}$$

Note that $f(\theta)$ is the scattering amplitude for *elastic scattering*, but when it enters the optical theorem, it is related to the total cross-section, which also includes the inelastic cross-section.

Modified Analytic Properties of Partial Wave Amplitudes

Since $|S_l| < 1$, the expression (12.254) for the wave function in terms of the Jost functions must be modified because it leads to Eq. (12.255), which implies $|S_l| = 1$. Let us replace $\mathcal{F}_l^*(k)$ in Eq. (12.254) by $\mathcal{F}_l(-k)$ and notice from Eq. (12.256), which defines the Jost function, that unlike Eq. (12.257), for a complex potential, $\mathcal{F}_l(-k) \neq \mathcal{F}_l^*(k)$. Accordingly, Eq. (12.258) must be modified,

$$S_l(k) = \frac{\mathcal{F}_l(-k)}{\mathcal{F}_l(k)} = \beta_l e^{2i\delta_l(k)} \equiv e^{2i(\delta_l(k) - i\chi_l(k))}, \tag{12.681}$$

where $\chi_l(k) \equiv \log[\beta_l(k)]/2 < 0$. The poles of the S matrix in the upper half of the complex k plane are no longer pure imaginary as they appear in Fig. 12.9. This can be understood from Eq. (12.256), since for a complex potential, the RHS of this equation does not vanish for $k = i\kappa$, $\kappa > 0$, hence $\mathcal{F}_l(i\kappa) \neq 0$. Equation (12.681) has an important consequence for the partial wave scattering amplitude $f_l(k)$:

$$f_l(k) = \frac{1}{2ik}(S_l - 1) = \frac{\mathcal{F}_l(-k) - \mathcal{F}_l(k)}{2ik\,\mathcal{F}_l(k)} \Rightarrow f_l(k) - f_l(-k) = 2ikf_l(k)f_l(-k). \tag{12.682}$$

12.8 Particles with Internal Degrees of Freedom

After a little algebra, we find that $\left[\frac{1}{f_l(k)+ik}\right] = \left[\frac{1}{f_l(-k)-ik}\right]$. Defining $g_l(k^2) \equiv \left[\frac{1}{f_l(k)+ik}\right]$, which is an even function of k, we find

$$f_l(k) = \frac{1}{g_l(k^2) - ik}. \tag{12.683}$$

Recall that in the case $|S_l| = 1$, we have $g_l(k^2) = k \cot \delta_l(k)$ as in Eq. (12.240). For $|S_l| < 1$, however, $g_l(k^2)$ is not real. Using the expression (12.681) for $|S_l|$ we have $g_l(k^2) = k \cot[\delta_l(k) - i\chi(k)]$. Consequently, Eq. (12.260) is modified:

$$\delta_l(-k) = -\delta_l(k) \pmod{2n\pi}, \quad \chi_l(-k) = -\chi_l(k). \tag{12.684}$$

It should be stressed that only quantities with positive k are physically measurable.

Modified Effective Range Theory

Let us now consider s-wave scattering ($l=0$) at low energy ($k \to 0$) (we omit the ubiquitous angular momentum subscript). From Eq. (12.684), we infer that $\chi(0) = 0$ and $\delta(0) = n\pi$. Let us define the complex function $z(k) \equiv \delta(k) - i\chi(k)$. Following Eq. (12.684), we have $z(k) \sim k$ as $k \to 0$. Hence, $g(k^2) = k \cot z(k)$ is expected to be finite, albeit complex. At low-energy ε (small k), we can define a *complex scattering length*, $a = a_R + ia_I$, and *complex effective range*, $r_0 = r_{0R} + ir_{0I}$,

$$\lim_{k \to 0} k \cot z(k) = -\frac{1}{a} + \frac{1}{2} r_0 k^2 + O(k^4). \tag{12.685}$$

We can now express the low-energy cross-sections for s-wave scattering in terms of the real and imaginary parts of the complex scattering length, $a = a_R + ia_I$. From Eqs. (12.677)–(12.679), adapted for s-wave scattering,

$$\sigma_e = \frac{\pi}{k^2}|1 - S|^2, \quad \sigma_{in} = \frac{\pi}{k^2}(1 - |S|^2), \quad \sigma_t = \frac{2\pi}{k^2}(1 - \text{Re}[S]). \tag{12.686}$$

Problem 12.29

Express a_R and a_I in terms of $\delta(k)$ and $\chi(k)$.

Answer: Use Eq. (12.685) and employ trigonometric identities to get

$$a = a_R + ia_I = -\frac{1}{k}\tan[\delta(k) - i\chi(k)] \approx -\frac{1}{k}[\tan \delta(k) - i\tanh \chi(k) + O(k^3)].$$

It is crucial to note that $\chi(k) < 0$, which implies that $a_I < 0$. The three main results are then formulated as follows:
(1) σ_e is finite as $k \to 0$. To show this, use Eq. (12.681), and Eq. (12.684), $z(k) \sim k$ as $k \to 0$. Therefore, $|1 - S|^2 = |e^{iz(k)} \sin z(k)|^2 \sim k^2$. Using Eq. (12.686), we see that $f(k) = \frac{S-1}{2ik} = \frac{1}{k \cot z(k) - ik} \to -a$; expressing σ_e in terms of scattering length,

$$\boxed{\sigma_e = 4\pi |a|^2 = 4\pi \left(a_R^2 + a_I^2\right).} \tag{12.687}$$

(2) σ_{in} diverges as k^{-1} for $k \to 0$. Intuitively, this seems at odds with the second relation in Eq. (12.686) because $|S| < 1$. However, this relation does not exclude $1 - |S|^2 \sim k$, as shown below. Starting from the expression for the S matrix in Eq. (12.688), we find

$$S \to \frac{1 - ika}{1 + ika}, \quad \Rightarrow \quad |S|^2 = \frac{(1 + ka_I)^2 + ka_R^2}{(1 - ka_I)^2 + ka_R^2}, \quad (a_I < 0). \tag{12.688}$$

Therefore, $1 - |S|^2 = \frac{-4ka_I}{(1-ka_I)^2 + (ka_R)^2} \to -4ka_I$, and expressing σ_{in} in terms of scattering length,

$$\sigma_{in} = \frac{\pi}{k^2}(1 - |S|^2) = -\frac{4\pi a_I}{k} \quad (12.689)$$

as claimed. Note the importance of the sign $a_I < 0$. If $a_I > 0$, $|S| > 1$ and $\sigma_{in} < 0$, which does not make sense.

(3) **The optical theorem**: Let us take the imaginary part of the s-wave scattering amplitude given in Eq. (12.687) and divide it into its elastic and inelastic parts. After some algebra, we find $\text{Im}[f(k)] = k|a|^2 - a_I$ for $k \to 0$; hence

$$\sigma_t = \frac{4\pi}{k}\text{Im}[f(k)] = 4\pi|a|^2 - \frac{4\pi a_I}{k}. \quad (12.690)$$

Thus, elastic scattering occurs even when $a_R = 0$, but there is no inelastic scattering if $a_I = 0$.

Finally, writing the complex potential as $w_{ll;k^2}(r) = (2m/\hbar^2)(V_R + iV_I)$, ignoring the energy dependence of V_I, and denoting by $u(r)$ the (dimensionless) s-wave solution of the radial Schrödinger equation, we state (without proof) the relations between the scattering potential $V = V_R + iV_I$ ($V_I > 0$) and the s-wave inelastic cross section and a_I:

$$\sigma_{in} = \frac{4\pi}{kE}\int_0^\infty dr V_I |u(r)|^2, \quad a_I = -\lim_{E\to 0}\frac{1}{E}\int_0^\infty dr V_I |u(r)|^2. \quad (12.691)$$

12.8.10 SCATTERING FROM A COLLECTION OF IDENTICAL PARTICLES

Numerous scattering experiments involve scattering of a particle from a system of N identical particles (usually $N \to \infty$) whose positions $\mathbf{r}_i$ vary slowly relative to the speed of the scattered particle or are almost fixed in space. Scattering of a fast electron off such a system composed of identical heavy particles is an example of the first scenario, while scattering of electrons (or neutrons or X-rays) from the atoms of solid is an example of the second scenario, where the (average) position vectors $\mathbf{r}_i$ are fixed and form a lattice. These scenarios are schematically illustrated in Fig. 12.24. In both cases, it is assumed that the mass of the target particles is much larger than the projectile mass m, so that recoil of the scatterers is neglected and the reduced mass equals m. The statistics of the target particles does not play any significant role in these experiments.

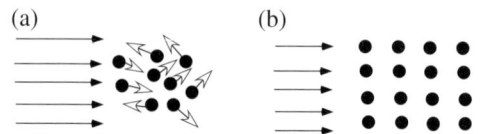

FIG 12.24 Illustration of scattering from a system of particles: (a) Scattering from a system of slowly moving particles whose positions are random (such as liquid or gas). (b) Scattering from a piece of solid material where the particles are fixed in space and form a lattice.

A complete analysis of such scattering is rather complicated. Here, we present a brief description based on the Born approximation (assuming that it can be justified) [190]. According to Eq. (12.123), the Born approximation expresses the scattering amplitude $f(\mathbf{k}', \mathbf{k})$ in terms of the Fourier transform of the potential at momentum transfer $\mathbf{q} = \mathbf{k}' - \mathbf{k}$. In addition to the Born approximation, the basic assumption is that the scattering potential $V(\mathbf{r})$ can be written as a sum of the potentials $v(\mathbf{r} - \mathbf{r}_i)$ of the separate scattering centers

$$V(\mathbf{r}) = \sum_{i=1}^N v(\mathbf{r} - \mathbf{r}_i). \quad (12.692)$$

12.8 Particles with Internal Degrees of Freedom

This approximation is justified when the distances $|\mathbf{r}_i - \mathbf{r}_j| \gg d$, where d is the range of the potential v. According to Eq. (12.123), the scattering amplitude in the Born approximation is

$$f_B(\theta,\phi) = -\frac{m}{2\pi\hbar^2} \sum_{i=1}^{N} \int d\mathbf{r}\, e^{-i\mathbf{q}\cdot\mathbf{r}} v(\mathbf{r}-\mathbf{r}_i)$$

$$= -\frac{m}{2\pi\hbar^2} \int d\mathbf{r}\, e^{-i\mathbf{q}\cdot\mathbf{r}} v(\mathbf{r}) \sum_{i=1}^{N} e^{-i\mathbf{q}\cdot\mathbf{r}_i} \equiv -\frac{\sqrt{2\pi}\,m}{\hbar^2} \tilde{v}(\mathbf{q}) S(\mathbf{q}), \quad (12.693)$$

where (θ,ϕ) are the spherical angles of $\mathbf{q}$, $\tilde{v}(\mathbf{q}) = (2\pi)^{-3/2} \int d\mathbf{r}\, e^{-i\mathbf{q}\cdot\mathbf{r}} v(\mathbf{r})$ is the Fourier transform of $v(\mathbf{r})$, and the *structure factor*, defined as

$$S(\mathbf{q}) \equiv \sum_{i=1}^{N} e^{-i\mathbf{q}\cdot\mathbf{r}_i}, \quad (12.694)$$

encodes the information on the geometry of the system of scatterers. The differential cross-section is

$$\frac{d\sigma}{d\Omega}(\theta,\phi) = 2\left[\frac{\pi m}{\hbar^2}\right]^2 |\tilde{v}(\mathbf{q})|^2 |S(\mathbf{q})|^2. \quad (12.695)$$

It is then interesting to analyze the modulus squared of the structure factor in some detail and, in particular, to distinguish between the two extreme cases according to whether the positions $\mathbf{r}_i$ are completely random or form a crystal.

If the positions are random, it is useful to separate the double sum in $|S(\mathbf{q})|^2 = S^*(\mathbf{q})S(\mathbf{q})$ as

$$|S(\mathbf{q})|^2 = N + \sum_{i\neq j=1}^{N} e^{-i\mathbf{q}\cdot(\mathbf{r}_i-\mathbf{r}_j)}. \quad (12.696)$$

In terms of the *average density of scatterers*

$$\rho(\mathbf{r}) = \frac{1}{N} \sum_{j=1}^{N} \delta(\mathbf{r}-\mathbf{r}_j), \quad (12.697)$$

we can use the randomness property and write

$$\sum_{j\neq i=1}^{N} e^{-i\mathbf{q}\cdot(\mathbf{r}_i-\mathbf{r}_j)} = N \int d\mathbf{r}\, \rho(\mathbf{r}) e^{-i\mathbf{q}\cdot\mathbf{r}}. \quad (12.698)$$

Thus,

$$\frac{d\sigma}{d\Omega}(\theta,\phi) = 2\left[\frac{\pi m}{\hbar^2}\right]^2 N|\tilde{v}(\mathbf{q})|^2 [1 + (2\pi)^{\frac{3}{2}} \tilde{\rho}(\mathbf{q})], \quad (12.699)$$

i.e., the cross-section is proportional to N.

Consider, on the other hand, the situation where the scatterers form a Bravais lattice (see Chapter 9, Sec. 9.3.1),

$$\mathbf{r}_i = n_i \mathbf{a} + m_i \mathbf{b} + l_i \mathbf{c}, \quad (12.700)$$

where $\mathbf{a}, \mathbf{b}$, and $\mathbf{c}$ are the vectors of the *unit cell* and n_i, m_i, and l_i are integers between 0 and $M-1$ such that $M = N^{1/3}$. Then, $S(\mathbf{q})$ is computed as a product of three geometric progressions and its modulus square is

$$|S(\mathbf{q})|^2 = \prod_{\gamma=a,b,c} \left[\frac{\sin(\frac{M\mathbf{q}\cdot\gamma}{2})}{\sin(\frac{\mathbf{q}\cdot\gamma}{2})}\right]^2, \quad (12.701)$$

so

$$\frac{d\sigma}{d\Omega}(\theta,\phi) = 2\left[\frac{\pi m}{\hbar^2}\right]^2 |\tilde{v}(\mathbf{q})|^2 \prod_{\gamma=a,b,c}\left[\frac{\sin(\frac{M\mathbf{q}\cdot\gamma}{2})}{\sin(\frac{\mathbf{q}\cdot\gamma}{2})}\right]^2. \quad (12.702)$$

For large enough M, the function $f(x) = \frac{\sin(Mx/2)}{\sin(x/2)}$ has sharp maxima at $x_n = n\pi$ (n an integer), i.e., $f(x_n) = M^2$, and it is negligible away from these points. Thus, as a function of $\mathbf{q}$, the differential cross-section is appreciable only for vectors such that

$$\mathbf{q}\cdot\gamma = 2n_\gamma \pi \quad (n_{\gamma=a,b,c} \text{ integers}). \quad (12.703)$$

In Problem 12.30 you are asked to determine $\mathbf{q}$. Thus, the analysis of the peaks obtained in particle scattering from a crystal yields structural information about the reciprocal lattice of the crystal (see Sec. 9.3.2).

Problem 12.30

Show that the equality (12.703) is satisfied if $\mathbf{q}$ is a vector in the *reciprocal lattice*, $\mathbf{q} = N_A \mathbf{A} + N_B \mathbf{B} + N_C \mathbf{C}$, where

$$\mathbf{A} = 2\pi \frac{\mathbf{b}\times\mathbf{c}}{\mathbf{a}\cdot\mathbf{b}\times\mathbf{c}}, \quad \mathbf{B} = 2\pi \frac{\mathbf{c}\times\mathbf{a}}{\mathbf{b}\cdot\mathbf{c}\times\mathbf{a}}, \quad \text{and} \quad \mathbf{C} = 2\pi \frac{\mathbf{a}\times\mathbf{b}}{\mathbf{c}\cdot\mathbf{a}\times\mathbf{b}}.$$

12.9 SCATTERING IN LOW-DIMENSIONAL SYSTEMS

So far, we have considered scattering in three dimensions (most experiments involving two colliding particles are carried out in 3D). We have also studied transmission and reflection formulated strictly in 1D in Chapter 1, mostly for pedagogical purposes. Recently, enormous progress in fabrication techniques has enabled the fabrication of systems confined to lower dimension. Hence, it is useful to develop the basic formalism for studying scattering in *low-dimensional systems*.

When a system is fabricated such that the motion is strongly confined along one direction and not confined along the other two directions, a 2D system results; such a 2D system is called a *quantum well*. When two directions are confined, a 1D system or *quantum wire* results, and when all three directions are confined, we have a zero-dimensional system or a *quantum dot*.

The physics of low-dimensional systems has become increasingly important with the discovery of physical phenomena in low dimensions. For example, integer and fractional quantum Hall effects in 2D have been discovered (see Sec. 9.5.8), interacting electrons in quantum wires have been shown to exhibit unusual Luttinger liquid properties (see Sec. 18.15.3 linked to the book web page), and Coulomb blockade (see Sec. 13.3.2), and Kondo effect (see Sec. 18.14) have been discovered in quantum dot systems. In the following sections, we introduce the basic tools required for analyzing scattering in low-dimensional systems.

12.9.1 SCATTERING IN TWO DIMENSIONS

2D scattering by a potential $V(\mathbf{r})$ results when the confinement in the z dimension is so strong that only one mode in the z dimension is populated; therefore, this degree of freedom is frozen out. Cartesian $\mathbf{r} = (x, y)$ or a polar $\mathbf{r} = (r, \theta)$ coordinates can be used. Assume that a scattered particle propagates initially to the right along the x axis (starting with $x < 0$ and very far from the origin $r = 0$). After scattering off the potential centered at the origin, it is observed by a detector situated very far from the origin and with polar coordinates (r, θ). The definitions of particle flux and scattering cross-section must be slightly modified compared with 3D scattering [see the discussion relating to Eqs. (12.52) and (12.167)]: cross-sections now have the units of length instead of area. In polar coordinates, the 2D Laplacian is

$$\nabla^2 = \frac{\partial^2}{\partial r^2} + \frac{1}{r}\frac{\partial}{\partial r} + \frac{1}{r^2}\frac{\partial^2}{\partial \theta^2}, \quad (12.704)$$

12.9 Scattering in Low-Dimensional Systems

and the Schrödinger equation for scattering from a potential $V(\mathbf{r})$, scaled as $v(\mathbf{r}) = \frac{2m}{\hbar^2} V(\mathbf{r})$, is

$$\left[\nabla^2 + k^2 - v(r,\theta)\right] \psi_{\mathbf{k}}^{+}(\mathbf{r}) = 0. \tag{12.705}$$

The asymptotic form of the wave function at large r [the 2D analog of Eq. (12.65)] is given by

$$\psi_{\mathbf{k}}^{+}(\mathbf{r}) \xrightarrow[r \to \infty]{} \frac{1}{2\pi} \left[e^{ikx} + f(\theta) \frac{e^{ikr}}{\sqrt{r}} \right]. \tag{12.706}$$

The direction of the incoming wave is along the x axis, $\mathbf{k} = k\hat{\mathbf{x}}$, θ is the scattering angle and $f(\theta)$ is the two-dimensional scattering amplitude, which has the dimensions of square root of length. The 2D current density,

$$\mathbf{J}(\mathbf{r}) = \frac{1}{m} \psi_{\mathbf{k}}^{+*}(\mathbf{r})[-i\hbar \nabla] \psi_{\mathbf{k}}^{+}(\mathbf{r}), \tag{12.707}$$

has units of (number of particles) per unit length per unit time, and $\mathbf{J}(\mathbf{r}) \cdot d\mathbf{s}$ gives the number of scattered particles per unit time per unit length that cross an infinitesimal circular arc, $d\mathbf{s} = r\, d\theta\, \hat{\boldsymbol{\theta}}$, centered at $\mathbf{r}$, where $d\mathbf{s}$ is perpendicular to $\mathbf{r}$. This is the local particle current, which in 2D is the number of particles crossing the arc $r\, d\theta$ per unit time, $dI(\mathbf{r}) = \mathbf{J}(\mathbf{r}) \cdot d\mathbf{s} = k|f(\theta)|^2 d\theta$. Note that the $r^{-1/2}$ in the denominator on the RHS of Eq. (12.706) is cancelled with the factor r appearing in the arc length. Dividing the current by the incoming flux k and by $d\theta$, we obtain the differential cross-section

$$\frac{d\sigma}{d\theta} = |f(\theta)|^2. \tag{12.708}$$

The total cross-section is given by

$$\sigma = \int_0^{2\pi} d\theta |f(\theta)|^2. \tag{12.709}$$

The formalism developed for studying potential scattering in 3D is easily transformed to the 2D case. The momentum kets $|\mathbf{k}\rangle$ corresponding to the 2D momentum vector $\mathbf{k} = (k_x, k_y)$, are normalized as $\langle \mathbf{k}' | \mathbf{k} \rangle = \delta^{(2)}(\mathbf{k}' - \mathbf{k})$ and the 2D plane wave is $\langle \mathbf{r} | \mathbf{k} \rangle = (2\pi)^{-1} e^{i\mathbf{k} \cdot \mathbf{r}}$. In analogy with Eq. (12.77) valid in 3D, we identify the Green's function of the free Schrödinger equation in 2D in momentum space as

$$\langle \mathbf{k}' | g_0^{+}(\varepsilon) | \mathbf{k} \rangle = \frac{\delta^{(2)}(\mathbf{k}' - \mathbf{k})}{k^2 - \varepsilon - i\eta}. \tag{12.710}$$

The definition holds off the energy shell, $k^2 \neq \varepsilon$, and the small positive imaginary part of the energy ε, $i\eta$, $\eta > 0$, guarantees that the Green's function in position space, $g_0(\mathbf{r}, \mathbf{r}'; k^2)$, has outgoing wave boundary conditions and satisfies,

$$[\nabla^2 + k^2] g_0(\mathbf{r}, \mathbf{r}'; k^2) = \delta^{(2)}(\mathbf{r} - \mathbf{r}'). \tag{12.711}$$

It can be obtained by taking the Fourier transform of $\langle \mathbf{k}' | g_0^{+}(\varepsilon) | \mathbf{k} \rangle$ to obtain

$$g_0^{+}(|\mathbf{r}' - \mathbf{r}|; k^2) = -\frac{i}{4} H_0^{(1)}(k|\mathbf{r}' - \mathbf{r}|), \tag{12.712}$$

which is the 2D analog of Eq. (12.80). Here $H_m^{(1)}(z)$ and $H_m^{(2)}(z)$ are the first and second Hankel functions of order m. In analogy with the discussion of the Lippmann–Schwinger equation, Eq. (12.79), we can write the equation for the wave function $\psi_{\mathbf{k}}^{+}(\mathbf{r})$ appropriate for 2D scattering,

$$\psi_{\mathbf{k}}^{\pm}(\mathbf{r}) = \frac{1}{2\pi} e^{i\mathbf{k} \cdot \mathbf{r}} + \int d\mathbf{r}' \, g_0^{+}(|\mathbf{r} - \mathbf{r}'|; k^2) v(\mathbf{r}') \psi_{\mathbf{k}}^{\pm}(\mathbf{r}'), \tag{12.713}$$

which is equivalent to the Schrödinger equation (12.705), together with the asymptotic boundary conditions (12.706).

To apply Eq. (12.713) for the calculation of the scattering amplitude, we employ the asymptotic form of the first Hankel function [see text following Eq. (1.7) and Eq. (B.23)] and use the expansion (12.81) in the aysmptotic form (B.23)

$$g_0^+(|\mathbf{r} - \mathbf{r}'|; k^2) \xrightarrow[r \to \infty]{} -\frac{e^{i\frac{\pi}{4}}}{4}\sqrt{\frac{2}{\pi k r}} e^{ikr} e^{-i\mathbf{k}' \cdot \mathbf{r}'}, \tag{12.714}$$

where $\mathbf{k}' = k\hat{\mathbf{r}}$. Substituting into Eq. (12.713) and comparing the result with the definition of $f(\theta)$ in terms of $\psi_k^+(\mathbf{r})$ in Eq. (12.706), we obtain the key relation between the scattering amplitude $f(\theta)$ and the scattering potential $v(\mathbf{r})$ in 2D [the analog of Eq. (12.87)]

$$\boxed{f(\theta) = -\sqrt{\frac{i\pi}{2k}} \int d\mathbf{r}' \, e^{-i\mathbf{k}' \cdot \mathbf{r}'} v(\mathbf{r}') \psi_k^+(\mathbf{r}').} \tag{12.715}$$

The Born approximation to the scattering amplitude is obtained by replacing the exact function $\psi_k^+(\mathbf{r}')$ in the integrand by the plane wave, $(2\pi)^{-1} e^{i\mathbf{k} \cdot \mathbf{r}}$,

$$f_B(\theta) = -\sqrt{\frac{i}{8\pi k}} \int d\mathbf{r}' \, e^{-i\mathbf{k}' \cdot \mathbf{r}'} v(\mathbf{r}') \, e^{i\mathbf{k} \cdot \mathbf{r}'} = -\sqrt{\frac{i\pi}{2k}} \tilde{v}(\mathbf{q}), \tag{12.716}$$

where $\tilde{v}(\mathbf{q})$ is the Fourier transform of $v(\mathbf{r})$ and $\mathbf{q} = \mathbf{k}' - \mathbf{k}$ is the (2D) momentum transfer, so the differential cross-section in the Born approximation is $\frac{d\sigma}{d\theta} = \frac{\pi}{2k} |\tilde{v}(\mathbf{q})|^2$.

Coulomb Scattering in Two Dimensions

Consider an incoming beam of particles of charge q_1 with energy $E > 0$ moving in the x, y plane along the x axis from left to right, and scattered by point charge q_2 that is fixed at the origin. The discussion below applies to $q_1 q_2 > 0$ ($q_1 q_2 < 0$) where the Coulomb potential $q_1 q_2 / r$ is repulsive (attractive). A collector located at very large r and at an angle θ relative to the x-axis measures the current of the scattered particles. Denoting $k^2 = \frac{2mE}{\hbar^2}$, $A \equiv \frac{2m q_1 q_2}{\hbar^2}$, $\eta = \frac{m q_1 q_2}{\hbar^2 k}$, the Schrödinger equation reads,

$$\left[-\nabla^2 + \frac{A}{r}\right] \psi(\mathbf{r}) = k^2 \psi(x, y). \tag{12.717}$$

Let us work out the solution of the problem by using plane parabolic coordinates and select the solution appropriate to the physical situation. The parabolic coordinates are defined as $\xi = r - x$ and $\zeta = r + x$, in terms of which the two-dimensional Laplacian reads

$$\nabla^2 = \frac{4}{\xi + \zeta} \left[\xi^{\frac{1}{2}} \frac{\partial}{\partial \xi} \left(\xi^{\frac{1}{2}} \frac{\partial}{\partial \xi} \right) + \zeta^{\frac{1}{2}} \frac{\partial}{\partial \zeta} \left(\zeta^{\frac{1}{2}} \frac{\partial}{\partial \zeta} \right) \right]. \tag{12.718}$$

Substituting into Eq. (12.717) and letting $\psi(\mathbf{r}) = e^{ikx} \Phi(\xi)$ yields the following equation for $\Phi(\xi)$

$$\xi \Phi'' + \left(\frac{1}{2} - ik\xi\right) \Phi' - \eta \Phi = 0, \tag{12.719}$$

whose solution is $N F(-i\eta, \frac{1}{2}, ik\xi)$, where N is a normalization constant and F is the hypergeometric function (see Chapter 5, Morse and Feshbach [26]). Letting $\xi \to \infty$ yields

$$\psi(\mathbf{r}) \to N \frac{\sqrt{\pi} e^{\pi \eta/2}}{\Gamma(\frac{1}{2} + i\eta)} \left[e^{ikx + i\eta \log k\xi} + \frac{e^{ikr - i\eta \log k\xi - i\pi/4}}{\sqrt{k\xi}} \frac{\Gamma(\frac{1}{2} + i\eta)}{\Gamma(-i\eta)} \right]. \tag{12.720}$$

12.9 Scattering in Low-Dimensional Systems

Restoring plane polar coordinates r, θ and identifying the scattering amplitude as the coefficient of $e^{ikr - \eta \log r}/\sqrt{r}$, we find

$$f(\theta) = -\frac{\eta e^{i\frac{\pi}{4}}}{\sqrt{2k \sin^2 \frac{\theta}{2}}} \frac{\Gamma(\frac{1}{2} + i\eta)}{\Gamma(1 - i\eta)} e^{-2i\eta \log \sin \frac{\theta}{2}}. \tag{12.721}$$

Using the identities $|\Gamma(\frac{1}{2} + i\eta)|^2 = \pi/\cosh \pi \eta$, and $|\Gamma(1 - i\eta)|^2 = \pi \eta/\sinh \pi \eta$, we obtain the differential cross-section

$$\frac{d\sigma}{d\theta} = |f(\theta)|^2 = \frac{\eta}{2k \sin^2 \frac{\theta}{2}} \tanh \pi \eta = \frac{|q_1 q_2|}{4E \sin^2 \frac{\theta}{2}} \tanh \frac{\pi m |q_1 q_2|}{\hbar^2 k}. \tag{12.722}$$

Except for the factor $\tanh \pi \eta$, the result coincides with the classical one. This is a beautiful example of quantum versus classical behavior. The differential cross-section for Coulomb scattering in 3D, Eq. (12.304), does not include $\hbar$ and, in fact, is equal to the classical result in Eq. (12.10). This is not the case in 2D since the factor $\tanh \pi \eta$ contains $\hbar$. When $\hbar \to 0$, the factor $\tanh \pi \eta \to 1$ and the classical result is recovered.

Partial Wave Expansion

The analysis leading to Eq. (12.715) is valid for a general potential $V(\mathbf{r}) = V(r, \theta)$ that falls off faster than r^{-2} as $r \to \infty$. In particular, no circular (or cylindrical) symmetry is required. If we limit our discussion to scattering potentials $V(\mathbf{r}) = V(r)$ with cylindrical symmetry, we can employ this symmetry through the technique of partial wave expansion that is now introduced for 2D scattering. In plane polar coordinates r, θ, the free Schrödinger equation has solutions $\psi_m(r, \theta)$, which are eigenstates of L_z with eigenvalue m,

$$\psi_m(r, \theta) = R_m(r) e^{im\theta},$$
$$L_z \psi_m(r, \theta) = \hbar m \psi_m(r, \theta). \tag{12.723}$$

To proceed, the following steps are carried out:

(1) The wave function, the plane wave, and the scattering amplitudes are expanded in the angular function basis $e^{im\theta}$. The partial wave expansion of the wave function reads

$$\psi_\mathbf{k}^+(\mathbf{r}) = \sum_{m=0}^{\infty} R_m(r) e^{im\theta}, \tag{12.724}$$

the expansion of the plane wave is

$$e^{ikr \cos \theta} = \sum_{m=-\infty}^{\infty} i^m J_m(kr) e^{im\theta}, \tag{12.725}$$

and the expansion of the scattering amplitude is given by

$$f(\theta) = \sum_{m=-\infty}^{\infty} a_m e^{im\theta}. \tag{12.726}$$

(2) Expansion (12.724) is inserted into the Schrödinger equation (12.705) and the orthogonality of the angular functions is used to obtain the radial equation

$$\left[\frac{d^2}{dr^2} + \frac{1}{r} \frac{d}{dr} - \frac{m^2}{r^2} + k^2 - v(r) \right] R_m(r) = 0. \tag{12.727}$$

To obtain the asymptotic form of $R_m(r)$, we note that when the potential in Eq. (12.727) is switched off, the equation reduces to the Bessel equation, whose solution, $J_m(kr)$, has the asymptotic form

$$J_m(kr) \xrightarrow[kr \to \infty]{} \sqrt{\frac{2}{\pi kr}} \cos\left[kr - \left(m + \frac{1}{2}\right)\frac{\pi}{2}\right] = -\sqrt{\frac{2}{\pi kr}} \sin\left[kr - \left(m - \frac{1}{2}\right)\frac{\pi}{2}\right]. \tag{12.728}$$

When the $v(r)$ is switched back on, the asymptotic form of $R_m(r)$ becomes,

$$R_m(r) \to -\sqrt{\frac{2}{\pi kr}} A_m \sin\left[kr - \frac{\pi}{2}\left(m - \frac{1}{2}\right) + \delta_m(k)\right], \tag{12.729}$$

where $\delta_m(k)$ is the partial wave phase shift and A_m is a dimensionless constant to be determined. The scattering information is encoded in the set of phase shifts $\delta_m(k)$, $m = 0, \pm 1, \pm 2, \ldots$. As in our discussion for the 3D partial wave expansion in Sec. 12.5.3, all radial wave functions are dimensionless.

(3) Now we substitute all three expansions (12.724), (12.725) and (12.726) into Eq. (12.706), take $r \to \infty$, write the trigonometric function $-\sin[kr - (\pi/2)(m - 1/2) + \delta_m(k)]$ in terms of complex exponentials, and equate the coefficients of e^{ikr} and e^{-ikr} separately to zero (real and imaginary parts). After integrating over angle, this gives four equations for the real and imaginary parts of the unknown coefficients A_m and a_m for each m, which are to be expressed in terms of the phase shift $\delta_m(k)$. The solutions are

$$A_m = e^{i\delta_m(k)}, \quad a_m = \frac{1}{\sqrt{k}} e^{i\delta_m(k)} \sin \delta_m(k). \tag{12.730}$$

(4) The partial wave expansion for the scattering amplitude is then

$$f(\theta) = \frac{1}{\sqrt{k}} \sum_{m=-\infty}^{\infty} e^{i\delta_m} \sin \delta_m e^{im\theta}. \tag{12.731}$$

The total cross-section in terms of phase shifts is given by

$$\boxed{\sigma = \int_0^{2\pi} d\theta |f(\theta)|^2 = \frac{2\pi}{k} \sum_{m=-\infty}^{\infty} \sin^2 \delta_m(k),} \tag{12.732}$$

and the optical theorem reads

$$\boxed{\sigma = \frac{2\pi}{\sqrt{k}} \operatorname{Im} f(0).} \tag{12.733}$$

One of the important applications of scattering in 2D is the scattering of electrons in graphene. We will treat this topic in Sec. 13.6.7.

Example: A Two-Dimensional Square Well

Consider scattering at energy $E = \frac{\hbar^2 k^2}{2m}$ from a two-dimensional (attractive) square well potential,

$$V(r) = \begin{cases} -V_0 & r < \rho, \quad (\rho > 0, \; V_0 > 0). \\ 0, & r > \rho, \end{cases} \tag{12.734}$$

12.9 Scattering in Low-Dimensional Systems

The wavenumber inside the well is $\kappa = \sqrt{k^2 + \frac{2mV_0}{\hbar^2}}$. The wave function of partial wave m outside the well $r > \rho$ is the analogous expression of Eq. (12.171), while inside the well it is the analogous expression of Eq. (12.205),

$$\psi_m(r) = \begin{cases} AJ_m(\kappa r), & (r < \rho), \\ H_m^{(2)}(kr) + S_m H_m^{(1)}(kr), & (r > \rho). \end{cases} \tag{12.735}$$

Following the procedure of matching logarithmic derivative that leads to Eq. (12.210) for scattering from a 3D square well, we arrive at an expression for the phase shift in terms of Bessel and Neumann functions that is analogous to the one displayed in problem (12.13). Denoting $\beta = k\rho, \zeta = \kappa\rho$, we find

$$\tan \delta_m(k) = \frac{kJ'_m(\beta)J_m(\zeta) - \kappa J'_m(\zeta)J_m(\beta)}{kN'_m(\beta)J_l(\zeta) - \kappa J'_m(\zeta)N_m(\beta)}. \tag{12.736}$$

Threshold Scattering

At very low energy, $\beta \ll 1$, only s-wave scattering occurs, hence, only the $m = 0$ phase shift is relevant. We may approximate $\kappa \approx \kappa_0 \equiv \sqrt{\frac{2mV_0}{\hbar^2}}$ and $\zeta \approx \zeta_0 = \kappa_0\rho$. We now use the following approximations for the Bessel functions with small arguments $z \ll 1$ that are valid up to the order z^2, see Abramowitz and Stegun [27], Eqs. (9.1.12) and (9.1.13).

$$J_0(z) = 1 + O(z^2), \quad N_0(z) = \frac{2}{\pi}\left[\ln(\tfrac{1}{2}z) + \gamma\right]J_0(z) + O(z^2),$$

where $\gamma = 0.5772157\ldots$ is Euler's constant. Moreover, we employ the recurrence relation in Eq. (9.1.27) in Abramowitz and Stegun [27],

$$J'_0(z) = -\frac{1}{2}J_1(z) \xrightarrow[z\to 0]{} 0.$$

Substitution into Eq. (12.736) then yields the approximate relation,

$$\tan \delta_0 \approx \frac{\pi \zeta_0 J_1(\zeta_0)}{2[\zeta_0 J_1(\zeta_0) \ln(y\beta) + J_0(\zeta_0)]}, \tag{12.737}$$

where $y = e^\gamma/2$.

The definition of scattering length is more subtle than for three-dimensional scattering. For $m = 0$, denoting $\delta(k) = \delta_0(k)$, the s-wave scattering amplitude is

$$f_0(k) = \frac{e^{i\delta} \sin \delta}{\sqrt{k}} = \frac{1}{\sqrt{k}(\cot \delta - i)}. \tag{12.738}$$

Making a low-energy approximation for the Bessel functions in Eq. (12.736), it can be cast in the general form

$$\cot \delta = c_0 + \frac{2}{\pi} \ln \beta + c_2 \beta^2 + O(\beta^4), \tag{12.739}$$

where c_0 and c_2 are energy-independent constants. Letting $k \to 0$ in Eq. (12.738) does not make sense because the denominator vanishes. The quantity that is left finite as $k \to 0$ is the (dimensionless) constant c_0, i.e.,

$$c_0 = \lim_{k\to 0}\left[\cot \delta_0(k) - \frac{2}{\pi}\ln k\rho\right]. \tag{12.740}$$

As it encodes the low-energy behavior of the scattering amplitude, it can serve as the substitute for the scattering length.

Bound States in 2D Systems

In Sec. 1.3.11, we showed that in 1D, an attractive potential always has a bound state, regardless of how small the well depth or width is. A similar statement holds for 2D, as we shall now show.

Consider, an attractive 2D circular well of radius ρ and potential depth $-V_0 < 0$, in which a particle is bound at energy $-E_B$, where, in this notation $E_B > 0$. Our task is to calculate E_B. The wave function decays exponentially as $r \to \infty$ and is regular at $r \to 0$:

$$\psi_m(r) = \begin{cases} AJ_m(\kappa r), & (r < \rho), \\ H_m^{(1)}(iqr) & (r > \rho). \end{cases} \qquad (12.741)$$

Here $q = \sqrt{\frac{2mE_B}{\hbar^2}}$, $\kappa = \sqrt{\kappa_0^2 - q^2}$, and $\kappa_0 = \sqrt{\frac{2mV_0}{\hbar^2}}$. By matching logarithmic derivatives at $r = \rho$, one obtains a transcendental equation for q from which the binding energy can be found. We will be interested in weak $m = 0$ bound states. In this case, $E_B \ll V_0$, i.e., $q^2 \ll \kappa_0^2$ and we may replace $\kappa \to \sqrt{\kappa_0^2 - q^2} \approx \kappa_0$. The required calculation is outlined in the following problem. Its solution implies that an attractive circular well in 2D always has an $m = 0$ bound state, but that the binding energy vanishes exponentially when the well-depth shrinks. This statement is then generalized for arbitrary attractive potentials.

Problem 12.31

(a) Write the logarithmic derivatives of the wave function inside and outside the well and then match them at $r = \rho$. Using recurrence relations and other identities for the Bessel functions, demonstrate the equality

$$\frac{\zeta_0 J_1(\zeta_0)}{J_0(\zeta_0)} = \frac{iq\rho H_1^{(1)}(iq\rho)}{H_0(iq\rho)} = \frac{q\rho K_1(q\rho)}{K_0(q\rho)},$$

where $K_m(\cdot)$ is the second kind Bessel function.

(b) Assume that E_B is small that is $q\rho \ll 1$ and use the small argument approximation of $K_m(\cdot)$ to show that the RHS is approximately equal to $-[\ln yq\rho]^{-1}$ (where $y = e^\gamma/2$). Then, derive the first equality:

$$E_B = \frac{\hbar^2}{2m\rho^2 y^2} e^{-\frac{2J_0(\zeta_0)}{\zeta_0 J_1(\zeta_0)}} \approx \frac{\hbar^2}{2m\rho^2 y^2} e^{-\frac{4}{(\kappa_0 \rho)^2}}.$$

(c) For weak attractive potential, $\zeta_0 = \kappa_0 \rho \ll 1$, use the small argument approximation for the Bessel functions $J_0(\cdot)$ and $J_1(\cdot)$ and obtain the second expression on the RHS.

More generally, for any attractive 2D potential $V(r)$ of range ρ, assuming that the magnitude of the bound-state energy, $E_B = |E|$, is much smaller than the maximum depth of the potential, V_0, let us assume (this will be confirmed below) that we neglect E on the RHS of the Schrödinger equation, $\frac{1}{r}\frac{\partial}{\partial r}\left(r\frac{\partial}{\partial r}\psi\right) = \frac{2m}{\hbar^2}[V(r) - E]\psi$, within the well, $r \leq \rho$. Multiplying by r and integrating from $r = 0$ to $r = \rho$, we find

$$r\frac{\partial}{\partial r}\psi\bigg|_0^\rho = \frac{2m}{\hbar^2} \int_0^\infty dr\, r V(r)\psi(r), \qquad (12.742)$$

where we have extended the integral from ρ to ∞, despite the fact that the potential vanishes beyond $r = \rho$. For $r \geq \rho$, $\psi(r) = CH_0^{(1)}(i\kappa r)$ and for $\kappa r \ll 1$, $H_0^{(1)}(i\kappa r) \approx \ln(\kappa r)$, where $\hbar^2\kappa^2/(2m) = |E|$. Matching the log derivative of ψ at

12.9 Scattering in Low-Dimensional Systems

$r = \rho$, and using Eq. (12.742), we find, $1/[\rho \ln(\kappa\rho)] = \frac{2m}{\rho\hbar^2} \int_0^\infty dr\, r\, V(r)$, hence

$$|E| = \frac{2\hbar^2\kappa^2}{2m} = \frac{m}{\rho\hbar^2} e^{-(\hbar^2/m)\int_0^\infty dr\, r\, V(r)}. \quad (12.743)$$

Thus, an $m = 0$ bound state always exists, and its energy vanishes as $e^{-4/(\kappa_0\rho)^2}$. It has an essential singularity as the dimensionless parameter $\frac{2mV_0}{\hbar^2}\rho^2 \to 0$.

12.9.2 1D SCATTERING: S MATRIX

In this section, we elaborate upon the theory of quantum scattering in 1D. This analysis is applicable to study electron transport along metallic or semi-conducting wires, which are very narrow so that only one transverse mode is populated. The simplest examples of quantum scattering are problems involving particle reflection and transmission in 1D. Penetration through a barrier is one of the first exercises appearing in quantum mechanical textbooks (see Sec. 1.3.11). Here, however, we have in mind not only scattering from a single barrier but also from a scattering potential that extends along a large part of the system (see below for a more quantitative statement).

Electronic 1D systems have recently become of great interest due to the experimental feasibility of fabricating quantum wires (1D quantum systems), which reveal several remarkable phenomena. For example, interacting electrons in quantum wires constitute a unique physical system called a *Luttinger liquid*, which cannot be described, even approximately, in terms of noninteracting fermions. Its description goes beyond the scope of this chapter. Remarkably, even within the seemingly simple independent particle model, the pertinent physics is rich. The theory discussed below can be tested experimentally due to the insight of Landauer [191], who showed that the quantum mechanical transmission and reflection in 1D scattering devices are directly related to conductance and resistance, which are measurable quantities (see Sec. 13.2). Moreover, once a quantum scattering formalism is developed for strictly 1D systems, it is straightforward to extend it to include scattering in quasi-1D systems, where electrons are confined in the transverse direction but several transverse modes are open. Note that the physics of electrons in quasi-1D systems is similar to the physics of light in waveguides.

In our analysis below, we will also discuss the effect of potential disorder on the transmission and reflection. Strictly speaking, disorder is unavoidable due to imperfections and impurities. In 1D systems, disorder has a dramatic effect, since it localizes the electrons in the sense that the transmission decays exponentially with length. This purely quantum mechanical fundamental phenomenon is referred to as *Anderson localization* [192].

Basic 1D Formalism

Consider first the simplest case of spinless particle of mass m moving in 1D along the x axis, $-\infty < x < \infty$, subject to a bounded potential $V(x)$, which falls off as $|x| \to \infty$ faster than $|x|^{-1}$ when $|x| > X$ where $X \gg \ell$ is some large length scale (here ℓ is the mean free path defined in Chapter 9, Sec. 9.2.1). The time-dependent Schrödinger equation describing the particle dynamics is

$$i\hbar \frac{d\Psi(x,t)}{dt} = H\Psi(x,t) \equiv \left[-\frac{\hbar^2}{2m}\frac{d^2}{dx^2} + V(x)\right]\Psi(x,t). \quad (12.744)$$

The fact that $V(x)$ falls off sufficiently fast beyond $|X|$ enables the definition of asymptotic regions, where $|x| \gg X$ and the effect of the potential $V(x)$ is virtually negligible. The segment $|x| < X$ for which $V(x)$ is not negligible is the scattering region. Within the stationary formulation of scattering, $\Psi(x,t) = e^{-iEt/\hbar}\psi(x)$ is a stationary state at energy $E = \frac{\hbar^2 k^2}{2m} > 0$, and $\psi(x)$ satisfies the (properly scaled) time-independent Schrödinger equation

$$H\psi(x) = \left[-\frac{d^2}{dx^2} + v(x)\right]\psi(x) \equiv [H_0 + v(x)]\psi(x) = k^2\psi(x), \quad (12.745)$$

where $v(x) = \frac{2m}{\hbar^2} V(x)$. A scattering event in one dimension is described as follows: A particle at a given energy, $E > 0$, approaches the scattering region either from the left or from the right, and it is partially reflected and partially transmitted. In the asymptotic regions $|x| > X$, the basic solutions of the stationary Schrödinger equation are plane waves, which we normalize to carry unit current,

$$\psi_0^{\pm}(x) = \frac{e^{\pm ikx}}{\sqrt{k}}, \tag{12.746}$$

$$J^{\pm} = \mathrm{Re}\left[\psi_0^{\pm}(x)^*\left(-i\frac{d}{dx}\right)\psi_0^{\pm}(x)\right] = \pm 1. \tag{12.747}$$

The wave functions in the corresponding (left and right) asymptotic regions are given as linear combinations of these basic states. When the particle approaches the scattering region from the left, the asymptotic form of the wave function is

$$\psi(x) = \psi_0^+(x) + r(k)\psi_0^-(x), \quad (x \to -\infty), \tag{12.748a}$$

$$\psi(x) = t(k)\psi_0^+(x), \quad (x \to \infty). \tag{12.748b}$$

Likewise, when the particle approaches the scattering region from the left, the asymptotic form of the wave function is

$$\psi(x) = \psi_0^-(x) + r'(k)\psi_0^+(x), \quad (x \to \infty), \tag{12.749a}$$

$$\psi(x) = t'(k)\psi_0^-(x), \quad (x \to -\infty). \tag{12.749b}$$

Here, $r(k)$ and $t(k)$ are the reflection and transmission amplitudes for scattering from the left, while $r'(k)$ and $t'(k)$ are the reflection and transmission amplitudes for scattering from the right. Note that, unlike the situation encountered in 3D scattering, the asymptotic wave functions in Eqs. (12.748) and (12.749) are exact eigenfunctions of the kinetic energy operator $H_0 = -\frac{\hbar^2}{2m}\frac{d^2}{dx^2}$.

Transmission Through a 1D Coulomb Potential

As an example, let us consider the transmission through a one-dimensional Coulomb potential barrier,

$$V_C(x) = \frac{A}{|x|}, \quad (A > 0). \tag{12.750}$$

For technical reasons, we first consider transmission through the right part of the barrier,

$$V(x) = \begin{cases} 0 & (x < 0), \\ \frac{A}{x} & (x > 0, \quad A > 0), \end{cases} \tag{12.751}$$

and then use this solution to evaluate the transmission through the potential $V_C(x)$. The Schrödinger equation is

$$\begin{cases} \left[-\frac{\hbar^2}{2m}\frac{d^2}{dx^2}\right]\psi(x) = E\psi(x) & (x < 0), \\ \left[-\frac{\hbar^2}{2m}\frac{d^2}{dx^2} + \frac{A}{x}\right]\psi(x) = E\psi(x) & (x > 0), \end{cases} \tag{12.752}$$

where $E = \hbar^2 k^2/(2m)$. Assuming that the particle approaches the barrier from left to right, the boundary conditions are,

$$\psi(x) = \begin{cases} e^{ikx} + r(k)e^{-ikx}, & (x < 0), \\ t(k)H_0^{(+)}(kx) & (x > 0), \end{cases} \tag{12.753}$$

12.9 Scattering in Low-Dimensional Systems

where $H_0^{(+)}(kx)$ is the Coulomb analog of the first Hankel function. Using $\rho \equiv kx$ and $\eta \equiv \frac{mA}{\hbar^2 k}$,

$$H_0^{(\pm)}(\rho) = G_0(\rho) \pm iF_0(\rho) \xrightarrow[\rho \to \infty]{} e^{\pm i[\rho - \eta \ln 2\rho + \text{Arg}\,\Gamma(1+i\eta)]}, \quad (12.754)$$

which are combinations of the s-wave Coulomb wave functions defined in Eqs. (12.310) and (12.311). The choice of $H_0^{(+)}(kx)$ in Eq. (12.753) dictates a rightward propagating Coulomb wave. The quantities $r(k)$ and $t(k)$ in Eq. (12.753) are the reflection and transmission amplitudes for scattering from left to right, which are to be calculated.

It is tempting to solve the problem directly by matching the logarithmic derivative of $\psi(x)$ at $x = 0$; however, this procedure is problematic because the function $G'(\rho)$ is singular at $\rho = 0$. Specifically, the values of the functions and their derivatives at $\rho = 0$ are

$$F_0(0) = 0, \quad F'(0) = C_0(\eta), \quad G_0(0) = 1/C_0(\eta), \quad G_0'(0) = -\infty. \quad (12.755)$$

In order to circumvent this problem, we consider a potential that is truncated at some small but finite $\delta > 0$ and let $\delta \to 0$ at the end of the calculations (this process is referred to as *regularization*). The truncated right half-potential is

$$V_\delta(x) = \begin{cases} 0 & x \leq \delta, \\ \frac{A}{x} & x > \delta, \end{cases} \quad (12.756)$$

as displayed on the right side of Fig. 12.25. The scattering boundary conditions for the wave function of the truncated potential are

$$\psi(x) = \begin{cases} e^{ik(x-\delta)} + r(k,\delta)e^{-ik(x-\delta)} & x \leq \delta, \\ t(k,\delta)H_0(kx) & x > \delta. \end{cases} \quad (12.757)$$

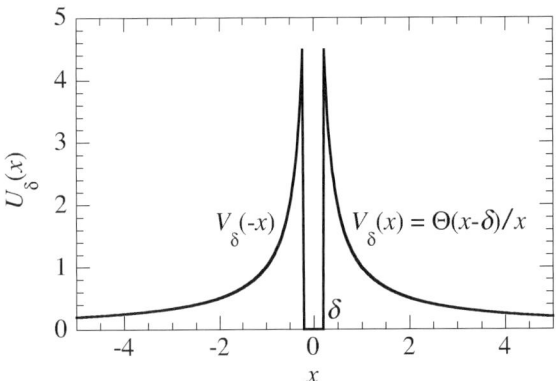

FIG 12.25 The potential $U_\delta(x)$ defined in Eq. (12.761). In the limit $\delta \to 0$ $U_\delta(x) \to A/|x|$, the 1D Coulomb potential $V_C(x)$, Eq. (12.750).

Matching at $x = \delta$ yields

$$1 + r(k,\delta) = t(k,\delta)H_0^{(+)}(k\delta),$$

$$1 - r(k,\delta) = -it(k,\delta)\dot{H}_0^{(+)}(k\delta), \quad (12.758)$$

where $\dot{H}_0^{(+)}(k\delta) \equiv \left[dH_0^{(+)}(\rho)/d\rho\right]_{\rho=k\delta}$. Hence,

$$t(k,\delta) = \frac{2}{H_0^{(+)}(k\delta) - i\dot{H}_0^{(+)}(k\delta)},$$

$$r(k,\delta) = \frac{H_0^{(+)}(k\delta) + i\dot{H}_0^{(+)}(k\delta)}{H_0^{(+)}(k\delta) - i\dot{H}_0^{(+)}(k\delta)}. \quad (12.759)$$

In view of Eq. (12.755), the limit as the truncation parameter $\delta \to 0$ are

$$t(k,\delta) \xrightarrow[\delta \to 0]{} 0, \quad r(k,\delta) \xrightarrow[\delta \to 0]{} -1. \quad (12.760)$$

Thus, the transmission through the half-Coulomb barrier (12.751) vanishes. In order to evaluate the transmission through the full Coulomb potential (12.750), we return to Eq. (12.759), just before taking the limit $\delta \to 0$, and compute the transmission $T(k,\delta)$ through the symmetric potential $U_\delta(x)$ (see Fig. 12.25)

$$U_\delta(x) \equiv V_\delta(x) + V_\delta(-x), \quad (12.761)$$

in terms of $t(k,\delta)$ and $r(k,\delta)$. Clearly, $\lim_{\delta \to 0} U_\delta(x) = V_C(x)$. Employing the technique of summing amplitudes of multiple reflected waves between the two parts of U_δ, the exact expression becomes

$$T(k,\delta) = \frac{t^2(k,\delta)}{1 - r^2(k,\delta)e^{2ik\delta}}. \tag{12.762}$$

Using expressions (12.755), one can show that $\lim_{\delta \to 0} T(k,\delta) = 0$, although this does not directly result from the fact that $t(k,\delta) \xrightarrow[\delta \to 0]{} 0$, because at the same time $r^2(k,\delta)e^{2ik\delta} \xrightarrow[\delta \to 0]{} -1$, and the denominator in Eq. (12.762) tends to be 0.

It is natural to try to apply the WKB approximation to this problem. Since $\int_{-A/b}^{A/b} dx \sqrt{A/x - b} = A\pi/\sqrt{b}$, the tunneling transmission coefficient $|T_{\text{WKB}}(k)|^2$ is finite, hence the WKB approximation fails to produce a correct result. The reason is that the condition for the validity of the WKB approximation,

$$\frac{\hbar}{\sqrt{2m|V_C(x) - E|}} \frac{V'_C(x)}{V_C(x)} \ll 1,$$

is violated as $x \to 0$, since the LHS becomes $\sim x^{-1/2} \to \infty$ as $x \to 0$. Thus, the 1D Coulomb barrier is impenetrable, unlike the case of the delta function potential for which the transmission remains finite (see Sec. 1.3.12). The peculiarities of the Coulomb potential in 2D and 3D scattering problems stem from its slow falloff at large r, but in 1D, the peculiarity is related to the behavior at the origin.

Channels and S matrix in 1D

Scattering in 1D can be regarded as a two-channel problem. Following the strict definition of channels, for the Hamiltonian $H = H_0 + V(x)$, where $V(x) \to 0$ as $x \to \pm\infty$, the channels are different orthogonal eigenstates of H_0, which are plane waves going either to the right, $|k\rangle$, or to the left, $|-k\rangle$, with $\langle -k|k\rangle = 0$. For elastic scattering, the on-shell condition should be obeyed. In this momentum representation, the elements of the S matrix are denoted as $S_{k,k}, S_{-k,k}, S_{k,-k}, S_{-k,-k}$. By convention, in this case, the channels are numbered as $1 = |k\rangle$ and $2 = |-k\rangle$. Historically, however, a somewhat different definition of channels is used in 1D scattering problems, where the two channels are determined by which side the scattered particle is located with respect to the scattering potential (left L or right R). In this scheme, that we call the left-right representation, the elements of the S matrix are denoted as $S_{L,L}, S_{R,L}, S_{L,R}, S_{R,R}$. By convention, in this case, the channels are numbered as $1 = L$ and $2 = R$. In both notations, the right index indicates the incoming (initial) wave and the left index indicates the outgoing (final) wave. Eventually, the decision of which choice to adopt is a matter of convenience.

In the absence of spin, the S matrix carries a channel index and hence it is a 2×2 matrix. This is in contradistinction to problems involving 3D scattering of spinless particles in central potential, for which the S matrix for a given angular momentum quantum number l is a single number $S_l = e^{2i\delta_l}$. Reflection and transmission amplitudes for an initial wave coming from the left (right) are denoted by r and t (r' and t'). Commensurate with our convention, the matrix element S_{ji} represents scattering amplitude from an initial channel i to a final channel j. It will be shown that for spinless particles, if the Hamiltonian is time-reversal invariant, $t = t'$, but we shall keep different notations to treat more general cases later on. Therefore, in the two schemes described above, the S matrix is written, respectively, as

$$S = \begin{pmatrix} t & r \\ r' & t' \end{pmatrix} \quad \text{or} \quad S = \begin{pmatrix} r & t' \\ t & r' \end{pmatrix}. \tag{12.763}$$

Note that in the first scheme, the S matrix reduces to the unit 2×2 matrix when the potential $V(x)$ is turned off, in accordance with the usual convention for S matrices. However, the second scheme is often used in the literature. In some cases, it is useful to express the solutions in the asymptotic regions on the left and right sides of the barrier without specifying the normalization of the incoming wave, i.e., $\psi(x) = A\psi_0^+(x) + B\psi_0^-(x)$ as $x \to -\infty$, and $\psi(x) = C\psi_0^-(x) +$

12.9 Scattering in Low-Dimensional Systems

$D\psi_0^+(x)$, as $x \to \infty$. The S matrix turns the vector of incoming amplitudes A, C into a vector of outgoing amplitudes B, D,

$$S\begin{pmatrix} A \\ C \end{pmatrix} = \begin{pmatrix} B \\ D \end{pmatrix}. \quad (12.764)$$

Problem 12.32

(a) Argue physically that $|t| = |t'|$ and then use the unitarity requirement $SS^\dagger = \mathbf{1}_{2\times 2}$ to show that $|r| = |r'|$.
(b) If time-reversal invariance is conserved, then as will be shown below, $t = t'$. In that case, we may write $t = t' = |t|e^{i\alpha}$ and $r = |r|e^{i\beta}$, $r' = |r|e^{i\beta'}$ to derive a relation between the three phases α, β, β'.

Answer: (b) $2\alpha - (\beta + \beta') = \pi$. Note that if the potential is symmetric, $V(x) = V(-x)$ then $r = r'$ and $\alpha - \beta = \frac{\pi}{2}$.

Symmetries of Transmission and Reflection Amplitudes

To check the symmetries of the scattering amplitudes, it is convenient to consider them within the framework of on-shell representation of the *S operator* denoted as $\hat{S}$, between momentum eigenstates $|\pm k\rangle$, whose configuration space representations are plane waves, $\langle x | \pm k \rangle = \psi_0^\pm(x)$, as defined in Eq. (12.746). Here, $k > 0$ and $\pm k$ refers to wave propagating from left to right ($+$) or from right to left ($-$). Thus,

$$t(k) = \langle k|\hat{S}|k\rangle, \quad t'(k) = \langle -k|\hat{S}| - k\rangle, \quad (12.765a)$$

$$r(k) = \langle -k|\hat{S}|k\rangle, \quad r'(k) = \langle k|\hat{S}| - k\rangle. \quad (12.765b)$$

Note that, here the only direction that the momentum can take is either left or right. Following the analysis of symmetries in three-dimensional scattering [see Eq. (12.547)], the following constraint is imposed by the time-reversal invariance,

$$t(k) = \langle k|\hat{S}|k\rangle = \langle -k|\hat{S}| - k\rangle = t'(k). \quad (12.766)$$

If the potential $V(x)$ is symmetric, $V(x) = V(-x)$, then the following constraint is imposed by invariance under parity,

$$r(k) = \langle -k|\hat{S}|k\rangle = \langle k|\hat{S}| - k\rangle = r'(k). \quad (12.767)$$

Problem 12.33

Show that invariance under parity also implies $t(k) = t'(k)$ even in the absence of time-reversal invariance.

Guidance: Time-reversal invariance reverse momentum directions and swaps initial and final states. Parity just reverses momentum directions. In one-dimensional systems, we see from Eq. (12.766) that the outcome is the same. However, note that the parity conservation holds only if $V(x)$ is symmetric.

Problem 12.34

The *transfer matrix M* turns the vector of the amplitudes (A, B) on the left into a vector of amplitudes (D, C) on the right,

$$M \begin{pmatrix} A \\ B \end{pmatrix} = \begin{pmatrix} D \\ C \end{pmatrix}. \quad (12.768)$$

> The order of components is such that the upper components on the LHS and RHS (A and D) correspond to leftward propagating waves, while the lower components (B and C) correspond to leftward propagating waves. Prove the following expression for M,
>
> $$M = \begin{pmatrix} t - r't'^{-1}r & r't'^{-1} \\ -rt'^{-1} & t'^{-1} \end{pmatrix}, \tag{12.769}$$
>
> then show that
>
> $$\det M = 1, \tag{12.770a}$$
> $$M^\dagger \sigma_z M = \sigma_z, \tag{12.770b}$$
>
> Demonstrate that Eq. (12.770b) results from current conservation (unitarity).

Green's Functions for 1D Scattering

Here, we will briefly develop Green's function formalism for 1D scattering problems [201], starting from the stationary Schrödinger equation (12.745). Green's function with the outgoing wave boundary condition satisfies the following requirements

$$\left(\frac{d^2}{dx^2} + k^2\right) g_0^+(x, x') = \delta(x - x'), \tag{12.771}$$

$$\lim_{x \to \pm\infty} g_0^+(x, x') = e^{\pm ikx} f(x'), \tag{12.772}$$

where $f(x')$ is a bounded function. This Green's function was already obtained in Problem (12.8),

$$g_0^+(x, x') = -\frac{e^{ik|x-x'|}}{2ik}, \tag{12.773}$$

with the "asymptotic" behavior,

$$\lim_{x \to \pm\infty} g_0^+(x, x') = \frac{1}{2ik} e^{\pm ikx} e^{\mp ikx'}. \tag{12.774}$$

The solution to the Lippmann–Schwinger equation derived from Eq. (12.745) with rightward propagating incoming wave is

$$\psi(x) = \frac{e^{ikx}}{\sqrt{k}} - \int dx' g_0^+(x, x') v(x') \psi(x'), \tag{12.775}$$

and the asymptotic form of the wave function is

$$\lim_{x \to \pm\infty} \psi(x) = \psi_0^+(x) - \frac{1}{2i} \psi_0^+(x) \int_{-\infty}^{\infty} dx' \psi_0^{+*}(x') v(x') \psi(x'). \tag{12.776}$$

Comparing Eqs. (12.748a) and (12.748b), we find

$$t = 1 - \frac{1}{2i} \int_{-\infty}^{\infty} dx' \psi_0^{+*}(x') v(x') \psi(x'), \quad r = -\frac{1}{2i} \int_{-\infty}^{\infty} dx' \psi_0^{-*}(x') v(x') \psi(x'). \tag{12.777}$$

12.9.3 1D SCATTERING: ANDERSON LOCALIZATION

In amorphous materials, electrons are subject to potentials that include many small attractive or repulsive centers located at random positions and having random strengths. The electrons are said to be affected by a *disordered potential*. The methods for handling scattering problems with disordered potentials are distinct from those used for treating ordinary potential scattering problems, and must involve a combination of quantum mechanics and statistical analysis of random potentials.

Disorder is often unavoidable. It may result from the presence of impurities or structural defects that usually occupy a sizable portion of the material. Disorder plays a crucial role in contemporary condensed matter physics and is responsible for numerous phenomena, such as disorder-induced localization and metal-insulator transitions. As the term metal–insulator transition implies, the effect of disorder is readily experimentally observable in conductance measurements. If disorder is strong (in the sense explained below), the material will be an insulator at zero temperature. The physics of how disorder and space dimensionality d affect localization of wave functions, and metallic or insulating properties of materials, is very rich; it is referred to as the theory of *Anderson localization* [192–194]. In this section, we analyze the problem of electron scattering from a disordered potential [195, 196] in one dimension. Strictly speaking, a disordered potential as a static distribution of impurities at random positions and random strengths is an over-simplified picture. Roughly speaking, impurities and other imperfections can move randomly from one place to another, but this occurs very slowly and the motion is then characterized by a very large relaxation time τ. During short enough time intervals, $\Delta t \ll \tau$, their positions can be regarded as "frozen"; the parameters determining the strength of disorder are called *quenched variables*. The actual value of a quenched variable (e.g., the electrostatic potential felt by an electron at a given point x) depends on the manner in which the sample was prepared. In another sample, prepared under similar conditions, the same variable may assume a different value. The system is then said to have (quenched) disorder. Each sample might have a different configuration of random impurities or different disorder realizations. The mathematical tool for dealing with such systems is an analysis of the Schrödinger equation with a static disordered potential, but as a result of randomness, the analysis must also include statistics.

Statistical Analysis of a Random Potential

The analysis of scattering from a disordered potential is quite different from that of scattering from a fixed potential due to the introduction of statistics into the problem. The interest is focused not just on a specific solution of a scattering problem for a given potential, but also on the *distribution* of observables calculated within the solution such as reflection and transmission coefficients. In this context, the statistical properties of the disordered potential play an important role. Consider the stationary Schrödinger equation (12.745) and suppose that the potential $v(x)$ is disordered (random). For a *random potential* $v(x)$ there is, in general, no definite functional relation $x \to v(x)$. Rather, there is a set of random potentials $\{v(x_i)\}$ generated by a set of random variables $\{V_i\}$ with joint distribution such that the probability that, at a given set of points x_i, the potential will assume a set of (real) values between $v(x_i)$ and $v(x_i) + dv(x_i)$, that is $P[v(x_1)v(x_2)\ldots]\prod_i d[v(x_i)]$. The Schrödinger equation can formally be solved once a set of values $\{v(x_i)\}$ is specified (this is referred to as a *realization of the random potential*). However, a numerical value of an observable quantity O (e.g., electrical conductance), derived from such a solution, assumes another value once a different realization is employed. In other words, O fluctuates as $\{v(x_i)\}$ fluctuates. Such sample to sample fluctuations are distinct from thermal or quantum fluctuations of a given realization. Since we are interested in disorder-averaged quantities, it is useful to introduce two kinds of averaging procedures, quantum statistical and disorder averaging:

$$\langle O \rangle = \frac{\text{Tr}[Oe^{-\beta H}]}{\text{Tr}[e^{-\beta H}]}, \quad \overline{O} = \int \prod_i d[v(x_i)]\, O[v(x_1)v(x_2)\ldots]P[v(x_1)v(x_2)\ldots]. \tag{12.778}$$

Here, H is the Hamiltonian for a fixed realization of the disordered potential and $\beta = (k_B T)^{-1}$.

Problem 12.35

For a given observable, O, explain the following variances,

$$\sigma_1^2 \equiv \overline{\langle O^2 \rangle} - \overline{\langle O \rangle^2}, \quad \sigma_2^2 \equiv \overline{\langle O \rangle^2} - \left[\overline{\langle O \rangle}\right]^2. \tag{12.779}$$

Answer: σ_1^2 is the variance of quantum fluctuations, while σ_2^2 is the variance due to disorder.

In this section, we focus on quantities whose fluctuations are solely due to disorder. In most physical situations of interest, it is legitimate to consider the random variables $\{V_i\}$ as independent identically distributed random variables, i.e., with equal probability distributions $p[v(x_i)]$. Moreover, the central limit theorem[13] suggests that this distribution is Gaussian. It is also convenient to shift the average value of $v(x_i)$ to zero. Thus, we arrive at the definition of *Gaussian white-noise random potential*,

$$P[v(x_1)v(x_2)\ldots] = \prod_i p[v(x_i)], \tag{12.780a}$$

$$p(v) = \frac{1}{\sqrt{2\pi\sigma^2}} e^{-\frac{v^2}{2\sigma^2}}, \tag{12.780b}$$

$$\overline{v(x)} = 0, \quad \overline{v^2(x)} = \sigma^2, \quad \overline{v(x)v(x')} = 0 \text{ if } x \neq x', \tag{12.780c}$$

where the constant σ^2 represents the degree of potential fluctuations.

Problem 12.36

A random potential is drawn from the distribution $P[v(x)] = \frac{2a^3}{\pi(v(x)^2 + a^2)^2}$, where a is constant. After verifying that $P[.]$ is normalized, calculate $\overline{v(x)}$ and $\overline{v(x)v(x')}$.

Answer: $\overline{v(x)} = 0, \overline{v(x)v(x')} = 0$, for $x \neq x'$ and $\overline{v(x)v(x')} = \frac{5}{4\pi a}$, for $x = x'$.

Tight-Binding Formulation of a Random Potential

Potential scattering problems can be formulated within the tight-binding approximation, a method closely related to a model used in solid-state physics, (see Sec. 9.4.7). In 1D, this approximation is obtained by replacing the continuum variable x by a lattice of equally spaced points $x_n = na$, where a is the lattice constant. The physical motivation for this approach is that in many realistic problems in solid-state physics, the particles (usually electrons) are tightly bound to the atoms composing a Bravais lattice, yet occasionally they hop between atoms. Thus, there is a natural length-scale in the problem, i.e., the lattice constant a, which should be employed in order to construct the tight-binding approximation. In the simplest version of the approximation, the hopping occurs between nearest neighbors only. Technically, this is obtained by replacing the Laplacian in the Schrödinger equation by a three-term difference approximation according to the prescription,

$$\frac{d^2}{dx^2}f(x) \approx \frac{f(x+\Delta x) - 2f(x) + f(x-\Delta x)}{\Delta x^2}. \tag{12.781}$$

---

[13] The central limit theorem gives conditions under which the mean of a sufficiently large number of independent random variables, each with finite mean and variance, are normally distributed.

12.9 Scattering in Low-Dimensional Systems

and taking $\Delta x = a$. Thus, the particle can hop between lattice sites $x_n = na$, with $n = 0, \pm 1, \pm 2, \ldots$. Applying the kinetic energy operator on the wave function then yields

$$-\frac{\hbar^2}{2m}\frac{d^2}{dx^2}\psi(x) \to -\frac{\hbar^2}{2ma^2}[\psi_{n+1} - 2\psi_n + \psi_{n-1}], \qquad (12.782)$$

where $\psi_n = \psi(x = na)$.

In addition to the kinetic energy, the potential energy term is included in the tight-binding approximation by the replacement $V(x_n)\psi(x_n) \to v_n\psi_n$, where v_n are random numbers drawn from a given distribution $p(v_n)$ as explained above. Written in the tight-binding scheme, Eq. (12.745) takes the form,

$$(\mathcal{H}\psi)_n = -[\psi_{n+1} + \psi_{n-1}] + v_n\psi_n = \varepsilon\psi_n, \qquad (12.783)$$

where ε is the scattering energy and the term $-2\psi_n$ on the RHS of Eq. (12.782) is absorbed into the energy. The Hamiltonian so defined is referred to as the *Anderson Hamiltonian* [192]. At this stage, our first task is to solve the scattering problem for a fixed realization of disorder, where $\{v_n\}$ is a fixed set of N real numbers. This procedure is carried out within the *transfer matrix* formalism. Once a solution is obtained and observables are calculated, the averaging procedure can be worked out.

The form of the wave function ψ_n in the different domains is illustrated in Fig. 12.26. For $n < 1$ and $n > N$, the wave function ψ_n is a combination of plane waves,

$$\psi_n = \frac{1}{\sqrt{\sin ka}}\left[Ae^{ikna} + Be^{-ikna}\right], \quad (n < 1), \qquad (12.784a)$$

$$\psi_0 = \frac{1}{\sqrt{\sin ka}}[A + B], \quad \psi_{-1} = \frac{1}{\sqrt{\sin ka}}\left[Ae^{-ika} + Be^{ika}\right], \qquad (12.784b)$$

$$\psi_n = \frac{1}{\sqrt{\sin ka}}\left[Ce^{ik(n-1-N)a} + De^{-ik(n-1-N)a}\right], \quad (n > N), \qquad (12.784c)$$

$$\psi_{N+1} = \frac{1}{\sqrt{\sin ka}}[C + D], \qquad (12.784d)$$

$$\psi_{N+2} = \frac{1}{\sqrt{\sin ka}}\left[Ce^{ika} + De^{-ika}\right], \qquad (12.784e)$$

where the energy is related to k as $\varepsilon = -2\cos ka$, and the current carried by a plane wave in the tight-binding approximation is $\sin ka$, so that the plane waves are current normalized. Solution of the scattering problem for a fixed realization requires the evaluation of the 2×2 transfer matrix M_N relating the complex coefficients A, B on the left to C, D on the right according to Eq. (12.768). The subscript N is to remind us that the disordered potentials occupies N points. In some cases, the limit $N \to \infty$ is examined.

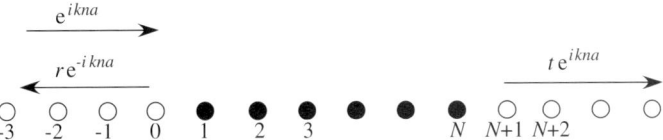

FIG 12.26 Scattering in a 1D system within the tight-binding approximation. Particle positions are represented on a lattice points $x_n = na$, where a is the lattice constant. The wave function satisfies the second-order difference equation (12.783). The scattering region consists of N sites (solid dots) for which $v_n \neq 0$. To its left and right, the potential (site energies) vanishes and the wave function is a combination of plane waves as detailed in the text.

Equation (12.783) can be recast in a transfer matrix form,

$$\begin{pmatrix} \psi_{n+1} \\ \psi_n \end{pmatrix} = \begin{pmatrix} \varepsilon - v_n & -1 \\ 1 & 0 \end{pmatrix} \begin{pmatrix} \psi_n \\ \psi_{n-1} \end{pmatrix} \equiv q_n \begin{pmatrix} \psi_n \\ \psi_{n-1} \end{pmatrix}, \quad (12.785)$$

where the matrices q_n are 2×2 independent *random matrices*. From Eq. (12.785), we obtain

$$\begin{pmatrix} \psi_{N+1} \\ \psi_N \end{pmatrix} = \prod_{n=1}^{N} q_n \begin{pmatrix} \psi_1 \\ \psi_0 \end{pmatrix} \equiv Q_N \begin{pmatrix} \psi_1 \\ \psi_0 \end{pmatrix}. \quad (12.786)$$

These relations can be used for evaluating the transfer matrix M_N by noting that

$$\begin{pmatrix} \psi_1 \\ \psi_0 \end{pmatrix} = \begin{pmatrix} \varepsilon & -1 \\ 1 & 0 \end{pmatrix} \begin{pmatrix} \psi_0 \\ \psi_{-1} \end{pmatrix} = \frac{1}{\sqrt{\sin ka}} \begin{pmatrix} \varepsilon & -1 \\ 1 & 0 \end{pmatrix} \begin{pmatrix} 1 & 1 \\ e^{-ika} & e^{ika} \end{pmatrix} \begin{pmatrix} A \\ B \end{pmatrix}, \quad (12.787a)$$

$$\frac{1}{\sqrt{\sin ka}} \begin{pmatrix} e^{ika} & e^{-ika} \\ 1 & 1 \end{pmatrix} \begin{pmatrix} C \\ D \end{pmatrix} = \begin{pmatrix} \psi_{N+2} \\ \psi_{N+1} \end{pmatrix} = \begin{pmatrix} \varepsilon & -1 \\ 1 & 0 \end{pmatrix} \begin{pmatrix} \psi_{N+1} \\ \psi_N \end{pmatrix}. \quad (12.787b)$$

Using these equations, and taking into account the required order [see remark after Eq. (12.768)], we arrive at the final expression for the transfer matrix

$$M_N = \frac{1}{2i \sin ka} \begin{pmatrix} -1 & e^{ika} \\ 1 & -e^{-ika} \end{pmatrix} \begin{pmatrix} \varepsilon & -1 \\ 1 & 0 \end{pmatrix} Q_N \begin{pmatrix} \varepsilon & -1 \\ 1 & 0 \end{pmatrix} \begin{pmatrix} 1 & 1 \\ e^{-ika} & e^{ika} \end{pmatrix}, \quad (12.788)$$

where $\det M_N = 1$. The calculation of the transfer matrix M_N for a given disorder realization reduces to the problem of evaluating the matrix,

$$Q_N = \prod_{n=1}^{N} q_n = \prod_{n=1}^{N} \begin{pmatrix} \varepsilon - v_n & -1 \\ 1 & 0 \end{pmatrix}. \quad (12.789)$$

For finite N, the product defining Q_N can be easily calculated if N is not too large. Once the transfer matrix is known, the reflection and transmission amplitudes can be deduced by inspecting the elements of M_N and using the expression (12.769). Alternatively, one can show that the transmission coefficient $|t(k)|^2$ is directly related to the transfer matrix M_N via the expression,

$$|t(k)|^2 = 2 \text{Tr} \{ M_N M_N^\dagger + [M_N M_N^\dagger]^{-1} + 2 \}^{-1}. \quad (12.790)$$

As will be explained below, the transmission coefficient is directly related to the electrical conductance via the Landauer formula [191]. Therefore, after proper disorder averaging, the electrical conductance of such a system can be calculated. In most cases, however, the procedure of obtaining $\overline{|t(k)|^2}$ through averaging over disorder, as defined in Eq. (12.778), cannot be carried out in closed form, and numerical calculations should then be used.

The case $N \to \infty$ is of special interest, since it is related to the phenomenon of *Anderson localization*. The product of N independent random matrices defining Q_N according to Eq. (12.789) tends to diverge as mathematically formulated in the following statement [197]. The asymptotic behavior of a product of large number of random matrices q_n generated by independent identically distributed random variables is characterized by the existence of the following limit,

$$\gamma = \lim_{N \to \infty} \frac{1}{N} \log \text{Tr} \prod_{n=1}^{N} q_n > 0. \quad (12.791)$$

For any energy and any non-trivial disorder distribution $p(v_n)$ (not necessarily Gaussian), one can show that $\gamma > 0$. Roughly speaking, the trace of the product diverges as $e^{\gamma N}$. The number γ defined through the limit in Eq. (12.791) is referred to as the *Lyapunov exponent* of the product (of random matrices). Note that the limit does not depend on the realization and is the same for any infinite sequence $\{v_n\}$ of random numbers. The Lyapunov exponent is hence *self-averaging* in the sense that the averaging is achieved by performing the limit on an arbitrary given realization of disorder and dividing by the length of the system. In the language of statistical mechanics, $\log \text{Tr} \prod_{n=1}^{N} q_n$ is an *extensive quantity*.

The consequence of the above theorem is rather profound. Recall from Eq. (12.768) that the transfer matrix satisfies

$$M_N \begin{pmatrix} 1 \\ r_N \end{pmatrix} = \begin{pmatrix} t_N \\ 0 \end{pmatrix}, \tag{12.792}$$

where the subscript N indicates that the amplitudes pertains to a system of length N. Since $\det M_N = 1$, Eq. (12.791) implies that in the large N limit, M_N can be written as

$$M_N = V_N^{-1} \begin{pmatrix} e^{\gamma N} & 0 \\ 0 & e^{-\gamma N} \end{pmatrix} U_N, \tag{12.793}$$

where U_N and V_N are 2×2 unitary matrices. Since the transmission amplitude t_N is bounded (by unitarity), the eigenvalue $e^{\gamma N}$ must not appear in its expression, implying that the reflection amplitude r_N must be so chosen that the upper element of $U \begin{pmatrix} 1 \\ r_N \end{pmatrix}$ vanishes. This means that

$$t_N = C_N e^{-\gamma N}, \tag{12.794}$$

where C_N is an algebraic function of the elements of V_N and U_N. Note that t_N is not a self-averaging quantity. Although it decays exponentially with length, it fluctuates depending on realization. On the other hand, $\log t_N$ *is self-averaging*. Since the wave function for $n > N$ is a product of t_N and a plane wave, the wave function on the RHS of the potential is exponentially small. This is one facet of *Anderson localization*, which, for 1D, states that in a disordered system, all states are localized in the sense that they decay exponentially at large distance from a certain reference point at which their amplitude is finite.

12.9.4 SCATTERING IN QUASI-ONE-DIMENSIONAL SYSTEMS

A quasi-1D system has the geometry of a slab. Particles can propagate along the longitudinal direction of the slab, which we take to be the x axis, while the wave function is confined and square integrable in the other directions by a confining potential. To avoid complicated expressions, we illustrate the main ingredients of single-particle scattering in quasi-1D system employing the geometry of a 2D strip, $-\infty < x < \infty$, $-W/2 \leq y \leq W/2$, where W is the width of the strip. The confinement of the wave function $\Psi(x, y)$ in the lateral direction is implied by imposing Dirichlet boundary conditions $\Psi(x, \pm W/2) = 0$. The scattering potential $V(x, y)$ is active within a finite region around $x = 0$, and falls off faster than $1/|x|$ for large $|x|$, so

$$\lim_{|x| \to \infty} |xV(x, y)| = 0. \tag{12.795}$$

The Schrödinger equation for $\Psi(x, y)$ reads

$$\left[-\frac{d^2}{dx^2} - \frac{d^2}{dy^2} + v(x, y) \right] \Psi(x, y) = k^2 \Psi(x, y), \tag{12.796}$$

where $v(x, y) = \frac{2m}{\hbar^2} V(x, y)$. Far away from the scattering region where $v(x, y)$ can be neglected, separation of variables is employed. The corresponding asymptotic states, which are eigenstates of H_0, are

$$\Phi_n^{\pm}(x, y) = \frac{1}{\sqrt{k_n}} e^{\pm i k_n x} \sqrt{\frac{2}{W}} \sin\left[n\pi (y + W/2)/W \right], \tag{12.797}$$

and these define the various channels. These states are products of current normalized plane waves propagating along x and standing waves in y. The standing wave energy is $\varepsilon_n = \frac{n^2 \pi^2}{W^2}$, when $\varepsilon_n < k^2$, a finite energy remains for the longitudinal motion and the wave can propagate with longitudinal momentum

$$k_n \equiv \sqrt{k^2 - \varepsilon_n}. \tag{12.798}$$

Since the channel states are current normalized, the current carried along the x direction by a state $\Phi_n^\pm(x, y)$ is ± 1. For a given energy $E = \frac{\hbar^2 k^2}{2m}$, the number N of channels is the maximal number n for which k_n is real,

$$N = \left[\frac{kW}{\pi}\right], \tag{12.799}$$

where $[O]$ is the integer value of O. For $n > N$, $\varepsilon_n > k^2$ and k_n in the exponent should be replaced by $i\sqrt{\varepsilon_n - k^2} \equiv i\kappa_n$, $\kappa_n > 0$. The states (12.797) with $n > N$ are referred to as *evanescent waves*. The evanescent waves should not be regarded as channels (although sometimes they are called *closed channels*), since they decay at large distance from the scattering region. Thus, they do not appear in the S matrix. Yet, they need to be included in the expansion of the wave function close to the scattering region.

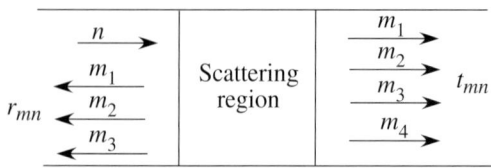

FIG 12.27 Schematic illustration of scattering in a quasi-1D system. The modes labeled m_i on the left and on the right are not the same.

A scattering event in a quasi-1D system is then described as follows: A particle at a given energy $E > 0$ approaches the scattering region (either from the left or from the right) in a given channel n and then it is partially reflected and partially transmitted into all possible channels (see Fig. 12.27). When the particle approaches the scattering region from the left, the asymptotic form of the wave function is

$$\Psi_n(x, y) = \Phi_n^+(x, y) + \sum_m \Phi_m^-(x, y) r_{mn}, \quad (x \to -\infty), \tag{12.800a}$$

$$\Psi_n(x, y) = \sum_m \Phi_m^+(x, y) t_{mn}, \quad (x \to \infty). \tag{12.800b}$$

Likewise, when the particle approaches the scattering region from the left, the asymptotic form of the wave function is

$$\Psi_n(x, y) = \Phi_n^-(x, y) + \sum_m \Phi_m^+(x, y) r'_{mn}, \quad (x \to \infty), \tag{12.801a}$$

$$\Psi_n(x, y) = \sum_m \Phi_m^-(x, y) t'_{mn}, \quad (x \to -\infty). \tag{12.801b}$$

When $|x|$ is large enough, all evanescent modes decay, and the sum in the above equations runs on all open channels. In the strip geometry, the number of open channels on the left and on the right is identical, so that $n = 1, 2, \ldots, N$. There are situations where this is not the case, but we will not discuss them here. The reflection and transmission amplitudes are $N \times N$ matrices. Thus, t_{mn} is the transmission amplitude for scattering of a wave propagating from the left in channel n and partially transmitted to the right in channel m. Similar definitions apply to the matrices t', r, r'. The transmission coefficient from initial channel n to all outgoing channels m is

$$T_n = \sum_m |t_{mn}|^2. \tag{12.802}$$

Since the plane waves are normalized to have unit current, the transmission amplitudes are current normalized, as can be proved by the following exercise.

12.9 Scattering in Low-Dimensional Systems

Problem 12.37

Show that the x components of the current carried by the states (12.800a) and (12.800b) across the strip of width L read

$$J_x^n = \int_{-\frac{W}{2}}^{\frac{W}{2}} dy \, \text{Re}\left[\Psi_n(x,y)^* \left(-i\frac{d}{dx}\right) \Psi_n(x,y)\right] \tag{12.803a}$$

$$= 1 - \sum_m |r_{mn}|^2, \quad (x \to -\infty), \tag{12.803b}$$

$$J_x^n = \sum_m |t_{mn}|^2, \quad (x \to \infty). \tag{12.803c}$$

Guidance: Use expansion (12.797) and employ the orthogonality of the transverse (trigonometric) functions. Current conservation requires that the current is independent of x, hence the two expressions must be equal,

$$\sum_m (|t_{mn}|^2 + |r_{mn}|^2) = 1 \quad \text{for any initial channel } n. \tag{12.804}$$

Summing over all channels $n = 1, 2, \ldots, N$ one gets,

$$\text{Tr}\,[t^\dagger t] + \text{tr}[r^\dagger r] = N. \tag{12.805}$$

The analysis of the 1D scattering problem carried out from Eqs. (12.763) to (12.770b) is also applicable here, but with a few obvious modifications. The S matrix is a $2N \times 2N$ unitary matrix, which has the same form as in Eq. (12.763) with $N \times N$ reflection and transmission matrices. Instead of Eqs. (12.748) and (12.749), it is useful to write the wave function in the asymptotic regions $x \to \pm\infty$ in its most general form

$$\Psi(x,y) = \sum_n \left[a_n \Phi_n^+(x,y) + b_n \Phi_n^-(x,y)\right], \quad (x \to -\infty), \tag{12.806a}$$

$$\Psi(x,y) = \sum_n \left[c_n \Phi_n^-(x,y) + d_n \Phi_n^+(x,y)\right], \quad (x \to \infty). \tag{12.806b}$$

The vectors a, b, c, d with $a^T = (a_1, a_2, \ldots, a_N)$, etc., respectively, define incoming left, outgoing left, incoming right, and outgoing right amplitudes, respectively. Equations (12.764) and (12.768) then hold with the $2N \times 2N$ transfer matrix M as defined in Eq. (12.769), where the order of matrices is important. Although the determinant of M equals unity as in Eq. (12.770a), the conservation of current Eq. (12.770b) now takes the form

$$M^\dagger \Sigma_z M = \Sigma_z, \quad \Sigma_z \equiv \sigma_z \otimes \mathbf{1}_{N \times N} = \begin{pmatrix} \mathbf{1}_{N \times N} & 0 \\ 0 & -\mathbf{1}_{N \times N} \end{pmatrix}. \tag{12.807}$$

Finally, it is worth pointing out that the quasi-1D formalism can also be formulated within the tight-binding approximation. The electron hops on a square lattice of integer pairs (n,m) and the Schrödinger equation reads

$$-t[\psi_{n+1,m} + \psi_{n-1,m} + \psi_{n,m+1} + \psi_{n,m-1}] + v_{nm}\psi_{nm} = E\psi_{nm}, \tag{12.808}$$

where $n = 0, \pm 1, \pm 2, \ldots$ and $m = 0, 1, 2, \ldots, M, M+1$. The hard wall boundary conditions require that $\psi_{n,0} = \psi_{n,M+1} = 0$.

Problem 12.38

For free motion, $v_{nm} = 0$, show that an appropriate solution is and the corresponding energy eigenvalue are, $\psi_{nm} = e^{ik_x n} \sin k_y m$ where $k_y = \frac{2\pi}{M+1}$, and $E(k_x, k_y) = -2t(\cos k_x + \cos k_y)$.

Effect of a Perpendicular Magnetic Field: Edge States

When the 2D strip is subject to a perpendicular magnetic field, we can treat the scattering rather simply using the *Landau gauge*,

$$\mathbf{B} = B\hat{\mathbf{z}} = \nabla \times \mathbf{A} = \nabla \times By\hat{\mathbf{x}}. \tag{12.809}$$

The first task is to find the magnetic analogs of the plane waves (12.797) in a 2D strip ($-\infty < x < \infty$, $-W/2 \leq y \leq W/2$). Recall the definition of the *magnetic length* $l \equiv \sqrt{\frac{\hbar c}{eB}}$, the Schrödinger equation at a given scattering energy $E = \frac{\hbar^2 \varepsilon}{2m}$ and the corresponding boundary conditions read

$$\left[\left(-i\frac{d}{dx} - \frac{y}{l^2}\right)^2 - \frac{d^2}{dy^2}\right]\Psi(x,y) = \varepsilon \Psi(x,y), \tag{12.810a}$$

$$\Psi(x, -W/2) = \Psi(x, W/2) = 0. \tag{12.810b}$$

Translational invariance along x enables the separation of variables as

$$\Psi(x,y) = e^{ikx} \psi_k(y). \tag{12.811}$$

Equation (12.810a) turns into a 1D harmonic oscillator in the variable y centered at $y_c \equiv kl^2$ with hard wall boundary conditions at $y = \pm W/2$,

$$\left[-\frac{d^2}{dy^2} + \frac{1}{l^4}(y_c - y)^2\right]\psi_k(y) = \varepsilon \psi_k(y), \tag{12.812a}$$

$$\psi_k(-W/2) = \psi_k(W/2) = 0. \tag{12.812b}$$

Although the wave function is separable as in Eq. (12.811), the energy cannot be written as a sum of longitudinal and transverse contributions. The scattering energy ε is just the Fermi energy of the scattered electron, and our task is to find the channel momenta $k_n(\varepsilon)$ in analogy with Eq. (12.798). This can be achieved by considering Eqs. (12.812a) and (12.812b) as an eigenvalue problem with k dependent eigenvalues $\varepsilon_n(k)$. Then, the channel momenta are obtained as solutions of the implicit equations $\varepsilon_n(k) = \varepsilon$.

The eigenvalue problem implied by Eqs. (12.812a) and (12.812b) is not simple to solve. Indeed, in the absence of boundaries, the system extends from $-\infty < y < \infty$, so the boundary conditions on $\psi_k(y)$ are $\psi(\pm \infty) = 0$ and the eigenvalues are simply the Landau energies, $\varepsilon_n = (2n+1)/l^2$, independent of $y_c = kl^2$. However, the presence of boundaries changes this infinite degeneracy and the energies become dependent on $y_c = kl^2$. The general solution of Eq. (12.812a) is a superposition of parabolic cylinder functions, and the eigenvalues $\varepsilon_n(k)$ are obtained by requiring that this superposition vanishes at $y = \pm W/2$. The corresponding eigenfunctions $\psi_{kn}(y)$ are orthogonal to each other. When $|y_c - \frac{W}{2}| < l$, the oscillator center is $y_c = kl^2$ is located within a magnetic length from the edge, so that $\psi_{kn}(y)$ are localized near the edge of the strip at $y = \pm W/2$. Hence, these states are referred to as *edge states*, and they play a fundamental role in the theory of the quantum Hall effect [122, 198, 200], and also in the description of recent experiments with electron interferometers. An approximate formula for $\varepsilon_n(k)$ has recently been proposed based on the semiclassical (WKB) approximation. The first few eigenvalues are displayed in Fig. 12.28.

12.9 Scattering in Low-Dimensional Systems

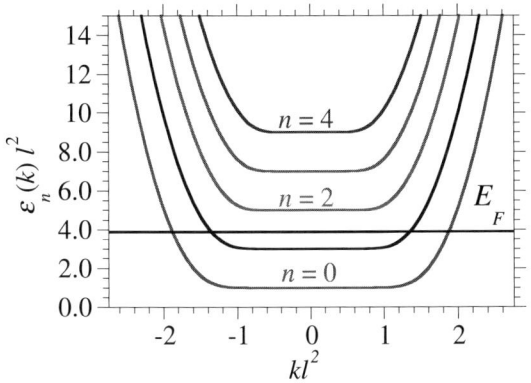

FIG 12.28 Eigenvalues $\varepsilon_n(k)$ for Eqs. (12.812a) and (12.812b) that model an electron in an infinite strip subject to a perpendicular magnetic field corresponding to a magnetic length $l^2 = 20$. Using the magnetic length as unit, the width of the strip is $W = 2.5l$. The dimensionless quantities $\varepsilon_n(k)l^2$ are displayed as a function of kl^2 for $n = 0, 1, 2, 3, 4$ from bottom up. The values of k corresponding to the boundaries are $k = \pm 1.25l^{-1}$. Near these values of k, the corresponding states are *edge states* propagating along the boundary of the strip and hardly scattered by impurities. For k close to 0, the eigenfunctions almost do not feel the edges and the corresponding energies are the degenerate Landau levels $\varepsilon_n = (2n+1)/l^2$.

Returning now to the problem of determining the channel momenta $k_n(\varepsilon)$, draw a horizontal line of height ε (the Fermi energy of the scattered electron) and determine the $k_n(\varepsilon)$ for which this line intersects the energy curves. It is evident from this procedure that if $k_n(\varepsilon)$ is a solution, so is $-k_n$. This symmetry is due to the reflection symmetry of the strip. The fact that the momenta have different signs is robust and means that edge states propagate on both edges in opposite directions. The asymptotic form of the wave function in the presence of a perpendicular magnetic field at a given scattering energy ε is given by

$$\Psi_n^{\pm}(x, y) = e^{\pm i k_n(\varepsilon) x} \psi_{\pm k_n(\varepsilon) n}(y). \tag{12.813}$$

The number of channels is equal to the number of bulk Landau levels below the Fermi energy ε,

$$N = \text{Int}\left[\frac{l^2 \varepsilon + 1}{2}\right]. \tag{12.814}$$

Problem 12.39

Find the current of the state $\Psi_n^+(x, y)$ along the x direction.

Solution: Following the definition of the current in the presence of a vector potential

$$J_n = \text{Re}\left[\int_{-\frac{W}{2}}^{\frac{W}{2}} dy\, \Psi_n^+(x,y)^* \left(-i\frac{d}{dx} - \frac{y}{l^2}\right) \Psi_n^+(x,y)\right] = k_n - \frac{1}{l^2}\langle\psi_{k_n}|y|\psi_{k_n}\rangle,$$

since $\psi_{k_n}(y)$ is real and normalized to 1 on the interval $[-\frac{W}{2}, \frac{W}{2}]$. Note that the current operator contains a contribution from the magnetic field.

The expression for the current as presented in the problem can also be derived by noting the relations

$$\varepsilon_n(k) = \langle\psi_{kn}| -\frac{d^2}{dy^2} + \left(k - \frac{y}{l^2}\right)^2 |\psi_{kn}\rangle, \tag{12.815a}$$

$$\frac{\partial \varepsilon_n(k)}{\partial k} = \varepsilon_n(k) \frac{\partial \langle\psi_{kn}|\psi_{kn}\rangle}{\partial k} + \langle\psi_{kn}|2\left(k - \frac{y}{l^2}\right)|\psi_{kn}\rangle = 2\left[k - \frac{1}{l^2}\langle\psi_{kn}|y|\psi_{kn}\rangle\right]. \tag{12.815b}$$

Applying the Hellman–Feynman theorem (see Sec. 11.3.3) on relation (12.815a) leads to the useful expression

$$J_n = \frac{1}{2}\left[\frac{\partial \varepsilon_n(k)}{\partial k}\right]_{k=k_n(\varepsilon)}. \tag{12.816}$$

Qualitatively, the value of the derivative can be deduced from Fig. 12.28. The current is appreciable near the edges and virtually vanishes at the bulk because $\varepsilon_n(k)$ is flat there.

The Role of Spin in Quasi-1D Scattering

This study is useful to analyze the propagation of electrons in solid materials with the appropriate geometry. In these systems, there are spin-dependent interactions that should be taken into account. The first example of spin-dependent interaction enters through a Zeeman term

$$v_Z = -g\mu_B \mathbf{s} \cdot \mathbf{B}, \tag{12.817}$$

which breaks time-reversal invariance. Here, $g \approx 2$ is the anomalous g factor, μ_B is the Bohr-magneton, and $\mathbf{s}$ is the electron spin operator. As a second example, we note that in many semiconductors, spin–orbit interaction is present, as we discussed in Sec. 9.6.6. Therefore, a short discussion on transmission and reflection of electrons in regions where spin–orbit interaction is appreciable is useful. Unlike the interactions involving the presence of magnetic field, the spin–orbit interaction conserves time-reversal invariance. In semiconductors, it might assume various forms, most notably *Rashba* or *Dresselhouse* interaction, defined by

$$v_R = \alpha_R(p_x\sigma_y - p_y\sigma_x), \tag{12.818a}$$
$$v_D = \alpha_D(p_x\sigma_x - p_y\sigma_y), \tag{12.818b}$$

where α_R and α_D are the corresponding interaction strength constants.

If the spin–orbit potential is active within a finite domain, say $|x| < X$ then for $|x| > X$ spin is conserved and the channel wave functions are written as $\Phi_{n\sigma}^{\pm}(x,y) = \Phi_n^{\pm}(x,y)\chi_\sigma$, where $\Phi_n^{\pm}(x,y)$ are defined in Eq. (12.797) and χ_σ is a spin function. The quantum numbers defining channels then include the energy quantum number n and the spin projection $\sigma = \pm 1/2$ along a given axis (usually z perpendicular to the plane of the strip). Thus, $t_{m\sigma'n\sigma}$ is the transmission amplitude for the scattering of a wave incoming from the left in channel $|n\sigma\rangle$ and leaving rightward in channel $|m\sigma'\rangle$. In particular, the case $\sigma' = \bar{\sigma} = -\sigma$ corresponds to transmission involving spin flip.

Symmetries of Transmission and Reflection Amplitudes

Now, we are in a position to analyze the symmetries of the transmission and reflection amplitudes under discrete symmetry operations. Our system is composed of an electron in a strip subject to a potential $v(x, y)$ that might depend both on space and spin. This dependence is due to scattering centers inside the material and/or to the application of perpendicular magnetic and/or electric fields. As far as space inversion symmetry is concerned, it is hardly expected that the scattering potential will have any kind of space symmetry, so that what is left is time-reversal symmetry, which is broken by the magnetic field.

In analogy with Eqs. (12.765), the relation of reflection and transmission amplitudes at energy ε to the S operator for quasi-1D scattering from an initial channel $|n\sigma\rangle$ into a final channel $|m\sigma'\rangle$ (here $\sigma, \sigma' = \pm\frac{1}{2}$) can be written as,

$$t_{m\sigma'n\sigma} = \langle k_m\sigma'|\hat{S}|k_n\sigma\rangle, \quad t'_{m\sigma'n\sigma} = \langle -k_m\sigma'|\hat{S}|-k_n\sigma\rangle, \tag{12.819a}$$
$$r_{m\sigma'n\sigma} = \langle -k_m\sigma'|\hat{S}|k_n\sigma\rangle, \quad r'_{m\sigma'n\sigma} = \langle k_m\sigma'|\hat{S}|-k_n\sigma\rangle. \tag{12.819b}$$

Consider first the situation where there is no magnetic field and time reversal is conserved. We have already shown that the invariance of the S operator under time-reversal implies the following constraint,

$$\langle \mathbf{q}\sigma'|\hat{S}|\mathbf{k}\sigma\rangle = (-1)^{\sigma-\sigma'}\langle -\mathbf{k}-\sigma|\hat{S}|-\mathbf{q}-\sigma'\rangle. \tag{12.820}$$

12.9 Scattering in Low-Dimensional Systems

Employing the relation (12.820) for the relations (12.819a), (12.819b) leads to the following symmetry relations,

$$t_{m\sigma'n\sigma} = (-1)^{\sigma-\sigma'} t'_{n\bar{\sigma}m\bar{\sigma}'}, \quad r_{m\sigma'n\sigma} = (-1)^{\sigma-\sigma'} r_{n\bar{\sigma}m\bar{\sigma}'}, \quad (12.821)$$

where $\bar{\sigma} = -\sigma$. An immediate consequence of the second relation is that there is no reflection involving spin-flip at the same channel,

$$r_{n\sigma n\bar{\sigma}} = 0. \quad (12.822)$$

When a magnetic field $\mathbf{B}$ (not necessarily homogeneous) is present, time-reversal invariance is broken. The eigenfunctions of the system Hamiltonian $H(\mathbf{B})$ are transformed by the time-reversal operation into those of $H(-\mathbf{B})$. Consequently, relations (12.821) should be modified as,

$$t_{m\sigma'n\sigma}(\mathbf{B}) = (-1)^{\sigma-\sigma'} t'_{n\bar{\sigma}m\bar{\sigma}'}(-\mathbf{B}),$$
$$r_{m\sigma'n\sigma}(\mathbf{B}) = (-1)^{\sigma-\sigma'} r_{n\bar{\sigma}m\bar{\sigma}'}(-\mathbf{B}), \quad (12.823)$$

which is a simple consequence of the Onsager relations, derived in Sec. 7.9.3. Further constraints might appear if the scattering potential, the magnetic field, and the geometry of the system have additional space symmetries, but this is not always experimentally achievable.

Transmission and reflection in quasi-1D disordered systems can be analyzed in a fashion similar to the strictly one-dimensional formulation. However, in a quasi-1D system, some of the waves are evanescent and cannot propagate very far. This happens for mode n when $\varepsilon_n > k^2$ and the momentum k_n defined in Eq. (12.798) is pure imaginary, as discussed following Eq. (12.799). The transmission problem then includes the propagation of waves from one interval to another, including evanescent modes. On the other hand, unitarity relations and current calculations (12.803a) should be formulated solely in terms of propagating modes for which k_n is real, i.e., the S matrix involves only open channels. This makes the use of the transfer matrix method for the solution of the transmission problem in quasi-1D systems rather cumbersome. However, in disordered systems, the fact that the potential is random implies that the transfer matrices are random as well, and one can derive many useful results based on the theory of *random transfer matrices* [209]. This is discussed in Sec. 13.4.2.

Electron localization occurs also in quasi-1D systems. In particular, it is possible to formulate the problem within a tight-binding approximation, such that the number M of transverse modes (propagating and evanescent) is finite. In the absence of spin degrees of freedom, the transfer matrices are of $2M \times 2M$ order and the Lyapunov exponent exists as defined in Eq. (12.791). Exceptions might occur at the band center, but these are artifacts of the tight-binding approximation.

Low-Dimensional Quantum Systems

13

The focus of condensed matter physics was, until recently, on bulk 3D systems. This focus has broadened during the last few decades to include the properties of low-dimensional systems having electrons (and holes and excitons) confined to move in a plane ($d = 2$), along a wire ($d = 1$), or within a quantum dot ($d = 0$). Low-dimensional systems have been realized in semiconductors and in ultra-cold atomic systems confined in optical lattices.[1] Experimental and theoretical progress have gone hand in hand with new revelations, and surprising results are rapidly emerging. The progress in experimental techniques has made it possible to fabricate novel semiconductor reduced dimensional structures with submicron resolution. This has led to the understanding of fundamental physical concepts in transport in low-dimensional structures that are markedly distinct from their analogs in 3D (bulk) systems.

Low-dimensional systems hold great promise in the quest for smaller and faster electronic and spintronic devices. They are natural candidates for miniaturization of solid-state devices whose behavior will be based on quantum mechanics. Such structures are expected to be the building blocks for the next generation of electronic and optoelectronic devices ranging from ultrafast optical switching devices to ultra-dense memory devices.

Low-dimensional systems can be classified according to the dimension of confinement. When electrons are confined within a very small segment of, say, the *z*-axis, but are free to move in the *x-y* plane, the structure is referred to as a *quantum well*. Electrons or holes confined in a quantum well form a *2D electron gas* or hole gas. When electrons are confined in two directions (say *y* and *z*) but are free to move along *x*, a *quantum wire* results. When the motion of electrons is confined in all three directions, a *zero-dimensional system*, or *quantum dot*, results. The single-particle energy levels of charge carriers in quantum dots are quantized; therefore, quantum dots are sometimes referred to as *artificial atoms*.

This chapter begins with an introduction to mesoscopic systems, a topic that has emerged as an important branch of condensed matter physics. Section 13.1 also presents a brief description of quantum wells, quantum wires, quantum dots, heterostructures, superlattices, and quantum point contacts. Section 13.2 introduces the *Landauer formula*, which establishes a crucial relation between quantum mechanical scattering in quasi-1D systems and the experimentally measured conductance. It was originally developed (and will be presented here) for noninteracting systems (its extension to interacting systems requires *many-body Green's functions*). The multi-port extension of the Landauer formula is also introduced in this section. Conductance of Aharonov-Bohm interferometers, quantum point contacts and three-port systems is calculated using the Landauer formula. Section 13.3 presents quantum dots and covers, among other topics, the phenomena of *Coulomb blockade* and spin physics. The treatment of the electron–electron interaction responsible for Coulomb blockade is introduced on a classical level according to which the quantum dot behaves as a charged capacitor. Disorder effects in mesoscopic systems and quantum dots in particular, are discussed in Sec. 13.4, and random matrix theory is used to predict their behavior of such systems. The Kondo effect in quantum dots is described in Sec. 13.5 within a first quantization formulation. A more elegant description requires the formalism of second quantization employed for studying the many-body problem, as detailed in Sec. 18.13 which is linked to the book web page. Section 13.6 presents graphene (a monolayer of graphite having a 2D honeycomb lattice of carbon atoms); this section will focus mainly on its electronic properties. The discovery and fabrication of graphene was an important development in low-dimensional systems. Graphene has remarkable electronic properties. It has a very high electron mobility, $\mu \approx 2 \times 10^4$ cm^2/(Vs). Under favorable conditions, the dynamics of electrons in graphene are governed by a 2D Dirac Hamiltonian. The Dirac electron theory is presented within this section. Finally, in Sec. 13.7, we briefly survey some novel phenomena that occur in low-dimensional systems, including Wigner crystals, exotic concepts associated with the fractional quantum Hall effect (composite fermions and fractional statistics), high temperature superconductivity, 1D spin systems, and finally, the spin Hall effect.

---

[1] This subfield, which belongs to both condensed matter physics and atomic physics, is one of the most rapidly growing areas of contemporary physics.

Quantum Mechanics with Applications to Nanotechnology and Information Science
Copyright © 2013 Elsevier Ltd. All rights reserved.

13.1 MESOSCOPIC SYSTEMS

A *mesoscopic system* is one whose linear size L ranges from a few Angstroms to a few micrometers, and *quantum coherence* is maintained throughout the entire system. A quantum mechanical wave function is coherent only as long as the phase of the wave is well defined (and therefore, it can interfere with other waves). However, spatial coherence cannot be maintained over infinite length. At some length scale (which depends on temperature T and space dimension d), coherence is lost, and the behavior of the system ceases to be purely quantum mechanical. The existence of a finite *coherence length*, denoted by ℓ_φ, leads to the distinction between coherent and incoherent regimes of propagation. For microscopic systems such as atoms and molecules, coherence is typically maintained, and the physics is governed fully by quantum mechanics. Macroscopic systems (e.g., a piece of solid material) generally contain a huge number of particles, and their linear size is much larger than the coherence length of the system. Hence, the existence of a coherence length is relevant in condensed matter physics, and in particular, for electrons propagating in solids. Coherence and decoherence will be extensively discussed in Chapter 17 to be linked to the book web page; these are fundamental concepts in quantum mechanics that have great bearing on mesoscopic physics.

The study of mesoscopic systems has become an integral part of condensed matter physics. We will focus below on the application of scattering theory in quasi-1D and -2D systems and to the calculation of experimental observables that can be measured in mesoscopic devices. The theoretical treatment of these processes is based on the relation between quantum mechanical transmission and electrical conductance, which has been derived by several authors, most notably, Rolf Landauer (for additional material beyond our presentation, the reader can consult Refs [131, 202, 203]).

There are at least two reasons for the recent interest in mesoscopic systems. First, there has been significant progress in fabrication techniques. An important achievement in this context is the possibility of attaching electrodes and measuring the electrical conductance of micro- and nano-sized systems at very low temperature where coherence is maintained and the conductance is determined almost entirely by the laws of quantum mechanics. One of the intriguing results of these measurements is that electrical conductance at the quantum level does not follow the classical laws of addition of resistance, neither in series nor in parallel. A second important experimental feature of mesoscopic systems is that the density of electrons in the system can be controlled by gate potentials and the effect of external fields can be traced very precisely. Therefore, mesoscopic systems are ideal experimental and theoretical sandboxes to test fundamental concepts in quantum mechanics, such as decoherence, quantum dissipation, and Aharonov–Bohm effects, edge-state properties, and electron–electron interaction effects. Moreover, disorder plays an important role in mesoscopia. Performing measurements on different samples of the same material with the same geometry yields different results that fluctuate around a certain average, a property referred to as *sample-to-sample fluctuations*. Plays an important role in mesoscopia. Performing measurements on different samples of the same material with the same geometry yields different results that fluctuate around a certain average, a property referred to as *sample-to-sample fluctuations*.

Length Scales in Mesoscopic Systems

There are a number of length scales that play a role in mesoscopic physics. An important one is the coherence length ℓ_φ. It is determined by various scattering mechanisms of the itinerant electrons. For example, inelastic scattering is one of the mechanisms that determines the coherence length of an electron propagating in a metal or a semiconductor (see Sec. 13.7.2). The coherence length ℓ_φ is associated with a time scale τ_φ that takes an electron in a diffusive system to traverse a distance ℓ_φ. For reasons that will become clear later, τ_φ is called *dephasing time* and τ_φ^{-1} is the *dephasing rate*. If inelastic scattering is the dominant mechanism responsible for the loss of coherence, ℓ_φ is the mean distance between two inelastic collisions and τ_φ is the mean time between two inelastic collisions. The relation between ℓ_φ and τ_φ is given by $\ell_\varphi = \sqrt{D\tau_\varphi}$, where D is the *diffusion constant*, which is related to the conductivity σ through the *Einstein relation*, $\sigma = e^2 (dn/dE) D$, and n is the electron density. Both ℓ_φ and τ_φ decrease with increasing temperature. Nowadays, it is possible to fabricate very small solid materials and study their electronic properties. The size, L of such systems typically ranges from a few nanometers to a few microns. At low enough temperature, the coherence length ℓ_φ becomes very large and exceeds the system size L.

13.1 Mesoscopic Systems

The technological achievement of fabricating systems with $L < \ell_\varphi$ has far reaching consequences for nanotechnology. Mesoscopic systems possess interesting and novel physical properties, because they maintain coherence and therefore behave quantum mechanically, yet they are much larger than the size of a typical 'nanoscopic' (atomic) system. In some sense, they behave as big molecules, having a low-lying discrete electronic spectrum, and electron orbitals that can be readily identified. However, they differ from natural atoms or molecules first, because they can be manipulated and because the relevant energy level spacing is of the order of meV, much smaller than the spacing between the electronic spectral lines of atoms.

Another important length scale for mesoscopic systems is the *elastic mean free path* l_{el}, i.e., the average distance an electron travels between two consecutive elastic collisions. Its range of values is anywhere between a few a (the distance between atoms) to the linear size of the system, L. A mesoscopic system is referred to as *diffusive* if $a \ll l_{el} \ll L < \ell_\varphi$. Finally, in considering the *diffusive metallic regime* (or metallic grain), we require that the system size L is much less than the localization length ξ (see Secs. 9.5.2 and 12.9.3). A mesoscopic system is said to be in the *diffusive regime* if

$$a \ll l_{el} \ll L < \ell_\varphi, \xi. \tag{13.1}$$

The quantum mechanical conductance in diffusive systems is finite but less than the quantum mechanical conductance unit, $2e^2/h$ per mode [see discussion related to Eq. (13.15) below, where the quantum mechanical unit of conductance is introduced].

The length-scale relations (13.1) defining diffusive system properties should be slightly modified in very clean mesoscopic samples, where an electron can travel the length of the entire system, undergoing very few elastic scattering events, if any. In this case, the length-scale relation becomes, $a \ll \ell_{el} \sim L < \ell_\varphi$, which defines the *ballistic regime*. In a ballistic mesoscopic system, the quantum mechanical conductance is close to $2e^2/h$ per open mode.

Sensitivity to Boundary Conditions

Electronic wave functions in mesoscopic systems show sensitivity to variations in the boundary conditions. Consider a mesoscopic system of linear size L and compare two possible boundary conditions imposed on the wave function: periodic boundary conditions $\psi(x+L) = \psi(x)$ and *twisted boundary conditions* $\psi(x+L) = e^{i\alpha}\psi(x)$, where $\alpha > 0$ is the twisting parameter. In particular, the case $\alpha = \pi$ yields *antiperiodic boundary conditions*. Comparing periodic and twisted boundary conditions is a good theoretical tool for investigating an important property of wave functions, their degree of localization. The single-particle energies, $\varepsilon_n(\alpha)$, vary smoothly with α. If $\varepsilon_{n_F}(0) \approx \varepsilon_F$, and $\Delta\varepsilon$ denotes a typical single-particle level spacing, a dimensionless quantity that reflects the sensitivity to boundary conditions is the *Thouless number*. It can be defined as

$$g_{Th} \equiv \frac{|\varepsilon_{n_F}(0) - \varepsilon_{n_F}(\pi)|}{\Delta\varepsilon}. \tag{13.2}$$

The numerator in Eq. (13.2), $E_{Th} \equiv |\varepsilon_{n_F}(0) - \varepsilon_{n_F}(\pi)|$, is one way to define the *Thouless energy* [see also (Eq. 13.3)]. The identification of $g_{Th} = E_{Th}/\Delta\varepsilon$ with the conductance of the system in units of the quantum unit of conductance e^2/h was first suggested by Thouless. For an Anderson insulator (see Sec. 9.9.2) where all states are localized, it is expected that the effect of changing boundary conditions will be negligible, $g_{Th} \approx 0$. On the other hand, a strong sensitivity of the wave function to boundary conditions implies that the wave functions are extended over the entire system and can carry current.

The Thouless energy is connected by the uncertainty relation to the *diffusion time* τ_D required for an electron to travel through the system of length L,

$$E_{Th} = \frac{\hbar}{\tau_D} = \frac{\hbar D}{L^2}, \tag{13.3}$$

where D is the diffusion constant. A somewhat finer definition of the Thouless energy can be obtained by inspecting the curvature of the energies $\varepsilon_n(\alpha)$ near the Fermi energy,

$$E_{Th} = \left[\frac{\partial^2 \varepsilon_{n_F}(\alpha)}{\partial \alpha^2}\right]_{\alpha=0}. \tag{13.4}$$

13.1.1 LOW-DIMENSIONAL NANOSTRUCTURES

Confining structures such as quantum wells (2D sheets), quantum wires (1D wires), and quantum dots are important tools in the quest for new electronic and optical devices. These low-dimensional systems can exhibit novel electric, magnetic, mechanical, chemical, and optical properties. Here, we consider a number of low-dimensional nanostructures. We briefly describe confinement effects that arise in quantum wells, quantum wires, quantum dots, heterostructures, superlattices, and quantum point contacts.

If the thickness of a nanostructure is small compared to the electron de Broglie wavelength, $\lambda = h/(m_e^* v_F)$, the energy momentum dispersion relation for the bulk material is no longer applicable. The energy of an electron in the conduction band of the quantum well of thickness d is

$$E_e = E_g + \frac{\hbar^2 (n\pi/d)^2}{2m_e^*} + \frac{\hbar^2 (k_y^2 + k_z^2)}{2m_e^*}, \tag{13.5}$$

where E_g is the gap energy (see Fig. 13.1). In a quantum wire stretched along the z-axis having small transverse dimensions d_x and d_y, the single-particle electron energy is given by

$$E_e = E_g + \frac{\hbar^2 [(n_x \pi/d_x)^2 + (n_y \pi/d_y)^2]}{2m_e^*} + \frac{\hbar^2 k_z^2}{2m_e^*}, \tag{13.6}$$

and in a quantum dot of dimensions d_x, d_y, and d_z, the electron energy is given by

$$E_e = E_g + \frac{\hbar^2 [(n_x \pi/d_x)^2 + (n_y \pi/d_y)^2 + (n_z \pi/d_z)^2]}{2m_e^*}. \tag{13.7}$$

Figure 9.4 schematically shows the geometry of quantum wells, wires and dots, and their density of states, and Fig. 13.1 shows further details of the geometry of a quantum well and the energy levels and energy-momentum dispersion relations for such a device. Heterojunctions, quantum wells, quantum wires, and quantum dots are widely used for lasers. Quantum dot lasers exhibit performance that is less temperature dependent than existing semiconductor lasers, and they do not degrade at elevated temperatures. Other benefits include further reduction in threshold currents and an increase in differential gain, i.e., more efficient laser operation.

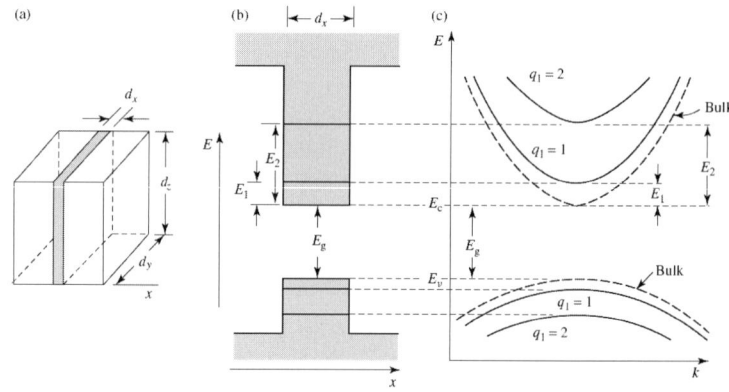

FIG 13.1 (a) Geometry of a quantum well structure. (b) Energy levels of electrons and holes in a quantum well. (c) Energy-momentum dispersion relations. The quantum numbers q_i in the figure correspond to the quantum numbers n in the text. (Reproduced from Figs 15.1–21 in Saleh and Teich [204].)

13.1.2 QUANTUM WELLS

Quantum wells are heterostructures in which a thin layer of one semiconductor is sandwiched between two layers of a different semiconductor material, thereby forming a *heterojunction*. An important requirement is that the two semiconductors have different energy gaps (and for optical applications, also different refractive indices). Materials are chosen so that electrons available for conduction in the middle layer have lower energy than those in the outer layers, creating an energy dip (or well) that confines the electrons in the middle layer [see Fig. 13.2(a)]. This is possible since differences in

13.1 Mesoscopic Systems

the energy gap permit spatial confinement of electrons and holes injected into the middle layer. Quantum well devices are often fabricated using molecular beam epitaxy. The most intensively studied and thoroughly documented semiconductor materials for heterostructures are GaAs and $Al_xGa_{1-x}As$, but several other III-V (InGaAs, GaN, and AlN) and IV-VI (PbS, PbTe, and GeTe) systems are also used.

The thickness of the middle layers (a few nanometers) is comparable with the wavelength of the electrons confined in it, thereby modifying electron behavior, since they are now confined within a plane (2D), rather than in a bulk (3D) sample. The energies of electrons in the middle layer exhibit quantized thresholds. When the middle layer is attached to leads and a small voltage is applied, a current flows. The current will be large when the energy of injected electrons matches the electron energy level in the middle layer, and small otherwise. The characteristics of the heterojunction can be tuned using the layer thickness and the material composition. Further tuning is achieved by building multiple quantum wells, made from alternating layers of semiconductor materials. This structure is referred to as *superlattice*, a structure comprising alternating layers of different materials.

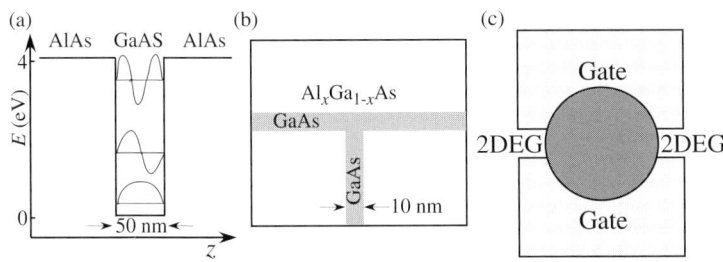

FIG 13.2 A quantum well, wire, and dot. (a) Schematic diagram of an AlAs/GaAs quantum well. The electrons are confined along z (showing the three lowest energy confined eigenstates) and free to move in the x-y plane. (b) A quantum wire is formed at the intersection of the T-shaped (shaded) region formed by two 10 nm GaAs type I quantum wells, confined by $Al_xGa_{1-x}As$ barriers. (c) Application of gates on a 2DEG to confine electrons into a lateral quantum dot, typically of size of a few tens of nanometer.

Quantum well systems can be used to create compact, fast computer chips, highly efficient microscopic lasers, and optoelectronic devices; they form the basis of lasers in CD players and microwave receivers. Blue light semiconductor lasers use quantum wells.

13.1.3 QUANTUM WIRES

Quantum wires are extremely narrow structures where electron transport is possible only in a very few transverse modes (with energies less than the Fermi energy). Quantum wires can be used as electron waveguides. Semiconductor quantum wires have been used to make switchable high-speed lasers. Quantum wires can be fabricated by an appropriate arrangement of metallic gates on top of a 2D electron gas. The electron gas beneath a negative gate voltage is depleted, slicing the quantum well into two, as shown in Fig. 13.2(b). Nanotubes can also be used as quantum wires.

Interesting physics emerges when quantum wires are used in transport experiments. Quantum wires are easily attached to electrical contacts that are formed within the underlying 2D electron gas, and this allows transport measurements. The length and width of a quantum wire can be controlled during an experiment. By varying the width, the number of transverse propagating modes that contribute to the conductance of the wire is changed. The Landauer formula, described below, predicts that the conductance of a perfectly clean wire equals $2e^2/h$ times the number of transverse modes below the Fermi energy (assuming noninteracting electrons). As the width of the wire is varied, the conductance as function of gate volatge is expected to display a step structure with plateaus quantized in units of $2e^2/h$. This has been experimentally observed. However, in some recent experiments, there are plateaus at $0.7e^2/\hbar$. The origin of this phenomenon is still under study, but it is well accepted that the reason is related to the interaction between the electrons (and more concretely, to the Kondo effect, that is discussed below).

The effects of electron–electron interaction become more pronounced when the transverse size of the quantum wire is of the order of the electron wavelength, and there is a single propagating mode. The motion of electrons along the longitudinal dimension is virtually 1D, and this leads to novel phenomena. The reason is that in 1D, the interaction between electrons is only weakly screened and interaction plays a central role. Some of the interesting physical phenomena related to interacting electrons in 1D will be discussed in Sec. 13.7.4.

13.1.4 QUANTUM DOTS

Roughly speaking, a *quantum dot* is a zero-dimensional system. It is a small, isolated region within a semiconductor or metallic material. The size L of a quantum dot is typically much smaller than the coherence length ℓ_φ, so that coherence is maintained throughout, and this often requires low temperatures. A quantum dot can be formed by cleaving a quantum wire and then overgrowing with high bandgap material. Another technique is to confine the electrons in a 2DEG by applying gates, as schematically shown in Fig. 13.2(c). This is especially useful since it enables the design of the quantum dot as part of an electric circuit between two leads. Other methods for fabricating quantum dots include: (1) Strained epitaxial growth, where the dot is "self-assembled" by depositing a semiconductor material with larger lattice constant onto a semiconductor with a smaller lattice constant. Examples include Ge on Si and InAs on Ga. (2) Quantum dots can be chemically grown in a liquid. (3) Quantum dots can also be formed in large molecules such as the fullerene molecule (sometimes called Buckminsterfullerene, named after Richard Buckminster Fuller, who developed the geodesic dome). When the Fullerene molecule is doped, electrons can move freely on its surface, hoping from atom to atom, thereby turning the system into an isolated quantum dot. A single quantum dot is the mesoscopic analog of a single atom and can be considered an artificial atom, since electron confinement results in a quantized energy spectrum. Hence, several neighboring quantum dots can be thought to constitute the mesoscopic analog of a molecule (this analogy is far from perfect—molecules have vibrational and rotational degrees of freedom, while a system of quantum dots does not vibrate or rotate).

The feasibility of spin control in a system of a few quantum dots has important ramifications for quantum information and quantum computing. For example, a two-level system in a double quantum dot can serve as the basic elementary qubit that is a key ingredient in quantum computation devices.

Several quantum dots can be integrated into an electric circuit of resistors and capacitors. This enables the construction of rich structures designed to study many interesting phenomena, such as resonant tunneling, the two-channel Kondo effect, and the Kondo–Fano effect. Some examples are shown in Fig. 13.3.

13.1.5 HETEROJUNCTIONS AND SUPERLATTICES

Heterojunctions can be grown epitaxially, one material on top of another, using molecular beam epitaxy, liquid phase epitaxy, or vapor phase deposition [e.g., metalorganic chemical vapor deposition (MOCVD)]. The semiconductor materials most often used are GaAs and the ternary compound $Al_x Ga_{1-x}As$. The combination of multiple *heterojunctions* in a device is often called a *heterostructure*, although the two terms are commonly used interchangeably. In 2000, the Nobel Prize in physics was awarded jointly to Herbert Kroemer and Zhores I. Alferov for developing semiconductor heterostructures.

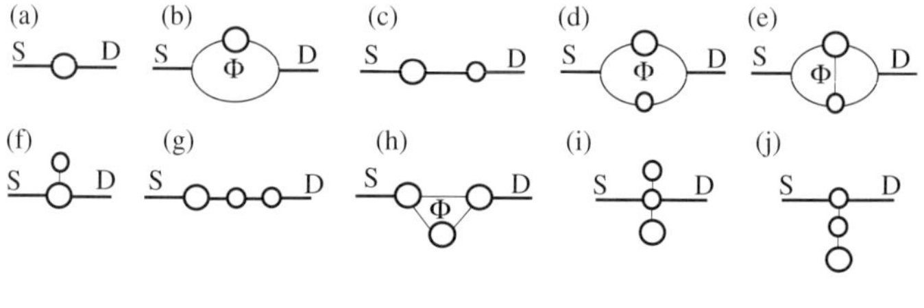

Heterojunctions are used in many solid-state device applications, including semiconductor lasers, solar cells, and transistors. The electronic energy bands in heterojunctions are engineered for specific purposes. For example, heterojunction devices are often used in field-effect transistors where, by proper control of the bandgaps and the impurity concentrations of the two semiconductor materials, a conductive channel can be formed at the interface of the two semiconductors. Because of the high conductivity in the channel, a large current can flow from source to drain.

FIG 13.3 A few quantum dots (QDs) assembled and integrated within electrical circuits between source (S) and drain (D) electrodes. (a) Single QD as a resistor, (b) QD in an Aharonov–Bohm loop, (c) two QDs in series, (d) two disjoint QDs in parallel within an Aharonov–Bohm loop, (e) two connected QDs in parallel within an Aharonov–Bohm loop, (f) T-shaped two QD configuration, (g) three QDs in series, (h) three QDs in a triangle within an Aharonov–Bohm loop, (i) cross-shaped three QD configuration, and (j) T-shaped three QD configuration.

13.2 The Landauer Conductance Formula

Using heterojunctions, barriers for charge carriers can be created, which prevent charges from entering undesired regions or confining charge carriers to a desired region of space. Moreover, energy discontinuities can accelerate carriers in specific regions. This is used to selectively enhance impact ionization in avalanche photodiodes. Semiconductors with different bandgap type (direct or indirect) can be introduced into specific regions to control whether light is emitted (or not emitted), since indirect bandgap materials do not efficiently emit light. Furthermore, the different indices of refraction of the heterostructure materials can be used to waveguide light in devices.

13.1.6 QUANTUM POINT CONTACTS

A *quantum point contact* (QPC) is a constriction located between two electron reservoirs. An experimental realization of a QPT can be obtained by depleting a 2DEG with the help of a number of tunable gates [see Fig. 13.4]. Tuning these gates enables variation of the width of the channel where electrons can flow, since depletion of electrons is equivalent to making the channel narrower. If the size of the constriction is smaller than the mean free path, the motion of electrons in the constriction occurs almost freely, and the collisions occur only at the boundaries of the QPC or outside it. In this case, the QPC is very similar to a waveguide (i.e., a photonic waveguide) and *ballistic transport* through the QPC ensues. The width of the constriction is often taken to be comparable to the Fermi wavelength (a few units of $1/k_F$, where k_F is the Fermi wavenumber). The point contact consists of a slit (formed by electrostatic gates) between two regions of 2DEG.

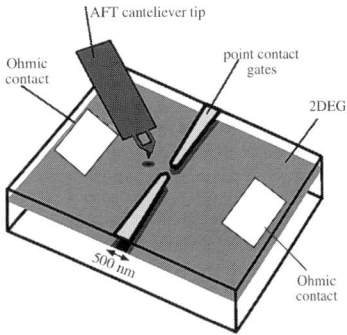

FIG 13.4 A quantum point contact with an AFM canteliever tip that can be used to measure its properties. Placing the tip so that it interrupts the flow in particular modes lowers the conductance of the point contact.

13.2 THE LANDAUER CONDUCTANCE FORMULA

We are familiar with electrical resistance in a system of linear size L at temperatures such that $L \gg \ell_\varphi(T, d)$. In this regime, the resistance is largely due to inelastic collisions of the itinerant electrons with lattice vibrations and impurities, and the resistance is well described by the classical Drude model (Ohm's Law), where the rate of collisions, $1/\tau$, increases with temperature (see Sec. 9.2.1). However, at very low temperatures, the behavior of the resistance is very different. In a mesoscopic system with linear size $L < \ell_\varphi(T, d)$, coherence is maintained through the entire system and the quantum mechanical nature of the resistance is manifest. At even lower temperatures, when $L \ll \ell_\varphi$, electrons can propagate the length of the entire system without undergoing a single inelastic collision.

This somewhat ideal description is not realized in actual experiments. To measure the resistance of a mesoscopic system, it must be connected to electrodes. The resistance is then due to both elastic collisions inside the system and elastic and inelastic collisions in the contacts and the leads. As shown by Landauer [191], the quantum mechanical resistance in such systems is determined by the rate of elastic scattering inside the system. To appreciate the importance of the *Landauer formula* for the resistance (conductance) of a low-temperature system, we note that one of the central challenges in studying the transport of electrons in solid materials is to relate experimentally measurable quantities, such as electrical conductance of fully coherent systems (for which the laws of quantum mechanics prevail), to transmission and reflection amplitudes that are calculated within quantum mechanics.

We will apply the theoretical tools developed in Chapter 12 to determine the electrical conductance of such systems, and derive a fundamental relation between transmission and conductance that is called the Landauer formula. For additional discussions, the reader is referred to the book of Imry [131].

Consider Fig. 13.5, which schematically illustrates a system consisting of a scattering region located between two ideal conductors that are connected to electron reservoirs. This kind of quantum scattering problem can be formulated using Eqs (12.800) and (12.801). In ideal conductors, the solution of the Schrödinger equation for a single electron at energy E is the product of standing waves $\varphi_n(y)$ in the transverse direction y with energy ε_n and plane waves $e^{\pm ik_n x}$ for the longitudinal motion, where $k_n = \sqrt{\frac{2m}{\hbar^2}(E - \varepsilon_n)}$ is the longitudinal wavenumber,

$$\psi_n^{\pm}(x, y) = \frac{1}{\sqrt{k_n}} \varphi_n(y) e^{\pm ik_n x}. \qquad (13.8)$$

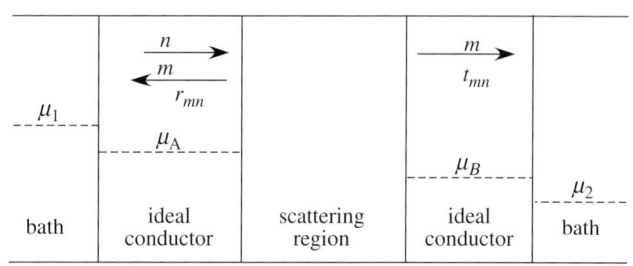

FIG 13.5 Schematic illustration of the Landauer construction that relates transmission and reflection amplitudes to conductance for scattering in a quasi-1D system. An input wave is incident from the left in channel n, and output waves emerge to the right and left in channels labeled by m. The chemical potentials in the left and right leads, and in the ideal conductors surrounding the scattering region (the quantum point contact region), are marked by dashed lines.

The normalization factor ensures that for real k_n, the state $\psi_n^{\pm}(x, y)$ carries 1 unit of current in the positive (+) or negative (−) longitudinal direction. These basic solutions define the physical channels for real k_n. An incoming wave $\psi_n^+(x, y)$ in channel n in the left ideal conductor propagates rightward toward the scattering region, as indicated by the corresponding arrow in Fig. 13.5. It is partially reflected backward into a wave $\psi_m^-(x, y)$ in channel m with reflection amplitude $r_{mn}(E)$ and partially transmitted into the wave $\psi_m^+(x, y)$ in the ideal conductor on the right with transmission amplitude $t_{mn}(E)$. A similar scenario holds for waves propagating from right to left, defining reflection and transmission amplitudes r'_{mn} and t'_{mn}, respectively. The reflection and transmission coefficients from channel n to channel m are $|r_{mn}|^2$ and $|t_{mn}|^2$, respectively, and the transmission coefficient for channel n is

$$T_n = \sum_m |t_{mn}|^2. \qquad (13.9)$$

In Sec. 12.9.4, we considered scattering in quasi-1D systems and defined the corresponding S matrix. For N_L channels on the left and N_R channels on the right, the S matrix can be written in terms of the reflection and transmission matrices $r(N_R \times N_R)$, $t(N_L \times N_R)$, $r'(N_R \times N_R)$, and $t'(N_L \times N_R)$ as

$$S = \begin{pmatrix} r & t' \\ t & r' \end{pmatrix}. \qquad (13.10)$$

This is the generalization of Eq. (12.763) for quasi-1D scattering. The S matrix is a $(N_L + N_R) \times (N_L + N_R)$ unitary matrix.

The Landauer formula relates the transmission coefficients T_n in Eq. (13.9), to the experimentally measured conductance. The latter can be measured by applying a voltage V across the system and measuring the current I. The chemical potentials μ_1 and μ_2 for the electrons in the two reservoirs 1 (on the left) and 2 (on the right) are related to the applied voltage as

$$eV = \mu_1 - \mu_2, \qquad (13.11)$$

We assume that each bath is in thermal equilibrium at temperature T, so the occupation of electrons in the bath is given by the corresponding Fermi functions $f(E - \mu_1)$ and $f(E - \mu_2)$, where E is the energy of electrons in the reservoirs. The difference in chemical potentials results in a current I between the reservoirs. This causes a difference in the chemical potentials of the ideal conductors, $\mu_A > \mu_B$ (see Fig. 13.5), but this consideration does not enter at this level of description (self-consistency of this nonequilibrium process is not maintained). The conductance is given by $G = I/V$. To be "counted" in a conductance measurement apparatus, an electron must cross from the left reservoir into the right reservoir, so that the conductance includes also the effect of the boundary between the ideal conductors and the electron baths. The electron reservoirs feed the ideal conductors with electrons in all channels that reach the scattering region from both

13.2 The Landauer Conductance Formula

left and right. To calculate the current, we first consider a channel n and calculate its contribution to the current. Due to current conservation, this can be calculated at any point. Here, it will be calculated in the ideal conductor to the right of the scattering region. The current contributed between energies E and $E + dE$ from an incoming wave from the left at channel n is

$$dI_n(L \to R) = \rho_n v_n T_n f_1(E - \mu_1) dE = \frac{e}{\pi\hbar} T_n f_1(E - \mu_1) dE. \tag{13.12}$$

Here, v_n is the initial velocity and $\rho_n = e/(\pi\hbar v_n)$ is the 1D charge density, see Eq. (9.26), with spin included. At the same time, the reservoir on the right feeds the ideal conductor with electrons in channel n that propagate leftward and are reflected by the scattering potential, contributing a term $\frac{e}{\pi\hbar}(1 - \sum_m |r'_{mn}|^2)|dE = T_n f_2(E - \mu_2)$ to the current. Within linear response,

$$f_2(E - \mu_2) - f_1(E - \mu_1) \approx -\left(-\frac{\partial f(E)}{\partial E}\right)(\mu_1 - \mu_2),$$

and summing over channels, the current between reservoirs is given by

$$I = \frac{\mu_1 - \mu_2}{\pi\hbar} \int dE \left(-\frac{\partial f(E)}{\partial E}\right) \sum_{mn} |t_{mn}(E)|^2. \tag{13.13}$$

Finally, dividing by the voltage $V = (\mu_1 - \mu_2)/e$, one arrives at the *multichannel Landauer conductance formula*,

$$\boxed{G = 2\frac{e^2}{h} \int dE \left(-\frac{\partial f(E)}{\partial E}\right) \text{Tr}[t^\dagger(E)t(E)] \xrightarrow{T \to 0} 2\frac{e^2}{h} \text{Tr}[t^\dagger(E)t(E)].} \tag{13.14}$$

The factor 2 originates from spin degeneracy. If the conductance depends on spin projections σ for the initial channel n and σ' for the final channel m, the transmission and reflection amplitudes are labeled as $t_{m\sigma';n\sigma}$ and $r_{m\sigma';n\sigma}$, respectively, and the Landauer formula can be used to calculate spin transmission coefficients $G_{\sigma'\sigma}$ expressing the conductance from a state prepared at spin σ into a final state at spin σ'. These quantities are difficult to resolve experimentally. A more accessible quantity is the conductance $G_\sigma = \sum_{\sigma'} G_{\sigma'\sigma}$, expressing the conductance for a given initial spin state.

- As emphasized above, the quantity

$$G_0 \equiv \frac{e^2}{h} \approx 3.874 \times 10^{-5} \, \Omega^{-1} \tag{13.15}$$

is the *quantum unit of conductance*. Its inverse, $h/e^2 \approx 26 \, \text{K}\Omega$, is the *quantum unit of resistance*.
- Equation (13.14) expresses the ratio of the current and the potential difference between the electron reservoirs. G includes the contact resistance between the ideal conductors and the reservoirs, and hence remains finite even if the barrier is transparent and $\text{Tr}[t^\dagger(E)t(E)]$ equals the number of open channels N. At zero temperature, the number of channels depends on the Fermi energy according to Eqs (12.799) and (12.814). When the Fermi energy is increased, new channels open, and for weak scattering, the conductance plotted as function of the Fermi energy displays step structures (see Fig. 13.11).
- The symmetries derived in Eq. (12.821) for the transmission amplitudes apply for the conductance as well. Specifically, in the presence of an external magnetic field **H**, the conductance is independent of the sign of the magnetic field,

$$G(\mathbf{H}) = G(-\mathbf{H}). \tag{13.16}$$

13.2.1 AHARONOV–BOHM INTERFEROMETER

As an example of the use of the Landauer formula, we calculate the conductance of an Aharonov–Bohm (AB) interferometer, to highlight some fundamental concepts in quantum mechanics such as interference, coherence, gauge invariance, and Onsager relations. From the point of view of using the Landauer formula, the analysis is straightforward. There is a single incoming channel and a single outgoing channel, and therefore, according to the Landauer formula, the conductance in units of G_0 is equal to the transmission coefficient. The nontrivial aspects arise because (1) between the initial and final ideal conductors, there are two modes that carry current, and (2) a magnetic flux threads the interferometer; hence, the AB effect must be properly treated (see Sec. 9.5.2).

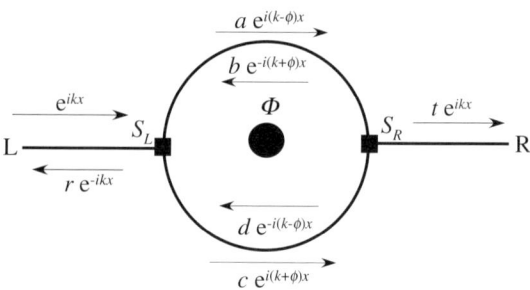

FIG 13.6 The schematic illustration of scattering through an Aharonov–Bohm interferometer.

In its simplest version, an AB interferometer (see Fig. 13.6) consists of two 1D ideal conductors L and R (for left and right) connected by three port contacts S_L and S_R, sometimes referred to as *splitters*, to an ideal ring-shaped conductor of radius ρ threaded by a central magnetic flux Φ. A convenient choice of gauge takes the tangential vector potential along the polar angle in cylindrical coordinates to be

$$\mathbf{A}(\rho,\theta,z) = \frac{\Phi}{2\pi\rho}\hat{\boldsymbol{\theta}}. \tag{13.17}$$

The magnetic field $\mathbf{H} = \nabla \times \mathbf{A}$ resulting from this vector potential is 0 everywhere except at the origin $r = 0$. As discussed in Sec. 9.5.2, the electrons stay on the wires of the interferometer and do not reach the central point where the magnetic field is present. Nevertheless, the transmission (and hence the conductance) depends on the magnetic flux. This is a purely quantum mechanical effect with no classical analog.

Problem 13.1

Show that the magnetic field on the ring is $\mathbf{H} = \nabla \times \mathbf{A} = \frac{\Phi}{2\pi r}\delta(\rho)\hat{\mathbf{z}}$.

Hint: Substituting Eq. (13.17) into Eq. (C.46) of Appendix C yields $\mathbf{H} = \mathbf{0}$, except for point $\rho = 0$. There, $H_z(\mathbf{r}) = \frac{1}{\rho}\frac{\partial(\rho A_\theta)}{\partial \rho}$ is not well defined since $1/\rho \to \infty$ when $\rho \to 0$. So, to determine $H_z(\mathbf{r})$ at $\rho = 0$, use Stokes' theorem.

The incoming plane wave with energy $E = \hbar^2 k^2/(2m)$ propagates toward the interferometer from the left and is reflected leftward with reflection amplitude r and transmitted through the ring with transmission amplitude t. The goal is to calculate the transmission and hence the conductance $G(k,\Phi) = 2\frac{e^2}{h}|t|^2$ via the Landauer formula. The external flux is a convenient control parameter, and the dependence of $G(k,\Phi)$ on energy and magnetic flux can tell us about the effects of quantum interference and gauge invariance. Note that the system cannot be regarded as purely one dimensional since part of it (the ring with the flux) is not simply connected. The calculation requires finding the two linearly independent solutions on every 1D conductor in the system and matching the solutions at the two splitters. The motion of the electron on the ring is governed by the Schrödinger equation in the presence of magnetic flux.

It is useful to introduce dimensionless variables using the radius ρ of the ring as the unit of length. The dimensionless coordinate is x (in units of ρ), the wavenumber is k (in units of $1/\rho$), the dimensionless energy is $\varepsilon = \frac{2mE\rho^2}{\hbar^2} = k^2$, and the dimensionless magnetic flux (the magnetic flux in units of the quantum unit of flux) is

$$\phi = \frac{\Phi}{\Phi_0}, \quad \text{where} \quad \Phi_0 \equiv \frac{hc}{e}. \tag{13.18}$$

The variable x in Fig. 13.6 assumes values $-\infty < x \leq 0$ on L, $0 \leq x < \infty$ on R, and $0 \leq x \leq \pi$ on either arm of the ring, where $x = 0$ at S_L. Gauge invariance requires that all measurable quantities are periodic functions of ϕ with period 1, and Onsager reciprocity relations require that the conductance is symmetric with respect to ϕ (see Eq. 13.22).

13.2 The Landauer Conductance Formula

Thus, we have

$$-\frac{d^2}{dx^2}\psi(x) = k^2\psi(x), \quad (x < S_L, \quad x > S_R), \tag{13.19a}$$

$$\left(-i\frac{d}{dx} + \phi\right)^2 \psi(x) = k^2\psi(x), \quad (S_L \le x \le S_R). \tag{13.19b}$$

The solutions corresponding to an incoming wave from the left (see Fig. 13.6) are

$$\psi(x) = e^{ikx} + re^{-ikx} \quad (x < S_L), \tag{13.20a}$$

$$\psi(x) = te^{-ikx} \quad (x > S_R), \tag{13.20b}$$

$$\psi(x) = ae^{i(k-\phi)x} + be^{-i(k+\phi)x} \quad (S_L \le x \le S_R, \text{ upper arm}), \tag{13.20c}$$

$$\psi(x) = ce^{i(k+\phi)x} + de^{-i(k-\phi)x} \quad (S_L \le x \le S_R, \text{ lower arm}). \tag{13.20d}$$

The junctions at points S_L and S_R are referred to as splitters. The precise relations between the incoming and outgoing amplitudes at each splitter depend on the detailed structure of the device. The only restriction is that current should be conserved at each splitter. To satisfy this restriction, we characterize the splitters S_L and S_R by two unitary 3×3 $\mathcal{S}$ matrices $\mathcal{S}_L$ and $\mathcal{S}_R$ relating the three amplitudes of the incoming waves in the junction to the three amplitudes of the outgoing waves at this junction. The form of these $\mathcal{S}$ matrices is then a property of the contacts and depends on the experimental setup. Current conservation at the two junctions yields,

$$\mathcal{S}_L \begin{pmatrix} 1 \\ b \\ d \end{pmatrix} = \begin{pmatrix} r \\ a \\ c \end{pmatrix}, \quad \mathcal{S}_R \begin{pmatrix} 0 \\ ae^{i(k-\phi)\pi} \\ ce^{i(k+\phi)\pi} \end{pmatrix} = \begin{pmatrix} t \\ be^{-i(k+\phi)\pi} \\ de^{-i(k-\phi)\pi} \end{pmatrix}. \tag{13.21}$$

These are six linear inhomogeneous equations for the six unknowns a, b, c, d, r, and t, which can be solved once the unitary matrices $\mathcal{S}_L$ and $\mathcal{S}_R$ are given. The solution must obey the current conservation constraint $|r|^2 + |t|^2 = 1$. Another test is that the conductance through the interferometer $G(\phi) = 2\frac{e^2}{h}|t(\phi)|^2$ should be a symmetric and periodic function of ϕ with period 1,

$$G(\phi) = G(-\phi), \quad G(\phi) = G(\phi+1). \tag{13.22}$$

The first equality was discussed in connection with Eq. (13.16). It is a special case of a more general set of constraints resulting from the combination of time-reversal invariance and magnetic field reversal, known as *Onsager relations* (see Sec. 7.9.3). The second relation, the periodicity of $G(\phi)$, is a consequence of gauge invariance and is referred to as the *Byers–Yang theorem*. Roughly speaking, it states that every physical quantity that is measured in a system which is not simply connected and threaded by a magnetic flux Φ is a periodic function of Φ with period given by the flux quantum $\Phi_0 = hc/e$.

As an example, we can take a specific form for the $\mathcal{S}$ matrices $\mathcal{S}_L$ and $\mathcal{S}_R$. To keep the number of parameters at a minimum yet avoid a trivial case, we take the matrices $\mathcal{S} = \mathcal{S}_L = \mathcal{S}_R$ to be real and to depend only on two parameters $0 < x_1, x_2 < 1$ as specified in Problem 13.2. This form is just for illustration purposes; it should not be considered as describing a specific realization of a splitter designed in a laboratory. Figure 13.7 plots the dimensionless conductance $G(k, \phi) = |t(k, \phi)|^2$ versus k and ϕ for $x_1 = 1/2$ and $x_2 = 1/2$.

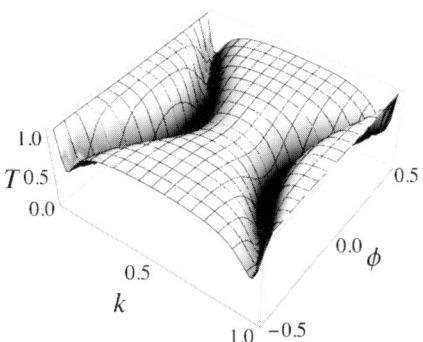

FIG 13.7 The dimensionless conductance $g(k, \phi) = |t(k, \phi)|^2$ versus k and ϕ for $x_1 = 1/2$ and $x_2 = 1/2$.

Problem 13.2

(a) Show that the matrix

$$S = \begin{pmatrix} \sqrt{1-x_2} & \sqrt{(1-x_1)x_2} & \sqrt{x_1 x_2} \\ -\sqrt{x_2} & \sqrt{(1-x_1)(1-x_2)} & \sqrt{(1-x_2)x_1} \\ 0 & -\sqrt{x_1} & \sqrt{1-x_1} \end{pmatrix}$$

is unitary if $0 < x_1, x_2 < 1$.

(b) Assuming $S_L = S_R = S$, solve Eqs. (13.21) and get a closed expression for the dimensionless conductance $g(k,\phi)$.

Answer: $g(k,\phi) = |t(k,\phi)|^2$, where

$$t(k,\phi) = \frac{e^{i\pi(k-\phi)}(e^{2i\pi k} - 1)x_2\sqrt{1-x_1}}{1 + 2e^{2i\pi k}x_1\sqrt{1-x_2}\cos 2\pi\phi - e^{2i\pi k}(1-x_1)(2-x_2) + e^{4i\pi k}(1-x_2)}.$$

13.2.2 MULTIPORT LANDAUER FORMULA

The Aharonov–Bohm interferometer shown in Fig. 13.6 is an example of a two-port (or two-lead) device, where current enters the device via one port (say, from the left) and leaves via the second port. There are many physical devices with more than just two ports where the current enters the device from a single port and can leave via several ports. For example, the Hall bar, used in the experimental discovery of the quantum Hall effect, is a six-port device (see Fig. 9.55). A schematic device with four ports is shown in Fig. 13.8(a). Each port k is composed of a wire carrying (for simplicity) a single mode and is attached to an electron reservoir whose chemical potential is μ_k. Hence, the potential difference between ports k and l is $V_{lk} = (\mu_l - \mu_k)/e$. The current supplied (or removed) at port i is denoted as I_i and the sum of all currents is 0 (current conservation).

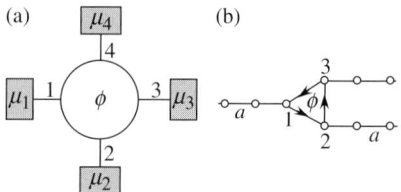

FIG 13.8 (a) Schematic illustration of a multiport device with an Aharonov–Bohm flux. (b) A three-port device is treated in the text within the tight-binding model. The arrows indicate the direction of the vector potential along the links.

The elements of the real *resistance tensor*,

$$\mathcal{R}_{ji,lk}(\phi) = \frac{e(I_j - I_i)}{\mu_l - \mu_k}, \qquad (13.23)$$

can be determined experimentally by measuring the potential drop and the currents at the appropriate ports. The indices in each pair are distinct, but those in the first pair need not be different from those in the second pair. The quantum mechanical formalism of electron transport in multiport devices was developed by Büttiker [205]. In a system with N ports, an electron reaching the device at port i is partially reflected with amplitude r_{ii} (reflection coefficient $R_{ii} = |r_{ii}|^2$) and partially transmitted into other ports $j \neq i$ with corresponding amplitudes t_{ji} (transmission coefficients $T_{ji} = |t_{ji}|^2$). The amplitudes form an $N \times N$ unitary (flux dependent) S matrix, whose elements are

$$S_{ii}(\phi) = r_{ii}(\phi), \quad S_{j\neq i}(\phi) = t_{ji}(\phi), \quad SS^\dagger = S^\dagger S = \mathbf{I}_{N\times N}. \qquad (13.24)$$

Extension of Eq. (12.821) (constraints imposed on the S matrix elements by time-reversal symmetry) to a multiport geometry system is straightforward:

$$S_{ji}(\phi) = S_{ij}(-\phi), \quad \Rightarrow \quad R_{ii}(\phi) = R_{ii}(-\phi), \; T_{ji}(\phi) = T_{ij}(-\phi). \qquad (13.25)$$

These are *Onsager reciprocal relations* (see Sec. 7.9.3). In the two-port geometry, the relation (13.16) is a special case of Eq. (13.25). Using current conservation and reciprocal relations, Büttiker was able to relate the elements of the resistance

13.2 The Landauer Conductance Formula

tensor (13.23) directly to the transmission and reflection coefficients T_{ji} and R_{ii}. The pertinent expressions for $N = 2, 3, 4$ are presented in Ref. [205], which become more and more complex for larger N. In the special case for $N = 2$, the expression reduces to the Landauer formula $1/\mathcal{R}_{21,21} = (e^2/h)|T_{21}|^2$. For $N = 3$, there are three different configurations of current voltage arrangements,

$$\mathcal{R}_{ji,jk}(\phi) = \frac{h}{e^2} \frac{T_{ki}}{T_{ij}T_{ik} + T_{ij}T_{jk} + T_{ki}T_{ik}} \equiv \frac{h}{e^2} \frac{T_{ki}}{D} = \mathcal{R}_{jk,ji}(-\phi). \tag{13.26}$$

$$\mathcal{R}_{ji,ki}(\phi) = \frac{h}{e^2} \frac{T_{kj}}{D}, \quad \mathcal{R}_{ji,ji}(\phi) = \frac{h}{e^2} \frac{T_{ki} + T_{kj}}{D}, \quad k \neq i \neq j \neq k. \tag{13.27}$$

These three resistance tensors refer to the potential and current configurations displayed in Fig. 13.9.

Example: Three-port Device.

Let us work out a solvable tight-binding model for the three-port device shown in Fig. 13.8(b). The ring is represented by an equilateral triangle and the wires by discrete chains. For triangular symmetry, we impose a vector potential $(e/\hbar c)aA = \phi/3$ along each link of the triangle [see arrows in Fig. 13.8(b)], where a is the length of each link. By Ampere's law, the total flux is ϕ. The incoming wave approaches the device from port 1, so that along the wires, the wave functions are

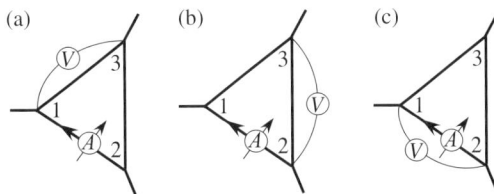

FIG 13.9 Current and voltage settings corresponding to the elemental resistance tensors defined in Eqs (13.26) and (13.27) for (a) $\mathcal{R}_{12,13}$, (b) $\mathcal{R}_{12,32}$, and (c) $\mathcal{R}_{12,12}$.

$$\psi_1(n) = e^{ikn} + r_{11}e^{-ikn}, \quad (n \leq 0), \quad \psi_2(n) = t_{21}e^{ikn}, \quad \psi_3(n) = t_{31}e^{ikn}, \quad (n \geq 0), \tag{13.28}$$

where k is related to the scattering energy as $\varepsilon = -2\cos k$. The values of the wave function on vertices are denoted as f_1, f_2, and f_3. Together with r_{11}, t_{21}, and t_{31}, we have six unknowns that should be evaluated from the solution of the Schrödinger equation applied to the wave functions at the vertices and at the sites $n = 0$ on wires 1, 2, and 3. In the simplest version of the tight-binding model [see discussion near Eq. (12.781)], the Schrödinger equation for a wave function $\psi(\mathbf{n})$ defined on lattice sites $\{\mathbf{n}\}$ reads,

$$-\sum_{\mathbf{q_n}=1}^{Z_\mathbf{n}} c_{\mathbf{q_n}} \psi(\mathbf{n}+\mathbf{q_n}) = \varepsilon \psi(\mathbf{n}), \tag{13.29}$$

where ε is the scattering energy and $\{\mathbf{q_n}\}$ are vectors connecting site $\mathbf{n}$ to its $Z_\mathbf{n}$ nearest neighbors (in a Bravais lattice, $Z_\mathbf{n}$ is the coordination number). The coefficients $c_{\mathbf{q_n}}$ are *hopping matrix elements* reflecting the degree of overlap of the wave function on sites $\mathbf{n}$ and $\mathbf{n}+\mathbf{q_n}$. Assuming constant hopping strength, and taking account of the magnetic flux, we use

$$c_{\mathbf{q_n}} = e^{i\frac{e}{\hbar c}A_{\mathbf{q_n}}a}, \tag{13.30}$$

where $A_{\mathbf{q_n}}$ is the component of the vector potential along the link $\mathbf{q_n}$. Along the chains, $A_{\mathbf{q_n}} = 0$ and $c_{\mathbf{q_n}} = 1$, while along the links of the triangle, $c_{\mathbf{q_n}} = e^{\pm i\phi/3}$, where the $\pm$ sign depends on orientation.

Application of the above formalism to the three-port system described in Fig. 13.8(b) results in the following six equations,

$$-[\psi_1(0) + e^{i\frac{\phi}{3}}f_2 + e^{-i\frac{\phi}{3}}f_3] = \varepsilon f_1, \quad -[\psi_1(-1) + f_1] = \varepsilon \psi_1(0), \tag{13.31a}$$

$$-[\psi_2(0) + e^{i\frac{\phi}{3}}f_3 + e^{-i\frac{\phi}{3}}f_1] = \varepsilon f_2, \quad -[\psi_2(1) + f_2] = \varepsilon \psi_2(0), \tag{13.31b}$$

$$-[\psi_3(0) + e^{i\frac{\phi}{3}}f_1 + e^{-i\frac{\phi}{3}}f_2] = \varepsilon f_3, \quad -[\psi_3(1) + f_3] = \varepsilon \psi_3(0), \tag{13.31c}$$

Writing the values of the wave functions on the chains in terms of transmission and reflection amplitudes given by Eq. (13.28), we end up with six inhomogenous linear equations for the six unknowns, with relatively simple expressions for r_{11}, t_{21}, and t_{31}. The reflection and transmission coefficients are

$$R_{11} = \frac{(\cos k + \cos \phi)^2}{D}, \quad T_{21} = \frac{2\sin^2 k[1 + \cos(k+\phi)]}{D}, \quad T_{31} = \frac{2\sin^2 k[1 + \cos(k-\phi)]}{D},$$

$$D = (\cos k + \cos \phi)^2 + 4\sin^2 k(1 + \cos k \cos \phi). \tag{13.32}$$

> **Problem 13.3**
>
> (a) Use the geometrical symmetry of the device shown in Fig. 13.8(b) to show that
>
> $$R_{22} = R_{33} = R_{11}, \quad T_{12} = T_{23} = T_{31}, \quad T_{21} = T_{13} = T_{32}. \tag{13.33}$$
>
> (b) Verify the symmetry relations (13.25).
> (c) Write an expression for the element resistance tensor when the current is fed at port 3 and withdrawn at port 1, while the voltage is measured between ports 2 and 1. For a given value $k = 0.65$, plot it as function of $0 \le \phi \le 1$.
>
> **Answer:** (a) The triangular symmetry implies that the enumeration of ports as $1, 2, 3$ can start at any point. Therefore, $R_{ii} = R_{jj}$. On the other hand, the occurrence of flux introduces left–right asymmetry at each port. Therefore, $T_{i+1i} = T_{ii-1}$ and $T_{ii+1} = T_{i-1i}$, but generically $T_{i+1i} \ne T_{i-1i}$ (here, the port indices are considered modulo 3). (b) Using the analytic expressions (13.32), we see that $D(\phi) = D(-\phi)$, and since the numerator of $R_{11}(\phi)$ is symmetric in ϕ, we get $R_{11}(\phi) = R_{11}(-\phi)$. Due to the symmetry established in (a), we get $R_{ii}(\phi) = R_{ii}(-\phi)$. The two expressions for the transmission amplitudes in Eq. (13.32) written for $i = 1$ imply $T_{i+1i}(\phi) = T_{i-1i}(-\phi)$. Using the symmetry relations for T_{ij} proved in (a), we get the second part of the symmetry relations (13.25).
> (c) The desired quantity is $\mathcal{R}_{13,12}$, and according to Eq. (13.26), we have $i = 3, k = 2$, and $j = 1$, so $\mathcal{R}_{13,12} = \frac{h}{e^2} \frac{T_{23}}{T_{31}T_{32}+T_{31}T_{12}+T_{23}T_{32}}$. All the transmission amplitudes can be obtained from Eq. (13.32) and the symmetry relations derived in part (a).

13.2.3 CONDUCTANCE OF QUANTUM POINT CONTACTS

A quantum point contact (QPC) is a long quasi-1D electron waveguide built on a semiconductor substrate that narrows and broadens at what is referred to as the *contact point* (see Fig. 13.4 and Sec. 13.1.6). The narrow region is referred to as the *constriction*. The passage from the broader waveguide and the constriction may be smooth or sharp. The ratio between the width of the constriction and its length is the *aspect ratio*.

A schematic illustration of a QPC is shown in Fig. 13.10. If the distance between the walls, $2w(x)$ (the width of the contact), changes slowly in comparison with the wavelength, transverse and longitudinal motion can be approximately separated, i.e., the dynamics is adiabatic in the same sense that flow of fluid in a pipe is adiabatic if the width $w(x)$ of the pipe changes sufficiently slowly as a function of the longitudinal coordinate x. A convenient way to describe the degree of adiabaticity in 2D is to use a parameterization of the form,

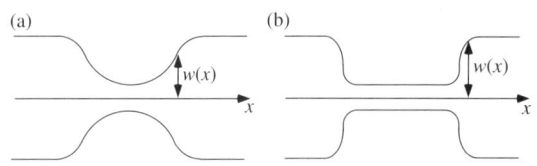

FIG 13.10 Geometries of a quantum point contact in the hard-wall boundary condition model. (a) Smooth QPC with large aspect ratio. (b) Sharp QPC with small aspect ratio.

$$w(x) = \pm w_0 - \frac{\Delta w}{2}\left\{\tanh\left[\frac{x+x_1}{\sigma_w}\right] - \tanh\left[\frac{x-x_1}{\sigma_w}\right]\right\}, \tag{13.34}$$

13.2 The Landauer Conductance Formula

which yields a symmetric QPC with minimum width Δw at $x = 0$, with crossover region proportional to σ_w. As shown in Fig. 13.4, it is possible to attach electrodes to the quantum point contact and integrate it within an electric circuit and control the gate voltage on the contact (which determines the scattering energy E of the electrons traversing the conctact). This was one of the first experimental demonstrations of mesoscopic physics. The conductance through a quantum point contact reveals several surprising effects. Remarkably, when plotted as function of gate voltage, it is approximately quantized in units of the quantum mechanical conductance $G_0 = e^2/h$ (see Fig. 13.11).

The conductance can be calculated by solving the Schrödinger equation with the appropriate boundary conditions and using the Landauer formula. For ballistic transport through a sharp constriction of length L and width w located between two leads of width $w_0 > w$, as shown in the inset of Fig. 13.11, the calculations are carried out following the discussion in Sec. 13.2. We choose a 2D Cartesian coordinate system such that the constriction is defined by $0 \le x \le L$, $-w_0/2 \le y \le w_0/2$, while the left and right leads are defined as $-\infty < x \le 0$ (left lead), $L \le x < \infty$ (right lead), and $-w/2 \le y \le w/2$ (both leads). For a given scattering energy, $E = \frac{\hbar^2 \varepsilon}{2m}$, we look for a solution of the Schrödinger equation,

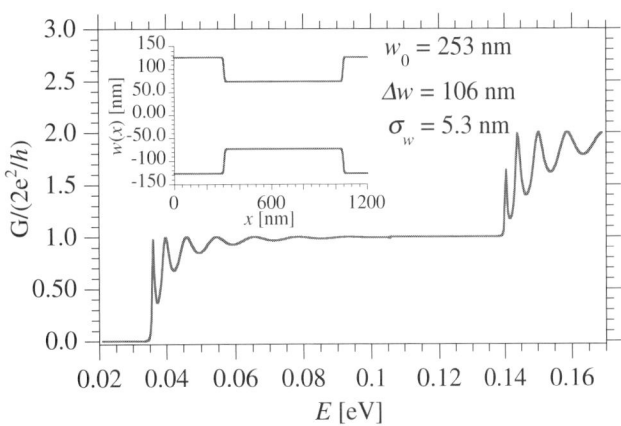

FIG 13.11 Conductance of a quantum point contact, in units of $G_0 = e^2/h$ (per spin), versus energy (controlled by a gate voltage). The inset shows a schematic quantum point contact with nearly sharp boundaries. The conductance displays a step structure, where the step occurs whenever a new transverse mode enters the constriction. Between steps, the conductance is approximately quantized in units of G_0. The resonance oscillations are due to diffraction around the corners. They are absent for adiabatic quantum point contacts [see Fig. 13.10(a)].

$$-\nabla^2 \Psi(x,y) = \varepsilon \Psi(x,y), \tag{13.35}$$

that vanishes on the boundaries and satisfies scattering boundary conditions along the propagation direction x. More specifically, we define the solutions

$$\psi_n^\pm(x,y) = \frac{e^{\pm i k_n x}}{\sqrt{k_n}} \sqrt{\frac{2}{w_0}} \sin \frac{n\pi(y + \frac{w_0}{2})}{w_0} \quad \text{(in leads)}, \tag{13.36a}$$

$$\varphi_n^\pm(x,y) = \frac{e^{\pm i q_n x}}{\sqrt{q_n}} \sqrt{\frac{2}{w(x)}} \sin \frac{n\pi(y + \frac{w_0}{2})}{w(x)} \quad \text{(in constriction)}, \tag{13.36b}$$

where $k_n = \sqrt{\varepsilon - \frac{n^2 \pi^2}{w_0^2}}$ and $q_n = \sqrt{\varepsilon - \frac{n^2 \pi^2}{w^2}}$. Scattering boundary conditions corresponding to an incoming wave from the left in channel n lead to the following solution of Eq. (13.35):

$$\Psi_n(x,y) = \begin{cases} \psi_n^+(x,y) + \sum_m \psi_m^-(x,y) r_{mn}, & \text{(in the left lead)}, \\ \sum_m a_m \varphi_m^+(x,y) + b_m \varphi_m^-(x,y) & \text{(in the constriction)}, \\ \sum_m \psi_m^-(x,y) t_{mn}, & \text{(in the right lead)}. \end{cases} \tag{13.37}$$

Problem 13.4

Prove that the functions $\psi_n^\pm$ are current normalized, i.e., the total current carried by $\psi_n^\pm(x,y)$ across the lead is
$\text{Re}\left[\int dy\, \psi_n^{\pm*}(x,y)[-i\frac{\partial}{\partial x}]\psi_n^\pm(x,y)\right] = \pm 1$.

Answer: $[-i\frac{\partial}{\partial x}]\psi_n^\pm(x,y) = \pm k_n \psi_n^\pm(x,y)$, $\int dy\, \psi_n^{\pm*}(x,y)\psi_n^\pm(x,y) = k_n^{-1}$.

The matching at the interfaces $x = 0$ and $x = L$ yields the coefficients appearing in Eq. (13.37). For proper matching, the sum must also include *evanescent waves* for which k_m are imaginary and $\varepsilon < m^2\pi^2/w_0^2$. The evanescent waves decay exponentially far away from the constriction. The quantities to be computed are the transmission and reflection matrices t_{mn} and r_{mn}, respectively, between the propagating (non-evanescent) waves. The number N of propagating waves (or channels) is determined by the inequality,

$$\frac{N^2\pi^2}{w_0^2} < \varepsilon < \frac{(N+1)^2\pi^2}{w_0^2}. \tag{13.38}$$

Since, as shown in Problem 13.4, the functions ψ_n are normalized to unit current, the conductance is given by the Landauer formula $G = G_0 \text{Tr}[tt^\dagger]$.

Problem 13.5

Let $x_1 \ll 0$ and $x_2 \gg L$. Calculate the total current contributed from mode n at x_1 and x_2 using $\Psi_n(x,y)$, making use of the orthogonality of the transverse functions.

Answer: $1 - \sum_{m=1}^{N}|r_{mn}|^2$ at x_1 and $\sum_{m=1}^{N}|t_{mn}|^2$ at x_2.

The formalism above is useful when the conductance needs to be calculated exactly. However, the step structure can be understood on a more qualitative level. A step occurs each time the energy is capable of accommodating one more propagating mode into the constriction. Thus, a step number j will occur when the scattering energy is such that $\varepsilon \simeq j^2\pi^2/w^2$. In the region $j^2\pi^2/w^2 < \varepsilon < (j+1)^2\pi^2/w^2$, the conductance is nearly quantized at jG_0, independent of energy.

13.3 PROPERTIES OF QUANTUM DOTS

Semiconductor quantum dots are small fabricated regions, typically of order of a few tens to 100 nanometers. The number of atoms in such dots is of the order of a few hundred to a few million (which reflects on the number of electrons). However, almost all the electrons are tightly bound to the atomic nuclei, and the number of free electrons (not bound to atoms) is extremely small and can be controlled to be from one to a few hundred. The size of the dot is comparable to the de Broglie wavelength of the electrons, and the single-particle energy levels are discrete. The shape of the dot plays a role in determining the properties of the single-particle spectrum. If the dot is highly symmetric (e.g., a circular dot in 2D), the spectrum, within a single-particle picture, can be calculated analytically and reveals a shell structure, as in atoms. If the dot is not symmetric (e.g., if it is in the form of a stadium), the classical dynamics may be *chaotic*, and there may not be any conserved quantities (except energy). Then, the particle trajectory $\mathbf{r}(t)$ can have a finite probability of being in any finite region inside the cavity. The classical system is then referred to as a *non-integrable system*. When a classically chaotic system is quantized, the system displays *quantum chaos*. Quantum dots can serve as an experimental tool for studying quantum chaos. The absence of conservation laws (hence classical nonintegrability and quantum chaos) can be caused not only by an asymmetric geometrical shape but also by the presence of a disordered potential such as resulting from localized impurities. We will discuss this in Sec. 13.4.1, but first we concentrate on the electrostatic aspects of electron transport in quantum dots.

The Coulomb interaction between free electrons in a dot is very important. This interaction is characterized by the *charging energy*, the energy required to add or remove an electron from the dot (analogous to the ionization energy

13.3 Properties of Quantum Dots

and electron affinity of an atom). Quantum dots are probed mainly by studying their transport properties and also by their optical properties (like atoms). Electron transport in quantum dots can be easily controlled. Charge transfer through a quantum dot is possible only when the repulsive Coulomb energy is compensated by a gate voltage. Otherwise the motion of electrons through the dot is blocked by Coulomb repulsion, a phenomena referred to as *Coulomb blockade*.

13.3.1 EQUILIBRIUM PROPERTIES OF QUANTUM DOTS

Figure 13.12 schematically shows a quantum dot in the shape of a disk. It is connected on the left to a source electrode held at potential V_s and on the right to a drain electrode held at potential V_d. These potentials control the chemical potentials μ_s and μ_d, respectively, of the electrons in the leads, $\mu_s - \mu_d = -e(V_s - V_d)$. When $\mu_s > \mu_d$, a current might flow across the dot. The current is sensitive to the energy spectrum of electrons in the dot. This spectrum can be controlled by a gate voltage V_g. Quantum dots are usually formed from extrinsic

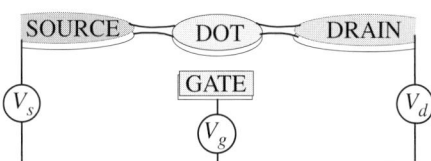

FIG 13.12 Schematic structure of a lateral quantum dot coupled with source and drain electrodes and controlled by a gate. The quality of the links between the dot and the source and drain electrodes determines the dot's physical properties. These links are quantum point contacts, whose conductances g_s and g_d can be controlled by separate gates.

semiconductors (see Sec. 9.6.3), which have fully ionized donors. With N_0 donors in the dot, the number of electrons free to propagate within a neutral dot is N_0, and a positive background charge of eN_0 originating from the donor ions remains in the heterostructure. The number N of electrons in the dot can be tuned by varying a gate potential on the dot. When $N \neq N_0$, the dot is negatively charged. Electron transport between the source and the drain electrodes, and its control by gate voltage, makes the quantum dot system an ideal tool for investigating the properties of quantum systems of nanoscopic size that contain a few mobile electrons.

Before analyzing electron transport through quantum dots, let us study their equilibrium properties. When the contact between the dot and the leads is severed, the dot is a cavity in which electrons interact with each other and with external potentials, such as the gate potential V_g, as well as with the confining potential. Often, the dot is assembled on a substrate, and the electrons in the dot interact with substrate atoms. The similarities between this system and an atom are strong in that both contain a finite number of confined interacting electrons that can move through the system. However, there are important differences. The size of the dot ($L_{\text{dot}} \approx$ thousands of Angstroms) is much larger than the size of an atom ($L_{\text{atom}} \approx$ a few Angstrom); hence, their energy level spacing is much smaller. If the dot is fabricated as a 3D cavity, the ratio of energy spacings between two adjacent levels in the dot is smaller by a factor $(L_{\text{atom}}/L_{\text{dot}})^2$. The typical level spacing in quantum dots, Δ, is of the order of a few millielectron volt. Another difference is that the electrons in an atom are subject to a central (spherically symmetric) Coulomb potential. In quantum dots, the degree of symmetry depends on the symmetry of the confining potential and the presence or absence of disorder. Most experimentally realizable quantum dots are fabricated on 2D substrates, and the symmetry (if it exists) is typically cylindrical. Perhaps, the most important distinction between quantum dots and atoms pertains to controllability: It is very difficult to manipulate single atoms, but quantum dots can be manipulated. The main tools for manipulating quantum dots in equilibrium are gates (electrical fields), and the application of magnetic and electromagnetic fields. We briefly discuss the constant-interaction model for electron–electron interactions, few-electron quantum dot systems, and various aspects of spin physics and optical transitions in quantum dots.

The Constant-Interaction Model

The classical *constant-interaction model* of quantum dots provides a simple approximate description of the electronic energy levels and the environment which typically consists of two leads (a source S and a drain D) and a gate G, held at potentials V_s, V_d, and V_g, respectively. It is adequate when the detailed electronic structure is not relevant, e.g., in the Coulomb blockade regime. The repulsive electron–electron interaction is accounted for through the charging energy of the dot. In semiconductor nanostructures, space quantization and Coulomb blockade effects are important because the Fermi wavelength in semiconductor heterostructures is much larger than in metallic systems, due to the relatively small electron density in semiconductors. The constant-interaction model accounts for both the Coulomb blockade effect and the energy spectrum of electrons in the quantum dot.

The two assumptions of the constant-interaction model are as follows: (1) The magnetic field dependent single-particle energy levels $\varepsilon_n(H)$ in the dot are independent of the number N of electrons in the dot and (2) the electron–electron interaction is determined by the capacitive couplings between the dot and its environment, C_s, C_d, and C_g; the total capacitance is $C = C_s + C_d + C_g$. Here, $s, d,$ and g stand for source, drain, and gate, respectively (see Fig. 13.12). The ground-state energy of a dot with N electrons is

$$E(N) = \frac{[e(N_0 - N) + C_s V_s + C_d V_d + C_g V_g]^2}{2C} + \sum_{n=1}^{N} \varepsilon_n(H). \tag{13.39}$$

Problem 13.6

(a) Show that the chemical potential of the dot with N electrons, defined as $\mu(N) \equiv E(N) - E(N+1)$, is

$$\mu(N, H) = \left(N - N_0 - \frac{1}{2}\right)\frac{e^2}{C} - \frac{e}{C}(C_s V_s + C_d V_d + C_g V_g) + \varepsilon_N(H). \tag{13.40}$$

(b) Show that the gate voltage for which the ground states of the dot with N and $N+1$ electrons are degenerate is

$$V_g = \frac{2C\varepsilon_n + e^2[2(N - N_0) - 1] - 2e(C_d V_d + C_s V_s)}{2eC_g}.$$

Guidance: (b) Set $\mu(N) = \mu(N+1)$ and solve for V_g.

The chemical potential determines the energy required to obtain the ground state of the N electron system from that of the $N-1$ electron system. It depends linearly on the voltages V_s, V_d, and V_g, and this dependence is the same for all values of N. A related quantity is the electrochemical potential of transition between successive ground states, referred to as addition energy,

$$E_{add} \equiv \mu(N+1) - \mu(N) = \frac{e^2}{C} + \varepsilon_{N+1}(H) - \varepsilon_{N-1}(H). \tag{13.41}$$

When $\mu(N)$ exhibits a jump (related to magic electron numbers, see below), its derivative, E_{add}, displays a peak. This is shown schematically in Fig. 13.13.

Problem 13.7

Consider a quantum dot with harmonic 2D confinement potential $V(r) = m\omega^2 r^2/2$. Find $E(N)$, $\mu(N)$, and E_{add}.
Answer: $E(N) = N^2\hbar\omega$, $\mu(N) = 2N\hbar\omega$, and $E_{add} = 2\hbar\omega$.

Few-Electron States in a Quantum Dot

We already noted that electron confinement in quantum dots results in a quantized electronic energy spectrum, similar to atoms. The spectrum can be indirectly probed via electronic transport if the quantum dots are fabricated between the source and the drain electrodes, since current-voltage measurements are sensitive to the

FIG 13.13 Chemical potentials and addition energies in a circular quantum dot containing a few electrons numbered 1,2, ..., 13. The ticks indicate the location of the chemical potentials $\mu_1, \mu_2, \ldots, \mu_{13}$ and the differences between the ticks correspond to the addition energies. The bold ticks correspond to "magic" numbers.

13.3 Properties of Quantum Dots

energy spectrum. The number of electrons on the dot is controlled by tuning the gate voltage. By varying the energy $E(N)$ and the chemical potential $\mu(N)$, it is possible to scan through the entire "periodic table" of the dot. This will be shown for a simple example, where the energy spectrum of a planar quantum dot with cylindrical symmetry, containing a few noninteracting electrons subject to a perpendicular magnetic field, is calculated.

First, recall some features of the few-electron spectra of atoms [206]. In atoms, particularly, stable electronic configurations occur when shells are completely filled. The corresponding atomic *magic numbers* are 2, 10, 18, 36, ... The degeneracy of an energy level in a given shell is partially lifted by electron–electron interaction and by spin–orbit coupling. The prescription of shell filling by electrons is summarized in Hund's rules (see Sec. 9.5.9).

Unlike in atoms, where the confining potential is spherically symmetric, the confinement potential of planar dots has the form of a 2D cavity with soft boundaries, which to some extent can be controlled. Here, we consider the highest possible symmetry, i.e., cylindrical symmetry. The repulsive confinement potential will be approximated by a harmonic potential. This has several consequences for the energy spectrum and relaxation times appearing in the decay of an excited state. In the absence of an external magnetic field, the energy spectrum has a shell structure with magic numbers 2, 6, 12, 20, ..., as shown in Fig. 13.14.

The energetics of adding electrons is displayed in Fig. 13.14(a) and depends on the number of electrons each orbit can accommodate. The first orbit can hold two electrons. Putting the first electron does not cost energy but adding the second electron costs an energy e^2/C, and that closes the first shell, with a magic number 2. The second orbit can accommodate up to four electrons. Putting the third electron (the first electron in the second orbit) costs an energy $e^2/C + \Delta\varepsilon_2$ and adding the other three electrons to the second orbit cost e^2/C each. This completes the closing of the second shell, with magic number 6. To add the seventh electron (the first on the third orbit) again costs an extra energy $e^2/C + \Delta\varepsilon_2$. These energetic considerations lead to the formation of a periodic table for the 2D quantum dot with cylindrical symmetry as displayed in Fig. 13.14(b). The magic numbers are shown in the shaded squares on the rightmost column. When the addition energy E_{add} defined in Eq. (13.41) is plotted against N, it displays peaks at the magic numbers. There are also half magic numbers resulting from Hund's rules, where E_{add} displays smaller peaks. Hund's rules in atomic physics were formulated in Sec. 9.5.9; an atomic shell is first filled with electrons with parallel spins until the shell is half full, then filling of the shell continues with spin in the opposite direction. In quantum dots with cylindrical symmetries, the second shell is half filled when $N = 4$. Half filling of the third and fourth shells occur for $N = 9$ and 16, respectively. These half magic numbers appear in the weakly shaded squares in Fig. 13.14(b).

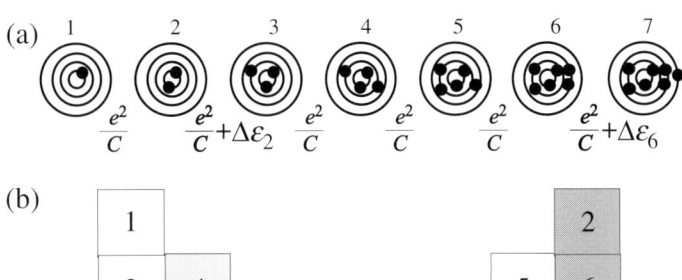

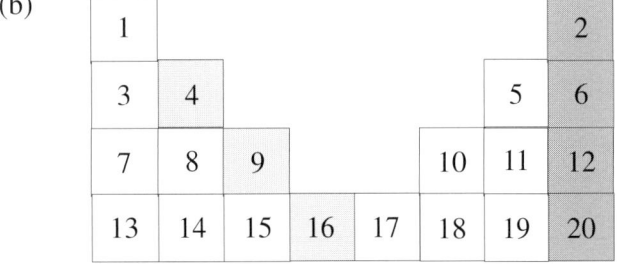

FIG 13.14 (a) Schematic illustration of electron orbits in a circular quantum dot with a harmonic oscillator confining potential. (b) The energetics of electron addition is summarized in a "periodic table." Each row corresponds to an orbit and the number of squares in a row is the number of electrons that this orbit can accommodate.

Spin does not play a direct role in energetics when no external magnetic field is present, and when spin–orbit effects are neglected, yet it enters indirectly through the Pauli principle and Hund's rules due to exchange energy between electrons.

The analysis above applies in the absence of a magnetic field. Application of a strong magnetic field H (e.g., such that the flux through the dot of area A is $HA = \Phi \gtrsim hc/e$) significantly modifies the eigenenergies and eigenfunctions.

A useful energy scale for systems in the presence of a magnetic field is the cyclotron energy $E_H = \hbar\omega_c \equiv \hbar e H/(m^*c)$, where m^* is the effective mass. In GaAs, $m^* = 0.067 m_e$, so $E_H/H \approx 1.8$ meV/T.

Problem 13.8

(a) Consider a circular quantum dot of radius r subject to a perpendicular magnetic field such that the magnetic flux through it is a single-flux quantum $\Phi_0 = hc/e$. What is the magnitude of the magnetic field.

(b) What should be the strength of the magnetic field acting on an atom such that the flux through an electronic orbit of radius 5 Å equals $10^{-3}\Phi_0$?

Answer: (a) $\pi r^2 H = \Phi_0, \Rightarrow H = \frac{\Phi_0}{\pi r^2}$. (b) In Gaussian units, $H = \frac{10^{-3}hc/e}{\pi(5\times 10^{-8})^2} = 527 \times 10^4$ Gauss ($= 5.27$ T). In SI units, $\Phi_0 = h/e$, and substituting e and h in SI units yields $H = 5.27$ T.

Problem 13.9

Calculate the magnetic Zeeman energy of an electron in GaAs $E_Z = g\mu_B H$, where μ_B is the Bohr magneton and $g_{GaAs} = -0.44$. In particular, show that $E_Z/E_H \approx 3 \times 10^{-2}$.

Single-Particle States: The Fock–Darwin Spectrum

Consider a charged particle in a plane confined by a 2D harmonic potential $V(r) = m^*\omega^2 r^2/2$ and subject to a perpendicular magnetic field H. In the absence of confining potential, the system exhibits the Landau levels as discussed in Sec. 9.5.4. When the harmonic potential is included, it is convenient to use the symmetric gauge for the vector potential, as in Eq. (9.216b). Using plane polar coordinates r, θ and using cylindrical symmetry, the z component $\hbar \ell_z$ of the orbital angular momentum is conserved ($\ell_z = 0, \pm 1, \pm 2, \dots$). The Hamiltonian for the radial motion is

$$H = -\frac{\hbar^2}{2m^*}\left[\frac{d^2}{dr^2} + \frac{1}{r}\frac{d}{dr} - \frac{\ell_z^2}{r^2}\right] - \frac{1}{2}\hbar\omega_c \ell_z + \frac{1}{2}m^*\left(\frac{1}{4}\omega_c^2 + \omega^2\right)r^2, \qquad (13.42)$$

which is composed of a 2D harmonic oscillator with frequency $\Omega \equiv \sqrt{\omega_c^2/4 + \omega^2}$ shifted by $-\hbar\omega_c \ell_z/2$.

Problem 13.10

Derive the Hamiltonian (13.42).

Guidance: In the symmetric gauge, the kinetic energy operator is

$$T = \frac{1}{2m^*}\left(\mathbf{p} - \frac{e}{2c}\mathbf{r}\times\mathbf{H}\right)^2.$$

Expand the RHS and use the definition of the angular momentum operator $L_z = (\mathbf{r}\times\mathbf{p})_z$. In 2D, the Hamiltonian of Eq. (13.42) results after adding the harmonic oscillator part.

The energy eigenvalues of Eq. (13.42) are

$$\varepsilon_{n\ell_z} = (2n + |\ell_z| + 1)\hbar\Omega - \frac{1}{2}\ell_z \hbar\omega_c. \qquad (13.43)$$

This is the *Fock–Darwin* spectrum, shown in Fig. 13.15(a). The wave functions are identical to 2D harmonic oscillator wave functions, with Ω replacing ω. From this figure, the chemical potential, with spin degeneracy and charging energy e^2/C included, has the following properties: (1) The level crossing is slightly avoided and (2) between two magic numbers, the curves are separated by e^2/C. The energy of adding the first electron above a magic number equals $e^2/C + \hbar\omega$ [see Fig. 13.15(a)].

13.3 Properties of Quantum Dots

Spin States in Few Electron Quantum Dots

It is typically not easy to manipulate the spin of a single electron in a condensed phase material. However, this is somewhat easier to do in a small quantum dot containing a few electrons, since the spin states of a few electrons in a quantum dot are determined by the orbital levels and the strength of the applied magnetic field.

Single-Electron States: Suppose there is a single electron in the dot. Let us denote by $E_n(H)$ the Fock–Darwin energies, counted from below with $E_0(H) < E_1(H) < E_2(H)\ldots$, as shown in Fig. 13.15(a). When the Zeeman energy E_Z is taken into account, the single-electron energies in the dot are

$$E_{0\uparrow}, \quad E_{0\downarrow} = E_{0\uparrow} + E_Z, \quad E_{1\uparrow}, \quad E_{1\downarrow} = E_{1\uparrow} + E_Z, \ldots \quad (13.44)$$

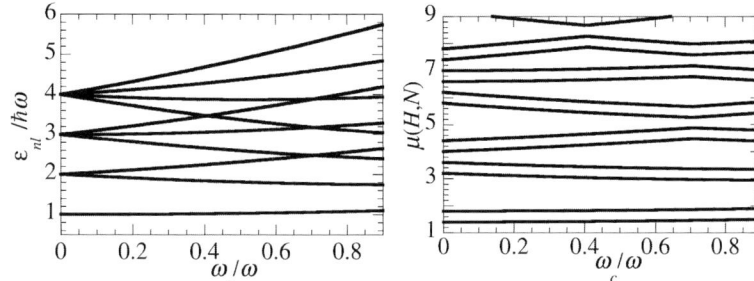

FIG 13.15 (a) Fock–Darwin spectrum [see Eq. (13.43)]. The scaled energies $\varepsilon_{n\ell_z}/(\hbar\omega)$ are plotted as function of ω_c/ω. The quantum numbers of the energy curves are assigned at low field (before level crossing) from the bottom up as $(n, \ell_z) = (0,0), (0,1), (0,-1), (0,2), (1,0), (0,-2), (0,3), (1,1), (1,-1), (0,-3)$. The numbers at $\omega = 0$ are the orbit numbers. When they are multiplied by 2 (for spin degeneracy) and summed, they give the magic numbers in the right column in Fig. 13.14. (b) Chemical potential $\mu(H, N)$, Eq. (13.40), with spin degeneracy and charging energy taken into account versus ω_c/ω. These results are obtained from (a) in two steps: (1) imposing avoided crossings and (2) putting two electrons on each level and separating their energies by the charging energy e^2/C.

Two-Electron States: As we have seen in Secs 4.7 and 8.4, the spin state of two electrons can be either an antisymmetric spin singlet state (then the spatial wave function will be symmetric) or any of the triplet states (then the spatial wave function will be antisymmetric):

$$|S\rangle = \psi_S(\mathbf{r}_1, \mathbf{r}_2) \frac{1}{\sqrt{2}}(|\uparrow\downarrow\rangle - |\downarrow\uparrow\rangle),$$

$$|T^+\rangle = \psi_{AS}(\mathbf{r}_1, \mathbf{r}_2)|\uparrow\uparrow\rangle,$$

$$|T^-\rangle = \psi_{AS}(\mathbf{r}_1, \mathbf{r}_2)|\downarrow\downarrow\rangle,$$

$$|T^0\rangle = \psi_{AS}(\mathbf{r}_1, \mathbf{r}_2) \frac{1}{\sqrt{2}}(|\uparrow\downarrow\rangle + |\downarrow\uparrow\rangle). \quad (13.45)$$

Both electrons can occupy the same orbital wave function ϕ_n with energy $E_n(H)$ if the total spin state of the two electrons is a singlet, and the charging energy cost is $E_c = e^2/C$. In a triplet state, the two electrons must occupy different orbits, say, n and $n' > n$. This costs an orbital energy $\Delta_{\text{orbit}} = E_{n'}(H) - E_n(H)$ but reduces the interaction energy by the exchange energy E_{ex}. Thus, the triplet–singlet energy difference is

$$\Delta_{TS} = \Delta_{\text{orbit}} - E_{\text{ex}}. \quad (13.46)$$

In the absence of magnetic field, the three triplet states are degenerate; in the presence of a magnetic field, this degeneracy is lifted, and $E(T^+) < E(T^0) = E(T^+) + E_Z < E(T^-) = E(T^0) + E_Z$. Restricting our discussion to the ground state, only levels $n = 0, 1$ are populated, and the singlet and triplet energies are

$$E(S) = E_{0\uparrow} + E_C + E_Z,$$
$$E(T^+) = 2E_{0\uparrow} + \Delta_{TS} + E_C, \quad E(T^-) = E(T^+) + 2E_Z,$$
$$E(T^0) = E(T^+) + E_Z. \quad (13.47)$$

Because population of the dot is controlled by a gate potential, we may regard the operations of adding or removing electrons from the dot as transitions between states of different numbers of electrons that are induced by the gate potential.

We will develop expressions for the energies required to add an electron in a given spin state to a dot that is already populated with a single electron in a given spin state. By definition, these are the corresponding chemical potentials. They depend on the spin of the two-electron states (S, T^+, T^0, T^-) and the one-electron spin states ($|\uparrow\rangle, |\downarrow\rangle$). Note that some combinations are blocked, for example, if the single electron in the dot is in a state $\uparrow$, the addition of a second electron cannot lead to a state T^- of the two-electron system. Thus, transitions $T^+ \leftrightarrow \downarrow$ or $T^- \leftrightarrow \uparrow$ are not included here because they require removal of one electron and spin-reversal of the second electron. Specifically, we will write expressions for:

$\mu(S, \uparrow) \equiv E[\text{two electrons in } S \text{ state}] - E[\text{one electron in spin-up state}]$.
$\mu(T^+, \uparrow) \equiv E[\text{two electrons in } T^+ \text{ state}] - E[\text{one electron in spin-up state}]$.
$\mu(T^0, \uparrow) \equiv E[\text{two electrons in } T^0 \text{ state}] - E[\text{one electron in spin-up state}]$.
$\mu(S, \downarrow) \equiv E[\text{two electrons in } S \text{ state}] - E[\text{one electron in spin-down state}]$.
$\mu(T^0, \downarrow) \equiv E[\text{two electrons in } T^0 \text{ state}] - E[\text{one electron in spin-down state}]$.
$\mu(T^-, \downarrow) \equiv E[\text{two electrons in } T^- \text{ state}] - E[\text{one electron in spin-down state}]$.

Thus, combining Eqs (13.44) and (13.47), we have

$$\mu(S, \uparrow) = E(S) - E_{0\uparrow} = E_{0\uparrow} + E_C + E_Z,$$
$$\mu(T^+, \uparrow) = E(T^+) - E_{0\uparrow} = E_{0\uparrow} + \Delta TS + E_C,$$
$$\mu(T^0, \uparrow) = E(T^0) - E_{0\uparrow} = E_{0\uparrow} + \Delta TS + E_C + E_Z,$$
$$\mu(S, \downarrow) = E(S) - E_{0\downarrow} = E_{0\uparrow} + E_C,$$
$$\mu(T^0, \downarrow) = E(T^0) - E_{0\downarrow} = E_{0\uparrow} + \Delta TS + E_C,$$
$$\mu(T^-, \downarrow) = E(T^-) - E_{0\downarrow} = E_{0\uparrow} + \Delta TS + E_C + E_Z. \quad (13.48)$$

The discussion above is highlighted in Fig. 13.16.

13.3.2 CHARGE TRANSPORT THROUGH QUANTUM DOTS

Transport experiments are an important tool for studying quantum dots. Transport occurs when the system under study is out of equilibrium. Theoretical tools for studying physical systems out of equilibrium are introduced in Chapter 18 (linked to the book web page). Fortunately, for small bias potential, these tools are not required, since the Landauer formula is adequate for calculating the conductance of quantum dot structures within linear response.

First, we consider Coulomb blockade effects on transport using the constant-interaction model. Classically, electric transport in bulk materials at finite temperature is characterized by the conductivity, σ, which is a local property of the material. The longitudinal conductance, G, of a finite sample of length L and cross-section area A is $G = \sigma A/L$. Since the constant-interaction model is essentially based on classical considerations, it overlooks important quantum interference effects

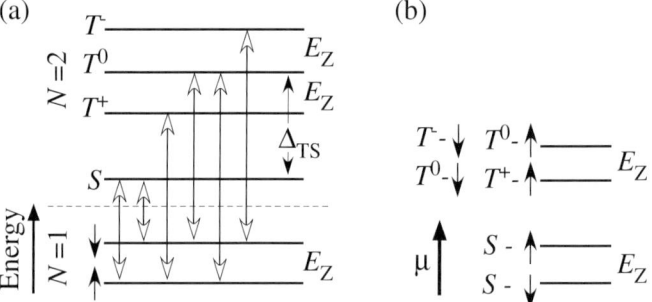

FIG 13.16 (a) Energies of one- and two-electron spin states in a simple dot and possible transitions between these states. (b) Chemical potentials for these transitions. The heights of the chemical potential lines are commensurate with the transition energies specified by the double white arrows in (a). The highest chemical potentials are $\mu(T^0, \uparrow) = \mu(T^-, \downarrow)$, followed by $\mu(T^0, \downarrow) = \mu(T^+, \uparrow)$ and the lowest ones are $\mu(S, \uparrow) > \mu(S, \downarrow)$.

13.3 Properties of Quantum Dots

that need to be considered at low temperatures. If the sample is small and the temperature is low, such that $\ell_\varphi > L$, the conductivity does not follow the classical Drude picture. In this regime, quantum corrections to the conductivity are essential. We briefly consider these features at the end of this section.

Coulomb Blockade at Small Bias

In Fig. 13.12, a potential difference $V_{sd} = V_s - V_d$ is maintained between the source and drain electrodes. As a result, the electrochemical potential on the left (source) is higher than that on the right (drain), so that

$$eV_{sd} = (\mu_L - \mu_R). \qquad (13.49)$$

First, let us assume that eV_{sd} is much smaller than $\mu(N)$ defined in Eq. (13.40). The energetics of quantum dot with $\mu_L \simeq \mu_R$ is displayed in Fig. 13.17.

The dot is located between two metallic leads whose conduction bands on the left (L) and right (R) (black rectangles) are filled up to chemical potentials μ_L and μ_R, respectively. The dot is separated from the leads by high tunneling barriers (white rectangles) and accommodates an integer number of electrons. When the dot contains N electrons, its chemical potential $\mu_D(N)$ is given in Eq. (13.6), and the addition energy $E_{add} = \mu_D(N+1) - \mu_D(N)$ is given in Eq. (13.41). The chemical potential of the dot can be

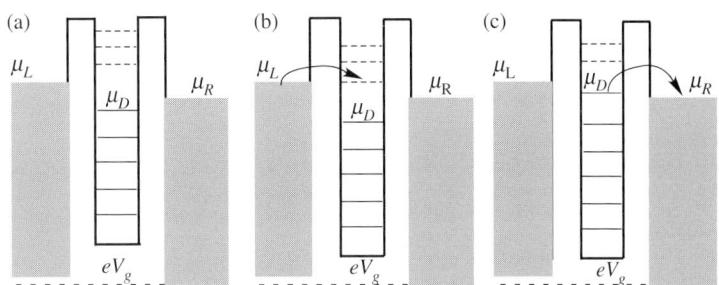

FIG 13.17 Three stages of energy profile in a simple quantum dot (see Fig. 13.12). (a) The dot contains N electrons such that $\mu_D(N) < \mu_R$ and $\mu_D(N+1) > \mu_L$. The tunneling current vanishes, and the dot is in a Coulomb blockade state. (b) $\mu_D(N+1) = \mu_L$. The dot states with N and $N+1$ are degenerate. (c) $\mu_L > \mu_D(N+1) > \mu_R$, the current through the dot is maximal.

shifted up and down by an appropriate gate voltage V_g. A small chemical potential difference $\mu_L - \mu_R > 0$ is maintained across the dot. The tunneling current, I, is strongly dependent on the position of $\mu_D(N)$. N electron levels below μ_R (full lines) are filled up to the dot chemical potential $\mu_D(N)$. The levels above μ_L are empty (dashed line). When $\mu_D(N) < \mu_R$ and $\mu_D(N+1) > \mu_L$ as shown in Fig. 13.17(a), the tunneling current is blocked and there is a *Coulomb blockade*. Application of gate voltage shifts the levels downward. When $\mu_D(N+1) = \mu_L$ as shown in Fig. 13.17(b), dot states with N and $N+1$ electrons are degenerate, and tunneling current can flow through the dot. The current has a peak when $\mu_R < \mu_D(N+1) < \mu_L$, as shown in Fig. 13.17(c). Thus, by changing the gate voltage, electrons enter the dot one by one. This is why a quantum dot is sometimes referred to as *single electron transistor*. The size of the current I through the dot depends on the spectrum of the dot and the conductances g_s and g_d between the dot and the source and drain. Since the links between the dot and the leads are quantum point contacts, g_s and g_d can be controlled by separate gates. The quantities,

$$\Gamma_{s,d} = g_{s,d}\Delta, \quad \Gamma = \Gamma_s + \Gamma_d, \qquad (13.50)$$

specify the partial widths and total width (broadening) of the upper dot level. They occur because an electron in this level is not perfectly bound inside the dot and can tunnel to either a source or a drain lead. Therefore, the corresponding level is not sharp; it is broadened by Γ. Equivalently, it has a relaxation time equal to $\hbar/\Gamma$.

The measurable quantity is the *differential conductance*,

$$G(V_{sd}; V_g) = \frac{dI}{dV_{sd}}, \qquad (13.51)$$

that depends on source drain and gate voltages. Following our analysis above, a plot of $G(V_{sd}; V_g)$ as function of V_g for small and fixed bias displays a series of sharp peaks occurring when $\mu_R < \mu_D(N+1) < \mu_L$, as shown in Fig. 13.18. The distance between peaks is then equal to $E_{add} = \delta\varepsilon_n + \frac{e^2}{C}$, Eq. (13.41). When the dot does not have a geometrical symmetry, the single-particle levels are not degenerate, and the level spacing $\Delta \equiv \varepsilon_{n+1} - \varepsilon_n$ is assumed to be independent of the single-particle energy quantum number n. Since each single-particle level accommodates two electron with opposite spin directions, we can replace Eq. (13.41) by

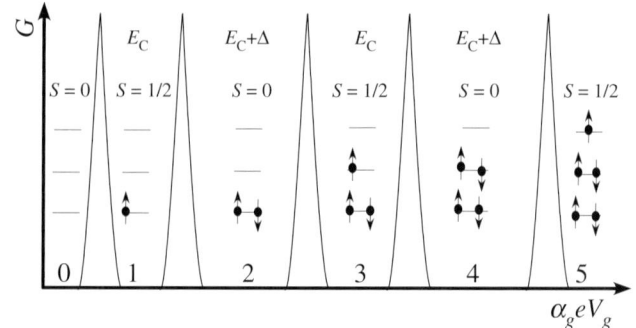

FIG 13.18 Schematic illustration of the differential conductance of a quantum dot versus $\alpha_g eV_g$, where $\alpha_g \equiv C_g/C$, displaying Coulomb blockade peaks. The energy difference between two adjacent peaks alternates between E_C and $E_C + \Delta$, and the ground-state spin of the few-electron system alternates between $S = 1/2$ and $S = 0$.

$$E_{add}(N) = \mu_D(N+1) - \mu_D(N) = \begin{cases} E_c, & \text{(if } N \text{ is odd)}, \\ E_c + \Delta, & \text{(if } N \text{ is even)}. \end{cases} \quad (13.52)$$

A Quantum Dot at Finite Bias: Coulomb Diamonds

In the discussion above, it was assumed that the bias voltage V_{sd} is small in the sense that only a single level in the dot is included between $e\mu_d$ and $e\mu_s$, as illustrated in Fig. 13.17(c). Experimentally, the bias can be made higher and several levels in the dot can be found between $e\mu_d$ and $e\mu_s$. In this situation, the system is driven way out of equilibrium, and an exact treatment requires special techniques explained in Chapter 18. However, within the classical approach the constant-interaction model is capable of treating a quantum dot at finite bias, either positive or negative (the notation V_{sd} will be used in either case). As we have noticed above, for $V_{sd} \approx 0$, the source (left) and drain (right) chemical potentials are nearly equal, $\mu_L \approx \mu_R$, and the Coulomb blockade peaks are very sharp because they occur only for a narrow interval of the gate voltage such that $\mu_L > \mu_D > \mu_R$ for $V_{sd} > 0$ or $\mu_L < \mu_D < \mu_R$ for $V_{sd} < 0$. For finite bias, the borders of these domains recede, and the interval at which the differential conductance is large becomes wider, while the regions of Coulomb blockade shrink. This can be deduced from Fig. 13.17(c). When $\mu_L - \mu_R$ is increased, the gate voltage can vary on a larger interval and still the inequality $\mu_L > \mu_D > \mu_R$ is maintained. From Fig. 13.17(a), we can see that μ_L and μ_R can vary on a finite interval and still keep the Coulomb blockade effective, as long as $\mu(N+1) > \mu_L$ and $\mu_D = \mu(N) < \mu_R$.

For a quantitative description of these features, we need to know the border lines in the V_g–V_{sd} plane that separate the regions with small differential conductance. In the former case, there is no energy level of the dot located between μ_L and μ_R, whereas in the latter case, at least one dot level exists between μ_L and μ_R. Let μ_0 be the common chemical potential of the source and drain leads in the absence of bias, and assume a symmetric application of the bias, i.e., $\mu_L = \mu_0 + eV_{sd}/2$ and $\mu_R = \mu_0 - eV_{sd}/2$. Then the quantum dot with N electrons is stable (in a state of Coulomb blockade) if

$$\mu(N+1) > \mu_0 + eV_{sd}/2, \quad \mu(N) < \mu_0 - eV_{sd}/2 \quad (V_{sd} > 0), \quad (13.53)$$

$$\mu(N) < \mu_0 + eV_{sd}/2, \quad \mu(N+1) > \mu_0 - eV_{sd}/2, \quad (V_{sd} < 0). \quad (13.54)$$

13.4 Disorder in Mesoscopic Systems

Using Eq. (13.40), we obtain two linear relations between V_{sd} and V_g that determine the border lines,

$$V_g = \frac{1}{2C_g e}[A(N) + (C + C_d - C_s)eV_{sd}], \quad (13.55)$$

$$V_g = \frac{1}{2C_g e}[A(N+1) - (C - C_d + C_s)eV_{sd}], \quad (13.56)$$

where

$$A(N) \equiv 2C(\varepsilon_n - \mu_0) - 2(C_d + C_s)\mu_0 + e^2[2(N - N_0) - 1]. \quad (13.57)$$

When N is used as a parameter, relations (13.55) and (13.56) represent two families of straight lines $V_g(V_{sd})$ with positive and negative slopes, respectively. For a symmetric dot, the slopes are of different sign. These two families define a set of parallelograms that tile the (V_{sd}, V_g) plane as shown in Fig. 13.19. This pattern is referred to as a *Coulomb diamond*.

13.4 DISORDER IN MESOSCOPIC SYSTEMS

A perfectly ordered crystal is an idealization; *disorder* is typically unavoidable. Disorder can take on different forms: point defects, interstitial atoms, dislocations, magnetic disordering, or the presence of impurity atoms. The degree of disorder can vary from very weak, for a metal with a very low impurity density, to very strong, for amorphous materials.

Disorder plays a crucial role in many branches of physics, chemistry, and engineering. In low-dimensional systems, its effects can be dramatic. For example, in space dimension $d \leq 2$, in absence of spin-orbit coupling, any amount of disorder leads to electron localization, and the system becomes an Anderson insulator (provided its size L exceeds the localization length ξ). This kind of disorder will be discussed further in Chapter 19, which is linked to the book web page. In this section, we focus on the effect of disorder in mesoscopic systems, where $L \ll \ell_\varphi, \xi$.

13.4.1 DISORDER IN QUANTUM DOTS

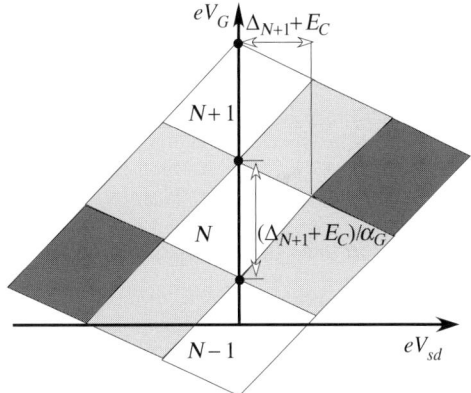

FIG 13.19 Coulomb diamond pattern for a quantum dot in the (eV_{sd}, eV_g) plane. For $V_{sd} = 0$, the black points along the eV_G axis indicate the positions of the Coulomb peaks discussed in connection with Fig. 13.18. For finite bias, Coulomb blockade is present in the bright areas where no level of the dot occurs between μ_L and μ_R. In the gray areas, a single dot level lies between μ_L and μ_R, and the dot serves as a single-electron transistor for finite range of V_G. In the dark areas, a couple of dot levels lie between μ_L and μ_R, and the dot serves as two-electron transistor for finite range of V_G.

Disorder in quantum dots can result from a variety of causes. If the boundaries of the dot are corrugated, and/or if there are imperfections in the semiconductor material on which the quantum dot is fabricated, or if impurities are present in the dot itself, quenched disorder can occur (see Sec. 12.9.3). A quantum dot with disorder can be regarded as a confined region which will give rise to chaotic classical dynamics: A point particle moving in the dot, subject to the disordered potential and reflected from its walls, will not possess any conserved quantities except energy. Chaotic motion will also occur in cases where there are no impurities and the boundaries of the dot are smooth, but the geometrical shape has low symmetry. The motion of the particle in the dot will be *ballistic*, but the classical dynamics can be chaotic. Examples of chaotic classical dynamics are dots with impurities, dots in the form of a stadium, or in a form of a square billiard with a circular block at one of its corners (or in the middle of the square). Figure 13.20 shows these examples of dots with classical trajectories that are chaotic.

Disorder, and classical chaos, can greatly affect electron transport through a quantum dot. For weak disorder, the motion of electrons in the dot is *diffusive* in the sense that $\ell_{el} \ll L \ll \ell_\varphi$. In diffusive mesoscopic systems, the dimensionless conductance displays significant sample-to-sample fluctuations. The main question of physical interest in such systems is not related to the conductance (or other physical properties) of a specific system, rather to the statistical behavior of transport quantities. For example, in a collection of quantum dots $\{D_i\}$ with slightly different shapes, Fermi energy, or applied magnetic field, the dimensionless conductances $\{g_i\}$ correspond to a random variable g with a probability distribution $P(g)$. For quantum dots in the Coulomb blockade regime, other observables have interesting statistical properties in addition to the conductance. Within a given ensemble of quantum dots, the distance between two Coulomb blockade peaks will fluctuate, as well as the peak heights.

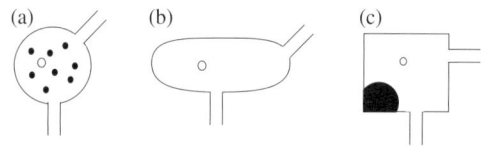

FIG 13.20 A particle (empty circle) moves in a quantum dot with two ports. (a) Quenched disorder due to impurities (filled circles). (b) Ballistic motion in a stadium. (c) Ballistic motion in a rectangular domain with a blocked corner (the somewhat more familiar *Sinai billiard* has a disk removed from the center of the rectangle).

13.4.2 DISORDERED SYSTEMS AND RANDOM MATRICES

In mesoscopic systems with disorder (or with classical chaos), the statistical properties of physical observables have interesting characteristics. A mathematical formalism geared to obtaining the statistical properties of the eigenvalues of measurable operators for these systems is *random matrix theory* (RMT) [131, 207, 209, 228]. RMT cannot be fully covered in this book; however, some results derived within RMT will be presented here without proof.

Random Matrix Theory: Gaussian and Circular Ensembles

The concept of random matrices was introduced into physics by Wigner to model systems having very complicated spectra, such as highly excited states of heavy nuclei.

To introduce the concept of random matrices, let us consider an ensemble of real symmetric matrices $\{M\}$ of dimension $N \times N$ that have the following properties:

1. M is symmetric.
2. Each entry of M is a random variable.
3. $\{M_{ij}\}$ are independent, subject to symmetry constraints.
4. The elements $\{M_{ij}\}$ have the same distribution $\forall\, i, j$.
5. $\langle M_{ij} \rangle = 0\ \forall\, i, j$, where $\langle \cdot \rangle$ refers to expectation with respect to the distribution.
6. $\langle M^2_{i \neq j} \rangle = 1$ and $\langle M^2_{ii} \rangle = 2$.
7. All the moments of M_{ij} are finite.

Let us cite a few results regarding the distribution of eigenvalues of M for such matrices of large dimension. Using a theorem by C. Tracy and H. Widdom, it is possible to show that the probability that the largest eigenvalue of M is smaller than $2\sqrt{N} + 13N^{-1/6}$ tends to 1 as $N \to \infty$. Therefore, due to symmetry, the spectrum of M is a set of N eigenvalues lying "mostly" in the interval $[-2\sqrt{N}, 2\sqrt{N}]$. For this reason, it is sometimes useful to consider the normalized matrix $\bar{M} = M/\sqrt{N}$ whose spectrum lies mostly within the interval $[-2, 2]$. A natural question is, how are the eigenvalues $\{\lambda_i\}$ of $\bar{M}$ distributed on the interval $[-2, 2]$? For large N, we can consider a distribution $\rho_N(E)$ such that $\rho_N(E)dE$ is the number of eigenvalues in the interval $[E, E + E]$. For finite N, $\rho_N(E)$ consists of a series of delta functions, and the integrated density of states, $\int_{-2}^{E} dx\, \rho_N(x)$, is a monotonic function of E, varying between 0 and N. For $N \to \infty$, the

13.4 Disorder in Mesoscopic Systems

Wigner semicircle law states that $\rho_N(E)$ weakly converges to

$$\rho(E) = \frac{1}{2\pi}\sqrt{4-E^2}. \tag{13.58}$$

Wigner–Dyson Matrix Ensembles: There are many physical systems that exhibit some kind of randomness, such as the spectra of heavy nuclei, disordered electronic systems, photon mode frequencies in irregular cavities, and scattering matrices for scattering from impurities. The latter requires an extension of the requirements imposed on real symmetric matrices to other classes of matrices (unitary matrices), as developed by Wigner and Dyson, see below.

In the Wigner–Dyson RMT formalism [207], one considers an ensemble $\mathbb{H}$ of matrices $\{H\}$ that is related to a physical system that displays some sort of randomness. The relevant physical system possesses (or lacks) special symmetries such as time-reversal and/or spin-rotation invariance. The elements h_{ij} of a matrix $H \in \mathbb{H}$ are random numbers with a certain distribution, constrained by the symmetries of the system. A matrix H is drawn from $\mathbb{H}$ with a certain distribution $\mathcal{P}(H)$. The probability to sample a matrix from the ensemble and obtain a result between H and $H + dH$ is $\mathcal{P}(H)dH$, where $dH = \prod_{i<j} dh_{ij} \prod_i dh_{ii} = \prod_{i<j} d\text{Re}[h_{ij}]\, d\text{Im}[h_{ij}] \prod_i dh_{ii}$.

Gaussian Ensembles: Anticipating the matrix H to correspond to the Hamiltonian $\mathcal{H}$ of a physical system, Wigner and Dyson suggested a specific form for $\mathcal{P}(H)$ in the case that $\mathbb{H}$ is an ensemble of Hermitian $N \times N$ matrices,

$$\mathcal{P}(H) = C_\beta\, e^{-\beta\, \text{Tr}[HH^\dagger]}, \tag{13.59}$$

where C_β is a normalization constant (its dependence on N will not be explicitly indicated), and the parameter β counts the number of degrees of freedom in a typical matrix element of H. We will encounter ensembles of real symmetric matrices, complex Hermitian matrices, real quaternionic matrices (see below for the precise definition), and unitary matrices, $\{S\}$. Each one of the first three ensembles represents a disordered system whose Hamiltonians have definite symmetry properties with respect to time-reversal and spin rotation, whereas the last ensemble represents an S matrix describing scattering from a disordered system. For the first three ensembles, β takes on values $\beta = 1, 2, 4$, respectively, and the corresponding three matrix ensembles are collectively referred to as *Gaussian Ensembles*. For this reason, β is also referred to as the *symmetry parameter*. A technical remark might be in order: Once a matrix derived from a physical system is considered (for example, the Hamiltonian of a disordered quantum system), some care must be exercised in order that the exponent in Eq. (13.59) is dimensionless. This can always be achieved by dividing H with some appropriate energy scale of the system.

Joint Eigenvalue Distribution: Let us write the Hermitian matrix as $H = U^\dagger \Lambda U$, where U is the unitary (or orthogonal) matrix of eigenvectors diagonalizing H and $\Lambda = \text{diag}\{\lambda_i\}, i = 1, 2, \ldots, N$, is the diagonal matrix composed of the eigenvalues of H. Then $\mathcal{P}(H)$ is independent of U, which means the U is uniformly distributed over the corresponding unitary group of matrices that diagonalize H. Then, one can write $dH = J\, d[U] \prod_{i=1}^N d\lambda_i$, where $d[U]$ is the measure of the corresponding unitary group (the so-called *Haar measure*), and $J = \prod_{i<j=1}^N |\lambda_i - \lambda_j|^\beta$ is the Jacobian resulting from the variable transformation $dH \to d[U] \prod_{i=1}^N d\lambda_i$. After integrating over $[U]$, we obtain the *joint distribution of eigenvalues*, $P(\boldsymbol{\lambda})$, that specifies the probability that the N eigenvalues $\boldsymbol{\lambda} \equiv (\lambda_1, \lambda_2, \ldots, \lambda_N)$ of H are located in the infinitesimal interval $[\boldsymbol{\lambda}, \boldsymbol{\lambda} + d\boldsymbol{\lambda}]$ [207],

$$P(\lambda_1, \lambda_2, \ldots, \lambda_N) \equiv P(\boldsymbol{\lambda}) = C_\beta \prod_{i<j} |\lambda_i - \lambda_j|^\beta e^{-\beta \sum_{i=1}^N \lambda_i^2}. \tag{13.60}$$

Note that the probability distribution vanishes when two levels coincide; this is referred to as *level repulsion* and is a hallmark of chaotic systems. The eigenvalues of the random matrix H cannot fluctuate independently and approach each other arbitrarily closely. In other words, they are correlated. As we shall see below, this feature has been experimentally verified for physical systems.

Once we know the joint distribution of eigenvalues, we are able to compute statistics of observables. Any function $\mathcal{O}(\lambda_1, \lambda_1, \ldots, \lambda_N)$ that depends only on the eigenvalues of H can be averaged as,

$$\langle \mathcal{O} \rangle = \int d\boldsymbol{\lambda}\, \mathcal{O}(\boldsymbol{\lambda}) P(\boldsymbol{\lambda}). \tag{13.61}$$

In particular, the averaged density of eigenvalues, $\rho(E)$, is

$$\rho(E) = \langle \rho(E;\lambda) \rangle, \quad \text{where} \quad \rho(E;\lambda) \equiv \sum_{i=1}^{N} \delta(E - \lambda_i). \tag{13.62}$$

Another important spectral property is the density–density correlation function,

$$C(E_1, E_2) \equiv \langle \rho(E_1;\lambda)\rho(E_2;\lambda) \rangle. \tag{13.63}$$

Circular Ensembles (Random Unitary Matrices): For the study of the statistics of transport properties in quantum dots attached to two narrow leads using RMT, we do not use the ensemble $\mathbb{H}$ of random Hermitian matrices, but rather random unitary matrices. This is because the physical properties of scattering from a disordered system that we typically seek is determined by the unitary S matrix, Eq. (13.10), so we need to consider an ensemble $\mathbb{U}$ of random unitary $(N_L + N_R) \times (N_L + N_R)$ matrices, referred to as a *Circular Ensemble* (CE). Here, N_L and N_R are the number of open channels in the left and right leads, respectively. There are three circular ensembles, characterized by the value of the symmetry parameter $\beta = 1, 2, 4$. The circular ensembles are characterized by the distribution $P(S) = $ constant. The scattering matrix is uniformly distributed in the unitary group $U(N_L + N_R)$, subject to the constraints imposed by the presence or absence of time-reversal and spin-rotation symmetries. Unlike for the Hermitian matrix ensembles, where $\mathcal{P}(H)$ contains the symmetry parameter β, see Eq. (13.59); here, β enters only when the joint distribution of eigenvalues is considered. Because S is unitary, its eigenvalues are unimodular complex numbers $\lambda_n = e^{i\phi_n}$, $n = 1, 2, \ldots, N_L + N_R$, and the angles $\{\phi_i\}$ are called *eigenphases*. The joint distribution of eigenphases of the S matrix is derived from $P(S) = $ constant as

$$P(\phi_1, \phi_2, \ldots, \phi_{N_L+N_R}) = D_\beta \prod_{m<n} \left| e^{i\phi_m} - e^{i\phi_n} \right|^\beta, \tag{13.64}$$

where D_β are normalization constants for the circular ensembles.

The Symmetries of the Three Gaussian Ensembles

The relevant symmetries in disordered systems are time-reversal and spin-rotation invariance. If the system conserves time-reversal invariance, $\mathcal{T}\mathcal{H}\mathcal{T}^{-1} = \mathcal{H}$, and $\mathcal{T}^2 = 1$, then all matrices in the ensemble can be chosen to be real and symmetric. A real symmetric matrix can be diagonalized by an orthogonal matrix transformation, hence the corresponding ensemble is referred to as a *Gaussian orthogonal ensemble* (GOE). This corresponds to the symmetry parameter $\beta = 1$. The physical system for $\beta = 1$ corresponds to a disordered system of spinless particles without the presence of a magnetic field. If time-reversal symmetry is not present, e.g., for systems with an external magnetic field, the Hamiltonian matrix is complex, and the corresponding ensemble consists of random complex Hermitian matrices. A complex Hermitian matrix can be diagonalized by a unitary matrix. Therefore, the corresponding ensemble is referred to as a *Gaussian unitary ensemble* (GUE). This corresponds to symmetry parameter $\beta = 2$. Finally, if the system is time-reversal invariant, $\mathcal{T}\mathcal{H}\mathcal{T}^{-1} = \mathcal{H}$, but $\mathcal{T}^2 = -1$, the system describes particles with half-integer spin, e.g., electrons when spin–orbit coupling is present (so that spin rotation symmetry is not present), in the absence of an external magnetic field (so that time-reversal symmetry is present). The corresponding random matrices have even dimension, $2N \times 2N$, and in addition to being Hermitian, they can be written as N^2 blocks of 2×2 matrices of the form $a\mathbf{1} - i\mathbf{b} \cdot \boldsymbol{\sigma}$, where the scalar a and the vector $\mathbf{b}$ are real and $\boldsymbol{\sigma}$ is the vector of Pauli matrices. Such a 2×2 matrix is called a *real quaternionic matrix*.[2] To expose the structure of the Hamiltonian containing blocks of real 2×2 quaternionic matrices that are time-reversal invariant, recall from Sec. 4.4 that for spin 1/2 particles, $\mathcal{T} = -iS_y K$, where K is the complex conjugation operator. For a $2N \times 2N$ Hamiltonian, $-iS_y = -(i\hbar/2)\sigma_y \otimes \mathbf{1}_{N \times N}$. In analogy with our discussion of the GOE and the GUE ensembles, we now look for a matrix that diagonalizes the $2N \times 2N$ Hamiltonian matrix that has the structure of N^2 blocks of

---

[2] Note that the elements of the real quaternionic matrix are not necessarily real; a and $\mathbf{b}$ are real.

13.4 Disorder in Mesoscopic Systems

2×2 real quaternionic matrices. Consider a $2N \times 2N$ real and unitary matrix M with the property, $M^T S_y M = S_y$, which reduces to

$$M^T \begin{pmatrix} 0 & 1_{N \times N} \\ -1_{N \times N} & 0 \end{pmatrix} M = \begin{pmatrix} 0 & 1_{N \times N} \\ -1_{N \times N} & 0 \end{pmatrix}.$$

Such a matrix is called a *symplectic matrix*. The set of symplectic matrices $\{M\}$ form a group called the *symplectic group*, denoted as $S_p(N)$. The ensemble of Hermitian random matrices $\{H\}$ composed of N^2 blocks of real quaternionic matrices is referred to as a *Gaussian symplectic ensemble* (GSE) and corresponds to symmetry parameter $\beta = 4$. A theorem from quaternion algebra states that a matrix H in the GSE can be diagonalized by a symplectic matrix M,

$$H = M \begin{bmatrix} \lambda_N & 0 \\ 0 & \lambda_N \end{bmatrix} M^T,$$

where $\lambda_N = \text{diag}(\lambda_1, \lambda_2, \ldots, \lambda_N)$ is the set of eigenvalues of H. Therefore, every eigenvalue is at least doubly degenerate (Kramers degeneracy).

Problem 13.11

Consider the 4×4 matrix H of 2×2 block form,

$$H = \begin{pmatrix} 0 & a + i\mathbf{b} \cdot \boldsymbol{\sigma} \\ a - i\mathbf{b} \cdot \boldsymbol{\sigma} & 0 \end{pmatrix},$$

where $a = 1$ and $\mathbf{b} = (1, 2, 3)$. Find its eigenvalues and verify the Kramers degeneracy.
Answer: The eigenvalues are $\pm\sqrt{15}$, where each is twofold degenerate.

Problem 13.12

Show that a 2×2 real quaternionic matrix $a - i\mathbf{b} \cdot \boldsymbol{\sigma}$ with real scaler a and vector $\mathbf{b}$ is invariant under time reversal.
Hint: Use $\mathcal{T} \mathbf{S} \mathcal{T}^{-1} = -\mathbf{S}$ and $\mathcal{T} i \mathcal{T}^{-1} = -i$.

The ensemble nomenclature, Orthogonal, Unitary, and Symplectic, correspond to $\beta = 1, 2, 4$, respectively, and this nomenclature is also used for the circular ensembles, referred to as circular Orthogonal, circular Unitary, and circular Symplectic ensembles (COE, CUE, and CSE).

13.4.3 APPLICATION OF RANDOM MATRIX THEORY TO DISORDERED MESOSCOPIC SYSTEMS

Random matrix theory can be used for analyzing statistical properties of spectra and transport observables. In the first case, the starting point is identification of the system's Hamiltonian, $\mathcal{H}$, with a random matrix H belonging to one of the Gaussian ensembles according to the symmetries of the system. In the second case, the starting point is the identification of the scattering matrix, $\mathcal{S}$, with a random unitary matrix that belongs to the circular ensemble possessing the symmetries of the system. We briefly present a few examples.

Level Correlations and Nearest Level Spacings

The fact that the Hamiltonian $\mathcal{H}$ of a disordered physical system is distributed according to one of the Wigner–Dyson Gaussian ensembles requires justification. For statistics of energy levels of heavy nuclei, justification was established by

Wigner and Dyson [207]. For a quantum dot in the diffusive regime (where the dot is sometimes called a *metallic grain*), this justification was established on energy scales below the Thouless energy defined in Eq. (13.4); see Refs [131, 209, 228] for more details.

The spectral properties of isolated dots with no contact to leads (the openings shown in Fig. 13.20 are assumed closed) can be investigated within RMT. The main calculational effort in RMT is the evaluation of averages and correlations, similar to those in Eqs (13.62) and (13.63). Powerful methods have been developed in Ref. [209] to perform this kind of stochastic integration. Much progress has been made, but there are still many quantities that cannot be obtained in closed form.

For example, the density–density correlation function (13.63) was calculated for the GUE at small energy difference $E_2 - E_1 \ll (E_N - E_1)$ [207] to be

$$R_2(\omega) \equiv \frac{C(E + \frac{\omega}{2}, E - \frac{\omega}{2})}{\rho^2(E)} = \delta(s) + 1 - \frac{\sin^2 \pi s}{(\pi s)^2}. \tag{13.65}$$

Here, $s = \omega/\Delta(E)$ denotes the level spacing in units of the mean level spacing at energy E, $\Delta(E) = 1/(\rho(E)N)$.

Level Fluctuations and Nearest Neighbor Spacing Distribution: The study of level fluctuations in disordered physical systems is a powerful tool for identifying the symmetry class and the degree of chaos. Level fluctuations can be obtained directly once the spectrum is measured. A spectrum of two systems A and B that are different only in the configuration of disorder, or a spectrum of two chaotic systems, where the shapes of their boundaries are slightly different, show energy level fluctuations, $\lambda_n(A) \neq \lambda_n(B)$. But the information gained from the fluctuation of each level separately is not very rich. A much more interesting quantity that can be measured in the same type of experiment is the fluctuation in the nearest level spacing distribution, i.e., fluctuation of the differences $q_n \equiv \lambda_{n+1} - \lambda_n$. This quantity introduces correlations among all levels because if $\lambda_2 - \lambda_1$ fluctuates, it automatically affects the fluctuation of $\lambda_3 - \lambda_2$, and so on. A physically relevant question in this respect is, what is the probability of find two levels whose spacing q_n is such that $q \leq q_n \leq q + dq$. This probability $P(q)dq$ is very sensitive to the symmetries of the single-particle Hamiltonian. Before proceeding, we need to solve a technical problem.

Spectral Unfolding: In many physical systems, the density of levels is not uniform. For example, in atomic or nuclear systems, the lower levels are well separated from each other, whereas at higher energy, the levels become more and more dense. Comparing the fluctuations of level spacings in both energy domains does not make sense. The study of level fluctuations of, say, $\lambda_2 - \lambda_1$ and $\lambda_{1002} - \lambda_{1001}$ on the same footing is meaningless, if the first energy difference is orders of magnitude larger than the latter. We are interested in a spectrum where the *average* spacing between two adjacent levels is identical on all energy scales, and our interest is focused on their fluctuations. There is a special procedure to transform a spectrum with energy-dependent averaged density of states to a spectrum with constant average density of states $\rho(E) = \rho$ in which the *mean level spacing* between two adjacent level is constant. This procedure is called *unfolding*. It does not affect the nature of level fluctuations. The technical details of the unfolding procedure will not be described here. It is pictorially illustrated in Fig. 13.21. From now on, when discussing level fluctuations, we assume that the average density of states is uniform, and so is the mean level spacing $\Delta = 1/(N\rho)$, which will then serve as an energy scale for the study of level fluctuations.

How does the spacing between two adjacent eigenvalues fluctuate around its mean value after the spectrum is unfolded? Let us assume that the eigenvalues $\{\lambda_n\}$ of a random matrix $H \in \mathbb{H}$ are ordered as $\lambda_1 < \lambda_2 < \ldots < \lambda_N$ and define a random variable for the level spacing as,

$$s_n \equiv \frac{\lambda_{n+1} - \lambda_n}{\Delta} \geq 0. \tag{13.66}$$

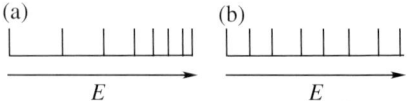

FIG 13.21 (a) Spectrum before unfolding. The average density of states and the mean level spacing are energy dependent. The low-lying states are widely separated, and the definition of mean level spacing is useless there. (b) Spectrum after unfolding. The average density of states is constant and so is the mean level spacing. But the level spacing fluctuates around the mean. This is where Eq. (13.66) is applicable.

13.4 Disorder in Mesoscopic Systems

Note that s is dimensionless. An example of nearest level spacing fluctuations in a matrix whose spectrum is unfolded is shown in Fig. 13.22, where we plot the eigenvalues $\{\lambda_n\}$ and the nearest level spacings $\{s_n\}$ of some random matrix as function of n. In this figure, we see that the average density of states is constant, but the nearest level spacings fluctuate around their mean value that, for the data chosen, is $\Delta = 1$. Therefore, we define the nearest level spacing distribution $P_\beta(s)$ of a given random matrix ensemble characterized by symmetry parameter β (and with unfolded spectrum) as follows: The probability that there are two adjacent levels whose spacing lies between s and $s+ds$ is $P_\beta(s)ds$. The constraints imposed by normalization and mean level spacing of unity requires the equalities,

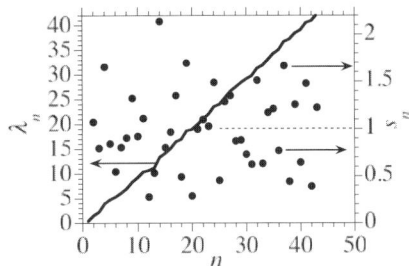

FIG 13.22 Energy spectrum λ_n (left ordinate), shown as a solid curve and level spacings s_n (right ordinate), shown as dots, as function of n. The overall dependence of λ_n on n is linear, indicating that the average density of states is constant. Yet $\{s_n\}$ fluctuate around the mean level spacing, $\Delta = 1$ (dashed line). (Reproduced from Castro Neto, *et al.* [214].

$$\int_0^\infty ds\, P_\beta(s) = \int_0^\infty ds\, s P_\beta(s) = 1. \tag{13.67}$$

The calculation of $P_\beta(s)$ directly from the joint probability distribution requires N point correlations, much higher than the two-point correlation defining $C(E_1, E_2)$ in Eq. (13.61). For $N=2$, using Eq. (13.61), with $\mathcal{O}(\lambda_1, \lambda_2) = \delta(s - |\lambda_1 - \lambda_2|)$, the required integral has the form

$$P_\beta(s) = C_\beta \int_{-\infty}^\infty d\lambda_1 d\lambda_2 |\lambda_1 - \lambda_2|^\beta e^{-\beta(\lambda_1^2 + \lambda_2^2)} \delta(s - |\lambda_1 - \lambda_2|) = C_\beta s^\beta e^{-A_\beta s^2}, \tag{13.68}$$

where the constants C_β is fixed by the normalization constraint [the first equality in Eq. (13.67)] and A_β results from the integral (you will calculate it in Problem 13.13). The physical meaning of A_β is that it controls the dependence of nearest level fluctuations at large distances. On the other hand, the level repulsion at short distance (vanishing of $P_\beta(s)$ as $s \to 0$) is controlled by the powers s^β before the Gaussian factor. Expression (13.68) is referred to as the *Wigner surmise*. It was conjectured to hold also for random matrices of arbitrary size (and this was numerically confirmed). Based on the constants calculated in Problem 13.13, the nearest level spacing distributions for the three Gaussian ensembles are plotted in Fig. 13.23.

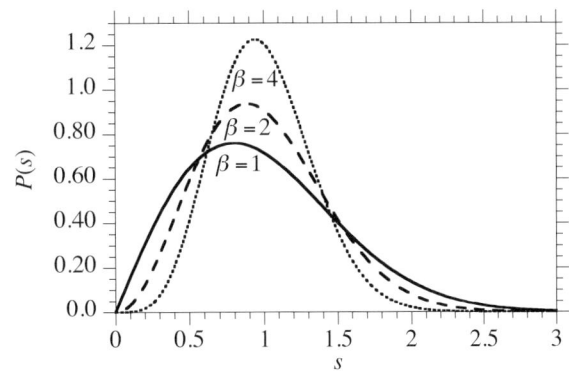

FIG 13.23 Nearest level spacing distributions $P(s)$ for the three Gaussian ensembles, GOE, GUE, and GSE ($\beta = 1, 2,$ and 4).

Problem 13.13

Use the two normalization constraints and unit average level spacings specified in Eq. (13.68) to find C_β and A_β for $\beta = 1, 2, 4$. Use the following relations, $\int_0^\infty dx\, x^n e^{-ax^2} = a^{-1/2-n/2} \Gamma[(1+n)/2]/2 = \frac{\sqrt{\pi}}{2\sqrt{a}}, \frac{1}{2a}, \frac{\sqrt{\pi}}{4a^{\frac{3}{2}}}, \frac{1}{2a^2}, \frac{3\sqrt{\pi}}{8a^{\frac{5}{2}}}, \frac{1}{a^3},$

for $n = 0, 1, 2, 3, 4, 5$.
Answer: $C_1 = \frac{\pi}{2}$, $A_1 = \frac{\pi}{4}$, $C_2 = \frac{32}{\pi^2}$, $A_2 = \frac{4}{\pi}$, $C_4 = \frac{2^{18}}{729\pi^3}$, $A_4 = \frac{64}{9\pi}$.

The Wigner surmise holds also for the level spacing distribution calculated for chaotic quantum dots such as the stadium or the Sinai billiard displayed in Fig. 13.20(b) and (c) (albeit without leads).

The quantity $P(s)$ is sensitive to the symmetry parameter β, as is clear from (Eq. 13.68). Moreover, it is a good indicator of whether a system is disordered (or chaotic) or integrable. It was shown by M. V. Berry and M. Tabor that for a generic integrable system, $P(s) = e^{-s}$, i.e., the distribution is that of a *Poisson process*. The main difference between the Poisson statistics and the Wigner surmise is that in the latter case, there is a level repulsion at small s, implying that levels are correlated and interact with each other, whereas in the former, there is no level repulsion at small s and levels are completely independent. The Wigner surmise result for nearest level fluctuations, and level repulsion has been experimentally verified in a series of experiments, see for example Ref. [210].

Universal Conductance Fluctuations

Consider the measurement of electrical conductance, $G = 2e^2 g/h \equiv G_0 g$, performed on several disordered mesoscopic systems that differ from each other only in their disorder configurations (otherwise, they have identical geometry and symmetry). Here, $G_0 = 2e^2/h$ is the quantum unit of conductance, and the factor 2 results from spin degeneracy. The measured (dimensionless) conductance g will vary from sample to sample, a phenomenon referred to as *sample-to-sample fluctuations*. Sample-to-sample fluctuations result from quantum interference and occurs in disordered systems. Sample-to-sample fluctuations, and the dependence on the symmetry parameter β, can be analyzed within RMT.

The statistical consequence of sample-to-sample fluctuations is that the dimensionless conductance g will obey a certain distribution $P(g)$. In several cases, RMT is capable of predicting the first two moments of $P(g)$, i.e., the averaged conductance $\langle g \rangle$ and its variance $\text{Var}(g)$, defined as,

$$\langle g \rangle \equiv \int dg \, g P(g), \quad \text{Var}(g) \equiv \langle g^2 \rangle - \langle g \rangle^2. \tag{13.69}$$

From the analysis of the Landauer formula, we expect that $\langle g \rangle$ is of order N_c, the number of open channels. What is less expected is that $\text{Var}(g) \simeq 1$, and its precise value depends mainly on the symmetries of the system under time reversal and spin rotation, and not on the detailed form of the disordered potential (it might depend weakly on the shape of the conductor). This phenomenon is referred to as *universal conductance fluctuations* (UCF). In particular, $\text{Var}(g)$ decreases by a factor of 2 when time-reversal symmetry is broken (e.g., by a magnetic field). This result was originally proven using the technique of *diagrammatic perturbation theory*, where the terms in the perturbation expansion are represented by *Feynman diagrams*. It was shown that breaking time-reversal symmetry suppresses one (of two) classes of diagrams, hence the factor 2. For a conductor in a form of a long wire, the expression $\text{Var}(g) = 2/(15\beta)$ was derived using RMT. Additional results pertaining to quantum dots will be discussed below.

Average Conductance: Having listed some general statements about the conductance fluctuations in general, it is instructive to present some examples where RMT is elegantly employed to yield analytical results that can be experimentally tested. For this purpose, we consider a chaotic quantum dot shown as in Fig. 13.20(b) with two ports supporting N_L and N_R propagating channels through which electrons can enter and leave. Classically, a point particle executes chaotic motion which implies that the transmission and reflection probabilities are equal. Quantum mechanically, the transmission probability is slightly smaller than the reflection probability, due to weak localization that results from constructive interference of pairs of time-reversed trajectories. The deviation of the quantum transmission coefficient from the classical one is referred to as the *weak localization correction to the conductance*.

13.4 Disorder in Mesoscopic Systems

Because the distribution $\mathcal{P}(S)$ for all three circular ensembles, COE, CUE, and CSE, is constant, the distribution of all its squared matrix elements can be evaluated within RMT. The result is

$$\langle |S_{mn}|^2 \rangle = \frac{\beta + (2-\beta)\delta_{nm}}{\beta(N_L + N_R - 1) + 2}, \quad n, m = 1, 2, \ldots, N_L + N_R. \tag{13.70}$$

To get the conductance according to the Landauer formula, we have to sum over $1 \leq n \leq N_L$ incoming modes and $N_L + 1 \leq m \leq N_L + N_R$ outgoing modes, which yields

$$\langle g \rangle = \sum_{m=N_L+1}^{N_L+N_R} \sum_{n=1}^{N_L} \langle |S_{mn}|^2 \rangle = \frac{\beta N_L N_R}{\beta(N_L + N_R - 1) + 2}. \tag{13.71}$$

From Eq. (13.71), we see that the conductance of disordered (or chaotic) system is increasing as a function of magnetic field strength. The increase of $\langle g \rangle$ as β changes from 1 to 2 is related to the phenomenon of *negative magnetoresistance* in 2D systems, one of the hallmarks of weak localization phenomena described in Sec. 13.7.2. An intuitive explanation is as follows. Assume that an electron moves in a 2D sample starting at some point A. What is the probability that it will return to this point after time t? For describing this process, we imagine a closed curve starting and ending at A as in Fig. 13.45. For every such closed curve, there is an identical closed curve, the *time-reversed curve*, that the electron can take on its way from A to A. In the absence of a magnetic field, the phase accumulated by the wave functions along these two curves is identical, and there is constructive interference at A leading to *enhanced backscattering*. When a magnetic field is applied, the phases gained by the electron along the two time-reversed paths are different, and there is no strong backward interference (see Sec. 13.7.2).

Higher conductance for $\beta = 4$ is due to a different mechanism leading to the absence of enhanced backward scattering. It occurs when the magnetic field is absent, but there is a strong spin–orbit interaction. From our discussion of the Aharonov–Casher effect in Sec. 9.5.2, we conclude that a spinor going along a closed loop acquires an $SU(2)$ phase factor [see Eq. (9.206)]. Because of this, the two paths along time-reversed closed loops starting and ending at a given point A interfere destructively, which leads to a higher conductance. This phenomenon is referred to as *weak antilocalization* induced by spin–orbit coupling.

Problem 13.14

Assume that the classical conductance through each port is proportional to the number of propagating channels and that the classical resistance of the dot can be calculated in series, $R = R_L + R_R$. (a) Calculate the classical dimensionless conductance g_{class}.
(b) Calculate the weak localization correction to the conductance.
Answer: (a) $g_{\text{class}} = 1/R = 1/(R_L + R_R) = 1/(1/N_L + 1/N_R) = N_L N_R/(N_L + N_R)$. (b) $\delta g = \langle g \rangle - g_{\text{class}}$.

Variance of the conductance: To calculate the variance, note that

$$g^2 = \left[\sum_{m=N_L+1}^{N_L+N_R} \sum_{n=1}^{N_L} |S_{mn}|^2 \right]^2,$$

where the expression inside the square brackets is the conductance according to the Landauer formula. Performing the integration of g^2 over the measure of the unitary group and subtracting $\langle g \rangle^2$ yields, for large values of channel numbers $N_L, N_R \gg 1$,

$$\text{Var}(g) = \frac{2(N_L N_R)^2}{\beta(N_L + N_R)^4} = \frac{1}{8\beta} \quad (\text{for } N_L = N_R). \tag{13.72}$$

This shows that fluctuations are universal. The calculations presented above are carried out for chaotic systems, as shown in Fig. 13.20(b) and (c), and show that the conductance fluctuations are independent of the shape of the dot and that the fluctuations are proportional to β^{-1}. These physical consequences are valid also for a quantum dot with impurities, as shown in Fig. 13.20(a). In this case, we consider many samples having the same average impurity disorder but different realizations of impurity configurations. As long as we are in the diffusive regime [see definition in Eq. (13.1)], there is neither dependence on the details of disorder, nor on the geometry of the dot, nor of its size. The conductance fluctuations depend only on the symmetry of the Hamiltonian with respect to time-reversal and spin-rotation operations. Experimental tests of universal conductance fluctuations and further discussion are found elsewhere [131]. One way to observe such fluctuations is to apply a weak magnetic field on a disordered system and vary the magnetic field. As long as the magnetic field remains weak, this variation is equivalent to keeping the magnetic field fixed and changing the impurity configurations in the sample (this kind of equivalence is referred to as *ergodicity*, we shall not discuss it here). An example of results of a typical experiment is displayed in Fig. 13.24.

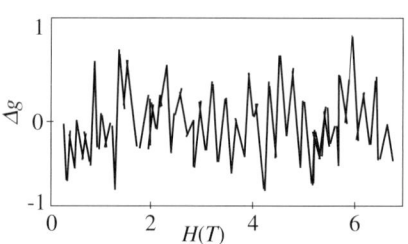

Conductance Distribution, $P(g)$: We have indicated after Eq. (13.69) that $\langle g \rangle \simeq N_c$ and $\text{Var}(g) \simeq 1$. For large N_c, the conductance of the dot has a Gaussian distribution. However, for small $N_c \simeq 1$, or more precisely, when the fluctuation of g is of the order of its mean, the conductance distribution $P(g)$ is not Gaussian. Within RMT, the distribution of the transmission coefficients $\{T_n\}$, defined in Eq. (13.9), can be derived. If there is only one channel, then $g = T_1$ and therefore $P(g) = P(T_1)$. In this case, RMT encodes the distribution of the conductance, and yields [209]

FIG 13.24 A typical experimental result showing the dimensionless conductance of a narrow wire, shifted to have zero mean, versus magnetic field. The conductance displays universal conductance fluctuations that are time independent and completely reproducible. The root mean square is close to $\sqrt{1/15}$ in accordance with the RMT prediction for a wire, $\text{Var}(g) = 2/(15\beta)$ with $\beta = 2$ (the unitary ensemble is appropriate due to the presence of the magnetic field).

$$P(T) = \frac{1}{2}\beta T^{\frac{\beta}{2}-1}, \quad T \in (0,1). \quad (13.73)$$

This expression is sensitive to the value of the symmetry parameter β. When time-reversal invariance is broken ($\beta = 2$), the ensemble is CUE and $P(g) = 1$ is uniform (since for 1 channel, $g = T$). When time-reversal and spin rotation invariance are both conserved, the relevant ensemble is COE and $P(T)$ is strongly peaked at $g = 0$. Finally, for the CSE (time reversal conserved with $T^2 = -1$ $P(T)$ is peaked at $T = 1$). The distribution of the transmission coefficient for a disordered wire with a single propagating mode for the three symmetry classes is shown in Fig. 13.25.

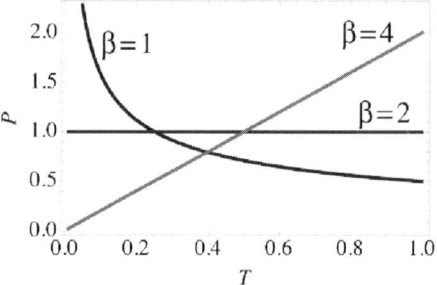

FIG 13.25 The RMT prediction of the distribution $P(T)$ of the transmission coefficient for a disordered wire with a single propagating mode for the three symmetry classes. For the orthogonal ensemble, $\beta = 1$, $P(T)$ is peaked at $T = 0$, for the unitary ensemble, $\beta = 2$, $P(T)$ is uniform, and for the symplectic class, $\beta = 4$, $P(T)$ is peaked at $T = 1$.

Problem 13.15

Calculate $\langle g \rangle$ and $\text{Var}[g]$ for the three-symmetry classes based on Eq. (13.73).

Answer:

$$\langle g \rangle = \int_0^1 dg\, g P(g) = \begin{cases} \frac{1}{4} & (\beta = 1) \\ \frac{1}{2} & (\beta = 2), \\ \frac{2}{3} & (\beta = 4) \end{cases}, \quad \text{Var}[g] = \begin{cases} \frac{11}{80} & (\beta = 1) \\ \frac{1}{12} & (\beta = 2). \\ \frac{5}{18} & (\beta = 4) \end{cases}$$

13.5 KONDO EFFECT IN QUANTUM DOTS

The Kondo effect, discovered in the 1930s, is a many-body phenomenon at the heart of a fundamental problem in the physics of dilute magnetic alloys. Consider the conductivity of a metal (e.g., copper) when it contains a very low concentration of *magnetic impurity atoms*, e.g., Cr. The magnetic impurity atoms are strongly localized in the host metal, hence, the underlying physics involves *localized magnetic moments*. The localized moments are coupled to the conduction electrons and affect the low-temperature electrical conductivity, since they act as scattering centers and tend to increase resistivity. Experiments showed that at low temperature, the resistivity of metals has a shallow minimum at a temperature near 10 K, below which it increases and eventually saturates at $T = 0$. In 1964, it was shown by Jun Kondo that this phenomenon is due to antiferromagnetic exchange scattering between the itinerant electrons and the localized magnetic moments. The shallow minimum represents a competition between the phonon contribution to resistivity, which diminishes as T^5 as $T \to 0$, and the magnetic scattering due to the exchange interaction, $J\mathbf{s} \cdot \mathbf{S}$, between the spin $\mathbf{s}$ of itinerant electrons and the magnetic impurity spin $\mathbf{S}$. This phenomenon pertains to the *Kondo effect in bulk metals*. In 1998, it was found that the Kondo effect also occurs in quantum dots, where the role of the localized moment is played by an electron trapped in the dot. This *Kondo effect in quantum dots* will be discussed in this section. It has some salient features that makes it different from the Kondo effect in bulk metals. A full discussion requires familiarity with theoretical tools presented in Sec. 18.14, linked to the book web page.

The constant-interaction model, while successful in encoding the energetics of charge transfer through quantum dots, ignores the important effects of quantum coherence and the complexities that result due to exchange interactions arising from Coulomb repulsion between the electrons in the dot and the electrons in the leads. Strong Coulomb repulsion makes it energetically unfavorable for the addition of an electron to the dot if the dot is in a Coulomb blockade valley. If the number of electrons in the dot is odd, the single unpaired electron left in the dot behaves as a localized magnetic moment. The derivation of the exchange interaction is qualitatively explained below and further worked out in Sec. 18.13.13 (Chapter 18 is linked to the web page of the book). These many-body aspects must be taken into account at low temperatures, where they affect the behavior of the conductance in a dramatic way. The electron–electron interaction, combined with the Pauli exclusion principle suggests a description of a quantum dot in terms of the Anderson impurity model, which is discussed in Sec.18.13. Some important differences between the Kondo effect in bulk metals and the Kondo effect in quantum dots are as follows.

(1) **The Local Moment Regime.** In a metal containing magnetic impurity atoms (e.g., Cr atoms in copper), the impurity atoms are localized. Achieving the local moment regime in quantum dots requires special conditions. The role of localized magnetic moment is played by an electron trapped in the quantum dot. The number of electrons in the dot must be odd for it to have a nonzero spin, as shown in Fig. 13.18. Moreover, the probability for the electron to escape from the quantum dot, or for other electrons to jump into the dot, should be very small. This is the case if the Coulomb repulsion between electrons on the dot is much larger than the width of the energy levels of the electrons inside the dot. The unpaired electron occupies the upper energy level, and the ground state of the isolated dot is a spin doublet.

(2) **Measurable Quantities.** The Kondo effect in bulk metals can be observed in measurements of both equilibrium properties (such as the magnetic susceptibility χ^M) and response functions (such as the electric conductivity σ). The experimental observables in the quantum dot system are related to transport measurements, in particular, the electric conductance G.

(3) **Low Temperature Physics.** For the Kondo effect in bulk metals, the resistivity increases from a shallow minimum at small T to a larger value at $T = 0$. In quantum dots, the temperature dependence is different. As discussed below, the density of states exhibits a resonance peak at the Fermi energy [the Abrikosov–Suhl resonance, see Eq. (13.78)

and Fig. 13.26]. This leads to a conductance maximum at $T = 0$.

13.5.1 THE DOT KONDO HAMILTONIAN

The first problem is to identify the relevant Hamiltonian that includes the metallic electrodes attached to the dot, the local moment in the dot, and the weak tunneling of electrons in and out of the dot. Consider a quantum dot in the configuration shown in Fig. 13.17(a), with almost zero bias, $\mu_L \approx \mu_R$ equal to the Fermi energy $\varepsilon_F = 0$ in the leads (recall that at low temperature, the chemical potential in metals approximately equals the Fermi energy). Within the Coulomb blockade picture, the state of the dot is deep in the valley between two Coulomb blockade peaks, and the conductance is almost 0.

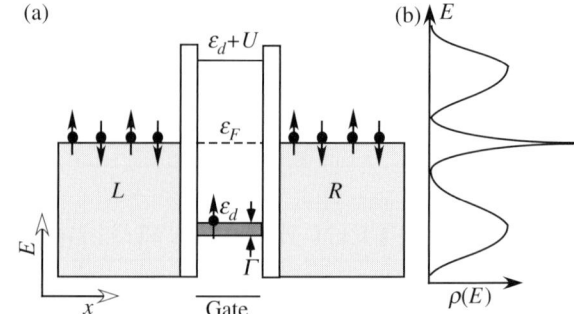

FIG 13.26 (a) Quantum dot as an Anderson impurity. The dot chemical potential $\mu_D = \varepsilon_d$ is tuned by the gate voltage to be deep in the Coulomb blockade valley with very small level broadening Γ, corresponding to Fig. 13.17(a), with equal chemical potentials $\mu_L = \mu_R = \varepsilon_F = 0$. The number of electrons in the dot is odd (only the single electron in the upper level is shown), see Fig. 13.18. (b) Density of states of the dot electron has broad maxima at ε_d and $\varepsilon_d + U$ and a narrow peak at $\varepsilon_F = 0$ (referred to as the Abrikosov–Suhl peak).

The quantum dot shown in Fig. 13.17(a) has the following properties. Electrons from the lead can hop in and out of the dot. A single-particle level in the dot can accommodate two electrons only if they have opposite spins. Using the Fermi energy as reference (taking $\varepsilon_F = 0$) and denoting the upper energy of the dot by $\varepsilon_d < 0$, the dot can accommodate 0, 1, and 2 electrons with corresponding energies 0, ε_d, and $U + 2\varepsilon_d$, where $U = E_c$ is the charging energy. A model for quantitatively describing such a system is the celebrated *Anderson impurity model*, for which the 1D Hamiltonian operator that operates on the N electron wave function $\Psi(n_1\sigma_1, n_2\sigma_2, \ldots, n_N\sigma_N)$ is

$$H = -\sum_{n_i\sigma_i} \Delta_{n_i} - \varepsilon_d \sum_{n_i\sigma_i} \delta_{n_i 0} + V \sum_{i\sigma_i} \delta_{n_i 0} \left(|n_i, \sigma_i\rangle\langle n_i \pm 1, \sigma_i| + |n_i \pm 1, \sigma_i\rangle\langle n_i, \sigma_i| \right) + \frac{1}{2} U \sum_{i \neq j} \delta_{n_i n_j} \delta_{\sigma_i \bar{\sigma}_j}. \qquad (13.74)$$

Here, n_i is the site occupied by electron i, and the attractive single-particle potential is designed to trap electrons in the dot located at site 0. Electrons on sites $n_i \neq 0$ are referred to as lead electrons, either left lead ($n_i < 0$) or right lead ($n_i > 0$). An electron trapped in the dot acts as an impurity that scatters the other electrons. Putting two electrons, electron i and j, in the dot is possible only if their spin projections are opposite $\sigma_i = -\sigma_j = \bar{\sigma}_j$ and costs a "charging" energy $U > 0$. Here, Δ_{n_i} is the second-order difference operator, and the term with V is a nonlocal operator, expressed in terms of projection operators, that lead to tunneling between sites ± 1 and the dot located at 0. When $U \gg |\varepsilon_d|$, the charging energy is too costly and the dot may contain either a single electron or none (the former case is energetically favorable). The possible tunneling of electrons from the dot to the leads means that the level ε_d has a finite lifetime $\tau = \hbar/\Gamma$, where the width Γ given by Eq. (13.50) is the broadening of the level. The parameters of this model are ε_d, U, Γ_s, and Γ_d. Figure 13.26(a) illustrates the quantum dot as an Anderson impurity problem.

Figure 13.26 suggests that the relevant energy scales for the occurrence of a local moment in a quantum dot with odd electron number is

$$U \gg |\varepsilon_d| \gg \Gamma. \qquad (13.75)$$

Under these inequalities, the dot electron is virtually immobile and its only degree of freedom is its spin, encoded by its spin operator **S**. The original Andesron impurity model (13.74) is mapped onto that of an electron gas whose electrons are subject to an exchange interaction induced by a localized quantum magnetic moment. This is the *Kondo model*.

13.5 Kondo Effect in Quantum Dots

The mapping is carried out using the *Schrieffer–Wolf transformation* (see Sec. 9.6.7). A quantitative treatment of the Anderson model, the Schrieffer–Wolf transformation, and the Kondo model in bulk 3D systems will be presented in Sec. 18.13.3, linked to the book web page. For a qualitative understanding of the Kondo Hamiltonian, consider a system of N electrons, for simplicity in 1D, interacting with a quantum spin $\mathbf{S}$ localized at $x = 0$. Denoting by $(x_i, \mathbf{s}_i)$ the space coordinate and spin operator for electron i, and by $T(x_i)$ the kinetic energy operator for electron i, the Hamiltonian is

$$H = \sum_{i=1}^{N}[T(x_i) + J\delta(x_i)\,\mathbf{s}_i \cdot \mathbf{S}], \tag{13.76}$$

where $J > 0$ is the *exchange constant* that can be expressed in terms of U, Γ, and ε_d (in the present notation, it has the dimension energy×length). Despite the apparent simplicity of Hamiltonian (13.76), it is a genuine many-body problem (see Problem 13.16).

Problem 13.16

A student tried to solve the Schrödinger equation $H|\Psi\rangle = E|\Psi\rangle$, where H is given by Eq. (13.76), by defining N single-particle Hamiltonians $h_i \equiv T(x_i) + J\delta(x_i)\mathbf{s}_i \cdot \mathbf{S}$ and employing separation of variables after writing $H = \sum_{i=1}^{N} h_i$. He argued that this procedure is legitimate because the different single-particle Hamiltonians h_i correspond to different electrons. Where is the error in this reasoning?

Hint: Check the commutation relations $[h_i, h_j]$.

13.5.2 THE DOT KONDO TEMPERATURE

One of the central features that distinguishes the Kondo effect in quantum dots from that in bulk 3D systems concerns the behavior of the conductance as function of temperature. To clarify this, we introduce the *dot Kondo temperature*, T_K. For a quantum dot attached to metallic leads described by the Anderson model, the dot Kondo temperature is expressible in terms of the parameters of the Anderson model [211],

$$k_B T_K \approx \sqrt{U\Gamma}\, e^{-\frac{\pi\varepsilon_d(\varepsilon_d+U)}{2\Gamma U}}. \tag{13.77}$$

The significance of the Kondo temperature as an energy scale and as a border-line between two temperature regimes with $T > T_K$ (the weak coupling regime) and $T < T_K$ (the strong coupling regime) will be clarified below.

13.5.3 ABRIKOSOV–SUHL RESONANCE AND KONDO CONDUCTANCE

A remarkable feature of the Kondo effect is that the density of states of the dot electron displays a narrow peak at the Fermi level $\varepsilon_F = 0$, as shown in Fig. 13.26(b). This is referred to as the *Abrikosov–Suhl resonance* [212]. One of the consequences of the Abrikosov–Suhl resonance is that the quantum dot conductance resulting from the Kondo effect rises slowly as a function of decreasing temperature and saturates at $T = 0$, as shown in Fig. 13.27. This is in contrast with the low-energy Kondo physics in bulk 3D systems, where the resistance has a shallow minimum at finite temperature and then slowly rises and saturates as $T \to 0$.

The conductance through the dot in the Kondo limit (unpaired electron localized on the upper dot level, hence only spin degrees of freedom are present) at finite temperature, in terms of the

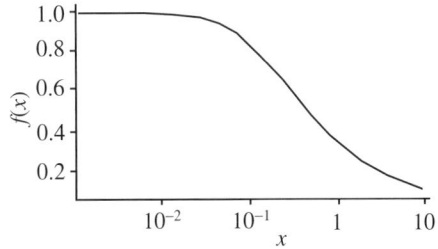

FIG 13.27 The scaling function $f(x)$, $x = T/T_K$ [see Eq. (13.78)]. The behavior at low temperature reflects the Kondo conductance within the Fermi liquid theory developed by Nozières.

dimensional conductances g_s and g_d of the point contacts (see Fig. 13.12), is

$$G = \frac{8e^2}{h} \frac{g_s g_d}{(g_s + g_d)^2} f\left(\frac{T}{T_K}\right), \qquad (13.78)$$

where $f(x)$ is only weakly dependent on the details of the model as long as the local moment behavior holds. Thus, the conductance is given by a "universal function" of the dimensionless variable $x \equiv T/T_K$. The region $T > T_K$ is called the *weak coupling regime*; the perturbation expansion in the exchange constant $J > 0$ is reasonable here and $f(x)$ has the form $f(x) = 1/[1 + J\rho(\varepsilon_F) \ln x]$, where $\rho(\varepsilon)$ is the electronic density of states in the leads. Perturbation theory fails for $T < T_K$ where the logarithm becomes negative. In this case, a non-perturbative calculation of $f(x)$ is required. The function $f(x)$ has been calculated using the formalism of *numerical renormalization group* developed by K. G. Wilson, who received the Physics Nobel prize in 1982. From its structure, shown in Fig. 13.27, it is evident that the conductance increases as T decreases below T_K ($x < 1$). This is the *strong coupling limit*. The conductance shoots up as $T < T_K$ and then saturates, indicating that as $T \to 0$, the scattering becomes weaker. Because the exchange constant J is antiferromagnetic, $J > 0$, the spin of the electrons tend to screen the spin of the localized moment, thereby forming a spin-screening cloud, and the many-body ground state becomes a singlet. When this happens, the rest of the electrons are effectively not scattered (there is no spin-flip), and the conductance increases and reaches the maximal value, $G = 2e^2/h$, at $T = 0$. If the geometry [left lead]-dot-[right lead] is strictly one dimensional, the S matrix $\begin{pmatrix} r & t' \\ t & r' \end{pmatrix}$ has dimension 4, where r and t are 2×2 matrices in spin space. At $T = 0$, there is neither reflection nor spin-flip, $r = r' = 0$ and $t = t' = \text{diag}(t_{\uparrow\uparrow}, t_{\downarrow\downarrow})$. Moreover, since the transmission is perfect, we may write $t_{\uparrow\uparrow} = t_{\downarrow\downarrow} = e^{i\delta}$, where δ is the *Friedel phase*.[3] The Friedel phase is a many-body quantity that expresses charge conservation. Roughly speaking, if a particle of unit charge is screened by a cloud of oppositely charged particles and a scattering event through the particle-cloud system occurs, the *Friedel sum rule* states that the (effective) charge of the screened particles is given by δ/π.

At zero temperature, the occupation of the dot is unity, $n_d = 1$, therefore, $\delta = \frac{\pi}{2}$ per spin. The limit in which $G = 2e^2/h$ and $\delta_\sigma = \pi/2$ is called the *unitary limit*. The low-temperature region, $T \ll T_K$, can be described by Nozières–Fermi liquid theory [160]. The first correction to the unitary limit is [see Eq. (13.78)]

$$f\left(\frac{T}{T_K}\right) \approx 1 - \frac{\pi^2 T^2}{T_K^2}, \quad T \ll T_K. \qquad (13.79)$$

Figure 13.28 illustrates the basic processes contributing to the conductance at finite and at zero temperature.

The consequences of the Kondo effect on the behavior of the low-energy conductance at $T < T_K$ are dramatic, both at zero bias, $V_{sd} \approx 0$, and at finite bias. At zero bias and at low temperature, the Kondo effect is realized only at the valleys between Coulomb blockade peaks, where N is odd (see Fig. 13.18). The upshot is that as $T < T_K$, the conductance

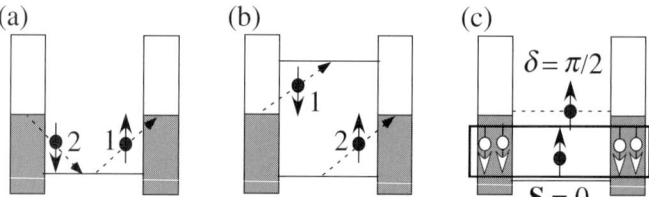

FIG 13.28 Processes contributing to the Kondo conductance of a quantum dot. (a) Direct cotunneling spin-flip occurs at finite temperature. The localized electron in the dot hops to the right lead, and then, an electron from the left lead with opposite spin projection hops into the dot. (b) Indirect cotunneling spin-flip occurs at finite temperature. An electron from the left lead with opposite spin projection hops into the dot and virtually occupies a higher level due to charging energy cost. Then, the localized electron in the dot hops to the right lead and its place is taken by the first electron. (c) At $T = 0$, the magnetic moment of the dot electron is fully screened by a cloud of opposite spin electrons in the leads, and the electron + cloud system is in a singlet state. Electrons in the left lead can pass without being affected by exchange interaction, but they undergo potential scattering. The Friedel sum rule requires that the scattering phase shift is $\delta_\uparrow = \delta_\downarrow = \pi/2$.

---

[3] Identification of the phase of the transmission amplitude with the Friedel phase is justified only if the reflection amplitude vanishes. Otherwise, the Friedel phase is the sum of the two eigenphases of the S matrix.

13.6 Graphene

in valleys with *odd* electron number rises as $T \to 0$, while the conductance in valleys with even electron number decreases slightly. This remarkable even-odd behavior, shown in Fig. 13.29, has been verified experimentally.

At finite bias, the local moment condition is slightly compromised because the depth of ε_d is measured from the lower chemical potential in either the left or the right lead [see Fig. 13.30(a)]. More importantly, the Abrikosov–Suhl resonance splits off into two peaks, one for μ_L and one for μ_R, as shown in Fig. 13.30(b). Note that the integrated density of states must be the same, independent of bias. When the gate voltage is fixed at the center of an odd N Coulomb blockade valley, the *differential conductance* dI/dV_{sd}, measured at low temperature $T \ll T_K$, decreases as function of source-drain voltage V. It has a peak at zero bias and decreases on either side, $V > 0$ or $V < 0$, as shown in Fig. 13.30(c). This zero-bias *anomaly* has been observed experimentally. The width of the peak is proportional to $k_B T_K$.

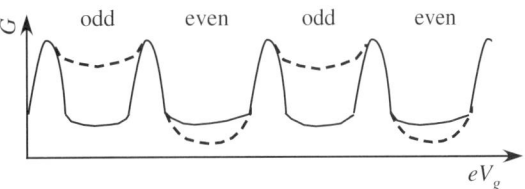

FIG 13.29 The schematic illustration of the Kondo effect in a quantum dot, which occurs at Coulomb blockade valleys with odd number of electrons in the dot. The conductance is displayed as a function of gate voltage at temperature $T > T_K$ (sold line). As the temperature decreases such that $T \ll T_K$, the conductance (dashed line) increases in valleys with odd electron number and decreases in valleys with even electron number.

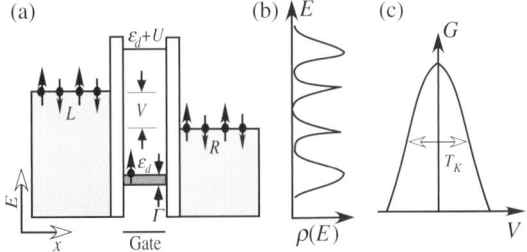

FIG 13.30 (a) Quantum dot in the local-moment configuration at finite bias V [compare with Fig. 13.26(a)]. (b) Density of states of dot electron at finite bias. Compared with Fig. 13.26(b), the Abrikosof–Suhl resonance is split into two peaks, one for each lead at its own chemical potential. (c) Low-temperature conductance of the quantum dot as function of V. It has a peak at zero bias with width $K_B T_K$. This is the zero-bias anomaly.

13.6 GRAPHENE

Graphene is a flat monolayer of carbon atoms arranged in a honeycomb lattice (see Fig. 13.31, top left). It is the thinnest known material and the strongest ever observed. The discovery of graphene, a natural planar material only one atom thick, came as a surprise, since there are predictions that an infinite 2D sheet of this type would not be stable against fluctuations that buckle the planar material. The discovery of graphene has heightened the recent interest in low-dimensional physics.

The basic chemistry of graphene is as follows. Carbon atoms have two $2s$ and two $2p$ occupied orbitals, i.e., $1s^2 2s^2 2p^2$. In graphene, these atomic orbitals hybridize into sp^2 orbitals; one s orbital and two in-plane p orbitals participate in strong in-plane covalent bonding, forming σ bonding orbitals and σ^* anti-bonding orbitals. The remaining p_z orbital oriented perpendicularly to the plane of graphene forms π (valence) and π^* (conduction) bands that participate in electronic conduction. The electronic structure is illustrated in Fig. 13.32. This electronic structure also explains how graphene forms the building block for other allotropes of carbon-based materials (see Fig. 13.31), such as three-dimensional graphite, quasi-one-dimensional carbon nanotubes [216] (see Sec. 13.6.1), and zero-dimensional fullerenes [213].

Graphene received early theoretical attention in the 1940s and 1950s [217] and was first isolated experimentally in 2004 by A. K. Geim and K. S. Novoselov (who were awarded the 2010 physics Nobel prize) and their collaborators [218]. The technique for fabricating graphene involves drawing with a piece of graphite (e.g., with a pencil) and repeated peeling with adhesive tape until the thinnest flakes are found. The problem is that graphite flakes of 10–100 layers thick are typically found because graphene crystallites left on a substrate are extremely rare and hidden among the thousands of thick graphite flakes. A critical ingredient for success was that graphene becomes visible in an optical microscope

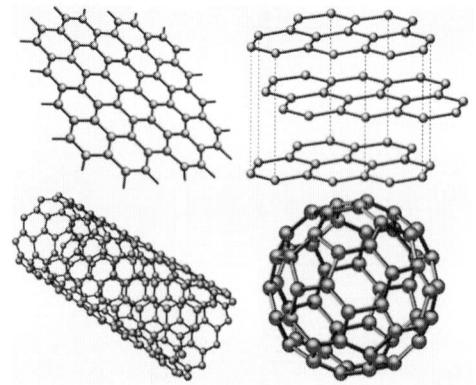

FIG 13.31 Some of the main materials composed only of carbon atoms. Top left: Graphene, a honeycomb lattice of carbon atoms. Top right: Graphite built from graphene layers shifted with respect to each other. Bottom left: Carbon nanotube, obtained by rolling graphene into a cylinder. Bottom right: C_{60} molecules (fullerene), 60 carbon atoms forming a structure with 12 pentagons and 20 hexagons [213]. (Reproduced from Ref. [214]).

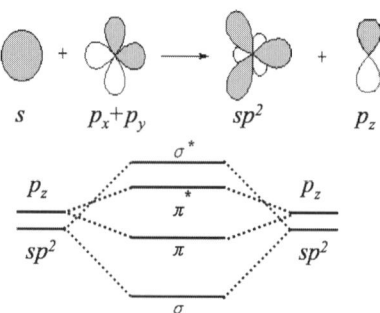

FIG 13.32 Electronic structure of graphene.

when placed on top of a silicon wafer with a carefully chosen SiO_2 thickness, due to an interferometric contrast with the silicon wafer. A mere 5% difference in silicon thickness (315 nm instead of the standard 300 nm) can make graphene completely invisible. More recently, graphene was found to have a clear signature in Raman microscopy [219], and this Raman technique is now used to categorize the crystallites obtained via optical microscopy.

Basic Properties of Graphene

Graphene is a very good conductor of both heat and electricity. It behaves in some ways like both a metal and a semiconductor. If the electrodes are placed at either end of a sheet, and a gate voltage is applied across the surface, the electrical conductance along the sheet is different for different values of the gate voltage, as is the case for semiconductors. But the conductance does not go to 0 when the gate voltage drops below a certain value, and in this sense, it behaves like a metal. The charge carriers exhibit very high intrinsic mobility and can have zero effective mass. They can propagate without scattering for micrometers at room temperature. Graphene can sustain current densities six orders of magnitude higher than that of copper and has record thermal conductivity and stiffness.

Graphene is an ideal material for exploring unusual properties of the 2D electron gas because, as we will see below, electrons in graphene behave as massless relativistic particles. If the graphene sheet is supported by a substrate material, the ballistic transport of charge carriers can be perturbed by interaction of the itinerant electrons with the substrate. Remarkably, this perturbation can be circumvented. It has proved experimentally feasible to fabricate suspended graphene sheets with a carrier mobility of $\mu \approx 2 \times 10^5$ cm^2/Vs and an electron density of $n \approx 2 \times 10^{11}$ cm^{-2}. The observed mobility depends only weakly on temperature [220]. This implies ballistic transport on a submicron scale even at room temperature. For comparison, in un-doped semiconductors (with very low carrier density), the highest mobility achieved is about 7.7×10^4 cm^2/Vs.

13.6.1 CARBON NANOTUBES

The most studied nanotubes are made of carbon, but inorganic nanotubes have also been synthesized, including boron nitride, silicon titanium oxide, tungsten disulfide, molybdenum disulfide, copper, and bismuth. Carbon nanotubes (CNTs)

13.6 Graphene

were discovered by Sumio Iijima in 1991 [216]. They were first synthesized as a byproduct in arc discharges used in the synthesis of fullerenes. Currently, CNTs are prepared by a variety of methods, including arc discharge, laser ablation, and catalytic decomposition of hydrocarbons. CNTs are light weight, have extremely high mechanical strength (larger tensile strength than steel), in the absence of oxygen, they withstand extreme heat (2000°C), and emit electrons efficiently when subjected to an electrical field. A single-walled CNT is graphene rolled into a cylinder with radius on the order of a nanometer. The length-to-radius ratio of single-walled CNT can exceed 10,000. The small diameter and the crystalline perfection of the atomic network make these ideal carbon-based 1D structures. See Figs 13.33 and 13.34 for illustrations of various carbon nanotube configurations.

Single-walled CNTs are often labeled in terms of the planar graphene basis vectors $\mathbf{a}_1$ and $\mathbf{a}_2$, with $|\mathbf{a}_1| = |\mathbf{a}_2| = 0.246$ nm $\equiv \sqrt{3}a$, where $a = 0.142$ nm is the carbon–carbon bond distance, as shown in Fig. 13.33. In this figure, we have taken the basis vectors of the hexagonal honeycomb lattice to be $\mathbf{a}_{1,2} = \frac{a}{2}(3, \pm\sqrt{3})$. The *chiral vector*, $\mathbf{C}_{n_1,n_2} = n_1\mathbf{a}_1 + n_2\mathbf{a}_2$, with integers n_1 and n_2, determines the circumference of the tube, hence, it is sometimes referred to as the *circumferential vector*. Therefore, the radius of the nanotube is

$$r = \frac{|\mathbf{C}_{n_1,n_2}|}{2\pi} = \frac{a}{2\pi}\sqrt{n_1^2 + n_1 n_2 + n_2^2}. \quad (13.80)$$

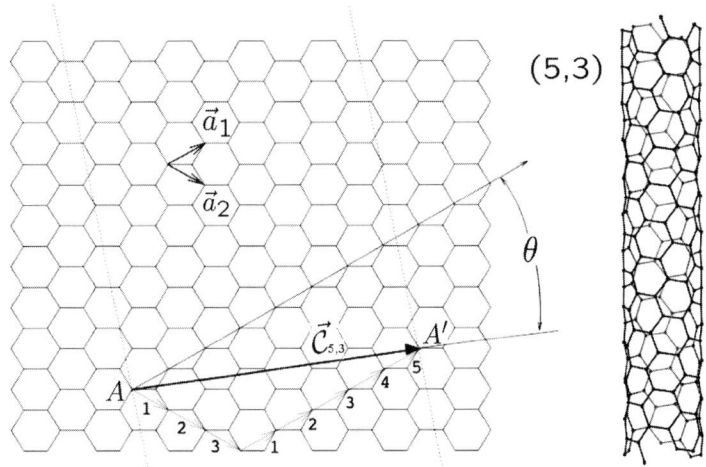

FIG 13.33 Left: Graphene honeycomb with lattice vectors $\mathbf{a}_1$ and $\mathbf{a}_2$. The horizontal (vertical) boundaries are armchair (zigzag) type. The chiral vector $\mathbf{C}_{5,3} = 5\mathbf{a}_1 + 3\mathbf{a}_2$ represents a possible wrapping of the two-dimensional graphene sheet into a tube. The direction perpendicular to $\mathbf{C}_{n_1,n_2}$ is the tube axis. The chiral angle is defined by the $\mathbf{C}_{n_1,n_2}$ vector and the $\mathbf{a}_1$ zigzag direction of the graphene lattice. Right: A 5,3 nanotube is constructed and the resulting tube is illustrated. (Adapted from Charlier et al. [221].)

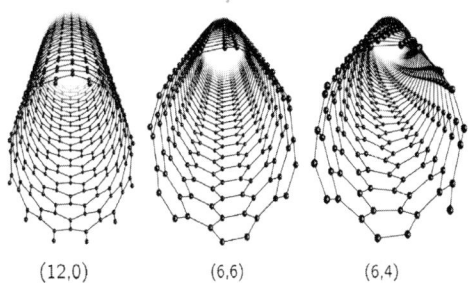

FIG 13.34 Atomic structures of (12, 0) zigzag, (6, 6) armchair, and (6, 4) chiral nanotubes. (Reproduced from Ref. [221].)

The chiral angle θ, i.e., the angle between $\mathbf{C}_{n_1,n_2}$ and $\mathbf{a}_1$ is

$$\cos\theta = \frac{\mathbf{C}_{n_1,n_2} \cdot \mathbf{a}_1}{|\mathbf{C}_{n_1,n_2}||\mathbf{a}_1|} = \frac{2n_1 + n_2}{2\sqrt{n_1^2 + n_1 n_2 + n_2^2}}. \quad (13.81)$$

The chiral angle θ also determines the tilt angle of the hexagons with respect to the direction of the nanotube axis. The nanotubes with $(n_1, n_2) = (n, 0)$, i.e., $\theta = 0$, are called zigzag tubes, because they exhibit a zigzag pattern along the circumference. These tubes have carbon–carbon bonds parallel to the nanotube axis. Nanotubes with $(n_1, n_2) = (n, n)$, i.e., $\theta = 30$ degrees, are called armchair tubes, because they exhibit an armchair pattern along the circumference and have carbon–carbon bonds perpendicular to the nanotube axis. Both zigzag and armchair nanotubes are achiral. Figure 13.34 shows the atomic arrangement of three-carbon nanotubes, zigzag, armchair, and $(n_1, n_2) = (6, 4)$. Carbon nanotubes can be either metals or semiconductors, depending on their diameters and helical arrangements. The $\mathbf{k} \cdot \mathbf{p}$ method introduced in Sec. 9.6.7 can be used to analytically describe electronic states of carbon nanotubes. It shows, for example, that the bandgap of a semiconducting carbon nanotubes is inversely proportional to the diameter because of a linear dispersion of the bands. It is also suitable for describing the electronic structure in external perturbations due to electric and magnetic fields.

13.6.2 LATTICE STRUCTURE AND DIRAC CONES

Consider electrons in a 2D honeycomb lattice, assuming that the graphene sheet is perfectly planar. This is an idealization, since 2D graphene crystals gain stability via reduction of elastic energy by forming shallow ripples along the perpendicular dimension, which leads to a small correction that will not be taken into account in our discussion. The physics of charge carriers (electrons or holes) in graphene is related to its lattice structure [214, 215]. Here, we discuss the lattice structure in real and reciprocal space, identify the Brillouin zone and mark its important symmetry points. Then, we present a qualitative description of the energy dispersion, i.e., its linear form, its gapless feature, and its relation to Dirac's relativistic theory of the electron (for treating the physics of charge carriers in graphene, a quantitative analysis requires a basic knowledge of Dirac's theory in 3D and in 2D).

Lattice Structure

Graphene is not a Bravais lattice; it is composed of two triangular Bravais lattices A and B, as shown in Fig. 13.35. The two triangular lattices are shifted with respect to each other to form a honeycomb lattice. Alternatively, graphene can be regarded as a single-triangular lattice with two atoms per unit cell. We shall analyze the geometry in direct and reciprocal space with the help of Fig. 13.36.

For the Bravais sublattice A, the basis vectors in the direct and reciprocal spaces are

$$\mathbf{a}_{1,2} = \frac{a}{2}(3, \pm\sqrt{3}), \quad \mathbf{b}_{1,2} = \frac{2\pi}{3a}(1, \pm\sqrt{3}), \tag{13.82}$$

respectively, where $a \approx 1.42$ Å is the distance between two neighboring carbon atoms. The three carbon atoms of sublattice B closest to the atom of sublattice A located at the origin are positioned at

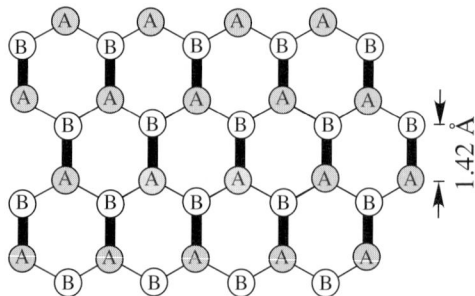

FIG 13.35 Honeycomb 2D lattice consisting of two shifted triangular lattices, denoted as A and B. In the combined structure, they are referred to as the A and B sublattices. The lattice is invariant under $120°$ rotations around any lattice site. Alternatively, the honeycomb lattice can be regarded as a triangular Bravais lattice A with two carbon atoms A and B per unit cell, joined by a thick line. The shape of the horizontal edges is zigzag, whereas that of the perpendicular edges is armchair.

$$\mathbf{d}_{1,2} = \frac{a}{2}(1, \pm\sqrt{3}), \quad \mathbf{d}_3 = a(-1, 0). \tag{13.83}$$

The Brillouin zone is a hexagon, as shown in Fig. 13.36(b). If the corners are ordered from 1 to 6, then corners with odd numbers are the corners of the Brillouin zone of the direct lattice A, denoted as $\mathbf{K}$, while the corners with even numbers are the corners of the Brillouin zone of direct lattice B, denoted as $\mathbf{K}'$. For example, the two adjacent apexes in Fig. 13.36(b) are given by,

$$\mathbf{K} = \frac{2\pi}{a}\left(\frac{2}{3}, \frac{1}{3\sqrt{3}}\right), \quad \mathbf{K}' = \frac{2\pi}{a}\left(\frac{2}{3}, -\frac{1}{3\sqrt{3}}\right). \tag{13.84}$$

13.6 Graphene

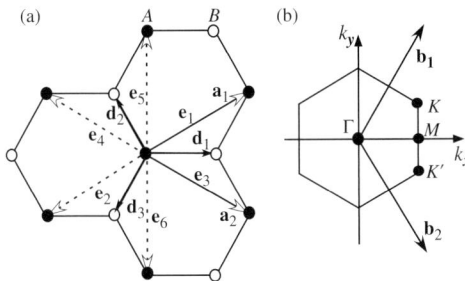

FIG. 13.36 (a) Direct lattice of graphene. The vectors $\mathbf{a}_1$ and $\mathbf{a}_2$ are the basis vectors for the Bravais sublattice A, and the vectors $\mathbf{d}_{1,2,3}$ connect points in sublattice A with their nearest neighbors in sublattice B. The vectors $\mathbf{e}_1, \ldots, \mathbf{e}_6$ connect points in sublattice A with their nearest neighbors in the same sublattice. (b) Brillouin zone in reciprocal lattice. $\mathbf{b}_{1,2}$ are the basis vectors in the reciprocal lattice corresponding to the direct lattice A. Similar basis vectors can be constructed for sublattice B. The bisectors of these reciprocal lattice vectors are the borders of the Brillouin zone. The two adjacent points K and K' are not equivalent (they are not connected by a reciprocal lattice vector). They are called Dirac points and play a crucial role in the physics of graphene. (Reproduced from Ref. [214].)

The points $\mathbf{K}$ and $\mathbf{K}'$ play a crucial role in the Bloch theory of electrons in graphene. As we show below, for clean graphene, the gap between valence and conduction bands is closed at only these points. Moreover, close to these points, the energy spectrum is linear, $\varepsilon(\mathbf{k}) = \pm C|\mathbf{k} - \mathbf{K}|$.

Dirac Cone Dispersion: Qualitative Description

The Bloch wave function for electrons on the two-dimensional honeycomb lattice can be obtained within the tight-binding model where electrons hop from site to site (see Sec. 13.6.4). Since the honeycomb lattice is a triangular lattice with two electrons per unit cell, the Schrödinger equation couples the wave function components on both sublattices. This leads to an eigenvalue problem for the energy $\varepsilon(\mathbf{k})$, which appears in an expression involving $\varepsilon^2(\mathbf{k})$. This is similar (but not identical) to what happens in the relativistic theory of the electron developed by Dirac more than eight decades ago. In Dirac theory, the energy of an electron of mass m_0 and momentum $\mathbf{p} = \hbar \mathbf{k}$ is given by $\varepsilon^2(\mathbf{k}) = (\hbar c \mathbf{k})^2 + m_0^2 c^4$, where c is the speed of light. Dirac's original theory was developed for electrons in 3D. Electrons in graphene are similar to relativistic electrons in quantum electrodynamics in two-space and one-time dimensions (referred to as 2 + 1 dimensions), but with c approximately equal to the speed of light divided by 300. The Dirac electron energy dispersion for graphene has the form $\varepsilon(\mathbf{k}) = \pm t\sqrt{3 + f(\mathbf{k})}$, where t is an energy scale. The function $f(\mathbf{k})$, whose precise expression will be specified in Eq. (13.111), is such that the quantity $3 + f(\mathbf{k})$ vanishes quadratically at $\mathbf{K}$ and $\mathbf{K}'$, hence,

$$\varepsilon(\mathbf{k}) \approx \pm C|\mathbf{k} - \mathbf{K}|, \quad |\mathbf{k} - \mathbf{K}| \ll \frac{1}{a}, \quad \varepsilon(\mathbf{k}) \approx \pm C|\mathbf{k} - \mathbf{K}'|, \quad |\mathbf{k} - \mathbf{K}'| \ll \frac{1}{a}, \quad (13.85)$$

where C is a constant that will be determined in Eq. (13.117). Near $\mathbf{K}$ and $\mathbf{K}'$, the bands $\varepsilon(k_x, k_y)$ form conical surfaces that touch, as shown in Fig. 13.37; these regions are referred to as *Dirac cones*. It should be stressed that, in the original Dirac theory, the electron mass results in a quadratic spectrum with a gap. The linear spectrum without a gap is obtained from Dirac theory only if the mass is 0. The regions around $\mathbf{K}$ and $\mathbf{K}'$ are referred to as *valleys*. There are only two valleys in each Brillouin zone, shown in Fig. 13.36. The other points shown in Fig. 13.37 are obtained by moving the first two by reciprocal lattice vectors, hence, they belong to different zones.

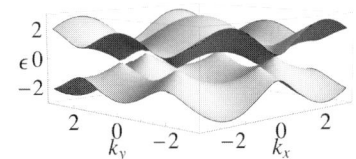

FIG. 13.37 Energy versus momentum of the π and π^* bands showing Dirac cone structures at $\mathbf{K}$ and $\mathbf{K}'$.

The Fermi energy in undoped graphene lies where the upper (conduction) band and the lower (valence) band meet at the Dirac points. It is experimentally possible to add or subtract electrons or holes by implanting (doping) donors or acceptors. Undoped graphene is a zero-gap semiconductor, in which low-energy quasi-particles (electrons or holes) within each valley $\mathbf{K}$ or $\mathbf{K}'$ are characterized by linear dispersion, as in Eq. 13.85. Adding more electrons to the sample (doping) moves the Fermi energy away from the Dirac points and changes the dispersion relation near the Fermi energy, making it quadratic.

Elucidating the behavior of electrons in graphene within the theory of massless Diracs fermions requires some background. The two concepts that are to be employed are the Dirac theory and the tight-binding model. We start with a short review of Dirac's electron theory in 3 + 1 dimensions, and then study massless Dirac fermions in 2 + 1 dimensions. Finally, within the tight-binding formulation, we will derive the energy dispersion and obtain the corresponding Dirac Hamiltonian near **K** and **K**′.

Before introducing the Dirac theory of the electron, we note that the relation between the physics of electrons in graphene and the Dirac equation can be described as follows: We know that a quadratic spectrum, $\varepsilon(\mathbf{k}) \propto k^2$, is associated with the kinetic energy operator that is proportional to the Laplacian operator. A linear spectrum $\varepsilon(\mathbf{k}) \propto |\mathbf{k}|$ is associated with the gradient operator in configuration space, $\mathbf{k} \to -i\hbar \nabla_\mathbf{r}$, i.e., an operator proportional to a first-order derivatives with respect to space, compared with the Laplacian, which has second-order derivatives. As we shall soon find out, the kinetic energy operator in the Dirac theory of the electron is also expressed in terms of the gradient operator (and not the Laplacian operator). Therefore, the cones defined in Eq. 13.85 are referred to as *Dirac cones*. However, it should be stressed that in the Dirac electron theory, the expression relating energy and momentum $\mathbf{p} = \hbar \mathbf{k}$ is $\varepsilon^2(\mathbf{k}) = (\hbar c \mathbf{k})^2 + m_0^2 c^4$. This spectrum does have a gap; hence, there are no Dirac cones. In particular, for $|\hbar \mathbf{k}| \ll m_0 c$, one has $\varepsilon_\pm(\mathbf{k}) \approx \pm m_0 c^2 (1 + \frac{(\hbar k)^2}{2 m_0 c^2})$. This is quadratic dispersion with a gap, $\varepsilon_+(0) - \varepsilon_-(0) = 2 m_0 c^2$. Linear dispersion with vanishing gap are obtained either if $m_0 = 0$ or in the ultrarelativistic limit $\hbar |\mathbf{k}|/(m_0 c) \gg 1$. Hence, electrons in graphene are said to behave as *massless Dirac fermions in two dimensions* because vanishing mass in the Dirac's theory implies a vanishing gap.

13.6.3 DIRAC EQUATION AND ITS RELEVANCE TO GRAPHENE

One of the hallmarks of graphene is its unusual electronic spectrum. Electron states with wave numbers **k** near **K** or **K**′ are described by a Dirac-like equation, similar to the Dirac equation for relativistic electrons. Electrons moving in graphene do not move relativistically, but their interaction with the periodic potential of the honeycomb lattice gives rise to quasi-particles with a linear dispersion (energy proportional to |**k**|). Instead of the speed of light c appearing in the Dirac equation, the speed for electrons in graphene is the Fermi velocity $v_F \approx 10^6$ m/s. With the experimental breakthrough in forming graphene sheets, the quantum electrodynamics of particles moving at much lower speed than c, and the electronic properties of graphene, can be probed.

We briefly present the basic ingredients of the original Dirac electron theory for electrons with mass m_0 moving in 3 + 1 dimensions (3 for space and 1 for time). Then, we treat the case of graphene, where electrons obey the Dirac equation for massless particles in 2 + 1 dimensions. The reader familiar with the Dirac equation can skip the next two sections and move to Sec. 13.6.4.

A Brief Overview of Dirac Electron Theory

The usual theory of electrons in atoms, molecules, and condensed matter physics is governed by the Schrödinger equation. For free electrons, the Hamiltonian consists of the kinetic energy $H_0 = -\frac{\hbar^2}{2m} \nabla^2$, which leads to a second-order differential equation for the wave function. Energy eigenfunctions are plane waves $e^{i\mathbf{k} \cdot \mathbf{r}}$ with energies $E_\mathbf{k} = \frac{\hbar^2}{2m} k^2$. Charge carriers in graphene are described by a different equation called the *Dirac equation* [222]. The Dirac equation combines the principles of quantum mechanics and special relativity. One of its properties is that space and time appear on equal footing, in the sense that differential operators are of first degree only, as opposed to the time-dependent Schrödinger equation, where the time derivative appears in first order, while space derivatives appear in second order.

In an attempt to fulfill the requirement of first-order space and time derivatives, we write a time-dependent equation for an electron of mass m in three-space dimensions,

$$i\hbar \frac{\partial \psi(\mathbf{r}, t)}{\partial t} = \left[c \sum_{k=1}^{3} \alpha_k p_k + \beta m c^2 \right] \psi(\mathbf{r}, t) \equiv H_D \psi(\mathbf{r}, t), \qquad (13.86)$$

13.6 Graphene

where H_D is *the Dirac Hamiltonian*, $p_k = -i\hbar \partial/\partial x_k$ (where $k = 1, 2$, and 3), c is the speed of light, and the dimensionless coefficients α_k and β are to be determined. Equation (13.86) is referred to as Dirac equation in $3 + 1$ dimensions. The space $\mathbf{r}$ and time t variables are now on an "equal footing"; they appear in terms of first-order derivatives. It is reasonable to require that if the Dirac Hamiltonian is applied twice, it should not contain mixed derivatives $p_i p_j$ with $i \neq j$. To satisfy this, the coefficients $\alpha_j, j = 1, 2, 3$, and β cannot be scalars, but rather must be (Hermitian) 4×4 matrices (or larger dimension) with appropriate algebraic properties. Otherwise, Eq. (13.86) does not have the correct physical interpretation (see below). Hence, the wave function $\psi(\mathbf{r}, t)$ must have four components. The basic properties of the matrices α, β are

$$\{\alpha_i, \alpha_j\} \equiv \alpha_i \alpha_j + \alpha_j \alpha_i = 0 \quad (i \neq j), \tag{13.87a}$$

$$\{\alpha_i, \beta\} = 0, \tag{13.87b}$$

$$\alpha_i^2 = \beta^2 = 1_{4 \times 4}. \tag{13.87c}$$

A collection of matrices obeying Eqs (13.87) is said to satisfy a *Clifford algebra*. Equation (13.86) with the requirements (13.87) on the matrices is one form of the Dirac equation, describing a free electron of spin 1/2, consistent with the basic principles of quantum mechanics and special relativity. This formalism is called the *Dirac relativistic electron theory*. The connection with spin 1/2 becomes more transparent once a specific representation of the 4×4 matrices α_k and β is given. For this purpose, define a three-component vector of 4×4 matrices, $\boldsymbol{\alpha} = (\alpha_x, \alpha_y, \alpha_z)$, in terms of the Pauli spin matrices $\boldsymbol{\sigma}$ and the 4×4 matrix β as follows [Eq. (13.87)]:

$$\boldsymbol{\alpha} = \begin{pmatrix} 0 & \boldsymbol{\sigma} \\ \boldsymbol{\sigma} & 0 \end{pmatrix}, \quad \beta = \begin{pmatrix} 1 & 0 \\ 0 & -1 \end{pmatrix}. \tag{13.88}$$

Problem 13.17

Use Eq. (13.88) to prove the anti-commutation relations (13.87).

If the four-component wave function $\psi(\mathbf{r}, t)$ is written in terms of a pair of two-component spinor wave functions, $\psi(\mathbf{r}, t) = \begin{pmatrix} \mathcal{U}(\mathbf{r}, t) \\ \mathcal{V}(\mathbf{x}, t) \end{pmatrix}$, the Dirac equation (13.86) takes the form,

$$i\hbar \frac{\partial}{\partial t} \begin{pmatrix} \mathcal{U}(\mathbf{r}, t) \\ \mathcal{V}(\mathbf{r}, t) \end{pmatrix} = \begin{pmatrix} mc^2 & c\boldsymbol{\sigma} \cdot \mathbf{p} \\ c\boldsymbol{\sigma} \cdot \mathbf{p} & -mc^2 \end{pmatrix} \begin{pmatrix} \mathcal{U}(\mathbf{r}, t) \\ \mathcal{V}(\mathbf{r}, t) \end{pmatrix}. \tag{13.89}$$

Stationary solutions of Eq. (13.89) are plane waves,

$$\psi(\mathbf{r}, t) = \begin{pmatrix} \mathcal{U}(\mathbf{r}, t) \\ \mathcal{V}(\mathbf{x}, t) \end{pmatrix} = e^{i(\mathbf{k} \cdot \mathbf{r} - \frac{E}{\hbar} t)} \begin{pmatrix} u_\lambda \\ v_\lambda \end{pmatrix}, \tag{13.90a}$$

$$\begin{pmatrix} mc^2 & c\boldsymbol{\sigma} \cdot \mathbf{p} \\ c\boldsymbol{\sigma} \cdot \mathbf{p} & -mc^2 \end{pmatrix} \begin{pmatrix} u_\lambda \\ v_\lambda \end{pmatrix} = E_\lambda \begin{pmatrix} u_\lambda \\ v_\lambda \end{pmatrix}, \tag{13.90b}$$

where $u_\lambda(\mathbf{k})$ and $v_\lambda(\mathbf{k})$ are two-component spinors, and $\lambda = 1, 2, 3, 4$ enumerates the solutions of the *algebraic* 4×4 eigenvalue problem obtained when $\mathbf{p} \to \mathbf{k}$ in the matrix multiplying $\begin{pmatrix} u_\lambda \\ v_\lambda \end{pmatrix}$. The eigenenergies are

$$E_{1\mathbf{k}} = E_{2\mathbf{k}} = c\sqrt{\hbar^2 k^2 + m^2 c^2} \equiv E_+, \tag{13.91a}$$

$$E_{3\mathbf{k}} = E_{4\mathbf{k}} = -c\sqrt{\hbar^2 k^2 + m^2 c^2} \equiv E_- = -E_+. \tag{13.91b}$$

The two degenerate solutions corresponding to $\lambda = 1, 2$ are *positive energy solutions*, and the two degenerate solutions corresponding to $\lambda = 3, 4$ are *negative energy solutions*. The eigenvectors are

$$\lambda = 1 \Rightarrow u_1 = 1, \quad u_2 = 0, \quad v_1 = -\frac{ck_z}{E_+ + mc^2}, \quad v_2 = -\frac{c(k_x + ik_y)}{E_+ + mc^2},$$

$$\lambda = 2 \Rightarrow u_1 = 0, \quad u_2 = 1, \quad v_1 = -\frac{c(k_x - ik_y)}{E_+ + mc^2}, \quad v_2 = \frac{ck_z}{E_+ + mc^2},$$

$$\lambda = 3 \Rightarrow u_1 = \frac{ck_z}{-E_- + mc^2}, \quad u_2 = \frac{c(k_x + ik_y)}{-E_- + mc^2}, \quad v_1 = 1, \quad v_2 = 0,$$

$$\lambda = 4 \Rightarrow u_1 = \frac{c(k_x - ik_y)}{-E_- + mc^2}, \quad u_2 = -\frac{ck_z}{-E_- + mc^2}, \quad v_1 = 0, \quad v_2 = 1.$$

(13.92)

There are several consequences of the dispersion relations (13.91).

- Each energy is doubly degenerate. This is a consequence of Kramers degeneracy resulting from time-reversal invariance. Kramers degeneracy at the same value of $\mathbf{k}$ can occur only for particles with spin, hence, the Dirac formalism implements the spin of the particle in a natural way.
- To each solution with positive energy, there corresponds a solution with negative energy of the same magnitude. A positive energy solution is referred to as a *particle* (p), whereas a negative energy solution is referred to as a *hole* (h).
- For $\mathbf{k} = 0$, there is an energy gap $2mc^2$ between the highest hole energy and the lowest particle energy.
- In the *ultra-relativistic limit*, $\hbar|\mathbf{k}| \gg mc$, the particle and hole energies can be approximated as $E_k \approx \pm \hbar c|\mathbf{k}|$, i.e., the dispersion relation is linear.
- In the *non-relativistic limit*, $\hbar|\mathbf{k}| \ll mc$, the approximate particle and hole energies are $E_k \approx \pm(mc^2 + \frac{\hbar^2 k^2}{2m})$. The quantity mc^2 is the *rest mass energy* of the electron whose *rest mass* is m. Subtracting the rest mass energy from the total energy yields the *quadratic dispersion* of the nonrelativistic Schrödinger equation. In this limit, for E_+, we have the inequality, $|u_1|^2 + |u_2|^2 \gg |v_1|^2 + |v_2|^2$, while for E_-, the inequality reads, $|u_1|^2 + |u_2|^2 \ll |v_1|^2 + |v_2|^2$. Accordingly, $\mathcal{U}(\mathbf{r}, t)$ is the "big" solution for positive energy and the "small" solution for negative energy, whereas $\mathcal{V}(\mathbf{r}, t)$ is the "small" solution for positive energy and the "big" solution for negative energy.
- $\mathcal{U}(\mathbf{r}, t)$ and $\mathcal{V}(\mathbf{r}, t)$ are two-component spinors. The two components of $\mathcal{U}(\mathbf{r}, t)$ and $\mathcal{V}(\mathbf{r}, t)$ are the amplitudes for the particle to be spin-up and spin-down.

A magnetic field $\mathbf{H}$ can be included in the Dirac equation, Eq. (13.86), by the replacement, $\mathbf{p} \to \mathbf{\Pi} \equiv \mathbf{p} + \frac{e}{c}\mathbf{A}$, where $\mathbf{A}$ is the vector potential such that $\mathbf{H} = \nabla \times \mathbf{A}$. If, in addition, the electron is subject to an electric field $e\mathbf{E} = -\nabla V(\mathbf{r})$, the Dirac Hamiltonian becomes,

$$H_D = c\boldsymbol{\alpha} \cdot \mathbf{\Pi} + V(\mathbf{r}) + \beta mc^2.$$

(13.93)

The presence of a potential breaks translation invariance and $\mathbf{k}$ is no longer a good quantum number. However, as long as $|V(\mathbf{r})| < 2mc^2$, the assignment of particles and holes makes sense and λ remains a relatively good quantum number. When the sign of the energy is specified (say $E > 0$), and in the nonrelativistic limit, the small component $\mathcal{V}(\mathbf{r}, t)$ can be dropped, leaving only the large one, $\mathcal{U}(\mathbf{r}, t)$, to arrive at a set of two equations for the large component spinor $\mathcal{U}(\mathbf{r}, t)$. The price of reducing the number of equations from four to two is that the resulting two equations are more complicated.

In Eq. (13.89), the two components of the spinor $\mathcal{V}(\mathbf{r}, t)$ are related to the two components of the spinor $\mathcal{U}(\mathbf{r}, t)$ by a factor of the order of $x \equiv \epsilon/mc^2$, where ϵ can be any energy in the original equation, such as V or $\mathbf{p}^2/2m$. At low energies, $x \ll 1$, and the expansion in powers of x up to first order is justified, resulting in the time-dependent equation for $\mathcal{U}(\mathbf{r}, t)$, referred to as the *Pauli equation*,

$$i\hbar \frac{\partial \mathcal{U}(\mathbf{r},t)}{\partial t} = \left[mc^2 + V + \frac{\mathbf{\Pi}^2}{2m} + 2\frac{e\hbar}{2mc}\left(\frac{\boldsymbol{\sigma}}{2}\right) \cdot \mathbf{H} + \lambda\,(\mathbf{\Pi} \cdot \boldsymbol{\sigma} \times \mathbf{E} + \boldsymbol{\sigma} \times \mathbf{E} \cdot \mathbf{\Pi}) + \lambda \hbar \nabla \cdot \mathbf{E} \right] \mathcal{U}(\mathbf{r},t),$$

(13.94)

where $\lambda \equiv \frac{e\hbar}{8m^2c^2}$ determines the spin–orbit strength and $\mathbf{E} = -\boldsymbol{\nabla} V/e$ is the electric field. This expansion is valid when $\lambda\hbar \left|\frac{\boldsymbol{\nabla} V}{V}\right| \ll 1$. Equation (13.94) contains the non-relativistic reduction of the Dirac equation that yields all the terms in the Schrödinger equation that involve spin and magnetic effects to order v^2/c^2, where v is the electron velocity. Thus, the electron spin emerges from the Dirac equation in a natural way and there is no need to introduce it *ad hoc*. The first term on the RHS of the Pauli equation is just the rest mass energy. The next two terms are the potential and kinetic energies, familiar from the non-relativistic Schrödinger equation for a spinless particle in a magnetic field. The fourth term is the Zeeman Hamiltonian, $H_Z = -\boldsymbol{\mu} \cdot \mathbf{H} = (g\mu_B/\hbar)\, \mathbf{S} \cdot \mathbf{H}$, where $\mathbf{S} = \hbar\boldsymbol{\sigma}/2$ is the electron spin operator, $g = 2$ is the electron g-factor, and $\mu_B = \frac{e\hbar}{2mc}$ is the Bohr magneton. The Dirac (and Pauli) equation yields $g = 2$; this is one of the achievements of Dirac electron theory. The magnetic moment of the electron is $\boldsymbol{\mu} = -g(\mu_B/\hbar)\,\mathbf{S}$. The g-factor g is not exactly equal to 2, because of radiative corrections that are higher order in $\alpha = e^2/(\hbar c)$. The small deviation of the magnetic moment resulting from the small deviation of g from 2 is referred to as the *anomalous magnetic moment* of the electron. The fifth term in Eq. (13.94) is the *spin–orbit* Hamiltonian. For a central potential $V(r)$, it reduces to the familiar form, $H_{\mathrm{so}}(\mathbf{r}) = \frac{e\hbar}{4m^2c^2 r}\frac{dV(r)}{dr}\mathbf{L}\cdot\mathbf{S}$. Finally, the last term on the RHS of Eq. (13.94) is called the Darwin term; it can be added as a small correction to the potential energy V.

Current and Density in the Dirac Formalism

In non-relativistic quantum mechanics, the current density and particle density are defined as [see Eq. (1.100)]

$$\mathbf{J}(\mathbf{r},t) = \mathrm{Re}\left[\psi^\dagger(\mathbf{r},t)\frac{\mathbf{p}}{m}\psi(\mathbf{r},t)\right], \tag{13.95a}$$

$$n(\mathbf{r},t) = \psi^\dagger(\mathbf{r},t)\psi(\mathbf{r},t), \tag{13.95b}$$

and the continuity equation,

$$\boldsymbol{\nabla}\cdot\mathbf{J} + \frac{\partial n}{\partial t} = 0, \tag{13.96}$$

results from the time-dependent Schrödinger equation. In Eq. (13.95), $\psi(\mathbf{r},t)$ can be taken as a two-component spinor and $\psi^\dagger(\mathbf{r},t)$ as its conjugate. Carrying out the operations within the continuity equation requires a second-order derivative in space (one from the momentum operator in the definition (13.95) of the current and the other due to the continuity equation). On the other hand, the Dirac equation is a first-order equation in space and time, so the definition of the current must be modified (keeping the continuity equation intact). A proper definition of the current in Dirac theory does not include the momentum operator.

Problem 13.18

Consider the four-component wave function $\psi(\mathbf{r},t)$ which satisfies the time-dependent Dirac equation (13.86). Show that the Dirac current and density, defined as,

$$\mathbf{J}_D(\mathbf{r},t) \equiv c\psi^\dagger(\mathbf{r},t)\boldsymbol{\alpha}\psi(\mathbf{r},t), \qquad n_D(\mathbf{r},t) \equiv \psi^\dagger(\mathbf{r},t)\psi(\mathbf{r},t), \tag{13.97}$$

satisfy the continuity equation, $\boldsymbol{\nabla}\cdot\mathbf{J}_D + \partial n_D/\partial t = 0$.

The proper definition of the current for the Pauli Hamiltonian (13.94) in the presence of the spin–orbit interaction $V_{\mathrm{so}} = \frac{e\hbar}{8m^2c^2}(\boldsymbol{\Pi}\cdot\boldsymbol{\sigma}\times\mathbf{E} + \boldsymbol{\sigma}\times\mathbf{E}\cdot\boldsymbol{\Pi})$ is somewhat subtle. We seek an analogy with the current that applies in the presence of a magnetic field, derived from a vector potential $\mathbf{A}$. The kinetic energy operator is $T = \frac{\Pi^2}{2m} \equiv m\mathbf{v}^2/2$, where $\mathbf{v} = \boldsymbol{\Pi}/m$ is the velocity operator and the (non-relativistic) current operator is $\mathbf{J} = \boldsymbol{\Pi}/m = (\mathbf{p}+\frac{e}{c}\mathbf{A})/m = \mathbf{v}$. In analogy with the definition of the kinetic energy operator T in the presence of a magnetic field alone, we define the kinetic energy operator $\bar{T} = T + V_{\mathrm{so}}$ as the kinetic energy operator in the presence of both magnetic and electric fields.

> **Problem 13.19**
>
> Show that the velocity operator derived from the modified kinetic energy $\bar{T}$ of the Pauli Hamiltonian $\bar{T} = T + V_{so}$ is
>
> $$\mathbf{v} = \dot{\mathbf{r}} = \frac{1}{i\hbar}[\mathbf{r}, H] = \frac{1}{m}[\mathbf{\Pi} - \lambda \boldsymbol{\sigma} \times \mathbf{E}]. \quad (13.98)$$

The corresponding current operator is $\mathbf{J} = \mathbf{v} = [\mathbf{\Pi} - \lambda\boldsymbol{\sigma}\times\mathbf{E}]/m$, where the role of the term $\lambda\mathbf{E}\times\boldsymbol{\sigma}$, which is due to spin–orbit interaction, is similar to that of $e\mathbf{A}/c$, related to the electromagnetic field. When $\lambda = 0$, the kinetic energy operator is simply $T = \frac{1}{2}m\mathbf{v}^2 = \frac{\mathbf{\Pi}^2}{2m}$. This analogy between the roles of $e\mathbf{A}/c$ and $\lambda\mathbf{E}\times\boldsymbol{\sigma}$ in relation to covariant momenta is true up to order λ^2, because when $\lambda \neq 0$, we have $\bar{T} = \frac{1}{2m}[\mathbf{\Pi} - \lambda\mathbf{E}\times\boldsymbol{\sigma}]^2 - \frac{1}{2m}(\lambda\mathbf{E}\times\boldsymbol{\sigma})^2$. Since λ is small, we can approximate $\bar{T} \approx m\mathbf{v}^2/2$ in analogy with the expression for T. Within this approximation, the Pauli equation has $U(1) \otimes SU(2)$ symmetry, $U(1)$ for electromagnetism and $SU(2)$ for spin rotation [223]. More explicitly, both $\mathbf{A}$ and $\mathbf{A}_{SU(2)} \equiv \boldsymbol{\sigma} \times \mathbf{E}$ contribute to the kinetic energy operator $\bar{T} \equiv \bar{\mathbf{P}}^2/(2m)$, which involves the square of a *covariant momentum* operator $\bar{\mathbf{P}} \equiv \mathbf{p} + \frac{e}{c}\mathbf{A} - \lambda\mathbf{A}_{SU(2)}$. The (slightly modified) stationary Pauli equation $\bar{T}\Psi(\mathbf{r}) = E\Psi(\mathbf{r})$ is invariant under the gauge transformations,

$$\Psi(\mathbf{r}) \to e^{-i\chi(\mathbf{r})}\Psi(\mathbf{r}) \quad \text{and} \quad \mathbf{A}(\mathbf{r}) \to \mathbf{A}(\mathbf{r}) + \nabla\chi(\mathbf{r}),$$

$$\Psi(\mathbf{r}) \to g(\mathbf{r})\Psi(\mathbf{r}) \quad \text{and} \quad \mathbf{A}_{SU(2)}(\mathbf{r}) \to \mathbf{A}_{SU(2)}(\mathbf{r}) + g(\mathbf{r})\nabla[g(\mathbf{r})]^{-1}.$$

The first equation is a $U(1)$ gauge transformation in which $\chi(\mathbf{r})$ is an arbitrary real valued function of space. Since $\mathbf{A}$ is a vector of c-numbers, the transformation is referred to as an *Abelian gauge transformation*. The second equation is an $SU(2)$ gauge transformation in which $g(\mathbf{r})$ is a spatially dependent $SU(2)$ matrix. Since $\mathbf{A}_{SU(2)}$ is a vector of $SU(2)$ matrices, the transformation is referred to as a *non-Abelian gauge transformation*. By gauge invariance we mean that the Pauli equation has the same form whether expressed in terms of the original or transformed variables $\Psi, \mathbf{A}$, and $\mathbf{A}_{SU(2)}$.

Dirac Equation in Two Dimensions

So far, our discussion of Dirac theory was carried out in $3 + 1$ dimensions, where there are three Dirac matrices α_k, $k = x, y$, and z, and the minimal dimension of the matrices α_k and β is 4. For graphene, we work in $2+1$ dimensions (two space and one time). For the Dirac equation in two-space dimensions (and in the absence of an external magnetic field), only two Dirac matrices, α_x, α_y, are required, and the minimal number of components in the wave function for satisfying the Clifford algebra, Eq. (13.87), is 2. These two matrices, together with the matrix β, can be taken as the Pauli matrices,

$$\boldsymbol{\alpha} = \boldsymbol{\sigma} = (\sigma_x, \sigma_y), \qquad \beta = \sigma_z. \quad (13.99)$$

For a free spin 1/2 particle in the $\mathbf{r} = (x, y)$ plane, the stationary Dirac equation in 2D is

$$H_D\psi(\mathbf{r}) \equiv (c\boldsymbol{\sigma}\cdot\mathbf{p} + \sigma_z mc^2)\psi(\mathbf{r}) = E\psi(\mathbf{r}), \quad (13.100)$$

where $\psi(\mathbf{r})$ is a two-component spinor. The plane wave solutions for wavenumber $\mathbf{k} = (k_x, k_y)$, and the corresponding energies, are

$$\psi_{\lambda=\pm,\mathbf{k}}(\mathbf{r}) = e^{i\mathbf{k}\cdot\mathbf{r}} u_{\lambda=\pm,\mathbf{k}} = e^{i\mathbf{k}\cdot\mathbf{r}} \frac{1}{\sqrt{2|E|}} \begin{pmatrix} \sqrt{|E_\lambda + mc^2|} \\ \lambda\sqrt{|E_\lambda - mc^2|}e^{i\theta_\mathbf{k}} \end{pmatrix}, \quad (13.101a)$$

$$E_{\lambda=+} = \sqrt{(\hbar c\mathbf{k})^2 + m^2c^4} = |E|, \quad E_{\lambda=-} = -|E|. \quad (13.101b)$$

where $\tan\theta_\mathbf{k} = k_y/k_x$. The energies $E_\pm$ versus (k_x, k_y) are plotted in Fig. 13.38.

13.6 Graphene

> **Problem 13.20**
>
> Check which properties listed after Eq. (13.92) for the 3+1 Dirac problem are valid for the Dirac problem in 2 + 1 dimensions.
>
> **Answer:** All the listed properties are valid. However, there is no non-relativistic analog leading to the Pauli equation (13.94) because the number of components is two, and not four, as in the original Dirac equation.

For $m \neq 0$, the spectrum has a gap $E_g = 2mc^2$, and for small $|\mathbf{k}|$, the dependence of the energy on wavenumber is quadratic. As we shall see below, electrons in graphene with momentum $\mathbf{k}$ near the Dirac points satisfy the 2D Dirac equation for particles without mass, and the spectrum is linear in $|\mathbf{k}|$.

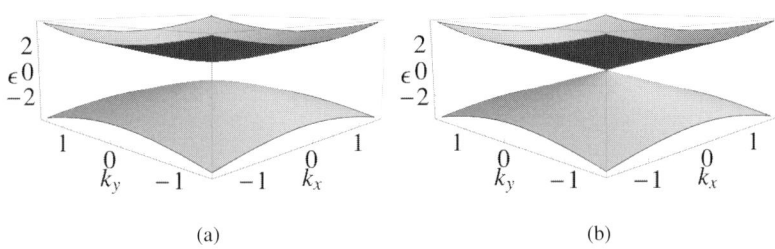

FIG 13.38 The energies $E_\pm$ defined in Eq. (13.101) are drawn as functions of the wave numbers (k_x, k_y). (a) For $\hbar k \ll mc$, the spectrum is quadratic and there is a gap $2mc^2$. (b) For $\hbar k \gg mc$, the spectrum is linear, the gap closes, and a Dirac cone is formed.

Finally, in analogy with expressions (13.97) derived for the Dirac theory in 3 + 1 dimensions, the corresponding expressions for the current and the density for the Dirac theory in 2 + 1 dimensions are

$$\mathbf{J}_D(\mathbf{r},t) = c\psi^\dagger(\mathbf{r},t)\boldsymbol{\sigma}\psi(\mathbf{r},t), \quad (13.102a)$$

$$n_D(\mathbf{r},t) = \psi^\dagger(\mathbf{r},t)\psi(\mathbf{r},t), \quad (13.102b)$$

which satisfy the continuity equation,

$$\nabla \cdot \mathbf{J}_D + \partial n_D/\partial t = 0.$$

13.6.4 TIGHT-BINDING MODEL FOR GRAPHENE

The tight-binding formulation of the Schrödinger equation in 1D systems has been discussed in Sec. 12.9.3. Extension from 1D to higher dimensions requires identification of the kinetic energy operator that replaces the Laplacian in the continuum theory. The details of this procedure depend on the structure of the lattice upon whose sites the wave function resides and enables several variants that are natural for tight-binding models, although they do not have clear and transparent continuum analogs.

Consider first a Bravais lattice where the wave function $\Psi(\mathbf{R})$ is defined on sites $\{\mathbf{R}\}$, and assume that the coordination number is Z, so that $\mathbf{R}$ has Z *nearest neighbors* (NN) located at sites $\mathbf{R}_\alpha$ $\alpha = 1, 2, \ldots, Z$. Then, the simplest variant, referred to as NN hopping, corresponds to the following replacement,

$$-\frac{\hbar^2}{2m}\Delta\Psi(\mathbf{R}) \to -t\sum_{\alpha=1}^{Z}\Psi(\mathbf{R}_\alpha). \quad \text{(NN hopping)}. \quad (13.103)$$

For graphene, $Z = 3$, and the constant t reflects the degree of itinerancy of the electrons for hopping from one site on sublattice A to its nearest neighbor site on sublattice B. It is obtained by calculating the overlap of wave functions localized at the pertinent sites, but, practically, it is often taken as a free parameter. In addition to Z NNs, each point $\mathbf{R}$ has Z' next nearest neighbors (NNNs), $\mathbf{R}_\beta$. For example, in a 2D square lattice of constant a, the coordination number is $Z = 4$ NNs at a distance a from $\mathbf{R}$ and $Z' = 4$ NNN at a distance $a\sqrt{2}$ from $\mathbf{R}$. In some cases, NNN hopping is also included, with a constant $|t'| < |t|$. The inclusion of both NN and NNN means the replacement,

$$-\frac{\hbar^2}{2m}\Delta\Psi(\mathbf{R}) \to -t\sum_{\alpha=1}^{Z}\Psi(\mathbf{R}_\alpha) - t'\sum_{\beta=1}^{Z'}\Psi(\mathbf{R}_\beta) \quad \text{(NN and NNN hopping)}. \quad (13.104)$$

Let us now apply the above formalism to graphene. Within the NN variant, the structure of the kinetic energy operator is similar to that of Eq. (13.103), except that the NN points do not belong to the lattice to which $\mathbf{R}$ belongs, but to the second lattice. In other words, there is a NN hopping between sites $\mathbf{R}_A \in A$ and sites $\mathbf{R}_B \in B$. It can be implemented by a two-component representation of the electron wave function

$$\Psi = \begin{pmatrix} \psi_A(\mathbf{R}_A) \\ \psi_B(\mathbf{R}_B) \end{pmatrix}, \quad (13.105)$$

one for each triangular sublattice, where $\mathbf{R}_A$ and $\mathbf{R}_B$ are full and empty circles in Fig. 13.36(a) connected by the vectors $\mathbf{d}_{1,2,3}$. Following Fig. 13.36, and recalling Eqs (13.82) and (13.83), we have, in this notation, $\mathbf{R}_A + \mathbf{d}_i \in B$ and $\mathbf{R}_B - \mathbf{d}_i \in A$. In most cases, the NN variant is sufficient, but we present the NNN variant as well, since it shows the richness of the tight-binding picture compared with the continuous version. Following Fig. 13.36(a), we denote by $\mathbf{e}_j$ the six vectors connecting a point on the triangular lattice to its six neighbors (belonging to the *same* lattice), and, recalling Eq. (13.82),

$$\mathbf{e}_{1,2} = \pm \mathbf{a}_1, \quad \mathbf{e}_{3,4} = \pm \mathbf{a}_2, \quad \mathbf{e}_{5,6} = \pm(\mathbf{a}_1 - \mathbf{a}_2). \quad (13.106)$$

The *tight-binding equations* for an electron in a clean graphene sheet are

$$-t \sum_i \psi_B(\mathbf{R}_A + \mathbf{d}_i) - t' \sum_i \psi_A(\mathbf{R}_A + \mathbf{e}_i) = E \psi_A(\mathbf{R}_A), \quad (13.107a)$$

$$-t \sum_i \psi_A(\mathbf{R}_B - \mathbf{d}_i) - t' \sum_i \psi_B(\mathbf{R}_B + \mathbf{e}_i) = E \psi_B(\mathbf{R}_B). \quad (13.107b)$$

The solution of Eqs (13.107) is worked out in two steps. First, like in the 1D tight-binding model, the wave functions are expressed in terms of their Fourier components, each one on its own Bravais lattice,

$$\psi_A(\mathbf{R}_A) = \sum_{\mathbf{k}} e^{i \mathbf{R}_A \cdot \mathbf{k}} \psi_A(\mathbf{k}), \quad (13.108a)$$

$$\psi_B(\mathbf{R}_B) = \sum_{\mathbf{k}} e^{i \mathbf{R}_B \cdot \mathbf{k}} \psi_B(\mathbf{k}). \quad (13.108b)$$

This leads to a set of two coupled *linear algebraic eigenvalue equations* for the corresponding Fourier components,

$$g(\mathbf{k}) \psi_A(\mathbf{k}) + h(\mathbf{k}) \psi_B(\mathbf{k}) = E(\mathbf{k}) \psi_A(\mathbf{k}), \quad (13.109a)$$
$$g(\mathbf{k}) \psi_B(\mathbf{k}) + h(\mathbf{k})^* \psi_A(\mathbf{k}) = E(\mathbf{k}) \psi_B(\mathbf{k}), \quad (13.109b)$$
$$g(\mathbf{k}) = -t' \sum_j e^{i \mathbf{k} \cdot \mathbf{e}_j} = g^*(\mathbf{k}), \quad h(\mathbf{k}) = -t \sum_j e^{i \mathbf{k} \cdot \mathbf{d}_j}. \quad (13.109c)$$

In the second step, the 2×2 matrix,

$$\mathcal{H}(\mathbf{k}) \equiv \begin{pmatrix} g(\mathbf{k}) & h(\mathbf{k}) \\ h(\mathbf{k})^* & g(\mathbf{k}) \end{pmatrix}, \quad (13.110)$$

is diagonalized, yielding the energy dependence on the momentum $\mathbf{k} = (k_x, k_y)$,

$$E_\pm(\mathbf{k}) = \pm t \sqrt{3 + f(\mathbf{k})} - t' f(\mathbf{k}),$$

$$f(\mathbf{k}) = 2 \cos(\sqrt{3} k_y a) + 4 \cos\left(\frac{\sqrt{3}}{2} k_y a\right) \cos\left(\frac{3}{2} k_x a\right), \quad (13.111)$$

as noticed by Wallace [217].

13.6 Graphene

> **Problem 13.21**
>
> Discuss the difference between the dispersion (13.111) assuming only NN coupling, i.e., $t' = 0$, and the dispersion for the square lattice lattice $E(\mathbf{k}) = -2t(\cos k_x a + \cos k_y a)$ corresponding to the tight-binding energy of an electron hopping on a square two-dimensional lattice.
>
> **Answer:** The main differences are as follows: (1) For each $\mathbf{k}$, there is only one energy $E(\mathbf{k})$ for the square lattice but two energies $E_\pm(\mathbf{k})$ for the honeycomb lattice. In the latter case, there are both particles and holes. (2) Near the extrema, the square lattice energy spectrum is quadratic, while the honeycomb lattice energy spectrum is linear.

The solutions $\psi_\pm(\mathbf{k}) = \begin{pmatrix} \psi_{A\pm}(\mathbf{k}) \\ \psi_{B\pm}(\mathbf{k}) \end{pmatrix}$ corresponding to positive (+) or negative (−) energies are the particle and hole wave functions, respectively,

$$\psi_\pm(\mathbf{k}) = \frac{1}{\sqrt{2}} \begin{pmatrix} e^{-i\frac{\theta_\mathbf{k}}{2}} \\ \pm e^{i\frac{\theta_\mathbf{k}}{2}} \end{pmatrix}, \quad \text{with} \quad \tan \theta_\mathbf{k} = \frac{\sum_j \sin \mathbf{k} \cdot \mathbf{d}_j}{\sum_j \cos \mathbf{k} \cdot \mathbf{d}_j}. \tag{13.112}$$

Dirac Points. Here, we consider the NN variant ($t' = 0$). At every point $\mathbf{k}$ for which the argument of the square root in Eq. (13.111) fails to vanish, there is an *energy gap* $E_+(\mathbf{k}) - E_-(\mathbf{k}) = 2E_+(\mathbf{k})$. There are two Dirac points $\mathbf{K}_\pm$,

$$\mathbf{K}_\pm = \frac{2\pi}{a}\left(\frac{2}{3}, \pm\frac{1}{3\sqrt{3}}\right), \tag{13.113}$$

located at the corners of the Brillouin zone of graphene [denoted as $\mathbf{K}, \mathbf{K}'$ in Fig. 13.37 and Eq. (13.84)], for which the argument of the square root in Eq. (13.111) vanishes,

$$E(\mathbf{K}_\pm) = 0. \tag{13.114}$$

For the *NN + NNN* variant ($t, t' \neq 0$), which we will not analyze here), the positions of the Dirac points are shifted.

Let us consider the dispersion relation (13.111) when $\mathbf{k}$ is close to $\mathbf{K}_\pm$, so that the difference vector $\mathbf{q} \equiv \mathbf{k} - \mathbf{K}_\pm$ is small. To first order in $\mathbf{q}$, the expression (13.109) reduces to,

$$h(\mathbf{k}) \approx -t \sum_j e^{i\mathbf{K}_\pm \cdot \mathbf{d}_j}(1 + i\mathbf{q} \cdot \mathbf{d}_j) = -it \sum_j e^{i\mathbf{K}_\pm \cdot \mathbf{d}_j} \mathbf{q} \cdot \mathbf{d}_j = \frac{3ta}{2}(\pm q_x + iq_y) + O(q^2), \tag{13.115}$$

since, by symmetry, $\sum_j e^{i\mathbf{K}_\pm \cdot \mathbf{d}_j} = 0$.

Let us denote the Dirac points by their *valley quantum numbers* $\eta \equiv \pm$. Near $\mathbf{K}_\eta$, the matrix $\mathcal{H}(\mathbf{k})$ defined by Eq. (13.110) takes the form,

$$\mathcal{H}_\eta(\mathbf{q}) = \frac{3ta}{2}\begin{pmatrix} 0 & \eta q_x + iq_y \\ \eta q_x - iq_y & 0 \end{pmatrix}. \tag{13.116}$$

> **Problem 13.22**
>
> Prove that $\{\mathcal{H}_\eta(\mathbf{q}), \sigma_z\} = 0$, i.e., $\mathcal{H}_\eta(\mathbf{q})$ anticommutes with σ_z. This property is referred to as *Chiral symmetry*. See the discussion below on the helicity quantum number.

Diagonalization of $\mathcal{H}_\eta(\mathbf{q})$ yields the energy dispersion as a function of the vector $\mathbf{q} \equiv \mathbf{k} - \mathbf{K}_\eta$, where the general relation (13.111) reduces to a linear one,

$$E(\mathbf{q}) = \pm \hbar v |\mathbf{q}| \quad \text{(independent of } \eta\text{)}, \tag{13.117}$$

where

$$v = \frac{3ta}{2\hbar} \qquad (13.118)$$

is a constant of dimension velocity. When $\varepsilon_F = E(\mathbf{q})$, v is the Fermi velocity of the electrons near the corresponding Dirac point $\mathbf{K}_\eta$; experimentally, $v \approx 10^6$ m/sec. A plot of $E(\mathbf{q})$, Eq. (13.117) for small q, reveals the same structure as in the Dirac theory for mass-less fermions, shown in Fig. 13.38(b). Unlike the quadratic dispersion relation, $E = \frac{\hbar^2 k^2}{2m}$, valid for an electron in free space, the dispersion relation for low-energy electrons in clean graphene is linear.

Graphene Hamiltonian in Second Quantization: Equations (13.107) encode the Schrödinger equation for graphene in a first quantization formalism within the tight-binding approximation. Often, the use of second quantization is used for writing down the graphene Hamiltonian. The kinetic energy operator for electron hopping on the sites of graphene in second quantization, based on the geometry of Fig. 13.36, is

$$H_0 = -t \sum_{\langle i,j \rangle} \left(a_i^\dagger b_j + h.c. \right) - t' \sum_{\langle\langle i,j \rangle\rangle} \left(a_i^\dagger a_j + b_i^\dagger b_j \right) \qquad (13.119)$$

where a_i, ($a_i^\dagger$) annihilates (creates) an electron on site $\mathbf{R}_i \in A$ and b_j, ($a_j^\dagger$) annihilates (creates) an electron on site $\mathbf{R}_j \in B$. The notation $\langle i,j \rangle$ implies NN hopping, i.e., for each site, $\mathbf{R}_i \in A$ the sum over $j = 1, 2, 3$ runs on the three vectors $\mathbf{R}_i + \mathbf{d}_j \in B$ with NN hopping energy $t \approx 2.8$ eV. Similarly, the notation $\langle\langle i,j \rangle\rangle$ implies NNN hopping; for each site, $\mathbf{R}_i \in A$, the sum over $j = 1, 2, \ldots, 6$ runs on the six vectors $\mathbf{R}_i + \mathbf{e}_j \in A$, with NNN hopping energy $t' \approx 0.1$ eV. Translation invariance can be used to express the operators a_j, b_j and $a_j^\dagger, b_j^\dagger$ via their Fourier transforms, in terms of creation and annihilation operators $a_\mathbf{k}, b_\mathbf{k}$ and $a_\mathbf{k}^\dagger, b_\mathbf{k}^\dagger$ corresponding to definite wave numbers. The resulting Hamiltonian is

$$H_0 = \sum_\mathbf{k} (a_\mathbf{k}^\dagger, b_\mathbf{k}^\dagger) \mathcal{H}(\mathbf{k}) \begin{pmatrix} a_\mathbf{k} \\ b_\mathbf{k} \end{pmatrix}, \qquad (13.120)$$

where $\mathcal{H}(\mathbf{k})$ is given in Eq. (13.110). This 2×2 Hamiltonian can be easily diagonalized.

13.6.5 CONTINUUM THEORY

The tight-binding model for graphene introduced through Eqs (13.107) is based on the assumption that electrons in graphene are tightly bound to the carbon atoms. However, in many cases, we need to examine its continuous version as well. First, notice that all the discussion so far was limited to a clean system. Once there are scattering events (either by impurities or through electron–electron or electron–phonon interaction), the continuum formulation might be more appropriate (see Sec. 13.6.7). Second, an external magnetic field and the quantum Hall effect are more elegantly treated within a continuum theory. Third, the continuum theory is that of massless Dirac fermions in two dimensions. The fact that the Fermi velocity plays the role of the speed of light enables us to test relativistic effects in solid-state physics. The quantum Hall effect and Klein tunneling in graphene (both discussed below) have been observed experimentally.

Our goal is to construct the effective low-energy Hamiltonian of an electron in graphene close to the Dirac points. This can be carried out by starting from the Hamiltonian $\mathcal{H}_\mathbf{q}$ in Eq. (13.116), and representing the momenta as derivative operators, $q_x = -i\hbar\partial_x$, $q_y = -i\hbar\partial_y$. Consider the two sublattices A and B as two (pseudo)-spin states, so that the wave function is a spinor in this pseudo-spin space. The corresponding pseudo-spin operator is $\boldsymbol{\sigma} = (\sigma_x, \sigma_y, \sigma_z)$. It should be kept in mind that the electron is treated as a spinless particle here. The effective low-energy Hamiltonians near $\mathbf{K}_\eta$ is

$$\mathcal{H}_0(\mathbf{K}_+) = v\boldsymbol{\sigma} \cdot \mathbf{p} = -i\hbar v \left(\sigma_x \frac{\partial}{\partial x} + \sigma_y \frac{\partial}{\partial y} \right), \qquad (13.121)$$

$$\mathcal{H}_0(\mathbf{K}_-) = \sigma_y \mathcal{H}_0(\mathbf{K}_+) \sigma_y = -i\hbar v \left(-\sigma_x \frac{\partial}{\partial x} + \sigma_y \frac{\partial}{\partial y} \right). \qquad (13.122)$$

13.6 Graphene

The two components of the wave function $\psi(\mathbf{r})$ [$\mathbf{r} = (x, y)$] for electrons in graphene with momentum near the Dirac points satisfy the Dirac equations,

$$-i\hbar v \left(\sigma_x \frac{\partial}{\partial x} + \sigma_y \frac{\partial}{\partial y} \right) \psi(\mathbf{r}) = E\psi(\mathbf{r}) = \pm \hbar v q \psi(\mathbf{r}), \quad (q = |\mathbf{k} - \mathbf{K}_+| \text{ small}), \quad (13.123)$$

$$-i\hbar v \left(-\sigma_x \frac{\partial}{\partial x} + \sigma_y \frac{\partial}{\partial y} \right) \psi(\mathbf{r}) = E\psi(\mathbf{r}) = \pm \hbar v q \psi(\mathbf{r}), \quad (q = |\mathbf{k} - \mathbf{K}_-| \text{ small}). \quad (13.124)$$

Hence, the electrons near the Dirac point in graphene obey the 2D Dirac equation for massless particles [cf., Eq. (13.100)]. In analogy with Eq. (13.102), the matter current in graphene is

$$\mathbf{J} = (J_x, J_y) = v\psi(\mathbf{r})^\dagger \boldsymbol{\sigma} \psi(\mathbf{r}). \quad (13.125)$$

Helicity

The relativistic form of the equations for the electrons in graphene has many interesting consequences. Here, we briefly consider the *helicity*, i.e., the component of the spin of the electron along its direction of motion, which plays an important role in the physics of relativistic particles. First, we discuss helicity in general and then analyze its role in the physics of graphene.

Definition of the Helicity Operator: The helicity formalism applies to particles of arbitrary spin. Consider first a particle with real spin $\mathbf{S}$ (as opposed to a pseudo-spin). The helicity operator is defined as $h = \mathbf{S} \cdot \hat{\mathbf{p}}$ and the physical meaning of its eigenstates is that instead of specifying states according to their spin projections along a fixed direction in space, we specify them according to helicity, the spin projection along its motion direction. For particles with nonzero mass, $m \neq 0$, there are $2S + 1$ linearly independent eigenstates of h for each momentum $\mathbf{p}$ with definite helicity λ. If $m = 0$, there are only two eigenstates $\lambda = \pm S$. For particles with spin $\mathbf{S} = \frac{\hbar}{2}\boldsymbol{\sigma}$, one can define a dimensionless *chirality operator* (also denoted as h and called helicity operator for convenience),

$$h \equiv \boldsymbol{\sigma} \cdot \hat{\mathbf{p}} = \sigma_x \hat{p}_x + \sigma_y \hat{p}_y, \quad (13.126)$$

where $\hat{\mathbf{p}} = \mathbf{p}/|\mathbf{p}|$. Its expectation value in a given spinor state gives the averaged projection of the spin along the momentum in this state (in units of $\hbar/2$).

> **Problem 13.23**
>
> A particle has angular momentum $\mathbf{J} = \mathbf{L} + \mathbf{S}$, where $\mathbf{L}$ is the orbital angular momentum operator and $\mathbf{S}$ is its real (physical) spin operator. Show that the helicity operator of the particle is $\mathbf{J} \cdot \hat{\mathbf{p}}$. Show that the helicity is the component of the spin angular momentum of a particle along the direction of its motion. Hint: Since $\mathbf{J} = \mathbf{L} + \mathbf{S}$, it remains to show that $\mathbf{L} \times \mathbf{p} = 0$.

Helicity plays an especially important role for massless particles. If a massless particle has a spin J, the only allowed helicity eigenvalues are $\pm J$. This is the case for the photon, which is a massless particle with spin $S = 1$. Although a spin 1 particle can also have $S_z = 0$, but because it has zero mass, this state is not possible. Thus, the wave functions of massless particles are eigenstates of the helicity operator with eigenvalues $\lambda = \pm 1$ with corresponding eigenvectors $|\phi_R\rangle$ (the right-handed positive helicity eigenstate with spin parallel to $\hat{\mathbf{p}}$) and $|\phi_L\rangle$ (the left-handed negative helicity eigenstate with spin antiparallel to $\hat{\mathbf{p}}$).

Helicity in Graphene: In Problem 13.22, you showed that the graphene Hamiltonian anticommutes with the z-component of the pseudo-spin operator, σ_z; this is referred to as chiral symmetry. This property enables the introduction of a chirality (or pseudo-helicity) operator $h = \boldsymbol{\sigma} \cdot \hat{\mathbf{p}}$ that commutes with the Hamiltonian. The reason for the notion of "pseudo" is that

σ is a pseudo-spin operator coupling the two sublattices A and B and not a real spin. The origin of the word "chirality" is that particles with spin parallel to the momentum (having spin angular momentum in the direction of $\mathbf{p}$) are said to have positive chirality, while particles with spin antiparallel to the momentum are said to have negative chirality. Here, we explain some of the properties of this pseudo-chirality operator.

So far, in our discussion of electrons in graphene, the spin of the electron has not yet been taken into considerations and that will continue to be the case in our analysis below. However, the structure of the graphene Hamiltonian is such that the pseudo-spin operator σ in Eq. (13.121) plays the same role as the physical spin operator, but here it is related to the two-sublattice structure of graphene. In the absence of an external magnetic field or spin–orbit coupling, the role of pseudo-spin is physically similar to that of real spin. For example, the expression for the current, defined in Eq. (13.125) in terms of the pseudo-spin operator σ, is identical with the expression (13.102) for the current in the 2D Dirac theory for massless fermions with real spin. In both cases, the currents are physically measurable.

We now define a (pseudo-)helicity operator and then employ the helicity formalism in studying some of its properties in graphene, bearing in mind that we deal with a pseudo-spin and not a real spin. The Hamiltonians $\mathcal{H}_0(\mathbf{K}_\eta) = \mathcal{H}_0(\mathbf{K}_\pm)$ in the vicinity of the $\mathbf{K}_\pm$ points as defined in Eqs (13.121) and (13.122) commute with the 2D pseudo-helicity operator. The corresponding eigenvalue equation reads,

$$h\psi(\mathbf{r}) \equiv \sigma \cdot \hat{\mathbf{p}}\,\psi(\mathbf{r}) = \lambda\psi(\mathbf{r}). \tag{13.127}$$

We already know that h has two eigenstates with eigenvalues $\lambda = \pm 1$. Since each one of the Hamiltonian operators $\mathcal{H}_0(\mathbf{K}_\eta)$ has two eigenvectors with energies $\pm\hbar v q$, the helicity eigenvalues $\lambda = \pm 1$ mark the sign of the energy. This sign is different for $\mathcal{H}_0(\mathbf{K}_+)$ and $\mathcal{H}_0(\mathbf{K}_-)$. To clarify this point, let us consider the three quantum numbers: the *valley index* for $\mathbf{K}_\eta$, $\eta = \pm 1$, the helicity quantum number, $\lambda = \pm 1$, and the sign of the energy, $\gamma = \pm 1$. The relation which connects these quantum numers is $\lambda = \gamma\eta$. These quantum numbers and their relation to the Dirac cones at the points $\mathbf{K}_\pm$ is shown in Fig. 13.39. The plane wavesolutions of Eqs (13.123) and (13.124) are spinors with definite helicities,

$$\psi_{[\mathbf{q}=\mathbf{k}-\mathbf{K}_+]\lambda}(\mathbf{r}) = \frac{e^{i\mathbf{q}\cdot\mathbf{r}}}{\sqrt{2}}\begin{pmatrix}1\\ \lambda e^{i\theta_\mathbf{q}}\end{pmatrix} \equiv e^{i\mathbf{q}\cdot\mathbf{r}} u_{\eta\lambda}(\mathbf{q}), \quad (\eta = 1), \tag{13.128}$$

$$\psi_{[\mathbf{q}=\mathbf{k}-\mathbf{K}_-]\lambda}(\mathbf{r}) = \frac{e^{i\mathbf{q}\cdot\mathbf{r}}}{\sqrt{2}}\begin{pmatrix}-\lambda e^{-i\theta_\mathbf{q}}\\ 1\end{pmatrix} \equiv e^{i\mathbf{q}\cdot\mathbf{r}} u_{\eta\lambda}(\mathbf{q}), \quad (\eta = -1). \tag{13.129}$$

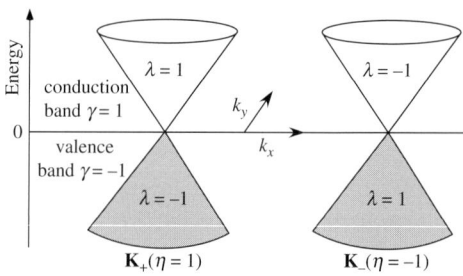

FIG 13.39 Dirac cones for graphene at the points $\mathbf{K}_\pm$ and the definitions of the quantum numbers η, λ, and γ.

where $\theta_\mathbf{q} = \arctan(q_y/q_x)$.

Berry Phase of Helicity Eigenstates: The Berry phase defined in Eq. 7.206 is related to the Berry connection [the vector potential in Eq. (7.203)],

$$\mathbf{A} = i\langle u_{\eta\lambda}(\mathbf{q})|\nabla_\mathbf{q}|u_{\eta\lambda}(\mathbf{q})\rangle = -\frac{\eta\lambda}{2}\nabla_\mathbf{q}\theta_\mathbf{q}. \tag{13.130}$$

Integrating $\mathbf{A}$ along a closed curve C encircling the corresponding Dirac point $\mathbf{K}_\eta$, we obtain

$$\gamma_{\eta\lambda} = \oint_C d\mathbf{k}\cdot\mathbf{A} = -\eta\lambda\pi. \tag{13.131}$$

Despite being defined as an integral of a gradient over a closed contour, $\gamma_{\eta\lambda} \neq 0$ is due to the fact that the Berry curvature $\mathbf{B} = \nabla \times \mathbf{A}$ is singular at $\mathbf{q} = \mathbf{K}_\eta$. $\mathbf{A}$ is a gradient only locally, and the singularity of $\mathbf{B}$ at $\mathbf{q} = \mathbf{K}_\eta$ is due to the degeneracy of $E_\lambda(\mathbf{q} = \mathbf{K}_\eta) = E_{-\lambda}(\mathbf{q} = \mathbf{K}_{-\eta})$ [cf., calculations of the Chern number in the IQHE, see Eq. (9.545) and also Eq. (7.211)].

13.6 Graphene

Problem 13.24

(a) Following the definition of current in Eq. (13.125), show that the current carried by the plane wave $\psi(\mathbf{r})$ in Eq. (13.128) is $\mathbf{J} = v\psi^\dagger(\mathbf{r})\sigma\psi(\mathbf{r}) = \lambda v \hat{\mathbf{q}}$.
Answer: For example, $J_x = vu_\lambda(\hat{\mathbf{q}})^\dagger \sigma_x u_\lambda(\hat{\mathbf{q}}) = \lambda v \cos\theta_\mathbf{q} = \lambda v \frac{q_x}{|q|}$.
(b) Prove that the plane waves defined in Eqs (13.128) and (13.129) are eigenstates of the helicity operator.
(c) Show that the energies $E_{[\mathbf{q}=\mathbf{k}-\mathbf{K}_\pm],\lambda}$ are $E_{[\mathbf{q}=\mathbf{k}-\mathbf{K}_\pm],\lambda} = \pm\lambda\hbar vq$.
(d) Show that the Berry curvature $\mathbf{B} = \nabla \times \mathbf{A}$, where $\mathbf{A}$ in Eq. (13.130) is $\mathbf{B} = -\eta\lambda\pi\delta^{(2)}(\mathbf{q})$.

Near the $\mathbf{K}_+$ valley, positive energy states have positive helicity, whereas negative energy states have negative helicity. Near $\mathbf{K}_-$, the signs of energies and helicities are opposite.

In addition to plane wavesolutions, we are interested in spherical spinor wave solutions with definite helicity. Consider Eqs (13.123) and (13.124) in plane polar coordinates $\mathbf{r} = (r,\phi)$ (the polar angle ϕ should not be confused with the angle $\theta_\mathbf{q}$).

Problem 13.25

Consider the dimensionless orbital angular momentum, (pseudo) spin and total angular momentum operators,

$$l_z = -i\frac{\partial}{\partial\phi}, \quad s = \frac{1}{2}\sigma, \quad j = l_z + s_z. \tag{13.132}$$

Prove the commutation relations, $[l_z, H_0] = -[s_z, H_0] = iv\sigma \times \mathbf{p}$, $[j, H_0] = 0$.

In analogy with the solution of the Dirac equation in spherical coordinates, we seek a solution of the two-component Schrödinger equation (13.123) in the circular wave form,

$$\psi_{m\lambda}(\mathbf{r}) = \begin{pmatrix} e^{im\phi} f_m(r) \\ \lambda e^{i(m+1)\phi} g_m(r) \end{pmatrix}. \tag{13.133}$$

Operating with H_0^2 yields equations for the radial functions $f_m(r)$ and $g_m(r)$,

$$\left(-\frac{d^2}{dr^2} - \frac{1}{r}\frac{d}{dr} + \frac{m^2}{r^2}\right) f_m(r) = q^2 f_m(r),$$

$$\left(-\frac{d^2}{dr^2} - \frac{1}{r}\frac{d}{dr} + \frac{(m+1)^2}{r^2}\right) g_m(r) = q^2 g_m(r), \tag{13.134}$$

whose solutions, regular at the origin, are Bessel functions. Taking into account the normalization and the fact that there are two solutions, one for each helicity, the solution of Eq. (13.123) in polar coordinates is,

$$\psi_{m\lambda}(\mathbf{r}) = \frac{1}{\sqrt{2}}\begin{pmatrix} e^{im\phi} J_m(qr) \\ \lambda e^{i(m+1)\phi} J_{m+1}(qr) \end{pmatrix}. \tag{13.135}$$

Problem 13.26

(a) Prove that the state $\psi_0(\mathbf{r})$ in Eq. (13.135) is an eigenstate of j_z with eigenvalue $m + \frac{1}{2}$.
Answer: Applying l_z to $\psi_0(\mathbf{r})$ "brings down" an m in the upper component and an $m+1$ in the lower component. Applying s_z to $\psi_0(\mathbf{r})$ yields a $+1/2$ in the upper component and a $-1/2$ in the lower component.

(b) Prove that the parity of $\psi_{m\lambda}(\mathbf{r})$ defined in Eq. (13.135) is $(-1)^m$. Note that for spinors in two dimensions, the parity operator yields $\mathcal{P}\psi_\lambda(\mathbf{r}) = \gamma_0 \psi_\lambda(-\mathbf{r})$, where the "Dirac matrix" γ_0 reduces to σ_z. Inversion $\mathbf{r} \to -\mathbf{r}$ in two dimensions yields $\phi \to \phi + \pi$.

13.6.6 LANDAU LEVELS IN GRAPHENE

Graphene reveals new physics in the integer quantum Hall effect. In graphene, the Landau levels are proportional to $\sqrt{n}$ (where n is the Landau level index), instead of the usual relation $E_n = (n + 1/2)\hbar\omega_c$ for the Landau energies in semiconductors. Moreover, since $n = 0, 1, 2 \ldots$, the quantum Hall effect in graphene has a Landau level at zero energy. There are at least two experimental distinctions of the Hall effect in graphene compared with the Hall effect in semiconductors. First, the separation between the $n = 0$ and $n = 1$ Landau level is much larger in graphene due to the $\sqrt{H}$ dependence of the Landau energy. Second, electrons in graphene have high mobility. As a consequence, the quantum Hall effect in graphene can be detected even at room temperature.

To analyze the effect of magnetic field, we concentrate on Eq. (13.123) near $\mathbf{K}_+$ and account for the application of a magnetic field by the substitution $\mathbf{p} \to \mathbf{p} + \frac{e}{c}\mathbf{A}$, where $\mathbf{A}$ is the vector potential such that the magnetic field $\mathbf{H} = H\hat{\mathbf{z}}$ is $\mathbf{H} = \nabla \times \mathbf{A}$. We will use the *magnetic length* $\ell_H \equiv \sqrt{\frac{\hbar c}{eH}}$ as a length scale and $b = \ell_H^{-2}$ as the scaled magnetic field. Dividing both sides of Eq. (13.123) by $\hbar v$, the energy $\varepsilon = \frac{E}{\hbar v}$ has the dimension of inverse length and can be expressed in units of ℓ_H^{-1}. In the Landau gauge, $\mathbf{A} = (0, -Hx)$, the translational invariance along y is still maintained, but translational invariance along x is lost. This lets us write the two-component wave function as

$$\begin{pmatrix} \Psi_A(x,y) \\ \Psi_B(x,y) \end{pmatrix} = e^{iqy} \begin{pmatrix} \psi_A(x) \\ \psi_B(x) \end{pmatrix}. \tag{13.136}$$

In a magnetic field, the 2D Dirac equation (13.123) is replaced by the following set of coupled equations:

$$\begin{pmatrix} 0 & -i\frac{d}{dx} + bx + iq \\ -i\frac{d}{dx} + bx - iq & 0 \end{pmatrix} \begin{pmatrix} \psi_A(x) \\ \psi_B(x) \end{pmatrix} = \varepsilon \begin{pmatrix} \psi_A(x) \\ \psi_B(x) \end{pmatrix}. \tag{13.137}$$

When $\varepsilon \neq 0$, it is possible to express $\psi_B(x)$ in terms of $\psi_A(x)$ as

$$\psi_B(x) = \frac{1}{\varepsilon}\left(-i\frac{d}{dx} + bx - iq\right)\psi_A(x), \tag{13.138}$$

and thereby get a second-order differential equation for $\psi_A(x)$,

$$\left[-\frac{d^2}{dx^2} + (q + bx)^2\right]\psi_A(x) = (\varepsilon^2 - b)\psi_A(x). \tag{13.139}$$

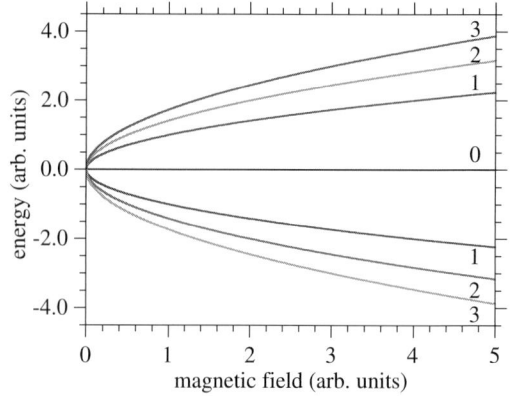

FIG 13.40 Landau energies in graphene versus magnetic field strength.

Equation (13.139) is the familiar Schrödinger equation for the harmonic oscillator problem. In the present units, the eigenvalues are $(2n+1)b$. In the non-relativistic case, restoring natural units, this is translated into $(n + 1/2)\hbar\omega_c$, where $\omega_c = \frac{eH}{mc}$ is the cyclotron frequency. However, for graphene, there is no mass. Equating the eigenvalues $\varepsilon^2 - b = (2n + 1)b$ and translating into natural units gives Landau energies $E_n = \pm \hbar v \sqrt{(n+1)b}$, $n = 0, 1, 2 \ldots$. So far it was assumed that

13.6 Graphene

$\varepsilon \neq 0$. It is not difficult to check that Eqs (13.137) also has a solution for $\varepsilon = 0$. Therefore, the Landau energies for graphene (see Fig. 13.40) are

$$E_n = \pm \hbar v \sqrt{nb} = \pm \frac{\hbar v}{\ell_H} \sqrt{n}, \quad n = 0, 1, 2 \ldots \tag{13.140}$$

Problem 13.27

(a) What is the energy difference between the first two Landau levels for the QHE in semiconductors and graphene.
(b) Compare the corresponding numerical values of the Landau level energies in semiconductors and graphene for $H = 10$ T, $v = c/100$, and $m^* = 0.1\, m_0$ (m^* is the effective mass of the electron in the semiconductor and m_0 is the mass of a free electron).

Answer (a): $\Delta E = \hbar \omega_c = \frac{\hbar^2}{m^* \ell_H^2}$ in semiconductors, $\Delta E = \frac{\hbar v}{\ell_H}$ in graphene.

The difference between the pattern of Landau levels in graphene compared with the corresponding pattern in semiconductors (i.e., the non-relativistic theory of 2D electrons subject to a perpendicular magnetic field) is twofold. First, the Landau energy for graphene grows as $\sqrt{n}$ and not as n. More striking is that in graphene, there is a Landau level at zero energy.

Integer Quantum Hall Effect in Graphene

The integer quantum Hall effect is peculiar due to the zero energy Landau level. We consider an infinite graphene sheet with weak disorder that leads to broadening of Landau levels. As in the ordinary IQHE, states on the Landau level energy are extended, and at these energies, ρ_{xx} and σ_{xx} are peaked, and σ_{xy} is not quantized. States between Landau levels are localized, hence, σ_{xy} is quantized and $\rho_{xx} = \sigma_{xx} = 0$. Recall that in graphene, the peaks are not equally spaced, since $\varepsilon_n = \sqrt{bn}$.

Let us follow the Laughlin argument in Sec. 9.5.8 and roll the graphene sheet into a CNT. One can ask, how many edge states are crossed at the Fermi energy in analogy with the argument presented in Fig. 9.56 pertaining to the integer quantum Hall effect in semiconductors? The edge state pattern is illustrated in Fig. 13.41(a). The longitudinal resistivity ρ_{xx} and Hall conductivity σ_{xy} are shown in Fig. 13.41(b). As explained in the caption, the Hall conductivity in graphene is quantized as $\sigma_{xy} = (2n+1)e^2/h$ per spin. There is no plateau at zero energy because it is the center of a Landau level, where states are extended and $\sigma_{xx} \neq 0$ (it is local maximum).

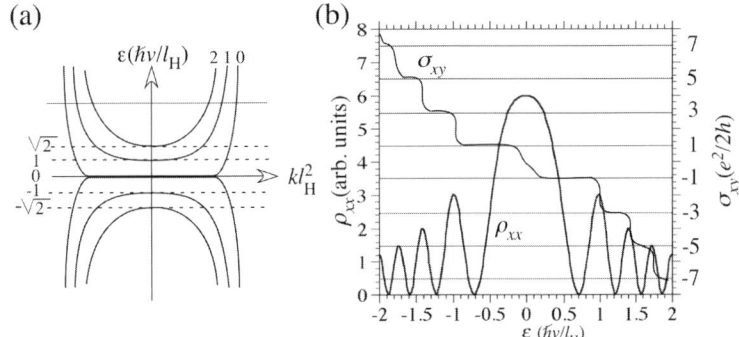

FIG 13.41 (a) Edge states in graphene rolled into a cylinder (CNT), as in the Laughlin gedanken experiment. Edge states with positive (negative) energies refer to particles (holes). Edge states with Landau level numbers $n \neq 0$ are doubly degenerate, one for each Dirac cone. The edge state with $n = 0$ is not degenerate because it is shared by the two Dirac cones. Therefore, on each edge, the Fermi energy between two Landau levels $\varepsilon_n < \varepsilon_F < \varepsilon_{n+1}$ crosses $2n + 1$ edge states, hence, $\sigma_{xy} = (2n+1)e^2/h$ per spin. (b) Longitudinal resistivity ρ_{xx} and Hall conductivity σ_{xy} for bulk graphene as function of Fermi energy. The peaks are the centers of Landau levels.

13.6.7 POTENTIAL SCATTERING IN GRAPHENE

Scattering from impurities in graphene is important because, experimentally, impurities are typically unavoidable. The scattering of massless 2D Dirac particles can be used to understand scattering in graphene at low energies.

From the Dirac Equation to Cross Sections

Consider an electron scattering at energy E from a potential $V(\mathbf{r})$ that does not mix sublattices A, B because potential does not depend on pseudo-spin. For concreteness, we consider scattering of states with small positive energy $E > 0$ close to $\mathbf{K}_+$. Adding the potential to the kinetic energy operator in Eq. (13.121), we obtain

$$[-i\boldsymbol{\sigma} \cdot \nabla + U(\mathbf{r})]\Psi_\mathbf{q}^+(\mathbf{r}) = q\Psi_\mathbf{q}^+(\mathbf{r}),$$
$$U(\mathbf{r}) = V(\mathbf{r})/(\hbar v), \quad q = E/(\hbar v) \quad (q > 0). \tag{13.141}$$

The asymptotic condition on the wave function consists of an incoming plane wave spinor in the $\hat{\mathbf{x}}$ direction (initial momentum $\mathbf{q}$) and a circular outgoing spinor wave whose amplitude depends on the angle of the final direction $\mathbf{q}'$, i.e., $\theta_{\mathbf{q}'}$ such that $\cos\theta_{\mathbf{q}'} = \hat{\mathbf{q}} \cdot \hat{\mathbf{q}}'$ (all vectors are two dimensional and energy conservation implies $|\mathbf{q}| = |\mathbf{q}'| = q$). Explicitly,

$$\Psi_\mathbf{q}^+(\mathbf{r}) \to \frac{1}{2\pi}\left[e^{iqx}u(\hat{\mathbf{q}}) + \frac{e^{iqr}}{\sqrt{r}}\begin{pmatrix}f_A(\theta_{\mathbf{q}'})\\f_B(\theta_{\mathbf{q}'})\end{pmatrix}\right], \tag{13.142}$$

where $u(\hat{\mathbf{q}})$ is defined in Eq. (13.128). The currents of the incoming wave along the direction $\hat{\mathbf{q}}$ and of the scattered wave along the direction $\hat{\mathbf{q}}'$ are

$$\mathbf{I}_\mathbf{q} = \text{Re}\left[e^{-iqx}u(\hat{\mathbf{q}})^\dagger \boldsymbol{\sigma} e^{iqx}u(\hat{\mathbf{q}})\right] = q\hat{\mathbf{x}}, \tag{13.143a}$$

$$\mathbf{I}_{\mathbf{q}'} = \text{Re}\left\{\frac{e^{-iqr}}{\sqrt{r}}(f_A(\theta_{\mathbf{q}'})^*, f_B(\theta_{\mathbf{q}'})^*)\boldsymbol{\sigma}\frac{e^{iqr}}{\sqrt{r}}\begin{pmatrix}f_A(\theta_{\mathbf{q}'})\\f_B(\theta_{\mathbf{q}'})\end{pmatrix}\right\} = \frac{|f_A(\theta_{\mathbf{q}'})|^2 + |f_B(\theta_{\mathbf{q}'})|^2}{r}\mathbf{q}'. \tag{13.143b}$$

The *differential cross-section* is equal to the number of particles crossing the arc of length $rd\theta_{\mathbf{q}'}$ per unit time and per unit angle divided by the incoming current,

$$\frac{d\sigma}{d\theta_{\mathbf{q}'}} = |f_A(\theta_{\mathbf{q}'})|^2 + |f_B(\theta_{\mathbf{q}'})|^2. \tag{13.144}$$

The *total cross-section* is then

$$\sigma = \int_0^{2\pi} d\theta_{\mathbf{q}'}[|f_A(\theta_{\mathbf{q}'})|^2 + |f_B(\theta_{\mathbf{q}'})|^2]. \tag{13.145}$$

Green's Function and the Lippmann-Schwinger Equation

The Lippmann–Schwinger equation is an integral equation for the wave function that incorporates the Dirac equation (13.141) and the asymptotic conditions (13.142). We will build on the free Green's function, from the free Hamiltonian $H_0 = -i\boldsymbol{\sigma} \cdot \nabla$, which can be written as a momentum or configuration space representation of the operator,

$$g_0^+(q) = (q + i\eta - H_0)^{-1} = (q + i\eta + i\boldsymbol{\sigma} \cdot \nabla)^{-1}. \tag{13.146}$$

In momentum space, this task is straightforward: multiply numerator and denominator by $(q + H_0)$ to obtain,

$$\langle \mathbf{k}|g_0^+(q)|\mathbf{k}'\rangle = \delta(\mathbf{k} - \mathbf{k}')\frac{\varepsilon + \boldsymbol{\sigma} \cdot \mathbf{k}'}{(q + i\eta)^2 - k^2}. \tag{13.147}$$

13.6 Graphene

Usually, this function is required off-energy shell, $k = k' \neq q$. The momentum coordinates $\mathbf{k}, \mathbf{k}'$ should not be confused with the momentum vectors $\mathbf{q}$ (initial direction) and $\mathbf{q}'$ (final direction), which are on shell, $q = q' = \varepsilon$. The representation of $g_0^+(q)$ in r space requires carrying out the Fourier transform of $\langle \mathbf{k}' | g_0^+(\varepsilon) | \mathbf{k} \rangle$ (in 2D). By construction, this function satisfies the inhomogenous equation,

$$[-i\boldsymbol{\sigma} \cdot \nabla + U(r)]\langle \mathbf{r} | g_0^+(q) | \mathbf{r}' \rangle = -\delta(\mathbf{r} - \mathbf{r}'). \tag{13.148}$$

The calculation of the Green's function is straightforward but tedious. For the analysis of the kinematics, we use Fig. 13.42.

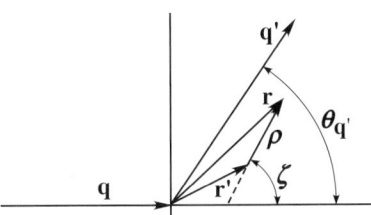

FIG 13.42 Kinematics of scattering from an impurity in graphene [see discussion of the Green's function defined in Eq. (13.149)].

Denoting $\boldsymbol{\rho} \equiv \mathbf{r} - \mathbf{r}'$ and $\cos \zeta \equiv \hat{\boldsymbol{\rho}} \cdot \hat{\mathbf{x}}$,

$$\langle \mathbf{r} | g_0^+(\varepsilon) | \mathbf{r}' \rangle = -\frac{iq}{\pi} \begin{pmatrix} H_0^{(1)}(q\rho) & -ie^{-i\zeta} H_1^{(1)}(q\rho) \\ -ie^{i\zeta} H_1^{(1)}(q\rho) & H_0^{(1)}(q\rho) \end{pmatrix}. \tag{13.149}$$

Thus, unlike the free Green's function in 2D, Eq. (12.712), the free Green's function for graphene, in addition to being a 2×2 matrix, depends not only on $|\boldsymbol{\rho}| = |\mathbf{r} - \mathbf{r}'|$ but also on the angle ζ, which $\boldsymbol{\rho}$ forms with the x axis. The solution to the Lippmann–Schwinger equation,

$$\Psi_\mathbf{q}^+(\mathbf{r}) = \frac{1}{2\pi} e^{iqx} u(\hat{\mathbf{q}}) + \int \langle \mathbf{r} | g_0^+(q) | \mathbf{r}' \rangle U(\mathbf{r}') \Psi_\mathbf{q}^+(\mathbf{r}') d\mathbf{r}', \tag{13.150}$$

constitutes a solution of Eq. (13.141) supplemented by the asymptotic boundary condition (13.142). To evaluate the scattering amplitude, we have to examine the asymptotic form of the Green's function as $r \to \infty$, where $|\mathbf{r} - \mathbf{r}'| \approx r - \hat{\mathbf{r}} \cdot \mathbf{r}'$ and the direction of $\boldsymbol{\rho}$ approaches that of $\mathbf{q}'$, so that $\zeta \to \theta_{\mathbf{q}'}$. Using Eq. (13.149), and employing the asymptotic expansion of the Hankel functions,

$$\langle \mathbf{r} | g_0^+(q) | \mathbf{r}' \rangle \to -\frac{iq}{\pi} \sqrt{\frac{\pi}{qr}} e^{iqr - i\frac{\pi}{4}} e^{-i\mathbf{q}' \cdot \mathbf{r}'} \begin{pmatrix} 1 & e^{-i\theta_{\mathbf{q}'}} \\ e^{i\theta_{\mathbf{q}'}} & 1 \end{pmatrix}. \tag{13.151}$$

Inserting Eq. (13.151) into Eq. (13.150) leads to the asymptotic expression,

$$\Psi_\mathbf{q}^+(\mathbf{r}) \to \frac{1}{2\pi} e^{iqx} u(\hat{\mathbf{q}}) - \frac{e^{iqr}}{\sqrt{r}} e^{i\frac{\pi}{4}} \sqrt{\frac{q}{\pi}} \begin{pmatrix} 1 & e^{-i\theta_{\mathbf{q}'}} \\ e^{i\theta_{\mathbf{q}'}} & 1 \end{pmatrix} \int e^{-i\mathbf{q}' \cdot \mathbf{r}'} U(\mathbf{r}') \Psi_\mathbf{q}^+(\mathbf{r}') d\mathbf{r}'. \tag{13.152}$$

Comparing this with Eq. (13.142), the connection between the wave function and the scattering amplitudes is established. If the two components of the spinor part of $\Psi_\mathbf{q}^+(\mathbf{r}')$ are equal at $\theta_{\mathbf{q}'} = \pi$, the amplitudes $f_A(\pi) = f_B(\pi) = 0$. For example, in the Born approximation, $\Psi_\mathbf{q}^+(\mathbf{r}')$ in the integrand is replaced by the incoming spinor $e^{iqx} u(\hat{\mathbf{q}}) = \frac{e^{i\mathbf{q} \cdot \mathbf{r}}}{\sqrt{2}} \begin{pmatrix} 1 \\ 1 \end{pmatrix}$. Therefore, the scattering amplitudes in the Born approximation are

$$f_A(\theta_{\mathbf{q}'}) = -e^{i\frac{\pi}{4}} \sqrt{\frac{q}{2\pi}} (1 + e^{-i\theta_{\mathbf{q}'}}) \tilde{U}(\mathbf{q}' - \mathbf{q}),$$

$$f_B(\theta_{\mathbf{q}'}) = -e^{i\frac{\pi}{4}} \sqrt{\frac{q}{2\pi}} (1 + e^{i\theta_{\mathbf{q}'}}) \tilde{U}(\mathbf{q}' - \mathbf{q}), \tag{13.153}$$

where $\tilde{U}(\mathbf{q}' - \mathbf{q})$ is the Fourier transform of $U(\mathbf{r})$ at momentum transfer $\mathbf{p} = \mathbf{q}' - \mathbf{q}$. In the Born approximation, both amplitudes vanish at $\theta_{\mathbf{q}'} = \pi$, i.e., there is no backward scattering. We will check below whether this *absence of backscattering in graphene* is valid to arbitrary order.

Partial Wave Expansion and Phase Shifts

Now we assume that the potential $U(\mathbf{r}) = U(r)$ is central. Therefore, j [defined according to Eq. (13.132)] is a good quantum number, and the Dirac Eq. (13.141) can be analyzed in terms of spinor circular waves as in Eq. (13.133), assuming $\lambda = 1$ for positive energy particles. It is convenient to use quantum number j instead of m, with $m = j - 1/2$ and $m + 1 = j + 1/2$. Recall that j is a magnetic quantum number (short notation for j_z) so that it is summed over all half-integers, positive and negative, $j = \pm 1/2, \pm 3/2, \ldots$. A spinor circular wave of definite j quantum number is

$$\Psi_j(\mathbf{r}) = \begin{pmatrix} e^{i(j-\frac{1}{2})\phi} f_j(r) \\ e^{i(j+\frac{1}{2})\phi} g_j(r) \end{pmatrix}. \tag{13.154}$$

The radial functions $f_j(r)$ and $g_j(r)$ satisfy the coupled radial equations [224, 225]

$$\frac{df_j}{dr} - \frac{j-\frac{1}{2}}{r} f_j + (\varepsilon - U) g_j = 0,$$

$$\frac{dg_j}{dr} + \frac{j+\frac{1}{2}}{r} g_j - (\varepsilon - U) f_j = 0. \tag{13.155}$$

Note the symmetry of Eqs (13.155) for any central potential $U(r)$:

$$f_{-j}(r) = g_j(r), \qquad g_{-j}(r) = -f_j(r). \tag{13.156}$$

Thus, if $(f_j(r), g_j(r))$ is a solution for a given j, then $(g_{-j}(r), -f_{-j}(r))$ is also a solution. The asymptotic forms of the radial functions read,

$$f_j(r) \to \frac{1}{\sqrt{r}} \cos\left(kr - j\frac{\pi}{2} + \delta_j^A\right), \tag{13.157}$$

$$g_j(r) \to \frac{1}{\sqrt{r}} \sin\left(kr - j\frac{\pi}{2} + \delta_j^B\right). \tag{13.158}$$

The symmetry (13.156) implies the relation between phase shifts,

$$\delta_j^A = \delta_{-j}^A, \qquad \delta_j^B = \delta_{-j}^B. \tag{13.159}$$

The rest of the formalism follows Sec. 12.5.3 and includes the following steps: (1) The wave function on the LHS of Eq. (13.142) and the spinor plane wave on the RHS of Eq. (13.142) are expanded in partial wave spinors (13.154), and the asymptotic forms of f_j, g_j and the Bessel functions are used. (2) When $r \to \infty$, the direction of $\mathbf{r}$ coincides with that of the final momentum $\mathbf{q}'$, so that the polar angle ϕ approaches $\theta_{\mathbf{q}'}$. (3) By expressing the trigonometric functions in terms of exponentials $e^{\pm iqr}$, it is possible to obtain a combination of two terms $C(\theta_{\mathbf{q}'}) \frac{e^{iqr}}{\sqrt{r}} + D(\theta_{\mathbf{q}'}) \frac{e^{-iqr}}{\sqrt{r}} = 0$, where $C(\theta_{\mathbf{q}'})$ and $D(\theta_{\mathbf{q}'})$ are spinors expressed in terms of the phase shifts and the angular exponential functions $e^{i(j\pm\frac{1}{2})\theta_{\mathbf{q}'}}$. This combination can vanish only if $C(\theta_{\mathbf{q}'}) = D(\theta_{\mathbf{q}'}) = 0$, and these two conditions yield the expansion coefficients and the expression for the scattering amplitude,

$$f_A(\theta_{\mathbf{q}'}) = \frac{1}{\sqrt{2iq}} \lim_{n \to \infty} \sum_{j=-n-\frac{1}{2}}^{n+\frac{1}{2}} (e^{2i\delta_j^A} - 1) e^{i(j-\frac{1}{2})\theta_{\mathbf{q}'}}. \tag{13.160a}$$

$$f_B(\theta_{\mathbf{q}'}) = \frac{1}{\sqrt{2iq}} \lim_{n \to \infty} \sum_{j=-n-\frac{1}{2}}^{n+\frac{1}{2}} (e^{2i\delta_j^B} - 1) e^{i(j-\frac{1}{2})\theta_{\mathbf{q}'}}. \tag{13.160b}$$

13.6 Graphene

Due to the symmetry (13.159), the contributions from positive and negative j can be grouped as follows:

$$e^{i(j-\frac{1}{2})\theta_{\mathbf{q}'}} + e^{i(-j-\frac{1}{2})\theta_{\mathbf{q}'}} = e^{i(j-\frac{1}{2})\theta_{\mathbf{q}'}}(1 + e^{-2j\theta_{\mathbf{q}'}}). \quad (13.161)$$

Because j is half-integer, the backward scattering amplitude vanishes [225],

$$\boxed{f_A(\theta_{\mathbf{q}'} = \pi) = f_B(\theta_{\mathbf{q}'} = \pi) = 0.} \quad (13.162)$$

The absence of backward scattering is related to the destructive interference of time-reversed paths of helicity eigenstates and the fact that the Berry phase accumulated after a 2π rotation is $e^{i\pi} = -1$ [see discussion after Eq. (13.129)].

The total cross-section is

$$\sigma = \int_0^{2\pi} \left[|f_A(\theta_{\mathbf{q}'})|^2 + |f_B(\theta_{\mathbf{q}'})|^2 \right] d\theta_{\mathbf{q}'}$$
$$= \frac{2\pi}{q} \sum_j \left[\sin^2 \delta_j^A + \sin^2 \delta_j^B \right] = \frac{2\pi}{q} \operatorname{Im}[f_A(0) + f_B(0)], \quad (13.163)$$

which is the optical theorem for scattering from an impurity in graphene.

Klein Tunneling

Since electrons in graphene satisfy a Dirac equation in 2D, where the role of the velocity of light c is played by the Fermi velocity, v_F, observation of relativistic effects in the regime of relatively low velocities is possible. One such effect is the *Klein paradox*, the scattering of Dirac particles from a high potential barrier. If the potential height V exceeds the particle rest energy, mc^2, the transmission probability is much larger than that obtained using the non-relativistic (Schrödinger) formalism and approaches unity in the limit $V \to \infty$. The reason is that a strong repulsive potential for electrons is attractive for holes in the Dirac theory. There are hole states inside the barrier whose energy match the electron energy states outside the barrier, and this affects scattering properties. Observation of the Klein paradox requires a steep potential gradient of the order of mc^2 over the Compton length scale, $\hbar/mc$, which is equivalent to a huge electric field ($\approx 10^{16}$ V/cm for electrons). However, in graphene, the particles satisfy the massless 2D Dirac equation, and no lower bound on the electric field is required. Reasonable estimates suggest that the required field for realizing the Klein paradox is $\approx 10^5$ V/cm.

The Klein scattering geometry is illustrated in the inset of Fig 13.43. Consider electrons in the vicinity of the Dirac point, $\mathbf{K}_+$, subject to the free Hamiltonian $\mathcal{H}_0(\mathbf{K}_+)$, Eq. (13.121), and a potential barrier $V(x) = V_0 \Theta(x) \Theta(a - x)$. Then, q_y is a good quantum number, and we can solve the barrier penetration problem using the plane waves in Eq. (13.128). Outside

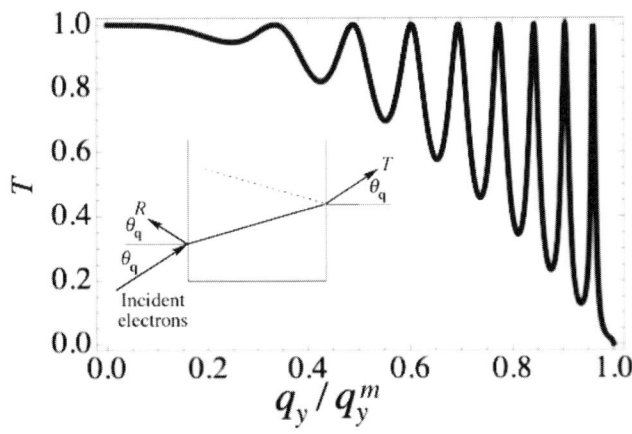

FIG 13.43 Transmission T versus transverse momentum, q_y/q_y^m where $q_y^m = E/(\hbar v_F)$. The geometry for scattering off the 2D barrier is shown in the inset. Note that $\sin \theta_{\mathbf{q}} = q_y/q_m$.

the barrier, the plane waves have scattering energy $E = v_F q = v_F \sqrt{q_x^2 + q_y^2}$, whereas inside the barrier, they have kinetic energy $E - V_0$. For a given energy E and q_y, the longitudinal momenta outside and inside the barrier are

$$q_x = \sqrt{[E/(\hbar v_F)]^2 - q_y^2},$$

$$\bar{q}_x \equiv \sqrt{[(E - V_0)/(\hbar v_F)]^2 - q_y^2}. \quad (13.164)$$

The details of the matching of the two components of the wave function at $x = 0$ and $x = a$ are easy to work out. For high potential barrier, $|E| \ll V_0$, and for q_y such that $\bar{q}_x$ is real, the transmission coefficient is given approximately by

$$T = \frac{\cos^2 \theta_\mathbf{q}}{1 - \cos^2(\bar{q}_x a) \sin^2 \theta_\mathbf{q}}, \quad (13.165)$$

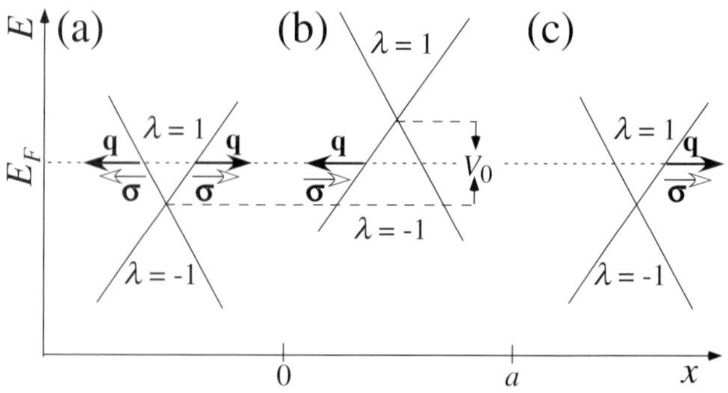

FIG 13.44 Klein tunneling in graphene for electrons near the Dirac point $\mathbf{K}_+$. (a) For $E_F > 0$, outside the barrier, electrons belonging to the right branch move toward the barrier with $\mathbf{q}$ (black arrow) and $\boldsymbol{\sigma}$ (white arrow) points rightward. Some of them are reflected and populate the left branch with $\mathbf{q}$, with $\boldsymbol{\sigma}$ pointing to the left. (b) Electrons that enter the barrier are converted into holes in the valence band. The holes move leftward, and to guarantee the helicity constraint $\lambda = -1$, the pseudo-spin points rightward. (c) Leaving the barrier after transmission, the electrons reappear on the right branch moving away from the barrier with $\mathbf{q}$ and, $\boldsymbol{\sigma}$ points rightward.

where $\tan \theta_\mathbf{q} = q_y/q_x$. Thus, despite the fact that the barrier is high, $|V_0| \gg |E|$, the transmission coefficient tends to unity at normal incidence $\theta_\mathbf{q} = 0$ and oscillates as the incidence angle is lowered (see Fig 13.43).

The reason for this kind of behavior is as follows. Suppose $E > 0$, so that outside the barrier the conduction band is partially filled. Since $\lambda = 1$, electrons moving to the right ($\partial E/\partial q_x > 0$) have $\boldsymbol{\sigma}$ pointing to the right, whereas electrons moving to the left ($\partial E/\partial q_x < 0$) have $\boldsymbol{\sigma}$ pointing to the left. Now suppose that $E - V_0 < 0$, so that under the barrier, the valence band is not filled and the helicity is -1. Here, we have holes with opposite velocities. To keep the helicity at -1, the electron that moved to the right outside the barrier with energy E is matched to a hole that moves to the left, so that the pseudo-spin vector of the hole continues to point to the right, i.e., it matches that of the electron. This matching leads to enhanced transmission. Figure 13.44 illustrates these scattering scenarios.

13.7 INVENTORY OF RECENTLY DISCOVERED LOW-DIMENSIONAL PHENOMENA

In this section, we present a few hallmarks and highlights of low-dimensional systems to expose the reader to important phenomena that have recently come to the forefront of research and to review some of the intriguing phenomena that have been discovered in this field. The presentation is qualitative, yet, hopefully conveys the depth and breadth of the research in this area during the last few decades, as well as challenges that remain. In Sec. 13.7.1, we discuss persistent currents in mesocsopic systems. Then, we discuss weak localization phenomena in Sec. 13.7.2, decoherence phenomena in Sec. 13.7.2, shot noise in Sec. 13.7.3, and the physics of strongly correlated systems in Sec. 13.7.4, Wigner crystals in Sec. 13.7.5, and the fractional Hall effect in Sec. 13.7.6 where we discuss Laughlin many electron wave functions and their low-lying excitations as well as the Janes idea of composite Fermions. Section 13.7.7 briefly considers the huge topic of High T_c superconductivity and 2D magnetism, and in Sec. 13.7.8, the basic notions of 1D interacting fermionic systems are introduced. Another important 1D system, the quantum spin chain, is introduced in Sec. 13.7.9, where we also mention the Haldane gap conjecture in the $S = 1$ spin chain. Finally, in Sec. 13.7.10, we briefly discuss the quantum spin Hall effect that occurs in systems conserving time-reversal invariance with strong spin–orbit coupling.

13.7 Inventory of Recently Discovered Low-Dimensional Phenomena

13.7.1 PERSISTENT CURRENTS

In Sec. 9.5.2, we discussed the Aharonov–Bohm and Aharonov–Casher effects, and showed that persistent currents are a consequence of the Aharonov–Bohm effect. In a conducting (metal or semiconductor) ring of size $L < \ell_\varphi$, the ground state of the itinerant electron system does not carry current in the absence of magnetic field. But when the ring is threaded by a magnetic flux, Φ, time-reversal invariance is broken. The wave function of an electron gains an Aharonov–Bohm (or Berry) phase ϕ when moving along the ring in one direction and $-\phi$ when moving in the opposite direction. Here, $\phi = \Phi/\Phi_0$, where $\Phi_0 = hc/e$ (the unit of flux quantum). Due to this *chirality*, the ground state carries a current $I(\phi)$. As a result of gauge invariance, the current is a periodic function of ϕ with period 1. In addition, the current is an antisymmetric function of ϕ. Persistent current is an equilibrium property: It is nondissipative and it is a property of all the states below the Fermi energy.

13.7.2 WEAK LOCALIZATION

In Secs 9.9 and 12.9.3, we discussed Anderson localization, where the eigenstates of an electron in a disordered potential become spatially localized, which is at the origin of the phenomenon known as the Anderson metal-insulator transition, and the occurrence of Anderson insulators. Strictly speaking, Anderson localization is a property of an infinite system or of finite systems of size $L \gg \xi$, where ξ is the localization length. In mesoscopic systems, we encounter another facet of localization, *weak localization*. Weak localization is different from Anderson localization and deals with relatively small variations of the resistance in a disordered conductor, but these variations do not turn it into an insulator. It has two facets that occur in weakly disordered 2D mesoscopic systems. The first is exposed when time-reversal symmetry is broken by a weak magnetic field [203, 228]. The second is exposed in the absence of a magnetic field when strong spin–orbit coupling is present. Both were already mentioned briefly in connection with Eq. 13.71.

Loss of Phase Coherence in a Magnetic Field: One of the hallmarks of weak localization is an enhanced magnetoconductance in weak magnetic fields [131]. Imagine, as illustrated in Fig. 13.45, an electron in a 2D disordered system that contributes to the current by moving to the right from an initial point A. Let P_A denote the probability for the electron to return back to the initial point A. The smaller P_A, the higher the contribution to the current. The figure shows two paths for which the electron returns to A after a sequence of collisions with randomly located impurities. The clockwise path is indicated by black arrows and the counter-clockwise path, also referred to as the time-reversed path, by white arrows. Along both paths, in the absence of magnetic field, the electron wave function accumulates the same phase, which in the WKB approximation can be written as $\oint \mathbf{k}(\mathbf{s}) \cdot d\mathbf{s}$, where $\mathbf{s}$ is an element of length along the closed loop and $\mathbf{k}(\mathbf{s})$ is the local wavevector. Hence, the two contributions, one from the clockwise trajectory and the other from the anti-clockwise trajectory, interfere constructively. This leads to enhanced backscattering, which increases the resistance of the system.

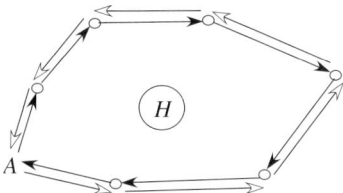

FIG 13.45 An example of weak localization: Two time-reversed paths that contribute to the return amplitude of an electron traveling in a 2D sample from a point A and return back to A. In the absence of a magnetic field, they interfere constructively at A, leading to enhanced backscattering, while in the presence of a perpendicular weak magnetic field, the two paths have different phases and the interference is not constructive. In other words, application of magnetic field reduces the probability of backscattering. This leads to negative magnetoresistance.

If a weak perpendicular magnetic field is applied, the two interfering components have different phases (see Sec. 9.5.2). Suppose the total magnetic flux through the closed loop is $\Phi = \phi \Phi_0$, where $\Phi_0 = hc/e$ is the flux quantum and $\phi = \Phi/\Phi_0$ is dimensionless. Along the counter-clockwise trajectory, the wave function accumulates a phase $2\pi\phi$ and along the clockwise trajectory, it accumulates a phase $-2\pi\phi$. Generically, phase coherence at the initial point is lost, therefore, the interference is not constructive and P_A is reduced. Thus, the application of a weak perpendicular magnetic field on a weakly disordered 2D system reduces its resistance, a phenomenon referred to as *negative magneto-resistance*. The correction can be shown to be a quadratic function of the magnetic field strength H,

$$\Delta R(H) = \frac{R(H) - R(0)}{R(0)} = -CH^2, \tag{13.166}$$

where $C > 0$ is a constant.

Weak Antilocalization due to Spin–Orbit Coupling: In the previous section, we discussed attenuation of probability of backward scattering due to an external magnetic field, where time-reversal invariance is broken. Here we will show that destructive interference at the backward direction occurs also due to spin-orbit interaction, i.e., time-reversal invariance symmetry is present, yet the origin of attenuation is internal. Suppose no external magnetic field is present but there is a strong Rashba type spin-orbit coupling. To simplify the illustration we assume that the closed loop in Fig. 13.45 is a circle and that an electric field $\mathbf{E}$ causes spin-orbit coupling which is generated by a 1D charged wire of charge density λ_c per unit length that is perpendicular to the 2D plane and passes through the center of the circular loop. This is precisely the geometry used to analyze the Aharonov-Casher effect in Sec. 9.5.2. The qualitative result regarding destructive interference in the backward direction remains valid for more realistic spin-orbit potentials, such as Rashba or Dresselhaus potentials discussed in Sec. 9.6.7). The electrons satisfy the Pauli equation (9.205), where the Aharonov-Casher $SU(2)$ flux $\Phi_{AC} \equiv \frac{g\mu_B^* \lambda_c}{2e}$ has units of magnetic flux. Here $\mu_B^* = e\hbar/(2m^*c)$ with m^* being the effective mass (in semiconductors, usually $m^* \ll m$). Because S_z is conserved, Eq. (9.205) can be decomposed into two separate equations, one for each spin component. The equation for the spin-up component is

$$\frac{1}{2m}\left(p + \frac{e}{c}\frac{\Phi_{AC}}{L}\right)^2 \psi_\uparrow = (E + \bar{E})\psi_\uparrow. \tag{13.167}$$

This equation is identical to that of a spinless charged particle moving on a ring subject to a magnetic flux $\phi = \phi_{AC}$ (in units of Φ_0). Therefore, we can use the results of the previous discussion. Consider an electron at point A that moves along a closed loop, either clockwise (black arrows in Fig. 13.45) or anti-clockwise (white arrows in Fig. 13.45). The corresponding two components, $\psi_\uparrow$ (clockwise) and $\psi_\uparrow$ (anti-clockwise), acquire phases $\mp 2\pi\phi_{AC}$, respectively, and interfere destructively on arriving at the initial point A. Similarly, the spin-down component equation is

$$\frac{1}{2m}\left(p - \frac{e}{c}\frac{\Phi_{AC}}{L}\right)^2 \psi_\downarrow = (E + \bar{E})\psi_\downarrow. \tag{13.168}$$

This equation is identical to that of a spinless charged particle moving on a ring subject to a magnetic flux $\phi = -\phi_{AC}$. The corresponding two components $\psi_\downarrow$ (clockwise) and $\psi_\downarrow$ (anti-clockwise) will acquire the corresponding phases $\pm 2\pi\phi_{AC}$ and interfere destructively on arriving back to the initial point A. The upshot is that, due to spin–orbit coupling, the probability of returning to A is reduced by a factor $\cos^2(2\pi\phi_{AC})$, compared with the situation where spin–orbit coupling is absent. Reduction of return probability means lower resistance. The conductance is slightly enhanced, a phenomenon referred to as *weak antilocalization*.

Dephasing

Once there is no definite relation between the phases of the components of a wave function due to mechanisms that are referred to as dephasing, coherence is suppressed. Low-dimensional systems constitute an effective testbed for studying dephasing. An extensive discussion of the mechanisms and effects of dephasing, and its importance for mesoscopic systems is given in Ref. ([131]). Here we will briefly discuss some of the main ideas and present a few results.

Consider a simple Fabry-Perot interferometer wherein the two components of a wave function that was initially in a pure state at one point in the interferometer can interfere at a later time at another point of the interferometer, and thereby the phase relation between the two components can be determined. In order to describe the loss of phase coherence,

13.7 Inventory of Recently Discovered Low-Dimensional Phenomena

we consider a procedure involving a transition of an electron from an initial (pure) state at $t = 0$ to a final state at $t > 0$ that undergoes some kind of perturbation as it propagates. The procedure is organized in steps as follows: (1) An external field induces a transition of the initial state to another quantum state. This transition is encoded by a quantum amplitude with a modulus and a phase $\varphi(t)$. The latter can be measured, e.g., in a Fabry-Perot interferometer. (2) Because we cannot control the external field (it is not perfectly deterministic), the phase contains a random component $\delta\varphi(t)$ that might be time-dependent. By "random" we mean that it displays sample-to-sample fluctuations. (3) For $t = \tau_\varphi$ such that $\delta\varphi(\tau_\varphi) \approx 2\pi$, interference is "averaged out" and thereby lost. τ_φ is the dephasing time and $1/\tau_\varphi$ is the dephasing rate discussed in Sec. 13.1.

The dephasing rate, τ_φ^{-1}, is intimately related with the accuracy in which energy of a system can be measured. It can be shown that, if one were able to measure the energy transferred as function of time between a system and an environment coupled to it, the accuracy of such a measurement would be of order $\hbar/\tau_\varphi$. This suggests the idea that depahsing rates can be determined using energy transfer between a system and its environment. In fact, other scenarios that do not involve energy transfer such as weak localization, or persistent currents, are insufficient for extracting information on dephasing rates because every event of time-reversal violation, such as application of an external magnetic field, destroys constructive interference. On the other hand, mesoscopic fluctuations involve energy transfer and are not destroyed by a magnetic field (recall from Eq. (13.72) that the application of a magnetic field reduces fluctuations by a factor 2 but does not destroy them). Thus, energy transfer often plays a role in dephasing.

Let us briefly mention some mechanisms leading to dephasing. The most familiar mechanism is *inelastic dephasing* caused by phonons, electron–electron interaction, magnons, etc. Another important mechanism for dephasing is the presence of magnetic impurities. In this case, energy transfer takes place between the itinerant electrons mediated through the impurities having an exchange interaction with the itinerant electrons. Since time-reversal invariance is violated, dephasing occurs. As $T \to 0$, the main source of dephasing in this case is Kondo scattering.

The question of whether dephasing can result from purely elastic scattering is not fully answered. For electrons in mesoscopic systems, the situation can be summarized as follows: Elastic scattering modifies the phase of the electronic wave function but does not destroy it. The phase is, in some sense, reproducible no matter how complicated it appears after a series of elastic scattering events. On the other hand, inelastic scattering introduces phase uncertainty which smears quantum interference effects.

The dephasing rate, τ_φ^{-1}, increases with increasing temperature. The question of whether it tends to zero or saturates at a finite value as $T \to 0$ was under debate for a long time, but it is now accepted that the dephasing time τ_φ diverges as a negative power of the temperature, i.e., $\tau_\varphi \propto T^{-p}, p > 0$. Consequently, no dephasing occurs at zero temperature. The value of p depends on the dimensionality d and the dephasing mechanism. When the dominant dephasing mechanism is due to electron–electron scattering, it was found that $p = 2/(4-d)$.

13.7.3 SHOT NOISE

When a current $I(t)$ is measured as function of time, it reveals fluctuations around a smooth average $\bar{I}(t)$. The fluctuations are called *current noise* [232]. Schottky found in 1918 that in ideal vacuum tubes, there are two kinds of noise. The first is called thermal noise, and is nowadays referred to as *Johnson–Nyquist noise*, is due to thermal motion of the charge carriers. The second is *shot noise*, stemming from the fact that charge is not continuous but discrete. Experiments on nanoscale conductors in the quantum mechanical regime indicate that shot noise contains information on dynamics and correlations of charge carriers. Let us denote the time-dependent fluctuation of the current by $\Delta I(t) \equiv I(t) - \bar{I}(t)$, and the power spectrum of the noise by

$$S(\omega) = 2 \int_{-\infty}^{\infty} dt \langle \Delta I(t+\tau) \Delta I(\tau) \rangle e^{i\omega t}, \qquad (13.169)$$

where $I(t)$ is the measured current at time t and $\langle \ldots \rangle$ stands for ensemble average (if the system is disordered) or averaging over the initial time τ. The prefactor 2 is due to spin degeneracy. Consider the thermal noise that occurs when $V = 0$

(no bias voltage) and $T > 0$. This noise is related to the conductance G by the fluctuation-dissipation theorem (see Sec. 7.9.4),

$$S_{\text{thermal}} = 4k_B T G, \quad (\text{for } \hbar\omega \ll k_B T). \tag{13.170}$$

Consequently, S_{thermal} at low frequency contains no new information beyond the conductance.

For $V > 0$ and $T = 0$, the remaining noise is shot noise, which reflects the correlation between charge carriers at different times (information not contained in G). In many electronic devices, such as p-n junctions, Schottky barriers, and tunnel junctions, the electrons are transmitted randomly and independently of each other. The number of electrons transmitted per unit time follows a Poisson distribution, and the low frequency ($\approx$ 10 KHz) noise is proportional to the average current,

$$S_{\text{Poisson}} = 2e\bar{I}. \tag{13.171}$$

When the transported electrons are correlated, the shot noise S is suppressed compared with S_{Poisson}. The stronger the correlation, the lower the shot noise. One source of correlation is the interaction between electrons, but in metals, it is effectively screened. In the absence of interactions, a significant source of correlations is the Pauli principle: If one electron is transmitted into a given quantum state, a second electron cannot occupy that same state. A dimensionless quantity that quantifies the correlations is the *Fano factor*,

$$F \equiv \frac{S}{S_{\text{Poisson}}} = \frac{S}{2e\bar{I}} \leq 1. \tag{13.172}$$

Within the single-particle approximation, the scattering problem in quasi-1D systems developed in Eqs (12.800b), (12.801b), and (12.802), the shot noise can be expressed in terms of the transmission coefficients,

$$T_n = \sum_m |t_{mn}|^2, \tag{13.173}$$

where t_{mn} is the transmission amplitude from initial channel n to final channel m. At zero temperature,

$$S(T=0) = 2e\bar{I} \frac{\sum_n T_n(1-T_n)}{\sum_n T_n}, \quad \Rightarrow \quad F = \frac{\sum_n T_n(1-T_n)}{\sum_n T_n}. \tag{13.174}$$

At finite temperature, both Johnson–Nyquist and shot noise are present,

$$S(T) = S(T=0) \coth\left(\frac{eV}{k_B T}\right) + 4k_B T G. \tag{13.175}$$

In disordered systems with many channels, the summation over n is replaced by integration,

$$\sum_n f(T_n) \rightarrow \int_0^1 dT f(T) P(T), \tag{13.176}$$

where $P(T)$ is the distribution of transmission coefficients. The distribution $P(T)$ in disordered systems is bimodal, with peaks at transmission $T = 0$ (closed channels) and $T = 1$ (for open channels). An explicit expression for $P(T)$ can be derived, leading to a Fano factor $F = 1/3$ [232].

Shot noise in interacting systems may be used to determine the charge of the carriers. In some interacting systems, such as the fractional quantum Hall effect, the basic unit of charge is not e but νe, where ν is some appropriate filling fraction, e.g., 1/3 (see discussion in Secs 13.7.4 and (9.5.8)). In analogy with Eq. (13.171) for S_{Poisson}, we write $S = 2q\bar{I}$, where q is the quantity of charge of a quasi-particle carrying the current. Therefore, the Fano factor is

$$F = \frac{S}{2e\bar{I}} = \frac{q}{e}. \tag{13.177}$$

Thus, measurement of noise as a function of the current reveals the ratio of the charge of the carrier to the charge of the electron. The fraction $q = e/3$ in the fractional quantum Hall effect has been measured in a shot noise experiment. This information extracted from shot noise cannot be obtained with measurement of the averaged current $\bar{I}$.

13.7.4 STRONGLY CORRELATED LOW-DIMENSIONAL SYSTEMS

Electron–electron interaction plays a central role in low-dimensional systems. Theoretically, when electron–electron interaction is omitted, electronic and thermal transport properties in systems with confined geometries are often well understood. For example, the integer quantum Hall effect, which is one of the most striking phenomena related to electron confinement in low dimensions ($d = 2$) under strong perpendicular magnetic field, is adequately explained in terms of the Landau level quantization, as discussed in Sec. 9.5.8. Inclusion of electron–electron interaction significantly complicates calculations, and makes the physics much richer. Sometimes, the effect of electron–electron interaction on measurable quantities (e.g., conductance) is rather dramatic. In 2D, electron–electron interaction is responsible for the fractional quantum Hall effect (see Sec. 9.5.8) in which the Hall conductance is quantized as $\sigma_H = \nu e^2/h$ where the filling factor ν are rational numbers. The fractional Hall effect has led to many new concepts such as fractional statistics, composite quasi-particles (bosons and fermions), and braid groups. Electron–electron interaction in 1D systems leads to new physical concepts such as Tomonaga–Luttinger liquids (a manifestation of the deviation from Fermi liquid behavior). Finally, electron–electron interaction in zero-dimensional systems underlies the Coulomb blockade, spin blockade, and the Kondo effect in quantum dots.

In Chapter 14, we will see that some interacting electron systems can be treated within the Fermi liquid formalism, which leads to a single-particle picture, whereas some cannot. Interacting electron systems for which the description within Fermi liquid theory is inadequate are referred to as *strongly correlated electron systems*. Its analysis requires the introduction of new mathematical techniques [212], some of which will be encountered in Chapters 14 and 18. Traditional many-body perturbation theory, which is developed in Sec. 18.2, linked to the book web page, is sometimes inadequate for studying strongly correlated electron systems in low-dimensions, due to lack of an appropriate small parameter. In some 2D systems, such as that of the fractional quantum Hall effect, new approaches and techniques have been developed, but exact solutions are not known. In 1D, there are several models of interacting systems whose ground-state can be calculated exactly. These include: (1) the Heisenberg spin 1/2 chain, (2) the 1D Bose gas with delta-function interaction, (3) the 1D Hubbard model (see Sec. 18.15.3 linked to the book web page), (4) the Kondo model (see Sec. 18.14).

13.7.5 WIGNER CRYSTALS

The ratio U/K between the electron–electron interaction energy and the kinetic energy increases as the electron density decreases. Consider a system of electrons occupying a volume $\mathcal{V}$ in d dimensions. Denote by n the density of electrons, and let $a \sim n^{-1/d}$ be the average distance between two adjacent electrons. For Coulomb interactions, the typical scale of the potential energy is $U = e^2/a$. From the properties of the free electron gas, the kinetic energy is of order $K \sim \hbar^2/(2ma^2)$ (see Sec. 9.1). For low densities, a is large, so $U/K \sim a$. Hence, electrons at low densities are a strongly correlated system in the sense that the interaction energy dominates the kinetic energy.

Achievement of low density is an experimental challenge. Recall the definition (9.11) of the Wigner–Seitz radius r_s in 3D, as the radius of a sphere containing a single electron such that if N free electrons occupy a volume $\mathcal{V}$, then $\mathcal{V} = N 4\pi r_s^3/3$. The density is $n_{3D} = N/\mathcal{V} = [4\pi r_s^3/3]^{-1}$. In metals and semiconductors, the effective mass m^* and the dielectric constant ε should be accounted for. Denoting by a_0 the Bohr radius for a free electron gas and by $a_0^* = \varepsilon \hbar^2/(m^* e^2)$ the effective Bohr radius in the material, we define the dimensionless quantity $\rho_s \equiv r_s/a_0$. The critical density for achieving Wigner crystallization in 3D is $n_{3D} = [(4\pi/3)(\rho_s a_0^*)^3]^{-1}$, which for real metals is too high to realize. On the other hand, in 2D, the inequality $a_0^* \gg a_0$ can be realized in many semiconductors, and with $\rho_s \approx 37$, the low density limit might be attainable.

A system of interacting electrons at low density might form a crystal to minimize its interaction energy. In 1934, Wigner predicted this phenomenon, referred to as a *Wigner crystal* [233]. In 3D, it would be a body-centered cubic lattice (although the required low-density limit has not been attained). 3D Wigner crystals have not been observed experimentally. In addition to the difficulties in achieving low density, a Wigner crystal, once created, is not stable against various perturbations. For example, once disorder is present, electrons minimize their energy by binding to attractive impurities.

Wigner crystallization has been sought at the surface of liquid Helium, where there is no disorder. A Wigner crystal has been observed in semiconductor heterostructures, where one can create a 2D electron gas at the surface of the semiconductor. The concentration of 2D electrons can be controlled by a gate. The Wigner crystals in these systems have been observed only in strong magnetic fields (the strong magnetic field suppresses the kinetic energy of the electrons, and Wigner crystallization is possible at higher densities). Recently, the occurrence of Wigner crystals in 1D has been analyzed theoretically.

13.7.6 THE FRACTIONAL QUANTUM HALL EFFECT

In Sec. 9.5.8 we discussed the quantum Hall effect, which occurs in a 2D electron gas at low temperature subject to a strong perpendicular magnetic field. The integer quantum Hall effect (IQHE) can be explained within the single-particle picture, while in the fractional quantum Hall effect (FQHE), the interaction between electrons plays a crucial role. Here, we briefly elaborate on the FQHE within the context of strongly correlated electron systems in 2D and mention a number of unusual physical properties that were not considered in Sec. 9.5.8. The origin of these properties is that in a system of interacting electrons in 2D subject to a strong magnetic field, the electrons occupy the lowest Landau level. For a clean system, the only energy scale is due to interaction between electrons. Thus, there is no small dimensionless parameter in the theory. This is different from electrons in metals, where the ratio between interaction and kinetic energies is small.

Recall that the distinction between the IQHE and the FQHE is determined according to whether the filling factor $\nu = N/(2N_L)$, defined in Eq. (9.241), is an integer or belongs to a certain class of rational numbers. Here, N is the number of electrons (or holes) in the sample and N_L is the degeneracy of the Landau levels. In both cases, the Hall conductance σ_H, defined as the ratio of the current through the sample and the voltage drop across the sample in the direction perpendicular to the current flow, is quantized to an extremely high precision as $\sigma_H = \nu \frac{e^2}{h}$. The IQHE occurs when ν is a positive integer and the system is also subject to a disordered potential. When σ_H is quantized, the longitudinal resistance ρ_{xx} vanishes. The IQHE is understood in terms of the Landau-level structure of the noninteracting electron spectrum (see Sec. 9.5.8). When the Landau level is filled, the system is gapped. In the FQHE, σ_H is quantized as $\sigma_H = \nu e^2/h$ and the system is gapped when the filling factor ν takes on certain rational numbers.

Ground State Wave Function and Fractional Charge: In 1983, Robert Laughlin [150] explained the occurrence of the simplest FQHE states having $\nu = 1/q$ with odd q, by suggesting the following many-body wave function for the ground state [see Eq. (9.297)],

$$\Psi_0(\mathbf{r}_1, \mathbf{r}_2, \ldots \mathbf{r}_N) = \Psi_0(z_1, z_2, \ldots z_N) = \prod_{i<j=1}^{N} (z_i - z_j)^q \, e^{-\sum_{k=1}^{N} |z_k|^2 / 4\ell_H^2}, \tag{13.178}$$

where $z_i = x_i + iy_i$ are the complex coordinates of electron i in the 2D x-y plane and ℓ_H is the magnetic length. At the special fractions, $\nu = 1/q$ with q odd, the 2D electron system has a gap of order $10^{-2} e^2/\ell_H$ above its ground state. This gap scenario is similar to that in the IQHE, where quantization is due to the energy gap between Landau levels. For $\nu = 1/q$, the degeneracy of the Landau level is q times higher than for $\nu = 1$, $N_L(1/q) = qN_L(1)$. The question in this context is how the physics of N electrons that *fully* occupy a Landau level with $N_L(1)$ places is similar to that of N electrons that *partially* occupy a Landau level with $N_L(1/q) = qN_L(1)$ places, as in both cases, there is a gap above the ground state. Laughlin's idea is to consider the system of N electrons occupying a fraction $1/q$ of the available $N_L(1/q)$ places as qN quasi-particles occupying *all* $N_L(1/q)$ places. To conserve the total charge, $Q = Ne$, each quasi-particle

13.7 Inventory of Recently Discovered Low-Dimensional Phenomena

must have a *fractional charge* $e/q = \nu e$. Quasi-particles of charge $e/3$ have been indeed observed by a shot noise experiment [see Eq. (13.177) and the discussion below it].

Low-Lying Excitations: Fractional Statistics: The Laughlin construction can be extended to include low-lying excitations above the ground state. The excitations, whose energy is above the gap, are obtained by either adding a quasi-particle (quasi-particle excitations) or by removing a quasi-particle (quasi-hole excitations). The quasi-particle or quasi-holes are localized at points $z_a, z_b, \ldots$. The wave functions for a single quasi-particle and quasi-hole excitations localized at z_0 are

$$\psi_p(z_0) = N_p \prod_{i=1}^{N} \left(\frac{\partial}{\partial z_i} - \frac{z_0}{\ell_H^2} \right) \Psi_0, \quad \psi_h(z_0) = N_h \prod_{i=1}^{N} (z_i - z_0) \Psi_0, \tag{13.179}$$

where N_p and N_h are normalization factors.

This construction can yield the statistics of the quasi-particles or quasi-holes [234]. Starting from a wave function describing excitation of, e.g., two holes, we assume that exchange operation endows the wave function with a phase factor $e^{i\theta}$ that is not necessarily equal ± 1, as for bosons ($+$) or fermions ($-$). Thus,

$$\psi_h(z_a, z_b) = N_{ab} \prod_{i=1}^{N} (z_i - z_a)(z_i - z_b) \Psi_0, \quad \psi_h(z_b, z_a) = P_{ab} \psi_h(z_a, z_b) = e^{i\theta} \psi_h(z_a, z_b). \tag{13.180}$$

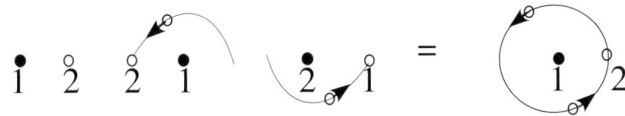

FIG 13.46 Two exchange operators, P_{ab}^2, in 2D: Starting from the initial configuration on the left, the first exchange operation moves particle 2 to the left of particle 1 on the upper curve. The second exchange operation moves particle 2 back to the right of particle 1 on the lower curve. The net operation is equivalent to moving particle 2 in a closed curve that encircles particle 1. This construction only makes sense in 2D.

To calculate the phase θ, consider the product of two exchange operators, P_{ab}^2. In 2D, and *only* in 2D, the product of two exchange operators is equivalent to adiabatically moving one particle in a closed curve around the other one, as shown in Fig. 13.46.

Therefore, starting from $\psi_h(z_a, z_b)$, we consider adiabatic motion of $z_b(t)$ along a closed loop parametrized by a real parameter $t \in [0, T]$ such that $z_b(0) = z_b(T) = z_b$ and inspect the Berry phase $\gamma(t)$ accumulated by the wave function at point $z_b(t)$ (see Sec. 7.8.5). Then, we relate the phase θ to the Berry phase $\gamma(T)$ [235]. Using Eq. (7.200), we obtain

$$\frac{d\gamma(t)}{dt} = i \left\langle \psi_h(z_a, z_b(t)) \left| \frac{\partial \psi_h(z_a, z_b(t))}{\partial z_b(t)} \right. \right\rangle \frac{dz_b(t)}{dt}. \tag{13.181}$$

Using the definition (13.180), the rest of the calculation of the accumulated phase $\gamma(T) = \int_0^T \frac{d\gamma(t')}{dt'}$ is not difficult and we will skip it. The result of the integration contains a term that depends on the flux encircled by the curve $z_b(t)$, in which we are *not* interested. To get rid of it, the phase γ is calculated along two identical curves, C_1 and C_2, such that C_1 encircles z_a and C_2 does not. The difference $\gamma_1(T) - \gamma_2(T)$ does not depend on flux and is a generic property of quasi-hole statistics. The final result is

$$\theta = \frac{1}{2} [\gamma_1(T) - \gamma_2(T)] = \pi \nu. \tag{13.182}$$

Quasi-holes (and quasi-particles) in the FQHE with filling factor $\nu = 1/q$ with q odd obey *fractional statistics* in the sense that following an exchange of two quasi-particles, the wave function is multiplied by a phase $e^{i\theta} = e^{i\pi\nu}$. This statistics is intermediate between Bose–Einstein ($\theta = 0$) and Fermi–Dirac ($\theta = \pi$) statistics. Fractional statistics has been confirmed experimentally.

Composite Fermions: An idea due to J. K. Jain [236] makes it possible to transform the original system of strongly interacting electrons in a strong magnetic field B into a system of weakly interacting composite fermions in a weaker effective magnetic field

$$B^* = B - 2pn_{2D}\Phi_0, \qquad (13.183)$$

where p is an integer, n_{2D} is the electron density in 2D, and $\Phi_0 = hc/e$ is the quantum flux unit. The idea is to replace each electron by another particle called a *composite fermion*. A composite fermion is an electron with a massless magnetic solenoid attached to it, carrying a flux $2p\Phi_0$ pointing antiparallel to B. This bound state of an electron and an even number of flux quanta is itself a fermion. The wave function of composite fermions is antisymmetric under the exchange of two composite fermions. Consequently, the attached flux has no observable consequence. The new problem, formulated in terms of composite fermions, is identical to the one with which we began. The advantage of this new formulation is that in the mean field picture, the magnetic field carried by the composite fermions, $\bar{B} = 2pn_{2D}\Phi_0$, acts oppositely to the original field B, and this leads to the field B^* of Eq. (13.183). Equivalently, if Eq. (13.183) is written as $B = B^* + 2pn_{2D}\Phi_0$, then each electron strips $2p$ flux units from the original field, thereby decreasing it to B^*. This is shown in Fig. 13.47.

Problem 13.28

Show that if the filling factor in the original problem is ν, the filling factor in the transformed system is

$$\nu^* = \frac{\nu}{1 - 2p\nu}, \quad \Rightarrow \quad \text{if } \nu = \frac{1}{2p+1} \quad \text{then} \quad \nu^* = 1. \qquad (13.184)$$

Problem 13.29

Show that if a composite fermion encircles another composite fermion and returns to its initial position, its wave function remains unchanged.

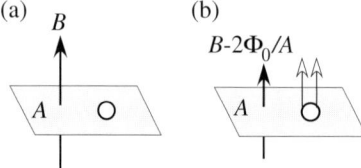

FIG 13.47 The transformation from electrons to composite fermions. (a) An electron in a 2D system of area A subject to a perpendicular magnetic field B. (b) The electron strips the magnetic field from two flux units and becomes a composite fermion [i.e., $p = 1$ in Eq. (13.183)]. The composite fermion contributes a field $2\Phi_0/A$, while the original field is reduced to $B^* = B - 2\Phi_0/A$.

What has been gained by this procedure? In the absence of interaction, all states on the lowest Landau level have the same energy, and the degeneracy N_D of the many-electron ground state at $\nu = 1/q$ is

$$N_D = \binom{2N_L}{N} = \binom{2N_L}{\left[\frac{2N_L}{q}\right]}. \qquad (13.185)$$

Equation (13.184) shows that after the transformation, the filling factor $\nu^* > \nu$, and by properly choosing p, it may be very close to 1, so that $q^* \gtrsim 1$ and the degeneracy is significantly reduced. This reduced degeneracy suggests that a starting point where composite fermions are treated as independent particles is quite reasonable. For $2p = 1/\nu$, one has $B^* = 0$ and the composite fermions fill a fermi sea. But since $1/\nu = q$ is odd, they fill the lowest Landau level corresponding to B^*. The interaction between the composite fermions remains the Coulomb repulsion, and similar to the Laughlin wave function, it is encoded in the *Janes wave function* for

N electrons at filling factor $\nu < 1$ on the lowest Landau level,

$$\Psi_\nu(\mathbf{r}_1, \mathbf{r}_2, \ldots \mathbf{r}_N) = \prod_{i<j=1}^{N} (z_i - z_j)^{2p} D_{\nu^*}(\mathbf{r}_1, \mathbf{r}_2, \ldots, \mathbf{r}_N), \quad (13.186)$$

where $D_{\nu^*}(\mathbf{r}_1, \mathbf{r}_2, \ldots, \mathbf{r}_N)$ is the Slater determinant for N noninteracting electrons occupying the lowest Landau level in a field B^* and filling factor ν^* (the electrons are assumed polarized, and spin degrees of freedom are omitted). $\Psi_\nu(\mathbf{r}_1, \mathbf{r}_2, \ldots \mathbf{r}_N)$ is an excellent approximations for the N electron ground state. Even more significantly, it clarifies the intuitive physics of composite fermions. The factors in the product preceding the Slater determinant tell us that every electron regards every other electron as a center of $2p$ vortices. Equivalently, when an electron encircles another electron, it acquires a phase $2\pi \times 2p$, which is equivalent to attaching $2p$ vortices to each electron in the Slater determinant for noninteracting electrons.

13.7.7 HIGH-TEMPERATURE SUPERCONDUCTIVITY AND 2D MAGNETISM

The physics of superconductivity is presented in Sec. 18.10 which is linked to the book web page. There, the focus is on the traditional theory, although new developments and the discovery of high-temperature superconductivity (HTSC) is briefly considered, in connection with the $t - J$ model (see Sec. 18.15.2 therein). A very brief qualitative account of this topic within the context of low-dimensional strongly correlated systems is presented below.

In 1986, a new class of oxides superconductors with high critical temperatures (up to 130 K) was discovered by George Bendorz and Karl Alex Müller [237]. There are hundreds of HTSC compounds known today. The precise mechanism responsible for high-temperature superconductivity is not fully established, but the superconducting state is similar to that in conventional superconductors, where electrons (or holes) attract each other and form a condensed state. Unlike ordinary superconductors, the pairing potential $\Delta(\mathbf{k})$ has d-wave symmetry (see Sec. 18.10.3 in the link on the book web page).

HTSC is obtained by doping certain parent compounds that, before doping, are Mott insulators (see Sec. 9.9). For example, La_2CuO_4 is a Mott insulator with antiferromagnetic ground state. Doping it into $La_{2-x}Sr_xCuO_4$ removes electrons, allowing in-plane conduction by holes. Hence, it is said to be *hole doped*. The quantity $0 \leq x < 1$ is the degree of doping. HTSC develops for $0.025 \leq x \leq 0.27$. Similarly, the parent compound Nd_2CuO_4 can be electron doped into $Nd_{2-x}Ce_xCuO_4$. This scenario is displayed schematically in the phase diagram of Fig. 13.48.

The combined appearance of HTSC and antiferromagnetism is interesting in itself. As a hole moves around, the surrounding spins must constantly readjust themselves to lower their antiferromagnetic exchange energy. It turns out that the exchange energy can be kept lower if the holes move together. In this context, antiferromagnetism helps hole pairing. The physics of other phases on the hole doped side are not yet elucidated. In the pseudogap region, there is still a gap in the density of states, but it is not a bona fide superconductor. It is not even clear whether the normal states labeled as "metal(?)" in Fig. 13.48 are Fermi liquids or non-Fermi liquids. The temperature dependence of the resistivity in this phase is not yet fully explained [238].

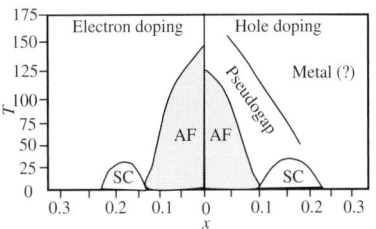

FIG 13.48 Schematic phase diagram of HTCS, showing hole doping on the right and electron doping on the left, starting from two different Mott insulators with antiferromagnetic ground state. For small doping, the ground state remains antiferromagnet, but its Neel temperature decreases. For higher doping, antiferromagnetism diminishes and superconductivity develops. The other phases on the hole-doped side are discussed in the text.

Another family of superconducting materials, based on Fe and As, was discovered in 2008 [239]. An example is $LaFeAsO_{1-x}F_x$ with $x \sim 0.11$, with $T_c \approx 26$ K. The crystal structure contains FeAs layers with Fe atoms in a square planar lattice arrangement, and these layers alternate with LaO layers along the c-axis (hence, the relevance of low dimensionality). We shall not discuss the physical origin of HTSC in these materials.

The relation between HTSC and antiferromagnetism of the CuO planes is related to 2D magnetism, usually described by the Heisenberg model, which we will encounter in Secs 18.15.2 and 18.16 (see also Sec. 9.8.3). The Heisenberg model is the basic tool for elucidating the physics of magnetic materials. It encodes the dynamics and the thermodynamics of a magnetic system including the magnetization curve $M(T)$ and the Curie temperature T_c. In general, it does not admit an exact solution and some approximations or numerical solutions are required to obtain the desired results.

According to the Mermin–Wagner theorem [241], mentioned in Sec. 18.15.4, 1D and 2D magnetic systems cannot have any long-range magnetic order at finite temperature $T > 0$. Thus, magnetism in the 2D Heisenberg model exists, if at all, at zero temperature only. The question of whether there is long-range order at zero temperature depends on the parameters in the Hamiltonian. By varying the parameters in the Hamiltonian, the quantum ground state can change from paramagnetic to antiferromagnetic and vice versa. This *quantum phase transition* will be discussed in Chapter 18, which is linked to the book web page. It is analogous to a thermodynamic phase transition, which is controlled by varying the temperature, but here the temperature is zero, and other parameters vary. There exist 2D magnets with finite (albeit low) Curie temperature; the main reason is that magnetocrystalline anisotropy and certain terms in the Heisenberg spin Hamiltonian referred to as *dipolar interactions* violate the conditions of the Mermin–Wagner theorem.

13.7.8 1D CORRELATED ELECTRON SYSTEMS

One-dimensional strongly correlated systems involve interacting particles or spins. Examples are the 1D Hubbard model (discussed in Sec. 18.15, which will be linked to the book web page) and the Heisenberg spin model in 1D. We have stressed that, in many cases, a system of interacting fermions can be described in terms of quasi-particles that behave as almost free particles, a concept developed by L. D. Landau and often referred to as *Fermi liquid Theory* (and sometimes as Landau–Fermi liquid theory). The properties of interacting particles in 1D systems are distinct from those of noninteracting ones, and often the Fermi liquid picture is not adequate to describe such systems. Instead, the eigenstates of a 1D system of bosons, fermions, or spins form what is referred to as a Luttinger liquid. Although the bare system of electrons in a quantum wire consists of fermions, the low-energy elementary excitations in a Luttinger liquid are bosons, that are sound waves of the 1D electron gas. This property is at the heart of a powerful technique called *bosonization*, which in some cases can be used to arrive at exact solutions [242]. The main idea is to recast the physics not in terms of wave functions but, rather, in terms of local densities, $\rho_L(x)$ and $\rho_R(x)$, of particles moving from left to right and from right to left. Here, the fact that the motion is strictly constrained in 1D is crucial. The interacting electron gas behaves, in some respects, similar to a liquid, where the dynamical variables include charge density, spin density, and current.

An important property of electronic systems in 1D is *spin-charge separation*. The state of an electron injected into an interacting system is decomposed into two elementary excitations: charge and spin density modes, which propagate at different velocities, u_c and u_s, respectively. This feature is absent in higher dimensions when the physics is described by Landau–Fermi liquid theory. Luttinger liquids display exotic behavior, e.g., the conductivity is extremely sensitive to disorder, and at $T = 0$, it vanishes if the wire contains a single impurity, no matter how weak. The conductivity has unusual behavior as function of temperature, voltage, and frequency.

A Luttinger liquid could arise in a quantum wire. However, due to the extreme sensitivity of conductance to disorder (and the electron–electron interaction), the wire needs to be extremely clean, and this is difficult to guarantee. Alternatively, consider edge states in a Hall bar. In the IQHE, when the Fermi energy is between the n and $n + 1$ Landau levels, there are n edge states at the Fermi energy on each side of the strip. Edge states on different edges move in opposite directions, a property referred to as *chirality* [243]. In the FQHE, where interaction plays a key role, the edge states form a 1D system of interacting quasi-particles. The number of edge states in the FQHE is ν^*, defined in Eq. (13.184), assuming the composite fermion analysis leads to an integer ν^*. The chirality of the edge states in the FQHE constitutes a realization of *chiral Luttinger liquids*.

13.7.9 1D SPIN CHAINS

In addition to the 1D electronic systems, there is a great interest in 1D spin chains [244, 245]. They can be realized in 3D compounds such as CsNiCl$_3$ and RbNiCl$_3$, and attract theoretical interest because they display remarkable quantum effects and define new quanta (magnons) that occur in higher dimensions as well. There are a few exactly solvable spin chain models using Bethe ansatz, bosonization, or transfer-matrix techniques. They can be efficiently studied numerically using *Monte Carlo algorithms*. The most familiar 1D spin chain system is the Heisenberg model [see Eq. (9.515)]. In its simplest version, the Hamiltonian is

$$H = -\frac{J}{N}\sum_{n=1}^{N} \mathbf{S}(n) \cdot \mathbf{S}(n+1). \tag{13.187}$$

where S_n is a quantum spin operator attached to lattice point $n = 1, 2, \ldots, N$, and J is an exchange constant, which can be ferromagnetic ($J > 0$) or antiferromagnetic ($J < 0$). The dimensions are chosen such that $J\hbar^2$ is an energy. Periodic boundary conditions $S_{N+1} = S_1$ are adopted, and the normalization factor $1/N$ in Eq. (13.187) is useful in the limit $N \to \infty$. The simplifying assumptions are that exchange interaction is effective only between nearest neighbors and that there is no external magnetic field. In $d > 1$ dimensions, the spins are localized at Bravais lattice points $\mathbf{R}$ surrounded by q nearest neighbor lattice points $\mathbf{R} + \mathbf{r}_i$, $i = 1, 2, \ldots q$, where q is the coordination number, so,

$$H = -\frac{J}{N}\sum_{\mathbf{R}}\sum_{i=1}^{q} \mathbf{S}(\mathbf{R}) \cdot \mathbf{S}(\mathbf{R} + \mathbf{r}_i). \tag{13.188}$$

Although this section deals with 1D systems, a few words about higher space dimensions is in order. For antiferromagnetic exchange, the role of space dimensionality d and value of $S = 1/2, 1, 3/2, \ldots$ is especially intriguing. For $d = 2$, on a square lattice, it is believed that the ground state is *Neel ordered*, i.e., the ground-state wave function $|GS\rangle$ has spin-up at points $\mathbf{R}$ and spin-down at points $\mathbf{R} + \mathbf{r}_i$. This implies a *long-range order* in the sense that, at zero temperature, as $|\mathbf{R} - \mathbf{R}'| \to \infty$,

$$\langle GS|\mathbf{S}(\mathbf{R}) \cdot \mathbf{S}(\mathbf{R}')|GS\rangle \to \text{Const.} \tag{13.189}$$

In the limit $N \to \infty$, spin excitations above this ground state are gapless.

The situation is somewhat different for $d = 1$. First, as discussed in Sec. 9.8.3, for $S = \frac{1}{2}\hbar$, the ground state for $d = 1$ is not Neel ordered. Rather, it is the Bethe–Hultein solution derived by the Bethe ansatz technique. Second, as we know from the Mermin–Wagner theorem, there is no long-range order in $d = 1$ even at $T = 0$. For $S = \frac{1}{2}\hbar$, it is found that the spin–spin correlation for $d = 1$ falls off as a power law,

$$\langle GS|\mathbf{S}_n \cdot \mathbf{S}_m|GS\rangle \sim |n - m|^{-\alpha}. \tag{13.190}$$

Third, and somewhat unexpectedly, the spectrum is gapless *only for half-integer spin* $S/\hbar = 1/2, 3/2, \ldots$ but there are strong indications that it is gapped for integer $S/\hbar = 1, 2, 3, \ldots$ [246]. The *Haldane gap* has been verified experimentally.

Problem 13.30

Consider the Heisenberg Hamiltonian for a 1D chain of three sites with periodic boundary conditions, $H = J(\mathbf{S}_1 \cdot \mathbf{S}_2 + \mathbf{S}_2 \cdot \mathbf{S}_3 + \mathbf{S}_3 \cdot \mathbf{S}_1)$, $(J > 0)$.

(a) Show that $[S^2, H] = 0$, where $\mathbf{S} = \mathbf{S}_1 + \mathbf{S}_2 + \mathbf{S}_3$ is the total spin.
(b) Find the energy spectrum for $S = \hbar/2$ and discuss its degeneracy.
(c) Explain why the method you used to solve (a) does not work for more than three spins, e.g., $H = J\sum_{i=1}^{4} \mathbf{S}_i \cdot \mathbf{S}_{i+1}$, $(\mathbf{S}_{4+1} = \mathbf{S}_1)$.

Answer: (a) Because $H = \frac{J}{2}(S^2 - \sum_{i=1}^{3} S_i^2) = \frac{J}{2}(S^2 - 9\hbar^2/4)$ it is clear that $[S^2, H] = 0$.
(b) For $S = (3/2)\hbar$ we have $E = (J/2)(S^2 - 9/4) = 3J\hbar^2/4$ for all three spins being parallel, which is the highest energy for $J > 0$. This level is two-fold degenerate because the three parallel spins might point either "up" or "down". For $S = (1/2)\hbar$ we have $E = -3J\hbar^2/4$. This is the ground-state energy and corresponds to one "up" electron and two "down" electrons. This level is six-fold degenerate.
(c) S^2 contains terms with exchange interaction between spins that are not nearest neighbors, such as $\mathbf{S}_1 \cdot \mathbf{S}_3$. Such terms are not present in H. Hence $[S^2, H] \neq 0$.

13.7.10 QUANTUM SPIN HALL EFFECT

The quantum spin Hall effect (QSHE) [247] occurs in topological insulators with time reversal invariance symmetry (see Sec. 9.9.2). It will also be discussed in Chapter 19, linked to the web page of the book. It is briefly mentioned here, since low dimensionality plays a crucial role. A system without edges is a bulk system. Usually, it is an infinite system in all directions, but for actual calculations, we may consider a finite system with periodic boundary conditions, such as a 2D system with a torus geometry. The phrase "bulk system is opened" means that there is at least one edge, such as in half a plane, half a cylinder, a cylinder, or a ribbon. When a bulk gapped system possessing a nontrivial topological quantum number is opened, it has gap-closing edge states that carry current along the edges. By this we mean that, for any energy E in the gap of the bulk system, there is at least one current-carrying state $|\psi_E\rangle$ with energy E moving close to the edge of the open system.

In the IQHE, the bulk topological quantum number is the first Chern number, defined in Chapter 19, and the edge states are chiral, i.e., on one edge, they are moving in one direction, whereas on the other edge, they are moving in the opposite direction.

For systems with time-reversal invariance and without spin-rotation symmetry (e.g., for systems with spin–orbit coupling without the presence of a magnetic field), the edge states (if they exist) come in pairs due to Kramers degeneracy. One edge state in such a pair can be obtained by operating on the other edge state of the pair with the time-reversal operator $\mathcal{T}$, and the two edge states are said to form a *Kramers pair*. The two states that form a Kramers pair move in opposite directions on the same edge of the sample.

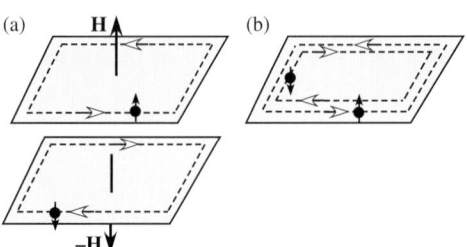

FIG 13.49 Toy model for the quantum spin Hall effect. (a) Two copies of the IQHE with opposite magnetic fields. In each copy, time reversal is violated due to the external magnetic field, and the edge state has a definite orientation (hence, it is a chiral edge state). In the upper copy, the (first) Chern number is $n = 1$, whereas in the lower copy, $n = -1$. (b) When the two copies are combined, the total magnetic field vanishes and time-reversal invariance is maintained. The first Chern number is $1 + (-1) = 0$. The edge states have opposite orientations, since they form a Kramers pair. There is spin current but no charge current along the edge. Due to conservation of S_z, the spin Hall conductance is quantized.

A toy model exemplifies the relation between edge states in the IQHE (a system that violates time-reversal invariance) and edge states in spin Hall systems (a system that conserves time-reversal invariance) [248]. Consider two copies of an open system displaying one edge state in the IQHE, one for spin-up and one for spin-down electrons, subject to equal but opposite magnetic fields, as displayed in Fig. 13.49. Then, as detailed in the caption, these two copies can be combined to form a spin Hall system with a single Kramers pair.

In generic 2D models, where time-reversal invariance is conserved and spin currents along the edges are present [similar to Fig. 13.49(b)], S_z is not always conserved and the edge states are not necessarily specified by the spin projection of the electron. The two edge states still form a Kramers pair in the sense that one is obtained by applying the time-reversal operator $\mathcal{T}$ on the other, up to a constant (see Sec. 9.6.7). The corresponding integer topological number (the analog of the Chern number in the IQHE) is denoted by $\mathbb{Z}_2$.

13.7 Inventory of Recently Discovered Low-Dimensional Phenomena

Whereas the definition (9.544) of the first Chern number in terms of Bloch functions is rather simple, and is directly related to the Kubo formula for the calculation of the Hall conductivity σ_{xy}, the definition of $\mathbb{Z}_2$ in terms of Bloch function is not so simple, and will be skipped. Like the Chern number, $\mathbb{Z}_2$ is calculated in the bulk system, but it is related to the open system and its edge states. The parity of $\mathbb{Z}_2$ tells us whether the number of Kramers pair edge states is even or odd [in Fig. 13.49(b) it is odd, $\mathbb{Z}_2 = 1$]. Only an odd number of Kramers pairs on the edges leads to a nontrivial physics [249]. To show this, assume there are two Kramers pairs on the edges, (ψ_1, ψ_2) and (ϕ_1, ϕ_2). Then, two states of the same parity under $\mathcal{T}$, such as ψ_1 and ϕ_1, can be scattered by an impurity potential V that is invariant under $\mathcal{T}$, that is, $\langle\psi_1|V|\phi_1\rangle \neq 0$. Thus, when there is an even number of Kramers pairs, the system of edge states is not robust against weak disorder. In summary, a 2D system displays the quantum spin Hall effect if (1) its Hamiltonian is invariant under time reversal and all states are twofold Kramers degenerate. (2) The bulk system has a gap and the topological quantum number $\mathbb{Z}_2$ is odd. (3) When the system is opened, there is an odd number of Kramers doublets on each edge at the Fermi energy. The spin Hall effect has been observed experimentally.

Now consider applying Laughlin's idea discussed in Sec. 9.5.8 on the Quantum Hall effect. Assume the system has the geometry of a cylinder, and let the flux $\Phi(t)$ along the axis change adiabatically from 0 to Φ_0. There will be a spin transfer between the left (L) and the right (R) edges at a rate,

$$\frac{d\langle S_z\rangle}{dt} = \sigma^s_{xy} \frac{d\Phi}{dt}, \qquad (13.191)$$

where $\langle S_z\rangle$ is the expectation of the operator S_z in a given edge state, and σ^s_{xy} is the spin Hall conductance. At the end of the adiabatic change, we have

$$\sigma^s_{xy} = \frac{e}{h} \left(\langle S_z\rangle_R - \langle S\rangle_L\right)|_{E_F}. \qquad (13.192)$$

S_z is not conserved in the presence of spin–orbit interaction; consequently, σ^s_{xy} is *not quantized*. Several models exhibiting the quantum spin Hall effect have been developed, but the details of these models go beyond the scope of this book.

Many-Body Theory 14

Quantum many-body theory deals with systems of many particles that interact with each other. Often, the systems treated contain (groups of) identical particles. A familiar example is that of electrons in a solid which interact with each other (in addition to interacting with the lattice ions). In systems with identical particles, be they bosons (e.g., photons, phonons or a degenerate gas of ^{4}He atoms) or fermions (e.g., electrons, or a degenerate gas of ^{3}He atoms), a crucial issue is particle statistics. The symmetry of the state of the system under exchange of identical particles must be properly incorporated in many-body theory. This requirement will be assured within the formalism developed here. Many-body physics deals with both equilibrium properties, such as energy, compressibility, magnetization, etc., and non-equilibrium properties such as conductance, heat transfer, noise, optical response and other non-equilibrium phenomena. The quantum many-body problem is one of the most thoroughly-studied subjects in quantum mechanics. The underlying theory is strongly based on, and intimately related to, the general framework of *quantum field theory*. Many-body theory is frequently used to treat systems containing a macroscopic number of identical particles. Although exact solutions for some many-body problems are known, in most cases a sophisticated scheme of approximations is required. Quantum field theory is designed so that it can also treat physical systems with infinitely many degrees of freedom.

The breadth and scope of quantum many-body physics is enormous. It requires the mathematical framework of second quantization and employs it to elucidate the physics of a myriad of interacting-particle physical systems. These include the electron gas, strongly correlated electrons, superconductivity, superfluidity, magnetism, spin systems, Fermi liquids, ultracold atomic gases, and low dimensional systems. The theory is capable of treating physical systems in equilibrium and out of equilibrium, at zero and finite temperatures. In this chapter, the basic ideas of second quantization are presented and then a few important representative physical systems are presented. A companion Chapter entitled *Field Theory Applications*, Chapter 18, can be regarded as part II of this chapter; it will be linked to the book web page. Many books devoted to many-body physics using the methods of quantum field theory are available, e.g., Refs. [250–256].

This chapter is organized as follows. The basic techniques required to handle the many-body problem are introduced in Sec. 14.1, starting with the fundamental concept of second quantization, construction of Fock space, and creation and annihilation operators. The Hamiltonian in Fock space is developed in Sec. 14.1.5. Section 14.1.6 introduces the field operators of quantum field theory. Quantization of the radiation field, and the concept of photons, is presented in Sec. 14.1.7, followed by the quantization of crystal vibrations in terms of phonons in Sec. 14.1.8. Systems containing two kinds of particles are discussed in Sec. 14.1.9. Section 14.2 presents the formulation of quantum statistical mechanics within the second quantization formalism. As an example of the use of the tools developed in this chapter, an approximation for the ground-state energy of an electron gas with Coulomb inter particle interactions is presented in Sec. 14.3. Finally, mean-field theory, which is a central ingredient in the quest for nonperturbative approximations for the solution of the many-body problem, is presented in Sec. 14.4.

14.1 SECOND QUANTIZATION

In this section, we introduce the formalism of second quantization for bosons and fermions. This formalism is designed to treat systems of identical particles in a natural way, with particle statistics being properly incorporated. We start by writing the Hamiltonian of a many-body system containing one kind of interacting particle in first quantized form, and then develop the second quantized form step by step. Later, we show how to treat systems containing more than one kind of particle within second quantization.

14.1.1 INTERACTING IDENTICAL PARTICLES: A GENERIC MANY-BODY PROBLEM

As an example of a many-body problem, consider a closed system containing N identical particles of mass m occupying a volume $\mathcal{V}$. Often one is interested in letting N and $\mathcal{V}$ go to infinity in such a way that the average density $\bar{\rho} \equiv N/\mathcal{V}$ is constant. This limiting procedure is referred to as the *thermodynamic limit* (although thermodynamics does not enter the definition per se). The Hamiltonian operator for the system is

$$H = \sum_{i=1}^{N}[T(\mathbf{r}_i) + U(\mathbf{r}_i)] + \frac{1}{2}\sum_{i\neq j=1}^{N} V(\mathbf{r}_i - \mathbf{r}_j) \equiv H_0 + V, \quad (14.1)$$

where $\mathbf{r}_i$ is the position vector of particle i ($i = 1, 2, \ldots N$). The terms on the RHS of Eq. (14.1) are the kinetic energy operator, $T(\mathbf{r}_i) = -\frac{\hbar^2}{2m}\nabla_i^2$, the single-particle external potential for particle i, $U(\mathbf{r}_i)$, and the two-body interaction between particles i and j, $V(\mathbf{r}_i - \mathbf{r}_j)$. The latter is often spin-independent and invariant under translations and rotations [as is the case for the Coulomb potential between electrons, $V(\mathbf{r}_i - \mathbf{r}_j) = e^2/|\mathbf{r}_i - \mathbf{r}_j|$]. The operators $T(\mathbf{r}_i)$ and $U(\mathbf{r}_i)$ act only on particle i, hence they are referred to as *single-particle operators*. We denote the sum of the single-particle operators by

$$H_0 = \sum_i [T(\mathbf{r}_i) + U(\mathbf{r}_i)] \equiv \sum_i h(\mathbf{r}_i). \quad (14.2)$$

In general, both the space coordinate $\mathbf{r}_i$ and the spin projection σ_i of particle i need to be specified in the many-body wave function Ψ. For this purpose we will use the notation $\mathbf{x}_i = (\mathbf{r}_i \sigma_i)$ and refer to $\mathbf{x}_i$ as *space-spin coordinate* of particle i. The many-body wave function will be denoted by

$$\Psi(\mathbf{r}_1\sigma_1, \mathbf{r}_2\sigma_2, \ldots, \mathbf{r}_N\sigma_N) = \Psi(\mathbf{x}_1, \mathbf{x}_2, \ldots, \mathbf{x}_N). \quad (14.3)$$

For electrons, $\sigma_i = \pm 1/2$, whereas for spin 1 particles $\sigma_i = 1, 0, -1$. This notation is required to account for the symmetry of the many-body wave function under particle exchange. The eigenstates of H_0 can be written as sum of products of single-particle spin orbitals $\langle q|\mathbf{x}\rangle = \phi_q(\mathbf{x})$ satisfying the single-particle Schrödinger equation,

$$h\phi_q(\mathbf{x}) \equiv [T + U(\mathbf{r})]\phi_q(\mathbf{x}) = \varepsilon_q \phi_q(\mathbf{x}). \quad (14.4)$$

Here, q represents a set of single-particle quantum numbers, including spin, necessary for characterizing the single-particle states, corresponding to a complete set of operators that commute with $h = T + U$. If $U(\mathbf{r})$ does not depend on spin, then $[h, \boldsymbol{\sigma}] = 0$, and $q = (\varepsilon_\alpha, \sigma)$ represents the set of eigenvalues of $h(\varepsilon_\alpha)$ (the energies $\{\varepsilon_\alpha\}$) and σ_z (the spin projections $\pm 1/2$). Note that the spin of the particle appears in $\mathbf{x}$ as a coordinate, whereas the eigenvalue of the spin projection appears as a quantum number in q. The spin orbital $\phi_q(\mathbf{x}) = \psi_\alpha(\mathbf{r})\eta_\sigma$ is then separable (the space and spin degrees of freedom are un-entangled). For example, if $U(\mathbf{r})$ is a central potential (see Chapter 3), then for discrete energy states, $q = nlm\sigma$, where n is an energy quantum number, l is an orbital angular momentum, and m is its projection.[1] The operators $V(\mathbf{r}_i - \mathbf{r}_j)$ act on particles i and j and hence are referred to as *two-body operators*. The stationary many-body Schrödinger equation reads,

$$H\Psi_\lambda(\mathbf{x}_1, \mathbf{x}_2, \ldots, \mathbf{x}_N) = E_\lambda \Psi_\lambda(\mathbf{x}_1, \mathbf{x}_2, \mathbf{x}_3, \ldots, \mathbf{x}_N). \quad (14.5)$$

The energies E_λ are usually arranged in ascending order, $E_0 < E_1 < E_2 \ldots$. Of special importance are the ground-state energy E_0 and the ground-state wave function $\Psi_0(\mathbf{x}_1, \mathbf{x}_2, \ldots, \mathbf{x}_N)$. The eigenfunctions $\Psi_\lambda(\mathbf{x}_1, \mathbf{x}_2, \mathbf{x}_3, \ldots, \mathbf{r}_N)$ must have the right symmetry under particle exchange, i.e.,

$$P_{ij}\Psi_\lambda(\mathbf{x}_1 \ldots \mathbf{x}_i \ldots \mathbf{x}_j, \ldots \mathbf{x}_N) = \Psi_\lambda(\mathbf{x}_1 \ldots \mathbf{x}_j \ldots \mathbf{x}_i, \ldots, \mathbf{x}_N) = \pm \Psi_\lambda(\mathbf{x}_1 \ldots \mathbf{x}_i \ldots \mathbf{x}_j, \ldots \mathbf{x}_N). \quad (14.6)$$

---

[1] We will see below that for a proper handling of fermion statistics, the set of single-particle quantum numbers q needs to be ordered.

14.1 Second Quantization

The operator P_{ij} exchanges particles i and j, i.e., it exchanges $(\mathbf{x}_i = \mathbf{r}_i\sigma_i)$ and $(\mathbf{x}_j = \mathbf{r}_j\sigma_j)$. The $+$ sign is for bosons and the $-$ sign for fermions (see Chapter 8). The many-body wave function of Bose (Fermi) particles that satisfies Eq. (14.6) with a plus (minus) sign is said to have the *appropriate symmetry*.

Equation (14.5) is the stationary many-body Schrödinger equation. Many problems sometimes require knowledge of the dynamics of many-particle systems, i.e., the solution of the time-dependent Schrödinger equation,

$$i\hbar\frac{\partial}{\partial t}\Psi(\mathbf{x}_1,\mathbf{x}_2,\ldots,\mathbf{x}_N,t) = H(t)\Psi(\mathbf{x}_1,\mathbf{x}_2,\ldots,\mathbf{x}_N,t), \tag{14.7}$$

with initial condition

$$\Psi(\mathbf{x}_1,\mathbf{x}_2,\ldots,\mathbf{x}_N,0) = \chi(\mathbf{x}_1,\mathbf{x}_2,\ldots,\mathbf{x}_N), \tag{14.8}$$

where χ is the time-independent initial wave function. This is the case, for example, if the system is acted upon by external time-dependent fields, so the Hamiltonian depends explicitly on time. Another nonequilibrium situation occurs when the Hamiltonian does not depend on time but the initial state is not an eigenstate of the Hamiltonian. When the two-body interaction potentials $V(\mathbf{r}_i - \mathbf{r}_j)$ are present, a closed form analytic solution of the many-body Schrödinger equation [either Eq. (14.5) or (14.7)] is rarely available. One is unavoidably obliged to resort to various approximation schemes.

A central challenge to be addressed at the outset is how to incorporate the permutation symmetry (14.6) in a compact and efficient scheme. The quest for an elegant formulation of the quantum mechanical many-body problem which takes into account the correct permutation symmetry of the many-body wave function right at the onset is achieved using the *second quantization formalism*. The reason for calling it second quantization will be explained below.

The basic idea of second quantization is to map the Hilbert space, whose elements are states of N (identical) particles, onto another space, called the *Fock space*, where the basis kets are determined by the number of particles occupying each single-particle state. This mapping induces a transformation of the original Hamiltonian into a new Hamiltonian with the same spectrum. The details of this mapping and the structure of the new Hamiltonian depend crucially on whether the N particle states are symmetric or antisymmetric under particle exchange, corresponding to bosons or fermions, respectively. The mapping procedure requires several steps as detailed below.

The formalism of second quantization is not just an elegant technique for handling systems of identical particles. It enables the description of systems with infinite number of degrees of freedom and establishes an intimate relation with the many-body physics, statistical mechanics, and field theory. It is capable of describing physical processes that cannot be described within the formalism of first quantization, especially those that do not conserve the number of particles.

14.1.2 A BASIS FOR MANY-BODY WAVE FUNCTIONS

An N-body wave function for identical fermions or bosons must satisfy the exchange symmetry constraint, Eq. (14.6). Construction of a basis for the N-body problem consists of finding a set of (time-independent) functions $\Phi_\alpha(\mathbf{x}_1,\mathbf{x}_2,\ldots,\mathbf{x}_N)$ with the appropriate symmetry. Here, α is a set of quantum numbers for the N-particle system (see below). The basis should be complete and orthogonal in the sense that it spans the Hilbert space of functions with the appropriate symmetry. Any time-dependent N-body wave function $\Psi(\mathbf{x}_1,\mathbf{x}_2,\ldots\mathbf{x}_N;t)$ with the appropriate symmetry should be expandable as a linear combination,

$$\Psi(\mathbf{x}_1,\mathbf{x}_2,\ldots\mathbf{x}_N;t) = \sum_\alpha A_\alpha(t)\Phi_\alpha(\mathbf{x}_1,\mathbf{x}_2,\ldots\mathbf{x}_N), \tag{14.9}$$

where the time dependence enters solely through the set of coefficients $\{A_\alpha(t)\}$. The choice of the basis set functions Φ_α is dictated by convenience. First, the basis functions Φ_α should be constructed, and second, the matrix elements $H_{\alpha\beta} \equiv \langle\Phi_\alpha|H|\Phi_\beta\rangle$ should be calculated (both tasks require computational resources). For example, one can choose $\Phi_\alpha(\mathbf{x}_1,\mathbf{x}_2,\ldots\mathbf{x}_N)$ as eigenfunctions of H_0 or of the N particle kinetic energy operator $T \equiv \sum_{i=1}^N T(\mathbf{r}_i)$. The appropriate symmetry of $\Phi_\alpha(\mathbf{x}_1,\mathbf{x}_2,\ldots\mathbf{x}_N)$ is achieved by forming Slater determinants (for fermions) or permanents (for bosons) [see Chapter 8].

Ordering Single-Particle Quantum Numbers

We need to construct the N-body basis states $\Phi_\alpha(\mathbf{x}_1, \mathbf{x}_2, \ldots \mathbf{x}_N)$ (with the appropriate symmetry) using sums of products of single-particle wave functions (spin orbitals) $\phi_q(\mathbf{x})$, defined by Eq. (14.4). Recall that each q is a collection (group) of eigenvalues of a complete set of commuting hermitian operators for the underlying single-particle problem. Examples are $q = (k_x k_y k_z \sigma)$ (free particle in a box) or $q = (nlm\sigma)$ (particle in a central field). It is assumed that the spin orbitals $\phi_q(\mathbf{x})$ are easily calculated and that the set $\{q\}$ of all groups of single-particle quantum numbers is generically infinite but discrete. Since every infinite discrete set of cardinality $\aleph_0$ can be mapped onto the set of positive integers $\mathbb{Z}_+$, and the mapping is one-to-one, we can assign an integer $N(q) \in \mathbb{Z}_+$ to every group q of single-particle quantum numbers, such that $N(q) \neq N(q') \Leftrightarrow q \neq q'$. To simplify notation, we use the same notation q for $N(q)$ and regard $q \in \mathbb{Z}_+$ as an integer representing a collection of single-particle quantum numbers. In this way, the infinite set $\{q\}$ of groups of single-particle quantum numbers is automatically ordered. In the simple case of spinless particles where states are characterized solely by their energy E_n, q represents a single quantum number, so by taking $q = n$, we understand that q represents the quantum number E_n and the corresponding wave function is $\phi_n(\mathbf{r})$ satisfying Eq. (14.4). In this simple case, the ordering scheme is trivial. For example, in a 1D harmonic oscillator problem, the energies are $E_n = (n + 1/2)\hbar\omega$ and the quantum numbers $q = n$ are naturally ordered. If single-particle states are specified by more than one quantum number, one can adopt a lexicographic procedure of ordering. Thus, for a spinless particle in a central potential, we can order the set $\{q\}$ composed of groups of three single-particle quantum numbers according to the rule, $(nlm) > (n'l'm')$ if $n > n'$, or, in case $n = n', l > l'$, or in case $(nl) = (n'l')$, $m > m'$. Thus, $(nlm) = (000), (100), (11\bar{1}), (110), (111)$ correspond, respectively, to $q = 1, 2, \ldots, 5$ (here, $\bar{1} = -1$). If the particle has spin 1/2, the single-particle states in a central potential are determined in terms of the quantum numbers $(nlm\sigma_z)$ and the order can be defined by adopting the relation $(nlm \uparrow) > (nlm \downarrow)$. The upshot is that the set $\{q\}$ composed of groups of single-particle quantum numbers is unambiguously represented by the set of (positive) integers $\{q\} \subset \mathbb{Z}_+$ which is naturally ordered.

Problem 14.1

A free spinless particle in a box obeys the Schrödinger equation, $-\frac{\hbar^2}{2m}\nabla^2 \phi(x, y, z) = E\phi(x, y, z)$, with periodic boundary conditions, $\phi(x + L_x, y, z) = \phi(x, y + L_y, z) = \phi(x, y, z + L_z) = \phi(x, y, z)$.

(a) Determine the quantum numbers represented by q.
(b) Find a lexicographic order for determining when $q > q'$. Specify the quantum numbers corresponding to $q = 1, 2, 3, 4, 5$.
(c) For a particle in a central potential, show that in this ordering, the third state has $(n, l, m) = (1, 1, -1)$, hence, $\phi_3(r, \theta, \phi) = R_{11}(r)P_1(\cos\theta)e^{-i\phi}$.

Answer: (a) $q \leftrightarrow (k_x, k_y, k_z)$, with $k_x = \frac{2\pi}{L}n_x$, $n_x = 0, 1, 2, \ldots$, and similarly for k_y and k_z.
(b) For $q > q'$, if $|\mathbf{k}| > |\mathbf{k}'|$, while if $|\mathbf{k}| = |\mathbf{k}'|$, $\theta_\mathbf{k} > \theta_{\mathbf{k}'}$, while if $|\mathbf{k}| = |\mathbf{k}'|$ and $\theta_\mathbf{k} = \theta_{\mathbf{k}'}$ then $\phi_\mathbf{k} > \phi_{\mathbf{k}'}$, where $(\theta_\mathbf{k}, \phi_\mathbf{k})$ are the polar angles of $\mathbf{k}$. This ordering is reasonable because it insures that if $q \geq q'$, $\varepsilon_q \geq \varepsilon_{q'}$.

Construction of $\Phi_\alpha(\mathbf{x}_1, \mathbf{x}_2, \ldots \mathbf{x}_N)$

Following our discussion above, we conclude that an ordered sequence of nonnegative integers $\{q_1, q_2, \ldots, q_N\}$ that represents N sets of single-particle quantum numbers [e.g., $q_i = (k_{ix}, k_{iy}, k_{iz})$ for a free electron $i = 1, 2, \ldots N$ in a box] determines the N-body quantum number α. This is compactly written as,[2]

$$\alpha \leftrightarrow (q_1, q_2, \ldots, q_N). \tag{14.10}$$

---

[2] Strictly speaking the sequence should be written, $\{q_{i_1}, q_{i_2}, \ldots, q_{i_N}\}$, $i_k \in \mathbb{Z}_+$, $i_k < i_{k+1}$. For simplicity, we write this as $\{q_1, q_2, \ldots, q_N\}$, keeping in mind that this applies to any ordered sequence of N integers.

14.1 Second Quantization

The sequence $\{q_1, q_2, \ldots, q_N\}$ uniquely determines $\Phi_\alpha(\mathbf{x}_1, \mathbf{x}_2, \ldots \mathbf{x}_N)$ appearing in Eq. (14.9) and specifies which spin orbitals are used to construct it. The prescription for constructing $\Phi_\alpha(\mathbf{x}_1, \mathbf{x}_2, \ldots \mathbf{x}_N)$ from $\{q_1, q_2, \ldots, q_N\}$ is the familiar procedure for imposing the appropriate symmetry (see Chapter 8),

$$\Phi_\alpha(\mathbf{x}_1, \mathbf{x}_2, \ldots \mathbf{x}_N) = \begin{cases} C \sum_P (-1)^p P \prod_i \phi_{q_i}(\mathbf{x}_i) & \text{(for fermions)} \\ C \sum_P P \prod_i \phi_{q_i}(\mathbf{x}_i), & \text{(for bosons)} \end{cases} \qquad (14.11)$$

where C is a normalization constant. The subscripts q_i could be either the corresponding groups of single-particle quantum numbers or the integers that represent them (so far the ordering is immaterial). The summation over permutation operators P runs over all $N!$ permutations of the space-spin coordinates $(\mathbf{x}_1, \mathbf{x}_2, \ldots, \mathbf{x}_N)$, and p denotes the parity of the permutation P. The permutation P is expressible as a product of two-particle exchange operators P_{ij} swapping positions of space-spin coordinates $\mathbf{x}_i$ and $\mathbf{x}_j$ as indicated in Eq. (14.6). Note also that, as defined, P does not affect the sequence of quantum numbers $(q_1, q_2, \ldots, q_N)$, as it acts only on the space-spin coordinates $\{\mathbf{x}_i\}$. However, from the mathematical structure of Eq. (14.11), it is evident that we can replace the exchange operators P_{ij} exchanging space-spin coordinates $\mathbf{x}_i \leftrightarrow \mathbf{x}_j$ by an exchange operator $\mathcal{P}_{ij}$ exchanging quantum numbers $q_i \leftrightarrow q_j$. Indeed,

$$P_{ij} \phi_{q_i}(\mathbf{x}_i) \phi_{q_j}(\mathbf{x}_j) = \phi_{q_i}(\mathbf{x}_j) \phi_{q_j}(\mathbf{x}_i), \quad \mathcal{P}_{ij} \phi_{q_i}(\mathbf{x}_i) \phi_{q_j}(\mathbf{x}_j) = \phi_{q_j}(\mathbf{x}_i) \phi_{q_i}(\mathbf{x}_j), \qquad (14.12)$$

and the two operations lead to the same result. Note that for bosons, the case $q_i = q_j$ is allowed, while for fermions, it is forbidden.

The construction (14.11) of $\Phi_\alpha(\mathbf{x}_1, \mathbf{x}_2, \ldots \mathbf{x}_N)$ for bosons and fermions has already been detailed in Chapter 8. For fermions, $\Phi_\alpha(\mathbf{x}_1, \mathbf{x}_2, \ldots \mathbf{x}_N)$ is a *Slater determinant*, defined in Eq. (8.20), whereas for bosons, it is a *permanent* defined in Eqs. (8.23) and (8.43). From Problem 14.2(a), we see that the integers $q_1, q_2, \ldots, q_N$ defining the N-body quantum number α in Eq. (14.11) must be distinct for fermionic states, while for bosons, some or all of them may be equal. Thus, we have

$$0 \leq q_1 < q_2 \ldots < q_N, \quad \text{(for fermions)}, \qquad (14.13)$$
$$0 \leq q_1 \leq q_2 \ldots \leq q_N \quad \text{(for bosons)}. \qquad (14.14)$$

Problem 14.2

(a) Show that if $q_i = q_j$, the N fermion wave function vanishes.
(b) If in Eq. (14.11), $\mathcal{P}$ acts on the sequence of quantum numbers $(q_1, q_2, \ldots, q_N)$ (instead of $\mathcal{P}$ acting on the coordinates $(\mathbf{x}_1, \mathbf{x}_2, \ldots, \mathbf{x}_N)$, show that the resulting N body function is the same.
(c) Consider the 1D Hamiltonian, $H = \sum_{i=1}^{2} \left[-\frac{\hbar^2}{2m} \frac{d^2}{dx_i^2} + \frac{1}{2} m\omega^2 x_i^2 \right] + \lambda \delta(x_1 - x_2)$ for $N = 2$ identical particles of mass m. Denote by $\psi_n(x)$ ($n = 0, 1, \ldots$) the eigenfunctions of the single-particle Harmonic oscillator ($q = n$ is the single quantum number corresponding to the energy). Construct the two-body quantum numbers α and symmetrized basis functions $\Phi_\alpha(x_1, x_2)$ for the six lowest energy boson and fermion states.
(d) Calculate the matrix elements $\langle \Phi_\alpha | H | \Phi_\beta \rangle$.

Answer: (c) Since the energy is linear in n, the states $\psi_{n_1}(x_1) \psi_{n_2}(x_2)$ with lower sum $n_1 + n_2$ have the lower energy. Following Eq. (14.10), the six sequences α and the six energies ε_α (in units of $\hbar\omega$) are $\alpha = (q_1, q_2) = (0, 0), (0, 1),$ $(0, 2), (1, 0), (1, 1),$ and $(2, 0)$, and $\varepsilon_{\alpha=(q_1, q_2)} = 1, 2, 3, 2, 3, 3$, respectively.
For bosons, $\Phi_\alpha^B(x_1, x_2) = \left(\frac{N_{q_1}! N_{q_2}!}{N!} \right)^{1/2} \text{Perm}[\psi_{q_1}(x_1) \psi_{q_2}(x_2)]$, $q_1 = 0, 1, 2$, and $q_2 = 0, 1, 2$. The factor $\left(\frac{N_{q_1}! N_{q_2}!}{N!} \right)^{1/2}$ properly normalizes $\Phi_\alpha^B(x_1, x_2)$.
For fermions, $\Phi_\alpha^F(x_1, x_2) = |\psi_{q_1}(x_1) \psi_{q_2}(x_2)|$, $q_1 = 0, 1, 2$, and $q_2 = 0, 1, 2$, $q_2 \neq q_1$. See Eq. (8.39).
(d) $\langle \Phi_\alpha | H | \Phi_\beta \rangle = \varepsilon_\alpha \delta_{\alpha\beta} + \lambda \int_{-\infty}^{\infty} dx\, \Phi_\alpha^*(x, x) \Phi_\beta(x, x)$.

14.1.3 MAPPING ONTO FOCK SPACE

In the next step, the N-body wave functions $\Phi_\alpha(\mathbf{x}_1, \mathbf{x}_2, \ldots \mathbf{x}_N)$ with the appropriate symmetry ($P_{ij}\Phi_\alpha = +\Phi_\alpha$ for bosons and $P_{ij}\Phi_\alpha = -\Phi_\alpha$ for fermions) are mapped onto an abstract Hilbert space called *Fock space*, $\mathcal{F}_N$, for N identical particles. A basis vector in $\mathcal{F}_N$ has an infinite number of components; most of them are 0, except those in locations corresponding to the integers, $\{q_i\}$, $i = 1, 2, \ldots, N$. This is where we use the construction with symbols $\{q_i\}$ that form an ordered set of integer numbers representing groups of single-particle quantum numbers. The entry in location q_i is a nonnegative integer number that tells us how many particles in Φ_α have the same quantum number q_i. This is the *occupation number* in location q_i that plays a central role in the present formalism. Following the solution of Problem 14.2, we know that for fermions, the occupation number is 0 or 1, while for bosons, it can be any nonnegative integer $\leq N$. The mapping is based on the fact that the sequence $q_1, q_2, \ldots, q_N$ unambiguously determines the basis function $\Phi_\alpha(\mathbf{x}_1, \mathbf{x}_2, \ldots \mathbf{x}_N)$ via Eq. (14.11). Usually, the number of terms in the expansion (14.9) is infinite. When translated into a sum over the integers q_i, it involves the sum over all possible sequences $q_1, q_2, \ldots q_N$ subject to the constraints (14.13). The Hilbert spaces $\mathcal{F}_N$ for fermions and bosons are distinct in that, as we have noticed, the basis states of $\mathcal{F}_N$ contain occupation numbers that are either 0 or 1 for fermions but are nonnegative integers less than or equal to N for bosons. The sum of all occupation numbers must equal the number N of particles.

Basis States in Fock Space for Fermions

We are now in a position to construct a basis in the Hilbert space $\mathcal{F}_N$. Dirac notation will be extremely useful in what follows. For fermions, the Fock space $\mathcal{F}_N$ is spanned by an infinite basis of vectors (kets) $|\alpha\rangle \leftrightarrow |q_1, q_2, \ldots, q_N\rangle$ with N distinct quantum numbers $q_1 < q_2 \ldots < q_N$ singly occupied. More explicitly,

$$|\alpha\rangle = |000\ldots 1_{q_1}00\ldots 1_{q_2}00\ldots 1_{q_N}00\ldots\rangle \equiv |q_1, q_2, \ldots, q_N\rangle, \tag{14.15}$$

where the subscripts denote the locations q_i in the infinite sequence that correspond to single-particle quantum numbers q_i that are occupied. The occupation numbers 1 indicate that the locations q_i are singly occupied, so that each spin orbital $\phi_{q_i}(\mathbf{x})$ appears just once in the product defined in Eq. (14.11). The second identification on the RHS of Eq. (14.15) is a compact notation where only the occupied locations (quantum numbers) are specified, bearing in mind that for fermions, the corresponding occupation numbers are 1. In a more heuristic interpretation, we say that the basis state $|\alpha\rangle$ is obtained by "occupying" or "putting" one particle in a single-particle state whose set of quantum numbers are represented by q_1 and so on. Once this is conceptually understood, one may prefer to use the quantum numbers themselves as indices, bearing in mind that they should specify the location of a certain component of a vector in Fock space. For example, in the study of electrons in a lattice, we will use the indices $n\mathbf{k}\sigma = n(k_x, k_y, k_z)\sigma$ of band index, lattice momentum, and spin projection keeping in mind that this set of five quantum numbers corresponds to a single integer q that specifies a location (a component number) in the vector $|\alpha\rangle \in \mathcal{F}_N$.

Basis States in Fock Space for Bosons

The corresponding procedure of constructing a basis in $\mathcal{F}_N$ for bosons is similar to that for fermions, but there are some important differences. For bosons, some (or even all) of the numbers q_i might be equal. As an example, for $N = 6$, a sequence of six quantum numbers q_i might actually be $(q_1, q_1, q_2, q_2, q_2, q_3)$. It is then natural to require that the only nonzero components of the corresponding basis state $|\alpha\rangle \in \mathcal{F}_N$ are located in (q_1, q_2, q_3) with nonzero occupation numbers equal to 2, 3, and 1, according to the number of equal q_i appearing in each group. In our example, we have

$$|\alpha\rangle = |000\ldots 2_{q_1}00\ldots 3_{q_2}00\ldots 1_{q_3}00\ldots\rangle. \tag{14.16}$$

Recalling our heuristic interpretation, the basis state $|\alpha\rangle$ is obtained by putting two particles in the single-particle quantum number q_1, three particles in the single-particle quantum number q_2, and one particle in q_3. The sum of the numbers in the nonzero components (i.e., the sum of the boson occupation numbers) should of course be equal to N ($N = 6$ in the present example).

14.1 Second Quantization

Occupation Number Representation of States

Equations (14.15) and (14.16) represent the N-body state vector $|\alpha\rangle$ in terms of the corresponding occupation numbers (0 or 1 for fermions, nonnegative integers for bosons) of single-particle quantum numbers $q_i, i = 1, 2, \ldots K$. Therefore, the state $|\alpha\rangle$ is uniquely determined by a set of K single-particle quantum numbers $q_i, i = 1, 2, \ldots K$ ($K = N$ for fermions and $K \leq N$ for bosons) and by K occupation numbers n_{q_i}. Thus, we write,

$$|\alpha\rangle = |n_{q_1}, n_{q_2}, \ldots, n_{q_K}\rangle, \tag{14.17}$$

where $q_i \neq q_j$ and n_{q_i} is the nonzero occupation number in mode (i.e., single-particle level) q_i. The latter is well defined once the system of single-particle quantum numbers are ordered as explained above. In this notation, the components with occupation number 0 are not specified. Equation (14.17) extends the notation introduced in Eq. (14.15) to be applicable for both bosons and fermions.

Orthogonality Relations and Wave Functions

Two basis vectors $|\alpha\rangle \in \mathcal{F}_N$ and $|\beta\rangle \in \mathcal{F}_N$ are different, $|\beta\rangle \neq |\alpha\rangle$, if they differ by at least one component. For example, for $N = 2$, the two basis vectors $|\alpha\rangle = |1_3 1_5\rangle = |00101\ldots\rangle$ and $|\beta\rangle = |1_3 1_4\rangle = |0011\ldots\rangle$ are distinct. An inner product between two basis vectors is defined to be zero if they are different and 1 if they are identical,

$$\langle\alpha|\beta\rangle = \delta_{\alpha\beta}. \tag{14.18}$$

With this definition, the Hilbert space $\mathcal{F}_N$ is unambiguously defined, since a complete basis, $\{|\alpha\rangle\}$, spanning it has been introduced together with an inner product. A general normalized vector in this space is defined in an analogy with expansion (14.9) as,

$$|\Psi(t)\rangle = \sum_{\alpha} A_\alpha(t)|\alpha\rangle, \quad \sum_{\alpha} |A_\alpha(t)|^2 = 1. \tag{14.19}$$

The amplitudes $A_\alpha(t)$ are complex numbers. An operator O acting on the vector $|\Psi(t)\rangle$ is defined in terms of its operation on the basis vectors $|\alpha\rangle$,

$$O|\Psi(t)\rangle = \sum_{\alpha} A_\alpha(t) O|\alpha\rangle. \tag{14.20}$$

Thus, we have defined the N-particle *Fock space* or *space of occupation numbers*, pertaining to N identical particles, for fermions and bosons. This Hilbert space $\mathcal{F}_N$ is defined in terms of its basis vectors $\{|\alpha\rangle\}$ and the scalar product (14.18).

The set of basis vectors $\{|\alpha\rangle\}$ is in one-to-one correspondence with the set of wave functions $\{\Phi_\alpha(\mathbf{x}_1, \mathbf{x}_2, \ldots \mathbf{x}_N)\}$ spanning the Hilbert space $\mathcal{H}_N$ of N-body wave functions with the appropriate symmetry. Formally, we may write,

$$\Phi_\alpha(\mathbf{x}_1, \mathbf{x}_2, \ldots \mathbf{x}_N) = \langle \mathbf{x}_1, \mathbf{x}_2, \ldots \mathbf{x}_N|\alpha\rangle. \tag{14.21}$$

In this notation, the concept of transformation between the Hilbert space $\mathcal{H}_N$ spanned by the functions $\{\Phi_\alpha(\mathbf{x}_1, \mathbf{x}_2, \ldots, \mathbf{x}_N)\}$ and the Fock space $\mathcal{F}_N$ spanned by the states $\{|\alpha\rangle\}$ is more transparent, as the expression on the RHS serves as an element of the transformation matrix. This definition is based on the completeness relations

$$\sum_{\alpha} |\alpha\rangle\langle\alpha| = 1_{\mathcal{F}_N}, \quad \sum_{\{\mathbf{x}_i\}} |\mathbf{x}_1, \mathbf{x}_2, \ldots, \mathbf{x}_N\rangle\langle\mathbf{x}_1, \mathbf{x}_2, \ldots, \mathbf{x}_N| = 1_{\mathcal{H}_N}. \tag{14.22}$$

The Complete Fock Space

So far we have defined the Fock space $\mathcal{F}_N$ pertaining to a fixed number N of identical particles. We can formally construct the total Fock space by performing a direct sum,

$$\mathcal{F} = \oplus_{N=0}^{\infty} \mathcal{F}_N. \tag{14.23}$$

Of particular interest is the subspace $\mathcal{F}_{N=0}$, which contains no particles at all. It is spanned by a single ket, all of whose components are zero,

$$|\alpha\rangle = |0000\ldots\rangle \equiv |0\rangle. \tag{14.24}$$

This is referred to as the vacuum state or simply the *vacuum*.

One might argue that building $\mathcal{F}$ from $\{\mathcal{F}_N\}$ is nothing more than a mathematical construction devoid of physics, since classically, particle number is conserved, and there are no transitions between subspaces of $\mathcal{F}$ with different number of particles. However, as we will encounter later on, in quantum mechanics, there are situations where particle number is not conserved. Particles can be created or annihilated, and subspaces $\mathcal{F}_N$ and $\mathcal{F}_{N'}$ corresponding to different particle numbers may be coupled. Examples are systems of electrons and phonons. This is an aspect of quantum mechanics that has no classical analog. The mathematical apparatus of second quantization developed in this section enables the inclusion of such cases. The procedure is effected in terms of operators creating or annihilating particles, thereby coupling $\mathcal{F}_N$ and $\mathcal{F}_{N\pm 1}$.

14.1.4 CREATION AND ANNIHILATION OPERATORS

Having constructed the Fock spaces $\mathcal{F}_N$ and $\mathcal{F}$, it is now possible to define operators that act on vectors in these spaces. All the operations on vectors in Fock space are fully determined by their operation on the basis vectors $\{|\alpha\rangle\}$, see Eq. (14.20). As will be demonstrated, it is sufficient to define just two kinds of operators, *creation and annihilation operators* (defined below). All operators of physical interest, such as the Hamiltonian or the current, are obtained by taking appropriate sums of products of creation and annihilation operators.

Definition of Creation and Annihilation Operators

The most elementary operation on a basis vector in $\mathcal{F}$ is to change an occupation number at a single location q by one unit. Consider the annihilation and creation operators, a_q and $a_q^\dagger$, that affect only the occupation number n_q of α at component number q. For simplicity of notation, we concentrate only on this component, writing,

$$|\alpha\rangle = |\ldots n_q \ldots\rangle, \tag{14.25}$$

recalling once again that $n_q = 0$ or 1 for fermions and that n_q is a nonnegative integer for bosons. The result of applying the *annihilation operator* a_q on $|\alpha\rangle$ is to subtract 1 from n_q, whereas the result of applying the *creation operator* $a_q^\dagger$ on $|\alpha\rangle$ is to add 1 to n_q. Explicitly,

$$a_q|\alpha\rangle = C(n_q)|\ldots n_q - 1\ldots\rangle, \quad a_q^\dagger|\alpha\rangle = \bar{C}(n_q)|\ldots n_q + 1\ldots\rangle, \tag{14.26}$$

where $C(n_q)$ and $\bar{C}(n_q)$ are numerical constants, which can be chosen to be real. The creation and annihilation operators do not conserve particle number. If the RHS of Eq. (14.26) are not 0, then, formally, $|\alpha\rangle \in \mathcal{F}_N \Rightarrow a_q|\alpha\rangle \in \mathcal{F}_{N-1}$, whereas $a_q^\dagger|\alpha\rangle \in \mathcal{F}_{N+1}$. The statistics of particles (Bose or Fermi) and the normalization requirements determine the numerical values of the constants $C(n_q)$ and $\bar{C}(n_q)$. They also determine the commutation or anticommutation relations between annihilation and creation operators (see below). We will prove below that

$$C(n_q) = \begin{cases} \sqrt{n_q} & \text{(bosons)}, \\ \sqrt{n_q}\phi(n_q) & \text{(fermions)}, \end{cases}$$

$$\bar{C}(n_q) = \begin{cases} \sqrt{1+n_q} & \text{(bosons)}, \\ \sqrt{1-n_q}\phi(n_q) & \text{(fermions)}, \end{cases}$$

$$\phi(n_q) = (-1)^{\sum_{p=1}^{q-1} n_p}. \tag{14.27}$$

14.1 Second Quantization

> The result of applying a fermion annihilation (a_q) or creation ($a_q^\dagger$) operator in Fock space to a state $|\alpha\rangle$ depends on the parity of the sum of the occupation numbers in the locations smaller than q. This is the reason for the necessity to order the single-particle quantum numbers.

Special cases of the expressions (14.27) should be obvious already at this stage, e.g., if $n_q = 0$, then $C(0) = 0$ because the entries of a basis ket cannot be negative. Furthermore, if $n_q = 1$ and the particles are fermions, then $\bar{C}(1) = 0$, because fermion occupation numbers are either 0 or 1. They cannot be greater than 1.

Let us now consider the relation between a_q and $a_q^\dagger$. For $n_q > 0$ (i.e., $n_q = 1$ for fermions), $C(n_q) \neq 0$, and we can apply the bra $\langle n_q - 1|$ to the left of the equation $a_q|n_q\rangle = C(q)|n_q - 1\rangle$ to obtain $\langle n_q - 1|a_q|n_q\rangle \neq 0$. Taking the complex conjugate, we then find,

$$\langle n_q - 1|a_q|n_q\rangle^* = \langle n_q|[a_q]^\dagger|n_q - 1\rangle, \quad (14.28)$$

which can be nonzero only if $[a_q]^\dagger = a_q^\dagger$ is a creation operator as defined independently in Eq. (14.26). In other words, the creation operator $a_q^\dagger$ defined in the second equation of (14.26) is the Hermitian conjugate of the annihilation operator a_q defined in the first equation of (14.26).

Since boson and fermion creation and annihilation operators have different properties (they act in different spaces), it is useful to denote them by different letters. We shall use b_q and $b_q^\dagger$ for bosons and c_q and $c_q^\dagger$ for fermions. Following our definitions above, we schematically illustrate the action of fermion and boson creation and annihilation operators on basis vectors in Fock space in Fig. 14.1.

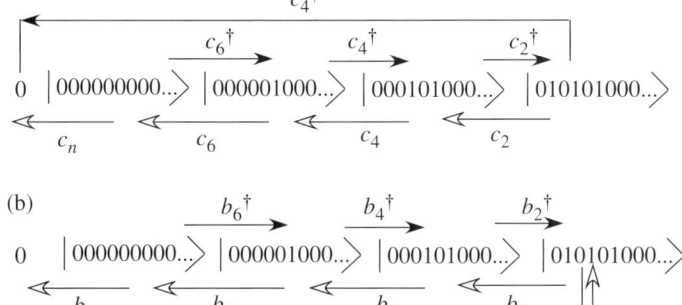

FIG 14.1 Fermion and boson creation operators. (a) The action of fermion creation operators $c_q^\dagger$ (black arrows) and annihilation operators c_q (white arrows) on basis vectors in Fock space. The leftmost vector with all components equal to 0 is the vacuum state $|0\rangle$. When an annihilation operator c_q ($q = 1, 2, \ldots$) is applied on $|0\rangle$, the vacuum state is destroyed. Starting from the vacuum state, particles are consecutively added by operating with $c_6^\dagger$, $c_4^\dagger$, and $c_2^\dagger$. They can be subtracted by application of the corresponding annihilation operators. If we apply a creation operator on a filled location (such as $c_4^\dagger$), the state is immediately killed. (b) Boson creation operators. The main difference from (a) is that application of a creation operator on a filled location (such as $b_4^\dagger$) does not kill the state but rather adds a second particle to the same location.

From the preceding discussion, we conclude that the concept of *second quantization* can be described as follows: A quantum mechanical problem involving N interacting particles defined by its Hamiltonian (14.1) acting in the Hilbert space $\mathcal{H}$ of N-particle wave functions with the appropriate symmetry is mapped onto a new problem, such that the underlying Hilbert space is the Fock space $\mathcal{F}_N$, and the pertinent physical operators (Hamiltonian, current, etc.) are expressible in terms of creation and annihilation operators. In this case, the problem is said to be posed in the formalism of second quantization. We have already encountered a simple example of this construction in our study of the harmonic oscillator in Sec. 2.7.2. The treatment of quantum mechanics starting from a Hamiltonian such as Eq. (14.1) is then referred to as *first quantization*.

To complete the construction of the second quantization formalism we need to determine the properties of the creation and annihilation operators and map the original Hamiltonian (14.1) onto a new Hamiltonian acting in Fock space. These two tasks are carried out separately for bosons and fermions. Henceforth, we shall occasionally use indices such as $i, j, k, \ldots$ rather than $q_1, q_2, \ldots, q_N$.

Algebra of Boson Creation and Annihilation Operators

Having defined the boson creation and annihilation operators $b_i^\dagger$ and b_i, respectively, through Eqs (14.26) and (14.27), we now consider products of such operators and derive the algebra that they obey. First, consider the commutation relation of two creation operators $b_i^\dagger$ and $b_j^\dagger$ or two annihilation operators b_i and b_j. To do so, we consider the wave function (14.21) and, following the procedure exemplified in Eq. (14.12), perform the exchange operation not on the coordinates as in Eq. (14.6) but rather on the quantum numbers α as in Eq. (14.11), replacing $q_i \leftrightarrow q_j$. Since the N boson wave function is symmetric under exchange, we conclude that two boson creation operators commute, $[b_i^\dagger, b_j^\dagger] = 0$, as do the two annihilation operators, $[b_i, b_j] = 0$.

Now we study products of creation and annihilation operators. For example, consider a ket state $b_i^\dagger b_j |\alpha\rangle$ for $i \neq j$ and operate on it with $\mathcal{P}_{ij}$. As $|\alpha\rangle$ and $|\beta\rangle \equiv b_i^\dagger b_j |\alpha\rangle$ are symmetric, $\mathcal{P}_{ij}|\alpha\rangle=|\alpha\rangle$ and $\mathcal{P}_{ij}|\beta\rangle=\mathcal{P}_{ij}b_i^\dagger b_j|\alpha\rangle =b_i^\dagger b_j|\alpha\rangle$. Thus,

$$b_i^\dagger b_j |\alpha\rangle = \mathcal{P}_{ij} b_i^\dagger b_j |\alpha\rangle = \mathcal{P}_{ij} b_i^\dagger b_j \mathcal{P}_{ij}^{-1} \mathcal{P}_{ij} |\alpha\rangle = \mathcal{P}_{ij} b_i^\dagger b_j \mathcal{P}_{ij}^{-1} |\alpha\rangle = b_j b_i^\dagger |\alpha\rangle.$$

Therefore, we find that $[b_i^\dagger, b_j] = 0$ ($i \neq j$).

Next, let us consider the case $i = j$. The fact that $[b_i, b_i^\dagger] \neq 0$ is easily seen when it is applied on the vacuum state $|0\rangle$. The defining equations (14.26) imply,

$$[b_i, b_i^\dagger]|0\rangle = \bar{C}(0)C(1)|0\rangle - b_i^\dagger b_i|0\rangle = \bar{C}(0)C(1)|0\rangle, \tag{14.29}$$

because application of b_i on the vacuum state appearing in the second term destroys it (see Fig. 14.1). Since $[b_i, b_i^\dagger]$ is the only nonvanishing commutator, and since it is a bilinear combination, we can safely scale the operators b and $b^\dagger$ by a real number, such that the product $\bar{C}(0)C(1) = 1$ without affecting the other (vanishing) commutation relations. Thus, we have $[b_i, b_i^\dagger]|0\rangle = |0\rangle$. We also require that the commutation relation $[b_i, b_i^\dagger]$ is independent of the state on which it operates; hence, we conclude that $[b_i, b_i^\dagger] = 1$. This is consistent with the result obtained in our discussion of the harmonic oscillator in Sec. 2.7.2, where the commutator $[x, p] = i\hbar$ yields the commutator $[b, b^\dagger] = 1$. Collecting the above results, we arrive at the key equation expressing commutation relations between boson operators,

$$\boxed{[b_i, b_j] = 0, \quad [b_i^\dagger, b_j^\dagger] = 0, \quad [b_i, b_j^\dagger] = \delta_{ij}.} \tag{14.30}$$

Problem 14.3

Prove the following commutation relations using Eq. (14.30),

$$[b_i^\dagger b_i, b_i] = -b_i, \quad [b_i^\dagger b_i, b_i^\dagger] = b_i^\dagger. \tag{14.31}$$

Guidance: $[b_i^\dagger b_i, b_i] = b_i^\dagger b_i b_i - b_i b_i^\dagger b_i$, and $b_i^\dagger b_i = b_i b_i^\dagger - 1$.

The *number operators* $\hat{n}_i$ are bilinear combinations of creation and annihilation operators,

$$\boxed{\hat{n}_i \equiv b_i^\dagger b_i,} \tag{14.32}$$

that play a special role in the formalism (we temporarily use $\hat{n}_i$ to avoid confusion with the *occupation numbers* n_i). Some of their properties are as follows: (1) $\hat{n}_i$ is a nonnegative operator. Indeed, the expectation value of $\hat{n}_i$ for any ket $|\phi\rangle$ (not necessarily a basis ket $|\alpha\rangle$) is equal to the norm squared of the state $b_i|\phi\rangle$, which is either positive or 0 (if $|\phi\rangle = |0\rangle$). (2) If a ket $|n_i\rangle$ is an eigenstate of the number operator $\hat{n}_i$ with an eigenvalue n_i, then $b_i|n_i\rangle$ is also an eigenstate of n_i with an eigenvalue $n_i - 1$. To show this, employ Eq. (14.31) in the form, $\hat{n}_i b_i |n_i\rangle = (-b_i + b_i \hat{n}_i)|n_i\rangle$

14.1 Second Quantization

$= (n_i - 1)b_i|n_i\rangle$, where the last equality follows because $\hat{n}_i|n_i\rangle = n_i|n_i\rangle$. By repeated application of b_i on $|n_i\rangle$, we find that $b_i^n|n_i\rangle$ is an eigenstate of the number operator $\hat{n}_i$ with eigenvalue $(n_i - n)$. If n_i is not an integer, then $(n_i - n)$ becomes negative for $n > n_i$, but this contradicts (1). Hence, the eigenvalues n_i of $\hat{n}_i$ must be nonnegative integers. The reason that $\hat{n}_i$ is referred to as the number operator is now clear, because its eigenvalue give the number of particles in quantum state i. Similar considerations lead to an analogous expression combining $\hat{n}_i$ and $b_i^\dagger$, as you will demonstrate in Problem 14.4.

Problem 14.4

Use Eq. (14.31) to show that if $\hat{n}_i|n_i\rangle = n_i|n_i\rangle$, then

$$\hat{n}_i b_i^\dagger |n_i\rangle = (\hat{n}_i + 1) b_i^\dagger |n_i\rangle.$$

We are now in a position to verify the normalization constants defined in Eq. (14.26) for bosons in the first row of Eqs (14.27):

$$C(n_i)^2 = ||b_i|n_i\rangle||^2 = \langle n_i|\hat{n}_i|n_i\rangle = n_i, \tag{14.33}$$

$$\bar{C}(n_i)^2 = ||b_i^\dagger|n_i\rangle||^2 = \langle n_i|1 + \hat{n}_i|n_i\rangle = n_i + 1. \tag{14.34}$$

The above analysis is summarized in the following key relations,

$$\hat{n}_i|n_i\rangle = n_i|n_i\rangle, \quad n_i = 0, 1, 2, \ldots \tag{14.35}$$

$$b_i|n_i\rangle = \sqrt{n_i}\,|n_i - 1\rangle, \quad b_i^\dagger|n_i\rangle = \sqrt{n_i + 1}\,|n_i + 1\rangle, \tag{14.36}$$

$$[b_i^\dagger]^{n_i}|0\rangle = \sqrt{n_i!}\,|n_i\rangle. \tag{14.37}$$

Note that this is an extension of the formalism describing the harmonic oscillator raising and lowering operators. Equation (14.36) proves Eq. (14.27) for bosons.

Finally, before moving on to analyze the analogous expressions for fermion operators, we define the *normal ordering* of an operator product composed of annihilation and creation operators. In a normal ordered product, creation operators appear to the left of all annihilation operators. Thus, the product $b_i^\dagger b_i$ is a normal ordered product, while $b_i b_i^\dagger$ is not. Commutation relations can be used to rearrange any operator product $\mathcal{O}$ and turn it into its normal form denoted as $:\mathcal{O}:$. For example,

$$:b_i b_i^\dagger: = 1 + \hat{n}_i. \tag{14.38}$$

Problem 14.5

(a) Show that $:b_i^2[b_i^\dagger]^2: = 1 + 4n_i + [b_i^\dagger]^2 b_i^2$.
(b) Construct the state $|\alpha\rangle = |030020004000\ldots\rangle$ by application of powers of $b_2^\dagger$, $b_5^\dagger$, and $b_9^\dagger$ on the vacuum $|0\rangle$. (Do not ignore the coefficients and the normalization requirement.)

Answer: Using Eq. (14.37), $|\alpha\rangle = |030020004000\ldots\rangle = \frac{1}{\sqrt{3!2!4!}}[b_2^\dagger]^3[b_5^\dagger]^2[b_9^\dagger]^4|0\rangle$.

Algebra of Fermion Creation and Annihilation Operators

The fermion creation and annihilation operators, $c_i^\dagger$ and c_i, are defined as in Eq. (14.26). To determine the properties of operator products, let us construct a Fock space vector $|\alpha\rangle = |\ldots 1_{q_1} \ldots 1_{q_2} \ldots 1_{q_N} \ldots\rangle$ by applying fermion creation operators on the vacuum, guided by Fig. 14.1. It proves convenient to start from $c_{q_N}^\dagger$ and finish with $c_{q_1}^\dagger$, i.e.,

$$|\alpha\rangle = |\ldots 1_{q_1} \ldots 1_{q_2} \ldots 1_{q_N} \ldots\rangle = c_{q_1}^\dagger c_{q_2}^\dagger \ldots c_{q_N}^\dagger |0\rangle. \tag{14.39}$$

The occupation numbers are equal to 0 in all dotted locations. The absence of a phase factor in (14.39) is due to our choice of the order of application of the creation operators. Using this form for $|\alpha\rangle$ in Eq. (14.21), we find,

$$\Phi_\alpha(\mathbf{x}_1, \ldots, \mathbf{x}_N) = \langle \mathbf{x}_1, \ldots, \mathbf{x}_N | \alpha \rangle = \langle \mathbf{x}_1, \ldots, \mathbf{x}_N | c_{q_1}^\dagger c_{q_2}^\dagger \ldots c_{q_N}^\dagger |0\rangle. \tag{14.40}$$

The antisymmetry of $\Phi_\alpha(\mathbf{x}_1, \mathbf{x}_2, \ldots \mathbf{x}_N)$ under the exchange $\mathbf{x}_i \leftrightarrow \mathbf{x}_j$ is guaranteed by Eq. (14.11), which also implies antisymmetry under the swap $c_{q_i}^\dagger \leftrightarrow c_{q_j}^\dagger$. Explicitly, for $N = 2$,

$$\langle \mathbf{x}_1, \mathbf{x}_2 | c_{q_1}^\dagger c_{q_2}^\dagger |0\rangle = -\langle \mathbf{x}_2, \mathbf{x}_1 | c_{q_1}^\dagger c_{q_2}^\dagger |0\rangle = \langle \mathbf{x}_2, \mathbf{x}_1 | c_{q_2}^\dagger c_{q_1}^\dagger |0\rangle = -\langle \mathbf{x}_1, \mathbf{x}_2 | c_{q_2}^\dagger c_{q_1}^\dagger |0\rangle, \tag{14.41}$$

i.e., $\langle \mathbf{x}_1, \mathbf{x}_2 | c_{q_1}^\dagger c_{q_2}^\dagger + c_{q_2}^\dagger c_{q_1}^\dagger |0\rangle = 0$. In other words, two fermion creation operators and, by Hermitian conjugation, two fermion annihilation operators satisfy *anticommutation relations*,

$$\{c_i^\dagger, c_j^\dagger\} \equiv c_i^\dagger c_j^\dagger + c_j^\dagger c_i^\dagger = 0, \quad \{c_i, c_j\} = 0, \tag{14.42}$$

which include the following equalities as special cases,

$$(c_i^\dagger)^2 = (c_i)^2 = 0. \tag{14.43}$$

In analogy with the definition of boson number operators in Eq. (14.32), we define the *fermion number operator*,

$$\hat{n}_i \equiv c_i^\dagger c_i. \tag{14.44}$$

Clearly, $\hat{n}_i$ is Hermitian. Using the same analysis as that which follows Eq. (14.32), it is possible to show that $\hat{n}_i$ is nonnegative. Moreover, if we apply $\hat{n}_i$ to a basis vector $|i\rangle \equiv |00 \ldots 1_i 000 \ldots\rangle$, using Eq. (14.26), we obtain $\bar{C}(0)C(1)|i\rangle$. Employing the normalization $|\bar{C}(0)C(1)|^2 = 1$ and positivity, we have $\bar{C}(0)C(1) = 1$. As in the analogous discussion for bosons, we can choose,

$$\bar{C}(0) = C(1) = 1, \quad \bar{C}(1) = C(0) = 0, \tag{14.45}$$

where the second set of equations results because (a) a given state cannot be occupied by two fermions and (b) the lowering operator destroys an unoccupied state,

$$c_i^\dagger | \ldots 1_i \ldots \rangle = 0, \quad c_i | \ldots 0_i \ldots \rangle = 0, \tag{14.46}$$

as schematically indicated in the upper part of Fig. 14.1. Consider now the application of $\hat{n}_2 = c_2^\dagger c_2$ on the state vector $|1100 \ldots\rangle = c_1^\dagger c_2^\dagger |0\rangle$. This is done by first swapping c_2 and $c_1^\dagger$, incurring an as yet unknown phase ϕ and then swapping $c_2^\dagger$ and $c_1^\dagger$, such that $c_2^\dagger$ is moved to the left of c_2 incurring a phase -1. Since the eigenvalue of $\hat{n}_2$ is 1 in this case, we get $\phi = -1$. Explicitly,

$$c_2^\dagger c_2 c_1^\dagger c_2^\dagger |0\rangle = \phi c_2^\dagger c_1^\dagger c_2 c_2^\dagger |0\rangle = \phi c_2^\dagger c_1^\dagger |0\rangle = -\phi c_1^\dagger c_2^\dagger |0\rangle = c_1^\dagger c_2^\dagger |0\rangle. \tag{14.47}$$

The upshot is then $c_2 c_1^\dagger = -c_1^\dagger c_2$. On the other hand, using arguments similar to those used in Eq. (14.29), it is easy to show that

$$\{c_i, c_i^\dagger\}|0\rangle = (c_i c_i^\dagger + c_i^\dagger c_i)|0\rangle = |0\rangle - 0. \tag{14.48}$$

14.1 Second Quantization

Combining this result with Eq. (14.42), we arrive at the fermionic analog of Eq. (14.30),

$$\boxed{\{c_i, c_j\} = \{c_i^\dagger, c_j^\dagger\} = 0, \quad \{c_i, c_j^\dagger\} = \delta_{ij}.} \tag{14.49}$$

Finally, we need to determine what results when an annihilation or creation operators acts on a basis state $|\alpha\rangle = |n_1 n_2 n_3 \ldots\rangle$, where we recall that for fermions, $n_i = \pm 1$. Since we adopt the order convention as in Eq. (14.39) and the normalization (14.45), the discussion around Eq. (14.47) implies,

$$c_i |n_1 n_2 \ldots n_i \ldots\rangle = \sqrt{n_i}(-1)^{\sum_{j=1}^{i-1} n_j} |n_1 n_2 \ldots n_i - 1 \ldots\rangle,$$

$$c_i^\dagger |n_1 n_2 \ldots n_i \ldots\rangle = \sqrt{1 - n_i}(-1)^{\sum_{j=1}^{i-1} n_j} |n_1 n_2 \ldots n_i + 1 \ldots\rangle, \tag{14.50}$$

where the coefficients $\sqrt{n_i}$ and $\sqrt{1 - n_i}$ reflect the constraints of Eq. (14.45). The above equations are the analogs of their bosonic counterparts, Eqs (14.36) and (14.37), where the square root is essential because n_i may be greater than unity for bosons. Equation (14.50) substantiates Eq. (14.27).

Problem 14.6

(a) Show that $\{n_i, c_i\} = c_i n_i$ and $\{n_i, c_i^\dagger\} = c_i^\dagger n_i$.

(b) Show that $:c_i c_j c_i^\dagger c_j^\dagger: = -1 + n_i + n_j + c_i^\dagger c_j^\dagger c_i c_j$.

14.1.5 THE HAMILTONIAN IN FOCK SPACE

The aim of this section is to map the Hamiltonian H of Eq. (14.1) onto the corresponding Hamiltonian operator $\hat{H}$ acting in Fock space. The latter will be expressed in terms of sum of products of creation and annihilation operators. After specifying the general framework, we separate the discussion for fermions and bosons.

Consider the bare N particle Hamiltonian in Eq. (14.1). Its matrix elements between basis states $\Phi_\alpha(\mathbf{x}_1, \mathbf{x}_2, \ldots \mathbf{x}_N)$ and $\Phi_\beta(\mathbf{x}_1, \mathbf{x}_2, \ldots \mathbf{x}_N)$ are explicitly given by,

$$\langle \Phi_\alpha | H | \Phi_\beta \rangle = \int d\mathbf{x}\, \Phi_\alpha^*(\mathbf{x}_1, \mathbf{x}_2, \ldots \mathbf{x}_N) H \Phi_\beta(\mathbf{x}_1, \mathbf{x}_2, \ldots \mathbf{x}_N), \tag{14.51}$$

where $\int d\mathbf{x} = \int \prod_{i=1}^{N} d\mathbf{r}_i \sum_{\sigma_i}$. We already defined a mapping between the basis functions $\Phi_\alpha(\mathbf{x}_1, \mathbf{x}_2, \ldots \mathbf{x}_N)$ and the basis states $|\alpha\rangle$ in Fock space through Eq. (14.21). An explicit evaluation of the matrix elements in (14.51) in terms of single-particle wave functions will be given below.

The guideline for identifying the Hamiltonian operator $\hat{H}$ in Fock space is the equality of the matrix elements,

$$\langle \alpha | \hat{H} | \beta \rangle = \langle \Phi_\alpha | H | \Phi_\beta \rangle. \tag{14.52}$$

Equation (14.52) provides the crucial link between the description of a physical system within the formalism of *first quantization* (RHS) and *second quantization* (LHS). Since $\hat{H}$ acts in Fock space, it should be written in terms of creation and annihilation operators. The procedure of finding the precise combination of operators and the appropriate coefficients will be worked out explicitly for a two-particle system. Since the operators appearing in the Hamiltonian are either one-body (such as the kinetic energy) or two-body operators (the two-body interaction potential), it is evident that a generalization to arbitrary number N of identical particles is straightforward.

Second Quantized Hamiltonian: Fermions

First, consider a two fermion system with basis functions (Slater determinants)

$$\Phi_\alpha(\mathbf{x}_1, \mathbf{x}_2) = \frac{1}{\sqrt{2}} \begin{vmatrix} \phi_m(\mathbf{x}_1) & \phi_m(\mathbf{x}_2) \\ \phi_n(\mathbf{x}_1) & \phi_n(\mathbf{x}_2) \end{vmatrix}, \quad \Phi_\beta(\mathbf{x}_1, \mathbf{x}_2) = \frac{1}{\sqrt{2}} \begin{vmatrix} \phi_p(\mathbf{x}_1) & \phi_p(\mathbf{x}_2) \\ \phi_q(\mathbf{x}_1) & \phi_q(\mathbf{x}_2) \end{vmatrix}. \qquad (14.53)$$

Here, $m \neq n$ and $p \neq q$ are integers designating sets of single-particle quantum numbers as discussed in detail in connection with Eq. (14.11). Without loss of generality, we can assume that $m > n$ and $p > q$. The two-body Hamiltonian is,

$$H = T_1 + T_2 + U(\mathbf{r}_1) + U(\mathbf{r}_2) + V(|\mathbf{r}_1 - \mathbf{r}_2|), \qquad (14.54)$$

where $T_i = -\frac{\hbar^2}{2m}\nabla_{\mathbf{r}_i}^2$ is the kinetic energy operator for particle i. To compute $\langle \Phi_\alpha | H | \Phi_\beta \rangle$, as in Eq. (14.51), let us define the following quantities,

$$t_{ij} \equiv \int d\mathbf{x}\, \phi_i^*(\mathbf{x}) \left[-\frac{\hbar^2}{2m} \nabla_{\mathbf{r}}^2 + U(\mathbf{r}) \right] \phi_j(\mathbf{x}), \qquad (14.55a)$$

$$v_{ijkl} \equiv \int d\mathbf{x}d\mathbf{x}'\, \phi_i^*(\mathbf{x})\phi_j^*(\mathbf{x}') V(\mathbf{r} - \mathbf{r}') \phi_k(\mathbf{x}) \phi_l(\mathbf{x}'). \qquad (14.55b)$$

Recall that $\int d\mathbf{x} = \int d\mathbf{r} \sum_\sigma$. When U and V are spin independent, the spin summation asserts that initial and final spin states are equal.

Problem 14.7

Verify the relations: $t_{ji} = t_{ij}^*$, $v_{jikl} = v_{ijkl}$, $v_{klij} = v_{ijkl}^*$.

In terms of these integrals, the calculation of the matrix element (14.51) is straightforward albeit tedious,

$$\langle \Phi_\alpha | H | \Phi_\beta \rangle = t_{mp}\delta_{nq} + t_{nq}\delta_{mp} - t_{mq}\delta_{np} - t_{np}\delta_{nq} + \frac{1}{2}(v_{mnpq} + v_{nmqp} - v_{mnqp} - v_{nmpq}). \qquad (14.56)$$

Now let us find the second quantized Hamiltonian $\hat{H}$ guided by Eq. (14.52). It is useful to distinguish the kinetic and single-particle potential terms T_i and $U(\mathbf{r}_i)$ from the interaction potential $V(\mathbf{r}_1 - \mathbf{r}_2)$. The former are one-body operators O_1, while the latter are two-body operators O_2.

Consider two basis vectors in $\mathcal{F}_2$, $|\alpha\rangle$ and $|\beta\rangle$. If $|\alpha\rangle$ differs from $|\beta\rangle$ by more than a single component, the matrix elements $\langle \alpha | O_1 | \beta \rangle$ of a single-particle operator O_1 vanish identically. Similarly, if $|\alpha\rangle$ differs from $|\beta\rangle$ by more than two components, the matrix elements $\langle \alpha | O_2 | \beta \rangle$ of a two-body operator O_2 vanish. When expressed in terms of annihilation and creation operators, single-particle operators are quadratic (any product contains two operators), while two-body operators are quartic (any product contains four operators).

Since we assumed $m > n$ and $p > q$, according to the phase convention (14.39), we have $|\alpha\rangle = c_n^\dagger c_m^\dagger |0\rangle$ and $|\beta\rangle = c_q^\dagger c_p^\dagger |0\rangle$. The single-particle operators associated with t_{ij} can be written as,

$$\hat{H}_0 \equiv \sum_{i=m,n} \sum_{j=p,q} t_{ij} c_i^\dagger c_j, \qquad (14.57)$$

hence,

$$\langle \alpha | \hat{H}_0 | \beta \rangle = \langle 0 | (c_m c_n) [t_{mp} c_m^\dagger c_p + t_{mq} c_m^\dagger c_q + t_{np} c_n^\dagger c_p + t_{nq} c_n^\dagger c_q] (c_q^\dagger c_p^\dagger) | 0 \rangle. \qquad (14.58)$$

14.1 Second Quantization

Consider the first term, $\langle 0|(c_m c_n)t_{mp}c_m^\dagger c_p(c_q^\dagger c_p^\dagger)|0\rangle$. Since $m \neq n$ and $p \neq q$, we can swap $c_m^\dagger$ with c_n and c_p with $c_q^\dagger$, incurring two sign changes. With the help of the analysis of the previous section, this gives, $t_{mp}\langle 0|c_m c_m^\dagger c_n c_q^\dagger c_p c_p^\dagger|0\rangle = t_{mp}\langle 0|c_n c_q^\dagger|0\rangle = t_{mp}\delta_{nq}$.

> **Problem 14.8**
>
> Work out the second term in Eq. (14.58) and show that it is equal to $t_{nq}\delta_{mp}$. In this case, there is no need to move the operators.

Now consider the third term, $\langle 0|(c_m c_n)t_{np}c_n^\dagger c_p(c_q^\dagger c_p^\dagger)|0\rangle$. Here, c_n and $c_n^\dagger$ are already adjacent and can be removed. On the other hand, we need to move c_p across $c_q^\dagger$ to put it near $c_p^\dagger$, and this results in a sign change and yields, $-t_{np}\langle 0|c_m c_q^\dagger|0\rangle = -t_{nq}\delta_{mq}$. Similarly, the fourth term yields $-t_{nq}\delta_{mp}$. Thus, we have shown that

$$\underbrace{\langle \alpha | \hat{H}_0 | \beta \rangle}_{\text{Second quantization}} = \underbrace{\langle \Phi_\alpha | H_0 | \Phi_\beta \rangle}_{\text{First quantization}} . \quad (14.59)$$

We now turn to the two-body operators, which will be constructed so that their matrix elements yield the last four terms in Eq. (14.56).

> **Problem 14.9**
>
> Show that $\langle \alpha | c_m^\dagger c_n^\dagger c_q c_p | \beta \rangle = \langle 0|(c_m c_n)c_m^\dagger c_n^\dagger c_q c_p (c_q^\dagger c_p^\dagger)|0\rangle = 1$.

Using this result, the first of these terms in Eq. (14.56) is obtained as,

$$\langle \alpha | v_{mnpq} c_m^\dagger c_n^\dagger c_q c_p | \beta \rangle = v_{mnpq}.$$

Note that the order of operators implied by their subscripts $mnqp$ does not follow the order of indices $mnpq$ appearing on the potential matrix element. The other terms are obtained similarly using the operators $v_{ijkl}c_i^\dagger c_j^\dagger c_l c_k$ (remember that for fermions $i \neq j$ and $k \neq l$). According to this prescription, the third term (which is an exchange term) becomes,

$$\langle \alpha | v_{mnqp} c_m^\dagger c_n^\dagger c_p c_q | \beta \rangle = v_{mnqp}\langle 0|(c_m c_n)c_m^\dagger c_n^\dagger c_p c_q (c_q^\dagger c_p^\dagger)|0\rangle = -v_{mnqp}.$$

The last equality is due to the fact that only one swap (between c_n and $c_m^\dagger$) is required to bring the matrix element to unity, and it costs a single minus sign.

Although our discussion referred specifically to a two-particle system, $N = 2$, the result is general, because $\mathcal{H}$ consists of only one- and two-body operators, and the matrix elements of $\hat{H}$ operating in $\mathcal{F}_N$ between states $|\alpha\rangle$ and $|\beta\rangle$ are calculated exactly as for the two-particle system. Combining this result with Eq. (14.57), we arrive at the central result,

$$\boxed{\hat{H} = \sum t_{ij} c_i^\dagger c_j + \frac{1}{2} \sum_{i \neq j; k \neq l} v_{ijkl} c_i^\dagger c_j^\dagger c_l c_k \equiv \hat{H}_0 + \hat{V}.} \quad (14.60)$$

Once again we stress the different orders of indices in the second term involving the two-body interaction: $ijkl$ for v and $ijlk$ for the operator product. The first part on the RHS of Eq. (14.60) is the single-particle term, that is quadratic (it contains products of two-particle operators, compared with the second term that contains products of four-particle operators, hence, it is quartic).

A quadratic Hamiltonian can always be diagonalized by performing a unitary transformation on the creation and the annihilation operators. Denoting the column array of the annihilation operators by $\mathbf{c}$, the row vector of the creation operators by $\mathbf{c}^\dagger$, and the hermitian matrix with elements $\{t_{ij}\}$ by $\mathbf{t}$, we use the unitary matrix U that diagonalizes $\mathbf{t}$ to perform the transformation,

$$\mathbf{c} = U\boldsymbol{\gamma}, \quad \mathbf{c}^\dagger = \boldsymbol{\gamma}^\dagger U^\dagger. \tag{14.61}$$

The new creation and annihilation operators, $\gamma_i^\dagger$ and γ_i, do not represent the physical particles that are represented by the operators $c_i^\dagger$ and c_i but they represent fictitious fermions (or quasi-particles) because the anticommutation relations are preserved,

$$\{\gamma_i, \gamma_j^\dagger\} = \{c_i, c_j^\dagger\} = \delta_{ij}, \tag{14.62}$$

as can be proven. Any transformation that preserves commutation relations for boson or anticommutation relations for fermions is referred to as *canonical transformation*. The single-particle term on the RHS of Eq. (14.60) is then written as,

$$\sum t_{ij} c_i^\dagger c_j = \sum_i \lambda_i \gamma_i^\dagger \gamma_i, \tag{14.63}$$

where $\{\lambda_i\}$ are the (real) eigenvalues of the matrix $\mathbf{t}$. In this form, the single-particle Hamiltonian has been diagonalized.

Problem 14.10

For $N = 3$ particles, specify the nonzero matrix elements of H_0 and V between states, $|\alpha\rangle = |11100\ldots\rangle$ and $|\beta\rangle = |0011100\ldots\rangle$.

A note on notation:
After the procedure for ordering single-particle quantum numbers is taken care of, and a given ordering scheme is established, it is still simpler and more transparent to index annihilation and creation operators in terms of the relevant physical quantum numbers instead of the integer numbers that represent them. For example, consider an electron in a 3D box with periodic boundary conditions. The quantum numbers are $\mathbf{k}\sigma = (k_x, k_y, k_z, \sigma)$. This group of quantum numbers is represented by some integer q that indicates the location of this state in a Fock space basis vector. But it is much more reasonable to use the notation $c_{\mathbf{k}\sigma}^\dagger$ and $c_{\mathbf{k}\sigma}$ instead of $c_q^\dagger$ and c_q.

Example 1: Three Spinless Fermions

Consider three spinless fermions of mass m confined to move in one dimension in the segment $-\frac{L}{2} \leq x \leq \frac{L}{2}$. The wave function $\Psi(x_1, x_2, x_3)$ is required to be antisymmetric and periodic with period L. As a basis set in the Hilbert space $\mathcal{H}$, it is natural to use the eigenstate of the kinetic energy operator for constructing Slater determinants. (a) If there is no interaction between the particles, write the ground-state wave function $\Psi_0(x_1, x_2, x_3)$ as a Slater determinant and find the ground-state energy. (b) Assume a two-body interaction $V(|x_i - x_j|) = V_0 L \delta(x_i - x_j)$. (Here, V_0 is a constant, and the factor L is inserted to guarantee that V_0 has a dimension of energy). Calculate the matrix element V_{ijkl} as in Eq. (14.55). (c) Write the Hamiltonian H in second quantized form.

Solution: (a) The kinetic energy part of the Hamiltonian is $H_0 = -\frac{\hbar^2}{2m} \sum_{i=1}^3 \frac{\partial^2}{\partial x_i^2}$. The normalized periodic single-particle wave functions are eigenfunctions of the kinetic energy operator $-\frac{\hbar^2}{2m} \frac{\partial^2}{\partial x^2}$, and the corresponding energies are

$$\psi_k(x) = \frac{1}{\sqrt{L}} e^{ikx}, \quad k = \frac{2\pi}{L} n_k, \quad n_k = 0, \pm 1, \ldots \quad \varepsilon_k = \frac{\hbar^2}{2m} k^2. \tag{14.64}$$

14.1 Second Quantization

The three lowest single-particle energies correspond to $n_k = 0 \pm 1$. We order the spectrum (somewhat arbitrarily) as $n_k = 0 \to n = 1, n_k = 1 \to n = 2$, and $n_k = -1 \to n = 3$; hence,

$$\phi_1(x) = \frac{1}{\sqrt{L}}, \quad \phi_2(x) = \frac{1}{\sqrt{L}} e^{ikx}, \quad \phi_3(x) = \frac{1}{\sqrt{L}} e^{-ikx}. \tag{14.65}$$

Thus, the ground-state wave function Ψ_0 and ground-state energy E_0 are

$$\Psi_0(x_1, x_2, x_3) = \frac{1}{\sqrt{3!}} \begin{vmatrix} \phi_1(x_1) & \phi_1(x_2) & \phi_1(x_3) \\ \phi_2(x_1) & \phi_2(x_2) & \phi_2(x_3) \\ \phi_3(x_1) & \phi_3(x_2) & \phi_3(x_3) \end{vmatrix}, \quad E_0 = 2\frac{\hbar^2}{2m}k^2. \tag{14.66}$$

(b) From Eq. (14.55), we find,

$$V_{k_1 k_2 k_3 k_4} = \frac{V_0}{L} \int_{-\frac{L}{2}}^{\frac{L}{2}} e^{-i(k_1 x_1 + k_2 x_2)} \delta(x_1 - x_2) e^{i(k_3 x_1 + k_4 x_2)} dx_1 dx_2$$

$$= \frac{V_0}{2} \delta_{k_3 + k_4 - k_1 - k_2}. \tag{14.67}$$

The Kronecker delta function in Eq. (14.67) reflects the translation invariance of the two-body potential.
(c) Momentum conservation yields the Kronecker delta function, and we can set $k_3 = k_1 + q$ and $k_4 = k_2 - q$ and thereby obtain

$$\hat{H} = \sum_k \varepsilon_k c_k^\dagger c_k + \frac{V_0}{2} \sum_{k_1 k_2 q} c_{k_1}^\dagger c_{k_2}^\dagger c_{k_2 - q} c_{k_1 + q}. \tag{14.68}$$

Example 2: Electrons with Screened Coulomb Interaction

A system of N electrons (charge $-e$, mass m, and spin $1/2$) occupy a box of volume $\mathcal{V} = L^3$. The many-electron wave function is required to be antisymmetric and periodic with period L in all three directions. First, we consider free (noninteracting) electrons, calculate the ground-state energy for $N = 14$ electrons, and write the corresponding Slater determinant. Second, assuming a *screened Coulomb interaction*,

$$V(\mathbf{r}_i - \mathbf{r}_j) = \frac{e^2}{|\mathbf{r}_i - \mathbf{r}_j|} e^{-\kappa |\mathbf{r}_i - \mathbf{r}_j|}, \tag{14.69}$$

write the second quantized Hamiltonian $\hat{H}$ and discuss the limit $\kappa \to 0$.
Solution: (a) Let us write the Hamiltonian in first quantization,

$$H = -\frac{\hbar^2}{2m} \sum_{i=1}^N \nabla_i^2 + \frac{e^2}{2} \sum_{i \neq j=1}^N \frac{e^{-\kappa |\mathbf{r}_i - \mathbf{r}_j|}}{|\mathbf{r}_i - \mathbf{r}_j|}. \tag{14.70}$$

We choose a basis of eigenfunctions of the single-particle Hamiltonian $h = -\frac{\hbar^2}{2m}\nabla_i^2$. A single-particle basis wave function is a product of a plane wave and a two-component spinor. The single-particle function for particle number i is

$$\phi_{\mathbf{k}\sigma}(\mathbf{r}_i) = \frac{1}{L^{3/2}} e^{i\mathbf{k} \cdot \mathbf{r}_i} \eta_\sigma(i). \tag{14.71}$$

Here, the momenta $\mathbf{k}$ are quantized as $\mathbf{k} = \frac{2\pi}{L}(n_x, n_y, n_z)$, and η_σ ($\sigma = \uparrow, \downarrow$) is a spin function. In two-component spinor notation, $\eta_\uparrow = \binom{1}{0}$ and $\eta_\downarrow = \binom{0}{1}$. The single-particle wave functions $\phi_{\mathbf{k}\sigma}(\mathbf{r}_i)$ introduced in Eq. (14.71) are orthogonal and complete:

$$\langle \phi_{\mathbf{k}\sigma}|\phi_{\mathbf{k}'\sigma'}\rangle = \delta_{\mathbf{k}\mathbf{k}'}\delta_{\sigma\sigma'}, \quad \sum_{\mathbf{k}\sigma}\phi^*_{\mathbf{k}\sigma}(\mathbf{r})\phi_{\mathbf{k}\sigma}(\mathbf{r}') = \delta(\mathbf{r}-\mathbf{r}'). \tag{14.72}$$

(b) Since the Hamiltonian does not depend on spin, the two states $\phi_{\mathbf{k}\uparrow}(\mathbf{r})$ and $\phi_{\mathbf{k}\downarrow}(\mathbf{r})$ are degenerate. The lowest energy is obtained by identifying the seven smallest combinations $n_x^2 + n_y^2 + n_z^2$ and putting two electrons in each of them, one with spin up and one with spin down. Clearly, this is satisfied by the following set of triples, $(n_x, n_y, n_z) = (0,0,0,), (\pm 1,0,0), (0,\pm 1,0), (0,0,\pm 1)$. Denoting the seven corresponding momenta by $\mathbf{k}$ as $\mathbf{k}_i, i = 1, 2, \ldots, 7$, the Slater determinant corresponding to the ground state is composed of 14 rows and columns with

$$\text{Odd numbered rows} = \frac{1}{\sqrt{14!}}\frac{1}{L^{3/2}} e^{i\mathbf{k}_i\cdot\mathbf{r}_j}\eta_\uparrow(j),$$

$$\text{Even numbered rows} = \frac{1}{\sqrt{14!}}\frac{1}{L^{3/2}} e^{i\mathbf{k}_i\cdot\mathbf{r}_j}\eta_\downarrow(j),$$

$$i = 1, 2, \ldots, 7, \quad j = 1, 2, \ldots, 14. \tag{14.73}$$

The ground-state kinetic energy is contributed from the 12 electrons occupying the states $\mathbf{k}_i, i = 2, 3, \ldots, 7$, i.e., $E_0 = 12\,\varepsilon_{\mathbf{k}} = 6\hbar^2 k^2/m$. Now, we calculate the matrix elements of the screened potential,

$$V_{k_1\sigma_1 k_2\sigma_2 k_3\sigma_3 k_4\sigma_4} = \frac{e^2}{L^6}\int d\mathbf{r}_1 d\mathbf{r}_2\, e^{-i\mathbf{k}_1\cdot\mathbf{r}_1}e^{-i\mathbf{k}_2\cdot\mathbf{r}_2}\frac{e^{-\kappa|\mathbf{r}_1-\mathbf{r}_2|}}{|\mathbf{r}_1-\mathbf{r}_2|}e^{i\mathbf{k}_3\cdot\mathbf{r}_1}e^{i\mathbf{k}_4\cdot\mathbf{r}_2}\langle\eta_{\sigma_1}(1)|\eta_{\sigma_3}(1)\rangle\langle\eta_{\sigma_2}(2)|\eta_{\sigma_4}(2)\rangle. \tag{14.74}$$

The electron–electron interaction is translationally invariant since it depends only on the distance between the two electrons ($|\mathbf{r}_1 - \mathbf{r}_2|$). This property is employed to reduce the number of integration variables. Defining $\mathbf{r} = \mathbf{r}_2$ and $\mathbf{y} = \mathbf{r}_1 - \mathbf{r}_2$ yields,

$$V_{k_1\sigma_1 k_2\sigma_2 k_3\sigma_3 k_4\sigma_4} = \delta_{\sigma_1\sigma_3}\delta_{\sigma_2\sigma_4}\frac{e^2}{L^6}\int d\mathbf{r}\, e^{i(\mathbf{k}_3+\mathbf{k}_4-\mathbf{k}_1-\mathbf{k}_2)\cdot\mathbf{r}}\int d\mathbf{y}\, e^{i(\mathbf{k}_3-\mathbf{k}_1)\cdot\mathbf{y}}\frac{e^{-\kappa y}}{y} = \frac{4\pi e^2}{L^3}\delta_{\sigma_1\sigma_3}\delta_{\sigma_2\sigma_4}\frac{\delta_{\mathbf{k}_3+\mathbf{k}_4,\mathbf{k}_1+\mathbf{k}_2}}{(\mathbf{k}_1-\mathbf{k}_3)^2+\kappa^2}. \tag{14.75}$$

Using a similar notation as before (14.68), we finally obtain,

$$\hat{H} = \sum_{\mathbf{k}\sigma}\varepsilon_{\mathbf{k}}c^\dagger_{\mathbf{k}\sigma}c_{\mathbf{k}\sigma} + \frac{2\pi e^2}{L^3}\sum_{\mathbf{k}_1\mathbf{k}_2\mathbf{q}}\sum_{\sigma_1\sigma_2}\frac{1}{q^2+\kappa^2}c^\dagger_{\mathbf{k}_1\sigma_1}c^\dagger_{\mathbf{k}_2\sigma_2}c_{\mathbf{k}_2-\mathbf{q}\sigma_2}c_{\mathbf{k}_1+\mathbf{q}\sigma_1}. \tag{14.76}$$

In this notation, the spin and orbital indices appear separately [see remark after Eq. (14.60)].
(c) On inspecting the second term in Eq. (14.76), we see that when $\kappa \to 0$, the term with $q = 0$ is singular. This is due to the long-range nature of the Coulomb interaction and is called an *infrared divergence* (long distance is equivalent to small momentum, hence low energy). This divergence can be avoided if it is assumed that there is a background positive charge $Q = Ne$, which is distributed uniformly in the volume $\mathcal{V} = L^3$. The presence of this background guarantees charge neutrality and leads to screening of the electron–electron interaction. The upshot is that the $q = 0$ term in the second term in Eq. (14.76) should be omitted. With this proviso, Eq. (14.76) is the basis for approaching the problem of the interacting electron gas within second quantization.

Second Quantized Hamiltonian: Bosons

Having worked out the second quantized Hamiltonian for identical fermions in Eq. (14.60), it is straightforward to repeat the same procedure for identical bosons. The two differences are (1) the boson operators commute (whereas fermion operators anti-commute) and (2) the number of bosons occupying a given quantum state can be arbitrary. These

14.1 Second Quantization

considerations lead to a form similar to that in Eq. (14.60) but without the restriction $i \neq j$ and $k \neq l$. Explicitly,

$$\hat{H} = \sum t_{ij} b_i^\dagger b_j + \frac{1}{2} \sum_{ij;kl} v_{ijkl} b_i^\dagger b_j^\dagger b_l b_k. \tag{14.77}$$

Since $[b_l, b_k] = 0$, their order is immaterial. As discussed after Eq. (14.60), the indices should incorporate spin quantum numbers as well, since bosons might have nonzero (albeit integer) spin. The impressive progress recorded in the physics of cold atoms enables the creation of optical lattices where bosonic atoms with nonzero integer spin S move in a periodic lattice exactly as Bloch electrons do in metals. Therefore, within the formalism of second quantization, it is useful to write the corresponding operators as $b_{\mathbf{k}\sigma}$ and $b_{\mathbf{k}\sigma}^\dagger$ with $\sigma = -S, -S+1, \ldots, S$.

Once the second quantized Hamiltonian $\hat{H}$ is obtained, the formulation of the problem in terms of second quantization formalism is complete. Thus, we have specified the general steps that should be taken to construct the many-body problem in a second quantized form.

14.1.6 FIELD OPERATORS

For a system of identical particles in 3D, the creation operators $c_{\mathbf{k}\sigma}^\dagger$ ($b_{\mathbf{k}\sigma}^\dagger$) create a fermion (boson) of momentum $\mathbf{k}$ and spin projection σ. Similarly, the annihilation operators $c_{\mathbf{k}\sigma}$ ($b_{\mathbf{k}\sigma}$) annihilate a fermion (boson) of momentum $\mathbf{k}$ and spin projection σ. These operators are defined in the momentum representation, which together with the spin projection defines a complete set of single-particle quantum numbers. The choice $\mathbf{k}\sigma$ as a complete set is of course not unique. Instead of using the momentum representation, a position representation could be used. Thus, we can speak of an operator that creates (or annihilates) a particle at space point $\mathbf{r}$ with spin projection σ. The corresponding creation and annihilation operators are usually denoted as $\hat{\psi}_\sigma^\dagger(\mathbf{r})$ and $\hat{\psi}_\sigma(\mathbf{r})$, respectively.[3] These operators are called *field operators*. The mathematical framework for treating field operators in Hilbert space is *Quantum Field Theory*.

Once the creation and annihilation operators in the momentum representation, $c_{\mathbf{k}\sigma}^\dagger$ ($c_{\mathbf{k}\sigma}$) for fermions or $b_{\mathbf{k}\sigma}^\dagger$ ($b_{\mathbf{k}\sigma}$) for bosons, are defined, creation (annihilation) field operators are easily constructed using Fourier expansion. The coefficients are the single-particle wave functions $\phi_{\mathbf{k}\sigma}^*(\mathbf{r})$ and $\phi_{\mathbf{k}\sigma}(\mathbf{r})$. Some caution is required in handling summations over quantum numbers. For a particle in a box, the momentum quantum number $\mathbf{k}$ is discrete and the construction of Fock space vectors is straightforward. On the other hand, the quantum number $\mathbf{r}$ is continuous and this introduces a mathematical subtlety, e.g., how to define an operation as specified in Eq. (14.39) for a continuous set of single-particle quantum numbers. In most cases, however, this can be circumvented as seen below. Alternatively, one can use a tight-binding formalism where space coordinates are discrete.

Construction of Field Operators

The procedure for constructing the field operators for fermions and bosons is virtually identical; hence, there is no need for a separate discussion of each. We will denote annihilation and creation operators for either fermions or bosons in the momentum representation by $a_{\mathbf{k}\sigma}$ and $a_{\mathbf{k}\sigma}^\dagger$. However, note that the algebraic relations are distinct (commutation relations for bosons and anticommutation relations for fermions). Let us then consider, within the formalism of first quantization, a system composed of N identical particles in a box of volume $\mathcal{V}$ with the Hamiltonian (14.1). To be definite, we assume that the single-particle wave functions to be used for the construction of the field operators are eigenfunctions of the *momentum operator* $\mathbf{p}$ (i.e., plane waves) and the spin operator S_z. Note that $\phi_{\mathbf{k}\sigma}(\mathbf{r})$ has a definite spin quantum number; hence, it is not a spin orbital. The annihilation field operator and its Hermitian adjoint, the creation field operator, are obtained from the operators $a_{\mathbf{k}\sigma}$ and $a_{\mathbf{k}\sigma}^\dagger$ by the expansions,

$$\hat{\psi}_\sigma(\mathbf{r}) = \sum_{\mathbf{k}} \phi_{\mathbf{k}\sigma}(\mathbf{r}) a_{\mathbf{k}\sigma}, \quad \hat{\psi}_\sigma^\dagger(\mathbf{r}) = \sum_{\mathbf{k}} \phi_{\mathbf{k}\sigma}^*(\mathbf{r}) a_{\mathbf{k}\sigma}^\dagger. \tag{14.78}$$

---

[3] The notation used for these operators is not $c_{\mathbf{r}\sigma}, c_{\mathbf{r}\sigma}^\dagger$.

Alternatively, instead of a plane-wave basis, we may use a complete set of single-particle wave functions $\{\phi_{j\sigma}(\mathbf{r})\}$ that solve the single-particle eigenvalue problem, $\left[-\frac{\hbar^2}{2m}\nabla^2 + U(\mathbf{r})\right]\phi_{j\sigma}(\mathbf{r}) = \varepsilon_j\phi_{j\sigma}(\mathbf{r})$ and construct the field operators as in Eqs (14.78),

$$\hat{\psi}_\sigma(\mathbf{r}) = \sum_j \phi_{j\sigma}(\mathbf{r})a_{j\sigma}, \quad \hat{\psi}_\sigma^\dagger(\mathbf{r}) = \sum_{j\sigma} \phi_{j\sigma}(\mathbf{r})^* a_{j\sigma}^\dagger. \tag{14.79}$$

One easily verifies that $\hat{\psi}_\sigma(\mathbf{r})|0\rangle = 0$, where $|0\rangle$ is the particle vacuum (all occupation numbers are 0 in any representation). On the other hand, when the creation operator $\hat{\psi}_\sigma^\dagger(\mathbf{r})$ acts on the vacuum $|0\rangle$, it creates a particle at the position $\mathbf{r}$ with spin projection σ. The corresponding vector in Fock space is the infinite combination,

$$\hat{\psi}_\sigma^\dagger(\mathbf{r})|0\rangle = \sum_k \phi_{k\sigma}^*(\mathbf{r})a_{k\sigma}^\dagger|0\rangle. \tag{14.80}$$

In analogy with Eq. (14.42), the anticommutation relations for fermion field operators are

$$\{\hat{\psi}_\sigma(\mathbf{r}), \hat{\psi}_{\sigma'}^\dagger(\mathbf{r}')\} = \delta_{\sigma\sigma'}\delta(\mathbf{r}-\mathbf{r}'), \quad \{\hat{\psi}_\sigma(\mathbf{r}), \hat{\psi}_{\sigma'}(\mathbf{r}')\} = \{\hat{\psi}_\sigma^\dagger(\mathbf{r}), \hat{\psi}_{\sigma'}^\dagger(\mathbf{r}')\} = 0, \tag{14.81}$$

whereas the commutation relations for boson field operators are

$$[\hat{\psi}_\sigma(\mathbf{r}), \hat{\psi}_{\sigma'}^\dagger(\mathbf{r}')] = \delta_{\sigma\sigma'}\delta(\mathbf{r}-\mathbf{r}'), \quad [\hat{\psi}_\sigma(\mathbf{r}), \hat{\psi}_{\sigma'}(\mathbf{r}')] = [\hat{\psi}_\sigma^\dagger(\mathbf{r}), \hat{\psi}_{\sigma'}^\dagger(\mathbf{r}')] = 0. \tag{14.82}$$

The proofs of relations (14.81) and (14.82) is completed by substituting the expansions (14.78) for the field operators on the LHS, using the commutation relations (14.30) and the anticommutation relations (14.49) together with the completeness relations of the single-particle wave functions. For spin 1/2 particles, one defines two-component spinor field operators,

$$\hat{\psi}(\mathbf{r}) = \begin{pmatrix} \hat{\psi}_\uparrow(\mathbf{r}) \\ \hat{\psi}_\downarrow(\mathbf{r}) \end{pmatrix}, \quad \hat{\psi}^\dagger(\mathbf{r}) = \begin{pmatrix} \hat{\psi}_\uparrow^\dagger(\mathbf{r}) & \hat{\psi}_\downarrow^\dagger(\mathbf{r}) \end{pmatrix}. \tag{14.83}$$

Problem 14.11

Consider a system of fermions in 1D confined to an interval $0 \leq x \leq L$, such that the single-particle wave functions satisfy periodic boundary conditions, $\phi(x+L) = \phi(x)$. Denote the creation and annihilation operators in momentum space corresponding to wave number $k = 2\pi n_k/L$ and spin projection σ, where $n_k = 0, \pm 1, \pm 2, \ldots$, by $c_{k\sigma}^\dagger$ and $c_{k\sigma}$, respectively.

(a) Write expressions for the creation and annihilation field operators $\hat{\psi}_\sigma^\dagger(x)$ and $\hat{\psi}_\sigma(x)$.
(b) Express the matrix element of the momentum operator $\langle \hat{\psi} | -i\hbar\frac{d}{dx} | \hat{\psi} \rangle$ in terms of $c_{k\sigma}^\dagger$ and $c_{k\sigma}$ (note that $|\hat{\psi}\rangle$ is a spinor as defined in Eq. (14.83)).
(c) Consider spin $s = 1/2$ particles subject to a Zeeman Hamiltonian $H_Z = \mathbf{h}\cdot\mathbf{s}$. Express the matrix element $\langle \hat{\psi} | H_Z | \hat{\psi} \rangle$ in terms of $c_{k\sigma}^\dagger$ and $c_{k\sigma}$.

Answer: (a) The properly normalized single-particle wave functions are $\phi_{k\sigma}(x) = \frac{1}{\sqrt{L}}e^{ikx}|\sigma\rangle$. Using Eq. (14.79) we obtain,

$$\hat{\psi}_\sigma^\dagger(x) = \sum_k \frac{1}{\sqrt{L}}e^{-ikx}\langle\sigma|a_{k\sigma}^\dagger, \quad \hat{\psi}_\sigma(x) = \sum_k \frac{1}{\sqrt{L}}e^{ikx}|\sigma\rangle a_{k\sigma}.$$

(b) $\langle\hat{\psi}|(-i\hbar\frac{d}{dx})|\hat{\psi}\rangle = \hbar\sum_{k\sigma} k c_{k\sigma}^\dagger c_{k\sigma}$. (c) Taking $\mathbf{h}$ along the z axis, we get, $\langle\hat{\psi}|H_Z|\hat{\psi}\rangle = h\sum_{k\sigma}(c_{k\uparrow}^\dagger c_{k\uparrow} - c_{k\downarrow}^\dagger c_{k\downarrow})$.

14.1 Second Quantization

Constructing Physical Operators Using Field Operators

In Eq. (14.76), we express the Hamiltonian of fermions in terms of creation and annihilation operators in momentum space. In the following, we show how to express operators in terms of field operators. Let $\mathcal{O}(\mathbf{r})$ denote a one-particle (spin-independent and local) operator defined within the first quantization formalism. For example, $\mathcal{O}(\mathbf{r})$ may be the kinetic energy of a particle $T(\mathbf{r})$ or a single-particle potential $U(\mathbf{r})$. When this function of $\mathbf{r}$ is integrated with field operators, we get an operator defined in Fock space,

$$\hat{\mathcal{O}}_\sigma = \int d\mathbf{r}\, \hat{\psi}_\sigma^\dagger(\mathbf{r})\mathcal{O}(\mathbf{r})\hat{\psi}_\sigma(\mathbf{r}). \tag{14.84}$$

Some care should be exercised when we consider two-particle (local and spin independent) operators $\mathcal{O}(\mathbf{r} - \mathbf{r}')$ such as the two-body potentials $V(\mathbf{r} - \mathbf{r}')$. The desired expression is,

$$\hat{\mathcal{O}}_{\sigma\sigma'} = \int d\mathbf{r} d\mathbf{r}'\, \hat{\psi}_\sigma^\dagger(\mathbf{r})\hat{\psi}_{\sigma'}^\dagger(\mathbf{r}')\mathcal{O}(\mathbf{r} - \mathbf{r}')\hat{\psi}_{\sigma'}(\mathbf{r}')\hat{\psi}_\sigma(\mathbf{r}), \tag{14.85}$$

where the order of $\mathbf{r}$ and $\mathbf{r}'$ is swapped in the field operators to the right of the operator. Some important physical operators expressed in terms of field operators are listed below. For definiteness, they are written for fermion fields, but the analogous expressions for boson fields are virtually identical.

Hamiltonian: For definiteness, we will assume a translation invariant system without single-particle potential. Employing Eq. (14.84) for the kinetic energy part of the Hamiltonian, $-\frac{\hbar^2}{2m}\nabla^2$, and using Eqs (14.78) and the orthogonality of the single particle wave functions, we get,

$$\hat{T} = \int d\mathbf{r} \underbrace{\sum_{\mathbf{k}\sigma} \phi_{\mathbf{k}\sigma}(\mathbf{r})^* a_{\mathbf{k}\sigma}^\dagger}_{\hat{\psi}^\dagger(\mathbf{r})} \left[-\frac{\hbar^2}{2m}\nabla^2\right] \underbrace{\sum_{\mathbf{k}'\sigma'} \phi_{\mathbf{k}'\sigma'}(\mathbf{r}) a_{\mathbf{k}'\sigma'}}_{\hat{\psi}(\mathbf{r})} = \sum_{\mathbf{k}\sigma} t_\mathbf{k} a_{\mathbf{k}\sigma}^\dagger a_{\mathbf{k}\sigma}, \tag{14.86}$$

where $t_\mathbf{k} = \hbar^2 k^2/(2m)$. Similarly, employing Eq. (14.85) for the two-body interaction part of the Hamiltonian, $V(\mathbf{r}-\mathbf{r}')$, we get,

$$\hat{V} = \frac{1}{2}\int d\mathbf{r}d\mathbf{r}' \sum_{\mathbf{k}_1,\mathbf{k}_2,\mathbf{q}} \sum_{\sigma_1,\sigma_2} \left[\phi_{\mathbf{k}_1\sigma_1}(\mathbf{r})^* a_{\mathbf{k}_1\sigma_1}^\dagger\right]\left[\phi_{\mathbf{k}_2\sigma_2}(\mathbf{r}')^* a_{\mathbf{k}_2\sigma_2}^\dagger\right]$$

$$\times V(\mathbf{r}-\mathbf{r}')\left[\phi_{\mathbf{k}_2-\mathbf{q},\sigma_2}(\mathbf{r}')a_{\mathbf{k}_2-\mathbf{q},\sigma_2}\right]\left[\phi_{\mathbf{k}_1+\mathbf{q},\sigma_1}(\mathbf{r})a_{\mathbf{k}_1+\mathbf{q},\sigma_1}\right]$$

$$= \frac{1}{2}\sum_{\mathbf{k}_1,\mathbf{k}_2,\mathbf{q}}\sum_{\sigma_1,\sigma_2} v_{\mathbf{k}_1\mathbf{k}_2\mathbf{k}_2-\mathbf{q}\mathbf{k}_1+\mathbf{q}} a_{\mathbf{k}_1\sigma_1}^\dagger a_{\mathbf{k}_2\sigma_2}^\dagger a_{\mathbf{k}_2-\mathbf{q},\sigma_2} a_{\mathbf{k}_1+\mathbf{q},\sigma_1}, \tag{14.87}$$

where the matrix element $v_{\mathbf{k}_1\mathbf{k}_2\mathbf{k}_2-\mathbf{q}\mathbf{k}_1+\mathbf{q}}$ is defined in Eq. (14.55b). We have also used the translation invariance of the interaction potential, $V(\mathbf{r},\mathbf{r}') = V(\mathbf{r}-\mathbf{r}')$ and its spin independence. When the single-particle potential $U(\mathbf{r}) \neq 0$, the formalism is similar except that there is no translation invariance and the field operators are defined as in Eq. (14.79). With this modification, the many-body Hamiltonian, which in first quantization has the form of Eq. (14.1), has the second-quantized form

$$\boxed{\hat{H} = \sum_\sigma \int d\mathbf{r}\, \hat{\psi}_\sigma^\dagger(\mathbf{r})\left[-\frac{\hbar^2}{2m}\nabla^2 + U(\mathbf{r})\right]\hat{\psi}_\sigma(\mathbf{r}) + \frac{1}{2}\sum_{\sigma,\sigma'}\int d\mathbf{r}d\mathbf{r}'\, \hat{\psi}_\sigma^\dagger(\mathbf{r})\hat{\psi}_{\sigma'}^\dagger(\mathbf{r}')V_{\sigma\sigma'}(\mathbf{r}-\mathbf{r}')\hat{\psi}_{\sigma'}(\mathbf{r}')\hat{\psi}_\sigma(\mathbf{r}).} \tag{14.88}$$

The form of the Hamiltonian (14.88) is the same for both fermions and bosons and is equivalent to Eq. (14.60) for fermions or Eq. (14.77) for bosons. The Schrödinger equation (14.1), when written in the first quantization formalism, contains the Hamiltonian operator that operates on wave functions. In Eq. (14.88), the Hamiltonian operator operates on Fock space vectors rather than wave functions. The origin of the name *second quantization* may come from the fact that the Hamiltonian in Eq. (14.88) looks like the expectation value of the "first quantized" Hamiltonian taken between wave functions.

Gauge Invariance: The Hamiltonians (14.60) and (14.77) have the following property. If the annihilation and creation operators are multiplied by phases,

$$c_j \to e^{-i\alpha} c_j, \quad c_j^\dagger \to e^{i\alpha} c_j^\dagger, \tag{14.89}$$

(similarly for the bosonic operators b_j and $b_j^\dagger$), the Hamiltonian remains unchanged. This property is referred to as *global gauge invariance*. Similarly, the Hamiltonian (14.88) is gauge invariant under the transformation

$$\hat{\psi}_\sigma(\mathbf{r}) \to e^{-i\alpha} \hat{\psi}_\sigma(\mathbf{r}), \quad \hat{\psi}_\sigma^\dagger(\mathbf{r}) \to e^{i\alpha} \hat{\psi}_\sigma^\dagger(\mathbf{r}). \tag{14.90}$$

We will encounter situations where, after implementing some approximations to the original Hamiltonian, the approximate Hamiltonian is no longer gauge invariant. Although this might seem to be an artifact of an approximation scheme, there are situations where interacting many-body systems do lose their gauge invariance and this situation is not an artifact of some approximation, rather it has a physical origin. This type of situation is called a *broken gauge symmetry*. In some systems, broken gauge symmetry is required by reality and is accompanied by dramatic effects. Examples are superconductivity and Bose–Einstein condensation, subjects that will be discussed later on in Chapter 18 which is linked to the book web page. A new quantity called an *order parameter* emerges, which in the case of a Bose condensate is the expectation value of the field operator $\hat{\psi}_\sigma(\mathbf{r})$ in the ground state $|G\rangle$,

$$\Phi_\sigma(\mathbf{r}) = \langle G | \hat{\psi}_\sigma(\mathbf{r}) | G \rangle. \tag{14.91}$$

It has a definite phase and magnitude that encode the underlying many-body physics.

Kinetic energy in the presence of an external magnetic field: The kinetic energy operator in a (static) magnetic field $\mathbf{B}(\mathbf{r}) = \nabla \times \mathbf{A}(\mathbf{r})$, assuming the particles have charge q, is given by

$$\hat{T} = \sum_\sigma \int d\mathbf{r}\, \hat{\psi}_\sigma^\dagger(\mathbf{r}) \left[\frac{1}{2m} \left(-i\hbar \nabla - \frac{q}{c} \mathbf{A}(\mathbf{r}) \right)^2 \right] \hat{\psi}_\sigma(\mathbf{r}). \tag{14.92}$$

Particle density: The density operator for spin projection σ at point $\mathbf{r}$ is,

$$\rho_\sigma(\mathbf{r}) = \int d\mathbf{r}'\, \hat{\psi}_\sigma^\dagger(\mathbf{r}') \delta(\mathbf{r} - \mathbf{r}') \hat{\psi}_\sigma(\mathbf{r}') = \hat{\psi}_\sigma^\dagger(\mathbf{r}) \hat{\psi}_\sigma(\mathbf{r}), \tag{14.93}$$

which counts the number of identical particles (fermions or bosons) with spin projection σ located at point $\mathbf{r}$ in space. The total density operator is,

$$\rho(\mathbf{r}) = \sum_\sigma \rho_\sigma(\mathbf{r}). \tag{14.94}$$

Problem 14.12

(a) Show that $\rho(\mathbf{r}) = \frac{1}{V} \sum_{\mathbf{k}\mathbf{q}\sigma} c_{\mathbf{k}\sigma}^\dagger c_{\mathbf{k}+\mathbf{q}\sigma} e^{i\mathbf{q}\cdot\mathbf{r}}$ (the volume appears in the definition (14.71) of the plane-wave spinors).

(b) Show that the Fourier transform of $\rho(\mathbf{r})$ is

$$\rho(\mathbf{q}) = \sum_{\mathbf{k}\sigma} c_{\mathbf{k}\sigma}^\dagger c_{\mathbf{k}+\mathbf{q}\sigma}. \tag{14.95}$$

14.1 Second Quantization

(c) Evaluate the density–density commutation relations $[\rho(\mathbf{r}), \rho(\mathbf{r}')]$ and $[\rho(\mathbf{q}), \rho(\mathbf{q}')]$.

Answer: $[\rho_\sigma(\mathbf{r}), \rho_\sigma(\mathbf{r}')] = [\rho_\sigma(\mathbf{q}), \rho_\sigma(\mathbf{q}')] = 0.$

Matter current density: The matter current density operator for particles of spin projection σ at space point $\mathbf{r}$ is denoted as $\mathbf{J}_\sigma(\mathbf{r})$. It is derived from the classical expression $\mathbf{J} = n(\mathbf{r})\mathbf{v}$, where $\mathbf{v} = (\mathbf{p} - (q/c)\mathbf{A})/m$ is the velocity operator that contains the vector potential $\mathbf{A}$ if there is an external electromagnetic field present. The Hermitian current density operator is

$$\mathbf{J}_\sigma(\mathbf{r}) = \frac{1}{2}\int d\mathbf{r}'\,\hat{\psi}_\sigma^\dagger(\mathbf{r}')\{\delta(\mathbf{r}-\mathbf{r}'), \mathbf{v}\}\hat{\psi}_\sigma(\mathbf{r}')$$

$$= \frac{\hbar}{2im}\left[\hat{\psi}_\sigma^\dagger(\mathbf{r})\left(\nabla\hat{\psi}_\sigma(\mathbf{r})\right) - \left(\nabla\hat{\psi}_\sigma^\dagger(\mathbf{r})\right)\hat{\psi}_\sigma(\mathbf{r})\right] - \frac{q}{mc}\mathbf{A}(\mathbf{r})\hat{\psi}_\sigma^\dagger(\mathbf{r})\hat{\psi}_\sigma(\mathbf{r})$$

$$\equiv \mathbf{J}_\sigma^\mathrm{p}(\mathbf{r}) + \mathbf{J}_\sigma^\mathrm{A}(\mathbf{r}). \tag{14.96}$$

Problem 14.13

(a) Show that the Fourier transform of $\mathbf{J}_\sigma^\mathrm{p}(\mathbf{r})$ is

$$\mathbf{J}_\sigma^\mathrm{p}(\mathbf{q}) = \frac{\hbar}{m}\sum_\mathbf{k}\left(\mathbf{k} + \frac{1}{2}\mathbf{q}\right) c_{\mathbf{k}\sigma}^\dagger c_{\mathbf{k}+\mathbf{q}\sigma}. \tag{14.97}$$

(b) Show that the Fourier transform of $\mathbf{J}_\sigma^\mathrm{A}(\mathbf{r})$ is

$$\mathbf{J}_\sigma^\mathrm{A}(\mathbf{q}) = -\frac{q}{mc}\mathbf{A}(\mathbf{q})\sum_\mathbf{k} c_{\mathbf{k}\sigma}^\dagger c_{\mathbf{k}+\mathbf{q}\sigma}. \tag{14.98}$$

Spin Density: Let us denote the spin density operator as $\mathbf{S}(\mathbf{r})$. There are many situations where the potentials appearing in the Hamiltonian depend on spin. This is the case, for example, when there is an external magnetic field (through the Zeeman effect, see Chapter 4) or when there is a strong spin–orbit interaction (see Chapter 9, Subsec. 9.6.6). In that case, an important quantity is the local spin density, which gives the magnetic moment per unit volume at space point $\mathbf{r}$. The corresponding operator is given by [recall that the field operator $\hat{\psi}(\mathbf{r})$ is a two-component spinor, see Eq. (14.83)],

$$\mathbf{S}(\mathbf{r}) = \frac{\hbar}{2}\hat{\psi}^\dagger(\mathbf{r})\,\boldsymbol{\sigma}\,\hat{\psi}(\mathbf{r}) = \frac{\hbar}{2}\sum_{\mathbf{kq}}\sum_{\mu,\nu} e^{i\mathbf{q}\cdot\mathbf{r}} c_{\mathbf{k}\mu}^\dagger [\boldsymbol{\sigma}]_{\mu\nu} c_{\mathbf{k}+\mathbf{q}\nu}, \tag{14.99}$$

where $\boldsymbol{\sigma} = (\sigma_x, \sigma_y, \sigma_z)$ is the vector of Pauli matrices.

Problem 14.14

(a) Prove the second equality in Eq. (14.99).
(b) Calculate the commutator $[S_x(\mathbf{r}), S_y(\mathbf{r}')]$.

Answer: (b) $[S_x(\mathbf{r}), S_y(\mathbf{r}')] = i\hbar\delta(\mathbf{r}-\mathbf{r}')S_z(\mathbf{r}).$

Spin Current Density: The spin current density operator will be denoted as $\mathbf{J}_\alpha^{(s)}(\mathbf{r})$. In analogy with the charge current defined in Eq. (14.96), we can formally define a spin current density operator,

$$\mathbf{J}_\alpha^{(s)}(\mathbf{r}) = \frac{\hbar}{4im}\left\{\hat{\psi}^\dagger(\mathbf{r})[\sigma_\alpha\nabla\hat{\psi}(\mathbf{r})] - [\sigma_\alpha\nabla\hat{\psi}^\dagger(\mathbf{r})]\hat{\psi}(\mathbf{r})\right\}, \tag{14.100}$$

where $\alpha = x, y, z$. The spin current density $\mathbf{J}^{(s)}$ is a tensor (it has two Cartesian indices, $J^{(s)}_{\alpha,i}(\mathbf{r})$, one for the direction of flow, i, and the other for the direction of spin projection, α). An important difference between the current density and the spin current density is that charge is conserved; hence, there is a continuity equation relating charge current and charge density, but spin is not always conserved, hence, the existence of a continuity equation relating spin and current densities is not necessarily guaranteed.

Time-Dependence of Operators

In the Schrödinger representation, operators are usually time-independent (the case of a potential that is time-dependent clearly an exception). An operator $\mathcal{O}$ in the Schrödinger representation is transformed into an operator $\mathcal{O}_H(t)$ in the Heisenberg representation by

$$\mathcal{O}_H(t) = e^{iHt/\hbar}\mathcal{O}e^{-iHt/\hbar}, \tag{14.101}$$

where H is the Hamiltonian of the system (which is here assumed to be time independent). The operator $\mathcal{O}_H(t)$ appearing on the LHS is given in the *Heisenberg representation*, whereas the operator $\mathcal{O}$ appearing on the RHS is written in the *Schrödinger representation* or *Schrödinger picture*. When the Hamiltonian of a system is time independent, no confusion may arise if we use either $O_H(t)$ or simply $O(t)$ to denote an operator in the Heisenberg representation. The main properties of this transformation can be derived directly from the definition.

(1) If $[\mathcal{O}, H] = 0$, then $\mathcal{O}_H(t) = \mathcal{O}$ (e.g., when $\mathcal{O} = H$).
(2) The operator $\mathcal{O}_H(t)$ satisfies the *Heisenberg equation of motion*,

$$i\hbar\frac{d\mathcal{O}_H(t)}{dt} = [\mathcal{O}_H(t), H]. \tag{14.102}$$

A useful tool for evaluating transformations like in Eq. (14.101) is the *Baker–Hausdorff formula*: If G is Hermitian operator and A is any operator, then

$$e^{iG\lambda}Ae^{-iG\lambda} = A + i\lambda[G, A] + \frac{(i\lambda)^2}{2!}[G, [G, A]] + \cdots + \frac{(i\lambda)^n}{n!}[G, [G, [G, \ldots [G, A]]] + \cdots \tag{14.103}$$

Problem 14.15

Let $H = \varepsilon b^\dagger b + \eta(b + b^\dagger)$. Find $b_H(t)$ in terms of $b_H(0) = b$.
Answer: $b_H(t) = \frac{\eta}{\varepsilon} + e^{-i\varepsilon t/\hbar}[b - \frac{\eta}{\varepsilon}]$.

Field Operators Within the Tight-Binding Formalism

A useful method for avoiding the mathematical subtleties that arise in the case of continuous quantum variables, such as the position variable $\mathbf{r}$ within a field operator $\hat{\psi}_\sigma(\mathbf{r})$, is to discretize space. This has physical justification in solid-state systems where electrons are tightly bound to the atoms. Hence, this approach is referred to as tight-binding formalism. Let us imagine, for simplicity, a 1D system such that the position coordinate x is restricted to the lattice,

$$x_n = na, \quad a = \text{lattice constant}, \quad n = 0, \pm 1, \pm 2, \ldots \tag{14.104}$$

14.1 Second Quantization

The notation for field operators, potential energies, etc. [see Eq. (14.1)], are modified as follows:

$$\hat{\psi}_\sigma(\mathbf{r}) \to \hat{\psi}_{n\sigma}, \quad \hat{\psi}_\sigma^\dagger(\mathbf{r}) \to \hat{\psi}_{n\sigma}^\dagger,$$
$$U(\mathbf{r}) \to v_n, \quad V(|\mathbf{r} - \mathbf{r}'|) \to V_{|n-m|},$$
$$-\frac{\hbar^2}{2m}\nabla^2 \hat{\psi}_\sigma(\mathbf{r}) \to -\frac{\hbar^2}{2ma^2}(\hat{\psi}_{n+1\sigma} - 2\hat{\psi}_{n\sigma} + \hat{\psi}_{n-1\sigma}). \quad (14.105)$$

Defining $t \equiv \frac{\hbar^2}{2ma^2}$, we arrive at the second quantized form of the so-called *tight-binding Hamiltonian*. In terms of these field operators the Hamiltonian is given by

$$H_{TB} = -t\sum_{n\sigma}\left(\hat{\psi}_{n\sigma}^\dagger \hat{\psi}_{n+1\sigma} + \hat{\psi}_{n+1\sigma}^\dagger \hat{\psi}_{n\sigma}\right) + \sum_{n\sigma} v_n \hat{\psi}_{n\sigma}^\dagger \hat{\psi}_{n\sigma} + \frac{1}{2}\sum_{nm\sigma_1\sigma_2} V_{|n-m|}\hat{\psi}_{n\sigma_1}^\dagger \hat{\psi}_{m\sigma_2}^\dagger \hat{\psi}_{m\sigma_2}\hat{\psi}_{n\sigma_1}. \quad (14.106)$$

Moreover, the algebra of field operators summarized in Eqs (14.81) and (14.82) is repeated here with $\delta(\mathbf{r} - \mathbf{r}') \to \delta_{nn'}$. Let us solve the tight-binding Schrödinger equation for spinless free electrons ($v_n = V_{|n-m|} = 0$) for the cases of (1) a finite 1D system between $n = 0$ and $n = N - 1$, with periodic boundary conditions, and (2) for a finite 1D system with open boundary conditions. The solution is as follows.

(1) We look for a solution of the eigenvalue problem

$$H_0|\phi\rangle = -t\sum_n\left(\hat{\psi}_n^\dagger \hat{\psi}_{n+1} + \hat{\psi}_{n+1}^\dagger \hat{\psi}_n\right)|\phi\rangle = \varepsilon|\phi\rangle, \quad (14.107)$$

where the eigenfunction $|\phi\rangle$ is a vector in Fock space. By *periodic boundary conditions* for a system of N sites ($x_n = na$, $n = 0, 1, 2, \ldots, N - 1$), we mean

$$\hat{\psi}_{n+N} = \hat{\psi}_n, \quad \hat{\psi}_{n+N}^\dagger = \hat{\psi}_n^\dagger. \quad (14.108)$$

Now we use translation invariance and consider transformation of the field operators into a new set of operators c_k and $c_k^\dagger$,

$$\hat{\psi}_n = \frac{1}{\sqrt{N}}\sum_k c_k e^{ikn}, \quad \hat{\psi}_n^\dagger = \frac{1}{\sqrt{N}}\sum_k c_k^\dagger e^{-ikn}. \quad (14.109)$$

Periodic boundary conditions require quantization of k as, $k = \frac{2\pi n_k}{N}$, where $n_k = 1, 2, \ldots, N$. Note that this is a *canonical transformation* since it leaves the anticommutation (and commutation) relations unchanged. Substituting into Eq. (14.107) and using the orthogonality relation of the discrete set of plane waves, we obtain the diagonal form of the Hamiltonian,

$$H = \sum_k (-2t\cos k) c_k^\dagger c_k. \quad (14.110)$$

Hamiltonian operators written in the form appearing on the RHS of Eq. (14.110) are already diagonal, and the coefficients of the number operators $c_k^\dagger c_k$ are the eigenenergies. The explicit solution of Eq. (14.107) is,

$$|\phi_k\rangle = c_k^\dagger|0\rangle = \frac{1}{\sqrt{N}}\sum_n e^{-ikn}\hat{\psi}_n^\dagger|0\rangle, \quad \varepsilon_k = -2t\cos k. \quad (14.111)$$

The second equation specifies the dispersion relation. When k varies continuously, say between $-\pi$ and $+\pi$, then the allowed energies form a band,

$$-2t \leq \varepsilon_k \leq 2t. \quad (14.112)$$

(2) By *free boundary conditions* for a finite system of sites ($x_n = na$, $n = 0, 1, 2, \ldots, N + 1$), we mean the following: suppose we expand the wave function,

$$|\phi\rangle = \sum_{n=0}^{N+1} \alpha_n \hat{\psi}_n^\dagger |0\rangle, \quad (14.113)$$

where $\{\alpha_n\}$ is a set of (possibly complex) numerical coefficients whose knowledge is equivalent to the knowledge of $|\phi\rangle$. As will be shown below, this will result in a set of linear equations for the coefficients $\{\alpha_n\}$. The solution must satisfy the conditions $\alpha_0 = \alpha_{N+1} = 0$. The requirement of free boundary conditions breaks translation invariance, and the canonical transformation implied in Eq. (14.109) is not applicable. Substitute the expansion (14.113) into the Schrödinger equation (14.107) and recall that

$$\hat{\psi}_n^\dagger \hat{\psi}_{n+1} \hat{\psi}_m^\dagger |0\rangle = \delta_{m,n+1} \hat{\psi}_n^\dagger |0\rangle,$$
$$\hat{\psi}_n^\dagger \hat{\psi}_{n-1} \hat{\psi}_m^\dagger |0\rangle = \delta_{m,n-1} \hat{\psi}_n^\dagger |0\rangle, \tag{14.114}$$

to obtain

$$-t \sum_{n=1}^{N} (\alpha_{n+1} + \alpha_{n-1}) \hat{\psi}_n^\dagger |0\rangle = \varepsilon \sum_{n=1}^{N} \alpha_n \hat{\psi}_n^\dagger |0\rangle, \tag{14.115}$$

with $\alpha_0 = \alpha_{N+1} = 0$. If we now apply $\langle 0|\hat{\psi}_n$ on the left to obtain a set of equations for the coefficients,

$$-t(\alpha_{n+1} + \alpha_{n-1}) = \varepsilon \alpha_n, \tag{14.116}$$

with $\alpha_0 = \alpha_{N+1} = 0$. This eigenvalue problem is solved by eigenvectors $\boldsymbol{\alpha}_k = (\alpha_{1k}, \alpha_{2k}, \ldots, \alpha_{Nk})$ and eigenvalues ε_k with $(k = 1, 2, \ldots, N)$, which are given explicitly as

$$\alpha_{nk} = \sqrt{\frac{2}{N}} \sin \frac{n\pi k}{N+1}, \quad \varepsilon_k = -2t \cos \frac{\pi k}{N+1}, \tag{14.117}$$

with $n, k = 1, 2, \ldots, N$. If we substitute $n = 0$ or $n = N + 1$ into the relation for α_{nk} above, we get the requirement of free boundary conditions, $\alpha_0 = \alpha_{N+1} = 0$.

14.1.7 ELECTROMAGNETIC FIELD QUANTIZATION: PHOTONS

The dual nature of particles and waves suggests that the electromagnetic field, which classically takes the form of electromagnetic waves, can also be represented as particles, or light quanta, which are called *photons*. The formalism of second quantization treats photons as bosonic particles that are created and annihilated in terms of photon creation and annihilation operators that act on the Fock space states that are labeled by photon occupation numbers. The formulation worked out below for the quantization of the electromagnetic field resembles the construction of raising and lowering operators for the quantum harmonic oscillator. Since photons can be absorbed and emitted, this is a remarkable instance where the advantage of the formalism of second quantization shows up in systems that do not conserve the number of particles. The representation of the electromagnetic field in terms of photons is referred to as *quantum electrodynamics*. It forms the basis for quantum description of the interaction of light with matter. Quantum electrodynamics is the most stringently tested theory in physics, and its predictions are the most accurate and precise available in the physical sciences.

The starting point is Maxwell's equations. The electric and magnetic fields can be written in terms of the scalar and the vector potentials, $\phi(\mathbf{r}, t)$ and $\mathbf{A}(\mathbf{r}, t)$,

$$\mathbf{E} = -\nabla \phi - \frac{1}{c} \frac{\partial \mathbf{A}}{\partial t}, \quad \mathbf{H} = \nabla \times \mathbf{A}. \tag{14.118}$$

Using gauge freedom, the vector potential can be chosen to satisfy the *Coulomb gauge* condition, $\nabla \cdot \mathbf{A} = 0$. If no charges and currents are present in the spatial region of interest, the electric field $\mathbf{E}$ and magnetic field $\mathbf{H}$ satisfy the Faraday and Ampère equations in vacuum (in Gaussian units),

$$\nabla \times \mathbf{E} = -\frac{1}{c} \frac{\partial \mathbf{H}}{\partial t}, \quad \nabla \times \mathbf{H} = \frac{1}{c} \frac{\partial \mathbf{E}}{\partial t}. \tag{14.119}$$

14.1 Second Quantization

These equations can be cast as a wave equation for $\mathbf{A}$,

$$\nabla^2 \mathbf{A} - \frac{1}{c^2} \frac{\partial^2 \mathbf{A}}{\partial t^2} = 0, \tag{14.120}$$

supplemented by the Coulomb gauge condition $\nabla \cdot \mathbf{A} = 0$. The wave equation for $\mathbf{A}(\mathbf{r},t)$ suggests an expansion in terms of traveling waves $e^{i(\mathbf{k}\cdot\mathbf{r} - \omega_k t)}$ with wave vector $\mathbf{k}$ and frequency $\omega_k = ck$. If the field is contained in a box of volume $\mathcal{V} = L^3$ and periodic boundary conditions are imposed then the wave number is quantized, such that $\mathbf{k} = \frac{2\pi}{L}\mathbf{n}$, where $\mathbf{n} = (n_x, n_y, n_z)$ (integers), and a real solution of Eq. (14.120) can be written in the form,

$$\mathbf{A}_{\mathbf{k}\lambda}(\mathbf{r},t) = \frac{c}{\sqrt{\mathcal{V}}} \hat{\boldsymbol{\epsilon}}_\lambda(\mathbf{k}) \left(A_{\mathbf{k}\lambda}(t) e^{i\mathbf{k}\cdot\mathbf{r}} + A^*_{\mathbf{k}\lambda}(t) e^{-i\mathbf{k}\cdot\mathbf{r}} \right), \tag{14.121}$$

where

$$A_{\mathbf{k}\lambda}(t) = A_{\mathbf{k}\lambda} e^{-i\omega_k t}. \tag{14.122}$$

The amplitudes $A_{\mathbf{k}\lambda}$ are constants, and the two real and orthogonal unit *polarization vectors* $\hat{\boldsymbol{\epsilon}}_\lambda(\mathbf{k})$ with $\lambda = 1, 2$, determine the polarization of $\mathbf{A}(\mathbf{r},t)$ given the Coulomb gauge condition. The field expansion takes the form,

$$\mathbf{A}(\mathbf{r},t) = \sum_{\mathbf{k}\lambda} \mathbf{A}_{\mathbf{k}\lambda}(\mathbf{r},t), \tag{14.123}$$

and the Coulomb gauge condition implies,

$$\nabla \cdot \mathbf{A} = 0 \implies \hat{\mathbf{k}} \cdot \hat{\boldsymbol{\epsilon}}_\lambda(\mathbf{k}) = 0. \tag{14.124}$$

The three unit vectors $\hat{\mathbf{k}}, \hat{\boldsymbol{\epsilon}}_1(\mathbf{k}), \hat{\boldsymbol{\epsilon}}_2(\mathbf{k})$ are mutually orthogonal. In classical electrodynamics, the field energy in volume $\mathcal{V}$ is

$$H_{\text{EM}} = \frac{1}{8\pi} \int d\mathbf{r}\, (\mathbf{E}^2 + \mathbf{H}^2) = \frac{1}{8\pi} \int d\mathbf{r} \left[\frac{1}{c^2} \dot{\mathbf{A}}^2 + (\nabla \times \mathbf{A})^2 \right]. \tag{14.125}$$

In analogy with the quantum harmonic oscillator, we define the analogs of oscillator coordinates and momenta,

$$Q_{\mathbf{k}\lambda}(t) = \frac{\sqrt{4\pi}}{c}[A_{\mathbf{k}\lambda}(t) + A^*_{\mathbf{k}\lambda}(t)], \quad P_{\mathbf{k}\lambda}(t) = -i\frac{\omega_k \sqrt{4\pi}}{c}[A_{\mathbf{k}\lambda}(t) - A^*_{\mathbf{k}\lambda}(t)]. \tag{14.126}$$

Substituting the expansion (14.123) into expression (14.125) for the energy, integrating over the volume $\mathcal{V}$, and using the orthogonality of the plane waves yields the expression for the energy as a sum of independent oscillators,

$$H_{\text{EM}} = \frac{1}{2} \sum_{\mathbf{k}\lambda} [P^2_{\mathbf{k}\lambda} + \omega_k^2 Q^2_{\mathbf{k}\lambda}]. \tag{14.127}$$

Hamilton's equations give the following relations:

$$\dot{Q}_{\mathbf{k}\lambda} = \frac{\partial H_{\text{EM}}}{\partial P_{\mathbf{k}\lambda}} = P_{\mathbf{k}\lambda}, \quad \dot{P}_{\mathbf{k}\lambda} = -\frac{\partial H_{\text{EM}}}{\partial Q_{\mathbf{k}\lambda}} = -\omega_k^2 Q_{\mathbf{k}\lambda}. \tag{14.128}$$

This shows that the coordinates $Q_{\mathbf{k}\lambda}$ and $P_{\mathbf{k}\lambda}$ are conjugate to each other. Quantization is formally achieved by imposing the commutation relation,

$$[Q_{\mathbf{k}\lambda}, P_{\mathbf{k}\lambda}] = i\hbar. \tag{14.129}$$

Following the discussion in Sec. 2.7.2, creation and annihilation operators $a_{\mathbf{k}\lambda}^\dagger(t)$ and $a_{\mathbf{k}\lambda}(t)$, respectively, in the Heisenberg picture are taken, such that

$$Q_{\mathbf{k}\lambda}(t) = \sqrt{\frac{\hbar}{2\omega_\mathbf{k}}}(a_{\mathbf{k}\lambda}(t) + a_{\mathbf{k}\lambda}^\dagger(t)), \quad P_{\mathbf{k}\lambda}(t) = -i\sqrt{\frac{\hbar\omega_\mathbf{k}}{2}}(a_{\mathbf{k}\lambda}(t) - a_{\mathbf{k}\lambda}^\dagger(t)), \tag{14.130}$$

where

$$a_{\mathbf{k}\lambda}(t) = a_{\mathbf{k}\lambda}\, e^{-i\omega_\mathbf{k} t}, \quad a_{\mathbf{k}\lambda}^\dagger(t) = a_{\mathbf{k}\lambda}^\dagger\, e^{i\omega_\mathbf{k} t}. \tag{14.131}$$

The proof is given in the answer to Problem 14.17(c). The time dependence shows that the operators are expressed in the Heisenberg picture.

Problem 14.16

Express the creation and annihilation operators $a_{\mathbf{k}\lambda}^\dagger$ and $a_{\mathbf{k}\lambda}$, respectively, in terms of $Q_{\mathbf{k}\lambda}$ and $P_{\mathbf{k}\lambda}$ and prove that $[a_{\mathbf{k}\lambda}, a_{\mathbf{k}\lambda}^\dagger] = 1$.

Answer: $a_{\mathbf{k}\lambda} = \tfrac{1}{2}\left(\sqrt{\tfrac{2\omega_\mathbf{k}}{\hbar}}Q_{\mathbf{k}\lambda} + i\sqrt{\tfrac{2}{\hbar\omega_\mathbf{k}}}P_{\mathbf{k}\lambda}\right)$, $a_{\mathbf{k}\lambda}^\dagger = \tfrac{1}{2}\left(\sqrt{\tfrac{2\omega_\mathbf{k}}{\hbar}}Q_{\mathbf{k}\lambda} - i\sqrt{\tfrac{2}{\hbar\omega_\mathbf{k}}}P_{\mathbf{k}\lambda}\right)$.

The quantum "excitation" associated with the operators $a_{\mathbf{k}\lambda}^\dagger$ and $a_{\mathbf{k}\lambda}$ is called the *photon*. In this representation, a photon has a wave vector $\mathbf{k}$ and one of two possible polarizations, $\lambda = 1, 2$. According to the result of Problem 14.16, the photon is a boson. In terms of the operators $a_{\mathbf{k}\lambda}^\dagger$ and $a_{\mathbf{k}\lambda}$, the second quantized operators $\mathbf{A}(\mathbf{r}, t)$ and the Hamiltonian H_{EM} are then expressed as,

$$\mathbf{A}(\mathbf{r}, t) = \frac{c}{\sqrt{V}} \sum_{\mathbf{k}\lambda=1,2} e^{i\mathbf{k}\cdot\mathbf{r}} \mathbf{A}_\lambda(\mathbf{k}, t), \tag{14.132}$$

$$\mathbf{A}_\lambda(\mathbf{k}, t) = \sqrt{\frac{\hbar}{2\omega_\mathbf{k}}}\left[a_{\mathbf{k}\lambda}\, e^{-i\omega_\mathbf{k} t} + a_{-\mathbf{k}\lambda}^\dagger\, e^{i\omega_\mathbf{k} t}\right] \hat{\epsilon}_\lambda(\mathbf{k}), \tag{14.133}$$

$$H_{\mathrm{EM}} = \sum_{\mathbf{k}\lambda} \hbar\omega_\mathbf{k}\left(a_{\mathbf{k}\lambda}^\dagger a_{\mathbf{k}\lambda} + \tfrac{1}{2}\right). \tag{14.134}$$

The photon energy $\hbar\omega_\mathbf{k}$ is independent of polarization. As defined in Eq. (14.133), the operator $\mathbf{A}_\lambda(\mathbf{k}, t)$ is time dependent. Following the discussion in Sec. 14.1.6, this means that it is given in the Heisenberg representation derived from the Hamiltonian H_{EM}. At $t = 0$, we obtain the *Schrödinger representation*:

$$\mathbf{A}_\lambda(\mathbf{k}) = \sqrt{\frac{\hbar}{2\omega_\mathbf{k}}}\left[a_{\mathbf{k}\lambda} + a_{-\mathbf{k}\lambda}^\dagger\right] \hat{\epsilon}_\lambda(\mathbf{k}). \tag{14.135}$$

Problem 14.17

(a) Show how to obtain the wave equation in Eq. (14.120) from Eq. (14.119) and the Coulomb gauge condition.
(b) Verify that the dimension of $e\mathbf{A}(\mathbf{r}, t)/c$ is that of momentum.
(c) Prove that $\mathbf{A}_\lambda(\mathbf{k}) = e^{-iH_{\mathrm{EM}}t/\hbar}\mathbf{A}_\lambda(\mathbf{k}, t)e^{iH_{\mathrm{EM}}t/\hbar}$.

Guidance for (c): In the Baker–Hausdorff formula, Eq. (14.103), put $\lambda = 1$ and $G = H$, where H is given in Eq. (14.134) and $A = a_{\mathbf{k}\lambda}$ or $a_{\mathbf{k}\lambda}^\dagger$. Then use Eq. (14.31) to show that $[H, a_{\mathbf{k}\lambda}] = -\hbar\omega_\mathbf{k} a_{\mathbf{k}\lambda}$ and $[H, a_{\mathbf{k}\lambda}^\dagger] = \hbar\omega_\mathbf{k} a_{\mathbf{k}\lambda}^\dagger$. The series obtained from the Baker–Hausdorff formula yield Eq. (14.131) from which the result follows.

14.1 Second Quantization

Electron–Photon Interaction

The electron–photon interaction is derived from the Hamiltonian of N electrons subject to an electromagnetic field derived from a vector potential $\mathbf{A}$,

$$H = \frac{1}{2m} \sum_{i=1}^{N} \left(\mathbf{p}_i + \frac{e}{c} \mathbf{A}(\mathbf{r}_i, t) \right)^2, \tag{14.136}$$

and expressing all the operators in their second quantized form. When the brackets are opened, there are two terms contributing to the electron–photon interaction. The first one is given by,

$$H_1 \equiv \frac{e}{2c} \sum_i \mathbf{J}^{\mathrm{p}}(\mathbf{r}_i) \cdot \mathbf{A}(\mathbf{r}_i, t) + \mathbf{A}(\mathbf{r}_i, t) \cdot \mathbf{J}^{\mathrm{p}}(\mathbf{r}_i), \tag{14.137}$$

where $\mathbf{J}^{\mathrm{p}}(\mathbf{r}_i)$ is the matter current defined in Eq. (14.96). It is more convenient to write this term in momentum space using Eqs (14.132) and (14.97):

$$H_1 = \frac{e}{\sqrt{V}} \sum_{\mathbf{q}\lambda} \mathbf{J}^{\mathrm{p}}(\mathbf{q}) \cdot \mathbf{A}_\lambda(\mathbf{q},t) = \frac{e\hbar}{m\sqrt{V}} \sum_{\mathbf{q}\lambda} \mathbf{A}_\lambda(\mathbf{q},t) \cdot \sum_{\mathbf{k}\sigma} \left(\mathbf{k} + \frac{1}{2}\mathbf{q} \right) c^\dagger_{\mathbf{k}+\mathbf{q}\sigma} c_{\mathbf{k}\sigma}. \tag{14.138}$$

Substitution of $\mathbf{A}_\lambda(\mathbf{q},t)$ from Eq. (14.135) completes the second quantized representation of H_1. The second contribution H_2 to the electron–photon interaction comes from the square of the vector potential. It can be written as,

$$H_2 = \frac{e^2}{2mc^2} \sum_{i=1}^{N} A^2(\mathbf{r}_i, t) = \frac{e^2}{2m} \sum_{\mathbf{k}\mathbf{q}\lambda\lambda'} \hat{n}(\mathbf{q}) \mathbf{A}_\lambda(\mathbf{q},t) \cdot \mathbf{A}_{\lambda'}(\mathbf{q} - \mathbf{k}, t), \tag{14.139}$$

where $\hat{n}(\mathbf{q})$ is the particle density operator defined in Eq. (14.95). Substituting the expression for $\mathbf{A}_\lambda(\mathbf{k},t)$ in terms of photon operators $a_{\mathbf{k}\lambda}$ and $a^\dagger_{-\mathbf{k}\lambda}$ in Eq. (14.133) into $H_1 + H_2$ yields the second quantized form of the electron–photon interaction.

14.1.8 QUANTIZATION OF LATTICE VIBRATIONS: PHONONS

The second quantized description of the electromagnetic field, as described above, led to the concept of photons, i.e., a quanta of the electromagnetic field having definite energy, wave vector, and polarization. Analogously, the second quantized description of the oscillations of crystals leads to the concept of *phonons*, quanta of lattice vibrations. Consider a Bravais lattice with a basis consisting of p atoms per unit cell. The constituents of the lattice (atoms or ions) oscillate around their equilibrium positions. When the amplitudes of these oscillations are small, the theory of small oscillations in classical mechanics (see Sec. 16.9, linked to the book web page) can be employed and leads to a description of crystal vibrations in terms of *normal modes*, which are equivalent to a system of independent harmonic oscillators. As described in Sec. 9.8.1, the energy $\hbar\omega_s(\mathbf{k})$ of each oscillator is determined by the wave vector $\mathbf{k}$ in reciprocal space and a *branch index* $s = 1, 2, \ldots, 3p$. Of these $3p$ branches, three are referred to as *acoustic branches* that vanish linearly with $\mathbf{k}$ as $k \to 0$. The other $3p - 3$ branches do not vanish as $k \to 0$ and are referred to as *optical branches*. $\hbar\omega_s(\mathbf{k})$ is periodic as a function of $\mathbf{k}$ in reciprocal lattice space. The Hamiltonian in first quantization is already diagonal,

$$H = \sum_{\mathbf{k}s} \left(n_{\mathbf{k}s} + \frac{1}{2} \right) \hbar\omega_{\mathbf{k}s}, \quad (n_{\mathbf{k}s} = 0, 1, 2, \ldots). \tag{14.140}$$

Passage to second quantization is carried out simply by turning the number of quanta $n_{\mathbf{k}s}$ into a number operator $\hat{n}_{\mathbf{k}s} = \hat{b}^\dagger_{\mathbf{k}s} \hat{b}_{\mathbf{k}s}$ and interpreting the operators $b^\dagger_{\mathbf{k}s}$ and $b_{\mathbf{k}s}$ as creation and annihilation operators, respectively, for lattice vibrations with wave vector $\mathbf{k}$ and branch s. The operators $b^\dagger_{\mathbf{k}s}$ and $b_{\mathbf{k}s}$ are the creation and annihilation operators for phonons, similar to the raising and lowering operators for the quantum harmonic oscillator (see Sec. 2.7.2); their bosonic commutation relations are given in Eq. (2.122). Hence, the second quantized form of lattice vibrations (phonons) Hamiltonian is,

$$\hat{H}_{\text{ph}} = \sum_{\mathbf{k}s} \left(\hat{b}^\dagger_{\mathbf{k}s} \hat{b}_{\mathbf{k}s} + \frac{1}{2} \right) \hbar \omega_{\mathbf{k}s}, \quad ([\hat{b}_{\mathbf{k}s}, \hat{b}^\dagger_{\mathbf{k}'s'}] = \delta_{\mathbf{k}\mathbf{k}'} \delta_{ss'}). \tag{14.141}$$

This is the phononic part of the Hamiltonian required to treat lattice vibrations. The Phonons are coupled to electrons moving in the crystal. Our goal here is to derive the interaction between lattice vibrations and electrons, i.e., the *electron–phonon interaction*, within the formalism of second quantization. This interaction is central to the description of numerous fundamental physical phenomena such as sound propagation and superconductivity. The electron–phonon interaction originates from the fact that a small lattice vibration in which ions change their position from a Bravais lattice point $\mathbf{R}$ to a nearby point $\mathbf{R} + \mathbf{u}$ destroys the lattice translation symmetry and leads to scattering of electrons with Bloch wave number $\mathbf{k}$ to a new wave number $\mathbf{k} + \mathbf{q}$. This is schematically depicted in Fig. 14.2. In an idealized situation [Fig. 14.2(a)], the atoms of a Bravais lattice (filled circles) are fixed at lattice sites $\mathbf{R}_i$, and the electron at point $\mathbf{r}$ (empty circle) feels a periodic potential $U(\mathbf{r}) = \sum_i V(\mathbf{r} - \mathbf{R}_i)$, where V is the electron–atom interaction potential. The electron is in a Bloch state of crystal wave number $\mathbf{k}$. In reality, the atoms execute small oscillations around their equilibrium points $\mathbf{R}_i$. In the harmonic oscillator description of small oscillations, this motion is due to the zero point energy (at $T = 0$). Atom i is then found at position $\mathbf{R}_i + \mathbf{u}_i$ as shown in Fig. 14.2(b). The frequency of oscillations is small compared with the inverse of the time scales associated with the electron motion, so that the problem can be treated adiabatically. The electron is affected by the potential of the slowly vibrating crystal, $U(\mathbf{r}) = \sum_i V(\mathbf{r} - \mathbf{R}_i - \mathbf{u}_i)$. It is scattered from an initial state of momentum $\mathbf{k}$ into a final state of momentum $\mathbf{k} + \mathbf{q}$. Since $|\mathbf{u}_i|$ are small, the expansion

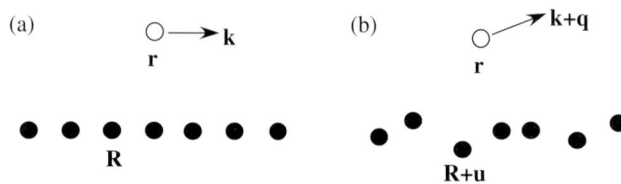

FIG 14.2 (a) Electron (empty circle) in a perfect Bravais lattice is in a Bloch state with wave vector $\mathbf{k}$. (b) The lattice atoms oscillate around their equilibrium points, and the electron scatters into a state characterized by slightly different wave number $\mathbf{k} + \mathbf{q}$.

$$V(\mathbf{r} - \mathbf{R}_i - \mathbf{u}_i) \approx V(\mathbf{r} - \mathbf{R}_i) - \nabla V(\mathbf{r} - \mathbf{R}_i) \cdot \mathbf{u}_i, \tag{14.142}$$

is justified. To first order in $\mathbf{u}_i$, the electrostatic energy of the lattice electrons due to the shift of the atoms $\mathbf{R}_i \to \mathbf{R}_i + \mathbf{u}_i$ is,

$$H_{\text{electron–lattice}} = \int d\mathbf{r} \rho(\mathbf{r}) U(\mathbf{r}) \approx V_{\text{Bloch}} + V_{\text{e–ph}}, \tag{14.143}$$

where $\rho(\mathbf{r})$ is the density of electrons. The energy of electrons in the perfectly periodic lattice is,

$$V_{\text{Bloch}} = \int d\mathbf{r} \rho(\mathbf{r}) \sum_i V(\mathbf{r} - \mathbf{R}_i), \tag{14.144}$$

while the interaction potential with the lattice vibrations is,

$$V_{\text{e–ph}} = -\int d\mathbf{r} \rho(\mathbf{r}) \left[\sum_i \nabla V(\mathbf{r} - \mathbf{R}_i) \cdot \mathbf{u}_i \right]. \tag{14.145}$$

Our goal is to cast $V_{\text{e–ph}}$ in a second quantized form. For convenience, it is assumed that the Bravais lattice is confined in a box of volume $\mathcal{V}$ containing N atoms and that periodic (Born–von Karman) boundary conditions are imposed on the electron wave functions. The wave numbers are quantized and summed within the first Brillouin zone.

14.1 Second Quantization

Problem 14.18 will help facilitate the derivation of the electron–phonon interaction in second quantization (see also Sec. 9.8.4).

Problem 14.18

Consider Problem (9.19) where an electron is affected by a sequence of 1D potential barriers centered at $X_n = n(s+d)$, $U(x) = \sum_{n=-\infty}^{\infty} V(x-X_n)$, $V(x-X_n) = V_0 \Theta(x - X_n + \frac{s}{2}) - \Theta(X_n + \frac{s}{2} - x)$. The Bloch wave function $\psi_k(x) = e^{ikx} u_k(x)$, $u_k(x+L) = u_k(x)$ was calculated exactly, but here it is assumed to be known.

(a) Write the Fourier series of the density $\rho_k(x) = |\psi_k(x)|^2$ in terms of the plane waves of period L.
(b) Use this expansion in Eq. (14.145) to evaluate the electron–phonon interaction for an electron with crystal momentum k.

Answer: (a) Since k is fixed, we drop it below for convenience and expand the periodic function as $u(x) = \frac{1}{\sqrt{L}} \sum_q f_q e^{iqx}$, $q = \frac{2\pi}{L} n_q$, where $n_q = 0, \pm 1, \pm 2, \ldots$. Therefore, $\rho(x) = |u(x)|^2 = \frac{1}{L} \sum_{qp} f_q f^*_{p+q} e^{ipx}$.

(b) $U'(x) = V_0 \sum_{n=-\infty}^{\infty} [\delta(x - X_n + \frac{s}{2}) - \delta(x - X_n - \frac{s}{2})]$.

Before substitution into Eq. (14.145), we introduce the expansion of the small displacement u_n as, $u_n = \frac{1}{\sqrt{L}} \sum_p u_p e^{ipX_n}$. Collecting, we obtain, $V_{e-ph} = \frac{2iV_0}{\sqrt{L}} \sum_{pq} f_q f^*_{p+q} \sin(ps/2) u_p$.

The quantization procedure is carried out by the following five steps:

(1) The small shifts $\mathbf{u}_i$ about $\mathbf{R}_i$ are expanded in normal coordinates $u_{\mathbf{k}s}$ as,

$$\mathbf{u}_i = \frac{1}{\sqrt{N}} \sum_{\mathbf{k}s} u_{\mathbf{k}s} \hat{\boldsymbol{\epsilon}}_s(\mathbf{k}) \, e^{i\mathbf{k}\cdot\mathbf{R}_i}, \tag{14.146}$$

where $\hat{\boldsymbol{\epsilon}}_s(\mathbf{k})$ is a polarization vector determining the direction of the normal coordinate of mode $\mathbf{k}s$. Each normal coordinate $u_{\mathbf{k}s}$ vibrates with frequency $\omega_{\mathbf{k}s}$.

(2) $u_{\mathbf{k}s}$ is quantized as in Eq. (14.130), [see also discussion after Eq. (2.121)],

$$u_{\mathbf{k}s} = \sqrt{\frac{\hbar}{2M\omega_{\mathbf{k}}}} (b_{\mathbf{k}s} + b^\dagger_{-\mathbf{k}s}), \tag{14.147}$$

where M is the atomic mass.

(3) The potential in the gradient term $\nabla V(\mathbf{r}-\mathbf{R}_i)$ appearing in Eq. (14.145) is Fourier expanded, and the gradient brings down an $i\mathbf{k}$ factor from the plane wave exponent. Summation over $\mathbf{k}$ is conveniently divided into sum over $\mathbf{q}$ in the first Brillouin zone and sum over reciprocal lattice vectors $\mathbf{G}$,

$$\nabla V(\mathbf{r} - \mathbf{R}_i) = -i \sum_{\mathbf{qG}} (\mathbf{q} + \mathbf{G}) V_{\mathbf{q}+\mathbf{G}} \, e^{-i(\mathbf{q}+\mathbf{G})\cdot(\mathbf{r}-\mathbf{R}_i)}. \tag{14.148}$$

(4) The electronic density $\rho(\mathbf{r})$ is written in terms of field operators using Eq. (14.94), which is then expanded in terms of $c_{\mathbf{k}\sigma}$ and $c^\dagger_{\mathbf{k}\sigma}$ [see Problem 14.12 and Eq. (14.95)],

$$\rho(\mathbf{r}) = \sum_\sigma \hat{\psi}^\dagger_\sigma(\mathbf{r}) \hat{\psi}_\sigma(\mathbf{r}) = \frac{1}{V} \sum_{\mathbf{k}p\sigma} e^{-i\mathbf{p}\cdot\mathbf{r}} c^\dagger_{\mathbf{k}+\mathbf{p}\sigma} c_{\mathbf{k}\sigma}. \tag{14.149}$$

(5) Now we substitute Eq. (14.147) in Eq. (14.146) to express $\mathbf{u}_j$ in terms of the phonon operators $b_{\mathbf{k}s}$ and $b^\dagger_{\mathbf{k}s}$ and insert it together with expressions (14.148) and (14.149) into Eq. (14.145) and integrate over $\mathbf{r}$ employing the orthogonality of the plane waves to get,

$$V_{\text{e-ph}} = \frac{1}{\sqrt{\mathcal{V}}} \sum_{\mathbf{k}\mathbf{q}\mathbf{G}} \sum_{\sigma s} M_s(\mathbf{q}, \mathbf{G}) c^\dagger_{(\mathbf{k}+\mathbf{q}+\mathbf{G})\sigma} c_{\mathbf{k}\sigma} (b_{\mathbf{q}s} + b^\dagger_{-\mathbf{q}s}), \qquad (14.150)$$

where we have defined,

$$M_s(\mathbf{q}, \mathbf{G}) \equiv i \sqrt{\frac{\hbar}{2nM\omega_{\mathbf{q}s}}} (\mathbf{q} + \mathbf{G}) \cdot \hat{\boldsymbol{\epsilon}}_{\mathbf{q}s}(\mathbf{q}) \, V_{\mathbf{q}+\mathbf{G}}, \qquad (14.151)$$

which represents the strength of the electron–phonon interaction.

The interaction as derived in Eq. (14.150) has a simple intuitive interpretation. The operator $c^\dagger_{(\mathbf{k}+\mathbf{q}+\mathbf{G})\sigma} c_{\mathbf{k}\sigma} b_{\mathbf{q}s}$ represents a process of *phonon absorption*. Reading from right to left, in the initial state (before the operator is applied), there is a phonon of momentum $\mathbf{q}$ and mode s and an electron of momentum $\mathbf{k}$ and spin projection σ. The phonon is annihilated, and the electron is scattered into a new state having momentum $\mathbf{k} + \mathbf{q} + \mathbf{G}$ and spin σ. This is consistent with momentum conservation, up to an inverse lattice vector $\mathbf{G}$, due to the lattice periodicity. The strength of this process is determined by $M_s(\mathbf{q}, \mathbf{G})$. A process with $\mathbf{G} = 0$ is referred to as *normal scattering*, whereas one with $\mathbf{G} \neq 0$ is referred to as an *umklapp scattering*. In most cases, the latter is much smaller in magnitude than the normal ones, but there are situations that umklapp processes cannot be neglected. The two fundamental processes of (normal) phonon absorption and emission are depicted in Fig. 14.3. These processes reveal the power of the second quantization formalism as it accommodates realistic physical situations where particle number is not conserved (here, the processes of phonon absorption and phonon emission do not conserve phonon number). Such a description is unimaginable within the formalism of first quantization.

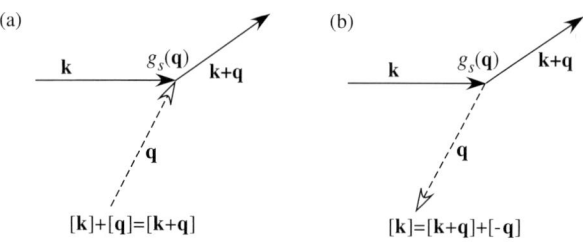

FIG 14.3 (a) A phonon of momentum $\mathbf{q}$ is absorbed by an electron of momentum $\mathbf{k}$ resulting in an electron of momentum $\mathbf{k} + \mathbf{q}$. (b) An electron of momentum $\mathbf{k}$ emits a phonon of momentum $-\mathbf{q}$ and scattered into a final state of momentum $\mathbf{k} + \mathbf{q}$.

Problem 14.19

(a) Verify that the dimension of $V_{\text{e-ph}}$ is energy.

(b) Explain why the operator $c^\dagger_{(\mathbf{k}+\mathbf{q}+\mathbf{G})\sigma} c_{\mathbf{k}\sigma} b^\dagger_{-\mathbf{q}s}$ represents *phonon emission* and identify how momentum conservation in this process is insured.

14.1.9 SYSTEMS WITH TWO KINDS OF PARTICLES

The Hamiltonian written in Eq. (14.60) governs the physics of a system containing a single group of identical particles. We often encounter systems containing more than one group of identical particles, most notably, the case of two groups A and B. The electron–phonon interaction in Eq. (14.150) is an obvious example (here, electrons and phonons). Another example is the Hamiltonian for a gas containing two kinds of atoms A and B. We may also consider a system consisting of electrons in a solid belonging to two different bands A and B in which there is an

14.2 Statistical Mechanics in Second Quantization

interaction between the two bands. In these cases, the second quantization formalism turns out to be very effective and compact. From a mathematical viewpoint, the Hilbert space of such a system is an outer product $\mathcal{H} = \mathcal{H}_A \otimes \mathcal{H}_B$ of the corresponding Hilbert spaces of systems A and B. A pure state $|\Psi\rangle$ can always be written as a linear combination of product states, $|\psi_A\rangle \otimes |\psi_B\rangle$, and operators that are a tensor product, $O_A \otimes O_B$, act on such product states. The two groups of identical particles need not have the same statistics, thus, in the electron–phonon case, they are fermions (electrons) and bosons (phonons). On the other hand, in the case of two band systems, they are both fermions.

Problem 14.20

Consider the Hamiltonian, $H_0 = \sum_{i\geq 1}[\varepsilon_i^F c_i^\dagger c_i + \varepsilon_i^B b_i^\dagger b_i]$, where $\varepsilon_{i+1}^{F,B} > \varepsilon_i^{F,B}$, $c_i^\dagger$ (c_i) creates (annihilates) a fermion on level ε_i^F and $b_i^\dagger$, (b_i) creates (annihilates) a boson on level ε_i^B.

(a) For a system of four particles, two fermions and two bosons, find the ground-state energy ε_1 and the ground-state wave function $|\psi_1\rangle$.

(b) The energy ε_2 and the wave function $|\psi_2\rangle$ of the first excited state depend on the energies ε_i^F and ε_i^B. Find the conditions under which $|\psi_2\rangle$ is a fermion or boson excited state.

(c) A perturbation $V = V_0(c_3 c_2^\dagger b_1 b_1^\dagger)$+h.c. is added to H_0. Calculate the matrix element $\langle \psi_1|V|\psi_2\rangle$, where $|\psi_2\rangle$ corresponds to fermion excitation.

Answer: (a) Two fermions cannot occupy the same level, so they must occupy levels ε_1^F and ε_2^F. On the other hand, the two bosons can occupy the lowest level ε_1^B. The two-fermion and two-boson ground state and its energy are
$|\psi_1\rangle = c_1^\dagger c_2^\dagger [b_1^\dagger]^2 |0\rangle$, $\varepsilon_1 = \varepsilon_1^F + \varepsilon_2^F + 2\varepsilon_1^B$.
(b) $\varepsilon_3^F - \varepsilon_2^F < \varepsilon_2^B - \varepsilon_1^B \Rightarrow |\psi_2\rangle = c_1^\dagger c_3^\dagger (b_1^\dagger)^2 |0\rangle$, $\varepsilon_2 = \varepsilon_1^F + \varepsilon_3^F + 2\varepsilon_1^B$. $\varepsilon_3^F - \varepsilon_2^F > \varepsilon_2^B - \varepsilon_1^B \Rightarrow |\psi_2\rangle = c_1^\dagger c_2^\dagger b_1^\dagger b_2^\dagger |0\rangle$, $\varepsilon_2 = \varepsilon_1^F + \varepsilon_2^F + \varepsilon_1^B + \varepsilon_2^B$. The first case is a fermion excitation and the second case is a boson excitation.
(c) $\langle \psi_1|V|\psi_2\rangle = -V_0$.

Problem 14.20 already suggests the general form of the Hamiltonian for such system, which can be written as,

$$H = H_A + H_B + H_I, \quad (14.152)$$

where H_A and H_B are given by Eq. (14.60) while for the case that the number of particles in each group is conserved,

$$H_I = \sum_{\alpha\alpha',\beta\beta'} V_{\alpha\alpha',\beta\beta'} a_\alpha^\dagger b_\beta^\dagger b_{\beta'} a_{\alpha'}. \quad (14.153)$$

14.2 STATISTICAL MECHANICS IN SECOND QUANTIZATION

Statistical mechanics is designed to treat physical systems whose detailed description is beyond reach because they have too many degrees of freedom to be treated exactly, such as many-body systems. Hence, it is necessary to formulate statistical mechanics within second quantization. In statistical mechanics, we deal with a statistical ensemble, i.e., a collection of many similar systems [119, 120], and our interest is focused on the average properties of the ensemble, instead of on an exact many-body wave function for each element of the ensemble. Technically, this is achieved as follows: The system is divided into the subsystem of particular interest and the rest of the system, which is referred to as a *reservoir* or a *bath* or an *environment*. The coupling between the quantum system of interest and the bath is such that energy exchange (and in the case of the grand canonical ensemble, also particle exchange) between the two can occur. If we assume that the reservoir is in thermal equilibrium at temperature T and the Hamiltonian of the quantum system is H, with eigenstates $|n\rangle$ having energies E_n, $H|n\rangle = E_n|n\rangle$, then under appropriate conditions (see Refs. [119, 120]), the probability that the quantum system will be in state $|n\rangle$ is given by the *Boltzmann distribution*:

$$P(E_n) = \frac{e^{-\beta E_n}}{\mathcal{Z}}, \quad \mathcal{Z} = \sum_{n'} e^{-\beta E_{n'}}, \qquad (14.154)$$

where $\beta = 1/k_B T$. The quantity $\mathcal{Z}$ defined above is called the *partition function*. The density operator ρ for the Boltzmann distribution is

$$\rho = \frac{e^{-\beta H}}{\mathcal{Z}} = \frac{\sum_n |n\rangle e^{-\beta E_n} \langle n|}{\mathcal{Z}}. \qquad (14.155)$$

With these definitions, the constraint $\text{Tr}[\rho] = 1$ is satisfied. The *quantum thermodynamic expectation* of an operator, $\mathcal{O}$, is defined as,

$$\langle \mathcal{O} \rangle = \frac{1}{\mathcal{Z}} \sum_n \langle n | e^{-\beta E_n} \mathcal{O} | n \rangle = \text{Tr}[\rho \, \mathcal{O}]. \qquad (14.156)$$

The trace appearing in Eq. (14.156), and employed in Problem 14.21, is evaluated under the assumption that the number N of particles in the quantum system is fixed, i.e., there is no particle exchange between the quantum system and the bath. This is referred to as a *canonical ensemble*.

Problem 14.21

(a) Show that when $T \to 0$, the quantum thermodynamics average reduces to the expectation value in the ground state $|G\rangle$, namely $\langle \mathcal{O} \rangle \to \langle G|\mathcal{O}|G\rangle$.

(b) A quantum Hamiltonian is given by $H = \varepsilon_1 a_1^\dagger a_1 + \varepsilon_2 a_2^\dagger a_2 + V(a_1^\dagger a_2 + a_2^\dagger a_1)$. Find the canonical ensemble partition function $\mathcal{Z}$ for the following cases: (1) The system contains a single particle ($N=1$). (2) The system contains two identical bosons. (3) The system contains two identical (spinless) fermions.

Answer: (b) The Hamiltonian is easily diagonalized by writing it as
$H = (a_1^\dagger a_2^\dagger) \begin{pmatrix} \varepsilon_1 & V \\ V & \varepsilon_2 \end{pmatrix} \begin{pmatrix} a_1 \\ a_2 \end{pmatrix}$ and diagonalizing the matrix by a unitary matrix U, thereby defining new quasi-particles, $\begin{pmatrix} \alpha_1 \\ \alpha_2 \end{pmatrix} = U \begin{pmatrix} a_1 \\ a_2 \end{pmatrix}$. The statistics of the new particles is the same as the old ones. U is said to be a *canonical transformation* if it conserves the commutation or anticommutation relations. In the new basis, $H = \sum_{i=1}^{2} \lambda_i \alpha_i^\dagger \alpha_i$, where λ_i are the two eigenvalues of the matrix H.

(1) For a single particle, Pauli statistics does not enter and $\mathcal{Z} = e^{-\beta \lambda_1} + e^{-\beta \lambda_2}$.
(2) There are three ways to accommodate two identical bosons. They can occupy the lowest level, the upper level, and one on each level. Hence, $\mathcal{Z} = e^{-2\beta \lambda_1} + e^{-2\beta \lambda_2} + e^{-\beta(\lambda_1 + \lambda_2)}$.
(3) The only way to accommodate two identical fermions is to put one on each level, hence, $\mathcal{Z} = e^{-\beta(\lambda_1 + \lambda_2)}$.

In many realistic situations, however, the quantum system is allowed to exchange particles with the reservoir, albeit keeping its average number of particles constant. The statistical mechanical term for this type of ensemble is the *grand canonical ensemble* (GCE). In the GCE formalism, the particle number operator [see Eq. (14.32)],

$$\hat{N} = \sum_n a_n^\dagger a_n, \qquad (14.157)$$

fluctuates around its average N. The constraint $\langle \hat{N} \rangle = N$ is implemented in the GCE formalism by introducing a chemical potential μ, which, in mathematical terms, is a *Lagrange multiplier*. The corresponding density operator, partition

14.2 Statistical Mechanics in Second Quantization

function, and quantum thermodynamic averaging are defined as

$$\rho_{GCE} = \frac{e^{-\beta(H-\mu\hat{N})}}{\mathcal{Z}}, \quad \mathcal{Z} = \text{Tr}[e^{-\beta(H-\mu\hat{N})}], \quad \langle\mathcal{O}\rangle = \text{Tr}[\rho_{GCE}\mathcal{O}]. \tag{14.158}$$

Here, the trace is performed over all states (even states with different particle number). Note that the value of the chemical potential μ is not known in advance. Once we require that $\langle\hat{N}\rangle = N$, we get an implicit equation for μ through the relation,

$$\langle\hat{N}\rangle = k_B T \frac{\partial \ln \mathcal{Z}}{\partial \mu} = N. \tag{14.159}$$

It is common to call $H - \mu\hat{N}$ the "Hamiltonian" in the GCE. If this is done, Eq. (14.158) looks just like the canonical density operator (14.155).

Once a partition function is defined and computed, all important statistical functions familiar from classical statistical mechanics can be obtained. These include the energy, $E - \mu N = \frac{\partial \ln \mathcal{Z}}{\partial \beta}$, the free energy F (for the canonical ensemble), or the grand potential Ω (for the GC ensemble),

$$F = -k_B T \ln \mathcal{Z}, \quad \Omega = -k_B T \ln \mathcal{Z}, \tag{14.160}$$

and the entropy S, which is related to the F and $E = \langle H \rangle$, or Ω and E, as,

$$F = E - TS, \quad \Omega = E - TS - \mu N. \tag{14.161}$$

Problem 14.22

Refer to Problem 14.21(b) and consider the GCE Hamiltonian, $\hat{H} = \varepsilon_1 a_1^\dagger a_1 + \varepsilon_2 a_2^\dagger a_2 + V(a_1^\dagger a_2 + a_2^\dagger a_1) - \mu\hat{N}$. Find the chemical potential μ for a system of fermions with constraints $\langle\hat{N}\rangle = 1$ and $\langle\hat{N}\rangle = 2$.

Answer: Once the particle number can fluctuate, then, in calculating the trace, we must include states of both $N = 1$ and $N = 2$ (not more because we have only two levels). The Pauli statistics enters even in the case $\langle\hat{N}\rangle = 1$. In the new fermion operators, we have $H - \mu\hat{N} = \sum(\lambda_i - \mu)\alpha_i^\dagger \alpha_i$. Thus, in the GCE formalism, we have $\mathcal{Z} = e^{-\beta(\lambda_1-\mu)} + e^{-\beta(\lambda_2-\mu)} + e^{-\beta(\lambda_1+\lambda_2-2\mu)}$. Therefore, under the constraints $\langle\hat{N}\rangle = N = 1, 2$, we get μ as the solution of $k_B T \frac{\partial \ln \mathcal{Z}}{\partial \mu} = \frac{1}{\mathcal{Z}}\left[e^{-\beta(\lambda_1-\mu)} + e^{-\beta(\lambda_2-\mu)} + 2e^{-\beta(\lambda_1+\lambda_2-2\mu)}\right] = N$.

The Fermi–Dirac Distribution Function

It is instructive to carry out the quantum thermodynamic averaging procedure in Eq. (14.158) for the fermion single-level number operator $\hat{n}_{\mathbf{k}\sigma}$ for a free electron gas. The free electron "Hamiltonian" $\hat{H} - \mu\hat{N}$ in the GCE is,

$$\hat{H}_0 = \sum_{\mathbf{k}\sigma}(\varepsilon_{\mathbf{k}} - \mu)c_{\mathbf{k}\sigma}^\dagger c_{\mathbf{k}\sigma} \equiv \sum_{\mathbf{k}\sigma} \xi_{\mathbf{k}} \hat{n}_{\mathbf{k}\sigma}, \tag{14.162}$$

where

$$\hat{n}_{\mathbf{k}\sigma} = c_{\mathbf{k}\sigma}^\dagger c_{\mathbf{k}\sigma} \tag{14.163}$$

is the fermion number operator and

$$\xi_{\mathbf{k}} \equiv \varepsilon_{\mathbf{k}} - \mu. \tag{14.164}$$

Since the Hamiltonian $\hat{H}_0$ commutes with $\hat{n}_{\mathbf{k}\sigma}$ and the states can be eigenstates of $\hat{n}_{\mathbf{k}\sigma}$ with eigenvalues $n_{\mathbf{k}\sigma} = 0, 1$, the calculation is simple,

$$\langle \hat{n}_{\mathbf{k}\sigma} \rangle = \frac{\text{Tr}[\rho_{\text{GCE}} \hat{n}_{\mathbf{k}\sigma}]}{\mathcal{Z}} = \frac{\sum_{n_{\mathbf{k}\sigma}=0}^{1} n_{\mathbf{k}\sigma} e^{-\beta \xi_{\mathbf{k}} n_{\mathbf{k}\sigma}}}{\sum_{n'_{\mathbf{k}\sigma}=0}^{1} e^{-\beta \xi_{\mathbf{k}} n'_{\mathbf{k}\sigma}}}. \tag{14.165}$$

The sums are easily performed to give the *Fermi–Dirac distribution*,

$$\boxed{n_{\text{FD}}(\xi_{\mathbf{k}}) \equiv \langle n_{\mathbf{k}\sigma} \rangle = \frac{1}{e^{\beta \xi_{\mathbf{k}}} + 1}.} \tag{14.166}$$

$n_{\text{FD}}(\xi_{\mathbf{k}})$ is the average occupation of the quantum state with energy $\xi_{\mathbf{k}}$.

Boson Occupation Number: Bose–Einstein Distribution Function

Unlike the fermion number operator $\hat{n}_{\mathbf{k}\sigma}$ that has eigenvalues of either 0 or 1, the boson occupation number operator $\hat{n}_{\mathbf{k}\sigma} = \hat{b}^{\dagger}_{\mathbf{k}\sigma} \hat{b}_{\mathbf{k}\sigma}$ can take any nonnegative integer value, $n_{\mathbf{k}\sigma} = 0, 1, 2, \ldots$. Thus, instead of Eq. (14.165), we now have,

$$\langle \hat{n}_{\mathbf{k}\sigma} \rangle = \frac{\text{Tr}[\rho_{\text{GCE}} \hat{n}_{\mathbf{k}\sigma}]}{\mathcal{Z}} = \frac{\sum_{n_{\mathbf{k}\sigma}=0}^{\infty} n_{\mathbf{k}\sigma} e^{-\beta \xi_{\mathbf{k}} n_{\mathbf{k}\sigma}}}{\sum_{n'_{\mathbf{k}\sigma}=0}^{\infty} e^{-\beta \xi_{\mathbf{k}} n'_{\mathbf{k}\sigma}}}. \tag{14.167}$$

The infinite series in the denominator and the numerator can be written and summed up in terms of the *fugacity* $\lambda_{\mathbf{k}} \equiv e^{-\beta \xi_{\mathbf{k}}}$ as,

$$\sum_{n_{\mathbf{k}\sigma}=0}^{\infty} \lambda_{\mathbf{k}}^{n_{\mathbf{k}\sigma}} = \frac{1}{1 - \lambda_{\mathbf{k}}}, \quad \sum_{n_{\mathbf{k}\sigma}=0}^{\infty} n_{\mathbf{k}\sigma} \lambda_{\mathbf{k}}^{n_{\mathbf{k}\sigma}} = \frac{\lambda_{\mathbf{k}}}{(1 - \lambda_{\mathbf{k}})^2}. \tag{14.168}$$

The latter equation can be obtained from the former by differentiation with respect to $\lambda_{\mathbf{k}}$. These sums converge only if $0 \leq \lambda_{\mathbf{k}} < 1$, i.e., $\xi_{\mathbf{k}} > 0$. Hence, the *Bose–Einstein distribution* is

$$\boxed{n_{\text{BE}}(\xi_{\mathbf{k}}) \equiv \langle \hat{n}_{\mathbf{k}\sigma} \rangle = \frac{1}{e^{\beta \xi_{\mathbf{k}}} - 1}.} \tag{14.169}$$

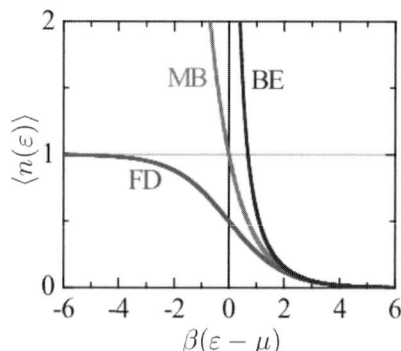

FIG 14.4 Comparison of the Maxwell–Boltzmann distribution $n_{\text{MB}}(\varepsilon_{\mathbf{k}} - \mu) = e^{-\beta(\varepsilon_{\mathbf{k}} - \mu)}$, the Fermi–Dirac distribution $n_{\text{FD}}(\varepsilon_{\mathbf{k}} - \mu) = \frac{1}{e^{\beta(\varepsilon_{\mathbf{k}} - \mu)} + 1}$, and the Bose–Einstein distribution $n_{\text{BE}}(\varepsilon_{\mathbf{k}} - \mu) = \frac{1}{e^{\beta(\varepsilon_{\mathbf{k}} - \mu)} - 1}$.

Note that both the Fermi–Dirac and the Bose–Einstein distribution functions defined in Eqs. (14.166) and (14.169) are the average occupation of quantum states and approach the classical Maxwell–Boltzmann distribution $n_{\text{MB}}(\varepsilon_{\mathbf{k}} - \mu) \equiv \langle \hat{n}_{\mathbf{k}} \rangle = e^{-\beta(\varepsilon_{\mathbf{k}} - \mu)}$ when $\beta(\varepsilon_{\mathbf{k}} - \mu) \gg 1$, as shown in Fig. 14.4. This is the high-energy limit, $\varepsilon_{\mathbf{k}} - \mu \gg k_B T$, where the energy quantum numbers are very high.

14.3 THE ELECTRON GAS

We now apply the second quantization formalism to an *electron gas*, i.e., a system of interacting electrons moving in a background of positive charge, such that the system is electrically neutral. The second quantization formalism proves to be natural and elegant for this task. The formalism employed is an elaboration of what we discussed in Example 2 in Sec. 14.1.5, where the Hamiltonian is written in Eq. (14.70) in first quantization and Eq. (14.76) in

14.3 The Electron Gas

second quantization language. The goal is to elucidate the equilibrium physics at zero and at finite temperatures. A systematic study is made easier by employing Feynman diagram techniques, to be presented in Chapter 18, which is linked to the book web page. Here, we will be content with analysis of the interacting electron gas to first order in the electron–electron interaction strength scaled as a dimensionless parameter. The analysis that goes beyond first-order perturbation is presented in Sec. 18.2. The system of interacting electrons that we study here is assumed to be confined in a box, and the electrons are assumed to be free (there is no single-particle potential $U(\mathbf{r})$ but rather, a positive background charge that keeps the system electrically neutral); this is referred to as *the jellium model*. The single-particle wave functions from which the basis states will be constructed are plane waves, and the analysis extends the discussion started in relation with Eq. (14.69). In the jellium model, the background positive charge is represented as a medium with constant density. A more realistic situation to be discussed later is that with a background charge consisting of positive ions arranged on a lattice. Such a system, with Bloch single-particle wave functions, is more difficult to handle. An approximation in which the Bloch functions are replaced by plane waves but the mass of the electrons in the lattice is replaced by an effective mass (the *effective mass approximation*) is often adequate.

14.3.1 ELECTRONS IN THE JELLIUM MODEL

Consider a system of N electrons moving in a medium of volume $\mathcal{V} = L^3$ with constant positive charge density $\rho = en$, where $n = N/\mathcal{V} = N/L^3$ is the particle density. No single-particle potential confines the electrons, but the spectrum is discrete since the system occupies a finite volume. A useful dimensionless parameter specifying the density of the electron gas is $r_s = r_0/a_0$, where $a_0 = \hbar^2/me^2$ is the Bohr radius and r_0 is such that N electrons each occupying a sphere of radius r_0 fill the volume $\mathcal{V}$,

$$\frac{\mathcal{V}}{N} = \frac{4\pi r_0^3}{3} \Rightarrow r_s = \frac{1}{a_0}\left(\frac{3}{4\pi n}\right)^{\frac{1}{3}}. \tag{14.170}$$

The smaller $r_s = r_0/a_0$, the higher the density. In metals, $2 \leq r_s \leq 6$. The quantity $e^2/(2a_0) \approx 13.6$ eV is a convenient energy unit called 1 Rydberg (Ry). Alternatively, one may adopt the atomic unit of energy e^2/a_0 (the Hartree).

Imposing periodic boundary conditions on the wave function, the single-particle wave functions $\phi_{\mathbf{k}\sigma}(\mathbf{r}) = \langle \mathbf{r} | \mathbf{k}\sigma \rangle$ are defined in Eq. (14.71), and the single-particle energies are $\varepsilon_{\mathbf{k}} = \frac{\hbar^2 k^2}{2m}$. The wave numbers are quantized as $\mathbf{k} = \frac{2\pi}{L}(n_x, n_y, n_z)$ with $n_x, n_y, n_z = 0, \pm 1, \pm 2, \ldots$. Following our quantum number ordering procedure, the single-particle quantum numbers are assumed to be ordered, so that for any two sets of quantum numbers, either $(\mathbf{k}\sigma) > (\mathbf{k}'\sigma')$ or vice versa. The particle number operator is $\hat{N} = \sum_{\mathbf{k}\sigma} \hat{n}_{\mathbf{k}\sigma} = \sum_{\mathbf{k}\sigma} c^\dagger_{\mathbf{k}\sigma} c_{\mathbf{k}\sigma}$. We will occasionally replace summation over $\mathbf{k}$ by integrals, guided by the rule, $\frac{1}{\mathcal{V}} \sum_{\mathbf{k}} f(\mathbf{k}) \to \frac{1}{(2\pi)^3} \int d\mathbf{k} f(\mathbf{k})$.

> **Problem 14.23**
>
> Discuss the degeneracy of the single-particle energy levels. For $\mathbf{k} = \frac{2\pi}{L}(1, 5, 5)$, how many energies are equal to $\varepsilon_{\mathbf{k}}$? *Answer:* 48.

Noninteracting Electrons at Zero Temperature

Let us first consider noninteracting electrons at zero temperature, where the Hamiltonian includes only the kinetic energy operator, whose second quantized expression is given in Eq. (14.86), where the field operators are introduced in Eq. (14.78),

$$T = \sum_\sigma \int d\mathbf{r}\, \hat{\psi}^\dagger_\sigma(\mathbf{r}) \left[-\frac{\hbar^2}{2m}\nabla^2\right] \hat{\psi}_\sigma(\mathbf{r}) = \sum_{\mathbf{k}\sigma} \varepsilon_{\mathbf{k}} c^\dagger_{\mathbf{k}\sigma} c_{\mathbf{k}\sigma}. \tag{14.171}$$

The ground state has the N lowest energy levels occupied [recall the order of the $\mathbf{k}$ vectors worked out in Problem 14.1],

$$|G\rangle = c^\dagger_{\mathbf{k}_1\downarrow} c^\dagger_{\mathbf{k}_1\uparrow} c^\dagger_{\mathbf{k}_2\downarrow} c^\dagger_{\mathbf{k}_2\uparrow} \cdots c^\dagger_{\mathbf{k}_{N/2}\downarrow} c^\dagger_{\mathbf{k}_{N/2}\uparrow} |0\rangle, \tag{14.172}$$

where $|0\rangle$ is the vacuum. The magnitude of the highest wave number is the *Fermi wave number* corresponding to energy ε_F,

$$k_F \equiv |\mathbf{k}_{N/2}|, \quad \varepsilon_F \equiv \varepsilon_{\mathbf{k}_{N/2}} = \frac{\hbar^2 k_F^2}{2m}. \tag{14.173}$$

Problem 14.24

Prove the equality $\hat{n}_{\mathbf{k}\sigma}|G\rangle = \theta(k_F - |\mathbf{k}|)|G\rangle$.

Answer: If we order the single-electron quantum numbers according to their magnitudes, we have $|G\rangle = \prod_{|\mathbf{k}|<k_F,\sigma} c^\dagger_{\mathbf{k}\sigma}|0\rangle$, where $|0\rangle$ is the particle vacuum. Applying the number operator $n_{\mathbf{k}\sigma} = c^\dagger_{\mathbf{k}\sigma} c_{\mathbf{k}\sigma}$ for $|\mathbf{k}| > k_F$ results in application of an annihilation operator on an unoccupied state, yielding zero. For $|\mathbf{k}| < k_F$, the eigenvalue of $n_{\mathbf{k}\sigma}$ just counts the number of electrons in an occupied state and yields 1.

Using the solution of Problem 14.24, we find,

$$N = \langle G|\hat{N}|G\rangle = \sum_{\mathbf{k}\sigma} \langle G|\hat{n}_{\mathbf{k}\sigma}|G\rangle = 2\frac{\mathcal{V}}{(2\pi)^3} \int d\mathbf{k}\, \theta(k_F - |\mathbf{k}|), \tag{14.174}$$

where the factor 2 comes from summation over spin. The integral simply yields the volume $\frac{1}{3} 4\pi k_F^3$ of the Fermi sphere, so

$$n = \frac{N}{\mathcal{V}} = \frac{k_F^3}{3\pi^2} \Rightarrow \varepsilon_F = \frac{\hbar^2}{2m}(3\pi^2 n)^{\frac{2}{3}}. \tag{14.175}$$

When the second equality is used at any positive energy ε, we get the *density of states*, see Eq. (9.15)

$$g_F(\varepsilon) = \frac{dn}{d\varepsilon} = \frac{1}{2\pi^2}\left(\frac{2m}{\hbar^2}\right)^{\frac{3}{2}} \varepsilon^{\frac{1}{2}}. \tag{14.176}$$

where we use g_F instead of g to emphasize that it applies for fermions. Manipulations similar to those of Eq. (14.174) yield the ground-state energy,

$$E_0 = \sum_{\mathbf{k}\sigma} \langle G|\varepsilon_\mathbf{k} \hat{n}_{\mathbf{k}\sigma}|G\rangle = 2\frac{\mathcal{V}}{(2\pi)^3}\frac{\hbar^2}{2m} \int d\mathbf{k}\, \theta(k_F - |\mathbf{k}|) k^2 = \frac{\hbar^2 \mathcal{V}}{10\pi^2 m} k_F^5 = \frac{3}{5} N\varepsilon_F. \tag{14.177}$$

Problem 14.25

Assume a simple cubic lattice of monovalent atoms with lattice constant $a = 10^{1/3}$ Å.

(a) Find the density n of free electrons.
(b) For an effective electron mass $m = 10^{-31}$ kg, calculate k_F and ε_F.
(c) Calculate the density of states at the Fermi energy ε_F.
(d) For a fixed volume $\mathcal{V}$, write the ground-state energy in the form $E_0 = AN^\alpha$ and determine A and α.
(e) Show that $E_0 = \frac{e^2}{2a_0} N \frac{2.21}{r_s^2}$.
(f) Prove the equality $\frac{E_0}{\mathcal{V}} = \int_0^{\varepsilon_F} d\varepsilon\, \varepsilon g_F(\varepsilon)$.

14.3 The Electron Gas

> **Answers:** (a) $n = 10^{29}/m^3$, (b) $k_F = \left[3\pi^2 n\right]^{1/3}$ and $\varepsilon_F = \frac{\hbar^2}{2m}k_F^2$, (c) Use Eq. (14.176) with $\varepsilon \to \varepsilon_F$, (d) Use Eq. (14.177).

Noninteracting Electrons at Finite Temperature

The electronic system at finite temperature is considered within the grand canonical formalism of quantum statistical mechanics. The system is allowed to interact with a reservoir containing many electrons in thermal equilibrium at temperature T. The system can exchange electrons with the reservoir, and the number of electrons in the system is not fixed. The relevant parameter in this case is the chemical potential μ, which is defined as the energy required to remove a single electron from the system. The probability that a level $\varepsilon_\mathbf{k}$ is occupied is given by the Fermi–Dirac distribution $n_F(\xi_\mathbf{k})$, see Eq. (14.166) and Fig. 14.4.

> **Problem 14.26**
>
> Demonstrate that $n_{FD}(\xi_\mathbf{k}) \xrightarrow[T\to 0]{} \theta(-\xi_\mathbf{k})$ and $-\frac{\partial n_{FD}(\xi_\mathbf{k})}{\partial \xi_\mathbf{k}} \xrightarrow[T\to 0]{} \delta(-\xi_\mathbf{k})$.

The electron density $n(T, \mu)$ at temperature T and chemical potential μ is

$$n(T, \mu) = \int d\varepsilon\, g_F(\varepsilon) n_{FD}(\varepsilon). \tag{14.178}$$

Assuming that the density is independent of temperature implies a temperature dependence of the chemical potential,

$$0 = \frac{\partial n(T, \mu)}{\partial T}, \quad \Rightarrow \mu(T) = \varepsilon_F \left[1 - \frac{\pi^2}{12}\left(\frac{k_B T}{\varepsilon_F}\right)^2 + \ldots\right], \tag{14.179}$$

> **Problem 14.27**
>
> Calculate the ratio $k_B T : \varepsilon_F$ for $T = 300K$ and for ε_F of Problem 14.25(b).
>
> **Answer:** $k_F = (3\pi^2 n)^{1/3} = 1.46 \times 10^8$ cm, $\varepsilon_F = \frac{\hbar^2 k_F^2}{2m} = 1.29 \times 10^{-11}$ erg, $k_B \times 300 = 4.14 \times 10^{-14}$ erg, $k_B T/\varepsilon_F = 3.21 \times 10^{-3}$.

Interacting Electrons: First-Order Perturbation Theory

The starting point for calculating first-order contribution to the ground-state energy is the Hamiltonian (14.76). As noted in comment (c) below Eq. (14.76), the divergent term resulting from $\mathbf{q} = 0$ is equal in magnitude and opposite in sign to the background positive energy when the parameter $\kappa \to 0$. Therefore, our starting point is the Hamiltonian,

$$\hat{H} = \sum_{\mathbf{k}\sigma} \varepsilon_\mathbf{k} c^\dagger_{\mathbf{k}\sigma} c_{\mathbf{k}\sigma} + \frac{2\pi e^2}{L^3} \sum_{\mathbf{k}_1 \mathbf{k}_2 \mathbf{q}\neq 0} \sum_{\sigma_1 \sigma_2} \frac{1}{q^2} c^\dagger_{\mathbf{k}_1 \sigma_1} c^\dagger_{\mathbf{k}_2 \sigma_2} c_{\mathbf{k}_2 - \mathbf{q}\sigma_2} c_{\mathbf{k}_1 + \mathbf{q}\sigma_1} \equiv \hat{H}_0 + \hat{V}. \tag{14.180}$$

The unperturbed energy E_0 was calculated in Eq. (14.177), and now the task is to calculate

$$E_1 \equiv \langle G|\hat{V}|G\rangle = \frac{2\pi e^2}{L^3} \langle G| \sum_{\mathbf{k}_1 \mathbf{k}_2 \mathbf{q}\neq 0} \sum_{\sigma_1 \sigma_2} \frac{1}{q^2} c^\dagger_{\mathbf{k}_1 \sigma_1} c^\dagger_{\mathbf{k}_2 \sigma_2} c_{\mathbf{k}_2 - \mathbf{q}\sigma_2} c_{\mathbf{k}_1 + \mathbf{q}\sigma_1} |G\rangle. \tag{14.181}$$

For a given term in the summand, a nonzero matrix element occurs if the occupations are such that

$$n_{\mathbf{k}_1\sigma_1} = n_{\mathbf{k}_2\sigma_2} = n_{\mathbf{k}_1+\mathbf{q}\,\sigma_1} = n_{\mathbf{k}_2-\mathbf{q}\,\sigma_2} = 1, \tag{14.182}$$

and the quantum numbers are such that

$$(\mathbf{k}_1, \sigma_1) = (\mathbf{k}_1 + \mathbf{q}, \sigma_1) \quad \text{and} \quad (\mathbf{k}_2, \sigma_2) = (\mathbf{k}_2 - \mathbf{q}, \sigma_2), \tag{14.183}$$

or

$$(\mathbf{k}_1, \sigma_1) = (\mathbf{k}_2 - \mathbf{q}, \sigma_2) \quad \text{and} \quad (\mathbf{k}_2, \sigma_2) = (\mathbf{k}_1 + \mathbf{q}, \sigma_1). \tag{14.184}$$

Equation (14.183) cannot be satisfied because $\mathbf{q} = 0$ is excluded from the sum in Eq. (14.181). The contribution of this *direct term* to the energy in first order then vanishes. Taking the constraints in Eq. (14.184) into account, the nonzero matrix element is,

$$\delta_{\sigma_1\sigma_2}\langle G|c^\dagger_{\mathbf{k}_1\sigma_1}c^\dagger_{\mathbf{k}_2\sigma_2}c_{\mathbf{k}_1\sigma_1}c_{\mathbf{k}_2\sigma_2}|G\rangle = -\delta_{\sigma_1\sigma_2}\langle G|\hat{n}_{\mathbf{k}_1\sigma_1}\hat{n}_{\mathbf{k}_2\sigma_2}|G\rangle, \tag{14.185}$$

where $\hat{n}_{\mathbf{k}\sigma}$ is the Fermion number operator defined in Eq. (14.163). The contribution of this *exchange term* to the energy in first order is negative, as we show below. To proceed, we use the equality $\hat{n}_{\mathbf{k}\sigma}|G\rangle = \theta(k_F - |\mathbf{k}|)|G\rangle$ proved in Problem 14.24 and change variables $(\mathbf{k}_1, \mathbf{k}_2) \to (\mathbf{p} + \tfrac{1}{2}\mathbf{q}, \mathbf{p} - \tfrac{1}{2}\mathbf{q})$. Replacing summation by integration in Eq. (14.181), we find,

$$E_1 = -\frac{4\pi e^2 V}{(2\pi)^6} \int \frac{d\mathbf{q}}{q^2} \int d\mathbf{p}\, \theta\left(k_F - \left|\mathbf{p} + \tfrac{1}{2}\mathbf{q}\right|\right) \theta\left(k_F - \left|\mathbf{p} - \tfrac{1}{2}\mathbf{q}\right|\right). \tag{14.186}$$

Integration over $\mathbf{p}$ equals the volume obtained as a union of two spheres in k space of radius k_F and centers separated a distance $|\mathbf{q}|$ as explained in Fig. 14.5.

Problem 14.28

Let $x = \frac{q}{2k_F}$, and prove that $\int d\mathbf{p}\, \theta(k_F - |\mathbf{p} + \tfrac{1}{2}\mathbf{q}|)\, \theta(k_F - |\mathbf{p} - \tfrac{1}{2}\mathbf{q}|) = \frac{4\pi k_F^3}{3}\left[1 - \tfrac{3}{2}x + \tfrac{1}{2}x^3\right]\theta(1-x)$.

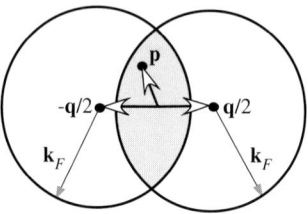

FIG 14.5 Integration region over $\mathbf{p}$ in Eq. (14.186). Draw two spheres of radius k_F centered at $\pm\tfrac{1}{2}\mathbf{q}$. The integration over $\mathbf{p}$ yields the volume of the overlap (shaded) region.

Following the solution of the above problem, the integration over $\mathbf{q}$ is simple, involving only q. The result, expressed in terms of a_0 and r_s, is,

$$E_1 = -\frac{2e^2 V k_F^4}{3\pi^3}\int_0^1 dx\left[1 - \frac{3}{2}x + \frac{1}{2}x^3\right]$$

$$= -\frac{e^2 V k_F^4}{4\pi^3} = -\frac{e^2}{2a_0}N\frac{0.916}{r_s}. \tag{14.187}$$

Combined with the expression for E_0 derived in Problem 14.25(e), we have the energy up to first order in perturbation theory. Anticipating higher order terms, we write the energy per particle in the form

$$E_g \equiv \frac{E}{N} = \frac{e^2}{2a_0}\left[\frac{2.21}{r_s^2} - \frac{0.916}{r_s} + Q_c\right], \tag{14.188}$$

14.3 The Electron Gas

where the *correlation energy* $E_c \equiv \frac{e^2}{2a_0} Q_c$ includes terms that are less singular at the *high density limit* $r_s \to 0$ than the first two terms. Using perturbation theory based on the Green's function formalism (see Sec. 18.2), one can calculate a few terms,

$$Q_c = -0.094 + 0.0622 \ln(r_s) + O(r_s), \tag{14.189}$$

but for the time being, we concentrate on $E_0 + E_1$. The result (14.188) will be re-derived below within the mean-field (Hartree–Fock) formalism. The exchange energy E_1 is negative; moreover, there is a region of r_s, where $E_1(r_s) + E_2(r_s) < 0$, namely, the exchange energy overcomes the positive kinetic energy E_0. When $r_s^* = 4.83$, the function $E(r_s)$ has a stable minimum, $E(r_s^*) = -1.29$ eV, as shown in Fig. 14.6. The experimental result for electrons in Na is $E/N = -1.13$ eV. The result obtained above is just $\langle G|H|G \rangle$. Regarding $|G\rangle$ as a variational function, it means that the exact ground-state energy might be lower, so that stability of the electron gas is valid exactly.

In addition to the ground-state energy, other quantities of interest include the pressure,

$$P = -\left(\frac{\partial E}{\partial V}\right)_N = -\frac{dE}{dr_s}\frac{dr_s}{dV} = n\frac{e^2 r_s}{6a_0}\left[\frac{2 \times 2.21}{r_s^3} - \frac{0.916}{r_s^2}\right], \tag{14.190}$$

and the bulk modulus,

$$B = -V\left(\frac{\partial P}{\partial V}\right)_N = n\frac{e^2}{9a_0}\left[\frac{5 \times 2.21}{r_s^2} - \frac{2 \times 0.916}{r_s}\right]. \tag{14.191}$$

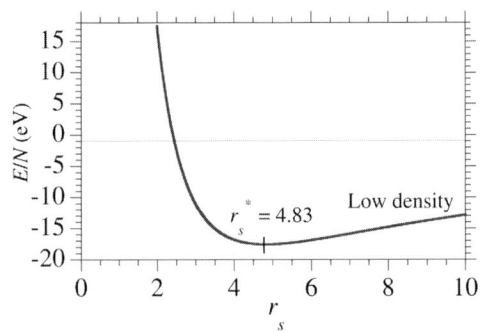

FIG 14.6 First-order estimate of the energy per particle for an interacting electron gas as function of r_s in the jellium model given by Eq. (14.188). To this order, the electron gas is stable at $r_s = r_s^* = 4.83$.

It should be stressed that approximation (14.188) is a high-density (small r_s) expansion, so it cannot be used for the large r_s region. In 1938, Wigner argued that a lower energy than $E(r_s^*)$ can be achieved at large r_s if the electrons form a lattice with negligible zero-point kinetic energy. Using an asymptotic expansion at large r_s, he obtained what is today called the "Wigner cyrstal" (see Sec. 13.7.5) with energy per particle given by

$$E_g = \frac{E}{N} = \frac{e^2}{2a_0}\left[-\frac{1.79}{r_s} + \frac{2.66}{r_s^{3/2}} + \ldots\right]. \tag{14.192}$$

Problem 14.29

(a) Show that

$$\frac{E_0 + E_1}{N} = \left[2.21\, a_0^2 \left(\frac{4\pi n}{3}\right)^{2/3} - 0.916\, a_0 \left(\frac{4\pi n}{3}\right)^{1/3}\right]\frac{e^2}{2a_0}.$$

Guidance: In Eq. (14.188), use (14.170) to write r_s in terms of n.

(b) Consider a gas of spin 1/2 particles interacting through a two-body central (Yukawa) potential $V(|\mathbf{r}-\mathbf{r}'|) = V_0 \frac{e^{-\kappa r}}{\kappa r}$. Determine how Eq. (14.181) should be modified and then calculate E_1.

(c) In a system with N electrons, the numbers N_+ and N_- of spin up and spin down electrons are fixed. Calculate E_g to first order in the interaction and check under what conditions $E(N_+ \neq N_-) < E(N_+ = N_-)$.

Answer: (b) In Eq. (14.181), $e^2 \to V_0/\kappa$ and $q^2 \to (q^2 + \kappa^2)$. There is no need to impose background charge because the term with $\mathbf{q} = 0$ is now allowed. Therefore, the summation over $\mathbf{q}$ in Eq. (14.181) includes $\mathbf{q} = 0$. Dividing the summation to $\mathbf{q} = 0$ and $\mathbf{q} \neq 0$, the second contribution is calculated as before, leading to Eq. (14.187) with the modifications $e^2 \to V_0/\kappa$ and $dx \to x^2 dx/(x^2 + [\kappa/(2k_F)]^2)$. Finally, the calculation of the RHS of Eq. (14.181) with the replacement $e^2/q^2 \to V_0/\kappa^3$ is straight forward.

(c) Expressing E_0 from Eq. (14.177) and E_1 from Eq. (14.187) in terms of k_F and a_0, we obtain $\frac{E_g}{N} \approx \frac{e^2}{2a_0}[\frac{3}{5}(ka_0)^2 - \frac{3}{2\pi}ka_0]$, by assuming that every level is occupied by two electrons with opposite spin projections. Having $N_+ \neq N_-$ implies that electrons with spin-up and spin-down can fill up levels to different wave numbers $k_\uparrow$ and $k_\downarrow$. Is this energetically worthwhile? From Eq. (14.185) we see that the exchange interaction acts only between electrons of parallel spins. Therefore, $E_g = \sum_\sigma N_\sigma E_\sigma$, $E_\sigma = \frac{e^2}{2a_0}[\frac{3}{5}(k_\sigma a_0)^2 - \frac{3}{2\pi}k_\sigma a_0]$, $n = n_\uparrow + n_\downarrow = \frac{k_\uparrow^3}{6\pi^2} + \frac{k_\downarrow^3}{6\pi^2} = \frac{k_F^3}{3\pi^2}$. Now, assume $n_\uparrow = xn$, and $n_\downarrow = (1-x)n$, $0 \leq x \leq 1$, to obtain $k_\uparrow(x), k_\downarrow(x)$ and calculate $E_g(x)$. Then check under what condition it has a minimum at $x \neq 1/2$.

A Gas of Bloch Electrons: Effective Mass Approximation

For electrons in a lattice, the single-particle wave functions are the solutions of the stationary Schrödinger equation,

$$H_{\text{Bloch}} \phi_{n\mathbf{k}\sigma} = \varepsilon_{n\mathbf{k}\sigma} \phi_{n\mathbf{k}\sigma}, \quad \phi_{n\mathbf{k}\sigma}(\mathbf{r}) = e^{i\mathbf{k}\cdot\mathbf{r}} u_{n\mathbf{k}}(\mathbf{r}) \chi_\sigma, \tag{14.193}$$

where n = band index, $\mathbf{k} \in$ FBZ, and $u_{n\mathbf{k}}(\mathbf{r}+\mathbf{R}) = u_{n\mathbf{k}}(\mathbf{r})$. Here, FBZ stands for "First Brillouin zone" and $\mathbf{R}$ is a lattice vector in the direct lattice. All functions of $\mathbf{k}$ are periodic with period $\mathbf{K}$, where $\mathbf{K}$ is any reciprocal lattice vector. It is customary to confine the system in a volume $\mathcal{V} = N^3 \mathbf{a} \times \mathbf{b} \cdot \mathbf{c}$, where $\mathbf{a}, \mathbf{b}$, and $\mathbf{c}$ are the vectors defining the primitive unit cell in the direct lattice. Employing Born–von Karman periodic boundary conditions leads to quantization of $\mathbf{k}$, such that its components along the reciprocal lattice axes are $\mathbf{k} = \frac{2\pi}{N}(\frac{n_a}{a}, \frac{n_b}{b}, \frac{n_c}{c})$ with $n_{a,b,c}$ integers. In constructing the second quantized Hamiltonian, it is possible, in most cases, to restrict oneself to single band n. The operators $c^\dagger_{\mathbf{k}\sigma}$ and $c_{\mathbf{k}\sigma}$ then create and annihilate an electron in a state of Bloch momentum $\mathbf{k}$, respectively. The difficult task is to calculate the matrix element of the Coulomb potential as in Eq. (14.55b) between Bloch functions instead of plane waves. Fortunately, for many applications, the Bloch electron wave function can be approximated by plane waves if at the same time the electronic mass m is changed into a material-dependent effective mass m^*. Explicitly,

$$\phi_{n\mathbf{k}\sigma} \to \frac{1}{\sqrt{\mathcal{V}}} e^{i\mathbf{k}\cdot\mathbf{r}}, \quad m \to m^*, \tag{14.194}$$

where in the plane wave representation $\mathbf{k}$ is not restricted to the FBZ.

14.4 MEAN-FIELD THEORY

In the previous sections, we presented the second quantization formalism for treating interacting identical particles in quantum mechanics. Now, we want to solve some nontrivial problems using this formalism. As has been stressed, solving the Schrödinger equation for an interacting system of particles is, in most cases, virtually impossible. Hence, it is useful to treat a second quantized Hamiltonian such as written in Eq. (14.60) within an approximation scheme. Since problems governed by the noninteracting part H_0 are often soluble, the first approach that comes to mind is using perturbation theory using an expansion in a small dimensionless parameter. Unfortunately, such a small parameter does not always exist (one may think of defining such a parameter as the ratio of the strength of the interaction and kinetic energies, however, often, these energies are comparable and one must go beyond standard perturbation theory). Impressive progress in elucidating the physics of many-body systems have been obtained via development of powerful approximation methods. The Green's function approach discussed in Chapter 18 is a perturbation expansion based on the technique of Feynman diagrams. Yet, it is capable of going beyond perturbation theory where some classes of diagrams can be summed to all orders.

14.4 Mean-Field Theory

In this section, we discuss the *Mean-Field Approximation* (MFA), which is not based on perturbation expansion. The main idea behind the MFA is rather simple. The potential energy felt by a particle in a system of interacting particles is computed as a sum of contributions from all other particles. When this sum is replaced by an averaged field, the many-body problem is reduced to one involving noninteracting particles. The solution of this problem generates a new average field, and the problem has to be solved again with the new field as input. This procedure terminates when self-consistency is achieved and the newly generated average field coincides with the old input field. The MFA formalism in first quantization is presented in Sec. 10.6. Here, we will work it out in second quantization formalism.

14.4.1 MEAN-FIELD EQUATIONS IN SECOND QUANTIZATION

For definiteness, we first consider a system of interacting fermions. The modifications required for treating a bosonic system will be specified at the end, as they are minuscule. As a second example, we will study a system of two kinds of particles. In the first example, we assume $T = 0$ for simplicity, and recall that in that case, the symbol $\langle O \rangle$ indicates expectation of the operator O in the many-body ground state. The second example is worked out at finite temperature. Because we do not necessarily consider systems with translational invariance, creation and annihilation operators are denoted as $c_\alpha^\dagger$ and c_α, respectively, where α is a set of single-particle quantum numbers that include spin projection. Using Eq. (14.60), the second quantized Hamiltonian in the GCE formalism is,

$$H = \sum_\alpha \xi_\alpha c_\alpha^\dagger c_\alpha + \frac{1}{2} \sum_{\alpha\alpha'\beta\beta'} V_{\alpha\beta,\alpha'\beta'} c_\alpha^\dagger c_\beta^\dagger c_{\beta'} c_{\alpha'} \equiv H_0 + V, \quad (14.195)$$

where $\xi_\alpha = \varepsilon_\alpha - \mu$ is the single-particle energy relative to the chemical potential as defined in Eq. (14.164). In second quantization language, the basic approximation of MFA is that the deviation of a product $c_\alpha^\dagger c_{\alpha'}$ from its average $\langle c_\alpha^\dagger c_{\alpha'} \rangle$ is small. This suggests the following decomposition,

$$c_\alpha^\dagger c_{\alpha'} = \langle c_\alpha^\dagger c_{\alpha'} \rangle + [c_\alpha^\dagger c_{\alpha'} - \langle c_\alpha^\dagger c_{\alpha'} \rangle] \equiv \langle c_\alpha^\dagger c_{\alpha'} \rangle + \Delta_{\alpha\alpha'}. \quad (14.196)$$

When this is inserted into the interaction terms, that include products of four operators, terms of order $\Delta_{\alpha\alpha'}^2$ are assumed to be small, hence, they are dropped. To take advantage of this hypothesis, we need to rearrange the interaction and move operators in the original four operator product $c_\alpha^\dagger c_\beta^\dagger c_{\beta'} c_{\alpha'}$, rewriting it, not in its normal order form, but rather in creation-annihilation-creation-annihilation order form. There are two ways to do this, and we need to consider both. First, $c_{\alpha'}$ can hop to the left once over $c_{\beta'}$ and once over $c_\beta^\dagger$. Taking into account the fermion anticommutation relations, the result is $c_\alpha^\dagger c_{\alpha'} c_\beta^\dagger c_{\beta'} - \delta_{\alpha'\beta} c_\alpha^\dagger c_{\beta'}$. Second, c_β can jump over $c_\beta^\dagger$ to the left, yielding $-c_\alpha^\dagger c_{\beta'} c_\beta^\dagger c_{\alpha'} + \delta_{\beta\beta'} c_\alpha^\dagger c_{\alpha'}$. The Hamiltonian (14.195) is then rewritten as,

$$H = H_0' + V',$$

$$H_0' = H_0 + \frac{1}{2} \left\{ \sum_{\alpha\alpha'} \left[\sum_\beta V_{\alpha\beta\alpha'\beta} \right] c_\alpha^\dagger c_{\alpha'} - \sum_{\alpha\beta'} \left[\sum_\beta V_{\alpha\beta\beta\beta'} \right] c_\alpha^\dagger c_{\beta'} \right\},$$

$$V' = \frac{1}{2} \sum_{\alpha\alpha'\beta\beta'} V_{\alpha\beta,\alpha'\beta'} [c_\alpha^\dagger c_{\alpha'} c_\beta^\dagger c_{\beta'} - c_\alpha^\dagger c_{\beta'} c_\beta^\dagger c_{\alpha'}]. \quad (14.197)$$

So far no approximation has been made, we just rearranged terms. The first term in the expression for V' is the *direct contribution*, while the second term is the *exchange contribution*. When this procedure is worked out for bosons, the result is almost the same, except that the sign in front of the exchange contribution that is negative for fermions will be positive for bosons. Note that although the order of operators is modified, the order of indices on the interaction matrix elements is unchanged.

Hartree–Fock Formalism

The Hamiltonian (14.197) is still exact, since it is identical with the initial Hamiltonian (14.195). The modified single-particle Hamiltonian H_0' is quadratic and can be diagonalized. The remaining task is to perform the procedure as in Eq. (14.196) on V' and employ the assumption that terms of order Δ^2 can be neglected. As a result, the interaction becomes quadratic and composed of the Hartree term (from the direct contribution) and the Fock term (from the exchange contribution). The corresponding interactions are

$$V_H = \frac{1}{2} \sum_{\alpha\alpha'\beta\beta'} V_{\alpha\beta,\alpha'\beta'}[\langle c_\alpha^\dagger c_{\alpha'}\rangle c_\beta^\dagger c_{\beta'} + \langle c_\beta^\dagger c_{\beta'}\rangle c_\alpha^\dagger c_{\alpha'} - \langle c_\alpha^\dagger c_{\alpha'}\rangle \langle c_\beta^\dagger c_{\beta'}\rangle], \quad (14.198)$$

$$V_F = \frac{1}{2} \sum_{\alpha\alpha'\beta\beta'} V_{\alpha\beta,\alpha'\beta'}[\langle c_\alpha^\dagger c_{\beta'}\rangle c_\beta^\dagger c_{\alpha'} + \langle c_\beta^\dagger c_{\alpha'}\rangle c_\alpha^\dagger c_{\beta'} - \langle c_\alpha^\dagger c_{\beta'}\rangle \langle c_\beta^\dagger c_{\alpha'}\rangle]. \quad (14.199)$$

The Hartree–Fock mean-field Hamiltonian is then,

$$H^{HF} = H_0' + V_H \mp V_F. \quad (14.200)$$

where $(-)$ is for fermions and $(+)$ is for bosons.

The Hartree–Fock Hamiltonian is quadratic (it contains products of at most two operators); hence, it can be diagonalized. Its ground state $|G\rangle$ depends on all the averages $\langle c_\alpha^\dagger c_\beta\rangle$, which are not known, because to determine them requires knowledge of $|G\rangle$. The way out of this conundrum is achieved by iteration. In the kth iteration, the averages are denoted by $\langle\ldots\rangle^{(k)}$. They serve to generate the kth iteration of the ground state $|G^{(k)}\rangle$. Then we have

$$\langle c_\alpha^\dagger c_\beta\rangle^{(k+1)} = \langle G^{(k)}|c_\alpha^\dagger c_\beta|G^{(k)}\rangle, \quad (14.201)$$

and the iterations stop when $\langle c_\alpha^\dagger c_\beta\rangle^{(k+1)} = \langle c_\alpha^\dagger c_\beta\rangle^{(k)}$.

Example: The Free-Electron Gas

The self-consistent equations (14.201) are rather complicated, and the number of unknown averages $\langle c_\alpha^\dagger c_\beta\rangle$ is M^2, where M is the number of orbitals. In some cases, a given symmetry helps to reduce it. For example, if the Hamiltonian is translationally invariant, then $\langle c_{\mathbf{k}}^\dagger c_{\mathbf{k}'}\rangle = n_\mathbf{k}\delta_{\mathbf{k}\mathbf{k}'}$ and the number of unknowns is reduced to M. This happens for an interacting electron gas in a box.

Consider the free-electron gas governed by the Hamiltonian (14.76) in the absence of screening ($\kappa \to 0$) taking into account the comment after Eq. (14.76). The contribution of the product $c_{\mathbf{k}_1\sigma_1}^\dagger c_{\mathbf{k}_2\sigma_2}^\dagger c_{\mathbf{k}_2-\mathbf{q}\sigma_2} c_{\mathbf{k}_1+\mathbf{q}\sigma_1}$ is easily calculated using equalities such as

$$\langle c_{\mathbf{k}_1\sigma_1}^\dagger c_{\mathbf{k}_1+\mathbf{q}\sigma_1}\rangle c_{\mathbf{k}_2\sigma_2}^\dagger c_{\mathbf{k}_2-\mathbf{q}\sigma_2} = \delta_{\mathbf{q}0}\, n_{\mathbf{k}_1\sigma_1} c_{\mathbf{k}_2\sigma_2}^\dagger c_{\mathbf{k}_2\sigma_2}.$$

The sum of all contributions with $\mathbf{q} = 0$ gives the total electron classical energy, which is cancelled by the positive charge background as asserted by the comment just after Eq. (14.76). Thus, the Hartree term does not appear in the mean-field Hamiltonian of the free-electron gas. The contribution to the Fock term is calculated based on the relation,

$$\langle c_{\mathbf{k}_1\sigma_1}^\dagger c_{\mathbf{k}_2-\mathbf{q}\sigma_2}\rangle c_{\mathbf{k}_2\sigma_2}^\dagger c_{\mathbf{k}_1+\mathbf{q}\sigma_1} = \delta_{\mathbf{q},\mathbf{k}_2-\mathbf{k}_1}\delta_{\sigma_1\sigma_2} c_{\mathbf{k}_2\sigma_2}^\dagger c_{\mathbf{k}_2\sigma_2}.$$

Combining these results, the upshot is that for the free-electron gas,

$$H^{MF} = H^{HF} = \sum_{\mathbf{k}\sigma}\left[\xi_\mathbf{k} - \frac{2\pi e^2}{L^3}\sum_{\mathbf{k}'}\frac{n_{\mathbf{k}'\sigma}}{(\mathbf{k}-\mathbf{k}')^2}\right] c_{\mathbf{k}\sigma}^\dagger c_{\mathbf{k}\sigma}. \quad (14.202)$$

14.4 Mean-Field Theory

In other words, H^{HF} is already diagonal, and the spectrum $E_\mathbf{k}$ is given by the expression in the square brackets. Since $n_{\mathbf{k}'\sigma} > 0$, the exchange (Fock) contribution to the energy is negative. At zero temperature, $n_{\mathbf{k}'\sigma} = \Theta(|\mathbf{k}'| - k_F)$. The sum over $\mathbf{k}'$ can be replaced by an integral using

$$\frac{1}{V} \sum_{|\mathbf{k}'|<k_F} f(\mathbf{k}') \to \frac{1}{(2\pi)^3} \int_{|\mathbf{k}'|<k_F} d\mathbf{k}'\, f(\mathbf{k}').$$

Problem 14.30

Prove that $\frac{2\pi e^2}{(2\pi)^3} \int_{|\mathbf{k}'|<k_F} \frac{d\mathbf{k}'}{|\mathbf{k}-\mathbf{k}'|} = \frac{e^2 k_F}{\pi} F(k/k_F)$, where $F(x) = \frac{1}{2} + \frac{1-x^2}{4x} \ln\left|\frac{1+x}{1-x}\right|$.

Thus, the Hartree–Fock energy (per spin) of an electron of wave number $\mathbf{k}$ in an electron gas interacting through the Coulomb interaction is

$$\varepsilon_\mathbf{k}^{HF} = \frac{\hbar^2 k^2}{2m} - \frac{e^2 k_F}{\pi} F(k/k_F). \tag{14.203}$$

Mean-Field Formalism for Two-Species Hamiltonian

It is instructive to develop the mean-field formalism for a system consisting of two species of particles with the Hamiltonian introduced in Eqs (14.152) and (14.153). It is also assumed that the temperature is finite, $T > 0$, so the averages $\langle \ldots \rangle$ refer to quantum statistical averaging of the corresponding products. Thus,

$$a_\alpha^\dagger a_\beta = \langle a_\alpha^\dagger a_\beta \rangle + [a_\alpha^\dagger a_\beta - \langle a_\alpha^\dagger a_\beta \rangle], \tag{14.204a}$$

$$b_\alpha^\dagger b_\beta = \langle b_\alpha^\dagger b_\beta \rangle + [b_\alpha^\dagger b_\beta - \langle b_\alpha^\dagger b_\beta \rangle]. \tag{14.204b}$$

The product of two small deviations,

$$[a_\alpha^\dagger a_\beta - \langle a_\alpha^\dagger a_\beta \rangle][b_\alpha^\dagger b_\beta - \langle b_\alpha^\dagger b_\beta \rangle] \approx 0, \tag{14.205}$$

is small and can be neglected. Substituting the decompositions (14.204) into the interaction term (14.153), and using (14.205), we obtain the *mean-field potential*,

$$V_{MF} = \sum_{\alpha\alpha',\beta\beta'} V_{\alpha\beta,\alpha'\beta'} \left[a_\alpha^\dagger a_{\alpha'} \langle b_\beta^\dagger b_{\beta'} \rangle + b_\beta^\dagger b_{\beta'} \langle a_\alpha^\dagger a_{\alpha'} \rangle - \langle a_\alpha^\dagger a_{\alpha'} \rangle \langle b_\beta^\dagger b_{\beta'} \rangle \right]. \tag{14.206}$$

Compared with the interacting fermion mean-field Hamiltonian, Eqs (14.198) and (14.198), we see that there is no exchange term here, because in the original two-species Hamiltonian, there is no quadratic terms in the fermion operators. The averages $\langle b_\beta^\dagger b_{\beta'} \rangle$ and $\langle a_\alpha^\dagger a_{\alpha'} \rangle$ are just numbers; hence, V_{MF} is quadratic in creation and annihilation operators, compared with H_I [Eq. (14.153)] which is quartic. Assuming H_A and H_B to be already diagonal, we arrive at the *mean-field Hamiltonian*,

$$H_{MF} = \sum_\alpha \varepsilon_\alpha^A a_\alpha^\dagger a_\alpha + \sum_\beta \varepsilon_\beta^B b_\beta^\dagger b_\beta + V_{MF}. \tag{14.207}$$

Problem 14.31

Use the definition (14.206) to write an expression for $\langle V_{MF} \rangle$ and discuss the role of the last term $-\langle a_\alpha^\dagger a_{\alpha'} \rangle \langle b_\beta^\dagger b_{\beta'} \rangle$.

Answer: From Eq. (14.206) we obtain, by averaging and taking account of cancellation of equal terms, $\langle V_{MF} \rangle = \sum_{\alpha\alpha',\beta\beta'} V_{\alpha\beta,\alpha'\beta'} \langle a_\alpha^\dagger a_{\alpha'} \rangle \langle b_\beta^\dagger b_{\beta'} \rangle$. Consequently, the role of $-\langle a_\alpha^\dagger a_{\alpha'} \rangle \langle b_\beta^\dagger b_{\beta'} \rangle$ is to prevent double counting.

To construct the MF Hamiltonian, we need to know $\langle b_\beta^\dagger b_{\beta'}\rangle$ and $\langle a_\alpha^\dagger a_{\alpha'}\rangle$. From Eq. (14.158), it is clear that this requires knowledge of the eigenstates of the full Hamiltonian. This brings us back again to the iteration scheme. The averages at iteration k, denoted as $\langle a_\alpha^\dagger a_{\alpha'}\rangle^{(k)}$ and $\langle b_\beta^\dagger b_{\beta'}\rangle^{(k)}$, are served to define the mean-field Hamiltonian pertaining to iteration k, denoted as $H_{\text{MF}}^{(k)}$. The dependence of $H_{\text{MF}}^{(k)}$ on $\langle a_\alpha^\dagger a_{\alpha'}\rangle^{(k)}$ and $\langle b_\beta^\dagger b_{\beta'}\rangle^{(k)}$ is found from Eq. (14.207) where on the RHS, the averages are known from the kth iteration.

In the previous discussion, we wrote the iteration equation at zero temperature [see Eq. (14.201)]. At finite temperature, the iteration equations are,

$$\langle a_\alpha^\dagger a_{\alpha'}\rangle^{(k+1)} = \frac{\text{Tr}[e^{-\beta H_{\text{MF}}^{(k)}} a_\alpha^\dagger a_{\alpha'}]}{\text{Tr}[e^{-\beta H_{\text{MF}}^{(k)}}]}, \qquad (14.208)$$

$$\langle b_\beta^\dagger b_{\beta'}\rangle^{(k+1)} = \frac{\text{Tr}[e^{-\beta H_{\text{MF}}^{(k)}} b_\beta^\dagger b_{\beta'}]}{\text{Tr}[e^{-\beta H_{\text{MF}}^{(k)}}]}. \qquad (14.209)$$

The iteration procedure stops when $\langle\ldots\rangle^{(k+1)} = \langle\ldots\rangle^{(k)}$. These are the self-consistent equations for the two-species problem.

Density Functional Theory 15

Most ground-state electronic structure calculations of atoms, molecules, and solids are carried out today using the Hohenberg–Kohn–Sham density functional theory (DFT) method. DFT focuses its attention on the density of systems rather than the wave function as in the usual Schrödinger equation method. The computational effort required with DFT rises as a power law in the number N of electrons as N^α, with $\alpha \approx 2\text{--}3$, rather than the much higher power law of wave function methods such as configuration interaction. Therefore, DFT can handle larger systems. DFT is now incorporated in most electronic structure codes, including GAUSSIAN, GAMESS, MOLPRO, etc., and it is basically the only method that can be used for really big electronic structure calculations, e.g., proteins, enzymes, and nucleic acids.

In standard quantum mechanics bound state calculations, one writes down the Schrödinger equation and solves for the energy eigenvalues and eigenfunctions ψ. The average values of observables are then calculated by taking expectation values of the observable operators with this wave function, typically using some approximation scheme. In DFT, the approach is quite different, as we shall see.

For electronic structure, the Hamiltonian appearing in the Schrödinger equation is comprised of the kinetic energy operator, $\hat{T}$, the interaction energy operator, $\hat{U}$, and the external potential energy operator, $\hat{V}$, $\hat{H} = \hat{T} + \hat{U} + \hat{V}$. The interaction potential of the particles comprising the system is given by a sum over pairs of particles of the form

$$\hat{U} = \sum_{i<j}^{N} U_{ij} = \frac{1}{2} \sum_{i,j}^{N} U_{ij}. \tag{15.1}$$

It is convenient to use the language of second quantization discussed in Chapter 14 to introduce DFT.[1] In second-quantized language, the Hamiltonian for identical particles takes the form [see Eq. (14.88)],

$$\hat{T} = \int d\mathbf{r}\, \hat{\Psi}^\dagger(\mathbf{r}) \left(-\frac{\hbar^2}{2m}\nabla^2\right) \hat{\Psi}(\mathbf{r}), \tag{15.2a}$$

$$\hat{U} = \int d\mathbf{r} \int d\mathbf{r}'\, \hat{\Psi}^\dagger(\mathbf{r})\hat{\Psi}^\dagger(\mathbf{r}')U(\mathbf{r},\mathbf{r}')\hat{\Psi}(\mathbf{r}')\hat{\Psi}(\mathbf{r}), \tag{15.2b}$$

$$\hat{V} = \int d\mathbf{r}\, \hat{\Psi}^\dagger(\mathbf{r})v_{ex}(\mathbf{r})\hat{\Psi}(\mathbf{r}). \tag{15.2c}$$

Here, $v_{ex}(\mathbf{r})$ is the external potential, which in the case of atoms or molecules includes the Coulomb potential between the nuclei and the electrons and any additional external potential, e.g., the potential arising due to external electric fields applied to the system. Expectation values of these operators yield the appropriate average energies, e.g., the external potential energy is given by

$$V = \left\langle \psi \left| \int d\mathbf{r}\, \hat{\Psi}^\dagger(\mathbf{r})v_{ex}(\mathbf{r})\hat{\Psi}(\mathbf{r}) \right| \psi \right\rangle = \int d\mathbf{r}\, n(\mathbf{r})v_{ex}(\mathbf{r}), \tag{15.3}$$

where $|\psi\rangle$ is the state of the system, and we have made use of the fact that $n(\mathbf{r}) = \langle \psi|\hat{n}(\mathbf{r})|\psi\rangle$ with $\hat{n}(\mathbf{r}) \equiv \hat{\Psi}^\dagger(\mathbf{r})\hat{\Psi}(\mathbf{r})$ [$= \sum_{j=1}^{N} \delta(\mathbf{r}-\mathbf{r}_j)$] in first quantization being the density operator.

Many powerful methods for solving the Schrödinger equation exist. However, these methods demand large computational resources, and it is impossible to apply them efficiently to large and complex systems. DFT provides a viable

---

[1] If you have not yet read Chapter 14, just ignore those parts of the equations below that contain the field operator $\hat{\Psi}(\mathbf{r})$; you should still be able to follow the gist of the discussion. Then, after completing this chapter you may want to read Chapter 14 and again look over this chapter.

alternative. It prescribes an algorithm for dealing with the kinetic energy operator $\hat{T}$ and the interaction potential operator $\hat{U}$ in such a way as to turn the quantum many-body problem into a single-particle problem. The key quantity in DFT is the particle density

$$n(\mathbf{r}) = N \int d\mathbf{r}_2 \int d\mathbf{r}_3 \ldots \int d\mathbf{r}_N \, \psi^*(\mathbf{r}, \mathbf{r}_2, \ldots, \mathbf{r}_N) \, \psi(\mathbf{r}, \mathbf{r}_2, \ldots, \mathbf{r}_N). \tag{15.4}$$

The DFT approach can be summarized as follows:[2]

$$n_0(\mathbf{r}) \Longrightarrow v_{ex}(\mathbf{r}) \Longrightarrow \psi_0(\mathbf{r}_1, \ldots, \mathbf{r}_N). \tag{15.5}$$

The particle density specifies the external potential $v_{ex}(\mathbf{r})$, which in turn specifies the many-body ground-state wave function, ψ_0 and, therefore, also specifies all other observables. In DFT, we say that the many-body ground-state wave function is a functional of the density, and we use the notation $|\psi_0[n]\rangle$. For a discussion of functionals, see Sec. 16.10.1 linked to the book web page.

The Hohenberg–Kohn–Sham DFT approach to electronic structure was introduced in the 1960s [259, 260]. It provides both a rigorous conceptual framework and a set of practical tools for calculating the ground-state properties of interacting electron systems. Reviews of DFT are available in Refs. [261–264], and the books by Dreizler and Gross [265], Parr and Yang [266], and Sholl and Steckel [267] are devoted to this subject. DFT has been generalized in many ways, e.g., to treat systems at finite-temperature [268], in time-dependent external fields, in superconducting electronic systems, and in systems as diverse as nuclei, classical fluids, spin density waves, and superfluid liquid He. The finite-temperature version of DFT can be viewed as a fundamental thermodynamic representation of the free energy, with the zero-temperature DFT obtained in the $T \to 0$ limit [270]. DFT, and in particular, time-dependent DFT (see Sec. 15.6) is still a developing field, with a bright future ahead. In this chapter, we provide an introduction to this field. This introduction should suffice to bring the reader to a level, where he/she can access research articles on DFT, and use the DFT computer codes discussed in Sec. 15.7 to carry out DFT calculations.

We begin our description of DFT by introducing the Hohenberg–Kohn theorems as they were introduced in Ref. [259].

15.1 THE HOHENBERG–KOHN THEOREMS

The ground-state wave function, $|\psi_0\rangle$, is a unique functional of the external potential $v_{ex}(\mathbf{r})$, i.e., $|\psi_0[v_{ex}]\rangle$, because the Schrödinger equation with a specific external potential can in principle be solved (for a given interaction potential between the particles comprising the system), to obtain the ground-state wave function. This seems pretty obvious. The Hohenberg–Kohn theorem, which is far from obvious, states

> $v_{ex}(\mathbf{r})$ is a unique functional of the ground-state density $n_0(\mathbf{r})$.

This theorem shows that for the ground state, Eq. (15.4) can be inverted, i.e., given a ground-state density $n_0(\mathbf{r})$, it is possible, at least in principle, to determine the external potential and, therefore, the corresponding ground-state wave function $\psi_0(\mathbf{r}_1, \mathbf{r}_2, \ldots, \mathbf{r}_N)$. Hence, all ground-state observables are also functionals of n_0. If ψ_0 can be determined from n_0 and vice versa, both contain exactly the same information. This seems truly amazing. How can a function of one vector variable, $\mathbf{r}$, be equivalent to a function of N vector variables $\mathbf{r}_1, \ldots, \mathbf{r}_N$? Yet, somehow, it is!

The original proof of the Hohenberg–Kohn theorem [259] is by *reductio ad absurdum*. It goes as follows. Let us assume that the ground state is nondegenerate and that another potential, $\hat{V}'$ [i.e., $v'_{ex}(\mathbf{r})$], has ground state $|\psi'_0\rangle$ that gives rise to the same density $n_0(\mathbf{r})$. Clearly, $|\psi'_0\rangle$ cannot be equal to $|\psi_0\rangle$ because they satisfy different Schrödinger equations, unless $v'_{ex}(\mathbf{r}) - v_{ex}(\mathbf{r}) = $ constant. Denoting the Hamiltonians and ground-state energies associated with $|\psi_0\rangle$ and $|\psi'_0\rangle$ by

---

[2] In what follows, we shall exclusively consider the *ground state electron density*, $n_0(\mathbf{r})$ (at least until Sec. 15.6 which deals with time-dependent DFT). Sometimes we will use the notation $n_0(\mathbf{r})$, but often it will be more convenient to denote the ground state density by $n(\mathbf{r})$.

15.1 The Hohenberg–Kohn Theorems

$\hat{H}, \hat{H}'$ and E_0, E_0', the minimum property of the ground state of the Hamiltonian $\hat{H}'$ yields the condition,

$$E_0' = \langle \psi_0' | \hat{H}' | \psi_0' \rangle < \langle \psi_0 | \hat{H}' | \psi_0 \rangle. \tag{15.6}$$

Letting $\hat{H}' = \hat{H} + (\hat{V}' - \hat{V})$ on the RHS of Eq. (15.6), we find that

$$E_0' < E_0 + \int d\mathbf{r}\, n_0(\mathbf{r})[v_{ex}'(\mathbf{r}) - v_{ex}(\mathbf{r})]. \tag{15.7}$$

Similarly, interchanging primed and unprimed quantities, we find,

$$E_0 < E_0' + \int d\mathbf{r}\, n_0(\mathbf{r})[v_{ex}(\mathbf{r}) - v_{ex}'(\mathbf{r})]. \tag{15.8}$$

Adding Eqs. (15.7) and (15.8) gives

$$E_0 + E_0' < E_0 + E_0', \tag{15.9}$$

which is clearly wrong. Hence, $v_{ex}(\mathbf{r})$ must be a unique functional of n_0. This completes the proof. Since $v_{ex}(\mathbf{r})$ completely specifies the Hamiltonian, for a given interaction between particles, the ground-state wave function, ψ_0, is a unique functional of n_0, i.e., $\psi_0(\mathbf{r}_1, \mathbf{r}_2, \ldots, \mathbf{r}_N) = \psi[n_0(\mathbf{r})]$.

Several alternative proofs of the Hohenberg–Kohn theorem have been developed, e.g., the constrained-search proof of the Hohenberg–Kohn theorem, given independently by Levy and Lieb, and the "strong form of the Hohenberg–Kohn theorem", which is based on the inequality

$$\int d\mathbf{r}\, \Delta n(\mathbf{r}) \Delta v_{ex}(\mathbf{r}) < 0, \tag{15.10}$$

where $\Delta v_{ex}(\mathbf{r})$ is a change in the external potential and $\Delta n(\mathbf{r})$ is the resulting change in the density. Equation (15.10) shows that a change in the potential must result in a change of the density. This is another way of stating that there cannot be two external potentials with the same ground-state density because a change in the potential necessarily yields a change in the ground-state density. The restriction to nondegenerate ground states is not essential.

Hohenberg and Kohn [259] went on and relate the energy functional, $E[n]$, for a given external potential, $v_{ex}(\mathbf{r})$, to an auxiliary quantity, the internal energy functional, $F[n]$, which is a universal functional of the density,

$$E[n] = \min_{\psi \to n} \langle \psi | \hat{T} + \hat{U} | \psi \rangle + \int d\mathbf{r}\, n(\mathbf{r}) v_{ex}(\mathbf{r}) \equiv F[n] + V[n], \tag{15.11}$$

where $\psi \to n$ means that ψ must be such that it yields the correct ground state density. The energy functional is given by $E[n] = T[n] + U[n] + V[n] = F[n] + V[n]$, where the kinetic energy functional is given by [see Eq. (14.88)]

$$T[n] = \left\langle \psi \left| \int d\mathbf{r}\, \hat{\Psi}^\dagger(\mathbf{r}) \left(-\frac{\hbar^2}{2m} \nabla^2 \right) \hat{\Psi}(\mathbf{r}) \right| \psi \right\rangle, \tag{15.12}$$

the external potential function is

$$V[n] = \int d\mathbf{r}\, n(\mathbf{r}) v_{ex}(\mathbf{r}) = \left\langle \psi \left| \int d\mathbf{r}\, \hat{\Psi}^\dagger(\mathbf{r}) v_{ex}(\mathbf{r}) \hat{\Psi}(\mathbf{r}) \right| \psi \right\rangle, \tag{15.13}$$

and for a system of particles interacting with the Coulomb potential, the interaction operator is

$$\hat{U} = \frac{e^2}{2} \int d\mathbf{r} \int d\mathbf{r}'\, \frac{\hat{\Psi}^\dagger(\mathbf{r}) \hat{\Psi}^\dagger(\mathbf{r}') \hat{\Psi}(\mathbf{r}') \hat{\Psi}(\mathbf{r})}{|\mathbf{r} - \mathbf{r}'|}, \tag{15.14}$$

so the interaction energy functional is

$$U[n] = \langle \psi | \hat{U} | \psi \rangle = \sum_{i<j} \frac{e^2}{|\mathbf{r}_i - \mathbf{r}_j|}. \tag{15.15}$$

The internal energy functional, $F[n]$, plays a central role in DFT. But for systems with electron-electron interactions, it is convenient to separate out the classical Coulomb energy and define the functional

$$G[n] \equiv F[n] - U_H[n], \tag{15.16}$$

where $U_H[n]$ is the Hartree energy,

$$U_H[n] \equiv \frac{e^2}{2} \int d\mathbf{r} \int d\mathbf{r}' \frac{n(\mathbf{r})n(\mathbf{r}')}{|\mathbf{r} - \mathbf{r}'|}. \tag{15.17}$$

It is convenient to define the Hartree potential $v_H[n]$ by

$$v_H[n] \equiv e^2 \int d\mathbf{r}' \frac{n(\mathbf{r}')}{|\mathbf{r} - \mathbf{r}'|}, \tag{15.18}$$

so that $U_H[n] = \frac{1}{2}\int d\mathbf{r}\, n(\mathbf{r}) v_H[n(\mathbf{r})]$ and $\frac{\delta U_H}{\delta n(\mathbf{r})} = v_H[n(\mathbf{r})]$. Sometimes, it is an adequate approximation to estimate the interaction energy by the Hartree energy, $U[n] \approx U_H[n]$ (this will give the Hartree approximation). In terms of $G[n]$, the energy functional can be written as

$$E[n] = \int d\mathbf{r}\, n(\mathbf{r}) v_{ex}(\mathbf{r}) + U_H[n] + G[n]. \tag{15.19}$$

Thus, $G[n] = T[n] + (U[n] - U_H[n])$.

Hohenberg and Kohn then provided a variational principle for the ground-state energy; this variational principle is often called the *second Hohenberg–Kohn theorem*. It states:

For any given non-negative trial density, $n(\mathbf{r})$, that integrates to the correct number of electrons, N, i.e., $\int d\mathbf{r}\, n(\mathbf{r}) = N$, the true ground-state energy, E_0, satisfies the relation: $E_0 \leq E[n]$.

That is, by the variational principle, we know that the energy functional is a minimum for the correct ground-state wave function $E_0 = E[\psi_0] = \langle \psi_0 | \hat{H} | \psi_0 \rangle < \langle \psi' | \hat{H} | \psi' \rangle = E[\psi']$; therefore,

$$E[n_0] = F[n_0] + \int d\mathbf{r}\, n_0(\mathbf{r}) v_{ex}(\mathbf{r}) < F[n] + \int d\mathbf{r}\, n(\mathbf{r}) v_{ex}(\mathbf{r}). \tag{15.20}$$

The kinetic energy can similarly be broken up into the kinetic energy of noninteracting particles of density n, $T_{ni}[n]$, and one that represents the remainder, denoted $T_c[n]$. The subscripts ni and c stand for "noninteracting" and "correlation," respectively. In much of the DFT literature, T_{ni} is denoted as T_s, where s could stand for "single-particle" or "Slater-determinant." T_{ni} is the expectation value of the kinetic energy operator $\hat{T}$ with a Slater determinant $\Phi[n] = |\phi_1(\mathbf{r}_1) \ldots \phi_N(\mathbf{r}_N)|$ (for fermions and permanent for bosons), with density n, $T_{ni}[n] = \langle \Phi[n] | \hat{T} | \Phi[n] \rangle$, whereas the full kinetic energy is defined as $T[n] = \langle \psi[n] | \hat{T} | \psi[n] \rangle$. Note that the orbitals $\phi_i(\mathbf{r}_i)$ in $\Phi[n]$ are noninteracting orbitals that satisfy a mean-field equation that will be presented below. Hence, $T[n] = T_{ni}[n] + T_c[n]$, and

$$E[n] = T[n] + U[n] + V[n] = T_{ni}[\{\phi_i[n]\}] + U_H[n] + E_{xc}[n] + V[n], \tag{15.21}$$

where $E_{xc}[n] = T_c[n] + (U[n] - U_H[n])$. The quantity $E_{xc}[n]$ is called the *exchange-correlation* (xc) energy and is composed of the sum of the differences $T - T_{ni}$ (i.e., T_c) and $U - U_H$. The exchange-correlation energy is often decomposed as $E_{xc} = E_x + E_c$, where the exchange energy E_x is due to the Pauli principle (exchange energy) and the correlation energy E_c is due to correlations (meaning all other possible contributions to the energy, even ones that are not understood), with T_c taken to be a part of E_c. The quantity $T_{ni}[n]$ plays an important role in setting up an effective single-body *formally exact* theoretical framework for solving the quantum many-body problem, i.e., setting up the Kohn–Sham equations.

Another way of understanding the correlation energy is to note that the operator representing the Coulomb interaction,

$$\hat{U} = \frac{e^2}{2} \int d\mathbf{r} \int d\mathbf{r}' \, \frac{\hat{n}(\mathbf{r})\hat{n}(\mathbf{r}') - \hat{n}(\mathbf{r})\delta(\mathbf{r}-\mathbf{r}')}{|\mathbf{r}-\mathbf{r}'|}, \quad (15.22)$$

where $\hat{n}(\mathbf{r}) = \hat{\Psi}^\dagger(\mathbf{r})\hat{\Psi}^\dagger(\mathbf{r})$ is the density operator and the term with the delta function subtracts out the interaction of a charged particle with itself, is equivalent to Eq. (15.14). Equation (15.22) can be derived by using the anticommutation relations for fermionic field operators. In contrast, in the Hartree term (15.17), this expectation value of a product is replaced by a product of expectation values, each of the form $n(\mathbf{r}) = \langle \psi | \hat{n}(\mathbf{r}) | \psi \rangle$. This mean-field approximation neglects quantum fluctuations about the expectation values. By writing $\hat{n} = n + \delta \hat{n}_{\text{fluc}}$ and substituting into Eq. (15.22), it becomes clear that the difference between $\langle \psi | \hat{U} | \psi \rangle$ and the Hartree term (15.17) is due to the fluctuations $\delta \hat{n}_{\text{fluc}}$ and the self-interaction correction to the Hartree term. Hence, quantum fluctuations about the expectation value are the origin of quantum correlations between interacting particles.

15.2 THE THOMAS–FERMI APPROXIMATION

An approximate semiclassical method for calculating the electron density of atoms, molecules, as well as condensed phase materials, that is also independent of wave function methods was developed by Llewellyn Thomas in 1927 and Enrico Fermi in 1928. The Thomas–Fermi (TF) method turns out not to be accurate; e.g., it predicts monotonically decreasing electron density in an atom (i.e., no spatial oscillation of the charge density in an atom due to the shell structure). The Kohn–Sham method, to be discussed in the next section, is a generalization of the TF method and does yield quantitative results, but it is useful to first understand the TF method. Let us begin by considering the expression for the number of states that can be filled by electrons up to a given energy E in terms of the integral over the phase space density [see Sec. 9.1, and specifically, Eq. (9.16)]:

$$N(E) = \frac{2}{(2\pi\hbar)^3} \int \int d\mathbf{p} \, d\mathbf{r}, \quad (15.23)$$

where the integral over momentum is up to momentum states with energy E. The factor of 2 in the numerator is due to electron occupation of spin-up and spin-down states for every spatial mode. One defines the local Fermi momentum $p_F(\mathbf{r})$ at a given coordinate point $\mathbf{r}$ to be such that

$$\frac{p_F^2(\mathbf{r})}{2m} + V(\mathbf{r}) = E_F, \quad (15.24)$$

where E_F is the Fermi energy. Assuming that the momentum distribution of the electrons is spherically symmetric, so $d\mathbf{p} = 4\pi \, dp \, p^2$, we find the local density of particles at $\mathbf{r}$ is

$$n(\mathbf{r}) = \frac{2}{(2\pi\hbar)^3} \int d\mathbf{p} = \frac{2}{(2\pi\hbar)^3} 4\pi \int_0^{p_F(\mathbf{r})} dp \, p^2. \quad (15.25)$$

Carrying out the integral on the RHS of Eq. (15.25), we find

$$n(\mathbf{r}) = \frac{1}{3\pi^2 \hbar^3} p_F^3(\mathbf{r}). \quad (15.26)$$

The integral over volume of $n(\mathbf{r})$ must give the number of particles, $\int d\mathbf{r} \, n(\mathbf{r}) = N$. We can calculate the local kinetic energy per unit volume,

$$t_{\text{TF}}(\mathbf{r}) = \frac{2}{(2\pi\hbar)^3} \int d\mathbf{p} \, \frac{p^2}{2m} = \frac{2}{(2\pi\hbar)^3} \frac{4\pi}{2m} \int_0^{p_F(\mathbf{r})} dp \, p^4$$

$$= \frac{2}{(2\pi\hbar)^3} \frac{4\pi}{10m} p_F^5(\mathbf{r}) = \frac{3\hbar^2}{10m} (3\pi^2)^{2/3} n^{5/3}(\mathbf{r}), \quad (15.27)$$

where we have used Eq. (15.26). Hence, the TF kinetic energy functional is

$$T_{TF}[n] = \int d\mathbf{r}\, t_{TF}(\mathbf{r}) = \frac{3\hbar^2}{10m}(3\pi^2)^{2/3} \int d\mathbf{r}\, n^{5/3}(\mathbf{r}). \tag{15.28}$$

Sometimes this is written in terms of the kinetic energy per unit density:

$$\tau_{TF}(\mathbf{r}) = \frac{3\hbar^2}{10m}(3\pi^2 n(\mathbf{r}))^{2/3}, \quad T_{TF}[n] = \int d\mathbf{r}\, n(\mathbf{r})\, \tau_{TF}(\mathbf{r}). \tag{15.29}$$

Adding the Thomas–Fermi kinetic energy, the Hartree energy, and the external potential energy, one obtains the TF approximation to $E[n]$,

$$E_{TF}[n] = T_{TF}[n] + V[n] + U_H[n] = \int d\mathbf{r}\, n(\mathbf{r}) \left[\frac{3\hbar^2}{10m}(3\pi^2 n(\mathbf{r}))^{2/3} + v_{ex}(\mathbf{r}) + v_H[n(\mathbf{r})] \right]. \tag{15.30}$$

The variational equation, $\frac{\delta E_{TF}[n]}{\delta n} = 0$, yields

$$\frac{\hbar^2}{2m}(3\pi^2 n(\mathbf{r}))^{2/3} + v_{ex}(\mathbf{r}) + \frac{e^2}{2}\int d\mathbf{r}'\, \frac{n(\mathbf{r}')}{|\mathbf{r} - \mathbf{r}'|} = 0. \tag{15.31}$$

For an atom with nuclear charge Ze, the external potential is $v_{ex}(\mathbf{r}) = -Ze^2/r$, whereas for a molecule with several nuclear charges eZ_α located at $\mathbf{R}_\alpha$, $v_{ex}(\mathbf{r}) = -\sum_\alpha \frac{e^2 Z_\alpha}{|\mathbf{r}-\mathbf{R}_\alpha|}$. The integral equation (15.31) for $n(\mathbf{r})$ can be turned into a differential equation by applying ∇^2 to both sides and noting that

$$\nabla^2 |\mathbf{r} - \mathbf{r}'|^{-1} = -4\pi \delta(\mathbf{r} - \mathbf{r}'). \tag{15.32}$$

Note that the *Poisson equation* tells us that ∇^2 applied to the quantity $[v_{ex}(\mathbf{r}) + v_H[n(\mathbf{r})]]$ appearing in Eq. (15.31) is given in terms of the total change density (due to the electrons and the nuclei),

$$\nabla^2 [v_{ex}(\mathbf{r}) + v_H[n(\mathbf{r})]] = -4\pi e \rho_{tot}(\mathbf{r}), \tag{15.33}$$

where $\rho_{tot}(\mathbf{r})$ is the total charge density due to the electrons and nuclei, $\rho_{tot}(\mathbf{r}) = (-e)n(\mathbf{r}) + e\sum_\alpha Z_\alpha \delta(\mathbf{r} - \mathbf{R}_\alpha)$.

Problem 15.1

What is the total Thomas–Fermi energy of a neutral atom with nuclear charge Ze in terms of the electron density $n(\mathbf{r})$?

Answer: $E_{TF}[n] = \int d\mathbf{r} \left[\frac{3\hbar^2 n(\mathbf{r})}{10m}(3\pi^2 n(\mathbf{r}))^{2/3} - \frac{Ze^2 n(\mathbf{r})}{r} + \frac{Ze^2}{2} n(\mathbf{r}) \int d\mathbf{r}'\, \frac{n(\mathbf{r}')}{|\mathbf{r}-\mathbf{r}'|} \right].$

Problem 15.2

Obtain the variational solution for the minimum of the energy functional
$\tilde{E}[n(\mathbf{r})] = \tilde{T}[n(\mathbf{r})] + \tilde{U}_H[n(\mathbf{r})] + \tilde{E}_{xc}[n(\mathbf{r})] + \tilde{V}_{ex}[n(\mathbf{r})]$ by taking the variation with respect to $n(\mathbf{r})$, when
$\tilde{T}[n(\mathbf{r})] = C_T \int d\mathbf{r}\, n^{5/3}(\mathbf{r})$, $\tilde{U}_H[n(\mathbf{r})] = \frac{e^2}{2}\int d\mathbf{r}\, \frac{n^2(\mathbf{r})}{r}$, $\tilde{E}_{xc}[n(\mathbf{r})] = -C_{xc} \int d\mathbf{r}\, n^{4/3}(\mathbf{r})$, and
$\tilde{V}_{ex}[n(\mathbf{r})] = \int d\mathbf{r}\, v_{ex}(\mathbf{r}) n(\mathbf{r})$, for a system with N electrons, i.e., $\int d\mathbf{r}\, n(\mathbf{r}) = N$.

15.2 The Thomas–Fermi Approximation

Answer: $\frac{5}{3}C_T n^{2/3}(\mathbf{r}) + \frac{e^2}{r}n(\mathbf{r}) - \frac{4}{3}C_{xc}n^{1/3}(\mathbf{r}) + v_{ex}(r) - \lambda = 0$, where λ is the Lagrange multiplier for the constraint on the number of electrons. The solutions are comprised of the three roots of the cubic equation for $n^{1/3}(\mathbf{r})$ at each r, and the minimum value of $\tilde{E}[n(\mathbf{r})]$ at each r must be determined from one of these solutions. The value of the Lagrange multiplier is determined by ensuring that $\int d\mathbf{r}\, n(\mathbf{r}) = N$.

A major source of inaccuracy in the Thomas–Fermi approach is due to the procedure of approximating the kinetic energy as the density functional in (15.28). Moreover, TF also neglects correlations, i.e., it assumes that $U[n] = U_H[n]$. As a result of these approximations, the TF density of atoms has no shell structure. Furthermore, within TF, molecules are unstable, i.e., the energy of isolated atoms is lower than the bound molecule they comprise. These errors will be corrected in the Kohn–Sham approach. Nevertheless, it is of interest to understand the TF approximation to see how the Kohn–Sham approach overcomes the shortcomings inherent in the TF approximation. In Problem 15.2, you will work out how to determine the TF density for atoms. Figure 15.1 shows the function $\chi(x)$ of Problem 15.2, from which the total potential $v(\mathbf{r}) = v_{ex}(\mathbf{r}) + v_H(\mathbf{r})$ and the electron density $n(\mathbf{r})$ are obtained; clearly, no shell structure is present.

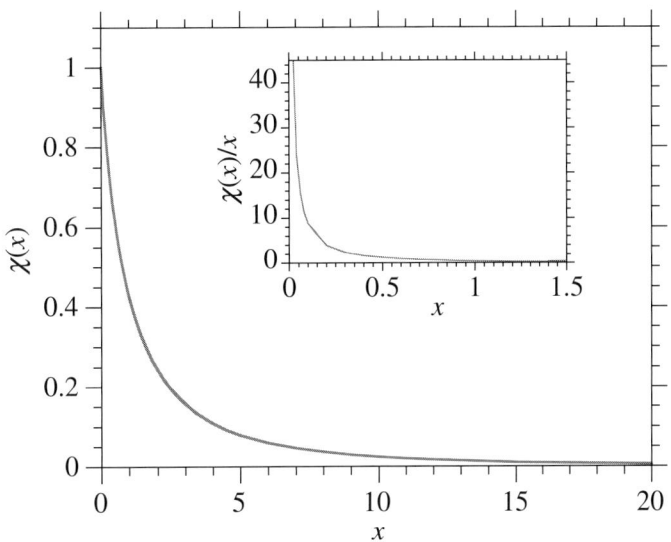

FIG 15.1 The function $\chi(x)$ of Problem 15.2 versus x. Inset shows $\chi(x)/x$. The density is proportional to $(\chi(x)/x)^{3/2}$.

For the exchange energy, $E_x[n]$ can be written in terms of the exchange energy density $e_x(n)$, $E_x[n] = \int d\mathbf{r}\, e_x(n(\mathbf{r}))$. For the homogeneous electron gas with wave functions given by plane waves, the exchange energy density was originally calculated by Dirac and is given by [265, 266]

$$e_x(n) = -\frac{3e^2}{4}\left(\frac{3}{\pi}\right)^{1/3} n^{4/3}, \quad E_x[n] = -\frac{3e^2}{4}\left(\frac{3}{\pi}\right)^{1/3} \int d\mathbf{r}\, n^{4/3}(\mathbf{r}). \quad (15.34)$$

The Dirac exchange energy $E_x[n]$ can also be written in terms of the *exchange energy per particle*, $\epsilon_x[n] = -\frac{3e^2}{4}\left(\frac{3}{\pi}\right)^{1/3} n^{1/3}$, as $E_x[n] = \int d\mathbf{r}\, n(\mathbf{r})\epsilon_x(\mathbf{r})$. We can append the exchange energy $E_x[n]$ of Eq. (15.34) to the TF energy to obtain the so-called *Thomas–Fermi-Dirac approximation* to $E[n]$. Multiplying $E_x[n]$ in Eq. (15.34) by an adjustable parameter α, one obtains the so-called $X\alpha$ approximation to $E_{xc}[n]$ (applying the uniform-electron gas approximation in different places leads to different values of α [265, 266]).

Attempts to fix the problem of approximating the kinetic energy in the TF approximation have been developed. The local Fermi momentum is given by Eq. (15.26) as $p_F(\mathbf{r}) = [3\pi^2\hbar^3 n(\mathbf{r})]^{1/3}$ and the local kinetic energy per unit density is given by Eq. (15.28) as $\tau_{TF}(\mathbf{r}) = \frac{3\hbar^2}{10m}(3\pi^2)^{2/3} n^{2/3}(\mathbf{r})$. But these local quantities do not really account for the variation of gradient of the density with position. The *gradient expansion approximation* (GEA) for the kinetic energy can correct this. A generalized functional $\mathcal{G}[n]$ that includes a gradient expansion can be written in the form:

$$\mathcal{G}[n] = \int d\mathbf{r}\, g(n(\mathbf{r}), \nabla n(\mathbf{r})) = \int d\mathbf{r}\, \Big\{ g_1(n(\mathbf{r})) + g_2(n(\mathbf{r}))|\nabla n(\mathbf{r})|^2 \\ + g_3(n(\mathbf{r}))[\nabla^2 n(\mathbf{r})]^2 + g_4(n(\mathbf{r}))\nabla^2 n(\mathbf{r})|\nabla n(\mathbf{r})|^2 + \ldots \Big\}. \quad (15.35)$$

The GEA for the kinetic energy of a noninteracting electron gas is given by $T_{ni}[n] = \int d\mathbf{r}\, n(\mathbf{r})\tau(\mathbf{r})$, where to lowest order in $\nabla n(\mathbf{r})$,

$$\tau(\mathbf{r}) = \frac{3\hbar^2}{10m}(3\pi^2 n(\mathbf{r}))^{2/3} + \frac{1}{72}\frac{|\nabla n(\mathbf{r})|^2}{n^2(\mathbf{r})}. \quad (15.36)$$

The $|\nabla n(\mathbf{r})|^2$ term is called the Weizsäcker correction [265, 266].

Problem 15.3

Determine the self-consistent potential, $v(r)$, and electron density, $n(r)$, for an atom with nuclear charge Ze, using the Thomas–Fermi approximation, $\tau_{TF}(\mathbf{r}) = \frac{3\hbar^2}{10m}(3\pi^2 n(\mathbf{r}))^{2/3}$, the local energy equation $\tau_{TF}(\mathbf{r}) + v(\mathbf{r}) = \epsilon$, and Poisson's equation, $\nabla^2 v = -4\pi e \rho(\mathbf{r}) = -4\pi e[Ze\, \delta(\mathbf{r}) + (-e)n(\mathbf{r})]$.

Answer: Substituting the kinetic energy functional into the local energy equation, we find $v(\mathbf{r}) = \epsilon - \frac{3\hbar^2}{10m}(3\pi^2)^{2/3} n^{2/3}(\mathbf{r})$, which can be inverted to yield $n(\mathbf{r}) = \frac{(10m/3)^{3/2}}{3\pi^2 \hbar^3}[v(\mathbf{r}) - \epsilon]^{3/2}$. Substituting this expression into the RHS of Poisson's equation, we obtain

$$\nabla^2 v = 4\pi e^2 \left(-Z\delta(\mathbf{r}) + \frac{(10m/3)^{3/2}}{3\pi^2 \hbar^3}[v(\mathbf{r}) - \epsilon]^{3/2} \right).$$

Writing $v - \epsilon = \frac{Ze^2}{r}\chi(r)$ and using Eq. (15.32), we find $\frac{d^2\chi}{dr^2} = \frac{4Z^{1/2}e^3}{3\pi\hbar^3}(10m/3)^{3/2}r^{-1/2}\chi^{3/2}$. Changing variables by defining a dimensionless position variable x such that $r = ax$ to eliminate the constants in the previous equation yields, i.e., using $a = \left(\frac{4Z^{1/2}e^3}{3\pi\hbar^3}\right)^{2/3}(10m/3)$, we find $\frac{d^2\chi}{dx^2} = x^{-1/2}\chi^{3/2}(x)$. This *Thomas–Fermi equation* can be solved numerically with boundary conditions, $\chi(0) = 1$ and $\chi(\infty) = 0$ (see Fig. 15.1). We thereby obtain $v(r) = \frac{Ze^2}{r}\chi(r) - \epsilon$, and $n(\mathbf{r}) = \frac{(10m/3)^{3/2}}{3\pi^2\hbar^3}[\frac{Ze^2}{r}\chi(r)]^{3/2}$ is in hand. In order for $v(r)$ to go asymptotically to zero, we require $\epsilon = 0$.

15.3 THE KOHN–SHAM EQUATIONS

We already mentioned in the paragraph containing Eq. (15.21) that a more accurate scheme than TF for treating the kinetic energy functional of interacting electrons, $T[n]$, is based on decomposing it into one part that represents the kinetic energy of noninteracting particles of density n, and another part that comes from correlation. That is, $T[n]$ is given by the sum of the quantity we have called $T_{ni}[n]$, and the remainder, denoted $T_c[n]$, where the subscripts ni and c stand for noninteracting and correlation,

$$T[n] = T_{ni}[n] + T_c[n]. \quad (15.37)$$

T_{ni} is defined as the expectation value of the kinetic energy operator $\hat{T}$ with a Slater determinant arising from noninteracting electron wave functions yielding density n, i.e., $T_{ni}[n] = \langle \Phi[n]|\hat{T}|\Phi[n]\rangle$ and $n(\mathbf{r}) = |\Phi[n(\mathbf{r})]|^2$. Similarly, the full kinetic energy is defined as $T[n] = \langle \psi[n]|\hat{T}|\psi[n]\rangle$, where $|\psi[n]\rangle$ is the full N-electron wave function with density n. All consequences of antisymmetrization (exchange) are incorporated by using a determinantal wave function in defining T_{ni}. The difference between T and T_{ni}, i.e., T_c, is solely due to correlation. $T_{ni}[n]$ is an unknown functional of n. Using a

15.3 The Kohn–Sham Equations

local density approximation (LDA) to approximate it, leads back to the TF approximation (which is an LDA). Instead, expressing T_{ni} in terms of the single-particle orbitals ϕ_i of a noninteracting system with density n leads to the expression

$$T_{ni}[n] = -\frac{\hbar^2}{2m}\sum_i^N \int d\mathbf{r}\, \phi_i^*(\mathbf{r})\nabla^2\phi_i(\mathbf{r}), \qquad (15.38)$$

For the noninteracting particles, the total kinetic energy is just the sum of the individual kinetic energies of the particles. Since all ϕ_i are functionals of n, this expression for T_{ni} is NOT an explicit orbital functional but an implicit density functional, $T_{ni}[n] = T_{ni}[\{\phi_i[n]\}]$.

As noted in Eq. (15.21), the exact energy functional can be written as

$$E[n] = T[n] + U[n] + V[n] = T_{ni}[\{\phi_i[n]\}] + U_H[n] + E_{xc}[n] + V[n], \qquad (15.39)$$

where by definition E_{xc} contains the differences $T - T_{ni}$ (i.e., T_c) and $U - U_H$, so clearly, a part of the correlation energy E_c is due to the difference T_c between the noninteracting and the interacting kinetic energies. Equation (15.39) is formally exact, but E_{xc} is unknown. However, the HK theorem guarantees that it is a density functional. The exchange-correlation functional, $E_{xc}[n]$, is often decomposed as $E_{xc} = E_x + E_c$, where E_x is due to the Pauli principle (i.e., the exchange energy) and E_c is due to correlation. The exchange energy can be written explicitly in terms of the single-particle orbitals as a Fock-type integral,

$$E_x[\{\phi_i[n]\}] = -\frac{e^2}{2}\sum_{jk}\int d\mathbf{r}\int d\mathbf{r}'\,\frac{\phi_j^*(\mathbf{r})\phi_k^*(\mathbf{r}')\phi_j(\mathbf{r}')\phi_k(\mathbf{r})}{|\mathbf{r}-\mathbf{r}'|}, \qquad (15.40)$$

but no general exact expression for E_x in terms of the density is known. This expression differs from the exchange energy used in the Hartree–Fock approximation only in the substitution of Hartree–Fock orbitals $\phi_i^{HF}(\mathbf{r})$ by the Kohn–Sham orbitals $\phi_i(\mathbf{r})$, which are to be introduced below.

Since T_{ni} is written as an orbital functional, we cannot directly minimize Eq. (15.39) with respect to n. Instead, we use a scheme suggested by Kohn and Sham [260] for performing the minimization indirectly. We write the minimization $\frac{\delta E[n]}{\delta n} = 0$ as,

$$\frac{\delta T_{ni}[n]}{\delta n} + \frac{\delta V[n]}{\delta n} + \frac{\delta U_H[n]}{\delta n} + \frac{\delta E_{xc}[n]}{\delta n} = \frac{\delta T_{ni}[n]}{\delta n(\mathbf{r})} + v_{ex}(\mathbf{r}) + v_H[n(\mathbf{r})] + v_{xc}[n(\mathbf{r})] = 0. \qquad (15.41)$$

The term $\frac{\delta U_H}{\delta n(\mathbf{r})} = v_H[n(\mathbf{r})]$ is the Hartree potential, introduced in Eq. (15.18), and the term $\frac{\delta E_{xc}}{\delta n(\mathbf{r})} = v_{xc}[n(\mathbf{r})]$ can be calculated explicitly once an approximation for E_{xc} is known. The quantity $v_{xc}[n]$ is called the *exchange-correlation potential*.

To put this into perspective, consider a system of noninteracting particles moving in the potential $v_{ni}(\mathbf{r})$. For this system, the minimization condition is $\frac{\delta E_{ni}[n]}{\delta n} = \frac{\delta T_{ni}[n]}{\delta n} + \frac{\delta V_{ni}[n]}{\delta n} = 0$, i.e.,

$$\boxed{\frac{\delta T_{ni}[n]}{\delta n(\mathbf{r})} + v_{ni}(\mathbf{r}) = 0.} \qquad (15.42)$$

No Hartree and exchange-correlation terms are present in the absence of interactions. The density obtained by solving this Euler equation is denoted by $n_{ni}(\mathbf{r})$. By comparing Eq. (15.42) with Eq. (15.41), we note that both minimizations have the same solution $n_{ni}(\mathbf{r}) \equiv n(\mathbf{r})$, if v_{ni} is taken to be

$$v_{ni}[n(\mathbf{r})] = v_{ex}(\mathbf{r}) + v_H[n(\mathbf{r})] + v_{xc}[n(\mathbf{r})]. \qquad (15.43)$$

Hence, the density of the interacting many-body system in a potential $v_{ex}(\mathbf{r})$ can be calculated by solving the noninteracting system in a potential $v_{ni}[n(\mathbf{r})]$ given by Eq. (15.43). Equations (15.42) and (15.43) must be solved self-consistently. In the absence of $v_{xc}[n(\mathbf{r})]$, this would be accomplished by solving the self-consistent Hartree equations (10.35). Hence, with $v_{xc}[n(\mathbf{r})]$ present, it is reasonable to require the solution of the following equations:

$$\left\{-\frac{\hbar^2}{2m}\nabla^2 + v_{ni}[n(\mathbf{r})]\right\}\phi_i(\mathbf{r}) = \varepsilon_i\phi_i(\mathbf{r}), \quad (15.44a)$$

$$v_{ni}[n(\mathbf{r})] = v_{ex}(\mathbf{r}) + v_H[n(\mathbf{r})] + v_{xc}[n(\mathbf{r})]. \quad (15.44b)$$

These equations are called the *Kohn–Sham self-consistent eigenvalue equations*. A detailed derivation of these equations can be found in Appendix A of Ref. [262] using functional differentiation. Note that the density-dependent effective potential $v_{ni}[n(\mathbf{r})]$ is a functional of the density due to the presence of both the Hartree potential $v_H[n(\mathbf{r})]$, as given by Eq. (15.18), and the exchange-correlation potential $v_{xc}[n(\mathbf{r})]$. The total effective potential is called the *Kohn–Sham noninteracting potential* and is given by $v_{ni}[n(\mathbf{r})] = v_{ex}(\mathbf{r}) + v_H[n(\mathbf{r})] + v_{xc}[n(\mathbf{r})]$. The total electron density is given by $n(\mathbf{r}) = \sum_i |\phi_i(\mathbf{r})|^2$, where the sum is over all occupied orbitals.

In atoms or molecules, the external potential $v_{ex}(\mathbf{r})$ is due to both nuclei and any external potential applied to the system (e.g., external electric or magnetic fields). The electron density, $n(\mathbf{r})$, is obtained from the occupied orbitals, which are usually doubly occupied for spin-singlet states,

$$n(\mathbf{r}) = 2\sum_{i=1}^{N/2} |\phi_i(\mathbf{r})|^2. \quad (15.45)$$

The electrostatic potential for a system of electrons and nuclei, which includes the external potential from the nuclei plus the Hartree potential, is given by

$$v_{ex}(\mathbf{r}) + v_H[n(\mathbf{r})] = e^2\left[-\sum_{\alpha=1}^{N_n} \frac{Z_\alpha}{|\mathbf{r}-\mathbf{R}_\alpha|} + \int d\mathbf{r}' \frac{n(\mathbf{r}')}{|\mathbf{r}-\mathbf{r}'|}\right]. \quad (15.46)$$

This potential can be obtained by numerical solution of the Poisson equation, (15.33), if the electron density and the nuclear positions and charges are specified.

The total ground-state electronic energy $E[n]$ can be obtained by noting that

$$\sum_i \varepsilon_i = \sum_i \langle\phi_i|-\frac{\hbar^2}{2m}\nabla^2 + v_{ni}(\mathbf{r})|\phi_i\rangle = T_{ni}[n] + \int d\mathbf{r}\, n(\mathbf{r})v_{ni}(\mathbf{r}). \quad (15.47)$$

Hence, using Eqs (15.39) and (15.44b), we find

$$E[n] = \sum_i \varepsilon_i - U_H[n] + E_{xc}[n] - \int d\mathbf{r}\, n(\mathbf{r})v_{xc}(\mathbf{r}). \quad (15.48)$$

Problem 15.4

Derive Eq. (15.46) using Eqs (15.47), (15.39), and (15.44b).
Hint: Note that $U_H[n] = \frac{1}{2}\int d\mathbf{r}\, n(\mathbf{r})v_H[n(\mathbf{r})]$.

Much effort has been directed in the derivation of an accurate exchange-correlation potential by gradient expansions and hybrid methods that include Hartree–Fock exchange contributions. Using these modifications, results of chemical accuracy[3] can be obtained using DFT for all ground-state properties of electronic systems.

---

[3] Chemical accuracy is the accuracy required to make realistic chemical predictions, which is considered to be about 4 kJ/mol.

15.3 The Kohn–Sham Equations

Given an exchange-correlation potential, self-consistent solution of the Kohn–Sham equations [Eqs (15.44a)–(15.44b)] for fixed nuclear locations is conceptually straightforward. An initial guess is made for the orbitals. This yields an electron density from which the effective noninteraction potential v_{ni} is constructed by solution of the Poisson equation and generation of the exchange-correlation potential. The eigenvalue equation is solved with the current effective potential, Eq. (15.44b), resulting in a new set of orbitals. The process is repeated until the total energy [or density] change is smaller than some desired tolerance. A flow chart of the process is shown in Fig. 15.3.

15.3.1 LOCAL DENSITY APPROXIMATION

If the system is such that the density is slowly spatially varying, the exchange-correlation potential can be taken as a local function of the density, i.e., the LDA can be made:

$$E_{xc}[n] = \int d\mathbf{r}\, n(\mathbf{r})\epsilon_{xc}[n(\mathbf{r})], \tag{15.49}$$

where $\epsilon_{xc}[n(\mathbf{r})]$ is the exchange-correlation energy per particle in a uniform electron gas of density n. The exchange-correlation potential is then given by

$$v_{xc}[n(\mathbf{r})] = \epsilon_{xc}[n(\mathbf{r})] + n(\mathbf{r})\frac{\delta\epsilon_{xc}[n]}{\delta n(\mathbf{r})}. \tag{15.50}$$

The total energy in the LDA becomes

$$E[n] = \sum_i \varepsilon_i - \frac{e^2}{2}\int d\mathbf{r}\int d\mathbf{r}'\,\frac{n(\mathbf{r})n(\mathbf{r}')}{|\mathbf{r}-\mathbf{r}'|} + \int d\mathbf{r}\, n(\mathbf{r})\{\epsilon_{xc}[n(\mathbf{r})] - v_{xc}[n(\mathbf{r})]\}. \tag{15.51}$$

The simplest LDA uses the form

$$\epsilon_{xc} = -\frac{9}{8}\left(\frac{3}{\pi}\right)^{1/3}\alpha\, n^{1/3}. \tag{15.52}$$

With $\alpha = 2/3$, Eq. (15.52) corresponds to the exchange energy, Eq. (15.34). This yields an exchange potential

$$v_{xc} = -\frac{3}{2}\left(\frac{3}{\pi}\right)^{1/3}\alpha\, n^{1/3}. \tag{15.53}$$

For arbitrary α, this corresponds to the $X\alpha$ approximation mentioned after Eq. (15.34). Figure 15.2 plots ϵ_{xc} versus n for $\alpha = 1$.

Several other approximations have been developed for the exchange-correlation energy, including the Gunnarsson and Lundqvist form

$$\epsilon_{xc,GL} = -\frac{0.458}{r_s} - G\left(\frac{r_s}{11.4}\right), \tag{15.54}$$

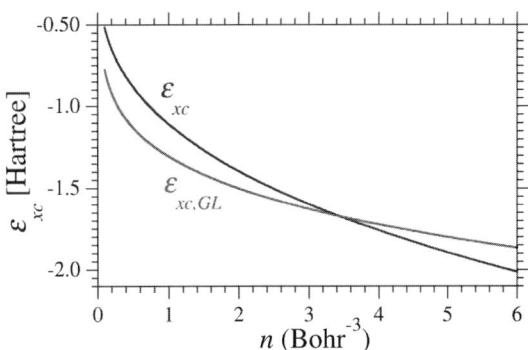

FIG 15.2 Exchange-correlation energy per particle, ϵ_{xc}, given by Eq. (15.52) with $\alpha = 1$ in a uniform electron gas versus density n, and the Gunnarsson–Lundqvist exchange-correlation energy per particle, $\epsilon_{xc,GL}$ given in Eq. (15.54).

where the Wigner–Seitz radius r_s is defined by the relation, $n^{-1} = \frac{4\pi}{3}r_s^3$, and the function $G(x)$ is defined by $G(x) = \frac{1}{2}[(1+x^3)\ln(1+1/x) - x^2 + x/2 - 1/3]$ (see Fig. 15.2). Ceperley and Alder [274] numerically determined the exchange-correlation energy for a uniform electron gas versus density via Monte Carlo simulation. Their result has been parameterized and used in many calculations. A (spin) LDA (see Sec. 15.4.1) that is more commonly used today is one that was developed by Perdew and Wang [275].

15.4 SPIN DFT AND MAGNETIC SYSTEMS

The most widely used form of DFT is *spin DFT*. It is necessary because the ground state of many atoms and molecules are not spin-singlet states, e.g., the ground state of the oxygen molecule is a triplet state, and many atoms and molecules have an odd number of electrons so not all spins can be paired. Moreover, in the presence of an external magnetic field, $\mathbf{H}(\mathbf{r})$, the system ground state has a net magnetization. Furthermore, DFT can be used to treat ferromagnetic systems, wherein a finite magnetization is present even without the presence of an external magnetic field. To treat such systems within DFT, we must consider two densities, $n_\uparrow(\mathbf{r})$ for spin-up and $n_\downarrow(\mathbf{r})$ for spin-down electrons. The spin DFT formulation is not very different from the spinless DFT presented earlier.

In the presence of an external magnetic field, the external potential depends on the density of the spin components,

$$V[n_\uparrow, n_\downarrow] = \int d\mathbf{r}' \, [n(\mathbf{r})v_{ex}(\mathbf{r}) - \mathbf{H}(\mathbf{r}) \cdot \mathbf{m}(\mathbf{r})], \tag{15.55}$$

where the total electron density is given by

$$\mathbf{n}(\mathbf{r}) = \sum_{i,m_s} \phi^*_{i,m_s}(\mathbf{r})\phi_{i,m_s}(\mathbf{r}), \tag{15.56}$$

and the magnetization vector is given by

$$\mathbf{m}(\mathbf{r}) = -\frac{g\mu_B}{2} \sum_{i,m_s,m'_s} \phi^*_{i,m_s}(\mathbf{r}) \sigma_{m_s,m'_s} \phi_{i,m'_s}(\mathbf{r}). \tag{15.57}$$

Here, σ_{m_s,m'_s} is the m_s, m'_s matrix element of the Pauli spin vector. For a uniform magnetic field, we can take the the z-axis to be along the field direction. The density is the sum of the spin-up and spin-down densities $n(\mathbf{r}) = n_\uparrow(\mathbf{r}) + n_\downarrow(\mathbf{r})$, and the magnetization, $m(\mathbf{r}) = -\frac{g\mu_B}{2}[n_\uparrow(\mathbf{r}) - n_\downarrow(\mathbf{r})]$, is proportional to the difference. There are now two external "potentials" present, $\{v_{ex}(\mathbf{r}), \mathbf{H}(\mathbf{r})\}$, and these fields couple to $\{n(\mathbf{r}), \mathbf{m}(\mathbf{r})\}$ in the external potential $V[n, \mathbf{m}] \equiv V[n_\uparrow, n_\downarrow]$.

We now show that, in direct analogy with the previous sections, the external potentials $\{v_{ex}(\mathbf{r}), \mathbf{H}(\mathbf{r})\}$ uniquely determine the ground state "density" $\{n(\mathbf{r}), \mathbf{m}(\mathbf{r})\}$. The energy functional is now given by

$$\begin{aligned}E[n, \mathbf{m}] &= T[n, \mathbf{m}] + U[n, \mathbf{m}] + V[n, \mathbf{m}] \\ &= T_{ni}[\{\phi_i[n, \mathbf{m}]\}] + U_H[n, \mathbf{m}] + E_{xc}[n, \mathbf{m}] + V[n, \mathbf{m}],\end{aligned} \tag{15.58}$$

where all the functionals depend on $\{n(\mathbf{r}), \mathbf{m}(\mathbf{r})\}$, or, $n_\uparrow(\mathbf{r})$ and $n_\downarrow(\mathbf{r})$. Again, it is convenient to separate the kinetic energy into a noninteracting part and the remainder, $T[n, \mathbf{m}] = T_{ni}[n, \mathbf{m}] + T_c[n, \mathbf{m}]$, where T_{ni} is defined as the expectation value of the kinetic energy operator $\hat{T}$ with the Slater determinant, $T_{ni}[n, \mathbf{m}] = \langle \Phi[n, \mathbf{m}] | \hat{T} | \Phi[n, \mathbf{m}] \rangle$. Expressing T_{ni} in terms of the single-particle orbitals $\phi_{i,m_s}(\mathbf{r})$ of a noninteracting system with density n and magnetization $\mathbf{m}$ leads to the expression

$$T_{ni}[n, \mathbf{m}] = -\frac{\hbar^2}{2m} \sum_{i,m_s} \int d\mathbf{r} \, \phi^*_{i,m_s}(\mathbf{r}) \nabla^2 \phi_{i,m_s}(\mathbf{r}). \tag{15.59}$$

An explicit expression for the exchange-correlation functional, $E_{xc}[n, \mathbf{m}] = T[n, \mathbf{m}] + U[n, \mathbf{m}] - U_H[n, \mathbf{m}]$, is not known. Nevertheless, we can define the quantities

$$v_{xc}(\mathbf{r}) = \frac{\delta E_{xc}[n, \mathbf{m}]}{\delta n(\mathbf{r})}, \quad \mathbf{H}_{xc}(\mathbf{r}) = \frac{\delta E_{xc}[n, \mathbf{m}]}{\delta \mathbf{m}(\mathbf{r})}, \tag{15.60}$$

15.4 Spin DFT and Magnetic Systems

By taking the variation of the energy functional with respect to $\{n(\mathbf{r}), \mathbf{m}(\mathbf{r})\}$ one can derive the Kohn–Sham equations for spin DFT:

$$\sum_{m'_s}\left[\left(-\frac{\hbar^2\nabla^2}{2m} + v_{in}\right)\delta_{m_s,m'_s} - \mathbf{H}_{ni}\cdot\mathbf{m}(\mathbf{r})\right]\phi_{i,m'_s}(\mathbf{r}) = \varepsilon_{i,m_s}\phi_{i,m_s}(\mathbf{r}), \quad (15.61a)$$

$$v_{in}[n(\mathbf{r}),\mathbf{m}(\mathbf{r})] = v_{ex}(\mathbf{r}) + v_H[n(\mathbf{r})] + v_{xc}[n(\mathbf{r}),\mathbf{m}(\mathbf{r})], \quad (15.61b)$$

$$\mathbf{H}_{ni}[n(\mathbf{r}),\mathbf{m}(\mathbf{r})] = \mathbf{H}(\mathbf{r}) + \mathbf{H}_{xc}[n(\mathbf{r}),\mathbf{m}(\mathbf{r})]. \quad (15.61c)$$

If the external magnetic field is homogeneous and only has a z component, $H = H_z$, the magnetization vector of Eq. (15.57) reduces to $m_z(\mathbf{r}) = -\frac{g\mu_B}{2}[n_\uparrow(\mathbf{r}) - n_\downarrow(\mathbf{r})]$ and $H_{ni} \equiv \frac{\delta E_{xc}[n,\mathbf{m}]}{\delta m(\mathbf{r})}$. In any case, the total density is given by $n(\mathbf{r}) = n_\uparrow(\mathbf{r}) + n_\downarrow(\mathbf{r})$ irrespective of whether the magnetic field is uniform or not.

Ferromagnetic systems, and spin unpaired systems, can be treated by taking a very small external magnetic field, which serves to break the symmetry between $n_\uparrow$ and $n_\downarrow$. One can verify that the magnetic field is small enough by checking that the resultant electronic structure does not change on halving the external magnetic field. This method of including a field that breaks a symmetry [here, spin rotation symmetry, $SU(2)$], and eventually setting the field equal to zero, can also be used to treat other systems where the ground-state does not have the symmetry of the bare Hamiltonian. Examples are Bose-Einstein condensation and superconductivity. In these systems, the bare Hamiltonian is gauge invariant but a mean-field approximation breaks gauge invariance, leading to macroscopic condensation in the ground state (Cooper pairs in superconductors and a Bose-Einstein condensate in bosonic systems). The occurrence of macroscopic condensates is quantified by a nonzero expectation value of certain operators (magnetization in ferromagnetic systems, $\Phi(\mathbf{r}) \equiv \langle\hat{\Psi}(\mathbf{r})\rangle$ in the case of a single component BEC, or the pairing parameter $\Delta(\mathbf{r}) \equiv \langle\hat{\Psi}_\downarrow(\mathbf{r})\hat{\Psi}_\uparrow(\mathbf{r})\rangle$ in the case of superconductivity). Such nonzero expectation values that persist even when the symmetry breaking field is turned off are referred to as an *order parameters*. A nonzero order parameter indicates that the ground state does not have the symmetry of the bare Hamiltonian. This symmetry is spin rotation in ferromagnets and gauge invariance in BEC and superconductors (see Secs. 9.5.10, 9.9, 18.10, 18.11, and 18.13.1, which are linked to the book web page).

An extension of spin DFT to include spin–orbit coupling and other relativistic effects has also been developed.

15.4.1 SPIN DFT LOCAL DENSITY APPROXIMATION

If $\{n(\mathbf{r}), \mathbf{m}(\mathbf{r})\}$ are both slowly varying with position, one can make a *local spin density approximation* and expand the exchange-correlation energy in the form

$$E_{xc}[n,\mathbf{m}] = \int d\mathbf{r}\, n(\mathbf{r})\epsilon_{xc}[n(\mathbf{r}),\mathbf{m}(\mathbf{r})]. \quad (15.62)$$

The exchange-correlation potential and magnetic field are then given by

$$v_{xc}(\mathbf{r}) = \epsilon_{xc}[n,\mathbf{m}](\mathbf{r}) + n(\mathbf{r})\frac{\delta\epsilon_{xc}[n,\mathbf{m}]}{\delta n(\mathbf{r})}, \quad (15.63)$$

$$\mathbf{H}_{xc}(\mathbf{r}) = n(\mathbf{r})\frac{\delta\epsilon_{xc}[n,\mathbf{m}]}{\delta\mathbf{m}(\mathbf{r})}. \quad (15.64)$$

The total energy in the spin LDA (SLDA) becomes

$$E[n,\mathbf{m}] = \sum_{i,m_s}\varepsilon_{i,m_s} + U_H[n(\mathbf{r})]$$

$$+ \int d\mathbf{r}\, n(\mathbf{r})\{\epsilon_{xc}[n(\mathbf{r}),\mathbf{m}(\mathbf{r})] - v_{xc}[n(\mathbf{r}),\mathbf{m}(\mathbf{r})]\}. \quad (15.65)$$

The Kohn–Sham equations are solved recursively (see Fig. 15.3), just as discussed for the Hartree–Fock equations, until the resulting energies $\varepsilon_{i,m_s}^{(j+1)}$ (and wave functions) no longer change with increasing j. The total energy is then formed using Eq. (15.65).

15.5 THE GAP PROBLEM IN DFT

Any property based solely on the ground-state density is identical in the noninteracting Kohn–Sham system and the physical system under study. For example, the ground-state electronic dipole moment depends solely on the electronic density and, therefore, is correctly obtained in DFT. But most physical properties computed with DFT are not identical to the actual properties of physical system, e.g., the ground-state energy, the kinetic energy, and polarizability. Perhaps surprisingly, the physical and noninteracting KS systems do have exactly the same ionization potential [273]. Since the ionization potential of the KS system is equal to the negative of its highest orbital energy, one obtains the *ionization potential theorem* (see also Koopman's theorem, Sec. 10.7, the negative of the HOMO energy equals the ionization potential of the real system): the negative of the Kohn–Sham energy eigenvalue of the outermost electron orbital equals the corresponding many-body ionization energy (including electronic relaxation) [273].

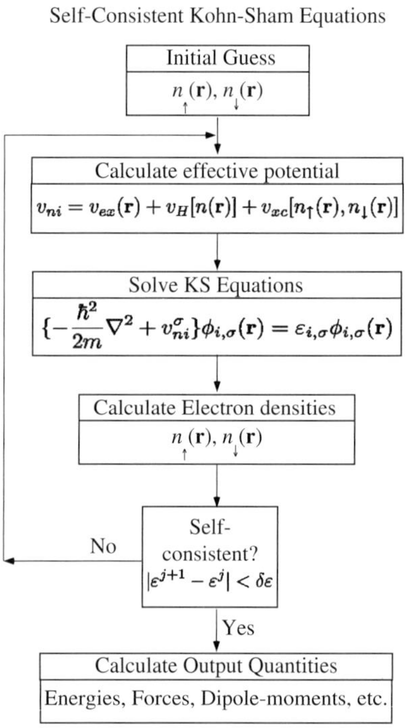

FIG 15.3 Schematic flow diagram of the DFT method.

Let us consider using DFT to compute the *fundamental gap* of an electronic system, defined as the difference between the ionization potential and electron affinity (see Sec. 10.7). The fundamental gap is an example of a "quasi-particle excitation energy"; these are important quantities in molecular systems, especially in the context of molecular electronics and photovoltaics. Quasi-particle excitation energies include the energy necessary to create a hole in the molecule (i.e., the ionization potential) and the energy to create an electron in a molecule (i.e., the electron affinity). It is through these levels that molecules conduct electrons or holes. The fundamental gap is the energy difference of the lowest hole energy and the lowest electron energy, i.e., the difference between the ionization potential and electron affinity. There are several different options available for calculating the gap. One option involves three separate DFT calculations: calculate the ground-state energy of the anion, the neutral, and the cation. Using the differences, one can obtain the fundamental gap. Another option requires only two DFT calculations, one for the anion and one for the neutral. Using the ionization potential theorem, the fundamental gap is equal to the difference between the HOMO energy of the KS system corresponding to the anion (since the ionization potential of the anion is the electron affinity of the neutral) and the HOMO energy of the KS system corresponding to the neutral.

It is somewhat unsatisfying that we need two different KS systems to compute the quasi-particle levels. It would be advantageous if the quasi-particle energies were equal to the orbital energies of the KS system, then one KS calculation could be used to compute electric conduction. Indeed, it is not uncommon to see computational studies using KS–DFT orbital energies as quasi-particle energies and in particular the LUMO–HOMO gap as an approximation for the fundamental gap. However, this is different from the above exact prescription and for KS approaches to DFT, this is usually not a very good approximation, i.e., the negative of the LUMO energy is not a good approximation to the electron affinity. For a compelling example, consider the F atom. Experimentally, one finds that the fundamental gap of F is approximately 14 eV. Let us determine the KS orbital gap for this system. The KS fictitious system contains nine noninteracting electrons occupying in pairs the first four KS orbitals, whereas the fifth orbital, the HOMO, is half-filled. Adding a 10th electron will release exactly the negative of the HOMO energy. Thus in this case, the electron affinity of the fictitious system is

15.6 Time-Dependent DFT

exactly equal to its ionization potential and the LUMO–HOMO orbital gap, i.e., the KS fundamental gap is zero, which is very different from the experimental fundamental gap.

The fact that the electron affinity of the fictitious system is usually much larger than the electron affinity of the physical system renders the Kohn–Sham orbital gap much smaller than the physical fundamental gap. This is true for most molecules and solids, and it should serve as a warning for users of KS–DFT: do not attribute to the LUMO–HOMO gap any quantitative meaning. In particular, the use of KS approaches in molecular electronics is unreliable for this reason.

The above reasoning still seems to be problematic, and it leads to a "paradox" in KS–DFT. Suppose we place a hydrogen atom at a large but finite distance from the F atom, i.e., we consider the H...F system. Because the ionization potential of H is much larger than the electron affinity of F, the ground-state density of H...F is that of two neutral atoms, almost exactly equal to the sum of displaced densities of the separate atoms. Based on this density, what should the KS potential be? We expect it to be just the sum of displaced KS potentials for the separate atoms. However, this seems to be wrong. The combined system has 10 electrons and since the IP of H is higher than that of F, so that the HOMO energy of F is lower than the HOMO energy of H, all 10 electrons would occupy the 5 lowest orbitals of F and none would occupy the HOMO of H. The KS system would then necessarily exhibit a density corresponding to $H^+...F^-$, which is very different from the correct experimental ground-state density.

This "paradox" can be avoided if we assume that the combined system KS potential is not the sum of displaced atomic potentials. But we cannot assume this, for the Hohenberg–Kohn theorem guarantees that the density uniquely determines the potential for each atom. This predicament is resolved if we remember that the Hohenberg–Kohn theorem does not really determine the potential uniquely. It determines it only up to a constant. Thus we can allow the potential of one atom to be displaced by a constant. The conclusion is that in the presence of the H atom, the potential of the F atom jumps up by the exact amount needed to make the HOMO energies of H and F to be equal. Once this equality of HOMO energies is achieved, the system can allocate the noninteracting electrons such that the two atoms are neutral, which conforms to the physical situation.

The inevitable jump of the F atom KS potential in the presence of the H atom shows that the exchange-correlation (XC) energy of the F atom must change nonanalytically when the electron density on F is slightly elevated as it "senses" the tunneling tail of the electron on the H. This nonanalyticity appears as a discontinuous jump in the XC potential, i.e., in the functional derivative of the XC energy with respect to the potential, and is called a *derivative discontinuity* [273].

KS DFT uses approximate XC functionals, such as the LDA. This energy functional is an analytic functional of the density and does not have a derivative discontinuity. LDA, applied within the KS procedure, will align the HOMO energy of the well-separated H and F atoms, but will do so by forcing the H and F atoms to partially "share" charge: the density of the H...F system wihin LDA will be of the form $H^{+0.2}...F^{-0.2}$. This nonphysical spurious long-range covalency is a result of the missing derivative discontinuity in the LDA.

Note that a different brand of DFT, not based on the KS approach but rather on the "generalized Kohn-Sham" approach using a range-separated hybrid functional, has recently shown good fundamental gaps for atoms, molecules, clusters, and solids [276].

15.6 TIME-DEPENDENT DFT

Time-dependent DFT applies the DFT philosophy to time-dependent problems. It replaces the many-body time-dependent Schrödinger equation by a set of time-dependent single-particle equations whose orbitals yield the same time-dependent density $n(\mathbf{r}, t)$. Time-dependent DFT is implemented in some quantum-chemical DFT program packages (see above). It is now used in solid-state physics, atomic and molecular physics, and optical processes [278].

For example, one can consider a system of N electrons in a potential $v_{ex}(\mathbf{r}, t)$ which includes the interaction with an electromagnetic field composed of a pulse of light centered around central frequency ω, so the external potential would, to a good approximation, include

$$v_{ex,EM}(\mathbf{r}, t) = -\frac{E_0(t)}{2} \sum_i \left(e^{-i\omega t} + e^{i\omega t}\right) \hat{\boldsymbol{\varepsilon}} \cdot \mathbf{r}_i, \qquad (15.66)$$

Another example would be the case of a vibrating molecule or a collision of two atoms, where in both cases the external potential arising from the nuclei is time dependent if we consider the motion of the nuclei as classical.

No variational principle exists for time-dependent quantum problems. However, a quantity analogous to the energy is the quantum mechanical action

$$\mathcal{A}[\psi] = \int_{t_0}^{t_1} dt \, \langle \psi(t) | i\hbar \partial_t - H(t) | \psi(t) \rangle, \qquad (15.67)$$

where $\psi(t)$ is a N-body wave function. Equating the functional derivative of the action with respect to $\psi^*(t)$ to zero, we obtain the time-dependent Schrödinger equation. Therefore, one can solve the time-dependent problem by calculating the stationary point of the functional $\mathcal{A}[\psi]$. The wave function $\psi(t)$ that makes the functional stationary is the solution of the time-dependent Schrödinger equation. However, there is no variational theorem that ensures that the energy is minimum, but only a stationary principle. Note that the action is always zero for the wave function that is a solution of the time-dependent Schrödinger equation, $\mathcal{A}[\psi] = 0$. Also, the solution of the time-dependent Schrödinger equation requires an initial condition at time t_0 on the wave function. These factors make time-dependent DFT more difficult than the time-independent theory.

15.6.1 THE RUNGE–GROSS THEOREM

The *Runge–Gross theorem* [277] is a time-dependent generalization of the Hohenberg–Kohn theorem:

> A given time-dependent density $n(\mathbf{r}, t)$ can arise from at most one time-dependent external potential $v_{ex}(\mathbf{r}, t)$, given an initial wave function, $|\psi(t_0)\rangle$, particle statistics (i.e., Bose or Fermi) and interaction potential, i.e., the time-dependent potential uniquely determines the time-dependent density.

More explicitly, if two potentials, $v_{ex}(\mathbf{r}, t)$ and $v'_{ex}(\mathbf{r}, t)$, differ by more than a purely time-dependent function $c(t)$ (i.e., the difference is spatially dependent), they cannot produce the same time-dependent density. Thus, given external potentials such that

$$v_{ex}(\mathbf{r}, t) \Longrightarrow n(\mathbf{r}, t) \quad v'_{ex}(\mathbf{r}, t) \Longrightarrow n'(\mathbf{r}, t), \qquad (15.68)$$

$$\text{then } n(\mathbf{r}, t) \neq n'(\mathbf{r}, t) \quad \text{if} \quad v'_{ex}(\mathbf{r}, t) \neq v_{ex}(\mathbf{r}, t) + c(t). \qquad (15.69)$$

The theorem considers external potentials that can be expanded as a Taylor series with respect to the time around the initial time t_0,

$$v_{ex}(\mathbf{r}, t) = \sum_{k=0}^{\infty} c_k(\mathbf{r})(t - t_0)^k, \qquad (15.70)$$

where $c_k(\mathbf{r}) = \frac{1}{k!} \frac{\partial^k}{\partial t^k} v_{ex}(\mathbf{r}, t)\big|_{t=t_0}$. If the two potentials $v_{ex}(\mathbf{r}, t)$ and $v'_{ex}(\mathbf{r}, t)$ differ by more than a time-dependent constant, then the quantity defined by

$$u_k(\mathbf{r}) = \frac{\partial^k}{\partial t^k} [v_{ex}(\mathbf{r}, t) - v'_{ex}(\mathbf{r}, t)]_{t=t_0}, \qquad (15.71)$$

is nonvanishing, and the corresponding coefficients $c_k(\mathbf{r})$ differ by more than a constant. The first step in the proof is to show that if $v'_{ex}(\mathbf{r}, t) \neq v_{ex}(\mathbf{r}, t) + c(t)$, the resulting current densities are also different. The current in first quantized language is [see (1.100)]

$$\mathbf{J}(\mathbf{r}, t) = \frac{\hbar}{2im} \left\{ \psi^*(\mathbf{r}, t) \nabla \psi(\mathbf{r}, t) - [\nabla \psi^*(\mathbf{r}, t)] \psi(\mathbf{r}, t) \right\}. \qquad (15.72)$$

15.6 Time-Dependent DFT

The current can be written in terms of a current operator; in the Schrödinger representation,

$$\hat{\mathbf{J}}(\mathbf{r}) = \frac{\hbar}{2im} \left[\hat{\Psi}^\dagger(\mathbf{r}) \nabla \hat{\Psi}(\mathbf{r}) - \nabla \hat{\Psi}^\dagger(\mathbf{r}) \hat{\Psi}(\mathbf{r}) \right], \tag{15.73}$$

where $\mathbf{J}(\mathbf{r}, t) = \langle \psi(t) | \hat{\mathbf{J}}(\mathbf{r}) | \psi(t) \rangle$. The time derivative of the current in the primed and unprimed systems is

$$i\hbar \frac{\partial}{\partial t} \mathbf{J}(\mathbf{r}, t) = \langle \psi(t) | [\hat{\mathbf{J}}(\mathbf{r}), \hat{H}(t)] | \psi(t) \rangle, \tag{15.74a}$$

$$i\hbar \frac{\partial}{\partial t} \mathbf{J}'(\mathbf{r}, t) = \langle \psi'(t) | [\hat{\mathbf{J}}(\mathbf{r}), \hat{H}'(t)] | \psi'(t) \rangle. \tag{15.74b}$$

Note that the wave function in the Schrödinger representation in the primed system equals the unprimed at the initial time, $|\psi(t_0)\rangle = |\psi'(t_0)\rangle$. Taking the difference of the two equations in Eq. (15.74) at the initial time yields

$$i\hbar \frac{\partial}{\partial t} [\mathbf{J} - \mathbf{J}']_{t=t_0} = \langle [\hat{\mathbf{J}}(\mathbf{r}), v_{ex}(\mathbf{r}, t_0) - v'_{ex}(\mathbf{r}, t_0)] \rangle_{t_0} = i\hbar n(\mathbf{r}, t_0) \nabla [v_{ex}(\mathbf{r}, t_0) - v'_{ex}(\mathbf{r}, t_0)]. \tag{15.75}$$

If Eq. (15.71) is satisfied for $k = 0$, the two potentials differ at t_0. Hence, the derivative on the LSH of Eq. (15.75) differs from zero, and the two current densities differ for $t > t_0$. If $k > 0$, then differentiating Eq. (15.75) k times yields

$$\frac{\partial^{k+1}}{\partial t^{k+1}} [\mathbf{J}(\mathbf{r}, t) - \mathbf{J}'(\mathbf{r}, t)]_{t=t_0} = n(\mathbf{r}, t_0) \nabla u_k(\mathbf{r}, t_0). \tag{15.76}$$

The RHS of Eq. (15.76) is nonzero, so $\mathbf{J}(\mathbf{r}, t) \neq \mathbf{J}'(\mathbf{r}, t)$ for $t > t_0$. For the second step, we must prove that if $\mathbf{J}(\mathbf{r}, t) \neq \mathbf{J}'(\mathbf{r}, t)$, then $n(\mathbf{r}, t) \neq n'(\mathbf{r}, t)$. This follows from the continuity equation, $\frac{\partial n(\mathbf{r}, t)}{\partial t} + \nabla \cdot \mathbf{J}(\mathbf{r}, t) = 0$, see Eq. (1.99), applied to the primed and unprimed system

$$\frac{\partial}{\partial t} [n(\mathbf{r}, t) - n'(\mathbf{r}, t)] = -\nabla \cdot [\mathbf{J}(\mathbf{r}, t) - \mathbf{J}'(\mathbf{r}, t)]. \tag{15.77}$$

Taking the $k + 1$ derivative with respect to time of Eq. (15.77), we find, using Eq. (15.76),

$$\frac{\partial^{k+2}}{\partial t^{k+2}} \left[n(\mathbf{r}, t) - n'(\mathbf{r}, t) \right]_{t=t_0} = -\nabla \cdot [n(\mathbf{r}, t_0) \nabla u_k(\mathbf{r}, t_0)]. \tag{15.78}$$

Hence, <u>if</u>, the RHS of Eq. (15.78) is nonvanishing, $n(\mathbf{r}, t) \neq n'(\mathbf{r}, t)$. The proof that the RHS of Eq. (15.78) is nonvanishing follows by *reductio ad absurdum*. Assume that $\nabla \cdot [n(\mathbf{r}, t_0) \nabla u_k(\mathbf{r}, t_0)] = 0$ with $u_k(\mathbf{r}, t_0)$ not constant in coordinate space, and consider the integral (at t_0)

$$\int d\mathbf{r}\, n(\mathbf{r}) [\nabla u_k(\mathbf{r})]^2 = \int d\mathbf{r}\, u_k(\mathbf{r}) \nabla \cdot [n(\mathbf{r}) \nabla u_k(\mathbf{r})] + \int d\mathbf{S} \cdot \nabla u_k(\mathbf{r})\, n(\mathbf{r}) u_k(\mathbf{r}). \tag{15.79}$$

This equation is obtained using Green's theorem, $\int d\mathbf{r} \nabla \cdot \mathbf{A}(\mathbf{r}) = \int d\mathbf{S} \cdot \mathbf{A}(\mathbf{r})$. The first term on the RHS is zero by assumption, while the second term vanishes if the density and the function $u_k(\mathbf{r}, t_0)$ decays sufficiently quickly at large r, as is ensured for finite systems. Moreover, the integrand is always positive. These conditions can only be satisfied if either the density n or u_k vanishes identically. The density cannot vanish everywhere, while the second possibility contradicts the initial assumption that $u_k(\mathbf{r}, t_0)$ is not constant in coordinate space. Thus, the proof of the Runge–Gross theorem is complete.

15.6.2 TIME-DEPENDENT KOHN–SHAM EQUATIONS

The time-dependent Kohn–Sham equations describe the evolution of N noninteracting electrons in the time-dependent Kohn–Sham potential $v_{ni}(\mathbf{r},t)$ but produce the same density $n(\mathbf{r},t)$ as that of the interacting system of interest are given by the time-dependent Kohn–Sham equations:

$$i\hbar \frac{\partial \phi_j(\mathbf{r},t)}{\partial t} = \left(-\frac{\hbar^2 \nabla^2}{2m} + v_{ni}[n](\mathbf{r},t) \right) \phi_j(\mathbf{r},t), \tag{15.80a}$$

$$v_{ni}[n](\mathbf{r},t) = v_{ex}[n](\mathbf{r},t) + \int d\mathbf{r}' \frac{n(\mathbf{r}',t)}{|\mathbf{r}-\mathbf{r}'|} + v_{xc}[n](\mathbf{r},t). \tag{15.80b}$$

The time-dependent density is given by $n(\mathbf{r},t) = \sum_{j=1}^{N} |\phi_j(\mathbf{r},t)|^2$ and is identical to that of the actual system. The exchange-correlation potential, $v_{xc}[n](\mathbf{r},t)$, is in general not known, but it is a functional of the entire history of the density, $n(\mathbf{r},t)$, the initial interacting wave function $\psi(t_0)$, and the initial Kohn–Sham determinantal wave function, $\Phi(0)$. This functional is rather complicated, much more than the ground-state functional. Knowledge of it implies solution of all time-dependent electron Coulomb interacting problems. The time-dependent Kohn–Sham equations must be solved as an initial value problem, with the system in the initial state specified by the Kohn–Sham orbitals $\{\phi_i(\mathbf{r})\}$ of the ground state of the system at $t = t_0$.

15.6.3 ADIABATIC LDA

In contrast with time-independent DFT, where good xc functionals exist, approximations to $v_{xc}[n](\mathbf{r},t)$ are still not well explored. The simplest approximation is the adiabatic LDA, related to the LDA of time-independent DFT. The adiabatic time-dependent xc potential is given by

$$v_{xc}^{ad}[n](\mathbf{r},t) = v_{xc}[n(\mathbf{r})]|_{n=n(t)}, \tag{15.81}$$

where the quantity on the RHS of Eq. (15.81) is the ground-state xc density functional evaluated at each time with the density $n(\mathbf{r},t)$. Hence, the adiabatic LDA uses the homogeneous gas xc potential (15.53)

$$v_{xc}^{ad}[n](\mathbf{r},t) = -\frac{3}{2} \left(\frac{3}{\pi} \right)^{1/3} \alpha n^{1/3} \bigg|_{n=n(t)}. \tag{15.82}$$

This quantity is based on a ground-state property, so we expect the adiabatic approximation to work only where the temporal dependence is small, i.e., when the system is locally close to equilibrium.

15.7 DFT COMPUTER PACKAGES

Time-dependent DFT has become popular as a method for calculating electronic excited-state energies. Often the same density functional approximations from ground-state DFT can be used for time-dependent DFT. Moreover, time-dependent DFT yields predictions for an enormous variety of phenomena, e.g., optical response of atoms, molecules and solids, electronic transport through single molecules, high harmonic generation in strong laser fields, and multiphoton ionization [278].

For more details on time-dependent DFT, see http://www.tddft.org.

Below we list a number of computer codes for calculating the electronic structure of atoms, molecules, and periodic structures using DFT. Note that this list is not complete. Most of the codes listed are not commercial and can be freely obtained (at least for academic use).

Gaussian – A commercial package for computing the electronic structure of atoms and molecules using local basis sets (http://www.gaussian.com/index.htm).

15.7 DFT Computer Packages

GAMESS – General Atomic and Molecular Electronic Structure System. Uses local basis sets (http://www.msg.chem.iastate.edu/GAMESS).

Octopus – DFT and TD-DFT using real-space grid discretization (no basis sets) (http://www.tddft.org/programs/octopus/wiki/index.php/Main_Page).

PARSEC – Real-space, pseudopotential DFT and TD-DFT for atoms, molecules and periodic systems (http://parsec.ices.utexas.edu/index.html).

Siesta – Ab initio structure and molecular dynamics simulations of molecules and solids using local basis sets (http://www.icmab.es/siesta).

VASP – Ab initio electronic structure code using pseudopotentials and a plane wave basis set (http://cms.mpi.univie.ac.at/vasp).

ABINIT – Computes the total energy, charge density and electronic structure of atoms, molecules and periodic structure using pseudopotentials and a plane wave basis (http://www.abinit.org).

CPMD Program – Plane wave/pseudopotential implementation of DFT, particularly designed for ab initio molecular dynamics (http://www.cpmd.org).

Quantum-Espresso (PWSCF) – DFT based on plane wave and pseudopotentials (http://www.quantum-espresso.org).

FPLO – Full-potential local-orbital minimum-basis code to solve the Kohn–Sham equations on a regular periodic lattice using the local spin density approximation (http://www.fplo.de).

FHI-aims – Full-potential local-orbital code for atoms, molecules and periodic systems (https://aimsclub.fhi-berlin.mpg.de).

Q-CHEM – A (gaussian) local-orbital code for atoms and molecules (http://www.q-chem.com).

Linear Algebra

The basic mathematical structure of quantum mechanics involves state vectors and linear operators that transform one vector into another. The vectors, which represent quantum states, are members of a complex vector space, or more specifically, a complex inner product space, which is a vector space having an associated inner product (sometimes called a scalar product). The inner product allows determination of the length of state vectors and the extent of their overlap with other state vectors. In this appendix, we review the key elements of linear algebra required for quantum mechanics. Textbooks treating linear algebra and inner product spaces include Refs. [57, 279–281]. If you have never studied linear algebra, you will have to do so before continuing with this appendix, which is meant only as a review.

A.1 VECTOR SPACES

We begin by defining a *vector space* over the field $\mathbb{C}$ of complex numbers as a set V of elements, called *vectors*, along with two operations "+" and "·" called *vector addition* and *scalar multiplication* satisfying the following properties:

1. If $\mathbf{u}, \mathbf{v} \in V$, their *vector sum* $\mathbf{u} + \mathbf{v}$ is an element of V.
2. If $\mathbf{u}, \mathbf{v} \in V$, then $\mathbf{u} + \mathbf{v} = \mathbf{v} + \mathbf{u}$.
3. If $\mathbf{u}, \mathbf{v}, \mathbf{w} \in V$, then $(\mathbf{u} + \mathbf{v}) + \mathbf{w} = \mathbf{u} + (\mathbf{v} + \mathbf{w})$.
4. There is a *zero vector* $\mathbf{0} \in V$ such that $\mathbf{u} + \mathbf{0} = \mathbf{u}$ for all $\mathbf{u} \in V$.
5. Each vector $\mathbf{v} \in V$ has an *additive inverse* $\mathbf{w} \in V$ such that $\mathbf{u} + \mathbf{w} = \mathbf{0}$. The inverse of a vector $\mathbf{v}$ is often denoted by $-\mathbf{v}$.
6. If $r \in \mathbb{C}$ and $\mathbf{u} \in V$, then $r \cdot \mathbf{u} \in V$. Henceforth, this operation of scalar multiplication will be written simply as $r\mathbf{u}$.
7. If $r, s \in \mathbb{C}$ and $\mathbf{u} \in V$ then $(r+s)\mathbf{u} = r\mathbf{u} + s\mathbf{u} \in V$. Here, the $+$ on the left hand side (LHS) of the equation is addition in $\mathbb{C}$, and the $+$ on the RHS is vector addition in the vector space V.
8. If $r \in \mathbb{C}$ and $\mathbf{u}, \mathbf{v} \in V$, then $r(\mathbf{u} + \mathbf{v}) = r\mathbf{u} + r\mathbf{v}$.
9. If r and s are any scalars, and $\mathbf{u} \in V$, $(rs)\mathbf{u} = r(s\mathbf{u})$.
10. $1\mathbf{u} = \mathbf{u}$, and $0\mathbf{u} = \mathbf{0}$.

The simplest examples of vector spaces come from plane and 3D Euclidean geometry. All vectors lying in the plane that originate from the same point (say, the origin) form a two-dimensional Euclidean vector space over the field of real numbers. A similar construction holds in three (and higher) dimensions.

It is obvious that if $\mathbf{v}_1, \mathbf{v}_2 \in V$ and $c_1, c_2 \in \mathbb{C}$ then $\mathbf{w} \equiv c_1\mathbf{v}_1 + c_2\mathbf{v}_2 \in V$; $\mathbf{w}$ is then said to be a *linear combination* of $\mathbf{v}_1$ and $\mathbf{v}_2$ with coefficients c_1 and c_2. This construction is easily extended to form linear combinations of n vectors, where n is a positive integer. The N vectors $\mathbf{v}_1, \mathbf{v}_2, \ldots, \mathbf{v}_N \in V$ with $\mathbf{v}_i \neq \mathbf{0}$ for all ($\forall$) i are said to be *linearly independent* if and only if (iff)

$$\sum_{i=1}^{N} c_i \mathbf{v}_i = \mathbf{0}, \quad \Leftrightarrow \quad c_i = 0 \; \forall \, i. \tag{A.1}$$

That is, it is not possible to express $\mathbf{0}$ as a linear combination of linearly independent $\{\mathbf{v}_i\}$ except when all complex coefficients $c_i \in \mathbb{C}$ are 0. If N is maximal in the sense that there is at least one set of N linearly independent vectors $\mathbf{v}_1, \mathbf{v}_2, \ldots, \mathbf{v}_N \in V$ with $\mathbf{v}_i \neq \mathbf{0}$ $\forall i$, but there is no set of $N + 1$ linearly independent vectors, then V is said to be a vector space of dimension N, or equivalently, N is the dimension of V; this is denoted by $V^{(N)}$. A maximal set of linearly independent vectors can be used to represent every vector in $V^{(N)}$. Any maximal set of linearly independent vectors $\mathbf{v}_1, \mathbf{v}_2, \ldots, \mathbf{v}_N \in V^{(N)}$ can be considered as "coordinate system" and is called as *basis*, i.e., a *complete set of basis vectors*. In this basis, we can write an arbitrary vector **u** as $\mathbf{u} = \sum_{j=1}^{N} c_j \mathbf{v}_j$, and it can be represented by

$$\mathbf{u} = \sum_{j=1}^{N} c_j \mathbf{v}_j := \begin{pmatrix} c_1 \\ \vdots \\ c_N \end{pmatrix}, \quad (A.2)$$

where := means "can be represented as." A basis is not unique; generally, there are an infinite number of bases that can be used. It should be stressed that vectors can be defined without specifying a particular basis.

Problem A.1

Write the two-dimensional vector $\mathbf{u} = 2\hat{\mathbf{x}} + 4\hat{\mathbf{y}} = \begin{pmatrix} 2 \\ 4 \end{pmatrix}$ in terms of the basis functions $\mathbf{v}_1 = \hat{\mathbf{x}} + \hat{\mathbf{y}}$ and $\mathbf{v}_2 = \hat{\mathbf{x}} - \hat{\mathbf{y}}$, where $\hat{\mathbf{x}}$ and $\hat{\mathbf{y}}$ are unit vectors along the Cartesian axes in the plane.
Answer: $\mathbf{u} = 3[\hat{\mathbf{x}} + \hat{\mathbf{y}}] - 1[\hat{\mathbf{x}} - \hat{\mathbf{y}}] = \begin{pmatrix} 3 \\ -1 \end{pmatrix}$.

Problem A.2

Show that if $\mathbf{v}_1, \mathbf{v}_2, \ldots, \mathbf{v}_N \in V^{(N)}$ with $\mathbf{v}_i \neq 0 \ \forall i$ are linearly independent, then an arbitrary vector $\mathbf{u} \in V^{(N)}$ can be written as a linear combination $\mathbf{u} = \sum_{i=1}^{N} c_i \mathbf{v}_i$.
Answer: If the statement is false then the $N + 1$ vectors $\mathbf{u}, \mathbf{v}_1, \mathbf{v}_2, \ldots, \mathbf{v}_n$ are linearly independent. Hence, N is not maximal as there are $N + 1$ linearly independent vectors.

A.1.1 DIRAC NOTATION

In quantum mechanics, *Dirac notation* is often used as a powerful tool to treat vector spaces. The use of Dirac notation is not limited to quantum mechanics. Among its other virtues, it significantly simplifies manipulations in vector spaces with inner products, such as Hilbert spaces (see below), which form the mathematical basis for quantum mechanics. In this sense, Dirac notation is much more than just notation and serves as a conceptual framework for dealing with state space in quantum mechanics.

Vectors in Dirac notation are written as $|u\rangle$, $|v\rangle$ (instead of $\mathbf{u}$, $\mathbf{v}$ or $\vec{u}$, $\vec{v}$), $|\psi\rangle$, $|\phi\rangle$, etc., and are called *ket vectors*. Note that Greek letters are often used in representing state-vectors of a quantum system. When Dirac notation is used, the vector space V is also called a *ket space* (the dimension N is often not explicitly specified). Any vector $|\psi\rangle \in V$ is expressible as a linear superposition of basis vectors $\{|\phi_j\rangle\} \subset V$. Thus, in Dirac notation, Eq. (A.2) is written:

$$|\psi\rangle = \sum_{j=1}^{N} c_j |\phi_j\rangle := \begin{pmatrix} c_1 \\ \vdots \\ c_N \end{pmatrix}, \quad (A.3)$$

where, as in Eq. (A.2), the symbol := means that the ket $|\psi\rangle$ is represented as a column vector of the coefficients $\{c_j\}$ (in quantum mechanics, these are sometimes called amplitudes) that multiply the basis kets $\{|\phi_j\rangle\}$.

To properly understand Dirac notation, we need to define another vector space, the *dual space*, or the *bra space*, $V^\dagger$, that is directly related to V. Vectors in $V^\dagger$ are called *bras* and are written as $\langle u|$, $\langle v|$, $\langle \chi|$, $\langle \psi| \ldots$. The two vector spaces V and $V^\dagger$ have the same dimension N and basically have identical structure. Vectors and operations in V are in one-to-one correspondence with those in $V^\dagger$ [in mathematical parlance, one says that the two spaces are isomorphic, meaning that there is a mapping (called an isomorphism — in Greek, iso means equal and morphosis means to shape) between elements of the two spaces that preserves the structure of the vector space]. At this stage, the Hermitian adjoint superscript † is

A.1 Vector Spaces

used just to distinguish between the ket space V and the dual or bra space $V^\dagger$, and to map kets onto bras and vice versa. Thus, $|\psi\rangle \in V \Leftrightarrow \langle\psi| \in V^\dagger$ is compactly written, $|\psi\rangle^\dagger = \langle\psi|$. Note that $(c|\psi\rangle)^\dagger = c^*\langle\psi|$, and the correspondence between linear combinations in V and $V^\dagger$ is

$$(c_1|\phi_1\rangle + c_2|\phi_2\rangle)^\dagger = c_1^*\langle\phi_1| + c_2^*\langle\phi_2|. \tag{A.4}$$

The expansion of a bra vector $\langle\psi| \in V^\dagger$ in a given basis $\{\langle\phi_j|\}$ and its representation in terms of the coefficients (the image of Eq. (A.3) in $V^\dagger$) is given by

$$\langle\psi| = \sum_{j=1}^N \langle\phi_j|c_j^* := \begin{pmatrix} c_1^* & \cdots & c_N^* \end{pmatrix}. \tag{A.5}$$

Hence, bra vectors are represented as row vectors whose coefficients (amplitudes) are complex conjugated.

In Dirac notation, the dual space is used to define the inner product of two vectors, a topic we now address.

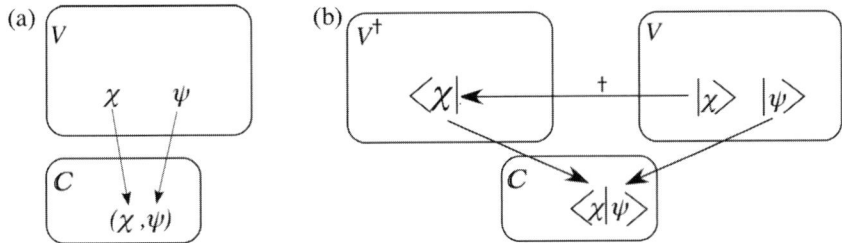

FIG A.1 An inner product associates a complex number $\in \mathbb{C}$ to any ordered pair of vectors in a vector space V. (a) Inner product without using Dirac notation. (b) Inner product using Dirac notation.

A.1.2 INNER PRODUCT SPACES

Just as in Euclidean geometry, in quantum mechanics, we need to specify the notion of length of a vector (also referred to as *norm*) and the notion of projection of a vector onto another vector. This requires the introduction of a new binary operation on the vector space V called an *inner product* (or sometimes a *scalar product*, not to be confused with multiplication by a scalar), which associates a complex number to any ordered pair of vectors in V. If Dirac notation is not used, the inner product of the vectors $\chi, \psi \in V$ is written, $(\chi, \psi) \in \mathbb{C}$ [see Fig. A.1(a)]. When the Dirac notation is used, the inner product of the kets $|\chi\rangle, |\psi\rangle \in V$ is written, $\langle\chi|\psi\rangle \in \mathbb{C}$. Note that $\langle\chi| = |\chi\rangle^\dagger \in V^\dagger$. In Dirac notation, this binary operation can be viewed as follows: first map the ket $|\chi\rangle \in V$ onto its bra image $\langle\chi| \in V^\dagger$ using $|\chi\rangle^\dagger = \langle\chi|$. Then associate a complex number $\langle\chi|\psi\rangle \in \mathbb{C}$ with $\langle\chi|$ and $|\psi\rangle$. The inner product $\langle\chi|\psi\rangle$ is called a *bracket*, a composition of bra and ket. This view of inner product is pictorially illustrated in Fig. A.1(b).

The inner product has the following attributes:

1. $\langle\chi|\psi\rangle$ is a complex number, independent of the basis in which the vectors are expanded.
2. $\langle\chi|\psi\rangle = \langle\psi|\chi\rangle^*$.
3. For any complex numbers c_1 and c_2,[1]

$$\langle\chi|(c_1|\psi_1\rangle + c_2|\psi_2\rangle) = c_1\langle\chi|\psi_1\rangle + c_2\langle\chi|\psi_2\rangle,$$
$$(c_1^*\langle\psi_1| + c_2^*\langle\psi_2|)|\chi\rangle = c_1^*\langle\psi_1|\chi\rangle + c_2^*\langle\psi_2|\chi\rangle.$$

4. $\langle\psi|\psi\rangle \geq 0$, with equality if and only if $|\psi\rangle$ is the zero vector, $|\psi\rangle = |0\rangle$.

---

[1] To facilitate the notation, one could write the vector $(c_1|\psi_1\rangle + c_2|\psi_2\rangle)$ as $|c_1\psi_1 + c_2\psi_2\rangle$. In this "augmented Dirac notation", the first equation reads, $\langle\chi|c_1\psi_1 + c_2\psi_2\rangle = c_1\langle\chi|\psi_1\rangle + c_2\langle\chi|\psi_2\rangle$, and the second reads, $\langle c_1\psi_1 + c_2\psi_2|\chi\rangle = c_1^*\langle\psi_1|\chi\rangle + c_2^*\langle\psi_2|\chi\rangle$. This "augmented Dirac notation" can be dangerous, and will be avoided, despite its convenience (see Problem A.5).

Properties of Inner Products

Two kets $|\psi\rangle$ and $|\chi\rangle$ are said to be *orthogonal* if

$$\langle \chi | \psi \rangle = 0. \tag{A.6}$$

The square of the length of the vector $|\psi\rangle$ is defined as the inner product of the vector with itself, $\|\psi\|^2 = \langle \psi | \psi \rangle$, i.e., the length (norm) of $|\psi\rangle$ is

$$\|\psi\| \equiv \sqrt{\langle \psi | \psi \rangle} \geq 0. \tag{A.7}$$

The length of any vector is real and non-negative, by virtue of the property, $\langle \chi | \psi \rangle = \langle \psi | \chi \rangle^*$. Only the null vector has zero length.

Just like for ordinary vectors in 3D Euclidean space, the *triangle inequality* holds for vectors in an inner product space,

$$\|\psi + \phi\| \leq \|\psi\| + \|\phi\|, \tag{A.8}$$

with equality if and only if one of the vectors is a non-negative scalar multiple of the other one. The *Cauchy–Schwarz inequality* holds,

$$|\langle \phi | \psi \rangle| \leq \|\psi\| \|\phi\|, \tag{A.9}$$

with equality only for vectors that are scalar multiples of one another.

It is instructive to prove the Cauchy–Schwarz inequality in order to see how it arises. Let us consider a ket $|f\rangle$ composed of a superposition of two vectors $|\phi\rangle$ and $|\psi\rangle$ and defined as follows:

$$|f\rangle \equiv \langle \phi | \psi \rangle |\phi\rangle - \langle \phi | \phi \rangle |\psi\rangle. \tag{A.10}$$

The bra $\langle f|$ is then given by $\langle f| = \langle \phi | \langle \psi | \phi \rangle - \langle \psi | \langle \phi | \phi \rangle$ and $\langle f | f \rangle \geq 0$. After some algebra, we arrive at the following expression for the inner product:

$$\langle f | f \rangle = \langle \phi | \phi \rangle (\langle \phi | \phi \rangle \langle \psi | \psi \rangle - |\langle \phi | \psi \rangle|^2). \tag{A.11}$$

Since $\langle f | f \rangle \geq 0$ and $\langle \phi | \phi \rangle \geq 0$, we conclude that $(\langle \phi | \phi \rangle \langle \psi | \psi \rangle - |\langle \phi | \psi \rangle|^2) \geq 0$, hence

$$\langle \phi | \phi \rangle \langle \psi | \psi \rangle \geq |\langle \phi | \psi \rangle|^2. \tag{A.12}$$

This concludes the proof.

Other useful identities include the *parallelogram* and *polarization identities*,

$$\|\phi + \psi\|^2 + \|\phi - \psi\|^2 = 2(\|\phi\|^2 + \|\psi\|^2), \tag{A.13}$$

$$\langle \phi | \psi \rangle = \frac{1}{4}(\|\phi + \psi\|^2 - \|\phi - \psi\|^2) - \frac{i}{4}(\|\phi + i\psi\|^2 - \|\phi - i\psi\|^2). \tag{A.14}$$

Problem A.3

Prove the triangle inequality (A.8) using the Cauchy–Schwarz inequality.

Given two ket vectors, $|\psi\rangle$ and $|\chi\rangle$, written in terms of the same basis $\{|\phi_j\rangle\}$, and represented as [see (A.3)]

$$|\psi\rangle = \begin{pmatrix} a_1 \\ \vdots \\ a_N \end{pmatrix}, \quad |\chi\rangle = \begin{pmatrix} b_1 \\ \vdots \\ b_N \end{pmatrix}, \tag{A.15}$$

A.1 Vector Spaces

their inner product is given by the complex number,

$$\langle \chi | \psi \rangle = \sum_{i,j=1}^{N} b_i^* a_j \langle \phi_i | \phi_j \rangle. \tag{A.16}$$

Basis vectors $\{|\phi_j\rangle\}$ are *orthonormal* if they satisfy the conditions

$$\langle \phi_i | \phi_j \rangle = \delta_{ij} = \begin{cases} 1 \text{ for } i = j \\ 0 \text{ for } i \neq j \end{cases}, \tag{A.17}$$

where δ_{ij} is called the *Kronecker delta function*. For an orthonormal basis,

$$\langle \chi | \psi \rangle = \begin{pmatrix} b_1^* & \cdots & b_N^* \end{pmatrix} \begin{pmatrix} a_1 \\ \vdots \\ a_N \end{pmatrix} = \sum_{j=1}^{N} b_j^* a_j. \tag{A.18}$$

If two kets $|\psi\rangle$ and $|\chi\rangle$ are orthogonal, and the basis functions are orthonormal, Eq. (A.6) reads, $\langle \chi | \psi \rangle = \sum_{j=1}^{N} b_j^* a_j = 0$. Moreover, using Eq. (A.18), $\|\psi\| = \sqrt{\sum_{j=1}^{N} |a_j|^2}$. Orthonormal bases are almost always used in quantum mechanics.

Problem A.4

(a) Given the vector $|\phi_1\rangle = \begin{pmatrix} \sqrt{2} \\ i \\ \frac{-1}{\sqrt{2}} \end{pmatrix}$ in an orthonormal basis, find the normalization constant, $\mathcal{N}$, such that $\mathcal{N}|\phi_1\rangle$ has unit length.

(b) Find $\langle \phi_2 | \phi_1 \rangle$, where $|\phi_1\rangle$ is given in (a) and $|\phi_2\rangle = \begin{pmatrix} 2 \\ 1 \\ i/\sqrt{2} \end{pmatrix}$.

Answers: (a) $\mathcal{N} = \sqrt{\frac{2}{7}}$. (b) Since $\langle \phi_2 | = \begin{pmatrix} 2 & 1 & -i/\sqrt{2} \end{pmatrix}$, we find $\langle \phi_2 | \phi_1 \rangle = 2\sqrt{2} + 3i/2$.

Problem A.5
Using "augmented Dirac notation",

(a) Show that $\langle c_1\phi_1 + c_2\phi_2 | b_1\psi_1 + b_2\psi_2 \rangle = c_1^* b_1 \langle \phi_1 | \psi_1 \rangle + c_2^* b_1 \langle \phi_2 | \psi_1 \rangle + c_1^* b_2 \langle \phi_1 | \psi_2 \rangle + c_2^* b_2 \langle \phi_2 | \psi_2 \rangle$.
(b) Show that $\langle b_1\psi_1 + b_2\psi_2 | c_1\phi_1 + c_2\phi_2 \rangle = \langle c_1\phi_1 + c_2\phi_2 | b_1\psi_1 + b_2\psi_2 \rangle^*$.
(c) To point out the danger in "augmented Dirac notation", rewrite $\langle \mathbf{p}|\mathbf{x}\rangle + \langle \mathbf{p}|-\mathbf{x}\rangle$ using it (see Sec. 1.3.8 for the interpretation of this inner product).

Answer: (c) $\langle \mathbf{p}|\mathbf{x} - \mathbf{x}\rangle$. This could be easily misinterpreted as $\langle \mathbf{p}|0\rangle$.

The expansion coefficients c_j in Eq. (A.3) can be computed by taking the inner product of Eq. (A.3) with a complete set of basis vectors $|\phi_i\rangle$:

$$\langle \phi_i | \psi \rangle = \sum_j c_j \langle \phi_i | \phi_j \rangle \; \forall \, i. \tag{A.19}$$

Equation (A.19) can be inverted to find the coefficients c_j. If the basis vectors $|\phi_j\rangle$ are *othonormal*, as is the case if these vectors are eigenvectors of a Hermitian (or self-adjoint) operator [see below and Secs. (1.3.2) and (2.2)], we obtain

$$c_i = \langle \phi_i | \psi \rangle, \qquad (A.20)$$

and thus we can write Eq. (A.3) as

$$|\psi\rangle = \sum_j \langle \phi_j | \psi \rangle |\phi_j\rangle. \qquad (A.21)$$

Sometimes, it is convenient to write Eq. (A.21) as $|\psi\rangle = \sum_j |\phi_j\rangle \langle \phi_j | \psi \rangle$ because this makes clear the insertion of the unit operator in the basis set expansion [see Eq. (A.32)].

Given two arbitrary vectors $|\phi_1\rangle$ and $|\phi_2\rangle$ that are not orthogonal, they can be orthogonalized using the *Gram–Schmidt orthogonalization* procedure. Adding $-\frac{\langle \phi_1 | \phi_2 \rangle}{\langle \phi_1 | \phi_1 \rangle} |\phi_1\rangle$ to the second vector, $|\phi_2\rangle$, results in a new vector, let us call it $|\psi_2\rangle = |\phi_2\rangle - \frac{\langle \phi_1 | \phi_2 \rangle}{\langle \phi_1 | \phi_1 \rangle} |\phi_1\rangle$, which is orthogonal to $|\psi_1\rangle \equiv |\phi_1\rangle$, as can easily be verified. The idea behind this procedure is to form the vector $|\psi_2\rangle \equiv |\phi_2\rangle + c|\psi_1\rangle$ and choose the coefficient c such that $\langle \psi_1 | \psi_2 \rangle = 0$. Solving for c yields $c = -\frac{\langle \psi_1 | \phi_2 \rangle}{\langle \psi_1 | \psi_1 \rangle}$. Generalizing the Gram–Schmidt procedure to orthogonalize k vectors $\{|\phi_n\rangle\}$, $n = 1, \ldots, k$, (taking $|\psi_j\rangle \equiv |\phi_j\rangle + \sum_n^{j-1} c_n |\psi_n\rangle$), we find:

$$|\psi_1\rangle = |\phi_1\rangle, \quad |\psi_2\rangle = |\phi_2\rangle - \frac{\langle \psi_1 | \phi_2 \rangle}{\langle \psi_1 | \psi_1 \rangle} |\psi_1\rangle, \ldots, |\psi_j\rangle = |\phi_j\rangle - \sum_{n=1}^{j-1} \frac{\langle \psi_n | \phi_j \rangle}{\langle \psi_n | \psi_n \rangle} |\psi_n\rangle, \ldots, |\psi_k\rangle = |\phi_k\rangle - \sum_{n=1}^{k-1} \frac{\langle \psi_n | \phi_k \rangle}{\langle \psi_n | \psi_n \rangle} |\psi_n\rangle.$$

Note that the new vectors $\{|\psi_n\rangle\}$ with $n = 1, \ldots, j-1$ (not the old vectors $\{|\phi_n\rangle\}$) appear on the RHS of the equation for $|\psi_j\rangle$. If so desired, the orthogonal vectors $\{|\psi_n\rangle\}$ can be normalized to obtain a set of *orthonormal vectors*.[2]

Problem A.6

(a) Given vectors $|v_1\rangle = \begin{pmatrix} 2 \\ 0 \\ 0 \end{pmatrix}$, $|v_2\rangle = \begin{pmatrix} 1 \\ i \\ 0 \end{pmatrix}$, $|v_3\rangle = \begin{pmatrix} 1 \\ 1 \\ -1 \end{pmatrix}$, use the Gram–Schmidt procedure to find orthogonal vectors $|u_1\rangle$, $|u_2\rangle$, and $|u_3\rangle$, with $|u_1\rangle = |v_1\rangle$.

(b) Normalize the vectors obtained in (a).

(c) Repeat (a) and (b) using $|u_1\rangle = |v_2\rangle$.

Answer: (a) $|u_1\rangle = \begin{pmatrix} 2 \\ 0 \\ 0 \end{pmatrix}$, $|u_2\rangle = \begin{pmatrix} 0 \\ i \\ 0 \end{pmatrix}$, $|u_3\rangle = \begin{pmatrix} 0 \\ 0 \\ -1 \end{pmatrix}$. (b) $|\tilde{u}_1\rangle = \begin{pmatrix} 1 \\ 0 \\ 0 \end{pmatrix}$, $|\tilde{u}_2\rangle = \begin{pmatrix} 0 \\ 1 \\ 0 \end{pmatrix}$, $|\tilde{u}_3\rangle = \begin{pmatrix} 0 \\ 0 \\ 1 \end{pmatrix}$.

Hilbert Spaces

As we have already stressed, the mathematical formulation of quantum mechanics is based on Hilbert space, where vectors correspond to quantum states. For finite dimension N, a Hilbert space is just an inner product (vector) space. However, the mathematical description of quantum systems often requires the structure of infinite dimensional inner product spaces. In such cases, proper convergence and limiting procedures should be taken into consideration.

---

[2] Gram–Schmidt can be used to show that any matrix A with linearly independent columns (e.g., a matrix with columns given by vectors $\{|\phi_n\rangle\}$) can be decomposed into the product of two matrices, $A = QR$, where the columns of Q are orthonormal (the vectors $\{|\psi_n\rangle\}$), and R is upper triangular and invertible. This is called *QR decomposition* or *QR factorization*.

A.1 Vector Spaces

The definition of Hilbert space is as follows: a Hilbert space $\mathcal{H}$ is an inner product space with two additional properties denoted here as I and II. If N is finite, these properties can be proved from the previous definitions of an inner product space, but for an infinite dimensional space they should be considered as axioms.

I. $\mathcal{H}$ is complete. If an infinite sequence $|\psi_1\rangle, |\psi_2\rangle, \ldots, \in \mathcal{H}$ satisfies the Cauchy convergence criterion [for each $\varepsilon > 0$ there exists a positive integer $M(\varepsilon)$, such that for $m, n > M(\varepsilon)$, $|||\psi_n\rangle - |\psi_m\rangle|| < \varepsilon$], then the sequence is convergent, i.e., it possesses a limit $|\psi\rangle$ such that $\lim_{n\to\infty} |||\psi_n\rangle - |\psi\rangle|| = 0$.

II. $\mathcal{H}$ is separable: that is, there is a sequence $|\psi_1\rangle, |\psi_2\rangle, \ldots, \in \mathcal{H}$, which is everywhere dense in $\mathcal{H}$.[3] Roughly speaking, elements of a dense set come arbitrary close to any element in $\mathcal{H}$. The property of separability is equivalent to the statement that there is a countably infinite complete orthonormal set $\{|\phi_n\rangle\}$ such that every vector $|\psi\rangle \in \mathcal{H}$ can be expanded as,

$$|\psi\rangle = \sum_{n=1}^{\infty} \langle\phi_n|\psi\rangle |\phi_n\rangle \Leftrightarrow \lim_{N\to\infty} \left\| \sum_{n=1}^{N} \langle\phi_n|\psi\rangle |\phi_n\rangle - |\psi\rangle \right\| = 0. \quad (A.22)$$

A necessary and sufficient condition for convergence is $\sum_{n=1}^{\infty} |\langle\phi_n|\psi\rangle|^2 < \infty$. Hence, this sum equals $\langle\psi|\psi\rangle$.

In summary: A Hilbert space is an inner product vector space which is also complete and separable [28, 282]. For practical applications, it is useful to have a represention of the Hilbert space. Let $|\psi\rangle$ be a ket in the Hilbert space $\mathcal{H}$ of dimension N ($N = \infty$ is possible). Its expansion in a given basis $\{|\phi_i\rangle\}$ is given by $|\psi\rangle = \sum_{i=1}^{N} |\phi_i\rangle \langle\phi_i|\psi\rangle$, and its representation as a row vector by Eq. (A.21),

$$|\psi\rangle := \begin{pmatrix} \langle\phi_1|\psi\rangle \\ \langle\phi_2|\psi\rangle \\ \cdot \\ \cdot \\ \cdot \\ \langle\phi_N|\psi\rangle \end{pmatrix}. \quad (A.23)$$

The set of N coefficients forming the vector on the RHS of Eq. (A.23) is the representation of $|\psi\rangle$ in the basis $\{|\phi_j\rangle\}$. The inner product between two kets $|\psi\rangle$ and $|\chi\rangle$ is given by Eq. (A.16) [or, for an orthonormal basis, by Eq. (A.18), which is simply the scalar product of the two corresponding vectors of coefficients]. Note that the inner product does not depend on the representation used. Note also that the bra $\langle\psi|$ can be expanded as $\langle\psi| = \sum_{i=1}^{N} \langle\psi|\phi_i\rangle \langle\phi_i|$, and it is represented by a column vector with components $\langle\psi|\phi_i\rangle = \langle\phi_i|\psi\rangle^*$ [see Eq. (A.18)].

Hilbert Space of Functions

One can define a Hilbert space whose elements (vectors) are complex functions $\psi(\mathbf{x}), \chi(\mathbf{x}), \ldots$ (or, $\psi_1(\mathbf{x}), \psi_2(\mathbf{x}), \ldots$) defined on $X = \mathbb{R}^k$, i.e., $\mathbf{x} = (x_1, x_2, \ldots, x_k)$ with real x_i. A *Hilbert space $\mathcal{H}$ of functions* satisfies the following:

1. There exists an operation $+$ and an operation of multiplication by a (complex) scalar satisfying *all* the requirements of a linear vector space. In particular, if $\psi(\mathbf{x}), \chi(\mathbf{x}) \in \mathcal{H}$, so is any function obtained as a linear combination, $\eta(\mathbf{x}) = a\psi(\mathbf{x}) + b\chi(\mathbf{x})$, where a, b are complex numbers.
2. The inner product in $\mathcal{H}$ is defined as $(\psi, \chi) = \int d\mathbf{x}\, \psi(\mathbf{x})^* \chi(\mathbf{x})$. It has the same properties as the inner product $\langle\psi|\chi\rangle$ of kets. The norm of a function $\psi(x) \in \mathcal{H}$ is $||\psi|| = \sqrt{(\psi, \psi)}$.
3. Functions $\{f_n(\mathbf{x}) \in \mathcal{H}\}$, $n = 1, 2, \ldots, m$, are said to be linearly independent if the equality $\sum_{n=1}^{m} a_n f_n(\mathbf{x}) = 0$, $\forall\, \mathbf{x} \in X$, implies $a_n = 0$ for $n = 1, 2, \ldots, m$. (3$_N$) $\mathcal{H}$ is said to be of finite dimension, $0 < N < \infty$, if there are at most N linearly independent functions in $\mathcal{H}$. We write it as $\mathcal{H}_N$.

---

[3] An example of a dense sequence is the rational numbers which are dense in $\mathbb{R}$.

(3_∞) $\mathcal{H}$ is said to be of infinite dimension, if there are an infinite number of independent functions in $\mathcal{H}$. We write it as $\mathcal{H}_\infty$. Hilbert spaces of functions of a continuous variable are typically infinite-dimensional, however, in practical calculations a judicious choice of a finite basis often suffices.

4. $\mathcal{H}$ is complete. For $\mathcal{H}_N$, this means that an orthonormal basis of functions $\{f_n(\mathbf{x})\}, n = 1, 2, \ldots$, exists in $\mathcal{H}_N$ such that any function $\psi(\mathbf{x}) \in \mathcal{H}_N$ can be expanded as $\psi(\mathbf{x}) = \sum_{n=1}^{N} (f_n, \psi) f_n(\mathbf{x})$. For $\mathcal{H}_\infty$, the definition of completeness is as follows. If an infinite sequence of functions $\psi_1(\mathbf{x}), \psi_2(\mathbf{x}), \ldots, \in \mathcal{H}_\infty$ satisfies the Cauchy convergence criterion [for each $\varepsilon > 0$, there exists a positive integer $M(\varepsilon)$, such that for $m, n > M(\varepsilon)$, $\| \psi_n(\mathbf{x}) - \psi_m(\mathbf{x}) \| < \varepsilon$, $\forall \mathbf{x} \in X$], then the sequence is convergent, i.e., it possesses a limit $\psi(\mathbf{x})$.

5. $\mathcal{H}_\infty$ is separable, i.e., there is a sequence $\psi_1(\mathbf{x}), \psi_2(\mathbf{x}), \ldots, \in \mathcal{H}_\infty$, which is everywhere dense in $\mathcal{H}_\infty$. Equivalently, there is a countable infinite complete orthonormal set $\{f_n(\mathbf{x})\}$ of functions $f_n(\mathbf{x}) \in \mathcal{H}_\infty$ such that every function $\psi(\mathbf{x}) \in \mathcal{H}_\infty$ can be expanded as,

$$\psi(\mathbf{x}) = \sum_{n=1}^{\infty}(f_n, \psi)\phi_n(\mathbf{x}) \Leftrightarrow \lim_{N \to \infty} \left\| \sum_{n=1}^{N}(f_n, \psi)f_n(\mathbf{x}) - \psi(\mathbf{x}) \right\| = 0, \forall \mathbf{x} \in X. \tag{A.24}$$

A necessary and sufficient condition for convergence is $\sum_{n=1}^{\infty} |(f_n, \psi)|^2 < \infty$. Hence, this sum equals $\langle \psi | \psi \rangle$.

Problem A.7

(a) Given the two real functions $f_1(x) = x$ and $f_2(x) = x^2$, orthogonalize f_2 to f_1 over the interval $x \in [-1, 1]$, i.e., obtain the function $\tilde{f}_2(x) = f_2(x) + cf_1(x)$ such that $\langle f_1 | \tilde{f}_2 \rangle \equiv \int_{-1}^{1} dx f_1(x) \tilde{f}_2(x) = 0$.

(b) Given the function $f_3(x) = x^3$, obtain the function $\tilde{f}_3(x) = f_3(x) + c_1 f_1(x) + c_2 f_2(x)$ such that $\int_{-1}^{1} dx f_1(x) \tilde{f}_3(x) = 0$ and $\int_{-1}^{1} dx \tilde{f}_2(x) \tilde{f}_3(x) = 0$, i.e., such that $\langle f_1 | \tilde{f}_3 \rangle = 0$ and $\langle \tilde{f}_2 | \tilde{f}_3 \rangle = 0$.

A.2 OPERATORS AND MATRICES

An *operator* $\hat{O}$ applied to a state $|\psi\rangle$ yields another state $|\chi\rangle$: $|\chi\rangle \equiv \hat{O}|\psi\rangle$. A *linear operator* transforms states so that

$$\hat{O}(c_1|\psi_1\rangle + c_2|\psi_2\rangle) = c_1 \hat{O}|\psi_1\rangle + c_2 \hat{O}|\psi_2\rangle, \tag{A.25}$$

for any states $|\psi_1\rangle, |\psi_2\rangle$ and complex numbers c_1, c_2. The *expectation value of an operator* $\hat{O}$ in the state $|\psi\rangle$ is denoted as $\langle \psi | \hat{O} | \psi \rangle$ and is defined to be $\langle \psi | \hat{O} | \psi \rangle = \langle \psi | (\hat{O} | \psi \rangle)$.

Given an orthonormal basis $\{|\phi_j\rangle\}$, an operator $\hat{O}$ can be written as a sum of the form,

$$\hat{O} = \sum_{ij} |\phi_i\rangle o_{ij} \langle \phi_j|, \tag{A.26}$$

where $o_{ij} = \langle \phi_i | \hat{O} | \phi_j \rangle$. If the basis is N dimensional, the operator $\hat{O}$ can be represented by the $N \times N$ matrix

$$\mathcal{O} = \begin{pmatrix} o_{11} & \cdots & o_{1N} \\ \vdots & & \vdots \\ o_{N1} & \cdots & o_{NN} \end{pmatrix}. \tag{A.27}$$

Moreover, in the $\{|\phi_j\rangle\}$ basis, the expectation value of the operator $\hat{O}$ in the state $|\psi\rangle$ is given by

$$\langle \psi | \hat{O} | \psi \rangle = \sum_{ij} a_i^* o_{ij} a_j, \tag{A.28}$$

A.2 Operators and Matrices

where $|\psi\rangle = \sum_i a_i |\phi_i\rangle$. Furthermore, in the $\{|\phi_j\rangle\}$ basis, the *matrix element* $\langle \chi | \hat{O} | \psi \rangle$ of the operator $\hat{O}$ is given by the (in general complex) number $\langle \chi | \hat{O} | \psi \rangle = \sum_{ij} b_i^* o_{ij} a_j$, where $|\chi\rangle = \sum_i b_i |\phi_i\rangle$. Sometimes, we shall use Einstein summation notation and not explicitly write the sum in expressions containing a repeated index, e.g., we can write the RHS of Eq. (A.28) simply as $a_i^* o_{ij} a_j$, and $\langle \chi | \hat{O} | \psi \rangle = b_i^* o_{ij} a_j$.

A.2.1 OUTER PRODUCT

In quantum mechanics, the *outer product* of two vectors is a special kind of *tensor product* of the vectors. In linear algebra, the tensor product of two vectors $\mathbf{u} = (u_1, \ldots, u_m)$ and $\mathbf{v} = (v_1, \ldots, v_n)$ is the $m \times n$ matrix

$$\mathbf{u} \otimes \mathbf{v} \equiv \begin{pmatrix} u_1 v_1 & \cdots & u_1 v_n \\ \vdots & & \vdots \\ u_m v_n & \cdots & u_m v_n \end{pmatrix}. \tag{A.29}$$

Using Dirac notation, if the two vectors are $|\psi\rangle$ and $|\chi\rangle$, their outer product is given by $|\psi\rangle\langle\chi|$, which is an operator that acts on ket vectors. Specifically, when $|\psi\rangle\langle\chi|$ operates on the state vector $|\varphi\rangle$, one obtains the vector $|\psi\rangle \langle\chi|\varphi\rangle$.

For a finite dimensional vector space ($N < \infty$), the outer product can be represented, using an orthonormal basis, by an $N \times N$ matrix:

$$|\psi\rangle\langle\chi| = \begin{pmatrix} a_1 \\ \vdots \\ a_N \end{pmatrix} \begin{pmatrix} b_1^* & \cdots & b_N^* \end{pmatrix}$$

$$= \begin{pmatrix} a_1 b_1^* & \cdots & a_1 b_N^* \\ \vdots & & \vdots \\ a_N b_1^* & \cdots & a_N b_N^* \end{pmatrix}. \tag{A.30}$$

In terms of the outer product, the notion of completeness of an orthonormal basis $\{|\phi_j\rangle\}$ has a particularly simple and compact form,

$$\sum_{j=1}^{N} |\phi_j\rangle\langle\phi_j| = \mathbf{1}. \tag{A.31}$$

Here, $\mathbf{1}$ is the unit $N \times N$ matrix. Then, Eq. (A.21) can be written as, $|\psi\rangle = \sum_j |\phi_j\rangle \langle\phi_j|\psi\rangle$, which just implies the tautology $|\psi\rangle = \mathbf{1}|\psi\rangle$, where the unit operator in Eq. (A.31) acts on $|\psi\rangle$. To reiterate, the important properties of completeness and orthogonality can be written as

$$\left[\sum_j |\phi_j\rangle\langle\phi_j| \right] = \mathbf{1}, \tag{A.32a}$$

$$\langle\phi_i|\phi_j\rangle = \delta_{ij}. \tag{A.32b}$$

It is also possible to formally consider matrices with infinitely many rows and/or columns. Infinite matrices arise upon considering operators in infinite dimensional Hilbert spaces (see Sec. A.1.2), where convergence, completeness and continuity issues require a bit more care. Here, we note that for position states $|q\rangle$, completeness reads, $\int dq \, |q\rangle \langle q| = \mathbf{1}$, and for momentum states, it reads $\int dp \, |p\rangle \langle p| = \mathbf{1}$, i.e., the integrals replace the finite sum in Eq. (A.31), and these states form a complete basis (see Sec. 1.3.8). Moreover, the states $|q\rangle$ and $|q'\rangle$ with $q \neq q'$ are orthogonal, $\langle q'|q\rangle = 0$, and similarly, $\langle p'|p\rangle = 0$ if $p \neq p'$.

A.2.2 DETERMINANTS AND PERMANENTS

The *determinant* of a square $N \times N$ matrix A is a scalar, denoted by $\det(A)$ or $|A|$, defined by $\det(A) = \sum_P (-1)^P P \prod_{i=1}^N a_{i,P(i)}$, where P are permutation operators that permute the indices (see, e.g., Sec. 8.1 and 8.3). It is often written in the form

$$\det(A) = \begin{vmatrix} a_{11} & a_{12} & a_{13} & \cdots & a_{1N} \\ a_{21} & & & & a_{2N} \\ a_{31} & & \ddots & & a_{3N} \\ \vdots & & & & \vdots \\ a_{N1} & a_{N2} & a_{N3} & \cdots & a_{NN} \end{vmatrix}. \quad (A.33)$$

For a 2×2 matrix,

$$\begin{vmatrix} a_{11} & a_{12} \\ a_{21} & a_{22} \end{vmatrix} = a_{11}a_{22} - a_{21}a_{12}, \quad (A.34)$$

and for a 3×3 matrix,

$$\begin{vmatrix} a_{11} & a_{12} & a_{13} \\ a_{21} & a_{22} & a_{23} \\ a_{31} & a_{32} & a_{33} \end{vmatrix} = a_{11} \begin{vmatrix} a_{22} & a_{23} \\ a_{32} & a_{33} \end{vmatrix} - a_{12} \begin{vmatrix} a_{21} & a_{23} \\ a_{31} & a_{33} \end{vmatrix} + a_{13} \begin{vmatrix} a_{21} & a_{22} \\ a_{31} & a_{32} \end{vmatrix}, \quad (A.35)$$

and expanding we find, $\det(A)_{3 \times 3} = a_{11}(a_{22}a_{33} - a_{32}a_{23}) - a_{12}(a_{21}a_{33} - a_{31}a_{23}) + a_{13}(a_{21}a_{32} - a_{31}a_{22})$. For a 4×4 matrix,

$$\begin{vmatrix} a_{11} & a_{12} & a_{13} & a_{14} \\ a_{21} & a_{22} & a_{23} & a_{24} \\ a_{31} & a_{32} & a_{33} & a_{34} \\ a_{41} & a_{42} & a_{43} & a_{44} \end{vmatrix} = a_{11} |3 \times 3| - a_{12} |3 \times 3| + a_{13} |3 \times 3| - a_{14} |3 \times 3|,$$

where the $|3 \times 3|$ are the determinants of the appropriate 3×3 cofactor matrix. For a general $N \times N$ matrix, Eq. (A.33) can be reduced to $\det(A) = \sum_j (-1)^{1+j} a_{1j} C_{1j}$, where C_{1j} is the cofactor matrix formed by deleting the first row and jth column of A, or alternatively, $\det(A) = \sum_j (-1)^{i+j} a_{ij} C_{ij}$.

Determinants have the following properties, from which the general formula for the determinant can be derived. (1) If two columns or rows of a determinant are exchanged, then the value of the determinant changes sign. (2) If two columns or rows of a determinant are the same, or proportional, then the value of the determinant is 0. (3) Multiplying any column or row by a constant c, results in the determinant being multiplied by c. (4) Adding any two rows (i.e., replacing $a_{i,j}$ by $a_{i+k,j}$) or columns leaves the determinant unchanged.

The determinant of the product of two square matrices A and B of the same size is the product of their determinants, $\det(AB) = \det(A) \det(B)$. Hence, $\det(A^k) = [\det(A)]^k$. The determinant of the transpose of a square matrix equals the determinant of the matrix, $\det(A^t) = \det(A)$. Since $AA^{-1} = \mathbf{1}$, and $\det(AA^{-1}) = \det(A) \det(A^{-1})$, we conclude that $\det(A^{-1}) = [\det(A)]^{-1}$. A matrix is invertible only if its determinant is non-zero.

The *permanent* of a square $N \times N$ matrix A is a scalar, denoted by $\text{perm}(A)$ defined by $\text{perm}(A) = \sum_P P \prod_{i=1}^N a_{i,P(i)}$. Note the similarity with the definition of the determinant of a matrix; only the factors $(-1)^P$ are missing.

A.2 Operators and Matrices

A.2.3 NORMAL, UNITARY, HERMITIAN-CONJUGATE AND HERMITIAN OPERATORS

In Sec. 1.3.2, we defined the *Hermitian conjugate* of an operator $\hat{O}$, $\hat{O}^\dagger$, to be the operator that satisfies the following equation for any vectors $|\psi_1\rangle$ and $|\psi_2\rangle$:

$$\langle\psi_2|\hat{O}^\dagger|\psi_1\rangle = \langle\psi_1|\hat{O}|\psi_2\rangle^*. \qquad (A.36)$$

In a basis, the Hermitian conjugate matrix is given by $\mathcal{O}^\dagger = \text{transpose}(\mathcal{O}^*)$, i.e., the *ij* element of the Hermitian conjugate matrix is given by the complex conjugate of the *ji* element of $\mathcal{O}$, $(\mathcal{O}^\dagger)_{ij} = o_{ji}^*$. The Hermitian conjugate of the product of two matrices satisfies the property $(\mathcal{O}_1\mathcal{O}_2)^\dagger = \mathcal{O}_2^\dagger \mathcal{O}_1^\dagger$.

When the Hermitian conjugate of a state vector is taken, the dual space vector is obtained, $|\psi\rangle^\dagger = \langle\psi|$; moreover, $(|\psi\rangle\langle\chi|)^\dagger = |\chi\rangle\langle\psi|$. The Hermitian conjugate of $a\mathcal{O}$ is given by $(a\mathcal{O})^\dagger = a^*\mathcal{O}^\dagger$, the Hermitian conjugate of $\alpha|\psi\rangle$ is $\alpha^*\langle\psi|$ and

$$(\alpha|\psi\rangle + \beta|\chi\rangle)^\dagger = \alpha^*\langle\psi| + \beta^*\langle\chi|. \qquad (A.37)$$

An operator, $\hat{O}$, is *normal*, if it satisfies $\hat{O}\hat{O}^\dagger = \hat{O}^\dagger\hat{O}$. Operators are diagonalizable if they are normal. For a normal operator $\hat{O}$, one can always find an orthonormal basis $\{|\phi_j\rangle\}$ such that

$$\hat{O} = \sum_j |\phi_j\rangle\lambda_j\langle\phi_j|. \qquad (A.38)$$

The scalars λ_j are called the eigenvalues of the operator, and the vectors $|\phi_j\rangle$ are the eigenvectors of the operator; they satisfy the equation

$$\hat{O}|\phi_j\rangle = \lambda_j|\phi_j\rangle. \qquad (A.39)$$

The eigenvalues λ_j of a normal operator need not be real. A matrix $\mathcal{O}$ is *normal* if it satisfies $\mathcal{O}\mathcal{O}^\dagger = \mathcal{O}^\dagger\mathcal{O}$. Normal matrices are unitarily equivalent to diagonal matrices, i.e., $\mathcal{O} = \mathcal{U}\Lambda\mathcal{U}^\dagger$, where the matrix $\mathcal{U}$ is *unitary* (meaning $\mathcal{U}\mathcal{U}^\dagger = 1$, i.e., $\mathcal{U}^{-1} = \mathcal{U}^\dagger$), with columns that are the eigenvectors of $\mathcal{O}$, and Λ is a diagonal matrix with matrix elements λ_j. Non-normal matrices can still be spectrally resolved using singular value decomposition (see Sec. 5.2.2).

In Sec. 1.3.2, we also defined a Hermitian operator $\hat{\mathcal{H}}$ to be one that is equal to its Hermitian conjugate:

$$\hat{\mathcal{H}} = \hat{\mathcal{H}}^\dagger. \qquad (A.40)$$

A Hermitian operator or Hermitian matrix is sometimes called *self adjoint*. The eigenvalues of a Hermitian matrix are real, and its eigenvectors can always be orthonormalized, i.e.,

$$\mathcal{H}|\phi_j\rangle = h_j|\phi_j\rangle, \qquad (A.41)$$

with real eigenvalues, $h_j^* = h_j$, and orthonormal eigenvectors, $\langle\phi_i|\phi_j\rangle = \delta_{ij}$. To prove this, we first note that $\langle\phi_j|\mathcal{H}^\dagger = h_j^*\langle\phi_j|$ as can be easily seen by taking the Hermitian conjugate of Eq. (A.41). Taking the inner product of this equation with $|\phi_i\rangle$, we find

$$\langle\phi_j|\mathcal{H}^\dagger|\phi_i\rangle = h_j^*\langle\phi_j|\phi_i\rangle, \qquad (A.42)$$

and now re-writing Eq. (A.41) as $\mathcal{H}|\phi_i\rangle = h_i|\phi_i\rangle$ and multiplying from the left by $\langle\phi_j|$, we find

$$\langle\phi_j|\mathcal{H}|\phi_i\rangle = h_i\langle\phi_j|\phi_i\rangle. \qquad (A.43)$$

Subtracting Eqs. (A.42) and (A.43), and using the fact that $\hat{\mathcal{H}} = \hat{\mathcal{H}}^\dagger$, we obtain

$$(h_j^* - h_i)\langle\phi_j|\phi_i\rangle = 0. \qquad (A.44)$$

For $i = j$, we find that $h_i^* = h_i$, since $\langle\phi_i|\phi_i\rangle$ does not vanish, i.e., the eigenvalues are real. For $h_j \neq h_i$, we find that $\langle\phi_j|\phi_i\rangle = 0$. Degenerate eigenvectors (i.e., eigenvectors belonging to eigenvalues that appear more than once, e.g., if there are two eigenvectors, $|\phi_i\rangle$ and $|\phi_j\rangle$ with exactly the same eigenvalue, $h_j = h_i$) can be orthogonalized using Gram–Schmidt.

The expectation value of a Hermitian operator is real. To prove this, consider the matrix element $\langle\chi|\hat{\mathcal{H}}|\psi\rangle$. Because $\hat{\mathcal{H}}$ is Hermitian, $\langle\chi|\hat{\mathcal{H}}|\psi\rangle = \langle\chi|\hat{\mathcal{H}}^\dagger|\psi\rangle = \langle\psi|\hat{\mathcal{H}}|\chi\rangle^*$. Now let $|\chi\rangle = |\psi\rangle$, to obtain $\langle\psi|\hat{\mathcal{H}}|\psi\rangle = \langle\psi|\hat{\mathcal{H}}|\psi\rangle^*$. This completes the proof.

An *anti-Hermitian operator* is one that satisfies the relation $\hat{A} = -\hat{A}^\dagger$. The expectation value of an anti-Hermitian operator is imaginary. To prove this, consider the matrix element $\langle\chi|\hat{A}|\psi\rangle$. Because $\hat{A}$ is anti-Hermitian, $\langle\chi|\hat{A}|\psi\rangle = -\langle\chi|\hat{A}^\dagger|\psi\rangle = -\langle\psi|\hat{A}|\chi\rangle^*$. Now let $|\chi\rangle = |\psi\rangle$, to obtain $\langle\psi|\hat{A}|\psi\rangle = -\langle\psi|\hat{A}|\psi\rangle^*$. This completes the proof. Note that the product of an anti-Hermitian operator and an imaginary number results in a Hermitian operator, e.g., multiplying an anti-Hermitian operator by i yields a Hermitian operator, and visa versa.

Two $N\times N$ matrices, A and B are *similar*, if there exists an nonsingular (i.e., invertible) matrix $N\times N$ X such that $A = XBX^{-1}$. The *spectra* (i.e., the eigenvalues) of similar matrices are equal, as you will prove in Problem A.8.

Problem A.8

Prove that the eigenvalues of two similar matrices A and B are equal.

Answer: The eigenvalues of a square matrix A can be determined by calculating the roots λ of the *characteristic polynomial*, $\det(A - \lambda\mathbf{1})$. Now, $\det(A - \lambda\mathbf{1}) = \det(XBX^{-1} - X\lambda\mathbf{1}X^{-1}) = \det[X(B - \lambda\mathbf{1})X^{-1}]$. Using the properties of determinants, this equals $\det(X)\det(B - \lambda\mathbf{1})\det(X^{-1}) = \det(B - \lambda\mathbf{1})$.

Problem A.9

(a) Calculate the eigenvalues/vectors of $H = \hbar\begin{pmatrix} \Delta & \Omega/2 \\ \Omega/2 & 0 \end{pmatrix}$.

Answer: $|H - E\mathbf{1}| = \begin{vmatrix} \hbar\Delta - E & \hbar\Omega/2 \\ \hbar\Omega/2 & -E \end{vmatrix} = 0$, so $E_\pm = \frac{\hbar}{2}\left(\Delta \pm \sqrt{\Delta^2 + \Omega^2}\right)$. $|\psi_+\rangle = \begin{pmatrix} \Delta + \sqrt{\Delta^2 + \Omega^2} \\ \Omega \end{pmatrix}$, $|\psi_-\rangle = \begin{pmatrix} \Delta - \sqrt{\Delta^2 + \Omega^2} \\ \Omega \end{pmatrix}$.

A.2.4 TRACE AND PROJECTION

The trace of an operator (and of a matrix) is the sum of its diagonal elements,

$$\text{Tr}(\hat{O}) \equiv \sum_j \hat{O}_{jj}. \tag{A.45}$$

The trace is independent of the choice of basis, as long as the basis vectors are complete. An operator whose trace vanishes is said to be traceless. The trace of the product of two matrices satisfies the property

$$\text{Tr}(\hat{A}\hat{B}) = \text{Tr}(\hat{B}\hat{A}), \tag{A.46}$$

since $\hat{A}_{ij}\hat{B}_{ji} = \hat{B}_{ij}\hat{A}_{ji}$, where Einstein summation notation has been used.

A *projection operator* is a Hermitian operator that satisfies the property $\hat{P}\hat{P} = \hat{P}$. This property is called *idempotence*. The eigenvalues of $\hat{P}$ are either 0 or 1. The complement of a projection operator, $\hat{\mathbf{1}} - \hat{P}$, is also a projector. The operator $|\phi\rangle\langle\phi|$ for any normalized ket $|\phi\rangle$ is clearly a projection operator.

A.2 Operators and Matrices

A.2.5 ANTILINEAR AND ANTIUNITARY OPERATORS

Antilinear operators arise in quantum mechanics when dealing with a symmetry called time-reversal (which actually describes motion reversal). An antilinear operator transforms a linear combination of two states such that

$$\hat{A}(c_1|\psi_1\rangle + c_2|\psi_2\rangle) = c_1^* \hat{A}|\psi_1\rangle + c_2^* \hat{A}|\psi_2\rangle, \tag{A.47}$$

for any states $|\psi_1\rangle$, $|\psi_2\rangle$ and complex numbers c_1, c_2. The product of two antilinear operators $\hat{A}_2$ and $\hat{A}_1$, where $(\hat{A}_2\hat{A}_1)|\psi\rangle = \hat{A}_2(\hat{A}_1|\psi\rangle)$ for any state $|\psi\rangle$, is a linear operator, since the second operation of complex conjugation undoes the first. An operator $\hat{A}$ is antiunitary, if it is antilinear, if its inverse $\hat{A}^{-1}$ exists, and it satisfies

$$\| \hat{A}|\psi\rangle \| = \||\psi\rangle\|, \tag{A.48}$$

for all $|\psi\rangle$. This definition of an antiunitary operator implies that if $|\phi_1\rangle = \hat{A}|\psi_1\rangle$ and $|\phi_2\rangle = \hat{A}|\psi_2\rangle$, then $\langle\phi_1|\phi_2\rangle = \langle\psi_1|\psi_2\rangle^*$. We shall see in Sec. 2.9.3 that the time reversal operator $\hat{T}$ is antiunitary. Any antiunitary operator $\hat{A}$ can be written as the product of a unitary operator $\hat{U}$ and the complex conjugation operator $\hat{K}$, $\hat{A} = \hat{U}\hat{K}$. A word of caution is in order: Dirac notation is perhaps ideal for linear operators, but it can be a bit confusing for antilinear operators (as we shall see in Sec. 2.9.3).

Problem A.10

Prove that for an antilinear operator $\hat{A}$, Eq. (A.48) implies that $\langle\hat{A}\psi_1|\hat{A}\psi_2\rangle = \langle\psi_1|\psi_2\rangle^*$.

Hint: The proof is sketched out in Sec. 2.9.3, see Eq. (2.175) and the surrounding text. Using this equation, fill in the details of the proof.

Some Ordinary Differential Equations B

In this appendix, we discuss some ordinary differential equations (ODEs) whose solutions are required in the various chapters of this book. This is not a tutorial on methods to solve differential equations; we would not be able to do justice to the subject of ODEs in a short appendix. Readers interested in the latter are referred to Refs. [283–286].

Quantum mechanics is linear, therefore most of the ODEs we shall encounter are linear ODEs. A linear differential operator $\mathcal{L}$ of order n which operates on a function $y(x)$ takes the form

$$\mathcal{L} = \sum_{j=0}^{n} p_j(x) \left(\frac{d}{dx}\right)^j, \tag{B.1}$$

where the $p_j(x)$ are functions of x, and the highest function, p_n, must be nonzero.

Homogeneous ODEs take the form $\mathcal{L}y(x) = 0$, and *inhomogeneous ODEs*, sometimes called driven ODEs, take the form $\mathcal{L}y(x) = f(x)$ with nonvanishing $f(x)$. The general solution to an inhomogeneous ODE takes the form of the sum of a particular solution $y_p(x)$ to the driven equation, $\mathcal{L}y(x) = f(x)$, and a solution $y_h(x)$ to the homogeneous ODE, i.e., $y(x) = y_p(x) + y_h(x)$.

A first-order ODE has the highest order $n = 1$, e.g.,

$$\left((2x^2 + 3)\frac{d}{dx} + (4x - 1)\right) y(x) = 0.$$

A second-order ODE has highest order $n = 2$, e.g., $\frac{d^2 y}{dx^2} - xy = e^{-x^2}$, which is an inhomogeneous differential equation, or, $-\frac{d^2 \psi}{dx^2} + \frac{2m}{\hbar^2}(V(x) - E)\psi = 0$, which is homogeneous. An example of a nonlinear differential equation is $\frac{dy}{dx} + y^2 = 0$.

Initial conditions (or boundary conditions) are necessary to completely specify the solution to differential equations. First-order ODEs require that an initial condition be imposed on the solution of the form $y(x = x_0) = y_0$, where x_0 and y_0 are constants. Second-order ODEs require initial conditions of the form $\{y(x = x_0) = y_0, y'(x = x_0) = y'_0\}$ to completely specify the solution, or the value of the dependent variable y can be specified at two x points. Higher order ODEs require additional initial conditions to fully specify the solution.

ODEs with *constant coefficients* take the form

$$\sum_{j=0}^{n} A_j \left(\frac{d}{dx}\right)^j y(x) = f(x), \tag{B.2}$$

where the A_j are constants. The solution of the homogeneous version of such ODEs is given by $y(x) = e^{zx}$, where z satisfies the nth degree polynomial equation $\sum_{j=0}^{n} A_j z^j = 0$.

Below we discuss some specific differential equations and their solutions that are encountered in our study of quantum mechanics.

The decay of the time-dependent quantity $y(t)$ with decay rate α is described mathematically by the differential equation,

$$\frac{dy}{dt} = -\alpha y. \tag{B.3}$$

The general solution to this equation is [see Eq. (7.131)]

$$y(t) = y_0 \, e^{-\alpha t}. \tag{B.4}$$

If the initial condition at time $t = 0$ is $y(0)$, then the coefficient y_0 is such that $y_0 = y(0)$. Exponential growth is obtained in the differential equation,

$$\frac{dy}{dt} = \alpha y, \tag{B.5}$$

whose general solution is $y(t) = y_0 \, e^{\alpha t}$.

The second-order differential equation

$$\frac{d^2 y}{dx^2} = -k^2 y, \tag{B.6}$$

is sometimes called the *Helmholtz equation*, or the *spatial wave equation* (see Sec. 1.3.10). It has solutions of the form

$$y(x) = a \sin(kx) + b \cos(kx). \tag{B.7}$$

The two constants a and b can be determined if boundary conditions for $y(x)$ are given for two points $x = x_1$ and $x = x_2$ or if boundary conditions on $y(x)$ and dy/dx are given at a point $x = a$. To contrast Eq. (B.6), the ODE,

$$\frac{d^2 y}{dx^2} = K^2 y, \tag{B.8}$$

has a general solution that is a linear combination of exponentially increasing and decreasing functions,

$$y(x) = a \, e^{Kx} + b \, e^{-Kx}. \tag{B.9}$$

This differential equation describes tunneling through a constant height potential barrier.

For tunneling through a linearly varying potential, the Schrödinger equation can be reduced to the form:

$$\frac{d^2 y}{dz^2} = zy. \tag{B.10}$$

The solutions to this ODE are called *Airy functions*, after the astronomer George Biddell Airy. The two linearly independent Airy functions are denoted by $\text{Ai}(z)$ and $\text{Bi}(z)$, and the general solution is,

$$y(z) = a \, \text{Ai}(z) + b \, \text{Bi}(z). \tag{B.11}$$

The asymptotic expansions of the Airy functions is given by [27],

$$\text{Ai}(z) \xrightarrow[z \to \infty]{} \frac{1}{2\pi^{1/2} z^{1/4}} \exp(-\zeta), \tag{B.12}$$

$$\text{Ai}(-z) \xrightarrow[z \to \infty]{} \frac{1}{\pi^{1/2} z^{1/4}} \sin(\zeta + \pi/4), \tag{B.13}$$

$$\text{Bi}(z) \xrightarrow[z \to \infty]{} \frac{1}{\pi^{1/2} z^{1/4}} \exp(\zeta), \tag{B.14}$$

$$\text{Bi}(-z) \xrightarrow[z \to \infty]{} \frac{1}{\pi^{1/2} z^{1/4}} \cos(\zeta + \pi/4), \tag{B.15}$$

where $\zeta = \frac{2}{3} z^{3/2}$. The regular and irregular Airy functions, Ai and Bi, are plotted in Fig. 1.24. The integral representation of the Airy function can be obtained by considering its Fourier transform:

$$\text{Ai}(z) = \int_{-\infty}^{\infty} d\xi \, e^{i\left(\frac{1}{3}\xi^3 + z\xi\right)}, \tag{B.16a}$$

$$\text{Bi}(z) = \int_0^{\infty} d\xi \left[e^{i\left(-\frac{1}{3}\xi^3 + z\xi\right)} + \sin\left(\frac{1}{3}\xi^3 + z\xi\right) \right], \tag{B.16b}$$

APPENDIX B Some Ordinary Differative Equations

The following ODE arises in the quantum solution of the harmonic oscillator (see Sec. 1.3.15):

$$\left(-\frac{1}{2}\frac{d^2}{dy^2}\psi(y) + \frac{1}{2}y^2\right)\psi(y) = \mathcal{E}\psi(y). \tag{B.17}$$

The solutions to this ODE which vanish as $y \to \pm\infty$ are given by

$$\psi_n(y) = H_n(y)\exp\left(-\frac{y^2}{2}\right), \tag{B.18}$$

when $\mathcal{E} = (n + 1/2)$. The functions $H_n(y)$ are called *Hermite polynomials* and are given by the generating function (1.127). A normalization constant is necessary to normalize these solutions to unity. The lowest six harmonic oscillator wave functions are shown in Fig. 1.26.

ODEs can often be solved by assuming a *series solution* $y(x) = \sum_n a_n x^n$, and plugging this expression into the ODE to obtain a recursion formula for the a_n term, and then writing the series expansion in terms of the a_n. The *Frobenius method* uses a generalized expansion of the form $y(x) = x^r \sum_n a_n x^n$. After substitution, equating the a_0 term in the ODE to 0 will produce the so-called *indicial equation*, which gives the allowed values of r in the series expansion. For example, substituting the generalized expansion into the *Bessel differential equation*,

$$x^2 y'' + xy' + (x^2 - l^2)y = 0, \tag{B.19}$$

yields

$$\sum_{n=0}^{\infty}[(r+n)(r+n-1) + (r+n) - l^2]a_n x^{r+n} + \sum_{n=2}^{\infty} a_{n-2} x^{r+n} = 0. \tag{B.20}$$

The a_0 term is $[r(r-1) + r - l^2]a_0 = 0$, so $r = \pm l$. Ignoring the $r = -l$ case and taking $r = l$, for example, gives, after some algebra, the series solution that yields the *Bessel function of the first kind*, $y(x) = J_l(x)$ (see Fig. B.1). The Bessel differential equation is often written as

$$\left[\frac{d^2}{dz^2} + \frac{1}{z}\frac{d}{dz} - \frac{l^2}{z^2} + 1\right] J_l(z) = 0. \tag{B.21}$$

The Bessel functions of the first kind, $J_l(z)$, are regular at the origin. The lowest order Bessel functions, $J_l(z)$, are plotted in Fig. B.1(a) for $l = 0, 1, 2, 3, 4$. The limiting forms of the Bessel functions for small argument, $z \to 0$, are $J_l(z) \sim (z/2)^l/\Gamma(l+1)$, where Γ is the gamma function. At large z, $J_l(z) \sim \sqrt{2/(\pi z)}\cos(z - l\pi/2 - \pi/4) = -\sqrt{2/(\pi z)}\sin(z - (l-1/2)\pi/2)$. Bessel functions of noninteger order, $l \to \nu \neq$ integer, are also commonly encountered. Bessel functions of

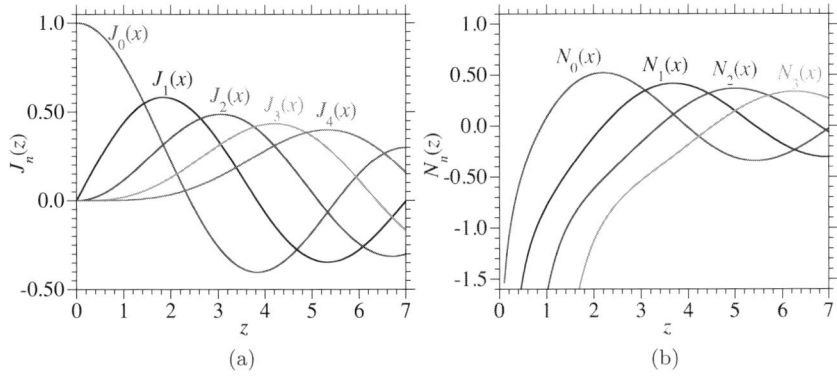

FIG B.1 Bessel functions, (a) $J_l(z)$, for $l = 0, 1, 2, 3, 4$, (b) $N_l(z)$, for $l = 0, 1, 2, 3$.

the second kind, $N_\nu(z)$ (often given the symbol $Y_\nu(z)$ in the mathematics literature), sometimes called Neumann functions, also satisfy the Bessel equation, $\left[\frac{d^2}{dz^2} + \frac{1}{z}\frac{d}{dz} - \frac{\nu^2}{z^2} + 1\right]N_\nu(z) = 0$, as do Bessel functions of the third kind (see below). They have a singularity at the origin, as shown in Fig. B.1(b). The Hankel functions (often called Bessel functions of the third kind)

$$H_l^{(1)} = J_l + iN_l, \quad H_l^{(2)} = J_l - iN_l, \tag{B.22}$$

are also commonly defined and satisfy Eq. (B.21); their aysmptotic forms are

$$H_l^{(1)}(z) \xrightarrow[z\to\infty]{} \sqrt{\frac{2}{\pi z}} e^{i(z - l\pi/2 - \pi/4)}, \quad H_l^{(2)}(z) \xrightarrow[z\to\infty]{} \sqrt{\frac{2}{\pi z}} e^{-i(z - l\pi/2 - \pi/4)}. \tag{B.23}$$

The Bessel functions of integer order are the asymptotic solutions to two-dimensional scattering problems [see Eq. (12.727)].

The radial part of the plane wave in spherical coordinates, $R_l(r) = \psi(r)/r$, is such that $\psi(r)$ satisfies the ODE

$$\frac{d^2\psi}{dr^2} = \frac{\ell(\ell+1)}{r^2}\psi - k^2\psi. \tag{B.24}$$

Defining the new variable $z \equiv kr$ and writing the ODE in terms of this variable, Eq. (B.24) becomes

$$\frac{d^2\psi(z)}{dz^2} = \frac{\ell(\ell+1)}{z^2}\psi(z) - \psi(z). \tag{B.25}$$

The solutions to this equation are called the *Riccati–Bessel functions*, $\hat{j}_l(z) = zj_l(z)$ and $\hat{n}_l(z) = zn_l(z)$, which are given in terms of the *spherical-Bessel functions* $j_l(z)$ and $n_l(z)$ [27]. Since the spherical-Bessel functions can be written in terms of the *Bessel functions* $J_l(z)$ and $N_l(z)$,

$$j_l(z) = \sqrt{\frac{\pi}{2z}} J_{l+1/2}(z), \quad n_l(z) = \sqrt{\frac{\pi}{2z}} N_{l+1/2}(z), \tag{B.26}$$

we find that

$$\hat{j}_l(z) = zj_l(z) = \sqrt{\frac{\pi z}{2}} J_{l+1/2}(z), \quad \hat{n}_l(z) = zn_l(z) = \sqrt{\frac{\pi z}{2}} N_{l+1/2}(z). \tag{B.27}$$

Note that in the literature, sometimes the Neumann functions n_l and N_l are denoted by y_l and Y_l, respectively. The power series expansion of the spherical-Bessel functions near the origin is

$$j_l(z) = \frac{z^l}{(2l+1)!!}\left(1 - \frac{z^2/2}{1!(2l+3)} + \frac{(z^2/2)^2}{2!(2l+3)(2l+5)} - \cdots\right), \tag{B.28a}$$

$$n_l(z) = -\frac{(2l-1)!!}{z^{l+1}}\left(1 - \frac{z^2/2}{1!(1-2l)} + \frac{(z^2/2)^2}{2!(1-2l)(3-2l)} - \cdots\right). \tag{B.28b}$$

The asymptotic form of these functions as $z \to \infty$ are

$$j_l(z) \xrightarrow[z\to\infty]{} \frac{\sin(z - l\pi/2)}{z}, \tag{B.29a}$$

$$n_l(z) \xrightarrow[z\to\infty]{} -\frac{\cos(z - l\pi/2)}{z}. \tag{B.29b}$$

APPENDIX B Some Ordinary Differential Equations

Rayleigh's formulas for the spherical-Bessel functions are $j_l(z) = z^n \left(-\frac{1}{z}\frac{d}{dz}\right)^n \frac{\sin z}{z}$ and $n_l(z) = -z^n \left(-\frac{1}{z}\frac{d}{dz}\right)^n \frac{\cos z}{z}$. The general solution to Eq. (B.24) is given by a linear combination of the regular and irregular Riccati–Bessel functions:

$$\psi(r) = a\hat{j}_l(kr) + b\hat{n}_l(kr). \tag{B.30}$$

The lowest order Riccati–Bessel functions are: $\hat{j}_0(z) = \sin z, \hat{j}_1(z) = z^{-1}\sin z - \cos z, \hat{j}_2(z) = (3z^{-2} - 1)\sin z - 3z^{-1}\cos z$, and $\hat{n}_0(z) = -\cos z, \hat{n}_1(z) = -\sin z - z^{-1}\cos z, \hat{n}_2(z) = -3z^{-1}\sin z - (3z^{-2} - 1)\cos z$. Figure 3.3 shows the regular and irregular Riccati–Bessel functions, $\hat{j}_l(z)$ and $\hat{n}_l(z)$, plotted versus z for $l = 0, 1, 2$. Rayleigh's formulas for the Riccati–Bessel functions are

$$\hat{j}_l(z) = z^{n+1}\left(-\frac{1}{z}\frac{d}{dz}\right)^n \frac{\sin z}{z}, \tag{B.31a}$$

$$\hat{n}_l(z) = -z^{n+1}\left(-\frac{1}{z}\frac{d}{dz}\right)^n \frac{\cos z}{z}. \tag{B.31b}$$

One can define the so-called spherical-Bessel functions of the third kind ($j_l(z)$ is the first kind, $n_l(z)$ is the second kind) as follows:

$$h_l^{(+)}(z) = j_l(z) + in_l(z), \tag{B.32a}$$

$$h_l^{(-)}(z) = j_l(z) - in_l(z). \tag{B.32b}$$

The asymptotic form of these functions is clear from (B.29a–B.29b). The Riccati–Bessel functions $\hat{h}_l^{(+)}(z) \equiv zh_l^{(+)}(z)$ and $\hat{h}_l^{(-)}(z) \equiv zh_l^{(-)}(z)$ are also often referred to. Clearly, $\hat{h}_l^{(+)}(z) \xrightarrow{z\to\infty} -i\,e^{i(z-l\pi/2)}$ and $\hat{h}_l^{(-)}(z) \xrightarrow{z\to\infty} i\,e^{-i(z-l\pi/2)}$.

The differential equation

$$\left(z(1-z)\frac{d^2}{dz^2} + [c - (a+b+1)z]\frac{d}{dz} - ab\right)f(z) = 0 \tag{B.33}$$

has a particular solution known as the *hypergeometric function*

$$f(z) = F(a,b,c;z). \tag{B.34}$$

This function is also sometimes denoted as $_2F_1(a,b,c;z)$; it has the power series

$$_2F_1(a,b,c;z) = 1 + \frac{ab}{1!c}z + \frac{a(a+1)b(b+1)}{2!c(c+1)} + \cdots = \sum_{n=0}^{\infty} \frac{(a)_n(b)_n}{n!(c)_n}z^n, \tag{B.35}$$

where $(a)_n = a(a+1)\ldots(a+n)$. There are many special cases of the coefficients a, b, c that yield other well-known special functions (e.g., Chebyshev, Legendre, Gegenbauer, and Jacobi polynomials). Hypergeometric functions can be generalized to *generalized hypergeometric functions*, $_nF_m(a_1,\ldots,a_n,b_1,\ldots,b_m;z)$. A special case is the function of the form, $_1F_1(a,b;z)$, called a *confluent hypergeometric function* of the first kind, which satisfies the differential equation

$$\left[z\frac{d^2}{dz^2} + (b-z)\frac{d}{dz} - a\right]f(z) = 0. \tag{B.36}$$

The *associated Legendre polynomials* [27], $P_l^m(z)$, are special functions that satisfy the differential equation

$$\frac{d}{dz}\left[(1-z^2)\frac{dP_l^m(z)}{dz}\right] + \left[l(l+1) - \frac{m^2}{(1-z^2)}\right]P_l^m(z) = 0. \tag{B.37}$$

Here, $l = 0, 1, 2, \ldots$, and $l \geq |m|$. For $m = 0$, $P_l^m(z) \equiv P_l(z)$, where P_l is called the *Legendre polynomial* of order l; the lowest few Legendre polynomials are

$$P_0(z) = 1, \quad P_1(z) = z, \quad P_2(z) = \frac{1}{2}(3z^2 - 1).$$

The *generating function* for the Legendre polynomials is

$$g(z,t) \equiv (1 - 2zt + t^2)^{-1/2} = \sum_{l=0}^{\infty} P_l(z) t^l, \tag{B.38}$$

the *Rodrigues formula*, which allows for the calculation of the Legendre polynomials via differentiation, is given by

$$P_l(z) = \frac{1}{2^l l!} \left(\frac{d}{dz}\right)^l (z^2 - 1)^l, \tag{B.39}$$

and the recurrence relation, which can be used to obtain higher order Legendre polynomials from lower ones, is given by

$$P_{l+1}(z) = \frac{1}{l+1} \left[(2l+1) z P_l(z) - l P_{l-1}(z)\right]. \tag{B.40}$$

The associated Legendre polynomials with $m > 0$ can be generated from the Legendre polynomials as follows:

$$P_l^m(z) = (1 - z^2)^{|m|/2} \left(\frac{d}{dz}\right)^{|m|} P_l(z) \quad \text{for} \quad m > 0. \tag{B.41}$$

Substituting Eq. (B.41) into the LSH of Eq. (B.38) yields the Rodrigues formula for P_l^m for all m, $-l \leq m \leq l$:

$$P_l^m(z) = \frac{1}{2^l l!} (1 - z^2)^{m/2} \left(\frac{d}{dz}\right)^{l+m} (z^2 - 1)^l. \tag{B.42}$$

The Legendre polynomials take the following values at $z = \pm 1$:

$$P_l(1) = 1, \quad P_l(-1) = (-1)^l, \quad P_l^m(1) = P_l^m(-1) = 0 \quad \text{if} \quad m \neq 0. \tag{B.43}$$

Recurrence relations and the generating function for the P_l^m can be developed using Eq. (B.42). The associated Legendre polynomials obey the orthogonality relations,

$$\int_{-1}^{1} dz\, P_{l'}^m(z) P_l^m(z) = \frac{(l+m)!}{(l-m)!} \frac{2}{2l+1} \delta_{l,l'}. \tag{B.44}$$

The *spherical harmonics* are obtained in terms of the associated Legendre polynomials as follows:

$$Y_{lm}(\theta, \phi) = (-1)^{m+|m|} \left[\frac{(2l+1)}{4\pi} \frac{(l-m)!}{(l+m)!}\right]^{1/2} P_l^m(\cos\theta) e^{im\phi}. \tag{B.45}$$

It is easy to verify by direct substitution that the spherical harmonics satisfy the differential equations,

$$-i\hbar \frac{\partial}{\partial \phi} Y_{lm}(\theta, \phi) = \hbar m Y_{lm}(\theta, \phi), \tag{B.46}$$

$$-\hbar^2 \left[\frac{1}{\sin\theta} \frac{\partial}{\partial \theta}\left(\sin\theta \frac{\partial}{\partial \theta}\right) + \frac{1}{\sin^2\theta} \frac{\partial^2}{\partial \phi^2}\right] Y_{lm}(\theta, \phi) = \hbar^2 l(l+1) Y_{lm}(\theta, \phi). \tag{B.47}$$

APPENDIX B Some Ordinary Differential Equations

There are a number of methods used to find solutions for ODEs. Linear ODEs with constant coefficients are easily solved analytically by algebraic methods. Series solutions are rather easy to obtain (see, e.g., the series solution to the radial Schrödinger equation in Sec. 3.2.6), Green's function methods are a powerful tool for inhomogeneous ODEs, and Fourier and Laplace methods are also very important tools. The references provided at the beginning of this appendix should be consulted to learn more about these methods.

Vector Analysis

Vector analysis is the mathematical framework associated with differentiation and integration of vector fields. In this Appendix, we shall define a vector field, then discuss scalar and vector products, differential operators that act on scalar and vector fields, present the Divergence and Stokes theorems, and then introduce curvilinear coordinates. We closely follow the discussion in Ref. [18].

If to each point of a region in coordinate space $\mathbb{R}^3$ there corresponds a vector $\mathbf{A}(x,y,z)$, then $\mathbf{A}$ is a vector function of position, and we say that the vector field $\mathbf{A}(x,y,z)$ has been defined in $\mathbb{R}^3$. Any vector $\mathbf{A}$ in three dimensions can be represented in a Cartesian coordinate system as a sum of its rectangular components, (A_x, A_y, A_z), as

$$\mathbf{A}(x,y,z) = A_x(x,y,z)\mathbf{i} + A_y(x,y,z)\mathbf{j} + A_z(x,y,z)\mathbf{k}, \tag{C.1}$$

where $\mathbf{i}$, $\mathbf{j}$, and $\mathbf{k}$ are the rectangular unit vectors having the directions of the positive x, y, and z axes, respectively.

C.1 SCALAR AND VECTOR PRODUCTS

Let us begin by reminding the reader about scalar and vector products of vectors. The scalar (or dot or inner) product of two vectors, $\mathbf{A}$ and $\mathbf{B}$, is a scalar denoted as $\mathbf{A} \cdot \mathbf{B}$ and is defined as the product of the magnitudes of $\mathbf{A}$ and $\mathbf{B}$, and the cosine of the angle θ between them. In symbols,

$$\mathbf{A} \cdot \mathbf{B} = AB\cos\theta, \quad 0 \leq \theta \leq \pi. \tag{C.2}$$

In a Cartesian coordinate system, with $\mathbf{B} = B_x\mathbf{i} + B_y\mathbf{j} + B_z\mathbf{k}$, the scalar product $\mathbf{A} \cdot \mathbf{B}$ is given by

$$\mathbf{A} \cdot \mathbf{B} = A_xB_x + A_yB_y + A_zB_z. \tag{C.3}$$

The cross (or vector) product of $\mathbf{A}$ and $\mathbf{B}$ is a vector, $\mathbf{C} = \mathbf{A} \times \mathbf{B}$. The magnitude of the vector is the product of the magnitudes of $\mathbf{A}$, $\mathbf{B}$, and the sine of the angle θ between them, $C = AB\sin\theta$. The direction of the vector $\mathbf{C} = \mathbf{A} \times \mathbf{B}$ is perpendicular to the plane made by $\mathbf{A}$ and $\mathbf{B}$ and such that $\mathbf{A}$, $\mathbf{B}$, and $\mathbf{C}$ form a right-handed coordinate system:

$$\mathbf{A} \times \mathbf{B} = AB\sin\theta\,\mathbf{u}, \quad 0 \leq \theta \leq \pi, \tag{C.4}$$

where $\mathbf{u}$ is a unit vector indicating the direction of $\mathbf{A} \times \mathbf{B}$. The magnitude of $\mathbf{A} \times \mathbf{B}$ equals the area of a parallelogram with sides $\mathbf{A}$ and $\mathbf{B}$. The cross product of $\mathbf{A}$ and $\mathbf{B}$ can be written as a determinant,

$$\mathbf{A} \times \mathbf{B} = \begin{vmatrix} \mathbf{i} & \mathbf{j} & \mathbf{k} \\ A_x & A_y & A_z \\ B_x & B_y & B_z \end{vmatrix}. \tag{C.5}$$

Problem C.1

(a) Show that

$$\mathbf{A} \cdot (\mathbf{B} \times \mathbf{C}) = \begin{vmatrix} A_x & A_y & A_z \\ B_x & B_y & B_z \\ C_x & C_y & C_z \end{vmatrix}. \tag{C.6}$$

(b) Show that

$$\mathbf{A} \cdot (\mathbf{B} \times \mathbf{C}) = \mathbf{B} \cdot (\mathbf{C} \times \mathbf{A}) = \mathbf{C} \cdot (\mathbf{A} \times \mathbf{B}). \tag{C.7}$$

C.2 DIFFERENTIAL OPERATORS

Let us consider a scalar function $\Phi(x, y, z)$ defined in a region of $\mathbb{R}^3$, and let $\Phi(x, y, z)$ be differentiable at each point (x, y, z) in the region (i.e., Φ defines a differentiable scalar field). Then the *gradient* of Φ, written $\nabla \Phi$ or grad Φ, is defined by

$$\nabla \Phi = \frac{\partial \Phi}{\partial x}\mathbf{i} + \frac{\partial \Phi}{\partial y}\mathbf{j} + \frac{\partial \Phi}{\partial z}\mathbf{k}. \tag{C.8}$$

$\nabla \Phi$ defines a vector field. The component of $\nabla \Phi$ in the direction of a unit vector $\mathbf{a}$ is given by $\nabla \Phi \cdot \mathbf{a}$ and is called the directional derivative of Φ in the direction $\mathbf{a}$. Physically, this is the rate of change of Φ [at position (x, y, z)] in the direction $\mathbf{a}$.

The vector differential operator *del* or *nabla*, written ∇, is defined by

$$\nabla \equiv \mathbf{i}\frac{\partial}{\partial x} + \mathbf{j}\frac{\partial}{\partial y} + \mathbf{k}\frac{\partial}{\partial z}. \tag{C.9}$$

The vector operator ∇ transforms as a vector under rotations and inversion. It is useful in defining the gradient of a scalar or vector function, the divergence of a vector function, and the curl of a vector function, as we now explain.

Let $\mathbf{A}(x, y, z)$ be differentiable at each point (x, y, z) in a certain region of space (i.e., $\mathbf{A}$ defines a differentiable vector field). Then the *divergence* of $\mathbf{A}$, written $\nabla \cdot \mathbf{A}$ or div $\mathbf{A}$, is

$$\nabla \cdot \mathbf{A} = \left(\mathbf{i}\frac{\partial}{\partial x} + \mathbf{j}\frac{\partial}{\partial y} + \mathbf{k}\frac{\partial}{\partial z}\right) \cdot (A_x\mathbf{i} + A_y\mathbf{j} + A_z\mathbf{k}) = \frac{\partial A_x}{\partial x} + \frac{\partial A_y}{\partial y} + \frac{\partial A_z}{\partial z}. \tag{C.10}$$

Note the analogy with $\mathbf{A} \cdot \mathbf{B} = A_xB_x + A_yB_y + A_zB_z$. Note also that $(\nabla \cdot \mathbf{A}) \neq (\mathbf{A} \cdot \nabla)$ [the latter defines an operator that acts on a function of (x, y, z)].

If $\mathbf{A}(x, y, z)$ is a differentiable vector field, then the *curl* or *rotation* of $\mathbf{A}$, written $\nabla \times \mathbf{A}$, curl $\mathbf{A}$ or rot $\mathbf{A}$, is defined by

$$\nabla \times \mathbf{A} = \left(\mathbf{i}\frac{\partial}{\partial x} + \mathbf{j}\frac{\partial}{\partial y} + \mathbf{k}\frac{\partial}{\partial z}\right) \times (A_x\mathbf{i} + A_y\mathbf{j} + A_z\mathbf{k}) = \begin{vmatrix} \mathbf{i} & \mathbf{j} & \mathbf{k} \\ \frac{\partial}{\partial x} & \frac{\partial}{\partial y} & \frac{\partial}{\partial z} \\ A_x & A_y & A_z \end{vmatrix}, \tag{C.11}$$

which can be expanded to yield

$$\nabla \times \mathbf{A} = \left(\frac{\partial A_z}{\partial y} - \frac{\partial A_y}{\partial z}\right)\mathbf{i} + \left(\frac{\partial A_x}{\partial z} - \frac{\partial A_z}{\partial x}\right)\mathbf{j} + \left(\frac{\partial A_y}{\partial x} - \frac{\partial A_x}{\partial y}\right)\mathbf{k}. \tag{C.12}$$

Note that in the expansion of the determinant, the operators $\frac{\partial}{\partial x}$, $\frac{\partial}{\partial y}$, and $\frac{\partial}{\partial z}$ must precede A_x, A_y, and A_z, respectively.

If $\Phi(x, y, z)$ and $\mathbf{A}(x, y, z)$ have continuous second partial derivatives, the following relations are valid (see, for example, Refs. [287, 288]). The curl of the gradient of any scalar field is zero,

$$\nabla \times (\nabla \Phi) = \mathbf{0}. \tag{C.13}$$

The divergence of the curl of any vector field is zero, i.e.,

$$\nabla \cdot (\nabla \times \mathbf{A}) = 0. \tag{C.14}$$

In addition,

$$\nabla \times (\nabla \times \mathbf{A}) = \nabla(\nabla \cdot \mathbf{A}) - \nabla^2 \mathbf{A}, \tag{C.15}$$

where ∇^2 is called the *Laplacian* operator,

$$\nabla^2 \Phi \equiv \nabla \cdot (\nabla \Phi). \tag{C.16}$$

C.3 Divergence and Stokes Theorems

In a Cartesian coordinate system, ∇^2 is given by

$$\nabla^2 = \frac{\partial^2}{\partial x^2} + \frac{\partial^2}{\partial y^2} + \frac{\partial^2}{\partial z^2}. \tag{C.17}$$

Formulas for the gradient, divergence, curl, and Laplacian in non-Cartesian coordinate systems can be found in Sec. C.4 (see also Refs. [34, 287, 288]).

It is convenient to introduce the Levi–Civita symbol (also called the permutation symbol) to treat vector products, curls and antisymmetric tensors. In three dimensions, the Levi–Civita symbol is defined by ε_{ijk}, with $\varepsilon_{123} = 1$, an even number of permutations of the indices 123 in ε_{123} also yield 1, an odd number permutations of the subscripts 123 yield -1, e.g., $\varepsilon_{132} = -1$, and if any of the subscripts ijk are equal, $\varepsilon_{ijk} = 0$. Another way of saying this is, even permutations of the indices 123 yield unity, odd permutations yield -1, and otherwise the Levi–Civita symbol vanishes. (The Levi–Civita symbol can be generalized to n dimensions.)

The cross product of two vectors can be written in terms of the Levi–Civita symbol as follows: $(\mathbf{A} \times \mathbf{B})_i = \varepsilon_{ijk} A_j B_k$, and $\mathbf{C} \cdot (\mathbf{A} \times \mathbf{B}) = \varepsilon_{ijk} C_i A_j B_k$, where the summation notation wherein repeated indices are summed over is intended. Similarly, $(\nabla \times \mathbf{A})_i = \varepsilon_{ijk} \partial_j A_k$, where we have used the shorthand notation $\partial_j = \frac{\partial}{\partial x_j}$. An important property involving Levi–Civita symbols that is extremely useful in carrying out manipulations is given by

$$\varepsilon_{ijk} \varepsilon_{ilm} = \delta_{jl} \delta_{km} - \delta_{jm} \delta_{kl}. \tag{C.18}$$

This identity can be used to easily prove the cross product and curl relationships,

$$(\mathbf{A} \times (\mathbf{B} \times \mathbf{C}))_i = B_i (\mathbf{A} \cdot \mathbf{C}) - C_i (\mathbf{A} \cdot \mathbf{B}), \tag{C.19}$$

$$(\nabla \times (\nabla \times \mathbf{A}))_i = \nabla_i (\nabla \cdot \mathbf{A}) - \nabla^2 A_i. \tag{C.20}$$

Problem C.2

(a) Verify Eq. (C.18).
(b) Prove Eq. (C.19) by writing

$$(\mathbf{A} \times (\mathbf{B} \times \mathbf{C}))_i = \varepsilon_{ijk} \varepsilon_{klm} A_j B_l C_m, \tag{C.21}$$

rewriting the product of the Levi–Civita symbols in the form $\varepsilon_{ijk} \varepsilon_{klm} = \varepsilon_{kij} \varepsilon_{klm}$ and using Eq. (C.18).
(c) Using the same method as in (b), prove Eq. (C.20). (Carefully consider the ordering of terms.)
(d) Show that $\varepsilon_{ijk} \varepsilon_{ijl} = 2\delta_{kl}$ and that $\varepsilon_{ijk} \varepsilon_{ijk} = 6$.

Problem C.3

(a) Prove that $\nabla^2 (1/r) = 0$ for $\mathbf{r} \neq 0$.
(b) Prove that $\nabla \cdot (\phi \mathbf{A}) = (\nabla \phi) \cdot \mathbf{A} + \phi \nabla \cdot \mathbf{A}$.
(c) Prove that $\nabla \cdot (\mathbf{r}/r^3) = 0$.
(d) Prove Eq. (C.14), $\nabla \cdot (\nabla \times \mathbf{A}) = 0$.

C.3 DIVERGENCE AND STOKES THEOREMS

Two important theorems of vector analysis that are useful in the mathematical formulation of electrodynamics and solving electrodynamic problems are the Gauss divergence theorem and Stokes theorem.

The divergence theorem of Gauss states that if V is the volume bounded by a closed surface S and $\mathbf{A}$ is a vector field with continuous derivatives, then

$$\int_V \nabla \cdot \mathbf{A} \, dV = \int_S \mathbf{A} \cdot \mathbf{n} \, dS, \tag{C.22}$$

where **n** is the positive (outward drawn) unit vector normal to S. In other words, the volume integral of the divergence of a vector field **A**, taken over any finite volume, is equal to the surface integral of **A** of the surface of that volume. Using this theorem, the integral form of the divergence can be obtained (Ref. [287, 288]):

$$\nabla \cdot \mathbf{A} = \lim_{\Delta V \to 0} \frac{\int_S \mathbf{A} \cdot \mathbf{n}\, dS}{\Delta V}. \tag{C.23}$$

Physically, the divergence of a vector **A** is the ratio of the flux of the vector **A** through the surface of a small volume element ΔV (i.e., $\int \mathbf{A} \cdot \mathbf{n}\, dS$) to the volume of the element.

Stokes' theorem states that if S is an open, two-sided surface bounded by a closed, nonintersecting simple closed curve C, then if **A** has continuous derivatives,

$$\oint_C \mathbf{A} \cdot d\boldsymbol{\ell} = \int_S (\nabla \times \mathbf{A}) \cdot \mathbf{n}\, dS, \tag{C.24}$$

where the line integral over path C on the LSH is traversed in the positive sense. The direction of C is called positive if an observer walking on the boundary of the surface S in the positive direction, with his head pointing in the direction of the positive normal to S, has the surface on his left (his traversal is in the sense given by the right hand rule). In other words, Stokes' theorem states that the surface integral of the curl of a vector field **A**, taken over any finite surface, is equal to the line integral of **A** around the boundary of that surface. If P is any point in S and **n** is a unit normal to S at point P, then at P:

$$(\nabla \times \mathbf{A}) \cdot \mathbf{n} = \lim_{S \to 0} \frac{\oint_C \mathbf{A} \cdot d\boldsymbol{\ell}}{S}, \tag{C.25}$$

where the limit is taken in such a way that S shrinks to P. Since $\oint_C \mathbf{A} \cdot d\boldsymbol{\ell}$ is called the circulation of **A** about C, the normal component of the curl can be interpreted physically as the limit of the circulation per unit area (thus the synonym "rotation of **A**" [rot **A**] for curl**A**).

C.4 CURVILINEAR COORDINATES

Geometrical considerations, or the nature of boundary conditions, often make it useful to consider coordinate systems other than Cartesian. A general curvilinear coordinate system in N dimensions can be defined in terms of a transformation T involving the Cartesian coordinates $(x_1, \ldots, x_N)$ and curvilinear coordinates $(u_1, \ldots, u_N)$,

$$x_i = T_i(u_1, \ldots, u_N), \tag{C.26}$$

where T_i are given functions of the arguments. The transformation $T \equiv \{T_i\}$ must be invertible, i.e., T^{-1} must exist. Moreover, it is convenient to assume that at every point **u** in the N-dimensional space, T is continuously differentiable. We define the jth column of the Jacobian matrix to be the vector $\mathbf{c}_j = \frac{\partial \mathbf{x}}{\partial u_j}$ (i.e., $c_{ij} = \frac{\partial T_i}{\partial u_j}$). This is the tangent vector to the curvilinear coordinate formed by allowing the jth coordinate of the vector **u** to vary. The *tangent vectors* $(\mathbf{c}_1, \ldots, \mathbf{c}_N)$ exist in the Cartesian coordinate space, and they form a local coordinate axis system, but they may not be orthogonal. The scalar product of these vectors define a matrix called the *metric tensor* of the transformation,

$$g_{ij} = \mathbf{c}_i \cdot \mathbf{c}_j. \tag{C.27}$$

Because $\mathbf{c}_i \cdot \mathbf{c}_j = \mathbf{c}_j \cdot \mathbf{c}_i$, the metric tensor is symmetric, $g_{ij} = g_{ji}$.

For example, let us consider the two-dimensional case and define arbitrary curvilinear coordinates (u, v) related to the Cartesian coordinates (x, y) by

$$\begin{pmatrix} x \\ y \end{pmatrix} = \begin{pmatrix} x(u, v) \\ y(u, v) \end{pmatrix}. \tag{C.28}$$

C.4 Curvilinear Coordinates

The Jacobian of the coordinate transformation, $\frac{\partial(x,y)}{\partial(u,v)}$ is

$$\begin{pmatrix} \frac{\partial x}{\partial u} & \frac{\partial x}{\partial v} \\ \frac{\partial y}{\partial u} & \frac{\partial y}{\partial v} \end{pmatrix}. \tag{C.29}$$

The tangent vectors are

$$\mathbf{c}_1 = \begin{pmatrix} \frac{\partial x}{\partial u} \\ \frac{\partial y}{\partial u} \end{pmatrix}, \quad \mathbf{c}_2 = \begin{pmatrix} \frac{\partial x}{\partial v} \\ \frac{\partial y}{\partial v} \end{pmatrix}, \tag{C.30}$$

and the metric tensor is given by

$$g_{11} = \left(\frac{\partial x}{\partial u}\right)^2 + \left(\frac{\partial y}{\partial u}\right)^2, \tag{C.31a}$$

$$g_{12} = g_{21} = \frac{\partial x}{\partial u}\frac{\partial x}{\partial v} + \frac{\partial y}{\partial u}\frac{\partial y}{\partial v}, \tag{C.31b}$$

$$g_{22} = \left(\frac{\partial x}{\partial v}\right)^2 + \left(\frac{\partial y}{\partial v}\right)^2. \tag{C.31c}$$

The differential of the vector $\mathbf{x}$, $d\mathbf{x}$, can be expressed in terms of the differential of the curvilinear coordinate vector $\mathbf{u}$ as follows:

$$d\mathbf{x} = \begin{pmatrix} dx \\ dy \end{pmatrix} = \begin{pmatrix} \frac{\partial x}{\partial u}du + \frac{\partial x}{\partial v}dv \\ \frac{\partial y}{\partial u}du + \frac{\partial y}{\partial v}dv \end{pmatrix}. \tag{C.32}$$

The squared differential element of length, $d\ell^2 = dx^2 + dy^2$, is

$$d\ell^2 = g_{11}du^2 + 2g_{12}dudv + g_{22}dv^2, \tag{C.33}$$

and the volume element is

$$dV = g^{1/2}\,dudv, \tag{C.34}$$

where g denotes the determinant of the matrix g_{ij}.

In an N-dimensional curvilinear coordinate system [287, 288], $g_{ij} = \frac{\partial x_\alpha}{\partial u_i}\frac{\partial x_\alpha}{\partial u_j}$,

$$d\ell^2 = \sum_{i,j=1}^{N} g_{ij}\,du_i du_j, \tag{C.35}$$

$$dV = g^{1/2}\prod_{i=i}^{N} du_i. \tag{C.36}$$

If the vectors $(\mathbf{c}_1,\ldots,\mathbf{c}_N)$ are *orthogonal*, only the diagonal elements of g_{ij} can be nonzero and $g_{ij} = 0$ for $i \neq j$. Curvilinear coordinates with orthogonal tangent vectors $(\mathbf{c}_1,\ldots,\mathbf{c}_N)$ are much simpler than the nonorthogonal case, and such coordinates are called *orthogonal curvilinear coordinates*. In this case, it is customary to define functions $h_i(u_1, u_2,\ldots, u_N)$, $i = 1,\ldots N$, as follows: $h_i = |\mathbf{c}_i| = |g_{ii}|^{1/2}$. These functions are called *scale factors*. Note that the $h_i(u_1, u_2,\ldots, u_N)$ are functions of the coordinates (only in Cartesian coordinates are $h_i = 1$ for all i). It is also useful to define the orthonormal unit basis vectors $\hat{\mathbf{e}}_i = h_i^{-1}\mathbf{c}_i$ for $i = 1,\ldots N$. In an *orthogonal* curvilinear three-dimensional coordinate system, with coordinates u_1, u_2, and u_3, we have $\mathbf{r} = \mathbf{r}(u_1, u_2, u_3)$, and

$$d\mathbf{r} = \frac{\partial \mathbf{r}}{\partial u_1} du_1 + \frac{\partial \mathbf{r}}{\partial u_2} du_2 + \frac{\partial \mathbf{r}}{\partial u_3} du_3 = \hat{\mathbf{e}}_1 h_1 du_1 + \hat{\mathbf{e}}_2 h_2 du_2 + \hat{\mathbf{e}}_3 h_3 du_3. \tag{C.37}$$

The squared differential element of length is, therefore, given by

$$d\ell^2 = d\mathbf{r} \cdot d\mathbf{r} = h_1^2 du_1^2 + h_2^2 du_2^2 + h_3^2 du_3^2, \tag{C.38}$$

and the volume element is given by

$$dV = |(\hat{\mathbf{e}}_1 h_1 du_1) \cdot [(\hat{\mathbf{e}}_2 h_2 du_2) \times (\hat{\mathbf{e}}_3 h_3 du_3)]| = h_1 h_2 h_3 \, du_1 du_2 du_3. \tag{C.39}$$

The operations, ∇, $\nabla \cdot$, $\nabla \times$, and ∇^2 can be expressed in terms of the h_i functions as follows [287, 288]:

$$(\nabla \Phi)_i = \frac{1}{h_i} \frac{\partial \Phi}{\partial u_i}, \tag{C.40}$$

$$\nabla \cdot \mathbf{A} = \frac{1}{h_1 h_2 h_3} \sum \frac{\partial}{\partial u_1} (h_2 h_3 A_1), \tag{C.41}$$

where the summation is over cyclic interchanges of the subscripts 1, 2, and 3,

$$(\nabla \times \mathbf{A})_1 = \frac{1}{h_2 h_3} \left(\frac{\partial}{\partial u_2} (h_3 A_3) - \frac{\partial}{\partial u_3} (h_2 A_2) \right), \tag{C.42}$$

where the remaining components of $\nabla \times \mathbf{A}$ are obtained by cyclic interchange of the subscripts, and

$$\nabla^2 \Phi = \frac{1}{h_1 h_2 h_3} \sum_{\text{cyclic interchange}} \frac{\partial}{\partial u_1} \left(\frac{h_2 h_3}{h_1} \frac{\partial \Phi}{\partial u_1} \right). \tag{C.43}$$

In cylindrical coordinates, ρ, θ, z, $d\ell^2 = d\rho^2 + \rho^2 d\theta^2 + dz^2$, i.e., the scale factors are $h_\rho = 1$, $h_\theta = \rho$, and $h_z = 1$. The following formulas apply:

$$\nabla \Phi = \hat{\mathbf{e}}_\rho \frac{\partial \Phi}{\partial \rho} + \hat{\mathbf{e}}_\theta \frac{1}{\rho} \frac{\partial \Phi}{\partial \theta} + \hat{\mathbf{e}}_z \frac{\partial \Phi}{\partial z}, \tag{C.44}$$

$$\nabla \cdot \mathbf{A} = \frac{1}{\rho} \frac{\partial (\rho A_\rho)}{\partial \rho} + \frac{1}{\rho} \frac{\partial A_\theta}{\partial \theta} + \frac{\partial A_z}{\partial z}, \tag{C.45}$$

$$\nabla \times \mathbf{A} = \hat{\mathbf{e}}_\rho \left(\frac{1}{\rho} \frac{\partial A_z}{\partial \theta} - \frac{\partial A_\theta}{\partial z} \right) + \hat{\mathbf{e}}_\theta \left(\frac{\partial A_\rho}{\partial z} - \frac{\partial A_z}{\partial \rho} \right) + \hat{\mathbf{e}}_z \frac{1}{\rho} \left(\frac{\partial (\rho A_\theta)}{\partial \rho} - \frac{\partial A_\rho}{\partial \theta} \right), \tag{C.46}$$

$$\nabla^2 \Phi = \frac{1}{\rho} \left(\frac{\partial (\rho \frac{\partial \Psi}{\partial \rho})}{\partial \rho} \right) + \frac{1}{\rho^2} \frac{\partial^2 \Phi}{\partial \theta^2} + \frac{\partial^2 \Phi}{\partial z^2}. \tag{C.47}$$

Notice that $\frac{1}{\rho} \left(\frac{\partial (\rho \frac{\partial \Phi}{\partial \rho})}{\partial \rho} \right) = \left(\frac{\partial^2}{\partial \rho^2} + \frac{1}{\rho} \frac{\partial}{\partial \rho} \right) \Phi$. Note also that the expressions for these operators in 2D polar coordinates can be obtained by neglecting the z degree of freedom.

C.4 Curvilinear Coordinates

In spherical coordinates, r, θ, ϕ, $d\ell^2 = dr^2 + r^2 d\theta^2 + r^2 \sin^2\theta d\phi^2$, i.e., $h_r = 1$, $h_\theta = r$, and $h_\phi = r\sin\theta$. The following formulas apply:

$$\nabla\Phi = \hat{\mathbf{e}}_r \frac{\partial\Phi}{\partial r} + \hat{\mathbf{e}}_\theta \frac{1}{r}\frac{\partial\Phi}{\partial\theta} + \hat{\mathbf{e}}_\phi \frac{1}{r\sin\theta}\frac{\partial\Phi}{\partial\phi}, \tag{C.48}$$

$$\nabla\cdot\mathbf{A} = \frac{1}{r}\frac{\partial(rA_r)}{\partial r} + \frac{1}{r\sin\theta}\frac{\partial A_\theta}{\partial\theta} + \frac{1}{r\sin\theta}\frac{\partial A_\phi}{\partial\phi}, \tag{C.49}$$

$$\nabla\times\mathbf{A} = \hat{\mathbf{e}}_r \frac{1}{r\sin\theta}\left[\frac{\partial(\sin\theta A_\phi)}{\partial\theta} - \frac{\partial A_\theta}{\partial\phi}\right] + \hat{\mathbf{e}}_\theta \left[\frac{1}{r\sin\theta}\frac{\partial A_r}{\partial\phi} - \frac{1}{r}\frac{\partial(rA_\phi)}{\partial r}\right]$$

$$+ \hat{\mathbf{e}}_\phi \left(\frac{1}{r}\frac{\partial(rA_\theta)}{\partial r} - \frac{\partial A_r}{\partial\theta}\right), \tag{C.50}$$

$$\nabla^2\Phi = \frac{1}{r^2}\left(\frac{\partial(r^2\frac{\partial\Phi}{\partial r})}{\partial r}\right) + \frac{1}{r^2\sin\theta}\frac{\partial(\sin\theta\frac{\partial\Phi}{\partial\theta})}{\partial\theta} + \frac{1}{r^2\sin^2\theta}\frac{\partial^2\Phi}{\partial\phi^2}. \tag{C.51}$$

Note that

$$\frac{1}{r^2}\left(\frac{\partial(r^2\frac{\partial\Phi}{\partial r})}{\partial r}\right) = \left(\frac{\partial^2}{\partial r^2} + \frac{2}{r}\frac{\partial}{\partial r}\right)\Phi = \frac{1}{r}\frac{\partial^2(r\Phi)}{\partial r^2} = \left(\frac{1}{r}\frac{\partial}{\partial r}r\right)\left(\frac{1}{r}\frac{\partial}{\partial r}r\right)\Phi. \tag{C.52}$$

In *nonorthogonal* curvilinear coordinate systems, the expressions for gradient, divergence, curl, and Laplacian are more complicated. See, for example, Spiegel [287], Synge and Schild [288], and Landau and Lifshitz [24] for details.

We conclude this section on curvilinear coordinate systems by noting that if the kinetic energy of a dynamical system is written in the form, $T = \frac{1}{2}m_{ij}\dot{u}_i\dot{u}_j$, where the u_i are the coordinate components in an N-dimensional curvilinear coordinate system, then the Laplacian operator can be written as

$$\nabla^2\Phi = \frac{1}{\sqrt{g}}\frac{\partial}{\partial u_k}\sqrt{g}\,m^{kl}\frac{\partial}{\partial u_l}\Phi, \tag{C.53}$$

where the metric g and the tensor m^{kl} are given by

$$g = \det|m_{ij}|, \quad m^{kl} = \frac{\text{minors of } m_{kl}}{g}. \tag{C.54}$$

By Eq. (C.36), the volume element is $dV = g^{1/2}du_1 du_3 \ldots du_N$, and the tensor m^{kl} satisifies $m^{ik}m_{il} = \delta_{kl}$. Equation (C.53) can be used to write the Schrödinger equation in curvilinear coordinates.

Fourier Analysis D

Fourier analysis is a method for analyzing functions by expanding them in terms of a complete orthogonal set of trigonometric functions (sine, cosine, or complex exponentials). Joseph Fourier (1768–1830) developed what is now called Fourier series and Fourier transforms to model heat-flow problems. These techniques can be employed to find the frequency components of simple or complex functions (wave forms), i.e., to break up complex wave forms into their fundamental frequencies, and to solve differential and integral equations by turning them into algebraic equations. Consequently, many problems in applied mathematics, mathematical physics, science, and engineering (and even economics and finance) can be effectively solved in terms of Fourier analysis. In quantum mechanics, Fourier analysis is a ubiquitous and indispensable tool for treating wave propagation problems by expanding the wave function $\psi(\mathbf{r}, t)$ in terms of traveling waves $e^{i(\mathbf{k}\cdot\mathbf{r}-\omega t)}$. Moreover, much of modern-day spectroscopy is carried out by using Fourier transform methods, e.g., Fourier transform infrared (FTIR) spectroscopy and Fourier transform NMR techniques.

Fourier expansions can be applied to functions of space, of time, or of both. The details of the Fourier expansion depend on whether the function to be expanded is defined on a finite domain or on an infinite domain.

Consider a function $f(\mathbf{x})$ of spatial variables, $\mathbf{x} = (x_1, x_2, \ldots, x_n)$. In the finite domain case, the function $f(\mathbf{x})$ is defined in an N-dimensional rectangular box $a_i \leq x_i \leq b_i$, for $i = 1, 2, \ldots, N$, of volume $V = \prod_{i=1}^{N}(b_i - a_i)$, is assumed to be periodic outside the domain. The expansion is achieved in terms of a *Fourier series* of the form

$$f(\mathbf{x}) = C \sum_{\mathbf{k}} F(\mathbf{k}) e^{i\mathbf{k}\cdot\mathbf{x}}, \tag{D.1}$$

where C is a constant and $F(\mathbf{k})$ are the expansion coefficients, sometimes called *Fourier amplitudes*, which depend on the real N-dimensional vector $\mathbf{k} = (k_1, k_2, \ldots, k_N)$ whose components are such that $k_i = \frac{2\pi n_i}{L_i}$, $n_i = 0, \pm 1, \pm 2, \ldots$, and $L_i \equiv b_i - a_i$. The Fourier amplitudes can formally be determined by multiplying both sides of Eq. (D.1) by $e^{-i\mathbf{q}\cdot\mathbf{x}}$ and integrating over $\mathbf{x}$. On using the orthogonality of the plane wave functions,

$$\int_V d\mathbf{x}\, e^{-i(\mathbf{q}-\mathbf{k})\cdot\mathbf{r}} = V\delta_{\mathbf{k}\mathbf{q}}, \tag{D.2}$$

one immediately obtains

$$F(\mathbf{k}) = \bar{C} \int d\mathbf{x}\, f(\mathbf{x}) e^{-i\mathbf{k}\cdot\mathbf{x}}, \tag{D.3}$$

where $\bar{C}C = 1/V$. The choice of C is somewhat arbitrary; the two most common choices are $C = 1/V$ and $\bar{C} = 1$, and the symmetric version, $C = \bar{C} = 1/\sqrt{V}$. Thus, in a Fourier series, the set of trigonometric functions (complex exponentials) participating in the expansion is discrete, and for each coordinate x_i, the functions have the same basic periodicity $L_i = b_i - a_i$. The expansion in Eq. (D.1) is referred to as a *Fourier series*, and Eq. (D.3) determines the Fourier amplitudes appearing in the Fourier series. The function $f(\mathbf{x})$ can be defined outside the finite domain through the requirement of periodicity.

In the infinite domain case, $\mathbf{x} = (x_1, x_2, \ldots x_n) \in \mathbb{R}^N$, $-\infty \leq x_i < \infty$, for $i = 1, 2, \ldots N$, the expansion involves an integral, rather than a sum,

$$f(\mathbf{x}) = C \int d\mathbf{k}\, F(\mathbf{k}) e^{i\mathbf{k}\cdot\mathbf{x}}, \tag{D.4}$$

where C is a normalization constant. The set of trigonometric functions (complex exponentials) participating in the expansion is now continuous, i.e., $\mathbf{k} = (k_1, k_2, \ldots k_n) \in \mathbb{R}^N$, $-\infty \leq k_i < \infty$, for $i = 1, 2, \ldots, N$. The Fourier amplitudes

$F(\mathbf{k})$ are obtained [by multiplying both sides of Eq. (D.4) by $e^{-i\mathbf{k}'\cdot\mathbf{x}}$ and integrating over $\mathbf{x}$ using $\int_{-\infty}^{\infty} d\mathbf{x}\, e^{i(\mathbf{k}-\mathbf{k}')\cdot\mathbf{x}} = (2\pi)^N \delta(\mathbf{k} - \mathbf{k}')$, see Eq. (1.75)] as,

$$F(\mathbf{k}) = \bar{C} \int d\mathbf{x} f(\mathbf{x}) e^{-i\mathbf{k}\cdot\mathbf{x}}, \tag{D.5}$$

where $\bar{C}$ is another constant, such that $C\bar{C} = (2\pi)^{-N}$. The most commonly used choices for the Cs are $C = 1$, $\bar{C} = (2\pi)^{-N}$, and the symmetric form $C = \bar{C} = (2\pi)^{-N/2}$.

Eqations (D.4) and (D.5) are examples of an *integral transform* given in terms of an integral operator. These equations are referred to as *Fourier integral transformations* or simply a *Fourier transform* (FT), which, when applied to $F(\mathbf{k})$, results in the function $f(\mathbf{x})$, and vice versa. Other integral transforms that are important in physics include the Laplace, Hankel, and Mellin transforms (they will not be discussed here). Some general properties of the Fourier integral transform will be discussed at the end of this chapter, but first, let us become acquainted with the Fourier expansion technique.

For the sake of clarity, it is useful to first discuss Fourier analysis for a function $f(x)$, which depends on a single space variable x. Extension to more than one-space dimension is then clear cut. The order of presentation will be as follows: (1) Fourier analysis for functions of a single-spatial variable defined on a finite domain. In this part, we also include Fourier analysis of functions defined on a discrete set of equally spaced points (sites), which is useful for the study of tight-binding models. Numerical techniques using Fourier transforms is of necessity of this form. (2) Fourier analysis for functions of a single-space variable defined on an infinite domain. (3–4) Fourier analysis for functions defined in three-space dimensions (finite and infinite domains). (5) Fourier analysis for functions depending on time. (6) Fourier expansion of operators. (7) The convolution property. (8) General properties of Fourier integral transforms (in one variable). Following this discussion we present in parts (9–10) several examples showing how the method of Fourier transform is used in the solutions of differential and integral equations.

Books having deatiled chapters on Fourier analysis include Refs. [25, 26]. Integral transforms and their applications are presented in Ref. [289]. Numerical Fourier analysis based on the fast Fourier transform (FFT) method (not discussed here) is presented in detail in Ref. [51].

D.1 FOURIER SERIES

Consider a function $f(x)$ defined on the finite interval $[0, L]$. If it has a finite number of discontinuities and only a finite number of extrema in $[0, L]$, $f(x)$ can be represented as the complex Fourier series,

$$f(x) = \frac{1}{\sqrt{L}} \sum_{n=-\infty}^{\infty} F(k_n) e^{ik_n x}, \quad k_n \equiv \frac{2\pi n}{L}, \tag{D.6}$$

where the symmetric C form has been adopted. The expansion uses the complex exponentials, which form a complete orthogonal set on the interval $0 \leq x \leq L$. The expansion coefficients are

$$F(k_n) = \frac{1}{\sqrt{L}} \int_0^L dx\, e^{-ik_n x} f(x). \tag{D.7}$$

The function $f(x)$ derived within this representation is periodic, with period L, i.e., $f(x) = f(x \pm L) = f(x \pm 2L)$, etc. The expansion is based on the fact that the complex exponentials form a complete orthogonal set on the interval $0 \leq x \leq L$.

A similar expansion can also be carried out in terms of trigonometric functions, $\sin\frac{2n\pi x}{L}$ ($n = 1, 2, \ldots$) and $\cos\frac{2n\pi x}{L}$ ($n = 0, 1, 2, \ldots$), which form a complete set. Traditionally, a somewhat different (the nonsymmetric) normalization is used in this case. The trigonometric expansion is given by

$$f(x) = \frac{1}{2}a_0 + \sum_{n=1}^{\infty} \left[a_n \cos\frac{2n\pi x}{L} + b_n \sin\frac{2n\pi x}{L} \right]. \tag{D.8}$$

D.1 Fourier Series

Using the normalization and orthogonality relations of the trigonometric functions, the expressions for the Fourier coefficients are found to be

$$a_n = \frac{2}{L} \int_0^L dx f(x) \cos \frac{2n\pi x}{L}, \quad b_n = \frac{2}{L} \int_0^L dx f(x) \sin \frac{2n\pi x}{L}. \tag{D.9}$$

An example of the convergence of Fourier expansion (D.8) to a function with discontinuous derivatives is shown in Fig. D.1.

Problem D.1

(a) Derive the following results,

$$\int_0^L dx \sin \frac{2n\pi x}{L} \sin \frac{2m\pi x}{L} = \int_0^L dx \cos \frac{2n\pi x}{L} \cos \frac{2m\pi x}{L} = \frac{L}{2} \delta_{nm}, \quad n, m \neq 0$$

$$\int_0^L dx \sin \frac{2n\pi x}{L} \cos \frac{2m\pi x}{L} = 0.$$

(b) Explain the factor 1/2 appearing before the coefficient a_0 in Eq. (D.8).
(c) Using these relations, derive Eq. (D.9).

So far, the specific boundary conditions at $x=0,L$ were not specified, but in some cases, they can help in removing (zeroing) some of the expansion coefficients. For example, if $f(0)=f(L)=0$, then $a_n=0$ (a sine function expansion is sufficient), whereas if $f'(0)=f'(L)=0$, then $b_n=0$.

It is easy to consider a function in the interval $[a,b]$, instead of the interval $[0,L]$, simply by making the transformation $y = \frac{b-a}{L}x + a$, and rewriting the Fourier series equations with the substitution $x \to \frac{L}{b-a}(y-a)$. For example, Figure D.2 shows the expansion of the function $f(x) = x^2 - x$ in the interval $x \in [-\pi, \pi]$, using $N=4$ and $N=50$ exponentials. Note the periodicity, $f(x) = f(x \pm \pi) = f(x \pm 2\pi)$, etc., of the function.

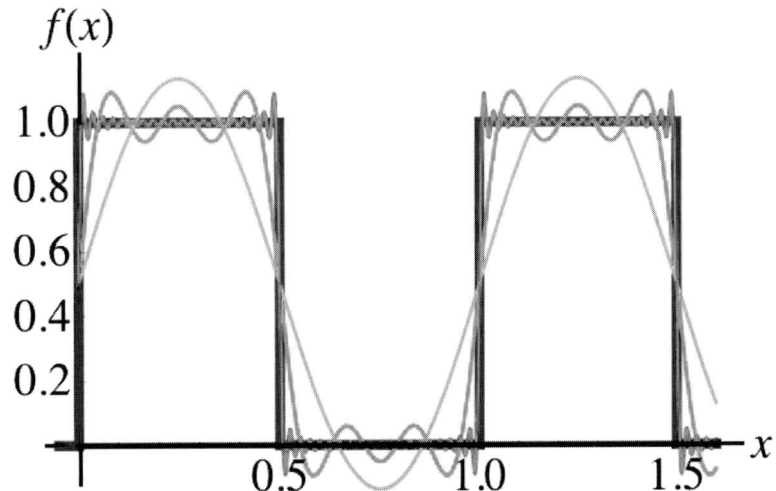

FIG D.1 Convergence of Fourier expansion (D.8) to the square-wave function $f(x) = [\theta(2x) - \theta(2x-1) + \theta(2x-2)]\theta(3-2x)$ (blue lines). Orange, green, and red curves are approximations with $N=1, 5$, and 40 terms, respectively. (For the color version of this figure, please refer the color plate section at the end of book.)

Problem D.2

(a) Carry out the details of the transformation of the Fourier series for functions defined on the interval $[a, b]$.
(b) Simplify the result obtained in (a) by letting $L = b - a$.

Problem D.3

(a) Calculate the Fourier series expansion of the function $f(x) = x$ on the interval $[-\pi, \pi]$.
(b) Similarly for the function $f(x) = x^2$.
(c) Generalize your results in (a) and (b) to the interval $[-\frac{L}{2}, \frac{L}{2}]$.

Hint: Use symmetry and note that $\int_{-\pi}^{\pi} dx \, x \sin nx = \frac{2\pi(-1)^{n+1}}{n}$, $\int_{-\pi}^{\pi} dx \, x^2 \cos nx = \frac{4\pi(-1)^n}{n^2}$.

D.1.1 FOURIER SERIES OF FUNCTIONS OF A DISCRETE VARIABLE

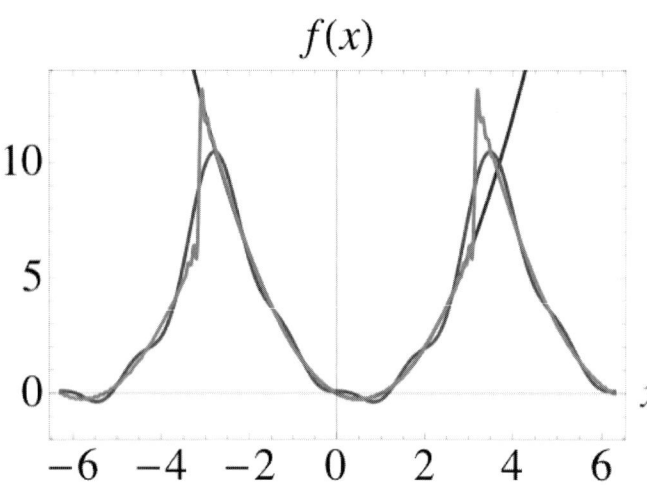

FIG D.2 Fourier series expansion of the function $f(x) = x^2 - x$ in the interval $[-\pi, \pi]$ using $N = 4$ and $N = 50$ terms.

A useful tool in the study of electronic properties of solids is the tight-binding model, in which the wave function $\psi(x)$ is defined on a discrete set of equally spaced points x_n (referred to as sites), and the continuous Schrödinger equation is replaced by a difference equation. In its simplest form, the set of points are the integers, and in a single-space dimension, the relevant (eigenvalue) equation is

$$-t[\psi(n+1) + \psi(n-1)] + v_n \psi(n) = \varepsilon \psi(n), \quad \text{(D.10)}$$

where the parameter t is the analog of $\frac{\hbar^2}{2m}$, which multiplies the Laplacian operator in the Schrödinger equation, and v_n is the potential at site n. If the number of sites N is finite, the boundary conditions are either periodic $\psi(n+N) = \psi(N)$ or of Dirichlet form, e.g., $\psi(0) = \psi(N+1) = 0$; in either case, the values of ψ at N sites are required. The corresponding Fourier expansion equations for the function ψ defined on the N sites $x_n = n = 1, 2, \ldots, N$, with periodic or Dirichlet boundary conditions, respectively, are

$$\psi(n) = \frac{1}{\sqrt{N}} \sum_{k_m} \phi(k_m) e^{ik_m n}, \quad \phi(k_m) = \frac{1}{\sqrt{N}} \sum_n \psi(n) e^{-ik_m n}, \quad \text{(D.11)}$$

$$\psi(n) = \sqrt{\frac{2}{N}} \sum_{q_m} \phi(q_m) \sin q_m n, \quad \phi(q_m) = \sqrt{\frac{2}{N}} \sum_n \psi(n) \sin q_m n. \quad \text{(D.12)}$$

where $k_m = \frac{2\pi m}{N}$, $m = 1, 2, \ldots, N$, and and $q_m = \frac{m\pi}{N+1}$, $m = 1, 2, \ldots, N$. Note that here m can be interpreted as a quantum number that corresponds to a given "energy" ε, but n is just the coordinate of the site.

D.2 Fourier Integrals

> **Problem D.4**
>
> Assume that $v_n = 0$ in Eq. (D.10).
>
> (a) For periodic boundary conditions show that the functions, $\phi(k) \equiv \frac{1}{\sqrt{N}} e^{ikn}$, are solutions of Eq. (D.10) and determine the values of k and the corresponding energies.
>
> (b) For Dirichlet boundary conditions show that the functions, $\phi(q) \equiv \sqrt{\frac{2}{N}} \sin qn$, are solutions of Eq. (D.10) and determine the values of k and the corresponding energies.
>
> (c) Prove the orthogonality of the discrete exponential and sine functions,
>
> $$\frac{1}{N} \sum_{n=1}^{N} e^{i(k_m - k_{m'})n} = \delta_{mm'}, \quad \frac{2}{N} \sum_{n=1}^{N} \sin q_m n \sin q_{m'} n = \delta_{mm'}.$$
>
> **Answers:** (a) $k_m = \frac{2m\pi}{N}$, $m = 1, 2, \ldots N$, $\varepsilon_m = -2t \cos k_m$.
> (b) $q_m = \frac{m\pi}{N+1}$, $m = 1, 2, \ldots N$, $\varepsilon_m = -2t \cos q_m$.

D.2 FOURIER INTEGRALS

Fourier analysis on finite interval leads to Fourier series, but on an infinite interval, Fourier integrals are required. To pass from a Fourier series to a Fourier integral, it is useful to consider a symmetric interval $-\frac{L}{2} \leq x \leq \frac{L}{2}$ and substitute Eq. (D.7) for F_n into Eq. (D.6) to obtain

$$f(x) = \frac{1}{L} \sum_{n=-\infty}^{\infty} \left[\int_{-\frac{L}{2}}^{\frac{L}{2}} d\xi \, f(\xi) e^{-ik_n \xi} \right] e^{ik_n x}. \tag{D.13}$$

As $L \to \infty$, the interval between two successive wavenumbers shrinks to zero, $dk = k_{n+1} - k_n = \frac{2\pi}{L} \to 0$, and the sum over n can be turned into an integral. Substituting $\frac{1}{L} = \frac{dk}{2\pi}$ for the prefactor, Eq. (D.13) takes the form,

$$f(x) = \frac{1}{2\pi} \int_{-\infty}^{\infty} dk \left[\int_{-\infty}^{\infty} d\xi \, f(\xi) e^{-ik\xi} \right] e^{ikx}. \tag{D.14}$$

The term in square brackets is the Fourier amplitude $F(k)$ (except perhaps for a multiplicative factor – see footnote 1), and Eq. (D.14) is then equivalent to the following relations[1]:

$$f(x) = \frac{1}{\sqrt{2\pi}} \int_{-\infty}^{\infty} dk F(k) e^{ikx}, \quad F(k) = \frac{1}{\sqrt{2\pi}} \int_{-\infty}^{\infty} dx f(x) e^{-ikx}. \tag{D.15}$$

---

[1] We have used a symmetrized version of the definition of the Fourier integrals. As stated in the introduction of this Appendix, the Fourier integrals are sometimes defined as

$$f(x) = \int_{-\infty}^{\infty} dk \, F(k) e^{ikx}, \quad F(k) = \frac{1}{2\pi} \int_{-\infty}^{\infty} dx f(x) e^{-ikx}.$$

Table D.1 *Fourier amplitudes $F(k)$ of selected functions $f(x)$.*

Function name	$f(x)$	$\sqrt{2\pi}F(k)$ [as defined in Eq. (D.15)]
Constant	1	$2\pi\delta(k)$
Cosine	$\cos k_0 x$	$\pi[\delta(k-k_0)+\delta(k+k_0)]$
Sine	$\sin k_0 x$	$\frac{\pi}{i}[\delta(k-k_0)-\delta(k+k_0)]$
Delta function	$\delta(x-x_0)$	e^{-ikx_0}
Exponential	$e^{-|k_0|x}$	$\frac{2k_0}{k^2+k_0^2}$
Gaussian	e^{-ax^2}	$\sqrt{\frac{\pi}{a}}e^{-\frac{k^2}{4a}}$
|Inverse Root|	$\frac{1}{\sqrt{|x|}}$	$\sqrt{\frac{2\pi}{|k|}}$
Step	$\Theta(x)$	$\frac{1}{i}\lim_{\varepsilon\to 0^+}\frac{1}{k-i\varepsilon}$
Inverse function	$\mathcal{P}\frac{1}{x}$	$\frac{\pi}{i}[1-2\Theta(-k)]$
Lorenzian	$\frac{\Gamma}{(x-x_0)^2+\Gamma^2/4}$	$2\pi\, e^{-ikx_0-\frac{\Gamma}{2}|k|}$
Ramp function	$R(x)=x\Theta(x)$	$i\pi\delta'(k)-\frac{1}{k^2}$

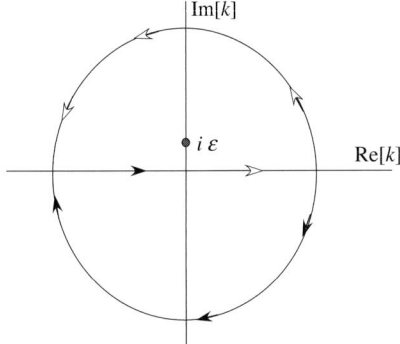

FIG D.3 Contour integrations for Eq. (D.16). The contour with upper half circle (white arrows) is used for $x>0$, whereas the contour with lower half circle (black arrows) is used for $x<0$.

The Fourier amplitudes for several commonly used functions are given in Table D.1. With the definition of $F(k)$ in Eq. (D.15), the tabulated results must be divided by $\sqrt{2\pi}$ (for the definition in the footnote, the tabulated results must be divided by 2π).

As an example of the use of Fourier integrals, let us calculate the Fourier amplitude of the step function $\Theta(x)$, $\Theta(k)$, and then calculate the step function $\Theta(x)$ from its Fourier amplitude $\Theta(k)$ using Eq. (D.15). From Eq. (D.15), $\Theta(k) = \frac{1}{\sqrt{2\pi}}\int_{-\infty}^{\infty}dx\,\Theta(x)e^{-ikx} = \frac{1}{\sqrt{2\pi}}\int_0^{\infty}dx\,e^{-ikx} = \frac{1}{\sqrt{2\pi}}\frac{1}{i}\lim_{\varepsilon\to 0^+}\frac{1}{k-i\varepsilon}$. For the inverse transformation, we need to show that

$$\Theta(x) = \frac{1}{\sqrt{2\pi}}\frac{1}{\sqrt{2\pi}}\frac{1}{i}\int_{-\infty}^{\infty}dk\,\frac{1}{k-i\varepsilon}e^{ikx}. \quad \text{(D.16)}$$

Note that on differentiating with respect to x, one obtains $\delta(x)$ on both sides; the singularity in the denominator on the RHS is cancelled. For $x>0$, the integral can be performed on a closed contour composed of the real k axis and an upper semicircle, $k=Re^{i\theta}$, as shown by the white arrows in Fig. D.3. This latter contribution vanishes in the $R\to\infty$ limit since the exponent decays as $e^{-\mathrm{Im}[k]x}$. The contour encloses the only singularity at $k_0=i\varepsilon$, and the use of the residue theorem[2] for its evaluation yields $e^{-x\varepsilon} \to 1$ as $\varepsilon \to 0$. On the other hand, for $x < 0$, we use the lower semicircle where the exponent decays as $e^{|\text{Im}[k]|x}$. There is no singularity here and the integral gives zero. Thus, we have shown that the integral indeed equals $\Theta(x)$.

D.3 FOURIER SERIES AND INTEGRALS IN THREE-SPACE DIMENSIONS

A typical example where Fourier analysis in three-space dimension is employed is provided by one of the most studied problems in solid-state physics, namely, that of an electron gas. A realistic theoretical model for its treatment starts by assuming that the electrons are confined within a three-dimensional rectangular box $0 < x < L_x, 0 < y < L_y, 0 < z < L_z$ of volume $V = L_x L_y L_z$. [The many electron wave function is constructed by sum of products of single-electron wave functions $\psi(\mathbf{r})$ (e.g., the Slater determinants, see Sec. 8.3).]. The functions $\psi(\mathbf{r})$ are required to satisfy periodic boundary conditions,

$$\psi(\mathbf{r} + L_x\hat{\mathbf{x}}) = \psi(\mathbf{r} + L_y\hat{\mathbf{y}}) = \psi(\mathbf{r} + L_z\hat{\mathbf{z}}) = \psi(\mathbf{r}). \tag{D.17}$$

In this case, the Fourier expansion reads,

$$\psi(\mathbf{r}) = \frac{1}{\sqrt{V}} \sum_{\mathbf{k}} a_{\mathbf{k}} e^{i\mathbf{k}\cdot\mathbf{r}}, \quad (k_x, k_y, k_z) = 2\pi \left(\frac{n_x}{L_x}, \frac{n_y}{L_y}, \frac{n_z}{L_z}\right), \tag{D.18}$$

where n_x, n_y, and n_z are integers. As functions of the discrete variable $\mathbf{k}$, the coefficients $a_{\mathbf{k}}$ constitute functions that are called the (three dimensional) *Fourier coefficients* of $\psi(\mathbf{r})$. Explicitly,

$$a_{\mathbf{k}} = \frac{1}{\sqrt{V}} \int_V d\mathbf{r}\, e^{-i\mathbf{k}\cdot\mathbf{r}} \psi(\mathbf{r}). \tag{D.19}$$

These expressions are easily obtained using the relations,

$$\frac{1}{V}\sum_{\mathbf{k}} e^{i\mathbf{k}\cdot\mathbf{r}} = \delta(\mathbf{r}), \quad \int \frac{1}{V} \int_V d\mathbf{r}\, e^{-i\mathbf{k}\cdot\mathbf{r}} = \delta_{\mathbf{k}0}. \tag{D.20}$$

D.3.1 3D FOURIER INTEGRALS

In this case, the variable $\mathbf{k}$ becomes continuous, and the summation in Eq. (D.18) turns into an integral over $\mathbf{k}$. Following the analysis of Eqs (D.15), the prescription in an arbitrary space dimension d is

$$\frac{1}{V}\sum_{\mathbf{k}} f_{\mathbf{k}} \to \frac{1}{(2\pi)^d} \int d\mathbf{k}\, f(\mathbf{k}). \tag{D.21}$$

Consequently, the analogous equations to Eqs (D.18) and (D.19) are (for $d = 3$),

$$\psi(\mathbf{r}) = \frac{1}{(2\pi)^{3/2}} \int d\mathbf{k}\, a(\mathbf{k}) e^{i\mathbf{k}\cdot\mathbf{r}}, \quad a(\mathbf{k}) = \frac{1}{(2\pi)^{3/2}} \int d\mathbf{r}\, e^{-i\mathbf{k}\cdot\mathbf{r}} \psi(\mathbf{r}), \tag{D.22}$$

and equations analogous to Eq. (D.20) are

$$\frac{1}{(2\pi)^3} \int d\mathbf{k}\, e^{i\mathbf{k}\cdot\mathbf{r}} = \delta(\mathbf{r}), \quad \frac{1}{(2\pi)^3} \int d\mathbf{r}\, e^{-i\mathbf{k}\cdot\mathbf{r}} = \delta(\mathbf{k}). \tag{D.23}$$

---

[2] The residue theorem of complex analysis, sometimes called Cauchy's Residue Theorem, states that the line integral over a closed curve in the complex z plane of a function $f(z)$ that is analytic inside the curve, except for a finite number of poles, can be evaluated as follows: $\oint_\gamma dz f(z) = 2\pi i \sum_{k=1}^{n} \text{Res}(f, z_k)$. Here, $\text{Res}(f, z_k)$ denotes the residue of function $f(z)$ defined at the *simple* poles z_k, $\text{Res}(f, z_k) = \lim_{z \to z_k}(z - z_k) f(z)$. See Refs. [25, 26].

D.4 FOURIER INTEGRALS OF TIME-DEPENDENT FUNCTIONS

We have already mentioned that the Fourier expansion of functions of space and time, $\psi(\mathbf{r}, t)$, is carried out in terms of traveling waves $e^{i(\mathbf{k}\cdot\mathbf{r}-\omega t)}$. Let us first assume that $\psi(\mathbf{r}, t)$ satisfies periodic boundary conditions in both space and time variables [in the latter case, it is defined for $0 \leq t \leq T$ and can be extended to be defined outside this interval through the requirement $\psi(\mathbf{r}, t+T) = \psi(\mathbf{r}, t)$]. The series expansion then reads,

$$\psi(\mathbf{r}, t) = \frac{1}{\sqrt{VT}} \sum_{\mathbf{k}_n, \omega_m} a(\mathbf{k}_n, \omega_m) e^{i(\mathbf{k}_n\cdot\mathbf{r}-\omega_m t)}, \quad \omega_m = \frac{2\pi m}{T}, \quad m = 0, \pm 1, \pm 2, \ldots \quad (D.24)$$

This can be rewritten as

$$\psi(\mathbf{r}, t) = \frac{1}{\sqrt{V}} \sum_{\mathbf{k}_n} \phi(\mathbf{k}_n, t) e^{i\mathbf{k}_n\cdot\mathbf{r}}, \quad \phi(\mathbf{k}_n, t) = \frac{1}{\sqrt{T}} \sum_{\omega_n} a(\mathbf{k}_n, \omega_m) e^{-i\omega_n t}. \quad (D.25)$$

Thus, for each fixed wavenumber $\mathbf{k}_n$, the second equation above is a Fourier series of a time-dependent function, which can be studied separately from its space part, and we shall now do so.

If $\phi(t)$ is periodic in time with period T, i.e., $\phi(t+T) = \phi(t)$, then its Fourier analysis is very similar to that of a function $f(x)$ depending on one-space variable as discussed above. The only difference is that the sign before the i is opposite due to the traveling wave form $a(\mathbf{k}\omega)e^{i(\mathbf{k}\cdot\mathbf{r}-\omega t)}$. The Fourier expansion then reads,

$$\phi(t) = \frac{1}{\sqrt{T}} \sum_{\omega_n} b_n e^{-i\omega_n t}, \quad b_n = \frac{1}{\sqrt{T}} \int_0^T dt\, e^{i\omega_n t} \phi(t). \quad (D.26)$$

If $\phi(t)$ is defined for $-\infty < t < \infty$, then it is expanded in terms of time variable Fourier integral,

$$\phi(t) = \frac{1}{\sqrt{2\pi}} \int_{-\infty}^{\infty} d\omega\, b(\omega) e^{-i\omega t}, \quad b(\omega) = \frac{1}{\sqrt{2\pi}} \int_{-\infty}^{\infty} dt\, e^{i\omega t} \phi(t), \quad (D.27)$$

where we have used the relations,

$$\frac{1}{2\pi} \int_{-\infty}^{\infty} d\omega\, e^{-i\omega t} = \delta(t), \quad \frac{1}{2\pi} \int_{-\infty}^{\infty} dt\, e^{i\omega t} = \delta(\omega). \quad (D.28)$$

D.5 CONVOLUTION

Let $\psi(\mathbf{r})$ be a periodic function defined by a *convolution integral*,

$$\psi(\mathbf{r}) = \int d\mathbf{r}'\, \phi(\mathbf{r}-\mathbf{r}')\gamma(\mathbf{r}'). \quad (D.29)$$

Denote by $f_\mathbf{k}$ and $g_\mathbf{k}$ the Fourier amplitudes of $\phi(\mathbf{r})$ and $\gamma(\mathbf{r})$, respectively. The Fourier expansion of $\psi(\mathbf{r})$ given by (this result is called the *faltung theorem* — faltung means 'folding' in German)

$$\psi(\mathbf{r}) = \int d\mathbf{r}'\, \phi(\mathbf{r}-\mathbf{r}')\gamma(\mathbf{r}') = \frac{1}{V} \int d\mathbf{r}' \sum_{\mathbf{k}\mathbf{k}'} f(\mathbf{k})e^{i\mathbf{k}\cdot(\mathbf{r}-\mathbf{r}')} g(\mathbf{k}')e^{i\mathbf{k}'\cdot\mathbf{r}'} = \sum_{\mathbf{k}} f_\mathbf{k} g_\mathbf{k} e^{i\mathbf{k}\cdot\mathbf{r}}. \quad (D.30)$$

D.7 Fourier Transforms

Hence, in the Fourier expansion of $\psi(\mathbf{r})$, the coefficient of $e^{i\mathbf{k}\cdot\mathbf{r}}$ is just the product of the Fourier coefficients of the two convoluted functions. Note, however, that according to Eq. (D.19), the Fourier amplitude of $\psi(\mathbf{r})$ is

$$\psi_{\mathbf{k}} = \frac{1}{\sqrt{V}} \int d\mathbf{r}\, e^{-i\mathbf{k}\cdot\mathbf{r}} \psi(\mathbf{r}) = \sqrt{V} f_{\mathbf{k}} g_{\mathbf{k}}. \tag{D.31}$$

As a special case of Eq. (D.30), we can set $\mathbf{r} = 0$ and thereby obtain an expression for an integral of a product of functions in terms of their Fourier coefficients. For example, a scalar product $\langle \phi | \gamma \rangle$ can be expanded as,

$$\int d\mathbf{r}\, \phi^*(\mathbf{r}) \gamma(\mathbf{r}) = \sum_{\mathbf{k}} f_{\mathbf{k}}^* g_{\mathbf{k}}. \tag{D.32}$$

This analysis has an obvious extension for the case of infinite domain and will not be detailed here.

D.6 FOURIER EXPANSION OF OPERATORS

Often, operators are given in configuration space representation, $\langle \mathbf{r} | \mathcal{O} | \mathbf{r}' \rangle = \mathcal{O}(\mathbf{r}, \mathbf{r}')$. The Fourier integral takes the form of a double Fourier expansion,

$$\mathcal{O}(\mathbf{r}, \mathbf{r}') = \frac{1}{(2\pi)^3} \int d\mathbf{k}\, d\mathbf{k}'\, e^{i\mathbf{k}\cdot\mathbf{r}} \mathcal{O}(\mathbf{k}, \mathbf{k}') e^{-i\mathbf{k}'\cdot\mathbf{r}'}. \tag{D.33}$$

In particular, for translationally invariant systems, the operators are expected to be functions of $\mathbf{r} - \mathbf{r}'$ [e.g., for the two-body interaction potential, $v(\mathbf{r} - \mathbf{r}')$]. Using simple manipulations, it is easy to see that the expansion becomes a single-variable Fourier transformation,

$$\mathcal{O}(\mathbf{r} - \mathbf{r}') = \frac{1}{(2\pi)^{3/2}} \int d\mathbf{k}\, e^{i\mathbf{k}\cdot(\mathbf{r}-\mathbf{r}')} \mathcal{O}(\mathbf{k}), \quad \mathcal{O}(\mathbf{k}) = \frac{1}{(2\pi)^{3/2}} \int d\mathbf{r}\, e^{-i\mathbf{k}\cdot\mathbf{r}} \mathcal{O}(\mathbf{r}). \tag{D.34}$$

Problem D.5

Find the Fourier coefficient of the Coulomb potential,

$$v(\mathbf{r} - \mathbf{r}') = \frac{e^2}{|\mathbf{r} - \mathbf{r}'|} = \frac{1}{(2\pi)^{3/2}} \int d\mathbf{k}\, e^{i\mathbf{k}\cdot\mathbf{r}} v(\mathbf{k}), \quad v(\mathbf{k}) = \frac{1}{(2\pi)^{3/2}} \int d\mathbf{r}\, e^{-i\mathbf{k}\cdot\mathbf{r}} \frac{e^2}{|\mathbf{r}|}.$$

Guidance: Choose $\mathbf{k}$ along the z axis and use spherical coordinates (k, θ, ϕ). The integration on ϕ just gives 2π and the volume element $r^2 dr \sin\theta\, dr\, d\theta$ will remove the r in the denominator yielding,

$$\int d\mathbf{r}\, e^{-i\mathbf{k}\cdot\mathbf{r}} \frac{e^2}{|\mathbf{r}|} = 2\pi e^2 \int_0^\infty dr \int_0^\pi d\theta\, e^{ikr\cos\theta} r \sin\theta.$$

D.7 FOURIER TRANSFORMS

In the discussion above, we introduced the basic definitions and techniques required for application of Fourier analysis to physical problems. We used the terms Fourier expansion, Fourier series, and Fourier integrals. But the Fourier integral formulas (D.15) can be regarded as a mathematical structure, which involves mapping of functions by integral operators, as studied in the important mathematics subfield referred to as *the theory of integral transforms*. The notion of *Fourier transform* (FT) is used to underline the operator content of Fourier analysis and applies mainly to the case where the functions are defined on infinite domains without any specific periodicity. We briefly introduce this important and useful notion. For the sake of clarity, the discussion is for the case of a one-dimensional domain (with variables denoted by x and k).

Consider a function $f(x)$, which is absolutely integrable, $\int_{-\infty}^{\infty} |f(x)|dx < \infty$. The Fourier transform of $f(x)$ is denoted by $\mathcal{F}[f(x)] = F(k)$ and is defined as,

$$\mathcal{F}[f(x)] \equiv \frac{1}{\sqrt{2\pi}} \int_{-\infty}^{\infty} dx f(x) e^{-ikx} = F(k). \tag{D.35}$$

This is also called the *forward Fourier transform*; it has $-i$ in the exponent. The *Inverse FT* or *backward FT* (with $+i$ in the exponent) is denoted by $\mathcal{F}^{-1}[F(k)] = f(x)$, which is defined by,

$$\mathcal{F}^{-1}[F(k)] \equiv \frac{1}{\sqrt{2\pi}} \int_{-\infty}^{\infty} dk F(k) e^{ikx} = f(x). \tag{D.36}$$

Some properties of FT are listed below for $\mathcal{F}$. With appropriate modifications, they apply also to $\mathcal{F}^{-1}$ as well.

Linearity $\quad \mathcal{F}[af(x) + bg(x)] = a\mathcal{F}[f(x)] + b\mathcal{F}[g(x)]$, (D.37)

Unitarity (Parceval's relation) $\quad ||f|| = ||F||$, (D.38)

Symmetry $\quad F(k) = \mathcal{F}[f(x)] \to F(-k) = \mathcal{F}[f(-x)]$, (D.39)

Shift $\quad \mathcal{F}[f(x-y)] = e^{-2\pi i y} F(k)$, (D.40)

Translation $\quad \mathcal{F}[e^{iqy} f(x)] = F(k-q)$, (D.41)

Scaling $\quad \mathcal{F}[f(\lambda x)] = \frac{1}{|\lambda|} F\left(\frac{k}{\lambda}\right)$, (D.42)

Conjugation $\quad \mathcal{F}[f^*(-x)] = \mathcal{F}[f^*(x)]$, (D.43)

Duality $\quad \mathcal{F}[F(x)] = f(-k)$, (D.44)

Riemman-Lebesgue Lemma $\quad \lim_{|k|\to\infty} |F(k)| = 0$. (D.45)

Moreover, the convolution property discussed above can be presented in a more compact form. Denoting the convolution of two functions as

$$[f * g](x) \equiv \int_{-\infty}^{\infty} dy f(x-y) g(y), \tag{D.46}$$

we obtain the following transforms[3]:

$$\mathcal{F}[f*g] = \sqrt{2\pi} \mathcal{F}[f] \mathcal{F}[g], \tag{D.47}$$

$$\mathcal{F}[fg] = \frac{1}{\sqrt{2\pi}} \mathcal{F}[f] * \mathcal{F}[g], \tag{D.48}$$

---

[3] With the alternative Fourier transform definition of footnote 1, the factors of $\sqrt{2\pi}$ are not present in Eqs (D.47)–(D.51).

D.8 FT for Solving Differential and Integral Equations

$$\mathcal{F}^{-1}[\mathcal{F}[f]\mathcal{F}[g]] = \frac{1}{\sqrt{2\pi}} f*g, \tag{D.49}$$

$$\mathcal{F}^{-1}[\mathcal{F}[f]*\mathcal{F}[g]] = \sqrt{2\pi} fg. \tag{D.50}$$

An important relationship between the convolution $f^* * f$ and the Fourier transform $F(k)$ is known as the *Wiener-Khinchin theorem*,

$$\mathcal{F}^{-1}\left[|F(k)|^2\right] = \frac{1}{\sqrt{2\pi}} \int_{-\infty}^{\infty} dy\, f^*(x-y) f(y). \tag{D.51}$$

If $\lim_{x \to \pm\infty} |f(x)| = 0$, then it is possible to express the FT of the derivatives of $f(x)$ in terms of the FT of $f(x)$ by using partial integration because the boundary terms vanish. Then,

$$\mathcal{F}\left[f^{(n)}(x)\right] = (ik)^n \mathcal{F}[f(x)] = (ik)^n F(k). \tag{D.52}$$

Problem D.6

If $F(k)$ is the FT of $f(x)$, show that the nth moment of $f(x)$ is given by $\int dx\, x^n f(x) = \frac{1}{(-i)^n} F^{(n)}(k=0)$.

D.8 FT FOR SOLVING DIFFERENTIAL AND INTEGRAL EQUATIONS

Let us consider the inhomogeneous differential equation,

$$-\frac{d^2y}{dx^2} + a^2 y = f(x), \tag{D.53}$$

with boundary conditions

$$y(x_1) = Y_1, \quad y(x_2) = Y_2, \tag{D.54}$$

Such linear inhomogeneous differential equations can be solved by the FT method, for any function $f(x)$. The corresponding homogeneous equation is obtained by replacing $f(x)$ on the RHS of Eq. (D.53) by 0 [see Eq. (B.8)],

$$-\frac{d^2y}{dx^2} + a^2 y = 0. \tag{D.55}$$

It is easy to show by direct substitution that any solution $y(x)$ of the inhomogeneous equation (D.53) can be written as the sum of the general solution $y_0(x)$ of the homogeneous equation (D.55) and a particular solution $y_p(x)$ of the inhomogeneous equation (D.53),

$$y(x) = y_0(x) + y_p(x). \tag{D.56}$$

We shall see below how the homogeneous solution of Eq. (D.55),

$$y_0(x) = Ae^{ax} + Be^{-ax}, \tag{D.57}$$

is used to account for the boundary conditions (D.54), but first let us find a particular solution to the inhomogeneous equation. By using Eq. (D.52) and taking the FT of both sides, we get the following expression for $Y(k) \equiv \mathcal{F}[y(x)]$,

$$Y(k) = \frac{F(k)}{k^2 + a^2}. \tag{D.58}$$

This is of product form and therefore can be inverted by using the convolution theorem,

$$y(x) = \frac{1}{\sqrt{2\pi}} \int_{-\infty}^{\infty} d\xi \, f(\xi) g(x-\xi), \tag{D.59}$$

where $g(x) = \mathcal{F}^{-1}\left[\frac{1}{k^2+a^2}\right] = \frac{1}{a}\sqrt{\frac{\pi}{2}} e^{-a|x|}$. Thus, the particular solution is

$$y(x) = \frac{1}{2a} \int_{-\infty}^{\infty} d\xi \, f(\xi) e^{-a|x-\xi|} \equiv y_p(x). \tag{D.60}$$

Returning now to the boundary conditions, we substitute into Eq. (D.56) using the solutions $y_0(x)$ from Eq. (D.57) and $y_p(x)$ from Eq. (D.60) and obtain the boundary conditions,

$$A e^{a x_1} + B e^{-a x_1} + y_p(x_1) = Y_1, \quad A e^{a x_2} + B e^{-a x_2} + y_p(x_2) = Y_2, \tag{D.61}$$

from which the constants A and B can be found. The problem is thereby completely solved.

Let us consider the following typical form for linear integral equation for an unknown function $f(x)$,

$$f(x) + \lambda \int_a^b dy \, K(x,y) f(y) = g(x), \tag{D.62}$$

where the inhomogeneous term $g(x)$ and the *integral kernel* $K(x,y)$ are known, and λ is a constant. If the kernel is translationally invariant, $K(x,y) = K(x-y)$, the integral is a convolution, and the FT technique is quite effective in obtaining a solution. As an example, consider the integral equation,

$$f(x) + 4 \int_{-\infty}^{\infty} dy \, e^{-a|x-y|} f(y) = g(x). \tag{D.63}$$

Application of the FT yields

$$F(k) = \frac{a^2 + k^2}{a^2 + k^2 + 8a} G(k), \tag{D.64}$$

hence,

$$f(x) = \frac{1}{\sqrt{2\pi}} \int_{-\infty}^{\infty} dk \, \frac{a^2+k^2}{a^2+k^2+8a} G(k) e^{ikx}, \tag{D.65}$$

where $G(k) = \mathcal{F}[g(x)]$.

Problem D.7

Find the solution of Eq. (D.63) for $a = 1$ and $g(x) = e^{-|x|}$.

Guidance: First show that $G(k) = \sqrt{\frac{2}{\pi}} \frac{1}{k^2+1}$. You will need to determine the integral of $I(x) = \frac{1}{\pi} \frac{e^{ikx}}{k^2+9}$, which can be evaluated by completing the integration contour (including the real line) into an upper or lower semicircle, depending on the sign of x, with radius $R \to \infty$ [see discussion near Eq. (D.16)]. The integral can be evaluated using the residue theorem.

Symmetry and Group Theory

Symmetry is all around us, from the shapes of butterflies to the form of snow crystals. Symmetries are often expressed and quantified by group theory. Furthermore, group theory allows simplification of physical problems possessing symmetry, e.g., the hydrogen atom, the benzene molecule, and the properties of electrons in crystals. It makes the symmetry properties of systems easier to recognize and understand, and identifies conserved quantities and invariances. Here, we present the rudiments of group theory with an eye to use in quantum problems. Textbooks on group theory and its application to the physical sciences include Refs. [99, 100, 124, 290, 291].

E.1 GROUP THEORY AXIOMS

A group is defined as a set of elements, G, and a binary operator (sometimes called the group multiplication operator) • that maps $G \times G \to G$, with the following properties. Let $\{O_i\}$ be a set of elements of group G. The following four conditions must be satisfied:

1. For any $O_i, O_j \in G$, $O_i \bullet O_j = O_k \in G$ (Group Property).
2. $O_i \bullet (O_j \bullet O_k) = (O_i \bullet O_j) \bullet O_k$ (Associativity).
3. There exists an element $I \in G$ called the *unit element* or *identity*, such that for any element $O_i \in G$, $I \bullet O_i = O_i \bullet I = O_i$. The unit element is sometimes denoted by e.
4. There exists an element, $O_i^{-1} \in G$ for every $O_i \in G$, such that $O_i^{-1} \bullet O_i = O_i \bullet O_i^{-1} = I$. This element is called the *inverse of* O_i.

The inverse of the product of two group elements $O_i \bullet O_j$ is $(O_i \bullet O_j)^{-1} = O_j^{-1} \bullet O_i^{-1}$. Similarly, for the product of any number of group elements, e.g., $(O_i \bullet O_j \bullet O_k)^{-1}$ is $O_i^{-1} \bullet O_j^{-1} \bullet O_k^{-1}$, as can readily be shown using the associative law.

The *order of a group* is the number of elements in the group.

E.2 GROUP MULTIPLICATION TABLES

Let us consider the group named D_{2h} which contains four elements, which we shall call I, 2_x, 2_y, and 2_z. The order of the group D_{2h}, i.e., the number of elements in the group, is 4. The *group multiplication table* specifies how to "multiply" elements of the group. A group multiplication table lists the group elements in the first row and the first column of the table. The remaining entries of the table are obtained by multiplying the element in the first row with the element in the first column, so the *ij* element of the table corresponds to the product $O_i \bullet O_j$. The table completely specifies the group elements and the group multiplication operator of the group. See Table E.1 for the multiplication table of group D_{2h} and Table 11.1 for group C_{2v} (the names used for the point groups is explained below). In these two cases, the groups are abelian, meaning that the order of multiplication of the group elements is irrelevant (see below). However, for the nonabelian groups, the order does matter. For example, Table E.2 shows a group multiplication table for a nonabelian group isomorphic to C_{3v}.[1] In this table, we arbitrarily use the convention for multiplication: [(row) (column) = product], e.g., $AB = D$.

---

[1] An isomorphism is a map that preserves sets and relations among elements.

Table E.1 *Group multiplication table for the group D_{2h}.*

D_{2h}	I	2_x	2_y	2_z
I	I	2_x	2_y	2_z
2_x	2_x	I	2_z	2_y
2_y	2_y	2_z	I	2_x
2_z	2_z	2_y	2_x	I

Table E.2 *Group multiplication table for a group isomorphic to C_{3v}.*

C_{3v}	E	A	B	C	D	F
E	E	A	B	C	D	F
A	A	E	D	F	B	C
B	B	F	E	D	C	A
C	C	D	F	E	A	B
D	D	C	A	B	F	E
F	F	B	C	A	E	D

E.3 EXAMPLES OF GROUPS

Abelian Groups are groups whose group multiplication operator is commutative, i.e., $O_i \bullet O_j = O_j \bullet O_i$. This is not typical groups; e.g., the elements of the symmetry group (i.e., a group based upon the symmetry of an object) of rotations in 3D do not commute.

Cyclic groups are groups wherein all elements can be generated from powers of a single generating element. For example, a rotation by $2\pi/n$ generates the group C_n. An example is the group C_6, which has the group elements C_6, C_6^2, C_6^3, C_6^4, C_6^5, $C_6^6 = I$. The group multiplication operator has the following properties: $C_6^n \bullet C_6^m = C_6^{(m+n)\bmod 6}$. For example, $C_6^2 \bullet C_6^3 = C_6^5$, $C_6^2 \bullet C_6^5 = C_6^1$, etc.

Symmetry groups can be finite or continuous. Finite symmetry groups (groups whose order is finite) include (a) point groups (groups leaving a point fixed), which contain group elements of finite rotations, reflections, and inversion; (b) lattice groups, which contain translations by finite specific lattice vectors (c) space groups that combines elements of point groups and space groups; and (d) the permutation group of n objects, S_n. The permutation groups are discussed in Sec. 8.1. Continuous symmetry groups (groups whose elements can be characterized by a continuous parameter or parameters) include the rotation groups [rotations in 2D, rotations in 3D that can be thought of as rotations about any 3D unit vector by an arbitrarily angle (see Sec. 3.3)], the group of translations, the unitary groups, the general linear groups, and the symplectic groups (see Sec. 13.4.2).

Many groups are relevant for quantum mechanics, including:

1. O(n) — the group of all orthogonal matrices of size $n \times n$. SO(n) is group of orthogonal $n \times n$ matrices with determinant equal +1. For example, the group O(3) is the group of all orthogonal 3×3 matrices. The group SO(3), where S stands for special, is the group of all proper rotations,[2] for which the determinant of the matrices representing the group elements is unity.
2. U(n) — the group of all unitary $n \times n$ matrices. SU(n) is the group of all unitary $n \times n$ matrices with determinant equal to unity (again special means $\det U = 1$). For example, SU(2) is the special unitary group of 2×2 unitary matrices with determinant equal to unity. Rotations of spins are carried out mathematically using matrices that belong to SU(2).
3. GL(n, C) and GL(n, R) — the general linear group of n-dimensional matrices (complex and real). SL(n, C) and SL(n, R) are the special general linear group of n-dimensional complex and real matrices, respectively (see Sec. 13.4.2).
4. S_n — the group of permutations of n objects.
5. Point groups in 2D and 3D (see below).

---

[2] An improper rotation is a linear combination of a (proper) rotation and a reflection in a plane perpendicular to the axis of rotation.

E.3 Examples of Groups

6. Space groups (see below).
7. Lorentz group — the group of linear space-time preserving transformations of (t, x, y, z) which leave $t^2 - x^2 - y^2 - z^2$ invariant.

E.3.1 POINT GROUPS

A point group is a group of geometric symmetries that leaves a point fixed, i.e., the term 'point group' indicates symmetry elements that leave at least one point fixed upon applying any of the group elements. Different collections of symmetry operations are organized into different groups. There are an infinite number of discrete point groups in each number of dimensions (e.g., 1, 2, 3, ...), but only a finite number (32) of point groups are possible for crystals in 3D; these are called the *crystallographic point groups*. Point groups can be used to classify molecules in terms of their internal symmetry. Molecules can be subjected to symmetry operations that result in indistinguishable configurations.

Eight symmetry elements can belong the to crystallographic point groups:
E (identity element),
C_2 (rotation diad),
C_3 (rotation triad),
C_4 (rotation tetrad),
C_6 (rotation hexad),
i (inversion),
σ_v (reflection in a plane passing through an axis of symmetry), and
σ_h (reflection in a plane perpendicular to an axis of symmetry).

The *crystallographic restriction theorem* proves that only the 32 groups listed below, with $n = 1, 2, 3, 4,$ and 6, are compatible with translational symmetry. Point groups can be categorized according to *Schoenflies notation* or *International notation* (sometimes called Hermann-Mauguin notation). In Schoenflies notation, the point groups are as follows:

C_1 — contains only the identity element. Example molecule: CHBrClF.
C_s — contains only a reflection plane (and identity). Example molecule: CH_2BrCl
C_i — contains only a center of symmetry. Example molecule: 1,2-difluoro-1,2-dichloroethane.
C_n — contains only a C_n center of symmetry. An example molecule of C_2 is hydrogen peroxide, H_2O_2.
C_{nv} — contains only n-fold axis and n vertical (or dihedral) mirror planes. An example of a C_{2v} molecule is water; and example of a C_{3v} molecule is ammonia.
C_{nh} — contains only an n-fold axis, a horizontal mirror plane, a center of symmetry or an improper axis ($\sigma_h C_n$). Example: $B(OH)_3$ belongs to C_{3h}.
S_{2n} — contains only a S_{2n} rotation–reflection axis. Example: 1,3,5,7-tetramethylcyclooctatetraene belongs to S_4.
D_n — contains only a C_n and C_2 perpendicular to it. See the web pages below for examples for this and other groups.
D_{nd}: — contains a C_n, n perpendicular C_2 and n dihedral mirror planes colinear with the principal axis.
D_{nh} — contains a C_n and a horizontal mirror plane perpendicular to C_n. An example of D_{6h} is the benzene molecule.
T, T_d, T_h — the tetrahedral groups. A tetrahedron has symmetry T.
O, O_h — the octahedral groups. A cube has the symmetry O_h.
I, I_h — the icosahedral groups (sometimes denoted Y, Y_h). These groups do not occur as symmetry groups for molecules in nature.

For molecules, other values of n, such as $n = 5$ and $n = 7$ are possible. For a discussion of the point groups and their properties, see Chapter 12 of Landau and Lifshitz, *Quantum Mechanics* [2], and Refs [124, 290]. For a web-based list of point groups, see
http://mathworld.wolfram.com/CrystallographicPointGroups.html
http://mathworld.wolfram.com/CharacterTable.html.
Examples of molecules having the symmetry of the point groups are beautifully illustrated on the web page
http://www.staff.ncl.ac.uk/j.p.goss/symmetry/Molecules_pov.html. It is recommended that you have a

look at this site. The site has links to many other group theoretical topics. The home directory of this site is
http://www.staff.ncl.ac.uk/j.p.goss/symmetry/index.html
Another lovely site on point group symmetry is
http://symmetry.otterbein.edu/index.html.

E.3.2 SPACE GROUPS

To classify crystals, we also need to consider translational symmetry. An infinite crystal containing atoms or molecules has translational symmetry elements that can be combined into 230 combinations of elements called *space groups*. The set of all equivalent lattice points in the crystal (i.e., the lattice points that can be brought into coincidence by the translational symmetry operations of the crystal) is called a Bravais lattice. A Bravais lattice in 3D (see Sec. 9.3.1) is the set of all points $\mathbf{R}_{n_1 n_2 n_3} \in \mathbb{R}^3$ that can be written as $\mathbf{R}_{n_1,n_2,n_3} = n_1 \mathbf{a}_1 + n_2 \mathbf{a}_2 + n_3 \mathbf{a}_3$, where $\mathbf{a}_i$ are three independent *primitive vectors* and n_i are integers.

A grouping of several classes of space groups is called a crystal system. There are seven crystal systems: Cubic Hexagonal, Trigonal (Rhombohedral), Tetragonal, Orthorhombic, Monoclinic, and Triclinic. There are 14 types of crystal space lattices based on the seven simple geometric groupings of the lattice points within the crystal. Section 9.3.1 discusses many aspects of space groups.

E.3.3 CONTINUOUS GROUPS

Sophus Lie was the first to study the continuous transformation groups, which is now called *Lie groups*. A circle has a continuous group of symmetries; one can rotate a circle an arbitrarily small angle and it looks the same. This is in contrast to the hexagon, where only rotations that are multiples of one-sixth of a full turn are symmetries. The group of rotations about a given axis is an example of a continuous group that can be parameterized by one parameter, the angle θ by which an object is turned about the given axis. Generally, Lie groups have a more complicated group structure. Examples include the group of orthogonal $n \times n$ matrices, O(n), the group of unitary $n \times n$ matrices, U(n), or the general linear group GL(n) of invertible $n \times n$ matrices. It is clear that O(n) is a group because the product of two orthogonal matrices is orthogonal, and the other group requirements (an inverse for all elements, a unit element, etc.) are satisfied. Similarly, it is clear that the product of two unitary matrices is unitary. As we have already noted, if the $n \times n$ orthogonal matrices have determinant $+1$, and the group is called SO(n). If the $n \times n$ unitary matrices are unimodular, the group is called SU(n).

E.4 SOME PROPERTIES OF GROUPS

If there exists a subset of elements, H, of a group G, such that the subset itself forms a group, then the group H is called a *subgroup* of the group G. The group consisting of the identity element I itself and the whole group itself are called improper subgroups. All other subgroups are "proper" subgroups.

The order h of any subgroup H must be a divisor of the order g of the main group G, i.e., g/h is a positive integer.

Two elements A and B of a group are said to be *conjugate elements* if $A = PBP^{-1}$ where P is also some element of the group. If A is conjugate to B, and B to C, then A is conjugate to C, since if $B = P^{-1}AP$, $C = Q^{-1}BQ$, it follows that $C = (PQ)^{-1}A(PQ)$. Hence, we can speak of sets of conjugate elements of a group. Such a set is called a *class of conjugate elements* or simply *class*.

Each class is completely determined by any one element A of the class, since, given A, we obtain the whole class by forming the products HAH^{-1}, where H is successively every element of the group (of course, this may give each element of the class several times). Thus, we can divide the whole group into classes; each element of the group can clearly appear in only one class. The unit element of the group is a class by itself, since for every element of the group, $AIA^{-1} = I$. If a group is abelian, each of its elements is a class by itself.

Let us consider a group $\mathcal{A}$ with n elements $A, A', A'', \ldots$, and a group $\mathcal{B}$ with m elements $B, B', B'', \ldots$. We can obtain a new group, called the *direct product* of the groups $\mathcal{A}$ and $\mathcal{B}$, denoted by $\mathcal{A} \times \mathcal{B}$, that is of order nm, as follows. We multiply every element of group $\mathcal{A}$ by every element of group $\mathcal{B}$, to obtain a set of nm elements which forms a group. For any two elements of this set, we have $AB \bullet A'B' = AA' \bullet BB' = A''B''$, i.e. another element of the set of elements of the direct product group. This is similar to the outer product of vector spaces (see Sec. A.2.1).

All the elements of a group can be divided into complexes called *cosets* of the group. There are left cosets and right cosets. The right coset of a subgroup H of a group G is the set of elements formed by taking combinations of the subgroup with the remaining elements, writing the latter on the right. For a subgroup, H, of order h: $HO_k \equiv \{H_1 O_k, H_2 O_k, \ldots, H_h O_k\}$ for all $O_k \notin H$. Every element is either in the subgroup or one of its cosets. No element is in both subgroup and its cosets. Another way of saying this is that every element of G belongs to one and only one coset. H itself is the only coset that is a subgroup. We have $HO_k = H$ if and only if O_k is an element of H. Since H is a subgroup, it must contain the identity, i.e, the identity is only in one coset. The order of a subgroup of a finite group is always a divisor of the order of the whole group, i.e., the order h of any subgroup H must be a divisor of the order g of the main group G, hence g/h is a positive integer. This result is called *Lagrange's theorem*.

E.5 GROUP REPRESENTATIONS

One of the most useful topics in group theory for quantum mechanics applications is the topic of group representations. A representation of a group G is a group action of G on a vector space V by invertible linear maps, i.e., group elements are represented by square non-singular matrices that operate on vectors in a vector space. The group multiplication operation is then represented by matrix multiplication, so if A and B are two group elements, the element $C = A \bullet B$ is represented by the matrix $C_{ki} = A_{kl} B_{li}$, where Einstein summation notation is intended and where A_{kl} and B_{li} are matrices that represent the group elements A and B. The *dimension of the representation* is the dimension of the vector space that the matrices of the representation operate on. If the representation is n dimensional, the matrices are $n \times n$ matrices.

For example, consider the group SO(3) of rotations in 3D. The generators of rotations are the angular momentum operators $\mathbf{L}$. The group elements of SO(3) are given by $\mathcal{R} = e^{-i\boldsymbol{\varphi} \cdot \mathbf{L}/\hbar}$ [see Eq. (3.114)]. The rotation operators (in Sec. 3.3.1 we specify the rotations in terms of Euler angles α, β, γ and the rotation operators are denoted $\mathcal{R}_{\alpha\beta\gamma}$) form a group called SO(3). We can apply the group elements to coordinate vectors $\mathbf{r}$. A group element $\mathcal{R}$ transforms $\mathbf{r} \to \mathbf{r}' = \mathcal{R} \mathbf{r}$, and this transformation is represented by the 3×3 matrix, $\mathcal{R}_{ij}$, i.e., the three-dimensional representation of the group SO(3) is composed of 3×3 orthogonal matrices. Rotation about the z-axis by an angle φ is given by the 3×3 rotation matrix, $\mathcal{R}_z(\varphi)$:

$$\begin{pmatrix} x' \\ y' \\ z' \end{pmatrix} = \mathcal{R}_z(\varphi) \begin{pmatrix} x \\ y \\ z \end{pmatrix} = \begin{pmatrix} \cos\varphi & \sin\varphi & 0 \\ -\sin\varphi & \cos\varphi & 0 \\ 0 & 0 & 1 \end{pmatrix} \begin{pmatrix} x \\ y \\ z \end{pmatrix}, \tag{E.1}$$

Similarly, rotation about the x-axis by an angle θ is given by the 3×3 rotation matrix $\mathcal{R}_x(\theta)$,

$$\begin{pmatrix} x' \\ y' \\ z' \end{pmatrix} = \mathcal{R}_x(\theta) \begin{pmatrix} x \\ y \\ z \end{pmatrix} = \begin{pmatrix} 1 & 0 & 0 \\ 0 & \cos\theta & \sin\theta \\ 0 & -\sin\theta & \cos\theta \end{pmatrix} \begin{pmatrix} x \\ y \\ z \end{pmatrix}, \tag{E.2}$$

The matrices representing the group elements have the multiplication properties specified by the group, and the product matrix represents a group element, e.g., $\mathcal{R}_x(\pi) \mathcal{R}_z(\pi) = \mathcal{R}_y(\pi)$. These 3×3 matrices form a three-dimensional representation of the group SO(3).

If reflections and inversion are added to the group SO(3), the group is denoted by the symbol O(3). The 3×3 matrices representing the group elements are orthogonal and have determinant $+1$ or -1. If reflections and inversion are not included, then the matrices of the group have determinant $+1$, and the group is denoted SO(3).

Problem E.1

For the group O(3), construct the 3×3 matrices representing (a) inversion i ($\mathbf{r} \to -\mathbf{r}$), (b) reflection in the x-z plane, σ_{xz}, and (c) reflection in the y-z plane, σ_{yz}.

Answers:

$$i = \begin{pmatrix} -1 & 0 & 0 \\ 0 & -1 & 0 \\ 0 & 0 & -1 \end{pmatrix}, \quad \sigma_{xz} = \begin{pmatrix} 1 & 0 & 0 \\ 0 & -1 & 0 \\ 0 & 0 & 1 \end{pmatrix}, \quad \sigma_{yz} = \begin{pmatrix} -1 & 0 & 0 \\ 0 & 1 & 0 \\ 0 & 0 & 1 \end{pmatrix}$$

The rotation group SO(3) has higher dimensional representations. Irreducible representations (see below) of dimension $(2j+1)$ are the matrices $D_{m'm}^{(j)}$, discussed in Sec. 3.3.2, where the basis of the representation for integer j are the $2j+1$ spherical harmonics $\{Y_{jm}\}$, $m = -j, \ldots, j$.

E.5.1 IRREDUCIBLE REPRESENTATIONS

Suppose that all the matrices A representing the group elements of a group G are transformed into matrices A' by applying a linear transformation matrix R, and we find that the matrices A' are block-factored matrices. For example, if

$$A' = R^{-1}AR = \begin{pmatrix} A'_1 & 0 & 0 \\ 0 & A'_2 & 0 \\ 0 & 0 & A'_3 \end{pmatrix}, \tag{E.3}$$

for *all* matrices A of the group, then these matrices form a *reducible representation* of G. The dimensions of the matrices A'_1, A'_2, A'_3 are in general not equal (and are distinct for all matrices A) but their sum equals the dimension of the matrices A for all $A \in G$. If the matrices A cannot be reduced to block-diagonal form, the representation is called *irreducible*.

For finite groups, every representation is equivalent to a unitary representation, i.e., the matrices A'_1, A'_2, etc., can be made to be unitary.

Let us consider a group representation of dimension f where the f-dimensional matrices operate on a set of functions which is spanned by basis functions $\{\psi_k\}$, $k = 1, \ldots, f$. For example, the vectors $\hat{\mathbf{x}}, \hat{\mathbf{y}}, \hat{\mathbf{z}}$ form a basis for the 3-dimensional representation of O(3) [see the beginning of Sec. E.5]. If, as a result of a suitable linear transformation R, $\psi'_k = R\psi_k$, it is possible to divide the basis functions of the representation into sets of $f_1, f_2, \ldots$ functions, with $f_1 + f_2 + \ldots = f$, such that, when any element of the group acts on them, the functions in each set are transformed only into combinations of themselves (and do not involve functions from other sets), this f dimensional representation is said to be *reducible*. A representation of a group G onto a vector space V is *irreducible* if it cannot be broken up into smaller representation spaces.

Problem E.2

Prove that if V is a vector space of dimension larger than unity, then, if there exists a vector $v \in V$ such that $\Phi(A)v = v$ for all elements A of a finite group G, the matrices $\Phi(A)$ form a reducible representation of G.

Answer: Without loss of generality, we can take $v = (1, 0, \ldots, 0)^\dagger$. Then the matrices $\Phi(A)$ have the form

$$\Phi(A) = \begin{pmatrix} 1 & A_{12} & \ldots & A_{1n} \\ 0 & A_{22} & \ldots & A_{2n} \\ 0 & & \ddots & \\ 0 & A_{n2} & \ldots & A_{nn} \end{pmatrix}.$$

Since the representations of a finite group are unitary, $A_{12} = \ldots = A_{1n} = 0$. Hence, $\Phi(A)$ are block diagonal and the representation is reducible.

Any reducible representation can be decomposed into irreducible ones. Thus, by the appropriate linear transformation, the base functions divide into several sets, of which each is transformed by some irreducible representation when the

E.5 Group Representations

elements of the group act on it. For example, the $2L+1$ spherical harmonic functions $Y_{L,M}$, $M = -L, \ldots, L$, are transformed into linear combinations of one another when the coordinate system is rotated. Hence, the spherical harmonic functions $\{Y_{L,M}\}$ for a given L is an irreducible representation. The set of hydrogenic wave functions corresponding to a given principle quantum number n divide up into several sets, each of which is an irreducible representation of the rotation group. As another example, consider identical particles of spin 1/2 (or more generally, any spin). The Young tableau encountered in Sec. 8.1.2 specifies how to obtain the irreducible representations of the permutation group (see Sec. 8.1 for a discussion of this group). Any reducible representation can be decomposed into irreducible ones. Hence, by the appropriate linear transformation, the base functions divide into several sets, each of which is transformed by some irreducible representation when the elements of the group act on it. Perhaps several different sets transform by the same irreducible representation. If this occurs, this irreducible representation is said to be contained so many times in the reducible one.

Irreducible representations play a crucial role in all quantum mechanical applications of group theory. The following theorems regarding irreducible representations are called *Schur's lemmas*.

LEMMA 1. Let $\{\Phi(A)\}$ and $\{\Psi(A)\}$ denote two irreducible representations of a group G of dimension n and m with $n \neq m$, and let B be a rectangular matrix of dimension $n \times m$. If $\Phi(A)B = B\Psi(A)$ for all $A \in G$, then $B = \mathbf{0}$.

LEMMA I'. If $\{\Phi(A)\}$ and $\{\Psi(A)\}$ are irreducible representations of the group G having the same dimension, and if $\Phi(A)B = B\Psi(A)$ for all $A \in G$, then either $\Phi(A)$ and $\Psi(A)$ are equivalent (there exists a unitary matrix U such that $U\Phi(A)U^\dagger = \Psi(A) \, \forall \, A \in G$), or $B = \mathbf{0}$.

LEMMA II. Let $\{\Phi(A)\}$ be an irreducible representation of a group G. If $\Phi(A)B = B\Phi(A)$ for all $A \in G$, then $B = $ constant $\mathbf{1}$, i.e., a nonzero matrix that commutes with all the matrices of an irreducible representation of a group G is proportional to the unit matrix.

E.5.2 GROUP ORTHOGONALITY THEOREM

The properties of group representations that are important in dealing with matrix elements can be derived from the *group orthogonality theorem* concerning the elements of the matrices, which constitute the irreducible representations of a group. We introduce this theorem here.

The matrices of irreducible representations satisfy several orthogonality relations. Let us denote the matrix of irreducible representation α for group element A belonging to group G by $A_{ij}^{(\alpha)}$ (rather than $\Phi(A)_{ij}$ as we did in Schur's lemmas). The matrices of two different irreducible representations α and β satisfy the orthogonality condition, $\sum_A A_{ij}^{(\alpha)} \left(A_{kl}^{(\beta)}\right)^* = 0$, where the summation is taken over all the elements of the group. Moreover, for any irreducible representation α, $\sum_A A_{ij}^{(\alpha)} \left(A_{kl}^{(\alpha)}\right)^* = \frac{g}{f_\alpha} \delta_{ik} \delta_{jl}$, where g is the order of the group G and f_α is the dimension of the αth irreducible representation; hence, $\sum_A |A_{ij}^{(\alpha)}|^2 = \frac{g}{f_\alpha}$. Combining these results, we obtain a relation that is called the *group orthogonality theorem* for matrix elements of irreducible representations:

$$\sum_A A_{ij}^{(\alpha)} \left(A_{kl}^{(\beta)}\right)^* = \frac{g}{f_\alpha} \delta_{\alpha\beta} \delta_{ik} \delta_{jl}. \tag{E.4}$$

This is a central result in the theory of group representations. It holds for finite groups; there is a generalization for compact groups (a topological group whose topology is compact).

E.5.3 CHARACTERS AND CHARACTER TABLES

A *character* of an element of a group is the trace of the matrix corresponding to the group element. Given element A of group G, the character of the element A is denoted by $\chi(A)$, where $\chi(A) \equiv \sum_i A_{ii}$. Characters depend on the representation of the group, but don't change upon change of basis, $A \to SAS^{-1}$, as can be proved using the properties of the trace of a matrix. The character of the unit element I of a group equals the dimension f of the representation, since the matrix representing I is the $f \times f$ unit matrix, i.e., the character $\chi(I)$ is just the dimension of the representation, f.

If the representation is the αth irreducible representation, then $\chi^{(\alpha)}(I) = f_\alpha$, where we have denoted the characters of the various irreducible representations by attaching superscripts.

The number of different irreducible representations of a group is equal to the number n_c of classes in the group. The characters of the matrices of the element A in the representations are $\chi^{(1)}(A), \chi^{(2)}(A), \ldots, \chi^{(n_c)}(A)$.

If we take the trace of the matrix representations in the group orthogonality theorem, Eq. (E.4), we find the character orthogonality condition

$$\sum_A \chi^{(\alpha)}(A) \left(\chi^{(\beta)}(A)\right)^* = g\delta_{\alpha\beta}. \tag{E.5}$$

Setting $\alpha = \beta$, we obtain

$$\sum_A |\chi^{(\alpha)}(A)|^2 = g. \tag{E.6}$$

The sum in Eq. (E.5) contains only n_c independent terms, since the characters of elements of the same class are equal. Hence, Eq. (E.5) can be written as

$$\sum_c g_c \chi^{(\alpha)}(c) \left(\chi^{(\beta)}(c)\right)^* = g\delta_{\alpha\beta}, \tag{E.7}$$

where the sum is over the classes in the group and g_c is the number of elements in the class c.

A trivial irreducible representation is given by a single-base function invariant under all group transformations. This 1D representation is called the unit representation, and all its characters are unity. If one of the representations in Eq. (E.7) is the unit representation, the other obeys the relation

$$\sum_A \chi^{(\alpha)}(A) = \sum_c g_c \chi^{(\alpha)}(c) = 0. \tag{E.8}$$

Hence, the sum of the characters of all the group elements is zero.

Any reducible representation can be decomposed into irreducible representations. Denoting $\chi(A)$ the characters of the reducible representation of dimension f, and $n^{(\alpha)}$ the number of times irreducible representation, α appears in the reducible representation, so that

$$\sum_\alpha n^{(\alpha)} f_\alpha = f, \tag{E.9}$$

then the characters $\chi(A)$ are given by

$$\chi(A) = \sum_\alpha n^{(\alpha)} \chi^{(\alpha)}(A). \tag{E.10}$$

Multiplying Eq.(E.10) by $\left(\chi^{(\beta)}(A)\right)^*$, and summing over A, we find, using Eq. (E.5),

$$n^{(\beta)} = \frac{1}{g} \sum_A \chi(A) \left(\chi^{(\beta)}(A)\right)^*. \tag{E.11}$$

A particular representation of dimension g, called the *regular representation*, is obtained by operating on a function ψ of coordinates with the operators A corresponding to the group elements, such that the g functions $A\psi$ obtained from ψ are linearly independent. The operators A can be thought of as the matrices representing the group elements. None of the matrices of this representation contain any diagonal elements, except the matrix corresponding to the unit element I, i.e., $\chi(A) = 0$ for all elements except I, and $\chi(I) = g$. Decomposing the regular representation into irreducible representations, we have from Eq. (E.11), $n^{(\alpha)} = f_\alpha$. Hence, the irreducible representations are contained in the regular

E.5 Group Representations

representation as many times as their dimension. Substituting this result into Eq. (E.9), we find the relation

$$\sum_\alpha f_\alpha^2 = g, \qquad (E.12)$$

the sum of the squared dimensions of the irreducible representations of a group is equal to its order.

> **Problem E.3**
>
> Prove that the irreducible representations of abelian groups are of one dimensional.
>
> **Hint:** Use Schur's Lemma.

Character tables are often used for summarizing the symmetry aspects of groups. Character tables can be thought of as shorthand versions of matrix representations. Table E.3 shows the character table for the group C_{3v} and Table E.4 shows the character table for the group C_6. The rows in a character table are labeled by the irreducible representation, and the columns are labeled by the classes. Mulliken symbols are used for the irreducible representations as explained in Sec. 11.1.2. The numbers in front of the symbols for the classes of the group show the numbers of elements in the corresponding classes. The classes are named according to the kind of symmetry elements that are contained in them. Often, in tabluations of group characters, isomorphic groups, which have the same representations, are given together. In quantum mechanics, wave functions transforming according to a degenerate representation, such as representation E of the group C_{3v}, or representations E_1 and E_2 of the group C_6, are degenerate. In these particular cases, the representations are doubly degenerate, but higher degeneracy is possible in other groups.

The characters of irreducible representations can be added together to produce a character table of an reducible representation, just like the result produced from adding the matrices, because they are shorthand matrix representations.

Given a representation of a group, for example, the regular representation, one can construct a character table for the group as follows. Take the traces of matrices for each element. The set of characters also forms a representation for the group. This is a reducible representation since it is a combination of irreducible representations. The matrices for the symmetry operations can be broken down into smaller matrices along the diagonal with all other elements equal to zero, i.e., they can be block diagonalized.

Table E.3 Character table for the group C_{3v}.

C_{3v}	I	$2C_3$	$3\sigma_v$
A_1	1	1	1
A_2	1	1	−1
E	2	−1	0

Table E.4 Character table for the group C_6. Here $e \equiv e^{2\pi i/6}$.

C_6	I	C_6	C_3	C_2	$(C_3)^2$	$(C_6)^5$
A	1	1	1	1	1	1
B	1	−1	1	−1	1	−1
E_1	1	e	$-e$	−1	$-e$	e^*
	1	e	$-e$	−1	$-e$	e
E_2	1	$-e^*$	$-e$	1	$-e^*$	$-e$
	1	$-e$	$-e$	1	$-e$	$-e$

The complex representations E_1 and E_2 are doubly degenerate and can be viewed as two dimensional.

For a web based list of character tables, see

http://symmetry.jacobs-university.de
http://www.webqc.org/symmetry.php
http://mathworld.wolfram.com/CharacterTable.html
http://www.staff.ncl.ac.uk/j.p.goss/symmetry/Molecules_pov.html.

E.5.4 CONSTRUCTING IRREDUCIBLE REPRESENTATIONS

The decomposition of a regular representation into its irreducible parts is accomplished by means of the projection operator $P_i^{(\alpha)} \equiv \frac{f_\alpha}{g} \sum_A \left(A_{ii}^{(\alpha)}\right)^* A$. In quantum mechanics, these irreducible representation projection operators can be used to project out the irreducible parts of wave functions or vibrational modes of molecules,

$$\psi_i^{(\alpha)} = P_i^{(\alpha)} \psi = \frac{f_\alpha}{g} \sum_A \left(A_{ii}^{(\alpha)}\right)^* A \psi. \tag{E.13}$$

The functions $\psi_i^{(\alpha)}$, for $i = 1, 2, \ldots, f_\alpha$, obtained in Eq. (E.13) are transformed by the group elements according to

$$A \psi_i^{(\alpha)} = \sum_k A_{ki}^{(\alpha)} \psi_k^{(\alpha)}. \tag{E.14}$$

They form a basis of the αth irreducible representation. There are f_α different base functions $\psi_i^{(\alpha)}$ for the irreducible representation α. Equation (E.13) shows that any function ψ may be written as a sum of functions transformed by the irreducible representations of the group

$$\psi = \sum_\alpha \sum_i \psi_i^{(\alpha)}, \tag{E.15}$$

with

$$\psi_i^{(\alpha)} = \frac{f_\alpha}{g} \sum_A \left(A_{ii}^{(\alpha)}\right)^* A \psi. \tag{E.16}$$

To project a function on irreducible representation α, one can use the projection operator $P^{(\alpha)} \equiv \frac{f_\alpha}{g} \sum_A \left(\chi^{(\alpha)}\right)^* A$, where $\chi^{(\alpha)}$ is the character,

$$\psi^{(\alpha)} = \frac{f_\alpha}{g} \sum_A \left(\chi^{(\alpha)}\right)^* A \psi. \tag{E.17}$$

BIBLIOGRAPHY

[1] L.I. Schiff, *Quantum Mechanics*, (McGraw Hill, 1955).
[2] L.D. Landau and E.M. Lifshitz, *Quantum Mechanics Non-relativistic Theory*, (Pergamon, 1977).
[3] C. Cohen-Tannoudji, B. Diu, and F. Lalo, *Quantum Mechanics*, Volumes 1 and 2, (John Wiley, 1977).
[4] H. Bethe and E.E. Salpeter, *Quantum Mechanics of One and Two Electron Atoms*, (Plenum, 1977).
[5] A.M. Dirac, *The Principles of Quantum Mechanics*, (Oxford University Press, 1982).
[6] D.J. Griffiths, *Introduction to Quantum Mechanics*, (Prentice Hall, 2004).
[7] E. Merzbacher, *Quantum Mechanics*, (Wiley, 1997).
[8] J.J. Sakurai, *Modern Quantum Mechanics*, (Addison Wesley, 1994).
[9] L. Ballentine, *Quantum Mechanics – A modern Development*, (World Scientific, 1998).
[10] A. Messiah, *Quantum Mechanics*, (Dover, 2000).
[11] R.B. Griffiths *Consistent Quantum Theory*, (Cambridge University Press, 2002).
[12] D.P. Bertsekas and J.N. Tsitsiklis, *Introduction to Probability*, (Athena Scientific, 2002).
[13] H. Hsu, *Schaum's Outline of Probability, Random Variables, and Random Processes*, (McGraw-Hill, 1996).
[14] Yakov G. Sinai and D. Haughton, *Probability Theory: An Introductory Course*, (Springer, 1992).
[15] Y.A. Rozanov, *Probability Theory: A Concise Course*, (Dover, 1977).
[16] A good source of information on probability theory on the internet is the review by A. Maleki and T. Do, http://cs229.stanford.edu/section/cs229-prob.pdf.
[17] The transcript of the classic talk that Richard Feynman gave at the 1959 annual meeting of the American Physical Society at the California Institute of Technology was published in the February 1960 issue of Caltech's *Engineering and Science*, which owns the copyright. It is available on the web at
http://www.zyvex.com/nanotech/feynman.html.
[18] Y.B. Band, *Light and Matter: Electromagnetism, Optics, Spectroscopy and Lasers*, (Wiley, 2006).
[19] L. Pauling, *General Chemistry*, (Dover, 1970).
[20] A. Einstein, N. Rosen, and B. Podolsky, "Can Quantum-Mechanical description of physical reality be considered complete?", Phys. Rev. **47**, 777–780 (1935).
[21] H. Goldstein, C.P. Poole, and J.L. Safko, *Classical Mechanics*, 3rd Ed., (Addison Wesley, 2002).
[22] L.D. Landau and E.M. Lifshitz, *Mechanics*, (Permagon, 1994).
[23] J.B. Marion, *Classical Dynamics of Particles and Systems*, (Academic Press, 1970).
[24] L. D. Landau and E. M. Lifshitz, *The Classical Theory of Fields*, (Pergamon, 1962).
[25] G.B. Arfken and H.J. Weber, *Mathematical Methods For Physicists*, (Academic Press, 2005).
[26] P.M. Morse and H. Feshbach, *Methods of Theoretical Physics*, (McGraw-Hill, 1953).
[27] M. Abramowitz and I.A. Stegun, *Handbook of Mathematical Functions*, (Dover, 1965).
[28] J. von Neumann, *Mathematische Grundlagen der Quantenmechanik*, (Springer 1932). English translation, by Robert T. Beyer, *Mathematical Foundations of Quantum Mechanics*, (Princeton University Press, 1955).
[29] W.H. Zurek, "Decoherence and the transition from quantum to classical", Phys. Today **44** (10), 36 (1991); /quant-ph/0306072.
[30] M.O. Scully and M.S. Zubairy, *Quantum Optics*, (Cambridge University Press, 1997).

[31] K. Wodkiewicz and J.H. Eberly, "Coherent states, squeezed fluctuations, and the SU(2) and SU(1,1) groups in quantum-optics applications", J. Opt. Soc. Am. **B 2**, 458 (1985).

[32] M. Kitagawa and M. Ueda, "Squeezed spin states", Phys. Rev. **A 47**, 5138 (1993).

[33] I. Tikhonenkov, E. Pazy, Y.B. Band, and A. Vardi, "Matter-wave squeezing and the generation of SU(1,1) and SU(2) coherent matter-states, via Feshbach resonances in Bose and Fermi quantum gases", Phys. Rev. **A 77**, 063624 (2008).

[34] J.D. Jackson, *Classical Electrodynamics*, 3rd Ed., (J. Wiley, 1998).

[35] W. Grotrian, *Graphische Darstellunger der Spectren* (Graphical Description of Spectrum), (Springer, 1928).

[36] G. Gabrielse and Y.B. Band, "Coherent-State multipole moments: source of important scattering information", Phys. Rev. Lett. **39**, 697 (1977).

[37] Y.B. Band, "Coherent-State multipole moments in electron-hydrogen impact excitation", Phys. Rev. **A 19**, 1906 (1979).

[38] A.R. Edmonds, *Angular Momentum in Quantum Mechanics*, (Princeton University Press, 1968).

[39] D.M. Brink, G.R. Satchler, *Angular Momentum*, (Oxford University Press, 1994).

[40] D. Delande and J.C. Gay, "Quantum chaos and statistical properties of energy levels: Numerical study of the hydrogen atom in a magnetic field", Phys. Rev. Lett. **57**, 2006 (1986).

[41] A. Abragam and B. Bleaney, *Paramagnetic Resonance of Transition Metal Ions*, (Oxford University Press, 1970).

[42] C.P. Slichter, *Principles of Magnetic Resonance*, (Springer, 1996); A. Abragam, *Principles of Nuclear Magnetism*, (Oxford University Press, 1983); E.M. Haacke, R.W. Brown, M.R. Thompson, and R. Venkatesan, *Magnetic Resonance Imaging: Physical Principles and Sequence Design*, (Wiley-Liss, 1999).

[43] R. Landauer, "Irreversibility and heat generation in the computing process", IBM J. Res. Dev. **5**, 183 (1961); C.H. Bennett and R. Landauer, "The fundamental physical limits of computation", Scientific American **253**(1), 38 (1985).

[44] P. Benioff, "The computer as a physical system: A microscopic quantum mechanical hamiltonian model of computers as represented by Turing machines", J. Stat. Phys. **22**, 563 (1980); P. Benioff "Quantum mechanical hamiltonian models of Turing machines", J. Stat. Phys. **29**, 515 (1982).

[45] R.P. Feynman, "Simulating physics with computers", Int. J. Theor. Phys. **21**, 467–488 (1982); R. P. Feynman, "Quantum mechanical computers", Found. of Phys. **16**, 507 (1986).

[46] M.A. Nielsen and I.L. Chuang, *Quantum Computation and Quantum Information*, (Cambridge University Press, 2000).

[47] P. Lambropoulos and D. Petrosyan, *Fundamentals of Quantum Optics and Quantum Information*, (Springer-Verlag, 2007).

[48] P.W. Shor, "Polynomial time algorithms for prime factorization and discrete logarithms on a quantum computer", SIAM J. Comput. **26**, 1484 (1997).

[49] A.M. Turing, "On computable numbers: With an application to the Entscheidungsproblem", Pro. Lond. Math. Soc., ser. 2, 42 (1936).

[50] C.E. Shannon, *A Mathematical Theory of Communication*, (University of Illinois Press, 1949).

[51] W.H. Press, B.P. Flannery, S.A. Teukolsky, and W.T. Vetterling, *Numerical Recipes*, (Cambridge University Press, 1986).

[52] D.M. Greenberger, M.A. Horne, and A. Zeilinger, "Going beyond Bell's theorem", in *Bell's theorem, quantum theory and conceptions of the universe*, M. Kafatos, ed., (Kluwer Academic, Dordrecht) 73–76 (1989); D.M. Greenberger, M. Horne, A. Shimony, and A. Zeilinger, "Bell's theorem without inequalities", Am. J. Phys. **58**, 1131 (1990).

[53] J.S. Bell, "On the Einstein-Podolsky-Rosen paradox", *Physics* **1**, 195 (1964).

[54] J.S. Bell, "On the problem of hidden variables in quantum theory", Rev. Mod. Phys. **38**, 447 (1966).

[55] J.S. Bell, *Speakable and Unspeakable in Quantum Mechanics*, (Cambridge University Press, 1987).

[56] E. Knill, R. Laflamme, and G.J. Milburn, "A scheme for efficient quantum computation with linear optics", Nature (London) **409**, 46 (2001).

[57] G. Strang, *Linear Algebra and Its Applications*, (Brooks Cole, 2005).

Bibliography

[58] R.F. Werner, "Quantum states with Einstein-Podolsky-Rosen correlations admitting a hidden-variable model", Phys. Rev. **A 40**, 4277 (1989).

[59] A. Peres, "Separability criterion for density matrices", Phys. Rev. Lett. **77**, 1413 (1996).

[60] R. Horodecki, P. Horodecki, M. Horodecki, and K. Horodecki, "Quantum entanglement", Rev. Mod. Phys. **81**, 865 (2009).

[61] M. Horodecki, P. Horodecki, and R. Horodecki, "Inseparable two spin-1/2 density matrices can be distilled to a singlet form", Phys. Rev. Lett. **78**, 574 (1997).

[62] D. Deutsch, "The Church-Turing principle and the universal quantum computer", Proc. R. Soc. Lond. **A 400**, 96 (1985); "Quantum computational networks", **425**, 73 (1989).

[63] C.H. Bennett and S.J. Wiesner, "Communication via one- and two-particle operators, Rev. Lett. **69**, 2881 (1992).

[64] R. Jozsa and B. Schumacher, "A new proof of the quantum noiseless coding theorem", J. Mod. Opt. **41**, 2343 (1994).

[65] B. Schumacher, "Quantum coding", Phys. Rev. **A 51**, 2738 (1995).

[66] C.H. Bennett, G. Brassard, C. Crépeau, R. Jozsa, A. Peres, and W.K. Wootters, "Teleporting an unknown quantum state via dual classical and Einstein-Podolsky-Rosen channels", Phys. Rev. Lett. **70**, 1895 (1993). C.H. Bennett, "Quantum information and computation", Phys. Today **48** (10), 24 (1995).

[67] Y. Aharonov, D.Z. Albert, and L. Vaidman, "How the result of a measurement of a component of the spin of a spin-1/2 particle can turn out to be 100", Phy. Rev. Lett. **60**, 1351(1988).

[68] C.M. Caves *et al.*, "On the measurement of a weak classical force coupled to a quantum-mechanical oscillator. I. Issues of principle", Rev. Mod. Phys. **52**, 341 (1980).

[69] L.K. Grover, "Quantum mechanics helps in searching for a needle in a haystack", Phys. Rev. Lett. **79**, 325 (1997).

[70] J. von Neumann, *Probabilistic Logic and the Synthesis of Reliable Organisms from Unreliable Components*, Automata Studies, (Princeton University Press, 1956).

[71] P. Shor, "Scheme for reducing decoherence in quantum computer memory", Phys. Rev. **A 52**, R2493 (1995).

[72] A. Steane, "Multiple-Particle interference and quantum error correction", Proc. Roy. Soc. Lond. **A 452**, 2551 (1996).

[73] A.Yu. Kitaev, "Fault-tolerant quantum computation by anyons", Ann. Phys. **303**, 2 (2003).

[74] J.I. Cirac and P. Zoller, "Quantum computations with cold trapped ions", Phys. Rev. Lett. **74**, 4091 (1995).

[75] G.K. Brennen, C.M. Caves, P.S. Jessen, and I.H. Deutsch, "Quantum logic lates in optical lattices", Phys. Rev. Lett. **82**, 1060 (1999).

[76] D. Jaksch *et al.*, "Entanglement of atoms via cold controlled collisions", Phys. Rev. Lett. **82**, 1975 (1999).

[77] S. Haroche and J.-M. Raimond, *Exploring the Quantum: Atoms, Cavities and Photons*, (Oxford University Press, 2006).

[78] N.A. Gershenfeld and I.L. Chuang "Bulk spin-resonance quantum computation", Science **275**, 350 (1997).

[79] R. Hanson, *et. al.*, "Spins in few-electron quantum dots", Rev. Mod. Phys. **79**, 1217 (2007).

[80] Y. Nakamura, Yu.A. Pashkin, and J.S. Tsai, "Coherent control of macroscopic quantum states in a single-Cooper-pair box", Nature **398**, 786 (1999).

[81] D.P. DiVincenzo, "The physical implementation of quantum computation", Fortschr. Phys. **48**, 771 (2000).

[82] P.J. Lee *et al.*, "Phase control of trapped ion quantum gates", J. Opt. B: Quantum Semiclass. Opt. **7**, S371 (2005).

[83] O. Mandel, *et al.*, "Controlled collisions for multi-particle entanglement of optically trapped atoms", Nature (London) **425**, 937 (2003).

[84] D. Vager, B. Segev, and Y.B. Band, "Engineering entanglement: The fast-approach phase gate", Phys. Rev. **A 72**, 022325 (2005).

[85] D. Loss and D.P. DiVincenzo, "Quantum computation with quantum dots", Phys. Rev. **A 57**, 120 (1998).

[86] G. Burkard, D. Loss and D.P. DiVincenzo, "Coupled quantum dots as quantum gates", Phys. Rev. **B 59**, 2070 (1999).

[87] R.J. Epstein, F.M. Mendoza, Y.K. Kato, and D.D. Awschalom, "Anisotropic interactions of a single spin and dark-spin spectroscopy in diamond", Nat. Phys. **1**, 94 (2005).

[88] D. Bohm and Y. Aharonov, "Discussion of experimental proof for the paradox of Einstein, Rosen, Podolsky", Phys. Rev. **108**, 1070 (1957).
[89] J.F. Clauser, M.A. Horne, A. Shimony, and R.A. Holt, "Proposed experiment to test local hidden-variable theories", Phys. Rev. Lett., **23**, 880 (1969).
[90] A. Aspect, J. Dalibard, and G. Roger, "Experimental test of Bell inequalities using time-varying analyzers", Phys. Rev. Lett. **49**, 1804 (1982).
[91] A. Peres, "Unperformed experiments have no results", Am. J. of Phys. **46**, 745 (1978).
[92] D. Harrison, "Bell's inequality and quantum correlations", Am. J. Phys. **50**, 811 (1982).
[93] G. Lindblad, "On the generators of quantum dynamical semigroups", Commun. Math. Phys. **48**, 119 (1976).
[94] Q. Xie and W. Hai, "Analytical results for a monochromatically driven two-level system", Phys. Rev. **A 82**, 032117 (2010).
[95] See the Nobel Lectures by S. Chu, C. Cohen-Tannoudji and W. D. Phillips published in Rev. Mod. Phys. **70**, No. 3, pp. 685, 707, and 721 respectively, (1998).
[96] J.M. Radcliffe, "Some properties of coherent spin states", J. Phys. (Paris) **A 4**, 313 (1971).
[97] P. Lemonde *et al.*, "Cold-Atom clocks on earth and in space", in *Frequency Measurement and Control*, (Springer-Verlag, Berlin, 2001); Topics Appl. Phys. **79**, 131–153 (2001). A. Bauch, "Caesium atomic clocks: Function, performance and applications", Meas. Sci. Technol. **14**, 1159 (2003).
[98] D.J. Wineland, J.J. Bollinger, W.M. Itano, and D.J. Heinzen, "Squeezed atomic states and projection noise in spectroscopy", Phys. Rev. **A 50**, 67 (1994).
[99] M. Hamermesh, *Group Theory and Its Application to Physical Problems*, (Dover, 1989).
[100] H.J. Lipkin, *Lie Groups for Pedestrians*, (Dover, 2002).
[101] M. Lewenstein, B. Kraus, J.I. Cirac, and P. Horodecki, "Optimization of entanglement witnesses", Phys. Rev. **A 62**, 052310 (2000).
[102] L. Mandel and E. Wolf, *Optical Coherence and Quantum Optics*, (Cambridge University Press, 1995).
[103] D. F. Walls and G.J. Milburn, *Quantum Optics*, (Springer-Verlag, 1995).
[104] E.J. Heller, "Time-dependent approach to semiclassical dynamics", J. Chem. Phys. **62**, 1544 (1975).
[105] S.A. Rice, M. Zhao, *Optical Control of Molecular Dynamics*, (Wiley, 2000).
[106] M. Shapiro, P. Brumer, *Principles of the Quantum Control of Molecular Processes*, (Wiley, 2003).
[107] D.J. Tannor, *Introduction to Quantum Mechanics: A Time-Dependent Perspective*, (University Science Books, 2006).
[108] W. Zhu, J. Botina, and H. Rabitz, "Rapidly convergent iteration methods for quantum optimal control of population", J. Chem. Phys. **108**, 385 (1998); W. Zhu and H. Rabitz, "A rapid monotonically convergent iteration algorithm for quantum optimal control over the expectation value of a positive definite operator", J. Chem. Phys. **109**, 385 (1998).
[109] W.P. Schleich, *Quantum Optics in Phase Space*, (Wiley-VCH, 2001).
[110] D. Meshulach and Y. Silberberg, "Coherent quantum control of multiphoton transitions by shaped ultrashort optical pulses", Nature **396**, 239 (1998).
[111] J.J. Sakurai, *Advanced Quantum Mechanics*, (Addison Wesley, 1967).
[112] B. Misra and E.C.G. Sudarshan, "The Zeno's paradox in quantum theory", J. Math. Phys. **18**, 756 (1977).
[113] M. Loy, "Observation of population inversion by optical adiabatic rapid passage", Phys. Rev. Lett. **32**, 814 (1974).
[114] H. Bateman, *Tables of Integral Transforms and Higher Transendental Functions*, Volume II, (MacGraw-Hill, 1953).
[115] M.V. Berry, "Quantal phase factors accompanying adiabatic changes", Proc. R. Soc. Lond. **A 392**, 45 (1984).
[116] R. Kubo, "Statistical-Mechanical theory of irreversible processes. I. General theory and simple applications to magnetic and conduction problems", J. Phys. Soc. Japan **12**, 570 (1957); M.S. Green, "Markoff random processes and the statistical mechanics of time-dependent phenomena. II. Irreversible processes in fluids", J. Chem. Phys. **20**, 1281 (1952); "Comment on a paper of mori on time-correlation expressions for transport properties", Phys. Rev. **119**, 829 (1960).
[117] D. Förster, *Hydrodynamic Fluctuations, Broken Symmetry and Correlation Functions*, (Westview Press, 1995).
[118] H.B. Callen and T.A. Welton, "Irreversibility and generalized noise", Phys. Rev. **83**, 34 (1951).

[119] F. Reif, *Fundamentals of Statistical and Thermal Physics*, (McGraw-Hill, 1965).
[120] L.D. Landau and E.M. Lifshitz, *Statistical Mechanics, Part 1*, (Pergamon, 1970).
[121] L.D. Landau, "Theory of Fermi-liquids", Sov. Phys. JETP **3**, 920 (1957); "Oscillations in a Fermi-liquid", *ibid*, **5**, 101 (1957); "On the theory of the Fermi-liquid", *ibid*, **8**, 70 (1959).
[122] N.W. Ashcroft and N.D. Mermin, *Solid State Physics*, (Holt, Rinehart and Winston, 1976).
[123] J.L. Birman *Theory of Crystal Space Groups and Lattice Dynamics: Infra-red and Raman optical processes of insulating crystals*, (Springer Verlag, 1984).
[124] F.A. Cotton, *Chemical Applications of Group Theory*, 3rd Ed., (Wiley-Interscience, 1990).
[125] C. Kittel, *Introduction to Solid State Physics*, 6th Ed., (John Wiley, 1986).
[126] D. Shechtman, I. Blech, D. Gratias, and J.W. Cahn, "Metallic phase with long-range orientational order and no translational symmetry", Phys. Rev. Lett. **53**, 1951 (1984).
[127] J.I. Pankove, *Optical Processes in Semiconductors*, (Dover, 1971).
[128] R. de L. Kronig and W.G. Penney, "Quantum mechanics of electrons in crystal lattices", Proc. Roy. Soc. (London) A**130**, 499 (1931).
[129] Y. Aharonov and D. Bohm, "Significance of electromagnetic potential in quantum theory", Phys. Rev. **115**, 485 (1959); M. Peshkin and A. Tonomura, *The Aharonov Bohm Effect*, (Springer, 1989).
[130] R.G. Chambers, "Shift of an electron interference pattern by enclosed magnetic flux", Phys. Rev. Lett. **5**, 3 (1960).
[131] Y. Imry, *Introduction to Mesoscopic Physics*, (Oxford University Press, 1997).
[132] M. Büttiker, Y. Imry, and R. Landauer, "Josephson behavior in small normal one-dimensional rings", Phys. Lett. **96A**, 365 (1983).
[133] S. Datta, *Electronic Transport in Mesoscopic Systems*, (Cambridge University Press, 1997).
[134] T.T. Wu and C.N. Yang, "Concept of nonintegrable phase factors and global formulation of gauge fields", Phys. Rev. D **12**, 3845 (1975).
[135] N. Byers and C.N. Yang, "Theoretical considerations concerning quantized magnetic flux in superconducting cylinders", Phys. Rev. Lett. **7**, 46 (1961).
[136] B.L. Altshuler *et al.*, "The Aharonov-Bohm effect observation in metallic cylinders", JETP Lett. **35**, 588 (1982); B.L. Altshuler, Y. Gefen, and Y. Imry, "Persistent differences between canonical and grand canonical averages in mesoscopic ensembles: Large paramagnetic orbital susceptibility", Phys. Rev. Lett. **66**, 88 (1991).
[137] L.P. Lévy, G. Dolan, J. Dunsmuir, and H. Bouchiat, "Magnetization of mesoscopic copper rings: Evidence for persistent currents", Phys. Rev. Lett. **64**, 2074 (1990).
[138] Y. Aharonov and A. Casher, "Topological quantum effects for neutral particles", Phys. Rev. Lett. **53**, 319 (1984).
[139] L. Onsager, "Reciprocal relations in irreversible processes. I.", Phys. Rev. **37**, 405 (1931); L. Onsager, "Reciprocal relations in irreversible processes. II.", Phys. Rev. **38**, 2265 (1931).
[140] L. Onsager, "Interpretation of the de Haas-van Alphen effect", Phil. Mag. **43**, 1006 (1951).
[141] D. Yoshioka, B.I. Halperin, and P.A. Lee, "Ground state of two-dimensional electrons in strong magnetic fields and $\frac{1}{3}$ quantized hall effect", Phys. Rev. Lett. **50**, 1219 (1983).
[142] W.J. de Haas and P.M. van Alphen, "The dependence of the susceptibility of diamagnetic metals Amsterdam Ac. **33**, 1106 (1930).
[143] L.D. Landau, "Diamagnetismus der Metalle (Diamagnetism in metals)", Z. Physik, **64**, 629 (1930).
[144] P.M.C. Rourke and S.R. Julian, "*Numerical extraction of de Haas - van Alphen frequencies from calculated band energies*", arXiv:0803.1895.
[145] W. Kohn, "Cyclotron resonance and de Haas-van Alphen effect of an interacting electron gas", Phys. Rev. **123**, 1242 (1961).
[146] L.Shubnikov, W.J. de Haas, "magnetische wiederstandsvergrösserung in einkristallen von wismut bei tiefen tempretaure", Leiden Commun. **N 207**, 3 (1930); L. Shubnikov and W.J. de Haas, "A new phenomenon in the change of resistance in a magnetic field of single crystals of bismuth", Nature **126**, 500 (1930).
[147] K. von Klitzing, G. Dorda, and M. Pepper, "A new method of high accuracy determination of the fine structure constant based on quantized Hall resistance", Phys. Rev. **B 25**, 2185 (1982).

[148] R.B. Laughlin, "Quantized Hall conductivity in two dimensions", Phys. Rev. **B 23**, 5632 (1981).
[149] D.C. Tsui, H.L. Störmer, and A.C. Gossard, "Two-dimensional magnetotransport in the extreme quantum limit", Phys. Rev. Lett. **48**, 1559 (1982).
[150] R.B. Laughlin, "Anomalous quantum hall effect: An incompressible quantum fluid with fractionally charged excitations", Phys. Rev. Lett. **50**, 1395 (1983).
[151] R.E. Peierls, *Quantum Theory of Solids*, (Oxford University Press, 1955), pp. 144–149.
[152] H.-A. Engel, E.I. Rashba, and B.I. Halperin, "Theory of spin hall effects in semiconductors", In *Handbook of Magnetism and Advanced Magnetic Materials*, H. Kronmuller and S. Parkin (Eds.), pp 2858–2877 (John Wiley, 2007).
[153] S. Glutsch, *Excitons in Low-Dimensional Semiconductors, Theory, Numerical Methods, Applications*, (Springer, Solid-State Scieces, 2003).
[154] R. Winkler, *Spin-Orbit Coupling Effects in Two-Dimensional Electron and Hole Systems*, Volume 191, (Springer Tracts in Modern Physics, 2003).
[155] N. Nagaosa *et al.* " Anomalous Hall effect", Rev. Mod. Phys. **82**, 1539 (2010).
[156] I. Zutic, J. Fabia, and S. Das Sarma, "Spintronics: Fundamentals and applications", Rev. Mod. Phys. **76**, 323 (2004).
[157] J. Fabian *et al.*, "Semiconductor spintronics", Acta Phys. Slov. **57**, 565 (2007).
[158] S. Datta, and B. Das, "Electronic analog of the electro-optic modulator", Appl. Phys. Lett. **56**, 665 (1990).
[159] D. Pines, *Elementary Excitations in Solids*, (Benjamin, 1968).
[160] P. Nozières and D. Pines, *The Theory of Quantum Liquids*, (Perseus Books, 1999).
[161] Giamarchi, *Quantum Physics in One Dimension*, (Oxford University Press, 2003).
[162] S. Sachdeev, *Quantum Phase Transitions*, (Cambridge University Press, 2001).
[163] D.J. Thouless, M. Kohmoto, M.P. Nightingale, and M. den Nijs, "Quantized hall coductance in a Two-Dimensional periodic potential", Phys. Rev. Lett. **49**, 405 (1982).
[164] F. Gebhard, *The Mott Metal-Insulator Transition: Models and Methods*, (Springer, 1997).
[165] I.N. Levine, *Quantum Chemistry*, 6th Ed., (Pearson Prentice Hall, Upper Saddle River, 2009).
[166] A. Szabo and N.S. Ostlund, *Modern Quantum Chemistry: Introduction to Advanced Electronic Structure Theory*, (Dover, 1996).
[167] L. Piela, *Ideas of Quantum Chemistry*, (Elsevier, 2007).
[168] J.C. Slater, *Quantum Theory of Matter*, 2nd ed., (McGraw-Hill, 1968).
[169] R. Weinstock, *Calculus of Variations*, (Dover, 1974).
[170] E.U. Condon and G.H. Shortley, *The Theory of Atomic Spectra*, (Cambridge University Press, 1953).
[171] J. Killingbeck, "Microcomputer calculations in physics", Rep. Prog. Phys. **48**, 54 (1985); G. Hadinger, M. Aubert-Frécon, and G. Hadinger, "The Killingbeck method for the one-electron two-centre problem", J. Phys. **B 22**, 697 (1989).
[172] I. Tuvi and Y.B. Band, "Modified Born-Oppenheimer basis for nonadiabatic coupling: Application to the vibronic spectrum of HDD^+", J. Chem. Phys. **111**, 5808 (1999); I. Tuvi and Y.B. Band, "Nonadiabatic coupling using a corrected Born-Oppenheimer basis: The vibronic spectrum of HD^+", Phys. Rev. **A 59**, 2680 (1999).
[173] J.M. Hollas, *Modern Spectroscopy*, (Wiley, 2004).
[174] I.N. Levine, *Molecular Spectroscopy*, (Wiley, 1975).
[175] G. Herzberg, *Molecular Spectra and Molecular Structure, Vol. I. Spectra of Diatomic Molecules*, (Van Nostrand, 1950); *Molecular Spectra and Molecular Structure, Vol. II. Infrared and Raman Spectra of Polyatomic Molecules*, (Van Nostrand, 1960); *Molecular Spectra and Molecular Structure, Vol. III. Electronic Spectra and Electronic Structure of Polyatomic Molecules*, (Van Nostrand, 1966).
[176] T.E. Sharp, "Potential-energy curves for molecular hydrogen and its ions", Atomic Data **2**, 119 (1971).
[177] L Veseth, "Hund's coupling case (c) in diatomic molecules. I. Theory", J. Phys. **B 6**, 1473 (1973); "Hund's coupling case (c) in diatomic molecules. II. Examples", J. Phys. **B 6**, 1484 (1973).
[178] R.S. Berry, S.A. Rice, and J. Ross, *Physical Chemistry*, (Wiley, 1980).
[179] M.L. Goldberger and K.M. Watson, *Collision Theory*, (Dover, 2004).

[180] J.R. Taylor, *Scattering Theory*, (J. Wiley, 1972).
[181] L.S. Rodberg and R.M. Thaler, *Introduction to the Quantum Theory of Scattering*, (Academic Press, 1967).
[182] R.G. Newton, *Scattering Theory of Waves and Particles*, (Dover, 2002).
[183] H. Feshbach, *Theoretical Nuclear Physics*, (Wiley, (1992).
[184] E. Timmermans, P. Tommasini, M. Hussein, and A. Kerman, "Feshbach resonances in atomic bose einstein condensates", Phys. Rep. **315**, 199 (1999).
[185] T. Wasak, P. Szańkowski, M. Trippenbach, Y. Avishai, and Y.B. Band, "Surprising Results of Solvable Feshbach Resonance Models" (to be published).
[186] U. Fano, "Effects of configuration interaction on intensities and phase shifts", Phys. Rev. **124**, 1866 (1961). The first attempt at a theory was conducted as part of Fano's Ph.D. thesis with Enrico Fermi. See U. Fano, "On the absorption spectrum of noble gases at the arc spectrum limit", Nuovo Cimento **12**, 154 (1935).
[187] M. Brillouin, "La mécanique ondulatoire de Schrödinger: une méthode générale de resolution par approximations successive (Schrödingier wave mechanics: A general solution using the method of successive approximations)", J. de Physique, **3**, 65 (1922); H. Jeffreys, "On certain approximate solutions of linear differential equations of the second order", Proc. Lond. Math. Soc. **23**, 428 (1924); G. Wentzel, "Eine Verallgemeinerung der Quantenbedingungen für die Zwecke der Wellenmechanik (Generalized quantum conditions for the purposes of analyzing wave mechanics)", Z. Physik, **38**, 518 (1926); H.A. Kramers, "Wellenmechanik und halbzählige Quantisierung (Wave mechanics and half-integer quantization)", Z. Physik, **39**, 828 (1926).
[188] K.M. Watson, "Multiple scattering and the many-body problem: Applications to photomeson production in complex nuclei", Phys. Rev. **89**, 575 (1953).
[189] R.J. Glauber, *Lectures in Theoretical Physics*, "Theory of high-energy scattering", W.E. Brittin and L.G. Dunham, (Eds.) (Interscience Publishers, 1959); Volume I, p. 315; R.J. Glauber and V. Schomaker, "The theory of electron diffraction", Phys. Rev. **89**, 667 (1953).
[190] L. Marchildon, *Quantum Mechanics: From basic Principles to Numerical Methods and Applications*, (Springer, 2002).
[191] R. Landauer, "Electrical resistance of disordered one-dimensional lattices", Phil. Mag. **21**, 863 (1970).
[192] P.W. Anderson, "Absence of diffusion in certain random lattices", Phys. Rev. **109**, 1492 (1958).
[193] D.J. Thouless, "Electrons in dirordered systems and the theory of localization", Rep. Prog. Phys. **13**, 93 (1974).
[194] P.A. Lee and T.V. Ramakrishnan, "Disordered electronic systems", Rep. Mod. Phys. **57**, 287 (1985).
[195] I.M. Lifshitz, S.A. Gredeskul, and L.A. Pastur, *Introduction to the Theory of Disordered Systems*, (Wiley, 1988).
[196] J-M. Luck, *Systeme Deśordone's Unidimensionnels* (One dimensional disordered systems), (Aléa Saclay, 1992).
[197] H. Furstenberg and H. Kesten, "Products of random matrices", Ann. Math. Stat. **31**, 457 (1960); H. Furstenberg, "Noncommuting random products", Trans. Am. Math. Soc. **108**, 377 (1963); V.I. Oseledec, "A multiplicative ergodic theorem", Trans. Moscow. Math. Soc. **19**, 197 (1968).
[198] B.I. Halperin, "Quantized Hall conductance, current-carrying edge states, and the existence of extended states in a two-dimensional disordered potential", Phys. Rev. **B 25**, 2185 (1982); Y. Avishai and G. Montambaux, "Semiclassical analysis of edge state energies in the integer quantum Hall effect", Eur. Phys. J. **B 66**, 41 (2008).
[199] A.H. MacDonald and P. Streda, "Quantized Hall effect and edge currents", Phys. Rev. **B 29**, 1616 (1984).
[200] M. Büttiker, "Absence of backscattering in the quantum Hall effect in multiprobe conductors", Phys. Rev. **B 38**, 9375 (1988).
[201] Y. Avishai and Y.B. Band, "One dimensional density of states and the phase of the transmission amplitude", Phys. Rev. **B32**, R2674 (1985).
[202] T. Ihn, *Electronic Quantum Transport in Mesoscopic Semiconductor Structures*, (Springer-Verlag, 2004).
[203] E. Akkermans and G. Montambaux, *Mesoscopic Physics of Electrons and Photons*, (Cambridge University Press, 2007).
[204] B.E.A. Saleh and M.C. Teich, *Fundamentals of Photonics*, (J. Wiley, 1991).
[205] M. Büttiker, "Symmetry of electrical conduction", IBM J. Res. Develop. **32**, 317 (1988).
[206] L.P. Kouwenhoven, D.G. Austig, and S. Tarucha, "Few-electron quantum dots", Rep. Prog. Phys. **64**, 701 (2001).

[207] M.L. Mehta, *Random Matrices*, (Academic, 1991); F.J. Dyson, "Statistical theory of the energy levels of complex systems, I–IV", J. Math. Phys. **3**, 140 (1962); *ibid* **3**, 157 (1962). *ibid* **3**, 1191 (1962); *ibid* **3**, 1199 (1962).

[208] B.L. Altshuler and B.I. Shklovskii, "Repulsion of energy levels and the conductance of small metallic samples", Sov. Phys. JETP **64**, 127 (1986).

[209] C.W.J. Beenakker, "Random-matrix theory of quantum transport", Rev. Mod. Phys. **69**, 73 (1997).

[210] H. Alt *et al.*, "Superconducting billiard cavities with chaotic dynamics: An experimental test of statistical measures", Phys. Rev. **E 50**, R1 (1994).

[211] I.L. Aleiner, P.W. Brouwer, and L.I. Glazman "Quantum effects in coulomb blockade", Phys. Reports **358**, 309 (2002).

[212] P.W. Anderson, *Condensed Matter Physics, the Continuous Revolution*, Physics World **12**, 37 (1995); P. Coleman, *Introduction to Many Body Physics* available at
http://www.physics.rutgers.edu/users/coleman/620/mbody/pdf/bk.pdf.

[213] H.W. Kroto, J.R. Heath, S.C. O'Brien, R.F. Curl, and R.E. Smalley, "C_{60}: Buckminsterfullerene", Nature **318**, 162 (1985).

[214] A.H. Castro Neto, *et al.*, "The electronic properties of graphene", Rev. Mod. Phys. **81**, 109 (2008).

[215] A.K. Geim and K.S. Novoselov, "The rise of graphene", Nature Mater. **6**, 183 (2007).

[216] S. Iijima, Nature **354**, 56 (1991).

[217] P.R. Wallace, "The band theory of graphite", Phys. Rev. **71**, 622 (1947); J.W. McClure, "Diamagnetism of graphite", Phys. Rev. **104**, 666 (1956).

[218] K.S. Novoselov, *et. al.*, "Electric field effect in atomically thin carbon films", Science **306**, 666 (2004).

[219] A.C. Ferrari, *et al.*, "Raman spectrum of graphene and graphene layers", Phys. Rev. Lett. **97**, 187401 (2006); A. Gupta *et al.*, "Raman scattering from high-frequency phonons in supported n-graphene layer films", Nano. Lett. **6**, 2667 (2006).

[220] K.I. Bolotin, *et al.*, "Ultra high electron mobility in suspended graphene", Solid State Communications **146**, 351 (2008).

[221] J.-C. Charlier, X. Blase and S. Roche, "Electronic and transport properties of nanotubes", Rev. Mod. Phys. **79**, 000677 (2007).

[222] P.A.M. Dirac, "The quantum theory of the electron", Proc. R. Soc. **A117**, 610 (1930).

[223] J. Fröhlich and U.M. Studer, "Gauge invariance and current algebra in nonrelativistic many-body theory", Rev. Mod. Phys. **65**, 733 (1993).

[224] D.P. DiVincenzo and E.J. Mele, "Self-consistent effective-mass theory for intra-layer screening in graphite intercalation compounds", Phys. Rev. **B 29**, 1685 (1984).

[225] N.H. Shon and T. Ando, "Quantum transport in two-dimensional graphite system", J. Phys. Soc. Japan **67**, 2421 (1998).

[226] H. Suzuura and T. Ando, "Crossover from symplectic to orthogonal class in a two-dimensional honeycomb lattice", Phys. Rev. Lett. **89**, 266603 (2002).

[227] D.S. Novikov, "Scattering theory and transport in graphene", Phys. Rev. **B 76**, 245435 (2007).

[228] B.L. Altshuler, P.A. Lee, and R.A. Webb (eds.), *Mesoscopic Phenomena in Solids*, (North-Holland, 1991).

[229] R.P. Feynman and F.L. Vernon, "Theory of quantum system interacting with a linear dissipative system", Ann. Phys. NY **24**, 118 (1963).

[230] A.O. Caldeira and A.J. Leggett, "Influence of dissipation on quantum tunneling in macroscopic systems", Phys. Rev. Lett. 46, 211 (1981); A.O. Caldeira and A.J. Leggert, "Path integral approach to quantum brownian motion", Physica **121A**, 587 (1983).

[231] A.O. Caldeira and A.J. Leggett, "Influence of damping on quantum interference: An exactly soluble model", Phys. Rev. **A 31**, 1059 (1985).

[232] Ya.M. Blanter and M. Büttiker, "Shot noise in mesoscopic conductors", Phys. Rep. **336**,1 (2000).

[233] E. Wigner, "On the interaction of electrons in metals", Phys. Rev. **46**, 11 (1934).

[234] G. Moore and N. Read, "Nonabelions in the fractional Quantum Hall effect", Nucl. Phys. **B 360**, 362 (1991).

[235] D. Arovas, J.R. Schrieffer, and F. Wilcheck, "Fractional statistics and the Quantum Hall effect", Phys. Rev. Lett. **53**, 722 (1984).
[236] J.K. Jain, *Composite Fermions*, (Cambridge University Press, 2007).
[237] J.G. Bendorz and K.A. Müller, "Possible high T_c superconductivity in the Ba-La-Cu-O system", Z. Phys. **B 64**, 189 (1986).
[238] P.A. Lee, N. Nagaosa, and X-G Wen, "Doping a Mott insulator: Physics of high-temperature superconductivity", Rev. Mod. Phys. **78**, 17 (2006).
[239] H. Takahashi, K. Igawa, K. Arii, Y. Kamihara, M. Hirano, and H. Hosono, "Superconductivity at 43 K in an iron-based layered compound $LaO_{1-x}F_xFeAs$", Nature **453**, 376 (2008).
[240] M.R. Norman, "Trend: High-temperature superconductivity in the iron pnictides", Physics **1**, 21 (2008).
[241] N.D. Mermin and H. Wagner, "Absence of ferromagnetism or antiferromagnetism in one- or two-dimensional isotropic heisenberg models", Phys. Rev. Lett. **17**, 1133 (1966).
[242] A.O. Gogolin, A.A. Nersesyan, and A.M. Tsvelik, *Bosonization and Strongly Correlated Systems*, (Cambridge University Press, 1998).
[243] X.G. Wen, "Chiral luttinger liquid and the edge excitations in the fractional quantum hall states", Phys. Rev. **B 41**, 12838 (1990).
[244] R.J. Baxter, *Exactly Solved Models in Statistical Mechanics*, (Academic Press, 1982).
[245] A. Auerbach, *Interacting Electrons and Quantum Magnetism*, (Springer, 1994).
[246] Ian Affleck, "Quantum spin chains and the Haldane gap", J. Phys.: Condens. Matter. **1**, 3047 (1989).
[247] M. König *et al.*, "The quantum spin Hall effect: theory and experiment", J. Phys. Soc. Jpn **77**, 310 (2008) [arXiv:0801.0901].
[248] F.D.M. Haldane, "Model for a quantum Hall effect without Landau levels: Condensed-matter realization of the 'parity anomaly'", Phys. Rev. Lett. **61**, 2015 (1988).
[249] C.L. Kane and E.J. Mele, "Quantum spin hall effect in graphene", Phys. Rev. Lett. **95**, 226801 (2005).
[250] A.L. Fetter and J.D. Walecka, *Quantum Theory of Many Particle Systems*, (McGraw-Hill, 1971).
[251] G.D. Mahan, *Many Particle Physics*, (Plenum Press, 1993).
[252] H. Bruus and K. Flensberg, *Many-Body Quantum Theory in Condensed Matter Physics: An Introduction*, (Oxford University Press, 2004).
[253] A.M. Zagoskin, *Quantum Theory of Many-Body Systems: Techniques and Applications*, (Springer, 1998).
[254] A.A. Abrikosov, L.P. Gorkov, and I.M. Dzyaloshinsky, *Methods of Quantum Field Theory in Statistical Physics*, (Dover, 1975).
[255] E. Fradkin, *Field Theory of Condensed Matter Systems*, (Addison-Wesley, 1991).
[256] Alexei M. Tsvelik, *Quantum Field Theory in Condensed Matter Physics*, (Cambridge University Press, 1996).
[257] N. Nagaosa, *Quantum Field Theory in Condensed Matter Physics*, (Springer, 2010).
[258] T-K. Ng, *Introduction to Classical and Quantum Field Theory*, (Willey, 2009).
[259] P. Hohenberg and W. Kohn, "Inhomogeneous electron gas", Phys. Rev. **B 136**, 864 (1964).
[260] W. Kohn and L.J. Sham, "Self-Consistent equations including exchange and correlation effects", Phys. Rev. **140**, A1133 (1965).
[261] W. Kohn, "Nobel Lecture: Electronic structure of matterwave functions and density functionals", Rev. Mod. Phys. **71**, 1253 (1999).
[262] W. Kohn and P. Vashishta, "General density functional theory", In *Theory of the Inhomogeneous Electron Gas*, S. Lundqvist and N.H. March, (Ed.), (Plenum Publishing Corp., 1983), p. 79–147, Chapter 2.
[263] K. Capelle, "A bird's-eye view of density-functional theory", Braz. J. Phys. **36**, 1318 (2006) [cond-mat/0211443].
[264] K. Burke, *The ABC of DFT*, http://dft.rutgers.edu/kieron/beta/b4.pdf.
[265] R.M. Dreizler and E.K.U. Gross, *Density Functional Theory*, (Springer, 1990).
[266] R.G. Parr and W. Yang, *Density-Functional Theory of Atoms and Molecules*, (Oxford University Press, 1989).
[267] D.S. Sholl and A. Steckel, *Density Functional Theory, A Practical Introduction*, (Wiley, 2009).
[268] N.D. Mermin, "Thermal properties of the inhomogeneous electron gas" Phys. Rev. **137**, A1441 (1965).

[269] M.A.L. Marques, C.A. Ullrich, F. Nogueira, A. Rubio, K. Burke, and E.K.U. Gross, *Time-Dependent Density Functional Theory*, (Springer, 2006).
[270] N. Argaman and G. Makov, "Density functional theory: An introduction", Am. J. Phys. **68**, 69 (2000); N. Argaman and G. Makov, "Thermodynamics as an alternative foundation for zero-temperature density-functional theory and spin-density-functional theory", Phys. Rev. **B 66**, 052413 (2002).
[271] N. Argaman and Y.B. Band, "Finite-temperature density-functional theory of bose-einstein condensates", Phys. Rev. **A 83**, 023612 (2011).
[272] M. Levy, "Electron densities in search of hamiltonians", Phys. Rev. **A 26**, 1200 (1982); E.H. Lieb, "Density Functionals for Coulomb Systems", in *Density Functional Methods in Physics*, R.M. Dreizler and J. da Providencia, (Eds.), (Plenum, 1985).
[273] J.P. Perdew, R.G. Parr, M. Levy, and J.L. Balduz, "Density functional theory for fractional particle number: Derivative discontinuities of the energy", Phys. Rev. Lett. **49**, 1691 (1982).
[274] D.M. Ceperley and B.J. Alder, "Ground state of the electron gas by a stochastic method", Phys. Rev. Lett. **45**, 566 (1980).
[275] J.P. Perdew and Y. Wang, "Accurate and simple analytic representation of the electron-gas correlation energy", Phys. Rev. **B 45**, 13244 (1992).
[276] T. Stein, H. Eisenberg, L. Kronik, and R. Baer, "Fundamental gaps of finite systems from the eigenvalues of a generalized Kohn-Sham method", Phys. Rev. Lett. **105**, 266802 (2010); H.R. Eisenberg and R. Baer, "A new generalized Kohn-Sham method for fundamental band-gaps in solids", Phys. Chem. Chem. Phys. **11**, 4674 (2009).
[277] E. Runge and E.K.U. Gross, "Density-Functional theory for Time-Dependent systems", Phys. Rev. Lett. **52**, 997 (1984).
[278] Y.B. Band, S. Kallush, and R. Baer, "Rotational aspects of short-pulse population transfer in diatomic molecules", Chem. Phys. Lett. **392**, 23 (2004).
[279] K. Jänich, *Linear Algebra*, (Springer, 1994).
[280] S. Lang, *Introduction to Linear Algebra*, (Springer, 2005).
[281] P. O. Löwdin, *Linear Algebra for Quantum Theory*, (Wiley-Interscience, 2004).
[282] P.R. Halmos, *A Hilbert Space Problem Book*, 2nd Ed., (Springer, 1982); see also P.R. Halmos, *Introduction to Hilbert Space*, (Chelsea Pub. Co., 1987).
[283] E.L. Ince, *Ordinary Differential Equations*, (Dover, 1956).
[284] M. Tenenbaum, H. Pollard, *Ordinary Differential Equations*, (Dover, 1985).
[285] V.I. Arnold, *Ordinary Differential Equations*, (MIT Press, 1978).
[286] Daniel Zwillinger, *Handbook of Differential Equations*, (Academic Press, 1998).
[287] M.R. Spiegel, *Theory and Problems of Vector Analysis and an Introduction to Tensor Analysis*, (Schaums's Outline Series, McGraw Hill, 1959).
[288] J.L. Synge and A. Schild, *Tensor Calculus*, (Dover, 1978).
[289] L. Debnath, *Integral Transforms and their Applications*, (CRC Press, 2000).
[290] M. Tinkham, *Group Theory and Quantum Mechanics*, (Dover, 2003).
[291] H. Weyl, *The Theory of Groups and Quantum Mechanics*, (Dover, 1950).

INDEX

Page numbers followed by "f" indicate figures, "t" indicate tables, and "n" indicate footnotes.

A

A. *See* Ampère
a_0. *See* Bohr
Abrikosov-Suhl resonance, 785–787, 787f
Absorption, 273–275. *See also* Photoabsorption
    coefficients, 274–275, 514f
    detuning *versus*, 275f
    exciton, 514–515
    infrared, 599f
    phonon, 856
    photon, 277
    spectra, 175, 513–514
    two-photon, 293, 319
    wavelengths, of semiconductors, 513t
AC. *See* Alternating current
Addition theorem for spherical harmonics, 120
Adiabatic limit, 273–273
Adiabatic passage, 336–337. *See also* Stimulated Raman adiabatic passage
Adiabatic theorem, 335
Adleman, Leonard, 203
AFM. *See* Atomic force microscopy
Aharonov, Yakir, 436
Aharonov-Bohm effects, 169, 435–443, 437f, 441f, 451, 760f
Aharonov-Casher effects, 435, 441–442, 441f, 504, 781
Airy function form, 307f
Airy functions, 54, 55f, 906
Alignment, 147
Alkali-metals
    absorption spectra of, 175
    Zeeman splitting of, 182f
Allan standard deviation, 287
Allotropes of carbon, 20, 785
Alpha particle decay, 14f
Alternating current (AC), 395–396
Ammann, Robert, 412
Ammann-Beenker tiling, 412, 413f
Ampère (A), 124
    equations, 396

Ampère's law, 761
Analytic functions, 453
Anderson impurity model, 784, 784f
Anderson localization, 441, 731, 737–741
Anderson metal-insulator transition, 464
Angular momentum, 32–36, 110–115
    addition, 139–143, 166
    coupling, 156
    decomposition of phase waves, 119–120
    eigenstate of, 373
    eigenvectors, 135
    electronic, 595
    internal, of photons, 14
    lowering operators, 107–110
    matrices, 109f
    operators, 105–106
    orbital, 106
    quantization, 12–13
    quenching, 473
    raising operators, 107–110
    rotations and, 132–139
    selection rules, 152
    in spherical coordinates, 111–112
    spin, 106, 159–160, 172
    standard deviation of, 283
Anharmonicity constant, 122
Annihilation operators, 832–837
Anomalous Hall effect (AHE), 512
Anticommutation relations, 793
Antiferromagnetism, 468
Antiferromagnets, 475, 475f, 478t
Anti-Hermitian operators, 66–67
Antilinear operators, 101
Antilocalization, weak, 781, 812
Antiparallel configuration, 480
Antiresonance, 270
Anti-Stokes transition, 327
Antisymmetric spin function, 665
Antisymmetrization operators, 369, 553
Anti-Zeno effect, 331
Apparatus-environment interaction, 79
Approximations. *See also specific approximation methods*
    adiabatic, 335–349

Born, 632, 683–684, 704
    effective range, 651
    eikonal, 690–692
    Hartree, 551–552
    Hartree-Fock, 572–574
    Heitler-London, 476
    mean-field, 414, 651
    methods, 682–692
    reflection, 603, 603f
    semiclassical, 304–309
    single-particle, 381
    sudden, 333–335
    tight-binding, 739f
Arithmetic coding, 197
Ashkin, Arthur, 278
Asymptotic states, 695–696
Atalla, John, 14
Atom spectroscopy, 152
Atomic clocks, 285n10
Atomic energies, splitting of, 415f
Atomic force microscopy (AFM), 18, 19f
Atomic magic numbers, 767
Atomic radii, 560
Atomic units (a.u.), 124–127, 127t
Atoms
    artificial, 747
    band structure of, 415f
    closed shell, 470
    Doppler cooling of, 275–278
    in electromagnetic field, 273–275
    electron affinity of, 559
    hydrogen, 124–132
    ionizatin energy of, 559
    incommensurate chain of, 413–414
    magnetic impurity, 783
    neutral, in optical lattices, 245–246
    optical trapping of, 278–279
    with partially filled shell, 470–473, 471f
    radii of, 560
    spin-orbit interaction in, 175–178
    term symbols, 175n5, 547
    as two-level system, 259

B

Babbage, Charles, 194
Backscattering, enhanced, 781
Baker-Hausdorff theorem, 94, 848
Balanced function, 227
Ball-and-stick model, 580f
Ballistic transport, 755
Balmer formula, 128n2
Balmer transitions, 131, 132f
Band. *See also* Energy bands
    bending, 495
    conduction, 50, 416, 416f
    formation, 416–417
    gap, 415–416, 417f, 482t
    index, 422
    insulators, 416, 543
    potential, 416–417
    valence, 50, 416, 416f
Band, Yehuda, 328
Band structure, 414–415
    of atoms, 415f
    electron transport and, 415–416
    semiconductor, 482–483, 483f
    silicon, 483f
Bandwidth, 415
Barrier potential, wave function for, 47f
Basis, 400
Basis states, 206
    adiabatic, 344
    expansions, 64–65, 303–304
    in Fock space, 830–832
Basis vectors, 61–62
BEC. *See* Bose-Einstein condensate
Bell, John S., 27, 251
Bell states, 207–213, 217
Bellman, Richard, 300
Bell's inequalities, 251–258, 254f
    aspects of, 256–258
    CHSH inequalities and, 255–256
Bendorz, George, 819
Bennett, Charles, 219, 223
Bennett-Brassard 1984 (BB84), 222, 222f
Benzene, 571f
Bernoulli's numbers, 390
Berry, Michael V., 344, 780
Berry phase, 344–349, 802–803
Bessel functions, 117, 119, 637–638, 907–908, 907f
Beta decay, 159
Bethe ansatz equations, 538
Biexcitons, 515
Binnig, Gerd, 15, 18
Bio-nanotechnology, 21
Bipartite systems, 296
Bipolar junction transistors (BJTs), 19
Bir-Aronov-Perel mechanism, 523, 524f
Blackbody radiation, 6–8, 7f
Bloch, Felix, 247
Bloch equations, 167, 188–189, 265
    with decay terms, 268
    optical, 271
Bloch function, 417–421
Bloch oscillations, 431–432
Bloch sphere, 262–266
    coherent states and, 284f
    N-two-level system, 282–284
    representation of, 264f
    representation of Ramsey fringe spectroscopy, 287f
Bloch theorem, 418–419
Bloch vectors, 263, 263f
Bloch walls, 480
Bloch wave functions, 417–421
BOA. *See* Born-Oppenheimer approximation
Bohm, David, 436
Bohr, Niels, 5, 251
Bohr (a_0), 125–126
Bohr magneton, 166–167, 469
Bohr quantization condition, 307
Bohr velocity, 125, 126
Bohr-Sommerfeld quantizing condition, 456
Boltzmann, Ludwig, 385n1
Boltzmann distribution, 858
Boltzmann equation, 84, 309
Born, Max, 589
Born-Oppenheimer approximation (BOA), 589–590
Born-Oppenheimer (BO) potentials, 591–592
Born-von Karman periodic conditions, 419–420, 429
Bose, S., 99
Bose-Einstein condensate (BEC), 246
Bose-Einstein distribution, 380, 388, 860
Bose-Einstein statistics, 99
Bosonic states
    symmetric, 373
    wave functions for, 379
Bosons, 98, 372
    annihilation operators, 834–835
    basis states in Fock space for, 830
    creation operators, 833f, 834–835
    identical, with antisymmetric spin function, 665
    identical spinless, 665, 666f
    occupation number, 860
    second quantized Hamiltonians and, 842–843
    spatial degrees of freedom of, 378
    wave functions, 99
    zero-temperature occupation of, 99f
Bound states, 4–5
    energies, 48f
    Green's function and, 627
    parity symmetry and, 48
    in potential well, 47–49
    resonance scattering and, 666–667
    in 2D systems, 730–731
    WKB approximation and, 304
Boundary conditions, 447
    antiperiodic, 751
    asymptotic, 610, 617, 622, 624, 635, 660, 680, 698, 710, 725
    free, 849–850
    hard-wall, 465f
    periodic, 849
    retardation, 719
    sensitivity to, 751
    softening, 451
    twisted, 751
Bragg, William H., 11, 409
Bragg, William L., 11, 409
Bragg condition, 409–411
Bragg peak, 411
Bragg scattering, 11–12, 409, 417
Bra-ket notation, 22, 61–63, 506, 892–893
Branch index, 853
Bras, 62, 146
Bravais lattices, 401–405, 401f, 854
    FCC (face-centered-cubic), 406f
    with motif, 411
    points, 401
    reciprocal lattice of, 406
    two-dimensional, 402, 403f
    unit cells of, 405f
    vectors, 401
Breakup reactions, 693, 715
Breakup threshold, 716
Breit-Wigner formula, 668–669
Brillouin, Leon, 304
Brillouin function, 472, 472f
Brillouin theorem, 577
Brillouin transitions, 326–329
Brillouin zone, 407–408, 407f
    first, 407
    lattice structure and, 790f
    magnetic zone, 543
Brownian motion, 362, 366
Buckminsterfullerine (Buckyballs), 20
Bulk metals, 783
Bulk pictures, 464–465
Bulk systems, IQHE in, 464–465
Byers-Yang theorem, 759

C

Callen, Herbert, 362
Campbell-Baker-Hausdorff theorem, 94

Atypical sequences, 199
a.u. *See* Atomic units
Aufbau principle, 99, 383
Avoided level-crossing, 424

Index

Canonical ensemble, 448, 858
Canonical transformation, 840
Carbon, 788f
 allotropes, 20
 chemical shifts, 191f
Carbon nanotubes (CNts), 788–790, 788f, 789f
Carrier densities, 444
 extrinsic semiconductor, 489–490
 impurities and, 488
 across p-n junctions, 494f
Casimir forces, 18
Cauchy principle, 351
Cauchy's residue theorem, 655
Cauchy-Schwarz inequality, 68, 894
Causality, 350, 655
Cayley's theorem, 370
Center of mass (CM), 619, 620f
Central limit theorem, 738n13
Central potentials, 635–666
Centrifugal force, 103, 117
CG coefficients. See Clebsch-Gordan coefficients
Chadwick, James, 159
Channels
 closed, 681, 695, 742
 internal degrees of freedom and, 693–695
 open, 681
 S matrix in 1D and, 734–735
Character tables, 581–582, 939–942, 945t
Charge carriers, 393f
 density, 484–485
 majority, 494
 minority, 494
Charge density waves, 527
Charge distribution
 Coulomb energy of, 151
 of surface plasmon, 533f
Charge transport, through quantum dots, 770–773
Charged particles
 Hamiltonian for, 169
 in magnetic field, 169–172
Charging energy, 764–765
Chemical accuracy, 880n3
Chemical potential, 381f9.1, 388, 482, 483, 484t9.4, 519f9.90
Chemical shifts, 189–190, 191f
Chern number, 544, 822–823
Chiral vector, 789
Chirality, 465, 801, 820
CHSH inequalities, 252, 255–256, 292–293
Chu, Steven, 278
Church-Turing principle, 194
CI. See Configuration interaction
Ciphers, 202

Ciphertext, 202
Cirac, Ignacio, 244
Circular ensemble (CE), 774–776
Circumferential vector, 787
Classical computation, 194–205
Classical electron radius, 126
Classical information, 194–205
Classical limit, 97–98
Classical mechanics
 Euler angles in, 133f
 failure of, 1
 Hamiltonians in, 29–30
Clebsch-Gordan (CG) coefficients, 140–143, 326, 373, 635
Clebsch-Gordan series, 143
Clifford algebra, 796
Closed channel potential, 669n6
Cluster operators, 578
CM. See Center of mass;
Codebreaking, 202
Coercive field H_C, 481
Coherences, 24, 71
 Bloch vectors as, 263
 first-order spatial, 76
 length, 751
Coherent states, 96–97, 96f
 Bloch sphere and, 284f
 ideal, 283
Cold atom systems, 651
Collisions
 direct, 703
 inverse, 703
 rearrangement, 715
Commutation relations, 159
Commutators, 39–40
Compatible operators, 66–67
Completeness, 61
Complex-conjugation operators, 101
Compression rate, 197
Compton wavelength, 126
Computational complexity, 204–205
Condon, Edward, 601
Condon energy, 602
Condon points, 601
Conductance. See also Magnetoconductance
 ballistic, 751
 differential, 771–772, 772f, 787
 dimensionless, 759f
 distribution, 782
 generalized, 355–356
 heat, 528
 Kondo, 785–787, 785f, 786f
 Kubo formula for, 360
 Landauer formula for, 755–764
 mismatch, 518
 of QPCs, 763f

quantum unit of, 463, 757
 semiconductor, 484
 spin Hall, 511–512
 universal, 780–782
 weak localized corrections to, 780–781
Conducting ring
 electron in, 348f
 near magnetic moment, 348–349
Conduction band, 50, 416, 416f
Conductivity
 AC, 395–396
 Drude, 392–396, 393f, 444
 electrical, 356–358, 394
 elementary theories of, 391–400
 of nonuniform systems, 359
 spin, 517
 tensor, 443, 542
 thermal, of metals, 397–399
 using generalized forces, 360
Configuration interaction (CI), 574–576
Configuration mixing (CM), 574–576
Configuration space representation, of Green's function, 642
Confluent hypergeometric function, 660
Conical intersections, 591
Conservation laws, 98–103
Constant energy surface, 384f
Constant function, 227
Constant-interaction model, 765–766
Constriction, 762
Contact points, 762
Continuity equation, 45
Continuous-variable systems, 295–296
Continuum theory, 800–803
Convergence, 302
Convolution, 928–929
Cook, Stephen A., 204
Cooper pairs, 249
Coordinate system, 103
Coordinates
 curvilinear, 916–919
 cylindrical, 918
 spherical, 919
Coordination number, 402
Coriolis force, 103
Corrections
 electron energy, 541
 inclusion of, 651–653
 weak localized, 780–781
Correlation
 classification of, 290–293
 electron, 574–578
 energy, 559
 level, 777–778
 retarded current-current, 359

Correlation functions, 75–76
  first-order, 76
  Fourier transform of, 365
  Gaussian, 365
  Kubo formula and, 352
  pair, 76
  second-order, 76
  spin-spin, 478
  temporal, 364
  typical, 364f
Correspondence principle, 97–98, 304
Cöster, Fritz, 577
Coulomb blockade, 747, 765
  at small bias, 771–772
  state, 771f
  tunneling through, 686f
Coulomb interaction
  exciton binding and, 500
  fQHE and, 467
  screened, 841–842
Coulomb potential
  expansion of, 150
  instantaneous, 321
  1D, 732–734, 733f
  partial wave expansion in, 661–662
  repulsive, 14, 14f
  scattering from, 659–664
Coulomb repulsion
  integral, 550, 553
  Pauli exclusion principle and, 185
Coulomb (C), 124
  diamonds, 772–773
  energy of charge distribution, 151
  exchange integral, 553
  gauge, 170, 321, 357
  ion traps and, 244
  peaks, 772f
  phase shift, 130–131, 662
  radial wave function, 128–131
  scattering, 726–731
  screening, 501–502
  unit of length, 660
Counter rotating, 270
Counterfeit protection, 193
Coupled cluster (CC) method, 577–578
Coupling limit, strong, 786
Covariant momentum, 436, 453, 796
Critical temperature T_C, 395
Cross-sections, 617–618
  calculating, 709–710
  differential, 606, 618
  Dirac equations and, 806
  partial wave, 639
  resonance, 668–669
  scattering, 607
  to wave functions, 622–623

Cryptanalysis, 202
Cryptography, 193, 222–223
  classical, 202–204
Cryptosystems, 202
  private key, 203
  public key, 203–204
Crystallography, 411
Crystals. *See also* Quasicrystals
  field splitting, 473
  modified definition of, 414
  molecular, 502
  momentum, 421
  structure, 400–414, 401f
  systems, 404t
  translations, 405
  Wigner, 815–816
  X-ray scattering from, 409–410
Curie law, 472–473
Curl, Robert, 20
Current. *See also* Spin current
  Hall, 444
  charge, 496
  density, 392–393, 432
  diffusion, 496
  fluctuating, 366
  generation, 495, 496f
  matter, 496, 847
  noise, 813–815
  persistent, 441, 811
  spin current, 510–511
  spin polarization, 517
Curvilinear coordinates, 916–919
Cyclotron frequency, 445

D

D (rotation) functions, 134–137
Data compression, 198–200
  in classical information theory, 199–200
  of quantum information, 220–221
Datta-Das spin transistors, 525–526, 526f
DC limit, 358
De Broglie, Louis, 11
De Broglie wavelength, 126, 752
De Haas, W. J., 459
De Haas-van Alphen effect, 435, 450, 454, 459–461, 460f, 461f
Debye model of specific heat, 529–530
Decay
  alpha particle, 14f
  beta, 159
  exponential, 331
  nonexponential, 331
  terms, 268
  width, 329
Decision problems, 204–205

Decoherence, 24, 70, 240
  time, 267
  two-level systems and, 266–268
Decoupling, 588
Degeneracy, 66–67, 447
  doubling, 589
  Kramers, 173
  spin, 131f, 757
Dehmelt, H., 275
Delta functions
  derivative of, 37
  Dirac, 36–38
  Kronecker, 895
  potential, 52–53
  3D, 37
Dense coding, 219
Density functional theory (DFT), 884f
  computer packages, 888–891
  gap problem in, 884–885
  Hohenberg-Kohn-Sham, 871–872
  spin, 882–883
  time-dependent, 885–888
Density matrices, 27
  components of, 147f
  of continuous-variable systems, 295
  correlated spins and, 279–281
  decohered, 78
  formalism of, 72, 294
  formulation, 70–75
  irreducible representations of, 146–148, 148f
  Liouville-von Neumann, 294
  multipole expansion of, 150
  in Pauli spin matrices, 264
  for qutrit, 288
  reduced, 76–77
  RWA, 271
  spatial degrees of freedom of, 378
  three-level bipartite, 288
  two-qubit, 291
  unitary transformation of, 78
  Werner, 291f
Density operators, 71, 146
Dephasing, 810
Depletion layer, 491, 493–494
Detailed balance, 703
Determinants, 374–375, 900
Determinism, 255
Deterministic polynomial time (P), 204
Detuning, 275f
Deutsch algorithm, 225–228, 227f
Deutsch-Jozsa algorithm, 225–228
DFT. *See* Density functional theory
Diamagnetism, 169–170, 468–475
  Landau, 473–475
  Larmor, 470

Index

Diatomic electronic states, 582–583, 584f
Diatomic molecules, 582–583
  dissociation energy of, 122
  Franck-Condon principle for, 602f
  hyperfine interactions in, 589
  photodissociation of, 603f
  potential, 123
  selection rules for, 597–600, 598t
Differential equations, 931–932
Diffraction, 6n4
  of electromagnetic waves, 409
  of electrons, 409
  of neutrons, 409
  patterns, 11f
  quasicrystal, 412f
Diffuse transitions, 176f
Diffusion
  constants, 497
  current, 496
  lengths, 499, 517
  time, 751
Diffusive metallic regime, 751
Digital communication, 194
Digital signatures, 193
Dimensionless anharmonicity constant, 594
Diophantic equation, 238
Dipolar interactions, 480–481, 820
Dipole force, 276, 278
Dipole moments, 179f, 312
Dirac, Paul, 61, 98–99, 159, 323
Dirac cones, 790–792, 791f, 797f, 802f
Dirac electron theory, 792–795
Dirac equation, 159, 792–797
  cross-sections and, 806
  in two-dimensions, 794
  relativistic, 128n2
Dirac notation, 22, 61–63, 892–893
Disorder, 773–782
Disordered systems, 774–777
Dispersion, 273–275
  curves, for light, 534f
  relation, 7, 655–656
Dissipation coefficient, 356
Dissipative force, 276, 278
Dissipative processes, 264n2
Dissociation energy, of diatomic molecules, 122
Divergence theorem, 915–916
Domain walls, 480
Doping, 481, 486, 819
Doppler cooling
  of atoms, 275–278
  temperature, 244
Doppler shift, 181, 330
Dot Kondo Hamiltonians, 784–785
Dot Kondo temperature, 785

Double barrier resonance structures, 50–51
Double-slit experiment, 8–10, 9f
Doublets, 175, 176f
Dresselhaus interaction, 747
Dresselhaus potentials, 507–508
Drift currents, 497
Drift velocity, 516–517
Drude, Paul, 393
Drude conductivity, 392–396, 393f, 444
Drude model, 755
Dyakonov-Perel mechanism, 523, 524f

E

Edge pictures, 464–465
Edge states, 464–465, 744–746
  propagation, 465f
  topological insulators and, 543
Effective mass, 425–426
Effective range
  analysis, 650
  approximation, 651
  complex, 721
  modified theory of, 721–722
Efimov states, 651
Ehrenfest, Paul, 13, 40
Ehrenfest theorem, 39–40, 86, 309
Eigenfunctions, 25
  reality of, 174–175
  rigid-rotor, 137–139
  zero-order, 310
Eigenphases, 640, 776
Eigenstates
  adiabatic, 345
  of angular momentum, 373
  helicity, 802–803
  nondegenerate, 174
  of spin, 160
  time dependence of, 65
Eigenvalue-eigenvector equation, 422
Eigenvalues, 25
  adiabatic, 337f
  basis-state expansions and, 64
  generalized, 332
  of Hermitian operators, 29
  hydrogenic energy, 128n2
  joint distribution of, 775–776
  lowest energy, for particle-in-a box, 44f
  nondegenerate, 48
  of parity operators, 100
  parity symmetry and, 48
  of periodic potentials, 52
  in two-level system, 260f
Einstein, Albert, 7, 99
Einstein model of specific heat, 529–530
Einstein notation, 62

Einstein Podolsky Rosen (EPR) paradox, 27, 207, 250–253
Einstein summation convention, 162
Einstein-Smoluchowski relation, 366
Electric breakdown, 430
Electric dipole radiation, 324–326
Electric fields
  effects, 430–434
  perturbative, 312–313
  semiclassical motion in, 457
  spin manipulation in, 516
  static, 495–496
  of surface plasmon, 533f
Electric quadrupole
  moment, 182
  radiative transition, 325
Electromagnetic fields, 6
  atoms in, 273–275
  dynamics, 320–329
  intensity, 321
  oscillating, 276
  quantization, 323, 850–853
Electromagnetic waves
  diffraction of, 409
  dipole-allowed, 154
  in metals, 396
Electron gas, 381, 860–866, 865f
  free, 381–391, 386f
  2D, 448–454, 449f, 747
Electron magnetic resonance (EMR), 185
Electron paramagnetic resonance (EPR), 185
Electron spin resonance (ESR), 185
Electron systems, strongly correlated, 541
Electron-empty levels, 433
Electronic band spectrum, 601f
Electronic devices, 481–482
Electronic states, diatomic, 582–583, 584f
Electronic transitions, 599
Electron-phonon interactions, 538, 854
Electron-photon interaction, 853
Electrons
  affinity, 559
  amplitude for scattering of, 74
  band, spin Hall conductance of, 511
  Bloch, 460, 866
  carriers, 495–496
  classical radius of, 126
  closed orbit, 455f
  in conducting ring, 348f
  conduction band, 50
  contribution, 530
  correlation, 574–578
  diffraction of, 409
  diffusion current, 496
  diffusion lengths, 499
  dressed, 392

energy band diagram, 496f
energy correction, 541
energy splitting for, 168
exciton binding and, 500
generation current, 495, 496f
independent model of, 414
interacting, 863–866
in jellium model, 861–866
level filling of, 383f
lifetimes, 498
in magnetic field, 166–172, 454–459
magnetic moment, 181f
mass, 5, 426, 486t
in metals, 51–52
noninteracting, 861–863
orbits, 767f
in periodic potential, 414–434, 454–459
in photoelectric effect, 15–17
relativist theory of, 159
screening, 632
in semiconductor conduction, 484
spin, 13, 110, 167
transport, 391–392, 415–416
tunneling, 55f
velocity, 395
wave properties of, 10
Electrostatic unit, 124
Elliot-Yafet mechanism, 523, 524f
Emission. See also Photoemission
  exciton, 514–515
  spectra, 5f, 513–514
EMR. See Electron magnetic resonance
Encryption transformation, 202–203
Energy bands, 52, 52f, 53f, 417f
  Bloch wave functions and, 418–421
  diagrams, 416f, 496f
Energy current density, 432
Energy gap, 249, 413–418, 422–428, 480t9.3
Energy quantization, 2–6
Energy splitting, for electron, 168
Energy surface, curvature of, 426
Enigma Rotor machine, 202
Ensemble of pure states, 71
Entangled states, 27
Entanglement, 26–27, 207–213
  classification of, 290–293
  dilution, 213
  distillation, 213
  entropy, 208, 210–211
  as information resource, 221
  mixed-state, 211–212
  Peres-Horodecki, 212, 290–291
  witness operators, 292–293
Entropy
  entanglement, 208, 210–211
  information and, 194–196

joint, 198
  relative, 197
  Shannon, 196–198
  von Neumann, 196n1, 211
EPR. See Electron paramagnetic resonance
EPR paradox. See Einstein Podolsky Rosen paradox
Equilibrium properties, 461
Ergodicity, 782
Error correction, 199–200. See also Quantum error correction
Error syndrome, 242
ESR. See Electron spin resonance
esu. See Statcoulomb
Ethanol, NMR spectrum of, 191f
Euclid's algorithm, 235
Euler, Leonhard, 6, 133
Euler angles, 133–134, 133f
Euler's equation for hydrodynamic fluid flow, 309
Evanescent waves, 742, 764
Evolution operators, 28, 35, 93, 612
Exchange coupling, 185, 476
Exchange integrals, 476, 553
Exchange interactions, 475–477
Exchange operators, 817f
Exchange reactions, 693
Exchange symmetry, 367, 371–373
  many-particle, 378–380
  two-level systems, 377–378
Exchange-correlation energy, 874
Excitations
  collective, 527f
  electron-hole, 526
  elementary, 501, 526
  helium, 557–558
  low-energy collective, 526–541
  low-lying, 817
  plasmon, 531f
  spectrum lineshape, 682
  SPR, 535f
Excitons, 50, 500–504
  absorption, 514–515
  binding, 500
  charge transfer, 501
  emission, 514–515
  Frenkel, 500, 500f, 502–504, 502f
  states, 515f
  Wannier-Mott, 500–502, 500f, 502t
Expansion coefficients, time-dependent, 315
Expectation value. See Mean value
Extensive quantity, 741

F

Face-centered cubic (FCC), 406f
Factorization, 204, 235–236

Faltung theorem, 355
Fano factor, 814
Fano lineshapes, 680, 680f
Fano resonance, 605, 658, 677–682
Faraday effect, 259
Faraday equations, 396
Far-off resonance force, 276, 278
Far-off resonance traps (FORTs), 278, 279f
Fast Fourier transform (FFT), 232, 922
Fast multipole method, 553
FBZ. See First Brillouin zone
FCC. See Face-centered cubic
Fermi, Enrico, 99, 180, 318, 383, 875
Fermi contact term, 179–180
Fermi energy, 383, 383f, 426
  fixed, 448–450, 450f
Fermi Golden Rule, 318, 323, 361
Fermi liquid theory, 414, 820
Fermi momentum, 383
Fermi statistics, 383
Fermi surface, 459, 461f
Fermi velocity, 383
Fermi-Dirac distribution, 380, 388–391, 388f, 859–860, 860f
Fermi-Dirac statistics, 99
Fermi-Dirac velocity distribution, 399
Fermionic states
  antisymmetric, 373
  wave functions for, 99, 379
Fermions, 98, 372
  annihilation operators, 835–837
  basis states in fock space for, 830
  composite, 818–819, 818f
  creation operators, 833f, 835–837
  Dirac, 795
  identical, 378
  identical, with antisymmetric spin function, 665
  identical, with symmetric spin function, 665, 666f
  number operator, 836
  second quantized Hamiltonians and, 838–842
  spinless, 840–841
  statistics, 383, 826
  wave functions, 99, 379
  zero-temperature occupation of, 99f
Ferrimagnets, 475, 475f
Ferromagnetic-nonmagnetic (F/N) contact, 516f, 517–518
Ferromagnets, 468, 475, 475f, 478t
  hysteresis in, 480f
  magnetization of, 479f
  semiconductors, 518

Index

Feshbach resonance, 658, 669–677, 670f, 682
  models, 674–677, 674f
  scattering, 670f
  scattering formalism, 670–674
  scattering length for, 677f
  in two-atom scattering, 670f
  for two-channel square well, 675f
FET. *See* Field-effect transistor
Feynman, Richard, 1n1, 328, 592
Feynman diagrams, 327f, 328, 780
FFT. *See* Fast Fourier transform
Field operators, 843–850
Field-effect theories, 14
Field-effect transistor (FET), 19
Filling factor, 450
Fine structure constant, 5, 124, 126
Finite bias, 772–773
Finite depth square well, 42f
First Brillouin zone (FBZ), 866
Fluctuating currents, 366
Fluctuation-dissipation theorem, 362–366, 392
F/N contact. *See* Ferromagnetic-nonmagnetic contact
Fock, Vladimir, 90, 548
Fock operators, 556
Fock space, 830–832, 837–843
Fock states, 90
Fock-Darwin spectrum, 768, 769f
Fokker-Planck equation, 268
Fortrat diagram, 601f
FORTs. *See* Far-off resonance traps
Fourier amplitudes, 921, 930t
Fourier expansion, 923, 923f, 924f
  of operators, 929
  of SVE, 320
Fourier integrals, 925–928
Fourier series, 405, 922–924, 927
Fourier transform (FT), 37n13, 922, 929–936. *See also* Fast Fourier transform
  algorithms, 224
  of correlation function, 365
  faltung theorem for, 355
  NMR, 190–192
  quantum, 232–234, 233f, 238
Fourier transform infrared (FTIR) spectroscopy, 921
FQHE. *See* Fractional quantum Hall effect
Fractional charges, 816–817
Fractional quantum Hall effect (FQHE), 462, 466–467, 816–819
Fractional statistics, 817
Franck, James, 6, 601
Franck-Condon principle, 600–603, 602f

Fraunhofer, Josef, 6
Free phonon wave functions, 540
Fresnel, Augustin Jean, 6
Friction coefficient, 356
Friction force, 276, 278, 393–394
Friedel sum rule, 786
Frozen-core approximation, 576
FT. *See* Fourier transform
FTIR spectroscopy. *See* Fourier transform infrared spectroscopy
Fundamental gap, 559

G

Galilean transformations, 33–36, 102–103
Gamow, George, 13
Gamow factor, 686
Gap problem, 884–885
Gates, 200t
  Cirac-Zoller CNOt, 244, 245f
  CNOt, 201f, 201t, 216f
  controlled, 215
  controlled-NOT, 244
  controlled-phase, 215, 233
  discrete, 217
  faulty, 241
  Fredkin, 217
  Hadamard, 216f
  one-qubit, 246, 247
  quantum, 213–218
  in quantum Fourier transform, 234
  single-qubit, 214–215
  SWAP, 216
  three-qubit, 217
  Toffoli, 201, 201f, 202t, 217
  two-qubit, 215–216
  U_f, 225–227
  universal quantum, 217–218
Gauge freedom, 435–436
Gauge invariance, 435
Gauge theories, 442
Gauge transformation, 346, 438f
Gaussian ensembles, 775, 776–777
Gaussian orthogonal ensemble (GOE), 776
Gaussian symplectic ensemble (GSE), 777
Gaussian type orbitals (GtOs), 546–547
Gaussian unitary ensemble (GUE), 776
Gaussian units, 124–127
  conversion table, 125t
  Fourier transform of, 81
GCE. *See* Grand canonical ensemble
GEA. *See* Gradient expansion approximation
Geim, A. K., 787
Gell-Mann matrices, 288
Generalized displacements, 343–344
Generalized eigenvalue problem, 332
Generalized forces, 343–344, 360

Generalized functions, 36
Generalized velocities, 343
Generating functions, 56–57, 113
Geometrical phase, 345
Geometrical structure factor, 411
Gerber, Christoph, 18
GF matrix method, 595
g-factor, 167, 170
GHZ state. *See* Greenberger-Horne-Zeilinger state
Giant magnetoresistance (GMR), 524
Gibbs inequality, 197
GMR. *See* Giant magnetoresistance
GOE. *See* Gaussian orthogonal ensemble
Goudsmit, Samuel A., 13
Gouy phase, 275n7
Gradient expansion approximation (GEA), 877–878
Gram-Schmidt orthogonalization, 896
Grand canonical ensemble (GCE), 389, 858–859
Graphene, 402f, 787–810
  Absence of backscattering, 805
  Dirac cones for, 805f
  direct lattice of, 791f
  Hamiltonians, 800
  honeycomb, 789f
  Klein tunneling, 807
  Landau levels in, 804–805, 804f, 805f
  properties of, 788
  scattering potential in, 806–810, 807f
  structure of, 788f
Greatest common divisor, 235
Green, Melville S., 352
Greenberger-Horne-Zeilinger (GHZ) state, 208–209
Green-Kubo formulas. *See* Kubo formulas
Green's function, 623–628
  configuration space representation of, 642
  Lippmann-Schwinger equation and, 806–807
  many-body, 749
  for 1D scattering, 736
Grotrian diagrams, 131, 132f
Group multiplication tables, 933, 937t
Group orthogonality theorem, 135, 939
Group theory, 135
  axioms, 933
  molecular orbitals and, 580–581
Groups, 934–940. *See also* Point groups
  continuous, 936
  properties, 936–937
  representations, 936–937
  space, 403, 405, 936
Grover, Lov, 228
Grover search algorithm, 228–231, 231f
GSE. *See* Gaussian symplectic ensemble

GTOs. *See* Gaussian type orbitals
GUE. *See* Gaussian unitary ensemble

H

Haar measure, 775
Hadamard transformation, 227
Hall angle, 444
Hall bar geometry, 458, 459*f*
Hall effect, 395, 435, 443–444, 443*f*, 450. *See also* Anomalous Hall effect; Fractional Quantum Hall effect; Quantum Hall effect; Spin Hall effect
    anomalous, 714
    in 3D periodic systems, 457–459
Hall resistance, 443
Hall transition, 464
Hall voltage, 444
Hamiltonians, 23. *See also* Zeeman Hamiltonian
    Anderson, 739
    anisotropic Heisenberg spin, 185
    center-of-mass, 124
    for charged particle, 169
    in classical mechanics, 29–30
    diagonalized, 98, 339*f*
    Dirac, 793
    dot Kondo, 784–785
    energy and, 32
    exact, 549
    factoring, 89
    in fock space, 837–843
    graphene, 800
    Hartree-fock, 556
    Heisenberg spin, 184–185, 476–477, 535–536
    helium, 548
    of hydrogen atom in magnetic field, 171
    hyperfine interaction, 180
    Luttinger, 507
    mean-field, 869
    multielectron system, 545–546
    nuclear spin, 189
    1D, 244
    Pauli, 442, 795
    in periodically driven two-level systems, 268
    perturbation, 310
    polaron, 539
    in quantum simulation, 239–240
    rigid-rotor, 137
    second quantized, 838–843
    spin-dipolar, 184
    spin-orbit, 176, 795
    Stark, 312
    tight-binding, 849
    time independent, 28
    time-dependent, 299–300
    in two-level system, 260
    two-species, 869–870
    vibrational-rotational, 594
    Zeeman, 247, 311
    zero-order, 65
Hamilton-Jacobi expansion, 309
Hamilton-Jacobi theory, 305
Hankel functions, spherical, 637–638
Hanle effect, 519, 519*f*
Hänsch, t. W., 275
Hard-wall boundary conditions, 382, 384*f*
Harmonic functions, 112
Harmonic oscillators, 55–58
    dissipative, 366
    raising-lowering operators, 88–92
    3-D, 121
    wave functions, 58*f*
Hartree, Douglas, 548
Hartree (E_0), 125–126
Hartree-Fock approximation, 572–574
Hartree-Fock formalism, 868
Hartree-Fock method, 548–550, 552–558, 555*f*
Hartree-Fock potential, 556
Hat notation, 63
Heat transfer, 397*f*
Heisenberg, Werner, 65
Heisenberg equation of motion, 848
    for electron spin, 167
    for nuclear magnetic moment, 167
Heisenberg matrix mechanics, 65, 92
Heisenberg models, 477
Heisenberg representations, 85–97, 353
Heisenberg Uncertainty Principle, 67–69
Heitler-London-Slater-Pauling method, 567–569
Helicity, 801–803
Helium, 548–550
    excitation, 557–558
    Hartree-fock method and, 557–558
    lowest few states of, 558*f*
Hellman, Hans, 592
Hellman-Feynman theorem, 343, 592, 747
Helmholtz equations, 396, 906
    multipole expansion of, 151
    one-dimensional, 41
    vector, 41
Hermite polynomials, 56–57, 907
Hermitian conjugates, 25, 901
Hertz, Gustav, 6
Hertz, Heinrich Rudolf, 16
Heterojunctions, 752, 754–755
Heterostructures, 754
Heun confluent functions, 269–270
Hidden subgroup problem, 239
Hidden variables, 253–256
Highest occupied molecular orbital (HOMO), 559, 884
High-temperature superconductivity (HTSC), 818–819, 819*f*
Hilbert spaces, 61–63, 627, 896–898
Hohenberg-Kohn theorems, 872–875
Holes, 433, 433*f*
    conduction, 434
    diffusion lengths, 499
    doping, 819
    exciton binding and, 500
    heavy, 505
    lifetimes, 498
    light, 505
    masses, 486*t*
    in semiconductor conductance, 484
    spin Hall conductance of, 512
Holonomy, 347
HOMO. *See* Highest occupied molecular orbital
HTSC. *See* High-temperature superconductivity
Hubble's law, 330
Hückel approximation, 569–572, 570*f*
    benzene, 571*f*
    1D solid state, 572, 572*f*
Huffman coding, 197–198
Hund, Friedrich, 562
Hund's coupling, 586–589, 587*f*
Hund's rules, 470, 547, 560–562, 767
    exchange interaction and, 475
    generalized, 561, 562*f*
Huygens, Christian, 6
Hybridization, 566
Hydrocarbons, cyclic, 569*f*
Hydrodynamic fluid flow, 309
Hydrogen
    absorption spectra of, 175
    atom, 124–132
    atom spectrum, 131–132
    emission spectrum of, 5*f*
    energy levels, 131*f*
    Grotrian diagram for, 132*f*
    hyperfine transition in, 325
    ion MOs, 563–564
    in magnetic field, 171–172, 171*f*
    masers, 181
    molecule, 565–569
    potential curves, 566*f*
    radial wave functions, 129*f*
    Zeeman splitting of, 182*f*
Hydrogenic energy eigenvalues, 128*n*2
Hyperfine interactions, 178–182
    in diatomic molecules, 589
    electric quadrupole, 182

Index

Hyperfine splitting, 178–179
Hyperfine transition, in hydrogen, 325
Hysteresis, 480–481, 480f

I

Identical particles
  interacting, 826–827
  scattering from collection of, 722–724, 722f
  scattering of two, 367f, 664–666
  with spin, 711
IE. *See* Ionization energy
Impact parameter, 606
Impurities, 486
  acceptor, 486–489, 487f
  carrier densities and, 488
  donor, 486–489, 487f, 488f
  magnetic, 783
  n-type, 487
  p-type, 487
Incident beam, 606, 706
Independent electron model, 414
Infinite square well, 42f, 43
Infinite volume, 381
Information, 193. *See also* Classical information; Quantum information
  content, 195
  entanglement, 221
  entropy and, 194–196
  source, 195–196
  technology, 17–21
  theory, 193
Infrared absorption spectrum, 599f
Infrared divergence, 574, 842
Inner product spaces, 893–898
Insulators, 541–544
  Anderson, 544
  band, 416, 543
  classification of, 543–544
  magnetic susceptibility of, 468–469
  Mott, 544
  Peierls, 544
  topological, 543
Integer quantum Hall effect (IQHE), 462–466, 805
Integral equations, 931–932
  Lippmann-Schwinger equation, 806
Integration contour, 658f
Intelligent states, 283
Intensity ratios, 190
Interaction representations, 87–88
  operators in, 353
  scattering within, 615–616
Interatomic potentials, 585–586
Interface potentials, 52f
Interference, 6n3, 9f
Interferometers, 677–678, 749, 757–759, 758f

Interleaving theorem, 333
Internal degrees of freedom, 693–724
International classification symbols, 405
International System of Units (SI), 124–127, 125t
International Union of Crystallography, 414
Internuclear distances, 123, 585–586
Invariance, 98
  gauge, 846
  rotational, 152, 708
  under space inversion, 701
  time-reversal, 98, 173–175, 702
Inverse scattering problem, 647
Inversions
  about average operator, 229–230
  space, 701, 702f
  symmetry, 100, 152
  symmetry operations and, 403
Ion cooling, 244
Ion traps, 243–245
Ionization energy (IE), 559, 559f
Ionization potential theorem, 884
IQHE. *See* Integer quantum Hall effect
Irreducible representations, 938–939
  construction, 942
  of density matrix, 146–148, 148f
Irreducible tensor operators, 144–145

J

Jeffreys, Harold, 304
Jellium model, 861–866, 865f
J-J coupling, 561
Johnson, John, 362
Johnson-Nyquist noise, 366, 813
Jordan, Pascual, 65
Josephson devices, 268
Josephson junction, 249
Jost functions, 653–654
  dispersion relation for, 655–656
  zeros of, 654, 655f
JWKB approximation. *See* WKB approximation

K

Kamerlingh-Onnes, H., 395
Karp, Richard, 204
Kerr nonlinearity, 248
Kets, 28
  channel, 694
  density operators and, 146
  Hilbert spaces and, 61
Key distribution problem, 203, 222
Kinematics, relativistic, 619
Kinetic energy operator, 846
Kitaev, Alexei, 241
Klein paradox, 809

Kohn-Sham equations, 878–881, 888
Kohn, Walter, 871, 872–875
Kondo effect, 783–787, 787f
Koopman's theorem, 558–560
Kosterlitz-thouless-Berezinsky transition, 479
Kramers, Hendrik Anthony, 173, 304
Kramers degeneracy, 173
Kramers doublet, 505
Kramers pair, 822
Kramers-Kronig relations, 351–352, 360–362, 655
Kronecker delta function, 895
Kronig-Penney potential, 417, 417f
Kroto, Harold, 20
Kubo, Ryogo, 352, 362
Kubo formulas, 349–350, 352–360, 542
Kümmel, Hermann, 577
Kummer functions, 662

L

Laboratory frame, 619–621, 621f
Lagrange multiplier, 332, 554, 688
Lagrange's theorem, 937
Laguerre polynomials, 123
Lamb-Dicke criterion, 244
Landau, Lev, 72, 339, 414, 459–460
Landau diamagnetism, 473–475
Landau energies, 446, 446f, 447f
Landau levels
  broadening, 461
  in graphene, 804–805, 805f
  IQHE and, 463–464
Landau quantization, 445–447
Landauer formula, 19, 749, 756f
  for conductance, 755–764
  multichannel, 757
  multiport, 760–762
Landau-fermi liquid theory, 392
Landau-Zener transition, 339–343, 340f, 341f
Landé factor, 471. *See also* g-factor
Landés interval rule, 561
Laplace's equation, 112
Larmor diamagnetism, 470
Larmor formula, 5, 324
Larmor frequency, 167
Lasers
  cooling, 278–279
  fixed frequency, 335
  ion cooling, 244
  polarization for, 279
  quantum dot, 752
  semiconductor, 481
  spatial interference patterns, 246f
Lattices, 401. *See also* Bravais lattices
  classes, 403
  classification of, 402–404
  constants, 401

direct, 406–407, 791f
optical, 245–246
parameters, 401
periods, 401
reciprocal, 405–411
structure, 790–792, 790f
translation operations, 405
2D honeycomb, 402f
vectors, 401
vibrations, 527–528
Laughlin, Robert, 464, 465, 816
Laughlin wave function, 465–466
LCAO. *See* Linear combination of atomic orbitals
LDA. *See* Local density approximation
LEDs. *See* Light-emitting diodes
Lee, T. D., 100
Legendre equation, 113
Legendre polynomials, 113–114, 909–910
Lempel-Ziv algorithm, 198
Lennard-Jones, John, 123, 562
Lennard-Jones potential, 123–124, 123f
Letokhov, V. S., 278
Level correlations, 777–778
Level fluctuations, 778
Level repulsion, 775
Level spacing, 779f
  mean, 779
  nearest, 777–778, 779f
Levi-Civita symbol, 106, 159, 915
Levin, Leonid, 204
Levinson theorem, 656–658
Light, 6–8
  dispersion curve for, 534f
  Doppler shift of, 330
  holes, 505
  slow, 294
  wave properties of, 10
Light and Matter (Band), 328
Light-emitting diodes (LEDs), 481
Linear combination of atomic orbitals (LCAO), 563, 567f, 574
Linear polyenes, 569f
Linear potential, 54–55
Linear response hypothesis, 354
Linear response theory, 349–366, 392
Liouville equation, 84, 309
Liouville theorem, 432, 432f
Liouville-von Neumann equation, 71–72, 83
Lippmann-Schwinger equation
  Born approximation and, 683
  Green's function and, 806–807
  operator, 629, 634
  partial wave, 641
  for radial function, 643–646

Local density approximation (LDA), 881
  adiabatic, 888
  spin DFT, 883
Local moment regime, 783
Local quantum operations and classical communication (LOCC), 213
Local realism theory, 251
Localization
  length, 441
  weak, 811–813, 811f
LOCC. *See* Local quantum operations and classical communication
Logarithmic derivatives, 647
Longitudinal relaxation, 188
Long-range order, 412
Lorentz, Hendrik A., 166
Lorentz factor, 330
Lorentz force, 449
Lorentz lineshape, 329, 330f
Lorentz number, 400
Lorentz transformations, 33
Low equations, 632–633
Low temperature physics, 783–784
Low-dimensional systems, 19–20, 810–823
  scattering in, 724–748
  strongly correlated, 815
Low-energy collective excitations, 526–541
Lowering operators, 58, 89, 107–110
Lowest unoccupied molecular orbital (LUMO), 559, 884
Loy, M., 335
L-S coupling, 561
LUMO. *See* Lowest unoccupied molecular orbital
LUMO-HOMO gap, 884–885
Luttinger Hamiltonian, 507
Luttinger liquids, 20, 731, 820
Lyapunov exponent, 741
Lyman transitions, 131, 132f

M

Magnetic breakdown, 430
Magnetic dipole radiative transition, 325
Magnetic fields
  charged particle in, 169–172
  crown-shaped, 348f
  effects, 434–481
  electrons in, 166–172, 454–459
  hydrogen atom in, 171–172, 171f
  inverse, 460f
  from magnetic moments, 179
  perpendicular, 448–454, 449f, 744–746, 745f
  perturbative effects, 311–312
  resonance energy tuning, 673
  semiclassical motion in, 457

spin manipulation in, 516
static, 168
2D electron gas in, 448–454, 449f
Magnetic impurities, 783
Magnetic moments, 475
  anomalous, 795
  conducting ring near, 348–349
  electron, 181f
  interaction of, 179
  localized, 783
  magnetic fields from, 179
  nuclear, 167
  orbital, 169
  of particles, 13
  proton, 181f
Magnetic order, 475–481
Magnetic susceptibility, 468–469
Magnetic tunnel junctions (MTJs), 520, 521f, 524
Magnetically ordered structures, 477–478
Magnetization, 168
  atomic, 469–473
  dimensionless, 263
  equilibrium, 188
  of ferromagnets, 479f
  magnetic susceptibility and, 468–469
  remnant, 480–481
  saturation, 480
  spontaneous, 477
  vector, 468
Magnetoconductance, 461, 463
Magneto-optical traps (MOts), 278–279
Magnetoresistance, 443–444, 520. *See also* Giant magnetoresistance; tunneling magnetoresistance
  IQHE and, 463
  negative, 781, 811–812
Magnets
  antiparallel configuration in, 480
  bar, 480f
  rare earth, 477–478
  types, 478t
Magnons, 527, 535–538
Many-body perturbation theory (MBPt), 576
Many-body problem, 381, 826–827
Many-body theory, 466–467
Many-particle systems, 75–76
Markov processes, 268
Markovian Lindblad form, 268
Masers, 181
Maslov index, 456
Mass, effective, 425–426
Mass action, law of, 485
Massive quantum parallelism, 225–226
Mathieu equations, 423–426
Matrices, 898–903. *See also specific matrix types*

Index

Matrix elements
   bound-free, 602–603
   hopping, 761
   of operators, 173–174
   reduced, 154
   of spin-dipolar interactions, 183
   of tensor operators, 153
   vector potential, 345–346
Matter-waves, 8
Mauborgne, Joseph, 203
Maxwell-Boltzmann velocity distribution function, 399–400
MBPt. See Many-body perturbation theory
Mean free path, 350, 395
Mean value, 24, 88
Mean-field approximation, 414, 651, 867
Mean-field equations, 867–870
Mean-field zone, 866–870
Measurements
   demolition, 79
   of first register, 238
   multi-qubit, 242
   nondemolition, 224
   of polarization states, 248
   POVM, 79
   problem, 69–70, 78–79
   quantum computing despite, 224
   qubit collapse after, 241
   syndrome, 242
   von Neumann, 79
   weak, 79, 224
Mermin-Wagner theorem, 820
Mesoscopic conductors, 360
Mesoscopic systems, 18–19, 438, 750–755, 773–783
Metallic diffusive regime, 751
Metallic grain, 778
Metal-oxide-semiconductor field-effect transistors (MOSFEts), 14, 14f
Metals. See also Alkali-metals
   AC conductivity of, 395–396
   electromagnetic waves in, 396
   electrons in, 51
   Hartree-fock approximation for, 572–574
   Sommerfeld theory of transport in, 392, 399–400
   thermal conductivity of, 397–399
Metal-semiconductor interfaces, 51–52, 52f
Metal-vacuum interfaces, 51–52
Meters kilogram second Ampère (MKSA), 124–125
Meters kilogram second (MKS) systems, 124
Microscopic causality, 655
Miller indices, 408–409, 408f, 409
Minimum uncertainty state, 95
Mixed states, 70–75

MKS systems. See Meters kilogram second systems
MKSA. See Meters kilogram second Ampère
Mobility, 394
Modular arithmetic, 234
Modular equality, 235
Molecular orbitals (MOs), 562, 574
   diatomic electronic states and, 583
   group theory and, 580–581
   hybrid, 567f
   hydrogen ion, 563–564
   method, 565–567
   molecular symmetries and, 579
Molecules. See also Diatomic molecules
   asymmetric top, 593
   electronic structure of, 562–572
   hydrogen, 565–569
   polyatomic, 591
   rotational structure of, 593–595
   spherical top, 593, 593f, 597, 597f
   symmetric top, 593
   symmetries, 579–582
Möller operators, 612–615, 614f, 633–634, 696–699, 697f
Möller-Plesset perturbation theory (MPPT), 577
Möller-Plesset theorem, 577
Momentum, 32–36. See also Angular momentum
   crystals, 421
   representations, 63–64
   states, 38–39
   transfer, 410, 631
Momentum units (p_0), 126
Monte Carlo algorithms, 821
Moore, Gordon, 201
Moore's law, 201
Morse oscillators, 122–123
Morse potential, 122f
MOs. See Molecular orbitals
MOSFETs. See Metal-oxide-semiconductor field-effect transistors
Motif, 400, 411
MOTs. See Magneto-optical traps
Mott skew scattering, 509
MPPT. See Moller-Plesset perturbation theory
Müller, Karl Alex, 819
Mulliken, Robert S., 562
Mulliken symbols, 581–582
Multichannel scattering theory, 694, 716
Multielectron systems, 545–546, 551–552
Multipole expansions, 150–151
Multipole moments, 147, 150–151
Multiqubit states, 207

N

Nanoparticles, 534–535, 535f
Nanostructures, low-dimensional, 752
Nanotechnology, 17–21
Nanotubes, 19–20
Nearest neighbors (NN), 797
   points, 402
   spacing distribution, 778
Neutrinos, 159
Neutrons
   diffraction of, 409
   discovery of, 159
Newton, Isaac, 6
Newtonian mechanics, 605
Newton's laws, 1
Nine-qubit-code, 241
Nitrogen-vacancy (NV) centers, 250
NMR. See Nuclear magnetic resonance
NN. See Nearest neighbors
No-cloning constraints, 241
No-cloning theorem, 218–219
Noether, Emmy, 98, 607
Noether's theorem, 607
Noise, 366, 738, 813–815
Nonabelian gauge terms, 592
Non-central potentials, 635
Non-crossing rule, 590, 591n
Nondeterminism, 204n3
Nondeterministic polynomial time (NP), 204
Nonintegrable systems, 764
Nonrelativistic limit, 794
Nonuniform systems, conductivity of, 359
Nonzero singular values (NSVs), 290
Noon state, 209n5
Normalization coefficient, 138
Novoselov, K. S., 787
NP. See Nondeterministic polynomial time
NP-complete (NPC), 204–205
NSVs. See Nonzero singular values
Nuclear derivative coupling, 591–592
Nuclear magnetic resonance (NMR), 185–187
   Fourier transform, 190–192
   liquid-state, 247
   spectrum of ethanol, 191f
   systems, 247–250
   Zeeman Hamiltonian for, 187
Nuclear shell model, 715
Nuclei
   alpha particle decay of, 14f
   magnetic moments, 167
   spin, 167, 189
Number operator
   bosons, 834
   fermions 836
Number states, 90

Number theory, 234
Numerical renormalization group, 786
NV centers. *See* Nitrogen-vacancy centers
Nyquist, Harry, 362

O

Observables, 22, 28
Occupation number
  bosons, 860
  representation of states, 831
Octopus project, 302
ODEs. *See* Ordinary differential equations
Off-shell conditions, 694n10
Ohm, 463
Ohm's law, 349, 393, 396, 542
1D formalism, 731–732
1D systems
  correlated electron, 820
  density of states in, 386–387, 387f
  scattering in, 731–741
Onsager, Lars, 185, 362, 460
Onsager analysis, 459
Onsager reciprocal relations, 360–362, 760–761
Onsager relations, 444, 757–758, 759
On-shell conditions, 694n10
Operators, 898–903. *See also specific operator types*
  anti-Hermitian, 902
  antilinear, 903
  antiunitary, 903
  average, 229–230
  creation, 832–837, 833f
  destruction, 832–837
  differential, 914–915
  Fourier expansion of, 929
  Hermitian, 24, 29, 66–67, 901
  in interaction representation, 353
  matrix elements of, 173–174
  non-commuting, 67
  number, 90, 836
  parity, 100
  Pauli spin, 160
  physical, 845–848
  projection, 279, 671, 707, 902
  resolvent, 623
  rotation, 132–133
  squeezing, 96
  statistical, 71
  symmetrization, 207n4, 368
  transformation, 32
  transition, 628–633
  unitary, 35
  unitary rotation, 105
  wave, 578
  witness, 292–293

Oppenheimer, Robert, 589
Optical molasses, 278–279
Optical susceptibility, 274
Optical theorem, 634, 639, 700, 722
Optical transitions, 596–600
Optical trapping, of atoms, 278–279
Optical tweezers, 278
Optics, 6
Optimal control theory, 246
Optimal entanglement witness, 292
Oracles, 228–229
Orbital effects, 169–172
Orbits. *See also* Molecular orbitals
  atomic, 414–415, 546–547
  classical, 464f
  closed, 455, 455f, 456f, 459
  electron, 767f
  on fermi surface, 461f
  skipping, 464
Order parameters, 479
  in many-body physics, 846
Order-finding problem, 235
Ordinary differential equations (ODEs), 905–911
Orthogonal vectors, 23
Orthogonality, 61
  relations, 114, 831
  of scattering states, 699–700
Oscillator strength, 324
Outer products, 899

P

P. *See* Deterministic polynomial time
p_0. *See* Momentum units
Pair annihilation, 496
Parabolic coordinates, 659
Parabolic cylinder functions, 342
Parabolic wells, 260f
Parallel transition, 599
Paramagnetism, 170, 468–475
  contribution, 358
  Curie law of, 472–473
  Pauli, 473, 474f
  Van Vleck, 470
Parity, 48
  bits, 200
  conservation, 100
  operators, 100
  symmetry, 48, 98
Partial representation, 695–696
Partial traces, 73
Partial wave
  amplitudes, 720–721
  analysis, 636–647, 717–718
  cross-section, 639
  Lippmann-Schwinger equation, 641

  S matrix for, 640, 717–718
  scattering amplitude, 640
  Schrödinger equations, 705
Partial wave expansion, 636–640, 719, 727–729
  in Coulomb potential, 661–662
  phase shifts and, 808–809
Particle-in-a-box, 42–52, 44f
Particles. *See also* Charged particles; Identical particles
  ballistic, 773–774
  composite, 693
  density, 846
  with internal degrees of freedom, 693–724
  magnetic moment of, 13
  with spin, 704–715
  spin 1/2, 711–712
  spin 0, 711–712
Partition function, 72, 353, 858
Partitions, 370
Paschen-Back regime, 182
Pauli equations, 794–795
Paul traps, 244
Pauli, Wolfgang, 98, 159
Pauli exclusion principle, 98–99, 185, 372, 382–383
Pauli spin
  matrices, 161–164, 264
  operators, 160
Pauling, Linus, 567
PDF. *See* Probability distribution function
Penalties, 301
Penrose, Roger, 412
Penrose tiling, 413f
Peres-Horodecki entanglement, 212, 290–291
Periodic potential
  band potential and, 416–417
  eigenvalues of, 52
  electrons in, 414–434, 454–459
Periodic table, 767f
Periodicity, 412
Permanents, 374–375, 900
Permutations
  cyclic, 368
  even, 368
  odd, 368
  operators, 367
  products, 368f
  symmetry, 367–369
  two-cycle, 367
Perturbation
  Hamiltonians, 310
  harmonic, 318–319, 361
  nearly harmonic, 320–321
  piecewise-constant, 317–318
  of spin-orbits, 311

Index

Perturbation theory, 309–321. *See also* Many-body perturbation theory; Moller-Plesset perturbation theory
  degenerate, 313–314
  diagrammatic, 780
  first-order, 863–865
  k · p, 506–509
  Moller-Plesset many-body, 576–577
  nondegenerate, 310–313
  Rayleigh-Schrödinger, 310
  second-order, 318
  time-dependent, 315–321
Phase breaking length, 360
Phase change errors, 243
Phase coherence, loss of, 811–812
Phase shifts
  analysis, 647–659
  Coulomb, 130–131, 662
  eigen, 640
  Jost functions and, 654
  at low energy, 649–650
  numerical evaluation of, 646
  partial wave expansion and, 808–809
  partial wave S matrix and, 640
  for penetrable sphere, 645
  photonic, 248
  resonance, 679
  scattering, 118
  WKB approximation for, 686–688
Phase transitions
  first-order, 479
  Kosterlitz-Thouless-Berezinsky, 477
  quantum, 543, 820
  second-order thermodynamic, 479
Phase-dependent interaction, nonlinear, 96
Phonon-electron interaction, 856f
Phonons, 527–531, 853–856
  absorption, 856
  acoustic, 528, 528f
  contribution, 528–530
  density of states, 530–531
  optical, 528
Photoabsorption, in semiconductors, 513f
Photodissociation, of diatomic molecule, 603f
Photoelectric effect, 5, 15–17, 16f, 17f
Photoemission, 5, 513f
Photons, 850–853
  absorption, 277
  induced processes, 512–515
  internal angular momentum of, 13
  phase-shifting of, 248
  polarization states of, 248
  qubits, 248f
  as two-level system, 259

Photosynthesis, 21
Photovoltaic effect, 492
Piecewise-constant potentials, 42–52, 42f
p-i-n junction, 492, 492f
Plaintext, 202
Planck, Max, 6, 7f
Planck blackbody radiation law, 388
Planck constant, 2–4
Planck distribution, 389f
Plane waves, 34, 37
Plasma frequency, 396, 531
Plasmons, 527, 531–538
  excitation, 531f
  in nanoparticles, 534–535, 535f
  surface, 532–535, 533f, 534f, 535f
p-n junctions, 490–500, 492f
  carrier densities across, 494f
  current generation in, 496f
  energy parameters of, 493f
  in equilibrium, 492–495, 495f
  out of equilibrium, 496–498
  qualitative description of, 491–492
  schematic of, 491f
  in static electric field, 495–496
Poincaré, Jules Henri, 265
Poincaré sphere, 265
Point groups, 403, 404t, 935–936
  decision tree, 580f
  symmetry, 581f
Poisson distribution, 93
Poisson process, 780
Polaritons, 541
Polarization
  Bell's inequalities and, 257
  for lasers, 279
  Poincaré sphere and, 265
  rotation, 248
  spin current, 517
  states, of photons, 248
  vectors, 851
Polarons, 538–541, 538f
Polynomial time, 204
Pontryagin, Lev, 300
Population inversion, 263, 286f
Position representations, 38–39, 63–64
Positive Operator-Value Measure (POVM), 79, 224
Potential barriers, transmission through, 684–686, 684f
Potential energy crossings, 590–591, 591f
Potential wells, 47–49
POVM. *See* Positive Operator-Value Measure
Power broadening, 272, 275
Principal axes coordinate system, 426
Principal quantum number, 5

Principal transitions, 176f
Principal value integral, 634
Probability amplitudes, 23, 28–29
Probability density, 44–45, 129
Probability distribution function (PDF), 905–908
Probability distributions, 22–24, 97, 97f
Probability flux vector, 44–45
Probability theory, 258
Projections, 73, 902
  operators, 279, 671, 707, 902
  theorem, 155
Proteins, 21
Protons, 159
  chemical shifts, 191f
  magnetic moment, 181f
Pseudocrossings, 590–591
Pump pulse frequencies, 338
Purcell, Edward, 247
Pure states, 289
Purity, 77, 77f

Q

QED. *See* Quantum electrodynamics
QHE. *See* Quantum Hall effect
QPCs. *See* Quantum point contacts
QSHE. *See* Quantum spin Hall effect
Quadratic dispersion, 794
Quantization. *See also* Second quantization
  angular momentum, 12–13
  electromagnetic field, 323, 850–853
  first, 833, 837
  of lattice vibrations, 527–528
  procedure, 855–856
Quantized wavenumbers, 447
Quantum averages, 276
Quantum chaos, 764
Quantum circuits, 205, 223–224, 223f
Quantum coherence, 750
Quantum computing, 193, 224–240
  all-optical, 248
  cavity based, 246
  despite measurement, 224
  fault-tolerant, 241
  scalable, 240
Quantum dots, 19–20, 724, 749, 754, 754f
  charge transport through, 770–773
  circular, 766f
  constant-interaction model of, 765–766
  differential conductance of, 772f
  diffusive, 774
  disorder in, 773–774
  equilibrium properties of, 765–770
  few electron, 766–767, 769–770
  at finite bias, 773–774
  Kondo conductance of, 786f

Kondo effect in, 783–787, 787f
lasers, 752
lateral, 765f
potential, 50
properties of, 765–773
qubits, 249–250
selection rules, 121
spherical, 120–121
Quantum electrodynamics (QED), 167, 246, 850
Quantum entanglement. See Entanglement
Quantum error correction, 240–243
Quantum field theory, 843
Quantum unit of flux, 435
Quantum unit of conductance, 460
Quantum gates, 205
Quantum Hall effect (QHE), 20, 462–467
  detecting of, 463f
  SHE versus, 510f
Quantum information, 205–213. See also Information
  data compression of, 220–221
  entanglement as, 221
  redundant, 240
Quantum mechanics
  basics of, 21–59
  definition of, 2–17
  early history of, 2, 3–4t
  postulates of, 28–29
  relativistic, 619
  semiclassical, 304
Quantum money, 193
Quantum numbers
  magnetic, 142
  principal, 128
  single-particle, 828
  topological, 543
Quantum optimal control theory, 300–302
Quantum point contacts (QPCs), 755, 755f, 762–764, 762f, 763f
Quantum propagation, 29–30
Quantum simulation, 224, 239–240
Quantum spin Hall effect (QSHE), 544, 822–823, 822f
Quantum states, 22–24, 61
Quantum teleportation. See teleportation
Quantum theory, 5
Quantum thermodynamic expectation, 858
Quantum units, 463, 757
Quantum wells, 19–20, 724, 749, 752–753, 753f
  potential, 50
  structure, 752f
Quantum wires, 19–20, 387, 724, 753, 753f
Quasicrystals, 412–414, 412f
Quasi-momentum, 421

Quasi-one-dimensional systems, 741–747, 742f
Quasi-particles, 500–501, 526
Quasi-probability distribution, 80
Qubits, 17, 205–206
  collapse, 241
  interacting, 241
  oracle, 229, 230f
  photon, 248f
  quantum dot, 249–250
  solid-state, 248–250
  space, 236–237
  superconducting, 249
  swap of two, 234f
  teleportation, 222f
Quenched disorder, 440–441
Qutrits, 205
  density matrix for, 288
  pure states, 289

R

Rabi frequencies
  adiabatic limit and, 272
  coupling, 294
  generalized, 261
  time-dependent, 268
  vector, 263n1
Rabi oscillations, 262, 262f
Racah, Giulio, 141
Radial function, 643–646
Radial momentum operators, 116
Radiation
  blackbody, 6–8, 7f
  electric dipole, 324–326
  multipole, 324–326
  pressure, 278
  spontaneous emission of, 323–324
  stimulated emission of, 323–324
Raising operators, 58, 89, 454
  angular momentum, 107–110
Raising-lowering operators, 88–92
Raman scattering, 293, 319, 326–329, 327f
Raman transitions, 600
Ramsauer-Townsend effect, 658–659
Ramsey, Norman, 285
Ramsey fringe spectroscopy, 285–287, 286f, 287f, 294
Random matrix theory (RMT), 777–782
  conductance distribution and, 782
  prediction, 782f
Random numbers, 236
Random potentials, 737–738
  Gaussian white-noise, 738
  tight-binding formulation of, 738–741
Rashba energy, 511
Rashba interaction, 747

Rashba potentials, 507–508
Rayleigh range, 275n7
Rayleigh transitions, 600
Rayleigh-Jeans limit, 7
Rayleigh-Schrödinger perturbation theory, 310
Real quaternionic matrix, 776
Realism, 251, 253
Reciprocal lattice space, 422–423
Recombination current, 496
Rectification, 495–496
Recurrence relations, 114, 128
Recursion relation, 57
Reflections
  amplitudes, 685, 735, 747–748
  approximation, 603, 603f
  coefficients, 46, 47f
  probability, 46
  symmetry operations and, 403
Refractive index, 274–275, 275f
Regge poles, 646
Relative velocities, 181
Relaxation time, 188, 267, 350
Resistance
  quantum unit of, 463, 757
  tensor, 760, 761f
Resonance, 50–51. See also specific resonance types
  cross-sections, 668–669
  energy tuning, 673
  Fano, 677–682
  Feshbach, 605, 658, 669–677
  magnetic, 185–192
  phase shift, 679
  scattering, 666–682
  shape, 668, 668f
Resonant tunneling diodes, 51
Response functions, 349
Response kernels, 349
Riccati-Bessel functions, 117–118, 118f, 637, 662
Riccati-Hankel functions, 637, 719
Riemann sheet, 657, 668
Riemann zeta function, 390
Rigid-rotor eigenfunctions, 137–139
Rivest, Ronald, 203
RMT. See Random matrix theory
Rodrigues formula, 113
Rohrer, Heinrich, 15
Rotating-wave approximation (RWA), 187, 270–271, 320
Rotation
  angular momentum and, 132–139
  constants, 593–594
  D functions and, 134–137
  function, 134

Index

infinitesimal, 144
matrix, 133–134
operators, 132–133
polarization, 248
of spin, 136
of spinors, 164–165
structure, 593–595, 593f
symmetry operations and, 403
transitions, 597
unitary, 105
of vector field, 148–149
RSA algorithm, 203–204
Runge-Gross theorem, 886–887
Runge-Lenz vector, 131
Russell-Saunders coupling, 470, 561
Rutherford formula, 633
Rutherford scattering, 607–608, 608f
RWA. See Rotating-wave approximation
Rydberg constant, 132f
Rydberg states, 171, 296

S

S matrix, 612–615, 633–634
    multichannel, 699–703
    1D scattering and, 731–736
    for partial wave, 640, 717–718
    symmetries of, 701
Sample-to-sample fluctuations, 748, 780
Saturation parameter, 272, 275f
Scalar potential, 169
Scalar products, 913
Scanning tunneling microscopy (STM), 14–15, 15f, 18, 54
Scattering. See also Bragg scattering; Raman scattering
    in abstract Hilbert space, 627
    amplitude, 74, 610–611, 645, 646f
    amplitude, partial wave, 640
    angle, 606
    average density of, 723
    classical, 605–608, 606f
    from collection of identical particles, 722–724, 722f
    Compton, 328
    Coulomb, 726–731
    from Coulomb potential, 659–664
    Coulomb-modified, 663
    cross-sections, 607
    differential, 618f
    elastic, 411, 616–617, 693
    electron-atom, 659
    electron-electron, 659
    of electrons, 74
    Feshbach resonance, 670f
    Feynman diagrams for, 327f
    formal theory of, 627–634
    formalism, 670–674
    of identical particles with spin, 711
    from impenetrable sphere, 643–644
    inelastic, 693, 715–722
    initial state of, 611–612
    within interaction representation, 615–616
    inverse, 647
    length, 651, 677f
    length, for Feshbach resonance, 677f
    in low-dimensional systems, 724–748
    low-energy, 667f
    multichannel, 694, 716
    nonrelativistic, 619
    normal, 856
    1D, 731–741
    of particles with spin, 704–715
    from penetrable sphere, 644–646
    phase shifts, 118
    potential, 617f
    potential, in graphene, 806–810, 807f
    problems, 44
    process, 702f
    quantum, 609–616
    in quasi-one-dimensional systems, 741–747, 742f
    Rayleigh, 319, 326–329, 327f
    reactions, 715–722
    resonance, 666–682
    Rutherford, 607–608, 608f
    by short-range potentials, 662–664
    single-channel formalism, 718
    skew, 509, 509f, 714
    of spin 1/2 particles, 711–712
    of spin 0 particles, 711–712
    states, orthogonality of, 699–700
    stationary, 616–627
    s-wave, 650, 680–682
    Thomson, 326–329
    threshold, 729
    time-dependent, 609–610, 612–616
    time-independent, 610–611
    total cross-section, 607
    transition probability for, 329
    in two dimensions, 724–731
    of two identical particles, 367f, 664–666
    from two potentials, 703–704
    two-atom, 670–674
    umklapp, 856
    von Laue analysis of, 410f
    X-ray, 409–410, 409f, 411
SCF calculation. See Self-consistent field calculation
Schawlow, A. L., 275
Schmidt decomposition, 209–211
Schönflies notation, 405
Schrieffer-Wolff transformation, 507, 784–785
Schrödinger equations, 23
    classical limit of time-dependent, 309
    partial wave, 705
    for piecewise-constant potentials in 2D, 49–50
    in quantum simulation, 239
    radial, 116–117, 120, 308
    in reciprocal lattice space, 422–423
    with spin-orbit interaction, 712–715
    time-dependent, 29–31, 39, 300
    time-independent, 29–31
    time-independent 1D, 304
Schrödinger representations, 85–97, 353
Schrödinger wave function method, 92
Schrödinger's cat paradox, 24
Schumacher quantum noiseless channel coding theorem, 220
Schwarz Inequality. See Cauchy-Schwarz Inequality
Search algorithms, 224, 230f. See also Grover search algorithm
Second quantization, 825–857
    concept of, 833
    formalism, 827, 837
    graphene Hamiltonian in, 800
    of Hamiltonians, 838–843
    mean-field equations in, 867–870
Second register, measurement of, 237–238
Seebeck effect, 398–399, 398f
Selection rules, 121, 152
    for diatomic molecules, 597–600, 598t
    electric dipole radiation, 326
    multipole radiation, 326
    for optical transitions, 596–600
    Raman, 600
    Rayleigh, 600
    in Wigner-Eckart theorem, 154
Self-adjoint, 24
Self-averaging, 741
Self-consistent field (SCF) calculation, 562
Self-energy, 672
Semiconductors, 481–515
    absorption coefficients, 514f
    absorption wavelengths of, 513t
    band gap energy of, 482t
    band structure, 482–483, 483f
    charge carrier density of, 484–485
    conductance, 484
    direct gap, 482
    extrinsic, 481, 486–490
    ferromagnetic, 518
    indirect gap, 482
    inhomogenous, 490–500
    intrinsic, 481
    lasers, 481
    materials, 481–482

n-doped 3D, 508
n-type, 490
  photoabsorption in, 513f
  photoemission in, 513f
  photon induced processes in, 512–515
  p-type, 490
  spin injection into, 518
  spin relaxation in, 524f
  spintronics, 518
  zinc-blende, 505, 506f
Semiconductor-vacuum interfaces, 51–52
Semimetal systems, 416
Separability, 211–212
Separation of variables method, 31
Shamir, Adi, 203
Shannon, Claude Elwood, 194
Shannon entropy, 196–198
Shannon's noiseless channel coding
  theorem, 199
Shechtman, Dan, 412
Shockley, William, 14
Shockley equation, 498–499
Shor, Peter, 21, 232, 241
Shor factorization algorithm, 234–239
Shor's quantum error correction algorithm,
  242–243
Short-range potentials, 662–664
Shot noise, 366, 813–815
Shubnikov-de Haas effect, 435, 454,
  459–462, 462f
SI. *See* International System of Units
Si(CH$_3$)$_4$. *See* tetramethylsilane
Side jump, 509, 509f
Silicon band structure, 483f
Sinai billiard, 774f
Single-channel formalism, 718
Single-electron states, 769, 770f
Single-mode field, 322
Single-particle
  approximation, 381
  density of states, 384
  energy levels, 386f
  operators, 826
  states, 768
Singlet state, rotational invariance of, 164
Singular value decomposition, 209, 210f
Sinusoidal potential, 423–426
$6j$ coefficients, 156–157
6-12 potential, 123
Size consistency, 576
Slater, John, 374, 567
Slater determinants, 98–99, 374–375, 467,
  829
Slater type orbitals (STOs), 546–547
Slowly varying envelope (SVE), 320
Slusher, R. E., 96

Smalley, Richard, 20
Sommerfeld expansion, 390
Sommerfeld parameter, 660
Sommerfeld theory, 392
  experiment compared with, 400
  of metals, 392, 399–400
Source coding, 198. *See also* Data
  compression
Space-fixed coordinate system, 103
Spatial degrees of freedom, 378
Spatial interference patterns, 246f
Specific heat
  Debye model of, 529–530
  Einstein model of, 530
  phonon contribution to, 528–530
Spectral decomposition, 364
Spectral function, 627
Spectral lines
  broadening of, 330
  splitting of, 312
Spectral representation, 359
Spectral unfolding, 778–779, 778f
Spheres. *See also* Bloch sphere
  impenetrable, 643–644
  penetrable, 644–646
  Poincaré, 265
Spherical coordinates, 111–112, 121
Spherical harmonics, 112–115, 910
  addition theorem for, 120
  Clebsch-Gordan series and, 143
  normalization of, 129
  real linear combinations of, 115f
  spinor, 149–150, 166, 706
  vector, 148–149
Spherical tops, 138
Spherically symmetric systems, 116–132
Spin. *See also* Pauli spin
  angular momentum, 106, 159–160, 172
  antisymmetric, 665
  basis functions, 149
  Bell's inequalities and, 257
  central potentials and, 635
  chains, 1D, 821
  conductivity, 517
  correlated, 253, 279–282
  degeneracy, 131f, 757
  dephasing time, 523
  detection, 515
  DFT, 882–883
  diffusion length, 517
  dynamics, 261–262, 518
  eigenstates of, 160
  electron, 13, 110, 167
  Hall conductance, 511–512
  Heisenberg, 184–185, 476–477, 535–536
  history of, 159

  identical particles with, 711
  injection, 515–518, 516f
  local density, 510
  manipulation tools, 516–520
  models, 536n7
  nuclear, 167, 189
  particle current, 517
  particles with, 704–715
  particles with internal degrees of freedom
    and, 693
  Pauli, 110
  polarization, 517
  precession, 509
  in quasi-1D scattering, 747
  relaxation, 188–189, 515–516, 522–523,
    523f, 524f
  relaxation time, 523
  rotation of, 136
  states, in few electron quantum dots,
    769–770
  symmetric, 665
  torque, 511, 525, 525f
  transport mechanisms, 509
  uncoupling, 588
  valves, 524
  wave functions, 371
  waves, 477, 527, 537
  Young tableau for, 371f
  Zeeman Hamiltonian for single, 186
Spin current
  definition of, 510–511
  density operator, 847–848
  operators, 510
  polarization, 517
Spin density
  operator, 847
  steady state, 519
Spin Hall effect (SHE), 510–512, 510f
Spin 1/2 particle, 711–712
  representation of, 264f
  Zeeman energy of, 168, 168f
  Zeeman splitting of, 186f
Spin 0 particle, 711–712
Spin-charge separation, 820
Spin-dipolar interactions, 183–185
Spin-orbit, 165–166
  extrinsic potential, 509
  gap, 505
  Hamiltonian, 176, 795
  Hartree-Fock method and, 554
  Hund's coupling, 586–589
  interactions in atoms, 175–178
  intrinsic, 505
  lines, 178
  perturbations of, 311

Index

Schrödinger equation with, 712–715
splitting, 177, 177f
Spin-orbit coupling, 165
   constant, 176
   intrinsic, 507–509
   in solids, 504–506
   spin transport mechanisms and, 509
   weak antilocalization due to, 812
Spinors, 110, 160–161
   fields, 149–150
   rotation of, 164–165
   spherical harmonics, 149–150, 166, 706
   spin-orbitals and, 165
   time-reversal properties of, 172–175
Spin-spin relaxation time, 188
Spin-states, 26–27
Spin-transfer torque, 525
Spintronics, 159, 515–526
   devices, 523–526
   Hanle effect, 519
   semiconductor, 518
Split-off gap, 505
Splitters, 758–759
SPR. *See* Surface plasmon resonance
Square barrier potential, 42f
Square step potential, 42f
Square wells
   potential, 48f, 645
   systems, 260f
   two-channel, 675f
   two-dimensional, 728–729
Squeezed states, 95–97, 96f
Stabilizer codes, 241
Stark, Johannes, 312
Stark shift, 335
Statcoulomb (esu), 124
States
   distance between, 77–78
   purity, 77
   vectors, 35
Statistical analysis, of random potentials, 737–738
Statistical mechanics, 857–860
Steane, Andrew, 241
Stefan-Boltzmann constant, 7–8
Step functions, 350
Step potential, 45f, 45–46
Stern, Otto, 159
Stern-Gerlach experiment, 12–13, 12f, 27, 159, 252
Stimulated Raman adiabatic passage (StIRAP), 249, 293–294, 338
STM. *See* Scanning tunneling microscopy
Stokes parameter, 266
Stokes theorem, 346–347, 915–916
Stokes transition, 327

STOs. *See* Slater type orbitals
Strongly correlated electron systems, 541
Structure constants, 288
Stueckelberg, Ernst, 339
Sudden approximation, 333–335
Superconductivity, 249, 395. *See also* High-temperature superconductivity
Superlattices, 754–755
Superposition
   incoherent, 73
   principle, 23
   state, 23
Surface plasmon resonance (SPR), 534, 535f
Susceptibilities, 349–352, 360–361
   electric, 350
   magnetic, 350, 468–469
   optical, 274
   tensor, 468
SVE. *See* Slowly varying envelope
s-wave scattering, 650, 680–682
Symmetric gauge, 435, 445, 452
Symmetric group S_N, 369–370
Symmetric spin function, 665
Symmetrization postulate, 372
Symmetry, 98–103. *See also* Exchange symmetry
   appropriate, 827
   axial, 635–636
   considerations, 151–157
   inversion, 100, 152
   molecular, 579–582
   operations, 403
   parameter, 775
   parity, 48, 98
   permutation, 367–369
   point group, 581f
   of S matrix, 701
   selection rules and, 152
   spherical, 606, 635
   spin, 665
   between susceptibilities, 360–361
   time-reversal, 100–102, 153, 708
   of transmission, 735, 747–748
   vibrational modes and, 595–596
Symplectic matrix, 777
Syndrome measurements, 242

T

t matrix, 628–633
   partial wave analysis, 717–718
   transition, 698
   wave function and, 642
Tabor, M., 780
Taylor expansion, 426
Teleportation, 221, 222f
Tensor operators, 144–151

   irreducible, 144–145
   matrix elements of, 153
Tensors
   conductivity, 443, 542
   dielectric, 350
   effective mass, 426
   electric susceptibility, 350
   magnetic susceptibility, 350
   multipole moments, 150–151
   polarizability, 312
   resistance, 443, 760, 761f
   spin Hall conductance, 511
   susceptibility, 468
Term symbols
   atomic, 175n5, 547
   Russell-Saunders, 560
Tetramethylsilane ($Si(CH_3)_4$), 191f
Thermal conductivity coefficient, 397
Thermal current density, 397
Thermodynamic limit, 381, 826
Thermoelectric effect, 398–399
Thermopower, 398
Thomas, L. H., 176
Thomas, Llewellyn, 159, 875
Thomas precession, 159, 176
Thomas-Fermi approximation, 875–878, 877f
Thomas-Fermi formalism, 632
Thomas-Reiche-Kuhn sum rule, 324
Thouless energy, 778
Thouless number, 751
3D periodic systems, 457–459
$3j$ symbols, 140–143
Three-electron systems, 375–376
Three-level systems, 287–289, 288f
   dressed, 293f
   dynamics, 293–294
   optically coupled, 293f
   two or more, 288–289
   two-particle, 289
Threshold theorem, 241
Tight binding equations, 798
Tight-binding formalism, 738–741, 848–850
Tight-binding model, 427–428
Time
   averages, 276
   decoherence, 267
   dependence, of eigenstates, 65
   diffusion, 751
   evolution, of state vectors, 35
   quantum propagation in, 29–30
Time reversal, 100–102
   invariance, 98, 173–175, 702
   properties of spinors, 172–175
   symmetry, 153, 708
Time-ordered product, 316, 616

Time-reversed curve, 781
TMR. *See* tunneling magnetoresistance
Tomonaga-Luttinger liquids, 815
Torque method, 460, 460f
Trace distance, 77–78
Traces, 902
Transfer matrix, 685, 735, 739
Transformation
  coefficients, 595
  operators, 32
  unitary, 78
Transistors. *See also* Metal-oxide-semiconductor field-effect transistors
  Datta-Das spin, 525–526, 526f
  field-effect, 14
  magnetic spin, 524–525, 525f
  single electron, 20, 771
Transition probability, for scattering, 329
Translation generators, 33
Translation operations, 405
Translation-invariant media, 350–351
Transmission
  amplitude, 685
  coefficients, 46, 47f, 756
  through 1D Coulomb potential, 732–734
  through potential barriers, 684–686, 684f
  symmetries of, 735, 747–748
Triangle inequalities, 139, 894
Tunneling, 13–15
  through Coulomb barrier, 686f
  through double barriers, 50–52
  electron, 55f
  Klein, 809–810, 810f
  Landau-Zener, 343
  linear potential and, 54–55
  states, 304
Tunneling magnetoresistance (TMR), 520–524, 521f
Turing, Alan, 194
Turning points, 306, 307f
2D systems
  bound states in, 730–731
  Coulomb scattering in, 726–731
  density of states in, 386–387, 387f
  with fixed fermi energy, 448–450, 450f
  with fixed number of particles, 450, 451f
  scattering in, 724–731
  symmetric gauge and, 452
Two-body collisions, 618–621
Two-body operators, 826
Two-electron states, 769–770
Two-electron systems, 375–376, 547–550
Two-level dynamics, 261–262
Two-level systems, 259–261, 260f
  Berry phase in, 347–348
  decoherence and, 266–268

exchange symmetry, 377–378
N, 282–284
periodically driven, 268–273, 269f
representations of, 264f
Two-particle systems, 856–857
Two-qubit states
  classification of, 291f
  density matrix, 291
Typical sequences, 199

U

Uhlenbeck, George E., 13
Ultrarelativistic limit, 794
Unbound states, 4–5
Uncertainty principle, 67–69
  squeezed states and, 95–97
  wave packet dynamics and, 297–298
Unfolding, 778–779
Unit cells, 401–402
  of Bravais lattices, 405f
  primitive, 402
Unitarity, 708
Unitary limit, 786
Universal conductance fluctuations (UCF), 780–782
Universal sets, 217
Universal Turing machine, 194
Unoccupied states, 433

V

Vacuum state, 832
Valence bond
  approximation, 476
  method, 567–569
Valley index, 802
Van Alphen, P. M., 459
Van der Waals, Johannes, 123–124
Van der Waals (VdW) units, 123–124
Van Hove singularities, 425, 530–531, 531f
Van Vleck paramagnetism, 470
Vanishing potential, 31
Variables
  hidden, 253–256
  random, 195–196
  separation of, 31
Variational method, 331–333
Variational principle, 688–690
Variational theorem, 333
VdW units. *See* Van der Waals units
Vector addition, 891
Vector coupling coefficients, 139
Vector fields, 148–149
Vector potential, 169, 345–346, 435
Vector products, 913
Vector space, 29, 891–898

Vernam, Gilbert, 203
Vibrational modes, 595–596
Vibrational structure, 593–595
Vibrational transitions, 598, 599f
Vibrational-rotational coupling, 594
Virial theorem, 87
Vlasov equation, 84
Von Laue, Max, 11, 409
Von Laue analysis, 410f
Von Laue condition, 410–411
Von Neumann, John, 72
Von Neumann entropy, 77

W

Walsh-Hadamard transformation, 226–227, 237
Wannier, Gregory H., 429
Wannier functions, 427, 429–430, 502–503
Watson transform, 646
Wave functions, 22
  angular, 706
  for barrier potential, 47f
  Bloch, 417–421
  boson, 99
  for bosonic states, 379
  bound-state, 128
  Coulomb radial, 128–131
  cross-sections to, 622–623
  electronic, 375
  even parity, 48
  fermionic, 99, 379
  free phonon, 540
  ground state, 816–817
  harmonic oscillators, 58f
  Landau, 446
  Laughlin, 465–466
  many-body, 827–830
  near band-edge, 424–425
  nonholonomic, 347
  notation, 29
  odd parity, 48
  orthogonality relations and, 831
  plane waves and, 34
  radial, 116, 129f
  Schrödinger method, 92
  spin, 371
  for step potential, 45f
  t matrix and, 642
  temporal dependence of, 298
  time-dependent, 31
  variational, 688
Wave packets, 53–54
  dynamics, 297–299
  normalized 1D Gaussian, 297
  spatial width of, 299
  spreading, 298f
Wave-particle duality, 8–12
Waves, 6–8. *See also specific wave types*
Weizsäcker correction, 878

Index

Well depth, 122
Welton, Theodore, 362
Wentzel, Gregor, 304, 318
Werner density matrix, 291f
White noise, 366, 738
Whittaker equations, 662
Wiedmann-Franz law, 392, 397, 528
Wien tail, 7
Wiesner, Stephen, 219
Wigner, Eugene P., 135, 152
Wigner crystals, 815–816
Wigner functions, 135, 295
Wigner representation, 79–85
Wigner solid, 865
Wigner surmise, 779–780
Wigner $3j$ symbols. *See* $3j$ symbols
Wigner transformation, 82–83
Wigner-Dyson matrix ensembles., 775
Wigner-Eckart theorem, 152–155, 326, 471
Wigner-Seitz cell, 407–408, 407f
Wigner-Seitz radius, 384, 394
Wilson, W., 307
Wineland, D., 275
WKB approximation, 304–308, 684–688
   1D Coulomb potential and, 734
   validity, 691
WKB connection formulas, 306–307
Wronskian, 648

X

X-ray scattering, 409–410, 409f, 411
X-ray spectrometry, 11f

Y

Yang, C. N., 100
Young, Alfred, 370
Young, Thomas, 6, 8
Young tableaux, 370–371, 370f, 371f
Yukawa potential, 632, 641, 683

Z

Zavoisky, Y. K., 185
Zeeman, Pieter, 166
Zeeman energy, 166, 168f, 177f
Zeeman Hamiltonian, 166, 169–170
   for NMR, 186
   for single spin, 186
Zeeman splitting, 159, 182, 182f, 247
   of alkali-metals, 182f
   of spin 1/2 particle, 186f
Zener, Clarence, 339
Zeno effect, 331
Zero phonon state, 540
Zero-bias anomaly, 787
Zero-dimensional systems, 747
Zero-energy resonance, 658
Zero-point energy, 57, 528–529
Zero-temperature occupation, 99f
Zinc-blende, 406f, 505, 506f
Zoller, Peter, 244

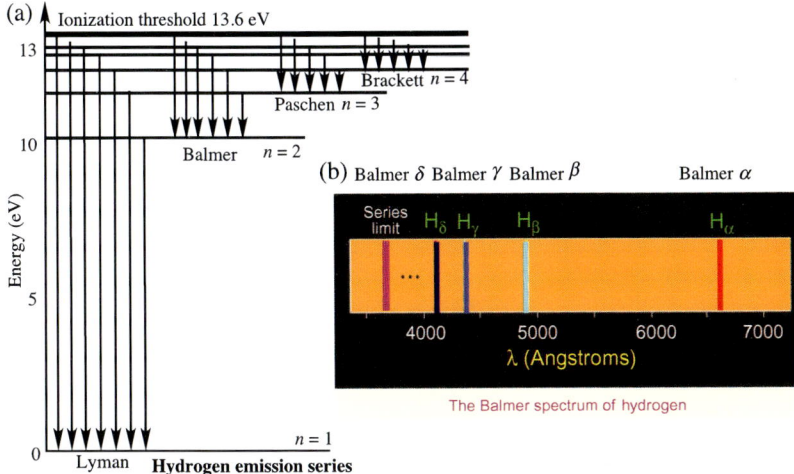

PLATE 1 (Figure 1.1 on page 5 of this book)

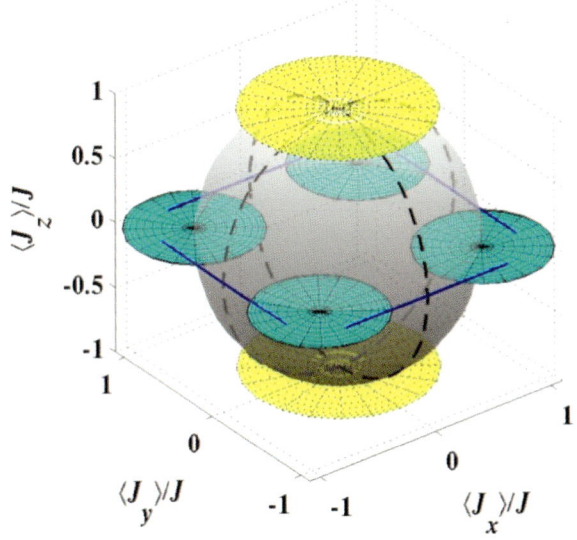

PLATE 2 (Figure 6.10 on page 284 of this book)

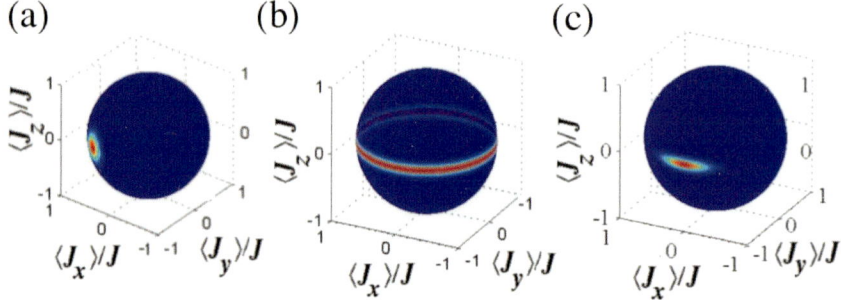

PLATE 3 (Figure 6.11 on page 284 of this book)

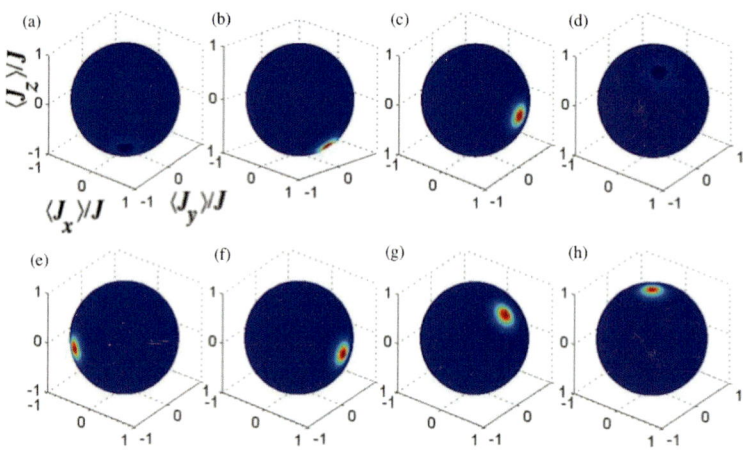

PLATE 4 (Figure 6.14 on page 287 of this book)

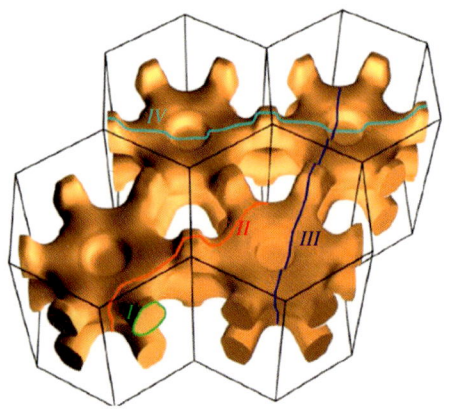

PLATE 5 (Figure 9.50 on page 461 of this book)

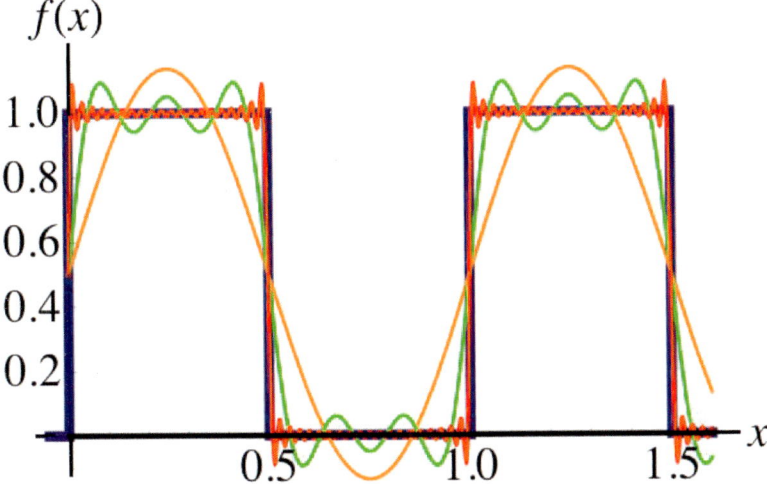

PLATE 6 (Figure D.1 on page 923 of this book)